Lecture Notes in Computer Science 3042

Commenced Publication in 1973
Founding and Former Series Editors:
Gerhard Goos, Juris Hartmanis, and Jan van Leeuwen

Springer-Verlag Berlin Heidelberg GmbH

Nikolas Mitrou   Kimon Kontovasilis
George N. Rouskas   Ilias Iliadis
Lazaros Merakos (Eds.)

NETWORKING 2004

Networking Technologies,
Services, and Protocols;
Performance of Computer
and Communication Networks;
Mobile and Wireless Communications

Third International IFIP-TC6 Networking Conference
Athens, Greece, May 9-14, 2004
Proceedings

 Springer

Volume Editors

Nikolas Mitrou
National Technical University of Athens
School of Electrical and Computer Engineering, Computer Network Laboratory
9 Heroon Polytechneiou str., 157 80, Athens, Greece
E-mail: mitrou@softlab.ntua.gr

Kimon Kontovasilis
National Center for Scientific Research "Demokritos"
Institute of Informatics and Telecommunications
P.O. Box 60228, 15310 Agia Paraskevi Attikis, Greece
E-mail: kkont@iit.demokritos.gr

George N. Rouskas
North Carolina State University, Department of Computer Science
Box 7534, 461 EGRC, 2410 Campus Shore Drive, Raleigh, NC 27695-7534, USA
E-mail: rouskas@csc.ncsu.edu

Ilias Iliadis
IBM Research, Zurich Research Laboratory
Säumerstrasse 4, 8803 Rüschlikon, Switzerland
E-mail:ili@zurich.ibm.com

Lazaros Merakos
University of Athens, Department of Informatics and Telecommunications
Panepistimiopolis, 15784 Athens, Greece
E-mail: merakos@di.uoa.gr

Library of Congress Control Number: 2004104505

CR Subject Classification (1998): C.2, C.4, H.4, D.2, J.2, J.1, K.6, K.4

ISSN 0302-9743
ISBN 978-3-540-21959-0   ISBN 978-3-540-24693-0 (eBook)
DOI 10.1007/978-3-540-24693-0

springeronline.com

©2004 Springer-Verlag Berlin Heidelberg

Originally published by IFIP International Federation for Information Processing, Hofstraße 3, 2361
Laxenburg, Austria in 2004.

Typesetting: Camera-ready by author, data conversion by PTP-Berlin, Protago-TeX-Production GmbH
Printed on acid-free paper   SPIN: 11007159   06/3142   5 4 3 2 1 0

Preface

This book contains the refereed proceedings of the 3rd International IFIP-TC6 Networking Conference, *Networking 2004*. Conferences in the *Networking* series span the interests of several distinct, but related, TC6 working groups, including Working Groups 6.2, 6.3, and 6.8. Reflecting this, the conference was structured with three Special Tracks: (i) Networking Technologies, Services, and Protocols; (ii) Performance of Computer and Communication Networks; and (iii) Mobile and Wireless Communications.

However, beyond providing a forum for the presentation of high-quality research in various complementary aspects of networking, the conference was also targeted to contributing to a unified view of the field and to fostering the interaction and exchange of fruitful ideas between the various related (and overlapping) specialized subcommunities therein. Towards this second objective, more than a few conference sessions (and thematic sections in this book) 'cut across' the Special Tracks, along more generic or fundamental concepts.

Networking 2004 was fortunate to attract very high interest among the community, and the conference received 539 submissions from 44 countries in all five continents. These figures correspond to a remarkable increase in submissions from the previous very successful events (roughly, a 156% increase over *Networking 2000* and 71% over *Networking 2002*), and indicate that *Networking* conferences are progressively becoming established as worldwide reference events in the field.

The high number of submissions provided an excellent opportunity for a high-quality program but also called for a very demanding and laborious paper evaluation process. The 99 members of the Technical Program Committee, together with more than 750 additional experts engaged as external reviewers, worked efficiently and responsibly under tight time constraints to produce a total of some 1600 reviews, towards responding to the authors with insightful, detailed feedback and providing the basis for the final paper selection.

As a result of this process, 103 papers were finally selected from the 539 submissions (at an acceptance rate below 20%) for presentation during the conference's technical sessions. Additionally, to allow the conference participants to benefit from further worthwhile and stimulating research results, 40 papers were accepted for presentation in a poster session; these papers appear in the proceedings in short form. Overall, the final program was excellent and diverse, and covered both fundamental issues in networking as well as emerging topics. It should be noted that, given the many submissions and the tight constraints, many good papers could not be accommodated.

The main program of *Networking 2004* covered three days and included streams of up to three parallel sessions, plus a poster session for the presentation of the short papers. In addition to these components, the program was enriched by three stimulating invited talks offered by world-renowned researchers in the field, specifically: Jim Roberts on "From ATM to IP and back again: the label switched path to the converged Internet, or another blind alley?", Leandros Tassiulas on "Cross-layer design issues for quality of service provisioning in wireless networks", and Ian Akyildiz on "Research challenges in wireless sensor and actuator networks". The main conference program was complemented by a diverse set of high-quality tutorials and workshops.

The final result wouldn't have been possible without the dedication and hard work of many colleagues. Special thanks are due to the members of the Technical Program Committee and to all external referees for their invaluable help with the paper reviews and their overall sense of responsibility and responsiveness under very tight deadlines in a particularly busy period. Another word of thanks goes to the Webmaster Manolis Solidakis and to the General Chair Nikolas Mitrou, who worked hard to ensure that flexible and effective technical support was available during the paper evaluation and the final paper selection. Finally, we are grateful to all authors who honored *Networking 2004* by trusting their work to the conference.

March 2004   Kimon Kontovasilis
                                                          George Rouskas
                                                          Ilias Iliadis
                                                          Lazaros Merakos

Message from the General Chair

No doubt the world economy is being globalized at a rapid pace, while, at the same time, it is becoming more and more information intensive. The joint evolution and convergence of these fundamental trends towards the global, knowledge-based society leans upon a bold scientific and technological objective: pervasive information processing and ubiquitous networking. Clearly, technological advances in all aspects of networking are central to this objective.

In such a context, *Networking 2004* strove for a distinct and valuable contribution to the field. Organized jointly by the Institute of Communication and Computer Systems of the National Technical University of Athens and the Institute of Informatics and Telecommunications of the NCSR (National Centre for Scientific Research) "Demokritos", it was the third event in a series of IFIP-TC6 networking conferences, launched in 2000 and organized every 2 years thereafter. In common with its predecessors, *Networking 2004* aimed at reviewing progress, presenting recent advances, identifying trends and refreshing vision in this hot field, by bringing together active and proficient members of the networking community from both academia and industry. It covered all aspects of networking, focusing especially on three main areas (Special Tracks): (a) networking technologies, services, and protocols, (b) performance of computer and communication networks, and (c) mobile and wireless communications. The conference lasted six days and included the main program (three days), tutorials (one and a half days) and thematic workshops attached to the main event (one day).

The organization of such an ambitious event required a considerable effort and commitment by very many people. I therefore wish to express my deep appreciation to all of the Executive Committee members for their outstanding work.

The heart of the conference was, of course, its main program. The dramatic increase in the number of submitted papers (more than 70%, compared to the previous event in 2002) clearly indicates the increasing interest of the community in our IFIP-TC6 networking conferences. Diligence in the reviewing and selection process was necessary, in order to thoroughly handle such a large number of submissions and, hopefully, to hand over to the next organization a conference standard even higher than that inherited. A special appreciation is due to the Technical Program Chair, Kimon Kontovasilis, for his deep devotion to this extremely delicate and demanding task. Sincere thanks are addressed also to the three Special Track Chairs, George Rouskas, Ilias Iliadis and Lazaros Merakos, who worked diligently towards a high-level conference program. The TPC members, as well as more than 750 additional external experts/referees, are also warmly thanked for their excellent work.

Within the main program, keynote speeches are "tracer bullets" in the unexploited sky of new trends and ideas, or critical retrospections on fundamental aspects of networking. Here I have to thank the Invited Speaker Co-chairs Leonidas Georgiadis and Michael Paterakis and, of course, our distinguished invited speakers.

The tutorial program enriched the main conference program with thorough half-day presentations, covering theoretical and practical aspects of recent advances in networking. The tutorials were mainly targeted to young researchers and practitioners in the field who wish to foster their skills and background knowledge. Many thanks to the Tutorial Co-chairs Christos Douligeris, Gunnar Karlsson and Dimitris Pendarakis for organizing an interesting set of tutorials.

The last day of the conference was devoted to thematic workshops focusing on areas of special interest or of remarkable evolution dynamics. Four such workshops were scheduled: *Next Generation Networking Middleware* (*NGNM04*, organized by Nikos Anerousis and George Kormentzas), *Data Processing and Storage Networking: Towards Grid Computing* (*DPSN04*, organized by Fotini-Niovi Pavlidou, John Soldatos and Evangelos Vayias), *Streaming Media Distribution over the Internet* (*SMDI04*, organized by Charalampos Patrikakis, Marco Combetto and Constantinos Boukouvalas) and *Wireless Internet and Reconfigurability* (*4th ANWIRE*, organized by Nikos Passas). The workshop chairs did an excellent job to achieve quality workshops, commensurate with the high level of the rest of the conference.

Our Web engineer, Manolis Solidakis, did his best to provide user-friendly on-line facilities during all preparatory phases: paper submission and review, registration, program dissemination. I thank him for his commitment to that.

Jointly with the Technical Program Chair, we would like to address our final word of thanks to the administration of the organizing institutes within NTUA and NCSR "Demokritos" for their support, as well as to the IFIP-TC6 management, who entrusted us with the conduct of this important event.

Not too long ago, we received the staggering news that Olga Casals, our good friend and colleague, an active researcher and TPC member of previous *Networking* conferences (intended as a TPC member of this conference too) suddenly passed away. The Best Paper Award of *Networking 2004*, presented to its recipient on the last day of the conference, was devoted to her memory.

March 2004   Nikolas Mitrou

Organization

Conference Executive Committee

General Chair
Nikolas Mitrou, National Technical University of Athens, Greece

Technical Program Chair
Kimon Kontovasilis, National Centre for Scientific Research "Demokritos", Greece

Special Track Chair for Networking Technologies, Services, and Protocols
George Rouskas, North Carolina State University, USA

Special Track Chair for the Performance of Computer and Communication Networks
Ilias Iliadis, IBM Research, Zurich Laboratory, Switzerland

Special Track Chair for Mobile and Wireless Communications
Lazaros Merakos, University of Athens, Greece

Tutorial Program Co-chairs
Christos Douligeris, University of Piraeus, Greece
Gunnar Karlsson, KTH, Sweden
Dimitris Pendarakis, IBM, USA

Invited Speakers Co-chairs
Leonidas Georgiadis, Aristotle University of Thessaloniki, Greece
Michael Paterakis, Technical University of Crete, Greece

Workshop Chairs

Next Generation Networking Middleware (NGNM04)
Nikos Anerousis, IBM Research, USA
George Kormentzas, University of the Aegean, Greece

Data Processing and Storage Networking: Towards Grid Computing (DPSN04)
Fotini-Niovi Pavlidou, Aristotle University of Thessaloniki, Greece
John Soldatos, Athens Information Technology, Greece
Evangelos Vayias, Intracom SA, Greece

Streaming Media Distribution over the Internet (SMDI04)
Charalampos Patrikakis, National Technical University of Athens, Greece
Marco Combetto, Microsoft Research, University Relations, Cambridge, UK
Constantinos Boukouvalas, OTE, S.A., Greece

Wireless Internet and Reconfigurability (4th ANWIRE)
Nikos Passas, University of Athens, Greece

Publicity Chair
Demetres Kouvatsos, University of Bradford, UK

Steering Committee Chair
Harry Perros, North Carolina State University, USA

Steering Committee Members
Augusto Casaca, IST/INESC, Portugal
Erol Gelenbe, University of Central Florida, USA
Harry Perros, North Carolina State University, USA
Guy Pujolle, University of Paris 6, France
Harry Rudin, Switzerland
George Stassinopoulos, National Technical University of Athens, Greece

Finance Chair
Athanasios Drigas, National Centre for Scientific Research "Demokritos",
Greece

Electronic Submission/Web Designer
Emmanuel Solidakis, National Technical University of Athens, Greece

Local Organizing Committee
Haralampos Skianis, National Centre for Scientific Research "Demokritos",
Greece
George Kormentzas, University of the Aegean, Greece

Technical Program Committee

Special Track for Networking Technologies, Services and Protocols

Kevin Almeroth, University of California, Santa Barbara, USA
Andrea Bianco, Politecnico di Torino, Italy
Ken Calvert, University of Kentucky, USA
Claudio Casetti, Politecnico di Torino, Italy
Laurie Cuthbert, Queen Mary and Westfield College, UK
Jordi Domingo-Pascual, Universitat Politecnica de Catalunya, Spain

Constantinos Dovrolis, Georgia Institute of Technology, USA
Eylem Ekici, Ohio State University, USA
Mihalis Faloutsos, University of California, Riverside, USA
Serge Fdida, Université Pierre et Marie Curie, France
Luigi Fratta, Politecnico di Milano, Italy
Maurice Gagnaire, École Nationale Supérieure des Télécommunications, France
Leonidas Georgiadis, Aristotle University of Thessaloniki, Greece
David Hutchison, Lancaster University, UK
Admela Jukan, National Science Foundation, USA
Mohan Kumar, University of Texas at Arlington, USA
G. Manimaran, Iowa State University, USA
Ibrahim Matta, Boston University, USA
Prasant Mohapatra, University of California, Davis, USA
Ioanis Nikolaidis, University of Alberta, USA
Peng Ning, North Carolina State University, USA
Jaudelice C. de Oliveira, Drexel University, USA
Symeon Papavasileiou, New Jersey Institute of Technology, USA
Michael Paterakis, Technical University of Crete, Greece
Guido Petit, Alcatel, Belgium
Chiara Petrioli, University "La Sapienza", Rome, Italy
Nineta Polemi, University of Piraeus, Greece
Martin Potts, Martel GmbH, Switzerland
Erwin Rathgeb, Universität Duisburg-Essen, Germany
Luigi Rizzo, University of Pisa, Italy
Hussein Salama, Cisco Systems International, Egypt
Saswati Sarkar, University of Pennsylvania, USA
Dimitris Serpanos, University of Patras, Greece
Mihai Sichitiu, North Carolina State University, USA
Efstathios Sykas, National Technical University of Athens, Greece
Andras Valko, Ericsson, Sweden
Iakovos Venieris, National Technical University of Athens, Greece
Lars Wolf, Braunschweig University of Technology, Germany
Zhi-Li Zhang, University of Minnesota, USA

Special Track for the Performance of Computer and Communication Networks

Ron Addie, University of Southern Queensland, Australia
Eitan Altman, INRIA, France
Miltiades Anagnostou, National Technical University of Athens, Greece
Andrea Baiocchi, University "La Sapienza", Rome, Italy
Chris Blondia, University of Antwerp, Belgium
Pierre Boyer, France Telecom R&D, France
Tosten Braun, University of Bern, Switzerland
Herwig Bruneel, University of Ghent, Belgium
Wojciech Burakowski, Warsaw University of Technology, Poland

Werner Bux, IBM Research, Zurich Laboratory, Switzerland
Mariacarla Calzarossa, University of Pavia, Italy
Constantinos Courcoubetis, Athens University of Economics and Business, Greece
Khaled Elsayed, Cairo University, Egypt
Guenter Haring, University of Vienna, Austria
Peter Harrison, Imperial College London, UK
Krishna Kant, Intel Corporation, USA
Peter Key, Microsoft Research Ltd., Cambridge, UK
Demetres Kouvatsos, University of Bradford, UK
Emilio Leonardi, Politecnico di Torino, Italy
Michela Meo, Politecnico di Torino, Italy
Edmundo Monteiro, University of Coimbra, Portugal
Ilkka Norros, VTT, Finland
Andreas Pitsillides, University of Cyprus, Cyprus
Ramon Puigjaner, Universitat de les Illes Balears, Spain
Otto Spaniol, Aachen University of Technology, Germany
George Stamoulis, Athens University of Economics and Business, Greece
Yutaka Takahashi, Kyoto University, Japan
Don Towsley, University of Massachusetts, USA
Jorma Virtamo, Helsinki University of Technology, Finland

Special Track for Mobile and Wireless Communications

Ian Akyldiz, Georgia Institute of Technology, USA
Nancy Alonistioti, University of Athens, Greece
Roberto Battiti, University of Trento, Italy
Luciano Bononi, University of Bologna, Italy
Azzedine Boukerche, University of Ottawa, Canada
Marco Conti, National Research Council, Italy
Luis Correia, Technical University of Lisbon, Portugal
Panagiotis Demestichas, University of Piraeus, Greece
Anthony Ephremides, University of Maryland, USA
Laura Feeney, SICS, Sweden
Gabor Fodor, Ericsson, Sweden
Jerome Galtier, France Telecom R&D and INRIA, France
Silvia Giordano, ICA-DSC-SUPSI, Switzerland
Enrico Gregori, National Research Council, Italy
Gerald Maguire, Royal Institute of Technology, Sweden
Takis Mathiopoulos, National Observatory of Athens, Greece
Stephan Olariu, Old Dominion University, USA
Sergio Palazzo, University of Catania, Italy
Nikos Passas, University of Athens, Greece
Jorge Pereira, European Commission, Belgium
George Polyzos, Athens University of Economics and Business, Greece
Guy Pujolle, Université Pierre et Marie Curie (Paris 6), France

Kimmo Raatikainen, University of Helsinki, Finland
Ivan Stojmenovic, University of Ottawa, Canada
Violet Syrotiuk, University of Arizona, USA
Michael Theologou, National Technical University of Athens, Greece
Bernhard Walke, Aachen University of Technology, Germany

Referees

Finn Arve Aagesen	Enzo Baccarelli	Chris Blondia
Samuli Aalto	Andrea Baiocchi	Fernando Boavida
Alhussein Abouzeid	Mario Baldi	Gregor von Bochmann
John Adams	George Balis	Eliane Bodanese
Ron Addie	Frank Ball	Rene Boel
Anjali Agarwal	Roland Balmer	Raffaele Bolla
Sharad Agarwal	Simonetta Balsamo	Alberto Bononi
Rui Aguiar	Albert Banchs	Luciano Bononi
Ozgur Baris Akan	Chadi Barakat	Eleonora Borgia
Nail Akar	Mario Barbera	Oliver Bott
Basheer Al-Duwairi	Antonio Barili	Abdelmadjid Bouabdallah
Tarek Al-Gizawi	Dhiman Barman	Azzedine Boukerche
Timo Alanko	Stefano Baroni	Christos Bouras
Guido Albertengo	Novella Bartolini	Raouf Boutaba
Kevin Almeroth	Stephen Barton	Onno Boxma
Nancy Alonistioti	Stefano Basagni	Jacqueline Boyer
Nicola Altan	Anup Basu	Pierre Boyer
Eitan Altman	Roberto Battiti	Adam Bradley
Paolo Ameigeiras	Daniel Bauer	Jeremy Bradley
Marcelo Dias de Amorim	Florian Baumgartner	Michael Bradshaw
Miltiadis Anagnostou	Buyurman Baykal	Arnold Bragg
Giuseppe Anastasi	John Bellardo	Christof Brandauer
Stergios Anastasiadis	Paolo Bellavista	Torsten Braun
Matthew Andrews	Mats Bengtsson	Tonnes Brekne
Vladimir Anisimov	Hans van der Berg	Erik van Breusegem
Tricha Anjali	Marco Bernardo	Mauro Brunato
Panayotis Antoniadis	Christian Bettstetter	Herwig Bruneel
George Apostolopoulos	Bharat Bhargava	Raffaele Bruno
John Apostolopoulos	Bobby Bhattacharjee	Roland Bueschkes
Martin Arlitt	Giuseppe Bianchi	Wojciech Burakowski
Jesus Artalejo	Andrea Bianco	Werner Bux
Gerald Ash	Hannes Birck	Juan Caballero
Sasikanth Avancha	Jozsef Biro	Tom van Caenegem
Husam Awadalla	Erez Biton	Gruia Calinescu
Irfan Awan	Danilo Blasi	Franco Callegati
Dimitrios Axiotis	Nicola Blefari-Melazzi	Rita de Caluwe
Fulvio Babich	Holger Bleul	Ken Calvert

Mariacarla Calzarossa
Pietro Camarda
Srjdan Capkun
Narcis Cardona
Yannick Carlinet
Loren Carrasco
Iacopo Carreras
Graca Carvalho
Emiliano Casalicchio
Vicente Casares
Ramon Casellas
Claudio Casetti
Marco Casole
Luca Casone
Erdal Cayirci
Nedo Celandroni
Isabella Cerutti
Ram Chakka
Anirban Chakrabarti
Rajiv Chakravorty
Shueng-Han Gary Chan
Soamsiri Chantaraskul
Hakima Chaouchi
Martin Chapman
Prasanna Chaporkar
Joachim Charzinski
Bensong Chen
Chao Chen
Jiangzhuo Chen
Peter Chen
Trista Chen
Yinong Chen
Yuchung Cheng
Jan Cheyns
Sundaram Chinthamani
Sunghyun Choi
Ishtiaq Ahmed Choudhry
Charilaos Christopoulos
Chrysost. Chrysostomou
Thomas Clausen
Jorge Cobb
Denis Collange
Jeff Considine
Marco Conti
Chris Cooper

Antonio Corradi
Olivier Corre
Luis Correia
Americo Correia
Jose Costa-Requena
Constantinos Courcoubetis
Piergiorgio Cremonese
Paolo Cremonesi
Mark Crovella
Juan Carlos Cruellas Ibarz
Pedro Cuenca
Jun-Hong Cui
Minghao Cui
Marilia Curado
Laurie Cuthbert
Marc Danzeisen
Vasilios Darlagiannis
Dai Davies
Panagiotis Demestichas
Dee Denetree
Wolfgang Denzel
Evaggelos Dermatas
Andrea Detti
Peter Deutsch
Ashay Dhamdhere
Eric Diehl
Antonis Dimakis
George Dimitrakopoulos
Nick Dingle
Jana Dittmann
Tien V. Do
Jordi Domingo-Pascual
Jeff Donahoo
Yingfei Dong
Yixin Dong
Yesid Donoso
Satish Doraiswamy
Christos Douligeris
Elias Doumith
Constantinos Dovrolis
Manos Dramitinos
Falko Dressler
Wenliang Du
Zhenhai Duan
Parijat Dube

Frederich Ducatelle
Michael Dueser
Tolga Duman
Adam Dunkels
Sandra Dykes
Elias Efstathiou
Gerald Eichler
Eylem Ekici
Karim El-Khazen
Abdulmotaleb El Saddik
Khaled Elsayed
Anthony Ephremides
Alberto Escudero-Pascual
Do Young Eun
Tony Eyers
Ramon Fabregat
Sonia Fahmy
Mihalis Faloutsos
Andras Farago
Julien Fasson
Serge Fdida
Laura Feeney
Zongming Fei
Guillem Femenias
Markus Fiedler
Ioannis Fikouras
Joe Finney
Jorge Fiñochietto
Eric Fleury
Gabor Fodor
Olivier Fourmaux
Jean-Michel Fourneau
Roberta Fracchia
Chuck Fraleigh
Mirko Franceschinis
Gerald Franzl
Vassilis Freiderikos
Mario Freire
Timur Friedman
Norihito Fujita
Andrea Fumagalli
Maurice Gagnaire
Dominique Gaiti
Giulio Galante
Clemente Galdi

Jose Ruela
Zhao Rui
Silvia Ruiz
Selwyn Russell
Mikalai Sabel
Dan Sadot
David Safford
Amit Sahoo
Hussein Salama
Apostolis Salkintzis
Elio Salvadori
Petros Sampatakos
Dimitris Sampson
Sergio Sanchez-Lopez
Kamil Sarac
Lambros Sarakis
Susana Sargento
Saswati Sarkar
Stelios Sartzetakis
Anna Satsiou
Eric Scharf
Matthias Scheidegger
Giovanni Schembra
Enrico Schiattarella
Marc Schinnenburg
Rudiger Schollmeier
John Schormans
Bruno Richard Schulze
Caterina Scoglio
Mirela Sechi-Notare
Srinivasan Seetharaman
Amit Seghal
Ahmet Sekercioglu
Sidi-Mohammed Senouci
Dimitrios Serpanos
Bartomeu Serra
Sanjay Shakkottai
Srinivas Shakkottai
Jonathan Shapiro
Vinod Sharma
Jang-Ping Sheu
Xiyu Shi
Ali Shoja
Peter Sholander
Karen Shoop

Rajeev Shorey
David Shrimpton
Stavroula Siachalou
Mihai Sichitiu
Nikolaos Sidiropoulos
Matthias Siebert
Manolis Sifalakis
Georgos Siganos
Dimitra Simeonidou
David Simplot-Ryl
Robby Simpson
Narendra Singhal
Prasun Sinha
Vasilios Siris
Krishna Sivalingam
Anders Sjogren
Harry Skianis
Dimitris Skyrianoglou
Avril Smith
Paul Smith
Peter Soerensen
John Soldatos
Sergios Soursos
Kathleen Spaey
Otto Spaniol
Alexander Sprintson
Adarsh Sridhar
Ashwin Sridharan
K. Sriram
Rolf Stadler
Yiannis Stamatiou
George Stamoulis
Panagiotis Stathopoulos
Brikena Stavovci-Halimi
Ioannis Stavrakakis
Kiriakos Stefanidis
Cliff Stein
Burkhard Stiller
Ivan Stojmenovic
Aaron Striegel
Peter Stuckmann
Marinos Stylianou
Weilian Su
Dhanant Subhadrabandhu
Kyoungwon Suh

Zhi-Li Sun
Min-Ho Sung
Irini Sygkouna
Efstathios Sykas
Violet Syrotiuk
Robert Szabo
Janos Sztrik
Abd-Elhamid Taha
Yutaka Takahashi
Alessandro Tarello
Leandros Tassiulas
Y. C. Tay
Jing Teng
Daniele Tessera
Michael Theologou
Petia Todorova
Birger Toedtman
Samir Tohme
Tiziana Toniatti
Corinne Touati
Don Towsley
Alex Trifunovic
Axel Truong
Konstantinos Tsagaris
Sofia Tsekeridou
Stavros Tsiakkouris
Orestis Tsigas
Masato Tsuru
Tuna Tugcu
Zoltan Turanyi
Damla Turgut
Filip de Turck
Shelley Unger
Alessandro Urpi
Francesco Vacirca
Kiran Vadde
Rui Valadas
Andras Valko
Srivatsan Varadara
Anujan Varma
Manos Varvarigos
Constantinos Vassilakis
Vasos Vassiliou
Jon-Olov Vatn
Evaggelos Vayias

Teresa Vazao
Hector Velayos
Andrea de Vendictis
Iakovos Venieris
Roberto Verdone
Dimitrios Vergados
Evangelos Vergetis
Giacomo Verticale
Rolland Vida
Attila Vidacs
Paulo Cesar Salgado Vidal
Laurent Viennot
Manuel Villen-Altamirano
Jorma Virtamo
Guillaume Vivier
Danny de Vleeschauwer
Raimund Vogl
Thiemo Voigt
Milan Vojnovic
Vinod Vokkarane
Artemios Voyiatzis
Stavros Vrontis
Can Vuran
Krzysztof Wajda
Marcel Waldvogel
Bernhard Walke
Jia Wang
Lan Wang

Lili Wang
Quanhong Wang
Tao Wang
Wenye Wang
Xin Wang
Xudong Wang
Zhipeng Wang
Gill Waters
Richard Weber
Steven Weber
Klaus Wehrle
Erik Weiss
Markus Werner
Attila Weyland
Jeff Wieselthier
Carey Williamson
Craig Wills
Larry Wilson
Lars Wolf
Tilman Wolf
Lloyd Wood
Ningning Wu
Christos Xenakis
Xing Xie
Yufeng Xin
Bagnan Xu
Kenan Xu
Kuai Xu

Lisong Xu
Shouhuai Xu
George Xylomenos
Stavros Xynogalas
Baijian Yang
Sruthi Yellamraju
Jian Yin
Y. Ying
Hongnian Yu
Lige Yu
Yinzhe Yu
Stathis Zachos
Paolo Zaffoni
Alberto Zanella
Hui Zang
Daniel Zappala
Djamal Zeglache
Soraya Zertal
Evangelos Zervas
Honggang Zhang
Jingyuan Zhang
Qingyu Zhang
Zhi-Li Zhang
Sencun Zhu
Yong Zhu
Artur Ziviani
Voula Zoi
Vasilis Zorkadis

Table of Contents

Wavelength Management

Multicast I

Wireless Network Performance I

Inter-domain Routing

Packet Classification and Scheduling

Services and Monitoring

Admission Control

MPLS and Related Technologies

Flow and Congestion Control

Performance of IEEE 802.11

Authentication and DOS Prevention

Multicast II

Energy Aspects of Wireless Networks

Optical Network Access

TCP Performance II

Routing in Ad Hoc Networks

Fault Detection, Restoration, and Tolerance

QoS Metrics, Algorithms, Architectures

Content Distribution, Caching, and Replication

A New Digital Signature and Certificate Architecture with Shortest Certification Path

Sung Keun Song[1], Hee Yong Youn[1], and Kang Shin Lee[2]

[1] School of Information and Communication Engineering
Sungkyunkwan University, 440-746, Suwon, Korea
kkskk103@skku.edu, youn@ece.skku.ac.kr
[2] Critical Information Infrastructure Protection Division
Korea Information Security Agency, 138-803, Seoul, Korea
kslee@kisa.or.kr

Abstract. The existing certificate architecture has two problems in terms of security and authentication. One is that there exists some possibility of certificate forgery by exploiting the collision problem associated with the hash algorithm used for signing. The other is that certification path complicates user authentication because it increases according to the distance between communicating users. In this paper we propose a new digital signature scheme and certificate architecture that solve the problems. It is achieved by using two-public key cryptography and a new certificate architecture. The proposed approach can be used without reconstructing the structure of the existing PKI system and always allows a certification path whose length is 1 regardless of the distance between the users. This is because each user confirms only the digital signature of root CA in the combined hierarchical-network infrastructure.

1 Introduction

Recently, user authentication has become an important issue in communication and e-commerce. The present target in authentication is implementation in PKI (Public Key Infrastructure) environment so that illegal connection and information leak can be prevented. Here the public key certificate of user is a crucial part, while the main concerns with the certificate are security and certification path. If a certificate is counterfeited by a malicious third party, catastrophic results may occur in the system. Also, the longer a certification path becomes, the larger authentication time and inconvenience to users are [1-4].

A public key certificate is based on digital signature. A certificate is known to the public after signed by a certificate authority (CA). The existing certificate architecture has two problems in terms of security and authentication. One is that possibility of certificate forgery exists. The other is that certification path complicates user authen-

This work was supported in part by 21C Frontier Ubiquitous Computing and Networking, Korea Research Foundation Grant (KRF - 2003 - 041 - D20421) and the Brain Korea 21 Project in 2003. Corresponding author: Hee Yong Youn

N. Mitrou et al. (Eds.): NETWORKING 2004, LNCS 3042, pp. 1–12, 2004.

tication. The reason why forgery of a certificate is possible is forgery of a digital signature of the certificate is possible. Digital signature algorithms have the collision problem of hash algorithm used for signing. Even though it is not easy for an adversary to attack a certificate by taking advantage of the hash collision problem, it is still possible that an adversary counterfeits the certificate. Here, an important problem is that if a certificate is counterfeited, one cannot prove illegality of the counterfeited certificate except the CA published the certificate. This hash collision problem may cause a devastating result especially when the counterfeited certificate is used for some important services. It is also a pivotal point to optimize the certification path of a certificate in PKI. A certification path is decided according to the distance between communicating users and user authentication depends on it in the existing PKI architecture; hierarchy, network, combined hierarchy-network. Therefore, the longer the certification path becomes, the more user authentication is complicated [1-10].

In this paper we develop the methods solving these problems by proposing a new digital signature scheme and certificate architecture using it. It is based on two-public key cryptography, and the proposed approach can be employed without reconstructing the structure of the existing digital signature scheme. Therefore, we can flexibly select the new scheme or the existing one according to the required degree of security. The proposed certificate architecture always allows a certification path whose length is 1 regardless of the distance between the users because each user confirms only the digital signature of root CA in the combined hierarchical-network infrastructure. Therefore, user authentication can be finished quickly. We provide detail of the proposed signature scheme and analyze its security.

The rest of the paper is organized as follows. Section 2 presents a brief description of digital signature and PKI architecture. Section 3 investigates the vulnerability of digital signature due to collision problem of the hash algorithm and fragile certification path of current PKI architecture. Section 4 proposes a new digital signature scheme and certificate architecture, and security of the scheme is evaluated. Finally, we conclude the paper in Section 5.

2   The PKI Architecture

2.1 Digital Signature

A digital signature is a pair of large numbers represented as strings of binary digits. Digital signature is computed using a set of rules and parameters with which identity of the signatory and integrity of the data can be verified. An algorithm is used to provide the way how to generate and verify the signature. The signature generation process makes use of a private key to generate a digital signature, while the signature verification process makes use of a public key corresponding to the private key. Each user possesses a private and public key pair. Public keys are known to the public using the certification of CA in general. Private keys are never shared. One can verify the signature of a user by using the user's public key. Only the possessor of a private key can generate signatures as long as the key has not been revealed [1].

A hash algorithm is used in the signature generation process to obtain a condensed version of message, called a message digest. The message digest is then input to the digital signature algorithm to generate a digital signature. The digital signature is sent to the intended verifier along with the message. The verifier of the message and signature verifies the signature using the sender's public key [1].

The same hash algorithm as the one used by the sender must be used in the verification process. The hash algorithm is specified in a separate standard, the Secure Hash Standard, FIPS 180-1 [2]. FIPS approved several digital signature algorithms implemented with the Secure Hash Standard. Similar procedures may be used to generate and verify signatures for stored as well as transmitted data.

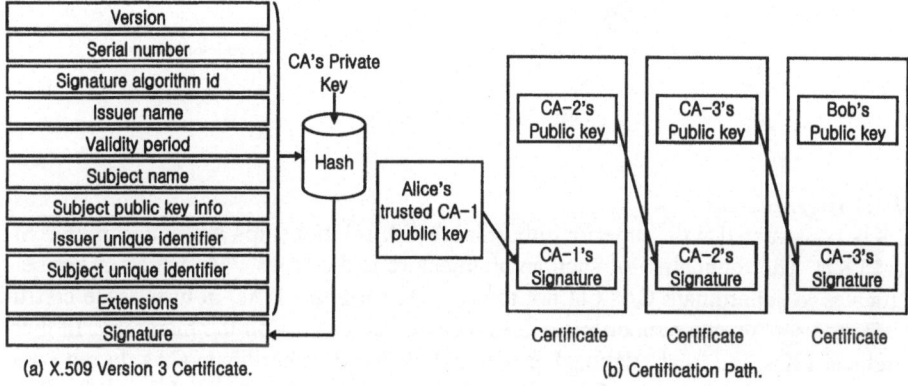

Fig. 1. Certificate architecture and certification path.

2.2 PKI Architecture

Figure 1(a) illustrates the X.509 v3 certificate. A certificate includes the issuer name, the subject name and the subject's public key, and is signed with the issuer's private key. For example, if Alice has Bob's certificate and knows the issuing CA's public key, she can verify Bob's certificate and then use Bob's public key to verify Bob's signature in any document. Certification path is a chain of certificates that use trust relationship between the CAs to determine if a certificate signed by a CA is trusted. This is illustrated in Figure 1(b); Bob has been issued a certificate by CA-3, which has been issued a certificate by CA-2, which in turn has been issued a certificate by CA-1. If Alice trusts CA-1 and knows its public key, she can verify each certificate in the certification path until she reaches Bob's certificate and verifies it. At that point, Alice knows Bob's public key and can verify his signature. CAs can certify each other in a systematic manner to form a PKI. A CA may be issued a certificate by another CA. Two CAs may issue each other a certificate; this is known as cross-certification, and the pair together is a cross-certificate.

PKI architectures fall into three configurations: hierarchy, network, combined hierarchy-network. Each configuration is characterized by the number of CAs, the trust relationship between the CAs, and where the PKI users place their trusts [1-4].

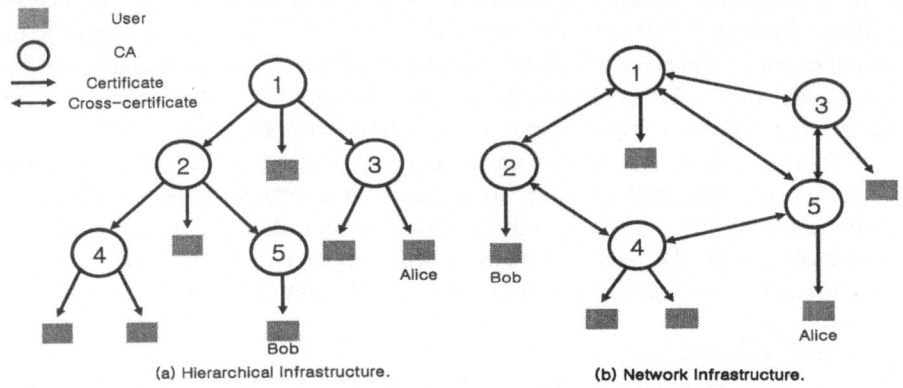

(a) Hierarchical Infrastructure.   (b) Network Infrastructure.

Fig. 2. PKI architecture.

2.2.1 Hierarchical PKI

PKIs constructed with superior-subordinate CA relationships are called hierarchical PKIs. The foundation of such an architecture is the "root" CA, which issues certificates to subordinate CAs but not to users. Subordinate CAs, in turn, issue certificates to users or other subordinate CAs located below them in the hierarchy. In hierarchical PKIs, the trust relationship is one-directional; subordinate CAs do not issue certificates to their superior CAs. Figure 2(a) shows an example of a hierarchical PKI with the root CA numbered as 1. The superior CAs impose rules governing the types of certificates their subordinates can issue; applications leverage these conditions to identify acceptable certificates. Since certificate suitability is implied by issuing the CA's identity, application-specific information can be omitted from the certificates. Every user knows the public key of the root CA, and any user's certificate may be verified by verifying the certification path that leads back to the root CA. For example, Alice verifies Bob's certificate, issued by CA-5, then CA-5's certificate issued by CA-2, and then CA-2's certificate issued by CA-1, the root, whose public key she knows.

The hierarchical PKI architecture has some advantages. The structure of many organizations such as government is largely hierarchical and trust relationships are frequently aligned with the organizational structure. A hierarchical PKI may be aligned with hierarchical directory names, and the search strategy of certification path is straightforward. Each user has a certification path back to the root; the user can provide this path to other users and every user can verify the path since all users know the root's public key.

Meanwhile, the hierarchical certification path architecture has some disadvantages. It is improbable that there will be a single root CA in the world, and therefore cross-certificates must exist at some level. Also, certification path verifiers must be able to cope with the topologies that are not entirely hierarchical. Moreover, compromise of the root private key is catastrophic because every certification path is compromised and recovery requires secure "out-of-band" distribution of the new public key to every user [1-4].

2.2.2 Network PKI

The traditional alternative to hierarchical PKIs is to create a network PKI or web of trust to connect CAs via P2P relationships. A CA in a network PKI can be a trust anchor, although users generally consider the CA issued their certificates as their trust anchor. In this architecture, CAs issue certificates to each other, and a pair of certificates describes a bi-directional trust relationship. Specific P2P CAs can specify any limitation of trust in the certificates they exchange between them. Figure 2(b) illustrates a network PKI. A user knows the public key of a CA near itself, generally the local CA that issued its certificate, and verifies the certificates by verifying the certification path leading back to the trusted CA. For example, Alice knows the public key of CA-5. There are several certification paths that lead from Bob to Alice, but the shortest path requires Alice to verify Bob's certificate issued by CA-2, then CA-2's certificate issued by CA-4, and finally CA-4's certificate issued by CA-5. CA-5 is Alice's CA, and she trusts CA-5 and knows its public key.

The network PKI architecture has the advantages that it is flexible and facilitates adhoc associations and trust relationships, and readily reflects bilateral trust relationships. It is likely that a national or worldwide PKI will evolve in an adhoc fashion involving isolated CAs, and this can be more easily accommodated in a network than a hierarchy. The CAs widely spread out but supporting the users working together with a high degree of trust can be directly cross-certified under a high trust policy that is higher than would be practical through a long, hierarchical chain of certificates. The CAs whose users communicate frequently can cross-certify directly, which can reduce certification path processing.

Perhaps the most compelling argument for a network PKI is that it is more convenient and natural for a certificate holder to place its trust in the local CA issued its certificate rather than a remote root CA, and make this the foundation of all trust relationships. Moreover, this simplifies the out-of-band secure distribution of the public key of CA. Also, recovery from the compromise of any CA's private key requires only that the new public key is securely distributed to the holders of the certificates issued from that CA and new certificates are generated for them. The network PKI has at least two disadvantages though [1-4]:

• Search of efficient certification path is complex.
• A user cannot provide a single certification path guaranteeing verification of its signatures by all other users of the PKI.

2.2.3 Combined Hierarchical-Network Federal PKI

The hierarchical and network PKI architecture are not mutually exclusive. Figure 3 illustrates a combined hierarchical-network federal PKI. There is a hierarchical path of certificates leading from the root CA to its subordinate CAs, and from each of these CAs to their subordinates, and so on, until every Federal end user is issued a certificate with a certification path from the root CA. Each Federal CA will have a single parent. There is one or more instances of the directory attribute certificate for the certificates issued by the parent. There is only one hierarchical path to the root

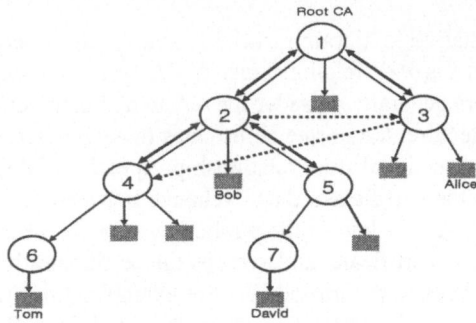

Fig. 3. Combined hierarchical-network architecture.

CA based on the directory attribute certificate. Other certificates held by a CA, which was generated by another issuer, will be posted in the directory in a cross-certificate pair. In parallel to the certificates hierarchically linking CAs to the root cross-certificate pair attributes also link those CAs. These parallel cross-certificate pairs are shown in Figure 3 as solid double-headed arrows. This allows client applications to operate with any Federal CA that perform certification path verification from the verifier's parent CA using the cross-certificate pair directory attribute. Federal CAs may cross-certify each other along the paths not parallel in the hierarchy. Optional cross-certificate pairs are shown in Figure 3 as dotted double-headed arrows. If Alice wishes to verify Bob's signature, she can find either a certification path that relies on her trust in her parent CA, CA-3, or Bob's certification path back to the root. In general, Federal PKI clients and applications may choose to follow either a certification path verification strategy that leads to the root CA, or back to their own CA. Because of the hierarchical cross-certificates, a certification path is guaranteed to exist from a client's own CA to every Federal certificate through the root CA, but there may also be much shorter paths [2].

3 Vulnerabilities of PKI

3.1 Digital Signature

As explained earlier, digital signature algorithms have two connoted hazardous factors in terms of security. They are the inherent security limitation of a digital signature algorithm and the collision problem of hash algorithm used for signing. The security of a digital signature algorithm depends on the security of public key cryptography. The collision problem of the hash algorithm, the second hazardous factor, is another factor limiting the security of digital signature.

A hash algorithm maps an arbitrary-length message to a fixed-length hash value, which must be a fast operation. On the other hand, the hash algorithm must be collision-resistant, i.e. it must be computationally infeasible to find a collision, which is a pair of different messages with the same hash value. However, collision cannot be avoided. MD5, SHA, and RIPEMD-160 are representative hash algorithms [7-9].

Many of the existing hash algorithms follow a design principle of Merkle-Damgard [10] shown in Figure 4. Essentially, this model simplifies the management of large inputs and production of a fixed-length output by using a function F, which is usually called a compression function. Given a compression function, a hash algorithm can be defined as repeated applications of the function until the entire message has been processed.

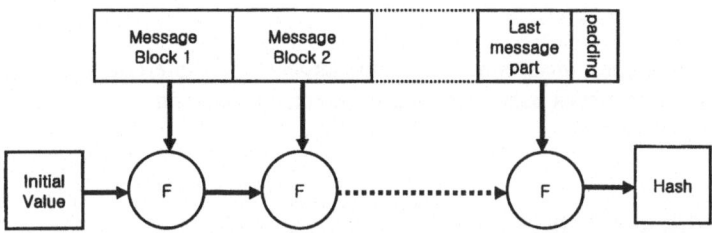

Fig. 4. Merkle-Damgard model.

In this process a message of arbitrary length is broken into blocks whose length depends on the compression function, and padded so that the size of the message becomes a multiple of the block size. The blocks are then processed sequentially, taking the result of hashing so far and the current message block as input, with the final output being the hash value for the entire message. The hash function is repeatedly applied to the message block and hash value of the previous blocks. The security of this scheme rests on the security of the F function. Note that the more the message size increases, the more the number of collisions per hash value increases exponentially. For example, we assume that one message block is 512 bits and the F function returns a 128-bit output. When a message needs 1 block, the number of collision is

$$\frac{1 \times 2^{512}}{2^{128}} = 2^{384}$$

When a message needs 2 blocks, the number of collision is

$$\frac{2^{128} \times 2^{384} \times 2^{512}}{2^{128}} = 2^{896}$$

When a message needs 3 blocks, it is

$$\frac{2^{128} \times 2^{896} \times 2^{512}}{2^{128}} = 2^{1408}$$

In general, for a message of n blocks,

$$\frac{2^{128} \times 2^{512n-640} \times 2^{512}}{2^{128}} = 2^{512n-128}$$

Namely, $2^{512n-128}$ messages per hash value have a same value. Because of this property, a third party can counterfeit a certificate signed by a CA. It is due to the collision problem of the hash algorithm, and many critical problems may occur in communication and e-commerce if that happens.

We can classify the attacks taking advantage of the collision problem of a hash algorithm into three types.

- An attacker researches the structural weakness of the hash algorithm to identify collision.

- An attacker accumulates digital signatures corresponding to each hash value for the life time of a public key of a target. For counterfeits, the attacker finds a digital signature from the database whose hash value is equal to that of the counterfeited message to a legitimate target message.
- An attacker counterfeits the message by modifying the counterfeited message until the hash value of it becomes same as that of the target message.

Among these types, the second case is the largest for counterfeiting a certificate. This is because a CA issues many certificates and an attacker can easily accumulate digital signatures of the CA.

3. 2 Certification Path

As explained in Section 2.2, the existing PKI architectures have a disadvantage in terms of certification path. It is that a certification path increases according to the distance between communicating users. This problem may complicate user authentication and give vexation to the users.

4 The Proposed Certificate Architecture

We have explained vulnerability of PKI. This section proposes a new digital signature scheme and certificate architecture solving the problem. The new digital signature scheme uses a cryptographic algorithm employing two different public keys. In this paper we call it "two-public key cryptography". The basic idea is to hide the hash value of a certificate an issuer signed using the two-public key cryptography. In the case of users the validity of the digital signature of a certificate is confirmed by the digital signature of root CA and public key of root CA. In the case of CAs the validity of the digital signature of a certificate is confirmed by the semipublic key of the issuer. First, we explain the two-public key cryptography. Then, we propose the new digital signature scheme.

4. 1 New Digital Signature Scheme

4.1.1 The Two-Public Key Cryptography

Figure 5 shows the structure of the proposed two-public key cryptography. Note that if the private key is used to encrypt something using Algorithm-B, only public key-2 can decrypt it. That is, the public key that can decrypt the message varies according to the algorithm used for encryption.

We show an example of two-public key cryptography using the RSA and the El-Gamal scheme, the two representative public key cryptography algorithms. First, we review the two.

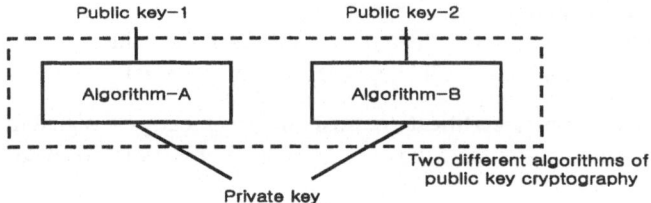

Fig. 5. The structure of two public key cryptography.

The RSA cryptography, named after its inventors R. Rivest, A. Shamir, and L. Adleman, is the most widely used public-key cryptography. It may be used to provide both secrecy and digital signatures, and its security is based on the intractability of the integer factorization problem. Each user creates an RSA public key and a corresponding private key. The users do the following [11]:
1. Generate two large random (and distinct) primes p and q, each roughly the same size.
2. Compute $n=pq$ and $\phi=(p-1)(q-1)$
3. Select a random integer e, $1< e < \phi$, such that $\gcd(e, \phi)=1$.
4. Use the extended Euclidean algorithm to compute the unique integer d, $1< d < \phi$, such that $ed\equiv 1(\bmod \phi)$.
5. The public key is (n, e); private key is d.

The ElGamal public-key encryption scheme can be viewed as Diffie-Hellman key agreement in the key transfer mode. Its security is based on the intractability of the discrete logarithm problem and the Diffie-Hellman problem. Each user creates a public key and a corresponding private key. The users do the following [12]:
1. Generate a large random prime p and a generator α of the multiplicative group Z_p of the integers modulo p.
2. Select a random integer a, $1\le a \le p-2$, and compute $y= \alpha^a \bmod p$.
3. The public key is (p, α, y) private key is a.

In above, we can recognize that if the prime p of the ElGamal and the ϕ of the RSA have a same value, the private key of the RSA and ElGamal are same. If the a of the ElGamal is denoted by the d of the RSA, public key-1 is (n, e), and public key-2 is (p, α, y), and the common private key is d in the proposed two-public key cryptography. In this way, we can construct two-public key cryptography using the RSA and ElGamal scheme. Of course, we can easily construct various two-public key cryptography scheme using any two different public key cryptographies.

4.1.2 The New Digital Signature Scheme with the New Certificate Architecture

P_A: a public key of algorithm-A; be known to all objects of PKI
P_B: a semipublic key of algorithm-B; be known only to the CA's
P_{AB}^{-1}: a private key of the two-public key cryptography
$\{ \} AP_{AB}^{-1}$: encrypt or decrypt the private key using algorithm-A

C: a certificate including a digital signature of a CA

H: a hash function that extends the input regardless of the value

h: a hash function that reduces the input regardless of the extent

The issue process of a certificate consists of two processes; signature generation process and certificate process. Signature generation process is executed by an issuer and certificate process is executed by root CA.

The issue process of a certificate handled by a general CA is as follows. First, the CA calculates a hash value of the random number (RN), H(RN). Here, the extent of the H(RN) has a fixed block size. The CA calculates a hash value where the H value is added to the certificate, h({M}K, H(RN)). When h is calculated, the H value is put on a specific block of the certificate that the CA selected. The CA generates a digital signature by encrypting the h value, the block position, and the random number using algorithm-B and its own private key. Digital signature of the CA is as follows.

$$\text{Digital signature: } \{h(\{M\}K, H(RN))\|\text{block position}\|RN\}\, BP_{CA\text{-}A}^{-1}$$

Thereafter, the CA requests certificate signature from root CA by sending the certificate to it. Figure 6(a) shows the signature generation process.

The certificate process handled by root CA is as follows. The root CA searches a semipublic key of CA-A, P_B, from a database using the ID of CA-A, and then decrypts the digital signature to obtain the block position and RN. Thereafter, the root CA calculates a hash value, h({M}K, H(RN)), by using the block position and the RN, and then compares the hash value with the h({M}K, H(RN)) value which is part of the decrypted digital signature of CA-A. If the values are same, the root CA calculates a hash value, h({C}K). The root CA generates a digital signature by encrypting h({C}K) by its own private key and algorithm-A for certifying the digital signature of CA-A. After the root CA calculates h(M), the value and digital signature of itself are attached to the certificate of CA-A. The root CA sends the certificate signed by itself to CA-A. As soon as CA-A receives the certificate from root CA, CA-A issues a certificate to the user.

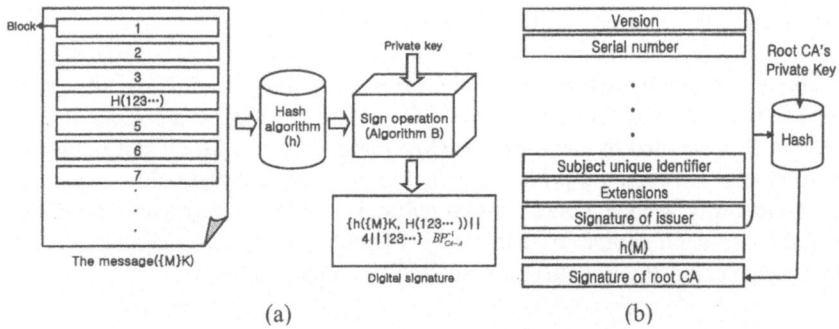

(a)   (b)

Fig. 6. The signature generation process and the new certificate architecture

4.2 Optimal Certification Path

The new certificate architecture that we propose is shown in Figure 6(b). The new certificate architecture is obtained by adding h(M) and digital signature of root CA to the existing certificate architecture. The system with the new certificate architecture needs additional certificate of the semipublic key of each CA. Certificates of semipublic keys are commonly used only between the CAs. We assume that all CAs share each certificate of semipublic key of all CAs. Then, all certificates have an optimal certification path regardless of the distance between the users in the combined hierarchical-network infrastructure. For example, in Figure 3, the length of certification path between Tom and David based on the existing certificate architecture is 5 for CA 6 and CA 7 building a cross certificate. If they have a cross certificate, the length of certification path is 2 with the existing certificate architecture. However, whether a cross certification exists or not between CA 6 and CA 7, the length of certification path is always 1 with the new certificate architecture. For all other cases the length is still 1 because each user in the PKI confirms only the hash value, h(M), and the digital signature of root CA. Therefore, the length of certification path is always 1 with the new certificate architecture. In case a user doubts other's certificate, the user can request verification of the certificate to a CA near itself, generally the parent CA issued its certificate. Then, the CA confirms the certificate using the semipublic key of the issuer of the certificate.

4.3 Security of the Proposed Certificate Architecture

The new proposed certificate architecture solves the counterfeit problem presented in Section 3. That is, any user of PKI cannot counterfeit a certificate by taking advantage of the collision problem of the hash algorithm in the new certificate architecture. Without the proposed scheme a malicious user can counterfeit a certificate using the collision problem. With our scheme, a malicious user cannot perfectly counterfeit a certificate since the user is not able to know the RN and block position of the signature of the certificate. The part where forgery is possible in the new certificate architecture by taking advantage of the collision problem of the hash algorithm is the digital signature of root CA. For this reason, the hash value, h(M), is added to the certificate. It is difficult to counterfeit the certificate such that h(M) and h(C) may be valid. However, because of the possibility of forgery, in case a user doubts a certificate, the user must request verification of the certificate to a CA near itself.

We need to compare security of two-public key cryptography and earlier public key cryptography. Note that security of any cryptographic algorithm is influenced by many factors such as difficulty of the mathematical problem of the cryptographic algorithm, complexity of the cryptographic algorithm, and key length, etc. If the securities of the two different cryptographic algorithms employed in the two-public key cryptography are similar, the security of the two-public key cryptography will be similar to the security of each of the two cryptographic algorithms since each of them is based on different problem of mathematics. Therefore, a system designer must design the two-public key cryptography using two different public key cryptographies of the same level of security.

5 Conclusion

In this paper we have proposed a new certificate architecture and digital signature scheme solving the collision problem of hashing required in the existing digital signature algorithms. As a result, the security of the new certificate architecture is not limited by the hash algorithm. The new digital signature scheme applied to the new certificate architecture can use a hash algorithm, which allows fast operation while providing high security. The new digital signature scheme can also be used without reconstructing the structure of the existing digital signature scheme.

The new certificate architecture allows an optimal certification path regardless of the distance between the users in the combined hierarchical-network infrastructure. If the e-commerce and communication system share certificates of the semipublic key of all CAs, user authentication can be done effectively with which illegal access is impossible. Therefore, we anticipate that the new certificate architecture can significantly activate e-commerce by increasing the security of transactions and effectively processing user authentication. In the future we plan to investigate the performance of the proposed scheme using various combinations of public key cryptographies.

References

1. William, T., Nelson E., Polk, Hastings, Ambarish Malpani.: Public Key Infrastructures that Satisfy Security Goals, IEEE Internet Computing. (2003)
2. William, E., Burr, Noel, A., Nazario and W. Timothy Polk.: A Proposed Federal PKI Using X.509 V3 Certificates. NIST. http://csrc.nist.gov/nissc/1996/papers/NISSC96/paper042/pkipap1.pdf
3. Adams, C., Cain, P., Pinkas, D., Zuccherato, R.: Internet X.509 Public Key Infrastructure Time Stamp Protocol. draft-ietf-pkix-time-stamp-00.txt. (1998)
4. Housely, R., Ford, W., Polk, W., Solo, D.: Internet X.509 Public Key Infrastructure. IETF RFC 2459. (1999)
5. National Institute of Standards and Technology (NIST).: Digital Signature Standard. FIPS PUB 186-2. (2000). http://csrc.nist.gov/publications/fips/fips186-2/fips186-2-change1.pdf
6. National Institute of Standards and Technology (NIST).: Secure Hash Standard. FIPS PUB 180-1. (1995). http://www.itl.nist.gov/fipspubs/fip180-1.htm
7. Dobbertin, H.: The status of MD5 after a recent attack. RSA Laboratories. CryptoBytes, 2(2). (1996)
8. Eastlake, D. 3rd, Jones, P.: US Secure Hash Algorithm 1 (SHA1). RFC 3174. (2001) http://www.faqs.org/rfcs/rfc3174.html
9. Keromytis, A., Provos, N.: The Use of HMAC-RIPEMD-160-96 within ESP and AH. RFC 2857. (2000)
10. Damgard, I.B.: A design principle for hash functions. Advances in Cryptology-Crypto '89, Lecture Notes in Computer Science, vol. 435, Springer-Verlag, (1990) 416-427
11. Rivest, R., Shamir, A., Adleman, L.: A Method for Obtaining Digital Signatures and Public Key Cryptosystems. Communications of the ACM, (1978) 120-126
12. ElGamal, T.: A public key cryptosystem and a signature scheme based on discrete logarithms. IEEE Trans. Info. Theory, IT-31, No. 4, (1985) 469-472

Inter/Intra Core Network Security with PKI for 3G-and-Beyond Systems

Georgios Kambourakis, Angelos Rouskas, and Stefanos Gritzalis

Department of Information and Communications Systems Engineering
University of the Aegean, Samos 83200, Greece
{gkamb, arouskas, sgritz}@aegean.gr

Abstract. With a large number of different heterogeneous network technologies (e.g. UMTS, WLAN, HIPERLAN) and operators expected in the future mobile communications environment, that should frequently and seamlessly interwork with each other and a constantly increasing population of communication parties, capturing the full benefits of open channel key transfers and scaling public key methods requires Public Key Infrastructure (PKI). In this paper, we discuss and investigate different ways to take advantage of a proposed PKI system. Focusing on UMTS Release 6 IP multimedia subsystem, we analyze the ongoing 3GPP specifications and its limitations and examine how PKI can provide robust security solutions to both 3G-and-beyond inter/intra core network and the mobile user. Public key security mechanisms to protect operator's core networks seem to gain ground and protocols like IPsec and SSL, seconded by PKI, can support the continuous growth of diverse technologies and solve inter-operator many-to-many modeled trust relationships. From the user's side we present solutions, which far enhance authentication procedures and end-to-end communication model trust. We argue that PKI can become a promising candidate, which offers the competitive framework to overcome symmetric key based security inefficiencies and provide powerful solutions to protect both network core signalling and user's data from potential intruders.

Keywords: PKI; Mobile Networks; UMTS; Network Domain Security; SSL/ TLS; Ipsec.

1 Introduction

An identified weakness in 2G systems security architecture is the absence of security in the core network. For instance, cipher keys are used to protect the traffic on the radio interface, but those keys are themselves transmitted unprotected between different networks. Originally and up to UMTS Release 99, this was not a problem, since 2G Signalling System Number 7 (SS7) networks were closed networks with very little interworking among different 2G operators and between 2G operators and the Internet.

Nevertheless, in a future wireless communication environment, like 3G and beyond, that will require frequent interworking of many different network technologies and providers there will also be a greater need for advanced security protection. Moreover, the introduction of IP, used not only for signalling traffic, but also for user

N. Mitrou et al. (Eds.): NETWORKING 2004, LNCS 3042, pp. 13–24, 2004.

traffic, as the network layer in the GPRS backbone network and later in the UMTS network domain (Figure 1), raises further reasons to worry about. Although this does not mean that inter/intra core network signalling would be carried over open connections, the involvement of many more "players" certainly brings a shift towards easier access to core network traffic.

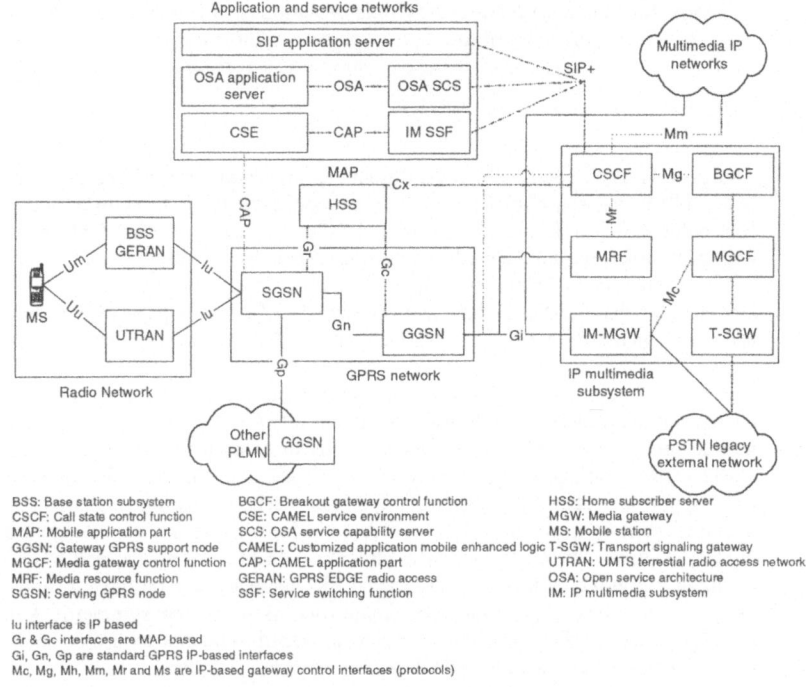

Fig. 1. UMTS Rel. 6 all-IP network architecture

Current mobile network standards perform user authentication, signalling and data encryption, as well as message integrity protection, by utilizing only symmetric key methods. However, as communication is envisaged to change from second generation (2G) person-to-person model to fourth generation (4G) machine-to-machine model, there is greater demand to provide more flexible, reconfigurable and scalable security mechanisms that can advance in a many-to-many trust relationship model.

PKI is gradually being introduced in the market, and its adaptation in future mobile networks will substitute long-term symmetric key relationships, with a flexible, reconfigurable and scalable public key based system. This will not only provide the appropriate level of inter/intra operator trust, but it will also offer solutions that far enhance user-to-network confidence and end-to-end security options.

The rest of this paper is organized as follows. In Section 2, we provide an overview of the current 3G-inter/intra security options and explain how PKI can adapt to existing architecture. Section 3 deals with PKI-proposed solutions that provide inter/intra operator trust, while Section 4 discusses user-to-network and end-to-end security. The paper is concluded in Section 5.

2 3G Core Network Security Specifications and PKI

2.1 Outline of 3G Inter/Intra UMTS Network Security

Global Mobile System (GSM) and Universal Mobile Telecommunication System (UMTS) networks, (Figure 1) use Mobile Application Part (MAP) protocol for the exchange of signaling messages between network Elements (NEs). User profile exchange, authentication, and mobility management are performed using MAP. MAP runs typically over the SS7 protocol stack. For instance, the signaling between the mobile, Serving GPRS node (SGSN) and Gateway GPRS support node (GGSN) to the Home Subscriber Server (HSS), and also the SMS message centre all consist of SS7 signaling.

3rd Generation Partnership Project (3GPP) has also defined a mechanism for protecting the MAP protocol at the application layer [1],[2]. MAP may also be protected at the network layer when IP is used as the transport protocol. However, when internetworking with networks using SS7-based transport is necessary, protection at the application layer shall be used. For this reason a new protocol header has been developed to protect MAP operations, much in the same way as the Encapsulating Security Payload (ESP) protocol protects IP packets. This new protocol is called MAPsec. In protection mode 2 of MAPsec, both confidentiality and integrity are protected, while in protection mode 1, only integrity is protected. When protection mode 0 is used there is no protection. While MAP runs over SS7, MAPsec and Internet Key Exchange (IKE) always run over IP. Therefore, it is assumed that nodes implementing MAPsec always have IP connectivity in addition to SS7 connectivity. In the 3GPP architecture MAPsec is typically running between two different network operators and the same Security Associations (SAs) are shared by a number of NEs. The necessary MAPsec-SAs between networks are negotiated between the respective Key Administration Centres (KACs) of the networks.

On the other hand, for native IP protocols, as in the GPRS backbone network, security shall be provided at the network layer. The security protocols to be used are the IETF defined IPsec suite [3]. The UMTS network domain control plane is sectioned into security domains, which typically coincide with operator borders. The borders between the security domains are protected by Security Gateways (SEGs) as shown in Figure 2.

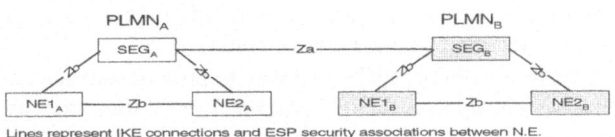

Lines represent IKE connections and ESP security associations between N.E.

Fig. 2. Network Domain architecture for IP-based protocols

All network domain traffic shall pass though a SEG before entering or leaving the security domain. Consequently, IP Network Domain Security (NDS/IP) will only support tunnel mode IPsec SAs, ESP and main mode. SEGs shall offer capabilities for secure storage of long term keys used for IKE authentication, so NDS/IP will only support Internet Security Association and Key Management Protocol (ISAKMP) SAs with pre-shared keys [4].

Only the inter-security domain SA IKE negotiations over the Za interface shall be mandatory, while the Zb interface is optional. Concluding, there is normally no NE-to-NE direct interface for NE belonging to different security domains.

2.2 PKI and Mobile Networks: A Viable Perspective

It was mentioned that in 3GPP's proposals and technical specifications the basic tool in protection of 3G-network domain traffic is IPsec protocol. The critical issue is key management: how to generate, exchange and distribute keys needed by algorithms that are used to provide confidentiality and integrity protection. Currently, agreements on keys and security associations are carried out on a bilateral basis between operators. However, as the number of network elements of each operator increases and the interworking between a high number of networks of different technologies will be frequent, a more scalable solution would be to replace those relationships with a PKI [5],[6]. This means that secure communications can be achieved without having to generate and distribute long-term secret keys.

PKI technology is gradually being introduced in the market. Projects like ASPeCT [7] and USECA [8], Third Generation Partnership Project (3GPP) discussion papers especially for UMTS R6 [9] as well as other papers [10] foresee that evolution. The eNorge 2005 strategy calls for a shared PKI for Norway, while advanced standards such MexE, WAP and i-mode from NTT DoCoMo have moved forward to introduce public key methods. Successful wireless PKI implementations and solutions from companies like Sonera Smarttrust, Lucent Technologies and Entrust, strengthens the assertion that PKI has become an acknowledged and promising component of standards.

Nevertheless, weighting up an asymmetric key system against a symmetric one, we note the following:

- The number of keys needed in a symmetric key system with n network elements communicating with each other is $O(n^2)$. On the other hand, in a public cryptosystem, the corresponding need for keys is $O(n)$. Therefore, when n increases, the costs in terms of key generation and distribution associated with the introduction of a new network element are quite different. In the symmetric model, we need to establish n new secret keys, while in the asymmetric case we only need 2 new keys (private + public) for any new network element.
- Pre-shared secrets are a rather inflexible way to provide authentication. A properly designed PKI, which supports digital certificates, will offer more dynamic, flexible and scalable mechanisms to issue certificates for new network elements and to revoke certificates that are no longer valid.
- One basic requirement and assumption in both GSM and UMTS, is that the Home Network has to trust the Serving Network, e.g. for the Authentication and Key Agreement (AKA) procedure. However, in future systems, where many different technologies, owned by different network operators, must frequently and seamlessly interwork, this is no longer the case. By introducing a Trusted Third Party (TTP) the requirement for bilateral trust is reduced.

- PKI can be used for authentication and symmetric key encapsulation and transport procedures, while derived symmetric session keys can be used to support confidentiality. Thus, we can by-pass the known public key cryptosystem disadvantages of key lengths and computational load.
- From the user scope, the implementation of public key algorithms in Mobile Stations (MSs) had been considered to be resource demanding. However, the increased processing requirements of IP capable terminals have driven towards high power computational platforms, which are now becoming ordinary in mobile devices.
- Furthermore, as IP-based networks are introduced to serve a large variety of applications, that may involve many and different network/service operators, complex and flexible communication relationships are necessary, which in turn demand a complex trust model. In many cases, the communication parties may not have pre-arranged security agreements. So, if unknown partners wish to perform mutual authentication and establish session keys, a public key based digital signature that is supported by a PKI will satisfy security needs. For example, a Session Initiation Protocol (SIP[1]) registration server, either proxy or redirect,[11] may not share any symmetric key with the User Equipment (UE). Instead, a digital signature may be an appropriate way to authenticate the proxy server.

2.3 Adaptation of PKI in Mobile Networks

Certainly, the support of asymmetric key services by a mobile network requires the adaptation of some PKI elements, which are not necessarily part of the current 3G-network core. Integration between 3G mobile systems and PKI has not been yet standardized, although, most recently 3GPP discussion papers deal with that particular subject [9],[12]. Figure 3 depicts the necessary PKI elements that should be included in the UMTS architecture. More specifically, we assume the following:

- There is some sort of Certification Authority (CA) per Public Land Mobile Network (PLMN) Operator, which issues and revokes certificates. Likewise, a presumable Attribute Authority (AA) can issue short-lived Attribute Certificates (ACs) [13] for the subscribers.
- There is at least one digital certificate database, which stores all the digital certificates and is being managed by the PLMN's CA.
- There is at least one revoked certificates database (CRL-database), which is being managed by the PLMN's CA and is accessible from all network elements that belong to the network core.
- Web servers or FTP servers can be used to store certificates and CRLs. Certificate revocation can be periodic or Online Certificate Status Protocol (OCSP) based. Revocation is generally a hard problem to run into. However, in that case, certificate revocation for core network elements can be handled manually as this proce-

---

[1] SIP is an application-layer, text-based, client-server control protocol that can establish, modify, or terminate user sessions. It has been chosen by 3GPP as the protocol for multimedia application in 3G mobile networks. See IP multimedia subsystem in Figure 1, where CSCF represent a SIP server. Upon registration every user is given a SIP URL of the format *sip:username@domainname*.

dure will happen infrequently. Moreover, subscriber's certificates revocation can be handled by International Mobile Subscriber Identity (IMSI).

- CAs which belong (or collaborate with) to different PLMN's issue (off-line) cross-reference certificates for inter-PLMN trust relationships. For instance in the case of two PLMNs with the corresponding certification authorities CA_a & CA_b, CA_a issues $Cert(CA_a)CA_a$[2] (the root certificate) and $Cert(CA_b)CA_a$. Respectively, CA_b issues $Cert(CA_b)CA_b$ and $Cert(CA_a)CA_b$.
- Cross-Reference certificates are cached in local Security Gateways SEGs (which probably implement firewall policies among other things) on the borders of IP security domains. Every PLMN can use one or more SEG, in order to balance inter network traffic.
- Every network element possesses a key pair (private, public), and the corresponding digital certificate (intra-operator trust). NE's private key and the public key of the local CA are stored locally in a secure manner.

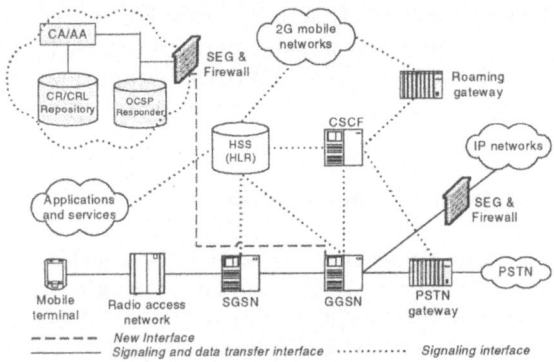

Fig. 3. General UMTS architecture Rel. 5 and PKI

If we are planning to extend PKI usage to the user, primarily for authentication, symmetric key encapsulation and support of certificates, we can assume the following:

- The User Services Identity Module (USIM) smart card should be a crypto-card with good pseudo-random (or random) generation capabilities and in-built crypto accelerator chip.
- Every subscriber possesses an asymmetric key pair and his private key is stored in his USIM card. The keys are associated with the user at registration time. IMSI handles revocation of subscriber's certificates, while attribute certificates are short-lived and will therefore not need to be included in any CRL.
- Furthermore, the USIM card is pre-loaded with all the CA's (root) public keys, which exist in the particular PLMN.

---

[2] $Cert(X)_y$ = Public key certificate of X with format X.509v3 (or subset) issued by Y.

3 PKI-Based Intra/Inter Network Domain Security

3.1 IPsec, IKE, and SAs Establishment

With network domain security we mainly mean secure communications between network elements. Thus, by introducing a PKI to a future wireless network we can use powerful protocols to protect signaling and user traffic both between inter-network and intra network elements.

Two connections have to be protected as shown in Figure 2:
– Za or SEG-to-SEG (inter-operator security),
– Zb or SEG-to-NE and NE-to-NE (intra-operator security).

One candidate for this task is IPsec [14],[15] and IKE in particular. As we already mentioned, 3GPP currently uses pre-shared secrets for IKE phase I. This means that each NE has to be configured with a password that is associated with the remote system's IP address being authenticated. Note, however, that the keys to be used for encryption and authentication (SKEYID_*), after the completion of phase I, have been generated solely based on the peer's IP address [15]. So, in scenarios where the IP address is dynamic, the responder cannot maintain pre-shared secrets indexed by an IP address that may not be known at that time. Remote access solutions are an example where the initiator's IP address may be different for each connection (road-warrior cases and IKE's phase I main mode). Additionally, the main drawback in pre-shared secret key authentication is the lack of a secure and scalable mechanism for exchanging pre-shared secret keys. That is appropriate only in a rather small-scale environment with a restrained number of systems, in which the set of peers is known in advance. However, if a pre-shared secret key is compromised, there is no universal method to alert the peer and launch a replacement.

An alternative solution based on PKI, can overcome these shortcomings. In that case, IKE is used for key exchange over the Za, Zb interfaces, while the authentication could be based on digital signatures with certificates instead of pre-shared secrets. Consequently, the generation of the keys to be used for encryption and message authentication is based solely on the peer's nonce and Diffie-Hellman key value (SKEID = Pseudo_random_function (Nonce_i | Nonce_r, DH_Key). For system authentication, a certificate request can be included to obtain the public key of the peer if the initiator does not already have it. The peer must have the other's public key to validate the signature and authenticate the peer in the third exchange (IKE messages 5 & 6). Also, the use of certificates in such a scheme can provide for non-repudiation in key exchange e.g. when attribute certificates are used [15].

3.2 Introducing SSL/TLS

Another solution, which benefits by the incorporation of PKI, is the use of SSL/TLS to protect communications between security gateways and probably between NEs. Authentication for the corresponding NEs during the handshake procedure is mutual, and is performed by exchanging their certificates. SSL/TLS has many of the advantages of IPsec and the successful introduction of the protocol in the wired Internet has proved its usability and effectiveness. Likewise, SSL/TLS can be part of an all-IP mobile environment, as it runs above TCP/IP and below higher-level protocols such as HTTP or FTP and consequently the TCP header is not encrypted.

For instance, using Performance Enhancing Proxies[3] (PEPs) in 3G in parallel with IPsec, end-to-end security can be compromised, as the PEP module must decode the encrypted IP protocol headers [16]. So either the packets bypass the PEP module and are directed to mobile hosts, in that case the connection will not benefit from any performance enhancement, or the user should trust the PEP in the middle (it is part of the IPsec's security association). In general however, the end system cannot trust PEPs.

3.3 UMTS IP Multimedia (IM) Subsystem

The case of application-level registration in UMTS Release 6, discussed below, proves even more the necessity for flexible and scalable public key security mechanisms. In an all-IP network, MS conducts two types of registration [11],[17].

Bearer-level registration (and authentication), where the MS registers with the GPRS network following the standard UMTS routing area update or attach procedures [18]. During that procedure the MS obtains an IP address and discovers (sending a DNS query) the Proxy CSCF (P-CSCF). The user may be at his home or at foreign network. The P-CSCF provides basic multimedia session support as well as functioning as a firewall to the IP multimedia (IM) subsystem.

Application-level registration (and authentication), where a Serving CSCF (S-CSCF) is assigned to MS. MS sends a REGISTER message to the P-CSCF, and this is relayed to an Interrogating CSCF (I-CSCF) in the home network (the home network can be found by the P-CSCF using the IMSI or SIP URL of the user). Thus I-CSCF acts as a gateway for serving networks. The I-CSCF in the home network communicates with HSS and retrieves the user's data from HSS's IM database (HSS is an HLR with new capabilities added to support IM subsystem. For example HSS may generate, store and manage security data and policies used in the IM subsystem). Then it selects an S-CSCF to deal with the requested service, as the latter has access to the resources needed to create services, such as video servers and media gateways.

We notice that a variety of network elements take part in application-level registration, especially when P-CSCF resides in a visited network. Moreover, the data being transferred are important to both the user and the network. Among others (e.g. CSCF-MS security parameters sent from the HSS to the I-CSCF), SIP messages may contain information a user or server wishes to keep private. For example, the headers can reveal information about the communication parties, or other confidential information. The SIP message body may also contain user information (media type, addresses, codec and ports, etc.) that should not be exposed.

Security should aim to keep network and user data private and prevent SIP sessions from being set up or modified by others masquerading the identity of the real user. As the confidentiality and integrity protection of SIP signaling is provided in a *hop-by-hop* fashion and SIP does not provide specific security features for that, protection relies on network level (IPsec) or transport-level (SSL) security. Note that hop-by-hop mechanisms are needed because intermediate elements may perform SIP processing by reading and/or writing some parts of SIP messages.

---

[3] PEPs improve the performance of wireless TCP connections between the core network and mobile hosts. Usually PEPs are implemented in the Radio Network Controller (RNC).

It is worth discussing the aforementioned Za and Zb interfaces in the case of SIP. 3GPP uses IPsec to secure communication between SIP entities that have preconfigured and thus have quite static security associations and policies. On the other hand, consider a roaming user who wishes to connect to a P-CSCF that resides in an Internet Telephony Service Provider (ITSP). No guarantee that secure transport will be used on the entire end-to-end path can be provided to the user.

To the best of our knowledge, the most recent version of the SIP specification [19] defines a way to indicate that a resource (e.g. a server or a user) should be reached securely using SSL. In that case a new type of URI (for example *sips:test@secure.com*) designates the use of SSL. This is well suited to architectures in which hop-by-hop security is required between hosts with a more dynamic and flexible security association using public key mechanisms. The incorporation of PKI can solve such problems, providing a scalable model, when interworking among different operators and diverse technologies (e.g. UMTS and IEEE 802.11) is required.

4 PKI and Mobile User Enhancements

From the user's side, a PKI can support the appropriate reconfigurable infrastructure, which offers great flexibility and scalability in an all-IP wireless environment. In this fashion, authentication and end-to-end security solutions can be provided to far enhance user's trust in a continuously evolving environment.

It is still a common misbelieve, that mobile devices are not ready for 'expensive', in terms of memory and processing power, public key computations. However, that is partially true, since contemporary mobile devices are featuring advanced architectures with processors up to 400 MHz, memory capacities of 64MB RAM and 32MB ROM, support for applications and strong operating systems. Besides that, these trends has also driven smart cards toward more advanced architectures, all the way to where we are beginning to see 32-bit RISC-based ARM processors in smart cards. These cards based on such modern chips from companies like Atmel and Infineon are just appearing in the market, and they can effectively store and protect the subscriber's private key, generate good pseudo-random values and take over of symmetric key (un)wrapping functions. The mobile's device processor can efficiently take over the rest of the calculations, needed by protocols like IPsec and SSL.

4.1 Providing Public Key Based Mutual Authentication

In an IP-enabled mobile device with the aforesaid characteristics, IPsec, can effectively secure signalling and user traffic, therefore, providing a secure end-to-end channel. Once again, IKE with authentication based on digital certificates will be used instead of pre-shared secrets. Road-warrior cases can also be effectively authenticated using this scheme.

For example, consider the following scenario. A business employee has IPsec-based Virtual Private Network (VPN) client software installed on his laptop, which is connected to his wireless network provider, via his mobile phone. Also assume that the employee is roaming to a foreign (serving) network. Upon connection, the employee is being authenticated by IKE sending its digital certificate and receiving

SGSN's certificate and cross-reference certificate. When IPsec's SAs have been created, VPN client filters the traffic, watching for IP packets destined to the employee's head office. It allows any traffic not going to the head office to pass unprotected. When however the client spots a packet that is addressed to the head office intercepts it. It then uses IPsec services to transmit the packet securely and to assure that all traffic back from the head office to him is also secure.

The first thing that the VPN client does is to establish a bi-directional IPsec Security Association (SA) with the head office server. IKE (ISAKMP) [20] defines the framework how the VPN client and server set up security associations. It does however require the use of digital signatures within the authentication section. This means that the VPN client and server must have IPsec public key certificates to be able to establish a security association.

Taking into account the aforesaid technological trends, SSL can provide for user authentication and end-to-end security [21],[22]. Until now performance considerations in using SSL in a resource-constrained environment drove wireless designers to choose a different, incompatible and insecure gateway oriented security protocol for their mobile clients, like in the case of WAP. The ASPeCT project has demonstrated that public-key authentication is possible and GSM and UMTS applications can co-exist on a single smart card. A recent study has also shown the feasibility of SSL in handheld wireless devices [22] while relevant work showed that SSL's handshake protocol time could be improved up to 5.7X times [23].

4.2 Support of Subscriber's Certificates

Another "added value" PKI service could be the support of subscribers' certificates [12]. Figure 4 depicts a presumable network architecture where a subscriber can obtain (attribute) certificates [13] regardless of the network (home or serving) he is connected to. Other alternative architectures are also possible for implementation; e.g. direct connection of CA/AA with GGSN or SGSN or even CSCF, although the proposed solution has minimum effects on existing 3G core network nodes.

The AAA server in the subscribers home network, provide the actual authentication for the subscriber. Extensible Authentication Protocol - Authentication and Key Agreement (EAP-AKA) [24] is an appropriate, directly applicable in UMTS, access-independent, user authentication method to support this architecture, as it provides a way to exchange AKA authentication messages encapsulated within the EAP protocol. Note that 3GPP has chosen EAP-AKA to support authentication of a subscriber who access WLAN subsystems. Of course SSL and/or IKE connections can be used to support such architecture. In either case, MS has to support new authentication mechanisms e.g. EAP-AKA or SSL.

In the first case, MS has to discover the appropriate (visited or home) Certificate Gateway (CG) and send a certificate request to it using EAP-AKA. CG acts as a certificate-provisioning gateway for the MS. EAP messages will be routed to an AAA server in the subscriber home network probably through a local AAA proxy. CG will wait for a reply from AAA server indicating successful or abortive authentication and other possibly required subscriber data, retrieved from HSS. If certificate issuing for this subscriber is endorsed, CA or AA generates and signs the certificate, updates its database and returns the certificate back to CG.

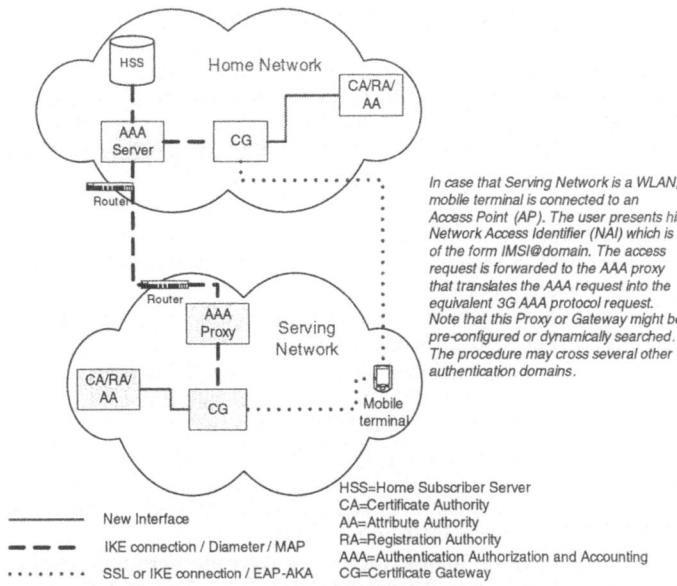

In case that Serving Network is a WLAN, mobile terminal is connected to an Access Point (AP). The user presents his Network Access Identifier (NAI) which is of the form IMSI@domain. The access request is forwarded to the AAA proxy that translates the AAA request into the equivalent 3G AAA protocol request. Note that this Proxy or Gateway might be pre-configured or dynamically searched. The procedure may cross several other authentication domains.

—————— New Interface
— — — IKE connection / Diameter / MAP
· · · · · · · SSL or IKE connection / EAP-AKA

HSS=Home Subscriber Server
CA=Certificate Authority
AA=Attribute Authority
RA=Registration Authority
AAA=Authentication Authorization and Accounting
CG=Certificate Gateway

Fig. 4. Network Architecture to support subscriber's certificates

5 Conclusions

The constantly increasing population of users expect from mobile operators to provide features that will protect their data while in transit, safeguard their billing and customer information, and offer availability and quality comparable to that of the wired services. Thus, more flexible, dynamic and scalable security mechanisms are necessary in order to support on-demand services and all-IP end-to-end solutions in a many-to-many trust model integrated with the Internet environment. In this paper, we proposed several alternative procedures based on PKI infrastructure and public key enabled protocols introduced in the mobile network architecture. The ultimate challenge is the invigoration of future inter/intra mobile core network security, enhancing authentication procedures and end-to-end communication model trust. We showed that PKI can be a competitive player, offering the appropriate framework to overcome symmetric key based security inefficiencies and providing powerful solutions to protect both 3G-and-beyond network core signaling and user's data.

References

1. 3GPP Tech. Spec. , "MAP Application Layer Security", (TS 33.200 v. 5.1.0), Dec. 2002.
2. Arko, J. and Blom, R., "The MAP Security Domain of Interpretation for Internet Security Association and Key Management Protocol", <draft-arkko-map-doi-07.txt>, May 2002.

3. Kent, S. & Atkinson, R., "Security Architecture for the Internet Protocol", IETF RFC 2401, Nov. 1998.
4. 3GPP Tech. Spec., "IP Network Layer Security", (TS 33.210 v.5.3.0), Mar 2003.
5. 3GPP TSG, "Using PKI to provide network domain security", Discussion Document S3-010622 SA WG3 Security – S3#21, Nov. 2001.
6. 3GPP TSG, "Security Services using Public Key Cryptography", Discussion Document S3z000025 SA WG3 Security – S3#15bis, Nov. 2000.
7. ASPeCT Proj, Securing the Future of Mobile Comm., www.esat.kuleuven.ac.be/cosic/aspect.
8. USECA Project, UMTS Security Architecture: Intermediate report on PKI architecture for UMTS, Public Report, July 1999.
9. 3GPP TSG, "Architecture proposal to support subscriber certificates", Discussion and Approval document, Tdoc S2-022854, Oct. 2002.
10. Kambourakis G., Rouskas A., Gritzalis S., "Introducing PKI to enhance Security in Future Mobile Networks", in the Proc. of the IFIPSEC'2003 18th IFIP Int'l Information Security Conf., pp.109-120, Athens, Greece May 2003.
11. 3GPP Tech. Spec., "Access security for IP-based services", (TS 33.203 v.5.2.0), June 2002.
12. 3GPP TSG, "Support of certificates in 3GPP security Architecture", Discussion Document S3-010353 SA WG3 Security – S3#19, July 2001.
13. Oppliger, R., Pernul, G. & Strauss, C., "Using Attribute Certificates to Implement Role Based Authorization and Access Control Models", In the Proc. of 4. Fachtagung Sicherheit in Informationsystemen (SIS 2000), pp. 169 – 184, Oct. 2000.
14. Frankel, S., Demystifying IPsec Puzzle, Artech House, 2001.
15. Tiller, J., A Technical Guide to IPsec Virtual Private Networks, Auerbach CRC Press, 2000.
16. Assaf, N. et. al., "Interworking between IP Security and Performance Enhancing Proxies for Mobile Networks", IEEE Comm. Mag., pp.138-144, May 2002.
17. Lin, Y., & Pang, A., "An All-IP Approach for UMTS Third-Generation Mobile Networks", IEEE Network, pp. 8-19, Sept./Oct. 2002.
18. 3GPP Tech. Spec., "Security Architecture", (TS 33.102 v.5.1.0), December 2002.
19. Rosenberg, J. et al., "SIP: Session Initiation Protocol", IETF RFC 3261, June 2002.
20. Maughan, D., et al, "Internet Security Association and Key Management Protocol (ISAKMP)", IETF RFC 2408, Nov. 1998.
21. Kambourakis G., Rouskas A., & Gritzalis S., "Using SSL/TLS in Authentication and Key Agreement Procedures of Future Mobile Networks", In the Proc. of the 4th IEEE Int'l Conf. on Mobile and Wireless Comm. Networks. (MWCN), Stockholm, pp. 152-156, Sep 2002.
22. Gupta V. & Gupta S., "Experiments in Wireless Internet Security", In the Proc. of IEEE Wireless Comm. & Networking Conf. (WCNC 2002),no. 1,pp. 859-863,March 2002.
23. Nachiketh, P., Srivaths, R., Anand, R. & Ganesh, L., "Optimizing Public-Key Encryption for Wireless Clients", In the Proc. of the IEEE Int'l Conf. On Communications (ICC 2002), no 1, pp. 1050 – 1056, April 2002.
24. Arkko, J. and Haverinen, H., "EAP-AKA Authentication", <draft-arkko-pppext-eap-aka-10.txt>, June 2003.

Improving Distributed Firewalls Performance through Vertical Load Balancing

Olivier Paul

GET/INT, LOR Department,
91000 Evry, France
Olivier.Paul@int-evry.fr

Abstract. In this paper we present an extension to an existing hash based packet classification technique in order to improve its performance in a distributed network access control environment. We show that such architecture can be modified so that flow states can be kept in a distributed fashion thus reducing the space needed for packet filtering in each component of the architecture. We also show how such approach can, in some cases, improve the overall time complexity of packet filtering operations by reducing the number of packet classification operations.

1   Introduction

As network architectures become more complex and interconnected, the number of interconnection devices regularly increases. At the same time, the number of network devices with security features in general and access control capabilities in particular is also rising. These capacities are widely used in companies in order to partition networks and limit the ability of users to interact with each others. The problem of automating and optimizing the configuration of such distributed access control architectures has recently raised a lot of interest. However most techniques have either focused on the optimization of distributed access control policies independently from packet classification schemes or focused on optimizing packet classification schemes independently from the distributed nature of distributed firewalls. In this paper we explore the optimization of a specific (while popular) packet classification scheme [4] in the context of a distributed access control architecture. We show that such a scheme can widely benefit from a distributed implementation.

Section 2 provides an overview of existing schemes for access control performance improvements. Section 3 presents our proposal to improve an hash based stateful packet classification scheme in a distributed access control environment. We address dimensioning issues in section 4 and perform a comparison with the non distributed approach. We also briefly present an implementation of our architecture with the *ns* simulator. Finally section 5 summarizes the pro and cons of our proposal and highlights possible extensions.

N. Mitrou et al. (Eds.): NETWORKING 2004, LNCS 3042, pp. 25–37, 2004.

2   Related Work

The management of distributed access control architectures has generated a lot of work in the last few years [1][2]. The goal is usually to automate the configuration of a set of access control devices under a single administrative domain. An interesting aspect regarding these techniques is that they acknowledge the fact that current access control architectures are usually made of several "layers" thus leading to the traversal of multiple redundant filters for most communications. Although this fact is not completely new, we believe that the integration of access control functions in a wide range of network devices as well as a better understanding of insider risks has led to a wide increase in this redundancy over the last decade. Our architecture takes advantage of this redundancy to distribute parts of the access control functions over several filters.

Load balancing is another well known technique to share processing load between several components. However the load is usually shared by a front device (e.g. DNS server, load balancer) among several components. Compared to such approaches our architecture may be considered as a load balancing scheme where the load balancing occurs between "serial" devices instead of occurring between "parallel" devices as usual, the distribution function being implemented in each component. Consequently our scheme does not necessitate more devices than those available in a network while providing similar benefits in term of processing performance. On the other hand load balancing architectures often provide fault resilience which is not our case.

Tunneling techniques can be used to bypass packet filters by short-cutting packet filtering functions or by obfuscating necessary filtering information. A similar result may be reached by disabling access control or state maintenance functions or limiting their use to specific communications. However they require some sort of signaling in order to set up a context in each device which can induce strong performance penalties in particular in the case of short flows. Moreover choosing the right source and destination for tunnels in order to bypass selected filtering components and making sure that the filtering policy is still enforced can be a difficult task.

3   A State Distribution Scheme

Stateful packet classification [3] is usually performed in two phases. The "dot" classification phase occurs first. The goal is to find if the packet can be matched to an existing connection. When it can, the action specified for the packet can be directly retrieved from the connection context. In the other case, the whole classification policy has to be searched for a relevant rule using "rectangle" classification. "dot" classification is distinguished from "rectangle" classification because it is usually much faster since it deals with specific values while "rectangle" classification deals with ranges thus leading to more complex treatments. As a result keeping a state about connection brings benefits from security and performance points of view. However "dot" classification is only possible if a connection context has been previously set up using "rectangle" classification. Several methods exist today to store such a context, however the method we focus on here is an hash based answer. The hash function is involved in two operations.

- After "rectangle" classification, a hash value is computed over a limited number of fields (i.e. protocol, source and destination addresses, source and destination ports). This hash value points to a bucket that will be used to store the connection state.
- For "dot" classification, the same fields are extracted from each packet and used to compute a hash value that points to the bucket used to store the connection state.

In the case where no corresponding connection is found (either because the bucket is empty or because no matching connection is found), "rectangle" classification has to be performed in order to fill the hash table with the corresponding connection.

Keeping a state in a single hash table can result in high collision rates thus leading to a large number of states kept in a single bucket and an increased "dot" classification time. A way to solve this problem is to use a multiple hash functions [4]. The state hash table of size m is divided into n smaller hash tables of size m/n where each table uses a separate hash function. n hash values are now computed over the packet content and point to a specific bucket in each hash table. In order to limit the number of concurrent states in each bucket, one selects the hash table including the bucket with the lowest depth. This approach however has two main drawbacks:

- n hash functions have now to be computed for each packet after rectangle classification and for hash classification.
- When performing hash classification, one bucket in each n tables has now to be searched in order to find the relevant state.

Another point we want to highlight is that contexts are usually memory consuming. For example keeping one state in pf requires roughly 150 bytes of memory which means that keeping states for a 10Gbits/s Ethernet links would require several gigabytes (We later detail how we reach these numbers).

3.1 State Distribution Problem

Let's now consider a distributed filtering architecture where several filters are available on the path between any source and destination. Let's also assume that each filter implements the basic hash classification process described earlier. Each filter has to compute at least one hash value over the packet content in order to perform access control functions. Our goal is to take advantage of these hash functions to simulate the behavior of a single packet filter implementing the multiple hash functions scheme. This would allows us to keep the improvements brought by a multiple hash function scheme while avoiding the corresponding drawbacks.

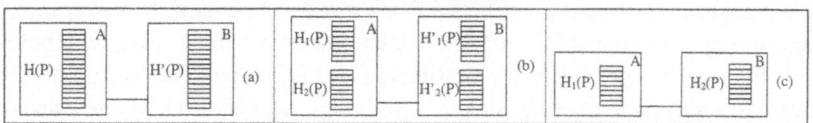

Fig. 1. Various distributed hash based filtering scenario. (a) represents two single hash filters. (b) represents two multiple hash filters. (c) represents our proposal

Figure 1 illustrate the difference between our proposal and other hash based "dot" classification methods in a distributed environment. (c) represents the suggested approach in which tables that were previously included in a single device are now scattered over several devices thus reducing the memory requirements for each filter.

Although this approach may seem superior it bears several problems:

1. The selection algorithm cannot be used as is since there is no way for device A to know how many packets are stored in device B. Packets stored in B may come from filters others than A. Additionally, A and B may have different resources.
2. When receiving a packet P, B doesn't know if P has been already filtered by A. Packets with spoofed addresses may reach B, be considered as already treated without actually going through any access control check.
3. If a corresponding state cannot be found in A when receiving a packet, it is impossible for A to know if the state is missing because no packet has been previously received for this connection or because the state is located on B.

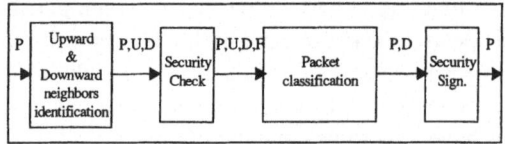

Fig. 2. Overall filtering process

We address these problems in the three following sections. The overall filtering process is presented in Figure 2. Given a packet P, we first identify the upward and downward neighbor filtering components U and D. We then check that the packet is not fraudulent (F) using the upward filtering component identity. Given this information, we execute the distributed packet classification algorithm. Finally, we sign the packet using the downward filter identity and send the packet to its destination.

3.2   Neighboring Filters Identification

The goal of stateful filtering being to match packets coming back and forth, symmetric routing is mandatory for the operation of our architecture. As a result we expect that firewalls [1] or at least the firewall management architecture [2] will be able, for each couple (source, destination) to define which filtering components are on the path. As routing tables may hold a limited view of the network topology (e.g. devices using a default route) this information may have to be provided to filtering components so that upward (U) and downward (D) components can be identified.

Classical routing structures such as a Patricia trie can be used. However in order to store filtering components information, we extend each routing entry with the downstream filtering component to the destination of the packet. The lookup process includes two addresses structure lookups providing the addresses of the upward and downward components if any.

3.3   Buckets State Estimation

Bucket State Transmission

As mentioned earlier we expect symmetric routing to be used among our distributed filtering components. Our approach to estimate the state of a bucket is to transmit this information between filtering components. In order to do so we define an experimental DSCP value using the first three bits (a,b,c) of the TOS field. This new field indicates the percentage of state space in use in 1/8th as indicated in Figure 3. Bit f indicates if the packet has already been filtered by an upstream component. Finally, the next two bits indicate an experimental DSCP as specified in [5].

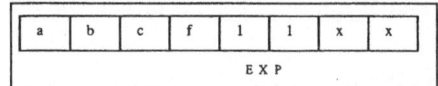

Fig. 3. TOS Field

Filtering components maintain a state table indicating the number of states used for each bucket in each adjacent filtering component. We say that two filtering components A and B are adjacent for a communication (Co) and a routing topology when there is no other filtering component C that Co would cross after going through A and before going through B and reciprocally. Let's consider the set made of the n filtering components adjacent to A, $S = \{B_1..B_n\}$. A maintains a state table $TB_i[1..SB_i]$ where $TB_i[j]$ indicates the proportion of states used in bucket j for the state table in B_i and SB_i represents the number of buckets used in B_i.

When a packet P is received, the DSCP value is retrieved after the packet is validated and used to update the adjacent component B_j, TB_j. This DSCP value is set to the value in TA for packet P before the packet gets forwarded.

Packet Treatment

When receiving a packet P, A extracts the set of invariant fields used to build an invariant descriptor P'. Note that the way P' is built depends on the direction of the packet. Depending on P' three cases may happen:
1. P belongs to a flow that is handled locally in which case a bucket in TA should include the corresponding state.
2. P belongs to a flow that is handled remotely. This flow can be either treated by an upstream component (2.1) or by a downstream component (2.2).
3. P belongs to a new flow that is not yet treated.

Using the local and adjacent filtering component identifiers A and B_i, A generates hash values H=H(P', A) and H'=H(P', B_i). In order to test case (1) we lookup the state in table TA[1..SA] by going through the states located in bucket TA[H%SA].

Using TB_i is unfortunately not sufficient to differentiate cases (2) and (3). We therefore need an additional table to distinguish flows that are handled remotely from unknown flows. This table has to allow us to clean connections that have timed out and therefore has to allow us to keep a timer for each flow. In order to do so we build a "temporal" bloom filter BF[1..SF] in which each component of the couple (BF[H%SF], BF[H'%SF]) can take two types of value.

- 00 indicates that the corresponding flow is not present in the filter.
- (01, 10, 11) indicates the value of the timer the last time a packet belonging to the flow was received.

Table 1. Meaning of timer values

(xx,yy) value	Meaning
(00,00)	No flow.
(00,xx), (xx,00)	Partial collision.
(xx,xx)	No collision, Total collision with same timer values.
(xx,yy), (yy,xx)	Partial collision, Total collision with different timer values.

The couple (BF[H%SF], BF[H'%SF]) can therefore indicate four types of events as indicated in Table 1. Note that partial collisions are not problem here since half timing information is still accurate. Total collisions are more problematic since they can transform a non existing or timed out flow into a valid one. We later show how the probability for these collisions can be controlled by rightly designing the bloom filter.

For a timeout value T we maintain a timer CT that is increased every T/3 and can take values (01,10,11). Before increasing CT we set to "00" entries that hold a value equal to the upcoming CT. Note that timers are not kept here for security purposes but only to limit the number of "foreign" states that have to be stored locally.

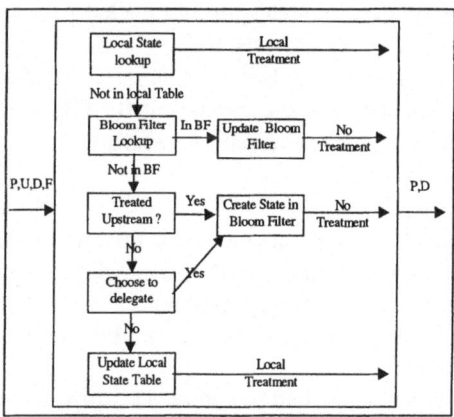

Fig. 4. Detailed packet treatment

If (3) applies to P (i.e. BF[H]=00 or BF[H']=00), we follow the multiple hash function algorithm [4] to decide where the state should be stored. To do so we compare TA[H%SA] and TB$_i$[H'%SB$_i$]

- If TA[H%SA] > TBi[H'%SBi] we store the packet value in the bloom filter (i.e. we set BF[H%SF] and BF[H'%SF] to CT).
- If TA[H%SA] <= TBi[H'%SBi] we perform classical packet classification operations and store the corresponding state in TA. We also mark the TOS f bit.

If (2) applies to P the action depends on the value of BF[H] and BF[H']. If both carry the same value, we update the bloom filter with the current timer. In the other case, we update the oldest identifiers. This is aimed to make sure that only one connection gets updated in the case of a partial collision.

Figure 4 provides the whole process followed to distinguish case 1, 2.1, 2.2 and 3.

3.4   Packet Checking

As mentioned earlier, our scheme does not provide a way for B, when receiving a packet, to know if it has been previously filtered and accepted by A or if it reaches B because of the state distribution technique. In order to solve this problem we suggest a weak authentication by overloading the ID field in the IP packet header. The usage of this field for operations unrelated to packet fragmentation as been suggested for packet tracking operations one of the reason being the small proportion (i.e. less than 0.5% on average) of fragmented traffic on the Internet [7]. Our scheme is also based on the ability for adjacent filtering components to share a common secret key K.

When a packet P reaches filtering component U, we use the downward filtering component identity D to retrieve the corresponding secret key K_{DU}. We then use the invariant part of the packet P' and K_{DU} to compute a HMAC value H" and store H" in the ID field of packet P before sending it toward its destination. When receiving packet P, a filtering component first identifies the neighboring filter U previously crossed by P. As routing is expected to be symmetric, this can be performed with an extension to the neighbor identification. It then retrieves the corresponding key K_{DU}, the invariant part P' of P, computes H"' using P' and K_{DU} and checks if H"' matches the content of the ID field. If the two values match, we perform previously specified operations. In the other case, we mark the F selector so that P is treated locally.

As the size of the authentication field makes it vulnerable to brute force attack we need a mechanism to discourage such attempts. The mechanism we present here does not prevent a determined attacker to find the key but instead gives the security officer enough time to identify such attempts. To do so, we maintain a counter C_{DU} for each key K_{DU} indicating the number of failed authentication operations since the last key change. When the number of tests performed by the attacker passes over 0.1% of the possible key space (i.e. roughly 65 packets), we reconfigure the filtering process so that every packet supposedly going through U is treated locally. In order to stop this local processing, we maintain a counter P_{DU} indicating the number of packets supposedly coming from U. When P_{DU} passes over 30.R where R is the maximum number of packets that D can handle per second, we perform a new key exchange and reset counters. As there is no difference in the filtering process once all packets are filtered, an attacker cannot learn anything from the system after the first 65 packets. He then has to wait for at least 30 seconds before being able to make another attempt. The probability for him to succeed to find the key evolves slowly from 10% after 2 hours, 50% after 5 hours to 90% after 20 hours.

Finally, we mention that this weak authentication process can be avoided in some cases when all communications received by a filtering component go through another filter first. This is in particular the case for filtering components that are collocated within a single device.

4   Dimensioning Issues and Evaluation

Bucket State/DSCP Code Relationship

As mentioned earlier two filtering components can use hash tables with different sizes, as a result the DSCP value is not absolute. We therefore need to define a relation between the number of states kept in a specific bucket, the number of states kept by the filtering component and the DSCP value. Given a worst case state lookup speed, and the fluid limit simulation results provided in [4], we can bound the maximal number of states expected for a bucket in the state table (Smax). Using this limit and the number of states S located in bucket TA[H%SA], we define the (a,b,c) DSCP bits as: $DSCP = TA[H\%SA] = |(S/ Smax) . 2^3|$ where $|x|$ represents the integer value for x.

Bloom Filter Dimensions

In order to dimension the bloom filter, we first select the bandwidth for which we want to dimension the filter (i.e. 10Gbps in our case) and consider that the bloom filter will store most of the traffic. We then compute the maximal and average number of new flows treated during a few selected lengths of time. To evaluate the likelihood of each flow duration we use an approximation of the flow duration probability distribution provided in [6]. This approximation is provided in Table 2.

Table 2. Distribution of flows

Flow dur. (s)	% flows in class	Flows/s (worst/avg)
[c0=0;c1=1]	p0=90	[28125.10^3; 351562]
[c1=1;c2=10]	p1=9	[28125.10^2; 35156]
[c2=10;c3=100]	p2=0.9	[281250; 3515]
[c3=100;c4=1000]	p3=0.09	[28125; 352]
[c4=1000;c5=10000]	p4=0.01	[3125; 39]

For the worst case scenario we consider single packet flows where the size of each packet is minimal (i.e. 40 bytes). For the average case, we use the internet average packet size (i.e. 320 bytes) and an average of 10 packets per flow. Although we expect our architecture to be subjected to the worst case scenario during short period of time (i.e. DDOS attacks), we do not expect these periods to exceed a few seconds as external mechanisms should be put in place to avoid such traffic conditions.

A point however not considered in Table 2 is the timeout value for the bloom filter. This value must be sufficiently low to avoid high collision rates and sufficiently high to prevent active connections to get cleared. Consequently we compute maximal collision probabilities for various bloom filter sizes and timeout values. To do so we first evaluate the number of flows in the filter at time t, Mf(t).

$$Mf\ (t) = (\ N \cdot To\ ) + Nr\ (t)) \tag{1}$$

Where N represents the number of new flows received each second, To represents the clearing timeout value and Nr(t) represents the number of flows from previous timeout periods that are still active at time t. Nr(t) can be computed as follows:

$$Nr\ (t) = N \cdot \sum_{i=1}^{i=t} (1 - (\ p_0 + \sum_{j=1}^{j=\min(\ 4.|\log_{10}(i)|+1)} \frac{(\min(\ i, c_{j+1}) - c_j)}{(c_{j+1} - c_j)} \cdot p_j\ ) \tag{2}$$

Mf(t) is maximized when t > 10000 so that:

$$Mf\ \max \approx N \cdot (To + 2) \tag{3}$$

As we expect most of the states to be in the current CT time period, we compute a collision rate as if we were using a classical bloom filter. We give here the maximal collision probability (Cmax) in a bloom filter of size 2^m after storing Mfmax flows.

$$C\ \max\ = (1 - (1 - 1/\ m)^{2 \cdot Mf\ \max})^2 \tag{4}$$

Figure 5 provides the maximal false positive probability for several bloom filters sizes (16Mb, 64Mb), flow rates (average and worst case scenarios) and timeout values. A 64Mbytes bloom filter with a 30 seconds timeout seems appropriate.

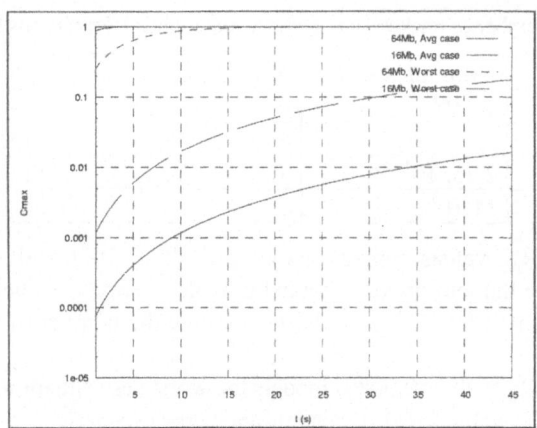

Fig. 5. False positive probability for various bloom filter sizes, flow rates and timeout values

Independently from the timeout value, flows identifiers can be overwritten by newer flows. In our case, flows under 100 seconds are unlikely to get overwritten (around 6%). On the other hand flows over 400 seconds are very likely to be overwritten in the filter (more than 50%). By being overwritten, a terminated flow will appear as still active for the duration of the overwriting flow. However, as shown in Table 2, the likelihood of the overwriting flows to have a long duration is very small.

As a result overwriting operations will have a limited impact (around 4 seconds on average) on the evaluation of long flows.

Another point is that packets generating a collision do not bypass the access control architecture. As each packet goes through several filtering components the overall probability to reach the last filtering component and being misclassified in each one decreases exponentially with the number of filtering component crossed.

Finally if some packets reach the last filter without being treated we expect the identification process to set mandatory filtering (F) for remaining packets.

Neighbor States Tables Storage
In order to evaluate the size for these tables we need to set a limit to the number of concurrent states that can be stored in a bucket. In order to do so we could have used formula (1). We instead choose to use indicators computed from real-life packet traces which allows us to evaluate state numbers and assess the validity of Mf(t) at the same time. We therefore use indicators of average link utilization (U_{avg}), average number of active flow entries (E_{avg}) and maximum number of active flow entries (E_{max}) computed over 51 long packet traces totaling to more than 900 hours of network monitoring. These indicators were computed by the Sprint IPMON project [9] on traces captured in 2002 and 2003 at several locations within Sprint operational network. We use these indicators to define ratios $R_{avg} = E_{avg}/U_{avg}$ (average number of states per bps) and $R_{max} = E_{max}/U_{avg}$ (maximum number of states per bps).

Table 3. Number of buckets and memory requirements for various bucket depths

Scenario	Num. of buckets	Max. Num. of states	Memory required
Worst Case	10.10^6	9	5Mbytes
	40.10^6	6	15Mbytes
Average Case	$2.75.10^6$	9	1.3Mbytes
	11.10^6	6	4.1Mbytes

In our case R_{avg} values remain within $[4.10^{-4}; 40.10^{-4}]$ with an average value $R_{avg}=11.10^{-4}$ (Note that our previous model provides a similar value) while R_{max} values remain within $[7.10^{-4}; 80.10^{-4}]$. In order to compute the number of states kept we use two indicators:
- The maximal R_{max} value (0.0080) models the worst case situation.
- The average R_{avg} value (0.0011) models an average scenario.

Table 4. Memory requirements and gains per filtering component depending on the average number of filtering components between a source and a destination

Num. filt. comp	Max memory req.	Avg memory req.	Gain per filter
Original scheme	12 GB	1.6 GB	1.0/1.0
2	6.2 GB	905 MB	1.9/1.8
5	2.6 GB	425 MB	4.6/3.8
10	1.4 GB	265 MB	8.6/6.1

Using these two indicators, we can compute the average and maximal number of flows. Combined with fluid limit simulation results provided in [4], we estimate the storage requirement for a component depending on several bucket depths (Table 3).

Assuming an average number of 10 neighbors we estimate the total storage to lie between 13 and 150 Mbytes depending on the maximum number of states per bucket. As a consequence the total storage required to implement our scheme stands between 77 and 214Mbytes. Table 4 provides memory requirements, overhead per filter and gain per filter depending on the number of filters crossed by a communication.

Processing Requirements
In order to evaluate our scheme, we compare the processing requirements to the original classification scheme [4]. Table 5 provides the number of CPU cycles used by both approaches in each filter where C represents the cost for "rectangle" classification, H the cost for a hash over 40 bytes, M the cost for an HMAC over 40 bytes, L the cost for an address lookup and N the number of packets per flow.

Table 5. Operations performed by Classical and Distributed architectures

Filter #	Classical	Distributed
1	C+2HN	C+N(2H+2M+2L)
2,3,...	C+2HN	N(2H+2M+2L)

Figure 6 shows the difference of time complexity in processor cycles between both schemes using C=10000 (linear search through 200 rules using 5 fields [3]), H=196, M=272 (UMAC performance tests [8]), L=250 and several N values. The distributed scheme is clearly more interesting for short flows (i.e. less than 6 packets per flow) while the regular scheme is more interesting for long flows (i.e over 15 packets per flow). It should however be noted that a faster packet classification scheme would render our scheme less attractive from a processing requirement point of view.

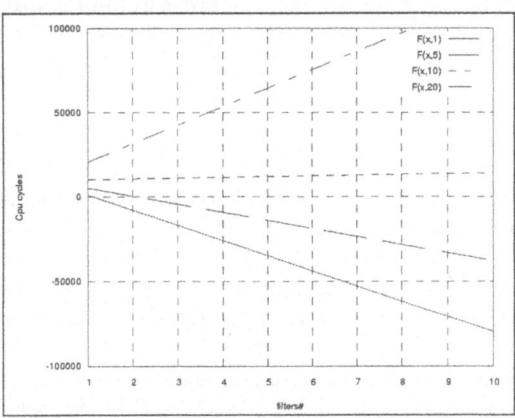

Fig. 6. Difference in term of processing requirements between distributed and regular approaches depending on the number of packet filters

Implementation and Tests

This architecture has been implemented under the *ns* simulator. The goals for the implementation were twofold, we wanted to check the correct behavior of the architecture and test its behavior under real life traffic. The neighbor identification module as well as the classification algorithm were implemented as an extension to the current address classifier class. As *ns* does not support real IP addresses in packets, we also extended IP packets allowing us to transport the packet header information required by our scheme as well as protocol specific information. A few modifications to other parts of the simulator were also performed in order to facilitate our tests. At the time of this writing only the protocol behavior was tested using generated traffic.

5   Conclusion

We present in this paper an architecture for connection states maintenance that was designed to reduce memory requirements as well as limit processing requirements in some specific cases. Our main contribution is to show how an existing state maintenance scheme can be distributed and that such a distribution can bring large improvements. Although this paper only focuses on a single state maintenance technique, we believe that such distribution could apply to other techniques.

On the other hand, our approach makes the implementation of functions like flow monitoring more difficult and renders functions like diffserv or fragmentation impossible. Additionally our scheme may generate flow interruptions in the case of route changes when the filter holding the state of the rerouted flows is longer part of the path between the source and the destination. However, this problem would also happen with a regular access control architecture.

This work is currently continued in two directions. We currently work on a real life implementation that would allow us to perform tests with real life traffic. This would allows us to compare results with traffic models used in this paper. Another direction is to explore improved bloom filter structures in order to limit collisions by performing a separation of short and long flows.

References

1. Joshua D. Guttman. Filtering Postures: Local Enforcement for Global Policies. IEEE Symposium on Security and Privacy. Oakland. May 1997.
2. Yair Bartal, Alain Mayer, Kobbi Nissim, Avishai Wool. Firmato, A Novell Firewall Management Toolkit. IEEE Symposium on Security and Privacy. Oakland. May 1999.
3. Daniel Hartmeier. Design and Performance of the OpenBSD Stateful Packet Filter (pf). Usenix Annual Technical Conference. June 2002.
4. Andrei Broder, Michael Mitzenmacher, Using Multiple Hash Functions to Improve IP Lookups, In proceedings of IEEE Infocom 2001. Anchorage, Alaska, April 2001.
5. K. Nichols and al. Definition of the Differentiated Services Field (DS Field) in the IPv4 and IPv6 Headers (RFC 2474). December 1998.

6. Yin Zhang and al. On the Characteristics and Origins of Internet Flow Rates. ACM SIGCOMM 2002. Pittsburgh, USA. August 2002.
7. Colleen Shannon and al.. Beyond Folklore: Observations on Fragmented Traffic". IEEE/ACM Transactions on Networking. December 2002.
8. John Black et al. UMAC: Fast and Secure Message Authentication. Advances in Cryptology - CRYPTO '99. Lecture Notes in Computer Science, vol. 1666, Springer-Verlag, 1999.
9. Sprint Labs. IP Monitoring Project. Available at http://ipmon.sprint.com/ipmon.php.

Performance Modelling and Evaluation of Firewall Architectures for Multimedia Applications

Utz Roedig[1] and Jens Schmitt[2]

[1]Mobile & Internet Systems Laboratory (MISL), University College Cork, Ireland
utz@cs.ucc.ie
[2]Distributed Computer Systems Lab (DISCO), University of Kaiserslautern, Germany
jschmitt@informatik.uni-kl.de

Abstract. Firewalls are a well-established security mechanism to restrict the traffic exchanged between networks to a certain subset of users and applications. In order to cope with new application types like multimedia applications, new firewall architectures are necessary. The performance of these new architectures is a critical factor because Quality of Service (QoS) demands of multimedia applications have to be satisfied. We show how the performance of firewall architectures for multimedia applications can be determined. A model is presented which can be used to describe the performance of multimedia firewall architectures. This model can be used to dimension firewalls for usage with multimedia applications. In addition, we present the results of a lab experiment, used to evaluate the performance of a distributed firewall architecture and to validate the model.

1 Motivation and Introduction

Within a global networked environment, security aspects have become more and more important and access control at network borders is considered essential. For this purpose firewalls are used. As an integral part of the network infrastructure, firewalls are strongly affected by the development and deployment of new communication paradigms and applications. Recently, there has been a rise in the use of multimedia applications which, from the perspective of firewalls, differ in many aspects from "traditional" applications. One of the most important aspects is the difference in performance requirements. Existing firewalls are not able to support multimedia applications in an efficient and secure manner [1]. In particular, a traditional firewall may not be able to support the QoS requirements of a multimedia application.

To overcome these deficiencies, new firewall architectures are currently discussed and proposed. Besides many other facets - e.g. security, maintainability, flexibility - these are intended to optimize firewall performance. Of course, all these characteristics have to be optimized simultaneously to meet the given requirements.

Currently, appropriate methods and tools to evaluate the performance of multimedia firewall architectures are missing. Hence, ascertained performance parameters of proposed firewall architectures are also unavailable. To solve these problems the following topics are covered in this paper:

N. Mitrou et al. (Eds.): NETWORKING 2004, LNCS 3042, pp. 38–51, 2004.
© IFIP International Federation for Information Processing 2004

(i)   Analysis of performance bottlenecks in multimedia firewall architectures;

(ii)   Performance modelling of multimedia firewall architectures;

(iii) Experimental performance evaluation and model validation.

In the remaining paragraphs of this section the terms "multimedia application" and "firewall architecture" are described in detail as they are used in the context of this paper. In Section 2, the parameters which characterize the performance of a multimedia firewall are defined. Further, performance bottlenecks in firewall architectures are analyzed. In Section 3, the performance model is introduced. In Section 4, the lab experiment is described, including measurement methods and tools that were used. In Section 5, the experimental results are compared with the model and the model is validated. Section 6 reviews related work. In the last section, our findings are summarized.

Multimedia Applications. Multimedia applications use a combination of continuous and discrete media data, with the continuous media usually being audio and/ or video streams. The discrete media often consist of control data streams for the audio and video data streams and additional information.

In order to describe communication scenarios, the following terms to distinguish the granularity at which an application's data stream is considered are defined. A *flow* is a single data stream, identified by a tuple of characteristic values (e.g. source address, source port, destination address, destination port, protocol number). A *session* describes the association of multiple flows which together constitute an application's data stream.

Firewall Types and Architectures. A firewall examines all network traffic between connected networks. Only data that is explicitly allowed to, as specified by a security policy, is able to pass through it. The tasks of a firewall are well defined, but there are many possible firewall architectures to fulfil them. Firewalls may consist of different firewall components, e.g. filters, stateful filters or proxies. In addition, the applications may interact explicitly with a firewall to support it to fulfil its task.

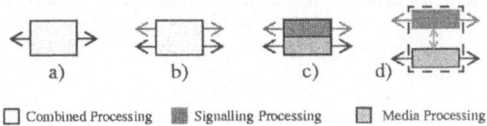

a)   b)   c)   d)

☐ Combined Processing   ▦ Signalling Processing   ▨ Media Processing

Fig. 1. Firewall Types

To select a useful architecture for the usage in conjunction with multimedia applications the following basic evolution of firewall types - illustrated by Figure 1 - has to be taken into account [1]. Figure 1a) abstractly describes the behavior of a "standard firewall". All traffic is sent through the firewall component which is responsible to apply the security functionality. In this case the specific characteristics of multimedia applications' traffic are not taken into account. If these specific characteristics (as shown in Figure 1b)) are regarded it is obvious that the same firewall component has to take care of different traffic types of the different traffic flows (control and media flows). In this case, it is not possible to adapt the one firewall component to the needs of the two different flow types. This results in many problems, in particular performance problems [2]. To overcome this weakness, two different firewall components for the processing of the two different flows can be used (Figure 1c)) [1]. This additional degree of freedom allows specific component optimizations for the different flow types. To maintain session state within the firewall, information exchange between the components is necessary. If the separation between signalling and media processing is further extended by even physically distributing them (Figure 1d) additional optimizations are possible [1], [3]. In this case the information exchange between the components has to be realized by an appropriate network protocol [4]. The implementation of the useful firewall types shown in Figure 1c) and Figure 1d) lead to different multimedia firewall architectures which are currently proposed. The focus is on these architectures in the remaining paper:

- **Architecture AI** (implementation of firewall type c)): The firewall consists of a single computer system containing a signal and media flow processing component. Well known firewalls following this design principle include firewall products like CISCO's PIX and Checkpoint's Firewall-1.

- **Architecture AII** (implementation of firewall type d)): The firewall consists of several computers. A well defined interface between signalling and media processing component(s) is used. A practical implementation of such an architecture is the Netscreen 500   firewall for SIP based IP-telephony applications [5].

- **Architecture AIII** (implementation of firewall type d)): In this case, the available signalling processing component within multimedia applications in end systems is used. By choosing this architecture, the need of centralized signalling processing components is avoided. These systems are not used today, but theoretical work exists [4].

To select one of the architectures, one has to consider the advantages and disadvantages and rate how important they are in the considered target scenario. Independent from these considerations, the firewall system has to be dimensioned to meet the QoS requirements of multimedia applications. It is necessary to know how many signalling and media processing units are necessary and what capacity they should have.

2   Firewall Performance

To determine the performance of a multimedia firewall architecture it is necessary to define the term performance in this context first. The performance of a firewall, respectively of a firewall architecture, is defined by:

(i)   its influence on applications' QoS parameters

(ii) its total capacity

The influence on QoS parameters of multimedia applications by a firewall within the communication path should be low and predictable. The maximum possible throughput, its capacity, should be as high as possible.

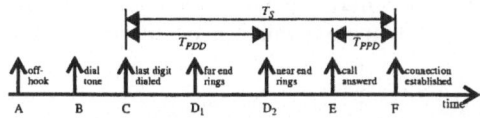

Fig. 2. Session setup steps

2.1 Quality of Service Parameters

To be able to rate the performance of a multimedia firewall, useful quality parameters have to be defined. These quality parameters should allow the objective validation of a firewall's performance. In the following, the necessary QoS parameters of multimedia applications are described. From these parameters quality parameters necessary to rate firewalls are derived.

Signalling Flow. The quality of the signalling plane is mostly influenced by the session setup delay. If the necessary time for a session setup is too long, a user of a multimedia application will feel disturbed or will regard the connection's quality unacceptable. The following definition is used:

*The **session setup time** T_s is the time from the setup of the control flow till the start of the first media flow.*

The determination of boundary values and an exact definition depends on the type of investigated application. The session setup time can also be divided in substeps, which might be subject to different requirements. The requirements for the session setup time for IP-telephony applications are described below, because these applications are used in the experiment described in Section 4. Figure 2 describes the substeps within the session setup as used in H.323 based IP-telephony applications [7]. In this case, the session setup time is given by $T_s = F - C$. In addition, the post dial delay $T_{PDD} = D_2 - C$ and post pickup delay $T_{PPD} = F - E$ can be defined. The post pickup delay is particulary critical. If the latter value is too high, the first words of the conversation are lost because the media channels are not yet established. Boundary values can be derived from values given for ISDN networks [6]. The post dial delay should be between 2 and 7 seconds, the post pickup delay should be between 0.75 and 2 seconds.

Media Flow. The media flows also have to meet specific requirements. Possible effects if specific bounds are violated might be for example echo or noise. The characteristic parameters to describe the quality of a media stream are delay $T_{D\,(i)}$, jitter $T_{J\,(i)}$ and loss L. As the experiments described in this paper target the control plane, we refer to [1] for a detailed definition and explanation of theses parameters.

Quality Index. Firewall quality indices can be derived from the previously described QoS parameters of multimedia applications. The following definition for quality indices is used:

The quality index G_X defines the percentage of the upper bound of a QoS parameter X of a specific multimedia application that is consumed by the firewall.

The different quality indices may depend on the number of similar active application sessions that are handled by the firewall. The quality indices are then given by:

$$G_X(n) = \frac{\Delta X(n)}{X_{max}} \qquad X \in \{T_S, T_D, T_J, L\} \tag{1}$$

with $\Delta X(n)$ describing the value consumed by the firewall and X_{max} representing the selected upper bound of the investigated QoS parameter.

2.2   Capacity of Firewall Architectures

The capacity of a firewall can be determined by the definition of upper bounds G_{Xmax} for the four different quality indices. The capacity is defined as:

The capacity N of a multimedia firewall is given by the number of concurrent active sessions such that

$$G_X(n) \leq G_{Xmax} \qquad \forall n \leq N, \forall X \tag{2}$$

In the following section bottlenecks in firewall architectures, their influence on the capacity and also the resulting impact on the dimensioning of firewalls is discussed.

Filter Bandwidth. The media flow processing within the firewall architectures described in Section 1 is normally implemented as a packet filter. For these filters, the maximum bandwidth B, which normally depends on the packet size s, is known. If the number of media flows r that are used for a specific multimedia application and the bandwidth b of these flows is also known, the upper bound on the capacity N_B of the firewall can be calculated:

$$N_B = \frac{B(s)}{r \cdot b} \tag{3}$$

The bandwidth used for the signalling and media control flows are not taken into account because they are small compared to the bandwidth of the media flows. In addition, it is assumed that the quality indices are within the boundary values according to equation (2).

Session Setup. The component used to process the signalling flow is limited in the amount of packets that can be processed in a certain time period. Therefore, a limit on

$$N_S = \mu \cdot T \tag{4}$$

the amount of session setups per second that can be handled exists. If it is assumed that all applications have duration T and further that the session setups are uniformly distributed, the upper bound on the capacity is given by:

In firewalls used today, the capacity of a firewall is mainly constrained by the signalling processing component, not by the available filter bandwidth ($N_B \gg N_S$). To overcome this shortage, several signalling processing components might be used. For each additional component the gain might be reduced (given by the parallelization efficiency α) due to the distribution overhead:

$$N_S = p \cdot \alpha(p) \cdot \mu \cdot T \qquad\qquad (5)$$

Summary. As shown, it is necessary to regard both factors, filter bandwidth and session setup, to determine the capacity of a multimedia firewall architecture. Especially in firewalls used today (implemented according to architecture AI) the session setup factor is not taken into account. This might lead, depending on multimedia applications characteristics, to a waste of resources and a lower than expected performance.

3   Performance Model

Today the performance behavior of media processing components (e.g. packet filters) is generally well understood and manageable. For the various available components, characteristic curves for the media flows related quality indices are directly or indirectly available. It is also possible to obtain media processing components suited for high bandwidths, so that desired bounds for the media flow quality indices can be met.

In contrast, the performance of signalling processing in multimedia firewalls has not been investigated in depth. To be able to state and predict the performance behavior of the signalling processing, a generic performance model is necessary. In this section, a performance model for the signalling processing using queuing theory is developed. In Section 4, a lab experiment is carried out which is used to verify the developed performance model.

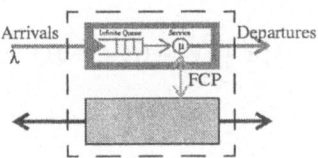

Fig. 3. Queueing System

3.1   Modelling of Multimedia Firewall Types

The signalling processing component of the multimedia firewall types (described in Section 1, Figure 1) can be modelled as shown in Figure 3.

With a rate of λ and a certain statistical distribution, new sessions arrive at the signalling processing component. In order to keep the model tractable but also due to many empirical studies on session arrival characteristics it is assumed that the session inter-arrival time is exponentially distributed. The queue is assumed to be infinite. This means the space (available memory) for waiting sessions is assumed to be sufficient at all times. The processor within the signalling processing component is able to process session setups with a rate of μ. The service time has a general distribution with average $1/\mu$ and variance σ^2. The variance of the service time is caused by the necessary communication between signalling and media processing components using a Firewall Control Protocol (FCP). The necessary processing time T_p for each session setup is composed of the following time segments:

$$T_p = i \cdot T_{sig} + k \cdot T_{FCP} \tag{6}$$

First T_p comprises the necessary and constant processing time T_{sig} for the i exchanged signalling messages used for session setup, second the time T_{FCP} necessary to submit and process FCP messages (e.g. containing flow specifications) is included. T_{FCP} might have a statistical distribution if T_{FCP} is strongly influenced by queueing effects within the FCP message handling in the signalling or media processing component or by the characteristics of the network used to transport the FCP messages.

The resulting queueing system to model the behavior of one signalling component is therefore an M/G/1 queue according to Kendall's notation. If p signalling processing components are used, the arrival rate λ' for each queueing system is:

$$\lambda' = \frac{\lambda}{p} \tag{7}$$

If T_{FCP} can be considered small compared to T_{sig} or shows little fluctuations, the service time can be assumed to be constant. In this case, the resulting queueing system to model the behavior of one signalling component is an M/D/1 queue.

To be able to predict the session setup time, the expected queueing delay (= expected session setup time introduced by the firewall) for the queueing system has to be calculated. The expected queueing delay in an M/G/1 system is given by [8]:

$$E(\Delta T_S) = \frac{1}{\mu \cdot (1 - \rho)} \cdot \left(1 - \frac{\rho}{2}(1 - (\mu^2 \cdot \sigma^2))\right) \text{ with } \rho = \frac{\lambda}{\mu} \tag{8}$$

For the special case of a deterministic (constant) service time, the variance of the service time is zero ($\sigma^2 = 0$). In this case, (8) gives the expected queueing delay in an M/ D/1 system.

3.2   Performance Models for Firewall Architectures

To model the architectures presented in Section 1, the number of signalling processing components has to be taken into account.

Architecture AI. For the hybrid architecture, where only one processing component is available, (8) can be used directly to give a model for the firewall's session setup time ΔT_s in relation to the number of sessions . If the duration of the sessions is assumed to be constant, $\lambda = n/T$ is obtained. With (8) the following model is obtained:

$$\Delta T_{S_{AI}}(n) = \frac{1}{\left(\mu - \frac{n}{T}\right)} \cdot \left(1 - \frac{n}{2 \cdot \mu \cdot T}(1 - (\mu^2 \cdot \sigma^2))\right) \tag{9}$$

Architecture AII. To model the locally distributed architecture, p signalling processing components have to be taken into account. Each signalling processing component comprises an M/G/1 queue. Therefore the arrival rate is split among the processing components and depends on their number p. Using (8) the following model is obtained:

$$\Delta T_{S_{AII}}(n) = \frac{1}{\left(\mu - \frac{n}{T \cdot p}\right)} \cdot \left(1 - \frac{n}{2 \cdot \mu \cdot T \cdot p}(1 - (\mu^2 \cdot \sigma^2))\right) \tag{10}$$

Architecture AIII. Within the totally distributed firewall architecture, for each session a distinct signalling processing component is available. Using (10) and $p = n$ the following model results:

$$\Delta T_{S_{AIII}}(n) = \frac{1}{\left(\mu - \frac{1}{T}\right)} \cdot \left(1 - \frac{1}{2 \cdot \mu \cdot T}(1 - (\mu^2 \cdot \sigma^2))\right) \tag{11}$$

Therefore, $T_{SAIII}(n)$ is constant and does not depend on the number of concurrent active sessions. If it is also assumed, that the session duration T is long and the service rate μ is high, $\Delta T_{SAIII} \approx 1/\mu$ is obtained.

Summary. With $\sigma^2 = 0$ all three analytical models can also be adapted to the assumption of constant service times (M/D/1). Using equation (1), the session setup time given by the models can be used to determine the firewall's session setup quality index.

4   Performance Evaluation

To gain realistic performance numbers for the session setup quality index of multimedia firewall architectures, a lab experiment has been conducted. The results of the experiment are used in Section 5 to validate the performance models developed above.

4.1   Measurement Tool

To be able to determine the quality indices of firewalls, the traffic generator and a measurement tool *KOMtraffgen* [1] is used.

Core. The KOMtraffgen tool can be used to generate traffic of concurrently running multimedia applications. The exact behavior, control and media flows of each individual application, is modeled. The software is divided into two parts, the core and the application specific part. The core carries the generic parts, e.g. measurement facility, timer and hooks to include the application specific parts. The application specific part carries the state machine (client or server side) of the emulated application.

Application. To carry out the experiments, an application with IP-telephony like characteristics was implemented (see Figure 4).

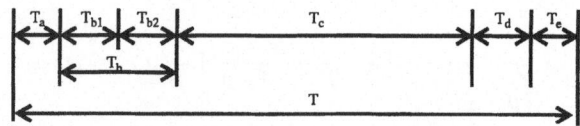

Fig. 4. Test application – Time chart

At the beginning of the communication a TCP control flow between both endpoints is set up (T_a). On the control channel, the parameters for the subsequent audio communication are negotiated $(T_b; T_{b1}$ is the post dial delay, T_{b2} is the post pickup delay). Then the audio flows are initiated and media packets are exchanged. The session setup time as well as the media QoS parameters are measured. When the session time is exceeded (T_c), the session teardown is initiated. Appropriate messages are exchanged on the control channel and the media channels are closed (T_d), finally the control flow is closed (T_e). The session setup time according to the definition in Section 2.1 is given by: $T_s = T_a + T_{b1} + T_{b2}$

Configuration. The *KOMtraffgen* system has to be configured by specifying the number n of concurrent active application sessions and the session duration T_c. Also the specification of the media flows has to be given (packet rate, packet size).

The time between session setups is exponentially distributed which generates a Poisson process of session setups. The setup rate is implicitly specified by: $\lambda = n/ T_c$.

Calibration. Before the quality index of the firewall can be determined, a calibration measurement without any firewall intervention is necessary. Two computers, one running the client part of *KOMtraffgen*, the other one running the server part of *KOMtraffgen* are connected via a 100 Mbit Ethernet switch and an intermediate router (see Figure 5). Then the session setup times for different setup rates are measured. The setup rate λ is adjusted by varying T_c with a fixed n. The calibration curves are later used to determine the difference in the session setup time introduced by the analysed firewall.

4.2   Experiment Setup

For the experiment, two different firewall systems - shown in Figure 5 - have been used. The first firewall system (FWa) is an implementation of architecture AI (see Section 1), the second firewall system (FWb) is an implementation of architecture AII.

Both firewall systems are based on firewall components, called the *KOMproxyd* system implemented by ourselves [1]. Our own firewall implementation was necessary for two reasons. First, a locally distributed firewall with several signalling processing units (according to FWb) is not available. Second, it is necessary to be able to compare the measurement results of the two firewall systems. This is only possible if both systems only differ in the interaction between signalling and media processing. If both systems are internally structured differently it is nearly impossible to determine performance differences caused by the architectural changes.

In the first scenario (FWa), the interaction between the signalling and media processing component is implemented as I/O-controls. In the second scenario, the exchanged information between the components is transported by a reliable UDP-based Firewall Control Protocol (FCP). All machines used are PIII 450 MHz with a FreeBSD 4.5 operating system. All links are 100 MBit full dupex switched Ethernet.

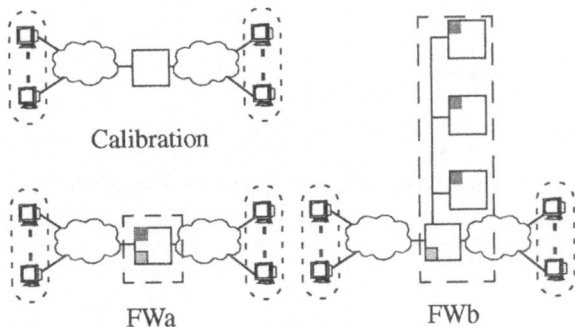

Calibration

FWa   FWb

Fig. 5. Experiment Setup

4.3   Experiment Results

First the monolithic and centralized firewall system FWa is tested. *KOMtraffgen* is parameterized with $n = 50$ concurrent sessions. The setup time required for small session setup rates is nearly constant with $T_{SFW_q} = 24$ *ms*. As the load increases, the setup times rise steeply. If $T_{S_{max}} = 6$ *s* is defined (according to boundary conditions for telephony calls as stated in Section 2.1) and a quality index of $G_{T_s}(n) \leq 2.5\%$ is recommended, we obtain using (1) with $T_c = 180$ s (standard phone call duration) a total capacity of $N_{SFWa} = 7112$.

Second, the firewall system FWb with 1, 2 and 3 processing units is tested. For the measurement of FWb with one processing unit ($p = 1$), *KOMtraffgen* is parameterized with $n = 50$ concurrent sessions. For the measurement of FWb with $p = 2$ resp. $p = 3$ processing units $n = 100$ resp. $n = 150$ concurrent sessions are used.

Using (1), the calibration measurements and the measurement results, the quality indices $G_{T_s}(n)$ as shown in Figure 6 result.

The quality index for the FWb system with one processing unit ($p = 1$) is always higher than the quality index of FWa. The setup time required for small session setup rates is nearly constant with $T_{SFWbp1} = 31$ *ms*. The difference between FWa and FWb$_{p=1}$

is caused by the difference in the communication between signalling and media processing component. The transportation of necessary information (e.g. flow specifications to adjust the filter configuration of the media processing component) over the network accounts for an additional 7 ms. Therefore, the total capacity is $N_{SFWbp1} = 6188$. For the measurements with multiple signalling processing components the following values have been obtained: $N_{SFWbp2} = 13073$, and $N_{SFWbp3} = 20077$.

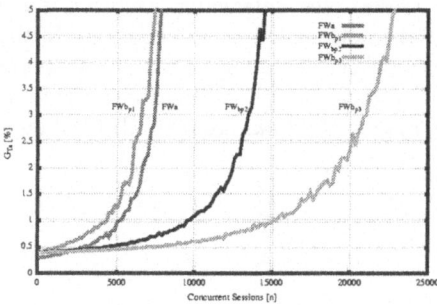

Fig. 6. Quality index $G_{TS}(n)$ for Fwa and FWb

4.4   Discussion

The experiment shows, that the distributed firewall architecture (AII) with $p > 1$ signalling components can be used to overcome the limits of a hybrid system (AI). Therefore, the trend towards distributed firewalls as currently discussed is justified.

Example. The measurement results obtained for the session setup delay can be used to dimension a firewall system. If an application with $r = 2$, $b = 87.2$ $Kbit/2$, and a media processing component with $B = 2$ $Gbit/2$, $N_D = 11468$ is assumed, architecture AII with $p > 1$ as used in the experiment is necessary to be able to fully utilize the available media processing capacity.

Comparison. If the total capacity of FWa and FWbp1 is compared, we see that 12.9% of the processing capacity of the signalling component has to be spent to implement the FCP communication. Therefore, architecture AII is only useful regarding performance optimization if used with $p > 1$.

Efficiency. If the total capacity of the firewall system FWb is compared using equation (5) we obtain:

$\alpha(p = 1) = 1$ $\alpha(p = 2) = 1.06$; $\alpha(p = 3) = 1.08$

At first glance it is surprising that the efficiency is slightly greater than 1 and that this factor is nearly independent from the degree of distribution. Yet, according to the performance model introduced in Section 3, this behavior has to be expected. A detailed comparison of the model and the experimental results is given in the next section.

5   Comparing Model and Experiment

For the comparison of the experimental results and the models introduced in Section 3, values for the variables and reflecting the experiment have to be determined.

Adaptation. To determine the service rate μ and the variance of the service time σ the appropriate model curve is fitted to the measurement curve using:

$$\begin{bmatrix} \mu \\ \sigma \end{bmatrix} = \underset{0 \le \mu,\, \sigma \le \infty}{\arg min} \left\{ \sum_{\aleph \le N} (\Delta T_{S_A}(n, \mu, \sigma) - \Delta T_{S_{FW}}(n))^2 \right\} \qquad (12)$$

For the model AI fitted to FWa $\mu = 42.5$ and $\sigma = 0.01$ is obtained. If the model of AII is fitted to the measurement curve of FWbp1 $\mu = 37.9$ with $\sigma = 0.01$ is obtained. In both cases, the variance of the service time σ is very close to 0. Thus, this gives evidence that the investigated firewalls process sessions with nearly constant service time. Therefore, the simplified models based on a M/D/1 queue are applicable (see Section 3.1).

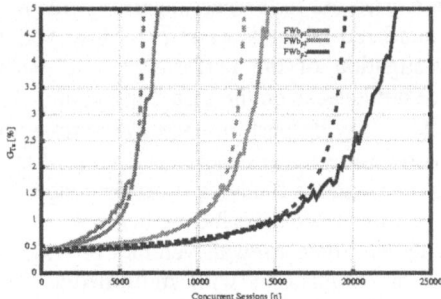

Fig. 7. Quality index $G_{T_s}(n)$ FWb

Comparison. For the comparison, the values μ and σ obtained from the fitting described before are used. Again the quality index $G^{All}_{T_s}(n)$ for the session setup delay is determined. With $T_{S_{max}} = 6\ s$ and $T_C = 180s$ using (1) the results shown in Figure 7 are obtained. Figure 7 also shows the measurement results $G^{FWb}_{T_s}(n)$ of Section 4.3.

Quality of Prediction. The model for FWa and FWbp1 can be used to calculate the FCP communication overhead. This number can be compared with the communication overhead determined by the experiment (Section 4.4). For the measurement an overhead of 12.9%, for the model an overhead of 10.8% is obtained (16% deviation).

If the model to determine the total capacity of the system assuming a recommended quality index of $G^{All}_{T_s}(n) \le 2.5\%$ is used, the results shown in Table 1 are obtained.

As it can be seen (Figure 7), the prediction of the the model regarding the total capacity tends to be more precise in the area where the signalling processing components are not stressed by heavy load. Compared with the experiments described in Section 4 the model allows us to predict the quality index curve (with $G^{All}_{Ts}(n) \leq 2.5\%$) with a deviation of at most 8%.

Table 1. Total Capacity for $G^{All}_{Ts}(n) \leq 2.5\%$

Measurement	Model	Deviation
$N_{S_{FWbp1}} = 6188$	$N_{S_{AIIp1}} = 6156$	0.05 %
$N_{S_{FWbp2}} = 13073$	$N_{S_{AIIp2}} = 12312$	5.8 %
$N_{S_{FWbp3}} = 20077$	$N_{S_{AIIp3}} = 18468$	8.0 %

6   Related Work

The performance of firewalls has always been a critical issue. Therefore, much research work has been carried out in the past regarding this topic. For basic firewall performance tests, standardized methods exist [9]. However, none of the previous work covered the investigation of the performance of *multimedia* firewalls and especially of perfomance bottlenecks on the signalling path.

Many firewall vendors provide performance evaluations of their firewalls (e.g. [10]). These evaluations do not give an exact description of the performed measurements. In addition, these evaluations focus on other protocols like HTTP or FTP and so the results cannot be transferred to describe the behavior of a firewall in interaction with multimedia applications. Some firewall vendors provide information about the performance evaluation in conjunction with multimedia applications resp. UDP processing [5]. Yet, these investigations only cover the media processing and make no statements about the signalling processing.

Beside the performance evaluation of firewalls, performance evaluations of multimedia components are available (e.g. performance evaluation of IP-telephony components [11]). These results also cannot directly be transferred to firewall architectures.

7   Summary

The work presented allows a rating and selection of firewall architectures for multimedia applications regarding performance issues. Therefore, the work clarifies many questions regarding firewall architectures that had been recently discussed (e.g. in the IETF). The contributions of the paper can be summarized as follows.

Bottlenecks. In the paper bottlenecks of multimedia firewalls were identified and analytically described. Lab experiments verified their existence. In particular, bottlenecks caused by the signalling processing component of a multimedia firewall were investigated.

Evaluation. Measurement methods that can be used to rate the performance of multimedia firewalls were developed and described. In addition, publicly available measurement tools are provided that can be used to perform firewall performance evaluation.

Modelling. In the paper a queueing model to describe the performance behavior of multimedia firewalls was introduced. This model was validated by a lab experiment.

Application. The above summarized results of the presented work allow two main applications. First, it is possible to use the analytical model to dimension multimedia firewalls. With the now available methods an unnecessary waste of resources can be avoided. Second, the model can be used to integrate a firewall actively in a network providing some form of QoS assurances. The model can be used to predict the behavior of a firewall and thus allows the derivation of information necessary for a dynamic admission control in a QoS-supporting network.

References

[1]   U. Roedig. Firewall Architectures for Multimedia Applications. PhD thesis, Darmstadt University of Technology, November 2002.

[2]   R. Knobbe, A. Purtell, and S. Schwab. Advanced security proxies: an architecture and implementation for high performance network firewalls. In Proceedings of DARPA information survivability conference and exposition 2000, pages 140–148, 2000.

[3]   P. Srisuresh, J. Kuthan, J. Rosenberg, A. Molitor, and A. Rayhan. Middlebox communication architecture and framework. Internet Engineering Task Force, RFC 3303, August 2002.

[4]   U. Roedig, M. Görtz, M. Karsten, and R. Steinmetz. RSVP as Firewall Signalling Protocol. In Proceedings of the 6th IEEE Symposium on Computers and Communications, Hammamet, Tunisia, pages 57–62. IEEE, July 2001.

[5]   NetScreen. NetScreen-500 System Product Description. P.Num.: 2002.6.50.1.500, 2002.

[6]   International Telecommunication Union. Network grade of service parameters and target values for circuit-switched services in the evolving ISDN. Recommendation E.721, Series E: Overall Network Operation, Telephone Service, Service Operation and human factors. Telecommunication Standardization Sector of ITU, Geneva, Switzerland, 1999.

[7]   European Telecommunications Standards Institute. End-to-End Quality of Service in TIPHON Systems; Part 2: Definition of speech Quality of Service (QoS) classes. Draft, Telecommunications and Internet Protocol Harmonization over Networks, ETSI, 2000.

[8]   L. Kleinrock and R. Gail. Queueing Systems: Problems and Solutions. John Wiley & Sons, 1996.

[9]   B. Hickman, D. Newman, S. Tadjudin, and T. P. Martin. Benchmarking Methodology for Firewall Performance. Internet Engineering Task Force, RFC 3511, April 2003.

[10] The Tolly Group. Test summary NetScreen-5200 versus Nokia IP740 and Cisco Systems Inc. PIX 535. Document No. 202121, March 2002.

[11] T. Eyers and H. Schulzrinne. Predicting Internet Telephony Call Setup Delay. In Proceedings of the 1st IP-Telephony Workshop (IPtel 2000), Berlin, Germany, April 2000.

A Multi-level TCP Model with Heterogeneous RTTs

Pasi Lassila[1] and Michel Mandjes[2,3]

[1] Helsinki University of Technology, P.O.Box 3000, FIN-02015-HUT, Finland.
Pasi.Lassila@hut.fi
[2] CWI, P.O. Box 94079, 1090 GB, Amsterdam, the Netherlands.
Michel.Mandjes@cwi.nl
[3] University of Twente, P.O. Box 217, 7500 AE Enschede, the Netherlands.

Abstract. This paper studies TCP performance focusing on the mean flow transfer delay and the average number of concurrent flows in the system. This is done for a dynamic population of users, rather than a static number of permanent flows. The modeling approach relies on using idealized processor sharing models as a starting point and modifying their properties to take into account some of TCP's non-ideal behavior. To this end, a model is derived that incorporates the effect of packet losses and RTTs on the goodput, as well as limited sending rates. Also, the unequal sharing of bandwidth between flows with different RTTs can be treated. The delay estimates take into account the initial slow start. Extensive ns2 simulations are used to verify the accuracy of the models.

1 Introduction

TCP data transfers account for most of the traffic volume in the Internet. TCP has been designed to support efficient and reliable transmission of data that tolerates variations in the throughput. An important task in TCP performance analysis is to study the dependence of throughput and file transfer delays as functions of the traffic parameters and network parameters. Traditional approaches to modeling TCP performance can be roughly grouped into two classes, which we call (i) flow-level models, and (ii) packet-level models.

Flow-level models are idealized models that include flow level dynamics, but ignore the impact of packet losses and round trip times (RTTs) on TCP performance. So-called processor sharing (PS) queues can be used in this context, see, e.g., [1] and [2]. PS queueing models and a number of their generalizations are attractive due to their insensitivity properties. Packet-level models capture more details of the system (RTTs, buffer size, etc.), but assume a constant number of persistent flows. Here an important result has been the "square-root-p" formula (see, e.g., [3]), which captures the impact of RTT on the throughput, for a given packet loss probability. By noticing that the throughput affects, in return, the packet loss probability, fixed-point models can be used to compute both throughput and loss probability, see, e.g., [4], [5] and [6]. The work in [7]

N. Mitrou et al. (Eds.): NETWORKING 2004, LNCS 3042, pp. 52–63, 2004.
© IFIP International Federation for Information Processing 2004

on modeling the conditional mean file transfer delay is also a packet-level model as no flow-level dynamics are considered.

Work on combined flow/packet-level modeling is done by Gibbens et al. [5]. To take into account the fluctuating number of concurrent flows N, they assume a priori N to be either Poisson or geometrically distributed. In Lassila et al. [8], a model is presented that combines both levels by computing first, at the packet level, the conditional throughput given the number of flows. These are then used as input to the flow level model by using a model that is sometimes called Generalized Processor Sharing (GPS) [9]. Hence the distribution of N results from the model. Both approaches apply a packet-level queueing model to obtain the throughput (as a fixed point). In Massoulié and Roberts [2], a GPS model is utilized, as well, but no queueing model is used to obtain the packet loss probabilities. Bu and Towsley [10] also study a single congested link and use the idealized unequal processor sharing model in [11] to provide a model for the mean flow transfer delays. In Ayesta et al. [4], a model for short TCP flows with stochastic flow arrivals has been given in, where the conditional mean-delay model is parameterized by the packet-level estimate for packet loss. To estimate the mean number of flows in the system, an $M/G/\infty$ approximation is used.

In this paper we develop a combined flow/packet-level model for studying TCP performance under user-heterogeneity, in the sense of users having different RTTs. We first consider a group of TCP users with identical RTTs, sharing a single congested link equipped with a finite buffer, and then generalize the model to heterogeneous RTT user groups with an access-rate limitation. Our contribution is that our model, while being tractable, covers all above features, whereas previous models included them only partially. The model in [10] considers user heterogeneity, but not the impact of limited access rates, nor finite buffers. On the other hand, the model in [4] considers buffer sizes and limited access rates, but not the impact of bandwidth sharing. The GPS model in [8] captures finite buffers and limited access rates, but can not be easily extended to heterogeneous RTTs. Our approach is based on deriving a system of differential equations representing the expected change per time unit in the packet sending rate (throughput) and the number of flows in the system. Thus, it is a generalization of the model in [6], where the number of flows is constant. Here we add a flow-level equation to the system. From the model we are able to obtain the mean delay and mean number of flows in the system. To increase the accuracy of the mean-delay estimates, additional heuristics are given to take into account the effect of initial slow starts. The models are validated through ns2 simulations.

2 TCP Model with Stochastic Flow Arrivals

We model the ideal behavior of stochastically arriving Reno-type TCP flows in the so-called congestion avoidance phase sharing the capacity of a single bottleneck link. Thus, timeouts are ignored and it is assumed that the sending rate of a TCP source increases linearly as long as acknowledgements are received and that the rate is halved for each packet loss. To this end a model is derived which consists of two parts: a flow-level model and a window-level model. At the packet level, we make a stationarity assumption and use a finite-buffer M/G/1 queue model (finite M/M/1 and M/D/1 models are used in the numerical results). The model is first derived for TCP users with the same RTT (homogeneous users) and then the model is generalized to the case of heterogeneous RTTs.

2.1 Homogeneous TCP Users

Consider TCP flows sharing a link with capacity C (pkts/sec) and buffer size K packets. Flows arrive according to a Poisson process with rate ν and the file lengths are assumed to be exponentially distributed with mean $1/\mu$ packets. The aggregate mean packet sending rate of the TCP population is denoted by $\lambda(t)$ at time t. All flows are assumed to have the same RTT, $R(\cdot)$,

$$R(t) = R_0 + d(t) \ ,$$

where R_0 represents the constant packet transmission and propagation delays and $d(t)$ is the mean delay in an M/G/1/K buffer with arrival rate $\lambda(t)$. The mean window size of the TCP population is denoted by $\bar{W}(t)$, and thus, the mean sending rate is given by $\lambda(t) = \bar{W}(t)/R(t)$. Correspondingly, the mean sending rate of a single TCP flow, that we give index i, is $\lambda_i(t) = \bar{w}_i(t)/R(t)$, where $\bar{w}_i(t)$ is the mean window size of flow i.

Flow-level equation: Consider a small time interval Δt. At the flow level, during Δt either a flow arrives or a flow departs. As flows arrive at rate ν, a flow arrival occurs with probability $\nu \Delta t$. Given that there are flows present in the system, a single flow in the system sends at mean rate $\lambda_i(t)$, but the goodput is $\lambda_i(t)(1 - P(t))$, where $P(t)$ is the packet loss probability. Thus, the probability of flow i completing its transfer during Δt equals $\mu\lambda_i(t)(1-P(t))\Delta t$. Assuming that the flows are independent, the probability that any flow finishes is $\mu\lambda(t)(1 - P(t))\Delta t$, where $\lambda(t) = \sum_i \lambda_i(t)$. The change in the mean number of flows, $\bar{N}(t)$, during Δt can be expressed as

$$\Delta\bar{N}(t) = \nu\Delta t - \mu\lambda(t)(1 - P(t))(1 - \pi_0(t))\Delta t \ ,$$

where $(1 - \pi_0(t))$ is the probability that there are flows present in the system. Letting $\Delta t \to 0$ we obtain

$$\frac{d\bar{N}(t)}{dt} = \nu - \mu\lambda(t)(1 - P(t))(1 - \pi_0(t)) \ . \tag{1}$$

The dynamics of $\pi_0(t)$ are unknown, but at the flow level, we assume that $N(t)$, the process of the number of flows present in the system at time t, behaves as in an ordinary PS queue. Even then we do not have an exact expression for the dynamics of $N(t)$, but we can use a similar quasi stationarity approximation as is done at the queue level in [6]. In equilibrium, π_0 is a function of the load ρ, $\pi_0 = 1 - \rho$ and $\bar{N}$ is also a function of ρ, $\bar{N} = \rho/(1 - \rho)$. Eliminating ρ from these gives $1 - \pi_0 = \bar{N}/(1 + \bar{N})$. Thus, the differential equation (1) becomes

$$\frac{d\bar{N}(t)}{dt} = \nu - \mu\lambda(t)(1 - P(t))\frac{\bar{N}(t)}{1 + \bar{N}(t)} \ . \tag{2}$$

Note that in the above $\lambda(t)(1 - P(t))$ represents the aggregate goodput of the flows, conditional on the event that there is a positive number of flows in the system. In an ordinary PS system this goodput is equal to C, the bandwidth of the bottleneck, implying that the flows can utilize the capacity fully without packet losses. In reality, TCP only approximates this ideal behavior, and packet losses and RTTs affect the goodput of the flows. To capture this effect, we next derive the equation describing the change in the expected value of the conditional mean sending rate $\lambda(t)$. In effect, the idea here is to compensate for the non-ideal performance of TCP compared to PS by assuming the system to behave as a PS system only with a smaller goodput than in the ideal PS system.

Window-level equation: In [6], for a fixed number, say n, of TCP flows, the change in Δt in $\lambda(t)$ has been shown under some approximations to be

$$\Delta\lambda(t) = \frac{\Delta \bar{W}(t)}{R(t)} = \frac{n}{R(t)} \left((1 - P(t)) \frac{1}{\bar{w}_i(t)} - P(t) \frac{\bar{w}_i(t)}{2} \right) \lambda_i(t) \Delta t , \qquad (3)$$

where $\lambda_i(t)\Delta t$ is the probability of a packet arrival from flow i and the term in brackets represents the change to the aggregate window size if a packet is accepted ($1/\bar{w}_i$ term) or lost ($\bar{w}_i/2$ term). In the setting of the present paper, n is a random variable which evolves according to a stochastic process $N(t)$. However, assuming that (3) holds approximately for any n, averaging it over the distribution of $N(t)$ gives

$$\Delta\lambda(t) = \frac{\bar{N}(t)}{R(t)} \left((1 - P(t)) \frac{1}{\bar{w}_i(t)} - P(t) \frac{\bar{w}_i(t)}{2} \right) \lambda_i(t) \Delta t .$$

Doing so means that the arrival and departure of flows are not modelled at the window level. Instead, we assume the time scale of changes in $N(t)$ to be much slower than that of the changes to the window sizes, i.e., that the file transmission times are long (cf. the separation of time scales principle). Thus, upon a change in $N(t)$, the window of a new flow quickly reaches the new stationary value and the other flows also adapt to the situation quickly (similarly for flow departures). Then, by noting that $\lambda_i(t) = \bar{w}_i(t)/R(t)$ and that $\lambda_i(t) = \lambda(t)/\bar{N}(t)$, and letting $\Delta t \to 0$, we obtain

$$\frac{d\lambda(t)}{dt} = (1 - P(t)) \frac{\bar{N}(t)}{R(t)^2} - P(t) \frac{\lambda(t)^2}{2\bar{N}(t)} . \qquad (4)$$

At the packet level in the queue, it is assumed that the window level always observes packet losses and queuing delays resulting from a stationary $M/G/1/K$ system with arrival rate $\lambda(t)$.

Complete model and steady-state solution: Combining (2) and (4), the complete model is given by

$$\begin{cases} \dfrac{d\bar{N}(t)}{dt} = \nu - \mu\lambda(t)(1 - P(t)) \dfrac{\bar{N}(t)}{1 + \bar{N}(t)} , \\ \dfrac{d\lambda(t)}{dt} = (1 - P(t)) \dfrac{\bar{N}(t)}{R(t)^2} - P(t) \dfrac{\lambda(t)^2}{2\bar{N}(t)} , \end{cases} \qquad (5)$$

where $R(t) = R_0 + d(t)$, $d(t)$ is the mean delay in an $M/G/1/K$ system with arrival rate $\lambda(t)$ and $P(t)$ equals the loss probability in an $M/G/1/K$ system with arrival rate $\lambda(t)$.

The steady-state solution of (5) is obtained by setting its right hand side equal to zero. From the second equation for $\lambda(t)$, we get $\bar{N} = \alpha R \lambda$, where $\alpha = \sqrt{P/(2(1-P))}$. Inserting this in the first equation results in a second order equation, for which the only positive solution equals

$$\lambda = \frac{\nu}{2\mu(1-P)} \left(1 + \sqrt{1 + \frac{4\mu(1-P)}{\nu\alpha R}} \right). \tag{6}$$

Note that to solve the above one needs to solve a fixed point equation since by (5) both P and R depend on the value of λ. Although confirmed by extensive numerical experiments, we have not succeeded in formally proving uniqueness of the fixed point.

The mean-delay model: The above model would already enable us to model the mean delay by simply using Little, i.e., the mean delay equals $\bar{N}/\nu$. However, this is not a very accurate model of reality as it assumes that the mean goodput rate of a single flow $\lambda(1-P)/\bar{N}$ is available instantly, whereas in reality there is a certain time how long it takes for the TCP's sending rate (window size) to grow up to the estimated steady-state rate.

Our basic idea is to approximate the total mean delay of a file transfer, $\bar{D}_{tot}$, by taking into account how much of the file is sent during the time it takes to reach the estimated steady-state goodput rate (initial slow start), $\bar{D}_{ss}$, and the remaining file size is then sent at the steady-state goodput rate, $\bar{D}_{eq}$, i.e., $\bar{D}_{tot} = \bar{D}_{ss} + \bar{D}_{eq}$. Observe that often it can also happen that the file size is so small and/or the bandwidth-delay product is so large that the predicted goodput rate is never reached, and the whole file is transmitted during the initial slow start. From our above model, the following performance measures are obtained: packet loss probability P, total goodput rate $\lambda(1-P)$, and mean RTT including the queuing delay R. The mean goodput rate equals approximately $\lambda(1-P)/\bar{N}$, except for cases where the mean number of flows in the system is so low (i.e., load is low) that $\lambda(1-P)/\bar{N} > C$, in which case we simply approximate that the goodput rate per flow equals C. Thus, the goodput rate of the TCP flows, $\bar{r}$, equals $\bar{r} = \min(C, \lambda(1-P)/\bar{N})$, which corresponds to a window size $\bar{w} = \bar{r} \cdot R$.

On the other hand, the mean time m (expressed in the number of RTTs) to send out a file of size $1/\mu$ assuming that the entire file is sent during slow start, is obtained from $\sum_{i=0}^{m} 2^i = 1/\mu$ yielding $m = \lceil \log_2(1/\mu - 1) \rceil - 1$. Note that the numbering of rounds starts from index 0 (actual number of rounds equals $m+1$). If $2^m \leq \bar{w}$ the file is sent during the slow start before the system reaches equilibrium, i.e., $\bar{D}_{ss} = (n+1)R$. If $2^m > \bar{w}$, the flow reaches equilibrium in $a = \lfloor \log_2 \bar{w} \rfloor$ rounds (indexing starts from 0 again) during which $(1 - 2^{a+1})/(1 - 2) = 2^{a+1} - 1$ packets are sent. The time to transmit the file equals the sum of the mean delay from slow start, $\bar{D}_{ss} = (a+1)R$, and the equilibrium delay $\bar{D}_{eq} = (1/\mu - 2^{a+1} + 1)/\bar{r}$ (time it takes to send the remaining packets at the equilibrium goodput rate). Putting the above pieces together gives us the following model for the mean transfer delay of files of size $1/\mu$,

$$\bar{D}_{tot} = \begin{cases} \bar{D}_{ss} = (m+1)R\,, & \text{if } 2^m \leq \bar{w}\,, \\ \bar{D}_{ss} + \bar{D}_{eq} = (a+1)R + (1/\mu - 2^{a+1} + 1)/\bar{r}\,, & \text{otherwise}\,, \end{cases} \quad (7)$$

where $m = \lceil \log_2(1/\mu - 1) \rceil - 1$ and $a = \lfloor \log_2 \bar{w} \rfloor$.

2.2 Heterogeneous TCP Users and Limited Access Rates

Here we present heuristics to extend the model to the case of M heterogeneous user groups, each with their own flow arrival rate ν_k and constant link delays $R_{0,k}$, for $k = 1, \ldots, M$. Each TCP user group may also have a limitation in their sending rate, $\lambda_{\max}^k$, caused, e.g., by an access link. The GPS model [9] does not apply under user heterogeneity and hence the approach in [8] can not be used. An idealized model for unequal bandwidth sharing is the so-called DPS model [11] (and applied in [10]). However, in its full generality it allows the mean file size to be different among the user groups and this model cannot be easily modified to take into account the effect of packet losses on goodput. Here we utilize the properties of the DPS model under the natural assumption of a common mean file length for all user groups and aim to give a simple approximate model that captures both the effect of a limited access rate, packet losses and different RTTs.

To model the above, observe that when the load of the system is low enough such that the number of flows in the system rarely exceeds the limit at which the bottleneck fills up, the system behaves as an $M/G/\infty$ system where the flow is only constrained by its access link rate and the behavior of TCP slow start. When the load is high enough to fill the bottleneck link, the system becomes a processor sharing system where the flows share the bandwidth. In this case, the total arrival rate of flows equals $\sum_k \nu_k$ and the aggregate goodput is $\sum_k \lambda_k(t)(1 - P(t))$, where $\lambda_k(t)$ is the mean sending rate of population k at time t. Now it is easy to see that for any work conserving service discipline (such as DPS), assuming exponential file lengths with a common mean $1/\mu$ for all classes, the total number of flows in the system is a Markov process with the same properties as the number of users in an ordinary PS system. Hence, we can use the same model for the time evolution of the mean total number of fl0ows in the system $\bar{N}(t)$ as earlier,

$$\frac{d\bar{N}(t)}{dt} = \sum_k \nu_k - \mu \sum_k \lambda_k(t)(1 - P(t)) \frac{\bar{N}(t)}{1 + \bar{N}(t)}\,. \quad (8)$$

For each TCP user group, the sending rate $\lambda_k(t)$ is determined by

$$\frac{d\lambda_k(t)}{dt} = (1 - P(t)) \frac{\bar{N}_k(t)}{R_k(t)^2} - P(t) \frac{\lambda_k(t)^2}{2\bar{N}_k(t)}\,, \quad (9)$$

where $R_k(t) = R_{0,k} + d(t)$ and $N_k(t)$ is the mean number of population k flows.

Given (8) and (9), we need a model for $\bar{N}_k(t)$ that captures the unequal sharing of the bottleneck bandwidth among heterogeneous users. In our system, the load caused by population k equals $\nu_k/(C \cdot \mu)$. However, according to our model, each population obtains a goodput equalling $\lambda_k(t)(1 - P(t))$, which includes the effects of packet loss and RTT. Thus, the *effective* load caused by population k is given by $\nu_k/(\mu \cdot \lambda_k(t)(1 - P(t)))$, and we approximate $\bar{N}_k(t)$ by dividing $\bar{N}(t)$ in proportion to the effective load of each class,

$$\bar{N}_k(t) = \frac{\nu_k/(\mu \cdot \lambda_k(t)(1 - P(t)))}{\sum_k \nu_k/(\mu \cdot \lambda_k(t)(1 - P(t)))} \bar{N}(t) = \frac{\nu_k/\lambda_k(t)}{\sum_k \nu_k/\lambda_k(t)} \bar{N}(t) . \quad (10)$$

Complete model and steady-state solution: Combining (8) and (9), the complete model is given by

$$\begin{cases} \dfrac{d\bar{N}(t)}{dt} = \sum_k \nu_k - \mu_k \sum_k \lambda_k(t)(1 - P(t))\dfrac{\bar{N}(t)}{1 + \bar{N}(t)} , \\ \dfrac{d\lambda_k(t)}{dt} = (1 - P(t))\dfrac{\bar{N}_k(t)}{R_k(t)^2} - P(t)\dfrac{\lambda_k(t)^2}{2\bar{N}_k(t)} , \quad k = 1, \ldots, M, \end{cases} \quad (11)$$

where $\bar{N}_k(t)$ is given by (10).

The steady-state values are again obtained by setting the right hand sides of (11) equal to zero. By solving $\bar{N}$ from the first equation as a function of λ_k, and solving from the rate equation for λ_k, the steady-state solution for λ_k can be expressed in the form

$$\lambda_k = N_k(\lambda_1, \lambda_2, \ldots) \cdot \sqrt{\frac{2(1 - P(\sum_k \lambda_k))}{P(\sum_k \lambda_k)}} \cdot \frac{1}{R(\sum_k \lambda_k)} ,$$

i.e., in the form of a fixed point equation. To prove the uniqueness of the fixed point is not easy but numerical experiments indicate a unique solution exists.

The mean-delay model: The model for the mean delay developed in the previous section can be used in this case with a simple modification. The model above provides estimates of the following performance measures: packet loss probability P, total goodput rate $\lambda_k(1 - P)$ of each population, and mean RTT including the queuing delay R_k. At the equilibrium, the goodput rate of a single flow is limited either by its access link rate, $\lambda_k^{\max}$, or, if the mean number of flows in the system is high enough, by the sharing of the bottleneck link. Thus, the goodput rate of the TCP flows of population k, $\bar{r}_k$, equals $\bar{r}_k = \min(\lambda_k^{\max}, \lambda_k(1 - P)/\bar{N}_k)$. With this modification the rest of the model for the file transfer delay is the same as given by (7).

The mean number of flows in the system: Here we can use similar ideas as in [4]. In the case, where the bottleneck load is high enough, the model above provides an estimate of the mean number of flows, $\bar{N}_k$. However, when the load is below the threshold for sharing, the system operates as an $M/G/\infty$ system with arrival rate equal to ν_k and mean service time equal to $\bar{D}_{tot,k}$, as given by (7). In an $M/G/\infty$ system, the distribution of the number of flows obeys a Poisson distribution with parameter $\nu_k \bar{D}_{tot,k}$, and the mean also equals $\nu_k \bar{D}_{tot,k}$. Thus, the mean number of flows is obtained from $\bar{N}_{k,final} = \max(\bar{N}_k, \nu_k \bar{D}_{tot,k})$.

3   Numerical Results

Here we present numerical results obtained from our models, and compare them against simulation results, which have been produced using the ns2 simulator version 2.1b9a. We experiment with different TCP variants (TCP Reno and TCP SACK), buffer sizes, RTTs and flow size distributions. In the simulations, the packet size for all TCP sources is 500 bytes, and the mean file size is 500 packets. We also experiment with different queuing models, namely $M/M/1/K$ and $M/D/1/K$, which are used as part of our packet-level model.

3.1   Examples with Homogeneous TCP Users

We first consider tests with TCPs having identical RTTs and illustrate the impact of the access rate limitation and the scaling of bottleneck capacity on the performance. By introducing an access rate limit on the flows the packet arrivals can be made less bursty, hence making our Poisson assumption more plausible. By scaling the bottleneck capacity our system becomes more like an $M/G/\infty$ system with constant mean delays and linearly increasing mean number of flows. This is illustrated in Figure 1 for the mean delays and in Figure 2 for the mean number of flows. In all figures, results are shown as a function of the load of the bottleneck link. Each figure has two sets of curves corresponding to results for two different RTTs ($R_0 = 20$ ms and $R_0 = 200$ ms) to assess the effect of the bandwidth delay product. In each set of curves, solid lines represent solutions to our analytical models and dashed lines are simulation results.

In Figure 1 (left), the bottleneck capacity $C = 10$ Mbps, the access links of the flows have the same capacity, $\lambda^{max} = 10$ Mbps, and the buffer size $K = 10$. To evaluate the effect of the TCP variant, results for TCP Reno and TCP SACK sources are shown. The system is, in a sense, a processor sharing system where any flow can utilize the full link capacity. This produces rather bursty packet arrivals and to compensate for this we have used the $M/M/1/K$ model as our packet level model (as opposed to $M/D/1/K$, which would be the more 'realistic' model). As seen from the results, TCP SACK is able to avoid time outs more effectively than TCP Reno (and thus has lower delays), especially at low loads. Thus, in the following we only use TCP SACK sources. Results for $R_0 = 200$ ms

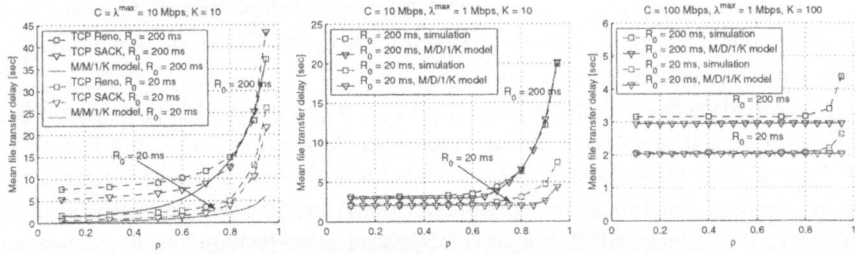

Fig. 1. Mean file transfer delays for three different scenarios.

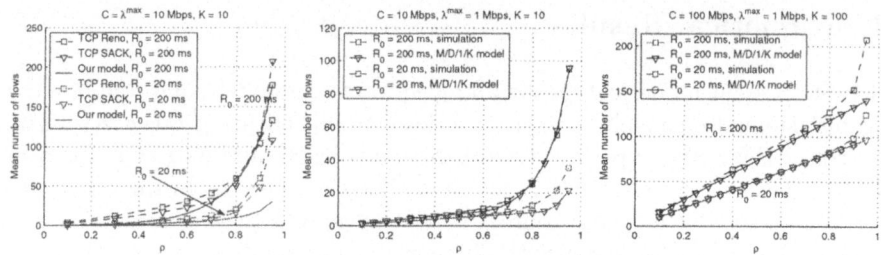

Fig. 2. Mean number of flows for three different scenarios.

are acceptable but for $R_0 = 20$ ms the model underestimates the delays. One reason for the inaccuracy is the very bursty nature of the packet arrivals.

Next we introduce an access rate limitation to smoothen the packet arrival process, and hence our Poisson assumption at the packet level should be more appropriate. As TCP actually sends constant size packets, the queue model is $M/D/1/K$ (using $M/M/1/K$ overestimates the packet losses). The result can be seen in Figure 1 (center), where $C = 10$ Mbps, access rate $\lambda^{max} = 1$ Mbps and $K = 10$. In the figure it can be seen how the results become more accurate. Finally, we scale the bottleneck capacity to $C = 100$ Mbps and also increase the buffer size to $K = 100$ in an attempt to have the system behave as an $M/G/\infty$ system with constant mean delays. The result is shown in Figure 1 (right). It can be seen that our model, indeed, predicts that the mean delays stay constant but in the high load region the delays in the simulated system are not quite constant and our model underestimates the simulation results somewhat.

Figure 2 contains the results for the mean number of flows in the system for the three cases described above. The accuracy of the results is similar to those of the mean delays. The right figure nicely shows how the system approximates the $M/G/\infty$ system with linearly increasing mean number of flows.

In general the accuracy depends on the numerous parameters of the system, but due to lack of space a systematic evaluation of the accuracy of the model can not be presented. Shortly, the dependencies are such that smaller buffer sizes give more accurate results; for larger buffers the model underestimates the loss probabilities. For smaller file sizes, our model does not estimate the available goodput for flows accurately, as our assumptions on the time scale decomposition between the rate adaptation time and flow interarrival times does not hold. Moreover, our TCP model does not take into account timeouts, which are more important the smaller the file sizes are.

System dynamics and effect of different file size distributions: Here we explore the insensitivity of the steady-state solution and the effect of the distribution on the dynamics. The main emphasis here is on the results concerning system dynamics under different distributions. Similar results on the insensitivity of the steady state have appeared elsewhere in the literature, see, e.g., [8], and they are shown here mainly for the sake of completeness. We study first the effect of different file size distributions on the mean file transfer delays.

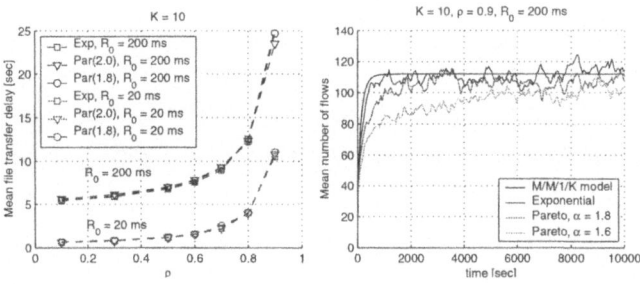

Fig. 3. Mean steady-state file transfer delays for $R_0 = \{20, 200\}$ ms (left) and the length of the transient period for different distributions (right).

The distributions that are used are: exponential, Pareto with shape parameter $\alpha = 2.0$ and $\alpha = 1.8$. In the simulation, the constant parameters are: $K = 10$ and $C = \lambda^{max} = 10$ Mbps. The varied parameters are the bottleneck link load and $R_0 = \{20, 200\}$ ms. As can be seen in Figure 3 (left), the steady-state mean file transfer delays are hardly affected by the distribution. In Figure 3 (right) the accuracy of our dynamical model is illustrated for exponential file lengths. Additionally, we show the impact of the type of the distribution on the time to reach stationarity. In the figure, the transient evolution for mean number of flows in the system is plotted in a system where $R_0 = 200$ ms, $K = 10$ and bottleneck load $\rho = 0.9$. Three file size distributions are considered: exponential and Pareto with shape parameters 1.8 and 1.6. The averages have been obtained by averaging over 50 sample paths for each case. As can be seen, the dynamics of the simulation with exponential file sizes nicely match the results of our analytical model. Regarding the effect of the distribution on the dynamics, for Pareto with shape parameter 1.8 the transient time is still manageable, though longer than for exponential. However, for Pareto with shape parameter 1.6 the transient to reach stationarity is much longer than for exponential.

3.2   Experiments with Heterogeneous RTTs

Next we experiment with four TCP populations with different RTTs that share a single bottleneck. Only results for the mean delays are presented due to lack of space (results for the mean number of flows are similar in accuracy). The bottleneck link bandwidth $C = 10$ Mbps and each TCP source has an access rate $\lambda^{max} = 1$ Mbps, and the link delays were chosen such that $R_{0,i} = \{30, 70, 150, 230\}$ ms. The bottleneck buffer size is either $K = 10$ or $K = 100$ and for the analytical results the queue model is M/D/1/K. To explore the impact of the distribution, the file length distributions were exponential and Pareto with shape parameter 2.0. The results for the mean delays are given in Figure 4. We can observe a rather good match between the simulation and analytical results, especially for the small buffer cases. For larger buffers the analytical model gives results that somewhat underestimate the simulated results. Note that the distribution type does not greatly affect the mean delays.

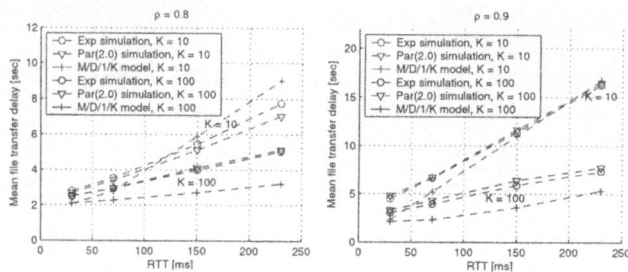

Fig. 4. Mean transfer delays with different RTTs for $\rho = 0.8$ (left) and $\rho = 0.9$ (right).

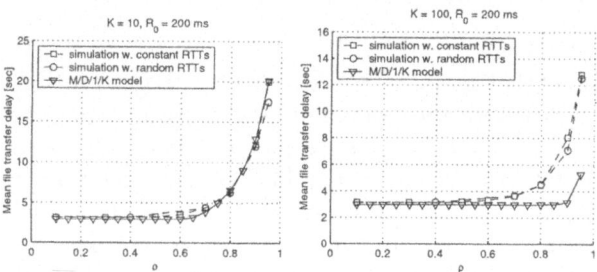

Fig. 5. Mean transfer delays with random RTTs for $K = 10$ (left) and $K = 100$ (right).

Tests with random RTTs: Here we present results from experiments where the delay on access link of each flow is drawn from a distribution separately for each new flow. We consider the same scenario as earlier with access rates being equal to 1 Mbps and the bottleneck link rate 10 Mbps. The bottleneck link is set to have a constant two-way delay of 10 ms. The two-way delay of the access link is drawn independently for each file transfer from a uniform distribution in the range [0, 390] ms. Hence, the mean RTT without queuing delay equals 200 ms and the variation in the RTTs of flows is significant. Two different buffer sizes are studied, $K = 10$ and $K = 100$ packets, respectively. We compare the results against simulation results with a constant two-way delay of 200 ms and the results from our model with the $M/D/1/K$ queuing model. The results are shown as a function of the bottleneck load in Figure 5 for the mean delays. As can be seen, perhaps even slightly surprisingly, the means are not really affected by the randomness of the access link delays (variability did increase, though). This suggests that from the point of view of just modeling overall mean delays, a model accounting just for the mean RTT seems enough.

4   Conclusions

This paper studies TCP performance, with a specific focus on the mean flow transfer delay and the average number of concurrent flows in the system. This is done for a dynamic population of users, rather than the situation of a static number of permanent flows. The modeling approach relies on using idealized PS models as a starting point and modifying their properties accordingly to take into account some of TCP's non-ideal behavior. A model has been derived that can incorporate the effect of packet losses and RTTs on the obtained goodput, as well as limited sending rates. Also, the unequal sharing of bandwidth between flows with different RTTs can be treated. The delay estimates additionally consider the impact of the initial slow start. Ns2 simulations have been used to verify the accuracy of the models. In general, the accuracy is better for systems where the ratio of the access link rate to the bottleneck link rate is relatively small, which corresponds to a higher degree of multiplexing on the bottleneck. It is also under such circumstances that the assumption of Poisson arrivals (at the packet level) may be assumed to be more applicable. However, the accuracy is in general dependent on the parameters; small buffer sizes usually give more accurate results. Future research topics include the derivation of more accurate packet-level models, modeling the effect of Web mice that share the capacity with longer flows, and extending the models to a multi-hop context.

References

1. Bonald, T., Proutiere, A., Régnie, G., Roberts, J.: Insensitivity results in statistical bandwidth sharing. In: Proc. of 17th International Teletraffic Congress, Bahia da Salvador, Brazil (2001) 125–136
2. Massoulié, L., Roberts, J.W.: Arguments in favour of admission control for TCP flows. In: Proc. of 16th International Teletraffic Congress, Edinburgh, UK (1999) 33–44
3. Mathis, M., Semke, J., Mahdavi, J., , Ott, T.: The macroscopic behavior of the TCP congestion avoidance algorithm. Computer Communication Review **27** (1997) 67–82
4. Ayesta, U., Avrachenkov, K., Altman, E., Barakat, C., Dube, P.: Multilevel approach for modeling short TCP sessions. In: Proc. of 18th International Teletraffic Congress, Berlin, Germany (2003) 661–670
5. Gibbens, R.J., Sargood, S.K., Eijl, C.V., Azmoodeh, F.P.K.H., Macfadyen, R.N., Macfadyen, N.W.: Fixed-point models for the end-to-end performance analysis of IP networks. In: Proc. of 13th ITC Specialist Seminar, Monterey, CA, USA (2000)
6. Kuusela, P., Lassila, P., Virtamo, J., Key, P.: Modeling RED with idealized TCP sources. In: Proc. of IFIP Conference on Performance Modeling and Evaluation of ATM & IP networks, Budapest, Hungary (2001) 155–166
7. Cardwell, N., Savage, S., Anderson, T.: Modeling TCP latency. In: Proc. of IN-FOCOM 2000, Tel Aviv, Israel (2000) 1742–1751
8. Lassila, P., van den Berg, H., Mandjes, M., Kooij, R.: An integrated packet/flow model for TCP performance analysis. In: Proc. of 18th International Teletraffic Congress, Berlin, Germany (2003) 651–660
9. Cohen, J.W.: The multitype phase service network with generalized processor sharing. Acta Informatica **12** (1979) 245–284
10. Bu, T., Towsley, D.: Fixed point approximations for TCP behavior in an AQM network. In: Proc. of SIGMETRICS 2001, Cambridge, MA, USA (2001) 216–225
11. Fayolle, G., Mitrani, I., Iasnogorodski, R.: Sharing a processor among many job classes. Journal of the ACM **27** (1980) 519–532

Improving the Performance of TCP in the Presence of Interacting UDP Flows in Ad Hoc Networks

Vikram Gupta, Srikanth V. Krishnamurthy, and Michalis Faloutsos

Department of Computer Science and Engineering,
University of California, Riverside,
Riverside, CA 92521, USA
{vgupta, krish, michalis} @ cs.ucr.edu

Abstract. In this paper, we study how fairness affects the performance of TCP over ad hoc networks with IEEE 802.11 at the MAC layer. The problem addressed is that the throughput of TCP flows degrades severely in the presence of heavily loaded UDP flows. Our contribution is twofold. First, we identify the factors affecting the TCP throughput by providing a micro-analysis of the performance at a level of detail that is not seen in previous studies. The intuition obtained from the first part leads us to our second contribution. We propose and study the use of per flow fairness through a mechanism we call **backpressure**. Backpressure improves the performance of TCP flows in the presence of heavy UDP flows. In fact, in some cases, this increased TCP throughput does not affect the throughput of UDP flows. We find that backpressure can increase the TCP throughput by as much as 95%. An advantage of backpressure is that it does not require any changes to the existing TCP or IEEE 802.11 protocols.

1 Introduction

In this paper, we study the effect of fairness on the throughput of TCP flows in the presence of UDP flows in ad hoc networks. The absence of a congestion control mechanism in UDP makes such flows monopolize the available bandwidth and drive TCP flows to starvation. In the Internet, this problem is partially addressed by intelligent buffer management policies such as the Random Early Drop (RED) and priority dropping of UDP packets [1, 2]. In ad hoc networks, the above schemes cannot be applied since congestion is manifested as channel access delays instead of queue build ups. In ad hoc networks, channel access delays at a node are dependent not only on its queue but also on that of its neighbors and possibly nodes further away. In this paper, we demonstrate the adverse effects of UDP flows on TCP performance and propose simple, yet effective mechanisms to overcome these effects.

It is well known that TCP performs poorly over IEEE 802.11 in multi-hop wireless networks [3]. The fundamental cause for this is the inability of the IEEE 802.11 MAC protocol to provide short-term fairness in medium access to certain nodes. In [4, 5], it has been shown that such short-term unfairness can lead to long-term unfairness with respect to goodput achieved by competing TCP flows. In [6], it has been shown that in the presence of UDP based flows the goodput of TCP flows is significantly reduced.

N. Mitrou et al. (Eds.): NETWORKING 2004, LNCS 3042, pp. 64–75, 2004.
© IFIP International Federation for Information Processing 2004

Previous efforts to improve TCP performance in ad hoc networks follow different approaches from the one presented here. Research falls into two categories: a) modifying TCP, or b) replacing the IEEE 802.11 MAC protocol. Most of the TCP-based proposals attempt to make TCP distinguish between packet losses due to mobility and congestion. To this effect, some researchers propose explicit link failure notifications (ELFN) [8] [7]. Significant work has been done at developing novel MAC layer schemes. In [9], Yu et al propose a hybrid scheme where senders as well as receivers are allowed to initiate MAC transmissions. Their results show that in certain scenarios significant improvements in fairness can be achieved without sacrificing throughput. Protocols other than the IEEE 802.11 MAC scheme have also been proposed [10, 11]. In [5] it is shown that intelligently tuning the parameters used in TCP and the IEEE 802.11 protocols improves performance. While these proposals may provide efficient solutions for improving TCP performance over the IEEE 802.11 MAC protocol, it is difficult to change the existing standards for TCP or IEEE 802.11 Distributed Co-ordination Function. Moreover, to the best of our knowledge, none of these solutions is specifically targeted at improving TCP performance in the presence of UDP flows.

In this work, we study: a) how MAC-layer unfairness degrades TCP performance, and b) how fairness could improve TCP performance in the presence of UDP flows. In more detail, we provide an elaborate microscopic study by creating specific scenarios that elucidate the effects of UDP flows on TCP traffic and show that providing fairness alleviates these effects to a large extent. We then propose **backpressure**, a scheme that significantly improves TCP performance in the presence of UDP flows. In a nutshell, our scheme provides burst regulation at the flow level at each node. A forwarding node refuses to accumulate large number of packets from a flow. Once the buffering quota is reached, the node will not accept more packets from that flow before it can forward some of the buffered packets. We show that our scheme can provide a coarse control over bandwidth allocation to TCP and UDP streams. Finally, an advantage of our scheme is that it can be implemented on top of the IEEE 802.11 MAC and does not require any changes to TCP.

The paper is organized as follows. In section 2, we build the background and summarize previous work relevant to the problem. In section 3, we explain the simulation set-up and present results demonstrating the adverse affect of UDP flows on TCP traffic. We also show that medium access control needs to be supported by fair queuing mechanisms, possibly at the network layer to alleviate the problem. In section 4, we present and evaluate backpressure. Finally, we conclude in section 5.

2 Background and Previous Work

In this section, we look into the poor performance of TCP over IEEE 802.11 multi-hop wireless networks. In particular, we first examine the limitations of the IEEE 802.11 MAC protocol when used in a multi-hop wireless environment. We then explain the implications on the performance of higher layer protocols. Due to space constraints, we have not covered certain well-known phenomenon in depth. We refer the reader to [12, 3] for more details on the section below.

One of the configurable parameters of the IEEE 802.11 DCF is the number of attempts made by a node to transmit a particular frame (retry limit). A node, wishing to transmit a frame to its neighbor, makes repeated attempts governed by the above parameter.[1] Upon failure to transmit, the link is assumed to be broken, i.e., the neighbor is assumed to have moved away. The duration between successive attempts to transmit a packet grows exponentially in accordance to the binary exponential back-off [13] algorithm. This algorithm is known to favor the last successful node [13]. Thus, once a node is successful in medium access contention, it is likely to send a large number of packets from its queue before losing the rights for medium access to another node. Nodes that are interfered by this transmission can falsely presume the links to be broken (termed as false link failures, see [3] for details). In general, with the default setting of parameters, frame sizes smaller than 1500 bytes can be transmitted without causing excessive false link failures.[2] However, in general, due to the capture effect [3, 5], a large number of failures occur. In particular, UDP flows are able to starve TCP flows in this scenario [5].

Solutions proposed for the above phenomenon propose changes to either TCP or the IEEE 802.11 DCF. In [5], Jiang et al show that increasing the retry limits leads to an increase in goodput of TCP connections. However, with this solution, it takes a long time to detect real link failures (typically due to mobility). In [14] the authors propose many changes to the IEEE 802.11 MAC such as a less aggressive back-off mechanism and an additional message to prevent false link failures. The false link failures described are known to lead to poor performance of TCP [3]. We have used the Ad hoc On demand Distance Vector (AODV) [15] as the routing protocol in our simulation studies. AODV provides an alternative to mechanism to determine link connectivity by periodically broadcasting Hello messages [16]. As the use of Hello messages has been shown to reduce false link failures, we use these messages in our simulations.[3]

We have utilized a simple fair-queuing scheme in developing our solution. To appreciate the novelty of our work in the presence of other literature on fair-queuing [22 23], consider that any fair-queuing scheme requires 3 policies [20] for choosing:
1) Which packet (queue) will be transmitted? For this, we use simple round-robin.
2) When is the packet transmitted?
3) Which packet(s) will be dropped in case of congestion?

The fundamental innovation of this work lies in combining the last 2 policies to control belligerent flows. Specifically, by restricting the queue size for a flow at each node in the network we are able to prevent a monopoly by belligerent flows. It should be noted that similar concepts have been proposed in the past, albeit for improving TCP congestion control [24] rather than for stemming UDP flows.

In summary, a UDP based flow can build up large queues on nodes on its route. This queue build-up, and the subsequent medium capture [5], creates congestion in the neighborhood of the path taken by the TCP flow. In such a scenario, the TCP flows that intersect[4] with the heavy UDP flow suffer delays and drops. TCP interprets this loss to be a mark of congestion and acts accordingly [17]. In this work, we first show the extent to which fair-queuing improves TCP performance in such a scenario. Later, we improve upon the gains achieved by fair-queuing through back-pressure.

---

[1] Depending on the frame-length, a transmitting node uses one of two different parameters [3].
[2] With default ns2 settings of 802.11, average time spent in repeated back-offs is more than the transmission time for a 1500 byte frame.
[3] This helps us in concentrating on the loss in goodput of TCP in the presence of UDP flows.
[4] Share medium or buffer

3   Tuning of the System Parameters and Simulation Scenarios

In this section, we describe the simulation scenarios and discuss the results. Although the conclusions are drawn from averages computed over many simulation runs, we perform a microscopic study to draw meaningful conclusions; towards this objective we set up specific scenarios and focus on individual traces.

Table 1. Parameters varied in simulations

Protocol	Parameter/Mode	Range Observed	Optimal Settings
TCP (FTP)	Max. segment size	200-1460 bytes	1460 bytes
UDP(CBR)	Data generation rate	50-800 KB/sec	800 KB/sec
	Packet size	200-2920 bytes	1460 bytes
	Num of application	1-2	2
	Time of start	+/-20 sec w.r.t. TCP start time	12 seconds after TCP
	Flow hop length	2-10	8
AODV	Local link maintenance mode	Link layer feedback or Hello Messages	Hello Messages
	Messages interval	0-α	9 Seconds
	Route repair wait duration	0-α	6 Seconds
802.11	Retry limits	(7,4) to (21,12)	21,12

3.1   Simulation Set-Up

We have used NS2 [18] for our simulations. A random topology or the incorporation of mobile nodes makes a microscopic performance analysis extremely difficult if not impossible. Thus, we test the various scenarios in a 13 x 13 static grid topology. Each node is separated from its neighbor by 200 meters. The transmission range of each node is fixed at 250m. A representative example of the topological structure of the network is shown in figure 1a. In this grid, the clients are placed at the corner nodes (0, 12, 156 and 168, refer figure 1a) and mid-way nodes (6, 78, 91 and 162, refer figure 1a) along the edges. The eight connections are labeled as Connection 1 to Connection 8, respectively. An FTP server is hosted at the node at the center of the grid (denoted by S). We use FTP application clients in NS2 to establish TCP based connections with the FTP server. Each client sends packets of fixed size to the server once a connection is established. The metric for performance is the goodput achieved by the 8 TCP clients. To simulate UDP flows, we place application agents that generate Constant Bit Rate (CBR) traffic at certain nodes (refer figure 1a). The traffic generated by these applications is transported through UDP. Further, the shortest hop path between the CBR source and destination passes through the FTP server. We simulated various CBR rates, but only a few have been reported due to space constraints. All simulations were run for fixed durations of 150 seconds.

3.2   Model Used

Due to a wide range of values for parameters in various protocols we faced the difficult task of simulating many possible scenarios. First, we observed that the performance of TCP in the presence of UDP flows is poor with the default settings.[5] We then observed the TCP performance variation with different parameter values. Table 1 lists these parameters. Our strategy was to choose the parameters such that TCP goodput in the *absence* of UDP flows (initial goodput) is high.[6] This can be observed in our choice of values (see column 4 of table 1). It should be noted that using protocols like DSR [19] did not qualitatively affect the results reported in this paper.

3.3   TCP Performance in Presence of UDP Flows

Stabilized Routing. The performance of TCP connections in the presence of UDP flows was found to be poor. Specifically, we observed that in presence of UDP streams, the aggregate goodput of the TCP clients was reduced to 10% of the aggregate goodput achieved in the absence of the UDP flows (initial conditions). The actual throughput achieved is dependent on the routes taken by the UDP streams. To explain this, we analyze two different cases.

Case 1. UDP streams encompass the TCP server. The scenario is shown in figure 1a. Note that once the UDP flows are established, the TCP server is isolated from the clients. Even if the TCP data packets are delivered to the server, the TCP-ACK packets sent by the server still face MAC congestion and are likely to be dropped.

Case 2. UDP streams do not isolate the server. The scenario is shown in figure 1b. Figure 2 compares the performance of various clients for case 1 and 2. Clearly, connections 6 and 7 (node 156 and 162 in the grid) are able to achieve a reasonable goodput in case 2. From these studies (more results are available in [21]) we concluded that:

1) UDP flows load the nodes on the route, leading to medium capture along the paths.
2) It is difficult for TCP based flows to "cross" such a heavily loaded path.

Stabilized Routing and Increased Persistence for MAC Transmission. A large number of MAC transmission failures result in poor TCP performance. We increase the persistence of the 802.11 MAC protocol by increasing the retry limit parameter from the default 7 to 21. We observed that the aggregate goodput of TCP clients improves significantly. Specifically, in the presence of UDP streams aggregate goodput for clients is approximately 33% of the initial goodput, as opposed to the 10% observed previously. Increasing the retry limit is not an acceptable solution, especially in case of mobility. However, for a static topology, this increase seems to

---

[5] Here, default settings refer to the use of default settings of IEEE 802.11 MAC (DHSS), AODV (with link layer detection) and TCP (Reno) as used in ns2 [18].

[6] These settings provide the highest throughput amongst the simulations carried out.

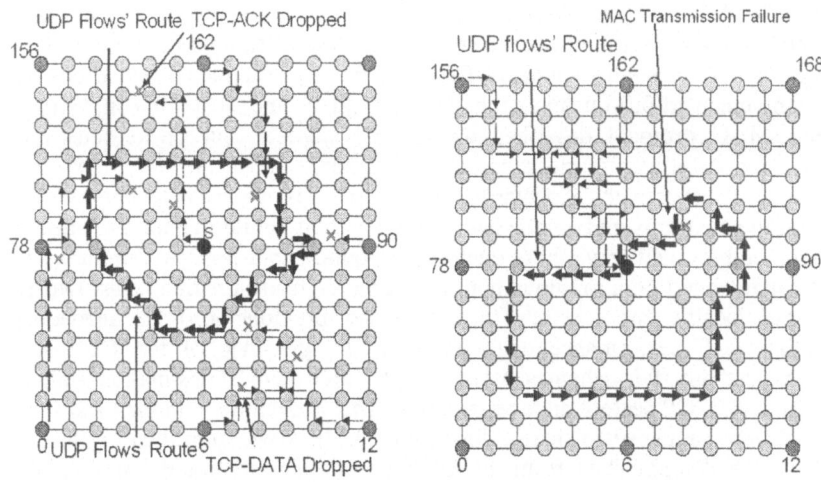

Fig. 1. a. When the TCP Server is encompassed by the UDP streams, all TCP clients suffer **b.** When TCP server is not isolated some TCP clients achieve normal throughput

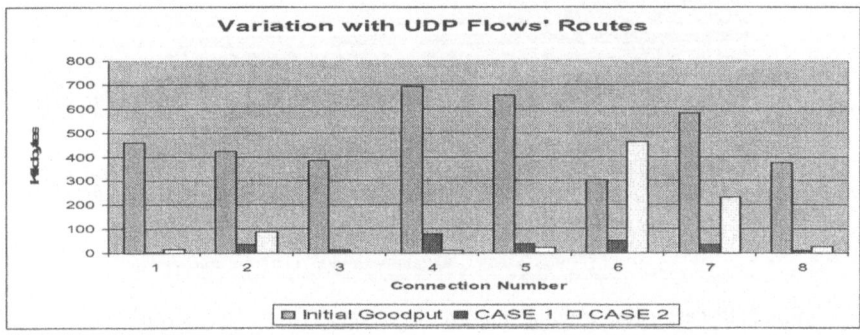

Fig. 2. *Column 1:* Initial Goodput, *Column 2:* Case 1 (all clients suffer), *Column 3:* Case2 (clients 6 and 7 are able to access the server)

benefit TCP clients. Hence, we use this case as a base for comparison with other schemes.

MAC and Network Layer Fairness. One might expect that a fair medium access control scheme would alleviate the effects discussed thus far. One could also expect such improvements by enforcing fairness at the network layer; however for doing so different policies may be chosen. In order to understand the effects of enforcing fairness, we consider the following scenarios for study:

1) We replaced the IEEE 802.11 with a fair MAC protocol. A Time Division Multiple Access scheme was implemented for a 5x5 grid. Four clients were placed on the corners and the server was again at the center. The time slot was chosen such that a MAC frame carrying single TCP packet could be transmitted in a time-slot. Although this scheme is unrealistic, we performed this experiment in order to understand the

effects of MAC layer fairness on TCP performance. The performance of TCP was still found to be poor. The primary cause for this degradation was the use of the FIFO queue at the interface which allows UDP based flows to fill queues on various nodes.
2) We implemented fair-queuing at the interface queue and used it in conjunction with the fair MAC protocol described above. We studied four packet classification schemes for enforcing fairness based on IP-source, IP-destination, next-hop and previous-hop respectively. To clarify the schemes, if we wish to enforce fairness based on the next-hop, and a node had K neighbors, it would maintain K queues, one for each neighbor and serve these queues in a round-robin fashion. Similarly, queues may be maintained on a previous hop, IP-source or IP-destination basis. Our results show that locally fair schemes (next-hop and previous-hop) are inadequate in preventing TCP performance degradation in the presence of the UDP streams. However, globally fair schemes based on IP-source and IP-destination address are able to prevent TCP performance degradation. This is because with the local schemes, if a TCP flow happened to share a link with a UDP flow, from then on, its packets would not be distinguished from the UDP flows' packets (for detailed results see [21]). Having understood the desirable effects of network fairness, we investigate its utility with the IEEE 802.11.

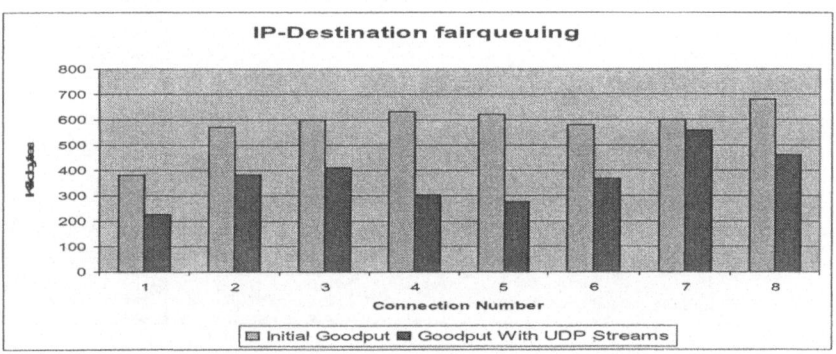

Fig. 3. IP-destination queuing improves aggregate TCP goodput significantly

Network Layer Fair-Queuing with the IEEE 802.11 MAC Scheme. In the scenario simulated, the TCP clients send data packets to the server. For this scenario, IP-destination based queuing is found to provide the best results. This is because TCP ACKs originating from the server are placed in separate queues by nodes on the return path to various clients. Thus, the TCP ACKs can receive up to 80% of the share of the bandwidth. This reduces the round trip times for various TCP connections, thus, increasing their goodput. If instead the server is sending data packets, then IP-source based queuing would be better. Figure 3 shows the performance of the TCP connections with IP-destination based queuing. We observe that:
1) Under the initial conditions without UDP traffic, with fair queuing the average performance of the TCP clients improves marginally (5%).

2) In the presence of UDP based flows, the TCP clients can now achieve about 64% of the initial goodput. Previously, with the increased MAC persistence this value was

about 33% of that achieved under initial conditions). For reasons explained above, for IP-source based fair-queuing this value was 42%.

3) The improvement comes at a price for UDP connections. The average reduction in UDP goodput is about 418 Kbytes. However, the TCP goodput increases by about 1501 Kbytes.

4) The UDP based flows were no longer able to create partitions in the network as observed earlier (Fig 2).
Fair-queuing schemes require each node to maintain state information with regards to the flows passing through the node. However, we argue that unlike in the Internet, this is less of a concern in ad hoc networks wherein most of the nodes would relay only a limited number of flows.

4 Back-Pressure

4.1 Overview

Our objective is to prevent an aggressive UDP source from injecting packets at a rate higher than what the network can afford. We achieve this by restricting the allocated buffer space for each particular flow at every node that the flow traverses. Consider a node that suddenly starts serving incoming flows at a lower rate[7]. Gradually, the buffer occupancy of incoming flows at the node is bound to increase. With the use of back-pressure, gradually, the rate at which the flows send the data to the node will also reduce. The effect travels *backwards* all the way to the source[8]. The source is then forced to adapt its rate to conform to the available bandwidth for the flow. Furthermore, if the congested node is able to serve the flows at a higher rate, then sources are automatically able to send packets at a higher rate.

4.2 Implementation

To implement back-pressure, we set a threshold, referred to as the back-pressure-threshold that restricts the buffer allocation to a particular IP-source (or IP-destination) at any node. Then, using the promiscuous mode of operation[9], a node keeps track of the number of packets in a downstream neighbor's queue. Upon receiving a MAC frame, the nodes operating in a promiscuous mode can determine if a neighbor has transmitted a packet belonging to a particular IP-source (or destination). Thus, for a flow, each upstream node is aware of the queue size at a downstream neighbor to which it forwards packets from the flow. Once the node recognizes that the back-pressure-threshold is reached at a downstream neighbor, it stops transmitting packets from the flow to that neighbor. Subsequently, the node's back-pressure limit is reached which would prevent its previous upstream relay from

---

[7] This can be due to medium access delays or from additional flows passing through the server.
[8] Hence the name back-pressure is adopted for the scheme.
[9] In promiscuous mode, nodes can overhear information broadcast on the wireless channel.

sending further packets from that flow. This effect is propagated all the way to the source of the flow.

It should be noted that the interface queue is a passive queue, i.e., it needs stimulus with respect to de-queuing of the packet. In our scheme the de-queuing of a packet is dependent on buffer of the neighbor. At present, our objective is to measure the gains possible by use of the backpressure. Thus, we have simplified the implementation by utilizing a virtual, globally accessible array that dynamically records the queue lengths for each flow at each node in the network.

4.3 Simulation Results

We observe that back-pressure prevents continuous medium occupation by nodes on the path of UDP flows. TCP goodput increases with this availability of the medium and reduction in transmission failures and buffer overflow. The improvements with IP-Destination based queuing combined with back-pressure are similar and are shown in figure 4. We observe that:

1) With a back-pressure threshold value of 1, the average gain in goodput over IP-Source Address based fair-queuing scheme for the TCP clients is 95%. (82% over the IEEE 802.11 MAC scheme as compared to the 42% achieved with fair-queuing).

2) The gain in TCP does not bring down the goodput of the UDP flows. In fact, we witness a gain of 4.5% in throughput with UDP back-pressure.

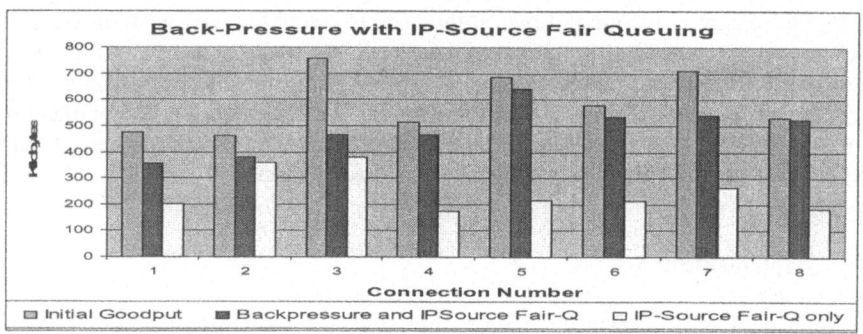

Fig. 4. Back-pressure Improves Performance Significantly in comparison to fair-queuing

4.4 Cause of Goodput Improvement

In figure 5, we plot the packets sent and received by UDP and TCP agents for IP-Source fair-queuing and UDP back-pressure. Notice that with back-pressure, the number of packets injected into the network by the UDP source is almost one-fifth (22%) of that with the simple IP-source fair-queuing. However, the number of packets actually delivered by the UDP flows is almost equal in both cases (back-pressure in fact results in the delivery of 4.5% more packets). With the increased medium availability TCP clients achieve a better throughput as seen in figure 5. This rate adaptation of the TCP source also leads to a reduction in MAC and Interface queue (IFQ) related drops for the TCP connections as well as for UDP flows.

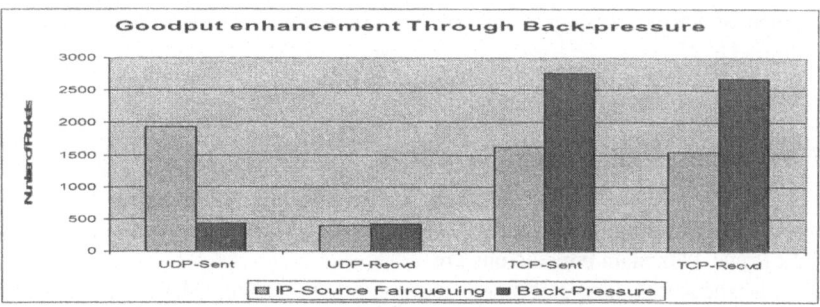

Fig. 5. With back-pressure number of packets injected by the UDP sources in the network is reduced by 80%, while the number of packets received by UDP destination remains the same as with fair-queuing, i.e. source adapts to delivery rate at a destination. Number of TCP packets sent by the clients increases by 95% and so does the packets reaching the TCP server

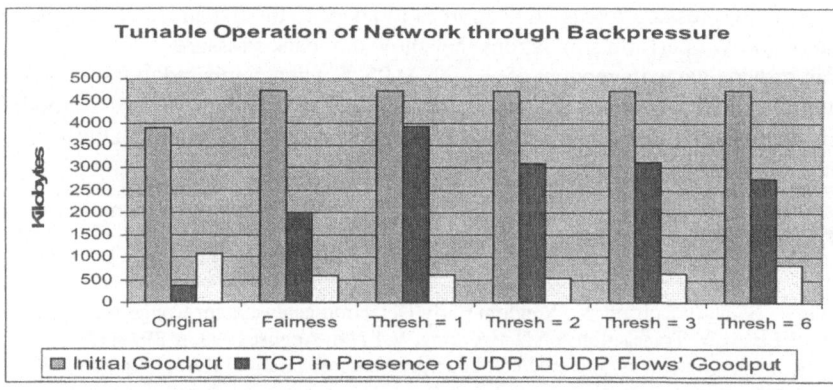

Fig. 6. Tunable operation of Network can be achieved by varying back-pressure threshold value. With Threshold =1, TCP achieves high goodput. Increasing the threshold increases the UDP goodput and reduces TCP goodput. For large value of Threshold, the results are similar to simple fairness

Another significant benefit of backpressure is that it reduces the jitter in the end-to-end delay of the UDP flows' packets. Further, we also witness an increase in jitter with increase in backpressure threshold value (see [21] for details).

4.5   Tunable Operation through UDP Back-Pressure

Figure 6 shows the performance of backpressure with different threshold values. By tuning the backpressure threshold, we are able to offer variable TCP and UDP goodputs. For example, if the threshold value is tuned to 1, the TCP clients achieve 82% of the goodput under initial conditions. Tuning the threshold to a value of 6 provides greater UDP throughput at the cost of TCP goodput. In general, a large threshold value will provide an advantage to UDP flows and a smaller threshold provides an advantage to TCP flows. Detailed results regarding the trade off between the choice of various back-pressure thresholds, retry limits and frame sizes can be found in [21].

5 Conclusions and Future Work

In this work we highlighted the poor performance of TCP connections in the presence of UDP flows. Our main conclusions are:
1) Simple measures like stabilized routing or increasing MAC persistence help achieve a 10% to 33% in the TCP goodput in the presence of interacting UDP flows.
2) A fair MAC protocol alone is insufficient in improving the performance of TCP in the presence of UDP flows.
3) Fair-queuing improves the TCP performance significantly. (33% to 42%, and 33% to 64% for IP-source and IP-destination based queuing respectively).
We propose a new scheme called back-pressure that provides hop-by-hop flow control. Back-pressure forces UDP sources to adapt to the dynamics of congestion in the network. We perform simulations that show that back-pressure
a) Can provide gains in goodput of as high as 95 % when compared to fair queuing.
b) Improves TCP performance significantly without sacrificing UCP flows' goodput.
c) Can be tuned for coarse allocation of bandwidth amongst TCP and UDP flows.

References

1.   Floyd, S., and Jacobson, V.: Random Early Detection gateways for Congestion Avoidance, in IEEE/ACM Transactions on Networking, V.1 N.4, August 1993, p. 397-413
2.   http://www.cisco.com/univercd/cc/td/doc/product/software/ios112/ios112p/gsr/wred _gs.pdf
3.   Xu, S., Saadawi, T.: Does the IEEE 802.11 MAC protocol work well in multihop wireless ad hoc networks?, in IEEE Communications Magazine, 39(6), Jun. 2001
4.   Gerla, M., Tang, K., Bagrodia, R.: TCP performance in wireless multi-hop networks, in 2nd IEEE Workshop on Mobile Computing Systems and Applications (WMCSA'99), Feb. 1999
5.   Jiang, R., Gupta, V., Ravishankar, C.V.: Interactions Between TCP and the IEEE 802.11 MAC Protocol, in DARPA Information Survivability Conference and Exposition (DISCEX), Volume I, 2003
6.   Gupta, V., Krishnamurthy, S., Faloutsos, M.: Denial of Service Attacks at the MAC Layer in Wireless Ad Hoc Networks, in Proceedings of IEEE Milcom, 2002
7.   Monks, J.P., Sinha, P., Bharghavan, V.: Limitations of TCP-ELFN for ad hoc networks, in Proc. of The 7th Int'l Workshop on Mobile Multimedia Communications MoMuC 2000
8.   Holland, G., Vaidya, N.: Analysis of TCP performance over mobile ad hoc networks, in 5th annual ACM/IEEE International Conference on Mobile Computing and Networking, Aug. 1999, pp. 219--230
9.   Wang, Y., Garcia-Luna-Aceves, J.J.: Throughput and Fairness in a Hybrid Channel Access Scheme for Ad Hoc Networks, in Proc. of the IEEE Wireless Communication and Networking Conf. (WCNC 2003)
10. Luo, H., Medvedev, P., Cheng, J., Lu,S.: A Self-Coordinating Approach to Distributed Fair Queueing in Ad Hoc Wireless Networks, in IEEE INFOCOM 2001

11. Kanodia, V., Li, C., Sabharwal, A., Sadeghi, B.,Knightly, E.: Distributed Multi-Hop Scheduling with Delay and Throughput Constraints, in proceedings of ACM MOBICOM 2001
12. http://standards.ieee.org/getieee802/802.11.html
13. Almes, G.T., Lazowska, E.D.: The Behavior of Ethernet-like Computer Communication Networks, Technical Report 79-05-01, University of Washington, 1979
14. Bharghavan, V., Demers, A., Shenker, S., Zhang, L.: MACAW: A Media Access Protocol for Wireless LAN's, in Proc. ACM SIGCOMM 9, pp. 212-25, London, UK, 1994
15. Perkins, C.E., Royer, E.M.: Ad-hoc On Demand Distance Vector Routing, in 2nd IEEE Workshop on Mobile Computing Systems and Applications (WMCSA'99)
16. Chakeres, I.D., Belding-Royer, E.M.: The Utility of Hello Messages for Determining Link Connectivity, in Proceedings of the 5th International Symposium on Wireless Personal Multimedia Communications (WPMC) 2002, Honolulu, Hawaii, October 2002
17. Mathis, M., Mahdavi, J., Floyd, S., Romanow, A.: TCP Selective Acknowledgement Options. RFC 2018, 1996
18. http://www.isi.edu/nsnam/ns/
19. Johnson, D.B., Maltz, D.A., Broch, J.: DSR: The Dynamic Source Routing Protocol for Multi-Hop Wireless Ad Hoc Networks. in Ad Hoc Networking, Addison-Wesley, 2001
20. Demers, A., Keshav, S., Shenker, S.: Analysis and simulation of a fair queuing algorithm, in Journal of Internetworking Research and Experience, vol. 1, no. 1, pp. 3--26, Sept. 1990
21. Gupta, V., "Denial of Service attacks in Wireless Ad Hoc Networks", Masters Thesis. University of California, Riverside, Department of Electrical Engineering
22. Lu, S., Bharghavan, V., Srikant, R.: Fair Scheduling in Wireless Packet Networks, in IEEE/ACM Trans. Networking, vol. 7, no. 4, pp. 473--489, Aug. 1999
23. Vaidya, N.H, Bahl, P, Gupta, S.: Distributed fair scheduling in a wireless LAN, in Sixth Annual International Conference on Mobile Computing and Networking, Boston (2000)
24. Pazos, C.M., Sanchez Agrelo, J.C, Gerla, M.: Using BackPressure to Improve TCP Performance with Many Flows, in IEEE INFOCOM'99, New York, NY, USA, March 1999

The Sensitivity of TCP to Sudden Delay Variations in Mobile Networks

Michael Scharf, Marc Necker, and Bernd Gloss

Institute of Communication Networks and Computer Engineering
University of Stuttgart, Germany*
{scharf,necker,gloss}@ikr.uni-stuttgart.de

Abstract. This paper studies the impact of variable transmission delays on the Transmission Control Protocol (TCP). Sudden delay variations, which are not uncommon in mobile networks, may degrade the performance since they may cause spurious TCP timeouts. The most important parameter in this context is the TCP retransmission timer. In this paper, we analyze TCP's round-trip time estimation for bulk data traffic over wireless links. The main contribution is a new analytical model that accurately predicts the timeout duration from given network parameters. As a first result, the model shows that the round-trip time sampling rate has a significant impact on the timer characteristics. Therefore, the standardized estimation algorithm does not harmonize well with timestamp-based measurement. Second, we quantify the risk of spurious TCP timeouts triggered by changing round-trip times, in particular long off periods. We conclude that delay variations are only critical when they are on the order of seconds.

1 Introduction

The Transmission Control Protocol (TCP) [1] is widely used in the Internet as a reliable end-to-end transport protocol. Mobile Internet access will be one of the key features in networks like the General Packet Radio Service (GPRS) or the Universal Mobile Telecommunication System (UMTS). Their link layer shields the transport layer from data corruption in the radio channel by using forward error correction (FEC) and automatic repeat request (ARQ) mechanisms. However, this error protection comes at the cost of higher latencies and jitter. Further effects in cellular networks are handovers, link outages and radio resource preemption caused by high-priority voice traffic. Thus, the data transmission may suffer from sudden delay variations or so-called "delay spikes" [2,3].

TCP relies on the retransmission timeout (RTO) to detect packet loss. The sender measures the round-trip time (RTT) and dynamically adapts the RTO value as specified in RFC 2988 [4], taking into account both the low-pass filtered RTT samples and the observed delay variance. Sudden delay variations are problematic since they may trigger a spurious timeout even if no packets are

---

* The research work of this paper was done in cooperation with Alcatel SEL AG, Research and Innovation Department, Lorenzstr. 10, 70435 Stuttgart, Germany.

N. Mitrou et al. (Eds.): NETWORKING 2004, LNCS 3042, pp. 76–87, 2004.

lost. This affects TCP performance in two ways: First, the TCP sender retransmits outstanding packets due to the *Go-back-N* mechanism; and second, the congestion window and thereby the sending rate are unnecessarily reduced. As a consequence, TCP may waste scarce bandwidth or underutilize the available resources. Both effects have already been studied extensively [5,6,7,8,9,10].

The most important parameter in this context is the TCP retransmission timeout. However, due to the complexity and the cumulative nature of the RTT estimation, this particular aspect of TCP has rarely been addressed so far. In this paper we give a detailed insight into the characteristics of the RTT estimation, both with and without timestamp-based measurement. We determine upper and lower bounds for the RTO duration as a function of the network configuration. This allows us to quantify the sensitivity of TCP to sudden delay variations. Unlike previous studies [11,12], we use an analytical approach and focus on networks with rather low data rates and high latencies. To the best of our knowledge, this is the first analytical model of the RTT estimator in TCP.

The remainder of the paper is organized as follows. In Section 2, we develop a delay model based on the saw-tooth behavior of bulk data traffic and describe the round-trip time measurement in TCP. Section 3 investigates the performance of the RTO estimation for this delay model using digital signal processing theory. In Section 4, we verify our analysis by simulation and discuss the impact of several parameters. Based on our model, Section 5 evaluates the risk of spurious timeouts. Finally, Section 6 concludes the paper and summarizes our results.

2   An End-to-End Delay Model

2.1   TCP in Networks with a High Bandwidth-Delay Product

Since significant delay variations usually come along with transient events like handovers, we first focus on the rather stationary situation between such events in order to determine the expected RTO duration. An analytical analysis of bulk data TCP traffic over a single bottleneck link has been presented in [13]. The underlying model is illustrated in Fig. 1. Some assumptions are: (1) There is one dedicated drop-tail buffer of fixed size B in front of the bottleneck link, (2) it is not necessary to consider two-way traffic, (3) all data packets are of equal size L_{MTU} ($=1500$ byte), and (4) all delays except for the service time and the queuing delay at the bottleneck link can be lumped together to single latency τ. When accessing the Internet through a cellular network, the radio link is likely to be the bottleneck. We thus argue that this model reflects the characteristics of 2.5 G and 3 G networks, in particular in downlink direction, as long as both the service rate μ and the latency τ remain at an approximately constant level.

In the network model depicted in Fig. 1, the congestion window W_C of a single TCP connection with greedy source follows a saw-tooth pattern because of the congestion control [14]. The sender goes into the *Congestion Avoidance* after having probed the available bandwidth during an initial *Slow Start*. In *Congestion Avoidance*, W_C is gradually increased until the capacity of the path

$$C = \lfloor B + 1 + \mu\tau \rfloor \tag{1}$$

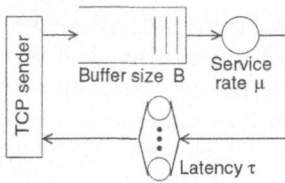

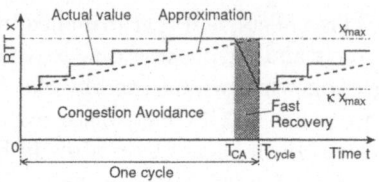

Fig. 1. End-to-end path model   **Fig. 2.** Typical RTT evolution

is exceeded and a single packet is dropped. Both B and W_C are here counted in segments. When detecting this packet loss, the sender enters the *Fast Recovery*, retransmits the missing segment, and approximately halves W_C from $W_{\max} = C + 1$ to $\frac{1}{2} W_{\max}$. On receiving new acknowledgments (ACKs), W_C is increased again. The congestion window thus follows a regular pattern that can be divided into "cycles" [13]. Note that the saw-tooth behavior may be prevented by a small receiver-advertised window W_R. However, W_R is unlikely to be a limiting factor if the *window scale option* [15] is used, as recommended in [16].

The evolution of W_C in one cycle is determined in [13]. We extend this analysis in two aspects. First, the assumption that B is smaller than the minimal bandwidth-delay product $\mu\tau + 1$ contradicts current buffer overprovisioning in mobile networks [2,3]. As a consequence, we rather assume $B > \mu\tau + 1$. Second, we do not neglect the *delayed acknowledgments* in TCP [14]: A TCP receiver should only generate an ACK for every second full-sized segment. Further on, ACKs must not be delayed for more than a certain duration d, which many TCP implementations set to 200 ms [17]. Thus, if $\mu < 1/d$, there is one ACK per received segment ($b = 1$). This corresponds to a data rate $r = \mu L_{MTU} < 60$ kbps. The delay d affects the total latency τ by an additional component $\tau_{DelAck} = d$. For higher data rates, only every second segment is acknowledged, i.e., $b = 2$.

The raise of W_C from $\frac{1}{2} W_{\max}$ to $W_{\max}$ during *Congestion Avoidance* can be described by the differential equation $\frac{dW_C}{dt} = \frac{dW_C}{da} \cdot \frac{da}{dt}$, where $\frac{dW_C}{da} \approx \frac{1}{W_C}$ denotes the rate of window growth per ACK, and $\frac{da}{dt} \approx \frac{\mu}{b}$ is the ACK arrival rate (see [13]). W_C in the first phase of one cycle ($0 \leq t \leq T_{CA}$) is thus given by

$$W_C(t) = \sqrt{\frac{1}{4}(C+1)^2 + \frac{2\mu t}{b}} \ . \tag{2}$$

2.2 Modeling the Round-Trip Time Measurement

While W_C raises, the buffer gradually fills up. At the same time the queuing delay increases and so does the round-trip time x. More precisely, the RTT is always incremented by $1/\mu$ when the congestion window advances by a full-sized segment. Provided that jitter can be neglected, x is a step-wise increasing function [18]. As sketched in Fig. 2, it roughly follows the square-root law given by (2). Apparently, the maximum value is defined by $x_{\max} = C/\mu$. When entering the *Fast Recovery* at the end of the cycle, W_C is halved and the buffer drains. This reduces the RTT by a factor of two.

The evolution of the round-trip time as shown in Fig. 2 is not a very useful function. In the following, we ignore it and instead use a linear approximation: If there are N samples in one cycle, we assume these samples $x(n)$ as

$$x(n) = \kappa\, x_{\max} + \hat{\kappa}\, x_{\max} \frac{n}{N-1} \tag{3}$$

for $0 \leq n \leq N - 1$. As already discussed, the ratio $\kappa\ (= 1 - \hat{\kappa})$ between the smallest and the largest RTT can be set to $\frac{1}{2}$.

The missing parameter in (3) is the number of RTT samples N. TCP may use two different measurement methods: By default, the TCP sender records the time when a segment has been sent. On arrival of an ACK comprising this segment, the RTT is obtained from the difference of the system time to this stored variable. This means that there is always only one ongoing measurement. A more frequent measurement is possible if the timestamp option [15] is included in the TCP header. The sender can then derive a sample from every new ACK.

From an abstract point of view, both methods differ in the sampling rate. In the first case, one measurement is taken per RTT. This corresponds to

$$N_{\text{default}} \approx b\left\lceil \frac{1}{2}\left(C+1\right) + 2 \right\rceil \tag{4}$$

samples in one cycle. In the case $b = 1$, this formula is straightforward: $\frac{1}{2}(C+1)$ is the number of increments in W_C occurring about once per RTT, just like the RTT measurement. The total number of samples is increased by two in order to account for the *Fast Recovery* and the transition back to *Congestion Avoidance*. For $b = 2$, only every second segment triggers an ACK, and $W(t)$ increases at a slower rate. This approximately doubles the number of samples.

With timestamps, every segment is timed during *Congestion Avoidance*. By solving (2) for $W_C(T_{\text{CA}}) = W_{\max}$, one obtains $T_{\text{CA}} = \frac{3b}{8\mu}(C+1)^2$. During this time, μT_{CA} segments are transmitted. Taking further into account the retransmitted segment and one additional window of data, which is sent before the packet loss is detected, the number of RTT samples can be approximated by

$$N_{\text{timestamps}} \approx \left\lceil \frac{3}{8}\left(C+1\right)^2 + \frac{C}{b} + 1 \right\rceil . \tag{5}$$

Note that for $b = 2$ the number of measurements is approximately halved because there is only one ACK for every other segment.

3   Performance of the Round-Trip Time Estimator

3.1   Mathematical Analysis

Setting the retransmission timeout is a challenging issue. On the one hand, an aggressive algorithm tends to trigger frequent spurious timeouts. On the other hand, a conservative timer may cause long idle times in case of packet loss. TCP tries to achieve a trade-off between both extremes by calculating the RTO duration $R(n)$ for every new sample $x(n)$ out of exponentially weighted moving averages for the smoothed round-trip time $s(n)$ and the RTT variation $v(n)$:

$$v(n) = \hat{\beta}\, v(n-1) + \beta \cdot \big| s(n-1) - x(n) \big| \tag{6}$$

$$s(n) = \hat{\alpha}\, s(n-1) + \alpha\, x(n) \tag{7}$$

$$R(n) = \max\big(s(n) + \gamma\, v(n), m\big) \ . \tag{8}$$

According to RFC 2988, the weighting factors should be set to $\alpha = 1 - \hat{\alpha} = \frac{1}{8}$, $\beta = 1 - \hat{\beta} = \frac{1}{4}$ and $\gamma = 4$. Furthermore, the lower bound m should be one second.

Given that the typical round-trip times in fixed networks are significantly lower than one second, $R(n)$ usually is dominated by m, while it is virtually unaffected by the other parameters [11]. In cellular networks, however, the RTT is on the order of several hundred milliseconds or even larger [2,3]. This makes it important to understand the performance of Equations (6)–(8) in detail, in particular if the clock granularity of the operating system is small.

In principle, the computation of the RTO value from RTT samples is a digital signal processing problem, where $s(n)$ is a linear low-pass filter. By applying (3) as input function to (7), the smoothed RTT value can be calculated as

$$s(n) = x(n) + x_{\max}\left(c_1\, \hat{\alpha}^{n+1} - c_2\right) \tag{9}$$

for $0 \le n \le N - 1$, e.g. using the well-known Z-transform. Note that $s(n)$ differs from $x(n)$ by a diminishing component, which results from the step at the beginning of the cycle, and by a constant offset. Utilizing the periodicity $s(n) \equiv s(n+N)$ we obtain:

$$c_1 = c_4 - \kappa + \frac{\hat{\kappa}}{\alpha\,(N-1)} \tag{10}$$

$$c_2 = \frac{\hat{\kappa}\,\hat{\alpha}}{\alpha\,(N-1)} = c_3\,\hat{\alpha} \tag{11}$$

$$c_4 = \frac{1}{1 - \hat{\alpha}^N}\left(1 - c_2 - (\kappa - c_3)\,\hat{\alpha}^N\right) \ . \tag{12}$$

The smoothed RTT variance $v(n)$ is a non-linear function because of the absolute value in Equation (6). Nevertheless, $v(n)$ can be analyzed similarly with the help of $y(n) = \big| s(n-1) - x(n) \big|$ being piecewise defined as follows:

$$\frac{y(n)}{x_{\max}} \approx \begin{cases} c_1\,\hat{\alpha}^n - c_3, & \text{if } 0 \le n < k \ , \\ c_3 - c_1\,\hat{\alpha}^n, & \text{if } k \le n \le N-1 \ . \end{cases} \tag{13}$$

Therein, $k = \ln\frac{c_3}{c_1}/\ln\hat{\alpha}$ is given by the solution of $y(k) = 0$. Applying (13) to $v(n) = \hat{\beta}\, v(n-1) + \beta\, y(n)$ yields the result

$$\frac{v(n)}{x_{\max}} = \begin{cases} -c_3 + (c_5 + c_3)\hat{\beta}^{n+1} + c_1\,\beta\,\dfrac{\hat{\alpha}^{n+1} - \hat{\beta}^{n+1}}{\hat{\alpha} - \hat{\beta}}, & \text{if } n < k \ , \\[2mm] c_3 + \big(v(l) - c_3\big)\hat{\beta}^{n-l} - c_1\,\hat{\alpha}^k\,\beta\,\dfrac{\hat{\alpha}^{n-l} - \hat{\beta}^{n-l}}{\hat{\alpha} - \hat{\beta}}, & \text{if } n \ge k \ , \end{cases} \tag{14}$$

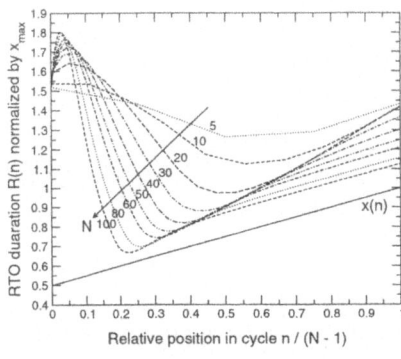
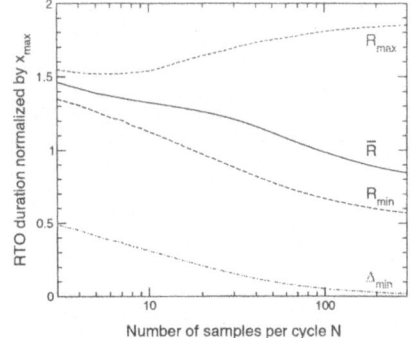

Fig. 3. Timeout duration in one cycle   **Fig. 4.** Characteristic RTO values

where $l = k - 1$ is used as abbreviation. Again, c_5 can be calculated by taking into account $v(n) \equiv v(n + N)$:

$$c_5 = \frac{1}{1 - \hat{\beta}^N} \left(c_3 \left(1 - 2\,\hat{\beta}^{N-k} + \hat{\beta}^N \right) + c_1 \frac{\beta}{\hat{\alpha} - \hat{\beta}} \left(2\,\hat{\alpha}^k\,\hat{\beta}^{N-k} - \hat{\alpha}^N - \hat{\beta}^N \right) \right) .$$

(15)

Finally, we can compute $R(n)$ by inserting (9) and (14) into (8).

3.2 Numerical Results

In Fig. 3, the function $R(n)$ is plotted for different values of N, under the assumption that it is not affected by the lower bound m. The shape of $R(n)$ depends on the sampling rate. If the number of measurements per cycle is small, $R(n)$ remains at an almost constant level and is always greater than the maximum RTT $x_{\max}$. For higher values of N, $R(n)$ steeply increases at the beginning of the cycle. This peak is due to the fact that the drop in the RTT from $x_{\max}$ to $\kappa\,x_{\max}$ turns into a sudden increase of the RTT variation $v(n)$. Subsequently, $R(n)$ decays to a value close to $x(n)$. Here, the exponential moving average does not "remember" information about the previous cycle and the maximum value $x_{\max}$. As a consequence, $s(n)$ converges to a value close to $x(n)$, and $v(n)$ converges to zero. The larger the sampling rate, the more noticeable is this effect.

In order to get a simpler view on the set of curves in Fig. 3, we define the following metrics in addition to the mean value $\overline{R}$:

$$R_{\max} = \max_{0 \leq n < N} \left(R(n) \right) \tag{16}$$

$$R_{\min} = \min_{0 \leq n < N} \left(R(n) \right) \tag{17}$$

$$\Delta_{\min} = \min_{0 \leq n < N} \left(R(n) - x(n) \right) . \tag{18}$$

These metrics are depicted in Fig. 4 as a function of N. Because of the peak, the maximum $R_{\max}$ increases with N. Contrary to this, both the mean $\overline{R}$ and the

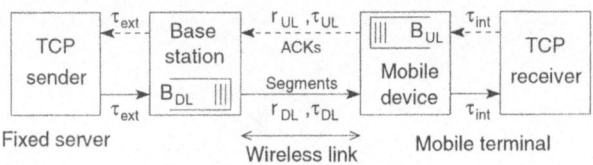

Fig. 5. Simulation topology

minimum R_{min} decrease. This also holds for Δ_{min} that gets very close to zero for $N \gg 10$. This means that the TCP sender is more aggressive when RTT measurement is based on timestamps.

The problem lies in the estimator weights α, β, and γ. They have been defined for the default RTT measurement taking at most one sample per RTT. Choosing adequate weights for the timestamps is an open research issue. One proposed solution is the *Eifel Retransmission Timer* [18] that adapts the parameters to the sampling rate. Also, newer versions of the Linux operating system deploy an algorithm that differs from the TCP specification [17]. Even though these empirical modifications address the issue of timestamp-based RTT measurement, a detailed analysis of their characteristics is still pending.

4   Model Validation

4.1   Simulation Setup

In order to compare our analysis with simulation results, we have modeled uni-directional TCP bulk data traffic over a mobile network using an object-oriented simulation tool [19]. Figure 5 shows the simulation setup.

The important simulation parameters are summarized in Table 1. We use several configurations both concerning the RTT estimator and the network model. The data rates in uplink and downlink direction are chosen from a set of values that are typical for cellular networks. As the link layer is supposed to be highly persistent, packet loss because of data corruption is not taken into account. In this section the radio channel is assumed to be rather ideal, i. e., it does not introduce a significant amount of jitter. B_{DL} is set to an integer value out of a range from 2 to 16 packets, which is equal to 3000–24,000 byte. In uplink direction we

Table 1. Simulation parameters

TCP parameter	Value	Netw. param.	Value
Algorithm	NewReno	(r_{DL}, r_{UL})	(13.4, 13.4); (53.6, 13.4);
L_{MTU} [byte]	1500	[kbps]	(64, 64); (128, 64); (384, 64)
Delayed ACK d [ms]	200	τ_{ext} [ms]	10; 100
Timestamps	disabled; enabled	$\tau_{DL} = \tau_{UL}$ [ms]	100
Granularity G [ms]	10; 200	τ_{int} [ms]	10
Min. RTO m [ms]	200; 1000	B_{DL} [pkts]	2–16
W_R [byte]	65536	B_{UL} [pkts]	unlimited

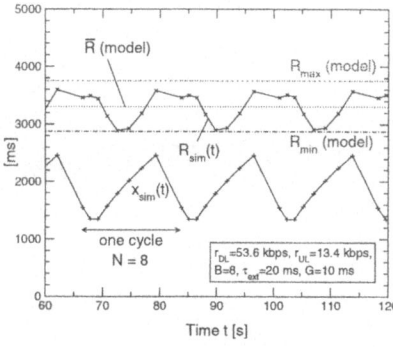 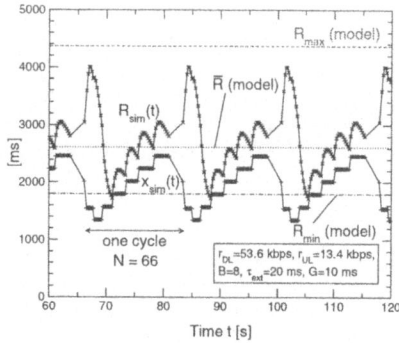

Fig. 6. Default RTT measurement   **Fig. 7.** Timestamp-based sampling

neglect the impact of finite buffer space because of the small size of acknowledgments $L_{ACK} = 40$ byte (52 byte with timestamp option). The corresponding parameters of the analytical model can easily be derived: Obviously, the service rate is $\mu = L_{MTU}/r_{DL}$, and B is equal to B_{DL}. Summing up all latencies yields $\tau = L_{ACK}/r_{UL} + \tau_{DL} + \tau_{UL} + 2\,(\tau_{ext} + \tau_{int}) + \tau_{DelAck}$.

4.2 Simulation Results and Discussion

In the following, we first discuss the accuracy of our analytical model for two selected sample configurations. Figure 6 compares the RTO trace obtained from a simulation with the results of the analytical model. x_{sim} has approximately the expected saw-tooth characteristic with $N = 8$ samples per cycle, which corresponds to $N_{default}$ as given by (4). R_{sim} remains within the bounds R_{max} and R_{min} determined from the analytical model, even though our model slightly overestimates the maximum. This value depends very much on the shape of the RTT reduction at the beginning of the cycle. From Fig. 6 it follows that, instead of immediately dropping from x_{max} to $\kappa\, x_{max}$, as assumed in our analysis, there may be a sample in between. A detailed investigation reveals that this RTT sample is taken immediately after having completed the *Fast Recovery* and measures the RTT of segments released during the *Fast Recovery*.

A further example in Fig. 7 shows the same scenario with RTT measurement based on timestamps. Obviously, the number of RTT samples increases and changes the characteristics of the RTT estimator. Again, $N_{timestamps} = 66$ from (5) perfectly matches the simulation results, and so does the minimum R_{min}. Regarding the maximum, there is a similar effect as in the previous example. However, the most important difference is that the measured RTT x_{sim} only jumps up if W advances by one. Every increment affects the RTT variation $v(n)$ and thus results in a small peak. Since we do not consider this phenomenon in our linear approximation, we may underestimate the mean round-trip time.

In order to validate our model for a larger parameter space, we performed the same analysis for all parameter combinations of r_{DL}, r_{UL}, B_{DL}, and τ_{ext}, as listed in Table 1. R_{max}, $\overline{R}$, R_{min}, and Δ_{min} were measured in the "steady state",

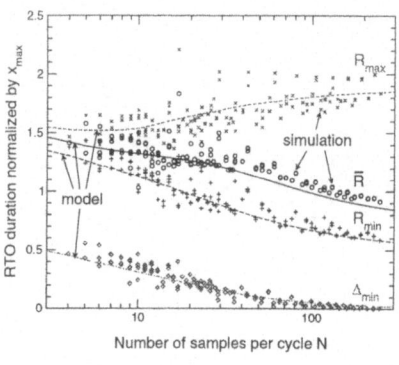 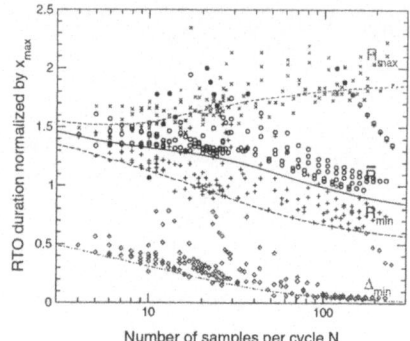

Fig. 8. Comparison model/simulation for $G = 10\,\text{ms}$ and $m = 200\,\text{ms}$

Fig. 9. Comparison model/simulation for $G = 200\,\text{ms}$ and $m = 1\,\text{s}$

i. e., after an initial transient phase had been passed. A separate simulation was started for every configuration with $B > \mu\tau + 1$. We determined N from (4) or (5) and used it as horizontal axis in the diagram. Furthermore, we normalized the measured values by the theoretical $x_{\text{max}} = C/\mu$.

Figure 8 shows the simulation results in a scatter-plot for a fine-granular timer and a rather small m. The comparison with the analytical results reveals a quite good match, in particular for R_{min} and Δ_{min}. As expected, our analysis underestimates the actual value of $\overline{R}$ when N is large, but only slightly. The largest discrepancies can be observed for R_{max}. As already discussed, R_{max} highly depends on the shape of the RTT reduction at the beginning of the cycle. Nevertheless, our model is a good approximation for a large parameter range. Besides, it is almost independent of the TCP variant as long as the *Additive Increase Multiplicative Decrease* mechanism is used.

Many TCP implementations use coarse-grained clocks. Unfortunately, a large granularity G introduces noise and degrades the model accuracy. Furthermore, our results are likely to be biased if R takes values below m. The simulation results in Fig. 9 reveal that for $m = 1\,\text{s}$ model and simulation do not match for a number of configurations. Concerning R_{max} and R_{min}, one can take this effect into account by setting $R'_{\text{max}} = \max(R_{\text{max}}, m)$ and $R'_{\text{min}} = \max(R_{\text{min}}, m)$.

Generally speaking, our analysis applies best if $R(n) > m$ for all n. This, as well as our modeling assumption of a single bottleneck link, is most likely fulfilled by high-latency wireless links, while it may not hold in other networks. We also do not consider jitter, e. g. caused by retransmissions of radio blocks in the link layer. As long as such jitter is on the order of tens of milliseconds, it hardly will trigger spurious timeouts (see next section). But the volatility affects the smoothed variance $v(n)$ and increases $R(n)$. To some extent, one can account for such random fluctuations by replacing μ and τ with their average values $\overline{\mu}$ and $\overline{\tau}$ (see [8]). Our model then provides lower bounds for the RTO duration and therefore is some kind of worst case analysis.

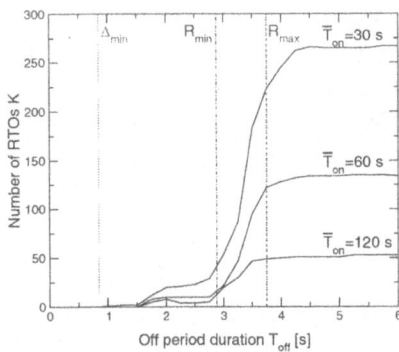 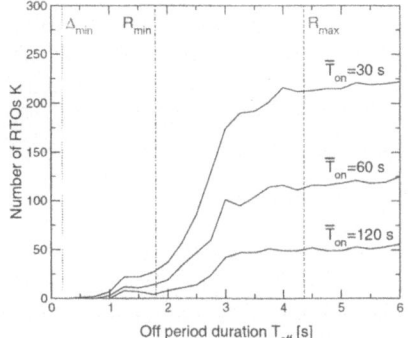

Fig. 10. Number of timeouts for default RTT measurement   **Fig. 11.** Number of timeouts for timestamp-based sampling

5   Quantifying the Sensitivity of TCP to Delay Variations

5.1   Spurious Timeout Probability

Based on the knowledge about the duration of the retransmission timeout, we can calculate the probability of spurious timeouts in the presence of delay variations. We assume the uplink and downlink transmission to be interrupted for a duration T_{off}. The time between such events $T_{on} \gg T_{off}$ is determined from a negative-exponential distribution. During the "off" period, no packets are transmitted but they remain buffered in the base station and the mobile terminal. The question we are interested in is whether such a delay spike triggers a timeout, which would be spurious because no packets are lost. Taking into account that the retransmission timer is restarted whenever an ACK acknowledges new data [4], this requires the inter-arrival time between two successive ACKs to exceed the current RTO duration. Or, in other words, a timeout should occur if $T_{off} > R$.

We have verified this by performing simulations for various values of T_{off}. The total number of retransmission timeouts K during a data transfer of about two hours is shown in Fig. 10 and Fig. 11. The configurations correspond to Fig. 6 and Fig. 7, respectively. As a reference, we also added Δ_{min}, R'_{min}, and R'_{max} from our theoretical model, which are calculated assuming no delay variations. This implicitly assumes that T_{on} is longer than the duration of one cycle.

Obviously, K depends on the frequency of delay spikes. The comparison with the total number of interruptions reveals that there is virtually always a (spurious) timeout if T_{off} is large, e. g. six seconds. In contrast, short interruptions do not trigger any timeout. Figure 10 and Fig. 11 show that the critical interval is between R'_{min} and R'_{max}, which are determined from the analytical model. This motivates us to define a *spurious timeout probability* as follows:

$$P_{TO}(T_{off}) = \begin{cases} 0, & \text{if} \quad T_{off} < R'_{min} \ , \\ \frac{T_{off} - R'_{min}}{R'_{max} - R'_{min}}, & \text{if} \quad R'_{min} \le T_{off} < R'_{max} \ , \\ 1, & \text{if} \quad T_{off} \ge R'_{max} \ . \end{cases} \tag{19}$$

This formula computes the probability that an interruption of duration T_{off} triggers a timeout. As presented, the parameters R'_{max} and R'_{min} can be determined from the maximum path capacity C and the bottleneck service rate μ, provided that it is known whether the TCP sender uses timestamps or not.

5.2 Evaluation

The *spurious timeout probability* quantifies the risk of inefficient interactions between TCP and delays, e. g. caused by the handover management or by resource allocation mechanism. Nevertheless, recall that R'_{min} is on the order of the maximum RTT x_{max} or m, whichever is larger. Assuming $m = 1\,s$, the data transmission must be interrupted for at least one second to trigger a timeout. Thus, a TCP-friendly system design should avoid such delay spikes. If this is not feasible, one might think about optimization techniques such as transport layer protocol helpers [7,8] or TCP enhancements like the *Eifel-Algorithm* [5].

Enabling the timestamp option in mobile networks is recommended by [16], claiming that this "reduces the risk of spurious timeouts". The results both of our analysis and our simulation disagree with this statement: Figure 10 and Fig. 11 show that smaller delay spikes trigger more spurious timeouts for timestamp-based measurement. Therefore, one must carefully balance the increased timer aggressiveness against the faster feedback which timestamps are able to provide.

Of course, (19) is only an approximation. A more detailed look at Fig. 10 and Fig. 11 reveals some timeouts for $T_{off} < R'_{min}$. They occur for two reasons. First, the retransmission timer is not restarted when no new ACKs arrive, e. g. during the *Fast Recovery*. In this case, a timeout can be triggered by any $T_{Off} > \Delta_{min}$. And second, bursty ACK arrival may cause multiple packet drops at the buffer in front of the bottleneck. TCP recovers from this by *real* timeouts. But such buffer overflows are only an issue if the buffer is almost completely filled, and they could be avoided by active queue management techniques.

Finally, it should be mentioned that a spurious TCP timeout may not only be caused by a complete interruption of the data transmission, but also by other delay variations. In particular, a sudden decrease of the service rate from μ_H to μ_L may have the same effect. An analysis of this scenario in [9] concludes that the retransmission timer will not expire if $R > 1/\mu_L$. With $\mu = \mu_H$ we can determine this value from our model and thus refine this condition: A bandwidth oscillation can only trigger a spurious timeout if $\mu_L < 1/R'_{min}$.

6 Conclusion

In this paper, we study the characteristics of the TCP retransmission timer for bulk data traffic over low-bandwidth paths. The main contribution is an analytical model to predict the RTO duration from given network parameters. We model both the standard RTT measurement and timestamp-based sampling. The results show that the sampling rate has a significant impact on the timer performance. From this we conclude that an RFC 2988 compliant timer does not harmonize well with timestamps. Given the complexity of underlying TCP mechanisms, we found that our model and simulation results match well over a wide range of scenarios. Based on the theoretical results, we are able to quantify the sensitivity of TCP to sudden delay variations. Our focus are longer off periods. We determine the probability of spurious timeouts depending on the off period duration, the path characteristics, and the TCP parametrization. As a rule of thumb, delay variations are critical whenever they are on the order of one second.

References

1. J. B. Postel (Editor), "Transmission control protocol," RFC 793, Sept. 1981.
2. A. Gurtov, M. Passoja, O. Aalto, and M. Raitola, "Multilayer protocol tracing in a GPRS network," in *Proc. IEEE VTC 2002 Fall*, Vancouver, Canada, Sept. 2002.
3. R. Chakravorty, J. Cartwright, and I. Pratt, "Practical experience with TCP over GPRS," in *Proc. IEEE GLOBECOM*, Taipei, Taiwan, Nov. 2002.
4. V. Paxson and M. Allman, "Computing TCP's retransmission timer," RFC 2988, Nov. 2000.
5. R. Ludwig and R. H. Katz, "The Eifel algorithm: Making TCP robust against spurious retransmissions," *ACM SIGCOMM Computer Communications Review*, vol. 30, no. 1, pp. 30 – 36, 2000.
6. A. Gurtov, "Effect of delays on TCP performance," in *Proc. IFIP Personal Wireless Communications*, Lappeenranta, Finland, Aug. 2001.
7. J. Schüler, S. Gruhl, T. Schwabe, and M. Schwiegel, "Performance improvements for TCP in mobile networks with high packet delay variations," in *Proc. 17th Int. Teletraffic Congress*, Salvador, Brazil, Sept. 2001.
8. M. C. Chan and R. Ramjee, "TCP/IP performance over 3G wireless links with rate and delay variation," in *Proc. ACM MOBICOM*, Atlanta, USA, Sept. 2002.
9. M. Yavuz and F. Khafizov, "TCP over wireless links with variable bandwidth," in *Proc. IEEE VTC 2002 Fall*, Vancouver, Canada, Sept. 2002.
10. A. A. Abouzeid and S. Roy, "Stochastic modeling of TCP in networks with abrupt delay variations," *Wireless Networks*, vol. 9, no. 5, pp. 509 – 524, 2003.
11. M. Allman and V. Paxson, "On estimating end-to-end network path properties," in *Proc. ACM SIGCOMM*, Seattle, USA, Sept. 1999.
12. M. Allman and J. Griner, "TCP behavior in networks with dynamic propagation delay," in *Proc. IEEE GLOBECOM*, San Francisco, USA, Dec. 2000.
13. T. Lakshman and U. Madhow, "The performance of TCP/IP for networks with high bandwidth-delay products and random loss," *IEEE/ACM Transactions on Networking*, vol. 5, no. 3, pp. 336 – 350, June 1997.
14. M. Allman, V. Paxson, and W. Stevens, "TCP congestion control," RFC 2581, Apr. 1999.
15. V. Jacobson, R. Braden, and D. Borman, "TCP extensions for high performance," RFC 1323, May 1992.
16. H. Inamura, G. Montenegro, R. Ludwig, A. Gurtov, and F. Khafizov, "TCP over second (2.5G) and third (3G) generation wireless networks," RFC 3481, Feb. 2003.
17. P. Sarolahti and A. Kuznetsov, "Congestion control in Linux TCP," in *Proc. USENIX Annual Technical Conference*, Monterey, California, USA, June 2002.
18. R. Ludwig and K. Sklower, "The Eifel retransmission timer," *ACM SIGCOMM Computer Communications Review*, vol. 30, no. 3, pp. 17 – 27, 2000.
19. "IKR Simulation Library," http://www.ikr.uni-stuttgart.de/IKRSimLib/.

Lost Retransmission Detection for TCP Part 2: TCP Using SACK Option[*]

Beomjoon Kim[1], Yong-Hoon Choi[2], Jaiyong Lee[3], Min-Seok Oh[1], and
Jin-Sung Choi[1]

[1] Standardiztion & System Research Group (SSRG), Mobile Communication
Technology Research Lab., CTO, LG Electronics Inc.,
LG R&D Complex, 533, Hogye1-Dong, Dongan-Gu, Anyang-City, Kyongki-Do,
431-749, Korea, {beom,minoh,jinsungc}@lge.com
[2] RAN S/W Group, System S/W Dept., System Research Lab., Telecommunication
Equipment & Handset Company, LG Electronics Inc.,
LG R&D Complex, 533, Hogye1-Dong, Dongan-Gu, Anyang-City, Kyongki-Do,
431-749, Korea, dearyonghoon@lge.com
[3] Department of Electrical & Electronic Engineering, Yonsei University, 134
Shinchon-Dong Seodaemun-Gu, Seoul, 120-749, Korea, jyl@nasla.yonsei.ac.kr

Abstract. The performance of transmission control protocol (TCP) is
largely dependent upon its loss recovery. Therefore, whether packet losses
may be recovered without a retransmission timeout (RTO) or not is a
very important issue. Although TCP using selective acknowledgement
(SACK) option can recover multiple packet losses in a window, it cannot avoid RTO if a retransmitted packet is lost again. In order to cope
with this problem, we propose a simple change to TCP SACK, which
is called TCP SACK+ in simple. We use a stochastic model to evaluate
the performance of TCP SACK+, and analyze its performance comparatively in terms of loss recovery probability. Numerical results evaluated
by simulations show that TCP SACK+ can improve the loss recovery
performance of TCP SACK significantly in presence of random losses.

1   Introduction

Since the specification of Transmission Control Protocol (TCP) is released [1],
implementations of TCP have been enhanced with several mechanisms such as
congestion control [4], [5]. TCP congestion control provides a function to detect
and recover packet losses by retransmissions, which is called loss recovery in
short. If the loss recovery is successful, packet transmission continues without
a retransmission timeout (RTO). The loss recovery function of TCP operates
using two basic algorithms of fast retransmit and fast recovery [2]–[5]. Because
overall TCP performance has close relation to loss recovery efficiency, it has been
a common focus to decrease the number of RTOs invoked if it is failed to recover packet losses by retransmissions [6]. Unnecessary RTOs can be divided into

[*] This work was supported by grant No.R01-2002-000-00531-0 from the interdiscipli-
nary research program of the KOSEF.

N. Mitrou et al. (Eds.): NETWORKING 2004, LNCS 3042, pp. 88–99, 2004.
© IFIP International Federation for Information Processing 2004

three classes. First, if multiple packets are lost in a window at the same time, then all of these packet losses cannot be recovered without RTO frequently. Recently, selective acknowledgement (SACK) option[1] [7], [8] is proposed to avoid performance degradation for this reason. Using SACK option, the sender can be informed about all the packets that have arrived successfully, so the sender need retransmit only the packets that have been lost. Second, if a packet is lost in a small window, then three duplicate ACKs may not be received to trigger a fast retransmit. Using Limited Transmit (LT) [10], most of RTOs corresponding to this class can be avoided. The third class of RTOs is caused by retransmission losses. All of existing TCP implementations including TCP SACK cannot avoid RTO when a retransmission is lost again [6], [11]. In order to avoid RTOs belonging to this class, we propose a simple algorithm that makes it possible for a TCP SACK sender to detect a lost retransmission and recover it by a retransmission. The proposed algorithm requires simple changes only to TCP implementation at the sender and is perfectly consistent with TCP specification such as additive increase multiplicative decrease (AIMD) principle [5]. According to [11], only 4% of the timeouts are due to lost retransmissions. However, it contributes to make TCP SACK more robust and perfect in that the final cause of unnecessary RTOs, which is remaining unsolved, can be avoided with no large modifications.

In order to evaluate the performance of the proposed algorithm, we model the sender's behavior during loss recovery. From the results of the modeling, the exact conditions for successful loss recovery can be derived. For derivation of loss recovery probability and stationary distribution of TCP window in steady-state, we mainly adopt the analysis using Markov process in [12]. Consequently, the improvement of the proposed algorithm can be comparatively analyzed with existing TCP versions in terms of the loss recovery probability that is normalized to the stationary distribution.

The rest of this paper is organized as follows. Section 2 provides a detailed presentation of the proposed algorithm with simulation results. In Section 3 the detailed analysis of loss recovery behaviors of TCP SACK is presented from the aspect of the loss recovery probability. Section 4 contains the numerical results and their discussion. Finally, some conclusions are summarized in Section 5.

2 Description of the Proposed Algorithm

In this paper, our focus is limited to TCP implementation using SACK option.[2] For TCP SACK, the proposed algorithm works for lost retransmission detection

---

[1] In the rest of this paper, we denote TCP implementation using SACK option by TCP SACK in simple. We consider "Sack1" presented in [6] as TCP SACK. Recently, the detailed loss recovery behaviors of Sack1 are addressed in terms of maintaining AIMD principle in [9].

[2] We have already proposed an algorithm called duplicate acknowledgement counting (DAC) which can be applied to TCP Reno and NewReno for lost retransmission detection. It cannot be included in this paper due to page limitation, and will appear in another publication as a Part 1 of this paper. Unlike DAC, because SACK information provides the information of well-transmitted packets, there is no need to count the number of duplicate ACKs.

on the basis of the packets transmitted after a retransmission. In the rest of this paper, TCP SACK using the proposed algorithm is indicated by a plus sign such as TCP SACK+.

2.1 Description of TCP SACK+

For TCP SACK+ operations, the sender keeps a variable per packet loss when its retransmission is performed. In the variable, the highest sequence number of the packets that are outstanding is stored when a lost packet is retransmitted. We denote the highest sequence number at the time of the retransmission of the hth packet loss by S_h. If the retransmission of the hth packet loss is successful, the right edge of the first block within SACK information is always smaller than or equal to S_h. During fast recovery, every time a duplicate ACK is received, the sender checks the right edge of the first block. If the sender receives any duplicate ACK for the hth lost packet indicating that the right edge of the first block is greater than S_h, it determines that the retransmission is lost again and retransmits it immediately. Consequently, if at least a new packet is transmitted successfully after a retransmission, then the sender can detect whether its retransmission is lost or not. From the aspect of *conservativeness*, TCP SACK+ is perfectly consistent with the AIMD principle specified in [5]. As described above, a lost retransmission is detected based on the packets transmitted after the retransmission. It means that if congestion is so heavy that no packets should be transmitted, TCP SACK+ does not transmit additive packets after detecting a lost retransmission because the packets after the retransmission would be likely to be lost as well in such a congested situation. However, a lost retransmission detected assures that there was a quite heavy congestion so that it should be taken as two indications of congestion, which leads to decreasing *cwnd* twice as specified in [5].

2.2 Simulations

Using *ns* simulations, we implement TCP SACK+ and show the loss recovery behaviors of a TCP SACK+ sender in fig. 1 and fig. 2. In these figures, the well-transmitted packet is indicated by a blank square, lost packet by a black-filled square, and an ACK packet by +. In the simulations, several specific packets are forced to be dropped as in [6].

In fig. 1, we compare the loss recovery behaviors of TCP SACK+ with TCP SACK when a single packet and its retransmission are lost. At about 0.7 second, packets 7–14 are transmitted with *cwnd* of 8 and packet 7 is dropped. When the sender receives three duplicate ACKs at about 0.9 second, the sender retransmits packet 7 and sets *cwnd* to $4(= \lfloor 8/2 \rfloor)$ and pipe to $5(= 8 - 3)$. The last duplicate ACK decrease pipe to one so that three new packets, packet 15, 16, and 17, are transmitted after the retransmission. Because the retransmission of packet 7 is lost again, as can be seen in fig. 1-(a), the sender receives the same repetitive duplicate ACKs per every round-trip time (RTT) until RTO occurs. At about 2.3 second, it can be seen that the sender restarts to transmit in slow-start mode.

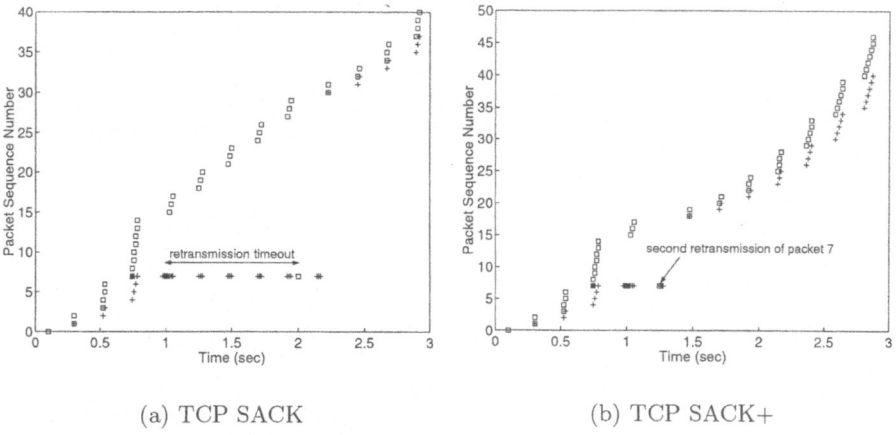

(a) TCP SACK   (b) TCP SACK+

Fig. 1. Comparison between TCP SACK and SACK+ for a single packet loss and lost retransmission. (+ indicates an ACK packet.)

For TCP SACK+ shown in fig. 1-(b), the sender sets S_1 to 14 when it retransmits packet 7. The eighth duplicate ACK includes SACK information indicating that the first block starts with packet 8 and ends with packet 15. At this time, because the right edge of the block is greater than the stored value in S_2, the sender transmits the second retransmission of packet 7 at about 1.3 second. After the retransmission, the sender halves *cwnd* again to be $2(= \lfloor 4/2 \rfloor)$. Therefore, even if two more duplicate ACKs by packet 16 and 17 decreases **pipe** to be 2, no packets cannot be transmitted. At about 1.5 second, the second retransmission delivers an ACK that acknowledges packets up to packet 17, which brings the sender out of fast recovery and congestion avoidance starts with *cwnd* of 2.

We perform the same simulation for two packet losses and the results are shown in fig. 2. In this simulation, two packets 7 and 12 are lost and the retransmission of packet 12 is lost again. After fast retransmit of packet 7, the fifth duplicate ACK decreases **pipe** to be 3 so that one packet transmission is allowed. At this time, the sender can be informed by SACK information in the duplicate ACK that packet 12 is also lost. The final duplicate ACK decreases **pipe** again and packet 15 is transmitted. After a RTT, a partial ACK and duplicate ACK for packet 12 are received, which decrease **pipe** by three, and three new packets, packet 16, 17, and 18 are transmitted. However, the retransmission of packet 12 is lost again, the sender cannot complete fast recovery but RTO occurs eventually at about 2.3 second. For TCP SACK+ shown in fig. 2-(b), the sender sets $D_1 = D_2 = 14$ when it retransmits packet 7 and 12. A partial ACK after a RTT means that the retransmission of packet 7 is well-transmitted so that S_1 is cleared and two new packets, packet 16 and 17, are transmitted. After that, the seventh duplicate ACK by packet 15 is received indicating that the first block starts with packet 13 and ends with packet 15, which is greater than S_2. Therefore, the sender transmits the second retransmission of packet 12

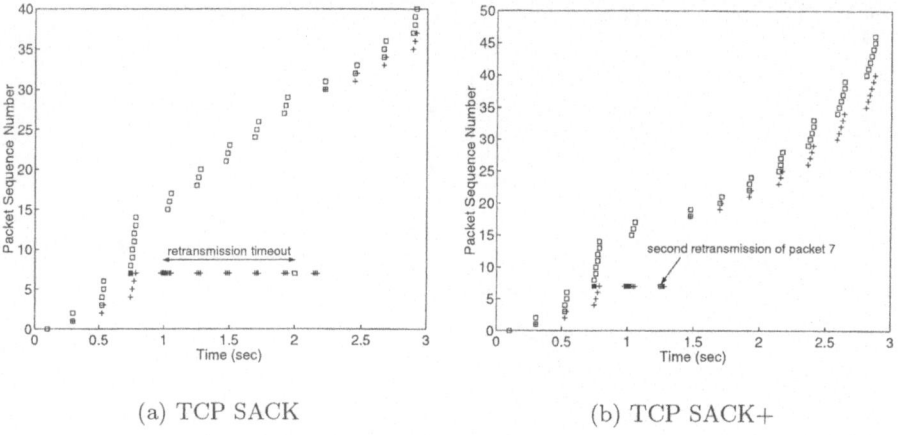

(a) TCP SACK   (b) TCP SACK+

Fig. 2. Comparison between TCP SACK and SACK+ for two packet losses and a single lost retransmission. (+ indicates an ACK packet.)

instead of packet 18 at about 1.3 second. An ACK that acknowledges packets up to packet 17 brings the sender out of fast recovery and congestion avoidance starts with *cwnd* of 2.

3   Modeling and Probabilistic Analysis

We adopt the concept of 'loss window' and 'round' defined in [12] and [13], respectively. If we denote a loss window by Ω and the ith packet loss in Ω by l_i, the first packet that Ω includes is always l_1. Additionally, we define Φ_k as the number of new packets that are transmitted in the kth round in loss recovery period. For n packet losses in Ω of u packets, Φ_0 is always equal to $u - n$. For modeling the evolution and obtaining stationary distribution of TCP congestion window, we mainly follow the procedures presented in [12] under the same assumptions such as fixed packet size, random packet losses, no ACK loss, and infinite packet transmission. We also follow some notations in [12] such as W_{max} for receiver's advertised window and K for *slow-start-threshold*.

3.1   TCP Reno

The recovery probability of TCP Reno is derived in terms of the number of packet losses in a loss window in [12]. In [12], it is assumed that the fast recovery of TCP Reno succeeds only if at most two packets are lost in a loss window. For $\Omega = u$ and a single packet loss, l_1 can be recovered without RTO if $\Phi_0 \geq K$ and the retransmission is not lost. Therefore, for $u \geq K + 1$, its probability is given by

$$R_R^{(1)}(u) = (1-p)^{\Phi_0}(1-p) = (1-p)^u. \tag{1}$$

In the same way, l_2 can be recovered if $\Phi_1 \geq K$ and there is no packet loss during fast recovery. Since Φ_1 is equal to the number of new packets transmitted after the fast retransmission of l_1, $\Phi_1 = \lfloor u/2 \rfloor + \Phi_0 - u = \lfloor u/2 \rfloor - 2.^3$ For $\Omega = u$, when it contents $\lfloor u/2 \rfloor - 2 \geq K$, the recovery probability of l_2 is given by

$$R_R^{(2)}(u) = \binom{u-1}{1}p(1-p)^{u-2}(1-p)^{\Phi_1}(1-p)^2 = (u-1)p(1-p)^{u+\Phi_2}. \tag{2}$$

Using LT, a packet loss can be fast retransmitted if only a single duplicate ACK can be received. Therefore, even if Ω does not content $u \geq K+1$, a single packet loss may be recovered by a retransmission. For $\Omega = u$ and $2 \leq u \leq 3$, its probability, $R_L(u)$, is given by

$$R_L(u) = (1-p)^{u-1}(1-p)^2(1-p) = (1-p)^{u+2}. \tag{3}$$

It means that exactly one packet is lost out of u packets and three packets including two new packets transmitted by LT and the retransmission itself should not be lost. Note that LT works for only a single packet loss or the first packet loss in case of multiple packet losses per window [14]. Therefore, the recovery probability of TCP Reno using LT is given by $R_{RL}(u) = R_R(u) + R_L(u)$.

3.2   TCP SACK

For loss recovery of TCP SACK, packet transmission may be stalled in the second round due to misbehavior of pipe. Suppose that n packet losses are included in a loss window of u packets. After the sender receives all duplicate ACKs for l_1, it sets pipe to $n(= u - \Phi_0)$. If n is equal to or greater than $cwnd(= \lfloor u/2 \rfloor)$, no new packet can be transmitted in the first round. When the sender receives a partial ACK by the retransmission of l_1, it decreases pipe by two so that it is equal to $n - 2$. Again, if $n - 2 \geq \lfloor u/2 \rfloor$, then no more packet can be transmitted but a RTO occurs in the end. Therefore, for successful loss recovery of TCP SACK, the number of packet losses is restricted by $n \leq \lfloor u/2 \rfloor + 1$. Consequently, the loss recovery probability of TCP SACK is given by

$$R_S(u) = \sum_{n=1}^{\mu_1} \binom{u-1}{n-1}p^{n-1}(1-p)^{u-n}(1-p)^n \tag{4}$$

where $\mu_1 = min(u - K, \lfloor u/2 \rfloor + 1)$. When LT is used for TCP SACK, its recovery probability, $R_{SL}(u)$, can be simply derived by replacement $(u - K)$ in μ_1 with $(u - 1)$.

---

^{3} According to our previous work presented in [14], three packet losses may be recovered without a RTO under the strict condition that a loss window size u is large enough to content $\lfloor u/4 \rfloor - 3 \geq K$ and there are at least $u - \lfloor u/4 \rfloor + (K - 1)$ packets between l_1 and l_2. However, its probability is so small that we do not include the case in our derivation of the loss recovery probability of TCP Reno.

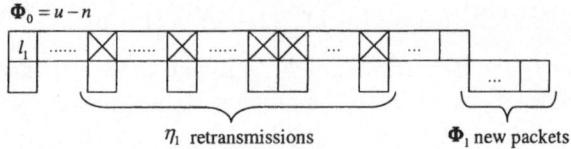

Fig. 3. Loss recovery behaviors of TCP SACK+ when the retransmission of n packet losses is completed in the first round.

3.3 TCP SACK+

We limit the number of lost retransmissions that the proposed algorithm can recover in a loss window to one and the same retransmission loss cannot be recovered twice. Therefore, the recovery probability of TCP SACK+ is given by

$$R_{S+}(u) = R_S(u) + p \cdot \Delta_S(u) \tag{5}$$

where $\Delta_S(u)$ is the probability that a lost retransmission is recovered. The term $p \cdot \Delta_S(u)$, therefore, means the probability that a retransmission is lost again and it is recovered by the proposed algorithm. When SACK option is used, the duration of loss recovery period is dependent on the position as well as the number of packet losses. For simplicity in the derivations of $\Delta_S(u)$, we limit our consideration to the cases that its loss recovery period is completed within a round or two.

First of all, as shown in fig. 3, we consider the case that loss recovery is finished at the first round. In this case, all of packet losses are retransmitted before a partial ACK is received by the retransmission of l_1. For n packet losses in Ω, if we denote the number of retransmissions sent by the decrement of pipe in the kth round by η_k, then we have

$$n = \eta_1 + 1. \tag{6}$$

When all of duplicate ACKs for l_1 are received, the values of cwnd and pipe are equal to $\lfloor u/2 \rfloor$ and n, respectively. Therefore, for $\Omega = u$, total number of packets that can be transmitted during the first round is given by

$$\eta_1 + \Phi_1 = \lfloor u/2 \rfloor - n. \tag{7}$$

For simplicity, as long as pipe permits, a retransmission is assumed to be always transmitted first regardless of its position in a loss window. Then, a retransmission loss out of n retransmissions can be detected if $\Phi_1 \geq 1$. From (6) and (7), the condition is given by

$$1 \leq n \leq \lfloor u/4 \rfloor. \tag{8}$$

If the last packet in a loss window is lost, even if the above condition is satisfied, the retransmission of n packet losses cannot be completed in the first round. Hence, we do not include this case in the derivation of $\Delta_S(u)$. As a consequence, the recovery probability of a lost retransmission in the first round, $\Delta_S^{(1)}(u)$, is

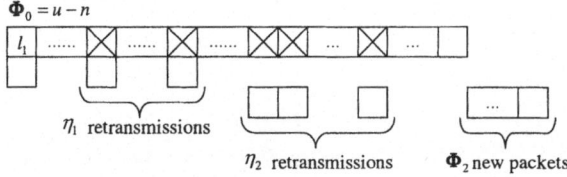

$\Phi_0 = u - n$

η_1 retransmissions

η_2 retransmissions Φ_2 new packets

Fig. 4. Loss recovery behavior of TCP SACK+ when the retransmission of n packet losses is completed in the second round.

given by

$$p \cdot \Delta_S^{(1)}(u) = p \cdot \sum_{n=1}^{\lfloor u/4 \rfloor} \binom{u-2}{n-1} \binom{n}{1} p^{n-1}(1-p)^{u+\Phi_1} \qquad (9)$$

where it reflects the following two facts:

i) $(n-1)$ packets are lost out of $(u-2)$ packets which do not include the last packet in Ω.

ii) A retransmission is lost out of n retransmissions, and Φ_1 packets and a retransmission by SACK+ should not be lost.

If $n \geq \lfloor u/4 \rfloor + 1$, the recovery period continues after the first round as shown in fig. 4. If it is assumed that the retransmission of n packet losses is to be completed in the second round, then we have

$$n = 1 + \eta_1 + \eta_2. \qquad (10)$$

Recalling the assumption that a retransmission is always transmitted first, it can be inferred that all of packets transmitted during the first round are retransmissions; i.e., $\Phi_1 = 0$. Therefore, the sender is going to receive $(\eta_1 + 1)$ partial ACKs in the second round, so that total number of packets that the sender is allowed to transmit in the second round is given by

$$\eta_2 + \Phi_2 = 2(\eta_1 + 1) = 2(\lfloor u/2 \rfloor - n + 1). \qquad (11)$$

Note that a partial ACK decrements **pipe** by two [6]. From (10) and (11), the condition for $\Phi_2 \geq 1$ is given by

$$n \leq 0.75\lfloor u/2 \rfloor + 0.5. \qquad (12)$$

As a consequence, for $\lfloor u/4 \rfloor + 1 \leq n \leq 0.75\lfloor u/2 \rfloor + 0.5$, the recovery probability of a retransmission loss in the second round is given by

$$p \cdot \Delta_S^{(2)}(u) = p \cdot \sum_{n} \binom{u-2}{n-1} \binom{\eta_2}{1} p^{n-1}(1-p)^{u+\Phi_2} \qquad (13)$$

where it reflects the following three facts:

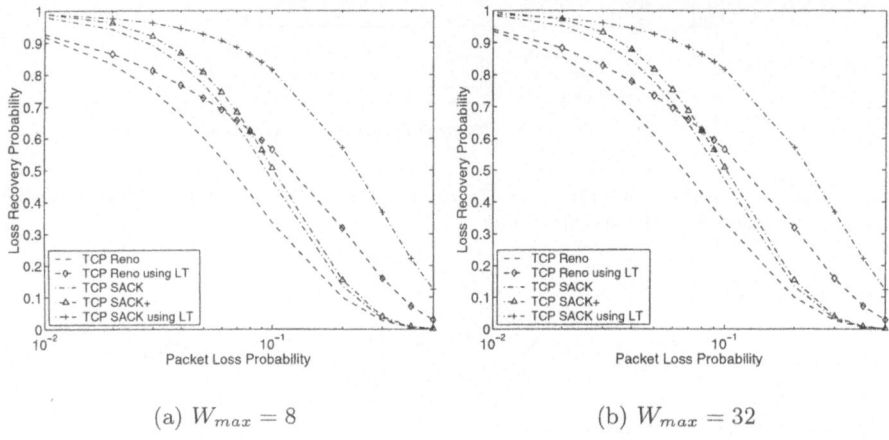

(a) $W_{max} = 8$ (b) $W_{max} = 32$

Fig. 5. Comparison of the loss recovery probability predicted by the developed model.

i) $(n-1)$ packets are lost out of $(u-2)$ packets which do not include the last packet in Ω.

ii) $(\eta_1 + 1)$ retransmissions should not be lost.

iii) A retransmission is lost out of η_2 retransmissions, and Φ_2 packets and a retransmission by SACK+ should not be lost.

From (5), (9), and (13), the total loss recovery probability of TCP SACK+ can be derived.

4   Results and Discussion

We calculate the loss recovery probability[4] that is normalized to the stationary distribution of the window, which is obtained from Markov process of the window evolution for packet loss probability as presented in [12]. The x-axis of each graph indicates packet loss probability, which corresponds to p. We assume that K is always three. In fig. 5, the loss recovery probability of each TCP is compared for two different values of W_{max}. When LT is not used, the loss recovery probability of all TCP starts to drop rapidly when packet loss probability exceeds 10^{-2}. For TCP SACK, the recovery of multiple packet losses would likely be successful if only the first lost packet can be recovered by fast retransmit. Therefore, the drop of the loss recovery probability for packet loss probabilities exceeding 10^{-2} can be explained by the fact that fast retransmit of the first lost packet cannot be triggered well due to lack of duplicate ACKs by the small congestion window. As can be seen in this figure, when LT is adopted, the loss recovery probability

---

[4] Since it has been already proved in [12] that TCP performance such as throughput or goodput is directly proportional to the loss recovery probability, we do not include the results of throughput comparison. Also, it can be inferred intuitively by simulation results presented in Section 2.

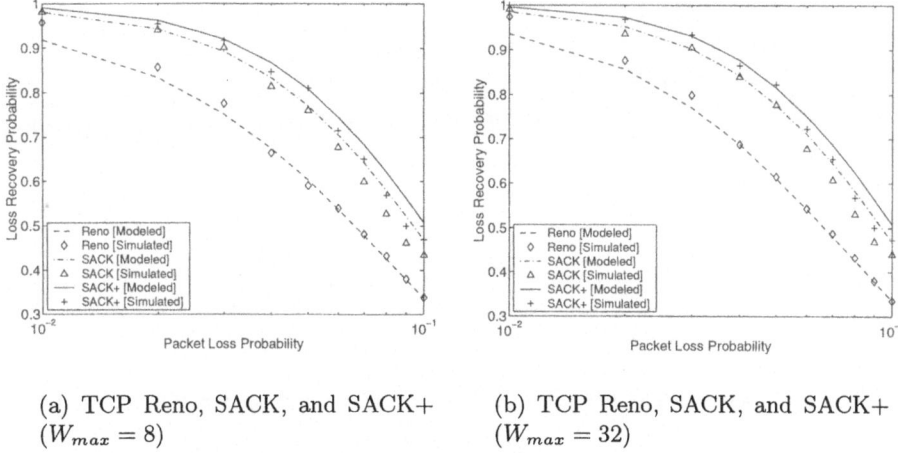

(a) TCP Reno, SACK, and SACK+ ($W_{max} = 8$)

(b) TCP Reno, SACK, and SACK+ ($W_{max} = 32$)

Fig. 6. Comparison of the loss recovery probability predicted by the developed model to the simulated results when the proposed algorithm is applied.

shows a slow decline with the increment of packet loss probability. For packet loss probability over 10^{-1}, the recovery probability of Reno using LT is rather higher than that of TCP SACK and SACK+. It is because, for such a large packet loss probability, the sender cannot keep its size large enough due to the frequent loss recovery events so that the capability of recovery for multiple packet losses makes little difference to the recovery probability. Since LT increases the likelihood of the first fast retransmit, when it is used with SACK option, the sender has more chances to recover multiple packet losses by retransmissions. It is the reason for the higher improvement when LT is used for TCP SACK than Reno.

There is a considerable improvement in the loss recovery probability of TCP SACK compared with TCP Reno. The difference between two lines of SACK and Reno reflects the capability to handle multiple packet losses without RTO or not. The slight difference between TCP SACK and SACK+ reveals the problem of lost retransmissions. As packet loss probability increases, more packet losses may occur and their retransmissions also tend to be lost again. On the other hand, as mentioned earlier, the congestion window is not large enough that multiple packet losses are not likely to be included in a window. That is, the probability for an event that a retransmission is lost again is quite low, which is the reason that the overall values of difference in loss recovery probability between SACK and SACK+ are not quite significant. According to the results measured in the real Internet presented in [11], the largest portion, about 85%, of timeouts are caused by small window, about 11% by multiple packet losses, and 4% by retransmission losses. The results obtained from our model verifies it again in that the largest improvement is made by LT, the second largest by SACK option, and the last by SACK+.

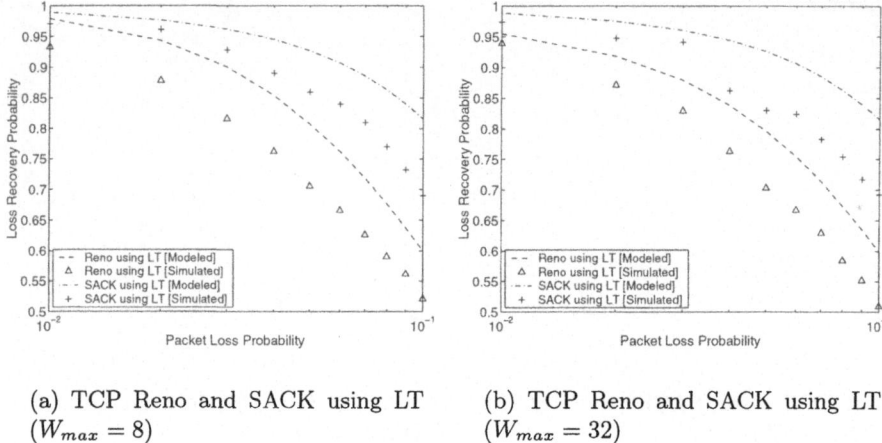

(a) TCP Reno and SACK using LT ($W_{max} = 8$)

(b) TCP Reno and SACK using LT ($W_{max} = 32$)

Fig. 7. Comparison of the loss recovery probability predicted by the developed model to the simulated results when LT is applied.

As shown in fig. 5, further increment of W_{max} to 32 makes no significant differences to the loss recovery probability. For low packet loss probability, it is unlikely that there are more than two packet losses in a window. Note that a single packet loss can be retransmitted if only three duplicate ACKs are received, namely, the window is equal to or greater than four. Even if multiple packet losses occur, they can be retransmitted if the first fast retransmit is successful. Consequently, a large W_{max} may benefit the throughput of TCP but not its loss recovery performance.

In fig. 6, we evaluate the loss recovery probability of TCP SACK and SACK+ using *ns* simulations. A sender and a receiver are connected with a long-fat link of 10Mbps and 100msec where packets are dropped in random. Using FTP, the sender transmits data consist of 10^4 packets whose size is 1kbytes, and the congestion window can grow up to W_{max}. The loss recovery probability is defined as the ratio of the number of the packets recovered by retransmissions to the total number of packet losses. It can be seen that the simulated values and the calculated values of our developed model fit well.

In fig. 7, we also evaluate the loss recovery probability when LT is used. There is a remarkable difference between the modeled results and the simulated results. It is because only the number of duplicate ACKs are considered as the condition for successful loss recovery in our modeling process. However, even if three duplicate ACKs are to be received, a RTO occurs if it expires before the third duplicate ACK arrives. If we suppose the worst case that the second packet is lost in a window of two in congestion avoidance, it takes about four RTTs to receive the third duplicate ACK. Therefore, the practical improvement of LT may be much smaller than the modeled results.

5   Conclusions

In this paper, we have proposed a simple algorithm that enables a TCP sender using SACK option to detect and recover retransmission losses without a RTO. The analysis in this paper assumes an environment where packets are dropped in random with probability p and the losses are independent to each other. Therefore, although the proposed algorithm can recover almost all of retransmission losses, its improvement is not significant since a retransmission loss itself is not such a common event in this loss model. However, according to queue management schemes such as droptail and with changing levels of congestion, the likelihood of a retransmission loss may have some variation. If it may occur more frequently than in our model, the change proposed in this paper is of a great benefit to loss recovery performance of TCP.

References

1. J. Postel: Transmission Control Protocol. RFC 793, (1981)
2. V. Jacobson: Congestion Control and Avoidance. ACM SIGCOMM'88, (1988)
3. V. Jacobson: Modified TCP Congestion Avoidance Algorithm. note sent to end2end-interest mailing list, (1990)
4. W. Stevens: TCP Slow Start, Congestion Avoidance, Fast Retransmit, and Fast Recovery Algorithms. RFC 2001, (1997)
5. M. Allman, V. Paxson, and W. Stevens: TCP Congestion Control. RFC 2581, (1999)
6. K. Fall and S. Floyd: Simulation-based Comparisons of Tahoe, Reno, and SACK TCP. ACM Computer Communication Review, vol. 26. (1996) 5–21
7. M. Mathis, J. Mahdavi, S. Floyd, and A. Romanow: TCP Selective Acknowledgement Options. RFC 2018, (1996)
8. S. Floyd, J. Madavi, M. Mathis, and M. Podolsky: An Extention to the Selective Acknowledgement (SACK) Option for TCP. RFC 2883, (2000)
9. E. Blanton, M. Allman, K. Fall, and L. Wang: A Conservative Selective Acknowledgement (SACK)-based Loss Recovery Algorithm for TCP. RFC 3517, (2003)
10. M. Allman, H. Balakrishan, and S. Floyd: Enhancing TCP's Loss Recovery Using Limited Transmit. RFC 3042, (2000)
11. Dong Lin and H. T. Kung: TCP Fast Recovery Strategies: Analysis and Improvements. IEEE INFOCOM'98, (1998) 263–271
12. Anurag Kumar: Comparative Performance Analysis of Versions of TCP in a Local Network with a Lossy Link. IEEE/ACM Transactions on Networking, vol. 6. (1998) 485–498
13. J. Padhye, V. Firoiu, D. F. Towsley, and J. F. Kurose: Modeling TCP Reno Performance: A Simple Model and Its Empirical Validation. IEEE/ACM Transactions on Networking, vol. 8. (2000) 133–145
14. Beomjoon Kim and Jaiyong Lee: Analytic Models of Loss Recovery of TCP Reno with Packet Losses. LNCS 2662, (2003) 938–947

High Coverage Broadcasting for Mobile Ad Hoc Networks

Dave E. Cooper, Paul Ezhilchelvan, and Isi Mitrani

School of Computing Science University of Newcastle, NE1 7RU, UK
{d.e.cooper, paul.ezhilchelvan, isi.mitrani}@ncl.ac.uk

Abstract. A family of message propagation protocols for highly mobile ad-hoc networks is defined, and is studied analytically and by simulation. The coverage of a message (the fraction of nodes that receive it), can be made arbitrarily close to 1, at a moderate cost of extra message traffic. Under certain simplifying assumptions, it is shown that a high coverage is achieved by making a total of $O(n \ln n)$ broadcasts, where n is the number of nodes, and the time to propagate a message is $O(\ln n)$. The effect of various parameters on the protocol performance is examined.

Keywords: Ad-hoc networks, mobility patterns, message propagation, broadcasting.

1   Introduction

Recent advances in the technologies of mobile devices and wireless communication have given rise to an increasingly popular form of networking, called Mobile Ad-hoc networking. A Mobile Ad-hoc network (MANET) consists of small, versatile and powerful mobile computing devices (nodes). It is typically formed at short notice and does not make use of any fixed networking infrastructure. A distinguishing feature of a MANET is that the nodes are not just the sources of message traffic but also engage in forwarding messages to final destinations; given that the nodes can be highly mobile, a MANET is a dynamic network characterized by frequent and hard-to-predict topological changes.

An application of a mobile network usually involves user collaboration towards achieving a common goal, in situations where access to base stations is unavailable or unreliable (e.g., command and control or disaster relief). The success of such collaborative undertakings depends to a large extent on the provision of reliable multicast [4]. That is, a message originating at any node should reach all other nodes within a reasonably short period of time. Unfortunately, both the nature of the devices (limited memory and power), and their mobility, imply that a guaranteed reliable multicast is not normally achievable in a MANET. Our objective, therefore, is to devise and evaluate multicast protocols which aim to maximize the probability of delivering a message to all nodes, while keeping the propagation time as low as possible.

N. Mitrou et al. (Eds.): NETWORKING 2004, LNCS 3042, pp. 100–111, 2004.

Existing work in this area has concentrated on minimizing the number of broadcasts carried out while propagating a message. Several protocols have been proposed, where the nodes maintain pro-actively, or construct on demand, distributed state information about the network topology. That state information is then used for the purpose of improving coverage with small overhead (see [2, 9,3,1]). When the degree of mobility is low, these protocols perform well, but when it is high, the network state information can become out-of-date quickly and the coverage achieved (i.e., the fraction of nodes that receive a message) can be poor [8,10].

A topology-independent and stateless protocol that seems to work better in highly mobile networks is 'flooding'. Every node broadcasts every message once, either immediately upon receipt or after a random interval (for a study of basic flooding, see Ho et al [5]; an optimized version was examined in Ni et al [7]). The coverage achieved by flooding depends not only on the mobility pattern, but also on the 'density' of nodes (usually defined as the average number of nodes within a disc of radius equal to the wireless range). When the density is low, the flooding coverage tends to be poor.

We propose, and study, a family of protocols which preserve the topology-independent nature of flooding, while being able to achieve coverage levels arbitrarily close to 1, for any node density. Of course a specific high coverage cannot be guaranteed in any given instance, but can be expected with high probability. These protocols are based on a notion of 'encounter', and are controlled by an 'encounter threshold' parameter. The cost paid for a high coverage is an increase in the message traffic, since messages are broadcast more than once by each node. Under certain simplifying assumptions, it is shown that to achieve a coverage close to 1 in a network with n nodes, the total average number of broadcasts per message is on the order of $O(n \ln n)$. This is a moderate increase on the $O(n)$ broadcasts carried out in flooding. The propagation time of a message is on the order of $O(\ln n)$. Various aspects of the protocols' performance are examined by simulation.

The model, and the message propagation protocols, are described in section 2. Some analytical results concerning the propagation time and the number of broadcasts are obtained in section 3. The outcomes of a number of simulation experiments are presented in section 4, while section 5 summarizes the results obtained and outlines avenues of further enquiry.

2   The Model

The system under consideration consists of n mobile nodes which move within a given terrain. The nodes communicate with each other using wireless technology, but without any fixed network infrastructure support. That is, the nodes themselves are the sources as well as the forwarders of the message traffic, and thus form a mobile ad-hoc network. Each node has a unique identifier (MAC or IP address). It is assumed that nodes do not fail; however, due to their mobility, they may become disconnected, and reconnected, as they move out of and into each other's wireless range. Thus, the structure of the network can change with time in an unpredictable manner. For simplicity, assume that the wireless ranges of all nodes are equal and remain constant during the period of interest.

The movement of each node is governed by some 'mobility pattern', which controls its current speed and direction. It is assumed that the n nodes are statistically identical, i.e. the rules of their mobility patterns are the same, and any random variables involved have the same distributions for all nodes.

We shall define a protocol whose principal objective is to deliver a message, originating at any node, to all other nodes with high probability. A secondary objective is to minimize, as far as possible, the memory requirements at each node. In fact, what will be defined is not a single protocol, but a family of protocols depending on an integer parameter, τ.

Node i $(i = 1, 2, \ldots, n)$ advertises its presence by broadcasting, at regular intervals, a signal carrying its identifier and saying, essentially, 'hello, this is node i'. It also listens for similar signals from other nodes and maintains a list, $\{j_1, j_2, \ldots, j_k\}$, of the nodes, other than itself, that it can hear. That list is called the 'current neighbourhood' of node i. At any moment in time, any current neighbourhood may be empty, or it may contain any number of other nodes.

The current neighbourhood of node i changes when a node which was in it, say j_1, moves out of range, or when a node which was not in it, say j_{k+1}, moves into range. The latter event is called an 'encounter'; that is, node i is said to encounter node j_{k+1}. Note that, since 'hello' signals are not assumed to be synchronized among the nodes, if node i encounters node j, node j does not necessarily encounter node i at the same time. Also note that, if node j leaves the current neighbourhood of node i and at some later point enters it again, then that entry constitutes an encounter. Nodes do not maintain a history of their current neighbourhoods, in order to keep their memory requirements low.

Now consider a message propagation protocol where each node behaves as follows:

1. Upon receiving or originating a new message, m, store it, together with an associated counter, $c(m)$, which is set to zero. Add the sending node to the current neighbourhood, unless already present. If the current neighbourhood contains nodes other than the sending one, broadcast m and increment $c(m)$ by 1.
2. At every encounter thereafter, if $c(m) < \tau$, broadcast m and increment $c(m)$ by 1.
3. When $c(m) = \tau$, remove m from memory (but keep its sequence number in order to remember that it has been handled).

Thus, every node receiving a message broadcasts it at τ consecutive encounters (one of which may be the message arrival), and then discards it. There are no acknowledgements. The integer τ is called the 'encounter threshold'. The above protocol, with encounter threshold τ, will be referred to as 'τ-propagation'.

When $\tau = 1$, the 1-propagation protocol behaves like flooding (except that the broadcast is delayed until the next encounter if the current neigbourhood contains only the sender). At the other extreme, if $\tau = \infty$, we have an ∞-propagation protocol whereby messages are kept forever and broadcast at every encounter. Assuming that the mobility pattern is such that every node eventually encounters every other node, ∞-propagation achieves coverage 1. Of course, ∞-propagation is not a practical option, but we shall see in section 3 that it can provide some useful insights.

It should be pointed out that τ-propagation trades memory capacity and probability of reaching all nodes against message traffic. Because past histories are not kept and exchanged, messages may be sent again to nodes who have already received them. By increasing the value of τ, the coverage can be made to approach 1, at the cost of having to store more messages for longer periods, and making more broadcasts.

In this paper, we place greater emphasis on evaluating the ability of τ-propagation to achieve high coverage, than on minimizing the message traffic overheads. That is why we assume the following:

- The overheads of collision resolution are negligible.
- Hello signals are sent and monitored at the MAC level; the information necessary to maintain the neighbourhood list is obtained at no extra cost to the higher level protocol.
- Encounters last long enough for a message to be received, i.e. the processing and propagation times of hello and broadcast messages are small enough for the encountered node to remain in the range of the encountering node.

The performance measures of interest are:

(i)   The average response time of τ-propagation, defined as the interval between the arrival (origin) of a message and the moment when no node can propagate it further.
(ii) The average propagation time of a message, defined as the interval between its arrival and the moment when either all nodes have received it, or no node can propagate it further.
(iii) The coverage of a message, i.e. the fraction of nodes that have received it by the end of its propagation time.

All of these performance measures are stated in terms of averages. However, the simulation results reported in section 4 provide some indication of the corresponding variances, by repeating each experiment 10 times with different random number streams. For example, observing a coverage of 1 implies that *all 10 runs* achieved a coverage of 1.

It is important to be able to choose the value of τ so as to achieve high coverage, without unduly increasing the response and propagation times. This question will be addressed in the following sections.

3   Analytical Approximation

Consider an idealized system with n mobile nodes who never cease to propagate the messages they receive (∞-propagation). Let T be the random variable representing a message propagation time, i.e., the interval between the origin of a message at some node, and the first instant thereafter at which all nodes have received it. If messages are not discarded, and every node eventually encounters every other node, T is finite with probability 1. It is then of interest to estimate its average value, $E(T)$. That quantity will also be used in choosing a suitable value for τ, when designing a practicable τ-propagation protocol.

An estimate for $E(T)$ will be obtained under the following simplifying assumptions:

(a) Each node experiences encounters at intervals which are exponentially distributed with mean ξ.
(b) At each encounter, a node meets one other node.
(c) The node encountered is equally likely to be any of the other nodes; that is, the probability that node i will next encounter node j, $j \neq i$, is equal to $1/(n-1)$, regardless of past history.

Assumption (a) can be justified by remarking that the interval until the next encounter experienced by a given node — say node 1 — is the smallest of the intervals until its next encounters with node 2, node 3, ..., node n. Some of these intervals may in fact be of length 0 with a positive probability. Nevertheless, it is reasonable (e.g., see [6]) to assume that the interval until the first of many random occurrences is approximately exponentially distributed. The value of ξ depends on the density of nodes, on the speed with which they move, and on the mobility pattern. It may be difficult to determine ξ analytically, but in practice it can be estimated by monitoring the system and taking measurements.

Assumption (b) is deliberately pessimistic, in order to give the estimate the character of an upper bound. If a node encounters more than one other node at the same time, then the propagation will proceed faster. In fact, it will be seen in the experiments that at high densities this assumption is *very* pessimistic.

Assumption (c) is loosely based on the fact that all nodes are statistically identical, and move independently of each other. If the starting positions of the nodes are uniformly distributed, the assumption is justifiable at the first encounter, although it may well be violated in subsequent ones. However, this assumption provides the simplification necessary for analytical tractability. Its effect on the performance measures will be evaluated in the simulation experiments.

Let $X = \{X(t); t \geq 0\}$ be the Markov process whose state at any given time is the number of nodes that have already received the message. The initial state of X is $X(0) = 1$ (only the originating node has received it; again, this is a pessimistic simplification since the the original neighbourhood may in fact contain other nodes). The random variable T is the first passage time of X from state 1 to state n.

Suppose that X is in state k, i.e. k nodes have received the message and $n-k$ have not. If any of the former k nodes encounters any of the latter $n-k$, the process will jump to state $k+1$. Since each node experiences encounters at rate $1/\xi$, and the probability of encountering any other node is $1/(n-1)$, the transition rate of X from state k to state $k+1$, $r_{k,k+1}$, is equal to

$$r_k = \left[\frac{k}{\xi}\right]\left[\frac{n-k}{n-1}\right] . \tag{1}$$

In other words, the average time that X remains in state k is

$$\frac{1}{r_k} = \frac{(n-1)\xi}{k(n-k)} . \tag{2}$$

Hence, the average first passage time from state 1 to state n is given by

$$E(T) = (n-1)\xi \sum_{k=1}^{n-1} \frac{1}{k(n-k)} . \tag{3}$$

This last expression can be simplified by rewriting the terms under the summation sign in the form

$$\frac{1}{k(n-k)} = \frac{1}{n}\left[\frac{1}{k} + \frac{1}{n-k}\right] .$$

The two resulting sums are in fact identical. Therefore,

$$E(T) = \frac{2(n-1)\xi}{n} \sum_{k=1}^{n-1} \frac{1}{k} = \frac{2(n-1)\xi H_{n-1}}{n} , \tag{4}$$

where H_n is the nth harmonic number. When n is large, the latter is approximately equal to

$$H_n \approx \ln n + \gamma ,$$

where $\gamma = 0.5772...$ is Euler-Mascheroni's number. Also, when n is large, $(n-1)/n \approx 1$ and $\ln(n-1) \approx \ln n$.

We have thus arrived at the following estimate, valid under assumptions (a), (b) and (c):

Proposition 1 *In a large mobile network where messages are not discarded, the average propagation period for a message is approximately equal to*

$$E(T) \approx 2\xi(\ln n + \gamma) . \tag{5}$$

An immediate corollary of Proposition 1 is that, during the propagation period T, the originating node experiences an average of $2(\ln n + \gamma)$ encounters. Other nodes, who receive the message later on, tend to experience fewer encounters. Thus, choosing the encounter threshold, τ, to have the value

$$\tau = 2\lceil \ln n + \gamma \rceil , \tag{6}$$

should ensure that, when the protocol terminates, most nodes will have received the message. This suggestion will be tested experimentally.

Note 1. An attractive aspect of equation (6) is that the only parameter appearing in it is the number of nodes, n. The mobility pattern and the node density do not matter, as long as assumptions (a), (b) and (c) are satisfied reasonably well.

Note 2. Since, under τ-propagation, every node that receives a message broadcasts it τ times, the total number of broadcasts per message is on the order of $O(n\tau)$. Hence, if τ is chosen according to (6), the total number of broadcasts per message is on the order of $O(n \ln n)$.

4 Experimental Results

A number of simulation experiments were carried out, aimed at evaluating the effect of various parameters on the performance of τ-propagation. The following factors were kept fixed:

The terrain is a square of dimensions $(1000\ m) \times (1000\ m)$.

The wireless range of a node is $50\ m$ (deliberately taken small compared to the size of the terrain).

The interval between 'hello' signals for each node is $25\ ms$.

The mobility pattern is 'Random Waypoint': Initially, the nodes are distributed uniformly on the square; thereafter, each node chooses a random destination (also uniformly distributed on the square) and moves towards it at a given speed; upon reaching the destination, the node pauses for a given interval ($1\ ms$ in our case), selects a new random destination and so on.

The speed, node density and encounter threshold were varied and the performance measures — average response time, average propagation time and coverage — were evaluated. Each run starts at time 0 with a message originating at node 1, and terminates when no node can propagate the message further. For each set of parameter values, the simulation ran 10 times, with different random number seeds, and the performance observations were averaged.

Figures 1 – 4 show the coverage achieved as a function of the encounter threshold, τ, for node densities ranging between 0.5 and 6.5, and speeds ranging between 20 ms^{-1} and 100 ms^{-1} (these values are not intended to represent any realistic application; they are chosen merely as illustration). In fact, only the density has a significant effect on the coverage function; the node speed is, on the whole, immaterial. The figures quantify the extent to which the coverage can be improved by increasing τ: at low densities, where flooding performs poorly ($\tau = 1$), the improvement is very considerable; at high densities, flooding performs well and the gain of increasing τ is correspondingly smaller.

Consider the analytical predictions concerning τ. For the assumed terrain area and wireless range, the densities 0.5, 1, 3.5 and 6.5 correspond to values of n equal to 64, 128, 446 and 828, respectively. For these numbers of nodes, the encounter thresholds given by equation (6) are $\tau = 10$, $\tau = 12$, $\tau = 14$ and $\tau = 16$, respectively, and the figures indicate that they do, indeed, achieve coverages close to 1. In fact, when the density is high, the threshold provided by equation

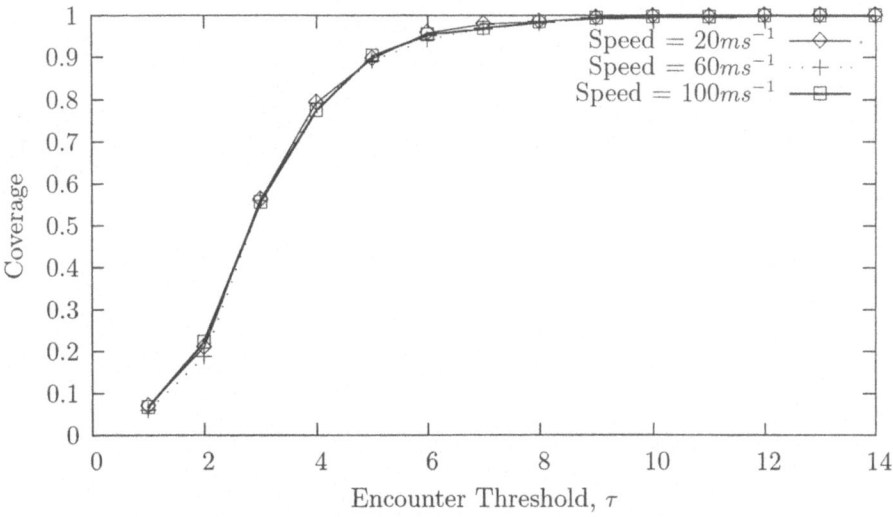

Fig. 1. Coverage achieved by τ-propagation: density 0.5

(6) is rather conservative. This is because, for those densities, assumption (b) in section 3 is too pessimistic.

Figure 5 shows the average response time and the average propagation time as functions of τ, for a particular density, 3.5 ($n = 446$), and node speeds 20 ms^{-1}, 60 ms^{-1} and 100 ms^{-1}. A noteworthy aspect of the figure is that, while the response time keeps increasing with τ (as expected), the propagation time increases up to a point ($\tau = 5$), and then decreases. To explain that behaviour, note that when the threshold is 5 or less, the coverage is less than 1 and therefore the propagation time is equal to the response time. When the threshold is 6 or more, a coverage of 1 is reached, and the propagation time completes, before nodes have stopped broadcasting. Moreover, further increases in τ tend to speed up the propagation, but prolong the response time.

Similar behaviour is observed at other densities.

The observed average intervals between encounters for density 3.5 and speeds 20 ms^{-1}, 60 ms^{-1} and 100 ms^{-1}, are $\xi = 0.96$, $\xi = 0.40$ and $\xi = 0.29$, respectively. According to equation (5), the corresponding limiting average propagation times (for $\tau = \infty$) should be 12.9, 5.4 and 3.9, respectively. These values agree quite well with the propagation times reached at $\tau = 14$.

The process of propagating a message among the nodes in a network where the speed (60 ms^{-1}) and threshold ($\tau = 14$) are fixed, while the density is varied in the range 0.5 – 6.5, is illustrated in figure 6. The graphs show how the rate of propagation changes as more and more nodes are covered. At high densities, it takes longer to cover the last 5% of the nodes than the first 95%. This phenomenon is due to the fact that some nodes on the periphery of the terrain can be relatively more difficult to reach than the others. It is less pronounced at

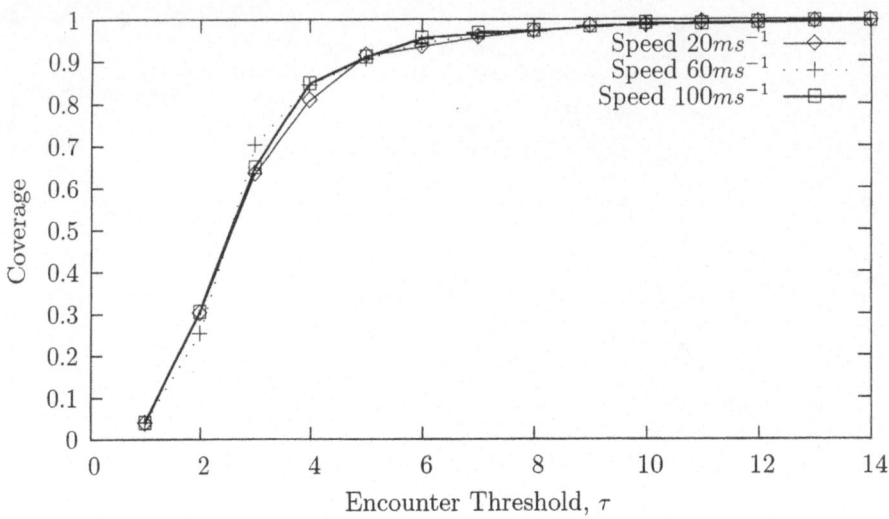

Fig. 2. Coverage achieved by τ-propagation: density 1

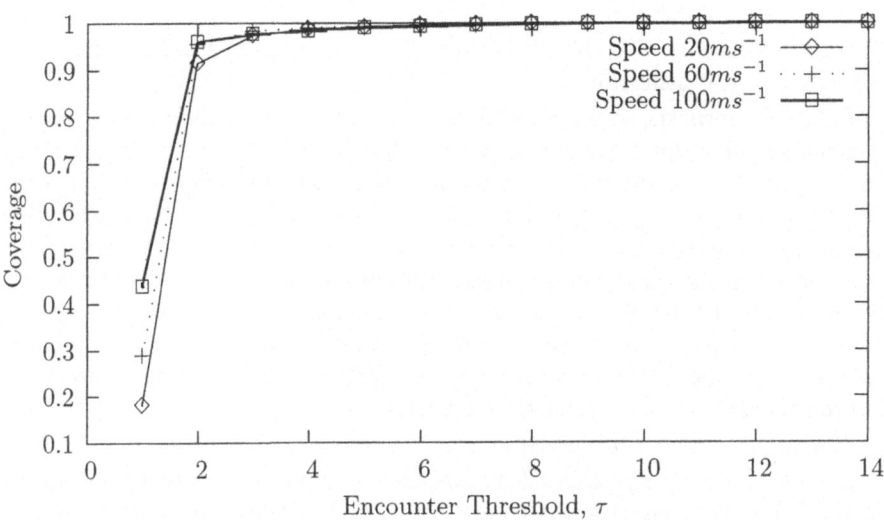

Fig. 3. Coverage achieved by τ-propagation: density 3.5

lower densities, but is still in evidence: the last 20% of the nodes take about as long to cover as the first 80%.

5   Conclusions

The main contributions of this paper can be summarized as follows:

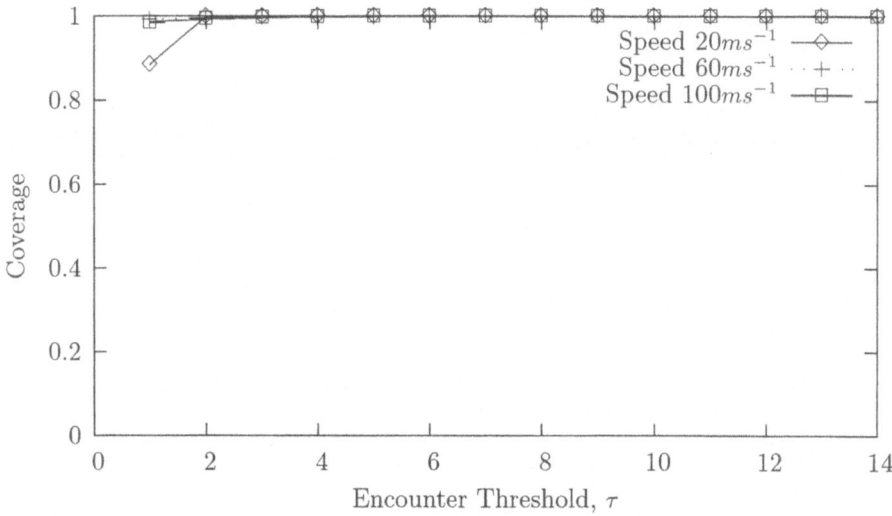

Fig. 4. Coverage achieved by τ-propagation: density 6.5

1. Introduction of the τ-propagation family (section 2).
2. Mobility-independent estimate for the value of τ that achieves high coverage (equation (6)).
3. Quantitative performance results obtained by experimentation (section 4).

It would clearly be desirable to relax some of the assumptions that were made in order to simplify the model, and evaluate the resulting changes. For example, if the 'hello' signals are not handled 'free of charge' by the MAC layer, one could not afford to broadcast them too frequently, encounters would be missed, and the response and propagation times would increase. Similarly, if messages can be lost because nodes do not remain within range long enough to receive them, the performance of the protocols would suffer. We intend to implement a more realistic simulation of the protocol using the GloMoSim tool. This will allow us to take into account the effects of collision, contention, congestion and varying quality of radio links.

A more adaptable family of propagation protocols may be designed by introducing a FIFO buffer for messages. Messages would be kept in the buffer, and re-broadcast, until either they are displaced by new messages or they reach an encounter threshold. The number of times a message is broadcast by a node would then change dynamically in response to changing conditions. That number could also be adjusted by keeping track of repeated receptions of the same message. A time-out interval can be introduced, to force the discarding of a message if the node does not experience a sufficient number of encounters. In addition, the encounter threshold may be controlled by the number of nodes already encountered, and possibly by the mobility pattern. All these are worthy topics for future research.

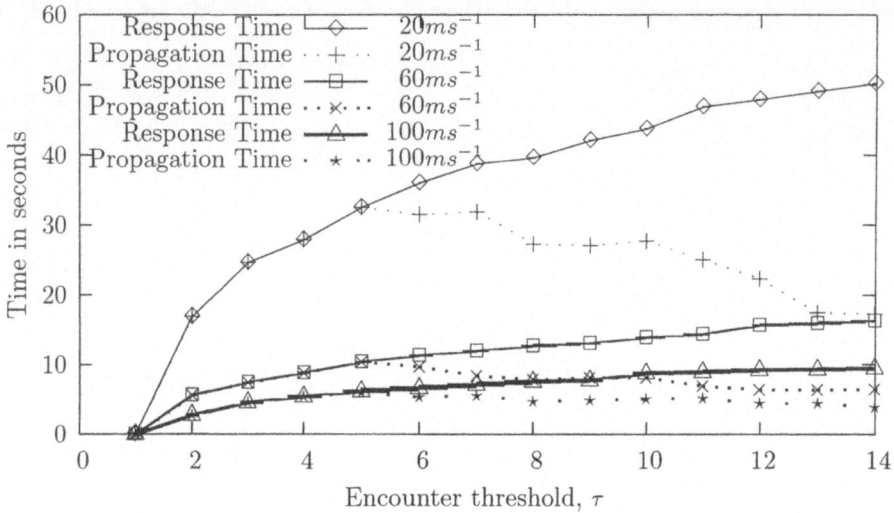

Fig. 5. Protocol times at different speeds: density 3.5

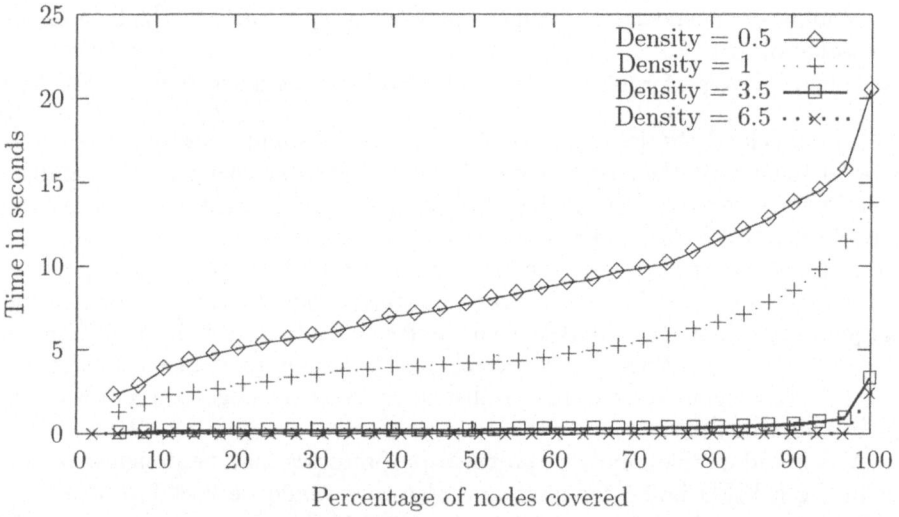

Fig. 6. Process of propagation: speed $= 60 \ ms^{-1}$; $\tau = 14$

Acknowledgement. This work was carried out as part of the research project PACE (Protocols for Ad-hoc Collaborative Environments), funded by the UK Engineering and Physical Sciences Research Council. We are also grateful to Einar Vollset, for some helpful discussions.

References

1. C. Chiang and M. Gerla, "On-Demand Multicast in Mobile Networks", Procs., *IEEE ICMP*, pp. 260-270, Austin, 1998.
2. J. Garcia-Luna-Aceves and E. Madruga, "A Multicast Routing Protocol for Ad-hoc Networks", Procs., *Infocom*, pp.784-792, New York, 1999.
3. T. Gopalsamy, M. Shinghal, D. Panda and P. Sadayappan, A Reliable Multicast Algorithm for Mobile Ad-hoc Networks", Procs., *ICDCS*, Vienna, 2002.
4. V. Hadzilacos and S. Tueg, "Fault-Tolerant Broadcasts and Related Problems", in *Distributed Systems* (Ed. S. Mullender), Addison-Wesley, 1993.
5. C. Ho, K. Obraczka, G. Tsudik and K. Viswanath, "Flooding for Reliable Multicast in Multihop Ad-hoc Networks", Procs., *Discrete Algorithms and Methods for Mobile Computing (DIAL-M)*, pp. 64-71, Seattle, 1999.
6. I. Mitrani, *Probabilistic Modelling*, Cambridge University Press, 1998.
7. S. Ni, Y. Tseng, Y. Chen and J. Sheu, "The Broadcast Storm Problem in a Mobile Ad-hoc Network", Procs., *ACM Int. Conf. on Mobile Computing and Networking (MOBICOM)*, pp. 151-162, Seattle, 1999.
8. E. Pagani and G.P. Rossi, "Reliable Broadcast in Mobile Multihop Packet Networks", Procs., *ACM Int. Conf. on Mobile Computing and Networking (MOBICOM)*, pp. 34-42, Budapest, 1997.
9. P. Sinha, R. Sivakumar and V. Bharagavan, "MCEDAR: Multicast Core Extraction Distributed Ad-hoc Networking", Procs., *Wireless Communications and Networking Conference (WCNC)*, pp. 1313-1317, New Orleans, 1999.
10. B. Williams and T. Camp, "Comparison of Broadcasting Techniques for Mobile Ad-hoc Networks", Procs., *ACM Int. Symp. on Mobile Ad-hoc Networking and Computing (MOBIHOC)*, Lausanne, 2002.

Combining Source- and Localized Recovery to Achieve Reliable Multicast in Multi-hop Ad Hoc Networks

Venkatesh Rajendran[1], Katia Obraczka[1], Yunjung Yi[2], Sung-Ju Lee[3], Ken Tang[4], and Mario Gerla[2]

[1] Computer Engineering Department, University of California, Santa Cruz
[2] Computer Science Department, University of California, Los Angeles
[3] Mobile & Media System Laboratory, Hewlett-Packard Laboratories
[4] Scalable Networks Technologies

Abstract. This paper proposes a novel reliable multicast transport protocol for multi-hop, wireless ad hoc networks (or MANETs). To recover from the different types of losses that may occur in MANETs, our Reliable Adaptive Congestion-controlled Transport protocol, or ReACT, combines source-based congestion- and error control with receiver-initiated localized recovery. While the latter attempts to recover localized losses (e.g., caused by transmission errors), the former is invoked only for losses and congestion that could not be recovered locally (e.g., caused by global congestion). Loss differentiation is an important component of ReACT and uses medium access control (MAC) layer information to distinguish between different types of losses. Through extensive simulations, we evaluate ReACT's performance under a variety of MANET scenarios, including different offered load and mobility conditions, and compare it against a strictly end-to-end (i.e., no localized recovery) scheme. Our results show that ReACT is the best performer in terms of reliability. Our results also showcase the effect of ReACT's local recovery mechanism which quickly corrects error- and path breakage induced losses and thus manages to prevent the source from reducing its rate unnecessarily , thus achieving significant throughput improvement with lower overhead when compared to the strictly end-to-end protocol.

1   Introduction

A multi-hop, wireless ad hoc network (or MANET) [1] operates without any fixed infrastructure. Hosts communicate with each other through wireless packet radios. Because of the limited radio propagation range, routes can often be multi-hop. Hence, every host may act as a packet forwarder as well as source or destination of traffic. Because of their ease of deployment, MANETs are an attractive choice for scenarios where the fixed network infrastructure is non-existent or unusable. Example applications include search and rescue, disaster recovery, digital battlefield, and covert military operations.

N. Mitrou et al. (Eds.): NETWORKING 2004, LNCS 3042, pp. 112–124, 2004.

Both unicast- and multicast routing in MANETs have been well-studied and, as a result, a number of protocols have been proposed [2,16]. Several research efforts have also focused on transport-layer approaches to achieve end-to-end reliable point-to-point communication. This includes the work on improving TCP performance in "last-hop" wireless networks and MANETs [3,4,5,6].

However, the types of scenarios targeted by MANETs make group-oriented services such as data dissemination and teleconferencing a key application domain. In particular, the mission-critical characteristics of a number of these applications (e.g., emergency response, special civilian or military operations) call for efficient **reliable** multi-point communication protocols for MANETs. Undoubtedly, "network-supported" multicast communication is an efficient means of supporting group-oriented applications. This is especially true in MANETs where nodes are energy- and bandwidth limited. In these resource-constrained environments, reliable point-to-point protocols (that may be viable in wired networks) can get prohibitively expensive: the convergence of multiple requests to a single node typically causes intolerable congestion, violating the reliability and time constraints of a critical mission, and may drain the node's battery, cutting short the network's lifetime. Despite the fact that it is a key enabling technology for mission critical applications in MANETs, surprisingly very few efforts to date focus on reliable multicast transport.

The Reliable Broadcast Protocol [7] addresses the problem of reliable atomic delivery of messages. While this protocol may work well in stable networks with low mobility and low failure rates, its performance will likely degrade in dynamic MANET scenarios where topology changes are frequent. Anonymous Gossip (AG) [8] recovers from losses by having pairs of multicast group participants exchange information on messages they have received or lost. AG uses solely local recovery from nearby members for error control. In our experiments, we compare the performance of ReACT against AG [1]. As expected, mainly due to the fact that AG does not implement congestion control, we observe that its performance deteriorates under heavy load.

In our previous work we demonstrated the importance of congestion control in improving reliability. Congestion-controlled Adaptive Lightweight Multicast (CALM) [10] is a multicast transport protocol that tries to achieve reliable delivery strictly through congestion control. The Reliable Adaptive Lightweight Multicast protocol [11] uses a congestion control scheme similar to that of CALM and recovers from losses using source-based retransmissions. It requires multicast group member information to perform congestion control and error recovery. In an extended version of RALM [12], we do away with the need to maintain group membership information at the source.

Several features unique to MANETs make the design of MANET reliable multicast transport mechanisms quite challenging. Among these features, we highlight: (1) MANET's heterogeneous loss characteristics due to factors such as mobility, node density, time-varying channel conditions, (2) effects of lower layer protocols, e.g., inherent unfairness and unreliability of contention-based

---

[1] These results that are not presented here due to space limitations can be found in [9]

medium access control protocols (e.g., IEEE802.11 [13] uses plain CSMA when broadcasting packets and thus do not provide reliable broadcast delivery), and (3) MANET's extreme sensitivity to offered load.

These MANET features render design choices used in reliable multicast protocols for wired networks not at all applicable to MANET environments. Based on observations from our prior work [10], we argue that multicast reliability in MANETs cannot be achieved solely by retransmission of lost packets as is typically done in wired networks with protocols such as Scalable Reliable Multicast (SRM) [14]. Our premise is that, besides error control, effective reliable multicast delivery in MANETs must also perform congestion control. As demonstrated in our previous studies [10,12,11], a simple congestion control scheme results in significant increase in delivery guarantees.

MANETs' complexity also calls for revisiting the layered system design argument which claims that, in a system, the design and implementation of each one of its layers should not be exposed to higher layers. We argue that in MANETs, information obtained from lower layers of the protocol stack is crucial for adequate performance at higher layers.

This motivated us to explore cross-layer mechanisms to achieve efficient reliable multicast transport. More specifically, we use information from lower layer protocols (in particular the MAC layer) to perform loss differentiation addressing MANETs' heterogeneous loss characteristics. Thus, some of the distinguishing features of ReACT are that (1) it combines source-based rate control with local error recovery and (2) uses loss differentiation to trigger either source-based control or local recovery. The goal is to recover from localized losses (e.g. due to node mobility, link quality, channel contention) using nearby group members, while congestion losses are reported to the source, triggering error- as well as congestion recovery.

Through extensive simulations, we evaluate ReACT's performance under a wide range of MANET conditions. In order to demonstrate the benefits of ReACT's loss differentiation and local recovery mechanisms, we also compare its performance against a strictly source-based control scheme (RALM [12]). In our experiments, as the underlying routing mechanism, we use a mesh-based multicast protocol, more specifically the On-Demand Multicast Routing Protocol (ODMRP) [15].

The remainder of this paper is organized as follows. Section 2 presents a detailed description of ReACT's source-based and local recovery mechanisms. Performance evaluation and simulation results follow in Section 3. Section 5 presents our concluding remarks and directions for future work.

2   ReACT

2.1   Overview

Our premise when designing ReACT is that in wireless environments losses may be caused by various factors and should be handled differently. For example losses caused by transmission errors (e.g., due to factors such as noise, interference, etc.) or hidden terminal collisions may be affecting only a small number of nodes in a neighborhood and thus can be recovered locally using a (non-congested) nearby member, i.e., without the involvement of the source. There is no need to trigger congestion control and slow down the source because these losses are not indicative of "global" congestion. Furthermore, by recovering locally, feedback and retransmissions are kept in the affected neighborhood and do not add to traffic destined to the source, hence improving protocol efficiency. On the other hand, congestion losses should be reported to the source triggering reduction of the sending rate as well as error recovery. However, special care should be taken as local recovery can exacerbate congestion if the network neighborhood performing recovery is already congested.

ReACT performs receiver-based loss differentiation to distinguish congestion- from local losses. A multicast receiver samples its MAC queue to detect congestion building up. Receivers also detect congestion building up anywhere on the path from the source by having intermediate nodes set a "congestion" flag in multicast data packets they forward. The congestion flag is set by any intermediate node whose MAC queue grows beyond a certain fraction of the maximum MAC queue size. By detecting incipient congestion (instead of waiting to take action until actual packet drops occur), ReACT tries to avoid persistent congestion conditions.

ReACT ensures that only multicast members that are situated in a non-congested area will be used to perform local recovery. This avoids contributing to congestion in an already congested neighborhood. The remainder of this section describes ReACT in detail by presenting its two main components, namely source-based (error and congestion) control and receiver-based error recovery.

2.2 Source-Based Control

ReACT employs a rate-based congestion control scheme that has two main modes of operation: initial rate set-up (i.e., determining the initial sending rate) [2], and congestion control. ReACT tries to determine the appropriate sending rate in order to avoid (1) initial bandwidth under utilization by starting too low, and (2) congestion by starting at too high of a rate.

One approach at setting up the initial rate is to probe the entire network and then decide on the rate based on the aggregate network condition. Though this approach can provide information on the overall state of the network, it is not scalable. Alternatively, in ReACT we establish the initial rate based on the set of members that are directly connected to the source. This provides the source with an estimate of its neighborhood's current conditions. The rate is decided so as to satisfy the worst receiver in this neighborhood.

---

[2] In our previous approaches [10,11,12], we start at the application sending rate and then react to congestion based on receiver feedback. Our experiments indicate that setting an initial rate too high may lead to extreme (sometimes unrecoverable) congestion and thus numerous packet losses (e.g., if the feedback path from the receivers to the source gets blocked.)

The first data packet sent by the multicast source serves as a probe packet and each directly connected member replies with a *PROBE_REPLY* packet.After sending the first packet, the source waits for *PROBE_WAIT_TIME* to receive replies to its probe. *PROBE_WAIT_TIME* is set based on the network diameter (*NET_DIAMETER*) and an estimate of average time to traverse one hop (*NODE_TRAVERSAL_TIME*) accounting for queuing and transmission delays (similar to the route reply timeout of AODV [17]). If the source does not hear from any receiver in response to the probe packet, it will continue to send (probe) packets every *PROBE_WAIT_TIME* interval. The source then computes the inverse of the largest round-trip time reported during the initial probing period and uses that as its initial sending rate.

The rate is periodically updated once every *PROBE_INTERVAL* by directly connected neighbors. If no feedback has been sent to the source for the last *PROBE_INTERVAL* seconds, the receiver generates an explicit *PROBE_REPLY* packet. Receivers only send an update to the source if they detect significant changes to the time it takes them to get packets from the source. *PROBE_INTERVAL* is set sufficiently large to prevent oscillations in the source sending rate and also to reduce feedback overhead due probing. The source continues to send at this rate, until it hears a negative acknowledgment (NACK) from any receiver experiencing congestion. In that case, it reverts to congestion control.

ReACT's congestion control works as follows. The source initially multicasts data packets at the rate decided using initial probing as described above. Upon reception of a NACK, the source adds the NACK sender to its *Receiver List* and enters loss recovery. The missing sequence numbers reported by the NACK are added to a global retransmission list, which is an aggregate of lost sequence numbers from all reporting receivers. This list is updated whenever the source retransmits a packet to prevent duplicate retransmissions. In addition, the source keeps track of the end-to-end latency between itself and each receiver that sent NACKs.

The source initiates loss recovery by selecting a receiver from the *Receiver List*, which we call *Feedback Receiver*. The source then retransmits a lost packet requested by the *Feedback Receiver* or multicasts a new packet (e.g., if all lost packets requested by that receiver had already been retransmitted). The packet header includes information instructing the *Feedback Receiver* to reply via "unicast" with a (positive) acknowledgment (ACK) indicating that all packets have been successfully received or specifying the sequence number(s) of packets that are still missing. All other receivers process the packet without replying to the source.

The source then responds by retransmitting the requested packets one at a time until the *Feedback Receiver* receives all packets (i.e., send-and-wait). The design philosophy behind retransmitting one packet at a time is to slow down the source when congestion is detected. Since only the *Feedback Receiver* replies to the source, the ACK/NACK implosion problem is avoided. NACKs are rate-limited to prevent excessive feedback overhead. The mechanism for controlling NACK generation is described in Section 2.3.

Once the *Feedback Receiver* obtains all packets, it unicasts an ACK to the source indicating successful reception of all packets. Upon reception of the ACK, the source removes the node from the *Receiver List*, chooses a new *Feedback Receiver* in a round robin fashion, and repeats this process until the *Receiver List* is empty.

When the *Receiver List* is empty, the source reverts to the latest sending rate decided based on periodic probe packets. If, however, the source does not receive a NACK or ACK from the *Feedback Receiver* within the time interval given by the measured round-trip time from the source to the *Feedback Receiver*, the source backs off and tries again up to a maximum number of times (which is three in our simulations) before removing it from the *Receiver List*. The removed receiver may later re-synchronize with the source through the normal NACK mechanism.

The round-robin send-and-wait approach does not require retransmissions of the same lost packets multiple times to each receiver. In the best-case scenario, lost packets are retransmitted only once by the source since retransmissions are multicast. For instance, if a set of receivers lost the same packet, it is retransmitted only once assuming the retransmitted packet is received by all the receivers. In the worse-case scenario, each receiver experiences different packet losses. In this case, all lost packets must be retransmitted to each receiver.

2.3   Receiver-Based Error Recovery

The main goal of ReACT's receiver-based recovery mechanism is to detect losses that can be recovered locally avoiding source involvement (and hence avoid triggering congestion control). Congestion losses, however, should be reported to the source so that it knows to slow down.

In order to recover from losses "locally", nodes must obtain information about other group members as potential *Recovery Nodes*. Our scheme gathers member information using multicast data packets as they get forwarded over the multicast tree or mesh. Hence it is independent of the underlying multicast routing protocol. We are only interested in *Recovery Nodes* that are in the forwarding path from the source. More specifically, ReACT only uses immediate upstream member node(s) for recovery.

The way recovery requests and replies (or retransmissions) are routed has significant impact on the overall performance of the reliable multicast mechanism. If the underlying unicast routing protocol does not have a valid path for the recovery request and performs flooding for route discovery, significant additional load may result. Our simulation study indicates that local recovery based protocols that do not address this problem (e.g., AG) suffer from congestion even at moderate loads. ReACT employs a source routing approach that makes use of valid cached paths. The main advantage of this approach is that it makes ReACT independent of the underlying unicast routing protocol. The tradeoff is the overhead involved in maintaining source routes, especially in highly mobile environments. ReACT restricts the maximum distance between a member and

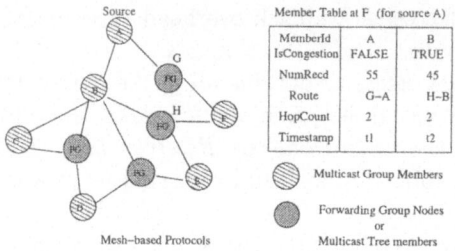

Fig. 1. Member table maintained for local recovery

its *Recovery Node* to *LR_ROUTE_LEN* hops to reduce the failure probability (e.g., due to node mobility) of source route.

Every node maintains a *Member Table* that stores information about current *Recovery Nodes*. To account for route volatility, *Member Table* entries are assigned an expiration time (*LR_VALID_TIME*). Additionally, each node maintains a metric of *reliability*, i.e., the rate at which it receives multicast data packets from *Recovery Nodes*. This information is used in selecting a *Recovery Node* if multiple ones exist. Each entry also has a flag to indicate if the path to the *Recovery Node* is congested. This flag is set if any intermediate node on the path to the *Recovery Node* has MAC queue size beyond the *CONGESTION_THRESHOLD*.

The IP option fields in the multicast data packet is used to carry route, hop count and congestion information. These fields are updated as the packet is forwarded to the group. The *route* field contains the path traversed by the packet from the upstream multicast group member. The *hopCount* field carries the length of the path. The *isCongestion* field denotes if any of the node in the path is congested. Whenever a node decides to perform local recovery, it selects a non-congested member that has the highest receive rate, lowest hop-count, and latest timestamp. Figure 1 shows a sample *Member Table* maintained by member node *F* when using either a tree- or mesh-based protocol. Mesh-based protocols may yield more than one upstream member because of path redundancy. As tree-based protocols also use broadcast for delivery, it is possible that a receiver might receive a packet from a node other than its parent node. Selecting an upstream *Recovery Node* based on its reliability and proximity increases the likelihood of successful local packet recovery.

Feedback generation is rate-limited to once every *MIN_FEEDBACK_INTERVAL* seconds to prevent excessive feedback overhead. Thus, every *MIN_FEEDBACK_INTERVAL*, receivers check if they need to perform error recovery by sending a NACK to a near-by member. There is a tradeoff in setting *MIN_FEEDBACK_INTEVAL*. When smaller intervals are used, we observe higher packet delivery ratios at the expense of higher overhead and lower throughput.

A NACK packet consists of an request array that is filled with the node's missing sequence numbers. The NACK is then sent to the selected *Recovery Node* if losses are found to be localized. Nodes use cached source routes to com-

Table 1. Simulation parameters

ReACT Parameters	Value
LR_ROUTE_LEN	3
LR_VALID_TIME	3 s
MIN_FEEDBACK_INTERVAL	5 s
NODE_TRAVERSAL_TIME	50 ms
NET_DIAMETER	35
PROBE_INTERVAL	50 s

municate with the *Recovery Node*. On the other hand, if losses are due to "global congestion" or if the node finds that it is experiencing congestion, then it sends the NACK to the source using the underlying unicast routing protocol.

A node checks if losses are due to "global congestion" by examining the paths to *Recovery Nodes*. If all paths are congested, then all valid *Recovery Nodes* in the *Member Table* will have the *isCongestion* flag set. Additionally, the node also examines its queue to check if it is congested. If any of the above conditions is true, then losses are classified as due to "global" congestion and feedback is sent to the source directly triggering congestion control.

Besides missing sequence numbers, a NACK packet destined to the source also includes the average delay multicast packets take to reach the node from the source. Receivers update the average delay to a multicast source every time they receive a data packet from that source. The average delay is computed as an exponential average with more weight to recent measurements. Receivers sending NACKs to the source are placed in the *Receiver List*. The source then transmits to each *Receiver List* receiver based on its reported delay using a send-and-wait approach as described in Section 2.2.

3   Experimental Setup

As our simulation platform, we use the QualNet network simulator [19]. ODMRP [18] and AODV [17] are used as the underlying multicast and unicast routing protocols, respectively. The transmission range for the radio is $447.807m$ with a data rate of $2Mbps$. The MAC protocol used is IEEE 802.11 DCF [13].

We evaluate the performance of ReACT in comparison with plain Reliable Adaptive Lightweight Multicast (RALM) [12], a strictly source-based control scheme. We evaluate ReACT's performance subject to a wide range of network conditions. We are particularly interested in how ReACT performs under various offered loads, and what is the impact of node mobility. Table 1 shows the values of the parameters used by ReACT.

As we target applications that require the highest possible delivery guarantees, protocol reliability is a critical performance metric. We define *Reliable Delivery Ratio* as the fraction of packets successfully (or reliably) delivered to ALL receivers over the total number of packets sent. We also measure *Reliable Goodput* defined as the throughput of packets reliably delivered, i.e., packets that are received by all members. Finally, we measure the overhead incurred by the protocols. To account for control packets sent by underlying unicast/multicast protocols, we measure the total number of packets sent by each node at the MAC

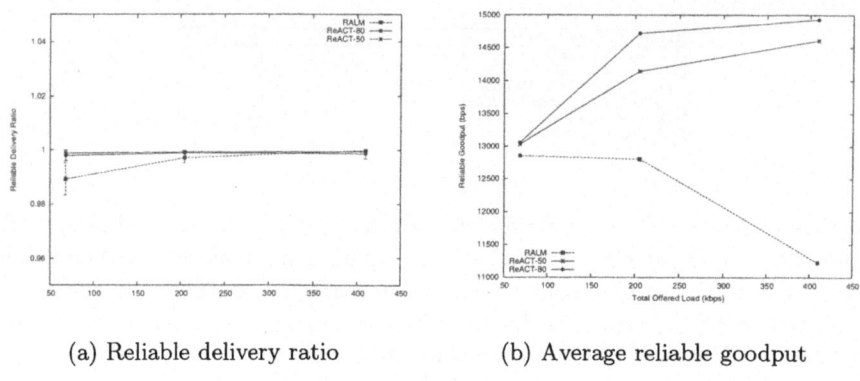

(a) Reliable delivery ratio   (b) Average reliable goodput

Fig. 2. Effect of congestion

layer. *Normalized Overhead* is thus computed as the ratio of total packets sent at the MAC layer to total data packets delivered to all members. This measures the total number of packets transmitted to successfully deliver one data packet to all members.

First, we study the importance of congestion control by simulating a scenario with multiple sources generating different traffic loads and then we analyze the impact of node mobility. For these sets of experiments, 50 nodes are placed randomly in a $1500m \times 1500m$ field and 10 randomly chosen nodes join the multicast group. These group nodes join at the start of the simulation and stay subscribed to the group till the end of the simulation. Five randomly selected members continuously send CBR traffic throughout the whole duration of the simulation with payload size of 512 bytes. The results are averaged over several runs and presented with 95% confidence.

4   Simulation Results

4.1   Effect of Congestion

Figure 2(a) shows reliable delivery ratio under different loads. The error bars correspond to reliable delivery ratio's 95% confidence intervals. As the load increases, so does packet loss due to congestion and hidden terminal collisions. Both RALM and ReACT perform error recovery and congestion control and hence they achieve very high reliability. RALM employs strictly source-based error recovery using NACKs. NACKs are generated when lost sequence numbers are detected upon arrival of a new data packet. Hence, if a node stops receiving data from a particular source, it will never generate NACK to recover lost packets. This leads to the reduced reliability of RALM at low loads and also higher reliability variance when compared to ReACT. On the other hand, ReACT achieves perfect reliability under various loads due to its robust error recovery mechanism.

We show results for two different versions of ReACT: one that detects congestion when queues grow above 80% of their maximum size, while the other version
uses 50% as the queue threshold indicating congestion. As observed from Figure
2(a), both versions deliver almost perfect reliability.

Figure 2(b) illustrates the impact of ReACT's combined local- and source-
based recovery mechanisms on goodput. We observe that ReACT achieves considerably higher (reliable) goodput when compared with RALM. Furthermore,
ReACT is able to keep its goodput steady even at higher loads, while RALM
suffers severe degradation at higher traffic rates. This is mainly because local
recovery prevents the source from backing off its rate when packet losses are
recovered locally. It should also be noted that RALM starts sending at the application rate and then performs congestion control when it receives feedback
from the receivers. This aggressive behavior can potentially lead to severe congestion preventing NACKs from receivers to reach the source. This is one of the
reasons for RALM's reliable goodput degradation as we increase the load. On
the other hand, ReACT's initial setting of the sending rate also contributes to
its high reliable goodput.

Figure 2(b) also illustrates the effect of *CONGESTION_THRESHOLD*,
which is set to 80% and 50% of the maximum MAC queue size. As expected,
goodput is lower at 50% as ReACT becomes more conservative, generates feedback sooner and thus causes more frequent rate decreases.

Normalized overhead for RALM and ReACT are depicted in Figure 4(a).
ReACT incurs significantly lesser overhead than RALM due to its local recovery mechanism at higher loads. It prevents unnecessary source retransmissions
(which are multicast) for errors that are recoverable locally. Source route caching
used by local recovery also helps to reduce the overhead incurred by local recovery. Otherwise, route discovery flooding by the underlying unicast routing
protocols can significantly increase the total overhead. However, at low loads,
probe replies sent by receivers for updating the source sending rate and the
corresponding route discovery initiated by AODV slightly increases the overall
normalized overhead.

4.2   Effect of Node Mobility

In these experiments, we use the random-way-point mobility model with no
pause time and $0m/s$ minimum speed. We vary maximum speed from $5m/s$ to
$20m/s$. The total network load injected in these mobile scenarios is $200Kbps$.

Figure 3(a) shows the reliable delivery ratio achieved by RALM and ReACT
for different node velocities. Both RALM and ReACT are able to achieve perfect
reliability even at high mobility. As previously discussed, the slight variation
in RALM's reliable delivery ratio is due to the NACK generation mechanism
driven by received data. As expected, Figure 3(b) shows that both protocols
exhibit degradation in goodput as we increase node mobility. In ReACT, the
sending rate is updated based on the measured delay reported by the probed
set of receivers. As we increase node mobility, the delay experienced by nodes
becomes highly variable. ReACT uses the highest delay reported to update the

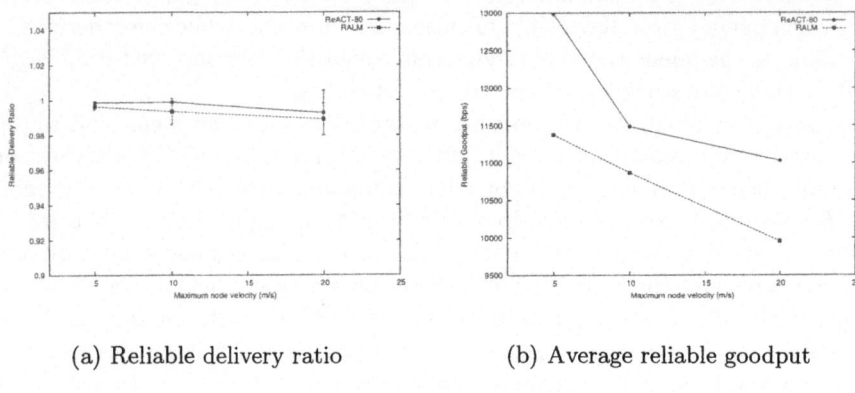

(a) Reliable delivery ratio   (b) Average reliable goodput

Fig. 3. Effect of mobility

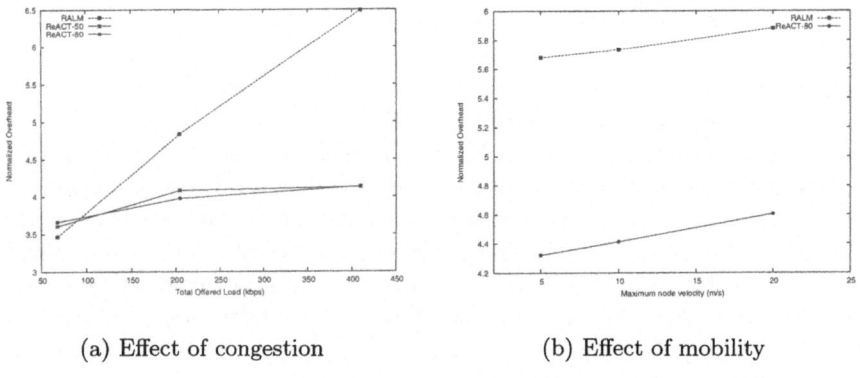

(a) Effect of congestion   (b) Effect of mobility

Fig. 4. Normalized overhead

sending rate, which can substantially reduce throughput. Thus, probing more frequently can improve the goodput with increased node mobility, at the expense of increased overhead due to probe replies. ReACT's local recovery mechanism is able to recover locally some mobility-induced losses, and thus achieves higher goodput than RALM.

As shown in Figure 4(b), ReACT's overhead is significantly lesser than RALM for all mobility conditions. This is mainly due to ReACT's local recovery mechanism which recovers mobility losses locally. As we increase mobility, Re-ACT's overhead increases as its local recovery effectiveness decreases: mobility leads to frequent timeouts of source routes maintained in the *Member Table*. This invalidates potential *Recovery Nodes* and forces receivers to resort to the source for error recovery. As previously discussed, this effect also contributes to the reduction in goodput with increased mobility. In our future work, we plan to overcome this problem by expanding cross-layer interaction and using more information from the lower layers (e.g., unicast routing).

5   Conclusion and Future Work

In this paper, we presented ReACT, an adaptive, congestion-controlled multicast transport protocol for reliable and timely multicast delivery in MANETs. One of ReACT's main distinguishing feature is its combination of source-based control and local recovery. ReACT's source-based control includes initial rate setup and congestion recovery which adjusts the sending rate using a simple stop-and-wait mechanism based on receivers' feedback.

Through simulations, we evaluated ReACT's performance and compared it with RALM, a strictly source-based protocol. Our results show that ReACT significantly improves both goodput and packet delivery with lower overhead. By way of its congestion control mechanism, ReACT is able to deliver perfect reliability under a wide range of conditions. We also demonstrate the benefits of ReACT's local recovery mechanism which prevents the source from reducing its rate unnecessarily and restricts the scope of receiver feedback yielding reduced protocol overhead.

In our future work, we will focus on improving the efficiency of ReACT at higher node mobility scenarios by extending the interaction with the underlying unicast routing layer. As a follow-on to our initial study on layer interaction, we will also investigate the interactions between the transport- and the MAC layer. In particular, we will investigate the synergy between ReACT and MAC protocols that provide link-level reliability for broadcast/multicast data.

References

1. Perkins, C.E.: Ad Hoc Networking. Addison Wesley (2001)
2. Broch, J., Maltz, D.A., Johnson, D.B., Hu, Y.C., Jetcheva, J.: A performance comparison of multi-hop wireless ad hoc network routing protocols. In: Proceedings of the Fourth Annual ACM/IEEE International Conference on Mobile Computing and Networking. (1998)
3. Kim, D., Toh, C.K., Choi, Y.: TCP-BuS: Improving TCP performance in wireless ad hoc networks. Journal of Communications and Networks **3** (2001)
4. Sun, D., Man, H.: ENIC - an improved reliable transport scheme for mobile ad hoc networks. In: Proceedings of IEEE GLOBECOM 2001, San Antonio, TX (2001)
5. Liu, J., Singh, S.: ATCP: TCP for mobile ad hoc networks. IEEE Journal on Selected Areas in Communications **19** (2001) 1300–1315
6. Sundaresan, K., Anantharaman, V., Hsieh, H.Y., Sivakumar, R.: ATP: A reliable transport protocol for ad-hoc networks. In: Proceedings of ACM MobiHoc, Annapolis, MD (2003)
7. Pagani, E., Rossi, G.P.: Reliable broadcast in mobile multihop packet networks. In: Proceedings of the third annual ACM/IEEE international conference on Mobile computing and networking, ACM Press (1997) 34–42

8. Chandra, R., Ramasubramanian, V., Birman, K.P.: Anonymous Gossip: Improving multicast reliability in mobile ad-hoc networks. International Conference on Distributed Computing Systems (2001) 275–283

9. Rajendran, V.: Reliable multicasting in ad hoc networks. Master's thesis, University of California (2003)

10. Tang, K., Obraczka, K., Lee, S.J., Gerla, M.: Congestion controlled adaptive lightweight multicast in wireless mobile ad hoc networks. Proceedings of IEEE ISCC (July 2002)

11. Tang, K., Obraczka, K., Lee, S.J., Gerla, M.: A reliable, congestion-controlled multicast transport protocol in multimedia multi-hop networks. In: Proceedings of IEEE WPMC 2002. (2002)

12. Tang, K., Obraczka, K., Lee, S.J., Gerla, M.: Reliable adaptive lightweight multicast protocol. In: Proceedings of IEEE ICC 2003. (2003)

13. IEEE: Wireless LAN medium access control (MAC) and physical layer specifications. ANSI/IEEE Standard 802.11, 1999 Edition (1999)

14. Floyd, S., Jacobson, V., Liu, C.G., McCanne, S., Zhang, L.: A reliable multicast framework for light-weight sessions and application level framing. IEEE/ACM Transactions on Networking 5 (1997) 784–803

15. Lee, S.J., Su, W., Gerla, M.: On-demand multicast routing protocol for multi-hop wireless mobile networks. ACM/Kluwer Mobile Networks and Applications 7 (2002) 441–453

16. Lee, S.J., Su, W., Hsu, J., Gerla, M., Bagrodia, R.: A performance comparison study of ad hoc wireless multicast protocols. In: INFOCOM (2). (2000) 565–574

17. Perkins, C.E., Royer, E.M.: Ad-hoc on-demand distance vector routing. In: Proceedings of IEEE WMCSA, New Orleans, LA (1999) 90–100

18. Lee, S.J., Gerla, M., Chiang, C.C.: On-demand multicast routing protocol. Proceedings of IEEE WCNC (1999)

19. Scalable Networks: http://www.scalble-networks.com.

A Lightweight Content Replication Scheme for Mobile Ad Hoc Environments

Vineet Thanedar, Kevin C. Almeroth, and Elizabeth M. Belding-Royer

Department of Computer Science
University of California, Santa Barbara
Santa Barbara, CA, USA 93106
{vineet, almeroth, ebelding}@cs.ucsb.edu

Abstract. The mobile, wireless, and self-organizing features of ad hoc networks pose many challenges with respect to continuous availability and accessibility of data. In such a dynamic environment, there are many advantages in replicating a data item so there are multiple copies, including reduced response times and higher data availability. Also, if done efficiently, replication can help reduce energy usage. In this paper, we propose the Expanding Ring replication strategy for pull-based information dissemination environments. One of our primary objectives is the development of a lightweight scheme for mobile nodes. We evaluate the performance of our scheme with respect to a number of parameters and compare it to a system without replication. Our results show a reduction in the average response times and the message processing overhead on nodes. The scheme also does well when both, the overall willingness of nodes to cache data and their individual caching capabilities vary.

1   Introduction

There has been considerable recent interest in the area of information dissemination in a wireless mobile environment. These environments are characterized by the presence of an information server that, through a public wireless access point, delivers useful information to a population of interested mobile clients. This information can be delivered in two ways: 1) a "push-based" approach where the server periodically broadcasts data items and client nodes access this data by monitoring the broadcast channel and waiting until the next broadcast of the data item, and 2) "pull-based" access methods where mobile nodes query the server for data they need. Since both push-based and pull-based systems have their own pros and cons, an efficient, integrated push-pull environment may be a better solution. Such systems have been explored previously [2], [7].

Previous work has addressed pieces of the problem but never in the context of or for the applications we consider. Research efforts in the past have mainly focused on push-based information delivery in a wireless mobile environment. The "Broadcast Disks" approach by Acharya et. al. [1], on-line scheduling algorithms by Vaidya and Hameed [8], [12], and caching strategies proposed by

N. Mitrou et al. (Eds.): NETWORKING 2004, LNCS 3042, pp. 125–136, 2004.

Hara [9] explore push-based systems in detail. Little work has been done in pull-based mobile environments. Also, the focus has mainly been on the scheduling of data items to broadcast in pull-based broadcast systems. Some of the work in pull-based systems include Aksoy et. al.'s scheduling algorithm for large scale on-demand broadcasting [3], [4], and Karakaya and Ulusoy's [11] scheduling algorithm based on an approximate version of the *Longest Wait First* heuristic. Replication schemes for ad hoc networks have been proposed before by Hara [10], and Chen and Nahrstedt [6]. However, these schemes are mainly directed towards replication within a group of mobile hosts. Replication of the requested data items in a pull-based environment has not been significantly explored.

In this paper, we propose the Expanding Ring replication scheme to replicate content in a mobile ad hoc network in the presence of a pull-based information delivery environment. It is worthwhile to mention that although our method is pull-based, the replication strategy is orthogonal to a system where a server is periodically broadcasting content. Hence, our solution can be deployed alongside a push-based system. Given the energy constraints in a mobile ad hoc network, our focus has been on developing a scheme that reduces the burden on a mobile node. Our proposed strategy is elegant and lightweight. The decision to replicate data is handled entirely by the server, thus relieving the nodes of this overhead. In our scheme, the server replicates *in-demand* data in a unique fashion that increases the likelihood of a node finding the required content within its neighborhood. The analysis and simulation results show that our scheme considerably reduces the energy spent by nodes in terms of the messages processed.

The remainder of the paper is organized as follows. In Section 2 we discuss the motivation behind our scheme. The system design and the replication strategy are discussed in Section 3. We present the simulation results of our scheme in Section 4 and concluding remarks and directions for future work in Section 5.

2   Motivation

The motivation behind our work is developing a lightweight replication scheme for a class of application environments that, we believe, will play a major role in the future. These are environments that consist of a server delivering publicly accessible and useful information. A few examples of these include:

- **Public Transportation Areas**: We are already seeing access points being deployed at airports and subway stations that serve information such as up-to-date schedules, last-minute ticket deals, hotel reservations, and car rental information. In addition, because travellers spend a significant amount of time waiting at such places, headline news updates, entertainment news, or similar data may be in high demand.
- **Large Events**: In the future, large sporting events such as the Olympic games will see the deployment of access points that deliver information such as up-to-date games schedules, real-time scores, medals tally, interesting trivia about the games and so on. Visitors carrying mobile devices can then request this information from the server.

We now present a formal analysis of a scheme with no replication, i.e., where each request travels to the server. The motivation for our scheme is the poor performance of this system with respect to the overhead on nodes.

2.1   Analysis of Zero Replication

In this section, we analyze a purely pull-based system with no replication of content in the network. We intend to show that such a model is not very efficient and can be improved by replicating content in the network. We also show that even when content is replicated in the network so that a fraction of content requests are satisfied by other nodes multiple hops away, there is considerable energy overhead on nodes in the network.

Say n items of data, $d_1, d_2, ..., d_n$ are being served by the information server. Consider a m-node one-hop network that generates M requests, $k_1, k_2, ..., k_m$ ($M = \sum_{j=1}^{m} k_j$) for content d_i in time period T. Here k_j is the number of requests generated by node j for d_i in time T. Say the server is an average of H hops away from this one-hop network. The requests then have to travel H hops to be fulfilled. This implies that an average of $2 * H * M$ extra messages will be processed by the intermediate nodes for all the requests to be satisfied. Thus, the energy consumption overhead, O_{norep}, on intermediate nodes for a single data item in terms of the additional messages processed per unit time, per hop of requests generated is:

$$O_{norep} = MP_{add} = \frac{2 \times H \times M}{T} \tag{1}$$

As an example, let $H = 3$, $M = 25$ generated by $m = 6$ nodes in the one-hop network, and $T = 1$ min. Substituting these values in the above equation we can see that 150 extra messages are processed in the intermediate hops for 25 content requests, i.e., 6 times more messages are processed in the network for requests generated just 3 hops away from the server. When applied to all the requests generated in the entire network for all the data items, we can see the large amount of energy consumed (in terms of the messages processed) in processing *extra* messages.

In the above analysis, we looked at the case where each request travels to the central server. This is an extreme option. A better scheme would be where some of the requests are satisfied by intermediate nodes rather than the requests travelling all the way to the server. Say M' out of M messages are satisfied by intermediate nodes and say they take an average of H' hops to find the cached data. In such a scenario, $2 * H * (M - M')$ messages are processed by intermediate nodes for requests going all the way to the server and $2 * H' * M'$ messages are processed by intermediate nodes for all other requests. The energy consumption overhead, $O_{partialrep}$, now in terms of additional messages processed per unit time, per hop is:

$$O_{partialrep} = MP'_{add} = \frac{2 \times H' \times M' + 2 \times H \times (M - M')}{T} \tag{2}$$

Substituting the same set of values as before with $H' = 1$, we see that the number of additional messages processed are 90, or almost 4 times the number of content requests generated. We can see that even if all the requests did not travel to the server, we could still have a significant message processing overhead on the intermediate nodes. This can discourage node participation in the network.

Our goal in this paper is to present our replication strategy, which replicates data in a manner such that requests are satisfied within a node's one-hop network. This considerably reduces the energy spent by nodes in processing content requests and their responses for other nodes.

3   Expanding Ring Replication Scheme Design

In this section, we first describe the core for our replication scheme. The data accessibility mechanism used to access information in the network is then discussed. Finally, our proposed replication scheme is described in Section 3.3.

3.1   Data Agent

Our scheme involves the presence of an agent on each mobile device, called the *Data Agent*. The data agent performs tasks related to the exchange of data between mobile devices. These include support for the replication scheme, management of data stored on the device, and resource monitoring to determine whether the device can participate in the replication mechanism. We term the information delivered by the server as *remote* data items. The data agent marks these items acquired from the server as one of the following two types: Shared Remote and Non-shared Remote data items. Shared remote data items are remote data items that are made available for sharing with other nodes. The node may have acquired these for its own purposes or may hold them as a result of replication. A node services requests for any shared remote data items. If the node lacks the resources to fulfill requests, the agent marks them as non-shared.

3.2   Data Accessibility

In our scheme, each node advertises to its neighbors only the kind of information it needs or in which it is interested. These advertisements, called *interest advertisements*, essentially contain a description of the information required. Once an interest advertisement is created, it is broadcast to all nodes within a distance of one hop. Neighboring nodes do not further broadcast the interest advertisement. This is to prevent a flood of broadcast requests in the network as well as a potential flood of responses. If any of the neighboring nodes has a copy of the requested information, it unicasts the data item back to the advertising node. The data agent on the advertising node sets a timeout period referred to as the *ad response time* for the request to be satisfied by its neighbors. If the request is not satisfied within this time period, the node initiates an information request to the server with the original version of the requested information.

One obvious question that could arise is why the advertisement is not forwarded to nodes beyond the one hop neighborhood. The rationale behind this is to reduce the energy expended by nodes, as well as the network traffic load. Since the request is not forwarded beyond the node's neighborhood, considerable energy is saved by other nodes in the network since they do not have to expend their battery power processing interest advertisements for nodes that may be several hops away. For example, consider a network of 3 hops. If each of these hops has about 5 nodes, then an interest advertisement, when forwarded to nodes beyond one hop, is processed by 12 nodes. On the other hand, only 6 nodes (4 neighbors + 2 additional nodes) will process the advertisement if it is broadcast only within the neighborhood and if not found, requested from the server. Also, nodes may shy away from responding to advertisements not from their neighbors due to a number of reasons such as selfishness, unwillingness to cooperate, or disinterest towards requests from non-neighbors. Hence, we do not forward the request by broadcasting it multiple hops. This also reduces the amount of network traffic.

3.3 Expanding Ring Replication Mechanism

In the previous section we discussed the data accessibility mechanism where the node initiates a content request to the server if it does not find a data item within its one-hop neighborhood. In our replication scheme, the data server monitors demand for each data item and replicates it in the network when demand crosses a certain threshold. The data server keeps track of the demand for each data item by measuring the frequency of requests for various data items. If n is the number of data requests it receives for a data item d_i in time T, then the access frequency f_i for d_i is n/T. This time period, T, is called the *Window Period*. If f_i exceeds a threshold value, λ_i, set by the server for d_i, then the server decides to replicate the data on one or more *capable* nodes in the network. An obvious question here is the accuracy of this method in estimating content access frequencies. For example, if a number of mobile hosts request a data item such that the item is eventually replicated near them and then do not request it again from the server since it is cached nearby, how does the server know whether the items are not being requested or are serviced by nearby nodes? We argue that if such a situation arises, it would be a success of the replication scheme. Also, if requests are not being generated for an item, then the server need not replicate such data items as they are either already replicated or are not being requested.

In the above discussion, we used the term *capable* node. Here, the capability of a node is a measure of a node's ability to service requests for the replicated data items. The data agent on the node computes the capability by applying a capability function. The capability function considers parameters such as available memory space, remaining battery power, and processing power[6]. We assume that such a function exists and it gives a good measure of the node's ability to cache data and service requests.

Node capability is an issue that has been ignored for the most part by research efforts in the past. In this paper, we designate a percentage of the nodes initially as capable nodes. Only these nodes have the ability to cache replicated

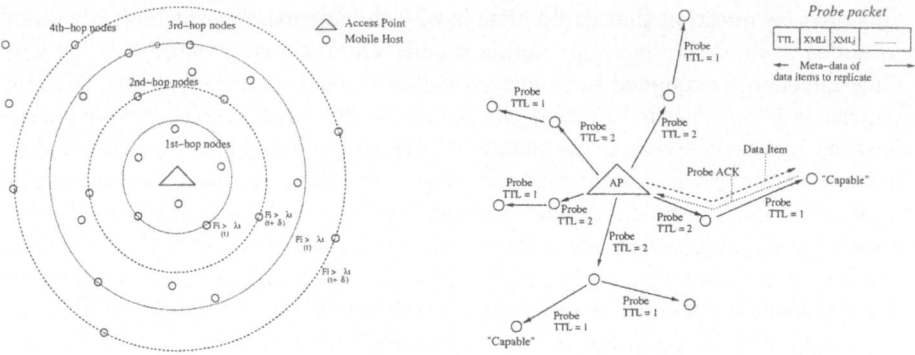

Fig. 1a. Expanding Ring Replication   **Fig. 1b.** Expanding Ring Replication

data items and respond to advertisements for the same. Furthermore, in our simulations we also vary this set of capable nodes. After each window period, we toggle the capabilities of a fraction of the nodes in the network. Thus, some fraction of the capable nodes are *turned* into incapable nodes and an equal number of incapable nodes are made capable. This ensures a practical, dynamic simulation environment.

Our replication strategy to some extent resembles the expanding ring multicast query technique that is based on a Time-To-Live (TTL) search. In our scheme, the data server maintains a set of hop count values, S_i, for each data item, d_i. These hop count values represent the number of hops from the server where the data should be replicated. For example, the set of hop count values could be: $S_i = 1, 3, 5, etc.$ Each time the access frequency for a data item exceeds the threshold, the server selects the current set of hop count values for the data item. The server then attempts to replicate the data item on capable nodes in each of the hops in the set. Upon successfully replicating data, the server sets the values in the set, S_i, to the alternate hops which will now be: $S_i = 2, 4, 6, etc.$

The difference between successive hop count values is no less than two hops, i.e., data is not replicated on nodes nearer than alternate hops from each other. Typically, the transmission range of a wireless access card is around 150 feet in an indoors/crowded environment. Thus, if data is cached on a host, devices within a 300 feet diameter have access to it. This is the primary reason why the server picks alternate hops. The replication scheme is illustrated in Figure 1a.

To replicate the data in the network, the server probes the nodes $h \in S$ hops into the network, soliciting their capabilities to replicate the data item. This is done by the server through broadcast of a small *probe* packet with the current set of hops, S, included in the packet. The TTL value of the probe packet is set to the maximum of the hop numbers in the current set, i.e., $TTL = max(h_l)$, where $h_l \in S$. To reduce the broadcast traffic generated, the server does not broadcast a separate probe packet for every data item. Meta-data information about multiple data items is included in a single probe packet. Intermediate nodes decrease the TTL value by one and further broadcast the packet. This is depicted in Figure 1b. Since the probe packet is broadcast by the server and the nodes, the possibility of duplicate probes reaching a node arises. We detect duplicate probe packets using sequence numbers and ignore them. When the probe packet reaches a host $h \in S$ hops away, the data agent on the host computes the capability of the mobile device to replicate a data item. If the device is capable, the data agent replies back to the server, sending a *probe acknowledgement* packet. The probe acknowledgement packet contains the list of items the node is interested in or has the memory space to cache. The server then directly unicasts the data items to the hosts from which it receives a probe acknowledgement as shown in Figure 1b. The data agent on the mobile host then marks the data item as *shared replicated* and services any requests for the data item.

3.4 Formal Analysis

Consider the same parameters used in Section 2.1. For content to be available to the nodes in the one-hop network, it has to be replicated by the server on at least one of the nodes in the one-hop network. To replicate the content, the server probes nodes h hops away soliciting their capabilities. The *capable* nodes then reply back to the server indicating their interests of data items to cache. The server then directly sends the data items to the *capable* nodes. Since the one-hop network under consideration is an average H hops away from the server, the number of *probe* messages processed by intermediate nodes will be $H * m$ in the worst case, i.e., if each of the m nodes in the one-hop network receives a probe packet on a different route. In the best case, all the nodes in the one-hop network may receive the probe packet on the same route; the number of messages processed by intermediate nodes in that case will be only H. In our analysis, we consider the worst case to demonstrate the effectiveness of the scheme. Say $1/p$ of the nodes are capable of caching the replicated data. The number of messages then processed by intermediate nodes to transmit replies to the server and content from the server will be $2 \times 1/p \times m \times H$ in the worst case. The number of additional messages then processed by intermediate nodes per unit time, per one-hop network, is

$$MP''_{add} = \frac{H \times m + 2 \times 1/p \times m \times H}{T} \tag{3}$$

As before, let $H = 3$ hops and $m = 6$ nodes. Let the percentage of capable nodes be 50%, i.e., $1/p = 1/2$. We then have $MP''_{add} = 36$ messages by intermediate nodes in the network. This is the overhead that the replication scheme has on intermediate nodes in the worst case. As we see, this is only one fifth of the overhead of the pure pull-based replication-less model and close to one third of the overhead of the improved replication model. We can imagine that this value would be considerably smaller in a better case where each node does not receive a probe packet on a different route, i.e., where some of the routes overlap.

Let r be the rate of replication, i.e., the number of times the server picks a set of hops $h_1, h_2, ..., h_l$ for replication in time T. The general equation for the overhead imposed by our scheme per unit time in terms of the messages processed by network nodes is,

$$MP''_{add} = r \times \frac{\sum_{i=1}^{i=l} m \times (h_i + 2 \times h_i \times 1/p)}{T} \tag{4}$$

Note here that the index, i, increments in steps of 2. We can see from the above equation that the scheme is independent of the demand for the data item, i.e., the number of content requests generated in each hop of the network. Although the scheme depends on the number of nodes per hop of the network, we argue that the scheme will work better even in case of high-density networks. This is because in a high density network, the number of content requests generated will also be high and this will impact the performance of schemes that depend on requests going to the server or being forwarded on multiple hops in the network.

Our simulation results, which we now discuss, show that such a strategy efficiently replicates data in the network by gradually replicating the data in the network as the demand for data increases.

4   Evaluation

4.1   Simulation Model

We use the GloMoSim discrete event simulator to evaluate the performance of our proposed scheme. Our simulation environment is shown in Table 1, which shows a listing of the simulation parameters, the range of values tested, and the nominal value for each parameter when not varied. To model data requests by mobile hosts, each node throughout the simulation period generates random requests for data items with a random delay between any two subsequent requests. The random delay is the *Inter-request time* shown in Table 1. The server evaluates the access frequency for each data item once in every *Window Period*. In our simulations we model data requests as an approximate Zipf distribution [5]. We have three categories of data items: highly popular data items, moderately popular, and less popular data items. The probability of a request for a data item depends on the category to which it belongs and therefore to its popularity. All categories have an equal number of data items and the probability of occurrence of a data item within a category is uniformly random. We also vary the cache size of each node in the simulations to model the capability of a node. The cache size i.e., the number of replicated data items a node can hold, is modelled in terms of a fraction of the total number of items that the server delivers. In our simulations, we also vary the total number of nodes in the network that are capable of participating in the replication mechanism from a low 25% to a high of 75% of the total number of nodes.

4.2   *Response Time Simulation Results*

We first examine the effect of the number of nodes in the network on the data request response time. The Response Time, R_t, is measured as the time delay

Table 1. Simulation parameters and values

Parameter	Range of values simulated	Nominal value
Network area	500mx500m, 1000mx1000m, 1500mx1500m	1000mx1000m
Simulation time	1200s - 1800s	1800s
Mobility model	-	Random waypoint
Mobility	0-10m/s, 30s pause time	0-5m/s, 30s pause time
Number of Nodes	10-100	60
Percentage of capable nodes	25-75%	50%
Number of data items	10-50	15
Cache Size	10%-100% of total items	30% of total items
Inter-request time	-	20-40s
Ad-response time	-	100ms
Window period	-	60s
Number of Seed values	-	5

experienced by a node to receive a data item from the time it issues an interest advertisement. Thus, $R_t = T_{data\_received} - T_{interestad\_broadcast}$.

The average of the response times of all requests made within the simulation period is plotted against the number of nodes in the network. The graph of the simulation results is shown in Figure 2. From Figure 2 we can see that when there is no data replication, the average response time increases as the number of nodes in the network increases. This is because the increased number of data requests results in longer response times at the server. In our scheme, for a small number of nodes in the network, the average response time is higher than without replication. This is because of two reasons: (1) there are not sufficient nodes in the network to replicate data, and (2) the node density is very low and hence the probability of a node finding content within its neighborhood is greatly reduced. Since the node initially does a local one hop broadcast to search for content in its neighborhood but does not find it, this additional time results in higher average response times for fewer nodes. However, as the number of nodes in the network increase, we see that the average response times decrease appropriately. Due to a higher number of nodes in the network, the number of content requests rises. This results in the access frequency crossing the threshold at the server more often. The server then frequently attempts to replicate data in the network. As the number of data item copies increases in the network, more requests are satisfied within one hop of a node's range resulting in shorter response times.

4.3 *Message Processing Overhead Simulation Results*

In this section we evaluate the performance of our scheme with respect to the overhead it imposes on nodes in the network. We measure this overhead in terms of the number of *extra* messages processed by the nodes, as explained earlier in

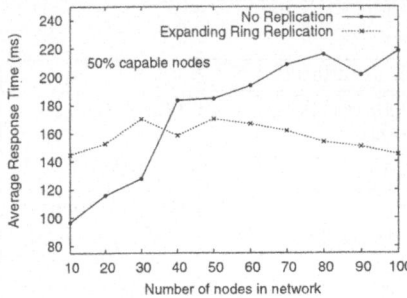
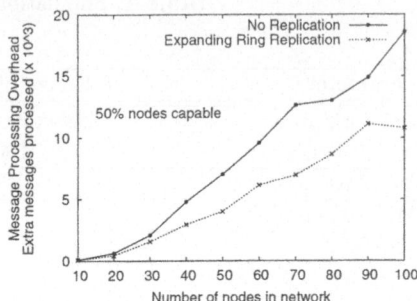

Fig. 2. Average response time versus number of nodes in the network

Fig. 3. Control message overhead versus number of nodes in the network

the formal analysis of the scheme. We vary the number of nodes in the network and measure the additional messages processed in each case. This also gives us an indication of the scalability of our scheme with respect to the population of hosts in the network. The graph depicting this is shown in Figure 3.

We can see from the graph that without replication, there is a steep linear rise in the number of messages processed as the network gets more dense. In our scheme, although there is a rise in the overhead, there is a significant difference between the extra messages processed in our scheme and one without replication. From the graph we can see that even in a dense network of 100 nodes, the number of additional messages processed in our scheme is on the order of 11,000 whereas without replication, the number is on the order of 19,000 messages. Thus there is a considerable decrease in the overall energy usage in terms of the additional messages processed in the network with our scheme.

4.4   Capable Node Percentage Simulation Results

We now evaluate the performance of our scheme when the actual number of capable devices can vary. To test this aspect, in each simulation run, only a percentage of the total number of nodes are capable. Therefore, when the server probes nodes h hops away, only some of the nodes respond to the probe packet. We evaluate this aspect with respect to the one-hop hit percentage, i.e., the fraction of the total number of times that a requested data item is found in a node's neighborhood or is present in the node's cache as a result of replication. The results of the simulation are shown in Figure 4.

It is clear from the graph that as the percentage of nodes capable and willing to cache data items increases, the hit ratio increases significantly. Also, as the number of nodes in the network increases, the one-hop hit ratio increases. From the graph we can also see that beyond a point (50% nodes capable), the increase in the one-hop hit ratio is not as rapid. This is because even if a higher number of nodes are capable of caching the data, only one of a nodes' neighbors needs to cache the data item for a node requesting a data item to find it within its neighborhood. Thus, we argue that the scheme will work well even if less than

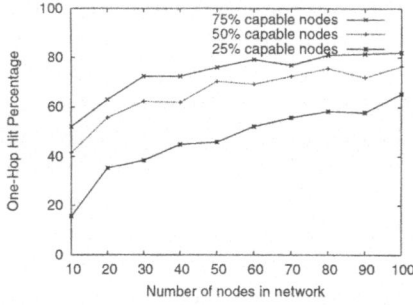
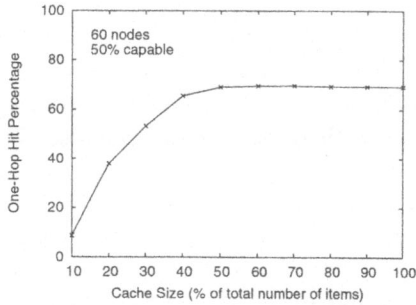

Fig. 4. Hit ratio versus number of nodes   **Fig. 5.** Hit ratio versus cache size

100% of nodes in the network are capable of participating in the replication scheme.

4.5 Cache Size Simulation Results

Since a mobile device in general has less memory, we evaluate the effect of the cache size on the one-hop hit ratio. We vary the number of data items that a node is capable of caching in its *shared replicated* segment in terms of the percentage of the total number of data items offered by the server. The graph depicting the simulation result for one-hop hit percentage is shown in Figure 5. We can see that initially as the cache size increases, the one-hop hit ratio increases as more data items can be cached on a node and hence a greater number of requests are satisfied within a node's neighborhood. However, increasing the cache size beyond a point does not make a significant difference in the hit percentage because it then depends on the data items requested, i.e., the data request distribution pattern. Since the popular data items may have been already cached on one or more neighboring nodes, even if the cache size is increased, no major improvement is noticed. Hence, our scheme does not require an excessively large cache size. It depends on the cache size only to an extent where some nodes in each neighborhood can, in a combined manner, cache a few data items. The server replicates data items on these nodes based on their demand such that requests are satisfied within the one-hop neighborhood of a node. Also, since the server replicates a data item on multiple capable nodes in a hop, the caching capacity of a single node does not significantly affect the scheme.

5 Conclusions and Future Work

In this paper, we have proposed a lightweight replication mechanism for ad hoc networks by utilizing the communication channel in a pull based information delivery system. Since connectivity can be intermittent in such an environment, a node's best source for the data it needs may indeed be the nodes in its neighborhood. Thus, a mechanism that replicates data in a manner such that the likelihood of finding data within the neighborhood is high, can effectively reduce the response time as well as the network traffic generated due to data requests.

The simulation results show that our scheme performs well in a variety of scenarios. To summarize, the features of our replication scheme are, 1) it is scalable with respect to the number of nodes in the network, 2) it is lightweight in terms of message processing overhead for the mobile device, 3) it does not depend excessively on the percentage of *capable* mobile devices in the network, and 4) it does not depend on a single mobile device's cache size to a large extent.

In our scheme we assume that if a node has the resources to cache a data item and service requests for it, then the node behaves as a cooperative member in the network. This unfortunately may not be true in a real ad hoc network setting. Hence, if an incentive scheme can be added to the system so that a node is encouraged to cache data items and share them with others, the effectiveness of the replication scheme can be greatly increased.

References

1. Acharya, S., Franklin, M., Zdonik, S.: Dissemination-based data delivery using broadcast disks. IEEE Personal Communications, 2(6) (1995)
2. Acharya, S., Franklin, M., Zdonik, S.: Balancing push and pull for data broadcast. In Proc. of the ACM SIGMOD (1997) 183-194
3. Aksoy, D., Franklin, M.: Scheduling for large-scale on-demand data broadcasting. In Proc. of the IEEE INFOCOM (1998) 651-659
4. Aksoy, D., Franklin, M.: Rxw: A scheduling approach for large-scale on-demand data broadcast. IEEE/ACM Transactions on Networking, Vol. 7 (1999) 846-860
5. Breslau, L., Cao, P., Fan, L., Phillips, G., Shenker, S.: Web caching and zipf-like distributions: Evidence and implications. In Proc. of the IEEE INFOCOM (1999) 126-134
6. Chen, K., Nahrstedt, K.: Cross-layer design for data accessibilty in mobile ad hoc networks. In Proc. of the fifth World Multiconference on Systemics, Cybernetics and Informatics (2001) 315-320
7. Deolasee, P., Katkar, A., Panchbudhe, A., Ramamritham, K., Shenoy, P.: Adaptive push-pull: disseminating dynamic web data. In the tenth International World Wide Web Conference on World Wide Web (2001) 265-274
8. Hameed, S., Vaidya, N.H.: Efficient algorithms for scheduling data broadcast. Wireless Networks, Vol. 5 (1999) 183-193
9. Hara, T.: Cooperative caching by mobile clients in push-based information systems. In Proc. of the eleventh International Conference on Information and Knowledge Management (2002) 186-193
10. Hara, T.: Effective replica allocation in ad hoc networks for improving data accessibility. In Proc. of the IEEE INFOCOM (2001) 1568-1576
11. Karakaya, M., Ulusoy, O.: Evaluation of a broadcast scheduling algorithm. Lecture Notes in Computer Science (2001)
12. Vaidya, N.H., Jiang, S.: Data broadcast in asymmetric wireless environments. In Proc. of the first International Workshop on Satellite-based Information Services (WOSBIS) (1996)

Scalable Service Discovery in Mobile Ad Hoc Networks

Uday Mohan[1], Kevin C. Almeroth[1], and Elizabeth M. Belding-Royer[1]

Department of Computer Science, University of California, Santa Barbara,
CA 93106-5110, Ph: 805-893-7520, Fax: 805-893-8553
{udaym, almeroth, ebelding}@cs.ucsb.edu

Abstract. Recent advances in hardware technology and wireless networking have made it possible to deploy large scale ad hoc networks[1]. As these networks begin to grow in size, an efficient mechanism is needed to locate services distributed within them. Protocols such as the Service Location Protocol (SLP)[2] and Universal Plug and Play (UPnP)[3] exist for service discovery in static networks. These protocols are based on a centralized server where services can register themselves and clients can query for them. However, maintaining such a server in an ad hoc network is difficult as nodes randomly join and leave the network. In this paper we study the problems associated with service discovery by first simulating two well known service discovery techniques and investigating their limitations for large network sizes. We then combine the best features of each approach to present an innovative, scalable mechanism. We simulate and analyze this mechanism and show it to scale well for large network sizes without increasing the latencies in locating a service.

Keywords: Ad hoc network, Service Discovery, Simulations

1 Introduction

A mobile ad hoc network (MANET) is a network formed by a group of wireless devices with limited power and transmission range[4]. These networks do not need any existing infrastructure but can form a network on the fly, with each device acting as a relay to forward packets for other nodes. With recent advances in hardware technology and wireless networking, it is now possible to deploy large scale MANETs[1]. As these networks begin to grow in size, an efficient mechanism is needed to locate services distributed within them. A *service* may be a computation, storage, a communication channel to another user, a software filter or a hardware device[5] that can be used by a person or a software program.

There are currently a number of existing protocols for service discovery [2, 6,3,7,8]. These protocols are centralized, registration-oriented protocols, with an assumption that a centralized database of services can be maintained and accessed by every node. However, these existing strategies do not work well for MANETs because of several reasons. First, in such networks a centralized server

N. Mitrou et al. (Eds.): NETWORKING 2004, LNCS 3042, pp. 137–149, 2004.

is difficult to maintain as the nodes can join or leave the network at random and therefore no node is part of the network permanently. Second, because of the dynamic nature of the network, every time a service leaves or joins the network it has to inform the centralized server about its presence and this presents a scalability issue. Considering these and other factors, a decentralized approach to service discovery is desirable in ad hoc networks.

Several intuitive ways to solve this problem come to mind. A straightforward solution is a Push-based solution in which *services* advertise themselves in the network by periodically broadcasting packets. A second method to locate services is a Pull-based method where *clients* actively broadcast requests into the network. In this paper we evaluate these two well known discovery techniques for MANETs. We then propose a model that uses the best features of each, and add mechanisms to provide scalability.

The rest of the paper is organized as follows. In Section 2 we describe a motivating scenario for service discovery in MANETs. Section 3 describes the push-based and pull-based techniques to discover services in a MANET and identifies their limitations. In Section 4 we present our Adaptive Service Discovery Model. Section 5 presents simulation and analysis of our model and Section 6 offers our conclusions.

2 Motivation

To understand resource discovery and its requirements, we now discuss in detail a motivating example. Consider a shopping mall, where a MANET is formed by people carrying mobile devices. Some of those devices may be in need of traditional services such as printing, scanning or access to the Internet, while others may be looking for new services such as spare CPU cycles. There will be companies advertising their services for which people may want to pay, e.g. McDonalds offering their lunch time specials or banks offering stock quotes. Some people may be running a peer-to-peer system such as Gnutella[9] to exchange files or multimedia content. There may be teenagers wanting to locate friends to chat with, or parents monitoring their children. In essence, there is a dynamic network with a wide selection of resources available, and there are devices that constantly need to locate these resources. As the number of devices entering the network grows, an efficient mechanism will be required to discover these resources. In such a network,there is a high probability that a large percentage of the nodes will take part in service discovery.

A number of existing protocols for resource discovery exist like the Service Location Protocol(SLP)[2,6], Jini[7], Salutation Consortium[8], and Universal Plug and Play (UPnP)[3]. All these protocols have similar architectures, the common feature being a centralized database that keeps track of all available services in the network. This feature works well in static networks where a centralized database can be maintained. However, it does not work well in MANETs because of several reasons.

First, maintaining a centralized database is difficult in MANETs because no node is a permanent member of the network. Nodes join and leave the network at random. There may be no pre-existing base stations, no well-known servers and no guaranteed Internet connectivity. This makes the centralized approach difficult.

Second, even if such a centralized database is somehow maintained, the nodes are mobile and the same service may enter and leave the network multiple times. Every time it disconnects and then connects, it has to contact the database and send updated information. Therefore, the centralized database has to keep track of every service in the network and has to do it every time the service joins or leaves the network. As the number of services in the network increases, this method does not scale well.

Third, several nodes may be providing the same service in the network and these nodes may enter at different times. An existing node might need updated information of all services in the network at all times. For example, there may be several nodes in a network providing Internet connectivity, with different costs associated to them. A node looking for an Internet connection will prefer to be connected to the server with the least cost of service. Hence, this node has to periodically poll the centralized database for updated information about any new server with better costs that might have entered the network at a later time. This continuous polling creates a serious scalability problem.

3 Architectural Considerations

We have described several scenarios where resource discovery is critical. The network used in each of these scenarios can be reduced to a generic architecture. This architecture is shown in Figure 1 and consists of mobile devices that fall into three categories.

1. Mobile devices that can provide services or have content that can be exchanged. We call these devices *servers*.
2. Mobile devices looking for such services. We call these devices *clients*.
3. Nodes that are not part of this resource discovery architecture but just members of the MANET helping in routing packets. We call them *forwarders*.

There are two intuitive methods to locate resources in the above architecture and are discussed next.

3.1 The Push and Pull Models

A simple way to enable service discovery is for services to advertise their presence in the network (Push model)[5]. Another method to locate services is for clients to actively query for them (Pull model)[5]. To understand these two models and create the motivation for our better, more scalable solution, we next simulate them and study their characteristics and performance impact on the network.

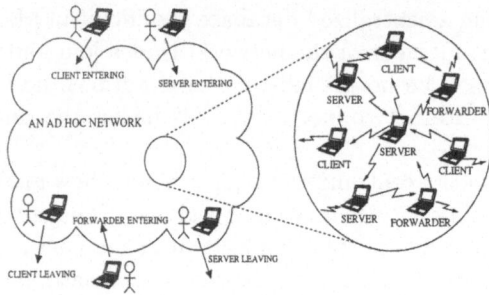

Fig. 1. A generic architecture for a Mobile Ad hoc Network.

Simulation of the Push and Pull models. Simulations were performed using the GloMoSim Network Simulator[10] developed at UCLA. In particular, we wanted to study the traffic generated and the latencies in locating services when the number of servers and clients varies.

To simulate a server or a client entering the network, we start it at a random node. A server provides only one service. Each client requests only one service. There can be several servers providing the same service and several clients interested in the same service. We keep the total number of nodes fixed to 400 but vary the percentage of servers and clients.

To simulate the **Push Model**, each server advertises its service by broadcasting packets containing the *source address*, the *advertised service type*, and a *sequence number*. Each client passively waits for a particular service to be advertised. Upon receiving the advertisement, it can determine the address of the service and can contact the service directly. The latency is the difference between the time a client starts looking for a service and the time it receives the first advertisement packet from a server.

For the **Pull model**, the servers listen passively for service requests. When a client wants a particular service, it periodically broadcasts a request packet containing the *node address*, the *required service*, and a *sequence number*. When a server receives a packet with a matching service type, it broadcasts a reply back to the client with its address. In our model, we assume that several clients may be interested in the same service and accordingly a broadcast is preferable over unicast. The latency here is the difference between the time a client sent out its first query packet and the time it got the first packet with the service address.

An important point to note in the Pull model is that a client may broadcast request packets even after it has located a server. It does this because several nodes may be providing the same service, and may join and leave the network at different times. By continuously broadcasting query packets, the client has up-to-date information about all services in the network. An optimization here could be to reduce the rate of querying after locating the first server. In our model, we have implemented a simplified approach in which we do not reduce the querying rate but keep it the same even after locating a service.

The number of nodes used for the simulations was 400. These nodes were initially placed randomly within a fixed-size of a 1500 x 1500 m^2 field. We choose this field size as it could be a typical shopping mall or a disaster rescue area. The random waypoint mobility model[11] was chosen with speeds between 0 and 5 meters/second and a rest period from 0 to 10 seconds. The range of each communicating node is 250 meters and the link bandwidth is 2 Mbits/second. The protocol used is UDP with each advertisement or query packet having a size of 512 bytes. We ran a number of simulations by varying the values of each parameter. Had we chosen a higher or lower value, the shape of the graphs would be slightly different but would still yield the same conclusion. Hence, the results using the above parameters are representative of the entire set of simulations we conducted.

Evaluation. Figure 2 shows the latency and overhead of the Push and Pull models. The left side of the x-axis has 0% servers and 100% clients and it varies linearly to 100% servers and 0% clients. The total number of nodes is kept constant at 400.

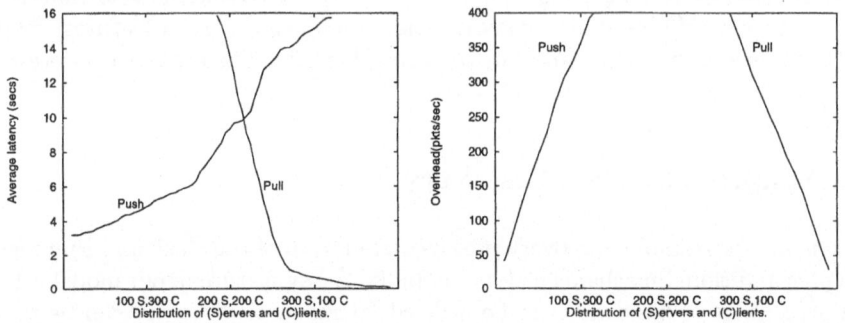

Fig. 2. Average latency and overhead for the Push and Pull models with varying distributions of Servers and Clients.

From Figure 2 it can be observed that the average latency in the Push model increases as the number of servers in the network increases. This can be explained by the overhead graph in Figure 2. As the number of servers increases, there is a corresponding rise in advertisement traffic. When the traffic increases beyond a certain limit, collisions start to occur and packets are lost. This results in the observed rise in latency to locate services. For example, suppose a server advertises a service every 10 seconds. If a packet is lost due to congestion, a new client entering will hear another advertisement only after 20 seconds.

One consideration then is how often should services advertise their presence. Advertising too often just increases the network traffic and redundant advertisements are flooded through the network. Advertising too infrequently results in high latencies. There is a fine balance between the number of services in the network and their frequency of advertisement so as to keep the number of collisions minimum. Achieving this balance is nearly impossible because the topology and membership of the network is dynamic.

Similarly, from Figure 2 it can be observed that the average latency in the Pull model increases as the number of clients in the network increases. This can also be explained from the overhead graph in Figure 2. When the number of clients in the network increases, the control traffic increases proportionately. This rise in control traffic causes congestion, and control packets are lost. This results in latencies becoming higher. For example, suppose a client queries the network every 10 seconds. If a packet gets lost due to congestion, the client will not get a reply. Clients then have to issue a request multiple times before they receive a response.

Like in the server Push case, the challenge is finding the appropriate interval for clients to query the network for a service. After sending out a query packet, how should the client decide when to query again? Sending it after a long interval of time will result in high latencies. Querying too often will result in more traffic being generated. There is a fine balance between the number of clients and their frequency of querying so as to keep the number of collisions minimum.

The simulations so far show that neither of the above methods is scalable to large networks. We need a mechanism that combines the good features of both methods, is scalable, and can limit the control traffic. The next section presents our scalable model for resource discovery.

4   Adaptive Service Discovery Model

Our model for resource discovery uses a combination of the Push and Pull methods plus additional mechanisms for scalability. A key feature in our model is that the percentage of bandwidth to be utilized for resource discovery can be set so as not to exceed a given upper bound.

In our model, a server and a client both actively try to locate each other. The goal of a server is to periodically advertise its services while the goal of the client is to query for a needed service. A client can locate a service in two ways. It can either receive a periodic advertisement by a server, or can receive a response to a query. In either case, this is done in an adaptive manner in order to keep the control traffic below a threshold limit. Every time a server is scheduled to advertise, it first listens for other control packets in the network. It listens for a time interval to estimate the number of members that advertised or queried within that period. If this number is below a threshold, the server sends an advertisement; otherwise it exponentially backs off. This is shown in Figure 3. At the end of the backoff period, it again senses the network and repeats the process. If it has backed off a maximum number of times, it ceases backing off and advertises *forcefully*. This forceful advertisement is special in that it forces others to back off, and lets it advertise. This is used to prevent starvation. This happens because other members also sense the traffic before advertising, and hear the forceful advertisement and back off themselves.

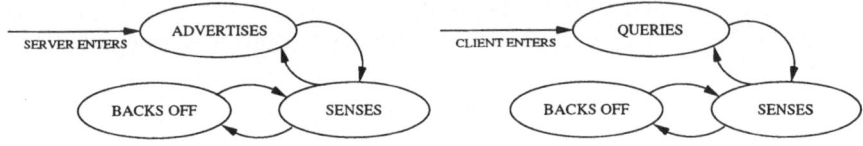

Fig. 3. State diagram for the server and client.

Along with advertising its presence, a server also listens for queries from clients. If it hears a query and can provide the requested service, it sends a reply immediately. This reply could be unicast to the client or could be a general broadcast into the network. In our model, the server broadcasts a reply to the client. This is because we assume that several clients may be interested in locating the same service and so a broadcast is preferable. To limit response scaling problems, a server only replies to a query if it has not heard from the client in one scheduled advertisement interval[1].

Similarly when a new client wants to locate a service, it first transmits a query packet to announce its presence. Before it broadcasts the next packet into the network, it listens to the control traffic and estimates the number of members in the network. It accordingly queries for services or backs off similar to the way a server adapts its rate. This is shown in Figure 3.

We now describe our model more formally. The parameters used are described in Table 1. The method to select an appropriate value for each parameter will be discussed in Section 5.2 and 5.3.

Table 1. Parameters used in our simulations.

$\mathcal{T}$	is the period of time that a server or client listens for other control packets in the network. This should be at least equal to the NET_TRAVERSAL_TIME[12], which is a conservative estimate of how long it should take a message to traverse the entire MANET.
$\mathcal{N}_{Threshold}$	is the maximum number of members that can advertise or query in $\mathcal{T}$ time. The amount of bandwidth taken up by the resource discovery protocol depends on this variable.
$\mathcal{S}_{Advertisement}$	is the number of advertisements packets a member hears from unique sources.
$\mathcal{C}_{Query}$	is the number of query packets a member hears from unique sources.
β	is the number of times a member has backed off.
β_{Max} & β_{Min}	is the maximum and minimum values of the backoff counter.

---

[1] This, and other security concerns will be dealt with in Section 5.3.

Model

1. When a new server (or client) enters the network, it advertises (or queries) once to announce its presence in the network[2]. It sets β to β_{Min}.
2. It listens for a time $\mathcal{T}$ to the control traffic to estimate the number of members $\mathcal{N}$ that advertised within that time period. $\mathcal{N}$ is calculated as

$$\mathcal{N} = \sum(\mathcal{S}_{Advertisement} + \mathcal{C}_{Query})$$

   – IF $\mathcal{N} <= \mathcal{N}_{Threshold}$ then
     Advertise (or Query).
     Set β to β_{Min}
   ELSE
       IF $\beta < \beta_{Max}$
          Increment β by one.
          Backoff for 2^β x $\mathcal{T}$ seconds
       ELSE
          Advertise(or Query) forcefully.
          Set β to β_{Min}

3. Goto step 2.

5   Simulation and Analysis

We simulate our Adaptive model to achieve several goals. First, we compare the performance of our model with both the Push and Pull models. Through this comparison we evaluate the improvements provided by our model. Second, we investigate the performance of our model in different scenarios to understand its characteristics and determine possible optimizations.

5.1   Simulation Environment

The Adaptive model is simulated in GloMoSim and works as described in Section 4. Servers and clients enter the network at random times advertising or querying for a service. The advertisement or query packets contain fields similar to the fields used in the Push and Pull models. Each server (client) then listens for $\mathcal{T}$ seconds to sense the channel. It advertises (queries) if it senses the number of members below $\mathcal{N}_{Threshold}$; otherwise it backs off exponentially. Every server (client) backs off a maximum of β_{Max} times. After these many unsuccessful attempts, it forcefully broadcasts a packet to tell others to back off and let it advertise (query).

The range of parameters used for the simulation of the Adaptive model is given in Table 2. The nominal values are the default value used in the simulations when the particular parameter is not varied.

---

[2] The problem of traffic explosion, i.e. when a large number of members suddenly enter the network at the same time, may occur but the impact is not significant. The reason for this will be explained in Section 5.2.

Table 2. Range of parameters for the Adaptive model.

Parameter	Min	Max	Nominal
Number of nodes	10	400	400
$N_{Threshold}$	2	10	5
T(ms)	150	1000	1000
β_{Min}	0	0	0
β_{Max}	3	6	5

5.2   Results

Our simulations begin by comparing the *latency* and *overhead* of the Adaptive model to the Push and Pull models. We then fine tune our model to determine the optimum range of values used for the various parameters.

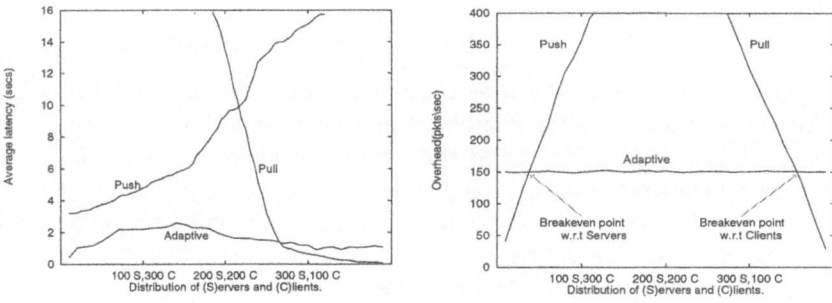

Fig. 4. Average latency and overhead for the Push, Pull and the Adaptive models with varying distributions of Servers and Clients.

To compare the latency and overhead, we run two different sets of simulations. In the first set, we keep the total number of nodes fixed at 400 but vary the percentage of servers and clients in the network. In the second set, we run simulations by varying the number of nodes from 10 to 500 and observe how overhead increases for the three models.

Figure 4 shows the same result as in Figure 2 but with additional lines for the Adaptive model. The distribution of clients and servers varies as explained in Section 3.1.

The Adaptive model performs significantly better in terms of latency and bandwidth utilization. Due to better bandwidth utilization, collisions are minimized and this results in lower latencies to locate services. However, this comes at a certain cost. The tradeoff is that the overhead in using our model for a 400 node network is greater than the push model if the number of servers is less than 40 servers. Similarly the overhead in our model is greater than the pull model if the number of clients is less than 35. This happens because the traffic is low when there are few clients in the Pull model and few servers in the Push model and there is a lot of free bandwidth available. In the Adaptive model, servers and clients both advertise and query, hence more traffic is generated.

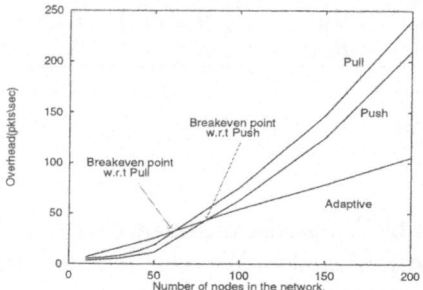

Fig. 5. Overhead of the Pull, Push and the Adaptive models with varying number of nodes.

Figure 5 shows the overhead as the number of nodes in the network is increased. The x-axis shows the number of nodes where each run consists of 50% servers and 50% clients. We varied the number of nodes to 500; however Figure 5 only presents our results till 200 nodes. This is because nothing unusual is observed after 200 nodes and the trend continues. We wanted to emphasize the breakeven points and hence, we only present simulation results till 200 nodes. Initially the overhead for the Adaptive model is slightly greater than for the Push and Pull model. However, when there are more than 60 nodes, the overhead for the Pull model becomes greater than the overhead for the Adaptive model. This is because when the number of nodes in the network exceeds 60, the overhead for 30 servers advertising *every 10 seconds* becomes higher than 60 members advertising or querying *adaptively*. Similarly, for the Push model this limit is 81 nodes.

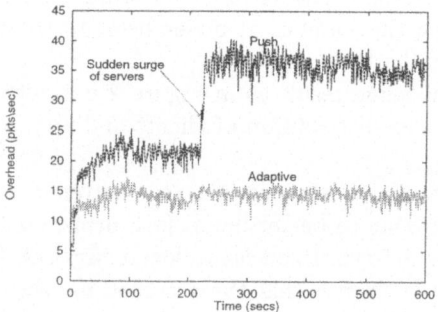

Fig. 6. Overhead with a sudden burst of servers at approximately 220 seconds.

Figure 6 shows the variation of overhead with time when a number of servers suddenly enter the network. In this simulation we start 50 servers at uniform times in the network. At around 220 seconds we simulate a sudden burst of 20 new servers entering the network. In the push model these servers immediately start advertising and hence there is a sharp rise in overhead. The adaptive model performs better under the same scenario as it is able to keep this overhead under control. This is because when new servers enter and advertise the first time, existing members in the network backoff to let these new servers advertise. This happens because in our model, every member listens for an interval $\mathcal{T}$ for advertisements or queries in the network. Existing members hear the new members and back off. If members suddenly enter the network within a time period of $\mathcal{T}$ seconds, there will be a noticeable rise in control traffic in the adaptive model. Hence, the value of $\mathcal{T}$ should be set to an interval such that the probability of several members suddenly entering the network within that interval is very low. Our model performs in a similar manner when there is a sudden burst of clients entering the network because clients follow the same mechanism to limit the overhead. Hence, our model is able to limit the overhead when there is a sudden burst of *servers or clients* entering the network.

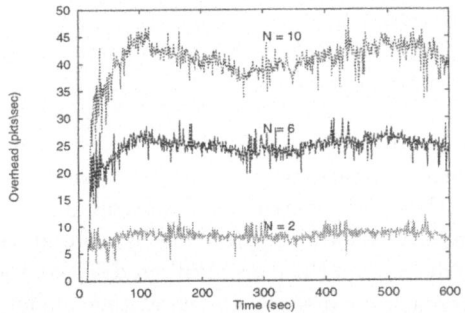

Fig. 7. Variation of overhead with time for different values of N=$\mathcal{N}_{Threshold}$.

Figure 7 shows the variation in overhead as $\mathcal{N}_{Threshold}$ is varied. $\mathcal{N}_{Threshold}$ is the number of members allowed to advertise or query in $\mathcal{T}$ seconds. Setting $\mathcal{N}_{Threshold}$ to a high value results in more members being able to advertise or query and hence more traffic is generated. As the value of $\mathcal{N}_{Threshold}$ is reduced, the control traffic decreases. The amount of bandwidth utilized by our resource discovery model depends on this variable and is an essential feature in our model.

5.3 Discussion

The Adaptive Service Discovery Model performs significantly better than the Push and Pull models. However, there are several practical considerations that need to be dealt with before our model can be deployed. First, how should parameters such as $\mathcal{N}_{Threshold}$ and $\mathcal{T}$ be communicated to new members entering the network? A simple mechanism to do this is to piggyback these two parameters in the dynamic address configuration packet which assigns an IP address to the node[13] when it first enters the network. Another method to set $N_{Threshold}$ is to set it to be a fixed percentage of the maximum bandwidth available in the MANET[14]. This maximum bandwidth value will be unambiguous and known to every member in the network. To set the value of $\mathcal{T}$, it should at least be equal to the NET_TRAVERSAL_TIME [12] and can be experimentally determined by a node in the network.

Second, there are several security concerns in our model. A server immediately replies with an advertisement when it receives a request packet. A malicious node can flood the network with query packets and force a server to keep replying, thereby causing a denial-of-service attack[15]. To prevent this, every server replies only to the first query it receives from a node within its scheduled advertisement interval. It ignores duplicate queries that it receives between two of its periodic advertisements. Another security concern is if a node cheats and advertises or queries at a higher rate. A node can do this by increasing its value of $N_{Threshold}$. There has to be a mechanism for neighboring nodes to sense this discrepancy and suppress the additional traffic generated by the cheating node. Techniques such as *Watchdog* and *Pathrater*[16] are likely to be useful if customized for this scenario.

6   Conclusions

As MANETs begin to grow in size, an efficient mechanism is needed to locate services in them. There are currently a number of existing protocols for service discovery. However, these are centralized protocols and do not scale well for MANETs. In this paper we study the problems associated with service discovery for MANETs by simulating the Push and Pull models. We then combine the best features of each approach to present an innovative model. Simulation results show that this model scale wells for large network sizes without increasing the latencies in locating a service. A further improvement can be achieved by setting priorities for services. Generally, 10% of services are used by 90% of clients. This 90-10 rule can be used to further reduce overhead and latency by setting up higher priorities for those 10% services used more often. We also need a formal model to define services in a network in order to let clients accurately define the services they require. All of these are interesting issues that need further research.

References

1. Intel Team: Largest Tiny Network. http://today.cs.berkeley.edu/800demo/ (2001)
2. J. Veizades, E. Guttman, C. Perkins, S. Kaplan: SLP: Service Location Protocol. Internet Engineering Task Force : RFC 2165 (1997)
3. Consortium Members of the UPnP Forum: Universal Plug and Play Device Architecture. Version 0.91. http://www.upnp.org (2000)

4. S. R. Das, C. E. Perkins, E. M. Belding-Royer: Performance comparison of two on-demand routing protocols for ad hoc networks. In: Proceedings of the IEEE Conference on Computer Communications (INFOCOM), Tel Aviv, Israel. (2000)
5. L. Cheng: Service advertisement and discovery in mobile ad hoc networks. In: Workshop on Ad hoc Communications and Collaboration in Ubiquitous Computing Environments. (2002)
6. C. E. Perkins, H. Harjono: Resource discovery protocol for mobile computing. In: IFIP World Conference on Mobile Communications. (1996)
7. K. Arnold, A. Wollrath, B. O'Sulliva, R. Scheifler, J. Waldo: The Jini specification. Addison-Wesley,Reading, MA, USA (1999)
8. Members of the Salutation Consortium: Salutation Consortium Homepage. http://www.salutation.org (2000)
9. M. Ripeanu: Peer-to-peer architecture case study: Gnutella network. In: Proceedings of International Conference on Peer-to-peer Computing. (2001)
10. X. Zeng, R. Bagrodia, M. Gerla: GloMoSim: A library for parallel simulation of large-scale wireless networks. In: Workshop on Parallel and Distributed Simulation. (1998)
11. T. Camp, J. Boleng, V. Davies: A survey of mobility models for ad hoc network research. In: Wireless Communications and Mobile Computing (WCMC): Special issue on Mobile Ad Hoc Networking: Research, Trends and Applications. (2002)
12. C. E. Perkins, E. M. Belding-Royer, S. R. Das: IP broadcast in ad hoc mobile networks. IETF Internet Draft, draft-ietf-manet-bcast-02.txt,(Work in Progress) (2001)
13. N. H. Vaidya: Weak duplicate address detection in mobile ad hoc networks. In: The Third ACM International Symposium on Mobile Ad Hoc Networking and Computing(MOBIHOC). (2002)
14. H. Schulzrinne, S. Casner, R. Frederick, V. Jacobson: RTP: A transport Protocol for Real-Time Applications. Internet Engineering Task Force : RFC 1889 (1996)
15. F. Lau, S. H. Rubin, M. H. Smith, L. Trajovic: Distributed denial of service attacks. In: IEEE International Conference on Systems, Man, and Cybernetics. (2000)
16. S. Marti, T. J. Giuli, K. Lai, M. Baker: Mitigating routing misbehavior in mobile ad hoc networks. In: 6th Annual International Conference on Mobile Computing and Networking(MOBICOM). (2000)

Fiber Cost Reduction and Wavelength Minimization in Multifiber WDM Networks

Christos Nomikos[1], Aris Pagourtzis[2], Katerina Potika[2], and Stathis Zachos[2,3]

[1] Department of Computer Science, University of Ioannina,
45110, Greece, cnomikos@cs.uoi.gr
[2] Computer Science, ECE, National Technical University of Athens,
15780, Greece, {pagour,epotik,zachos}@cs.ece.ntua.gr
[3] CIS Department, Brooklyn College, CUNY, NY, US

Abstract. Motivated by the increasing importance of multifiber WDM networks we study two routing and wavelength assignment problems in such networks:

- *Fiber Cost Minimization*: the number of wavelengths per fiber is given and we want to minimize the cost of fiber links that need to be reserved in order to satisfy a set of communication requests; we introduce a generalized setting where network pricing is *non-uniform*, that is the cost of hiring a fiber may differ from link to link.
- *Wavelength Minimization*: the number of available parallel fibers on each link is given and we want to minimize the wavelengths per fiber that are needed in order to satisfy a set of communication requests.

For each problem we consider two variations: undirected, which corresponds to full-duplex communication, and directed, which corresponds to one-way communication. Moreover, for rings we also study the problem in the case of pre-determined routing. We present exact or constant-ratio approximation algorithms for all the above variations in chain, ring, star and spider networks.

1 Introduction

All-optical networks make it possible to transmit data at very high speed. The technology that enables transmitting more than one signal along a single optical fiber is called *Wavelength Division Multiplexing (WDM)*; many signals can be simultaneously carried over the same physical link by light beams of different wavelengths. Recent developments make it possible to use multiple fibers on each link, allowing any signal to switch fiber at any node; however, it is preferred for each signal to remain on the same wavelength from transmitter to receiver, in order to avoid wavelength conversion.

A multifiber network can be described by a graph $G = (V, E)$ and a function $\mu : E \to \mathbb{N}$ that defines the multiplicity of fibers on each link. The set of requests $\mathcal{R}$ is a set of pair of nodes. A routing and path multicoloring[1] for $\mathcal{R}$ (w.r.t. $\mu(e)$)

---

[1] Color collisions between paths that use the same edge are allowed, so we use the term "path multicoloring", as opposed to classical "path coloring" where paths that share an edge must receive different colors.

N. Mitrou et al. (Eds.): NETWORKING 2004, LNCS 3042, pp. 150–161, 2004.

is valid w.r.t. μ (or simply valid) if all requests of $\mathcal{R}$ are satisfied, i.e. there is a colored path for each request, and for each edge e any color is used at most $\mu(e)$ times among paths that pass through e. The function μ may be given in advance, representing the number of available fibers on each link, or may be sought, representing the number of fibers that should be reserved on each link in order to satisfy a set of connection requests.

In the first part of this paper, we deal with the case where $\mu(e)$ is sought. Here we follow a more general setting where fiber costs are not the same everywhere; we call such a situation *non-uniform* pricing, as opposed to *uniform pricing* where the cost of a fiber on any link is the same. We consider networks where each fiber has a limited bandwidth (number of wavelengths) w and each link has a cost, representing the cost of using a fiber on this link for a certain time period T. For a given set of communication requests with duration at most T, we want to satisfy all requests minimizing the total cost of active fibers in the network. The number of fibers needed between two adjacent nodes of the network is the maximum number of connections that use the same wavelength and pass through the link between the two nodes. An example with two different solutions is shown in Figure 1 (left and right).

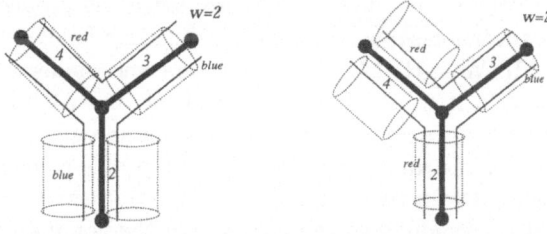

Fig. 1. An instance of *Fiber Cost Minimization Path Multi-Coloring* with 2 colors per fiber, a solution with cost 11 (left) and a solution with cost 13 (right).

We formalize this problem as the MINIMUM FIBER COST ROUTING AND PATH MULTI-COLORING (MINFIBCOST-RPMC) problem: *Given an undirected graph $G = (V, E)$, a cost function $c : E \to \mathbb{N}$, a set of requests $\mathcal{R}$ and w wavelengths (colors), assign paths to requests and colors to paths, so that the objective function $\sum_{e \in E} c(e) \cdot \mu(e)$ is minimized, where $\mu(e)$ is the maximum multiplicity of any color on edge e.*

In the second part of this paper we study the MINIMUM WAVELENGTHS ROUTING AND PATH MULTI-COLORING (MINWAV-RPMC) problem. This problem describes the situation where the number of available fibers is given and the goal is to minimize the number of wavelengths needed to satisfy all requests. Two examples are shown in Figure 2.

Formally, the problem MINWAV-RPMC is defined as follows: *Given a graph $G = (V, E)$, a function $\mu : E \to \mathbb{N}$ and a set of requests $\mathcal{R}$, find a valid routing and path multicoloring such that the number of colors used is minimized.*

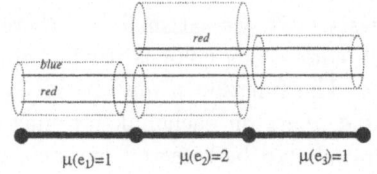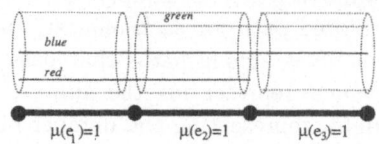

Fig. 2. Two instances of *Minimum Wavelengths Path Multi-Coloring*, the minimum number of colors needed is $w = 2$ for the left one and the minimum number of colors needed is $w = 3$ for the right one.

We also consider, for both problems, the variation in which the routing is pre-determined, i.e. a set of paths is given instead of a set of requests. The variations are called MINIMUM FIBER COST PATH MULTI-COLORING (MINFIBCOST-PMC) and MINIMUM WAVELENGTHS PATH MULTI-COLORING (MINWAV-PMC) respectively. Since any optimal routing must use simple paths these version make sense only in topologies where it is possible to route requests in more than one ways, e.g. ring, mesh, etc. In acyclic topologies there is a unique path between any two nodes, hence the problems MINFIBCOST-RPMC and MINFIBCOST-PMC coincide (as well as MINWAV-RPMC and MINWAV-PMC).

MINFIBCOST-RPMC in rings is NP-hard, since the problem with uniform costs, which is a special case, is NP-hard [9]; the same holds for MINFIBCOST-PMC in rings. MINWAV-RPMC in rings, stars and spiders is also NP-hard (since it is a generalization of the classical routing and path coloring problem which is NP-hard for such topologies [11]); this is also true for MINWAV-PMC in rings as well as for the directed version of both problems in rings.

We distinguish between two types of models: undirected and directed. The undirected model corresponds to the case where the communication for every request is two-way and signals in both directions must use the same set of links and the same wavelength (full-duplex communication). One-way communication can be modeled by using directed requests and paths; the corresponding problem variations have the same names, preceded by the word "DIRECTED". Note that in the directed case, color collisions may occur only between paths that pass through the same edge in the same direction.

In this paper we present constant-ratio approximation or exact algorithms for MINFIBCOST-RPMC in rings with or without pre-routed requests and for MINWAV-RPMC in chains, rings, stars and spiders. We also present appropriate adaptation of our algorithms for the directed versions of the problems. All the proposed algorithms run in polynomial time. A comprehensive table of the results is given in section 4.

1.1   Related Work

The problem of minimizing the number of active fibers in multifiber networks with *uniform fiber costs* was introduced in [9], where polynomial-time solvability was shown for chains and 2-approximation algorithms were given for the undirected problem in ring and star networks. Their results for chains and stars extend to MINFIBCOST-RPMC. Moreover an exact algorithm for DIRECTED MINFIBCOST-RPMC in chains and stars is implicit. In [13] they also study the undirected problem with uniform fiber costs for chains and give a new polynomial-time algorithm for this class of graphs; they also define other variations and show them NP-hard. A 2-approximation algorithm for MINFIBCOST-RPMC in spiders is given in [8]; this algorithm yields an exact algorithm for the directed case.

The problem MINWAV-RPMC was introduced in [6,7] for the special case where $\mu(e)$ is the same for all edges of the network; the more general definition that we use here was first given in [5].

Multifiber tree networks have been studied only recently. For the problem MINFIBCOST-RPMC with uniform fiber costs two approximation algorithms, with ratios $1 + 4|E| \log |V|/OPT$ and 4, are presented in [5]; these results can be immediately extended to the case of non-uniform fiber costs. For MINWAV-RPMC a 4-approximation algorithm is presented in [1].

A lot of work has been done on minimization and maximization routing and path coloring problems for single-fiber networks (see e.g. [11], [10] and references therein).

Other related work includes traffic grooming. In this approach we can combine low speed traffic components onto high speed channels in order to minimize the network cost. Traffic grooming for path, star and tree networks is studied in [4]; in [2] they consider the problem for ring networks.

1.2   Technical Preliminaries

A *chain* is a graph that consists of a single path, while a *ring* is a graph that consists of a single cycle. A *star* is a tree with one internal node. A *spider* is a star of chains, i.e. a star whose edges have been replaced by chains (also called *legs*).

Given a network $G = (V, E)$ and a set of requests $\mathcal{R}$ we denote by n the number of nodes, and by m the number of requests. A *routing* of the requests $\mathcal{R}$ is a set of paths $\mathcal{P}$, each connecting the endpoints of a request. For a set of paths $\mathcal{P}$ and an edge e we denote by $L(e, \mathcal{P})$ the *load* of edge e w.r.t. $\mathcal{P}$, i.e. the number of paths in $\mathcal{P}$ that pass through e.

Let a ring G consist of n nodes labeled clockwise from v_0 to v_{n-1}. We denote the path from u to v in clockwise direction by $\langle u, v \rangle$ and we say that it *begins* at u and it *ends* at v.

An algorithm A for a minimization problem Π is a ρ-approximation algorithm if for every instance I of Π, A runs in time polynomial in $|I|$ and delivers a solution with value $SOL \leq \rho \cdot OPT$, where OPT denotes the value of an optimal solution for I.

2   Minimizing Fiber Cost

In this section we deal with the problem of minimizing the cost of active fibers needed in order to satisfy all requests with a given number of wavelengths. We present approximation algorithms for ring networks. Recall that in rings we may consider two versions, depending on whether the routing is pre-determined (MINFIBCOST-PMC) or not (MINFIBCOST-RPMC).

Our algorithms make use of an algorithm for MINFIBCOST-RPMC in chains that gives optimal solutions in polynomial time. Such an algorithm was described in [9] for uniform fiber costs. That algorithm works for non-uniform costs too, as observed in [8].

Once a routing $\mathcal{P}$ is determined (or unique, or given in advance), each edge contributes at least cost $\lceil L(e_i, \mathcal{P})/w \rceil \cdot c(e_i)$, because at least $\lceil L(e_i, \mathcal{P})/w \rceil$ fiber units are needed for this edge. Summarizing over all edges in E we get $OPT \geq \sum_{e \in E} \lceil L(e_i, \mathcal{P})/w \rceil \cdot c(e_i)$. Note also that for the directed version the sum must be taken over both directions.

2.1   MinFibCost-PMC in Rings (Pre-routed Requests)

Without loss of generality we assume that all edges of G are used by some path (otherwise we would eliminate an unused edge and obtain a chain instance, which can be solved optimally using the algorithm for chains). Therefore, at least one fiber per edge is needed, thus the total cost of an optimal solution is at least $OPT \geq C = \sum_{e \in E} c(e)$.

We denote by $\mathcal{P}_v$ the set of paths in $\mathcal{P}$ that contain v as an internal node. Let $\mathcal{P}'_v$ be the set of paths that results from splitting paths in $\mathcal{P}_v$ at node v. Paths in $\mathcal{P}'_v$ are called v-clockwise if they contain edge (v, u), where u is the neighbor of v in clockwise direction; the remaining paths in $\mathcal{P}'_v$ are called v-counterclockwise. Consider the longest v-clockwise path and the longest v-counterclockwise path in $\mathcal{P}'_v$; let $p(v)$ be the one of the two using edges with minimum sum of costs. We define the $tare$ $t(v)$ of v to be the sum of edge-costs of $p(v)$ and the $span$ $s(v)$ of v to be the length of $p(v)$ (the number of its edges). If $\mathcal{P}_v$ is empty, then $t(v) = 0$ and $s(v) = 0$. Let v_0 be the node with minimum tare; let also $t = t(v_0)$ and $s = s(v_0)$. W.l.o.g. we may assume that $p(v_0)$ is v_0-clockwise (if not we may consider a 'mirror' instance instead). Our algorithm for MINFIBCOST-PMC in rings first selects node v_0 as above. The complete algorithm follows.

Algorithm for MINFIBCOST-PMC *in rings*

Input: $I = (G, c, \mathcal{P}, w)$; $G = (V, E)$ is a ring network, c is the edge-cost function,
   $\mathcal{P}$ is a set of paths and w is the number of colors.
Output: A multicoloring of paths in $\mathcal{P}$.

---

1. Find node v_0 with minimum tare and reindex nodes accordingly.
2. Transform the given ring instance to a chain instance $(G', c, \mathcal{P}', w)$ as follows:
   a. The chain graph G' consists of $n + s + 1$ nodes, namely $v'_0, \ldots, v'_{n+s}$.
      Set $c(e'_i) = c(e'_{i+n}) = c(e_i)$.
   b. For each path $\langle v_i, v_j \rangle \in \mathcal{P}$, add a path to $\mathcal{P}'$:
         if $i < j$ add $\langle v'_i, v'_j \rangle$ to $\mathcal{P}'$, otherwise add $\langle v'_i, v'_{j+n} \rangle$ to $\mathcal{P}'$.
3. Call Algorithm for MINFIBCOST-PMC in chains [9] on instance $(G', c, \mathcal{P}', w)$.
4. Color each path in $\mathcal{P}$ with the color of the corresponding path in $\mathcal{P}'$.

Theorem 1. *The algorithm for* MINFIBCOST-PMC *in rings computes a multicoloring with cost at most* $OPT + t$.

Proof. Let us abbreviate edges in G by $e_i = (v_i, v_{(i+1) \bmod n}), (0 \leq i \leq n-1)$, and edges in G' by $e'_i = (v'_i, v'_{i+1}), (0 \leq i \leq n+s-1)$. It is easy to see that if a path in $\mathcal{P}$ uses edge e_i and $0 \leq i \leq s-1$, then the corresponding path of $\mathcal{P}'$ uses either edge e'_i or edge e'_{i+n}. Thus, for $0 \leq i \leq s-1$, the load of e_i in G is split into two parts in G': $L(e_i, \mathcal{P}) = L(e'_i, \mathcal{P}') + L(e'_{i+n}, \mathcal{P}')$. Notice that $L(e_i, \mathcal{P}) = L(e'_i, \mathcal{P}')$ for $s \leq i \leq n-1$. Due to the optimality of the chain algorithm, the number of repetitions of any color on an edge e'_i is at most $\mu(e'_i) = \lceil L(e'_i, \mathcal{P}')/w \rceil$.

Hence the cost of the solution computed by our algorithm is:

$$
SOL \leq \sum_{i=0}^{n+s-1} \mu(e'_i) \cdot c(e'_i) = \sum_{i=0}^{n+s-1} \lceil \frac{L(e'_i, \mathcal{P}')}{w} \rceil \cdot c(e'_i)
$$

$$
= \sum_{i=0}^{s-1} (\lceil \frac{L(e'_i, \mathcal{P}')}{w} \rceil + \lceil \frac{L(e'_{i+n}, \mathcal{P}')}{w} \rceil) \cdot c(e'_i) + \sum_{i=s}^{n-1} \lceil \frac{L(e'_i, \mathcal{P}')}{w} \rceil \cdot c(e'_i)
$$

$$
= \sum_{i=0}^{n-1} \lceil \frac{L(e_i, \mathcal{P})}{w} \rceil \cdot c(e_i) + \sum_{i=0}^{s-1} c(e_i) \leq OPT + t \qquad \square
$$

The approximation ratio is at most $1 + \frac{t}{OPT}$ which is smaller than 2 and gets close to 1 for instances with heavy communication traffic.

The computation of each tare and span and the transformation can be performed in $O(m + n)$ time. The complexity of algorithm for MINFIBCOST-PMC in rings is determined by that of the chain algorithm, which is $O((m + n \cdot w) \log w)$ [9].

2.2 MinFibCost-RPMC in Rings

We now propose an algorithm for MINFIBCOST-RPMC in rings, i.e. the routing is also sought. Our algorithm uses lightest-path routing: each request is routed along the path with minimum cost (sum of edge costs along path p) between the two alternative complementary paths.

Algorithm for MINFIBCOST-RPMC *in rings*
Input: $I = (G, c, \mathcal{R}, w)$; $G(V, E)$ is a ring network, c is the cost function, $\quad \mathcal{R}$ is a set of requests and w is the number of colors.
Output: A routing $\mathcal{P}$ for $\mathcal{R}$ and a multicoloring of paths in $\mathcal{P}$.
1. Perform a lightest-path routing obtaining a set of paths $\mathcal{P}$.
$\quad$ Call Algorithm MINFIBCOST-PMC in rings (pre-routed requests) on $(G, c, \mathcal{P}, w)$
$\quad$ to multicolor the set of paths $\mathcal{P}$.
2. For each edge $e \in E$: route all requests in $\mathcal{R}$ avoiding e obtaining set of paths $\mathcal{P}_e$.
$\quad$ Call Algorithm for MINFIBCOST-PMC in chains on instance $(G, c, \mathcal{P}_e, w)$.
3. Choose the best solution among the one found in step 1 and those found in step 2.

The selection of lightest paths minimizes the quantity $\sum_{i=0}^{n-1} L(e_i) \cdot c(e_i)$, and decreases the upper bound for t to $t \leq C/2$, where $C = \sum_{i=0}^{n-1} c(e_i)$. A bound for the cost of the solution computed by this algorithm is given by the following theorem:

Theorem 2. *The algorithm for* MINFIBCOST-RPMC *in rings computes a multicoloring with cost at most* $OPT + C + t$.

Proof. We prove the claim for the solution returned by step 1 of the algorithm (in fact step 2 is only needed for the case in which an optimal solution completely avoids an edge).

Let $\mathcal{P}$ be the set of paths selected by our algorithm for MINFIBCOST-RPMC in rings and $\mathcal{P}^*$ be the set of paths in an optimal solution. We denote by e_i the edge between nodes i and $(i+1) \bmod n, 0 \leq i \leq n-1$. Note that $OPT \geq \lceil L(e_i, \mathcal{P}^*)/w \rceil \cdot c(e_i)$. Since $\mathcal{P}$ consists of lightest paths it holds:

$$\sum_{i=0}^{n-1} L(e_i, \mathcal{P}) \cdot c(e_i) \leq \sum_{i=0}^{n-1} L(e_i, \mathcal{P}^*) \cdot c(e_i) \tag{1}$$

The following properties of ceilings hold for all $n \in \mathbb{N}^+, a_i \in \mathbb{R}, 0 \leq i < n$:

$$\sum_{i=0}^{n-1} \lceil a_i \rceil \leq \lceil \sum_{i=0}^{n-1} a_i \rceil + n - 1 \quad \text{and} \quad \lceil a_i \rceil \cdot n - n + 1 \leq \lceil a_i \cdot n \rceil \tag{2}$$

From (1) and (2) we get:

$$\sum_{i=0}^{n-1} (\lceil \frac{L(e_i, \mathcal{P}) \cdot c(e_i)}{w} \rceil) \leq \sum_{i=0}^{n-1} \lceil \frac{L(e_i, \mathcal{P}^*) \cdot c(e_i)}{w} \rceil + n \Rightarrow$$

$$\sum_{i=0}^{n-1} (\lceil \frac{L(e_i, \mathcal{P})}{w} \rceil \cdot c(e_i)) - C \leq OPT$$

By an inequality used in the proof of Theorem 1 and the above inequality, the cost of the approximate solution returned by the algorithm is at most

$$SOL(I) \leq \sum_{i=0}^{n-1} (\lceil \frac{L(e_i, \mathcal{P})}{w} \rceil \cdot c(e_i)) + t \leq OPT + C + t \qquad \square$$

If $OPT \geq C$ then the algorithm for MINFIBCOST-RPMC in rings achieves approximation ratio 5/2, using the fact that $t \leq C/2$.

If $OPT < C$, then it must be the case that paths in the optimal solution do not pass through some edge, say e. In step 2, the algorithm considers, among others, the (unique) routing in which all requests avoid e. Algorithm MINFIBCOST-RPMC in rings then uses the Algorithm for MINFIBCOST-RPMC in chains, which returns an optimal solution for the corresponding chain instance. Hence, the solution returned is optimal.

As about the complexity, the most costly step is step 2, which employs n calls to algorithm for MINFIBCOST-PMC in chains. The overall cost is thus $O(n(m + nw) \log w)$.

2.3   Directed Fiber Cost Minimization

In the directed version the requests are directed, while the underlying graph is considered bidirected. We assume that the cost of an edge is the same in both directions.

- For DIRECTED MINFIBCOST-PMC in rings (pre-routed requests) we obtain the same approximation as for the undirected case, because we can split the instance into one instance of clockwise direction and one of counterclockwise direction and solve the two instances separately as undirected ones.
- For DIRECTED MINFIBCOST-RPMC in rings we obtain a 4-approximation algorithm by first performing lightest path routing and then applying the above algorithm for the problem with pre-routed requests.

3   Minimizing the Number of Wavelengths

In this section we present exact and approximate algorithms for the wavelength minimization problem in chains, rings, stars and spiders. In this problem, the multiplicity of fibers on each edge is given and the goal is to find a valid routing and path multicoloring using a minimum number of colors. This number is denoted by w_{opt}. Note that, once a routing $\mathcal{P}$ is determined (or unique, or given in advance), $w_{opt} \geq w_{lb} = \max_{e \in E} \lceil \frac{L(e,\mathcal{P})}{\mu(e)} \rceil$. In the directed version this maximum is taken over all edges in both directions.

We can solve MINWAV-PMC in chains using exactly w_{lb} colors, which is optimal. This can be done as follows: Call algorithm MINFIBCOST-PMC in chains for the same requests, unit edge cost everywhere, and w_{lb} available colors. As shown in [9] this call returns a multicoloring that uses exactly $\mu'(e) = \lceil L(e, \mathcal{P})/w_{lb} \rceil \leq \mu(e)$ fibers on each edge e. Hence, this is a valid path multicoloring.

3.1   MinWav-PMC in Rings (Pre-routed Requests)

For solving MINWAV-PMC in rings we observe that every instance of the problem falls in exactly one of the following three categories:

1. $\forall e \in E : \mu(e) \geq 2$.
2. There exists at least one edge $e_i \in E$ with $\mu(e_i) = 1$ and no edges of multiplicity 0 exist.
3. There exists at least one edge $e_i \in E$ with $\mu(e_i) = 0$.

Instances that fall in category 3 are actually chain instances and can be solved optimally. An algorithm that copes with instances in categories 1 and 2 is presented below.

Algorithm for MINWAV-PMC *in rings*

Input: $I = (G, \mathcal{P}, \mu)$; $G(V, E)$ is a ring network, $\mathcal{P}$ is a set of paths
and $\mu : E \to \mathbb{N}$ is the edge multiplicity function
Output: A valid multicoloring of paths in $\mathcal{P}$.

---
if $\forall e \in E : \mu(e) \geq 2$ **then** (*category 1*)
    Set $w = \max_{e \in E} \lceil \frac{L(e, \mathcal{P})}{\mu(e) - 1} \rceil$
    Call Algorithm MINFIBCOST-PMC in rings on $(G, 1, \mathcal{P}, w)$
else (*category 2*)
    Choose an edge e_i with $\mu(e_i) = 1$. Set of paths $\mathcal{P}_i$: paths in $\mathcal{P}$ passing through e_i.
    Set $\mathcal{P}' = \mathcal{P} \setminus \mathcal{P}_i$. Remove edge e_i from G, let this graph be G'.
    Call Algorithm for MINWAV-PMC in chains on instance $(G', \mathcal{P}', \mu)$.
    Color paths in $\mathcal{P}_i$ using $|\mathcal{P}_i|$ new colors.

Theorem 3. *Algorithm* MINWAV-PMC *in rings is a 2-approximation algorithm.*

Proof. Instance in Category 1: $\forall e \in E : \mu(e) \geq 2 \Rightarrow \mu(e) - 1 \neq 0$ and $\mu(e) - 1 \geq \frac{\mu(e)}{2}$. Consider an edge e^* for which $\lceil \frac{L(e^*, \mathcal{P})}{\mu(e^*) - 1} \rceil = \max_{e \in E} \lceil \frac{L(e, \mathcal{P})}{\mu(e) - 1} \rceil$.

The number of colors (w) used by the algorithm is:

$$w = max_{e \in E} \lceil \frac{L(e, \mathcal{P})}{\mu(e) - 1} \rceil = \lceil \frac{L(e^*, \mathcal{P})}{\mu(e^*) - 1} \rceil \leq \lceil \frac{2 \cdot L(e^*, \mathcal{P})}{\mu(e^*)} \rceil \leq 2 \cdot \lceil \frac{L(e^*, \mathcal{P})}{\mu(e^*)} \rceil$$

$$\leq 2 \cdot max_{e \in E} \lceil \frac{L(e, \mathcal{P})}{\mu(e)} \rceil \leq 2 \cdot w_{opt}$$

Instance in Category 2: The algorithm uses $|\mathcal{P}_i|$ colors for the paths passing through e_i. It is $|\mathcal{P}_i| \leq w_{opt}$, because any optimal solution would need at least $|\mathcal{P}_i|$ colors for paths passing through edge e_i.
The algorithm multicolors the remaining paths in $\mathcal{P}'$ $(\mathcal{P}' = \mathcal{P} \setminus \mathcal{P}_i)$. All paths in $\mathcal{P}'$ avoid edge e_i, thus we can remove e_i from G and get G', which is a chain network. Algorithm MINWAV-PMC in chains returns a solution using a number of colors $w = \max_{e \in E} \lceil \frac{L(e, \mathcal{P}')}{\mu(e)} \rceil \leq w_{opt}$. Hence, we use $w + |\mathcal{P}_i| \leq 2 \cdot w_{opt}$ colors in total. $\square$

3.2 MinWav-RPMC in Rings

We now turn to the problem in rings where the routing is also sought. Our algorithm is based on the idea of routing the requests in such a way that the edge with minimum number of available fibers is completely avoided.

Algorithm for MINWAV-RPMC *in rings*

Input: $I = (G, \mathcal{R}, \mu)$; $G(V, E)$ is a ring network, $\mathcal{R}$ is a set of requests
and $\mu : E \to \mathbb{N}$ is the edge multiplicity function.
Output: A routing $\mathcal{P}$ for $\mathcal{R}$ and a valid multicoloring of paths in $\mathcal{P}$.

---
1. Pick an edge e_0 with minimum fiber multiplicity $\mu(e_0)$.
2. Route all requests in $\mathcal{R}$ so that the corresponding paths avoid edge e_0.
    Let $\mathcal{P}$ denote the resulting set of paths. Remove edge e_0 from G, call the new graph G'.
3. Call Algorithm for MINWAV-PMC in chains on instance $(G', \mathcal{P}, \mu)$.

Theorem 4. *Algorithm* MINWAV-RPMC *in rings is a 2-approximation algorithm.*

Proof. Let $\mathcal{P}_{opt}$ denote the set of paths in an optimal solution, that uses w_{opt} colors. Let also w_{sol} denote the number of colors used by our algorithm for MINWAV-RPMC in rings.

First, we observe that $\mathcal{P}_{opt}$ can be seen as a transformation of $\mathcal{P}$ in which some paths have been replaced by their complementary paths (that necessarily use edge e_0). Therefore, for any edge $e \neq e_0$ it holds:

$$L(e, \mathcal{P}) \leq L(e, \mathcal{P}_{opt}) + L(e_0, \mathcal{P}_{opt})$$

Dividing by $\mu(e)$ and taking into account that $\mu(e) \geq \mu(e_0)$ we obtain:

$$\lceil \frac{L(e, \mathcal{P})}{\mu(e)} \rceil \leq \lceil \frac{L(e, \mathcal{P}_{opt})}{\mu(e)} \rceil + \lceil \frac{L(e_0, \mathcal{P}_{opt})}{\mu(e_0)} \rceil$$

Since the above holds for any edge e, it also holds for the edge e^* with maximum load/multiplicity ratio w.r.t. routing $\mathcal{P}$, which is equal to the number of colors used by Algorithm MINWAV-PMC in chains when applied to $(G', \mathcal{P}, \mu)$. On the other hand, each of the two quantities on the right side of the above inequality is a lower bound for the number of colors used by the optimal solution. Altogether:

$$w_{sol} = \max_{e \in E} \lceil \frac{L(e, \mathcal{P})}{\mu(e)} \rceil = \lceil \frac{L(e^*, \mathcal{P})}{\mu(e^*)} \rceil \leq \lceil \frac{L(e^*, \mathcal{P}_{opt})}{\mu(e^*)} \rceil + \lceil \frac{L(e_0, \mathcal{P}_{opt})}{\mu(e_0)} \rceil$$

$$\leq 2 \cdot \max_{e \in E} \lceil \frac{L(e, \mathcal{P}_{opt})}{\mu(e)} \rceil \leq 2 \cdot w_{opt} \qquad \square$$

3.3   MinWav-PMC in Stars and Spiders

We now propose a 3/2-approximation algorithm for MINWAV-PMC in stars, which is based on a transformation of the problem to edge coloring of a multigraph H. For the sake of brevity, we only point out few details: each node of H corresponds to a group of at most w_{lb} paths that use the same edge in the original graph G. There is an edge in H for each path p in $\mathcal{P}$, connecting the two groups that contain p. Multigraph H can be edge-colored using at most $3/2 \cdot w_{lb} \leq 3/2 \cdot w_{opt}$ colors [12]; it is not hard to see that assigning to each path p in $\mathcal{P}$ the color of the corresponding edge in H, we obtain a valid path multicoloring. The above idea can be extended to spiders (generalized stars), at a cost of at most w_{lb} additional colors (for paths that do not pass through the center), giving a valid path multicoloring with at most $5/2 \cdot w_{lb} \leq 5/2 \cdot w_{opt}$ colors. Hence the following is true:

Theorem 5. MINWAV-PMC *can be approximated within 3/2 in stars and 5/2 in spiders.*

3.4   Directed Wavelength Minimization

For the directed version of MINWAV-PMC and MINWAV-RPMC we assume
that fiber multiplicity is symmetric, i.e. the number of fibers in two opposite
edges (v_i, v_j) and (v_j, v_i) is the same. We briefly explain how to adapt the algorithm of this section to obtain algorithms for the version of DIRECTED MINWAV-
RPMC and DIRECTED MINWAV-PMC.

- For chain networks we can use the minimum possible number of wavelengths,
  by computing an optimal solution for each direction independently; our solution is the maximum of the two solutions. This gives exactly the above
  lower bound.
- For DIRECTED MINWAV-PMC in rings (pre-routed requests) we can easily
  obtain the same approximation ratio (2 in the worst case) as for the undirected case (subsection 3.1); we can split the instance into one of clockwise
  direction and one of counterclockwise direction and solve them independently
  as undirected ones. Our solution is the maximum of the two solutions.
- For DIRECTED MINWAV-RPMC in rings we can obtain an algorithm by
  modifying Algorithm MINWAV-RPMC in rings (subsection 3.2). This gives
  an approximation algorithm with ratio 2. In the analysis of the algorithm
  we use the fact that for the clockwise direction and for each edge $e \neq e_0$:
  $L(e, \mathcal{P}_+) \leq L(e, \mathcal{P}_+^*) + L(e_0, \mathcal{P}_-^*)$. A similar inequality holds for $L(e, \mathcal{P}_-)$.
  We use $\mathcal{P}_+$ $(\mathcal{P}_-)$ to denote the set of paths of our solution that are oriented
  clockwise (counterclockwise respectively); $\mathcal{P}_+^*$ $(\mathcal{P}_-^*)$ are defined analogously.
- For DIRECTED MINWAV-RPMC in stars our algorithm gives an optimal
  solution, due to the fact that the multigraph H is now bipartite and it is
  known that it can be edge-colored with exactly w_{lb} (degree of H) colors (see
  e.g. [3]). Similarly we obtain a 2-approximation algorithm for DIRECTED
  MINWAV-RPMC in spiders.

4   Summary of Results – Conclusions

We studied two up-to-date optimization problems: fiber cost minimization and
wavelength minimization in multifiber WDM networks. Both problems deal with
limited resources: in the former the number of wavelengths is given and the goal
is to minimize the cost of fiber usage; in the latter it is the number of fibers that
is given and we aim at minimizing the number of necessary wavelengths. We
remark that for MINWAV-RPMC we follow a very recently introduced model [5]
under which the number of fibers may differ from link to link; previous models
were based on the rather restrictive assumption that the number of fibers is
uniform [6,7]. We follow the same assumption for MINFIBCOST-RPMC.

We summarize our algorithms in the following table, where the approximation
ratio of each of them is shown (algorithms giving optimal solutions are referred to
as "exact" and the term "pre-rings" stands for "rings with pre-routed requests").
Note that our new results are shown in boldface; we also mention algorithms from
[9] (MINFIBCOST-RPMC in chains and stars) and [8] (MINFIBCOST-RPMC
in spiders) in order to obtain a complete picture.

Network	MinFibCost-RPMC		MinWav-RPMC	
	Undirected	Directed	Undirected	Directed
chains	exact	exact	**exact**	**exact**
pre-rings	**2-approx.**	**2-approx.**	**2-approx.**	**2-approx.**
rings	**5/2-approx.**	**4-approx.**	**2-approx.**	**2-approx.**
stars	2-approx.	exact	**3/2-approx.**	exact
spiders	2-approx.	exact	**5/2-approx.**	**2-approx.**

The proposed algorithms are easy to implement and we have proven for all of them a guaranteed approximation ratio. We anticipate that they will prove even better in practice. In particular, it can be shown that for heavily loaded instances the approximation ratio gets close to 1. This is due to the fact that the cost of our solutions differ from the cost of an optimal solution by an additive term only, which is usually very small.

References

1. C. Chekuri, M. Mydlarz, and F. B. Shepherd. Multicommodity demand flow in a tree. In *Proc. Automata, Languages and Programming, 30th International Colloquium, ICALP 2003*, pages 410–425, 2003.
2. A. L. Chiu and E. Modiano. Traffic grooming algorithms for reducing electronic multiplexing costs in WDM ring networks. *Journal of Lightwave Technology*, 18(1):2–12, 2000.
3. R. Cole, K. Ost, and S. Schirra. Edge-coloring bipartite multigraphs in $O(ElogD)$ time. *Combinatorica*, 21(1):5–12, 2001.
4. R. Dutta, S. Huang, and G. N. Rouskas. Traffic grooming in path, star, and tree networks: complexity, bounds, and algorithms. In *Proc. of the 2003 ACM SIGMETRICS*, pages 298–299. ACM Press, 2003.
5. T. Erlebach, A. Pagourtzis, K. Potika, and S. Stefanakos. Resource allocation problems in multifiber WDM tree networks. In *Proc. of the 29th Workshop on Graph Theoretic Concepts in Computer Science*, LNCS 2880, pages 218–229, 2003.
6. G. Li and R. Simha. On the wavelength assignment problem in multifiber WDM star and ring networks. *IEEE/ACM Transactions on Networking*, 9(1):60–68, 2001.
7. L. Margara and J. Simon. Wavelength assignment problem on all-optical networks with k fibres per link. In *Proc. Automata, Languages and Programming, 27th International Colloquium, ICALP 2000*, pages 768–779, 2000.
8. C. Nomikos, A. Pagourtzis, K. Potika, and S. Zachos. Path multi-coloring in weighted graphs. In *Proc. of the 8th PCI*, volume I, pages 178–186, 2001.
9. C. Nomikos, A. Pagourtzis, and S. Zachos. Routing and path multicoloring. *Information Processing Letters*, 80(5):249–256, 2001.
10. C. Nomikos, A. Pagourtzis, and S. Zachos. Minimizing request blocking in all-optical rings. In *Proc. INFOCOM 2003*, San Francisco, CA, 2003.
11. P. Raghavan and E. Upfal. Efficient routing in all-optical networks. In *Proc. of the twenty-sixth annual ACM STOC*, pages 134–143. ACM Press, 1994.
12. V. Vizing. On an estimate of the chromatic class of a p-graph (in russian). *Diskret. Analiz.*, 3:23–30, 1964.
13. P. Winkler and L. Zhang. Wavelength assignment and generalized interval graph coloring. In *Proc. of the 14th Annual ACM-SIAM SODA*, pages 830–831, Baltimore, MD, January 2003.

Static WDM Network Planning with TDM Channel Partitioning*

Achille Pattavina[1], Massimo Tornatore[1], Alessandro De Fazio[2], Guido Maier[2], and Mario Martinelli[2]

[1] Dept. of Electronics and Information,
Politecnico di Milano
P.za Leonardo da Vinci,32 - 20133 Milan, Italy
{pattavina, tornator}@elet.polimi.it
[2] CoreCom
Via Colombo, 81 - 20133 Milan, Italy
maier, martinelli@corecom.it

Abstract. In this paper a new solution for the routing, fiber, wavelength and time slot assignment problem (RFWTA) is proposed for a static WDM network. A set of electronic requests characterized by different bit-rates are groomed on lightpaths by a TDM-scheme, designing the logical topology so that the lightpath bandwidth is efficiently exploited. An efficient TDM-based mapping of low granularity electronic connection requests achieves savings in fiber deployment to accommodate a given static traffic. Our analysis comprises two different network layers. First, all the electronic requests are mapped on the logical topology and groomed to obtain the set of lightpath connection requests. Secondly, such requests are routed and fibers and wavelengths are jointly assigned. We adopt the concept of Multihop Lightpath to constrain the number of Time Slot Switching (TSS) operations for each connection request. After introducing our design approach, we discuss the optimization of two case-study networks under various conditions and we analyze the results.

1 Introduction

Nowadays, the Optical Transport Network (OTN) based on the Wavelength Division Multiplexing (WDM) technology is the well established primary system to carry traffic of the ever-growing Internet and of the other advanced telecommunication applications. An open question is how low-bandwidth digital data flows can efficiently access transmission trunks such as the lightpaths, having very large (10 Gbps) or huge (40 Gbps) capacity. The trivial solution of assigning an entire wavelength-channel capacity to each electronic connection would cause network oversizing with waste of expensive resources.

* This work has been partially supported by the Italian Ministry of Education, University and Research (MIUR) under the FIRB project TANGO.

N. Mitrou et al. (Eds.): NETWORKING 2004, LNCS 3042, pp. 162–173, 2004.

Traffic grooming is a new research topic for optical networks, trying to provide methods to design OTNs in which lightpath-bandwidth fractioning is performed in a cost-effective way. The work we are presenting in this paper deals jointly with grooming and physical resource optimization in OTNs.

The paper is organized as follows. In section 2 basic concepts concerning OTN and grooming are reported. Section 3 introduces and explains our heuristic optimization approach to solve the problem. Finally, section 4 is dedicated to the analysis of the results obtained by applying this approach to some case-study networks.

2   WDM-Network Model Definition

Let us describe the network environment we are referring to.

At the physical layer, an OTN is composed by WDM transmission links and Optical Cross Connect (OXC) switching nodes arranged according to the *physical topology*. An OTN offers an optical-circuit connectivity service by setting up lightpaths. Each lightpath is a point-to-point connection between a source and a destination OXC: it can cross other transit OXCs and it requires a wavelength channel per crossed WDM link. The set of all the lightpaths is called *logical topology*, and it can be represented by a graph having the termination OXCs as nodes and the lightpaths as edges. Resource allocation (WDM channels) to lightpaths is performed by mapping the logical over the physical topology. Since we are considering multifiber OTNs, this operation comprises Routing and Fiber and Wavelength Assignment (RFWA) for all the lightpaths.

The client layer of the OTN is an electronic circuit-switched network whose digital connections have to be hosted by the lightpaths.

The issue of access-to-bandwidth and channel-granularity adaptation between the electronic and optical flows has been taken into account by the International Communication Union (ITU) defining the recent OTN standard [1]. A three-level hierarchy has been introduced to classify the capacity of both lightpaths and client-layer electronic connections. The levels, denoted by j, correspond to bit-rates around 2.5 ($j = 1$), 10 ($j = 2$) and 40 Gbps ($j = 3$). A lightpath of a given level can accommodate digital tributaries of equal or lower level via Time Division Multiplexing (TDM). The basic client-bandwidth quantum (≈ 2.5 Gbps) corresponds to a time-slot in a lightpath payload. Thus, e.g. a 10 Gbps digital connection can be carried by a lightpath with $j \geq 2$ by allocating 4 time-slots. The problem of mapping the set of digital connections on the top of the logical topology thus involves time-slot assignment.

Let us now define the problem we intend to solve in this work. The physical topology and the physical link lengths are known: each WDM link is a multifiber cable and each fiber carries a given number of WDM channels. The fiber number on each link is a variable of the problem and has to be dimensioned in order to support the given set of static digital connections characterized by their bit-rate.

The first step to solve it consists in mapping the connections on the logical topology: we suppose in the present work that the logical topology is a full-mesh, i.e. a lightpath could be established between any couple of nodes; anyway this approximation does not affect the genarlity of our algorithm. Two grooming strategies have been jointly exploited to obtain an efficient mapping: TDM (following the previously described scheme) and the Time Slot Switching. Each node is equipped to support TSS. TDM technology is used to make a better use of free time slots within wavelengths. Some wavelengths could be already exploited by other connection requests, but not totally filled, so that residual time slots can be exploited by other requests. TSS technology allow the digital request to exploit multihop lightpaths, i.e. to reach their destination by a sequence of edges belonging to the logical topology (let us observe that between two lightpaths composing the multihop path of a connection an opto electronic conversion has to be performed to process different electronic tributaries). We call this first step step *TS-mapping*. In other words, we can say that TS-mapping is an aspect of the more general problem called traffic grooming [2,3] and comprises the necessity of solving the Logical Topology Design (LTD) problem [4].

The second part of the problem consists in mapping the logical connection requests on a physical topology to minimize the total number of fibers used to transport the requested traffic volume. This is the RFWA (Routing, Fiber, Wavelength Assignment) part of the problem that concerns the physical network planning [5]. In the present work we will consider Virtual Wavelength Path (VWP) networks, in which every node is fully equipped with converters (opaque OXC) so that any incoming optical signal can always be converted on any idle output wavelength [6]. A solution of the planning problem is found only when resources are chosen and allocated for all the optical connections derived by TS-mapping step (rejecting or blocking a connection request is not considered).

Some proposals in literature face the problem from a point of view slightly similar to our approach: in [7] the concept of Super-Lightpath is introduced to reach more than one destination node exploiting the TDM bandwidth of a single lightpath; in [8] the complete problem is efficiently solved in a dynamic scenario, and the authors referred to it as Routing, Fiber, Wavelength Assignment RWTA problem; in [9] and in [10] the problem is solved by a single step approach, anyway no remarks on physical resources utilization are taken into account.

3   Heuristic Optimization Method for RWTA Problem

We solve the RWTA problem using two different planning approaches: a novel two-step algorithm, that represents the proposal of this work, and a simple and intuitive algorithm, that will be useful as a term of comparison for our approach.

Our algorithm is divided in two steps: the former routes electronic connection requests on a logical topology (assumed to be full-mesh) providing an efficient mapping of time slots into lightpath bandwidth; the second solves the RFWA part of the problem. Each edge of logical topology corresponds to a lightpath. From now on we will refer to these approach as TOOL B and its distinct steps will be called respectively TS mapping phase and RFWA phase. A second approach has been implemented just to provide a comparison term: this approach does not distinguish two different operations, but the algorithm simply routes an optical connection for each electronic request; then it packs the requested time-slot into channel bandwidth using a first fit strategy, finally it eliminates the empty lightpaths and optimizes the fiber number. We will refer to this second approach as TOOL A in the results section.

Electronic connections can be mapped on the logical topology following two paradigm: *single hop* or *multihop* From now on, the term *lightpath* will be used to indicate a single edge in the logical topology, while the term *multihop lightpath* will refer to a sequence of edges in the logical topology.

3.1   TOOL B: TS-Mapping Phase

Let us consider a simple four-node ring to show TS-mapping phase of TOOL B. First of all, logical topology (in this example the associated full-mesh four-node network) is replicated to form a layered graph (we set a high number of replicated layer, compatible with the amount of offered traffic, so that no request will be refused). Obviously, path conversion between different layer is allowed thanks to vertical edges linking the same nodes of the different planes.

Once layered graph has been built, electronic connection requests have to be routed. The first operation is the analysis of connection bit-rate: the different bit-rate have to be converted in term of basilar time-slot granularity. As previously seen, 2.5 Gbps is the basilar unit, then, as far as connection requiring more than one time-slot unit are concerned, we impose all the 2.5 Gbps units associated to the same electronic request to follow the same route. At the end of the process we could choose the fittest bit-rate (2.5, 10 or 40) on that lightpath. This limitation on the set of admissible transmission bit-rates reflects the actual situation of network operators, that presumably have deployed a set of transmission equipment with fixed bit-rates. So the algorithm have to route electronic connection characterized by 3 distinct admissible granularity in term of time slots: 1, 4 and 16. Now the algorithm has to route this time slot connections

TS mapping metric. The routing on the layered logical topology is carried on using two well-known strategies: the Shortest Path Routing (SPR) and the Least Loaded Routing (LLR)[5].

SPR: Shortest Path Routing routes a connection on the path having the minimum number of hops (so trying to avoid excessively long multihop path). Applying the shortest path algorithm each connection request is routed on the replicated logical topology; if we hypothesize the case of full-mesh logical topology, this step simply assigns the direct link between each couple of nodes to the connection requested by the node couple and corresponds to a ceiling operation on the number of direct lightpaths needed to support direct traffic.

LLR: Least Loaded Routing strategy routes a connections choosing, among the different alternatives, the path having as most congested link the link with minimal load. This approach allows us to distribute the traffic among the network with the aim of avoiding excessively congested links.

We will show in the result section the performance of this optimization phase using respectively SPR and LLR metric. Let us observe that when two or more paths are characterized by the same weight with respect to the chosen metric, the algorithm routes the requested connection on the path minimizing another metric, the (SPR-L) metric, built with the aim to discourage utilization of lightpaths that would probably use more fiber mileage in the dimensioning step.

It's worth noting that in case of a link of logical topology with residual capacity, the next connection to be routed on that link can: 1) find a sufficient amount of available capacity (in terms of time slots), so exploiting the residual capacity of a pre-existing lightpath, 2) need the allocation of a new lightpath to support its bit-rate.

Minimizing the optical requests. After pre-optimization phase, the algorithm applies an optimization cycle whose aim is to minimize the number of used lightpaths (that is to say the number of links in the logical topology). The sequence of instructions in the optimization phase is the following:

1. Set the integer variable $\mathcal{L}$ to 1
2. Identify among the allocated lightpaths (logical topology links), those having $\mathcal{L}$ spare time slot
3. For each of the previous lightpath:
   - temporary delete the previous lightpaths
   - try to re-route all the connections owing to the deleted lightpath on the residual capacity of the remaining lightpaths
   - if re-routing succeeds for all connections, then delete definitively the lightpath, otherwise restore it
4. increase $\mathcal{L} = \mathcal{L} + 1$;
5. if $\mathcal{L}$ is larger than $T - 1$ exit, otherwise go to step 2 .

T is the maximum number of time slot supported by a single lightpath. During the re-routing phase, we have also introduced an input parameter K that sets the maximum number of TSS a connection request can cross. Since a logical link corresponds to a lightpath to be physically routed, it is of great importance to take under control the maximum number of hops. The number of hops a multihop lightpath can cross is constrained by the maximum number of Time Slot Switching (in fact $K=H$-1, i.e. with N admissible Time slot switching operations, a multihop lightpath composed by at most $N + 1$ hops can be built). A single TSS operation requires the demultiplexing of WDM signal in its electronic tributaries, the switching of distinct time slots contained in WDM channel, multiplexing of time slots in a new single WDM channel: it is a complex and expensive operation not to be abused.

The cost reduction introduced by this optimization phase thanks to TDM bandwidth utilization can be easily observed in the example shown in Fig 1, where, rerouting the connection b along the residual capacity of lightpaths pre-allocated to support connections a and c, we are able to route two electronic connections exploiting two (and not three) lightpaths in the physical network.

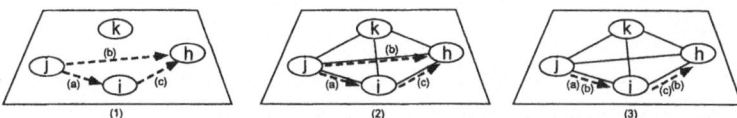

Fig. 1. Re-routing of an electronic tributary (time slots)

Let us assume that up to 16 time slots (according to G.709, 2.5 Gbps × 16= 40 Gbps) can be multiplexed on each lightpath. The offered electronic traffic is that shown in Fig. 1(1):

- *tributary (a): 1 connection request from node j to node i at 2.5 Gbps*
- *tributary (b): 1 connection request from node j to node h at 10 Gbps*
- *tributary (c): 1 connection request from node i to node h at 2.5 Gbps*

If we refer to SPR metric, during the first phase of TS-mapping algorithm, connection request are routed on the shortest path, which corresponds in this logical full-mesh topology to the direct link between source and destination node Fig.1(2). So this phase does not involve any TSS operation. Then optimization re-routes electronic flow mapped on $j - h$ lightpath (Fig.1(2)) along the hops $j - i$ and $i - h$ so giving origin to a multihop lightpath as shown in Fig. 1(3). This process allows us to remove a lightpath from optical traffic load, only at cost of the introduction of a TSS operation in node i.

Optimization runs are carried on varying the constraint on the number of admissible TSS (K). The increase of admissible cascaded TSS is expected to improve the grooming capability of the algorithm, so influencing the global resource utilization. As a matter of fact, larger values of K provide a wider spectrum of lightpath re-routing alternatives. On the other hand an enlarged set of possibilities is associated to an increase of the network cost, because additional TSS operation costs have to be taken into account.

3.2   TOOL B: RFWA Phase

After the optimization phase, the logical topology design is given and provides an optical offered traffic matrix as an input to solve the physical topology design (RFWA). In this matrix the bit-rates are not specified anymore: every lightpath assigned in the logical topology has to be mapped as an optical connection onto physical topology. The algorithm to solve RFWA problem is the same presented in [5], that has proven to provide good performance. It is worth noting that multihop lightpaths will be interpreted as a series of distinct single hop lightpath. For example in Fig. 1(3) the logical topology provides such a optical traffic load:

- *1 connection request from node j to node i*
- *1 connection request from node i to node h*

Concluding, the aim of this two-step study (TS mapping associated and followed by RFWA) consists in analyzing the effect of TDM channels partitioning in the mapping of electronic tributaries onto WDM channel, and secondarily in evaluating on its effect on the physical resource utilization.

4   Results

In this section we present and discuss the results we obtained by performing dimensioning experiments on two case-study networks, the EON and the NSFNET. The National Science Foundation Network has 14 nodes and 22 links; the offered traffic matrix (360 connection requests distributed on 108 node couples) is taken from [11]. The EON (European Optical Network) is constituted by 19 nodes and 39 links. The offered traffic matrix (1380 connection requests distributed on 342 source-destination node couples) is taken from [12].

Actually, in order to run our experiments, we need an electronic traffic matrix, while in literature we have found traffic matrix listing lightpath requests. So we have adopted the following conversion: each optical connection request is translated in a 10 Gbps unit of bandwidth. This conversion allows us to obtain the value of total bandwidth request per each node-couple requiring connectivity, then we have subdivided randomly the total request between connections characterized by 2.5 and/or 10 Gbps bit-rates. Each fiber is equipped with 16 WDM channels and there are no constraints on the number of fibers a link can support (coherently with our dimensioning approach).

4.1   ILP Comparison

First of all, we report the comparison between the performance obtained solving TS-mapping problem by an ILP formulation of the problem and by our algorithm presented in section 3. We have carried on our experiments on a very simple network case to allow the computationally intensive ILP approach to achieve optimal results in reasonable computational time (seconds). We solve the problem of mapping a dozen of electronic requests on a six-node network: without TSS operation ($K=0$) the two strategies return the same number of lightpaths needed to support the electronic requests (12); allowing $K=1,2,4$ TSS operations the difference between the number of lightpaths is just one (respectively 9 vs 8, 8 vs 7, 7 vs 6). Heuristic results are strictly near to the optimal ones and follow the same trend.

4.2   SPR vs. LLR in TS-Mapping

The objective of the TS-mapping in this work consist in minimizing the number of lightpaths necessary to support electronic traffic. We have exploited two distinct approaches to route TS on logical topology before the optimization by means our algorithm. Figures 2(a) and (b) allow us to appreciate the clear performance distance between the two metrics: SPR outperforms LLR for any values of K parameter and in both networks. We have reported results obtained with $T = 16$, anyway a variation on the number of available time slot per frame does not affect this conclusions. It is also possible to notice the positive effect due to introduction of TSS capability on the number of lightpaths needed to support electronic traffic. Let us point out the decrease in the number of connection due to TDM utilization compared with the 360 (1380) electronic requests on NSFNET (EON) case.

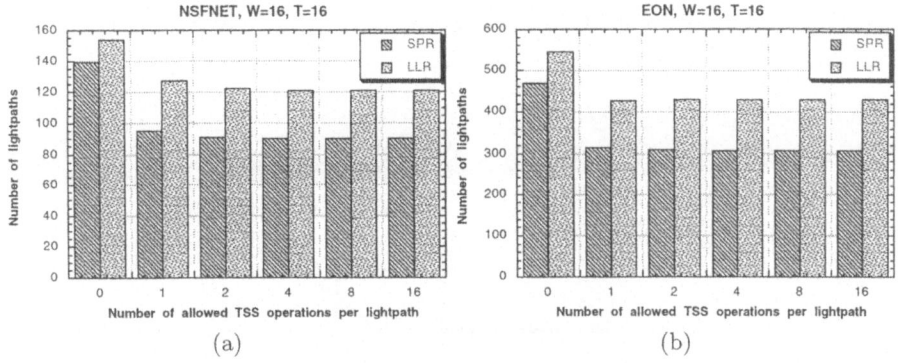

Fig. 2. Number of lightpaths returned by TS-mapping algorithm varying the number of allowed cascaded TSS on NSFNET (a) and EON (b) topology.

4.3   T and K: Two Key Parameters

As previously said in section 3, the K parameter sets an upper bound on the number of admissible TSS along a multihop lightpath. In our experiments this parameter assumes the values 0, 1, 2, 4, 8, 16, where this last value has been set high enough to consider the TS mapping problem solved without any constraint on the number of allowed TSS. Varying this parameter we can investigate the effective trade-off between the additional network equipment complexity and a better resource utilization achievable while allowing an increasing number of admissible cascaded TTS operations.

In Fig.3 we have plotted the average number of allocated time slots on a single lightpath, varying the number of available time slots on a single channel. Clearly the maximal bandwidth utilization could be represented by a line with unitary slope. We keep in consideration also the RWTA approach referred as TOOL A (see section 3), i.e. the absence of a preliminary and separated TS-mapping phase. The first consideration that rises observing this graphic is the relevant advantage in term of network resource usage deriving from the adoption of multihop lightpaths. For increasing values of allowed TSS operation (strictly related to the number of hops of multihop lightpaths), the curves come closer to the ideal one. Then, also this graphic confirm that the practical bound on the number of cascaded TSS to obtain an efficient utilization of channel bandwidth is $K=4$, although even $K=2$ provides quite good performances. Besides, Fig. 3 allows us to compare the TOOL A approach with the TOOL B experiments in case of $K = 0$. Both these strategies do not support multihop lightpaths, because TSS operations are not taken in considerations. Nevertheless, TOOL B achieves a better utilization degree than TOOL A. Therefore, we can conclude that the largest gain in term of efficient channel occupation is due to TSS capability, but we have not to neglect that a small contribute is associated to the utilization of a separated TS-mapping phase. This phase provides a preliminary packing of time-slots in wavelength bandwidth that outperforms the on-line packing of time slots provided by TOOL A approach, even in the case of absence of TSS capability.

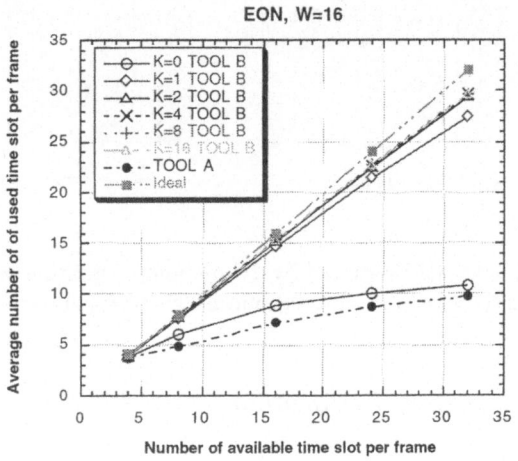

Fig. 3. Average number of allocated time slots per frame varying the number of available time slots, on EON topology.

4.4   TDM Effects on Fiber Dimensioning

In this paragraph we show the results obtained thanks to the RFWA step of our analysis. We refer to the number of fibers employed to satisfy all the connection requests in order to provide an estimation of the effect of the introduction of TDM technique on the physical network cost. In Fig.4(a), for the EON network, we have drawn the variation of the total fiber number after dimensioning, varying the number of available time slot per frame, for different values of parameter K. The graphic shows that the case without TSS capabilities is much more expensive in terms of fiber cost than the cases in which TSS operations are allowed. A key improvement can be observed between $K = 0$ and $K = 1$, while there is a modest gap between $K = 1$ and $K = 2$ and then a further increase of K does not imply any relevant cost reductions. Finally in Fig. 4(b) we plot the dependence of the total fiber number on the number of allowed TSS operations, keeping constant the number of time slot per frame. From this slightly different point of view, it is possible to appreciate the relevant decrement of total fiber number, when (also a small number as 1 or 2) TSS operations are allowed. In particular this graphic points out that TSS utilization is a key-factor especially when the frame supports an high number of time slots. In fact, with a larger number of time-slots per frame, there is a higher probability that request re-allocation succeeds, being available more residual capacity for the optimization algorithm described in the paragraph 3.1. Analogous results has been obtained in NSFNET case.

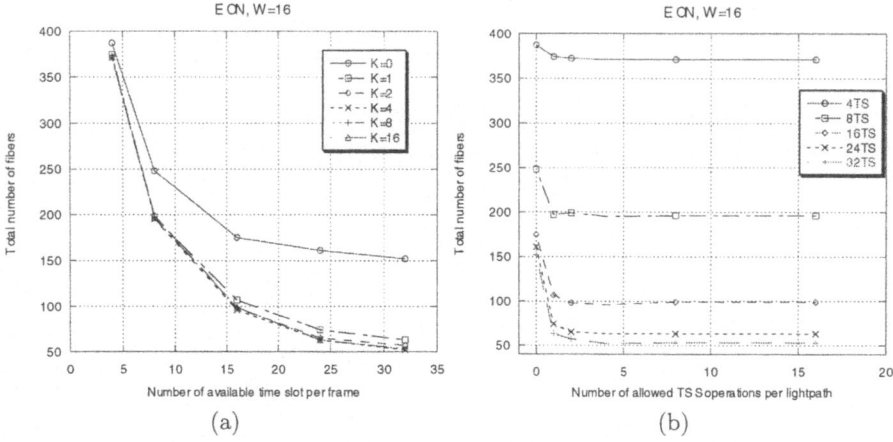

Fig. 4. (a)Total fiber number varying available time slots per frame, for different values of K and (b)total fiber varying admissible cascaded TSS operation, for different values of T, on EON topology.

4.5   Network Cost Analysis

In this last paragraph we propose a cost estimation that includes TDM technology equipment in order to provide a cost comparison with network scenario without TDM capabilities (Wavelength Routed, WR).

Let us now introduce a new cost function, alternative to the fiber or channel number: the graphics in Fig. 5(a) and 5(b) have been obtained using the simple cost function in equation (1),

$$C = F \cdot T \tag{1}$$

where F is a variable that expresses the number of employed fibers, while T is the number of time slot per frame.

The line WR identifies the cost of dimensioning under wavelength routed hypothesis, i.e. when for each electronic connection request, independently of the bandwidth, a wavelength channel is assigned to the request.

It is not difficult to argue that, when the number of time slot per frame increases (and so the capacity of a single wavelength channel), there is a lower fiber requirement, but at the same time the global network cost is affected by an increment due to the cost of TDM technology and of TSS equipment (i.e. there is an actual cost dependency also by T). Let us focus our attention on the curve with $K=0$, that shows an increasing trend much steeper than all the other curves, which on the contrary lie in the same order of magnitude. This leads to an important conclusion: TDM technology offers cost performance competitive with traditional WR approach only if TSS capabilities are implemented, providing a technology platform to efficiently exploit multihop lightpaths.

Then, it is worth noting as, in EON case, a network with higher connectivity degree and a larger offered traffic than in NSFNET, the TDM approach results

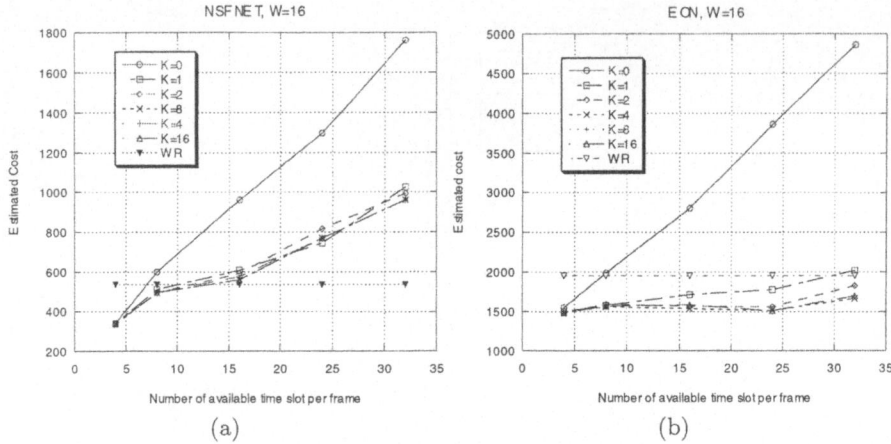

Fig. 5. Cost function varying available time slots per frame, for different values of T, on NSFNET (a) and on EON topology (b).

more cost-efficient thanks to the considerable reduction of fibers, even accepting a unique admissible TSS per connection request. In NSFNET case instead, costs show a steady state till $T = 16$ (except the case with $T = 0$ whose cost grows rapidly), then they exceed WR threshold reaching higher costs. We can argue that cost convenience in applying TDM channel partitioning strategy to map electronic requests on a WDM networks is in dependence of offered traffic amount (with a higher traffic load, it is possible to save a larger amount of network resource) and connectivity degree.

This work assumes that the number of TSS does not affect the global network cost, so that we observe that cost decreases when the number of admissible TSS increase thanks to an efficient grooming phase. This is a simplifying approximation, but it can be partially acceptable if we consider networks in which every nodes have TSS capabilities already installed, so that additional TSS operation cost is negligible. Besides, this hypothesis has been introduced, because in the grooming phase we impose a constraint on the number of sequential (cascaded) TSS, but we can not decide a priori in which nodes the grooming activity is performed: so all the nodes are assumed to perform TSS. A further development of this work would include an efficient strategy to choose the nodes where TSS capability could be located. Concluding, a cost function, which suitably describes the trade-off between physical resources and TDM processing, is difficult to identify and it follows different criteria according to network features. In this work we have fixed some of these network features, putting in evidence the dependence on network topology (number of nodes, connectivity degree), the volume of offered traffic, time slots per frame and admissible cascaded TSSs.

5   Conclusions

In this work we have developed a new heuristic method aimed at optimizing the number of lightpaths needed to support a given electronic traffic by means of TDM technique. Then we have extended our analysis on the effect of TDM channel partitioning on physical resources. This analysis has been carried out referring to a static traffic matrix and matching the TDM features with the requirements of G.709 ITU-T.

TDM is a key technique to obtain an efficient utilization of wavelength bandwidth. A relevant performance improvement has been achieved exploiting just one or two TSS operations, i.e. allowing the utilization of lightpath multihop to map electronic connections, particularly for high values of time-slots per frame. An approximated cost study, carried out to evaluate the convenience of TDM technique, has highlighted a positive dependence on the traffic load and connectivity degree.

References

1. Architecture of Optical Tranport Networks (OTN). G.709/Y.1331, Amendment 1. ITU-T International Communication Union (2001)
2. Thiagarajan, S., Somani, A.K.: Traffic grooming for survivable WDM mesh networks. SPIE Optical Network Magazine (2002)
3. Zhu, K., Mukherjee, B.: Traffic grooming in an optical WDM mesh networks. IEEE Journal on Selected Areas in Communications **20** (2002) 122–133
4. Dutta, R., Rouskas, G.N.: A survey of virtual topology design algorithms for wavelength routed optical networks. SPIE Optical Network Magazine **1** (2000)
5. Dacomo, A., Patre, S.D., Maier, G., Pattavina, A., Martinelli, M.: Design of static resilient WDM mesh-networks with multiple heuristic criteria. In: Proceedings, IEEE INFOCOM '02. (2002)
6. S. Chaudhuri, E.B., Ellinas, G.: Addressing transparency in DWDM mesh survivable networks. OFC 2001 **2** (2001) TuO5–1–TuO5–3
7. Mellia, M., Leonardi, E., Feletig, M.: Exploiting OTDM technology in WDM networks. In: IEEE INFOCOM 2002. (2002)
8. B.Wen, Sivalingam, K.: Routing, Wavelength and Time-Slot Assignment in Time Division Multiplexed Wavelength-Routed Optical WDM Networks. In: IEEE INFOCOM 2002. (2002)
9. Zhu, H., Zang, H., K.Zhu, Mukherjee, B.: A Novel Generic Graph Model for Traffic Grooming in Heterogeneous WDM Mesh Networks. IEEE/ACM Transactions on Networking **11** (2003) 1–16
10. Fumagalli, A., Cerutti, I.: Multi-Rate and multi-Hop Hybrid Optical Networks with Arbitrary Topology. In: Proceedings, HPSR '03. (2003)
11. Miyao, Y., Saito, H.: Optimal design and evaluation of survivable WDM transport networks. IEEE Journal on Selected Areas in Communications **16** (1999) 1190–1198
12. Fumagalli, A., Cerutti, I., Tacca, M.: Survivable Networks Based On Optimal Routing and WDM Self-Healing Rings. In: Proceedings, IEEE INFOCOM '99. (1999)

Traffic Grooming in WDM Ring Networks with the Min-Max Objective*

Bensong Chen, George N. Rouskas, and Rudra Dutta

Department of Computer Science, North Carolina State University, Raleigh, NC, USA
{bchen,rouskas,dutta}@eos.ncsu.edu

Abstract. We consider the problem of traffic grooming in WDM ring networks. Previous studies have focused on minimizing *aggregate* representations of the network cost. In this work, we consider a *Min-Max objective*, in which it is desirable to minimize the cost at the node where this cost is maximum. Such an objective is of practical value when dimensioning a network for unknown future traffic demands and/or for dynamic traffic scenarios. We prove that traffic grooming with the Min-Max objective is NP-Complete even when wavelength assignment is not an issue. We also present a new polynomial-time algorithm for Min-Max traffic grooming. Experiments with a wide range of problem instances demonstrate that our algorithm produces solutions which are always close to the optimal and/or the lower bound, and which scale well to large network sizes, large number of wavelengths, and high loads.

1 Introduction

Wavelength division multiplexing (WDM) technology has the potential to satisfy the ever-increasing bandwidth needs of network users on a sustained basis. In WDM networks, nodes are equipped with *optical cross-connects* (OXCs) or *optical add-drop multiplexers* (OADMs), devices which can optically switch wavelengths, thus making it possible to establish *lightpath* connections between pairs of network nodes. The set of lightpaths defines a *logical topology*, which can be designed to optimize some performance measure for a given set of traffic demands. The logical topology design problem has been studied extensively in the literature. Typically, the traffic demands have been expressed in terms of whole lightpaths, while the metric of interest has been the number of wavelengths, the congestion (maximum traffic flowing over any link), or a combination of the two. The reader is referred to [1] for a survey and classification of relevant work.

With the deployment of commercial WDM systems, it has become apparent that the cost of network components, especially *line terminating equipment* (LTE), is one of the dominant costs in building optical networks, and is a more meaningful metric to optimize than, say, the number of wavelengths. Furthermore, in order to utilize bandwidth more effectively, new models of optical networks allow several independent traffic streams to *share* the capacity

---

* This work was supported in part by the NSF under grant ANI-0322107.

N. Mitrou et al. (Eds.): NETWORKING 2004, LNCS 3042, pp. 174–185, 2004.

of a lightpath. These observations give rise to the concept of *traffic grooming*, a variant of logical topology design, which is concerned with the development of techniques for combining lower speed components onto wavelengths in order to minimize network cost. A comprehensive survey and classification of traffic grooming strategies can be found in [2].

Typically, the goal of traffic grooming studies is to minimize an *aggregate* representation of the LTE cost. While such a metric is important, not imposing a bound on the cost of individual nodes may result in a solution in which some nodes end up with a (very) large amount of LTE while some others with only a small amount of LTE. Such a solution has a number of undesirable properties. First, a node that requires a large amount of LTE may be too expensive or impractical to deploy (e.g., due to high interconnection costs, high power consumption, or space requirements). Second, the network can be highly heterogeneous in terms of the capabilities of individual nodes, making it difficult to operate and manage. Third, a solution minimizing the total LTE cost can be extremely sensitive to the assumptions regarding the traffic pattern, as previous studies [3] have demonstrated. Specifically, a solution that is optimal for a given set of traffic demands may be far away from optimal for a different such set. Since LTE involve expensive hardware devices that cannot be moved from one node to another on demand, an approach that attempts to minimize total LTE cost may not be appropriate for dimensioning a network unless the network operator has a clear picture of traffic demands far into the future *and* these traffic demands are unlikely to change substantially over the life of the network.

In this work, we consider a variant of the traffic grooming problem in which the objective is to minimize the LTE cost at the node where this cost is maximum. We believe that a Min-Max objective is of high practical value to network designers and operators. In particular, an approach that minimizes the maximum LTE cost at any network node is likely to be attractive because all the network nodes are likely to be provisioned with identical equipment. Effectively, all nodes will have a cost that is dictated by the node with the maximum LTE. Such a homogeneous network is easier to operate, manage and maintain, and is likely to be less expensive than a heterogeneous one due to the economies of scale that can be achieved when all nodes are subject to identical specifications. Furthermore, such an optimization approach can be of great importance to dimensioning the network for *unknown and/or dynamic future demands*. Specifically, the network designer may solve the optimization problem for a wide range of traffic scenarios, and equip each node with an amount of LTE equal to the highest solution obtained (plus a certain *fudge factor* for making the solution future-proof). We also note that a similar approach was taken in [4] in a different context, namely for routing and wavelength assignment in the presence of converters. Specifically, an algorithm was developed for distributing a number of converters uniformly across the ring nodes rather than placing them at a single hub node.

The paper is organized as follows. In Section 2, we review the general traffic grooming problem, we introduce the Min-Max objective we consider, and we present new theoretical results regarding the complexity of the problem. In

Section 3, we present a polynomial-time algorithm for rings to minimize the maximum LTE cost at any ring node. We present numerical results in Section 4, and we conclude the paper in Section 5.

2   Problem Definition and Complexity Results

New models of optical networks allow several independent traffic streams to share the bandwidth of a lightpath. If the multiplexing and demultiplexing of lower-rate traffic components is performed at the boundaries of the network only (i.e., at edge routers), and the aggregate traffic transparently traverses the optical network, this problem is equivalent to the *routing and wavelength assignment* (RWA) problem. It is known that wavelength assignment to minimize the number of wavelengths can be solved in polynomial time in paths and stars, but that it is NP-hard in rings, trees, and networks of a general topology. However, it is in general impossible to set up lightpaths between every pair of edge routers. Therefore, it is natural to consider optical networks in which nodes have both optical and electronic switching capabilities. Such nodes let some lightpaths pass through transparently, while they may terminate others. Traffic on terminating lightpaths may then be electronically switched (*groomed*) onto new lightpaths towards the destination node. Introducing some amount of electronic switching within the optical network has several advantages: it enhances the degree of *virtual connectivity* among the edge routers, which otherwise is limited by the number of optical interfaces at each router; it increases bandwidth utilization; and it may also reduce the wavelength requirements within the optical network for a given traffic demand. The trade-off is an increase in network cost due to the introduction of expensive active components (i.e., optical transceivers and electronic switches). These observations motivate us to define the following traffic grooming problem.

Let C be the capacity of each wavelength, expressed in units of some arbitrary rate (e.g., OC3); we will refer to parameter C as the *grooming factor*. Let W be the number of wavelengths that each fiber link in the network can support. We represent a traffic pattern by a demand matrix $T = [t^{(sd)}]$, where integer $t^{(sd)}$ denotes the number of traffic streams (each of unit demand) from node s to node d. Given matrix T, the traffic grooming problem involves the following conceptual subproblems (SPs):

1. *logical topology SP:* find a set R of lightpaths,
2. *lightpath routing and wavelength assignment SP:* solve RWA on R, and
3. *traffic routing SP:* route each traffic stream through the lightpaths in R.

The first and third subproblems together constitute the grooming aspect of the problem. Also, the number W of wavelengths per fiber link is taken into consideration as a constraint rather than as a parameter to be minimized.

The goal we consider in this paper is to minimize the *maximum number of lightpaths originating from or terminating at any node*. In our cost model, one unit of cost is incurred for each lightpath that terminates at, or originates from, a network node. Thus, this cost metric accurately reflects the amount of LTE needed at each network node. Note also that this objective is equivalent to minimizing the *maximum nodal degree* in the logical topology.

In this work, we study the traffic grooming problem in a unidirectional ring $\mathcal{R}$ with N nodes; this problem is NP-Complete since the RWA subproblem in rings is NP-Complete. In this section, we prove that the traffic grooming problem in path networks is also NP-Complete. Since the RWA problem can be solved in linear time in path networks, our results demonstrate that traffic grooming with the Min-Max objective is itself an inherently difficult problem, and that it remains NP-Complete in rings or other general topologies *even when* full wavelength conversion is available at the network nodes.

Let us consider a unidirectional path network $\mathcal{P}$ with N nodes. There is a single directed fiber link from node i to node $i+1$, for each $i \in \{1, 2, \cdots, N-1\}$. An instance of the traffic grooming problem is provided by specifying a number N of nodes in the path, a traffic matrix $T = [t^{(sd)}], 1 \leq s < d \leq N$, a grooming factor C, a number of wavelengths W, and a goal F. The problem asks whether a valid logical topology may be formed on the path and all traffic in T routed over the lightpaths of the logical topology so that the number of incoming or outgoing lightpaths at any node in the path is less than or equal to F.

We distinguish two problem variants depending on whether bifurcated routing of traffic is allowed or not. In the former variant, a traffic component $t^{(sd)}$ is allowed to be split into various subcomponents which may follow different routes (i.e., different lightpath sequences). The bifurcation is restricted to integer subcomponents. In the latter variant, for any pair (s, d) such that $t^{(sd)} \leq C$, we require that all $t^{(sd)}$ traffic units be carried on the *same* sequence of lightpaths from s to d. On the other hand, if $t^{(sd)} > C$, it is not possible to carry all the traffic on the same lightpath. In this case, we allow the traffic demand to be split into $\lfloor \frac{t^{(sd)}}{C} \rfloor$ subcomponents of magnitude C and at most one subcomponent of magnitude less than C, and the no-bifurcation requirement applies to each subcomponent independently. We then have the following two theorems; their proofs are omitted due to space constraints, and can be found in [5].

Theorem 1. *The decision version of the grooming problem in unidirectional paths with the Min-Max objective (bifurcated routing not allowed) is NP-complete.*

Theorem 2. *The decision version of the grooming problem in unidirectional paths with the Min-Max objective (bifurcated routing allowed) is NP-complete.*

Because of the construction of the proofs, we have the following corollary, which demonstrates that, even when solutions to the first two subproblems of the traffic grooming problem (see Section 2) are provided, the problem remains NP-Complete by virtue of the third subproblem (traffic routing). Therefore, traffic grooming is inherently more difficult than the well-known RWA problem.

Corollary 1. *The decision version of the traffic grooming problem in unidirectional paths with the Min-Max objective (bifurcated routing allowed or not) is NP-Complete even when a logical topology is provided.*

3   Min-Max Traffic Grooming Algorithm for Rings

We now present an algorithm for traffic grooming in WDM rings to minimize the maximum nodal degree (indegree or outdegree) in the logical topology. Rather than solving all three subproblems simultaneously (see Section 2), we decouple the logical topology and traffic routing subproblems from the RWA subproblem and tackle them independently. Specifically, the algorithm consists of the following steps which are explained in more detail in the following subsections:

- **Step 1.** Solve the logical topology and traffic routing subproblems on the ring network using the algorithm in Section 3.1. The result of this step is a set R of lightpaths and a routing of the traffic demands over the lightpaths in R that minimize the maximum nodal degree.
- **Step 2.** Use the algorithm in Section 3.2 to color the lightpaths of set R. The result of this step is a wavelength assignment that does not use more than W wavelengths. However, at the end of this step, the degree at one node, say, node i, of the ring may increase beyond the corresponding value after Step 1, by an amount equal to some value Δ.
- **Step 3.** Use the algorithm presented in [4] to distribute the additional Δ LTE at node i to other nodes in the ring network.

3.1   Min-Max Traffic Grooming Algorithm

We present a polynomial-time algorithm for the logical topology and traffic routing subproblems of the traffic grooming problem. Our algorithm attempts to minimize the maximum nodal degree by creating long lightpaths that bypass intermediate nodes whenever possible. Because of the results in Section 2, the traffic grooming subproblem is itself NP-Complete, and hence our polynomial-time algorithm will terminate without necessarily finding an optimal solution. However, numerical results to be presented later indicate that the solutions we obtain are close to the optimal and/or the lower bound.

Before we proceed, we introduce the concept of *reduction* of a traffic matrix. Specifically, we reduce the matrix T so that all elements are less than the capacity C of a single wavelength, by assigning a whole lightpath to traffic between a given source-destination pair that can fill it up completely. The available wavelengths on the links of the path segment from the source to the destination node are also decremented by the number of lightpaths thus assigned. Since breaking such lightpaths would increase the amount of LTE at some intermediate nodes of the path, this procedure does not preclude us from reaching an optimal solution, nor does it make the problem inherently easier or more difficult.

After the reduction, we initialize the logical topology to one in which a sufficient number of single-hop lightpaths is formed on each link of the ring network to carry the traffic using this link. We note that this initial solution is a feasible solution to the logical topology and traffic routing subproblems, in that it does not use more than W lightpaths on any link. However, this initial topology yields a large value of the maximum nodal degree F, which is equal to the number of single-hop lightpaths in the most congested link. Our approach, then, is to improve on this initial solution by joining short lightpaths to form longer ones, thus lowering the degrees at intermediate nodes. In the following, we summarize our algorithm for joining short lightpaths.

Let us define the relationship $i \prec j$ between ring nodes to denote that node i "precedes" node j in the direction of traffic flow; similarly, we will use the notation $i \preceq j$ to denote that node i precedes, or may be the same as, node j. The main idea of our algorithm is to consider the node with the maximum degree, and to attempt to decrease its degree by one at each iteration; this process repeats until no more improvement is possible. Let m be the node with the maximum degree. The algorithm searches for a pair of nodes $(i, j), i \prec m \prec j$, such that there exist lightpaths (i, m) and (m, j). The objective is to shift all the traffic from lightpaths (i, m) and/or (m, j) to either an existing or a new lightpath (i, j) in order to decrease the maximum of the indegree and outdegree of node m by one. If the traffic can be shifted entirely to an existing lightpath (i, j), then this procedure is always possible, since no wavelength limit constraints are violated, and also the degree of nodes i and/or j may also decrease in the process. However, if a new lightpath (i, j) must be created, the above procedure is carried out only if the wavelength limit constraint is not violated and the degrees of nodes i and j do not increase above the current maximum degree minus one (the minus one is necessary to ensure that the algorithm will not get into an infinite loop). A more detailed description is provided in Figure 1.

We now argue that at the end of each iteration of the **repeat** loop, the algorithm produces a solution to the logical topology and traffic routing subproblems that is feasible (i.e., no link carries more than W lightpaths – recall that we are not concerned with wavelength assignment at this point), and the maximum nodal degree is no larger (but possibly smaller) than that of the logical topology at the beginning of the iteration. At each iteration of the **repeat** loop, the algorithm tries to replace the lightpaths (i, m) and (m, j) for some nodes $i \prec m \prec j$ with a (possibly) new and longer lightpath (i, j) so as to decrease the nodal degree of node m. However, no action is taken if replacing the two lightpaths with the longer one would violate any wavelength constraints or would increase the degrees of i or j to more than the maximum at the start of the iteration, as we can see at Step 15. Since the initial topology at Step 2 of the algorithm is feasible, we conclude that the topology at the end of each iteration will be feasible and will not increase the maximum nodal degree.

The running time complexity of our algorithm is determined by the main iteration between Steps 6 and 21. In turn, the complexity of the iteration is determined by the two **for** loops, the inner **for** loop from Step 11 to 14, and the outer loop from Step 8 to 20. Each of these loops takes time $O(N^2)$ in the worst case, where N is the number of nodes in the ring. Therefore, each iteration through the **repeat** loop from Step 6 to 21 takes $O(N^4)$ time in the worst case. The main iteration of the algorithm (i.e., the **repeat** loop) will be executed at most $N\delta$ times, where δ is the maximum decrease in the degree of any node. Since $\delta < W$ always, the worst-case complexity of the algorithm is $O(WN^5)$.

Min-Max Traffic Grooming Algorithm
Input: A ring network with N nodes, W wavelengths each of capacity C, and reduced traffic matrix $T = [t^{(sd)}]$.
Output: The number of lightpaths b_{ij} from i to j, and the traffic routing quantities $t_{ij}^{(sd)}$ (which indicate the amount of the component $t^{(sd)}$ routed over a lightpath from i to j), so that the solution does not violate any wavelength constraints and minimizes the maximum number of lightpaths terminating at, or originating from, any node.

1. **begin**
2.   Initialize the logical topology to one with only single-hop lightpaths
        and initialize all b_{ij} and $t_{ij}^{(sd)}$ accordingly
3.   **for all** i,j **do** $r_{ij} \leftarrow$ capacity unused on the (i,j) direct lightpath in the
        current topology (residual capacity)
4.   **for all** j **do** $I_j \leftarrow$ indegree of j in the current topology
5.   **for all** j **do** $O_j \leftarrow$ outdegree of j in the current topology
6.   **repeat**   // Main iteration
            $m \leftarrow$ some node s. t. $\max\{I_m, O_m\}$ is maximum in the ring
            //Our objective is to replace a lightpath (i,m) and/or (m,j) with a
            //direct lightpath (i,j) in order to reduce $\max\{I_m, O_m\}$ by one
8.   **for** each pair (i,j) such that $i \prec m \prec j$ **do**
9.   **if** $I_m == O_m$ **then** $TotalToShift \leftarrow \max\{C - r_{im}, C - r_{mj}\}$
                else if $I_m > O_m$ **then** $TotalToShift \leftarrow C - r_{im}$
                else $TotalToShift \leftarrow C - r_{mj}$
                // $TotalToShift$ is the amount of traffic to be shifted to other
                // wavelengths in order to reduce $\max\{I_m, O_m\}$ by one
10. $TrafficToShift \leftarrow 0$
11.   **for** each pair (s,d), $s \preceq i, j \preceq d$ **do**
12. $TrafficToShift \leftarrow TrafficToShift + \min(t_{im}^{(sd)}, t_{mj}^{(sd)})$
13.   **if** $TrafficToShift > TotalToShift$ **then** break
                    // No more (s,d) pairs needed
14.   **endfor**   // of the (s,d) loop
15.   **if** $TrafficToShift < TotalToShift$ **then** break
                // Cannot replace lightpaths (i,m) or (m,j); continue w/ next (i,j) pair
                else if replacing lightpaths (i,m) or (m,j) would violate any
                wavelength limit constraints **or** would create new maximum nodal
                degrees at i or j **then** break
                else
16.   Remove lightpaths (i,m), (m,j) // $\max\{I_m, O_m\}$ decreases by 1
17.   **for** all pairs (s,d) contributing to $TrafficToShift$ **do**
18.   Reduce $t_{im}^{(sd)}$, $t_{mj}^{(sd)}$ by contributing amount, add equal amount to $t_{ij}^{(sd)}$
                        // This step reflects the new routing over the lightpath (i,j)
19.   **endif**
20.   **endfor**   // of the (i,j) loop
21.   **until** no decrease in the max degree at any node is possible
22. **end**   // of the algorithm

Fig. 1. Algorithm for logical topology and traffic routing

In practice, however, the algorithm runs much faster than the above worst-case analysis indicates, taking on the order of few tens of milliseconds for any problem instance with $N = 16$ nodes and $W = 128$ wavelengths.

3.2   An Algorithm for Wavelength Assignment

The output of the above algorithm is a set of lightpaths R between pairs of ring nodes, and a routing of the traffic elements $\{t^{(sd)}\}$ over these lightpaths. While the algorithm guarantees that the resulting logical topology is such that no link carries more than W wavelengths, it may not be possible to color the lightpaths in R using no more than W wavelengths. We now present a polynomial-time algorithm to perform wavelength assignment with at most W colors; the tradeoff in ensuring that the number of wavelengths does not exceed W is a modification of the logical topology (i.e., the set R) which may result in an increase in the degree of some node in the ring. Consequently, the objective F of our optimization problem may increase. Therefore, we then refine the new logical topology to decrease the objective F.

Let us start by describing how to assign wavelengths to the lightpaths of set R. Our approach is based on the observation that, while the wavelength assignment problem is hard for ring networks, it is solvable in linear time in paths. Consider some node m of the ring. Let R_1 denote the lightpaths in R which *optically bypass* node m, and let $R_2 = R - R_1$ be the set of remaining lightpaths. The lightpaths in set R_2 can be viewed as the logical topology on a path network, and thus, can be colored using no more than W wavelengths. Now consider all the lightpaths in set R_1. It may be possible to color some of them without violating any wavelength continuity constraints; in general, however, there may be some lightpaths in this set that cannot be colored without the need for additional wavelengths. In this case, we break such a lightpath $(x, y), x \prec m \prec y$ into two lightpaths (x, m) and (m, y). The new lightpaths do *not* bypass node m, and thus, can be colored along with the lightpaths in set R_2 using no more than W wavelengths. While breaking such a lightpath will increase the indegree and outdegree of node m by one, this approach guarantees a coloring of the new set of lightpaths that satisfies the wavelength constraints. The following steps describe our algorithm in more detail.

1. Let $R_1 \subset R$ be the set of lightpaths that optically bypass node m. Let $R_2 = R - R_1$ be the subset of remaining lightpaths.
2. Sort the lightpaths in R_2 in increasing order of their length.
3. Use the first-fit policy to color the lightpaths in R_2. This step is always possible since it corresponds to a first-fit wavelength assignment for a path.
4. Sort the lightpaths in R_1 in decreasing order of their length.
5. Use the first-fit policy to color the lightpaths in R_1. If lightpath $l = (x, y), y \prec x$, cannot be colored, then: break l into two lightpaths, $l_1 = (x, m)$ and $l_2 = (m, y)$ which do not bypass node m; increment the indegree and outdegree of node m to accommodate the new lightpaths; and repeat from Step 1 with $R_2 \leftarrow R_2 \cup \{l_1, l_2\}$ and $R_1 \leftarrow R_1 - \{l\}$.

Note that we arbitrarily select the node m above. An improvement would be to consider all N nodes, run the algorithm for each of them, and then select the solution with the least maximum nodal degree. However, we have found that the incremental improvement from such an approach does not justify the increase in running time, and thus we run the above algorithm just once.

Let Δ denote the increase in the nodal degree of node m after the termination of the above wavelength assignment algorithm. This increase is due to the fact that Δ lightpaths which optically bypassed node m under the initial logical topology defined by the set R, have now been broken into two lightpaths each. Of the 2Δ new lightpaths, Δ terminate at node m and Δ originate from it. Therefore, node m needs an additional Δ pairs of LTE, one for each of the original Δ lightpaths that used to bypass the node. Consequently, this increase of the objective F by Δ at node m increases the LTE cost in the ring network.

We now refine the new logical topology at the end of the wavelength assignment to improve on the objective F. Our approach is based on the observation that each additional pair of LTE at node m, one for an incoming and one for an outgoing lightpath, can be thought of as a *wavelength converter*. Indeed, consider, one of the Δ lightpaths that initially bypassed the node. This lightpath was broken by the algorithm into two shorter lightpaths that terminate at and originate from the node, respectively. This action was taken in Step 6 of the algorithm because it was not possible to assign the two shorter lightpaths on the two links in either side of node m the same color. Therefore, the additional pair of LTE at node m acts as a converter, changing the wavelength of the new incoming short lightpath to the wavelength of the new outgoing lightpath.

Consider a logical topology and corresponding feasible wavelength assignment on a ring network that requires a number Δ of converters at some node m. The recent study in [4] showed that it is possible to modify the wavelength assignment such that 2Δ converters are uniformly distributed across all N ring nodes (i.e., each node has at most $\lceil 2\Delta/N \rceil$ converters). Therefore, we use this algorithm to distribute the Δ pairs of LTE (i.e., "converters") at node m to the other ring nodes. As a result, the maximum degree at all the ring nodes will increase, but the maximum nodal degree of the network (i.e., of node m) will decrease, resulting in a new logical topology with a smaller value for the objective F. For the details of this algorithm, the reader is referred to [4].

4   Numerical Results

We present experiments to demonstrate the performance of our traffic grooming algorithm. The experiments are characterized by the following parameters: the traffic pattern, the number N of nodes in the ring, the number W of wavelengths per link, the capacity C of each wavelength, and the load L on the link carrying the most traffic. The maximum amount of traffic that can flow through a link is WC, hence we express the load L as a percentage of WC. For each experiment (i.e., each set of values for the above parameters), we generate

50 problem instances, and we compare the following values for each instance: the value of the objective F (i.e., the maximum nodal degree) returned by our algorithm; the *lower bound* F^l on the objective F, which can be obtained as: $F^l = \max_s \left(\max \left(\left\lceil \sum_d t^{(sd)}/C \right\rceil, \left\lceil \sum_d t^{(ds)}/C \right\rceil \right) \right)$; and the value F^e of the objective for a network using *all-electronic routing*, which corresponds to the number of wavelengths needed to carry the traffic on the link with the heaviest traffic load: $F^e = \max_l \left\lceil \sum_{(x,y) \in B(l)} t^{(sd)}/C \right\rceil$. We consider three traffic patterns in our study: uniform, random, and locality; for details on how these patterns are generated, refer to [5].

For small rings (i.e., $N \le 8$), we have obtained the optimal solution by using CPLEX to solve an ILP formulation (see [5]) of the min-max traffic grooming problem. We have found that our algorithm produces solutions at, or very close to, the optimal, for all traffic patterns we considered. Unfortunately, CPLEX fails to terminate for instances with $N > 8$ even after running for several days.

We now present results for rings with $N = 16$ and $W = 128$; such networks are of practical interest because the maximum size of a SONET ring is 16 nodes, and also because WDM links supporting 128 wavelengths are becoming commercially available. We note that our algorithm needed less than a second to find a solution for each problem instance that we present in this section.

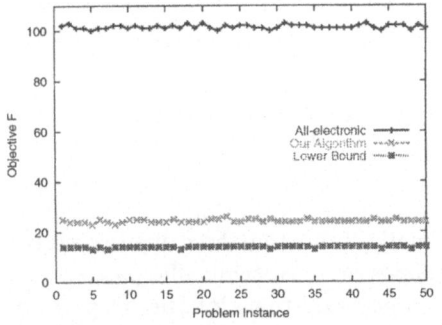 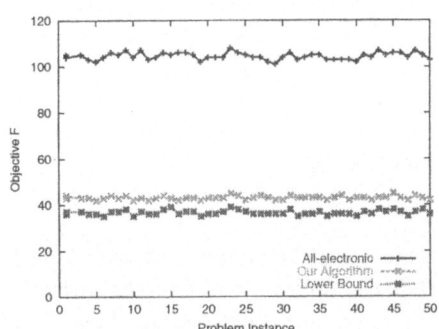

Fig. 2. Uniform pattern, $N = 16$, $W = 128$, $C = 12$, $L = 80\%$

Fig. 3. Locality pattern, $N = 16$, $W = 128$, $C = 12$, $L = 80\%$

Figures 2, 3, and 4 present the results of 50 different problem instances for each of the uniform, locality, and random traffic patterns, respectively. As we can see, the values of the objective F returned by our algorithm are close to the lower bound, and track it well across the different traffic patterns and problem instances within each pattern. The all-electronic solutions, on the other hand, are quite high with values at around LW. This result indicates that our approach of minimizing the maximum LTE cost in the network can produce significant cost savings. Another related observation is that, by using a Min-Max objective, we can ensure that the cost of any individual network node (as well as the overall network cost) is determined by the traffic demands of the node, and will not

184   B. Chen, G.N. Rouskas, and R. Dutta

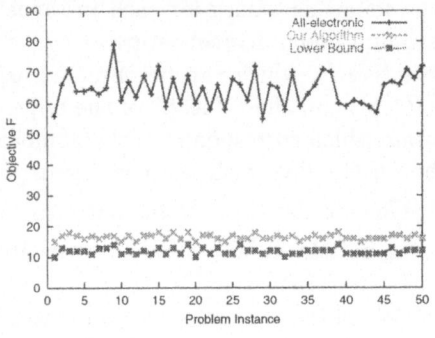

Fig. 4. Random pattern, $N = 16$, $W = 128$, $C = 12$, $L = 40 - 60\%$

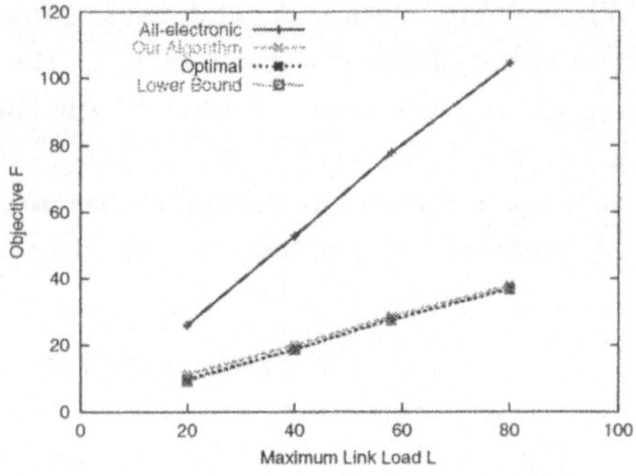

Fig. 5. Uniform pattern, $N = 6$, $W = 128$, $C = 12$, various loads

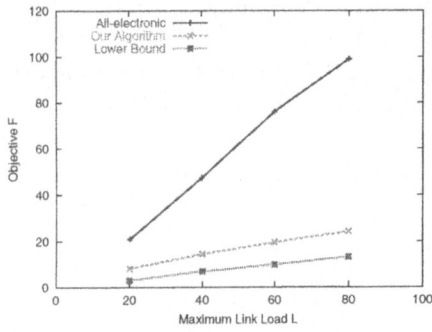

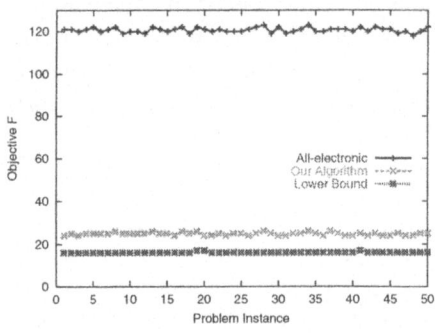

Fig. 6. Uniform pattern, $N = 16$, $W = 128$, $C = 12$, various loads

Fig. 7. Uniform pattern, $N = 16$, $W = 128$, $C = 12$, $L = 95\%$

scale with the number of wavelengths. Finally, from the three figures we see that the optimal value of the objective depends on both the traffic pattern and the maximum link load L. This observation suggests that, for the purposes of dimensioning the network for unknown future demands, the network designer should run our algorithm for a variety of traffic patterns and loads, and then equip each node with an mount of LTE determined by the highest value of F.

In Figures 5 and 6, we plot the values of F^e, F^l, and F against the load L for a uniform traffic pattern; each point in the figures is the average over 50 problem instances for the given value of L (the optimal values F^* are also plotted for the small network in Figure 5). The experiments presented in the two figures are identical in all respects except that Figure 5 is for 6-node rings, while Figure 6 is for 16-node rings. We have obtained similar results for the other two patterns as well. As we can see, F^e, F^l, and F all increase linearly with the load L. However, the curve corresponding to the all-electronic solution has a slope much steeper than that of the curves corresponding to the lower bound and our solution. This behavior demonstrates that the cost benefit of our optimization approach increases with the load of the network. This property is an important one, and it implies that network operators will be able to operate the network at high loads with only an incremental increase in cost.

Another important property of the Min-Max optimization approach becomes evident once we compare the curves of Figure 5 to those of Figure 6. Specifically, we note that the curves corresponding to the all-electronic solutions have similar slopes in both figures. On the other hand, the curves corresponding to the lower bound and our solution have a steeper slope in Figure 5 (which corresponds to a 6-node network) than in Figure 6 (for a 16-node network). Equally important, for a given load L, the actual values of the lower bound and our solution are *lower* for the larger network than the respective values for the smaller network; this is true despite the fact that link loads of both networks are similar and that the 16-node network carries a much larger aggregate amount of traffic than the 6-node network. Based on this observation, we conclude that our solution approach scales well not only with the load and number of wavelengths, as we explained above, but also with the size of the network.

Finally, in Figure 7, we present results for problem instances with $N = 16$ nodes, a uniform traffic pattern, and a high load $L = 95\%$. Let us compare Figure 7 to Figure 2 which plots results for the same values of the system parameters, except that the load $L = 80\%$ in Figure 2. As we can see, the all-electronic solution F^e is about 20% higher in Figure 7, as expected, and the lower bound is also somewhat higher. However, there is little difference in the values of F returned by our algorithm when the load increases from 80% to 95%, reinforcing our earlier conclusion that the benefit of Min-Max optimization increases with the traffic load, enabling cost-effective network operation at very high loads.

5   Concluding Remarks

We studied a variant of traffic grooming in rings, where the objective is to minimize the maximum LTE cost at any node. We presented an algorithm for grooming and wavelength assignment, and we showed that it performs well under a wide range of traffic patterns and system parameter values. Our algorithm can be used to dimension networks for unknown and/or dynamic traffic demands.

References

1. Dutta, R., Rouskas, G.N.: A survey of virtual topology design algorithms for wavelength routed optical networks. Optical Networks 1 (2000) 73–89
2. Dutta, R., Rouskas, G.N.: Traffic grooming in WDM networks: Past and future. IEEE Network 16 (2002) 46–56
3. Dutta, R., Rouskas, G.N.: On optimal traffic grooming in WDM rings. IEEE Journal on Selected Areas in Communications 20 (2002) 110–121
4. Chen, L.W., Modiano, E.: Efficient routing and wavelength assignment for reconfigurable WDM networks with wavelength converters. In: IEEE INFOCOM. (2003)
5. Chen, B., Rouskas, G.N., Dutta, R.: Traffic grooming in WDM rings with the Min-Max objective. Technical report, North Carolina State Univ., Raleigh, NC (2003)

A Case for Mesh-Tree-Interaction in End System Multicasting*

Anirban Chakrabarti and Govindarasu Manimaran

Dept. of Elecrical and Computer Engineering, Iowa State University
{anirban,gmani}@iastate.edu

Abstract. End System Multicasting (ESM) is fast becoming a feasible alternative to IP multicasting. ESM approaches can be broadly classified into two main categories: (i) Tree first approaches, where an overlay tree is constructed on the physical network, (ii) Mesh first approaches, where a mesh is constructed on the physical network and then a tree is created on the constructed mesh. In this paper, we propose a generic Mesh Tree Interaction (MTI) mechanism, which combines the management efficiency of the mesh first approaches and the performance benefits of the tree first approaches. To achieve this, MTI uses the concept of mesh and enables interactions between the mesh and the underlying multicast tree. Our simulation studies show that MTI results in significant improvement in the quality (average delay metric) of the multicast tree.

1 Introduction

Multicasting has been the most popular mechanism for supporting group communication. In a multicast session, the sender transmits only one copy of each message that is replicated at appropriate routers inside the network and delivered to multiple recipients. For this reason, multicasting typically requires less total bandwidth than separately unicasting messages to each receiver. In order to determine whether to implement multicasting at IP or application layer, implementation complexity vs. performance trade-offs need to be considered. Several prototypes have been developed and IP multicasting has been added as a feature in many commercial routers. In spite of the advancements in the field of IP multicasting, IP multicasting suffers from scalability problem, as each router needs to store group specific information. Also, implementing higher level features like congestion control, flow control, reliability and security have been shown to be more difficult in IP multicasting, than in the unicasting case.

As an alternative to IP multicasting, researchers have proposed the End System Multicasting (ESM) approach [4,5,6,7], wherein the complex multicasting features like replication, group membership management and multicast routing are implemented at the application layer, assuming only the end-systems or hosts are responsible for multicasting. As all the complexities are handled at the hosts rather than at the routers, it offers some distinct advantages over its IP counterpart. (i) ESM is easier to implement, as there is no complexity required at the routers, (ii) Complex functionalities like congestion control, reliable data transfer are handled separately at the unicast level, and

---

* This research was supported in part by the NSF under grant ANI-0240433

N. Mitrou et al. (Eds.): NETWORKING 2004, LNCS 3042, pp. 186–199, 2004.

therefore manageable, (iii) Adding security features to multicasting is easier as routers are not involved.

In spite of these advantages, ESM has some issues which need future research attention. (i) The quality of the multicast tree produced using ESM is worse than that produced using IP multicasting, (ii) Since each node in the ESM tree is a host, therefore the nodes have limited capability in terms of bandwidth and processor capabilities, (iii) The multicast sessions are unreliable as they depend on the hosts for data transmission.

Multicast trees in ESM can be constructed using two approaches: (a) Tree-first approaches and (b) Mesh-first approaches. In Tree-First approach, members directly select their upstream neighbors from among the known members [3]. In Mesh-first type of approach, a mesh is constructed on the physical network. Narada [4] and NICE [5] are examples of this approach.

2   Problem Statement and Motivation

In this section we formally define the ESM tree management problem, and then provide motivation for the approach taken in this paper to solve the problem.

Given an undirected network $N = (V, E)$, where V is the set of vertices or nodes, and E is the set of edges or links. Let e_{ij} be an edge between nodes i and j, such that $e_{ij} \in E \; \forall i, j \in V$. $D_{i,j}$ be the delay associated with the edge e_{ij}. Let S be the set of all shortest paths in N, and s_{ij} is the shortest path between nodes i and j, such that $s_{ij} \in S$ $\forall i, j \in V$. Let M be the set of members in a multicast session, such that $M \subseteq V$. Let F_i be the fanout constraint of each member i. The problem is to construct a multicast tree $T = (V_M, S_M)$, $S_M \subseteq S$ spanning all members so that the average delay to all members is minimized such that $d_i \leq F_i \; \forall \; i \in M$, where d_i is the degree at node i.

We call the above problem as ESM tree management problem. In this paper, whenever we refer to quality of a multicast tree we use average delay as the metric. The ESM tree management problem can be tackled using two methods. The first method creates a degree constrained spanning tree on a fully connected virtual graph. As shown in [8,9], the problem is NP-Complete. Tree-first techniques use this approach. The second approach, is through construction of a degree-constrained K-spanner on the fully connected virtual graph. A degree-constrained K-spanner is a subset of the fully connected virtual graph such that, each node satisfies the degree constraint and the shortest path between any two node in the K-spanner is not more than K times the shortest path in the fully connected virtual graph. As shown in [10], this problem is also NP-Complete. In this approach, after the construction of the K-spanner, a spanning tree is constructed on the K-spanner. Mesh-first techniques use this approach. In this paper, we propose a technique called Mesh Tree Interaction (MTI), which combines the management ability of the mesh-first approaches, and the performance benefits of the tree-first approaches.

As mentioned earlier, both Tree-first and Mesh-first approaches are based on NP-Complete problems. Therefore, both the approaches use approximations to construct spanning tree and K-spanner respectively. Independent mesh and tree, though result is simplicity in mesh management, result in creation of low-quality tree as mesh construction is done without taking the actual tree construction into account. Therefore, mesh construction may result in creation of mesh links which do not contribute to the improvement of the quality of the multicast tree. It is to be noted that mesh provides redundancy, however it is the multicast tree which is used for actual data dissemination. Therefore, quality of multicast tree is absolutely critical for group communication. In this paper, we refer to average delay as the 'quality' of the multicast tree.

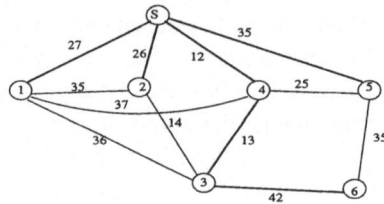

Fig. 1. An Example Mesh

In Figure 1, an example mesh is shown which is a subset of a fully connected virtual graph. Each link in the figure is the shortest path between the nodes. The number shown with each link indicates the delay of the shortest path between the two nodes. The links which are part of the multicast tree are indicated in the figure using darker lines. Let S be the source of the multicast communication; shortest path from node 6 to node 4 has a delay of 12. However, link $(6-4)$ is not part of the mesh, as shown in the figure. Let us assume the fanout limit for each node in this example is 4. Therefore, node 6, if connected to node 4 can provide a better delay path for itself. However, node 4 has reached its fanout limit (in this case 4). It is to be noted that in mesh-first approaches, each node independently tries to satisfy the fanout constraint and find the best neighbor at the mesh level. Since the Mesh-first protocols have no way to identify that link $(6-4)$ if added to the mesh, it will result in a better tree, will not add link $6-4$ to the mesh. This example shows that there is a need for interaction among the constructed mesh and tree so that "unimportant" mesh links can be removed to eventually produce better quality tree. Referring back to the above example, if node 4 had somehow realized, during link addition itself, that links $(1-4)$ and $(4-5)$ are "unimportant" links, or links which will not result in a better quality tree, then one of these links could be removed in this case and the link $(6-4)$ can be accommodated such that the quality of the overall multicast tree is improved. In other words, a continuous interaction between mesh and tree is needed to construct a better quality mesh, which eventually leads to the construction of better quality tree. In this paper we propose a mesh-tree interaction approach which achieves the above. MTI identifies whether a link is important (part of the tree) or not (part of the non-tree mesh), and takes action based on the information. MTI technique achieves the following objectives: (a) MTI achieves a better "quality" tree than other Mesh-first protocols. (b) MTI is easily deployable, as group management is still controlled at the mesh level, instead of tree level in case of Tree-first protocols. (c) MTI can be used in isolation, as well as in conjunction with any of the existing Mesh-first protocols like Narada and NICE.

The rest of the paper is organized as follows: In Section 3, an overview of the MTI approach is provided with important definitions to be used for the rest of the paper. In

Section 4, the different steps of MTI are described in detail. A Restricted MTI (R-MTI) approach is outlined in Section 5. Finally, in Sections 6 and 7, simulation results and some concluding statements are provided respectively.

3   Mesh-Tree Interaction (MTI) Overview

Mesh-tree Interaction (MTI) is a mechanism to create "good" quality multicast tree through the improvement of the mesh in an iterative manner. While all the mesh-first protocols create the tree from the mesh, MTI improves the quality of the mesh based on the constructed tree, which in turn improves the quality of the tree. On an abstract level, the main difference between a standard mesh-first approach and MTI lies in the inherent understanding of the nature of the multicast tree, which is used to construct a better mesh. The basic difference is illustrated in Figure 2(a). While in Mesh-first approaches quality of the tree depends enormously on the quality of the underlying mesh, MTI uses an iterative process as tree structure influences the mesh, which in turn influences the tree structure.

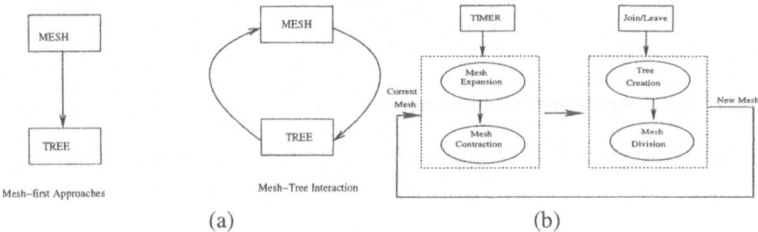

(a)   (b)

Fig. 2. (a)Mesh-first Approaches vs. MTI, (b) Different steps of MTI

To implement MTI, the mesh is divided into Primary Mesh (PM) and Secondary Mesh (SM), where PM contains all the links that are part of the multicast tree and SM contains all the links that are not part of the multicast tree. The second difference between Mesh-first approaches and MTI is the selection of the "best" neighbors for mesh optimization. Goodness of neighbors are identified by a parameter called the Upstream Correlation Factor (ρ), which determines how "good" the upstream neighbor of a node is. Mesh-tree Interaction has three main steps which differentiate the approach from the traditional mesh-first and tree-first approaches:

Mesh Division: This is the first step where the mesh is divided into Primary Mesh and Secondary Mesh. The intuition behind this is to differentiate between tree links and non-tree links within a mesh. Mesh division also helps MTI to keep track of the important links which constitute the Primary Mesh, and unimportant links which constitutes the Secondary Mesh. This prioritization helps MTI to accommodate links which eventually results in the construction of a better multicast tree.

Mesh Expansion & Contraction: In mesh expansion, links are added to the secondary mesh which eventually leads to the improvement of the quality of the mesh. Mesh contraction takes place when certain mesh links are deleted to make way for mesh expansion. The interaction between the different steps is illustrated in Figure 2(b).

Tree Creation: Tree is created using shortest path on the total mesh. Tree creation step leads to the mesh division. Tree creation step is not much different from the tree creation protocol described in [4].

Property 1: $-1 \leq \rho \leq 1$

Property 2: If $\rho_{ij}^s = \delta$, then $\rho_{ji}^s = -\delta$.

Property 3: $\rho_{is}^s = 1$.

Property 4: Let i, j and k are three nodes and source is s, $\rho_{ij}^s > 0$ and $\rho_{jk}^s > 0$, then $\rho_{ik} \geq \frac{\rho_{ij}^s \cdot \Delta_{ij} + \rho_{jk}^s \cdot \Delta_{jk}}{\Delta_{ij} + \Delta_{jk}}$.

Property 5: If i_j is the best upstream neighbor of i_{j-1} $\forall j = 1, 2...n$, and $\rho_{i_{j-1},i_j}^s > 0$, then $i_1, i_2 ... i_n$ cannot form a loop.

Fig. 3. Properties of ρ

3.1   Upstream Correlation Factor (ρ)

To understand Upstream Correlation Factor (ρ), we define the following terms:
Shortest Path Delay (Δ_{ij}): Δ_{ij} determines the delay of the shortest path between nodes i and j.
Upstream Neighbor (η_i^s): η_i^s indicates the upstream neighbor of node i with respect to source s.
Upstream Correlation Factor (ρ_{ij}^s): ρ_{ij}^s determines the quality of the upstream neighbor j of node i, where s is the source of the multicast tree. Mathematically,

$$\rho_{ij}^s = \frac{\Delta_{is} - \Delta_{js}}{\Delta_{ij}} \tag{1}$$

Properties of ρ are shown in Figure 3.

ρ or the Upstream Correlation Factor is an interesting and important metric to determines the quality of the upstream neighbor. $\rho_{ij}^s = 1$ indicates that the shortest path of i goes through j, and $\rho_{ij}^s = -1$ indicates that the shortest path of j goes through i. Higher the value of ρ, lower is the delay of the path through the upstream node. The reason ρ is such an important metric in ESM context is that it determines the quality of the upstream neighbor without actually knowing anything about the actual path. Therefore, the metric can be measured by sending ICMP packets to different nodes and measuring the delay experienced by the packets. Therefore, the metric does not violate the basic premises of the ESM architectures.

Let us illustrate the usefulness of ρ with the help of an example. Let node A and node B have shortest delay path to these source as 10 and 8 respectively. The current delay offered by the two nodes are 10 and 12 respectively. This is a possible scenario, as the current delay path depends on the quality of the underlying mesh. Now, a node C

has shortest delay path to nodes A and B as 2 and 3 respectively. Therefore, the current delay offered to node C, if node A is C's upstream neighbor is 12, while that offered if node B is the upstream neighbor is 15. However, node B has the capability of offering a delay path of 11 to node C. Therefore, it is a "better" upstream neighbor. ρ value reflects this as $\rho_{CA} = 0.5$ and $\rho_{CB} = 1$. From Property 1, ρ_{CB} is the maximum ρ value possible for any neighbor of C. Therefore, B is the best upstream neighbor.

4 Different Steps of MTI

As mentioned earlier, MTI consists of three steps: (a) Mesh Division, (b) Mesh Expansion, (c) Mesh Contraction and (d) Tree Construction. We describe the steps in detail, in the following subsections. In this paper, we only discuss about the first three steps as MTI is flexible, and any tree construction algorithm can be employed.

Mesh Division: In this step, the mesh is divided into Primary Mesh (PM) and Secondary Mesh (SM). PM consists of the links which are part of the multicast tree, and SM consists of all the links which are not part of the multicast tree. Each mesh consists of a two lists. SM consists of two lists SM^+ and SM^-, while PM consists of PM^+ and PM^-. List SM^+ consists of all links having positive ρ value among the SM links, sorted in descending order such that the head of the list contains the link having the maximum ρ value. On the other hand, SM^- consists of all links having negative ρ values among the SM links, arranged in ascending order such that the head of the list contains the link having minimum ρ value. The head and tail of SM^+ are called SM Maximum+ (Γ_{SM}^+), SM Minimum+ (γ_{SM}^+) respectively. The head and tail of SM^- are called SM Minimum- (γ_{SM}^-) and SM Maximum- (Γ_{SM}^-) respectively. It is to be noted that, the Minimum and Maximum of SM^+ and SM^- are reversed. The reason behind this is that, the "importance" of a link is higher if the ρ value is lower, if the link has negative ρ value. The links in PM are also arranged in PM^+ and PM^- in a similar way. Let us illustrate the mesh division concept based on the example mesh shown in the Figure 1. For Node 4. PM^+ has only one link having ρ value of 1.0. PM^- also has only one link $(4-3)$, having ρ value of -0.85. SM^+ is empty. SM^- has two links $(4-1)$ and $(4-5)$ having ρ values of -0.41 and -0.52 respectively. $\Gamma_{PM}^+ = \gamma_{PM}^+$, and $\Gamma_{PM}^- = \gamma_{PM}^-$, as there is only one link each in PM^+ and PM^-. Link $(4-1)$ is γ_{SM}^- and link $(4-5)$ is Γ_{SM}^-.

Mesh Expansion and Contraction: Mesh expansion forms the second step of MTI which may or may not lead to mesh contraction. Under mesh expansion each node proactively tries to expand the mesh by adding a neighbor to its mesh, which is better than at least one of its current neighbors. The "goodness" of a neighbor is measured by the ρ value mentioned earlier. Higher is the ρ value of the neighbor, better is the neighbor. The main principle behind mesh expansion is that each node (say i) attempts to find its Best Upstream Neighbor ($\mu+$). To identify $\mu+$ each node searches for the neighbor having the maximum ρ. Each searching node (i) searches for a set of candidate neighbors and calculates the ρ value for each of them. The candidate neighbor having the highest ρ value is selected as the best candidate neighbor (say n). If both i and n

have enough resources (fanout is less than the limit), the link is accommodated. In this case, mesh contraction is not called. Otherwise, mesh contraction is called.

In case of mesh contraction, some candidate links are found connected to both i (mentioned as $ReplaceLink_i$) and n (mentioned as $ReplaceLink_n$). If the resource constraint at node i is violated (the fanout at node i exceeds the limit) because of the addition of the link $(i - n)$, $ReplaceLink_i$ is deleted from the mesh to accommodate for link $(i - n)$. Similarly, if the resource constraint at node n is violated because of the addition of the link $(i - n)$, $ReplaceLink_n$ is deleted from the mesh. This step is part of the mesh contraction, as mentioned before. To search for $ReplaceLink_i$, firstly SM^+ of i is searched. The reason behind this is that, all links which are present in SM^+ are not part of the tree. If a link is found (say j) which has lower ρ than ρ_{in} (ρ value of link $i - n$), then j is identified as the $ReplaceLink_i$. If SM^+ is empty, γ_{SM}^- is identified as the $ReplaceLink_i$. If both SM^+ and SM^- are empty, then $ReplaceLink_i$ is identified from PM^+ if ρ value of the link is PM^+ is less than ρ_{in}. To identify $ReplaceLink_n$, if n is not the source same sequence is followed, only this time first SM^- is searched, then SM^+ and then PM^- as ρ changes sign in n (Property 2). However, if n is a source then $ReplaceLink_n$ is identified as a link (say j) in the PM^+ of n, if $\Delta_j > \Delta_{in}$. The reason for using Δ instead of ρ is because $\rho_{in} = 1$, in this case as n is the source (Property 3). The pseudo-code of the mesh is described in the Appendix.

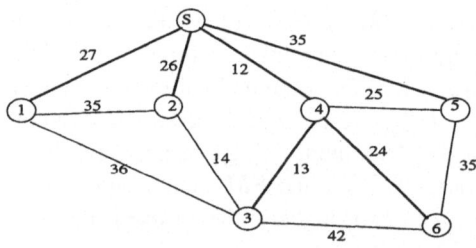

Fig. 4. The New mesh

To illustrate the above algorithm with the help of an example, we refer back to the Figure 1. Let us assume that $\Delta_{6S} = 36$ and $\Delta_{64} = 27$. This means that the shortest distance from node 6 to the source is 36, and the shortest delay from node 6 to node 4 is 27. From the figure, the shortest distance from source to node 4 is 12 $i.e.$ $\Delta_{4S} = 12$. Therefore, $\rho_{64}^S = 0.9$ and $\rho_{46}^S = -0.9$. $\gamma_{SM}^- = (4 - 1)$, and $\rho_{41}^S = -0.41$. Node 6 does not violate its fanout constraint by accommodating the link. However, node 4 does. Therefore, mesh contraction algorithm needs to be called at node 4. There exists at least one link in the SM^- of node 4 having lesser importance than link $(4 - 6)$. Therefore, γ_{SM}^+ $i.e.$ link $(4 - 1)$ is removed from the mesh to accommodate link $(4 - 6)$. Hence, link $(4 - 1)$ is removed from the SM and link $(4 - 6)$ is added to the SM^- of node 4. After tree creation, this link will be added to the tree. After the expansion/contraction of the mesh, the mesh looks like Figure 4. After the mesh addition and tree creation, the average delay improves from 32.17, in the first case to 27.33.

5   Restricted MTI (R-MTI)

Comparing between MTI and any other flat mesh based ESM protocol (Narada for example), the following points need to be considered:

Message Complexity: Message complexity of MTI and any mesh-first protocol is comparable. In both the cases each node need to send $O(n)$ messages, where n is the number of nodes in the mesh.

Message size: In case of standard mesh-first approaches (like Narada), probe messages calculate the distance of the probing node to the potential neighbors. In case of MTI, the probe message should also include the distance of the potential neighbor to the source in addition to the distance information between the two nodes. This requires 4 bytes of extra information in the probing message.

Computational Complexity: In case of ESM, since all multicasting activities are handled at the end systems, therefore computational complexity assumes important proportions. The computational complexity in case of any flat mesh-first protocol is $O(n)$, where n is the number of nodes in the mesh. MTI increases the computational complexity to $O(flogf + fn)$, where f is the fanout limit of the nodes in the mesh.

Since the message and computational complexity of MTI increases linearly, then the scalability of the protocol suffers for high number of nodes in the mesh. A message complexity of $O(n)$ has the potential of message explosion, if the number of nodes in the mesh increases. Therefore, there is a need to device means to reduce or restrict the number of messages transmitted. Reduction of message complexity motivates the development of Restricted MTI (R-MTI). In R-MTI, the potential neighbor search is only restricted to the neighbors in either SM^- and PM^-. The ρ value of neighbors of SM^+ and PM^+ are calculated based on Property 4. R-MTI helps to restrict the worst-case message complexity from $O(n)$ under normal MTI, to $O(f^2)$, where f is the fanout limit of the nodes. Under normal case, it will be still less because search will be restricted to only neighbors having negative ρ. Similarly, the computational complexity is reduced from $O(flogf + fn)$ to $O(flogf + f^2)$. Though R-MTI has low computational and message complexity, it produces lower "quality" tree as the search space is restricted. Therefore, R-MTI introduces a trade-off between message and computational complexity with the "quality" of the multicast tree. In the simulation section R-MTI is studied vis-a-vis MTI and Narada to quantify this trade-off.

6   Simulation Studies

In order to evaluate the effectiveness of our MTI model, we conducted extensive simulation studies using ns [11]. In our simulation studies, we compared our MTI model with Narada as well as several Centralized algorithms. The various inputs for the simulation studies were generated as follows: (a) Random network topologies were generated based on a given input parameter "graph density." This parameter determines the average node degree and hence the connectivity of the network. The higher the value, the denser the topology. (b) The selection of receivers for a given multicast session were uniformly distributed from the node set. (c) Members join and leave the multicast group, and the mesh is reorganized assuming a Distance Vector protocol running on the mesh level. (d) For each point in the graph, an average of 10 simulation runs were conducted.

The *default parameters:* are (i) Total number of nodes = 1000, (ii) 20% of all nodes are end hosts, (iii) Average Node degree = 4, (iv) Average number of members = 100, (v) Average link bandwidth = $15Mbps$, (vi) Average link delay = $12.5ms$ (vii) Member join/leave inter-arrival time = $100ms$, (viii) Average fanout of the nodes = 4.0.

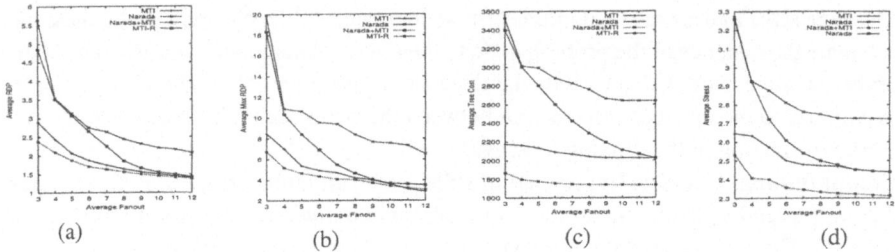

(a)   (b)   (c)   (d)

Fig. 5. Variation of (a) ARDP, (b) AMRDP, (c) Average Tree Cost and (d) Average Stress with varying Fanout Limit

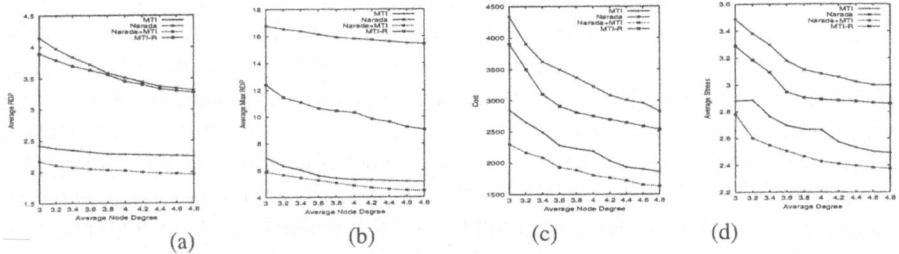

(a)   (b)   (c)   (d)

Fig. 6. Variation of (a) ARDP, (b) AMRDP, (c) Average Tree Cost and (d) Average Stress with varying Average Network Density

In order to compare the effectiveness of the various models, we evaluated the models according to the following performance metrics: (a) **Average Relative Delay Penalty (ARDP):** Relative Delay Penalty (RDP) is defined as the ratio of the delay provided by the current multicast tree to that provided by unicast averaged for all the members. (b) **Average Maximum Relative Delay Penalty (AMRDP):** Maximum Relative Delay Penalty (MRDP) is defined as the maximum RDP suffered by a node, among all the nodes currently in session. (c) **Average Stress:** Stress is defined as the average number of unicast flows per tree link. (d) **Average Tree Cost:** The average cost of multicast tree averaged out over time were compared to evaluate the effectiveness of the algorithms.

In order to evaluate the effectiveness of different approaches, we studied the effects of the following parameters: (a) Fanout constraint, (b) Group Dynamics, (c) Network Density and (d) Group Size.

Effect of Fanout Limit: In this set of experiments, the fanout limit of each node participating in the ESM session was varied and its effect was studied on different performance metrics for different approaches. The results are shown in Figure 5. In Figure 5(a), the

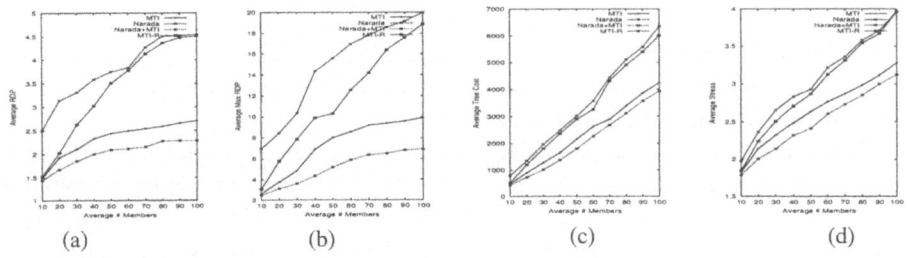

Fig. 7. Variation of (a) ARDP, (b) AMRDP, (c) Average Tree Cost and (d) Average Stress with varying Average Group Size

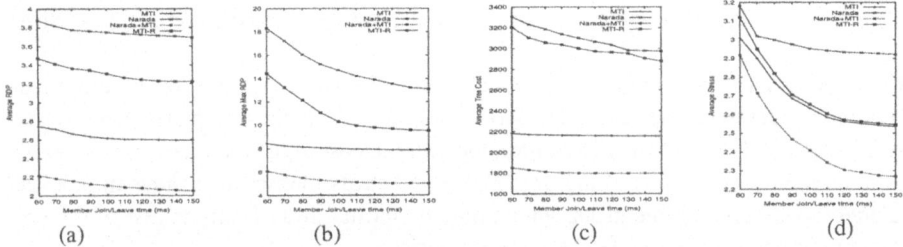

Fig. 8. Variation of (a) ARDP, (b) AMRDP, (c) Average Tree Cost and (d) Average Stress with varying Average Join/Leave Time

average RDP (ARDP) is studied with varying fanout limit. With the increase of fanout limit, each node in the multicast group can accommodate more mesh links, and therefore reduces the average delay of the overall tree. This trend can be observed for all the approaches. Comparing the relative performances of different approaches, the ARDP of the trees constructed using MTI is $50 - 60\%$ of that of the Narada trees. Narada and MTI combination reduces the ARDP value further by $10 - 15\%$, justifying that MTI and Narada can be combined in practice to get a better quality without losing the inherent easy maintenance of the mesh-first approaches. ARDP of the MTI-R approach, is lower (better) than that of Narada and gets closer to MTI with increase in fanout limit. The reason behind this is that, with increase in fanout limit, more links nodes are searched for the identification of better upstream neighbor.

Though ARDP is the primary metric used in MTI, the unique method of selecting best neighbors using ρ value reduces AMRDP, Average Tree Cost and Stress also. This point is justified in Figures 5(b), (c) and (d). The trends exhibited by each of these metrics is similar to that exhibited by the ARDP metric. Narada has the highest Average Tree Cost, MTI and Narada combination has the lowest. MTI and MTI-R lies somewhere in between. Trees produced by the MTI-R approach is similar to the MTI approach with increase in fanout limit.

Effect of Network Density: In this set of experiments the node degree of the physical network is varied and its effect is studied on the four different performance metrics mentioned above. Higher the average node degree, denser is the physical topology. Increasing the density of the nodes in the physical network has significant effect on the trees created on the overlay. Figure 6 illustrates the effect. As the network becomes denser, the chances of two paths on the overlay going through the same physical links gets substantially reduced. This effect gets reflected the Average Stress metric (shown in Figure 6(d)), and in turn on all the other metrics used for performance analysis. Average Stress decreases with increasing node degree.

Figure 6(a) shows the variation of ARDP with average node degree of the physical network. As the network becomes denser, then there are more options to go from one node to another, as a result the chances of getting a better delay path increases. This phenomenon is reflected in Figure 6(a). Among the different approaches, combination of MTI and Narada has the lowest ARDP value. ARDP value in case of MTI is nearly 50% less than that of Narada. R-MTI is in between Narada and MTI and ARDP of most R-MTI trees are around 10% lower than that of Narada, while $20-30\%$ more than MTI. Similar trends are also witnessed for AMRDP, Average tree cost and Average Stress metrics.

Effect of Group Size: In this set of experiments, the average group size of a multicast session is increased. The results are shown in Figure 7. As the group size increases the size of the overlay increases. Therefore, the average cost of the multicast tree increases, as more members are part of the tree. Delay and stress metrics also show an increase, as more members join the group which may be farther away from the rest of the tree, resulting in increase in these performance metrics.

In Figure 7(a), the variation of ARDP metrics is shown with average group size. At low group size, (≤ 60), R-MTI performs similar to MTI and its ARDP is nearly 50% less than that of Narada. The combination is 10% less than MTI. With increase in group size, the ARDP value increases for all approaches as the distance between nodes increases. The increase of R-MTI is maximum, as the ARDP value is nearly equal to that of Narada for group size ≥ 160. Increase of Narada is approximately linear with group size, while that of MTI and the combination is sub-linear. Therefore, ARDP of MTI is approximately 80% less than Narada at higher values of group size (≥ 160). Variation of AMRDP is shown in Figure 7(b), which is similar to RDP.

In Figure 7(c) and (d), the variation of Average Tree Cost and Stress is shown with average group size. Both the parameters increase approximately linearly with average group size. The relative performance of the approaches match that of ARDP.

Effect of Group Dynamics: In Figure 8 variation of group dynamics is shown on the performance metrics. Group dynamics is measured by the join/leave time. Higher the join/leave time interval, lesser dynamics is the group. All the performance metrics increase with the increase in the group dynamics. The reason for this is that, the multicast tree and the mesh size gets bigger at a faster rate than the optimization when the group dynamics is very high.

MTI is more immune to group dynamics, as the ARDP, Average Cost and AMRDP increase approximately $3-5\%$ for MTI, while $5-7\%$ for Narada.

7   Conclusion

In this paper, we have proposed a mesh-tree interaction (MTI) approach which does not compromise on the inherent simplicity in management of the mesh first approaches, however builds a much better quality multicast tree than the mesh-first approaches. The main principle behind the MTI approach is that the "quality" of the mesh is improved based on the underlying multicast tree, which in turn improves the quality of the tree itself. Thus, by keeping the mesh structure, the management simplicity of mesh-first approaches is maintained, and the iterative tree-building mechanism improves the quality of the tree dramatically. In this paper, we have carried out extensive simulation studies illustrating the MTI approach. In comparison to other mesh-first approaches like the Narada, MTI improves the ARDP metric by nearly $40 - 50\%$ and cost of the multicast tree improves by $20 - 30\%$ for lower fanout constraints (4-5). MTI can also be applied in conjunction with other mesh management techniques like Narada, which further improves ARDP by $10 - 15\%$. Future work includes: (i) Extending MTI to hierarchical mesh management techniques like NICE. (ii) Interoperability of MTI tree management techniques with that of IP tree management techniques.

References

1. S. Deering, "Multicast Routing in Internetworks and extended lans," in *Proc. SIGGCOM*, pp. 55-64, Aug. 1988.
2. C. Diot, B.N. Levine, B. Lyles, H. Kassem, and D. Balensiefen, "Deployment issues for the IP multicast service and architecture," *IEEE Network*, pp.78-88, Jan./Feb. 2000.
3. P. Francis, "Yoid: Your own Internet Distribution," *www.aciri.org/yoid*, Apr. 2000.
4. Y.-H. Chu, S. G. Rao, S. Seshan, and H. Zhang, "Enabling Conferencing Applications on the Internet using an Overlay Multicasting Architecture," in *Proc. SIGGCOM*, Aug. 2001.
5. S. Banerjee, B. Bhattacharya, and C. Kommareddy, "Scalable Application Layer Multicast," in *Proc. SIGGCOM*, Aug. 2002.
6. D. Pendarakis, S. Shi, D. Verma, and M. Waldgovel, "ALMI: An Application Level Multicast Infrastructure," in *Proc. USENIX Symp. on Internet Technologies and Systems*, Mar. 2001.
7. M. Castro, P. Druschel, A. -M. Kermarrec, and A. Rowstron, "SCRIBE: A Large-scale and decentralized application-level multicast infrastructure," in *IEEE JSAC*, vol. 20, no. 8, Oct. 2002.
8. N. Deo and S. L. Hakimi, "The shortest Generalized Hamiltonian Tree," in *Proc. Annual Allerton Conference*, pp. 879-888, 1968.
9. G. Zhou and M. Gen, "Application to Degree Constrained Minimum Spanning Tree Problem using Genetic Algorithm," in *Engineering Design and Automation*, vol. 3, no. 2, pp. 157-165, 1997.
10. G. Korsatz and D. Pelleg, "Generating Low-Degree 2-Spanners," in *SIAM Journal on Computing*, vol. 27, no. 5, pp. 1438-1456, Oct. 1998.
11. UCB/LBNL/VINT Network Simulator - ns (version 2), Available at www.isi.edu/nsnam/ns.

Appendix

Property 1: $-1 \leq \rho \leq 1$

Proof: Let us assume that $\rho_{ij}^s > 1$. Then, from Equation 1 we get,

$$\Delta_{is} > \Delta_{ij} + \Delta_{js} \tag{2}$$

Equation 2 shows that there is an alternate path through j which is shorter than the shortest path $i - s$. This leads to contradiction, therefore

$$\rho_{ij}^s \leq 1 \tag{3}$$

To prove the lower bound, let us assume that $\rho_{ij}^s < -1$. Therefore, from Equation 1, we get

$$\Delta_{js} > \Delta_{ij} + \Delta_{is} \tag{4}$$

Since the network is undirected, $\Delta_{ij} = \Delta_{ji}$. Therefore, Equation 4 leads to a contradiction as there exists a shorter path than the shortest from j to s through i. Therefore,

$$\rho_{ij}^s \geq -1 \tag{5}$$

Equations 3 and 5 prove the property.

Property 2: If $\rho_{ij}^s = \delta$, then $\rho_{ji}^s = -\delta$.

Proof: From Equation 1,

$$\rho_{ij}^s = \frac{\Delta_{js} - \Delta_{is}}{\Delta_{ji}} = -(\frac{\Delta_{is} - \Delta_{js}}{\Delta_{ij}}) = -\rho_{ij}^s \tag{6}$$

Equation 6 proves the Property.

Property 3: $\rho_{is}^s = 1$.

Proof: The property can be proved by substituting $\Delta_{ss} = 0$ in Equation 1.

Property 4: Let i, j and k are three nodes and source is s, $\rho_{ij}^s > 0$ and $\rho_{jk}^s > 0$, then $\rho_{ik} \geq \frac{\rho_{ij}^s \times \Delta_{ij} + \rho_{jk}^s \times \Delta_{jk}}{\Delta_{ij} + \Delta_{jk}}$.

Proof: From Equation 1,

$$\rho_{ik} = \frac{\Delta_{is} - \Delta_{ks}}{\Delta_{ik}} \geq \frac{\Delta_{is} - \Delta_{ks}}{\Delta_{ij} + \Delta_{jk}} \tag{7}$$

The Property can be proved by substituting the values of Δ_{is} and Δ_{js} from Equation 1 to Equation 7.

Property 5: If $i_j = \eta_{i_{j-1}}^s \ \forall j = 1, 2...n$, and $\rho_{i_{j-1}, i_j}^s > 0$, then $i_1, i_2 ... i_n$ cannot form a loop.

Proof: Let us assume that i, j and k are nodes such that $j = \eta_i^s$, $k = \eta_j^s$ and $i = \eta_k^s$ i.e., i, j and k form a loop. Also, $\rho_{ij}^s, \rho_{jk}^s, \rho_{ki}^s > 0$. From Equation 1, we get $\Delta_{is} > \Delta_{js}, \Delta_{js} > \Delta_{ks}$ and $\Delta_{ks} > \Delta_{is}$. This leads to contradiction, therefore such a loop cannot occur. This argument can be extended to prove the Property $\forall \ k = i_1, i_2...i_n$

Mesh Expansion & Contraction

1. If Current Fanout of i is less than the fanout limit of i goto 9. Therefore, if i can accommodate the link it will as long as n can also accommodate the link.
2. If SM^+ of i is empty goto 4. To accommodate the link, some link of i need to be removed, since SM^+ is empty, links from SM^- are searched.
3. If $\rho_{in} > \rho_{\gamma_{SM}^+}$
   a) This means that link $(i - n)$ is "better" than at least one link in, SM^+
   b) Set $ReplaceLink_i = \gamma_{SM^+}$ i.e. γ_{SM^+} is chosen as the likely candidate for removal from the mesh.
   c) Goto 9

4. Otherwise Goto 18 *i.e* current link is "good" enough, therefore the link is not added to the mesh.

5. If SM^- of i is empty Goto 7 *i.e.* SM is empty, therefore PM is searched.

6. If SM^- of i is non-empty
   a) set $ReplaceLink_i = \gamma_{SM^-}$ *i.e.* γ_{SM^-} is a likely candidate for removal.
   b) Goto 9 *i.e.* check whether n can accommodate the link.

7. If $\rho_{in} > \rho_{\gamma^+_{PM}}$
   a) Current Link is better than the link in PM^+, therefore PM^+ is the likely candidate for removal.
   b) Set $ReplaceLink_i = \gamma_{PM+}$ *i.e.* γ_{PM+} is a likely candidate of removal from the mesh.
   c) Goto 9

8. Otherwise Goto 18, *i.e.* no candidate can be found. Therefore, the link is not "good" enough.

9. If current fanout of n is less than the fanout limit of n Goto 17 *i.e.* the link can be added without removing any current link in n.

10. If n is a source node Goto 17.

11. If SM^- of n is empty Goto 13 *i.e.* n is not a source and SM^+ for candidates.

12. If $\rho_{in} < \rho_{\gamma^-_{SM}}$
    a) Current link is "better" than at least one link in SM^- of n.
    b) set $ReplaceLink_n = \gamma_{SM^-}$ *i.e.* γ_{SM^-} is the likely candidate for removal.
    c) Goto 18

13. If SM^+ of n is empty Goto 15

14. If SM^+ of n is non-empty
    a) set $ReplaceLink_n = \gamma_{SM+}$ *i.e.* γ_{SM+} is a likely candidate of removal from the mesh.
    b) Goto 18

15. If $\rho_{in} < \rho_{\gamma^-_{PM}}$
    a) Current link is "better" than at least one link in PM^- of n, therefore γ^-_{PM} is the likely candidate candidate for removal.
    b) set $ReplaceLink_n = \gamma_{PM-}$ *i.e.* γ_{PM-} is a likely candidate of removal from the mesh.
    c) Goto 18

16. Otherwise Goto 19

17. If $\Delta_{in} < \Delta_{jn}, j \in PM^+$
    a) set $ReplaceLink_n = j$ *i.e.* j is a likely candidate of removal from the mesh. Here Δ is used as a parameter instead of ρ because, in this case $\rho_{in} = 1$ (Property 3).
    b) Goto 18

18. The current link is added
    a) Link $(i - n)$ is added to the secondary mesh
    b) Links $ReplaceLink_i$ and $ReplaceLink_n$ are removed from the mesh of i and n respectively.

19. Exit

On Core Selection Algorithm for Reducing Delay Variation of Many-to-Many Multicasts with Delay-Bounds*

Moonseong Kim[1], Young-Cheol Bang[2], and Hyunseung Choo[1]

[1] School of Information and Communication Engineering
Sungkyunkwan University
440-746, Suwon, Korea +82-31-290-7145
{moonseong,choo}@ece.skku.ac.kr
[2] Department of Computer Engineering
Korea Polytechnic University
429-793, Gyeonggi-Do, Korea +82-31-496-8292
ybang@kpu.ac.kr

Abstract. With the proliferation of multimedia group applications, the construction of multicast trees satisfying the quality of service (QoS) requirements is becoming a problem of the prime importance. In this paper, we study the core selection problem that should produce the improved delay-bounded multicast tree in terms of the delay variation that is known to be NP-complete [8]. A solution to this problem is required to provide decent real-time communication services such as on-line games, shopping, and teleconferencing. Performance comparison shows that our proposed scheme outperforms that of DDVCA [18] that is known to be most effective so far in any network topology. The enhancement is up to about 11.1% in terms of normalized surcharge for DDVCA. The time complexity of our algorithm is $O(mn^2)$.

1 Introduction

New communication services involving multicast communications and real time multimedia applications are becoming prevalent. In multicast communications, messages are sent to multiple destinations that belong to the same multicast group. These group applications demand a certain amount of reserved resources to satisfy their quality of service (QoS) requirements such as end-to-end delay, delay jitter, loss, cost, throughputs, and etc. Since resources for multicast tree are reserved along a given path to each destination in a given multicast tree, it may fail to construct a multicast tree to guarantee the required QoS if a single link cannot support required resources. Thus an efficient solution for multicast communications includes the construction of a multicast tree that has the best chance to satisfy the resource requirements [1,4,9,10,11,12,19,21,22].

* This paper was supported in part by Brain Korea 21 and University ITRC project. Dr. H. Choo is the corresponding author.

N. Mitrou et al. (Eds.): NETWORKING 2004, LNCS 3042, pp. 200–210, 2004.

Algorithms for the tree construction in multicast protocols can be categorized as follows: source-based algorithms (SBA) and core-based algorithms (CBA) [20]. SBA constructs a tree rooted at source that originates and sends messages to each destination in the multicast group. SBA is currently being used as the tree construction algorithm for Distance Vector Multicast Routing Protocol (DVMRP) [14], Protocol Independent Multicast Dense Mode (PIM-DM) [7], and Multicast Open Shortest Path First (MOSPF) [13]. On the other hand, CBA that is used for many-to-may multicasts selects a core node as a root of the multicast tree we want to determine at the first step. Then a tree rooted at the core node is constructed to span all members in the multicast group. Thus it is very important to select the best core node as much as possible. To send messages originated at source, messages are sent to the core and distributed to destinations along the path to the core node. Once messages are reached at the core node, messages are sent to remaining destinations. Multicast protocols that use CBA as a tree construction algorithm include Protocol Independent Multicast Sparse Mode (PIM-SM) [5,7] and the Core-Based Tree (CBT) protocol [2,3].

For multicast communications such as a teleconference, it is very critical that the current speaker must be heard by all participants simultaneously, but otherwise the communication may lose the feeling of an interactive face-to-face discussion. Another similar dispute can be easily found in on-line video games. These are all related to the multicast delay variation problem [17]. In this paper, we introduce a novel core selection algorithm that can improve the delay and delay variation constraint algorithm (DDVCA) known to be the best algorithm [18]. In the DDVCA, the selection of a core node over several candidates (possible core nodes) is overlooked in the a core node is randomly selected among candidates. Meanwhile we investigate candidate nodes to select the better node with the same time complexity of DDVCA. Our algorithm with the tree construction part of DDVCA was empirically compared with DDVCA, and evaluation determined that the delay variation of the tree constructed by our algorithm is smaller than that of the tree constructed by DDVCA. The enhancement is up to about 3.6%~11.1% in terms of the normalized surcharge for DDVCA. Our main contribution in this research is as follows. We propose a new core selection algorithm that produces the multicast tree with the better delay variation in comparison with DDVCA. The time complexity of our algorithm is the same as that of the DDVCA.

The rest of the paper is organized as follows. In section 2, we study related works, and section 3 presents details of the proposed algorithm. Then, in section 4, we evaluate the proposed algorithm by the simulation model. Section 5 concludes this paper.

2   Related Works

We consider a computer network represented by a directed graph $G = (V, E)$ with n nodes and l links or arcs, where V is a set of nodes and E is a set of links (arcs), respectively. Each link $(i, j) \in E$ is associated with delay $d_{(i,j)}$. The delay of a link is the sum of the perceived queueing delay, transmission delay, and propagation delay over that link. We assume that the delay on each arc is asymmetric in general. Given a network G, we define a path as sequence of nodes u, i, j, $\ldots$, k, v, such that (u, i), (i, j), $\ldots$, (k, v), belongs to E. Let $P(u, v) = \{(u, i), (i, j), \ldots, (k, v)\}$ denote the path from node u to node v. If all elements of the path are distinct, then we say that it is a simple path. We define the length of the path $P(u, v)$, denoted by $n(P(u, v))$, as a number of links in $P(u, v)$. Let $\preceq$ be a binary relation on $P(u, v)$ defined by $(a, b) \preceq (c, d) \leftrightarrow n(P(u, b)) \leq n(P(u, d))$, $^\forall (a, b)$, $(c, d) \in P(u, v)$. $(P(u, v), \preceq)$ is a totally ordered set. For a given source node $s \in V$ and a destination node $d \in V$, $(2^{s \Rightarrow d}, \infty)$ is the set of all possible paths from s to d.

$$(2^{s \Rightarrow d}, \infty) = \{ \ P_k(s, d) \mid \text{all possible paths from } s \text{ to } d, \ ^\forall s, \ d \in V, \ ^\forall k \in \Lambda \ \}$$

where Λ is a index set. The delay of arbitrary path P_k is assumed to be real function from $(2^{s \Rightarrow d}, \infty)$ to nonnegative real number $\mathcal{R}^+$. Since $(P_k, \preceq)$ is a totally ordered set, if there exists a bijective function f_k then P_k is isomorphic to $\mathcal{N}_{n(P_k)}$.

$$P_k = \{(u, i), (i, j), \ldots, (m, v)\} \xrightarrow{f_k} \mathcal{N}_{n(P_k)} = \{1, 2, \ldots, n(P_k)\}$$

We define,

$$\text{function of delay along the path } \phi_D(P_k) = \sum_{r=1}^{n(P_k)} d_{f_k^{-1}(r)} , \ ^\forall P_k \in (2^{s \Rightarrow d}, \infty) \ .$$

$(2^{s \Rightarrow d}, supD)$ is the set of paths from s to d for which the end-to-end delay is bounded by $supD$. Therefore $(2^{s \Rightarrow d}, supD) \subseteq (2^{s \Rightarrow d}, \infty)$. For multicast communications, messages need to be delivered to all receivers in the set $M \subseteq V \setminus \{s\}$ which is called multicast group, where $|M| = m$. The path traversed by messages from the source s to a multicast receiver, m_i, is given by $P(s, m_i)$. Thus multicast routing tree can be defined as $T(s, M) = \bigcup_{m_i \in M} P(s, m_i)$, and messages are sent from s to destination of M using $T(s, M)$.

In the following we now introduce two important qualities of service metrics in multicast communications [17]. The multicast end-to-end delay constraint, $supD$, represents an upper bound on the acceptable end-to-end delay along any path from the source to a destination node. This metric reflects the fact that the information carried by the multicast messages becomes stale $supD$ time units after its transmission at the source. The multicast delay variation, δ, is the maximum difference between the end-to-end delays along the paths from the source to any two destination nodes.

$$\delta = max\{ \ |\phi_D(P(s, m_i)) - \phi_D(P(s, m_j))|, \ ^\forall m_i, m_j \in M, \ i \neq j \ \}$$

The issue first defined and discussed in [17] is of minimizing multicast delay variation under multicast end-to-end delay constraint. The authors referred to this problem as Delay- and Delay Variation-Bounded Multicast Tree (DVBMT) problem. The DVBMT problem is to find the tree that satisfies

$$min\{ \ \delta_\alpha \ | \ ^\forall m_i \in M, \ ^\forall P(s, m_i) \in (2^{s \Rightarrow m_i}, supD), \ ^\forall P(s, m_i) \subseteq T_\alpha, \ ^\forall \alpha \in \Lambda \ \}$$

where T_α denotes any multicast tree spanning $M \cup \{s\}$, and is known to be NP-complete [17]. There are two well known approaches to construct multicast tree for the DVBMT problem. One is DVMA (Delay Variation Multicast Algorithm) [17] and the other DDVCA [18]. The algorithm DDVCA proposed by Pi-Rong Sheu and Shan-Tai Chen is based on the Core Based Tree (CBT) [2,3]. It has been shown that DDVCA outperforms DVMA in terms of the delay variation of the constructed tree. Moreover, the time complexities of algorithms are $O(mn^2)$ and $O(klmn^4)$ for DDVCA and DVMA, where k and l is the number of paths at line 3 and line 11 in Fig. 3, Ref. [17], respectively; m represents the number of destination nodes, while n represents the number of nodes in the computer network.

3   Description of the Proposed Algorithm

In this section, we describe our proposed novel algorithm to construct multicast trees that is superior to DDVCA. In order to define a multicast tree, the basic idea of the proposed algorithm is based on CBT [2,3]. The method used in CBT for the establishment of a multicast tree is first to choose some core routers which compose the backbone. We also select a core router addressed as a core node.

3.1   The Basic Concept

The goal of this paper is to propose an algorithm which produces multicast trees with low multicast delay variation. In this subsection, we present our proposed algorithm. The proposed algorithm consists of a core node selection part and the multicast tree construction part. Hence we take an interest in a core node selection. When candidate of core node is several nodes, the DDVCA randomly choose a core node among candidates but the proposed algorithm presents lucid solution. To implement the our idea, the following data structures are employed.

- *Input* : A directed graph $G(V, E)$, M is the multicast group with $m = |M|$, a source node s, a end-to-end delay bound $supD$

- *Output* : The multicast tree T such that $\phi_D(P(s, m_i)) \leq supD$, $^\forall P(s, m_i) \subseteq T$, $^\forall m_i \in M$, and has a small multicast delay variation

- *candidate* : the candidates of core node

- *compare* : the max difference delay between core nodes and visited destinations

- $Dij(m_k, v_i)$ = Calculate the minimum delay between m_k and v_i

- $pass(s, v_i, m_k)$: $Dij(s, m_k)$ when any destination node m_k is visited in the path from s to v_i

- $max_i = max\{Dij(v_k, v_i) \mid {}^\forall v_k \in M\}$, ${}^\forall v_i \in V$

- $min_i = min\{Dij(v_k, v_i) \mid {}^\forall v_k \in M\}$, ${}^\forall v_i \in V$

- $diff_i = max_i - min_i$

In selecting such a core node, we use the minimum delay path algorithm [6]. Fig. 1 checks whether any destination node is visited in the path from source node to each other node. If any destination node is visited, then the proposed algorithm records in 'pass' data structure.

For $\forall l \in \{$the minimum delay path from s to $v_i\}$ Do
$/ * l$: the nodes in minimum delay path form s to v_i $* /$
    If $l = m_k$, $\forall m_k \in M$
      then $pass(s, v_i, m_k) = Dij(s, m_k)$
      else $pass(s, v_i, m_k) = 0$
    If $diff_i < diff_{min}$ and $Dij(s, v_k) + max_i \le supD$
      then $diff_{min} = diff_i$; $c = i$

Fig. 1. A partial amendment of the DDVCA - I

Fig. 2 conforms $supD$ and select nodes with the minimum delay variation as the candidates of core node. Next, our algorithm chooses the core node with $min\{ \phi_D(P(s, c_i)) - min\{ pass(s, c_i, m_j) \} \}$ in Fig. 3.

For $\forall v_i \in V$ Do
    If $diff_i = diff_c$ and $Dij(s, v_i) + max_i \le supD$
      then $candidate = candidate \cup v_i$

Fig. 2. A partial amendment of the DDVCA - II

As you shown in Fig. 4, our algorithm overcomes DDVCA's weaknesses. In construction of a multicast tree, we follow the DDVCA. The time complexity of the proposed algorithm is $O(mn^2)$, which matches the complexity of the DDVCA.

For $\forall c_i \in candidate$ Do

    If $pass(s,c_i,m_k) = 0$, for every $m_k \in M$

      then $compare_i = Dij(s,c_i) - min\{pass(s,c_i,m_j) | \text{positive and } \forall m_j \in M\}$

$c = min\{i | compare_i\}$

Fig. 3. A partial amendment of the DDVCA - III

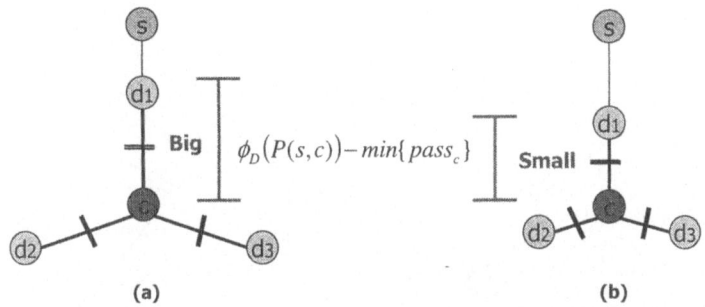

Fig. 4. The basic concept of the proposed algorithm

3.2 An Illustration

In this subsection, we illustrate an example with explanation of the algorithm. Fig. 5 (a) shows a given network topology with link delays specified on each link. Suppose that the multicast end-to-end delay constraint $supD$ is 11. Fig. 5 (b) represents the ultimate multicast tree obtained by the DDVCA. Fig. 5 (c) shows the path constructed by the proposed algorithm.

Table 1. The method by which proposed algorithm selects a core node

		v_1	v_2	v_3	v_4	v_5	v_6	v_7	v_8
source	v_1	0	2	8	8	4	9	3	6
pass	v_5	0	0	0	4	4	4	0	4
	v_6	0	0	0	0	0	9	0	0
destination	v_5	4	2	5	4	0	5	2	2
	v_6	9	7	6	3	5	0	7	3
max_i		9	7	6	4	5	5	7	3
min_i		4	2	5	3	0	0	2	2
$diff_i$		5	5	1	1	5	5	5	1
$compare_i$					4				2

From Table 1, we know that nodes with the smallest multicast delay variation are v_3, v_4, and v_8. However, since we must consider the delay bound $supD$, the node v_3 is ignored. The DDVCA randomly selects the node v_4, but the proposed

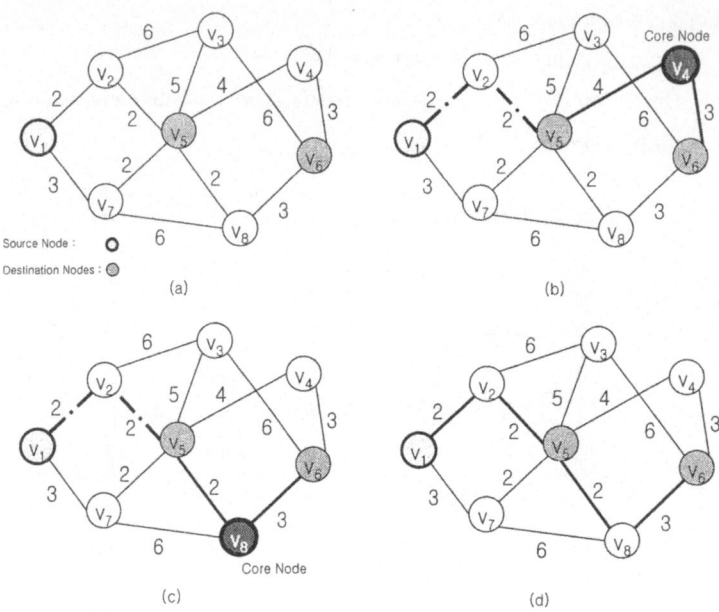

Fig. 5. (a) A Given network $G(V, E)$ and link delays are shown to each link, (b) DDVCA and $\delta_{DDVCA} = 7$, (c) Proposed Algorithm and $\delta_{Proposed} = 5$, (d) Optimal tree and $\delta_{Optimum} = 5$

algorithm selects the node v_8 as a core node. Because the proposed algorithm calculates the minimum among $compare_{v_4} = 8 - 4 = 4$ and $compare_{v_8} = 6 - 4 = 2$ in Fig. 3, we take the node v_8 as core node. Finally, the DDVCA's multicast delay variation is 7, but the proposed algorithm's multicast delay variation is 5. Fig. 6

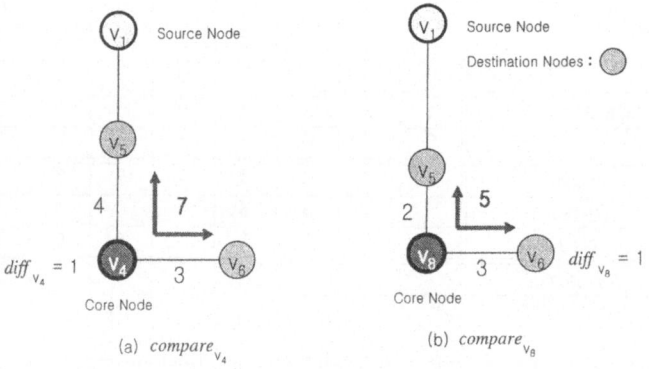

Fig. 6. The candidate nodes

shows that proposed algorithm chooses v_8 in Fig. 6 (b) in case of the same $diff_i$ value when it selects a core node.

4   Experimental Results

We compare our proposed algorithm with the DDVCA in terms of multicast delay variation. We describe the generation of random network topologies for the evaluation and the simulation results based on the network topology generated.

4.1   Random Graph Generation and Simulation Parameters

The details of the generation for random network topologies are as follows. The method uses parameters n - the number of nodes in networks, and P_e - the probability of edge existence between any node pair [15,16]. Let us remark that if a random graph models a random network then this graph should be connected. Hence, the graph should contain at least a spanning tree. So, firstly a random spanning tree is generated. As we know, we consider cases for $n \geq 3$. A tree with 3 nodes is unique, and thus we use this as an initial tree. And we expand to a spanning tree with n nodes. After adjusting the probability P_e, we generate other non-tree edges at random for the graph based network topology. Let us calculate the adjusted probability P_e^a. By $Prob\{event\}$ denote a probability of the event. Suppose e is a possible edge between a couple of nodes, then we have

$$P_e = Prob\{\ e \in spanning\ tree\ \} + Prob\{\ e \notin spanning\ tree\ \} \cdot P_e^a$$
$$P_e = \frac{n-1}{n(n-1)/2} + (1 - \frac{n-1}{n(n-1)/2}) \cdot P_e^a$$
$$\therefore\ P_e^a = \frac{nP_e - 2}{n-2}\ .$$

Let us describe a pseudo code for random network topologies. Here A is an incident matrix, r is a simple variable, and $random()$ is a function producing uniformly distributed random values between 0 and 1.

Graph Generation Algorithm
  Begin
  $A_{1,2} = A_{2,1} = A_{2,3} = A_{3,2} = 1$
  For $i = 4$ to n **Do**
      $r = (i - 1) \times random() + 1$
      $A_{r,i} = A_{i,r} = 1$
  For $i = 1$ to $(n - 1)$ **Do**
      For $j = (i + 1)$ to n **Do**
          If $P_e > random()$ **Then** $A_{i,j} = A_{j,i} = 1$
  End Algorithm.

4.2 Discussion of Results

We now describe some numerical results with which we compare the performance of the proposed scheme. The proposed algorithm is implemented in C++. We randomly selected a source node. We generate 10 different networks for each size of given 50, 100, and 200. The destination nodes are picked uniformly from the set of nodes in the network topology (excluding the nodes already selected for the destination). Moreover, the destination nodes in the multicast group will occupy 10, 20, 30, 40, 50, and 60% of the overall nodes on the network, respectively. We randomly choose $supD$ such that $supD \geq min\{\phi_D(P_k(s,m)) \mid s$ is source node, $\forall m \in M, \forall k \in \Lambda\}$. Delays are uniformly random integer values between 0 and 10. We simulate 1000 times $(10 \times 100 = 1000)$ for each $|V|$ and $P_e = 0.3$.

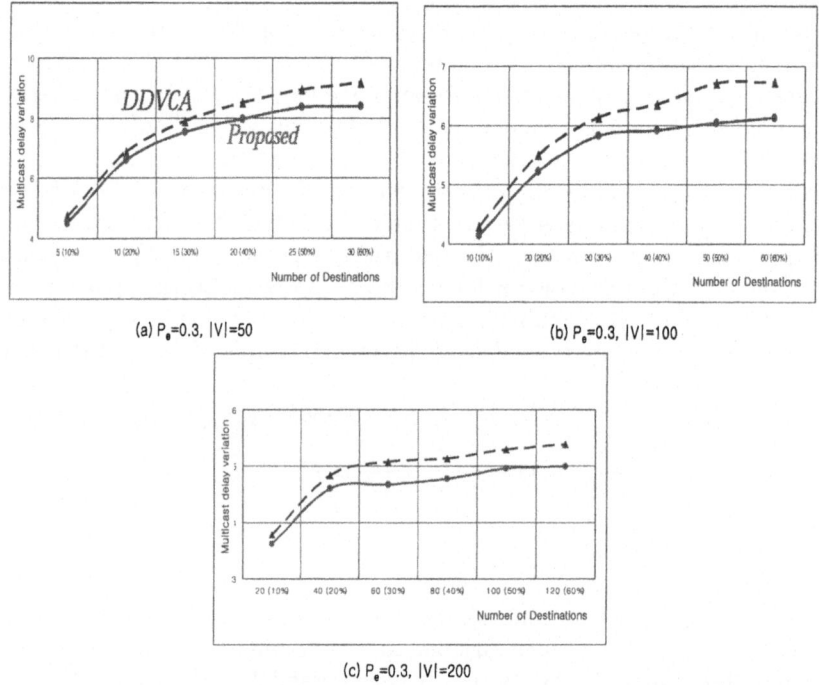

(a) P_e=0.3, $|V|$=50   (b) P_e=0.3, $|V|$=100

(c) P_e=0.3, $|V|$=200

Fig. 7. The multicast delay variations of the three different networks

For the performance comparison, we implement the DDVCA in the same simulation environment. We use the normalized surcharge, introduced in [11], of the algorithm with respect to our method defined as follows:

$$\bar{\delta} = \frac{\delta_{DDVCA} - \delta_{Proposed}}{\delta_{Proposed}}.$$

In our plotting, we express this as a percentage, *i.e.*, $\bar{\delta}$ is multiplied by 100. Fig. 7 (a), (b), and (c) show the simulation results of multicast delay variations.

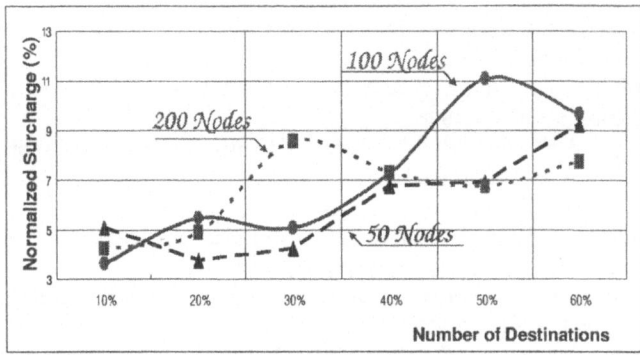

Fig. 8. Normalized Surcharges versus number of nodes in networks

As indicated in Fig. 8, it is easily noticed that the proposed algorithm is always better than the DDVCA. The enhancement is up to about 3.6%~11.1% in terms of normalized surcharge for the DDVCA.

5   Conclusion

In this paper, we consider the transmission of a message that guarantees certain bounds on the end-to-end delays from a source to a set of destinations as well as on the multicast delay variations among these delays over a computer network. There are two well known approaches for constructing a multicast tree with the DVBMT problem, which is known to be NP-complete. The one is the DVMA [17]. Although it provides smart performance in terms of the multicast delay variation, its time complexity is as high as $O(klmn^4)$. As we all know, a high time complexity dose not fit in large scale high speed networks. The other is the DDVCA [18]. It has been shown that the DDVCA outperforms the DVMA slightly in terms of the multicast delay variation for the constructed tree. Moreover, the time complexity of the DDVCA is $O(mn^2)$. In the meantime, the time complexity of the proposed algorithm is $O(mn^2)$, which is the same as that of the DDVCA. Furthermore, the comprehensive computer simulation results show that the proposed scheme obtains the better minimum multicast delay variation than the DDVCA.

Future work is that we examine the problem of constructing minimum cost multicast trees which guarantees certain bounds on the end-to-end delay from the source to the destination nodes and the inter-destination delay variations between paths from the source to the destination nodes.

References

1. K. Bharath-Kumar and J. M. Jaffe, "Routing to multiple destinations in computer networks," IEEE Trans. Commun., vol. COMM-31, no. 3, pp. 343-351, March 1983.

2. T. Ballardie, P. Francis, and J. Crowcroft, "Core based trees (CBT) : An architecture for scalable inter-domain multicast routing," Computer Commun. Rev., vol. 23, no. 4, pp. 85-95, 1993.
3. A. Ballardie, B. Cain, and Z. Zhang, "Core Based Trees (CBT Version 3) Multicast Routing," Internet draft, 1998.
4. Y.-C. Bang and H. Choo, "On multicasting with minimum costs for the Internet topology," Springer-Verlag Lecture Notes in Computer Science, vol. 2400, pp.736-744, August 2002.
5. S. Deering et al., "Protocol Independent Multicast-Sparse Mode (PIM-SM): Motivation and Architecture," Internet draft, 1998.
6. E. W. Dijkstra, "A note on two problems in connexion with graphs," Numerische Mathematik, vol. 1, pp. 269-271, 1959.
7. D. Estrin et al., "Protocol Independent Multicast (PIM) Sparse Mode/Dense Mode," Internet draft, 1996.
8. M. R. Garey, R. L. Graham, and D. S. Johnson, "The complexity of computing steiner minimal trees," SIAM J. Appl. Math., vol. 32, no. 4, pp. 835-859, June 1977.
9. E. N. Gilbert and H. O. Pollak, "Steiner minimal tree," SIAM J. Appl. Math., vol. 16, 1968.
10. S. L. Hakimi, "Steiner's problem in graphs and its implication," Networks, vol. 1, pp. 113-133, 1971.
11. V. P. Kompella, J. C. Pasquale, and G. C. Polyzos, "Multicast routing for multimedia communication," IEEE/ACM Trans. Networking, vol. 1, no. 3, pp. 286-292, June 1993.
12. L. Kou, G. Markowsky, and L. Berman, "A fast algorithm for steiner trees," Acta Informatica, vol. 15, pp. 141-145, 1981.
13. J. Moy, "Multicast Extension to OSPF," Internet draft, 1998.
14. T. Pusateri, "Distance Vector Routing Protocol", draft-ietf-idmr-dvmrp-v3-07, 1998.
15. A.S. Rodionov and H. Choo, "On generating random network structures: Trees," Springer-Verlag Lecture Notes in Computer Science, vol. 2658, pp. 879-887, June 2003.
16. A.S. Rodionov and H. Choo, "On generating random network structures: Connected Graphs," International Conference on Information Networking 2004, Proc. ICOIN-18, pp. 1145-1152, February 2003.
17. G. N. Rouskas and I. Baldine, "Multicast routing with end-to-end delay and delay variation constraints," IEEE JSAC, vol. 15, no. 3, pp. 346-356, April 1997.
18. P.-R. Sheu and S.-T. Chen, "A fast and efficient heuristic algorithm for the delay-and delay variation bound multicast tree problem," Information Networking, Proc. ICOIN-15, pp. 611-618, January 2001.
19. H. Takahashi and A. Matsuyame, "An approximate solution for the steiner problem in graphs," Mathematica Japonica, vol. 24, no. 6, pp. 573-577, 1980.
20. B. Wang and J. C. Hou, "Multicast Routing and its QoS Extension: Problems, Algorithms, and Protocols," IEEE Networks, Jan./Feb, 2000.
21. B. W. Waxman, "Routing of multipoint connections," IEEE JSAC, vol. 6, no. 9, pp. 1617-1622, December 1988
22. Q. Zhu, M. Parsa, and J. J. Garcia-Luna-Aceves, "A source-based algorithm for near-optimum delay-constrained multicasting," Proc. IEEE INFOCOM'95, pp. 377-385, March 1995.

Overlay Multicast Tree
Minimizing Average Time Delay

Hwangjun Song and Dong Sup Lee

School of Electrical Engineering, Hongik University,
72-1 Sangsudong Mapogu Seoul, Korea 121-791.
hwangjun@wow.hongik.ac.kr

Abstract. In this work, we present an overlay multicast tree constructing algorithm to minimize the average time delay from the sender to end-systems. At the same time, the proposed algorithm considers the computing power and the network condition of each end-system as a control variable and thus we can avoid the disastrous case that loads are concentrated to only several end-systems. The multicast tree is constructed by clustering technique and modified Dijkstra's algorithm in two steps, i.e. tree among proxy-senders and tree in each cluster. By the experimental results, we show that the proposed algorithm can provide an effective solution.

1 Introduction

Recently, Internet plays an important role in multimedia communication area and becomes an important way to obtain the information. It can be understood that the number of end-systems (or hosts) connected to the Internet has been exponentially increasing. Furthermore, the demand of various multimedia services through the Internet has been increasing very fast. Internet architectures can be traditionally divided into two entities, i.e. end-systems called hosts and the network consisting of routers and switches. Generally speaking, QoS (quality of service) and multicast are the most important features that should be added to support various multimedia services and increase network utilization [1]. Since QoS functionality can not be serviced without the help of IP layer, many research efforts have been devoted to QoS of IP layer. So far, various Internet protocols such as DiffServ, IntServ and RSVP have been proposed. Multicast has gained a large amount of interests after IP multicast [2] was proposed. Many research efforts have been focused on IP multicast, that is, IP router supporting multicast function. However, IP multicast has not been widely employed so far since most of current routers do not identify the class D of IP addresses. To fully support IP multicast, all routers in the worldwide Internet must maintain the multicast function. It is unlikely that IP multicast will be widely supported soon or later.

As an alternative of IP multicast, overlay multicast has recently proposed to realize the multicast over the current IP network that does not support multicast functional-

N. Mitrou et al. (Eds.): NETWORKING 2004, LNCS 3042, pp. 211–222, 2004.

ity. While overlay multicast does not need any new additional modification in IP routers, some of end-systems in the multicast group have to replace the multicast function of IP routers.

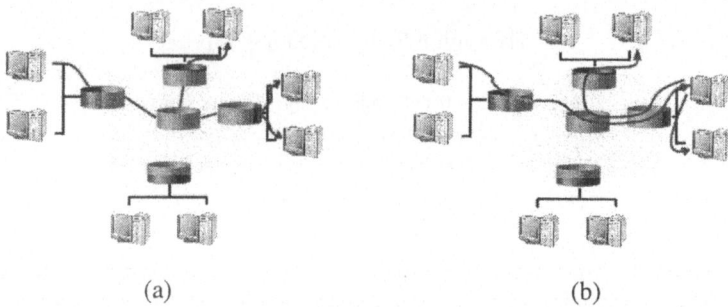

(a)   (b)

Fig. 1. Comparison between IP multicast and overlay multicast: (a) IP multicast and (b) overlay multicast.

Recently, overlay multicast becomes more feasible since the computing power of end-system becomes very strong due to the fast development of VLSI hardware technology and the network condition of end-system has been rapidly improved due to the fast progress of digital communication and network technology. If the end-system's resources such as the computing power, the storage devices, and network condition are efficiently used, the utilization of the Internet can be improved. The architectures of IP multicast and overlay multicast are shown in (a) and (b) of Figure 1, respectively.

As shown in (a) of Figure 1, router plays an important role to realize the multicast. Routers classify incoming multicast packets and transmit them to routers and end-systems participating in the multicast group only a time. Thus, we don't need to transmit the same packet several times. Many IP multicast protocols can be found in the literature [10]. On the other hand, overlay multicast does not need any modification in routers as shown in (b) of Figure 1. Instead, some end-systems participating in the multicast replace IP routers. So far, many algorithms have been proposed to construct and manage the overlay multicast tree. Y. Chu and et al [1] proposed Narada that constructs an overlay structure among participating end-systems in a self-organizing and fully distributed manner, Liebeherr and Nahas [3] proposed application-layer multicast with delaunay triangulations that each application locally derive next hop routing information without the need for a routing protocol in the overlay, Mathy and et al [4] studied a method to build a hierarchy of nodes, based on the notion of proximity, in a distributed and scalable way, and Park and et al proposed a realistic scheme that is based on the unicast transport from a remote sender to a local subnet and the multicast forwarding to receivers within the subnet in [6]. In addition, many effective algorithms have been proposed [7, 8, 9, 12].

While overlay multicast has many advantages as mentioned previously, there are still several problems to be considered. First of all, overlay multicast can not perform as well as IP multicast, that is, more redundant traffics over physical link are inevitable compared with IP multicast. But, it is still smaller than multiple unicasts. Thus, it may be acceptable. Secondly, more serious problem is the time delay. The time delay is increased compared with IP multicast since traffic must traverse several end-systems until it arrives at the final destination [1]. The increased time delay can be an obstacle for the real-time media delivery. In this paper, we consider an overlay multi-cast tree constructing algorithm to minimize the average time delay from the sender to the multicast members for the effective real-time media delivery. This paper is organized as follows. Problem formulation and effective multicast tree constructing algorithm are proposed in Section 2, experimental results are provided in Section 3 to show the superior performance of the proposed algorithm, and finally concluding remarks are presented in Section 4.

2   The Proposed Overlay Multicast Tree Constructing Algorithm

As mentioned earlier, we study the overlay multicast tree constructing algorithm to minimize the average time delay from sender to end-systems. RTT (round trip time) is employed as a measure of time delay, and we treat the computing capability and network condition of each end-system as a control variable, that is, the maximum number of streams that each end-system can support is determined considering its limited resources. It is one of the unique features of the proposed algorithm. First of all, we make the following assumptions.

① RTT values among multicast group members are known by using ping commands.
② The number of streams handled by each end-system is advertised to all members.

Under these assumptions, it is overlay multicast impossible to find the optimal multicast tree to minimize the average time delay from sender to end-systems since we have to investigate the full mesh cases since any two end-systems can be connected over IP layer and the number of possible multicast trees increases exponentially in terms of the number of end-systems.

2.1   Problem Formulation

In this scenario, end-systems participating in the multicast are divided into several clusters based on RTT values, and then one of end-systems in each cluster is selected as a proxy-sender that plays a role as a IP router and other end-systems in the cluster receive data from the proxy-sender. The reason is that it is more efficient for a proxy-sender to receive a stream from the remote original sender (or parent proxy-sender) and then distribute it to others when they are closely located than every end-system receives a stream from the remote sender with finite capability. After then, the effective tree among proxy-senders and tree in each cluster are constructed to minimize the time delay (An example constructed by the proposed algorithm is shown in Figure

2.). In this case, the RTT values experienced by a proxy-sender and an end-system can be described as follows.

$$RTT_{p_i}^s = \begin{cases} RTT_{p_i}^s \text{ if a proxy sender directly received a stream from sender,} \\ RTT_{p_1}^s + RTT_{p_2}^{p_1} + \cdots + RTT_{p_k}^{p_{k-1}} + RTT_{p_i}^{p_k} \text{ otherwise,} \end{cases}$$

Where $RTT_{p_i}^s$ is the RTT of the proxy-sender p_i and $RTT_{p_k}^{p_{k-1}}$ is the RTT between two adjacent proxy-senders on the path from the sender and the final proxy-sender. And the RTT experienced by an arbitrary end-system is described by

$$RTT_j^s = RTT_{p_i}^s + RTT_j^{p_i},$$

where $RTT_j^{p_i}$ is the RTT between the j_{th} end-system and the proxy-sender p_i in the i_{th} cluster. Then, the total sum of RTT values is

$$\sum_{i=1}^{n_p} RTT_{p_i}^s + \sum_{i=1}^{n_p} \sum_{j=1}^{m_i-1} \left(RTT_{p_i}^s + RTT_j^{p_i} \right),$$

where n_p is the number of clusters and m_i is the number of end-systems in the i_{th} cluster. Hence, the average RTT value can be expressed by the following Eq. 1. Now, we can formulate our problem to minimize the average RTT as follows.

Problem Formulation: Determine the number of clusters and trees to minimize

$$\frac{1}{N}\sum_{i=1}^{n_p} m_i \cdot RTT_{p_i}^s + \frac{1}{N}\sum_{i=1}^{n_p}\sum_{j=1}^{m_i-1} RTT_j^{p_i}, \qquad (1)$$

subject to $s_s \le ST_s$ and $s_i \le ST_i$,

where N is the number of end-systems participating the multicast ($N = \sum_{i=1}^{n_p} m_i$), s_s and s_i are the numbers of streams handled by the sender and the i_{th} end-system re-

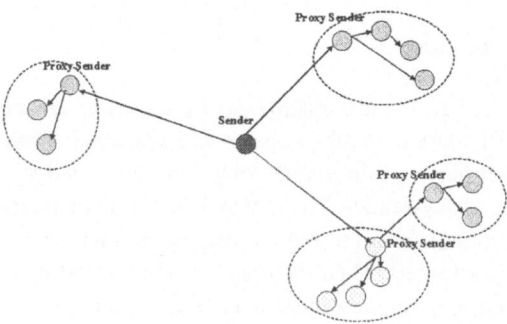

Fig. 2. An example of overlay multicast tree under consideration.

spectively, and ST_s and ST_i are the maximum numbers of streams that can be supported by the sender and the i_{th} end-system, respectively. We have to investigate the full meshes among all end-systems to find the optimal solution. The number of possible trees increases exponentially in terms of the number of end-systems, thus it is overlay multicast impossible to find the optimal solution. To reduce the required computational complexity, we neglect the dependency between two terms in Eq. 1. Actually, the dependency could be negligible if $RTT_{p_i}^s \gg RTT_j^{p_i}$. Under this assumption, we can simplify the above problem as follows.

Simplified Problem Formulation: Determine the number of clusters and a tree among clusters to minimize

$$\frac{1}{N}\sum_{i=1}^{n_p} m_i \cdot RTT_{p_i}^s \text{ subject to } s_s \le ST_s, \tag{2}$$

and then determine tree in each cluster to minimize

$$\sum_{j=1}^{m_i-1} RTT_j^{p_i} \text{ subject to } s_i \le ST_i, \text{ for } 1 \le i \le n_p. \tag{3}$$

It means that the multicast tree is constructed by two steps, i.e. a tree among proxy-senders minimizing the average RTT value weighted by the number of end-systems in each cluster and trees minimizing average RTT value in all clusters. To obtain the optimal solution of simplified problem, the followings must be taken into account: how to cluster the end-systems, how to choose the proxy-sender in each cluster, how to construct among proxy-senders, and how to construct the trees in the clusters. They are described in detail in the followings.

2.2 Clustering and Tree Constructing Algorithm

In this section, we describe clustering algorithm, selection of proxy-sender in each cluster, and the modified Dijkstra's algorithm to construct effective trees among proxy-senders and in each cluster in detail.

2.2.1 Clustering Algorithm and Proxy-Sender Selection

How to cluster the multicast members and select a proxy-sender in each cluster is greatly related to the performance of the proposed algorithm. The *k-mean clustering* algorithm [9], one of the most popular techniques in digital image processing, is employed in this paper. At the same time, we have to consider the number of streams that can be supported by sender when the number of clusters is determined. In this paper, the number of clusters is determined by iterative approach as shown in Figure 5.

To find the optimal selection of proxy-sender, we have to consider the two terms in Eq. 1 simultaneously. Thus, it is difficult to select a proxy-sender in each cluster to minimize the average of RTT values since the number of possible cases is too large. During the experiment, it is observed that it is optimal that an end-system with the minimum RTT from sender and maximum number of streams is chosen as a proxy-sender, and it is very close to the optimal solution that an end-system that is the closest to the sender and can support as many streams as other end-systems is selected as a proxy-sender. In this paper, we choose the end-system closest to sender/parent proxy-sender as a proxy-sender under the assumption that all members can support overlay multicast same number of streams.

2.2.2   Tree Construction by Using the Modified *Dijkstra's* Algorithm

To construct the tree to minimize the average RTT, *Dijkstra's* algorithm [8] can be employed, which finds the shortest path from a source node to all other nodes in a network. However, it can not be directly applicable to our problem since the number of streams may be greater than the number of streams that can be handled by an end-system. Furthermore, the link cost between sender and proxy-sender must be adjusted by the cluster size since the sum of RTT values in a cluster with m_i end-systems increases by $m_i \cdot \Delta RTT$ if the RTT value of proxy-sender is increased by ΔRTT.

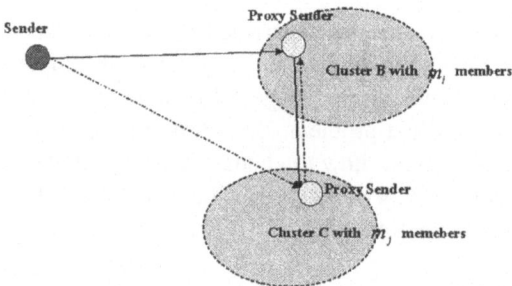

Fig. 3. The optimal selection of proxy-sender when two clusters are located closely.

Therefore, the RTT values between sender and proxy-sender must be weighted by $1/m_i$ before the *Dijkstra's* algorithm is applied. Basically, the proxy-sender with the minimum weighted RTT is connected to sender or parent proxy-sender to minimize the sum of RTT values. However, it is not always the best choice. More consideration is sometimes required when two clusters are closely located as shown in Figure 3. Under the assumption that $\dfrac{RTT_{p_i}^s}{m_i} < \dfrac{RTT_{p_j}^s}{m_j}$ and the RTT between two proxy-senders is $RTT_{p_j}^{p_i}$ (It is assumed that $RTT_{p_i}^{p_i}$ is equal to $RTT_{p_j}^{p_i}$.), we consider which proxy-sender must be chosen to minimize the average RTT value. The sum of RTT values of the solid line case in Figure 3 is $m_i \cdot RTT_{p_i}^s + m_j \cdot \left(RTT_{p_i}^s + RTT_{p_j}^{p_i}\right)$ while that of the dotted

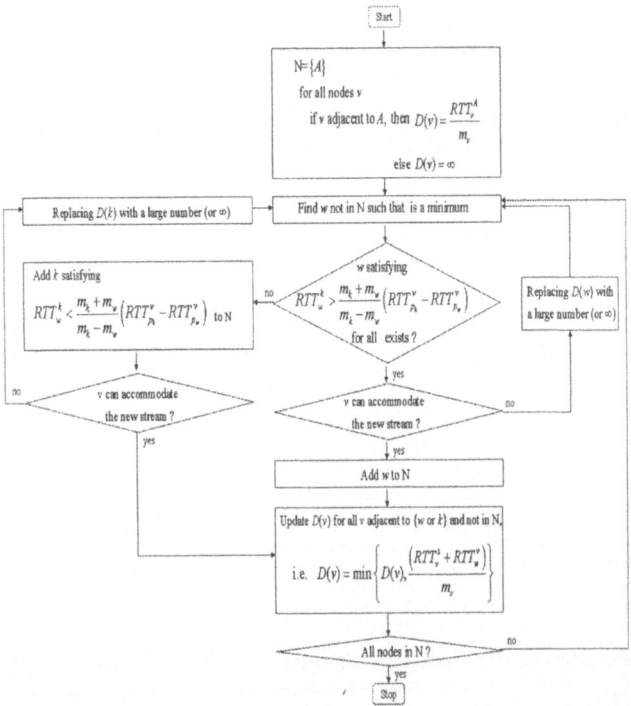

Fig. 4. Flow chart of the proposed Dijkstra's algorithm: where N is the set of nodes whose least RTT path from the sender is definitely known, RTT_v^A is the RTT from A to v, m_v is the number of end-systems in the v cluster, and $D(v)$ is the weighted RTT of the path from sender to destination v that has currently the least weighted RTT.

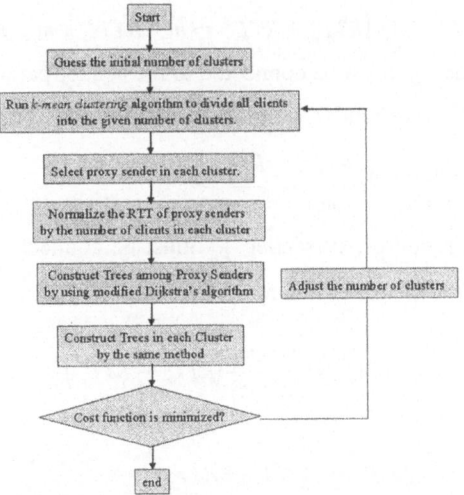

Fig. 5. Flow chart of the proposed algorithm.

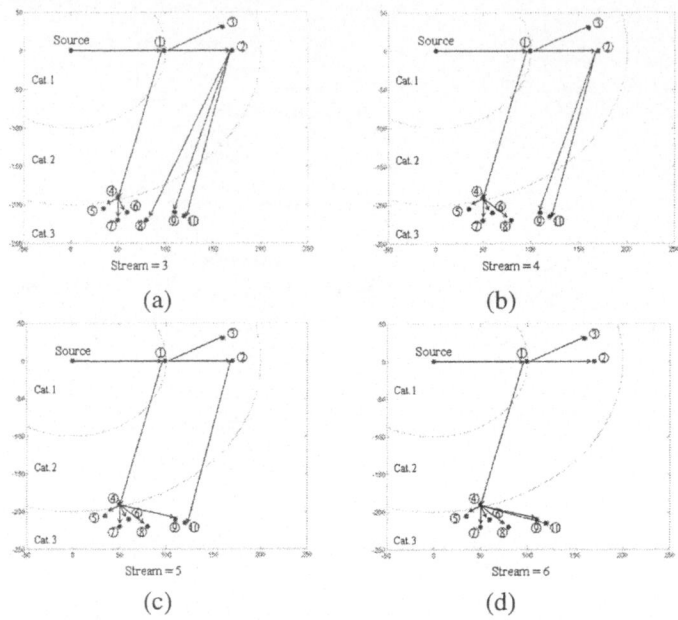

Fig. 6. Tree generated by expanded ring algorithm: (a) the number of streams that each end-system can support is 3, (b) the number of streams that each end-system can support is 4, (c) the number of streams that each end-system can support is 5, and (d) the number of streams that each end-system can support is 6.

line case in Figure 3 $m_j \cdot RTT_{p_i}^s + m_i \cdot \left(RTT_{p_i}^s + RTT_{p_i}^{p_j} \right)$. Now we can get the following rules based on the above two equations.

- If $m_i \cdot RTT_{p_i}^s + m_j \cdot \left(RTT_{p_i}^s + RTT_{p_i}^{p_i} \right) < m_j \cdot RTT_{p_j}^s + m_i \cdot \left(RTT_{p_j}^s + RTT_{p_i}^{p_j} \right)$, then proxy-sender p_i must be connected to the sender/parent proxy-sender.

- Otherwise proxy-sender p_j must be connected to the sender/parent proxy-sender even though $\dfrac{RTT_{p_i}^s}{m_i} < \dfrac{RTT_{p_j}^s}{m_j}$.

Thus, the condition that proxy-sender p_j must be connected to the sender/parent proxy-sender instead of the proxy-sender p_i with the minimum weighted RTT is

$$RTT_{p_i}^{p_j} < \frac{m_i + m_j}{m_i - m_j} \left(RTT_{p_j}^s - RTT_{p_i}^s \right).$$

It can be easily shown that $\dfrac{m_i + m_j}{m_i - m_j} \left(RTT_{p_j}^s - RTT_{p_i}^{p_j} \right) > 0$ by using $\dfrac{RTT_{p_i}^s}{m_i} < \dfrac{RTT_{p_j}^s}{m_j}$ and the fact that the proxy-sender p_i has the minimum weighted RTT. In the following, the modified *Dijkstra's* algorithm is summarized in Figure 4.

By the same procedure, the effective multicast tree can be constructed in each cluster. The summarized flow chart is given in Figure 5.

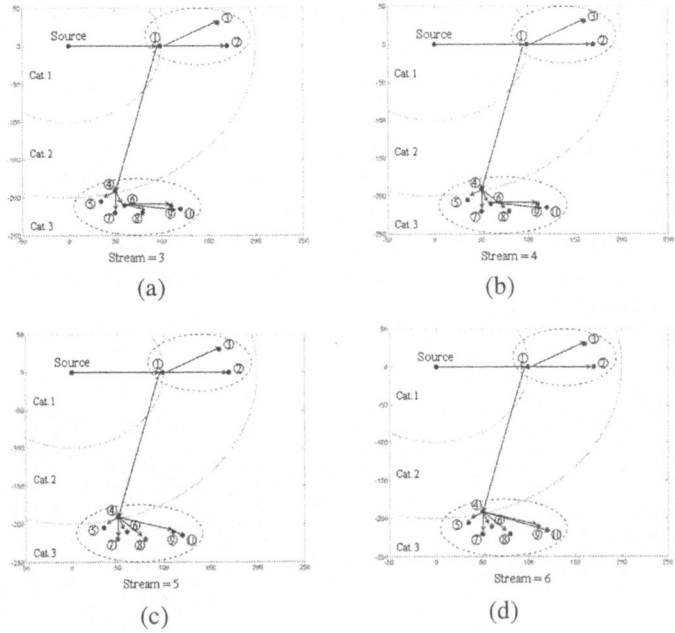

Fig. 7. Tree generated by the proposed algorithm when end-systems are densely located: (a) the number of streams that each end-system can support is 3, (b) the number of streams that each end-system can support is 4, (c) the number of streams that each end-system can support is 5, and (d) the number of streams that each end-system can support is 6.

3   Experimental Results

In this section, we compare the performance of the proposed algorithm in terms of tree structure, average time delay, and computational complexity. That is, average RTT value, CPU time and the number of streams assigned to each end-system are used as performance measures. For the simplicity, we assume that the members have the same computing power and network condition and thus they can support the same number of streams. However, it is still not easy to search other algorithms in order to directly compare the performance since they consider the different network conditions. In this paper, the proposed algorithm is compared with the expanded ring algorithm studied in [12] since they consider the relatively similar situations and also use average RTT as performance measure. The resulting trees of the proposed algorithm and the expanded ring algorithm are given in Figure 6, 7 and 8, and the performance comparison is provided in Figure 9 and 10. No matter how end-systems are located, the expanded ring algorithm give the same structure multicast tree shown in Figure 6. On the other hand, the proposed algorithm makes different trees according to the distribution of end-systems. When end-systems are relatively densely located and can support less than 5 streams, the proposed algorithm connect end-systems 9 and 10 to the closest other end-systems while the expanded ring algorithm connect them to the remote end-system 2 as shown in Figure 6 and 7, and thus the average RTT can be reduced. However, both algorithms give the same trees as shown in (d) of Figure 6 and 7 when end-systems can support more than 6 streams. Secondly, the difference is much more obvious when end-systems are sparsely located. That is, a stream from sender is given to end-system 4 instead of end-system 1 due to the numbers of end-systems in two clusters and the RTT between two proxy-senders. In this case, the proposed algorithm can significantly reduce the average RTT value compared to the expanded ring algorithm as shown in Figure 10. However, the required computational complexity and the number of ping commands may be greater than those of the expanded ring algorithm.

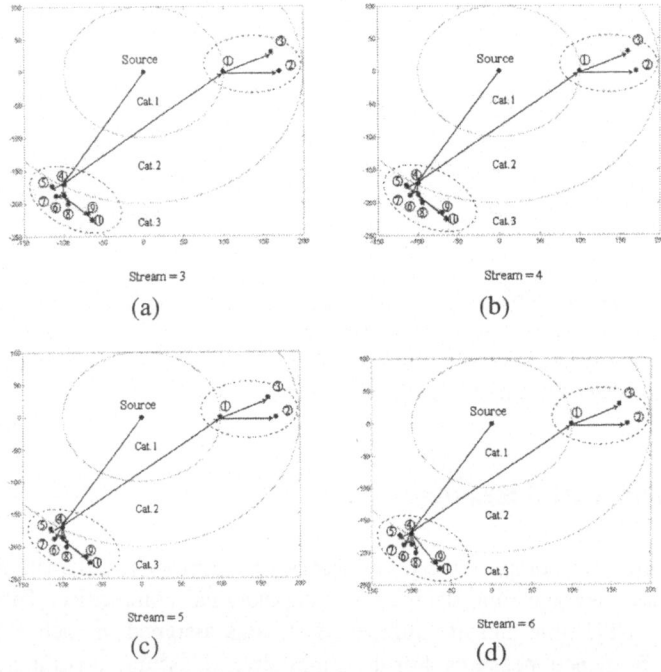

Stream = 3

(a)

Stream = 4

(b)

Stream = 5

(c)

Stream = 6

(d)

Fig. 8. Tree generated by the proposed algorithm when end-systems are sparsely located: (a) the number of streams that each end-system can support is 3, (b) the number of streams that each end-system can support is 4, (c) the number of streams that each end-system can support is 5, and (d) the number of streams that each end-system can support is 6.

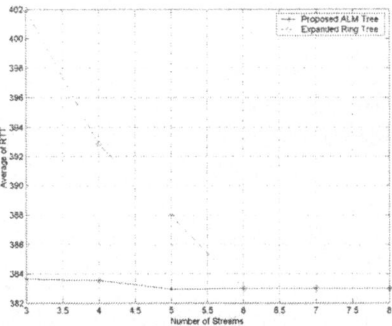

Fig. 9. Average RTT comparison between the proposed algorithm and the expanded ring algorithm when end-systems are densely located.

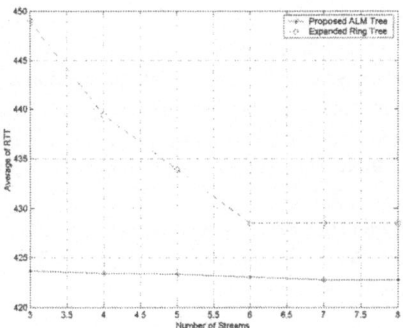

Fig. 10. Average RTT comparison between the proposed algorithm and the expanded ring algorithm when end-systems are sparsely located.

4   Conclusion

In this paper, we have presented an effective overlay multicast tree constructing algorithm minimizing the average RTT value of end-systems for real-time media delivery. Furthermore, the proposed algorithm has considered the computing power and network condition of each end-system as a control parameter to avoid the disastrous situation that loads are concentrated to several end-systems. The multicast tree has been constructed by clustering algorithm and modified Dijkstra's algorithm in two steps, i.e. a tree among proxy-senders and trees in every cluster. By the experimental results, we have showed that the proposed algorithm can provide an effective solution. For a complete solution, dynamic tree maintenance is needed when a new member joins or leaves the multicast group. Actually, it is under the current investigation.

References

1. Y. Chu, S. Rao, S. Seshan and H. Zhang, "A case for end-system multicast," ACM SIGMETRICS, Santa Clara, June 2000.
2. S. Deering, "Host Extensions for IP Multicasting," RFC1112, August 1989.
3. J. Liebeherr and M. Nahas, "Application-layer Multicast with Delaunay Triangulations," IEEE GLOBECOM, 2001.
4. L. Mathy, R. Canonico, S. Simpson, and D. Hutchison, "Scalable adaptive hierarchical clustering," IEEE Communication Letter, Vol. 6, No. 3, March 2002.
5. D. Pendarakis, S. Shi, D. Verma and M. Waldvogel, "OVERLAY MULTICASTI: An Application level multicast infrastructure," Proceedings of 3rd Usenix Symposium on Internet Technologies & Systefeeder (USITS 2001), San Francisco, Mar. 2001.
6. J. Park, S. J. Koh, S. G.. Kang, D. Y. Kim, "Multicast Delivery Based on Unicast and Subnet Multicast," IEEE Communications Letters, Vol. 5 No. 4 April 2001.
7. M. Castro, P. Druschel, A. Kermarrec, and A. Rowstron, "Scribe: A large scale and decentralized application-level multicast infrastructure," IEEE Journal on Selected Areas in Communications, Vol. 20, No. 8, Oct. 2002.
8. S. Y. Shi and J. S. Turner, "Multicast routing and bandwidth dimensioning in overlay network," IEEE Journal on Selected Areas in Communications, Vol. 20, No. 8, Oct. 2002.
9. S. Banerjee and B. Bhattacharjee, "Scalable secure group multicast over IP multicast," IEEE Journal on Selected Areas in Communications, Vol. 20, No. 8, Oct. 2002.
10. J. F. Kurose and K. W. Ross, Computer Networking: A Top-down approach featuring the Internet, Addison Wesley, 2001.
11. W. K. Pratt, Digital Image Processing, Wiley Interscience, 1991.
12. C. K. Yeo, B. S. Lee and M. H. Er, "A framework for multicast video streaming over IP networks," Journal of Network and Computer Applications, pp. 273-28, Vol. 26, 2003.

Maximum Throughput Analysis in Ad Hoc Networks

Bernardo A.M. Villela and Otto Carlos M.B. Duarte

Grupo de Teleinformática e Automação – PEE/COPPE-Poli/UFRJ
Universidade Federal do Rio de Janeiro
CP 68504 - 21945-970 - Rio de Janeiro, Brazil.
Phone: +55-21-2260-5010 x. 240 - Fax: +55-21-2290-6626
{bernardo,otto}@gta.ufrj.br

Abstract. The IEEE 802.11 standard for ad hoc networks uses a distributed access mechanism in the shared medium that attempts to avoid collisions by performing carrier sensing, inter-frame spaces, and a backoff mechanism. This paper aims at deriving an analytical expression for the maximum throughput of a communication between two nodes. The communication is achieved in a multi-hop scenario and the path from the source to the destination consists of a chain of nodes. We analyze the multi-path communication and show that the maximum throughput can be increased with the simultaneous use of two paths: the shortest path and an appropriate alternative path that takes into account the interference problem. We also derive the constraints for this alternative path. The use of multiple alternative paths can be considered in order to achieve a load and energy balancing.

1 Introduction

Nowadays, the IEEE 802.11 [1,2] is the most famous standard in local wireless networks. In the ad hoc mode, it uses the distributed medium access mechanism called DCF (Distributed Coordination Function), which applies the CSMA/CA (Carrier-Sense Multiple Access/Collision Avoidance) access method. The collision avoidance is carried out by carrier sensing, inter-frame spaces and a backoff mechanism that uses a contention window. Furthermore, after each successful transmission, the receiver must send an ACK to the sender indicating the success of the operation. That is necessary because in wireless networks, due to the significant difference between the transmitting and the receiving signal power, only the receiver is able to identify a collision.

The hidden terminal problem is a classical challenge in wireless networks. This problem arises because the carrier sensing is accomplished at the sender, but the transmission success is observed at the receiver. Therefore it is possible that a node senses the medium free and starts transmitting, but in the receiver point of view the medium was busy, which means that there was already a transmission that could not be noticed by the last sender due to the long distance between the two senders. To solve the hidden terminal problem, the DCF proposes the use of RTS (Request To Send) and CTS (Clear To Send) frames. By means of an RTS frame, the sender shows all his neighbors the intention to transmit and the receiver allows the transmission by sending a CTS frame, showing that its neighborhood is free. The DCF mechanism derives another benefit from these frames, making it possible for implementing a virtual carrier sense, by indicating in RTS and CTS frames how long the medium will be busy.

N. Mitrou et al. (Eds.): NETWORKING 2004, LNCS 3042, pp. 223–234, 2004.

Xu *et al.* [3] analyzed the effectiveness of the RTS/CTS handshake and showed that it cannot completely solve the hidden terminal problem, due to the effect of the interference. This mechanism assumes that all nodes that could interfere with the frame reception, which will be called hidden nodes, are able to receive the CTS too. Xu *et al.* [3] derived an expression showing that the interference range is a variable range depending on the distance from the sender to the receiver and the signal to interference relation at the receiver. Moreover, they showed that when the distance from the sender to the receiver is longer than a threshold value, the interference range becomes greater than the transmission range. That happens because the interfering signal power is much less than the signal power required for a correct reception. As a consequence, we cannot assume that the CTS frame is received by a hidden node.

Saadawi *et al.* [4] considered the performance of the IEEE 802.11 medium access control (MAC) protocol in multi-hop ad hoc networks, through the analysis of a TCP (Transmission Control Protocol) traffic. Although the IEEE 802.11 can support ad hoc networks, they argued that the multi-hop connectivity poses serious problems, degrading its performance.

Gupta *et al.* [5] and Li *et al.* [6] analyzed the capacity of ad hoc networks. They showed that the capacity clearly depends on some local radio parameters, the MAC protocol used, the network size, and the traffic patterns. These characteristics directly affect the efficiency and scalability of the network. If the communications take place distant from each other, then the spatial reuse of the bandwidth increases. In contrast, if neighboring nodes want to send a frame concurrently, they compete for the medium, according to the medium access mechanism. In addition, it is worth mentioning that the average number of hops in the communications also plays an important role.

Li *et al.* [6] analyzed the capacity of a chain of nodes considering the interference range. They showed the maximum utilization achievable for a fixed interference range. Considering that the interference range is actually variable, in this paper we generalize this result for any signal to interference relation required for a successful reception and any distance of neighboring nodes.

In addition, we use multiple paths to increase the throughput of a communication. We show that the maximum throughput is achieved with only two paths to the destination: the shortest path and an appropriate alternative path that takes into account the interference problem. We present the constraints that should be respected by this alternative path. Besides we generalize an expression for the maximum utilization achievable using multiple paths for any signal to interference relation required for a successful reception and any distance of neighboring nodes.

At last, we consider the implementation of the interference aware alternative path. Since it relies on some location information, it is suggested the use of geographic routing. If the nodes remain always static, as in a rooftop network, or they have a slow moving dynamics, as in sensor networks, then finding a good alternative path is easier. The faster the nodes in the network move, the more difficult becomes the task of finding a convenient alternative path. In this case, attempting to accomplish this task, we suggest the implementation of an anchored path, as it is done by the Terminodes routing [7,8], listing in each packet a list of geographic fixed regions strategically chosen to guide the way a packet is supposed to travel until it gets to the destination.

The remainder of this paper is organized as follows. In Section 2, we present the interference effect. In Section 3, we analyze the maximum utilization of a chain of nodes and derive a generic expression. In Section 4, we show that the use of an alternative path can improve the throughput experienced by the source. It is shown how this alternative path should be and it is calculated the maximum utilization provided by this method. In Section 5, we consider the use of multiple alternative paths. In Section 6, we take into account the implementation of the alternative path. Section 7 presents our conclusions.

2 The Interference Effect

Let us consider a transmission in an IEEE 802.11 network. In order to correctly receive a packet, the signal power at the receiver must be strong enough. Hence, the signal to noise plus interference relation must be greater than the minimum value specified for the receiver equipment. Due to the signal attenuation in the air, increasing the distance from the sender to the receiver results in the reduction of the signal power at the receiver, which means that nodes more distant from the receiver can become hidden nodes.

Let d be the distance from the sender to the receiver, r be the distance from the receiver to a third node that might want to transmit, and SIR_{TH} the minimum value for the signal to interference relation (SIR) required for a successful reception. Xu *et al.* [3] argued that the thermal noise can be ignored when compared to the interference signal; accordingly, we also ignored it in our analysis in this paper. Then Xu *et al.* derived the Equation 1, which implies that every node separated by less than $d\sqrt[4]{SIR_{TH}}$ meters from the receiver can indeed interfere with its reception.

$$SIR = \left(\tfrac{r}{d}\right)^4 < SIR_{TH}$$

$$r < d\sqrt[4]{SIR_{TH}} \tag{1}$$

Let us define R_{Tx} as the transmission range. It can be easily shown [3] that when the sender and the receiver are more than $\frac{1}{\sqrt[4]{SIR_{TH}}}R_{Tx}$ meters away from each other, the RTS/CTS handshake does not solve the hidden terminal problem.

Let us assume that the nodes N_S and N_R are the sender and the receiver of a transmission, respectively. In Figure 1, the A_{RTS} and A_{CTS} areas are the regions reached by the RTS and CTS frames, respectively sent by N_S and N_R. In Figure 1, the SIR_{TH} is such as $1 < SIR_{TH} < 16$. The A_{Int} area represents the region where hidden nodes might be located. In Figure 1, the sender and the receiver are separated by the maximum distance (R_{Tx}). In this case, the interference range and the $A_{Int} - (A_{RTS} \bigcup A_{CTS})$ area are maximized. Other ad hoc nodes, represented by N_A, N_B, and N_C, might want to transmit during the transmission of N_S. It can be seen at Figure 1 that N_A and N_B would not transmit simultaneously with N_S, since they previously received the RTS and CTS frames, respectively, of the handshake implemented by N_S and N_R. However if N_C, which is not aware of this handshake, wants to send a frame, then it would proceed to its transmission, interfering with the reception of N_R.

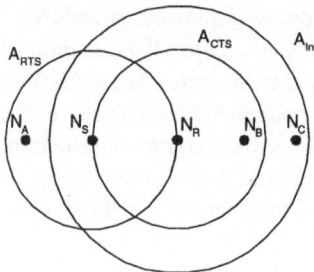

Fig. 1. The interference effect

3   Capacity Analysis of a Chain of Nodes

The highest throughput can be achieved when the destination is in the transmission range of the source, called "direct communication", and the medium is always free. Li *et al.* [6] showed that, for the IEEE 802.11 at 2Mbps, the maximum throughput is 1.7Mbps for 1500 bytes of packet size. The data rate reduction is due to the overhead added by the RTS/CTS/ACK exchange and the inter-frame timings. Furthermore the data rate depends on the packet size. The same considerations can be used in order to calculate the maximum throughput at other rates, such as 11Mbps or 54Mbps. The scenario with only two neighboring nodes can be considered as the simplest chain case and can be taken as a baseline for comparison, giving an upper bound for the throughput of a chain. Hence, in the following analysis, the maximum utilization of a general case is a fraction of this reference value. The general case is a "multi-hop communication" where, due to constraints in the consumption of power, bandwidth, and energy, the nodes need to cooperate forwarding packets from the source to the destination. We will analyze the behavior of a multi-hop chain of nodes transmitting a flow of packets.

Li *et al.* [6] analyzed the capacity of a chain of nodes aligned in a row and separated by 200m from each neighbor. They also simulated it in NS. In the NS, as a means of achieving a realistic model, the physical radio characteristics of each mobile node [9], such as the transmit power, the antenna gain, and receiver's sensitivity, are based on the Lucent WaveLAN model. This way, the transmission range is set to 250m and the SIR_{TH} is set to 10. In the analysis, Li *et al.* assumed a simplification: they considered the interference range fixed at 550m, instead of variable. Besides they considered the carrier sensing range equal to the transmission range; such consideration is also assumed in this paper. Figure 2 presents a chain of 7 nodes, separated by 200m from each neighbor. Since the transmission range is 250m, a packet sent by the node P_S is able to reach the destination P_D, by passing sequentially through the nodes P_1, P_2, P_3, P_4, and P_5.

Li *et al.* [6] argued that the maximum utilization achieved by their chain is $\frac{1}{4}$. As they remarked, nodes P_S and P_1 cannot transmit a frame together, since P_1 cannot send and receive at the same time; and the nodes P_S and P_2 cannot transmit simultaneously, because P_1 cannot correctly receive the frame of P_S if P_2 is transmitting together. Concluding the explanation of a maximum utilization of $\frac{1}{4}$, they justified that P_S and P_3 cannot transmit at the same time, because the transmission of P_3 would interfere

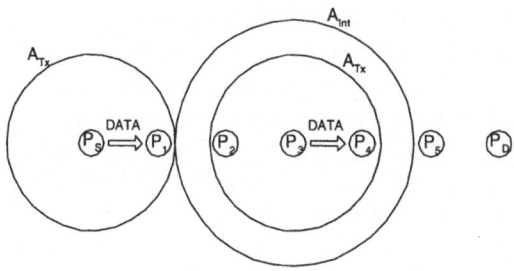

Fig. 2. Chain of 7 nodes separated by 200m from each neighbor. The interference range is according to Equation 1

with P_1's reception of the frame sent by P_S. This argument was based on the fixed interference range set to 550m.

Nevertheless, Xu *et al.* [3] showed that the interference range is variable, depending on the distance from the sender to the receiver and the signal to interference relation required for a correct reception. Accordingly, we consider this range variable in this work. Since the distance between two consecutive nodes is 200m, the Equation 1 shows that the interference range is only 356m. Then a transmission from P_3 does not disturb the transmission of P_S because P_3 is separated from P_1 by a distance greater than the interference range, as illustrated in Figure 2. So, as shown by Xu *et al.* [3], the maximum utilization achievable is $\frac{1}{3}$.

Let us define the maximum utilization as U_{max}. Now, we will analyze the maximum utilization of the chain when the distance d between neighbors in the chain is such as $\frac{R_{Tx}}{2} < d \leqslant R_{Tx}$. Our point is that if $d > R_{Tx}$ then it is impossible getting to the destination and if $d \leqslant \frac{R_{Tx}}{2}$ then each node would not send a packet to its closest neighbor, instead each node would send it to its neighbor closest to the destination, as a means of minimizing the number of hops. The arguments which justified that P_S cannot transmit simultaneously with P_1 or P_2 remain valid for any value of d and SIR_{TH}. Then $\frac{1}{3}$ is an upper bound for U_{max}.

Let K be the number of nodes which are in the interference range of the source. Therefore $K = \lfloor \frac{R_{Int}}{d} \rfloor = \lfloor \sqrt[4]{SIR_{TH}} \rfloor$. Then analyzing the successors of P_1, its reception of a frame can be interfered by all its successors until P_{K+1}. In addition, P_{K+2} and its successors cannot interfere with the transmission of P_S. So U_{max} is given by Equation 2 for $\frac{R_{Tx}}{2} < d \leqslant R_{Tx}$:

$$U_{max} = \frac{1}{K+2} \tag{2}$$

The maximum utilization depends on the SIR_{TH}. It can be remarked that the SIR_{TH} must be greater than 1. In this paper, we derive generic equations, but we evaluate $K = 1$ or $K = 2$, constraining the SIR_{TH} value to $1 < SIR_{TH} < 81$ that is a realistic case. For example, if $SIR_{TH} = 10$, then $K = 1$, which implies $U_{max} = \frac{1}{3}$, and if $SIR_{TH} = 20$, then $K = 2$, which implies $U_{max} = \frac{1}{4}$.

Let us assume P_S is transmitting a frame to P_1. Then the RTS/CTS handshake successfully forbids P_1 and P_2 from starting to transmit together. If $SIR_{TH} < 16$, this handshake performs well for achieving the maximum utilization, as P_3 and P_S can indeed send frames concomitantly. But if $SIR_{TH} > 16$, it does not perform as well. The nodes from P_3 until P_{K+1} do not listen to the RTS and CTS frames sent by P_S and P_1, so they will not avoid simultaneous transmissions with P_S. Therefore, increasing the frame rate at P_S beyond U_{max}, the chain will experience collisions, which will force retransmissions and implementation of longer backoff waiting times. Ultimately, it will reduce the utilization and the delivery rate.

4   The Use of an Alternative Path

We propose a way of improving the flow utilization. We define P_S and P_D as the source and the destination of a flow of packets. As illustrated in the Figure 3, let us suppose there is a second path $(P_S, A_1, A_2, \ldots, P_D)$ to the destination. This path is longer, so a single path routing protocol, executing a minimum hop metric, will keep on using the path $P_S, P_1, P_2, \ldots, P_D$.

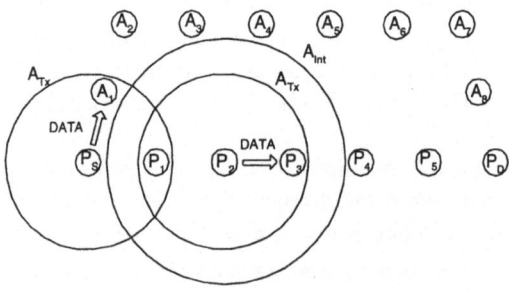

Fig. 3. A default chain and an alternative path

In the chain case, we argued that $\frac{1}{3}$ is an upper bound for the maximum utilization. We argue that we can increase the throughput, taking another way to get to the destination. The flow can be split in two parts using two different paths, as illustrated in Figure 3: the default path, through nodes $P_S, P_1, P_2, \ldots, P_D$ and an alternative path, through the nodes $P_S, A_1, A_2, \ldots, P_D$. The source injects frames into the network, alternating each frame through each path. In this case, we argue that $\frac{1}{2}$ is an upper bound for the maximum utilization, as it will be explained. If P_1 is transmitting, then P_S cannot send a frame, since P_S is able to listen to the transmission and senses the medium busy. In addition, if it is implemented the RTS/CTS handshake then P_S has received the RTS from P_1. But, if P_2 is transmitting, then P_S does not listen to the transmission and also has not listened to the handshake originated by P_2, if the RTS/CTS handshake is used. Therefore, P_S can send a frame simultaneously with P_2. P_S cannot send it to P_1, since P_1 can listen to both P_S and P_2 and will not correctly receive the frame sent by the source. In addition, if the RTS/CTS handshake is applied, then P_1 will not send a CTS in response to the RTS issued by P_S, forbidding the transmission from the source. However if the source has a neighbor, which is not in the interference range of P_2, then P_S can effectively send a frame to this node. This way, in Figure 3, where $1 < SIR_{TH} < 16$, P_2 can send a frame to P_3 while P_S send the next frame to A_1.

Again, let us assume that each node is separated from its neighbors in each path by a distance d, such as $\frac{R_{Tx}}{2} < d \leqslant R_{Tx}$. Let the function $dist(P_A, P_B)$ denote the distance from the node P_A to the node P_B. Again, let K be $\lfloor \sqrt[4]{SIR_{TH}} \rfloor$, which is the number of nodes that are in the interference range of the source. In the chain case, we allowed the concomitant transmission of P_S and P_{K+2} or any of its successors, but we forbade the simultaneous transmission of P_S and P_{K+1}, since $dist(P_1, P_{K+1}) \leqslant R_{Int}$. Aiming at increasing the throughput, now we allow this concomitant transmission, if P_S can send a frame to a neighbor A_1, such as $dist(A_1, P_{K+1}) > R_{Int}$. An equivalent law should be respected by the nodes P_1 and A_{K+1}. An analogous procedure is recommended at the end of the process, when the two paths get closer to each other. And in the middle of the chain, we recommend that a distance greater than R_{Int} should be guaranteed. This procedure, which will be called $Proc_{Alt}$, is implemented as an example in Figure 3.

In fact, allowing the simultaneous transmission of P_S and P_{K+1} with $Proc_{Alt}$ is the best option we have, since the simultaneous transmission of P_S and P_K would imply a lot of collisions, degrading the performance. Since $dist(P_S, P_K) \leqslant R_{Int}$, if the nodes are not synchronized (and we cannot assume such synchronism in the DCF mechanism of IEEE 802.11), then both P_S and P_K will often experience problems at accomplishing the DATA/ACK (or the RTS/CTS/DATA/ACK) handshake. Not receiving the ACK means to the sender a failure and forces the retransmission of the DATA frame. And if the RTS/CTS handshake is implemented, not receiving the CTS forces the retransmission of the RTS; in addition, it forbids the transmission of the DATA frame until the correct reception of the CTS. At last, if we allow the simultaneous transmission of P_S and P_{K-1} or any other node in the chain between P_S and P_{K-1}, then it is impossible finding a neighbor of P_S, which is not interfered by the concomitant transmission of this distant node. In addition, P_S is also in its interference range.

$Proc_{Alt}$ allows nodes P_S and P_{K+1} (and also nodes P_S and A_{K+1}) to transmit simultaneously. Then Equation 3 gives the maximum utilization achievable by $Proc_{Alt}$.

$$U_{max} = \frac{1}{K+1} \tag{3}$$

From Equations 2 and 3, we notice that the achievable gain is $100 \left(\frac{K+2}{K+1} - 1 \right)$ in percentage, where $K \geqslant 1$. If $1 < SIR_{TH} < 16$, then $Proc_{Alt}$ gives a 50% gain, achieving a maximum utilization of $\frac{1}{2}$, which is already the upper bound for the utilization. And if $16 \leqslant SIR_{TH} < 81$, then $Proc_{Alt}$ achieves a gain of 33% and a maximum utilization of $\frac{1}{3}$.

In Figure 4, it is presented the beginning of a default and an alternative path. The two paths are respecting $Proc_{Alt}$, so they will end in an analogous way. Let us assume that the nodes $P_S, A_1, \ldots, A_{K+1}$ of the alternative path are also aligned as a row. Let the angle $P_S \angle P_1, A_1$ be α. The triangles $\triangle P_S, A_1, P_{K+1}$ and $\triangle P_S, P_1, A_{K+1}$ are

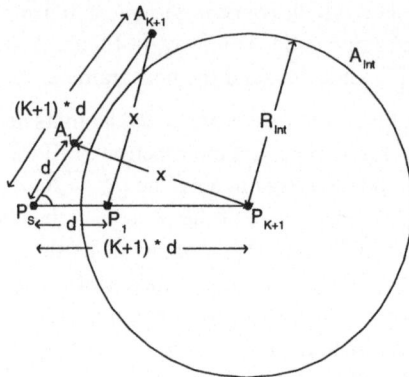

Fig. 4. Angle used to choose an alternative path with $Proc_{Alt}$

equivalent because

$$
\begin{aligned}
dist(P_S, P_1) &= dist(P_S, A_1) = d; \\
dist(P_S, P_{K+1}) &= dist(P_S, A_{K+1}) = (K+1)d; \\
P_S \angle A_1, P_{K+1} &= P_S \angle P_1, A_{K+1} = \alpha.
\end{aligned}
$$

Therefore $dist(P_1, A_{K+1}) = dist(A_1, P_{K+1})$, which we define as x. Let us define β as $P_S \angle A_1, P_{K+1}$ when $x = R_{Int}$. Then, by the cosine law, Equation 4 gives the value of β. To increase the throughput, avoiding collisions due to the interference, we have to choose α, such as $x > R_{Int}$, then $\beta < \alpha < 360 - \beta$. $d^2 + (K+1)^2 d^2 - 2d(K+1)d \cos \beta = d^2 \sqrt{SIR_{TH}}$

$$
\beta = \arccos \left(\frac{(K+1)^2 + 1 - \sqrt{SIR_{TH}}}{2(K+1)} \right) \tag{4}
$$

Figure 5 presents the value of β when the SIR_{TH} is varied. The lower is β, the greater is the neighborhood area of P_S which is not interfered by P_{K+1}, and also the greater is the range of values that α can assume. If the nodes are not previously ranged in a convenient way, instead they are randomly set in the scenario, then it is important having lower values of β (specially for not dense networks), because it results in a greater probability of finding an alternative path respecting $Proc_{Alt}$. As expected, in each interval limited by the fourth power of two integer values, the value of β increases, when the SIR_{TH} gets greater. Just after switching to the next interval, the node P_{K+1}, that can transmit simultaneously with P_S, gets one hop farther from the source. Therefore the maximum utilization gets lower, but, at this moment, the neighborhood area of the source disturbed by the interference is also reduced, explaining the reduction of β. In the extreme case when the SIR_{TH} is the fourth power of an integer value, β gets equal to zero. Accordingly, P_1 is the only neighbor of the source, which is in the interference range of P_{K+1}.

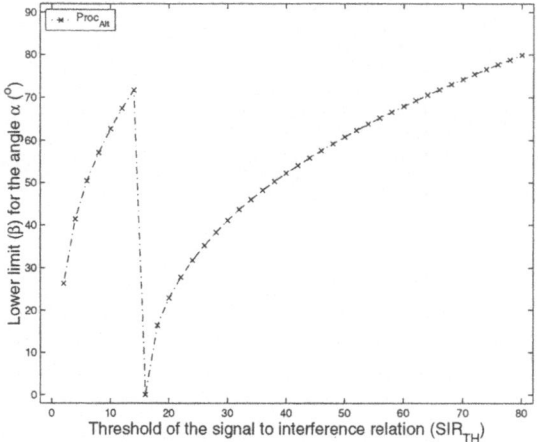

Fig. 5. The lower limit (β) for the angle α

5   Multiple Alternative Paths

As $dist(P_S, P_{K+1}) > R_{Int}$ and $dist(P_S, A_{K+1}) > R_{Int}$, it results that β is always less than 90°, in accordance with Figure 5. Then it may be implemented more than one alternative paths. In fact, it can be generalized that it is possible to use a maximum of $\lceil \frac{360}{\beta} - 1 \rceil - 1$ alternative paths, if we have enough nodes located in a convenient way.

Assuming that the paths begin and end as aligned rows, to implement the interference awareness between the paths, the angle between each row should be greater than β. In Figure 6, where β is assumed less than 45°, it is presented four alternative paths (ALT_1, ALT_2, ALT_3, and ALT_4), using $Proc_{Alt}$. All of them begin and end with a direction shifted by at least 45° from each other and the default path. And, in the middle, it is guaranteed a distance greater than R_{Int} between all the paths.

Using multiple alternative paths allows a better load and energy balancing, which are two scarce resources in wireless networks. With a more equal consumption of energy by the ad hoc nodes, the network can last much longer. Supposing a network with different communications taking place simultaneously, it may be the case that the flow of traffic is concentrated in a region or a group of regions. Then it may be very interesting taking alternative paths, trying to bypass the overloaded paths. Nevertheless, if the network area is approximately equally loaded, then the use of only one alternative path is enough for getting the maximum utilization. That is because the moment when the source can switch from the first alternative path to a second one is also the moment when it can go back to the default path, which is attractive, since it uses less nodes. In this paper, we calculated the maximum utilization achieved by a source, when there is no other source of traffic. Then, in this case, the load balancing with the use of more than one alternative paths is not necessary.

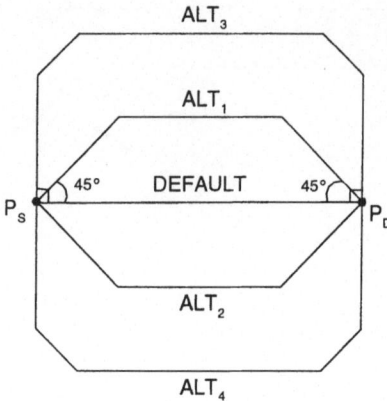

Fig. 6. Multiple alternative paths to the destination

6   Implementation of the Alternative Path

The implementation of the alternative path with an interference range aware procedure depends on some location knowledge. Hence, we suggest it should be based on a geographic routing paradigm. Implementing it in a distributed way is not trivial. It is made easier for static networks. For example, in an ad hoc rooftop network, used when the infra-structure is not properly working, the nodes remain still. We can even assume that a node previously knows the position of the other nodes, making the whole process easier. In this case, a source routing can be applied, where the source indicates in the packet all the nodes in the path to the destination, just like DSR [10] (Dynamic Source Routing) does. In an urban area (e.g. Manhattan), it rather be quite easy to find an alternative path able to improve the utilization. In addition, in a sensor network, it is common having slow moving nodes and a large density of nodes, then the process of finding an appropriate alternative path is also more easily accomplished.

We can also attempt to apply this procedure for rather faster nodes and evaluate its performance. In this case we suggest the use of an anchored path, as it is implemented by the Terminodes routing [7,8]. Instead of listing all nodes in the path to the destination, as it is done by the source routing paradigm of DSR, it can be listed some geographic fixed regions, called anchors, that will guide the way a packet is supposed to travel bypassing the interference.

In Figure 7, P_S and P_D are the source and the destination of a flow of packets, respectively; $DEFAULT$ and ALT are the default and the alternative paths respectively; $Anch_1$ and $Anch_2$ are two anchor regions strategically chosen. The anchor regions are calculated based on the angle α, which is relevant to the beginning and the end of the alternative path, and the necessity of respecting a distance greater than R_{Int} between the two paths in the middle of them. This way, in Figure 7, the alternative path, which passes by $Anch_1$ and $Anch_2$ before getting to the destination, is respecting the $Proc_{Alt}$ procedure. In the header of all the packets supposed to travel through the alternative path, it will be specified that the packet should go firstly to a point in the $Anch_1$ region, after it should go to a point in the $Anch_2$ region, and, at last, it will go to the destination. Between each of these areas, the packet will be routed in a greedy manner, trying to approach the desired position as fast as possible. In the header of the packets that go through the default path, it is specified only the position of the destination. They find their way in a greedy basis, with no need of anchors. In Figure 7, it is illustrated a simplified case where all the nodes are conveniently aligned. In this case, the greedy routing is able to get to each desired position in straight lines.

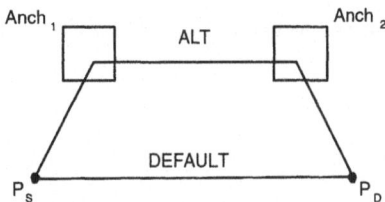

Fig. 7. Implementation of the anchored path to the destination

The faster the nodes move, the faster the links will break. But if we use anchors and we have a sufficiently high density of nodes well distributed over the total area, since geographic areas do not move, the path will be much more stable. Moreover the anchors are very appropriate to implement the bypassing of the default path.

7 Conclusions

In ad hoc wireless networks, the bandwidth is a scarce resource. In this work, we aimed at increasing the throughput verified by a source of traffic.

Analyzing the symptoms experienced by a chain of nodes is relevant, since the forwarding of packets is a frequent activity in ad hoc networks, where the nodes are expected to cooperate. If the source is transmitting, then its next two successors in the chain cannot transmit concomitantly. That reduces the maximum utilization of a chain to $\frac{1}{3}$ in the optimum case. If the signal to interference relation required for a good reception (SIR_{TH}) is greater or equal to 16, then, due to the hidden terminal problem, this optimum utilization cannot be achieved. We analytically derived Equation 2, which gives the maximum utilization in terms of the SIR_{TH}.

We analytically showed with Equation 3 that the simultaneous use of a default and an appropriate alternative path can increase the utilization of the chain. This way, if $1 < SIR_{TH} < 16$, then the maximum utilization gets $\frac{1}{2}$ and if $16 \leqslant SIR_{TH} < 81$, then the maximum utilization gets $\frac{1}{3}$. That represents a gain of 50% and 33%, respectively, over the single path case.

It is presented which kind of alternative path is able to bypass the interference problem. Since finding an appropriate alternative path depends on the knowledge of the position of the nodes, we suggested the use of the geographic routing paradigm. The process is easier accomplished for networks composed by static nodes, such as rooftop networks, or slow moving nodes, such as sensor networks. For networks composed by rather faster nodes, we suggested the use of an anchored path, as it is done by the Terminodes routing.

It is shown that the use of more than one alternative path does not increase the maximum utilization. Nevertheless, its use can be considered if the goal is load or energy balancing, which might be necessary since both bandwidth and energy are scarce resources in wireless networks.

Acknowledgment. This work has been supported by CNPq, CAPES, COFECUB, and FAPERJ.

References

1. IEEE Standard 802.11: Wireless LAN medium access control (MAC) and physical layer (PHY) specifications - Part 11. (1999)
2. Crow, B.P., Indra Widjaja, F., Kim, J.G., Sakai, P.T.: IEEE 802.11 wireless local area networks. IEEE Communications Magazine **35** (1997) 116–26
3. Xu, K., Gerla, M., Bae, S.: How effective is the IEEE 802.11 RTS/CTS handshake in ad hoc networks? In Proc. IEEE Globecom (2002)
4. Xu, S., Saadawi, T.: Does the IEEE 802.11 MAC protocol work well in multihop wireless ad hoc networks? IEEE Communications Magazine (2001) 130–137
5. Gupta, P., Kumar, P.R.: The capacity of wireless networks. IEEE Transactions on Information Theory **46** (2000)
6. Li, J., Blake, C., De Couto, D.S., Lee, H.I., Morris, R.: Capacity of ad hoc wireless networks. Proceedings of the 7th ACM International Conference on Mobile Computing and Networking (2001) 61–69
7. Blazevic, L., Giordano, S., Boudec, J.Y.L.: Self organized terminode routing. In Journal of Cluster Computing **5** (2002)
8. Blazevic, L., Giordano, S., Boudec, J.Y.L.: Anchored path discovery in terminode routing. Proc. of The Second IFIP-TC6 Networking Conference (Networking 2002) (2002)
9. Broch, J., Maltz, D.A., Johnson, D.B., Hu., Y.C., Jetcheva, J.: A performance comparison of multi-hop wireless ad hoc network routing protocols. In Proceedings of the Fourth Annual ACM/IEEE International Conference on Mobile Computing and Networking (1998)
10. Maltz, D.A., Johnson, D.B., Hu., Y.C., Jetcheva, J.G.: The dynamic source routing protocol for mobile ad hoc networks. Internet Draft, draft-ietf-manet-dsr-09.txt (2003)

Performance of Wireless Ad Hoc Networks under Balanced Fairness

Aleksi Penttinen* and Jorma Virtamo**

Networking laboratory, Helsinki University of Technology, P.O. Box 3000,
FIN-02015 HUT, Finland
{Aleksi.Penttinen, Jorma.Virtamo}@hut.fi

Abstract. Balanced fairness is a new resource sharing concept recently introduced by Bonald and Proutière. We extend the use of this notion to wireless networks where the link capacities at the flow level are not fixed but depend on the scheduling of transmission rights to interfering nodes on a faster time scale. The balance requirement together with the requirement of maximal use of the network's resources jointly determine both a unique state-dependent scheduling and bandwidth sharing between the contending flows. The flow level performance under the resulting scheme is insensitive to detailed traffic characteristics, e.g., flow size distribution. The theoretical and computational framework is formulated and illustrated by two examples for which the performance in terms of average flow throughputs in a dynamic system is explicitly worked out.

1 Introduction

As in fixed networks, also in wireless ad hoc networks the performance perceived by the users sending elastic traffic mainly manifests itself on the flow level. A flow of elastic traffic typically comprises a transfer of a document, file or message such that the transmission can use all the bandwidth that is available but can also adapt the transmission speed to the congestion and share the bandwidth with other concurrent flows. The performance, such as the average duration of transfer of a document of a given size, clearly depends on dynamic behavior of the system and on how the bandwidth is shared between different flows. Thus it is necessary to study the system in a dynamic setting where new flows arrive at the network, are transferred across the network, and upon completion depart from the system. As far as we are aware, no analysis of this type has been done for ad hoc networks previously.

In order to facilitate the analysis, a certain degree of abstraction is necessary. In particular, we ignore the detailed packet level behavior although the actual communication in an ad hoc network consists of a sequence of packet level operations of channel access and data forwarding over a multihop route. From a conceptual point of view we distinguish two layers of operations. On the lowest level, which we call scheduling, one basically determines which transmitters

---

* Partly funded by Finnish Defence Forces Technical Research Center.
** Supported by the Academy of Finland (grant n:o 80424)

N. Mitrou et al. (Eds.): NETWORKING 2004, LNCS 3042, pp. 235–246, 2004.

are allowed to send data at any given time (because of the interference not all transmitters are allowed to operate simultaneously). As the set of permissions is switched on a fast time scale, the resulting network appears at flow level as a virtual network with links with capacities that depend on the schedule. So the schedule determines the capacity to send data on different links. On the upper level of operation, which is referred to as bandwidth sharing or bandwidth allocation one determines how the capacity of the virtual network is shared between the flows, i.e. viewed from a lower level, what data is sent when a sender has a permission to send.

The main difference of the analysis of flow level performance of ad hoc networks from that in fixed networks stems indeed from the fact that by scheduling one can, in certain limits, shift capacity of the network from one link to another, and that this degree of freedom can be used to improve the performance.

In the flow level abstraction we assume that the flow durations are long compared with the time scale of the operation of a schedule as well as with the time it takes for whatever protocol or flow control mechanism is used for the bandwidth sharing to find a steady state. Thus we assume that as soon as a flow arrives or departs, a new steady state, with a new virtual network and its resource sharing, is reached instantaneously.

Fairness of the bandwidth sharing has been recognized an important consideration for fixed networks and different fairness concepts have been introduced, cf. [1]. In ad hoc networks this issue has received attention only recently. In particular the application of the notion of max-min fairness in a static setting has been studied in [2,3,4]. The problem is again more complex than in fixed networks because it is entangled with the lower layer scheduling. Using closely related network models and scheduling constraints to those in the present paper, the authors of the mentioned papers have presented centralized and distributed methods that achieve max-min fair rates for given set of flows. However, the fact that an allocation is optimal in the sense of an utility function does not necessarily guarantee that the system converges to a steady state that is optimal [5]. Also, an analysis of the performance of max-min fair resource sharing in a dynamic setting would be prohibitively difficult.

A new concept of balanced fairness (BF) has recently been introduced by Bonald and Proutière [6,7]. This is a very interesting notion on two accounts. First and foremost, it leads to a network performance which does not depend on the traffic characteristics except the traffic intensity on the different paths, in other words, the performance under BF is insensitive. Secondly, BF often allows an explicit analysis of the performance of simple systems in the dynamic setting.

It should be noted that BF does not represent a solution to an utility optimization problem or guarantee Pareto efficient use of the resources. However, studies of fixed networks have shown that in many cases the performance of a network under BF is similar to that under max-min fairness. BF provides therefore a useful approximation tool for evaluating network performance.

In this paper we introduce a natural extension of the notion of balanced fairness to take into account the fact that the capacities of the virtual network can be changed by the schedule. The resulting BF resource sharing problem is a joint problem of determining both the state-dependent schedule and bandwidth allocation to maximize the use of the network resources while at the same time retaining the balance and the related insensitivity properties. We demonstrate the analysis of the performance under BF by two examples illuminating different computational approaches. Furthermore, the paper contributes to general wireless network optimization problems by establishing a condition under which so called clique constraints provide sufficient conditions for the feasibility of a given set of link capacities of the virtual network.

The paper is organized as follows. Section 2 gives a formal description of the balanced fairness concept with an extension to variable link capacities. It also illustrates the main principles involved in this paper by carefully examining a simple example. These principles are then extended to general ad hoc networks in Sect. 3, while Sect. 4 brings forward an alternative approach to allow performance analysis without the need to explicitly solve the scheduling. Section 5 concludes the paper.

2   Extension of Balanced Fairness

Consider a network consisting of J unidirectional links and carrying N classes of flows using pre-defined routes. Let $\mathcal{F}_j$ denote the set of flows using link j. The network state is represented by a vector $x = (x_1, \ldots, x_N)$ in which x_i is the number of class-i flows in progress. Let $\phi_i(x)$ be the bandwidth allocation to class i in state x. This bandwidth is equally shared by all flows of class i. An allocation is said to be balanced if it holds that

$$\frac{\phi_i(x - e_j)}{\phi_i(x)} = \frac{\phi_j(x - e_i)}{\phi_j(x)} \qquad \forall i,j,\ x_i > 0,\ x_j > 0\ ,$$

where e_i is an N-vector with 1 in the ith component and 0 elsewhere. It can be shown [6] that an allocation is balanced if and only if the allocations can be expressed in terms of a so-called balance function $\Phi(x)$ as

$$\phi_i(x) = k\ \Phi(x - e_i)\ , \tag{1}$$

where the proportionality constant is $k = 1/\Phi(x)$. Conversely, any positive function $\Phi(x)$ defines a balanced allocation by (1).

Balanced fairness as defined by Bonald and Proutière [6,7] refers to the most efficient balanced allocation in a fixed network in the sense that in each state of the system at least one of the links is saturated. This leads to a unique allocation: the balance function is uniquely defined by the recursion,

$$\Phi(x) = \max_j \frac{1}{C_j} \sum_{i \in \mathcal{F}_j} \Phi(x - e_i)\ . \tag{2}$$

The seed of the recursion can be arbitrarily set, e.g. $\Phi(0) = 1$. The extended balanced fairness principle just says that, in each state x, the proportionality constant $k = 1/\Phi(x)$ is chosen as large as allowed by whatever constraints the system is subject to. In the case of fixed routes and fixed link capacities this leads to (2).

Now assume that we have fixed routes but the link capacities $C_j(p)$ depend on the schedule p, i.e. the capacities are related to a virtual network defined by schedule p. Then the constraints are

$$\sum_{i \in \mathcal{F}_j} \phi_i(x) \leq C_j(p), \ \forall j \ , \qquad \Rightarrow \qquad k \leq \min_j \frac{C_j(p)}{\sum_{i \in \mathcal{F}_j} \Phi(x - e_i)} \ ,$$

and the maximizing k is

$$k = \max_p \min_j \frac{C_j(p)}{\sum_{i \in \mathcal{F}_j} \Phi(x - e_i)} \ ,$$

that is, one finds the most constraining link for any schedule p and then makes this constraint as loose as possible by changing p. Accordingly, the balance function $\Phi(x) = 1/k$ of the balanced fairness is now uniquely determined by the recursion

$$\Phi(x) = \min_p \max_j \frac{1}{C_j(p)} \sum_{i \in \mathcal{F}_j} \Phi(x - e_i) \ . \qquad (3)$$

For each state x, this recursion defines both the balance function $\Phi(x)$ and the schedule $p = p(x)$. Recursion (2) is, of course, a special case of this, when the capacities are fixed and no scheduling alternatives are available.

Assume that the flows are generated by sessions, each session being composed of a random number of flows separated by think times. Flow sizes and think time durations can be arbitrarily distributed and need not to be independent. If balanced resource allocation is used and the sessions arrive as a Poisson process then, as shown in [7,5], the steady state distribution of the network state is given by

$$\pi(x_1, \dots, x_N) = \frac{1}{G(\rho)} \Phi(x_1, \dots, x_N) \rho_1^{x_1} \dots \rho_N^{x_N} \ , \qquad (4)$$

and depends on the traffic characteristics only through the traffic loads ρ_i of different routes. Load ρ_i is the product of the flow arrival rate and mean flow size on route i.

In (4), $G(\rho)$ is the normalization constant

$$G(\rho) = \sum_{x_1=0}^{\infty} \dots \sum_{x_N=0}^{\infty} \Phi(x_1, \dots, x_N) \rho_1^{x_1} \dots \rho_N^{x_N} \ , \qquad (5)$$

which depends on the traffic load vector $\rho = (\rho_1 \dots \rho_N)$. The normalization constant $G(\rho)$ is an important quantity as the performance measures can be derived from it. Under a specific condition detailed in [8] the constant can be calculated recursively directly without even solving $\Phi(x)$. In the examples presented in this paper the condition is satisfied and the normalization constant can indeed be obtained in a simple way.

Fig. 1. Example 1

A key performance measure for class-i flows is the throughput γ_i, defined as the ratio of the mean flow size to the mean flow duration. By Little's result this is equal to $\rho_i/\mathrm{E}[x_i]$. The denominator can be obtained by derivation yielding

$$\gamma_i = \frac{G(\rho)}{\frac{\partial}{\partial \rho_i} G(\rho)} \ . \tag{6}$$

Balanced fairness has the very desirable property that the performance of the network is insensitive to traffic details. Moreover, it allows us to evaluate the performance by (3), (5) and (6). Next we illustrate the concept of extended balanced fairness by a very simple example.

2.1   Example 1

The system consists of three nodes (A,B,C) and two radio links (1,2), each of nominal capacity 1. Node A cannot reach node C directly. There are two flow classes: class-1 flows from B to C utilize link 1 only while class-2 flows from A to C use both link 1 and link 2. Figure 1 illustrates the system.

Assume that the two links interfere and cannot be used at the same time. The schedule is now defined by a single parameter p defining which portion of time is scheduled for link 1. The effective link capacities are thus,

$$C_1(p) = p \ , \qquad C_2(p) = 1 - p \ ,$$

and the recursion (3) reads

$$\Phi(x) = \min_p \max\{\frac{\Phi(x - e_1) + \Phi(x - e_2)}{p}, \frac{\Phi(x - e_2)}{1 - p}\} \ .$$

The minimum with respect to p is obtained when the two expressions are equal,

$$p = \frac{\Phi(x - e_1) + \Phi(x - e_2)}{\Phi(x - e_1) + 2\Phi(x - e_2)} \ , \qquad 1 - p = \frac{\Phi(x - e_2)}{\Phi(x - e_1) + 2\Phi(x - e_2)} \ , \tag{7}$$

leading to the recursion for $\Phi(x)$ and its solution

$$\Phi(x) = \Phi(x - e_1) + 2\Phi(x - e_2) \qquad \Rightarrow \qquad \Phi(x) = \binom{x_1 + x_2}{x_1} 2^{x_2} \ . \tag{8}$$

The normalization constant can be easily calculated,

$$G(\rho) = \frac{1}{1 - \rho_1 - 2\rho_2} \ .$$

This is recognized to be the same as the normalization constant of a single fixed link of unit capacity shared by three flow classes with loads ρ_1, ρ_2 and ρ_2 (or two flow classes with loads ρ_1 and $2\rho_2$). Indeed, we have a single resource, the time slot, contended by three transmissions (a class-1 transmission from B to C and class-2 transmissions from A to B and from B to C). From the normalization constant one readily obtains the throughputs using (6),

$$\gamma_1 = 1 - \rho_1 - 2\rho_2 \ , \quad \gamma_2 = \frac{1}{2}(1 - \rho_1 - 2\rho_2) \ . \tag{9}$$

The throughputs are greater than those resulting from the basic BF using any fixed scheduling. Note that for negligible loads the throughputs are 1 and $\frac{1}{2}$, as they should, and also that for any loads the class-2 throughput is always half of that of class 1.

Finally, note that from the balance function (8) the explicit expressions for the state-dependent allocations and schedules can be obtained by (1) and (7), respectively, yielding

$$\begin{cases} \phi_1(x) = \dfrac{x_1}{x_1 + x_2} \ , \\[2mm] \phi_2(x) = \dfrac{\frac{1}{2}x_2}{x_1 + x_2} \ , \end{cases} \qquad \begin{cases} p(x) \ = \dfrac{x_1 + \frac{1}{2}x_2}{x_1 + x_2} \ , \\[2mm] 1 - p(x) = \dfrac{\frac{1}{2}x_2}{x_1 + x_2} \ . \end{cases}$$

These equations can be interpreted at the scheduling level as follows. The time slots are equally shared by all active flows, e.g. on a rotational basis. A class 2 flow needs two time slots for an end-to-end transmission; every second time slot assigned to it is used for transmission from A to B and every second time slot for transmission from B to C.

We reiterate that when the network is operated under this scheme, the performance (9) is insensitive to any detailed traffic characteristics. In this example, the scheme is also Pareto efficient in the sense that no resources are wasted; $p(x) = \phi_1(x) + \phi_2(x)$ and $1 - p(x) = \phi_2(x)$.

3 Scheduling under Interference Constraints

In the previous example, the two links interfered with each other and could not be used at the same time, resulting in effective capacities of the links that depend on scheduling. The same principle extends to general wireless networks as follows. One can define a set of permissible simultaneous transmissions, a transmission mode, consisting of the directed links that can be used at the same time. It suffices to consider only the links used by some flow class and the maximal transmission modes, i.e. those which are not contained in another mode. Denote the set of directed links in a maximal transmission mode by τ and the set of all maximal transmission modes by $\mathcal{T}$.

In this general setting, we define a schedule as the vector $p = \{p_\tau, \tau \in \mathcal{T}\}$, with the meaning that in each time slot one of the τ's is used in some order such that, on average, transmission mode τ is used the portion of time p_τ.

As in Example 1, the effective capacities of links can be defined. We assume that the flow and scheduling time scales are well separated, i.e. the duration of a typical flow is much longer than the time slot. Then, given the schedule p, we have on the flow level a virtual network with 'fixed pipes'. Link j of this fixed network has the effective capacity

$$C_j(p) = C_j \sum_{\tau \in \mathcal{T}: j \in \tau} p_\tau , \qquad (10)$$

where C_j is the nominal capacity (bandwidth) of the radio channel on link j. With these capacities one can solve the recursion (3) numerically or, as in Example 1, analytically. In general, we call a set of link capacities d_j, $j = 1, \ldots, J$, feasible if a schedule p exists such that $d_j \leq C_j(p)$ for all j.

3.1   Modeling the Interference

Interference determines which links can transmit simultaneously. Here we go briefly through different alternatives for modeling interference in ad hoc networks and define the concept of link graph.

Elementary interference models set the following constraints to the links in the network. A node may not transmit and receive simultaneously and it cannot transmit or receive more than one packet at a time. In other words, all the links connected to a given node belong to different transmission modes. Such models are often justified by the assumption that other transmissions in the vicinity of the node can operate without conflict using locally distinct frequencies [9,2].

A more detailed model would entail that no two links can be simultaneously active if either of the receiving ends is interfered by the other transmission. In the model presented in [10] a transmission can prevent reception everywhere within the transmission range, whereas in the widely applied protocol model [11] the interference depends on the locations of the transmitting node so that the closest (with a selected margin) transmission can be successfully received.

Note that the above models define interference as a pairwise property of the links. Generally, the successful reception depends on the signal to interference ratio calculated at each receiver, where the interference depends on all other links in use. This, however, complicates the search of transmission modes and in this paper we restrict ourselves to the pairwise models. No further restrictions are made on model selection.

Pairwise interference can be described using a link graph. Given a network and a set of flows with their routes, the corresponding link graph is constructed as follows: each active directed link in the network is mapped to a vertex and an edge connects two link graph vertices if the corresponding links interfere with each other. By definition, each τ corresponds to a maximal independent set in the link graph, i.e. a maximal subset of the vertices such that no two vertices in the subset represent an edge of the link graph. Therefore, in principle, one can find $\mathcal{T}$ by enumerating all independent sets of the link graph.

3.2   LP-Formulation for the Recursion Step

In some simple cases, such as the one in Example 1, one can find the optimal schedule p and bandwidth allocation analytically. In general, however, one has to resort to numerical analysis. In this case it is useful to formulate the recursion step and the corresponding schedule optimization as an LP-problem. We outline the approach, though it is not utilized in this paper.

Let $y = (p, k)^T$ be the decision vector and let $a = (0, ..., 0, 1)$. If the $\Phi(x - e_i)$ are known for all i, we obtain $\Phi(x) = 1/k$ and the corresponding schedule $p(x)$ from y by finding the maximum in the LP-problem:

$$\text{max } ay$$

subject to

$$My \leq (\bar{0}, 1)^T , \qquad y \geq 0 ,$$

where $\bar{0}$ is a J-vector of zeros. M is the constraint matrix given by

$$M = \begin{pmatrix} -A & B(x) \\ \bar{1} & 0 \end{pmatrix} ,$$

where $\bar{1}$ is a $|\mathcal{T}|$-vector of ones, A is the $J \times |\mathcal{T}|$ matrix $A_{j,\tau} = 1_{j \in \tau}$ and $B(x)$ is the column vector $B_j(x) = C_j^{-1} \sum_{j \in \mathcal{F}_j} \Phi(x - e_i)$. Note that only the element $B(x)$ in the matrix M needs to be updated in the recursion.

4   Maximal Clique Constraints

In Example 1, the bandwidth shares after schedule optimization could be interpreted to be limited solely by the fact that the three transmissions contend for the same time slot.

In a general scheduling and bandwidth sharing problem we can similarly identify one or several sets of transmissions such that transmissions in a given set contend for a common time slot. Such transmissions constitute a clique in the corresponding link graph. Each clique q imposes a necessary condition on the bandwidth allocation. The most stringent set of conditions is set by the maximal cliques, i.e. cliques that are not a subset of another clique. Thus, we have the necessary conditions for a feasible bandwidth allocation

$$\sum_{j \in q} \frac{1}{C_j} \sum_{i \in \mathcal{F}_j} \phi_i(x) \leq 1 , \qquad \forall q \in \mathcal{Q} , \tag{11}$$

where $\mathcal{Q}$ denotes the set of all maximal cliques. The maximal cliques can be enumerated, e.g. by an algorithm from [12].

An interesting question is whether the maximal clique constraints (11) also give sufficient conditions for a feasible bandwidth allocation, i.e. whether there exists a schedule that allows attaining the link capacities required by an allocation. By counter examples one can easily see that this is not generally true.

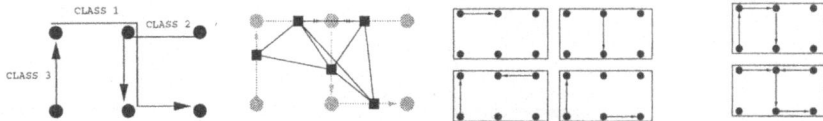

Fig. 2. Flows, link graph, maximal independent sets and cliques of Example 3.

In Appendix A, however, we prove a lemma stating that the maximal clique constraints do provide necessary and sufficient conditions for the feasibility of an allocation if the link graph is a perfect graph[1].

When this is the case the bandwidth allocation is only limited by conditions (11) which are of the same type as the link capacity constraints in a fixed network. Note, however, that now there is one constraint for each clique as opposed to one constraint for each link in a fixed network, and also that a bandwidth allocation ϕ_i can appear in a condition several times as the flows in class i can traverse several links in the same clique. The balance function can then be calculated using the recursion

$$\Phi(x) = \max_{q \in \mathcal{Q}} \sum_{j \in q} \frac{1}{C_j} \sum_{i \in \mathcal{F}_j} \Phi(x - e_i) \ . \tag{12}$$

While this greatly simplifies the task of evaluating the flow level performance as there is no need to explicitly consider the scheduling, it indeed leaves the schedule indetermined. Therefore, when needed, the scheduling must be worked out separately.

4.1 Example 2

Consider now a slightly more complicated example consisting of six nodes, three flow classes and five active unidirectional links of unit capacity, as shown in Fig. 2. The interferences shown in the link graph result from the protocol model (see Sect. 3.1) and the link graph has four maximal independent sets and two maximal cliques. Note that the link graph is triangulated (i.e. it contains no induced cycles other than triangles) and thus a perfect graph (cf. [13]) and the maximal clique constraints apply. These can be written as

$$\begin{cases} 3\phi_1(x) + 2\phi_2(x) & \leq 1 \ , \\ 2\phi_1(x) + \phi_2(x) + \phi_3(x) \leq 1 \ , \end{cases}$$

which results in the recursion

$$\Phi(x) = \max\{3\Phi(x - e_1) + 2\Phi(x - e_2), 2\Phi(x - e_1) + \Phi(x - e_2) + \Phi(x - e_3)\}$$

$$= 2\Phi(x - e_1) + \Phi(x - e_2) + \max\{\Phi(x - e_1) + \Phi(x - e_2), \Phi(x - e_3)\} \ .$$

---

[1] A graph G is called perfect if the chromatic number $\chi(H)$ every induced subgraph $H \subseteq G$ equals the maximum clique size $\omega(H)$ of the subgraph. A conjecture by Berge says that a graph G is perfect if and only if neither G nor its complement $\bar{G}$ contains an odd cycle of length at least 5 as an induced subgraph [13].

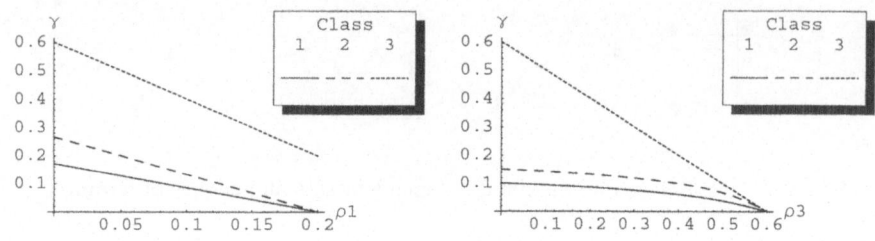

Fig. 3. Throughputs of the flow classes in Example 3. Left; $\rho_2 = 0.2$, $\rho_3 = 0.2$. Right; $\rho_1 = 0.1$, $\rho_2 = 0.2$

It can be shown by induction that when $x_3 > 0$ the latter expression realizes the maximum. Thus,

$$\Phi(x) = \begin{cases} 3\Phi(x - e_1) + 2\Phi(x - e_2) , & x_3 = 0 , \\ 2\Phi(x - e_1) + \Phi(x - e_2) + \Phi(x - e_3) , & x_3 > 0 . \end{cases}$$

As the recursion has a given form throughout a coordinate plane as well as in a positive part of the state space, the method of [8] can be applied to the calculation of the normalization constant in parts. For brevity we omit the details and only give the result

$$G(\rho) = \frac{1 - 2\rho_1 - \rho_2}{(1 - 3\rho_1 - 2\rho_2)(1 - 2\rho_1 - \rho_2 - \rho_3)} .$$

In the same way as in the first example, throughputs of different classes can be calculated by (6). The results are shown in Fig. 3 for two different scenarios. In the first, ρ_2 and ρ_3 are kept fixed while ρ_1 is varied. In the second, ρ_3 is varied while the two others are fixed.

5   Conclusions

This paper studied flow-level dynamics of ad hoc networks and illustrated how the concept of balanced fairness can be extended to allow the derivation of performance measures of a wireless multihop network. The resource allocation is defined by a balance function which can be computed recursively using, e.g. one of the three methods presented in this paper: One can either explicitly write down the capacities of the links as a function of schedule or solve the integrated recursion step and scheduling as an LP-problem. The third approach, applicable under special circumstances, is to use the maximal clique constraints that often yield a simple recursion for the balance function. When the link graph is perfect this approach is feasible and especially suitable for performance analysis since the actual schedule need not be worked out.

A noteworthy feature of the scheme is that the performance is insensitive to traffic details. This potentially allows one to develop simple and robust provisioning rules that depend only on traffic intensities.

A   Necessity and Sufficiency of Maximal Clique Constraints

Denote the set of maximal cliques of a link graph G by $\mathcal{Q}$. Let d_j denote the capacity of link j in the virtual network. As discussed in Sect. 3, the capacities $\{d_j\}$ are feasible if there exists a conflict-free schedule such that each link j is scheduled for transmission the fraction of time d_j/C_j.

Lemma 1. *If the link graph G of an ad hoc network is perfect then both a necessary and a sufficient condition for the feasibility of capacities $\{d_j\}$ is*

$$\sum_{j\in q}\frac{d_j}{C_j} \leq 1 \ , \qquad \forall q \in \mathcal{Q} \ , \tag{13}$$

Proof. The condition is necessary since d_j/C_j is the fraction of the time that has to be scheduled for transmission on link j and for all $j \in q$ the transmissions must be non-simultaneous.

To prove sufficiency, we assume that the d_j/C_j can be written in the form

$$\frac{d_j}{C_j} = \frac{n_j}{N} \ , \qquad \forall j \ ,$$

where the n_j and N are integers (by choosing N large enough these relations can be satisfied to any desired accuracy). Our task now is to show that if

$$\sum_{j\in q} n_j \leq N \ , \qquad \forall q \in \mathcal{Q} \ , \tag{14}$$

then there exists a conflict-free schedule in which each link j is given the fraction of time n_j/N. To this end, consider a frame of arbitrary duration and divide it into N time slots. Now, in order to realize the capacities $\{d_j\}$, one should be able to assign n_j time slots in the frame to each link j so that no conflicts occur.

If $n_j = 1$ for all j then finding a conflict free schedule is equivalent to the graph coloring problem on G in the following way. Let each vertex j represent a one-slot transmission on link j, which has to be scheduled to one of the N time slots in the frame, the position of which we refer to as the "color" of the one-slot transmission. That no two interfering transmissions can use the same time slot translates to that two vertices of G cannot have the same color if they are connected with an edge. As we assumed G to be perfect, the number of colors required equals the maximum clique size, which by (14), with $n_j = 1$ for all j, is no greater than N the number of available colors. So a conflict-free schedule for the case where each link has one one-slot transmission does exist.

We can extend this consideration to any numbers n_j satisfying (14). Make n_j copies of each vertex $j \in G$, each representing one of the n_j one-slot transmissions on link j. All replicas of vertex j have to be connected by an edge to each other and to all other nodes and their replicas in the same clique, as the corresponding transmissions interfere with each other. Such a graph can be constructed by expanding vertices one-by-one until each link j is represented by n_j vertices. A result of graph theory (Lemma 5.5.4 in [13]) states that a graph obtained from a perfect graph by expanding a vertex is again perfect. Thus, the final graph G^*, resulting from the $(n_j - 1)$-fold expansion of each link j, is perfect. Consequently, the number of colors needed for coloring G^* equals its maximum clique size, $\max_{q \in \mathcal{Q}} \sum_{j \in q} n_j$, which again by (14) is no greater than the number of available colors N.

Corollary 1. *A flow allocation $\{\phi_i(x)\}$ satisfying*

$$\sum_{j \in q} \frac{1}{C_j} \sum_{i \in \mathcal{F}_j} \phi_i(x) \leq 1 , \qquad \forall q \in \mathcal{Q} , \tag{15}$$

is feasible if the corresponding link graph G is perfect.

Proof. Setting $d_j = \sum_{i \in \mathcal{F}_j} \phi_i(x)$ the result follows directly from the lemma.

References

1. Mo, J., Walrand, J.: Fair end-to-end window-based congestion control. IEEE/ACM Transactions on Networking **8** (2000) 556–567
2. Tassiulas, L., Sarkar, S.: Maxmin fair scheduling in wireless networks. In: Proc. IEEE INFOCOM 2002. (2002) 764–772
3. Huang, X., Bensaou, B.: On max-min fairness and scheduling in wireless ad hoc networks: analytical framework and implementation. In: Proc. ACM MobiHOC 2001. (2001) 221–231
4. Luo, H., Lu, S., Bharghavan, V.: A new model for packet scheduling in multihop wireless networks. In: Proc. ACM MOBICOM 2000. (2000) 76–86
5. Bonald, T., Proutière, A.: Insensitive bandwidth sharing in data networks. Queuing Systems **44** (2003) 69–100
6. Bonald, T., Proutière, A.: Insensitivity in processor-sharing networks. Performance Evaluation **49** (2002) 193–209
7. Bonald, T., Proutière, A.: On performance bounds for balanced fairness. Performance Evaluation **55** (2004) 25–50
8. Bonald, T., Proutière, A., Roberts, J., Virtamo, J.: Computational aspects of balanced fairness. In Charzinski et al., eds.: Providing Quality of Service in Heterogeneous Environments. Elsevier Science (2003) 801–810
9. Hajec, B., Sasaki, G.: Link scheduling in polynomial time. IEEE Transactions on Information Theory **34** (1988) 910–917
10. Arikan, E.: Some complexity results about packet radio networks. IEEE Transactions on Information Theory **IT-30** (1984) 681–685
11. Gupta, P., Kumar, P.R.: The capacity of wireless networks. IEEE Transactions on Information Theory **46** (2000) 388–404
12. Bron, C., Kerbosch, J.: Algorithm 457: Finding all cliques of an undirected graph. Communications of the ACM **16** (1973) 575–578
13. Diestel, R.: Graph Theory. Springer-Verlag (2000) Electronic Edition 2000, http://www.math.uni-hamburg.de/home/diestel/books/graph.theory/.

Cell Throughput Analysis of the Proportional Fair Scheduling Policy

Jin-Ghoo Choi and Saewoong Bahk

School of Electrical Engineering & Computer Science
Seoul National University
Rm 314, Bldg 132, San 56-1, Shillim-dong, Kwanak-ku, Seoul 151-744, Korea
{cjk, sbahk}@netlab.snu.ac.kr
TEL: 82-2-880-1813
FAX: 82-2-880-8214

Abstract. The fairness concept has been widely studied in the area of data networks. The most well-known fairness criterion, max-min fairness, gives priority to the minimum rate session. Kelly questioned its appropriateness in his works on the bandwidth sharing among the end-to-end flows and proposed another fairness criterion preferring short distance flows to enhance the overall throughput, which is called the proportional fairness (PF). A simple scheduler achieving this objective was introduced in wireless access networks and revealed that it can achieve a good compromise between cell throughput and user fairness. Though it has received much attention for some time, research on its performance mainly depended on computer simulations. In this paper, we analyze the PF scheduler to obtain the cell throughput which is a primary performance metric.

Keywords: Wireless networks, proportional fair scheduler, cell throughput

1 Introduction

Traditionally wireless access networks, especially cellular networks, have provided only voice service and were considered as an extension of wireline telecommunication systems. However, as demands for ubiquitous data services arose with the rapid growth of the Internet, wireless access networks and requirements placed upon them are evolving to support various high data rate services including multimedia. To accommodate these requirements, 3rd generation cellular systems such as wideband code division multiple access (W-CDMA) and cdma2000 have standardized enhanced radio access, and their core networks which currently consist of circuit switches and mobility supporting devices, are expected to be replaced by IP enabled components including routers [1].

In wireless data networks an important tool for resource management is the packet scheduler. Several works have shown that the scheduling policy can significantly affect system performances such as throughput, delay, fairness and

N. Mitrou et al. (Eds.): NETWORKING 2004, LNCS 3042, pp. 247–258, 2004.
© IFIP International Federation for Information Processing 2004

loss rate in wireline and wireless domains [2]-[3]. Schedulers in wireless networks need to consider the unique characteristics of time-varying and location dependent channel conditions when contrasted with wired networks. Though many scheduling policies are proposed previously [4], they have the common weak point of assuming the channel to have only two states, "good" and "bad" states.

To know the channel state exactly, a BS requires fast feedback from mobile terminals about more detailed channel state information, which incurs a cost of increased signaling burden. For instance, the channel state can be divided into 2^m levels by using m bits per feedback. This structure of wireless link state has been realized in the IS-856 standard for the downlink direction, which is known as the High Data Rate (HDR) system of Qualcomm [5].

Several schedulers for this architecture have been proposed. [6] maximizes cell throughput under the constraint that the normalized average throughput for each user is equal, where the normalization factor reflects users' required quality of service (QoS). The scheme proposed in [7] achieves the same objective with a different constraint where the allocated time portion of each user is assumed given. This scheduler can be shown to maximize the objective function (for example, based on utility) if it depends only on the signal-to-noise ratio (SNR). [6] and [7] use adaptive schemes with time-varying parameters, which can lead to convergence problems when implemented in real systems.

We consider schedulers exploiting the time-varying channel condition without explicit constraints. The maximum cell throughput can be obtained by serving the user with the best channel condition, however, it raises a serious fairness problem. The packet scheduler should meet a reasonable tradeoff between throughput and fairness. A scheduler achieving the PF criterion was introduced in [5], [8]. It has low complexity, which facilitates its implementation in real systems. Though it has received much attention for some time, the cell throughput obtainable by the PF scheduling policy has not been analyzed yet. Cell throughput is defined as the sum of each user's average throughput and is considered a primary performance metric for schedulers in wireless networks. In this paper, we analyze the PF scheduler and derive an approximate expression for cell throughput.

This paper is organized as follows. In Section II, we introduce the system and channel model. Section III analyzes the PF scheduler by adopting the linear and logarithmic models for relating the feasible rate and the SNR. Then we compare the simulation results with numerical analysis in Section IV, followed by concluding remarks in Section V.

2   Model Description

2.1   System and Channel

We consider the downlink channel of a single cell wireless access network where a BS serves N mobile terminals (or users). The downlink structure is very similar to that of the IS-856 system. It is a single broadband channel shared by all users in the time division multiplexing manner. The BS exploits the pilot signal, which is pre-defined by the protocol, in the specified position of each time slot, and every mobile measures it to obtain the channel gain. The BS receives the fed back signal from all the users to collect the current channel status. Based on the channel information, the radio frequency scheduler selects a user among the active[1] ones to be served in the next slot. Though various schedulers are proposed in this context, among them we concentrate on the PF scheduler. A packet from the selected user is transmitted with the BS's full power, which is assumed fixed for all users.

When the transmission power of the BS is P_t, the receiving power of mobile user i is given by $P_i = |h_i|^2 P_t$, where h_i is the channel gain written as

$$h_i = \sqrt{cd_i^{-\alpha} s_i}\, z_i. \tag{1}$$

c is a constant incorporating the transmission and receiving antenna gains, d_i is the distance from the BS to user i, α is the path loss exponent (estimated to be about 4.0 in typical urban environments), s_i is a random variable for the shadow fading effect, and z_i represents the phasor sum of the multipath components. The shadow fading effect s_i is known to follow the log-normal distribution with zero-mean and the variance σ_s^2 (dB) in the log-scale. The multipath fading effect z_i is modeled as the second-order chi-square or exponential random variable with mean 1.0, which represents the Rayleigh fading channel.

Since we consider the single cell scenario, there is no inter-cell interference. Therefore, we can represent the SNR of user i as $Z_i = P_i/P_n$ where P_n is the background noise power which includes the thermal noise and other Gaussian interferences. [9] defined the median SNR at the cell edge, l, to represent the noise level of the wireless environment considered. For $l = cD^{-\alpha} P_t/P_n$ where D is the radius of the cell, we obtain the average[2] SNR of user i as $E\{Z_i\} = \overline{Z}_i = l(\frac{D}{d_i})^{\alpha} s_i$. Note that the received signal level follows the exponential distribution and the noise power is constant so that the SNR can also be modeled by the exponential random variable.

2.2 PF Scheduler

We briefly explain the operation of the PF scheduler. The average throughput of each user is tracked by an exponential moving average. At the beginning of each time slot, each user feeds back the channel state (or the feasible rate) to the BS. The BS calculates the ratio of the feasible rate to the average throughput for each user, which is defined as the *preference metric* and is the key selection criterion. The user with the maximum preference metric will be selected for transmission at the next coming slot.

This is described formally as follows. In time slot n, the feasible rate of user k is $R_k[n]$ and its moving average is denoted by $\tilde{R}_k[n]$. Then, user $k^* =$

---

[1] It means that its downlink queue is already backlogged.

[2] In this paper, we use both the notation $E\{\cdot\}$ and the overline to represent the average of a random variable.

$\arg\max_k R_k[n]/\tilde{R}_k[n]$ is served in time slot n, and the average throughput of each user is updated by

$$
\tilde{R}_k[n+1] = \begin{cases} \left(1 - \frac{1}{t_c}\right)\tilde{R}_k[n] + \frac{1}{t_c}R_k[n], \ k = k^* \\ \left(1 - \frac{1}{t_c}\right)\tilde{R}_k[n], \qquad\quad k \neq k^* \end{cases}
\tag{2}
$$

where t_c is the time constant for the moving average. It is clear that the PF scheduler affects *relative* preference to users with good channels as opposed to *absolute* preference.

In fast fading environments, the channel state of a user fluctuates around an average level in a random manner. When the number of users is small and all the users happen to be in bad state, then throughput in that slot will be low regardless of whoever is scheduled. On the other hand, when the number of users is large, some users will be in the good state with high probability. Therefore, we can obtain additional cell throughput gain by scheduling them first to utilize the characteristics of fast fading channels, which is called *multi-user diversity*.

3   Analysis

3.1   Assumptions

In this section we derive an approximate expression for the cell throughput under the PF scheduling policy. We make the following assumptions.

1. Users are distributed uniformly throughout the entire cell area.
2. Every session (or user) is always active in the downlink direction. We ignore the throughput loss due to the lack of data to be transmitted.
3. The distribution of the channel gain of user i, $h_i[n]$, does not depend on slot n and it is constant for the slot duration.
4. Our model uses the ratio of the SNR to the average SNR as the preference metric instead of the original PF metric of the ratio of the feasible rate to the average rate (or throughput). That is, in time slot n, the user k^* is served with $\arg\max_k Z_k[n]/\tilde{Z}_k[n]$ where $\tilde{Z}_k[n]$ denotes the exponential moving averaged SNR. Though this criterion is not exactly the same as that of the PF scheduler, they share the important characteristic of allocating almost the same portion of time slots to each user and allowing *relative* preference to users with good channels.
5. The feasible rate is a strictly monotonic increasing function of the SNR. It is also continuous while, in real systems, it takes values from a discrete set of supported rates. For example, the IS-856 system defines 11 feasible rates [5].
6. The average throughput and the average SNR are obtained by the time average rather than the moving average. For the throughput, $\tilde{R}_k[n+1] = \frac{1}{n}\sum_{i=1}^{n} R_k[i]I_k[i]$, where the indicator function $I_k[i]$ is 1 if the user k is scheduled in slot i and 0 otherwise. The procedures to obtain the average SNR are the same.

We add some remarks on assumptions 4 and 5. The feasible transmission rate depends on various factors such as the SNR, the degree of signal distortion during the transmission, the modulation scheme, the channel coding scheme, the hardware structure of the receiver, etc. Among these, the most important factor is the SNR, so we express the feasible rate as a function of the SNR aggregating other factors into a specific form.

We consider two popular models in this paper. They are the linear model and the logarithmic model. The linear model was widely adopted to describe CDMA systems though it is not accurate in very high SNR regions. In this model, our modified preference metric completely matches that of the PF scheduler. In the case of the logarithmic model, it does not coincide with the genuine preference metric of the PF scheduler and introduces errors in the analysis. This is evaluated by comparing the analysis with simulations.

3.2   Cell Throughput of the PF Scheduler

The redefined preference metric of the user i is $\Gamma_i[n] = Z_i[n]/\tilde{Z}_i[n]$ in time slot n. When n approaches infinity, we assume $\tilde{Z}_i[n]$ converges to a stationary value and denote it as T_i. Considering $Z_i[n]$ is independent of the slot time, we omit the time index n. Therefore, the preference metric is rewritten as

$$\Gamma_i = \lim_{n\to\infty} \frac{Z_i[n]}{\tilde{Z}_i[n]} = \frac{Z_i}{T_i}, \tag{3}$$

which follows the exponential distribution.

We can write the long-term average throughput of user i, Φ_i, which is the primary intermediate result to get the cell throughput, the product of the probability of scheduling user i and its expected feasible rate when scheduled. We watch the "scheduled probability" of the user i first. The scheduled user sees that its preference metric is larger than that of any other user. By denoting the maximum preference metric of all the users except i as Γ_{i-}, we can write the scheduled probability as $\Pr\{\Gamma_i > \Gamma_{i-}\}$. Since $\Gamma_{i-} = \max\{\Gamma_1, ..., \Gamma_{i-1}, \Gamma_{i+1}, ..., \Gamma_N\}$ for N users within the cell, its cumulative density function (CDF) $F_{i-}(t)$ is given as

$$F_{i-}(t) = \Pr\{\Gamma_{i-} \le t\} = \prod_{j\neq i} F_j(t), \tag{4}$$

where $F_j(t)$ is the CDF of Γ_j.

The feasible rate strongly depends on the SNR although the concrete relationship varies depending on the underlying technologies of the physical and link layers used. Since the SNR Z_i is given as $T_i \cdot \Gamma_i$, the feasible rate R_i is a function of Γ_i, which is denoted by $R_i = \xi(\Gamma_i)$. Since $\xi(\cdot)$ is a strictly monotonic increasing function of Γ_i, it always has the inverse function $\Gamma_i = \xi^{-1}(R_i) = \zeta(R_i)$. Then, the long-term average throughput of user i is

$$E\{\Phi_i\} = \Pr\{\Gamma_i > \Gamma_{i-}\} \cdot E\{R_i | \Gamma_i > \Gamma_{i-}\}$$

$$= \int_0^{\hat{R}} r \frac{d}{dr} \Pr\{R_i \leq r \text{ and } \Gamma_i > \Gamma_{i-}\}dr$$

$$= \int_0^{\hat{R}} r \frac{d}{dr} \Pr\{\Gamma_i \leq \zeta(r) \text{ and } \Gamma_i > \Gamma_{i-}\}dr,$$

where $\hat{R}$ indicates the supported maximum feasible rate. After some manipulation, we obtain

$$E\{\Phi_i\} = \int_{\zeta(0)}^{\zeta(\infty)} \xi(t) f_{\Gamma_i}(t) F_{i-}(t)dt, \tag{5}$$

where $f_{\Gamma_i}(t)$ denotes the probability density function (PDF) of Γ_i.

Consider the case where the SNR of user i in time slot n can be represented as the product of the constant c_i and the random variable $C_i[n]$. That is, $Z_i[n] = c_i C_i[n]$. T_i/c_i is known to be the same for all i's when the following conditions are satisfied [10]: i) random variable $C_i[n]$'s are i.i.d. regardless of users and time slots, ii) $C_i[n]$'s are independent of the transmission power, and iii) the preference metric is a linear function of the SNR.

Since our model corresponds to this case with $c_i = \overline{Z}_i$ and[3] $C_i[n] = |z_i|^2$, $T_i/c_i = T_i/\overline{Z}_i = \overline{\Gamma}_i$ is the same for all i's. Deleting the subscript i, we can see that, from (3), the PDF of Γ_i is given as $f_{\Gamma_i}(t) = \frac{1}{\overline{\Gamma}}\exp(-\frac{t}{\overline{\Gamma}})$ for $t \geq 0$ and, from (4), $F_{i-}(t) = \left(1 - \exp(-\frac{t}{\overline{\Gamma}})\right)^{N-1}$ for $t \geq 0$. Then, (5) can be rewritten as

$$E\{\Phi_i\} = \int_{\zeta(0)}^{\zeta(\infty)} \xi(t) \frac{1}{\overline{\Gamma}} \exp(-\frac{t}{\overline{\Gamma}}) \left(1 - \exp(-\frac{t}{\overline{\Gamma}})\right)^{N-1} dt. \tag{6}$$

Next, we consider the function $R_i = \xi(\Gamma_i)$ in the linear and logarithmic forms.

Linear model. In this model the feasible rate is linearly proportional to the SNR. Past works adopted this model to describe CDMA systems and used it as an approximation to represent relations containing the small variation of SNR. When the relation is given as $R_i = \beta W Z_i$, where β is a constant and W is system bandwidth, we obtain the average throughput of user i from (6) as follows.

$$E\{\Phi_i\} = \frac{\beta W \overline{Z}_i}{N} \cdot N \int_0^\infty t e^{-t}(1 - e^{-t})^{N-1} dt, \tag{7}$$

where $\frac{\beta W \overline{Z}_i}{N} = \frac{1}{N}E\{R_i\}$ represents the average throughput of user i when the round robin scheduler is used. The second term is the auxiliary gain by using the PF scheduling and we define it as the multi-user diversity factor $M(N)$. After some manipulation, we have

$$M(N) = N \sum_{m=0}^{N-1} \binom{N-1}{m} \frac{(-1)^m}{(m+1)^2}. \tag{8}$$

---

[3] Note that $|z_i|^2$ is the exponential random variable with the mean 1.0 regardless of i.

The multi-user diversity factor increases rapidly with the increase in the number of users. This is owing to the very high SNR occasionally experienced during fast fading. Recognizing that the linear model overestimates the feasible rate in this region, we can see that the result has limited meaning in practical environments.

We derived a user's long-term average throughput assuming that the user's location is fixed or the shadowing effect is given. Now it needs to be averaged for all possible shadow fading and for the entire cell area and, then, multiplied by the number of users to obtain the cell throughput. If we denote the expectation of the shadowing by $E_S\{\cdot\}$,

$$E_S\{\overline{\Phi_i}\} = \frac{\beta W}{N}M(N)E_S\{\overline{Z_i}\} = \frac{\beta W}{N}M(N)E_S\{s_i\}l\left(\frac{D}{d_i}\right)^\alpha. \qquad (9)$$

Representing the average over the entire area by $E_A\{\cdot\}$, we can express the cell throughput $\hat{T}_{cell}$ as

$$\hat{T}_{cell} = N \cdot E_A\{E_S\{\overline{\Phi_i}\}\} = \beta W M(N) E_S\{s_i\} \cdot \Omega_A^{-1} \int_A l\left(\frac{D}{d_i}\right)^\alpha dA, \qquad (10)$$

where A and Ω_A represent the cell and its area, respectively.

In this case, we can see that the cell throughput does not have a finite value. This means that the linear model is not accurate for very high SNR and produces unreasonably high data rates. Consequently, the extremely high throughput of users close to the BS gives rise to an infinite cell throughput. We can obtain the finite cell throughput by enforcing the distance between the BS and users to be larger than $D_0 = \eta D(\eta < 1)$. This reflects the fact that the transmission antenna of BS is, typically, apart from the ground. With this constraint, we can express the cell throughput as

$$\hat{T}_{cell} = W\frac{4l\beta}{2-\alpha}\frac{1-\eta^{2-\alpha}}{1-\eta^2}\exp\left(\left(\frac{\ln 10}{10\sqrt{2}}\sigma_s\right)^2\right)M(N) \qquad (11)$$

by using $E_S[s_i] = 2\exp\left(\left(\frac{\ln 10}{10\sqrt{2}}\sigma_s\right)^2\right)$. Henceforth, we use the *normalized* cell throughput $T_{cell} = \hat{T}_{cell}/W$ to make it independent of system bandwidth, which is the spectral efficiency of the cell in unit of bps/Hz.

Logarithmic model. Adaptive modulation schemes have been introduced to perform link adaptation. Among those, M-level quadrature amplitude modulation (M-QAM) is promising where M is typically set to the power of 2. With the increase of M, both the information bit rate and the bit error rate for the given SNR increase. So, the large M is used in good channel condition while the small M in bad condition. It was verified that, with M-QAM, the feasible transmission rate is related to the SNR in the logarithmic manner [9], [11]. In this paper, we adopt the form proposed in [9],

$$R_i = W\log_2\left(1 + \frac{Z_i}{K}\right), \qquad (12)$$

where K is a constant depending on the system design and the target bit error rate required for reliable transmission. We call K the system efficiency factor since the feasible rate for the same SNR varies according to K. When $K = 1.0$, (12) becomes the well-known Shannon capacity formula.

Substituting $\xi(\Gamma_i) = \frac{W}{\ln 2}\ln(1+\frac{T_i}{K}\Gamma_i)$ into (6), we obtain the long-term average throughput of user i as follows.

$$
E\{\Phi_i\} = \frac{W}{\ln 2} \int_0^\infty \ln(1 + \frac{\overline{Z_i}}{K}t)e^{-t}(1 - e^{-t})^{N-1}dt
$$

$$
= \frac{W}{\ln 2} \sum_{m=0}^{N-1} \binom{N-1}{m}(-1)^m \int_0^\infty \ln\left(1 + \frac{\overline{Z_i}}{K}t\right)e^{-(m+1)t}dt.
$$

As $\int_0^\infty \ln(1 + \mu t)e^{-\kappa t}dt = \frac{1}{\kappa}\exp(\frac{\kappa}{\mu})\text{expint}(\frac{\kappa}{\mu})$ in [12], we[4] can rewrite (13) as

$$
E\{\Phi_i\} = \frac{W}{\ln 2} \sum_{m=0}^{N-1} \binom{N-1}{m}\frac{(-1)^m}{(m+1)} \exp\left(\frac{K}{\overline{Z_i}}(m+1)\right)\text{expint}\left(\frac{K}{\overline{Z_i}}(m+1)\right). \quad (13)
$$

In [9], $\exp(t)\text{expint}(t)$ is closely approximated as $\nu_1 \ln 2 \cdot \ln(1 + \nu_2 t^{-1})$ for a wide range of t, where ν_1 and ν_2 are set to 1.4 and 0.82, respectively.

Applying this approximation to (13), we obtain

$$
E\{\Phi_i\} \approx W\nu_1 \sum_{m=0}^{N-1} \binom{N-1}{m}\frac{(-1)^m}{(m+1)} \ln\left(1 + \frac{\nu_2}{K(m+1)}\overline{Z_i}\right). \quad (14)
$$

Taking the expectation of the shadow fading, we can approximate the above as

$$
E_S\{\overline{\Phi_i}\} \approx W\nu_1 \sum_{m=0}^{N-1} \binom{N-1}{m}\frac{(-1)^m}{(m+1)} \cdot E_S\left\{\ln\left(1 + \frac{\nu_2}{K(m+1)}\overline{Z_i}\right)\right\}. \quad (15)
$$

Using $\overline{Z_i} = l(\frac{D}{d_i})^\alpha s_i$, we need to calculate the form $E_S\{\ln(1+\mu s)\}$. We omit the subscript i here for notational simplicity. In [9], $\Pr\{Y = \ln(1+\mu s) \leq y\}$ is shown to be close to $1 - \frac{1}{2}\text{erfc}(\frac{y - \ln\mu}{\frac{\ln 10}{10}\sqrt{2}\sigma_s})$. This indicates that $E_S\{Y\} = E_S\{\ln(1+\mu s)\}$ is well approximated by the Gaussian random variable with the mean $\ln\mu$ and the standard deviation $\frac{\ln 10}{10}\sigma_s$. [9] also pointed out that it is valid only for $y \gg 1$. Since we are concerned with the mean rather than the CDF itself, we can use this approximation for our purpose.

However this gives an unsatisfactory result for small μ. Especially, when μ is less than 1.0, we get a negative mean by the approximation while Y is always larger than 0. It motivates us to derive another approximation.

$$
E_S\{\ln(1 + \mu s)\} = \ln(1 + \mu) + \nu_3, \quad (16)
$$

---

[4] expint(t) is defined as $t\int_1^\infty \ln x \cdot e^{-tx}dx$ and called the exponential integral function.

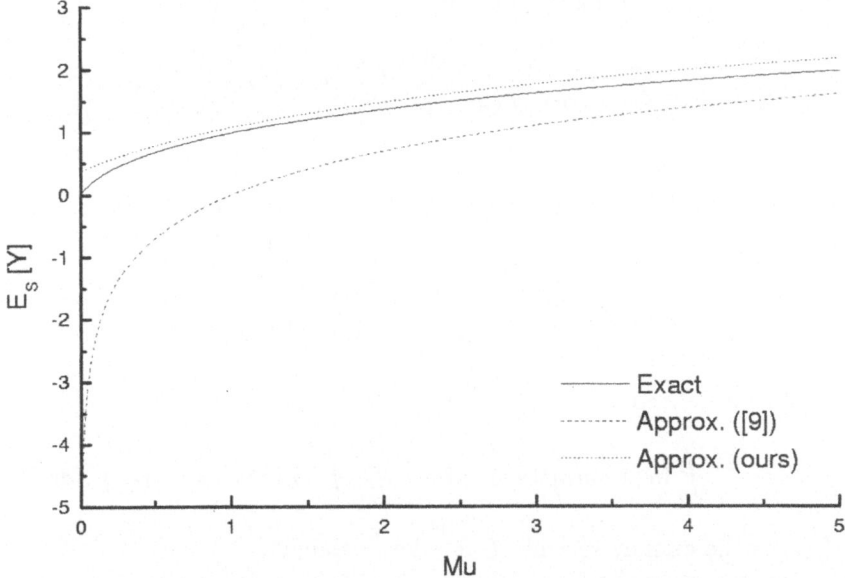

Fig. 1. Accuracy of the approximation for the average of the $\ln(1 + \mu s)$.

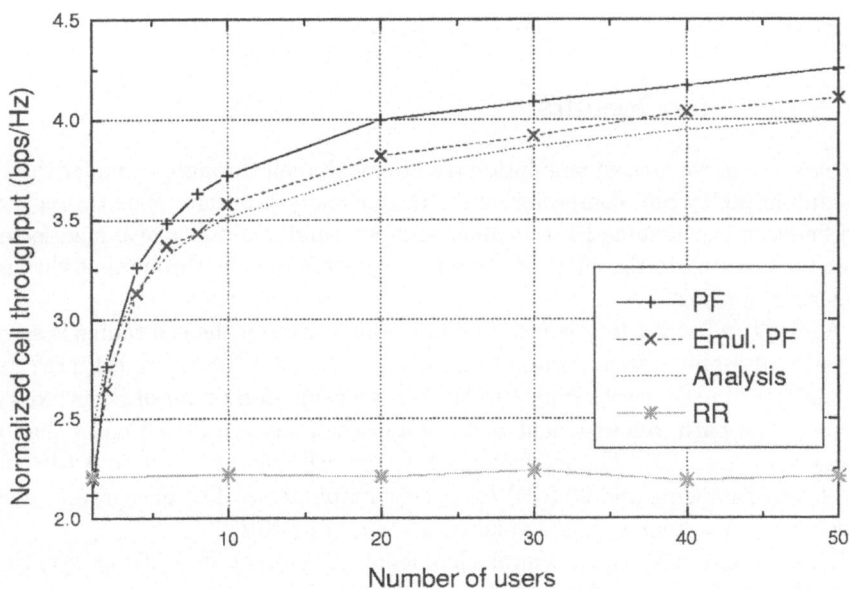

Fig. 2. Comparison of the analysis with simulations: $l = 0$ (dB) and $K = 8$ (dB).

where $\nu_3 = \frac{\ln 10}{10} \cdot \frac{\sigma_s}{\sqrt{2\pi}}$. See the Appendix for the detailed derivation. Its accuracy is depicted in Fig. 1. Using this approximation, (15) becomes

$$E_S\{\overline{\Phi_i}\} \approx W\nu_1 \sum_{m=0}^{N-1} \binom{N-1}{m} \frac{(-1)^m}{(m+1)} \cdot \left(\ln \left(1 + a(m)(\frac{D}{d_i})^\alpha\right) + \nu_3 \right), \quad (17)$$

where $a(m) = \frac{\nu_2 l}{K(m+1)}$.

Finally, taking the average for the entire cell area, we can write the normalized cell throughput as

$$T_{cell} \approx N\nu_1 \sum_{m=0}^{N-1} \binom{N-1}{m} \frac{(-1)^m}{(m+1)} (A(m) + \nu_3), \quad (18)$$

where $A(m)$ is defined as

$$A(m) = \Omega_A^{-1} \int_A \ln \left(1 + a(m)(\frac{D}{d_i})^\alpha\right) dA = \frac{2}{D^2} \int_0^D r \ln \left(1 + a(m)(\frac{D}{r})^\alpha\right) dr. \quad (19)$$

We changed the dummy variable d_i to r for readability.

Now we integrate the form $\int x \ln(1 + \mu x^{-\alpha}) dx$ to obtain $A(m)$ but this is difficult to do except in some special cases. $A(m)$ can be exactly calculated for the integer $\alpha \geq 3$. We present the result for $\alpha = 4$, which is the most interesting case:

$$A(m) = \ln(1 + a(m)) + 2a(m)^{\frac{1}{2}} \cdot \arctan a(m)^{-\frac{1}{2}}. \quad (20)$$

Other values of α can be similarly handled by referring to [12].

4   Simulation Results

In this section, we present simulation results for the cell throughput under the PF scheduling policy and compare them with our analysis results. The throughput gap between the genuine PF scheduler and our emulated one is also manifested. Then we investigate the effect of various parameters on the throughput through numerical analysis.

A single cell scenario is considered in simulations and the cell radius is set to 1 (km). All mobile users are uniformly distributed over the entire cell area and the number of users varies from 1 to 50. The transmission power of BS is fixed at 10 (W). The path loss exponent α is 4.0 and the standard deviation of shadow fading σ_s is 8.0 (dB). The median SNR at the cell edge, l, is set to 0 (dB) for hostile environments and 20 (dB) for good environments. The background noise power and the antenna gains are adjusted according to l.

The slot length for the downlink channel is 1.67 (msec). Each simulation runs for 1,000,000 slots and 1,000 outcomes of the same simulation runs are averaged to represent a single point of the graph. We only consider the logarithmic model for the relation between the feasible rate and the SNR since the linear model has only limited meaning in practical situations.

Fig. 2 compares the analysis result with simulations where $l = 0$ (dB) and $K = 8$ (dB). The system efficiency factor $K = 8$ (dB) represents the feasible rate of the typical M-QAM system.[5] We obtained very similar graphes for different l's and K's also, though they are not presented due to the lack of space. We observe that the simulation results for the PF scheduler are very close to those for the emulated PF scheduler[6].

There is a tendency that the throughput of the PF scheduler is slightly higher than that of the emulated one. The analysis always tracks well the throughput of the emulated PF scheduler. Strictly speaking, our analysis slightly underestimates the throughput of the PF scheduler. The throughput of the RR scheduler is constant regardless of the number of users while that of the PF scheduler increases logarithmically with the increase of the number of users. Obviously the RR scheduler shows the same performance as the PF scheduler when $N = 1$. The throughput gap between the two schedulers comes from the effect of the multi-user diversity for $N > 1$.

5   Conclusion

In legacy wireless access networks, user fairness was not a pressing consideration. Several newly proposed systems, however, seriously consider user fairness and operate with packet schedulers that treat users differently according to the channel condition to achieve enhanced performance goals. The PF scheduler is one of the strong candidates targeted for balancing the cell throughput against user fairness by considering the channel condition.

In this paper we analyzed the PF scheduler to obtain an analytic expression for the cell throughput. We verified the accuracy of our analysis by comparing it with simulation results.

References

1. W. Mohr and W. Konhäuser: Access network evolution beyond third generation mobile communications. *IEEE Commun. Mag.* vol. 38, no. 12 (2000) 122–133
2. H. Zhang: Service disciplines for guaranteed performance service in packet-switching networks. *Proc. IEEE* vol. 83, no. 10 (1995) 1374–1396
3. A. K. Parekh and R. G. Gallager: A generalized processor sharing approach to flow control in integrated services networks: the single-node case. *IEEE/ACM Trans. Networking* vol. 1, no. 3 (1993) 344–357
4. Y. Cao and Victor O. K. Li: Scheduling algorithms in broad-band wireless networks. *Proc. IEEE* vol. 89, no. 1 (2001) 76–87
5. P. Bender, P. Black, M. Grob, R. Padovani, N. Sindhushyana, and S. Viterbi: CDMA/HDR: a bandwidth efficient high speed wireless data service for nomadic users. *IEEE Commun. Mag.* vol. 38, no. 7 (2000) 70–77

---

[5] On the other hand, when K is 0 (dB) the feasible rate model follows the Shannon capacity formula and becomes an ideal system.

[6] Again the preference metric of the emulated PF scheduler is the ratio of the SNR to the average SNR, which was adopted for the facility of the analysis.

6. S. Borst and P. Whiting: Dynamic rate control algorithms for HDR throughput optimization. in *Proc. IEEE INFOCOM* (2001) 976–985

7. X. Liu, E. K. P. Chong, and N. B. Shroff: Opportunistic transmission scheduling with resource-sharing constraints in wireless networks. *IEEE J. Select. Areas Commun.* vol. 19, no. 10 (2001) 2053–2064

8. A. Jalali, R. Padovani, and R. Pankaj: Data throughput of CDMA-HDR a high efficiency-high data rate personal communication wireless system. in *Proc. IEEE VTC* (2000) 1854–1858.

9. S. Catreux, P. F. Driessen, and L. J. Greenstein: Data throughputs using multiple-input multiple-output (MIMO) techniques in a noise-limited cellular environment. *IEEE Trans. Wireless Commun.* vol. 1, no. 2 (2002) 226–234

10. J. M. Holtzman: Asymptotic analysis of proportional fair algorithm. in *Proc. IEEE PIMRC* (2001) 33–37

11. A. J. Goldsmith and S.-G Chua: Variable-rate variable-power MQAM for fading channels. *IEEE Trans. Commun.* vol. 45, no. 10 (1997) 1218–1230

12. I. S. Gradshteyn and I. M. Ryzhik: Tables of Integrals; Series and Products. 4th ed. New York: Academic (1995)

Appendix

We show that $E_S\{\ln(1+\mu s)\} \approx \ln(1+\mu) + \nu_3$, $\nu_3 = \frac{\ln 10}{10} \cdot \frac{\sigma_s}{\sqrt{2\pi}}$, which is, in fact, the upper bound. Let X be the log-normal random variable with the mean 0 (dB) and the variance σ_s^2 (dB) in log-scale. Since $Y = 10\log X$ follows the Gaussian distribution with the mean zero and the variance σ_s^2, from $f_X(x) = f_Y(y)\frac{dy}{dx}$, the PDF of X is given as

$$f_X(x) = \frac{c}{\sqrt{\pi}} \cdot \frac{1}{x}\exp(-(c\ln x)^2), \quad c = \frac{10}{\ln 10 \sqrt{2}\sigma_s}. \tag{21}$$

Then,

$$
\begin{aligned}
E_S\{\ln(1+\mu s)\} &= \int_0^\infty \ln(1+\mu x)f_X(x)dx \\
&< \int_0^1 \ln(1+\mu)f_X(x)dx + \int_1^\infty \ln((1+\mu)x)f_X(x)dx \\
&= \ln(1+\mu)\int_0^\infty f_X(x)dx + \int_1^\infty \ln x \cdot f_X(x)dx \\
&= \ln(1+\mu) + \frac{\ln 10}{10} \cdot \frac{\sigma_s}{\sqrt{2\pi}}.
\end{aligned}
$$

On Detection of Anomalous Routing Dynamics in BGP

Ke Zhang[1], Amy Yen[1], Xiaoliang Zhao[3], Dan Massey[3], S. Felix Wu[1], and Lixia Zhang[4]

[1] University of California, Davis, CA, U.S.A
{kezhang,ahyen,sfwu}@ucdavis.edu
[2] USC/ISI, Arlington, VA, U.S.A
{xzhao,masseyd}@isi.edu
[3] University of California, Los Angeles, CA, U.S.A
lixia@cs.ucla.edu

Abstract. BGP, the de facto inter-domain routing protocol, is the core component of current Internet infrastructure. BGP traffic deserves thorough exploration, since abnormal BGP routing dynamics could impair global Internet connectivity and stability. In this paper, two methods, signature-based detection and statistics-based detection, are designed and implemented to detect BGP anomalous routing dynamics in BGP UPDATEs. Signature-based detection utilizes a set of fixed patterns to search and identify routing anomalies. For the statistics-based detection, we devise five measures to model BGP UPDATEs traffic. In the training phase, the detector is trained to learn the expected behaviors of BGP from the historical long-term BGP UPDATEs dataset. It then examines the test dataset to detect "anomalies" in the testing phase. An anomaly is flagged when the tested behavior significantly differs from the expected behaviors. We have applied these two approaches to examine the BGP data collected by RIPE-NCC servers for a number of IP prefixes. Through manual analysis, we specify possible causes of some detected anomalies. Finally, comparing the two approaches, we highlight the advantages and limitations of each. While our evaluation is still preliminary, we have demonstrated that, by combining both signature-based and statistics-based anomaly detection approaches, our system can effectively and accurately identify certain BGP events that are worthy of further investigation.

1 Introduction

As the size, complexity, and connectivity of the Internet increase, the analysis of operational BGP dynamics becomes more and more challenging. First, because a huge amount of BGP UPDATE traffic is generated in a single domain everyday, operators are not able to conduct thorough analysis on the whole logged BGP dataset. Second, even for a single BGP event, the root cause analysis could be extremely hard. Sometimes, an experienced network administrator needs to, if possible, access the information in the core of the service networks, even from

N. Mitrou et al. (Eds.): NETWORKING 2004, LNCS 3042, pp. 259–270, 2004.
© IFIP International Federation for Information Processing 2004

different administrative domains, in order to identify potential faults or config- uration problems. Since the process of problem and fault analysis can be highly expensive, it is critical to put our focus on a small set of valuable network events. In other words, given a large set of BGP update messages, can we accurately cat- egorize them as "normal" or "abnormal"? With this categorization, we can then spend our precious resources mostly on the "abnormal" ones. Two criteria jointly define an anomaly. One criterion is related to BGP performance. For example, slow convergence for a router to reach a stable view of the Internet's available routes [1] belongs to this type, because the router announces many invalid routes to downstream BGP routers during the convergence process. The other criterion refers to a statistical anomaly (also called "relative anomaly")—the significant deviations of current routing behavior from expected routing behavior.

However, in practice, to our best knowledge, we do not have a systematic approach to consistently label a set of BGP events as normal or abnormal. Borrowing the techniques from intrusion detection area, we develop two ap- proaches to detect BGP anomalous routing dynamics—signature-based detec- tion and statistics-based anomaly detection. For signature-based detection, we devise a set of anomalous BGP routing update patterns to search for matching incidents in BGP UPDATEs data. For statistics-based anomaly detection, the long-term historical BGP UPDATEs datasets are used to train the detector to learn the statistical properties. Thereafter, we perform the anomaly detection on the short-term testing UPDATEs datasets. Following detection, we examine the anomalous routing incidents and explain why they should be categorized as anomalies and specify possible root causes of some incidents.

Analyzing the root causes for BGP dynamics is a very challenging task. Our work moves the first step towards this problem by providing the approaches to automatically locate the anomalous routing updates. Due to limited rout- ing information we can acquire, the anomalies discussed in the paper are still speculative ones. However, results from manual examination show that these anomalies are worthy of further investigation. Thus, we believe the approaches are valuable in that they drastically reduce the search space from a large amount of BGP data to a small set of "abnormal" BGP events. Moreover, the signatures and statistics developed in these approaches can be used to analyze BGP data and quantitatively evaluate the "normality" of each BGP UPDATE.

The rest of the paper is organized as follows. Section 2 introduces the concepts of signature-based detection and statistics-based detection, and briefly reviews related work. Section 3 describes the BGP UPDATE dataset that we have used in the experiments. Section 4 and 5 present signature-based detection and statistics based detection respectively. Section 6 compares these two approaches, followed by conclusions in section 7.

2   Related Work

Signature-based detection and statistics-based detection are two major ap- proaches in modern intrusion detection area. Signature based detection systems, such as the Snort IDS, report an attack when a set of symptoms corresponding to a predefined attack signature is observed. Statistics-based intrusion detection flags as attacks any traffic that is unusual for that system. Several research has been carried out on BGP routing behavior. Labovitz et al. [2] showed that around 1996 unstable and pathological routing behaviors dominated the Internet. Later, they presented possible explanations for these anomalies [3]. Other BGP routing problems, such as slow convergence [1], persistent MED oscillation [4,5], have also been well examined. Also concerning abnormal BGP route changes, Wang et al. [6] proposed a path-filtering approach to validate the correct route changes for DNS prefixes. Teoh et al. developed an interactive visualization process to explore BGP data [7]. These works are complementary to our approaches described in this paper.

3   Dataset

The dataset we examine in the paper consists of BGP UPDATE messages collected by the Routing Information Service of RIPE [8]. The collector has multi-hop BGP sessions with 9 peer ASes located at different countries.

We examine BGP updates for a set of networks (IP prefixes) in different peers. Different prefixes may show different behaviors. Even for a single prefix, routing behaviors may be different from different observation points. As an initial step, we choose the following prefixes as samples. We choose 8 prefixes of either root DNS servers or gTLD servers because of the critical role of DNS service. 4 prefixes from Korean and China are selected because we attempt to examine the impact of SQL worm attack. Similar to previous work [9,10], we also select 4 prefixes of popular destinations and 4 prefixes from Department of Defense (DoD).

In addition, we removed duplicated updates due to known implementation problem in some vender's router [3], although duplicate updates are anomalous.

4   Signature-Based Detection

4.1   Patterns of Anomalous BGP Dynamics

A route announced by a BGP router is generally the best route at that moment. Comparing the consecutively announced routes, we can infer the route changes in that router's BGP routing table. In order to compare the consecutively announced routes, we assign a value corresponding to the preference of each route based on BGP route selection process, which is described in [11]. Based on the relative preference values of two consecutive routes, we define four terms. [1]

---

[1] Since the BGP updates are collected through the EBGP sessions, we cannot acquire information on the following four aspects: LOCAL_PREFERENCE, source of the route(EBGP or IBGP), the IGP cost to the NEXT_HOP, and router ID. We can assign relative preference value by comparing the AS_PATH length, origin type and MED value of each consecutive route announcement.

Table 1. Signatures of BGP Update Burst

TYPE	Pattern	Examples	Indication
B	A sequence of updates with WD in the middle	<D,D,W,U,U> <D,W,U,W,U>	Transient failure followed by fast fail-over
C	A sequence of updates with only one preference fluctuation	<U,U,D,D,F> <D,D,U,F,U> <D,D,U,U,D>	Transient failure followed by fast fail-over OR Normal route changes
F	A sequence of updates with same preference	<F,F,F,F,F>	Anomaly in community attributes or aggregation or same length AS path oscillation OR Normal route changes

UP: if the second route is more preferable than the previous one, we label the second route as UP.

DOWN: if the second route is less preferable than the previous one, we label the second route as DOWN.

FLAT: if two routes have the same preference value, we label the second route as FLAT.

WD: if the second announcement is a route withdrawal, we label the second route as WD.

We define BGP update burst as a sequence of updates within a short time window. Formally, BGP update burst is K consecutive updates for the same prefix that space close together. The time interval between update messages is less than T and the average update rate $> \alpha$. In the experiments, we empirically set $K = 4, T = 240s$, and $\alpha = 1/90$. Examining different parameters is part of our future study.

Given a BGP update burst, we map the updates into a {UP, DOWN, FLAT, WD} sequence. We define 7 patterns for update sequences. Due to the limited space, we only show 3 types in the following. Other types are described in [12].

Type-B: If the update burst has a WD in the middle, it indicates a transient failure followed by a fast fail-over.

Type-C: If the update burst does not consist of WD and has only one <UP, DOWN> or <DOWN, UP> in the middle (the preference fluctuation only happens once in the sequence of updates), it indicates either a transient failure (or congestion) followed by a fast fail-over, or normal route changes.

Type-F: If the update burst consists of all the routes with the same preferences, it might be anomalous. These routes have the same length AS_PATH, the same origin types and the same MED values. Difference might lies in the content of AS_PATH or other attributes, such as community attributes, ATOMIC_AGGREGATE and AGGREGATOR. Since we cannot get the local preference of each route, these route changes might be legitimate. However, from discovered incidents(presented later in this paper), we believe this type of incident is speculative.

4.2   Experiment Results for Signature Detection

We have performed signature detection on 20 prefixes over the period from Feb. 2002 to Jan. 2003. We demonstrate here detected anomalies for four represen-

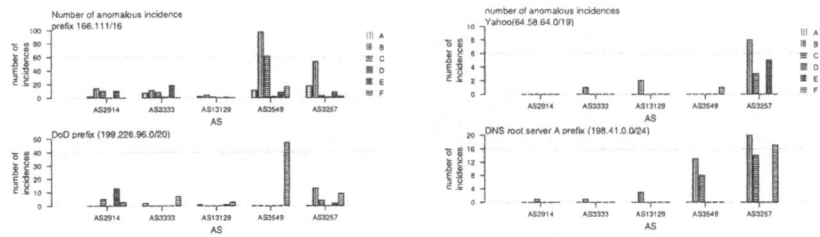

Fig. 1. Anomalies detected by Signature Detector

tative prefixes. Other prefixes have similar results. For each prefix, we choose to present results from 5 ASes (AS2914, AS3333,AS13129, AS3549, AS3257) out of 9 observed ASes. These four prefixes are the prefix for Yahoo.com, the DNS root Server-A prefix, a prefix for Department of Defense and a prefix from a University of China.

Figure 1 plots the number of the each type of incidents at five ASes. We notice that B,C,F are three major types of abnormal incidents. Through further analysis, we identify three different speculative incidents in type F.

We notice that the DoD prefixes has a total of 69 type-F incidents observed from 5 ASes. The pattern capturing these special sequences is frequent substitutions of AGGREGATOR. Due to limited information, we cannot verify whether or not this special behavior is normal BGP operation. However, the high rate of AGGREGATOR substitution, once per minute on average, deserves more attention from the operators. Since local AGGREGATOR changes should be restricted in the local area, and not be propagated to the outer networks. These anomalies indicate BGP operation in DoD networks violates this desired property.

AS3549 and AS3257 have more type-F sequences than other three ASes. We find that some Type-F update bursts for prefix 166.111/16 are due to community attribute changes. The change rate is very high. For example, in one case, AS 3549 changed the community attribute 4 times in 6 seconds. These frequent changes of community attribute generate a lot of BGP updates in a short period of time. If the downstream BGP router performs BGP route flap damping, the route announced by AS3549 would be suppressed. We performed BGP route flap damping (using default CISCO router's damping parameters) on the Jan. 2003 updates data. We find that 174 out of 250 effective updates would be suppressed, and the total suppression time in that month is 7.3 hours.

In addition, from AS3257, we observe that the DNS root server-A prefix has 17 type-F incidents, 7 out of which are oscillations of two routes. The two routes, {AS3257, AS1, AS10913, AS19836} and {AS3257, AS3356, AS10913, AS19836} have the same AS path length, the same origin type and the same MED value. They replace each other at least three times in a short time window. The possible root cause might be link flap, or transient link congestion or even other unknown reasons. Although we do not know the root cause, we believe

Table 2. Five Measures

Intensity Measure	BGP Updates Message Arrival Frequency
	Number of AS paths in a period
Categorical Measure	BGP Updates Type
	AS path Occurrence Frequency
Counting Measure	AS path Difference

this kind of incident should be anomalous because the frequent route changes can degrade packet forwarding.

5   Statistics-Based Anomaly Detection

We apply a statistics-based anomaly detection method, NIDES/STAT [13]. The NIDES/STAT algorithm monitors a subject's behavior on a computer system, and raises alarm when the subject's current (short-term) behavior deviates significantly from its expected behavior, which is described by its long-term profile. A subject's behavior is described by a set of detection measures. For each individual measure, there is a corresponding Q statistic. The historical profile records the frequency distribution of Q. For each measure, the corresponding S, derived from Q, is indicative of the degree of abnormality of the behavior with respect to that measure. T^2 summarizes the abnormality of many measures, reflecting the degree to which recent behavior is similar to the historical profile. Large values indicate abnormal behavior.

5.1   Measures

Like NIDES/STAT, we define 3 types of measures listed in Table 2.

BGP updates message arrival frequency (M1). This measure is one of activity intensity measures. It measures the inter-arrival time of BGP update messages sent by a router for a single prefix. We devise this measure to detect BGP update burst. BGP update burst most likely indicates abnormal operations. Moreover, the burst itself may impair the network because a huge number of update messages can occupy the overall resource of a BGP router, or even cause a router crash.

For this measure, the Q value corresponding to the current update message represents the number of update messages that have arrived in the recent past. In exponentially weighed sums scheme, whenever a new update arrives, the system will assign a Q value based on the following formula.

$$Q_n = 1 + Q_{n-1} * 2^{-r*\Delta t}$$

where r is the decay factor, Δt is the inter-arrival time between the current and the previous update.

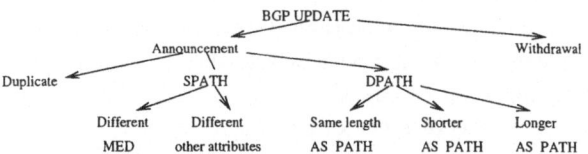

Fig. 2. BGP Update Class Hierarchy

Number of AS paths (M2). This measure is another intensity measure. Due to link failure or router crash, BGP will suffer slow convergence problem. During convergence process, BGP router may receive a number of potential AS paths that are seldom seen in the past. Therefore, the number of AS path in that period may drastically increase. This measure is devised to monitor the variation of the number of AS paths. The Q value is calculated by the following formula

$$Q_n = N_{new\_aspaths} + Q_{n-1} * 2^{-r*\Delta t}$$

where current Q is the number of new AS paths detected in the current audit record plus decayed previous Q.

BGP update type (M3). Similar to [14], we classify BGP update messages into 7 types in a hierarchical structure (Fig 2). At the top of the class hierarchy are two major classes: announcements and withdraws. Announcement is further classified into three sub-classes. Duplicate announcement indicates that the consecutive updates contain exactly the same information. (However, note that because we remove all the duplicate updates, this category does not exist in the experiments.) If the new route contains the same AS path as the current route, it is labeled as SPATH. Due to MED oscillation problem, we further distinguish SPATH by checking if the MED value is different. DPATH indicates that the current route is replaced by a different AS path. Because the length of AS path is a key factor in the BGP route selection process, we divide this sub-class into 3 more specific groups: same length AS path, longer AS path, Shorter AS path. The leaf nodes in this classification tree are the types of BGP update messages. Currently, we use these six types. If necessary, we can still sub-classify these types.

AS path occurrence frequency (M4). According to the observation that only a small number of different AS paths are announced, we define a categorical measure to capture the frequency distribution of AS paths occurrence. Each individual category within this measure is a different AS path. We calculate the frequency of each AS path occurrence. Since a new AS path will appear in the future, we utilize the "new path" category to denote the new path.

The Q computation for the categorical measure is:

$$Q_n = \sum_{m=1}^{M} [(g_{m,n} - f_m)^2 / V_m]$$

where

f_m = the relative frequency with the mth AS path has occurred in the history.

$g_{m,n}$ = the relative frequency with which the mth AS path has occurred in the recent past (which ends at the nth received UPDATE message).

V_m = the approximate variance of the $g_{m,n}$

For detailed computation of these variables, please refer [13].

AS path difference (M5). In order to compare the current AS path with historical dominant AS path , we employ this measure. We use a $Simi(path1, path2)$ function to calculate the difference between two AS paths. First, we define the AS path as a string in which each character is an AS number. Then we calculate the edit distance of two strings. In $Simi$ function, $path1$ is the current AS path, $path2$ is the historical dominant AS path which is usually the most stable path. The edit distance of two paths denotes their difference. The larger the distance, the greater the difference.

Combination of five measures. The NIDES/STAT algorithm defines another variable S which is "normalizing" transformation of Q statistics so that the degree of abnormality for different measures can be added on a comparable basis. S has a half-normal distribution. Since each individual measure has a S value for each BGP update message, the anomaly detector can generate a single score value T^2 by the following formula:

$$T^2 = (S_1^2 + S_2^2 + \cdots + S_n^2)/n$$

The details of transformation from Q to S can be found in [12].

Because the value of S ranges from 0 to 3.9, T^2 can range from 0 to 15.2 theoretically. In practice, we set the threshold of T^2 to be 2.5, since chances are very small for T^2 to have greater values based on our past experience.

5.2   Experiments for Statistics-Based Anomaly Detection

Experiments overview. Our experiments consist of two major parts, historical profile training and detecting process. Long term historical profile training is the process by which the anomaly detector learns the past behaviors of a subject. Detecting process examines the testing data by comparing current routing behaviors with the historical behaviors. If the deviation score is above the predefined threshold, a warning will be flagged. Otherwise, the data will be considered normal and incorporated into historical profile.

Experiments parameters. The decay factor has a significant impact on our detector. According to [1], most convergence time is about 3 minutes. Thus, in the case of inter-arrival time measure, the decay factor is set to be 1/300, which corresponds to the half-life of 300 seconds. Please note that convergence time is a function of the topology, MRAI timer, route flap damping, and routing policy. We cannot prove that this decay value is optimal. However, based on the distribution of the inter-arrival time for each prefixes we observed, most of inter-arrival time is less than 300 seconds or greater than 3000 seconds. We

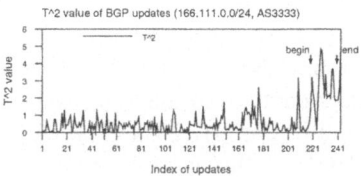

Fig. 3. 166

Time	AS_PATH
01:51:37	3333 3356 1239 9405 4538
02:08:07	3333 3356 1239 9407 9407 4538
02:09:55	3333 286 209 1239 9407 9407 4538
02:10:22	3333 12859 13237 1299 701 1239 9407 9407 4538
02:11:17	Path Withdrawal

Fig. 4. Anomalous Update Sequence

choose 300 seconds as half-life value to capture the frequent route changes. For the categorical measure, we set the half-life decay to 20 BGP update messages, corresponding to an $r = 0.05$.

Experiment results. We test our statistical anomaly detector on the BGP UPDATE data to see whether it is able to effectively detect the BGP routing anomalies and whether it can help users analyze BGP routing dynamics. We conduct our test in two directions. On one hand, we perform the test on the BGP UPDATEs data during SQL worm attack. Although SQL worm does not intent to attack BGP protocol, BGP has been impacted during worm attack. We test our detector to see if the detector can find out the anomalies in that period. On the other hand, in order to compare the results with those from signature detection, we perform the test on the same prefixes.

Table 3. S and T^2 values

Prefixes	S_{M1}	S_{M2}	S_{M3}	S_{M4}	S_{M5}	T^2
166.111.0.0/16	1.032996	2.809773	2.610673	1.220799	2.659919	4.868608
203.250.84.0/24	1.607696	1.967813	2.554678	0.212015	2.046884	3.235723
199.226.96.0/20	2.497571	1.907853	2.366445	0.773497	2.313003	4.285221

Detecting SQL worm. SQL worm attacked the Internet on Jan 25, 2003. Although SQL worm did not intent to attack the Internet routing architecture, a large increase of the number of BGP routing updates have been observed during that period. We apply the anomaly detection on the BGP updates and find that the warnings have been flagged for some prefixes from DoD, Korean and China, while the prefixes for popular destination and root server appear normal. Through the comparison of the prefixes from DoD, Korean and China, we can infer that their abnormal behaviors are similar in essence, while their BGP update sequences are different. Table 3 lists the S and T^2 values of the most abnormal update for each prefix. From the table, we observe some similarities in S distribution among these prefixes– all three assigned S values (S_{M2}, S_{M3}, and S_{M5}) are abnormally large (greater than 1.96, indicating that the probability for the update message occurrence is less than 5%). This similarity leads us to further manually examine the BGP traffic of the three prefixes.

Figure 3 plots the T^2 values for the prefix 166.111/16 that is the address block for a university in China. The X-axis denotes the index of each update. All updates were recorded from Sept 1, 2002 to Jan 31, 2003. The highlighted updates (between the arrow "begin" and "end" in the figure) were recorded on Jan 25, 2003 when SQL worm attacked the whole Internet. Observing many UP-DATEs with large T^2 values (statistically significant deviation) on that day, we can easily infer that BGP routing has produced many highly abnormal behaviors. One of BGP update sequences was shown in Fig 4. The last BGP withdrawal message flags a warning. The corresponding S and T^2 values are listed in the first row of table 3. The large S_{M2} and S_{M5} value were due to the previous two updates. Large S_{M2} value indicates the arrival of new pathes (third and fourth AS_PATH). large S_{M5} value indicates that the pathes are significantly different from the dominant path. This is a classical case of slow convergence [1]. Possible root cause could be that huge amount of traffic generated by SQL worm congested the link between AS1239 and AS9407 and tore down the BGP session. This example demonstrates that our detector can detect BGP slow convergence effectively, and the statistical information learned from the detector can help analyze what causes a anomaly.

Comparison with signature detection. We apply statistics-based anomaly detection on the same prefixes examined by signature-based anomaly detection. Compared with signature detection, the number of anomalous incidents identified by statistics-based detection is much smaller.

For example, we examined BGP updates for the prefix 166.111/16 from AS3257 in the dataset of Jan. 2003. The detector did not flag warning during the SQL worm attack, because of the imperfection of our training dataset. However, it did flag a warning for one type B incident that is worthy of more investigation than the incidents caused by the worm. The abnormal in that type B incident is a brand new fail-over path, {3257, 3356, 12013, 3681, 20080, 11537, 9405, 4538}. This path has never appeared before and remains for only 500 seconds. In addition, the new path is very different from the primary path {3257,1239,9405,4538}. Moreover, the three transit ASes (3681,12013 and 20080) are ASes of two universities in Florida. It is highly abnormal that the networks of two American universities provide transit service for a university in China. Thus, we believe this incident is more interesting than other type B incidents because it may indicate some routing policy misconfiguration.

6   Discussion

For signature-based BGP detection, if the patterns of anomalies are well defined and persistently updated, this method should be very efficient in terms of false rate. However, in the BGP scenario, it is very hard to accurately define signatures. In this paper, the parameters ($K = 4, T = 240s$) are set empirically.

We may miss some anomalous incidents with only three close updates, or may incorrectly treat two consecutive updates as two different events.

Statistics-based BGP anomaly detection does not require knowledge about patterns of anomalies in advance. It can assign BGP updates with different deviation scores, providing an objective measure telling which incident is more abnormal and deserves more attention. In addition, statistics-based detector can provide information on the detected anomalies and help network operators investigate what may have triggered a warning. In our detection system, whenever an alarm is raised, detector provides both expected distributions and observed distributions (an example is shown in [12]). With this additional information, operators might be able to speculate what could have accounted for the statistically significant deviation.

Limitations also exist in statistics-based method. In our experiments, anomaly detector appears to have a relatively higher false rate compared to signature-based detector, because we do not have a clean training dataset in advance. The expected behaviors learned by the detector may have included problematic BGP UPDATE sequences.

Through comparison, we find that while both approaches are capable of identifying BGP routing anomalies to some extent, the list of detected anomalies is not exhaustive. Experiments demonstrate that combination of the two approaches can generate more comprehensive results.

At the current stage, we are unable to evaluate the identified anomalies, because evaluation is based on root cause analysis which is still an open question. The major barrier for root cause analysis is that we cannot acquire the necessary information from real operational network. Further, root cause analysis for BGP anomalies may need cooperations among ASes, because under some circumstances, identification of certain causes is extremely hard for an individual AS. Thus, the goal of our current work is not to provide accurate root causes analysis for speculative anomalies. In stead, we aim to devise an approach to identify possible anomalies, which is the first step towards solving the root cause analysis problem.

7 Conclusion and Future Work

In this paper, two approaches, signature-based detection and statistics-based detection, are proposed to search for anomalous BGP routing dynamics. The value of our work lies in the following aspects: First, we develop two systematic approaches to detect abnormal BGP UPDATE traffic. In current network management, they can help operators and researchers to filter out the trivial events and focus mainly on the most important BGP events. Second, through our experiments, we identify advantages and limitations of both methods. A feasible way to overcome the weakness of each is through combination of both. Third, these two detection approaches can be further used in monitoring and analyzing the real-time BGP traffic. In particular, statistics-based approach can quantitatively measure the "abnormality" of each BGP UPDATE.

The limitation of our work lies in the lack of information from real BGP run-time environment. At this stage, we are not able to thoroughly evaluate the identified anomalies. Root causes for most of the anomalies are still our conjectures. However, in our future work, we plan to attack this problem by building a large scale BGP testbed [15]. In this simulated BGP operational environment, we can generate various fully-controlled network failures and attacks. Applying the detection approaches to examine the simulated BGP traffic, we can provide a more extensive evaluation of the detectors' performance.

Acknowledgment. We would like to thank Chen-Nee Chuah for valuable comments. We also thank Yifei Zhu for proofreading.

References

1. C. Labovitz, A. Ahuja, A. Bose, and F. Jahanian. Delayed Internet Routing Convergence. In *Proceedings of ACM Sigcomm*, August 2000.
2. C. Labovitz, G. Malan, and F. Jahanian. Internet Routing Instability. In *Proceedings of ACM Sigcomm*, September 1997.
3. C. Labovitz, F. Jahanian, and G.R.Manlan. Origin of Internet Routing Stability. In *Proceedings of the IEEE INFOCOM*, June 1999.
4. A. Basu, C. Ong, A Rasala, F. Shepherd, and G. Wilfong. Route Oscillations in I-BGP with Route Reflection. In *Proceedings of ACM Sigcomm*, August 2002.
5. T. Griffin and G. Wilfong. Analysis of the MED Oscillation Problem in BGP. In *Proceedings of ICNP*, November 2002.
6. L. Wang, X. Zhao, D. Pei, R. Bush, D. Massey, A. Mankin, S. Wu, and L. Zhang. Protecting BGP Routes to Top Level DNS Servers. In *Proceedings of the ICDCS 2003*, 2003.
7. S.T Teoh, K.L. Ma, and S.F. Wu. A Visual Exploration Process for the Analysis of Internet Routing Data. In *Proceedings of IEEE Visualization*, 2003.
8. The RIPE Routing Iformation Services. http://www.ris.ripe.net.
9. J. Rexford, J. Wang, Z. Xiao, and Y. Zhang. BGP Routing Stability of Popular Destinations. In *Proceeds of Internet Measurement Workshop*, November 2002.
10. X. Zhao, M. Lad, D. Pei, L. Wang, D. Massey, A. Mankin, S. Wu, and L. Zhang. Understanding BGP Behavior through a Study of DoD Prefixes. In *Proceedings of the IEEE DISCEX III*, 2003.
11. B. Halabi. *Internet Routing Architectures*. Cisco Press, second edition, 2001.
12. Ke Zhang, Amy Yen, Xiaoliang Zhao, Dan Massey, S.Felix Wu, and Lixia Zhang. On Detection of Anomalous Routing Dynamics in BGP. Technical Report CSE-2004-2, UCDAVIS, 2004.
13. H.S. Javitz and A. Valdes. The NIDES Statistical Components: Description and Justification. Technical report, SRI Network Information Center, March 1993.
14. L. Wang, X. Zhao, D. Pei, R. Bush, D. Massey, A. Mankin, S. Wu, and L. Zhang. Observation and Analysis of BGP Behavior under Stress. In *Proceedings of the ACM IMW 2002*, October 2002.
15. R. Bazjscy, T. Benzel, M. Bishop, B. Braden, C. Brodley, S. Fahmy, S. Floyd, W. Hardaker, G. Kesidis, K. Levitt, B. Lindell, P. Liu, D. Miller, R. Mundy, C. Neuman, R. Ostrenga, V. Paxson, P. Porras, C. Rosenberg, S. Sastry, D. Sterne, and S.F. Wu. Cyber Defense Technology: Experimental Research Network and Evaluation Methods. under submission.

Multi-objective Egress Router Selection Policies for Inter-domain Traffic with Bandwidth Guarantees

Kin-Hon Ho, Ning Wang, Panos Trimintzios, and George Pavlou

Centre for Communication Systems Research, University of Surrey,
Guildford, Surrey, England, United Kingdom GU2 7XH
{K.Ho, N.Wang, P.Trimintzios, G.Pavlou}@surrey.ac.uk

Abstract. The next generation Internet is designed to accommodate flows that span across multiple domains with quality of service guarantees, in particular bandwidth. In this context, destinations for inter-domain traffic may be reachable through multiple egress routers within a domain. In this paper, we formulate a bandwidth guaranteed egress router selection problem. The objective is to, for each aggregated inter-domain traffic flow, select an egress router that satisfies the end-to-end bandwidth requirement while optimizing the network resource utilization by which we consider three objective functions: minimizing the total bandwidth consumption, improving intra-domain and inter-domain load balancing in the network. We propose a heuristic algorithm with five egress router selection policies to solve this problem. The evaluation of these egress router selection policies through simulation benefits ISPs by choosing the one that fits their target objectives.

1 Introduction

As the Internet has grown in size and diversity of applications, the next generation Internet is intended to accommodate flows with end-to-end Quality of Service (QoS) guarantees across multiple domains. To provide efficient end-to-end QoS guarantees, QoS routing and Traffic Engineering (TE) have become indispensable: the former selects a path that meets the QoS requirements while the latter optimizes resource utilization in order to be able to carry more traffic flows in the network. In the past decade, there has been a considerable amount of work on QoS routing and traffic engineering at the intra-domain level. However, only little attention has been given to the inter-domain problem. We consider that inter-domain QoS routing and traffic engineering should be addressed for the following reasons.

Inter-domain QoS: End-to-end QoS over the Internet covers the intra-domain and inter-domain QoS. Even though research in intra-domain QoS is mature, the lack of inter-domain QoS support hinders the deployment of end-to-end QoS. Thus, together with the current QoS-aware intra-domain routing, inter-domain QoS routing will facilitate an end-to-end QoS-based Internet, which will benefit Internet Service Providers (ISPs) and their customers. The current inter-domain routing protocol, Border Gateway Protocol (BGP), however, does not cater for QoS support.

N. Mitrou et al. (Eds.): NETWORKING 2004, LNCS 3042, pp. 271–283, 2004.

Inter-domain TE: Inter-domain traffic engineering [1] concerns forwarding traffic entering or exiting a network based on some optimization objectives. One of the inter-domain traffic engineering problems is to direct the inter-domain traffic flows to the 'best' egress router within a domain towards certain destination prefixes; this we call the "egress router selection" problem. The problem arises when a domain has multiple connections to neighboring domains, so that a situation can emerge that a destination prefix is reachable through multiple egress routers. In that context, selecting different egress routers for traffic flows can have diverse effects on network resource utilization. Addressing inter-domain TE is important because appropriate selection of egress routers for inter-domain traffic flows benefits ISPs by improving the network resource utilization. Inter-domain traffic engineering, however, is commonly applied today in a trial-and-error only fashion [1].

Based on this reasoning, we aim to develop a systematic approach to solve this inter-domain TE problem with end-to-end QoS guarantees, i.e. QoS guaranteed egress router selection. Assuming that most QoS requirements can be derived from bandwidth [2], our work only focuses on providing bandwidth guarantees. Thus, the problem we address becomes Bandwidth Guaranteed Egress Router Selection (BGERS).

Our goal is summarized as follows: *Given an aggregated customer traffic flow in an ISP network, select an egress router that satisfies the customer end-to-end bandwidth requirement while optimizing resource utilization in the network. Each customer traffic flow consists of a destination prefix that belongs to a remote domain and a bandwidth requirement. An egress router must be selected amongst egress routers that offer the guaranteed bandwidth to the destination prefix.*

With respect to optimizing network resource utilization, we consider three objectives: Minimizing the total bandwidth consumption, improving intra-domain and inter-domain load balancing in the network. We do not consider optimizing multiple objectives simultaneously since, for example, the objective of minimizing bandwidth consumption and load balancing are contradictory with each other.

Related work on inter-domain QoS routing and TE are as follows. Bonaventure [3] focuses on how to distribute flexible QoS information by BGP in different network scenarios. Cristallo and Jacqenet [4] propose a new attribute, the QoS_NLRI (Network Level Reachability Information), for the BGP UPDATE message to carry QoS information. Xiao [5] proposes a similar QoS extension to BGP to perform the bandwidth advertising and routing. On the other hand, research on egress router selection has only been done in the context of best-effort traffic. Bressoud [6] determines an optimal selection of outgoing links and associated border routers, where the selection optimizes the ISP's network resource utilization. The ISP, however, can only select an egress router based on prefix reachability and the egress link capacity information, without knowing whether the selected egress router can satisfy the traffic flow's end-to-end bandwidth requirement.

The key to solve the BGERS problem is the support for traffic engineering information (e.g. bandwidth) distributed within and between domains. In this paper, we propose a TE-enabled Internet architecture to achieve this, which includes traffic engineering extensions to the current intra-domain and inter-domain routing protocols. Our work extends the egress router selection problem presented in [6] by considering end-to-end bandwidth guarantees. We propose a heuristic algorithm to solve the BGERS problem. In addition to this, we propose five egress router selection policies to address the optimization objectives that we consider. By evaluating the behavior of those egress router selection policies through simulation, we can provide an answer to the fundamental question of how an egress router is selected in order to give the best performance with respect to which optimization objective. This evaluation gives ISP insight into the relation of egress router selection policy and network resource utilization, based on which they can configure their networks in order to realize their target objectives. To the best of our knowledge, this work is the first attempt at inter-domain traffic engineering using BGP policies to control inter-domain traffic flows with end-to-end bandwidth guarantees.

The rest of the paper is organized as follows. In section 2 we present a TE-enabled Internet architecture. In section 3 we formulate the BGERS problem and propose a heuristic algorithm with five egress router selection policies to solve it. Section 4 presents the evaluation of those egress router selection policies through simulation. Finally, we conclude our work and discuss future research directions in section 5.

2   Traffic Engineering Enabled Internet Architecture

To address the BGERS problem, the TE-enabled Internet architecture requires that the current intra-domain and inter-domain routing protocols are able to convey TE information such as bandwidth. We assume Traffic Engineering extensions to OSPF (OSPF-TE) [7] as the intra-domain link-state protocol, which disseminates bandwidth information within a domain. Moreover, we assume using Constrained Shortest Path First (CSPF) [8] with unit link cost to calculate a bandwidth constrained path between an ingress router and an egress router in a domain. An explicit path is then configured for the selected path along with bandwidth reservation. This can be done by RSVP extensions for MPLS TE [8].

On the other hand, the lack of TE information support in the current BGP hinders the deployment of BGERS. To make up this deficiency, it is necessary to record bandwidth information in the BGP UPDATE message, which represents the ability of a neighboring domain to provide the route with such available bandwidth. In [4], a new attribute, the QoS_NLRI, is proposed for this purpose. We assume that bandwidth information, which takes a single value, is conveyed through a similar attribute and we call the extended BGP the Traffic Engineering extensions to BGP (BGP-TE). We simply extend BGP for the purpose of bandwidth advertisement and TE but do not use bandwidth information on the usual BGP path selection.

The bandwidth information advertised by BGP-TE is guaranteed by a Service Level Agreement (SLA) established between neighboring domains in a management time-scale. Each domain is configured to make sure that sufficient bandwidth is provisioned for other domains, conforming to the established SLAs. The BGERS is one of the issues in this bandwidth provisioning. The outcome of bandwidth provisioning is Bandwidth Capability (BC), which is the bandwidth that has been allocated to a path between an ingress router and an egress router within a domain towards certain destination prefixes. From the ISP point of view, establishing SLAs with neighboring domains can extend its services to customers by reaching more destination prefixes that belong to the remote domains with bandwidth guarantees. Technically, this service extension is done by bandwidth capability binding: A domain binds its bandwidth capability to the bandwidth information advertised by the neighboring domains and forms an extended *BC* (*eBC*) as the basis for agreeing new SLAs with its customers. Bandwidth capability binding is unidirectional and is done by a simple algebraic method. The *eBC* or *BC* is advertised as bandwidth information to neighboring domains through BGP-TE.

We give a small example of bandwidth capability binding in Figure 1. The example also shows how ISP 1 provides bandwidth guarantees to the traffic flows of customer 1 destined to customer 2, conforming to the customer SLA established with customer 1. We denote by *BCX* the unidirectional bandwidth capability of ISP *X* towards destination prefixes that belong to customer 2. We assume that ISP 1 has established a provider SLA with ISP 2 for bandwidth guarantees to customer 2, thus ISP 2 will advertise *BC2* to ISP 1 through BGP-TE. When ISP 1 receives *BC2*, it performs bandwidth provisioning based on the customer SLA and then binds *BC1* to *BC2*. This binding forms a unidirectional *eBC*. The value of *eBC* is equal to the minimum of *BC1* and *BC2*. ISP 1 can then provide *eBC* bandwidth guarantees to the customer 1's traffic flows destined to customer 2, conforming to the customer SLA.

For a large scale Internet to provide bandwidth guarantees between edge domains, ISPs have to collaborate and provide bandwidth guarantees to other domains' traffic flows. The concatenation of SLAs between domains can ensure end-to-end bandwidth guarantees for end customers. In this context, the bandwidth information advertised by BGP-TE is the unidirectional cascaded effect of bandwidth capability binding between each two domains along a BGP path. It is also the concatenated bandwidth that is guaranteed starting from the downstream domain (i.e. the one which advertises the bandwidth information) until the destination domain. Each domain uses bandwidth information, provided by BGP-TE and OSPF-TE, to optimize its network resource utilization by selecting appropriate egress routers for inter-domain traffic flows with bandwidth guarantees.

3 Bandwidth Guaranteed Egress Router Selection

With the TE-enabled Internet architecture, we can provision bandwidth guarantees for inter-domain traffic flows. In this section, we formulate the BGERS problem and propose a heuristic algorithm with several egress router selection policies to solve it.

Figure 2 shows a general TE-enabled Internet architecture. The ISP domain has a set of border routers as well as a set of intra-domain and inter-domain links. An inter-domain link connects between border routers of the ISP and the downstream domain. Each border router may connect to multiple inter-domain links. We assume that the ISP has established SLAs with its downstream domains for bandwidth guarantees and border routers in the ISP and downstream domains support BGP-TE. Through BGP-TE, the border routers of the ISP receive route advertisements of destination prefixes associated with the bandwidth information from the downstream domains.

Each border router that connects to the downstream domains selects the best route for each destination prefix based on the usual BGP decision process and then distrib-

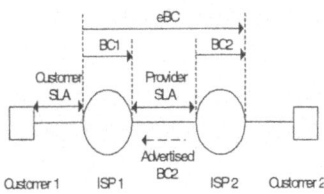

Fig. 1. BC binding

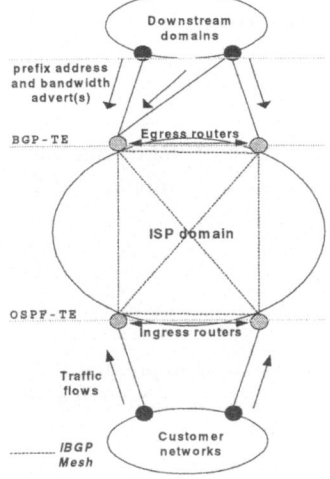

Fig. 2. A general TE-enabled
Internet architecture

Symbol	Description
E	A set of intra-domain links
K	A set of destination prefixes
I	A set of ingress routers
J	A set of egress routers
$t(i,k)$	The aggregated bandwidth requirement of customer traffic flows destined to destination prefix $k \in K$ at ingress router $i \in$
$Out(k)$	A set of egress routers that can reach destination prefix k
$NEXT_j$	A set of next hop addresses (addresses of border routers in downstream domains) that is connected to egress router $j \in J$
$f_k(j,n)$	True (1) / False (0); whether the prefix k can be reached through the inter-domain link between the egress router j and the next hop address $n \in Next_j$
C_{intra}^l	The capacity of intra-domain link $l \in E$
bW_{intra}^l	The current availability (unallocated bandwidth) on C_{intra}^l
$C_{inter}^{j,n}$	The capacity of the inter-domain link which is attached to egress router j and is connected to next-hop address n
$bW_{inter}^{j,n}$	The current availability (unallocated bandwidth) on $C_{inter}^{j,n}$
$p(k,j)$	The bandwidth advertised by BGP-TE on the egress router, to the destination prefix k after the BGP path selection
$bp(k,j)$	The current availability (unallocated bandwidth) on $p(k,j)$
$x_{(i,k)}^j$	True (1) / False (0); whether the customer traffic flow $t(i,k)$ has been assigned to the egress router j
$y_{(i,k)}^l$	True (1) / False (0); whether the customer traffic flow $t(i,k)$ has consumed bandwidth on the intra-domain link l
$d(i,j,k)$	Number of hops of the feasible shortest path (found by CSP between the ingress router i and the egress router j for $t(i,k)$
$bW_{intra}^{p_{i,j}}$	The bottleneck bandwidth of intra-domain path p between the ingress router i and the egress router j, i.e. $\underset{l \in p}{Min} \left(bW_{intra}^l \right)$

Fig. 3. Notations

utes the route to other border routers within the domain through Internal BGP (IBGP) mesh between border routers. It is possible that a border router receives multiple route advertisement with a common destination prefix through IBGP. Thus, an opportunity emerges to select the best among the appropriate egress routers for inter-domain traffic flows with bandwidth guarantees. The outcome of the BGERS can be realized by BGP policies such as using policy routing and manipulating BGP attributes. For the ISP's decisions on advertising bandwidth information to upstream domains such as how much bandwidth is advertised and where to advertise, we consider this as the subject of inbound inter-domain TE, which is out scope of this paper.

We assume that the traffic matrix of customer traffic flows is known through customer SLAs established in a management time-scale. Each customer traffic flow includes a bandwidth requirement to a destination prefix and the ingress router where it enters the ISP domain. Moreover, individual customer traffic flows are aggregated at each ingress router based on their destination prefixes. In the rest of this paper, we refer customer traffic flow as the one which is aggregated (including bandwidth) from those individual flows destined to the same destination prefix at the ingress router.

3.1 Problem Formulation

We formulate the BGERS as an integer programming problem. Figure 3 shows the notations that are used in the rest of this paper. A solution of BGERS should compute a set of allocation between traffic flows and egress routers, which yields the best value for one or more objective functions. In this paper, we consider three:

1)   *Minimizing the total bandwidth consumption:* The objective of minimizing the total bandwidth consumption in the network can be translated to the problem of minimizing the total number of hops that a traffic flow must traverse, i.e.

$$Minimize \sum_{k \in K} \sum_{i \in I} \sum_{j \in Out(k)} x^j_{(i,k)} \cdot d(i,j,k) \cdot t(i,k) \tag{1}$$

2)   *Improving intra-domain load balancing:* The objective of improving intra-domain load balancing can be approximated with the problem of minimizing the maximum link utilization within the network, i.e.

$$Minimize \quad Max_{l \in E}\left(1 - \frac{bw^l_{intra}}{c^l_{intra}}\right) \tag{2}$$

3)   *Improving inter-domain load balancing:* The objective of improving inter-domain load balancing can be approximated with the problem of minimizing the maximum link utilization among all the inter-domain links, i.e.

$$Minimize \quad Max_{j \in J, n \in NEXT_j}\left(1 - \frac{bw^{j,n}_{inter}}{c^{j,n}_{inter}}\right) \tag{3}$$

The BGERS is subject to the following constraints:

$$\sum_{k \in K} \sum_{i \in I} x^j_{(i,k)} \cdot t(i,k) \cdot f_k(j,n) \le c^{j,n}_{inter} \qquad \forall (j,n) \ where \ j \in J \ \& \ n \in NEXT_j \tag{4}$$

$$\sum_{k \in K} \sum_{i \in I} y^l_{(i,k)} \cdot t(i,k) \le c^l_{intra} \qquad \forall l \in E \tag{5}$$

$$\sum_{i \in I} x^j_{(i,k)} \cdot t(i,k) \le p(k,j) \qquad \forall (k,j) \ where \ k \in K, j \in J \tag{6}$$

$$x^j_{(i,k)}, y^l_{(i,k)} \in \{0,1\} \tag{7}$$

$$\sum_{j \in Out(k)} x^j_{(i,k)} = 1 \qquad \forall (i,k) \ where \ i \in I, k \in K \tag{8}$$

Satisfy customer bandwidth requirement $\forall t(i,k)$ (9)

Constraints (4)-(6) are the capacity constraint respectively on each inter-domain link, intra-domain link and advertised bandwidth towards certain destination prefixes; constraint (7) ensures that the discrete variables to assume binary values; constraint (8) ensures that only one egress router is selected for each customer traffic flow. Constraint (9) enforces that there is sufficient end-to-end bandwidth to accommodate each assigned customer traffic flow. We define the following criteria to determine whether the bandwidth requirement of customer traffic flow, $t(i,k)$, is satisfied:

1. There exists a feasible path p from the ingress router $i \in I$ to the selected egress router $j \in J$ such that $bw_{intra}^{p,i,j} \geq t(i,k)$

2. $bp(k,j) \geq t(i,k)$

3. $bw_{inter}^{j,n} \geq t(i,k)$ where $n = FindInterdomainLink(j,k)$

The function *FindInterdomainLink(j,k)* returns the next hop address n (i.e. the address of border router in the downstream domain) to which a specific inter-domain link at the egress router j is connected to reach the destination prefix k. The first criterion is an intra-domain bandwidth constraint, which ensures that a feasible path exists between the ingress router and the selected egress router, and the bottleneck bandwidth of the path is no less than the bandwidth requirement of the customer traffic flow. The second criterion ensures that the advertised bandwidth at the selected egress router is sufficient to accommodate the customer traffic flow. This implies that in each domain along the corresponding BGP path towards the destination prefix there is sufficient bandwidth for the customer traffic flow. The third criterion ensures that the inter-domain link, connected between the selected egress router and the downstream domain reaching the destination prefix, has sufficient bandwidth for the customer traffic flow. If all the above criteria are met, the constraint on customer bandwidth requirement is satisfied.

The objectives of load balancing defined by (2) and (3) are orthogonal to each other. However, the objective of minimizing total bandwidth consumption defined by (1) and load balancing may lead to contradictory solutions. Based on this reasoning, we do not consider optimizing both objectives simultaneously. Instead, we study the implication of our egress router selection policies on each of these objectives.

The work in [6] has formulated the egress router selection problem as Generalized Assignment Problem (GAP) [9] and proved that the problem is NP-complete. Due to the fact our work extends that of [6] and has the additional constraints (5) and (6) as an intra-domain and a cascaded inter-domain capacity constraint respectively, we consider the BGERS problem to be a variant of GAP, which is also NP-complete. Hence, we propose a heuristic algorithm to solve it.

3.2 Heuristic Algorithm

We propose a greedy-based heuristic algorithm, namely Greedy-cost heuristic, to solve the BGERS problem. The Greedy-cost heuristic takes the following steps:

Step 1: Sort customer traffic flows in descending order based on their bandwidth requirements and selects one, $t(i,k)$, at a time in that order. This sorting can be assumed that large customer traffic flows have higher priority to be considered.

Step 2: Identify all the feasible egress routers for the customer traffic flow. The egress router j is feasible if it meets two criteria: (1) $j \in Out(k)$, and (2) Satisfy the bandwidth requirement of the customer traffic flow; i.e. the constraint (9). The first criterion ensures that the egress router j has a route to reach the destination prefix k while the second criterion ensures that by selecting egress router j there is sufficient end-to-end bandwidth to provide bandwidth guarantees to the customer traffic flow.

Step 3: Compute the cost metric of selecting each feasible egress router. The cost metric encompasses two types of network information: (1) topology which includes the available bandwidth of each link, the number of hops and the bottleneck bandwidth on the path from the ingress to the egress router within a domain, and (2) resource at the egress router which includes the available bandwidth and the capacity of inter-domain links to which is connected. Among a set of feasible egress routers, the one that yields the best cost, determined by the adopted egress router selection policy which is discussed in section 3.4, is selected for the customer traffic flow.

Step 4: Once the egress router is selected, an explicit path is configured for the corresponding selected intra-domain path and the requested bandwidth is reserved on the explicit path, the corresponding selected inter-domain link and the advertised bandwidth capability for the assigned customer traffic flow.

Step 5: We consider the next customer traffic flow and repeat step 2 to step 5. The heuristic finishes when all the customer traffic flows have been considered.

3.3 Egress Router Selection Policies

With reference to step 3 of the Greedy-cost heuristic, we propose five egress router selection policies that use increasingly more network information to make a decision on egress router selection.

1)   *Random-egress*: select an egress router randomly. We consider this policy as the behavior of the current BGP for the BGERS. The current non-TE BGP will select an egress router with respect to bandwidth information completely at random.

2)   *Closest-egress-first*: selects the egress router which is the closest, in terms of number of hops, to the ingress router where the customer traffic flow enters the ISP domain. If there are several such egress routers, the selection tiebreak is in order of the maximum bottleneck bandwidth on the intra-domain path and the maximum available bandwidth on the inter-domain link.

3)   *Widest-egress-first*: selects the egress router which has the maximum bottleneck bandwidth on the path leading from the ingress to the egress router. If there are several such egress routers, the selection tiebreak is in order of the path with the least number of hops and the maximum available bandwidth on the inter-domain link.

4)   *Highest-availability-egress*: selects the egress router with which the selected inter-domain link, after being assigned the customer traffic flow, has the highest bandwidth availability. The bandwidth availability is the ratio of the available bandwidth to the capacity of a link.

5)   *Shortest-dist-egress*: This policy uses the network distance between the ingress and the egress router as the selection parameter. The scope of network distance covers the intra-domain path between the ingress and the egress router, and the inter-domain link associated with the egress router. We define two distance functions to quantify the available bandwidth on each intra-domain and inter-domain link, and then compute the network distance. The distance of intra-domain link l is approxi-

mated by (10) while the distance of inter-domain link that is selected for the customer traffic flow at the egress router j is approximated by (11). The parameter α in the distance functions represents the degree to which the greatest available bandwidth is favored over the least one. The network distance of selecting egress router j is defined by (12) where l is the link belongs to the selected intra-domain path p between the ingress router and the egress router j. The shortest-dist-egress selects the egress router with the minimum network distance. In the subsequent simulations, we use the notation *Shortest-dist-egress(α)* to denote this policy.

$$Dist_{intra}(l) = \frac{1}{\left[bw^l_{intra} - t(i,k) \right]^\alpha} \tag{10}$$

$$Dist_{inter}(j,n) = \frac{1}{\left[bw^{j,n}_{inter} - t(i,k) \right]^\alpha} \tag{11}$$

$$NetworkDist(j) = \sum_{l \in p} Dist_{intra}(l) + Dist_{inter}(j,n) \tag{12}$$

4   Performance Evaluation

We evaluate the Greedy-cost heuristic with the proposed egress router selection policies through simulation. Simulation results are based on 100-node transit domain topologies. The topologies are randomly generated by the Waxman's method [10]. The set of ingress and egress routers are disjoint. We set the number of ingress routers to 30, whereas the number of egress routers is a variable, as we will evaluate some effects by changing its value between 10 and 30. Each egress router is attached with a maximum of two inter-domain links. We assume that the inter-domain capacity is less than the intra-domain capacity, so the bottleneck resides at the former. The capacity of each intra-domain link is randomly generated between 400 and 500 while the capacity of each inter-domain link is randomly generated between 250 and 300.

Feamster [11] discovered that a typical default-free routing table may contain routes for more than 90,000 prefixes, but only a small fraction of prefixes are responsible for a large fraction of the traffic. Thus, we consider 1000 destination prefixes that belong to remote domains. As these prefixes are usually popular destinations, we assume that each egress router can reach all of them. This set of prefixes is randomly distributed on the inter-domain link(s) of each egress router. Each destination prefix is advertised with available bandwidth randomly generated between 200 and 250. For each customer traffic flow, the destination prefix and the ingress router are randomly generated and its bandwidth requirement is randomly generated between 10 and 40.

We evaluate the performance of the five egress router selection policies with respect to the three objective functions that we consider. Our evaluation consists of four scenarios. The first three scenarios evaluate the effects of selection policies on the total bandwidth consumption, intra-domain and inter-domain load balancing respectively. The last scenario studies the impact of α in the shortest-dist-egress on those objective functions. For the shortest-dist-egress, we use α=1.0 as the reference value.

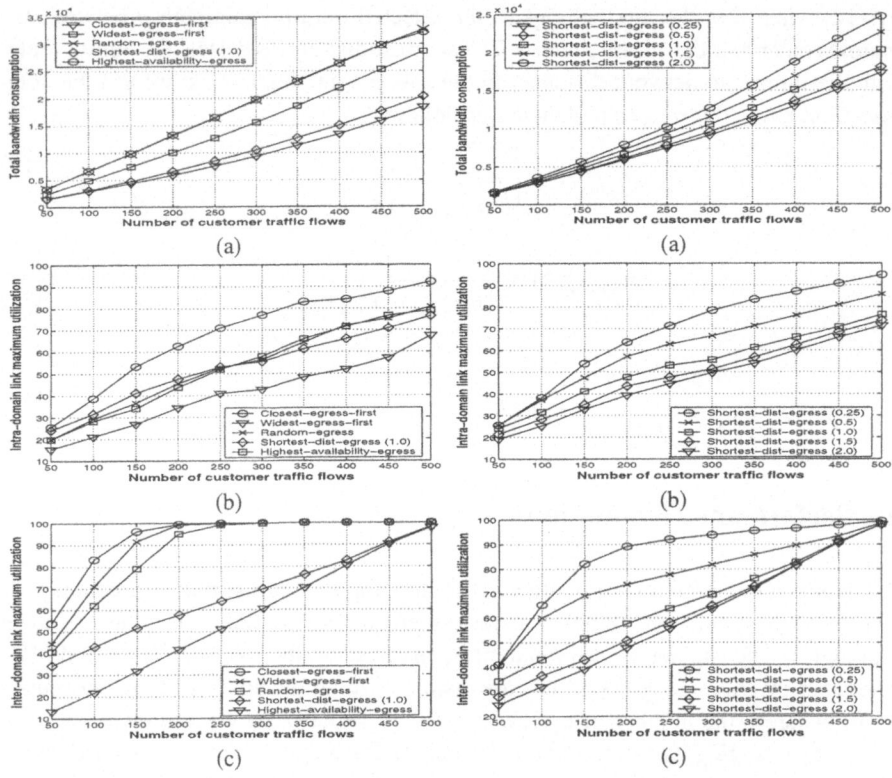

Fig. 4. Effects of egress router selection policies   **Fig. 5.** Impact of α

Figure 4(a) shows the total bandwidth consumption as a function of the total number of customer traffic flows under the five egress router selection policies. The closest-egress-first consumes the least total bandwidth because customer traffic flows have always been directed to the nearest egress routers, which attempts to minimize bandwidth consumption by using least-hop paths. The other selection policies consume more bandwidth because they take load balancing into consideration, which may result in using longer paths. Specifically, the selection policies such as the highest-availability-egress and the random-egress, which do not entirely consider the network distance to make the selection decision, consume relatively much more bandwidth. While the shortest-dist-egress jointly considers number of hops and path bandwidth availability, it consumes slightly more bandwidth than the closest-egress-first but better load balancing is achieved as shown in the subsequent simulation.

Figure 4(b) shows the performance of intra-domain load balancing under the five egress router selection policies. The closest-egress-first exhibits the worst performance. This is because customer traffic flows are always statically directed to the nearest egress routers using a small subset of intra-domain shortest paths, thus causing the links on these paths to be heavily loaded. The widest-egress-first performs the best in intra-domain load balancing because customer traffic flows are always distributed to the paths with the maximum bottleneck bandwidth. This minimizes the maximum utilization over all the links in the network. We find that an indirect way to achieve intra-domain load balancing is to evenly distribute traffic flows among all the egress routers. The random-egress and the highest-availability-egress are the examples. The rationale is that the more egress routers are evenly selected, the more different intra-domain paths are used to reach those egress routers. In general, this facilitates intra-domain load balancing. Since the shortest-dist-egress directly takes bandwidth availability as a factor to select egress routers, it performs better than all the other selection policies except the widest-egress-first especially under a heavily loaded network. Compared to the widest-egress-first, the shortest-dist-egress consumes much less bandwidth as this has been shown in the previous simulation.

Figure 4(c) shows the performance of inter-domain load balancing under the five egress router selection policies. Highest-availability-egress performs the best because customer traffic flows are distributed on the egress router with which the inter-domain link has the highest bandwidth availability. This minimizes the maximum utilization over all the inter-domain links. Selection policies such as the closest-egress-first, the widest-egress-first and the random-egress, which do not directly consider inter-domain link bandwidth availability to select egress routers, may direct traffic flows to a small subset of egress routers causing the inter-domain links at these egress routers to be heavily loaded while the other inter-domain links are barely used. This effect is more apparent for the closest-egress-first. As the shortest-dist-egress partly considers the inter-domain link bandwidth availability to select egress routers, it can achieve fairly good performance on inter-domain load balancing.

From the description of the shortest-dist-egress, α is an important selection parameter representing the degree to which the greatest available bandwidth is favored over the least one. The purpose of α is to balance the impact of path hops and path bandwidth availability on making the selection. This allows a selective adjustment between the objectives of minimizing the total bandwidth consumption and improving both the intra-domain and inter-domain load balancing. We investigate the impacts of α with respect to the three objective functions by adjusting its value. The simulation results are shown in Figure 5.

The graphs show that when α is small, the significance of path bandwidth availability to make egress router selection is reduced. As a result, the closest egress router is preferred and those heavily loaded intra-domain and inter-domain links may get higher chance to be selected. Thus, using small α would attempt to reduce bandwidth consumption at the expense of poor intra-domain and inter-domain load balancing. On the contrary, when α is big, the significance of path bandwidth availability to make the selection is amplified. This results in giving more emphasis on load balancing. In Figure 5, simulation results show that using large α would achieve better intra- and inter-domain load balancing at the expense of higher total bandwidth consumption. In summary, for the objective of minimizing the total bandwidth consumption and improving load balancing, small and big α are respectively used.

We summarize the implication of each egress router selection policy on the objectives that we consider as follows. The closest-egress-first, the widest-egress-first and the highest-availability-egress have the best performance with respect to the objective of minimizing the total bandwidth consumption, improving intra-domain and inter-domain load balancing respectively. The shortest-dist-egress has fairly good performance with respect to all the objectives. Specifically, it allows a selective adjustment between the objective of minimizing the total bandwidth consumption and improving load balancing by adjusting the parameter α. As minimizing the total bandwidth consumption and improving load balancing are contradictory, there is no policy that can achieve the best performance with respect to both objectives simultaneously.

5 Conclusions

In this paper, we presented the Bandwidth Guaranteed Egress Router Selection (BGERS) problem and solutions in the context of a TE-enabled Internet architecture. The architecture comprises traffic engineering extensions to the current intra-domain and inter-domain routing protocols. The objective of BGRES is that, for each customer traffic flow, select an egress router that satisfies the customer end-to-end bandwidth requirement while optimizing the network resource utilization by which we consider three objective functions: minimizing the total bandwidth consumption, improving intra-domain and inter-domain load balancing. We have developed a heuristic algorithm to solve the problem and studied the implication of egress router selection policies on the objective functions. This study benefits ISPs by choosing the egress router selection policy that fits their target objectives. As future work, we plan to extend the BGERS problem and solutions to accommodate the other QoS metrics, such as delay, with classes of services under the Differentiated Services network.

References

1. Awduche, D., Chiu, A., Elwalid, A., Widjaja, I., Xiao, X.: Overview and Prcinciples of Internet Traffic Engineering. RFC 3272 (2002)
2. Apostolopoulos, G., Williams, D., Kamat, S., Guerin, R., Orda A., Przygienda, T.: QoS Routing Mechanisms and OSPF Extensions. RFC 2676 (1999)
3. Bonaventure, O.: Using BGP to distribute flexible QoS information. Internet Draft <draft-bonaventure-bgp-qos-00.txt> (2001)
4. Cristallo, G., Jacquenet, C.: Providing Quality of Service Indication by the BGP-4 Protocol: the QoS_NLRI attribute. Internet Draft <draft-jacquenet-qos-nlri-04.txt> (2002)
5. Xiao, L., Lui, K.S., Wang, J., Nahrstedt, K.: QoS Extension to BGP. In the Proceedings of the 10th IEEE International Conference on Network Protocols. (2002) 100-109
6. Bressoud, T.C., Rastogi, R., Smith, M.A.: Optimal Configuration for BGP Route Selection. In the Proceedings of IEEE INFOCOM' 2003. (2003) 916-926

7. Katz, D., Kompella, K., Yeung, D.: Traffic Engineering Extensions to OSPF Version 2. RFC 3630 (2003)
8. Osborne E., Simha A.: Traffic Engineering with MPLS. Cisco Press (2002)
9. Martello, S.,Toth, P.: Knapsack Problems: Algorithms and Computer Implementations. John Wiley and Sons, New York (1990)
10. Waxman, B.M.: Routing of multipoint connections. IEEE Journal on Selected Areas in Communications. 6 (1988) 1617-1622
11. Feamster, N., Borkenhagen, J., Rexford, J.: Controlling the impact of BGP policy changes on IP traffic. AT&T Labs – Research, Technical Report HA173000-011106-02TM (2001)

On Properties of Internet Exchange Points and Their Impact on AS Topology and Relationship*

Kuai Xu[1], Zhenhai Duan[2], Zhi-Li Zhang[1], and Jaideep Chandrashekar[1]

[1] Department of Computer Science, the University of Minnesota,
200 Union St SE, Minneapolis, MN 55455, USA
{kxu, zhzhang, jaideepc}@cs.umn.edu
[2] Department of Computer Science, the Florida State University,
Tallahassee, FL 32306, USA
duan@cs.fsu.edu

Abstract. Internet eXchange Points (IXPs) are one of two primary methods for Autonomous Systems (ASes) to interconnect with each other for exchanging traffic and for global Internet reachability. This paper explores the properties of IXPs and their impact on the AS topology and AS business relations using Scriptroute and Skitter traceroute probes, BGP routing archives and other data. With these datasets we develop an algorithm to discover IXPs and infer ASes that participate at these IXPs. Using the discovered IXPs and their inferred AS participants, we analyze and characterize the properties of IXPs and their participants such as size, geographical locations. We also investigate the impact of IXPs on the global AS topology and business relations between ASes. Our study sheds light on the Internet interconnection practices and the evolution of the Internet, in particular, the potential role IXPs play in such evolution.

1   Introduction

The Internet is a collection of interconnected Autonomous Systems (ASes), which are independently administrated by various Internet Service Providers (ISPs), universities and other institutions. In general, two ASes interconnect with each other either via a dedicated circuit or through an IXP (See Fig. 1 for an illustration). In the former case, depending on their business agreement, either one party pays the full cost of the circuit for interconnection with the other party, or both parties pay half of the circuit cost to exchange traffic. Clearly via a dedicated circuit, only one pair of ASes can interconnect with each other. In contrast, an IXP is a *shared interconnection infrastructure*, where multiple ASes can interconnect with one another through switches and routers at the IXP. Each AS only bears the cost of a circuit from its own premises to the IXP, in

---

* This work was supported in part by the National Science Foundation under the grants ANI-0073819 and ITR-0085824. Any opinions, findings, and conclusions or recommendations expressed in this paper are those of the authors and do not necessarily reflect the views of the National Science Foundation.

N. Mitrou et al. (Eds.): NETWORKING 2004, LNCS 3042, pp. 284–295, 2004.

addition to facility fees paid for the IXP. More importantly, through the same infrastructure, an AS can interconnect with some or all of the other participants at the IXP, subject to mutual business agreements. Therefore, IXPs provide a cost-effective model for different ASes to interconnect. Indeed, hundreds of IXPs have been deployed around the world, and many ASes participate at such IXPs for Internet interconnection.

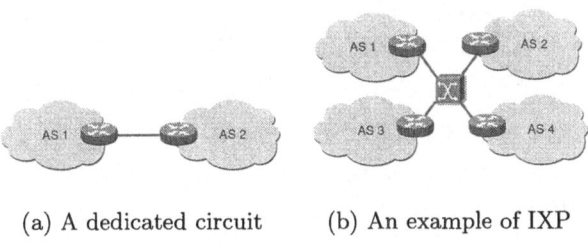

(a) A dedicated circuit   (b) An example of IXP

Fig. 1. AS interconnections

Although IXPs play an important role in the development of the Internet, many of their properties and their impact on the evolution of the Internet such as its topology and AS relationship, have not been systematically studied. For example, how are IXPs distributed on the Internet and how are they used in the different parts of the world? How do different ASes value and consequently utilize IXPs? Why do some ASes participate at an IXP, while others do not? How does an AS participating at an IXP decide which other AS to interconnect and what business relation to enter into? We believe that a better understanding of IXP properties and their impact on the global Internet will shed light on the Internet interconnection practices and the evolution of the Internet.

Towards this goal, we study the properties of IXPs such as geographical locations and size distributions, and investigate their impact on AS topology and relationship. To identify participants at an IXP, we design an inference algorithm based on traceroute probes and reverse DNS lookups. Using this algorithm, we find that IXPs appear to be more prevalent in Europe than in other continents. We also find that *transit* ASes (i.e., ASes with customer ASes) participating at IXPs have a higher percentage of peering relationship among themselves than, say, the average AS on the Internet. This reflects the fact that transit ASes generally peer more aggressively with other transit ASes to reduce transit cost, while stub ASes typically participate at an IXP primarily for transit service.

To investigate the impact of IXPs on AS topology and relationship, we decompose the AS topology into a *ring structure* based on the "distance" of an AS to the *Internet core* (or I-core in short). Using this ring classification, the (current) Internet AS topology can be decomposed into six rings. The I-core ASes normally only have a small percentage of IXP neighbors relative to their degree. This is not surprising, as the I-core ASes are known to prefer interconnection with their customer ASes via dedicated circuits for obvious economic reasons. However, a large percentage of transit ASes in Ring 2 and Ring 3, especially in Europe and Asia, have a significant number of IXP connections. These Ring 2 transit ASes more aggressively peer with each other to reduce transit costs and their reliance on the I-core, as is evidenced by the much higher percentage of peering relations among themselves.

The remainder of the paper is structured as follows. In Section 2, we describe the data sources and methodology used in the paper. We study the properties of IXPs in Section 3, investigate the impact of IXPs on AS topology and relationship in Section 4. The paper is concluded in Section 5.

2   Data Sources and Methodology

2.1   Data Sources

In order to have a more comprehensive set of *BGP route views*, we collected BGP routing tables from three vantage points, namely, the Route-views project [11], RIPE NCC [13], and a border router at the University of Minnesota and merged them to get an aggregated BGP routing table. We use the *public IXP database* available at Packet Clearing House (PCH) [12] for some basic information about IXPs. For each IXP, the database records geographical location (e.g., city, country), name, status (e.g., active, planning, defunct), IP address blocks, IP version (e.g., IPv4, IPv6), and some other attributes. In this paper, we mainly study the properties of *active layer-3 IXPs* supporting IPv4, where there are 148 such IXPs.

We use *traceroute* to discover IXPs and identify their AS participants. The data is collected from two sources, Scriptroute project [15] and Skitter project [1]. Scriptroute provides a measurement infrastructure for us to launch traceroute probes from more than two hundred nodes to a large set of sample destinations on the Internet, while Skitter project deploys about 20 monitors periodically sending out traceroute to end hosts on the Internet. The traceroute data were collected between 11/02/2003 and 01/16/2004.

2.2   Inferring IXP Participants

As observed in [9,7], if the forward path of a traceroute probe traverses an IXP, the IP addresses of the routers (or more precisely their interfaces) connecting to the IXP's switching facility are also recorded. Based on this intuition, we develop an IXP inference algorithm to infer the participants at an IXP. First of all, we filter out anomalous probes, such as those containing AS path loops. Then, we check if a forward IP path contains the IP address associated with an IXP by searching the IP address blocks in the IXP database. If a traceroute hop maps to an IXP, we consider the ASes before and after the hop as potential participants at the IXP. If the mapping fails, or if an IP address is mapped to multiple origin ASes, we do not consider them as IXP participants. Although [14] also describes a similar approach, it does not discuss several important issues such as multiple origin AS conflicts and AS loops [9] in traceroute results.

To obtain a larger set of IXP participants, we use another inference mechanism based on reverse DNS name lookups. The border router of an AS present at an IXP has an interface directly connected to the IXP switch facility. If the interface is associated with a DNS name, we will have a high probability to discover the participating AS from the domain name of the interface. Based on this, we design a simple scheme to identify more participants at an IXP. Given the IP address block of an IXP, we conduct brute-force reverse DNS lookups for the IP addresses within this block. If a reverse DNS lookup is successful, we extract the domain name from the host name. By looking up the domain name via DNS query, we obtain an IP address from the domain. We then map this IP address to an AS number, which is considered as a *potential* participant at the IXP. An AS inferred from this approach is labelled as a "true participant" at the IXP if the corresponding interface is alive by active probing using ping or traceroute.

2.3 Identifying IXP Edges and Inferring AS Relationship

An IXP provides a shared physical infrastructure for participating ISPs to interconnect. However, it is not necessary that every participant at an IXP will interconnect with every other. In other words, how a participant connects to others at the IXP depends on its mutual agreements, e.g., provider-customer or peering, with the others. We now briefly describe how to determine whether there is an interconnection between two ASes and how we infer their business relation. First we need to define some notation that will be used throughout the paper. We refer to the topology graph constructed *directly* from the aggregate BGP routing table as the AS topology graph, where each node represents an AS, and there is an edge between two ASes if they are neighbors on any AS path in the routing table. From the aggregate BGP routing table we constructed an AS topology graph with 16,014 nodes and 32,748 AS edges. Note that from traceroute measurements we may discover some ASes interconnections ("edges") at the IXPs that are not observed in the AS topology graph. We refer to them as *hidden (IXP) edges*. To infer the AS relation between two ASes, we apply the inference algorithm proposed by Gao [5] on the AS topology graph. In the remainder of the paper, we classify an AS as either a *transit* or *stub* AS based on its relations with its neighboring ASes. If an AS does not have any customer, it is a stub AS; otherwise, it is a transit AS.

3   Properties of IXPs and Their AS Participants

In this section we characterize the properties of IXPs in terms of their size, geographic location and hidden IXP edges. We further use three IXPs located in Asia, Europe and North America as representative examples to illustrate the relations between the IXPs and their AS participants and among the participants.

3.1   IXP Characteristics

Using the methodology described in the previous section, we discover 82 IXPs (referred to as *probed* IXPs) in the traceroute probes out of the 148 listed in the PCH database. We define the *size of an IXP* as the number of ASes participating at the IXP. Fig. 2(a) shows the *inferred* size distribution of the 82 IXPs, ranked by their size. Note that because traceroute probes may not traverse all participants of every IXP, the actual size of IXPs can be larger than what is reported here. In general, for IXPs close to traceroute monitors, their participants tend to be well discovered, as a large number of random probes generated by the traceroute monitors are more likely to traverse every single physical connection between the hosting ASes and their IXP neighbors. Hence, there is *bias* towards the IXPs close to traceroute monitors in terms of their inferred size [8]. To validate the coverage of the probed IXP participants, we use public participant lists released by six IXPs[1] on their Web sites. On average, 77.8% participants at these IXPs are discovered by our inference algorithm. Note that it is possible that we may not obtain such high coverage for the other IXPs due to limited traceroute probes.

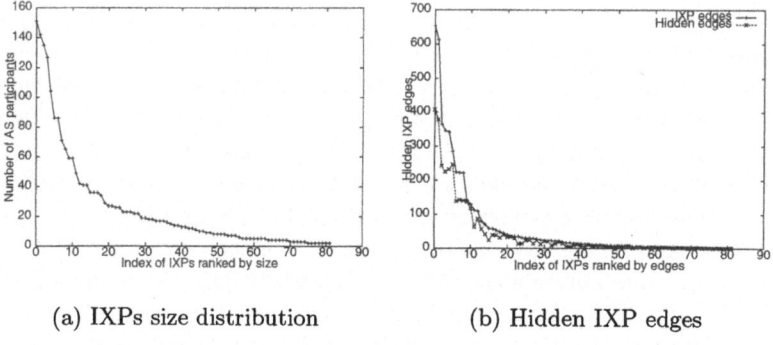

(a) IXPs size distribution   (b) Hidden IXP edges

Fig. 2. The size distribution of probed IXPs and hidden IXP edges

The 82 probed IXPs are located in five continents on the world: 7 in Asia (AS), 34 in Europe (EU), 35 in North America (NA), 3 in Oceania (OC), and 3 in South America (SA). It is not surprising to observe that North America has the largest number of IXPs as it contains more than 50% ASes on the Internet. However, 34 of 82 (41.5%) probed IXPs are from Europe, although it contains only 25% ASes on the Internet. These observations confirm the common belief that IXPs are a more prevalent method of Internet interconnection in Europe than other continents. Moreover, these statistics suggest that our inference algorithm is adequate to capture the characteristics of IXP distributions on the world.

---

[1] They are AMS-IX, DE-CIX, HKIX, JPIX, LINX-Extreme and LINX-Foundry.

Recall that we may discover some IXP edges that are not observed in the AS topology graph, and such an edge is called a *hidden* (or *missing*) edge. This may happen, for example, if ASes A and B have a *peering* relation at IXP X, and there is no BGP view data (i.e., BGP routing tables) from A, B and their respective customers. Fig. 2(b) shows the hidden IXP edges as well as the total number of IXP edges for the probed IXPs found in the traceroute dataset. We see that at 61 IXPs, 50% or more of the IXP connections among the AS participants are missing from the AS topology constructed from the BGP routing data alone. These results quantify the observation in [2] that a rich set of IXP edges are not shown in BGP-derived AS topology graph.

3.2   IXPs and Their AS Participants: A Detailed Look

In this section we study the role IXPs play in Internet interconnections and their influence on the relations among the AS participants by examining three representative IXPs in more detail. The three IXPs, PAIX, LINX-Foundry, and JPIX, are located in North America, Europe, and Asia, respectively. We classify AS participants at each IXP into *transit* and *stub* ASes. Table 1 lists the number of transit and stub ASes in these IXPs. We see that more than 70% of ASes participating at PAIX and LINX-Foundry are transit ASes, while JPIX has significantly more stub ASes (38%). These results suggest that these IXPs could potentially play different roles in AS interconnections among their AS participants by region, as the economics and interconnection structure are different and market conditions vary.

Table 1. AS types of participating ASes of selected IXPs

IXP Name	Size	# Transit ASes(percentage)	# Stub ASes(percentage)
PAIX	135	94 (70.0%)	41 (30.0%)
LINX-Foundry	151	119(78.8%)	32 (21.2%)
JPIX	71	44 (62.0%)	27 (38.0%)

We also study the business relations among IXP participants and observe that most transit ASes participate at IXPs to exchange traffic with other transit ASes, thus entering into a peering agreement with each other. Some are also there to provide transit service to stub (or other smaller transit) ASes. All stub ASes participate at IXPs to buy (less expensive) transit service from transit ASes at the IXPs. There are also slightly different roles for the IXPs: PAIX and LINX-Foundry are examples of a "peer traffic exchange" where mostly transit ASes participate to exchange traffic among themselves, with a relatively small number of ASes present to buy transit service. Whereas, JPIX seems to be an example of "transit service exchange" where many transit ASes not only participate to exchange traffic among themselves, but also provide transit service for stub (or smaller transit) ASes.

4 Impact of IXPs on AS Topology and Relationship

In this section, we first propose a top-down approach to decompose the Internet into a ring structure and classify the ASes into the "rings". For each ring, we further discuss the impact of IXPs on their AS interconnections and relationship.

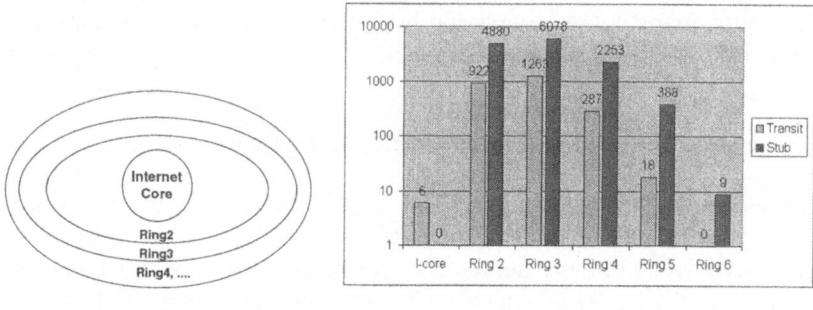

(a) The ring structure   (b) The size of Internet core and rings

Fig. 3. The ring structure of AS topology and their breakdown

4.1 IXPs and AS Topology

The Ring Structure of AS Topology. It is well-known that the Internet *AS topology* has an inherent *hierarchical* structure, arising from the interconnection relations among the ASes [6,17,16]. At the top, there is a collection of biggest ASes that *peer* with each other to exchange traffic with no monetary settlement. In other words, none of them carry transit traffic for another AS in the collection. Hence by definition, they form a full *peering* mesh. We will refer to this collection of ASes at the top of the AS hierarchy as the *Internet core*, or *I-core* in short. The rest of the ASes in the Internet rely on one or more ASes in the I-core to reach part of the Internet directly or indirectly.

In this study, we use the I-core identified in [3,10,18]. The I-core consists of the six largest ISPs that form a peering mesh, AS701(UUNET), AS1239(Sprint), AS7018(AT&T), AS3356(Level 3), AS209(Qwest) and AS3561(Cable & Wireless)[2]. We classify the rest of ASes based on their "distance" – the number of AS hops – from the I-core. We call the collection of ASes that are directly connected (i.e., one hop away) to one or more ASes in the I-core the *Ring 2*; the collection of ASes that are two hops away from the I-core (i.e., they can only reach an AS in the I-core via an AS in Ring 2) as *Ring 3*; and so forth. Using this ring

---

[2] The membership of the I-core has been changing as the Internet evolves. For instance, Genuity was used to be a member of the I-core. After being acquired by Level 3, it is no longer part of the I-core. The ASes in the I-core in our study are those identified during our data collection period.

classification, the Internet AS topology can be decomposed into six rings, as schematically depicted in Fig. 3(a).

To understand the role that IXPs play in the interconnections and business relations among ASes, we further divide each Ring i, $i = 2,3,4,5$, into stub ASes (referred to as *Ring i stub*) or transit ASes (referred to as *Ring i transit*). Fig. 3(b), with y-axis in log-scale, shows the number of transit ASes and stub ASes in each ring. The majority of ASes are in the Ring 2 (922 transit, 4880 stub) and Ring 3 (1263 transit, 6078 stub), comprising more than 80% of the total number of ASes in the current Internet. The I-core (transit ASes only) and Ring 6 (with 9 stub ASes) are the two smallest. Although this ring structure is very similar to various Internet topology models proposed in the literature, such as [6,17,16], there are two important difference between the ring structure and other models. First, the ring structure decomposes the Internet from two dimensions, namely, topology and business relations. Second, we choose the I-core based on the Internet peering practice [3,10,18], instead of simply picking a set of fully-connected "tier-1" ISPs[3].

IXPs and the I-core ASes. If an AS connects to an IXP, we refer to such a connection as an *IXP connection*, and the neighboring ASes thus connected via the IXP as *IXP neighbors*. By examining the IXP neighbors of the ASes in the I-core, we see that all the ASes are connected to a few of ASes via some IXPs. However, with the exception of AS3356, the number of ASes connected via IXPs relative to their degrees are fairly small, and the percentage of IXP neighbors are less than 2.5%. Hence the majority of ASes are connected to the I-core via "private peering", i.e., *dedicated circuits*. Although AS3356 has a relative high percentage (6.8%) IXP neighbors worldwide, nearly 75% of such IXP neighbors are located in Europe. In other words, Europe is the main location where AS3356 exchanges traffic with these neighbors through IXPs.

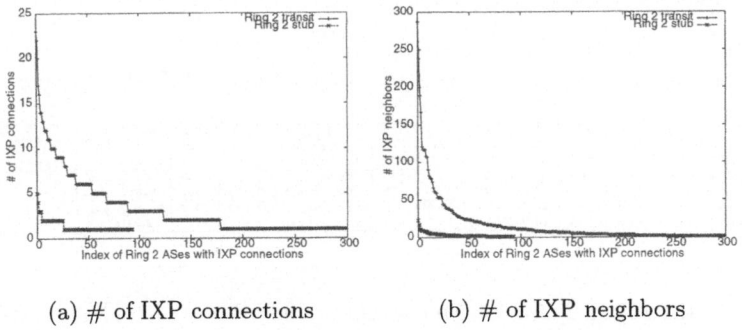

(a) # of IXP connections   (b) # of IXP neighbors

Fig. 4. IXP connections and neighbors of Ring 2 ASes

---

[3] Typically, the definition of "tier-1" is an ISP who does not buy any transit from others. Due to marketing purposes, some non tier-1 ISPs claim to be "tier-1".

IXPs and Ring 2 ASes. Using the traceroute probe data, we have found that a total of 1083 ASes are participating at at least one IXP. Among these 1083 ASes, 394 are from Ring 2 and 593 are from Ring 3. Among these 394 Ring 2 ASes, over 75% (300) are in Ring 2 transit, the rest (94) are in Ring 2 stub. Fig. 4 shows the number of IXP connections for the 300 ASes in Ring 2 transit and the 94 ASes in Ring 2 stub respectively, ranked by the decreasing number of IXP connections. We see that more than half of the ASes with IXP connections in Ring 2 transit are present at at least two IXPs, about 68 ASes are present at five or more IXPs. In contrast, only one out of 94 ASes with IXP connections in Ring 2 stub is present at five IXPs, while most of them are participating at only one IXP. This is consistent with the nature of transit and stub ASes, as transit ASes are more likely to participate at one or more IXPs to peer and exchange traffic with other transit ASes, while stub ASes participate at an IXP mostly for gaining transit service from one of the transit ASes at the IXP.

Table 2 shows the *geographic* break-down[4] of the ASes with IXP connections in Ring 2 as compared to that of all ASes in Ring 2. The large majority of ASes in Ring 2 are in North America, as most I-core ASes are in the United States. However, among these ASes, only a small percentage of the Ring 2 ASes are found to have IXP neighbors, i.e., connected to them via IXPs. In contrast, Europe, Asia as well as other regions have far fewer Ring 2 ASes, but they have a significantly higher percentage of Ring 2 ASes with IXP neighbors. Given that we tend to *underestimate* the number of IXPs and IXP neighbors in regions such as Asia where there are only a few traceroute monitors, due to the "biased" and partial data we have, the percentage of ASes with IXP neighbors can be considerably higher. These results illustrate that connection via IXPs is far more prevalent in Asia, Europe and other regions than in North America.

We also study the *degree distribution* of Ring 2 ASes and find out that Ring 2 ASes (especially transit ASes) *with* IXP connections tend to have higher degree than those *without* IXP connections. This finding indicates the strong influence of IXPs on the degree of ASes *with* and *without* IXP connections for ASes in Ring 2 transit and stub.

Table 2. Locations of Ring 2 and Ring 3 ASes with IXP neighbors

Ring	AF	AS	EU	NA	OC	SA	Total	Ring	AF	AS	EU	NA	OC	SA	Total
Ring2(R2)	5	118	406	5242	7	24	5802	Ring3(R3)	184	1385	2196	3098	227	251	7341
(R2 Transit)	(4)	(83)	(217)	(595)	(4)	(19)	922	(R3 Transit)	(14)	(299)	(458)	(387)	(54)	(51)	1263
R2 w/ IXPs	0	43	193	153	2	3	394	R3 w/ IXPs	0	158	261	136	17	21	593
(R2 Transit)	(0)	(40)	(148)	(107)	(2)	(3)	300	(R3 Transit)	(0)	(93)	(148)	(64)	(10)	(15)	330
Percentage (%)	0.0	36.4	47.5	2.92	28.6	12.5	6.8	Percentage (%)	0.0	11.4	11.9	4.4	7.5	8.4	8.1
(R2 Transit)	(0.0)	(48.2)	(68.2)	(18.0)	(50.0)	(15.8)	32.5	(R3 Transit)	(0.0)	(31.1)	(32.3)	(16.5)	(18.5)	(29.4)	26.1

IXPs and Ring 3 ASes. Among these 1083 ASes with IXP connections, 593 are from Ring 3. Table 2 also shows the geographic breakdown of ASes in Ring 3 and those with IXP connections. Relatively speaking, we see that a large portion of ASes in Asia, Europe and other continents are in Ring 3, most

---

[4] The AS to country and continent mapping is derived from CAIDA NetGeo database [1], which is updated every week.

of them in Ring 3 stub, as compared to North America. This is most likely due to the fact that most ASes, especially stub networks, in Asia, Europe and other continents typically connect to one or more regional transit ASes for global Internet reachability. Fig. 5(a) shows the number of IXP connections and IXP neighbors for Ring 3 *transit* vs. *stub* ASes that are present at IXPs. Similar to the case of Ring 2 ASes, Ring 3 transit ASes have relatively larger number of IXP neighbors than Ring 3 stub ASes, as transit ASes usually peer with others for economic reasons. Fig. 5(b) illustrates the degree distributions of Ring 3 transit and stub ASes with IXP connections. Furthermore, we also find that Ring 3 ASes (especially transit ASes) with IXP connections tend to have higher degree than those without. This again demonstrates the significant advantages and benefits of participating IXPs for Internet connections.

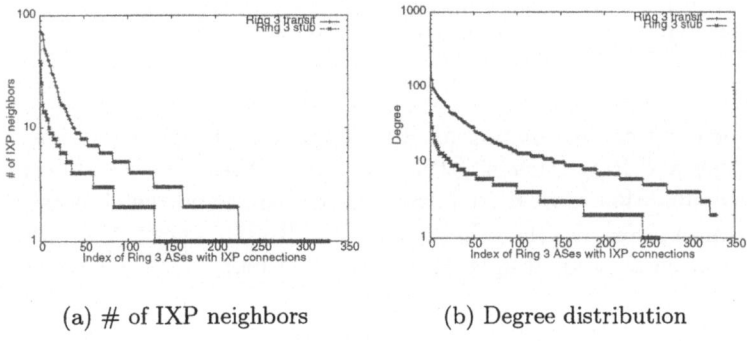

(a) # of IXP neighbors   (b) Degree distribution

Fig. 5. IXP neighbors and degree of Ring3 ASes w/ IXP connections

IXPs and Ring 4, 5, 6 ASes. Among the 2,955 ASes in Ring 4, Ring 5 and Ring 6, only 90 have IXP connections. Among them, 83 are in Ring 4, 7 in Ring 5 and none in Ring 6. Most of these ASes are stub ASes located in Asia and Europe, participating only at one IXP to buy transit service.

4.2   IXPs and AS Relationship

We now investigate the role of IXPs in influencing the business relations among ASes. To this end, we apply the AS relation inference algorithm to the aggregated BGP routing table. The results suggest that overall most dominant relations among ASes are of *provider-customer* type (93.9%), with only 6.1% inferred to be of *peering* relation type. However, among ASes that are participating at IXPs, the percentage of inferred peering relations are significantly higher (about 42.5%). This confirms the common belief that many (transit) ASes peer with each other at IXPs to exchange traffic was to reduce transit costs.

To gain a deeper understanding of how ASes interconnect at IXPs, we study the relations among ASes with IXP connections in the different rings of the AS topology. In general, we see that the I-core ASes participate at IXPs to provide transit service for other (Ring 2) ASes, *not* to peer with other I-core ASes. This is in contrast to Ring 2 transit ASes.

In Table 3, we examine how Ring 2 ASes form relations with their IXP neighbors. We present the number of provider-customer (PC) and peering (PP) relations and the percentage between Ring 2 and the following four groups: 1) I-core, 2) Ring 2 transit, 3) Ring 2 stub, and 4) Ring 3 ASes. The results suggest that most I-core ASes are providers of Ring 2 ASes, while most Ring 2 transit peer with each other. Most Ring 3 ASes are customers of Ring 2 transit.

Table 3. Relations: Ring 2 ASes and their IXP neighbors

Ring 2	I-core		Ring 2 Transit		Ring 2 Stub		Ring 3	
	PC (%)	PP (%)	PC (%)	PP (%)	PC (%)	PP (%)	PC (%)	PP (%)
transit	35 (72.9)	13 (27.1)	236 (41.0)	339 (59.0)	39 (67.2)	19 (32.8)	262 (62.5)	157 (37.5)
stub	8 (100)	0 (0.0)	39 (67.2)	19 (32.8)	0 (0.0)	6 (100)	4 (57.1)	3 (42.9)

We conduct a similar analysis for Ring 3 ASes and study the relations between Ring 3 ASes and Ring 2 transit/stub and Ring 4 ASes. The results confirm the previous finding that most Ring 3 ASes are customers and they buy transit from Ring 2 transit ASes. On the other hand, some Ring 3 transit ASes also provide Internet connectivity for Ring 4 ASes at some IXPs.

5   Conclusions

In this paper we have studied the properties of IXPs and investigated their impact on the AS topology and AS relationship, using traceroute results, BGP archives and other data sources. We developed an algorithm to discover IXPs and infer ASes that participate at these IXPs. Using the discovered IXPs and their inferred AS participants, we analyzed and characterized the properties of IXPs and their participants such as size, geographical location. We found that IXPs are a more prevalent method of Internet connections in Europe. Most participants at IXPs are transit ASes that more aggressively peer with each other, most likely to reduce transit costs. Furthermore, the percentage of peering relationship among transit ASes at IXPs are significantly higher than the average. We believe that our study is one of the first to perform a detailed analysis of the IXP properties and their impact on AS topology and AS relationship. Previous studies have focused on the properties of Internet topology at either AS-level or router-level [4], or on AS relationship [5,16]. The work [14] is most similar to ours, which also studies the properties of IXPs and their participants with only Skitter traces, but does not provide as detailed an analysis as we have done here, in particular, in terms of their impact on global AS topology and AS relationship. Our results indicate that it is interesting and important to take methods of Internet interconnections such as IXPs into account when analyzing the Internet topology and how it evolves. Furthermore, IXPs can also be used to provide additional information in inferring AS relationship. We believe that gaining a better understanding of IXPs and their role in Internet interconnections can shed light on the Internet interconnection practices and the evolution of the Internet, in particular, the potential role economics plays in such evolution. Due to limited data, our study constitutes only an initial step towards this goal.

Acknowledgments. We thank CAIDA, Route-views, RIPE, PCH for making the data available. We also thank Neil Spring for providing Scriptroute facility. We are grateful to David Farmer and University of Minnesota NTS for helping us set up measurement facilities. We are also grateful to Bill Woodcock for many helpful comments and discussions.

References

1. CAIDA. http://www.caida.org.
2. H. Chang, R. Govindan, S. Jamin, S. J. Shenker, and W. Willinger. Towards capturing representative AS-level Internet topologies. In *Proc. ACM SIGMETRICS*, 2002.
3. G. Cook. The Cook report on Internet protocol, technology, economics, policy: Economics of IP network interconnection peering transit and IXs, Nov. 2002.
4. M. Faloutsos, P. Faloutsos, and C. Faloutsos. On Power-law relationships of the Internet topology. In *Proc. ACM SIGCOMM*, Aug. 1999.
5. L. Gao. On inferring autonomous system relationships in the Internet. *IEEE/ACM Tran. on Networking*, Dec. 2001.
6. Z. Ge, D. Figueiredo, S. Jaiswal, and L. Gao. On the hierarchical structure of the logical Internet graph. In *SPIE ITCom*, 2001.
7. Y. Hyun, A. Broido, and k claffy. Traceroute and BGP AS path incongruities. In *Internetworking International Conference*, June 2003.
8. A. Lakhina, J. Byers, M. Crovella, and P. Xie. Sampling biases in IP topology measurements. In *Proc. IEEE INFOCOM*, Apr. 2003.
9. Z. M. Mao, J. Rexford, J. Wang, and R. Katz. Towards an accurate AS-level traceroute tool. In *Proc. ACM SIGCOMM*, Aug. 2003.
10. W. B. Norton. The evolution of the U.S. Internet peering ecosystem. Nov. 2003. http://www.equinix.com/pdf/whitepapers/PeeringEcosystem.pdf.
11. U. of Oregon. Routeviews archive project. http://archive.routeviews.org/.
12. PCH. IXPs database. http://www.pch.net/resources/data/exchange-points/.
13. RIPE. Routing information service raw data. http://data.ris.ripe.net/.
14. E. Silenok. Peering relationships through IX vs. direct peering, 2003. http://www.caida.org/~elena/projects.html.
15. N. Spring, D. Wetherall, and T. Anderson. Scriptroute: A public Internet measurement facility. In *USITS*, 2003.
16. L. Subramanian, S. Agarwal, J. Rexford, and R. H. Katz. Characterizing the Internet hierarchy from multiple vantage points. In *Proc. IEEE INFOCOM*, 2002.
17. S. L. Tauro, C. Palmer, G. Siganos, and M. Faloutsos. A simple conceptual model for the Internet topology. In *Global Internet*, 2001.
18. B. Woodcock. Internet topology and economics: How supply and demand influence the changing shape of the global network. Jan. 2003.

Scalable Packet Classification through Maximum Entropy Hashing

Lynn Choi[1], Jaesung Heo[1], Hyogon Kim[1], Jinoo Joung[2], and Sunil Kim[3]

[1] The Department of Electronics and Computer Engineering, Korea University,
Anam-Dong, Sungbuk-Ku, Seoul, Korea
{lchoi, jsheo, hyogon}@korea.ac.kr
Tel: +82-2-3290-3249
Fax: +82-2-921-0544
[2] i-Networking Laboratory, Samsung Advanced Institute of Technology
Giheung-Eup, Yongin-Shi, Gyeonggi-Do, Korea
jjoung@samsung.com
[3] The School of Information and Computer Engineering, Hongik University
72-1 Sangsu-Dong, Mapo-Gu, Seoul, Korea
skim@cs.hongik.ac.kr

Abstract. In this paper we propose a new packet classification algorithm, which can substantially improve the performance of a classifier by decreasing the rulebase lookup latency. The algorithm hierarchically partitions the rulebase into smaller independent sub-rulebases by employing hashing. By using the same hash key used in the partitioning a classifier only needs to look up the relevant sub-rulebase to which an incoming packet belongs. For an optimal partitioning of rulebases, we apply the notion of maximum entropy to the hash key selection. We performed the detailed simulations of our proposed algorithm on synthetic rulebases of size 1K to 500K entries using real packet traces. The results show that the algorithm can significantly outperform existing classifiers by reducing the size of a rulebase by more than four orders of magnitude with just two-levels of partitioning. Both the space and time complexity of the algorithm exhibit linearity in terms of the size of a rulebase, suggesting a good scalable solution for the packet classification with a large rulebase.

1 Introduction

Packet classification is one of the most fundamental building blocks in many networking functions such as Diff-Serv traffic conditioning, firewall, VPN, traffic accounting and billing, load-balancing, and policy-based routing. These functions need to track *flows* and give the same treatment to the packets in a flow. A rulebase stores classification rules, which define the flows and their corresponding treatments. Since a flow is defined by the header values of a packet, a classifier's duty is to examine the header and identify the corresponding flow.

Internet traffic is not only fast growing, but it is also diversifying both in applications and in protocols. New applications and protocols such as Internet telephony, security protocols, and peer-to-peer applications are being rapidly

N. Mitrou et al. (Eds.): NETWORKING 2004, LNCS 3042, pp. 296–307, 2004.
© IFIP International Federation for Information Processing 2004

deployed in addition to the traditional Internet applications such as Web, ftp, and email. As a result, the rulebase size is rapidly increasing. In a recent study [9], Woo argues that a rulebase with over a million entries is possible in future packet classification applications. From the classifier's viewpoint, this implies that for each packet the classifier must be able to find the matching rule with the highest priority amongst all the rules in the rulebase at the wire speed. Thus, there has been a renewal of interest [1, 2, 4, 5, 6, 7, 8, 10] in the scalability issue in terms of the size of a rulebase. Most of existing works, however, mainly focus on relatively small classifiers, e.g., with less than 20K rules [12]. To address this issue, we propose a new scalable packet classification algorithm that can scale well up to this size.

The motivation of our algorithm is based on the observation that a given packet matches only a few rules even in large classifiers [1]. This strongly implies that most of rules in any given rulebase are independent. Thus, we can partition the rulebase into many smaller independent sub-rulebases. As long as the matching sub-rulebase can be identified quickly, the performance of the rulebase lookup can be substantially improved since the lookup needs to be performed only in the final sub-rulebase. This is achieved by hierarchically decomposing the original rulebase into many smaller independent sub-rulebases based on the rules' definitions.

The algorithm is carried out in two phases: preprocessing and classification. First, during the preprocessing phase we hierarchically partition the original rulebase into many smaller independent sub-rulebases by hashing on the bit fields selected from the classification space. The degree of the partitioning depends on the density of a sub-rulebase in the classification space. The denser the sub-rulebase, the more partitioning is needed. This hierarchical partitioning stops until all the sub-rulebases are small enough. Then, during the classification phase a classifier inspects each incoming packet using the same hash key used in the preprocessing and identifies the sub-rulebase relevant to the packet. The search to find a matching rule is performed only in the final sub-rulebase where any existing lookup algorithm can be employed.

For an optimal partitioning of rulebases, we apply the notion of *entropy* in this paper, which guides us to choose the bits that most evenly divide the given rulebase. When the hash keys are selected to maximize the entropy, a rulebase is partitioned evenly into sub-rulebases under the smallest variance. As a consequence, we can achieve the smallest depth in the partitioning tree, which directly translates to the smallest number of hash table lookups. If the depth is small and the final sub-rulebase is small enough, we can achieve a low per-packet classification delay.

To evaluate the performance of our classification algorithm, we have applied our algorithm to real-life packet traces under synthetic rulebases of size 1K to 500K rules. The results show that the algorithm can reduce the size of the original rulebase by several orders of magnitude with only two-levels of partitioning, which requires only a couple of memory lookups. For example, a rulebase with 100K rules can be reduced to a sub-rulebase with only 7.6 rules on average and 258 rules in the worst case. In view of memory accesses, our algorithm requires 2 or 3 times less number of memory lookups compared to best classification algorithms known so far. Furthermore, the algorithm exhibits scalability in both its memory requirement and classification performance as we increase the size of a rulebase.

This paper is organized as follows. Section 2 defines the packet classification problem and presents the overall algorithm of our packet classification process. The section also introduces several partitioning algorithms including the notion of entropy and discusses the variations of the proposed algorithm to handle rule definitions with range and prefix mask description. Section 3 describes our experimentation methodology and summarizes the results. Section 4 concludes the paper.

2 The Proposed Classification Algorithm

2.1 Problem Definition

We can define the packet classification problem as follows. Given a *rulebase*, $R = \sum_{i=1}^{n} r_i$, which is a set of rules, a packet classifier needs to identify the rule that an incoming packet matches to by looking up one or more fields of the packet header. Each rule is specified by the range of values in one or more fields of a packet header. Specifically, in *d-dimensional packet classification*, each rule r_i is defined over d fields. Formally, r_i is defined by a tuple (C_i, A_i) where C_i is called a *classification space* and A_i is the associated action of rule r_i. The *classification space* is defined by the crossproduct, $C_i = F_1 \otimes F_2 \otimes \ldots F_d = \prod_{k=1}^{d} F_k^i$ where F_k^i is a range of values the field k must take. A rule r_i match a packet $p = \{b_1, b_2, \ldots b_d\}$ if for $\forall k, b_k \in F_k^i$ where b_k is a singleton. Multiple rules can match a packet. Thus, a classifier must identify the highest priority rule among all the matching rules. Intuitively, this requires the classifier to lookup the header fields of an incoming packet and to compare them against the rules in the rulebase one by one in order of decreasing priority. When n, i.e. the number of rules, is large or the arrival rate λ of incoming packets is high, this is a time-consuming serial process, which will limit the speed of the classifier. Thus, the essence of the problem is to find a fast yet scalable classification function both in time and in space.

2.2 Proposed Algorithm

Our classification algorithm is based on the conjecture that, in a rulebase, only a few rules have the possibility of matching a given packet. Let's look at the rulebase example of a typical firewall [15] shown in Table 1 where inner network serves several application services such as HTTP, telnet and FTP. Rules *R1, R2,* and *R3* represent grant of these connection requests while *R0* protects inner network against spoofing attacks. *D* is the default deny rule for all other communications. The protocol field in Table 1 suggests that a packet using UDP protocol can be matched only to *R0* or *D*. Thus, *R1*, *R2* and *R3* need not be matched against a UDP packet.

The algorithm consists of two phases: preprocessing and classification. The idea is to use divide-and-conquer approach. First, during the *preprocessing phase* we divide the original rulebase into many smaller independent sub-rulebases based on the values of classification fields where each rule is defined. Then, during the *classification phase* a classifier looks up the same header fields of an incoming packet and identifies the sub-rulebase where the relevant rules are stored. Thus, when the ratio (s/n) of the size of the sub-rulebase (s) over the size of the original rulebase (n) is small, then we can overcome the scalability issue by a single memory lookup to the hash table, which is constructed during the preprocessing phase. If a sub-rulebase is still large, then the sub-rulebase can be re-partitioned until the final sub-rulebase is small enough.

Table 1. A rulebase example of a firewall. †Inner side: protected local network by the firewall. ‡Outer side: network separated from inner side network by the firewall

Rules	Protocol	Src. Port	Dst. Port	Src. IP	Dst. IP	Action	Description
R0	*	*	*	Inner side†	Inner side†	Deny	Protection against Spoofing Attacks
R1	TCP	1024~65535	80	Outer side‡	Inner side†	Accept	HTTP Service
R2	TCP	1024~65535	23	Outer side‡	Inner side†	Accept	Telnet Service
R3	TCP	1024~65535	21	Outer side‡	Inner side†	Accept	FTP Service
D	*	*	*	*	*	Deny	Default Rule

According to [12], a rule $r_1 = (\prod_{k=1}^{d} F_k^1 , A_1)$ overlaps with a rule $r_2 = (\prod_{k=1}^{d} F_k^2 , A_2)$

if $\forall k\ F_k^1 \cap F_k^2 \neq \varnothing$. Intuitively, two rules overlap if there exists any instance of a packet that matches both rules. Since sub-rulebases differ at least in those bits that are selected as the hash key, a packet cannot match both sub-rulebases at the same time. Thus, the independence among sub-rulebases is guaranteed. Therefore, we need to look up only the relevant sub-rulebase after inspecting a packet on the same bit fields.

2.2.1 Preprocessing Phase: Rulebase Partitioning and Hash Table Construction

In preprocessing phase, we partition the original rulebase into many independent sub-rulebases. For example, the rules governing HTTP, FTP, and SMTP traffic can be partitioned into separate sub-rulebases. Then, by looking up the protocol field of an incoming packet, we only need to look up the sub-rulebase with the same protocol.

We can choose any of the bits in the classification fields as a hash key. If we select 8 bits, then we create a hash table with 2^8=256 entries, each of which points to a sub-rulebase. Intuitively, two rules may overlap if they map to the same sub-rulebase while rules mapped to different sub-rulebases would never overlap, which implies that they are independent. Sub-rulebases larger than a threshold value, such as 16 rules, can be repartitioned with another hash key, which must be different from the first hash key. This hierarchical partitioning stops until all the sub-rulebases are small enough. However, our experimentation results show that two levels of partitioning are enough for a rulebase under 500K rules.

Both the space and time complexity of our classification algorithm depend on the number of nodes and the depth of the partitioning hierarchy. To reduce the number of partitioning we need to partition a rulebase into sub-rulebases as evenly as possible so that the number of empty sub-rulebases is minimized and the number of rules in sub-rulebases must follow uniform distribution. This partitioning efficiency depends on the hash key selection algorithm, which we will discuss in detail in Section 2.3.

Table 2. A rulebase example and its hash tables. * denotes a don't care bit.

Classification space $(b_0b_1b_2b_3\ldots\ldots b_{103})$	Rule	Hash Key 8MSBs $(b_0b_1b_2b_3b_4b_5b_6b_7)$	Sub-rulebase	Hash Key b_3b_5	Sub-rulebase
0000 0110.............	R0	0000 0000	Null Entry	00	R3
		0000 0001	R3		
0000 0110...........	R1	:	Null Entry	01	R0, R1
		0000 0110	R0, R1		
0001 0001.............	R2	:	Null Entry	10	R2
		0001 0001	R2		
0000 0001.............	R3	:	Null Entry	11	Null Entry
**** ****.............	D	1111 1111	Null Entry		

2.2.2 Classification Phase

After we partition a rulebase and construct hash tables during the preprocessing stage, a classifier can narrow down the rulebase lookup by mapping an incoming packet into the corresponding sub-rulebase where the packet can be applied. The classifier looks up the hash table by using the hash key extracted from the packet header.

Let us consider the rulebase example shown in Table 2. Assume that rules *R0* to *R3* are listed in the decreasing order of priority. We assume 5-dimensional classification, which uses 104-bit fields from protocol (8), source port (16), destination port (16), source (32) and destination (32) IP addresses from the header.

In Table 2, we show only 8 most significant bits (MSBs) of a classification space, which may represent any header field such as the protocol. We partition the rulebase to 256 buckets by using the 8 MSBs and create the hash table as shown in Table 2. Rules in one sub-rulebase do not overlap with rules in other sub-rulebases. When a packet arrives, the classifier extracts the 8 MSBs from the header and uses it as an index to the hash table. If the hash table entry is not empty, then the classification is performed within the sub-rulebase. Otherwise, the default rule is the matching rule.

2.3 Hash Key Selection

2.3.1 The First-Level Partitioning

For the first-level partitioning, we only consider the protocol and port numbers as a hash key since these fields can naturally classify rules based on the Internet services governed by the rules. For example, HTTP service corresponds to protocol 6 and server port 80. The classification space at this level is comprised of protocol (8), source (16), and destination port (16) numbers. To limit the size of the hash table, we select a subset from the classification space as a hash key. In our implementation, we use a 17-bit hash key, which suggests a hash table with 128K entries. There is a tradeoff between the memory space and the depth of the partitioning hierarchy depending on the size of the hash key. Assuming each entry contains either a 32b address or NULL-pointer, the size of the table is 512Kbytes.

For the 17-bit hash key, we first select 6 bits from the protocol field using the entropy-maximizing key selection algorithm, which we will discuss in detail in Section 2.3.2. Since only two protocols, TCP and UDP, need to specify port numbers, we select up to 11 additional bits from the port numbers for these protocols. Since a port number is bi-directional, i.e. either source or destination, and specified by a range with upper and lower bounds, we select one of the port field by an additional bit to denote the direction and then select additional 10 LSBs or 6 MSBs from the port field by using the *precision directed grouping*, which we will describe in Section 2.4.2. Typically, a server port (dense area) designates a specific port number between 0 and 1023 while a client port (sparse area) uses a random port numbered from 1024 to 65536. Thus, lower 10 bits are used for a server port while upper 6 bits are used for a client port. Thus, a hash key is concatenated from [protocol field], [direction bit] and [10 LSBs | 6 MSBs in one of the port field]. Since a server port has a higher partitioning efficiency than a client port, we use the server port regardless of direction if a rule specifies a server port. If a rule specifies client ports in both source and destination ports, we use the 6 MSBs and spread the rules in both source and destination hash tables. This rule spreading is described in detail in Section 2.4.2.

2.3.2 The Second-Level Partitioning

The second level partitioning only applies to buckets larger than the threshold after the first level partitioning. Since the second-level hash key must be disjoint from the first-level hash key, we only consider source and destination IP addresses. To limit the size of the hash table, we only select a subset of the 64-bit fields as a second level hash key. In our implementation, we use a 16-bit hash key. We use the following four different hash key selection algorithms.

1. *MSB pattern* (represented as *MSB*): With this criterion, a 16-bit hash key is made by concatenating 8 MSBs from source and destination IP address fields. The idea is that most of prefix mask selects the first few significant bits from an IP address field. The time complexity of this key selection algorithm is $O(1)$.

2. *Exponential growing pattern* (*Exp*): With this criterion, a 12-bit hash key can be made by selecting the bit position corresponding to the exponential function of 2, namely $b_1 b_2 b_4 b_8 b_{16} b_{32}$ from both source and destination IP addresses. The idea is that the lower the position of a bit in an IP address field, the more likely to be masked out. We add extra two bits $b_6 b_{11}$ to create a 16-bit hash key. The time complexity of this key selection algorithm is also $O(1)$.

3. *Mask distribution pattern* (*Mask*): The basic idea of this heuristic method is that don't care bits in a classification space do not provide any information. Thus, each bit b_i in the classification space has the information in inverse proportion to the number of don't care bits in the bit position of rule definitions. The procedure of finding this key is as follows. For each bit position b_i we sum the number of non-don't care bits in all the rules in a rulebase, and accumulate them from the MSB to the LSB. For a k-bit hash key, we select a bit if the accumulated value of the bit position is the multiple of the total accumulated value divided by k. The time complexity of this key selection algorithm is $O(kn)$ assuming that the total number of rules in a rulebase is n. In our experimentation, k is 16.

4. *Entropy-maximizing pattern* (*Ent*): To find a good hash key we use the notion of *entropy*, which is used in information theory [11]. As known widely, the entropy is maximized when all the entries have the same probability of occurrence. Thus, we can find a good hash key through the calculation of entropy. Using the notion of entropy, a hash key K_σ of length σ can be expressed recursively by $\mathbf{K_\sigma} = \mathbf{K_{\sigma-1}}$ $\oplus$ **q**, where $\oplus$ is the concatenation operator and q is the bit from the

classification space that produces the maximum entropy. The algorithm starts by calculating the entropy for the hash key of length 1 and determines the bit position that produces the maximum entropy value. Then, the algorithm repeats this process for the hash key of length 2 and so on until the length of the hash key reaches σ or the entropy does not increase further. Based on this algorithm, we create a 16-bit hash key by selecting an 8-bit hash key from each IP address field. The time complexity this algorithm is $O(n \cdot [\frac{s}{2}(2w - s + 1)])$, where w is the length of classification space, s is the length of a hash key, and the n is the total number of rules in a rulebase. The detailed discussion of the entropy-maximizing key selection algorithm can be found in [16].

2.4 Adaptation of the Algorithm for Prefix Mask and Range Specification

So far our discussion implicitly assumed exact value matching for packet classification. However, a rule definition often includes field descriptions with prefix mask or range specification. In this section, we will discuss how our proposed algorithm can handle these different field specifications.

Table 3. A rulebase and its hash tables with two different hash keys of length 2.

Rule	Field Description $(b_0b_1b_2b_3b_4b_5b_6b_7)$	Index	Hash Key b_0b_1	Hash Key b_0b_2
R0	0000 0000	00	R0	R0
R1	0110 0000	01	R1	R1
R2	1000 0000	10	R2, R3	R2
R3	1*10 0000	11	R3	R3

2.4.1 Prefix Mask Field
This is commonly used to specify the range of an IP address field. Table 3 shows a rulebase example with prefix masks and its two different hash tables. The issue here is that we may need to duplicate a rule into multiple entries in a hash table if the rule contains a field specification with masks.

With b_0b_1 as a hash key, **R3** needs to be spread over two entries indexed by 10 and 11 since b_1 is don't care term in **R3**. We call this issue *rule spreading*, which may increase the size of the result hash table by duplicating rules. However, by selecting b_0b_2 as a hash key, the rule spreading can be avoided as. To avoid this rule spreading as much as possible, we need to modify the entropy-maximizing key selection algorithm such that when calculating the entropy of a bit, the algorithm must ignore a rule whose definition specifies don't care condition for the selected bit. This is simply because don't care bits do not add any information to the system in terms of entropy.

2.4.2 Range Field
This is commonly used to specify a TCP or a UDP port description. As described in Section 2.3.1, typically a server port designates a specific port number between 0 and 1023 while a client port uses a random port number between 1024 and 65536. The basic idea here is to transform the range specification to exact or prefix mask

specification. We can use *range to prefix conversion* [5], which splits a given arbitrary range to a group of prefix masks. For example, a 16-bit range [1024, 65535] can be split to six prefix masks such as 000001*, 00001*, 0001*, 001*, 01*, 1*. However, this method results in extensive rule spreading in our algorithm, which is not desirable. Alternatively, we propose *precision-directed grouping* in this paper.

It is straightforward to split a rule with a range specification into multiple rules with exact values. For example, a rule with a range [71, 74] can be split into four rules with exact values from 71 to 74. However, a rule with a wide range such as [49152, 65535] can create a huge number of rules (16,384). Fortunately, a TCP/UDP port description with a range is usually biased. For example, 80% of port numbers used in most rules are under 3,999 although the total number of port reserved is much higher (0 ~ 49,151) [13]. Thus, we can group a different number of rules depending on the density of the range. For the 16-bit port range, we use the 10 LSBs as a hash key for dense area [0, 1023], creating a single entry per port, while we use the 6 MSBs as a hash key for sparse area [1024, 65535], creating 63 entries, i.e. 1024 ports per entry.

3   Experimentation and Results

In this section, we demonstrate the performance of the proposed algorithms for 5-dimensional classification. Since it is difficult to obtain large real-life classification rulebases, we synthesized large rulebases from real-life packet traces. The packet traces were collected from PUBNET for five to eight hours during three days, 7/24/01, 12/14/01 and 12/17/01, respectively [14]. One-hour trace is nearly 70 million packets. The first trace is used to synthesize the rulebase and others are used as data for packet classification. To create a synthetic rulebase that resembles real-life rulebases, we carefully synthesized a rulebase by following the rulebase characteristics observed from real-life firewall applications [1, 2]. The detailed guidelines are described in [16]. All of our experimentation was performed in 1.7GHz Pentium IV system with 512MB of memory running Linux.

3.1   Rulebase Partitioning

3.1.1   First Level Partitioning
Figure 1 shows the results of the first level partitioning by displaying the average and maximum size of a sub-rulebase after the partitioning. By the partitioning we can reduce the average size of a rulebase substantially. For rulebases with 10K, 100K, and 500K rules, the reduction ratios are 0.0029, 0.0014, and 0.0014 respectively. This is very significant since we can reduce the size of a rulebase by more than two orders of magnitude by a single memory lookup to the corresponding hash table. However, as you can see from the maximum size of a sub-rulebase in the figure, rules are not evenly distributed in the partitioned rulebases. The largest sub-rulebase contains about 24% of rules of the original rulebase in all the rulebases tested. As we can predict, these rules are related to HTTP service, which corresponds to protocol 6 and port 80. The numbers of sub-rulebases over the threshold (16 rules per sub-rulebase) are 141, 192, and 1114 for 10K, 100K, and 500K cases. For these sub-rulebases we perform

the second-level partitioning. All of the first level partitioning is completed in less than one second in our experimentation platform. As a side effect of the first level partitioning, we observe that the partitioning more than doubles the total number of rules due to rule spreading. The actual inflation ratio is 2.42.

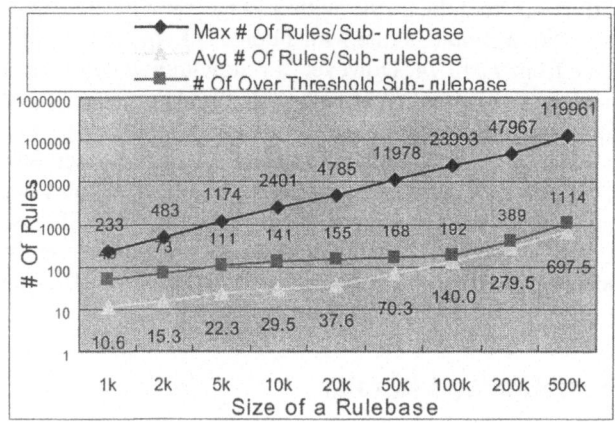

Fig. 1. The result of the first level partitioning

3.1.2 Second Level Partitioning

Figures 2 and 3 show the results of the second level partitioning with various hash key selection algorithms. Figure 2 shows the average number of rules per sub-rulebase while the Figure 3 shows the size of the largest rulebase. Assuming the entropy-maximizing key selection, the second-level partitioning further reduces the sub-rulebase by reduction ratios of 0.054, 0.054, and 0.052 for 10K, 100K, and 500K rulebases. When the first-level and the second-level partitioning are combined, 10K, 100K, and 500K rulebases are reduced to 1.6, 7.6, and 36.6 rules per sub-rulebase on average, which corresponds to reduction ratios of 0.00016, 0.000076, and 0.000073. This is very significant since we can reduce the size of a rulebase by more than four orders of magnitude by just two memory lookups to the hash tables.

The second level partitioning is also very effective in reducing the largest sub-rulebase, which contained 24% of the entire original rulebase after the first level partitioning. Assuming the entropy-maximizing key selection, with the second level partitioning we can reduce the size of the largest rulebase to contain 31, 258, and 1281 rules in 10K, 100K, and 500K rulebases respectively. This suggests that for a 100K rulebase we only need to compare a packet to those 258 rules in the worst case during classification phase. Note that the second-level partitioning is more effective than the first-level partitioning in reducing the largest rulebase. This is expected since the first-level partitioning partitions the original rulebase according to the Internet services. Thus, the largest sub-rulebase governing the HTTP service is grouped into a single sub-rulebase. On the contrary, the second-level partitioning can reduce this largest sub-rulebase substantially by considering the distribution of rules in the classification space with the entropy-maximizing key selection algorithm.

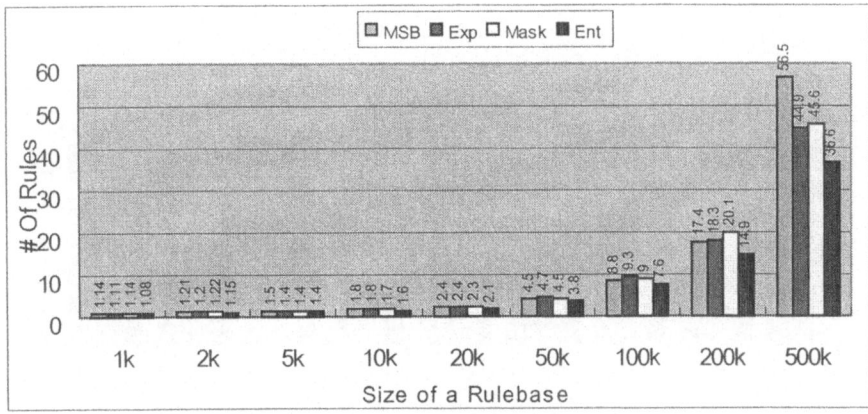

Fig. 2. The average size of a sub-rulebase with different key selection algorithms

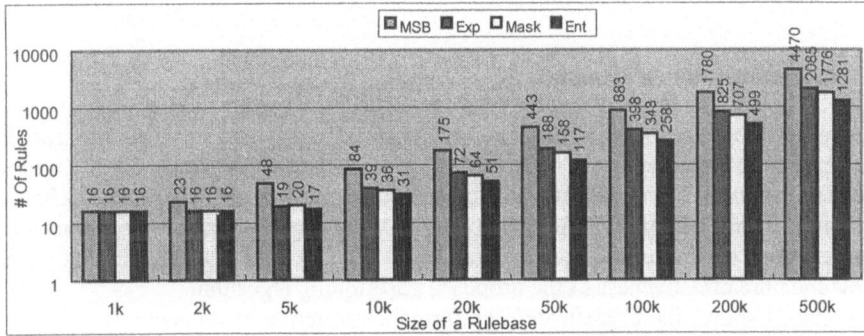

Fig. 3. The maximum size of a sub-rulebase with different key selection algorithms

Figures 2 and 3 also compare the effectiveness of different key selection algorithms. As expected, the entropy-maximizing key selection gives the best result. In particular, it can reduce the size of the largest sub-rulebase substantially compared to other heuristics. For 10K, 100K, and 500K rulebases, the entropy-maximizing key selection is more effective compared to MSB pattern key selection by factors of 2.38, 3.42, and 3.49. All the key selection algorithms show comparable results on the average size of a sub-rulebase. However, the entropy-maximizing key selection is the most effective in all the cases, especially for rulebases larger than 100K rules.

It is hard to judge the scalability of our rulebase partitioning from Figures 2 and 3 since the size of a rulebase does not increase linearly in the horizontal axis. But, we carefully scrutinized the numbers and could verify that both the maximum and the average size of a sub-rulebase grow linearly as we increase the size of rulebase. Note that these results assumed only two-levels of partitioning. If necessary, more levels of partitioning can be employed to significantly reduce the size of sub-rulebases. Likewise, we studied the memory requirement of our algorithm in terms of the growing rates of the total number of rules, the total number of buckets, and the total number of buckets over threshold [16]. We omit them here due to space limitation, but we find that all the numbers show a good scalability.

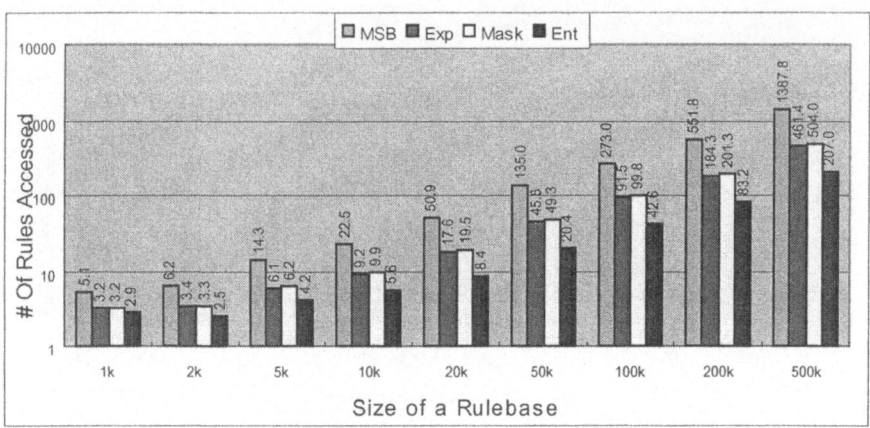

Fig. 4. The packet classification results

3.2 Classification Performance

To demonstrate the classification performance of our algorithm, we use real-life packet traces collected from PUBNET. The size of packet trace is 202 million packets, which is big enough to obtain the confidence of our experiment. After the first and second-level partitioning, we assume the worst-case search algorithm, i.e. a simple linear search, to find the matching rule in the final sub-rulebase so that we can show the pure effectiveness of the proposed partitioning algorithm.

Figure 4 shows the classification performance in terms of the average number of rules accessed to find a matching rule. Figure 4 reveals that our algorithm accesses only 5.6, 42.64, and 207.02 rules for 10K, 100K and 500K rulebases respectively. This is very promising since the best-known algorithm [1] requires at least 13 memory accesses for 5K rules. On the other hand, our partitioning algorithm needs to access only 5.6 rules in addition to 2 memory accesses required for hash table lookups although the worst-case linear search is used in the final sub-rulebase. By employing a more efficient algorithm [2, 5, 12] to find a matching rule in the final sub-rulebase, we can further improve the classification performance.

4 Conclusion

The growth of Internet in applications and protocols makes the size of a rulebase rapidly increasing. However, most of existing works mainly focus on relatively small classifiers, e.g., with less than 20K rules. Beyond this size, most of existing schemes may not scale either due to the memory explosion or due to the slowdown of classification. To address this issue, we propose a new classification algorithm that achieves the scalability by hierarchically partitioning a rulebase into many smaller independent sub-rulebases. By using the same hash key used in the partitioning a classifier can inspect an incoming packet and find its relevant sub-rulebase with a few memory lookups to the hash tables. Thus, the classification performance mainly depends on how much the final sub-rulebase can be reduced compared to the original rulebase. To make the reduction effective, we use the notion of entropy that finds a good hash key considering the distribution of rules in the classification space.

With synthesized rulebases of sizes ranging from 1K to 500K rules, we evaluate the effectiveness of the algorithm in both partitioning and classification. The experimental results show that two-levels of partitioning can substantially reduce the size of a rulebase. As a result, during classification a classifier needs to access only 4.2, 20.4, and 207 rules on average for rulebases with 5K, 50K, and 500K rules. This is very promising since one of the best-known algorithms [1] requires at least 13 memory accesses for 5K rules. Furthermore, we show that the proposed algorithm has the unique scalability both in space and in time as we increase the size of rulebase.

References

1. P. Gupta and N. McKeown, *"Packet Classification on Multiple Fields"*, In Proceedings of the ACM SIGCOM '99, Vol. 29, issue 4, August 1999.
2. F. Baboescu and G. Varghese, *"Scalable Packet Classification"*, In Proceedings of the ACM SIGCOM '01, Vol. 31, August 2001.
3. *Flow Analysis of Passive Measurement Data*, http://pma.nlanr.net/PMA/Datacube.html.
4. T. V. Lakshman and D. Stiladis, *"High-speed Policy-based Packet Forwarding using Efficient Multi-dimensional Range Matching"*, In Proceedings of the ACM SIGCOMM '98, Vol. 28, pp. 191-202, 1998.
5. V. Srinivasan, S. Suri, G. Varghese, and M. Valdvogel, *"Fast and Scalable Layer Four Switching"*, In Proceedings of the ACM SIGCOMM '98, Vol. 28, pp. 203-214, 1998.
6. V. Srinivasan, G. Varghese, and S. Suri, *"Packet Classification Using Tuple Space Search"*, In Proceedings of the ACM SIGCOM '99, Vol. 29, pp. 135-146, August 1999.
7. M. M. Buddhikot, S. Suri, and M. Waldvogel, *"Space Decomposition Techniques for Fast Layer-4 Switching"*, In Proceedings of the IFIP Sixth International Workshop on Protocols for High Speed Networks. Vol. 66, No. 6, pp. 277-283, August 1999.
8. A. Feldmann and S. Muthukrishnan, *"Tradeoffs for Packet Classification"*. In Gigabit Networking Workshop of the Proceedings of the IEEE INFOCOM '00. March 2000.
9. T. Woo, *"A Modular Approach to Packet Classification: Algorithms and Results"*, In Proceedings of the IEEE INFOCOM '00. March 2000.
10. P. Gupta and N. McKeown, *"Packet Classification using Hierarchical Intelligent Cuttings"*, In Proceedings of the Hot Interconnects VII, 1999.
11. Robert B. Ash, *"Information Theory"*, Dover Publications, 1ˢᵗ edition, November 1990.
12. H. Kim, J. Heo, L. Choi, and S. Kim, *"Taming Large Classifiers with Rule Reference Locality"*, In Proceedings of the ICOIN. Vol. 1 pp. 35-50, February 2003.
13. *IANA Port Number Assignment*, http://www.iana.org/assignments/port-numbers, 2002.
14. Korea Network Information Center, http://www.nic.or.kr/.
15. Elizabeth D. Zwicky et al. *"Building Internet Firewall"*, 2ⁿᵈ edition. O'Reilly, 2000.
16. Jaesung Heo, *"Scalable Packet Classification through Maximum Entropy Hashing"*, M.S. thesis, Department of Electronics and Computer Engineering, Korea University, 2003.

Diagonal Tuple Space Search in Two Dimensions

Mikko Alutoin and Pertti Raatikainen

VTT Information Technology
P.O. Box 1202, FIN-02044
Finland
{mikko.alutoin, pertti.raatikainen}@vtt.fi

Abstract. Due to the evolution of the Internet and its services, the process of forwarding packets in routers is becoming more complex. In order to execute the sophisticated routing logic of modern firewalls, multidimensional packet classification is required. Unfortunately, the multidimensional packet classification algorithms are known to be either time or storage hungry in the general case. It has been anticipated that more feasible algorithms could be obtained for *conflict-free* classifiers. This paper proposes a novel two-dimensional packet classification algorithm applicable to the conflict-free classifiers. It derives from the well-known tuple space paradigm and it has the search cost of $O(log\ w)$ and storage complexity of $O(n2^w\ log\ w)$, where w is the width of the protocol fields given in bits and n is the number of rules in the classifier. This is remarkable because without the conflict-free constraint the search cost in the two-dimensional tuple space is $\Theta(w)$.

1 Introduction

Traditional packet forwarding in the Internet is based on one-dimensional route look-ups: destination IP address is used as the key when the Forwarding Information Base (FIB) is searched for matching routes. The routes are stored in the FIB by using a network prefix as the key. A route matches a packet if its network prefix is a prefix of the packet's destination IP address. In the event that several routes match the packet, the one with the longest prefix prevails.

This well-know process does not inherently meet requirements of some of the new routing techniques. For example, in *firewalling, QoS based routing, programmable* and *active networking* [1] as well as in *application level routing* the forwarding decision is based on multiple protocol fields [2]. The forwarding is no longer based on just the destination IP address, but other attributes are considered as well. In firewalling, for example, the packet may be matched against a 5-tuple, composed of the source and destination IP address, source and destination port and the protocol field of the IP header. In application level routing, a URL can be used as an attribute when making the forwarding decision. In summary, all these new routing techniques require multidimensional packet classification [3, 4].

N. Mitrou et al. (Eds.): NETWORKING 2004, LNCS 3042, pp. 308–319, 2004.

When it comes to the methods of packet classification, linear search through the FIB is an option. In the linear search, all FIB entries are compared with the packet one by one in order to eliminate the non-matching entries. Among the matching entries, the one with the highest priority (e.g. the one with the longest network prefix in the case of one-dimensional packet classification) is the best matching entry. Unfortunately, the linear search is too time-consuming for backbone routers, which have to make tens of millions of forwarding decisions per second in order to keep up with the line speed. Thus packet classification algorithms are called for.

Efficient algorithms that facilitate wire-speed route look-ups have been developed for the problem of one-dimensional packet classification [5]. When the number of dimensions grows, so does the search complexity. A general k-dimensional ($k > 3$) packet classification algorithm has $O(log\ n)$ search complexity with $O(n^k)$ memory space or $O(log^{k-1}n)$ search complexity with $O(n)$ memory space, where n is the number of FIB entries [6]. This is impractical for a high-speed router. In [7], it has been suggested that more efficient packet classification algorithms could be developed for the conflict-free FIBs. To support this claim, a two-dimensional packet classification algorithm, which exploits the conflict-free constraint, has been provided. The algorithm is based on Tuple Space Search [8].

The work reported in [7] inspired us to study the subject more deeply. After a careful study, we came to the conclusion that the proposed algorithm does not work. However, by elaborating the ideas in [7] and by adding some new ones, it was possible to come up with an algorithm that makes use of the conflict-free constraint. The algorithm is shown to have the search cost of $O(log\ w)$ and the storage requirement of $O(n2^w\ log\ w)$. This is remarkable, because without the conflict-free constraint the number of search steps has been shown to be exactly 2w-1, i.e., $\Theta(w)$ [8].

Rest of the paper is organized as follows. Section 2 explains the concept of conflict-free constraint and section 3 describes the concept of tuple space. Section 4 introduces our contribution, the diagonal tuple space search in two dimensions, section 5 includes performance evaluation and section 6 concludes the paper.

2 Conflict-Free Constraint

Multiple FIB entries can match a packet at each look-up and thus some arbitration must be done to determine the best matching one. In the one-dimensional look-ups, the length of the prefix is used for this purpose. FIB entries with a longer network prefix get priority. In the multidimensional look-ups, the principles remain the same, i.e., the longer match gets priority. Nevertheless, it is not always that simple to determine which match is the longest one. From now on, a FIB entry is referred to as a *rule* - a commonly used term in packet classification [3]. Consider what happens if a FIB contains the following rules:

```
Rule1    From network a.b.c.* to network n.*.*.*
         DENY packets
```

Rule2   From network a.*.*.* to network n.b.c.*
          PERMIT packets

Let's suppose that a packet arrives from network a.b.c.* and its destination is in network n.b.c.*. Both rules match the packet, but which one is the longer match. It is impossible to say, because both are longer in one dimension but shorter in the other. So, the two rules are in conflict. Generally, two rules are in conflict when they *overlap* and neither one *encloses* the other [9]. Overlap means that all the prefixes of the two rules are non-disjoint. This is true in the example case, because a.b.c.* is a subset of a.*.*.* and n.b.c.* is a subset of n.*.*.*. Enclosure means that one of the rules is at least as specific as the other one in all dimensions. Clearly neither Rule1 or Rule2 encloses the other one.

Before a rule can be inserted into a FIB, all conflicts between the rule and the already inserted rules need to be detected [10] and resolved. There are two methods for resolving conflicts, i.e., *implicit conflict resolution* and *explicit conflict resolution* [9]. In the former case, the conflicting rules are assigned priorities that are used to arbitrate between the matching rules. In the explicit conflict resolution, a *resolving rule* is required for each conflict. A resolving rule specifies explicitly the action that prevails in the conflict region. Resolving rules are no different than the ordinary rules in the FIB, except that they are removed when either one of the conflicting rules is removed. A pseudo-code for computing prefixes of a resolving rule is given below. It is adapted from [9].

```
Function ResolvingRule(Rₐ,Rᵦ)
        for i = 1 to k do
                    Rᵨ[i]= Longer(Rₐ[i], Rᵦ[i])
        end for
        return (Rᵨ)
end Function
```

In the above example, the resolving rule would be:

Rule3   From network a.b.c.* to network n.b.c.*
          ACTION (= PERMIT packets or DENY packets)

Whenever a packet matches both Rule1 and Rule2, it will also match the resolving rule (Rule3). In such cases, the resolving rule is the best match, because it is always at least as specific as either one of the conflicting rules in every dimension. The action part of the resolving rule is decided by the entity that handles the conflict resolution.

The conflict-free constraint on the FIB means that there is a resolving rule for each conflicting rule pair in the FIB. This is a mandatory requirement for the algorithm which is put forward in this paper.

3 Tuple Space Search

The tuple space search [8] is a scheme proposed for multidimensional packet classification applications. Next, the basic features of the scheme and its major tools are explained.

3.1 Tuple Space Paradigm

In the tuple space search, rules are grouped based on their prefix length and a group is referred to as a *tuple* [8]. The groups are stored in hash tables and each group forms a separate hash table. In a k-dimensional FIB, the tuples are vectors of length k. For example, in a two-dimensional tuple space, rules $R_1 = (100^*, 11^*)$ and $R_2 = (001^*, 01^*)$ both map to tuple $T_1 = [3, 2]$, while $R_3 = (^*, 110^*)$ maps to tuple $T_2 = [0, 3]$.

The key idea is that rules are hashed by using the concatenation of the prefix strings. When a packet is being classified, each tuple is probed for a matching rule. The concatenation of bits from the packet's header fields forms the hash key. The tuple vector indicates the number of bits taken in each dimension. Note that a probe results in finding either one or none matching tuple entries.

The search complexity of this packet classification method is proportional to m, the number of tuples. This is an improvement to the basic linear search through the FIB, for which the search cost is proportional to N, the number of rules in the FIB. However, the worst case bound is still $O(N)$.

3.2 Markers and Pre-computation

Markers and pre-computation were introduced in [5] to carry out binary search for one-dimensional IP route look-ups. The FIB in [5] can be thought of as one-dimensional tuple space: the FIB entries are routes to networks and they are grouped to tuples by the length of the destination IP address prefixes. A hash key is generated for each tuple by taking as many most significant bits of the destination IP address as the hash table is wide. The basic linear search through the tuples has a search cost of $O(w)$, w being the width of the destination IP address in bits. However, a much better bound $O(log w)$ can be obtained by using binary search for the hash table probes. The binary search can be applied by employing *markers* and *pre-computation*.

Markers are used to direct the binary search to look for matching routes with even longer network prefixes. The idea is that adding a route that has a network prefix of length l will result not only in insertion of the route in the hash table of width l but also in insertion of a marker in each hash table of width shorter than l. For example, addition of route that has prefix 1101 will produce markers 110, 11 and 1, which are inserted in hash tables of width 3, 2 and 1, respectively. Thus an entry in a tuple can be associated with one route and one marker. Note that routes whose network prefixes start with the same l-bit sequence share markers in the hash tables that have width $\leq l$.

When a hash table (width l) is probed during a binary search, the other hash tables can be divided into two groups: longer half (width $> l$) and shorter half (width $\leq l$). If no matching marker is found, the longer half can be eliminated and the search focuses on the shorter half. This can be done because every matching route in the longer half would have left a matching marker in the probed hash table. If instead a matching marker is found, the binary search is directed to the longer half. Now, the shorter half cannot be dismissed straight away, because there is no guarantee that the longer half will eventually contain any matching routes. This situation is dealt with pre-computation.

The idea of pre-computation is that one can compute the best matching route in the shorter half for each marker beforehand and store it in the marker. In this way, one can dismiss the shorter half since the matching marker has already yielded the best matching route in that set. The algorithm must keep track of the current best matching route all along the search and update it each time a new matching marker is found. If the matching entry is associated with a route, but not with a marker, the search stops and that route is the best matching one.

To summarize, the markers and pre-computation can be used to trade off memory space and route/rule insertion time for faster look-up time.

3.3   Markers and Pre-computation in Multidimensional Tuple Space

To understand how markers and pre-computation work in the multidimensional tuple space, consider a tuple $T_i = [l_1, l_2, ..., l_k]$. The tuple space can be partitioned into three disjoint sets with respect to T_i, i.e., $Short(T_i)$, $Long(T_i)$ and $Incomparable(T_i)$ [8]. Set $Short(T_i)$ contains the tuples that are no longer than T_i in any dimension, i.e., tuple $T_j = [h_1, h_2, ..., h_k]$ belongs to set $Short(T_i)$ if and only if $h_i \leq l_i$ ($1 \leq i \leq k$) and $T_j \neq T_i$. Similarly, tuple $T_j = [h_1, h_2, ..., h_k]$ belongs to set $Long(T_i)$ if and only if $h_i \geq l_i$ ($1 \leq i \leq k$) and $T_j \neq T_i$. The rest of the tuple space belongs to set $Incomparable(T_i)$. Note particularly that if two overlapping rules R_i and R_j map to tuples T_i and T_j, respectively, and if tuple T_j belongs to set $Incomparable(T_i)$ then the two rules are in conflict.

Each rule that maps to tuple T_i can leave a marker in tuples in set $Short(T_i)$. Markers in T_i in turn contain their best matching rule, obtained by pre-computation, in set $Short(T_i)$. It follows that if tuple T_i is being probed and it does not contain a matching marker, set $Long(T_i)$ can be dismissed and the search can be restricted to sets $Short(T_i)$ and $Incomparable(T_i)$. Let us call the union of these sets as $Fail(T_i)$. If instead there is a matching marker in T_i then one can dismiss set $Short(T_i)$ by pre-computation and restrict the search to sets $Long(T_i)$ and $Incomparable(T_i)$. Let's call the union of these sets as $Success(T_i)$.

Can binary search work for k-dimensional tuple spaces? It turns out that it cannot, because set $Incomparable(T_i)$ is included both in $Success(T_i)$ and $Fail(T_i)$. Due to this overlap, the binary search cannot work. In fact, it has been proved in [8] that the best case search cost for any algorithm, which performs a search in k-dimensional tuple space ($k>2$), is $\Omega(w^{k-1})$. A related result has been provided in [11], where it has been stated that by deploying markers and pre-computation the worst case search cost is

$O(w^{k-1} \log w)$. For the special case of two-dimensional tuple space, the search cost has been shown to be exactly $2w-1$, i.e., $\Theta(w)$ [8].

Despite these rather disturbing results, it has been suggested in [7] that by imposing the conflict-free constraint on the FIB faster search algorithms can be obtained. To prove their claim the authors put forward an algorithm for two-dimensional packet classification for the conflict-free FIBs [7]. After a careful study, we came to the conclusion that the proposed algorithm does not work. However, by refining some of the given ideas and adding new ones, we were able to come up with an algorithm that seems to work. This algorithm is named as the diagonal tuple space search in two dimensions and indeed, it has the search cost lower than $O(w)$.

4 Diagonal Tuple Space Search in Two Dimensions

The diagonal tuple space search algorithm uses markers in a new and innovative way, i.e., markers are inserted diagonally. The pseudo-code below describes the procedure of inserting the markers. This procedure is executed each time a new rule is added into the FIB.

```
Function SetMarkers(Rule R)
/* Tuple T is initially the tuple
 * to which the rule maps to */
Tuple T = [|R[1]|, |R[2]|];

/* One marker is inserted in each iteration */
while (T ! = [0, 0])
if (T[1] > T[2])
        T[1] = T[1] - 1;
else
        if (T[2] > T[1])
                T[2] = T[2] - 1;
        else /* T[1] equals T[2] */
                T[1] = T[1] - 1;
                T[2] = T[2] - 1;
InsertMarkerAtTuple(T, R);
end while
end Function
```

An example of a two-dimensional FIB is shown in Fig. 1. The arrows describe the way the rules place their markers. For example, if a rule maps to the shadowed tuple $T_d = [2, 2]$ markers are inserted into tuples [1, 1] and [0, 0]. A rule mapping to tuple [0, 5] inserts markers to tuples [0, 4], [0, 3], [0, 2], [0, 1] and [0, 0]. Tuple [0, 0] is a virtual tuple which contains the default rule $R_{default} = (*,*)$. Fig. 1 also shows how the rest of the tuple space is divided into sets $Short(T_d)$, $Long(T_d)$ and $Incomparable(T_d)$

with respect to diagonal tuple $T_d = [2, 2]$. Definitions of the sets were given in the previous section.

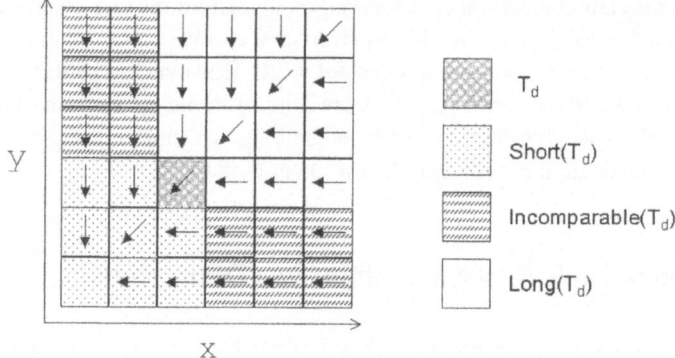

Fig. 1. Two-dimensional tuple space with respect to diagonal tuple [x, y]=[2, 2]

The algorithm starts with a binary search over the diagonal tuples in order to find the longest diagonal tuple T_d, which contains a matching entry. The following theorem states where the best matching rule resides with respect to tuple T_d.

Theorem 1

If the longest diagonal tuple, which contains a matching entry, is $T_d = [d, d]$ then the best matching rule resides in set Short(T_{d+1}) $\cup$ Incomparable(T_{d+1}), where $T_{d+1} = [d+1, d+1]$.

Proof.

Any matching rule in set $Long(T_{d+1})$ places a matching marker in the diagonal tuple T_{d+1}. If T_d is the longest diagonal tuple, containing a matching entry then set $T_{d+1} \cup Long(T_{d+1})$ contains no matching rules. Since $T_{d+1} \cup Long(T_{d+1}) \cup Short(T_{d+1}) \cup Incomparable(T_{d+1}) = 1$, it follows that the best matching rule resides in union $Short(T_{d+1}) \cup Incomparable(T_{d+1})$.

Recall that set $Short(T_d)$ is covered by pre-computation and T_d has already been probed. Thus the search can be restricted even further. Excluding set $T_d \cup Short(T_d)$ yields search area $(Short(T_{d+1}) \cup Incomparable(T_{d+1})) \cap (Long(T_d) \cup Incomparable(T_d))$. This remaining search area consists of two rectangles (see Fig. 2). Later we will show that if the matching entry in T_d, let this entry be E_d, is not associated with a marker, but a rule only, then that rule is the best matching rule. For now, suppose that a marker is associated with E_d in step 1 and the algorithm continues to step 2.

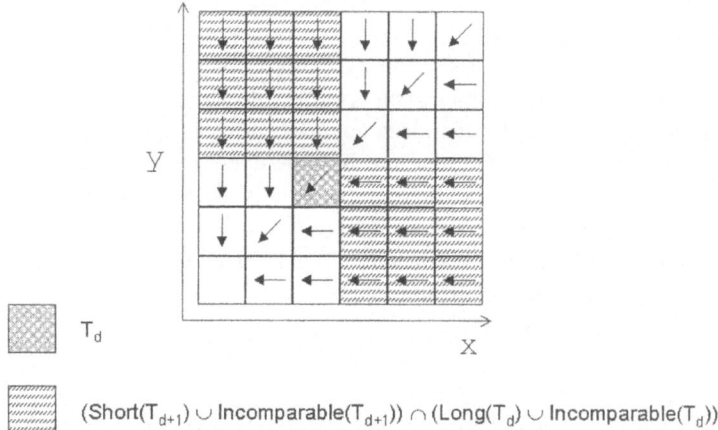

T_d

$(\text{Short}(T_{d+1}) \cup \text{Incomparable}(T_{d+1})) \cap (\text{Long}(T_d) \cup \text{Incomparable}(T_d))$

Fig. 2. Remaining search area after step 1 of the algorithm

In order to keep the search cost low, the algorithm uses a new technique that we call *mirroring*. Due to mirroring, only two additional binary searches are needed to conclude the packet classification. These two binary searches are performed on tuples $[d, y \geq d]$ and tuples $[x \geq d, d]$. This results in the search cost of $O(log\ w)$, because three binary searches are enough to classify a packet. The basic need for mirroring is that it must be ensured that every matching rule in any tuple $[x < d, y > d]$ is represented among the tuples $[d, y > d]$ and that every matching rule in any tuple $[x > d, y < d]$ is represented among the tuples $[x > d, d]$. This representation is achieved by using *mirror rules*.

Mirror rules are called for, when a conflict arises between a rule and a diagonal marker (i.e. a marker that resides in a diagonal tuple). The mirror rules are updated each time the FIB is changed. The condition part of a mirror rule is computed in the same way as the condition part of a resolving rule (see the `ResolvingRule` procedure in section 2). The action part of a mirror rule is the same as that of the rule, which is used to produce it. For example, if a rule $R_l = (*, 110*)$ has a conflicting marker $M = (00*, 11*)$ at tuple $[2, 2]$, a mirror rule $R_m = (00*, 110*)$ is produced in tuple $[2, 3]$ to represent R_l in column 2. If multiple rules produce a mirror in the same entry, the mirror that is produced by the longest rule prevails. For example, if there is another rule $R_2 = (0*, 110*)$, which is also in conflict with marker M, then the action part of mirror rule R_m is that of R_2.

Let us concentrate on explaining how mirroring works in the upper rectangle. From now on, the term *original rule* is used to refer to the rules which are not mirror rules. In other words, an original rule is either an ordinary rule or a resolving rule.

Step 2 of the algorithm is to perform binary search on tuples $[d, y \geq d]$. Suppose step 2 returns a matching entry that is in tuple $T_u = [d, y_u]$, $T_u \neq T_d$. Now, tuples $[x \leq d, y > y_u]$ can be dismissed from the search, because all the matching rules in that part of the tuple space are either in column d or have a corresponding mirror rule in column *d*. This is contradictory to the fact that [*d*, *y_u*] was found to be the longest tuple in column *d*. Hence, the tuples [*x* ≤ *d*, *y* > *y_u*] do not contain a matching rule.

At this point, it is clear that if the upper rectangle contains a matching rule, it will be in set $T_u \cup Short(T_u)$. If T_u contains an original rule, this is clearly the best matching rule in the upper rectangle. If it does not contain the original rule but contains a mirror rule, the mirror rule is the best match in the upper rectangle. If T_u contains no rules but only a matching marker, pre-computation is used to determine whether the upper rectangle contains matching rules at all.

The next theorem shows that if a matching rule is found in the upper rectangle, it is the best matching rule in the whole tuple space.

Theorem 2

If tuple T_m contains a matching rule and set $Long(T_m)$ contains no matching rules then no tuple in set $Incomparable(T_m)$ contains a matching rule.

Proof.

If two matching rules R_m and R_n reside in pair-wise incomparable tuples T_m and T_n then, by the conflict-free constraint, there is a third matching rule R_r which belongs to set $Long(T_m) \cap Long(T_n)$. Now, if set $Long(T_m)$ contains no matching rules then set $Long(T_m) \cap Long(T_n)$ contains no matching rules either. This is contradictory to the assumption that both tuples contain a matching rule. Thus if tuple T_m contains a matching rule and set $Long(T_m)$ contains no matching rules then no tuple in set $Incomparable(T_m)$ contains a matching rule.

The algorithm as a whole is as follows.

Step 1:   Perform binary search on the diagonal tuples in order to find the longest matching entry E_d among them. Let E_d reside in $T_d = [d, d]$. If E_d is associated with a marker proceed to step 2, otherwise return the rule which caused the match in T_d as the best matching rule.

Step 2:   Perform binary search on tuples [*d*, *y* ≥ *d*] to find the longest matching entry E_u among them. Let E_u reside in tuple T_u.

Step 3:   If $T_u \neq T_d$, go to step 4, else go to step 6.

Step 4:   If E_u is not associated with any rule (but a marker only), go to step 5.1, else go to step 5.2.

Step 5.1:   Find the best matching rule in set $Short(T_u)$ by pre-computation. If the rule is in rectangle [*x* ≤ *d*, *y* > *d*], return this rule, else go to step 6.

Step 5.2:   If E_u is associated with an original rule, return that rule, else return the mirror rule.

Step 6:   Perform binary search on tuples [*x* ≥ *d*, *d*] to find the longest matching entry E_l among them. Let E_l reside in tuple T_l.

Step 7:   If $T_l \neq T_d$, go to step 8, else go to step 10.

Step 8:   If E_t is not associated with any rule (but a marker only), go to step 9.1, else go to step 9.2.

Step 9.1:   Return the best matching rule residing in set $Short(T_t)$.

Step 9.2:   If E_t is associated with an original rule, return that rule, else return the mirror rule.

Step 10:   If T_d contains an original rule, return that rule, else return the best matching rule residing in set $Short(T_d)$ (and which is found by pre-computation).

Since the two rectangles are pair-wise incomparable, the search continues to the lower rectangle only when no matching rule is found in the upper rectangle. Due to the symmetrical nature of the problem, there is no need to explain steps 6 to 9 in detail. Step 10 is reached only if neither of the rectangles contains a matching rule.

To conclude the explanation, recall that the proof of step 1 was partly postponed to a later stage. Since Theorem 2 is now available, it is relatively easy to finalize the proof. The claim was that if E_d in step 1 is not associated with a marker, but with a rule only, then that rule is the best matching rule. Now, T_d obviously contains a matching rule while set $Long(T_d)$ contains no matching rules. Consequently, set $Incomparable(T_d)$ is also dismissed by Theorem 2.

As a final remark, a short explanation is provided why the algorithm in [7] does not work. The reasoning in [7] is based on the assumption that can be formulated as follows: *"If a matching marker resides in tuple T_m and a matching rule R_n resides in tuple T_n and tuples T_m and T_n are pair-wise incomparable then, by the conflict-free constraint, there is a matching resolving rule R_r in set $Long(T_m)$"*. This theorem does not hold, because a matching marker in tuple T_m does not guarantee that there is a matching rule in set $Long(T_m)$. Namely, it is possible for a rule to insert a matching marker even if the rule itself does not match. Our algorithm tackles this problem via mirroring.

5 Performance Evaluation

In this section, the search and storage complexities of the algorithm are evaluated. When it comes to other lookup algorithms [5, 7, 8, 11], which deploy hash tables, the search complexity/cost has been evaluated in terms of the asymptotic tight bound on the number of hash probes required to classify a packet. The storage complexity is generally evaluated by deriving asymptotic tight bound on the number of hash table entries needed to store the FIB and its associated data structures. These measures are used in the following analysis as well.

The search cost of the diagonal tuple space search in two dimensions is $O(log\ w)$, because three binary searches at most are needed to classify a packet. This looks very good, recalling that without the conflict-free constraint the theoretical best bound is $O(w)$. What can we say about the storage complexity? Recall that an original rule requires one mirror rule for each conflicting diagonal tuple. Within an incomparable diagonal tuple $T_d = [d, d]$ a rule that maps to tuple $[x < d, y > d]$ may have up to $2^{(d-x)}$

conflicting markers. At first glance, this yields the storage complexity of $O(n2^w w)$, where n is the number of rules in the FIB. However, the binary search on a column/row does not require that markers are created in all the tuples. It is enough to create them only in the tuples, which may be visited during the binary search [5]. Thus any rule leaves $log\ w$ markers at most. This gives the storage complexity of $O(n2^w\ log\ w)$. Table 1 contains a comparison between the search and storage complexities of packet classification algorithms, which are usable for two-dimensional FIBs.

Table 1. Comparison between two-dimensional packet classification algorithms

Algorithm	Search	Storage
Grid of tries	$O(w)$	$O(nw)$
Cross-producting	$O(log\ w)$	$O(n^2)$
Tuple space search	$O(w^2)$	$O(n)$
Rectangle search	$O(w)$	$O(nw)$
Diagonal tuple space search	$O(log\ w)$	$O(n2^w\ log\ w)$

The grid of tries and cross-producting have been described in [12]. Our algorithm exploits the conflict-free constraint in reducing the search cost dramatically, while the storage still remains linear with respect to the number of rules in the FIB. The downside is that the storage complexity grows drastically as the protocol fields get wider.

6 Conclusions

New routing techniques, such as firewalling and application level routing, require multidimensional packet classification in routers. Unfortunately, the general k-dimensional packet classification problem has been found to be either time or storage hungry. This fact has steered the research on packet classification algorithms towards hardware based as well as heuristic schemes. Nevertheless, it has recently been anticipated [7] that more efficient look-up algorithms could be achieved by imposing the conflict-free constraint on the Forwarding Information Base (FIB).

This paper proposes a novel search algorithm, named as the diagonal tuple space search in two dimensions, applicable to the conflict-free FIBs. The algorithm derives from the tuple space paradigm [8] and its search complexity is $O(log\ w)$. This is remarkable, because without the conflict-free constraint the number of search steps in a two-dimensional tuple space is known to be $2w-1$, i.e., $\Theta(w)$ [8].

The algorithm scales well with respect to the size of the FIB, because its storage complexity is $O(n2^w\ log\ w)$. Nonetheless, the worst case storage requirement grows drastically with respect to the width of the protocol fields. Yet, it has to be stated that the derived worst-case storage complexity is overly pessimistic and we believe that for real-life FIBs the scalability of the algorithm would be clearly better. It is for further study to develop estimates for the practical storage requirement.

Characteristics of the developed search algorithm support the claim that in some cases the conflict-free constraint can be leveraged in finding more efficient packet classification algorithms. At present, the algorithm is applied in two dimensions and further work concentrates on analyzing the implications of the conflict-free constraint for the tuple space in three or more dimensions.

References

[1]   Alutoin, M., Raatikainen, P.: Control Interface for Router Extension. Proceedings of 21st IASTED Conference on Applied Informatics (2003) 697-702

[2]   Gupta, P., McKeown, N.: Packet Classification on Multiple Fields. Proceedings of ACM SIGCOMM'99, vol. 29, no. 4 (1999) 147-160

[3]   Gupta, P., McKeown, N.: Algorithms for Packet Classification. IEEE Network, vol. 15, issue 2 (2001) 24-32

[4]   Chao, H.J.: Next Generation Routers. Proceedings of the IEEE, vol. 90, no. 9 (2002) 1518 – 1558

[5]   Waldvogel, M., Varghese, G., Turner, J., Plattner, B.: Scalable High Speed IP Routing Lookups. Proceedings of ACM SIGCOMM'97 (1997) 25-36

[6]   Overmars, M.H., van der Stappen, A.F.: Range Searching and Point Location Among Fat Objects. Journal of Algorithms, 21(3) (1996) 629-656

[7]   Warkhede, P., Suri, S., Varghese, G.: Fast Packet Classification for Two-Dimensional Conflict-Free Filters. Proceedings of 20th IEEE Infocom, vol. 3 (2001) 1434-1443

[8]   Srinivasan, V., Suri, S., Varghese, G.: Packet Classification using Tuple Space Search. Proceedings of ACM SIGCOMM'99 (1999) 135-146

[9]   Hari, A., Suri, S., Palkar, G.: Detecting and Resolving Packet Filter Conflicts. Proceedings of 19th IEEE Infocom, vol. 3 (2000) 1203-1212

[10] Baboescu, F., Varghese, G: Fast and Scalable Conflict Detection for Packet Classifiers, IEEE Computer Networks, vol. 42 (2003) 717-735

[11] Waldvogel, M.: Multi-dimensional Prefix Matching Using Line Search. Proceedings of IEEE Local Computer Networks (2000) 200-207

[12] Srinivasan, V., Varghese, G., Suri, S., M. Waldvogel: Fast and Scalable Layer Four Switching. Proceedings of ACM SIGCOMM'98 (1998) 191-202

An Evaluation of Delay-Aware Receiver Playout Strategies for VoIP Applications

Hugh Melvin and Liam Murphy*

Dept. of Computer Science, UCD, Ireland

Abstract. Previous work by the authors confirm the feasibility of implementing a delay-aware receiver buffering strategy based on synchronised time within VoIP applications. Our objective in this paper is to extensively test this delay-aware approach, termed the hybrid playout strategy over diverse networks using a combination of simulation and delay emulation. We utilise both measured delay data and derived delay models and quantify the significant performance gains through use of the ITU-T E-model. We also analyse recent Internet delay studies to further assess the hybrid's wider applicability.

Keywords: Periodic Network Measurements, Simulations, Network Emulation, Synchronised Time.

1 Introduction

A recent review of Internet telephony in [1] indicates that despite significant research, its deployment has been very limited mainly due to quality of service (QoS) concerns. In this paper we examine receiver-based buffering approaches to improving VoIP QoS.Adaptive buffering approaches generally have no knowledge of actual delays and adapt to trends in delay [2] [3] [4]. Non-adaptive approaches implement a simpler fixed delay at the expense of inflexibility.

In previous work [5] [6] we propose a hybrid playout strategy based on synchronised time, provided via the Network Time Protocol (NTP) and the Real-time Transport Protocol Control Protocol (RTCP) Sender Reports (SR). This delay-aware approach enables end-to-end delays to be determined on a per-packet basis and used to select an optimum playout strategy i.e. adaptive or fixed. The extent of testing in [6] however was very limited and in effect proved the technical feasibility of the approach rather than its applicability. In this paper we evaluate a revised version of the hybrid over a wide range of network conditions using both measured delays and derived delay models as advocated by [7]. As a basis of comparison, we use two of the baseline adaptive strategies, detailed in [2].

To facilitate trace-driven testing, extensive delay data was captured in tests within Ireland and to the UK, mainland Europe and the US. The relative performance of the various playout strategies was evaluated against this test data

---

* H. Melvin is contactable at hugh.melvin@nuigalway.ie

N. Mitrou et al. (Eds.): NETWORKING 2004, LNCS 3042, pp. 320–331, 2004.

using two different approaches. The first imposed the trace delay data on a simulated voice stream, and evaluated through simulation, the various playout strategies. The second approach utilized pre-recorded voice streams and a LAN-based testbed incorporating a delay emulator. Although the former approach facilitated rapid testing, the latter is useful in evaluating the various strategies in a real implementation environment and lends itself to subjective testing though various implementation issues arose which degraded testbed effectiveness.

In addition to the trace data, more general delay models were also developed.These models were tuned to present more severe network conditions than those experienced from trace data and relative performance was assessed via the simulated approach. A simplified yet conservative adaption of the ITU-T E-model was used in all cases to quantify performance gains.

Results show that the hybrid playout strategy performed better than the baseline adaptive approaches in over 99% of tests. The extent of gain in E-model R-factor varied up to 50% but averaged at about 10-15% relative to the two conventional adaptive approaches.

In general, we propose that although jitter remains a serious issue in *best-effort* Internet, actual delays are often well within the 150 msec M2E delay limit. In this context, a delay-aware playout strategy such as the hybrid algorithm can deliver significant voice quality improvements across Internet Wide Area Network (WAN) paths. We review recent delay studies by [9] [10]that support this position. Adaptive approaches, though useful where precise delay information is not available, will often result in unnecessary late packet loss and inherent voice distortion.

The remainder of the paper is organized as follows. Section 2 reviews the rationale for the hybrid algorithm and describes the additional dynamic features introduced in this paper. Section 3 deals with various test design issues including delay measurement, network emulation, general Internet delay models and also outlines the use and limitations of the ITU-T E-model.Section 4 outlines implementation details, section 5 summarizes results and finally section 6 concludes by describing current and future work plans.

2   Receiver Buffer Strategies

In order to cope with the Internet's *best-effort* service level, a great deal of research has focused on the optimal implementation of application delay or buffering delay within receivers. We evaluated many such adaptive approaches such as [2] [3] [4] and [9] in [6] and identified Alg. 1 and 4 from [2] (labelled Alg. A and B respectively in [6]) as common candidates for baseline comparison.

The following points summarize the key issues relating to the design of receiver buffer strategies for best-effort networks:

- Adaptive algorithms such as A & B that utilize TCP-like formula tend to react too slowly yet overestimate playout delays where delay variation is significant.

- The selection of tuning parameters is a non-trivial matter though is required to suit network characteristics. Such characteristics may however change with time.
- Adaptive approaches result in some degree of late loss.They also result in distortion of the original audio signal.
- Adaptive algorithms are most useful when receivers have no knowledge of actual one-way delays. If actual delays are known, and are within G.114 [8] requirements, a fixed playout delay will avoid many of the problems associated with adaptive algorithms. However a fixed playout delay imposed at the start and fixed for the duration of a session may quickly become inappropriate.

2.1 Hybrid Adaptive-Fixed Algorithm

In [5] and [6], we propose, describe and test to a limited degree, the hybrid approach. It combines the useful characteristics of both adaptive and fixed buffer strategies and requires end-host system clocks to be synchronized which is achieved through use of NTP and RTCP SR packets.

The following briefly describes the hybrid strategy, refer to [6] for more detail:

Session commences implementing an adaptive buffer algorithm. Synchronized time enables one-way packet delays to be precisely determined once the first RTCP SR packet is received. Each receiver generates and maintains a rolling delay histogram. In the following equation, est is the delay estimate value extracted from the histogram to meet target loss requirements, Wf is a weight factor applied in determining the fixed playout delay and Pkt refers to the packetisation delay. If est is within the G.114 limit, playout switches to fixed mode and the situation thereafter is monitored:

$$playout = est + (150 - Pkt - est) * Wf$$

Note that for the limited results reported in [6], both Wf and the frequency of re-evaluation of playout mode were fixed for the duration of a session. In this paper we introduce self tuning features to the hybrid.

Tuned Hybrid Algorithm. In [6], the authors test the hybrid algorithm over the limited Irish academic and research network and report significant performance gains. In this paper, we tune the hybrid as follows:

- *Dynamic Wf:* The weight factor Wf provides an extra safety margin in determining the fixed playout delay. Wf is initialized to 0.33 as before but subsequent values are set at *variation/median* with floor and ceiling values applied to maintain Wf in the range $0.1 - 1$. Essentially, this means that the safety margin applied is proportional to network jitter.

  As outlined by [9] and discussed in section 3.4, users are more sensitive to increased loss than to increased delay. This forms the basis for the development of the hybrid approach in that a higher (though fixed) delay is imposed, resulting in lower late losses and overall better quality.

– *Dynamic Estimation Interval:* The frequency at which playout mode is re-evaluated is set dynamically according to network conditions. A baseline value of 500 packets is used which equates to 16 sec with 32 msec packets but subsequent intervals are a function of $(150/median) * 500$. This ensures that where delays are running close to the G.114 limit, the operating mode is monitored more closely.

The core benefit of the hybrid algorithm is that an informed fixed playout is implemented whenever possible so that the integrity of speech both within and between talkspurts is maintained. Adaptive mode is thus used only when absolutely necessary.

3   Test Design Issues

As outlined in section 1, we utilised both measured delays and derived delay models for testing and regarding the former, used both a simulation-based and emulation-based approach. Model-driven testing was done through simulation only. Fig. 1 outlines the simulation model used and Fig. 2 illustrates the LAN-based emulator developed to evaluate the hybrid algorithm in a real implementation environment. The following sections briefly introduce and describe the network emulator, delay measurement issues, network delay and loss modelling and finally the ITU-T E-model.

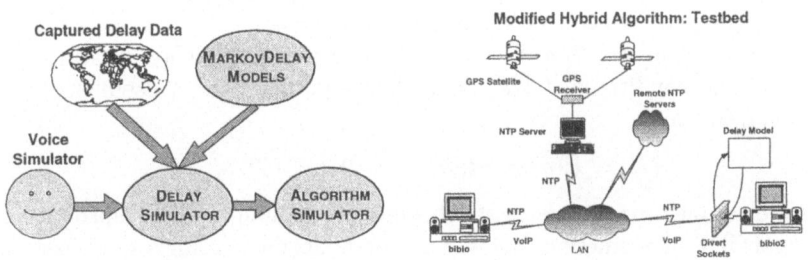

Fig. 1. Simulator Model   **Fig. 2.** LAN-based Emulator

3.1   Network Emulator

The network delay emulator is based on Linux Divert Sockets. A detailed description of its operation and use in testing applications over a variety of network models is given in [13]. Essentially it enables a user to intercept media packets at the socket level and divert them to a user process which reinjects them according to user specified rules of loss or/and delay.

3.2 Delay Measurement Issues

We gathered extensive delay data using the *ping* utility and estimated one-way delays assuming symmetric paths. Although one-way delay measurement is superior to round-trip-time (RTT) delays, RTT data is much easier to capture. Our objectives were to capture the general characteristics of diverse network paths and thus RTT data was considered sufficient. Bolot in his analysis [16] uses round-trip-times and Schulzrinne [14] though recording one-way delays suggests that round-trip-delays are generally sufficient.

Where precise one-way delay measurements are required, relying on RTT is unwise as outlined in RFC 2679 [11]. RFC 2330 [12] outlines that for delay measurements, periodic test packet generation can lead to synchronization issues or interference with network characteristics. For our tests, however the objective was somewhat different in that we wish to see the effect of simulated VoIP data streams on overall delay characteristics. In this regard, our objectives are similar to those of [16], [17], [14] detailed above and thus the more recent RFC 3432, on periodic sampling provided a more useful reference [15].

3.3 Internet Modelling

Internet loss and to a lesser degree delay modelling has attracted much research in recent years. In [7], the pitfalls of relying solely on trace-driven simulation are highlighted and a source-driven approach is also advised. As outlined in [14] and [16], network delay and loss both exhibit burstiness or temporal dependency and losses are often preceded by high delays which in a VoIP environment can lead to a final loss pattern of large loss bursts, comprising late loss and network loss. Our focus here is solely on delay modelling.

Schulzrinne et al in [17] and [14] examine the issue of delay burstiness and recommend the conditional CDF, defined in equation 1 as a useful metric to capture burstiness.

$$f(t) = P(d_i >= t | d_{i-l} >= t), l = 1, 2, 3, ... \qquad (1)$$

[17] also examines the degree to which different queuing models reproduce the conditional CDF phenomenon. Bolot et al [16] model delay using a single server, 2 input stream queue where one stream represents periodic media packets and the other represents all other Internet traffic. They report good correlation between analytical and experimental results. In [13], a Markov Modulated Poisson Process (MMPP) model is used to simulate zero-loss, high delay-variance in a wireless channel with retransmissions.

In analysing trace data and developing general delay models, we draw from a number of the above sources. We applied the conditional CDF method to the trace data to determine the extent of delay burstiness. We then developed general bursty delay models, by applying some characteristics of actual traces to a 2-state Markov model. Unlike [13] which used a Poisson process to model interarrival times (in a wireless environment), packet interarrival times within each state were centered on the mean, with a wider spread of values for the BAD state than for the GOOD state.

3.4   ITU E-model

Although the use of Mean Opinion Score (MOS) tests are the preferred mechanism for evaluating speech quality, it requires significant resources to yield useful results. The recently developed ITU-T E-model is a rapid planning tool for predicting how the average user rates the voice quality of a phone call with known transmission parameters. The model returns a transmission rating factor R (scale 0-100), defined as:

$$R = (R_o - I_s) - I_d - I_e + A$$

Technical details are specified in [19] including a mapping from R to MOS scales. The factors of interest here are I_d (delay impairment) and I_e (loss impairment). I_e includes the distortion caused by low bit rate codec operation as well as the effect of packet loss (both network and late loss). Due to its ease of deployment, many other researchers such as [20] [21] and [18] have adopted it for assessing speech quality.

In the context of evaluating real voice sessions, the E-model is however limited in that it delivers an instantaneous rating based on singular loss and delay figures. Other research has thus examined issues such as bursty versus random loss, recency [22] [9], perceived versus instantaneous quality [22] and rating of entire voice calls rather than segments [9]. For the purposes of this paper, we apply the same simplified yet conservative E-model analysis developed in our previous work [6]. As such, our analysis is aimed at extracting approximate relative values rather than detailed absolute values.

As such, the I_d impairment can be approximated by 10 units per 100 msec [9] ([21] uses 3 units/100 msec up to 170 msec). For the I_e factor, we base impairment on that for G.711, with PLC implemented, a packet size of 10 msec and bursty rather than random loss. Packet loss in the range 0-5% can therefore be approximated by a dual-slope curve set at 10 units per % above 3% and at 3 units per % below 3% [see G.113 [23] for details]. A critical factor for the hybrid algorithm is the lower sensitivity of I_d to increased delay than that of I_e to packet loss. As outlined in section 2.1, this justifies the use of the weight factor Wf in calculating the fixed playout delay from the delay estimate est.

4   Implementation Details

Implementation details for capturing trace data, for the simulator and for the LAN-based emulator are presented separately.

4.1   Trace Capture

Trace delay data was gathered from tests to a number of remote Internet machines, in Ireland, UK, mainland Europe and the US. The *ping* utility was configured to mimic actual data streams (eg. 256 byte packets every 32 msec mimics G.711 data packets) and round-trip-times were recorded. In all, over 1000 tests were conducted. Tests lasted between 20 seconds and three minutes and were repeated at various times in order to capture any diurnal variation.

4.2   Simulator

For simulating voice input, the distribution of talkspurts and silence periods was generated from actual voice streams. This was mapped to the delay values to simulate real voice traffic which was fed to the various playout algorithms.In [9], a scale of 1 to 6 is used to distinguish between conversations involving short, rapid exchanges and slower, more relaxed speech. We chose a *medium* scale voice session for tests. Both Alg. A and B had default settings from [2] (Alg. A:α=0.998002 and β=4, Alg. B:α=0.875, β=4). Spike threshold values were chosen based on network conditions and varied from 10-30 msec. This simulation-based approach facilitated extensive and rapid testing and is similar to that used by [2].

4.3   Emulator

- Both end hosts were situated on the same LAN with the divert sockets application running on one host. See Fig. 2.
- Trace delay data files were used as source data for the emulator.
- Pre-recorded voice streams were delivered across the testbed. As with the simulator, a *medium* scale ON/OFF speech pattern was used.

5   Results

We firstly analyze the subset of trace delay data used for algorithm evaluation. The relative performance of the various algorithms when tested with the simulator approach using both the trace-driven delays and general delay models is then assessed. The emulator was then used to confirm the relative performance using a subset of the trace-driven delay tests in a real VoIP environment. Finally, we summarize recent delay studies such as [9] [10]that support the wider applicability of the hybrid approach.

5.1   Measured Delay Characteristics

We present a representative sample of the total trace delay data captured. More extensive results are available from [24]. This sample focuses on tests to four remote servers located in Dublin (www.ucd.ie), London (www.cs.ucl.ac.uk), Munich (www.lkn.ei.tum) and Berkeley, Calif.(www.icir.org). Tests to each were carried out in blocks of 30 where test numbers correspond to the following times of day (GMT/UTC): Test 1-6 (17:00-18:00),Test 7-12 (22:00-23:00), Test 13-18 (09:00-10:00), Test 19-24 (10:00-11:00) and Test 25-30 (13:00-14:00).

The following points summarise trace delay results:

- Mean Delay & Diurnal Variation: Although the *mean* delay to each location varied little over the course of a day, diurnal variation was evident with greatest burstiness evident in the 17:00-18:00 window.

- Unconditional/Conditional CDF: Although there was evidence to support some temporal dependence of delay values, it was most evident in the 17:00-18:00 window GMT.
- Network Loss: Network loss rates were generally very low, though interestingly were highest for the most local path to UCD. This contrasts with the more usual findings that international paths suffer from higher loss rates.

General Delay Models. A number of 2-state Markov models were developed based loosely on captured traces to both Germany (www.ei.tum.de) and the US (www.icir.org). The mean delay required as a priming delay value was taken directly from the traces. The required Markov parameters such as the average high-delay burst length (i.e. BAD state) and overall extent of high delays were exaggerated from traces to increase burstiness. A range of different models were developed and tested; results presented are based on a high delay state probability of 0.05 and a mean high-delay burst of 320msec (10 packets). As outlined in section 4.3, the GOOD and BAD states were modelled using two different ranges of packet jitter, significantly higher than those recorded in traces (Compare Fig. 3 with Fig. 8).

5.2 Hybrid Performance Results

Simulator-based results using both the actual delay traces and the 2-state Markov models are presented separately.

Trace-Driven Testing. Fig. 3 illustrates for LKN, a single test performance of the hybrid algorithm along with Algorithm A/B (Adaptive/Spike). The spike threshold for Alg.B was set to 10 msec. As evident from these figures, the hybrid resulted in higher playout delays but significantly lower late losses.

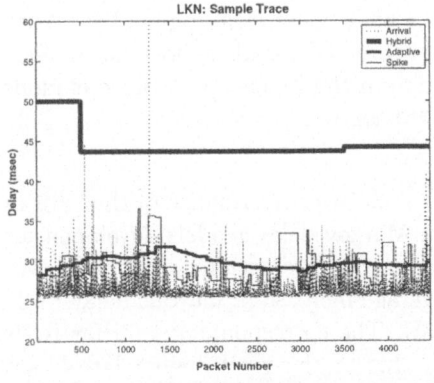
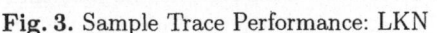
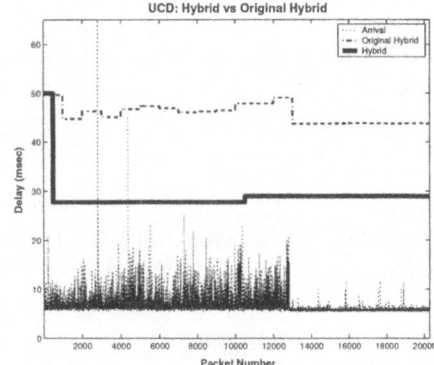

Fig. 3. Sample Trace Performance: LKN   **Fig. 4.** Self-tuning Hybrid

Figures 5, 6, and 7 outline the performance of the hybrid along with adaptive algorithms Alg. A and B over a single but representative days tests to UCL, ICIR and LKN respectively. The plots within each figure show the average late loss rate for all three algorithms, the additional delay introduced by the hybrid relative to Alg.A/B, and the net gain in R-factor.

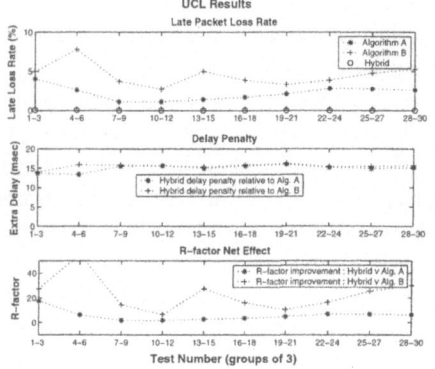

Fig. 5. Performance Comparison: UCL

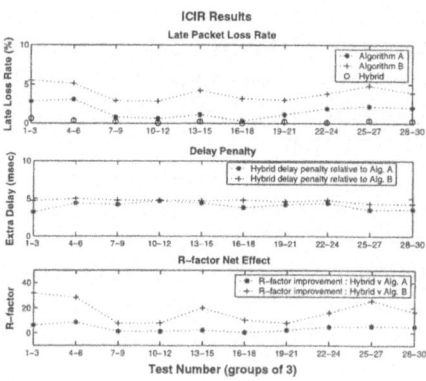

Fig. 6. Performance Comparison: ICIR

Of the 120 tests shown here, the hybrid returned a net gain in all but one. The extent of gain using our *conservative* E-model analysis varied up 50% and was highest in the test range 1-6 which corresponded to periods of highest jitter. Average gains relative to Alg. A/B were approximately 10/15%. Fig. 7 also shows that the R-factor gain is closely correlated with the standard deviation of packet delays. This reflects the fact, reported also by [9] that adaptive algorithms tend to track network conditions too closely resulting in unnecessary late loss when significant jitter occurs.

Although the primary objective of this paper was to assess the hybrid under more diverse network conditions, the performance improvements due to hybrid self-tuning features introduced in section 2.1 are evident from Fig. 4. This shows a decrease in playout delay (due to the variable W_f) with almost no increase in late loss along with a significant reduction in the frequency of playout mode re-evaluation (every 10000 versus every 500 packets).

General Delay Models. Fig. 8 shows the relative performance of the various algorithms when tested with the two-state Markov delay models developed for LKN from section 5.1. Relative to Alg. A, the hybrid resulted in a reduction of 2.8% (3.3 to 0.5) in late loss with an average increase in playout delay of 28 msec, resulting in a net R-factor gain of 8%. The corresponding gain for ICIR was higher at 35% (not shown). This was due to the much higher late losses in adaptive mode (6%) than in hybrid mode (2.4%). This reflects the fact that jitter which was set as a percentage of delays was significantly higher for ICIR,

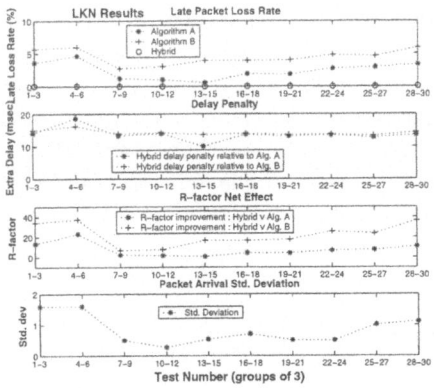

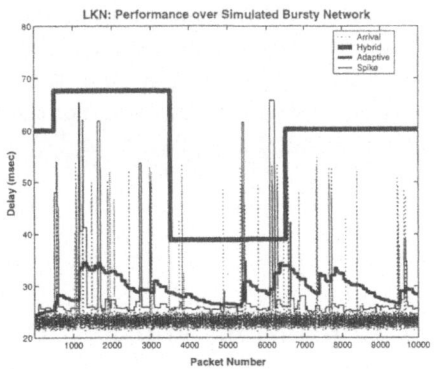

Fig. 7. Performance Comparison: LKN   **Fig. 8.** LKN: Bursty Delay Model

leading to poor adaptive performance. The late loss rate was higher in hybrid mode for the ICIR than for LKN model due to the proximity to the G.114 limit. Alg. B performed marginally better than A in both cases, due to the high degree of bursty delays. More generally, the hybrid consistently outperformed both Alg. A and B, once overall delay values were within the G.114 limit. See [24] for more detail.

5.3 Testbed Performance

The emulator enabled simulator-based results to be repeated in a real implementation environment.Although results correlated well with those from the simulator-based approach, the delay emulator suffered from delay uncertainty, caused principally by timer granularity limitations.This resulted in actual imposed delays differing slightly from the measured trace delays, making precise comparisons impossible. On a related note, [25] carry out black box testing of various VoIP endpoints to measure, amongst other things, M2E delay introduced solely by the endpoints. Their results show that the combination of OS, packetisation, buffering, and miscellaneous hardware delays can often accumulate to 150 msec and more which is a serious cause for concern. As such, although a primary advantage of the emulator approach is that it facilitates subjective testing, in practice, due to the emulator delay jitter and endpoint uncertainties, our performance gains reported are based solely on E-model analysis.

5.4 Internet Delay Studies

Our results confirm that the hybrid approach can deliver significant gains over both short and long distance Internet paths. In [10] Roychoudhuri et al carry out extensive testing (using a smoothed RTT metric) of US and International paths and report that though jitter remains a serious problem, delay rarely exceeds the acceptable limits for telephony. Finally in [9], testing of forty three (43)

US backbone paths was carried out. These paths were classified into five types according to fixed and variable delay components and were continuously tested over a 2.5 day period. For 39 of these paths, 98% of packets experienced delays less than or equal to 120 msec. For the remaining four, classified as long haul with high delay burstiness, 98% of packets experienced delays less than or equal to 170 msec. The above brief analysis confirms that delay-aware playout strategies such as the hybrid offer a viable alternative to conventional delay-unaware adaptive approaches.

6   Conclusions and Future Work

In this paper we extensively test the hybrid algorithm using both a trace-driven and derived delay model approach. Conventional adaptive algorithms often result in unnecessary late packet loss and voice distortion. Where M2E delays are within G.114 requirements, an informed fixed playout delay, though increasing overall end-to-end delay, can significantly reduce late packet loss and fully preserve speech integrity. This trade-off can often result in improved quality as users are more sensitive to increased loss than delay. Extensive delay measurements carried out in this paper coupled with an analysis of third party studies suggest that delay characteristics within the Internet make the hybrid algorithm a viable alternative. Ongoing work is examining the timing limitations of the delay emulator in order to facilitate more effective evaluation, including subjective testing. Further analysis of Internet delay studies such as the Active Measurement Project [26] is being undertaken. Finally, more analysis is required of the uncertainties caused by hardware/software within VoIP endpoints. This has particular implications for the hybrid algorithm but applies more generally to any Internet multimedia application.

References

1. S. Bradner, "Internet Telephony-progress along the road," *IEEE Internet Computing*, vol. 6,no. 3,May-Jun. 2002.
2. R. Ramjee, J. Kurose, D. Towsley, and H. Schulzrinne, "Adaptive playout mechanisms for packetized audio applications in wide-area networks," *Proc. Conf. Comp. Comm. (IEEE Infocom)*, IEEE CS Press, Los Alamitos, Calif., Jun. 1994.
3. S. Moon, J. Kurose, and D. Towsley, "Packet audio playout delay adjustment:performance bounds and algorithms," *ACM/Springer Multimedia Systems*, vol. 6, pp. 17-28, Jan. 1998.
4. Y. Liang, N. Farber, and B. Girod, "Adaptive playout scheduling using time-scale modification in packet voice communications," *Proc. of ICASSP 2001*.
5. H. Melvin and L. Murphy, "Time synchronization for VoIP Quality of Service," *IEEE Internet Computing*, vol. 6,no. 3,May-Jun. 2002.
6. H. Melvin and L. Murphy, "An evaluation of the potential of synchronized time to improve VoIP quality", *IEEE Intl Conf. on Comm. (ICC 2003)*,Anchorage, May 2003.

7. S. Floyd, "Difficulties in Simulating the Internet." *IEEE/ACM Trans. on Network-ing*,vol. 9, no. 4, Aug. 2001.
8. Recommendation G.114, "One way transmission time," ITU, May 2000.
9. A. Markopoulou, F. Tobagi, and M. Karam, "Assessment of VoIP quality over Internet backbones," *IEEE Proc of Infocom 2002*, NY, Jun. 2002.
10. L.Roychoudhuri, E.Al-Shaer, H.Hamed, and G.Brewster, "Audio Transmissions over the Internet:Experiments and Observations," *IEEE Intl. Conf. on Comm. (ICC 2003)*,Anchorage, Alaska, May 2003.
11. G. Almes, S. Kalidindi, M. Zekauskas, "A One-way Delay Metric for IPPM," *IETF RFC 2679*, Sept. 1999.
12. V.Paxson et al, "Framework for IP Performance Metrics," *IETF RFC 2330*, May 1998.
13. W.Kellerer, E. Steinbach, P. Eisert, and B. Girod, "A Real-Time Internet Stream-ing Media Testbed,"*Proceedings of the IEEE Int'l Conf. on Multimedia and Expo.*,Lausanne, Aug. 2002.
14. W. Jiang and H. Schulzrinne, "Modeling of packet loss and delay and their effect on real-time multimedia service quality," *NOSSDAV 2000*, Chapel Hill,NC, Jun. 2000.
15. V.Raisanen, G.Grotefield, and A.Morton, "Network performance measurement with periodic streams" *IETF RFC 3432*, Nov.2002.
16. J. Bolot, "End-to-End Packet Delay and Loss Behaviour in the Internet," *SIG-COMM 1993*, NY, Sept. 1993.
17. W. Jiang and H. Schulzrinne, "QoS measurement of Internet real-time multimedia services," *tech. report CUCS-05-99m, Columbia Univ.*, NY, Dec. 1999.
18. C.Boutremans and J.Boudec, "Adaptive Joint Playout Buffer and FEC Adjust-ment for Internet Telephony," *IEEE Proc. of Infocom 2003*, San Fran., Mar. 2003.
19. Recommendation G.107, "The E-model, a computational model for use in trans-mission planning," ITU, May 2000.
20. J. Janssen, D. De Vleeschauwer, M. Buchli, and G. Petit, "Assessing voice quality in packet-based telephony," *IEEE Internet Computing*, vol. 6,no. 3,May-June 2002.
21. W. Jiang and H. Schulzrinne, "Comparison and Optimization of Packet Loss Repair Methods on VoIP Perceived Quality under Bursty Loss," *NOSSDAV'02*, Miami, May 2002.
22. A. Clarke, "Modeling the effects of burst packet loss and recency on subjective voice quality," *Proc. of IP Telephony Workshop*, Mar. 2001.
23. Recommendation G.113, "Transmission impairments due to speech processing," ITU, Feb. 2001.
24. H. Melvin and L. Murphy, "Trace Driven Testing of the Hybrid Playout Strategy", *Dept. of IT, NUI,G, Tech.Report 0106003-2*, Jun. 2003.
25. W. Jiang and H. Schulzrinne,"QoS Evaluation of VoIP End-points," *IEEE Intl. Conf. on Comm. (ICC 2003)*,Anchorage, Alaska, May 2003.
26. Active Measurement project. See http://www.watt.nlanr.net/.

A Secure Web Services for Location Based Services in Wireless Networks*

Minsoo Lee[1], Jintaek Kim[1], Sehyun Park[1†], Jaeil Lee[2], and Seoklae Lee[2]

[1]School of Electrical and Electronics Engineering, Chung-Ang University,
221, HukSuk-Dong, DongJak-Gu, Seoul, Korea
lemins@wm.cau.ac.kr, groundiv@ms.cau.ac.kr, shpark@cau.ac.kr
http://bkmodem.cau.ac.kr/main.htm
[2] Korea Information Security Agency
78, Karak dong, Songpa-Gu, Seoul, Korea
{jilee, sllee}@kisa.or.kr
http://www.kisa.or.kr

Abstract. While Location Based Services (LBS) can make our lives more comfortable and productive, it may cause an invasion of privacy by disclosure and commercial use of location information. In this paper, we discuss privacy and security problems that may happen in the current LBS system and propose solutions. We propose a new secure Web services architecture for LBS in wireless network. Our architecture allows mobile users to create and enforce dynamic policy for safe and consistent LBS. We also describe some practical scenarios in which our architecture protects user's location privacy and security.

Keywords. Location Based Service, Privacy, Security, Interoperability

1 Introduction

With the development of mobile communication technologies, LBS are beginning to attract attention as a new research area of networking. LBS can offer much convenience to dynamic users in wireless network as well as provide significant revenue to mobile operators and content providers.

But the dark side of LBS, it also involves the threat of an invasion of privacy and security caused by indiscreet location tracking [1, 2, 21, 22, 23]. For example, if a company is using location tracking service to grasp where each employee is on duty hours, it must not observe their positions anymore off duty hours. If the tracking is continued, it will bring about a violation of privacy for the employees of the company. As another example, let's consider the case in which a LBS provider advertises to arbitrary users in a specific region. This may be also an incident of privacy violation if mobile users in the area do not want to receive these advertisements. Crimes

---

*This Research was supported by the Chung-Ang University Research Grants in 2003.
†The corresponding author

N. Mitrou et al. (Eds.): NETWORKING 2004, LNCS 3042, pp. 332–344, 2004.
© IFIP International Federation for Information Processing 2004

that use illegal location tracking may also be possible. So, LBS related groups [16, 17] are considering privacy problems in various aspects.

The problem which we have to consider next is location security. Since LBS are based on message exchange in wireless network, there are always security risks as location information could be stolen, lost, or modified. Therefore, we must concern the security mechanism for location information. The security mechanism must eliminate or minimize the potential for attacks against LBS entities and must reduce exposure of the user's identity and location.

The last point that we should discuss is interoperability problem. One of the concerns about national and global LBS roaming is ensuring the interoperability of LBS platforms. Most of LBS platform have optional and proprietary features that can interfere with interoperability. There is no guarantee that user's location privacy policies and authorization rules are observed through various LBS platforms.

In order to ensure a robust, consistent LBS environment, we need a secure architecture that is capable of supporting dynamic enforcement of user privacy policies, security mechanisms and convergence of services.

Therefore, this paper identifies these LBS problems and outlines the requirements for securing the LBS. And we propose a secure Web services architecture to protect the location information. We design LBS Policy Authority to resolve privacy problems and LBS Broker to solve authentication and authorization problems. The proposed architecture can overcome differences in LBS platforms, location information, and positioning technologies and network architectures. The architecture enhances interoperability among various LBS providers building Global LBS service on various platforms in the ubiquitous environment. Our model also guarantees Single Sign-On (SSO) among multi-vendor topologies by exchanging authentication and authorization information using Security Assertion Markup Language (SAML) token.

The rest of this paper is organized as follow. Section 2 identifies the problems of current LBS and requirements to solve them. Section 3 suggests a new web service architecture enhancing privacy, security and interoperability of LBS. Section 4 shows some of LBS scenarios applying the proposed architecture. Section In section 5, we discuss the simulation environment and the results. Finally, we conclude in Section 6.

2 Motivations and Requirements

2.1 Privacy Problems

LBS are considered as one of the main revenue generators for next generation wireless services. However, LBS do raise new privacy issues [1, 2, 27, 28, 29] integral to LBS. The major problem arises when location information is required in order to obtain a service and at the same time the user does not want to reveal more personal identifiable information. Users wish to have complete control over the visibility of their location, but in the most part of LBS scenarios users are not in full control.

After all, location privacy will have to be carefully managed and we need systematic method as well as technological method [3, 4]. One of the ways to solve the location privacy problem is to provide fine-grained privacy policies in user control. Policies that require service provider to adhere to strong privacy practices are needed to counterbalance the invisible nature of location collection in the wireless world.

In this paper, we present classified user profiles and location policy to access location information to cope with the privacy problem. The user's policy is created with the agreement of user and is effectively used by dynamic condition for access to user's location information.

2.2 Security Problems

Beside location privacy problems, there are some risks about location information itself. Location information may be sniffed, modified or stolen by attacker from communication channel between LBS entities. Security requirements and mechanisms must be addressed to ensure the safety of location information exchange among various location server that support different positioning methods. The mechanism should provide confidentiality against eavesdroppers and integrity to assure that the location information was not modified accidentally or deliberately in transit. The mechanism should provide mutual authentication guarantees that access to LBS applications is restricted to only those who can provide the appropriate proof of identity.

These requirements can be satisfied by using digital signature and encryption of location data because they concern how to protect communicated data. Apart from these mechanisms, we also have to consider the protection of location information so that only appropriate entities are allowed to access location information. Authorization process is required to decide whether or not the entity can access the particular location information. These cryptographic operations could create so many burdens of user's mobile terminal in LBS environments where network resources and computing power are usually limited.

To effectively perform these tasks, Agent or Broker could be deployed. The agent may provide combination features such as better communication facilities, high speed cryptographic engine and memory mechanism. In order to ensure a robust, consistent LBS environment, our model uses two agents on behalf of mobile terminal for enhancing security operation like secure key management, authentication and authorization.

2.3 Interoperability Problems

Another issue of LBS community is that LBS are challenged by the disparate location technology implemented by wireless infrastructure providers, service providers and equipment vendors. Most of LBS community are faced with having to support multiple, disparate location-determining technologies (LDT), and content implementations, and multiple data transport protocols. This is simply cost prohibitive. Thus the LBS technologies used must vary with the service context i.e. time constraints, location positioning method, network connection status.

The key to interoperability will be the development and adoption of a ubiquitous set of interconnected wireless communications and Internet location service standards. Open and scalable LBS architectures and common data structures are necessary for various types of location information. These common structures could be defined by XML. Location Inter-Operability Forum (LIF) developed the XML based Mobile Location Protocol (MLP) [15] standard, which is concerned with the integration of position or location information. And Open GIS Consortium (OGC) issues Request for Technology for Web Services Initiative [18] to provide interoperable Spatial Web Services.

However, in some complicated services, such as LBS roaming which includes service continuity and hand-off issues, more consistent security features should be partnered with these efforts for interoperability. In the future of LBS environment, LBS roaming scenario is likely widespread where many LBS service providers are used to implement functionality "behind the scenes." If a user does not know whether or not the location information is broadly secure in various LBS platform, when roaming across boundaries within interconnected wireless networks, LBS roaming may create new security and privacy challenges. Therefore, we should figure out how to seamlessly provide secure location information utilizing heterogeneous wireless networks without reauthenticating each time. In this paper, we consider these interoperable and consistent security needs as addressed by Web service security mechanisms, and map each of the requirements onto the construct of future global LBS environments.

3 Secure Web Services Architecture for LBS

In this section, we propose a secure Web Services architecture which is designed to meet the requirements in previous sections. The objective of the proposed architecture is to ease the development of secure LBS by providing customized privacy and security profiles which can be assembled to create concrete LBS applications. The Figure 1 shows proposed LBS privacy and security Enhanced Web Services architecture.

3.1 Enhanced Interoperability with Web Services

The need to integrate disparate LBS applications that run across the Internet on heterogeneous wireless networks, and the realization that proprietary approaches would not solve the integration problem, gave rise to use of Web Services for LBS. Web services are going to play a big role in the evolution of mobile business. A Web Services supports direct interactions with other software applications using XML based messages via internet-based protocols such as HTTP, SMTP, and FTP, including Simple Object Access Protocol (SOAP). For easier configuration, Web Services interfaces could be defined and modified by Web Services Description Language (WSDL). The defined Web Services can be registered and discovered at Universal Description, Discovery, and Integration (UDDI) registry.

Consequently, these advantages of Web Services could bring maximum efficiency and interoperability to the LBS in next generation wireless networks where loosely coupled and highly dynamic environments are expected. Global LBS [5] also could be provided by cooperation of LBS providers in different country.

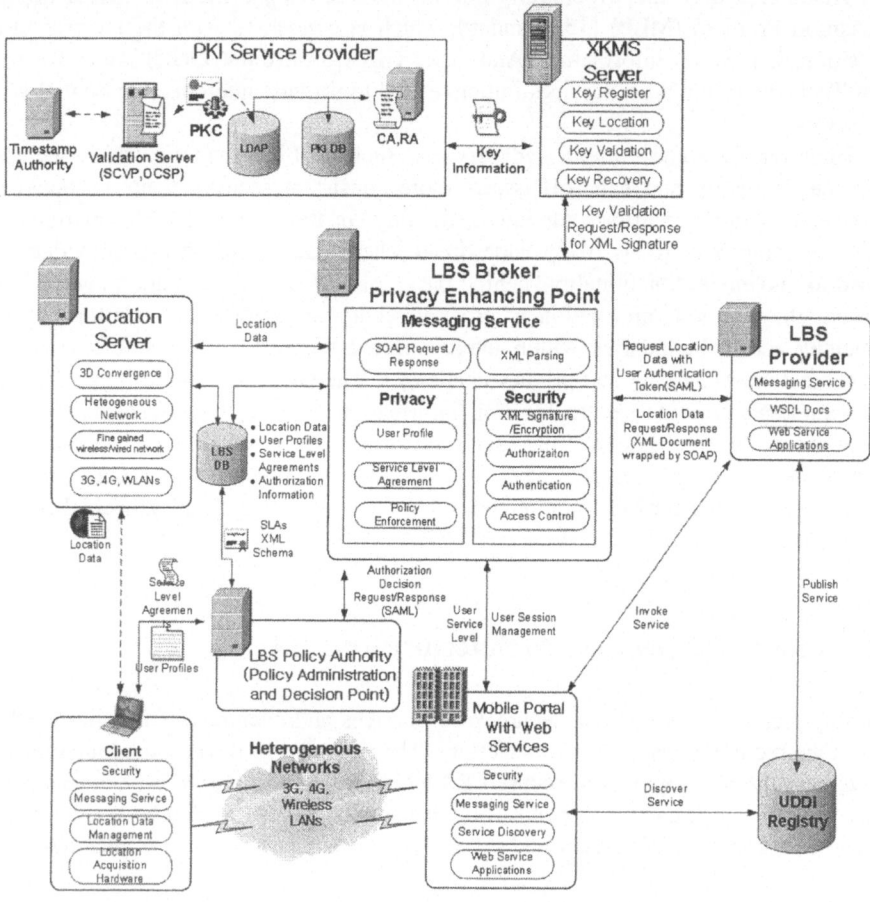

Fig. 1. LBS Privacy and Security Enhanced Web Service Architecture

3.2 Web Services Security for LBS

The Secure Socket Layer (SSL) is used to provide an encrypted means of data exchange between a web browser and a web server. Although SSL is widely treated as a standard, SSL is insufficient for Web Services Security in several ways. SSL only supports data in transit, not in storage. SSL does not support multi-party transactions and non-Repudiation. SSL is not granular enough because it encrypts everything.

To overcome the limited features of SSL, XML Signature [6] and XML Encryption [7] are used to forming a strong foundation for the development of secured web services by enabling partial signature and partial encryption respectively. These XML security specifications could provide authentication, encryption and non-repudiation in multiple participants from different location service domains. Additionally, distributed authorization and federated identity management like SSO are among the great challenges for LBS. Authorization policies for location information and other QoS parameters. To meet such security requirements, location information could be partnered with SAML [8] as addressed in [14]. SAML provide the basis for interoperable authentication, authorization and attributes among disparate systems including a SSO facility [9]. In our architecture, we took the advantage of incorporating location information with these Web Services security mechanisms to enhance the security and privacy of location based services.

3.3 LBS Broker

We design a LBS Broker to solve security problems of LBS. In some architecture, it may be useful to use a Broker to improve performance or security [19, 20]. LBS broker plays key role in protecting user's location information from unauthorized LBS service provider or malicious users.

The LBS broker act as a Policy Enforcement Point (PEP) that checks permission with the LBS policy authority, the Policy Decision Point (PDP) by exchanging SAML message before making decision and releasing the secured location information to the LBS service providers. LBS broker could provide users the greatest amount of control over their personal information, since the user is in control to choose whether their location is transmitted to the server for others to access. LBS broker supports XML signature and XML encryption to validating the signature of the SOAP messages. To validate the keys used in XML signature, it interacts with XML Key Management Specification (XKMS) [11] servers. XKMS helps to remove the complexity of working with PKI. SAML assertions are employed to for exchanging authentication and authorization token across different LBS entities like LBS brokers, LBS policy authorities, LBS service providers and mobile portals over Internet.

3.4 LBS Policy Authority for LBS Privacy

LBS Policy Authority acts as a Policy Administration Point (PAP) in LBS privacy agreement step and a PDP in LBS service step. As a PAP, it creates a LBS policy set to LBS Service Level Agreements (SLAs) with users using predefined XML Schema. The policy includes user profiles and other LBS service attributes.

3.4.1 Profiles for LBS Privacy
A simple set of location privacy rules is insufficient to enforce dynamic and consistent privacy when users roam. In Figure 2, we propose classified privacy and security profiles to accommodate more adaptive and optimal LBS environments. A key advantage that profiles offer is that LBS can be customized to fit user's specific needs.

Customization of LBS is performed through the classification of profiles. The classification of security needs to provide a wide scope for various LBS users. LBS Client adaptively modifies its profile for heterogeneous wireless networks. This mechanism could bring minimized leakage of privacy information that users wanted.

Profile Type			
Profile Type	Description	Fields	Usage
Basic Mobile Node Profile	• Basic properties of mobile equipment to use LBS • Basic network information	• Mobile type, power management Type, memory size, computation ability • Defined Max Data Rate, Sub-IP layer Type, Defined Mobility	• When minimum information are required • Basic group based service, basic roaming service
Basic User Profile	• Minimal privacy information • User registration information	• Preferred service type, role type, and protocol type • User ID, address	• Basic user information are required at LBS Policy Authority or Mobile Portal Service Provider
Extended Mobile Node Profile	• Advanced features for customized LBS • Detailed information of mobile equipment	• Minimum latency & throughput • Preferred location sensing type • Routing and roaming preference	• To envisage maximum performance • To enhance QoS features • Precise positioning • Faster roaming
Extended User Profile	• Advanced privacy and security features • Specific or optional contexts of user	• User Authentication Token • Security level (Signature Algorithm, Key Length) • Authorization Policy	• Advance group based service • Fine-grained access control • Single-sign-on • Secure global roaming

Profile Class							
Profile Class #	1	2	3	4	5	6	7
Profile composition	Basic Node	Basic User	Basic Node | Basic User	Basic Node | Ext. Node	Basic Node | Ext. Node | Basic User	Basic Node | Basic User | Ext. User	Basic Node | Ext. Node | Basic User | Ext. User

Fig. 2. Profile types and classes for LBS Privacy

3.4.2 Policy Setting

The policy model proposed in this paper provides direct control function to a user through policy decision procedures with LBS Policy Authority which performs effective policy enforcement. Figure 3 shows LBS SLAs procedures and examples of SAML message about user privacy.

4 Scenarios

4.1 A Secure LBS Push Scenario

The push scenario happens when LBS service provider requests user's location information for providing location services to user. When LBS service provider requests user's location information to LBS broker, Figure 4 presents a scenario in which validation of user's privacy, authentication and authorization are enforced. In the scenario, XML based protocol is used for interoperability between all type of system. LBS Policy Authority could prevent improper usages of location service in specific area, time or users. The model with LBS Broker can improve efficiency and performance of authorization and authentication validation.

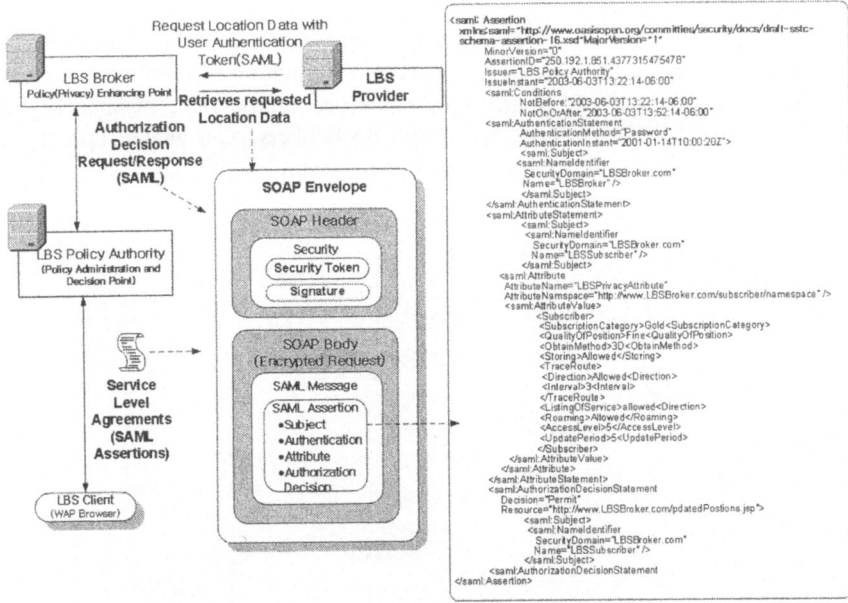

Fig. 3. Secure Messaging for LBS policy setting

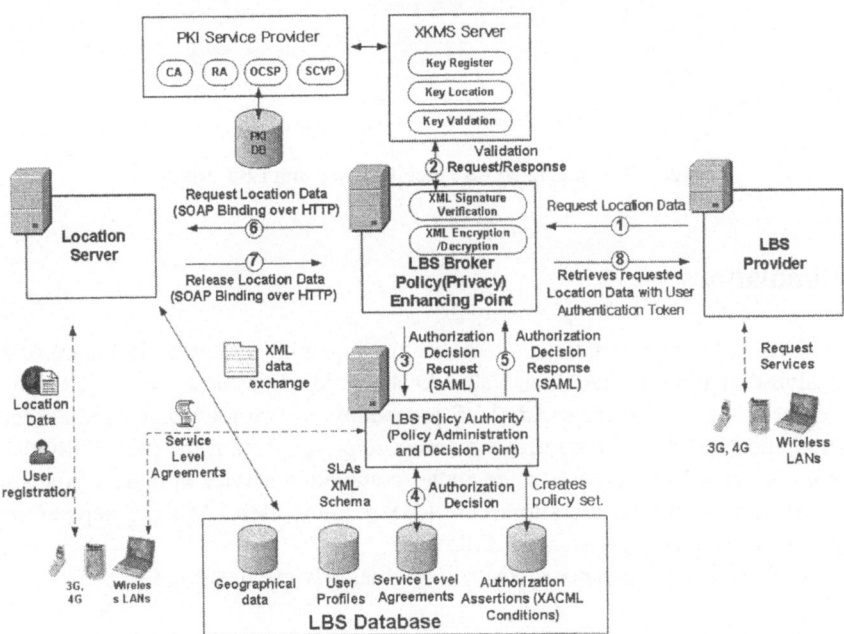

Fig. 4. A Secure LBS Web Service Scenario

4.2 A Convergence Model

In this scenario, we propose a convergence model for more consistent LBS environments. For future global LBS roaming, location information of users, which exist in various LBS platform, should be managed on secure interoperable manner. Figure 5 depicts the integrated LBS architecture using LBS Brokers. Our model supports multiple, disparate LDT and supports SSO functionality with LBS Brokers.

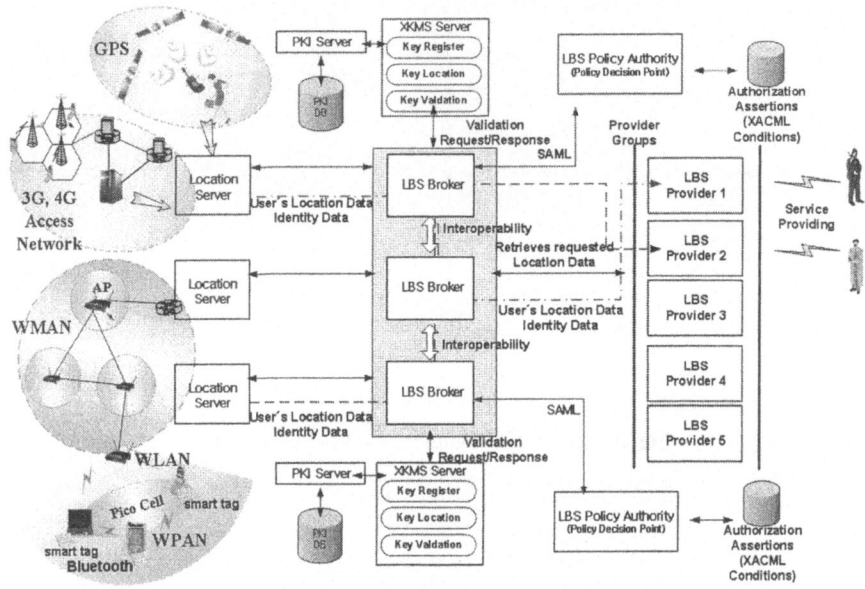

Fig. 5. The integrated LBS architecture using LBS Brokers

5 Simulations

We have modeled our architecture as a closed queuing system as in Figure 6, and we analyzed of approximate Mean Value Analysis (MVA) as described in [24, 25]. In the scenario of Figure 4, the secure LBS procedure has two job classes, initial secure location update step and secure LBS roaming step. $r_{im,jn}$ means the probability that a class m job moves to class n at node j after completing service at node i. And *ratio* represents a ratio of total users to secure LBS roaming users. Analyze steps of class switching closed queuing system are following.

Step1: Calculate the number of visits in original network by using (1)

$$e_{ir} = \sum_{j=1}^{K} \sum_{s=1}^{C} e_{js} r_{js,ir} \tag{1}$$

where K = total number of queues, C = total number of classes.

Step 2: Transform the queuing system to chain.

Step 3: Calculate the number of visits e_{iq}^{*} for each chain by using (2)

$$e_{iq}^{*} = \frac{\sum_{r \in \pi_q} e_{ir}}{\sum_{r \in \pi_q} e_{1r}}$$ (2)

where r = queue number in chain q, π_q = total queue number

Step 4: Calculate the scale factor α_{ir} and service times s_{iq} by using (3) with (1).

$$s_{iq} = \sum_{r \in \pi_q} s_{ir} \alpha_{ir} \ , \ \alpha_{ir} = \frac{e_{ir}}{\sum_{s \in \pi_q} e_{is}}$$ (3)

Step 5: Calculate the performance parameters for each chain using MVA.

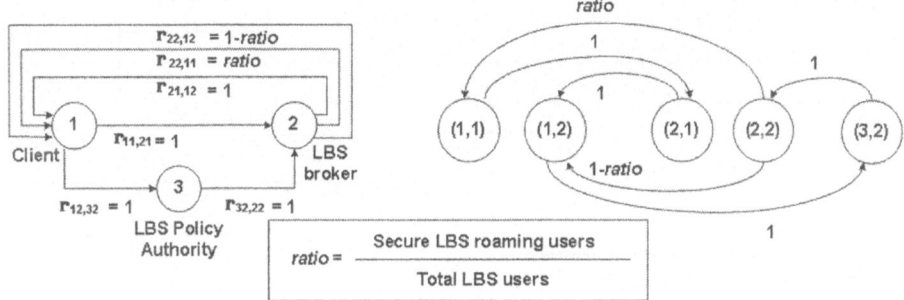

Fig. 6. Multiple class queuing system in the secure LBS push scenario

Table 1 summarizes the base parameter settings underlying the performance experiments. LBS Broker and LBS Policy Authority used Solaris 8 machine with Pentium III 933 MHz, 512 MB RAM. Mobile node used Pentium III 500 MHz, 128MB RAM, WindowsXP as operating system with Lucent Orinoco IEEE 802.11b wireless LAN card. The cryptographic library was Openssl 0.9.7a [12], and SAML Library was OpenSAML 0.9.1 [13]. Data size was 1KB in digital signature. Figure 7(a) shows average throughput at high security level when the roaming ratio r varies. As the initial secure location update needs more cryptographic operation, our secure Web service architecture show better performance in secure LBS roaming environments where users move fast. Figure 7(b) shows throughputs of secure location update with three security levels. Our secure Web service architecture could manage 12 users at high security level and up to 45 at one second. These simulation results could be useful to provide guidelines as to how the security level is set to meets the user needs. As we can see, the advantages of protecting privacy and security could far outweigh its overhead in specifying security assertions in XML.

Table 1. Base parameter settings of the queuing model

Entity	Operation in scenario	Description	Perform-ance
	Initial secure location update		
Mobile Node	Token Request with User's Private key	RSA with SHA-1 signature sign with a 512 bit key	5.5 ms
LBS Broker	Signature verification using User's Public key	RSA with SHA-1 signature verify with a 512 bit key	0.1 ms
LBS Broker	SCVP(OCSP) Request Message - signature of LBS Broker's Private Key	RSA with SHA-1 signature sign with a 1024 bit key	7.4 ms
XKMS with PKI	X.509 Certificate validation	Validate user certificate	30.3 ms
LBS Broker	OCSP Response Message validation	RSA with SHA-1 signature verify with a 1024 bit key	0.4 ms
LBS Broker	SAML Authorization Request	XML Parsing and RSA 1024 signature	27.4 ms
LBS Policy Authority	SAML Authorization Response	XML Parsing and RSA with SHA-1 1024 bit key signature verify	20.4 ms
LBS Policy Authority	SAML Authentication Token generation (and response to MN)	3DES Symmetric key encryption	7.702 MB/s
LBS Broker	Token Response with Location information	RSA encrypt on 512 bit keys	31.201 KB/s
LBS SP	Decrypt Token Response with Location Update Response	RSA decrypt on 512 bit keys	8.517 KB/s
	Secure LBS roaming		
Mobile Node	Location Request with Security Token	Average hand-off latency	30 ms
LBS Broker	Token verification	3DES Symmetric key decryption	1.090MB/sec
LBS Broker	Token Response with Location information	RSA encrypt on 512 bit keys	31.201 KB/s
LBS Provider	Decrypt Token Response with Location Update Response	RSA decrypt on 512 bit keys	8.517 KB/s

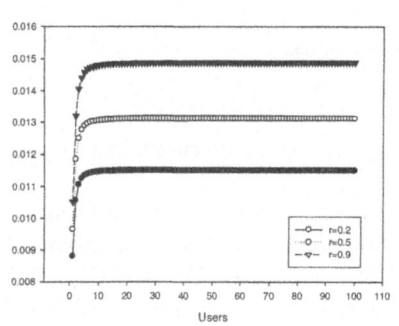

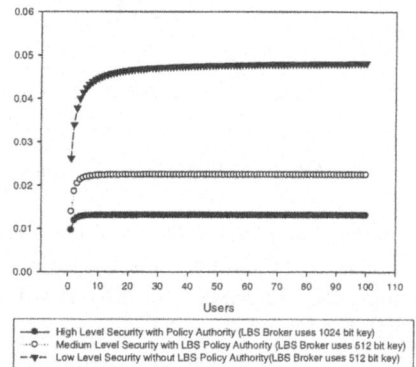

(a) Throughput of secure location update at high level security

(b) Throughputs of secure location update with various security levels

Fig. 7. Simulations results of the secure Web Services architecture

6 Conclusion

In this paper, we analyze privacy and security issues on location based service and give our view on the future prospects of LBS for the next generation wireless network. Users are increasingly concerned with the disclosure of location information to third parties and the potential consequences for their privacy. As the location privacy and security is emerging as one of the key issues that will have to manage before fulfilling the LBS revenue promise, we propose a secure Web Service architecture for location based service.

The architecture takes advantages of Web Services and is designed to maximize the efficiency and interoperability for the LBS in wireless networks where loosely coupled and highly dynamic environments are expected. We design a LBS Broker to effectively solve privacy, authentication and authorization problems. We introduce LBS Policy Authority with classified privacy and security profiles. Our model also guarantees SSO among LBS service providers by exchanging authentication and authorization information using SAML token. We also have shown some practical scenarios in which strong authentication and authorization are provided while preserving user's location privacy. And the simulation results could be useful to provide guidelines as to under which circumstances one security scheme may be used in preference to another.

References

1. Jorge Ceullar, John B. Morris, Deirdre Mulligan, Jon Peterson and James Polk, "Geopriv Requirements," draft-ietf-geopriv-reqs-0.3, 3, 2003-07-30
2. WLIA, "Adopted WLIA Privacy Policy (First Revision)," http://www.wliaonline.com/ indstandard/privacy.html
3. Dan Greening, "Location Privacy," location interoperability forum, 2002
4. Shereen Fink, "The Fine Line Between Location-Based Services & Privacy," http://www.sun.com/aboutsun/media/presskits/sp/
5. Ulf Leonhardt and Jeff Magee., "Security Considerations for a Distributed Location Service", Journal of Network and System Management, Vol 6(1):51-70, March 1998.
6. W3C Recommendation. D. Eastlake, J. Reagle, and D. Solo., "XML-Signature Syntax and Processing", February 2002.
7. W3C Recommendation T. Imamura, B. Dillaway, J. Schaad, E. Simon., "XML Encryption Syntax and Processing", December 2002.
8. OASIS Standard, Security Assertion Markup Language (SAML) 1.0, November 2002.
9. Ben Galbraith, et. al., *Professional Web Services Secuirty*, Wrox Press, 2002.
10. Alberto Escudero-Pascual, Gerald Q. Maguire Jr., "Role(s) of a proxy in location based services" 13th IEEE International Symposium on Personal, Indoor and Mobile Radio Communications. PIMRC2002. Lisbone. Portugal. September 2002.
11. W3C working draft, "XML Key Management Specification (XKMS) v 2.0", April 2003.
12. OpenSSL, http://www.openssl.org/
13. OpenSAML, http://www.opensaml.org/
14. Harsha Srivatsa, "Location, location, location-based services", IBM, November 2002.
15. Location Inter-operability Forum (LIF), Mobile Location Protocol (MLP), TS 101 Specification Version 3.0.0 6, June 2002.
16. Location Inter-operability Forum (LIF), "Privacy Guidelines", LIF TR 101 Report, 2002
17. 3GPP, Enhanced support for User Privacy in location services, TR 23.871,
18. Open GIS Consortium (OGC), "A Request for Technology In Support of an OGC Web Services Initiative," 2003

19. Michael Berger, et. al., "An Approach to Agent-Based Service Composition and Its Application to Mobile Business Processes", IEEE Transactions on Mobile Computing, VOL. 2, NO. 3, July-September 2003.
20. Aura Ganz, Se Hyun Park, and Zvi Ganz, "Security Broker for multimedia wireless LANs", Computer Communications, Vol.23, issue 5-6, pp. 588-592, March 2000.
21. Alberto Escudero-Pascual, Thijs Holleboom, and Simone Fischer-Hiibner, "Privacy for Location Data in mobile networks"
22. Euro Beinat, "Privacy and Location-based Stating the Policies Clearly", GEO Informatics, Volume 4, September 2001
23. Alastair R. Beresford and Frank Stajano, "Location Privacy in Pervasive Computing", PERVASIVE computing, January-March 2003.
24. Boudewijn R. Haverkort John, "Performance of Computer Communication Systems : A Model-Based Approach" , Wiley & Sons, October 1999.
25. Gunter Bolch, Stefan Greiner, Kishor Trevedi, "A Generalized Analysis technique for queueing networks with mixed priority strategy and class switching", Technical Report TR-I4-95-08, Oct. 1995.

Distributed Channel Monitoring for Wireless Bandwidth Aggregation

Puneet Sharma[1], Sung-Ju Lee[1], Jack Brassil[1], and Kang G. Shin[2]

[1] Hewlett-Packard Laboratories, Palo Alto, CA 94304
{puneet, sjlee, jtb}@hpl.hp.com
[2] University of Michigan, Ann Arbor, MI
kgshin@eecs.umich.edu

Abstract. Aggregating low-speed WAN links into a higher-speed logical link promises to improve data-transfer rates to collaborating communities of wireless mobile multi-homed devices. Such bandwidth aggregation systems must adapt to link dynamics as the number of links and the channel conditions vary with time due to mobility, power dissipation, and channel interference. A monitoring architecture that accurately measures the link dynamics and promptly feeds this information to the system is vital to realize significant bandwidth aggregation performance gains. We present various architectural design alternatives for such a monitoring system, and evaluate them using simulation. We show that a properly-designed monitoring system can accurately measure and quickly respond to changes in communication link performance while minimizing the control overhead.

1 Introduction

Users of wireless mobile computing devices seeking Internet connectivity in a public setting often face a choice between convenience and performance. One might locate, approach, and connect to a public wireless access point using a high-speed LAN such as IEEE 802.11x, or accept nearly ubiquitous but much slower access using a WAN such as a 2.5G or later-generation cellular network.

Although networks that provide high-speed access to mobile users are currently under development (e.g., *EvDO*, 4G cellular systems), they will not be widely available soon. To meet this need today, we have proposed an alternative, complementary solution to high-speed Internet access through collaborative resource sharing [13]. A group of multi-homed wireless, mobile computing and communication devices in close proximity dynamically form communities interconnected through their compatible high-speed LAN interfaces; we call these ad hoc groups *Mobile Collaborating Communities* (MC^2), though we will refer to them simply as *communities*. Each community member independently uses its WAN interface to create a communication *channel* to a remote inverse multiplexing or *aggregation* proxy, and optionally offers full or partial access to this channel to other community members. Each member volunteers to forward packets received on its WAN link to receiver(s) on the LAN. The set of channels connecting the participating community members to the proxy can be logically combined with an inverse multiplexing protocol to yield a higher-speed *aggregated channel* than is available from any one of the individual members. Hence, members using the aggregated

N. Mitrou et al. (Eds.): NETWORKING 2004, LNCS 3042, pp. 345–356, 2004.

channel enjoy higher bandwidth — and higher communication performance — than any one member alone could receive.

Striping data across multiple, parallel communication channels is a conventional communications technique used to improve system performance or reliability in varied but relatively static settings [3,16]. But due to end-device heterogeneity, mobility, and time-varying link transmission characteristics, an aggregated wireless channel is highly dynamic, and the challenge is to assemble, administer, and monitor its operation in a decentralized fashion.

In an earlier paper [13] we presented the initial design, simulation and implementation of a collaborative bandwidth aggregation system that is both practical and readily deployable. A key contribution of that work was to show that significant performance gains can be realized by adapting shared WAN link use to the specific application requirements of the flows sent over the aggregated channel. For a typical scenario, we demonstrated that the packet loss rate of a CBR video stream on an aggregated channel could be reduced by 71% by properly assigning packets to preferred links. But achieving these performance gains requires the aggregation system to be continuously aware of the communication characteristics of the constituent links.

In this paper we show that both WAN link communication performance as well as community membership dynamics must be accurately monitored and efficiently communicated to use an aggregated channel effectively. We explore the tradeoffs encountered in properly designing a decentralized monitoring system. We present a decentralized monitoring architecture and protocols designed to balance both system responsiveness and bandwidth efficiency. We also show how an inverse multiplexer should use measurements — possibly neither up-to-date nor consistent — to make decisions about proper channel use.

The rest of the paper is organized as follows. Sections 2 and 3 explore the requirements and issues associated with decentralized monitoring. Section 4 introduces a preferred monitoring architecture capable of meeting our system goals, and Section 5 presents simulation results exploring how effectively our proposed architecture balances the goals of responsiveness and bandwidth efficiency. Our conclusions are drawn in the final section.

2   Monitoring Requirements and Design Goals

2.1   Background

Prior to discussing the requirements and design goals of a monitoring architecture we briefly review the design and operation of a bandwidth aggregation system. Figure 1 shows a system that can be readily deployed by a network access provider, wireless telecommunication service provider, or a content distribution network operator. The specific implementation we have proposed has three principal components: a dedicated appliance providing aggregation proxy services, a standard LAN-based announcement and discovery protocol for mobile host community construction and maintenance, and standard protocol tunnels to facilitate both communication across shared links and packet forwarding at mobile hosts.

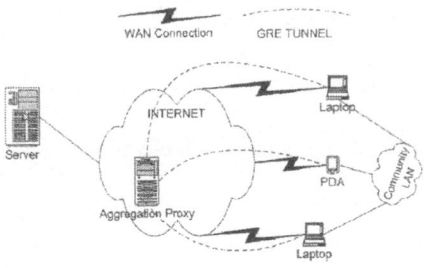

Fig. 1. A bandwidth aggregation service architecture.

The dedicated aggregation proxy performs inverse multiplexing at the application layer, intelligently striping downstream packets across available links to the community. Generic Routing Encapsulation [6] tunnels create channels between the proxy and participating MC^2 members, and support packet forwarding. This approach requires no modification to community members, as most operating systems (Linux, FreeBSD, Windows, etc.) today have built-in support for GRE tunnels. Each packet received by a member over the tunnel is automatically decapsulated and forwarded via the wireless LAN to the destination host. Since the destination is oblivious to which members forwarded the data packets, no additional data reassembly functionality is required at the receiver. Standard announcement and discovery protocols such as the Service Location Protocol are used for community and aggregated channel formation and management.

2.2 Challenges of Monitoring Systems

An aggregation proxy is responsible for assigning incoming traffic flows to available WAN channels. We refer to this function as *flow mapping* or *channel control*. A proxy might also be able to modify the incoming flows themselves (i.e., source control). The goal of monitoring communication dynamics is to provide a proxy's channel and traffic controllers with prompt and accurate information on the condition and number of WAN channels available between the proxy and the community. Only with this information can a proxy perform intelligent channel and source control in the face of rapid changes to the communication channels. One of the challenges for the flow mapper is how to use the measurement data it receives intelligently. For instance, frequently remapping flows to channels based on transient (fluctuating) channel quality measurements would not necessarily improve overall system performance.

We anticipate that both the availability and the quality of communication channels between a proxy and an MC^2 to vary with time. Community membership will change as mobile hosts join and leave the community, due to either end-system failures (e.g., power exhaustion) or simply moving out-of-range of LAN communications. Wireless WAN channel quality may change often and unpredictably because of fading, interference, and location-dependent coverage gaps. Delay and delay jitter will change as the heterogeneous, CPU-limited devices forwarding packets between WAN and LAN interfaces are subject to time-varying computing workloads. Hence, the parameters we expect our monitoring system to measure include:

- *Link quality*: raw and available bandwidth, delay, jitter, packet loss rate, signal strength
- *Community membership*: number of available WAN channels, participation time in system
- *Forwarding capability*: delay, jitter, available processing power

Beyond a channel's communication parameters, certain associated information might also be maintained — but not necessarily measured — by the monitoring system. This might include the 'cost' of a channel, or its expected departure time.

Though we anticipate that a community member will be capable of explicitly announcing its pending departure (from the community) to other members, one of the most difficult challenges our monitoring system faces is rapidly detecting sudden and *unannounced* leaves. We envision a LAN-based monitoring agent capable of tracking membership, including announced leaves and new members' joins. Such an agent would likely rely on an existing service discovery protocol, and a new member joining the MC^2 would register its identity and present available resource information. Such a system would likely have to be supplemented with an active mechanism to detect leaves. For example, the monitoring agent can periodically issue an echo request message (e.g., *ping* or *hello*) to active members and await a reply. The question of how often the monitoring agent should probe the members arises immediately. Clearly, there is a tradeoff between the probing overhead and the freshness of membership information. While we cannot afford to have excessive control message overhead in membership maintenance, we will typically assume that LAN bandwidth is a relatively plentiful resource.

To illustrate the importance of low latency in reporting WAN channel status to the aggregation proxy in improving the performance of an aggregated channel, we simulated an aggregation system with three community members. Each member offered a WAN channel with 20 kb/s bandwidth. Each channel has a time-varying packet loss rate (unknown to the proxy) that cycles as follows: a loss rate of 1% for 50 seconds, followed by a loss rate of 5% for 50 seconds, and then a loss rate of 10% for 50 seconds. The cycle is repeated multiple times during the lifetime of the session. The changes in loss rates across the three links are synchronized such that at any instant there is exactly one channel that has error rate of 1%, one channel with 5% and one channel with 10%. Thus, the total error rate is the same throughout the experiment.

An application-aware aggregation proxy [13] seeks to map hierarchically layer-coded [9] video to these three available channels. The simulated layered video consists of base layer (layer 0) and two enhancement layers (layers 1 and 2). Each layer is modeled as a 20 kb/s CBR stream. Using the channel loss rate as the reliability metric, the aggregation proxy maps each layer onto one of the three channels, ideally with higher layers assigned to increasingly less reliable channels; we referred to this flow assignment as the *Layer Priority Striping* (LPS) algorithm in [13]. Figure 2 shows the packet loss rate of each layer when the reporting latency (i.e., feedback delay) is varied. The feedback delay is defined as the time difference between the instant when the channel error rate changes and the time when the aggregation proxy remaps the flows onto the channels based on the newly-available information. As expected, the feedback delay decreases aggregated channel performance; the base layer is not transmitted over the most reliable channel during the feedback delay period following each loss rate transition event. In

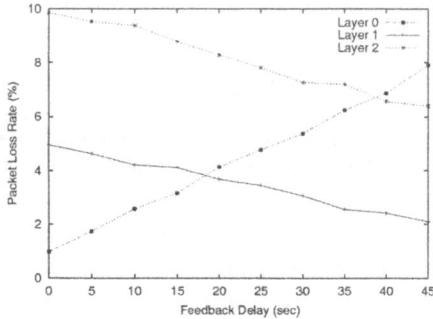

Fig. 2. The effect of reporting latency on aggregation performance.

fact, when the feedback latency is larger than 18 seconds, the loss rate of the base layer exceeds that of enhancement layer 1.

In general, the change in layer i's packet loss rate l_i is proportional to the feedback delay δ. Let the duration of a session be $N * T$ seconds where the link loss rate changes every T seconds. Let $P(i, j, k)$ be the packet loss rate of the channel during period k to which layer i has been assigned in period j. Then, layer i's packet loss rate can be written as

$$l_i = \frac{\sum_{j=1}^{N} P(i, j, j) * (T - \delta) + P(i, j - 1, j) * \delta}{N * T}.$$

In the above example we assumed that the aggregation proxy received *correct* measurements late. But measurement errors can also cause suboptimal mappings of application subflows to WAN channels. Hence, it is obviously important for a monitoring system to measure channel conditions accurately, and a tension exists between taking the time required for accurate measurements and keeping reporting latency short. In certain situations the system will tolerate even large measurement errors and continue to perform well. For instance, in the above example, even substantial errors in measuring link reliability would maintain the optimal channel ordering from most to least reliable.

In summary, our design goals for the overall monitoring system are:

- accurate measurement of link quality,
- low latency in reporting changes in link quality and community membership,
- low control message overhead,
- no or little software modification to the community members,
- minimal member performance degradation due to community participation,
- scalable design to support multiple aggregation proxies and large community memberships,
- scalable aggregation proxy capable of supporting a large number of communities simultaneously, and
- robustness to failures of members and their channels.

3   Design Choices

Designing an effective monitoring system forces us to answer key questions, including:

- *Architecture*: at which locations in the system should monitoring be performed? How do we design a scalable monitoring architecture capable of supporting both large community sizes and multiple proxies? What protocols should be used to feed the monitored information back to the aggregation proxy?
- *Measurement*: how should WAN channel communication performance and community membership dynamics be measured? Should measurement rely on active or passive techniques, or both? How do we minimize the burden of measurement placed on community members?
- *Configuration*: how do we dynamically set design parameters (e.g., proxy update interval, measurement intervals, active membership probing intervals) particularly as the community size and traffic changes? At what point should an aggregation proxy use measurement data it receives to decide to remap flows to available channels?

In the rest of this section we investigate design choices related to the above questions and discuss their strengths and weaknesses. This investigation will lead us to present a monitoring architecture in Section 4 which balances the many tradeoffs we must make.

The first and most important architectural issue we face is identifying the location of measurement points in the system. A *monitoring agent* will perform measurements at each of these points, and exchange information between themselves and the aggregation proxy. These agents may reside on one or more community members (i.e., mobile hosts), at the aggregation proxy, or both; we will exclude from our discussion the possibility that any type of dedicated equipment be required for monitoring, as that would preclude spontaneous formation of an MC^2.

3.1   Community Member-Based Monitoring

An agent may be located at one or more community members to monitor WAN channel condition and membership dynamics. Let us consider how such a system would operate. An arriving host seeking to participate in a pre-existing community discovers the community using a service discovery protocol (e.g., SLP) and registers with the monitoring agent(s). A member seeking to leave the community (i.e., an announced departure) broadcasts a departure notice to the community, and is deregistered by the monitoring agent(s). An active mechanism is used by monitoring agents to detect unannounced departures; an agent periodically probes the existence/condition of the community members. In such a case, the probing period is an important design parameter and must be determined by making a tradeoff between the probing overhead and the accuracy of the monitored information. On a high-speed LAN (e.g., IEEE 802.11x) the messaging overhead is not a significant issue, but the processing load and power consumption the agent imposes on a community member is an important issue. This is a particular concern if relatively few of the community members are providing monitoring services, such as when a single member is appointed or elected as the sole monitoring agent. The fact that a member serving as a monitoring agent consumes more power and processing than a regular member suggests that it is beneficial to have the agent's role rotated or shared among members. This also argues for power and processing availability at each node to be included in those parameters that are measured and maintained by monitoring agents.

The above sketch of system operation serves to highlight several of the advantages of deploying monitoring agents at a community member. A community member can quickly and easily track membership changes. But while a member can assess the quality of its own WAN channel to the proxy, it has very limited ability to assess other WAN channels. Moreover, a protocol must be established for identifying the members to serve as agents. Clearly, relying on a single (or even a few) monitor(s) can result in both a performance and reliability bottleneck.

This bottleneck problem can be solved by either replicating the monitoring agent or making every member the monitoring agent, i.e., distributed monitoring. We opt to use distributed monitoring which works as follows. Each member broadcasts its channel characteristics and associated information (e.g., communication costs and its energy balance) either periodically, or upon detection of an event, or when a certain threshold is exceeded. Each broadcast is timestamped. Upon receiving such a broadcast all the other members update the corresponding entry of their copy of the community communication status database. The aggregation proxy obtains a copy of the database in either of two ways. First, the proxy requests a copy of the database from any community member. Requests can be sent to a randomly-selected member, or a member identified by inspection of the most recent database the proxy has received. For example, an inquiry might be directed to a member with ample advertised available processing power, residual energy, or network bandwidth. A proxy might issue such an inquiry periodically, or be driven by an event such as the need to remap channels for a newly-arriving flow. The second way that a proxy obtains the database is simply by receiving an update report periodically or when a monitoring agent observes a significant local event (e.g., sudden channel failure).

Such a decentralized monitoring system is very attractive because it clearly improves overall system reliability and eliminates a potential bandwidth bottleneck. Note that each member's database need not be a perfect representation of current system state. Making each member a monitoring agent provides the best overall visibility of conditions of every channel.

3.2 Proxy-Based Monitoring

An alternative measurement architecture places a single monitoring agent at the location where the WAN channels terminate and the channel allocation is done. Depending on the link technology, a proxy may be able to detect an indication of a WAN channel failure rapidly. In other cases a proxy-based monitor might be able to infer failures over longer time periods. For example, a proxy observing a long duration flow using a transport protocol with end-to-end feedback (e.g., TCP) might conclude that a failure has occurred if traffic associated with that flow trickles to a halt. Here a proxy is using TCP as an implicit monitor of channel characteristics. Observing multiple coincident TCP rate decreases across multiple flows sharing a single channel would be a stronger indication of a failure.

A proxy-based monitoring system has the great advantage of simplicity; monitoring agents do not have to be deployed at members, no coordination is required, and no protocols need be defined. But the proxy's single vantage point provides low visibility to overall system state. Indeed, when a channel failure *is* detected a proxy is unlikely to know the cause, or other related effects.

3.3 Hybrid Proxy- and Member-Based Monitoring

It is clear that a combination of proxy- and member-based monitoring can be used to capture the most information about the current state of the system. As we demonstrated in Figure 2, providing the proxy with the most complete and up-to-date measurements improve channel allocation decisions and overall system performance. However, as the amount of measurement information that a proxy receives increases, the proxy is faced with ever more complicated decisions about how to allocate channels. An analysis of the proxy facing simple binary decision detailed in [12] illustrates this complexity.

3.4 Measurement Techniques

Though our monitoring system relies on the ability to measure channel characteristics, our focus is to identify appropriate existing measurement techniques, not invent them. There are numerous approaches to measuring and estimating link bandwidth and delay in the Internet [11]. Active probing schemes typically use pathchar to obtain link information [5]. The RTT of each hop is measured for variable packet sizes to calculate link bandwidth [8]. Packet pairing [4] is another popular technique for estimating link bandwidth. In this scheme, end-to-end path capacity is obtained from the dispersion between two packets of the same size transmitted one after the other. A centralized approach for measuring bandwidth and delay using tools such as SNMP and IP probes is proposed in [1].

Passive measurement schemes such as SPAND [15] do not use any probing messages and instead rely on observing traffic generated by the applications. In wireless networks radio signal-to-noise ratio (SNR) can be used to estimate hop-by-hop wireless link bandwidth [17]. SNR information can often be provided by a wireless network interface card (e.g., IEEE 802.11x card). A network service architecture that collects wireless channel conditions and provides them to the applications is proposed in [7].

4 Monitoring System for MC^2

We now describe a distributed monitoring architecture designed to meet the various system requirements and goals introduced earlier. The proposed architecture is decentralized; every community member participates in monitoring. Each member has a monitoring agent which joins a well-known multicast group G_m for exchanging community status information. Each monitoring agent broadcasts a *local report* R_l addressed to G_m on the LAN. Each local report contains information about the current state of the member and its offered WAN link(s).

Upon receiving a local report from member m_i, each member updates the information about member m_i in its locally-maintained community status database. In steady-state each member has up-to-date information about all community members. Each member issues a single packet containing the local report once every local reporting interval I_l. Though local report traffic grows linearly with the number of community members, this is not a concern as LAN bandwidth is plentiful, and report sizes are small. Though local report traffic grows linearly with the number of community members, this is not a concern for the following reasons. First, LAN bandwidth is plentiful, and report sizes are small. [1] Messaging overhead will be limited, and actions described below will help avoid redundant information exchange. Finally, in practical settings we anticipate that the size of most collaborating communities will be small, perhaps tens of members.

The collective information about the community members is sent to the inverse multiplexing proxy in *proxy reports* R_p. The community reports its current state to the proxy once every proxy reporting interval I_p. Instead of electing a particular member to send proxy reports, every member shares the responsibility. Each member sets a suppression timer for duration of $I_p + \delta$, where δ is a uniform random variable on $[0, S_d]$. Upon expiration of its suppression timer, member m_i sends R_p to the proxy via its local WAN link, and also multicasts the same report to the community on group G_m. Upon receipt of the multicast proxy report the members other than m_i cancel their suppression timers and report transmissions. At the same time, each member reschedules timers to send a proxy report for the next interval. Since R_p has the latest information about all the members, newly arriving members that have incomplete information about the community obtain complete system information quickly. Maintaining a distributed database is also advantageous for other reasons. Decentralization alleviates the potential problem of a control traffic bottleneck by spreading the traffic over multiple WAN links. Sharing responsibilities does not put an undue burden on any one node, provides fault-tolerance, and system reliability remains high even in a challenging 'high turnover' environment where members are arriving and departing at very high rates. A *soft-state* approach is used for maintaining member information in the monitoring databases. If the state is not periodically refreshed it is purged from the database. This approach also serves to purge the database of records of members who departed silently.

The system designer should configure the monitoring system to achieve high system responsiveness while limiting report traffic. Note that the maximum time between a state change and the proxy's knowledge of it is bounded by $I_l + I_p + S_d$. Increasing the reporting intervals I_p and I_l reduces both messaging traffic and responsiveness. Properly configuring these timers is challenging, as the optimal values depend upon community membership dynamics, the time-varying communication characteristics of WAN links, and the requirements and dynamics of the flows sent over the aggregated channel.

Where possible we opted to use passive methods for measuring channel characteristics. For example, it is reasonable to assume that each member has access to and monitors physical layer information such as the SNR of its wireless links. In some cases this information can also be used to estimate link quality parameters such as loss rate and bandwidth that are advertised in the local reports.

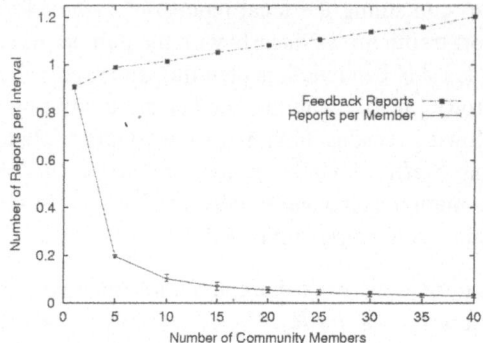

Fig. 3. The number of proxy reports issued per member per reporting interval I_p, and the average number of reports received by the proxy per reporting interval I_p.

5   Simulation Experiments

To explore the problems associated with monitoring system configuration we used *ns-2* based simulation. We set the value of both reporting intervals I_l and I_p to 1 second. Figure 3 shows the number of reports sent to the proxy in each proxy reporting interval I_p. As desired, the number of reports per reporting interval stays close to 1 even as the number of community members increases. The suppression algorithm is only slightly less effective in preventing multiple reports per interval in large communities (i.e., occasionally 2 reports are sent in one interval). If necessary, the number of instances of multiple reports can be reduced further by increasing the value for parameter S_d which controls the spread of the suppression timers. Figure 3 also plots the average number of proxy reports per interval sent by each member (R_p/I_p per member) with error bars showing the maximum and the minimum. As the community size increases the number of reports sent by each member declines as the reporting task is distributed across all community members. Note that the variability of the reports issued from member to member is very little; the reporting task is fairly equally split between all the members.

Though in our simulations all the members participated equally in the reporting process, in practice members will have differing capabilities (e.g., remaining battery life, compute power), so the system should permit different levels of participation by different members. Only members with sufficient memory and WAN bandwidth need to collect the information from the other members and share the load of informing the proxy. Biased suppression timers are one means of achieving this type of load balancing; more capable members can simply set shorter suppression timers (smaller value of S_d).

We also studied how the feedback latency varies with different settings of the reporting interval I_p. For this study, we generated a sequence of 100 events, each representing a change in the link state (such as bandwidth or loss rate) of a particular member. A member was chosen randomly from a 10-member community for each event. A change event occurs every 10 second period at a random time picked from a uniform distribu-

---

[1] Report sizes can be even smaller when schemes such as delta encoding are used.

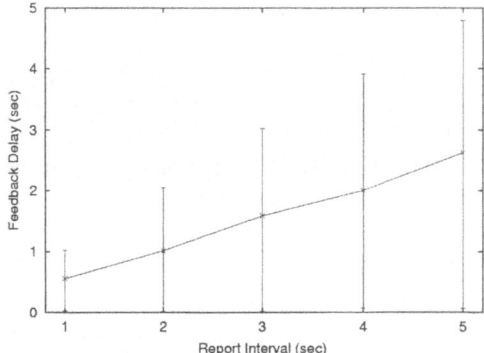

Fig. 4. The effect of reporting interval I_p on feedback latency.

tion on $[0, 10]$. The average feedback latency for this sequence of 100 events is shown in Figure 4 with the error bars showing the maximum and the minimum. As expected, the average feedback latency increases as I_p increases. We also observe that the maximum feedback latency is bounded by the reporting interval I_p. Although the feedback latency is low for small values of I_p, the amount of reporting traffic is large. This tradeoff between reporting overhead and reporting latency can have a significant effect on overall system performance because the WAN bandwidth between the agent and the proxy is relatively scarce, and the channel carries both data and control traffic. The reporting interval can be increased without greatly affecting the feedback latency by generating reports that are triggered by a significant event, e.g., a member departure, a measured channel characteristic exceeding a certain threshold.

6   Conclusion

Aggregating low-speed links to form a higher-speed logical link appears deceptively simple in principle. But as the communication characteristics of the underlying links grow increasingly erratic — as is the case in the challenging mobile setting we consider — potential performance improvements can vanish quickly. Hence a monitoring system that can accurately track communication link behavior and promptly inform a channel aggregator is crucial to achieving real performance gains in a practical bandwidth aggregation system.

We have designed and evaluated the performance of a decentralized channel monitoring system to support wireless bandwidth aggregation. An architecture that fairly distributes the burden of monitoring among community members can be made highly robust and responsive while limiting control message overhead. The monitoring architecture we have proposed in this paper is independent of the specific implementation of link aggregation, and can be used to support other aggregation and channel sharing systems [2,10,14]. We are also exploring use of monitored information in a multifuction proxy to perform joint channel and traffic control [12].

References

1. Y. Breitbart, C.-Y. Chan, M. Garofalakis, R. Rastogi, and A. Silberschatz, "Efficiently monitoring bandwidth and latency in IP networks," in Proceedings of *IEEE INFOCOM*, Anchorage, AK, Apr. 2001, pp. 933–942.
2. C. Carter and R. Kravets, "User device cooperating to support resource aggregation," in Proceedings of *IEEE WMSCA*, Callicoon, NY, June 2002, pp. 59–69.
3. P. M. Chen, E. K. Lee, G. A. Gibson, R. H. Katz, and D. A. Patterson, "RAID: Highperformance, reliable secondary storage," *ACM Computing Surveys*, vol. 26, no. 2, pp. 145–185, June 1994.
4. C. Dovrolis, P. Ramanathan, and D. Moore, "What do packet dispersion techniques measure?" in Proceedings of *IEEE INFOCOM*, Anchorage, AK, Apr. 2001, pp. 905–914.
5. A. B. Downey, "Using pathchar to estimate Internet link characteristics," in Proceedings of *ACM SIGCOMM*, Cambridge, MA, Sept. 1999, pp. 241–250.
6. D. Farinacci, T. Li, S. Hanks, D. Meyer, and P. Traina, "Generic routing encapsulation GRE," IETF, RFC 2784, Mar. 2000.
7. B.-J. Kim, "A network service providing wireless channel information for adaptive mobile applications: Proposal," in Proceedings of *IEEE ICC*, Helsinki, Finland, June 2001, pp. 1345–1351.
8. K. Lai and M. Baker, "Measuring link bandwidths using a deterministic model of packet delay," in Proceedings of *ACM SIGCOMM*, Stockholm, Sweden, Aug. 2000, pp. 283–294.
9. S. McCanne and M. Vetterli, "Joint source/channel coding for multicast packet video," in Proceedings of *IEEE ICIP*, Washington, DC, Oct. 1995, pp. 25–28.
10. M. Papadopouli and H. Schulzrinne, "Connection sharing in an ad hoc wireless network among collaborative hosts," in Proceedings of *NOSSDAV*, Florham Park, NJ, June 1999, pp. 169–185.
11. R. S. Prasad, M. Murray, C. Dovrolis, and K. Claffy, "Bandwidth estimation: Metrics, measurement techniques, and tools," *IEEE Network*, vol. 17, no. 6, pp. 27–35, Nov. 2003.
12. P. Sharma, J. Brassil, S.-J. Lee, and K. G. Shin, "Distributed channel monitoring for wireless bandwidth aggregation," HP Laboratories, Technical Report HPL-2003-171, Aug. 2003. [Online]. Available: http://www.hpl.hp.com/techreports/2003/HPL-2003-171.html
13. P. Sharma, S.-J. Lee, J. Brassil, and K. G. Shin, "Handheld routers: Intelligent bandwidth aggregation for mobile collaborating communities," HP Laboratories, Technical Report HPL-2003-37R1, May 2003. [Online]. Available: http://www.hpl.hp.com/techreports/2003/ HPL-2003-37R1.html
14. A. C. Snoeren, "Adaptive inverse multiplexing for wide area wireless networks," in Proceedings of *IEEE GLOBECOM*, Rio de Janeiro, Brazil, Dec. 1999, pp. 1665–1672.
15. M. Stemm, R. Katz, and S. Seshan, "A network measurement architecture for adaptive applications," in Proceedings of *IEEE INFOCOM*, Tel Aviv, Israel, Mar. 2000, pp. 285–294.
16. C. B. S. Traw and J. M. Smith, "Striping within the network subsystem," *IEEE Network*, vol. 9, no. 4, pp. 22–32, July/Aug. 1995.
17. J. Zhang, L. Cheng, and I. Marsic, "Models for non-intrusive estimation of wireless link bandwidth," in *Proceedings of PWC*, Venice, Italy, Sept. 2003.

Benefit of Admission Control in Aggregation Network Dimensioning for Video Services[*]

Gert Van Hoey[1], Danny De Vleeschauwer[1],
Bart Steyaert[2], Véronique Ingelbrecht[2] and Herwig Bruneel[2]

[1] Alcatel Bell, R&I / NSG, Fr. Wellesplein 1,
B-2018 Antwerp, Belgium
{Gert.Van_Hoey, Danny.De_Vleeschauwer}@alcatel.be
[2] University of Ghent, SMACS, Sint-Pietersnieuwstraat 41,
B-9000 Ghent, Belgium
{bs, vi, hb}@telin.ugent.be

Abstract. We derive a methodology to determine the effect of per service session admission control on the resource usage in an aggregation network for the delivery of video services. Given the network capacity and the characteristics of the service, we determine the number of subscribers that can be served in the scenarios with and without admission control. The results depend on the activity grade of the subscribers, and on the session blocking probability and/or the packet loss probability that can be tolerated. For a broad range of both probability parameters, the comparison shows a clear benefit of admission control, which permits an increase of the number of subscribers by up to 30%. The benefit of admission control decreases as the subscriber activity grade increases.

1 Introduction

An important problem encountered by network access providers (NAP) is the dimensioning of the aggregation network through which the NAP's customers can connect to Internet Service Providers (ISPs) or Application Service Providers (ASPs). An example is a DSL aggregation network, in which DSL access multiplexers (DSLAMs) aggregate traffic from typically hundreds of DSL subscribers. The traffic from the DSLAMs is then further aggregated by ATM- or Ethernet-based switched networks.

Dimensioning deals with determining the relationship between the number of customers attached to the aggregation network, the traffic characteristics of the services and applications used by these customers, and the network resources (link bandwidths, switch capacities) that are available or required in the network.

---

[*] This work was carried out within the framework of the project ARGUS sponsored by the Flemish Institute for the promotion of Scientific and Technological Research in the Industry (IWT).

N. Mitrou et al. (Eds.): NETWORKING 2004, LNCS 3042, pp. 357–368, 2004.

A straightforward approach for determining the required network resources is overdimensioning, which means that the network is dimensioned such that it can deal with the worst-case scenario that all customers are simultaneously on-line and using the service (voice-over-IP, video-on-demand (VoD), ...). In general, overdimensioning is not feasible because of limited aggregation network resources, which is particularly true for services requiring a relatively high bandwidth (e.g. video).

As a consequence, we need to take into account the fact that not all customers are consuming network resources simultaneously, and dimension such that the available resources, while considerably lower than in the overdimensioning case, still suffice to ensure users are fully satisfied. The user satisfaction is mainly determined by two factors. The first one is the availability of the subscribed service, which is the ability to initiate a service session at any time the customer wishes to do so. The second one is the quality of the transmitted information, which is determined by the extent to which packet delay and packet loss restrictions are met during an ongoing service session.

The dimensioning of the aggregation network will depend on whether or not the NAP performs admission control. Admission control (or call acceptance control, CAC) limits the number of simultaneously ongoing sessions (e.g. video-on-demand sessions) to a predetermined number, and thus, if needed, prevents a user from setting up a session at a moment when the maximum number of sessions is temporarily reached. The advantage of applying admission control is that it enables the NAP to guarantee the quality of transmitted information (packet loss, delay) because the maximum number of simultaneous sessions is known beforehand. The disadvantage is that from time to time, a customer has no access to the subscribed service (less than 100% availability). In a scenario without admission control, the availability of the service is 100%, but the possibility exists that temporarily, the amount of required network resources exceeds the provisioned resources, resulting in packet loss and excessive delay, and thus reduced service quality for all customers with ongoing sessions.

This paper compares the scenarios with and without CAC in terms of the network resources that need to be provisioned for a given customer population and service characteristics (or, equivalently, in terms of the number of subscribers that can be connected to the network, given the network resources). We focus on video services, which play a prominent role in emerging network-based applications and services. Because of the relatively high bandwidth requirements, the absolute differences in terms of required network capacity are expected to be substantial, and thus very relevant to network operators with limited resources in their aggregation network.

The remainder of the paper is organized as follows. Section 2 introduces the problem that is considered and describes the relevant parameters of the network, the users, and the service. The models on which this methodology relies, are introduced in section 3. Section 4 describes the methodology that is followed to determine the maximum number of VoD subscribers in the scenarios with and without CAC, while section 5 presents and discusses the results of the comparison between the two scenarios. Finally, section 6 concludes this paper.

2   Problem Statement

Figure 1 shows the considered aggregation network and the various network and traffic parameters that play a role in the dimensioning study of this paper. It shows a DSL aggregation network in which a group of N DSL subscribers are attached to the same DSLAM. The DSLAM's so-called "uplink" connects it with a switch in the aggregation network, that aggregates the traffic of several DSLAMs. Typical uplink capacities are 599 Mbps (STM-4, with 622 Mbps SDH bit rate) in the case of an ATM aggregation network, and 1 Gbps in the case of Ethernet. It should be noted that the methodology and results in this paper are equally applicable to aggregation networks other than DSL.

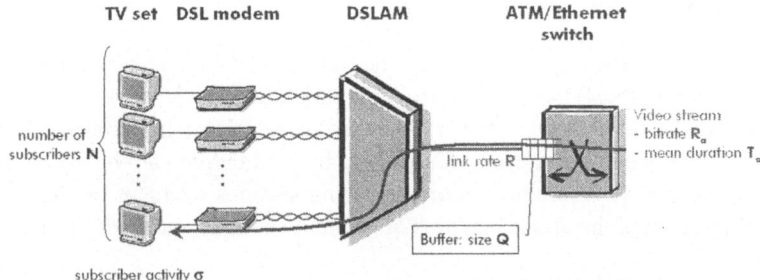

Fig. 1. Network setup

We assume that if a capacity bottleneck exists in the aggregation network, it is located in the aggregation segment (segment where resources are shared by multiple customers) closest to the customer side. In the DSL network of figure 1, this is the link between the DSLAM and the switch to which it is connected (the link modem-DSLAM is dedicated to a single subscriber, and hence not shared with other subscribers). In cable networks, the potential bottleneck would be located in the cable segment shared by a group of subscribers. Because the VoD service is considered in this paper, it is the network capacity in the downstream direction (from network towards user) that potentially forms a bottleneck. This means that the downstream link rate R and the and the size Q of the queue in the switch feeding the link towards the DSLAM, are the two parameters that determine to what extent packets of video streams can be lost. The packet loss probability will be denoted by P_{packet}.

Video streams are assumed to consist of 1500 byte packets and all have the same constant bit-rate (CBR), denoted by R_a (Mbps). The various video streams may have different durations, the average duration being denoted by T_a (sec). The parameter σ, the subscriber activity grade, indicates the probability that an individual VoD subscriber is actually watching a video stream at a randomly chosen moment in time.

In this paper, we take the point of view of a network operator that wants to determine the maximum number of VoD subscribers N that can be supported per DSLAM in his existing aggregation network. This means that we assume the network parameters R and Q as given, and that based on the values of R_a, T_a, and σ, the maximum N is determined that still guarantees a packet loss ratio below P_{packet}, and — in the case of admission control — a session blocking probability lower than $P_{session}$.

We confine ourselves to a network scenario with homogeneous (i.e., identical) sources, because a number of results in the literature suggest ([6]) that this serves as a worst-case to a heterogeneous scenario with comparable mean and peak rate. Hence, although many realistic traffic scenarios will include heterogeneous VoD sources, the homogeneous case can be used for dimensioning purposes.

3   Models for VoD Performance Assessment

3.1   Models to Assess the Packet Loss

The VoD quality will rapidly degrade when the packet loss becomes too high. It is well known (e.g. [5,7]) that under fairly general conditions the packet loss in a finite-capacity single-server queue of size Q can be accurately approximated by the probability that the queue length (represented here by the random variable u) in an infinite capacity exceeds K. Therefore, in the following sections we focus on the calculation of the P-quantile of the queue length in an infinite-capacity queue, defined as

$$Q_u(P) \triangleq \min_{Q}\{\Pr[u > Q] < P\} \quad , \tag{1}$$

This quantity calculated from the N×Geo/G/1 and N×MMBP/G/1 queueing model will be used to estimate the buffer size corresponding to a packet loss P in the associated finite-capacity queue.

When a subscriber is active (i.e., a session is ongoing), packets will be generated at a rate γ corresponding to R_a. The packet length (expressed in number of slots) is assumed to be generally distributed according to the probability generating function (pgf) $S(z)$. Let L_d (bytes) be the basic unit we use for expressing the length of the data packets and let S_l correspond to the slot length, then we clearly have that

$$S_l = 8L_d / R \quad ; \quad \gamma = S'(1)R / R_a \quad . \tag{2}$$

In a number of traffic scenarios it suffices to consider constant-length packets, in which case $S(z)=z$.

3.1.1   The N×Geo/G/1 Model
Let us consider the situation where N subscribers are active, each receiving packets at a rate γ. Since VoD sessions are typically very long (i.e., hundreds of seconds or more), it is reasonable to model the system performance under these circumstances by the steady-state behaviour of the discrete-time N×Geo/G/1 queue. If we define the queue length u as the amount of work in the queue (not including the packet being transmitted, if any), then it can be deduced that its steady-state pgf is given by ([1])

$$U(z) = (1-\rho)\left\{ \frac{x-1}{x-S(A(x))} \right\}_{x=S(z)} \quad , \tag{3}$$

where $\rho = N\gamma$ and $A(z)$ is the pgf describing the number of packet arrivals per slot and is therefore given by $A(z) = (1-\gamma+\gamma z)^N$.

From (1) it follows that we need an efficient way to calculate the tail distribution of u; in [4] it is shown that this quantity can be accurately approximated by using a dominant-pole approximation for $U(z)$, leading to

$$\Pr[u > Q] = \frac{-C}{z_0 - 1} z_0^{-Q-1} \quad , \tag{4}$$

where z_0 is the smallest pole of $U(z)$ and C the corresponding residue. In view of (3), it becomes clear that z_0 is the (smallest) real zero of $z=A(S(z))$ satisfying $z_0>1$ (we refer to [4] for further detail concerning these results).

Whenever the value of N becomes sufficiently high, $A(z)$ comes close to describing a Poisson arrival process, and the results of the M/G/1 model can be applied instead. This property will be exploited in the next sections.

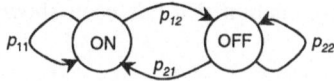

Fig. 2. The ON/OFF subscriber model

3.1.2 The NxMMBP/G/1 Model

Subscribers will not be watching VoD all the time, but will switch between active (during which a session is ongoing) and non-active periods. A popular approach in this case is to model each subscriber by a two-state ON/OFF source model in order to capture this type of behaviour. The ON state obviously corresponds to an ongoing VoD session during which packets are generated at a rate γ, while no packets are received during an OFF period. A graphic description of this subscriber conduct, in addition to the corresponding transition parameters, is shown in Fig. 2. The subscriber activity σ then equals $p_{21}/(p_{21}+ p_{12})$, and the transition probabilities can be derived from the VoD source characteristics in the following manner :

$$p_{12} = S_l / T_a \quad ; \quad p_{11} = 1 - p_{11}$$

$$p_{21} = \frac{1-\sigma}{\sigma} p_{12} ; \quad p_{22} = 1 - p_{21} \tag{5}$$

An important quantity in our source description is the 2x2 matrix $\mathbf{Q}(z)$, defined as

$$\mathbf{Q}(z) = \begin{bmatrix} p_{11}(1-\gamma+\gamma z) & p_{12} \\ p_{21}(1-\gamma+\gamma z) & p_{22} \end{bmatrix} \quad , \tag{6}$$

which fully captures the ON/OFF subscriber behaviour.

The packet arrival process described above is of the Markov Modulated Bernoulli Process (MMBP) type, and the analysis of an infinite-capacity single queue fed by this type of sources can be carried out by resorting to a pgf-approach ([8]). This leads to the following expression for the steady-state pgf $U(z)$ of the queue length u :

$$U(z) = (1-\rho) \sum_{m=0}^{N} \frac{(x-1)}{x - S(\lambda_1(x)^m \lambda_2(x)^{N-m})} \xi_m(x) \Bigg|_{x=S(z)} . \tag{7}$$

In this expression, $\lambda_j(z)$, $1 \leq j \leq 2$, represent the eigenvalues of $\mathbf{Q}(z)$. The functions $\xi_m(z)$ contain $N+1$ boundary probabilities that can be calculated by exploiting the analytic-

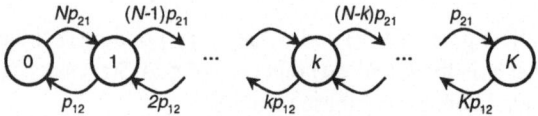

Fig. 3. The N×Geo/Geo/K/K state transition diagram

ity of $U(z)$ inside the complex unit disk, leading to a set of $N+1$ linear equations that needs to be solved. However, in order to keep the calculations as efficient as possible we prefer to use an approximation that has been shown to be sufficiently accurate, which yields

$$\xi_m(z) \cong \frac{N!}{m!(N-m)!} \left(\frac{\overline{\sigma}^T \mathbf{Q}(0) \overline{u}_1(z)}{\overline{\sigma}^T \mathbf{Q}(0) \overline{1}} \right)^m \left(\frac{\overline{\sigma}^T \mathbf{Q}(0) \overline{u}_2(z)}{\overline{\sigma}^T \mathbf{Q}(0) \overline{1}} \right)^{N-m} , \tag{8}$$

with $\overline{\sigma} \triangleq [\sigma \;\; 1-\sigma]^T, \overline{1} \triangleq [1 \;\; 1]^T$, and $\overline{u}_j(z)$ represent the column eigenvectors of $\mathbf{Q}(z)$ associated with $\lambda_j(z)$ satisfying $\overline{u}_1(1) + \overline{u}_2(1) = \overline{1}$. This approximation is based on the observation that the event of the buffer being empty at the beginning of a slot, implies that there were no packet arrivals during the preceding slot.

Once again, the tail distribution of the queue length is calculated by adopting a dominant-pole approximation. In this case however, it does not suffice to use an expression similar to (4) since a thorough evaluation of this approach has revealed that multiple poles of $U(z)$ can now contribute significantly to a tail probability. Hence we will use the following formula for tail probability calculations :

$$\Pr[u > Q] \cong \sum_{m \leq N} \frac{-C_m}{z_{0,m} - 1} z_{0,m}^{-Q-1} \tag{9}$$

where the pole $z_{0,m}$ of $U(z)$ is the real solution of $z = \lambda_1(S(z))^m \lambda_2(S(z))^{N-m}$ satisfying $z_{0,m} > 1$, and C_m its corresponding residue. Starting from $m=N$, decreasing values of m correspond to 'less dominant' poles, and for any given value of Q subsequent terms in (9) should be calculated, until the value of the term calculated last is negligible compared to the sum of the ones calculated before. For a wide range of parameter values, it has been found that this approach yields an efficient procedure for the calculation of queue length quantiles.

3.2 Models to Assess the Session Loss: The NxGeo/Geo/K/K Queueing System

When CAC is applied, we need some means to calculate the session blocking probability, for a population of N subscribers of which N_{stream} are allowed to be active simultaneously. The calculation of blocking probabilities is usually done by considering an appropriate server-loss (bufferless) model. Therefore, in view of the queueing model presented in section 3.1.2 for evaluation of the packet loss, the natural way to proceed is to calculate session blocking probabilities by means of a discrete-time NxGeo/Geo/K/K (where $K \equiv N_{stream}$ in our case) queueing model. In order to simplify the analysis, let us furthermore assume that the probability of two or more events (an event being an active subscriber terminating his session or a passive subscriber attempting to initiate a new session) occurring during a slot is negligible compared to the probability of a single event occurring. This assumption will yield accurate results when sessions (and silent periods between sessions) are typically very long, which is clearly the case for the VoD scenario under study. Under these circumstances, Fig. 3 represents the state transition diagram from which the steady-state probabilities of being in state k can be readily derived. We must point out that this diagram also describes the behaviour of the continuous-time M/M/K/K/N queueing system, also known as the Engset loss system ([1]).

The session blocking probability in the NxGeo/Geo/K/K model is therefore given by the Engset blocking probability. If we want to calculate the maximum number of subscribers that we can accommodate under the restriction that the session blocking probability may not exceed $P_{session}$, we must solve the following equation

$$\max_{N} \left\{ \frac{C_{N-1}^{K} \sigma^{K} (1-\sigma)^{N-K-1}}{\sum_{j=0}^{K} C_{N-1}^{j} \sigma^{j} (1-\sigma)^{N-j-1}} \right\} \leq P_{session} \tag{10}$$

i.e., we must calculate the largest value of N for which the inequality still holds.

3.3 A Numerical Example: Packet versus Burst Level Behaviour

Let us illustrate the typical buffer behaviour for R=450 Mb/s, R_a=2 Mb/s and T_a=5400 sec, and for constant-length VoD packets of 1500 byte (i.e., L_d=1500). In figure 4, we have plotted the 10^{-6} quantile of the queue length, obtained from the NxMMBP/D/1 model, versus the load ρ for various values of the subscriber activity σ. In this figure we express the quantile $Q_u(P)$ in seconds, because this immediately relates to the maximum delay that can be expected by a newly arriving packet.

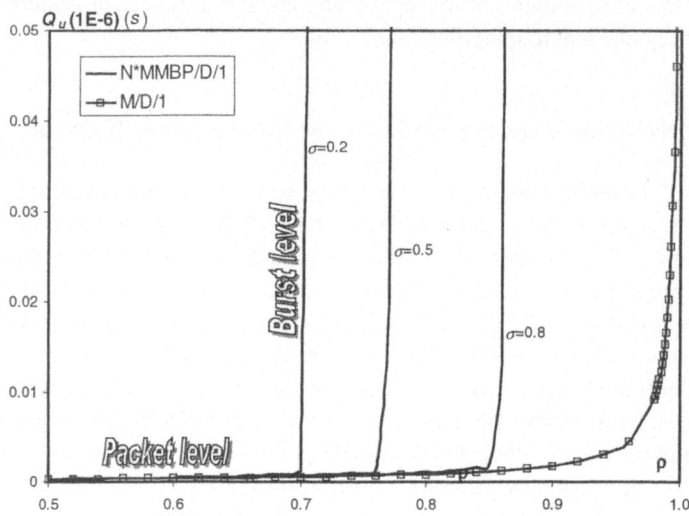

Fig. 4. Packet versus burst level buffer behaviour

Here we already observe the typical quantile behaviour that captures the essence of the VoD buffer performance. We can roughly distinguish two regions for the load ρ. On the one hand, an area of relatively small loads where the quantiles remain stable for increasing values of the load. This means that in this area packet loss can be avoided by assigning sufficient buffer space to the VoD buffer; the required buffer space however remains relatively small (i.e., ~5 *ms*), and the buffer performance can be regarded upon as being dominated by the *packet level* behaviour. On the other hand, for further increasing values of the load, one reaches a point where the quantiles suddenly explode : small increments of the load lead to huge increases of the quantiles. This point can be considered as the turning point where the buffer performance commences to be dominated by the *burst level* behaviour. In this region, large amounts of work can accumulate in the buffer, and whenever this occurs, the buffer will not be able to accommodate a newly arriving session. Note that, for these parameter values, a buffer of 0.01*s* can contain approximately 375 packets, while a VoD subscriber which is active during just 1 minute generates 10 000 packets on average. This explains why trying to buffer entire bursts becomes futile, especially if one wants to guarantee sufficiently small packet delays (and delay jitter), as well as small packet loss, as a measure of QoS.

When the packet level behaviour dominates the buffer performance, it suffices to have relatively small buffers in the switch-DSLAM (which also guarantees small packet delays) and the exact value of the buffer size Q is not important as long as it is 'high enough', which we assume to be the case in the remainder of this paper.

4   Methodology

4.1   Without Admission Control

In the scenario without admission control, the NxMMBP/G/1 model is applied to describe the switch-DSLAM buffer behaviour. Clearly, in view of the discussion in Section 3.3, from a network operator's point-of-view, the targeted working area is the range of ρ values where the packet level dominates, and the 'turning point' determines the maximum value of ρ that can be supported when no CAC is applied. For the given values of R (450 Mbps) and P_{packet} (10^{-6}) and the traffic parameters R_a (2 Mbps), T_a (5400 sec) and σ (0.2), the turning point is located at ρ=0.709. (cfr. $Q_u(P_{packet})$ in Fig. 4) Based on this value of ρ, the maximum number of subscribers N is then determined as $\lfloor R\rho/(\sigma R_a) \rfloor$, which yields N_{noCAC}=797.

4.2   With Admission Control

In the case of admission control, the maximum number of subscribers N is determined in two steps. First, the M/G/1 model is used to determine the maximum load ρ that can be tolerated on the switch-DSLAM link. In Fig. 4 we have plotted the 10^{-6} quantile of the M/D/1 queue, from which we can deduce that for P_{packet}=10^{-6} and if we allow packet delays up to $0.05s$, then the maximum tolerable load is nearly 1 (i.e. ρ=0.996), leading to N_{stream}=$\lfloor R\rho/R_a \rfloor$=224 (note that in most cases the value of ρ will be close to 1 leading to $N_{stream} \approx \lfloor R/R_a \rfloor$). Secondly, this value of N_{stream} is plugged into the NxGeo/Geo/K/K model (i.e., equation (10)) to determine the maximum number of subscribers that can be sustained for a given activity rate σ and blocking probability $P_{session}$ when CAC is applied. For $P_{session}$=10^{-4}, this model leads to N_{CAC}=915. This means that thanks to CAC, 915 subscribers can be supported instead of 797 in the case without CAC, which is an increase by 14.8%.

5   Results and Discussion

In this section, we quantify the number of VoD subscribers N that can be supported by an aggregation network, in the scenarios with (N_{CAC}) and without (N_{noCAC}) admission control, using the models and methodology described in the previous sections.

A video session is assumed to require a video codec R_a = 2 Mb/s and last on average T_a = 5400 s. We consider various values for the activity grade σ, (0.2, 0.5 and 0.8). A video packet has a size of 1500 byte. The aggregate bit rate that is available for the video service amounts to R = 450 Mb/s. This could, e.g., correspond with a situation with a 622 Mbps uplink on which a capacity of 172 Mbps needs to be available at any time for services other than VoD.

For the performance parameters we consider various values. For the tolerated packet loss P_{packet} we consider values from 10^{-10} to 10^{-3}, whereas for the session loss probability $P_{session}$ (or blocking probability) we consider values from 10^{-8} to 10^{-3}. The values that are tolerated depend on the type of service the user is requesting.

Video codecs are not very robust against packet loss (a value of the order of 10^{-6} is just bearable for MPEG codecs [3]). Moreover, packet loss may be very bursty: certainly in the NxMMBP/G/1 model there are indications that the system will slowly alternate from an under-load state in which no packets are lost at all to an overload state where the packet loss is very high (e.g. Fig. 4). Even though the overall (averaged) packet loss (the performance parameter that is resulting from the model studied) is low enough, it still could be that the loss during the (long-lasting) overload state would be detrimental. Therefore, the tolerated packet loss probability P_{packet} should be chosen very conservatively for video.

The choice of the tolerable session loss probability $P_{session}$ is set by the operator. A value of 10^{-3} seems reasonable, but the operator may choose a lower value.

In order to compare the scenario with AC with the one without AC, we define the gain due to AC as:

$$\frac{N_{CAC} - N_{noCAC}}{N_{noCAC}} \tag{11}$$

Table 1 gives the gain for various values of the tolerated packet loss and session loss for users that are only active during 20 % of the time on average. A considerable gain (in terms of the number of subscribers that can be supported) can be obtained by implementing an admission control mechanism, if we choose the tolerated P_{packet} conservatively (as we need to do for a video service) and as long as the session loss probability $P_{session}$ is not chosen too small. E.g., if we require the packet loss ratio to be below 10^{-8}, and put forward 10^{-4} as an acceptable blocking probability for the VoD service, the scenario with CAC enables the operator to have 20.7% more VoD subscribers attached to a single DSLAM than in the case without CAC. From Table 2 and Table 3 it can be seen that the benefit from implementing an admission control mechanism becomes gradually smaller as the activity grade of the user increases. The tables 1-3 further point out that for services that may be more tolerant w.r.t. packet loss and less w.r.t. blocking, the CAC case can turn out to enable less subscribers than the case without CAC.

In order to put the results of tables 1-3 in perspective, two important remarks should be made concerning the interpretation of the packet loss probability in the scenarios with and without CAC. First, in the case of CAC, P_{packet} has to be interpreted as the packet loss probability during time periods where the maximum number of VoD sessions (N_{stream}) are going on simultaneously. Because these time periods only represent a fraction of the total time, P_{packet} should be regarded as a "worst-case" packet loss ratio, and an "average" packet loss probability would have a smaller value. Secondly, in the case without CAC, packet loss is expected to be bursty. During long time intervals where the number of active users does not exceed $\lfloor R / R_a \rfloor$, the packet loss ratio will be very low. From time to time, however, the number of active users will exceed $\lfloor R / R_a \rfloor$, giving rise to a large packet loss ratio of approximately $1/\lfloor R/R_a \rfloor$. The time periods with high packet loss must be expected to last in the order of seconds or tens of seconds, given the fact that video streams of long duration are considered. This means that even if a low value of the "average" P_{packet} is put forward, time periods with high packet loss can occur in the case without CAC. From these considerations, it is clear that if models were available to quantify the discussed issues, the comparison between the scenarios with and without CAC would turn out more advantageous (or less negative) for CAC then currently indicated by the tables 1-3.

Table 1. The gain due to the activation of an AC mechanism for an activity grade σ of 0.2

σ = 0.2		$P_{session}$					
		1.E-08	1.E-07	1.E-06	1.E-05	1.E-04	1.E-03
P_{packet}	1.E-03	-8.1%	-5.5%	-2.5%	0.8%	4.8%	10.0%
	1.E-04	-5.0%	-2.3%	0.8%	4.3%	8.4%	13.7%
	1.E-05	-2.1%	0.7%	3.9%	7.4%	11.7%	17.2%
	1.E-06	0.6%	3.5%	6.8%	10.4%	14.8%	20.5%
	1.E-07	3.4%	6.3%	9.7%	13.4%	17.9%	23.7%
	1.E-08	5.8%	8.8%	12.3%	16.1%	20.7%	26.6%
	1.E-09	7.9%	11.0%	14.5%	18.4%	23.1%	29.2%
	1.E-10	10.6%	13.8%	17.4%	21.4%	26.2%	32.4%

Table 2. The gain due to the activation of an AC mechanism for an activity grade σ of 0.5.

σ = 0.5		$P_{session}$					
		1.E-08	1.E-07	1.E-06	1.E-05	1.E-04	1.E-03
P_{packet}	1.E-03	-6.2%	-4.3%	-2.2%	0.3%	3.2%	6.7%
	1.E-04	-3.9%	-1.9%	0.3%	2.8%	5.8%	9.4%
	1.E-05	-1.7%	0.3%	2.5%	5.1%	8.2%	11.9%
	1.E-06	0.3%	2.3%	4.6%	7.2%	10.4%	14.1%
	1.E-07	2.1%	4.1%	6.5%	9.1%	12.3%	16.1%
	1.E-08	3.9%	6.0%	8.4%	11.0%	14.3%	18.2%
	1.E-09	5.5%	7.6%	10.0%	12.7%	16.1%	20.0%
	1.E-10	7.1%	9.2%	11.7%	14.5%	17.8%	21.8%

Table 3. The gain due to the activation of a CAC mechanism for an activity grade σ of 0.8.

σ = 0.8		$P_{session}$					
		1.E-08	1.E-07	1.E-06	1.E-05	1.E-04	1.E-03
P_{packet}	1.E-03	-3.6%	-2.4%	-1.2%	0.0%	1.6%	3.2%
	1.E-04	-2.4%	-1.2%	0.0%	1.2%	2.8%	4.4%
	1.E-05	-1.2%	0.0%	1.2%	2.4%	4.1%	5.7%
	1.E-06	-0.4%	0.8%	2.1%	3.3%	4.9%	6.6%
	1.E-07	0.4%	1.7%	2.9%	4.1%	5.8%	7.5%
	1.E-08	1.3%	2.5%	3.8%	5.0%	6.7%	8.4%
	1.E-09	2.1%	3.4%	4.6%	5.9%	7.6%	9.3%
	1.E-10	3.0%	4.3%	5.5%	6.8%	8.5%	10.2%

Another consideration to take into is the effect of an increase in the value of the subscriber activity grade σ. In practice, the value of σ will be determined based on observations of the average number of subscribers that is simultaneously active during the busy hour of the service. It is possible, however, that at a certain moment in time, the demand for the service is significantly larger (e.g. caused by an unexpected important news item), such that the corresponding σ is larger than the value taken into account for dimensioning. The important distinction is that on the one hand, in the scenario with CAC, the higher σ will lead to an increased blocking probability $P_{session}$, while the packet loss guarantee still holds for all the service sessions that are in progress. On the other hand, in the scenario without CAC, the higher σ leads to a larger number of ongoing sessions, with all ongoing sessions experiencing increased packet losses. This increase in packet loss could be so severe that it renders the service practically useless for all ongoing sessions. It is difficult to point out which one of both effects, high blocking probability or excessive packet loss, is most undesirable. Still, a strong argument in favour of CAC is that it guarantees that a certain number (N_{stream}) of service sessions can always take place with the necessary packet loss guarantees.

6   Conclusion

The results of this study show that for a broad value range of the tolerated session blocking and packet loss probabilities, the application of CAC enables the service provider to serve considerably more subscribers. For a subscriber activity grade σ equal to 0.2, the difference amounts to up to 30%. This advantage of CAC decreases, however, with increasing subscriber activity. Due to a worst-case consideration concerning CAC, and the fact that a model of the packet loss burstiness in the case without CAC is not available, the quantitative comparison could be even more advantageous for CAC than our results indicate. Moreover, in case of exceptionally high service demands, CAC guarantees the quality and continuity of the service by refusing the excess of service requests, whereas in a scenario without CAC, the drastically increased packet losses may cause a complete break-down of the service.

References

1. Bruneel, H., Kim, B.G., "Discrete-time model for communication systems including ATM" (Kluwer Academic Publishers, Boston, 1993).
2. Cohen, J.W., "The generalised Engset formulae", Philips Telecommunication Review, 1957, vol. 18(4), pp. 158-170.
3. De Neve, P., et al.: A comprehensive quality performance study of MPEG streaming video over an IP-based network with packet loss. Submitted to ICC 2004, 20-24 June, 2004.
4. Bruneel, H., Steyaert, B., Desmet, E., Petit, G.H., "Analytic derivation of tail probabilities for queue lengths and waiting times in ATM multiserver queues", European Journal of Operational Research, 1994, vol. 76, pp. 563-572.
5. Kim, H., Shroff, N.B., "Loss Probability Calculations in a Finite Buffer Multiplexer", IEEE/ACM Trans. on Networking, vol. 9, no. 6, Dec. 2001, pp. 765–768.
6. Mao, G., Habibi , D., "Loss performance analysis for heterogeneous ON-OFF sources with application to connection admission control", IEEE/ACM Transactions on Networking, vol. 10 (1) (February 2002), pp. 125 - 138.
7. Steyaert, B., Bruneel, H., Analytic derivation of the cell loss probability in finite multiserver buffers from infinite-buffer results, Proceedings 2nd IFIP WG6.4 Workshop on Performance Modelling and Evaluation of ATM Networks (1994), pp. 18/1-18/11.
8. Steyaert, B., De Vleeschauwer, D., Queueing models for voice in packet-based networks, Presentation at the COST 257 Final Seminar "Impacts of new services on the architecture and performance of broadband networks" (Würzburg, 27-29 September 2000).

Call Admission Control for Voice/Data Integration in Broadband Wireless Networks

Majid Ghaderi and Raouf Boutaba

School of Computer Science
University of Waterloo, Waterloo N2L 3G1, Canada
{ghaderi, rboutaba}@uwaterloo.ca

Abstract. This paper addresses bandwidth allocation for an integrated voice and data broadband mobile wireless network. Specifically, we propose a new admission control scheme called EFGC, which is an extension of the well-known fractional guard channel scheme proposed for cellular networks supporting voice traffic. The main idea is to use two acceptance ratios, one for voice calls and the other for data calls in order to maintain the proportional service quality for voice and data traffic while guaranteeing a target handoff failure probability for voice calls. We describe two variations of the proposed scheme: EFGC-REST, a conservative approach which aims at preserving the proportional service quality by sacrificing the bandwidth utilization; and EFGC-UTIL, a greedy approach which achieves higher bandwidth utilization at the expense of increasing the handoff failure probability for voice calls. Simulation results show that our schemes satisfy the hard constraints on handoff failure probability and service differentiation while maintaining a high bandwidth utilization.

1 Introduction

With the expansion of wireless bandwidth, the next generations of mobile cellular networks are expected to support diverse applications such as voice, data and multimedia, demanding different quality of service (QoS) and bandwidth [1,2]. The bandwidth of the wireless links is inherently limited and is generally much smaller than that of wireline access links. Particularly, for integrated voice/data mobile networks it is necessary to develop mechanisms that can provide effective bandwidth management while satisfying the QoS requirements of both types of traffic.

Bandwidth allocation has been extensively studied in single-service (voice) wireless cellular networks. Hong and Rappaport [3] analyzed the famous *guard channel* (GC) scheme, which is optimal for minimizing an objective function of call blocking and dropping probabilities. Ramjee et al.[4] proposed a *fractional guard channel* scheme (FGC) which is optimal for minimizing call blocking probability subject to a hard constraint on call dropping probability. Instead of explicit bandwidth reservation as in GC, the FGC accepts new calls according to a randomization parameter called the *acceptance ratio*. One advantage of FGC

N. Mitrou et al. (Eds.): NETWORKING 2004, LNCS 3042, pp. 369–380, 2004.

over the GC is that it distributes the new accepted calls evenly over time which leads to a more stable control [5].

One of the challenges in moving to a multi-service system is that the limited bandwidth has to be shared among multiple traffics [2]. Epstein and Schwartz [6] investigated complete sharing, complete partitioning and hybrid reservation schemes for two classes of traffic, namely narrow-band and wide-band traffic. Haung et al.[1] proposed a bandwidth allocation scheme for voice/data integration based on the idea of *movable boundaries* (MB), however they completely neglected the prioritization of handoff calls over new calls and treated the two identically. Wu et al.[2] proposed a *dual threshold reservation* (DTR) scheme, which is static, i.e., the two reservation thresholds are fixed over time and it is too difficult to adjust them dynamically to satisfy the required service quality. Interested readers are referred to [7] for a comparison between DTR and MB schemes.

This paper introduces an *extended fractional guard channel call admission mechanism* (EFGC) for integrated voice/data mobile cellular networks that maximizes the wireless bandwidth utilization while satisfying a target call dropping probability and a relative voice/data service differentiation. The main idea is to use two acceptance ratios for voice and data according to the desired dropping probability of voice calls and relative priority of voice calls over data calls. Similar to [2,6,8], we assume that call dropping is not an important issue for data calls and treat handoff and new data calls in the same way. We define the extended MINBLOCK [4] problem as follows:

for a given cell capacity, maximize the bandwidth utilization subject to a hard constraint on the voice call dropping probability and relative voice/data call blocking probability.

To the best of our knowledge, extending the basic fractional guard channel scheme to address the extended MINBLOCK problem is a novel work. We follow an approach similar to the stable admission control algorithm proposed by Wu et al.[5] to derive the acceptance ratios for voice and data. In particular,

1. EFGC is dynamic, therefore, adopts to a wide range of system parameters and traffic conditions.
2. EFGC uses separate acceptance ratios for voice and data calls, therefore, it is very straightforward to enforce a relative or even strict service differentiation between voice and data traffic.
3. The control mechanism is stochastic and periodical to reduce the overhead associated with DCAC-based schemes.

The rest of the paper is organized as follows. Our system model, assumptions and notations are described in section 2. Section 3 is dedicated to the proposed admission control algorithm and presents the details of the analysis of the algorithm. Some numerical results and their analysis are presented in section 4. Finally, section 5 concludes this paper.

2   System Model

A cellular system which carries both voice and data traffic is considered. Let BU denote the smallest bandwidth unit that can be allocated to a call [2,6,8]. We assume that each voice call requires b_v BUs and each data call requires b_d BUs for the whole duration of the call. In this system, voice handoff calls have the highest priority then new voice calls are considered and lastly the new and handoff data calls.

Let random variables d and s denote the call duration and cell residency time of call X respectively. Following the convention (as in [1,2,3,4,5,6,7,8]), we assume that d and s have exponential distribution. In the real word, the cell residence time distribution may not be exponential but exponential distributions provide the mean value analysis, which indicates the performance trend of the system. Furthermore, the proposed admission control algorithm has a periodic control in which the length of the control period is set to less than the average cell residency time of a call in a cell to make the algorithm insensitive to this assumption.

The system under consideration is not necessarily uniform, i.e., each cell can experience a different load. Let $k = \{v, d\}$ denote the type of traffic, i.e., $k = v$ for voice and $k = d$ for data traffic. Below is the notation which will be used throughout this paper.

- Θ_i: the set of the adjacent cells of cell i
- C_i: the capacity of cell i in terms of BUs
- P_d: voice call dropping probability
- P_f: voice handoff failure probability
- P_{QoS}: target voice handoff failure probability
- λ_i^k: type-k new call arrival rate into cell i
- $1/\mu_k$: type-k average call duration
- $1/h_k$: type-k average cell residency
- r_{ji}: routing probability from cell $j \in \Theta_i$ to cell i
- a_i^k: type-k call acceptance ratio of cell i
- $E[z]$: the mean of random variable z
- $V[z]$: the variance of random variable z

Our calculations in this paper are based on the handoff failure probability P_f, which can be related to the call dropping probability P_d by the following relation (refer to [3] for more details):

$$P_f = \frac{P_d}{1 - P_d}\left(\frac{\mu_v}{h_v}\right). \tag{1}$$

Notice that exponential assumption is a necessary condition in deriving (1). Please refer to [9] for the handoff probability under a general call duration and cell residency distribution.

Next, we compute all the required probabilities for the rest of our discussion. Let $P_h^k(t)$ denote the probability that a type-k call hands off before time t and remains active until t, given that it has been active at time $t = 0$. Also, let $P_s^k(t)$ denote the probability that a type-k call remains active in its home cell until time t, given that it has been active at time $t = 0$.

$$P_h^k(t) = \Pr(s_k \leq t) \Pr(d_k > t) = \left(1 - e^{-h_k t}\right) e^{-\mu_k t} \tag{2}$$

$$P_s^k(t) = \Pr(s_k > t) \Pr(d_k > t) = e^{-(\mu_k + h_k)t}. \tag{3}$$

On average, for any call which arrives at time $t' \in (0, t]$, we can compute these probabilities as follows.

$$\hat{P}_h^k(t) = \frac{1}{t} \int_0^t P_h^k(t - t') \, dt' \tag{4}$$

$$\hat{P}_s^k(t) = \frac{1}{t} \int_0^t P_s^k(t - t') \, dt'. \tag{5}$$

Finally, let $P_{ji}^k(t)$ and $\hat{P}_{ji}^k(t)$ denote the probability that an active call in cell j at time $t = 0$ will be in cell i at time t, where $j \in \Theta_i$,

$$P_{ji}^v(t) = P_h^v(t) \, h_{ji} \tag{6}$$

$$\hat{P}_{ji}^v(t) = \hat{P}_h^v(t) \, h_{ji} \tag{7}$$

$$P_{ji}^d(t) = a_i^d \left[P_h^d(t) \, h_{ji} \right] \tag{8}$$

$$\hat{P}_{ji}^d(t) = a_i^d \left[\hat{P}_h^d(t) \, h_{ji} \right]. \tag{9}$$

In the next section we will use these probabilities to compute the maximal acceptance ratios for voice and data calls with respect to the specified call dropping probability and relative voice/data call blocking probability.

3   Admission Control Algorithm

Fig. 1 shows the state transition diagram of the EFGC scheme (call completions/handoffs are not shown). With the exception of state C, voice handoffs are always accepted with probability 1. In state C, the system is full and no more calls can be accepted. At each state there are two acceptance ratios for voice and data (one for voice and the other for data). Typically there is a service differentiation (priority) between voice and data calls which governs the relation between these two acceptance ratios. In this study, we assume that this relation is given and EFGC should maintain it given any traffic condition.

It is too expensive and impractical to repeat all the calculations necessary to determine the appropriate $a = (a_v, a_d)$ at each system state. Instead, our algorithm has a periodical control structure. At the beginning of each control interval of length T, each cell receives some information from its direct neighbor cells. Using this information and QoS constraints (handoff dropping probability and service differentiation), it will calculate an acceptance ratio $a = (a_v, a_d)$ for the current control interval. In addition to this, the call blocking probability in

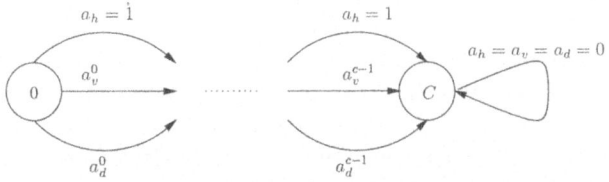

Fig. 1. Extended fractional guard channel.

cell k can be calculated by the acceptance ratio over multiple control periods as follows:

$$P_b^k = 1 - E[a_k]. \tag{10}$$

Therefore, we assume that the ratio of the voice blocking probability to the data blocking probability is given in terms of the relative acceptance probability of voice and data.

The admission control algorithm follows the pseudo-code in Fig. 2. In this algorithm, X is a type-k call where $b(X)$ and $a_i(X)$ denote the bandwidth requirement and corresponding acceptance ratios respectively. Also, $C_i^f(t)$ returns the amount of free bandwidth in cell i at time t and $\text{rand}(0,1)$ is the standard normal random generator function. In the next section, we will specify the acceptance ratio vector $a = (a_v, a_d)$ to complete the algorithm.

```
if (X is a voice handoff call) then
   if (b(X) ≤ Cᵢᶠ(t)) then
      accept call
   else
      reject call
   end if
else
   if (b(X) ≤ Cᵢᶠ(t)) ∧ (rand(0,1) < aᵢ(X)) then
      accept call
   else
      reject call
   end if
end if
```

Fig. 2. Call admission control algorithm in cell i.

3.1   Control Algorithm

We assume that during a control period each call experiences at most one handoff. This assumption is justified by choosing the length of control period T reasonably shorter than the average cell residency time (approximate equations are derived in [10] for appropriate control period length). Therefore, the immediate neighbors of cell i, i.e., Θ_i, are those which will affect the number of calls and consequently the bandwidth usage in cell i during a control period.

The number of calls in cell i at time t is composed of two elements: (1) the number of background calls which are already in cell i or its adjacent cells, and (2) the number of new calls which will arrive in cell i and its adjacent cells during the period $(0, t]$ $(0 < t \leq T)$. Let $g_i^k(t)$ and $n_i^k(t)$ denote the number of background and new type-k calls in cell i at time t respectively. Background calls have a Binomial distribution but new calls have a joint Poisson-Binomial distribution [11]. To have a closed form formula, we approximate this joint distribution by a Binomial distribution.

We then define the following quantities:

$$V_s^k(t) = P_s^k(t)\left(1 - P_s^k(t)\right) \tag{11}$$

$$V_{ji}^k(t) = P_{ji}^k(t)\left(1 - P_{ji}^k(t)\right) \tag{12}$$

$$\hat{V}_s^k(t) = \hat{P}_s^k(t)\left(1 - \hat{P}_s^k(t)\right) \tag{13}$$

$$\hat{V}_{ji}^k(t) = \hat{P}_{ji}^k(t)\left(1 - \hat{P}_{ji}^k(t)\right) \tag{14}$$

which show the variance of handoff/stay probabilities and are obtained with respect to the properties of Binomial distribution. Then the mean number of type-k active calls in cell i at time t is given by

$$E[N_i^k(t)] = E[g_i^k(t)] + E[n_i^k(t)] \tag{15}$$

where,

$$E[g_i^k(t)] = N_i^k(0)P_s^k(t) + \sum_{j \in \Theta_i} N_j^k(0)P_{ji}^k(t) \tag{16}$$

$$E[n_i^k(t)] = (a_i^k \lambda_i^k t)\hat{P}_s^k(t) + \sum_{j \in \Theta_i}(a_j^k \lambda_j^k t)\hat{P}_{ji}^k(t) \tag{17}$$

and similarly the variance is

$$V[N_i^k(t)] = V[g_i^k(t)] + V[n_i^k(t)] \tag{18}$$

where,

$$V[g_i^k(t)] = N_i^k(0)V_s^k(t) + \sum_{j \in \Theta_i} N_j^k(0)V_{ji}^k(t) \tag{19}$$

$$V[n_i^k(t)] = (a_i^k \lambda_i^k t)\hat{V}_s^k(t) + \sum_{j \in \Theta_i}(a_j^k \lambda_j^k t)\hat{V}_{ji}^k(t). \tag{20}$$

Knowing the bandwidth requirement of each type of calls, the mean and variance of bandwidth usage in cell i at time t, $C_i^u(t)$, are given by

$$E[C_i^u(t)] = b_v E[N_i^v(t)] + b_d E[N_i^d(t)] \tag{21}$$

$$V[C_i^u(t)] = b_v^2 V[N_i^v(t)] + b_d^2 V[N_i^d(t)]. \tag{22}$$

As we mentioned before, the cellular system considered in this paper is a broadband wireless system with capacity of several Mbps. In practice, 3G systems and beyond can be considered as broadband wireless systems (for example a UMTS system can support up to 2 Mbps). With this range of cell capacity it is reasonable to apply the central limit theorem. Then, the bandwidth usage in each cell can be approximated by a normal distribution:

$$C_i^u(t) \approx \mathcal{N}(E[C_i^u(t)], \ V[C_i^u(t)]). \tag{23}$$

Therefore, the original problem of maintaining a target handoff failure probability P_f is reduced to maintaining the bandwidth usage below the available capacity C_i at any point in time $t \in (0, T]$, or equivalently maintaining the overload probability below P_f. Hence, the time-dependent handoff failure probability $P_f(t)$ can be computed as follows:

$$P_f(t) = \Pr(C_i^u(t) > C_i) \tag{24}$$

therefore,

$$P_f(t) = \frac{1}{2} \operatorname{erfc} \left(\frac{C_i - E[C_i^u(t)]}{\sqrt{2 V[C_i^u(t)]}} \right) \tag{25}$$

where $\operatorname{erfc}(c)$ is the complementary error function. Then, similar to [5], the average handoff failure probability over a control period is given by

$$\widehat{P}_f = \frac{1}{T} \int_0^T P_f(t) \, dt. \tag{26}$$

To solve (26) for $a = (a_v, a_d)$ we need one more equation. This equation can be derived with respect to the required service differentiation. Given the service condition $a_d = \psi(a_v)$, the acceptance ratio vector $a = (a_v, a_d)$ can be found by numerically solving (26). Function ψ is such that $0 \leq \psi(a_v) \leq 1$ and $\psi(0) = 0$. In addition to this, ψ is uniformly increasing over $[0, 1]$. The boundary condition is that $a \in [0, 1] \times [0, 1]$, hence if $\widehat{P}_f < P_{QoS}$ than P_f for $a_v = 1$ then $a = (1, \psi(1))$. Similarly, if $\widehat{P}_f > P_{QoS}$ even for $a_v = 0$, then $a = (0, 0)$.

Furthermore, we have implemented two versions of this basic algorithm. EFGC-REST (or simply REST) is a restricted version which aims at satisfying the specified priority function ψ over time. In other words, REST always uses the acceptance ratio $a = (a_v, \psi(a_v))$ regardless of the congestion situation. It is obvious that in some states of the system it is possible to increase the acceptance ratio of data calls beyond the limit returned by the service differentiation function. For example when the network is not congested (at light traffic loads), we found that by increasing the priority of data traffic the overall utilization of the precious wireless bandwidth is increased while the handoff failure remains almost at the same level. This relaxed version is called EFGC-UTIL (or simply UTIL) due to its greedy approach for maximizing the utilization.

To estimate the control parameters of the EFGC, we simply use an exponentially weighted moving average technique. Each cell i estimates the average new call arrival rate of its neighbors based on their previous acceptance ratio:

$$(a_j^k \lambda_j^k) \leftarrow (1 - \epsilon) \, N_j^k / T + \epsilon (a_j^k \lambda_j^k) \tag{27}$$

where N_j^k is the number of type-k new calls that were actually accepted in cell j during the previous control period and ϵ is a weighting factor that should be specified with respect to the sampled observations. In our simulations we found that $\epsilon = 0.3$ leads to good estimation of the average new call arrival rate.

4   Numerical Results

4.1   Simulation Parameters

Simulations were performed on a two-dimensional cellular system consisting of 19 hexagonal cells (see Fig. 3). Opposite sides wrap-around to eliminate the finite size effect.

The common parameters used in the simulation are as follows. All the cells have the same capacity $C = 5$ Mbps, which is equal to 160 BU assuming each BU is equal to 32 Kbps (encoded voice using ADPCM requires 32 Kbps). Target handoff failure probability for voice calls is $P_f = 0.01$, $T = 20\,s$ and all the neighboring cells have the same chance to be chosen by a call for handoff, i.e. $h_{ji} = 1/6$. The system is uniform and the input load is the same for every cell, although EFGC is designed to handle the nonuniform case as well. For each load, simulations were done by averaging over 8 samples, each for 10 hours of traffic. Load distribution between voice and data traffic is fixed over time. At any load, 60% of the load is due to voice calls and the remaining 40% is composed of data calls. Table 1 summarizes service and traffic parameters for both traffic types. In this table, *priority* refers to the relative priority of voice and data calls. It means that new voice calls have higher priority than data calls for the admission control algorithm. In particular, the probability of accepting a new voice call is at least twice the probability of accepting a data call (new/handoff) at any time and any load.

Table 1. Voice and data service parameters.

Type	Priority	$1/\mu$ (s)	$1/h$ (s)	BU	Load
voice	1	180	100	1	60%
data	0.5	1000	800	2	40%

As mentioned earlier, this relative priority can be any service differentiation function. In our simulations, for the sake of simplicity we have chosen a constant service differentiation function.

We have also implemented the double threshold reservation (DTR) [2] scheme for comparison purposes. Since DTR is designed for a static traffic pattern, the handoff failure probability increases rapidly with the network load when the guard channels for handoff are few, but remains too low when the guard channels are many. Here, we choose the two thresholds in such a way that DTR achieves its objectives when the network starts to get overloaded. Hence, the voice threshold is set to 155 BUs and the data threshold is set to 151 BUs. Using these thresholds at load 2, P_f and $p = a_v/a_d$ were found to be 0.01 and 2 respectively.

4.2 Gaussian Verification

When the network is not congested and each cell has only a few active calls, it is clear that Gaussian approximation is not good. On the other hand, at light loads the admission algorithm does not require a high precision estimation of the load since there is no congestion in the network. As the load increases the number of active calls in each cell increases rapidly until no more calls can be accepted. Due to the high capacity of a broadband system, it is expected to have enough active calls in each cell so that central limit theorem can be applied.

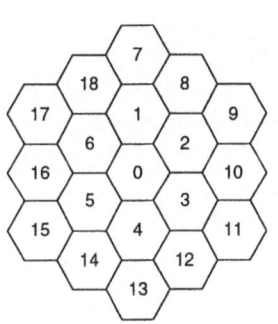

Fig. 3. Simulated cellular system.

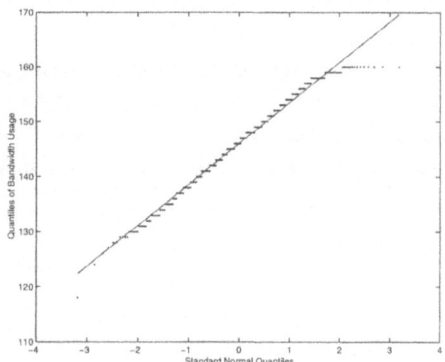

Fig. 4. QQ-plot of bandwidth usage.

We further investigated this issue in our simulation. At the beginning of each interval, the bandwidth usage at cell 0 is recorded until the end of simulation for load 2 (which is not a very high load). To verify the normality of these samples, we used the standard QQ-plot. Fig. 4 depicts the QQ-plot of the sample of the bandwidth usage at cell 0 versus the quantiles of the standard normal distribution. This plot clearly shows that Gaussian approximation of the bandwidth usage in each cell is satisfactory for our stochastic control.

4.3 Results and Analysis

Fig. 5 shows the handoff failure probability for the three schemes for a wide range of loads (interested readers are referred to [10] for more detailed simulation results which have not been presented in this paper due to space limitations). Both UTIL and REST maintain a constant failure probability independent of the load. For DTR, it grows very rapidly with the load (which was expected). With light loads (load < 2), DTR and REST have almost the same failure probability while UTIL has slightly higher P_f. But with high loads (load > 2) UTIL and REST are almost matched together while DTR has much higher failure probability. Fig. 8 shows that, although REST has better failure probability in light loads, this is accomplished by substantial sacrifice in the data call blocking probability. In addition to this, even in this region, UTIL has satisfied the target failure probability.

One of the objectives of EFGC is to maintain the relative service priority between voice and data calls. In our simulations, this relative priority is fixed and indicates that the acceptance probability of new voice calls should be twice the acceptance probability of new data calls. Fig. 6 depicts the ratio $p = a_v/a_d$ for different loads. It shows that EFGC maintains an almost constant service priority between the two types of traffic. Particularly, REST precisely preserves $p = 2$ for the whole range of loads while UTIL has $p = 1$ in light loads and $p = 2$ in high loads as expected because in light loads UTIL accepts as many data calls as there is enough bandwidth (without violating the target voice handoff failure probability). As the load increases, service priority of DTR increases rapidly. Fig. 8 shows that at high loads almost no data call is accepted. In other words, DTR is not fair and leads to starvation of data traffic. We have to mention that, although in this simulation the service differentiation is fixed, the EFGC can satisfy more complex functions such as state dependent functions.

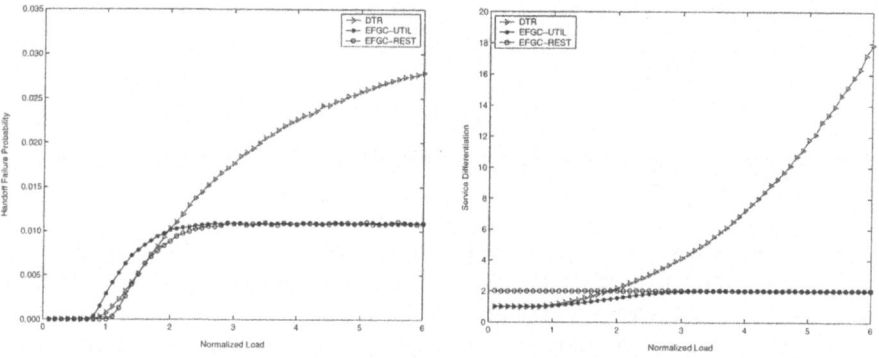

Fig. 5. Voice handoff failure probability. **Fig. 6.** Relative acceptance probability (p).

Figs. 7 and 8 show the new voice and new/handoff data call acceptance probability respectively. Again for high loads, UTIL and REST converge on the same result but the difference in their performance at light loads is significant. For data traffic at light loads the acceptance probability of UTIL is almost twice that of the REST. This explains why the utilization of UTIL is superior to REST. It can be seen that DTR has slightly higher acceptance probability for voice but much lower acceptance probability for data in comparison to UTIL and REST.

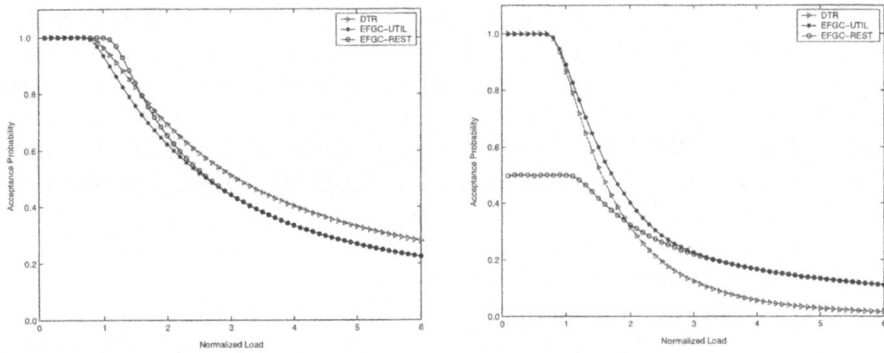

Fig. 7. Voice calls acceptance probability. **Fig. 8.** Data calls acceptance probability.

And finally, Fig. 9 depicts the wireless bandwidth utilization under the three bandwidth allocation mechanisms. Although DTR performs poorly in terms of handoff failure probability and service priority, its utilization is slightly better than EFGC. Interestingly, UTIL has exactly the same utilization as DTR at light loads and as expected, higher utilization than REST. In this simulation, voice traffic constitutes a larger portion of the total load. As the percentage of data traffic increases, the utilization of DTR is expected to drop (interested readers are referred to [10] for more detailed simulation results).

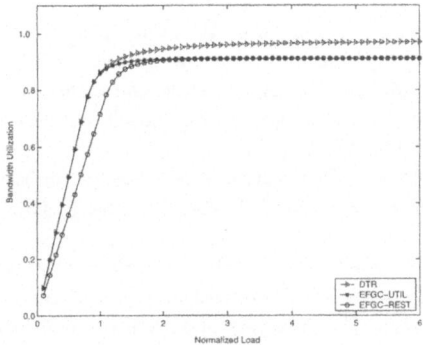

Fig. 9. Wireless bandwidth utilization.

5 Conclusion

In this paper, we developed a new admission control algorithm for voice/data integration in broadband wireless networks. Our algorithm is a natural extension of the well-known fractional guard channel proposed for voice cellular systems. EFGC always achieves the predetermined call dropping probability for voice calls while keeping the relative blocking probability of voice and data calls within a target threshold. We then described two versions of the EFGC, namely EFGC-UTIL and EFGC-REST. EFGC-UTIL follows a greedy approach to maximize the bandwidth utilization while EFGC-REST maintains the relative service priority. Both versions converged on the same result for high traffic loads. The major advantage of EFGC is its insensitivity to the network load. The dropping probability of voice and relative blocking probability of voice and data calls is maintained at a stable level over a wide range of traffic loads. According to the simulation results, we conclude that EFGC-UTIL is a good candidate for integrated voice/data cellular networks.

We are currently investigating the case of multiple classes of traffic where each class has its own QoS requirements in terms of call blocking and dropping probability. EFGC can readily support multiple classes of traffic by assigning a separate acceptance ratio to each class. However, computing these acceptance ratios in order to satisfy the desired QoS is not trivial.

References

1. Haung, Y.R., Lin, Y.B., Ho, J.M.: Performance analysis for voice/data integration on a finite-buffer mobile system. IEEE Trans. Veh. Technol. **49** (2000) 367–378
2. Wu, H., Li, L., Li, B., Yin, L., Chlamtac, I., Li, B.: On handoff performance for an integrated voice/data cellular system. In: Proc. IEEE PIMRC'02. Volume 5., Lisboa, Portugal (2002) 2180–2184
3. Hong, D., Rappaport, S.S.: Traffic model and performance analysis for cellular mobile radio telephone systems with prioritized and nonprioritized handoff procedures. IEEE Trans. Veh. Technol. **35** (1986) 77–92
4. Ramjee, R., Towsley, D., Nagarajan, R.: On optimal call admission control in cellular networks. ACM/Baltzer J. Wireless Networks **3** (1997) 29–41
5. Wu, S., Wong, K.Y.M., Li, B.: A dynamic call admission policy with precision QoS guarantee using stochastic control for mobile wireless networks. IEEE/ACM Trans. Networking **10** (2002) 257–271
6. Epstein, B., Schwartz, M.: Reservation strategies for multi-media traffic in a wireless environment. In: Proc. IEEE VTC'95. Volume 1., Chicago, USA (1995) 165–169
7. Li, B., Li, L., Li, B., Cao, X.R.: On handoff performance for an integrated voice/data cellular system. ACM/Baltzer J. Wireless Networks **9** (2003) 393–402
8. Li, B., Chanson, S., Lin, C.: Analysis of a hybrid cutoff priority scheme for multiple classes of traffic in multimedia wireless networks. ACM/Baltzer J. Wireless Networks **4** (1998)
9. Fang, Y., Chlamtac, I.: Analytical generalized results for handoff probability in wireless networks. IEEE Trans. Commun. **50** (2002) 396–399
10. Ghaderi, M., Boutaba, R.: Call admission control for voice/data integration in broadband wireless networks. Technical Report CS-2003-41, School of Computer Science, University of Waterloo (2003)
11. Naghshineh, M., Schwartz, M.: Distributed call admission control in mobile/wireless networks. IEEE J. Select. Areas Commun. **14** (1996) 711–717

Self-Admission Control for IP Telephony Using Early Quality Estimation

Olof Hagsand[1], Ignacio Más[1], Ian Marsh[2], and Gunnar Karlsson[1]

[1] Department of Microelectronics and Information Technology
Royal Institute of Technology (KTH)
S-16440 Kista, Sweden
[2] Swedish Institute of Computer Science
Box 1263
SE-164 29 Kista, Sweden

Abstract. If quality of service could be provided at the transport or the application layer, then it might be deployed simply by software upgrades, instead of requiring a complete upgrade of the network infrastructure. In this paper, we propose a self-admission control scheme that does not require any network support or external monitoring schemes. We apply the admission control scheme to IP telephony as it is an important application benefiting from admission control. We predict the quality of the call by observing the packet loss over a short initial period using an in-band probing mechanism. The quality prediction is then used by the application to continue or to abort the call. Using over 9500 global IP telephony measurements, we show that it is possible to accurately predict the quality of a call. Early rejection of sessions has the advantage of saving valuable network resources plus not disturbing the on-going calls.

1 Introduction

Quality of service in the Internet has been researched for the last twenty years, yet its introduction has been extremely slow. Differentiated services [1] was originally proposed in 1997 to overcome scalability problems of previous proposals. However, DiffServ is still not widely offered by Internet service providers, perhaps due to the required upgrade in network infrastructure. Our proposal offers a light QoS for multimedia stream traffic, by a regulated admission of sessions, rather than a regulation of the flow rate per session. In human terms it is better to block a call that has little chance of being completed with adequate quality rather than allowing it to start and potentially degrading the system. Therefore, admitted sessions gain by having a high probability of being completed with decent quality. All these properties can be successfully accomplished by using admission control.

The purpose of this paper is to devise an efficient and flexible admission control scheme for IP telephony. Although IP telephony is used as the example real-time application in this work, it should be clear that there are no inherent restrictions on the applicability of the admission control scheme.

The admission control can be performed without explicit support from the network [2]. The procedure is in–band probing [3], in which the first seconds of the voice

N. Mitrou et al. (Eds.): NETWORKING 2004, LNCS 3042, pp. 381–391, 2004.
© IFIP International Federation for Information Processing 2004

transmission are used as a probe stream. A new session is established only after estimating that the state of the network is acceptable. The receiver of the call measures the
packet loss ratio of the first few seconds and estimates the packet loss probability. This
estimated loss probability is compared to an acceptance threshold, which determines
whether the session should be established or not. Loss levels above the threshold result
in blocking of the new session and the sender should wait before establishing a new
session. Hence, ongoing calls are protected from new calls that could deteriorate the
overall quality to an unacceptable level by placing additional load on the network. The
admission control being proposed is related to the out-of-band probing scheme being
developed in our group [4,5,6,7].

   We claim that measurements can produce data useful for predicting future quality.
However, it is important to state we use only packet loss as the quality indicator of a VoIP
session in this work. Packets arrive at a receiver 50 times per second (assuming no loss)
in our VoIP scheme [8], so we have frequent sampling and observation of the network
state. The measured loss after an initial number of seconds (zero to ten) is compared
with the loss measured over the whole session. We use the correlation between the two
measurements to determine how accurate the estimation is. This is possible as we have
the whole session recorded at the receiver stored available for post processing[1].

   The structure of this paper is as follows. The next section gives some background
on how the empirical measurements were taken; we also describe how we measure the
packet loss ratio for one call. Section 3 shows the results for all considered calls and
offers a statistical analysis of the accuracy of the loss estimation for different initial time
intervals, as well as blocking and error probabilities. Section 4 gives some conclusions
of our work, some applications and pointers to future work. A preliminary version of
this work was published in [9].

2 Measurement Description

This paper uses the results of previous work where approximately 23000 VoIP calls
were measured between hosts at nine academic sites [10]. The locations of the sites are
shown in Figure 1. The sites were connected as a full-mesh, allowing us, in principle, to
measure the quality of 72 different Internet paths. These paths represent large differences
in timezones, hop counts and geographic distances.

   The measurements were performed over a period of 15 weeks in the following way:
A call between two hosts was initiated on an hourly basis between a sender and a
receiver. The sender transmitted a sequence of pre-recorded speech samples at 64 kbps
as a stream of RTP/UDP/IP datagrams. The receiver made a detailed log of the arrival
process, recording the reception time of each datagram. The complete details of the
measurements are described in our previous work [10].

2.1 Reducing the Sample Set

For the purposes of this paper, we needed a common basis for our analysis, and therefore
selected a subset of the 23000 calls. We only used calls that experience loss, since loss

---

[1] see also http://www.sics.se/~ianm/COST263/cost263.html

free calls do not provide any extra information for our analysis: both the probing and the total loss rate are zero giving perfect correlation. A large percentage of the calls are in fact loss free which reduces the sample set somewhat. We attribute the large number of loss free sessions to the fact that the sites are located on well provisioned (academic) networks. This restriction resulted in a subset of 9683 calls. Despite this reduction, all nine sites are represented in the subset.

2.2 Measuring a Single Call

Figure 2 shows the loss process of a sample call as observed by a receiver. The call was made between the Argentinian and Turkish sites. The figure shows a loss pattern that is representative of many other calls in the subset. The plot shows the number of lost packets on the y-axis versus time on the x-axis. It can be seen that the number of lost packets increases almost linearly as the call proceeds.

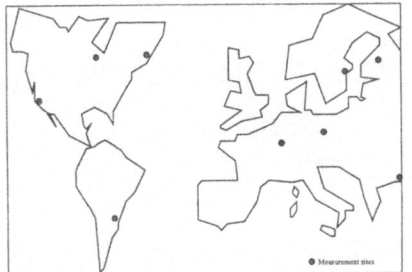

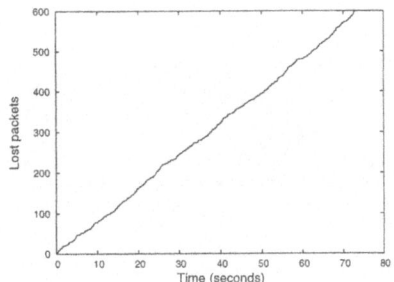

Fig. 1. Measurements were made between nine academic sites worldwide.

Fig. 2. Loss process of a single sample call between Turkey and Argentina.

Figure 3 shows the cumulative loss ratio for the same call. This ratio is defined as the number of lost packets divided by the number of sent packets. We show the cumulative plot to clarify how long we need to measure to obtain a good estimation of the final loss ratio. From the plot, we see that the final loss ratio for the complete call is approximately 18%.

In Figure 4 we show the first 20 seconds of the same call. From the figure, we see that the initial loss is approximately 14% after one second and 19% after ten seconds. These are early estimations of the final loss rate. We want to know how accurate such early estimations are. Therefore we need to study the relation between the loss ratio of an initial part of the call and the loss ratio of the whole duration.

3 Analysis

In the preceding section, only one call was considered. In Figures 5 and 6, the loss ratio for the whole call is plotted versus the loss ratio of an initial interval for all calls in the selected subset. In the figures, every point represents one call. The plots show that as the

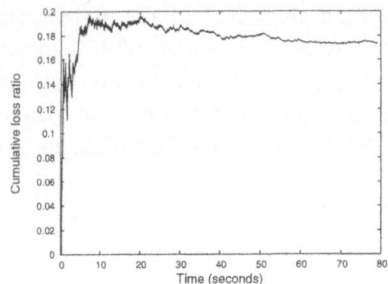

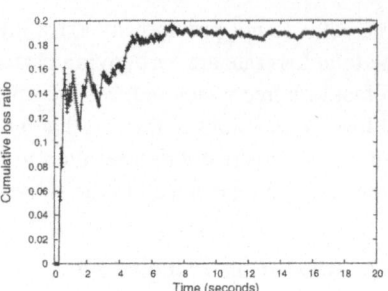

Fig. 3. Cumulative loss ratio of a single sample call between Turkey and Argentina.

Fig. 4. Cumulative loss ratio of the same session showing only the initial portion of the call.

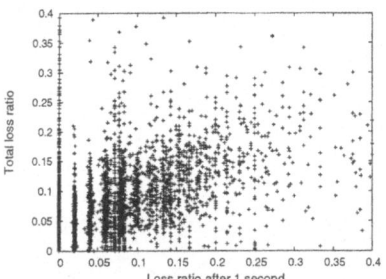

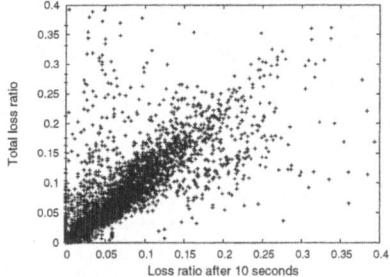

Fig. 5. Relation between the loss ratio after one second and the total loss ratio for all calls

Fig. 6. Relation between the loss ratio after ten seconds and the total loss ratio for all calls

initial interval increases, the points group closer around the line $y = x$. In other words, the correlation increases and the estimation improves.

The plots in Figures 5 and 6 give an intuitive measure of the correlation between the loss ratio of the initial interval and the total call. In order to evaluate more precisely the accuracy of the estimation, we computed the actual correlation factor as a function of the initial interval. The result is plotted in Figure 7.

Figure 7 shows that the correlation factor increases as the probing interval increases. From the figure, we can clearly see that the correlation stabilizes after four seconds. This is important, because after this point no further estimates are necessary.

The relation between the loss ratio of an initial interval, l_p, and the loss ratio of the total call, l_t can be further examined by forming an error function, such as $l_t - l_p$, and analyzing it as a stochastic variable ϵ.

Figure 8 shows histograms of ϵ for probing intervals of one, four and ten seconds. In the histograms, positive values represent calls where the initial loss rate is smaller than the total loss rate, i.e., $l_p < l_t$. In other words, those calls experienced a higher packet loss after the probing: the quality of the calls deteriorated after the initial interval. Likewise, negative values represent calls where the initial loss rate is greater than the total loss

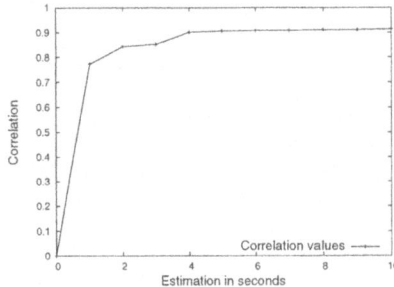

Fig. 7. The correlation factor as a function of the initial probing interval

rate, i.e., $l_p > l_t$. Note that the histograms represent probability density functions of ϵ that are not normally distributed.

Based on the values in the histograms we calculated the confidence intervals by counting the number of samples around $\epsilon = 0$ that sum up to the desired confidence level. The result is shown in Table 1.

Table 1. Table showing confidence levels and intervals of the error function $\epsilon = l_t - l_p$ for probing intervals one, four and ten seconds.

Level	interval (1 second)	interval (4 seconds)	interval (10 seconds)
0.75	$[-0.0288, 0.0336]$	$[-0.0141, 0.0187]$	$[-0.0087, 0.0151]$
0.80	$[-0.0424, 0.0436]$	$[-0.0180, 0.0260]$	$[-0.0124, 0.0213]$
0.85	$[-0.0608, 0.0568]$	$[-0.0252, 0.0344]$	$[-0.0183, 0.0299]$
0.90	$[-0.0848, 0.0752]$	$[-0.0416, 0.0496]$	$[-0.0280, 0.0424]$
0.95	$[-0.1768, 0.1144]$	$[-0.1200, 0.0800]$	$[-0.0888, 0.0696]$
0.99	$[-0.4000, 0.2536]$	$[-0.3200, 0.2216]$	$[-0.2480, 0.2144]$

Based on the table, we can express to what degree we can trust an initial observation. For example, if we measure the loss ratio l_p of a call after four seconds, we can be 80% certain that the total loss of the call will be in the interval $[l_p - 1.8\%, l_p + 2.6\%]$.

While the confidence intervals may be useful in themselves to express confidence in an observed value, forming the cumulative distribution function (cdf) of ϵ is more useful when an upper bound on the final loss value is of interest. This is typically the case in admission control scenarios, where we want to block calls that we believe will experience a loss higher than a certain threshold.

Table 2 shows the cdf of ϵ. Using the table, we can make statements such as: Given a probing loss and a confidence level, the final loss will be bounded by the probing loss plus a value given by Table 2. Figures 9, 10 and 11 show the cdf of ϵ in graphical form.

The cdf of ϵ can directly be used for admission control purposes. The table gives us the percentage of calls that have an error less or equal to the value of ϵ:

$$P(\epsilon < l_a - l_p) \geq confidence \quad level$$

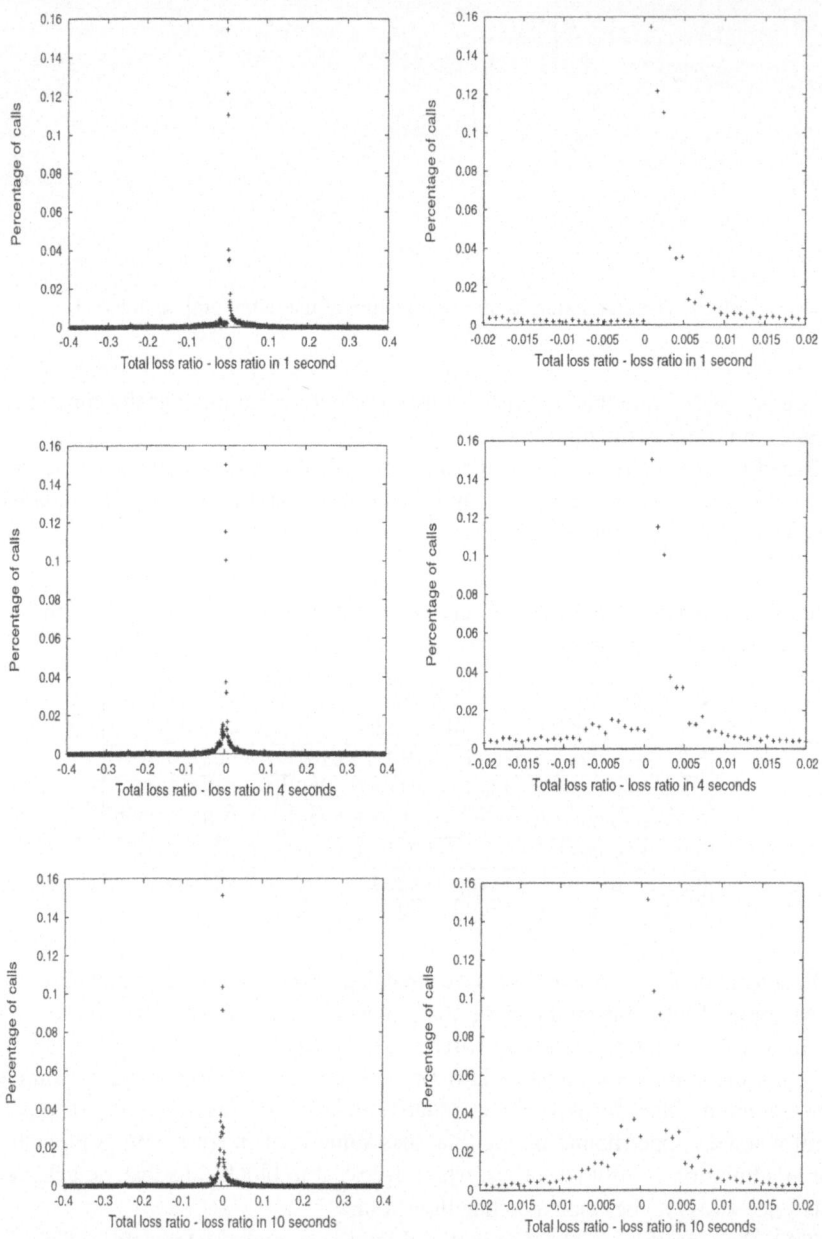

Fig. 8. Histograms of the error $\epsilon = l_t - l_p$ for initial probing intervals one, four and ten seconds. Each histogram is shown in full view on the left, while the right plot shows an enlarged region around $\epsilon = 0$.

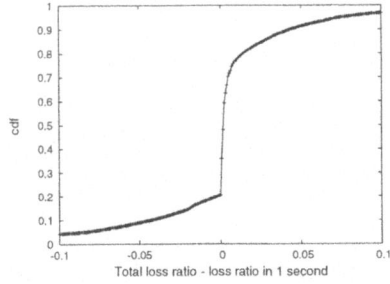

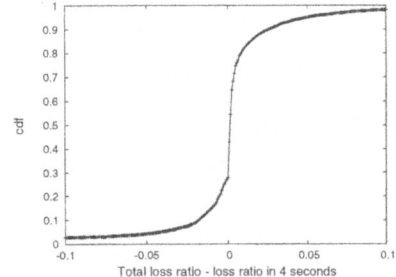

Fig. 9. Cumulative distribution function of $\epsilon = l_t - l_p$ for a probing interval of one second.

Fig. 10. Cumulative distribution function of $\epsilon = l_t - l_p$ for a probing interval of four seconds.

Table 2. Table showing cumulative values of the error function $\epsilon = l_t - l_p$ for probing intervals one, four and ten seconds.

Confidence level	1 second	4 seconds	10 seconds
0.05	−0.0848	−0.0416	−0.0280
0.1	−0.0424	−0.0181	−0.0101
0.2	−0.0016	−0.0056	−0.0038
0.3	–	–	−0.0008
0.4	–	–	–
0.5	0.0018	0.0014	0.0009
0.6	0.0025	0.0020	0.0017
0.7	0.0042	0.0038	0.0030
0.8	0.0144	0.0086	0.0069
0.9	0.044	0.0260	0.0212
0.95	0.0752	0.0496	0.0424

For example, suppose the aim of a strict admission control scheme using four seconds probing is to drop calls that have a higher risk than 10% to surpass a pre-established loss rate l_a. Retrieving the value of ϵ from Table 2 shows that $l_a - 2.6\%$ is a good threshold. A more relaxed policy could have the aim to reject all calls that have more than 90% risk to surpass l_a. In that case, again using Table 2, the threshold is $l_a + 1.81\%$. The strict and relaxed policies outlined above both have drawbacks. With a strict policy, most bad calls ($l_t > l_a$) will be blocked, along with a large number of good calls ($l_t < l_a$). A relaxed policy admits most good calls, while admitting many bad calls.

Table 12 shows a classification of calls with respect to an admission control strategy: classes AG and AB represent calls that were admitted while classes RG and RB represent calls that were blocked. Further, classes AG and RB represent categories where the admission control decision was correct. Classes AB and RG represent decisions that were wrong. An admission control policy based on probing, needs to consider the trade–off between classes.

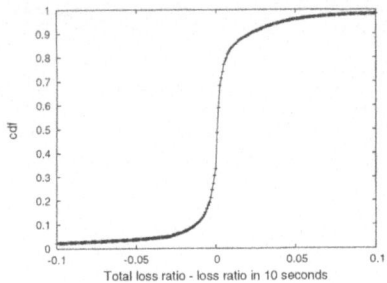

	Good calls $l_t < l_a$	Bad calls $l_t > l_a$
Admitted $l_p < l_\alpha$	AG	AB
Rejected $l_p > l_\alpha$	RG	RB

Fig. 11. Cumulative distribution function of $\epsilon = l_t - l_p$ for a probing interval of ten seconds.

Fig. 12. The table shows the different kinds of calls based on the initial estimation and the final outcome. l_α denotes an admission threshold applied after a probing interval, while l_a is the desired upper bound on the loss level.

If we return to the strict policy introduced above, it minimizes class AB while class RG is large, thus protecting on–going calls in a more successful manner whilst increasing the blocking probability. In the same way, the relaxed policy minimizes class RG, thus reducing the blocking probability at the risk of a higher number of bad calls.

To obtain absolute numbers on the number of calls in the classes, a real loss distribution has to be considered. By aiming at an upper bound of the loss rate and applying the *cdf* to that bound, it is possible to get absolute numbers of the different classes. The admission threshold can then be varied to find a desired optimum.

Figure 13 shows an example of a uniform loss distribution (calls can experience any packet loss rate between 0 and 100% with equal probability) with a desired upper bound on the loss rate l_a. The *cdf* for four seconds in Figure 10 has been superimposed[2] on the uniform loss distribution for two admission thresholds, strict policy and relaxed policy. The number of calls belonging to each class can be determined by the areas in the graph. The areas are bounded by l_α and the *cdf*. For example, it can be seen from the graph that area RG (rejected calls that turned out good) is large in the strict policy, but is small in the relaxed. Likewise, area AB (admitted calls that turned out bad) is small in the strict and large in the relaxed policies.

A uniform loss distribution is evidently unrealistic, but the same methodology can be applied for a real loss distribution. We have applied the method to the complete set of 9683 error-free calls in the measurements in the case of four seconds probing and calculated the percentage of calls that fall in each of the areas. The rest of this section deals with this case.

Figure 14 gives the blocking probability (RG+RB) for the complete sample space. From the figure it can be seen that rejecting calls that experience an initial loss rate equal or higher than 10% gives a blocking probability of around 15%, while a more stringent packet loss rate threshold would result in a rapidly increasing blocking probability. Note however, since the error-free calls are omitted, the blocking probability is overly pessimistic. We would expect a lower blocking probability with a factor of around three if the error free calls were included.

---

[2] Note, the *cdf* is reflected around x=0.

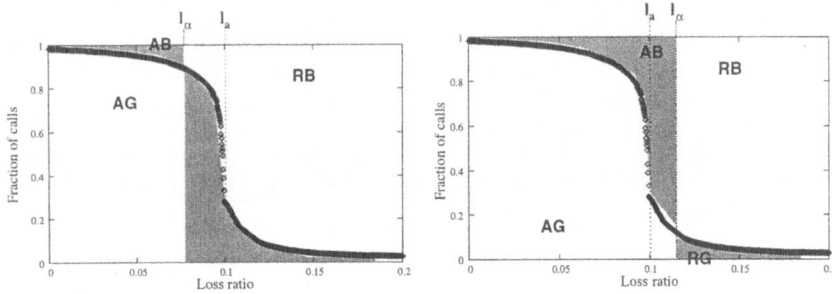

Fig. 13. Example showing the result of imposing admission control decision in the strict (left) and the relaxed (right) admission policy with a uniform loss distribution. The desired upper bound on packet loss is l_a and the imposed threshold is l_α.

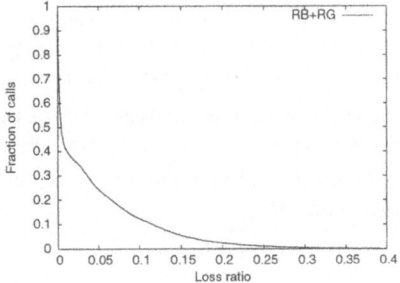

Fig. 14. Blocking probability as a function of the packet loss rate admission threshold.

As was previously noted, the accuracy of an admission control policy can be measured by counting the correct and incorrect decisions. Figure 15 shows the incorrect decisions for a packet loss rate target of 2%. The plot shows both kinds (AB and RG) as well as their sum.

The plot illustrates how the number of incorrectly admitted calls increases as the admission threshold is relaxed, while the incorrectly rejected calls decreases. The sum of the two functions has a minimum for a particular admission threshold at 1.8%, which can be considered as an optimum operating point. That is, a minimal number of incorrect decisions were made at this threshold.

Finally, Figure 16 illustrates the sum of incorrect decisions for different target loss rates as the acceptance threshold is varied. The results show a minimum close to the value of the target loss rate, as was intuitively expected.

The choice of an operating point for the admission control has to take into account many parameters. We can always increase the accuracy by measuring for a longer period. However, increasing the probing period reduces the advantages, since we are extending the period in which a bad call is disturbing the ongoing calls, reducing the overall quality in the process. Also, longer probing times increase the frustration in the case of a rejection.

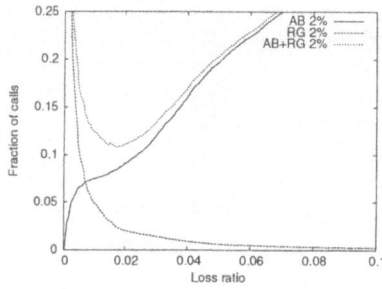
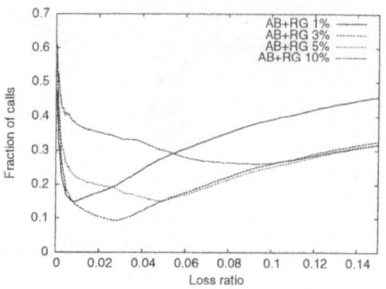

Fig. 15. Erroneous decisions as a function of the admission threshold for a 2% target loss rate

Fig. 16. Erroneous decisions as a function of the admission threshold for different target loss rates.

To summarize, if we use our measurements, we would probe for four seconds and use an admission threshold close to the targeted value. Assume that 2% packet loss is acceptable. In this case, the admission threshold should also be around 2%, which would give a blocking probability of 36%. The admission control decision would then have failed 11% of the time, the majority would be calls that were admitted although they turned out to be bad (9% of the total calls), a smaller fraction would be calls that were rejected but turned out to be good (2%).

4   Conclusions and Future Work

This paper proposes a quality differentiation scheme based on self-admission control without the need of infrastructure changes. The admission control is performed at the application layer and can provide statistical bounds on the packet loss rate that stream flows will experience in the network. We have shown how the admission control mechanism can be devised by blocking calls experiencing an initial loss rate exceeding an admission threshold. An initial admission threshold is motivated by two factors: (1) it makes sense to drop calls that will experience bad quality and thus reduce congestion in the network so that other calls may experience better quality; (2) an audio coder may have an upper bound on quality: exceeding a drop rate will result in unacceptable audio quality.

We have evaluated the admission control scheme by analyzing a large number of IP telephony calls that were made over the Internet. Based on this empirical data, we have shown that it is possible to predict the quality of a call by making an early measurement of the packet loss. From our particular data, we have shown that it is sufficient to make an estimation after four seconds. The analysis we have performed offers thresholds for call blocking probability and failure rates of the scheme.

From a practical point of view, the admission control scheme shown in the paper could be implemented using standard RTCP [11] receiver reports. A small adjustment of the rate that the receiver generates the reports would be enough for our probe-based admission control scheme.

One limitation with our method is that all calls in the experimental data are in fact admitted. The effects of dropping calls to the network as a whole has not been assessed. We claim that this observation is irrelevant in this study for two reasons: (1) all of the calls in the study were disjoint in time; (2) the effect could only be positive, thus our results can be seen as worst-case.

An interesting point is whether the results based on the measured data [10] are generally valid. This is a difficult question, and we cannot claim that the results hold for all network conditions. For example, one could claim differences in timescales (the measurements were made in 2001), networks (most data were made on academic networks), link technologies (no wireless access were available, etc). We hope that future work can help to get a larger understanding of such conditions.

References

1. S. Blake, D. Black, M. Carlson, E. Davies, Z. Wang, and W. Weiss, "An architecture for differentiated services," RFC 2475, IETF, December 1998.
2. J. H. Saltzer, D. P. Reed, and D. D. Clark, "End-to-end arguments in system design," *ACM Transactions on Computer Systems*, vol. 2, pp. 277–288, November 1984.
3. L. Breslau, E. W. Knightly, S. Shenker, I. Stoica, and H. Zhang, "Endpoint admission control: Architectural issues and performance," in *Computer Communication Review – Proc. of Sigcomm 2000*, vol. 30, (Stockholm, Sweden), pp. 57–69, ACM, August/September 2000.
4. G. Karlsson, "Providing quality for internet video services," in *Proc. of CNIT/IEEE ITWoDC 98*, (Ischia, Italy), pp. 133–146, September 1998.
5. V. Fodor (née Elek), G. Karlsson, and R. Rönngren, "Admission control based on end–to–end measurements," in *Proc. of the 19th Infocom*, (Tel Aviv, Israel), pp. 623–630, IEEE, March 2000.
6. I. Más Ivars and G. Karlsson, "PBAC: Probe–based admission control," in *Proc. of QoFIS 2001*, vol. 2156 of *LNCS*, (Coimbra, Portugal), pp. 97–109, Springer, September 2001.
7. I. Más, V. Fodor, and G. Karlsson, "The performance of endpoint admission control based on packet loss," in *Proc. of QoFIS 2003* [12].
8. O. Hagsand, I. Marsh, and K. Hansson, "Sicsophone: A low-delay internet telephony tool," in *Proc. of the 29th Euromicro Conference*, (Belek-Anatalya, Turkey), pp. 189–197, September 2003.
9. P. Biyani, O. Hagsand, G. Karlsson, I. Marsh, and I. Más, "Early estimation of voice over ip quality," in *Proc. of the 21st Nordunet network conference*, (Reykjavik, Iceland), August 2003.
10. I. Marsh, F. Li, and G. Karlsson, "Wide area measurements of voice over IP quality," in *Proc. of QoFIS 2003* [12].
11. H. Schulzrinne, S. Casner, R. Frederick, and V. Jacobson, "RTP: A transport protocol for real–time applications," RFC 1889, IETF, January 1996.
12. *Procedings of the 4th COST 263 International Workshop on Quality of Future Internet Services*, vol. 2856 of *LNCS*, (Stockholm, Sweden), Springer, October 2003.

Fair and Performance Guaranteed Methods
for Flat-Rate Unlimited Access Service Plan

Yeali S. Sun[1], Pei-Wen Chen[1], and Meng Chang Chen[2]

[1]Dept. of Information Management, National Taiwan University
sunny@im.ntu.edu.tw
[2]Institute of Information Science, Academia Sinica, Taiwan
mcc@iis.sinica.edu.tw

Abstract. Simplicity in administration and operation is the choice for production networks. Hence, flat-rate unlimited access service plan is the predominant form of retail pricing in Broadband Internet access services. However, this service plan can easily result in unfair resource sharing, abusive usage and poor performance. In this paper, we propose several fair and performance guaranteed methods to alleviate the problems. The working field trial project called Virtual Internet Pricing (VIP) project was deployed at the dormitory network of National Taiwan University with a total of 5355 users. In VIP, a quota-based priority control (QPC) scheme was proposed to resolve the problems. While it alleviated the problems, QPC however raised some issues such as chaotic periods, bandwidth stealing and weak performance guarantees for in-profile packets. Four methods were proposed to enhance the basic QPC scheme. The simulation results showed that the proposed methods significantly improved network performance and increased system stability.

1 Introduction

In the past, a number of Internet pricing models (such as [1,2,3,4]) have been proposed. The common argument is in favor of the model of charging users by actual usage as in many public utility services. Despite of the fact, today's Internet access services (e.g., dial-up, ADSL and cable modem) are mostly charged by flat rates probably for different access speeds or a fixed fee plus per unit time charge [5,6]. As pointed out in [6], the flat-rate unlimited-access service plan is economically inefficient. Because users do not face the true marginal cost of usage, it often results in over-usage, jeopardizing network performance. Studies also find that there is big usage difference of users between being served under the flat-rate unlimited access plan and being served under the usage-based charge plan. Hence, many service providers and network operators take the tactic of 50% or 70% threshold utilization as the rule of thumb in upgrading their resources/systems to avoid congestion.

We faced similar problems as do commercial networks in our university dormitory network. There were twelve dormitories with a total of 5355 students who paid a one-time network access fee for every semester. Each student was given a fixed IP address

N. Mitrou et al. (Eds.): NETWORKING 2004, LNCS 3042, pp. 392–404, 2004.

and a Fast Ethernet access to the Internet. The *unlimited* access service caused slow Internet access and frequent connection timeouts. After preliminary investigation and analysis, we found less than 10% of the total users contribute more than 90% of the daily traffic. When looking into individual user traffic, we found that this small group of *"heavy"* users had voluntarily contributed their computers to form various kinds of peer-to-peer networks for file downloading and content sharing. While large files were downloaded from these machines, largely 90% "regular" users in the dormitory network follow the typical internet access pattern.

Essentially, the problem we encountered is that, under the flat-rate unlimited service plan, users are not charged on the basis of how many packets are sent. Without control, user usage could become excessive and outrageous, causing severe network congestion, performance degradation and unfair resource sharing. Since network is not free, light users are indeed subsidizing heavy users. Similar problem also exist in today's intranets such as schools, companies, commercial buildings and residential community that share common links to the Internet. Those intranets are generally implemented with high-speed technology such as Gigabit/Fast Ethernet. But the bandwidth of the Internet access links are however much lower. Many of these network users experience congestion on the Internet access link. In some cases, the congestion period could last for almost entire day.

To address the poor Internet access performance and unfair resource sharing caused by excessive (selfish) use of a small number of users, we conducted an experimental field trial project called Virtual Internet Pricing (VIP) [7]. We consider performance incentive as an alternative to address the fairness and performance problems that result from the flat-rate unlimited access service plan. Although per-flow scheduling like Weighted Fair Queueing algorithm is widely considered as a good technique to enforce fairness and QoS guarantees in a link-sharing environment[8,9], not many network equipment implemented per-flow Quality of Service (QoS) due to its overhead. In a *Quota-based Priority Control* (QPC) system, each user is allowed to transmit no more than a maximum amount of high-priority traffic in each quota control period. (Note those packets are called *in-profile* packets.) We adopted QPC scheme to enforce per-user fairness and to relieve the congestion problem caused by a small group of selfish heavy users.

The QPC scheme however raises several interesting problems including chaotic periods, bandwidth stealing and weak performance guarantees for in-profile packets. When many users are backlogged at the beginning of the quota control period, undesirable congestion often occurs as soon as the period begins. Packet loss periods could last longer than thirty minutes with a loss rate of more than 10%. This phenomenon is referred to as the *chaotic period*.

Traffic metering is typically implemented in routers and the accounting is performed at the backend systems [11]. Accounting systems periodically collect user usage data from the metering routers. The accounting data collection interval is an important system parameter in quota control. To minimize the overhead, accounting interval is often set to the value much greater than packet transmission time (e.g., every 10 minutes). This way however makes the *bandwidth-stealing* possible. Because per-user account balance is checked per accounting interval, during the period a user may over use or steal the bandwidth. It directly affects the effectiveness of the QPC scheme in enforcing fairness and the performance guarantees to in-profile packets.

In QPC there are ways to overrun a bottleneck resource. For example, a user with a full quota account balance can legally introduce extremely bursty in-profile packets into the network within a short period of time. This makes the provision of performance guarantees to in-profile packets quite challenging. We will examine and discuss each of these problems in detail later in this paper.

1.1 Related Work on Internet Pricing

Works have been proposed to devise optimal pricing policy for optimal social welfare. Achieving this, a marginal congestion cost is charged. Congestion costs are the performance penalties incurred from imposing one user's traffic on other users. There are many ways to deal with congestion externalities, such as to establish social norms, to establish a rationing or quota system or to develop a pricing mechanism. In the literature, a group of researchers especially economists prefer using pricing to manage user behavior in dealing with the problems of network resource scarceness and congestion. The advantage of this approach is that one can effectively control network traffic as well as achieve economic profits. A good example is Shadow price [3]. However, there are three possible issues as pointed out in [4]. First, marginal congestion cost-based pricing may not produce sufficient revenue to fully recover actual costs. Second, congestion costs are difficult to characterize and obtain from the network, and therefore cannot reliably form the basis for pricing. Third, there are other more structural goals besides optimality.

Other works on pricing include smart market [2] and edge pricing [4]. In edge pricing, true congestion costs are approximated by replacing actual congestion conditions and the cost of actual paths with expected congestion conditions and the cost of expected paths. Under this model, charges depend only on source and destination pairs and therefore can be determined and assessed locally at the access point rather than computed in a distributed fashion along the entire path.

2 The VIP (Virtual Internet Pricing) Project

In addition to the unfairness and congestion problems, the university campus network administrators also face the problem that the total traffic emitted from the dormitory network constitutes more than 50% of the total traffic in then the ATM 155Mbps campus backbone. Initially, the university network administrators decided to impose a maximum rate of 54Mbps on the traffic from the dormitory networks. Unfortunately, the uplink became even more congested. As a result, the VIP project was initiated and an experimental field trial of virtual internet pricing was conducted in the NTU dormitory network to solve the problem. The traffic-metering device is a Cisco router in which Netflow collects IP usage data. The QoS router is a home-made Linux-based device implementing priority scheduling and IP packet classification. The meter-reading server performs usage data analysis. Per-user usage accounting, service charging are implemented on a server machine which is also responsible for sending traffic control commands to the QoS router to real-time configure individual user's priority levels in accordance with the account balance.

In the Virtual Internet Pricing (VIP) project, the dormitory network severed 5355 users. For the purpose of administrative and operational simplicity, the quota-based control combined with priority scheduling was used to achieve per-user fairness under the flat-rate unlimited service plan. There are two services: the *Regular* service and the *Custody* service, corresponding to the high and low priority queues, respectively. The default service is high-priority regular service. Each user is given an *account* of *virtual network dollars* (abbreviated as Net$). At the beginning of a *quota control period*, each account is credited with a fixed amount of Net$. Traffic sent through different priority queues is charged at different rates (e.g., Net$10/Mbits for high priority and no charge for low priority). During the period, the system records the amount of traffic (in bytes) each user transmits over the uplink and accordingly deducts service charges from the account. When a user's account balance becomes zero, the user is classified as *under custody* – subsequent traffic from the user is marked as low priority until the current quota control period ends.

There are two performance goals. The first goal is to satisfy the bandwidth demands of the majority users under max-min fairness. Assume users are numbered from 1 to N according to the increasing order of their traffic demands. To satisfy the demands of the first s of the total user population, the quota is set to be equal to or more than the demand of the sth user. Note that this quota assignment guarantees the minimum bandwidth to every user. During a quota control period, in-profile traffic is transmitted as high priority. Excess traffic will be marked as out-profile packets, served by using the remaining bandwidth in the low-priority queue.

The second goal is to guarantee a maximum average packet loss rate for in-profile packets in each quota control period. Given the user population and bottleneck link capacity, the performance of the high priority queue directly relates to the quota assignment. The larger the quota, the more the traffic ranked as high priority, and possibly, the greater packet loss in the queue. If, however, the quota amount is too small, most users' basic traffic demands will not be met.

3 Problems in Quota-Based Priority Control

Three problems – chaotic periods, bandwidth stealing and weak performance guarantees to in-profile packets obstructed the two performance goals.

A. Chaotic Periods

In the field trial, congestion occurred almost at the beginning of every quota control period. Because every user's account was deposited with full credits at the beginning of a control period, suddenly bursts of packets were injected into the high-priority queue and the queue quickly built up. The situation usually continued for half an hour and resulted in severe loss of in-profile packets until heavy users gradually used up quota. This congestion period is referred to as the chaotic period. We presented the simulation results in Figure 1 o illustrate the traffic load and packet loss rate dynamics during a chaotic period. In the simulation, the link capacity is 6Mbps, the quota amount is 6Mbits, the quota control period is 60 seconds, the accounting interval is per-packet, and there are 100 users. In the Figure, the chaotic period lasted for about 13 seconds until a sufficient number of users used up quotas and their packets were thereafter directed to the low-priority queue. After then, the high priority queue becomes stable and the performance of the in-profile packets was no longer affected by the out-profile packets in the low-priority queue. Besides quota-based control can achieve long-term fairness, it leaves users great flexibility in sending packets. Even we interleave user with different starting time of quota control period, chaotic period might still happens when some users transmit in-profile packets at very high transmission rate. In practice, due to the dynamics of user demands and behavior, it is difficult to predict when a chaotic period will take place and how long it will last. It therefore poses great challenges in resource planning to support performance guarantees such as maximum packet loss rate to in-profile packets under the basic QPC scheme.

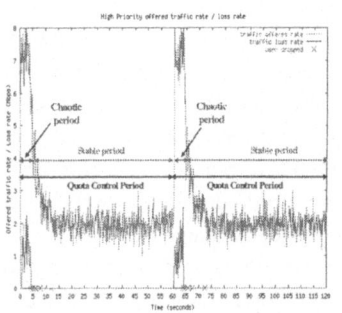 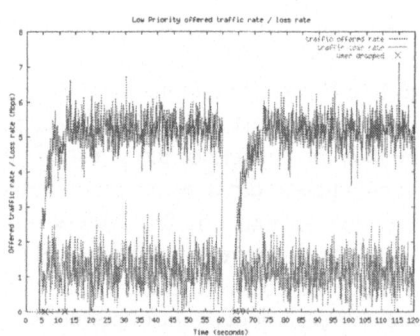

(a) Regular service (high priority queue).   (b) Custody service (low priority queue).

Fig. 1. Severe packet loss in the high-priority queue during a chaotic period.

B. Bandwidth Stealing

By taking into consideration the operational issues, such as feasibility, complexity and overhead, of this large-scale field trial, we decided to perform traffic data collection and usage accounting every 10 minutes. During the accounting interval, a user may generate more in-profile packets than allowed, i.e. over-use. Such *bandwidth-stealing* from heavy users happened in the field trial. It unfortunately worsened the congestion situation in the chaotic periods. It also degraded fairness in the quota-based control. A possible amendment is to reduce accounting interval with a penalty of extra computation resources needed.

C. Weak Performance Guarantees to In-Profile Packets

In the trials, through the use of smaller quota control period and smaller quota amount, the congestion problem in the NTU dormitory network was resolved and the average packet loss rate in the high priority queue was bounded during a quota control period. However, if any chaotic period took place, the packet loss rate in the chaotic period would be much higher than that in the stable period. To support more consistent performance guarantees, it is essential to control the duration of chaotic period.

We did not consider *credit-carry-over* in the trial (i.e. unused quota not carried over to the next period). The reason is that it would make the traffic demand in each control period even more uncertain and unpredictable. It becomes difficult to pre-plan and allocate sufficient resources to support performance guarantees to in-profile packets. In Section 6, we use simulations to show the relationship between the important operational parameters of the QPC scheme to provide performance guarantees.

4 Enhanced QPC Schemes

In view of the above-mentioned problems encountered in the field trials using the basic quota-based priority control (QPC) scheme, in this section we propose two different combinations of quota control and priority scheduling methods that achieve better fairness and the support of performance guarantees to in-profile packets using flat-rate unlimited access service plan.

A. Multiple Priority Levels

In a quota control system, the occurrence of chaotic periods is unavoidable. To shorten the duration of a chaotic period, we propose to divide per-user quota allotment into multiple partitions assigned to different priority levels. Each user starts from the highest priority level. As amount of traffic increases, a user is downgraded to one level lower. Depending on the quota amounts assignment to different priority levels, this method can quickly sort users into different usage groups to minimize the performance impact by heavy users to light users.

In *Uniform Quota Assignment* method, assume there are K+1 priority levels (K>1) and the lowest priority level is the best-effort service. The total quota amount Q is equally divided and assigned to the first K priority levels, i.e. each level has Q/K quota allotment. The best-effort service has no quota constraint. All users start service at the highest priority level. As usage accumulates, heavy users will move from the highest priority level to the lowest. The more priority levels, the better in separating different usage groups. As a result, the duration of chaotic periods at each level will be shortened, thus guaranteeing better performance to in-profile packets from light users and fairer sharing of resources. As the number of priority levels increases, the method approximates processor sharing-based scheduling [8].

In *Load-based Quota Assignment* method, we will take into account the distribution of user traffic demands in quota assignment to different priority levels. Consider that the total quota amount is 90 Mbits and there are four priority levels. If a uniform quota assignment is used, each of the first three priority level will have quota allotment of 30Mbits. Suppose regular users only have average 10Mbits usage, which is much smaller than the allocated quota for the highest priority level, the negative impact from heavy users remains severe. To address the problem, we propose to allocate quota amounts to meet regular users' demand, for instance 10Mbits for level 1, 15Mbits for level 2 and 35Mbits for level 3.

B. QoS Options

By having multiple priority levels, we are able to reduce the duration of and packet loss in chaotic periods by quickly differentiating heavy users from light users. However for users who only occasionally use the Internet still have a non-zero possibility of encountering a chaotic period. To address this issue, we propose best-fit and on-demand QoS-option service models.

1) Best-Fit QoS-Option Service Model

In the best-fit QoS-option service model, individual user submits an estimate of his/her expected traffic demand to the service provider before each quota control period begins. Consider K+1 service levels. The quality of service of level 1 is better than that of level 2, and so on. Assume non-increasing quota allotments, i.e. $Q_1 \leq Q_2 \leq ... \leq Q_{K+1}$, and non-decreasing charge rates, i.e. $p_1 \geq p_2 \geq ... \geq p_K$. For service level K+1, there is no charge because it is the best-effort service. The virtual Net\$ allotment is fixed and the same for each priority level denoted as M, i.e. $p_k \cdot Q_k = \text{M (Net\$)}, \forall k$. Each user i chooses a service level that best matches his/her expected offered load w_i, i.e. $Q_k \leq w_i$.

Once determined, a user's account is credited with Net\$ M. During a quota control period, when a user's account balance becomes zero, his/her access would be immediately downgraded to the lowest best-effort service.

In this method, users must follow the rule and choose a service level best for their needs to avoid a performance penalty. If a user cheats by giving a smaller than expected offered load and starting with a higher priority level, because the higher priority level has smaller quota allotment, this user will quickly use up the virtual money and be moved to the lowest best-effort queue, possibly experiencing poor performance for the rest of the control period. Directing overloading traffic to the best-effort queue when with empty account can be considered as a penalty to users for possible cheating. On the other hand, it is possible that a user may unintentionally underestimate the demand. This method can encourage users to accurately estimate their usage in order to receive good performance.

The design rationale of this method is to motivate users to better estimate their offered load in each quota control period. Since all users pay the same amount of service fee, if users want to receive better QoS, they must reduce their traffic demands to have higher priority. For users with large-usage demands, they will be served with lower priority to avoid penalty. Under this service model, light users will receive better performance for less use than heavy users. Moreover, the service provider is able to get more information about network load from user selections. This can greatly aid network capacity planning and traffic control. By proper choice of priority levels and quota allotments, performance guarantees to all levels except the best-effort service is possible. If all users are able to estimate well, the system will achieve good performance and fair resource sharing. In Section 5, simulation results show that users with wrong choice may experience 30% more packet loss.

2) On-Demand QoS-Option Service Model

In the best-fit service models, users are required to make good estimates about their access demands. However, some users may not be able to accurately forecast their requirements and possible incidental needs for higher quality of service. In [6], the authors reported that in their experiments, although the majority of users chose flat-rate unlimited-access service plans, almost every user purchased high quality service at least once. The objective of the on-demand QoS-option service model is to complement flat-rate service by allowing users to pay an extra fee in order to choose the level of priority service that best meets their QoS requirements. This approach is simple and more predictable than prior researches such as IntServ and optimal pricing policy [3,4]. Accordingly, network service providers may need to allocate extra resources and/or reconfigure the network to accommodate additional usages during a quota control period.

5 Performance Analysis

There are four important parameters in the basic QPC model in achieving fairness and performance guarantees: bottleneck link capacity, quota amount (in bytes), accounting interval (in seconds) and quota control period. In this section, we use Ns-2 [12] simulations to study the choices and relationship between these four parameters. Consider an environment with 100 users: 95 regular users and 5 heavy users. Packet arrivals of all users are Poisson processes with different rates. For the regular user group, the mean arrival rate is 20Kbps with 10Kbps standard deviation. For the heavy user group, the mean arrival rate is 1Mbps with 0.5Mbps standard deviation.

Given a six-hour quota control period and a per-packet accounting interval, Figure 2 compares the amount of successfully transmitted packets with and without a quota control. Users are indexed in the order of their offered loads. Each experiment is run thirty times, and the mean and standard deviation are taken. With a quota control, the throughput performance is significantly enhanced. The throughout line coincides with the offered load curve. For regular users, such performance improvement is owing to the effective reduction of packet loss in chaotic periods. This simulation result confirms the results in the field trials.

In practice, system administrators would like to do usage accounting as less frequently as possible. Figure 3 shows the packet loss rate in the high priority level under different accounting intervals for different quota assignments. The performance guarantees to in-profile packets is to maintain target maximum packet loss rate for the high priority class is 0.01. The packet loss rate performance curves exhibit a staircase shape when increasing the quota allotment. For each curve, the initial increase is due to packet loss occurring in a chaotic period that usually lasts longer for a larger accounting interval, resulting in greater packet losses. In this example, to achieve the target packet loss rate, the accounting interval cannot be longer than 20 minutes for quota less than 1800Mbits per a 6 hour quota control period

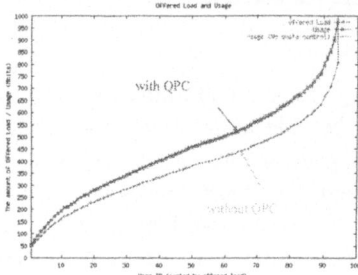

Fig. 2. Comparison of the throughputs with and without QPC scheme

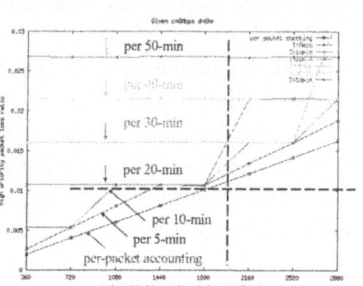

Fig. 3. Packet loss rate vs. quota assignment.

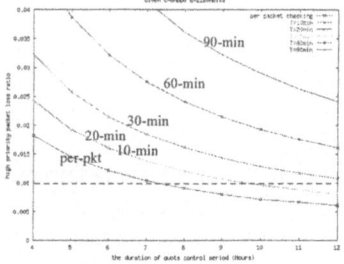

Fig. 4. Average packet loss rate of the high priority queue.

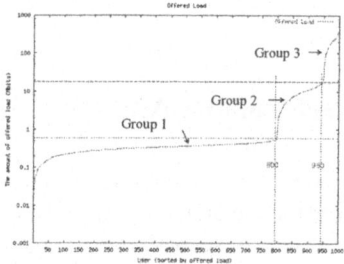

Fig. 5. The user traffic demand distribution.

In this experiment, we are interested to know the performance results under different quota control periods. Figure 4 shows the average packet loss rate for different durations of a quota control period. Because of high packet loss in the chaotic period that occurs at the beginning of a quota control period, a long enough quota control period is needed to average out the packet loss rate. The figure also shows that a shorter accounting interval results in better average packet loss rate.

Table 1 presents the comparison of average packet loss rate of each user group under different number of priority levels. Here, the bottleneck link capacity is 15Mbps, the total quota assignment is 14Mbits, the quota control period duration is 600 seconds and all assuming per-packet accounting interval. Compared with one single service level (i.e. only best-effort service), the packet loss performance of user group 1 is significantly reduced from 20.865% to 0.901% when having two priority levels. This again confirms the results in our field trial. As more levels of control or priority are added to the system, the packet loss rate performance is further improved. As long as a user stays in-profile, a good packet loss performance is guaranteed. For heavy users, packet loss rate becomes very high when entering the best-effort queue.

Table 1. Comparison of average packet loss rate for different user groups.

# of Priority Levels	User Group 1	User Group 2	User Group 3
Best Effort	20.865%	20.810%	19.496%
2	0.901%	1.622%	22.807%
3	0.438%	1.210%	22.879%
4	0.232%	1.062%	22.906%
5	0.135%	0.970%	22.922%
6	0.100%	0.924%	22.929%

Table 2. Performance comparison of two- and three-level of priority.

Quota Assignment Policy: Total Priority Levels	Priority Level	Duration of Chaotic Period (sec)	# of Remaining Users	Packet Loss Rate
Demand-based Quota Assign.: 3 levels	1	0.0 ~ 1.7	800	1.1 x10-5
	2	1.1 ~ 44.6	150	2.83x10-3
	3	N/A	50	0.230
Uniform Quota Assign.: 3 levels	1	0.0 ~ 14.9	849	4.38 x10-4
	2	11.2 ~ 28.2	82	1.21x10-2
	3	N/A	69	0.229
Uniform Quota Assign.: 2 levels	1	0.0 ~ 28.2	931	2.523x10-2
	2	N/A	69	0.228

In the previous experiment, 849 users use less than 7Mbits and remain in priority level 1 after the system becomes stable. Among them, 800 of them use less than 800kbits in total. The variance of usage within a priority level is quite large. To minimize the usage variance with a user group, quota allotment to different priority levels is proposed based on the demand distribution. 0.6Mbits and 19.4Mbps are taken as the quota allotments to priority level 1 and 2, respectively for the demand-based quota assignment. Performance results of each priority level are shown Table 2, compared with the results in two and three priority levels using uniform quota assignment. The demand-based quota assignment approach has the best result in guaranteeing packet loss rate performance to in-profile packets.

In the previous experiments, packet re-classification is based on user account balance. In the following experiments, we consider the Best-fit QoS-option service model with three priority levels. Let q_1=20Mbits, q_2=50Mbits, q_3=infinite and N=1000 users. The traffic demand distribution is shown in Figure 5. The quota control period is 600 seconds. Two experiments are conducted. In the first experiment, all users are assumed to make the right selection. There were 950 users choose service level 1 and the others choose service level 2. The throughput performance of each service level is shown in Figure 6. The x's at the x-axis denote the time instants at which users are reclassified to lower service levels. In this scenario, the traffic demands of the first 95% users are fully supported and there is no packet loss. The remaining 5% users received 0.2% to 30% packet loss depending on their traffic demands.

In the second experiment, assume a heavy user cheats – the 975th user purposely chooses service level 1 instead of level 2. As shown in Figure 7, before time 63 seconds, the throughput of level 1 is a bit higher than the case of no cheating as shown in Figure 6. At 63 seconds, the cheating user is moved to level 3. Thereafter the performance of level 1 remains stable. For the cheating user, his/her traffic will initially receive good performance but afterwards the performance will be very bad. If the user had not cheated, the average packet loss rate would have been around 22.1% instead of 29.8%. This is the performance penalty for the user. In summary, if all users properly choose their service levels, the overall performance for all users will be good.

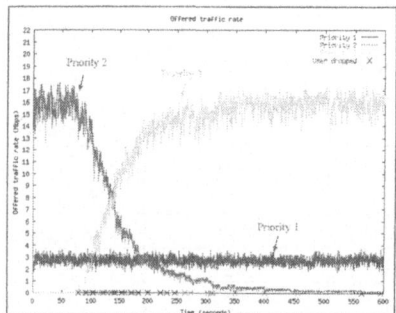

Fig. 6. Throughput performance using the Best-fit QoS-option service model assuming all users correctly select service level.

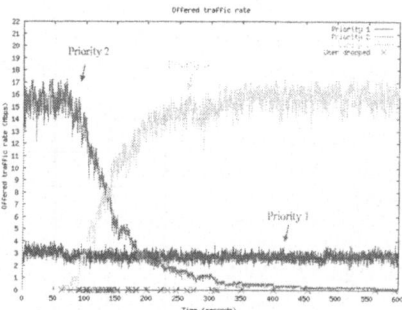

Fig. 7. Throughput performance when a heavy user cheats by choosing a higher service level.

6   Conclusion

The main appeal of the flat rate plan is its simplicity as it reduces risk and administrative costs. The unfortunate consequences of flat rate are congestion and no support of performance guarantees. Quality of service cannot be provided to those willing to pay for it. Furthermore, in largely best-effort service based Internet, excessive and abusive use of shared resources by selfish users may severely impair network performance and fairness between users. However, if the consumers regard the pricing structure as too complicated and service providers regard the implementation and administrative costs to be too high, flat rate plans will be preferred.

For the sake of operational simplicity and manageability, instead of using a complex per-flow scheduling approach, in the VIP project we combine a per-user quota control with a priority scheduling scheme. Measurement and performance results from the experiments demonstrate the benefits of this simple scheme. We discuss the problems of chaotic periods, bandwidth stealing, weak support of bandwidth guarantees and the difficulty of using quota incentives to change user network access behavior in quota-based control.

Several methods combining multiple priority levels with quota assignment are proposed to achieve fairer resource sharing, congestion control and support of performance guarantees to in-profile packets. By taking into account user traffic demand distribution in quota assignment to different priority levels, the method can quickly sort users into different usage groups, reducing the duration of chaotic periods and minimizing the performance impact from heavy users on light users. Better fairness and performance are achieved for the popular flat-rate unlimited access plan. Two QoS-option service models are also proposed. In the best-fit model, if all users can properly select the service levels best for their needs, good performance can be guaranteed to all users. This method motivates users to better estimate their offered load in each quota control period. Since all users pay the same amount of service fee, if users want to receive better QoS, they must reduce their traffic demands to have higher priority. For users with large-usage demands, they will be served with lower priority to avoid penalty. In the on-demand QoS-option service model, users can choose their level of priority service to meet their QoS requirements at any time. User can credit the account at any time to receive QoS service charges.

Simulation results of the proposed methods are presented to show the benefits of the simple quota control priority scheduling in resolving fairness, congestion and performance guarantees issues for service networks using flat-rate unlimited access service plan. It successfully provides a more predictive, affirmative service guarantees to the service users.

References

1. R. Cocchi, D. Estrin, S. Shenker and L. Zhang, "Pricing in Computer Networks: Motivation, Formulation, and Example", IEEE/ACM Transactions on Networking, 1993.
2. J. Mackie-Mason, H. Varian, "Pricing Congestible Network Resources", IEEE Journal on Selected Areas in Communications, September 1995
3. S. Shenker, D. D. Clark, D. Estrin, and S. Herzog, "Pricing in Computer Networks: Reshaping the Research Agenda", ACM Computer Communication Review 1996.
4. F. Kelly, "Charging and rate control for elastic traffic," European Transactions on Telecommunications, Volume 8, 1997, pp 33-37.
5. M.D. Biddiscombe, J.E. Midwinter and S. Sabesan, "Application of free-market principles to telecoms resource allocation," Electronics Letters, Vol. 35 No. 4, February 1999.
6. J. Altmann and K. Chu, "A Proposal for a Flexible Service Plan that is Attractive to Users and Internet Service Providers", INFOCOM 2001.

7. T. Lin, Y. Sun, S. Chang, S. Chu, Yi-Ting Chou and Mei-Wen Li, "Priority-Based Internet Access Control for Fairness Improvement and Abuse Reduction," *2nd International workshop on QoS in Multiservice IP Networks* (QoS-IP 2003), Italy, 2003.
8. A. K. Parekh and R. G. Gallager, "A Generalized Processor Sharing Approach to Flow Control in Integrated Services Networks: The Single-Node Case", IEEE/ACM Transactions on Networking, Vol. 1, No. 3, pp.344-357, June 1993.
9. S. Floyd and V. Jacobson, "Link-sharing and resource management models for packet networks," *IEEE/ACM* Trans. Networking, vol. 3 pp. 365-386, Aug. 1995.
10. S. Blake, et al., "An Architecture for Differentiated Services," IETF RFC2475, December 1998.
11. D. Estrin , and L. Zhang "Design Considerations for Usage Accounting and Feedback in Internetworks", ACM Comp. Commun. Rev., Vol. 20. No. 5. pp. 56-66, October 1990.
12. The Network Simulator, http://www.isi.edu/nsnam/ns/

Bounds on Benefits and Harms of Adding Connections to Noncooperative Networks

Hisao Kameda

University of Tsukuba, Tsukuba Science City 305-8573, Japan
kameda@is.tsukuba.ac.jp
http://www.osdp.is.tsukuba.ac.jp/~hkameda/

Abstract. In computer networks (and, say, transportation networks), we can consider the situation where each user has its own routing decision so as to minimize noncooperatively the expected passage time of its packet/job given the routing decisions of other users. Intuitively, it is anticipated that adding connections to such a noncooperative network may bring benefits at least to some users. The Braess paradox is, however, the first example of paradoxical cases where it is not always the case. This paper studies the bounds on the degrees of coincident cost improvement (benefits) and degradation (harms) for all users by adding connections to noncooperative networks. For Wardrop networks (noncooperative networks with infinitesimal users), the degree of benefits for all users can increase without bound by adding connections whereas no Wardrop network has been found for which the degree of harms can increase without bound for all users. In contrast, for Nash networks (noncooperative networks with a finite number of users), the degrees of both benefits and harms can increase without bound for all users. On the other hand, we see that, for some category of Wardrop networks, adding connections to them can bring neither benefits nor harms to all users, and that, for some homogeneous networks, adding connections to them can never bring benefits to all users under any static policy including cooperative and noncooperative ones.

Keywords: Braess paradox, Wardrop equilibrium, Nash equilibrium, coincident cost degradation and improvement, computer and communication networks, distributed computer systems, network routing, shortest path routing, source routing, static load balancing.

1 Introduction

There exist networks and systems wherein a number of independent users share and compete for resources. For example, communication networks like the Internet consist of a finite number of links, and of flows of packets each of which is issued by an independent user and flows through the links. Distributed computer systems like GRIDs [1] consist of computers widely distributed and interconnected by communication networks, and of flows of jobs each of which is issued by an independent user.

N. Mitrou et al. (Eds.): NETWORKING 2004, LNCS 3042, pp. 405–417, 2004.

We can consider communication networks where each infinitesimal user has its own routing decision so as to minimize the expected passage time of its packet from the origin and the destination given the routing decisions by other users. The situation where every infinitesimal user has attained its optimization given the decision of other users and would not unilaterally deviate its routing decision is called a *Wardrop equilibrium*, a Nash equilibrium with infinitesimal players. We call here such a noncooperative network in a Wardrop equilibrium a *Wardrop networks*. In computer networking, some shortest path routing protocols may bring about situations close to Wardrop equilibria. It would be anticipated that users' benefits would increase by adding connections to a network, and thus increasing the degree of freedom in choices to each user. This is not always the case, however, as first exemplified in the Braess paradox [2].

The famous Braess paradox shows that adding connections (links) to a network may sometimes degrade the cost for all users in a Wardrop equilibrium. The Braess paradox attracted the attention of many researchers and a lot of work has been accumulated, *e.g.*, [3, 4, 5, 6, 7, 8, 9, 10, 11, 12, 13, 14, 15, 16]. Coincident cost degradation for all users by adding connections to a network looks paradoxical, and is called the *paradox*. It seems that, since its discovery, the paradox has been studied mostly with respect to Wardrop equilibria. It also seems that the topologies of concrete networks examined with respect to the paradox have been similar to or centered on that of the Braess network.

On the other hand, we can consider networks where users are classified into a small number of groups, each of which optimizes its cost non-cooperatively. The situation where, in such a scheme, every user has attained its optimization given the decision of other users and would not unilaterally deviate its decision is called a *Nash equilibrium*. We call here such a noncooperative network in a Nash equilibrium a *Nash network*. In computer networking, some source routing protocols may bring about situations close to Nash equilibria. Examples of paradoxes similar to Braess's in a Nash equilibrium have been found not only for networks of topologies similar to the Braess one [17, 18], but also for a network of another topology [19].

It seems that, except [20, 21], few studies have emphasized the estimation of how harmful the paradox can be, *i.e.*, the worst-case degree of coincident cost degradation for all users by adding connections to a noncooperative network. Moreover, we have not seen the estimation of how beneficial the addition of connections to a noncooperative network can be, *i.e.*, the best-case degree of coincident cost improvement by adding connections to a noncooperative network.

We consider networks with fixed numbers of vertices. It has been shown that the degree of coincident cost degradation by adding connections to Wardrop networks with one origin-destination pair is bounded [20, 21]. On the other hand, it has been shown that the degree of coincident cost degradation by adding connections to Nash networks can increase without bound [22]. The present paper addresses the estimation of the best-case degrees of coincident cost improvement for all users by adding connections to Wardrop and Nash networks. We show that the degree of coincident cost improvement (benefits) by adding connections to both Wardrop and Nash networks can increase without bound. In contrast, for homogeneous distributed systems, no coincident cost improvement occurs by adding connections under any static load balancing policy including cooperative and noncooperative ones. Furthermore, in Wardrop equilibia for any distributed system, neither coincident cost degradation nor coincident cost improvement occurs by adding connections to the system.

Note, in passing, that each user of a network may have decisions on flow control, in addition to routing. In wireless networks, users may have power control. The concept of a Nash equilibrium is also discussed in flow control [23, 24, 25, 26, 27, 28] and in power control [29, 30, 31, 32, 33], both of which are not addressed in this paper.

The outline of this paper is as follows. Section 2 presents the concepts around Pareto superiority, Braess-like paradoxes, and their degrees of magnitude. Section 3 gives the description of the Wardrop networks investigated, and presents the results. On the other hand, Section 4 examines networks with a topology different from that of the Wardrop networks, in particular, with that of distributed systems. Section 5 concludes this paper.

2   A Measure Showing the Degree of the Paradox

This paper uses a single scalar measure that shows the degree of Pareto superiority of a system state before adding connections to that after doing so, or the degree of paradoxes in Nash equilibria.

The Pareto superiority is defined as follows. Consider a system consisting of n users (or players, decision makers), $1, 2, \ldots, n$. User i has its cost $C_i(S)$, in the system state S. Denote by S^a and S^b two different states of the system. $k_i \triangleq C_i(S^a)/C_i(S^b)$. S^b is *Pareto superior* to S^a iff $k_i > 1$ for some i and $k_j \geq 1$ for all other j. In particular, we say that S^b is *strongly Pareto superior* to S^a iff $k_i > 1$ for all i. A state to which some other state is Pareto superior is *Pareto inefficient*. Thus, the Pareto superiority depends on the vector $(k_1, k_2, \ldots, k_n)$. It may, however, be convenient to express the degree of Pareto superiority, using a single scalar measure. It is required that the measure should clearly reflect Pareto superiority. If $k_{\min} > 1$, the state S^b is (strongly) Pareto superior to S^a, and if $k_{\min} < 1$, the state S^b is Pareto indifferent or superior to S^a. Thus, the measure $k_{\min}$ may be used as a primary measure of Pareto superiority. In contrast, for example, a measure based on a certain average or on a product of all k_i cannot satisfy the above requirement, but may be used as a secondary measure for tie-breaking the case where $k_{\min} = 1$.

It would be anticipated that in a system state where each user has more freedom of choice than in another state, at least one user should enjoy higher utility than in the latter state. But, as the famous *Braess paradox* shows, it is not always the case for noncooperative systems. Thus, a Nash equilibrium of a system with less freedom may be *Pareto superior* to that with more freedom, which is called *paradox*. We, therefore, use the measure of Pareto superiority as the measure of the degree of the paradox, *i.e.*, the coincident cost degradation for all users by adding connections. The measure of the degree of coincident cost improvement for all users after due by adding connections can be considered in a similar way.

3   Wardrop Networks

3.1   Assumptions on Wardrop Networks

Wardrop networks considered here consist of one origin and one destination and some relay nodes, some pairs of which a one-way link connects. One of the simplest networks is the general Braess network (Fig. 1) discussed later. There are a number of paths each of which connects the origin and the destination through a different series of links. The cost of a path is the sum of the cost of each link in the path. Infinitely many infinitesimal users send their packets through the network. Each user chooses a path of the minimum cost. The choice of a single infinitesimal user has only a negligible impact on the cost of each link. The situation where no user can reduce his/her cost by unilaterally choosing another path is a Wardrop equilibrium, an infinitesimal-user version of a Nash equilibrium.

It is assumed that the cost of each link is a non-decreasing function of the total flow, $i.e.$, the rate of packets through the link. In a Wardrop equilibrium of the networks with only one pair of origin and destination, the costs of all users are identical. C_o and C_c, respectively, denote the costs of users of a Wardrop network before and after adding connections to the network. Define $k = C_c/C_o$. Then, k expresses the degree of cost change for all users by adding the connections. k_{min} of Section 2 reduces to k here.

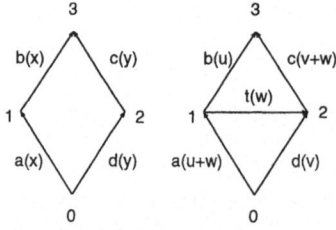

Fig. 1. General Braess network. **Left**: The network before link 1-2 is added. **Right**: The network after link 1-2 is added.

The Braess network consists of four nodes: one origin, one destination, and two relay nodes (Fig. 1). Each user flows through one of the paths. Before adding connections, the network has two paths, 0-1-3 and 0-2-3, each of which contains two links, the origin to one relay (0-1 or 0-2) and the relay to the destination (1-3 or 2-3), respectively. After adding connections, $i.e.$, a one-way link connecting two relays (1-2), the network has three paths including the new path (0-1-2-3) connecting the origin, one relay, the other relay, and the destination. Each user flows through one of the paths. In the original Braess network, the cost of each link is a linear function of the amount of the flow through the link [2].

This paper also considers the networks, called *general Braess networks*, that have nonlinear link cost functions. If we denote by η the flow of each link, the costs of links 0-1, 1-3, 2-3, 0-2, and 1-2 are, respectively, denoted by $a(\eta)$, $b(\eta)$, $c(\eta)$, $d(\eta)$, and $e(\eta)$. The total flow through the network is denoted by X. Denote the amounts of the flows through paths 0-1-3 and 0-2-3, respectively, by x and y before adding link 1-2. Denote the amounts of the flows through paths 0-1-3, 0-2-3, and 0-1-2-3, respectively, by u, v, and w after adding link 1-2. Denote by X the total flow, and, thus, $x+y = X = u+v+w$. The cost of links 0-1, 1-3, 2-3, and 0-2, are, respectively, $a(x)$, $b(x)$, $c(y)$, and $d(y)$ before adding link 1-2. The cost of links 0-1, 1-3, 2-3, 0-2, and 1-2 are, respectively, $a(u+v)$, $b(u)$, $c(v+w)$, $d(v)$, and $t(w)$ after adding the link.

For the original Braess network, $a(\eta) = c(\eta) = 10\eta$, $b(\eta) = d(\eta) = \eta + 50$, $t(\eta) = \eta + 10$, and $X = 6$, which result in $C_o = 83$ and $C_c = 92$, and thus $k = C_c/C_o = 1.1084\ldots$ [2]. Recall that k shows the degree of cost change by adding link 1-2. $k > 1$ means coincident cost degradation. In the above case, it is about 11 % degradation.

By *general Cohen-Kelly networks*, we mean a subset of general Braess networks for which the costs of links 0-1 and 2-3, are, respectively, $a(\eta) = \alpha/(a-\eta)$ and $c(\eta) = \alpha/(a-\eta)$ for $0 \le \eta < a$ and for which the costs of links 1-3 and 0-2, are, respectively, $b(\eta) = d(\eta) = b$ for any $\eta > 0$ with $t(\eta) = t$ (constant). Cohen and Kelly [12] considered a network of this type for which $\alpha = 1$, $a = \phi$, $b = 2$, $t = 1$, and $X = 2\lambda$. They showed that $C_o = 1/(\phi - \lambda) + 2 < 3 = C_c$, i.e., $1 < k < 3/2$, assuming that $2\lambda > \phi - 1 > \lambda > 0$, which is a paradox. In the above case, it is less than 50 % degradation.

As a general result on the Braess networks, it has been shown that the degree, k, of coincident cost degradation is bounded to 2 for the general Braess networks for which $a(\cdot)$ and $c(\cdot)$ are increasing and $b(\cdot)$, $d(\cdot)$, and $t(\cdot)$ are non-decreasing [20]. Furthermore, as a general result on the Wardrop networks, it has been shown that k is bounded to $\lfloor n/2 \rfloor$ for Wardrop networks that consist of n vertices and have link costs each of which is a nondecreasing function of the flow through the link [21].

We define more special Braess networks that will be used in the next section.

- [*Model Cohen-Kelly network*] Networks that are the same as general Cohen-Kelly networks with and $b \ge \alpha/(a - X) + t$. Thus, $u = v = 0$ and $w = X$. Then, the following relations hold.

$$C_o = \frac{\alpha}{a - X/2} + b, \tag{1}$$

$$C_c = \frac{2\alpha}{a - X} + t. \tag{2}$$

We consider the case of $t = 0$, in particular. Then, model Cohen-Kelly networks are described by the values of parameters α, a, b, and X that satisfy $0 < X < a$ and $0 < \alpha/(a - X) \leq b$.

- [*Reduced Cohen-Kelly network*] A subset of model Cohen-Kelly networks with $b = \alpha/(a - X) + t$. Thus, the following relation holds for C_c while C_o is given by (1).

$$C_c = \frac{2\alpha}{a - X} + t = 2b - t = \frac{\alpha}{a - X} + b. \tag{3}$$

We also consider the case of $t = 0$, in particular.

3.2 The Results

Proposition 1. *For every value of k, s.t. $0 < k < 2$, there exist model Cohen-Kelly networks for which the measure k is that value.*

[**Proof**] The outline of the proof is given as follows. For the details, see [34].

Step 1) shows that k depends only on ρ $(= X/a)$ and Z $(= b(a - X)/\alpha - 1)$, i.e., $k = k(\rho, Z)$.

Step 2) sees that $k = k(\rho, 0)$ monotonically increases in ρ with the range $1 < k < 2$ and the domain $0 < \rho < 1$. Thus, given k, s.t. $1 < k < 2$, the corresponding value of ρ can be obtained and thus, with $Z = 0$, the corresponding combinations of values of α, a, X, and b, can be obtained, which describe model Cohen-Kelly networks that are also reduced Cohen-Kelly networks.

Step 3) sees that, given ρ, s.t. $0 < \rho < 1$, $k = k(\rho, Z)$ monotonically decreases in Z with the range $0 < k \leq 1$ and the domain $Z_\rho \leq Z$ where $Z_\rho = \rho/(2 - \rho) > 0$. Thus, given k, s.t. $0 < k \leq 1$, the corresponding value of Z can be obtained and thus, with given ρ, the corresponding combinations of values of α, a, X, and b, can be obtained, which describe model Cohen-Kelly networks.

Steps 4) shows, by combining 2) and 3), that, for every value of k, s.t. $0 < k < 2$, the corresponding combinations of values of α, a, X, and b, can be obtained, which describe model Cohen-Kelly networks. Therefore, for every value of k, s.t. $0 < k < 2$, there exist model Cohen-Kelly networks that have the value of k.

From Proposition 1, we see that there exist Wardrop networks for which the degree of coincident cost improvement (benefits) for all users by adding connections to the networks can increase without bound. Therefore, we have the following property.

Theorem 1. *The degree of coincident cost improvement for all users by adding connections to Wardrop networks can increase without bound.*

4 Noncooperative Load Balancing in Distributed Systems

The previous section has shown that by adding connections to Wardrop networks with fixed numbers of vertices, the degree of coincident cost improvement can increase without bound, although the degree of coincident cost degradation cannot be over some bound.

This section gives a category of noncooperative networks [19], for which it has been shown that the degree of the Braess-like coincident cost degradation can increase without bound in Nash equilibria [22]. This category of networks presents models of distributed systems, like GRIDs [1]. Load balancing of jobs among nodes in distributed systems are regarded as routing in the equivalent networks [35, 36, 37].

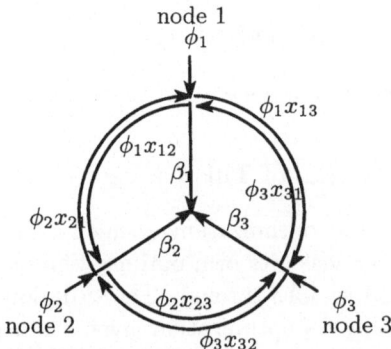

Fig. 2. The model of a distributed system for $m = 3$. The symbols near each arrow express the rate of jobs that flow through the arrow. $\beta_i = \phi_1 x_{1i} + \phi_2 x_{2i} + \phi_3 x_{3i}, i = 1, 2, 3$.

4.1 Assumptions on the Networks Equivalent to Distributed Systems

A network equivalent to a distributed system consists of m origins and one destination, with each origin being connected to the destination through one separate link, which is often called 'node' here. Denote the total flow of O-D pair i is ϕ_i.

Before adding connections, there is only one path for each O-D payer. After adding connections (see Fig. 2), some flow passes through the connections to forward to other nodes, out of the total flow ϕ_i that arrives at the origin i. Let the flow from origin i to node j be $\phi_i x_{ij}$, $i \neq j$. The amount of flow, $\phi_i x_{ii}$, goes to node i. Then, $0 \leq x_{ij} \leq 1$, $i, j = 1, 2, \ldots, m$, $\sum_p x_{ip} = 1$. Denote the vector $(x_{i1}, x_{i2}, \ldots, x_{im})$ by $\boldsymbol{x}_i$ and the vector $(\boldsymbol{x}_1, \boldsymbol{x}_2, \ldots, \boldsymbol{x}_m)$ by $\boldsymbol{x}$. $\boldsymbol{x}$ denotes the strategy profile. The load balancing policies considered determine (only) $\boldsymbol{x}$, and, thus, are static in nature. Denote by C the set of $\boldsymbol{x}$'s that satisfy the above constraints. The resulting flow β_i through node i, $i = 1, 2, \ldots, m$, is

$$\beta_i = \sum_p \phi_p x_{pi}. \tag{4}$$

Thus, the cost for the entire flow associated with the O-D pair i is

$$T_i(\boldsymbol{x}) = \sum_p x_{ip} T_{ip}(\boldsymbol{x}), \tag{5}$$

where $\quad T_{ii}(\boldsymbol{x}) = D_i(\beta_i), \quad$ and $\quad T_{ij}(\boldsymbol{x}) = D_j(\beta_j) + G_{ij}(\boldsymbol{x}),$ for $j \neq i.$ $\quad$ (6)

$D_i(\beta_i)$ denotes the cost of processing a packet/job at node i in the case where the load on node i is β_i. $G_{ij}(\boldsymbol{x})$ denotes the cost of sending a packet/job from origin i to node j in the case where the strategy profile is $\boldsymbol{x}$. It is assumed that, for all i, j, k $(i \neq j \neq k \neq i)$,

$$G_{ij}(\boldsymbol{x}) < G_{ik}(\boldsymbol{x}) + G_{kj}(\boldsymbol{x}). \tag{7}$$

Thus, a job forwarded from a node is not to be forwarded again to another node.

4.2   Wardrop Networks of This Category

We consider an individual optimization scheme for the networks of this category. Each infinitesimal user seeks its own optimization and chooses the path of the minimum cost to send its jobs through. The situation where every infinitesimal user attains unilaterally its optimization given the decisions of other users is a Wardrop equilibrium. We call the network at a Wardrop equilibrium a *Wardrop network* as before. A Wardrop network in question is characterized as follows: Infinitesimal users choose the path of the smallest cost for each O-D pair. Thus, the Wardrop equilibrium $\bar{x}$ after adding connections satisfies the following for all i, j:

$$T_{ij}(\bar{\boldsymbol{x}}) = \min_k\{T_{ik}(\bar{\boldsymbol{x}})\} \text{ for } \bar{x}_{ij} > 0, \tag{8}$$

$$T_{ij}(\bar{\boldsymbol{x}}) \geq \min_k\{T_{ik}(\bar{\boldsymbol{x}})\} \text{ for } \bar{x}_{ij} = 0. \tag{9}$$

That is, all the paths used have the equal cost for each O-D pair. We have the following property.

Theorem 2. *The costs of all users neither degrade nor improve coincidently by adding connections to any Wardrop network of this category.*

[**Proof**] Similarly as [35, 38, 37], after adding connections to a Wardrop network in question, nodes are one of the following:

(1) idle source (R_d): The node sends jobs and does not process any jobs. That is, $\beta_i = 0$.
(2) active source (R_d): The node sends jobs and does not receive any jobs. But, the node processes a part of the jobs that arrive at the node. That is, $\phi_i > \beta_i > 0$.
(3) neutral (N): The node processes jobs locally without sending or receiving jobs. That is, $\beta_i = \phi_i$.

(4) sink (S): The node receives jobs from other nodes but does not send out any jobs. That is, $\beta_i > \phi_i$.

That is, there does not exist such a node that both sends and receives jobs. Indeed, suppose that node i sends jobs to node j and receives jobs from node k. Then,

$$D_i(\beta_i) \geq D_j(\beta_j) + G_{ij}(\boldsymbol{x}), \tag{10}$$

$$D_k(\beta_k) \geq D_i(\beta_i) + G_{ki}(\boldsymbol{x}) = T_{ki}. \tag{11}$$

It is because, if (10) does not hold, processing jobs at node i incurs the smaller cost than sending jobs to node j from node i. Similarly for (11).

Then, from (10) and (7), we have $T_{ki} = D_i(\beta_i) + G_{ki}(\boldsymbol{x}) \geq D_j(\beta_j) + G_{ij}(\boldsymbol{x}) + G_{ki}(\boldsymbol{x}) > D_j(\beta_j) + G_{kj}(\boldsymbol{x}) = T_{kj}$. That is, $T_{ki} > T_{kj}$. Then, from (8) and (9), $x_{ki} = 0$. That is, node i does not receive jobs from node k, which contradicts the supposition.

Denote by $\bar{T}_i$ and $\hat{T}_i$, respectively, the cost of O-D pair i before and after adding connections to the network in question. Recall that D_i is nondecreasing for all i.

1) Assume that adding connections to a network in question brings about coincident cost degradation to all O-D pairs.

Suppose that there exists an idle or active node i after adding connections, then $\bar{T}_i = D_i(\phi_i) \geq D_i(\beta_i) \geq \hat{T}_i$. That is, the O-D pair i suffers no cost degradation. Thus, we see that there exists neither idle nor active node.

Then, since there exists neither idle node nor active node, there must exist no sink node. Only remain neutral nodes. That is, no coincident cost degradation occurs for all O-D pairs by adding connections to the network in question.

2) Assume that adding connections to a network in question brings about coincident cost improvement to all O-D pairs.

Suppose that there exists a sink node i, then $\bar{T}_i = D_i(\phi_i) \leq D_i(\beta_i) = \hat{T}_i$, which is no cost improvement for the O-D pair i. Thus, we see that there exists no sink node.

Then, since there exists no sink node, then there must exist neither idle node nor active node. Only remain neutral nodes. That is, no coincident cost improvement occurs for all O-D pairs by adding connections to the network in question. □

4.3   The Best-Case Degree of Coincident Cost Improvement by Adding Connections to Nash Networks

A Nash network in question is described as follows: For each O-D pair, there is one decision maker, or a player, that strives to minimize the cost for the O–D pair, i.e., decision maker i for O-D pair i, $i = 1, 2, \ldots, m$. Before adding connections, each decision maker has no choice since there is only one path for each O-D payer. After adding connections, decision maker i ($i = 1, 2, \ldots, m$) chooses the amount of the flow to pass through the connections to forward to other nodes, out of the total flow ϕ_i that arrives at the origin i. Thus, within these constraints, decision maker i ($i = 1, 2, \ldots, m$) chooses the values of x_{ij}, $j = 1, 2, \ldots, m$, to achieve optimization. Assume that each decision maker strives to optimize non-cooperatively the cost associated only with the corresponding O-D pair.

A Nash equilibrium is given by such $\tilde{x}$ as satisfies the following for all i,

$$T_i(\tilde{x}) = \min_{x_i} T_i(x_i; \tilde{x}_{-(i)}), \quad \text{such that } (x_i; \tilde{x}_{-(i)}) \in C,$$

where $(x_i; \tilde{x}_{-(i)})$ denotes an m-dimensional vector in which the element corresponding to $\tilde{x}_i$ has been replaced by x_i. $\tilde{x}$ is called a *solution* for the above non-cooperative optimization.

Theorem 3. *The degree of coincident cost improvement for all users by adding connections to Nash networks can increase without bound.*

[**Proof**] 1) We first show this property even if $D_i(\cdot)$ and $G_{ij}(\cdot)$ are nondecreasing, $i, j(\neq i) = 1, 2, \ldots, m$.

Consider the following network with $m = 2$, $\phi_1 = \phi$, $\phi_2 = \phi + \epsilon$ ($\phi > 0$, $\epsilon > 0$), $\delta/D = D/\Delta \triangleq \kappa$ ($0 < \delta < D < \Delta$),

$$D_1(\beta_1) = D \text{ (constant)},$$

$$D_2(\beta_2) = \begin{cases} \delta, & \text{for } 0 \le \beta_2 \le \phi, \\ \Delta, & \text{for } \beta_2 > \phi, \end{cases}$$

$$G_{12} = G_{21} = 0.$$

D_1, D_2, G_{12}, and G_{21} are nondecreasing. Then, clearly, $\tilde{x} = (\tilde{x}_{11}, \tilde{x}_{12}, \tilde{x}_{21}, \tilde{x}_{22}) = (0, 1, 1, 0)$ is a Nash equilibrium after adding connections, and

$$T_1(\tilde{x}) = \delta, \quad T_2(\tilde{x}) = D.$$

Note, however, that, before adding connections,

$$T_1(\tilde{x}) = D, \quad T_2(\tilde{x}) = \Delta.$$

Therefore, $k_1 = k_2 = \kappa < 1$, and both users 1 and 2 have coincident cost improvements by adding connections, and the best-case degree can increase without bound as $\kappa \to 0$.

2) We can easily show this property if we relax the condition that $D_i(\cdot)$ and $G_{ij}(\cdot)$ are nondecreasing, $i, j(\neq i) = 1, 2, \ldots, m$.

Consider the following network that is different from the above network only in the following. $0 < \delta < D = \Delta$,

$$D_1(\beta_1) = \begin{cases} D, & \text{for } 0 \le \beta_1 \le \phi, \\ \delta, & \text{for } \beta_1 > \phi, \end{cases}$$

$$D_2(\beta_1) = D \text{ (constant)}. \tag{12}$$

D_1, D_2, G_{12}, and G_{21} are nonincreasing. Then, clearly, $\tilde{\boldsymbol{x}} = (\tilde{x}_{11}, \tilde{x}_{12}, \tilde{x}_{21}, \tilde{x}_{22})$ $= (1, 0, 1, 0)$ is a Nash equilibrium after adding connections, and

$$T_1(\tilde{\boldsymbol{x}}) = \delta, \quad T_2(\tilde{\boldsymbol{x}}) = \delta.$$

Note, however, that, before adding connections,

$$T_1(\tilde{\boldsymbol{x}}) = D, \quad T_2(\tilde{\boldsymbol{x}}) = D,$$

Therefore, $k_1 = k_2 = \delta/D < 1$, and both users 1 and 2 have coincident cost improvements by adding connections, and the best-case degree can increase without bound as $\delta/D \to 0$.

3) Thus, we have seen that there exist Nash networks for which the degree of coincident cost improvement (benefits) for all users by adding connections to the networks can increase without bound. Therefore, we can say that the degree of coincident cost improvement for all users by adding connections to Nash networks can increase without bound. □

4.4   Homogeneous Distributed Systems

As previously noted, it has been shown that, for homogeneous distributed systems, the degree of coincident cost degradation by adding connections can increase without bound in Nash equilibria [22]. That is, for any value of the ratio k, s.t. $k > 1$, of coincident cost degradation by adding connections, there exists a homogeneous distributed system that has that value of the ratio in a Nash equilibrium. However, as to coincident cost improvement by adding connections to homogeneous distributed systems, we have the following property for all the static load balancing policies that determine $\boldsymbol{x}$.

Theorem 4. *No static load balancing policy in homogeneous distributed systems after adding connections can bring about coincident cost improvement for all users over the situations before adding connections, if node cost functions are nondecreasing and convex.*

[Proof] See Kameda [34]. □

Remark 1. From the above, we see that under any static load balancing policy (any policy that determines $\boldsymbol{x}$, not only noncooperatively but also cooperatively), it is impossible that all users have benefits coincidently from adding connections to homogeneous distributed systems. This result would give some insight into scheduling in GRID computing. □

5   Concluding Remarks

The present paper has examined Wardrop and Nash networks that have fixed numbers of nodes and links. The results imply the following: For Wardrop networks, the degree of coincident cost improvement (benefits) can increase without bound by adding connections whereas there has been found no Wardrop network for which the degree of coincident cost degradation (harms) can increase without bound. On the other hand, for Nash networks, the degrees of both benefits and harms can increase without bound. In contrast, we have seen that some Wardrop network can have neither cost improvement nor cost degradation coincidently for all users by adding connection to it, and that some homogeneous network cannot have cost improvement coincidently for all users by adding connection to it under any static policy including cooperative and noncooperative ones.

References

1. Foster, I., Kesselman, C., eds.: The Grid: Blueprint for a New Computing Infrastructure. Morgan Kaufmann (1998)
2. Braess, D.: Über ein Paradoxen aus der Verkehrsplanung. Unternehmensforschung **12** (1968) 258–268
3. Samuelson, P.A.: Tragedy of the open road: Avoiding paradox by use of regulated public utilities that charged corrected Knightian tolls. J. Int. and Comparative Econ. **1** (1992) 3–12
4. Cohen, J.E., Horowitz, P.: Paradoxial behaviour of mechanical and electrical networks. Nature **352** (1991) 699–701
5. Murchland, J.D.: Braess's paradox of traffic flow. Transpn. Res. **4** (1970) 391–394
6. Frank, M.: The Braess paradox. Mathematical Programming **20** (1981) 283–302
7. Taguchi, A.: Braess's paradox in a two terminal transportation network. J. Oper. Res. Soc. of Japan **25** (1982) 376–388
8. Frank, M.: Cost effective links of ladder networks. Methods of Operations Research **45** (1984) 75–86
9. Steinberg, R., Zangwill, W.I.: The prevalence of Braess's paradox. Transportation Science **17** (1983) 301–318
10. Dafermos, S., Nagurney, A.: Sensitivity analysis for the asymmetric network equilibrium problem. Mathematical Programming **28** (1984) 174–184
11. Dafermos, S., Nagurney, A.: On some traffic equilibrium theory paradoxes. Transpn. Res. B **18** (1984) 101–110
12. Cohen, J.E., Kelly, F.P.: A paradox of congestion in a queuing network. J. Appl. Prob. **27** (1990) 730–734
13. Calvert, B., Solomon, W., Ziedins, I.: Braess's paradox in a queueing network with state-dependent routing. J. Appl. Prob. **34** (1997) 134–154
14. Cohen, J.E., Jeffries, C.: Congestion resulting from increased capacity in single-server queueing networks. IEEE/ACM Trans. Networking **5** (1997) 1220–1225
15. Pas, E.I., Principio, S.L.: Braess's paradox: Some new insights. Transpn. Res. B **31** (1997) 265–276
16. Bean, N.G., Kelly, F.P., Taylor, P.G.: Braess' paradox in a loss network. Journal of Applied Probability **34** (1997) 155–159
17. Korilis, Y.A., Lazar, A.A., Orda, A.: Architecting noncooperative networks. IEEE J. Selected Areas in Communications **13** (1995) 1241–1251
18. Korilis, Y.A., Lazar, A.A., Orda, A.: Avoiding the Braess paradox in noncooperative networks. J. Appl. Prob. **36** (1999) 211–222
19. Kameda, H., Altman, E., Kozawa, T., Hosokawa, Y.: Braess-like paradoxes in distributed computer systems. IEEE Trans. Automatic Control **45** (2000) 1687–1691

20. Kameda, H.: How harmful the paradox can be in the Braess/Cohen-Kelly-Jeffries networks. In: Proc. IEEE INFOCOM 2002, New York (2002)
21. Roughgarden, T.: Designing networks for selfish users is hard. In: Proceedings of the 42nd Annual IEEE Symposium on Foundation of Computer Science. (2001) 472–481 (to appear in a special issue of Journal of Computer and System Sciences).
22. Kameda, H., Pourtallier, O.: Paradoxes in distributed decisions on optimal load balancing for networks of homogeneous computers. J. ACM **49** (2002) 407–433
23. Hsiao, M.T.T., Lazar, A.A.: Optimal decentralized flow control of Markovian queueing networks with multiple controllers. Performance Evaluation **13** (1991) 181–204
24. Parekh, A.K., Gallager, R.G.: A generalized processor sharing approach to flow control in integrated services networks: The single-node case. IEEE/ACM Trans. Networking **1** (1993) 91–10
25. Chakravorti, B.: Optimal flow control of an M/M/1 queue with a balanced budget. IEEE Trans. Automatic Control **39** (1994) 1918–1924
26. Korilis, Y.A., Lazar, A.A.: On the existence of equilibria in noncooperative optimal flow control. J. ACM **42** (1995) 584–613
27. Altman, E., Başar, T.: Multiuser rate-based flow control. IEEE Trans. Communications **46** (1998) 940–949
28. Ching, W.K.: A note on the convergence of asynchronous greedy algorithm with relaxation in a multiclass queueing environment. IEEE Communications Letters **3** (1999) 34–36
29. Famolari, D., Mandayam, N.B., Goodman, D., Shah, V.: A new framework for power control in wireless data networks: Games, utility and pricing. In Ganesh, Pahlavan, Zvonar, eds.: Wireless Multimedia Network Technologies. Kluwer Academic Publishers (1999) 289–310
30. Saraydar, C.U., Mandayam, N.B., Goodman, D.J.: Pricing and power control in a multicell wireless data network. IEEE JSAC **19** (2001) 1883–1892
31. Saraydar, C.U., Mandayam, N.B., Goodman, D.J.: Efficient power control via pricing in wireless data networks. IEEE Trans. Communications **50** (2002) 291–303
32. Ji, H., Huang, C.Y.: Non-cooperative uplink power control in cellular radio systems. Wireless Networks **4** (1998) 233–240
33. Sung, C.W., Wong, W.S.: A noncooperative power control game for multirate CDMA data networks. IEEE Trans. Wireless Communications **2** (2003) 186–194
34. Kameda, H.: Bounds on benefits and harms of adding connections to noncooperative networks. Technical Report ISE-TR-04-195, Institute of Information Sciences and Electronics, University of Tsukuba (2004)
35. Tantawi, A.N., Towsley, D.: Optimal static load balancing in distributed computer systems. J. ACM **32** (1985) 445–465
36. Kim, C., Kameda, H.: An algorithm for optimal static load balancing in distributed computer systems. IEEE Trans. Comput. **41** (1990) 381–384
37. Kameda, H., Li, J., Kim, C., Zhang, Y.: Optimal Load Balancing in Distributed Computer Systems. Springer (1997)
38. Zhang, Y., Kameda, H., Shimizu, K.: Parametric analysis of optimal load balancing in distributed computer systems. Journal of Information Processing (Info. Proc. Soc. of Japan) **14** (1992) 433–441

Fair Packet Forwarding in MANETs with Anonymous Stations: A Game-Theoretic Approach

Jerzy Konorski

Gdansk University of Technology
ul. Narutowicza 11/12, 80-952 Gdansk, Poland
jekon@eti.pg.gda.pl

Abstract. A station of a mobile ad-hoc network (MANET) may selfishly refuse to forward transit packets as it shortens the battery life and takes up a portion of the bandwidth that could be used for source packets. Due to a high degree of station anonymity, selfishness meets with little punishment. The well-known watchdog mechanism can be used to check if an adjacent station forwards packets. We point out that a watchdog may be unable to tell source from transit packets, which enables undetectable manipulation of local congestion controls in selfish stations. We allow each station to set its source packet admission threshold so as to maximise a throughput- and reputation-related payoff. The nature of possible Nash equilibria of the resulting noncooperative game are examined for a generic model of packet forwarding and symmetric traffic flows. A novel packet forwarding protocol called F^3T is proposed and the payoffs it yields are approximately analysed.

1 Introduction

A mobile ad-hoc network (MANET) consists of a number of mobile stations exchanging data packets over one or more wireless channels. MANETs rely on self-organisation rather than central administration. Owing to the falling prices of wireless network equipment as well as the advances in protocol design, MANETs have abandoned their traditional niche of military and emergency communications and increasingly enter the field of civilian data services [6]. Logically, a MANET can be visualised as a (time-varying) adjacency graph, station m being adjacent to station n if it remains within the latter's hearing range. Each station acts both as a mobile user terminal and a packet forwarder. As the former, it injects *source packets* into the network and absorbs *destination packets* therefrom; as the latter, it forwards *transit packets* on behalf of currently non-adjacent pairs of stations. Forwarding transit packets is a dual liability: it shortens the station's battery life and takes up a portion of the channel bandwidth it could use to transmit source packets. MANETs allow a high degree of station anonymity and so refusal to forward transit packets may meet with little punishment. One can therefore envisage various types of station misbehaviour; [11] presents a comprehensive taxonomy.

N. Mitrou et al. (Eds.): NETWORKING 2004, LNCS 3042, pp. 418–429, 2004.

We focus on *selfish* behaviour whereby stations try to reap some undue benefits (as distinct from *cooperative* behaviour and from *malicious* behaviour meant just to do some damage unto others). For a taxonomy of selfish behaviour in MANETs, see [10]. Cooperative behaviour cannot be enforced in a MANET other than by some incentive-based mechanisms. Buttyan and Hubaux [2] propose a virtual currency called *nuglets* that a station earns by forwarding transit packets and then uses to buy a similar service from other stations. Marti *et al.* [9] propose to equip a station with a *watchdog* mechanism which listens to adjacent stations' transmissions and checks if they perform forwarding. A number of recent papers adopts game theory, whereeach player (station) sets her own strategy at will, but the received payoff also depends on the other players' strategies. The play often reaches a Nash equilibrium (NE) from which no player wants to deviate [5]. Michiardi and Molva [10] incorporate a measure of reputation into the payoffs so that rational players forward transit packets to avoid being excluded from existing routing paths. Urpi *et al.* [14] and Srinivasan *et al.* [12] relate the payoffs to throughput efficiency and battery consumption. Zhong *et al.* [15] show that honesty in handling virtual currency can be made a payoff maximising strategy. In the approach of Felegyhazi *et al.* [4], each station sets its own level of cooperation based on its current perception of other stations' levels.

We examine the watchdog approach and argue that some of its weaknesses listed in [9] are not fundamental. However, we point to an unlisted one: being unable to tell source from transit packets, a watchdog is also unable to decide whether an adjacent station is misbehaving or it is backlogged due to heavy transit traffic. Undetectable selfish behaviour then consists in over-admittance of source packets. We address this issue in a game-theoretic framework. Next we propose a packet forwarding protocol called F^3T under which the NE of the underlying game prescribes fair and throughput-efficient settings of local congestion controls.

In Sec. 2 we formulate the network model and explain the nature of undetectable selfish behaviour. In Sec. 3 we define a noncooperative congestion control game and discuss its outcomes. In Sec. 4 we describe and approximately analyse the F^3T protocol. In Sec. 5 we discuss the relevance of proper configuration of F^3T from a game-theoretic perspective. Sec. 6 concludes the paper.

2 Network and Packet Forwarding Model

We assume that the stations use omnidirectional antennae so that the watchdogs can hear all adjacent stations' transmissions. The adjacency graph is assumed bidirectional. MAC and multihop routing protocols are not relevant to our considerations and will not be specified. MAC addresses need not be trustworthy. Data privacy and station anonymity can be achieved via a public-key cryptosystem, such as RSA (with off-line encryption and decryption), and a public hash function such as SHA-1 or MD-5 [13]. Public keys need not be permanent or unique per station. Packet forwarding can now be outlined as follows (Fig. 1):

- a pair of adjacent stations, n and m, establish a *neighbourhood relationship* by exchanging their public keys, key_n and key_m, and routing tables,
- to transfer a packet to a destination station d, a source station n first looks up the next-hop neighbour station m in the routing table, then uses key_d to encrypt the packet body along with key_n, next appends $h_n = hash(key_n)$, $h_m = hash(key_m)$ and $h_d = hash(key_d)$ and finally transmits the packet,
- if the packet is received error-free, station m acks it and compares h_n, h_m and h_d with locally stored hashes of public keys to check that it has a neighbourhood relationship with station n and to recognise itself as the receiver and (possibly) destination; if $m = d$ is detected, the packet body is decrypted using the private key key_m^{-1}, otherwise h_m is replaced by $h_l = hash(key_l)$ with l determined by station m's routing table, and the forwarding continues,
- if $m \neq d$, station n performs a *watchdog check*: upon reception of station m's transmission with h_l appended, it compares the packet body with a copy it has retained to check that the packet has indeed been forwarded by station m,
- based on the check statistics, a neighbourhood relationship may be terminated.

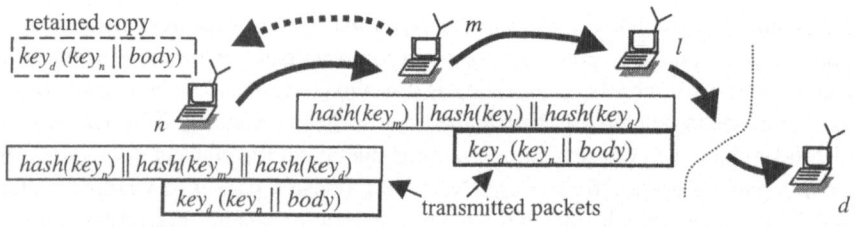

Fig. 1. Anonymous forwarding with public keys and a hash function

Packet collisions on a wireless channel may cause some ambiguity e.g., the watchdog at station n may not receive a packet being forwarded by station m or may be uncertain if the packet has been received at station l [9]. This danger is not serious if a powerful enough MAC protocol is employed e.g., CSMA/CA with RTS/CTS or multiple-channel CDMA. Also, most types of selfish behaviour can be countered under the above model. Refusal to forward a packet claiming transmission errors is counterproductive since the neighbour stations are likely to monitor the channel quality. Assumption of a new identity (new key_m) with a 'clean record' can be countered by requiring an initial silent period for each neighbourhood relationship. False deleting a neighbour from the routing table would not pay off if 'rich enough' routing tables were required to maintain a neighbourhood relationship. Finally, collusion between a pair of neighbours tolerating each other's misbehaviour is risky as either of them might use it for not forwarding the other's packets.

Note that the watchdog at station n must allow station m to transmit a number of packets prior to one being listened for, up to a public-knowledge deadline B beyond which a *failed check* occurs. Station m can legitimately refuse to receive a packet (via a Receive-Not-Ready frame) claiming a current backlog in excess of B. To prove its claim, station m appends to the RNR frame the hashes of B backlogged packets, to be subsequently listened for at station n. Pretending to have a large backlog implies appending fake hashes and subsequently transmitting dummy packets which compute to the same hashes. This could only be productive if the dummy packets were unusually short; an obvious remedy is to define some minimum packet size.

The encryption of a source station's public key (for reasons of privacy and anonymity) precludes the watchdog at station n from distinguishing between station m's source and transit packets. This leaves a possibility of undetectable selfish behaviour. Namely, the necessity to keep the backlog low mandates a local congestion control mechanism at each station. E.g., a simple Drop-and-Throttle (D&T) mechanism [7] permits a station to admit source packets only if its current backlog is below a, where a is called the *D&T threshold*. Since a is set locally, nothing stops station m from unrestrained admission of source packets and subsequent issuing of RNR frames. Such misbehaviour will go unnoticed as the watchdog at station n, listening to station m's transmissions, is unable to tell source packets from transit packets; legitimate refusal to receive packets is indistinguishable from selfish behaviour. A packet forwarding protocol is therefore needed that offers incentives to set a so as to achieve fair and efficient use of the channel bandwidth.

3 Game-Theoretic Model

In this section we describe a noncooperative 'D&T game' that arises when each station sets its D&T threshold so as to maximise the local source packet admission rate while keeping failed checks at neighbour stations tolerably rare. Following game theory [5], we presume that a set of selfish stations reach a Nash equilibrium. We show that even if the traffic flows are symmetric and the D&T thresholds are initially identical, the outcome at equilibrium may be unfair to some stations.

3.1 The D&T Game

Let us view all stations as players in a nonzero-sum game. Station n's feasible actions are the values of D&T threshold it sets locally. A *D&T threshold profile* has the form $[a_n (a)_{-n}]$, where a_n is station n's D&T threshold and $(a)_{-n} = (a_m, m \neq n)$ is the *opponent profile*. The payoff to any station n is determined by the D&T threshold profile. We define two payoff components:

- a throughput measure $S[a_n (a)_{-n}]$ – the local source packet admittance rate, and
- a deadline violation measure $V[a_n (a)_{-n}]$ – the station n-related rate of failed checks at a neighbour station i.e., the rate of reception of packets in whose presence station n will have transmitted at least B other packets.

The former component gives incentives to increase a_n, whereas the latter gives incentives to keep a_n moderate lest station n's selfish behaviour be detected. We take

$$V[a_n(a)_{-n}] \geq V* \tag{1}$$

as the condition of termination of all neighbourhood relationships involving station n, where $V*$ is a public-knowledge tolerance level. $V*$ should be set distinctly above the station malfunction rate. Also, it should be large enough for (1) to be detected with statistical credibility. $V*$ should be upper bounded in relation to the average route length; e.g., with 5-hop routes and $V* \leq 10^{-3}$, over 99.5% of traffic reaches destination. To include reputation effects in the payoffs we assume that

(i) Any station is interested in maintaining all its neighbourhood relationships for which (1) is false and none for which it is true.

Station n therefore wants to maximise

$$payoff[a_n(a)_{-n}] = \begin{cases} S[a_n(a)_{-n}], & \text{if } V[a_n(a)_{-n}] < V* \\ 0, & \text{otherwise} \end{cases} \tag{2}$$

A *Nash equilibrium* (NE) is a D&T threshold profile $[a_n^0(a^0)_{-n}]$ such that

$$payoff[a_n^0(a^0)_{-n}] \geq payoff[a_n(a^0)_{-n}] \text{ for all } n \text{ and any } a_n \tag{3}$$

i.e., no station has incentives to change its D&T threshold unilaterally.

The exact form of the payoffs (2) is determined by a packet forwarding protocol. The following assumption states the payoff structure needed for the D&T game to be nontrivial. The first two parts imply that when increasing its D&T threshold, station n faces a conflict between the increased $S[a_n(a)_{-n}]$ and $V[a_n(a)_{-n}]$; the third part implies that unrestrained admission of source packets does not pay off:

(ii) If D&T threshold profiles are ordered in the sense of vector inequality then
- $S[a_n(a)_{-n}]$ increases in a_n and decreases in $(a)_{-n}$,
- $V[a_n(a)_{-n}]$ increases in both a_n and $(a)_{-n}$, and
- $V[a_n(a)_{-n}] \geq V*$ for a large enough $[a_n(a)_{-n}]$.

In response to changes in $(a)_{-n}$, station n can adjust its D&T threshold. To capture the game dynamics while keeping the model simple, we assume that

(iii) The underlying D&T threshold adjustment mechanism is
- *locally adaptive* i.e., a_n is adjusted based on observed changes in $payoff[a_n(a)_{-n}]$,
- *gradual* i.e., a change of a_n by $\pm\Delta$ causes a payoff change to any other station equivalent of Δ consecutive changes of a_n by ± 1, and
- *prompt* i.e., any other station becomes aware of and can react to each of these Δ changes before the next one takes effect.

The first part reflects the distributed nature of ad-hoc network protocols, whereas the second part is motivated by the fact that $S[a_n(a)_{-n}]$ and $V[a_n(a)_{-n}]$ are medium-term statistics, so do not change abruptly. The third part implies unit changes of the D&T

thresholds: each station n moves sequentially; each move consists in changing a_n by ± 1 and immediately yields payoffs corresponding to the new D&T threshold profile; and no station lags more than one move behind the other stations.

3.2 Nash Equilibria of a Symmetric D&T Game

Suppose that the traffic flows are symmetric and a D&T threshold profile $[a_0 (a_0 ... a_0)]$ currently prevails, yielding fair nonzero payoffs. The following proposition categorises possible Nash equilibria of the D&T game (see [8] for proof and Sec. 5 for illustration).

Under assumptions (ii) and (iii), a symmetric D&T threshold profile $[a (a ... a)]$ will eventually be reached whereupon

- *each station receives a nonzero payoff and has no incentive to change its D&T threshold (a symmetric efficient NE), or*
- *all neighbourhood relationships will be terminated since each station receives a zero payoff, but has no incentive to change its D&T threshold (a symmetric inefficient NE), or*
- *a timing game (a 'war of preemption' or 'war of attrition') starts, leading to an asymmetric NE with unfair payoffs.*

Remarks:

- In a timing game, a player moves at most once and initially all players have incentives to move. In a 'war of preemption,' moving early yields higher payoffs than moving late or not at all, whereas in a 'war of attrition,' moving late or not at all yields higher payoffs.
- Of the above outcomes, the first one (a symmetric efficient NE) is the only desirable; in the following sections we will show how it can be attained by proper design and configuration of the packet forwarding protocol.
- Fairness is somewhat difficult to define in the case of asymmetric traffic flows; for the purpose of this paper we consider a packet forwarding protocol satisfactory if it yields a symmetric efficient NE for the symmetric D&T game.

4 F³T Protocol

Any packet forwarding protocol ought to include mechanisms of 1) provably legitimate refusal to receive packets (otherwise a station may be unduly punished for increased transit traffic), and 2) provably legitimate override of 1) (otherwise $V[a_n (a)_{-n}]$ need not increase in a_n or $(a)_{-n}$, contrary to assumption (ii)). We present a protocol called *Fair Forwarding with Forced Transmissions* (F³T) and approximately analyse the D&T game payoffs under F³T for a symmetric network model.

4.1 Protocol Description

Depending on its current backlog x, a station operates in the NORMAL mode (when $x < e$) and the CONGESTED mode (when $x \geq e$). The parameter e is public knowledge. In the CONGESTED mode, station n can legitimately refuse to receive packets. This it does by announcing the hashes of e backlogged packets, appended to an RNR frame or a transmitted packet. Let the current backlog at a neighbour station m be y. If $y < e$, station m suspends further packet transmissions to station n until $x < e$, as announced by the latter via a Receive-Ready frame or a suitable indication in a transmitted packet. On the other hand, if $y \geq e$ and station m has a packet ready for station n, it *forces* a packet transmission and appends to it the hashes of e backlogged packets as proof of the CONGESTED mode. Thus a CONGESTED station requests that inbound packet transmissions be suspended, which NORMAL neighbour stations comply with and CONGESTED ones disregard. Note that the protocol operation for $e > B$ is the same as for $e = 1$.

A backlogged packet at station n whose next-hop station is m remains enqueued if
- $x < e$ and $y \geq e$ i.e., a packet transmission to station m cannot be forced, or
- no channel is available for a packet transmission to station m, or
- some packet received prior to the packet in question remains enqueued.

Thus packets at a station form a common FIFO queue regardless of the selected next-hop stations. FIFO queuing is known to reduce channel utilisation; on the other hand it enables correct F^3T protocol operation in the following way:
- with per next-hop station queues it would be unclear whether a failed check is due to selfish behaviour or the fact that the neighbour station keeps transmitting packets from other queues,
- a packet received when $x = B$ is certain to cause a failed check; as such it can be immediately discarded instead of unproductively increasing the backlog, and
- a claim of the CONGESTED mode at a neighbour station can be verified within a definite time horizon by comparing the hashes of e subsequent packets transmitted by that station with the previously received hashes.

A neighbourhood relationship is terminated if a claim of the CONGESTED mode is not verified (the other condition is (1)). The goal is to configure B, V^* and e such that the D&T game has a symmetric efficient NE.

4.2 D&T Game Payoffs under F^3T

Assumption (iii) in Sec. 3.1 restricts our interest to payoffs to D&T threshold profiles of the form $[a' (a \ldots a)]$ with $|a' - a| \leq 1$. These will be calculated assuming that:
- (A1) all stations synchronise to fixed-size time slots, a slot accommodating a packet transmission along with related acks, RNR and RR frames,
- (A2) each station has M neighbour stations,
- (A3) the average source-to-destination path length is H hops,
- (A4) the next-hop station for a backlogged packet is selected at random,

(A5) the network operates under heavy load i.e., in each slot a station admits as many source packets as its D&T threshold permits, and

(A6) a station can simultaneously and error-free transmit to and receive from all its neighbour stations, at most one packet per slot per neighbour station.

Assumption (A4) simplifies the calculation; it reflects, in a somewhat exaggerated way, the path variability in ad-hoc networks. Assumption (A5) factors out traffic generation characteristics. Finally, assumption (A6) implies a powerful multipacket reception scheme e.g., CDMA. We stick to this assumption to avoid shifting the focus from packet forwarding to multiple access and physical transmission.

Following the 'isolated node' approach [1], we shall focus upon an arbitrarily chosen station n, where the current backlog x will be modelled as a homogeneous Markov chain. The transition probabilities depend on the current backlog $y_1, ..., y_M$ at the neighbour stations $n_1, ..., n_M$. The approximation consists in regarding $y_1, ..., y_M$ in each slot as drawn from the steady-state probability distribution $(p(x), 0 \leq x \leq B)$ of the above Markov chain. This leads to a fixed-point relationship of the form $p(\cdot) = f[p(\cdot)]$, which can be solved iteratively for $p(\cdot)$. Let $X^{(s)}$ and $Y_m^{(s)}$ denote the backlog at station n and n_m, respectively, at the start of the sth slot. The $Y_m^{(s)}$ will be treated as iid with respect to s and m; let $p_Y(y) = \Pr[Y_m^{(s)} = y]$ for $0 \leq y \leq B$ and $1 \leq m \leq M$. We will express the transition probabilities for $X^{(s)}$ through $p_Y(\cdot)$.

Given $X^{(s)} = x$ and $(Y_1^{(s)}...Y_M^{(s)}) = (y_1...y_M) = \mathbf{y}$, denote by $T|_{x,y}$ and $R_m|_{x,y}$ respectively the random number (between 0 and min[x, M]) of packet transmissions out of station n and the number (0 or 1) of non-destination packets received from station n_m in the sth slot. Let $P_Y(e) = \Sigma_{0 \leq y < e} p_Y(y)$. Recalling assumption (A6) and the conditions for a packet to remain enqueued (Sec. 4.1), one has

$$\Pr[T|_{x,y} \geq k] = \begin{cases} 0, \text{if } k > \min[x,M] \\ P_Y^k(e) \cdot \prod_{0 \leq j \leq k-1}(1-\frac{j}{M}), \text{if } k \leq \min[x,M] \text{ and } x < e \\ \prod_{0 \leq j \leq k-1}(1-\frac{j}{M}), \text{if } k \leq \min[x,M] \text{ and } x \geq e \end{cases} \quad (4)$$

$$\Pr[R_m|_{x,y} = 1] = \begin{cases} 0, \text{if } y_m < e \text{ and } x \geq e \\ (1-\frac{1}{H}) \cdot \frac{1}{M} \cdot \Sigma_{0 \leq j < \min[y_m,M]} \prod_{0 \leq i \leq j} P_Y^j(e), \text{otherwise} \end{cases} \quad (5)$$

The second part accounts for various positions ($1^{st}, ..., \min[y_m, M]^{th}$) the packet to be transmitted to station n may occupy in station n_m's FIFO queue. The random variable

$$I|_{x,y} = \Sigma_{1 \leq m \leq M} R_m|_{x,y} - T|_{x,y} \quad (6)$$

represents the net influx of transit packets at station n per slot. Its probability distribution is obtainable via (4) and (5) since $T|_{x,y}$ and $R_m|_{x,y}$ are independent. Unconditioning on **y** gives

$$\Pr[I|_x = i] = \sum_{0 \le y_1,\dots,y_M \le B} p_Y(y_1)\dots p_Y(y_M) \cdot \Pr[I|_{x,y} = i]. \qquad (7)$$

Note that there are $(B+1)^M$ summands in (7), each of which involves numerical inversion of a probability generating function for (6); this makes (7) the most tedious part of the calculation. Given $X^{(s)} = x$ one has

$$X^{(s+1)} = \min[B, x + (a-x)^+ + I|_x], \qquad (8)$$

where $(a-x)^+ = \max[0, a-x]$ is the number of source packets admitted in the sth slot. Hence, the calculation of the transition probabilities of interest as well as the steady-state probabilities $p(x) = \lim_{s\to\infty} \Pr[X^{(s)} = x]$ is straightforward, assuming that $p_Y(\cdot)$ is known. Since by symmetry $p(\cdot) = p_Y(\cdot)$, one can calculate $p(\cdot)$ based on an assumed $p_Y(\cdot)$, and in the next iteration substitute $p(\cdot)$ for $p_Y(\cdot)$ until the two differ insignificantly. Thereupon one obtains

$$S[a(a\dots a)] = \sum_{0 \le x \le a} p(x) \cdot (a-x)^+ \qquad (9)$$

$$V[a(a\dots a)] = \frac{1}{M} \cdot \sum_{0 \le x \le B} p(x) \sum_{i > B} (i-B)^+ \cdot \Pr[x + (a-x)^+ + I|_x = i] \qquad (10)$$

To calculate $payoff[a'(a \dots a)]$, put $a := a'$ in (9) and (10) while retaining $p(\cdot)$.

5   F³T Configuration and Performance

The approximate analysis outlined in Sec. 4.2 assumes a rather idealised network model and only yields to numerical calculation. Yet its results are instructive as they capture the possible outcomes of the D&T game under F³T. Provided that B, e and V^* are set properly, increasing a_n unilaterally backfires in terms of V: a larger backlog at station n initially reduces the local packet reception rate, but ultimately drives the neighbour stations CONGESTED and causes them to force packet transmissions into n. Without incentives to increase a, the adjustment mechanism mentioned in assumption (iii) of Sec. 3.1 remains dormant and a fair and throughput-efficient D&T threshold profile persists. The results also illustrate the difference in performance prediction compared to the classical cooperative paradigm.

A series of numerical experiments confirmed the validity of assumption (ii) of Sec. 3.1 under F³T. For various a and $a' = a-1$, a and $a+1$, $payoff[a'(a \dots a)]$ was calculated from (9) and (10) as a percentage of the maximum attainable value M/H. Sample results are depicted in Fig. 2a-c, assuming $B = 8$, $M = 3$, $H = 5$ and $V^* = 10^{-3}$.

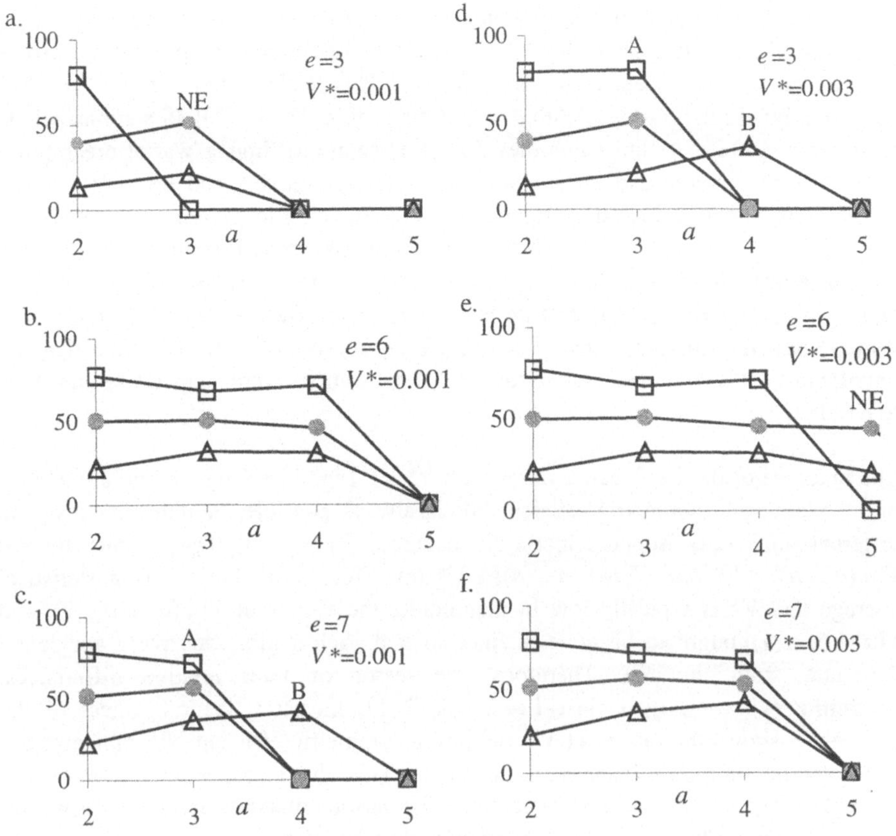

Fig. 2. *payoff*[*a'* (*a…a*)] (%) under F³T (□: *a'=a+1*, O: *a'=a*, Δ: *a'=a–1*)

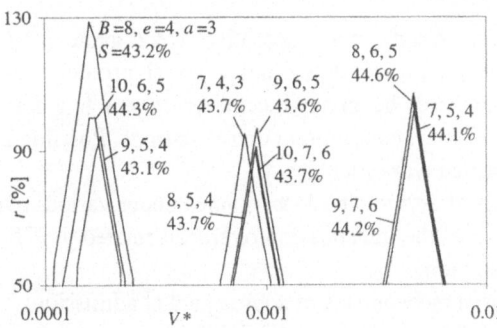

Fig. 3. Robustness of symmetric efficient Nash equilibria under F³T

One sees that $e = 3$ yields a symmetric efficient NE with $S[3 (3 ... 3)] = 51.2\%$: no station has incentives to increase or decrease its D&T threshold. With $e = 6$, at the profile $[4 (4 ... 4)]$ all stations have incentives to increase a, ending up at the profile $[5 (5 ... 5)]$ with no incentive to move further. This leads to the termination of all neighbourhood relationships. With $e = 7$, at the profile $[3 (3 ... 3)]$ all stations benefit from moving to $a = 4$. Supposing they do not do so at one time, a 'war of preemption' follows: early movers end up at point A, while late movers end up at B with no incentive to move further (since the payoff corresponding to $[3 (4 ... 4)]$ exceeds that corresponding to $[4 (4 ... 4)]$). Had all the stations moved from $a = 3$ to $a = 4$ simultaneously, they would have to consider retreating to $a = 3$, but early movers now end up at B and late movers end up at A. This is in effect a 'war of attrition.' The (unfair) payoffs thus may range from 71.4% (at A) to 41.8% (at B). Note that fair payoffs of 56.7% could be attained for $e = 7$ if the stations were cooperative and stuck to $a = 3$.

The outcome of the D&T game varies with V^*, cf. Fig. 2d-f. Given B and e, a choice of V^* yielding a symmetric efficient NE is always possible, though may result in different *robustness*. Suppose that a NE occurs at $[a (a ... a)]$. Then it must be that $V[a (a ... a)] < V^*$ and $V[a+1 (a ... a)] \geq V^*$ (cf. Fig. 2a,e). Since V is a statistical average and V^* is typically low in magnitude, the values of $V[a (a ... a)]$, V^* and $V[a+1 (a ... a)]$ ought to be quite distinct so that each station can avoid accidental departure from the NE. Therefore, the lesser of their relative differences, $r = \min\{V^*/V[a (a ... a)]-1, V[a+1 (a ... a)]/V^*-1\}$, measures the robustness of the NE. Fig. 3 shows the ranges of V^* yielding a symmetric efficient NE with $r \geq 50\%$ (the corresponding B, e, a and S are indicated). Thus that a proper setting of B and e guarantees a desirable and robust outcome for various magnitudes of V^*. However, the received payoffs are relatively invariant and under 45%.

6 Conclusion and Future Research

We have defined and analysed a noncooperative 'D&T game' played by anonymous MANET stations. For symmetric flows and a novel protocol called F^3T we have shown how the stations can be given incentives to reach a fair and throughput-efficient NE in terms of local congestion control settings. Possible improvements and extensions of the presented work include:

- simulative analysis of asymmetric D&T games under various dynamic scenarios,
- detailed calculation of the transmission overhead related to F^3T operation and the employed hash function,
- other e.g., rate-based mechanisms of source packet admission,
- single-channel MAC protocols e.g., IEEE 802.11,
- per next-hop station queueing to improve the channel utilisation (a suitable extension of F^3T is possible under source routing schemes e.g., DSR [3]), and
- individual negotiation of B, e and V^* for each neighbourhood relationship.

References

1. Agnew, G. B., Mark, J. W.: Performance Modeling for Communication Networks at a Switching Node. IEEE Trans. Comm. COM-32 (1984) 902-910
2. Buttyan, L., Hubaux, J. P.: Nuglets: A Virtual Currency to Stimulate Cooperation in Self-Organised Mobile Ad-Hoc Networks. Tech. Rep. DSC/2001/001, Swiss Federal Institute of Technology (2001)
3. Broch, J., Johnson, D., Maltz, D.: The Dynamic Source Routing Protocol for Mobile Ad Hoc Networks. IETF Internet Draft (1998)
4. Felegyhazi, M., Buttyan L., Hubaux, J. P.: Equilibrium Analysis of Packet Forwarding Strategies in Wireless Ad Hoc Networks – The Static Case. Tech. Rep. IC/2003/33, Swiss Federal Institute of Technology (2003)
5. Fudenberg, D., Tirole, J.: *Game Theory*. MIT Press, Cambridge Mass. (1991)
6. Goldsmith, A.J., Wicker, S.B.: Design Challenges for Energy-Constrained Ad Hoc Wireless Networks. IEEE Wireless Communications 4 (2002) 8-27
7. Kamoun, F.: A Drop-and-Throttle Flow Control Policy for Computer Networks. IEEE Trans. Comm. COM-29 (1981) 444-452
8. Konorski, J., Fair Packet Forwarding in Mobile Ad Hoc Networks: A Game-Theoretic Approach. Tech. Rep., Gdansk Univ. Tech. (2004)
9. Marti, S., Giuli, T. J., Lai, K., Baker, M.: Mitigating Routing Misbehavior in Mobile Ad Hoc Networks. In: Proc. 6th Annual Conf. on Mobile Computing and Networking MobiCom 2000 (2000) 255-265
10. Michiardi, P., Molva, R.: Making Greed Work in Mobile Ad Hoc Networks. Res. Rep. RR-02-069, Institut Eurecom, Sophia-Antipolis, France (2002)
11. Obreiter, P., Koenig-Ries, B., Klein, M.: Stimulating Cooperative Behaviour of Autonomous Devices – An Analysis of Requirements and Existing Approaches. Tech. Rep. 2003-1, Univ. of Karlsruhe, Germany (2003)
12. Srinivasan, V., Nuggehalli, P., Chiasserini, C. F., Rao, R. R.: Cooperation in Wireless Ad Hoc Networks. In: Proc. IEEE INFOCOM 2003 (2003)
13. Stallings, W.: Cryptography and Network Security: Principles and Practice. Prentice-Hall, Englewood Cliffs NJ (1999)
14. Urpi, A., Bonuccelli, M., Giordano S.: Modelling Cooperation in Mobile Ad Hoc Networks: a Formal Description of Selfishness. In: Proc. WiOpt 2003 Workshop on Modeling and Optimization in Mobile, Ad Hoc and Wireless Networks (2003)
15. Zhong, S., Chen, J., Yang, Y. R.: Sprite: A Simple, Cheat-Proof, Credit-Based System for Mobile Ad-Hoc Networks. In: Proc. IEEE INFOCOM 2003 (2003)

Pricing Differentiated Services:
A Game-Theoretic Approach

Eitan Altman[*1], Dhiman Barman[**2], Rachid El Azouzi[*3], David Ros[4], and
Bruno Tuffin[5]

[1] INRIA, BP 93, 06902 Sophia Antipolis Cedex, France
[2] 111 Cummington Street, Dept. of C.S., Boston University, Boston, MA 02215, USA.
[3] LIA/CERI, Univesité d'Avignon, Agroparc, BP 1228, 84911, Avignon, France
[4] GET/ENST Bretagne, 35567 Cesson Sévigné Cedex, France
[5] IRISA/INRIA, Campus Universitaire de Beaulieu, 35042 Rennes Cedex, France

Abstract. We consider TCP and real-time connections sharing a
bottleneck buffer using RED. The latter gives differentiated services to
the applications according to their choice of service class. The choice of
a service class of an application depends both on the QoS as well as on
the cost. We first study the performance of the system as a function of
the connections' parameters and choice of service classes. We then study
the decision problem of how to choose the service classes as a game.
We finally study how to choose prices so that the resulting equilibrium
would maximize the network's benefit.

Keywords: TCP, Buffer Management, RED/AQM, Nash equilibrium,
Pricing, Economics

1   Introduction

We consider TCP connections as well as CBR (Constant Bit Rate) connections
sharing a bottleneck link where a RED buffer management is used. We allow for
service differentiation between the connections through the rejection probability
which may depend on the connection. More specifically, we consider a buffer
management scheme that uses a single averaged queue length to determine the
rejection probabilities; for any given averaged queue size, packets belonging to
connections with higher priority have smaller probability of being rejected. To
obtain this differentiation we assume that the loss curve of RED is scaled by
a factor that represents the priority level of the application. We obtain various
performance measures of interest such as the throughput, the average queue size
and the average drop probability.

   We then address the question of the choice of priorities. Given utilities that
depend on the performance measures and on the cost, the users are faced with

---

* The work of these authors was supported by France Telecom R&D 001B001 contract.
** The work of this author was performed during internship at INRIA, financed by the
INRIA's PrixNet ARC collaboration project

N. Mitrou et al. (Eds.): NETWORKING 2004, LNCS 3042, pp. 430–441, 2004.

a non-cooperative game. We establish conditions for an equilibrium to exist. We further provide conditions for convergence to equilibrium from non equilibria initial states. We finally study numerically the pricing problem of how the network should choose prices so that the resulting equilibrium would maximize its benefit.

We briefly mention some recent work in that area. Reference [5] has considered a related problem where the traffic generated by each session followed a Poisson process, and the service time was exponentially distributed. The decision variables were the input rates and the goodput was taken as the performance measure. The paper restricts to symmetric users and equilibria and pricing was not considered. In this framework, with a common RED buffer, it was shown that an equilibrium did not exist. An equilibrium was obtained and characterized for an alternative buffer management. We note that in contrast to [5], since we also include in the utility of CBR traffic a penalty for losses we do obtain an equilibrium when using RED. For other related papers, see [1,3,7,9]. Our pricing problem formulation is related to Stackelberg equilibrium [4,6].

The structure of the paper is as follows: In Sec. 2 we describe the model and compute the performance for connections with given priority choices. Sec. 3 introduces the competition between connections at given prices, and concludes with the pricing problem. In Sec. 4 we study the game in the case of only CBR or only TCP connections. In Sec. 5 we provide an algorithm for computing the equilibrium for the symmetric case. Numerical examples are given in Sec. 6. We conclude the paper in Sec 7.

2   The Model and Computation of Throughput

RED is based on the following idea: there are two thresholds q_{min} and q_{max} such that the drop probability is 0 if the average queue length q is less than q_{min}, 1 if it is above q_{max}, and $p(i)(x - q_{min})/(q_{max} - q_{min})$ if it is x with $q_{min} < x < q_{max}$; the latter is the *congestion avoidance* mode of operation. See Fig. 1.

Consider a set $\mathcal{N}$ containing N TCP flows (or aggregate of flows) and a set $\mathcal{I}$ containing I real-time flows that can be differentiated by RED; they all share a common buffer yet RED treats them differently. They all have common q_{min} and q_{max} but each flow i may have a different $p(i)$, which is the value of the drop probability as the average queue tends to q_{max} (from the left). In other words, the slope t_i of the linear part of the curve in Fig. 1 depends on the flow i: $t_i = p(i)/(q_{max} - q_{min})$. Denote $\mathbf{t} = (t_i, i \in \mathcal{I} \cup \mathcal{N})$. We identify t_i as the priority class of a connection. The service rate of the bottleneck router is given by μ. We use the well-known relation for TCP rate:

$$\lambda_i = \frac{1}{R_i}\sqrt{\frac{\alpha}{p_i}}, \quad i \in \mathcal{N}, \tag{1}$$

where R_i and p_i are TCP flow i's round trip time and drop probability, respectively. α is typically taken as $3/2$ (when the delayed ack option is disabled) or $3/4$ (when it is enabled). We shall assume throughout the paper that the queueing delay is negligible with respect to R_i for the TCP connections.

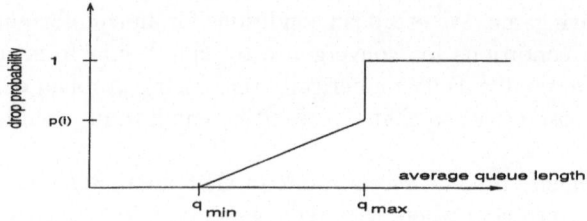

Fig. 1. Drop probability in RED as function q

In contrast, the rates λ_i, for $i \in \mathcal{I}$ are not controlled and are assumed to be fixed. If $\mathcal{N} = \emptyset$ we assume throughout the paper that $\sum_{j \in \mathcal{I}} \lambda_j > \mu$ (unless otherwise specified), otherwise the RED buffer is not a bottleneck. Similarly, if $\mathcal{I} = \emptyset$ we assume that TCP senders are not limited by the receiver window. In general, since the bottleneck queue is seen as a fluid queue, we can write $\sum_{j \in \mathcal{I} \cup \mathcal{N}} \lambda_j (1 - p_j) = \mu$. If we operate in the linear part of the RED curve then this leads to the linear equations:

$$\sum_{j \in \mathcal{I} \cup \mathcal{N}} \lambda_j (1 - p_j) = \mu, \qquad p_i = t_i (q - q_{\min}), \quad \forall i \in \mathcal{I} \cup \mathcal{N}$$

with $(N + I + 1)$ unknowns: q (average queue length), and p_i, $i \in \mathcal{I} \cup \mathcal{N}$, where λ_i, $i \in \mathcal{N}$ is given by (1). Substituting (1) and

$$p_i = t_i (q - q_{\min}) \ \forall i, \tag{2}$$

we obtain a single equation for q:

$$\sum_{j \in \mathcal{N}} \frac{1}{R_j} \sqrt{\frac{\alpha}{t_j (q - q_{min})}} (1 - t_j (q - q_{\min})) + \sum_{j \in \mathcal{I}} \lambda_j (1 - t_j (q - q_{\min})) = \mu. \tag{3}$$

If we write $x = \sqrt{q - q_{\min}}$, then (3) can be written as a cubic equation in x:

$$Z(x) = z_3 x^3 + z_2 x^2 + z_1 x + z_0 = 0 \text{ where} \tag{4}$$

$$z_3 = \sum_{j \in \mathcal{I}} \lambda_j t_j, \ z_2 = \sum_{j \in \mathcal{N}} \frac{1}{R_j} \sqrt{\alpha t_j}, \ z_1 = \mu - \sum_{j \in \mathcal{I}} \lambda_j, z_0 = - \sum_{j \in \mathcal{N}} \frac{1}{R_j} \sqrt{\frac{\alpha}{t_j}}.$$

The equation has a unique positive solution if there are only TCP or only real-time connections; in either case, it becomes a quadratic equation.

Proposition 1 *[2] Fix the values of t_j, $j \in \mathcal{I} \cup \mathcal{N}$. The cubic equation (4) has a unique real positive solution. Assume that the solution lies in the linear region of RED. Then the average queue size is given as, $q = q_{\min} + x^2$ where x is the unique positive solution of (4) and the loss probability for session i is given by $p_i = t_i (q - q_{\min})$.*

In the case of only CBR traffic operating in the linear region,

$$q = q_{\min} + \frac{\sum_{j \in \mathcal{I}} \lambda_j - \mu}{\sum_{j \in \mathcal{I}} \lambda_j t_j} \quad \text{and} \tag{5}$$

$$p_i = t_i \frac{\sum_{j \in \mathcal{I}} \lambda_j - \mu}{\sum_{j \in \mathcal{I}} \lambda_j t_j}. \tag{6}$$

In the case of only TCP connections operating in the linear region, we have

$$q = q_{\min} + \left(-\mu + \sqrt{\mu^2 + 4\alpha \sum_{j \in \mathcal{N}} \left(\frac{1}{R_j \sqrt{t_j}} \right) \sum_{j \in \mathcal{N}} \left(\frac{\sqrt{t_j}}{R_j} \right)} \right)^2 \Big/ 4\alpha \left(\sum_{j \in \mathcal{N}} \frac{\sqrt{t_j}}{R_j} \right)^2 \tag{7}$$

$$p_i = t_i \left(-\mu + \sqrt{\mu^2 + 4\alpha \sum_{j \in \mathcal{N}} \left(\frac{1}{R_j \sqrt{t_j}} \right) \sum_{j \in \mathcal{N}} \left(\frac{\sqrt{t_j}}{R_j} \right)} \right)^2 \Big/ 4\alpha \left(\sum_{j \in \mathcal{N}} \frac{\sqrt{t_j}}{R_j} \right)^2. \tag{8}$$

3   Utility, Pricing, and Equilibrium

We denote a strategy vector by $\mathbf{t}$ for all flows such that jth entry is t_j. By $(t_i, [\mathbf{t}]_{-i})$, we define a strategy where flow i uses t_i and all other flows $j \neq i$ use t_j from vector $[\mathbf{t}]_{-i}$. We associate to flow i a utility U_i. The utility will be a function of the QoS parameters and the price payed by flow i, and is determined by the actions of all flows. More precisely, $U_i(t_i, [\mathbf{t}]_{-i})$ is given by $a_i \lambda_i (1 - p(t_i, [\mathbf{t}]_{-i}))$ $- b_i p(t_i, [\mathbf{t}]_{-i}) - d(t_i)$ where the first term stands for the utility for the goodput, the second term stands for the dis-utility for the loss rate and the last term corresponds to the price $d(t_i)$ to be paid by flow i to the network.

In particular, we find it natural to assume that a TCP flow i has $b_i = 0$ (as lost packets are retransmitted anyhow, and their impact is already taken into account in the throughput). Moreover, since λ_i for TCP already includes the loss term $p_i(t_i, [\mathbf{t}]_{-i})$, the utility function of TCP is assumed to be $U_i(t_i, [\mathbf{t}]_{-i}) = a_i \lambda_i (1 - p(t_i, [\mathbf{t}]_{-i})) - d(t_i)$.

We assume that the strategies or actions available to session i are given by a compact set of the form: $t_i \in \left[t^i_{min}, t^i_{max} \right]$, $i \in \mathcal{I} \cup \mathcal{N}$.

Each flow of the network strives to find its best strategy so as to maximize its own objective function. Nevertheless its objective function depends upon its own choice but also upon the choices of the other flows. In this situation, the solution concept widely accepted is the concept of Nash equilibrium.

Definition 1 *A Nash equilibrium of the game is a strategy profile* $\mathbf{t} = (t_1, t_2, .., t_M)$ *where* $M = I + N$ *from which no flow have any incentive to deviate. More precisely, the strategy profile* $\mathbf{t}$ *is a Nash equilibrium, if the following holds true for any* i $t_i \in \arg\max_{\bar{t}_i \in [t^i_{min}, t^i_{max}]} U_i(\bar{t}_i, [\mathbf{t}]_{-i})$. t_i *is the best flow* i *can do if the other flows choose the strategies* $[\mathbf{t}]_{-i}$.

Note that the network income is given by $\sum_{i \in \mathcal{I} \cup \mathcal{N}} d(t_i)$. Since the $p_i(t_i, [\mathbf{t}]_{-i})$'s are functions of t_i and $[\mathbf{t}]_{-i}$, d can include pricing per volume of traffic successfully transmitted. In particular, we allow for d to depend on the uncontrolled arrival rates of real-time sessions (but since these are constants, we do not make them appear as an argument of the function d).

We shall sometimes find it more convenient to represent the control action of connection i as $T_i = 1/t_i$ instead of as t_i. Clearly, properties such as existence or uniqueness of equilibrium in terms of t_i directly imply the corresponding properties with respect to T_i.

Optimal pricing: The goal here is to determine the pricing that maximizes the network's benefit; we seek to obtain a function $c(\mathbf{t}^*) = \arg\max_d \sum_{i=1}^{I} d(t_i^*)$, where $\mathbf{t}^*$ is a Nash equilibrium which can be obtained when considering special classes of function d. For instance, consider the set of functions $d(t) = d/e^t$. We then obtain a system of equations that can be solved numerically (to get the t^* satisfying the Nash equilibrium). Then a numerical optimization over the parameter d can be obtained. An assumption of this optimization problem is that the network knows the number of flows and the parameters a_i, b_i and $R_i \forall i$. The optimal pricing will be done numerically in Section 6.

4   Equilibrium for Only CBR or Only TCP Connections

We assume throughout that $t_{max}^i \leq 1/(q_{max} - q_{min})$ for all connections. The bound for t_{max}^i is given so that we have $t_{max}^i(q_{max} - q_{min}) \leq 1$. From (2) we see that $p_i \leq 1$ with equality obtained only for the case $t_i = 1/(q_{max} - q_{min})$. [1]

In our analysis, we are interested mainly in the linear region. For only real-time sessions or only TCP connections, we state the assumptions and describe the conditions for linear region operations and we show the existence of a Nash equilibrium. Proofs are given in [2].

Theorem 1 *A sufficient condition for the system to operate in linear region is that for all i: (a) For only real-time connections:*

$$\lambda > \mu \text{ and } t_{min}^i > \frac{\lambda - \mu}{\lambda(q_{max} - q_{min})}. \tag{9}$$

(b) Define $\lambda = \sum_{j \in \mathcal{I}} \lambda_j$ and $\Delta q := q_{max} - q_{min}$. For only TCP connections:

$$t_{min}^i > \left(\frac{-\mu + \sqrt{\mu^2 + 4\alpha(\sum_{j \in \mathcal{N}} \frac{1}{R_j})^2}}{4\sqrt{\alpha \Delta q} \sum_{j \in \mathcal{N}} \frac{1}{R_j}} \right)^2. \tag{10}$$

---

[1] Note that if the assumption does not hold then for some value $q' < q_{max}$ we would already have for some i, $p_i = 1$ so one could redefine q_{max} to be q'. An important feature in our model is that the queue length beyond which $p_j = 1$ should be the same for all j.

Theorem 2 *[2, Appendix 10.2] Assume operation at the linear region of RED. Assume that the functions d are convex in $T_i := 1/t_i$. Then a Nash equilibrium exists.*

Supermodular Games: In Thm 3 (resp. Thm 5) we present alternative conditions that provide sufficient conditions for a supermodular structure for real-time connections (resp. for only TCP connections). This implies in particular the existence of an equilibrium. Another implication of supermodularity is that a simple, so-called tatônnement or Round Robin scheme, for best responses converges to the equilibrium. To describe it, we introduce the following asynchronous dynamic greedy algorithm (GA).

Greedy Algorithm: Assume a given initial choice $\mathbf{t}^0$ for all flows. At some strictly increasing times τ_k, $k = 1, 2, 3, ...$, flows update their actions; the actions t_i^k at time $\tau_k > 0$ are obtained as follows. A single flow i at time τ_{k+1} updates its t_i^{k+1} so as to optimize $U_i(., [\mathbf{t}^k]_{-i})$ where $[\mathbf{t}^k]_{-i}$ is the vector of actions of the other flows $j \neq i$. We assume that each flow updates its actions infinitely often. In particular, for the case of only real-time sessions, we update t_i^{k+1} as follows:

$$t_i^{k+1} = \begin{array}{c} \arg\max \\ t_i \in [t_{\min}^i, t_{\max}^i] \end{array} a_i \lambda_i (1 - p_i) - b_i p_i - d(t_i) \tag{11}$$

where p_i in (11) is given by (6). For TCP-only, we update t_i^{k+1} as follows:

$$t_i^{k+1} = \begin{array}{c} \arg\max \\ t_i \in [t_{\min}^i, t_{\max}^i] \end{array} \frac{a_i}{R_i} \sqrt{\frac{\alpha}{p_i}} (1 - p_i) - d(t_i) \tag{12}$$

where p_i in (12) is given by (8).

Remark 1 *For the case of real-time sessions, we could obtain a closed form solution for t_j^{k+1} with specific cost function $d(t_i)$ such as $\frac{d}{t_i}$ which will lead to update of t_i^{k+1} as follows,*

$$\delta_i^k = \frac{\sum_{j \neq i} \lambda_j t_j^k}{\sqrt{(a_i \lambda_i + b_i)(\sum_{j \in \mathcal{I}} \lambda_j - \mu)(\sum_{j \neq i} \lambda_j t_j^k) - \lambda_i \sqrt{d}}} \quad \text{where } \delta_i^k \text{ is such that } \left. \frac{\partial U_i}{\partial t_i} \right|_{t_i = \delta_i^k} = 0 \text{ and}$$

U_i *corresponds to utility function of real-time session i. Then t_i^{k+1} is given by :*

$$t_i^{k+1} = \begin{cases} t_{\min}^i & \text{if } \delta_i^k < 0, \\ t_{\max}^i & \text{if } \delta_i^k < t_{\min}^i, \delta_i^k \geq 0, \\ t_{\min}^i & \text{if } \delta_i^k > t_{\max}^i, \delta_i^k \geq 0, \\ \delta_i & \text{otherwise.} \end{cases}$$

The theorems below follow by establishing supermodularity of the games [8,10], see [2] for detailed proofs.

Theorem 3 *For the case of only real-time connections we assume that $\forall j$, $\lambda_{\min} \leq \lambda_j \leq \lambda_{\max}$, and $(I - 1)\lambda_{\min} t_{\min} \geq \lambda_{\max} t_{\max}$, where $t_{\min} = \min_{i \in \mathcal{I}} \{t_{\min}^i\}$*

and $t_{\max} = \max_{i \in \mathcal{I}}\{t^i_{\max}\}$. *Then there is smallest equilibrium $\underline{t}$ and largest equilibrium $\bar{t}$, and the* **GA** *dynamic algorithm converges to $\underline{t}$ (resp. $\bar{t}$) provided it starts with $t^j_{\min}$ for all j (resp. $t^j_{\min}$ for all j)*

Theorem 4 *For the case of only real-time connections, we assume that $\forall j, \lambda_{\min} \leq \lambda_j \leq \lambda_{\max}$, and $2t^3_{\min}\lambda^2_{\min} > t^3_{\max}\lambda^2_{\max}$. Under supermodular condition, the Nash equilibrium exists and is unique.*

Theorem 5 *Case of only TCP connections: assume $\forall j, t_{min} \leq t \leq t_{max}$ and*

$$(3 + p_i)\frac{\partial p_i}{\partial t_i}\frac{\partial p_i}{\partial t_j} \geq 2p_i(p_i + 1)\frac{\partial^2 p_i}{\partial t_i \partial t_j} \quad \forall i, j, \; i \neq j. \tag{13}$$

Then the game is super-modular and an equilibrium exists.

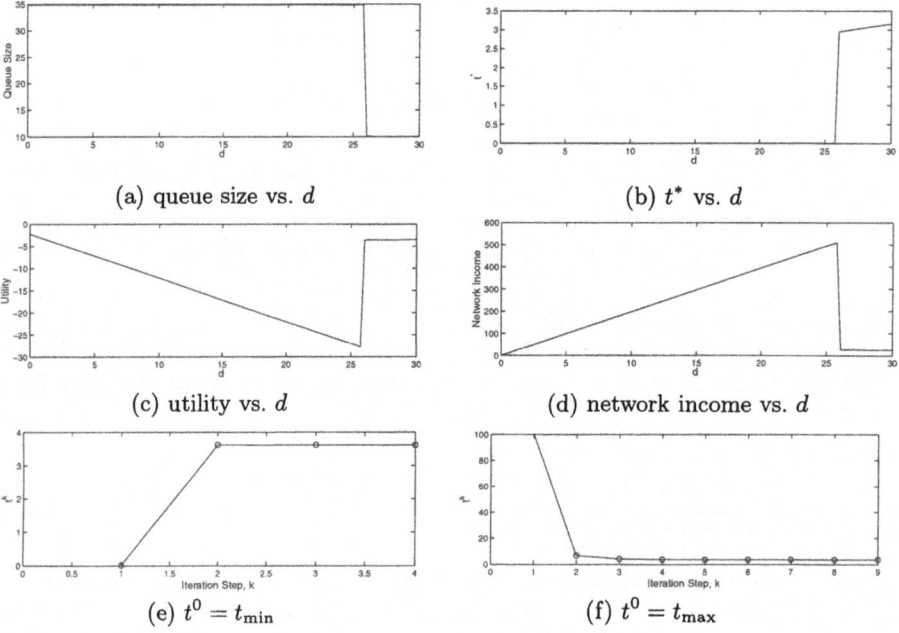

(a) queue size vs. d

(b) t^* vs. d

(c) utility vs. d

(d) network income vs. d

(e) $t^0 = t_{\min}$

(f) $t^0 = t_{\max}$

Fig. 2. Symmetric Real-Time flows: (a) queue size, (b) t^* vs. d, (c) utility, (d) network income vs. d. (e), and (f): Convergence to Nash equilibrium

5 Symmetric Users

In this section, we assume that all flows have the same utility function (for all i, $a_i = a$, $\lambda_i = \bar{\lambda}$ and $b_i = b$ for real-time sessions and $a_i = a$ and $R_i = R$ for TCP connections) and the same intervals for strategies ($t^i_{\min} = t_{\min}$ and $t^i_{\max} = t_{\max}$).

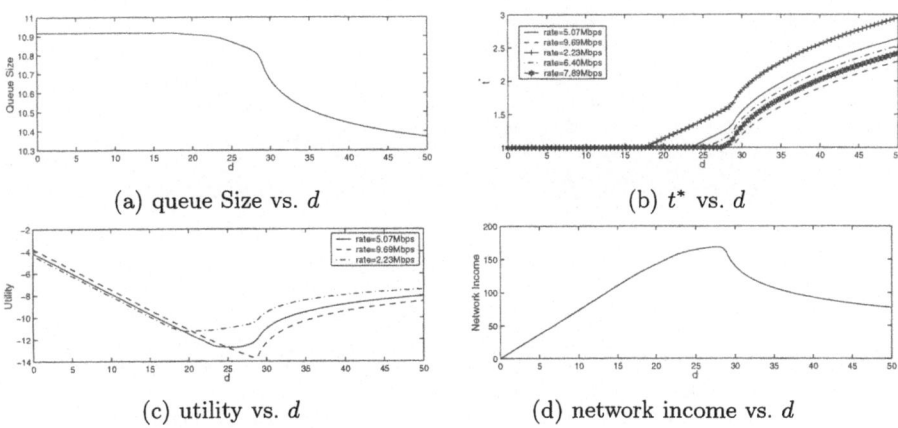

(a) queue Size vs. d

(b) t^* vs. d

(c) utility vs. d

(d) network income vs. d

Fig. 3. Non-symmetric Real-Time flows: (a) queue size, (b) t^* vs. d, (c) utility, and (d) network income vs. d

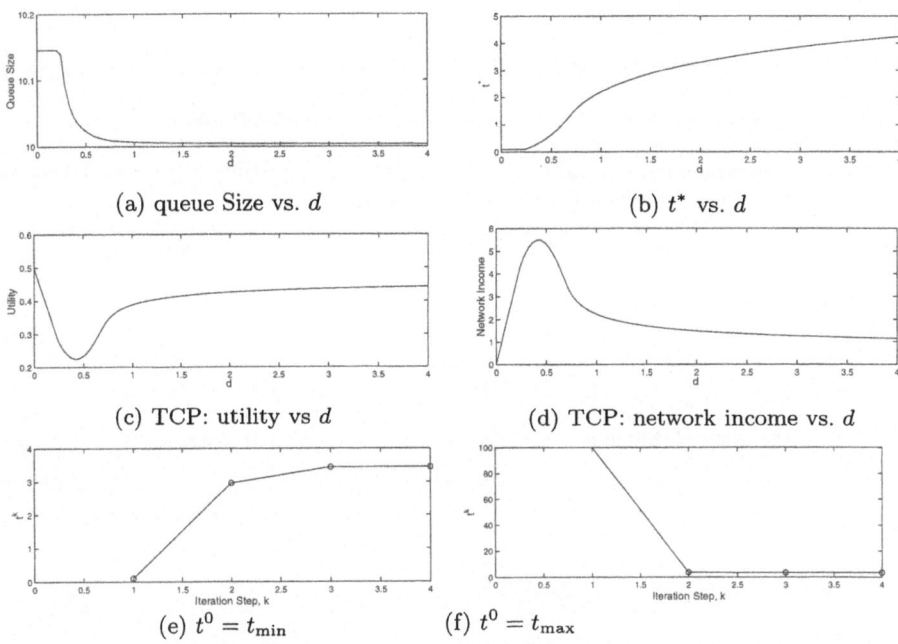

(a) queue Size vs. d

(b) t^* vs. d

(c) TCP: utility vs d

(d) TCP: network income vs. d

(e) $t^0 = t_{min}$

(f) $t^0 = t_{max}$

Fig. 4. Symmetric TCP flows: (a) queue size, (b) t^* vs. d, (c) utility, (d) network income vs. d, e) and f) convergence to Nash equilibrium

Algorithm for Symmetric Nash Equilibrium:

For symmetric Nash equilibrium, we are interested in finding a symmetric equilibrium strategy $\mathbf{t}^* = (t^*, t^*, .., t^*)$ such that for any flow i and any strategy t_i for that flow (real-time session or TCP connection),

$$U(\mathbf{t}^*) \geq U(t_i, [\mathbf{t}^*]_{-i}). \qquad (14)$$

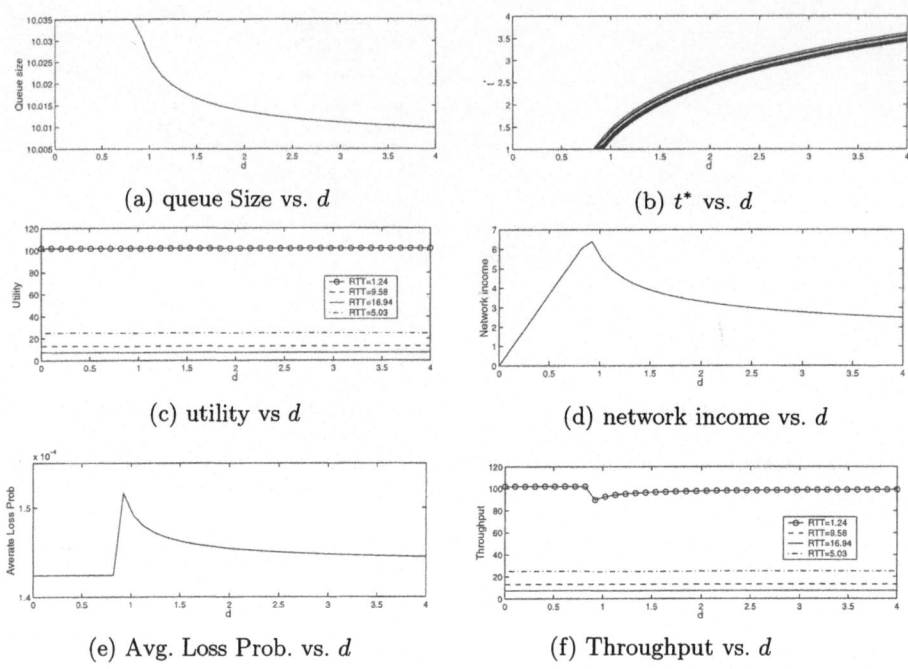

Fig. 5. Non-Symmetric TCP: (a) queue size, (b) t^* vs d; (c) utility, (d) network pricing vs. d, (e) Avg. Loss prob, and (f) throughputs vs. d

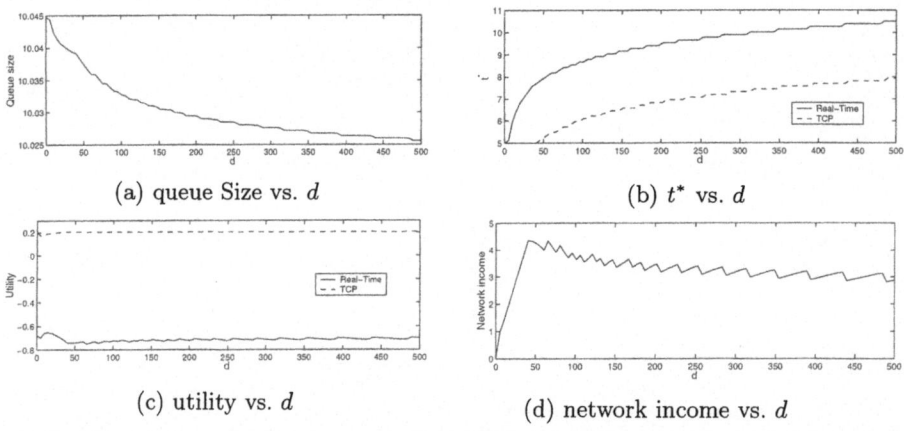

Fig. 6. Real-time and TCP: (a) utility and network income vs. d, b) t^* vs. d, c) utility vs. d, and d) network income vs. d

Next we show how to obtain an equilibrium strategy. Due to symmetry, to see whether $\mathbf{t}^*$ is an equilibrium it suffices to check (14) for a single flow. We shall thus assume that there are $L+1$ flows all together, and that the first L flows uses the strategy $\mathbf{t}^o = (t^o, ..., t^o)$ and flow $L+1$ use t_{L+1}. Define the set $\mathcal{Q}_{L+1}(\mathbf{t}^o) = \arg\max_{t_{L+1} \in [t_{\min}, t_{\max}]} \left(U(t_{L+1}, [\mathbf{t}^o]_{-(L+1)}) \right)$, where $\mathbf{t}^o$ denotes (with some abuse

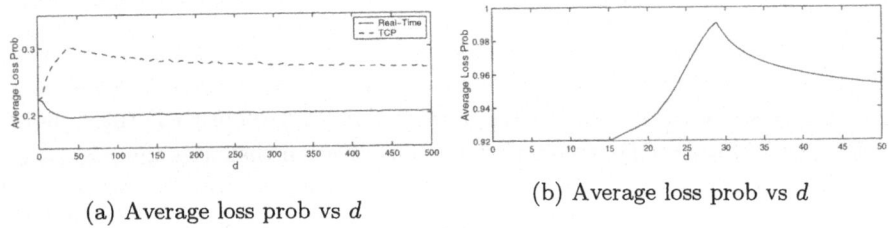

(b) Average loss prob vs d

(a) Average loss prob vs d

Fig. 7. (a) Real-time and TCP, (b) Non-Symmetric Real-Time flows

of notation) the strategy where all flows use t^o, and where the maximization is taken with respect to t_{L+1}. Then $\mathbf{t}^*$ is a symmetric equilibrium if $t^* \in \mathcal{Q}_{L+1}(\mathbf{t}^*)$.

Theorem 6 *[2] Consider real-time connections operating in linear region. The symmetric equilibrium t^* satisfies:* $T^* \left. \frac{\partial \hat{d}(T)}{\partial T} \right|_{T=T^*} = \frac{a\lambda+b}{(I\lambda)^2}$ *where $T^* = 1/t^*$ and $\hat{d}(T) = d(\frac{1}{T})$.*

6 Numerical Examples

In the following simulations, we obtain a unique Nash equilibrium for only real-time sessions or only TCP connections without satisfying the conditions in Thm 4. Moreover, the GA algorithm converges without satisfying the conditions of supermodularity. All the conditions of supermodular games (Thm 3 and Thm 5) and uniqueness of Nash equilibrium (Thm 2 and Thm 3) are only sufficient but not necessary as shown in the numerical results.

The pricing function that we use for player i throughout this section is $d/\exp(t_i)$. We shall investigate how the choice of the constant d will affect the revenue of the network.[2]

Symmetric Real-Time flows: In the following numerical evaluations, we show the variation of different metrics as function of d. Fig. 2 corresponds to a unique symmetric equilibrium in which all the real-time flows have $\lambda_i = 2$Mbps with $t_{min} = 0.01, t_{max} = 100, \mathcal{I} = 20, q_{min} = 10, q_{max} = 40, \mu = 30$Mbps. Here we set the values of parameters to ensure that the system operates in linear region such as $t_{min} > \frac{1}{\Delta q}(1 - \frac{\mu}{\sum_{j \in \mathcal{I}} \lambda_j}) = 0.0083$. The bound on t_{max} is needed only to limit the value of loss probability to 1. The value of d which maximizes the network revenue occurs at $d = 25.75$. All the flows attain a loss rate of 0.25. Note that for

---

[2] We note that it is desirable to have a "nontrivial" parameterized pricing function that leads to an optimal revenue for some parameter. We also tested other pricing functions that did turned out to be "trivial" in the sense that the benefit was always monotone in the parameter; an example of such a function is $\exp(-\beta t_i)$ and the network optimizes with respect to β.

real-time flows symmetric case, $p_i^* = (\sum\limits_{j \in \mathcal{I}} \lambda_j - \mu)/\sum\limits_{j \in \mathcal{I}} \lambda_j$ at the Nash equilibrium is a constant. The average queue size, given by $q_{\min} + p_i^*/t_i^*$, is shown in Fig. 2. We observe the value of t^* at which maximum network income is achieved is close to $t_{\min}$ while the system operates in the linear region of RED throughout.

We plot in Fig. 2 (e,f) sample paths of a connection that uses the Algorithm for symmetric users (Sec. 5) (the evolution for all connections is the same). The figure shows convergence to the same Nash equilibrium when t^0 started from $t_{\min}$ or $t_{\max}$. We plot it for $d = 41.525$. In Figure 2(e), the value of t^* is 3.6163, and in Figure 2(f), it is 3.6162.

Non-symmetric real-time flows: In the next experiment (asymmetric) rates λ_i are drawn uniformly from $[1, 10]$ Mbps with $t_{\min} = 1, t_{\max} = 100, q_{\max} = 40, q_{\min} = 10, \mathcal{I} = 20, \mu = 30$Mbps. Figures 3 (a-d) show how different metrics vary with d at unique Nash equilibrium. To ensure that the flows operate in linear region, we need $t_{\min} > \frac{1}{\Delta q} \geq \frac{1}{\Delta q}(1 - \frac{\mu}{\sum_j \lambda_j})$. We observe that $d = 27.27$ maximizes the network revenue. Figure 3(b) shows that values of t^* for flows having higher rates increase slower than that of flows having lower rates, i.e., higher rate flows experience lower loss rates. Figure 3(c) shows that flows having different rates gain similarly in their utility functions. We plot the average loss rate in Figure 7(b). We confirm in these experiments about uniqueness of Nash equilibrium, although the sample path of different connections will depend on the connection rates.

Symmetric TCP Connections: For symmetric TCP connections we have considered $R_i = R = 20$ms for all connections with $t_{min} = 0.1, t_{max} = 100, \mu = 30$Mbps, $N = 20$. Figures 4(a-f) show the corresponding figures. The maximum value of network revenue is found at $d = 0.4040$. In this symmetric case, the loss probability is given by

$$p^* = \frac{R^2}{3N^2}\left\{\mu^2 + \frac{3N^2}{R^2} - \mu\sqrt{\mu^2 + \frac{6N^2}{R^2}}\right\} = 0.0017$$

To ensure that the symmetric TCP flows operate in the linear region, we satisfy the condition on $t_{min} > \left(\dfrac{-\mu + \sqrt{\mu^2 + 4(\sum\limits_{j \in N} \frac{1}{R_j})^2}}{4\sqrt{\alpha\Delta q}\sum\limits_{j \in N} \frac{1}{R_j}}\right)^2 = 4.6271 \times 10^{-5}$.

We plot sample paths of a connection which show convergence to Nash equilibrium when t^0 started from $t_{\min}$ or $t_{\max}$. We plot it for $d = 3.821$. In Figure 4(e), the value of t^* is 3.4480, and in Figure 4(f), it is 3.4481.

Non-symmetric TCP connections: We present a non-symmetric case in Figures 5(a-f) in which R_is are drawn uniformly from $[1, 20]$ms with $t_{\min} = 1, t_{\max} = 100, \mu = 30$Mbps, $N = 20$. The value of d at which network revenue is highest is 0.9321. We ensure the operation in the linear region by setting

$$t_{\min} > \left(\frac{-\mu + \sqrt{\mu^2 + 4(\sum_{j \in \mathcal{N}} \frac{1}{R_j})^2}}{4\sqrt{\alpha \Delta q} \sum_{j \in \mathcal{N}} \frac{1}{R_j}} \right)^2 = 0.5476.$$

Real-time connections and TCP flows: In this experiment, we combine both real-time and TCP connections. We have $I = 15$, $N = 15$, $\mu = 13$Mbps, RTT=10ms, $t_{\min}^{real} = 5, t_{\max}^{real} = 11, t_{\min}^{TCP} = 5, t_{\max}^{TCP} = 11, \lambda = 1$Mbps, $q_{\min} = 10, q_{\max} = 40$. The highest network revenue is achieved at $d = 40.40, t^{real} = 7.69, t^{TCP} = 5$. In the simulations, we observe the values of $q < q_{\max}$ and since there is atleast one TCP flow i with throughput, $\lambda_i > 0$, it implies that the flow has loss probability, $p_i > 0$ and average queue length, $q > q_{\min}$. We conclude that system operates in linear region. Our objective in this set of experiments is to show that there exists a Nash equilibrium for both real-time and TCP connections.

7   Conclusions

We have studied a fluid model of RED buffer management with different drop probabilities applied to both UDP and TCP traffic. We first computed the performance for fixed drop policies. We then investigated how the drop policies are determined using game theory and obtained and characterized the equilibrium. The equilibrium depends on the pricing strategy of the network provider. We finally addressed the problem of optimizing the revenue of the network provider.

References

1. T. Alpcan and T. Basar, "A game-theoretic framework for congestion control in a general topology networks", *41st IEEE CDC*, Las Vegas, Nevada, Dec. 2002.
2. E. Altman, D. Barman, R. El Azouzi, D. Ros and B. Tuffin, "Pricing Differentiated Services: A Game-Theoretic Approach", INRIA Research report RR-4946, 2003.
3. E. Altman, T. Boulogne, R. El Azouzi, T. Jimenez and L. Wynter, "A survey on networking games", submitted. Available at
   http://www-sop.inria.fr/mistral/personnel/Eitan.Altman/ntkgame.html
4. T. Basar and R. Srikant, "A Stackelberg network game with a large number of followers", *J. Optimization Theory and Applications*, 115(3):479-490, 2002.
5. D. Dutta, A. Goel and J. Heidemann, "Oblivious AQM and Nash Equilibria", IEEE Infocom, 2003.
6. Y. A. Korilis, A. A. Lazar and A. Orda, "Achieving network optima using Stackelberg routing strategies", *IEEE/ACM ToN*, 5(1), pp. 161-173, 1997.
7. M. Mandjes, "Pricing strategies under heterogeneous service requirements", *Computer Networks* **42**, pp. 231-249, 2003.
8. D. Topkis, "Equilibrium points in nonzero-sum n-person submodular games", *SIAM J. Control and Optimization*, 17:773-787, Nov. 1979.
9. Y. Jin and G. Kesidis, "Nash equilibria of a generic networking game with applications to circuit-switched networks", *IEEE INFOCOM '03*.
10. D. D. Yao, "S-modular games with queueing applications", *Queueing Systems*, 21:449-475, 1995.

Performance Planning, Quality-of-Service, and Pricing under Competition

Corinne Touati*[1], Parijat Dube[2], and Laura Wynter[2]

[1] Institute of Information Sciences and Electronics, University of Tsukuba, 1-1-1
Tennoudai, Tsukuba-shi, Ibaraki 305 8573, Japan. `corinne@osdp.is.tsukuba.ac.jp`
[2] IBM T. J. Watson Research Center, Yorktown Heights, NY 10598, USA.
`{pdube,lwynter}@us.ibm.com`

Abstract. In this work we model the relationship between the capacity
and the Quality of Service (QoS) offered by the firm in a competitive
scenario of two firm's working to maximize their profits. Using simple
queueing theoretic models we study the sensitivity of a firm's market
share to price, capacity and market size. Our preliminary studies yield
important properties of the equilibrium solution which may further
provide important "engineering" guidelines for performance planning
and pricing strategies.

Keywords: Queueing theory, Nash equilibrium, e-commerce

1 Introduction

The pricing of electronic goods, network bandwidth, and the internet itself has received considerable attention in the literature in the past decade. In this paper we
are not concerned with pricing the internet, which generally involves discussions
of "best effort" classes versus paying customers, and often of shadow-price-based
schemes which assume marginal cost pricing [4,1]. Rather, we consider pricing
of, and more generally economic planning for e-commerce services, such as web
hosting, from the perspective of the major players in the market.

Many of the preoccupations are the same in modelling e-commerce markets
as in modelling the internet. Queueing theory and other stochastic relationships
are vital. Customer behavior, for example, is modelled through distributions, and
arrivals of customers may be assumed to be Poisson, exponential, etc. However,
we are not concerned with marginal cost pricing, or in ensuring that a best
effort (free) service remains in place. E-commerce services are by definition paid
services, and the motive of firms in the e-commerce marketplace is quite clearly
towards profit maximization, rather than towards public service, as much of the
internet is and will continue to be.

Nonetheless, pure profit maximization cannot be a representative model, as
the market allows for competition, and even very large providers can face shifts of

* The author would like to thank the INRIA for supporting this work. INRIA. Domaine
de Voluceau Rocquencourt - B.P. 105 - 78 153 Le Chesnay Cedex - France.

their clientèle depending upon what happens in the marketplace. In this respect, an equilibrium framework is appropriate for modelling firms' optimal choices. Indeed, the equilibrium framework allows us to compute *stable* price, capacity and QoS choices for the firm, in the presence of other firm(s) and a universe of customer demand that can shift across firms, as a function of the prices and QoS that each one offers. The paradigm that we employ is the Nash equilibrium concept, in mixed strategies; that is, the number of users is sufficiently large, that fractional quantities are quite justified, and can just as easily represent percentages of total demand levels. (Pure strategy equilibria would require each user to choose either provider 1 or 2, and the number of users choosing one or the other would be a natural number; this restriction on the strategy set leads to possible non-existence of a (pure strategy) equilibrium, and does not in our setting add any better insight into the model).

It is however of interest to develop models of pricing/QoS behavior of more than one provider in the electronic marketplace. Indeed, in the market for e-commerce services, other firms can adjust their price schedules rapidly in response to that of a competitor. Furthermore, in the *On Demand* paradigm, firms can augment their capacity/QoS levels instantaneously as well. Then, the question for any provider is no longer how to set prices or capacities when other firms' price choices are given, but rather whether the joint setting of prices by all providers will tend towards an equilibrium, and, in the affirmative, what are the properties of the equilibrium.

The basic formulation of the demand and the market, as well as the choice mechanisms of the users, is taken from [5]. In that reference, a two-firm market (which may represent one large firm, and the rest of the market as the second firm) is considered in a manner similar to that of [3], but with one very important difference. Namely, the *Quality of Service (QoS)* was introduced and along with it, a continuous distribution of price-QoS tradeoff parameters, to describe the *dispersion* of users' choices across the price-QoS frontier.

Indeed, the incorporation of QoS in the model is vital, and well understood: in the commerce of electronic goods, there is generally some product differentiation that is naturally present or can easily be introduced. While spatial factors do not play a role with respect to the Internet, other variations in the quality of service do exist, such as host server and network speeds or response times, availability, reliability etc.

However, if we assume that all users react in the same way to price-QoS tradeoffs, we would obtain seriously biased results in terms of the market share of each firm. Product differentiation allows firms to increase market share because the users are inherently different in their willingness to pay for different levels of quality. To use the internet as an example, some users will pay the higher price of DSL to have a faster, broadband access to the Web, whereas others will not be prepared to pay double the price of a telephone dialup carrier, and will experience usually slower service. The distinction is not necessarily binary; often DSL providers offer multiple service classes, higher QoS is accompanied by higher price. Assuming that the service choices are *Pareto optimal* for some user, then each price-QoS service offering will attract a different segment of the population, and each segment can be characterized by its own, unique *price-QoS tradeoff parameter*. We model these tradeoff parameters explicitly, as introduced in [5] and used in the context of strategic outsourcing in [2], by a continuous, general, random distribution. Depending on the particular distribution chosen, different results are obtained. It was argued in [5] that forms such as exponential (or Pareto, log-normal, etc) are most representative of these tradeoffs in practice.

In this paper, we extend the work of [5] by generalizing the notion of quality of service (QoS) ; that is, we concentrate on a particular characterization of QoS that is of importance in e-services, namely, response time, or delay, and model explicitly the dependence of delay on service capacity. The resulting model is significantly more complex than the capacity-independent versions of [5]. Indeed, it is a challenge to determine the feasible values of the parameters, price, capacity, and QoS (delay).

Our contribution in this paper is therefore to formulate this more complex model, and to derive an auxiliary problem whose solution gives feasible values of the QoS/market share for each firm. In this context, computing a Nash equilibrium becomes a complex numerical exercise that makes use of our derivations. We leave a study of particular Nash equilibrium, as a function of the input parameters, to a future research study, by ourselves or others.

In such an equilibrium setting, the paradigm would work as follows: Supplier 1 (for example) determines his capacity vector so as to maximize some objective (profit) as a function of prices of his own service and that of his competitor(s), and as a function of his competitor's capacity (which determines then the competition's delay, or QoS). Prices, however, are not fixed: for each value of capacity that supplier 1 considers, a vector of equilibrium prices (p_1, p_2) would be determined, using the Nash paradigm, described above. Depending on whether the overall profit of supplier 1 increases or decreases, he modifies his capacity, and so on, until reaching a stationary point (local optimum). This local optimum would represent a "good" capacity-price offering for supplier 1, given the market context, and the responsiveness of the competition (in its price(s)) and of the end user demand (in its patronage of supplier 1 or 2). This paradigm represents an instance of a *Stackelberg*, or leader-follower, game. While we do not compute values of the Stackelberg equilibrium here, we provide the necessary machinery to formulate and solve that important problem.

The structure of the paper is as follows. In Section 2, we recall the framework of [5], that is the price and QoS hypotheses, and price-QoS tradeoff parameters, and how they fit into a Nash equilibrium model. In Section 3 we model the explicit relationship between the QoS offered by a firm and its capacity and provide conditions for the existence of non-trivial market share of each firm. The model is studied for the special case of uniformly distributed price-QoS tradeoff parameter and an explicit closed form expression for each company's share is obtained in Section 4. We then study the sensitivity of the solution to different parameters which provides further insight. Finally we conclude in Section 5 and present directions for further research in this area.

2   The Price-QoS Market Model with Delay-Capacity Relations

Suppose that the e-service offered by firm $i = 1, \ldots I$ is characterized by a 2-tuple $(p_i, d_i(c_i))$ where p_i is the price charged for use of the service and $d_i(c_i)$ is some measure of the quality of service perceived by the customer. Here, as opposed to in [5,2], the QoS shall depend upon, among other things, the capacity held by firm i. Note that p_i is independent of the usage level of the customer, referred to as *flat* price in literature. (Usage-dependent prices are treated in [5,2]).

The quality of service will be taken in the remainder of this paper to be some measure of service performance, namely, the *expected* delay incurred on a typical request. Note that it is possible to extend this framework to more than two (possibly usage-dependent) service characteristics. For simplicity of analysis, however, we shall continue to refer only to the two QoS characteristics of price and capacity-dependent delay.

Each user is then characterized by a particular value of the variable α that models his willingness to pay for a higher quality of service. That is, α gives the user's own tradeoff between price and delay. We shall suppose that the user tradeoff parameter α is described by a random variable, distributed over the population of potential customers and taking values in $[0, 1]$. Let F be the distribution of α. Consider one potential customer n. Given his own value of the tradeoff parameter, α_n, the customer will optimize his choice of provider, among the I firms, by choosing the one that minimizes his combined cost:

$$i^* \in \arg\min_{i}\{\alpha_n p_i + (1 - \alpha_n)\gamma d_i(c_i)\}, \tag{1}$$

where $\gamma = 1$ and is introduced for dimension compatibility (e.g., if p_i is in dollars and d_i in minutes then the unit of γ is dollars/minutes). Observe that α is a dimensionless quantity. Taking α to be a random variable is a critical feature; we are in effect capturing the *universe of users' behaviors* with respect to the cost vs. quality tradeoff. For example, a user requiring low-priority service, for email or file transfer operations, would be characterized by a *high value of QoS, α, e.g., close to 1*, whereas a job requiring more bandwidth, faster service, etc. and for which the user is willing to pay for the better quality, would be characterized by a low value of α (e.g., close to 0). As has been observed in internet traffic as well as in the population in general, the percentage of low values of QoS is much higher than the percentage of high values, across users. This observation has an impact on the *form* of the distribution of the tradeoff parameters, α, as we shall discuss later in this paper.

Note that it is possible to have the tradeoff parameter be dimensionfull, call it w –in units of dollars per time, by defining a *generalized cost* of $p + wd(c)$. It is this latter definition that was used in [5,2]. Here, the use of a parameter that varies from 0 to 1 facilitates some of our computation and hence the price-QoS tradeoff parameter was normalized in the above manner.

We analyze the case of $I = 2$ providers. While [5] considered different price structures (linear, flat, etc. and the different possible combinations of those across providers), we simplify that part of the model here by letting all prices be flat, i.e., usage *in*dependent, and instead exploit the explicit dependence of QoS on the firm's capacity.

Thus a customer chooses provider 1 if

$$\alpha p_1 + (1 - \alpha)\gamma d_1(c_1) < \alpha p_2 + (1 - \alpha)\gamma d_2(c_2), \tag{2}$$

and chooses provider 2 otherwise.

Without loss of generality, we can suppose that $p_1 > p_2$. We can then note that if $d_1(c_1) \geq d_2(c_2)$ then no rational user will join the first firm. In other words, $\forall \alpha \in [0, 1], \alpha p_1 + (1 - \alpha)\gamma d_1(c_1) \geq \alpha p_2 + (1 - \alpha)\gamma d_2(c_2)$. Therefore, as we are interested in the scenario of competitive markets, we suppose that $d_1 > d_2$, that is, the supplier 1 offers a better quality of service (lower delay) but as such charges a higher subscription price. We will then denote in the following by d and p the delays and prices differences respectively:

$$d = d_2(c_2) - d_1(c_1) > 0 \text{ and } p = p_1 - p_2 > 0.$$

In the general setting of usage-based pricing there are thresholds for which one or the other supplier is cost-effective for a user. Since in our model, the customer pays a one-time subscription fee for both providers the threshold is only in α and can be written as $\alpha \leq \frac{\gamma d}{p + \gamma d}$ for choosing supplier 1. Indeed, since supplier 1 offers a better QoS (lower delay), users with lower price-QoS tradeoff parameters prefer supplier 1.

The threshold value of the price-QoS tradeoff parameter, $\hat{w} = \frac{\gamma d}{p + \gamma d}$ determines the split of users between the two providers. We also introduce the notation $\bar{F} = 1 - F$. Thus the profits of providers 1 and 2 can be expressed as follows:

$$\begin{cases} \Pi_1(p) = \lambda p_1 F(\hat{w}) - \xi_1 c_1, \\ \Pi_2(p) = \lambda p_2 \bar{F}(\hat{w}) - \xi_2 c_2, \end{cases} \tag{3}$$

where ξ_i are the marginal costs of providing capacity for each of the firms, $i = 1, 2$. This can represent e.g., the amount paid by the provider i to the bandwidth agent if he leases capacity.

3 Modeling Capacity-Related QoS Metrics

The arrival process of customers is a Poisson process with rate λ. To customer n we associate a vector (S_n, α_n) where S_n is the amount of work brought by user n and α_n is the preference parameter which reflects the customer's choice. The amount of work brought by a customer has some general distribution with mean $1/\mu$ and second moment σ^2. Each customer is processed at the server in a particular discipline, e.g. First-In-First-Out (FIFO), Last-In-First-Out (LIFO), Processor Sharing (PS), etc. An arriving customer joins the server which minimizes its disutility function which we take as a function of the QoS perceived by the user and the price paid by the user. Let us assume that $\{\alpha_n\}$ are i.i.d. random variables with distribution F.

We shall assume that both firms make use of the same service discipline. We then have (see e.g. [6]) the following expressions for average delay depending on the service disciplines at each firm's servers:

- *Case I* – FIFO/LIFO: Then each server can be modeled as an $M/G/1$ queue with FIFO/LIFO service. The mean delay at server i $(i = 1, 2)$, d_i is given by the classical *Pollaczek-Khinchin* formula:

$$D_1 = \frac{\lambda F(\hat{w})\sigma^2}{2c_1 \left(c_1 - \frac{\lambda F(\hat{w})}{\mu}\right)} \quad \text{and} \quad D_2 = \frac{\lambda \bar{F}(\hat{w})\sigma^2}{2c_2 \left(c_2 - \frac{\lambda \bar{F}(\hat{w})}{\mu}\right)}.$$

- *Case II* – PS or LIFO with pre-emption: The mean delay is insensitive to the service distribution and is same as the delay in an $M/M/1$ FIFO queue with mean service rate $= 1/\mu$. Thus:

$$D_1 = \frac{1}{\mu c_1 - \lambda F(\hat{w})} \quad \text{and} \quad D_2 = \frac{1}{\mu c_2 - \lambda \bar{F}(\hat{w})}. \tag{4}$$

Observe that (4) is implicit in d_i, as the right hand side is also a function of d_i, since $\hat{w} = \frac{\gamma d}{p + \gamma d}$. We shall next study the sensitivity of delay to capacity for some specific distributions for α. We restrict the analysis to the case where delays are given by (4).

3.1   Existence of Solutions

Consider a system in which the two competitors announce prices p_1 and p_2 and expected delay d_1 and d_2. The customers arrive and join the queue which minimizes their disutility function. Thus there is an independent splitting of the aggregate arrival process λ based on the two portions of the price-QoS tradeoff distribution, into $\lambda_1 = \lambda F(\hat{w})$ and $\lambda_2 = \lambda \bar{F}(\hat{w})$ where λ_i is the rate of Poisson arrivals at firm i, $i = 1, 2$. D_i are the true mean delays (given by (4)) and d_i are the announced delays. We do not consider here cases when the firms can *cheat* the customers by announcing a smaller delay but later not satisfying it, i.e., $D_i > d_i$, for any $i = 1, 2$ (because the capacity of the firm may not be sufficient to provide the announced delay to the customers). Also we are not interested in the case when $D_i < d_i$, for any $i = 1, 2$, because this will result in less revenue for the firm i. Thus our study is restricted to the scenario where $D_i = d_i$, $i = 1, 2$; in other words, we are interested in the study of the fixed point equations (4).

Proposition 1. *Let p_i and c_i be given, for all i. Then, for any CDF F, the system of fixed point equations (4) admits at most one solution.*

Proof. Let us assume there are two sets of solutions to (4), (d_1, d_2) and $(\tilde{d}_1, \tilde{d}_2)$. Also, let us suppose that $\tilde{d}_1 > d_1$. From equation (4), we can write that $d_1^{-1} + d_2^{-1} = \mu(c_1 + c_2) - \lambda = \tilde{d}_1 + \tilde{d}_2$ which is constant for given parameters. Therefore, we have $d_2 > \tilde{d}_2 > \tilde{d}_1 > d_1$ Thus $\tilde{d} = \tilde{d}_2 - \tilde{d}_1 < d = d_2 - d_1$, implying $\frac{\gamma \tilde{d}}{p+\gamma \tilde{d}} >$ $\frac{\gamma d}{p+\gamma d}$. Therefore $\hat{w} < \tilde{w}$ and finally, as F is non-decreasing: $d_1 = \frac{1}{\mu c_1 - \lambda F(\hat{w})} \geq \frac{1}{\mu c_1 - \lambda F(\tilde{x})} = \tilde{d}_1$ which is a contradiction. $\square$

3.2   A General Beta Distribution for Price-QoS Tradeoff α

We shall suppose throughout that α follows a Beta distribution with parameters a, b. The use of the Beta distribution on a random variable over the interval $[0,1]$ is very natural, and flexible. Indeed, depending on how one sets the two parameters, a and b, one can obtain a distribution approaching normal, exponential, uniform, etc. over the given, finite interval.

The probability density $f(x)$ and the cumulative distribution function $F(x)$ of the Beta distribution are characterized by (with $a, b > 0$):

$$f(x) = \frac{\Gamma(a+b)}{\Gamma(a)\Gamma(b)}(1-x)^{b-1}x^{a-1}, \tag{5}$$

$$F(x) = \frac{\Gamma(a+b)}{\Gamma(a)\Gamma(b)}\int_0^x u^{a-1}(1-u)^{b-1}du. \tag{6}$$

Remark 1. Most types of market scenarions can be captured by working with different values of a and b in the Beta distribution for α. For example, to characterize the price-QoS-queueing game when the value-of QoS tradeoff parameter is not uniform, one can choose parameters a and b so that the form of the Beta distribution is skewed towards the origin, much like a truncated log-normal distribution over $[0,1]$. This can capture the dynamics of a *quality dominant market*. Further with $a = 3, b = 2$, the distribution is skewed towards 1, making the market predominantly *price-dominant* and with $a = 3, b = 3$, the market is sort of an *average market* (For price-dominant market one gets $F(\hat{w}) = 12\int_0^{\hat{w}}(1-u)u^2 du = 12\hat{w}^3\left(\frac{1}{3}-\frac{1}{4}\hat{w}\right)$ and for an average market $F(\hat{w}) = 12\int_0^{\hat{w}}(1-u)^2u^2 du = 30\hat{w}^3\left(\frac{1}{3}-\frac{\hat{w}}{2}+\frac{\hat{w}^2}{5}\right).$)

3.3   Feasible Solutions

Having characterized the distribution of α we proceed to obtain the solution set d_i, $i = 1, 2$ of (4). From (4) we have

$$d = \frac{1}{\mu c_2 - \lambda \bar{F}(\hat{w})} - \frac{1}{\mu c_1 - \lambda F(\hat{w})}. \tag{7}$$

We now introduce a variable X to represent the fraction of users joining the second operator's system times λ. Thus $X = \lambda \bar{F}(\hat{w})$. Also define $A = \mu c_1 - \lambda$ and $B = \mu c_2$. Then from (7) we have:

$$d = d_1 - d_2$$
$$= \frac{1}{B-X} - \frac{1}{A+X}. \tag{8}$$

Solving for X, we obtain:

$$X = \frac{d(B-A) - 2 + \epsilon\sqrt{d^2(A+B)^2 + 4}}{2d}, \text{ with } \epsilon = \pm 1. \qquad (9)$$

The following Lemma allows to constrain the feasible values of X and assures that X needs to be greater than (B-A)/2 and hence $\epsilon = 1$:

Lemma 1 (Existence of a solution to (4)). *Any feasible solution X must satisfy*

$$\max(0, -A, \frac{B-A}{2}) \leq X \leq \min(\lambda, B). \qquad (10)$$

Thus, a necessary condition for the system of fixed point equations (4) to have at least one solution is $\max(0, -A, \frac{B-A}{2}) \leq \min(\lambda, B)$, or equivalently that :

$$\frac{\lambda}{\mu} \leq c_1 + c_2 \text{ and } c_2 - c_1 \leq \frac{\lambda}{\mu}. \qquad (11)$$

That is, together, the two providers can accommodate all the traffic, and the capacity of the second provider (with the higher price and lower delay) is not too much larger than that of the first.

Proof. The constraint $0 \leq X \leq \lambda$ is given by the definition of X . The constraint $\frac{B-A}{2} \leq X$ results from the fact that $d > 0$. Finally, the positivity of d_1 and d_2 implies that $-A \leq X \leq B$. □

Note that, from Lemma 1 we have $\frac{B-A}{2} \leq X$ and therefore $\epsilon = +1$ in equation (9). Let us assume that a and b are integers. Then from the definition of $F(.)$ in (6), we note that for any pair $(a, b) \in N^2$, F is a polynomial of order $a + b - 1$. From (9) we conclude that an acceptable X satisfies:

$$2\lambda\hat{w}\bar{F}(\hat{w}) = (B-A)\hat{w} - 2\frac{(1-\hat{w})}{p} + \sqrt{(A+B)^2\hat{w}^2 + \left(\frac{2(1-\hat{w})}{p}\right)^2}. \qquad (12)$$

Thus $\hat{w}$ can be solved as the solution of a $2(a+b)$ order polynomial from (12). Observe that atmost one solution of (12) is acceptable as the feasible solution $\hat{w}$. Indeed, it is the cut-off value-of-QoS parameter, $\hat{w}$, which allows us to compute the market share of each firm.

In the next section, we explicitly solve this quantity when the Beta distribution parameters are both equal to 1, thereby defining a uniform distribution of price-QoS tradeoffs on the interval $[0, 1]$.

4   Application: Uniformly Distributed Price-QoS Tradeoff Parameter α

We consider a special case of our model in which we let the distribution of α be uniform on the interval $[0, 1]$. Then, $a = 1, b = 1$ in (6) and $F(\hat{w}) = \hat{w}$, and we have $X = \lambda(1 - \hat{w})$, which gives $\frac{d}{p} = \frac{\lambda - X}{X}$. Equation (12) allows us to write X as a solution of a fourth order polynomial.

However, by substituting $\frac{d}{p} = \frac{\lambda - X}{X}$ in (8), we obtain X as the solution of the third order polynomial:

$$P(X) - R(X) = 0, \tag{13}$$

where $P(X) = p(A + X)(B - X)(\lambda - X)$ and $R(X) = X(A - B + 2X)$. Thus, X can be found by examining the intersection of P and R. As P is a third order polynomial, there is either one or three possible values for X. However, observe, from Lemma 1 that the solutions of (13) may not all be acceptable in our system.

Proposition 2. *In the case of uniformly-distributed price-QoS tradeoff parameters α, when the conditions (11) are satisfied, (13) has always a feasible solution, where feasibility means a solution satisfying (10).*

Proof. We must distinguish three cases, shown in Figs. 1-3. We can check that that for any values of c_1 and c_2, (13) always admits three *real* roots, that we denote by X_1, X_2 and X_3 with $X_1 < X_2 < X_3$. We can also check that the only acceptable solution to our system is X_2.

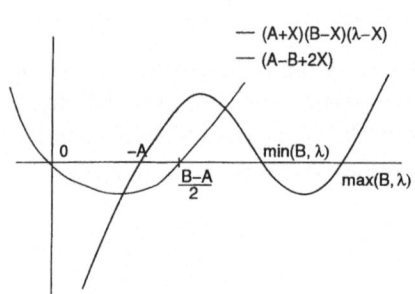

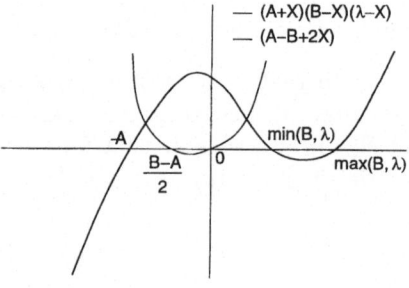

Fig. 2. Case 2: $\mu c_1 > \lambda$ and $\frac{\mu c_2 - \mu c_1 + \lambda}{2} < 0$

Fig. 1. Case 1: $\mu c_1 < \lambda$

- Case 1: $\mu c_1 < \lambda$ (i.e., $c_1 \geq \frac{\lambda}{\mu}$): the first operator does not have a capacity large enough to handle all the traffic. In that case, we recall that Lemma 1 imposes that $\frac{\mu c_2 - \mu c_1 + \lambda}{2} \leq \min(\lambda, \mu c_2)$. Also $\frac{\mu c_2 - \mu c_1 + \lambda}{2} \geq \lambda - \mu c_1$. Then the system has exactly one feasible solution (see Fig. 1).
- Case 2: $\mu c_1 > \lambda$ and $\frac{\mu c_2 - \mu c_1 + \lambda}{2} < 0$. As $\lambda - \mu c_1 \leq \frac{\mu c_2 - \mu c_1 + \lambda}{2}$, the system has exactly one solution (see Fig. 2).
- Case 3: $\mu c_1 > \lambda$ and $\frac{\mu c_2 - \mu c_1 + \lambda}{2} > 0$ (i.e., $c_1 - c_2 \geq \frac{\lambda}{\mu}$). Then, the result comes from the fact that $\frac{\mu c_2 - \mu c_1 + \lambda}{2} \leq \min(\mu c_2, \lambda)$ (see Fig. 3).

□

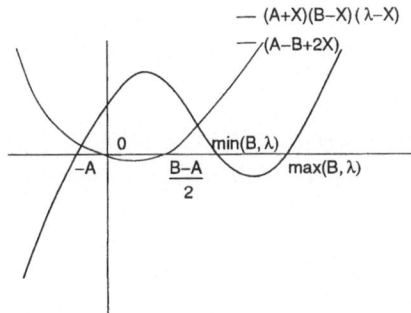

Fig. 3. Case 3: $\mu c_1 > \lambda$ and $\frac{\mu c_2 - \mu c_1 + \lambda}{2} > 0$

Proposition 3. *If the system satisfies the constraints of Lemma 1 then* Cardan's formula *gives the unique solution X as:*

$$X^* = -\frac{(\mu c_1 - \mu c_2 - 2\lambda)p - 2}{3p}$$

$$+ \frac{2}{3p} Cos\left[\frac{1}{3}(4\pi + ArcCos(\mathcal{Y})\right] \sqrt{2 + p\mathcal{G}^2 + 3p\mathcal{H} + (\mu c_1 - \lambda)\mathcal{I}}, \quad (14)$$

where, $\mathcal{G} = (\mu c_2 - \mu c_1 + 2\lambda))$, $\mathcal{H} = -\mu c_2(1 + p\lambda)$, $\mathcal{I} = (1 + p(\mu c_2 + \lambda))$, *and*

$$\mathcal{Y} = \frac{-27(\mu c_1 - \lambda)\mu c_2 p^3 \lambda + 2(2 + p\mathcal{G})^3 - 9p((\mu c_1 - \lambda)p - 1 - \mathcal{I})(\mathcal{H} + (\mu c_1 - \lambda)\mathcal{I})}{2\sqrt{(2 + p\mathcal{G}^2 + 3p\mathcal{H} + (\mu c_1 - \lambda)\mathcal{I})^3}}.$$

Remark 2. Obtaining a closed form equation for $X = \lambda \bar{F}(\hat{w})$ is of interest since the profit functions of each provider are linear in X (from Equation (3)). Observe further from (8) that d_i's can be directly obtained from X.

In this section, we have studied the case of a uniform distribution of the price-QoS tradeoff parameter. We have shown that if the providers choose their capacities so that they can accommodate all the traffic, and the capacity of the provider with the higher price and lower delay is not too much larger than that of the other then the the price-delay-capacity system admits a unique solution. Finally, we gave an analytical formulation of this solution. Additionally, we formulated X as the intersection point of two polynomials. In the next section we shall exploit this characterization to obtain qualitative results on the properties of X.

4.1   Sensitivity Analysis

We would like to determine the influence of the parameters c_1, c_2, λ and the price difference, p, on X, the market share of provider 2. One can show that:

Proposition 4. *The market share of the second provider $(X \equiv X^*)$ is increasing in c_2, the price difference p and the total arrival rate λ, and decreasing in c_1.*

Proof. Let us define $h_{c_1,c_2,\lambda,p}(X) = P(X) - R(X)$:

$$h_{c_1,c_2,\lambda,p}(X) = p(\mu c_1 - \lambda + X)(\mu c_2 - X)(\lambda - X) - X(\mu c_1 - \mu c_2 - \lambda + 2X). \tag{15}$$

As seen in the previous section, h is null and locally decreasing at $X = X^*$. The local behavior of X with the parameters c_1, c_2, λ and p is given by the sign of the quantities $h_{c_1',c_2,\lambda,p}(X)$, $h_{c_1,c_2',\lambda,p'}(X)$, $h_{c_1,c_2,\lambda',p}(X)$ and $h_{c_1,c_2,\lambda,p'}(X)$ respectively where $c_1' > c_1$, $c_2' > c_2$, $\lambda' > \lambda$ and $p' > p$. As each of these h function is locally decreasing and X is a continuous function of parameters c_1, c_2, p and λ, we conclude that if these quantities are positive, then X is an increasing function with their related parameter and decreasing otherwise.

One can show that: $h_{c_1,c_2,\lambda,p'}(X) = X(\mu c_1 - \lambda - \mu c_2 + 2X)\frac{p'-p}{p}$. Therefore if $p' > p$, then $h_{c_1,c_2,\lambda,p'}(X) \geq 0$ and X is an increasing function with p.

Similarly:
$$\begin{cases} \text{if } X \neq \mu c_2, & h_{c_1,c_2',\lambda,p}(X) = X(\mu c_1 + X)\frac{\mu c_2' - \mu c_2}{\mu c_2 - X}, \\ \text{else} & h_{c_1,c_2',\lambda,p}(X) = \mu c_2(\mu c_2' - \mu c_2). \end{cases}$$

And :
$$\begin{cases} \text{if } X \neq \lambda - \mu c_1, & h_{c_1',c_2,\lambda,p}(X) = X(\mu c_2 - X)\frac{\mu c_1 - \mu c_1'}{\mu c_1 + X} \\ \text{else} & h_{c_1',c_2,\lambda,p}(X) = -\mu c_2(\mu c_2 + \mu c_1' - \lambda). \end{cases}$$

We finally study the impact of λ. Let us suppose that $X \neq \lambda$ and $X \neq \lambda - \mu c_1$. We can write $h_{c_1,c_2,\lambda',p}(X) =$

$$X\left[\frac{(\lambda' - X)(\mu c_1 - \lambda' + X)}{(\lambda - X)(\mu c_1 - \lambda + X)}(\mu c_1 - \lambda - \mu c_2 + 2X) - (\mu c_1 - \lambda' - \mu c_2 + 2X)\right].$$

$$h_{c_1,c_2,\lambda',p}(X) = \frac{X}{(\lambda - X)(\mu c_1 - \lambda + X)}\begin{array}{l}[(\lambda' - X)(\mu c_1 - \lambda' + X)(\mu c_1 - \lambda - \mu c_2 + 2X) \\ -(\lambda - X)(\mu c_1 - \lambda + X)(\mu c_1 - \lambda' - \mu c_2 + 2X)]\end{array}$$

$$\geq \frac{X}{(\mu c_1 - \lambda + X)}\begin{array}{l}[(\mu c_1 - \lambda' + X)(\mu c_1 - \lambda - \mu c_2 + 2X) \\ -(\mu c_1 - \lambda + X)(\mu c_1 - \lambda' - \mu c_2 + 2X)]\end{array}$$

$$\geq \frac{X}{(\mu c_1 - \lambda + X)}[(\lambda' - \lambda)(\mu c_2 - X)].$$

$\square$

We note that, while the behavior of x as a function of c_1, c_2 and p is intuitive, the results obtained for λ is quite interesting. It states that in a competitive market, an increase in the total load benefits the provider having a higher delay or "poorer" service.

5   Conclusions and Suggestions for Further Research

We have presented an extension of a line of competitive market models of e-commerce services, such as web hosting, or the internet. The novelty of these models is that they employ a randomly-distributed value of tradeoff parameter, which captures the way different firms, or individuals, react to a palette of price-QoS tradeoffs. In this work, we included the explicit dependence of QoS on a system's capacity, through queueing models. This allows a good number of further generalizations to follow: capacity planning, hierarchical, or Stackelberg, equilibrium, Nash equilibrium models in terms of capacity, etc.

The underlying framework is, however, significantly more complex than without the explicit QoS-capacity relationships. Our contribution is to present the derivations needed to make use of this framework, since obtaining a single feasible point requires the solution of a complex fixed point equation. We provided a general representation of the price-QoS tradeoffs that uses the flexible Beta distribution, as well as an application to uniformly-distributed tradeoff parameters, which is a special case of the beta distribution.

It is clear that it would be of great value to make use of this framework and study the resulting Nash equilibrium, under various hypotheses. Indeed, several questions are of interest: does the resulting Nash system have a nontrivial solution, that is, one in which $p_i \neq 0$, $i = 1, 2$ for different assumptions on the forms of the distribution function F, and, if so, what are the properties of that equilibrium? For capacity planning, we must go one step further; supplier 1 is interested in optimally setting its capacity, given the capacity of its competitor(s) and the equilibrium prices. Therefore supplier 1 formulates a *bilevel program*, over c_1 and (p_1, p_2), where (p_1, p_2) are given by the Nash equilibrium problem across both suppliers. This formulation is also known as a Stackelberg equilibrium, in which supplier 1 represents the "leader" since he can set his capacity and predict the price responses of the competition. Preliminary studies that we have done indicate that, contrary to the constant-delay cases (see e.g., [5]), once capacity-delay relationships are explicitly taken into account, price wars may ensue in a Nash equilibrium. This very preliminary observation requires further study.

References

1. M. Bouhtou, M. Diallo, and L. Wynter. Capacitated Network Revenue Management through Shadow Pricing. *proc. of ICQT, Munich, Germany*, 2003.
2. P. Dube, Z. Liu, L. Wynter, and C. Xia. Outsourcing and price-QoS equilibrium for E-commerce and internet firms: IT *on demand. Proc of IEEE CDC*, Dec. 2003.
3. P. C. Fishburn and A. M. Odlyzko. Competitive pricing of information goods: Subscription pricing versus pay-per-use. *Economic Theory*, 13:447–470, 1999.
4. F.P. Kelly, A.K. Maulloo, and D.K.H. Tan. Rate Control for Communication Networks: Shadow Prices, Proportional Fairness and Stability. *Journal of the Operational Research Society*, 49:237–252, 1999.
5. Z. Liu, L. Wynter, and C. Xia. Usage-based versus Flat Prcing for E-business Services with Differentiated QoS. *Proc. of IEEE CEC '03*, June 2003.
6. R. W. Wolff. *Stochastic modeling and the theory of queues*. Prentice-Hall, Inc., 1988.

Delay Sensitivity Based Assignment of OVSF Codes for the Downlink DCH in WCDMA

Dimitrios N. Skoutas and Angelos N. Rouskas

Department of Information and Communication Systems Engineering,
University of the Aegean, Samos 83200, Greece
{d.skoutas, arouskas}@aegean.gr

Abstract. A significant problem that deteriorates the performance of
3G W-CDMA systems is code blocking. The complete elimination of this
phenomenon is accomplished only if a code reassignment procedure is
employed. However, the relatively high total signaling delay for a single
OVSF code reassignment combined with the strict delay requirements
of some services makes the reassignment of some OVSF codes impractical. In other words, that a reassignment procedure cannot always be
performed and therefore code blocking cannot be eliminated. In this paper, we introduce a new code selection scheme, named Delay Sensitivity
Based Assignment (DSBA), which takes into account the different delay requirements of the incoming calls. The performance of the proposed
scheme is evaluated through event driven simulation and the results show
a significant decrease in code blocking probability, especially for high rate
calls, compared to other previously proposed schemes.

1 Introduction

In third generation mobile communication systems, WCDMA has been selected
as the most promising technology to support high data rate and variable data
rate services with different QoS requirements. Multiple rate transmission is supported by using Orthogonal Variable Spreading Factor (OVSF) codes as channelisation codes [1],[2].

At the downlink of (UTRA FDD) WCDMA systems, a constant bit rate
service is transmitted through a Dedicated Channel (DCH) while a variable bit
rate service is transmitted through a combination of a DCH and a Downlink
Shared Channel (DSCH) or a DCH and a High Speed-DSCH channel. Each
channel is assigned an OVSF code, which in the case of the downlink DCH
channel is normally fixed for the duration of the connection [4].

On the other hand, due to code blocking, it is often necessary to reallocate a
downlink DCH from one OVSF code to another in order to accommodate a new
call. However, if the corresponding connection has very strict delay requirements
the downlink DCH may be "unmovable" and cannot be reallocated to another
OVSF code. Therefore, a reassignment procedure may not be successful due to
the unmovable DCH channels. Thus, code blocking phenomenon cannot always
be eliminated, even if a code reassignment procedure is employed along with an
OVSF code assignment scheme.

N. Mitrou et al. (Eds.): NETWORKING 2004, LNCS 3042, pp. 454–464, 2004.
© IFIP International Federation for Information Processing 2004

Previous works on the reduction of code blocking are attempting to keep the code tree less fragmented, but they do not differentiate between movable and unmovable DCH channels [3], [9], [10]. Consequently, they allow the spreading of the unmovable DCH channels across the OVSF code tree. That decreases the efficiency of the code reassignment procedure, which in turn increases the code blocking probability especially for high rate calls.

In this paper, we propose a new code selection scheme, named Delay Sensitivity Based Assignment (DSBA). DSBA tries to keep the allocation of the unmovable DCH channels as compact as possible at the OVSF code tree. Thus, a code reassignment procedure is more efficient and the code-blocking phenomenon is reduced compared to previously proposed schemes.

The rest of the paper is organized as follows. The system model is presented in Section 2. In Section 3, the problem statement is outlined. Section 4 presents an overview of previously proposed schemes. The DSBA scheme is described in Section 5. Section 6 presents the numerical results and Section 7 concludes our study.

2   System Description

A detailed description on OVSF code generation can be found in [1], [2]. The OVSF code tree, shown in Figure 1, is a binary tree where each node represents a channelisation code. Each code can be denoted as $C_{SF,k}$, where SF is the spreading factor of the code and k is the code number, $1 \leq k \leq SF$. The higher the spreading factor the lower the transmission rate supported by a code. Leaf codes have the maximum spreading factor (SF_{max}) and therefore the minimum data rate, which is denoted by R. The transmission rate R_S supported by an OVSF code with spreading factor SF is always a multiple of a power of two of the lowest available rate:

$$R_S = K \times R, K = \frac{SF_{max}}{SF} = 1, 2, 4, 8, \ldots \tag{1}$$

Two OVSF codes are orthogonal, if and only if, none of them is an ancestor of the other. Therefore, once a code is assigned, all of its ancestors, as well as all of its descendants are blocked and cannot be used until the code is released.

2.1   Code Blocking Condition

A difficulty in the assignment of OVSF codes is a condition called code blocking [3]. An example is shown in Figure 1. Suppose that codes C_{41} and C_{44} are already assigned. Due to the orthogonality constraint, all the descendant and ancestor codes of C_{41} and C_{44} respectively cannot be used. Although the system has adequate capacity to support an additional connection at rate 8R the only codes that can support rate 8R (C_{21} and C_{22}) are already blocked.

Code blocking reduces spectral efficiency and increases call-blocking probability especially for higher data rate users. A first possible countermeasure that reduces this phenomenon is the clever selection among possible candidate codes

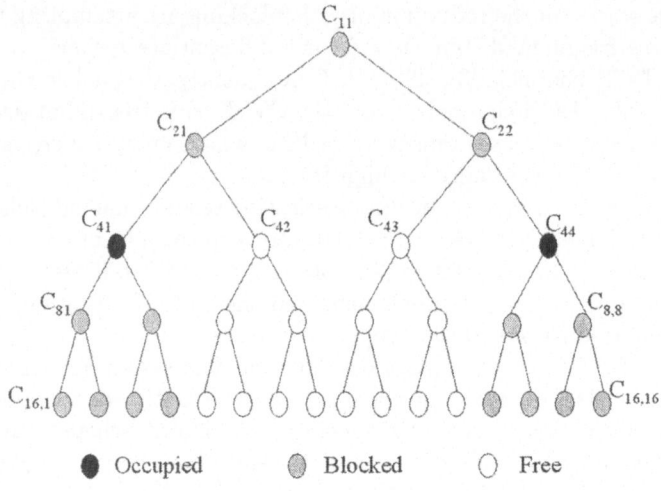

Fig. 1. Code blocking: although the total remaining capacity is 8R, a connection of rate 8R will be blocked

during the assignment process. However, because of the statistical nature of the departure process, the complete elimination of code blocking is accomplished only if a code reassignment procedure can be employed. A reassignment procedure reallocates ongoing calls to other codes so that a new call can always be supported if the system has enough free capacity to support the requested rate.

2.2   Transport Channels

In (UTRA FDD) WCDMA [4], there are three types of downlink transport channels:

1. Common channels are suited for the transmission of small data amounts such as signaling data or small IP packets. Therefore, common channels are mainly used during connection set up process.
2. Shared channels allow a single OVSF channelization code to be shared among several users. They are made for the transmission of medium or large data amounts. There are two types of downlink shared channels:
   a) The Downlink Shared Channel (DSCH) and
   b) The High Speed Downlink Shared Channel (HS-DSCH).
   DSCH and HS-DSCH may change transmission rate every 10ms or 2ms respectively. Both DSCH and HS-DSCH channels are associated with a DCH.
3. Dedicated channels are reserved for a single user only and support transmission rates from a few kbps up to 2Mbps. Each downlink DCH channel is assigned an OVSF code, which is normally fixed for the duration of the connection. Downlink DCH is used in the following cases [4], [5]:
   a) For conversational Real Time (RT) services that have fairly constant bit rate but high delay sensitivity.
   b) In association with a DSCH Channel.
   c) In association with a HS-DSCH Channel.

3   Problem Statement

Code blocking can be easily solved in the case of DSCH or HS-DSCH channels since for these channels a reassignment procedure can always be performed every 10ms or 2ms respectively. This is not the case for the downlink DCH channels: The total signaling delay for a single OVSF code reassignment, performed with an RRC layer procedure called Physical Channel Reconfiguration, is evaluated at 220ms [6]. Such a delay can not be tolerated by Real Time (RT) conversational services which in most cases have very strict delay limits (around 250ms) [7], [8] but can be tolerated by an interactive service such as Web- browsing.

We will refer to a DCH channel as "unmovable" if it is used by an RT conversational service, or if it is associated with a HS-DSCH, which is used by some service with very strict delay limit. Then in either of these cases, DCH cannot be reassigned. That is because such a reassignment would significantly increase the total delay, resulting in unacceptable quality of service.

On the other hand, we will refer to a DCH channel as "movable" if it is related with a delay insensitive service. For example, there is no restriction when reassigning a downlink DCH associated with a DSCH, as the applications served by the latter usually have very low delay sensitivity.

Concluding, the problem at hand is that the reassignment procedure is impossible when some of the involved OVSF codes cannot be reassigned, and hence code blocking cannot be eliminated. In the rest of the paper, we will refer to these codes, related to an unmovable DCH, as the "unmovable" codes. On the other hand, we will refer to the codes related to some service with low delay sensitivity, as the "movable" codes since these codes can be reassigned (e.g. codes assigned to a DCH associated with a DSCH).

4   Overview of Other Proposed Schemes

4.1   Dynamic Code Assignment (DCA)

A dynamic code assignment (DCA) scheme is proposed in [3]. DCA aims at minimizing the number of OVSF codes that must be reassigned in order to support a new call. The algorithm is based on the concept of assigning a cost function to each candidate branch and identifying a branch with a minimal cost. The cost of a branch is defined as the minimum number of code reassignments, which are necessary in order to empty the branch. This scheme does not specify a code allocation strategy when a new call arrives or when an ongoing call is reallocated. If there are more than one candidate codes for an incoming call then any of them can be assigned to the new call. The complexity of the DCA scheme depends on the type of the search algorithm used to locate the minimum cost branch.

4.2   Crowded-First Scheme

Crowded-first strategy is proposed in [9]. The main idea behind this scheme is to keep the code tree less fragmented by allocating the codes as compact as possible. This is achieved by always selecting the code, whose immediate upper layer subbranch will have the least free capacity after the allocation. In the case of ties the search continues one more layer up, until the tie is resolved. Crowded-first is very efficient and significantly reduces code-blocking probability provided that all the DCH channels are allocated to movable OVSF codes.

4.3   Fewer Codes Blocked (FCB) Scheme

In [10] the Fewer Codes Blocked (FCB) scheme is introduced. In this scheme the measure that differentiates the candidate codes during the selection process is the number of upper layer codes that are not already blocked due to some other existing code allocation and will be blocked if the candidate code is assigned. The code, which has the minimum value according to the FCB criterion, is the one to be selected. Ties are resolved by choosing the first, from the left, code of the OVSF code tree, which supports a rate equal to the requested rate (Leftmost criterion). FCB has the same efficiency as the crowded-first scheme, but it has less computational complexity. That is because FCB does not require a recursive search from the level of the code up to the root level during decision process.

During system operation it is inevitable that the OVSF code tree will become fragmented regardless of the code assignment scheme. Therefore, a code assignment scheme has always to be combined with a reassignment procedure in order to eliminate code blocking. Because FCB and Crowded-first schemes use the same criterion for the allocation of both movable and unmovable OVSF codes, the unmovable codes are scattered across the code tree. Consequently, a reassignment procedure may fail due to code blocking, and unmovable OVSF codes must be allocated as compact as possible at the OVSF code tree.

5   The Delay Sensitivity Based Assignment Scheme

Assuming that all the connections are served by DCH channels a code at the OVSF code tree may be a:

1. Free code: The code and all of its ancestors as well as all of its descendant codes are not assigned to a downlink DCH channel. (e.g. C_{82} in Figure 2)
2. Occupied movable code: The code is assigned to a movable downlink DCH channel. (e.g. C_{86} in Figure 2)
3. Occupied unmovable code: The code is assigned to an unmovable downlink DCH channel. (e.g. C_{81} in Figure 2)
4. Blocked movable code: One of its ancestors or one or more of its descendant codes is assigned to a movable downlink DCH channel. (e.g. C_{43} in Figure 2)
5. Blocked unmovable code: One of its ancestors or one or more of its descendant codes is assigned to an unmovable downlink DCH channel. (e.g. C_{41} in Figure 2)

The main idea behind the proposed Delay Sensitivity Based Assignment (DSBA) scheme is to avoid the scattering of the unmovable OVSF codes across the code tree. That is accomplished with clever selection among possible candidate codes during the assignment process. In that way, the use of a code reassignment procedure will be successful more frequently and code blocking phenomenon will be reduced. The proposed scheme consists of two sub-schemes namely M-FCB and U-FCB. More specifically, at the arrival of a new call requesting a rate of kR, where k is a power of two:

1. IF the system has enough capacity to support the rate requirement of the call then:
   a) IF the service has low delay sensitivity, it will be allocated to a movable code and the M-FCB scheme will be used.
   b) ELSE the service has high delay sensitivity, it will be allocated to an unmovable code and the U-FCB scheme will used.
2. ELSE the call is blocked

5.1   The M-FCB Scheme

Candidate codes to receive the new call are all the free codes that support a rate of kR.

1. IF there is one or more candidate codes: The criterion for the selection of a candidate code is the number of upper layer codes that are not already blocked by some other movable codes and will be blocked if the candidate code is assigned. The code that has the minimum value according to this criterion is the one to be selected. Ties are resolved by choosing the first, from the right, candidate code of the OVSF code tree, which supports a rate equal to the requested rate.
2. ELSE the Code Reassignment Scheme is applied. If the reassignment scheme fails due to code blocking, the call is blocked.

5.2   Code Reassignment Scheme

Code Reassignment Scheme is a heuristic algorithm, which aims at minimizing the number of OVSF codes that must be reassigned in order to support a new call. Candidate codes to receive the new call are those codes that are restrained from use because some of their descendants are occupied only by movable OVSF codes.

1. IF there is one or more candidate codes:
   a) Sort the candidate codes according to the number of their occupied descendant codes and choose the one with the smallest number. Resolve ties by selecting the code whose occupied descendant codes carry the smallest capacity.
   b) Once the subbranch to become empty is selected, each of the occupied descendant codes must be transferred to other subbranches. Each reassignment is treated like a new call. However, a reassignment may fail due to code blocking.

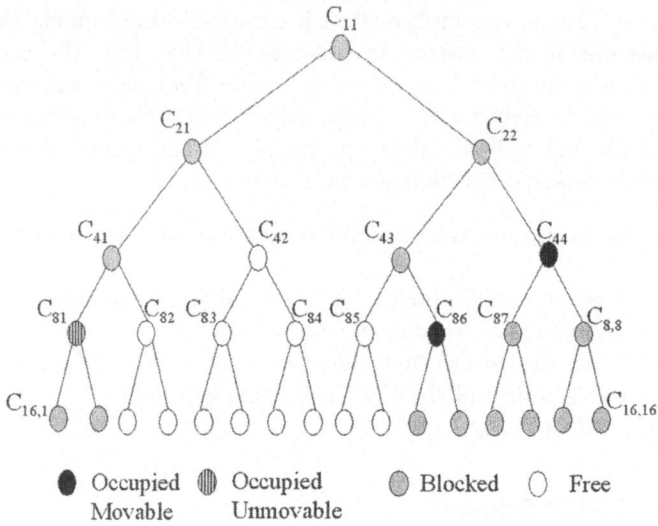

Fig. 2. Code assignment examples

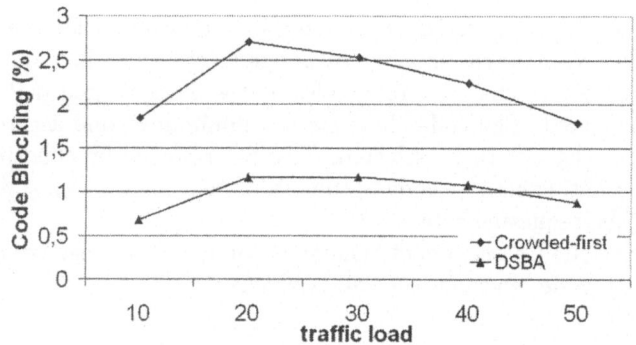

Fig. 3. Code blocking probability when percentage of unmovable codes is low

      i. IF all the reassignments are completed successfully the candidate code is assigned to the new call.

      ii. ELSE the candidate code is rejected and the procedure continues from step 1 for the rest of the candidate codes.

2. ELSE the call is blocked.

5.3 The U-FCB Scheme

From all the OVSF codes of rate equal to the requested rate, the candidates codes to receive the new call are:

- The free OVSF codes.
- The occupied movable OVSF codes.

– The OVSF codes which are blocked only by some other lower rate movable codes.

1. IF there is one or more candidate codes:
   a) The criterion for the selection of a candidate code is the number of its ancestor codes that will be blocked after the code assignment and they are not already blocked by some other unmovable codes. The code that has the minimum value according to this criterion is the one to be selected. Ties are resolved by choosing the first from the left candidate code at the OVSF code tree.
   b) If the chosen candidate code is already blocked or occupied by movable codes then the codes that cause the blocking must be transferred to other subbranches. If any of the reassignments fail due to code blocking then the candidate code is rejected and U-FCB criterion is again applied to the rest of the candidate codes. The code assignment process concludes when an appropriate code is found or when all the candidate codes are rejected.
2. ELSE the call is blocked

For example consider the code tree in Figure 2. Suppose a new call requesting a code of rate $2R$:

1. If the user's service is not delay sensitive then the OVSF code, which will be allocated to the call, will be a movable code. In this case the M-FCB scheme will be used in order to find the most appropriate OVSF code for the call. According to the M-FCB scheme candidate codes for the new call are C_{82}, C_{83}, C_{84} and C_{85}. The value of the M-FCB criterion for the C_{82}, C_{83}, C_{84} and C_{85} codes are respectively 2, 2, 2 and 0. Consequently code C_{85}, which has the minimum value according to M-FCB criterion is the one to be selected.
2. If the user's service is very delay sensitive then the OVSF code, which will be allocated to the call, will be an unmovable code. In this case the U-FCB criterion will be used in order to find the most appropriate OVSF code for the call. According to the U-FCB criterion candidate codes for the new call are C_{82}, C_{83}, C_{84}, C_{85} and C_{86} and the values of the U-FCB criterion are respectively 0, 1, 1, 2 and 2. Consequently code C_{82}, which has the minimum value of the U-FCB criterion will be selected.

Concluding we can say that the main idea behind the DSBA scheme is to allocate a movable code closer to other movable codes and an unmovable code closer to other unmovable codes. In that way, the scattering of unmovable codes along the code tree is avoided and therefore code blocking is reduced.

6   Numerical Results

The performance of the proposed scheme is evaluated through event driven simulation on a 7-layer OVSF code tree. Calls are assumed to arrive according to

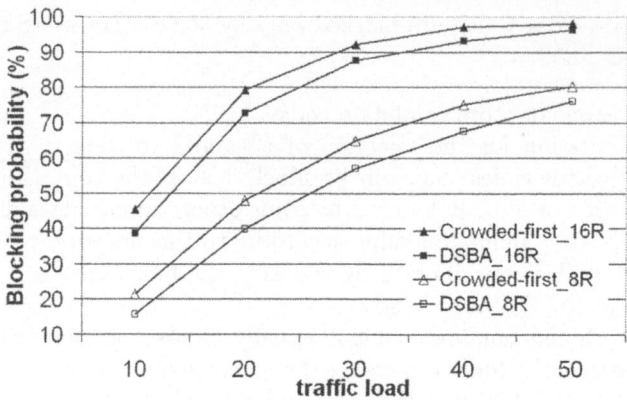

Fig. 4. Blocking probability of high rate calls when percentage of unmovable codes is low

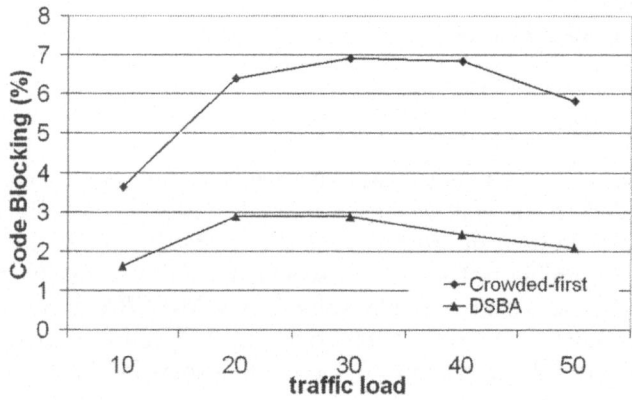

Fig. 5. Code blocking probability when percentage of unmovable codes is high

a Poisson process, while their duration is exponentially distributed with equal mean. We assume that all the connections are served by DCH channels. Each incoming call may request a rate of $R, 2R, 4R, 8R,$ or $16R$ and has a P_U probability to be related with a delay sensitive service with a very strict delay limit. In the last case, the incoming call will be allocated to an unmovable OVSF code. To ensure the stability of the results, each simulation run consisted of at least 100000 incoming calls.

The goal of the simulation is to study the code blocking performance of DSBA scheme in two different study scenarios. Crowded-first scheme combined with a Code Reassignment Scheme has also been evaluated for comparative purposes. The uniform rate distribution $(R : 2R : 4R : 8R : 16R = 20 : 20 : 20 : 20 : 20)$ is used in both scenarios while the P_U probability changes.

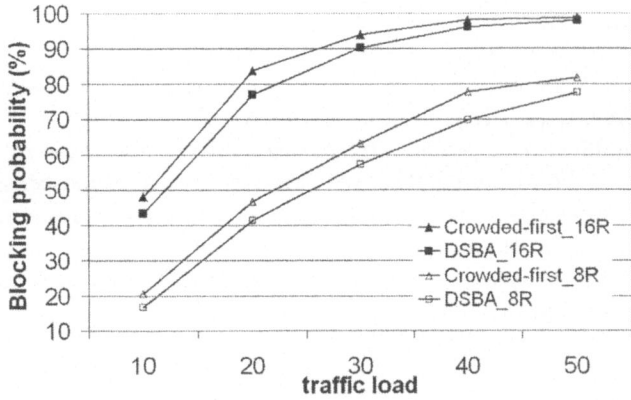

Fig. 6. Blocking probability of high rate calls when percentage of unmovable codes is high

6.1   Low Percentage of Unmovable Codes

At the first study scenario $P_U = 0, 4$. The code blocking probability for increasing traffic load is shown in Figure 3. For both schemes as the traffic load increases, the code blocking probability reaches its peak value and then, as the traffic load continues to increase, the code blocking probability decreases. That is because at high traffic loads the calls often cannot be served by the system due to lack of capacity (capacity blocking). Therefore, capacity blocking gradually replaces code blocking which is reduced. When DSBA is used, it can be observed that the code blocking probability is very low and ranges approximately between 0,7 and 1,2. This is not the case for the Crowded-first scheme. The scattering of the unmovable OVSF codes across the code tree leads to higher code blocking probability, which ranges approximately between 1,8 and 2,7. Calls requesting for high rates are the most affected by code blocking. By reducing code blocking, the blocking rate for such calls is reduced and therefore call acceptance becomes more fair. Figure 4 shows the blocking rate of calls requesting for an $8R$ or $16R$ rate under the DSBA and Crowded-fist schemes respectively. As we can observe the use of the DSBA scheme results in reduced blocking rate at all traffic loads.

6.2   High Percentage of Unmovable Codes

At the second study scenario the percentage of unmovable OVSF codes is increased to $P_U = 0, 6$. Figure 5 shows the code blocking probability for increasing traffic load. As we expected, the increased number of unmovable OVSF codes results in increased code blocking for both schemes. However, the code blocking probability for the DSBA scheme does not exceed 3 per cent while at the same time for the Crowded-first scheme the code blocking probability ranges approximately between 3.8 and 7. Finally, Figure 6 shows the blocking rate for calls requesting for an $8R$ or $16R$ rate. As in the first study scenario, DSBA scheme outperforms Crowded-first at all traffic loads.

7   Conclusions

We proposed a dynamic OVSF code allocation scheme namely Delay Sensitivity Based Assignment (DSBA), for the downlink DCH channel of W-CDMA 3G mobile communication systems. DSBA takes into account the different delay requirements of the incoming calls, reduces code blocking and consequently reduces the blocking rate especially for higher rate calls, which are most affected otherwise. Therefore, call acceptance becomes fairer. The simulation results have demonstrated the efficiency of the DSBA scheme compared to some other previously published approach.

References

1. F. Adashi, M. Sawahashi, and K. Okawa: Tree structured generation of orthogonal spreading codes with different lengths for forward link of DS-CDMA mobile radio. Electronic Letters, vol. 33, pp. 27-28, Jan 1997.
2. 3GPP, Technical Specification 25.213 V5.2.0 (2002-09), Spreading and Modulation (FDD).
3. T. Minn, K.Y. Siu: Dynamic Assignment of Orthogonal Variable Spreading Factor Codes in W- CDMA. IEEE Journal on Selected Areas in Communications, Vol. 18, No. 8, August 2000, pages 1429-1440.
4. Harri Holma and Antti Toskala: WCDMA for UMTS Radio Access For Third Generation Mobile Communications. John Wiley and Sons, 2001.
5. 3GPP, Technical Specification TR 25.848 V4.0.0 (2001-03), Physical layer aspects of UTRA High Speed Downlink Packet Access.
6. Paolo Goria, Claudio Guerrini, Alessandro Vaillant: Signaling delay of code allocation strategies. In Proc. of IST Mobile & Wireless Telecommunications Summit 2002.
7. 3GPP Technical Specification 22.105 V6.2.0 (2003-2006) Services and service capabilities.
8. 3GPP Technical Report 25.853 V4.0.0 (2001-2003) Delay Budget within the Access Stratum.
9. Y.-C. Tseng, C.-M. Chao, and S.-L. Wu: Code placement and replacement strategies for wideband CDMA OVSF code tree management. in Proc. of IEEE GLOBECOM, vol. 1, pp. 562-566, 2001.
10. A. Rouskas and D. Skoutas: OVSF code assignment and reassignment at the forward link of W- CDMA 3G systems. In Proc. of IEEE PIMRC, vol. 5, pp. 2404-2408, 2002

Interoperability Criteria and Mechanisms for Seamless Inter-working between UMTS-HSDPA and HIPERLAN/2 Networks Enhanced with MIMO Techniques

Tareq Al-Gizawi[1], Kostas Peppas[1], Fotis Lazarakis[2], Remy Pnitenet[3], and
Jeremy Gosteau[3]

[1] Institute of Communications and Computer Science, National Technical University of Athens,
9 Heroon Polytechniou, Zographou, 15773, Athens, Greece
tarek@telecom.ntua.gr
[2] Institute of informatics 7 telecommunications,
National center of scientific research "Demokritos", Athens, Greece
flaz@iit.demokritos.gr
[3] Motorola Labs, paris, France, Parc les algorithmes st. Aubin 91193
{remey.pintenet,jerem.gosteau}@motorola.com

Abstract. The concept 4G wireless communication systems is to provide a user with a rich range of services across different radio access technologies, while maintaining the service's minimum QoS requirement, independently on the coverage area, mobility conditions, and using a single mobile terminal. To achieve this goal the need for interoperable heterogeneous wireless networks emerges. In this article we identify the challenges that arise when enhancing interoperability functionality to two different radio access networks. Specifically we will focus on inter-working UTRA FDD HSDPA and WLANs networks, as two strong candidates for composing the 4G environment. We also propose an interoperable architecture to achieve seamless inter-working between the aforementioned networks. The article also introduces two interoperable criteria that triggers interoperability based on the use of Cost functions. Namely, these triggers are the initial user assignment to the optimal network and inter-system handover. Each of the two triggers initiates a respective interoperability algorithm. Finally, In order to characterize the inter-operability mechanisms behaviour and address an accurate QoS performance analysis, a software simulation platform has been developed. The platform is enhanced with MIMO transceivers and takes into consideration: network configuration, propagation conditions, fast fading, and service requirements.

Keywords: Heterogeneous HSDPA, HIPERLAN\2, interoperability, Cost function, MIMO , BS, AP

1 Introduction

The huge growth of the number of mobile subscribers world-wide, during the last decade, together with the increasing demand of higher transmission rates and flexible

N. Mitrou et al. (Eds.): NETWORKING 2004, LNCS 3042, pp. 465–475, 2004.

access to diverse services, motivated significant research standardization and development effort in the area of mobile communication systems. In Europe the transition from 2G to the 3G generation has started, and system designers and IST project (ex. FITNESS, METRA, ASILUM) are already working on the 4G technology. The main goals of 4G networks are to succeed a universal ubiquitous coverage across different radio technologies using a single terminal, and also to provide the 4G subscriber with rich range of services with variable bandwidth and QoS. These features demand connectivity across multiple networks with different radio technologies, over different geographic areas, as well as for accessing different types of services. This raises the need for interoperable networks.

(4G) communication systems will consist of multi-radio technologies and overlapping wireless networks. The wireless network can be organized in layered structures similar to hierarchical cell structure in cellular mobile radio systems as depicted in fig. 1

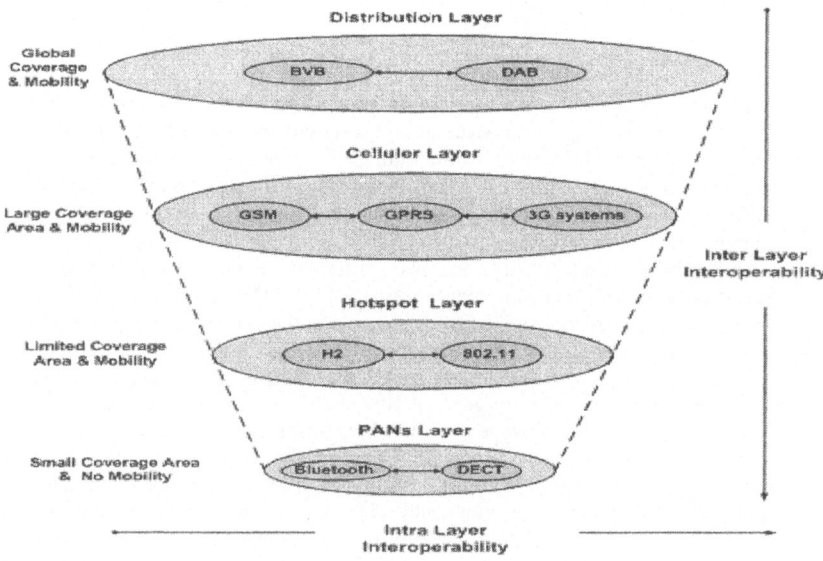

Fig. 1. 4G Layered Architecture

Interoperability can be defined as the capability of a heterogeneous network to support seamless mobility (roaming) between different access radio technologies, maintaining at least the users' minimum QoS requirements.

In this paper we investigate interoperability between High Speed Downlink Packet Access (HSDPA) [1] channel of UTRAN and HIPERLAN/2 [2]. These two radio access technologies can be seen as complementary: the first network supports user mobility while the second supports portability and provides users with higher throughput within limited coverage areas (i.e. hotspots). Introducing a seamless interworking platform would result in higher bandwidth, higher capacity and enhanced QoS. These benefits can be achieved only if inter-network (roaming) enables an efficient management of radio resources. As a result, users can access new services and also select the most appropriate network with an enhanced QoS with respect to the desired service requirements. The achievement of these enhancements depends on the integration architecture of the two technologies and on developing efficient interoperability mechanisms and criteria.

2   Interoperability Criteria and Mechanisms

In this section we address the challenges that arise when incorporating interoperability mechanisms and algorithms between HSDPA and WLAN networks. The inter-operability platform along with the interoperability criteria and mechanisms that the platform takes into account are also presented.

2.1   Interoperability Challenges

When enhancing a 4G network with interoperability capabilities two main challenges arise at the system level:

- *User assignment to the optimal access technology*: given that a user may be offered access from more than one wireless technology, one has to consider how the terminal and the network will choose the optimum access technology suitable for the required services. The decision could be taken based on a combination of criteria such as the measured link quality, users' profile, required service and terminal type

- *Mobility*: mobility is seen as a critical issue when enhancing interoperability in 4G networks. Specifically, at the IP layer the design of mobility enabled IP networking architecture is needed. This includes the exploitation of the existing mobile IP protocols (i.e. Mobile IPv4, Mobile IPv6, and cellular IP [3,4]) in order to support fast and seamless inter-system handover, QoS and accounting and security issues. A handover at the link layer could trigger a handover procedure at the IP layer in order to minimize IP handover delay. Finally, the strategy and criteria for interoperability at the link level should also be defined. This strategy will depend on the operators' system optimization criterion (i.e. traffic balancing, maximizing the number of satisfied users)

- *Re-configurability*: the link level algorithms used at the transceiver should be able to adapt to the changes in propagation environment, interference condition, users speed, etc.

2.2 Interoperability Platform the m Description (Criteria and Mechanisms)

The interoperability platform illustrated in Fig. 2 describes our approach to incorporating interoperability between HSPA and WLAN networks. The main concept is to use Cost functions for both the Initial User Assignment (IUA) and Inter-System Handover (ISH) [5]. Each of the two Cost functions monitors the essential parameters at the system level in order to trigger the respective interoperability algorithm. The weights assigned to the Cost functions parameters are tuned adequately. The user computes the relative Cost function. Upon the computed value an interoperability procedure can be triggered. Following the Cost function computation, the target network is also identified. A negotiation between network capabilities and QoS user requirements takes place, leading to a final decision. The weight values of the Cost function are adapted appropriately after a number of simulation runs in order to meet the network optimization criterion for different test cases.

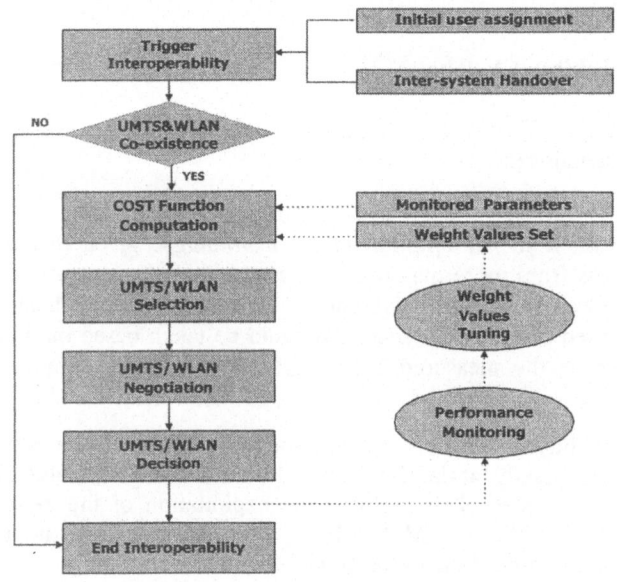

Fig. 2. General Interoperability platform

- *IUA Algorithm:* When a user is within the joint HSDPA/HIPERLAN/2 deployment the Cost function will be computed to make the decision to which of the two networks the user should initially request access. The parameters of the IUA Cost function are :

- *Terminal type*: two terminal types (PDAs and laptops) are considered. The weight value assigned to a laptop will contribute in giving higher priority for accessing the WLAN

- *Traffic specification*: some services are more adapted to a network due to their characteristics (maximum delay, minimum throughput, priority, security requirements). Hence we take into account the traffic specifications for assigning the service to the appropriate system– e.g. between FTP and web browsing, the former will be given a higher weight to access the WLAN

- *Speed:* stationary or users moving at walking speed are considered more appropriate to a WLAN, hence they are given a higher weight, which will contribute to giving preference access to the WLAN network

- *User preference*: depending on the users' preference (profile, given user previous experience, subscription specification or existing connections), a weight will be given in order to contribute to the users' final decision. Users that are not authorized to access either of the two networks may also be considered.

A Cost function can be expressed as a linear combination of the preference functions with the associated weights; hence the IUA Cost function can be expressed as in eq.1 [6]:

$$CF_{IUA} = \beta_{TT} \cdot P_f(TT) + \beta_{TS} \cdot P_f(TS) + \beta_{Sp} \cdot P_f(Sp) + \beta_{UP} \cdot P_f(UP) \quad (1)$$

Where:

$P_f(X)$: is the preference function associated to the criterion X,

β_X are the weights corresponding to the preference function $P_f(X)$ for the criterion X,

TT: the Terminal Type criterion,

TS: the Traffic Specification criterion

Sp: Speed criterion

UP: the User Preference criterion.

The Traffic Preference (*TP*) ranges from 1 to 4 and is defined as follows:

- *TP 1:* Requests a WLAN connection and cannot access the HSDPA network

- *TP 2:* Requests a WLAN connection but can be allocated to the HSDPA network as well

- *TP 3:* Requests an HSDPA connection but can be allocated to a WLAN network as well

- *TP 4:* Requests an HSDPA connection and cannot access the WLAN network

The output of the Cost function determines the traffic preference according to the following rule:

$$\begin{cases} 0 \le CF < 0.25 & \Rightarrow & \text{Traffic Preference 1} \\ 0.25 \le CF < 0.5 & \Rightarrow & \text{Traffic Preference 2} \\ 0.5 \le CF < 0.75 & \Rightarrow & \text{Traffic Preference 3} \\ 0.75 \le CF \le 1 & \Rightarrow & \text{Traffic Preference 4} \end{cases}$$

- **_ISH Algorithm:_** The inter-system handover algorithm aims at the successful switching from one network to another, maintaining the minimum QoS requirements for the considered on-going communication.

The proposed ISH Cost function takes into account the following parameters in order to make the decision on when an inter-system handover procedure must be triggered:

- *Average Users' throughput*: dropping below the minimum QoS requirements based on the average taken over one second of real time simulation.

- *Link quality*: if the received signal-to-noise ratio results in a Packet Error Rate (PER) that exceeds the upper bound defined by the minimum QoS requirements.

- *Mobility*: the departure of users from the area where WLAN and HSDPA coexist, and entering an exclusively HSDPA coverage area.

- *Traffic load*: if the load in one system exceeds a predefined threshold, while the other network has still resources available.

ISH triggering can be categorized as event triggered, periodic, or on demand. When the ISH is triggered, the user computes the Cost function with inputs provided from both user measurements and network signalling information as illustrated in eq. 2:

$$CF_{ISH} = \beta_{MR} \cdot P_f(MR) + \beta_{LQ} \cdot P_f(LQ) + \beta_{Lo} \cdot P_f(Lo) + \beta_{Pr} \cdot P_f(\text{Pr}) \quad (2)$$

Where:

$P_f(X)$: is the preference function associated to the criterion X,

β_x : are the weights corresponding to the preference function $P_f(X)$ for the criterion X

MR: the Mobility Requirement criterion

LQ: the Link Quality criterion

Lo: the Load criterion

Pr: the Prediction criterion

If the value of the Cost function computed by the user initiates a handover procedure then:

- In the case of a WLAN user, a request is made to the HSDPA network. If the contacted Node B cannot satisfy the connection QoS, the user maintains the connection with the current AP but requests repeatedly an access to the Node B until a predefined delay constraint is reached;

- In the case of an HSDPA user, the user requests access to the AP with the best signal-to-noise ratio. If the specific AP cannot satisfy the connection QoS requirements, it tries the AP with the second best received signal-to-noise ratio. The procedure continues until an AP accepts the user or a predefined delay constraint is reached.

In both cases, once the delay constraint is reached, the hand-over request is rejected and the user is dropped.

3 Scenarios and Simulation Assumptions Considered for the Interoperability Investigation

In this paper, two interoperability scenarios have been selected for evaluation via a simulation platform enhanced with Multiple Input Multiple Output (MIMO) transceivers.
- The first scenario considers the coexistence of the HSDPA macro deployment with an indoor airport HIPERLAN\2 hotspot deployment
- The second one represents the coexistence of the HSDPA macro deployment with an outdoor city center HIPERLAN\2 hotspot as illustrated in Fig. 2

To investigate interoperability, two common data switched services are considered for both networks, namely; FTP and web browsing with service activity factor of 20% and 80% for FTP and web browsing, respectively. Since both networks are enhanced through diversity techniques using multi-antennas, the platform incorporates the appropriate functionalities to evaluate the benefits of MIMO technology, and to examine seamless inter-working between the two networks. The developed simulation platform takes into consideration: network configuration, propagation conditions, fast fading, service requirements and interoperability algorithms. Finally, the full set of simulation parameters can be found in Appendix 1.

4 Performance Assessment and Simulation Results

In order to assess performance of the interoperability algorithms and mechanisms that we have introduced in this paper a system level simulation platform enhanced with MIMO transceivers have been developed. The results were collected from different simulation test cases. Moreover, the performance metric was observed under different traffic loads that correspond to (3, 6, 9 Mbps per Base Station \ Access Point). The MIMO enhancement of the simulation platform introduces complexity and high simulation time; hence each simulation run was of 30000 frames (i.e. 1 minute) duration.

To evaluate the interoperability algorithms we define an appropriate performance metric for both the initial user assignment and the inter-system handover algorithms. Finally, the performance of our algorithms is compared with test cases where the interoperability algorithms were not implemented.

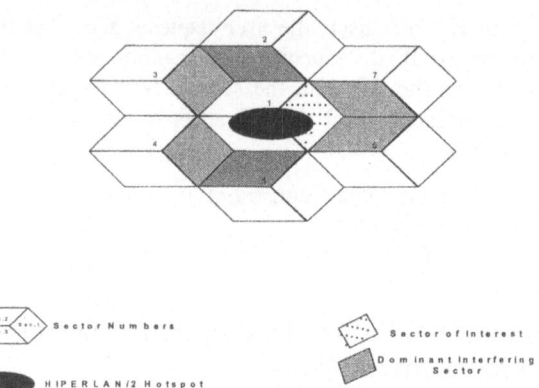

Fig. 3. The Joint HSDPA-HIPERLAN\2 Deployment

– (IUA) Simulation Results:
For the IUA algorithm, the following performance metric is defined:
- ***Percentage of user preference un-satisfaction (PU_unsat):*** is the percentage of users that are assigned to a network other than the one of their preference. Since the output of the introduced Cost function may assign a user to a network different from his original preference, PU_{-unsat} is considered appropriate for performance evaluation.

Table 1 illustrates a comparison between the PU_{-unsat} in the case where the initial user assignment algorithm is implemented and in the case where the algorithm is not taken into consideration. The results show the significant reduction (about 70%) of the mean PU_{-unsat} due to the algorithm implementation. Moreover, the issue of the IUA was found to be independent of the network load. This makes our algorithm robust and applicable to all traffic conditions

Table 1. PU_unsat Vs. Algorithm application

	With IUA algorithm	With no IUA
PU_{-unsat}	19.7%	64.6%

In Fig. 4, we illustrate the results of a test case where the PU_{-unsat} was computed for all possible combinations of weight distributions of the four parameters of the Cost function (with weight step= 0.05). Each one of the points (a,b,c,...) on the horizontal axis represent a weight combination that corresponds to a PU_{-unsat} value. These weight combinations are adressed in Table 2. For example, we can see that point (b) corresponds to the case where equal weights of 0.25 are assigned to all of the four parameters of the Cost function and results in a PU_{-unsat} = 27.5%. Based on Figure 5, service providers can determine the desired PU_{-unsat} target in their network and assign the respective weights to the four parameters.

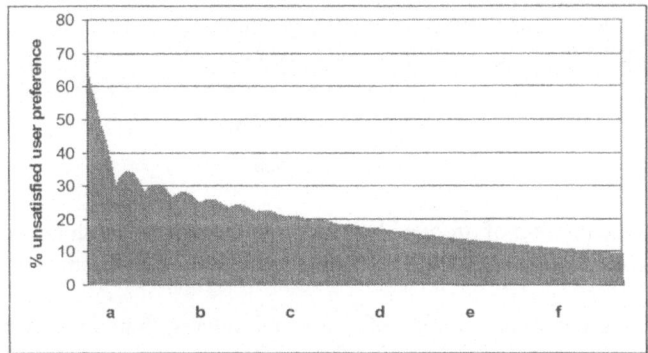

Fig. 4. PU_unsat Vs. Different weight contribution of the Cost function parametres

Table 2. Different combination of wieght values assigned to the Cost function parameters

	Terminal Type	Traffic Specification	Speed	User Preference
	W1	W2	W3	W4
a	0.1	0.85	0.65	0.65
b	0.25	0.25	0.25	0.25
c	0.45	0.15	0.8	0.75
d	0.6	0.5	0.4	0.4
e	0.75	0.5	0.55	0.45
f	0.9	0.75	0.75	0.85

Fig. 5 investigates the effect of each of the four parameters separately on the PU_{-unsat}. In this test case, the PU_{-unsat} was calculated for all possible weight values assigned to the parameter of interest, while the remaining three parameters were considered to have a weight equal to 0.1. We can see that all parameters follow similar behaviour; assigning higher weights results in a smaller PU_{-unsat}.

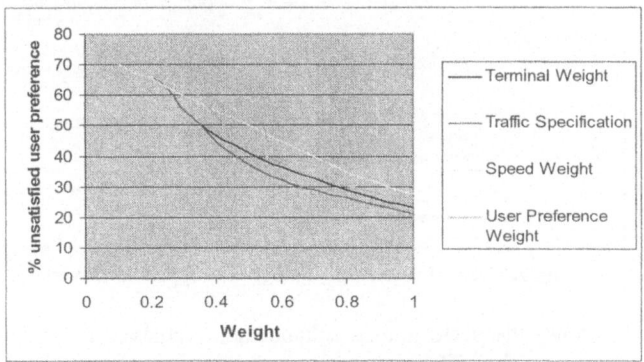

Fig. 5. Cost function parametres Vs. PU_unsat

– **ISH Simulation Results**

For the ISH algorithm, the following performance metric is identified:

• **The Handover Drop Rate** is identified as the appropriate performance metric for evaluating the inter-system handover algorithm. The *Handover Drop Rate* is defined as the percentage of dropped handovers as given below

$$Drop\_rate = \frac{N_{HO\_dropped}}{N_{HO\_tot}}$$

Where:

– $N_{HO\_dropped}$ is the number of dropped handovers in the system (both networks)
– N_{HO_tot} is the total number of handovers that occurred in the system

Fig. 6 shows the total number of intra and inter-system handovers that were observed during a simulation run for four different network loads. Since higher network loads correspond to a larger number of users within the deployment, the increase of the number of handovers is considered reasonable.

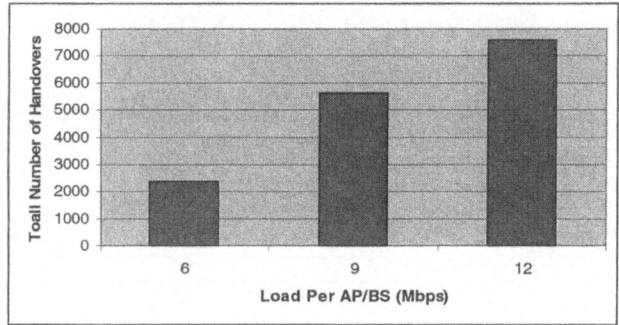

Fig. 6. Network Load Vs. Total number of handovers

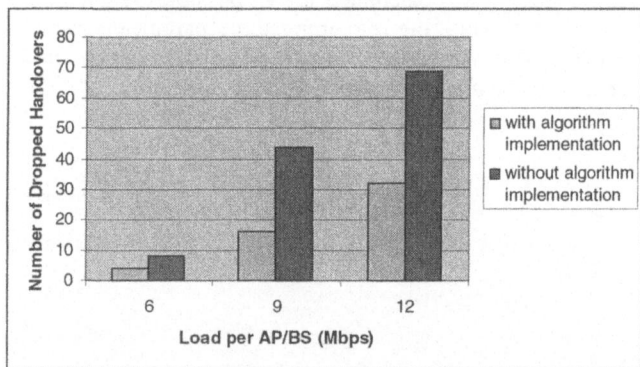

Fig. 7. Network Load vs. Number of dropped handovers

Finally, Fig. 7 shows the performance enhancement achieved by means of the inter-system handover algorithm. The test case compares the number of dropped handovers when inter-system handover is incorporated within the simulation platform vs. the case where the system supports only intra-system handover. The implementation of the algorithm achieves approximately 60% reduction in the handover drop rate for low traffic loads (3, 6) Mbps BS/AP and 40% reduction in high traffic load test cases (9) Mbps BS/AP.

5 Conclusions

This paper has proposed an approach to enable inter-operability between HSDPA and WLAN by introducing initial user assignment and inter-system handover mechanisms. Realistic scenarios have been developed to demonstrate the efficiency of the proposed mechanisms and criteria through system simulations. Performance characterization of the two inter-operability mechanisms was also presented and compared with the case where the algorithms were not implemented. The efficiency of the algorithms was proved since the initial user assignment algorithm distributes the users in both networks in a way that reduces the PU_{unsat} from 64.6% to 19.7% while the inter-system handover algorithm achieves approximately 60% reduction in the handover drop rate for low traffic loads (3, 6) Mbps BS/AP and 40% reduction in high traffic load test cases (9) Mbps BS/AP.

Acknowledgments. The authors wish to thank all contributors of the IST- FITNESS project within which these studies have been conducted. Deliverables are available on the project web site

References

1. 3GPP TR 22.934 V6.1.0 (2002-12): Feasibility study on 3GPP system to Wireless Local Area Network (WLAN) inter-working (Release 6)
2. ETSI/TR 101 683 V1.1.1 (2002-02): BRAN, HiperLAN Type 2, System Overview
3. http://www.ctr.columbia.edu/~andras/cellularip
4. Ramjee et al., "HAWAII: A Domain -based Approach for Supporting Mobility in Wide-area Wireless Networks", International Conference on Network Protocols, ICNP'99, June 1999
5. R.Pintenet, J.Gosteau, T. Al-Gizawi, F.Lazarakis, K. Peppas, A. Alexiou, VTC fall 2003, " Evaluation of interoperability mechanisms for Co-existing HSDPA and WLAN enhanced with MTMR techniques
6. http://www.ist-fitness.org

Efficient Location-Based Hard Handoff Algorithms for Cellular Systems

Antonis Markopoulos, Panagiotis Pissaris, Sofoklis Kyriazakos, and
Efstathios D. Sykas

National Technical University of Athens,
School of Electrical and Computer Engineering, Telecommunications Laboratory,
9 Heroon Polytechneiou Street, Zographou 15773, Athens, GREECE
Contact: antony@telecom.ntua.gr

Abstract. Mobile world's rapid growth has spurred development of new protocols and new algorithms to meet changing operation requirements- such as mobile networking, and quality-of-service support. Handoff is one of the most
critical procedures in cellular systems. Network operators give emphasis to optimize handover, since it is strongly related to dropped calls, network overload
and subsequently users' criticism. Handoff can be seen as a blind procedure, if
it is only based on the comparison of measurements, without the information of
location. Since signal propagation and pathloss are complex in nature, we can
expect unnecessary and wrong handoff executions. Both UMTS and those of
the second generation (GSM) systems will require redefined handoff algorithms
of active connections as the smooth mobility support and continuous connection are essential issues for obtaining high performance and increasing user
satisfaction. In this paper we present a set of intelligent algorithms using the
mobile terminal (MT) location information and area awareness to assist safe
handoff decisions. The implemented algorithms are validated by means of cellular network simulators that clearly show the impact of these techniques to
major system performance metrics.

1 Introduction

The paper mainly focuses on the improvement of network performance when MT
position is taken into account. For that purpose, we defined and implemented new
innovative handoff algorithms. We have also developed two network simulators, one
for GSM and one for UMTS, to validate those algorithms. The simulators were developed from scratch and are used to evaluate the performance of existing handoff
algorithms as well as to validate the simulator models of our new algorithms. To
produce safe results a large number of simulations were performed.

The paper is organized around 5 sections. Section 1 is the introduction. Section 2
introduces innovative handoff algorithms based on MT location and area information.
In section 3 we present the basic simulators layout and results of numerous simulations that show the network performance behavior for both systems. The results are
very optimistic since increased network performance under normal and high traffic
load situations is clearly shown. Finally in section 4 we sum up with the conclusions.

N. Mitrou et al. (Eds.): NETWORKING 2004, LNCS 3042, pp. 476–489, 2004.
© IFIP International Federation for Information Processing 2004

2   Location Aided Handover Algorithms

Stability and optimization of network performance in wireless 2G-3G will be the bedrock for network and service providers to maintain a smooth operation and flourish revenue results. Such providers hold the opinion that "user satisfaction and experience" has become an essential prospective to retain subscribers and increase new customers. Inaccessibility of network, low call quality, coverage and interference issues, block and drop of calls lead customers easily to change loyalties in a competitive wireless market [1]. A critical mechanism affecting network performance for existing and next generation systems is the hand-off procedure.

Handoff is the mechanism that transfers an ongoing call from one cell to another as a user moves through the coverage area of a cellular system. Each handover requires network resources to reroute the call to the new base station. Cellular networks recommendations do not specify an algorithm for the handover decision or target cell selection. Usually the vendor defines the handover algorithm. Network operators and manufacturers have hence the possibility of defining and using a handover algorithm based on the various parameters. The signal level of the mobile stations, which can be used to estimate the path loss of a radio signal, is one of the criteria for the handover initiation. Besides the signal strength, the measured bit-error ratio is also a very important indication of the signal quality of the uplink and downlink.

Some complementally intelligent software technologies, i.e. fuzzy logic [2] , neural networks and pattern recognition techniques [3]-[5] were proposed to optimize the handover process. Moreover, hardware solutions such as adaptive antennas were introduced to make the handovers "smart". However, it is worthwhile observing that intelligent handover based on mobile location information and cell RF propagation statistic could decrease the occurrence of unnecessary handovers and so the signaling for handover procedure can be reduced [6]. Triggered by that, we introduce a set of new intelligent handover algorithm (Location-Aided Handover, LAH) based on mobile phone location capability and area information.

LAH is a set of algorithms uses the information of instantaneous mobile location, data from the Mobile Network Geographic Information System (MGIS) [7] and area characteristics to make the decision of the most appropriate target base station for handover. MGIS database consists of all these essential performance metrics (see Section 2) per cell. MGIS is specified in [7]. The major benefits from the above described use of the MGIS data for efficient handover procedures are the decreasing of signaling traffic, avoidance of drop calls, increased speech quality, solution of several planning shortcomings and enabling seamless handovers for high data-rate users in UMTS by providing a mechanism for resource allocation in the target cell.

In the following sections follow a detailed description of Location Aided Handover algorithms for GSM and UMTS networks. These algorithms are:

- Ping-Pong avoidance (PPA)
- Towards the border (TTB)
- MGIS Data Resolution (MDR)
- Transmission Power and Interference Optimization (TPIO) for UMTS

2.1 Ping-Pong Avoidance (PPA) Algorithm

One very undesirable effect that occurs relatively frequently is the so-called ping-pong handoff. This is a handoff that takes place when the MT being at the cells' border oscillates from the current cell to a neighboring cell and back. A ping-pong handoff depends on the power budget criterion.

The hysteresis margin and the duration of averaging window length are two parameters that could be used to prevent the MT from such a situation. Previous research studied the handoff process in terms of the level crossing process of the difference between the two base stations' received signals, where the MTs move on the straight line that connects them [9]. At first the level crossing process was assumed to be stationary; later work [10] included the nonstationary case. A simulation study [11] used to evaluate the trade-off between delay and number (and probability) of unnecessary handoffs. Basic simulation parameters were the hysteresis level and the averaging window length. A work that has focused on the shape of the averaging window and considered both rectangular windows and exponential windows with various weights concludes that long windows reduce the number of handoffs but increase the delay [11].

A conclusion from these simulation studies is that none of these parameters is appropriate in actual operation. A high hysteresis value would practically eliminate a power budget handoff. A change in the boundary area of two cells will also be delayed far beyond the boundary. A high averaging length makes the handoff process too sluggish and leads to a situation in which a really necessary handoff sometimes is not carried out in time and the connection is then broken off. This is the reason why ping-pong handoff cannot be avoided with the current handoff algorithms and the modified variants, even if the appropriate parameter adjustments are used.

Our PPA algorithm on the other hand aims to minimize such undesirable handoffs using MT location and taking advantage the area information and mobility model to predict the user's movement. The objective is to avoid continuous unnecessary handoffs between two neighboring cells. The basis of the algorithm is the movement prediction of each user, who appears to have such a behavior. The major steps are to define when the algorithm should be triggered and how it could be applied more efficiently. For that purpose some new parameters are attached to the MT. A timer (for each user) is used to keep the time period from the last successful handoff. If that timer has not exceeded a specific threshold, which is an input for the PPA algorithm, (e.g. 10secs) and the user is again requesting a handoff to his previously serving cell, the algorithm is triggered. Given the user location info, the algorithm predicts his future move. Before applying this algorithm there was a need to validate the accuracy of user's movement prediction. How much does it converge to the real movement? The knowledge of area (Grid of pixels classified into building or street – Movement is only allowed to streets) combined to the exact mobility model (see Section 4.1 - Mobility model) give us a very high prediction accuracy. In Figure 1 the predicted movement versus the real movement is shown.

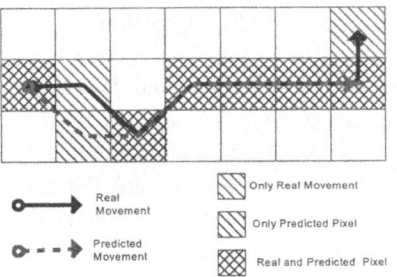

Fig. 1. Real and predicted movement congruency

As soon as the algorithm is triggered, the objective is to decide which cell appears to be the most appropriate for accommodating the MT, avoiding unnecessary repeated handoffs at the same time. In Figure 2, ping-pong phenomenon is depicted. We assume that the predicted movement is the one shown in that figure. In a short time, several handoffs are requested.

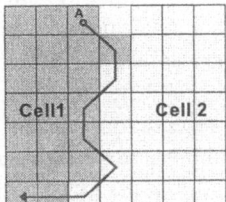

Fig. 2. Cell prediction example

Suppose that the algorithm is triggered at position A. The dominant cell in the pixel where the algorithm is triggered is C_A. In the specific example, C_A=Cell 1. A new parameter S is then attached to MT. This parameter will assist the algorithm to attach the user to the most proper cell. Assume that P is the number of pixels that MT will go through during its complete route, p_i $\forall i \in P$ is the current pixel and $Rx(p_i)$ is the received field strength in p_i from the dominant cell $C_d(p_i)$ in p_i. S_i is initialized in the pixel where the algorithm is triggered and evolves during the movement as follows:

$$S_i = \begin{cases} S_{i-1} + Rx(p_i) & \text{, if } C_d(p_i) = C_A \\ S_{i-1} - Rx(p_i) & \text{, if } C_d(p_i) \neq C_A \end{cases} \quad \forall i \in P, \quad S_0 = 0 \qquad (1)$$

If the final value $S_P \geq 0$, the most appropriate cell to cover MT is the C_A (Cell 1), otherwise if $S_P < 0$ Cell 2 should cover it. In that case a handoff will take place. Then, the MT is locked to the estimated most proper cell for a fixed time period. During this period, MT cannot request another handoff.

2.2   "Towards the Border" (TTB) Algorithm

PPA algorithm tries to avoid continuous handoffs between two cells whenever users move almost parallel to cell borders. What happens when a user moves almost vertically towards a border? In this case it is almost certain to expect a handoff from one cell to another. According to the mobility model, the direction of a user can change in every step. Despite the fact that the user might change direction of 360° or even 180°, the probability of this to happen is extremely low due to the fact that the mean of the normal distribution, that we choose to model the user's probability of changing direction, is small (10° to 20°). Practically, it is not common to dramatically change your direction within a few seconds when you are moving on a straight highway.

This additional information gives us a significant advantage. A user moving towards a different cell is almost certain that is going to request for a handoff within a small period of time. The idea is that whenever such behavior (i.e. vertical movement towards the border) is detected, the target cell needs to bind sources for a dynamically configured period of time. The most popular non-predictive channel reservation approach for handoff prioritization in cellular systems is the guard channel (GC) protocol. The logical channels that are set aside are often called guard channels. Initially, the guard channel scheme improves the handoff success rate; on the other hand it might increase the blocking probability, since it reserves channels for the handoff procedure. Predictive channel reservation scheme considers the probabilistic user's movement and can result in a higher network performance. The predictive channel reservation scheme (PCR) is based on the MT position, which can be either the result of an accurate location server, or prediction of the location. According to PCR, if the MT is moving towards a new cell, it transmits a channel reservation request to the new cell. At this time a counter is set in order to cancel the reservation in case the MT direction changes dramatically [12].

For example, assume that a user is inside Cell 1 and moves toward Cell 2 (Figure 3). The algorithm predicts that within a specified time period this user will reach the boundary.

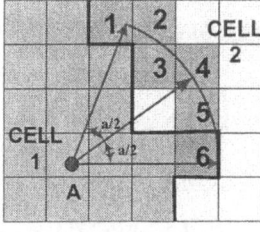

CELL 2 - Target Cell

Fig. 3. Prediction of movement towards the border

In such a case, Cell 2 binds resources for this user before his actual handoff request. When this request takes place the user utilizes the already bound resource to the effect that the call will not be blocked. This algorithm uses the position and velocity of each user in order to predict the new cell, which we call "Target Cell". The direction of the

user can be extracted by two sequential positions. We extend this direction by a selected angle. Taking into account the prediction time, user speed, angle   and the area
characteristics we have an estimate whether the user is moving towards another cell.
In the example of Figure 3 cell number 2 is the "Target Cell". The question that arises
is how the network would know whether the user approaches or actually enters another cell in order to trigger the execution of handoff. The user starts from position A,
and after a predicted time, with speed and possible angle   the user will end up in any
of six different destination pixels (1 to 6 in Figure 3). Cell 2 and only one by Cell 1
dominate five of these pixels. There is a high probability therefore for Cell 2 to cover
this user after time. Taking that as a fact, the network will be able to make the
handoff in a more efficient way by reserving appropriate resources to the "Target
Cell" thus avoiding undesirable events such as drop and block call. To improve the
prediction, angle   is initially selected to have the same value as the mean value of the
normal distribution used by the mobility model for the user direction changes (see
Section 4.1- Mobility Model). What happens when the movement prediction fails? In
this case the network releases the reserved resources after the timeout period.

2.3   MGIS Data Resolution (MDR)

Next algorithm uses the stored data collected by the Mobile Network Geographic
Information System server (MGIS server)[7]. The Set of cells is C and pixels is P.
We assume that every selected cell $j \in C$ has two extra features: a) mean value of
Drop Call Rate ($DCR_j \ \forall j \in C$) and b) mean value of Block Rate ($BR_j \ \forall j \in C$).
Both of them come from the constantly and remotely collected MGIS data. The above
features are attached to the cells as a total attribute. We have to map these to each
pixel $p_{i,j} \ \{\forall (i, j) \in (C \times P)\}$. A formula is used for this purpose (Equation 2) based
on these attributes and the signal strength $Rx(p_{i,j}) \ \{\forall (i, j) \in (C \times U)\}$. Each of the
components participates with a selected weight in the proposed cost function. Thus:

$$C_{i,j} = w_1 \cdot (1 - Rx_{IND}) + w_2 \cdot DCR_{IND} + w_3 \cdot BR_{IND} \quad 0 \le C_{i,j} \le 1 \qquad \forall j \in C \quad (2),$$

$$Rx_{IND} = \frac{Rx(p_{i,j})}{Rx_{MAX}} \qquad (3),$$

$$DCR_{IND} = DCR_j \cdot \frac{\sum\limits_{\forall i \in j} Rx(p_{i,j})}{\sum\limits_{\forall i \in j} P_{(i,j)}} \cdot \frac{1}{Rx(p_{i,j})} \qquad (4),$$

$$BR_{IND} = BR_j \cdot \frac{\sum\limits_{\forall i \in j} Rx(p_{i,j})}{\sum\limits_{\forall i \in j} P_{(i,j)}} \cdot \frac{1}{Rx(p_{i,j})} \qquad (5),$$

Where $w_1 + w_2 + w_3 = 1$.

This formula (Equation 2) calculates a cost $C_{i,j}$ at a specific pixel i for each cell j. This pixel is the reply of the Location Server (LS), which is used (as in previous algorithm) to estimate the real user position (location error).

The core of the algorithm is the calculation of the cost $C_{i,j}$, $\forall j \in C$ at a specific pixel p_i. If a user, while he is moving, comes to a pixel p_i where the cost of his serving $Cell_j$, $j \in C$ is above a selected threshold, we consider that the coverage area of $Cell_j$ is a malicious one. Before a handover execution, MDR algorithm makes the calculation $\forall j \in C$ and provides the system with an extra check. The cell of the best cost should not be the same with serving cell ($Cell_j$). Only if there is another cell with a better cost, a handover should take place. Therefore the basic criterion of executing the handover procedure based on MDR algorithm is that a user gets inside a "critical/problematic area/hot spot" and prevision for a better cell coverage.

However, if PPA algorithm is active and a user is "locked", as explained in previous section, a handover (even a "MGIS handover") is not permitted and user still remains locked.

As in PPA algorithm, there are also some critical parameters that need to be initiated and fine tuned through a large number of simulations. First and most important is the selection of the cost threshold that indicates whether a user is inside a "critical area" or not. Last but not least, the three weights used in cost calculation formula are needed to be fixed depending on which metric is more essential for the provider policy. Depending on what the provider consider more critical in defining cell cost (i.e. Rx Level, DCR or BR), we need to choose the right weights for the cost function.

2.4 Transmission Power and Interference Optimization (TPIO)

Handover is a critical feature in cellular systems and therefore handover algorithms are under extensive research. Problems in handover algorithm or its parameters may lead into call drops with a direct effect on user satisfaction. This is particularly critical for 3G systems, where high data rate users will be prime candidates for being dropped. Unnecessary handovers may lead to degraded call quality and waist of capacity in signaling. Since the cell-size is constantly decreasing, it is important for the handover algorithm to identify users with different mobility and data rate characteristics. The user position can affect in different ways the interference that is caused to serving and neighboring base stations (intracell and intercell interference). Moreover since in a multimedia system, there are different kinds of services (see Table 1), the users have different requirements in terms of bandwidth, maximum bit rate, bit rate availability, BER, delay tolerance, grade of service and so on [16]. For example, these users that demand high throughput (e.g. video conference) it would be more efficient to be close to a cell center. If they are quite far from the center, position near the cell boundary, it is possible to transmit with high power (due to power control by the base station), so that they will cause interference and probably be dropped. This algorithm copes with a problem, called Transmission Power and Interference Optimization

(TPIO). The core concept of the algorithm is based on the following: "Given the service area layout, mobile position, the services and their requirements, aim at optimizing the transmitted powers and therefore the interference levels in both uplink and downlink" [14].

The service area is split into a set of pixels, P. Each pixel $p \in P$ corresponds to a small part ($10 \times 10m$) of the area. The set of services is S. The QoS requirements of $s \in S$ are expressed through the minimum signal-to-interference ratio, SIR_s. It takes into account the bit-rate, service activity factor (SAF) and the minimum required E_b/I_0. The set of cells is C and the set of active users U. UTPIO algorithm uses a cost function applied not only for each new connection but also performed for each active user in every step since UMTS simulator investigates also mobility issues.

$$Q_i = w_1 \cdot [1 - fs(p_i)] + w_2 \cdot s_i + w_3 \cdot d_{i,j} \qquad (6)$$

Equation 6 denotes the cost function $Q_i \; \forall i \in U$, given field strength $fs(p_i)$ in specific pixel $p \in P$, requested service $s_i \in S$ and in distance $d_{i,j}$ from cell $j \in C$.

This algorithm should aim to define the following thresholds/targets according to the location (pixel) and service distribution of the offered load.

- Uplink interference threshold, $I_{up}^{thres}(c) \; \forall c \in C$.

- Total downlink transmitted power threshold, $P_{down}^{thres}(c) \; \forall c \in C$.

- Uplink power target level per pixel and service, $p_{up}^{thres}(p,s)$ $\forall p \in P, s \in S$.

- Downlink power target level per pixel and service, $p_{down}^{thres}(p,s)$

The above parameters are used for a) the admission control mechanism, [15]-[19], and b) the power control mechanism, [20][21]. In the first case, the admission of a new call or a handover depends on the interference it would have caused if accepted. If there is a feasible power level that can be assigned to the new connection so as to alleviate with the new interference conditions, in both uplink and downlink, the MT is admitted otherwise the call is blocked. In the second case, during the system operation, the TPIO algorithm assigns the optimum power levels for the MT/Node-B and the respective Node-B/MT gives the appropriate power commands in order to reach these levels. In any case, the algorithm should maintain the QoS levels required by the transmissions, namely bit rate and SIR levels, ensuring the compatibility of the assigned powers to the terminal capabilities.

3   The Simulation Models and Results

While measurement and experimentation provide a means for exploring the "real world", simulation is restricted to exploring a constructed, abstracted model of the world. Measurements are needed for a crucial "reality check". Experiments are frequently vital for understudying the behavior of otherwise intractable systems. However, measurements and experiments have limitations in that can only be used to explore the existing handover procedures. They cannot be used to explore different possible new handover procedures [22].

Simulations are not only complementary to analysis, but allow exploration of complicated scenarios that would be either difficult or impossible to analyze. Simulations can also play a vital role in helping researchers to develop intuition about the behavior of new handover procedures [22]. To simulate each one of the previously mentioned algorithms, two cellular system simulators were developed (GSM & UMTS) [23]. The objectives of those simulators were to investigate network performance metrics (mean blocking and dropping call rates, handover failure rate, and power allocation in UMTS, interference issues and QoS provisioning), existing handover procedures and simulating and validating the proposed above innovative algorithms. LAH simulators consist of several software modules. Each module is responsible for a set of tasks and communicates with other modules [23].

Each simulator characterized by:

1.   **For GSM Simulator**
   - Area definition: An urban geometry (Downtown, Athens, Greece) given area's low left point coordinates (WGS 84, UTM Zone 32, Southern Hemisphere), the dimensions of the area (given in pixels of 10X 10m and 3,2 Km X 3,2 Km totally) grid having 14 cells.
   - Cell radius: 200 m to 2 Km
   - Environment of application: outdoor (typically urban areas)
   - Main propagation mechanisms: forward and backward diffraction over rooftops
   - Location Accuracy - GPS solutions are assumed
   - Traffic model: Defines traffic model parameters (Erlangs per user, Mean call duration, Number of SDCCHs, Number of TCHs per cell, etc).
   - Mobility model: Medium and higher mobility. This model is used to move each user inside the selected area. Each user position corresponds to an exact pixel. Mobility is allowed only to pixel that corresponds to the Street class.

2.   **For UMTS Simulator**
   - Generic Principles of CDMA systems [24]
   - Area definition, Mobility model, Location Servers: The same as in GSM
   - Cell radius: 200m to 2 Km.
   - Traffic model: Defines traffic model parameters (Circuit and Packet Switched services, Erlangs per user, Mean call/session duration, Average number of users per service and cell).

- Services: The traffic model with the assumed service characteristics is summarized in Table 1
- Admission and Power Control: In conjugation with the thresholds defined by the TPIO algorithm

Table 1. Assumed UMTS service characteristics

	Voice Call		Videoconference		Interactive Data		Streaming Video	
	Uplink	Downlink	Uplink	Downlink	Uplink	Downlink	Uplink	Downlink
Bit Rate (kbps)	12.2	12.2	64	64	144	144	-	384
Eb/Io requirement (dB)	4	6.5	2	5	1.5	4	-	3
Service Activity Factor	0.67	0.58	1.0	1.0	1.0	1.0	-	1.0
Orthogonality factor	-	0.5	-	0.5	-	0.5	-	0.5

Several simulations lead us to the following results described in the next section. Results are categorized for GMS in the investigation of performance metrics such as Drop Call, Handover and Handover Failure Rates and for UMTS analyzing uplink power transmission and interference behavior depending on mobile distance from the cell and the type of service being requested.

3.1 Simulation Results

Figure 4 (left side) shows the handoff failure rate in cell level. The data were drawn from a real network and is a result of averaged measurements for a time period of two weeks.

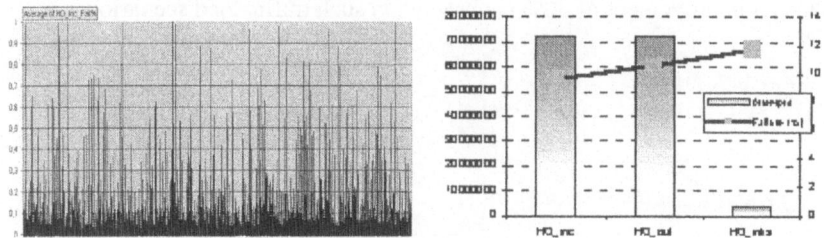

Fig. 4. Handover Failure and HO Statistics

The first observation is that the average handoff failure is around 10%, which is quite high. Considering that users making calls while moving may require several handoffs during a call, the call-drop probability increases. Figure 4 (right side) shows the number of handoff attempts in a period of one week in the complete coverage of a GSM operator. The failure rate is shown according to the type of the handoff procedure (Inc-incoming, out-outgoing, intra-intra handover).

The total number of around 150 million handoffs per week and the average handoff failure (around 10%) shows the need for handoff procedure enhancement. The reduction of the total number of handoffs in the whole system is of a high priority in all vendor's policies for network optimization.

PPA algorithm leads the system to a 5% fewer total handoffs (HoRate) as Figure 5 depicts. The right side of this figure depicts that the PPA algorithm achieves a significant reduction in the number of ping-pong handoffs. An estimated 40% reduction is achieved in all traffic load scenarios.

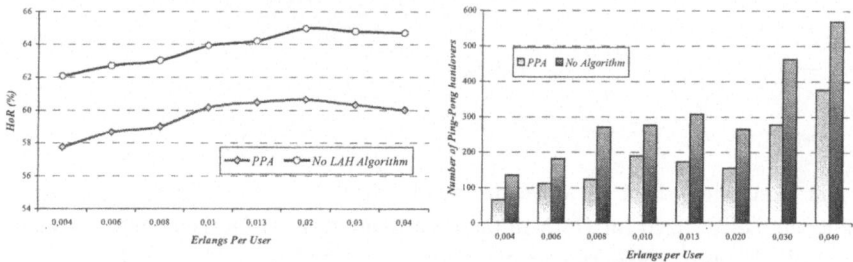

Fig. 5. HOR progress and improvement and No Of Ping-Pong HO

Figure 6 (left side) presents the Drop Call Rate progress when no algorithm is applied compared to the case when PPA, MDR and TTB algorithms are triggered. The improvement to the drop call rate clearly increases with the traffic load. A similar behavior can be observed in the same figure (right side) where the Handoff Block Rate is plotted. Here, the improvement of the HOBR is even better than the DCR in the previous figure. When traffic load reaches 0,06 Erlangs per user, more than a 50% reduction in blocked handoffs is achieved. When the "towards the border" algorithm is applied an improvement of 80% is observed in such traffic load scenarios.

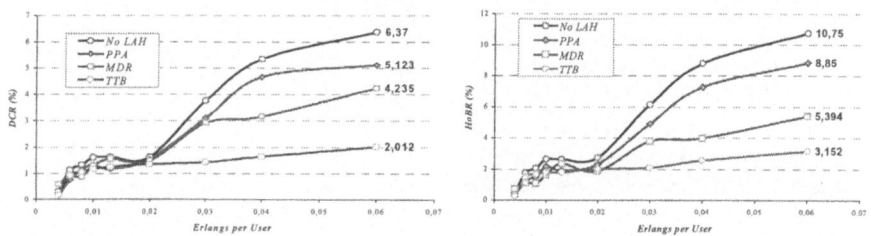

Fig. 6. DCR progress and improvement and HO progress and improvement

From Figure 7 (left side) we can conclude that the MDR algorithm can optimize the network performance since two major performance metrics namely the Handoff Block Rate and the Drop Call Rate are drastically reduced. The mean decline is about 40% for the HOBR and 30% for the DCR.

Similarly a major reduce of around 60% in Handoff Block Rate and 40% in Drop Call Rate can be observed in Figure 7 (right side)when the TTB algorithm is applied.

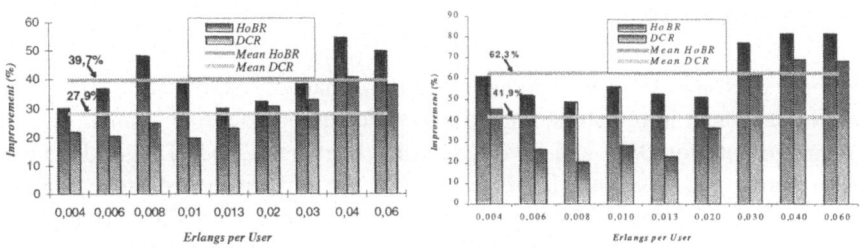

Fig. 7. DCR and HOBR improvement due to MDR and TTB respectively

In the case of the UMTS simulations each of the 10 cells is split into 4 zones around the respective NodeB. A zone consists of several pixels according to the distance from their controlling NodeB. Mobile terminals can transmit at maximum 300mW(25dBm) for all the provided services, while base stations can transmit at maximum 20W(43dBm). Chip rate is set to 3,84Mcps. The thermal noise density is -174dBm/Hz corresponding to a noise power of -108,1dBm. Figure 8 depicts the number of active downlink connections per service (columns) and the total power that should be transmitted by each NodeB in the forward link (row) with and without TPIO algorithm respectively.

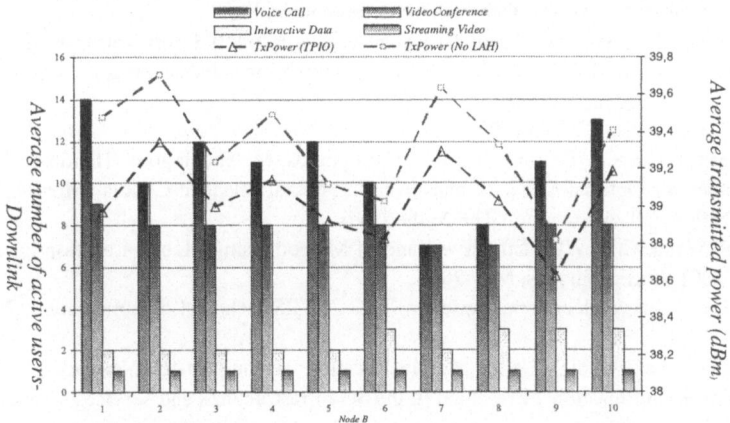

Fig. 8. Downlink load – Transmitted power per NodeB

The uplink/downlink power control mechanisms should drive the mobile/NodeBs, respectively, to transmit at these power levels in order to cope with the offered load and satisfy the QoS requirements of the requested services. The algorithm is capable of handling with different load situations and traffic variations and should be run at any time that re-engineering process is needed.

4   Conclusions

We have presented an investigation in the area of handoff optimization, aided by MT location and area information, which is an outcome of extensive simulative studies, performed both for GSM and UMTS systems. Intelligent handoff algorithms exploit MT location and area knowledge to increase the performance of several network metrics. PPA, TTB and MDR algorithms are implemented for GSM systems. Nevertheless the concept could be extended to UMTS with proper modifications as well. The objective is to eliminate all the existing problems in the handoff procedure, improve vital network performance metrics, and help systems to adapt to unforeseen situations such as critical areas and hot spots. Moreover, the TPIO algorithm addresses a hot issue for 3rd generation systems, which is the optimal transmission power allocation and interference optimization, aided by MT location and variant service characteristics.

References

1. Isaac Samuel, Kishore Arora, Bhuvarahamurthy Nasarimhan, "Location-Based Performance – Measuring Techniques in UMTS", Bell Labs Technical Journal 8(2), pp. 15-32, 2003 Lucent Technologies
2. http://www.comnets.rwth-aachen.de/report96/node36.html
3. M. Chiu, M. Bassiouni, "Predictive Schemes for Handoff Prioritization in Cellular Networks Based on Mobile Positioning", IEEE Journal on selected areas in communications, Vol. 18, No. 3, March 2000
4. 3GPP TR 23.920 V3.1.0 (1999-10)
5. J. Markoulidakis, J. Dermitzakis, G. Lyberopoulos, M. Theologou, "Handover Prioritised Schemes for Optimal Capacity and Overload Management in Cellular Mobile Systems", VTC 1999 Fall, Amsterdam, The Netherlands
6. S.S.P. Wang, C.H. WU "Effective Handoff Method Using Mobile Location Information" VTC2001 Spring, Rhodes, May 2001.
7. IST-Cello (www.telecom.ntua.gr/cello) D22 "MGIS technical documentation",November 2001
8. J. Lähteenmäki et al., "Using Mobile Location Techniques for Network Planning and Handover Optimisation", Presented in the 3G infrastructures and services symposium, 2-3 July, Athens, 2001.
9. R. Vijayan and J. M. Holtzman, "Analysis of Handover Algorithms Using Nonstationary Signal Strength Measurements," Proc. Globecom '92, Orlando, FL, paper 41.2, Dec. 6p;9, 1992.
10. G. E. Corazza, D. Giancristofaro, and F. Santucci, "Characterization of Handover Initialization in Cellular Mobile Radio Networks," Proc. Vehicular Tech. Conf. '94, Stockholm, Sweden, June 8p;10, 1994, pp. 1869-72.
11. N. Zhang and J. M. Holtzman, "Analysis of Handoff Algorithms Using Both Absolute and Relative Measurements," Proc. Vehicular Tech. Conf. '94, Stockholm, Sweden, June 8-p;10, 1994, pp. 82p;86.

12. M. Chiu, M. Bassiouni, "Predictive Schemes for Handoff Prioritization in Cellular Networks Based on Mobile Positioning", IEEE Journal on selected areas in communications, Vol. 18, No. 3, March 2000

13. Nikos Dimitriou, Rahim Tafazolli, Georgios Sfikas, "Quality of Service for Multimedia CDMA", IEEE Communications Magazine, pp 88-94, July 2000

14. P, Demestichas, K.Tsagkaris, G.Kotsakis, E.Tzifa, M.Theologou, " Uplink Transmission Power Allocation in the Design and Management of CDMA-based Cellular Networks", IEE Electronic Letters Journal, Vol. 39, No. 1, 2003

15. Dahlman, E.,Knutsson J., Ovesjo, F., Persson, M. and Roobol, C., "WCDMA - The Radio Interface for Future Mobile Multimedia Communications". IEEE Transactions on Vehicular Technology, Vol. 47. No. 4. November 1998. pp. 1105-1118.

16. Z. Liu and M. E. Zarki, "SIR-based call admission control for DS-CDMA cellular systems", IEEE Journal on Selected Areas in Communications, vol. 12, no. 4, pp. 638-644, May 1994.

17. Y. Ishikawa and N. Umeda, "Capacity Design and Performance of Call Admission Control in Cellular CDMA Systems", IEEE Journal on Selected Areas in Communications vol. 15, no. 8, 8 Otc. 1997

18. S. Shin, C.H. Cho and D.K. Sung "Interference Based Channel Assignment for DS-CDMA Cellular Systems," IEEE Trans. Vehic. Tech, vol. 48, no 1 Jan 1999

19. J.Laiho, A.Wacker, T.Novosad (Editors), "Radio network planning and optimisation for UMTS", J.Wiley&Sons

20. Roy D. Yates, "A Framework for Uplink Power Control in Cellular Radio Systems", IEEE Journal on Selected Areas in Communications, Vol. 13, No. 7, pp 1341-1348, Sept 1995

21. Fredrik Berggren, Student Member, IEEE, Seong-Lyun Kim, Member, IEEE, Riku Jantti, and Jens Zander, Member, IEEE, "Joint Power Control and Intracell Scheduling of DS-CDMA Nonreal Time Data", IEEE Journal on Selected Areas in Communications, Vol. 19, No. 10, October 2001

22. Sally Floyd,Vern Paxson "Difficulties in simulating the Internet" IEEE/ACm Transactions on Networking, Vol.9 No4. August 2001

23. A.Markopoulos, P.Pissaris, S.Kyriazakos, Combining Position Location Information and Network performance data for simulating Location Aided Handover , IST Summit 2002, Thessalonica , Greece

24. T. Ojampera, R. Prasad, "Wideband CDMA for third generation mobile communications", Artech House 1998

Priority Queuing for IP-Based Service Differentiation in the UMTS Radio Access Network

Abedellatif Samhat, Tijani Chahed, and Gerard Hébuterne

GET/Institut National des Télécommunications
9 rue Charles Fourier - 91011 Evry CEDEX - France
{abedellatif.samhat, tijani.chahed, gerard.hebuterne}@int-evry.fr

Abstract. In this work, we investigate service differentiation under IP for the transport of both real-time and non real-time user traffic in the UMTS Terrestrial Radio Access Network (UTRAN). Therein, stringent delay bounds are to be met for both types of traffic, albeit tighter for the voice traffic. For the sake of simplicity, we suggest, model and analyze the use of priority queuing, as an efficient way to implement service differentiation. Our results are validated empirically on a test-bed emulating the UTRAN transport functionalities. This is carried out following a mean value as well as a percentile analysis. Based on these results, we draw the proper dimensioning of the UTRAN so as to meet the target QoS requirements.

1 Introduction

The Universal Mobile Telecommunications System (UMTS) promises to enable a wide range of multimedia applications and seamless service delivery in multiple mobile environments while granting them Quality of Service (QoS). An intensive research activity is currently investigating the basic design of UMTS, and a significant amount of this activity focuses on the UMTS Terrestrial Radio Access Network (UTRAN) [1] and the transport technology that shall be used therein. Specific to UMTS are the stringent delay bounds for the transport of various types of user traffic real-time as well as non real-time, with various QoS needs, over the UTRAN. This is imposed by the WCDMA advanced radio control functions. The transport in the UTRAN should meet these requirements in a cost effective way in terms of efficiency and maximal utilization of the bandwidth. This latter is typically formed of E1 links. Early works on the UTRAN focused solely on AAL2/ATM, especially in the release 99 of IMT-2000 standards [1]. AAL2 offers an elegant way both to multiplex voice traffic and to differentiate between voice and data traffic in the UTRAN, as reported in some analytical works [2] [3] or others carried out by simulations [4] [5] [6].

With the advent of IP as a de facto networking technology and its presence in 3G core network, IP is making its way to the UTRAN. Its standardization is still under way and should be finalized in Release 6 of the 3GPP standards [7]. IP-based UTRAN is further supported by the Mobile Wireless Internet Forum (MWIF) [8]. Contrary to most works in this area, that use simulations, we developed in [9] an analytical model for the transport of real-time voice traffic over the UTRAN using IP adopting a similar approach than the one in [2] for the AAL2/ATM case. Our results show the feasibility of IP as a transport technology in the UTRAN as well as its efficiency.

N. Mitrou et al. (Eds.): NETWORKING 2004, LNCS 3042, pp. 490–501, 2004.

In this paper, we consider IP service differentiation between voice and data traffic in the UTRAN. The main difficulty lies in the fact that, in this case, both traffic types have stringent delay bounds to be met, albeit tighter for the real-time voice. We hence suggest priority queuing as a simple yet efficient way to meet these objectives. Data traffic shall however not be severely penalized owing to its elastic nature and the low bit rate, well controlled nature of voice traffic. We then validate our proposed model through an empirical work on a test-bed emulating the UTRAN transport functionalities. This is done in two fashions : mean value as well as percentile analysis. We eventually use these results to draw proper dimensioning of the UTRAN in order to meet the required performance.

The remainder of this paper is organized as follows. In Section II, we present the main features of the UTRAN and recall the issues related to QoS, mainly in terms of delay. In Section III, we present our model for the UTRAN which we analyze in Section IV. We consider both voice and data traffic and evaluate their delay performance. In Section V, we show empirically as well as numerically the benefits of priority queuing over a non-priority scheme for the transport of both traffic types. We then present in section VI the required dimensioning of the UTRAN in this case. Section VII eventually concludes the paper.

2   Radio Access Network in UMTS

2.1   Architecture

The UTRAN, as described in [1], interconnects the Uu interface (air interface) and the Iu interface. It contains the Node B, the Radio Network Controller (RNC) and their interconnection. The RNC is responsible for the control of radio resources of UTRAN. It plays a very important role in power control, handover control, admission control and load control. RNC interfaces the core network via Iu interface and uses Iub interface to control one Node B. The Iur interface between RNCs allows soft handover between RNCs. Node B is equivalent to the GSM base station (BS/BTS), and it is the physical unit for radio transmission and reception with cells. Node B performs the air interface processing, which includes channel coding, interleaving, rate adaptation and spreading. The connection with the user equipment is made via Uu interface, which is actually the WCDMA radio interface.

The user plane protocol stack in the UTRAN is shown in Figure 1. RLC (Radio Link Control) establishes the RLC connection between UE and RNC. The MAC layer deals with logical channels. It handles the mapping between the logical channels and the transport channels. Transport channels are categorized depending on the transmission format. The output of the MAC layer consists of sets of Transport Block (TB) periodically generated every Transmission Time Interval (TTI) of the transport channel. For each transport channel, FP layer assembles the bursts transmitted in one TTI into one FP frame which is transmitted to the transport network layer, the IP transport technology in our case.

Why IP? IP is already present in the 3G core network; it would make it easier both for users and operators to have an all-IP setting. Besides, IP as the common layer 3

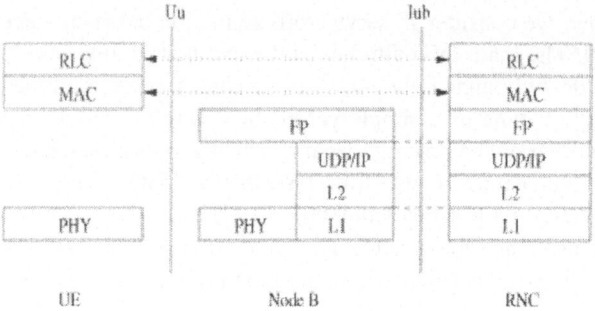

Fig. 1. Protocol stack

protocol in the UTRAN brings flexibility to an operator in choosing a Layer 1/2 backhaul technologies, including options of IP over synchronous optical network (SONET) or IP over wavelength-division multiplexing (WDM). Eventually, with IP, several possibilities exist to provide sufficient QoS, DiffServ for instance.

2.2 QoS Requirements and Service Differentiation

The real-time nature of voice-oriented applications and the WCDMA radio control functions impose rather stringent delay requirement on the UTRAN transport network for both types of traffic, real-time as well as non-real-time: 3GPP specifies 5 ms delay bound for real-time traffic in addition to minimal jitter. The delay requirement for data traffic can be as small as 10ms, possibly 50ms, owing to the radio functions of the outer-loop power control and soft-handoff control. This implies that challenging demands are imposed for the fulfillment of QoS requirements. Service differentiation should thus give voice traffic a higher priority while guaranteeing both delay targets for both traffic types, as explicited in the next section.

3   Model for UTRAN System

We propose to model the Iub interface of the UTRAN as shown in Figure 2. FP frames belonging to voice and data traffic are separately packetized and processed through a transfer unit. They however share a common output link that includes a scheduler that should enforce their respective QoS needs while optimizing the bandwidth utilization.

In this work, we suggest the use of Priority Queuing (PQ) as an efficient yet simple means of implementing service differentiation. With strict, non-preemptive PQ, IP voice packets are always served first since they have a tighter latency constraint; IP data packets are only served when the voice queue is empty. We note that the main factor influencing the jitter for voice traffic is the inability to pre-empt a data packet that has just begun receiving service when the voice packet arrives. To minimize this effect, data packet sizes should be kept low, for example on a 2Mbps link, a 1500 byte data packet introduces a 6ms link blocking, too high a value with regards to our voice delay budget. Service differentiation should also make sure that data traffic is not starved. In this case, proper dimensioning shall be implemented.

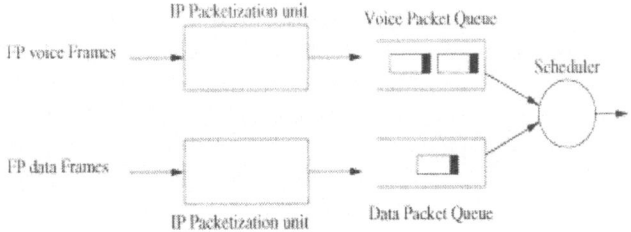

Fig. 2. IP model in the UTRAN

4 Analysis

4.1 High-Priority Voice Traffic

Voice traffic is generated by Adaptive Multi-Rate (AMR) codecs at 12.2 kbps. The transport channels carrying speech traffic shall be assigned a 20 ms TTI value. One DCH (Dedicated CHannel) is allocated to each user. Voice traffic consists of a succession of ON and OFF periods. When a user is in ON period, the MAC layer transfers a 31 bytes speech frame each 20ms; this 31 bytes speech frame is encapsulated into one FP voice frame by adding a 5-bytes header. The individual channel's FP frames arrive periodically, every TTI, at the IP packetization unit. As voice traffic is symmetric, the model holds for both uplink and downlink.

For voice case, the IP packetization unit (Figure 2) is a multiplexing or assembly unit where IP voice packets are filled with FP frames, each FP frame has an additional 3-byte as Multiplexed Header (MH) including one byte termed User IDentifier (UID) unique to each user. The maximum size for IP voice packets is s_v and the payload for this completely filled packet corresponds to n FP frames. A Timer Common Usage (TCU) is associated with the assembly unit so as to avoid unacceptably large packetization delay. Then some IP packets can be partially filled and the packetization delay is bounded by the TCU value.

For large number of simultaneously ON DCH channels N_v, i. e. large number of active users, the arrival process of FP voice frames is modeled by a Poisson process with mean rate r_v equal to $\frac{N_v}{20}$ FP frames/ms. The arrival process of the IP packets to the voice queue is affected by the use of the TCU in the packetization unit. In [9], we derive analytically the IP packet arrival rate taking into account the TCU effect; including all IP packet configurations : partially as well as completely filled ones. We now reproduce briefly our work (see [9]for details).

Let the time axis be discrete, scaled according to Time Units (TUs). Let λ be the mean arrival rate of frames to the packetization unit, in units of frames per TU. Since arrivals are Poisson with mean rate λ, the probability that k FP frames arrive in one TU in the assembly unit is given by $P_k = \frac{\lambda^k}{k!}e^{-\lambda}$.

Let Y be the process that describes the number of FP frames in the assembly unit and let the following matrices denote the dynamics of Y during one TU :

- Let A_0 be a $1 \times (n-1)$ transition matrix from $Y = 0$ (empty unit) to $Y > 0$ (non empty unit) during one TU without IP Packet generation. Then, $A_0 = (P_1\ P_2\ \dots\ P_{n-1})$.
- Let A_{00} be an $(n-1) \times (n-1)$ transition matrix from $Y > 0$ to $Y > 0$ during one TU without IP Packet generation.

$$A_{00} = \begin{pmatrix} P_0 & P_1 & P_2 & \dots & P_{n-2} \\ 0 & P_0 & P_1 & \dots & P_{n-3} \\ \vdots & \vdots & \vdots & \ddots & \vdots \\ 0 & \dots & \dots & \dots & P_0 \end{pmatrix} \qquad (1)$$

- Let A_{0i} be a $1 \times n$ transition matrix from $Y = 0$ to Y during one TU with the generation of i IP packets. Then, $A_{0i} = (P_{in}\ P_{in+1}\ \dots\ P_{in+n-1})$.
- Let A_i be an $(n-1) \times (n)$ transition matrix from $Y > 0$ to Y during one TU with the generation of i IP packets.

$$A_i = \begin{pmatrix} P_{in-1} & P_{in} & P_{in+1} & \dots & P_{in+n-2} \\ \vdots & \vdots & \vdots & \ddots & \vdots \\ P_{in-n+1} & P_{in-n+2} & \dots & \dots & P_{in} \end{pmatrix} \qquad (2)$$

- Let A_{+0} be an $(n-1) \times n$ transition matrix from $Y > 0$ to Y during one TU when TCU expires.

$$A_{+0} = \begin{pmatrix} \sum_{k=0}^{n-2} P_k & 0 & 0 & \dots & 0 \\ \vdots & \vdots & \vdots & \ddots & \vdots \\ 0 & & & & \\ \sum_{k=0}^{0} P_k & 0 & 0 & \dots & 0 \end{pmatrix} \qquad (3)$$

Just after the beginning of a TU, FP frames arrive and IP voice packet(s) is (are) immediately generated if there are enough frames to constitute them. Let Y_m be Y just after the mth generation of a group of IP packets and let t_m be the corresponding TU. Let T_m be the time between the mth and $(m+1)$th generation of a group of IP packets, i.e., $T_m = t_{m+1} - t_m$. Let S_m be the number of FP frames that arrive during T_m and let C_{m+1} be the number of IP packets generated in the $(m+1)$th group packet generation. We have,

$$Y_{m+1} = \begin{cases} 0 & \text{if TCU expires and } C_{m+1} = 1 \\ Y_m + S_m - nC_{m+1} & \text{otherwise} \end{cases} \qquad (4)$$

where $C_{m+1} = \lfloor \frac{Y_m + S_m}{n} \rfloor$. Let $r = (r(0), r(1), r(2), \dots\dots, r(n-1))$ be the corresponding stationary probability vector of Y, with $r_0 = r(0)$ and $r_+ = (r(1), r(2), \dots, r(n-1))$. For e a $n \times 1$ all-1 column vector, the mean number g of IP packets in a group and the mean inter-group generation time τ are given by

$$g = r_+ \left(\sum_{t=1}^{TCU} A_{00}^{t-1} \sum_{i=1}^{\infty} i A_i + A_{00}^{TCU-1} A_{+0} \right) e$$

$$+ r_0 \sum_{s=1}^{\infty} P_0^{s-1} \left(\sum_{t=1}^{TCU} A_0 A_{00}^{t-1} \sum_{i=1}^{\infty} i A_i + A_0 A_{00}^{TCU-1} A_{+0} + \sum_{i=1}^{\infty} i A_{0i} \right) e \quad (5)$$

$$\tau = r_+ \left(\sum_{t=1}^{TCU} t A_{00}^{t-1} \sum_{i=1}^{\infty} A_i + TCU A_{00}^{TCU-1} A_{+0} \right) e + r_0 \sum_{s=1}^{\infty} P_0^{s-1}$$

$$\left(\sum_{t=1}^{TCU} (s+t) A_0 A_{00}^{t-1} \sum_{i=1}^{\infty} A_i + (s+TCU) A_0 A_{00}^{TCU-1} A_{+0} + s \sum_{i=1}^{\infty} A_{0i} \right) e \quad (6)$$

The mean rate of IP packet generation at the assembly unit λ_v packets per TU, which is equal to the mean IP packet arrival rate to the voice queue, is given by $\lambda_v = \frac{g}{\tau}$.

At this point, we make a change in the time scale as follows. We assume that the service time is constant and equal to the time needed to serve an IP packet with s_v size at output link capacity C, denoted by Service Time Unit (STU). Normalizing by STU, the mean packet arrival rate in one STU is then equal to the server load ρ_v. $n\rho_v$ is the mean arrival rate of FP frames. This is an over-estimation as some IP packets can leave the assembly unit with less than n FP frames.

Using the independence approximation (see [2]) which states that FP frames arrive according to a Poisson process with mean rate $n\rho_v$, we calculate the probability θ_i that i IP packets arrive to the transmission queue in one STU. That is approximately equivalent to $i \times n$ FP frames arriving to the system in one STU. We have,

$$\theta_i = \begin{cases} \displaystyle\sum_{j=0}^{n-1} \frac{n-j}{n} \frac{(n\rho_v)^j}{j!} e^{-(n\rho_v)} & i = 0 \\[2em] \displaystyle\sum_{j=1}^{n-1} \frac{j}{n} \frac{(n\rho_v)^{(i-1)n+j}}{((i-1)n+j)!} e^{-(n\rho_v)} + \sum_{j=0}^{n-1} \frac{n-j}{n} \frac{(n\rho_v)^{in+j}}{(in+j)!} e^{-(n\rho_v)} & i = 1, 2, \dots \end{cases}$$

$$(7)$$

We now define the embedded Markov chain X to be the number of IP packets present in the transmission queue at the end of the service of one packet. The matrix P of transition probabilities $[p_{ij}]$ $(i, j = 0, 1, 2, \dots)$ is given by (see [11] page 178)

$$P = \begin{pmatrix} \theta_0 & \theta_1 & \theta_2 & \theta_3 & \cdots \\ \theta_0 & \theta_1 & \theta_2 & \theta_3 & \cdots \\ 0 & \theta_0 & \theta_1 & \theta_2 & \cdots \\ 0 & 0 & \theta_0 & \theta_1 & \cdots \\ \vdots & \vdots & \vdots & \vdots & \ddots \end{pmatrix} \quad (8)$$

The stationary probability that there are i IP packets in the voice queue is given by $\pi = \pi P$ where $\pi = (\pi(0), \pi(1), \pi(2), \dots)$. Taking into account that the mean arrival rate is ρ_v and the average number of IP packets in the queue is q, we apply Little's formula and obtain d_q, the average queuing time in the transmission unit

$$d_q = \frac{q}{\rho_v} = \frac{\sum_{i=1}^{\infty} i\pi(i)}{\rho_v} \quad (9)$$

Let M be the time it takes for n frames to arrive. As the frames arrival process is Poisson with rate r_v, M is distributed according to an Erlang$(n-1, r_v)$ distribution with mean $E(M) = \frac{n-1}{r_v}$. The mean delay d_v of individual FP frames at the multiplexer including packetization at the assembly unit and queuing at the transmission queue is given by

$$d_v = \begin{cases} d_q + E(M) & \text{if } E(M) \leq TCU \\ d_q + TCU & \text{if } E(M) > TCU \end{cases} \tag{10}$$

Owing to the presence of data traffic with packet size s_d, the total mean delay D_v on voice traffic is given by

$$D_v = d_v + \frac{s_d}{2C} \tag{11}$$

4.2 Low-Priority Data Traffic

Data traffic refers to various applications. It ranges from background traffic, such as mail, to interactive traffic, such as Web surfing. In this work, we focus on downlink Web browsing traffic, transported over a Downlink Shared CHannel (DSCH), shared by several users. By reference to the preceding voice traffic model over dedicated channels, the TTI value is 40 ms in this case. The DSCH channel bit rate can be 64 kbps, 144kbps or 384kbps. At the FP layer, and as is the case of voice, Web traffic over DSCH can be modeled as a succession of ON and OFF periods. In one ON period, for a 64 kbps channel, a frame of 320 bytes is generated every 40 ms and is encapsulated into one FP data frame by adding a 5-byte header.

We focus on the case where one FP data frame is encapsulated into one IP packet by adding UDP/IP header. The packet size s_d amounts to 353 bytes. Assuming that the time for forwarding IP packets from the packetization unit to the transmission queue is negligible, the system is then equivalent to the transmission queue only. For N_d simultaneously ON data channels, the arrival process of IP data packets to the data transfer queue can be modeled by a Poisson process and the queue itself is viewed as an $M/D/1$ queue with vacations. The vacations correspond to epochs when the server is busy serving higher-priority voice traffic. Let V be a random variable indicating the duration of vacations. The values of V correspond thus to the length of the busy period for the voice queue. Normalizing the time axis, both for arrival and service processes, to the fixed time needed to serve a data packet of size s_d at link capacity C, and using the Pollaczek-Khinchin formula for an $M/D/1$ queue, we calculate the mean queuing delay d_d for a data packet as

$$d_d = \frac{2 - \rho_d}{2(1 - \rho_d)} \tag{12}$$

where ρ_d is the data workload. The mean queuing delay for an $M/D/1$ queue with vacations is given by [12]

$$D_d = d_d + \frac{E[V^2]}{2E[V]} \tag{13}$$

where $E[V]$ and $E[V^2]$ are respectively the first and second moments of the vacation interval V and correspond to the first and second moments of the busy periods for the voice queue and are thus given by

$$E[V] = \sum_{i=1}^{\infty} (i * STU)\pi(i) \qquad (14)$$

$$E[V^2] = \sum_{i=1}^{\infty} (i * STU)^2 \pi(i). \qquad (15)$$

5   Numerical Results and Empirical Validation

In this section, we first describe the UTRAN test-bed that we use for the empirical validation of our results. We then validate our model through a comparison between the analytical and the empirical results. Our analytical work was based on a mean value analysis. A percentile analysis can give further insights into the problem. It is carried out empirically in this work and is presented in the third subsection. Note that throughout this section and for the case of voice traffic, we set n, the number of FP voice frames per IP packet equal to 8 and the TCU timer value equal to 2 ms. Those values yield an optimal link utilization, as shown in our work in Reference [9].

5.1   Emulating UTRAN Transport on a Local Test-Bed

Our test-bed emulating the UTRAN transport functionalities is based essentially on PCs operating with FreeBSD Operating System (OS). All elements of the test-bed are synchronized using GPS equipment, essential for accurately measuring delays at very small scales. The traffic at the FP layer is generated using the server-client UDP/IP model with socket programming. The assembly and the transmission units are implemented as follows: at the assembly unit, threads are implemented to execute the multiplexing algorithm including the assembly process and the timer process. The PQ algorithm is implemented through the use of the ALTQ tool [13] on FreeBSD system. Tcpdump and other locally-coded software are used to record desired parameters.

5.2   Mean Value Results

In Figure 3, we compare the analytical mean delay D_v for high-priority voice traffic and the empirical one obtained from experiments on the test-bed for a number of simultaneously active voice channels N_v=100 and in the presence of data traffic. PQ ensures service differentiation. The x-axis shows different values of the link capacity C. As expected, the mean delay decreases when the link capacity increases. One can see that the empirical mean delay is less than the analytical one, this is explained by the fact that analytically there is an over-estimation due to the assumption that all packets are completely filled with FP frames. This assumption leads to an analytical packetization delay higher than the empirical one.

Figure 4 compares the analytical and the empirical mean delay D_d for data traffic as a function of data load ρ_d. The three sets of curves correspond to three different values of ρ_v, the load of voice traffic : zero, medium and high. The output link capacity is fixed at 2 Mbps. Curves are truncated at values where the total load is strictly less than 1, i.e.,

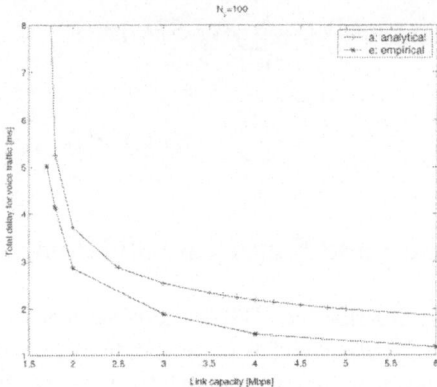

Fig. 3. Mean delay for voice traffic: analytical versus empirical

$\rho = \rho_v + \rho_d < 1$. Note that the empirical curves are generated through a Poisson arrival process. We note that the mean delay D_d increases as the data workload increases. When no voice traffic is present, the results obtained analytically match very closely with the empirical ones when the arrival process of data packets is Poisson. When the load of voice traffic is medium or high, the analytical and empirical curves differ slightly. This is mainly due to the terms related to vacations in the analytical case.

In Figure 5, we reproduce the same performance measures as in the previous one. The curves are however both empirical and correspond to two different traffic generation schemes: Poisson versus periodic. The latter represents best the real system as it reproduces the periodic nature of arrivals per TTI. The results, in terms of mean delay, show that the Poisson approximation is a good one. We note that for a low load of data traffic, i.e. a low number of data channels, the periodic traffic is less bursty than Poisson. When the number of data channels increases, the periodic traffic can be more bursty.

5.3   Percentile Results

Figures 6 and 7 confirm further the benefits of the use of service differentiation through a percentile analysis. In effect, those Figures draw the 95 percentile of three quantities :

1. Total delay of voice traffic including packetization, queuing and possible jitter in the presence of data traffic (denoted by total delay in the figures).
2. Queuing (or transmission) delay of voice traffic and possible jitter in the presence of data traffic (denoted by trans delay in the figures).
3. Delay of data traffic.

Figure 6 shows the results of the above mentioned delays with and without the use of PQ for ρ_v=0.36 and output link capacity equal to 2 Mbps. The x-axis shows the total load $\rho = \rho_v + \rho_d$. We observe that when no differentiation is implemented, the voice transmission delay and the data delay are almost equal. The total voice delay is higher due to the packetization delay component and violates the voice delay budget for almost

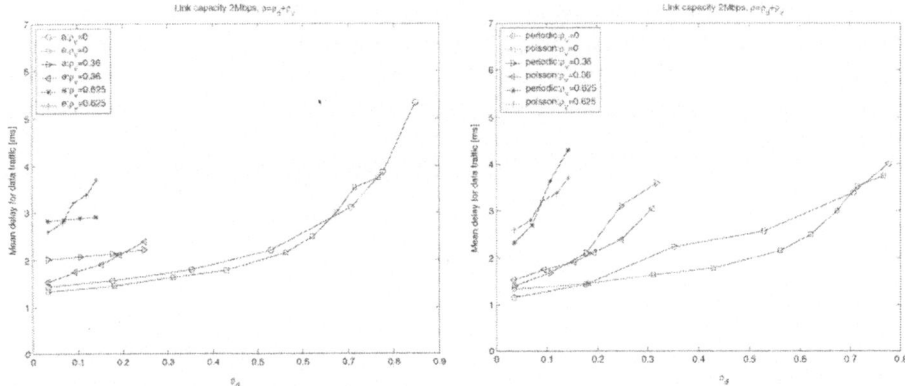

Fig. 4. Mean delay for data traffic: analytical versus empirical

Fig. 5. Mean delay for data traffic: Poisson versus periodic

all values of ρ_d. Using PQ, voice delays both total and transmission get less than the 5ms delay bound. At the expense of a large delay for data traffic, yet below the data delay budget. The same observations can be made for Figure 7 where $\rho_v = 0.5$. In this case however, the total load is less than the one in the previous case. This is due to the fact that voice traffic has an even tighter delay than the data one and gets more critical as its percentage increases with respect to the one of data in the total load.

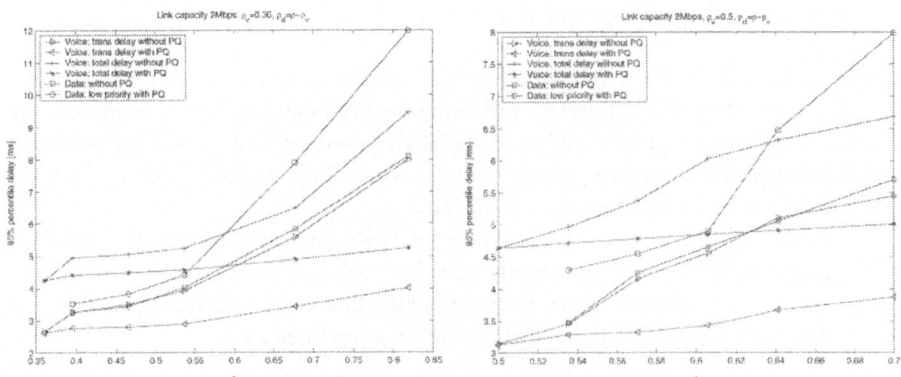

Fig. 6. 95 percentile delay of voice and data traffic with and without PQ

Fig. 7. 95 percentile delay of voice and data traffic with and without PQ

Figure 8 shows the Cumulative Distribution Function (CDF) of the total delay of voice traffic when $\rho_v = 0.36$ for different values of ρ_d. The difference between curves is the jitter due to the presence of data traffic. When no data traffic we note that 98.15 % of voice traffic has a delay less than 5 ms. When $\rho_d = 0.2$ and 0.36, this figure is about

96.75% and 95.53% respectively. Figure 9 shows the CDF of the delay of data traffic when ρ_d=0.2 for different values of ρ_v. We note that when the load of voice traffic is low or medium the data delay bound is respected. For high loaded systems, when ρ_v=0.625 or larger, one can observe that about 88% of the data traffic experiences a delay less than 10 ms .

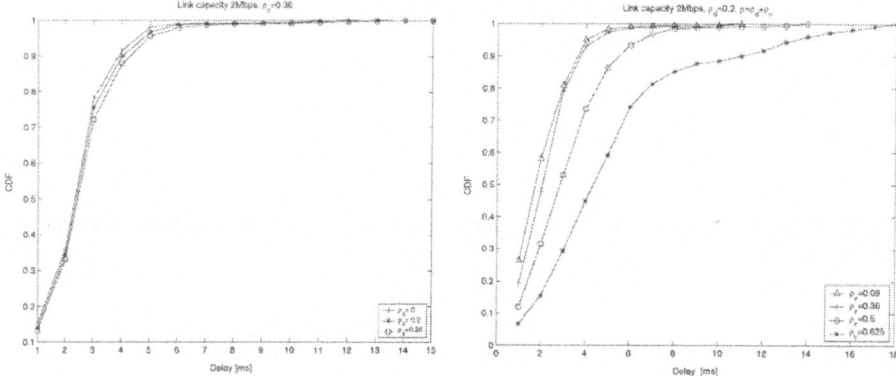

Fig. 8. Cumulative Distribution Function of the total delay of the voice traffic

Fig. 9. Cumulative Distribution Function of the delay of the data traffic

6   Dimensioning of the UTRAN

The volume of incoming traffic, be it voice or data or both, present in the UTRAN is typically dictated by the air interface following the availability of the air resources. The latter are scarce and expensive and so the UTRAN wired capacity should not be restrictive. Instead, proper ressource provisioning following accurate dimensioning should be implemented so as to accommodate the demand.

Consider traffic i, $i \in \{v, d\}$, standing for voice and data respectively and for a QoS criterion defined as the probability of delay violation by any type of traffic, i.e., $P(D_i > b_i) \leq \epsilon_i$, where b_i is the maximum delay that can be tolerated by traffic i in the UTRAN. We can, at this point, distinguish between three cases :

1. presence of voice traffic only in the UTRAN. In this case, b_v =5ms. Let ρ_{0v} denote the maximum load of voice traffic that the UTRAN can accomodate in this case.
2. presence of data traffic only. In this case, b_d =10ms. Let ρ_{0d} denote the maximum load of data traffic that the UTRAN can accomodate in this case.
3. presence of both voice and data traffic in the UTRAN. PQ is here to prioritize voice over data. In this case, in the worst case, the voice delay bound is $(b_v - \frac{s_d}{C})$ ms due to the non pre-emptive nature of PQ. Let ρ'_{0v} be the maximum load of voice traffic corresponding to this delay budget. As of data traffic, the delay bound is $(b_d - b_v)$; let ρ'_{0d} be the load of data traffic in this case.

For an offered load $\rho = \rho_v + \rho_d$, proper dimensioning should satisfy the following constraints:

- for $\rho_d = 0$, $\rho \leq \rho_{0v}$.
- for $\rho_v = 0$, $\rho \leq \rho_{0d}$.
- for $\rho_d > 0$ and $\rho_v > 0$, $\rho_v \leq \rho_{0v}'$ and $\rho_d \leq \rho_{0d}'$.

7 Conclusion

In this work, we developed an analytical model for IP service differentiation in the UTRAN where both real-time voice and non-real-time data traffic are to meet severe QoS constraints in terms of delay. We validated our model on a local test-bed emulating the UTRAN transport functionalities. Our results show that PQ is essential to make the voice traffic meet its delay requirements. PQ ensures not only service differentiation between voice and data traffic but also helps minimize jitter for voice traffic. We also drew proper dimensioning for the UTRAN so as to make voice and data traffic meet their delay constraints.

References

1. 3GPP TS 25.430 V3.6.0 (2001-06) 3rd GPP; Technical Specification Group Radio Access Network; UTRAN Iub Interface: General Aspects and Principles (Release 1999).
2. H. Saito, Performance Evaluation and Dimensioning for AAL2 CLAD, IEEE INFOCOM'99, New-York, 1999.
3. A-F. Canton, S.Tohmé, D. Zeghlache, T. Chahed, Performance analysis of ATM/AAL2 in UMTS Radio Access Network, in press, IEEE PIMRC 2002, Lisbon, 2002.
4. R. Makké, S. Tohmé, J-Y. Cochennec, S. Pautonnier , Performance of the AAL2 Protocol within the UTRAN, IEEE ECUMN 2002.
5. O. Isnard, et all, Handling Traffic Classes at AAL2/ATM layer over the logical Interfaces of the UMTS Terrestrial Radio Access Network, IEEE PIMRC, 2000.
6. S. Nananukul, S. Kekki, Simulation Studies of bandwidth Management for the ATM/AAL2 Transport in the UTRAN. Vehicular Technology Conference, 2002.
7. 3GPP TS 25.933 V1.7.1 (2002-01), IP Transport in UTRAN Work Task Technical Report.
8. Mobile Wireless Internet Forum IP in the RAN as a Transport Option in 3rd Generation Mobile Systems, Technical Report MTR-006, Release v2.0.0, Ratified June 18, 2001.
9. A. Samhat, T. Chahed, Performance Evaluation of IP in the UMTS Terrestrial Radio Access Network, In Proc. 18th International Teletraffic Congress (ITC), Berlin, September 2003.
10. A. Samhat et all, Transport in UMTS Radio Access Network: IP versus AAL2/ATM, 8th International Conference on Cellular and Intelligent Communications (CIC), Seoul, 2003.
11. L. Kleinrock, Queueing Systems, Volume I: Theory, John Wiley and Sons, 1975.
12. D. Bertsekas, R. Gallager, Data Networks, second edition, Prentice-Hall, 1992.
13. Home page of ALTQ project http://www.csl.sony.co.jp/ kjc/ projects.html.

Multi-hour Design of Dynamically Reconfigurable MPLS Networks*

Balázs Gábor Józsa[1,2], Dániel Orincsay[1,2], and Levente Tamási[2]

[1] Budapest University of Technology and Economics, Department of Telecommunication and Media Informatics, High Speed Networks Laboratory, H-1117 Magyar tudósok körútja 2, Budapest, Hungary
{jozsa, orincsay, tamasi}@tmit.bme.hu
[2] Ericsson Research Hungary, Traffic Analysis and Network Performance Laboratory, H-1300 POB 107, Budapest, Hungary
{Balazs.Jozsa, Daniel.Orincsay}@ericsson.com

Abstract. This paper proposes an algorithm for cost-efficient network design. The two main network components are the routers and the links connecting them. As these network elements have discrete capacity values, their costs are modeled by stepwise functions. In the current interpretation of the network design problem the subproblems of topological design, resource dimensioning, and routing of traffic flows are handled simultaneously. Using new network technologies (e.g., MPLS) the operators have the opportunity to reconfigure their networks dynamically. By using this function the periodic (e.g., daily, weekly) changes of traffic volumes and directions can be taken into consideration during network design resulting in lower deployment costs. The algorithm proposed for multi-hour design is based on an efficient, published method. The efficiency of the new algorithm is demonstrated by simulation on random and real network topologies with the help of two reference algorithms.

Keywords: MPLS, traffic engineering, network planning, multi-hour design.

1 Introduction

Nowadays, when the number of users of networking applications increases enormously, the management of network resources is a hot-topic issue. On the other hand, as technology develops, the capacities of network devices become larger and larger; a few years ago 155 Mbps links were general but now 10 Gbps links are also available. However, the need for bandwidth increases more rapidly than capacities of routers and links, since the bandwidth requirements of novel networking applications are higher and higher. Further, sometimes it is not worth extending the existing networks since the old technology that does not support novel applications or the extension is restricted by physical limitations. For these reasons the establishment of new networks using new technologies is essential. Thus, Internet service providers (ISPs) can make use of an algorithm that can solve the problem of network design aiming at reducing the deployment cost.

* This work was supported by the Ministry of Education, Hungary, under the reference No. IKTA-0092/2002. http://w3.ttt.bme.hu/ikta-2002

N. Mitrou et al. (Eds.): NETWORKING 2004, LNCS 3042, pp. 502–513, 2004.

An important issue considering the profitability of ISPs is how economically they can utilize their resources. In order to reach the optimal network performance, administrators should have full control over traffic flows. Fortunately, using novel network technologies, e.g., multiprotocol label switching (MPLS) [11], several traffic engineering [1] tools are available for the operator. For example, the paths of traffic flows can be given explicitly and can be reconfigured without the interruption of traffic [6]. The method of reconfiguration according to the daily and/or weekly traffic changes is called *capacity management* in a recommendation by the Telecommunication Standardization Sector of International Telecommunication Union (ITU-T) [4]. A possible approach of capacity management is multi-hour design (MHD). Multi-hour design—as opposed to single-hour design (SHD), which results in a network that is dimensioned for maximal (busy hour) traffic demands—takes the periodic change of traffic volumes and directions into account by partitioning the whole time scale into several intervals and calculating the maximal traffic demands separately for each interval. As the maximal demand between different node pairs may occur in different intervals, the capacities of network devices may be smaller than in the case of single-hour design, resulting in lower deployment cost. This design approach takes the ability of reconfiguration into account already at the network design phase.

In this study the relation between capacity and cost values tries to represent well the real situations, where network devices are built up from smaller modules with discrete capacity values. Thus, stepwise cost functions are used, which makes the complexity of the design task higher implying the need for a heuristic approach. This paper proposes an algorithm for multi-hour design problem which is based on the algorithm for single-hour design presented in [7]. Although MPLS is assumed as underlying network technology throughout this paper, the algorithm to be presented can be used for any technology supporting the traffic engineering features described above.

Similar problem for ATM networks was already discussed in the literature. In [2] the problem is divided into three sub-problems and during the iterative solution two of them are considered as static at one time. Medhi has several methods [8,9,10] for the problem, moreover he also investigated the issue of protection. The novelty compared to the previous approaches is that our algorithm uses stepwise cost functions and it does not restrict the paths between node pairs to pre-determined path sets.

The rest of the paper is organized as follows. The next section describes the applied models and gives the problem definition. In Section 3 the proposed algorithm is detailed. Then the simulation environment is presented and the numerical results are evaluated in Section 4. Finally, Section 5 contains the concluding remarks.

2   Problem Statement

This section presents the used interpretation of cost-optimal multi-hour network design problem. First, the network, traffic, and cost models are described. Then the problem formulation is given, finally the issue of complexity is discussed.

2.1   Network Model

The network is modeled by a directed graph where the routers and the links between them are represented by a set of nodes and a set of edges, respectively. Duplex links are supposed that are represented by oppositely directed edge pairs, where the two edges corresponding to a given link must have the same capacity values. For the easier handling of capacities of routers, the nodes are substituted by virtual links (see Fig. 1). A particular node n is decomposed into two nodes n_{in} and n_{out} belonging to the incoming and outgoing traffic of node n, respectively. Thus, n_{in} is connected only to the incoming links, while n_{out} is connected to the outgoing links, and node n is represented by the virtual link between n_{in} and n_{out} that the transit traffic must traverse. In this way the capacity constraint of nodes can be taken into consideration easily.

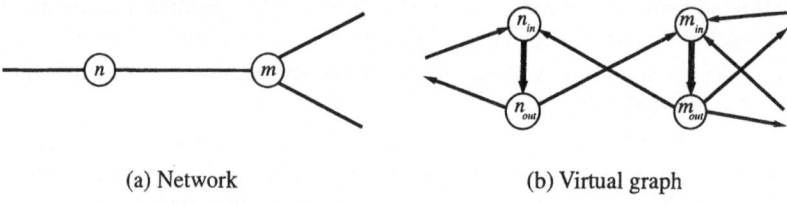

(a) Network   (b) Virtual graph

Fig. 1. Mapping the network onto a directed graph using virtual links.

2.2   Traffic Model

In the case of multi-hour design the time-scale is divided into several intervals. In the time intervals different traffic volumes and distributions can be given by different sets of traffic demands. A particular traffic demand is described by its source and destination routers, and its required capacity. Generally, one unsplittable traffic demand is assumed per node-pair, however, the model can handle more parallel demands for the same node-pair (e.g., representing more traffic classes) as well.

Since capacitated traffic demands are assumed, mapping them onto the network means that the required amount of capacity has to be reserved on each link and router along their previously computed paths. In MPLS an explicitly determined label switched path (LSP) should be established for each traffic demand using, e.g., the traffic engineering extension of the resource reservation protocol (RSVP-TE). Obviously, the sum of capacity reservations on a device cannot exceed its maximum (reservable) capacity value.

2.3   Cost Model

Basically, cost is one of the most important factors during the process of network design, therefore the applied cost model has large influence on the accuracy and applicability of the used approach. The cost/capacity dependencies of various network devices are generally described by cost functions. Various approaches co-exist in network design starting from the simplest linear cost model up to the complex *stepwise* cost model (see Fig. 2). When using linear functions the problem is simpler, however, this approximation of the cost/capacity relationship is very inaccurate. On the contrary, stepwise cost functions can model real situations well, however, the drawback of this approach is the much higher complexity. Since this study aims at developing a method that could also be applied in real-world design tasks, the more sophisticated model of stepwise cost functions is followed.

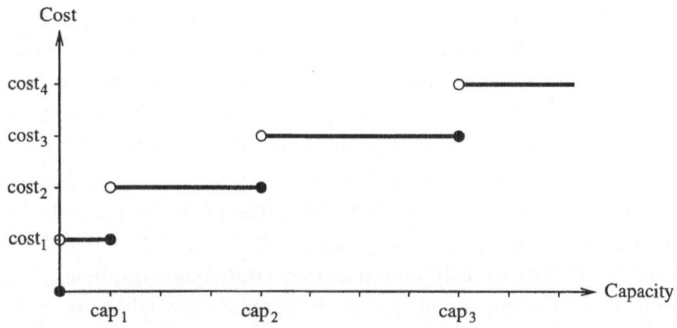

Fig. 2. Stepwise cost function example.

2.4   Problem Formulation

The formulation of the problem can be given by the sets of input and output data. The input sets are the following:

- locations of routers,
- possible links between the pairs of routers,
- individual cost functions for routers and links,
- traffic demand sets to be satisfied corresponding to different time intervals.

The following output data are determined as the result of the design process:

- capacity of each device,
- explicit path set for each traffic demand set.

2.5   Complexity

As it has been discussed in [7], the single-hour network design problem is already NP-hard [3] when using stepwise cost functions. Because of the more time intervals handled in multi-hour network design the problem is more difficult than in the case of single-hour network design. Since the latter one is a special case of the first one, the mathematical complexity of MHD is at least the same as that of SHD.

3   Multi-hour Core Network Design Algorithm (CND^MH)

In this section the new algorithm CND^MH proposed for cost-efficient multi-hour network design is detailed. First, the outline of the algorithm is described and then its three phases are presented.

3.1   Algorithm Outline

The multi-hour network design algorithm CND^MH has the same three phases as its single-hour design version CND presented in [7]. Naturally, the particular phases are extended due to the more sets of traffic demands corresponding to the different time intervals. The first phase is the initial capacity estimation (ICE^MH) that tries to foresee the approximate capacity needs. Then the iterative routing optimization (IRO^MH) follows, whose task is to provide a feasible network configuration using the output of ICE^MH. The last phase called posterior capacity refinement (PCR^MH) serves for finalizing the actual network configuration to achieve a cheaper solution. Note that IRO^MH is the main phase that can give a valid solution for the problem by itself, while the use of ICE^MH and PCR^MH is optional. However, it is worth using them as they contribute significantly to the quality of the solution. It is also important that in the special case when the number of time intervals equals to 1 all phases of CND^MH operate in the same way as their single-hour versions.

3.2   Initial Capacity Estimation (ICE^MH)

The goal of ICE^MH is to foresee the necessary capacities of network devices by analyzing the sets of traffic demands. This phase provides a partially dimensioned network for IRO^MH, therefore the remaining tasks are to finish dimensioning of the devices and to route the sets of traffic demands corresponding to the different time intervals. Although ICE^MH does not give a feasible solution, it specifies a good starting state based on a global view of the problem instance.

The operation of ICE^MH is based on the routing of traffic demands one by one in random order. All traffic demand sets are routed in a predefined number of rounds using shortest path routing. We suppose that if all the random accommodations needed a certain amount of capacity on a particular device, then it is probable that also the optimal accommodation needs so much capacity. Therefore, the capacity values of the devices are specified in such way that they get the lowest value of the arising ones in the above accommodations.

3.3   Iterative Routing Optimization (IRO^MH)

IRO^MH is the main phase of the algorithm, i.e., it can provide a full solution for the problem without the use of ICE^MH and/or PCR^MH phases. The base of this phase is an algorithm that is capable of routing a given set of traffic demands in a capacitated graph. In this paper—as in [7]—the algorithm proposed in [5] is used, which performs well in terms of *feasibility*, namely, it is very probable that it finds a solution for a given problem instance provided that a solution exists. However, this applied routing optimization algorithm can be substituted by any other one solving the same task.

IRO^{MH} is based on an iteration having two steps. In the first step the traffic demand sets are tried to be routed in the graph while taking the actual capacity constraints into account. If the algorithm terminates with success for all traffic demand sets IRO^{MH} finishes. Otherwise, in the second step the capacity of a particular device is increased by one capacity step, and the first step follows again. The selection of the device to be enlarged is done by routing the rest of the traffic demands without capacity constraints for each time interval and then the device having the maximal capacity excess—that is averaged for the different time intervals—is chosen.

3.4 Posterior Capacity Refinement (PCR^{MH})

Although a feasible solution is available after the IRO^{MH} phase, this result can be improved with the help of PCR^{MH}. This phase is based on a local search procedure, i.e., the process concentrates only on one part of the network at one time. The idea behind PCR^{MH} is to reduce the capacity of such devices whose *relative step utilization* is low. The relative step utilization refers to the ratio of necessary and total capacity values on the actual capacity step (corresponding to the actual device cost).

PCR^{MH} starts with the sorting of the devices by their relative step utilizations—that are averaged for the different time intervals—and then the first device is chosen. The capacity of the actual device is decreased by one step and the sets of traffic demands are tried to be accommodated by IRO^{MH} under these tighter capacity conditions. If all sets can be routed then PCR^{MH} restarts, otherwise the capacity of the current device is reset and the next device is tried to be shrunk. After the last device has been tried to be shrunk without success PCR^{MH} finishes.

4 Results

This section contains the performance evaluation of the proposed algorithm CND^{MH} which was carried out by means of simulation. First, the process of problem instance creation is described including the generation of network topologies, traffic demand sets, and cost functions. Then the reference algorithms used for the comparison are introduced. Finally, the numerical results and their analysis are presented.

4.1 Problem Instance Generation

To ensure the reliability of simulations many different problem instances were investigated. The automated process of problem instance generation was similar to the one applied in [7]. However, in case of traffic demand sets a more complex generation method had to be used since more time intervals are distinguished. Note that CND^{MH} is able to handle any kind of problem instance that can be described by the model introduced in Section 2.

Network Topology. Generally, the task of network design is to plan possible future networks, thus the main focus of examinations was on random topologies. Besides, we also considered two existing real-world topologies (see Fig. 3). To create random topologies the random graph generation method published in [5] was used that tries to model the characteristics of real networks. In this study we concentrated on 25-node networks, however, we present results for 15-node and 35-node network sizes as well. The initial average nodal degree of topologies was set to 5.

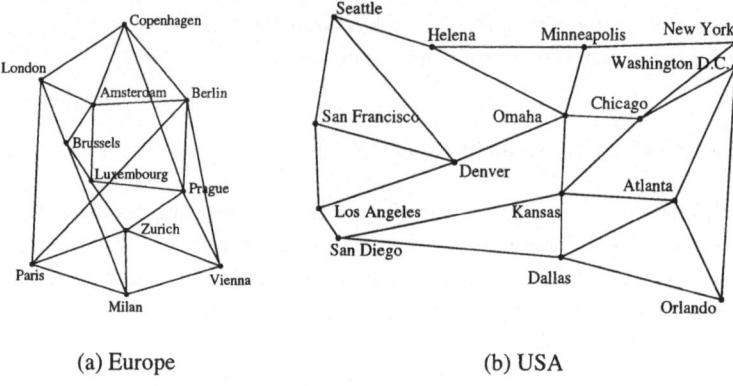

(a) Europe   (b) USA

Fig. 3. The investigated real network topologies.

Traffic Demand Sets. The main characteristic of the multi-hour design problem is that the bandwidth demand of traffic flows between node pairs is varying in time. In order to create more traffic demand sets we divided the initial network topology into certain number of regions in the following way. First, the center of the topology was determined as a circle shape area. Its radius was set based on the expected number of nodes per region value that is the total number of nodes divided by the number of regions. Then the remaining topology was divided into angle ranges around the center, and the out-of-center nodes within the same angle range were assigned to one certain region. In a particular time interval each region was matched with exactly one other region. One part of the sum outgoing traffic of a given region specified by parameter Δ flowed towards its dedicated region pair, while the remaining part of traffic was distributed evenly among all regions (including itself as well as its region pair). In this way, $\Delta = 0$ means that there was no extra traffic between the specified region pairs and $\Delta = 100\%$ refers to the case when the total traffic flowed between the specified region pairs. When calculating the traffic demands for a given time interval the original traffic generation method detailed in [7] was used as first step, then the values for a given region were scaled based on parameter Δ so that the sum of traffic did not change. The difference between time intervals was based on the varying matching of regions. In this paper both the number of time intervals and the number of regions were set to 6. Since we had no information available about realistic traffic distributions—moreover traffic modeling is a huge research area that is out of the scope of this paper—different possible cases were examined with the help of parameter Δ that was shifted from 0% up to 100% by steps of 25%. The traffic volume was varied within a wide interval with the help of an input parameter representing the required average traffic per link value (referred as 'traffic volume' in tables). Further, the traffic volume was constant in the different time intervals of a particular problem instance.

Cost Functions. During the investigation of CND^{MH} the same cost model was followed as in case of the original CND, in this way the same cost function generation was used (see [7]). In case of routers three capacity levels were distinguished specifying the maximum capacity to be handled (see Table 1). The cost functions of links were based on the standard capacity values of STM and it was supposed that the deployment of two parallel devices having the same capacity costs less than the device of the next capacity level (see Table 2). Note that the capacity values were modified randomly for each network device so that they would have unique cost functions.

Table 1. Costs of the routers.

capacity	cost (unit)
1 Gbps	5
10 Gbps	15
100 Gbps	60

Table 2. Costs of the links.

capacity	cost (unit)
155 Mbps	3
310 Mbps	6
622 Mbps	9
1244 Mbps	18
2.5 Gbps	27
5 Gbps	54
10 Gbps	81
20 Gbps	162

4.2 Reference Algorithms

Since information on the absolute optimal solutions of the problem instances were not available, two reference algorithms were used for evaluating the proposed algorithm CND^{MH}. Both the two references are simple extensions of CND. In this way the cost savings using CND^{MH} for multi-hour network design compared to CND can be investigated. After the reference algorithms determine the capacities of devices, the routes are calculated for each time interval with the help of global path optimizer [5] used in IRO^{MH} phase of CND^{MH}.

Traffic Bandwidth Maximizer (TBM). The main idea behind TBM is that the maximal traffic bandwidth requirement arising in the different traffic demand sets is considered between each node pair. Then the original CND is executed for the maximized traffic demand set. As the maximized traffic demands have at least the same bandwidth requirements as the individual traffic demands corresponding to the different time intervals, the designed network can surely accommodate every demand set. An important advantage of TBM—resulting from the bandwidth maximization—is that a feasible solution based on static routes can be provided for the case when path reconfiguration cannot be performed.

Link Capacity Maximizer (LCM). LCM handles the different time intervals separately, which means that it designs one particular network—with the help of CND—for each time interval independently from the other ones. After that the maximal arising capacity values are taken for every network device. In this way, the resulting network configuration can fulfill the capacity requirements of all traffic demand sets.

4.3   Numerical Results

This section presents the numerical results of the performed simulations. 5 different network topologies and 3 different traffic situations per topology were investigated for each problem class specified by network size, traffic volume, and traffic distribution value Δ. Due to the stepwise nature of the cost functions the size of the confidence interval of the results were relatively large, however, the tendencies were the same for each particular problem instance. Since the problem to be solved is cost optimization, the main focus of the investigations was on the total network cost. On the other hand, the running time was also examined to get a clear picture about the applicability of the algorithm in real situations.

The applied cost functions were based on cost units as it was discussed in Section 4.1, because we concentrate on the ratios of the cost values of different devices instead of the real commercial prices. When comparing the total cost values of the different design algorithms, the ratios were considered in percentage where 100% meant the best solution for the given problem class.

Table 3 shows the total cost values for 25-node networks. The results correspond to 10 different traffic volumes and 5 different Δ values (referring to the different traffic distributions). As one can see CND^{MH} gave the best solution almost in all cases. Exceptional cases can be found at $\Delta = 0$ setting, when TBM was slightly better than CND^{MH}. The largest differences between CND^{MH} and the reference algorithms can be observed at $\Delta = 100\%$: TBM and LCM were worse than CND^{MH} by 44% and 18% on average, respectively. Although we have no information about real traffic models, we suppose that real situations correspond to $25\% \leq \Delta \leq 75\%$ values, where the use of CND^{MH} results in a performance gain of 4–39% compared to the references. The ratios of solutions of different algorithms varied a bit as the traffic volume changed between 200 and 5500 Mbps, however, the tendency was clear.

In Table 4 the results (averaged for all traffic volumes) of different network sizes as well as real network topologies can be seen. The first observation is that CND^{MH} was the best for 15-node networks also at $\Delta = 0$. Another thing to note is that the performance gain attained by using CND^{MH} was slightly reduced when increasing the network size in general. In case of real networks—according to the results of the similar size random networks—CND^{MH} overperformed the references also in the case of evenly distributed traffic ($\Delta = 0$). The improvement gained by applying CND^{MH} varied between 7% and 35% compared to the references.

Considering the running time values—that were measured on a Sun Ultra Enterprise 420R with an Ultra II 450Mhz processor and 1GByte memory—it can be seen that for 15-node networks (as well as for the real networks) the time consumption is very low. Further, the worst running time value for 35-node networks is only about 1 hour, which is still moderate for an off-line network design tool. One can note that the running time of TBM is significantly lower than in case of the other two algorithms. The reason for this is that TBM performs the original CND only once, while LCM repeats CND for all

Table 3. Ratio of total network costs for 25-node networks (%).

traffic volume (Mbps)	algorithm	Δ				
		0%	25%	50%	75%	100%
200	CND^{MH}	100.10	100.00	100.00	100.00	100.00
	TBM	100.00	104.67	120.70	133.22	137.66
	LCM	105.62	105.31	107.85	110.48	115.09
300	CND^{MH}	101.35	100.00	100.00	100.00	100.00
	TBM	100.00	104.40	116.53	137.65	142.89
	LCM	105.75	107.68	106.69	117.31	119.47
500	CND^{MH}	101.16	100.00	100.00	100.00	100.00
	TBM	100.00	105.58	118.99	140.00	150.88
	LCM	109.28	109.18	109.80	117.80	121.07
800	CND^{MH}	100.00	100.00	100.00	100.00	100.00
	TBM	100.62	106.04	125.08	142.19	142.44
	LCM	108.66	108.70	108.80	113.03	117.25
1200	CND^{MH}	101.20	100.00	100.00	100.00	100.00
	TBM	100.00	107.33	121.61	133.52	138.99
	LCM	105.24	104.79	107.34	113.59	122.16
1700	CND^{MH}	101.57	100.00	100.00	100.00	100.00
	TBM	100.00	101.93	116.76	140.01	140.23
	LCM	106.86	104.20	108.36	115.67	120.27
2400	CND^{MH}	102.32	100.00	100.00	100.00	100.00
	TBM	100.00	104.36	119.32	138.53	150.51
	LCM	109.67	107.40	107.14	117.42	118.37
3200	CND^{MH}	100.00	100.00	100.00	100.00	100.00
	TBM	101.85	102.37	121.43	150.75	145.10
	LCM	106.98	108.57	108.69	115.31	120.04
4200	CND^{MH}	100.01	100.00	100.00	100.00	100.00
	TBM	100.00	104.10	116.24	143.43	146.03
	LCM	107.40	106.33	110.72	113.95	113.71
5500	CND^{MH}	101.45	100.00	100.00	100.00	100.00
	TBM	100.00	100.37	114.58	131.66	147.31
	LCM	102.89	107.25	110.70	114.14	114.60
Average	CND^{MH}	100.67	100.00	100.00	100.00	100.00
	TBM	100.00	104.11	119.12	139.09	144.20
	LCM	106.57	106.94	108.61	114.87	118.20

time intervals and CND^{MH} has more sub-processes that are multipled due to the more sets of traffic demands.

5   Concluding Remarks

This paper addressed the problem of multi-hour design of MPLS networks. In the current interpretation of the problem the time-scale is divided into several intervals, and different sets of traffic demands are given describing the traffic volumes and distributions for all time intervals. The cost/capacity relations of network elements such as routers

Table 4. Cost and running time results for different network sizes.

network size	algorithm	Δ					time (minute) min–max (avg)
		0%	25%	50%	75%	100%	
15 nodes	CND^{MH}	100.00	100.00	100.00	100.00	100.00	0–3 (< 1)
	TBM	100.17	107.32	123.10	145.80	152.23	0–1 (< 1)
	LCM	105.75	108.59	113.76	125.29	126.65	0–2 (< 1)
25 nodes	CND^{MH}	100.67	100.00	100.00	100.00	100.00	1–20 (7)
	TBM	100.00	104.11	119.12	139.09	144.20	1– 8 (2)
	LCM	106.57	106.94	108.61	114.87	118.20	1–37 (6)
35 nodes	CND^{MH}	104.83	100.00	100.00	100.00	100.00	1–61 (25)
	TBM	100.00	102.62	116.83	134.33	149.62	1–29 (10)
	LCM	108.25	106.80	107.79	112.34	121.88	1–58 (30)
Europe 11 nodes	CND^{MH}	100.00	100.00	100.00	100.00	100.00	0–1 (< 1)
	TBM	111.42	106.86	120.06	129.85	128.74	0–1 (< 1)
	LCM	126.74	122.54	127.83	128.40	115.19	0–1 (< 1)
USA 15 nodes	CND^{MH}	100.00	100.00	100.00	100.00	100.00	0–2 (< 1)
	TBM	108.08	109.14	115.12	126.67	134.96	0–1 (< 1)
	LCM	121.11	121.98	123.12	123.69	117.57	0–2 (< 1)

and links are modeled by stepwise functions. The advantage of this approach is that it can approximate real situations well. On the contrary, its drawback is the high mathematical complexity, which makes it reasonable to handle the problem with the help of a heuristic approach. We proposed an algorithm for the multi-hour network design problem combining the benefits of global and local search methods.

The performance of the novel algorithm was investigated with the help of simulations. Various size random networks as well as real world topologies were considered. In order to get a reliable picture, various traffic situations were examined starting from the evenly distributed one up to the fully polarized one. Further, two simple algorithms—based on simple extensions of the original single-hour design algorithm—were introduced as reference. Generally, the new proposal overperformed the reference algorithms. In exceptional situations, in the case of evenly distributed traffic combined with larger networks the performance of the new algorithm sometimes fell behind, compared to the first reference algorithm. However, this does not contradict our expectations since multi-hour design makes sense when significant difference can be observed between the traffic distributions of various time intervals. The running time of algorithms can be considered moderate also in the case of larger networks, regarding that they solve off-line network design task. In summary, we may say that the proposed multi-hour design algorithm can provide economical solutions with acceptable running times for the examined network situations.

Our future work can include the extension of the current algorithm so that it can handle backup paths as well. Further, in order to increase the reliability of simulation results more realistic traffic demand models should be used. Finally, a new reference algorithm optimized for multi-hour design should be investigated, which will possibly require adaptation in order to fit the problem definition applied (e.g., stepwise cost functions).

References

1. Awduche, D., Malcolm, J., Agogbua, J., O'Dell, M., McManus, J.: Requirements for Traffic Engineering Over MPLS. Internet Engineering Task Force, Request for Comments: 2702 (1999)
2. Bauschert, T.: Multihour design of multi-hop virtual path based widearea ATM networks. Proc of the 15th International Teletraffic Congress (ITC-15), Washington DC, USA (1997) 1019–1029
3. Garey, M.R., Johnson, D.S.: Computers and Intractability: A Guide to the Theory of NP-Completeness. Freeman, San Francisco, CA. (1979)
4. ITU-T: QoS Routing and Related Traffic Engineering Methods – Capacity Management Methods. Recommendation E.360.6, CA. (2002)
5. Józsa, B.G., Király, Z., Magyar, G., Szentesi, Á.: An Efficient Algorithm for Global Path Optimization in MPLS Networks. Optimization and Engineering, Vol. 2., No. 3. Kluwer Academic Publishers (2001) 321–347
6. Józsa, B.G., Makai, M.: On the Solution of Reroute Sequence Planning Problem in MPLS Networks. Computer Networks, Vol. 42., No. 2. Elsevier Science (2003) 199–210
7. Józsa, B.G., Orincsay, D., Kern, A.: On the Use of Routing Optimization for Virtual Private Network Design. In Proc. of the 7th IFIP Working Conference on Optical Network Design & Modelling (ONDM'2003), Budapest, Hungary (2003) 865–880
8. Medhi, D.: A Unified Approach to Network Survivability for Teletraffic Networks: Models, Algorithms and Analysis. IEEE Transactions on Communications, Vol. 42. (1994) 534–548
9. Medhi, D.: Multi-Hour, Multi-Traffic Class Network Design for VP-based Wide-Area Dynamically Reconfigurable ATM Networks. In Proc of IEEE INFOCOM'95, Boston, MA. (1995) 900–907
10. Medhi, D.: Some Approaches to Solving a Multi-Hour Broadband Network Capacity Design Problem with Single-Path Routing. Telecommunication Systems, Vol. 13. (2000) 269–291
11. Rosen, E., Viswanathan, A., Callon, R.: Multiprotocol Label Switching Architecture. Internet Engineering Task Force, Request for Comments: 3031 (2001)

A Tabu Search Heuristic for the Offline MPLS Reduced Complexity Layout Design Problem

Sergio Beker, Nicolas Puech, and Vasilis Friderikos

École Nationale Supérieure des Télécommunications,
Computer Science and Networks Department,
46, Rue Barrault, 75634 Paris CEDEX 13, France
{beker|npuech|freideri}@enst.fr

Abstract. MPLS paths can be calculated on-line as demands arrive or off-line for an estimate of the traffic demand over time. Off-line calculation has the advantage of allowing a globally optimal network design. From the operational standpoint, a layout design for large networks should consider minimizing the layout complexity, thus reducing the cost of operation. In this paper, we formulate an optimization problem whose objective is to minimize the number of required hops. An end-to-end path delay constraint provides quality of service (QoS) guarantees. The resulting Minimum Path Set and Flow Allocation Problem (MPSFAP), formulated as a Mixed Integer Non Linear Program (MINLP) is $\mathcal{NP}$-complete. As such, it is only possible to solve it exactly on small size networks. To overcome this limitation, we designed a Tabu Search (TS) based algorithm that computes good quality approximate solutions for the MPSFAP problem. Solutions for medium to large size networks in reasonable CPU time and within acceptable optimality tolerances are found using the proposed TS based algorithm.

Keywords: MPLS, Traffic Engineering, Layout Complexity, Multicommodity Flow, MINLP, Tabu Search, LSP.

1 Introduction

One of the main reasons for present market enthusiasm around MPLS architecture is its capability to implement evolved traffic engineering (TE) functionalities [1] (e.g. required to support the Next Generation Internet). In MPLS based IP networks, all packets in a Label Switched Path (LSP) between source and destination nodes follow a predefined route, enabling TE mechanisms such as load sharing and flow separation. This allows the network operator to offer differentiated services with Quality of Service (QoS) guarantees, while using the available resources cost-effectively. Planning and dimensioning the network under varying traffic conditions then becomes a central issue for the operator.

Given the physical topology, the operator has to design a layout or *virtual* topology (i.e. find an optimal set of paths and a flow distribution over the physical network) to meet a given demand, as well as to adapt the layout to varying

N. Mitrou et al. (Eds.): NETWORKING 2004, LNCS 3042, pp. 514–525, 2004.

traffic conditions. Additionally, each path must ensure some Quality of Service (QoS) requirements (e.g. end-to-end delay, throughput).

To design the MPLS layout, on-line or off-line approaches have been proposed. On-line methods calculate paths as demands arrive, either by updating the routing protocol metrics for each link [2], or by calculating the route at the source node using available network load information. As the resulting problem of finding one constrained path is usually $\mathcal{NP}$-complete [3], a number of algorithms have been proposed in order to find approximate solutions in times compatible with the on-line operation. Constrained Shortest Path First (CSPF) [4] provides Open Shortest Path First (OSPF) extensions to find the shortest path meeting one or a set of constraints. Minimum Interference Routing (MIRA) [5] proposes an algorithm to find a path avoiding the critically loaded links, and [6] improves MIRA by finding a set of k-minimum interferring paths with bounded length. On-line approaches have the advantage of adapting the network layout to changing load conditions. However, by taking into account the flow dynamics incrementally, an on-line method might produce a sub-optimal resource usage in the long run. Off-Line LSP layout design considers global information about the state of the network and traffic characteristics over time. The network can be optimally engineered and set up around a point of operation on a long term basis. Complementary, on-line decisions can be made as how to route arriving demands, in order to be able to accommodate traffic variations in the short term. Common objective functions for the off-line design aim at minimizing the maximum loaded link or minimizing the average cross network packet delay. However, the operator's main concern is to reduce the cost of operation while honoring the Service Level Agreements (SLAs). The cost of operation is directly related to the number of paths (hence to the layout complexity) to be monitored [7]. A main objective in the network design should then be to minimize the layout complexity, while meeting QoS guarantees.

In this paper we focus on the problem of obtaining an MPLS layout which is optimal w.r.t. the number of required hops in the set of LSP in the layout (indirectly driving to a reduction in the number of required paths). The Minimum Path Set and Flow Allocation Problem (MPSFAP) is formulated as a Mixed Integer Non-Linear Multicommodity Flow Problem (MINLP-MCFP). The objective is to obtain the minimum set of paths and the associated flow distribution under bounded end-to-end path delay for each Class of Service (CoS) for a given traffic demand. We first define the notation used through the paper in Section 2. In section 3 the general MPSFAP problem is set for multiple Classes of Service (CoS). Even in the case when we consider only one class of traffic, the model is intractable and may be solved exactly only for small size networks. This led us to develop algorithms that provide (hopefully good quality) approximate solutions to the considered problem. An adaptation of the Tabu Search (TS) heuristic applied to the MPSFAP problem is presented in section 4. The evaluation of numerical results for the MPSFAP problem with a single CoS are presented in Section 5. Results obtained from exact solvers for small size networks are compared to results obtained from the TS algorithm. The comparison shows that the approximate TS solutions are close to the exact ones, making the TS based algorithm a good candidate to solve MPSFAP on large problem instances. We then test the TS algorithm on real size networks to assess our approach. Finally, in Section 6 conclusions and future work are presented.

2   Model and Notation

2.1   Network

The physical network is represented by a directed graph $\mathcal{G} = (\mathcal{V}, \mathcal{E})$, where $\mathcal{V}$ is the set of vertices indexed $1, 2, \ldots, N$; $\mathcal{E}$ is the set of directed edges indexed $1, 2, \ldots, M$. The vertices represent the MPLS Label Switching Routers (LSR), and each edge represents a connecting link between two LSRs. Link i has capacity C_i. Let $\mathbf{C}$ be the single column matrix representing the M-dimensional capacity vector.

2.2   Paths

Let the node couples $q = (m, n)$, with $m \neq n$, be indexed $1, 2, \ldots, Q$ where $Q = N(N-1)$. Let the classes of service supported by the network be indexed $1, \ldots, L$. For each source-destination couple, a *commodity* $d_{q,l}$ describes the demand requested for service class l and couple q. The commodity $d_{q,l}$ may be routed via K_q different paths (or routes) binding the node couple q. We denote by K_q the total number of such paths, indexed $a_q^1, a_q^2, \ldots, a_q^{K_q}$. Any path a_q^k can be represented as a vector with entries $a_{q,i}^k = 1$ if path a_q^k lies on link i, and 0 otherwise, for $i = 1, 2, \ldots, M$. The paths are acyclic, i.e. no path traverses twice the same link or the same node. We can represent the $K = \sum_{q=1}^{Q} K_q$ single paths between all pair of nodes as the $M \times K$ arc-path incidence matrix $\mathbf{A}$:

$$\mathbf{A} = (A_1 | A_2 | \ldots | A_Q)$$

where :

$$A_q = \begin{pmatrix} a_{q,1}^1 & \cdots & a_{q,1}^{K_q} \\ a_{q,2}^1 & \cdots & a_{q,2}^{K_q} \\ \vdots & & \vdots \\ a_{q,M}^1 & \cdots & a_{q,M}^{K_q} \end{pmatrix}$$

for $q = 1, \ldots, Q$, and where the block A_q is the arc-path matrix corresponding to the K_q paths for the node couple q. Each path a_q^k has an associated weight $w_{q,l}^k$ to carry a portion of commodity $d_{q,l}$. Let $\mathbf{W}$ be the $K \times L$ weight matrix.

2.3   Demands and Flows

The demand matrix is represented by a $Q \times L$ dimensional matrix $\mathbf{D}$ whose entry $d_{q,l}$ stands for the amount of traffic requested for the service class l for the node couple q. Let be the flow assigned to path a_q^k for the service class l denoted by $b_{q,l}^k$. The $K \times L$ matrix $\mathbf{B}$ represents the flow distribution per class and path:

$$\mathbf{D} = \begin{pmatrix} d_{1,1} & \cdots & d_{1,L} \\ d_{2,1} & \cdots & d_{2,L} \\ \vdots & \vdots & \vdots \\ d_{Q,1} & \cdots & d_{Q,L} \end{pmatrix} \qquad \mathbf{B} = \begin{pmatrix} b_{1,1}^1 & \cdots & b_{1,L}^1 \\ \vdots & \vdots & \vdots \\ b_{1,1}^{K_1} & \cdots & b_{1,L}^{K_1} \\ \vdots & \vdots & \vdots \\ b_{Q,1}^1 & \cdots & b_{Q,L}^1 \\ \vdots & \vdots & \vdots \\ b_{Q,1}^{K_Q} & \cdots & b_{Q,L}^{K_Q} \end{pmatrix}$$

Let us define $\mathbf{R} = (r_{q,i})$ the $Q \times K$ commodity-path incidence matrix, where entry $r_{q,i} = 1$ if i lies in the range $[K_1 + K_2 + \ldots + K_{q-1} + 1, K_1 + K_2 + \ldots + K_{q-1} + K_q]$ and 0 otherwise, for $1 \leq q \leq Q$ and $1 \leq i \leq K$. Let x_i be the value of the total flow traversing link i, for $1 \leq i \leq M$, and $\mathbf{X}$ be the M dimensional single column matrix of link flows:

$$\mathbf{X} = \mathbf{A} \cdot \mathbf{B} \cdot \mathbb{1}_L \tag{1}$$

where $\mathbb{1}_L$ is the L-dimensional column matrix with all its entries set to 1.

3   Minimum Path Set and Flow Allocation Problem (MPSFAP)

As stated before, the cost of operation for large networks is related to layout complexity. We define layout complexity as the number of paths necessary to transport a given demand matrix $\mathbf{D}$. The MPSFAP problem can be formulated as a MINLP problem, whose objective is to minimize the total number of required paths:

Given:

$$\mathbf{A}, \mathbf{C}, \mathbf{D}, \Theta, \Delta, \epsilon$$

minimize:

$$\sum_{l=1}^{L} {}^t \mathbf{W}^{(l)} \cdot \mathbf{H}^{(l)} \tag{2}$$

subject to:

$$\mathbf{X} = \mathbf{A} \cdot \mathbf{B} \cdot \mathbb{1}_{\mathbf{L}} \leq (1 - \epsilon)\mathbf{C} \tag{3}$$

$$\mathbf{R} \cdot \mathbf{B} = \mathbf{D} \tag{4}$$

$$h_{q,l}^k \sum_{i=1}^{M} \frac{\lambda a_{q,i}^k}{C_i - x_i} \leq \theta_{q,l} \qquad l = 1, \dots, L; \quad k = 1, \dots, K_q; \quad q = 1, \dots, Q \tag{5}$$

$$0 \leq b_{q,l}^k \leq h_{q,l}^k \delta_{q,l} \qquad l = 1, \dots, L; \quad k = 1, \dots, K_q; \quad q = 1, \dots, Q \tag{6}$$

$$h_{q,l}^k \in \{0,1\} \qquad l = 1, \dots, L; \quad k = 1, \dots, K_q; \quad q = 1, \dots, Q \tag{7}$$

where:

- Θ is the $Q \times L$ matrix containing the maximum tolerable delay $\theta_{q,l}$ for each commodity $d_{q,l}$;
- Δ is the $Q \times L$ matrix containing the maximum acceptable flow $\delta_{q,l}$ on one path for the commodity $d_{q,l}$;
- $\epsilon > 0$, and λ is the average packet size.
- $\mathbf{H}$ is a $K \times L$ matrix, whose entries are binary variables taking the value $h_{q,l}^k = 1$ if path a_q^k is in use by commodity $d_{q,l}$, and 0 otherwise. $\mathbf{H}^{(1)}$ and $\mathbf{W}^{(1)}$ are the single columns corresponding to the lth CoS of $\mathbf{H}$ and $\mathbf{W}$ respectively.

Constraint (4) expresses that traffic demands must be met by allocated flows (no flow loss). Constraint (5) ensures that no packet belonging to class l traversing a path connecting the node pair q will experience a delay greater than $\theta_{q,l}$. The path weights $w_{q,l}^k$ can be used by the operator to fine tune the cost structure of the path layout. Hop-count based weights, for instance, account for the cost of signaling long paths:

$$w_{q,l}^k = \sum_{i=1}^{M} a_{q,i}^k \qquad l = 1, \dots, L; \quad k = 1, \dots, K_q; \quad q = 1, \dots, Q \tag{8}$$

According to the studied objective function, flows are allocated to the paths up to the limit given by the constraints (5) and (6).

The advantage of the flow distribution obtained with the objective function (2) is that commodity splitting is reduced. Its main drawback is that it usually leads to an important path delay difference between the various LSPs. This doesn't have a major impact if a flow-by-flow load sharing strategy is used (e.g. all packets in a TCP flow are forwarded on the same LSP).

In the case of packet-by-packet load sharing strategy, the objective function should include a term to distribute flows according to path delay to avoid significant packet reordering:

$$\alpha \sum_{l=1}^{L} {}^t\mathbf{W}^{(1)} \cdot \mathbf{H}^{(1)} + \beta \sum_{q=1}^{Q} \sum_{k=1}^{K_q} \sum_{l=1}^{L} h_{q,l}^k \sum_{i=1}^{M} \frac{\lambda a_{q,i}^k}{C_i - x_i} \tag{9}$$

where $0 \leq \alpha \leq 1$ and $0 \leq \beta \leq 1$ are factors used to balance the importance of each term in the objective function.

4 Tabu Search: A Metaheuristic Approach

The major drawback of multicommodity flow models is that they are computationally intractable even for small problem instances. Optimization algorithms based on metaheuristics have recently gained interest because they are able to cope with problem instances of large size. Among others, simulated annealing [8], genetic algorithms [9] and Tabu Search (TS) [10,11] have been proposed to tackle multicommodity flow problems. The computed solution however is not guaranteed to be optimal. That is why it is necessary to check the accuracy of the solutions provided by such a heuristic based algorithm and to compare them with exact solutions on small size problems.

A detailed description of the TS algorithm can be found in [10]. Briefly, TS consists of exploring the space of solutions until a number of iterations is reached or until a specific cost criterion is satisfied. The exploration starts with an initial solution computed by another algorithm (e.g. randomly generated). At each iteration, TS computes a set of solutions or *neighborhood* derived from the current solution via perturbations applied to it. All the solutions of the neighborhood are evaluated (see below) and the best one is selected as the new current solution. In order to prevent the algorithm from cycling along the same series of current solutions, a tabu list is maintained. It contains a number of last visited solutions, which cannot be chosen as long as they belong in this list. This allows the algorithm to choose a solution worse than the current one, allowing it to escape from the local minima encountered during the search. Three problem-specific elements must be defined in order to apply TS to the considered problems:

Initial Solution: An initial solution (i.e. layout) must be computed. It will constitute the starting point for the exploration of the solution space. The initial layout is the one obtained by choosing the shortest path for all the demands in the traffic matrix given as input.

Perturbation mechanism: A perturbation mechanism is necessary to generate a neighborhood of the current solution. Neighbor solutions are calculated by randomly choosing a source-destination pair. The current solution sets a given flow value along each path connecting this node pair. This flow distribution is randomly changed so that a new flow distribution meeting the flow demands but not necessarily meeting the link flow and path delay constraints is obtained. Hence, the neighbor generation process leads to new candidate solutions that may be considered as *valid* (solutions that meet all the constraints) or *invalid* (solutions that do not meet the link flow or path delay requirements). In particular, the flows in the paths belonging to the chosen node pair are rearranged in a way that current flow traversing each path is taken into account in order not to unnecessary make appear new paths.

A cost function: A cost function $f : S \rightarrow \mathbb{R}$ that maps elements in the solution space S to real numbers must be defined to allow for solution comparison. We used a cost function that evaluates distinctly valid and invalid solutions. For valid solutions, the TS cost function is the function defined by Equation 2 for MPSFAP and defined by Equation 10 for MTDFAP (see Section 5 below for full definition of MTDFAP problem). For invalid solutions, an extra threshold value is added to ensure that the evaluation of an invalid solution is always greater than for any valid solution, along with a term dependent on the overload. This term is a descendent function of the overload proportion on overloaded links, so a slope towards valid solutions is created in the solution landscape. A complete description of the TS algorithm can be found in [12].

5   Experimental Results

MPSFAP belongs to the class of $\mathcal{NP}$-complete problems, as it can be viewed as a constrained shortest path problem between all couples, which is itself a $\mathcal{NP}$-complete problem [3]. In order to find the TS parameters which lead the algorithm to the production of good results, and in order to show that our TS based algorithm provides acceptable approximate solutions, we used a couple of small sized networks, denoted by NET1 and NET2. NET1 is a 4-node, 8-link and 24-LSP network. NET2 is a 4-node, 10-link and 38-LSP network. The simplicity of these networks allowed us to obtain exact results from a numerical solver and to use them as a reference for the evaluation of the TS solutions.

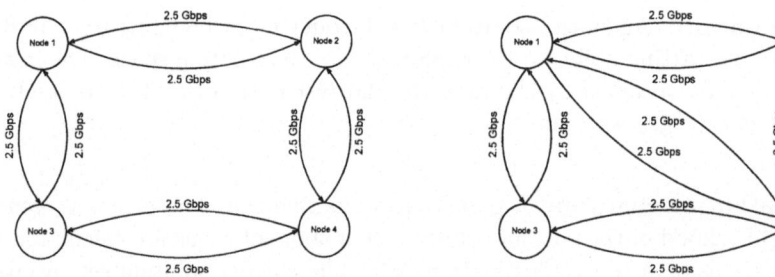

Fig. 1. NET1 and NET2 networks.

The TS algorithm was then tested on two real size networks denoted by VTHD and NSFNET. VTHD, consisting of 9 nodes 24 links and 550 LSPs, is a model version of a French network designed for research purposes, in the context of the VTHD (Vraiment Très Haut Débit) [13] research project. NSFNET is a 14-node, 42-link and 496-LSP network.

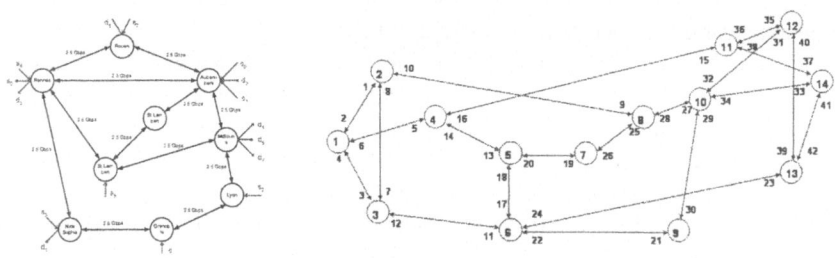

Fig. 2. VTHD and NSFNET networks.

We tested the algorithms in the particular case of a single service class. In order to derive results we had to consider the solutions obtained under various traffic conditions. We generated a series of 25 traffic matrices for the considered test networks. The matrices were obtained randomly according to the following method, which is derived from the method described in [14]. A certain percentage F of the demands for each source- destination node pair q is a random variable uniformly distributed within the interval $[0, \frac{C_q}{a}]$ and the rest $(100 - F)$ is uniformly distributed within the interval $[0, \frac{C_q Y}{a}]$, where C_q is the total capacity connecting the end nodes of couple q. The parameter a is an arbitrary integer which may be 1 or more, and Y is the ratio $\frac{max_q\{C_q\}}{min_q\{C_q\}}$. Parameter a is used to tune the traffic intensity on the network so that the produced matrices can lead to feasible MPSFAP problems, whereas Y allows a specific traffic distribution for some node pairs in the network. Mean packet size is assumed to be 128 bytes. In all cases link capacities are set to the same value of 2.5 Gbps. Results obtained for MPSFAP are compared with the results obtained with a reference objective function minimizing the total delay in the network, which we denote by minimum total delay flow allocation problem MTDFAP, and where the objective is to minimize:

$$\sum_{q=1}^{Q}\sum_{k=1}^{K_q}\sum_{l=1}^{L}h_{q,l}^k \sum_{i=1}^{M}\frac{\lambda a_{q,i}^k}{C_i - x_i} \tag{10}$$

subject to the same path-delay constraints. MTDFAP is also formulated as a MINLP problem. This objective function is extensively used through the literature and serves as a reference. All numerical results are obtained from the MINLP solver [15], available through the NEOS server on the Internet [16]. In all cases, solutions are obtained with $\epsilon = 0.0001$, and the flow limit for all commodities is $\delta_q = 2.5$ Gbps (i.e. the path flows can use up to the maximum available capacity in any link). Path delay limits are set to $\theta_q = 30$ μsec for NET1 and NET2 and $\theta_q = 50$ μsec for VTHD and NSFNET networks.

Table 1 shows the values obtained for the objective function with the Tabu Search heuristic algorithm and with the exact solver for both the MPSFAP and the MTDFAP problems. The presented values represent the range and the mean value of the objective function evaluations for 25 executions of each method (one execution for each traffic matrix). The chosen parameter values were $a = 4$, $Y = 1$ and $K = 24$ for the matrices generated for the tests with NET1 and $a = 5$, $Y = 1.33$ and $K = 38$ for the tests with NET2. Although the TS algorithm cannot always find the optimal value, the average evaluation difference between the approximate and the exact solution is not very large, since it reaches 13, 75% in the worst case. Results can be further improved by sacrificing more CPU time, i.e. by examining a greater number of possible solutions. Figure 3 shows the results obtained by the two algorithms for every matrix with MPSFAP for the NET1 and NET2 networks. Infeasible matrixes show no value in the figure.

Table 1. Comparison of the values (number of hops) obtained by the TS algorithm and the exact solver for NET1 and NET2.

	MPSFAP		MTDFAP ($\times 10^{-3}$)	
Network	Range	Average	Range	Average
	Tabu Search Results			
NET1	16-19	16.48	13.36-24.56	17.65
NET2	14-20	15.48	12.74-122.89	31.10
	Exact Solutions			
NET1	16-16	16	12.7-21.86	15.52
NET2	14-16	14.92	11.65-96.09	29

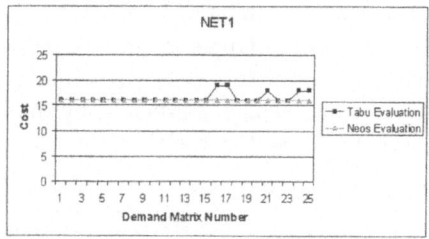
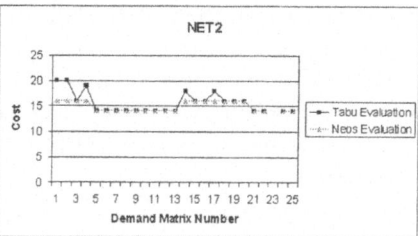

Fig. 3. MPSFAP evaluations obtained by Tabu Search and NEOS Server for NET1 and NET2 networks.

Table 2 presents the total number of paths required by MPSFAP and MTD-FAP to allocate the demand imposed by the 25 used traffic matrices. Despite the small size of the tested networks, we can see that the MPSFAP problem leads to solutions with a smaller set of paths than MTDFAP. MTDFAP tends to use more available paths to balance the network load, in order to lower the total cross network delay. The approximate solutions obtained with the TS algorithm are close to the exact ones in terms of used paths. Nevertheless we must keep in mind that the considered cost functions aim at minimizing the hop count, and as such the TS algorithm sometimes reaches better values than the exact solution in terms of used paths.

Table 2. Comparison of the number of used paths obtained by the TS algorithm and the exact solver for NET1 and NET2.

	MPSFAP		MTDFAP ($\times 10^{-3}$)	
Network	Range	Average	Range	Average
	Tabu Search Results			
NET1	12-13	12.20	12-14	12.52
NET2	12-14	12.64	12-14	12.96
	Exact Solutions			
NET1	12-14	12.32	12-14	12.44
NET2	12-14	12.66	12-14	12.72

From this first series of tests on small size networks NET1 and NET2, we observe that the approximate solutions computed by the TS algorithm are close to the exact ones. This makes our algorithm a good candidate to solve more complex problems. We performed a second series of experiments with TS on real size networks. This time, the deterministic solver is unable to compute (exact) solutions, so reference values will not be further available.

Exact solvers on powerful contemporary computers are not able to give a solution in an acceptable amount of time for these networks, whereas our TS algorithm is able to propose approximate solutions in a reasonable amount of time. The time of execution of the heuristic method has a primary role in the quality of the solutions, which can be clearly noticed in cases where a feasible solution can hardly been found for a particular problem instance.

As before, we generated random sets of 25 traffic matrices for the considered networks. More precisely, in order to take into account various parameter configurations, we generated two different matrix sets for the VTHD network (which leads to test1 and test2). The chosen parameter values were $a = 100$, $Y = 2.6$ and $K = 550$ (respectively $a = 120$, $Y = 2.6$ and $K = 550$) for the matrices generated for the first series (respectively second series) of tests with VTHD, and $a = 30$, $Y = 1.5$ and $K = 496$ for the tests with NSFNET. In the first set of VTHD traffic matrices (test1), parameter a is chosen to produce a quite heavy load in the network, while in the second set (test2) the choice $a = 120$ leads to milder traffic conditions. Table 3 shows the hop count on the network for two set of traffic matrices in VTHD and one in the NSFNET network. Under heavy traffic conditions, more paths are needed to allocate the demand, resulting in a greater number of required hops. Table 4 shows the corresponding number of required paths for the same sets of traffic matrices for VTHD and NSFNET networks. As seen before, MTDFAP solutions require additional paths with respect to MPSFAP in order to minimize the total delay in the network.

Table 3. Value of the cost function (number of hops in the case MPSFAP) by the TS algorithm for VTHD test1, VTHD test2 and NSFNET.

Network	MPSFAP		MTDFAP ($\times 10^{-3}$)	
	Range	Average	Range	Average
VTHD test1	145-227	162.67	149-214	169.35
VTHD test2	139-160	144	140-162	152.04
NSFNET	390-430	410.6	411-466	428.17

Table 4. Number of used paths computed by the TS algorithm for VTHD test1, VTHD test2 and NSFNET.

Network	MPSFAP		MTDFAP ($\times 10^{-3}$)	
	Range	Average	Range	Average
VTHD test1	71-88	75.76	73-89	78.1
VTHD test2	71-75	72.20	72-77	74.48
NSFNET	182-193	186.83	187-200	191.17

We used a personal computer equipped with a Pentium IV processor at 2.4 GHz and 512 MB of physical memory to perform our tests. The typical processing time for the above given solutions is 5-11min. for the VTHD topology and around 20min. for NSFNET.

6   Conclusions and Future Directions

Our main concern in this paper is to obtain low complexity layouts, thus reducing the cost of operation for large networks, while still providing QoS guarantees for several CoSs. Reducing reconfiguration complexity (i.e. the number of paths that must be changed) helps reducing service disruption times and resource overdimensioning during transition management. We propose a MINLP model to obtain low complexity layouts for the traffic demand, while keeping the QoS guarantees. This model is $\mathcal{NP}$-complete and as such solvers are able to compute (exact) solutions only for small size networks. To overcome this limitation, we designed a Tabu Search based algorithm that provides approximate solutions for the studied problem. We compared the solutions computed by our algorithm to the exact ones for small size problem instances. The comparison showed that the approximate solutions are close to the exact ones, hence making the TS based algorithm a good one to deal with the MPSFAP problem on real size problems. We presented the results obtained with TS for two different real size networks. The reasonable amount of CPU time required by the TS algorithm to solve these problems make it a good method to tackle the MPLS layout design problem.

Results obtained on small networks show that layout and reconfiguration complexity are significantly reduced when compared with classical approaches. The formulated optimization problems are $\mathcal{NP}$-complete. In order to compute optimal layouts for larger networks, we need to devise efficient algorithms providing approximate solutions. We are currently working on algorithms implementing heuristics which take advantage of the nature and knowledge of the underlying problem, and metaheuristic such as simulated annealing or tabu search.

References

[1] D. Adwuche, G. Malcom, J. Agogbua, M. O'Dell, and J. McManus. Requirements for traffic engineering over MPLS. RFC 2702, IETF, 1999.

[2] B. Fortz and M. Thorup. Internet traffic engineering by optimizing OSPF weights. In *Volume 2, p.519-528*, INFOCOM 2000, 2000. IEEE.

[3] R. Ahuja, T. Magnanti, and J. Orlin. *Network Flows: Theory, Alogorithms and Applications*. Prentice Hall Inc., 1993.

[4] D. Katz, D. Yeung, and K. Kompella. Traffic engineering extensions to OSPF version 2. Draft, IETF, 2002.

[5] M. Kodialam and T. Lakshman. Minimum interference routing with applications to MPLS traffic engineering. In *Volume 2, p.884-893*, INFOCOM 2000, 2000. IEEE.

[6] G. Banerjee and D. Sidhu. Path computation for traffic engineering in MPLS networks. ICN 2001, 2001. IEEE.

[7] S. Beker, D. Kofman, and N. Puech. Off-line reduced complexity layout design for MPLS networks. In *Proceedings of IP Operations and Management*, IPOM 2003, 2003. IEEE.

[8] S. Kirkpatrick, C.D. Jr. Gelatt, and M.P. Vecchi. Optimization by simulated annealing. *Science*, (4598), 1983.

[9] C. Gazen and C. Ersoy. Genetic algorithms for designing multihop lightwave network topologies. *Artificial Intelligence in Engineering*, 13:211–221, 1999.

[10] F. Glover and M. Laguna. *Tabu Search*. Kluwer Academic Publishers, Boston, MA, 1997.

[11] N. Puech, J. Kuri, and M. Gagnaire. Models for the logical topology design problem. In *Proceedings of the 2nd IFIP-TC6 Networking Conference*. Springer-Verlag, May 2002.

[12] S. Beker. *Thesis: Optimization Techniques for the Dimensioning and Reconfiguration of MPLS Networks*. Télécom Paris, 2004 (to appear).

[13] Réseau national de reherche en télécommunications. Technical report, Ministère de l'Economie, des Finances et de l'Industrie de France, 2000-2004. www.vthd.org.

[14] D. Banerjee and B. Mukherjee. Wavelength-routed optical networks: Linear formulation, resource budgeting tradeoffs, and a reconfiguration study. *Networking, IEEE/ACM Transactions on*, 8(5):598–607, 2000.

[15] R. Fletcher and S. Leyffer. Numerical experience with lower bounds for MIQP branch and bound. *SIAM Journal of Optimization*, 8(2):604–616, 1998.

[16] J. Czyzyk, M. Mesnier, and J. Moré. The Neos Server. *IEEE Journal on Computational Science and Engineering*, 5:68–75.

Self-Protecting Multipaths – A Simple and Resource-Efficient Protection Switching Mechanism for MPLS Networks

Michael Menth[1], Andreas Reifert[2]*, and Jens Milbrandt[1]

[1] Department of Distributed Systems, Institute of Computer Science,
University of Würzburg, Am Hubland, 97074 Würzburg, Germany
{menth,milbrandt}@informatik.uni-wuerzburg.de
[2] Institut für Kommunikationsnetze und Rechnersysteme (IKR),
University of Stuttgart, Pfaffenwaldring 47, D-70569 Stuttgart, Germany
reifert@ikr.uni-stuttgart.de

Abstract. In this paper we propose the concept of an end-to-end (e2e) Self-Protecting Multi-Path (SPM) as a protection switching mechanism that may be implemented, e.g., in Multiprotocol Label Switching (MPLS) networks. In case of local outages, resilient networks redirect the traffic from a failed link over an e2e backup path to its destination. In this case, Quality of Service (QoS) can only be provided if sufficient extra capacity is available. If backup capacity can be shared among different backup paths, multi-path routing allows for considerable savings regarding this extra capacity. The SPM consists of disjoint paths that carry the traffic both in normal operation mode and during local outages. If a partial path is affected by a network failure, the traffic is just distributed to the remaining working paths. This structure is easy to configure and the switching to failure mode operation is simple since no signalling is required. Based on analytical results, we show that load balancing of the traffic across the disjoint paths can reduce the required backup capacity significantly. The backup performance depends strongly on the network topology, and the SPM outperforms simple Open Shortest Path First (OSPF) rerouting by far.

1 Introduction

Carrier grade networks can not afford outages due to internal link or router failures that are visible to their customers. Therefore, they require mechanisms to deviate affected traffic aggregates around the outage location. In contrast to Open Shortest Path First (OSPF) rerouting, these mechanisms have to react fast and they have to provide control over redirected traffic. Fast failure detection is achieved by frequently exchanged "Hello" messages and fast reaction is done by switching the traffic onto pre-computed and pre-installed backup paths. This is called protection switching [1]. In contrast, rerouting denotes the convergence of routing protocols in a narrow sense, i.e., reachability information is exchanged and the routing tables are calculated anew. Since we focus only on path layout and load distribution and not on signaling details, we use the terms rerouting and protection switching synonymously in this work.

---

* This work was done while the author was with Inst. 1

N. Mitrou et al. (Eds.): NETWORKING 2004, LNCS 3042, pp. 526–537, 2004.

Traffic rerouting to maintain pure connectivity does not suffice in carrier grade networks since Quality of Service (QoS) must be maintained. Our objectives are resilient networks, i.e., the customer should not perceive an internal outage by service interruptions or degraded QoS due to bottlenecks on backup paths. Therefore, resilient networks need some extra capacity which is the difference between the required network capacity with and without resilience requirements. Extra capacity is needed for backup purposes, however, it is costly and should be small, so we take it as a performance measure in our study.

Many different rerouting approaches have been proposed in the literature [2,3]. For example, the traffic may be rerouted only locally or to a different end-to-end (e2e) backup path. However, the backup capacity has not been considered. An optimum path layout and load balancing that requires a minimum backup capacity is computed in [4,5] for a given network topology and traffic matrix. This optimal solution leads to complex multi-paths that may branch and join at interior nodes, i.e. they are hard to configure. Furthermore, it makes the reorganization of unaffected paths necessary in case of a network failure, which imposes heavy signaling load on the network in a critical situation.

The contribution of this paper is the proposal of a new and simple e2e protection switching mechanisms – called Self-Protecting Multi-Path (SPM) – that may be implemented by explicit routing mechanisms like MPLS. We take advantage of the load balancing potential of multi-path forwarding and minimize the required extra capacity by a polynomial-time optimization algorithm. Our multi-path structures are significantly simpler than general multi-paths since they consist only of disjoint paths. Only traffic shifting of affected traffic aggregates onto backup paths is needed. The minimization of the extra capacity is still very effective such that – depending on the network topology – 20% additional transmission capacity is sufficient to provide full resilience against all single node and link failures.

Given this result, resilience can be implemented at lower cost on the network layer than on the physical layer where fault tolerance is achieved by resource duplication. An exception is the concept of p-Cycles [6,7] which allows for a more economic protection. It also achieves savings in backup capacity by shared protection and implicit multi-path routing in failure cases. The path layout must adhere to physical layer restrictions and its optimization is more difficult than the one for SPM.

The paper is organized as follows. In Section 2 we point out the difference between other routing optimization approaches and our work. In Section 3 we explain the SPM together with its load balancing options to minimize the required extra capacity for network resilience. The numerical results in Section 4 demonstrate the performance of the SPM. Section 5 summarizes this work and gives some outlook on further work.

2   Related Work

This work is about routing optimization and load balancing in a very broad sense. To avoid any confusion, we delimit it from other network optimization studies.

2.1   Routing Paradigms

There are two major forwarding paradigms: destination based forwarding and connection oriented forwarding.

Destination Based Forwarding. In pure Internet Protocol (IP) technology, routers identify the corresponding output interface based on the destination address in the packet header according to their routing tables. The routes in IP forwarding are usually set up by means of routing protocols like the Open Shortest Path First (OSPF) protocol [8]. They exchange reachability information associated with link costs based on which the output ports for the shortest paths to certain destinations are computed. By manipulating the link costs, the routing can be influenced which gives room for traffic engineering. Load balancing over multiple paths is possible if several paths to the same destination have equal costs. This Equal Cost Multi-Path (ECMP) is implemented, e.g., in OSPF.

Connection Oriented Forwarding. MPLS is a connection oriented switching technology, i.e., traffic is forwarded along virtual connections that build an overlay network. Packets matching a set of attributes in a router create a Forwarding Equivalent Class (FEC). A so-called LSP Ingress Router (LIR) identifies them and groups them together into a single traffic aggregate by assigning the packets a common label on top of their header. This traffic aggregate is forwarded along a Label Switched Path (LSP) to the LSP Egress Router (LER) that pops the label. The intermediate routers of the LSP forward the packets by label swapping corresponding to the information in their label information base (LIB). The LIB holds a table about incoming LSPs that are identified by their ingress interface and their ingress label and maps them to their egress interface and their egress label. In contrast to routing tables, the information in the LIBs is provided at connection setup. At that occasion, the path of an LSP may be determined automatically by routing protocols or it may follow a pre-computed explicit route.

The routing granularity and the forwarding resolution in MPLS is much finer than in IP because the attributes of a FEC may be, e.g., source *and* destination address. Traffic to a same destination may be carried over different paths that have completely different costs by using explicit routes in MPLS. Explicit routing can be mimicked by source routing in IP technology but this is not advisable since it slows down the forwarding speed of routers considerably. In addition, explicit routing along multiple paths is restricted to ECMP. Therefore, connection oriented technologies like MPLS allow for more powerful traffic engineering than destination based forwarding.

2.2   Routing Optimization

A well investigated problem is routing optimization in the presence of limited link capacities to maximize the supportable traffic intensity whose e2e structure is given by a traffic matrix. This is a multi-commodity flow problem and its solution can be implemented, e.g., by LSPs. For IP routing, a similar approach can be done by setting the link cost appropriately such that all traffic is transported through the network and that the mean and maximum link utilization is minimized [9]. Pure IP and MPLS solutions may also be combined [10]. These approaches require the knowledge of the traffic matrix which is usually not known for best effort traffic. This problem is tackled by [11] presenting a stable closed loop solution using multi-path structures. Load balancing should be done on a per flow basis and not on a per packet basis to avoid packet reordering which has a detrimental effect on the TCP throughput. The hash based algorithm in [12] achieves that goal very well. The authors of [13] present an online solution for routing with resilience requirements. They try to minimize the blocking probability of successive path requests using suitable single-paths as primary paths and backup paths. The backup bandwidth may be shared or dedicated.

Routing with resilience requirements can also be considered under a network dimensioning aspect, i.e. the traffic matrix is given and the link capacities must be set. This problem is trivial without resilience requirements since a suitable bandwidth assignment for the shortest paths is already an optimum solution. It becomes an optimization problem if capacity sharing for backup paths is allowed. The routing must be designed and the capacity must be assigned such that primary paths and shared backup paths require minimal network capacity while the backup mechanisms provide full resilience for a given set of protected failure scenarios. This is fundamentally different from the above problem since both the routing and the link bandwidth are optimized simultaneously. Note that the results of such calculations depend on the capabilities of the applied restoration schemes. The results of [14] can be well implemented since this work applies only single-paths for both primary and backup paths and relocates only affected primary paths. However, they renounce on multi-path routing and load distribution for path restoration purposes. This is especially important in outage scenarios because traffic diverted over several different paths requires only a fraction of the backup capacity on detour links. If backup capacity sharing is allowed, this backup capacity may be used in different failure scenarios by different rerouted traffic aggregates, which leads to increased resource efficiency since less additional resources must be provisioned in the network. In [4,5] multi-path routing is used. The required network resources are minimized by calculating the optimum path layout and routing independently for each failure scenario. These backup solutions are too difficult for implementation but they present lower bounds for the required backup capacity.

2.3 Restrictions for Path Layout

We explain why the results in [4,5] can not be implemented as restoration mechanisms and derive technical side constraints for feasible backup solutions. The path layout and the load balancing is calculated for the normal operation mode and for each failure scenario independently and general multi-path structures are allowed. In an outage case, broken paths must be rerouted but aggregates that are not affected by the failure might also need to be shifted to implement the resource minimal solution.

Firstly, the knowledge of the specific location of the failure is required to apply the optimized path layout and load balancing. Therefore, the exact outage information must be propagated to all ingress routers to trigger protection switching for a specific outage scenario. This entails extensive signaling in a critical system state where the reachability is corrupted.

Secondly, the relocation of the paths can not be done simultaneously. Deflecting more paths than necessary might lead to transient overload on some network elements and can be avoided if only broken paths are redirected.

Thirdly, if each connection holds a backup path for each protected failure scenario, a large amount of paths must be pre-installed and administered. This makes the path configuration very complex and the large number of paths is a problem for the state maintenance of today's core network routers.

Fourthly, to keep the fault diagnostics and the reaction to failures simple, the ingress router should be able to detect a failure and to react locally by switching the traffic to another path. With general multi-path structures, paths may fork and join in transit routers. If a partial path fails, the entire multi-path looses some packets and can not be used anymore. Implementing general multi-paths as a superposition of overlapping single-paths prevents that problem because only some partial paths may fail in case of a local outage. However, this increases the number of parallel LSPs and makes the state management more complex. Hence, only disjoint paths should be used to achieve simple fault diagnostics for multi-path forwarding.

Another restriction for path layout are Shared Risk Link Groups (SRLGs) [15,16, 17] which group network elements together that may fail simultaneously with a high probability. For instance, all links originating at the same router fail if the router goes down. SRLGs are motivated by optical networking where a single optical fiber duct accommodates several logically separate links. In our work, we consider only the first scenario and the second one in a trivial way by excluding parallel links. However, we do not take general SRLGs into account because our focus is the performance evaluation of the basic SPM and not its adaptation to SRLGs.

3   Self-Protecting Multi-path for Simple Protection Switching

The experiments in [4] have shown that e2e protection mechanisms require less backup capacity than local detours because the traffic of failed paths is redirected early at the source avoiding bottlenecks or much backup capacity around the outage region. Therefore, we focus only on e2e protection switching. We use e2e multi-paths routing because it allows for load distribution in failure cases. As outlined above, only multi-path structures consisting of disjoint paths should be applied and only traffic from paths that are affected by a failure should be rerouted. The basic structure of an e2e Self-Protecting Multi-Path (SPM) for a single e2e aggregate d consists of parallel disjoint paths. We compute them using a k (link and node) Disjoint Shortest Paths (kDSP) algorithm [18, 19] whose calculation is fast. However, it does not take general SRLGs into account, which is a NP hard problem. The SRLGs are not focus of this work but they can be easily integrated into SPMs by substituting the kDSP heuristic by any other calculation yielding a link and node disjoint multi-path. The SPM sends traffic over all its partial paths. If the LIR recognizes that a partial path fails, it simply redistributes the traffic onto the working paths.

Fault tolerance depends on the set of considered failure scenarios $s \in S$ (s signifies failed links or routers), including the working mode, for which resilience is guaranteed. The LIR is not aware of the exact failure scenario s but loss of light (LoL) or missing keep-alive or "Hello" messages [1] for a partial path indicate a failure symptom $f_d(s)$. It consists of the failed and working paths of the SPM that carries the traffic aggregate d. For every aggregate d and for every failure symptom $f_d(s)$ a load balancing function l_d^f is configured. If the LIR diagnoses the failure symptom $f_d(s)$, it redistributes the traffic of d according to l_d^f on the working paths. For example, the equal distribution of the traffic aggregate d onto all working paths is a very simple load balancing function.

The traffic matrix specifies the rates for all traffic aggregates. They have to be supported in all protected scenarios S which entails a lower bound on the link bandwidths.

The sum of all link capacities is the required overall capacity. It should be as small as possible because it represents capital or operational costs. The load balancing function provides some degrees of freedom for the minimization of backup capacities. A simple optimization approach is the assignment of a large portion of d to short partial paths and of a small portion of d to long partial paths. Mathematically speaking, we distribute the rate of a traffic aggregate onto the working paths of an SPM reciprocally to the lengths of these paths. The load balancing function of an SPM can also be exactly optimized. In [20] we modelled technical constraints by linear equations and used linear programming as optimization method. As the solution for l_d^f consists of real values, the computation can be performed in polynomial time.

4 Backup Efficiency of Self-Protecting Multi-paths

In this section we evaluate the performance of the SPM both in example and random networks using homogeneous traffic matrices. The impact of heterogeneous traffic matrices is investigated in [21]. We determine the required network capacity, i.e. the sum of all link bandwidths, which is required to accommodate the traffic matrix without resilience if shortest path routing (OSPF) is used based on the hop count metric. We take it as a reference value since it is a lower bound for the required network capacity. Then we calculate the required capacity for a given protection scheme to meet the resilience requirements. The resulting extra capacity is the performance measure in our studies. Note that this extra capacity is not always used for backup purposes only because protection mechanisms require sometimes longer paths than the shortest paths for normal operation. However, we use the term extra capacity and backup capacity exchangeably since the extra capacity is required to provide resilience with the respective protection mechanism. For resilience purposes, we take all single link and router failures in the set of protected failure scenarios S into account.

The calculations for the routing and the load balancing were carried out on a Pentium IV 1.5 GHz standard PC and took some seconds for small networks and some minutes for large networks.

4.1 Impact of Path Layout and Load Balancing on the Required Backup Capacity

We investigate the impact of path layout and load balancing for SPM on the backup performance in two test networks. The Lab03 network in Figure 1(b) is taken from the testbed of the KING project [22]. It is a modification of the UUNET in 1994 where all nodes with a node degree of at most 2 are successively removed. The network in Figure 1(a) is the optical core of the infrastructure in the COST-279 project [23]. The project was part of the "European Co-operation in the Field of Scientific and Technical Research" and concentrated on ultra-high capacity optical transmission networks. We use both networks in our performance evaluation because they have different properties.

Figures 2(a) and 2(b) show the backup performance for different SPMs in the COST-239 and in the Lab03 network. The x-axis shows the parameter k for the kDSP calculation for the path layout. The load balancing options are given by different curves. The additional capacity for SPM with equal load balancing is marked by kSPM-E, for SPM

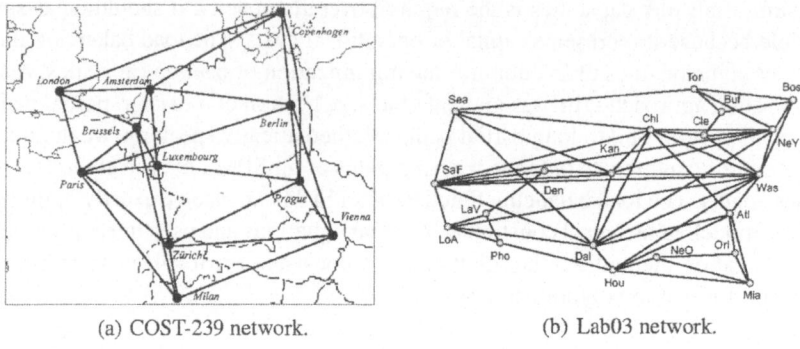

(a) COST-239 network.   (b) Lab03 network.

Fig. 1. Test networks.

with reciprocal load balancing it is marked by kSPM-R, and for SPM with optimized load balancing it is marked by kSPM-O. In addition, the backup capacity for OSPF is given. The SPM require clearly less capacity than OSPF rerouting. The kSPM-O is most economic and its efficiency increases with increasing k. As there are more disjoint backup paths available for larger k, the traffic can be better redistributed in a failure case and less extra capacity is required. The most articulate performance gain is achieved for taking $k=3$ instead of $k=2$ disjoint paths. Due to the network topological restrictions, only 4 disjoint paths can be found mostly even for $k=5$. Therefore, the backup capacity can not be arbitrarily reduced.

In the COST-239 network, the performance of kSPM-E and kSPM-R degrades for increasing k and more extra capacity is needed. The same effect can also be observed to a minor extent in the Lab03 network. If an SPM consist of more disjoint shortest paths, some of them are significantly longer than the shortest one. Their extensive use can not be avoided with kSPM-E or kSPM-R which leads to an increased required network capacity. Hence, SPM with simple load balancing schemes reveal only minor benefits and the optimization is worthwhile.

4.2   Impact of Network Topology Characteristics

To study the impact of the network topology in more detail, we conduct studies based on random networks and take for 5SPM-O as protection switching mechanism. At first, we describe our algorithm for the construction of random networks. Then we illustrate the impact of the network topology on the backup performance of SPMs both in absolute values and in comparison to the backup performance of OSPF rerouting.

Construction of Random Networks. We construct random networks and control some of their essential characteristics. One of them is the degree $deg(v)$ of a node v, which is the number of links v is connected with. We briefly explain our network construction method that incorporates features of the well know Waxman model [24,25]. It is an efficient algorithm that provides control over the minimum, the average, and the maximum node degree (deg_{min}, deg_{avg}, deg_{max}), and avoids loops and parallels.

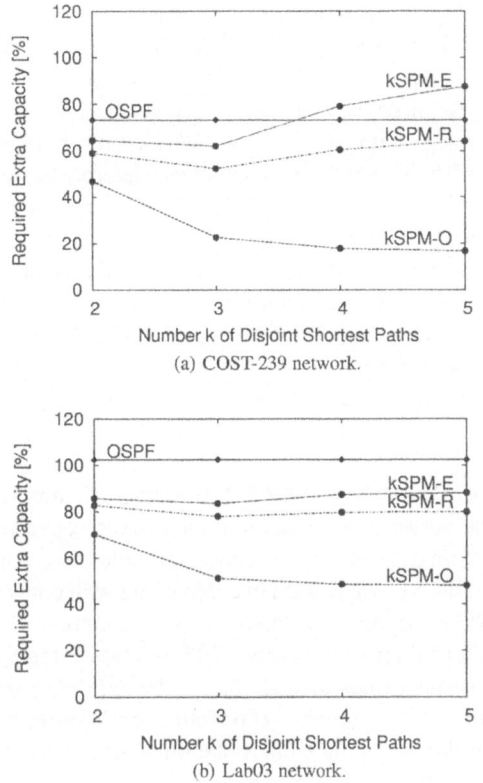

Fig. 2. Impact of multi-path routing and load balancing on the backup capacity of SPMs.

The algorithms starts with an empty link set $\mathcal{E} = \emptyset$ and defines a single arbitrary node $v_{start} \in \mathcal{V}$ connected. Then, $\frac{|\mathcal{V}| \cdot deg_{avg}}{2}$ links are added successively to $\mathcal{E}$ by connecting suitable nodes v_α and v_ω. An arbitrary node v_α is chosen from a set of preferred nodes $\mathcal{V}_\alpha$ with the following properties. All $v \in \mathcal{V}_\alpha$ are connected and have $deg(v) \le deg_{max}$. If a node $v \in \mathcal{V}$ exists with $deg(v) < deg_{min}$, all $v \in \mathcal{V}_\alpha$ must have $deg(v) < deg_{min}$. The set of potential neighbor nodes $\mathcal{V}_\omega$ obeys the following requirements: Loops and parallels must be avoided, i.e. $v_\alpha \notin \mathcal{V}_\omega$ and $(v_\alpha, v_\omega) \notin \mathcal{E}$. Furthermore, if an unconnected node $v \in \mathcal{V}$ exists, all $v \in \mathcal{V}_\omega$ must be unconnected. The node $v_\omega \in \mathcal{V}_\omega$ is chosen according to a probability distribution which depends on v_α and $\mathcal{V}_\omega$. Here, the Waxman model comes into play. Each node has a position in the plane. The Euclidean distance $d(v, w)$ induces a weight $P(v, w) = a \cdot e^{-\frac{d(v,w)}{b \cdot d_{max}}}$ with $d_{max} = \max_{v,w \in \mathcal{V}} d(v, w)$, and $P(v, w)$ produces the probability distribution $p_{v_\alpha}(w) = \frac{P(v_\alpha, w)}{\sum_{v \in \mathcal{V}_\omega} P(v_\alpha, v)}$. Given a maximum node degree deviation deg_{dev}^{max}, the minimum node degree is set to $deg_{min} = \max(deg_{avg} - deg_{dev}^{max}, 2)$ and the maximum node degree is set to $deg_{max} = deg_{avg} + deg_{dev}^{max}$.

Absolute Backup Performance. We investigate the required backup capacity for 240 random networks of different size, different average node degree deg_{avg}, and different maximum node degree deviation deg_{dev}^{max}. There are 5 random networks for each topology description. In Figure 3(a), the x-axis indicates the average number of disjoint parallel paths k^* that are found for all source–destination pairs in a network and the y-axis shows the required backup capacity. In general, we observe that the required backup capacity decreases with increasing k^*. We identify four clusters of networks that are marked by dashed lines which are least square interpolations among the points of these clusters according to an exponential function. It turns out that all networks of a cluster have the same average node degree deg_{avg}. The dashed lines make the clusters more visible, however, the extrapolation of those curves does not make sense since deg_{avg} is a trivial upper bound on k^*. Within a cluster, the network size n seems to be irrelevant. A small maximum deviation deg_{dev}^{max} of the node degrees $deg(v)$ from the average node degree deg_{avg} seems to increase k^*, and leads to more efficient backup solutions within a cluster. Therefore, resilience can be achieved at lower cost if the network topology is symmetric.

Backup Performance Relative to OSPF Rerouting. Figure 3(b) shows the backup capacity for the same networks in relation to the backup capacity for OSPF rerouting. The OSPF normalization dampens the influence of topological characteristics and shows clearly the benefits of the SPM approach in comparison with conventional rerouting. For all 5 random networks with the same topological characteristics, we build the mean of their k^* and the mean of their ratios of the SPM and OSPF rerouting backup capacity. The horizontal and vertical lines provide the 90% confidence intervals. The data are plotted on a logarithmic scale to make exponential trends better visible.

The dashed line is the least square interpolation of all experiments and the solid lines are the interpolations within a cluster of networks with the same average node degree deg_{avg}. The four clusters confirm the above observation that deg_{avg} of a network is strongly correlated with k^*. Increasing the average node degree deg_{avg} shifts the exponential trend slightly towards larger backup capacity. Again, we observe an exponential decay with regard to an increasing k^*, i.e., the superiority of the SPM over OSPF rerouting increases with a larger average number of disjoint paths k^* because SPM reduces the required backup capacity by multi-path forwarding.

5   Conclusion

In this paper we have proposed the Self-Protecting Multi-Paths (SPM) as an e2e protection switching mechanism for MPLS networks. It is used to achieve network resilience in a single autonomous system. We started with an overview of related work and argued for a simple backup solution like the SPM. The SPM carries the traffic of a single e2e traffic aggregate over disjoint parallel paths that may have different length. The traffic of an aggregated my be distributed over these paths according to a load balancing function. If a single partial path fails, the traffic is redirected to the other working paths according to another load balancing function. This action can be performed without signalling across the network because traffic aggregates are only shifted to parallel paths and only aggregates that are affected by a network failure are relocated.

The network capacity dimensioned for OSPF routing without fault tolerance is the reference case in our performance evaluation study. The performance measure for a cer-

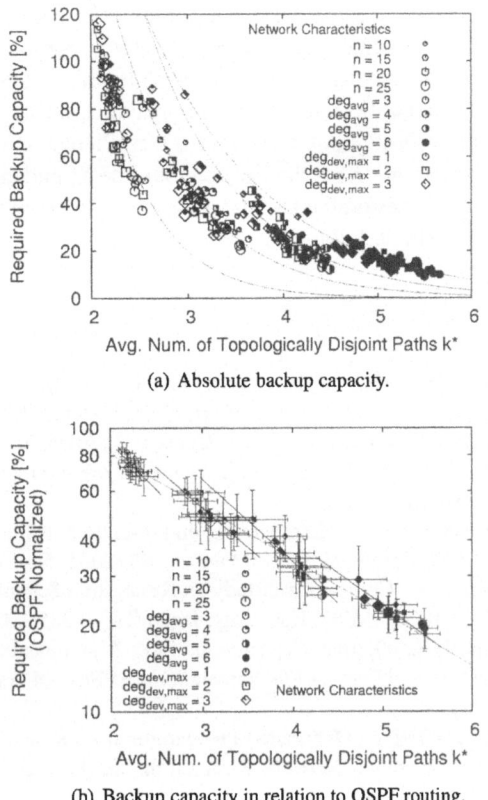

(a) Absolute backup capacity.

(b) Backup capacity in relation to OSPF routing.

Fig. 3. Required extra capacity for SPM in random networks.

tain protection switching mechanism is the additional network capacity that is required to achieve resilience. Our experiments showed that the backup performance benefits from the transmission over disjoint parallel paths. Optimized load balancing leads to enormous bandwidth savings for the SPM such that it can provide full resilience against all single link and node failures with less than 17% backup capacity in the COST239 network. This makes failure protection on the network layer significantly cheaper from a resource point of view than on the physical layer if resource doubling is applied.

We constructed random networks and controlled some of their fundamental network characteristics. The amount of required extra capacity depends on the network topology and, in particular, on the average number k^* of disjoint paths in the network whereas the network size has no influence on the required extra capacity. Since OSPF rerouting can also achieve network resilience, we compared the backup capacity required for SPMs and OSPF rerouting. Our simulations revealed that the amount of extra capacity for SPMs decays exponentially with k^* compared to OSPF routing. Only 20% of the OSPF extra bandwidth is needed in suitable networks. In addition, the reaction time of SPM is faster than OSPF rerouting because no reachability information must be exchanged if a failure occurs.

As a challenge remain, e.g., fast heuristics for the calculation of an optimized load balancing are needed for large networks. Suitable network structures are a prerequisite for cheap backup capacities and should be further identified. The optimization of the SPM must be adapted to networks with given link capacities and a structure of their traffic matrix to maximize their throughput while meeting resilience requirements. Moreover, the underlying path computation for the SPMs should be extended towards SRLGs and the impact of multiple failures on the QoS degradation is to be investigated in networks that are resilient against single failures.

References

1. Sharma (Ed.), V., Hellstrand (Ed.), F.: RFC3469: Framework for Multi-Protocol Label Switching (MPLS)-based Recovery. http://www.ietf.org/rfc/rfc3469.txt (2003)
2. Pan, P., Gan, D.H., Swallow, G., Vasseur, J.P., Cooper, D., Atlas, A., Jork, M.: Fast Reroute Extensions to RSVP-TE for LSP Tunnels. http://www.ietf.org/internet-drafts/draft-ietf-mpls-rsvp-lsp-fastreroute-03.txt (2003)
3. Autenrieth, A., Kirstädter, A.: Engineering end-to-end ip resilience using resilience-differentiated qos. IEEE Communications Magazine 40 (2002) 50–57
4. Murakami, K., Kim, H.S.: Comparative Study on Restoration Schemes of Survivable ATM Networks. In: IEEE INFOCOM'97, Kobe City, Japan (1997) 345 – 352
5. Murakami, K., Kim, H.S.: Optimal Capacity and Flow Assignment for Self–Healing ATM Networks Based on Line and End–to–End Restoration. IEEE/ACM Transactions of Networking 6 (1998) 207–221
6. Grover, W.D.: Cycle–Oriented Distributed Preconfiguration: Ring–like Speed with Mesh–like Capacity for Self–planning Network Restoration. In: Proceedings of IEEE ICC '98. (1998) 537–543
7. Gruber, C.G., Schupke, D.A.: Capacity–efficient Planning of Resilient Networks with p–Cycles. In: Proceedings of Networks 2002. (2002) 389–395
8. Moy, J.: RFC2328: OSPF Version 2. ftp://ftp.isi.edu/in-notes/rfc2212.txt (1998)
9. Fortz, B., Thorup, M.: Internet traffic engineering by optimizing OSPF weights. In: IEEE INFOCOM'00. (2000) 519–528
10. Köhler, S., Binzenhöfer, A.: MPLS traffic engineering in OSPF networks - a combined approach. In: 18th International Teletraffic Congress (ITC18), Berlin (2003)
11. Gojmerac, I., Ziegler, T., Ricciato, F., Reichl, P.: Adaptive Multipath Routing for Dynamic Traffic Engineering. In: GLOBECOM'03, San Francisco (2003)
12. Dittmann, G., Herkersdorf, A.: Network Processor Load Balancing for High–Speed Links. In: SPECTS 2002, San Diego, CA (2002) 727–735
13. Kodialam, M.S., Lakshman, T.V.: Minimum Interference Routing with Applications to MPLS Traffic Engineering. In: Proceedings of IEEE INFOCOM 2000. Volume 2. (2000) 884–893
14. Iraschko, R.R., MacGregor, M.H., Grover, W.D.: Optimal Capacity Placement for Path Restoration in STM and ATM Mesh-Survivable Networks. IEEE/ACM Transactions on Networking 6 (1998) 328 – 336
15. Strand, J., Chiu, A.L., Tkach, R.: Issues For Routing In The Optical Layer. IEEE Communications Magazine 39 (2001) 81–87
16. Rajagopalan, B., Luciani, J.V., Awduche, D.O.: IP over Optical Networks: A Framework. http://www.ietf.org/internet-drafts/draft-ietf-ipo-framework-05.txt (2003)
17. Kompella, K., Rekhter, Y.: Routing Extensions in Support of Generalized Multi–Protocol Label Switching. http://www.ietf.org/internet-drafts/draft-ietf-ccamp-gmpls-routing-09.txt (2003)

18. Suurballe, J.W.: Disjoint Paths in a Network. Networks **4** (1974) 125–145
19. Edmonds, J., Karp, R.M.: Theoretical Improvements in the Algorithmic Efficiency for Network Flow Problems. Journal of the ACM **19** (1972) 248–264
20. Menth, M., Reifert, A., Milbrandt, J.: Optimization of End-to-End Protection Switching Mechanisms for MPLS Networks. Technical Report, No. 320, University of Würzburg, Institute of Computer Science (2004)
21. Menth, M., Milbrandt, J., Reifert, A.: Self-Protecting Multipaths - A Simple and Resource-Efficient Protection Switching Mechanism for MPLS Networks. Technical Report, No. 322, University of Würzburg, Institute of Computer Science (2004)
22. Hoogendoorn, C., Schrodi, K., Huber, M., Winkler, C., Charzinski, J.: Towards Carrier-Grade Next Generation Networks. In: ICCT 2003, Beijing, China (2003)
23. Batchelor et al., P.: Ultra High Capacity Optical Transmission Networks. Final report of Action COST 239. http://barolo.ita.hsr.ch/cost239/network/ (1999)
24. Zegura, E.W., Calvert, K.L., Donahoo, M.J.: A Quantitative Comparison of Graph-Based Models for Internet Topology. IEEE/ACM Transactions on Networking **5** (1997) 770–783
25. Waxman, B.M.: Routing of Multipoint Connections. IEEE Selected Areas in Communications **6** (1988) 1617–1622

Integrated Dynamic Routing of LSPs in IP over WDM Networks: Full Protection and Partial Spatial-Protection

Qin Zheng and Mohan Gurusamy*

Department of Electrical and Computer Engineering
National University of Singapore, Singapore 117576
{engp1752, elegm}@nus.edu.sg

Abstract. We investigate the problem of Label Switched Path (LSP) protection using online integrated routing in IP over WDM networks. While certain mission- and time-critical applications require guaranteed 100% protection, other applications may have less stringent protection requirements. We consider these two kinds of protection scenarios and refer them as full protection (FP) and partial protection (PP), respectively. We first consider the full protection scenario and develop two integrated routing algorithms to select primary LSPs and backup LSPs, respectively. Next we consider the partial protection scenario, particularly, the partial spatial-protection (PSP) wherein the working traffic is unprotected against the failure of certain links along the primary LSPs according to the specified connection protection requirements. We develop an algorithm to determine the set of unprotected links with the objective of improving the backup resource sharing efficiency. We evaluate the performance of the proposed algorithms on the NSFNET and Pan-European optical networks.

1 Introduction

In IP/multi-protocol label switching (MPLS) over Wavelength Division Multiplexing (WDM) networks, IP/MPLS routers are directly connected to optical cross-connects (OXCs) which are interconnected via fiber links carrying multiple wavelength channels. IP/MPLS routers are also refereed to as Label Switched Routers (LSRs). End-to-end lightpaths are created across the optical core on the wavelength channels which in turn form the virtual topology to be used by the IP layer. A lightpath must use the same wavelength on all the links along its physical route. The virtual topology is then used by the IP layer for IP routing. Consequently, the IP layer paths can traverse multiple hops (lightpaths).

We consider a single link failure model in this paper and a connection is restorable by using a link-disjoint pair of an active LSP and a backup LSP. Both primary and backup LSPs may traverse a number of lightpaths which in turn

---

* This work was supported in part by the ONFIG-GMPLS project (NUS WBS No: R-263-000-231-593) funded by SERC, ASTAR, Singapore.

N. Mitrou et al. (Eds.): NETWORKING 2004, LNCS 3042, pp. 538–549, 2004.

are routed over a number of OXCs. The traffic are transmitted on the active (primary or working) LSPs during normal network operations and switched to the backup (protection) LSPs when failure occurs. Protection resources can be shared among multiple backup LSPs if their corresponding primary LSPs will not fail simultaneously.

As a variety of novel types of applications appear in Internet besides the traditional voice and data services, the ability of providing multiple levels of service performance becomes necessary. While voice traffic should have guaranteed 100% protection, other applications may require less stringent protection requirements [1]. Consequently, having various protection grades to satisfy the multi-level service requirements has received much attention recently [1,2,3]. Another motivation is the fact that since Internet traffic is often more sensitive to cost, it is desirable to have a range of protection services and cost [1].

1.1 Related Work

Integrated routing that incorporates the resource and topology information from both IP and optical layers has been proposed as a promising solution [4,5,6]. A salient feature of the integrated routing is that the path found can traverse IP logical (virtual) links in the IP layer and new wavelength channels at the optical layer which will lead to the creation of new virtual links. The integrated routing problem was first introduced in [5] and the impact of IP subnets on it was studied in [4]. The differentiated integrated routing considering o-e-o conversion constraints was studied in [7]. These approaches considered LSP provisioning without taking into account the survivability requirements. The problem of integrated routing of restorable connections was studied in [6]. The algorithms proposed for routing of working paths are essentially integrated physical-hop-based routing. The QoS requirements of working traffic and the wavelength resource constraints are not taken into account.

Next we describe the related work on partial protection. The problem of partial protection can be classified into three categories: partial traffic-protection, partial temporal-protection, and partial spatial-protection. In partial traffic-protection, the percentage of working traffic to be protected depends on the specified protection grade. Such a partial traffic-protection is considered in [1]. Partial temporal-protection is introduced in [2] where protection bandwidth can be shared with some working path which allows the connection to be unprotected during some periods of time. In partial spatial-protection, a connection is unprotected against some fiber link failures based on the survivability requirements. The differentiated reliability (DiR) problem studied in [3] belongs to this category.

We consider LSP-level partial spatial-protection in this paper wherein backup LSPs may not be available when certain links along working LSPs fail. We consider online integrated sub-λ LSP routing of dynamic requests that arrive one by one with no prior information. In [3], offline λ (lightpath) routing of static traffic was studied and lightpath-level partial spatial-protection was considered wherein some links along primary lightpaths are not protected by corresponding backup lightpaths. As an LSP can traverse one or more lightpaths, protection at the LSP level makes it possible to specify the end-to-end protection grade for each connection request at the IP/MPLS layer.

1.2 Contributions

First we consider full protection scenario and develop integrated routing algorithms to select the primary LSPs and backup LSPs taking into account the constraints at both the MPLS and optical layers. Next we consider the LSP-level partial spatial-protection following the single link failure assumption. A connection request specifies the required protection grade (pg) where $0 \leq pg \leq 1$. An equivalent term is maximum failure probability (MFP) where $MFP = 1 - pg$. The MFP of a connection defines the maximum allowable probability with which backup path is not available upon occurrence of a single link failure in the network. For a connection request with less stringent protection requirement, the backup LSP need not be available for every possible link fault that may occur along the primary LSP. It is thus possible to select a set of links along the primary LSP for which the connection is unprotected, given the required protection grade is guaranteed. These links are called *unprotected links* with respect to the primary LSP of interest. As a result, the backup bandwidth on lightpaths along a backup LSP can be shared by another backup LSP even if the corresponding primary LSPs of these two requests share a common link, if one of these two primary LSPs choose the shared link to be an unprotected link.

In this paragraph we illustrate how to determine the number of unprotected links for a connection request with protection grade *pg*. Suppose F is the set of unprotected links. The link failure probability (LFP) is defined as the probability that the given link fails upon a single link fault in the network. The LFP can be derived from normalized link downtime ratio and the length of the links in the network [3]. The following condition must be satisfied $\sum_{i \in F} LFP_{(i)} \leq MPF$ to select F for a specified *pg*. For simplicity, we assume that all the links have equal failure probabilities and thus the LFP for each link is $\frac{1}{L}$ for a network with L fiber links. The number of unprotected links allowed is denoted by F_m where F_m is the largest integer that satisfies the condition $\frac{1}{L} \times F_m \leq MFP$. The case where links have different failure probabilities can be dealt with using the first condition. After converting the protection grade to the number of unprotected links allowed, one crucial problem is to determine the set of unprotected links. We propose an algorithm which selects unprotected links with the objective of reducing the backup bandwidth needed.

2 An Example of LSP-Level Partial Spatial-Protection

Fig. 1 shows a network which comprises four IP routers connected to four OXCs through wavelength ports. The OXCs are interconnected by fiber links labeled a through e which carry multiple wavelength channels. Assume that at an instance of time, a new request which requires bandwidth b from router4 to router3

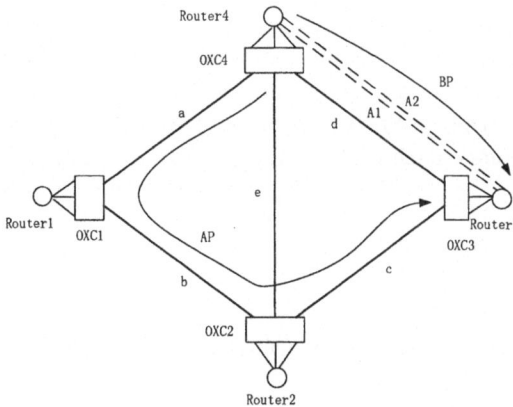

Fig. 1. Example of LSP-level partial spatial-protection

arrives. There exists a connection from router4 to router3 with bandwidth b1. The existing connection routes its active path (AP) and backup path (BP) on virtual links (dashed lines) A1 and A2, respectively. A1 traverses fiber links e,c and A2 traverses fiber link d. Suppose that the new request opens a new virtual link on fiber link a-b-c for the primary path and uses existing virtual link A2 in the backup path. Assume that the existing request has 100% protection requirement while the new request specifies the connection to be 80% restorable against any single link failure.

Now we compute the amount of protection bandwidth to be reserved on the backup path for the new request. As the primary paths of the two requests traverse one common fiber link c, the backup resources on arc A2 cannot be shared by the new request. As a result, $b1 + b$ amount of bandwidth needs to be reserved on arc A2 if the new request is to be provided with 100% protection. Next we consider the protection grade of specified 80% survivability for the new request. Since we assume that all the fiber links are equally likely to fail and there are five links in the network, it implies that the new request can be unprotected against the failure of one fiber link. As fiber link c is the common risk of the two requests which makes sharing impossible, we choose it to be unprotected. Consequently, $max(b1, b)$ amount of bandwidth needs to be reserved on arc A2. We note that choosing fiber link a or b to be unprotected will require $b1 + b$ amount of protection bandwidth on arc A2.

3   The Proposed Integrated Routing Algorithms

3.1   Network Model and Problem Statement

We consider a network of N nodes connected by L bidirectional fiber links each carrying W wavelength channels. We assume that each node comprises an OXC and a LSR. The OXCs have no wavelength conversion capability. We consider dynamic traffic where LSP requests arrive one-by-one with no prior information about future requests. An LSP-request is specified as $< s, d, b, pg >$ where s is the source node, d is the destination node, b is the amount of bandwidth required, and pg is the specified protection grade. For each connection request, a link-disjoint primary path and backup path must be found. The objective is to reduce the delay for working traffic and minimize the protection bandwidth needed on the backup path while satisfying the protection grade. We consider the case of $pg = 100\%$ in this section.

Notations
- l_j^W is the number of available wavelength channels on fiber link l_j at an instance of time. Initially, $l_j^W = W$.
- a_m is the unidirectional wavelength-switched path (lightpath) defined as an ordered vector of traversed fiber links $a_m = < l_1, l_2, ... l_{h_m} >$, where h_m denotes the physical length of a_m. Further, a_m represents the directed arc between two nodes in the virtual topology, with a fixed bandwidth denoted by B_m. We use the terms *link* and *arc* to refer to the edges in the physical topology and virtual topology.
- r_m^j is a binary variable which indicates whether link l_j is used in arc a_m.
- n_p^l denotes the number of LSRs traversed by the primary LSP.
- V_p^m is a binary variable which indicates whether the primary LSP traverses arc a_m.
- C_p^j is a binary variable which indicates whether the primary LSP traverses a free wavelength channel on link l_j. Note the path found by integrated routing can traverse arcs and wavelength channels which lead to the creation of new arcs.
- A_p^j is a binary variable which indicates whether the primary LSP traverses link l_j.
- A_p denotes the set of links traversed by the primary LSP.
- V_b^m is a binary variable which indicates whether the backup LSP traverses arc a_m.
- C_b^j is a binary variable which indicates whether the backup LSP traverses a free wavelength channel on link l_j.
- T_m is an ordered vector associated with arc a_m to record the backup bandwidth required to protect against each fiber link failure in the network. $T_m = < B_m^1, B_m^2, ..., B_m^j, ... B_m^L >$, where B_m^j is the amount of backup bandwidth needed on a_m when link l_j fails.
- T_m^B denotes the backup bandwidth reserved on arc a_m which is the maximum value in the vector T_m.
- T_m^l denotes the link corresponding to T_m^B. This information is used to determine the set of unprotected links as discussed in Section 4.1.
- b_m^a denotes the additional backup bandwidth needed on arc a_m to route the backup path for the current request.
- k_1, k_2 constants, $k_1 \gg k_2$ such that $k_1 x' > k_2 y'$, where x' is the smallest possible non-zero x-value and y' is the largest possible non-zero y-value in a function of the form $k_1 x + k_2 y$.

3.2   Algorithms

Primary Path Selection. The Minimum Delay Least Congestion integrated routing algorithm (MDLC-IRA) is used to select the primary path. MDLC-IRA chooses a path that traverses minimum number of LSRs, which attempts to minimize the global average queuing delay. In case of a tie, the path which creates lightpaths on fiber links with more available channels is preferred. The objective is to avoid saturating wavelength resources on certain links, thus increases the possibility of opening new lightpaths on these links. Consider a path p which traverses n_p^l number of LSRs. If l_j^W is the number of free wavelength channels on link j traversed by p, $W - l_j^W$ gives the number of occupied channels on it. Now the cost C of path p is defined as

$$C = k_1 n_p^l + k_2 max_{C_p^j = 1}(W - l_j^W) \tag{1}$$

The MDLC-IRA chooses the path that minimizes the cost C as the primary path. MDLC-IRA assigns edge weights as follows: each o-e-o edge is assigned weight k_1. For each link j, $k_2(W - l_j^W)$ is set as weight. A Dijkstra-like shortest path algorithm is used to compute the minimum cost path and the wavelength resources on links are dealt with to decide the bottleneck similar to the widest-shortest path selection.

Backup Path Selection. The Minimum Bandwidth Least Congestion integrated routing algorithm (MBLC-IRA) is used to route the backup path. MBLC-IRA minimizes the total amount of bandwidth that needs to be reserved on the backup path. The additional bandwidth needed on links traversed is b and that on existing arcs is given by b_m^a ($b_m^a \leq b$) which is determined by Equation (3) and (4). The tie is broken using a method as in MDLC-IRA. MBLC-IRA assigns edge weights in the following way: Each o-e-o edge is assigned weight ϵ. For each arc a_m, $k_1 h_m b_m^a$ is set as weight. For each link j, $k_1 b + k_2(W - l_j^W)$ is set as weight. A Dijkstra-like shortest path algorithm is used to compute the minimum cost path and the wavelength resources on links are dealt with to decide the bottleneck similar to the widest-shortest path selection.

$$C = k_1 \left(\sum_{C_b^j = 1} b + \sum_{V_b^m = 1} h_m b_m^a \right) + k_2 max_{C_b^j = 1}(W - l_j^W) \tag{2}$$

$$A_p^j = C_p^j \quad or \quad V_p^m r_m^j \tag{3}$$

$$b_m^a = max_{j=1}^L (B_m^j + A_p^j b) - T_m^B \tag{4}$$

Once the primary path is chosen, whether fiber link j is traversed by it can be determined from Equation (3). This helps to determine the set of fiber links traversed by the chosen primary path. The b_m^a value on arc a_m is calculated using Equation (4). It requires updates of the entries in T_m that correspond to the links traversed by the primary path. For each link j traversed by the primary path, the entry B_m^j in T_m associated with arc a_m is increased by b. The additional backup bandwidth needed b_m^a on arc a_m is the amount by which the maximum value T_m^B is increased. For the example in Fig. 1, the T_m^B value of arc A2 is b1 and b1+b before and after the new request is honored, respectively. The b_m^a value of arc A2 is b and T_m^l is link c.

4   LSP-Level Partial Spatial-Protection

In this section, we consider multiple levels of protection grades of connections and the objective is to satisfy these user-specific requirements to minimize the network resources. We first propose an algorithm to determine the unprotected links according to the protection grades. We explain how the algorithm can improve the backup sharing efficiency and in turn reduce the total amount of bandwidth required on the backup path. Finally, we discuss the actual *restorable probability* of each connection request which is defined as the probability that the backup LSP is available upon a single link failure.

4.1   Improved Backup Sharing Efficiency with PSP

Consider a connection request $< s, d, b, pg >$ where pg is the specified protection grade denoting the PSP requirement. We translate the protection grade into the permissible number of unprotected links, denoted by F_m. We recall that the failure probability of each of the L links is assumed to be the same and F_m is the largest integer number that satisfies $\frac{1}{L} \times F_m \leq MFP$ where $MFP = 1 - pg$. We emphasize that our routing algorithm can be easily modified to account for the case where links have different failure probabilities.

We first choose the primary path and backup path by using MDLC-IRA and MBLC-IRA, respectively. We then choose F_m number of links to be unprotected by using the following algorithm in a way to increase the backup sharing efficiency. The following pseudo code shows the steps taken place to determine the unprotected link set F.

The algorithm searches the existing arcs traversed by the current backup path in step1. If $b_m^a > 0$ (which means that the backup bandwidth on arc a_m is increased) and T_m^l is not in F, then T_m^l is added to F. The number of chosen links (F_l) is increased by 1. The idea is to combine the choice of unprotected links with the backup sharing on existing arcs along the backup path. Step1 continues if $F \subset A_p$ (the fiber link set traversed by the current primary path) and $F_l < F_m$. If the above condition still holds when all the existing arcs are searched, the algorithm chooses the unprotected links randomly from the remaining links in A_p but not in F in step2. We note that the unprotected links can also be selected from the remaining links based on the link criticality if such information is available in the network. The proposed algorithm only uses T_m^l information on the arcs traversed by the current backup path which makes the decision of unprotected links quickly.

For the example in Fig. 1, the proposed algorithm will choose link c as the unprotected link. As a result, the corresponding B_m^c value of arc A2 will remain unchanged. The T_m^B value of arc A2 is b1 and max(b1,b) before and after the new request is honored, respectively. The additional backup bandwidth needed on arc A2 is max(b1,b)-b1 to protect the new request.

4.2 Discussion on Connection Restorable Probability

Although connections are protected with specific survivability grades, they have higher probability to be restored against a single link failure. Generally, the unprotected links selected in step 2 can be restored as they are allowed to share the backup resources on a_m freely (without increasing T_m^B). Furthermore, we consider dynamic traffic and the values in T_m for each arc a_m keep changing whenever a new request is honored or an existing request terminates. In both cases, the B_m^j value for unprotected link j on arc a_m could be sufficiently lower than T_m^B to satisfy the condition $B_m^j + b \leq T_m^B$ which means that this arc can be used as backup even when unprotected link j fails. We study this effect in Section 5.3.

5   Performance Study

5.1   Simulation Model

We consider a dynamic network traffic model, and connections are setup and torn down dynamically. The traffic arrival at a node follows Poisson distribution with rate λ and the holding time of a connection is exponentially distributed with a mean of $1/\mu$. The destination node for a connection is selected using a uniform distribution among all the nodes except the source node. The traffic load per node is defined as λ/μ and expressed in Erlangs.

Simulation experiments are performed on two networks: NSFNET with 14 nodes and 21 links and the Pan-European optical network with 19 nodes and 38 links. We assume 8 wavelength channels on each fiber link in the two networks. The bandwidth requested by a connection is uniformly distributed in the range of (1, 6). The maximum capacity of a wavelength is assumed to be 10. The system parameter varied is the load per node. For the NSFNET, the load is varied from 2.0 to 8.0 Erlangs. For the Pan-European optical network, the load is varied from 2.0 to 12.0 Erlangs as it is denser than the NSFNET.

In the first set of experiments, we consider full protection (FP) for all requests and compare the performance of the proposed integrated routing algorithms MDLC-IRA and MBLC-IRA to the integrated IP-hop routing and integrated physical-hop routing algorithms. Both the integrated IP-hop routing algorithm and integrated physical-hop routing algorithm route a path on virtual links and wavelength channels. The integrated IP-hop routing finds the primary path and backup path based on the virtual hop counts. The integrated physical-hop routing simply finds both paths based on the physical hop counts. In the second set of experiments, the protection grades are taken into account and we show the improvements obtained using partial spatial-protection (PSP) compared to the full protection. We consider three classes of traffic which permit 0, 1 and 2 unprotected links in the primary path, respectively. We assume that the failures are uniformly distributed among all the fiber links in the network and each link has the equal probability to fail. Each request is randomly assigned to class 0, 1, or 2 with probability 0.4, 0.3 and 0.3, respectively.

The performance metrics considered are the *blocking probability* and *average restorable probability*. Each simulation experiment is run with a large number of connection requests on the order of 100000 per node. The experiment is repeated several times to achieve accurate results with a small confidence interval for a 95% confidence level.

5.2   Blocking Probability

Blocking probability is defined as the percentage of rejected connections among all the connection requests. The objective of online routing algorithms is to minimize this metric. Figure 2 and Fig. 3 show the blocking probability of different integrated routing algorithms in the two networks with full protection. We recall that, the MDLC-IRA and MBLC-IRA are used to route the primary path and backup path, respectively. In both figures we observe that the proposed routing algorithms MDLC-IRA and MBLC-IRA perform best and the integrated physical-hop routing algorithm is better than the integrated IP-hop routing algorithm. The integrated IP-hop routing algorithm performs poorly as it prefers paths traversing less virtual links. When there is no existing IP link with enough bandwidth between the ingress and egress nodes, the algorithm will try to open a new virtual link which leads to inefficient resource usage and higher blocking.

Figure 4 and Fig. 5 show the blocking probability of the proposed routing algorithms MDLC-IRA and MBLC-IRA with and without considering protection grades. In both figures we observe that the performance is much better when protection grades are taken into account. This is because the bandwidth on the backup paths are reduced by appropriately choosing the set of unprotected links.

5.3   Average Restorable Probability

The protection grades of class 0, 1, 2 traffic are 100%, $20/21 = 95.24\%$ and $19/21 = 90.48\%$, respectively in the NSFNET; and 100%, $37/38 = 97.37\%$ and $36/38 = 94.74\%$, respectively in the Pan-European network. Average restorable probability is defined as the average probability that a connection can be restored against any link failure in the network. As differentiated protection grades are provided to each class of traffic, this metric is important to measure whether the user-specific requirements can be met. In our experiments, we measure the restorable probability for each connection constantly at a time period 0.01 of the mean connection holding time. Then for each traffic load, these values of all the measuring periods are used to get the mean probability for each class.

Figure 6 and Fig. 7 show the average restorable probability for each class of traffic in the two networks. In the experiments, all the requests can satisfy their corresponding protection requirements. In both figures we observe that the

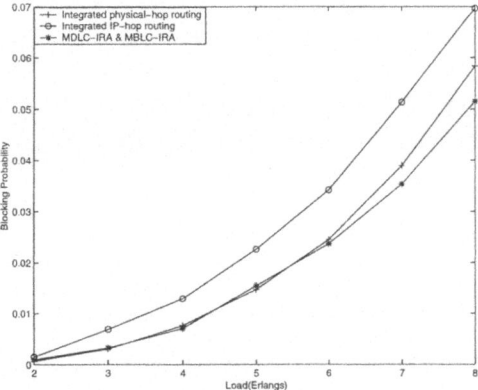

Fig. 2. Blocking probability with FP in NSFNET

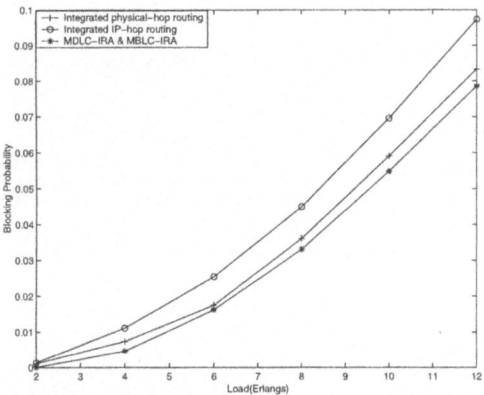

Fig. 3. Blocking probability with FP in Pan-European Network

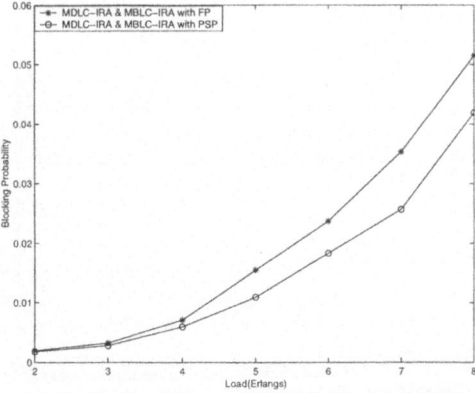

Fig. 4. Blocking probability with FP and PSP in NSFNET

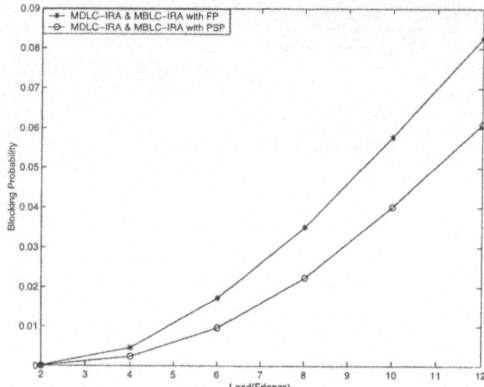

Fig. 5. Blocking probability with FP and PSP in Pan-European Network

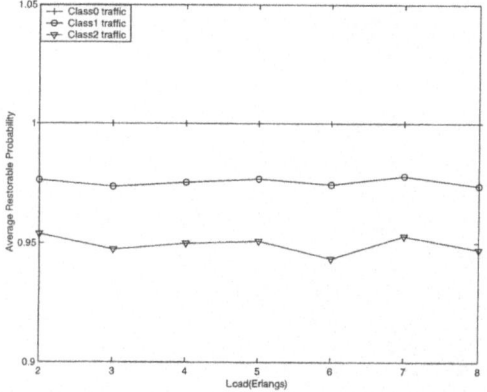

Fig. 6. Average restorable probability with PSP in NSFNET

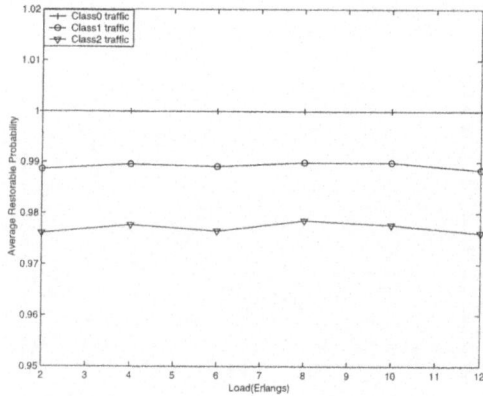

Fig. 7. Average restorable probability with PSP in Pan-European Network

average restorable probability is above the protection grades required. As explained in Section 4.2, although each connection has the number of unprotected links corresponding to the protection grade, backup resources could be available even when these links fail due to the backup sharing and dynamic nature of connection requests.

6   Conclusions

We considered the problem of LSP protection using integrated routing with the objective of providing different protection grades (full and partial). We developed integrated routing algorithms MDLC-IRA and MBLC-IRA to select the primary LSPs and backup LSPs taking into account the constraints at both the MPLS and optical layers. We proposed an algorithm to determine the set of unprotected links according to the specific spatial-protection requirements to reduce the bandwidth to be reserved on the backup path. We argued that although connections are provided with a specific protection grade, they have higher probability to survive from a single link failure due to backup sharing and the dynamic nature of traffic. We evaluated the performance of the proposed algorithms on the NSFNET and Pan-European optical networks. Through extensive simulations, we demonstrated that LSP full protection using the proposed integrated routing algorithms can reduce the connection blocking probability considerably. The blocking performance improves significantly due to the improved backup sharing efficiency when protection grades are considered. Further, connections in each class have higher probability to be restored than their protection requirements.

References

1. Gerstel, O., Sasaki, G.: Quality of protection (QoP): a quantitative unifying paradigm to protection service grades. Optical Networks Magazine Vol. 3 (2002)
2. Mohan, G., Somani, A.: Routing dependable connections with specified failure restoration guarantees in WDM networks. IEEE INFOCOM Vol. 3 (2000) 1761–1770
3. Fumagalli, A., Tacca, M., Unghvary, F., Farago, A.: Shared path protection with differentiated reliability. IEEE ICC Vol. 4 (2002) 2157–2161
4. Acharya, S., Gupta, B., Risbood, P., Srivastava, A.: IP-subnet aware routing in WDM mesh networks. IEEE INFOCOM Vol. 1 (2003) 1333–1343
5. Kodialam, M., Lakshman, T. V.: Integrated dynamic IP and wavelength routing in IP over WDM networks. IEEE INFOCOM Vol. 1 (2001) 358–366
6. Zheng, Q., Mohan, G.: An Efficient Dynamic Protection Scheme in Integrated IP/WDM Networks. IEEE ICC Vol. 2 (2003) 1494–1498
7. Cheng Tien E., Mohan, G.: Differentiated QoS routing in GMPLS-based IP/WDM Networks. IEEE Globecom Vol. 3 (2002) 2757–2761

A Receiver Based Single-Layer Multicast Congestion Control Protocol for Multimedia Streaming[*]

Miguel Rodríguez-Pérez, Manuel Fernández-Veiga, Sergio Herrería-Alonso,
Andrés Suárez-González, and Cándido López-García

E.T.S.E. Telecomunicación
Campus universitario s/n
36200 Vigo, Spain

Abstract. The transmission of multimedia content in a multicast environment is still a topic of active research. New congestion control protocols that can support the needs of real time transmission, while keeping the current Internet stability, are needed. While there already exist some protocols suitable for multimedia transmission in unicast communications, and there are also protocols capable of transmitting data to a large number of receivers simultaneously, our protocol is designed for achieving both goals at the same time. In this paper we present a novel protocol designed for multimedia streaming in multicast networks. Our simulation results show that the protocol is compatible with TCP flows, thus preserving the network stability, and is suitable for real-time multimedia transmissions, as it has low oscillations in its throughput and imposes no additional network delays that could increase the latency.

Keywords: Multicast, Congestion Control, Multimedia

1 Introduction

Multimedia streaming across the Internet to a large and disperse number of receivers has still to improve before being safely deployed. At the network level, the essence of the problem lies in devising a bandwidth-efficient transport protocol which does not jeopardize the stability of the current Internet.

In designing a TCP-compatible congestion control algorithm for multicast streaming, we have identified the following set of desirable features:

Smoothness. It is well known that the abrupt changes in the throughput (window) of a TCP connection, due to the loss of even a single packet, can severely degrade the subjective quality of a multimedia flow [1,2,3,4]. Therefore, it is imperative that any candidate protocol adjusts its transmission

---

[*] This work was supported by the project "TIC2000-1126 of the Plan Nacional de Investigación Científica, Desarrollo e Innovación Tecnológica" and by the grant PGIDT01PX132202PN of the "Secretaría Xeral de I+D da Xunta de Galicia."

rate at a slow pace. Slowly-responsive algorithms also allow the use of smaller buffers at the receivers which, in turn, help to reduce overall latency.

Low latency. This requirement is not as much important for one-way communications as it is for two-way communications. So, although not strictly needed, it is needed for the protocol to support interactive applications.

Scalability. The protocol should be deployable in multicast networks with a large number of receivers. This point poses some interesting design issues, especially the avoidance of the catastrophic feedback implosion phenomenon [5]. It is critical to find a way to collect enough state information about the receivers without saturating the network with their feedback messages.

Stability. Network conditions vary dynamically. After any change, the protocol must eventually reach a steady state where all other goals are preserved.

Intra-protocol fairness. The protocol must be fair to other instances of itself. Informally, if two sessions are experiencing similar end-to-end network conditions they should obtain equitable bandwidth.

Inter-protocol fairness. No other protocols should be mistreated when they compete for bandwidth along a network path or link. This compatibility requirement is as essential as the intra-protocol fairness, because it guarantees safe deployment on the network. In this paper, we limit ourselves to protocols fair against TCP, since this is the predominant traffic in the Internet [6, 7].

So as to fulfill the goals stated above, we ought to make some choices about the use of different techniques. The last years have witnessed important advances in this area, resulting in two broad approaches: the *multiple-layer* proposals and the *single-rate* approach. While it is better to use a multiple-layer approach if best-quality delivery to a very different set of receivers is the concern [8], single-rate protocols are a better fit for keeping the sender and the encoding of the data simpler [2,9]. In addition, single-rate protocols are generally easier to understand and analyze because of their similarity with unicast protocols. Hence, in this paper we focus on a single-rate algorithm.

There exist two main strategies aimed to avoid network congestion suffered with the single-rate multicast streaming protocols, both borrowed from unicast streaming solutions and based on the robust congestion control mechanisms embedded in TCP. One of the options is to rely on an equation-based algorithm, like in [2], that, by using an approximate throughput formula for TCP as a nonlinear control rule, tries to track the same transmission rate that a real TCP connection would have between the two endpoints. However the equation-based protocols react too conservatively to changes in the network conditions and do not efficiently use all the available bandwidth in dynamic conditions [10]. The alternative pursued by TCP *emulators* is to reproduce directly the behavior of TCP. In this case, and considering a multicast transmission paradigm, the main problem is getting enough feedback information (acknowledgments) from all receivers without flooding the sender. One possible solution is using complicated algorithms for aggregating feedback from the receivers like in [11]. Another possibility, chosen in our work, is to use a representative of all the receivers [9,

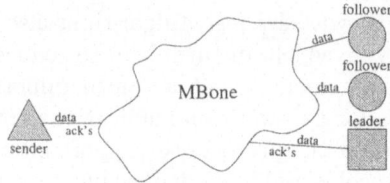

Fig. 1. Main elements present in a VLMCC transmission.

12]. This schema does not necessarily need any help from routers in order to be scalable and avoids completely the threat of *feedback implosion*. Furthermore, unlike [9,12] where non-representative receivers send a NACK packet whenever they detect a packet loss, in our proposed protocol these receivers only send feedback when they consider that they should take the role of the representative host, reducing, even more, the feedback traffic.

Another problem we faced in our design was devising a way for the protocol to be TCP-fair. To this end, we have preferred to emulate the TCP behavior between the sender and a representative host. This way, we can trust that, if the representative is well chosen, i.e., if it is the host with the worst connection among all, then the protocol will be TCP-fair along all the links. The specific TCP implementation chosen is TCP-Vegas [13], hence the name of our proposed protocol: VLMCC. TCP-Vegas was chosen for two main reasons: (1) It usually exhibits small variations in its sending rate, yielding a smooth throughput in the transmission; (2) With TCP-Vegas it is easier for the receivers to analyze the network conditions in a given instant. If the sending rate fluctuates amply, like in other TCP implementations, it becomes more difficult to know what the long term sending rate is, and this would force the receivers to delay their responses to congestion until a reliable measure of the sending rate had been obtained.

The rest of this paper is organized as follows. Section 2 presents the protocol architecture and its behavior. In Section 3 simulation results are shown. Our conclusions are exposed in Section 4.

2   Protocol Description

2.1   High Level Overview

VLMCC works by emulating a TCP-Vegas connection between the *sender*, represented with a triangle in Fig. 1, and the host experiencing the worst network conditions, called the *leader*. The rest of the hosts, the *followers*, are passive most of the time and they simply monitor the network conditions, asking the sender to become the new leader if they detect that their conditions are worse than those of the *leader*.

Unlike PGMCC [9] where the sender is in charge of detecting hosts in worse conditions than those the leader is experiencing, in VLMCC the receivers themselves are able to detect this situation and report it to the sender. This frees the

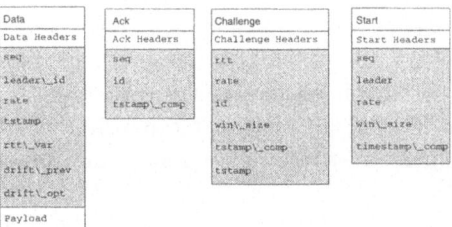

Fig. 2. Information interchanged between the sender and the rest of the hosts in the transmission. The area in grey holds the data needed by VLMCC.

server of some load and completely avoids the problem of the sender not electing the worst receivers because of NAKs suppression.

We have depicted in Fig. 2 the different packets interchanged between the sender and the rest of the nodes participating in the transmission. *Data* packets and *Start* packets are sent by the sender, *Ack* packets are sent by the session leader and *Challenge* packets can only be sent by follower nodes. In the following sections, we will show what the information conveyed in those packets is for.

2.2 The Sender-Leader Pair

The Sender-Leader pair is responsible for emulating the behavior of a TCP-Vegas connection running between those two nodes. For the sake of this discussion, we will assume that the leader has already been elected.

To accomplish their mission the Sender-Leader Pair uses the *Ack* packet and three fields of the *Data* packet (cf. Fig. 2): seq, leader_id and tstamp. The other fields are just information for the rest of the nodes. seq is a sequential number used to differentiate the packets. leader_id is the address of the current leader, and it is used by the receivers to know if they have been chosen as the leader responsible to emit an acknowledgment for the packet. The id field in the *Ack* packet is there just to prevent errors and contains the address of the node that emits the *Ack*.

The sender acts like a simplified TCP-Vegas server that does not provide any guarantee about delivery of the packets neither has to bother with retransmissions. In contrast, the job of the leader is much easier, for it only has to answer with an ack to every packet it receives. Using the variations in the RTT to detect incipient congestion, as TCP-Vegas does, VLMCC is able to react lowering its transmission rate before packet losses are likely to happen, avoiding drastic reductions in the window size. Therefore, to properly emulate the TCP-Vegas behavior, the sender must maintain a congestion window (W), and estimations of both the RTT and its variance ($\widehat{rtt}$ and $\hat{\sigma}^2_{\text{rtt}}$ respectively). Both $\widehat{rtt}$ and $\hat{\sigma}^2_{\text{rtt}}$ are estimated using exponential moving averages according to (1) and (2), where g and g_{σ^2} are gains and rtt is the value of the RTT observed by the last acknowledged packet. We have chosen, rather empirically, $g = g_{\sigma^2} = 0.1$.

$$\widehat{rtt} \leftarrow (1 - g)\,\widehat{rtt} + g \cdot rtt \tag{1}$$

$$\hat{\sigma}^2_{\text{rtt}} \leftarrow (1 - g_{\sigma^2})\,\hat{\sigma}^2_{\text{rtt}} + g_{\sigma^2}\left|\widehat{rtt} - rtt\right| \qquad (2)$$

For measuring the rtt, the sender could just store the time it sends every packet and compare it with the time it receives its corresponding ack, but in doing so, the time spent by the receiver generating the ack would be part of the measured rtt. To avoid this bias the sender timestamps the *Data* packet (tstamp field) and the leader echoes a corrected value back in the *Ack* packet (tstamp_comp field). Instead of just copying the tstamp into tstamp_comp, leaders add the time spent by the receiver processing the packet to it. This way, the sender obtains a more accurate measure of rtt that accounts only for the time the packet actually spent in the network.

Every $\widehat{rtt}$ the sender adjusts its congestion window according to the variations observed in the RTT. To this end, it compares the obtained throughput ($packetsInTravel/\widehat{rtt}$), in packets per RTT, with the maximum expected of ($W/base\_rtt$), where $base\_rtt$ is the minimum rtt measured in the transmission. If the difference is below a certain threshold α,[1] the sending rate can still be safely increased, and so W is increased in one packet. When the difference is considered too high, i.e., when it is above a certain level β, the connection is starting to cause congestion, and W is decreased in one packet. In any other case, W is kept constant.

The sender also modifies W whenever packet losses are detected (packets whose acks do not arrive in $\widehat{rtt} + 4\hat{\sigma}^2_{\text{rtt}}$ seconds after they have been sent[2]) during a RTT period. When such a loss is detected, the value of W is halved. Other packet losses occurring in the same RTT would be ignored, as the decrease in the congestion window size has not taken effect for those packets yet.

With all of this taken into account, a VLMCC sender can send packets to the network provided all of the following conditions hold true:

1. The sending application has pending packets to send.
2. The number of packets in travel is less than the congestion window size.
3. Some time has passed since the previous burst of packets was transmitted. This condition is a consequence of the fact that TCP-Vegas and VLMCC try to avoid sending long bursts of packets to the network. In our protocol, we have limited the burst length to just two packets, after which the sender has to wait the delay dictated by (3) before sending the next packet. In any case, a full window of packets can still be sent in just half the time of an RTT.

$$delay = \widehat{rtt} \cdot \frac{MaxSegment}{W} \qquad (3)$$

---

[1] The difference is first normalized dividing $\widehat{rtt}$ over $base\_rtt$, so α and β can be measured in bytes or packets and not in bytes per second or packets per second. This lets the configuration parameters do not depend on the physical capabilities of the underlying network, such as the links bandwidth or their length.

[2] This is the same value as the TCP retransmission timeout [6].

2.3   The Followers

The followers are nodes who are experiencing better network conditions than the *leader*. So, their job is to monitor network conditions and, provided such conditions worsen, warn the sender requesting to replace it. We call this *challenging the leader*, and the procedure involves sending a *Challenge* packet.

The tricky part is, of course, monitoring the network conditions, because the followers are forbidden to send any regular feedback as that would render the protocol unscalable. So, the followers estimate the network conditions by calculating the congestion window size a TCP-Vegas connection would have in use under the same circumstances.

Followers apply the same algorithm described in Section 2.2, but with one additional difficulty: since they cannot send regular data to the *sender*, they cannot know directly the value of *rtt*. The solution proposed here consists on asking the sender for a little help and not calculate *rtt* directly, but rather its variation (Δ_{rtt}) between two consecutive packets.

When a node joins a transmission it must get a first estimation of *rtt*. For this it sends a special timestamped packet to the sender that is always answered. Receivers use a *Challenge* packet for this, but only two fields are filled: id and tstamp. id carries the identity of the node that sends the *Challenge* while tstamp is used by the receiver to measure *rtt*. The answer comes in the form of a *Start* packet, that carries the following info: the seq number of the next expected packet, leader which is the address of the leader receiver, rate that is the current sending rate of the sender, win_size as the current congestion window size and finally timestamp_comp, which is used to measure *rtt*.

After this initial estimation, the follower adjusts *rtt* according to

$$rtt \leftarrow rtt + \Delta_{\text{rtt}} \tag{4}$$

every time two consecutive packets are received. With this *rtt* value, the estimation $\widehat{rtt}$ can be computed using (1). But how is Δ_{rtt} calculated? The sender just provides the needed information in every packet by inserting the time since the previous packet was sent in the drift_prev field of *Data* packets. So

$$\Delta_{\text{rtt}} = now - (lastArrivalTime + drift\_prev). \tag{5}$$

One remarkable side benefit of calculating Δ_{rtt} that way, is that only takes into account the variations suffered in the direction data is travelling. This makes the protocol robust against other connections causing congestion in the return path. If VLMCC reacted to congestion in the return path, not only it would not alleviate the congestion, but it would also affect the performance adversely.

With this knowledge, the follower can calculate a value for W that it will indirectly use for deciding when to request the leadership. In every packet sent by the sender there is information about the current server sending rate in the rate field ($W_{\text{sender}}/\widehat{rtt}_{\text{sender}}$). Followers use this information to calculate a moving average of the difference between their expected rate ($W_{\text{follower}}/\widehat{rtt}_{\text{follower}}$), and the rate announced by the sender, where W_{sender} is the window size calculated by the sender, $\widehat{rtt}_{\text{sender}}$ is its RTT estimation and W_{follower} and $\widehat{rtt}_{\text{follower}}$ are analogous values calculated by the follower host. A *Challenge* packet is generated by the followers every time this difference increases and it is larger than the equivalent of increasing W_{follower} with N_{extra} packets; that is, if

$$\frac{W_{\text{sender}}}{\widehat{rtt}_{\text{sender}}} > \frac{W_{\text{follower}} + N_{\text{extra}}}{\widehat{rtt}_{\text{follower}}}. \tag{6}$$

The N_{extra} value is a tradeoff between having a transmission that is completely fair in all the network links with competing flows, but with a high frequency of leader changes, and one that may take more bandwidth in some links, but is much more stable. In any case, the amount of extra bandwidth allocated is bounded by N_{extra}, so VLMCC can be adjusted to be as much fair as needed.

The *Challenge* packet is filled with information about the current $\widehat{rtt}$ (`rtt`) as measured by the receiver, the calculated rate (`rate`) and the follower's window size (`win_size`). If the sender accepts the challenge, it will use the follower's window size immediately as the new window size, and it will assume all packets sent before the new leader election happened have been acknowledged, even if this is not the case. It would make no sense to react to losses in those packets, when the sending rate, RTT and leader have changed.

In any case, it will be up to the sender to either accept the challenge or ignore it. For example, the sender can decide to ignore a challenge if a new leader has been elected in the last RTTs to let the situation stabilize before revoking that new leader. If the new sending rate would be too low, the sender can also decide not to attend the challenger demands. The challenger can resolve to abandon the transmission if it is not able to cope with the minimum required rate.

2.4 Leadership Loss

As stated in Section 2.1, VLMCC needs the presence of a host responsible of acknowledging every packet sent by the sender. If this leader host is missing, the sender must manage somehow to encounter a substitute. In case the protocol did not provide a mechanism for selecting a new leader, the congestion window size at the sender would drop to 1, and packets would only be sent due to timeouts. Moreover, as probably every host in the transmission would be able to admit higher throughput, no challenges would be produced, and the no-leader situation would perpetuate itself. To avoid this deadlock, the protocol needs to solve two problems, the first is detecting a leadership loss, and the second is being able to recover from it without causing *feedback-implosion*.

There is obviously no reliable way to know that the leader host is down or that it has dropped its connection, because in the first case it may not be able to warn the sender. For this reason VLMCC uses a heuristic to detect leadership loss. The sender assumes that the leader is not responsible when more than a whole window of packets is sent without hearing a single acknowledgment.

Once the leader loss has been detected the sender has to elect a new leader. For this, it ignores the fact that it is not getting acknowledgments and keeps on sending a whole window of packets during each RTT, so that the connection throughput does not degrade unnecessarily. As it is a fact that all the nodes were

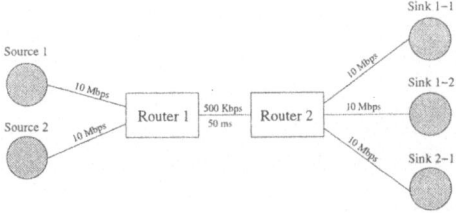

Fig. 3. Simulation topology (delays for links for which no delay is specified is just 0. Higher delays could be used, but we were interested in the bottleneck sharing behavior). All the routers use simple FIFO queues.

not experiencing worse network conditions than the leader, the sender increases the window size in one packet after each RTT to provoke a *challenge* from some host in the network. Once this *challenge* arises, the sender can safely choose the challenger host as the new leader.

3 Experimental Results

We have made several tests to our implementation of the VLMCC protocol with the help of the *ns-2* [14] network simulator. Throughout this section we will discuss the most illustrative experiments that we have performed. We will try to show tests for validating most of the desired properties listed in Section 1.

We have represented in Fig. 3 the topology used in the first simulation. We will use this topology to test some basic fairness properties of the protocol. The network shown consists of five end node hosts engaged in two different transmissions. One from *Source 1* to both *Sink 1-1* and *Sink 1-2* and the other one from *Source 2* to *Sink 2-1*. We will be interested in observing how the connections share the bandwidth of the bottleneck between both routers.

The first simulation presented here tries to examine the inter-protocol fairness properties of VLMCC; in particular we are interested in the fairness degree of VLMCC when confronted with TCP flows, as those represent the predominant traffic in the current Internet. To this end, we have run a simulation using *Source 1* as the VLMCC source and *Sink 1-1* and *Sink 1-2* as its receivers, and *Source 2* as the sender of an TCP flow with destination *Sink 2-1*.[3]

Fig. 4(a) shows the obtained results. The simulation has been repeated several times changing the relative start time of both flows, and the averaged throughput has been represented. The VLMCC session manages to equitably share the bottleneck bandwidth with the TCP flow. Only at session startup, when both flows have not still reached their steady state, there are little differences in the obtained bandwidth.

In Fig. 4(b) we show other interesting data obtained in the previous simulation. Instead of representing the received data, we depict the amount of data lost

---

[3] Unless otherwise noticed, the packet size for all simulations is 1 000 bytes and the TCP version employed is TCP Reno.

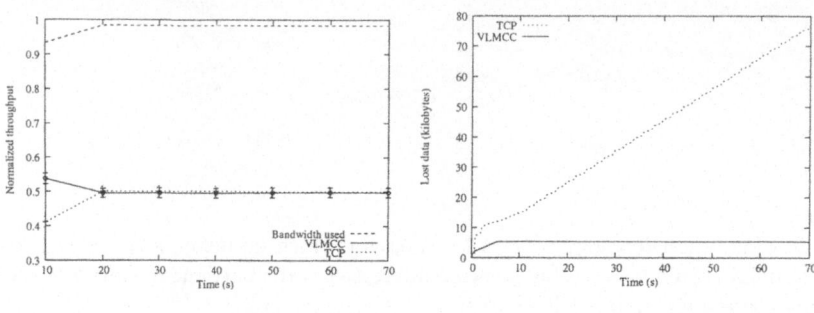

(a) Throughput obtained. Error bars for a 95% confidence interval.

(b) Accumulated data lost.

Fig. 4. One VLMCC connection and a TCP flow sharing a single bottleneck.

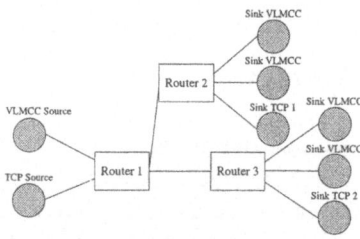

Fig. 5. Simulation topology used to test the stability of VLMCC. All the links have a bandwidth of 10 Mbps and 1 ms delay, except the links between the routers, that have 1 Mbps bandwidth. All the routers use simple FIFO queues.

by each flow. Because TCP uses packet losses as the means to detect congestion and adjust its sending rate, it suffers packet losses all along the simulation; meanwhile, VLMCC manages to find an appropriate sending rate after some seconds of operation, and does not suffer from any packet loss from that instant. Because VLMCC is designed with multimedia streaming requirements in mind, the fact that it suffers from less packet losses is greatly important, as that is directly related to the perceived transmission quality. In real-time streaming, packet losses can not usually be recovered, so it is important to avoid them before they happen. The pro-active window management algorithm employed by VLMCC manages to accomplish this goal fairly well, as shown.

The following simulation tests the ability of VLMCC to adapt quickly to changes in the available network resources. Fig. 5 shows the topology used in the test. During the simulation a VLMCC session is run originating from the node VLMCC *Source* and having all VLMCC *Sink* as destinations. A TCP connection is established between hosts TCP *Source* and *Sink* TCP *1* in the time intervals [10, 25] and [41, 60]. Finally, another TCP connection is established between nodes TCP *Source* and *Sink* TCP *2* from second 27 until second 45. The

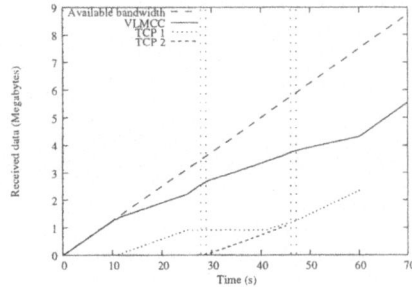

Fig. 6. Accumulated data for both a VLMCC connection and two TCP sessions.

purpose of this simulation is to test if, as the links between the routers become congested, the VLMCC session leader changes and VLMCC is able to adapt and share the bandwidth fairly in both links.

The results are shown in Fig. 6. The vertical lines represent the instants when a new leader is elected in the VLMCC session. The other curves show the amount of properly received data by each connection. The slope of these curves represents the obtained throughput. At the beginning of the simulation one of the VLMCC receivers connected to *Router 2* is selected as the leader, so, when the first TCP flow starts at second 10 no leader change is needed, as both the VLMCC leader and the TCP flow share the same bottleneck. In the graph it can be noticed how the throughput of the VLMCC session decays. At second 25 when the first TCP flow stops, the VLMCC throughput increases again, until second 27 when the second TCP flow starts both VLMCC receivers under *Router 3* try to become the leader, to be able to adapt the throughput to the bandwidth in the new bottleneck. When, at second 41, the first TCP flow restarts, VLMCC changes nothing, as both bottlenecks are in the same conditions, but, when the second TCP flow ends (second 45), one of the receivers under *Router 2* becomes the leader again to be able to share the bandwidth in the link *Router 1-Router 2* in a fair manner. As seen, VLMCC reacts quickly in dynamic scenarios moving the leader location to a host in the congested part of the network.

In the last simulation we want to show the behavior of the protocol in a realistic scenario (similar to one used in [15]), such as the one depicted in Fig. 7. Here, a VLMCC session is established between one long-lived VLMCC source and three different sinks. At the same time a TCP connection, some UDP traffic and some HTTP sessions populate the network. Fig. 8(a) shows that even in this complicated situation VLMCC is capable of behaving in a TCP-fair manner against the competing TCP flow. We have chosen to represent the accumulated data received in order to see more clearly the results in the long range, given that the instantaneous throughput is just too noisy.

In Fig. 8(b) we plotted the congestion window size of both the TCP flow and the VLMCC sender. Although both experience noticeable variations due to the congested state of the network, the oscillations of the VLMCC window are

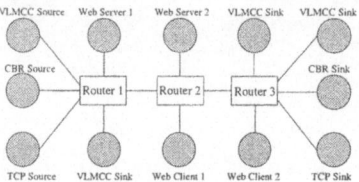

Fig. 7. Simulation topology for testing the VLMCC performance in a complex scenario. All the links have a bandwidth of 100 Mbps and 24 ms delay, except those that interconnect the routers, that have a bandwidth of 10 Mbps and 1 ms delay. All the routers use simple FIFO queues.

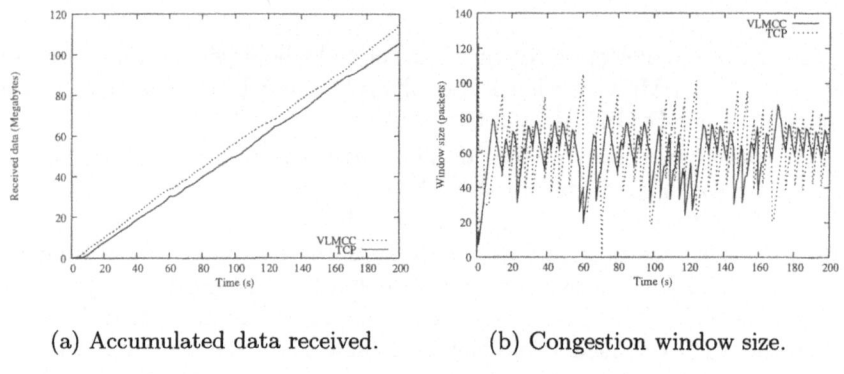

(a) Accumulated data received.   (b) Congestion window size.

Fig. 8. Bandwidth sharing between VLMCC and TCP sessions.

less severe than those of TCP. Although in isolation VLMCC would probably be able to reach a constant window size, the fact that the network is congested makes VLMCC suffer from packet losses and so its behavior is closer to that of TCP-Reno.

4   Conclusions

We have presented a multicast congestion control protocol suitable for the streaming of real-time multimedia data. The main innovations of the protocol are the use of the variations in the round trip time to detect congestion at the receivers, and the decentralization of the leader election process.

Because of the use of variations of the RTT to detect congestion VLMCC is able to react quickly to congestion. This is important, because slow reaction has been shown to loose throughput over faster ones. This is one of the main drawbacks of equation-based protocols [10].

The decentralization of the leader election process greatly reduces the amount of feedback that receivers send back to the sender, up to the point that there is no need to regulate it to avoid feedback-implosion. Unlike other schemata where the sender is in charge of selecting the leader, VLMCC receivers can detect when they need to become the leader, reducing the load suffered by the sender.

Simulation results show that the protocol behaves in a TCP-fair manner and thus it should be safe for deploying in the Internet. Finally all the configuration parameters of the protocol are time-independent, what makes unnecessary reconfiguring it for different link bandwidths or delays present in the network.

References

1. Tan, W., Zakhor, A.: Real-time internet video using error resilient scalable compression and TCP-friendly transport protocol. IEEE Transactions on Multimedia **1** (1999) 172–186
2. Widmer, B., Handley, M.: Extending equation-based congestion control to multicast applications. In: Proceedings of the 2001 conference on applications, technologies, architectures, and protocols for computer communications, ACM Press (2001) 275–285
3. Rejaie, R., Handley, M., Estrin, D.: RAP: An end-to-end rate-based congestion mechanism for realtime streams in the internet. In: Proceedings of the IEEE INFOCOM. Volume 3. (1999) 1337–1345
4. Feamster, N., Bansal, D., Balakrishnan, H.: On the interactions between layered quality adaptation and congestion control for streaming video. In: 11th International Packet Video Workshop. (2001)
5. Yang, Y.R., Lam, S.S.: Internet multicast congestion control: A survey. In: Proceedings of ICT 2000, Acapulco, Mexico, ICT 2000 (2000)
6. Jacobson, V.: Congestion avoidance and control. In: Symposium proceedings on Communications architectures and protocols, ACM Press (1988) 314–329
7. Floyd, S., Fall, K.: Promoting the use of end-to-end congestion control in the Internet. IEEE/ACM Transactions on Networking (TON) **7** (1999) 458–472
8. McCanne, S., Jacobson, V., Vetterli, M.: Receiver-driven layered multicast. In: ACM SIGCOMM. Volume 26,4., New York, ACM Press (1996) 117–130
9. Rizzo, L.: pgmcc: a TCP-friendly single-rate multicast congestion control scheme. In: Proceedings of the conference on Applications, Technologies, Architectures, and Protocols for Computer Communication, ACM Press (2000) 17–28
10. Bansal, D., Balakrishnan, H., Floyd, S., Shenker, S.: Dynamic behavior of slowly-responsive congestion control algorithms. In: Proceedings of the 2001 conference on Applications, technologies, architectures, and protocols for computer communications, ACM Press (2001) 263–274
11. Rhee, I., Balaguru, N., Rouskas, G.N.: MTCP: Scalable TCP-like congestion control for reliable multicast. Computer Networks: The International Journal of Computer and Telecommunications Networking **38** (2002) 553–575
12. DeLucia, D., Obraczka, K.: Multicast feedback suppression using representatives. In: INFOCOM (2). (1997) 463–470
13. Brakmo, L.S., O'Malley, S.W., Peterson, L.L.: TCP Vegas: New techniques for congestion detection and avoidance. ACM SIGCOMM Computer Communication Review **24** (1994) 24–35
14. NS: ns Network Simulator (2003) http://www.isi.edu/nsman/ns/.
15. Bansal, D., Balakrishnan, H.: TCP-friendly congestion control for real-time streaming applications. MIT Technical Report, MIT-LCS-TR-806 (2000)

The Role of Information Update in Flow Control

Eitan Altman[*][1], Tamer Başar[**][2], and Naceur Malouch[3]

[1] INRIA, BP 93, 06902 Sophia Antipolis Cedex, FRANCE
[2] University of Illinois, 1308 West Main Street, Urbana, IL 61801-2307, USA
[3] LIP6, Université Pierre et Marie Curie (Paris VI), 75015 Paris - FRANCE

Abstract. A common feature of congestion control protocols is the presence of information packets used to signal congestion. We study here the question of how frequently such protocols need to generate information packets in order to optimize their performance. Through a number of models, we identify and quantify different types of effects of the frequency of generating information packets. We consider both TCP-type protocols, in which controlling the frequency of information packets is done through static or dynamic delayed ACK options, as well as ATM type flow control, where the optimal time spacing between the generation of network management packets is computed. We show how the spacing between information packets influences the throughput and stability of the system.

Keywords: TCP, Delayed ACK, Throughput, Stability, Flow Control.

1 Introduction

Congestion and flow control protocols make use of special control packets to indicate to the traffic sources actions to be taken in order to adapt the transmission rates to the available bandwidth. In TCP/IP these are the acknowledgement packets; in the ABR (Available Bit Rate) class of ATM these are the resource management packets. Even when the size of these packets is significantly smaller than that of the data packets, they may still compete over network resources with the data packets and thus decrease the amount of resources available to them. In many cases, they may require an amount of resource much larger than what their size would suggest. For example, when using the IEEE 802.11 MAC protocol, each data as well as ACK packet of the TCP flow requires the same (large) overhead of three link layer packets (RTS, CTS and a link layer ACK). Even in the absence of such overheads, the processing time of an ACK at the destination might require an additional overhead that can be pretty large with respect to the transmission time when very high speed networks are considered. We therefore raise the question of with what frequency flow control protocols should send control packets.

[*] Work of this author was partially supported by the EuroNgi Network of Excellence.
[**] Research of this author was partially supported by the NSF Grant CCR 00-85917.

N. Mitrou et al. (Eds.): NETWORKING 2004, LNCS 3042, pp. 562–573, 2004.
© IFIP International Federation for Information Processing 2004

We should note that TCP/IP already has the "delayed ACK" option that allows it to reduce the ACK frequency from one ACK for every received packet to one ACK every $d = 2$ received packets. Note that ACKs could also be filtered within the network (see e.g. [4] and references therein). However we shall not investigate here the question of how ACKs should be thinned.

We introduce three frameworks in which to study the optimization of the frequency of control packets. The first is the case of TCP/IP traffic sources with routers using drop tail queues. A simple mathematical model is derived for optimizing the amount of ACK thinning at the destination so as to maximize the system's throughput. A simulation study validates the conclusions we obtained from the mathematical model. In the second framework, we study the dynamics of an AIMD (additive increase multiplicative decrease) flow control interacting with a RED type buffer. We model the system's dynamics through a system of delay-differential equations, and study the stability of the system as a function of the frequency of ACKs. We finally propose an abstract model for an optimal rate control with sampled delay information, similar to models used for rate control in the ABR class in ATM [1]. We adopt a linear-quadratic model and optimize the time between two successive packets in information feedback.

2   A Fixed-Point Method to Model ACK Thinning

In this section, we use the expression for TCP throughput developed in the literature and standard queueing models to quantify the impact of the delay factor d on the throughput. The throughput T of a TCP connection can be approximated by [7]

$$\frac{1}{RTT\sqrt{\frac{2\,d\,p}{3}} + RTO\,\min\left(1,\,3\sqrt{\frac{3\,d\,p}{8}}\right)p\left(1 + 32p^2\right)} \tag{1}$$

where p is the loss probability of TCP packets, RTT is the round-trip delay experienced by the TCP connection, and RTO is the retransmission timeout.

We start by modeling the network as a bidirectional link. Each direction of the link is modeled by a queue system. Two sets of N symmetric TCP sources send data from both end-points of the link. Each source of the first set connects to a receiver that does not belong to the second set and vice versa. Thus, in each queue TCP packets and ACK packets from different connections are multiplexed in the same queue and served by the same server. Using this model, we assume that losses that occur in the system are only due to congestion, i.e. buffer overflow.

We denote by α the "effective" size of ACK packets and by Z the size of TCP packets. The parameter α would not only model the actual size of ACK packets but also the eventual overheads introduced in the processing time. In order to develop a tractable model, we need a simple formula that relates the throughput at the buffers to the losses that will be experienced there. To that end we shall assume that the packet arrival process at each queue can be approximated by a Poisson process. (We note that the validity of this approximation in a similar context has been discussed and examined in [3].) Then, the packet loss probability is the loss probability of an $M/G/1/K$ system. Note that the loss probability seen by ACK packets or TCP packets is theoretically the same because of the PASTA property.

Below we propose two modeling approaches for the service time: the exponential service time which provides a simple expression for the losses but gives a rough approximation, and the deterministic service time (whose duration varies according to whether it is an ACK or a data packet) which gives a better approximation but with a more complex expression for the losses. In the deterministic model there are thus two possible values of service times: α/ZC and $1/C$, where C is the capacity of the link in TCP packets per unit of time.

The Exponential Service Time Case. In this case, the service time of packets is exponentially distributed. The loss probability is then given by the loss probability of an $M/M/1/K$ system:

$$p = \rho^K \frac{1-\rho}{1-\rho^{K+1}} \qquad (2)$$

where ρ is the load of the system and it is computed as follows:

$$\rho = \left(NT + \frac{NT}{d}\right)\left(\frac{1}{d+1}\frac{\alpha}{ZC} + \frac{d}{d+1}\frac{Z}{ZC}\right) = \frac{1}{C}(NT + NT\frac{\alpha}{dZ}) \qquad (3)$$

We use the fixed-point method to solve numerically the system of equations (1), (2) and (3). The advantage of this simple model is that we can compute the loss probability and thus the throughput for large values of K and ρ.

The Deterministic Service Time Case. Here, we modify only the assumption that the service times are exponentially distributed and thus only equation (2) is replaced by the expression of the loss probability of an $M/G/1/K$ queue [8]:

$$p = \frac{1 + (\rho - 1)f}{1 + \rho f}, \text{ where } f = \frac{1}{2\pi i}\oint_{D_r}\frac{1}{G(s)\,s^{K-1}}\,ds \qquad (4)$$

D_r is any circle in the complex plane with center 0 and radius r chosen small enough so that all the zeros of the function $G(s)$ are outside D_r, i.e. $r < |G(z)| \; \forall \; z$ such that $G(z) = 0$. The complex function $G(s)$ is defined as $G(s) = LST(b(\lambda(1-s))) - s$ where $LST(b())$ is the Laplace Stieltjes Transform of the service time distribution $b()$:

$$LST(b(s)) = \int_0^\infty b(t)\,e^{-st}\,dt = \underbrace{\frac{1}{d+1}e^{-(\alpha s/ZC)}}_{ACK} + \underbrace{\frac{d}{d+1}e^{-s/C}}_{TCP} \qquad (5)$$

The parameter λ is the total arriving rate at the entrance of each queue, which is equal to $(NT + NT/d)$. Computation of f is detailed in [2].

Again, we solve the three equations (1), (4) and (3) numerically, and we compare the results to those obtained using the exponential time distribution.

Numerical Results and Simulations. Figure 1 depicts plots of the throughput of TCP as a function of the delay factor d. We consider an "effective" size of TCP data packet of 500 bytes and various "effective" sizes of ACKs, ranging between $\alpha = 40$ and $\alpha = 250$. (as mentioned in the introduction, the difference between actual and effective size of a packet is that additional overhead may be added to its real size due to other protocols of other layers; in addition, processing an ACK at some nodes may take longer than its relative size with respect to a TCP data packet). The queue size is 20 packets and the round-trip time RTT is 200 ms. We set C to 125 TCP packets/s, and N to 1 since the throughput depends only on the fraction C/N. The plots were generated numerically using the more precise model of deterministic service times. The figure shows that

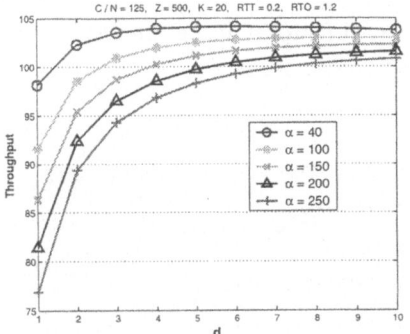
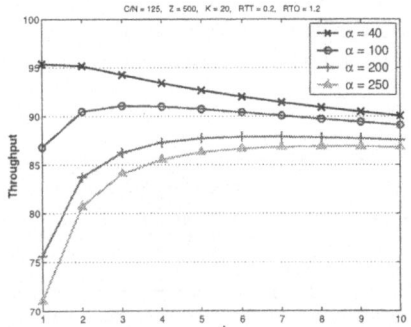

Fig. 1. Effect of d on the throughput for various ACK sizes — the deterministic service time case

Fig. 2. Effect of d on the throughput for various ACK sizes — the exponential service time case

for the small ACK size 40, spacing the ACK (using $d > 1$) results in a small improvement of the performance (6%). However, the throughput is maximized when $d = 4$. For higher values of "effective" ACK size ($\alpha \geq 100$), we see an improvement of 12% to 31%, with the optimum obtained for example at $d = 5$ for $\alpha = 100$. Using a value of $d > 2$, which is the default value in TCP, results in a relative gain of around 10%. We have observed similar trends with larger link capacities C (larger congestion window) and with much smaller buffer sizes K (higher loss probabilities). When $K \geq 60$, the throughput approaches the value obtained simply by solving $\rho = 1$, which gives $Thpt = \frac{C}{N(1+\alpha/(dZ))}$, and the relative gain can reach 50%.

Figure 2 depicts plots of the throughput vs. d with the same parameters as in the previous scenario, but using the exponential service time approximation. In this case, the packet size can represent the average of the "effective" packet size in the network. This model is useful when the packet size in the backbone is variable and the exact distribution is unknown. The figure shows that the trend of the throughput is almost the same, except for the smallest ACK size 40 where spacing the ACK results in deterioration of the performance for all values of

566   E. Altman, T. Başar, and N. Malouch

$d > 1$. For larger ACK sizes ($\alpha \geq 200$), once again, the gain of delaying ACKs goes beyond 20%.

Next, we perform two sets of simulations to study the robustness of the models described herein. First, we study the effect of the number of connections N. Then, we look at the effect of the receiver timeout[1] which is not considered by the two models. We use a similar network configuration as the one described in the model of Section 2, and we add N access links of capacity 1 Mbs each to the bottleneck link. The average round-trip propagation delay ($\approx RTT$) is set to 200 ms.

In the first set of simulations, we vary the number of connections N from 30 to 200, and we vary the bottleneck capacity C in order to keep the ratio C/N constant at 125. We fix the ACK size to $\alpha = 100$. In this set of simulations, we use the default value of the receiver timeout which is set to 100 ms. Figure 3 plots the average TCP throughput vs. the delayed factor d, using the two models and simulation traces for $N = 30$, 50, 100 and 200.

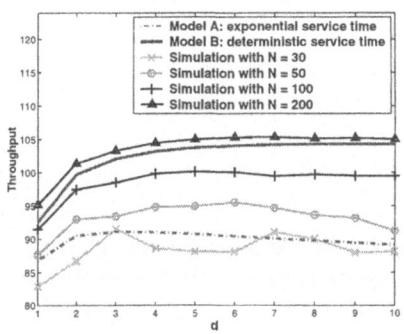

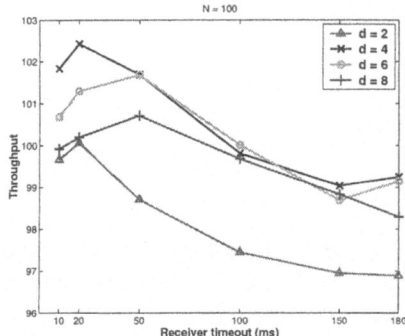

Fig. 3. Simulation results vs. numerical results

Fig. 4. Effect of the receiver timeout on the TCP throughput

The main observation we make is that as N increases, the relative error induced by the deterministic model decreases notably. This is because the resulting process of multiplexing the TCP connections approaches a Poisson process [3].

Next, we examine the impact of receiver timeout on TCP throughput. We use the same parameter values as in the previous simulations, and set $N = 100$. We vary the receiver timeout from $t_1 = 10$ ms to $t_2 = 180$ ms. Note that t_1 is just larger than the minimum inter-arrival time of two consecutive TCP packets sent in the same window, and t_2 is just less than RTT. Figure 4 depicts plots of the average TCP throughput as a function of the receiver timeout for various values of d. We see clearly that when the timeout is large, then the throughput

---

[1] The receiver timeout is used to send an ACK for arriving packets even before d packets arrived at the receiver if the time since the unacknowledged packet arrived exceeds the timeout.

is reduced. In the cases $d = 2$ and $d = 4$, the throughput is maximized when the timer is equal to 20 ms. For $d = 6$ and $d = 8$, the throughput is maximized at a timer of 50 ms.

Here, the gain in the throughput is relatively small ($\approx 3\%$), but this is because the average window size is also small (≈ 20). For larger window sizes, for example when the delay-bandwidth product is large, the gain is more significant. However, setting a small value for the timeout is risky since the packet inter-arrival time depends on the cross traffic along the path and hence could vary over time.

More generally, it is difficult to find a constant timeout that is adequate for all network scenarios. Another alternative is to compute dynamically the timeout using a similar method as the one used by the sender to compute the retransmission timeout.

To summarize, it is worth using a delay factor $d > 2$, particularly for long TCP connections and when the window size is large. In fact, the improvement obtained from reducing the number of ACKs in the network is more significant than the decrease in the throughput due to the lack of information update.

3   Control Model of TCP Traversing a RED Buffer

After focusing on the quantitative impact of the rate of information on the throughput, we next study how stability conditions are influenced by the rate of information packets. We assume that N symmetric persistent TCP connections share a bottleneck link of capacity C located close to the source, assumed to operate in the congestion avoidance regime. Let $W(t)$ be the window size of a connection at time t, and R the round trip delay (including queueing delay), which is assumed to be a constant (this assumption holds when queueing delays are much smaller than propagation delays). Let $p(t)$ be the loss (or marking) probability of the RED buffer at time t. Let $q(t)$ be the amount of buffered traffic at the queue at the bottleneck link. We consider a fluid approximation of the window size of TCP given by $\frac{dW}{dt} = \frac{1}{R} - \frac{W(t)W(t-R)}{2R}p(t-R)$, see [6]. This equation is obtained under the assumptions that (i) the delayed ACK mechanism is not used; and (ii) a new ACK is generated with each TCP packet that arrives at the destination, resulting in an increase of the window size by one unit every round trip time.

We shall now consider the possibility of using a (dynamic) delayed ACK approach in which an ACK is generated for every d TCP packets that arrive at the destination. d will be considered as a control variable and will thus be allowed to be time dependent. Both the increase rate as well as the decrease rate are divided by a factor d since the rate of ACKs that arrive at the sources is d times smaller. In particular indications for decreasing the rate (we assume that ACKs have marks indicating congestion) return less frequently. The window size evolution then becomes

$$\frac{dW}{dt} = \frac{1}{Rd(t-R)} - \frac{W(t)W(t-R)}{2Rd(t-R)}p(t-R).$$

The queue dynamics are given in [6] by $\frac{dq}{dt} = \frac{N}{R}W(t) - C$. We shall assume that not only the TCP packets have to queue but also the ACKs. To model the difference between the size of an ACK and that of a TCP packet, we assume that an ACK requires a fraction γ of buffer space required by a TCP packet. Under this assumption, the queue dynamics above are modified to $\frac{dq}{dt} = \frac{N}{R}W(t) - C + \frac{\gamma W(t-R)N}{d(t-R)R}$. The tradeoff that influences the choice of the control d will be the following: on one hand, when the window size is small, we may want d to be small so that the window size can grow quickly so as to achieve higher throughput. On the other hand, when the window size is large then we may want to increase d so as to limit the congestion due to ACKs in the bottleneck queue. We shall analyze in this section a linear control mechanism in which d has the form $d(t) = \alpha(1+\beta W(t))$, $\alpha > 0$, $\beta \geq 0$. Finally, we shall consider the RED buffer marking probability (ignoring the averaging of the queue size) in its linear operation regime: $p(t) = \eta_1 q(t) - \eta_2$. We now summarize the system's overall dynamics below:

$$\frac{dW}{dt} = \frac{1}{Rd(t-R)} - \frac{W(t)W(t-R)}{2Rd(t-R)}p(t-R) \tag{6}$$

$$\frac{dq}{dt} = \frac{N}{R}W(t) - C + \frac{\gamma W(t-R)N}{d(t-R)R} \tag{7}$$

$$d(t) = \alpha(1+\beta W(t)), \qquad p(t) = \eta_1 q(t) - \eta_2 \tag{8}$$

Equilibrium: The equilibrium point is obtained by equating to 0 the time derivatives in the previous equations. This leads to:

$$\frac{N}{R}W_o + \frac{N\gamma W_o}{d_o R} = C \Rightarrow W_o = \frac{CR}{N(1+\gamma/d_o)}$$

$$d_o = \alpha(1+\beta W_o), \qquad W_o^2 p_o = 2 \Rightarrow p_o = 2/W_o^2$$

$$p_o = \eta_1 q_o - \eta_2 \Rightarrow q_o = (p_o + \eta_2)/\eta_1$$

Hence the throughput is given by: $Thp = \frac{W_o}{R} = \frac{C}{N(1+\gamma/d_o)}$. The throughput is seen to increase with d_o.

A linearization of the dynamical system in a neighborhood of the equilibrium point yields

$$\frac{d\delta W}{dt} = -\frac{1}{RW_o d_o}\delta W(t) - \frac{1}{RW_o d_o}\delta W(t-R) - \frac{\eta_1 W_o^2}{2Rd_o}\delta q(t-R)$$

$$\frac{d\delta q}{dt} = \frac{N}{R}\delta W(t) + \frac{\alpha\gamma N}{d_o^2 R}\delta W(t-R),$$

where δ stands for the shifted version of variables in which the equilibrium value is subtracted (e.g. $\delta W := W - W_o$). Taking the Laplace transform of these equations we obtain

$$s\delta W(s) = -\frac{1}{RW_o d_o}\delta W(s) - \frac{e^{-sR}}{RW_o d_o}\delta W(s) - \frac{\eta_1 W_o^2}{2Rd_o}e^{-sR}\delta q(s)$$

$$s\delta q(s) = \left(\frac{N}{R} + \frac{\alpha\gamma N}{d_o^2 R}e^{-sR}\right)\delta W(s)$$

With $z = sR$, the stability condition is then given by requiring that all zeros of $g(z) = 0$ have strictly negative real parts, where $g(z)$ is given by

$$g(z) = z^2 + \frac{z}{W_o d_o}(1 + e^{-z}) + \frac{\eta_1 W_o^2 N}{2d_o}(1 + \frac{\alpha\gamma}{d_o^2}e^{-z})e^{-z} \tag{9}$$

The goal. One may identify two possible goals: (i) maximize the system throughput while keeping it stable, and (ii) for a given desired throughput, make the system "as stable as possible" by which we mean to choose the parameters so as to have the real part of the largest zero of $g(z)$ as negative as possible.

Note that whereas the throughput only depends on d_o, γ, N and C, and not directly on the values of $\alpha, \beta, \eta_1, \eta_2$, the stability regime does depend directly on α and η_1, and hence these parameters do enter into the optimization under goal (i). For the second goal, we may first compute d_o and then optimize stability with respect to the other parameters.

A numerical example. This numerical example is picked to show that the system could be unstable for meaningful choices of the values of the parameters: $\eta_1 = 0.001$, $\eta_2 = 0.02$, $\alpha = 1$, $d_0 = 3$. Further let $N = 5$, $\gamma = 250/500 = 0.5$, $RC = 100$ so that $RC >> q_o$. $W_o = \frac{CR}{N(1+\gamma/d_o)} = 100/(5 \times 1.166) = 17.142857$. Hence $p_o = 2/W_o^2 = 0.006806$, $\beta = (d_0/\alpha - 1)/W_o = 0.11666$ and $q_0 = \frac{p_0 + \eta_2}{\eta_1} = 26.80555$, which is indeed small with respect to RC. We can verify that $z = 0.09964430039 + 0.4704656808 * I$ is a zero of (9), and hence the system is unstable.

Using the Model: Impact on Stability. Examining the form of $g(z) = 0$ (whose solutions provide the stability condition), we observe the following:
- The stability condition is not a function of the round trip delay. This is seen directly from (6)-(8): if we scale time so that a time unit corresponds to a round trip time, we arrive at a new system of equations which do not involve R.
- From the form of the expression $g(z)$ we see that for the same values of $C, R, N, \gamma, \eta_1, \eta_2$ and for the same value d_o at equilibrium, the stability region can change according to the choice of the parameters α and β. $\beta = 0$ corresponds to a non-dynamic value of d (i.e. a value that does not change with W) in which case $\alpha = d_o$.
- We also see that for fixed $C, R, N, \gamma, \alpha, \beta$ and a fixed queue size q_o, at equilibrium, $g(z)$ will be influenced by η_1 (and hence the stability region). In fact $g(z)$ does not depend on η_2, but note that since we assume that R includes the mean queueing delay (which is proportional to q_o), this means that fixing q_o and η_1 already determines η_2.

In the following, we choose a scenario that shows clearly the impact of using a dynamic delayed ACK factor d on the stability of RED. We fix the parameters used in the numerical example. Then, we set $N = 14$, and vary α from 1 to 4. Since we fix d_0, β is deduced. Table 1 presents the results concerning the stability of the system.

Table 1. Impact of dynamic control on stability

$d0$	α	$\Rightarrow \beta$	The system is
3	1	0.33 ($W \nearrow d \nearrow$)	stable
3	2	0.08 ($W \nearrow d \nearrow$)	stable
3	3	0 (no control)	unstable
3	4	-0.04 ($W \nearrow d \searrow$)	unstable

The first observation we make is that when $\alpha = 3 (= d_0)$, the system is unstable, which means that when there is no dynamic variation of the delayed factor d, the system is unstable. However, for $\alpha = 1$ or 2, the system is stable. Besides, β is positive, which means that the linear control is correct ($d(t)$ is a non-decreasing function of $W(t)$). When $\alpha = 4$, β is negative and the system is unstable. These results corroborate the fact that with an increase of the TCP window, we should increase the delay factor d. Moreover, using an adequate adaptive control of d, we can improve the stability of the system.

4   Linear-Quadratic Approaches to Flow Control

We now study the problem of optimum choice of the time interval between successive transmission of information using a control theoretic approach within a linear-quadratic framework. Such frameworks have frequently been used for approximating rate-based flow control by ignoring the nonlinearity at queue boundaries (empty or full buffers). More precisely, we make four simplifying assumptions:

1. *Fluid approximation.* We replace a discrete number of packets by a continuous fluid.

2. *Linearized dynamics.* The network has linearized dynamics for the control of queue length; see (10) below: we neglect losses when the buffer is full, and we neglect the boundary effect of an empty queue. As discussed below, we set some desirable threshold on the queue length which we attempt to track as closely as possible so as to avoid large queues (which might lead to losses) or empty queues (which might result in loss of potential throughput).

3. *Bottleneck assumption.* All performance measures (such as throughput, delays, loss probabilities, etc.) are determined essentially by a bottleneck node.

4. *Information flow.* Information on the queue length is sent to the controller periodically, every $1/\lambda$ seconds.

We now introduce the model. Let $q(t)$ denote the queue length at a bottleneck link. Assume that the information packets use the same link and have priority over data packets. The link capacity available to data packets, C, is thus assumed to depend on λ: $C(\lambda) = C - \frac{a}{\lambda}$. We assume that other uncontrolled inputs share the buffer, and their total input rate is given by $C_1 + v(t)$ where C_1 is some constant and v represents a stochastic process with zero mean. The average rate of the controlled source is assumed to be $C_2 = C - C_1$. Thus we let $u(t) + C_2$ be the rate of the controlled input into the queue at time t. Then the queue length dynamics are given by

$$dq = udt + dv \qquad (10)$$

which is *idealized* because the end-point effects have been ignored. The objectives of the flow controller are *(i)* to ensure that the bottleneck queue size stays around some desired level $\overline{Q}$, and *(ii)* to minimize variations in the rates. The choice of $\overline{Q}$ and the variability around it have a direct impact on loss probabilities and throughput. We therefore define a shifted version of q: $x(t) := q(t) - \overline{Q}$, in view of which (10) now becomes $dx = udt + dv$. An appropriate local cost function that is compatible with the objectives stated above would be the one that penalizes variations in $x(t)$ and $u(t)$ around *zero* — a candidate for which is the weighted quadratic cost function: $x^2 + ku^2$. We seek an optimal control policy among those which choose $u(t)$ as a function of the queue length at the times when information is available: $u(t) = \mu(x(0), x(\lambda), ..., x(n\lambda))$, for $t \in [n\lambda, (n+1)\lambda)$, $\quad n = 0, 1, 2, ...$

LQG model. We assume that v is a zero-mean Brownian motion with incremental variance r. The expected average cost for a given policy μ and initial state x is defined as $J(x, \mu, \lambda) = \lim_{T \to \infty} \frac{1}{T} E_x^\mu \left[\int_0^T (x^2(t) + ku^2(t))dt \right]$ where E_x^μ is the expectation with respect to the probability measure induced by a policy μ and an initial state x. We first seek to obtain the optimal policy and value for a given λ: $J(x, \lambda) := \min_\mu J(x, \mu, \lambda)$.

Theorem 1. *[2] The optimal value of the LQG problem is independent of the initial state x and is given by $J(\lambda) = \frac{\lambda^2}{2} r + \sqrt{k} r$, and the unique policy that attains the minimum is given by*

$$u^*(t) = \mu^*(x) = -\frac{1}{\sqrt{k}} \exp\left(-\frac{t - n\lambda}{\sqrt{k}}\right) x(n\lambda) \text{ for } t \in [n\lambda, (n+1)\lambda). \qquad (11)$$

We thus conclude that as long as λ^2 is much smaller than $\sqrt{k}$, the value is quite insensitive to changes in the spacing λ. On the other hand, when it is much larger than $\sqrt{k}$, we see that the spacing of information packets has a huge impact on the performance: the cost grows quadratically in the spacing.

Optimal spacing of information for the LQG model. We obtained above the optimal control policy for a given parameter λ of spacing of information packets. Our next goal is to optimize λ. In addition to the cost $J(\lambda)$ which we have obtained, we assume that the flow control has some utility $U(\lambda)$ that can represent the utility for the average throughput $C - a/\lambda$ available for the data packets. We shall consider

1. A utility linear in the average throughput: $U(\lambda) = C - a/\lambda$.
2. A logarithmic utility: $U(\lambda) = \log(C - a/\lambda)$.

The cost to be minimized in both cases is $Z(\lambda) = J(\lambda) - \gamma U(\lambda)$. Straightforward minimization leads to the following:

Theorem 2. *The optimal information spacing is given by*
(i) linear case: $\lambda^* = \left(\frac{\gamma a}{r}\right)^{1/3}$ *(ii) logarithmic case:*

$$\lambda^* = \frac{1}{6}\left(108\beta + 8\alpha^3 + 12\sqrt{81\beta^2 + 12\beta\alpha^3}\right)^{1/3} + \frac{\alpha}{3}$$

$$+ \frac{2\alpha^2}{3\left(108\beta + 8\alpha^3 + 12\sqrt{81\beta^2 + 12\beta\alpha^3}\right)^{1/3}} \quad where \ \alpha = \frac{a}{C} \ and \ \beta = \frac{\gamma a}{rC}.$$

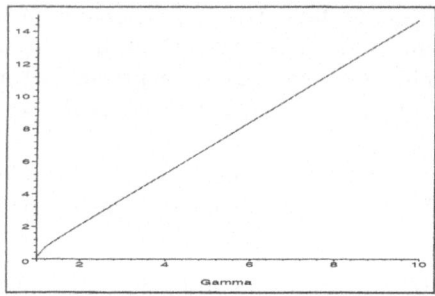

Fig. 5. Maximum spacing as a function of the attenuation Γ

An H^∞ approach. In this approach we do not make any statistical assumptions on the distribution of the noise process, and we adopt instead a robust approach that guarantees the best performance under the worst-case conditions. More precisely, we introduce

$$V(\mu, \lambda) = \sup_{\{v_n\}_{n=-\infty}^{\infty}} \frac{L(\mu, v, \lambda)}{\|v\|^2} \ \ where \ L(\mu, v, \lambda) = \int_{-\infty}^{\infty} (x^2(t) + ku^2(t))dt \quad (12)$$

and $\|v\|$ is the L^2 norm of v: $\|v\| = \sqrt{\sum_{n=-\infty}^{\infty}(v_n)^2}$. We seek a μ that minimizes $V(\mu, \lambda)$; denote its infimum over μ by $(\Gamma^*)^2$.

Define a soft-constrained cost function $L_\Gamma(\mu, v, \lambda) := L(\mu, v, \lambda) - \Gamma^2\|v\|^2$, and consider a two player game where L_Γ is to be minimized by Player 1 (controlling μ) and maximized by Player 2 (controlling v). If there exists some policy μ^* for the problem of minimizing $L(\mu, \lambda)$, then [5]: $\sup_v L_{\Gamma^*}(\mu^*, v, \lambda) = \inf_\mu \sup_v L_{\Gamma^*}(\mu, v, \lambda)$. The quantity above is the upper value of the zero-sum game with kernel L_{Γ^*}, which in fact equals zero. It can be shown that for any $\Gamma \geq \Gamma^*$, the upper value of the game with parameterized kernel L_Γ is zero, and for $\Gamma < \Gamma^*$, its upper value is infinite. Hence, Γ^* is the smallest positive scalar Γ for which the zero-sum game with kernel L_Γ has a finite upper value. Instead of obtaining μ^* defined above, we will solve a parameterized class of controllers, $\{\mu^\Gamma, \Gamma > \Gamma^*\}$, where μ^Γ is obtained from $\sup_v L_\Gamma(\mu^\Gamma, v, \lambda) = \inf_f \sup_v L_\Gamma(\mu, v, \lambda)$. The controller μ^Γ will clearly have the property that it ensures a performance level Γ^2 for the index adopted in (12), i.e. the attenuation is bounded by Γ, i.e., $V(\mu^\Gamma, \lambda) \le \Gamma^2$. It turns out that the limit $\lim_{\Gamma \to \infty} \mu^\Gamma =: \mu^\infty$ is a well-defined controller, and solves uniquely the control problem with the previous Gaussian model. Note that Γ^* will be a function of λ. Our goal is to determine the largest value of spacing for which there exists a policy that guarantees a given level of attenuation.

Theorem 3. *For a given level of attenuation Γ, the largest spacing λ of information for which there is a policy μ^Γ such that $V(\mu^\Gamma, \lambda) \le \Gamma^2$ is given by, for $\Gamma > k$, $\lambda = \frac{\pi}{2}\Gamma - \Gamma \arctan \frac{1}{\sqrt{\Gamma^2/k-1}}$.*

The proof follows from the material in [5, Chapter 5.3]. Figure 5 shows how the spacing grows as a function of the desired achievable attenuation for $k = 1$. The spacing (y-axis) is seen to be concave increasing in the desired attenuation level, and asymptotically its growth is linear in Γ.

5 Conclusions

We have studied various facets of the problem of determining how frequently information should be sent to the source in congestion control mechanisms. We have shown that the rate of information can have an impact on the throughput of the system, as conveying information requires resources that are then not available for data packet transmission. We have further identified an impact on the stability of congestion control. Finally, we have formulated two dynamic optimization (control) problems, one using a stochastic model for queue dynamics, and the other one using a deterministic worst-case model. In both cases, we have obtained apparent tradeoffs dictating the optimal choice of the spacing between successive information packets.

References

1. E. Altman and T. Başar, "Multi-user rate-based flow control", *IEEE Trans. on Communications*, pp. 940-949, 1998.
2. E. Altman, T. Başar and N. Malouch, "The Role of Information Update in Flow Control", INRIA Research report, 2004.
3. U. Ayesta, K. Avrachenkov, E. Altman, C. Barakat, P. Dube, "Simulation Analysis and Fixed Point Approach for Multiplexed TCP flows", *PFLDnet'03: Workshop on Protocols for Fast Long-Distance Networks*, CERN-Geneva, Switzerland, 2003.
4. C. Barakat and E. Altman, "On ACK filtering on a slow reverse channel", *QOFIS*, Berlin, Germany, September 2000.
5. T. Başar and P. Bernhard, *H∞-Optimal Control and Relaxed Minimax Design Problems: A Dynamic Game Approach*, Birkhäuser, Boston, MA, 1995 (2nd edt).
6. C. Hollot, V. Misra, D. Towsley and W.-B. Gong, "A control theoretic analysis of RED" *CMPSCI Technical Report TR 00-41*, July 2000.
7. J. Padhye, V. Firoiu, D. Towsley and J. Kurose, "Modeling TCP throughput: A simple model and its empirical validation", *ACM SIGCOMM'98*, 1998.
8. J. W. Cohen "The Single Server Queue", *North Holland publishing company*, 1982.

Stabilized Edge-to-Edge Aggregate Flow Control

Hyung-Keun Ryu, Jeong-Woo Cho, and Song Chong

Dept. EECS, KAIST, 373-1 Gusung-dong, Yusung-gu, Taejon, 305-701, Korea
{hkryu, ggumdol}@netsys.kaist.ac.kr, song@ee.kaist.ac.kr

Abstract. In this paper, we present a distributed flow control scheme which achieves weighted max-min fair bandwidth allocation among all source-destination pairs on a per-aggregate basis within its network. The motivation behind the scheme is the absence of per-aggregate flow control in the current Internet, resulting in inability to enforce a certain fairness on source-destination flows. In our scheme, the distributed algorithm to compute weighted max-min fair rates is based on PI control in feedback control theory. We mathematically prove the asymptotic stability of the algorithm in presence of aggregate flows with heterogeneous round-trip delays. Through simulations we demonstrate the effectiveness of the proposed scheme in controlling per-aggregate flows.

1   Introduction

The current Internet is a TCP-controlled network, hence, the Internet congestion control relies mostly on TCP, and TCP is an end-to-end protocol over which Internet service providers(ISPs) do not have any control. Consequently, ISPs are facing a big problem as they sell their bandwidth to customers but in the Internet core they have no tool to explicitly control or engineer bandwidth.

In order to tackle this problem, we present a distributed flow control scheme, in which traffic engineering is carried out by means of edge-to-edge aggregate flow control. More specifically, instead of relying on explicit admission control and/or explicit reservation [1,2], the edge-to-edge flow control on aggregate level plays a major role to ensure fair bandwidth sharing between aggregate flows. The proposed flow control scheme is hierarchical. In the upper layer, weighted max-min flow control is implemented and acting on a per-aggregate and edge-to-edge basis, and in the lower layer, TCP flows belonging to a source-destination flow share its per-aggregate bandwidth allocated by the upper layer in their normal way. Thus, the scheme does not require modification nor replacement of present TCP congestion control.

Several studies have been done in the area of aggregate flow control. [3] proposed an architecture to achieve a fair bandwidth allocation among individual flows without using per-flow state in the network core. However, [3] does not provide any minimum rate guarantees but also can cause a serious performance degradation of TCP, especially in a large delay-bandwidth network, because it achieves fair bandwidth allocation by probabilistically dropping packets in

N. Mitrou et al. (Eds.): NETWORKING 2004, LNCS 3042, pp. 574–587, 2004.
© IFIP International Federation for Information Processing 2004

network core. [4] proposed an overlay congestion control architecture for edge-to-edge traffic control, but it does not support minimum rate guarantee and differentiated service on a per-aggregate level. Some other studies [5,6,7,8] have addressed aggregate flow control by employing an aggregate TCP connection which multiplexes local TCP connections into a single, persistent TCP connection and is operated by TCP or its modified congestion control algorithm. The problem with these studies is that they still depend upon packet loss within the network to detect congestion in the bottleneck link. Compared to the previous works, the proposed scheme has the following features: 1) It achieves the stable target queue lengths at the bottlenecks and maximizes the network utilization. 2) It achieves almost no packet losses inside core network. Instead, it distributes the interior network congestion to the network edges. 3) It also guarantees minimum rates and supports weighted max-min fairness on a per-aggregate basis. 4) It imposes minimal complexity in the network core and is highly scalable, in that no per-aggregate state management are necessary at the network core.

In this paper we propose a distributed algorithm which is based on the rate-based closed loop control for the aggregate flow control. It is highly responsive and adaptive to available network bandwidth change and network congestion. There are extensive prior works on the design of distributed algorithms for rate-based flow control. However most of those works is not completely satisfactory for their complexity or for the lack of analysis from the point of view of stability. [9] derived analytically a control-theoretic fair rate allocation algorithm which allows for arbitrary control of the closed-loop performance, but its practical use is limited by high degree of implementation complexity as the round-trip delay increases. The work by [9,10,11] proposed distributed algorithms which adapt quickly to congestion while achieving max-min fairness, either with or without minimum rate guarantee, among competing flows. The work by [12] addressed a weighted max-min fair bandwidth sharing with minimum rate guarantee, but employs per-flow state management to calculate fair rate for each flow.

Our main objective is to develop a distributed algorithm to compute common fair rate in a weighted max-min fair sense, which is based on proportional and integral(PI) control in feedback control theory [13,14]. The algorithm is *scalable* in that the computational complexity imposed on each link is $O(1)$, i.e., independent of number of aggregate flows travelling through the link. It is *stable* in that it converges asymptotically to the desired equilibrium, and has explicit *link buffer control* in that buffer occupancy of every bottlenecked link in a path asymptotically converges to the desired value. Another objective is to derive an explicit and usable, sufficient and necessary condition to ensure asymptotic stability of the network employing the proposed distributed algorithm even in presence of aggregate flows with heterogeneous round-trip delays.

2   A Distributed Flow Control Scheme

The Internet can be thought of as a concatenation of heterogeneous network clouds. Our scheme is applied to a network cloud which consists of edge nodes at the network boundary and core nodes at the network interior. In each ingress edge node, incoming TCP flows having the same ingress-egress edge pair are classified and multiplexed into a single aggregate flow in a per-aggregate queue.

In the network, the proposed scheme runs a distributed and asynchronous algorithm to share available network bandwidth among competing aggregate flows. The distributed algorithm consists of two components, a link algorithm and a source algorithm.[1] The link algorithm, implemented at each outgoing link of edge nodes and core nodes, computes locally the *common fair rate* using the occupancy information of the link buffer. The common fair rate is same as the excess bandwidth normalized by the sum of pre-assigned weights of the aggregate flows sharing the link. The source algorithm, implemented in each edge node, computes the allowed source rate using feedback rate in its path.

For communication between sources and links, each source generates and inserts a control packet each time N_b bytes of data is transmitted. The control packet carries current source rate, pre-assigned weight, and minimum rate. The control packet is transmitted into the network together with data packets and travel along the forward path down to egress edge node, then returned to the source along the backward path, which may not be identical with the forward path in the current IP network. The control packets travelling in the forward and backward paths are called forward control packet(FCP) and backward control packet(BCP), respectively.

Each outgoing link along the forward path intercepts every FCP arriving at the link and updates the FCP's feedback rate field with the same result of the common fair rate computation no matter which aggregate flow it belongs to.

2.1   Link Algorithm

The proposed common fair rate computation at each outgoing link is as follows. The common fair rate is calculated periodically with an update interval T by

$$f[k] = -C_P(q[k] - q_T) - C_I \sum_{n=0}^{n=k}(q[n] - q_T) \qquad (1)$$

where $C_P > 0$ and $C_I > 0$ are the PI controller gains, $q[k]$ is the queue length at the link buffer, and q_T is the target queue length. The choice of C_P and C_I determines the convergence rate of the iteration as well as the stability of the distributed algorithm. The notable feature of this algorithm is that the common fair rate computation is virtually independent of the number of aggregate flows travelling through the link and thus highly scalable. Moreover, the proposed algorithm jointly controls rate allocation and link buffer control, meaning that as the iteration proceeds, it makes the link buffer occupancy converge to the target value, i.e., $\lim_{t \to \infty} q(t) = q_T$, while finding the weighted max-min fair rate as is proved in Section 3.

---

[1] In the proposed scheme, the edge nodes function as the effective sources and destinations of aggregate flows. More specifically, each ingress edge node has per-aggregate sources(say, virtual sources), each of which performs per-aggregate queueing and rate adaptation for an aggregate flow.

At each outgoing link, common fair rate allocation per aggregate flow is performed aperiodically upon arrival of the corresponding FCPs in forward path. That is, upon arrival of a FCP at time t, the common fair rate $f(t)$ computed locally by the link is compared with the feedback rate being carried by the FCP's feedback rate field, and the smaller value is written onto the field and delivered to the source. Note that $f(t)$ is the present value of $f[k]$.

2.2 Source Algorithm

Each aggregate flow has its own class of service, each of which is characterized by a 3-tuple consisting of weight w_i, minimum rate m_i, and peak rate p_i. Let f_i be the common fair rate notified by a BCP from the link to the source. Using the common fair rate f_i, each source calculates the *weighted fair rate* which supports minimum rate and excess bottleneck bandwidth proportional to a weight. Then it calculates allowed source rate from the minimum of the weighted fair rate and peak rate constraint.

Upon receipt of a BCP, allowed source rate a_i for aggregate flow i is computed as follows.

$$a_i = \min[w_i \cdot f_i + m_i, \ p_i] . \tag{2}$$

Now, the weighted max-min fair bandwidth allocation among competing aggregate flows is simply obtained by regulating the transmission rate of each aggregate flow using its allowed source rate.

3 Modelling and Analysis

3.1 Network Model

First, consider a network model in Fig. 1 where we model a single link explicitly and the other links implicitly to simplify the analysis. The link is an outgoing link with a FIFO queue and has N aggregate flows passing through it. We use a continuous-time fluid flow approximation to model the system.

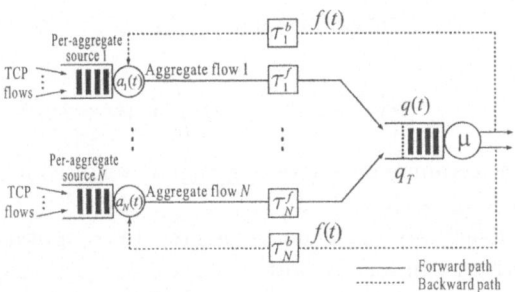

Fig. 1. Network model with a link of interest

Let AF_i denote an aggregate flow i and τ_i, τ_i^f, and τ_i^b denote the round-trip delay, forward-path delay, and backward-path delay of AF_i, respectively. We assume that $\tau_i = \tau_i^f + \tau_i^b$ is constant and the sources are *persistent* until the system reaches steady state. We also assume that the available bandwidth μ at the link is constant until the system reaches steady state and the buffer size at the bottleneck link is finite and larger than the target queue length.

Let $a_i(t)$ denote the rate at which the source i transmits data at the source time t, and denote p_i the peak rate constraint of AF_i. Next, let $f_i(t)$ denote the common fair rate allocated to AF_i by the link of interest and $b_i(t)$ be the latest minimum value of the common fair rates allocated to AF_i by other links along the AF_i's path. Moreover, let $f_i^w(t)$ and $b_i^w(t)$ denote the weighted fair rates of AF_i which are computed by source i at the source time t as follows:

$$f_i^w(t) = w_i f_i(t) + m_i, \quad f_i(t) = f(t - \tau_i^b), \quad \forall\, i \in N \tag{3}$$

and

$$b_i^w(t) = w_i b_i(t) + m_i, \quad \forall\, i \in N \tag{4}$$

where w_i and m_i denote a weight value and the minimum rate which the link is required to guarantee during the entire holding time of AF_i, respectively. We assume that $m_i \leq p_i$, $\forall\, i \in N$ and there exists an admission control which guarantees $\sum_{i \in N} m_i < \mu$.

The source behavior of AF_i can be modeled by

$$a_i(t) = \min[\, f_i^w(t),\ b_i^w(t),\ p_i\,], \quad \forall\, i \in N \tag{5}$$

where N denotes the set of all the aggregate flows whose route includes the bottleneck node of interest. This model implies that a source transmits data at the smallest value among the weighted fair rates allocated by the nodes along the route and the peak rate constraint of the aggregate flow.

By neglecting the buffer floor, the dynamics of the link buffer of interest is modelled in continuous time by

$$\dot{q}(t) = \sum_{i \in N} a_i(t - \tau_i^f) - \mu\,. \tag{6}$$

The common fair computation in equation (1) can be rewritten in continuous time by

$$f(t) = -C_P\{q(t) - q_T\} - C_I \int_0^t \{q(t) - q_T\}dt\,. \tag{7}$$

Note that $f(t)$ is the common part of per-aggregate weighted fair rate allocations, $f_i^w(t)$, $\forall\, i$, which implies that all the sources bottlenecked at the link are fed with the same feedback rate. Thus no per-aggregate computation is required.

Let Q denote the set of locally-bottlenecked aggregate flows, at a link, containing all those aggregate flows having common fair rate determined at the link in the steady state for a given network loading. In the same way, let $N - Q$ denote the set of remotely-bottlenecked aggregate flows, at a link, containing all those aggregate flows having common fair rate determined at some other link in the path, or having data transfer rate limited by their peak rate constraint. Let

$a_{is} = \lim_{t\to\infty} a_i(t)$, $f_{is}^w = \lim_{t\to\infty} f_i^w(t)$, and $b_{is}^w = \lim_{t\to\infty} b_i^w(t)$. Then Q at the link of interest is given by $Q = \{i | i \in N$ and $a_{is} = f_{is}^w\}$ and $N - Q$ at the link of interest is given by $N - Q = \{i | i \in N$ and $a_{is} = \min[b_{is}^w, p_i]\}$.

3.2 Steady State and Fairness

Suppose that the closed-loop dynamics have an equilibrium point at which the derivatives of the system variables are zero. Let $f_s = \lim_{t\to\infty} f(t) > 0$. Then, from (3), (5) and (7), we have

$$a_{is} = \min[f_{is}^w, \ b_{is}^w, \ p_i], \quad f_{is}^w = w_i f_s + m_i, \quad \forall \, i \in N \tag{8}$$

and $q_s = q_T$ where $q_s = \lim_{t\to\infty} q(t)$. Since $q_s = q_T > 0$, the buffer equation (6) implies that

$$\sum_{i\in N} a_{is} = \mu \ . \tag{9}$$

By combining the equations (8), (9), and the definitions of Q and $N - Q$, we obtain

$$f_s = \frac{\mu - \sum_{i\in N-Q} \min[b_{is}^w, p_i] - \sum_{i\in Q} m_i}{|Q|_w} \tag{10}$$

where $|Q|_w$ denote the weighted cardinality of Q, which is the weighted number of locally-bottlenecked aggregate flows, i.e., $|Q|_w = \sum_{i\in Q} w_i$. In addition, let $|Q|$ denote the cardinality of Q, which is the number of locally-bottlenecked aggregate flows. The following theorem summarizes the result.

Theorem 1. *For $\sum_{i\in N} m_i < \mu$ and $\min[b_{is}^w, p_i] \geq m_i$, there exists a unique equilibrium point at which* (i) *the buffer occupancy is equal to the target value* $(q_s = q_T)$, (ii) *the capacity at the link is fully utilized* $(\sum_{i\in N} a_{is} = \mu)$, *and* (iii) *individual minimum rates are guaranteed at the link and the unreserved portion of capacity, $\mu - \sum_{i\in N} m_i$, is allocated in the weighted max-min fair sense to the aggregate flows travelling through the link. That is,*

$$a_{is} = \begin{cases} \frac{w_i(\mu - \sum_{i\in N-Q} \min[b_{is}^w, p_i] - \sum_{i\in Q} m_i)}{|Q|_w} + m_i, & i \in Q \\ \min[b_{is}^w, p_i], & i \in N-Q \ . \end{cases} \tag{11}$$

3.3 Asymptotic Stability and Optimal Controller Gain

In this subsection, we study the local stability of the closed-loop system in the neighborhood of the equilibrium point in which the dynamics of the other links are in steady state, i.e., $b_i^w(t) = b_{is}^w$, $\forall \, i \in N$. Thus, (5) can be rewritten by

$$a_i(t) = \begin{cases} f_i^w(t), & i \in Q \\ \min[b_{is}^w, p_i], & i \in N-Q \ . \end{cases} \tag{12}$$

In a link, by combining (3), (6), and (12), we obtain

$$\dot{q}(t) = \sum_{i\in Q} w_i f(t - \tau_i) + \underbrace{\sum_{i\in Q} m_i + \sum_{i\in N-Q} \min[b_{is}^w, p_i] - \mu}_{constant} \ . \tag{13}$$

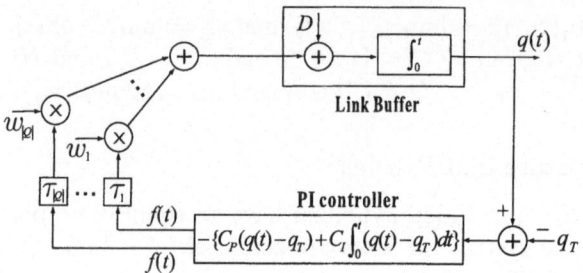

Fig. 2. The closed-loop system model

The constant part in (13) can be considered as an external disturbance, which is denoted by D. By substituting (7) for $f(t - \tau_i)$ in (13), we obtain the following closed-loop equation of the system.

$$\dot{q}(t) = D - \sum_{i \in Q} w_i \left[C_P \{ q(t - \tau_i) - q_T \} + C_I \int_0^{t - \tau_i} \{ q(t) - q_T \} dt \right] . \qquad (14)$$

Fig. 2 depicts this closed-loop system model.

We define the controller gains to be $(C_P, C_I) = (A/|Q|_w, B/|Q|_w)$, where A and B are some positive constants. The open-loop transfer function of the closed-loop system in Fig. 2 is then given by

$$F(s) = \left(\frac{A}{s} + \frac{B}{s^2} \right) \sum_{i \in Q} \rho_i e^{-\tau_i s} \qquad (15)$$

where $\rho_i = \frac{w_i}{|Q|_w} \geq 0$, $\forall i \in Q$ and $\sum_{i \in Q} \rho_i \leq 1$.

Now, the sufficient and necessary condition for the asymptotic stability of the closed-loop system is found in an usable form. For the page limitation, the detailed derivations of the theorems and corollary below are given in [16].

First, we consider a single source case, i.e., $|Q| = 1$ with round-trip delay τ and $\rho_1 = 1$ and $\rho_i = 0$, $\forall i > 1$. Note that this case is equivalent to the multiple source case with homogeneous delays τ. By letting $s = j\omega$, the open-loop transfer function becomes

$$F(j\omega) = \left(-\frac{B}{\omega^2} - j\frac{A}{\omega} \right) e^{-j\omega\tau} . \qquad (16)$$

Then, by appealing to the Nyquist stability criterion[15], we can find the stability condition for the single source system which is stated in the following theorem.

Theorem 2. *The closed-loop system with a single delay $\tau \geq 0$ is asymptotically stable if and only if the delay is bounded by*

$$0 \leq \tau < \frac{\arccos \left(\frac{B}{\bar{\omega}^2} \right)}{\bar{\omega}} \triangleq \tau^u \qquad (17)$$

where $\bar{\omega}$ is a unique $\omega > 0$ such that $F(j\omega) = 1$.

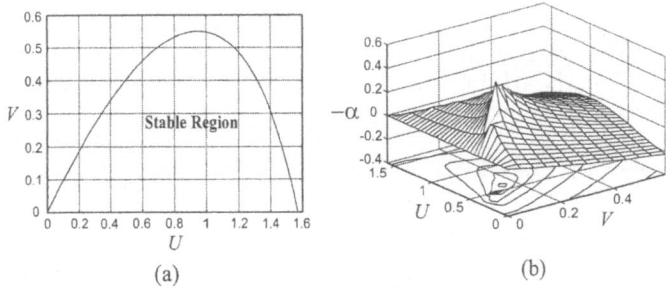

Fig. 3. (a) Stable region (b) Asymptotic decay rate $-\alpha$ as a function of U and V

We have found the upper bound of the round-trip delay for the single source system to be asymptotically stable. It is, however, difficult to apply the stability condition itself (17) to the design of a controller. We modify the condition into an usable form in the following corollary.

Corollary 1. *Let $U = A\tau$ and $V = B\tau^2$. Then the closed-loop system is asymptotically stable if and only if*

$$0 < U < \frac{\pi}{2} \text{ and } 0 < V < \omega_1^2 \cos \omega_1 \qquad (18)$$

where ω_1 is the unique solution of $U = \omega \sin \omega$ for $0 < \omega < \pi/2$.

We provide the stable region of U and V in Fig. 3(a). Now, we derived that the stable gain for the case of multiple sources with heterogeneous round-trip delays can be easily found from (18) by applying the theorem below.

Theorem 3. *The closed-loop system with heterogeneous delays is asymptotically stable for all $0 \leq \tau_i \leq \bar{\tau}$ and for all ρ_i satisfying $\sum_{i \in Q} \rho_i \leq 1$ if and only if the closed-loop system of the single-delay case with delay $\bar{\tau}$ is asymptotically stable.*

Consequently, once the upper bound of all the round-trip delays is known, the stable gain for the multiple source system can be obtained from $A = U/\bar{\tau}$ and $B = V/\bar{\tau}^2$ where U and V satisfies (18). In [16], we found the asymptotic decay rate or convergence speed of the closed-loop system numerically. Fig. 3(b) is the result of our numerical approach. The asymptotic decay rate is maximized approximately at $(U, V) = (0.5, 0.1)$. Hence, we can find a stable and optimal controller gain from $(A, B) = (0.5/\bar{\tau}, 0.1/\bar{\tau}^2)$. $\bar{\tau}$ is a possible maximum round-trip delay which can be obtained by off-line measurement in the network domain.

3.4 $|Q|_w$ Estimation

Based on the pair (A,B) found in the above subsection, we can find the controller gain as $(C_P, C_I) = (A/|Q|_w, B/|Q|_w)$ where $|Q|_w$ is obtained through the estimation of weighted number of locally-bottlenecked aggregate flows, $|\hat{Q}|$. We estimate $|Q|_w$ without doing per-aggregate accounting as follows. Suppose that the jth FCP arrives at a link at the link time t^j. If the jth FCP happens to be a control packet of AF_i, it carries the value $a_i(t^j - \tau_i^f)$, m_i, and w_i. The link monitors the FCP arrivals in a synchronous fashion over fixed-length intervals of W seconds. For the lth interval, the weighted number of locally-bottlenecked aggregate flows can be estimated by

$$|Q|_w^l = \sum_{t^j \in ((l-1)W, lW]} \frac{N_b + S_{CP}}{W \cdot a_i(t^j - \tau_i^f)} \cdot w_i \cdot 1\{a_i(t^j - \tau_i^f) - m_i \geq \delta \cdot w_i \cdot f(t^j)\}, 0 < \delta < 1 \quad (19)$$

where $1\{\cdot\}$ is the indicator function, S_{CP} is the byte size of a control packet, and $f(t^j)$ is the latest value of the common fair rate at time t^j. Note that all the rates have the values represented in terms of bytes per sec. Here δ is the margin to avoid the underestimation. Based on this estimate for each interval, the recursive estimate is computed at the end of every interval as follows.

$$|\hat{Q}|_w = \mathrm{sat}_1^{|N|_w}[\lambda|\hat{Q}|_w((l-1)W) + (1-\lambda)|Q|_w^l], \quad 0 < \lambda < 1 \quad (20)$$

where λ is an averaging factor and the saturation function ensures that $1 \leq |\hat{Q}|_w (t) \leq |N|_w$ for all t. We choose large λ at a value close to 1 in the hope that the averaging operation in (20) will effectively filter out the variability of $|Q|_w^l$.

In the implementation of $|Q|_w$ estimation, we introduce *virtual packet* concept to remove the impact of different packet size among aggregate flows. We first define a virtual packet with a fixed byte size, then a stream of variable-sized packets can be regarded as a stream of virtual packets. Now, the choice of N_b is obtained as $N_b = N_{CP} \cdot S_{VP}$, where N_{CP} is the number of virtual packets transmitted between two adjacent FCPs and S_{VP} is a virtual packet size.

Table 1. Recommended parameter values in the distributed flow control scheme($\bar{\tau} = \max\{\tau_i, i \in N\}$, Δ=one virtual packet transmission time)

Common fair rate computation			$|Q|_w$-Estimation				
A	B	T	W	δ	λ	N_{CP}	S_{VP}
$0.5/\bar{\tau}$	$0.1/\bar{\tau}^2$	30Δ	300Δ	0.9	0.98	30	1Kbytes

Table 2. The aggregate flow models and theoretical fair rates in the single bottleneck link configuration

AF#	m_i (Mbps)	w_i	Arrival (sec)	Departure (sec)	Fair rate(Mbps) 0~25	25~50	50~75	75~∞
AF1	0	1	0	∞	10.63	8.5	6.36	10.63
AF2	5	1.5	0	∞	20.94	17.75	14.54	20.94
AF3	0	2	25	75		17	12.73	
AF4	10	2.5	0	∞	36.56	31.25	25.91	36.56
AF5	0	3	0	∞	31.88	25.5	19.1	31.88
AF6	15	1	50	75			21.36	

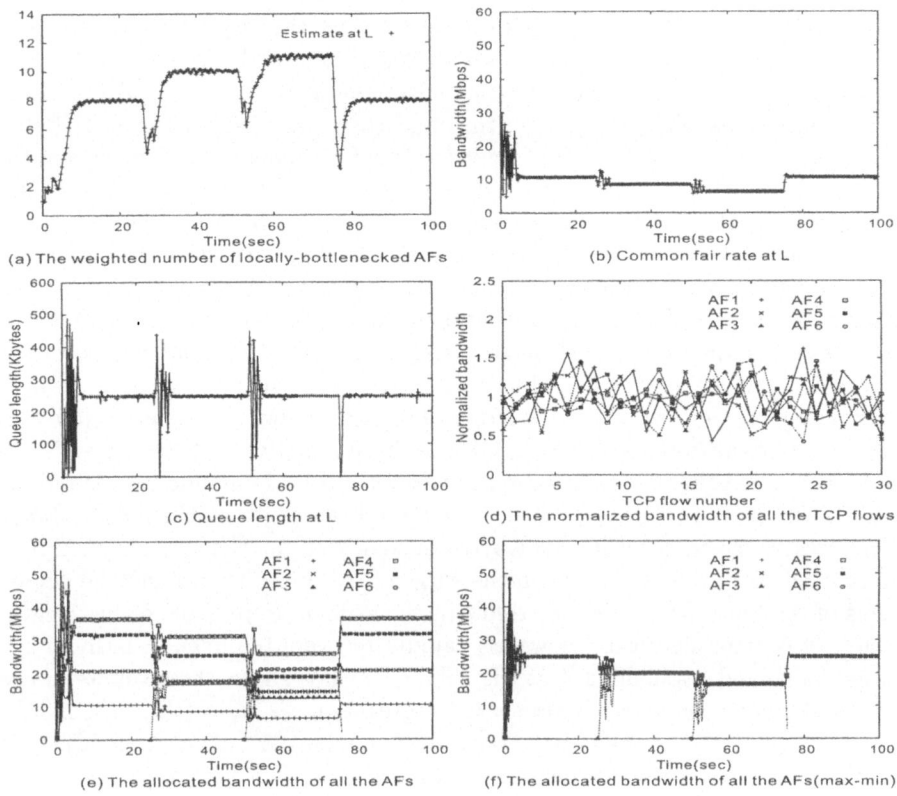

Fig. 4. Results in the single bottleneck link configuration

4   Simulation Results

In this section, we present simulation data to verify and demonstrate the performance of our scheme as described in the previous sections. All the simulation are performed in the ns-2[17] environment. We consider the single bottleneck link configuration, where the bottleneck is shared by six ingress-egress pairs and each ingress-egress pair establishes only one virtual path for an aggregate flow. The six aggregate flows have heterogeneous round-trip delays ranging from 20ms to 70ms. Thus the maximum round-trip delay($\bar{\tau}$) between ingress nodes and the egress nodes is about 70 ms. Each aggregate flow consists of 30 persistent TCP flows. In this section, we use simulations to verify and demonstrate the performance of our scheme as described in the previous sections. The simulations are performed in the ns-2[17] environment. We consider a typical configuration with a single bottleneck link, termed L, where the bottleneck link is shared by six ingress-egress pairs and each ingress-egress pair has an aggregate flow which consists of 30 persistent TCP flows. The six aggregate flows have heterogeneous round-trip delays ranging from 20ms to 70ms. Thus the maximum round-trip delay($\bar{\tau}$) between ingress nodes and the egress nodes is about 70 ms. In the configuration, the capacity of each outgoing link is equally set to 100Mbps, the propagation delay of the bottleneck link L is set to 10ms, the target queue length(i.e., q_T) of each outgoing link buffer is set to 256Kbytes, and the maximum buffer size of the per-aggregate queue in each edge node is set to 256 Kbytes. All the TCP sources use TCP Reno algorithm and their data packet size is 1Kbytes. In Table 1, we summarize recommended values for simulation parameters in the proposed scheme. The aggregate flow models used in this simulation are summarized in Table 2 and an aggregate flow i is denoted by AFi. The simulation results for weighted max-min fair bandwidth allocation are shown in Fig. 4. For comparison purpose, we have computed the theoretical fair rates for the given simulation scenario based on Proposition 1, and include the results in Table 2. The transmission rate of each aggregate flow in Fig. 4(e) exactly follows the theoretical fair rates given in Table 2 although there is a transient period whenever an aggregate flow arrives or leaves. Fig. 4(b) shows the common fair rate computed by the bottleneck link L. Observe from Fig. 4(e) that the transmission rate of AF1 is equal to the common fair rate since its minimum rate is 0Mbps and its weight is 1. Fig. 4(c) shows that the queue length at the link L always converges to the target value 256Kbytes in steady state. The arrivals of AF3 at 25 sec and AF6 at 50 sec result in the surge of the queue length and the departures of AF3 and AF6 at 75 sec result in the sudden drop of the queue length. The flow control algorithm, however, rapidly recovers the queue length to the target value and restabilizes it at the value. Fig. 4(a) shows the estimate of the weighted number of locally bottlenecked aggregate flows, $|\hat{Q}|_w(t)$, at the link L. Fig. 4(d) shows the normalized average throughputs(over a 20sec interval) of all the TCP flows belonging to each aggregate flow. The TCP flows track and share the bandwidth allocated to the aggregate flow in their normal way.

Next, we show that max-min fairness is achieved if all aggregate flows have same weight. In this simulation, each aggregate flow has the weight value of 1 without minimum rate guarantee. Observe from Fig. 4(f) that the bottleneck link capacity is divided equally among the aggregate flows.

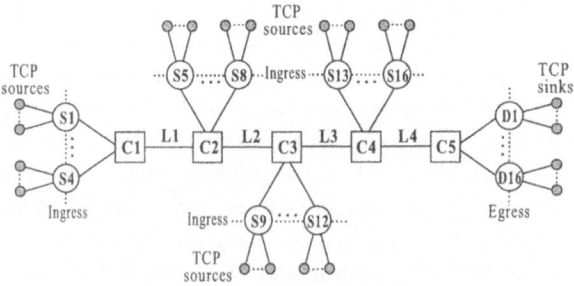

Fig. 5. Multiple bottleneck link configuration

Finally, we study the multiple bottleneck link configuration, shown in Fig. 5. In this simulation, we consider peak rate constraint. Each aggregate flow

Table 3. The aggregate flow models and the theoretical fair rates in the multiple bottleneck link configuration.

AF#	p_i (Mbps)	m_i (Mbps)	w_i	Fair rate (Mbps)	Bottlneck
AF1,AF5,AF9	100	0	1	5	L3
AF13	100	0	1	36.67	L4
AF2,AF6,AF10	100	0	2	10	L3
AF14	100	0	2	73.33	L4
AF3,AF7,AF11	20	10	1	15	L3
AF15	20	10	1	20	p_i
AF4,AF8,AF12,AF16	20	10	3	20	p_i

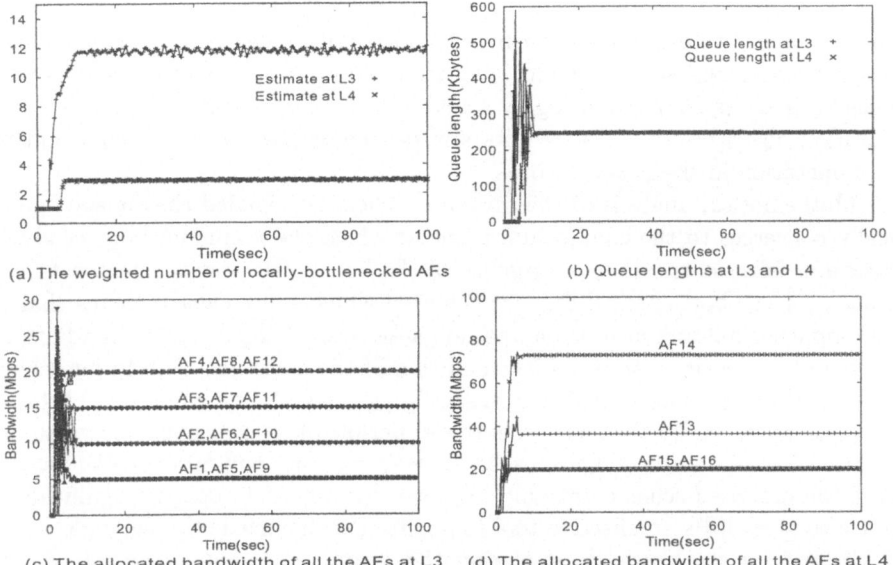

(a) The weighted number of locally-bottlenecked AFs

(b) Queue lengths at L3 and L4

(c) The allocated bandwidth of all the AFs at L3

(d) The allocated bandwidth of all the AFs at L4

Fig. 6. Results in the multiple bottleneck link configuration

contains 50 TCP flows. 16 aggregate flows with different edge node locations are contained and the capacities of the links between core nodes are set to 300 Mb/s, except that the link between C3 and C4 is 150 Mb/s. The link delays between core nodes are all 10ms and the other link delays are 1ms. All other parameters have the same values used in the previous experiment. The aggregate flow models used in this simulation configuration are summarized in Table 3.

For comparison purpose, we also computed the theoretical fair rates satisfying the weighted max-min fairness with minimum rate guarantee for the given simulation scenario. We also include the theoretical bottleneck location of each aggregate flow in the table, signifying the location at which each fair rate is determined. Fig. 6 shows the simulation results. The actual transmission rate of each aggregate flow in Fig. 6(c) and 6(d) exactly follows the theoretical fair rates given in Table 3, irrespective of their round-trip delays and the bottleneck locations. Thus weighted max-min fair bandwidth allocation among aggregate flows are achieved in multiple bottleneck links. In the given scenario, there are two congested links, L3 and L4. As expected, the queue length at these congested nodes converges to the target value, 256 Kbytes, which is shown in Fig. 6(b). Fig. 6(a) shows the estimate of the weighted number of locally bottlenecked aggregate flows, $|\hat{Q}|_w$ (t), at L3 and L4, respectively. We see that in the steady state the estimate stays around 12 and 3 at L3 and L4, respectively, which agrees with the data in Table 3.

5   Conclusions

In this paper, we propose a distributed flow control scheme for aggregate flows mainly concerning fair bandwidth allocation on a per-aggregate basis. The proposed scheme is simple and highly scalable because its common fair rate computation algorithm does not require any per-aggregate flow state management and operation in the network core.

Mathematical analysis of the proposed scheme concluded that it asymptotically converges to the equilibrium point at which the minimum plus weighted max-min fairness and the convergence of the link buffer occupancy to a target value at every bottlenecked link are achieved which subsequently means that it accomplishes full link utilization and no packet loss at steady state. In addition, we found the asymptotic stability condition of the controller gain in an usable form and the optimal controller gain satisfying the stability condition. Through simulations we verify that our scheme can perform the excellent bandwidth allocation for the aggregate flows based on weighted max-min fairness. We believe that the proposed scheme not only improves the Internet capacity significantly but also gives ISPs an effective tool to engineer traffic inside the network

In ongoing work, we are exploring the quantized common fair rate delivery using standard Explicit Congestion Notification(ECN) framework to remove control packet overhead in our scheme and thus making our algorithm readily implementable in current network status, hence, enabling easy deployment of our scheme.

References

1. Blake, S., Blake, D., Carlson, M., Davies, E., Wang, Z., Weiss, W.: An Architecute for Differentiated Services. IETF RFC 2475 (Dec. 1997)
2. Braden, R., Zhang, L., Berson, S., Herzog, S., Jamin, S.: Resource Reservation Protocol (RSVP). IETF RFC 2205 (Sept. 1997)
3. Stoica, I., Shenker, S., Zhang, H.: Core-Stateless Fair Queueing: Achieving Approximately Fair Bandwidth Allocations in High-Speed Networks. Proc. ACM SIG-COMM'98 Conference (Sep. 1998) 118-130
4. Harrison, D., Kalyanaraman, S.: Edge-To-Edge Traffic Control for the Internet. Technical Report ECSE-NET-2000-I, RPI ECSE Networks Laboratory (Jan. 2000)

5. Chapman, A., Kung, H.T.: Traffic Management for Aggregate IP Streams. Proc. CCBR'99 (Nov. 1999) 1-9
6. Lee, B.P., Balan, R.K., Jacob, L., Seah, W.K.G., Ananda, A.L.: TCP Tunnels: Avoiding Congestion Collapse. Proc. IEEE LCN'00 (Nov. 2000) 408-417
7. Nandy, B., Ethridge, J., Lakas, A., Chapman, A.: Aggregate Flow Control: Improving Assurances for Differentiated Services Network. Proc. IEEE INFOCOM'01 (Apr. 2001) 1340-1349
8. Pradhan, P., Chiueh, T., Neogi, A.: Aggregate TCP Congestion Control Using Multiple Network Probing. Proc. ICDCS'00 (Apr. 2000) 30-37
9. Benmohamed, L., Meerkov, S.M.: Feedback Control of Congestion in Packet Switching Networks: The Case of Single Congested Node. IEEE/ACM Trans. on Networking, Vol. 1. (Dec. 1993) 693-708
10. Charny, A., Clark, D., Jain, R.: Congestion Control with Explicit Rate Indication. Proc. IEEE ICC'95 (June 1995) 1954-1963
11. Kolarov, A., Ramamurthy, G.: A Control Theoretic Approach to The Design of Closed Loop Rate Based Flow Control for High Speed ATM Networks. IEEE INFOCOM'97 (Apr. 1997) 293-301
12. Hou, Y.T., Tzeng, H., Panwar S.S., Kumar, V.P.: A Generic Weight-Proportional Bandwidth Sharing Policy for ATM ABR Service. Performance Evaluation, Vol. 38. (Sep. 1999) 21-44
13. Aström, K.J., Wittenmark, B.: Computer Controlled Systems: Theory and Design. Prentice-Hall, Englewood Cliffs, New Jersey (1984)
14. Barmish, B.R.: New Tools for Robustness of Linear Systems. MacMillan (1994)
15. Franklin, G.F., Powell, J.D., Workman, M.L.: Digital Control Systems. Addison-Wesley, Reading, MA, New York (1990)
16. Ryu, H.K., Cho, J.W., Chong, S.: Stabilized Edge-to-Edge Aggregate Flow Control (extended version). http://netsys.kaist.ac.kr/ ~hkryu/networking2004.pdf
17. ns-2, Network Simulator (ver-2), Available at http://www.isi.edu/nsnam/ns/

MaxNet: Faster Flow Control Convergence

Bartek P. Wydrowski[1,2], Lachlan L.H. Andrew[1], and Iven M.Y. Mareels[1]

[1] The ARC Special Research Centre for Ultra-Broadband Information Networks.
Department of Electrical and Electronic Engineering, The University of Melbourne,
Australia. {b.wydrowski,l.andrew,i.mareels}@ee.mu.oz.au
[2] NetLab, Department of Computer Science, California Insitute of Technology

Abstract. MaxNet is a distributed congestion control architecture in which only the most severely bottlenecked link on the end-to-end path controls the source rate. This paper shows that the small-signal convergence speed of MaxNet is higher than that of conventional architectures, such as the current Internet or REM. It also shows that MaxNet decouples the control, so that each pole position depends only on parameters of one bottleneck link and of the sources controlled by that bottleneck, enabling optimal pole placement.

1 Introduction

Network flow control aims to control source rates so that link capacities are utilised. Internet-like networks, where links and sources can only have local information, must use fully distributed control.

Models of Internet-like networks control the source rate by a scalar feedback congestion signal which aggregates the congestion prices of links on the end-to-end connection. The signals, such as loss or delay, used by current flow control algorithms such as TCP, are implicitly summed over all links by the network. We refer to these networks as SumNets. In [1], we introduced MaxNet, where the congestion prices are aggregated by taking the maximum link price along the connection path, rather than the sum. In [1], we showed that MaxNet results in Max-Min fairness for sources with homogeneous demand functions.

Sufficient conditions for the stability of MaxNet are known [2], but its transient dynamics are yet to be studied. The convergence time of network flow control impacts on the Quality of Service of the network. A slow response results in long traffic transients which are responsible for packet delay, delay-jitter, under-utilisation and buffer-overflow. Reducing the duration and overshoot of transients improves these performance measures and makes smaller buffer sizes possible.

This paper compares the convergence times of MaxNet and SumNet, using a common framework given in Section 2. The first part of the investigation involves a local analysis for which we develop small-signal linearized models in Sections 3 and 4. Using these models, Section 6 finds the position of poles which determine the convergence time of MaxNet. Section 7 compares the convergence time of MaxNet with a lower bound of the performance of SumNet, and shows that a

N. Mitrou et al. (Eds.): NETWORKING 2004, LNCS 3042, pp. 588–599, 2004.

faster pole placement is possible with MaxNet than with SumNet. This makes it possible for MaxNet to achieve better QoS performance. In the second part of this paper, Section 8 investigates the global performance by simulating the full non-linear system, and relates the local analysis to these simulation results.

2 Control Architecture

The following is a brief overview of the MaxNet and SumNet control algorithms. For a fuller description, see [1,3]. The behaviour of source i is governed by an explicit demand function, $D_i(\cdot)$, such that its transmit rate is

$$x_i = D_i(q_i) \tag{1}$$

for a congestion signal q_i. In both MaxNet and SumNet, the congestion signal, q_i, summarises the prices, p_l, of all links, l, on the end-to-end path, L_i, of source i. In MaxNet the congestion signal is the maximum of all link prices,

$$q_i = \max\{p_l : l \in L_i\}. \tag{2a}$$

In contrast, SumNet uses the sum

$$\hat{q}_i = \sum_{l \in L_i} p_l. \tag{2b}$$

(Throughout this paper, variables with a hat pertain to SumNet, and the corresponding variables without a hat pertain to MaxNet.)

The Active Queue Management (AQM) algorithm in a router sets the price of an outgoing link according to the well studied integrator process [3]:

$$p_l(t+1) = p_l(t) + (y_l(t) - c_l)\varphi_l, \tag{3}$$

where $y_l(t) = \sum_{i:l \in L_i} x_i(t)$ is the aggregate traffic for link l at time t, φ_l is the control gain and c_l is the target capacity of link l which is related to its physical capacity, C_l by the target utilisation, $0 < \mu_l < 1$, such that $c_l = \mu_l C_l$.

3 MaxNet Control System Model

This section describes the MaxNet model from [2] which will be used in subsequent sections to investigate convergence time. This model makes a number of simplifications of the network. The first is to use a fluid-flow approximation of the packet based information flow. The second simplification is that the global non-linear system is linearized about its equilibrium point. MaxNet contains two sources of non-linearity. The first is the Max operation itself and the second is the non-linear demand function, D. These will be linearised separately.

Using these simplifications, the network is represented as a multi-variable control system, shown in Figure 1. Note that Figure 1, for illustration only, shows a large-signal source, and small-signal links and network. The interconnection of sources with links is piecewise linear, due to the Max operation. It is described in the Laplace domain by forward and backward routing matrices. The matrices specify the interconnection and the delay incurred in signal flow from source to link and vice versa. The forward routing matrix is

$$[\overline{R}_f(s)]_{l,i} = \begin{cases} e^{-\tau^f_{i,l}s} & \text{if source } i \text{ uses link } l \\ 0 & \text{otherwise}, \end{cases} \tag{4}$$

where $\tau^f_{i,l}s$ is the forward delay between source i and link l. Note that the bar notation in $\overline{R}_f(s)$ indicates it has a row for every link in the network. We later reduce this to a matrix representing only bottleneck links, without the bar.

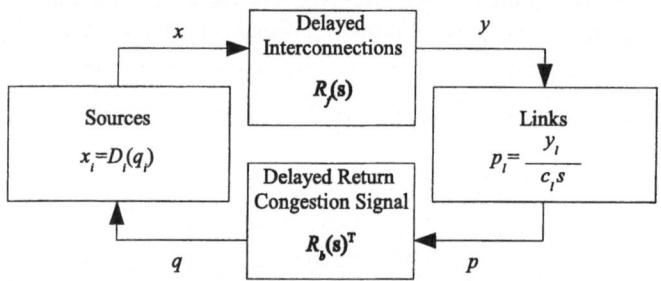

Fig. 1. Flow Control Structure.

Let n_i be the bottleneck link that controls source i (the link on L_i with the maximum price). Then the backward routing matrix depends on n, as

$$[\overline{R}_b(s;n)]_{l,i} = \begin{cases} e^{-\tau^b_{i,l}s} & \text{if } n_i = l \\ 0 & \text{otherwise}. \end{cases} \tag{5}$$

Note that the round-trip time of source i's connection is $\tau_i = \tau^f_{i,l} + \tau^b_{i,l}$. Let L be the number of links in the network. Without loss of generality, order the link prices such that

$$p_1 \ge p_2 \ge \ldots \ge p_L. \tag{6}$$

The backward routing matrix remains static over a period where the variations in link prices do not change the ordering of link prices (6). The overall multi-variable feedback loop in the configuration of Figure 1 is

$$\overline{y}(s) = \overline{R}_f(s)x(s) \tag{7}$$

$$q(s) = \overline{R}_b(s;n)^T \overline{p}(s). \tag{8}$$

We can construct a small signal model as in [3]. Consider small perturbations around equilibrium, $x = x_0 + \delta x$, $\overline{y} = \overline{y}_0 + \delta \overline{y}$, $\overline{p} = \overline{p}_0 + \delta \overline{p}$, $q = q_0 + \delta q$, where subscript 0 denotes a steady state value and prefix δ denotes a perturbation. Note that the bar notation still denotes variables that contain non-bottleneck links and $\delta \overline{p_l}$ is only non-zero for bottleneck links. Note also that when all link prices are distinct, the vector of bottlenecks, n, is unchanged by a sufficiently small perturbation. In this case, the small signal model does not explicitly involve n. This is the first linearisation. Form the vectors $\delta p(s)$, $\delta y(s)$ and the matrices R_f, R_b by eliminating the elements (or rows) corresponding to non-bottleneck links. This gives the reduced small signal model

$$\delta y(s) = R_f(s)\delta x(s) \tag{9}$$

$$\delta q(s) = R_b(s)^T \delta p(s) . \tag{10}$$

To achieve stable control for networks of arbitrary dimensions, it is sufficient that the gains that sources and links introduce be as follows [2]. The second linearisation replaces the demand function of a source by a small-signal gain, κ_i, between a perturbation in δq_i and the resulting perturbation in δx_i, given by

$$\kappa_i = D_i'(q_i) . \tag{11}$$

For robust stability, it is sufficient that this gain be scaled such that

$$\kappa_i = \frac{\alpha_i x_{0i}}{\tau_i} . \tag{12}$$

Here, $\alpha_i \in (0, 1)$ reflects the source's need for capacity, and the factor τ_i provides robustness to delay. To make stability robust to the number of sources, a gain x_{0i}/c_l is introduced in the closed-loop, with the x_{0i} component put into the source and the $1/c_l$ component in the link as $\varphi_l = 1/c_l$.

Note that (11) implicitly assumes a static demand function. As discussed in [3], the requirement (12) determines the shape of the static demand function. However, recent work in [4] provides dynamic source algorithms which allow arbitrary demand functions, whilst preserving the control gain required for robust stability. They separate the high-frequency gain AC from the DC gain.

In the Laplace domain, the integrator AQM of (3) with the required gain between the coupling of δp_l and δy_l is

$$\delta p_l = \frac{1}{c_l s}\delta y_l . \tag{13}$$

The open-loop transfer function for the small signal MaxNet model is

$$H(s) = \frac{1}{s}R_f(s)\mathcal{K}R_b(s)^T \mathcal{C} , \tag{14}$$

where

$$\mathcal{K} = \mathrm{diag}(\kappa_i), \qquad \mathcal{C} = \mathrm{diag}\left(\frac{1}{c_l}\right) . \tag{15}$$

4   SumNet Control System Model

This section describes the model from [3] for a SumNet network, by highlighting the difference from the MaxNet model. Recall that the hat symbol identifies SumNet variables which have a related variable in MaxNet.

The SumNet forward routing matrix is the same as that of MaxNet, $\hat{\overline{R}}_f = \overline{R}_f$. The backward routing matrix, which describes the flow of congestion information from each link back to sources, is independent of the current transmission rates:

$$[\hat{\overline{R}}_b(s)]_{l,i} = \begin{cases} e^{-\tau^b_{i,l}s} & \text{if source } i \text{ uses link } l \\ 0 & \text{otherwise} . \end{cases} \tag{16}$$

Thus (8) becomes

$$q(s) = \hat{\overline{R}}_b(s)^T \overline{p}(s). \tag{17}$$

The small signal variables also take on the hat notation: $\hat{x} = \hat{x}_0 + \delta\hat{x}$, $\hat{\overline{y}} = \hat{\overline{y}}_0 + \delta\hat{\overline{y}}$, $\hat{\overline{p}} = \hat{\overline{p}}_0 + \delta\hat{\overline{p}}$, $\hat{q} = \hat{q}_0 + \delta\hat{q}$.

For SumNet, the routing matrices can again be reduced to contain only bottleneck links. These reduced matrices are applicable so long as the bottlenecks remain the same throughout the perturbations. The reduced small-signal model has the same form as (9), (10), in the variables $\hat{R}_f = R_f$, $\hat{R}_b \, \delta\hat{p}(s)$ and $\delta\hat{y}(s)$.

To achieve stable control for networks of arbitrary dimensions, the gains that sources and links introduce need to satisfy the bounds detailed in [3]. For SumNet, a source i requires a gain $\hat{\kappa}_i$ of

$$\hat{\kappa}_i = \frac{\hat{\alpha}_i \hat{x}_{0i}}{M_i \tau_i} , \tag{18}$$

where M_i is the number of controlling bottleneck links on the end-to-end path, and $\hat{\alpha}_i \in (0,1)$ is again an adjustable parameter.

To maintain stability, SumNet must either estimate and communicate M_i to each source [3], or use a slow, conservative control policy. MaxNet has $M_i = 1$, eliminating these drawbacks.

The complete SumNet open loop small signal transfer function has the same form as (14), (15).

5   Root Loci

Despite their non-linear nature, the small signal convergence behaviour of MaxNet and SumNet can be characterised by the positions of the dominant poles of their linearisation. The MaxNet case is described; SumNet is analogous.

The closed-loop transfer function is

$$T(s) = G(s)(Is + G(s))^{-1} , \tag{19}$$

where $G(s) = sH(s)$. The poles of $T(s)$ are values of s satisfying either of the equivalent equations

$$\det(I + H(s)) = 0 \quad \text{or} \quad \text{eig}(H(s)) = -1 . \tag{20}$$

For non-zero poles, corresponding conditions are

$$\det(Is + G(s)) = 0 \qquad \text{or} \qquad \text{eig}(G(s)) = -s. \qquad (21)$$

The root loci of MaxNet and SumNet have many similarities, but some important differences. The open loop transfer function of each has L poles at zero. In MaxNet, these correspond directly to the sources controlled by the L links. In SumNet, there is intrinsic coupling between the links, and it is not helpful to think of poles as belonging to particular links.

For very small, but positive, loop gain, the poles at the origin move left on the real line. Meanwhile, L infinite sets of poles appear with real part $-\infty$, and with imaginary parts uniformly spaced [5]. These poles move right in the complex plane as the loop gains are increased. Importantly, L of these poles move along the real axis. For MaxNet, it is once again possible to associate each pole with a specific link, while for SumNet, the poles can only associated with eigenvalues of a less structured matrix.

The point at which the rightmost of the poles coming from infinity meets the leftmost of the poles coming from zero is called a breakpoint. At this point, the two poles become a complex-conjugate pair, and start moving at right angles to the real axis, before going right again to eventually cross the imaginary axis and cause instability. As the gains increase further, subsequent pairs of real poles will meet at their respective break points, and also eventually become unstable. Under MaxNet, the pairs of poles which meet at break points always belong to the same link.

The value of the maximum real pole is minimised at the break point, when two real solutions of (20) coincide. At that point, s_l^*, not only are the left and right hand sides equal, but their derivatives are also equal [6].

6   MaxNet Convergence Time

This section will derive bounds on the fastest possible convergence time of MaxNet; that is, the most negative value the real part of the dominant pole as the feedback gain is varied. These results hold for MaxNet networks with arbitrary topology, delay, number of sources and capacity.

Lemma 1. *For sufficiently small gain, each link, l, introduces a pair of real poles. The minimum value achieved (by increasing the gain) of the maximum of these poles is the break point, s_l^*, which lies between $-1/t_{\max_l}$ and $-1/t_{\min_l}$, where $t_{\max_l}$ and $t_{\min_l}$ are the maximum and minimum round trip times (RTTs) of all of the sources being controlled by link l.*

Proof. Since $G(s)$ is lower triangular under MaxNet, the eigenvalues are simply the diagonal elements, each of which corresponds to a particular link. Thus (21) decouples, and we get one equation per link. From (14), poles associated with link l satisfy

$$-s = \sum_{k \in m_l} \frac{a_i e^{-\tau_i s}}{c_l}, \qquad (22)$$

where

$$a_i = \frac{\alpha_i x_{0i}}{\tau_i}. \tag{23}$$

Each of these equations clearly has a real solution for sufficiently small a_i, establishing the first part of the lemma.

Differentiating (22) to find the break point, s_l^*, yields the condition

$$1 = \sum_{k \in m_l} \frac{a_i \tau_i e^{-\tau_i s_l^*}}{c_l} = \sum_{k \in m_l} b_i, \tag{24}$$

where

$$b_i = \frac{a_i \tau_i e^{-\tau_i s_l^*}}{c_l}. \tag{25}$$

Substituting (25) into (22) gives

$$-s_l^* = \sum_{k \in m_l} \frac{b_i}{\tau_i}. \tag{26}$$

Since $1 = \sum_{k \in m_l} b_i$, then (26) is a weighted sum of $1/\tau_i$. A weighted sum is between the maximum and minimum elements in the sum, giving

$$-\frac{1}{\tau_{\min_l}} \le s_l^* \le -\frac{1}{\tau_{\max_l}}. \tag{27}$$

Proposition 1. *At the break point, s_l^* is the dominant pole due to link l.*

Proof. Except for the pole at the origin, all poles of (19) start with infinitely negative real part for low loop gain. Thus it suffices to show that, as the loop gain is increased, no complex pole crosses the line $\mathrm{Re}(s) = s_l^*$ before the real pole starting at $-\infty$ does.

Substituting $s = -\sigma + j\omega$ into (22) yields the implicit equation for pole positions at link l

$$\sum_{k \in m_l} a_k e^{\sigma \tau_k} (\cos(\omega \tau_k) - j\sin(\omega \tau_k)) = \sigma c_l - j\omega c_l. \tag{28}$$

Taking the real part of (28) gives

$$\sum_{k \in m_l} a_k e^{\sigma \tau_k} \cos(\omega \tau_k) = \sigma c_l. \tag{29}$$

Consider a line on the complex plane where $\mathrm{Re}(s) = \sigma$. If we fix the operating point for parameters x_{0k} and τ_k, then by (28), a is element-wise minimized when $\omega = 0$. Since complex poles begin at negative infinity for $a = 0$, and for the minimum $a_{\min}$ that satisfies (28) there is only a real pole on the line $\mathrm{Re}(s) = \sigma$, it follows that the real pole is the first to cross this line as the gain is increased. Complex poles, with $\omega \neq 0$, that cross this line have an element-wise higher a, and are therefore to the left of the real pole when the gain is $a_{\min}$. Thus the real pole at the break point will be the dominant pole for that link, since no complex poles have crossed to its right.

Remark 1. A key conclusion from this analysis is that because the links are independent, it is possible to adjust the control gains such that all links are simultaneously at their break points. That implies that the fastest operation of MaxNet is governed by poles satisfying (27).

7   Bound on SumNet Convergence Time

This section will show that, at least for the specific case analysed, MaxNet has a faster transient response than SumNet.

Due to the complexity of the SumNet analysis, we will consider a two link SumNet network only, where all sources have a common round trip time, τ, and only one source traverses both links. The assumption of a common round trip time is expected to favour SumNet by reducing the coupling between link. Thus we have no reason to believe that any other SumNet will be able to achieve a faster transient response than the equivalent MaxNet. It is sufficient to consider only the real pole, even though there may be complex poles which are slower, since this gives a lower bound for the transient response time.

The SumNet system can be described by a 2×2 open-loop transfer function matrix, $\hat{H}$. Expanding the SumNet form of (14) gives the elements of $\hat{H}$ as

$$\hat{H}_{ij}(s) = \frac{1}{sc_j} \sum_{k \in U_i \cap U_j} e^{-(\tau_{ki}^F + \tau_{kj}^B)s} \hat{a}_k , \tag{30}$$

where U_i is the set of sources that uses link i and

$$\hat{a}_k = \frac{\hat{\alpha}_k x_{0k}}{M_k \tau_k} . \tag{31}$$

The following lemmas are proved in the appendix.

Lemma 2. *For a two link SumNet, where only one source traverses both links, and all sources have the same RTT τ, the unique break point is at $-1/\tau$.*

Lemma 3. *Unless $\hat{\alpha}_k = 0$ for all k, $\hat{G}(s)$ for a two link SumNet does not have a repeated eigenvalue for real s.*

Together, these two lemmas imply that there must be a pole to the right of $-1/\tau$. Therefore SumNet must have a slower transient response than MaxNet.

8   Numerical Results

In this section, the full non-linear SumNet and MaxNet networks are simulated to compare their transient response speeds. The results are evidence that the small signal linearized properties proven analytically in the previous sections are relevant to the practical non-linear system.

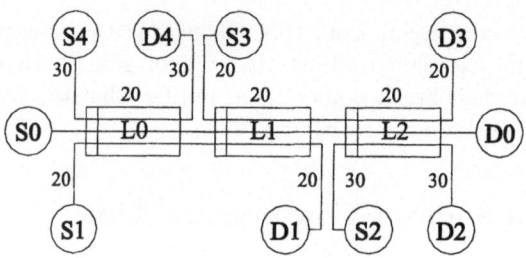

Fig. 2. Network Simulated.

The system simulated in this section is intended to reflect a physically realisable system. Whilst it may be possible to devise a control strategy where each source measures network properties and tunes its own gains (equation (18) for SumNet or (12) for MaxNet) to optimize transient speed, an online algorithm to achieve this is not trivial. In this paper we consider a practical strategy where all sources use the same demand function. We simulate sources with the same static demand function

$$x_i(t) = x_{\max} e^{-\rho q_i(t)} \qquad (32)$$

where ρ is a network wide parameter and $x_{\max}$ is the maximum transmission rate. A similar demand function was introduced in [3], and was shown to be able to satisfy the gain requirements (12). For MaxNet, the parameter ρ relates to the small-signal source gain (12) such that

$$\rho = \frac{\alpha_i x_{0i}}{\tau_i} \qquad (33)$$

and for SumNet the equivalent relationship is with (18)

$$\rho = \frac{\hat{\alpha}_i x_{0i}}{M_i \tau_i} \qquad (34)$$

Note that ρ may be tuned to improve transient performance. This strategy will in general not result in the fastest possible transient response for MaxNet or SumNet, as the poles are not necessarily placed at their closet position to the break-points. Nevertheless it allows us to demonstrate some important properties.

A small network of 5 sources and 3 links, shown in Figure 2, is simulated using both SumNet and MaxNet congestion signaling. Sources $S0 \ldots S4$ transmit to destinations $D0 \ldots D4$ respectively.

We model traffic by a fluid flow approximation, that is, the source transmission rate and congestion price are continuous. At each time step in this discrete time simulation, the flow rate values and price feedback move one unit along in the forward and backward delay paths between sources and links. Acknowledgements are assumed to traverse the same links in the reverse direction, and consume negligible bandwidth. The numbers near each line in Figure 2 represent delays, in simulation time step units. Note that every source has a RTT of 160 units, and for all sources x_{max} is set to 15. The MaxNet or SumNet link control law (3) is at the head of the link, represented by the rectangle inside each link in Figure 2.

In the simulation scenarios, we assume that the best-effort congestion controlled traffic is receiving only a portion of the link's physical capacity. This represents the situation of having higher-priority constant-bit-rate (CBR) traffic occupying some capacity. We simulate two scenarios with different proportions of CBR traffic and different link capacities.

In Scenario 1, the physical link capacities are $c_0 = 5$, $c_1 = 3$ and $c_2 = 5$. To generate a transient, we assume that initially the capacities available to best-effort traffic are 5, 3 and 2 at links $L0$, $L1$ and $L2$ respectively. A transient occurs when the CBR traffic source using $L2$ stops and the available capacities become 5, 3 and 5. Throughout the whole experiment the link gains are, as stipulated in [3], $1/c_l$ such that $\varphi_0 = 1/5$, $\varphi_1 = 1/3$ and $\varphi_2 = 1/5$.

Scenario 2 is the same as scenario 1 except that the physical link capacity of link 1 is $c_1 = 12$, and correspondingly $\varphi_1 = 1/12$. The available capacities again start at 5, 3 and 2, and link 2's available capacity increases to 5.

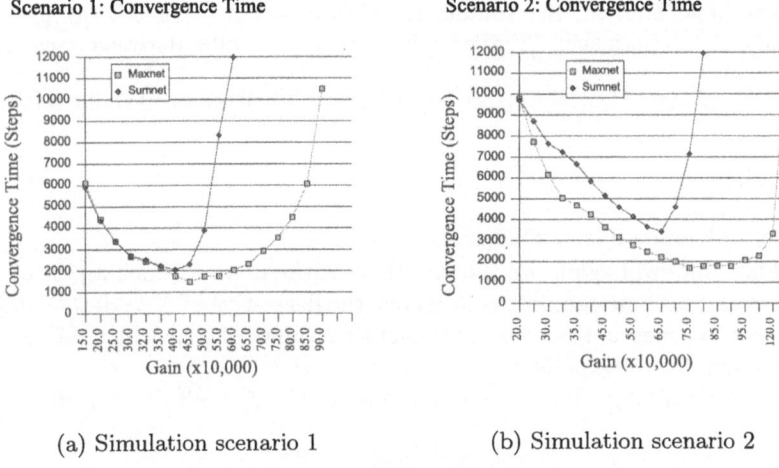

(a) Simulation scenario 1   (b) Simulation scenario 2

Fig. 3. SumNet and MaxNet Convergence Time.

The transient response metric used is the settle time, which is the time from the change of capacity to when the last source is within ±1% of its final value. The settle time is measured in simulation time steps. Figures 3(a) and 3(b) show this convergence time for both SumNet and MaxNet for gains $\rho = 0.0015$ to $\rho = 0.009$ for scenario 1, and $\rho = 0.002$ to $\rho = 0.012$ for scenario 2.

9   Conclusion

This paper has shown that MaxNet flow control has favourable convergence properties compared with traditional SumNet flow control. For small perturbations from the operating point, MaxNet permits a pole placement that has a faster transient response than that possible with SumNet. Numerical results for the complete nonlinear system confirm the conclusions drawn from the analysis of the linear model.

Acknowledgement. This work was funded by the Australian Research Concil.

References

1. Wydrowski, B., Zukerman, M.: MaxNet: A congestion control architecture for maxmin fairness. IEEE Commun. Lett. **6** (2002) 512–514
2. Wydrowski, B., Andrew, L.L.H., Zukerman, M.: MaxNet: A congestion control architecture for scalable networks. IEEE Commun. Lett. **7** (2003) 511–513
3. Paganini, F., Doyle, J.C., Low, S.H.: Scalable laws for stable network congestion control. In: Proc. IEEE Conf. Decision Contr. (CDC), Orlando, FL (2001) 185–90
4. Paganini, F., Wang, Z., Low, S., Doyle, J.: A new TCP/AQM for stable operation in fast networks. In: Proc. IEEE Infocom, San Francisco, CA (2003) 96–105
5. Marshall, J.E., Górecki, H., Walton, K., Korytowski, A.: Time-Delay Systems: Stability and Performance Criteria with Applications. Ellis Horwood, New York, NY (1992)
6. Ogata, K.: Modern Control Engineering. Prentice-Hall, Englewood Cliffs, NJ (1970)

Appendix

The proof of Lemma 2 is as follows.

Proof. This proof will again use the fact that, at the point a which a pair of real poles meet and become complex conjugates, the derivative of $X = \det(I + \hat{H}(s))$ with respect to s is zero. It also uses the fact that values of s for which $dX/ds = 0$ but $X \neq 0$ are not breakpoints.

Since the round trip time of each route is equal, $\tau^F_{ki} + \tau^B_{ki} = \tau$ for all k, (30) implies

$$\hat{H}_{ii}(s) = \frac{1}{sc_i} \sum_{k \in U_i} e^{-\tau s} \hat{a}_k = \frac{e^{-\tau s}}{s} R_i , \tag{35}$$

for $i = 1, 2$, where $R_i = \sum_{k \in U_i} \hat{a}_k / c_i$.

The fact that only one source traverses both links implies that, for $i \neq j$, the sum in (30) contains a single term. Without loss of generality, let that source be source 1. Then

$$\hat{H}_{12}(s)\hat{H}_{21}(s) = \frac{\hat{a}_1^2}{s^2 c_1 c_2} e^{-(\tau^F_{11} + \tau^B_{12} + \tau^F_{12} + \tau^B_{11})s}$$

$$= \frac{e^{-2\tau s}}{s^2} R_3 , \tag{36}$$

where $R_3 = \hat{a}_1^2 / (c_1 c_2)$.

From (20), the 2×2 SumNet poles are at $X = 0$, where

$$X = \hat{H}_{11}(s)\hat{H}_{22}(s) - \hat{H}_{12}(s)\hat{H}_{21}(s) + \hat{H}_{11}(s) + \hat{H}_{22}(s) + 1 .$$

Substituting (35) and (36) into this gives

$$X = \frac{e^{-2\tau s}}{s^2}(R_1 R_2 - R_3) + \frac{e^{-\tau s}}{s}(R_1 + R_2) + 1 . \tag{37}$$

Differentiating (37) to find the break point gives

$$\frac{dX}{ds} = (\tau s + 1)\left(\frac{2e^{-2\tau s}}{s^3}(R_1 R_2 - R_3) - \frac{e^{-\tau s}}{s^2}(R_1 + R_2)\right)$$

$$= (\tau s + 1)A\frac{2e^{-\tau s}(R_1 R_2 - R_3) - s(R_1 + R_2)}{s^3} , \tag{38}$$

where

$$A \equiv \frac{e^{-\tau s}}{s} .$$

This derivative, (38), is zero when $s = -1/\tau$. The root locus occupies the entire negative real axis, and thus $s^* = -1/\tau$ corresponds to an actual breakpoint. It remains to show that there are no other breakpoints.

Assume, with a view to obtaining a contradiction, that there is another breakpoint, s'. At s', the final factor of (38) must be zero. That implies

$$\frac{e^{-\tau s}}{s} = \frac{R_1 + R_2}{2(R_1 R_2 - R_3)} . \tag{39}$$

Substituting (39) into (37) gives

$$X = \frac{3(R_1 + R_2)^2}{4(R_1 R_2 - R_3)} + 1 .$$

But the left had side is positive, since R_3 is one of the terms in the positive-term sum $R_1 R_2$, and so $X \neq 0$. Thus s' is not a pole, and cannot be a breakpoint. This establishes the result.

The proof of Lemma 3 is as follows.

Proof. The eigenvalues of $\hat{G}(s)$ are

$$\hat{G}_{11}(s) + \hat{G}_{22}(s) \pm \sqrt{(\hat{G}_{11}(s) - \hat{G}_{22}(s))^2 + 4\hat{G}_{21}(s)\hat{G}_{12}(s)} . \tag{40}$$

Equating the two solutions to (40) gives the condition for poles being coincident as

$$0 = (\hat{G}_{11}(s) - \hat{G}_{22}(s))^2 + 4\hat{G}_{21}(s)\hat{G}_{12}(s) . \tag{41}$$

When s is real, $\hat{G}_{ij}(s)$ is also real. A real solution to (41) is only possible when $\hat{G}_{21}(s)\hat{G}_{12}(s) \leq 0$. For SumNet, $\hat{G}_{21}(s)\hat{G}_{12}(s) > 0$ for real s, unless $\hat{\alpha}_k = 0$ for all k. Thus (41) cannot be satisfied.

Improving MAC Performance in Wireless Ad-Hoc Networks Using Enhanced Carrier Sensing (ECS)

Zhifei Li[1], Sukumar Nandi[2], and Anil K. Gupta[1]

[1] School of Computer Engineering, Nanyang Technological University,
Singapore, 639798
{pg03802331, asgupta}@ntu.edu.sg
[2] Dept. of Computer Science & Engineering, Indian Institute of Technology,
Guwahati, India, 781039
sukumar@iitg.ernet.in

Abstract. In IEEE 802.11, whenever a node detects an erroneous frame (e.g., a sensing range frame) on the medium, it defers the transmission by a fixed duration (represented by EIFS). We show that this duration is sometimes smaller and sometimes larger than the desired period by which the transmission should be deferred, and it leads to substantial unfairness and throughput degradation. We propose an enhanced carrier sensing (ECS) scheme, which distinguishes among the type of the erroneous frames based on their lengths and defers the transmission accordingly. Simulation results show that the ECS improves the fairness as well as the throughput substantially.

1 Introduction

Recently, wireless ad-hoc networks have attracted considerable research interest as they are easy to deploy and maintain. The Distributed Coordination Function (DCF) in IEEE 802.11 [4] is a CSMA/CA based protocol and is popularly adopted as the MAC protocol for ad-hoc networks. In the IEEE 802.11-based multi-hop wireless ad-hoc networks, since all the nodes share a common medium, only one flow among the contending flows can transmit during a certain duration. Therefore, whenever a frame exchange sequence between two nodes is in progress, it is extremely important to guarantee that all the other nodes in the interference range should defer their own transmission. In IEEE 802.11, whenever a node detects a physical carrier on the medium, it does not transmit. This is known as the *physical carrier sensing*. Moreover, even after the medium becomes idle, the node may need to defer further to allow the transmission of the remaining frames in the same sequence. This is known as the *virtual carrier sensing* (VCS). Under VCS, every frame will carry a duration value indicating the time by which the overhearing nodes should defer. It is clear that the VCS requires that all the potentially interfering nodes should overhear the ongoing frame transmission clearly, which may not be true due to wireless transmission

N. Mitrou et al. (Eds.): NETWORKING 2004, LNCS 3042, pp. 600–612, 2004.

errors and due to the fact that the carrier sensing range is normally greater than the transmission range. Therefore, an important issue is that whenever a node detects an erroneous frame, how long should the node defer its transmission?

In the current IEEE 802.11, whenever a node detects an erroneous frame on the medium, it always defers the transmission by a fixed duration indicated by the Extended Inter-Frame Space (EIFS) constant. We show that in some situations, the EIFS value is too large compared to the desired value, while in some other situations the EIFS value is too small. Respectively, we refer to these two cases as (i) large-EIFS problem, and (ii) small-EIFS problem. The two problems, together, are referred as the imprecise-EIFS problem. When small-EIFS problem occurs, a node may begin its transmission even though some other nodes in the interference range are still transmitting, resulting in collisions. On the other hand, when the large-EIFS problem occurs, the medium may be unnecessarily idle and the node may experience unfairness. Therefore, the imprecise-EIFS problem leads to immense unfairness and throughput degradation. In order to solve the imprecise-EIFS problem, we propose an enhanced carrier sensing (ECS)[1] mechanism in which the EIFS value is made variable in an adaptive manner. The simulation results show that our ECS greatly improves the fairness as well as the capacity utilization.

In the literature, there have been a lot of research-work [1,2,9,10] focusing on improving fairness or throughput by modifying the contention avoidance (CA) or contention resolution (CR) in the CSMA/CA-based protocols. However, to the best of our knowledge, this is the first work that focuses on the carrier sensing (CS) part to improve the performance of the CSMA/CA-based MAC protocols.

The rest of the paper is organized as follows. The imprecise-EIFS problem and the ECS are discussed in sections 2 and 3, respectively. The performance of ECS is studied in Section 4. Section 5 concludes the paper.

2 Imprecise-EIFS Problem in IEEE 802.11

2.1 Preliminaries

A special characteristic of wireless propagation is the attenuation of the transmission power over the distance traversed by the signal. Based on the attenuation, two ranges are defined: the transmission range (TR) and the sensing range (SR). Normally, the SR range is greater than the TR range [11]. Correspondingly, we call a frame detected by a node within TR as the TR frame, and a frame detected by a node out of TR but within SR as the SR frame. If there is no collision or transmission error, a node can receive a TR frame correctly. On the contrary, an SR frame can be detected by the carrier sensing but it cannot be received correctly, and therefore will be treated as an error. If capture is not allowed, a collision occurs whenever there are multiple frames (which may be a mixture of TR and SR frames) around a receiving node and all the frames will

---

[1] The concept of ECS was introduced in [7], where only the unfairness is considered for specific topologies. In this paper, the ECS is extended and generalized.

be destroyed. On the contrary, if capture is allowed, one of the multiple frames may be captured (i.e., received correctly). In this work, we assume that capture is not supported.

To cope with the hidden-terminal problem, IEEE 802.11 defines a four-way handshaking, where a sequence of Request To Send (RTS), Clear To Send (CTS), Data, and Acknowledgement (ACK) frames, is transmitted for the transmission of every single data packet. For the convenience, we call the exchange of RTS/CTS/Data/ACK frames as a frame exchange sequence (FES). FES(X, Y) represents a FES between nodes X and Y, initiated by node X.

We now describe the simulation environment. NS-2 with CMU wireless extensions [3] is used for the simulations. For each single-hop flow, a Constant Bit Rate (CBR) traffic generates 200 packets per second. Each packet is 1000-bytes long, resulting in a traffic source rate of 1.6 Mbps. The raw bandwidth is set with 2 Mbps, leading to a maximum throughput about 1.4 Mbps due to the overhead in IEEE 802.11. Static routing is used. Mobility and transmission errors are not explicitly considered in the simulation. The sensing range (SR) is 550 meters whereas the transmission range (TR) is 250 meters. Other system parameters are set according to the Direct Sequence Spread Spectrum (DSSS) [4], i.e., Slot-Time: 20 μs; SIFS: 10 μs; DIFS: 50 μs; EIFS: 364 μs.

2.2   Large-EIFS Problem

To explain the large-EIFS problem, scenario shown in Figure 1 is used, where the distance between two neighboring nodes is 200 meters. Therefore, nodes A and C are out of TR but within the SR of each other. As there are two single-hop flows, we expect that each flow should get a bandwidth about 0.7 Mbps. However, in the simulation, we found that the flow from B to C gets about 1.15 Mbps whereas the other flow gets only about 0.25 Mbps, showing how much unfair the IEEE 802.11 is in such a simple scenario.

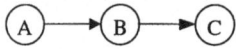

Fig. 1. 3-nodes with Two Single-hop Flows (Scenario-1)

We now explain the reason of the unfairness. Figure 2 demonstrates the process how a node defers its transmission while a FES between two other nodes is in progress. From Figure 2(a), it is easy to see that after a FES(A, B) is successfully completed, nodes A and B will start contending for the medium at about the same time. However, this is not true after the completion of a FES(B, C), as shown in Figure 2(b). Specifically, after node C sends an ACK to node B, node A will detect an SR frame and thus defer its transmission by EIFS rather than by DIFS as done at node B. Since EIFS is equal to *SIFS + TxTime (ACK) + DIFS* [4], the deferment at node A is certainly of much longer duration than that at node B, resulting in unfairness. We call this large-EIFS problem because the EIFS value is larger than it should be to reflect the state of the medium. In fact, the EIFS value should be equal to DIFS in this case. To explain why the throughputs of the two flows differ so much, an analytical model has been developed, which validates the results here. Please refer to [8] for the model.

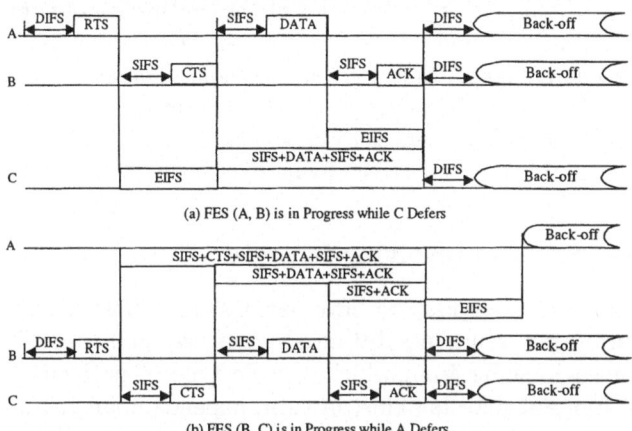

(a) FES (A, B) is in Progress while C Defers

(b) FES (B, C) is in Progress while A Defers

Fig. 2. Time Diagram Showing Frame Exchange Sequence

2.3   Small-EIFS Problem

Now we introduce the small-EIFS problem using the scenario of Figure 3. The distance between two neighboring nodes is again 200 meters. Our results show that the two flows share the bandwidth equally (in a long term). However, each of them only gets about 0.31 Mbps, resulting in an aggregate throughput about 0.62 Mbps, which is much smaller than the capacity (i.e., 1.4 Mbps).

Fig. 3. 4-nodes with Two Single-hop Flows (Scenario-2)

To explain why the throughput degrades so much, let us consider the situation where nodes C and D are deferring their transmission, while FES(A, B) is in progress. After node B sends a CTS to A, since node D is in the sensing range of node B, node D gets an SR frame and defers its transmission with EIFS. Then, node A begins to send its Data frame to B. However, since node D is out of the SR of A, node D cannot detect this Data frame. Moreover, since the EIFS value is much smaller than the transmission time of the Data frame, after deferring for the EIFS duration, node D may begin transmit a frame even though node A is still transmitting the Data frame to B. The two transmissions result in a collision at node B and thus node B will discard the Data frame. Likewise, due to symmetry in this topology, for a FES(D, C), node C is also likely to discard the Data frame, explaining why the throughputs degrade so much. We call this as the small-EIFS problem because the EIFS value is smaller than it should be to reflect the state of the medium. In fact, in this situation, the EIFS value should be large enough to allow for the complete transmission of the Data frame.

In the above discussion, we assumed that the capture is not supported. When the capture is supported, node B *may* capture the Data frame from node A if the SNR of the frame is higher than a given threshold. Therefore, the throughputs will be greatly improved. However, when the capture is supported, the small-EIFS problem leads to unfairness as discussed in [8]. Therefore, we should resolve the small-EIFS problem irrespective of whether or not the capture is supported.

2.4   General Scenario Involving Imprecise-EIFS Problem

So far using specific examples, we have demonstrated how the large-EIFS and small-EIFS problems result in substantial unfairness and throughput degrade. In fact, a node will suffer from which specific type of problem depends upon what kind of frame it detects. This, in turn, depends upon the location of the node with respect to the location of the two nodes between whom the FES is in progress. For example, if a node gets an SR frame corresponding to a CTS, it may suffer from the small-EIFS problem. If a node gets an SR frame corresponding to an ACK, it may suffer from the large-EIFS problem. However, if a node gets an SR frame corresponding to a Data frame, the imprecise-EIFS problem does not arise because the EIFS value caters for the time needed for the next frame (i.e., ACK) to pass through. This is also true when an SR frame corresponding to a RTS is detected, since the next frame is a CTS frame, and the length of the CTS is equal to an ACK. Figure 4 shows a generalized scenario with imprecise-EIFS problem where FES(A, B) is in progress. In the figure, T(frame) denotes a TR frame and S(frame) denotes an SR frame. We classify the entire region into seven areas. Nodes within different areas get different type of frames and thus may suffer from different type of problems. Note that in the area 6, a node will not suffer from the small-EIFS problem after detecting an SR frame of CTS, since the node can also detect the SR frame corresponding to the Data.

3   Enhanced Carrier Sensing (ECS)

EIFS based deferment, as mentioned, is designed to allow the next frame in a FES to pass through. Since the next frame can be of any type (i.e., CTS, Data, or ACK), the transmission time of this frame may differ substantially. However, the IEEE 802.11 does not distinguish among different SR frames and uses the same constant EIFS value in all the cases, resulting the imprecise-EIFS problem. Therefore, if somehow a node can distinguish among different type of SR frames and adopts different value of EIFS accordingly, the problem can be greatly reduced and thus the performance (i.e., fairness and throughput) can be improved. In the light of this discussion, we make two proposals: *(1)* Whenever

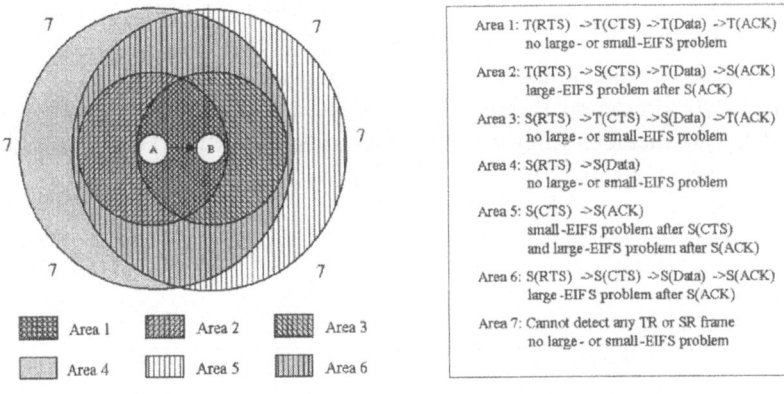

Fig. 4. Problems Arising in Different Areas while FES(A, B) is in Progress

a node detects an SR frame, it should try to identify the type of that frame; *(2)* EIFS value should be directly linked to the type of SR frame detected.

Since our CS mechanism tries to identify the type of an SR frame rather than just sensing its presence, we call it Enhanced Carrier Sensing (ECS).

3.1 Distinguishing among SR Frames

Basically, there are two methods in which we can distinguish among different type of SR frames. The first method is to use the recent history of frames observed on the medium to decide the type of the current SR frame as the frames are transmitted in a specific sequence. However, this method is very complex and the recent history may not provide very precise information as there may be multiple FESs in progress at any given time. The second method is to get the information from the frame itself. However, it is not trivial as the node cannot interpret the contents of an SR frame. We propose that the various type of frames should have different lengths, and based on the *length* of an SR frame observed on the medium, the type of the frame is identified.

We first need to differentiate the lengths of different type of frames. In IEEE 802.11 [4], the lengths corresponding to the control frames are as follows, RTS: 20 bytes, CTS: 14 bytes, ACK: 14 bytes. On the other hand, the header of a Data frame is 34 bytes, implying that the length of a Data frame must be greater than 34 bytes. In order to distinguish between CTS and ACK, the size of CTS should be increased by a few bytes. The reason why we increase the size of CTS rather than ACK is that the extra bytes in the CTS may be used by the receiver to add some receiver status information, which may be useful to the sender before it sends out the Data frame. What is an appropriate number of bytes that should be added to the CTS depends upon the trade-off between the sensitivity of the physical layer and the additional overhead introduced due to these bytes. In our implementation, we change the length of the CTS to 17 bytes.

When a node gets an SR frame, to identify the type of the frame, the node only needs to detect the length of that frame. Before discussing how to detect the length, we have to introduce the format of the physical layer frame in the IEEE 802.11 [4]. As presented in Figure 5, a Physical Layer Convergence Protocol (PLCP) frame includes three parts: PLCP Preamble, PLCP Header, and MAC Protocol Data Unit (PDU). The PLCP preamble is used by the receiver to synchronize, while the PLCP header contains information to help the receiver to decide the end of the frame. The MAC PDU corresponds to the MAC layer frame, e.g., a RTS frame.

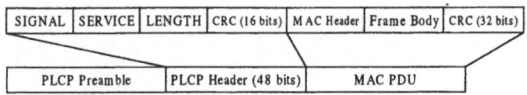

Fig. 5. PLCP Frame Format

Now we discuss how to detect the length (in terms of bytes) of an SR frame by considering two possibilities. The *first* possibility is that the PLCP header part of the frame can be correctly received by the physical layer but the MAC PDU part cannot be interpreted by the MAC layer. This is very likely due to the following two reasons. *(i)* In the IEEE 802.11, the PLCP header is always transmitted at the lowest transmission rate (i.e., 1 Mbps) while the MAC PDU in the same PLCP frame may be transmitted at a higher rate (e.g., 2 Mbps or even higher). Generally, the Bit Error Rate (BER) under the case of lower transmission rate should be smaller if the Signal to Noise Ratio (SNR) is assumed to be the same, which should be true during the transmission of an entire PLCP frame. Therefore, the BER for the PLCP header may be much smaller than that for the MAC PDU. *(ii)* Moreover, the length of the PLCP header (i.e., 6 bytes) is much smaller than that of the MAC PDU (e.g., 20 bytes for a RTS frame). Due to the above two reasons, it is very likely that the PLCP header is correctly received while the MAC PDU is not. In such a situation, the MAC frame length (and thus the type) can be easily calculated from the fields contained in the PLCP header. Note that the above discussion also applies for a TR frame containing transmission errors as the PLCP header of a TR frame should always be interpretable.

The *second* possibility is that, when an SR frame is detected, even the PLCP header part in the PLCP frame cannot be correctly received by the physical layer. In such a situation, to identify the type of the MAC frame, we can make use of the Clear Channel Assessment (CCA) mechanism [4]. Specifically, based on the length of the time duration that the CCA mechanism indicates a busy medium, if the transmission rate is fixed for all the frames, the MAC layer can calculate the length (in terms of bytes) and thus identify the type. When the transmission rate of the MAC PDUs varies, as discussed in [8], the type of an SR frame can also be identified with the help of CCA mechanism. However, for the simplicity, in our simulation the transmission rate of the MAC PDU is fixed.

While our ECS will get help from the physical layer to detect the type of a frame, it does not need any enhancement at the physical layer. Therefore, it can be easily incorporated into IEEE 802.11.

3.2 EIFS Values in ECS

Based on the type of the frame observed on the medium, an EIFS value should be chosen accordingly. The basic rule for deciding the EIFS value is that it should be large enough to allow the complete transmission of the *next* frame in the sequence. Based on this rule, five different values of EIFS are defined in Figure 6. For a RTS type SR frame, the EIFS value is equal to SIFS + TxTime (CTS). When a node detects a CTS type SR frame, since the next frame in the sequence is a Data frame whose length may be variable and cannot be detected from the CTS type SR frame, we simply set the EIFS according to the maximum length allowed for the Data frames, i.e. Max-Data-Length. For a Data type SR frame, the EIFS value is set to SIFS + TxTime (ACK). Lastly, for an ACK type SR frame, since this is the last frame in the sequence, the EIFS value is set to DIFS.

In wireless ad-hoc networks, due to the mobility and dynamic propagation characteristics (e.g., fading), when an SR frame is detected by a node, it may only be a part of a frame rather than being a complete frame. We call this partial-frame problem. Another type of problem may occur as follows. In the multi-hop scenario, spatial reuse of the bandwidth is very likely. Therefore, multiple frames may be there around a node at any given instant, resulting in a collision. As a result, the SR frame that a node detects may be an overlapped of several frames rather than a single frame. We call this overlapping-frames problem. In fact, it is very difficult for an MAC protocol to fully take care of these two problems. Therefore, whenever any of these two problems occur, as shown by the last line in Figure 6, we use the same value of EIFS as defined in the standard [4].

```
EIFS(RTS) = SIFS + TxTime(CTS);
EIFS(CTS) = SIFS + TxTime(Max-Data-Length);
EIFS(Data) = SIFS + TxTime(ACK);
EIFS(ACK) = DIFS;
EIFS(collision or unknown type) = Standard EIFS;
```

Fig. 6. Different EIFS Values in ECS

4 Simulation Results

In the performance evaluation of our Enhanced Carrier Sensing (ECS), the same simulation parameters are used as described in Section 2.

4.1   Performance for Scenarios with Two-Flows

Scenario-1: This scenario (Figure 1) has been used to demonstrate the large-EIFS problem in Section 2. The average throughputs under the scenario are presented in Table 1. The IEEE 802.11 is very unfair due to the large-EIFS problem at node A. However, under our ECS, the two flows share the bandwidth equally. Moreover, as shown in Table 1, the aggregate throughput under the ECS is slightly greater than that under IEEE 802.11. This can be explained as follows. Consider that after a successful transmission by node B, nodes A and B contend for the medium. Suppose that the node A generates a random back-off timer equal to 10 slots, while the node B generates a random back-off timer with 30 slots, the node A will certainly win the contention in this round. If the large-EIFS problem does not occur as in the case of ECS, the medium will be idle for only 10 slots and then node A begins to transmit. On the contrary, if the large-EIFS problem occurs as in the case of IEEE 802.11, the medium will be idle by an extra duration equal to $(EIFS - DIFS)$, i.e., 16 slots. In summary, since the ECS solves the large-EIFS problem, it reduces the idle time of the medium, explaining the throughput improvement in ECS. Moreover, the throughput improvement should be much higher if a high-rate physical layer is used.

Scenario-2: This scenario (Figure 3) has been used to demonstrate the small-EIFS problem in Section 2. The average throughputs are presented in Table 2. Under the IEEE 802.11, due to the small-EIFS problem, the aggregate throughput is very small (i.e., 0.621 Mbps). However, under our ECS, the aggregate throughput greatly improves (i.e., 1.334 Mbps) as the ECS solves the small-EIFS problem. Note that the results are obtained under static routing and without capture capability. In [6], by varying the factors including the capture ability and the routing protocols, we have conducted an extensive simulation study for this topology.

Table 1. Throughput under Scenario-1   **Table 2.** Throughput under Scenario-2

Throughput (Mbps)	IEEE 802.11	ECS
A to B	0.254	0.705
B to C	1.154	0.718
Aggregate	1.408	1.423

Throughput (Mbps)	IEEE 802.11	ECS
A to B	0.314	0.662
D to C	0.307	0.672
Aggregate	0.621	1.334

Scenario-3: Now we discuss the results for the scenario shown in Figure 7. This scenario is similar to scenario-2 except that the direction of the flows is reversed. It is easy to see that after node A sends back an ACK frame to node B, node C suffers from the large-EIFS problem if IEEE 802.11 is used. This is also true for node B after the node D sends back an ACK frame to node C. The average throughputs of the two flows are presented in Table 3. In contrast to scenario-1, the large-EIFS problem does not result in *long-term* unfairness in this scenario (i.e., otherwise the flows will have different average throughputs). However, it results in the *short-term* unfairness. To show this, in Figure 8, we present the Jain's index results using the sliding window method discussed in [5]. Generally, the higher the index, the fairer the protocol is. It is easy to see that the ECS greatly improves the fairness compared to IEEE 802.11. The reason that the large-EIFS problem leads to short-term unfairness is as follows. Consider that node B successfully transmits an upper-layer packet, and the chance that the node B gets control of the medium *again* is larger than that of node C, as node C suffers from the large-EIFS problem. On the other hand, once node C controls the medium, node C will also have high chance to transmit *consecutively*. Therefore, the large-EIFS problem results in short-term unfairness. However, since the large-EIFS problem occurs at nodes B and C with the same possibility, the long-term fairness between the two flows is ensured.

Fig. 7. 4-nodes with Two Single-hop Flows (Scenario-3)

Table 3. Throughput Comparison for Scenario-3

Throughput (Mbps)	IEEE 802.11	ECS
B to A	0.708	0.719
C to D	0.702	0.710
Aggregate	1.410	1.429

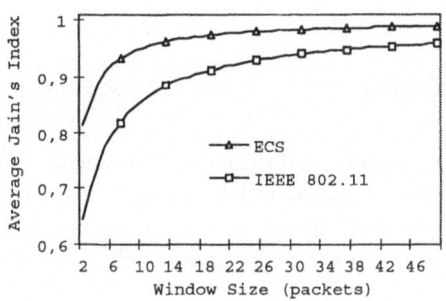

Fig. 8. Fairness Index Comparison for Scenario-3

Scenario-4: Now we discuss the results for the scenario shown in Figure 9, which is similar to scenario-2 except that the distance between nodes B and C is increased to 400 meters. Therefore, nodes B and C are out of the TR but within the SR of each other, and they suffer from the small-EIFS problem as indicated by area 5 in the generalized scenario (Figure 4). Table 4 presents the throughput results. The aggregate throughput under the IEEE 802.11 is 0.155 Mbps, which is much smaller than that under the ECS (i.e., 0.578 Mbps). However, even in the ECS, the aggregate throughput is still small compared to the medium capacity (i.e., about 1.4 Mbps). However, this is not due to the deficiency of ECS. For a detailed reason, please refer to [8].

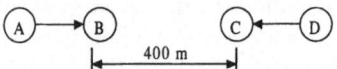

Fig. 9. 4-nodes with Two Single-hop Flows (Scenario-4)

Table 4. Throughput Comparison for Scenario-4

Throughput (Mbps)	IEEE 802.11	ECS
A to B	0.079	0.290
D to C	0.076	0.288
Aggregate	0.155	0.578

Scenario-5 and -6: Here we present the results for two more scenarios. The topologies are presented in figures 10 and 11, respectively, while the throughputs are presented in tables 5 and 6, respectively. It is noted that the ECS improves both the fairness and the throughput. For a detailed explanation of these results, please refer to [8].

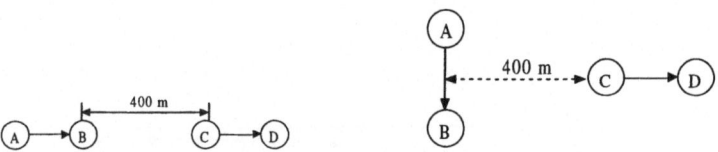

Fig. 10. Scenario-5   **Fig. 11.** Scenario-6

By referring to the generalized scenario discussed in Section 2.4, one can easily design some other scenarios (with two flows) to show the effects of the imprecise-EIFS problem and the advantage of our ECS [8]. However, due to space limitation, we do not present any more such scenarios here.

4.2   Performance for a Complex Scenario

Figure 12 shows the double-ring topology with 16 nodes. The distance between the sender and the receiver of each flow is 200 meters. The diameter of the inner circle is also 200 meters. Therefore, the diameter of the outer circle is 600 meters. The angle between any two neighboring flows is 45 degrees. The performance measures are presented in tables 7 and 8. The ID of a flow is the same as the node ID of the flow's sender. The ECS greatly improves the aggregate throughput in this topology. As for the fairness, both the standard deviation and Jain's index under ECS are slightly greater than those under IEEE 802.11,

Table 5. Throughput under Scenario-5   **Table 6.** Throughput under Scenario-6

Throughput (Mbps)	IEEE 802.11	ECS
A to B	0.0	0.075
C to D	1.398	1.338
Aggregate	1.398	1.413

Throughput (Mbps)	IEEE 802.11	ECS
A to B	1.161	0.672
C to D	0.254	0.766
Aggregate	1.415	1.438

which may convey very conflicting conclusions. If the standard deviation alone is considered, it seems that IEEE 802.11 is more fair than the ECS. On the other hand, the Jain's index is indicating that the ECS is more fair. Because the throughputs under the two schemes differ by an order of magnitude, a larger standard deviation does not imply that ECS is more unfair. Therefore, standard deviation is not always a good measure of fairness, while the Jain's index is.

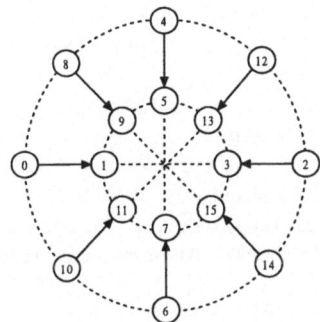

Fig. 12. Double-ring Topology: 16 Nodes with 8 Flows

Table 7. Throughput in Double-ring Topology   **Table 8.** Performance Measures

Flow ID	0	2	4	6	8	10	12	14
IEEE 802.11	0.0169	0.0214	0.0197	0.0195	0.0207	0.0206	0.0197	0.0209
ECS	0.1586	0.1587	0.1542	0.1517	0.1641	0.1627	0.1663	0.1483

Metrics	Aggregate	STD DEV	Jain's Index
IEEE 802.11	0.1594	0.0013	0.9957
ECS	1.2646	0.0059	0.9986

We also conducted the simulation for several more complex scenarios (e.g., chain and grid topologies) [8]. It is found that the ECS always improves the performance of IEEE 802.11.

5   Conclusions

In IEEE 802.11, whenever a node detects an erroneous frame, it will always defer by a fixed duration (represented by EIFS). Due to this fixed EIFS value, we showed that two problems arise: small- and large-EIFS problems, which lead to considerable unfairness and throughput degradation. To solve the problems, we have proposed an enhanced carrier sensing (ECS) mechanism. In the ECS, the lengths of the frames are made different. Based on the length of a frame observed on the medium, the type of the frame can be detected, and the node defers the transmission for a duration accordingly. In order to identify the length of an erroneous frame, the ECS utilizes the information provided by the physical layer. However, the ECS does not require any enhancement at the physical layer, and thus it can be easily incorporated into IEEE 802.11. The extensive simulation results have shown that the ECS eliminates the small- and large-EIFS problems to a great extent, and thus improves the fairness as well as throughput drastically.

References

1. V. Bharghavan, A. Demers, S. Shenker, and L. Zhang, "MACAW: A Media Access Protocol for Wireless LANs," ACM SIGCOMM, 1994.
2. F. Cali, M. Conti, E. Gregori, "Dynamic Tuning of the IEEE 802.11 Protocol to Achieve a Theoretical Throughput Limit," IEEE Journal on Selected Areas in Communications, December 2000, pp.785-799.
3. CMU Monarch Group. CMU Monarch Extensions to NS, http://www.monarch.cs.cmu.edu/.
4. IEEE, "Wireless LAN Medium Access Control (MAC) and Physical Layer (PHY) specifications," IEEE 802.11 standards, June 1999.
5. C.E. Koksal, H. Kassab, H. Balakrishnan, "An Analysis of Short-Term Fairness in Wireless Media Access Protocols," ACM SIGMETRICS, 2000.
6. Z.F. Li, S. Nandi, A.K. Gupta, "Study of IEEE 802.11 Fairness and its Interaction with Routing Mechanism," IFIP MWCN, 2003.
7. —— "Improving Fairness in IEEE 802.11 using Enhanced Carrier Sensing (ECS)," accepted by IEE Proceedings on Communications, Dec., 2003.
8. —— "Improving MAC Performance in Wireless Ad-hoc Networks using Enhanced Carrier Sensing (ECS)," full version of the paper, avaliable at http://www.ntu.edu.sg/home5/pg03802331/.
9. H. Lou, S. Lu, V. Bharghavan, "A New Model for Packet Scheduling in Multi-hop Wireless Networks," ACM MOBICOM, 2000.
10. T. Nandagopal, T. Kim, X. Gao, V. Bharghavan, "Achieving MAC Layer Fairness in Wireless Packet Networks," ACM MOBICOM, 2000.
11. K.X. Xu, M. Gerla, S. Bae, "How Effective is the IEEE 802.11 RTS/CTS Handshke in Ad Hoc Networks?", IEEE Globecom, 2002.

Modeling the Short-Term Unfairness of IEEE 802.11 in Presence of Hidden Terminals

Zhifei Li[1], Sukumar Nandi[2], and Anil K. Gupta[1]

[1] School of Computer Engineering, Nanyang Technological University,
Singapore, 639798
{pg03802331, asgupta}@ntu.edu.sg
[2] Dept. of Computer Science & Engineering, Indian Institute of Technology,
Guwahati, India, 781039
sukumar@iitg.ernet.in

Abstract. In this paper, using a simple hidden-terminal scenario, we show that IEEE 802.11 exhibits substantial short-term unfairness, though it provides long-term fairness. We analyze the short-term behavior using embedded-Markov chain method to answer the following two questions: *(i)* once a node gets control of the medium, what is the average number of packets this node can transmit consecutively without experiencing any collision, *(ii)* once a node loses its control of the medium, what is the average time the node has to wait before it gets control of the medium again. The first question reflects on how long a node can *capture* the medium, whereas the second question reflects on how long a node may be *starved*. The analytical model is validated by the simulation results. Our work is distinct from most of the work published in the literature in two aspects: we focus on the short-term behavior rather than the long-term, and the analytical method is adopted for the study.

1   Introduction

IEEE 802.11 [6] is the de facto standard for Wireless LANs, and it defines two MAC protocols: Point Coordination Function (PCF) and Distributed Coordination Function (DCF). However, only the DCF is popular. As DCF operates in a *distributed* manner, achieving fairness in accessing the medium is one of the most challenging issues. Based on the *length* of the time over which we observe the system, the fairness can be defined on a short-term basis and a long-term basis. The short-term fairness automatically gives rise to long-term fairness, but not the vice versa [7]. In particular, under certain scenarios, though the bandwidth allocation is fair in a long-term, it is very unfair if we view the system from a short-term viewpoint. The short-term fairness is important for the adaptive traffic (e.g., TCP traffic) and for the delay- or jitter-sensitive traffic [7]. In this paper, we aim to analyze the short-term unfairness of IEEE 802.11.

The *duration*, over which the short-term fairness should be measured, is difficult to define as it depends upon the requirements of applications as well as upon the channel bandwidth. To get around this problem, Jain's index [7] can

N. Mitrou et al. (Eds.): NETWORKING 2004, LNCS 3042, pp. 613–625, 2004.

be used to reflect fairness over different time scales. Though the index is useful in comparing fairness of two *different* protocols, the *absolute* value of the index for a *given* protocol does not express the fairness of the protocol very clearly. Therefore, we measure the short-term fairness in an alternate way by evaluating the following two metrics: *(i)* once a node gets control of the medium, what is the average number of packets this node can transmit *consecutively* without experiencing any collision, *(ii)* once a node loses its control of the medium, what is the average time the node has to wait before it gets control to the medium again.

In this paper, using an embedded Markov chain model, the above two metrics are measured based on the concepts of *'expected state holding time'* and *'expected first passage time'*. The analytical model is validated by the simulation results. Our results show that the IEEE 802.11 exhibits substantial short-term unfairness even in a very simple hidden-terminal scenario.

The remainder of the paper is organized as follows. In Section 2, using simulation method, we show that IEEE 802.11 exhibits considerable short-term unfairness in the hidden-terminal scenario. In Section 3, the Markov chain model is described. Section 4 presents the analytical and simulation results. Related work is reviewed in Section 5 and the paper is concluded in Section 6.

2   Short-Term Unfairness in IEEE 802.11

To show the short-term unfairness in IEEE 802.11, we simulate the well-known hidden-terminal scenario depicted in Figure 1. There are three nodes, A, B and C, with two single-hop flows: flow from A to B, and flow from C to B. Since nodes A and C cannot hear from each other, they may simultaneously try to communicate with a common node, i.e., node B, resulting in a collision. In such a situation, nodes A and C are referred as the hidden terminals of each other.

Fig. 1. Hidden-terminal Scenario

In the simulation, NS-2 with CMU wireless extensions [4] is used. For each single-hop flow, a Constant Bit Rate (CBR) traffic is adopted, where each packet is 1000-bytes long. The raw bandwidth is set with 2 Mbps, and the *maximum* throughput is about 1.4 Mbps due to the overhead of IEEE 802.11. The source rate of each single flow is made greater than the medium capacity, since unfairness occurs only when the system is overloaded.

The simulation results show that each flow gets an *average* throughput of about 0.7 Mbps, indicating that the two flows share the medium fairly on a *long-term* basis. However, if we compute the values of the two *short-term* metrics defined in Section 1, the protocol exhibits substantial unfairness. For metric *(i)*, we find that, on an average, once a node gets control of the medium, it can transmit about 6.4 packets *consecutively* without collision. For the metric *(ii)*, once a node (say node C) loses control of the medium, it has to wait for the other node (i.e., A) to transmit about 27 packets before it gets control of the medium again. However, this does not mean that node A can transmit 27 packets *consecutively* without collision. We illustrate this using the following example. Consider that after node C *loses* control of the medium, node A gets control of the medium and it transmits one or more packets consecutively. Then, one or more collisions occur. After the collision(s), node A again gets control and transmits one or more packets consecutively. In addition to the average values, we also observed the corresponding *maximum* values, which are equal to 35 and 160, respectively, for the two metrics. This shows how much short-term unfairness is ingrained in IEEE 802.11, which is unacceptable for jitter-sensitive traffic. In this section, we explain how hidden-terminal problem causes short-term unfairness.

2.1   Basic Techniques in IEEE 802.11

IEEE 802.11 adopts the well-known binary exponential back-off (BEB) algorithm as its contention resolution mechanism, which is described as follows. Every node maintains a Contention Window (CW) and a back-off timer. Before every transmission, the node first defers by a back-off timer, which is generated according to equation (1), unless the back-off timer already contains a non-zero value, in which case it is unnecessary to generate a new random back-off timer.

$$BackoffTime = Random() \times SlotTime \qquad (1)$$

The *SlotTime* is specified by the physical layer, and the *random* value is uniformly distributed over the range [0, CW]. For the *first* transmission attempt of a packet, the CW is set to CW_{min}. Whenever a retransmission is initiated, the CW is doubled until the CW_{max} is reached. After that, the CW remains at the maximum value until the retry limit (say, n) is reached. Once the retry limit is reached, the CW will be reset to CW_{min}. The CW is also reset to CW_{min} whenever a transmission attempt is successful. For the convenience, we call each retransmission attempt as a *stage*, whose number is in the range $[0, n - 1]$.

To combat with the hidden-terminal problem, IEEE 802.11 defines a four-way handshake, where a sequence of Request To Send (RTS), Clear To Send (CTS), Data, and ACK frames, is transmitted for the transmission of every single data packet. For the convenience, we call the exchange of RTS/CTS/Data/ACK frames as a *frame exchange sequence (FES)*. Moreover, *FES (X, Y)* represents a FES between nodes X and Y, *initiated* by node X, implying node X sends one *packet* successfully to node Y. Moreover, we call node X as the *transmitting* node and Y as the *receiving* node, while all the other nodes are called the *waiting* nodes. Note that the word '*packet*' implies the protocol data unit (PDU) of a higher layer whereas '*frame*' is the MAC layer PDU.

When a FES is in progress, the *waiting* node *freezes* its back-off timer. After the FES is successfully completed, all the nodes first defer for a DCF Inter-Frame

Space (DIFS) period. Then, the *transmitting* node generates a new random value from its CW and backs-off before it initiates another FES. On the other hand, the *waiting* node simply resumes to count down from its *frozen* back-off timer. It is easy to see that the *transmitting* node may transmit several packets *consecutively* before the *waiting* node's back-off timer is reduced to zero. Contrary to a successful transmission, when a collision occurs, *all* the *colliding* nodes will generate a new random value from their corresponding CWs.

2.2   Explanation for the Short-Term Unfairness

In the four-way handshake, once the RTS/CTS has been completed successfully, the hidden-terminal problem may not arise any more. For example, in Figure 1, once node B sends back a CTS to node A, node C overhears this CTS and thus defers its transmission, avoiding collision. The four-way handshake solves the hidden-terminal problem largely by introducing the RTS/CTS handshake before the real Data frame is transmitted. However, it cannot *eliminate* the problem completely as RTS/CTS cannot always be transmitted successfully.

Now let us derive the *condition* under which the RTS/CTS can be successful when two hidden nodes (A and C) contend for the medium (Figure 1). The condition is as follows: after a collision or a FES, the difference between the back-off timers at the two hidden nodes should be large enough for node B to send back a CTS to node A (C) before that node C (A) starts sending its RTS. The *minimum* time difference required is equal to the transmission time of RTS plus a Short Inter-Frame Space (SIFS). This can be expressed as:

$$| Z |> Len = TxTime(RTS) + SIFS \qquad (2)$$

where Z is the difference between the back-off timer. *Len* is equal to about 19 slots when the slot time is 20 μs for DSSS [6]. It is easy to see that the condition is difficult to satisfy when the CWs at the contending nodes are small (e.g., 31).

Now let us explain how the hidden-terminal problem causes short-term unfairness. Consider the situation that the CWs at nodes A and C are very small (e.g., 31). As discussed above, under such situation, the transmission of RTSs of nodes A and C may overlap partially, and as a result collide. The collision may occur several times until the CWs are large enough to allow either node (say, node A) to get control of the medium. Once the FES (A, B) is completed, node A resets its CW to CW_{min} and backs-off before initiating another FES. However, the remaining back-off timer at node C may be large compared to the back-off timer at node A, and thus nodes A and B may exchange several more FESs before node C's back-off timer reduces to a small value.

Whenever the back-off timer at node C reduces to a small value, node C contends for the medium. However, as the CW at node A is equal to CW_{min}, the contention is most likely to result in a collision again. After the collision, node A doubles its CW from CW_{min} whereas node C doubles its CW from a larger value (at least 63). Therefore, the CW at node C is greater than that at A, and node A is more likely to get control of the medium *again*. Moreover, this process (i.e., several packet transmissions by node A, followed by collisions, and then transmissions by node A again) may repeat several times, leading to starvation at node C for a long period (compared to the time needed for a FES).

However, several mechanisms incorporated in IEEE 802.11 prevent node C from starving *completely*, such as: *(i)* after every FES, node A will back-off before initiating another FES, which gives node C a chance to contend for the medium with node A; *(ii)* the CW at node C will be reset to CW_{min} after the retry limit n is reached. Once node C controls the medium, it can transmit consecutively in a similar way, and thus the long-term fairness between the two flows is ensured.

3 Analytical Modeling

In this section, we model the hidden-terminal scenario in Figure 1 using an *embedded* Markov chain.

3.1 Markov Chain Model

At any point of time, the *medium* is in one of the following five states: T_A, T_B, T_C, *Col* and *Idle*, where T_A means that node A is getting the control of the medium and transmitting its packet, i.e., FES(A, B) is in progress. Similarly, T_B and T_C correspond to nodes B and C, respectively. However, in our considered scenario, T_B does not arise, as node B does not have any data packets to send. State *Col* means that there is a collision on the medium, while state *Idle* means that there is no transmissions or collisions over the medium. As our objective is to analyze the fairness rather than the capacity utilization of the medium, we do not need to consider the *Idle* state. As a result, only three states of the medium are considered: T_A, T_C, and *Col*. When the medium is either in T_A or in T_C, we simply say that the medium is in a T state. Since the transition probabilities among these three states depend on the values of CWs at nodes A and C, which, in turn, are determined by the corresponding *stages*, the *system* can be modeled using three random variables: state of the *medium*, *stage* at node A, and *stage* at node C. Therefore, the system states are (T_A, k, l), (T_C, k, l), and (Col, k, l), where k and l denote the *stages* at nodes A and C, respectively. Obviously, $k, l = 0, ..., (n-1)$. Note that we must use *stages* rather than the values of CW to represent the system state, since once the CW reaches the maximum value, it remains unchanged before resetting. When the medium is in state T_A, it is easy to see that the stage at the *transmitting* node (i.e., node A) must be zero, that is, only $(T_A, 0, l)$ system states are possible. Similarly, only $(T_C, k, 0)$ system states are possible. Therefore, if the retry limit is n, there are n^2 number of *Col* states, and n number of states corresponding to *each* of T_A and T_C, and thus the number of all possible system states is:

$$N_{state} = n^2 + 2n \tag{3}$$

From a *Col* state, whenever a *transition* occurs, the system can enter anyone of the three kinds of states as shown in the leftmost diagram of Figure 2. If the *next* state is also a *Col* state, both of the stages, k and l, are incremented by one, except that the stage is reset to zero whenever the retry limit n is reached. On the other hand, if the next state is a T state, the stage at the *transmitting* node is reset while the other stage remains unchanged.

In state T_A, whenever node A transmits another *packet*, it is natural to view this event as a *self-transition*. However, if we model the system in such a way, the transition probabilities depend upon the remaining back-off timer at the waiting node (i.e., node C), which in turn, depends upon how many times this timer has been *frozen*, i.e., how many self-transitions have occurred in the T_A state. This requires *memorizing* the *history* to obtain the *current* transition probabilities, which violates the *memoryless* requirement of a Markov chain. Therefore, we do *not* treat this event as a state transition. Rather, whenever in state T_A, a state transition occurs *only* when the system enters a *Col* state or a T_C state. Therefore, whenever the system enters a T_A state, the time that the system will *remain* in that state depends on the number of packets that node A can transmit consecutively before that node C controls the medium or that a collision occurs. Similar explanation applies for the state T_C. The state transitions from a T state are illustrated in the remaining diagrams of Figure 2. It is easy to see that the chain obtained in such a way is an *embedded* Markov chain, modelling the underlying semi-Markov process.

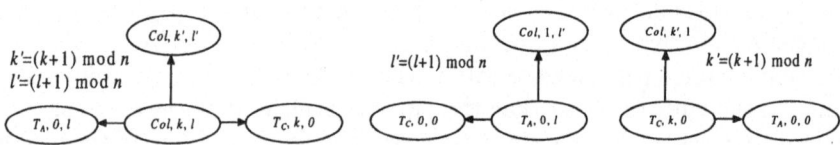

Fig. 2. State Transitions Diagrams

To illustrate the model clearly, we present the complete state transition diagram in Figure 3, where we assume $n=3$.

3.2   Basic Analysis of the Model

For the convenience, rather than representing the system states using three variables as above, we assign a *single* variable to represent the states, by ordering them as indicated in Table 1.

For the embedded Markov chain, we need to know the transition probability matrix, P. Once we get the matrix P, we can find the *steady state* probability vector, π, by solving the following equation:

$$\begin{cases} \pi = \pi \cdot P \quad \pi = [\pi_0, \pi_1, \pi_2, \ldots], \; P = [p_{ij}] \\ \sum_i \pi_i = 1 \quad i \in [0, n^2 + 2n - 1] \end{cases} \tag{4}$$

In a *discrete* Markov chain, if the *interval* between two *consecutive* transitions (including self-transition) is *identical*, the steady state probability π_i reflects the

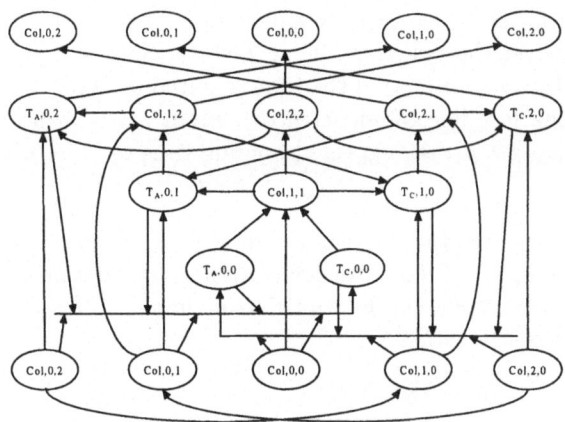

Fig. 3. State Transition Diagram for n=3

Table 1. Re-designating the State Variables

States	Single Variable Range
$(T_A, 0, 0)$ to $(T_A, 0, n\text{-}1)$	$[0, n\text{-}1]$
$(T_C, 0, 0)$ to $(T_C, n\text{-}1, 0)$	$[n, 2n\text{-}1]$
$(Col, 0, 0)$ to $(Col, n\text{-}1, n\text{-}1)$	$[2n, n^2\text{+}2n\text{-}1]$

proportion of the *time* that the system is in state i. However, this is not true in our model. For example, the interval between a T state and its next state is a *random* variable. Therefore, π_i can only tell us the probability that the system *enters* state i whenever a *transition* occurs [5]. To get the *time average* state probability of being in state i, we must first analyze the *holding time* of state i.

Let μ_{ji} denote the average time that the system will remain in state i once a *transition* from state j to i occurs. The μ_{ji} is given by:

$$\mu_{ji} = \begin{cases} 0 & p_{ji} = 0 \\ 1 & p_{ji} > 0 \,\&\, i \in [2n, n^2 + 2n - 1] \\ r \times num(j, i) & p_{ji} > 0 \,\&\, i \in [0, 2n - 1] \end{cases} \tag{5}$$

Whenever the transition probability p_{ji} is zero, the μ_{ji} must also be zero. The holding time in a *Col* state, since it does not depend upon the previous or present state, is assigned a *unit* value as in the second line of the above formula. To express the holding time of a T state, let us define $num(j, i)$, which denotes the average number of packets the transmitting node can transmit consecutively once the system reaches the T state (i.e., state i) from state j. Clearly, the holding time of a T state is proportional to $num(j, i)$. Moreover, we define the '*FES time*' as the time needed for a FES to be completed, and '*Col time*' as the time needed to detect the collision, while r is the *ratio* between the '*FES time*' and the '*Col time*'. The '*Col time*' is corresponding to the CTSTimeout interval defined in [6], which is independent from the length of the Data frame. Therefore, whenever the system enters a T state, the average holding time can be represented by $r \times num(j, i)$, if the time required to detect a collision is unity. This explains the last line of equation (5). From μ_{ji}, which corresponds to a *transition*, we can get μ_i, the expected holding time of state i, as follows:

$$\mu_i = \sum_j \frac{\pi_j \times p_{ji}}{\pi_i} \mu_{ji} \qquad (6)$$

It is easy to see that the μ_i corresponding to a *Col* state is always equal to *one*. Corresponding to a T state, we define $num(i)$ in equation (7), which denotes the average number of packets the *transmitting* node can transmit consecutively once the system reaches state i.

$$num(i) = \sum_j \frac{\pi_j \times p_{ji}}{\pi_i} num(j, i) \quad i \in [0, 2n - 1] \qquad (7)$$

As a result, formula (6) can be replaced by:

$$\mu_i = \begin{cases} num(i) \times r & i \in [0, 2n - 1] \\ 1 & i \in [2n, n^2 + 2n - 1] \end{cases} \qquad (8)$$

Now we can get ρ_i, which represents the *time average* state probability of state i [5]. Note that here we have ignored the time the meidum being idle.

$$\rho_i = \frac{\pi_i \times \mu_i}{\sum_j \pi_j \times \mu_j}; \ i, j \in [0, n^2 + 2n - 1] \qquad (9)$$

3.3   Derivation of the Metrics

We now obtain the two metrics defined in Section 1. To recall, the *first* metric is, once a node gets control of the medium, what is the average number of packets this node can transmit consecutively without any collision. In our Markov chain model, let us say for node A, it is *not* possible for the system to travel from one T_A state to another T_A state without visiting a T_C or *Col* state. Moreover, once the system enters a given T_A state (let us say, state i), the average number of units of '*FES time*' for which the system *remains* in that T_A state, is simply equal to $num(i)$, which is defined by equation (7). Therefore, the metric-1 can be obtained by taking average of all $num(i)$ corresponding to T_A states:

$$H_{metric-1} = \sum_{i=0}^{n-1} \frac{\pi_i}{\pi(A)} \times num(i); \ \ \pi(A) = \sum_{i=0}^{n-1} \pi_i \qquad (10)$$

Since the behavior is *identical* at nodes A and C, the metric obtained for node A are also applicable to node C.

Now, we recall that the *second* metric is, once a node *loses* its control of the medium, what is the average time the node has to wait before it gets control to the medium again. Since there are only two nodes (i.e., A and C) contending for the medium (Figure 1), the metric, let us say for node C, can be replaced by: once the medium is controlled by node A, what is the average number of packets that

A can transmit before the medium is controlled by node C. This simply implies that once the system enters a T_A state, what is the average number of units of 'FES time' that the system can remain in any T_A state (via visiting Col state in-between) before the system enters a T_C state. Note that the *second* metric allows a visit to a *Col* state in-between two T_A states, which is *not* permitted in the *first* metric.

Here we use the concept of "*expected first passage time*" [5], which means that if the system *starts* in a given T_A state (let us say, state i), what is the expected time after which the system will enter any T_C state for the *first* time. The expected first passage time, V_i, can be expressed as follows [5]:

$$V_i = \begin{cases} 0 & i \in [n, 2n-1] \\ R_i + \sum_j p'_{ij} V_j & i \in [0, n-1] \cup [2n, n^2 + 2n - 1] \end{cases} \quad (11)$$

where R_i is the *immediate reward* once the system enters state i. For a T_A state, it is equal to the corresponding $num(i)$. Since all T_C states are the *trapping* states, R_i is zero for these states [5]. We should also assign R_i with zero for all *Col* states, as our objective is to obtain the number of FESs, rather than the number of collisions. Therefore,

$$R_i = \begin{cases} num(i) & i \in [0, n-1] \\ 0 & i \in [n, n^2 + 2n - 1] \end{cases} \quad (12)$$

In equation (11), p'_{ij} is a *modified* value of the transition probability p_{ij}. Since the T_C states are considered as *trapping* states, the transition probabilities out of a T_C state is set to zero, whereas the *self-transition* probability for each T_C state is set to one. Other transition probabilities remain unchanged. Therefore,

$$p'_{ij} = \begin{cases} 0 & i \in [n, 2n-1] \& j \neq i \\ 1 & i \in [n, 2n-1] \& j = i \\ p_{ij} & i \in [0, n-1] \cup [2n, n^2 + 2n - 1] \end{cases} \quad (13)$$

Obviously, V_i corresponding to the *trapping* states (i.e., all T_C states) should be zero. For all the other states, V_i is equal to the 'immediate reward' R_i plus the expected reward earned from whatever state is entered *next*. This explains the equation (11).

The metric-2 can be obtained by taking average over all V_i values corresponding to the T_A states:

$$H_{metric-2} = \sum_{i=0}^{n-1} \frac{\pi_i}{\pi(A)} \times V_i; \quad \pi(A) = \sum_{i=0}^{n-1} \pi_i \quad (14)$$

To solve the above equations, we *only* need to know the transition probability matrix P and the state holding time $num(j, i)$. The calculation of the above values is not trivial. However, due to space limitation, we do not present it here. Please refer to [8] for a detailed analysis.

4   Numerical Results

Here we evaluate the equations derived in Section 3, and compare the analytical results with the simulation results. The simulation environment is the same as described in Section 2. The results correspond to the case when $CW_{min}=31$, $Len=19$, $CW_{max}=1024$, and $n=7$, which are typically used in IEEE 802.11. Since our main objective is to analyze the behavior at T states, and the behavior at nodes A and C is *identical*, we only present the results for T_A states.

State Probabilities: Table 2 presents the values of π_i and ρ_i. To recall, π_i represents the proportion of *transitions* entering state i, while ρ_i reflects the proportion of *time* spent in state i. The analytical results are quite close to the simulation results. We also notice that, though the sum of π_i is quite small (about 0.24), the sum of ρ_i is quite large (0.496). Since the T_C states also have the *same* values, the total fraction amount of *time* spent in T states is about 0.992, implying that only a very small amount of time is spent in the large number of *Col* states. The reason is that the ratio (i.e., r) between the '*FES time*' and the '*Col time*' is very large (i.e., 20 in our case). This shows the advantage of using the *short* RTS/CTS frames before the transmission of the *long* Data frame.

Table 2. State Probabilities Comparison

	From	$(T_A,0,0)$	$(T_A,0,1)$	$(T_A,0,2)$	$(T_A,0,3)$	$(T_A,0,4)$	$(T_A,0,5)$	$(T_A,0,6)$	Total
π	Model	0.025	0.031	0.037	0.038	0.037	0.035	0.032	0.236
	Simulation	0.025	0.038	0.039	0.039	0.037	0.034	0.031	0.243
ρ	Model	0.008	0.014	0.025	0.046	0.088	0.165	0.150	0.496
	Simulation	0.008	0.015	0.023	0.045	0.088	0.167	0.150	0.496

Expected State Holding Time: Table 3 presents the results of $num(i)$, which denotes the average number of packets node A can transmit consecutively once the system enters a given T_A state (i.e., state i). We see that the results match very closely. As the stage at the waiting node (i.e., node C) increases, the $num(i)$ also increases, which indicates that it becomes more unfair for node C.

Table 3. Expected Holding Time Comparison

num(i)	$(T_A,0,0)$	$(T_A,0,1)$	$(T_A,0,2)$	$(T_A,0,3)$	$(T_A,0,4)$	$(T_A,0,5)$	$(T_A,0,6)$
Model	1.006	1.450	2.143	3.830	7.496	14.858	14.893
Simulation	1.002	1.206	1.829	3.616	7.461	15.336	15.348

Expected First Passage Time: Table 4 presents the results of V_i, which represents, if the system starts in a given T_A state (i.e., state i), what is the expected number of FESs after which the system will enter any T_C state for the first time. Again, the results obtained from the model are quite close to those from simulation. When the stage at the waiting node (i.e., node C) increases (from 0 to 4), the V_i also increases. However, when the stage at the node C further increases (i.e., from 4 to 5, and then 6), the V_i decreases. First, let us explain why the V_i corresponding to the $(T_A, 0, 6)$ state is small. We recall that V_i is equal to the immediate reward (i.e., $num(i)$) plus the expected reward earned from whatever state is entered next. When the system departs from the $(T_A, 0, 6)$ state, the system is likely to enter state $(Col, 1, 0)$ where the stage at node C has been reset. From this Col state, the system is more likely to enter a T_C state, in comparison to, from other Col states. For instance, when the system transits to a Col state from the $(T_A, 0, 5)$ state, the stage at node C will not be reset, and the probability of transiting to a T_C state is small. This implies that after leaving $(T_A, 0, 6)$ state, the expected reward earned from the future states is smaller in comparison to that after leaving the state $(T_A, 0, 5)$. Therefore, the V_i corresponding to $(T_A, 0, 6)$ is small compared to the $(T_A, 0, 5)$ state. Now, we explain why the V_i corresponding to the $(T_A, 0, 5)$ state is smaller than that for the $(T_A, 0, 4)$ state. Since the state $(T_A, 0, 6)$ is a *future* state of $(T_A, 0, 5)$, a small V_i for $(T_A, 0, 6)$ will also affect the V_i for the $(T_A, 0, 5)$ state. However, the effect of the reset behavior decreases rapidly as the stage at the waiting node becomes smaller than 4. From the above discussion, it is clear that the *resetting CW mechanism* adopted in IEEE 802.11 improves the short-term fairness.

Table 4. Expected First Passage Time Comparison

V_i	$(T_A,0,0)$	$(T_A,0,1)$	$(T_A,0,2)$	$(T_A,0,3)$	$(T_A,0,4)$	$(T_A,0,5)$	$(T_A,0,6)$
Model	11.796	25.416	31.081	34.059	34.404	30.273	17.661
Simulation	10.413	23.584	29.998	34.031	34.896	31.430	18.555

Metrics: Table 5 presents the values of the two metrics defined earlier. We also present the corresponding maximum values. Again, the analytical results match the simulation results. Note that the values of the two metrics *differ* largely. The reason is that whenever the system departs from a T_A state, the probability that the system enters a Col state is very large. After the collision(s), the system is more likely to enter another T_A state (rather than a T_C state) as the CW at node A is smaller than that at node C. This may be repeated several times, resulting in such a large difference.

General Applications of the Model: The above results correspond to the case when $CW_{min}=31$, $Len=19$, $CW_{max}=1024$, and $n=7$. By varying these parameters, the *short-term* behavior for other scenarios can be obtained. For example, by making $Len=TxTime(Data)+SIFS$, we can model the *two-way* handshake in the presence of hidden terminals. On the other hand, by making $Len=0$, we can model *anyone* of the two handshakes *without* hidden terminals. By varying n and CW_{min}, we can model different physical layers, such as FSSS, DSSS and

Table 5. Comparison of the Metrics

From	metric-1	Max metric-1	metric-2	Max metric-2
Model	6.683	NA	27.379	NA
Simulation	6.413	35	27.090	160

IR [6]. Moreover, since the model can predict the *short-term* behavior *precisely*, it would also predict the *long-term* behavior accurately.

5   Discussion and Related Work

5.1   Future Work

We have presented a novel embedded-Markov model to study the short-term unfairness in a simple 3-nodes hidden-terminal case. We are extending the model to a more general scenario. However, it is necessary to mention that the modelling process described in Section 3 is quite general for the study of short-term behavior, especially the adoption of the *first passage time*.

Another focus is to propose a solution to cope with the short-term unfairness problem. From the results, we have already seen that the *resetting CW mechanism* improves the short-term fairness. Therefore, in addition to the standard resetting mechanism, we are of the opinion that the CW should also be reset whenever the short-term unfairness occurs that can be detected using dynamic measurements. Our preliminary results show that this method improves the fairness. However, the aggregate throughput may degrade.

5.2   Related Work

The fairness problem in wireless networks have been extensively addressed [9,10, 11]. However, most of these work do not consider the hidden-terminal problem *explicitly*. Also, they mainly consider *long-term* unfairness. More importantly, there is no analytical model predicting the unfairness in IEEE 802.11. In contrast, in this paper, we have developed an analytical model to explain and predict the short-term unfairness due to the hidden-terminal problem.

Several *analytical* models of IEEE 802.11 [1,2,3] have studied the *long-term* behavior (i.e., capacity) by *ignoring* many details of the protocol and adopting *simplified* assumptions. For example, the model in [1] overlooks the *resetting CW mechanism* and assumes a *constant* collision probability, which is clearly imprecise as shown in our results. While these models are able to predict the *long-term* behavior, they cannot be used to study the *short-term* behavior *accurately* because the required details are lost in their models. In contrast, we model the IEEE 802.11 in a *detailed* manner (e.g., by including the resetting mechanism and all the possible collision states) to predict the short-term fairness *precisely*.

The authors of [7] have studied the short-term fairness by first developing two fairness metrics and then applying the metrics in analyzing two MAC protocols: CSMA/CA and ALOHA. Though IEEE 802.11 is mainly based on CSMA/CA, it has many other features, and thus [7] cannot apply to IEEE 802.11. Moreover, they have not considered the hidden-terminal problem and they mainly focus on *developing* general fairness metrics, which are different from our work.

6   Conclusions

In this paper, we have presented an Embedded-Markov chain model for IEEE 802.11. Our model is novel in that it predicts the *short-term unfairness* of IEEE 802.11 very precisely, which is not available in the literature. The key concepts used in the model include the 'state holding time' and the 'first passage time'. Our results show that IEEE 802.11 exhibits substantial short-term unfairness in presence of hidden terminals. One important implication of our results is that the resetting of the contention window may improve the short-term fairness.

References

1. G. Bianchi, "Performance Analysis of the IEEE 802.11 Distributed Coordination Function," IEEE Journal on Selected Area in Communications, March 2000, pp.535-547.
2. F. Cali, M. Conti, E. Gregori, "Dynamic Tuning of the IEEE 802.11 Protocol to Achieve a Theoretical Throughput Limit," IEEE/ACM Transactions on Networking, December 2000, pp.785-799.
3. H.S. Chhaya, S. Gupta, "Performance modeling of asynchronous data transfer methods of IEEE 802.11 MAC protocol," ACM Wireless Networks, 3, 1997, pp.217-234.
4. CMU Monarch Group. CMU Monarch Extensions to NS, http://www.monarch.cs.cmu.edu/.
5. R.G. Gallager, "Discrete Stochastic Processes," Kluwer Academic Publishers, Boston,1996.
6. IEEE, "Wireless LAN Medium Access Control (MAC) and Physical Layer (PHY) specifications," IEEE 802.11 standards, June 1999.
7. C.E. Koksal, H. Kassab, H. Balakrishnan, "An Analysis of Short-Term Fairness in Wireless Media Access Protocols," in ACM SIGMETRICS, 2000.
8. Z.F. Li, S. Nandi, A.K. Gupta, "Modeling the Short-term Unfairness of IEEE 802.11 in Presence of Hidden Terminals," full version of the paper, avaliable at http://www.ntu.edu.sg/home5/pg03802331/.
9. H. Lou, S. Lu, V. Bharghavan, "A New Model for Packet Scheduling in Multi-hop Wireless Networks," in ACM MOBICOM, 2000.
10. T. Nandagopal, T. Kim, X. Gao, V. Bharghavan, "Achieving MAC Layer Fairness in Wireless Packet Networks," in ACM MOBICOM, 2000.
11. N.H. Vaidya, P. Bahl, S. Gupta, " Distributed fair scheduling in a wireless LAN," in ACM MOBICOM, 2000.

Analytical Modeling of TCP Clients in Wi-Fi Hot Spot Networks*

Raffaele Bruno, Marco Conti, and Enrico Gregori

Italian National Research Council (CNR) – IIT Institute
Via G. Moruzzi, 1 - 56100 Pisa, Italy
{firstname.lastname}@iit.cnr.it

Abstract. The channel utilization of 802.11-based wireless local area networks has been studied extensively from an analytical perspective. However, these studies consider wireless networks formed by *saturated* stations that transmit UDP-like traffic to randomly selected destinations. This model cannot be easily applied to hot spots since: *i*) the majority of Internet applications are based on the TCP protocol; *ii*) the TCP traffic is rate-controlled and the TCP flow control mechanisms impede the stations to saturate; and *iii*) the users' traffic is either sent towards or received from the access point. This paper is the first to provide an analytical model of the channel utilization in hot spots when there are M users which use TCP connections to download traffic from the Internet. Specifically, in this work we analytically investigate the complex interaction of the TCP flow control mechanisms and MAC collision avoidance techniques in 802.11b-based hot spot networks. We validate our model by means of simulations, and we exploit it to provide thorough reasons of the following counter-intuitive observations: *i*) the channel utilization is almost independent of the number of TCP connections; *ii*) on average the access point contends with few users for the channel bandwidth.

1 Introduction

The IEEE 802.11b standard [1], also named *Wi-Fi*, is the dominant technology in the market of network cards for wireless local area networks (*WLANs*). Now the attention of manufactures and service providers is turning to deploying Wi-Fi-based networks over *hot spots*, such as cafes, retail shops, convention centers, airports and other areas where people can benefit by a seamless public access to the Internet. As the number of users increases, the throughput obtained by each user is one of the most crucial concerns for the service providers and the hot spot operators. It is important to study the MAC protocol efficiency since the 802.11b standard uses a CSMA/CA-based random access protocol that allows unconstrained movement of mobile hosts, but doesn't provide guarantees on the

---

* This work was carried out under the financial support of the Italian Ministry for Education and Scientific Research (MIUR) in the framework of the Projects: FIRB-PERF and FIRB-VICOM

N. Mitrou et al. (Eds.): NETWORKING 2004, LNCS 3042, pp. 626–637, 2004.

stations' throughput. An extensive literature of analytical models of the channel utilization in 802.11-based networks can be found [2, 3, 4, 5]. However, these studies consider wireless networks formed by *saturated* stations, i.e., each station has immediately a packet available for transmission, which transmit UDP-like traffic to randomly selected destinations. This model cannot be easily applied to hot spots since: *i*) the majority of Internet applications are based on the TCP protocol; *ii*) the TCP traffic is rate-controlled and the TCP flow control mechanisms impede the stations to saturate; and *iii*) the users' traffic is either sent towards or received from the access point (AP). This paper aims at developing an analytical model that characterizes the channel utilization in Wi-Fi hot spots when there are M users, hereafter indicated as STAs, which use TCP connections to download traffic from the Internet. We consider only downstream data traffic because TCP-based downloads are the dominant part of the user traffic in typical local area networks [6]. The extension of our model to deal with upstream data traffic is left to future works. To the best of authors' knowledge this is the first attempt to model the interaction between the 802.11 MAC protocol and the TCP considering this increasingly important network architecture. The key approximation used in our model is the assumption that the devices access the wireless slotted channel adopting a p-persistent IEEE 802.11b protocol [5, 7]. This assumption is commonly used to study the binary exponential backoff adopted in the 802.11 MAC protocol [1]. However, our work significantly differs from previous studies because we integrate into the model the feedback-based behavior of the TCP protocol. Specifically, one of the major consequences of the flow control techniques used by the TCP protocol is that the amount of acknowledgment traffic the STAs have to send back to the AP mainly depends on the rate the AP delivers the TCP data packets to the STAs. In [8] we show that, even with *asymptotic* TCP connections, the STAs may not have a TCP ACK immediately available for transmission. Our model takes into consideration that the instantaneous number of STAs participating to the channel contention is variable. By differentiating the AP's and STAs' traffic arrival processes, we can model the way the TCP feedback-based behavior affects the MAC protocol operations. Our model provides an in-depth understanding of the MAC protocol operations in hot spot networks. By exploiting our analytical results, in this paper we show that: 1. even with a large number of TCP connections, on average there are less than two STAs contending with the AP for the channel bandwidth; 2. the channel utilization is almost independent of the number of downstream TCP connections; and 3. the MAC protocol efficiency depends mainly on the AP's capability to pump traffic towards the STAs. It is worth pointing out that observations 1.-3. are counter-intuitive results, because they indicate that the users' throughput is limited by the interaction of the MAC protocol with the TCP flow control mechanisms rather than the contention level of the network.

2   Model Definition

In our study, we have considered a hot spot with a fixed number M of users
(the STAs) that are associated to the AP. In the analysis we assume ideal channel conditions (i.e., no hidden terminals or packet losses). The stations access
the wireless channel adopting a p-persistent IEEE 802.11b MAC protocol [5,7],
which differs from the standard protocol only in the backoff selection: instead
of using the binary exponential backoff algorithm, the backoff interval is sampled from a geometric distribution of parameter p. Since the STAs and the AP
can suffer different number of collisions, we assume that the AP and the STAs
use different p values: p_{AP} for the AP, and p_{STA} for the STAs. Each station
performs a TCP-based download from remote Internet servers. Hence, the AP
delivers TCP data packets towards the STAs, while the STAs reply back with
TCP ACKs. The TCP sources are *asymptotic* connections, i.e., they have always a packet ready to be transmitted (ftp-like traffic). We assume that the AP
and the STAs have infinite size buffers, therefore neither TCP data packets nor
TCP ACKs can be lost due to buffer overflows. The key approximation of our
model is the p-persistent behavior of the MAC protocol. However, the dynamic
of the TCP protocol is complex and to make the problem analytically tractable
we have adopted further simplifying assumptions. Specifically, we assumed that
each TCP data packet is acknowledged separately, that is the delayed-ACK
mechanism [9] is disabled. This implies that the TCP receivers generate the
maximum offered load for the network, and this is the worst-case situation for
the TCP throughput performance and the contention on network resources. The
last assumption we have made is to consider that no more than one TCP ACK is
queued in the active STAs' buffers[1]. The major outcomes of this assumption are:
1) after a STA's successful transmission the STA's buffer is empty and the STA
becomes inactive; and 2) each AP's successful transmission activates a new STA.
As a consequence, during the analytical study we can assume that the number of
active STAs in the network is a discrete random variable that increments by one
every time the AP performs a successful transmission, and decrements by one
every time a STA performs a successful transmission. Preliminary simulations
results have substantiated this approximation [8, 10].

3   Model Analysis

In this section we outline the analytical study we have developed to compute
the channel utilization. Due to the space constraints we present the main results and discuss the line of reasoning we followed. The complete mathematical
development and the formal proofs are presented in the technical report [11],
available at the url http://cnd.iit.cnr.it/tech-report/2003/. To model
the MAC protocol behavior, we observe the channel between two consecutive
successful transmissions. From the p-persistent behavior it is straightforward to
derive that all the stochastic processes that define the occupancy pattern of the

---

[1] An *active* STA is a STA with a not empty transmission buffer.

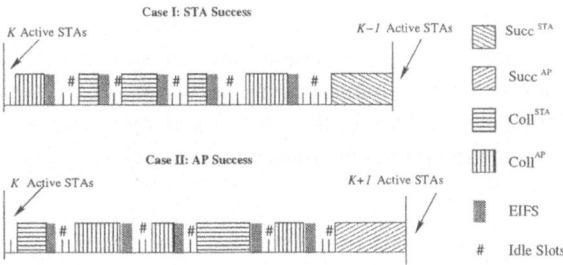

Fig. 1. Structure of channel events between to consecutive successful transmissions

channel (i.e., idle slots, collisions and successful transmissions) are regenerative, since they renew at the end of each successful transmission. In the following we will exploit the regenerative properties of the system and classical renewal theoretical arguments [12] to compute the time between two AP's successful transmissions, say T_v. Hereafter, we will refer to T_v also as *virtual transmission time*. During the development of the analysis we assume to know the steady state probability that there are K active STAs after the AP's successful transmissions, denoted as $\pi(K)$, with $K \leq M$. In section 3.1 we present approximated formulas to compute the $\pi(K)$. First of all, let us introduce the notation that will be useful in the following analysis:

- $E[X]_K$ is the expectation of the random variable X conditioned to having K active STAs;
- I_{AP} is a function that has value 1 when the AP performs a transmission attempt, 0 otherwise;
- $Idle\_p$ is the duration of the idle time that precedes a transmission attempt;
- *Coll* and *Succ* are, respectively, the duration of a collision and a successful transmission (including all the MAC protocol overheads due to interframe spaces and control frames);
- given the random variable X, we define $X^{AP} = \{X, I_{AP} = 1\}$ and $X^{STA} = \{X, I_{AP} = 0\}$;
- t_B, t_H and t_{ACK} are, respectively, the time needed to transmit a byte, the overhead added to the MAC data payload, and the MAC ACK frame;
- t_{SLOT} and τ are, respectively, the time slot duration and the maximum propagation delay over the wireless channel.

To derive the closed formula of $E[T_v]_K$, i.e., the average virtual transmission time conditioned to having K active STAs at the beginning of the T_v, it is useful to consider which events occur between two consecutive successful transmissions. Hereafter, we denote the time interval between two consecutive successful transmissions as T_s. Again, let us assume that the number of active STAs is K. In Fig. 1 we show the structure of the T_s. Collisions and idle periods may occur before a successful transmission. An idle period is a time interval in which the transmission medium remains idle due to the backoff algorithm. A collision may either involve the AP, or be caused only by STAs' concurrent transmission attempts. After the successful transmission, the hot spot contention level, that is the number of active STAs, changes. The analytical study is simplified by assuming that in the active STAs' buffers there is at most one queued TCP ACK. Hence, if an AP's successful transmission completes the T_s, the number of active STAs changes from K to $K+1$, since the AP's transmission causes an inactive STAs to generate a TCP ACK. On the other hand, if a STA's successful transmission completes the T_s, the number of active STAs changes from K to $K-1$, since the STA has sent the only TCP ACK it has. With reference to a generic T_s, N_c^{AP} indicates the number of collisions that involve the AP, and N_c^{STA} indicates the number of collisions that don't involve the AP. Considering the system behavior we can write that

$$E[T_s|Succ^{AP}]_K = E\Big[\sum_{i=1}^{N_c^{AP}} (Idle\_p_i + Coll_i^{AP} + \tau + EIFS) + \sum_{i=1}^{N_c^{STA}} (Idle\_p_i +$$

$$Coll_i^{STA} + \tau + EIFS) \Big]_K + E[Idle\_p_{N_c^{STA}+N_c^{AP}+1}]_K + E[Succ^{AP}]_K , \quad (1a)$$

$$E[T_s|Succ^{STA}]_K = E\Big[\sum_{i=1}^{N_c^{AP}} (Idle\_p_i + Coll_i^{AP} + \tau + EIFS) + \sum_{i=1}^{N_c^{STA}} (Idle\_p_i +$$

$$Coll_i^{STA} + \tau + EIFS) \Big]_K + E[Idle\_p_{N_c^{STA}+N_c^{AP}+1}]_K + E[Succ^{STA}]_K , \quad (1b)$$

where $T_s|Succ^{STA}$ is the Case I shown in Fig. 1, whereas $T_s|Succ^{AP}$ is the Case II shown in Fig. 1. The assumption on the p-persistent behavior of the backoff algorithm implies that the idle periods $\{Idle\_p_i\}$, and the collisions times $\{Coll_i^{AP}, Coll_i^{STA}\}$ are i.i.d. random variables. Thus (1a) and (1b) can be rewritten as

$$E[T_s|Succ^{AP}]_K = (E[N_c^{AP}|N_{tr} \geq 1]_K + E[N_c^{STA}|N_{tr} \geq 1]_K + 1) \cdot E[Idle\_p]_K +$$
$$E[N_c^{AP}|N_{tr} \geq 1]_K \cdot (E[Coll^{AP}]_K + \tau + EIFS) + E[N_c^{STA}|N_{tr} \geq 1]_K \cdot$$
$$(E[Coll^{STA}]_K + \tau + EIFS) + E[Succ^{AP}]_K , \quad (2a)$$

$$E[T_s|Succ^{STA}]_K = (E[N_c^{AP}|N_{tr} \geq 1]_K + E[N_c^{STA}|N_{tr} \geq 1]_K + 1) \cdot E[Idle\_p]_K +$$
$$E[N_c^{AP}|N_{tr} \geq 1]_K \cdot (E[Coll^{AP}]_K + \tau + EIFS) + E[N_c^{STA}|N_{tr} \geq 1]_K \cdot$$
$$(E[Coll^{STA}]_K + \tau + EIFS) + E[Succ^{STA}]_K . \quad (2b)$$

The average number of collisions that involve the AP during a T_s can be further expanded in

$$E[N_c^{AP}|N_{tr} \geq 1]_K = E[N_c^{AP}|I_{AP} = 1, N_{tr} \geq 1]_K \cdot Pr\{I_{AP} = 1|N_{tr} \geq 1\}_K ,$$

where $Pr\{I_{AP}=1|N_{tr} \geq 1\}_K$ is the probability that the AP transmits given that there is at least a transmission on the channel and there are K active STAs. Similarly, denoting with $Pr\{I_{AP}=0|N_{tr} \geq 1\}_K$ the probability that the AP isn't transmitting, given that there is at least a transmission on the channel and there are K active STAs, the average number of collisions that don't involve the AP during a T_s can be further expanded in

$$E[N_c^{STA}|N_{tr} \geq 1]_K = E[N_c^{STA}|I_{AP} = 0, N_{tr} \geq 1]_K \cdot Pr\{I_{AP} = 0|N_{tr} \geq 1\}_K .$$

The following Lemma provides closed formulas for the expressions in (2a) and (2b).

Lemma 1. *By assuming that: i) the STAs sample the backoff interval from a geometric distribution of parameter p_{STA}, while the AP samples the backoff interval from a geometric distribution of parameter p_{AP}; ii) the number of active STAs at the beginning of the T_s is K; and iii) the sequence of message lengths $\{l_i^{AP}\}$ ($\{l_i^{AP}\}$) of AP's transmissions (STAs' transmissions), expressed in bytes, are i.i.d. random variables with mean $\bar{l}_{AP}$ ($\bar{l}_{STA}$), it holds that*

$$Pr\{I_{AP} = 0|N_{tr} \geq 1\}_K = \sum_{n=1}^{K} \frac{\binom{K}{n}(1-p_{AP})\cdot p_{STA}^n(1-p_{STA})^{K-n}}{1-(1-p_{AP})(1-p_{STA})^K} , \quad (3a)$$

$$Pr\{I_{AP} = 1|N_{tr} \geq 1\}_K = \sum_{n=0}^{K} \frac{\binom{K}{n-1}p_{AP}\cdot p_{STA}^n(1-p_{STA})^{K-n}}{1-(1-p_{AP})(1-p_{STA})^K} , \quad (3b)$$

$$E[Idle\_p]_K = \frac{(1-p_{AP})(1-p_{STA})^K}{1-(1-p_{AP})(1-p_{STA})^K} \cdot t_{SLOT} , \quad (3c)$$

$$E[N_c^{AP}|I_{AP} = 1, N_{tr} \geq 1]_K = \frac{1-(1-p_{STA})^K}{(1-p_{STA})^K} . \quad (3d)$$

$$E[N_c^{STA}|I_{AP} = 0, N_{tr} \geq 1]_K = \frac{1-[(1-p_{STA})^K + K p_{STA}(1-p_{STA})^{K-1}]}{K p_{STA}(1-p_{STA})^{K-1}} , \quad (3e)$$

$$E[Coll^{AP}]_K = t_H + \frac{t_B}{1-(1-p_{STA})^K} \sum_{m=1}^{l_{MAX}} m\{F_{AP}(m)\{1-p_{STA}$$

$$[1-F_{STA}(m)]\}^K - F_{AP}(m-1)\{1-p_{STA}[1-F_{STA}(m-1)]\}^K -$$

$$[F_{AP}(m)-F_{AP}(m-1)](1-p_{STA})^K\} , \quad (3f)$$

$$E[Coll^{STA}]_K = t_H + \frac{t_B}{1-[(1-p_{STA})^K + K p_{STA}(1-p_{STA})^{K-1}]}$$

$$\sum_{m=1}^{l_{MAX}} m\{\{1-p_{STA}[1-F_{STA}(m)]\}^K - \{1-p_{STA}[1-F_{STA}(m-1)]\}^K$$

$$-[F_{STA}(m)-F_{STA}(m-1)]p_{STA}(1-p_{STA})^{K-1}\} , \quad (3g)$$

$$E[Succ^{AP}]_K \leq 2\tau + t_H + t_B \cdot \bar{l}_{AP} + SIFS + t_{ACK} + DIFS , \quad (3h)$$

$$E[Succ^{STA}]_K \leq 2\tau + t_H + t_B \cdot \bar{l}_{STA} + SIFS + t_{ACK} + DIFS , \quad (3i)$$

where $F_{AP}(m) = Pr\{l_{AP} \leq m\}$ and $F_{STA}(m) = Pr\{l_{STA} \leq m\}$.

Proof. See Appendix I in [11].

Using the $E[T_s|Succ^{AP}]_K$ and $E[T_s|Succ^{STA}]_K$ formulas, we are able to express the $E[T_v]_K$. The following Lemma defines a recursive algorithm to compute the $E[T_v]_K$.

Lemma 2. *By assuming that: i) the devices access the channel according to a p-persistent access scheme; ii) the number of active STAs at the beginning of the T_v is K; and iii) there is at most one ACK queued in the active STAs' buffers, it holds that*

$$
\begin{cases}
E[T_v]_K = \{E[T_s|Succ^{STA}]_K + E[T_v]_{K-1}\} \cdot Pr\{Succ^{STA}|Succ\}_K + \\
\qquad\qquad E[T_s|Succ^{AP}]_K \cdot Pr\{Succ^{AP}|Succ\}_K \\
E[T_v]_0 = E[Idle\_p]_0 + E[Succ^{AP}|Succ]_0
\end{cases}
$$

where $Pr\{Succ^{STA}|Succ\}_K$ and the $Pr\{Succ^{AP}|Succ\}_K$, are, respectively, the probability that a successful transmission belongs to either the Case I of Fig. 1 or Case II of Fig. 1.

Proof. In [11].

Finally, the $E[T_v]$ value is computed as

$$
E[T_v] = \sum_{K=1}^{M} E[T_v]_K \cdot \pi(K) . \tag{4}
$$

By exploiting the analytical formulas we have derived so far, it is straightforward to compute the average channel utilization achieved by the AP's transmissions, say ρ_{AP}. In fact, ρ_{AP} is the ratio between the average time the channel is occupied by the payload transmission of the AP's successful transmission and the average virtual transmission time, that is

$$
\rho_{AP} = \frac{t_B \cdot \bar{l}_{AP}}{E[T_v]} . \tag{5}
$$

To derive the average channel utilization achieved by the STAs' transmissions, say ρ_{STA}, we have to calculate the average number of STAs' successful transmissions, say $E[N_{succ}^{STA}]$, which occur during a T_v. Specifically, ρ_{STA} is the ratio between the average time the channel is occupied by the payload transmission of the STAs' successful transmissions and the average virtual time. Similarly to the reasoning used in Lemma 2, we start by deriving the $E[N_{succ}^{STA}]_K$, i.e., the average number of STAs' successful transmissions during a T_v, conditioned to having K active STAs at the beginning of the T_v. The following Lemma provides a recursive algorithm to compute $E[N_{succ}^{STA}]_K$

Lemma 3. *By assuming that: i) the devices access the channel according to a p-persistent access scheme; ii) the number of active STAs at the beginning of the T_v is K; and iii) there is at most one TCP ACK queued in the active STAs' buffers, it holds that*

$$
\begin{cases}
E[N_{succ}^{STA}]_K = \{1 + E[N_{succ}^{STA}]_{K-1}\} \cdot Pr\{Succ^{STA}|Succ\}_K \\
E[N_{succ}^{STA}]_0 = 0
\end{cases}
$$

Proof. In [11].

Similarly to (4), it follows that $E[N_{succ}^{STA}] = \sum_{K=1}^{M} E[N_{succ}^{STA}]_K \cdot \pi(K)$. Hence, it holds that

$$\rho_{STA} = \frac{t_B \cdot \bar{l}_{STA} \cdot E[N_{succ}^{STA}]}{E[T_v]} . \tag{6}$$

Finally, the overall channel utilization, say ρ, is computed as

$$\rho = \rho_{AP} + \rho_{STA} , \tag{7}$$

The maximum achievable ρ identifies the MAC protocol *capacity*, which provides a good indication of the overheads introduced by the MAC protocol to perform its coordination task among devices. The derivation of the couple of $\{p_{AP}, p_{STA}\}$ values that guarantees to attain the protocol capacity is left to further studies.

3.1   Estimation of $\pi(K)$

The analytical study we have developed so far requires the knowledge of the probability that there are K active STAs after an AP's successful transmission, denoted as $\pi(K)$. In this section we propose a simple model to derive the $\pi(K)$. It is worth pointing out that the analytical framework presented in the previous section is independent of the $\pi(K)$ formula. However, an accurate estimation of the $\pi(K)$ improves also the accuracy of the ρ computation. The p-persistent behavior of AP and STAs implies that each transmission attempt, regardless of the number of consecutive collisions experienced by the AP or the STAs, is a successful transmission with a constant and independent probability (see Lemma 1). Therefore, the stochastic process that describes the number of active STAs after an AP's successful transmission, say $S(n)$, can be modeled through an embedded Markov process. The time scale adopted is discrete: the index n refers to the beginning of the nth virtual transmission time T_v. The probability $\pi(K)$, for $K = 1, 2, \ldots, M$, is the steady state probability of the chain, that is $\pi(K) = \lim_{n \to \infty} Pr\{S(n) = K\}$. In Fig. 2 we depict the Markov chain that models the $S(n)$ process when the M value is finite. Henceforth, to indicate the transition probability from state i to state j, i.e., the $Pr\{S(n+1) = j|S(n) = i\}$, we use the notation $p(i,j)$. The following Lemma derives the $p(i,j)$ for the chain depicted in Fig. 2.

Lemma 4. *Under the assumption that: i) the AP and the STAs have the same probability to perform a successful transmission; and ii) no more than one TCP ACK is queued in the active STAs' buffers, it holds that*

$$\begin{cases} p(K_1, K_0) = \frac{1}{K_0+1} & K_0 = 1, 2, \ldots, M-1 \\ & K_1 = 1, 2, \ldots, K_0+1 \\ p(K_1, M) = \frac{1}{M+1} & K_1 = 1, 2, \ldots, M-1 \\ p(M, M) = \frac{2}{M+1} \end{cases} \tag{8}$$

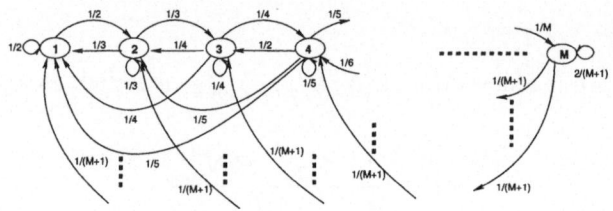

Fig. 2. Markov chain model of the $S(n)$ process for a finite M.

Proof. In [11].

Using the knowledge of the transition probabilities $p(i, j)$ given in (8), it is straightforward to find a closed formula for the $\pi(K)$ by writing the equilibrium at each node of the chain. Thus all the $\pi(K)$ are expressed in terms of $\pi(1)$ and of the K value. Finally, $\pi(1)$ is computed by imposing the normalization condition on the steady state probabilities (see [11] for the details). The $E[N_{STA}]$ can be computed as $E[N_{STA}] = \sum_{K=1}^{M} K \cdot \pi(K)$. In [11] we proved that $E[N_{STA}] \leq 2$. Thus, in hot spot networks the contention level is very low. In [11] it is considered also the case of an infinite number of STAs, that is $M \to \infty$, and it is proved the following proposition.

Proposition 1. *When $M \to \infty$, the average number of active STAs that are active at the beginning of a T_v is* $E[N_{STA}] = \sum_{K=1}^{\infty} K \cdot \pi(K) = \sum_{K=1}^{\infty} \frac{K}{e(K-1)!} = 2$.

This result is counter-intuitive because it demonstrates that, even when there are an infinite number of asymptotic TCP connections , the contention level in the network is very low. On average, only two STAs contend with the AP for the channel bandwidth. In [11] numerical results confirming the accuracy of the proposed $\pi(K)$ estimation are shown.

4   Model Validation

In this section we validate the correctness and accuracy of the analytical study performed in section 3 by means of realistic simulations. The simulation tool we have used is an extension of the one used in [5] and [7]. In addition to all the IEEE 802.11b MAC standard protocol details, it implements also the behavior of asymptotic TCP connections. The TCP version we have used is the TCP Reno [9] with the delayed-ACK mechanism disabled. The formulas we have derived in section 2 are applicable to a general payload distribution. In the following study we consider a constant distribution for the payload length. If not otherwise specified, the TCP senders delivers TCP data packets with a payload equal to 1500 bytes. As a consequence, we used $\bar{l}_{AP} = 1540$ bytes and $\bar{l}_{STA} = 40$ bytes[2] to compute the numerical results. As far as the p_{AP}

---

[2] The $\bar{l}_{AP}$ value characterizes a TCP data packet with payload equal to 1500 bytes, while the $\bar{l}_{STA}$ value is the typical size of a TCP ACK.

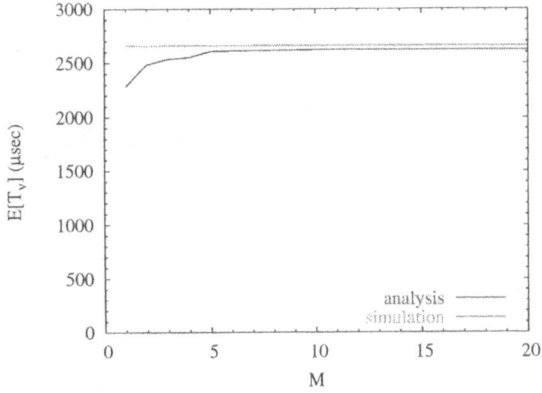

Fig. 3. $E[T_v]$: analysis versus simulation

and p_{STA} values, we use the values estimated through the iterative algorithm described in Appendix II of [11]. The constant overheads of the IEEE 802.11 MAC protocol are fully compliant with the standard and the values we used are the same adopted in [7]. All the results presented henceforth were obtained by performing simulation runs long enough to guaranteed a 99% confidence level with a precision lower that 1%. The first set of simulations was run to validate the accuracy of formula (4). In Fig. 3 we plot the average virtual time versus the number M of users, with $M \in [1, \ldots, 20]$, and we compare the analytical results against the behavior of the real system. Fig. 3 confirms that the p-persistent model provides a close approximation of the real behavior at least for $M \geq 5$. The analytical results are slightly lower than the simulation results for $M < 5$ because assuming that the active STAs have at most a TCP ACK to transmit is less accurate with a few stations. However, despite of this slight difference our model accurately captures the dynamic of the contention level in the hot spot network. Lemma 2 shows that the $E[T_v]$ depends on the $E[T_v]_K$ distribution. For this reason, we have not limited our study to the $E[T_v]$ computation, but we have also investigated if our model is able to match the $E[T_v]_K$ distribution. Fig. 4(a) and Fig. 4(b) show the $E[T_v]_K$ respectively for $M = 5$ and $M = 20$. These figures indicate that the analytical results are slightly lower than the simulation results, with an error that never exceeds the 10%. In Fig. 5 we show the channel utilization ρ for several M values, $M \in [1 \ldots 20]$. The numerical results show a close correspondence between the analysis and the simulation for $M \geq 5$. The differences when $M < 5$ are due to the underestimate introduced by our model in the $E[T_v]$ computation for small M values. Another important outcome of our analysis is that the channel utilization is almost independent on the number M of stations (see Fig. 5). This can be explained by considering the TCP feedback-based behavior. Specifically, the more data traffic the AP sends to the STAs, the more STAs become active. However, the larger is the number of active STAs the lower is the probability that the AP can perform a successful transmission

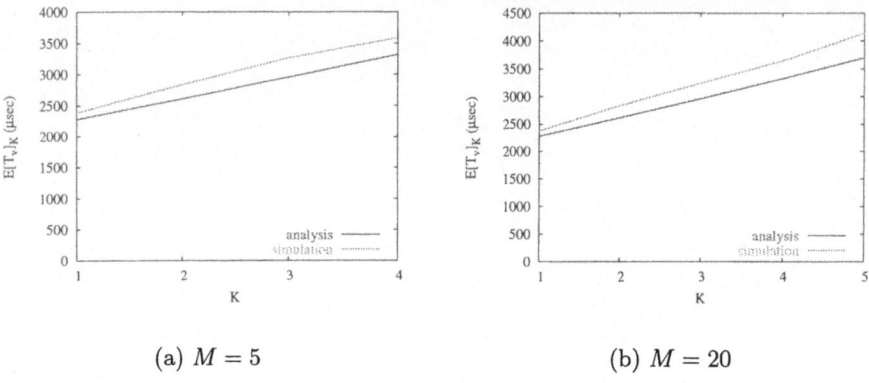

(a) $M = 5$ (b) $M = 20$

Fig. 4. $E[T_v]_K$ for various M: analysis versus simulation

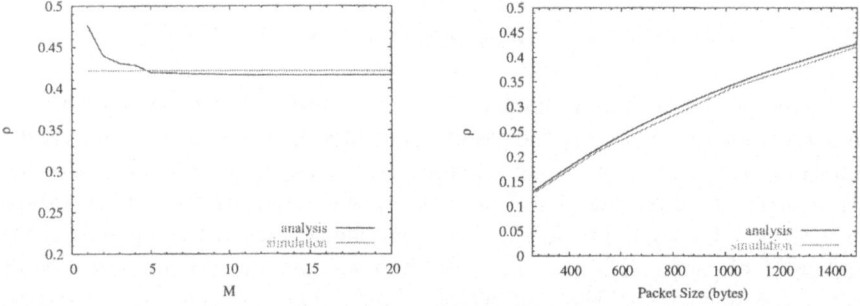

Fig. 5. ρ: analysis versus simulation   **Fig. 6.** ρ for different payload length and $M = 10$: analysis versus simulation

to activate new stations. Therefore the contention level in the network cannot increase by increasing the number M of users. A last set of simulations was run to investigate the impact of the payload length over both the throughput performance and the model accuracy. In Fig. 6 we show the channel utilization in a network with $M = 10$ by varying the $\bar{l}_{AP}$ value. Specifically, Fig. 6 shows the channel utilization ρ versus $\bar{l}_{AP} \in [256, \ldots, 1540]$ bytes, with $\bar{l}_{STA} = 40$ bytes as in the previous experiments. As expected ρ increases as the payload size increases because the fraction of channel time per successful transmission that is occupied by the protocol overheads reduces. The numerical results clearly confirm that the model accuracy is independent of the payload length.

5   Conclusions

To the best of authors' knowledge, this work is the first attempt of developing an analytical model to compute the channel utilization in hot spot networks.

Our model defines the *contention level* of the network as the number of STAs with a packet available for transmission. By considering i) the feedback-based behavior of the TCP protocol, and ii) the fact that the STAs' traffic has to be delivered through the AP, we were able to model how the contention level changes. This allowed us to derive an accurate characterization of the MAC protocol operations, with a good correspondence between the analytical results and the simulation results. By exploiting the developed analysis, we provided thorough reasons of the following counter-intuitive observations: i) the channel utilization is almost independent of the number of TCP connections; ii) on average the access point contends with few users for the channel bandwidth. This is significantly different from the results of previous analytical studies that considered UDP-like traffic. In fact, with asymptotic not-responsive traffic flows the channel utilization degrades as the number of users increases.

References

1. Part 11: Wireless LAN Medium Access Control (MAC) and Physical Layer (PHY) Specification/Amendment 2: Higher-speed Physical Layer (PHY) in the 2:4 GHz band (2001)
2. Kleinrock, L., Tobagi, F.: Packet Switching in Radio Channels: Part I - Carrier Sense Multiple-Access Modes and their Throughput-Delay Characteristics. IEEE Trans. Commun. **23** (1975) 1400–1416
3. Chhaya, H., Gupta, S.: Performance modeling of asynchronous data transfer methods on IEEE 802.11 MAC protocol. Wireless Networks **3** (1997) 217–234
4. Bianchi, G.: Performacs Analysis of the IEEE 802.11 Distributed Coordination Function. IEEE J. Select. Areas Commun. **18** (2000) 1787–1800
5. Calí, F., Conti, M., Gregori, E.: Dynamic Tuning of the IEEE 802.11 Protocol to Achieve a Theoretical Throughput Limit. IEEE/ACM Trans. Networking **8** (2000) 785–799
6. Tang, D., Baker, M.: Analysis of a Local-Area Wireless Network. In: Proc. of MobiCom 2000, Boston, MA (2000)
7. Bruno, R., Conti, M., Gregori, E.: Optimization of Efficiency and Energy Consumption in p-Persistent CSMA-Based Wireless LANs. IEEE Trans. Mob. Comp. **1** (2002) 10–31
8. Bruno, R., Conti, M., Gregori, G.: Throughput Evaluation and Enhancement of TCP clients in Wi-Fi Hot Spots. In: Proc. of WONS 2004, Madonna di Campiglio, Italy (2004) 73–86
9. Stevens, W.: TCP Illustrated, Volume 1: The Protocols. Addison-Wesley, New York, NY (2001)
10. Pilosof, S., Ramjee, R., Raz, D., Shavitt, Y., Sinha, P.: Understanding TCP fairness over wireless LAN. In: Proc. of IEEE Infocom 2003, San Francisco, CA (2003)
11. Bruno, R., Conti, M., Gregori, E.: Throughput Analysis of TCP Clients in Wi-Fi Hot Spot Networks. Technical report, IIT – CNR Pisa (2003)
12. Heyman, D., Sobel, M.: Stochastic Models in Operations Research. Volume 1. McGraw-Hill, New York, NY (1982)

On the Feasibility of Integrated MPEG Teleconference and Data Transmission, over IEEE 802.11 WLANs

Ioannis Broustis and Michael Paterakis

Information & Computer Networks Laboratory
Department of Electronic & Computer Engineering
Technical University of Crete
73100 Chania, Greece
{broustis, pateraki}@ telecom.tuc.gr

Abstract. The most widespread Wireless Local Area Networks (WLANs) are based today on the IEEE 802.11 standard and its various versions, especially the IEEE 802.11b. In this article we first briefly explain the IEEE 802.11 architecture and the possible implementations of the Data Link Layer. We then present and discuss the results from simulation experiments we performed in order to evaluate the protocol performance. We studied cases of integrated MPEG teleconference and computer data transmissions, over the IEEE 802.11b WLAN. Our simulation results clearly demonstrate the difficulty of the protocol to support time sensitive applications with a large number of wireless users under the same Access Point.

Keywords: IEEE 802.11b Wireless Local Area Networks, MPEG Teleconference, Video, Network Protocols, Performance Evaluation.

1 Introduction

A WLAN provides its connected users with the capability of communicating with each other, by sending packetized data, without using cables that connect their devices. The WLANs focus on a small geographical area, such as a building, an airport or a part of a university campus. Every time a mobile user wishes to connect to the Internet or to any other network from such place, the connection is almost always carried out via a WLAN. For this kind of access, the user's host is a node belonging to the network, and the LAN provides access to the Internet, via a router. The connected to the WLAN users transmit their data packets via a common wireless channel. Every time a mobile station transmits a packet, this packet "goes out in the open", so every other mobile user may receive it. However, a user usually sends a packet to one recipient only (unicast). In order for a station to be able to send a packet to a specific receiver, it must know the recipient's network address. As a result, we must correspond addresses to users, and transmitted packets must contain the receiver's WLAN address in their headers. By this way, a wireless station is able to know whether a sent packet is designated to it, or to some other station.

In the following sections we briefly explain the possible implementations of the IEEE 802.11b MAC layer, and we indicate the basic concepts concerning multimedia applications and their transmission problems. We especially focus on the integrated teleconference and data transmission over IEEE 802.11b. We perform an extensive

N. Mitrou et al. (Eds.): NETWORKING 2004, LNCS 3042, pp. 638–649, 2004.

simulation study on various scenarios of teleconference video and computer data traffic, and we present and explain results from these simulations.

2 Multimedia Applications and Video

During the last decade, a remarkable explosion of network-multimedia applications has taken place. Such applications have been developed so that users can send/receive audio and video, especially via the Internet. The IP telephony, the prerecorded or the real-time video transmission, the teleconference, the Internet radio and the network games are some of the most popular ones. We can generally distinguish these applications in three different classes: prerecorded audio and video streaming, real-time audio/video streaming, and interactive real-time audio/video streaming.

Multimedia has an important characteristic: it can be compressed. Without compression, multimedia consumes a large amount of disk storage and network bandwidth. A good compression algorithm must not only save disk space, but also avoid multimedia quality loss. Such a good and modern algorithm is MPEG-4 [9]. MPEG-4 manages to put by enough storage, without diminishing the reproduction quality. A 2-hour DVD movie, encoded using the MPEG-2 algorithm [10], consumes almost 7,5 - 8 Gigabytes. If the MPEG-4 algorithm is used, the same movie will now consume about 2,5 Gigabytes, that is 2/3 less disk space.

The multimedia applications differ from other standard widespread Internet applications (such as FTP, e-mail etc), in that they are delay sensitive, and tolerant to few packet losses. As a result, wireless networks that have been developed for fast data transfer, and which provide reliability and security, may not be appropriate for multimedia transmission, because of the various packet delays that are introduced. This time sensitivity generates many problems in multimedia transmission over some wireless media, because, if the packet delay becomes larger than a specific value (usually of the order of few tens of milliseconds), those packets will then be considered as old and will be dropped, as the receiver's application layer cannot wait more for these packets. However, multimedia applications are packet loss tolerant, and, according to how strict they are about the maximum percentage of packets that may be lost, we can determine a QoS *(Quality of Service)* level for the transmissions. Furthermore, if we do not lose packets very often, the gaps from the lost packets may be partially or completely covered (in some occasions), using various techniques [4].

3 The 802.11 Architecture

In order for a mobile station to be able to connect to an 802.11 WLAN, a wireless card must be available. The fundamental building block of the 802.11 is the BSS *(Basic Service Set)*. The BSS is a set of stations that use one of the following coordination functions: DCF *(Distributed Coordination Function)*, or PCF *(Point Coordination Function)*, which we briefly explain later in this section. The geographical area, covered by a BSS is called BSA *(Basic Service Area)* and corresponds to the notion of a "cell" that is used in mobile telephony. Stations that belong to the same BSS are able to communicate directly to each other, as in an Ad-Hoc network. In contrast to the Ad-Hoc network topology, *Infrastructure networks* have been developed to provide specific services and cover wider ranges. The basic equipment includes an AP *(Access Point)*, which corresponds to the BTS *(Base Transceiver Station)*, used in mobile telephony networks [3], [5]. Using APs we manage to interconnect stations outside the limits of a BSS. As a result, APs connect a BSS with the rest of the network world, so as to create an ESS *(Extended Service Set)*. An ESS consists of many interconnected BSSs through a DS *(Distribution system)*. The Distribution System may be considered as a *backbone* network, responsible for the data link layer transmissions [3].

3.1 MAC Layer

In 802.11, CSMA-CA *(Carrier Sense Multiple Access with Collision Avoidance)* is used: when a node wants to transmit, it senses the medium and if it does not detect any activity, it will wait for an additional time interval and will sense the medium again. If the medium is still idle, then the station will assume that nobody uses the medium, so it will send a data packet.

The mechanism that stations follow to gain access to the medium may alternate between two different modes: Contention-free and Contention. While in contention-free, *(CFP – Contention Free Period)*, access to the channel is determined by an AP *(Access Point)*, in that the AP decides which station has the right to transmit during a certain period of time. During this period, other stations only hear the medium and do not have authorization to transmit. In contrast to CFP, while in CP *(Contention Period)* the CSMA-CA algorithm is followed.

3.1.1 Distributed Coordination Function
DCF is the fundamental channel access method, based on the CSMA-CA protocol, and is used to support asynchronous data transfer. All stations are required to be able to operate in DCF, which may coexist with the PCF (described later), or operate solely. While in DCF, every station with a packet to transmit must contend to gain access to the medium. In every access, the wireless node may transmit one data frame only and, after the end of this transmission, it must re-contend, in order to transmit another frame from its queue [1], [2].

The main advantage of this algorithm is fairness: every station must re-contend for the channel after every packet transmission. All nodes have equal probability of gaining access to the medium after each *DCF-InterFrame Space* (DIFS) time interval. The basic disadvantage though, is that it does not guarantee a minimum access delay to stations running time-sensitive applications.

3.1.2   Point Coordination Function
PCF is an optional connection-oriented operation, based on a PC *(Point Coordinator)*, whose responsibility is to select the station that transmits during a specific time period. Wireless nodes do not have to contend with each other in order to gain access to the medium. The AP usually performs the role of the PC. The way stations are selected is based on an algorithm, chosen by the PC (e.g. Round Robin). PCF provides contention-free (CF) frame transfers and is usually used for time-bounded services.

PCF needs Access Points. All nodes obey the medium access rules determined by the AP/PC. Every time the AP contacts a station, giving it the order to transmit, the station may transmit only one frame, destined to any other station (inside the BSA or not), and not just to the AP/PC. PCF controls the CFP *(Contention Free Period)*. At the nominal start of every CFP, the AP/PC senses the medium. If the medium is considered to be idle for a *PCF_InterFrame Space* (PIFS) time period, it will broadcast a beacon frame to all nodes, informing them that CFP has began. If neither the stations nor the PC have packets to transmit, the PC may terminate the PCF immediately after the beacon frame transmission [1].

The entire channel operation, alternating between PCF and DCF periods, is shown in the following figure. Finishing with the brief protocol description, we notice that because of the alternating operation (PCF - DCF), we are able to transmit different kinds of data during each function. Thus, we may use CFP to transmit delay sensitive data, such as audio and video, and leave other kinds of data to be transmitted during the CP.

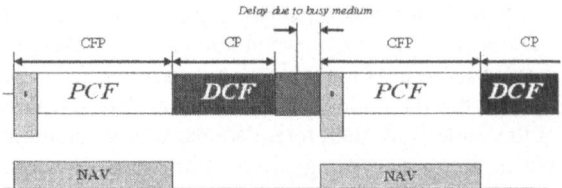

Fig. 1. Alternation between Point Coordination and Distributed Coordination Functions

Indeed, delay sensitive data packets cannot be efficiently transmitted, if they collide frequently, as additional delays are incurred in such cases. In contrast, data packets that can put up with some delay (such as FTP or email data) may wait more, so, they can be transmitted during the CP.

4   The Simulator

In order to determine the protocol's efficiency and the network capacity, we developed a software simulator. It is generally difficult to develop such simulator from scratch, because of the significant complexity of the protocol. Our simulator does not cover cases of noisy environment and hidden nodes. Some of protocol's parameters either could not be found, or are left to the implementer - administrator. Thus, we assumed certain values for these parameters, shown in Table 1, based on the relevant bibliography and the values that have been used in protocol simulations by others. We varied these values in a number of experiments and noticed that they do not significantly affect the results.

Table 1. Values of some important parameters, used in our simulations

DIFS	34 μsec
SIFS	16 μsec
PIFS	25 μsec
Slot time	9 μsec
DCF period	1000 μsec
PCF period	10000 μsec

The simulator views the wireless network as follows. Every single station maintains two separate queues: one for video packets and one for data packets. The teleconference application, assumed to run at each station, adds fragmented video frames into the station's video queue. The video frame lengths (in bytes) are being read from 1-hour playback duration trace files, corresponding to MPEG-4 teleconference movies [7]. Every node has its own trace file, which is different from the others', meaning that every 40 msecs, stations read a new different (for every station) video frame, which they fragment and insert into their video queues. Moreover, every station's data queue is being filled with ATM size packets, according to a Poisson arrival process with arrival rate λ. The video system is represented with a circular queue, nodes of which are the stations' separate video queues. In addition, the data system is represented with another circular queue, nodes of which are the separate stations' data queues. For the representation of the AP we follow a similar design: There is a circular queue, nodes of which are video queues, one for every destination station. Every time station A wants to send a video packet to station B, it sends the video packet to the AP. This packet is stored in B's queue at the AP. When the AP contacts station B, it will send packets to it from its queue, maintained at the AP. There is no need for a data circular queue at the AP, since we assume that data transfers (DCF) occur between stations of the same BSS, therefore, data packets are sent directly to the receivers. The simulator is time – triggering; we assume the existence of a minimum time period of 1 μsec, and represent all time values as multiples of this time period [8]. Finally, the simulator was developed in C++.

4.1 Scenario 1: Full Teleconference

In this scenario we assume that we have a number of users in a BSA, which use a common teleconference application to communicate, under the IEEE 802.11b wireless environment, with a transmission rate of 11Mbps. Every user has a video camera connected to a portable PC (laptop), and either can move inside the BSA, or may stay still at a certain point. We also assume that every station is running several other applications, among which is a data transfer application, used to send ATM size data packets to the other teleconference users. The video produced from the teleconference application is encoded using the MPEG4 algorithm, and the time elapsed between two video frame arrivals is 40 msecs. Furthermore, video packets have a maximum life time of 40 msecs in the system. The ATM data packets are generated according to a Poisson arrival process, with arrival rate λ. We also assume that the protocol alternates between CP and CFP, so both PCF and DCF are used. Stations exchange video packets during the PCF period and data packets during the DCF period of time.

In order for a station to participate in the teleconference, it must send its video packets to all the other teleconference users, and receive video packets from all other users. As a result, a packet that has been received by a node, has been sent to all the other teleconference nodes, except the one who sent it. Such functionality could be easily performed if the sender sent the packet to the AP, and the AP broadcasted or multicasted it to all other teleconference participants, by attaching a broadcast/multicast address corresponding to the set of them. However, in our scheme we do not select this broadcast/multicast functionality, since it is unreliable. Indeed, if a packet is broadcasted or multicasted to a set of nodes, then those nodes - recipients do not send acknowledgements back to the sender. That is, the sender does not have any clue about whether a video packet reached correctly all teleconference participants, so as to decide whether to retransmit it or not. Because of the latter, we avoid using the broadcast/multicast feature, even though it appears to be the easiest way to send the video packets to the teleconference participants. Thus, we have to find another way to reliably send video packets from a node to all the other teleconference nodes, before the 40 msecs lifetime deadline is exceeded. We assume that we have a transport protocol, responsible for the end-to-end packet delivery, such as UDP, which sets up the path that packets follow. As a result, every node has a separate UDP connection with every other teleconference node in the BSS. A solution to the problem of sending a packet to all other stations is to send that packet separately to each station (*"multicast through unicast"*). To do this, every video packet must be copied as many times as its recipients, that is all nodes minus one – the sender, and every copy must be sent to a recipient. There are two ways to implement this network operation:

1. The originator of the packet performs the copy procedure: When a station contacts the AP, it must send as many packet copies as the number of the other teleconference users. The problem here appears when the sender is given the authorization to transmit. It can send only one packet upstream. Thus, if we have, for example, 8 teleconference users, the sender must transmit 7 packet copies, one copy for every time it contacts the AP. So it will take seven round robin cycles until the packet has completely left the transmitter. This is a significant amount of time for the video packets waiting in queues.

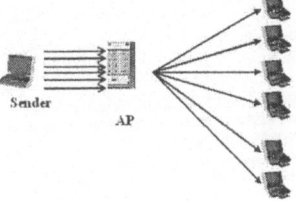

Fig. 2. The packet originator sends copies of the packet to the AP, which further delivers them

The 40 msecs lifetime deadline is frequently exceeded and video packets are dropped while inside the sender's queue - that is while in the process of being transmitted on the uplink.

2. The copy procedure is performed by a copy function behind the AP/PC: In this case, a packet copy (replication) function must operate behind the AP. Every time a video packet arrives at the AP, the copy function replicates the packet as many times as the number of destination stations. The AP stores the copies in its local buffers and sends these copies to the stations (one by one) every time it contacts them.

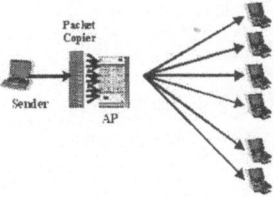

Fig. 3. A packet copier supplies the AP with copies of the packet sent by the originator

Even in this case, however, we observe packet loss, *on the downlink*, rather than on the uplink: the Access Point has stored many packet copies, which does not manage to transmit - at least not all of them, as their lifetimes expire, while still stored in AP's buffers. In the sequel, we assume the existence of a copy function behind the AP/PC.

Initially we assume that the teleconference application fragments video frames to packets with maximum length of 1000 bytes. The video *mean bit rate* is 42 Kbps and the *peak bit rate* is 690Kbps. Moreover, as we have mentioned before, the ATM data packets are transmitted during the DCF, they have a constant length of 53 bytes and the Poisson arrival rate is $\lambda = 4*10^{-4}$, that is 4 data packets every 10 msecs. Furthermore, we require a quite small video packet loss, specifically, we require: *Packet Video Drop (PVD)* 10^{-4}, where PVD stands for the number of the overall lost video packets, divided by the number of the overall video packets that have been generated. The following figures represent the PVD and the average video packet delay as we add more teleconference stations in the network, respectively:

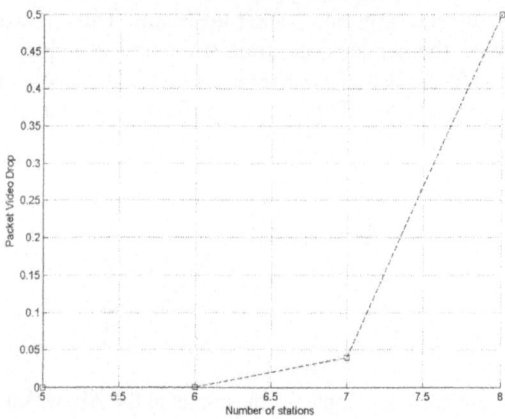

Fig. 4a. For 7 users or more, the system does not operate efficiently; the PVD increment is prohibitive for our QoS requirements

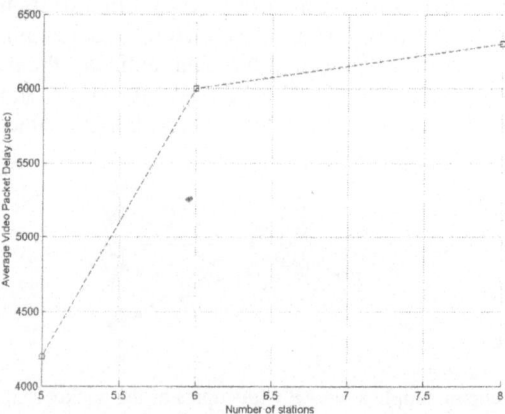

Fig. 4b. Average video packet delays are acceptable

From figures 4a and 4b we notice that the system operates efficiently with a maximum capacity of 6 stations. If we increase the number of stations to 7 or more, we observe a severe PVD increase, something unacceptable for the assumed QoS level. Moreover, as we expect, the average video packet delay increases too, since many more video packets are waiting to be transmitted now. Of course, a maximum number of 6 users may be a bit small for certain teleconference scenarios. This relatively small capacity is due to the protocol's significant time overheads. All those time safety valves consume quite large time that otherwise could have been used for actual packet transmissions. As a result the channel throughput decreases significantly. As we also notice, concerning the DCF, by adding more stations we observe large packet delays leading the system to be unstable, (something that we thoroughly explain later). For 6 users, however, packet delays and loss are acceptable, as we see from figures 4a, 4b and 5.

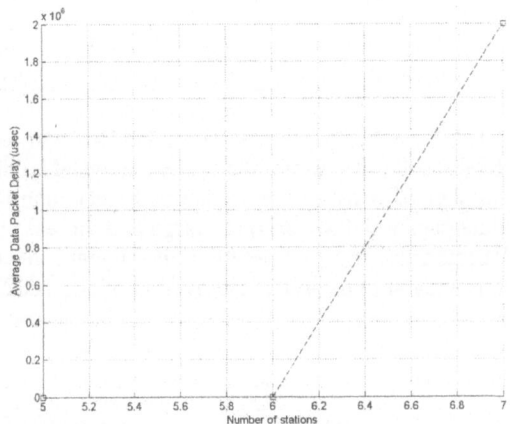

Fig. 5. Average data packet delays are acceptable

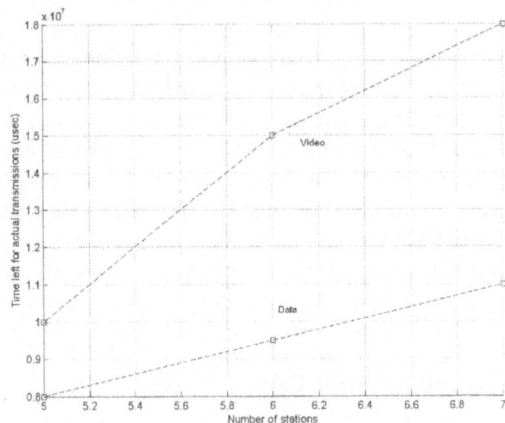

Fig. 6. The amount of time left for video and data packet transmissions is quite small

The channel time used for actual packet transmissions is quite small. From the simulations we see that for 6 stations and 100 secs of network operation, video packets are transmitted for 15 seconds (overall) only and data packets for 10 seconds only. The remaining 75 seconds constitute the time overhead and the CP periods. These numbers do not seem to change dramatically for a different number of wireless stations, as we see from figure 6 above. It is quite interesting to see how the average data packet delay and loss change, for different data packet arrival rates λ. In the simulation we assumed 6 teleconference users and a maximum video packet size of 1000 bytes (payload). *The values are aggregate, (i.e. for the packets of all stations).*

Table 2. Average data packet delays and data packet losses for various values of λ

λ	Arrivals	Losses	Average Data Packet Delay
10^{-4}	59908	0	1439,50 µsec
$2*10^{-4}$	120278	1	2968,09 µsec
$3*10^{-4}$	179301	16	4975,05 µsec
$4*10^{-4}$	240134	178	11054,0 µsec
$5*10^{-4}$	299446	1050	2868192 µsec

From the results in Table 2 we observe that, for $\lambda \geq 5*10^{-4}$ and for the maximum sustained number of 6 teleconference users, the system is unstable: data packet loss (due to experiencing a number of consecutive collisions higher than the protocol's threshold) and average data packet delays are quite large and increase as the simulation time is prolonged. We must also mention here that the changes in λ do not affect the *video packet loss*, while they slightly affect the average video packet delays, as shown in Table 3.

Table 3. Average video packet delays and video packet losses for various values of λ

λ	Losses	Average Video Packet Delay
10^{-4}	5	5781,70 µsec
$3*10^{-4}$	5	5969,13 µsec
$4*10^{-4}$	5	6008.20 µsec
$5*10^{-4}$	7	6010,20 µsec

This behavior is due to the following reason. For larger values of λ every data queue fills with more packets. This means that the medium is busy for a longer time during the DCF. It is thus likely that consecutive transmissions will be taking place, even at the time point that DCF ends. As a result, the protocol has to extend the DCF period, so that the current data packet transmission can be completed. This time extension acts burdensomely for the delays of the video packets waiting to be transmitted during the following CFP. We conclude by saying that the DCF period extensions generally increase the data and video packet delays. Respectively, the same conclusion stands for the PCF extension.

4.2 Scenario 2: Teleconference for Pairs of Users

This case differs from the previous one, in that every pair of users holds its own teleconference. We have as many teleconferences in progress, as the number of users divided by two (fig. 7). Of course we assume an even number of users. As before, we assume that, all parallel teleconferences begin at the same time, so, new video frames arrive at the same time to the wireless stations. Because of the nature of this scenario, a video packet sent by a node is destined to one node only, the teleconference partner of the transmitter. As a result, we do not have here to replicate packets destined to a number of receivers. As expected, the system capacity increases. From the results presented in table 4, we can see that the system can now support up to 16 users (8 pairs), that is 10 more users than in the previous scenario.

As we expect, if we have more teleconference user pairs, video and data packet delays are increased. As we add more users, we observe a severe increase in the average data packet delays, during the DCF period. This is because, in a system with many users, it is more difficult for a station to gain access to the medium through the CSMA-CA algorithm. A large number of users implies frequent packet collisions and thus, increased data packet delays.

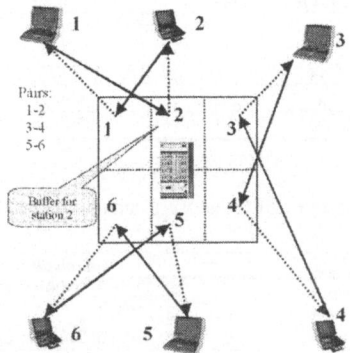

Fig. 7. Teleconference for pairs of users

Table 4. Average video packet delays and video packet losses for pairs of teleconference users

Users	Arrivals	Losses	Average Video Packet Delays
6	16252	0	2378,303 µsec
8	21654	0	3196,960 µsec
10	27098	0	3991,544 µsec
12	32766	0	4837,170 µsec
14	38207	0	5304,612 µsec
16	43640	0	6148,205 µsec
18	49060	372	6603,696 µsec

4.3 Scenario 3: Teleconference for Groups of Users

In the last scenario we assume groups of teleconference users under the same AP. For the groups with more than two users, we must use a copy function that replicates video packets behind the AP. These packets suffer higher delays (scenario #1). In contrast, for teleconference groups with two users only, we do not have to copy packets. As a result, video packets exchanged between these pairs of users do not suffer from time delays, at least not as much as packets in larger teleconference groups. Indeed, packet loss (because of lifetime deadlines) occurs more frequently in groups with many users. The advantage of the system is that, groups of different sizes do not interfere with each other. Large groups suffer from delays and losses, while small ones hold teleconference more efficiently. In the following tables we provide representative simulation results, supporting all of the above statements.

Table 5. One group of three users and another of two users

VIDEO	Groups	Losses	Arrivals	Average Delay
	0 – 1 – 2	0	0: 2709, 1: 2709, 2: 2709	6055.66 μsec
	3 – 4	0	3: 2708, 4: 2708	2055 μsec
DATA	λ	Losses	Arrivals	Average Delay
	0.0004	0	199460	1016.298 μsec

Table 6. One group of three users and two other groups of two users each

VIDEO	Groups	Losses	Arrivals	Average Delay
	0 – 1 – 2	0	0: 2707, 1: 2707, 2: 2707	8101.66 μsec
	3 – 5	0	3: 2708, 5: 2708	2821 μsec
	4 – 6	0	4: 2707, 6: 2707	3051.5 μsec
DATA	λ	Losses	Arrivals	Average Delay
	0.0004	91	279303	4447,46 μsec

Table 7. Two groups of three users each and a group of two users:

VIDEO	Groups	Losses	Arrivals	Average Delay
	0 – 1 – 2	0	0: 2709, 1: 2709, 2: 2709	9618.61 μsec
	3 – 4	0	3: 2708, 4: 2708	3145.7 μsec
	5 – 6 – 7	0	5: 2707, 6: 2707, 7: 2709	9668.8 μsec
DATA	λ	Losses	Arrivals	Average Delay
	0.0004	2164	319443	1479940.1 μsec

Table 8. Three groups of two users each and a group of five users:

VIDEO	Groups	Losses	Arrivals	Average Delay
	0 – 1 – 2 – 7 – 8	0:629, 1:615 2:622, 7:652, 8: 651	0: 2710, 1: 2710 2: 2710, 7: 2710 8: 2710	24721,56 μsec
	3 – 6	0	3: 2710, 6: 2710	4367,2 μsec
	5 – 9	0	5: 2710, 9: 2710	4750 μsec
	4 – 10	0	4: 2710, 10:2710	4383 μsec
DATA	λ	Losses	Arrivals	Average Delay
	0.0004	205	218578	7074,45 μsec

From tables 5 and 6 we can see that the simulation results are quite satisfactory and agree with intuition. Thus, in the first sub-scenario the group with the three nodes encounters larger time delays, because we are obliged to replicate all video packets at the AP. In contrast, the other two stations, involved in the second group, encounter smaller delays, as video packets are sent directly between partners. In the second sub-scenario we notice similar results. We should point here, that in this second case, packets exchanged between pairs of users in groups 2 and 3 encounter larger time delays from the packets exchanged between pairs of users in the first sub-scenario. This is because in the second case we have added two more users (which constitute a new teleconference pair) under the same AP. The following results are shown to further support our observations and conclusions.

From the results of the last sub-scenario we observe that the group with the five nodes suffers severe packet delays and loss. In contrast, the other groups hold teleconference quite efficiently.

5 Conclusions

The purpose of the work presented in this paper was to examine the capability of the widespread IEEE 802.11b protocol, to efficiently support time-sensitive applications. We chose the integrated MPEG teleconference and data exchange, and we examined scenarios of full teleconference, teleconference with pairs of users and teleconference with groups of different numbers of users. In the first scenario (full teleconference) the system may support up to 6 wireless stations. As it is expected, in the 3rd scenario (groups of teleconference users) we cannot transmit video and data efficiently, if we have a group with more than 5 users, as this will be an extension to the first scenario. In reality, according to the number of groups and the number of stations per group, the system may support a larger or a smaller number of parallel teleconferences. This is clear in the second scenario, in which every group consists of 2 stations only, thus the system may support 16 users. Potentially, a maximum number of wireless nodes in one group is 5, however if we have a few users per group, we may support more teleconferences. As for the data packet transmissions during DCF, the simulation results are as expected. By adding more stations we observe an increase in the average data packet delay and in the packet losses.

In conclusion, our simulation results clearly demonstrate the difficulty of the protocol to support time-sensitive applications with a large number of wireless users, under the same Access Point.

References

1. ANSI/IEEE 802.11 Standard, 1999 Edition.
2. IEEE 802.11b Standard, 1999, (Supplement to ANSI/IEEE 802.11 Standard, 1999 Edition).
3. Brian P. Crow, Indra Widjaja, Jeong Geun Kim, Prescott T. Sakai: "IEEE 802.11 Wireless Local Area Networks", IEEE Communications Magazine, 1997.
4. James F. Kurose, Keith W. Ross: Computer Networking: "A Top Down Approach Featuring The Internet", Addison – Wesley Publications, 2001.
5. F. Cali, M. Conti, E. Gregori: "IEEE 802.11 Wireless LAN: Capacity Analysis and Protocol Enhancement", Proceedings of the 1998 IEEE Infocom Conference.
6. Breeze Wireless Communications Ltd.: IEEE 802.11 Technical Tutorial, Israel 2000.
7. Video traces: http://www.tkn.ee.tuberlin.de/research/trace/trace.html
8. Averill M. Law, W. David Kelton: "Simulation modeling & Analysis", 2nd Edition, McGraw Hill International Editions, 1991.
9. Frank H. P. Fitzek, Martin Reisslein: "MPEG-4 and H.263 Video Traces for Network Performance Evaluation (Extended Version)", TU Berlin, Dept. of Electrical Engineering, Telecommunication Networks Group, Technical Report: TKN-00-06, October 2000, http://www.tkn.ee.tu berlin.de/research/trace/pub.html
10. P. N. Tudor: "MPEG-2 Video Compression", Tutorial, Electronics and Communications Journal, December 1995, http://www.bbc.co.uk/rd/pubs/papers/paper_14/paper_14.html

Differentiated Quality-of-Protection Provisioning in Optical/MPLS Networks*

Canhui (Sam) Ou and Biswanath Mukherjee

Department of Computer Science, University of California, Davis, CA 95616, USA
{ouc, mukherje}@cs.ucdavis.edu

Abstract. This paper investigates the problem of dynamic survivable lightpath provisioning against single node/link failures in optical mesh networks employing wavelength-division multiplexing (WDM). We present a new approach to provisioning lightpath requests according to their differentiated quality-of-protection (QoP) requirements. We focus on one of the most important QoP parameters, namely protection-switching time, since lightpath requests may have differentiated protection-switching-time requirements. For example, lightpaths carrying voice traffic may require 50-ms protection while lightpaths carrying data traffic may have a wide range of requirements. Numerical results show that, compared to shared-path protection, our approach achieves significant performance gain which leads to remarkable reduction in blocking probability. While our focus is on optical WDM network, the basic ideas of our approaches can be applied to multiprotocol label switching (MPLS) networks with appropriate variations, e.g., differentiated bandwidth granularities.

1 Introduction

In a wavelength-routed optical network, the failure of a network element can cause the failure of several lightpaths, thereby leading to large data and revenue loss. Protection, a proactive procedure in which spare capacity is reserved during lightpath setup [1,2,3,4,5], is essential for recovering from such failures in a short time period, e.g. 50 ms. Protection schemes can be classified by the type of routing used (link-based versus path-based) and by the type of resource sharing (dedicated versus shared). A path carrying traffic during normal operation is known as a *working* path[1]. When a working path fails, the lightpath is rerouted over a *backup* path. High bandwidth efficiency and short protection-switching time are two of the most important features of a protection scheme [6], where protection-switching time for a lightpath is the time period the lightpath takes to properly signal/configure the nodes along the backup path before switching traffic to the backup path after a failure occurs on the working path [5].

We consider the problem of dynamic survivable lightpath provisioning against single node (crossconnect) and single link (fiber) failures. Specifically, we focus on

---

* This work has been supported by NSF Grant No. ANI-98-05285.

[1] Working path is also referred to as primary path, active path, and service path.

N. Mitrou et al. (Eds.): NETWORKING 2004, LNCS 3042, pp. 650–661, 2004.
© IFIP International Federation for Information Processing 2004

shared protection (because of its desirable resource efficiency) with the assumptions that existing lightpaths cannot be disturbed and no knowledge of future arrivals is available at the time of provisioning the current lightpath request. While we consider full wavelength-convertible networks here, the extension to the wavelength-continuous case is straightforward.

Much work has been conducted on dynamic shared protection [7,8,9,10] in optical WDM networks and on dynamic routing of restorable bandwidth-guaranteed connections in MPLS networks [11,12,13,14]. A widely considered approach, called shared-path protection [5], is bandwidth efficient due to backup sharing. Consequently, how to increase backup sharing based on different cost models and route-computation techniques is of particular interest and has been reported in [12,15,16,17,18,19,20]. The complexity of shared-path protection is high as shown in [9,21] that it is NP-complete to find a working path and a backup path for a new lightpath request when backup sharing with existing backup paths is allowed. As a result, practical heuristics are usually employed.

One possible limitation of shared-path protection is that backup paths may sometimes become longer due to backup sharing [15]. Consequently, protection-switching time may increase because of longer backup paths. The relation between backup sharing and backup-path hop distance for path protection have been shown to be that one trades off another in [10,15].

Furthermore, lightpath requests may have differentiated protection-switching-time requirements. For example, lightpaths carrying voice traffic may require 50 ms protection while lightpaths carrying data traffic may require a wide range of protection-switching-time requirements. Due to the path-wise node-/link- disjoint nature of path protection, shared-path protection may not provision lightpath requests according to their protection-switching-time requirements effectively in practical-sized networks [22,23]. Clearly, proper mechanisms are needed to provision such lightpath requests in a resource-efficient manner.

Motivated by the above considerations, we present a new and effective approach to provisioning lightpath requests according to their protection-switching-time requirements while taking into account backup sharing. While our focus is on optical WDM network, in which the bandwidth requirement of a lightpath request is one wavelength, our approaches can also be directly applied to MPLS networks for provisioning restorable, bandwidth-guaranteed connections of differentiated bandwidth granularities with appropriate adjustments.

2   Provisioning Lightpaths of Differentiated Quality-of-Protection (QoP) Requirements

We focus on one of the most important QoP parameters, namely protection-switching time. The protection-switching time of a shared-path protected lightpath can be based on the hop count of the working/backup paths [5,23,24]. Therefore, we consider QoP in terms of hop count.

Below, we argue that new mechanisms are needed to provision lightpaths of differentiated QoP requirements and present a new approach to achieve the QoP.

2.1   Motivation

Lightpath requests may have differentiated protection-switching-time requirements. For example, lightpaths carrying voice traffic may require 50 ms protection while lightpaths carrying data traffic may need a wide range of protection-switching-time requirements. While some mission-critical lightpath requests can be dedicate protected, it is not economically viable to provide dedicated protection to each lightpath request due to its excessive resource requirement.

Below, we show that shared-path protection cannot provide the desired level of protection-switching time either. Let us consider a simple case in which the backup-path hop count of any lightpath cannot exceed a constant H_b (ignoring the constraint on working path for now). A lightpath request will be blocked if the computed backup path is longer than H_b hops. We simulate a dynamic network environment with the assumptions that the lightpath-arrival process is Poisson and the lightpath-holding time follows a negative exponential distribution. In every experiment, 10^6 lightpath requests are simulated; they are uniformly distributed among all node pairs; average lightpath-holding time is normalized to unity; the cost of any link is unity; and our example network topology with 16 wavelengths per fiber is shown in Fig. 1.

A commonly used route-computation heuristic for shared-path protection is as follows: compute a least-cost path as the working path and then compute as the backup path a link (or node) disjoint path of least additional cost [15]. In general, we can compute K candidate working paths, repeat the above heuristic for each candidate working path, and select the working and backup path pair of minimum cost.

Figure 2(a) plots the blocking probability of shared-path protection for $H_b = 6$ for the network shown in Fig. 1 with different values of ϵ. Figure 2(a) confirms the conclusion in [15,10] that a larger value of ϵ leads to shorter backup path but decreased backup sharing, and a smaller value of ϵ leads to increased backup sharing but longer backup path. Please note that $H_b = 6$ for this network is reasonably large since the average hop distance is 2.99. However, regardless of the values of ϵ, the blocking probability in Fig. 2(a) is quite high. The main reason is that some lightpath requests are blocked because their backup paths span more than $H_b = 6$ hops. Figure 2(b) shows the impact of H_b on shared-path protection. While the blocking probability drops significantly as H_b increases, the blocking for $H_b = 7$, which is quite large, is still unacceptable.

As network size grows, it is clear that shared-path protection cannot achieve reasonable blocking for practical values of H_b due to its fundamental limitation: the backup path has to be end-to-end node-/link- disjoint to the working path.

Consequently, new mechanisms, which can relax the end-to-end node-/link-disjointness, are needed. One such mechanism is segment protection, as described in the next section.

2.2   Segment Protection (SP)

Various forms of segment protection (SP) have been reported in [25,26,27,28]. The approaches proposed in [25] addressed single-link failures by dividing a

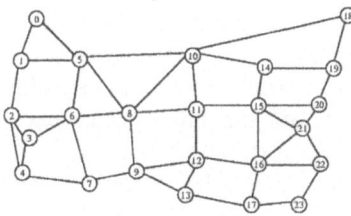

Fig. 1. A representative topology whose average hop distance is 2.99 and average nodal degree is 3.58.

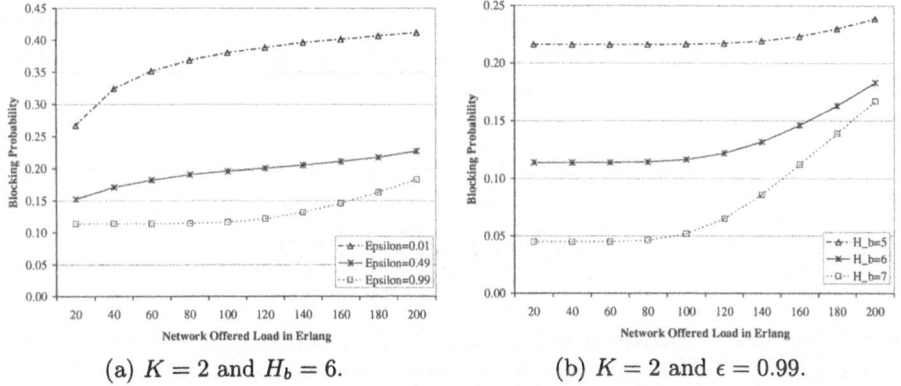

(a) $K = 2$ and $H_b = 6$.   (b) $K = 2$ and $\epsilon = 0.99$.

Fig. 2. Blocking probability of shared-path protection.

working path into a sequence of non-overlapping segments and protecting each such segment individually. Nodes were assumed to be robust in these approaches as consecutive non-overlapping segments share the same node failure.

Even though node failures are not as often as link failures, they need to be carefully treated because the impact of node failures is much more disastrous than that of link failures. The work in [26,27,28] addressed single-node/link failures by dividing a working path into a sequence of overlapping segments and protecting each such segment separately. As shown in Fig. 3, the lightpath from node s to node d is partitioned into two overlapping segments: one with working segment $\langle s, i, j, u \rangle$ and backup segment $\langle s, u \rangle$; another with working segment $\langle j, u, v, d \rangle$ and backup segment $\langle j, d \rangle$. (There are other nodes along the two backup segments $\langle s, u \rangle$ and $\langle j, d \rangle$. Those nodes are not shown in Fig. 3 to preserve clarity.)

The advantages of SP are numerous. The end-to-end protection entity is *a segment* in SP as opposed to *a path* in path protection. When a failure occurs along a working path (segment), the source node of that path (segment) switches to its backup. Since a segment is typically shorter than a path in terms of hop count, SP is expected to have shorter protection-switching time.

Meanwhile, two segments (or two lightpaths in path protection) can share backup wavelength links as long as their working segments (or working paths in path protection) do not share the same node/link failure. Since, in general, a segment is shorter than a path, the probability of two working segments sharing the same risk is typically lower than the probability of two working paths sharing the same risk. As a result, SP can have better backup sharing compared to shared-path protection.

Furthermore, SP has more flexibility in routing compared to path protection since path protection is a special case of SP in which every lightpath has exactly one segment. Apart from these differences, it is clear that the longer the working path is, the more difficult it is to find a node-disjoint backup path [29].

Based on SP, we present below an approach, called SP_QoP, which: (1) dynamically divides a judiciously selected working path into multiple overlapped working segments, (2) computes a backup segment for each working segment, (3) accommodates backup sharing, and (3) ensures the hop-count constraint on working/backup segments.

Fig. 3. Segment protection: node j and node u can be adjacent or separated by other nodes. (The solid lines from node s to node d represent the working path, and the dashed lines represent the backup segments. While only two segments are shown in these illustrations, in general, a path may employ many segments.)

2.3   SP_QoP Heuristic

Upon the arrival of a new lightpath request, the network management system needs to compute a working path l_w and a list of backup segments $\{l_b^i\}$, which divide the working path into overlapping segments $\{l_w^i\}$ such that l_w^i and l_b^i are node-/link- disjoint. New backup segments $\{l_b^i\}$ can share wavelength links with existing backup segments as well as among themselves. Unfortunately, it is NP-hard to determine if there exists an eligible solution as we have proved the NP-completeness of the existence version of shared-path-protection problem, which is a special case of SP with the number of segments being one in [9]. As a result, we resort to a heuristic.

A network is represented as a weighted, directed graph $G = (V, E, C, \lambda)$, where V is the set of nodes, E is the set of unidirectional fibers (referred to as links), $C : E \to R^+$ is the cost function for each link (where R^+ denotes the set of positive real numbers), and $\lambda : E \to Z^+$ specifies the number of wavelengths on each link (where Z^+ denotes the set of positive integers).

A conflict set is associated with a link to identify the sharing potential between backup segments[2]. The conflict set ν_e for link e defines the set of nodes traversed by such working segments whose backup segments utilize wavelengths

---

[2] The conflict set is similar to the conflict vector in [8], the aggregated square matrix in [17], and the "bucket" link metric in [19].

on link e. The conflict set ν_e for link e can be represented as an integer set, $\{\nu_e^u \mid \forall u \in V, 0 \leq \nu_e^u \leq \lambda(e)\}$, where ν_e^u specifies the number of working segments which traverse node u and are protected by link e (or, in other words, their corresponding backup segments traverse link e). The number of wavelengths reserved for backup segments on link e is thus $\nu_e^* = \max_{\forall u}\{\nu_e^u\}$. Clearly, the union of the conflict sets for all the links aggregates the per-segment-based information, and the size of the conflict set depends only on the number of nodes, not on the number of segments.

For a candidate working path l_w^k, our heuristic, called SP_QoP, performs the following recursive procedure to compute a list of eligible backup segments.

1. Starting from node s, compute a least-cost path to all the other nodes along l_w^k, where the cost function is C_1 defined in Algorithm 1.
2. Starting from node d and following the reverse direction of l_w^k, find the first node v which satisfies the constraint that the least-cost path from node s to node v is of at most H_b hops.
3. If node v is the destination node d, the heuristic succeeds and terminates; otherwise, starting from all the nodes between node s and node v (excluding nodes s and v) along l_w^k, recursively apply the above procedure. (If there is no node between node s and node v, the heuristic fails.)

Our SP_QoP heuristic is specified in Algorithm 1. For a node $u \in V$, $PC(u)$ denotes the cost of the least-cost path destined to node u; $HC(u)$ represents the hop count of the least-cost path; and $PH(u)$ records the previous hop along the least-cost path. For a path l_w, $Head(l_w)$ returns the first node along l_w.

We make the following remarks. 1) In Algorithm 1, the candidate working path l_w^k is given. This is just for the purpose of simplifying the presentation; in our implementation, we dynamically compute K candidate working paths based on Yen's K shortest loopless paths algorithm [30], execute Algorithm 1 for each candidate working path, and select the working path and the list of backup segments of minimal cost.

2) Backup sharing in this case is tricky as the situation shown in Fig. 4 can arise. In the presence of H_b, the path $\langle s, x, p, q, y, d \rangle$ is not valid when $H_b = 4$. However, the two segments $\langle s, x, p, q, u \rangle$ and $\langle j, p, q, y, d \rangle$ so formed are still valid. Our heuristic accommodates this type of backup sharing in Steps 3 and 7 of Algorithm 1 since the freshly reserved backup wavelengths for a newly computed backup segment is used for computing later backup segments for the same lightpath request.

3) Sometimes, it may be desirable that the hop count of any working segment plus the hop count of its backup segment is no more than some constant H. We can modify Step 5 to satisfy this constraint as follows. For any node v along path l_w, denote as $l_b^{i,v}$ the least-cost path destined to node v and denote as h_w^v the number of hops from $Head(l_b^{i,v})$ to node v along l_w. Starting from node d and following the reverse direction of l_w, find the first node v which satisfies the constraint $HC(v) + h_w^v \leq H$. Other constraints based on combinations of working and backup segment hop count also can be easily incorporated.

Algorithm 1 SP_QoP

Input: $G = (V, E, C, \lambda)$, $\nu = \{\nu_e \mid e \in E\}$, s, d, a candidate working path l_w^k

Output: a list of backup segments $\{l_b^i\}$, each of which spans no more than H_b hops and they collectively partition l_w^k into overlapped segments $\{l_w^{k,i}\}$ such that $l_w^{k,i}$ and l_b^i are node-/link- disjoint; otherwise, NULL if no such list is found.

1. $S \leftarrow \{s\}, L_b \leftarrow \phi, l_w \leftarrow l_w^k, i \leftarrow 0$
2. $V' \leftarrow V$; $\forall u \in S, PC(u) \leftarrow 0, HC(u) \leftarrow 0, PH(u) \leftarrow NULL$; $\forall u \in V \wedge u \notin S, PC(u) \leftarrow \infty, HC(u) \leftarrow \infty, PH(u) \leftarrow NULL$; $i \leftarrow i + 1$
3. define link-cost function $C_1(e)$, $e \in E$, with respect to l_w:

$$C_1(e) := \begin{cases} +\infty & \text{if } l_w \text{ traverses link } e, \text{ or } \nu_e^* \text{ is equal to } \nu_e^u \text{ for some node } u \text{ along} \\ & l_w \text{ and link } e \text{ does not have any free wavelength} \\ \epsilon \times C(e) & \text{if for any node } u \ (u \neq s, d) \text{ traversed by } l_w, \nu_e^u < \nu_e^* \\ C(e) & \text{otherwise} \end{cases}$$

4. while $(V' \neq \phi)$ do {
       $u \leftarrow \arg \min_{u \in V'} \{PC(u)\}$, $V' \leftarrow V' - \{u\}$
       if $(u = Head(l_w))$ or (l_w does not traverse u) {
           $\forall v \in V'$, s.t. $\langle u, v \rangle \in E$
           if $PC(v) > PC(u) + C_1(\langle u, v \rangle)$ then {
               $PC(v) \leftarrow PC(u) + C_1(\langle u, v \rangle)$
               $HC(v) \leftarrow HC(u) + 1$
               $PH(v) \leftarrow u$
           } // if
       } // if
   } // while
5. starting from node d and following the reverse direction of l_w, find the first node v which satisfies $HC(v) \leq H_b$
6. retrieve the least-cost path destined to node v by following $PH(v)$ and denote the path as l_b^i
7. allocate backup wavelengths along l_b^i: let l_w^i be the working segment starting from $Head(l_w)$ and ending at v along l_w (inclusively); for any link e that l_b^i traverses and for any node u along l_w^i (excluding the source and the destination nodes of l_w^i), $\nu_e^u \leftarrow \nu_e^u + 1$; if $\nu_e^u > \nu_e^*$, then reserve one more wavelength on link e and let $\nu_e^* \leftarrow \nu_e^u$
8. if v is d, then return $\{l_b^i\}$
9. $S \leftarrow$ all the nodes between $Head(l_w)$ and node v along l_w, excluding $Head(l_w)$ and node v; if S is empty, then undo any changes made to G in Step 7 and return NULL
10. $l_w \leftarrow$ the path starting from node v to node d along l_w
11. go to Step 2

Computational Complexity: The computational complexity of Algorithm 1 is $O(|V|^3 + |E|)$. In particular, the computational complexities for Steps 1-11 are $O(1)$, $O(|V|)$, $O(|E|)$, $O(|V|^3 + |E|)$, $O(|V|)$, $O(|V|)$, $O(|V|)$, $O(|E|)$, $O(|V|)$, $O(|V|)$, and $O(1)$, respectively. If we compute K candidate working paths and

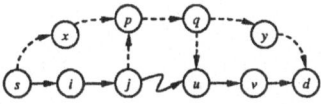

Fig. 4. Two backup segments, $\langle s, x, p, q, u \rangle$ and $\langle j, p, q, y, d \rangle$, of the same lightpath share the same wavelength link on link $\langle p, q \rangle$, assuming $H_b = 4$.

execute Algorithm 1 for each candidate working path, then the computational complexity is $O(K \cdot (|V|^3 + |E|))$.

3   Illustrative Numerical Results

We now quantitatively compare SP_QoP to shared-path protection under the simulation configuration as described in Section 2.1. For the illustrative results shown here, we use $K = 2$ as we found the performance improvement is marginal if we increase K to any larger value.

3.1   Blocking Probability under Different Values of ϵ

Figure 5(a) plots the blocking performance for $H_b = 6$ under $\epsilon = 0.01, 0.49$, and 0.99. We observe that our SP_QoP approach has significantly lower blocking probability than shared-path protection under the same ϵ. We further observe that large values of ϵ, e.g., $\epsilon = 0.49$ or $\epsilon = 0.99$, are preferable as both SP_QoP and shared-path protection have significantly lower blocking when ϵ has a large value. Later, we shall use large values of ϵ.

Figure 5(b) shows the *performance gain*, defined as the percentage of lightpath requests which are blocked in shared-path protection but can be accepted by SP_QoP. Performance gain can be calculated as follows. Whenever shared-path protection needs to block a lightpath request, we apply SP_QoP to check whether the same lightpath request can be provisioned under the same network state (but we do not set up the lightpath request even if it can be provisioned). We observe that SP_QoP has a remarkable performance gain (over 70% across all load regions). The huge performance gain results from the fact that SP_QoP relaxes the path-wise node-/link- disjointness to segment-wise node-/link- disjointness and computes segments with respect to H_b.

3.2   Blocking Probability under Different Values of H_b

Figure 6(a) examines the impact of H_b on both SP_QoP and shared-path protection with $\epsilon = 0.99$. We observe that: (a) SP_QoP has much lower blocking probability when load is not very high. (b) When H_b increases from 5 to 6, SP_QoP has noticeable reduction in blocking probability while the reduction is marginal when H_b further increases to 7. (c) As H_b increases, the blocking probability of shared-path protection drops significantly. However, shared-path

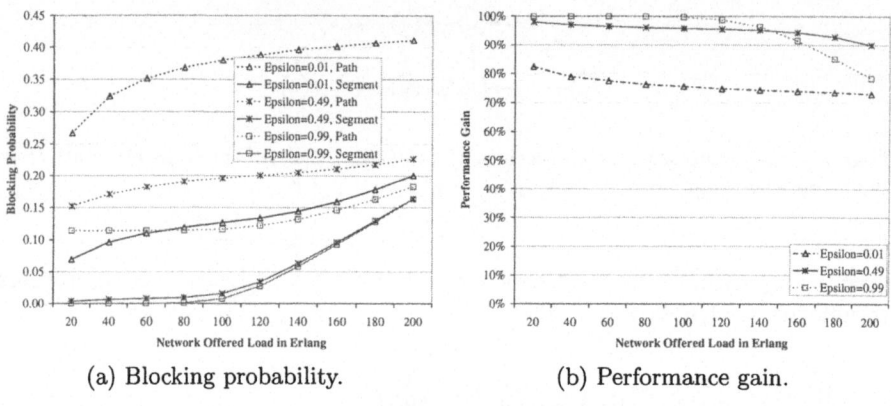

(a) Blocking probability.   (b) Performance gain.

Fig. 5. Impact of ϵ ($H_b = 6$).

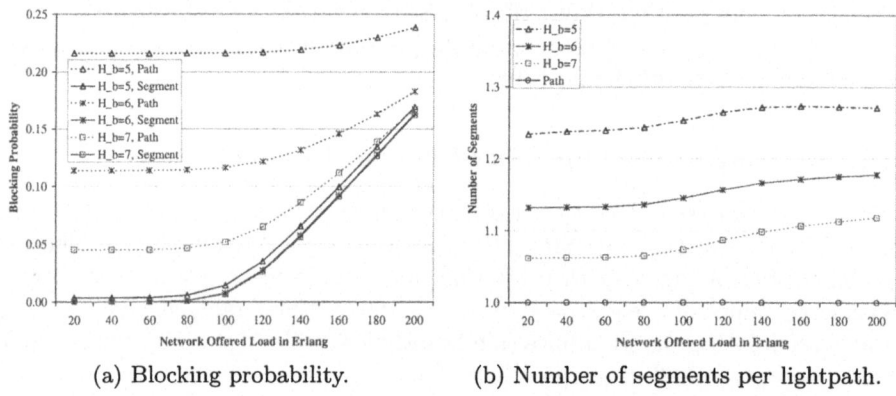

(a) Blocking probability.   (b) Number of segments per lightpath.

Fig. 6. Impact of H_b ($\epsilon = 0.99$).

protection still has remarkable blocking (above 4%) even when the network offered load is low, e.g., 20 Erlangs which translates to about 8.5% average link utilization. This is due to the path-wise end-to-end node-/link- disjoint nature of shared-path protection.

Figure 6(b) shows that more segments are needed for smaller value of H_b. However, the average number of segments per lightpath is still quite low, e.g., less than 1.3, even for $H_b = 5$. This implies that the control and management overhead due to segmentation is not very significant.

3.3   Blocking Probability for Lightpath Requests with Differentiated QoP Requirements

Different lightpath requests may have differentiated QoP requirements, as discussed earlier in Section 2.1. Figure 7 compares the performance of SP_QoP to shared-path protection under two types of traffic. The QoP of the lightpath re-

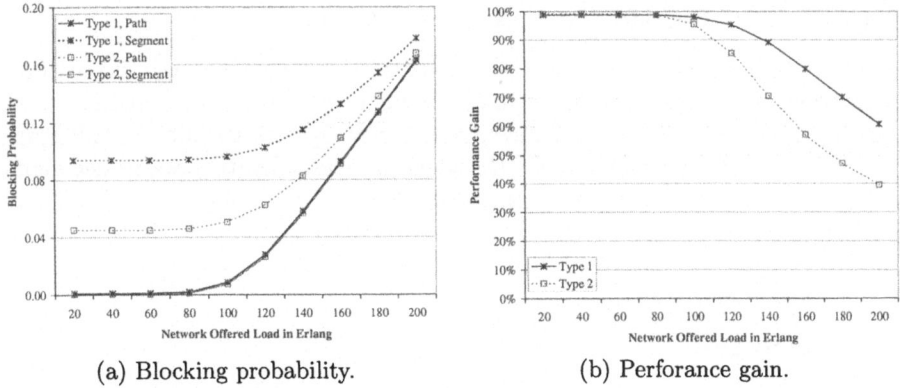

(a) Blocking probability.   (b) Perforance gain.

Fig. 7. Impact of differentiated QoP requirements ($\epsilon = 0.99$). In Type 1, H_b follows $5 : 6 : 7 : \infty = 30 : 20 : 10 : 40$; In Type 2, H_b follows $5 : 6 : 7 : \infty = 10 : 20 : 20 : 50$.

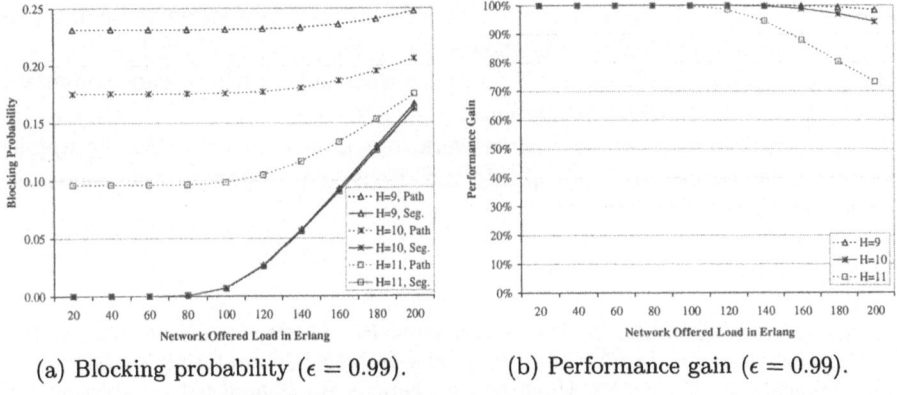

(a) Blocking probability ($\epsilon = 0.99$).   (b) Performance gain ($\epsilon = 0.99$).

Fig. 8. Impact of H ($\epsilon = 0.99$).

quests in terms of H_b follows the distribution $5 : 6 : 7 : \infty = 30 : 20 : 10 : 40$ in Type 1 and $5 : 6 : 7 : \infty = 10 : 20 : 20 : 50$ in Type 2.

SP_QoP has much lower blocking probability than shared-path protection, as shown in Fig. 7(a). For shared-path protection, the large difference between the blocking probability for the two types of traffic implies that shared-path protection cannot effectively provision lightpath requests based on their differentiated QoP requirements. However, the difference between the blocking probability for the two types of traffic in SP_QoP is very small. This indicates that SP_QoP can properly provision lightpath requests according to their differentiated QoP requirements.

As shown in Fig. 7(b), when load is modest or low, SP_QoP achieves close to 100% performance gain; even when load is high, SP_QoP still achieves more than 35% performance gain.

3.4   Blocking Probability for Different Values of H

Figures 8(a) and 8(b) examine the impact of H on SP_QoP and shared-path protection. The curves in Fig. 8 have similar trend to the ones in Fig. 6 and can be explained similarly. Meanwhile, since H applies to both working and backup segments, as opposed to H_b which applies only to backup segments, the H constraint is more stringent than the H_b constraint. As a result, the performance gain for different values of H is even higher, above 70% across all load regions, as shown in Fig. 8(b).

4   Conclusion

This paper considered the problem of dynamic survivable lightpath provisioning against single node/link failures in optical mesh networks. We presented a new approach to provisioning lightpath requests according to their differentiated quality-of-protection (QoP) requirements with focus on protection-switching time since lightpath requests may have differentiated protection-switching-time requirements. Numerical results showed that, compared to shared-path protection, our approach achieves significant performance gain which leads to remarkable reduction in blocking probability.

While our focus is on optical network, our approaches can also be applied to MPLS networks with appropriate variations, e.g., differentiated bandwidth granularities.

References

1. Doshi, B.T., Dravida, S., Harshavardhana, P., Hauser, O., Wang, Y.: Optical network design and restoration. Bell Labs Technical Journal 4 (1999) 58–84
2. Fumagalli, A., Tacca, M., Unghvary, F., Farago, A.: Shared path protection with differentiated reliability. In: Proc. IEEE ICC. (2002) 2157–2161
3. Lumetta, S.S., Medard, M., Tseng, Y.C.: Capacity versus robustness: a tradeoff for link restoration in mesh networks. IEEE J. Lightwave Technology 18 (2000) 1765–1775
4. Modiano, E., Narula-Tam, A.: Survivable lightpath routing: a new approach to the design of WDM-based networks. IEEE J. Selected Areas in Communications 20 (2002) 800–809
5. Ramamurthy, S., Sahasrabuddhe, L., Mukherjee, B.: Survivable WDM mesh networks. IEEE J. Lightwave Technology 21 (2003) 870–883
6. Koo, S., Subramaniam, S.: Trade-offs between speed, capacity, and restorability in optical mesh network restoration. In: Proc. OFC. (2002) 487–489
7. Hauser, O., Kodialam, M., Lakshman, T.V.: Capacity design of fast path restorable optical networks. In: Proc. IEEE INFOCOM. Volume 2. (2002) 817–826
8. Mohan, G., Murthy, C.S.R., Somani, A.K.: Efficient algorithms for routing dependable connections in WDM optical networks. IEEE/ACM Trans. Networking 9 (2001) 553–566
9. Ou, C., Zhang, J., Zang, H., Sahasrabuddhe, L., Mukherjee, B.: Near-optimal approaches for shared-path protection in WDM mesh networks. In: Proc. IEEE ICC. (2003) 1320–1324

10. Xiong, Y., Xu, D., Qiao, C.: Achieving fast and bandwidth-efficient shared-path protection. IEEE J. Lightwave Technology **21** (2003) 365–371
11. Kar, K., Kodialam, M., Lakshman, T.V.: Routing restorable bandwidth guaranteed connections using maximum 2-route flows. In: Proc. IEEE INFOCOM. (2002) 113–121
12. Kodialam, M., Lakshman, T.V.: Dynamic routing of bandwidth guaranteed tunnels with restoration. In: Proc. IEEE INFOCOM. Volume 2. (2000) 902–911
13. Kodialam, M., Lakshman, T.V.: Dynamic routing of locally restorable bandwidth guaranteed tunnels using aggregated link usage information. In: Proc. IEEE INFOCOM. Volume 1. (2001) 376–385
14. Qiao, C., Xu, D.: Distributed partial information management (DPIM) schemes for survivable networks – Part I. In: Proc. IEEE INFOCOM. (2002) 302–311
15. Bouillet, E., Labourdette, J.F., Ramamurthy, R., Chaudhuri, S.: Enhanced algorithm cost model to control tradeoffs in provisioning shared mesh restored lightpaths. In: Proc. OFC. (2002) ThW2
16. Li, G., Wang, D., Kalmanek, C., Doverspike, R.: Efficient distributed path selection for shared restoration connections. In: Proc. IEEE INFOCOM. (2002) 140–149
17. Liu, Y., Tipper, D., Siripongwutikorn, P.: Approximating optimal spare capacity allocation by successive survivable routing. In: Proc. IEEE INFOCOM. Volume 2. (2001) 699–708
18. Su, C., Su, X.: Protection path routing on WDM networks. In: Proc. OFC. Volume 2. (2001) TuO2–T1–3
19. Su, X., Su, C.: An online distributed protection algorithm in WDM networks. In: Proc. IEEE ICC. Volume 5. (2001) 1571–1575
20. Wang, H., Modiano, E., Medard, M.: Partial path protection for WDM networks: end-to-end recovery using local failure information. In: Seventh International Symposium on Computers and Communications (ISCC 2002). (2002) 719–725
21. Ellinas, G., Bouillet, E., Ramamurthy, R., Labourdette, J., Chaudhuri, S., Bala, K.: Routing and restoration architectures in mesh optical networks. SPIE Optical Networks Magazine **4** (2003) 91–106
22. Li, J., Park, H., Lee, H.: Shared sub-path protection with overlapped protection areas in WDM networks. In: Proc. OFC. (2003) FQ7
23. Ou, C., Zang, H., Mukherjee, B.: Sub-path protection for scalability and fast recovery in optical WDM mesh networks. In: Proc. OFC. (2002) 495–496
24. Assi, C., Ye, Y., Shami, A., Dixit, S., Ali, M.: Efficient path selection and fast restoration algorithms for shared restorable optical networks. In: Proc. IEEE ICC. (2003) 1412–1416
25. Anand, V., Chauhan, S., Qiao, C.: Sub-path protection: A new framework for optical layer survivability and its quantitative evaluation (Dept. of CSE, State University of New York at Buffalo, Tech. Report 2002-01, Jan. 2002)
26. Gummadi, K., Pradeep, M., Murthy, C.: An efficient primary-segmented backup scheme for dependable real-time communication in multihop networks. IEEE/ACM Trans. Networking **11** (2003) 81–94
27. Ho, P.H., Mouftah, H.: A framework for service-guaranteed shared protection in WDM mesh networks. IEEE Communication Magazine **40** (2002) 97–103
28. Ho, P.H., Mouftah, H.T.: A novel strategy for protection domain allocation in dynamic mesh WDM networks. In: Proc. IEEE ICC. (2003) CQ22–1
29. Liu, Y., Tipper, D.: Successive survivable routing for node failures. In: Proc. IEEE Globecom. (2001) 2093–2097
30. Yen, J.Y.: Finding the K shortest loopless paths in a network. Management Science (1971) 712–716

Hierarchical Routing with QoS Constraints in Optical Transport Networks*

Xavier Masip-Bruin, Sergio Sánchez-López, Josep Solé-Pareta
Jordi Domingo-Pascual, and Eva Marín-Tordera

Departament d'Arquitectura de Computadors, Universitat Politècnica de Catalunya
Avgda. Víctor Balaguer, s/n– 08800 Barcelona, Spain
{xmasip, sergio, pareta, jordid, eva} @ac.upc.es

Abstract. Optical Transport Networks (*OTN*) with automatic switching capabilities are named *ASON*. Hierarchical routing is required in the *ASON* recommendations to achieve scalability. Basically, hierarchical routing consists of three main components, an aggregation scheme, an update policy and a routing algorithm. This paper proposes a new network structure focusing on these three components. We propose a new aggregation scheme; an update policy based on a threshold value; and we also extend an already proposed routing mechanism to be applied to a hierarchical network. Main skill of this routing mechanism is to reduce the connection blocking increase because of selecting paths based on inaccurate routing information. This inaccuracy is introduced both by the aggregation process and by the update policy.

Keywords. Hierarchical optical routing, ASON, aggregation schemes, routing inaccuracy.

1   Introduction

In recent years the introduction of high capacity and reliable transport networks is being necessary in order to cover the needs of Internet traffic demands. New incoming Internet applications increasingly request greater capacity and guarantees of traffic delivery. Optical Transport Networks (*OTN*) with automatic switching capabilities (*ASON*, Automatically Switched Optical Networks) appear as a potential solution to cope with such a situation.

A single carrier optical network may consist of hundreds of network nodes, with each node containing more than one thousand physical ports. An appropriate *ASON* routing protocol has to be scalable enough for this network size. Some approaches may be used to achieve scalability, such as hierarchical routing, link bundling, and classification of link state information into static and dynamic information. *ASON* specifications [1] specifically recommend hierarchical routing to guarantee network

---

* This work was partially funded by the MCyT (Spanish Ministry of Science and Technology) under contract FEDER-TIC2002-04344-C02-02 and the CIRIT (Catalan Research Council) under contract 2001-SGR00226.

N. Mitrou et al. (Eds.): NETWORKING 2004, LNCS 3042, pp. 662–674, 2004.

scalability. The main advantage of hierarchical routing is to reduce large communication overhead while providing efficient routing.

ASON must include a Control Plane able to provide features such as Traffic Engineering. One of the essential components of this Control Plane is the routing protocol, which has to include both a mechanism to disseminate routing information throughout the network and a routing algorithm to select a suitable path. The routing algorithm dynamically computes paths supporting the QoS constraints required by the incoming call. Assuming source-based routing as the most commonly used QoS routing algorithms, routes are computes on the source nodes according to the routing information contained in their network state databases (named Traffic Engineering Database, *TED* when including QoS parameters). The Control Plane has to provide a flexible, fast and reliable mechanism to both disseminate and update the topology and the available resource information throughout the network. An Optical NNI (*O-NNI*) should be standardized for *ASON* to support such a mechanism.

For some time it seemed clear that such an *O-NNI* would be based on the *GMPLS* [2] paradigm. Solutions based on the *GMPLS* paradigm only consider typical Internet protocols such as link state based Interior Gateway Protocols (*IGPs*). Nevertheless, recently the idea of an *O-NNI* based on the *ATM PNNI* [3] paradigm is gaining support. There are some works in the recent literature focusing on adapting the *PNNI* to the Optical Network Control Plane [4].

The main goal of this paper is to provide *ASON* with a hierarchical routing in order to ensure the scalability for large worldwide networks. Hierarchical routing functions include two parts:

– Network information dissemination among nodes and networks.
– Constraint based path computation.

According to these main functions we decompose the hierarchical routing problem into three main issues: the aggregation process used to reduce disseminated information throughout the network; the update policy used to keep network state databases perfectly updated; and the lightpath selection process assuming the potential network state inaccuracy introduced by both the aggregation process and the update policy.

The remainder of this paper is organized as follows. Section 2 is devoted to define a hierarchical structure for *ASON*. In Section 3 we propose a new hierarchical structure including a new aggregation scheme, a new update policy and a new routing mechanism. After that, Section 4 evaluates the proposed mechanism and finally Section 5 concludes the paper.

2   Hierarchical Routing Issues

An *ASON* should be subdivided into routing areas (*RAs*), i.e., X1, X2, X3, Y1, Y2 and Y3 (see Fig.1 as an example), which have to contain physical nodes with similar features. The *RA* nodes should exchange topology and resource information among themselves in order to maintain an identical view of the *RA*. This information should be contained in a Routing Controller (*RC*) component, which will respond both to requests from connection controllers (*CC*) for path information needed to set up connections and to requests for topology information from hierarchical mechanisms.

Each *RA* should be represented by a *"Logical Routing Area Node"* (*LRAN*) in the next hierarchical level. The necessary functions to perform this role should be executed by a node called the *"Routing Area Leader"* (*RAL*). This node will receive complete topology state information from all *RA* nodes and will send information up to the *LRAN*. The propagated information should only include the information needed by the higher level.

As mentioned above, hierarchical routing consists of three main issues: an aggregation process, an update policy and a hierarchical routing algorithm. Next, we shortly introduce these issues.

2.1   Network Information Aggregation

There are many possible aggregation schemes. One possible adopted in [3] consists of a complex node representation, which represents the internal structure of each *RA*, called Peer Group, as a symmetric star topology with a uniform radius. B. Awerbuch et al. [5] compared the performance of the different aggregation schemes. They include a star with radius equal to half the cost of the network diameter (*DIA*), a star radius equal to half the average cost between nodes (*AVE*), minimum spanning Tree (*MST*), Random Spanning Tree (*RST*) and t-spanner. Other aggregation schemes, namely *Full Mesh Aggregation* scheme and *Asymmetric Simple Aggregation* scheme are presented in [6]. They are based on the aggressive mode defined in [3], which chooses the best value of each QoS parameter for every intra-group links. The QoS parameters considered by these schemes are delay, capacity and cost. However, being aware that QoS parameters are different in optical networks and in *ATM* networks, the QoS parameters considered for optical networks are the propagation delay that is proportional to the fiber distance between two nodes, and the number of wavelengths per link.

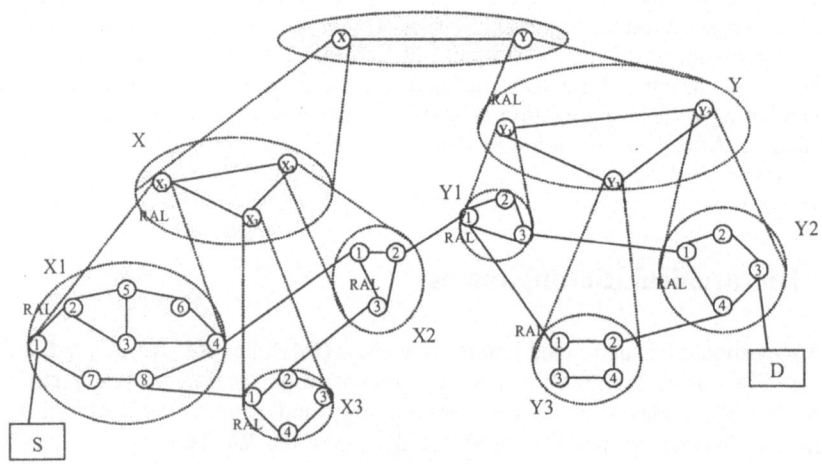

Fig. 1. Hierarchical QoS Routing

2.2   Update Policy

A key factor in the path decision process is the accuracy of the routing information used to compute paths. It has been demonstrated [7] that selecting paths under inaccurate network state information leads to have a significant connection blocking increment, known as the routing inaccuracy problem. There are two main factors in our hierarchical network structure contributing to the inaccuracy: the state aggregation process implemented to reduce the amount of information to be flooded, since information about physical nodes and links is not distributed; and the update policy implemented to reduce the huge number of update messages required to keep *TED* information perfectly updated. Generally speaking, existing update policies are based on updating by either a periodical refresh or a threshold value. The former is not suitable for large networks mainly owing to network dynamics. The latter is more appropriate since it takes into account the network congestion in the updating decision.

2.3   Hierarchical Routing Algorithm

New specific hierarchical routing algorithms should be generated to address the problem of selecting lightpaths under inaccurate network state information. Although the routing inaccuracy problem has been widely analyzed in an IP scenario, there are not many contributions coping with this problem in optical transport network. Main contributions can be found in [7], [8], [9], [10] and [11]. As stated before, in [7] authors only present the effect produced in the blocking probability because of having inaccurate routing information when selecting lightpaths. In [8] the routing inaccuracy problem is addressed by modifying the lightpath control mechanism, and a new distributed lightpath control based on destination routing with rerouting capabilities is suggested. In [9] authors propose a mechanism whose goal is to control the amount of signaling messages flooded throughout the network. The most recent and novel contribution is referred as *BYPASS Based Optical Routing* mechanism *(BBOR)* [10], [11]. The *BBOR* mechanism is a new adaptive source routing mechanism aiming to reduce the connection blocking probability due to perform routing and wavelength assignment decisions under inaccurate network state information. It can also be found in [11] a deeply description of the previously mentioned proposals.

To the authors' best knowledge there is not any proposal in the literature proposing a complete network structure addressing all the three aspects mentioned above and also assuming the routing inaccuracy problem in the lightpath selection process.

3   A Proposal for Hierarchical Routing in Optical Transport Networks

This Section proposes a source-based QoS hierarchical routing including a new aggregation scheme to reduce the volume of information to be disseminated throughout the network, an update policy to determine when update messages must be triggered and a routing algorithm to address the routing inaccuracy problem.

3.1   The Aggregation Scheme

Consider a network consisting of Q OXCs. Each node is assumed to have a fixed number of ports. According to the proposed hierarchical structure, an optical network is divided into M RAs connected by border OXCs, each one composed of a set of OXCs with similar characteristics. Let $G(Q,U)$ describe the given physical network, where Q is a set of OXCs and U is a set of links (i.e. fibers) connecting the nodes. Let $g(q,u)$ describe the given physical RA, where q is the set of nodes (OXCs) in the RA and u is a set of links connecting the nodes within the RA. Therefore, $g \in G$, $q \in Q$ and $u \in U$. Each fiber (link) supports c different wavelengths, i.e. from λ_1 to λ_c. Moreover, we consider that wavelength conversion does not exist in any OXC. Thus, an incoming call is associated to the same wavelength color along the lightpath.

Generally speaking, aggregation schemes work as follows: Firstly, a RA pre-computes all the lightpaths existing between all border nodes along with the QoS parameters allocated to each lightpath. Secondly, an aggregation scheme summarizes this information reducing the amount of data to be flooded throughout the physical network. Finally, the aggregate information from each RA is grouped in a topology database, which will be used by a source node to compute an end-to-end lightpath.

We propose the following network parameters for optical networks:

- D: Propagation delay in a link which is proportional to the fiber distance be-
       tween two nodes.
- W_p: Number of available wavelength of each color in a link

According to the aggregation schemes presented above and the QoS parameters proposed for optical networks, we propose a new aggregation scheme for ASON, namely Node Aggregation Scheme (NAS). The aggregation process performed in this scheme turns out two QoS parameters, the aggregated delay and the aggregated number of available wavelengths.

Node Aggregation Scheme (NAS). Node Aggregation Scheme associates the aggregated QoS parameters to each border node. The process is as follows:

Aggregated Delay (D_i):
1.   Compute all the lightpaths from node i to all border nodes.
2.   Add the propagation delay of each link for each lightpath.
3.   Select the minimum value among the values computed in the step 2.

Aggregated number of available wavelength (W_p^i):

1.   Compute all the lightpaths from node i to all border nodes.
2.   Select the minimum number of wavelength per color that is available on each
      path.
3.   Select the maximum value among the values computed in the step 2.

Formally, the aggregated delay and available wavelength are defined according to (1) and (2) as follows:

$$D_i = \min_{R_{ij}, \forall j \neq i}\left[\sum_{l \in R_{ij}} D(l)\right],$$

(1)

$$W_p^i = \max_{\substack{R_{ij},\forall j\ne i \\ p=1..c}} \left\{ \min_{l\in R_{ij}} \left[W_p(l) \right] \right\} . \tag{2}$$

Fig. 3 illustrates the effects of using the proposed aggregation scheme in terms of entries of the *Aggregated TED (ATED)*. It shows a comparison between the *ATED* size produced when applying the proposed aggregation scheme and the *TED* size produced when an aggregation scheme is not used. As expected the ATED size is shorter when applying the *NAS* scheme.

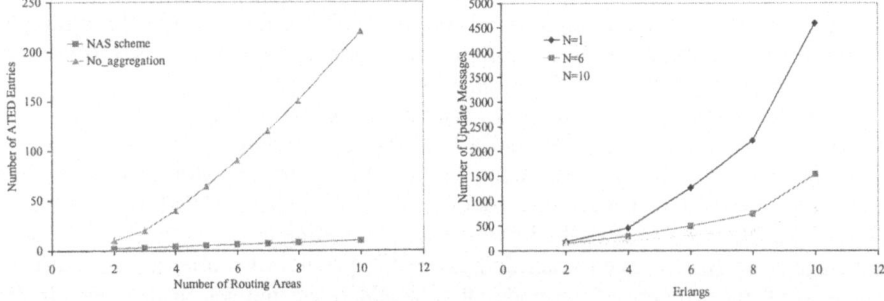

Fig. 3. Comparison with/without aggregation   **Fig. 4.** Number of update messages

3.2   Update Policy

The update policy is an important issue of hierarchical routing. Usually, an update policy is implemented by a triggering policy, which decides when update messages must be flooded throughout the network. Most triggering policies are time-based which are not valid for large dynamic networks since they do not represent traffic evolution. Moreover, the accuracy of aggregated information is dependent on the update interval. If the update is set shorter, the aggregate information is more accurate, but the overhead of re-aggregation and information distribution becomes higher. Another triggering policy may be based on sending an update messages whenever there is a change in the network state. This triggering policy guarantees a low inaccuracy but adds an important signaling overhead.

The *BBOR* mechanism proposed by the authors, introduces a new triggering policy based on a threshold value aiming to include network congestion (available network resources) in the triggering decision. In this way, a network node triggers an update message when a fixed number N of wavelengths changes their status, i.e. after a fixed number of N connections are established or released. This update policy is properly modified to be applied to hierarchical routing. Assuming that the update messages sent by the *RAL* nodes consist of aggregated information, i.e., information already reduced, the signaling overhead pending to be reduced is that produced into each *RA*. Hence, in hierarchical routing we only apply the update policy into each *RA*.

Fig.4 shows the reduction obtained in the quantity of update messages flooded throughout the hierarchical network topology of Fig.1, when increasing the values of

N. As expected, the larger the N the lower the number of update messages. Note that the case of $N = 1$ corresponds to a policy that triggers update messages whenever a change occurs.

3.3   Routing Algorithm

ASON recommendations do not specify a routing algorithm in order to compute routing paths. However, it defines a set of features that have to be supported by any routing algorithm running over the optical networks. One of them recommends path computation based on source routing. In this paper we extend the *ALG3* [11] routing algorithm inferred from the *BBOR* to be applied to hierarchical networks. A short description of such an algorithm is now presented.

When a source node is required to establish a new incoming call it selects the k-shortest paths to the destination and a free wavelength and sends a set-up message piggybacking the explicit route along the selected lightpath. However, since light-paths are selected under inaccurate routing information, the selected wavelength may be unavailable at the time of the lightpath set-up, leading to the rejection of the set-up message. The *BBOR* mechanism addresses this problem by offering an alternative route to all those intermediate nodes that would reject the set-up message. In fact, when an intermediate node detects that the explicitly routed output link has insufficient resources, i.e., there is not the same incoming wavelength to accommodate the new incoming call (no wavelength conversion is provided), it dynamically sends the set-up message along a pre-computed *bypass-path* which bypasses this link. Therefore, the wavelength availability is the critical parameter in deciding when a lightpath must be rerouted. Assuming that any link is a bundle of B fibers, the main *BBOR* performance can be detailed according to the next steps:

Defining wavelengths to be bypassed: Those wavelengths in a link that potentially might not be available are defined as *Obstruct-Sensitive-Wavelength (OSW)*. Being B the total number of a certain λ_i on a link, R the current number of available (not assigned to an already established lightpath) λ_i on this link we can say that according to the triggering policy defined by the *BBOR* mechanism, a wavelength λ_i is defined as *OSW*, namely λ^{os}_i, on a certain link, when R is lower or equal than a *threshold percentage* T_p of N, being N the number of changes established in the triggering policy to send an update message.

Selecting the lightpath: Once the OSW detection has finished, the lightpath is computed. This process consists of two basic steps, selecting the working path and selecting as many bypass-paths as wavelengths defined as OSWs . On one hand, a new algorithm named ALG3 is proposed when including the OSW in the path selection process. This is done by adding a new parameter named OSW_i (L, F) where L is the number of links where λ_i has been defined as OSW and F is the minimum value of available λ_i along the lightpath. This is done by adding a new parameter named OSW_i (L, F) where L is the number of links where λ_i has been defined as OSW and F is the minimum value of available λ_i along the lightpath. Hence, L represents the *degree of obstruction* and F the *degree of congestion* of the path. The weight associated to each link is represented by the factor L/F. This factor stands for a balance between the number of potentially obstructed links and the real congestion. Moreover, since longer paths than the shortest ones can be selected, the length of the path is also included in the path decision. Hence, in order to avoid those paths that are either widest (in terms of wavelength availability) but too long or shortest but too narrow, the weight factor of each path is modeled by F_p according to

$$F_p = H\left(\frac{L}{F}\right) ,$$

(3)

being H the path length. Then, *ALG3* selects that λ in the pre-selected k-shortest paths minimizing the F_p value. On the other hand, once the working path is selected a *bypass-path* must be computed for those wavelengths defined as *OSW* in this lightpath. Although other criteria could be used to compute the *bypass-paths*, such as minimizing the number of wavelengths defined as *OSW*, the shortest (number of hops) *bypass-paths* are selected. Summarizing, in order to explicitly distribute in the set-up message the *bypass-paths*, source nodes must perform both the detection of those wavelengths on a link that potentially cannot be available when establishing the path, and the computation of a *bypass-path* for each one of these wavelengths.

In this paper, *ALG3* is modified to be adapted to hierarchical routing turning out a new routing algorithm named *ALG3_H*. In this hierarchical structure, wavelengths can be defined as *OSW* on different hierarchical levels. This implies two main differences regarding a flat network structure. The former, the *OSW (L,F)* value can be computed on each hierarchical level. Hence, the expression used so far to model the weight factor F_p of each path must be adapted to these different *OSW (L,F)* values. The new weight factor is F_{p_H} and can be computed according to.

$$F_{p\_H} = \sum_{i=1}^{n} F_p^i ,$$

(4)

where n is the number of hierarchical levels and F_p^i is the parameter defined in (3) per each hierarchical level. The selected wavelength will be that minimizing F_{p_H}.

The latter, *bypass-paths* may be computed on each hierarchical level. Meanwhile the ingress node for each *RA* must compute the route and the required *bypass-paths* along its network, the source node receiving the call request must compute the path to the destination node and the required *bypass-paths* on each hierarchical level.

Bypass-paths usage: Once the working lightpath is computed the set-up message is sent along the selected route. Intermediate nodes can send this set-up message along either the working path, i.e., the selected wavelength is indeed available, or the *bypass-path*, i.e., the selected wavelength is not really available. It is worth noticing that *bypass-paths* are only used when the selected wavelength is really not available at the path set-up time in that link defined as *OSW*.

4   Performance Evaluation

Before evaluating our proposal, we present an example to illustrate its performance. The example is based on the topology shown in Fig.1. Considering that every *OXC* includes control functions with signaling capabilities, we assume 10 fibers per link and 4 wavelengths per fiber. Update messages are sent according to $N = 6$ and a wavelength is defined as *OSW* according to T_p = 50% (i.e., when the minimum number of available wavelengths on this link is lower than or equal to 3). These values are obtained from the simulations performed in [10] .Suppose that incoming call requests arrive between nodes *S* and *D*. Shaded areas in Fig.1 show the whole network perspective seen by nodes belonging to *Routing Area* X1. Each node has a topology database containing both complete topology information of the *RA* X1 and information summary of the rest of the network. This is represented in Fig.1 by three different types of links, which are grouped on either physical or logical links. For example, physical link X11-X12 stands for available resources in the link connecting nodes X11 and X12. The physical link X14-X2 stands for the available resources in the link connecting *RA* X1 to *RA* X2 through node X14. Finally, logical link X1-X2 stands for the aggregated information of *RA* X2 from the node directly connected to *RA* X1, i.e., X21. This aggregated information of the logical link X1-X2 has been obtained according to the information of the topology and available resources database shown in Table 1 and an aggregation process applied to that database, which is described as follows.

Table 1. Topology/Available Resource Database in X2

Link	λ_1	λ_2	λ_3	λ_4	D
X21-X22	2	3	4	7	1
X21-X23	1	5	3	2	1
X23-X22	4	1	6	5	1

Firstly, the *Node Aggregation Scheme* (*NAS*) summarizes the information of Table 1, according to the expressions (3) and (4). Feasible lightpaths from X21 to X22 and X23 are: a) X21-X22, b) X21-X23 and c) X21-X23-X22. Therefore, according to (2) the aggregated available wavelength for each color is represented in (5). Moreover, according to (1) the aggregated delay is also represented in (6).

$$W_1^1 = 2 ; W_2^1 = 3; W_3^1 = 4; W_4^1 = 7 , \qquad (5)$$

$$D_1 = \min_{a,b,c}\left[\sum_{l\in a,b,c} D(l) \right] = \min[1,1,2] = 1 . \qquad (6)$$

Secondly, X21 bundles its information and disseminates it throughout the *RA's* according to a flooding mechanism proposed in [4].

Finally, Table 2 depicts the topology database in X1 built according to the received dissemination messages. In particular, the topology database provides all the information required to compute a route from the given node to any reachable node.

When a call request from node *S* to node *D* reaches node X11, this node applies *ALG3_H* to select the lightpath based on the information represented in Table 2. Table 3 shows possible shortest paths from node *S* to node *D* as well as the *OSW* (*L,F*) value on each hierarchical level.

Table 2. Topology/Available Resource Database in X1

Link	λ_1	λ_2	λ_3	λ_4	D	Link	λ_1	λ_2	λ_3	λ_4	D
X11-X12	6	3	3	6	1	X17-X18	0	4	6	1	1
X12-X13	2	3	6	0	1	X18-X14	6	6	0	6	1
X13-X14	6	3	0	2	1	X14-X2	6	7	7	5	1
X12-X15	6	2	0	1	1	X18-X3	5	6	7	5	1
X15-X13	6	6	6	6	1	X1-X2	2	3	4	7	1
X15-X16	0	7	3	3	1	X1-X3	1	2	3	5	1
X16-X14	1	1	1	1	1	X3-X2	4	4	5	7	3
X11-X17	6	3	1	6	1	X-Y	4	3	2	3	5

Table 3. Routing Table in X11

Route 1	λ_1 λ_2 λ_3 λ_4	D	OSW (L,F)
X11-X12-X13-X14-X2	2 3 0 0	4	$\lambda_1(1,2)$, $\lambda_2(3,3)$
X1-X2	2 3 4 7	1	$\lambda_1(1,2)$, $\lambda_2(1,3)$
X-Y	4 3 2 3	5	$\lambda_2(1,3)$,$\lambda_3(1,2)$, $\lambda_4(1,3)$
Route 2	λ_1 λ_2 λ_3 λ_4	D	OSW (L,F)
X11-X17-X18-X3	0 3 1 1	3	$\lambda_2(1,3)$,$\lambda_3(1,1)$, $\lambda_4(1,1)$
X1-X3-X2	1 2 3 5	4	$\lambda_1(1,1)$,$\lambda_2(1,2)$, $\lambda_3(1,3)$
X-Y	4 3 2 3	5	$\lambda_2(1,3)$,$\lambda_3(1,2)$, $\lambda_4(1,3)$

Table 4. F_{p_H} value for Route 1 and Route 2

Route 1	$F_p(\lambda_1)$	$F_p(\lambda_2)$	Route 2	$F_p(\lambda_2)$	$F_p(\lambda_3)$	$F_p(\lambda_4)$
X11-X12-X13-X14-X2	2	4	X11-X17-X18-X3	1	3	3
X1-X2	0.5	0.33	X1-X3-X2	2	0.67	0
X-Y		1.67	X-Y	1.67	2.5	1.67
F_{p_H}	2.5	6	F_{p_H}	4.67	6.17	4.67

Once shortest paths are selected and the *OSW (L,F)* has been defined, *ALG3* must select the wavelength to be used. This is performed by computing the F_{p_H} value for each wavelength on both paths according to (4). Table 4 illustrates the F_{p_H} computation. *ALG3* selects that wavelength minimizing the F_{p_H} value. Therefore, the first route is selected and λ_1 will be used to transport the traffic.

Once the lightpath is selected node X11 must select the required *bypass-paths*. The selected wavelength λ_1 has been defined as *OSW* in two different levels therefore two *bypass-paths* must be computed. On the one hand, in the low level λ_1 is *OSW* in the link X12-X13. A *bypass-path* across X15, i.e. X12-X15-X13, may be used to bypass the link where the selected wavelength has been defined as *OSW*. On the other hand, λ_1 has also been defined as *OSW* in the logical link X1-X2. This means that logical node X1 should compute a *bypass-path* to bypass this link. There is only one option to bypass the logical link X1-X2, that made up of X1-X3-X2. The usage of this *bypass-path* must be carefully analyzed. It must be noticed that the edge node from X1 to X2 is different from that connecting X1 to X3. In fact, X14 and X18 connects X1 to X2 and to X3 respectively. Assuming that the primary path ends at X14, the *bypass-path* to X2 is made up of X14-X18-X3-X2. When node X14 receives the set-up message it forwards this message to X2 through X14-X2 if the selected wavelength is available on this link. At this point the set-up message reaches *RA* X2 through X21. If the selected wavelength is not available on the logical link X1-X2, node X21 must send the set-up message back to find the computed *bypass-path*. A crankback mechanism is implemented to send the message back to node X14. Unlike the traditional crankback mechanism where the alternative route is computed in all nodes in the reverse path, in our proposal the route must not be computed instead only first node of the *bypass-path* must be reached.

Once the proposed hierarchical network structure has been analyzed by the illustrative example presented above, we also use the network topology shown in Fig. 1 to evaluate our proposal. However, unlike the illustrative example, in this case we suppose a 5-fiber topology, with 16 wavelengths on all the fibers on all the bi-directional links. Call arrivals are modeled by a Poisson distribution and the connection holding time is assumed to be exponentially distributed.

In Fig.5 the impact on the blocking probability reduction obtained when applying the aggregation schemes and the update policy proposed in this paper is clearly presented. A value of N must be selected to implement the update policy. According to the results obtained in Fig. 4 and considering that the larger the N the lower the number of update messages but the larger the computational cost of the *BBOR* mechanism, a value of 6 is assumed for the N. Moreover, a value of the T_p parameter must also be considered. Again, $T_p = 50\%$ to reduce the cost of the *BBOR* mechanism. A detailed impact of the N and the T_p values on global network performance can be found in [10]. The routing algorithm used to select the routes is based on applying the *First-Fit (FF)* heuristic [12] on the pre-computed shortest path. Analyzing the results obtained in Fig.5 we can observe that compared to the case when there is not information aggregation only a minor blocking probability increase of 3.5% is obtained by the *NAS* scheme.

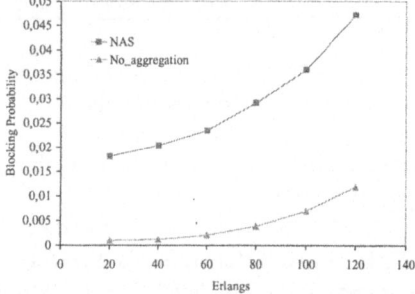 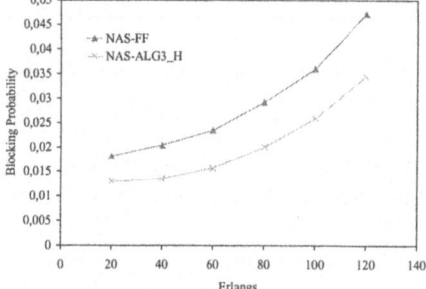

Fig. 5. Blocking probability with/without aggregation

Fig. 6. Blocking probability for the First-Fit and the ALG3_H

In Fig.6 we evaluate the impact on the blocking probability because of applying the *ALG3_H* inferred from the *BBOR* mechanism. In fact, *NAS* scheme performs better when the *ALG3_H* is also included. Compared to the *First-Fit* heuristic a reduction of 1.25% is obtained in the blocking probability under the higher network load by the *NAS* scheme.

5   Conclusions

In this paper a new source-based hierarchical routing mechanism is proposed to be applied to optical networks. The proposed routing mechanism deals with three main aspects related to hierarchical routing, namely the aggregation schemes, the update policies and the routing algorithms. We propose a new aggregation scheme to reduce the overhead produced by the routing information dissemination in large networks. We also propose an update policy to reduce the number of update messages needed to guarantee accurate network state information in the border nodes that select paths. Finally, a new routing algorithm is also presented to address the effects on global network performance because of having a hierarchical structure. Mainly this routing algorithm is oriented to reduce the negative effects on the blocking performance because of selecting paths under inaccurate network state information.

The source-based hierarchical routing proposed in this paper has been evaluated on a realistic network topology by simulation. Obtained results show a reduction in the required size of the *TED* (because of applying the suggested aggregation scheme), a reduction in the update messages flooded throughout the network (because of applying the suggested update policy) and a reduction on the blocking probability (because of applying the suggested routing algorithm).

References

[1]   ITU-T Rec. G8080/Y.1304, "Architecture for the Automatically Switched Optical Network (ASON)", November 2001.
[2]   P.Ashwood, et al., "Generalised MPLS – Signalling Functional description", draft-ietf-mpls-generalised-signalling-09.txt, February 2003.
[3]   ATM Forum, "Private Network to Network Interface Specification Version 1.1", af-pnni-0055.002, April 2002.
[4]   S. Sánchez-López, J. Solé-Pareta, J. Comellas, J. Soldatos, G. Kylafas, M. Jaeger, "PNNI Based Control Plane for Automatic Switched Optical Networks", Accepted for publication to IEEE Journal of Lightwave Technology, Special Issue on Optical Networks, vol.21, n°.11, November 2003.
[5]   Awerbuch B., et al., "Routing Through Network with Hierarchical Topology Aggregation", IEEE Symposium on Computer and Communications 1998.
[6]   Ben-Jye, et al., "Dynamic Update of Aggregated Routing Information for Hierarchical QoS Routing in ATM Network", ICPADS 2001.
[7]   J.Zhou, X.Yuan, "A Study of Dynamic Routing and Wavelength Assignment with Imprecise Network State Information", in Proceedings of ICPP Workshop on Optical Networks, Canada, August 2002
[8]   J.Zheng, H.T.Mouftah, "Distributed lightpath control based on destination routing in wavelength-routed WDM networks", Optical Networks Magazine, Vol.3, n°:4, pp.38-46, July/August 2002.
[9]   S. Darisala, A. Fumagalli, P. Kothandaraman, M. Tacca, L. Valcarenghi, "On the Convergence of the Link-State Advertisement Protocol in Survivable WDM Mesh Networks", in Proceedings of 7th IFIP Working Conference on Optical Network Design & Modelling, ONDM'03, pp. 433-447, Budapest, Hungary, February 2003.

[10] X. Masip-Bruin, R. Muñoz, S. Sànchez-López, J. Solé-Pareta, J. Domingo-Pascual, G. Junyent, "An Adaptive Routing Mechanism for Reducing the Routing Inaccuracy Effects in an ASON", in Proc. ONDM'03, pp.333-349, Budapest, Hungary, February 2003.

[11] X. Masip-Bruin, S. Sànchez-López, J. Solé-Pareta, J.Domingo-Pascual, D. Colle, "Routing and Wavelength Assignment under Inaccurate Routing Information in Networks with Sparse and Limited Wavelength Conversion", in Proc. IEEE GLOBECOM 2003, San Francisco, USA, December 2003.

[12] H. Zang, J.P. Jue, B. Mukherjee, "A Review of Routing and Wavelength Assignment Approaches for Wavelength-Routed Optical WDM Networks", Optical Networks Magazine, January 2000.

Analysis of a Finite Number of Deflections in Fully and Uniformly Loaded Regular Networks

Ireneusz Szcześniak

Institute of Theoretical and Applied Informatics
Polish Academy of Sciences
ul. Bałtycka 5, 44-100 Gliwice, Poland
Tel: +48 32 2317319, Fax: +48 32 2317026
iszczesniak@iitis.gliwice.pl

Abstract. This paper presents an analytical methodology to obtain steady state throughput of a uniformly and fully loaded regular network. The network operates using deflection routing under the condition that a data packet is allowed to experience a finite number of deflections. Unlike the published analytical methods, the proposed method allows the analysis of a final number of deflections a packet can experience by means of harnessing polynomials in a novel way. The analytical results of the network throughput agree with the simulation results with relative error of 1% on average. The analysis is presented in the context of the shufflenet with the node connectivity of two. The largest network considered in the article has 896 nodes.

Keywords: Deflection routing, performance evaluation, simulations.

1 Introduction

The all-optical technology [1] is being developed as the promise of future communication networks. Among various technical challenges, the most notable is the lack of optical memories [2], which eliminates store-and-forward routing. Therefore the all-optical switches cannot buffer packets but only delay, and then relay them using deflection routing. Thus deflection routing is currently essential to all-optical technology.

Deflection routing copes with the lack of optical memories by misdirecting a data packet (i.e. sending it using a wrong connection) due to the lack of the desired connection. Since this routing strategy is probabilistic, a packet can travel indefinitely long in the network (the livelock problem, [3]) and impede network performance. Moreover, unlimited deflection is undesirable, as the severely delayed packets are considered lost by the TCP protocol.

Barth et al. introduce in [4] the notion of mixed routing, a new method of ensuring the maximal delay of a packet, according to which a packet is forwarded with deflection routing up to the moment when it reaches some fixed number of deflections. After that the packet is relayed in the network according to the rules of convergence routing on an Euler cycle. In order to evaluate the performance

N. Mitrou et al. (Eds.): NETWORKING 2004, LNCS 3042, pp. 675–686, 2004.
© IFIP International Federation for Information Processing 2004

of their method, it is first necessary to evaluate the performance of deflection routing with a finite number of allowed deflections, which motivates this article.

We describe an analytical method to evaluate the performance of a regular network which operates using deflection routing under the condition of a finite number of deflections. After a packet reaches the limit of deflections, it is eliminated from the network and lost. The analysis applies to regular networks under a full and uniform traffic load.

The analysis is based on the approach of Acampora and Shah presented in [5], where they analyze uniformly loaded regular networks, but with no restrictions imposed on the number of deflections (packets can experience any number of deflections). We adopt their method to facilitate the analysis under the condition of a finite number of deflections. We encourage a reader unfamiliar with their method to read their article.

We extend their approach by the introduction of polynomials. Acampora and Shah express the probabilities of packet presence by real numbers, while we express these probabilities by polynomials of real coefficients. A polynomial serves as a tool to remember a number of deflections a packet undergoes.

2   Network Model

The network is homogeneous: every node of the network functions both as a routing node for packets in transit and as an access point where packets can enter or exit the network.

The network works synchronously, i.e. time is divided into time slots and is the same for every node. Packet transmissions take place at the beginning of time slots.

The load of the network is full and uniform. During every time slot at every node there are always enough packets waiting for admission, so that the network is constantly fully loaded. The probabilistic nature of packet arrivals before their admission is irrelevant to this analysis. A packet's source and destination are chosen uniformly from all the nodes. To ensure uniform traffic we allow a node to be both a packet's source and destination.

The analysis can be applied to a regular network or, using the graph theory nomenclature, to a node symmetric network. In a regular network every node is equivalent to every other node, which, together with the assumption of traffic uniformity, allows the analysis to be carried out for one node only.

2.1   Shufflenet

To ease the description of our analysis we shall present it for one regular network. For this purpose we choose the shufflenet with the node connectivity, also called node degree, of two: two inbound and two outbound links. The network has $N = n2^n$ nodes and $M = 2N$ unidirectional links, where n is the network size. There are n columns of nodes, where each column contains 2^n nodes. A sample network of size $n = 2$ is shown in Fig. 1. Nodes in the first column and the last column are the same nodes.

This network can operate with mixed routing since it has an Euler cycle. A necessary and sufficient condition for a network to have an Euler cycle is that the node connectivity of its every node is even. Thus, the considered network has an Euler cycle because the node connectivity of its every node is two.

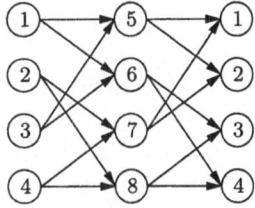

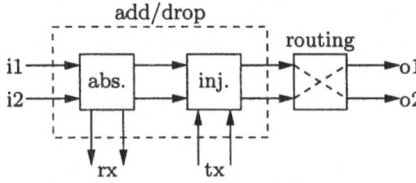

Fig. 1. A shufflenet of size $n = 2$.   **Fig. 2.** Logical structure of a node.

2.2 Deflection Routing

Deflection routing does not buffer packets as store-and-forward routing does but instead relays them during the next time slot after the time slot in which they arrived.

A packet at a node may have a *preferred link* which is a link that yields a shortest path journey to its destination. A link is not considered preferred when every link at the node offers a journey of the same length.

For a packet every node can be only a "care" node or a "don't care" node. At a "care" node the packet has a preferred link, as opposed to a "don't care" node where neither of the links is preferred.

At a node a packet is either deflected (it undergoes a deflection) or correctly routed. A packet is deflected only at "care" nodes when it is refused a preferred link, and it is correctly routed otherwise.

For instance, consider the network in Fig. 1 and a packet which is destined for node 2. Node 5 is a "care" node where the bottom link is preferred. There the packet experiences a deflection when it is sent along the upper link. At node 6 the packet does not possess a preferred link, is insusceptible to a deflection, and therefore node 6 is a "don't care" node.

Let an integer S be defined as the maximal number of deflections a packet can suffer. A packet is allowed to have at most $(S-1)$ deflections, and is eliminated from the network when it undergoes the Sth deflection. Every packet keeps a counter of its number of experienced deflections.

A packet in the network always either is delivered at the destination or is eliminated from the network. A successful packet experiences during its journey at most $(S-1)$ deflections and finally reaches the destination. Conversely, for an unsuccessful packet the number of deflections increases to S, and the packet is eliminated from the network.

2.3 Node Type

Each node is of the type shown in Fig. 2, which is a modified version of the type proposed in [6]. There are two major blocks: "add/drop" for receiving and injecting packets, and "routing" for directing packets to appropriate neighbor nodes. The "add/drop" block consists of two minor blocks, one absorbs packets, the other injects packets. The node connectivity is two (two inputs "i1, i2", and two outputs "o1, o2"), but the analytical method is not limited to this case only and should apply to any other node connectivity.

When a packet arrives at a node it is absorbed by the "abs." block on condition that this node is the packet's destination. In every time slot at most two packets can be absorbed, which are then relayed by the "rx" links. Packets are injected into the network by the "inj." block that can receive at most two packets from its four inbound links. First, the packets from the "abs." block are accepted. Next, new packets from the "tx" links are admitted if no packet or one packet arrived from the "abs." block. New packets are admitted even if they cause a contention.

The second major block is responsible for routing packets. Packets arrive along the block's inbound links, and are directed appropriately to the outbound links. If a contention takes place, then the winning packet uses the preferred outbound link, while the deflections counter of the losing packet is increased by one. The losing packet is eliminated from the network provided that its deflections counter reaches the value of S, or otherwise it is sent to the remaining outbound link.

A packet's elimination renders one output link (either "o1" or "o2") idle, which cannot be utilized to transmit any other packet. Therefore the number of links used by the eliminated packet during its journey is the packet's number of hops increased by one.

2.4 An Upper Bound on the Number of Used Links

In this subsection we obtain an upper bound on the number of links any packet can use, which is defined as an integer K. The upper bound is derived for any network, not only for the shufflenet. A successful packet, one that gets to its destination, makes at most K hops, whereas an unsuccessful packet, one that is eliminated from the network, uses at most K links.

To derive the value of K, consult Fig. 3, which presents the worst scenario of a packet journey. The packet enters the network at the source node A and is bound to the destination node B. The node A is at the greatest possible distance from the node B, i.e. at the distance equal to the network's diameter (the network's diameter is the length of the longest path of all shortest paths between any pair of nodes). The packet travels without a deflection from the node A to the node C using ($diameter - 1$) links. Right at the node C the packet experiences its first deflection, and is sent back to the node A using one link. Now the packet is again at the source node A where up to now it has used *diameter* links and experienced one deflection. This situation happens exactly $(S-1)$ times, after which the packet is at the node A with $(S-1) \cdot diameter$ used links and $(S-1)$ experienced deflections.

Next, the packet is at the node C after $(diameter - 1)$ hops. From here a packet either gets to the node B in one hop, or is eliminated and uses one link. Therefore the worst scenario requires a packet to use K links:

$$K = S \cdot diameter. \tag{1}$$

For instance, for the network from Fig. 1 and $S = 1$ (no deflections allowed), the value of K equals 3.

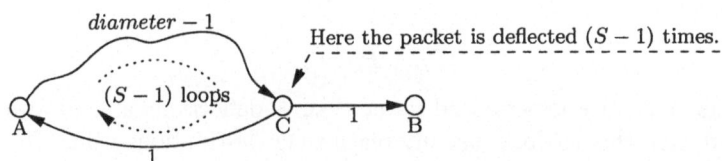

Fig. 3. The worst scenario of delivering a packet.

3   Analysis

The objective of the analysis is to obtain the steady state throughput λ of the system, i.e. the average number of packets that reach their destinations in a single time slot. Note that the number of packets admitted into the network is not equal to the number of packets arriving at their destinations because packets can be eliminated from the network.

The analysis in the following subsections is presented with the top-down approach: the throughput is derived using values that are described in detail in subsequent subsections.

3.1   Deriving λ

To obtain the network throughput first definitions of some probabilities are introduced. The probability that a packet is delivered to its destination using k links is denoted by:

$$P_D[k] = P[\text{a packet is delivered} \\ \text{using } k \text{ links}], k = 1, \ldots, K, \tag{2}$$

and the probability that a packet reaches its destination is:

$$P_D = \sum_{k=1}^{K} P_D[k]. \tag{3}$$

The probability that a packet is eliminated and used k links is denoted by:

$$P_E[k] = P[\text{a packet is eliminated and} \atop \text{used } k \text{ links}], k = 1, \ldots, K, \tag{4}$$

and the probability that a packet is eliminated is:

$$P_E = \sum_{k=1}^{K} P_E[k]. \tag{5}$$

Due to the limited number of deflections, every packet in the network eventually either reaches its destination or is eliminated from the network, and therefore:

$$P_D + P_E = 1. \tag{6}$$

Packets that have experienced fewer than S deflections and have not been eliminated from the network, usually reach their destinations using D_D links:

$$D_D = \frac{1}{P_D} \sum_{k=1}^{K} k P_D[k]. \tag{7}$$

Correspondingly, eliminated packets (every such packet has experienced exactly S deflections) use on average D_E links:

$$D_E = \frac{1}{P_E} \sum_{k=1}^{K} k P_E[k]. \tag{8}$$

Not all the packets present in the network reach their destinations, but only a fraction η of those packets do,

$$\eta = \frac{D_D P_D}{D_D P_D + D_E P_E}. \tag{9}$$

The number of packets present in the network is equal to the number M of the network links, since the network is fully loaded. However, the average number of packets that are present in the network and that reach their destinations is:

$$M' = \eta M. \tag{10}$$

Finally, from Little's law the network throughput is given by

$$\lambda = \frac{M'}{D_D}. \tag{11}$$

To compute λ we need not only the value of M, which is provided by the user, but more importantly the values of $P_D[k]$ and $P_E[k]$ for $k = 1, \ldots, K$. This is the crux of the analysis, discussed in detail in the following subsection.

3.2   Deriving $P_D[k]$ and $P_E[k]$

To obtain $P_D[k]$ and $P_E[k]$, the behavior of an average packet in the system is modeled by the behavior of the test packet. Since the network is regular and the traffic is uniform, the test packet is traced on its journey not to every node, but only to node 1.

We calculate the probabilities that the test packet resides at a particular node of the network during a particular time slot. These probabilities are stored in the vectors P_k for $k = 0, 1, 2, \ldots, K$, where K is the maximal number of time slots any packet can be present in the network. The probabilities of the test packet residency in the network during the kth time slot are expressed by the vector P_k:

$$
P_k = \begin{pmatrix} p_{1,k}(x) \\ \vdots \\ p_{i,k}(x) \\ \vdots \\ p_{N,k}(x) \end{pmatrix}.
\tag{12}
$$

with polynomial elements $p_{i,k}(x)$ of the form:

$$
p_{i,k}(x) = \sum_{j=0}^{S-1} p_{i,k}[j]x^j.
\tag{13}
$$

Each polynomial $p_{i,k}(x)$ provides information on the probability that the test packet resides at node i during the kth time slot. This probability is hereafter also referred to as *residency probability*. The polynomial is of the degree at most $(S-1)$, $\deg p_{i,k}(x) \leq S - 1$. The coefficient $p_{i,k}[j]$ provides the probability that not only the packet resides at node i during the kth time slot, but also that the test packet suffered j deflections on its journey up to this point. The polynomial $p_{i,k}(x)$ for $x = 1$ becomes: $p_{i,k}(1) = \sum_{j=0}^{S-1} p_{i,k}[j]$ and expresses the probability of the test packet presence at the ith node during the kth time slot regardless of the specific number of suffered deflections.

With the probability $1/N$ the test packet starts a journey at every node of the network, including the destination node (i.e. the 1st node) to preserve traffic uniformity. We realize that it is unrealistic for a packet to have the same node as the source and the destination. Nonetheless, under this assumption the presented analysis is more accurate. The vector P_0 is:

$$
P_0 = \left(\frac{1}{N}, \ldots, \frac{1}{N} \right)^T.
\tag{14}
$$

The average value of the probability of a packet deflection at a "care" node is denoted by d and called *the probability of deflection*. This value is assumed to be constant and equal for every node. Conversely, $p = (1-d)$ is the probability that the test packet at a "care" node is not misrouted, but assigned to its preferred outbound link.

Let us consider three examples of calculating residency probabilities. For each example the setting is identical: the considered network is shown in Fig. 1; during time slot 0 the test packet resides at every node with an equal probability; the test packet is destined to node 1. We are interested in the residency probabilities for nodes 3, 1 and 2 during time slot 1.

First, let us study the probability that the test packet resides at node 3. The packet can reach node 3 from nodes 6 and 8, which are "don't care" nodes with respect to node 1. From a "don't care" node the packet departs on any of the node's outbound links with an equal probability (the probability is $1/2$ for the node connectivity of two), and so: $p_{3,1}(x) = \frac{1}{2} \cdot p_{5,0}(x) + \frac{1}{2} \cdot p_{7,0}(x) = \frac{1}{8}$.

Second, consider the probability that the test packet arrives at node 1. There the test packet can arrive from nodes 5 and 7, which are both "care" nodes. Hence the test packet at these two nodes has to be routed according to its preference: $p_{1,1}(x) = p \cdot p_{5,0}(x) + p \cdot p_{7,0}(x) = \frac{p}{4}$.

The last example is most interesting. Node 2 can be reached by the test packet from nodes 5 and 7 under the condition of losing a contention at each of the two nodes. Losing a contention occurs with the probability d and thus: $p_{2,1}(x) = xd \cdot p_{5,0}(x) + xd \cdot p_{7,0}(x) = \frac{d}{4}x$. Multiplication by x expresses a deflection. Therefore the probability $\frac{d}{4}x$ provides the information that the packet experienced one deflection.

A very important fact to stress is that multiplication of a residency probability $p_{i,k}(x)$ by xd may result in a polynomial of the degree equal to S. As stated earlier, a polynomial can be of the degree at most $(S-1)$. When such an event arises, the polynomial's highest term is discarded, which corresponds to elimination of the test packet from the network.

To calculate the residency probabilities for every node during time slot 1, the matrix T_0 is introduced. The polynomial element $t'_{i,j}(x)$ of the matrix expresses the probability of the test packet transition from node j to node i, provided that the test packet resides at node j. Having defined T_0, the computation of the vector P_1, i.e. residency probabilities for every node, is elementary:

$$P_1 = T_0 P_0. \tag{15}$$

The following is the transition matrix T_0 for the network shown in Fig. 1:

$$T_0 = \begin{pmatrix} 0 & 0 & 0 & 0 & p & 0 & p & 0 \\ 0 & 0 & 0 & 0 & xd & 0 & xd & 0 \\ 0 & 0 & 0 & 0 & 0 & \frac{1}{2} & 0 & \frac{1}{2} \\ 0 & 0 & 0 & 0 & 0 & \frac{1}{2} & 0 & \frac{1}{2} \\ p & 0 & p & 0 & 0 & 0 & 0 & 0 \\ xd & 0 & xd & 0 & 0 & 0 & 0 & 0 \\ 0 & p & 0 & p & 0 & 0 & 0 & 0 \\ 0 & xd & 0 & xd & 0 & 0 & 0 & 0 \end{pmatrix}. \tag{16}$$

The elements of the transition matrix are deduced as follows. If a link between nodes j and i does not exist, then the test packet cannot make a transition, and for that reason $t'_{i,j}(x)$ equals 0. The test packet leaves a "don't care" node j by any of the two outbound links with an equal probability $t'_{i,j}(x) = 1/2$. Transition along a preferred link is represented by $t'_{i,j}(x) = p$, while $t'_{i,j}(x) = xd$ corresponds to a transition with a deflection. These rules are summarized below:

$$
t'_{i,j}(x) = \begin{cases} 0 & \text{for no transition} \\ \frac{1}{2} & \text{for routing without preference} \\ p & \text{for preferred routing} \\ xd & \text{for routing with a deflection.} \end{cases} \tag{17}
$$

The first time slot is special, because the test packet is allowed to leave the destination node, whereas during every other time slot the packet is absorbed by the destination. The form of the matrix T_0 reflects this exception by having two nonzero elements in the first column. For every other time slot the transition matrix is different and denoted by T:

$$
T = \begin{pmatrix} 0 & 0 & 0 & 0 & p & 0 & p & 0 \\ 0 & 0 & 0 & 0 & xd & 0 & xd & 0 \\ 0 & 0 & 0 & 0 & 0 & \frac{1}{2} & 0 & \frac{1}{2} \\ 0 & 0 & 0 & 0 & 0 & \frac{1}{2} & 0 & \frac{1}{2} \\ 0 & 0 & p & 0 & 0 & 0 & 0 & 0 \\ 0 & 0 & xd & 0 & 0 & 0 & 0 & 0 \\ 0 & p & 0 & p & 0 & 0 & 0 & 0 \\ 0 & xd & 0 & xd & 0 & 0 & 0 & 0 \end{pmatrix}. \tag{18}
$$

The following matrix equation serves to obtain the vectors P_k:

$$
P_k = TP_{k-1}, k = 2, \ldots, K. \tag{19}
$$

Up to this point the polynomial vectors P_k of probabilities have been derived, from which the values of $P_D[k]$ and $P_E[k]$ are calculated. Let us start with providing the easier one:

$$
P_D[k] = p_{1,k}(1), k = 1, \ldots, K. \tag{20}
$$

The harder one is the probability of a packet elimination $P_E[k]$, which is equal to the product of the probability of deflection d and the probability that the test packet which endured $(S-1)$ deflections is subject to routing at a "care" node during the kth time slot:

$$
P_E[k] = d \cdot \sum_{i=1}^{N} b_{i,k-1}[S-1], k = 1, \ldots, K, \tag{21}
$$

where the vector B_k, with elements $b_{i,k}$, represents the residency probabilities of the test packet being subject to routing at "care" nodes during the previous time slot. The vector B_k is defined as:

$$
B_k = (\alpha_1 e_{1,k}(x), \ldots, \alpha_i e_{i,k}(x), \ldots, \alpha_N e_{N,k}(x))^T, \tag{22}
$$

where $\alpha_i = 1$ if node i is a "don't care" node (with regard to node 1), $\alpha_i = 0$ if node i is a "care" node (with regard to node 1), and elements $e_{i,k}(x)$ belong to the vector E_k:

$$E_k = \begin{cases} P_0 & \text{for } k = 0 \\ (0, p_{2,k}(x), \ldots, p_{N,k}(x))^T & \text{for } k = 1, 2, \ldots, L. \end{cases} \tag{23}$$

The vector E_k expresses the probabilities that the test packet is subject to routing at the kth time slot, while the vector B_k expresses the probabilities that the test packet at the kth time slot is subject to routing at "care" nodes.

Now, the matrices T_0 and T depend on the value of d. Following [5] the value of d for a full and uniform load is:

$$d = \frac{1}{4}(1 - P_{dc}), \tag{24}$$

where P_{dc} is defined in the following subsection.

3.3   Deriving P_{dc}

P_{dc} is the probability of encountering a "don't care" node (with regard to node 1) by the test packet any time it arrives at a node during its entire journey. It is defined as follows:

$$P_{dc} = \frac{\sum_{k=1}^{K} \sum_{i=1}^{N} b_{i,k}(1)}{\sum_{k=1}^{K} \sum_{i=1}^{N} e_{i,k}(1)}. \tag{25}$$

In [5] the method for obtaining P_{dc} substantially differs from the method presented here. There the probability is calculated for the test packet when it reaches the destination, while our method takes into account not only delivered but also eliminated packets.

The probability P_{dc} depends on the vectors P_k. These vectors rely on the probability of deflection d, which in turn depends on P_{dc}. Thus a more precise value of P_{dc} is acquired by successive approximations.

The initial guess for the value of P_{dc} is the ratio of the number of "don't care" nodes to the number of all nodes, which is the first rough approximation. For the approximated value of P_{dc} the value of d is calculated from (24), new matrices T_0, T are generated, then vectors P_k are computed using (15), (19), and finally a new and refined value of P_{dc} is obtained from (25).

The process is repeated the necessary number of times until the desired precision of P_{dc} is reached. In our calculations the precision of the order of 10^{-4} was attained after a few iterations, as the successive values of P_{dc} converged smoothly and fast.

Let us evaluate the running time of the algorithm. One iteration, which produces the next approximation of P_{dc}, requires $O(KN^2)$ multiplications of two polynomials: K number of vectors P_k need to be obtained, and each of them costs $O(N^2)$ multiplications. A multiplication of two polynomials costs $O(S^2)$, and therefore the overall cost of an iteration is $O(KN^2S^2)$.

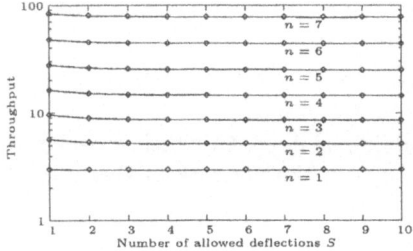

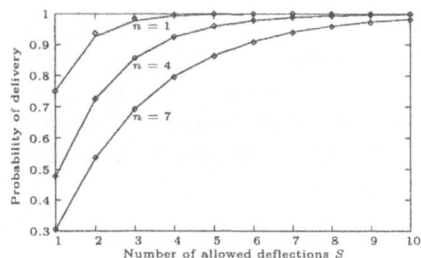

Fig. 4. Network throughput for various network sizes n.

Fig. 5. Probability of a packet delivery for several networks.

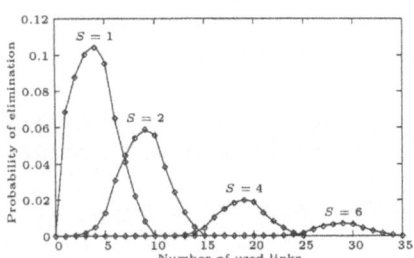

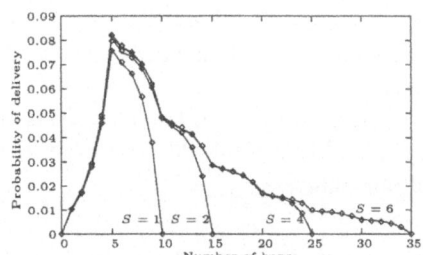

Fig. 6. Probability of a packet elimination for network size 5.

Fig. 7. Probability of a packet delivery for network size 5.

4   Results and Verification

This section presents our results of the analysis and their verification with simulations carried out within the framework of the *Omnet++* software [7]. The results are presented in Figures 4, 5, 6, 7 which display the analysis results as fine squares, whereas the discrete simulation results are presented as lines to distinguish them from the analysis results.

In comparison with simulations, the relative error of our analytical throughput was on average below 1%, and for the network of size $n = 1$ it equaled approximately 2%. The analysis of the network of size $n = 1$ was the most inaccurate, whereas the accuracy of the analysis was increasing as we were enlarging the size of the network.

Fig. 4 shows the network throughput λ as a function of the number of allowed deflections S for various network sizes n. The throughput is similar for different values of S. When the value of S is small, then packets do not live long in the network, and in their place new packets are admitted. As a result, even though many packets are eliminated (more than 50%), other packets are promptly admitted. When the value of S is large, then almost no packet is eliminated, but there are long living packets which block the network and lower the thoughput.

Fig. 5 presents the probability P_D with which a packet successfully arrives at the destination provided it is allowed to undergo S deflections. There are three series for $n = 1, 4, 7$. As expected, the probability increases in tandem with the number of allowed deflections.

The probability $P_E[k]$, that a packet in the network of size $n = 5$ is eliminated and that it used k number of links, is shown in Fig. 6 for $S = 1, 2, 4, 6$.

The last figure (Fig. 7) depicts the probability $P_D[k]$ with which a packet is delivered to its destination in k number of hops in the network of size $n = 5$. There are four series for $S = 1, 2, 4, 6$.

5   Conclusion

The article presented above discusses a novel methodology which incorporates polynomials of real coefficients to allow the analysis of a finite number of deflections. Simulations have confirmed that this method is accurate and serves its purpose.

Future work includes an analysis of mixed routing [4], to which our methodology should be applicable. Another future plan is to further increase the accuracy of our analysis.

Acknowledgment. The author wishes to thank his wife Arlena for assistance with English and the referees of this article for their constructive comments.

References

1. P. Green, "Progress in Optical Networking," *IEEE Commun. Mag.*, pp. 54-61, Jan. 2001.
2. L. Xu, H. G. Perros, G. Rouskas, "Techniques for Optical Packet Switching and Optical Burst Switching," *IEEE Commun. Mag.*, pp. 136-142, Jan. 2001.
3. J. T. Brassil, R. L. Cruz, "Bounds on Maximum Delay in Networks with Deflection Routing," *IEEE Transactions on Parallel and Distributed Systems*, vol. 6, no. 7, pp. 724-732, July 1995.
4. D. Barth, P. Berthomé, T. Czachórski, J. M. Fourneau, C. Laforest, S. Vial, "Mixed Deflection and Convergence Routing Algorithm: Design and Performance," in *Proc. Euro-Par 2002*, pp. 767-774, Aug. 2002.
5. A. S. Acampora, S. I. A. Shah, "Multihop Lightwave Networks: A Comparison of Store-and-Forward and Hot-Potato Routing," *IEEE Trans. Commun.*, vol. 40, pp. 1082-1090, June 1992.
6. A. Bononi, P. R. Prucnal, "Analytical Evaluation of Improved Access Techniques in Deflection Routing Networks," *IEEE/ACM Trans. Networking*, vol. 4, no. 5, pp. 726-730, Oct. 1996.
7. A. Varga, "The OMNeT++ Discrete Event Simulation System," in *Proc. European Simulation Multiconference (ESM'2001)*, Prague, Czech Republic, June 2001.

Improving Perceived Web Performance by Size Based Congestion Control

Thomas Ziegler, Hung Tuan Tran, and Eduard Hasenleithner

ftw. Telecommunications Research Center Vienna[*]
Donaucitystr.1, 1220 Vienna, Austria
{ziegler, tran, hasenleithner}@ftw.at

Abstract. Flow size based congestion control has the potential to improve user perceived Web performance due to the heavy tailed characteristic of file size distributions in the Web. After discussing the benefits and drawbacks of transport protocol and router based solutions for size based congestion control, guidelines for algorithm design are developed. Using these guidelines we find that size based congestion control needs to incorporate TCP models to avoid undesirable user incentives. Based on this insight we specify enhancements to TCP featuring size based congestion control and provide arguments for parameter settings. It is shown by simulation that our modified version of TCP significantly outperforms NewReno from a user perspective in scenarios using realistic models for Web traffic and topologies with multiple congested links.

1   Introduction

Heavy tailed distributions can be considered as one of the invariants when analyzing Internet performance. The evidence of heavy tails can for instance be found in traffic arrival patterns causing burstiness over multiple time scales and thus a variety of headaches in traffic analysis for researchers and network designers. By exploiting the fact that file size distributions on Web servers are heavy tailed this paper can be seen as an attempt to draw benefits out of the heavy-tail misery.

As shown first in [1] for scenarios having the computational power in web servers as the bottleneck, average response times for Web downloads can be dramatically reduced using process scheduling mechanisms giving priority to short flows. The rationale behind this finding lies in a property of Web traffic we will call the "heavy tailed property of file sizes" for the remainder of this paper: the majority of files is short (the so called web mice) and constitutes only a relatively small portion of the load; long files are less numerously but constitute the major portion of the load. Thus, because short flows are high in number, we may expect to reduce average response times giving preference to short flows. Additionally, we expect that average response time improvements won't negatively affect performance of long flows because of the minor load caused by short flows.

---

* This work is funded by the Austrian K*plus* research program.

N. Mitrou et al. (Eds.): NETWORKING 2004, LNCS 3042, pp. 687–698, 2004.

TCP congestion control currently adapts the congestion window independently of the size of the flow to be transferred. Additionally, it can be shown that TCP throughput for short flows is significantly smaller than for long flows. In other words, TCP does the contrary of what the heavy tailed property of file sizes would suggest for maximization of efficiency and additionally causes unfairness among short and long flows.

This paper proposes and investigates TCP Vienna, a new version of TCP employing flow size based congestion control. The basic motivation is to propose straightforward modifications of TCP alleviating unfairness against short flows and thus increasing efficiency. Additionally, it is of major importance to keep congestion control conservative and avoid incentives for misbehaving users to gain an unfair high share of the link capacity due to prioritization of short flows.

The paper is structured as follows. Section 2 reviews related work. The pros and cons of various flow size aware mechanisms to increase Web efficiency and fairness are discussed in Section 3. Section 4 explains the Vienna enhancements to TCP congestion control. After giving an overview on the simulation scenarios, performance evaluation results are shown in Section 5. Finally, Section 6 concludes this paper.

2   Related Work

[1], [2], and [3] have proposed and analyzed size based process scheduling in servers to improve Web performance by exploiting the heavy tailed property of file sizes. Using queueing theory and measurements it has been shown that the performance of overloaded Web servers can be improved by a factor four and more if Shortest Remaining Processing Time (SRPT) scheduling is employed. SRPT naturally favours short jobs having shorter remaining processing times than long jobs. The size of a Web object is not known a-priori in case of dynamic content. Thus [4] shows that similar web server performance improvements can be achieved by SRPT process scheduling of jobs with unknown duration.

Inspired by the work on web server performance improvements [5] aims at achieving similar goals in case the link bandwidth at an Internet router, and not the computational power at a web server, constitutes the bottleneck resource. The basic idea is to keep per flow state at the edge router and mark flows according to their length using a few DiffServ Codepoints. A core router can examine Codepoints and assign packet drop priorities accordingly. As a metric for the flow length a flow's number of bytes received by an edge router is used. For the reminder of the paper we will refer to this idea as "the Router Based Approach (RBA)". In [6] it is shown by analysis and simulation that RBA can improve response times by an order of magnitude in case of a heavy tailed flow size distribution. In case of exponentially distributed flow sizes (light tail) average response times can be improved slightly, however, this improvement comes at the cost of long flows.

[7] investigates bandwidth allocation criteria using flow size based differentiation. As a preliminary implementation of these criteria it is shown that an upgraded TCP Reno source setting the increase and decrease parameters of the congestion window during TCP's congestion avoidance phase according to the residual flow size can increase performance in the range of 30-40% compared to standard TCP Reno. With-

out argumentation for parameter dependency on flow size, round trip time and drop probability the multiplicative decrease parameter is varied between 0.01 and 1; the additive increase parameter is varied between 0.25 and 10. For the reminder of the paper we refer to this kind of approach as "the Transport protocol Based Approach (TBA)".

Other examples for the TBA approach to avoid a retransmission timeout and Slow-start in case of small congestion window sizes at the start of a TCP connection have been proposed in [8], [9], and [10]. [8] proposes to set the initial TCP congestion window size to a maximum of 4 segments (see section 4.2 for more details) increasing the performance of short flows. [9] proposes to chop TCP segments into smaller chunks increasing the congestion window size in units of packets and thus avoiding retransmission timeouts in case of small windows. The main idea in [10] is to allow the TCP data sender to transmit new segments already in response to the first and the second duplicated ACK, keeping the ACK clock going and thus avoiding Slowstart in case of small windows.

3   Router vs. Transport Protocol Based Approach

Router and transport protocol based approaches both have their benefits and drawbacks which shall be highlighted in subsequent high-level considerations.

Performance gains: RBA has the potential to provide higher performance gains than TBA. For instance, TBA exhibits limitations in increasing throughput for flows having a size of only a few packets (which is quite common in case of Web traffic [15]). RBA routers could schedule packets of short flows with strict priority queuing, which corresponds to a significantly stronger flow length based differentiation than achievable with TBA without violating the conservativeness of TCP congestion control. However, as shown in [7], strict priority queueing would be an undesirable policy for RBA due to discrimination of long flows in case flow sizes are not heavy tailed and due to undesirable bandwidth allocation effects. Thus a more conservative policy for flow length based discrimination has to be chosen anyway for RBA (see for instance [6]).

Incremental deployment: TBA only requires modification of the TCP data sender. It thus supports incremental deployment and would typically be implemented in Web server transport protocols. RBA does not support incremental deployment, it rather requires standardization as core routers need to be able to correctly interpret code-points set by edge routers. A full RBA implementation in an Autonomous System would require upgrading all routers potentially subject to congestion.

Scalability: RBA requires per microflow state at edge routers. It is questionable whether the performance gains by flow size based congestion control (although impressive) would balance the cost for per flow state in routers considering the fact the Web and TCP traffic is usually rather low priority traffic. We are not aware of scalability problems with TBA.

Application specific protocol design: Exploiting the heavy tailed property of Web file sizes per definition means violating the important paradigm of designing protocols for a broad spectrum of applications. In case of RBA this is problematic because edge routers are generally not able to distinguish between applications, e.g. if IPsec is used.

Thus RBA would be performed for all kinds of applications using TCP, independently of the characteristic of their flow size distribution. On the contrary, TBA would typically be deployed in Web servers where this problem does not exist. Of course, for RBA and TBA the argument that the Web is not "just another Internet application" holds.

Fairness and avoidance of undesirable incentives for misbehaving users: Another important problem that might come with TBA and RBA is an undesirable incentive for users to chop a long flow into many short flows in order to gain higher throughput. This incentive is created in case TBA or RBA were designed to provide short flows with higher throughput than long flows. A possible solution to this problem is that TBA and RBA like mechanisms need to be designed such that a short TCP flow achieves smaller or equal throughput than a long flow under the same network conditions, i.e. packet loss probability and round trip time (RTT). In other words, fairness should be a design goal not only for congestion control purposes but also to avoid undesirable user incentives. Estimating throughput of a long flow under the same network conditions is hard to achieve in case of RBA because routers have no straightforward possibility to measure path drop probabilities and RTTs. As will be shown in Section 4, estimation of drop probability is feasible with TBA; RTT estimation is already implemented in TCP. Based on drop probability and RTT estimation congestion control parameters can be set such that throughput of short flows approaches, but does not exceed throughput of long flows.

Summarizing above bullets, we observe the dominant advantages of the transport protocol based approach in terms of deployment, scalability, application specific protocol design and potential for avoidance of undesirable user incentives.

4   Enhancements to TCP Congestion Control

4.1 Design Guidelines and Basic Idea

Section 3 provides a design guideline for size based congestion control. Adapting TCP congestion control parameters as a function of the flow size such that throughput of short flows approaches but does not exceed throughput of long flows enables exploitation of the heavy tailed property while maximizing fairness and avoiding undesirable user incentives[1].

The basic idea behind TCP Vienna is to estimate the throughput a "long flow exhibiting typical TCP behavior" would have under the network conditions a flow of arbitrary size currently experiences. By network conditions we mean drop probability and RTT; a "long flow with typical TCP behavior" means a TCP Reno flow with delayed ACKs enabled having infinite flow length. Having estimated drop probability and RTT the throughput of such a typical long flow can be computed using an approved model

---

1. As shown in figure 1, Section 4, throughput as a function of the flow size is a monotonically increasing function in case of current versions of TCP. Thus long flows achieve a disproportionally high share of the bottleneck capacity compared to short flows.

of TCP [11]. Using the same model, congestion control parameters can be adjusted such that throughput of an arbitrary-length flow approaches throughput of a long flow under the same network conditions as closely as possible.

Model based adoption of TCP parameters enabling the improvement of Web performance while avoiding undesirable user incentives is the main innovation of TCP Vienna compared to the RBA and TBA solutions proposed so far in [6] - [10].

4.2 What TCP Parameters to Adjust

Having defined flow length independent fairness as a design guideline there remains an important constraint to be considered when increasing throughput for short flows. TCP congestion control must stay conservative in order to avoid congestion collapse in certain scenarios. Keeping this constraint in mind the TCP parameters to be adjusted according to the flow length and their bounds may be identified.

Additive increase, multiplicative decrease: the additive increase (α) and multiplicative decrease (β) parameters may be adapted according to the flow size during the congestion avoidance phase. Without argumentation the decrease factor is varied between 0.01 and 1 and the increase factor is varied between 0.25 and 10 in [7], dependent on the residual size of a flow. In TCP Vienna parameter adoption happens based on TCP models, more conservatively, and taking above design guidelines into account. An arbitrary sized flow should exhibit congestion control behavior at least as aggressively as a typical TCP Reno flow having infinite length to achieve a fair share of the throughput. A typical TCP Reno flow increases the congestion window (cwnd) by 1/cwnd at the receipt of an ACK and halves the congestion window at fast retransmit, fast recovery in case a packet is lost. Thus the decrease parameter is lower bounded by $\beta_{min} = 0.5$; the increase parameter is lower bounded by $\alpha_{min} = 1$. In order to stay conservative, and independently of a flow's size, we always want to decrease the window somewhat in case a packet is lost. Thus we use $\beta_{max} = 0.9$ as an preliminary upper bound for the multiplicative decrease. We set the upper bound of the increase parameter to $\alpha_{max} = 8$ in our simulations, a similar value as proposed in [7]. Note that while lower bounds for additive increase and multiplicative decrease are well argued in case of TCP Vienna, upper bounds require further substantiation by simulation experiments. Thus, as a first step, it is shown in Section 5 that the above parameter settings improve Web performance without exhaustively increasing loss rates at congested router output ports.

Window increase during Slowstart: a TCP flow having delayed ACKs disabled doubles the congestion window every RTT during Slowstart. A TCP flow having delayed ACKs enabled does so roughly every two RTTs due to the reduced ACK frequency. As TCP Slowstart behavior without Delayed ACKs is well known and has been identified as sufficiently conservative we may increase the congestion window more aggressively in case our TCP flow has delayed ACKs enabled in order to increase throughput of short flows. Consequently, we define the "Exponential Increase" parameter (γ), determining the congestion window increase in units of segments at the receipt of an acknowledgement during the Slowstart phase. In case of delayed ACKs disabled γ is fixed to 1*MSS (usual TCP without delayed ACKs; MSS denotes the Maximum Segment Size). In case of delayed ACKs enabled $\gamma_{max} = 2MSS$ (imitating TCP behavior with delayed ACKs disabled) and γ_{min} = 1MSS (TCP with delayed ACKs enabled) define upper and lower bounds for γ.

Initial window: RFC 3390 [8] recommends min(4*MSS,max(2*MSS,4380bytes)) as an upper bound for the initial window size of TCP and discusses the benefits and drawbacks of increasing the initial window. Obviously, an initial window higher than 1*MSS favors short flows and is thus in-line with our goal. Independently of the flow length the initial window is set as recommended in [8] for TCP Vienna.

4.3 Modeling TCP Throughput

Taking constraints on upper and lower bounds for the TCP parameters identified in Section 4.2 into account, these parameters need to be adapted dynamically such that throughput B of a flow with size d approaches throughput of an infinite-length flow having delayed ACKs enabled and experiencing the same network conditions in terms of drop probability p and *RTT*. Note that B is a function of α, β, γ, d, p, and *RTT*.

For computation of B the approved TCP model in [12] which is based on the model in [11] is employed. This model computes the expected steady-state TCP transfertime for a flow as $E(T) = E(T_f) + E(T_{ss}) + E(T_{ca}) + E(T_d)$; throughput $B = d / E(T)$. T_f models the time during connection setup, T_{ss} models the time spent during initial Slowstart, T_{ca} models the time spent in congestion avoidance, and T_d models additional delay introduced by delayed ACKs.

We generalize this model to incorporate arbitrary values for α and β instead of fixed parameter settings, as for instance $\beta = 0.5$. This generalization requires a modified derivation of the expression for the expected congestion window size $E(W)$ which is used to compute T_{ca} (see eq. 22 in [12]). Due to space limitations we only roughly sketch this derivation, based on the analysis in [11]. Let TDP_i denote the Triple Duplicate Ack Period i (the time period between two consecutive packet losses), W_i denote the congestion window size at the end of TDP_i, and X_i the length of a TDP_i in units of RTTs. We observe that during TDP_i the congestion window increases linearly with a slope of α/b between βW_{i-1} and W_i. Thus, modifying eq. 2.8 in [11] to incorporate arbitrary values for α and β, W_i can be written as follows.

$$W_i = \beta W_{i-1} + \alpha \frac{X_i}{b} \qquad (1)$$

In eq. 1, b equals 2 in case of delayed ACKs, else it equals 1. Eq. 1 needs to substituted into eq. 2.9 in [11] and simplified resulting in an expression for Y_i, the number of packets transmitted in TDP_i.

$$Y_i = \frac{X_i}{2}\left(W_{i-1}\left(2\beta - \frac{\alpha}{2}\right) + \alpha(W_i - 1)\right) + \kappa_i \qquad (2)$$

κ_i denotes the number of packets sent in the last RTT of TDP_i. From eq. 2 and equation 2.6 in [11] it follows an expression for $E(W)$ by solving the resulting quadratic equation (compare to eq. 2.14 in [11] or eq. 23 in [12], respectively):

$$E(W) = \frac{1 + \alpha\beta b}{db(4d + \alpha)} + \sqrt{\left(\frac{1 + \alpha\beta b}{db(4d + \alpha)}\right)^2 + \frac{2(1 - p)}{pb\beta(2\beta + \alpha/4)}} \qquad (3)$$

Figure 1 shows throughput B as a function of flow length calculated by the updated model for two settings of α and β. Other parameters: $\gamma = 1$, initial window = 1 MSS, $p = 0.1$, $RTT = 0.2$s.

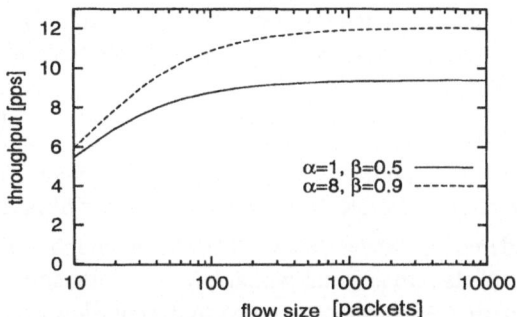

Fig. 1. TCP throughput according to model

4.4 The TCP Vienna Algorithm

Let α^* and β^* define k-dimensional vectors of increase and decrease values, and γ^* be a two dimensional vector. In correspondence with the guidelines presented in section 4.2 we set k = 5, $\alpha^* = (1,2,4,6,8)$, $\beta^* = (0.5,0.6,0.7,0.8,0.9)$, and $\gamma^* = (1,2)$. Additionally, let B_{max} denote the throughput of a typical TCP flow with delayed ACKs enabled and infinite length, where "infinity" equals a flow size of 10000 packets[2]. Given above definitions, the algorithm to compute size based congestion control parameters for a delayed acks enabled TCP data sender can be outlined as follows.

```
Initialization:
    Set α = αmax; β = βmax; γ = γmax

Every m segments sent to the data receiver:
    Estimate p and RTT
    Bmax = B(1,0.5,1,10000,p,RTT)
    Search for i,j:
        Bmax - B(αi*,βi*,γj*,d,p,RTT) is minimal and strictly positive
    Set α = αi*; β = βi*; γ = γj*
```

Fig. 2. Simplified TCP Vienna pseudo code

---

2. Figure 1 illustrates that TCP throughput stays constant for flow lengths above a certain limit. Using the model and simulations we have verified that 10000 packets is above this limit for all relevant settings of p and RTT. Thus it is reasonable to consider 10000 packets as infinite flow length.

The search operation consists of a simple loop scanning α^*, β^*, and γ^*. *RTT* equals the smoothed round trip time as computed in all standard versions of TCP. The distance *m* in terms of number of packets to be transmitted between executions of the computations shown in figure 2 is set as an exponentially increasing function of N, the total number of packets transmitted so far. The rationale behind this choice is the need of frequent throughput estimations for short lived flows to enable correct parameter settings. Flows having already transmitted many packets can base their estimations on a high number of samples thus *m* may be higher. The exponentially increasing function is bounded by a constant *M*, which is set to 600 packets in our simulations.

$$m = min(M, 2^N) \qquad (4)$$

For estimation of the drop probability *p* we employ a simple heuristic. Let *L* denote the number of triple duplicate ACK events plus the number of partial ACKs seen so far[3]. *L* can be considered as a lower bound for the actual number of losses because it does not take lost packets into account which have been transmitted directly after a packet for which a partial ACK has been received during Slowstart. Let *R* denote the total number of retransmitted packets as counted by TCP. *R* overestimates the actual number of losses because packets which have already been delivered correctly at their first transmission are counted twice in case they are retransmitted during Slowstart. We consider underestimation of the loss probability by *L* as less erroneous than overestimation by *R*, thus the drop probability *p* is computed as the weighted average of *L* and *R* divided by *N*, the total number of packets sent so far:

$$p=(a*L+(1-a)*R)/N, \ a > 0.5 \qquad (5)$$

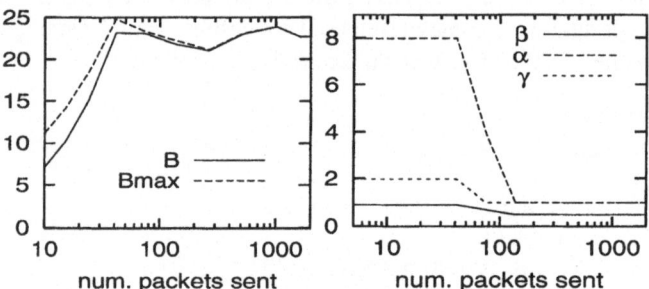

Fig. 3. Throughput and TCP parameter adoption

Figure 3 shows an example of our ns simulation [14] results with a single TCP Vienna flow over a single link topology to illustrate and evaluate the basic dynamics of the algorithm. The link enforces a uniformly distributed packet drop probability of 0.025; link propagation delay equals 100ms; the link bandwidth equals 10Mbps; initial

---

3. Triple duplicate ACK events (causing fast retransmit/recovery) and partial acks indicate packet losses. A partial ACK acknowledges new data but not the highest-sent packet before invocation of Slowstart or fast retransmit, respectively.

window = 4 MSS. The left part figure shows the typical evolution of B and B_{max} in units of packets per second; the right part figure shows α, β, γ parameters during the lifetime of a TCP Vienna flow. Comparing the initial phase in the left and right part of figure 3 we observe the effect of the upper bounds α_{max}, β_{max}, γ_{max} on α, β, γ and thus on throughput B. Although TCP Vienna parameters equal their upper bounds ($\alpha = 8$, $\beta = 0.9$, $\gamma = 2$), B is smaller than B_{max} for $N < 50$. In the course of the simulation B converges to B_{max} and α, β, γ are decreased to $\alpha = 1$, $\beta = 0.5$, $\gamma = 1$. In case of constant RTTs and drop probabilities B and B_{max} would both converge to a constant value. This is, however, not the case in our simulation due to RTT and drop probability variations.

Further simulations with different settings for drop probability and propagation delays show similar results. In case of shorter propagation delays the decrease of α, β, γ parameters is less steep and happens at a later point in time.

5 Performance Evaluations

Simulations are performed with ns-2 using the topology shown in figure 4 with n set to 1, 5, and 9. Links employ Drop Tail queue management; buffer sizes are set to 100 packets; packet sizes equal 500 octets. To ensure tight confidence intervals, sufficiently long simulation times (8000s) are selected.

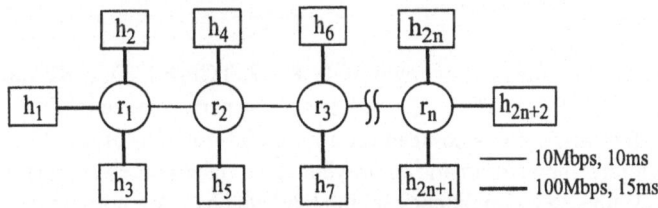

Fig. 4. Simulated network

Several aggregates of Web traffic generating load according to SURGE [15], a state of the art model for Web traffic based on real traffic traces, are simulated. SURGE models the object size as a mixed lognormal/pareto distribution (see [15] for details). We evaluate performance of an aggregate having Web servers at host h_{2n+2} and clients at h_1. Crosstraffic is created in forward direction by aggregates having clients at h_{2i-1} and servers at h_{2i+1} $(2<=i<=n)$ and in reverse direction by aggregates having clients at h_{2j+2} and servers at h_{2j}, where $1<=j<=n-1$. Load is varied by changing the number of users per aggregate. Considering a link between router r_i and r_{i+1}, the number of users in the evaluated aggregate equals the number of users in both crosstraffic aggregates sharing the link.

TCP Vienna has been implemented as a modified version of TCP NewReno; see Section 4 for details on the algorithm and parameter settings. The a parameter for averaging the drop probability is set to 0.75. Delayed ACKs are enabled. The flow length d is not known a-priori in case of dynamic Web content or HTTP1.1. Thus two versions of the algorithm are investigated. In TCP-V1 d corresponds to the total size of the flow (assuming a-priori knowledge); in TCP-V2 d corresponds to the number of bytes sent so far. These two versions are compared against TCP NewReno having an initial congestion window of 1 MSS (TCP-R1) and NewReno having an initial congestion window of 4 MSS (TCP-R2) according to [8].

In accordance with related work in [1]-[7], we measure mean slowdown, defined as the mean of response times divided by Web object sizes. Note that slowdown is best to measure user perception of Web performance because it takes into account that absolute reponsetime improvements are perceived more dramatically in case of small objects.

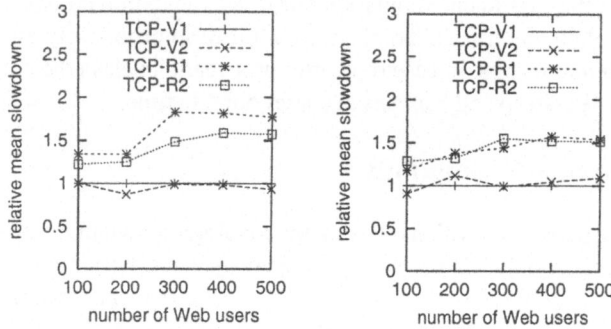

Fig. 5. Mean slowdown as a function of load for n = 1 (left part figure) and n = 9 (right part figure)

Figure 5 shows the mean slowdown of TCP-V2, TCP-R1, TCP-R2 normalized by the slowdown of TCP-V1. TCP-V2 exhibits similar performance to TCP-V1, thus we can conclude that there exists no need for a-priori information of the flow size in TCP Vienna. The figure shows dramatic improvements in user perceived performance with TCP Vienna. With slight dependency on the load we observe improvements of a factor 1.3 up to 1.8.

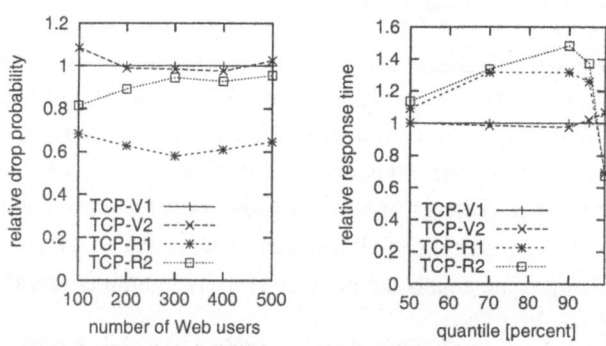

Fig. 6. Drop probabilities and quantiles of response time

The left part of figure 6 shows the drop probability as a function of load for n = 1. The dramatic performance improvements with TCP Vienna shown in figure 5 come to the expense of a slightly increased drop probability compared to TCP-R2, i.e.

NewReno with an initial congestion window set as recommended in [8]. TCP-R1 shows significantly smaller drop probabilities than the other versions of TCP as its congestion control is less aggressive than TCP-R2 and TCP Vienna.

The right part of figure 6 shows a typical example for the quantiles of the response time distribution of TCP-V2, TCP-R1, TCP-R2 normalized by the responsetime quantiles of TCP-V1. The selected scenario is n = 9 with a load of 300 Web users. As expected, both versions of TCP Vienna cause significantly shorter response times for the majority of short flows. This is indicated in the figure by showing that TCP Vienna causes smaller response times up to the 95% quantile. For the longest flows TCP Vienna causes higher responsetimes because short TCP Vienna flows grab a higher portion of the link capacity than TCP-NewReno in comparable scenarios. Thus for quantiles greater than 95% NewReno drops below TCP Vienna. Again, both versions of TCP Vienna exhibit similar performance.

Due to space limitations only a small subset of the simulations results shown in [13] can be presented in this paper. Simulations in [13] investigate the dependency of TCP Vienna performance on parameter setting. Furthermore, we find that it is beneficial for TCP Vienna if the TCP max_burst parameter, limiting the number of back-to-back segments sent in response to a single acknowledgement, is set to a value between 4 and 8. Additionally, comparing scenarios with active queue management and drop-tail in the routers, show that active queue management does not have a significant impact on the performance of TCP-Vienna in relation to TCP NewReno. Extensive simulations using topologies with multiple congested hops loaded with a realistic traffic mix of peer-to-peer and Web traffic show significant performance improvements of TCP-V1 and TCP-V2 compared to TCP-R2 in terms of response time quantiles and slowdown without significantly increasing drop probabilities or decreasing the goodput (throughput at user level) of the peer-to-peer traffic. This finding is in accordance with the "heavy tailed property of file sizes" (see section 1).

6   Conclusions

Based on high level considerations we find that transport protocol based solutions for size based congestion control provide significant advantages compared to router based solutions in terms of scalability, ease of deployment, and avoidance of undesirable user incentives. We emphasize that flow size based congestion control needs to stay conservative to avoid congestion collapse and must not have a bias against long flows but rather converge to a state of flow length independent fairness in order to avoid undesirable user incentives. The latter design guideline motivates us to enhance a well known TCP model for the requirements of size based congestion control. Subsequently, we identify the TCP parameters to be adapted as a function of the flow size, provide arguments for their parameter setting, and define flow size based enhancements to congestion control in TCP Vienna incorporating the enhanced model of TCP. Simulations illustrate the dynamics of parameter adaptations during the lifetime of a TCP Vienna flow.

Performance evaluations with realistic Web traffic over topologies with several congested hops show that TCP Vienna with and without a-priori information on the flow length outperforms TCP NewReno by a factor 1.3-1.8 in terms of user perceived performance. Additionally, we investigate the effect of TCP Vienna on drop probability and response time.

As a promising future research topic the investigation of TCP-Vienna in combination with TCP-SACK can be mentioned. Additionally, the proposals of [9] or [10] could easily be integrated into TCP Vienna. Subsequently to implementing TCP models based on table lookups for a variety of drop probability and RTT settings, a Web server implementation of TCP Vienna is planned.

References

[1]   M.E. Crovella, R. Frangioso, M. Harchol-Balter, "Connection Scheduling in Web Servers", USENIX Symposium on Internet Technologies and Systems (USITS '99), Boulder, Colorado, October '99

[2]   M. Harchol-Balter, M.E. Crovella, S. Park, "The Case for SRPT Scheduling in Web Servers", MIT-LCS-TR-767, October 1998.

[3]   N. Bansal and M. Harchol-Balter, "Analysis of SRPT Scheduling: Investigating Unfairness", Proceedings of ACM Sigmetrics 2001, Conference on Measurement and Modeling of Computer Systems, 2001.

[4]   M. Harchol-Balter, "Task Assignment with unknown Duration", Proceedings of ICDCS 2000

[5]   L. Guo and I. Matta, "The War between Mice and Elephants", Proc. 9th IEEE International Conference on Network Protocols (ICNP'01) , Riverside, CA, November, 2001.

[6]   L. Guo and I. Matta, "Scheduling Flows with Unknown Sizes: Approximate Analysis", Proceedings of ACM SIGMETRICS'02, poster session.

[7]   S.J. Yang and G.D. Veciana, `` Size Based Adaptive Bandwidth Allocation: Optimizing the Average QoS for Elastic Users", Proceedings of IEEE INFOCOM 2002.

[8]   M. Allman, S. Floyd, C. Partrige, "Increasing TCP's initial Window", RFC 3390, Oct. 2002

[9]   M. Mellia, M. Meo, C. CAsetti, "TCP Smart Framing: using smart segments to enhance the performance of TCP", Proceedings of Globecom 2001

[10]   M. Allman, H. Balakrishnan, S. Floyd, "Enhancing TCP's Loss Recovery Using Limited Transmit", RFC 3042, January 2001

[11]   J. Padhye et al.,"Modeling TCP Troughput: A simple Model and its empirical Validation", Proceedings of ACM SIGCOMM 1998, August 1998

[12]   N. Cardwell, S. Savage, T. Anderson, "Modeling TCP Latency", Proceedings of IEEE Infocom 2000, Tel Aviv, Israel, 2000

[13]   T. Ziegler, "Investigating size Based Congestion Control", techn. Report, Oct. 2003, www.userver.ftw.at/~ziegler

[14]   NS Simulator Homepage, http://www.isi.edu/nsnam/ns/

[15]   P. Barford, M.E. Crovella, "Generating Representative Web Workloads for Network and Server Performance Evaluation", Sigmetrics 1998.

An Adaptive AIMD Congestion Control Protocol for Communication Networks

Robert Shorten*, Douglas Leith**, and Peter Wellstead

Hamilton Institute, NUI Maynooth, Ireland

Abstract. We present a new adaptive TCP protocol for the control of congestion in communication networks. The key innovative idea in our protocol is to combine results from Frobenius-Perron theory with online adaptation to realise a TCP variant that: (i) allocates the network pipe fairly amongst competing flows; (ii) coexists with other TCP variants; (iii) responds rapidly to changes in available bandwidth; and (iv) which strives to achieve a high data rate through the bottleneck link. When implemented, the new protocol requires only sender-side modification of the standard window congestion control scheme and is transparent to network buffers and network sources operating other TCP-variants. Results from the network simulator NS are presented to illustrate the key features of our scheme.

Keywords: Adaptive congestion control; Network congestion control; Communication networks

1 Introduction

In this paper we propose a new congestion control protocol for networks of sources employing additive-increase multiplicative-decrease (AIMD) congestion control algorithms. We show that using this protocol it is possible to obtain networks that achieve: (i) fair allocation of the network pipe among competing flows; (ii) TCP-friendliness toward other conventional TCP sources; (iii) rapid allocation of available bandwidth among sources (network responsiveness); and (iv) an efficient use of the bottleneck link capacity. We have shown in an earlier paper [1] that for conventional AIMD networks operating under-provisioned links, high data throughput cannot be achieved without adversely affecting the responsiveness characteristics of the network. Roughly speaking, networks whose sources multiplicative-decrease (backoff) factors are close to zero are very responsive whereas those networks with backoff factors close to unity are highly efficient in terms of data throughput. In this paper we show that both of these objectives can be achieved simultaneously by means of a new type of AIMD algorithm. In a similar manner to TCP-Westwood [2], our key idea is to use available measurements to adapt the network backoff factors to reflect prevailing network conditions.

---

* Joint first author; robert.shorten@may.ie
** Joint first author; robert.shorten@may.ie

N. Mitrou et al. (Eds.): NETWORKING 2004, LNCS 3042, pp. 699–711, 2004.
© IFIP International Federation for Information Processing 2004

Crucially, we use results from positive matrices and Frobenius-Perron theory to ensure that our adaptation strategy for the AIMD parameters is constrained to result in fairness amongst competing operating conventional TCP-Reno.

The adaptive control ideas build upon the analysis of the dynamic properties of congestion control mechanisms presented in [3,1] for networks of synchronised sources. The results in these papers shows that the dynamic congestion avoidance properties of networks whose sources employ AIMD congestion control algorithms can be deduced from the study of a positive matrix [4]. At a more general level however, the use of adaptive control in congestion control is a quite natural sequel to Jacobson's original analysis [5]. In particular, the increase/backoff format of AIMD flows has a strong family resemblance to Dual Control concepts [6] of probing (to gather information concerning a system), combined with control (to effectively use a system). Many of these issues are particularly topical in the context of high-speed networks [7,8]) where a number of proposals for modifications to TCP are currently being studied (it seems likely that high-speed networks will have queues that are small relative to the window sizes of the flows that utilize the network due to cost and to reduce network latency and jitter).

This paper is structured as follows. An overview of the results derived in [3, 1] is given in Section 2. In Section 3 we discuss why adaption is required to achieve the three criteria, of fairness, efficiency and responsiveness. In Section 4, an adaptive scheme is proposed. The paper concludes with a series of examples of the adaptive method for various traffic and network scenarios.

2   Positive Matrices and the Internet

The algorithm presented in this paper is motivated by the insights obtained from the analysis framework for synchronised communication networks developed in [3,1][1]. A summary of the main results in these papers is presented here for completeness.

A communication network consists of a number of sources and sinks connected together via links and routers. We assume that these links can be modelled as a constant propagation delay together with a queue, and that all of the sources are operating a TCP-like congestion control algorithm. TCP (transmission control protocol) operates a window based congestion control algorithm. The TCP standard defines a variable *cwnd* called the congestion window. Each source uses this variable to track the number of sent unacknowledged packets that can be in transit at any time. When the window size is exhausted, the source must wait for an acknowledgement before sending a new packet. Congestion control is achieved by dynamically adapting the window size according to an additive-increase multiplicative-decrease (AIMD) law. The basic idea is for

---

[1] The problem of developing an analysis framework for designing communication networks has become topical in the context of internet congestion control [5,9,10,11, 12,13], but is also of relevance in a variety of problems where a number of devices compete for a shared resource.

a source to gently probe the network for spare capacity and rapidly back-off its send rate when congestion is detected as depicted in Figure 1. Each source is parameterized by an additive increase parameter and a multiplicative decrease factor, denoted α_i and β_i respectively. These parameters satisfy $\alpha_i \geq 1$ and $0 < \beta_i < 1 \; \forall i \in \{1, ..., n\}$. Then, under the assumptions of source synchronisation and uniform round-trip-time (RTT)[2] for all sources, a convenient network model can be found using elementary algebra.

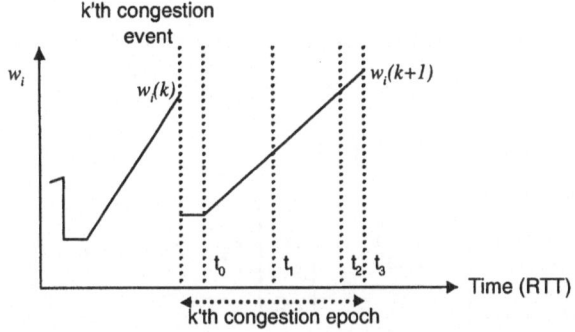

Fig. 1. Evolution of window size

Let $w_i(k)$ denote congestion window size of source i immediately before the kth network congestion event is detected by the sources. Over the kth congestion epoch three important events can be discerned: $t_a(k), t_b(k)$ and $t_c(k)$ in Figure 1. The time $t_a(k)$ is the time at which the number of unacknowledged packets in the pipe equals $\beta_i w_i(k)$; $t_b(k)$ is the time at which the pipe is full; and $t_c(k)$ is the time at which packet drop is detected by the sources, where time is measured in units of RTT. It follows from the definition of the AIMD algorithm that the window evolution is completely defined over all time instants by knowledge of the $w_i(k)$ and the event times $t_a(k), t_b(k)$ and $t_c(k)$ of each congestion epoch. We therefore only need to investigate the behaviour of these quantities.

We have that $t_c(k) - t_b(k) = 1$; namely, each source is informed of congestion exactly one RTT after the first dropped packet was transmitted. Also,

$$w_i(k) \geq 0, \sum_{i=1}^{n} w_i(k) = P + \sum_{i=1}^{n} \alpha_i, \; \forall k > 0, \tag{1}$$

where P is the maximum number of packets which can be held in the 'pipe'; this is usually equal to $q_{max} + BT$ where q_{max} is the maximum queue length of the congested link, B is the service rate in packets per second and T is the round-trip time when the queue is empty. At the $(k + 1)$th congestion event

$$w_i(k + 1) = \beta_i w_i(k) + \alpha_i[t_c(k) - t_a(k)]. \tag{2}$$

---

[2] One RTT is the time between sending a packet and receiving the corresponding acknowledgement when there are no packet drops.

and

$$t_c(k) - t_a(k) = \frac{1}{\sum_{i=1}^{n} \alpha_i}[P - \sum_{i=1}^{n} \beta_i w_i(k)] + 1. \tag{3}$$

Hence, it follows that

$$w_i(k+1) = \beta_i w_i(k) + \frac{\alpha_i}{\sum_{j=1}^{n} \alpha_i}[\sum_{i=1}^{n}(1 - \beta_i)w_i(k)], \tag{4}$$

and that the dynamics an entire network of such sources is given by

$$W(k+1) = AW(k), \tag{5}$$

where $W^T(k) = [w_1(k), \cdots, w_n(k)]$, and

$$A = \begin{bmatrix} \beta_1 & 0 & \cdots & 0 \\ 0 & \beta_2 & 0 & 0 \\ \vdots & 0 & \ddots & 0 \\ 0 & 0 & \cdots & \beta_n \end{bmatrix} + \frac{1}{\sum_{j=1}^{n} \alpha_i} \begin{bmatrix} \alpha_1 \\ \alpha_2 \\ \cdots \\ \alpha_n \end{bmatrix} \begin{bmatrix} 1 - \beta_1 & 1 - \beta_2 & \cdots & 1 - \beta_n \end{bmatrix}. \tag{6}$$

The matrix A is a positive matrix (all the entries are positive real numbers) and it follows that the synchronised network (5) is a positive linear system [4]. Many results are known for positive matrices and we will exploit some of these to analyse the properties of synchronised communication networks. In particular, from the viewpoint of designing communication networks the following properties are very important: (i) network fairness and TCP-friendliness; (ii) network stability; (iii) convergence rate; and (iv) and network throughput. Roughly speaking, window or pipe fairness refers to a steady state situation where n sources operating $AIMD$ algorithms have an equal number of packets in flights at each congestion event; stability refers to the existence of a unique fixed point to which the network dynamics converge; and throughput refers to the objective that the network operates at the bottleneck-link capacity. It is shown in [1] that these properties can be deduced from the network matrix A. We briefly summarise here the relevant results in these papers.

Theorem 1. *[3] Let A be defined as in Equation (6). Then, a Perron eigenvector of A is given by $x_p^T = [\frac{\alpha_1}{1-\beta_1}, ..., \frac{\alpha_n}{1-\beta_n}]$.*

The following corollary follows from Theorem 1 and properties of non-negative matrices [14,4].

Corollary 1. *[3] For a network of synchronised time-invariant AIMD sources: (i) the network has a Perron eigenvector $x_p^T = [\frac{\alpha_1}{1-\beta_1}, ..., \frac{\alpha_n}{1-\beta_n}]$; and (ii) the Perron eigenvalue is $\rho(A) = 1$. It follows that all other eigenvalues of A satisfy $|\lambda_i(A)| < \rho(A)$. The network possesses a unique stationary point $W_{ss} = \Theta x_p$, where Θ is a positive constant such that the constraint (1) is satisfied; $\lim_{k\to\infty} W(k) = \Theta x_p$, and the rate of convergence of the network to W_{ss} is bounded by the second largest eigenvalue of A ($\max|\lambda|, \lambda \neq 1 \in spec(A)$).*

The following facts are easily deduced from the above discussion.

(i) **Fairness and friendliness:** Window fairness is achieved when the Perron eigenvector is a scalar multiple of $x_p^T = [1, ..., 1]$. Consequently, if $\frac{\alpha_i}{1-\beta_i}$ does not depend on i then the matrix is symmetric and the system is fair. Further, since for conventional TCP-flows we have that

$$\alpha = 2(1 - \beta), \tag{7}$$

it follows that any new protocol operating an AIMD variant that satisfies $\frac{\alpha_i}{1-\beta_i} = 2$ will be TCP-fair at each congestion event.

(ii) **Network responsiveness:** It follows from the Corollary that the second largest eigenvalue of the matrix A determines the convergence properties of the entire network. It is therefore important to determine this eigenvalue. We show in [1] that the 95% network rise-time when measured in number of congestion epochs is bounded by $n_r = \frac{log_e(0.05)}{log_e(\lambda_{n-1})}$ where λ_{n-1} is the second largest eigenvalue of A. For the frequently encountered situation where $\beta_i = 0.5$ for all i, we show in [3] that $\lambda_{n-1} = 0.5$; hence $n_r \approx 4$. Note that n_r gives the number of congestion epochs until the network dynamics have converged to 95 % of the final network state: the actual time to reach this state depends on the length of the congestion epochs which is ultimately dependent on the α_i. We also show in [1] that all the eigenvalues of A are real and positive and lie in the interval $[\beta_1, 1]$ (where we assume that $\beta_1 \leq \beta_2 \leq .... \leq \beta_n$). In particular, the second largest eigenvalue is bounded above by β_n and below by β_{n-1}. Consequently, fast convergence to the equilibrium state (the Perron eigenvector) is guaranteed if the largest backoff factor in the network is small.

(iii) **Network throughput :** It follows that immediately before a congestion event the network bottleneck is operating at link capacity and the the total data throughput through the link is given by

$$R(k)^- = \frac{\sum_i^n w_i(k)}{T + \frac{q_{max}}{B}}, \tag{8}$$

where B is the link capacity, q_{max} is the bottleneck buffer size, T is a fixed delay and $T + q_{max}/B$ is the round-rip time when the queue is full. After backoff, the data throughput through the link is given by

$$R(k)^+ = \frac{\sum_i^n \beta_i w_i(k)}{T} \tag{9}$$

under the assumption that the bottleneck buffer empties. Evidently, if the sources backoff too much, data throughput will suffer as the link operates below its maximum rate. A simple method to ensure maximum throughput is to equate both rates yielding the following equation for the β_i:

$$\beta_i = \frac{T}{T + \frac{q_{max}}{B}} = \frac{RTT_{min}}{RTT_{max}}. \tag{10}$$

(iv) **Maintaining fairness** : Note that setting $\beta_i = \frac{RTT_{min}}{RTT_{max}}$ requires a corresponding adjustment of α_i to maintain network fairness. TCP fairness is ensured by adjusting α_i according to $\alpha_i = 2(1 - \beta_i)$.

In summary, if $\frac{\alpha_i}{1-\beta_i}$ does not depend on i then the matrix is symmetric and the system is fair. Since A is positive all the non-Perron eigenvalues are in the interior of the unit circle so the network has a unique stationary point (the Perron eigenvector) that is stable. More generally, it is shown in [1] that even when the matrix A is not symmetric, its eigenvalues are real and positive, and all the non Perron eigenvalues lie inside the unit circle and interlace the β_i's. Hence, the rate of convergence to the stationary solution, which depends on the second largest eigenvalue of A, is bounded from above by the largest β_i. Thus to obtain rapid convergence one is always interested in ensuring that the β_i's are uniformly small (close to zero). On the other hand the link utilisation is highest for β_i close to unity since this ensures that network buffers do not empty following a congestion event.

Finally we note that many of our results extend in a natural manner to the case of non-synchronised networks in the following way. Networks without synchronisation can be modelled by letting $\beta_i(k)$ be either β_i or 1 for each source; i.e. $\beta_i(k) = 1$ when the i'th source does not experience a packet drop at the k'th congestion event and $\beta_i(k) = \beta_i$ otherwise. This yields the following description of the network:

$$W(k + 1) = A(k)W(k), \tag{11}$$

where $A(k)$ is one of $2^n - 1$ possible matrices over each congestion epoch. Here the asymptotic properties of the matrix product $A(k)A(k-1)....A(0)$ as $k \to \infty$ determine network fairness, stability, convergence and throughput properties.

3   Adaptive Congestion Control

In this section we discuss how the results quoted in the previous section can be used to guide the design of an adaptive scheme that ensures efficient bandwidth utilisation and responsiveness to disturbances. Our basic strategy is to continuously adapt the AIMD flow parameters to ensure that the network is responsive and that the bottleneck link capacity is fully utilised and consists of two main components: (A) adaptation to ensure bandwidth utilisation; and (B) adaptation to ensure network responsiveness.

(A) Adaptation to Achieve Bandwidth Utilisation

In standard TCP congestion control the AIMD parameters are set as follows: $\alpha_i = 1$ and $\beta_i = 0.5$. These choices are reasonable when the maximum queue size in the bottleneck buffer is equal to the delay-bandwidth product, and backing off by a half should allow the buffer to just empty. However, is is generally impractical to provision a network in this way when, for example, each flow sharing a common bottleneck link has a different round-trip time. Moreover, in high-speed networks large high speed buffers are problematic for technical and cost reasons.

The solution is an adaptive backoff mechanism along the lines suggested in Item (iii) in Section 2 in which the provisioning of each TCP flow is estimated on-line and the backoff factor set such that the throughput on a per flow basis is matched before and after backoff. In ideal circumstances this should ensure that the buffer just empties following congestion and the link remains operating at capacity[3]. The parameters required for such an adaptive mechanism can be easily obtained at each flow by measuring the maximum and minimum round trip time. Since we have that:

$$RTT_{min,i} = T_i,$$
$$RTT_{max,i} = \frac{q_{max}}{B} + BT_i.$$

then the multiplicative backoff factor β_i that ensures efficient use of the link is $\beta_i = \frac{RTT_{min,i}}{RTT_{max,i}}$ where $RTT_{min,i}$ and $RTT_{max,i}$ are the minimum and maximum RTT's seen by the i'th source, and T_i is the minimum round trip time as seen by the i'th source when the queue is empty.

In summary: each source operates the adaptive backoff mechanism operates as follows.

(i) Estimate $\frac{RTT_{min,i}}{RTT_{max,i}}$.

(ii) Set the multiplicative backoff factor to be equal to $\frac{RTT_{min,i}}{RTT_{max,i}}$.

(iii) Adjust the corresponding additive increase parameter according to $\alpha_i = K(1 - \beta_i)$, $K > 0$, $K \in R$: $K = 2$ gives TCP fairness.

(iv) Monitor continuously the ratio $\frac{RTT_{min,i}}{RTT_{max,i}}$ to check for dynamic changes in the link provisioning.

Comment : In real networks, $RTT_{min,i}$ and $RTT_{max,i}$ are noisy quantities. It is therefore prudent for each source to estimate the ratio of these quantities indirectly; for example using filtered RTT measurements, or by matching throughput (averaged over an RTT) immediately before and immediately after backoff (see [15] and [16] for further details).

Comment : This adaptation strategy is similar TCP-Westwood. However, it differs from Westwood in a number of key areas. In particular: (i) we make no attempt to estimate the packet rate of the bottleneck link; (ii) our adaptation scheme is based upon easily obtained network measurements and does not require complex filtering strategies; and (iii) and network fairness properties are guaranteed in our scheme by controlling the Perron eigenvector (adjusting the parameters according to $\alpha_i = 2(1 - \beta_i)$ gives TCP fairness).

---

[3] In circumstances where the buffer does not empty the link will still operate at capacity.

(B) Adaptation to Achieve Responsiveness

The ratio $\frac{RTT_{min,i}}{RTT_{max,i}}$ may approach unity on highly under provisioned links. However values of β_i close to one will give slow convergence after a disturbance (e.g. traffic joining or leaving the route associated with the link, see examples below). It follows that we need a further adaptive mechanism which continuously adjusts the trade-off between network responsiveness and efficient link utilisation. This requires a network quantity that changes sensibly during disturbances and can be used to trigger an adaptive reset. One variable that does this is the minimum of the mean Inter Packet Time for each source ($IPT_{min,i}$), where the mean is taken over a round-trip time. The $IPT_{min,i}$ is a measure of the link bandwidth allocated to a particular flow. This in turn is determined by the link service rate B (which we assume is constant), the number of flows and the distribution of bandwidth among the flows. Thus as new flows join we expect the $IPT_{min,i}$ for 'our' flow to increase. On the other hand the value of $IPT_{min,i}$ will decrease when the traffic decreases. Thus by monitoring $IPT_{min,i}$ for changes it is possible to detect points at which the flows need to re-adjust and reset β_i to some suitable low value for a time.

In summary, an adaptive reset algorithm for each source is:

   (i) Continually monitor the value of $IPT_{min,i}$.
  (ii) When the measured value of $IPT_{min,i}$ moves outside of a threshold band, reset the value of β_i to β_{reset}.
 (iii) Once $IPT_{min,i}$ returns within the threshold band (e.g. after convergence to a new steady state, which might be calculated from β_{reset}), re-enable the adaptive backoff algorithm $\beta_i = \frac{RTT_{min,i}}{RTT_{max,i}}$.

The two adaptive mechanisms (backoff and reset) that comprise the adaptive control algorithm are shown schematically in Figure 3.

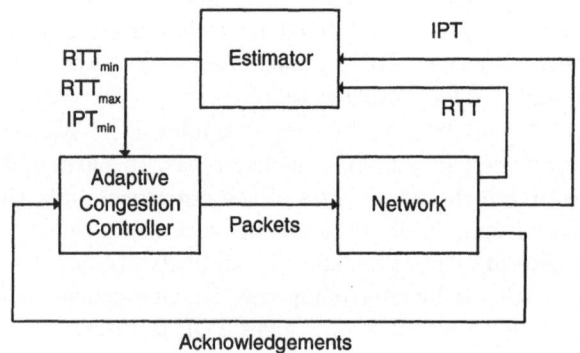

Fig. 2. Adaptive congestion control scheme

The figure depicts a classical adaptive control paradigm [17] in which an on-line estimator continuously estimates relevant system parameters and uses these to continuously update the settings in a control law. In this particular case we argue that the parameters which control the adaptation can be determined exactly. The value of $IPT_{min,i}$ may vary slightly due to other network mechanisms, therefore the band of variations that are allowed before reset must be selected by the user. At the moment we consider this to be a user 'tuning knob', although in future work we are considering sequential change detection methods which can automate this process.

4   Examples

In this section we present examples that illustrate some of the main points of the paper.

Example 1: Fair coexistence of flows. In this example we consider two flows, each with different choices of α_i and β_i. Both flows satisfy the formula $\alpha_i = 2(1 - \beta_i)$. NS simulation results are shown in Figure 3. Evidently, each of the flows coexist fairly at the congestion event, i.e. the have the same window size at each congestion event.

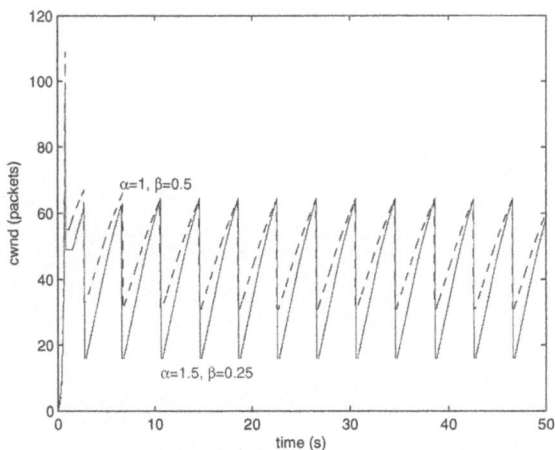

Fig. 3. Example of fair coexistence of two AIMD sources (network simulation parameters are: 10Mb bottleneck link, 100ms delay, maximum queue size is 40 packets).

Example 2: Dependence of convergence on backoff factor. Figure 4 shows NS simulation results for three flows operating the adaptive backoff mechanism $\beta_i = \frac{RTT_{min,i}}{RTT_{max,i}}$. To illustrate the behaviour of the adaptive backoff mechanism in the context of recent proposed modifications to TCP, we show results for the high speed variant of TCP proposed in [3] (this modifies the TCP AIMD algorithm to achieve a faster than linear rate of increase while ensuring backward compatibility with standard TCP on slower networks). For the choice of network parameters used, we have that $\beta_i = 0.77$. For this value of backoff factor, the foregoing positive matrix analysis indicates a convergence time (95% rise time) of 11 congestion epochs and it can be seen from the figure that this is in good agreement with packet-level simulation results.

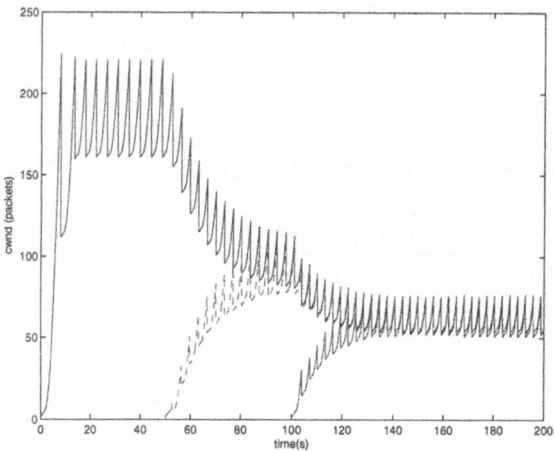

Fig. 4. Illustrating poor responsiveness with adaptive back off (network simulation parameters are: 20Mb bottleneck link, 100ms delay, maximum queue size is 50 packets).

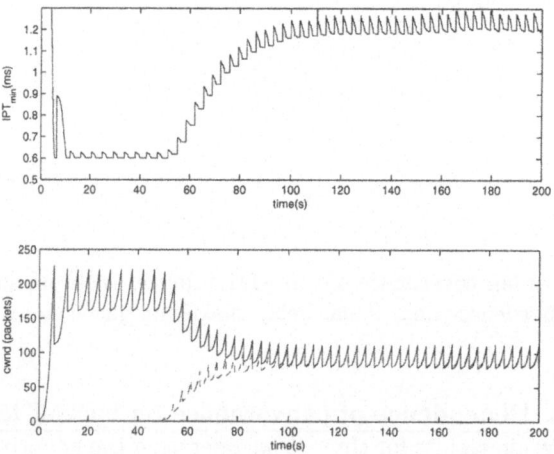

Fig. 5. Illustrating the change in IPT with traffic (network simulation parameters are: 20Mb bottleneck link, 100ms delay, maximum queue size is 50 packets).

Example 3: Variation of ITP_{min} with traffic. Figure 5 shows the $IPT_{min,i}$ time history of flow 1 in the previous example (the congestion window evolution is also shown for comparison). Notice that the IPT for the first flow increases as the second flow start to seize bandwidth.

Example 4: Adaptive congestion control. Figure 6 repeats the simulation of Example 2 but now with the both adaptive backoff and reset, as described in Section 3B. It can be seen that the backoff factor of flow 1 is reset to 0.5 temporarily when flow 2 starts, ensuring rapid convergence (in around 4 congestion epochs). This behaviour is repeated when a third flow starts.

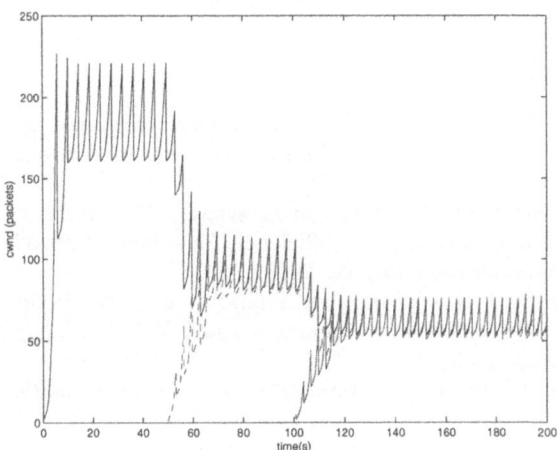

Fig. 6. Adaptive congestion control. Notice that the effective backoff is reset in response to new flows starting (network simulation parameters are: 20Mb bottleneck link, 100ms delay, maximum queue size is 50 packets).

Notice that the flows now converge quickly to the fair allocation, at which time the adaptive reset is disabled and the value of the β_i that utilises the link bandwidth effectively is used instead.

5   Discussion and Concluding Remarks

In this paper we have discussed the dynamic performance of communication networks where each source operates an AIMD congestion control algorithm. We have shown how adaptive control has a clear role in achieving the twin objectives of efficient bandwidth use and responsiveness. For rapid response to disturbances, low values of β_i are required. In the case of underprovisioned bottleneck links this may lead inefficient use of the link bandwidth. To overcome this problem we have proposed an adaptive congestion control protocol that improves both bandwidth utilisation and network responsiveness. Our protocol, which is based on adapting the backoff factor of each flow to reflect the prevailing network conditions, involves only sender side modification of TCP. Simulations are presented to illustrate the efficacy of the proposed algorithm.

Acknowledgements. This work was supported by Science Foundation Ireland grant 00/PI.1/C067 and the Walton Visitor Programme. This work was also partially supported by the European Union funded research training network *Multi-Agent Control*, HPRN-CT-1999-00107[4] and by the Enterprise Ireland grant SC/2000/084/Y. Neither the European Union or Enterprise Ireland is responsible for any use of data appearing in this publication.

References

1. A. Berman, R. Shorten, and D. Leith, "Positive matrices associated with synchronised communication networks." Submitted to Linear Algebra and its Applications, 2003.
2. M. Gerla, M. Sandaidi, M. Valla, and R. Wang, "TCP Westwood with adaptive bandwidth estimation to improve efficiency/friendliness tradeoffs." To appear in Computer Communication Journal, 2003.
3. R. Shorten, D. Leith, J. Foy, and R. Kilduff, "Analysis and design of synchronised communication networks," in *Proceedings of 12th Yale Workshop on Adaptive and Learning Systems*, 2003.
4. A. Berman and R. Plemmons, *Nonnegative matrices in the mathematical sciences.* SIAM, 1979.
5. V. Jacobson, "Congestion avoidance and control," in *Proceedings of ACM SIG-COMM*, pp. 314–329, 1988.
6. A. A. Feldbaum, "Dual control theory: Part I," *Automation and Remote Control*, vol. 21, no. 9, pp. 874–880, 1960.
7. S. Floyd, "High-Speed TCP for large congestion windows," tech. rep., Internet draft draft-floyd-tcp-highspeed-02.txt: Work in progress, February 2003.
8. T. Kelly, "On engineering a stable and scalable TCP variant," tech. rep., Cambridge University Engineering Department Report CUED/F-INFENG/TR.435, 2002.
9. S. Floyd and K. Fall, "Promoting the use of end-to-end congestion control in the internet," *IEEE/ACM Transactions on Networking*, vol. 7, no. 4, pp. 458–472, 1999.
10. Various authors, "Special issue on TCP performance in future networking environments," *IEEE Commuications magazine*, vol. 39, no. 4, 2001.
11. S. Low, F. Paganini, and J. Doyle, "Internet congestion control," *IEEE Control Systems Magazine*, vol. 32, no. 1, pp. 28–43, 2002.
12. Various authors, "Special issue on internet technology and convergence of communication services," *Proceedings of the IEEE*, vol. 90, no. 9, 2002.
13. J. Hespanha, S. Hohacek, K. Obrarzka, and J. Lee, "Hybrid model of TCP congestion control," in *Hybrid Systems: Computation and Control*, pp. 291–304, 2001.
14. R. Horn and C. Johnson, *Matrix Analysis.* Cambridge University Press, 1985.

---

[4] This work is the sole responsibility of the authors and does not reflect the European Union's opinion

15. R. Shorten, D. Leith, and P. Wellstead, "Adaptive congestion control of the internet." Submitted to Automatica, 2004.
16. D. Leith and R. Shorten, "H-TCP:TCP for high-speed and long-distance networks," in *Web-proceedings of the Second International Workshop on Protocols for Fast Long-Distance Networks (http://www-didc.lbl.gov/PFLDnet2004/)*, Argonne National Laboratory, Argonne,, Illinois USA, 2004.
17. P. E. Wellstead and M. B. Zarrop, *Self-Tuning Systems: Control and Signal Processing*. Wiley, 1995.

TCP-DCR: Making TCP Robust to Non-congestion Events* **

Sumitha Bhandarkar and A.L. Narasimha Reddy

Dept. of Electrical Engineering
Texas A & M University
{sumitha,reddy}@ee.tamu.edu

Abstract. In this paper, we propose and evaluate TCP-DCR. TCP-DCR makes simple modifications to the TCP congestion control algorithm to make it more robust to non-congestion events. The key idea here is to delay the congestion response of TCP for a short interval of time τ, thereby creating room for local recovery mechanisms to handle any non-congestion events that may have occurred. If at the end of the delay τ, the event is not handled, then it is treated as a congestion loss. We evaluate TCP-DCR through analysis and simulations. The evaluation is done for three scenarios - a wireless network with channel errors, a wired network with packet reordering and a network with zero non-congestion events. The simulation results show that significant performance improvements can be achieved by using TCP-DCR in the presence of non-congestion events with zero or marginal impact in the absence of non-congestion events. TCP-DCR remains fair to the native implementations of TCP that respond to congestion immediately after receiving three dupacks. TCP-DCR is a simple, effective scheme providing a unified solution to several problems with minimal implementation overhead.

1   Introduction

The strength of TCP lies in the fact that it tries to mitigate congestion in the network by reducing the sending rate in response to loss of packets. Historically, using packet loss as a measure for perceiving congestion has worked quite well. But in the recent past, the nature of the networks has changed significantly. As a consequence, severe penalty is paid in terms of degraded performance in networks where the reason for packet loss is not necessarily network congestion. Recent studies [1], [2] have shown that packet reordering is more prevalent in the current Internet than was assumed earlier rendering the wait of three dupacks used in TCP, an inefficient heuristic. While this in itself is a good reason for investigating the robustness of TCP to non-congestion events such as packet reordering, the authors of [3] present a more compelling reason - the taboo against packet reordering prevents or restricts the research and deployment of several new, beneficial schemes on the Internet for providing efficient routing or differentiated services. Another

---

* This work is supported in part by a grant from The Texas Higher Education Board, by NSF grant ANI-0087372 and by Intel Corp.
** An extended version of this paper is available in [22]

common situation that has spurred the interest in improving the robustness of TCP to non-congestion events is the ever increasing use of wireless networks. Wireless networks are characterized by higher channel error rates than wired networks. When TCP is used in wireless networks, the losses due to channel errors (non-congestion events) are mistaken for congestion losses and the sending rate is unnecessarily reduced, resulting in degraded performance [7]. Several different solutions have been proposed to improve the performance of TCP in the face of packet reordering or in wireless networks. In this paper, we aim to provide a single generalized solution that can be used to improve the robustness of TCP to all non-congestion events.

Our solution is intuitive and employs two simple ideas: (a) delay the congestion response of TCP for a short interval of time τ, creating room to handle any non-congestion events that may have occurred, and (b) employ "local recovery" techniques to recover from non-congestion events during this interval. If at the end of the delay τ the event has not been handled, then it is treated as a congestion event. This simple concept fits into the general philosophy of segregation between the different layers of the network model. The modifications to TCP do not handle the non-congestion event, but rather, rely on some lower layer mechanism to do local recovery, if necessary. To distinguish this flavor of TCP from the original, we call it the Delayed Congestion Response TCP (TCP-DCR for short). This is a general solution that can be extended to any network with non-congestion events and an underlying mechanism for recovering from them.

The rest of the paper is organized as follows. Section 2 provides intuition, analyses and discussion of the TCP-DCR modifications in general. This is followed by discussion and ns-2 [16] simulation results for three specific cases - wired networks with non-negligible packet reordering (Section 3), wireless networks with non-negligible channel errors (Section 4) and finally regular networks with no non-congestion events at all (Section 5). Section 6 concludes the paper by summarising the results and looking at the future work.

2   Delayed Congestion Response TCP

When a TCP receiver finds an intermediate packet missing, but subsequent packets are being received, it sends dupacks to the sender. The sender using the standard TCP algorithms treats the receipt of three consecutive dupacks as an indication that the intermediate packet is lost and responds by reducing the congestion window and triggering the fast retransmit/recovery algorithms. The sender using TCP-DCR is modified to wait for an interval of τ after receiving the first dupack to make room for local recovery of the packet, if possible. The study presented in [17] has shown that even in dynamic network conditions, slowly responding protocols are fair and safe for deployment. In this section we present the details of the proposed TCP-DCR modifications.

2.1   Choice of τ

The delay in responding to congestion determines the performance of TCP-DCR and the choice of τ is a critical aspect for the TCP-DCR modifications. Too large a delay would mean that the protocol responds too sluggishly to congestion in the network. Too small a delay would not allow the lower layer sufficient time to recover from the non-congestion events. In this section we provide guidelines for choosing reasonable bounds on the delay τ.

Consider first the wireless scenario. Fig. 1 shows a general case where the TCP receiver is connected to a base station over a wireless link. In this scenario, for the link layer to recover a packet lost due to channel error, τ should be atleast as large as the round trip time of the wireless link. On the other hand, to avoid an expensive timeout at the TCP-DCR sender, τ should be smaller than the retransmission timer (RTO) value. The RTO is usually set to RTT + 4 times the RTT variance, where RTT is the estimated end-to-end round trip time. A choice of one RTT for the value of τ allows the link layer sufficient time to recover the packet while at the same time avoiding an RTO. Same argument holds good even when the sender and receiver are connected directly over a wireless link, as in the case of an adhoc network.

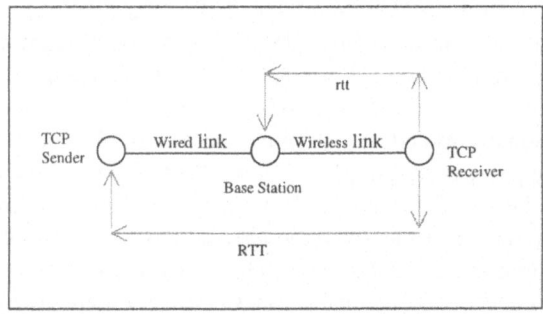

Fig. 1. Analysis of TCP-DCR in a Wireless Network with no Congestion Losses

In the case of packet reordering, the amount by which the packet is reordered could be highly variable - the time to recover the lost packet is the time that the reordered packet takes to reach the receiver. Hence there is no preset lower bound for the delay τ, that will facilitate the recovery of all reordered packets. However, the upper bound is still decided by the value of the RTO. So, a value of one RTT for τ is still a reasonable choice.

Based on the discussion above, we modify the heuristic for the wait after the sender starts receiving dupacks (τ) to one RTT. Setting τ to one RTT, rather than a fixed value, also provides inherent robustness to fluctuations in the queuing delays ensuring that we do not get into RTO timeout even during sudden changes in the network load.

2.2   Steady State Analysis of TCP-DCR

The steady state throughput of TCP-DCR with the assumption of uniform periodic loss probability model can be shown to be similar to that of TCP ($throughput \propto \frac{1}{\sqrt{p}}$). Detailed analysis has been omitted here due to lack of space. An interested reader may find the same in [22]. The main difference however, is that in the presence of both non-congestion events and congestion losses, for the standard TCP algorithm p is the sum of the non-congestion event rate and the loss rate due to congestion, but for TCP-DCR p is only the loss rate due to congestion. As a result, in the presence of both non-congestion events and congestion, TCP-DCR can achieve better throughput.

2.3 Receiver Buffer Requirement When TCP-DCR Is Used

When TCP-DCR is used, the receiver will need to have additional buffer space to accommodate the extra packets corresponding to the delay τ, when a packet is lost due to congestion. Having these extra buffers allows TCP-DCR to achieve the best performance. In the absence of the additional buffer space, the flow control mechanism of TCP limits the sending rate, preventing TCP-DCR from achieving the maximum performance improvement.

2.4 Local Recovery Mechanisms

The performance benefits to be gained from using the TCP-DCR modifications depend on the existence of an underlying scheme for recovering the losses due to non-congestion events. In case of packet reordering, nothing needs to be done explicitly to recover the reordered packet. In case of wireless networks, we assume that the underlying mechanism is a simple link level retransmission scheme, possibly NACK-based, that does not attempt in-order delivery. Some of the recent research in the area of networking for multimedia [18] also advocate the use of link level retransmission schemes that do not attempt in-order delivery. Alternatively, FEC (Forward Error Correction) schemes could also be used.

2.5 Summary of Modifications

The TCP-DCR modifications need to be applied only to the sender. The congestion response is delayed only during the congestion avoidance phase. During the congestion response delay, the congestion window continues to evolve using additive increase but only one new packet is transmitted in for each dupack (similar to the proposed standard limited transmit algorithm [15]). Thus, TCP-DCR remains ack-clocked during the congestion response delay period and the sending rate during τ remains at best, the same as when the first dupack was received.

If the congestion response delay timer expires, the fast retransmit/recovery algorithms are triggered. The *ssthresh* and the congestion window are set to half the current value of the congestion window just as it would be in a traditional implementation of TCP.

The sender can implement the delay either by using a timer or by modifying the threshold on the number of dupacks to be received before triggering the congestion recovery algorithms (*dupthresh*). The timer based implementation however depends on the clock granularity. To ensure that faulty implementation of the timer does not result in an RTO, for the timer-implementation the timer should be set to one RTT as indicated by the *smoothed_rtt* and for the dupack-based implementation, the new value for *dupthresh* should be scaled by the factor $(smoothed\_rtt)/(current\_instantaneous\_rtt)$.

The TCP-DCR modifications work with most flavors of the TCP protocol. However, in this paper we advocate the use of TCP-DCR with TCP-SACK [14], if the TCP-SACK option is available. When used with TCP-SACK, the only thing modified by TCP-DCR is the time at which the fast retransmit/recovery algorithm is triggered in response to the first loss within a window of packets. All subsequent losses within the same window (irrespective of whether they are due to congestion or non-congestion events) are handled in exactly the same way as TCP-SACK would in the absence of TCP-DCR modifications.

Use of delayed_acks will not intervene with the TCP-DCR modifications, provided that the implementation of delayed acks follow the guidelines in [19] that the dupacks (or SACKs) are not delayed.

3 TCP-DCR and Reordering Robustness

In current networks, packet reordering is observed to be not negligible [1],[2]. Also, many new design alternatives for routers or network architectures may benefit if there are no strict restrictions of zero packet reordering. Several different solutions have been proposed in literature to solve this problem. In [3] and [4] the authors present schemes for improving the reordering robustness of TCP that use DSACKs [5] or timestamps [6] to identify reordering and the possible amount of reordering. TCP-DCR on the other hand, aims to improves the reordering robustness of TCP without having to identify the exact amount of reordering in the network or using complex state or algorithms.

3.1 Simulation Topology

We evaluated the performance of TCP-DCR using the ns-2 simulator [16] (version 2.26). A simple dumbbell topology is used where *n* different sources are connected to *n* different receivers via a single bottleneck link between routers R1 and R2. The default values for the bandwidth and delay for the links between the routers and the end nodes is fixed at 10 Mbps and 1 ms respectively. The bandwidth and the delay for the bottleneck link is varied in accordance with the requirements of the experiment. Each source *i* performs bulk data transfer to the receiver *i* with a packet size of 1000 bytes. DropTail buffer management scheme is used at the routers and the queue size is set to 50 packets, unless otherwise specified. Packet reordering is simulated by modifying the errormodel object of ns-2 such that randomly selected packets can be delayed for a random amount of time.

The TCP-DCR agent is implemented by modifying the tcp-sack1 implementation of TCP-SACK agent in ns-2. Ack-based implementation is used for the congestion response delay. The TCPSink/Sack1 agent is used for the receivers. FTP sources start sending data at time 0 and are staggered to avoid synchronization. All simulations are run for 1100 seconds, but data is collected only after the first 100 seconds to ensure that steady state is reached. The receiver advertises a large window such that the sending rate is not limited by the receiver dynamics.

3.2   Performance at Varying Packet Delay Rate

One of the primary reasons for reordering in the network is that some of the packets get delayed more than others, and hence arrive out of order. In this experiment randomly selected packets are delayed, with a delay picked from a normal distribution with a mean of 25ms and a standard deviation of 8ms so that most chosen packets are delayed in the range 0 to 50ms. This simulates mild but persistent reordering. The bottleneck link bandwidth is set to 8Mbps and the delay to 50ms. There is no congestion in the network. The topology consists of a single flow. The experiment is first run with TCP-SACK and repeated for TCP-DCR. Fig. 2 shows the results

As can be seen from the graph, the performance of TCP-SACK degrades rapidly, since persistent reordering keeps the sender congestion window small, reducing the throughput drastically. TCP-DCR performs significantly better than TCP-SACK. Since there is no congestion in the network, and the packets are only mildly reordered, most packets are recovered during the delay in the congestion response.

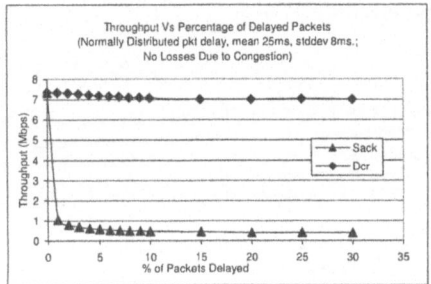

Fig. 2. Throughput Vs Percentage of Packets Delayed (With Single Flow)

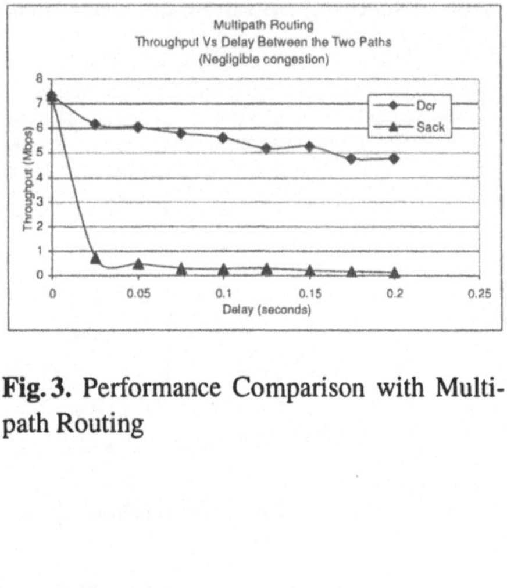

Fig. 3. Performance Comparison with Multi-path Routing

3.3   Performance Comparison with Multi-path Routing

One of the situations that can cause packet reordering is when packets are routed over different paths. Suppose, a router chooses between two different paths for load balancing. In the worst case, alternate packets get routed over the different routes, causing 50% of all packets to get delayed. In this simulation we examine such a situation. The x-axis shows the difference between the RTTs of the two routes. The link delay of the shorter route is fixed at 50ms. Fig. 3 shows the results. It can be seen from the graph that TCP-DCR performs significantly better than TCP-SACK. When the delay between the two paths becomes larger than the round trip time of the shorter path, the performance of TCP-DCR starts to degrade a little. However, the smoothed RTT estimate at the TCP sender will reflect the average round trip time of the link, and the congestion response delay is scaled by this value. As a result, the performance degradation is not drastic.

3.4   Performance Comparison with Congestion in the Network

One of the primary concerns with using TCP-DCR is the effect of delaying congestion response on other flows in the network. We study that in this experiment. The bottleneck link has a capacity of 10Mbps and a link delay of 10ms. The number of flows in the network is 12, with 6 of them using TCP-DCR and the other 6 using TCP-SACK. Congestion in the network is controlled by varying the buffer size at the router R1 for the link between R1-R2. Fig. 4 shows the results when 10% of the packets are delayed. Thus, packets are reordered as well as lost due to congestion in this simulation. When a packet is lost due to congestion, the sending rate is reduced both in the case of TCP-SACK and TCP-DCR. However, when a packet is reordered, the sending rate is reduced only in the case of TCP-SACK. As a result at low congestion levels, TCP-DCR flows utilize more link capacity than TCP-SACK flows and show better throughputs. When the congestion levels in the network increases, the link capacity is more and more equitably shared. It is to be noted that the reason for TCP-DCR realizing better throughputs (when packets are reordered) is not due to unfairness, but due to correctly recovering from the reordering events (without reducing the congestion window). We address the fairness issue in section 5 when we consider zero non-congestion events.

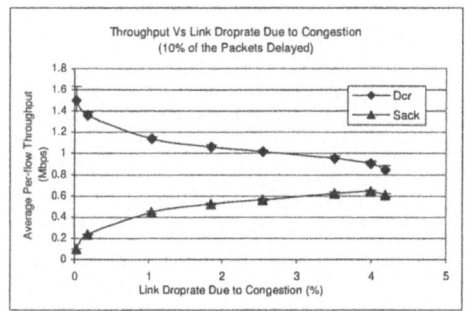

Fig. 4. Throughput Vs Link Droprate due to Congestion

4   TCP-DCR in a Wireless Network

Wireless networks are characterized by high channel error rates. When TCP is used in wireless networks, the losses due to channel errors are mistaken for congestion losses and the sending rate is unnecessarily reduced, resulting in degraded performance [7]. Several solutions have been proposed to improve the performance of TCP over wireless networks. These solutions fall in one of the following broad categories: (a) Split connection approaches (eg. [8]) (b) TCP-aware link layer protocols (eg. [9]) (c) Explicit loss notification approaches (eg. [10]) (d) Receiver-based approaches (eg. [11]) and (e) Modifications to TCP (eg. [12]) When both congestion losses and losses due to the transmission errors can occur, a simple solution would be to let the link layer mechanisms to recover from losses due to transmission errors, allowing the transport protocol to recover from congestion losses. When TCP-DCR is used in wireless networks, a simple link level retransmission scheme that is not aware of TCP semantics would suffice to recover from transmission errors without any explicit notification from the network regarding the type of the loss. Earlier work has shown that local recovery of channel errors is efficient [13].

4.1 Simulation Topology

The network topology used in these simulations is similar to that in the previous section, except that R2 is the Base station connected to the receivers via wireless links. The default values for the wired link bandwidth and delay is fixed at 100 Mbps and 5 ms respectively. The wireless link bandwidth is kept fixed at 1 Mbps and the delay is varied in accordance with the requirements of the experiment.

The TCP-DCR agent in these simulations uses the timer-based implementation of the congestion response delay. Link level retransmission is simulated by using the error model and the queue object provided by ns-2. The error model is exponential, and the corrupted packets are buffered at the base station and retransmitted after a delay corresponding to the round trip time of the wireless link, thus simulating link level retransmission. The packet to be retransmitted is added at the head of the queue that holds the packets awaiting transmission. The TCP/IP and MAC layer headers are ignored in the throughput calculations.

4.2 Performance at Different Channel Error Rates

First, we present the results for the simulation showing the performance improvement offered by TCP-DCR at various channel error rates in Fig. 5. The workload consists of a single flow in this case. The wireless link bandwidth and delay are set to 1Mbps and 45ms respectively, so that the total round trip time is comparable to the simulation in section 3.2. There is no congestion in the network. As can be seen from the graph, TCP-DCR performs better than TCP-SACK. Since there is no congestion in the network, most of the packet losses can be recovered using the link layer retransmission scheme making a window reduction unnecessary. Thus, the performance of TCP-DCR even at high channel error rates stays close to the performance that can be obtained when there is no channel errors at all. On the other hand, TCP-SACK treats the losses due to channel errors as congestion loss and hence the throughput stays below 0.5 Mbps.

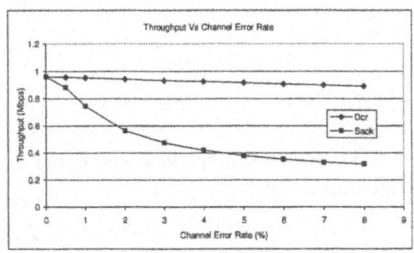

Fig. 5. Throughput Vs Channel Error Rate

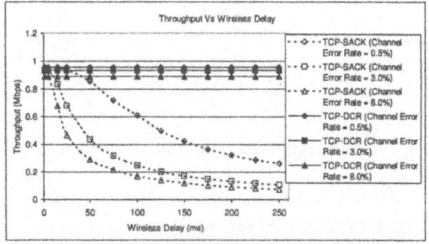

Fig. 6. Throughput Vs Wireless Link Delay

4.3 Performance at Different Wireless Delays

Wireless networks have highly varying delays ranging from few milliseconds to few tens of milliseconds for a LAN to several hundred of milliseconds for satellite links[20,21]. In this section we show the effect of the wireless delay on the performance of the different protocol flavors. The topology is similar to that in the previous section. Fig. 6 shows the results. It can be seen from the graph that as the wireless link delay is increased, the throughput of the TCP-SACK flows degrades significantly. This is because when the window is reduced incorrectly due to a packet lost by channel errors, it takes a longer time for the protocol to increase the window to the correct value again.

4.4 Performance with Congestion in the Network

In this set of simulations, the workload consists of 24 flows, half of which use TCP-DCR and the other half use TCP-SACK. The different levels of congestion are obtained by varying the buffersize at the router R1. The bottleneck link capacity is set to 10Mbps and the delay to 5ms. The wireless link bandwidth and delay are 1Mbps and 20ms. Fig. 7 shows the results. In the graph, congestion loss rates of less than 1% are labelled as low error, in the range of 2.5-3.5% are labelled as moderate congestion and greater than 3.5% are labelled as high congestion.

It can be seen from the figure that when the congestion loss rate is low, the average throughput of the TCP-DCR flows is far more than that of TCP-SACK flows. The throughput achieved by TCP-DCR flows is inversely proportional to the congestion loss rate in the network, whereas the throughput of the TCP-SACK flows is inversely proportional to the sum of the congestion loss rate and the channel error rate. So, as the congestion loss rate in the network increases, the difference in the average throughput of the TCP-DCR flows in the network compared to that of the TCP-SACK flows becomes narrower.

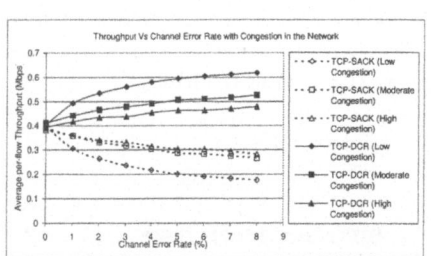
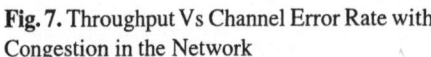
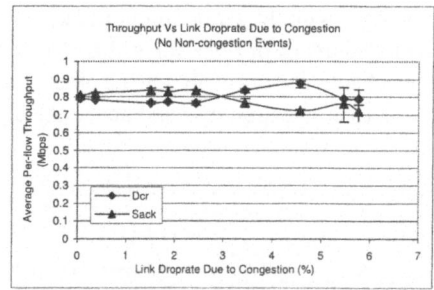

Fig. 7. Throughput Vs Channel Error Rate with Congestion in the Network

Fig. 8. Throughput comparison with Zero Non-congestion Events.

5   TCP-DCR with Zero Non-congestion Events

The earlier two sections have shown that TCP-DCR provides a simple, but effective mechanism for tolerating non-congestion events in networks that cause packet reordering or have significant channel errors. The natural questions that arise: what is the consequence of employing TCP-DCR in networks that do not experience any non-congestion events? Does TCP-DCR impact the throughput realized by individual flows? Is it fair to other flows that respond to congestion immediately? Does it impact queue lengths? We examine such question in this section.

5.1   Fairness

The results in section 3.4 and section 4.4 show that in the presence of non-congestion events, TCP-DCR utilizes the network bandwidth better than TCP-SACK flows at lower congestion, and the bandwidth is shared more equitably as congestion losses become the major contributing factor towards the total losses. In this section we evaluate the fairness of TCP-DCR when there are no non-congestion events at all in the network.

The simulation set up is similar to that in section 3.4. The graph shows the average throughput realized by DCR and SACK flows. From Fig. 8 we see that the average throughput achieved by the DCR flows is very close to the average throughput of the SACK flows, even at fairly high levels of congestion. The throughput of each individual flow does not vary too much from the average as indicated by the confidence intervals. These results indicate that TCP-DCR does not behave more aggressively than TCP-SACK, when τ is set to one RTT.

5.2   Packet Delivery Time and RTT Estimates

Since TCP-DCR delays the congestion response by one RTT, it takes a longer time to recover a packet lost due to congestion when compared to SACK. In order to evaluate the extent of the additional time taken by DCR, we conducted this experiment. For the network topology chosen, majority of the packets for both the flavors are delivered in 0.05seconds. For packets lost due to congestion and recovered using TCP retransmission, the average packet delivery time for case 1 (100% flows use TCP-DCR) is 207ms, for case 2 (100% flows use TCP-SACK) it is 178ms and for case 3 (50% of the flows use TCP-DCR and 50% use TCP-SACK), it is 201ms for TCP-DCR and 182ms for TCP-SACK. TCP-DCR does not affect the packet delivery time when there is no congestion. However, when a packet is lost due to congestion, the time to recover it could be higher by about one RTT. Also, according to Karn's algorithm used by most standard implementations of TCP, a retransmitted packet is not used in estimating the round trip time. Thus the delayed congestion response of TCP-DCR does not affect the rtt estimation of TCP. Our simulation results agree with the discussion above. An interested reader may find the detailed results in [22].

5.3   Response to Sudden Increase in Traffic

In this experiment we study the response of TCP-DCR to sudden increase in the traffic on the network. For this experiment, we first allowed six flows to run for 50 seconds until they reached steady state. At the end of 50 seconds, an additional six TCP-SACK flows were added. We compared the response of TCP-DCR with that of TCP-SACK for this sudden increase in traffic. Fig. 9 shows the results. It can be seen from the graph that the response of TCP-DCR is similar to that of TCP-SACK. The time to reach (55%, 45%) allocation for TCP-SACK was 3.1 seconds and for TCP-DCR, it was 3.67 seconds.

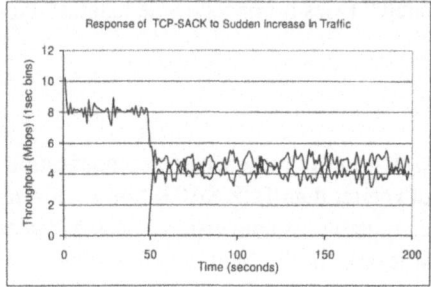 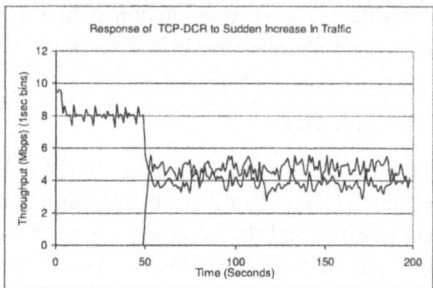

Fig. 9. Response to Sudden Increase in Traffic

5.4   Summary of Other Observations

We conducted several other experiments to evaluate the queue lengths, the timeouts, the perflow droprates, and the link utilization to understand the impact of DCR flows on the network characteristics. We summarize these results for three cases - case 1 (100% flows use TCP-DCR), case 2 (100% flows use TCP-SACK) and case 3 (50% of the flows use TCP-DCR and 50% use TCP-SACK) - in Table 10. As seen from these results, DCR flows do not drastically alter the observed network characteristics compared to a network with only SACK flows.

Case	Protocol of flow	Avg.Flow throughput Mbps	Avg.Flow Drop rate %	Avg.Flow timeouts %	Avg.que Length Pkts
1	DCR	0.801	2.72	0.0	44
2	SACK	0.801	2.27	0.009	44
3	DCR	0.748	2.77	0.0	45
	SACK	0.852	2.13	0.02	

Fig. 10. Summary of observations with Zero Non-congestion Events.

We also conducted several simulations with the sudden increase in background traffic due to several short term TCP flows simulating web-traffic. The results were similar to those in Fig. 9.

6   Conclusions and Future Work

In this paper, we proposed TCP-DCR that employs delayed congestion response and local recovery to recover from non-congestion events. We studied DCR's handling of non-congestion events in two specific scenarios, namely, packet reordering and wireless channel errors. In both the scenarios, results from simulations have shown that DCR offers significantly better performance by simply delaying congestion response for one RTT. We then studied the impact of employing DCR in networks with zero non-congestion events. Several other simulations were conducted the results of which have not been included here due to lack of space. An interested reader can find them in [22]. Our evaluation at multiple levels - individual flows, TCP characteristics and network characteristics - has shown that DCR does not significantly impact other flows or the network even when all the packet losses are due to congestion alone. Based on these results, DCR seems to offer a simple, unified solution to handle non-congestion events safely. We have also implemented DCR on a Linux platform and the preliminary results (available in [22]) look promising. We plan to continue with the tests under different scenarios.

Acknowledgements. Nauzad Sadry and Nitin Vaidya have contributed to the work reported in the wireless network section. Comments from Sally Floyd on an earlier draft have helped the paper.

References

1. Jon Bennett, Craig Partridge, and Nicholas Shectman, "Packet Reordering is Not Pathological Network Behavior," *IEEE/ACM Transactions on Networking*, December 1999.
2. Sharad Jaiswal, Gianluca Iannaccone, Christophe Diot, Jim Kurose, and Don Towsley, "Measurement and Classification of Out-of-Sequence Packets in a Tier-1 IP Backbone," *Proceedings of IEEE INFOCOM*, 2003.
3. M. Zhang, B. Karp, S. Floyd, and L. Peterson, "RR-TCP: A Reordering-Robust TCP with DSACK," *ICSI Technical Report TR-02-006*, Berkeley, CA, July 2002.
4. E. Blanton and M. Allman, "On Making TCP More Robust to Packet Reordering," *ACM Computer Communication Review*, January 2002.
5. Sally Floyd, Jamshid Mahdavi, Matt Mathis and Matt Podolsky, "An Extension to the Selective Acknowledgement (SACK) Option for TCP," *RFC 2883*, July 2000.
6. R. Ludwig and M. Meyer, "The Eifel Detection Algorithm for TCP," *RFC 3522*, April 2003.
7. H. Balakrishnan, V. Padmanabhan, S. Seshan, and R. H. Katz, "A Comparison of Mechanisms for Improving TCP Performance over Wireless Links," *IEEE/ACM Transactions on Networking*, 1997.
8. K. Brown and S. Singh, "M-TCP: TCP for mobile cellular networks," *ACM Computer Communications Review, vol. 27, no. 5*, 1997.
9. H. Balakrishnan, S. Seshan, E. Amir and R. Katz, "Improving TCP/IP performance over wireless networks," *Proc. of ACM MOBICOM*, Nov. 1995.
10. H. Balakrishnan and R. H. Katz, "Explicit Loss Notification and Wireless Web Performance," *Proc. of IEEE GLOBECOM*, Nov. 1998.

11. N. H. Vaidya, M. Mehta, C. Perkins and G. Montenegro, "Delayed Duplicate Acknowledgement: a TCP-unaware Approach to Improve Performance of TCP over Wireless," *Journal of Wireless Communications and Mobile Computing*, special issue on Reliable Transport Protocols for Mobile Computing, February 2002.
12. S. Mascolo, C. Casetti, M. Gerla, M. Sanadidi and R. Wang, "TCP Westwood: Bandwidth Estimation for Enhanced Transport over Wireless Links ," *Proceedings of ACM MOBICOM*, 2001.
13. D. Eckhardt and P. Steenkiste, "Improving Wireless LAN Performance via Adaptive Local Error Control," *Proceedings of IEEE ICNP*, Austin, TX, 1998.
14. M. Mathis, J. Mahdavi, S. Floyd and A. Romanow, "TCP selective acknowledgement options," *Internet RFC 2018*.
15. M. Allman, H. Balakrishnan, and S. Floyd, "Enhancing TCP's Loss Recovery Using Limited Transmit," *RFC 3042, Proposed Standard*, January 2001.
16. ns-2 Network Simulator. **http://www.isi.edu/nsnam/**
17. D. Bansal, H. Balakrishnan, S. Floyd and Scott Shenker, "Dynamic Behavior of Slowly Responsive Congestion Control Algorithms," *Proceedings of ACM SIGCOMM*, Sep. 2001.
18. R. Han and D.G. Messerschmitt, "A Progressively Reliable Transport Protocol For Interactive Wireless Multimedia", *ACM/Springer-Verlag Multimedia Systems Journal, vol. 7, no. 2*, March 199 9.
19. M. Allman, V. Paxson and W. Stevens, "TCP Congestion Control," *RFC 2581*, April 1999.
20. M. Allman, D. Glover and L. Sanchez, "Enhancing TCP Over Satellite Channels using Standard Mechanisms," *RFC 2488*, January 1999.
21. J. Border, M. Kojo, J. Griner, G. Montenegro and Z. Shelby, "Performance Enhancing Proxies Intended to Mitigate Link-Related Degradations," *RFC 3135*, June 2001.
22. Sumitha Bhandarkar, Nauzad Sadry, A. L. N. Reddy and Nitin Vaidya, "TCP-DCR: A Novel Protocol for Tolerating Wireless Channel Errors" *Tecnhical Report TAMU-ECE-2003-01*, February 2003.

An Enhanced Hybrid Key Management Protocol for Secure Multicast in Ad Hoc Networks

Mohamed Salah Bouassida[1], Isabelle Chrisment[1], and Olivier Festor[2]

[1] LORIA-UHP
[2] LORIA-INRIA
name.surname@loria.fr
MADYNES, Campus scientifique
B.P. 239, 54506 Vandœuvre-lès-Nancy Cedex - France
tel : +33-3-83-59-30-49 - fax : +33-3-83-41-30-79

Abstract. An ad hoc network is a collection of wireless mobile hosts forming a temporary network without the aid of any established infrastructure or centralized administration. This flexibility in space and time induces new challenges towards the security infrastructure needed to support secure unicast and multicast communications. Especially, traditional group key management architectures meant for wired networks are not appropriate in such environment due to high dynamics and mobility of nodes.

In this paper, we propose an enhanced hybrid key management protocol for secure multicast dedicated to operate in ad hoc networks. Built on a protocol called BAAL dedicated to key distribution in wired networks, our approach integrates threshold cryptography and the services of the AKMP protocol to deliver fast, efficient and mobility aware key distribution in a multicast service.

Keywords: Ad hoc networks, multicast security, group key management, threshold cryptography

1 Introduction and Motivation

The last decade saw the exponential deployment of wireless networks thanks to the emergence of new technologies and standards (e.g. the 802.11 series, Hiperlan [7], ...). The wireless networks can be used either in a base or in an ad hoc mode where hosts do not rely on any fixed infrastructure. The combined use of wireless ad hoc networks with wired gateways enables an easy network coverage extension at low cost. These networks also called hybrid networks are gaining more and more interest by the community including operators which see the clear advantages in deployment of these technologies. Those networks which are dynamic in space and time, offer great flexibility. However, this flexibility associated with the wireless connection vulnerability, requires an increased need in securing users and data.

The characteristics of ad hoc networks pose challenges in achieving the main security goals : authentication, confidentiality, non repudiation, integrity and

N. Mitrou et al. (Eds.): NETWORKING 2004, LNCS 3042, pp. 725–742, 2004.

availability. Using wireless links makes an ad hoc network vulnerable to link attacks ranging from passive eavesdropping to active impersonation, message replay and message distortion. Eavesdropping might give an adversary access to secret information, violating confidentiality. Active attacks might allow the adversary to delete messages, to inject erroneous messages, to modify messages and to impersonate a node, thus violating availability, integrity, authentication and non repudiation. The nodes in an ad hoc network are heterogenous and may have relatively poor physical protection, so they can be compromised. Therefore, we should not only consider malicious attacks from outside a network, but also take into account the attacks launched from within the network by compromised nodes. A major constraint placed on candidate security architectures in ad hoc networks is the absence of any possibility to put a centralized component in the solution. Security mechanisms must also be dynamic and efficient to adapt themselves to the dynamic and scalable nature of ad hoc network.

In parallel with the deployment of wireless ad hoc networks, multicast services gained acceptance through several applications like software distribution, multimedia conferencing, radio casting. The combination of an ad hoc infrastructure with multicast services which have to be operated induces new challenges towards the security infrastructure needed to enable acceptance and wide deployement of these multicast services.

In this article, we propose an approach to enable secure group communication within an ad hoc environment. Our solution focuses on group key management which is the major issue in group security. The principle of our approach is not to develop another new and specific solution but to adapt a protocol we have already tested and validated within wired networks.

Thus, our approach combines the functional architecture of BAAL [5], which is an group key management protocol in wired networks, with the dynamic support of AKMP [3]. And for ensuring secure generation of group keys, we use the threshold cryptography.

To present our solution, this paper is organized as follows. Section 2 identifies the multicast security challenges that emerge in ad hoc environment. In Section 3, we present related works concerning multicast security. Then, before describing our enhanced hybrid key management protocol for secure multicast in ad hoc networks, we give the main building blocks we have reused. Finally, we present the initial simulations and results obtained. Finally, we summarize the contribution and identify the directions for future work.

2   Multicast Security Challenges in an Ad Hoc Environment

Securing multicast communications is challenging because they present more opportunities for traffic interception. In addition, the identity and the addresses of multicast groups are known in a large scale, which help adversaries to orient theirs attacks. Multicast routing information can also be attacked, which can prevent a node from knowing the exact way to join an group (e.g. : wormhole problem).

Several other challenges raised by ad hoc network features. These challenges are :

- The lack of infrastructure that implies there is no central authority to be referenced for trust decisions about other parties within the network. The transient relationships do not help in building trust based on direct reciprocity and incite some malicious nodes to cheat.
- The size and dynamicity of multicast group which can be very high in ad hoc networks : the size cannot be controllable and, in the same way, dynamicity of members addition or removal.
- The mobility of nodes which has to be taken into account in security architecture. Some members move and still want to be able to receive multicast data. Nodes can also disappear without leaving the group (battery problem,...). Thus, when they appear again, they want to receive multicast flow. In this case, the multicast tree changes frequently but not the group members.
- The scalability, in the context of group communications in an ad hoc network, refers to the capacity of security mechanisms to cover great size multicast groups, without affecting the performances of the whole security system. The problem concerns the group key and security policies management and distribution.
- The trust model in an environment without fixed infrastructure concerns entities generating, distributing and managing cryptographic keys and security policies. Thus, we need one trust model to answer to the following questions : to which entities trust is granted for ensuring security services, which level of trust must be granted to them and which is the authority alive source.

3   Related Work

3.1   Secure Multicast Communication and State of the Art

IP multicast is an efficient communication mechanism for group-oriented applications, such as video conferencing, interactive group games and video on demand. IP multicast saves bandwidth by sending the source traffic on a multicast tree that spans all the members of the group. Group communication confidentiality requires that only valid users can decrypt the multicast data even if the data is broadcasted to the entire network. We assume that multicast data is encrypted using a symmetric cryptosystem ; the same key called Traffic Encryption Key (TEK) is used to encrypt and decrypt data.

The confidentiality requirements can be mainly translated into two key distribution rules :

- Forward confidentiality : users that left the group should not have access to any future key. This ensures that a member can not decrypt data after leaving the group.

- Backward confidentiality : a new user should not have access to any old key. This ensures that a member can not decrypt data sent before joining the group.

In order to meet the above requirements, a re-key process should be triggered after each Join/Leave procedure. It consists in generating a new TEK and distributing it to the members including the new one in case of a joining, or the residual members in case of a removal. We classify group key management proposals into three approaches:

- Approach A : all group members share a unique single symmetric key (TEK). This approach is mainly used within a centralized architecture where a single key server is responsible for generating and redistributing the new TEK whenever a member joins or leaves the group. This approach does not meet the scalability requirements since the number of transmitted messages to update TEK is proportional to n, the number of group members. This is known as the "1 affects n" phenomenon [12] where a single group membership changes results in a re-keying process that disturbs all group members to update TEK. In addition, the use of a single key server leads to a bottleneck problem during TEK distribution and suffers from a single point of failure. The BAAL protocol [5] belongs to this approach.
- Approach B : the multicast group is divided into multiple subgroups. Each subgroup shares a local TEK managed by a special entity : the subgroup controller. Protocols proposed within this approach are more scalable than centralized protocols, they also attenuate the "1 affects n" phenomenon. However, the drawback of this approach is that subgroups have different TEKs, multicast packets should be decrypted and re-encrypted by subgroup controllers whenever they pass from a subgroup to another. IOLUS [12] and AMAM [14] protocols belong to this approach.
- Approach C : To solve the "1 affects n" problem, without generating a great overhead due to the encryption/decryption process, this approach consists on merging the two preceeding approaches. The basic idea is to start a multicast session with centralized key management (approach A), and to divide the network dynamically in order to delegate key management to local controllers (approach B). AKMP [3](An Adaptive Key Management Protocol for secure Multicast) is proposed within this hybrid approach.

3.2 Secure Multicast Communication and Ad Hoc Environment

Few research results were published so far on how to secure multicast comunications within an ad hoc network.

The proposal of [15] defines the NTDR ad Hoc networks (Near Term Digital Radio). In NTDR architecture, there is a set of clusters, each containing a clusterhead, which when linked together form a routing backbone. A cluster has a single level consisting of nodes within one hop of a clusterhead. Intercluster communication is restricted to clusterheads only and intracluster communication between nodes that are within one hop of each other must traverse the clusterhead. This cluster based control structure promotes more efficient use of resources in controlling large dynamic networks, but generates more computing overhead due to the network clustering, the clusterheads election, and the establishment of the routing backbone.

For ensuring authentication, [15] makes use of public key systems and protocols and involves the use of certificates and certification authorities. Thus, all network participants have some time access to a public key infrastructure in their own fixed network domains. In this sense, [15] assumes that some hierarchical based PKI in fixed networks to which the participants have access to sometime before getting involved in a mobile ad hoc network. However, having an established PKI within an ad hoc environment suffering from lack of infrastructure is a very big challenge.

For ensuring secure group key management, [15] assumes that the clusterhead in an NTDR network could serve as a trusted entity to coordinate packet routing and manage security for the cluster. Thus, each clusterhead manages the cluster keys for its cluster and mediates all communication between its cluster and other clusters. Key generation and distribution is ensured by having two types of keys : cluster group key which is used to encrypt all cluster traffic to secure intra-cluster communication, and key encryption key which is a shared secret between a clusterhead and a node. This group key management is vulnerable because it is established around clusterheads which can be compromised. A computing overhead is also generated due to clusterheads movements, clusterheads deletion,...

[4] defines an authentication framework for hierarchical ad hoc Sensor networks. In this proposal, the sensor network consists of three tiers of devices with varying levels of computational and communication capabilities. This architecture is more scalable than flat ad hoc networks. The lowest tier consists of compute-constrained sensors that are unable to perform public key cryptography. Thus, [4] presents a new type of certificate, called a TESLA certificate. This certificate can be used by low-powered nodes to perform entity authentication. TESLA is more detailled in [8,2]. This three-tier architecture consists of three classes of wireless devices : high power access points that route packets received via radio links to the wired infrastructure, mobile medium-powered forwarding nodes that relay information from sensor nodes to access points, and low-powered mobile sensor nodes. [4] assumes that each forwarding node and access point has an RSA-key-pair along with its certificate. Like in [15], this assumption poses a very big challenge.

The proposal from [6] presents a new secure multicast communication approach using the measures from GPS units (latitude, longitude and altitude) and the Prüfer decoding algorithm. The group key is managed by using the Prüfer number and the Group Diffie Hellman key-exchange protocol (GDH). Any user in the multicast group can use the group key distributed by the source to securely receive multicast messages from the multicast source. This group key distribution model is efficient and robust but suffers from "1 affects n" phenomenon, and does not treat mobility factor. Computing Prüfer algorithm can also generate an overhead within ad hoc network nodes which have a relatively low power computing.

[10] proposes to reduce the communication and computation load on the source by having active group members which participate to the group security. This approach is based on IOLUS. The reliability is improved by allowing a node to maintain more than one link and the security is increased by requiring a joining node to authenticate with at least k members of the group. On the other hand, this approach generates a computing overhead due to the encryption/decryption process of IOLUS.

4   The Building Blocks

The aim of our approach is to solve the problems described in section 3.2 and to reduce the "1 affects n" phenomenon while limiting the computing overhead due to encryption-decryption process. For ensuring authentication within our network, we use threshold cryptography instead establishing a PKI infrastructure over an ad hoc environment. Finally, our approach treats mobility factor which is a big challenge to secure ad hoc networks.

In this section, we describe how we start from an existing group key management protocol we have already tested and validated over wired networks and how we adapt it to an ad hoc environnment by adding other building blocks.

Figure 1 shows the main three building bloks of our enhanced approach :

- the functionnal architecture from the BAAL protocol [5]
- the hybrid support from the AKMP protocol [3]
- the cryptography issue from the threshold cryptography [17]

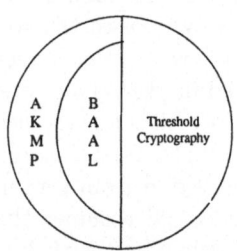

Fig. 1. Enhanced Approach Architecture

4.1   BAAL

BAAL is a group key management protocol which ensures access control, data confidentiality and authentication of group members within wired networks.

These services are achieved by distributing only one key K_{grp}, this makes BAAL belongs to the approach A presented in Section 3.

The three actors met in the BAAL architecture are :

1. The Global Controller (GC): may be an organizer of conferences, which can create one or several secured groups in the Internet. It holds a list of future participants in the group. This list can be created by others means like e-mail or fax. The GC creates the key group and distributes it to all the participants via the local controllers. Moreover, the GC has to re-key periodically or sometimes occasionally.
2. The Local Controller (LC) : is delegated by the GC. It receives the group key and distributes it to all the participants in its network, during the initial configuration of the group. The LC can create and distribute a new group key, accept or refuse a member and notify the others LCs in the case of group changes.
3. The Members of Group (MG) : member of the list of participants, or any member which joins the group later on.

The operations defined for group key management are :

- Group initialisation. Within this operation, K_{grp} is safely distributed to all elements of the participant list. The delegation of the LCs by the GC is also done, in order to guarantee the local access to the group and the cooperation with others controllers to manage the group key. The group initialisation is performed in two phases : invitation phase which is reserved to the invitation of all members of the participant list, and group key distribution phase which is dedicated to safely distribute the group key to group members and to delegate the LCs.
- Addition of a new entity. To join a multicast group, an entity must first be allowed to do so. This condition is verified by the LC which decides to add the entity to the group or not. In the affirmative case a rekey process must be triggered. If the LC is already delegated, it will generate and distribute a new key for all members of the group, else it will negociate with the GC to obtain permission for triggering the re-key process.
- Withdrawal of an entity. A member wanting to leave the group sends a Leave message in order to stop group traffic flow. If a member is detected as malicious, it will be automatically excluded. This is done by renewing the group key.
- Periodic re-key. Generally, cryptographic keys have a limited time to live and must be periodically renewed. This renewal can be done by the GC or by the LCs.

4.2 AKMP

The main idea of AKMP (Adaptive Key Management Protocol) is to meet approach A as long as no frequent membership change is depicted by group members, and to switch to approach B whenever members show a certain level of dynamicity. The decryption/re-encryption process is only restricted to sub-networks that are subject of high dynamicity.

The protocol begins with a single group that shares a unique TEK. This group is initially managed by one AKMP router. During the multicast secure session, if an AKMP router detects a local dynamicity, it initiates a subgroup with an independent local key. To do so, the concerned AKMP router generates and distributes the local key to the members in the constructed subgroup. This key is called a Downstream Key (DK). Then, the router decrypts received packets using its parent AKMP router key (called Upstream Key UK), and re-encrypts the packets using DK. The AKMP router has so switched from an inactive state to an active state. Thus, AKMP reduces decryption/re-encryption overhead to the minimum while attenuating the "1 affects n" phenomenon. Within each AKMP router, an evaluation function f_i is implemented and sets the AKMP router state according to the mcf: number of members changes per unit of time.

Each AKMP router holds DK_i and UK_i, upstream and downstream keys, and a pair of keys (public and private) allowing secured exchange between different AKMP routers. When an AKMP router i detects a high dynamicity within his sub network (f_i = true), it switches to active state, generates a new DK_i and distributes it to all local members. Then it must send its old DK_i to its AKMP router parent J so that J generates and distributes its new key DK_j for all its child local members in the case of $oldDk_i = DK_j$. In the case of (f_i = false), the AKMP router stays passive. Thus, when it detects a Join or a Leave event, it must notify its AKMP router parent in order to update and distribute DK_j.

4.3   Threshold Cryptography

BAAL uses a public key infrastructure which involves the presence of a CA (Certification Authority). But, within an ad hoc network, having only one CA presents a single point of failure. If the CA is unavailable, secured communications between nodes become impossible. The adversaries can also use this failure to compromise all the network. The CA duplication within ad hoc networks would bring more reliability but also more risks of malicious attacks since it also duplicates the possibility to compromise one CA.

Threshold cryptogpraphy [17] solves this problem. The new key management service, having $(n, t + 1)$ configuration, consists on having n special nodes called servers within the ad hoc network. Every server holds its pair keys, and public keys of all nodes in the network, particularly those of the others servers. This fact allows servers to communicate together securely. [16] proposes to distribute trust to nodes having a relatively hight physical security and a good computing power. The authors call these nodes MOCA (Mobile Certificate Authority). In the $(n, t + 1)$ configuration, the n servers share capability to sign certificates for the other nodes in the network. The private key of all the service (k) is divided into n secrecies $(s1, s2, ..., sn)$, each secrecy corresponding to one server.

Each server generates a partial signature of node cetificate and sends it to a combiner, which needs t+1 partials signatures to compute the complete signature. [17] assumes that $(n >= 3t + 1)$; t represents the maximum number of compromised servers. Thus, even if t servers are compromised, the combiner is able to generate the node signature.

The choice of t is detailed in [16], a great t offers more safety but at the same time, generates more traffic overhead.

The combiner, essential for the node signatures generation, can also be compromised. [9] proposes a replication of the combiner in many CA : a co-operative architecture of combiners, which can be able to be formed with stolen around the node, and generate for it its signature.

[16] presents a certification protocol, called MP (Moca Certification Protocol). The clients according to this protocol, broadcast messages Send Request (SREQ). Each MOCA server receiving this message, responds with a message Certif Response (CREP), containing a partial signature. When the client collects the $t + 1$ valid CREPs, it can constitute its signature. To reduce the flooding, a proposed solution is the B-unicast, which allows nodes to send by unicast to $t+1$ MOCAs, if it holds in its routing table enough information concerning their routes. Otherwise, the node will be obliged to broadcast its SREQ over all the network.

5   An Enhanced Hybrid Key Management Protocol for Secure Multicast in Ad Hoc Networks

The context of our approach is a set of ad hoc nodes, having capability of communicating in both unicast and multicast mode. This proposal has to be independent of used routing protocols, and thus, will start after construction of the multicast tree.

To ensure different security services for group communication in ad hoc networks : e.g. authentication, confidentiality, integrity and non repudiation, we need a group key management architecture.

Within our environment, we establish a threshold cryptography infrastructure which gives to each entity a public and private key (K_i , k_i). Group key generation is also achieved using threshold cryptography.

5.1   Approach Architecture

Like in BAAL, the main actors in our architecture are the global controller (GC), local controllers (LCs) and Group Members (GMs).

- The Global Controller (GC) is the group source. Initially, this entity holds group participants list named Participant_List. The GC is responsible for the generation and distribution of the group key, it also ensures periodic renewal of this key, and group security management (controlling local controllers and group members behavior). All the controllers (global and local) hold a same list named Recovery_List, which contains members excluded of the group, this list will be used at the time of Join and Leave of entities in the group.

– The Local Controller (LC). Every mobile node, belonging to the multicast tree, as a group member or a simple participant to the multicast tree, and having child nodes to which it conveys multicast flow, is considered as a passive local controller. Every local controller holds its local members list, named Local_Participant_List. It must convey multicast flow, sent by source, to all members of this list. If a passive LC is a group member, it must hold the same cryptographic key as its parent node. Otherwise, if the passive LC is a simple participant, it will not need the node-parent key. When the local dynamicity rate and the local members number reaches certain thresholds, the LC decides to switch to an active state. Thus, it generates a new local key, distributes it to all its members and starts a decryption/re-encryption process. We say that this active LC forms with its local members a new cluster. To decide on its state, every LC holds a dynamicity-evaluation function, described in what follows :

if (mcf > d1 or mn > d2) then {switch to dec/rec process}
    f_i = true;
else
    f_i = false;
end;
with mcf: number of members changes per unit of time,
d1 : predefined fequency threshold.
          mn : local members number, d2 : predefined member number.

To obtain mn, a LC counts its passive child nodes with their child nodes, and its active child nodes without their child nodes. Figure 2 gives an example of calculation of mn.

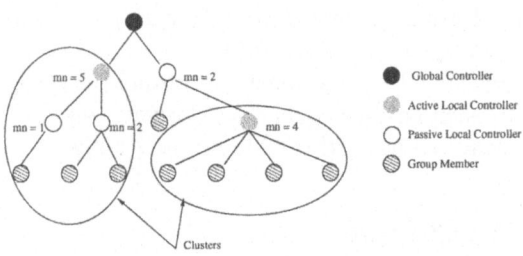

Fig. 2. Example of mn evaluation

This function differs from the AKMP evaluation function by taking into account not only the members-changes frequency but also their number. This is necessary to secure group communications in ad hoc networks for two reasons. First, all group members are also routers and so they can be considered as local controllers (if they have child nodes). Second, at the time of a leave, the active local controller is obliged to renew the local key and to

distribute it, in unicast, at all its cluster members. Thus,time necessary for this renewal is proportional to cluster-members number.
- Group Member(GM) is a member of the list Participant_List or a member joining the group later.

We now present the different operations we have extended in our approach.

5.2   Operations

Group initialization. The GC initializes the both lists Participant_List and Recovery_List. Then, it starts the phase of the group key generation and distribution. The group key generation is realized using threshold cryptography with B-Unicast as decrypted below. If an $(n, t+1)$ configuration is established, the GC consults its routing table, and checks whether it holds routes to t+1 servers. If it is the case, it will send them a Key_Query message, otherwise it will broadcast this message to the whole network. At the reception of one Key_Query message a server starts by authenticating the sender. If the authentication succeeds, it generates a partial signature and includes it within a Key_Resp message. This message will be sent to the GC encrypted with its private key. The GC remains on standby of t+1 valid partial signatures, sent by the t+1 first servers. On their arrivals, it combines them to obtain the key group. The servers are provided with algorithms allowing them to generate, randomly, partial signatures for other nodes. In order to secure communications during key distribution, the sent message includes the sender's signed token. Thus, the signed token is essential for the authentication process in our approach. It allows a receiver to check the message origin and the sender's identity. A token contains :

1. an identity of the sender, e.g. its IP address ;
2. a timestamp ;
3. a random number, used to protect receivers against the replay of messages.

Each token is included within the message, signed with the private key of its sender. The GC sends the following message to Participant_List members (M_i), encrypted with their respective public keys (K_i). This message contains the group key, the group identity, the identity and the signed token of the GC.

GC -> M_i : $\{K_{grp}$, IDG, ID_{GC}, [token_GC]'Prv_GC$\}$'K_i
with IDG : group identity, ID_{GC} : GC identity, [token_GC]'Prv_GC : GC signed token.

After receiving this message, each group member decrypts it, authenticates the GC and extracts the key group. Then, it sends to GC a Report message containing the group identity, its identity and its signed token, encrypted with GC public key.

M_i -> GC : $\{$IDG, ID_M_i, [token_M_i]'Prv_$M_i\}$'Pbk_GC
with Pbk_GC : GC public key, [token_M_i]'Prv_M_i : M_i signed token.

Addition of a new entity. An entity wanting to join the group, sends a Report message to its LC. This message contains its signed token, its identity and the group identity.

$M_i \rightarrow$ LC : {IDG, ID_M_i, [token_M_i]'Prv_M_i}'Pbk_LC
with LC : Local controller of the new entity.

After receiving this message, the LC authenticates the signed token. If the authentication succeeds, the LC checks whether the new entity does or not appear in the Recovery_List. At this stage, the LC calculates its dynamicity-evaluation function. Two cases are identified :

1. if (f_i = true) : the LC switches to active state. Thus, it must generate a new local key using threshold cryptography and distribute the new key K_i^{loc} to its local old members, encrypted with the old key old_K_i^{loc} , and send the same key to the new member encrypted with its public key. Moreover, the LC must send its old local key to its active parent node M_l in order to update the key K_l^{loc} if old_K_i^{loc} = K_l^{loc} (when the LC switches, at the first time, to active state, its key is the same as its active parent node key).

    for j:1..nb_old_attached_members
       LC $\rightarrow M_j$: {K_i^{loc}}'old_K_i^{loc}
    LC $\rightarrow M_i$: {K_i^{loc}}'K_i with M_i : new member
    LC $\rightarrow$ Active_Parent_Node M_l: {old_K_i^{loc}}'K_l where K_l :
    public key of the parent node M_l.

2. if (f_i = false) : the LC remains in passive state. Thus, it sends a request for key renewal to its active parent node, which will start generation and distribution of a new key to all its local members.

Withdrawal of a group entity. We distinguish two cases, voluntary withdrawal and expulsion. The first is realized when a member decides to leave the group. Thus, it sends a Leave message to its LC, containing its signed token. In this case, the LC removes the entity from its Local_Participant_List and starts the renewal key phase. The second case (expulsion) is named member revocation, which takes place when the member can place the safety of the group in danger. Thus, the LC adds this member in Recovery_List, removes it from the Local_Participant_List and starts the renewal key phase. Within the renewal key phase, which is basically the same as entity addition, two cases are identified :

1. If (f_i = true) : the LC switches to the active state. Thus, it must generate a new local key using threshold cryptography and distribute then new key K_i^{loc} to its local old members, while excluding the leaving member, encrypted with their respective public keys. Moreover, the LC must send its old local key to its active parent node M_l in order to update the key K_l^{loc} if old_K_i^{loc} = K_l^{loc}.

M_s member leaving the group,
for j:1..nb_attached_members, j different from s,
   LC -> M_j : $\{K_i^{loc}\}'K_j$
LC -> Active_Parent_Node M_l: $\{\text{old\_}K_i^{loc}\}'K_l$ with K_l :
public key of the parent node M_l.

2. If (f_i = false) : the LC remains in passive state. Thus, it sends a request for key renewal to its active parent node, which will start generation and distribution of a new key to all its local members.

Periodic renewal of the group key. Cryptographic keys have a limited time to live. Thus, they must be renewed periodically. The renewal period is determined according to the key length and the key generation algorithm. The periodic key renewal is done by the GC or all the active LCs. This renewal is operated in two stages : key generation using threshold cryptography and key distribution to the local members.

5.3   Mobility Processing

Our approach has to be adapted to mobile ad hoc network features. So we have studied some mobility scenarii involving the main actors of the key distribution architecture :

1. When an LC moves, all its local members will be unable to receive multicast data sent by the source. To solve this problem, these members must immediately connect themselves to another LC in order to continue to receive multicast flow. It remains to see how the transition between clusters is done and how much data members transiting between clusters will lose. We distinguish two cases :
   a) The LC moves with notification : this notification can be a message sent by the LC, through multicast, to all its local members. Thus, the local members will send Report messages to another LC in order to obtain its local key and continue multicast data reception.
   b) The LC moves without notification : the local members will realize, after a certain period of time, that the route to the source is not assured any more by their LC. They must join the group via another cluster.
2. When a group member moves outside its cluster, it will lose the multicast connectivity. It must choose another route to the group source by connecting itself to another LC and by authenticating itself to others LCs when it begins to move.

To notify of its movement, a node must detect as soon as possible changes in its multicast routing table. These solutions are dependent of security and QOS policies, established in the ad hoc network. According to these policies, we can allow or not a latency time necessary for a member, in movement, to find the multicast data reception.

6   Simulation and Results

In this section, we present simulations realized to define frequency and member number thresholds. Beyond these thresholds, a local controller switches from passive to active state. For the simulation, we have studied the 1 affects n behavior, using the threshold cryptography with (5,3) configuration. We have also measured the time necessary for the group key renewal, after a Join or a Leave message, according to the event frequency and the group members number.

These two thresholds give us a first evaluation of our solution. They garantee that the time necessary for a Join or a Leave cannot exceed an upper value. Having these threshold, we can also vary the threshold cryptography configuration, in order to enhance the processing within our architecture.

6.1   Simulation Model

To do the simulation, we used the NS simulator, version ns2.1b9a [11]. The simulated network is an ad hoc network, composed of 100 nodes, and using MAODV [13] as the multicast routing protocol. Nodes moving is generated randomly in order to take account the mobility factor of ad hoc networks. To generate realistic multicast sessions, we use the model presented by Almeroth [1], which suggests that member arrival follows a Poisson processus ($\lambda = 10$ arrivals by time unit) and the membership duration is an exponential distribution (in average $\mu = 145$ time units). This model is deduced from real multicast sessions observed on the Mbone.

Figure 3 shows the variation of group members number by time, for the simulated model.

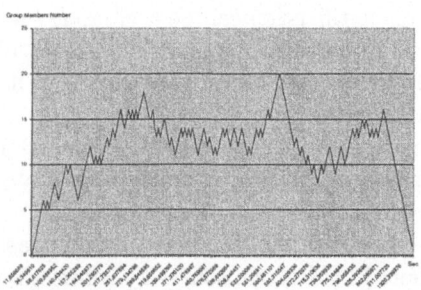

Fig. 3. Variation of group members number

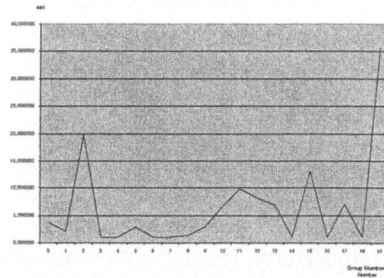

Fig. 4. Renewal Key Time following a Join by group members number

6.2   Simulation Results

Before starting simulation, we have calculated the time induced by a Join and a Leave procedure within the multicast group. For the Join event, the number of sent messages to renew the group key is 10, which are :

- 2 messages for new member authentication.
- 3 messages sent by the GC to 3 servers in order to generate the new group key, and 3 response messages.
- 1 message sent by the GC in multicast, to all old members, containing the new group key.
- 1 message sent by the GC in unicast, to the new member, containing the new group key.

The transmission of the new key needs operations of encryption/decryption, for which we chose the 3DES algorithm. Eight operations of encryption/decryption are necessary for a Join event:

- 3 operations of encryption realized by servers generating the new key.
- 3 operations of decryption realized by the GC while receiving the 3 partial signatures from servers.
- 1 operation of encryption of the new key while sending it in multicast, to all group members.
- 1 operation of decryption of the new key realized by the new member. We assume that old members will decrypt the key at the same time.

The average time to send a message in our network is $t = 0.002579s$ and the time to achieve one decryption or encryption operation, according to 3DES, is $d = 0.000712s$ (1500 bytes). Thus, we estimate the average cost for the renewal key during a member addition to $10 * t + 8 * d = 0.031486s$. This cost is added to the latency time between the Join query and the effective Join.

The same study is done for a Leave procedure: the number of sent messages for renewing a group key is $(7+n)$ with n : group-members number :

- 1 message of Leave query.
- 3 messages sent by the GC to 3 servers in order to generate the new group key, and 3 response messages.
- n messages sent by GC in unicast to residual members, containing the new group key.

To ensure key confidentiality after a Leave event, we need $(6+2n)$ encryption/decryption operations :

- 3 operations of encryption realized by servers generating the new key.
- 3 operations of decryption realized by the GC while receiving the 3 partial signatures from servers.
- n operations of encryption of the new key while sending it in unicast, to all residual group members.

– n operations of decryption realized by residual group members while receiving the new group key, using their private keys.

Figure 4 presents the time necessary for the rekey-process following a Join procedure by the group members number.

Considering that the generated cost for the renewal key is constant, the different variations shown in the curve are due to the fact the latency time between a Join query and the effective Join can vary due to the MAODV protocol and also to the location of the new member compared to the others group members. Thus, this curve does not permit to define the threshold of group members number in a cluster.

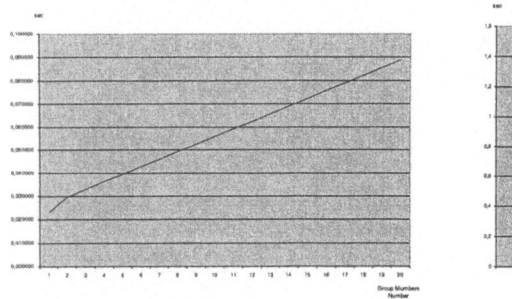 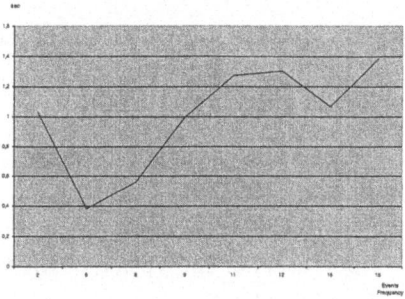

Fig. 5. Renewal Key Time following a Leave by group members number

Fig. 6. Renewal Key Time by events frequency

Figure 5 shows that the time necessary for the rekey-process following a Leave event, is proportional to the group members number. Thus, we can define the threshold of members number within a cluster. If we take, for example, as a constraint that the time necessary for the rekey-process following a Leave event cannot exceed 0.05s, the threshold will be 8 members per cluster.

To carry out Figure 6, we calculated the average time necessary for the rekey-process following a Join or a Leave procedure by events frequency calculated within equal-time intervals. If we take as a second constraint the fact that the time for the rekey-process by the frequence cannot exceed 1s, the curve shows that, once established the steady operation, the threshold of frequence is 9 events per unit of time.

7 Conclusion and Future Works

In this paper, we addressed the security of one service within an ad hoc infrastructure namely multicast.

We presented an enhanced architecture starting from BAAL which is a group key management protocol, already tested and validated over wired networks. To adapt BAAL to ad hoc environment, we use the hybrid support of AKMP in order to ensuring dynamicity and scalability. Our enhanced architecture ensures also secure generation of group keys by using the threshold cryptography.

Our approach meets some challenges posed by securing multicast communications within an ad hoc network, namely :

- The problem of lack of infrastructure to ensure group source and members authentication is solved by using the threshold cryptography. In fact, this method consists on sharing a central authority into many nodes which we call servers. These servers share the capacity to sign certificates for all the ad hoc network nodes.
- The support of AKMP ensures dynamicity and scalability of our approach. Indeed, AKMP solves the problem "1 affects n" met in BAAL while limiting the encryption/decryption computing overhead due to the clusterisation of the group in sub-groups. Thus, our enhanced architecture allows to cluster the multicast group, dynamically, according to the frequence of the events Join and Leave, and to the members-number per cluster.
- The mobility of nodes within our environment has been taken into account. In fact, we studied some mobility scenarios involving the main actors of the key distribution architecture.
- To solve the trust problem concerning securing cryptographic keys generation, we also used the threshold cryptography which allows the keys generation by servers having a great computing power and a good physical security.

We realized simulations to define frequency and member number thresholds. Beyond these thresholds, a local controller switches from passive to active state. Thus, these thresholds garantee that the time necessary for a Join or a Leave cannot exceed an upper value.

As future works, we plan to improve this approach by integrating service availability and securing routing informations within the network. We plan also to achieve the reliability of the keys distribution using acknowledgements.

References

1. K. Almeroth and M. Ammar. Collecting and modelling the join-leave behaviour of multicast group members in the mbone. In *The Symposium on High Performance Distributed Computing*, Syracuse NY, 1996.
2. M. Archer. Proving correctness of the basic tesla multicast stream authentication protocol with tame. In *Workshop on Issues in the Theory of Security, 2002*, 2002.
3. H. Bettahar, A. Bouabdallah, and Y. Challal. An adaptive key management protocol for secure multicast. In *11th International Conference on Computer Communications and Networks ICCCN*, Florida USA, October 2002.
4. M. Bohge and W. Trappe. An authentication framework for hierarchical ad hoc sensor networks. In *WISE'03, San Diego, California, USA*, September 2003.
5. G. Chaddoud, I. Chrisment, and A. Schaff. Baal : Sécurisation des communications de groupes dynamiques. In *The Proceedings of the 8th Colloque Francophone sur l'Ingénierie des Protocoles CFIP'2000*, Toulouse, France, October 2000.

6. T. Chiang and Y. Huang. Group keys and the multicast security in ad hoc networks. In *Proceedings of the 2003 International Conference on Parallel Processing Workshops (ICPP 2003 Workshops)*, 2003.

7. ETSI. High Performance Radio Local Area Network (Hiperlan), draft standard ETS 300652, March 1996.

8. T. Hardjono and L. Dondeti. *Multicast and Group Security*. Computer Security Series. Artech House, Librarie Eyrolles, 2003.

9. V. Legrand, F. Abdesselam, and S. Ubéda. Etablissement de la confiance et réseaux ad hoc - un état de l'art. Technical report, Laboratoire CITI - INRIA ARES, 2003.

10. G. Lin and G. Noubir. Secure multicast over multihop wireless ad hoc networks. In *Workshop on Mobile Ad Hoc Networking and Computing*, March 2003.

11. C. Lindemann. http://rul-www.cs.uni-dortmund.de/MobileP2P/mainE.html, 2003.

12. S. Mittra. Iolus: A framework for scalable secure multicasting. In *SIGCOMM*, pages 277–288, 1997.

13. E. Royer and C. Perkins. Multicast Ad hoc On-Demand Distance Vector (MAODV) routing, IETF Internet Draft: draft-ietf-manet-maodv-00.txt, 2000.

14. H. Sallay, A. Lahmadi, O. Festor, and I. Chrisment. Extension de l'architecture active amam pour le support des services de sécurité multicast. In *GRES'2003 Colloque Francophone sur la Gestion de Réseaux et des Services*, February 2003.

15. V. Varadharajan, M. Hitchens, and R. Shankaran. Securing ntdr ad-hoc networks. In *IASTED International Conference on Parallel and Distributed Computing and Systems 2001*, pages 593–598, Anaheim California, August 2001.

16. S. Yi and R. Kravets. Key management for heterogeneous ad hoc wireless networks. Report Research UIUCDCS-R-2002-2290, UILU-ENG-2002-1734, University of Illinois at Urbana-Champaign, Department of Computer Science, 1304 West Springfield Avenue, Urbana, IL 61801-2987 USA, July 2002.

17. L. Zhou and J. Haas. Securing ad hoc networks. *IEEE Network*, 13(6):24–30, 1999.

On Dynamic Subset Difference Revocation Scheme

Weifeng Chen[1], Zihui Ge[2], Chun Zhang[1], Jim Kurose[1], and Don Towsley[1]

[1] Department of Computer Science
University of Massachusetts at Amherst, MA 01002, USA
{chenwf, czhang, kurose, towsley}@cs.umass.edu
[2] AT&T Labs-Research, Florham Park, NJ 07932
gezihui@research.att.com

Abstract. Subset Difference Revocation (SDR) [8] has been proposed
to perform group rekeying in a stateless manner. However, statelessness
comes at a cost in terms of storage and message overhead when the
number of currently active members is much smaller than the number
of potential group members [3]. We propose a *dynamic* SDR scheme
to address these problems. Rather than maintaining a large static key
tree that accommodates all potential group members, we use a smaller
dynamic key tree for only currently active members. We dynamically
assign current members to the positions in the key tree rather than using
fixed pre-assignment. The smaller key tree requires less storage and
dynamic assignment achieves a smaller rekeying cost. Our evaluation
shows that the dynamic scheme significantly improves the performance
of SDR, reducing by half the rekey communication cost in the case
that the number of the currently active members is much less than the
total number of potential members. Compared to SDR in [8], dynamic
SDR does not need to know the maximum number of potential group
members in advance, a value that can be difficult to estimate in practice.

Keywords: Multicast security, Group rekeying, Subset Difference Revocation

1 Introduction

Membership-based applications, such as pay-per-view and specialized information services (e.g., stock price, live news), require that information content be delivered to (and only to) subscribed members. This is typically accomplished by encrypting data using a common *Traffic Encryption Key* (TEK) that is shared by all currently active members. When a member joins the group, the TEK must be changed to ensure that the newly joining member cannot decrypt previous communications (a requirement known as "backward confidentiality"). Similarly, the TEK must be changed when a member leaves the group to ensure that future messages cannot be decrypted by the departing member (a requirement known as "forward confidentiality"). The algorithms that manage the distribution, updating and revocation of the TEK are collectively known as *group key management*

N. Mitrou et al. (Eds.): NETWORKING 2004, LNCS 3042, pp. 743–758, 2004.

protocols. The IETF MSEC framework suggests using a *Group Controller and Key Server (GCKS)* for rekeying. Generally, the TEK is encrypted using *Key Encryption Keys* (KEKs) and then multicast by the GCKS. Several works have dealt with the group rekeying problem [2,6,7,8,9,10,11]. Subset Difference Revocation (SDR) [8] has been proposed as a "stateless" group rekeying algorithm. By stateless, it is meant that members do not need to keep track of rekeying messages in order to maintain their states. This desirable property comes as a price, however [3] - SDR can require high key storage at both the member and GCKS sides, and can generate a significant amount of messaging traffic during rekeying.

These two problems arise from the fact that SDR maintains a static key tree that is constructed in advance. This tree must be large enough to hold all potential members. Since key storage cost is determined by the size of the key tree [8], key storage costs can thus be large. Further, a member joining the group is attached to a pre-assigned position in the key tree. Assuming that member activity is independent of position, when the number of currently active members is much smaller than the number of potential members, the positions occupied by active members are likely to be *sparse*, i.e., there are many holes (positions not occupied by an active member) among the positions occupied by active members. The sparse distribution of these positions can cause SDR to perform as inefficiently as encrypting the TEK separately for each active member [3]. We will refer to the SDR proposed in [8] as *static* SDR in this paper.

In this paper, we propose a *dynamic SDR* scheme to reduce both the key storage cost and the rekey communication cost of static SDR. Dynamic SDR uses a key tree that is large enough to hold currently active members (as opposed to *all* members, both active and inactive) in order to reduce key storage. This is done by dynamically assigning a joining member a new position in the tree. Ideally, our goal will be to position active members adjacent in the key tree in order to reduce rekeying cost.

This paper has the following contributions. First, we design a new group rekeying algorithm, dynamic SDR. The algorithm is based on Subset Difference and uses a dynamic key tree to reduce both storage costs and messaging overhead. The algorithm is "multicast stateless" (i.e., it does not require nodes to maintain states when receiving the multicast rekeying messages, which distribute only TEKs) and does not require *a priori* knowledge of the number of potential members. Secondly, we propose and evaluate several enhancements on dynamic SDR. Our simulations show that dynamic SDR significantly reduces both the key storage cost and rekey communication cost when the number of currently active members is much less than the number of potential members. Finally, we investigate the tradeoff between the unicast and multicast costs in dynamic SDR.

The rest of the paper is organized as follows: In Section 2, we briefly overview the static Subset Difference Revocation algorithm. The dynamic SDR scheme is described in Section 3. Section 4 presents our evaluation. Section 5 discusses various properties of dynamic SDR. Related work is given in Section 6. Finally, we conclude the paper in Section 7.

2   Background: Subset Difference Revocation Algorithm

Static SDR [8] is a tree-based group key management protocol. For a finite set of potential members $\mathcal{N}$ ($N = |\mathcal{N}|$), the GCKS maintains a tree with N leaves and assigns a fixed position (a leaf in the key tree) to each distinct member[1]. For a node i in the key tree, let S_i be the set of the potential members which are descendants of i. Given two nodes i and j, where j is a descendant of i, subset $S_{i,j}$ is defined as $S_i \setminus S_j$. Each subset $S_{i,j}$ is initially assigned a long-lived key $K_{i,j}$ that is only known by members covered by $S_{i,j}$. Let $\mathcal{M} \subseteq \mathcal{N}$ be the set of members currently active in the group and $M = |\mathcal{M}|$. To distribute an updated TEK, the GCKS uses a *Subset-Cover* framework to partition $\mathcal{M}$ into disjoint subsets. More specifically, the GCKS divides $\mathcal{M}$ into s subsets such that $\bigcup_{d=1}^{s} S_{i_d, j_d} = \mathcal{M}$ and $\forall m \in \mathcal{M}$, m is covered by one and exact one resultant subset. The updated TEK is then encrypted using K_{i_d, j_d}. Consequently, only members in $\mathcal{M}$ can deduce the new TEK.

SDR is a *stateless algorithm* for the KEKs ($K_{i,j}$) are unchanged after initialization such that each member m in $\mathcal{M}$ is able to deduce the KEKs at each rekeying instance based on the *secret information*, denoted as I_m, received during initialization. I_m allows m to deduce the keys of *all* possible subsets to which m may belong.

Although the total number of possible subsets to which a member m may belong is $O(N)$, the size of I_m is $|I_m| = (\log^2 N + \log N)/2 + 1$ [3]. It is important to note that I_m is the only required information for member m to participate in the group communication for the statelessness of SDR. Also note that I_m is determined by the position of m in the key tree. As a consequence, once a leaf position in the key tree is assigned to a member m, that position cannot be assigned to any other member even when m is currently not in the group. That is the reason why the GCKS in static SDR needs to maintain a key tree large enough for $\mathcal{N}$. Typically, in static SDR, a *returned* member (a member joining the group again after leaving) is assigned to the position that the member was assigned last time.

The number of the resultant subsets of $\mathcal{M}$ is determined by the positions of members of $\mathcal{M}$ in the key tree of size N. Generally speaking, under the assumption that member activity is independent of position in the key tree, the larger the difference between N and M, the more likely that active members are sparsely distributed in the key tree, resulting in $O(M)$ disjoint subsets. On the other hand, the smaller the difference between N and M, the more adjacent the positions of active members in the key tree, resulting in $O(N - M)$ disjoint subsets. As pointed out by [3], given M, the expected number of resultant subsets

---

[1] In the rest of this paper, a leaf node in the key tree and the member assigned to that node are treated indistinguishable when there is no risk of ambiguity.

of $\mathcal{M}$ is $\Theta(\min(N - M, M))$. Moreover, static SDR requires $O(N \log N)$ storage overhead at the GCKS [3], and $O(\log^2 N)$ storage overhead at the member side.

3   Dynamic SDR

We have seen that static SDR generally requires a very large key tree to accommodate all unique members. As a consequence, currently active members, usually a small fraction of all of the unique members, are likely to be widely dispersed in the key tree space. Both of these factors decrease the performance of SDR. In this section, we propose a *dynamic SDR* approach that addresses these two inefficiencies of static SDR. First, we describe an observation on static SDR, based on which dynamic SDR is proposed.

In the binary key tree of SDR, a node i has a height of h if the subtree T_i rooted at i has 2^h leaves. A leaf itself has a height of 1. The heights of T_i and $S_{i,j}$ are also defined as the height of node i.

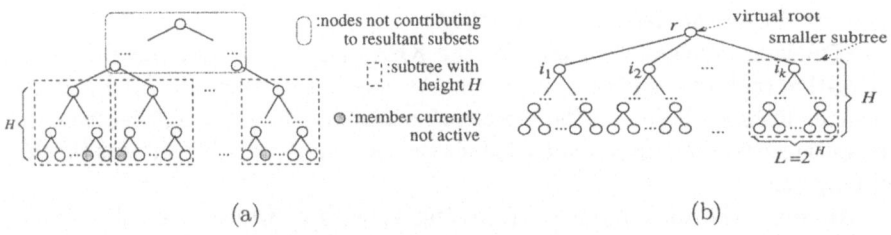

(a)   (b)

Fig. 1. Constructing dynamic SDR key tree

Proposition 1. *In the key tree T of static SDR, if there exists a set of disjoint subtrees $\{T_i\}$ with the same height H such that*
(1) leaves of $T = \bigcup_i \{$leaves of $T_i\}$;
(2) each T_i has at least one leaf in $\mathcal{N} \setminus \mathcal{M}$, i.e., a member currently inactive, then all resultant subsets of $\mathcal{M}$ have height $h \leq H$ (Fig. 1(a)).

Proposition 1 implies that, if such set of $\{T_i\}$ exists in the key tree of static SDR, all nodes with height $h > H$ will not contribute to the resultant subsets of $\mathcal{M}$. Thus the GCKS only needs to maintain a set of smaller subtrees satisfying Proposition 1 with an appropriate height H (we will address how to choose H shortly). The idea of dynamic SDR is to dynamically maintain such set of subtrees.

3.1   Scheme of Dynamic SDR

In dynamic SDR, the positions of members are not pre-assigned. Instead, the spaces in the key tree are dynamically allocated and reclaimed, adapting to the current set of active members. More specifically, the GCKS dynamically creates leaves when new member joins, or discards a subtree when all positions of the subtree are inactive. By doing this, the GCKS maintains active members in a dynamic key tree, rather than a large key tree constructed in advance.

For simplicity, we use set of subtrees $\{T_{i_k}\}$ as allocation units, one can view the subtrees connected to a virtual root r (Fig. 1(b)). Initially, the GCKS has a single subtree T_{i_1} connected to the virtual root r. When a member joins the group, regardless of being a new member or a returned member, the member is assigned to the next available position in the key tree (from left to right) and is *unicast* the secret information associated with the new position. The GCKS thereafter encrypts and *multicasts* the updated TEK to the current members in exactly the same way as in static SDR. If new positions are required, the GCKS creates a new subtree T_{i_k}. When a member, m, leaves the group, the position becomes empty and will never be used by any member (even m itself). And a new TEK is multicast to the members that remain in the group. If all positions of the leftmost subtree become empty, the GCKS discards that subtree. This process is detailed in [4].

The advantages of maintaining such a dynamic-membership key tree are two-fold. First, dynamic SDR may require a much smaller key tree of a size sufficient to accommodate the maximum number of concurrently active members. This helps reduce key storage cost, both at the members and at the GCKS. Second, by assigning members that arrive close in time to positions that are close in the key tree, the GCKS is likely to find a subset that can cover many adjacent members. This implies that the messaging overhead associated with rekeying is also reduced. Dynamic SDR achieves the advantages by introducing additional unicast, corresponding to deliver I_m to a joining member m. However, the overall communication cost (in bytes per second), can be reduced by more than 50% in comparison to that of static SDR, as we will see in Section 4.

Since the key tree of dynamic SDR can be extended arbitrarily, dynamic SDR does not require *a priori* knowledge of the size of total member population, N. This avoids the problem, which exists in static SDR, of estimating N. Overestimating N makes the static SDR key tree unnecessarily large, increasing both rekey communication cost and key storage cost, whereas, underestimating N may introduce the problem of having to reject members when all positions have been assigned.

In dynamic SDR, however, the size, $L = 2^H$, of SDR subtree T_{i_k} should be chosen properly, as we describe next.

3.2   Determining L

The choosing of L is a design tradeoff. Based on Proposition 1, one subset covers at most L members. Therefore, when L is small, more resultant subsets are required to cover $\mathcal{M}$, which consequently increases the multicast cost. When L is large, we may still encounter the space inefficiency of static SDR that active members disperse in the subtree.

We thus choose L as a reasonable value of $2^{\lceil \log E[M] \rceil}$, where $E[M]$ is the expected value of the number of concurrent members. Ideally, we want to put the concurrent members in one subtree.

3.3   Key Storage of Dynamic SDR

In dynamic SDR, the size of secret information, $|I_m|$, is reduced from $(\log^2 N + \log N)/2 + 1$ to $(\log^2 L + \log L)/2 + 1$ for the following Proposition.

Proposition 2. *In dynamic SDR, if a member, m, is assigned a position in subtree T_{i_k}, for any resultant subset $S_{i,j}$ covering m, i is a descendant of i_k.*

Although the key storage size required by a member is fixed when L is chosen, this is not the case for the GCKS, whose key storage is related to the number of subtrees T_{i_k}. Assuming that the GCKS sequentially assigns the available positions of the key tree from the left to the right to the joining members, we define S as the distance from the leftmost position occupied by an active member to the first available position at the right side. These S positions are referred as the *concurrent spaces*, which determine the key storage at the GCKS. To hold S concurrent spaces, at most $\lceil S/L \rceil + 1$ subtrees are required. Since the GCKS has $2L \log L + 1$ key storage for a subtree T_{i_k} of size L [3], it follows that the key storage at the GCKS is $(\lceil S/L \rceil + 1)(2L \log L + 1)$.

When members arrive according to a Poisson process with rate λ and the time that each active member stays in the group (which is referred as *lifetime*) is exponentially distributed with mean $1/\mu$, the expectation of S can be computed as follows [4], where $\rho = \lambda/\mu$:

$$E[S] = \rho e^{-\rho} \sum_{m=0}^{\infty} \frac{\rho^m}{m!} \sum_{i=1}^{m+1} \frac{1}{i} \tag{1}$$

Using Little's law, the average number of the concurrent members $E[M] = \rho = \lambda/\mu$, thus $E[S]$ can be viewed as a function of $E[M]$, as shown in Fig. 2(a). The top curve in the figure presents $E[S]$ as a function of $E[M]$ according to (1), which shows that $E[S]$ increases super-linearly with $E[M]$.

$E[S]$ also depends on the distribution of members' lifetime. If members' lifetime is deterministic (e.g., members are First-In-First-Out), $E[S]$ is identical to $E[M]$. Generally, the higher variance members' lifetime has, the larger $E[S]$ is.

A large value S results in an increased key storage at the GCKS side. Also, currently active members disperse in the key tree as S increases, incurring more resultant subsets and thus more rekeying messages. As a result, it is desirable to keep S small, a topic we address next.

3.4   Reducing S by Shifting

In this subsection, we propose a simple operation, namely *shifting*, to reduce S.

We define *shifting* as the operation of detaching the leftmost active member in the key tree and re-attaching the member to the next available position (for

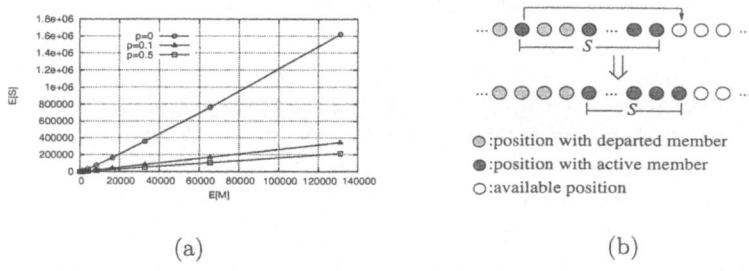

(a)   (b)

Fig. 2. Relationship between $E[S]$ and $E[M]$ (a); The shifting scheme (b)

new arrivals) in the key tree. When some holes (i.e., positions with departed members) are generated in the key tree, shifting the leftmost active member may reduce the concurrent spaces, S, and make active members more adjacent, as illustrated in Fig. 2(b).

When active members are shifted, they are delivered new secret information associated with the new positions by unicast. From the collusion-proof property of static SDR, we can show that shifting does not jeopardize the confidentiality [4]. There are many strategies for shifting. Here, we investigate two approaches, namely *probabilistic shifting* and *threshold-based shifting*.

Probabilistic shifting. Probabilistic shifting is defined as follows. When a member joins the group, that member is assigned to the next available position in the key tree. Meanwhile, with a probability p, the GCKS shifts the leftmost member to the position just to the right of the newly arrived member. Note that the shifting probability p affects the tradeoff between the shifting cost, associated with unicasting the secret information for the new position to the shifted member, and the value of S. In practice, the shifting probability p is a parameter set by the GCKS and can be application specific.

When members join the group according to a Poisson process with rate λ and a member's lifetime is exponentially distributed with mean $1/\mu$, we can compute the expected number of the concurrent spaces, $E(S)$, as a function of p, where $\rho = \lambda/\mu$:

$$E[S] = \rho e^{-\rho} \sum_{m=0}^{\infty} \frac{\rho^m}{m!} \left(\sum_{i=1}^{m+1} \frac{1}{i+p\rho} + p \sum_{i=1}^{m} \frac{1}{i+p\rho} \right) \qquad (2)$$

Fig. 2(a) presents the result of $E[S]$ for $p = 0, 0.1$ and 0.5 respectively. We observe that $E[S]$ decreases as a function of p. In particular, when the expected group size $E[M] = 10^5$, the space-member-ratio, $E[S]/E[M]$ is 12.1 for dynamic SDR without shifting, and reduces to 2.6 when $p = 0.1$, and further reduces to 1.6 when $p = 0.5$. Thus, by introducing a small shifting probability p, dynamic SDR can effectively reduce the average concurrent space S, therefore reducing the key storage cost at the GCKS and potentially the rekey cost as well.

We next consider a different kind of shifting strategy – threshold-based shifting.

Threshold-based shifting. We define the occupancy ratio γ as the number of active group members to the number of concurrent spaces, i.e., $\gamma = M/S$. Informally, the larger the occupancy ratio is, the more likely the members are adjacent to each other in the key tree, and thus can be covered by fewer subsets. To keep the rekey process efficient, the GCKS should keep the occupancy ratio high. A natural way to achieve this is to define a threshold $\Gamma < 1$; when a member leaves the group, the GCKS computes the occupancy ratio γ and compares it to the threshold Γ. If the occupancy ratio falls below the threshold (i.e., $\gamma < \Gamma$), the GCKS will keep shifting the leftmost member until $\gamma \geq \Gamma$. We refer to this strategy as threshold-based shifting.

As with the shifting probability p in the probabilistic shifting scheme, the threshold Γ is also an application-specific parameter affecting the tradeoff between the shifting cost (unicast) and the rekey cost (multicast), as discussed in Section 4.2.

3.5 Block Alignment

So far, we have treated the newly arrived members and the shifted members identically when assigning a member to an available position. However, a member that has been in the group for a long time and has become the leftmost member in the key tree may have very different characteristics in terms of the remaining service time,than that of a member who just joined the group.

Given the above considerations, we further propose an enhancement to the dynamic SDR with shifting scheme by allocating different blocks with size B in the key tree for new members and shifted members. Detailed procedure of block alignment is left in [4] for page limitation. B is an application-specific parameter similar to p and Γ. We also evaluate the effects of such block alignment through simulation in Section 4.2.

4   Evaluation

In this section, we evaluate the performance of the dynamic SDR through simulation. We will first describe the simulation model and the performance metrics, then present the results.

4.1 Simulation Model and Performance Metrics

We assume a fixed $N = 2^{17}$ to which a GCKS provides key management, i.e., the number of potential members is 2^{17}. Each member independently decides to join or leave the group. We approximate the members' arrival by a fixed-rate Poisson process and assume that the lifetimes are iid random variables.

We evaluate different lifetime distribution functions, which include exponential, lognormal and uniform distribution.

In the following, we will only present results for immediate-rekey scheme since the performance of SDR is insensitive to the rekey period [3].

The performance metrics of interest are the *key storage cost* (both at the member side and at the GCKS side) and *rekey communication cost*. The key storage cost is measured as the key storage described in the previous section. More specifically, we assume to use 3DES for encryption and each label has 128-bit of storage.

The rekey communication cost is measured as the number of unit-size rekey messages (assuming one message contains one 128-bit key) per unit time, which is further divided into *multicast cost* and *unicast cost*.

The multicast cost equals to the minimum number of subsets used to cover the active members in the key tree. Since the GCKS is performing immediate rekeying, and when the system is in steady state, the rate at which members depart the group should equal to the rate that members join the group, we can compute the overall multicast cost C_M as $C_M = 2\lambda \overline{N_{SD}}$, where λ is the arrival rate and $\overline{N_{SD}}$ is the average number of subsets that the GCKS uses for one TEK update.

The unicast cost includes the messages for delivering the secret information to a joining member or a shifted member. For static SDR scheme, the secret information is delivered to a user when that user joins the group for the first time. Since a member's position in the key tree is fixed, no additional unicast costs are incurred when the member returns to the group. To favor static SDR, we assume that the system has been running for long enough so that each member has received the secret information for its position. Thus, we count the unicast cost for static SDR as zero.

For dynamic SDR schemes, the unicast cost is computed as $C_U = (\lambda + v)N_K$, where v is the rate that members are shifted and N_K is the key storage at the member side and equals to $(\lceil \log E[M] \rceil^2 + \lceil \log E[M] \rceil)/2 + 1$ based on the analysis result in Section 3.

The overall rekey communication cost of dynamic SDR is the weighted sum of the multicast cost and the unicast cost. In the rest of the evaluation, we treat the cost of unicasting a message the same as that of multicasting, even though the unicast cost should be much lower than the multicast cost with respect to the number of links that the message travels. Some more discussion on the relative weight of the unicast cost and multicast cost is included in Section 4.2.

In summary, the overall rekey communication cost of static SDR is

$$C_s = 2\lambda \overline{N_{SD}^s} \qquad (3)$$

where $\overline{N_{SD}^s}$ is the average number of subsets using static SDR for one TEK update.

The overall rekey communication cost of dynamic SDR is

$$C_d = 2\lambda \overline{N_{SD}^d} + (\lambda + v)N_K \qquad (4)$$

where $\overline{N_{SD}^d}$ is the average number of subsets using dynamic SDR for one TEK update.

4.2   Simulation Results

Effects of group size. We first compare the performance of static SDR and dynamic SDR for different group sizes.

With a fixed total population ($N = 2^{17} = 131072$) and a fixed mean lifetime ($1/\mu = 100$), we use different value of the arrival rate (λ) to vary the expected group size ($E[M] = \lambda/\mu$) from 100 to 1.3×10^5. For now, we only consider the dynamic SDR scheme without shifting.

Fig. 3(a) compares the key storage of static SDR and dynamic SDR for different group size with exponentially-distributed lifetimes. The top two curves represent the key storage cost at GCKS and the bottom two curves represent the key storage cost at member side. Since static SDR maintains a fixed key tree whose size is determined by the total member population, its key storage costs, at both GCKS and member side, are invariant with different group size. For dynamic SDR, however, the key storage increases when the expected group size increases. Compared to static SDR, the member-side key storage cost of dynamic SDR is consistently lower, since M is always smaller than N. However the GCKS key storage cost of dynamic SDR begins to exceed that of static SDR as $E[M]$ increases. This is because the expected size of concurrent spaces $E[S]$, which includes both active members in the key tree and departed members in between, becomes larger than N when $E[M]$ is significant ($> 10\%$) compared to N.

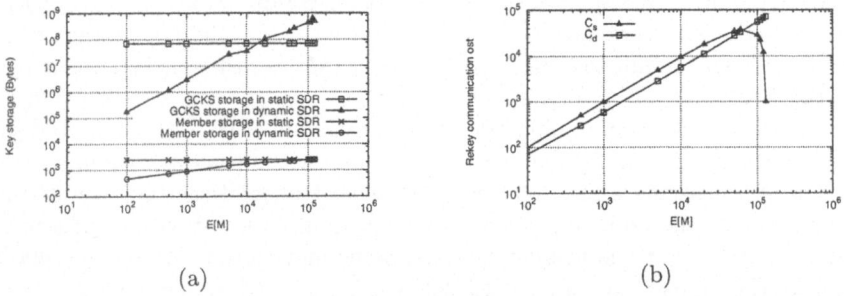

(a)   (b)

Fig. 3. Key storage (a) and rekey communication cost (b) of static SDR and dynamic SDR for different group size

We next compare the rekey communication cost of static SDR, C_s, and dynamic SDR, C_d, as described in (3) and (4) respectively (Fig. 3(b)). We observe that the rekey communication cost of static SDR increases as $E[M]$ increase, reaching a maximum when $E[M]$ is about $N/2$, and then starts to decrease when $E[M]$ gets close to N. This behavior matches well the reasoning in Section 2.

Furthermore, we observe that when $E[M] < 2 \times 10^4$, the rekey communication cost C_d is much lower, nearly half, compared to C_s. Only when $E[M]$ is greater than $N/2$, does C_s outperform C_d.

We have seen the benefit of dynamic SDR when $M \ll N$. In fact, many practical applications have this property. For example, the MBone STS-71 session has $M \approx 360$ while having $N \approx 4000$ [1]. Another example is in pay-per-view service: the number of people watching a movie at the same time, M, is usually orders of magnitude smaller than the total number of people having cable TV, N.

In the next, we will focus on the scenario where $M \ll N$ and evaluate the impact of shifting and block alignment.

Table 1. Parameters of the simulation configurations

Config. ID	Poisson arrival	Distr. of lifetime	Mean of lifetime	Var. of lifetime
EXP	$\lambda = 10$	exponential	100	10^4
UNI	$\lambda = 10$	uniform	100	3.3×10^3
LOG_l	$\lambda = 10$	lognormal	100	10^3
LOG_m	$\lambda = 10$	lognormal	100	10^4
LOG_h	$\lambda = 10$	lognormal	100	10^5

Impact of shifting. In this subsection, we study the performance of dynamic SDR with shifting. We consider the case where $E[M] = 1000(\ll N)$. Table 1 shows the five different configurations that we use to obtain the simulation results. We simulate probabilistic shifting and threshold-based shifting strategies for each configuration.

Figure 4(a) shows the communication cost, C_d, of dynamic SDR with probabilistic shifting for the five different configurations. We observe that, without shifting ($p = 0$), C_d, tends to be higher when members lifetime is of high variance.

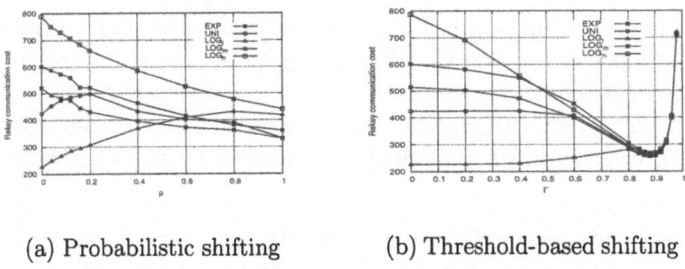

(a) Probabilistic shifting   (b) Threshold-based shifting

Fig. 4. Rekey communication cost of dynamic SDR combined shifting ($E[M] = 10^3$)

With shifting ($p > 0$), we find two different kinds of behaviors among the five configurations. For configurations EXP, LOG_m and LOG_h, which have high variance lifetime, increasing the shifting probability generally reduces C_d. This is because, with no shifting, S is large; while introducing shifting, although unicast cost is increased, S can be substantially reduced, which results in a reduced multicast cost.

For low variance configurations, UNI and LOG_l, when increasing the shifting rate, the communication cost C_d does not necessarily decrease. In these cases, S is close to M without shifting. Shifting cannot reduce S much. On the contrary, shifting might affect the distribution of members in the key tree, which may increase the number of subsets needed to cover the active members. As a result, the choice of optimal value of p depends on the lifetime distribution.

Fig. 4(b) shows the evaluation of threshold-based shifting scheme applied to the five configurations in Table 1. In the figure, it appears that, all configurations have a local minimum rekey communication cost when $\Gamma \approx 0.9$. When Γ goes beyond 0.9, there is a dramatically increase of the rekey cost. This is because, for a threshold $\Gamma > 0.9$, even though multicast cost may be reduced, unicast cost associated with frequent shifting becomes so high that the overall rekey cost is dramatically increased. Except for LOG_l, the local minimum rekey cost is also the global minimum. For configuration LOG_l, although the global minimum cost is achieved when $\Gamma = 0$, the local minimum rekey cost at $\Gamma \approx 0.9$ is very close to the global minimum. Thus choosing a proper value of the threshold parameter Γ is not as sensitive to members' lifetime distribution as in probabilistic shifting.

Enhancement with block alignment. We next evaluate the performance of block alignment as an enhancement to probabilistic shifting and threshold-based shifting for the five simulation configurations. Here we present the results of configuration EXP and LOG_h.

Fig. 5 plots the rekey communication cost when applying block alignment with $B = 0, 2, 16$ and 32 for probabilistic shifting (5(a) and 5(b)) and threshold-based shifting (5(c) and 5(d)). We observe that, for configuration EXP, introducing block alignment (with various B) does not have much improvement on reducing rekey communication cost (sometimes is even worse). This is due to the memoryless property of exponential distribution However, for configuration LOG_h, the improvement of block alignment is significant. Furthermore, for this particular group size, we find that increasing block size B beyond 16 does not provide much additional improvement.

From the figure, one can also see that, compared to threshold-based shifting, improvement of using block alignment is more evident in probabilistic shifting.

Tradeoff between unicast cost and multicast cost. As described in Section 3, dynamic SDR reduces multicast costs by introducing additional unicast, by which the secret information is delivered to shifted or returned members. Since the different strategies proposed in this paper have different parameters (p or Γ) to configure, each of which reflects the tradeoff between unicast cost and multicast cost. To compare different schemes, we study the tradeoff graph of these proposed schemes as shown in Fig. 5(e). In the tradeoff graph, a point

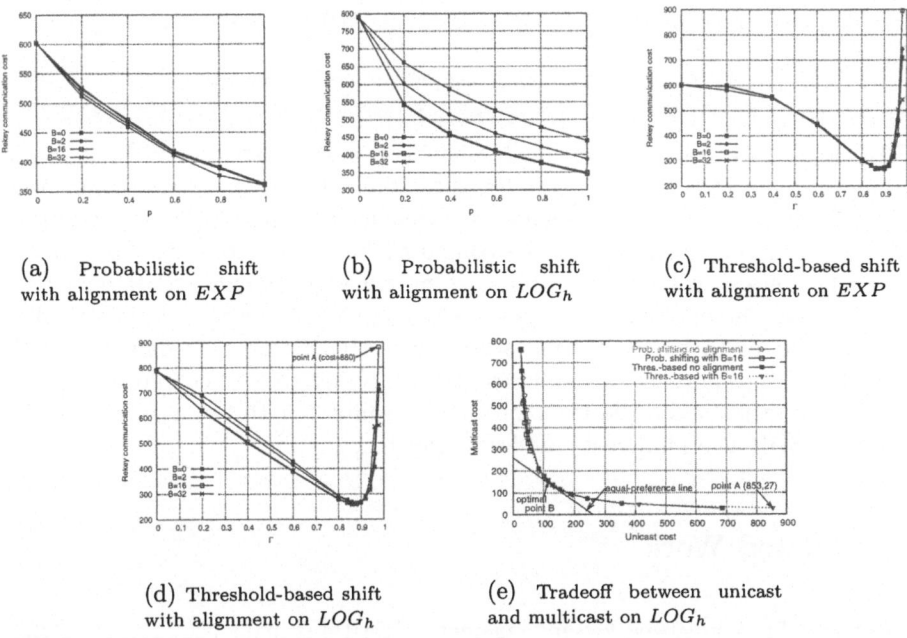

(a) Probabilistic shift with alignment on EXP

(b) Probabilistic shift with alignment on LOG_h

(c) Threshold-based shift with alignment on EXP

(d) Threshold-based shift with alignment on LOG_h

(e) Tradeoff between unicast and multicast on LOG_h

Fig. 5. Costs and tradeoff of dynamic SDR using shifting and alignment ($E[M] = 10^3$)

on a curve denotes the multicast and unicast cost for the corresponding strategy with a particular parameter. For example, point A in Fig. 5(e) is associated with unicast cost of 853 and multicast cost of 27, denoting the total cost of 880 for the threshold-based shifting combined alignment of $B=16$ with parameter $\Gamma = 0.98$ (point A in Fig. 5(d)).

From the figure, we observe that reducing multicast cost comes at a cost of increasing unicast cost, and vice versa. The relative weight of unicast cost and multicast cost affects the choice of the optimal schemes and the operating parameters. If we treat unicast cost as expensive as multicast cost, in the tradeoff graph, all points on a line with a slope -1 are equally preferable. While points on a line close to point (0,0) are preferred over points on lines far away. In this sense, the threshold-based shifting combined alignment with block size $B=16$ offers the best tradeoff among the algorithms considered, achieving an optimal value with multicast cost of 146 and unicast cost of 114 (point B in Fig. 5(e)). In general, if the relative weight of unicast cost and multicast cost is w, the equal-preference lines will have slope $-w$ in the tradeoff graph. In this case, the best approach and the optimal parameters may be different.

5 Discussion

Security. In this paper, we use three criteria to measure security: forward confidentiality, backward confidentiality and collusion problem. As we know, static

SDR maintains forward/backward confidentiality, and has no collusion problem. We find that those properties hold for dynamic SDR. Scratch proof is given out in [4].

Multicast Stateless. Among all group key management algorithms, static SDR belongs to the so-called stateless algorithms in that members do not need to keep track of history of rekeying. In dynamic SDR, KEKs (keys of resultant subsets) are long-lived and can be computed based on the secret information securely unicast to members. Multicast messages in dynamic SDR distribute only TEKs. The multicast messages are not required to be reliably delivered to members so as to maintain their states correctly. However, in dynamic SDR, the GCKS is required to reliably unicast a member the secret information for the new position when the member joins or shifts. For this reason, we classify dynamic SDR as a multicast stateless algorithm.

6   Related Work

Most scalable centralized key-management algorithms make use of a tree structure to manage members. These algorithms could be broadly divided into *stateful* algorithms and *stateless* algorithms depending on whether members need to track the communication history to participate in the group communication. In stateful algorithms ([2,6,10,11]), the active members are leaves of the tree. While in stateless algorithms ([8]), the potential members are leaves of the tree.

LKH is a stateful algorithm. In an LKH tree, there is a leaf node corresponding to each active member. There is a key associated with each node in the tree, and each member holds a copy of every key on the path from its corresponding leaf node to the root of the tree. Hence, the key corresponding to the root node is shared by all members, and serves as the TEK.

Static SDR is a stateless algorithm. In a static SDR tree, there is a leaf node corresponding to each potential member. Subsets are defined through the tree, and each member holds subset keys for all subsets to which it belongs.

The performance of key-management algorithms is mostly determined by the positions of concurrent members in the tree. [5,12] propose methods to improve the performance of LKH by adjusting the positions of members dynamically so as to balance the key tree and reduce the overall height. Our work aims to improve the performance of SDR using the similar methodology – dynamically adjusting the positions of members.

A performance comparison between static SDR and LKH is given in [3]. Both the key storage and the rekey communication cost are compared in different scenarios, e.g. immediate rekeying, periodical batch rekeying and membership batch rekeying.

7   Conclusion

Static Subset Difference Revocation (SDR) is the current state of the art in stateless group rekeying algorithms. However, it works inefficiently when the number of the active members in the group is much less than the number of potential members, which is the case in many practical applications.

In this paper, we have proposed a group rekeying algorithm, dynamic SDR, which still keeps multicast stateless without the requirement of estimating the number of all potential members. By dynamically constructing the key tree, dynamic SDR uses a smaller key tree sufficiently large for the currently active members rather than the potential members. The smaller key tree reduces both the key storage cost and rekey communication cost compared to static SDR. We also introduce some enhancements to further improve the performance of dynamic SDR. Our evaluation shows that dynamic SDR significantly improves the performance of static SDR, reducing by half the rekey communication cost in the case that the number of the currently active members is much less than the total number of potential members. Also, compared to static SDR, dynamic SDR does not need to know the maximum number of potential group members in advance, a value that can be difficult to estimate in practice.

Acknowledgments. This research has been supported in part by the NSF under grant awards UF-EIES-0205003-UMA and EIA-0080119. Any opinions, findings, and conclusions or recommendations expressed in this material are those of the author(s) and do not necessarily reflect the views of the National Science Foundation. The authors would also like to thank the anonymous reviewers for their helpful comments.

References

1. Almeroth, K.C., Ammar, M.H.: Collecting and Modeling the Join/Leave Behavior of Multicast Group Members in the MBone. In: Proc. of the Symposium on High Performance Distributed Computing. Syracuse, NY (1996) 209–216
2. Chang, I., Engel, R., Kandlur, D., Pendarakis, D., Saha, D.: Key Management for Secure Internet Multicast using Boolean Function Minimization Techniques. In: Proc. IEEE INFOCOM 1999, Vol. 2. New York, NY (1999) 689–698
3. Chen, W., Dondeti, L.R.: Performance Comparison of Stateful and Stateless Group Rekeying Algorithms. In: Proc. Fourth International Workshop on Networked Group Communication. Boston, MA (2002)
4. Chen, W., Ge, Z., Zhang, C., Kurose, J., Towsley, D.: On Dynamic Subset Difference Revocation Scheme. Technical Report UM-CS-2003-22, University of Massachusetts at Amherst. (2003)
5. Dondeti, L.R., Mukherjee, S., Samal, A.: DISEC: A Distributed Framework for Scalable Secure Many-to-Many Communication. In: Proc. Fifth IEEE Symposium on Computers and Communications, Antibes-Juan les Pins, France (2000)
6. McGrew, D.A., Sherman, A.T.: Key Establishment in Large Dynamic Groups Using One-Way Function Trees. Technical Report No. 0755, TIS Labs at Network Associates, Inc., Glenwood, MD (1998)

7. Mittra, S.: Iolus: A Framework for Scalable Secure Multicasting. In: Proc. ACM SIGCOMM 1997, Cannes, France (1997) 277–288
8. Naor, D., Naor, M., Lotspiech, J.: Revocation and Tracing Schemes for Stateless Receivers. In: Proc. CRYPTO 2001, Lecture Notes in Computer Scienc, Vol. 2139, (2001) 41–62
9. Setia, S., Koussiah, S., Jajodia, S., Harder, E.: Kronos: A Scalable Rekeying Approach for Secure Multicast. In: Proc. IEEE Symposium on Security and Privacy, Berkeley, CA (2000)
10. Wallner, D., Harder, E., Agee, R.: Key Management for Multicast: Issues and Architectures. IETF Informational RFC, (1999)
11. Wong, C.K., Gouda, M.G., Lam, S.S.: Secure Group Communications Using Key Graphs. In: Proc. ACM SIGCOMM 1998, Vancouver, Canada (1998) 68–79
12. Yang, Y., Li, X., Zhang, X., Lam, S.: Reliable Group Rekeying: A Performance Analysis. In: Proc. ACM SIGCOMM 2001, San Diego, CA (2001) 27–38

Nikolas Mitrou   Kimon Kontovasilis
George N. Rouskas   Ilias Iliadis
Lazaros Merakos (Eds.)

NETWORKING 2004

Networking Technologies,
Services, and Protocols;
Performance of Computer
and Communication Networks;
Mobile and Wireless Communications

Third International IFIP-TC6 Networking Conference
Athens, Greece, May 9-14, 2004
Proceedings

Springer

Volume Editors

Nikolas Mitrou
National Technical University of Athens
School of Electrical and Computer Engineering, Computer Network Laboratory
9 Heroon Polytechneiou str., 157 80, Athens, Greece
E-mail: mitrou@softlab.ntua.gr

Kimon Kontovasilis
National Center for Scientific Research "Demokritos"
Institute of Informatics and Telecommunications
P.O. Box 60228, 15310 Agia Paraskevi Attikis, Greece
E-mail: kkont@iit.demokritos.gr

George N. Rouskas
North Carolina State University, Department of Computer Science
Box 7534, 461 EGRC, 2410 Campus Shore Drive, Raleigh, NC 27695-7534, USA
E-mail: rouskas@csc.ncsu.edu

Ilias Iliadis
IBM Research, Zurich Research Laboratory
Säumerstrasse 4, 8803 Rüschlikon, Switzerland
E-mail:ili@zurich.ibm.com

Lazaros Merakos
University of Athens, Department of Informatics and Telecommunications
Panepistimiopolis, 15784 Athens, Greece
E-mail: merakos@di.uoa.gr

Library of Congress Control Number: 2004104505

CR Subject Classification (1998): C.2, C.4, H.4, D.2, J.2, J.1, K.6, K.4

ISSN 0302-9743
ISBN 978-3-540-21959-0   ISBN 978-3-540-24693-0 (eBook)
DOI 10.1007/978-3-540-24693-0

springeronline.com

©2004 Springer-Verlag Berlin Heidelberg

Originally published by IFIP International Federation for Information Processing, Hofstraße 3, 2361
Laxenburg, Austria in 2004.

Typesetting: Camera-ready by author, data conversion by PTP-Berlin, Protago-TeX-Production GmbH
Printed on acid-free paper   SPIN: 11007159   06/3142   5 4 3 2 1 0

Preface

This book contains the refereed proceedings of the 3rd International IFIP-TC6 Networking Conference, *Networking 2004*. Conferences in the *Networking* series span the interests of several distinct, but related, TC6 working groups, including Working Groups 6.2, 6.3, and 6.8. Reflecting this, the conference was structured with three Special Tracks: (i) Networking Technologies, Services, and Protocols; (ii) Performance of Computer and Communication Networks; and (iii) Mobile and Wireless Communications.

However, beyond providing a forum for the presentation of high-quality research in various complementary aspects of networking, the conference was also targeted to contributing to a unified view of the field and to fostering the interaction and exchange of fruitful ideas between the various related (and overlapping) specialized subcommunities therein. Towards this second objective, more than a few conference sessions (and thematic sections in this book) 'cut across' the Special Tracks, along more generic or fundamental concepts.

Networking 2004 was fortunate to attract very high interest among the community, and the conference received 539 submissions from 44 countries in all five continents. These figures correspond to a remarkable increase in submissions from the previous very successful events (roughly, a 156% increase over *Networking 2000* and 71% over *Networking 2002*), and indicate that *Networking* conferences are progressively becoming established as worldwide reference events in the field.

The high number of submissions provided an excellent opportunity for a high-quality program but also called for a very demanding and laborious paper evaluation process. The 99 members of the Technical Program Committee, together with more than 750 additional experts engaged as external reviewers, worked efficiently and responsibly under tight time constraints to produce a total of some 1600 reviews, towards responding to the authors with insightful, detailed feedback and providing the basis for the final paper selection.

As a result of this process, 103 papers were finally selected from the 539 submissions (at an acceptance rate below 20%) for presentation during the conference's technical sessions. Additionally, to allow the conference participants to benefit from further worthwhile and stimulating research results, 40 papers were accepted for presentation in a poster session; these papers appear in the proceedings in short form. Overall, the final program was excellent and diverse, and covered both fundamental issues in networking as well as emerging topics. It should be noted that, given the many submissions and the tight constraints, many good papers could not be accommodated.

The main program of *Networking 2004* covered three days and included streams of up to three parallel sessions, plus a poster session for the presentation of the short papers. In addition to these components, the program was enriched by three stimulating invited talks offered by world-renowned researchers in the field, specifically: Jim Roberts on "From ATM to IP and back again: the label switched path to the converged Internet, or another blind alley?", Leandros Tassiulas on "Cross-layer design issues for quality of service provisioning in wireless networks", and Ian Akyildiz on "Research challenges in wireless sensor and actuator networks". The main conference program was complemented by a diverse set of high-quality tutorials and workshops.

The final result wouldn't have been possible without the dedication and hard work of many colleagues. Special thanks are due to the members of the Technical Program Committee and to all external referees for their invaluable help with the paper reviews and their overall sense of responsibility and responsiveness under very tight deadlines in a particularly busy period. Another word of thanks goes to the Webmaster Manolis Solidakis and to the General Chair Nikolas Mitrou, who worked hard to ensure that flexible and effective technical support was available during the paper evaluation and the final paper selection. Finally, we are grateful to all authors who honored *Networking 2004* by trusting their work to the conference.

March 2004   Kimon Kontovasilis
                                                        George Rouskas
                                                          Ilias Iliadis
                                                      Lazaros Merakos

Message from the General Chair

No doubt the world economy is being globalized at a rapid pace, while, at the same time, it is becoming more and more information intensive. The joint evolution and convergence of these fundamental trends towards the global, knowledge-based society leans upon a bold scientific and technological objective: pervasive information processing and ubiquitous networking. Clearly, technological advances in all aspects of networking are central to this objective.

In such a context, *Networking 2004* strove for a distinct and valuable contribution to the field. Organized jointly by the Institute of Communication and Computer Systems of the National Technical University of Athens and the Institute of Informatics and Telecommunications of the NCSR (National Centre for Scientific Research) "Demokritos", it was the third event in a series of IFIP-TC6 networking conferences, launched in 2000 and organized every 2 years thereafter. In common with its predecessors, *Networking 2004* aimed at reviewing progress, presenting recent advances, identifying trends and refreshing vision in this hot field, by bringing together active and proficient members of the networking community from both academia and industry. It covered all aspects of networking, focusing especially on three main areas (Special Tracks): (a) networking technologies, services, and protocols, (b) performance of computer and communication networks, and (c) mobile and wireless communications. The conference lasted six days and included the main program (three days), tutorials (one and a half days) and thematic workshops attached to the main event (one day).

The organization of such an ambitious event required a considerable effort and commitment by very many people. I therefore wish to express my deep appreciation to all of the Executive Committee members for their outstanding work.

The heart of the conference was, of course, its main program. The dramatic increase in the number of submitted papers (more than 70%, compared to the previous event in 2002) clearly indicates the increasing interest of the community in our IFIP-TC6 networking conferences. Diligence in the reviewing and selection process was necessary, in order to thoroughly handle such a large number of submissions and, hopefully, to hand over to the next organization a conference standard even higher than that inherited. A special appreciation is due to the Technical Program Chair, Kimon Kontovasilis, for his deep devotion to this extremely delicate and demanding task. Sincere thanks are addressed also to the three Special Track Chairs, George Rouskas, Ilias Iliadis and Lazaros Merakos, who worked diligently towards a high-level conference program. The TPC members, as well as more than 750 additional external experts/referees, are also warmly thanked for their excellent work.

Within the main program, keynote speeches are "tracer bullets" in the unexploited sky of new trends and ideas, or critical retrospections on fundamental aspects of networking. Here I have to thank the Invited Speaker Co-chairs Leonidas Georgiadis and Michael Paterakis and, of course, our distinguished invited speakers.

The tutorial program enriched the main conference program with thorough half-day presentations, covering theoretical and practical aspects of recent advances in networking. The tutorials were mainly targeted to young researchers and practitioners in the field who wish to foster their skills and background knowledge. Many thanks to the Tutorial Co-chairs Christos Douligeris, Gunnar Karlsson and Dimitris Pendarakis for organizing an interesting set of tutorials.

The last day of the conference was devoted to thematic workshops focusing on areas of special interest or of remarkable evolution dynamics. Four such workshops were scheduled: *Next Generation Networking Middleware* (*NGNM04*, organized by Nikos Anerousis and George Kormentzas), *Data Processing and Storage Networking: Towards Grid Computing* (*DPSN04*, organized by Fotini-Niovi Pavlidou, John Soldatos and Evangelos Vayias*), *Streaming Media Distribution over the Internet* (*SMDI04*, organized by Charalampos Patrikakis, Marco Combetto and Constantinos Boukouvalas) and *Wireless Internet and Reconfigurability* (*4th ANWIRE*, organized by Nikos Passas). The workshop chairs did an excellent job to achieve quality workshops, commensurate with the high level of the rest of the conference.

Our Web engineer, Manolis Solidakis, did his best to provide user-friendly on-line facilities during all preparatory phases: paper submission and review, registration, program dissemination. I thank him for his commitment to that.

Jointly with the Technical Program Chair, we would like to address our final word of thanks to the administration of the organizing institutes within NTUA and NCSR "Demokritos" for their support, as well as to the IFIP-TC6 management, who entrusted us with the conduct of this important event.

Not too long ago, we received the staggering news that Olga Casals, our good friend and colleague, an active researcher and TPC member of previous *Networking* conferences (intended as a TPC member of this conference too) suddenly passed away. The Best Paper Award of *Networking 2004*, presented to its recipient on the last day of the conference, was devoted to her memory.

March 2004   Nikolas Mitrou

Organization

Conference Executive Committee

General Chair
Nikolas Mitrou, National Technical University of Athens, Greece

Technical Program Chair
Kimon Kontovasilis, National Centre for Scientific Research "Demokritos", Greece

Special Track Chair for Networking Technologies, Services, and Protocols
George Rouskas, North Carolina State University, USA

Special Track Chair for the Performance of Computer and Communication Networks
Ilias Iliadis, IBM Research, Zurich Laboratory, Switzerland

Special Track Chair for Mobile and Wireless Communications
Lazaros Merakos, University of Athens, Greece

Tutorial Program Co-chairs
Christos Douligeris, University of Piraeus, Greece
Gunnar Karlsson, KTH, Sweden
Dimitris Pendarakis, IBM, USA

Invited Speakers Co-chairs
Leonidas Georgiadis, Aristotle University of Thessaloniki, Greece
Michael Paterakis, Technical University of Crete, Greece

Workshop Chairs

Next Generation Networking Middleware (NGNM04)
Nikos Anerousis, IBM Research, USA
George Kormentzas, University of the Aegean, Greece

Data Processing and Storage Networking: Towards Grid Computing (DPSN04)
Fotini-Niovi Pavlidou, Aristotle University of Thessaloniki, Greece
John Soldatos, Athens Information Technology, Greece
Evangelos Vayias, Intracom SA, Greece

Streaming Media Distribution over the Internet (SMDI04)
Charalampos Patrikakis, National Technical University of Athens, Greece
Marco Combetto, Microsoft Research, University Relations, Cambridge, UK
Constantinos Boukouvalas, OTE, S.A., Greece

Wireless Internet and Reconfigurability (4th ANWIRE)
Nikos Passas, University of Athens, Greece

Publicity Chair
Demetres Kouvatsos, University of Bradford, UK

Steering Committee Chair
Harry Perros, North Carolina State University, USA

Steering Committee Members
Augusto Casaca, IST/INESC, Portugal
Erol Gelenbe, University of Central Florida, USA
Harry Perros, North Carolina State University, USA
Guy Pujolle, University of Paris 6, France
Harry Rudin, Switzerland
George Stassinopoulos, National Technical University of Athens, Greece

Finance Chair
Athanasios Drigas, National Centre for Scientific Research "Demokritos",
Greece

Electronic Submission/Web Designer
Emmanuel Solidakis, National Technical University of Athens, Greece

Local Organizing Committee
Haralampos Skianis, National Centre for Scientific Research "Demokritos",
Greece
George Kormentzas, University of the Aegean, Greece

Technical Program Committee

Special Track for Networking Technologies, Services and Protocols

Kevin Almeroth, University of California, Santa Barbara, USA
Andrea Bianco, Politecnico di Torino, Italy
Ken Calvert, University of Kentucky, USA
Claudio Casetti, Politecnico di Torino, Italy
Laurie Cuthbert, Queen Mary and Westfield College, UK
Jordi Domingo-Pascual, Universitat Politecnica de Catalunya, Spain

Constantinos Dovrolis, Georgia Institute of Technology, USA
Eylem Ekici, Ohio State University, USA
Mihalis Faloutsos, University of California, Riverside, USA
Serge Fdida, Université Pierre et Marie Curie, France
Luigi Fratta, Politecnico di Milano, Italy
Maurice Gagnaire, École Nationale Supérieure des Télécommunications, France
Leonidas Georgiadis, Aristotle University of Thessaloniki, Greece
David Hutchison, Lancaster University, UK
Admela Jukan, National Science Foundation, USA
Mohan Kumar, University of Texas at Arlington, USA
G. Manimaran, Iowa State University, USA
Ibrahim Matta, Boston University, USA
Prasant Mohapatra, University of California, Davis, USA
Ioanis Nikolaidis, University of Alberta, USA
Peng Ning, North Carolina State University, USA
Jaudelice C. de Oliveira, Drexel University, USA
Symeon Papavasileiou, New Jersey Institute of Technology, USA
Michael Paterakis, Technical University of Crete, Greece
Guido Petit, Alcatel, Belgium
Chiara Petrioli, University "La Sapienza", Rome, Italy
Nineta Polemi, University of Piraeus, Greece
Martin Potts, Martel GmbH, Switzerland
Erwin Rathgeb, Universität Duisburg-Essen, Germany
Luigi Rizzo, University of Pisa, Italy
Hussein Salama, Cisco Systems International, Egypt
Saswati Sarkar, University of Pennsylvania, USA
Dimitris Serpanos, University of Patras, Greece
Mihai Sichitiu, North Carolina State University, USA
Efstathios Sykas, National Technical University of Athens, Greece
Andras Valko, Ericsson, Sweden
Iakovos Venieris, National Technical University of Athens, Greece
Lars Wolf, Braunschweig University of Technology, Germany
Zhi-Li Zhang, University of Minnesota, USA

Special Track for the Performance of Computer and Communication Networks

Ron Addie, University of Southern Queensland, Australia
Eitan Altman, INRIA, France
Miltiades Anagnostou, National Technical University of Athens, Greece
Andrea Baiocchi, University "La Sapienza", Rome, Italy
Chris Blondia, University of Antwerp, Belgium
Pierre Boyer, France Telecom R&D, France
Tosten Braun, University of Bern, Switzerland
Herwig Bruneel, University of Ghent, Belgium
Wojciech Burakowski, Warsaw University of Technology, Poland

Werner Bux, IBM Research, Zurich Laboratory, Switzerland
Mariacarla Calzarossa, University of Pavia, Italy
Constantinos Courcoubetis, Athens University of Economics and Business, Greece
Khaled Elsayed, Cairo University, Egypt
Guenter Haring, University of Vienna, Austria
Peter Harrison, Imperial College London, UK
Krishna Kant, Intel Corporation, USA
Peter Key, Microsoft Research Ltd., Cambridge, UK
Demetres Kouvatsos, University of Bradford, UK
Emilio Leonardi, Politecnico di Torino, Italy
Michela Meo, Politecnico di Torino, Italy
Edmundo Monteiro, University of Coimbra, Portugal
Ilkka Norros, VTT, Finland
Andreas Pitsillides, University of Cyprus, Cyprus
Ramon Puigjaner, Universitat de les Illes Balears, Spain
Otto Spaniol, Aachen University of Technology, Germany
George Stamoulis, Athens University of Economics and Business, Greece
Yutaka Takahashi, Kyoto University, Japan
Don Towsley, University of Massachusetts, USA
Jorma Virtamo, Helsinki University of Technology, Finland

Special Track for Mobile and Wireless Communications

Ian Akyldiz, Georgia Institute of Technology, USA
Nancy Alonistioti, University of Athens, Greece
Roberto Battiti, University of Trento, Italy
Luciano Bononi, University of Bologna, Italy
Azzedine Boukerche, University of Ottawa, Canada
Marco Conti, National Research Council, Italy
Luis Correia, Technical University of Lisbon, Portugal
Panagiotis Demestichas, University of Piraeus, Greece
Anthony Ephremides, University of Maryland, USA
Laura Feeney, SICS, Sweden
Gabor Fodor, Ericsson, Sweden
Jerome Galtier, France Telecom R&D and INRIA, France
Silvia Giordano, ICA-DSC-SUPSI, Switzerland
Enrico Gregori, National Research Council, Italy
Gerald Maguire, Royal Institute of Technology, Sweden
Takis Mathiopoulos, National Observatory of Athens, Greece
Stephan Olariu, Old Dominion University, USA
Sergio Palazzo, University of Catania, Italy
Nikos Passas, University of Athens, Greece
Jorge Pereira, European Commission, Belgium
George Polyzos, Athens University of Economics and Business, Greece
Guy Pujolle, Université Pierre et Marie Curie (Paris 6), France

Kimmo Raatikainen, University of Helsinki, Finland
Ivan Stojmenovic, University of Ottawa, Canada
Violet Syrotiuk, University of Arizona, USA
Michael Theologou, National Technical University of Athens, Greece
Bernhard Walke, Aachen University of Technology, Germany

Referees

Finn Arve Aagesen	Enzo Baccarelli	Chris Blondia
Samuli Aalto	Andrea Baiocchi	Fernando Boavida
Alhussein Abouzeid	Mario Baldi	Gregor von Bochmann
John Adams	George Balis	Eliane Bodanese
Ron Addie	Frank Ball	Rene Boel
Anjali Agarwal	Roland Balmer	Raffaele Bolla
Sharad Agarwal	Simonetta Balsamo	Alberto Bononi
Rui Aguiar	Albert Banchs	Luciano Bononi
Ozgur Baris Akan	Chadi Barakat	Eleonora Borgia
Nail Akar	Mario Barbera	Oliver Bott
Basheer Al-Duwairi	Antonio Barili	Abdelmadjid Bouabdallah
Tarek Al-Gizawi	Dhiman Barman	Azzedine Boukerche
Timo Alanko	Stefano Baroni	Christos Bouras
Guido Albertengo	Novella Bartolini	Raouf Boutaba
Kevin Almeroth	Stephen Barton	Onno Boxma
Nancy Alonistioti	Stefano Basagni	Jacqueline Boyer
Nicola Altan	Anup Basu	Pierre Boyer
Eitan Altman	Roberto Battiti	Adam Bradley
Paolo Ameigeiras	Daniel Bauer	Jeremy Bradley
Marcelo Dias de Amorim	Florian Baumgartner	Michael Bradshaw
Miltiadis Anagnostou	Buyurman Baykal	Arnold Bragg
Giuseppe Anastasi	John Bellardo	Christof Brandauer
Stergios Anastasiadis	Paolo Bellavista	Torsten Braun
Matthew Andrews	Mats Bengtsson	Tonnes Brekne
Vladimir Anisimov	Hans van der Berg	Erik van Breusegem
Tricha Anjali	Marco Bernardo	Mauro Brunato
Panayotis Antoniadis	Christian Bettstetter	Herwig Bruneel
George Apostolopoulos	Bharat Bhargava	Raffaele Bruno
John Apostolopoulos	Bobby Bhattacharjee	Roland Bueschkes
Martin Arlitt	Giuseppe Bianchi	Wojciech Burakowski
Jesus Artalejo	Andrea Bianco	Werner Bux
Gerald Ash	Hannes Birck	Juan Caballero
Sasikanth Avancha	Jozsef Biro	Tom van Caenegem
Husam Awadalla	Erez Biton	Gruia Calinescu
Irfan Awan	Danilo Blasi	Franco Callegati
Dimitrios Axiotis	Nicola Blefari-Melazzi	Rita de Caluwe
Fulvio Babich	Holger Bleul	Ken Calvert

Mariacarla Calzarossa
Pietro Camarda
Srjdan Capkun
Narcis Cardona
Yannick Carlinet
Loren Carrasco
Iacopo Carreras
Graca Carvalho
Emiliano Casalicchio
Vicente Casares
Ramon Casellas
Claudio Casetti
Marco Casole
Luca Casone
Erdal Cayirci
Nedo Celandroni
Isabella Cerutti
Ram Chakka
Anirban Chakrabarti
Rajiv Chakravorty
Shueng-Han Gary Chan
Soamsiri Chantaraskul
Hakima Chaouchi
Martin Chapman
Prasanna Chaporkar
Joachim Charzinski
Bensong Chen
Chao Chen
Jiangzhuo Chen
Peter Chen
Trista Chen
Yinong Chen
Yuchung Cheng
Jan Cheyns
Sundaram Chinthamani
Sunghyun Choi
Ishtiaq Ahmed Choudhry
Charilaos Christopoulos
Chrysost. Chrysostomou
Thomas Clausen
Jorge Cobb
Denis Collange
Jeff Considine
Marco Conti
Chris Cooper

Antonio Corradi
Olivier Corre
Luis Correia
Americo Correia
Jose Costa-Requena
Constantinos Courcoubetis
Piergiorgio Cremonese
Paolo Cremonesi
Mark Crovella
Juan Carlos Cruellas Ibarz
Pedro Cuenca
Jun-Hong Cui
Minghao Cui
Marilia Curado
Laurie Cuthbert
Marc Danzeisen
Vasilios Darlagiannis
Dai Davies
Panagiotis Demestichas
Dee Denetree
Wolfgang Denzel
Evaggelos Dermatas
Andrea Detti
Peter Deutsch
Ashay Dhamdhere
Eric Diehl
Antonis Dimakis
George Dimitrakopoulos
Nick Dingle
Jana Dittmann
Tien V. Do
Jordi Domingo-Pascual
Jeff Donahoo
Yingfei Dong
Yixin Dong
Yesid Donoso
Satish Doraiswamy
Christos Douligeris
Elias Doumith
Constantinos Dovrolis
Manos Dramitinos
Falko Dressler
Wenliang Du
Zhenhai Duan
Parijat Dube

Frederich Ducatelle
Michael Dueser
Tolga Duman
Adam Dunkels
Sandra Dykes
Elias Efstathiou
Gerald Eichler
Eylem Ekici
Karim El-Khazen
Abdulmotaleb El Saddik
Khaled Elsayed
Anthony Ephremides
Alberto Escudero-Pascual
Do Young Eun
Tony Eyers
Ramon Fabregat
Sonia Fahmy
Mihalis Faloutsos
Andras Farago
Julien Fasson
Serge Fdida
Laura Feeney
Zongming Fei
Guillem Femenias
Markus Fiedler
Ioannis Fikouras
Joe Finney
Jorge Finochietto
Eric Fleury
Gabor Fodor
Olivier Fourmaux
Jean-Michel Fourneau
Roberta Fracchia
Chuck Fraleigh
Mirko Franceschinis
Gerald Franzl
Vassilis Freiderikos
Mario Freire
Timur Friedman
Norihito Fujita
Andrea Fumagalli
Maurice Gagnaire
Dominique Gaiti
Giulio Galante
Clemente Galdi

Laura Galluccio
Sebastia Galmes
Jerome Galtier
Atilio Gameiro
Ayalvadi Ganesh
Wilfried Gansterer
Christos Gantsidis
Javier Garcia
Anargyros Garyfalos
Michael Gastpar
Cristoph Gauger
Damianos Gavalas
Evangelos Gazis
Jim Gemmell
Leonidas Georgiadis
Yacine Ghamri
Nasir Ghani
Vittorio Ghini
Andrea Ghittino
Paolo Giaccone
Richard Gibbens
Silvia Giordano
Giodi Giorgi
Tolga Girici
Roch Glitho
Sylvain Gombault
Antonio Gomez Skarmeta
Jose-L. Gonzalez-Sanchez
Pawan Goyal
Enrico Gregori
Costas Grekas
Luigi Alfredo Grieco
Dimitris Gritzalis
Stefanos Gritzalis
Yu Gu
Isabelle Guerin Lassous
Ratul Guha
Chao Gui
Fabrice Guillemin
Mina Guirguis
Liang Guo
Minaxi Gupta
Prabhanjan Gurumohan
Mohan Gurusamy
Martin Gutbrod

Vincent Guyot
Stathis Hadjiefthymiades
George Hadjipollas
Hans Hall
Youngnam Han
Uli Harder
Khaled Harfoush
Guenter Haring
Fotios Harmantzis
Janelle Harms
Peter Harrison
Hannes Hartenstein
Henry Haverinen
Sugata Hazarika
Oliver Heckmann
Marc Heissenboettel
Jane Hillston
Helmut Hlavacs
Lester Ho
Petra Hoepner
Tom Hofkens
Y. Charlie Hu
Benny van Houdt
David Hutchison
Ren-Hung Hwang
Antonio Iera
Ilias Iliadis
Sandor Imre
Paola Iovanna
Tim Irnich
Cyril Iskander
Zabeer Jainullabudeen
Clark Jeffries
Jorjeta Jetcheva
Hao Jiang
Wenyu Jiang
Shudong Jin
Jorma Jormakka
Anupam Joshi
Jÿrki Joutsensalo
Jason Jue
Carlos Juiz
Admela Jukan
Alexandros Kaloxylos
Kostas Kalpakis

Ahmed Kamal
Thanassis Kanatas
S. Kannan
Krishna Kant
Rohit Kapoor
George Karagiannidis
Pradnya Karbhari
Holger Karl
Ahmed Karmouch
Jouni Karvo
Shoji Kasahara
Sneha Kumar Kasera
Sokratis Katsikas
Jonathan Katz
Andre Kaup
Kenichi Kawanishi
Takehiro Kawata
Jean-Marc Kelif
Tom Kelly
Lukas Kencl
Stephen Kent
Anne-Marie Kermarrec
George Kesidis
Csaba Keszei
Peter Key
Ibrahim Khalil
Aram Khalili
Kalevi Kilkki
Jorma Kilpi
Taehyun Kim
Yongdae Kim
Andreas Kind
Csaba Kiraly
Leila Kloul
Svein Knapskog
Geir Koien
Markku Kojo
N. Kokkalis
Fatih Koksal
Ramanarao Kompella
Kimon Kontovasilis
Thanasis Korakis
George Kormentzas
Mohamed Koubaa
Manolis Koubarakis

Dimitris Kouis
Christos Koukourlis
Giannis Koukoutsidis
George Koundourakis
Elena Kouri
Polychronis Koutsakis
Artemis Koutsopodi
Maria Koutsopoulou
Nikolaos Koutsouris
Demetres Kouvatsos
Ulas Kozat
Srikanth Krishnamurthy
Thomas Kuerner
Sandeep Kulkarni
Abhishek Kumar
Mohan Kumar
Zefir Kurtisi
Christian Kurz
Pirkko Kuusela
Persefoni Kyritsi
Koen Laevens
Paola Laface
Samer Lahoud
Anukool Lakhina
Maria Lambrou
Stijn van Langen
Bart Lannoo
Pasi Lassila
William Lau
Jean Yves Le Boudec
Guy Leduc
Sanghwan Lee
Franck Legendre
Alessandro Leonardi
Emilio Leonardi
Michah Lerner
Marios Lestas
Brian Neil Levine
Bo Li
Jian Li
Jiangbo Li
Jikai Li
Song Li
Yue Li
Yuhong Li

Yun Li
Zhi Li
Yi Liang
Francesco Licandro
Jorg Liebeherr
Noura Limam
Christoph Lindemann
Antionio Lioy
Donggang Liu
Enjie Liu
Peng Liu
Yong Liu
Catalina Llado
Renato Lo Cigno
Michael Logothetis
Stephane Lohier
Francesca Lo Piccolo
Francesco Lo Presti
Jean-Raymond Louvion
Steven Low
Thomas Luckenbach
Henrik Lundqvist
Jie Luo
Emil Lupu
Andrey Lyakhov
Joaquim Macedo
Edmundo Madeira
Damien Magoni
Gerald Maguire
Petri Mahonen
Matthias Malkowski
Anjlica Malla
Szabolcs Malomsoky
Marcel Mampaey
Luigi Mancini
Vincenzo Mancuso
Sotiris Maniatis
Demetres Manikis
G. Manimaran
Jukka Manner
Petteri Mannersalo
Ma Maode
Peter Marbach
Mario Marchese
Margaritis Margaritidis

Athina Markopoulou
Alan Marshall
Jim Martin
Ruediger Martin
Jean-Phil. Martin-Flatin
Fabio Martinelli
Peter Martini
Josep Luis Marzo
Ignacio Mas
Gaia Maselli
Xavier Masip-Bruin
Dan Massey
Laurent Massoulie
Hiroyuki Masuyama
Bertrand Mathieu
Takis Mathiopoulos
Laurent Mathy
Ibrahim Matta
Ravi Mazumdar
Gianluca Mazzini
Muriel Medard
Djamal Edine Meddour
Alberto Medina
Rob van der Mei
Marco Mellia
Tommaso Melodia
Paulo Mendes
Michela Meo
Madjid Merabti
Lazaros Merakos
Marcin Michalak
Ioannis Milis
Geyong Min
Gary Minden
Amitabh Mishra
Vishal Misra
Isi Mitrani
Nikolas Mitrou
Naoto Miyoshi
Nader Moayeri
Eytan Modiano
Prasant Mohapatra
Antonella Molinaro
Miklos Molnar
Sandor Molnar

Refik Molva
Raul Mondragon
Nicolas Montavont
Edmundo Monteiro
Paolo Monti
Giacomo Morabito
Nikos Moschopoulos
Sead Muftic
Billy Mullins
Maurizio Munafo
Anelise Munaretto
C. Siva Ram Murthy
Luca Muscariello
Yiannos Mylonas
Hassan Naser
Leandro Navarro
Srihari Nelakuditi
Dragos Niculescu
Jeffrey Nie
Georgios Nikolaidis
Ioanis Nikolaidis
Eugenia Nikolouzou
Dessislava Nikolova
Petros Nikopolitidis
Peng Ning
Zoltan Nochta
Christos Nomikos
Erik Nordstrom
Ilkka Norros
Antonio Nucci
Mario Nunes
Eeva Nyberg
Antonis Oikonomou
Stephan Olariu
Jaudelice C. de Oliveira
Jauvane C. de Oliveira
Ruy de Oliveira
Huw Oliver
Philippe Olivier
Giwon On
Dirk Ooms
Katsunori Ori
Joao Orvalho
Olav Osterbo
Sam Ou

Sara Oueslati
Philippe Owezarski
Sai Sudhir Padmanabhan
Elena Pagani
Fernando Paganini
Aris Pagourtzis
Sergio Palazzo
Georgios Papadimitriou
Christos Papadopoulos
Thanassis Papaioannou
Evangelos Papapetrou
Symeon Papavasileiou
Vasileios Pappas
Sarantis Paskalis
Nikos Passas
Michael Paterakis
Babis Patrikakis
Achille Pattavina
Giovanni Pau
Niovi Pavlidou
George Pavlou
Magdalena Payeras
Kiamal Pekmestzi
Aleksi Penttinen
Jorge Pereira
Rubem Pereira
Chiara Petrioli
Marinella Petrocchi
Dimitrios Pezaros
Chris Phillips
Chiara Piglione
Antonio Pinizzotto
Paulo da Fonseca Pinto
Andreas Pitsillides
Nineta Polemi
George Polyzos
Dario Pompili
Ana Pont
Dario Ponzio
Jonathan Poritz
Martin Potts
Niels Kjoelstad Poulsen
Leonid Poutievski
Ravi Prakash
George Prezerakos

Richard Price
Gregorio Procissi
Julian Proenza
Konstantinos Psounis
Spyros Psychis
Nicolas Puech
Alessandro Puiatti
Ramon Puigjaner
Guy Pujolle
Francesco Quaglia
Francisco-Jose Quiles
Juergen Quittek
Kimmo Raatikainen
Carla Raffaelli
S. Raghavan
Krishna Ramachandran
Ram Ramjee
Sanjeev Rampal
Ananth Rao
Erwin Rathgeb
Martin Reisslein
Hannu Reittu
Tianmin Ren
Simos Retalis
Laurent Reynaud
Jose Rezende
Injong Rhee
Fabio Ricciato
Julien Ridoux
Rudolf Riedi
Reinhard Riedl
Vincent Rijmen
Herve Rivano
Luigi Rizzo
James Roberts
Vincent Roca
Rui Rocha
Antonio Rodrigues
Jose Rolim
Sami Rollins
Simon Pietro Romano
Emilia Rosti
Angelos Rouskas
George Rouskas
Bart Rousseau

Jose Ruela
Zhao Rui
Silvia Ruiz
Selwyn Russell
Mikalai Sabel
Dan Sadot
David Safford
Amit Sahoo
Hussein Salama
Apostolis Salkintzis
Elio Salvadori
Petros Sampatakos
Dimitris Sampson
Sergio Sanchez-Lopez
Kamil Sarac
Lambros Sarakis
Susana Sargento
Saswati Sarkar
Stelios Sartzetakis
Anna Satsiou
Eric Scharf
Matthias Scheidegger
Giovanni Schembra
Enrico Schiattarella
Marc Schinnenburg
Rudiger Schollmeier
John Schormans
Bruno Richard Schulze
Caterina Scoglio
Mirela Sechi-Notare
Srinivasan Seetharaman
Amit Seghal
Ahmet Sekercioglu
Sidi-Mohammed Senouci
Dimitrios Serpanos
Bartomeu Serra
Sanjay Shakkottai
Srinivas Shakkottai
Jonathan Shapiro
Vinod Sharma
Jang-Ping Sheu
Xiyu Shi
Ali Shoja
Peter Sholander
Karen Shoop

Rajeev Shorey
David Shrimpton
Stavroula Siachalou
Mihai Sichitiu
Nikolaos Sidiropoulos
Matthias Siebert
Manolis Sifalakis
Georgos Siganos
Dimitra Simeonidou
David Simplot-Ryl
Robby Simpson
Narendra Singhal
Prasun Sinha
Vasilios Siris
Krishna Sivalingam
Anders Sjogren
Harry Skianis
Dimitris Skyrianoglou
Avril Smith
Paul Smith
Peter Soerensen
John Soldatos
Sergios Soursos
Kathleen Spaey
Otto Spaniol
Alexander Sprintson
Adarsh Sridhar
Ashwin Sridharan
K. Sriram
Rolf Stadler
Yiannis Stamatiou
George Stamoulis
Panagiotis Stathopoulos
Brikena Stavovci-Halimi
Ioannis Stavrakakis
Kiriakos Stefanidis
Cliff Stein
Burkhard Stiller
Ivan Stojmenovic
Aaron Striegel
Peter Stuckmann
Marinos Stylianou
Weilian Su
Dhanant Subhadrabandhu
Kyoungwon Suh

Zhi-Li Sun
Min-Ho Sung
Irini Sygkouna
Efstathios Sykas
Violet Syrotiuk
Robert Szabo
Janos Sztrik
Abd-Elhamid Taha
Yutaka Takahashi
Alessandro Tarello
Leandros Tassiulas
Y. C. Tay
Jing Teng
Daniele Tessera
Michael Theologou
Petia Todorova
Birger Toedtman
Samir Tohme
Tiziana Toniatti
Corinne Touati
Don Towsley
Alex Trifunovic
Axel Truong
Konstantinos Tsagaris
Sofia Tsekeridou
Stavros Tsiakkouris
Orestis Tsigas
Masato Tsuru
Tuna Tugcu
Zoltan Turanyi
Damla Turgut
Filip de Turck
Shelley Unger
Alessandro Urpi
Francesco Vacirca
Kiran Vadde
Rui Valadas
Andras Valko
Srivatsan Varadara
Anujan Varma
Manos Varvarigos
Constantinos Vassilakis
Vasos Vassiliou
Jon-Olov Vatn
Evaggelos Vayias

Teresa Vazao
Hector Velayos
Andrea de Vendictis
Iakovos Venieris
Roberto Verdone
Dimitrios Vergados
Evangelos Vergetis
Giacomo Verticale
Rolland Vida
Attila Vidacs
Paulo Cesar Salgado Vidal
Laurent Viennot
Manuel Villen-Altamirano
Jorma Virtamo
Guillaume Vivier
Danny de Vleeschauwer
Raimund Vogl
Thiemo Voigt
Milan Vojnovic
Vinod Vokkarane
Artemios Voyiatzis
Stavros Vrontis
Can Vuran
Krzysztof Wajda
Marcel Waldvogel
Bernhard Walke
Jia Wang
Lan Wang

Lili Wang
Quanhong Wang
Tao Wang
Wenye Wang
Xin Wang
Xudong Wang
Zhipeng Wang
Gill Waters
Richard Weber
Steven Weber
Klaus Wehrle
Erik Weiss
Markus Werner
Attila Weyland
Jeff Wieselthier
Carey Williamson
Craig Wills
Larry Wilson
Lars Wolf
Tilman Wolf
Lloyd Wood
Ningning Wu
Christos Xenakis
Xing Xie
Yufeng Xin
Bagnan Xu
Kenan Xu
Kuai Xu

Lisong Xu
Shouhuai Xu
George Xylomenos
Stavros Xynogalas
Baijian Yang
Sruthi Yellamraju
Jian Yin
Y. Ying
Hongnian Yu
Lige Yu
Yinzhe Yu
Stathis Zachos
Paolo Zaffoni
Alberto Zanella
Hui Zang
Daniel Zappala
Djamal Zeglache
Soraya Zertal
Evangelos Zervas
Honggang Zhang
Jingyuan Zhang
Qingyu Zhang
Zhi-Li Zhang
Sencun Zhu
Yong Zhu
Artur Ziviani
Voula Zoi
Vasilis Zorkadis

Table of Contents

Inter-domain Routing

Packet Classification and Scheduling

Services and Monitoring

Admission Control

Competition in Networks

3G/4G Wireless Systems

MPLS and Related Technologies

Flow and Congestion Control

Performance of IEEE 802.11

QoS Metrics, Algorithms, Architectures

Content Distribution, Caching, and Replication

Routing Theory and Path Computation

Wireless Network Performance II

Short Papers (Poster Presentations)

Efficient, Authenticated, and Fault-Tolerant Key Agreement for Dynamic Peer Groups

Li Zhou and Chinya V. Ravishankar

Department of Computer Science & Engineering
University of California, Riverside
Riverside, CA 92521, USA
{lzhou,ravi}@cs.ucr.edu

Abstract. We present an efficient authenticated and fault-tolerant protocol (*AFTD*) for tree-based key agreement. Our approach is driven by the insight that when a Diffie-Hellman blinded key is updated, in a tree-based method, it suffices to send the update to a small subset of the group, instead of entire group, as current methods require. Our scheme distributes each updated public key to a relatively small subgroup, called its *trust set*, greatly improving performance. Moreover, we use a threshold secret sharing method to distribute the function of the trusted authority across trust sets, thereby guaranteeing key authentication, enhancing fault-tolerance, and protecting our protocol from impersonation attacks. Our performance analysis suggests that our scheme significantly reduces the communication overhead and storage requirement.

Keywords: Secure Group Communication, Key Agreement, Key Authentication

1 Introduction

As a result of the increased popularity of group-oriented applications, such as pay-TV, distributed interactive games, teleconferencing and chat rooms, there is a growing demand for security services to achieve secure group communication. A common method is to encrypt messages with a group key, so that entities outside the group cannot decode them. A satisfactory group communication system would possess the properties of *group key security, forward secrecy, backward secrecy,* and *key authentication/integrity* [1,2,3]. In this paper, we focus on *Key authentication/integrity*, which ensures that public keys of group members cannot be modified by adversaries. There are two approaches for generating such group keys: centralized key distribution and distributed key agreement. Centralized key distribution uses a dedicated key server, resulting in simpler protocols. However, centralized methods fail entirely once the server is compromised, so that the central key server makes a tempting target for adversaries. In addition, centralized key distribution is not suitable for dynamic peer groups, in which all nodes play the same function and role, thus it is unreasonable to make one the

N. Mitrou et al. (Eds.): NETWORKING 2004, LNCS 3042, pp. 759–770, 2004.

key server, placing all trust in it. In contrast, distributed key agreement requires each member to contribute a share to generate the group key, resulting in more complex protocols.

The group key is updated on every membership change for forward and backward secrecy, a method called *group rekeying*. To reduce the number of rekeying operations, Wong et al. [4] proposed a logical data structure called a *key tree* that reduces the rekeying overhead from $O(n)$ to $O(\log n)$, where n is the group size. Based on this idea, Kim et al. proposed a tree-based key agreement protocol, *TGDH* [1], which is a combination of key tree and Diffie-Hellman key exchange to generate and maintain the group key.

Unfortunately, *TGDH* suffers from two drawbacks. As explained in Section 3, it remains prone to impersonation attacks, and uses more messages than necessary.

1.1   Our Work

In this paper, we propose a novel **A**uthenticated, **F**ault-tolerant **T**ree-based **D**iffie-Hellman key agreement protocol, *AFTD*, based on two key ideas. First, as explained in Section 3, it is gross overkill to broadcast updated public keys to all group members for recomputing the group key when a node n_i joins or leaves. It suffices to send each update to a much smaller subset of nodes in the tree, called its *trust set* $TS(n_i)$. Second, we achieve robust key authentication by distributing the function of trusted authority among the nodes in $TS(n_i)$, using a threshold cryptographic scheme. Any k members of a node's trust set can serve as its public key certificate authority. Our performance analysis shows this scheme can reduces the communication overhead from $O(n^2)$ to $O(n \log n)$ for initialization, and from $O(n \log n)$ to $O(n)$ for rekeying. It also reduces the storage requirement for blinded keys from $O(n)$ to $O(\log n)$. This feature is particularly useful when a broadcast channel is unavailable.

The rest of this paper is organized as follows. We survey related work in Section 2. Section 3 motivates our work and defines our intrusion model. We present our solution in Section 4 and demonstrate performance analysis and comparison in Section 5. Finally we make a conclusion in Section 6.

2   Related Work

Key trees [4,5] were first proposed for centralized key distribution, while Kim et al. [1] adapted it to a distributed key agreement protocol *TGDH*. Every group member create a key tree separately. Each leaf node is associated with a *real* group member, while each non-leaf node corresponds to a subgroup of group G, considered a *virtual member*. In Figure 3, virtual member V_4 corresponds to the subgroup that contains two real group members M_3 and M_4.

In *TGDH*, every node on the key tree has a Diffie-Hellman key pair based on the prime p and generator α, used to generate the group key. Secret-public key pair $\{K_{M_i}, BK_{M_i} = \alpha^{K_{M_i}} \bmod p\}$ is for real member M_i, and $\{K_{V_i}, BK_{V_i} =$

$\alpha^{K_{V_i}} \bmod p\}$ is for virtual member V_i. Public key BK_{M_i} is also called as *blinded key*. Consider a node M_v whose left child is M_{lv} and right child node is M_{rv} (to simplify the description, we do not distinguish real members from virtual members here). M_i's secret key can be computed in the usual Diffie-Hellman fashion as $K_{M_v} \equiv (BK_{lv})^{K_{rv}} \equiv (BK_{rv})^{K_{lv}} \bmod p$.

With all blinded keys well-known, each group member can compute the secret keys of all nodes on its key path, comprising the nodes from the leaf node up to the root. The root node's secret key K_{V_0} is known to all group members, and becomes the group key. In Figure 3, group member M_2 knows the key pairs of M_2, V_3, V_1 and V_0. V_0's secret key is the group key.

Steiner et al. [6,2] proposed a family of Group Diffie-Hellman (*GDH*) protocols for dynamic peer groups. Based on them, Ateniese et al. [7] proposed a new multiparty authenticated key agreement protocol, which offers key authentication/integrity, key confirmation, and non-repudiation of group membership. However, some flaws in this protocol have been found by Pereira et al. [8].

Lee et al. [9,10] have designed several tree-based distributed key agreement protocols, reducing the rekeying complexity by performing interval based rekeying. They also present an authenticated key agreement protocol. As the success of their scheme is partially based on a certificate authority, their protocol will encounter the same problems as centralized trust mechanisms.

In [11], Kong et al. provide robust and ubiquitous security support for mobile ad-hoc networks. In their scheme, they distribute the certificate authority functions through a threshold secret sharing mechanism, in which each entity holds a secret share, and multiple entities in a one-hop neighborhood jointly provide certificate services. Our distributed trust mechanism differs from theirs in two aspects. First, our goals are different. Second, the nodes that offer valid certificates are different [12].

3   Motivation and Attack Model

TGDH [1] is a simple and efficient approach for the establishment of ephemeral keys for group sessions, however, it suffers from two significant drawbacks.

First, although *TGDH* uses authenticated channels, it still seems vulnerable to impersonation attacks. To provide authenticated blinded keys, *TGDH* suggests that every protocol message be signed by its sender using some strong public signature method such as DSA or RSA, and then verified by all receivers using the sender's public key. However, *TGDH* is a session key generation protocol, and does not address the long-term security of DSA or RSA keys. Since adversaries can compromise those keys in the long run, these keys must be refreshed periodically from a trusted source that is available online. Our approach is to define a *trust set* for each member M_i, and distribute the function of trusted authority across trust sets so that any k members from this set can offer M_i's public key certificate, enhancing fault-tolerance. We distribute the function of trust authority in a manner similar to the scheme in [11], which is based on (k, n)-threshold secret sharing scheme. It is pointed out in [13] that scheme in [11] is not usable in a group with malicious members since it does not provide an important property known as *verifiability*. However, we focus on the correct and secure generation of group keys in the face of *outsider* attacks mounted by non-group members as in [14,15]. We do not address insider attacks mounted by malicious group members because they can always reveal their own private keys or the group key to non-group members, thus causing fraudulent membership events or compromising group communication. Thus our scheme is unaffected by the flaw pointed out in [13]. Detailed description of intrusion model appears in [12].

Second, *TGDH* involves excessive communication and storage overheads caused by broadcasting updated blinded keys. This problem becomes more serious if membership events are common. Storage requirement also becomes an important issue when resource limited devices, such as PDAs, are able to join the group as qualified group members. In our scheme, instead of broadcasting these updates, they are transmitted to a smaller subset of nodes (*trust set*), so that communication and storage overheads can be reduced significantly.

4   Our Solution

4.1   Overview

TGDH [1] observes that a node n_1 in the group needs to know the blinded key of another node n_2 only if n_2 is on its *co-path*, defined as the set of siblings of each node in the key path of n_1. However, we offer the key insight that even more is true. In fact, an update of a blinded key need be sent only to a small subset of the group, instead of entire group. Because of the way Diffie-Hellman key exchanges are used in a key tree to generate the group key, the blinded key of any node n_i is only needed by the leaf nodes of the subtree rooted at n_i's sibling. This group of nodes, which we call n_i's *trust set*, forms the basis for both improved efficiency as well as key authentication. We send each node's blinded keys only to its trust set. A node's trust set is also entrusted with the task of responding to requests for its public key, and provides key certificates using a threshold cryptographic scheme. This insight is missing from earlier work.

Key Management Phases. In our scheme, each group member construct a key independently. Each real group member M_i has two key pairs: a Diffie-Hellman key pair, $\{K_{M_i}, BK_{M_i} = \alpha^{K_{M_i}} \bmod p\}$, which is used to generate the group key, and an RSA secret-public key pair, $\{D_i, E_i\}$, which is used to provide source authentication. Non-leaf nodes V_i are virtual members, and have only a Diffie-Hellman key pair $\{K_{V_i}, BK_{V_i} = \alpha^{K_{V_i}} \bmod p\}$.

Group key management in our approach occurs in two phases: the *initialization phase* and the *rekeying phase*. Initialization is a one-time activity that distributes appropriate public-key certificates to trust sets. While such initialization may be done in many ways, we simplify our presentation by postulating that this function is performed by a trusted authority (TA), which subsequently goes offline. Offline here means the TA will not issue renewed public key certificates for existing group members during the process of group rekeying. New

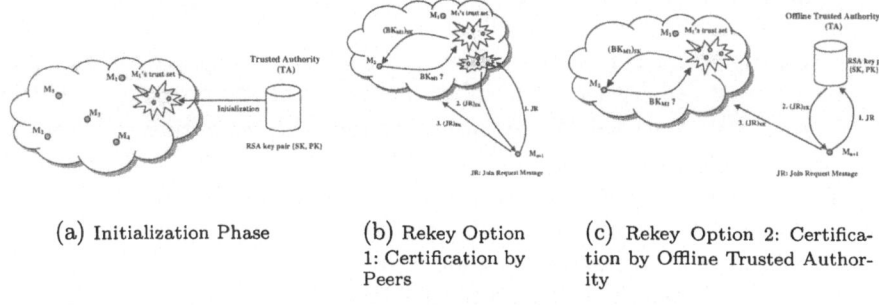

(a) Initialization Phase   (b) Rekey Option 1: Certification by Peers   (c) Rekey Option 2: Certification by Offline Trusted Authority

Fig. 1. *AFTD* overview

members wishing to join the group may obtain initial certificate from the TA at any time prior to join.

This TA uses an RSA secret-public key pair $\{SK, PK\}$, and establishes public key certificates for each group member M_i by signing M_i's public key with its secret key SK. M_i's public key certificate $\langle M_i, BK_{M_i}, E_i \rangle_{SK}$ is now distributed to its *trust set*. Since the public key PK is well known, any member of M_i's trust set can verify this certificate and obtain M_i's public key.

The initialization phase also distributes a secret share SK_j of the secret key SK to each group member M_j using Shamir's (k, n)-threshold sharing [16], which is used for creating partial public key certificates held by members of trust sets. A node M_j in M_i's trust set verifies the original certificate for M_i (signed using SK), and re-encrypts it with SK_j to create a partial certificate. Now any k members in the trust set of a given group member can offer that group member's public key certificate by group signing of certificates.

In the rekeying phase, *AFTD* includes protocols in support of three operations: join, leave and interval multicast.

4.2   Initialization Phase

Assume the group G has n real group members $M_1, M_2, \ldots, M_n$ initially. We describe how to distribute the function of the trusted authority to appropriate subgroups (*trust set*) so that any k member nodes in an appropriate subgroup can offer the corresponding valid certificate. Here "valid" means the certificate has been signed with the system secret key SK.

Distributing the system secret key shares SK_i. Our design uses Shamir's (k, n)-threshold scheme [16]. First, the TA randomly selects a $(k-1)$-degree polynomial $f(x) = SK + a_1 \cdot x + \cdots + a_{k-1} \cdot x^{k-1}$, such that the shared secret is $f(0) = SK$. Each group member obtains a secret share $SS_{M_i} = (f(M_i) \bmod m)$. For any k group members $\{M_1, M_2, \ldots, M_k\}$, Lagrange interpolation yields $SK \equiv \sum_{i=1}^{k}(SS_{M_i} \cdot l_{M_i}(0)) \equiv \sum_{i=1}^{k} SK_i \pmod{m}$, where $l_{M_i}(0)$ are the Lagrange coefficients[1].

Obtaining valid certificates: The certificate X for any node is served by the node's *trust set*, with each member in that trust set providing a partial certificate X^{SK_i}. With any k partial certificates, the requesting member can compute the valid certificate as $X^{SK_1} \cdot X^{SK_2} \dots X^{SK_k} = X^{(\sum_{i=1}^{k} SK_i)} = X^{SK}$ [11]. Thus, these k members can work like a trusted authority, and jointly offer the certificate. (We use the *t-bounded coalition offsetting* algorithm proposed in [11] to ensure that the above equation is valid.)

This approach has the nice feature that the system secret key SK is never revealed to any member node nor to any subset of member nodes. They can jointly reconstruct X^{SK}, but never SK itself. While this method can be unsafe if group members can be compromised [13], this difficulty does not arise in our case, as explained in Section 3.

Further, *AFTD* improves fault-tolerance, since Shamir's threshold scheme ensures that any set of $k-1$ or less secret shares cannot jointly obtain SK. Thus if any set of $k-1$ or less secret shares have been discovered, the system secret key SK is still safe from adversaries.

Defining Trust Sets. At the beginning, each group member is assigned a unique member ID and associated with a leaf node of the key tree in ascending order. To define trust sets, the group is first split into k-member clusters. The members in the last cluster may have more than k group members when n is not a multiple of k. The upper part of Figure 3 shows a 7-member group. When $k = 2$, the group is divided into 3 clusters, and the last one has three members.

Definition 1 (Trust Set). *The trust set of M_i or V_i (See Figure 2) is the set of nodes in the union of all clusters that contain one or more leaf nodes of the subtree rooted at M_i or V_i's sibling node, and represented as $TS(M_i)$ or $TS(V_i)$.*

In Figure 3, $TS(V_2)$ is the set of nodes in the first two clusters, which contain all leaf nodes $\{M_1, M_2, M_3, M_4\}$ of the subtree rooted at V_2's sibling node V_1.

When the number of clusters in $TS(M_i)$ is less than two, we improve fault tolerance by including in $TS(M_i)$ the first adjacent cluster formed by the leaves of the sibling of M_i.

Distributing the Certificates to Trust Sets. After distributing the system secret key shares to all group members, the trusted authority distributes the stored public key certificates to appropriate trust sets. Then the TA works offline and is used only to initialize new members joining the group. The system can provide key authentication service to renewed RSA keys without the help of the TA, since its function has been fully distributed to appropriate trust sets.

4.3   The Rekeying Phase

To achieve forward and backward secrecy (Section 1), the group key must be updated whenever new members join or old members leave. Each new member

---

[1] Defined as $l_{M_i}(x) = \frac{(x-M_1)\cdots(x-M_{i-1})(x-M_{i+1})(x-M_k)}{(M_i-M_1)\cdots(M_i-M_{i-1})(M_i-M_{i+1})(M_i-M_k)}$.

M_J must also obtain a system secret share SS_{M_J} and some certificates so that it can offer certificate services.

Localized Self-initialization. When an uninitialized new member M_J joins the group, a k-coalition of old members which are share holders $\{M_{J_1}, M_{J_2}, \cdots M_{J_k}\}$, can jointly offer a system secret share SS_{M_J} to M_J as follows:

$$SS_{M_J} \equiv \sum_{i=1}^{k} SS_{M_{J_i}} \cdot l_{M_{J_i}}(M_J) \equiv \sum_{i=1}^{k} SS_{J_i} \pmod{m}.$$

The Lagrange coefficients and SS_{J_i} are known by M_{J_i}, so $SS_{M_{J_1}}$ can be derived directly. Here we use the *shuffling* scheme presented in [11] to maintain the secrecy of M_{J_i}'s share.

The Join Protocol. Assume that a new member M_{n+1} wishes to join a n-member group which contains $\{M_1, M_2, \ldots, M_n\}$. M_{n+1} is required to authenticate itself by presenting a join request signed with SK. M_{n+1} may obtain a signature on its join request either by establishing credentials with the offline trusted authority, or by enlisting the cooperation of at least k nodes from the group G willing to recognize and certify M_{n+1}'s request.

When the other group members receive this request, they independently determine M_{n+1}'s insertion node [1] in the key tree, which is the shallowest rightmost node, or the root node when the key tree is well-balanced. They also independently determine a real member called *join sponsor* M_S [1] to take responsible for coordinating the join, which is the rightmost leaf node in the subtree rooted at the insertion node.

No keys change in the key tree at a join, except the blinded keys for nodes on the key path for the sponsor node. The sponsor simply recomputes the group key, and sends updates for blinded keys on its own key path to their corresponding trust sets. Each member M_j of the sponsor node's trust set creates a new partial share SS'_j of the secret key SK, and forwards it to M_{n+1}, which combines them to obtain its new secret share SK_{n+1}. The new node M_{n+1} also sends its signed certificate to the members of its trust set $TS(M_{n+1})$, and gets the public keys needed for generating the group key. The join works as shown in Algorithm 1.

Algorithm 1 Join Protocol in $AFTD$
___

1: The new member M_{n+1} broadcasts the signed join request to the group.
2: Group members determine the insertion point, and update their key trees by creating a new intermediate node and promoting it to become the parent of both the insertion node and M_{n+1}.
3: Each group member adjusts the clusters in its key tree by adding M_{n+1} to the smallest cluster adjacent to the insertion point, or to the cluster on its right one in case of a tie. If the size of the modified cluster goes up to $2k$, split it into two clusters.
4: The sponsor M_S computes the new group key, and sends the updated blinded keys of nodes on its key path to their corresponding trust sets. These messages are signed by the sponsor M_s.
5: The members in these trust sets request the sponsor M_S's certificate from $TS(M_S)$ to verify the updated blinded keys they received.
6: M_{n+1} obtains its secret share SK_{n+1} from $TS(M_S)$.
7: M_{n+1} sends its valid public key certificates to its trust set, and gets the public keys needed for generating the group key.

In Figure 3, M_8 joins a 7-member group, and $k = 2$. The join sponsor M_7 creates a new intermediate node V_6 in the key tree and promotes it to become

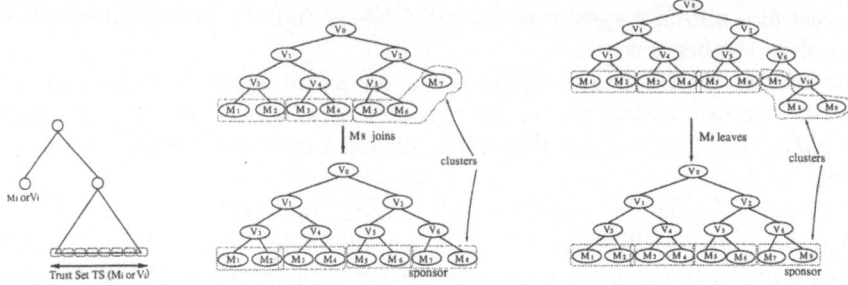

Fig. 2. Trust Set   **Fig. 3.** Join Process in AFTD   **Fig. 4.** Leave Process in AFTD
of M_i or V_i

the parent of M_7 and M_8. The sponsor M_7 computes the new group key, sending the updated BK_{V_6} and BK_{V_2} to their corresponding trust sets $\{M_5, M_6, M_7, M_8\}$ and $\{M_1, M_2, M_3, M_4\}$ respectively. Finally as the size of the third cluster is extended to $2k = 4$, it splits into two clusters: $\{M_5, M_6\}$ and $\{M_7, M_8\}$.

Leave Protocol. Assume that a member M_L wishes to leave a n-member group. First M_L initiates the leave protocol by sending a leave request. When the other group members receive the request, they independently determine the sponsor node, which is defined as in [1] to be the right-most leaf node of the subtree rooted at the leaving member's sibling node. The leave protocol works as shown in Algorithm 2.

Algorithm 2 Leave Protocol in *AFTD*

1: The former sibling node of M_L is promoted to replace M_L's parent node.
2: The size of the cluster that formerly contained M_L is decreased by one, and merges with an adjacent cluster if its size drops below k. The new cluster may split if its size is $2k$ or larger.
3: The sponsor M_S picks a new secret key K'_{M_S}, computes the new group key, and sends the updated blinded keys of nodes on its key path to their corresponding trust sets. These messages are signed by the sponsor M_s.
4: The members in these trust sets request M_S' certificate from $TS(M_S)$ to verify the updated blinded keys they received.

In Figure 4, M_8 leaves a 9-member group where $k = 2$. The sponsor M_9 picks a new secret key K_{M_9} and computes the new group key, sending updated BK_{M_9}, BK_{V_6} and BK_{V_2} to their corresponding trust sets $\{M_5, M_6, M_7, M_9\}$, $\{M_5, M_6, M_7, M_9\}$ and $\{M_1, M_2, M_3, M_4\}$ respectively.

Interval Multicast Protocol. *AFTD* can also realize secure interval multicast, in which a group member wants to send data to a subgroup of group G. This problem is discussed by Gouda et al. [17], who describe a new use of key trees. They are concerned about using the existing subgroup keys in the key tree to securely multicast data to different subgroups within the group. Unlike their approach, which depends on a centralized key server to maintain the unique key tree and manage all keys, *AFTD* solves this problem in a distributed fashion. For the detailed algorithm, please refer to [12].

Updating Secret Keys & Secret Shares. In *AFTD*, each group member is required to update its Diffie-Hellman keys before each group session, or during a session when it is selected as a sponsor on a member's leaving. Source authentication of the updated blinded keys is guaranteed by the sender's RSA signature. Further, to ensure the long-term secrecy of the RSA keys, *AFTD* requires each group member to renew its RSA key pair periodically, and send it to its trust set securely using its current RSA secret key. *AFTD* adopts the proactive secret share update algorithm in [18] to periodically update the system secret shares to invalidate compromised secret shares.

4.4 Security of Trusted Authority

The trusted authority, which may be distributed, is on-line during initialization, but remains offline subsequently. During initialization, the TA distributes valid key certificates and secret shares of its secret key *SK*, so that the function of key authentication can be realized and distributed across appropriate trust sets. Since the duration of initialization is relatively short, it is safe for us to use the TA at that time.

During the rekeying phase, the trusted authority may be approached by new group members for authentication and creation of valid key certificates for them. In this mode, the trusted authority works offline, in that it only communicates with new group members, making compromises of TA unlikely.

4.5 The Number of Rekeying Messages Received

On a rekeying event, all members in the trust sets of the nodes on the sponsor's key path will receive an updated blinded key. As in [1], we use the term *co-path* for the set of siblings of each node on the key path of member M_i. The nodes on a co-path have disjoint subtrees, so that the set of leaf nodes for these subtrees are also disjoint. Thus, the leaf nodes of each node in the co-path fall into clusters with minimal overlaps.

Because of the way they are defined(Section 4.2), the trust sets of the nodes on the key path of member M_i have small overlap. Consequently, each group member receives nearly the same number of rekeying messages in our scheme. For example, in Figure 3, V_6 and V_2 are on the key path of the sponsor M_7. V_5 and V_1 are on the co-path of M_7, which have disjoint subtrees, so that V_6 an V_2 have disjoint trust sets. V_6's trust set is $\{M_5, M_6, M_7, M_8\}$, and V_2's trust set is $\{M_1, M_2, M_3, M_4\}$. Each member of these trust sets will receive one updated blinded key.

5 Performance Analysis

5.1 Communication Overheads

The Initialization Phase. The communication overhead is measured by the number of the messages. In the initialization phase of *AFTD*, the trusted authority distributes the certificates of each node in the key tree to its trust set.

Since every node in each trust set receives a message, the overall communication overhead of the first phase is measured by the number of nodes in all trust sets.

If n is the group size and h_{V_i} is the height of V_i, there are at most $2^{\lg n - h_{V_i}}$ leaf nodes on the subtree of V_i's sibling node. Since the cluster size is k, the number of shares needed to reconstruct the certificate, these nodes will fall into at most $\lceil \frac{2^{\lg n - h_{V_i}}}{k} \rceil$ clusters. Hence, the size of V_i's trust set is no more than $\lceil \frac{2^{\lg n - h_{V_i}}}{k} \rceil \cdot k < 2^{\lg n - 1 - h_{V_i}} + 2k$ nodes. As the maximum number of nodes in each level h_{V_i} is $2^{h_{V_i}}$, the communication overhead of each level is at most $2^{h_{V_i}} \cdot (2^{\lg n - h_{V} M_i} + 2k)$. Therefore, we can compute the overall communication overhead in the initialization phase as $C_{Initial} = \sum_{i=1}^{\lg n} 2^i \cdot (2^{\lg n - i} + 2k) = O(n \log n)$.

The Rekeying Phase. In the rekeying phase of $AFTD$, when a new member joins or an old member leaves, keys for nodes on the sponsor node's key path must be updated and multicast. Hence for each level of the key tree, only one node's blinded key has been updated and must be multicast to its trust set. Since the size of the trust set of the nodes on level h_{V_i} is at most $\lceil \frac{2^{\lg n - h_{V_i}}}{k} \rceil \cdot k < 2^{\lg n - h_{V_i}} + 2k$ nodes, we can compute the overall communication in this stage as $C_{Rekey} = \sum_{i=1}^{\lg n} (2^{\lg n - i} + 2k) = O(n)$.

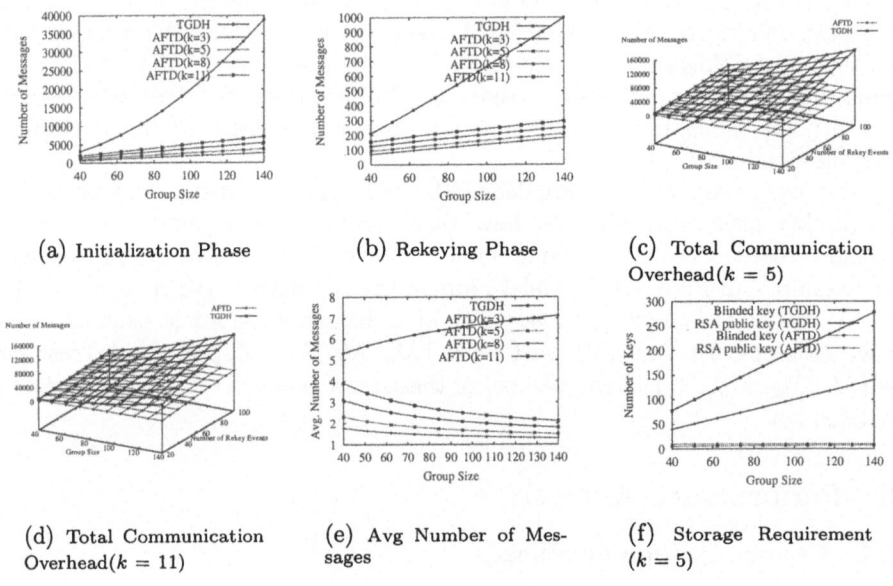

(a) Initialization Phase

(b) Rekeying Phase

(c) Total Communication Overhead$(k = 5)$

(d) Total Communication Overhead$(k = 11)$

(e) Avg Number of Messages

(f) Storage Requirement $(k = 5)$

Fig. 5. Communication & Storage Overheads

Table 1. Performance of *TGDH* and *AFTD* Compared

	Communication Overhead		Storage Requirement	
	Initialization Phase	Rekeying Phase	Blinded Key	RSA Public Key
TGDH	$O(n^2)$	$O(n \log n)$	$O(n)$	n
AFTD	$O(n \log n)$	$O(n)$	$O(\log n)$	$2k$

Figure 5(a) and (b) show the effects of different thresholds k on the communication overheads of *AFTD*. They also compare the communication overhead of our scheme to that of *TGDH*. Figure 5(c) and (d) compare the total communication overhead of *AFTD* with that of *TGDH*, where $k = 5$ and 11 separately. Total communication overhead is defined as the combined communication overhead of initialization phase and multiple rekeying events. Clearly, the communication overhead of our scheme is significantly smaller than that of *TGDH*, especially in large dynamic group scenarios.

The value of k is a system-dependent parameter, and represents a tradeoff between system security and fault-tolerance. Further, as seen from the above figures, the communication overhead of our scheme is insensitive to k.

Number of Messages Received on Rekeying. As explained in Section 4.5, each group member receives nearly the same number of rekeying messages in our scheme, because of the way trust sets are defined. Thus, the average number of messages on an rekeying event is a reasonable measure of communication overhead. Figure 5(e) compares the average number of messages received by each group member on an rekeying event. As the group size increases, the average number of messages received decreases to approximately one in our scheme, while it remains at $\lg n$ in *TGDH*. For example, when the group size is 128, each group member receives around two rekeying message in our scheme, but about seven in *TGDH*. This demonstrates the scalability of *AFTD* in terms of the load experienced by group members.

5.2   Computation Overhead and Storage Requirements

Table 1 compares the performance of *TGDH* and *AFTD*. Due to space limitation, readers are referred [12] for a detailed analysis of computation and storage requirements.

6   Conclusion

In this paper, we have presented *AFTD*, an efficient, authenticated and fault-tolerant tree-based key agreement protocol. Central to our technique is a threshold secret sharing based method to distribute the function of trusted authority to appropriate trust sets. Our performance analysis shows that our approach can significantly reduce the communication and storage overheads.

Acknowledgement. This work is supported in part by grants from Tata Consultancy Services, Inc., and the Fault-Tolerant Networks program of Defense Advanced Research Projects Agency, under contract F30602-01-2-0536.

References

1. Kim, Y., Perrig, A., Tsudik, G.: Simple and fault-tolerant key agreement for dynamic collaborative groups. In: Proceedings of the CCS'00. (2000)
2. Steiner, M., Tsudik, G., Waidner, M.: Key agreement in dynamic peer groups. IEEE TRANSACTIONS on Parallel and Distributed Systems **11** (2000)
3. Perrig, A.: Efficient collabortive key management protocols for secure automonomous group communication. In: Proceedings of CrypTEC'99. (1999)
4. Wong, C., Gouda, M., Lam, S.: Secure group communication using key graphs. In: Proceedings of the ACM SIGCOMM'98, Vancouver, Canada (1998)
5. Wallner, D., Harder, E., Agee, R.: Key management for multicast: Issues and architecture. In: Internet Draft, draft-wallner-key-arch-01.txt. (1998)
6. Steiner, M., Tsudik, G., Waidner, M.: Cliques: A new approach to group key agreement. In: Proceedings of the ICDCS'98, Amsterdam, Netherlands (1998)
7. Ateniese, G., Steiner, M., Tsudik, G.: New multiparty authentication services and key agreement protocols. IEEE Journal of Selected Areas in Communications **18** (2000)
8. Pereira, O., Quisquater, J.: A security analysis of the cliques protocols suites. In: Proceedings of the 14-th IEEE Computer Security Foundations Workshop. (2001)
9. Lee, P., Lui, J., Yau, D.: Distributed collaborative key agreement protcols for dynamic peer groups. In: Proceedings of the ICNP'02. (2002)
10. Lee, P., Lui, J., Yau, D.: Distributed collaborative key agreement protcols for dynamic peer groups. Technical report, Dept. of Computer Science and Engineering, Chinese University of Hong Kong (2002)
11. Kong, J., Zerfos, P., Luo, H., Zhang, L.: Providing robust and ubiquitous security support for mobile ad-hoc networks. In: Proceedings of the ICNP'01. (2001)
12. Zhou, L., C.V.Ravishankar: Efficient, authenticated, and fault-tolerant key agreement for dynamic peer groups. Technical Report 88, Dept. of Computer Science and Engineering, University of California, Riverside (2003)
13. Narasimha, M., Tsudik, G., Yi, J.H.: On the utility of distributed cryptography in p2p and manets: the case of membership control. In: Proceeding of the ICNP'03. (2003)
14. Amir, Y., Kim, Y., Nita-Rotaru, C., Tsudik, G.: On the performance of group key agreement protocols. In: Proceedings of the ICDCS'02. (2002)
15. Amir, Y., Nita-Rotaru, C., Stanton, J., Tsudik, G.: Scaling secure group communication systems: Beyong peer-to-peer. In: Proceedings of the DISCEX'03, Washington DC (2003)
16. Shamir, A.: How to share a secret. Communications of the ACM **22** (1979)
17. M.G.Gouda, Huang, C., E.N.Elnozahy: Key trees and the security of interval multicast. In: Proceedings of the ICDCS'02, Vienna, Austria (2002)
18. Herzberg, A., Jarecki, S., Krawczyk, H., Yung, M.: Proactive secret sharing or: How to cope with perpetual leakage. extened abstract, IBM T.J. (1995)

Proactively Detecting Distributed Denial of Service Attacks Using Source IP Address Monitoring

Tao Peng, Christopher Leckie, and Kotagiri Ramamohanarao

ARC Special Research Center for Ultra-Broadband Information Networks
Department of Computer Science and Software Engineering
The University of Melbourne
Victoria 3010, Australia
{tpeng, caleckie, rao}@cs.mu.oz.au

Abstract. In this paper, we propose a simple but robust scheme to detect denial of service attacks (including distributed denial of service attacks) by monitoring the increase of new IP addresses. Unlike previous proposals for bandwidth attack detection schemes which are based on monitoring the traffic volume, our scheme is very effective for highly distributed denial of service attacks. Our scheme exploits an inherent feature of DDoS attacks, which makes it hard for the attacker to counter this detection scheme by changing their attack signature. Our scheme uses a sequential nonparametric change point detection method to improve the detection accuracy without requiring a detailed model of normal and attack traffic. Furthermore, we show that with the combination of monitoring per flow speed, we can detect all types of DDoS attacks. We demonstrate that we can achieve high detection accuracy on a range of different network packet traces.

1 Introduction

A denial-of-service (DoS) attack is a malicious attempt by a single person or a group of people to cripple an online service. The impact of these attacks can vary from minor inconvenience to users of a website, to serious financial losses for companies that rely on their on-line availability to do business. Sophisticated tools to gain root access to other people's machines are freely available on the Internet. These tools are easy to use, even for unskilled users. Once a machine is cracked, it is turned into a "zombie" under the control of one "master". The master is operated by the attacker. The attacker can instruct all its zombies to send bogus data to one particular destination. Simultaneously, the resulting traffic can clog links, and cause routers near the victim or the victim itself to fail under the load. The type of DoS attack that causes problems by overloading the victim with useless traffic is known as a *bandwidth attack*. This paper focuses on curtailing bandwidth attacks.

A key problem to tackle when solving bandwidth attacks is *attack detection*. However, there are two challenges for detecting bandwidth attacks. The first

N. Mitrou et al. (Eds.): NETWORKING 2004, LNCS 3042, pp. 771–782, 2004.

challenge is how to detect malicious traffic close to its source. This is particularly difficult when the attack is highly distributed, since the attack traffic from each source may be small compared to the normal background traffic. The second challenge is to detect the bandwidth attack as soon as possible without raising a false alarm, so that the victim has more time to take action against the attacker.

Previously proposed approaches rely on monitoring the volume of traffic that is received by the victim [10][16][2]. A major drawback of these approaches is that they do not provide a way to differentiate flash crowds from DDoS attacks. Due to the inherently bursty nature of Internet traffic, a sudden increase of in traffic may be mistaken as an attack. If we delay our response in order to ensure that the traffic increase is not just a transient burst, then we risk allowing the victim to be overwhelmed by a real attack. Moreover, some persistent increases in traffic may not be attacks, but actually "flash crowd" events, where a large number of legitimate users access the same website simultaneously. Clearly, there is a need for a better approach to detecting bandwidth attacks.

A better approach is to monitor the number of new source IP addresses, rather than the local traffic volume. Jung et al. [8] have observed that during bandwidth attacks, most source IP addresses are new to the victim, whereas most source IP addresses in a flash crowd appeared at the victim before. Previously, this observation has been used as the basis for a mechanism to filter out attack traffic at the victim [12]. In this paper, we propose to monitor the number of new IP addresses in a given time period in order to detect bandwidth attacks. We demonstrate that this is a more sensitive variable for detecting bandwidth attacks than monitoring the total volume of incoming traffic. In addition, we present a method for detecting changes in our monitoring variable, based on the non-parametric Cumulative Sum (CUSUM) algorithm [3][15]. The CUSUM algorithm reduces the false positive rate, and has been shown to optimal in terms of detection accuracy and computing overhead for parametric model and have good performance for non-parametric model [3].

Our main contribution in this paper is a novel approach to detecting band-width attacks by monitoring the arrival rate of new source IP addresses. We show that this approach is much more effective than earlier schemes, especially when there are multiple attack sources and the attack traffic is highly distributed. We adapt the detection scheme proposed by Wang et al. [15], which is based on an advanced non-parametric change detection scheme, CUSUM, and demonstrate that this approach detects a wide range of simulated attacks quickly and with high accuracy.

The rest of the paper is organized as follows. Section 2 gives a detailed expla-nation of our solution to this problem. Section 3 explains CUSUM algorithm and the model we proposed for the bandwidth attack detection. Section 4 presents the simulation results of our detection mechanism. Section 5 analyzes possible attacks against our detection mechanism.

2   Our Solution: Source IP Address Monitoring

We propose a scheme called Source IP address Monitoring (SIM) to detect the Highly Distributed Denial of Service (HDDoS) attacks. This detection scheme uses an intrinsic feature of HDDoS attacks, namely the huge number of new IP addresses in the attack traffic to the victim. This novel approach has the advantage that it can detect attacks close to their sources in the early stages of the attack.

2.1   Overview of Source IP Address Monitoring

SIM contains two parts: *off-line training*, and *detection and learning*. The first part is the *off-line training*, where a learning engine adds legitimate IP addresses into an IP Address Database (IAD) and keeps the IAD updated by adding new legitimate IP addresses and deleting expired IP addresses. This is done off-line to make sure the traffic data used for training does not contain any bandwidth attacks. A simple rule can be used to decide whether a new IP address is legitimate or not. For example, a TCP connection with less than 3 packets is considered to be an abnormal IP flow. How to build an efficient IAD is discussed in detail in [12].

The second part is *detection and learning*. During this period, we collect several statistics of incoming traffic for the current time interval Δ_n. By comparing the IP addresses during Δ_n with the IAD, we can calculate how many new IP addresses have appeared in this time slot. If the rate per IP address is larger than a certain threshold, an alarm is set to indicate a bandwidth attack. This is used to detect some unsophisticated attacks that use a small number of source IP addresses. More importantly, by analyzing the number of new IP addresses, we can detect whether a HDDoS attack is occurring. If an attack is detected, the on-line-learning is suspended. Otherwise, on-line-learning proceeds in the same matter as off-line training.

2.2   The Choice of a Detection Feature

The key aspect of our detection scheme is that we choose a completely new detection feature compared to earlier detection proposals. We collect the IP addresses during each time slot Δ_n (*n=1, 2, 3, ...*), which determines the detection resolution. We assume $\Delta_1 = \Delta_2 = ... = \Delta_n$, which means the time slots are of equal length. The choice of Δ_n is a compromise between making Δ_n small so that the detection engine can quickly detect an attack, and making Δ_n large so that the detection engine has less computation load because it checks the traffic less often.

Let $\mathcal{T}_n$ represent the set of unique IP addresses and $\mathcal{D}_n$ represent the items of IP Address Database (IAD) at the end of the time interval Δ_n ($n = 1, 2, 3, ...$). As we discussed before, $|\mathcal{T}_n - \mathcal{T}_n \cap \mathcal{D}_n|$,which represents the number of new IP addresses in Δ_n, can be used to detect the DDoS attack. However, $|\mathcal{T}_n - \mathcal{T}_n \cap \mathcal{D}_n|$ varies according to the position of the network traffic monitoring point (NTMP) and different Δ_n. We can normalize this value by defining $X_n = \frac{|T_n - T_n \cap D_n|}{T_n}$, which will not be affected by the NTMP and Δ_n. Consequently, we use $\overset{\star}{X}_n$ for our detection mechanism.

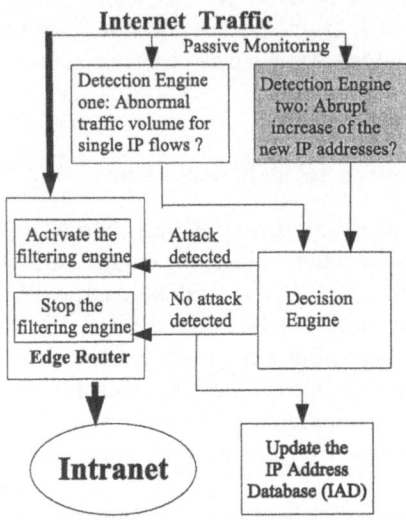

Fig. 1. DDoS attack detection and history-based IP source address filtering

2.3   Implementation of Our Source IP Address Monitoring (SIM) Scheme

Figure 1 provides an overview of our SIM scheme. The SIM scheme consists of three parts: *detection engine*, *decision engine*, and *filtering engine*. The *detection engine* analyzes the incoming traffic pattern to detect any abnormalities. The *decision engine* summarizes the results from the detection engine and decides whether an attack is occurring. The *filtering engine* filters the attack traffic according to the identified attack traffic pattern. Note that there are two detection engines. The first detection engine is used to detect non-distributed attacks from a single source, while the second detection engine is used to detect highly distributed denial of service attacks.

There are two steps in the detection engines. First, the detection engine sorts the incoming IP flows according to source IP addresses, and identifies whether there is an IP flow with an unusually large number of packets. If there is, we activate the *filtering engine* to block this abnormal IP flow. This step is very effective for defending against some naive DoS attacks launched from a single or small number of sources. The second step is the core technology of our SIM scheme, which is shown in the shadow part of Figure 1. This step is designed to defend against sophisticated DDoS attacks and is described in detail in the following sections. As we can see from Figure 1, the detection engine monitors the traffic through a passive (read-only) interface which is pre-configured with a non-routable IP address. This implementation feature can make the detection engine immune to the attacks since it is invisible to the attacker. When no attack is detected in the detection engine, a control signal is sent to the edge router [1] to stop the *filtering engine*.

3 Abrupt Change Detection

In order to detect a DDoS attack, we need to be able to detect changes in our detection feature over time. However, our detection feature is a random variable due to to the stochastic nature of Internet traffic. Consequently, before describing the proposed flooding detection mechanism, we discuss the details of the theoretical background of our detection algorithm.

3.1 Change Detection Modelling

Internet traffic can be viewed as a complex stochastic model and any traffic abnormalities, for example, a HDDoS attack, can lead the abrupt change of the model. Our goal is to detect the change in the number of new IP addresses. There are two approaches to detect this change. One is *fixed-size batch detection*, which monitors the change of mean value every fixed time period. Another is *sequential change-point detection*, which monitors the variables successively. The latter is designed to detect a change in the model as soon as possible after its occurrence, which meets the key design requirement for our detection engine. Thus, we can model our task as a sequential change point detection problem. Consider the illustrative example in Figure 3. For the random sequence $\{X_n\}$, there is a step change of the mean value at m from α to $\alpha + h$. We require an algorithm to detect changes of at least step size h and estimate m in a sequential manner so that the detection delay and false positive rate are both minimized. The random sequence $\{X_n\}$ can be formalized as follows:

$$X_n = \alpha + \xi_n I(n < m) + (h + \eta_n)I(n \geq m), \qquad (1)$$

where $\xi = \{\xi_n\}_{n=1}^{\infty}$, $\eta = \{\eta_n\}_{n=1}^{\infty}$ are random sequences such that $E(\xi_n) = E(\eta_n) \equiv 0$, $h \neq 0$. $I(\mathcal{H})$ is the indicator function, it equals "1" when the condition $\mathcal{H}$ is satisfied and "0" otherwise.

3.2 The CUSUM Algorithm

The CUSUM (Cumulative Sum) algorithm is a commonly used algorithm in statistical process control, which can detect the change of mean value of a statistical process. CUSUM relies on the fact that if a change occurs, the probability distribution of the random sequence will also change. Generally, CUSUM requires a parametric model for the random sequence so that the probability density

---

[1] We use the term *edge router* to refer to the router that provides access to the Internet for the victim's subnetwork that we are defending.

function can be applied to monitor the sequence. Unfortunately, the Internet is a very dynamic and complicated entity, and the theoretical construction of Internet traffic models is a complex open problem, which is beyond the scope of this paper. Thus, a key challenge is how to model $\{X_n\}$. Since non-parametric methods are not model-specific, they are more suitable for analyzing the Internet. In our experiment, we applied the non-parametric CUSUM (Cumulative Sum) method [3] in our detection algorithm. This general approach is based on the model presented in Wang et al. [15] for attack detection using CUSUM. The main idea behind the non-parametric CUSUM algorithm is that we accumulate values of X_n that are significantly higher than the mean level under normal operation. One of the advantages of this algorithm is that it monitors the input random variables in a sequential manner so that real-time detection is achieved.

Let us begin by defining our notation before we give a formal definition of our algorithm. As we mentioned in Sec 2.2, X_n represents the fraction of new IP addresses in the measurement interval Δ_n. The top graph in Figure 3 shows an illustrative example of $\{X_n\}$. In normal operation, this fraction will be close to 0, i.e. $E(X_n) = \alpha \ll 1$, since there is only a small proportion of IP addresses that are new to the network under normal conditions [8] [12]. However, one of the assumptions for the nonparametric CUSUM algorithm [3] is that mean value of the random sequence is negative during normal conditions, and becomes positive when a change occurs. Thus, without loss of any statistical feature, $\{X_n\}$ is transformed into another random sequence $\{Z_n\}$ with negative mean a, i.e. $Z_n = X_n - \beta$, where $a = \alpha - \beta$ (See the middle graph of Figure 3). Parameter β is a constant value for a given network condition, and it helps to produce a random sequence $\{Z_n\}$ with a negative mean so that all the negative values of $\{Z_n\}$ will not accumulate according to time. When an attack happens, Z_n will suddenly become large and positive, i.e. $h + a > 0$, where h can be viewed as a lower bound of the increase in Z_n during an attack. Hence, Z_n with a positive value ($h + a > 0$) is accumulated to indicate whether an attack happens or not (See the bottom graph of Figure 3). One thing worth noting is that h is defined as the minimum increase of the mean value during an attack and it is not the threshold for the bandwidth attack detection. The attack detection threshold N is used for the y_n, accumulated positive values of Z_n, which is illustrated in Figure 3. Our change detection is based on the observation of $h \gg \beta$. Now our detection problem is to find the abrupt change in the random sequence $\{Z_n\}$ which is described as follows:

$$Z_n = a + \xi_n I(n < m) + (h + \eta_n)I(n \geq m), \qquad (2)$$

where $a < 0$, $-a < h < 1$, and other conditions are the same as Eq. 1.

The formal definition of the non-parametric CUSUM algorithm is illustrated as follows:

$$y_n = S_n - \min_{1 \leq k \leq n} S_k, \qquad (3)$$

where $S_k = \sum_{i=1}^{k} Z_i$, with $S_0 = 0$ at the beginning, and y_n is our test statistic. In order to reduce the overhead for online implementation, we use the recur-

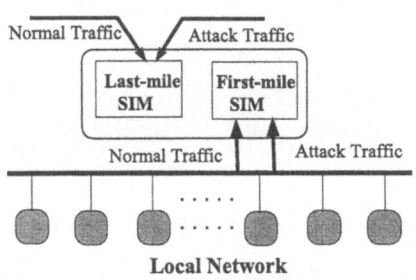

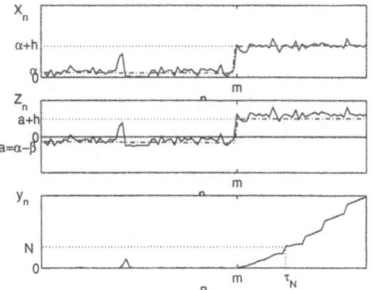

Fig. 2. The trace-driven simulation experiment

Fig. 3. The CUSUM algorithm

sive version of non-parametric CUSUM algorithm [1][3][2][15] which is shown as follows:

$$y_n = (y_{n-1} + Z_n)^+,$$
$$y_0 = 0, \qquad (4)$$

where x^+ is equal to x if $x > 0$ and 0 otherwise. A large y_n is a strong indication of an attack.

As we see in the bottom graph of Figure 3, y_n represents the cumulative positive values of Z_n. We consider the change to have occurred at time τ_N if $y_{\tau_N} \geq N$. The decision function can be described as follows:

$$d_N(y_n) = \begin{cases} 0 \text{ if } y_n \leq N; \\ 1 \text{ if } y_n > N. \end{cases}$$

N is the threshold for attack detection and $d_N(y_n)$ represents the decision at time n: '1' if the test statistic y_n is larger than N, which indicates an attack, and '0' otherwise, which indicates the normal operation (no statistical feature change for the random sequence $\{Z_n\}$).

4   Performance Evaluation

The CUSUM algorithm detects changes based on the cumulative effect of the changes made in the random sequence instead of using a single threshold to check every variable. Therefore, with the deployment of the CUSUM algorithm, the performance of our detection scheme will not be affected by whether the attack rate is bursty or constant. To evaluate the efficacy of our detection scheme SIM, we conducted the following simulation experiments. As shown in Figure 2, we created different types of DDoS attack traffic and merged them with the normal traffic. SIM was then applied to detect the attacks from the merged traffic. The normal traffic traces used in our study are collected at different times from two different sources. The first set was gathered at the University of Auckland [7]

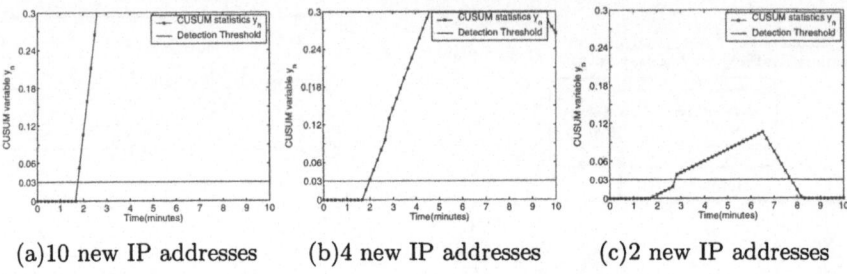

(a)10 new IP addresses   (b)4 new IP addresses   (c)2 new IP addresses

Fig. 4. The DDoS attack detection sensitivity in the first-mile router using the Auck-IV-out trace

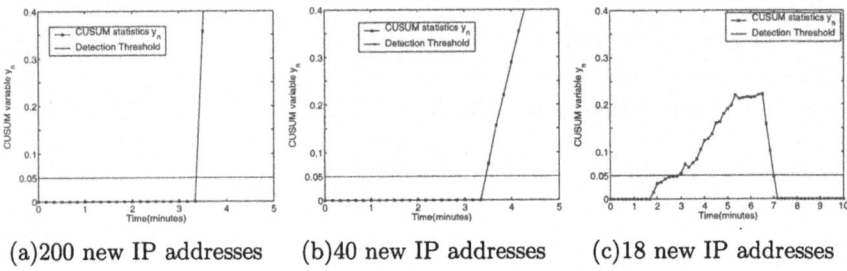

(a)200 new IP addresses   (b)40 new IP addresses   (c)18 new IP addresses

Fig. 5. The DDoS attack detection sensitivity for the last-mile router using the Auck-IV-in trace

with an OC3 (155.52 Mbps) Internet access link [6]. The second data trace was taken on a 9 MBit/sec Internet Connection in Bell Labs [13].

We use the first-mile SIM to monitor the traffic coming into the target network, and last-mile SIM to monitor the traffic going out from the target network. Hence, the outgoing traffic data traces of the University of Auckland can be used evaluate the performance of the first-mile SIM while the incoming traffic data traces can be used evaluate the performance of the last-mile SIM. For the simplicity of the experiment design, we assume the attack traffic rate to be constant. The attack period is set to be 5 minutes, which is a commonly observed attack period in the Internet [11]. The attack traffic rate for all the simulated DDoS attacks is set to be 1 Mbps. Since the network we are defending has the connection capacity of 155.2 Mbps and the average peak connection speed of about 6 Mbps, we define 1 Mbps as the minimum traffic rate to disrupt the network services. We set this conservative attack traffic rate, and aim to test the *detection sensitivity* of the SIM. Attack traffic with higher traffic volume should be easier to detect, and hence is not covered by our performance evaluation.

4.1   DDoS Detection Using Detection Engine Two

In an attempt to avoid detection by our scheme, attackers may try to constrain the number of spoofed IP addresses that they use. Let $\mathcal{W}$ represent the number IP addresses in the attack traffic which are new to the network. We tested

Table 1. Detection Performance of the first-mile router and the last-mile router

The first-mile router			The last-mile router		
W	Detection Accuracy	Detection Time	W	Detection Accuracy	Detection Time
2	99%	69.7s	15	90%	127.3s
4	100%	20.1s	18	100%	81.1s
6	100%	18.9s	40	100%	18.9s
8	100%	10s	60	100%	10s
10	100%	10s	200	100%	10s

different values of W in our simulation, and the detection performance for the first and last-mile routers are shown in Figure 4 and Figure 5 respectively. We repeated the attack detection under a variety of different network conditions, and listed both the average detection accuracy and detection time in Table 1.

As we can see from the simulation results, our detection algorithm is very robust in both the first-mile and last-mile routers. For the last-mile router, we can detect the DDoS attack with $W = 18$ within 81.1 seconds with 100% accuracy, and detect the DDoS attack with $W = 15$ within 127.3 seconds with 90% accuracy. Given the attack traffic length is no more than 5 minutes, only the attack traffic with $W < 18$ has the possibility of sometimes avoiding our detection. However, by forcing the attacker to use a small number of new IP addresses, we can detect the attack by observing the abrupt change of the number of packets per IP source address using the first detection engine which is described in Sec. 2.3.

For the first-mile router, we can achieve 99% detection accuracy even when there are only 2 new IP address in the attack traffic. The reason lies in the fact that the background traffic for the first-mile router is very clear. Generally, there will be very few IP addresses that are new to the network since all the valid IP packets originated from within the same network. Since the IP addresses in the *IP Address Database* (IAD) will expire and be removed after a certain time period, the IP addresses within the subnetworks which have not been used recently will be new to IAD. This is very similar to ingress filtering [5]. However, ingress filtering cannot detect the attack when the spoofed IP addresses are within the subnetworks. In contrast, our first-mile router detection algorithm can detect the spoofed IP addresses within the subnetworks if they are new to the IAD.

It is worth noting that we choose our detection interval $\Delta_n = 10s$ in our experiment, which is a conservative choice for a real implementation. If we decrease the detection interval by using more computing resources, we can reduce the detection time accordingly.

4.2   False Positives and Implementation Overhead

We define a *false positive* as an attack that is reported by the SIM during normal network operation. We use the following method to evaluate the false positive rate of the SIM. We use the data traces collected at the University of Auckland and Bell labs as the normal traffic input to the SIM. If any attack is detected, then a false positive is generated. We randomly choose 20 one-hour-worth normal traffic snapshots as inputs to the SIM. Each experiment runs independently and no false positive is found.

Our trace-driven experiments were run on a Linux machine with dual 900MHz Xeon and 512 MB RAM. We can achieve an average throughput of 10 Gbps. Therefore, the SIM will not be a bottleneck for network implementation. It is worth noting that we use a two-weeks of data traces to build the IAD due to the short of publicly available data traces. In practice, if we can build the IAD using traces of a longer period, we can expect better detection and reaction performances.

5   Discussion

5.1   Possible Attacks against the SIM

If the attackers know that the SIM is based on previous network connections, they could mislead the server to be included in the IP address database. For example, they can first use a certain group of IP addresses to do some reconnaissance before the real attack. The attackers can control the reconnaissance traffic to be sufficiently low so as not to trigger the IP packet filtering process. If the server considers the reconnaissance traffic to be part of the normal traffic, it will add the attacker's reconnaissance IP addresses into the IP address database. Therefore, the attacker can use the IP addresses which they used before to launch the DDoS attack. Since these IP addresses appear in the IP address database, the attack traffic can pass the filter easily, which constitutes a successful denial-of-service attack.

We can prevent this by increasing the period over which IP addresses must appear in order to be considered frequent. Moreover, we can randomize the learning time for the IAD and keep it secret to the attacker. Furthermore, we can ensure that we only include an IP address in our database if it has successfully completed a TCP connection. This prevents the attacker from using spoofed IP addresses for which no host exists. The attacker can only launch their attack using the real IP address of their computer, which makes it much easier to identify and block the source of the attack. We may also be able to use techniques from our previous work on scan detection [9] in order to identify IP addresses with unusual patterns of accesses. Moreover, we can combine additional rules for defining frequent IP addresses in order to improve the accuracy of the SIM. For example, the type of service accessed by the user and the length of each session may be useful measures for identifying frequent IP addresses.

5.2   Other Related Issues

With the deployment of Network Address Translation (NAT), Dynamic Host Configuration Protocol (DHCP) and proxy services, multiple users can share the same source IP address. Moreover, the source IP address can still represent some level of identity, for example, a group of users with geographically proximity. Since the IP addresses in our experimental data traces have been sanitized using one-to-one hash mapping, the network information in the IP address is lost. In practice, we can use network addresses, for example, a class C network address, to represent the user's identity. Moreover, the increasing implementation of IP.v6 [4] will strengthen the correlation between source IP address and user identity.

As high profile websites, such as Yahoo and CNN, will have visitors from all around the world, maintaining the IAD is a very challenging task. Fortunately, the deployment of Content Distribution Network (CDN) [14] has limited the users to their local CDN server. Hence, the users for each CDN server will keep consistent and we can build an IAD for each local CDN server separately.

6   Conclusion and Future Work

In this paper we proposed a scheme to detect distributed denial of service attacks by monitoring the increase of new IP addresses. We have also presented a sequential change point detection algorithm that can identify when an attack has occurred. We demonstrated the efficiency and robustness of this scheme by using trace-driven simulations. The experimental results in the Auckland traces show that we can detect DDoS attacks with 100% accuracy using as few as 18 new IP addresses in the last-mile router and DDoS attacks using as few as 2 new IP address in the first-mile router. Our online detection algorithm is fast and has a very low computing overhead. Our first-mile SIM has the advantage over ingress filtering [5] that it can detect attack traffic with spoofed source IP addresses within the subnetworks. Further, with the combination of two detection engines, all the DDoS attacks can be detected. Our future work will include combining other network traffic statistics to detect bandwidth attacks and using distributed detection to detect DDoS attacks.

Acknowledgement. We would like to thank the Waikato Applied Network Dynamics Research Group, and the Internet Traffic Research group in Bell Labs for making available their data traces. This work is funded by Australia Research Council.

References

1. M. Basseville and I. V. Nikiforov. *Detection of Abrupt Changes: Theory and Application*. Prentice Hall, 1993.
2. Rudolf B. Blažek, Hongjoong Kim, Boris Rozovskii, and Alexander Tartakovsky. A novel approach to detection of "denial-of-service" attacks via adaptive sequential and batch-sequential change-point detection methods. In *Proceedings of IEEE Systems, Man and Cybernetics Information Assurance Workshop*, June 2001.
3. B. E. Brodsky and B. S. Darkhovsky. *Nonparametric Methods in Change-point Problems*. Kluwer Academic Publishers, 1993.

4. S. Deering and R. Hinden. Internet protocol, version 6 (IPv6) specification. RFC 2460, December 1998.
5. P. Ferguson and D. Senie. *Network ingress filtering: Defeating denial of service attacks which employ IP source address spoofing.* RFC2267, IETF, January 1998.
6. Waikato Applied Network Dynamic Research Group. http://wand.cs.waikato.ac.nz/wand/wits/auck/4/.
7. Waikato Applied Network Dynamics Research Group. Auckland university data traces. http://wand.cs.waikato.ac.nz/wand/wits/.
8. Jaeyeon Jung, Balachander Krishnamurthy, and Michael Rabinovich. Flash crowds and denial of service attacks: Characterization and implications for CDNs and web sites. *Proceeding of 11th World Wide Web Conference*, May 2002. Honolulu, Hawaii, USA.
9. C. Leckie and R. Kotagiri. A probabilistic approach to detecting network scans. In *Proceedings of Eighth IEEE Network Operations and Management Symposium (NOMS 2002)*, Florence, Italy, 15-19 April 2002.
10. Ratul Mahajan, Steven M. Bellovin, Sally Floyd, John Ioannidis, Vern Paxson, and Scott Shenker. Controlling high bandwidth aggregates in the network. Technical report, AT&T Center for Internet Research at ICSI (ACIRI) and AT&T Labs Research, February 2001.
11. David Moore, Geoffrey M. Voeker, and Stefan Savage. Inferring Internet Denial-of-Service activity. In *Proceedings of the USENIX Security Symposium*, pages 9–22, August 2001.
12. Tao Peng, Christopher Leckie, and Kotagiri Ramamohanarao. Prevention from distributed denial of service attacks using history-based IP filtering. In *Proceeding of ICC 2003*, Anchorage, Alaska, USA, May 2003.
13. NLANR PMA and the Internet Traffic Research group. Bell Labs - I data set. http://pma.nlanr.net/Traces/long/bell1.html.
14. Dinesh C. Verma, Seraphin Calo, and Khalil Amiri. Policy-based management of content distribution networks. *IEEE Network*, 16(2):34 –39, March-April 2002.
15. Haining Wang, Danlu Zhang, and Kang G. Shin. Detecting SYN flooding attacks. In *Proceedings of IEEE Infocom'2002*, June 2002.
16. David K. Y. Yau, John C. S. Lui, and Feng Liang. Defending against distributed denial-of-service attacks with max-min fair server-centric router throttles. In *Proceedings of IEEE International Workshop on Quality of Service (IWQoS)*, Miami Beach, FL, May 2002.

Secure Name Service: A Framework for Protecting Critical Internet Resources*

Yingfei Dong[1], Changho Choi[2], and Zhi-Li Zhang[2]

[1] Dept. of Electrical Engineering University of Hawaii, Honolulu, HI 96822
yingfei@hawaii.edu
[2] Dept. of Computer Science and Engineering University of Minnesota,
Minneapolis, MN 55455
{choi,zhzhang}@cs.umn.edu

Abstract. We propose a novel Secure Name Service (SNS) framework for protecting critical Internet resources from unauthorized accesses, denial of service (DoS) and other attacks. The key idea is to enforce packet-origin authentication through *resource virtualization* and utilize dynamic name binding for protecting servers under attacks and improving service availability. Different from static network-level security schemes such as IPsec and VPN, SNS is able to dynamically bind the names of critical resources at the service level, which allows us to actively protect the service resources through a distributed filtering mechanism built on authenticated packet forwarding paths. Our prototype implementation of authenticated packet forwarding components on Pentium 4 Linux machines demonstrates that regular Linux platforms are sufficient to support SNS authenticated packet forwarding on 100Mbps or 1Gbps LANs.

1 Introduction

As we become more and more reliant on the Internet for a variety of networking services, the number of network security attacks with the aim to abuse or disrupt such services has also significantly increased. Furthermore, the sophistication of cyber attacks has also increased. The emergence of massive distributed denial-of-service (DoS) attacks is one such example. Unfortunately, because of the *decentralized* and *open* nature of the Internet, it is nearly impossible to protect the entire Internet from cyber attacks. In addition, the cost of such a solution will be economically prohibitive, due to the sheer size of the Internet. It is therefore important to *selectively* secure and protect Internet services that are *critical*, namely, those services that provide significant values. In this paper we propose a novel approach – *Secure Name Service (SNS)* – to protect critical Internet services from cyber attacks. The proposed SNS mechanism serves as a

---

* This work was supported in part by the National Science Foundation (NSF)under the Grant ITR-0085824. Any opinions, findings, conclusions or recommendations expressed in this paper are those of the authors and do not necessarily reflect the views the NSF.

N. Mitrou et al. (Eds.): NETWORKING 2004, LNCS 3042, pp. 783–794, 2004.

comprehensive "first-line of defense" against unauthorized accesses, intrusions as well as DoS attacks. SNS is built upon and an extension of the standard domain name service (DNS). The basic ideas behind the SNS approach are as follows: A critical Internet service and its associated resources (e.g., servers, databases, etc.) are placed within a (virtual) *secure zone* in the network domain of the service provider, and correspondingly the names of the service and its resources are placed within a *secure name space*, separate from the standard domain name space.

Unlike DNS, where in response to a query for a host name, the corresponding IP address of the host is returned, SNS only answers queries originated from *trusted* network domains, and returns a so-called *secure handle (SH)* instead of an IP address in response to a query for a secure name. In other words, the IP addresses of protected resources such as servers are always concealed from the requesters (even from a trusted domain), and the protected resources are in essence "virtualized" from both trusted and untrusted users. Consequently, a unauthorized user cannot gain access to a protected resource (say, a server) directly via IP address spoofing. Furthermore, legitimate packets from a trusted domain carry *security authenticators* – generated by the trusted domain based on secure handles – and are *verified* before they can enter the secure zone containing the protected resources.

In this paper we describe the proposed SNS architecture which is comprised of two major mechanisms: i) *secure name service* that consists of secure name servers that virtualize protected resources within secure zones, set up security associations (SAs) between domains, and perform secure name resolutions; and ii) *authenticated packet forwarding* that consists of *security checkpoints (SCs)* and *security gateways (SGs)*, which verify security authenticators, filter out illegitimate packets, and map secure handles to the IP addresses of protected resources. In addition to *proactive protection*, we also explicitly incorporate *active monitoring* and *rapid response* mechanisms into our proposed architecture for further securing critical services.

The remainder of this paper is organized as follows. In Section 2, we compare the proposed SNS framework with the related work. In Section 3 we describe the design of SNS naming scheme. We present the design of authenticated packet forwarding components and our experimental evaluation in Section 4. In Section 5, we devise two fast lookup schemes for secure name translation and evaluate their performance through analysis and simulation. We conclude the paper in Section 6.

2   Related Work and Discussion

In the SNS framework, we combine name service and network security into a unified framework. We briefly review the related work in naming security, traffic security, entity authentication, and proactive and reactive defense schemes. For naming security, DNSSEC [1,2] mostly focuses on protecting the authenticity and integrity of DNS databases and DNS responses. Although DNSSEC is indeed an effective way to avoid DNS forgery, it does not address the issue of protecting services under attacks. IPsec [3,4] supports traffic security at only the network layer with several limitations. First, IPsec is a rather heavy-duty mechanism which poses many preliminary requirements that hinder its deployment. Furthermore, the scalability of IPsec is a potential issue because an IPsec server needs to negotiate and maintain a security association for each client connection. Lastly, IPsec focuses on traffic security at the network layer, and does not address the issue of protection of service and active defense for improving service availability. Regular VPNs also suffer from this problem. Similarly, TLS [5] ensures the security at the transport layer and does not address the defense issue.

Kerberos [6] is designed for entity authentication that allows a client and a server to mutually authenticate each other across an insecure network. After the mutual authentication, they are able to negotiate a shared secret to exchange encrypted messages for privacy and data integrity. Kerberos does not address the issue of active defense for improving service availability. Existing mechanisms to deal with DoS attacks are often classified into proactive and reactive approaches. Proactive approaches eliminate packets with forged source addresses, such as ingress filtering, Secure Overlay Service (SOS) [7], Mayday [8], and VPN Shield [9]. Ingress filtering uses known unambiguous traffic information to filter out invalid packets at an ingress point, such as source addresses or destination addresses. Therefore, it is suggested for stub domains and low-rate ingress links, but not for transit domains and high-rate links. Ingress filtering does not preclude an attacker using a forged source address within a legitimate prefix filter range. SOS requires a wide-area overlay infrastructure with a large number of intermediate nodes to filter out attacking traffic. VPN Shield provides a limited capability of reacting to flooding attacks. However, it is built on the static IPsec and requires bandwidth reservation at the ingress links of secure domains.

Reactive approaches for DoS attacks include firewalls, IP traceback [10], link testing, input debugging [11], controlled flooding [12], logging [11], ICMP traceback [13], packet marking [12,10], aggregate-based congestion control, and so forth. They all require either the coordination of human administrators of related domains or the modification of intermediate routers. The complexity of the coordination and the slow error-prone human actions hinder the deployment of these approaches. Furthermore, these approaches only work when attacks have caused some damage, and are less useful to stop unknown attacks.

Compared to the related work, the proposed SNS shows several salient advantages. First, the SNS framework provides a comprehensive first-line of defense through resource virtualization and dynamic name binding, which allows us to apply different security policies at multiple levels and components to address different security threats. As a result, it enhances the service availability with low management costs. In particular, SNS distributes the security check load over security gateways (SGs) and security checkpoints (SCs) in authenticated packet forwarding, and therefore significantly reduces the security costs at critical servers. SCs are responsible for filtering out ingress attacking traffic, while

SGs mostly emphasize secure-packet translation. Consequently, critical servers can sustain their service performance under attacks. In contrast, existing approaches such as IPsec or TLS does not address this issue. As a result, a critical server could not sustain its performance under attacks because it has to devote itself to intensive security checking. Furthermore, SNS is incrementally deployable as it does not require to have a broad infrastructure in place, and it does not require to replace application software.

3   Secure Name Service (SNS)

The main functionalities of the SNS naming system are 1) to authenticate hosts, security gateways, and checkpoints in a domain, and manage corresponding security keys and IDs in order to ensure intra-domain packet authentication between hosts and security gateways (or between gateways and checkpoints); 2) to build security associations (SAs) between SNS servers. An SA includes the IP addresses of corresponding security gateways and secret keys for generating and verifying packet authenticators between domains; 3) to maintain a secure name database for secure name resolutions; 4) to resolve secure name queries from trusted hosts. To support these features, we design the SNS naming system consisting of SNS servers, SNS-aware DNS servers, SH managers at SGs, and stub resolvers at hosts. We refer readers to [14] for the details of the secure name service mechanism and components. In the SNS naming framework and forwarding mechanism, we add other three identities combining with an IP address to represent a host at different stages of packet forwarding, i.e., *Secure Handle (SH)*, *Host ID* and *External Identity*. We use a 32-bit secure handle (SH_X) in a response as the *SNS identity* to represent a destination host X when a packet is sent from a host to an SG. This SNS identity is viewed as a *virtual* IP address by applications, and it is used a forwarding label in the authenticated packet forwarding in a secure zone. When a packet is forwarded from an SG to a host, we use the host IP address to represent the host. Because we hide each host behind an SG, to distinguish each host, we assign a host identifier H_ID_X to a host X. In addition, we define (SG_IP_X, H_ID_X) as its *external identity* to represent X outside its home zone, where SG_IP_X is the IP address of the SG for this host X.

   A secure name resolution maps a secure name into an SNS identity (an SH). The basic process of resolving a secure name query is shown in Fig.1. An SNS stub resolver S_1 at a host recognizes an SNS query Q for the identity of a secure name X, and then forwards this query to its SNS. When this query arrives at SG_1, SG_1 authenticates this message and then forwards it to SNS_1. SNS_1 looks up its secure name database and finds the external identity of X, i.e., (SG_IP_X, H_ID_X). (If X is not in the database, SNS_1 will obtain the external identity of X by issuing a secure name query to SNS server SNS_2 that manages secure name X.) Then SNS_1 passes the external identity of X to SH manager M_1 at SG_1 in a response R'. Upon receiving R', M1 first checks if the external identity of X is in its SH database. If it is, M1 finds SH_X from the database;

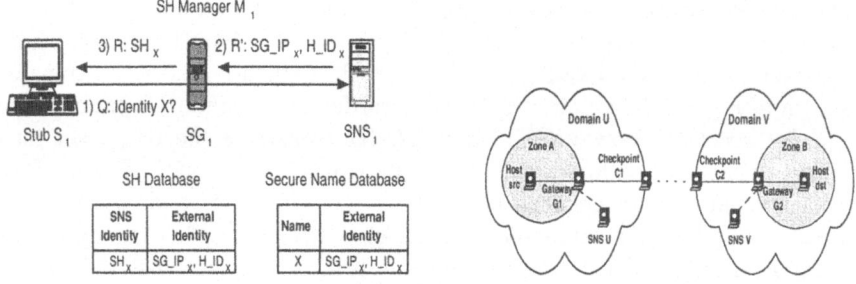

Fig. 1. Resolving a query by a local SNS.   **Fig. 2.** Two SNS-enabled Domains.

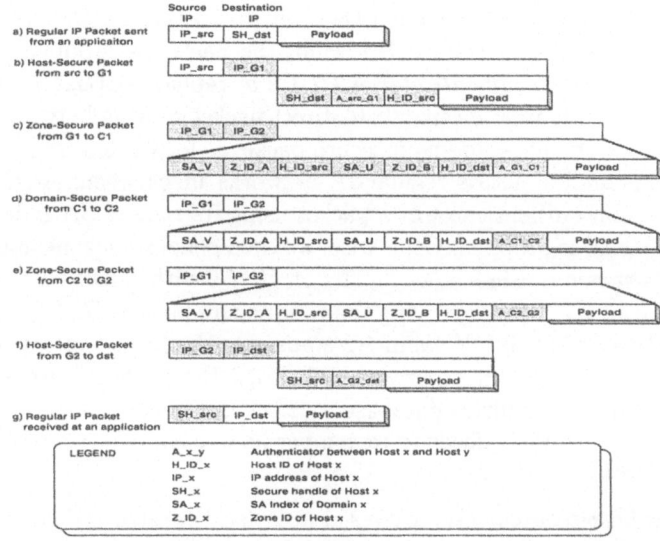

Fig. 3. Packets from Host *src* to Host *dst*.

otherwise, M_1 inserts an entry into the SH database for this external identity and obtains SH_X. Then, M_1 sends a response R to S_1 with the SH_X as the response to query Q.

4   Authenticated Packet Forwarding

The secure packet forwarding mechanism consists of secure IP layers at end hosts, security gateways (SGs) of secure zones, and security checkpoints (SCs) of secure domains. We use an example as shown in Fig.2 to explain how the SNS framework achieves the secure communication between Host *src* in Zone A of Domain U and Host *dst* in Zone B of Domain V, without revealing their IP addresses. Assume an application on host *src* first obtains a secure handle SH_dst of host *dst*, and it then constructs a regular IP packet using SH_dst as the destination address, as shown in Fig.3.a. Before this packet is passed the link layer at *src*, it is intercepted by the sIP layer at *src*. The sIP layer recognizes this packet by its secure handle, and then translates it into a host-secure packet, as shown in Fig.3.b. The packet is then forwarded as a regular IP packet. When the packet reaches gateway $G1$ of Zone A, $G1$ translates the IP packet into a zone-secure packet, and forwards it to checkpoint $C1$, as shown in Fig.3.c. Based on security parameters between $G1$ and $C1$, $G1$ generates and inserts a *zone authenticator* (A_G1_C1) into the packet. As shown in Fig.3.c, the destination host ID H_ID_dst and the remote zone ID Z_ID_B are also inserted into the packet to ensure this packet is correctly routed to the host *dst*. Moreover, the source host ID H_ID_src and the source Zone ID Z_ID_A are also inserted into the packet in order to provide sufficient routing information for return packets to be routed back to host *src* when they return to $G1$. At $C1$, we first check the zone authenticator A_G1_C1. If invalid, the packet is dropped. Otherwise, we compute a *domain authenticator* A_C1_C2 to replace A_G1_C1, as shown in Fig.3.d. We use BGP announcements to direct packet routing between domain U and V such that the above domain-secure packet is forwarded from Checkpoint $C1$ to Checkpoint $C2$ across regular IP networks in between. At $C2$, we first check the domain authenticator of a packet using its remote SA Index SA_U. If invalid, the packet is dropped. Otherwise, we then generate a zone authenticator A_C2_G2. As shown in Fig.3.e, we replace A_C1_C2 with A_C2_G2 in the packet and forward it to $G2$. Upon receiving the zone-secure packet, $G2$ first checks if its zone authenticator is valid. If valid, $G2$ translates the packet into a host-secure packet as shown in Fig.3.f; otherwise, $G2$ drops the packet. Furthermore, $G2$ looks up its remote IP address database to check if it needs to insert a new entry in the database because it needs to remember how to route a return packet from Host *dst* to Host *src*.

When the host-secure packet arrives at host *dst*, the secure IP layer recognizes it as a secure packet based on the protocol field in its IP header. It first translates the host-secure packet into a regular IP packet, and then puts this new packet into the IP input queue. Consequently, an application at Host *dst* receives a regular IP packet as shown in Fig.3.g.

We have implemented the prototypes of sIP layer, SG and SC on Linux kernel 2.4.20 using Linux Netfilter for evaluating authenticated packet forwarding of SNS. We refer readers to [15] for the details of the implementation and introduce the performance results in the following. Utilizing the time stamp counter (TSC) of Pentium CPUs to directly read CPU clock cycles, we can measure the delay at each step of our implementation in clock cycles. We use three Linux machines such as H1, H2, and H3. H1 and H2 houses a 2GHz Pentium 4 processor,512 MB memory, 8KB L1 cache, and 512KB L2 cache. H3 is a 2.8GHz Pentium 4 processor machine with 1GB memory, 8 KB L1 cache, and 512KB L2 cache. We summarize the delays at the components of authenticated packet forwarding in Table 1. For the testing of sIP layer, we send 10,000 UDP packets of 1024 bytes over a direct link between H1 and H2. We also use HMAC-MD5 for MAC generations. The overall delay of the sIP layer is 6879 cycles (3.44μs). We also

Table 1. Delays of Forwarding Components(in clock cycles)

	Authenticator Initialization	MAC Check	Secure Packet Translation	MAC Generation	Total	Effective Bandwidth
sIP	3067	-	-	3812	6879	291 MB
SG	-	4463	450	3587	8500	329 MB
SC	-	4455	-	3869	8324	337 MB

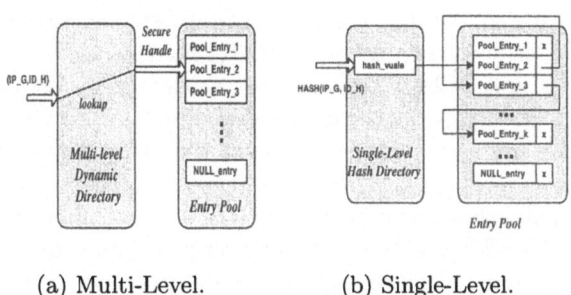

(a) Multi-Level.   (b) Single-Level.

Fig. 4. Two Types of Address Translation Table.

measure the effect of sIP on end-to-end bandwidth using *Iperf* from NLANR (www.nlanr.net). On a 100Mbps link, we can achieve a transmission rate of 93.9 Mbps over regular IP and a transmission rate of 91.9Mbps over sIP, which is 98% of the rate using IP.

To evaluate the performance of an SG, we measure the delays of packet authentication (MAC check), secure packet translation, and MAC generation, as shown in the second row of Table 1. We connect host H1 to H2 through H3, which acts as an SG. Again, we send 10,000 UDP packets of 1024 bytes from H1 to H2. We use the similar setting of SG to test H3 as an SC. The results are also shown in the third row of Table 1. The last column in Table 1 shows that our prototype can support a transmission rate around 300 MBps, which is sufficient for a LAN environment with a 100Mbps or 1Gbps link. The experimental measurements on the prototype of sIP layer, SG and SC on regular Linux machines have shown the feasibility of the SNS authenticated packet forwarding schemes.

5   Dynamic Table Management at an SG

In the process of secure address translation at an SG, we need to authenticate and translate an incoming secure packet based on its address pair (IP_G, ID_H) or an outgoing packet based on its SH, where IP_G is the 32-bit IP address of a remote security gateway and ID_H is a 16-bit remote host ID. To ensure the correct mapping in both incoming and outgoing directions, we need both an SH and a (IP_G, ID_H) pair of the same flow to point to the same entry in the address table. Different from traditional dynamic table mechanisms, which only access tables through a primary key, we need to use both a (IP_G, ID_H) pair and an SH to access an address entry. Therefore, we design a two-layer structure to address this issue. At the lower layer, we use an *Address Entry Pool* consisting of address entries, which allows us directly to access address entries using its indexes as SH's. At the upper layer, we build a *dynamic directory* for fast lookups based on a primary key, i.e., (IP_G, ID_H) pair. For fast lookups based on (IP_G, ID_H) pairs, we design a multi-level directory scheme and a single-level directory scheme described in the following. The corresponding structure of entry pools is shown in Figure 4. We first propose a *Multi-Level Directory Scheme*. Let us denote a 48-bit primary key, a (IP_G, ID_H) pair, as $k_{47}k_{46} \cdots k_0$. At the first level, we use the first 16 bits, $k_{47}k_{46} \cdots k_{32}$, as the index. We use the next 8-bit $k_{31}k_{30} \cdots k_{24}$ as the index of the second-level directory. Similarly, at level three, four and five, we use corresponding 8 bits as the index of subdirectories.

We also design a *Single-Level Hashing Scheme* to reduce potential delays and memory cost in the above scheme, because the total number of hosts is assumed to be smaller than 2^{32} and using 48 bits as a primary key may result in an uneven directory tree, which causes unnecessary delays in operations. In this scheme, we need to search through a list by comparing the primary keys of a list to find an SH, because we allow collisions on a table entry. We use hash value v to find the header of a list, where $v = H_1(IP\_G, ID\_H)$, and hash function H_1 is implemented using Knuth's multiplication method [16], which can be computed in less than 100 clock cycles on Pentium-4.

We analyze the performance of the above directory schemes in the following. Let us first define the traffic model used in evaluation. Assume we have N clients, each has an on-period T_i^{on} seconds with a rate of r_i packets/sec, and an off-period T_i^{off} seconds, where $1 \le i \le N$. Then the average number of active flows generated by clients will be $N_{active} = \sum_{i=1}^{N} \frac{T_i^{on}}{(T_i^{on} + T_i^{off})} \cdot N$.

For a packet j, the probability that it belongs to an existing flow i is $P[j \in flow\ i] = \frac{r_i}{\sum_{k=1}^{N_{active}} r_k}$. We assume that an address entry is expired after each on-period. Then we need to insert an address entry for a flow in each on-off cycle. The probability that packet j causes a table insertion for flow i is $P[j\ causes\ an\ insertion] = \frac{1}{T_i^{on} \cdot r_i}$. Therefore, for packet j, the probability that it causes an insertion for flow i is $P_{insert}^{(i)} = P[j \in flow\ i] \cdot P[j\ causes\ an\ insertion]$.

We first analyze the performance of the multi-level directory scheme under the above traffic model. Figure 5 shows the lookup algorithm that decides the action for a packet of flow i, whose address is fallen into directory entry e. Consider level l directory with 2^k entries, where $k = 16$ when $l = 1$, and $k = 8$, when $2 \le l \le 5$. Let N_l be the current flow population in level l and its sub-directories. We know $N_1 = N_{active}$. Assume client addresses are uniformly distributed across the whole directory, the expected population in the level l is $N_l = \frac{N_1}{2^{16+8 \cdot (l-2)}}$, $2 \le l \le 5$.

Assume packet j arrived at directory level l is fallen into an entry e with a uniform probability of $\frac{1}{2^k}$. Let $p_0^l = P^l[e = 0]$ be the probability that entry e is not occupied currently (i.e., flag $F = 0$); $p_1^l = P^l[e = 1]$ is the probability that entry e is currently occupied by a single flow (i.e., flag $F = 1$), and $p_2^l = P^l[e = 2]$

```
1. if (entry e is empty)
2.    INSERT(i); // insert client i into entry e
3.    return a secure handle;
4. else
5.    if (exact one client is in entry e)
6.       if (i is the same as the client in entry e)
7.          return a secure handle;
8.       else // collision
9.          EXPAND(); // expand a next-level directory
10.         INSERT(i); INSERT(i'); // insert both into the next level
11.         return a secure handle;
12.    else // at least two clients are in entry e
13.       step down into the next level directory.
```

```
1. if H_i(key) >= p
2.    index = H_i(key);
3. else
4.    index = H_{i+1} (key)
5. access the entry at the index;
6. search through a overflow list if necessary;
```

Fig. 5. Lookup of Multi-Level Directory   **Fig. 6.** Lookup in Linear Hashing.

is the probability that entry e is currently occupied by more than one flow (i.e., flag $F = 2$), and thus it is expanded into the next level $l + 1$ (for $l < 5$). Then we have $p_0^l = (1 - \frac{1}{2^k})^{N_l}$, $p_1^l = (1 - \frac{1}{2^k})^{N_l-1} \cdot \frac{1}{2^k}$, and $p_2^l = 1 - p_0^l - p_1^l$. Because of no collisions in the fifth level, we have $p_0^5 = 1$, $p_1^5 = 0$, and $p_2^5 = 0$. Therefore, the expected delay of inserting a new entry into a directory at level l and its sub-directories, denoted by D_{insert}^l, is given recursively by Equation 1.

$$D_{insert}^l = d_{flag} + p_0^l \cdot d_{insert} + p_1^l [d_{compare} + d_{expand} + E_{insert}^{l+1}(i, i')] \\ + p_2^l [d_{down} + D_{insert}^{l+1}] . \tag{1}$$

where d_{flag} is the delay to determine the flag value of a directory entry, d_{insert} is the delay to insert client information into an entry, $d_{compare}$ is the delay to compare the destination of a packet with that of an existing entry, d_{expand} is the delay to expand a sub-directory in the next level, d_{down} is the delay to step down into the next-level sub-directory, and $E_{insert}^{l+1}(i, i')$ is the delay to insert two distinct entries, i and i', into a newly-expanded sub-directory at level $l + 1$, as defined in Equation 2.

$$E_{insert}^l(i, i') = \frac{1}{2^{16+8\cdot(l-1)}} E_{insert}^{l+1}(i, i') + (1 - \frac{1}{2^{16+8\cdot(l-1)}}) \cdot 2 \cdot d_{insert} . \tag{2}$$

where $2 \leq l \leq 4$. For $E_{insert}^5(i, i') = 2 \cdot d_{insert}$ because no collision occurs at the fifth level. The expected delay of searching an entry at level l and its sub-directories, denoted by D_{lookup}^l, is given recursively by Equation 3.

$$D_{lookup}^l = d_{flag} + p_1^l \cdot d_{compare} + p_2^l [d_{down} + D_{lookup}^{l+1}] . \tag{3}$$

In summary, for the packets of flow i, the expected delay of an address insertion is D_{insert}^1, and the expected delay of an address lookup is D_{lookup}^1. Then the expected delay of a directory lookup/insertion is thus:

$$D(i) = P_{insert}^{(i)} \cdot D_{insert}^1 + (1 - P_{insert}^{(i)}) D_{lookup}^1 . \tag{4}$$

Now let us analyze the expected memory cost in the multi-level directory scheme. First, we always allocate the top level directory with 2^{16} entries. Then, for each collision on an entry, we allocate a sub-directory of 2^8 entries. For each flow i, it may cause an expansion of a sub-directory at level $l + 1$ if it is collided with another address entry at level l (i.e., when flag $F = 1$), $1 \leq l \leq 4$. The probability that flow i is collided with another entry at level l is $m(i, l) = (\prod_{k=1}^{l-1} p_2^k) \cdot p_1^l$. Therefore, the potential memory cost due to flow i is $m_i = \sum_{l=1}^{4} m(i, l)$. The potential memory cost of N_1 flows is denoted as M, where $M = \sum_{i=1}^{N_1} m_i$.

We now analyze the performance of the linear hashing directory scheme. Assume we initialize the directory with $\tilde{N}_0$ entries, say $\tilde{N}_0 = 2^8$. Assume we have a perfect hashing function, then the memory cost of the single-level directory for a population of N_1 is denoted as $M_{N_1} = \tilde{N}_0 \cdot 2^k$, where $k = \lfloor log_2 N_1 / \tilde{N}_0 \rfloor$, such that $2^{k-1} \cdot \tilde{N}_0 \leq N_1 \leq 2^k \cdot \tilde{N}_0$. We only expand the directory after $2^{k-1} \cdot \tilde{N}_0$ collisions.

For each packet, we need to first search the table to check if it has a corresponding entry there. If not, we then insert an address entry. The probability that the address of the packet is hashed into an empty directory entry is $p_0 = P[X = 0] = (1 - \frac{1}{2^k})^{N_1}$, while the probability that its address is hashed into an occupied directory entry is $p_1 = 1 - p_0$. The search procedure of linear hashing is shown in Figure 6.

$$D_{lookup} = d_{hash} + d_p + D_{list} .\tag{5}$$

where d_{hash} is the delay of computing the hashing function, d_p is the delay to compare with a splitting pointer p, and D_{list} is the expected delay of searching through the overflow list. For a good hashing function, we assume that the average length of the list is less than two. As a result, the upper bound of the delay of searching the list is $D_{list} \leq 1.5 \cdot d_{compare} + 0.5 \cdot d_{next}$, where $d_{compare}$ is the delay to compare the address of the packet with the address in a name entry, and d_{next} is the delay to access the next entry on a list. We then have

$$D_{insert} = p_0 \cdot d_{insert} + p_1 \cdot (d_{hash} + d_p + D_{list} + d_{insert}) .\tag{6}$$

And the expected lookup/insertion delay of packets of flow i is

$$D(i) = P_{insert}^{(i)} \cdot D_{insert} + (1 - P_{insert}^{(i)}) \cdot D_{lookup} .\tag{7}$$

We measure the delay of memory read/write and hashing computation in Linux kernel and plug in these parameters into our models. Figure 7 shows the comparison of the multi-level approach with a perfect linear hashing approach. For a uniform distribution of addresses, although the multi-level approach does well for a small population, its delay grows as the population increases. We also test the multi-level approach with a skewed input, in which all address entries are in a single directory entry at the first level and they are uniformly distributed below the first level. In this case, the delay of multi-level approach is increased significantly. While the linear hashing approach keeps a constant delay under the assumption of a perfect hashing function. In addition, the memory cost of the hashing approach is less compared with the multi-level approach, as shown in Figure 7.b.

We also conduct simulations to evaluate the two schemes. We use a multiplication approach for fast computing hash values, and generate a random set of

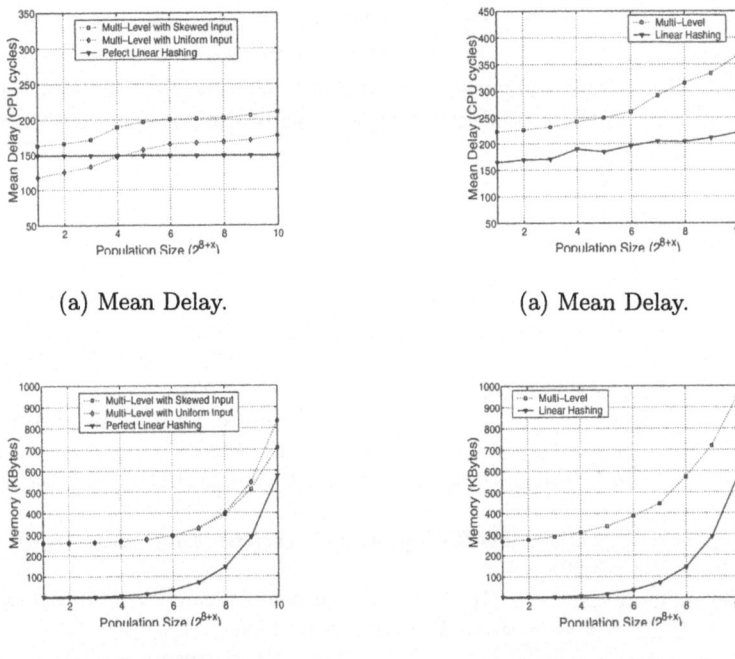

(a) Mean Delay.   (a) Mean Delay.

(b) Memory Cost.   (b) Memory Cost.

Fig. 7. Analytical Models.   **Fig. 8.** Simulations.

address lookups. Figure 8.a shows the mean delay of the hashing scheme is significantly better than the multi-level scheme. Figure 8.b shows that the memory cost of the hashing scheme is also better than the multi-level scheme.

6   Conclusion and Ongoing Work

We have proposed the SNS framework to protect critical resources from unauthorized accesses and DoS attacks. Through the *resource virtualization* of SNS, we build a distributed filtering scheme to enforce packet-origin authentication. We have described the basic design of the SNS framework, and addressed the performance bottleneck in its authenticated packet forwarding. Based on our prototype on Linux, we have shown the feasibility of implementing SNS on regular Linux machines. We have also designed two fast secure-handle schemes to address the scalability issue in fast address translation. To fully exploit the advantages of the SNS framework, we face several challenges in the design of the SNS framework, i.e., scalability, reliability, efficiency, and easy deployment. For reliability, we need to protect security gateways from attacks (such as packet replay and flooding) because these gateways are exposed to attackers. We will address this issue from two perspectives. First, we will evaluate the tradeoffs between computation costs and probabilities that invalid packets penetrate an ingress filtering mechanism using Bloom Filter [17]. Furthermore, we will investigate the effect of reconstructing dynamic packet forwarding paths to defeat attacks. Currently, we are working on these issues and implementing the complete SNS framework for further investigation.

References

1. R. Arends and et.al, "DNS security introduction and requirements," *Internet Draft, draft-ietf-dnsext-dnssec-intro-03, IETF*, Oct. 2002.
2. G. Ateniese and S. Mangard, "A new approach to DNS security (DNSSEC)," *ACM Conf. on Computer and Communications Security*, 2001.
3. S. Kent and R. Atkinson, "Security architecture for the internet protocol," *RFC2401, Internet Engineering Task Force*, Nov. 1998.
4. D. Harkins and D. Carrel, "The internet key exchange (IKE)," *RFC2409,Internet Engineering Task Force*, Nov. 1998.
5. E. Rescorla T. Dierks, "The TLS protocol," *Internet Draft, draft-ietf-tls-rfc2246-bis-02.txt*, Oct. 2002.
6. B. Neuman and T. Ts'o, "Kerberos: An authentication service for computer network," *IEEE Comminucation Magazine*, Sept 1995.
7. A. Keromytis, V. Misra, and D. Rubenstein, "SOS: Secure overlay services," *In Proc. of ACM SIGCOMM'02*, 2002.
8. David Aderson, "Mayday: Distributed filtering for internet services," *4th Usenix Symposium on Internet Technologies and Systems, Seattle, Washington*, March 2003.
9. R. Ramanujan and et. al., "Organic techniques for protecting virtual private network (vpn) services from access link flooding attacks," *International Conference on Networking'02*, 2002.
10. Stefan Savage, David Wetherall, Anna Karlin, and Tom Anderson, "Practical network support for IP traceback," *Proc. of the 2000 ACM SIGCOMM Conference, Stockholm, Sweden*, Aug., 2000.
11. R. Stone, "Centertrack: An IP overlay network for tracking DOS floods," *Proc. of 2000 USENIX Secuirty Symposium*, July, 2000.
12. H. Burch and B. Cheswick, "Tracing anonymous packets to their approximate source," *Unpublished Paper*, Dec. 1999.
13. "IETF ICMP traceback working group," *http://www.ietf.org/html.charters/itrace-charter.html*.
14. Y. Dong, C. Choi, and Z.-L. Zhang, "Design of secure name service," *Techincal Report, CS, UMN*, 2003.
15. C. Choi, Y. Dong, and Z.-L. Zhang, "Implementation of SNS authenticated packet forwarding mechanism," *Techincal Report, CS, UMN*, 2003.
16. Thomas Cormen, Charles Leiserson, and Ronald Rivest, "Introduction to algorithm," *MIT Press, ISBN 0262031418*, 1986.
17. B. Bloom, "Space/time trade-offs in hash coding with allowable errors," *Communications of the ACM, 13 (7). 422-426.*

Detecting and Blocking Unauthorized Access in Wi-Fi Networks

Haidong Xia and José Brustoloni*

University of Pittsburgh, Dept. Computer Science, 210 S. Bouquet St. #6135,
Pittsburgh, PA 15260, USA,
{hdxia,jcb}@cs.pitt.edu

Abstract. Academic and commercial 802.11 hotspots often use an SSL-secured captive portal to authenticate clients. Captive portals provide good usability and interoperability, but poor security. After a captive portal has authenticated a client, session hijacking and freeloading allow attackers to capture or use the client's session. Freeloading does not require special tools and, surprisingly, is strengthened by the (widely recommended) use of personal firewalls. We propose and evaluate novel defenses against these attacks, session id checking and MAC sequence number tracking, both of which are transparent to clients and do not require changes in client computers. Experiments demonstrate that the proposed defenses are effective against the mentioned attacks and have little overhead.

1 Introduction

Wi-Fi networks [1,2] provide an unprecedented combination of low cost, high bandwidth, and support for mobility. Consequently, they have become extremely popular, as evidenced by the recent 42%-a-year growth rate of the Wi-Fi chipset market [3]. Wi-Fi networks do have, however, a major weakness: poor security. Wi-Fi's original security scheme, WEP (Wired Equivalent Privacy), has been demonstrated to be easily broken [4,5,6,7,8].

Security has to be balanced with other requirements, such as usability and interoperability. In Wi-Fi networks, the relative importance of these requirements varies with application, as summarized in Table 1. *Enterprise* networks are installed in companies or government offices and therefore require a high level of security, with mutual authentication between mobile stations and network at connection time and encryption and authentication of all user traffic after that. On the other hand, an enterprise usually can provide technical support to users and owns all equipment connected to its network. Consequently, security solutions that end users find difficult to install or configure or that are proprietary can be used in such networks. *Home* networks differ from enterprise networks in

---

* This project was funded in part by the Pittsburgh Digital Greenhouse through a grant from the Commonwealth of Pennsylvania, Department of Community and Economic Development.

N. Mitrou et al. (Eds.): NETWORKING 2004, LNCS 3042, pp. 795–806, 2004.

Table 1. The relative importance of security, usability, and interoperability in Wi-Fi networks varies according to application. In some cases, users prize usability and interoperability more than security.

Wi-Fi Application	Requirement		
	Security	Usability	Interoperability
Enterprise	High	Medium	Medium
Home	High	High	Medium
Access	Medium	High	High
Open	Low	High	High

that they need to operate with little or no technical support. Therefore, security solutions for such networks have to be easy to install, configure, and use. *Access* networks using Wi-Fi are often deployed in locations such as universities and commercial hotspots. Their primary purpose is to provide Internet connectivity to users who are away from the respective offices or homes. Access networks typically need to block unauthorized users, while interoperating with a wide variety of user-owned equipment and providing little or no on-site technical support. Access networks usually do not attempt to secure user communication. At least in principle, local security would be either redundant, if the user uses end-to-end security protocols, such as IPsec [9] or SSL [10], or insufficient, if the user does not use such protocols (local security cannot prevent attacks from occurring elsewhere on the user's communication paths). *Open* networks need to interoperate with other equipment quickly, with minimal or no configuration or technical support, but do not provide any security. In theory, this mode exists only for initial equipment configuration and testing. However, nearly 70% of existing production Wi-Fi networks actually operate in this mode [11]. This fact is contrary to what a designer might imagine or prefer, and shows that users do, in many cases, prize ease-of-use and interoperability more than security.

There currently isn't a scheme that simultaneously provides high security, usability, and interoperability in Wi-Fi networks. Several solutions exist or are being developed for securing enterprise Wi-Fi networks, including IPsec, WPA (Wi-Fi Protected Access) [2], IEEE 802.11i [12], and a variety of proprietary alternatives. However, these solutions currently involve installation, configuration, or interoperation difficulties that make them poorly suited for home or access networks.

Access Wi-Fi networks typically use MAC address filtering or captive portals to authenticate users, as explained in Section 2. These schemes are easy to understand and use and do not require special client hardware or software. However, they can be defeated by MAC address spoofing, session hijacking, or freeloading attacks, as explained in Sections 2, 3, and 4. The latter attack was previously unreported. It requires no special tools, is strengthened by the (widely recommended) use of personal firewalls, and can easily go undetected.

This paper contributes novel mechanisms for detecting and blocking session hijacking and freeloading attacks in access Wi-Fi networks. The proposed mech-

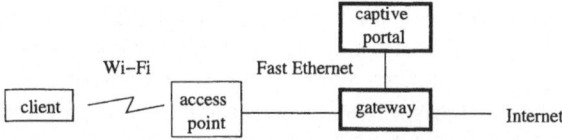

Fig. 1. Captive portals are widely used for user authentication at universities and commercial hotspots, but they are vulnerable to session hijacking and freeloading attacks.

anisms, session id checking and MAC sequence number tracking, are transparent to users and work well with default hardware and software configurations, which makes them well-suited for access applications. Experiments demonstrate that the proposed defenses are effective and impose little overhead.

The rest of this paper is organized as follows. Section 2 discusses previously proposed security schemes for Wi-Fi networks. Section 3 describes session hijacking attacks and a novel defense against them, session id checking. Session 4 characterizes freeloading attacks and presents MAC sequence number tracking, a novel defense against them. The new defenses are evaluated experimentally in Section 5 and discussed in Section 6. Finally, Section 7 concludes.

2   Related Work

MAC address filtering is a simple solution for authenticating users in an access Wi-Fi network. It consists in configuring access points so that they accept association only of computers using certain MAC addresses. Most access points allow this type of configuration, and no changes are necessary in client computers.

Unfortunately, this solution is very insecure. Attackers can simply sniff the network to find the MAC address of an approved computer. Applications for such sniffing are freely available from the Internet (e.g., ethereal). Attackers can then spoof their own MAC address to be that of an approved computer. In Windows, several Wi-Fi drivers support MAC address spoofing from the control panel (e.g., Dell TrueMobile). Even if the driver does not support it, spoofing can be performed using the registry; an application that automates the process (smac) is freely available from the Internet. In Linux, MAC address spoofing is enabled by the built-in command `ifconfig ethXX hw ether XX:XX:XX:XX:XX:XX`.

Many networks combine MAC address filtering with suppression of access points' 802.11 beacon messages. This technique has limited value because user network traffic can still be picked up by sniffers. Some networks combine MAC address filtering with WEP (using static keys). This technique also has limited value, since attackers often can obtain the key by social engineering or by using freely available tools, such as airsnort or WEPcrack.

Captive portals were first proposed in Stanford's SPINACH project [13]. They are widely used for user authentication in access Wi-Fi networks at universities and commercial hotspots. Captive portals require a special gateway between the Wi-Fi network and the rest of the network, as depicted in Figure 1. The gateway allows unauthorized clients to use DHCP, ARP, and DNS for initial configuration. The gateway also redirects any Web requests from unauthorized clients to the network's captive portal. The captive portal is an SSL-secured Web page where unauthorized users enter their id and password. The captive portal may use a variety of back-ends for user authentication, e.g. Kerberos, RADIUS, or LDAP. The gateway drops any other traffic of unauthorized clients. After authentication, the captive portal authorizes the client to access the rest of the network. Authorization is done by registering the client's MAC and IP addresses in the gateway. The captive portal typically also sends the client a *session management page* that contains a button for terminating the session. This page is usually displayed on a small pop-up window that is not used for browsing. Finally, the captive portal redirects the client to the page that the client had initially requested.

Captive portals do not require special hardware or software configuration in client computers: most contemporary computers already have a Web browser. The Web-based interface is also very intuitive. Captive portals are significantly more secure than is MAC address filtering because they resist simple MAC address spoofing attacks. However, captive portals are still vulnerable to session hijacking and freeloading attacks, as discussed in Sections 3 and 4.

IPsec was proposed for securing access Wi-Fi networks in [14]. However, IPsec configuration and interoperation continue to be problematic. Recent Microsoft operating systems include a VPN (Virtual Private Network) client that implements IPsec in a way that does not support nested secure connections (L2TP over IPsec transport mode). If this VPN client is used for local security, it is not available for secure end-to-end connections, which is unacceptable. Third-party IPsec clients can be used, but IPsec's excessive configuration options can be daunting. Moreover, IPsec currently omits several details that are needed for mobile access, such as methods for obtaining networking configuration parameters from a VPN gateway and support for legacy user authentication methods. These omissions are often supplanted by non-standard drafts or proprietary extensions. IKEv2 is a future version of IPsec's IKE protocol that hopefully will resolve interoperability problems. However, it will not address configuration problems, which continue to hamper the use of IPsec in access networks.

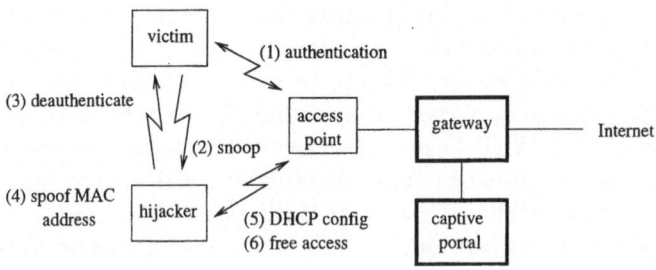

Fig. 2. Low-level 802.11 vulnerabilities allow a hijacker to gain control of a session of an authorized user.

PANS is a security scheme for access Wi-Fi networks that provides high security [15]. However, it requires installation of non-standard software in client computers. Non-standard schemes that provide high security are also available from several vendors, including Cisco (LEAP). Proprietary schemes can be easier to install and configure than is IPsec. However, it is highly desirable that access networks be able to interoperate with a wide range of client equipment; proprietary solutions preclude that, unless they become de facto standards.

IEEE 802.1x [16] is enabling significant improvements in native Wi-Fi security, including WEP with dynamic per-session keys (not a standard, but supported by Microsoft), WPA, and IEEE 802.11i. There are significant difficulties for adopting such schemes in access Wi-Fi networks. First, many details of these solutions have not yet been fully worked out and continue to change. Second, these solutions require the installation of new software, firmware, and possibly hardware in client computers, and these modifications need to implement the same draft versions as those adopted in the network's access points and authentication servers. Third, ongoing improvements are creating a multitude of configuration options that could be confusing for end users. For example, a variety of schemes may be used for initial authentication (e.g., pre-shared keys, EAP-TLS, or PEAP); if PEAP is used, many legacy user authentication schemes are possible (e.g., using passwords, one-time passwords, or tokens); and several packet encryption algorithms will soon be in use (e.g., WEP with dynamic keys, TKIP [2], and CCMP [12]). Resolution of these difficulties requires technical support at a level that typically is not offered in access Wi-Fi networks.

3   Session Hijacking and Session Id Checking

This section describes session hijacking and a new defense against it, session id checking.

Session hijacking is illustrated in Figure 2. The hijacker snoops on the victim's and the access point's MAC addresses. The hijacker then periodically sends to the victim disassociation or deauthentication notifications purported to come from the access point. According to the IEEE 802.11 standard, these notifications are not authenticated and must be obeyed. They cause denial of service to the victim. (The notifications need to be periodically repeated because the victim typically attempts to reauthenticate some time after the last notification.) The hijacker then spoofs his or her MAC address to be the same as the victim's and obtains the victim's networking configuration from the DHCP server. (Alternatively, the hijacker gleans this information by sniffing network traffic beforehand.) Using the victim's MAC and IP addresses and other configuration parameters, the hijacker can then access the rest of the network without being authorized to do so.

Session id checking detects and blocks session hijacking as follows. Captive portals usually send to authorized clients a session management page on a small pop-up window. Session id checking (1) associates with this page a secure non-persistent cookie containing a cryptographically random session id of

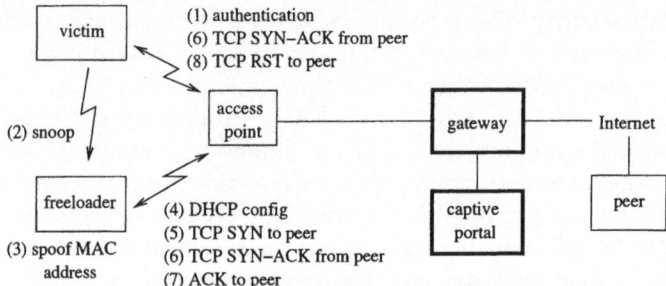

Fig. 3. A freeloader simply uses the MAC and IP addresses of an authorized victim. The victim also receives packets destined to the freeloader. These packets may elicit responses from the victim that disrupt the freeloader's communication, as shown. However, personal firewalls inhibit such responses and enable both freeloader and victim to communicate reliably.

sufficient length (e.g., 128 or more bits); (2) tags the page with the directive `http-equiv="refresh"` and a certain period; and (3) secures the page with SSL.

The refresh directive periodically causes the client's browser to request the captive portal to retransmit the page. Because the page has an associated cookie and is SSL-secured, each such request is accompanied by the cookie containing the client's session id and is also SSL-secured.

The captive portal detects that an authorized client's session has been hijacked when, after an entire period, the captive portal has not received a refresh request with the client's IP address and session id. A hijacker cannot guess the victim's session id because it is cryptographically random, and cannot capture it by eavesdropping because its transmission is SSL-secured. The captive portal then blocks the hijacker's communication by unregistering the victim's MAC and IP addresses in the gateway between the Wi-Fi network and the rest of the network.

4   Freeloading and MAC Sequence Number Tracking

This section describes freeloading and a new defense against it, MAC sequence number tracking.

The freeloading attack consists in spoofing a victim's MAC and IP addresses while the victim remains associated and communicating normally. In principle, freeloading would be expected to perform unreliably, because the victim also receives packets destined to the freeloader and may react in ways that disrupt the freeloader's communication. This problem is illustrated in Figure 3. In this example, the freeloader sniffs the MAC address of an authorized victim and spoofs the freeloader's MAC address to be the same as the victim's. The freeloader then obtains the rest of the victim's networking configuration from the DHCP

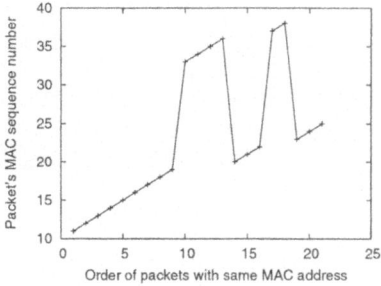

Fig. 4. Freeloading causes interleaving of different senders' MAC sequence number trend lines

server (alternatively, the freeloader gleans this information by sniffing the network beforehand). After that, the freeloader opens a TCP connection with the usual three-way handshake (send a SYN segment, receive a SYN-ACK segment, and send an ACK segment). However, the victim also receives the SYN-ACK segment destined to the freeloader. Given that the victim does not know about this connection, the victim, following the TCP standard, sends a RST segment to the segment's source (i.e., the freeloader's peer). This aborts the freeloader's connection.

If the victim and freeloader use personal firewalls, however, freeloading works reliably. Personal firewalls are widely recommended and quite common. For example, NIST recommends that all U.S. government employees use a personal firewall when on a Wi-Fi network [17]; the trade press makes similar recommendations to business users; and newer operating systems, such as Windows XP, often have a built-in personal firewall. A personal firewall typically allows the respective computer to respond only to packets that the firewall recognizes as part of a session initiated by that computer. Personal firewalls usually interpret other packets as attempts to fingerprint the respective computer's software or find vulnerable ports. Consequently, personal firewalls typically inhibit responses to packets that they do not recognize. In a victim computer, such packets include those that are actually destined to a freeloader. Thus, the victim and freeloader do not respond to or interfere with each other's communication. Because both freeloader and victim get access, freeloading may occur in collusion against the network's owner (e.g., commercial hotspot).

MAC sequence number tracking detects freeloading by observing that IEEE 802.11 frames contain a 12-bit sequence number that should increment by one for each new datagram sent, and remain the same in case of MAC-layer fragmentation or retransmission. Because of the tight timing constraints of MAC-layer acknowledgements and retransmissions, 802.11 sequence numbers are set and verified by network interface card (NIC) hardware or firmware. Host software typically cannot set such numbers.

Consequently, an access point or other device can detect freeloading by noticing that the MAC sequence numbers of successive packets with the same MAC

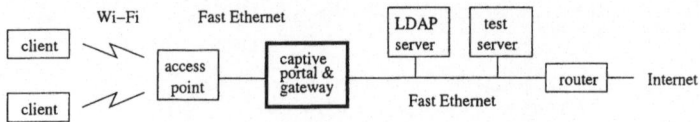

Fig. 5. Testbed used in the experiments

address form more than one trend line, as illustrated in Figure 4. One of the trend lines corresponds to the authorized client, while other trend lines correspond to freeloaders. The detecting device blocks freeloaders' communication by unregistering the victim's MAC address in the gateway between Wi-Fi network and the rest of the network.

Note that a simple jump in MAC sequence number is not a good criterion for detecting freeloading. Simple jumps occur also for other reasons, e.g. because a client has moved out of range and then back in range, or interference has caused several datagrams to be lost. For robustness, we detect freeloading when the MAC sequence number *returns* from one trend line to the previous trend line.

5 Experimental Evaluation

We performed experiments for evaluating the proposed defenses and report in this section the results obtained.

We used the testbed shown in Fig. 5. The access point is an IBM T30 1.8 GHz PC with built-in Intersil Prism 2.5-based 802.11b interface, running Linux 2.4.20 and HostAP driver. We modified HostAP to support MAC sequence number tracking. The captive portal/gateway and LDAP and test servers are Dell Dimension 4550 2.4 GHz PCs running Linux 2.4.20. We implemented session id checking on the captive portal, which also uses the Apache Web server. The LDAP server uses OpenLDAP. As clients, we used IBM T30, Dell, and Sony laptops, 5 Sharp Zaurus PDAs, and 10 Dell Dimension 8300 2.6 GHz PCs with a variety of Wi-Fi network interface cards (NICs), including Linksys, Netgear, D-Link, Proxim Orinoco Gold, and Cisco Aironet 350.

We tested session hijacking using a modified version of the airjack toolset, which is freely available from the Internet. We measured throughput using the ttcp benchmarking application between the test server and a test client. Delay was measured by pinging the test server from a test client and capturing packet times using Ethereal on the Wi-Fi network. Reported results are the average of five tries.

In the first experiment, we had one of the test clients hijack the session of another test client, after the captive portal authorized the latter client's access. We verified that session id checking promptly detects and blocks session hijacking. We varied the refresh period of the session management page between 1 s and 10 s, and verified that this period controls the latency for detecting and blocking a hijacked session.

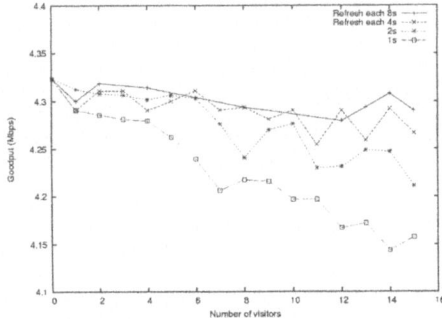

Fig. 6. Session id checking's impact on network throughput is quite low, especially at longer refresh periods

In the second experiment, we measured the throughput between the test server and a test client as the number of authorized clients varied, using session id checking but not MAC sequence number tracking, and for several different refresh periods. The results are shown in Figure 6. The curves show that session id checking decreases the network's throughput roughly in proportion to the number of authorized clients. This result is as expected, since each authorized client adds the same amount of overhead traffic for refreshing the session management page. The curves also show that the overhead increases as the refresh period decreases. This result is also expected, since a smaller refresh period requires the same amount of refresh traffic to occur in a shorter time interval. With 15 authorized clients and 1 s refresh period, the measured network throughput overhead of session id checking was approximately 4%.

In the third experiment, we measured the CPU utilization of the captive portal/gateway under the same conditions as the previous experiment, with a 1 s refresh period. The results are shown in Figure 7. The curve shows that the CPU overhead of session id checking increases roughly in proportion to the number of authorized clients, as expected. With 15 authorized clients and 1 s refresh period, the measured CPU overhead was approximately 5%.

In the fourth experiment, we measured the round-trip time (RTT) between a test client and the test server, under the same conditions as the previous experiment. The RTT remained steady at an average of 2.4 ms with standard deviation of 0.3 ms as the number of associated clients varied between 1 and 22. The same RTT was measured without session id checking. This shows that session id checking introduces negligible delay.

In the fifth experiment, we had one of the test clients freeload on the session of another test client, after the captive portal authorized the latter's access. We verified that MAC sequence number tracking detects and blocks freeloading as soon after the attack starts as the authorized client sends another packet.

In the sixth experiment, we measured the throughput between the test server and a test client, using MAC sequence number tracking but not session id check-

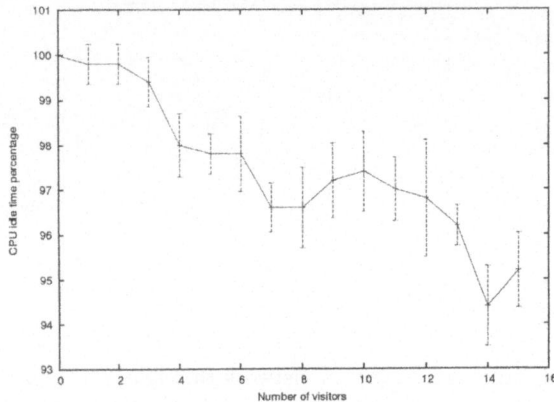

Fig. 7. Session id checking's impact on the captive portal's CPU utilization increases roughly proportionally to the number of authorized users (1 s refresh rate shown)

ing. The throughput remained steady at 4.42 Mbps with a standard deviation of 0.06 Mbps as the number of authorized clients varied between 1 and 8. These measurements show that MAC sequence number tracking has negligible impact on network throughput. This is expected, since this defense is passive and does not add traffic to the Wi-Fi network.

In the seventh experiment, we measured the round-trip time (RTT) between a test client and the test server, under the same conditions as the previous experiment. The RTT remained steady at an average of 2.4 ms with standard deviation of 0.3 ms as the number of associated clients varied between 1 and 8. The same RTT was measured without MAC sequence number tracking. This shows that MAC sequence number tracking introduces negligible delay.

In the eighth and final experiment, we used both session id checking and MAC sequence number tracking. We verified that the defenses interoperate well and continue to be effective against the mentioned attacks when used together. We also verified that they work well with all the Wi-Fi NICs mentioned above.

6   Discussion

It should be noted that neither of the proposed defenses can replace the other. Session id checking does not detect freeloading because the latter attack, unlike session hijacking, allows the victim to continue communicating and refreshing the session management page. Conversely, MAC sequence number tracking does not detect session hijacking because the latter attack, unlike freeloading, causes a simple jump in the MAC sequence number, without clearly delineating different trend lines.

The network throughput and CPU overheads measured in the experiments suggest that session id checking would scale, as shown, up to about 200 authorized users. In larger installations, it would probably be necessary to partition access points among several captive portals, or migrate session id checking to the access points, in order to handle the session id checking load.

Freeloading is ordinarily an access control violation. MAC sequence number tracking converts it into a denial-of-service attack. If access control is desired, this can be considered a good tradeoff: Much is gained by blocking unauthorized access and, given that Wi-Fi already has numerous other denial-of-service vulnerabilities for which no good defense is known (e.g., jamming Wi-Fi frequencies or broadcasting deauthentication notifications), little is lost by introducing yet another way to achieve denial-of-service. On the other hand, in networks that do not need access control, the proposed defenses probably should be disabled.

An attacker could attempt to thwart the proposed defenses by jumping from one victim to another as soon as the respective attack is detected. It is doubtful that such a strategy would be practical. First, the attacker would quickly run out of potential victims, given Wi-Fi's limited range and the typically low utilization of access Wi-Fi networks. Second, it would be difficult for such an attack to remain unnoticed, given the many victims that would suffer denial of service. Third, the attacker's IP address would keep quickly changing as the attacker jumps to each victim. It would therefore be difficult or impossible for the attacker to maintain connections, especially secure ones.

The proposed defenses do not detect attacks using rogue access points. However, there are several intrusion detection systems that can be used for such purpose and can be expected to interoperate well with the defenses proposed here. Commercially available examples of such systems include AirWave, Wavelink, AirMagnet, and AirDefense, among others.

7 Conclusions

Access Wi-Fi networks provide Internet connectivity to mobile users at locations such as universities and commercial hotspots. They need to block access by unauthorized users and to interoperate with a wide range of user-owned equipment, but usually cannot provide users much technical support. As a result of these usability and interoperability requirements, access Wi-Fi networks often forego security improvements that are being used in other applications (e.g., enterprise networks). Access Wi-Fi networks typically use a captive portal to authenticate users. Captive portals are intuitive and do not require special hardware or software. However, they are vulnerable to session hijacking and freeloading attacks. Freeloading is a previously unreported attack. It does not require special tools and can easily remain undetected. We proposed two novel defenses, session id checking and MAC sequence number tracking, against captive portals' vulnerabilities. These defenses are transparent to users and work with default client hardware and software configurations, and therefore are well-suited for use in access Wi-Fi networks. Our experiments show that the proposed defenses are effective against the mentioned attacks and impose modest overhead.

References

1. IEEE: Wireless LAN Medium Access Control (MAC) and Physical Layer (PHY) Specifications. 802.11 Std. (1999) [Online]
   http://standards.ieee.org/getieee802/download/802.11-1999.pdf
2. Wi-Fi Alliance: [Online] http://www.weca.net
3. Mackie, K.: Report Profiles Growth in Wi-Fi IC Shipments. Broadband Wireless Online, Dec. 19 (2002) [Online]
   http://www.shorecliffcommunications.com/magazine/news.asp?news=1105
4. Borisov, N., Goldberg, I., Wagner, D.: Intercepting Mobile Communications: The Insecurity of 802.11. In: Proc. Seventh Annual International Conference on Mobile Computing and Networking (MOBICOM), ACM (2001) 180–188
5. Fluhrer, S., Mantin, I., Shamir, A.: Weaknesses in the Key Scheduling Algorithm of RC4. In: Eighth Annual Workshop on Selected Areas in Cryptography. (2001)
6. Stubblefield, A., Ioannidis, J., Rubin, A.: Using the Fluhrer, Mantin, and Shamir Attack to Break WEP. Technical Report TD-4ZCPZZ, AT&T Labs (2001)
7. Arbaugh, W., Shankar, N., Wang, J.: Your 802.11 Network Has No Clothes. In: Proc. First IEEE International Conference on Wireless LANs and Home Networks. (2001)
8. Mishra, A., Arbaugh, W.: An Initial Security Analysis of the IEEE 802.1X Standard. Technical Report CS-TR-4328, University of Maryland (2002)
9. Kent, S., Atkinson, R.: Security Architecture for the Internet Protocol. IETF, RFC 2401 (1998) [Online] ftp://ftp.rfc-editor.org/in-notes/rfc2401.txt
10. Freier, A., Karlton, P., Kocher, P.: The SSL Protocol Version 3.0. [Online]
    http://wp.netscape.com/eng/ssl3/draft302.txt
11. Lemos, R.: Security: Open Networks Pose Dilemma. In: news.com, Feb. 5 (2003) [Online] http://news.com.com/2009-1033-982324.html?tag=rn
12. IEEE: Specification for Enhanced Security. (unapproved draft for 802.11i) [Online]
    http://standards.ieee.org/getieee802/new.html
13. Appenzeller, G., Roussopoulos, M., Baker, M.: User-Friendly Access Control for Public Network Ports. In: Proc. INFOCOM, IEEE, Mar. (1999) 699–707 [Online]
    http://mosquitonet.stanford.edu/publications/WebSpinach.ps
14. Brustoloni, J., Garay, J.: MicroISPs: Providing Convenient and Low-Cost High-Bandwidth Internet Access. In: Computer Networks **33** (2000) 789–802 [Online]
    http://www9.org/w9cdrom/249/249.html
15. Bahl, P., Venkatachary, S., Balachandran, A.: Secure Wireless Internet Access in Public Places. In: Proc. ICC, IEEE, June (2001) [Online]
    http://www.cs.ucsd.edu/users/abalacha/research/papers/ICC01.pdf
16. IEEE: Port-Based Network Access Control. 802.1x Std. (2001) [Online]
    http://standards.ieee.org/getieee802/download/802.1X-2001.pdf
17. Karygiannis, T., Owens, L.: Wireless Network Security — 802.11, Bluetooth and Handheld Devices. Special Publication 800-48, NIST (2002) [Online]
    http://csrc.nist.gov/publications/nistpubs/800-48/NIST_SP_800-48.pdf

Secure Reverse Communication in a Multicast Tree*

Josep Domingo-Ferrer, Antoni Martínez-Ballesté, and Francesc Sebé

Dept. of Computer Engineering and Maths,
Universitat Rovira i Virgili,
Av. Països Catalans 26,
E-43007 Tarragona, Catalonia, Spain
{jdomingo,anmartin,fsebe}@etse.urv.es

Abstract. Multicast content delivery can be expected to become a major source of revenue with the increase of private broadband fixed and mobile communications. Several multicast applications require the receivers to securely send some real-time information back to the source, which leads to a many-to-one communication scenario. Using unicast connections to send this reverse traffic results in a data implosion which may swamp the source with incoming communication. In order to avoid this problem, a method for aggregating the information as it is being sent back to the source is presented in this paper. Confidentiality and authentication are guaranteed at the symbol level for this reverse real-time traffic.

Keywords: Cryptography, Security and privacy, Multicasting.

1 Introduction

Multicast communication is a paradigm with one source and a group of receivers. Examples of multicast communication are cable TV, Internet live transmissions, pay-per-view video on demand, etc.

A multicast system can be represented as a tree structure where the root node is the content source, the leaves are users and the intermediate nodes are multicast routers. A multicast router receives the content from its parent node and forwards it to its child nodes.

Multicast applications usually result in one-to-many communication from the source to the users. Additionally, some multicast applications may require the source of the multicast tree to collect data from all users, which results in *many-to-one* communication [Mill99]. Many-to-one applications entail inherent scaling problems. Example data that may need to be collected by the source are network monitoring information, fee collection in case of pay-per-view or pay-as-you-watch transmissions, sensor information, resource discovery, etc. Too many

---

* This work has been partly supported by the Spanish Ministry of Science and Technology and the European FEDER fund under project TIC2001-0633-C03-01 "STREAMOBILE".

N. Mitrou et al. (Eds.): NETWORKING 2004, LNCS 3042, pp. 807–816, 2004.

simultaneous users sending data to the source can potentially overwhelm the latter, a situation usually known as implosion problem [Quin01]. In addition to requiring solutions to implosion, some many-to-one applications require secure and real-time transmission.

1.1   Contribution and Plan of This Paper

We propose in this paper a scalable and secure protocol for reverse many-to-one communication in a multicast tree. Our proposal is based on super-increasing sequences and probabilistic additive public-key privacy homomorphisms. Since our scheme is designed for real-time reverse traffic, it is assumed that information cannot be buffered but should be sent symbol by symbol as these are generated by the user. Therefore, security is provided at the symbol level and consists of communications secrecy and user authentication.

Section 2 gives some background on super-increasing sequences and probabilistic additive public-key privacy homomorphisms. Section 3 describes the mechanism we propose for many-to-one communication. Section 4 deals with the security of the proposed scheme. A performance analysis is reported in Section 5. Section 6 contains some conclusions and suggestions for future work.

2   Background Concepts

The basic tools used in our proposal are described next.

2.1   Super-Increasing Sequences and the Knapsack Problem

Given a sequence of positive integers $S = \{S_1, S_2, \ldots, S_{m-1}, S_m\}$ and a value T which is the sum of some elements of S, the knapsack problem [Merk78] consists of finding a subset $S' = \{S_a, S_b, \ldots, S_j\}$, of S whose sum equals T.

The general knapsack problem is known to be an NP-complete problem, but there are some cases in which the problem can be solved polynomially. This is the case when the sequence S is *super-increasing*.

A sequence of positive integers $S = \{S_1, S_2, \ldots, S_{m-1}, S_m\}$ is super-increasing if every term is greater than the sum of all previous terms, *i.e.*

$$S_k > \sum_{j=1}^{k-1} S_j$$

Given a super-increasing sequence, the knapsack problem can easily be solved using the following recursive procedure:

1. Initially, all values S_i greater than T are marked as not being part of the solution.
2. Take the largest unmarked element $S_j \in S$ and check whether $S_j \leq T$. If the check is positive, mark S_j as being part of the solution and let $T := T - S_j$. Otherwise mark S_j as not being part of the solution and leave T unaltered.

3. Solve the problem for the remaining unmarked elements in S.
4. The problem is correctly solved if and only if T equals 0 when all elements in S have been marked.

The scheme in this paper uses super-increasing sequences to aggregate integer values in a reversible way.

2.2   Additive Privacy Homomorphisms

Privacy homomorphisms (PHs) are encryption transformations mapping a set of operations on cleartext to another set of operations on ciphertext. Basically, PHs are encryption functions $E : CT \rightarrow CT'$ allowing a set F' of operations on a ciphertext domain CT' to be carried out without knowledge of the decryption function D. Knowledge of D allows the result of the corresponding set F of operations on a cleartext domain CT to be retrieved. A PH is called *additive* when its set F of cleartext operations contains addition. A PH is called *probabilistic* if the encryption algorithm E involves some random mechanism that chooses the ciphertext corresponding to a given cleartext from a set of possible ciphertexts.

Privacy homomorphisms that will be used in our proposal below must be additive, probabilistic and public-key. In [John02], it is shown that additive PHs are insecure in front of known-cleartext attacks if used for signing. Thus, in our scheme the public-key additive PH is only used for encryption. We next give an example of a probabilistic additive public-key PH.

Example 1. The Okamoto-Uchiyama [Okam98] probabilistic public-key cryptosystem (OUPH) has an additive homomorphic property. This probabilistic public-key cryptosystem is proven to be as secure as the intractability of factoring $n = p^2 p'$ against passive adversaries, where p and p' are two large primes.□

3   Secure Many-to-One Bit Transmission

This section describes our proposal for secure reverse many-to-one communication in a multicast context. A basic construction is first described which allows transmission of one binary or ternary symbol. A generalization is then presented for transmission of a q-ary symbol or a block of bits.

3.1   The Basic Construction

The construction consists of a set-up protocol to be run before any transmissions are started, and a transmission protocol to be run for each symbol transmission.

Protocol 1 (Set-up).

1. *The source chooses parameters l, u, where l will be used below and u is the number of users.*

2. *The source generates 2u intervals as follows:*

$$\mathbf{I}_1 = [1, 2^l - 1]$$

$$\mathbf{I}_j = [I_j^{min}, I_j^{max}] = [(2^j - 2)(2^l - 1) + 2^{j-1} - 1, (2^j - 1)(2^l - 1) + 2^{j-1} - 1]$$

for $j = 2$ to $2u$.

3. *The source generates u keys k_i, for $i = 1$ to u, corresponding to a block cipher (e.g. AES).*
4. *The source generates a key pair for a probabilistic additive public-key privacy homomorphism such that its cleartext space is $CT = \{0, 1, 2, \cdots, p-1\}$ where p should be larger than $2I_{2u}^{max} + 1$. After some manipulation, it can be checked that the lower bound on p is*

$$p > (2^{2u} - 1)(2^{l+1} - 1) \tag{1}$$

(E.g., for $u = 500$ users, p should have $O(10^3)$ bits.)
5. *The source multicasts the public key PK of the PH and I_{2i-1}^{min} and I_{2i}^{min} for $i = 1$ to u. In addition the source secretly sends k_i to each user U_i, who should keep this key confidential (storing it in a tamper-resistant device such as a smart card would seem appropriate).*

After set-up, the normal operation of the scheme consists of many-to-one transmissions of binary or ternary symbols. In order to collect a binary or ternary symbol from each user, the following four-step protocol is used:

Protocol 2 (Many-to-one binary or ternary transmission).

1. Transmission request. *A challenge message is multicast by the source to all users. This challenge contains a random value v.*
2. Message generation.
   (a) *When a user U_i receives the challenge message, she computes*

$$S_{2i-1} = I_{2i-1}^{min} + \mathcal{H}(v||k_i)$$

$$S_{2i} = I_{2i}^{min} + \mathcal{H}(v+1||k_i) \tag{2}$$

   where $\mathcal{H}$ is a one-way collision-free hash function yielding an l-bit integer as output. This condition on the output of $\mathcal{H}$ ensures that $S_{2i-1} \in \mathbf{I}_{2i-1}$ and $S_{2i} \in \mathbf{I}_{2i}$, which in turn guarantees that the sequence $S = \{S_j\}$ for $j = 1, \cdots, 2u$ is super-increasing. On the other hand, condition (1) ensures that no overflow in CT will occur when adding encrypted terms of the super-increasing sequence over the ciphertext space CT'.
   (b) *Next, each U_i generates her message as follows: i) If she wants to transmit a 0 bit value, she generates the message $M_i = E_{PK}(S_{2i-1})$ where $E_{PK}(\cdot)$ stands for the encryption function of the probabilistic additive public-key privacy homomorphism used. ii) If she wants to transmit a 1 bit value, she generates $M_i = E_{PK}(S_{2i})$). iii) In case of transmitting*

ternary symbols, a third symbol (other than 0 and 1) could be transmitted
if U_i sent $M_i = E_{PK}(S_{2i-1} + S_{2i})$.

Note that, since a probabilistic cryptosystem is being used, the same cleartext message can result in different encrypted messages. Finally U_i sends
M_i up to her parent router.

3. Message aggregation. *Intermediate routers receive messages from their child
routers/users and do the following:*
   (a) *Once all expected messages $\{M_i\}_i$ have been received, the router aggregates them as $M = \sum_i' M_i$, where $\sum'$ stands for the ciphertext operation
   of the privacy homomorphism corresponding to cleartext addition.*
   (b) *The router sends M up to its parent router.*
4. Symbol extraction. *When the previous process completes, the source finally
receives an aggregated message M, from which the transmitted symbols are
extracted as follows:*
   (a) *The source constructs the super-increasing sequence $S = \{S_j\}$ for $j = 1$
   to $2u$ using, for each user U_i, equations (2).*
   (b) *The source decrypts M using its private key to recover a value T which
   is used to solve the super-increasing knapsack problem and obtain the
   sequence $S' = \{S_a, S_b, \ldots, S_j\}$ that yields the symbols sent by the users.
   Specifically, there are four possible cases for each user U_i: i) If $S_{2i-1} \notin S'$
   and $S_{2i} \notin S'$, then U_i sent nothing. ii) If $S_{2i-1} \in S'$ and $S_{2i} \notin S'$, then
   U_i sent a bit value 0. iii) If $S_{2i-1} \notin S'$ and $S_{2i} \in S'$, then U_i sent
   a bit value 1. iv) If $S_{2i-1} \in S'$ and $S_{2i} \in S'$, then this is an error
   condition in a binary transmission. However, if using ternary symbols,
   this configuration can be used to send the third symbol.*

3.2   A Generalization for q-ary or Block Transmission

The basic construction given above can be generalized as follows to accomodate
transmission of q-ary symbols or blocks of bits:

1. During Protocol 1 generate and publish tu intervals $\mathbf{I}_j$ rather than $2u$ intervals. Condition (1) must be modified by replacing $2u$ with tu.
2. During Protocol 2:
   (a) Each user U_i takes t consecutive intervals $\mathbf{I}_{ti-t+1}, \cdots, \mathbf{I}_{ti}$ and generates
   t terms the super-increasing sequence as:

$$S_{ti-t+j} = I_{ti-t+j}^{min} + \mathcal{H}(v + j - 1 \| k_i)$$

   for $j = 1$ to t.
   (b) Now U_i can transmit $q = 2^t - 1$ different values by sending the encrypted sum of a subset chosen among the $2^t - 1$ non-empty subsets
   of $\{S_{ti-t+1}, \cdots, S_{ti}\}$ Note that the encrypted sum of the empty subset
   (*i.e.* $E_{PK}(0)$) cannot be used to encode a value in a secure transmission
   because anyone can send it (no authentication) or guess it (no confidentiality). Thus, U_i can either transmit a q-ary symbol or a block of $t - 1$
   bits (the first option is clearly less wasteful).

(c) Message aggregation in the generalization stays the same as in Protocol 2.
(d) Symbol extraction by the source in the generalization must use the same mapping between subsets and q-ary symbol values used by users during message generation. This mapping assumed to be public.

4   Security

We next state the security properties of our scheme, which are proven in the Appendix.

Property 1 (Confidentiality). *If a secure probabilistic additive public-key PH is used in which there is a negligible probability of obtaining the same ciphertext as a result of two independent encryptions of the same cleartext, then an intruder cannot determine the symbol transmitted by a user in Protocol 2.*

Property 2 (Authentication). *If a secure public-key PH and a one-way collision-free hash function with l-bit output are used, the following holds:*

1. *the probability of successfully impersonating another user when sending a bit value to the source is 2^{-l};*
2. *substituting a false message M' for a legitimate message $M \neq M'$ in the current transmission is at least as difficult as impersonation;*
3. *substituting a message M' for a legitimate message $M \neq M'$ in future transmissions using information from the current transmission is infeasible.*

5   Performance

Before presenting the performance comparison below, some preliminary remarks are required:

- The performance criterion considered is the bandwidth required by the reverse traffic.
- In order to benchmark the performance of our system, we will consider an alternative system based on unicast transmissions from each user to the source. Like in our system, the unicast transmissions in the benchmark system will be symbol-wise. We assume that the communication is real-time, so that symbols are transmitted as they are generated, rather than being buffered and transmitted in batches.
- We will require that each symbol transmission in the alternative unicast system has the same security properties as transmissions in our system.
- For the sake of concreteness, we will use OUPH as a privacy homomorphism in this section.

5.1   A Benchmark Unicast System

In order to avoid the need of public-key encryption for a user to send a confidential and authenticated symbol, we must assume that each user U_i shares with the source a key k_i corresponding to a block cipher (*e.g.* AES).

The message M containing the symbol b will thus look like

$$M = E_{k_i}(b||ts||ck), U_i$$

where $E_{k_i}(\cdot)$ stands for the encryption function of the block cipher, ts is a time-stamp, ck is a checksum and U_i is the identity of user U_i. Integrity is ensured by ck and ts (the time-stamp prevents replacing future transmissions with past transmissions).

5.2   Comparison

When u users simultaneously send their encrypted symbols with the benchmark unicast system, $u(B + \log_2 u)$ bits are received by the source, assuming that the B is the block bitlength of the block cipher and $\log_2 u$ is the bitlength of the user identifier U_i. We assume also that the bitlength of $b||ts||ck$ is less than or equal to B. For a block cipher such as AES, at least one has $B = 128$, so the previous assumption is reasonable.

When u users send their encrypted bits/symbols with our system, all symbol transmissions are eventually aggregated into a single message

$$M = \prod_i M_i \quad (\text{mod } n)$$

which is the only one reaching the source. M can be at most n, so its length is $\log_2 n$. Equivalently, the bitlength of M is

$$|M| = \log_2 n = \log_2(p^2 p') = 2\log_2 p + \log_2 p' = 3\log_2 p$$

where we have used that, in OUPH, $n = p^2 p'$ with $|p| = |p'|$. Now, already for a moderate number u of users, p can be chosen close to its lower bound (1) while remaining large enough for factoring of $n = p^2 p'$ to be hard, as required by OUPH. Therefore, if we use the generalized bound (1) with tu instead of $2u$, we have

$$|M| \approx 3\log_2[(2^{tu} - 1)(2^{l+1} - 1)] = 3\log_2(2^{tu} - 1) + 3\log_2(2^{l+1} - 1) \quad (3)$$

It can be seen that expression (3) is dominated by $3tu$ as the number of users grows. Therefore, if the number u of users is moderate to large and if the symbol bitlength is $t < (B + \log_2 u)/3$, the bandwidth $3tu$ required by our scheme is *less* than the bandwidth $u(B + \log_2 u)$ required by the benchmark unicast system. Since typical block sizes are as large as $B = 64, 128, 192$ or 256, the previous assumption on the symbol bitlength is reasonable.

Besides, our proposal only requires one reverse incoming connection to the source, whereas the unicast alternative requires u reverse connections to the source, which calls for allocation of additional overhead bandwidth not included in the above comparison.

Note 2. The primary aim of our proposal is reverse bandwidth reduction. It must be noticed that this is achieved without increasing the computational burden at the source. Symbol extraction during Protocol 2 requires the source to build tu terms of a super-increasing sequence and to solve a super-increasing knapsack problem. The computational cost of doing this is similar to the cost of the u block decryptions required by the unicast benchmark.

6   Conclusions and Example Applications

The thrust behind the design of the scheme in this paper was the need to securely send real-time reverse information in multicast scenarios, that is, information whose symbols should not be buffered but be sent as they are generated. This requirement not being exclusive of multicast, our scheme can be applied whenever a large number of users or devices must communicate in real-time with a single node and there is a risk that the incoming bandwidth available at the receiving node may be a bottleneck. Example applications include:

- *Secure multicast.* The scheme presented can be used for users to securely send keepalive messages to the source, who can keep track of who is logged on. A step further is to use the scheme for real-time pay-per-view multicast: the users send payment information back to the source, and stop receiving multicast contents if they stop sending payment.
- *Secure collection of control information.* A control center securely collects periodical status information from a large number of sensors or other devices. This is similar to sending reverse traffic in a multicast scenario.

The proposed scheme uses super-increasing sequences and probabilistic additive public-key homomorphic encryption to aggregate traffic. It guarantees confidentiality and authentication of the transmitted q-ary symbols. Thanks to aggregation, the source only needs to establish one incoming connection. In the special case where the Okamoto-Uchiyama PH is used, the required incoming bandwidth at the source for u users approximates $3tu$ bits when each user securely transmits one q-ary symbol at a time, with $q = 2^t - 1$. This is not so far from the $u \log_2 q \approx tu$ bits required for *insecure* transmission of u q-ary symbols. Achieving the same security properties using unicast transmissions would typically need Bu bits split in u user-source connections, where B is the block size of a block cipher.

References

[John02] R. Johnson, D. Molnar, D. Song and D. Wagner, "Homomorphic signature schemes", in *Topics in Cryptology - CT-RSA 2002*, ed. B. Preneel, LNCS 2271, Berlin: Springer Verlag, pp. 244-262, 2002.

[Merk78] R. C. Merkle and M. Hellman, "Hiding information and signatures in trapdoor knapsacks", in *IEEE Transactions on Information Theory*, vol. 24, no. 5, pp. 525-530, 1978.

[Mill99] C. K. Miller, *Multicast Newtorking and Applications*. Reading MA: Addison Wesley, 1999.

[MSEC03] Multicast Security Working Group (MSEC WG). http://www.securemulticast.org

[Okam98] T. Okamoto and S. Uchiyama, "A new public-key cryptosystem as secure as factoring", in *Advances in Cryptology - EUROCRYPT'98*, ed. K. Nyberg, LNCS 1403, Berlin: Springer-Verlag, pp. 308-318, 1998.

[Quin01] B. Quinn and K. Almeroth, "IP multicast applications: challenges and solutions", Internet RFC 3170, Sept. 2001. http://www.ietf.org.

Appendix

Proof (Property 1): Without loss of generality and to keep the proof simple, we can restrict ourselves to the basic construction given in Section 3.1, which corresponds to the case $t = 2$ of the general construction sketched in Section 3.2. Generalizing the proof to any t is cumbersome but straightforward. Now, assume the intruder captures a message M sent by U_i during Protocol 2. This message is either $E_{PK}(S_{2i-1})$, $E_{PK}(S_{2i})$ or $E_{PK}(S_{2i-1} + S_{2i})$. Decryption of M is not possible because the PH is secure and the intruder does not have access to the private key.

Exhaustive search of the cleartext carried out by M is the other attack strategy to be examined. Now, exhaustive search of the sequence values S_{2i-1} or S_{2i} by encrypting candidate values and comparing the result to M is not feasible because the PH is probabilistic and there is a negligible probability that two independent encryptions of the same cleartext yield the same ciphertext. Therefore, the above comparison (and thus exhaustive search) will fail with overwhelming probability. $\square$

Proof (Property 2): By the same argument as for Property 1, we need only prove this property for the basic construction. In the impersonation attack, an intruder who wants to impersonate user U_i tries to generate a message $E_{PK}(S_{2i-1})$, $E_{PK}(S_{2i})$ or $E_{PK}(S_{2i-1} + S_{2i})$, Now, the intruder needs to compute S_{2i-1} or S_{2i}. Each term S_j of the super-increasing sequence S is pseudo-randomly chosen within an interval $\mathbf{I}_j$ containing 2^l integer values. The choice is made using a one-way collision-free hash function of the challenge and the secret key k_i unknown to the intruder, as shown in equations (2). Thus, the probability of the intruder randomly hitting S_j is at most 2^{-l}. Remark that exhaustive search is not feasible, since there is no way of checking whether the right S_j has been hit (there is no way for the intruder to make sure whether the message generated with the candidate S_j is correct).

A substitution attack can be mounted in the current transmission or in future transmissions:

- In the current transmission, assume the intruder wants to substitute a false message M' for an authentic message M sent by U_i, with $M' \neq M$. Without loss of generality, let $M = E_{PK}(S_{2i})$; the intruder wants to transform M into $M' = E_{PK}(S_{2i-1})$ or $M' = E_{PK}(S_{2i-1} + S_{2i})$. This requires the following steps: i) recover S_{2i} from M; ii) compute S_{2i-1} with knowledge of S_{2i}; iii) compute M'. Thus, even if decrypting M at step i) was easy (which it is not), solving step ii) is as difficult as mounting a successful impersonation attack (see above).
- A second possibility is for an internal intruder to use information derived from a current transmission of a message by U_i to alter future messages sent by U_i. But this is infeasible, because in subsequent executions of Protocol 2, a different super-increasing sequence will be used to encode the messages which does not depend on the current super-increasing sequence (see equations (2)).

□

Stealth Multicast: A Novel Catalyst for Network-Level Multicast Deployment

Aaron Striegel

Systems & Software Laboratory
Dept. of Computer Science & Engineering
University of Notre Dame
Notre Dame, IN 46530 USA
striegel@cse.nd.edu

Abstract. While network-level multicast has tremendous potential for increasing the efficiency of group-oriented applications, the adoption of network-level multicast has been tepid at best. In this paper, we propose a novel concept entitled stealth multicast that allows for practical adoption of network-level multicast on a domain-wise basis rather than global scale. In the stealth multicast model, similar unicast packets are dynamically assembled into virtual groups for multicast transmission across the domain. At the edge of the domain, the packets are converted back to unicast, thus hiding the existence of stealth multicast from the external Internet. True to its namesake, stealth multicast operates in complete stealth, providing seamless interoperability without requiring any modifications to end-user applications nor requiring any inter-domain support. In this paper, we introduce the basic concepts of stealth multicast and show that the stealth multicast model can offer significant benefits in terms of bandwidth savings with minimal impact to the end-user QoS.

Keywords: Multicast Deployment, Routing, QoS

1 Introduction

While fundamental multicast concepts have been successfully deployed in the Mbone and exist in many commercially available routers, recent studies show a relatively lackluster adoption over the last decade [1, 2]. Furthermore, despite the large body of research on network-level multicast [3], recent trends in multicast research have shifted to application-level multicast (ALM) [4]. Although techniques such as ALM can offer near network-level multicast bandwidth savings, ALM can suffer from additional delay due to longer distribution trees and a dependence upon a rich end-user capacity to provide adequate downstream branching. Despite its weaknesses, ALM offers a compelling solution for bandwidth management as it avoids one of the key problems associated with network-level multicast, namely global network deployment.

Whereas much can be written about the tepid adoption of network-level multicast, the root of the problem arises from the fact that much of the benefit of

N. Mitrou et al. (Eds.): NETWORKING 2004, LNCS 3042, pp. 817–828, 2004.

network-level multicast only comes with complete global deployment. The challenge of global network-level multicast support appears especially daunting given that many other complex sub-issues also provide obstacles such as the support vs. development vs. demand dilemma, deployment complexity (billing, management), and ISP economic incentive [5]. Thus, our paper poses the following question, is it possible to offer a novel approach to multicast that allows for incremental deployment while avoiding the pitfalls that have plagued network-level multicast deployment (application adoption, ISP incentive, etc.)? This question provides the basis for the model proposed in our paper, *stealth multicasting*.

1.1 Stealth Multicast Overview

At its core, the stealth multicast model changes the context of the problem regarding multicast deployment. Rather than requiring participation on a global scale, stealth multicast abstracts multicast transport in an individual domain (Autonomous System) such that its entire presence operation is kept hidden from the outside world. At the edge of the domain, packets are dynamically converted to and from multicast, thus allowing for seamless interaction with existing unicast applications. The conversion for multicast is done only at inputs to the domain whereby maximum rewards can be gleaned through reduction to multicast transport. Hence, the target audience for stealth multicast is domains directly serving UDP applications that send out identical data streams to multiple users (i.e. streaming media, on-line games) that network characteristics that ALM cannot meet (minimal delay, reduced client capabilities, etc.).

The rest of our paper is organized as follows. In Section 2, we present the fundamentals of the stealth multicast model. Next, in Section 3, we present the MYDEKI model. Then, in Section 4 we conduct simulation studies regarding the performance of the MYDEKI model. Finally, in Section 5 we offer several concluding remarks and discuss our future work.

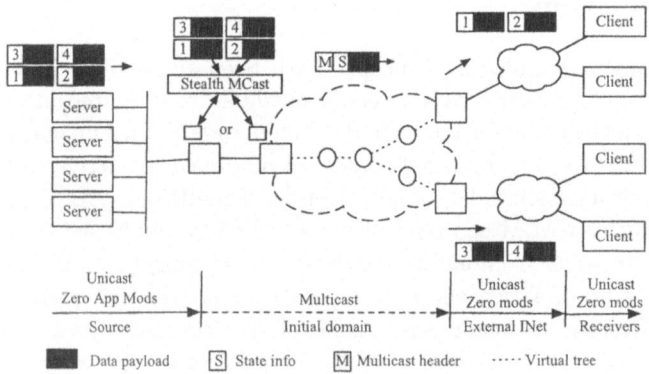

Fig. 1. The stealth multicast model

2   The Stealth Multicasting Model

In order to provide seamless interoperability with minimal end-system modification, the stealth multicast model relies on two governing principles that are as follows:

- *Externally transparent:* External unicast applications should not require any modifications in order to operate in the stealth multicast environment. The same applies for existing multicast applications (if present).
- *Negligible QoS impact:* The end-user should not experience a *noticeable* impact in QoS.

With the first principle of external transparency, the goal is to provide seamless functionality with existing networks. This principle is extremely critical due to the tremendous number of applications currently in use, many of which may be extremely difficult or even impossible to change. Note that this principle does not preclude future applications from taking advantage of the stealth multicast model, ALM, or or network-level multicast but rather ensures backwards compatibility with existing IP unicast applications.

The second principle ties into the first principle and into the stealth of the model itself. If the QoS of the user is significantly impacted in either the positive or the negative direction, the fact that stealth multicast is being employed may be discernible. A significant QoS change may impact the functionality of the applications utilizing the network as well. Although a positive QoS impact may not necessarily generate criticism, a negative impact on QoS will certainly cause issues with application functionality. However, this principle has an inherent amount of flexibility due to the fact that only a 'noticeable' QoS impact causes any issues. Due to the fact that QoS is subject to both the perception of the end user and the requirements of the application, it is the prerogative of the network administrator to determine what constitutes a noticeable QoS impact. For our paper, we define the term *noticeable QoS impact* to refer to the end user attributing the poor network performance to something other than the typical variations in Internet traffic behavior.

2.1   Stealth Multicast Operation

Figure 1 shows the overall concept of the stealth multicast model. The key component of the stealth multicast model is the Virtual Group Detection Module (VGDM) that is shown in Figure 2. The VGDM is placed at the edge of the domain and queues packets for assembly into *virtual groups* for multicast transport across the domain.

The stealth multicast process begins as a group-oriented application transmits packets via separate unicasts to multiple clients. The packets travel via the uplink to the domain and arrive at the edge router (see Figure 1). The packets are then transferred to the VGDM for virtual group consideration. A filter may be applied at the edge router to remove packets from consideration that should not or would never become part of a virtual group. Examples of such packets would include existing multicast packets, ICMP, and RSVP.

Depending upon the behavior desired and involvement of the source application, the VGDM may be placed along the path at various points in the network. While placement at the edge of an ISP's domain such as in Figure 1 would allow for maximum stealth and benefit solely for the ISP, the VGDM could alternatively be placed at the edge of the customer LAN to decrease the load on the uplink to the domain. Furthermore, the VGDM could also be placed on an incoming link from another domain. However, as will be discussed later, the stealth and utility of such placement may be reduced depending upon the nature of the underlying traffic. For conceptual purposes, the VGDM can be viewed as a collection of COTS hardware dedicated to serving an uplink whose traffic can benefit significantly from stealth multicast. Figure 2 shows the steps involved once the packet arrives at the VGDM which are discussed below.

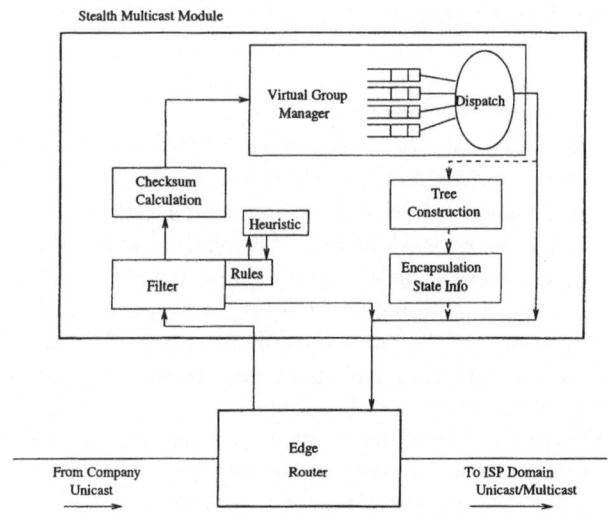

Fig. 2. Stealth multicast module - basic components

Initial Filter: Once a packet arrives at the VGDM, the first step is to apply a basic filter to the packet. The packet is filtered according to a set of rules as defined by the network administrator. The rules may also be generated by heuristics monitoring the incoming flows to optimize the candidates for consideration.

Checksum Calculation: If a packet passes the filter, it is uniquely categorized according to its data contents. The checksum calculation module creates a checksum that uniquely identifies the payload of the data packet. The checksum must be sufficient such that two unique payloads of the same size do not compute to the same checksum. The checksum module is only interested in the data payload as the remaining header information (IP, UDP) is handled by the virtual group manager. The checksum (digital signature) is computed using COTS hardware [6].

Virtual Group Manager: Next, the packet is passed to the virtual group manager for placement into virtual groups (queues). The virtual groups are uniquely identified by the shared checksum, packet size, source IP, and source port. The packets themselves are queued in the virtual group until an appropriate trigger causes the virtual group to be dispatched. The triggers determine both the performance (additional multicast efficiency) as well as the impact on end-user QoS introduced by the VGDM.

Tree Construction: After a sufficient stimulus has occurred (time, size, etc.), the packets are given to the dispatch mechanism for transmission onwards. Providing that the virtual group has sufficient membership to justify the use of multicast transport, the packet is given to the tree construction module. If the size of the virtual group is not sufficient, the packets are simply released as standard unicast packets back to the edge router.

Encapsulation of State Information: One of the inherent problems in stealth multicast is that virtual groups are constructed dynamically, i.e. the makeup of the end clients is not know a priori. In addition, since it is assumed that multicast does not exist outside of the ISP domain, the packets must be converted back to unicast at the edge of the domain. However, using standard IP multicast which contains only the source IP/ destination group address, the edge routers would not know whom is responsible for sending the unicast packet nor the unique portions of the headers (DS field, destination IP, UDP destination port). Hence, additional state information must be included in the packet or kept at the egress routers to identify how the packet should be converted back to unicast.

Transmission & Exit: After the packet has been modified for multicast transport (tree construction) and the state information has been added (encapsulation), the packet is given back to the edge router. The edge router then forwards the packet across the domain using the underlying multicast transport mechanism (broadcast, route-pinning, ALM, etc.). Once the packet reaches the edge of the domain, the egress router is responsible for converting the packet back into a unicast packet. Using the virtual group state information, the egress router reconstructs the appropriate unicast packets and forwards the replicated unicast packets onwards. The packets are identical to the packets that were originally seen at the VGDM with appropriate modifications for TTL and any other necessary fields. Once the packet leaves the domain, it is a standard unicast packet, indistinguishable from any other unicast packets that did not undergo multicast consolidation/transport across the domain.

To both the application sending the packets and the client receiving the packets, there is no difference in the contents of the packets nor a noticeable change in the QoS of the packet. However, there is certainly a noticeable impact for the domain. In addition to avoiding upstream bottlenecks, the general bandwidth requirements of the domain are alleviated due to the use of multicast. In fact, stealth multicast allows for a clear transition of deployment from the bare minimum (entirely stealthful - domain only) to full network-level multicast. Most notably, stealth multicast allows a domain to deploy multicast support and realize concrete benefits without waiting for application, customer, or global routing support.

3   The MYDEKI Architecture

The MYDEKI (Multicast and You Don't Even Know It) is an architecture based on the stealth multicast model. The MYDEKI architecture governs the undefined areas of stealth multicast which include virtual group management, multicast transport, and state management. The MYDEKI architecture is targeted towards medium-size (tens to hundreds of clients) UDP[1] group-oriented applications employing separate unicasts. These applications may either operate with or without knowledge of MYDEKI. The MYDEKI architecture offers three modes of deployment, two stealth modes (differing by the location of the VGDM), and an application-assisted mode.

3.1   MYDEKI: Stealth Mode (FullStealth and Local)

In the stealth mode of MYDEKI, all applications are unaware of the presence that stealth multicast is being employed. The first contact that the architecture has with the packet occurs when the candidate packet is presented to the VGDM (see Figure 2).

Dispatch Mechanism: The intuition behind the MYDEKI parameters is to allow for predictable tuning by the network administrator. At their core, the MYDEKI parameters capture the potential benefits of waiting (receiving another packet that can be put into a virtual group) versus the effect that the queuing (delay, buffer size) is having on the behavior of the packet flow. For optimal performance, a packet should be kept as a group candidate so as long as only to add more egress points to the multicast group and hence increase the efficiency of the virtual group. Thus, MYDEKI includes triggers that reward close proximity of matching packets (likely to have another match) while still putting a firm cap on the maximum delay that can be experienced by a packet. The following parameters govern how MYDEKI manages the triggers for release for its virtual groups (see Figure 3):

- *PSW - Packet Scan Width:* The number of packets to scan before the virtual group is released. In the event of a new addition to the virtual group, this count is reset.

---

[1] Although MYDEKI can be adapted for limited TCP support, such a topic is beyond the scope of this paper.

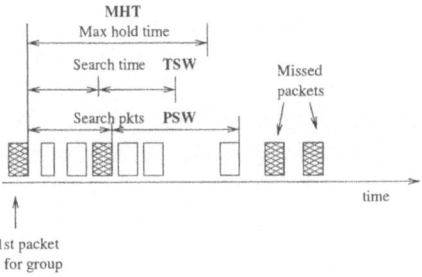

Fig. 3. MYDEKI - search settings

- *TSW - Time Scan Width:* The amount of time to scan before the virtual group is released. Similar to the PSW, this timer is reset upon the addition of a new packet to the virtual group.
- *MHT - Maximum Hold Time:* The maximum time that a virtual group can exist and hence the maximum time that a packet can sit inside a queue. This timer is set when the virtual group is started and places a maximum delay on the queuing time, regardless of additions to the virtual group.

Once a packet is triggered for release, it is given to the dispatch mechanism for dispatch to either the multicast modules or dispatch via standard unicast routing (no virtual group constructed). The MYDEKI group settings are discussed in more detail below:

- *MaxGS - Maximum Group Size:* This parameter forces a group to be dispatched when it passes a specific size.
- *MinGS - Minimum Group Size:* In order to be considered as a candidate for multicast, a group must meet a certain minimum membership level.

The *PSW* setting governs the search width from the perspective of flow aggregation / virtual group detection (i.e. how mixed is the packet in the incoming aggregate flow) whereas *TSW* reduces queuing time in the event of idle time on the link. Intuitively, the closer that the VGDM is to the source, the tighter the *PSW/TSW* values that can be employed. If the VGDM is located extremely close to the traffic source, there is a much better chance that virtual group packets will be located close together.

The *MHT* setting allows one to controllably affect the delay impact of virtual group detection. The *MHT* places a worst case bound on the virtual group detection while the actual delay experienced will depend upon the underlying traffic patterns and the *TSW/PSW* settings. Hence, the *PSW* and *TSW* parameters limit the effective search for a given packet and limit the queuing delay of packets even further by emptying out non-growing virtual groups.

The notion of the total number of groups introduces a significant factor for setting of the dispatch triggers in MYDEKI, the maximum number of concurrent virtual groups (i.e. buffer size). Since the number of virtual groups and the storage required for such groups and their packets cannot be considered infinite, the parameters should be set to minimize overflow conditions. Unlike traditional queue overflows where the packets are discarded, overflow packets are still forwarded onwards via unicast. Thus, overflow does not introduce catastrophic failure. However, once the capacity of virtual groups has been filled, no additional efficiency gains due to stealth multicast will occur until several of the existing virtual groups are dispatched.

Tree Construction & Multicast Transport: In order for a virtual group to be sent using multicast, it must first pass a minimum threshold ($MinGS$). Unlike standard IP multicast, the stealth multicast model introduces an overhead (transport information, state information) that must be offset by sufficient tree savings in order to offer an improvement in efficiency. The aspect of tree construction is encompasses how multicast packets are transported across the domain of the ISP. Since the destinations that make up the virtual group are not known a priori, we propose to use an encapsulation-based method for providing multicast transport across the domain. In short, an encapsulation-based approach includes the multicast tree inside the packet, thereby removing the need for multicast state along the routers in the multicast path. For MYDEKI, we have selected the DSMCast approach [7] as it was targeted towards multicast transport across a single domain. Although an encapsulation-based transport does introduce additional overhead due to the tree being encapsulated inside the packet, we believe the benefits of dynamic routing and resource management far outweigh the additional cost. If the minimum group size is set appropriately, the effects of the additional encapsulation overhead will be entirely offset by the multicast savings.

State Management: While the encapsulation-based transport allows for a stateless core to cross the domain, the unique portions of the packet (destination IP, destination port) must also be addressed. Furthermore, each unicast conversion must be appropriately associated with an egress point for the domain. Otherwise, it is extremely difficult for an egress point to accurately assess if it is responsible for converting a multicast packet for a specific destination.

Similar to how the encapsulation-based transport includes the tree information in the packet, the state information in MYDEKI is also bundled in the packet. In MYDEKI, each multicast packet includes three pieces of information for each client that is covered by the multicast packet, the egress point (who should convert the packet), the destination IP, and the destination port. Upon receiving a multicast packet, an edge (egress) router will inspect the packet to determine if it should appropriately convert/replicate the packet.

4   Simulation Studies

The simulations were developed using the ns-2 simulator and the GenMCast extension module for ns-2. The rationale behind our simulations was the following. In the network, Company X hosts an on-line gaming service with applications serving up to 120 clients. The application is hosted on a set of servers to users outside of the initial domain of the company's ISP. The parameters for the network simulation are summarized in Table 1.

Table 1. MYDEKI settings

Parameter	Setting	Parameter	Setting
Maximum Groups	50	ISP Core Nodes	32
Maximum Hold Time (MHT)	10 ms	ISP Edge Nodes	16
Time Search Width (TSW)	2 ms	Total Sources	40
Packet Search Width (PSW)	100	Avg Clients	32
Min Group Size (MinGS)	2	Avg Packet Rate	50 ms
Max Group Size (MaxGS)	200	Avg Packet Size	500 bytes

The primary purpose of the simulations is to evaluate the basic principles of the stealth multicast model (impact of queueing, predictability of control parameters, etc.). For our simulations, the MYDEKI model was evaluated according the following performance parameters:

- *Bandwidth utilization:* The bandwidth consumption of the server traffic on the uplink from Company X, the core of the domain, and the overall network were examined to determine the savings of the MYDEKI model.
- *End-user QoS:* The effects on end-user QoS and were examined to determine if the MYDEKI model was having a negligible or noticeable impact on the user QoS.

In our simulations, we compared the performance of five different distinct models under varying configurations:

- *Traditional Unicast:* In this model, no stealth multicasting is employed. This model is used as a base line for comparing the performance of the other two models.
- *MYDEKI-FullStealth:* In this model, the VGDM is placed at the edge router of the ISP. Traffic must first pass through the customer's uplink before being considered as a candidate for stealth multicasting.
- *MYDEKI-Local:* In this model, the group detection module is placed directly at the edge router of the company. The traffic can be considered for stealth multicasting before going on the customer's uplink to the ISP.
- *MYDEKI-AppAssist:* In this model, the application actively participates by submitting state information to the VGDM at the edge of the ISP domain [5].
- *ALM:* A generic version of ALM was used that is based on End System Multicast [8].

4.1   Effect of Client Subscriptions

The fundamental motivation for the stealth multicast model is to offer a significant bandwidth improvement in the core of the domain. Figure 4 plots the

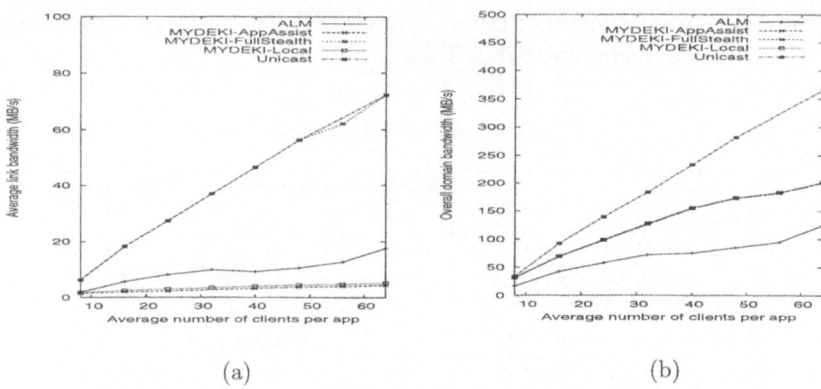

Fig. 4. Effect of average number of clients on (a) bandwidth - uplink (b) bandwidth - domain

performance as the average number of clients per server application is varied from 8 to 64. As would be expected, the unicast-only model offers the worst performance in all cases. However, the most notable aspect is the performance of ALM versus stealth multicast. Whereas ALM offers better performance in the core of the network, it achieves such a balance by pushing the bandwidth consumption out to the edge links (uplink, client links) as evidenced by the Figure 4(a).

Most notably, the actual queuing delay of the VGDM is quite minimal as shown in Figure 5(a). Unlike ALM which adds additional delay due to a longer distribution tree, stealth multicast adds a barely perceptible 1-2 milliseconds of delay to the end-to-end delay. Most important of all, stealth multicast can offer significant bandwidth improvements with zero modifications to the client or server applications ranging from 2x in the FullStealth case over the domain to over 10x for the uplink in the Local case.

4.2   Effect of MYDEKI – Packet Search Width

Figure 5(b) shows a negligble impact by the PSW on the performance of the various models. Due to the limited aggregation of the traffic (a single customer only), the packets frequently arrive in self-contained bursts rather than sporadically arriving due to aggregation with other flows. However, one can still discern the impact of an increased PSW in Figure 5(b) as an increase in PSW causes the VGDM to search a larger width and hence release the packet later from the queue. A inter-domain link employing a VGDM would experience a significantly increased impact of MYDEKI due to the TSW/PSW search widths being periodically reset rather than the burst of this scenario. Hence, the closer the VGDM can be placed to the customer, the less that MHT must be used to cap the maximum delay and the more that TSW/PSW can be relied upon (rewarding proximity).

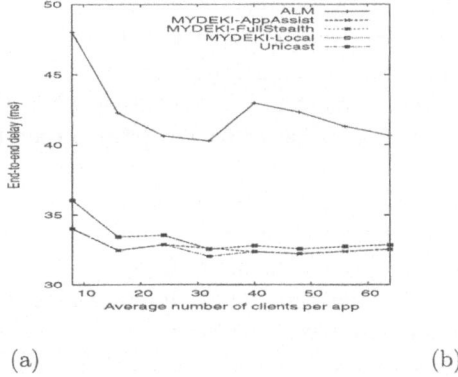

(a)   (b)

Fig. 5. Effect on end-to-end queuing delay of (a) average number of clients and (b) packet search width

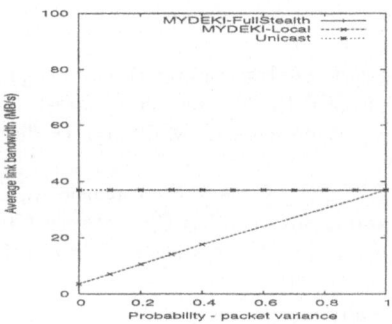

Fig. 6. Effect of packet variance on bandwidth – uplink

4.3   Effect of Aggregation in Client Packets

While the ideal case would be to only monitor traffic that will be part of a virtual group, such will most probably not be the case in practice. Hence, Figure 6 plots the performance of the various models as the probability of variance of client packets is varied from 0 to 1.0. In short, the probability of variance captures the chance that a given packet in a server transmission will be unique from all other packets (distinct checksum). As the variance is increased, more and more of the packets are sent out as unique packets that will become a 1-packet virtual group. The earlier plots of the number of clients had a variance of zero since variance represents intra-group distinct packets that neither AppAssist nor ALM models could handle (transmission to a partial subset of clients).

5   Related Work and Summary

While several works have base similarities with stealth multicast, the work in this paper is unique in that it is the first work to dynamically aggregate redundant payloads for multicast transmissions. The closest work is in [6] where packet-level caching was applied. Other works include ConCast [9] and GatherCast [10]. However, stealth multicast fundamentally differs from these two approaches in that ConCast provides a many to one multicast while GatherCast aggregates multiple small packets into a larger packet (still preserving the content of the packets).

The stealth multicast model offers a technically feasible approach for solving the issue of economic incentive for ISPs to deploy multicasting. Although the stealth multicast model does not ensure that multicast will be deployed multicast in a global sense, it addresses many of the fundamental issues that hinder multi-cast on even a domain-wise level. Furthermore, we believe the stealth multicast model offers a unique approach to multicast and provides an excellent platform for future research.

References

1. Almeroth, K.: A long-term analysis of growth and usage patterns in the multicast backbone. In: Proc. of IEEE INFOCOM'2000. (2000)
2. Beverly, R., K. Claffy: Wide-Area IP Multicast Traffic Characterization. IEEE Network (2003)
3. Diot, C., Levine, B., Lyles, B., Kassem, H., Balensiefen, D.: Deployment issues for IP multicast service and architecture. IEEE Network (2000) 78–89
4. El-Sayed, A., Roca, V., Mathy, L.: A survey of alternative group communication services. IEEE Network (2003)
5. Striegel, A.: Dynamically encapsulated trees for stealth multicast with mydeki. Technical Report TR-04-11, Univ. of Notre Dame Comp. Sci. and Engr. (2004)
6. Spring, N.T., Wetherall, D.: A protocol independent technique for eliminating redundant network traffic. In: Proc. of the 2000 ACM SIGCOMM Conference, Stockholm, Sweden (2000)
7. Striegel, A., Manimaran, G.: A scalable protocol for member join/leave in DiffServ multicast. In: Proc. of Local Computer Networks (LCN), Tampa, Florida (2001)
8. Chu, Y., Rao, S.G., Seshan, S., Zhang, H.: A case for end system multicast. (IEEE Journal on Selected Areas in Communication (JSAC), Special Issue on Networking Support for Multicast) To Appear.
9. Calvert, K.L., Griffioen, J., Mullins, B., Sehgal, A., Wen, S.: Concast: Design and implementation of an active network service. IEEE Journal on Selected Area in Communications (JSAC) (2001)
10. Badrinath, B., Sudame, P.: Gathercast: The design and implementation of a pro-grammable aggregation mechanism for the internet. In: Proc. of IEEE Int'l Conf. on Computer Communications and Networks (ICCCN). (2000)

AMBTS: A Scheme of Aggregated Multicast Based on Tree Splitting

Zhi-feng Liu, Wen-hua Dou, and Ya-jie Liu

School of Computer, National University of Defense Technology,
ChangSha, China, 410073
{liuzhif,liuyaj}@hotmail.com, wenhd@public.cs.hn.cn

Abstract. As the number of simultaneously active groups increases, IP multicast suffers from scalability problem. In backbone networks, the state scalability problem is even more serious. In this paper, we propose a novel scheme, called *AMBTS*, which uses the concept of aggregated multicast, and employs tree splitting before aggregating. We design an algorithm to divide the leaf nodes of backbone into different sub-groups, and then splitting the native multicast spanning tree. We propose a scheme to assign a set of aggregated multicast trees to deliver packets for a group. Analyses and simulations show that *AMBTS* can greatly improve state scalability: the number of aggregated trees is bounded in a small fixed number, and the multicast routing entries in transit nodes decreased dramatically.

1 Introduction

IP Multicast was first introduced in S. Deering's Ph.D. dissertation in 1988[1] and since then it has been an active area of research. However, the deployment of IP Multicast is limited and sparse due to a variety of technical and non-technical reasons. Among the issues which delay the deployment of IP multicast, state scalability is one of the most critical ones[2][3]. The state scalability concerns two aspects: scalability with regard to the number of receivers and scalability with regard to the number of simultaneously active multicast groups. In this paper, we only concern about the latter problem.

In traditional IP multicast, every on-tree node must maintain a forwarding state for each group (or even per-group/source), which increases linearly with the number of groups. On the Internet, the size of the multicast groups is limited only by the available address space: 2^{28} for IPv4 and 2^{120} for IPv6. Growing number of forwarding states means not only more memory requirement but also slower forwarding process since the forwarding action of each packet involves an address looking up. Thus, multicast will suffer from scalability problems when the number of simultaneously active multicast groups is very large.

In unicast, the hierarchical address allocation structure [4] can lead to significant reduction in the unicast forwarding table. Because the address prefixes reflect the physical proximity of the network nodes. However, in multicast one cannot make any assumption about the location of receivers of a group, it is

N. Mitrou et al. (Eds.): NETWORKING 2004, LNCS 3042, pp. 829–840, 2004.

impossible to aggregate multicast forwarding table entries of different multicast groups.

In recent years, researchers have proposed some techniques to solve the forwarding state scalability problem. We classify the different proposals into four categories. In the first category, the schemes aim at eliminating the forwarding states from the routers. Application layer multicast [5][6][7] belongs to this class. In these architectures, the complexity is pushed to the end points. A lot of obstacles that confronted by IP multicast can be conquered. But there are some other problems arise in those protocols: they are less efficient than IP multicast. In the second category, the forwarding states at non-branched routers are reduced [8][9]. These techniques mainly target networks with a large number of sparse groups. In the third category, researchers try to aggregate forwarding states in routers using the same techniques as unicast [10][11].These techniques depend heavily on multicast address allocation. In the last category, multiple different groups are forced to use the same deliver tree, so that the number of deliver trees and forwarding states in backbone can be reduced [12][13]. A.Fei and J.H.Cui et al. proposed an algorithm to assign aggregated trees to multicast groups[12] and extensive simulations proved that aggregated multicast was a promising direction for scalable transit domain multicast provisioning. However, if the size of groups is as different as possible and the receivers of groups spread as diffuse as possible, the basic aggregated multicast scheme can do little for states reduction. This phenomenon will be seen in the following sections.

The rest of this paper is organized as follows: Section 2 presents an algorithm based on tree splitting for group-tree matching. Section 3 analyses the performance of *AM* (Aggregated Multicast) and *AMBTS* (Aggregated Multicast Based on Tree Splitting). Section 4 provides extensive simulations to study the performance of *AM* and *AMBTS*. Section 5 concludes our work.

2 A Scheme of Aggregated Multicast Based on Tree Splitting

2.1 Overview of AMBTS

To improve the state scalability of multicast in backbone domains, we propose a novel scheme of aggregated multicast based on tree splitting (*AMBTS*). The main idea is that we can classify the leaf nodes of the backbone into different subgroups in advance. As a multicast group comes into the backbone, we can divide the leaf nodes of this group into different sub-trees following the dividing scheme defined beforehand. We assign an aggregated tree for each sub-tree, and then we can get a set of aggregated trees for the native multicast group tree.

Every multicast group has a *native* multicast tree, which we denote as $T_0(G)$, the native multicast tree can cover the entire group receivers and never forward packets to those nodes without receivers. The native tree $T_0(G)$ can be calculated through some routing algorithm, such as *PIM-SM* [14]and CBT[15]. A splitting of tree $T_0(G)$ can be defined as:

$$T_0(G) = \sum_{i=1}^{c} T_0^i(G) \,.$$

(1)

Where, c is the number of portions that the leaf nodes of backbone have been divided. If a set of aggregated trees $T_j, j = 1, 2, \ldots, c$, are used to replace the group's native sub-trees $T_0^j(G), j = 1, 2, \ldots, c$, the cost of bandwidth can be defined as:

$$\delta(G, \sum_{j=1}^{c} T_j) = \frac{C(\sum_{j=1}^{c} T_j) - C(T_0(G))}{C(T_0(G))} \,.$$

(2)

Where, $C(T_0(G))$ is the cost of tree $T_0(G)$. In intuition, $\delta(c_0, T)$ reflects the percentage of link overhead when we use aggregated sub-trees $T_0^j(G), j = 1, 2, \ldots, c$ to deliver packets for multicast group G.

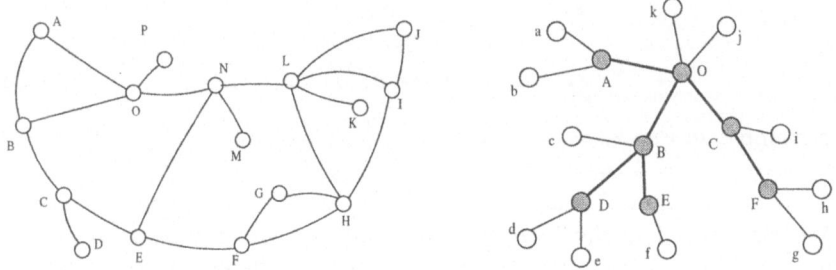

Fig. 1. vBNS backbone topology   **Fig. 2.** A spanning tree rooted at node O

In order to compare the performance of AMBTS and AM, we do our analyses and simulations with a real network topology, vBNS IP backbone, illustrated in Fig. 1. In vBNS backbone, there are 16 core routers (they will not be terminal nodes for any multicast group), and we assume each of these core routers is attached with an edge router. Thus, there are totally 32 routers in the target network. The 16 leaf routers can join any multicast session. So, the number of different spanning trees rooted at any core routers in backbone is $2^{16} - 1$.

If we only allow perfect match, as the number of simultaneously active groups increases, the number of aggregated trees will be very large, even if we adopt tree aggregation. But, if we classify all the leaf nodes into different parts in advance, then the potentially number of trees that the backbone should maintain can be dramatically decreased. For example, if we classify the leaf nodes rooted at a core router into 2 sub-groups, 8 leaf nodes in each group, the number of different spanning trees rooted at this core router can not exceed $2(2^8 - 1)$.

In conclusion, if we classify the leaf nodes into several groups beforehand and then aggregate the sub-trees based on the splitting scheme, we can get only a small number of aggregated trees. The number of aggregated trees can be bounded in a small fixed number when the simultaneously active groups become very enormous in backbone.

2.2 Classify Technique

In the following section, we will introduce a technique to split the leaf nodes. Using this technique, we will classify those leaf nodes into different sub-groups with similar size. It is not difficult to know that the number of different combinations of leaf nodes will be minimized if the size of each sub-group is equal.

Fig. 2. shows a spanning tree T built by PIM-SM or other core-based multicast routing protocol in a backbone network, the RP node is O. There are 6 core routers and 11 leaf routers. We use the term aggregated sub-tree leaf nodes (S_j) at router j to denote the entire set of leaf nodes served by all routers in the sub-tree rooted at j in T. The number of such aggregate sub-tree leaves, $s_j = |S_j|$ is given by:

$$s_j = \sum_{k \in children(j)} s_k$$
$$s_k = 1, \quad \text{if } k \in leaf(T) . \tag{3}$$

For example, in Fig. 2., $s_a = s_b = \ldots = s_j = s_k = 1$, $s_A = 2$, $s_B = 4$, $s_C = 3$, $s_D = 2$, $s_E = 1$, $s_F = 2$, $s_O = 11$.

In this example, node O has 5 direct children, i.e. there are 5 sub-trees rooted at O. We can set each of the 5 sub-trees into different sub-groups. If we classify the sub-trees into c sub-groups, the number of the different aggregated trees cannot exceed:

$$\sum_{i=1}^{c} (2^{n_i} - 1), \quad n_1 + \ldots + n_c = n \tag{4}$$

n_i is the number of leaf nodes in sub-group i.

The more parts we divide, the fewer trees we will maintain. On the other hand, the more parts that we divide, the more complexity that the RP node will involve. The reason is that each packet will encapsulate with different multicast address in RP node.

2.3 Algorithm Description

The deployment of aggregated multicast based on tree splitting includes two processes: foreclosing and group-tree matching. The foreclose process can be done offline. We describe the two processes in detail.

Foreclosing

1. Calculate a native multicast tree $T(r_i)$ rooted at some core router r_i. The spanning tree will cover all of the backbone leaf routers;

2. For each child node of r_i in tree $T(r_i)$, calculate the aggregated sub-tree leaves s_j, $j \in children(r_i)$;
3. Classify the leaf nodes into c sub-groups based on s_j, $j \in children(r_i)$, the number of leaf nodes in each sub-group should be as similar as possible;
4. For each potential root router r, repeat steps (1) to (3). After the foreclosing process, tree $T(r_i)$ is divided into c sub-trees: $T^1(r_i)$, $T^2(r_i)$, ..., $T^c(r_i)$.

Group-Tree Matching

1. According to foreclosing, we split the native tree $T_0(G)$ of group G into $T_0(G) = \sum_{i=1}^{c} T_0^i(G)$;
2. Find an appropriate match for $T_0^1(G)$, $T_0^2(G)$, ..., $T_0^c(G)$: for any sub-tree T_k that has the same root r in MTS(Multicast Tree Set), if $leaf(T_k) \supseteq leaf(T_0^j(G))$, $j = 1, 2, ..., c$, choose the one that can minimize $C(T_k) - C(T_0^j)$, $j = 1, 2, ..., c$, denoted as T_{j_min}, $j = 1, 2, ..., c$;
3. If there is not any sub-tree T_k rooted at r in MTS that can satisfy $leaf(T_k) \supseteq leaf(T_0^j(G))$, $j = 1, 2, ..., c$, $T_0^j(G)$ will be added to MTS;
4. Calculate the bandwidth overhead:

$$\delta(G, \sum_{j=1}^{c} T_{j\_min}) = \frac{C(\sum_{j=1}^{c} T_{j\_min}) - C(T_0(G))}{C(T_0(G))} \qquad (5)$$

if $\delta(G, \sum_{j=1}^{c} T_{j\_min}) > bth$, bth is the allowed bandwidth overhead threshold, we will choose a sub-tree T_{i_min} from T_{j_min}, $j = 1, 2, ..., c$ that can maximize $\delta(T_0^i(G), T_{i\_min})$. Then let the native sub-tree $T_0^i(G)$ be an aggregated tree, and add it to MTS, repeat step (4), until the set of sub-trees can satisfy $\delta(G, \sum_{j=1}^{c} T_{j\_min}) \leq bth$.

3   Performance Analysis of AMBTS

3.1   Performance Metrics

There are two natural metrics to evaluate aggregated multicast: the number of aggregated trees and the number of forwarding states in transit nodes.

Number of aggregated trees. (or "number of trees" for short) The number of aggregated trees is defined as $|MTS|$, where MTS is the current set of multicast trees maintained in backbone network. The more multicast trees, the more memory required and the more processing overhead involved.

Forwarding states in transit nodes. (or "transit states" for short). Forwarding states in terminal nodes can not be reduced in any multicast scheme, even in aggregated multicast. So we only measure the forwarding states in transit nodes.

3.2 Theoretical Analysis

To assist analyses. We list some parameters in table 1. In this section, we compare the performance between AM and $AMBTS$. We use NAT to represent the potential number of different aggregated multicast trees if perfect match allowed only. NAT is a stochastic variable, we can calculate the mean of it: $E(NAT)$.

Table 1. List of parameters

Parameters	Represent
l	the number of active groups
n	the number of leaf nodes
m	the number of RP routers
c	the number of divided portions of the backbone leaf nodes
p	session density
bth	bandwidth overhead threshold
L	the number of aggregated trees in MTS
N	the number of states in transit nodes

In AM, $E(NAT_{AM})$ can be calculated by formula (6):

$$E(NAT_{AM}) = m \sum_{i=1}^{c} C_n^i p^i (1-p)^{n-i} E(n,i)$$

$$E(n,i) = C_n^i \left[1 - \left(\frac{C_n^i - 1}{C_n^i} \right)^{n_i} \right]$$

$$n_i = \frac{l/(1-(1-p)^n)}{m} \times C_n^i p^i (1-p)^{n-i} \tag{6}$$

In $AMBTS$, $E(NAT_{AMBTS})$ can be calculated by formula (7):

$$E(NAT_{AMBTS}) = \sum_{k=1}^{m} \sum_{j=1}^{c_k} \sum_{i=1}^{n_{k,j}} C_{n_{k,j}}^i p^i (1-p)^{n_{k,j}-i} E(n_{k,j}, i)$$

$$E(n_{k,j}, i) = C_{n_{k,j}}^i \left[1 - \left(\frac{C_{n_{k,j}}^i - 1}{C_{n_{k,j}}^i} \right)^{n_{k,j}} \right]$$

$$n_{k,i} = \frac{l/(1-(1-p)^n)}{m} \times C_{n_{k,j}}^i p^i (1-p)^{n_{k,j}-i}$$

$$\sum_{i=1}^{c_k} n_{k,i} = n, \quad k = 1, 2, \ldots, m \tag{7}$$

From formula (6)., it is clear that $E(NAT_{AM})$ associate with the number of RP nodes, session density and the simultaneously active groups. We can simplify formula (6). as:

$$E(NAT_{AM}) = f(l, m, p) \tag{8}$$

From formula (7)., $E(NAT_{AMBTS})$ associate with the number of RP nodes, session density, the simultaneously active groups and the number of portions that the spanning tree have been divided into. We can simplify formula (7). as:

$$E(NAT_{AMBTS}) = g(l, m, p, c) \tag{9}$$

Table 2. $m=1$, $c=1$ the relationship of $E(NAT_{AM})$ with l and p

$p\backslash l$	500	1000	1500	2000	2500	3000	3500	4000	4500	5000
0.1	71	119	161	197	230	363	560	720	855	973
0.3	100	198	295	390	484	936	1772	2537	3247	3911
0.5	100	200	300	400	499	994	1971	2934	3882	4815
0.7	100	198	294	389	482	933	1767	2530	3238	3901
0.9	61	102	138	171	200	318	493	635	757	864

Table 3. $m=1$, $c=2$ the relationship of $E(NAT_{AMBTS})$ with l and p

$p\backslash l$	500	1000	1500	2000	2500	3000	3500	4000	4500	5000
0.1	54	76	92	104	112	144	184	206	222	236
0.3	132	202	250	284	312	388	448	472	484	492
0.5	168	280	356	406	440	500	510	510	510	510
0.7	132	202	250	284	310	388	448	472	484	492
0.9	48	70	84	96	104	134	172	196	212	224

We calculate $E(NAT)$ using formula (6). and (7) with different p, c, l and node B as RP node in Fig. 1. The result is shown in table 2. and 3. From these tables and formula (6).to(7)., we can see:

1. With the increase of the active groups, $E(NAT)$ increases;
2. $E(NAT)$ reaches maximum while p equals to 0.5. As p gets away from 0.5, $E(NAT)$ decreases;
3. $AMBTS$ superior to AM when l is a larger number;
4. For any m, p and c, we can find a critical number l_c, l_c is the critical performance point of AM and $AMBTS$:

$$f(l, m, p) < g(l, m, p, c) \quad l < l_c$$
$$f(l, m, p) = g(l, m, p, c) \quad l = l_c$$
$$f(l, m, p) > g(l, m, p, c) \quad l > l_c \tag{10}$$

l_c decrease as c increase if m and p unchanged.

4   Simulations and Analysis

Given the lack of large scale multicast experimental traces, we use the group
model developed in [12]: the random node-weighted model. In this model, each
node is assigned a weight representing the probability that the node belongs to a
group. Then we can control the size of the group, in other words, we can control
the session density [3]. We assume every node have the same probability to be in
each group and change this probability to find out the relevance between session
densities with the number of aggregated trees.

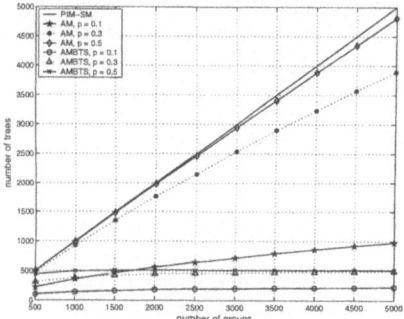 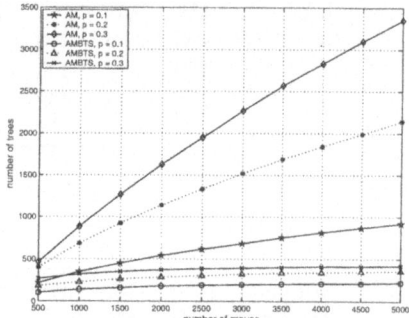

Fig. 3. L vs l, bth=0, c=2, m=1   **Fig. 4.** L vs l, bth=0.1, c=2, m=1

We design two set of experiments to compare *PIM-SM* and *AM* and *AMBTS*.
In the first group, there is only one core router selected as *RP* node in backbone.
In the second group, three core routers are used as *RP* nodes in order to achieve
better load balancing.

In the first set of experiments, the *RP* node is *B*, and the leaf routers in
backbone are divided into 2 groups according to the foreclosing process presented
in Sect.2., and the group-tree matching algorithm introduced in Sect.2. is used.
Correspondingly, the routing algorithm is *PIM-SM* like routing algorithm, which
uses unidirectional shared tree.

Fig. 3. shows the relevance of the number of aggregated trees with the simultaneously active groups when we only allow perfect-match. We can see that in
PIM-SM the number of trees is a linear function of the number of active groups.
If we use the basic *AM*, the number of aggregated trees is quite different as the
session density changes. In contrast, if we use *AMBTS*, although the number
of aggregated trees increases with the session density, the range is limited. As
the session density is 0.1 and the concurrent groups increase from 500 to 5000,
the number of aggregated trees only increases from 104 to 224. The influence of
session density is less than that in AM. When session density is 0.5 and active
groups are 5000, there are 510 aggregated trees in backbone, which is just as
analyzed before.

Fig. 4. compare the performance of *AM* and *AMBTS* when we allow pure-leaky match. Fig. 4. plots the relationship between the active groups and the aggregated trees when the bandwidth overhead threshold is 0.1. In this Fig., we can see that both *AM* and *AMBTS* can get more aggregation. In *AM*, when session density is 0.1 and the active groups are 5000, we can get 5.8% aggregation more than that in perfect-match. If the session density increases to 0.3, there are 4129 aggregated trees, decreased about 14.2% than that in perfect-match. In *AMBTS*, when session density is 0.1 and the number of active groups is 5000, we need to maintain 221 aggregated trees. When the session density increases to 0.3, there are only 358 aggregated trees, decreased about 29.8% compared to the perfect-match. The similar result can be attained if we change the bandwidth overhead threshold. In a word, increasing the bandwidth overhead threshold can improve the ability of aggregation.

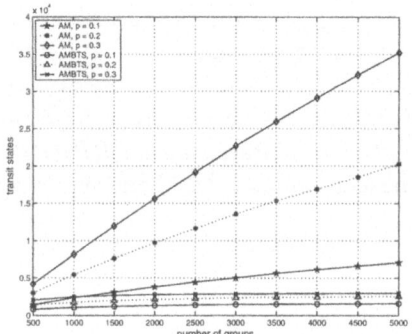 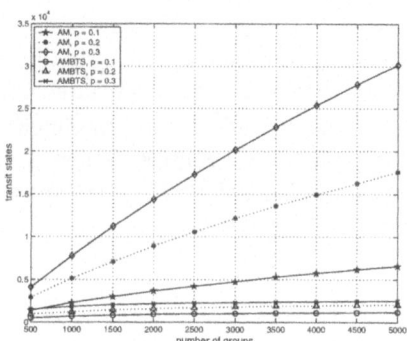

Fig. 5. N vs l, $bth{=}0$, $c{=}2$, $m{=}1$ **Fig. 6.** N vs l, $bth{=}0.1$, $c{=}2$, $m{=}1$

Fig. 5. shows the relevance of the number of forwarding entries in core routers and the number of active groups when we only allow perfect-match. From this Fig. we can see that if we use the *AMBTS*, the number of forwarding entries in core routers is much less than that in *AM*. There are two reasons: first, in *AMBTS*, there are less aggregated trees; second, the leaf nodes in aggregated trees of *AMBTS* are less than that in aggregated trees of *AM*. For these two reasons, *AMBTS* is superior to *AM* in aspect of the number of forwarding entries in core routers. In Fig. 5., when session density is 0.3 and the number of active groups is 5000, there are 35202 forwarding entries in core routers if we use *AM*. In contrast, if we use *AMBTS*, the number of forwarding entries in core nodes is only 2988, it is only 8.5% of *AM*'s.

Fig. 6. compares the performance of *AM* with *AMBTS* in the number of forwarding entries in transit routers when we allow pure-leaky match. We obtain the similar trends with Fig. 5. The more we allow the bandwidth overhead, the less number of forwarding entries we should maintain in transit routers. The reason is that more and more groups can use the same aggregated tree.

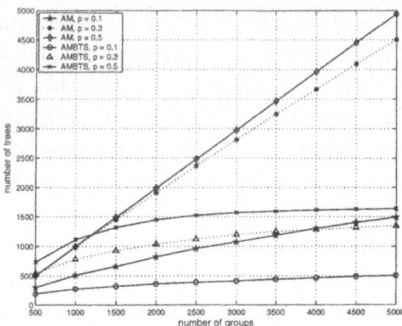
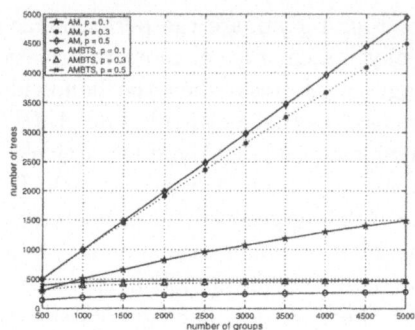

Fig. 7. L vs l, bth=0, c=2, m=3   **Fig. 8.** L vs l, bth=0, c=3, m=3

Compare to AM, $AMBTS$ can get even more aggregation. With the increase of simultaneously groups in backbone, the number of forwarding entries maintained in transit routers close to a smaller number, which is owing to the stability of the smaller number of aggregated trees.

In the second set of experiments, there are three core routers E, H and N chosen as RP routers. When a multicast session starts up, its RP node is randomly chosen from the 3 RP routers. The leaf routers in backbone are divided into 2 or 3 groups.

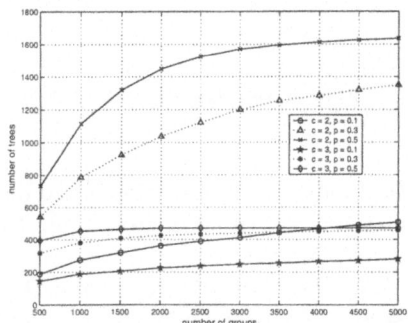

Fig. 9. L vs l, bth=0, c=2 or 3, m=3   **Fig. 10.** L vs l, bth=0.1 or 0.2, c=2, m=3

Fig. 7. shows the relevance of aggregated multicast trees with the number of active groups when the number of RP nodes great than 1. In this experiment, the leaf routers are divided into 2 groups. In this Fig., we can find that the limitation of $AMBTS$, the aggregated multicast trees may be small if the number of active groups is large, but there are possibly more aggregated trees than AM when the number of active groups in backbone is not large enough. For example, AM superior to $AMBTS$ when $l < 1200$, $p = 0.5$, $c = 2$ and $m = 3$. If we increase

c, the critical point l_c will decrease dramatically. In fig. 8., $p = 0.5$, $c = 3$ and $m = 3$, $AMBTS$ is superior to AM as l exceeds 500.

Fig. 9. shows the relevance between the number of divided portions with the number of aggregated trees. From this figure. we can see that increase the number of divided portions will decrease the number of aggregated trees in different group session density. When $c = 2$, $m = 3$, and the number of active groups is 5000 there are 1636 aggregated trees. However, if $c = 3$, and the other parameters keep same, there are only 471 aggregated trees. Namely, in the second case, there are only 28.7% of trees need to maintain compare to the first case.

Fig. 10. and Fig. 11. compares the performance of $AMBTS$ that in different bandwidth overhead threshold when the number of RP routers is 3. The trends are similar to the result of the first set of experiments. The difference is that there are more aggregated trees and more transit states in each circumstance.

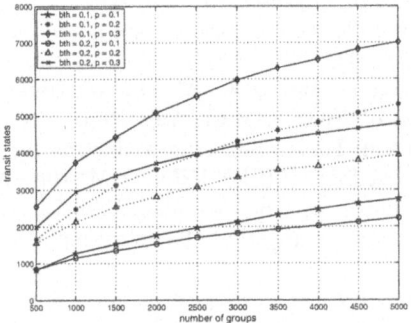

Fig. 11. N vs l, $bth = 0.1$ or $o.2$, c=2, m=3

5   Conclusions

In this paper, we propose a novel aggregated multicast scheme based on tree splitting, called $AMBTS$. This scheme can dramatically reduce the number of aggregated trees in backbone, and can take full advantage of aggregated multicast. In this paper, we design a tree splitting scheme, and a group-tree matching algorithm based on tree splitting. Through theoretical analysis and extensive simulations, we compare $AMBTS$ with traditional multicast scheme and the basic AM. The simulation results show that $AMBTS$ can obtain much more reduction in the number of aggregated trees and the number of forwarding state in core routers than the basic AM can get. The number of aggregated trees in backbone can be bounded in a small fixed number and this number can be calculated in advance, no matter how many simultaneous groups in the backbone. Thus, we can conclude that aggregated multicast based on tree splitting is a very promising scheme for transit domain multicast provision.

References

1. S. Deering: Multicast Routing in a Dategram Internetwork, Ph.D. thesis, Stanford University, 1991.
2. K. Almeroth: The evolution of multicast: From the MBone to inter-domain multicast to Internet2 deployment. IEEE Network, Jan./Feb. 2000.
3. T. Wong and R. Katz: An Analysis of Multicast Forwarding State Scalability, in Proceedings of the 8th IEEE International Conference on Network Protocols (ICNP 2000), Osaka, Japan, November 2000.
4. Y. Rekhter and C. Topolcic: Classless Inter-Domain Routing (CIDR). RFC1520, Septmember 1993.
5. D. Pendarakis, S. Shi, D. Verma, and M.Waldvogel.: ALMI: An application level multicast infrastructure. Proceddings of 3rd Usenix Symposium on Internet Technologies and Systems(USITS 2001), Mar. 2001.
6. M. Castro, P. Druschel, A-M. Kermarrec, and A. Rowstron.: SCRIBE: A large-scale and decentralized application-level multicast infrastructure. IEEE Journal on Selected Areas in communications(JSAC), 2002.
7. Y. Chu, S. Rao, and H. Zhang.: A case for end system multicast. Proceedings of ACM Sigmetrics, June 2000.
8. L. H. M. Costa, S. Fdida, and O. C. M. Duarte: Hop-by-hop multicastrouting protocol. Proceddings of SIGCOMM'01, Aug. 2001.
9. J. Tian and G. Neufeld: Forwarding state reduction for sparse mode multicast communications. Proceedings of IEEE INFOCOM, Mar. 1998.
10. P. I. Radoslavov, D. Estrin, and R. Govindan: Exploiting the bandwidth-memory tradeoff in multicast state aggregation. Technical report, USC Dept. of CS Technical Report 99-697 (Second Revision), July 1999.
11. D. Thaler and M. Handley: On the aggregatability of multicast forwarding state. Proceedings of IEEE INFOCOM, Mar. 2000.
12. A. Fei, J.-H. Cui, M. Gerla, and M. Faloutsos: Aggregated Multicast with inter-group tree sharing. Proceedings of NGC2001, Nov. 2001.
13. J.-H. Cui, J. Kim, D. Maggiorini, K. Boussetta, and M. Gerla: Aggregated Multicast — A Comparative Study. Accepted for publication in the special issue of Cluster Computing: The Journal of Networks, Software and Applications, Baltzer Science Publisher, 2003
14. D. Estrin, D. Farinacci, A. Helmy, D. Thaler, S. Deering, M. Handley, V. Jacobson, C. Liu, P. Sharma, and L. Wei: Protocol Independent Multicast-Sparse Mode (PIM-SM): Protocol Specification. IETF RFC 2362, June 1998
15. A. Ballardie: Core Based Trees (CBT version 2) multicast routing protocol specification. IETF RFC 2189, Sept. 1997

Cooperative Caching with Optimal Radius in Hybrid Wireless Networks

Huaping Shen, Sajal K. Das, Mohan Kumar, and Zhijun Wang

Center for Research in Wireless Mobility and Networking (CReWMaN)
Department of Computer Science and Engineering
The University of Texas at Arlington, Arlington, TX 76019, USA
{hpshen, das, kumar, zwang}@cse.uta.edu

Abstract. In this paper, we propose a novel caching scheme, called *energy efficient cooperative caching with optimal radius* (ECOR), to overcome the limitations of hybrid wireless networks such as high energy consumption, long access latency and poor load balance. In ECOR, each mobile user (MU) forms a *cooperation zone* with MUs in proximity to share cached data items. An analytical model of ECOR is developed to determine the optimal radius of the cooperation zone based on MU's location, data popularity and node density. Experimental results show that ECOR achieves significant performance improvement in power saving, network throughput and load balance in comparison with other existing approaches.

1 Introduction

The convergence of mobile wireless communications and the Internet is making pervasive computing a reality. The goal is to enable users to access information anytime, anywhere. For example, wireless mobile users may access stock quotes, news headline, video clips or games when in subway stations, campus or walking in the metropolitan area. However, sparse coverage of high speed wireless networks (e.g., IEEE 802.11 based WLAN) and limited battery power of mobile devices (e.g., PDA, Cellphone, Laptop, etc.) are among the major obstacles toward full realization of ubiquitous information access.

Given the current state of wireless technologies, we can classify mobile information access methods into three main categories. In the first approach, also called *infrastructure based*, there exists a fixed information access point (or base station) in the system, with which each mobile user (MU) interacts to retrieve the desired information. However, due to the expensive spectrum license fees and small coverage area of high speed wireless networks, the deployment cost of sufficient base stations to provide pervasive information service is prohibitive. The *ad hoc based* information access method, on the other hand, alleviates this problem by enabling MUs to cooperatively form a dynamic network topology without any fixed infrastructure. Each MU retrieves desired the information from other MUs in a peer-to-peer fashion. However, due to the absence of stable network connection, this approach has a low reliability. Recently, a *hybrid*

N. Mitrou et al. (Eds.): NETWORKING 2004, LNCS 3042, pp. 841–853, 2004.

based information access scheme [6][9] has been proposed to complement ad hoc and infrastructure based access methods. As shown in Figure 1, in the hybrid approach, the base stations (BSs) are assumed to have Internet connection. An MU within the transmission range of a BS can retrieve information from that BS directly. When an MU moves out of the BS transmission range, it can continue to access information by relaying data to other MUs along the path to the BS. The hybrid approach is an attractive architecture for pervasive data services as it significantly extends the wireless service area at low cost.

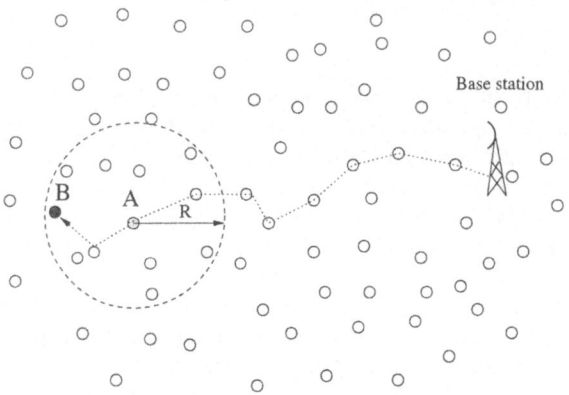

Fig. 1. A Hybrid Wireless Network

However, the hybrid wireless network is inadequate to provide ubiquitous information services due mainly to three problems. First, the battery power of the MUs that reside closer to the BS, exhaust faster than other MUs due to the frequent data exchange with the BS, which may result in a short service life of the whole network. Second, since all MUs retrieve data from the BS, the channel congestion or disconnection from the BS degrades the throughput of the network. Third, the MUs that are far away from the BS suffer long access latency for each data request. These problems motivate us to investigate energy efficient data caching schemes in hybrid wireless networks.

In order to reduce the access latency and power consumption, we propose a novel caching scheme called *energy efficient cooperative caching with optimal radius* (ECOR), to efficiently support data access in hybrid wireless networks. In ECOR, each mobile user has a cache to store the frequently accessed data items. The data items in the cache satisfy not only the MU's own requests but also the data requests passing through it from other MUs. Furthermore, to minimize bandwidth and energy cost for each data retrieval, the MUs in a proximity form a *cooperation zone* (CZ) by exchanging messages to share their cached data items. When a data request arrives, the MU first searches the data in its CZ before forwarding the request to the BS. We also develop an analytical model for the ECOR system to evaluate its performance. Based on this model, a function is derived to calculate the optimal radius of the cooperation zone of each MU.

Since the optimal radius calculation considers MU's location, data popularity, and network density, the ECOR system is expected to adapt to dynamic wireless and mobile environments. Extensive simulation experiments are conducted to validate our analytical results.

The rest of the paper is organized as follows. Section 2 gives an overview of the related work. In Section 3, a system model is developed for the proposed scheme, ECOR. Section 4 presents a detailed description of the proposed algorithms. The simulation model and experimental results of ECOR are discussed in Section 5. Section 6 concludes the paper.

2 Related Work

Caching technology has been widely studied in the context of wired networks to improve Web performance. Wessels and Claffy [14] introduced the Internet cache protocol (ICP) to support communication between caching proxies by using message exchange. Cache digests [13] and summary cache [3] enable proxies to exchange information about cached content. A cooperative hierarchical web caching architecture and fundamental design principles are investigated in [2]. However, the problem addressed in wired networks is significantly different from that in wireless mobile environments. These protocols and architectures usually assume fixed network topology and require high computation and cooperative communication capabilities. Due to the constrained resources (i.e., bandwidth, battery power and computing capacity) of portable devices along with node mobility in wireless networks, these techniques may not adapt well to the wireless network.

In the context of ad hoc wireless networks, Lau, et al. [7] proposed a cooperative caching architecture for supporting continuous media proxy caching. They introduced an application manager to transparently perform data location and session migration of continuous media streams among all proxy caches. To tolerate network partitions and improve the data accessibility, Hara [5] proposed several replica allocation methods for ad hoc networks. In these schemes, the replicated data are relocated periodically based on access frequency and overall network topology. Although data replication can improve data accessibility, the overhead for relocating replicas periodically is significantly high for mobile ad hoc networks. Nuggehalli, et al. [10] addressed the problem of optimal cache placement in ad hoc wireless networks and proposed a greedy algorithm, called POACH, to minimize the weighted sum of energy expenditure and access delay. Papadopouli and Schulzrinne [11] proposed the 7DS architecture, a peer-to-peer data sharing system, in which a couple of new protocols are defined to share and disseminate data among users that experience intermittent connectivity to the Internet. A cooperation concept was introduced in 7DS for data sharing among all mobile hosts. Similar to our proposed work, Lim, et al. [8] proposed a cooperative caching scheme for Internet based mobile ad hoc networks. A broadcast based simple search scheme is proposed to establish cooperation among all MUs in the network to share cached data items. Although the broadcast based data search scheme can locate the nearest requested data item, the energy and bandwidth cost of the flooding search is significantly high for a mobile ad hoc network. Sailhan and Issarny [16] limited the broadcast range in collaborative Web caching in ad hoc networks to minimize energy consumption and network load. The authors proposed a fixed broadcast range based on the underlaying routing protocol. However, the mobile users' location, data popularity and network density often change in a real mobile environment, so the fixed broadcast scheme is hard to adapt to real mobile applications. Yin and Cao [15] investigated cooperative caching algorithms in ad hoc networks to support data access. These algorithms mainly focus on the problem of choosing data item or data path for caching in the limited cache space of mobile devices. Their work can be considered as complementary to our work.

3   System Model

In hybrid wireless networks, as shown in Figure 1, the BS is a data source that contains a database of N items. A data request initialized by an MU is sent to the BS along with the routing path. Upon receiving the request the BS responds with the requested data item.

In ECOR, each MU maintains a *cooperation zone* (CZ) with radius R in its proximity by exchanging cooperation messages (CMs) with other MUs within its CZ. With reference to Figure 1, when a data request is initialized in an MU, say A, it first looks for the data item in its own cache. If there is a cache miss, A checks if the data item is cached in other MUs within A's CZ. If the item is cached in an MU, say B, in the CZ, then the request message is forwarded to B. Since the number of hops between A and B is at most R which is significantly less than that between A and the BS, the bandwidth and energy consumption to retrieve the data item are reduced. Now the question is: *What is the best size of CZ an MU is required to maintain in order to ensure optimal bandwidth and energy consumption of the whole network?* Intuitively, the MUs that are far away from the BS need to maintain a larger CZ than those residing closer to the BS, since the former MUs need more hops to retrieve the data from the BS. In the following, we develop an analytical model to optimize energy performance of the hybrid system. A function is derived from the model to calculate optimal radius of the cooperation zone.

3.1   Assumptions and Notations

In our analytical model, we assume that the data requests of each MU follow a Poisson process. Least Recently Used (LRU) cache replacement policy is deployed at each MU. All MUs have the same cache size and also all data items have the same size. A shortest path routing algorithm is deployed to route the data traffic in the network. According to [4], energy consumed by a mobile user for sending, receiving or discarding a message is given by the following linear equation:

$$\varepsilon(s) = m * s + p \tag{1}$$

where s is the message size, m denotes the incremental energy cost associated with a message, and p is the energy cost for the overhead of message. The parameters m and p are different for sending and receiving. The other notations used in the analytical model are listed in Table 1.

Table 1. Description of symbols

N	total number of data items in the system	C	cache size of each MU
s_{cm}	size of a cooperation message (CM)	λ_{ij}	access rate of MU_i for d_j
H_i	number of route hops between BS and MU_i	MU_i	mobile user i
H_{ik}	number of route hops between MU_i and MU_k	S	size of a data item
CZ_{ij}	cooperative zone of MU_i for data item d_j	PH_i	route path from MU_i to BS
s_{rm}	size of a data request message	f	CM exchange frequency
$\varepsilon(s_{rm})$	energy to relay a request message	$\varepsilon(s_{cm})$	energy to relay a CM
P_{ij}	probability of data item d_j cached in MU_i	R_{ij}	radius of CZ_{ij}
ρ	node density of the network	$\varepsilon(S)$	energy to relay a data item

3.2   Problem Formulation

In ECOR, each mobile user maintains a cooperation zone for each data item, so the total energy cost for MU_i to access a data item d_j can be calculated as:

$$E = E_1 + E_2 \qquad (2)$$

where E_1 is the energy cost to maintain CZ_{ij}, i.e., the energy consumed to exchange cooperation messages with other MUs; and E_2 is the energy cost used to retrieve d_j. In evenly distributed networks, the total number of MUs in the CZ_{ij} is $\rho \pi R_{ij}^2$ where R_{ij} is the radius of CZ_{ij}. We define the CM exchange frequency (f) as the number of exchanged cooperation messages among the MUs in CZ_{ij} in a unit time slot. Thus, E_1 is calculated as:

$$E_1 = \rho \pi R_{ij}^2 f \varepsilon(s_{cm}) \qquad (3)$$

In order to calculate E_2, three cases need to be considered. Case 1: data item d_j is cached by one of MUs in CZ_{ij}; Case 2: d_j is cached by one of the MUs in the routing path (PH_i) to the BS; and Case 3: d_j is retrieved from the BS. According to the algorithm of ECOR, in the case 1, when MU_i requests for d_j, it is always retrieved from the MU of smallest number of hops away within the CZ_{ij} (if multiple nearest MUs cache d_j, randomly pick one). Therefore, the energy cost for the case 1 (E_2^1) is calculated as follows:

$$E_2^1 = \sum_{r=1}^{R_{ij}} r (1 - \prod_{H_{ti}=r} (1 - P_{tj})) \prod_{H_{ki}<r} (1 - P_{kj}) \times [\varepsilon(s_{rm}) + \varepsilon(S)] \qquad (4)$$

If d_j is not cached in any MU within the CZ_{ij}, the request message is sent to the BS. In the case 2, d_j is retrieved from the nearest MU in the PH_i. Thus, the energy cost for the case 2 (E_2^2) is calculated as follows:

$$E_2^2 = \prod_{MU_t \in CZ_{ij}} (1 - P_{tj}) \sum_{h=1}^{H_i-1} hP_{qj} \prod_{MU_k \in PH_i,\ H_{ki} < h} (1 - P_{kj}) \times [\varepsilon(s_{rm}) + \varepsilon(S)]$$

$$(5)$$

where MU_q is the MU that resides in PH_i and h hops away from MU_i (i.e., $H_{qi} = h$). If d_j is neither cached in any MU within the CZ_{ij} nor cached in any MU in the PH_i, then d_j is retrieved from the BS. Thus, the energy cost for the case 3 (E_2^3) is calculated as follows:

$$E_2^3 = \prod_{MU_k \in CZ_{ij} \cup PH_i} (1 - P_{kj}) H_i \times [\varepsilon(s_{rm}) + \varepsilon(S)] \qquad (6)$$

Therefore, the energy cost (E_2) used to retrieve d_j is given as follows:

$$E_2 = E_2^1 + E_2^2 + E_2^3$$

In order to make the problem tractable and to simplify the system implementation, we use P_{ij} to estimate the probability of data item d_j cached in other MUs in CZ_{ij} and PH_i. Hence, E_2 can be approximated as:

$$E_2 \approx \left[\sum_{r=1}^{R_{ij}} r((1 - P_{ij})^{\rho\pi(r-1)^2} - (1 - P_{ij})^{\rho\pi r^2}) + (1 - P_{ij})^{\rho\pi R_{ij}^2} \sum_{k=1}^{H_i-1} (1 - P_{ij})^{k-1} P_{ij} k \right.$$
$$\left. + (1 - P_{ij})^{\rho\pi R_{ij}^2 + H_i - 1} H_i \right] \times [\varepsilon(s_{rm}) + \varepsilon(S)]$$

Thus, the total energy cost of any MU_i to access a data item d_j is expressed as follows:

$$E = \rho\pi R_{ij}^2 f\varepsilon(s_{cm}) + \left[\sum_{r=1}^{R_{ij}} r((1 - P_{ij})^{\rho\pi(r-1)^2} - (1 - P_{ij})^{\rho\pi r^2}) + (1 - P_{ij})^{\rho\pi R_{ij}^2} \right.$$
$$\times \frac{1 - (1 + H_i P_{ij} - P_{ij})(1 - P_{ij})^{H_i-1}}{P_{ij}} + (1 - P_{ij})^{\rho\pi R_{ij}^2 + H_i - 1} H_i \right] \times [\varepsilon(s_{rm}) + \varepsilon(S)] \qquad (7)$$

The probability (P_{ij}) of MU_i caching d_j is determined in terms of MU_i's data access probability for d_j, the cache replacement policy of MU_i, and the cache space of MU_i. Che, et al. [2] derived a function for P_{ij} under the LRU cache replacement policy. In order to minimize the energy consumption of the system, we need to find an optimal value for the radius R_{ij} such that the energy cost of each data access of MU_i is minimized. From Equation (7), the optimal value of R_{ij} can be calculated as follows:

$$\frac{\partial}{\partial R_{ij}} (E(R_{ij})) = 0 \quad for \quad 0 \leq R_{ij} \leq H_i \qquad (8)$$

4   Description of ECOR Algorithm

In ECOR, in order to minimize the transmission error of each CM exchange, we minimize the size of each CM. For each CM exchange, the MU only exchanges the ID information of the data items whose state have changed since the last CM exchange, i.e., new cached data items and evicted data items. We call the CM that includes only the IDs of data items that changed state since last exchange as the *asynchronous* CM (or ACM). In a mobile environment, the MUs frequently leave and join in a cooperation zone. To synchronize the information with the newly joining MUs, a CM that include the IDs of all cached data items is inserted in the sequence of ACM exchanges. We call the CM that includes the IDs of all cached data items as the *synchronous* CM (or SCM). In a CM exchange sequence, the ratio of the number of ACMs to SCMs will be defined as the *AS ratio*.

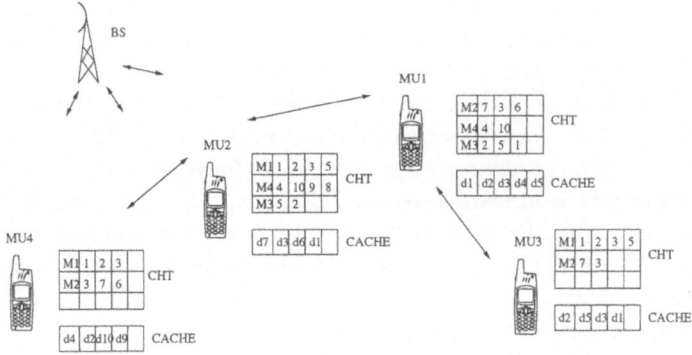

Fig. 2. A Simple Example of ECOR

With each data item's ID included in a CM, we associate a *time to live* (TTL) value to indicate the cooperation radius of the corresponding data item. The cooperation radius is calculated with the help of Equation (8). When an MU receives a CM from another MU, the TTL value of each data item is decreased by one. If the TTL value reaches zero, the corresponding data ID is deleted from the CM to indicate that the CM has reached the boundary of the cooperation zone for that item. In ECOR, each MU maintains a cache hint table (CHT) for the cache information of all MUs in its proximity. In order to handle mobility of MUs, each entry of CHT has an *aging timer* (AT) value. After receiving a CM from a source MU, the received MU needs to update the CHT entry of the source MU by resetting the AT to the initial value. If the AT reaches zero, the entry is deleted from the CHT, so that the MU that moved out of the cooperation zone will no longer participate in the cooperation. Additionally, Figure 2 illustrates the proposed ECOR algorithm. In this Figure, MU_1 is one hop away from MU_2 and MU_3, and 2 hops from MU_4. MU_1 caches $d_1,...,d_5$. When MU_1 constructs CM, the TTL value (i.e., optimal radius value) of d_4 is equal to zero, TTL value of d_5 is equal to one, and TTL values of d_1, d_2 and d_3 is equal to two. Thus, the cache hint entry of MU_1 in the CHT of MU_2 and MU_3, includes IDs of d_1, d_2, d_3, and d_5. IDs of d_1, d_2 and d_3 are included in the CHT of MU_4, since the ID of d_5 is deleted from the CM relay at MU_2 after the TTL value of d_5 reaches zero. It is also shown in Figure 2 that since the TTL values of all cached data items in MU_4 are less than 3, the cached data items' information is not included in the CHT of MU_3 which is 3 hops away from MU_4.

When MU_i initializes a data request, it first checks its local cache for the requested data item. In the case of a cache miss, MU_i searches its CHT. If the requested data item is cached by some MUs in the MU_i's cooperation zone, the request message is forwarded to the MU that is closest to MU_i. Otherwise, the request message is forwarded to the BS to retrieve the data item. As shown in Figure 2, if MU_1 requests for d_{10}, the request message is forwarded to MU_4. When an MU_i receives a request message from a neighboring MU, the MU_i first checks its local cache. If the requested data item is available at MU_i, the data item is replied to the source MU. If the requested data is not in MU_i's cache, the MU_i searches its cache hint table and forwards the request message to the MU that is closest and caches the requested data item. If MU_i finds itself as the destination of the request message, it means that the source MU of request message has inconsistent cache hint information of MU_i. For example, in Figure 2, the MU_4 receives a request message for d_8 from MU_2. In this case, the request message is forwarded to the BS to retrieve the requested data item. The inconsistency may be incurred by the loss of CM or the error of CM in the transmission. When the source MU receives the requested data item from other MUs or the BS, the LRU replacement policy is used to evict the existing data items to make space for the incoming item.

5   Performance Evaluation

In this section, we evaluate the performance of ECOR through simulation experiments. Three existing caching schemes are also implemented for comparison.

5.1   Simulation Model and System Parameters

We use the Network Simulator 2 (ns2) [17] with CMU wireless extension. In our simulations, an ad hoc network routing protocol, DSDV [12], is deployed to route the data traffic in the hybrid wireless network. We also use IEEE 802.11 as the MAC protocol and two-ray ground propagation as the radio propagation model. For the mobility model, we assume a service area of 1500m × 1500m. One fixed base station is located at (0m,750m). All mobile nodes moving in this area follow the random way point mobility model, in which the target location is also randomly chosen, and the moving speed is uniformly random chosen from $[0, v_{max}]$. After the mobile node reaches the destination, it pauses for a period of time (t_p) before continuing its next movement with the same pattern. The arrival requests for each mobile user is assumed to follow a Poisson process. After a request is sent out, if the MU does not receive the data item, it waits for an interval (t_w) before resending the same request message. We consider *Zipf-like* [1] distribution for the data popularity pattern of MU's access. In order to model the location based data access, the whole service area is equally divided into 5 × 5 square grids. Beginning with the left lowest grid, we name the grids as 1, 2, 3,..., 25 in a zig-zag diagonal-wise fashion. The MUs in the same grid have the same *Zipf* data popularity. MUs in different grids have different shift values for the *Zipf* pattern. For an MU in grid i, the *id* of data access is shifted by i so that $id = (id + i)$ mod N, where N is the total number of data items. The default settings of system parameters are given in Table 2. Some parameters may vary in the following experiments.

Table 2. System parameter setting

Parameters	default values	Parameters	default values
Number of mobile devices	100	Transmission range	200 m
Number of data items (N)	3000	Pause time (t_p)	2 sec
Data item size (S)	10K bytes	Max speed (v_{max})	2 m/s
Cache size (C)	500K bytes	Mean access rate (λ)	$1/30\ sec^{-1}$
Bandwidth	11M bps	AS ratio (AS)	10:1
Waiting interval (t_w)	10 sec	θ parameter of Zipf	0.8
Request message size (s_{rm})	10 bytes	CM message size(s_{CM})	2 bytes/item

5.2   Simulation Results

In the experiments, five metrics are used to fully evaluate the system performance: *hit ratio* (HR), *peer hit ratio* (PHR), *power per query* (PPQ), *average access delay* (AAD) and *throughput* (THU). We define PHR as the ratio of the number of data items retrieved from peer mobile users to the total number of requested data items of a mobile user. PPQ is the ratio of the total energy consumed in a given period to the total number of requests in the period. The throughput is the total number of data items successfully retrieved by all MUs in the network divided by the time. For performance comparison with ECOR, three other schemes (i.e., PathCC, BDCC and TPCC) are implemented. In the *path cooperative caching* (PathCC), only the nodes that reside in the route path from the source node to the BS participate in the cooperation to share the cache data for the data requests from the source node (i.e., cooperation radius of each MU is equal to zero). In [8], a broadcast search based cooperative caching scheme has been proposed for the Internet based mobile ad hoc networks, which is referred as *broadcast cooperative caching* (BDCC). In [16], a hybrid cooperative caching scheme is proposed for the web caching in ad hoc networks. In this scheme referred to as *terminal profile cooperative caching* (TPCC), each MU keeps a terminal profile for every other MU. In our experiments, the same data access patten, mobility model and LRU cache replacement policy are applied to all four schemes. Let us now summarize the results.

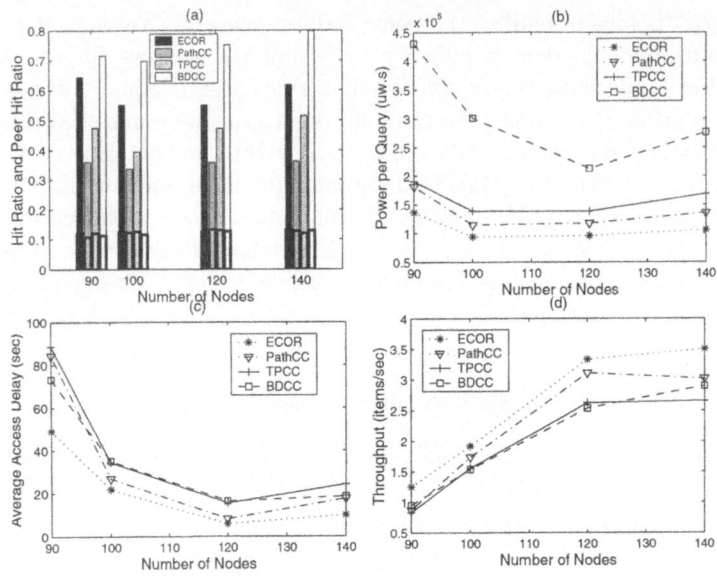

Fig. 3. Performance for different densities: (a)HR+PHR, (b)PPQ, (c)AAD, (d)THU

(1) Effect of Node Density

In this experiment, we vary the total number of MUs and fix the area of the
network to study the performance under different node densities. In Figure 3(a),
the low part of each column represents the hit ratio of the corresponding scheme,
and the upper part of each column represents the peer hit ratio of the scheme.
As shown in Figure 3(a), with identical node number, four schemes show almost
the same HR performance, because they use the same LRU cache replacement
policy. Figure 3(a) also shows that ECOR has a higher PHR at low node density
(i.e., 90 nodes) than at the median node density (i.e., 100, 120 nodes), and the
PHR improves at high node density (i.e., 140 nodes). When the density is low,
the BS has a high chance to be isolated from MUs, so that the MUs can only
retrieve data items from other peer MUs. When the density is high, the number
of MUs in a cooperation zone increases which also leads to a PHR improvement.
As shown in Figure 3(b), ECOR achieves the best PPQ performance among the
four schemes under different node densities. This is because: 1) ECOR uses the
optimal cooperation radius to share the cached data among all MUs, so that the
number of hops for one data item retrieval is minimized, and 2) asynchronous
piggyback CM exchange scheme minimizes the energy cost of CM exchanges. In
BDCC, the energy cost of the flooding search is huge for a wireless network, so
it has the worst PPQ performance. TPCC reduces the energy cost by constraint
flooding search in a local area (within two hops in the experiment). In each
scheme, the PPQ drops at the median node density and increases at the high
node density. In a low density, the isolation period of each node is long. Each
node needs to repeatedly send the request to retrieve a data item resulting in
high energy consumption. The number of resending messages decreases as node

density improves which results in PPQ dropping. However, if the network has too many nodes, the wireless channel congestion leads to high energy consumption of each data retrieval. As shown in Figure 3(c)(d), the AAD and THU performances show a similar behavior as the PPQ performance. It is shown in Figure 3(d) that the throughput improvement of ECOR over other three schemes increases for high density network. This is due to the good load balance property and choice of optimal cooperation radius of ECOR, which is further explained in the following experiment.

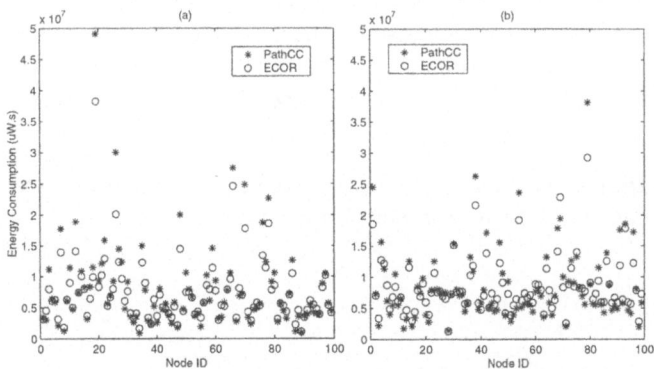

Fig. 4. Energy Consumption of Each Node: (a) $v_{max} = 2m/s$, (b) $v_{max} = 20m/s$

(2) Fairness in Energy Consumption and Load Balance

In this experiment, we investigate the energy consumption and load balance of ECOR and PathCC. We run the simulation for 4000 seconds and record the energy consumption of each MU under two mobility scenarios as $v_{max} = 2m/s$ and $v_{max} = 20m/s$. As shown in Figure 4(a), several nodes (e.g, nodes 19, 26, 66, 70 and 78) cost much more energy than other nodes. We call them as *key nodes*. Usually, the key nodes are the ones that reside in the transmission range of BS and relay the data traffic between the BS and other nodes. Since the MUs in PathCC retrieve most data items from BS, the energy consumption of each key node is high. In ECOR, the energy consumption of each key node is significantly reduced. This is because each MU in ECOR uses a cooperation zone to retrieve data items, and the radius of cooperation zone varies according to the distance from BS. The far-off MU has a bigger radius than the nearby MU. Thus, the data traffic is equally distributed in the whole network. If we consider the battery life of the key nodes as the indicator of the life time of the network, ECOR significantly improves the service life of the hybrid networks. Figure 4(b) shows that the difference of energy consumption of key nodes over other nodes decreases under a high mobility scenario. This is because the energy consumption of each node is more fair as the topology of the network changes fast.

6   Conclusion

Inheriting both advantages of infrastructure based and mobile ad hoc networks, the hybrid wireless networks are expected to become more and more important to provide ubiquitous data service with low cost and high reliability. However, long access latency, poor load balance and high energy consumption are among the major constraints for the data access in such networks. In this paper, we propose a novel cooperative caching scheme, called energy efficient cooperative caching with optimal radius (ECOR), to improve the network throughput and balance the data traffic while reducing the energy consumption. A system model is developed for ECOR to calculate the optimal radius of the cooperation zone of each mobile user. The experimental results demonstrate that ECOR outperforms existing approaches in the energy saving, access latency and network throughput.

Acknowledgement. This work is supported by a grant from Texas Advanced Research Program (TXARP) under Grant Number 14-771032.

References

1. L. Breslau, P. Cao, J. Fan, G. Phillips, and S. Shenker, "Web caching and Zipf-Like Distributions: Evidence and Implications," *In IEEE Proceedings of INFOCOM*, pp 126-134, 1999.
2. H. Che, Y. Tung, and Z. Wang,"Hierarchical Web Caching Systems: Modeling, Design, and Experimental Results," *IEEE Journal on Selected Areas in Communication*, 20(7), pp 1305-1315, 2002.
3. L. Fan, P. Cao, J Almeida, and A. Broder, "Summary cache: A Scalable Wide Area Web Cache Sharing Protocol," *ACM SIGCOMM*, pp. 254-265, 1998.
4. L. Feeney, and M. Nilsson, "Investigating the Energy Consumption of a Wireless Network Interface in an Ad hoc Networking Environment," *In IEEE Proceedings Of INFOCOM*, April 2001.
5. T. Hara, "Effective Replica Allocation in Ad Hoc Networks for Improving Data Accessibility," *IEEE INFOCOM*, 2001.
6. H.-Y Hsieh and R. Sivakumar, "On Using Peer-to-Peer Communication in Celluar Wireless Data Networks" *IEEE Tran. on Mobile Computing*, 3(1), pp. 57-72, 2004.
7. W.H.O. Lau, M. Kumar, and S. Venkatesh, "A Cooperative Cache Architecture in Supporting Caching Multimedia Objects in MANETs," *The Fifth International Workshop on Wireless Mobile Multimedia*, 2002.
8. S. Lim, W.C. Lee, G. Cao, and C. R. Das, "A Novel Caching Scheme for Internet based Mobile Ad Hoc Networks," *In IEEE Int'l Conf. on Computer Comminications and Networks (ICCCN)*, pp.38-43, October 2003.
9. H. Luo, R. Ramjee, P. Sinha, L. Li and S. Lu "UCAN: A Unified Cellular and Ad-Hoc Network Architecture" *In ACM Intl. Conf. on Computing and Networking (Mobicom)*, pp. 353-367, 2003.
10. P. Nuggehalli, V. Srinivasan, and C.F. Chiasserini "Energy-Efficient Caching Strategies in Ad Hoc Wireless Networks" *In Proceedings of ACM MobiHoc*, pp. 25-34, 2003.

11. M. Papadopouli, and H. Schulzrinne "Effects of power conservation, wireless coverage and cooperation on data dissemination among mobile devices" *In Proceedings of ACM MobiHoc*, pp. 117-127, 2001.
12. C. Perkins, and P. Bhagwat,"Highly Dynamic Destination-Sequenced Distance-Vector Routing (DSDV) for mobile computers," *ACM SIGCOMM*, pp 234-244, 1994.
13. A. Rousskov and H. Schulzrinne, "Cache Digests," *Computer Networks and ISDN Systems*, pp. 22-23, 1998.
14. D. Wssels, and K. Claffy,"ICP and the Squid Web Cache," *IEEE Journal on Selected Areas in Communication*, pp 345-375, 1998.
15. L. Yin, and G. Cao, "Supporting Cooperative Caching in Ad Hoc Networks," *IEEE INFOCOM*, March 2004.
16. F. Sailhan and V. Issarny, "Cooperative Caching in Ad Hoc Networks" *In Intl. Conf. on Mobile Data Management (MDM)*, pp. 13-28, 2003.
17. ns Notes and Documentation, "http://www.isi.edu/nsnam/ns/".

Power Adaptation Based Optimization for Energy Efficient Reliable Wireless Paths

Suman Banerjee[1] and Archan Misra[2]

[1] Computer Sciences Dept., University of Wisconsin, Madison, WI 53706, USA.
suman@cs.wisc.edu
[2] IBM T.J. Watson Research Center, Hawthorne, NY 10532, USA. archan@us.ibm.com

Abstract. We define a transmission power adaptation-based routing technique that finds optimal paths for minimum energy reliable data transfer in multi-hop wireless networks. This optimal choice of the transmission power depends on the link distance between the two nodes and the channel characteristics. Typical energy efficient routing techniques use a transmission power such that the received signal power at the destination minimally exceeds a desired threshold signal strength level. In this paper we argue that such a choice of the transmission power does not always lead to optimal energy routes, since it does not consider differences in the receiver noise levels.

We first analyze the optimal transmission power choices for both the ideal case from an information-theoretic perspective, and for realistic modulation schemes. Subsequently we define our technique for transmission power adaptation that can be used in existing routing protocols for multi-hop wireless networks. Our simulations show that current best-known schemes incur upto 10% more energy costs in low noise environments, and upto 6.67 times the energy costs in high noise environments compared to our proposed scheme.

1 Introduction

We define a distributed route computation technique that computes minimum-energy paths for reliable multi-hop wireless communication. This technique consists of two parts — (1) adaptation of the transmission power level for each wireless link, and (2) assignment of link costs to individual links which will be used by standard routing protocols to compute end-to-end paths. Both these mechanisms take into account the specific channel characteristics of the links and are necessary and sufficient to compute the optimal energy-efficient path for reliable communication. In this work we show that our technique is optimal, i.e. it computes the minimum-energy path for *reliable packet delivery* in a multi-hop wireless network.

Typical minimum-energy routing protocols for multi-hop wireless networks assign the transmission power required to sustain communication over a link as the link cost. Subsequently they use standard route computation techniques to obtain appropriate end-to-end paths. Due to the properties of wireless signal attenuation these algorithms observe that the total energy requirements for packet transfer over the entire path can be minimized by choosing a route consisting of a large number of small-distance hops over an alternative one with a small number of large-distance hops [6,12]. However, these

N. Mitrou et al. (Eds.): NETWORKING 2004, LNCS 3042, pp. 854–865, 2004.

algorithms do not necessarily yield minimum-energy paths for reliable packet delivery. This is because the the link metrics in these algorithms depend solely on the energy spent in a single transmission attempt and do not capture the additional energy expended on retransmissions in the presence of link errors. In [1] the authors had shown how link costs should be assigned to account for the total energy spent in reliable packet delivery, which also includes energy consumed in packet retransmissions. However the authors in [1] continue to make the assumption that the transmission power level on an individual link is determined solely by the corresponding link distance. In this work, we argue that such a choice of transmission power level is not optimal in finding minimum energy costs — appropriate adaptation of the transmission power based on the channel noise characteristics is a crucial component in reaching this objective.

2   Optimal Transmission Power for Individual Links

In our analytic study we first use an information-theoretic approach to show how the optimal transmission power and the associated minimum reliable transmission energy depends on both the link distance and the channel characteristics. Since the resulting bounds are essentially theoretical and not practically realizable due to severe buffering and delay constraints, we then apply our framework to practical channel models. Due to space constraints in this extended abstract, we only summarize the main results. Specific details of the analysis can be found in [2].

2.1   Information-Theoretic Bounds on Optimal Transmission Power

We utilize the information-theoretic bound on the maximum capacity of the well-known band-limited Gaussian channel in order to ascertain the existence of an optimal transmission power for energy efficient reliable data forwarding across a link. The Gaussian channel models an environment where the noise component (both thermal and due to interfering transmissions) at the receiver is assumed to have a Gaussian spectral distribution and is additive in nature. In our analysis we use Shannon's channel capacity theorem to show that if a node transmits at a power level P_t, it follows that the *normalized reliable transmission energy* (i.e. the energy needed per bit of reliable transfer) is related to its transmit power level, P_t, as :

$$E(P_t) = \frac{P_t}{W \times \log_2(1 + \frac{P_t}{D^\alpha \times \eta \times W})} \tag{1}$$

where W is the spectral width of the channel, η is the spectral noise density, and D is the distance between the transmitter and the receiver. This an increasing function of P_t.

Thus in theory, the smaller we make the transmission power, the more energy-efficient the communication process. The greatest energy efficiency (lowest cost per reliably transferred bit) is achieved as $P_t \downarrow 0$.

This minimum energy value can be obtained by applying the L'Hospital's rule to Equation 1, as is given by:

$$E_{opt} = \ln 2 \times \eta \times D^\alpha \tag{2}$$

Therefore, *every channel is associated with a fundamental theoretical (non-zero) lower bound on the minimum energy needed to reliably transfer a single bit.*

The above results show that maximum energy efficiency is achieved by transmitting at as low a power level as possible, and that a non-zero communication rate can be sustained even if the received power is much smaller than the channel noise. This is clearly not possible in any *practical* communication system with realistic bounds on the transfer latency. Indeed, Shannon's result is based on the use of asymptotically long coding sequences, resulting in unbounded transmission delays. Now we will consider a *practical* communication sub-system. We shall then see that there exists a non-zero optimal transmission power-level P_t^*: while smaller values of the transmission power result in a sharp increase in the total number of retransmissions needed, values larger than the optimum end up wasting unnecessarily large amounts of energy in a single transmission.

2.2 Optimal Transmission Power for Practical Modulation Schemes

Consider a wireless channel with a simple retransmission based error correction mechanism; if a data packet is lost in transmission across any link, it is retransmitted by the upstream node of that link. Assuming independent packet losses, the expected number of transmissions needed for the reliable transfer of one data packet across any link is $1/(1-p)$, where p is the packet error rate of the link. Let us assume that data packets are L bits long. Then for such a wireless channel with data transmission rate, f, the normalized reliable transmission energy (energy required per bit of reliable transfer) is given by:

$$E(P_t) = \frac{L}{f} \times P_t \times \frac{1}{1 - p(P_t)} \times \frac{1}{L} \tag{3}$$

The packet error rate depends on the chosen transmission power, P_t, and hence so does the normalized energy requirements. The packet error rate can be computed as:

$$p(P_t) = 1 - (1 - p_b(P_t))^L \tag{4}$$

where p_b is the bit error rate of the wireless channel. Consider Binary Phase Shift Keying (BPSK) as an example of a practical channel modulation scheme, where the bit error rate is given by:

$$p_b = 0.5 \times erfc(\sqrt{\frac{P_t \times W}{D^\alpha \times N_r \times f}}) \tag{5}$$

where D is the link distance, W is the channel bandwidth, N_r is the noise signal power, and and f is the data transmission rate. BPSK modulation is used in wireless environments, for example in the 1 Mbps version of the IEEE 802.11 wireless LAN standard.

In Figure 1 we plot the variation in $E(P_t)$ for a channel employing BPSK modulation as a function of the transmission power P_t and a packet size of 1000 bytes. We set the channel parameters to be representative of the 802.11b standard, with a bit rate of 1 Mbps and a noise bandwidth (post de-spreading) of 2 MHz. The link distance D is 100 meters and the spectral noise N_r is 4.0×10^{-11} W. We can see that the optimal transmission power for this channel exists, and is ≈ 20 mW.

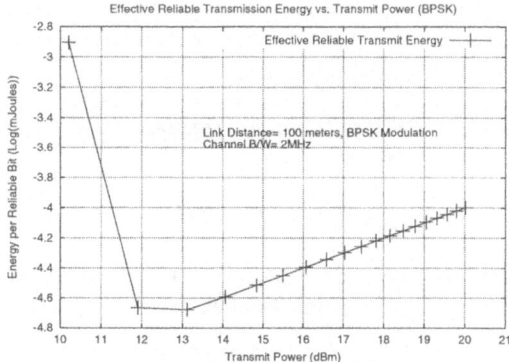

Fig. 1. Reliable Transfer Energy Behavior (BPSK)

In our analysis we have also explored how the optimal transmission power for a link varies with changes in link distance. Our results show that this optimal transmission power *increases at a slightly slower rate than D^α* where α is the constant is the signal attenuation model. Accordingly, conventional protocols, such as [1,6,12], which assign link costs to be proportional to D^α, penalize longer links more than needed. From Equations 3, 4, and 5 it also follows that the optimal transmission power depends on the spectral noise in the channel. Our study also showed that a policy of maintaining a "constant target Signal to Noise Ratio" at the receiver (by appropriately adjusting the transmission power) is a good informal rule that achieves "close to optimal" energy efficiency. In the full version [2] of this paper, we also show how the optimal energy efficiency of a practical modulation scheme compares with the theoretical lower bound provided by Equation 2.

3   Minimum Energy Routing — Assigning Link Costs

Consider a node that transmits a packet with transmit power P_t across a link. Let the corresponding energy required for this single transmission be E_t. (Assuming packets are of constant size, E_t differs from P_t by a proportionality constant.) Let E_{max} be the packet transmission energy corresponding to the maximum transmission power level at which the node can transmit. Similarly let E_{min} be the packet transmission energy corresponding to the minimum transmission power necessary to ensure that the signal strength at the receiver exceeds a desired threshold.

In this extended abstract we only describe the scenario where the wireless links implement link-layer retransmissions to recover from packet losses. However our proposed technique is equally applicable for alternative link layer reliability mechanisms like forward error correction. In another alternative scenario, if the wireless links do not implement any recovery mechanisms, data needs to be recovered using end-to-end retransmissions. Description of these alternative cases can be found in the full version of the paper [2].

Let $p(E_t)$ denote the packet error probability corresponding to the packet transmission energy, E_t. The optimal value of the energy required for reliable packet delivery across a single link is given by the solution to:

$$\frac{d}{dE_t}E_t(reliable, HHR) = 0 \tag{6}$$

$$\frac{d^2}{dE_t^2}E_t(reliable, HHR) \geq 0 \tag{7}$$

where $E_t(reliable, HHR) = E_t/(1 - p(E_t))$. It follows that this optimal value, E_t^*, that minimizes the energy cost for a link satisfies:

$$E_t^* . p'(E_t^*) - p(E_t^*) = 1 \tag{8}$$

where $p'(.)$ denotes the first derivate of $p(.)$ with respect to E_t. E_t^* can be computed using efficient numerical techniques. However, this optimal solution may exceed E_{max}. In such a case our choice of optimal transmission energy, $E_t^* = E_{max}$. Similarly if the computed E_t^* falls below E_{min}, we assign $E_t^* = E_{min}$. In the routing protocol, we assign each wireless link a cost which is given by energy required for reliable packet transmission across that link, given by $E_t^*/(1 - p(E_t^*))$. The end-to-end route can therefore be computed in a distributed manner by any standard routing protocol capable of computing minimum cost paths. It follows that shortest cost path found by the routing algorithm will be the *optimal energy-efficient route* for that end-to-end path.

4   Simulation Studies and Performance Evaluation

In this section, we report on extensive simulation-based studies on the performance impacts of our proposed modifications in the *ns-2* simulator. In these studies, we only consider the hop-by-hop retransmission scenario. We performed studies using both TCP and UDP traffic sources to study the effect of our routing schemes on these transport layer mechanisms. However, due to space constraints we report results for TCP experiments, performed using its NewReno variant. The details of the UDP experiments can be found in [2]. For all these simulation studies, we use link-layer retransmissions to recover from packet losses.

To study the performance of our suggested schemes, we implemented and observed three separate routing algorithms: (a) Energy-Aware (EA) scheme in which link costs are assigned based only the energy requirements for a single transmission attempt across the link. This scheme does not account for energy requirements for retransmissions and is equivalent to [6,12], (b) Re-transmission Aware (RA) scheme [1] in which link costs also account for energy required for retransmissions, and (c) our proposed Optimal Re-transmission Aware (RA-Opt) scheme which additionally performs optimal adaptation of transmission power.

4.1   Network Topology and Link Error Modeling

For our studies, we used different topologies having upto 100 nodes distributed over on a square region, to study the effects of various schemes on energy requirements and throughputs achieved.

1. *Grid topologies:* For comparison purposes, we first present results on the performance of the schemes on a 100 node grid topology similar to one used in [1]. The nodes on this topology were separated 100 units apart along each axis, and the maximum transmission radius of the node was limited to 150 units. Thus, each node has between 3 and 8 neighboring nodes on this topology [1].

2. *Random topologies:* We also present simulation results for randomly generated topologies. In the random topologies, the nodes were distributed uniformly at random in a 1000×1000 square grid. We experimented with different transmission radii for the nodes. In our random topology generator, we specified the desired number of links [2]. To avoid uni-directional links, we assigned the same transmission radii to all nodes. Note that a hop-by-hop retransmission scheme works only for bi-directional links. In the results presented in this section for the random topologies, we specified the number of wireless links to be one-eighth of a complete graph on these set of nodes. The consequent transmission radii for each node was about 210 units.

Each of the routing algorithms was then run on these topologies to derive the least-cost paths to each destination node. To simulate the offered traffic load typically of such ad-hoc wireless topologies, each of the corner node on the grid topology had 3 active flows, providing a total of 12 flows. In the random topology, we chose 12 random source-destination pairs from the entire set of nodes.

Since our objective was to study the transmission energies alone, we did not consider other factors such as link congestion, buffer overflow etc. Thus, each link had an infinitely larger transmit buffer; the link bandwidths for all links (point to point) was set to 2 Mbps. Each of the simulations was run for a fixed duration.

We choose BPSK as our representative modulation scheme and hence, use Equation 5 to derive the bit-error-rate. We varied the ambient noise to obtain different data points. For the non-adaptive transmission power algorithms (EA and RA) we chose a transmit power of 20 mW. The spectral noise for the different channels was chosen to vary between two configurable parameters, N_{min} and N_{max} corresponding to minimum and maximum noise respectively. less than a configurable parameter N_{max}.

We simulated two different environments:

1. *Low noise environment:* In this case, we chose N_{min} to be $1.8 \times 10^{-11}W$, while N_{max} was varied between $2.0 \times 10^{-11}W$ and $3.0 \times 10^{-11}W$. For the non-adaptive schemes (EA and RA) a maximum spectral noise of $2.0 \times 10^{-11}W$ leads to a corresponding channel packet error rate of 0.1 on a 100 unit link. Our adaptive

---

[1] Our-energy aware routing formulation does not directly define a transmission range. It is possible that a longer link with lower receiver noise may consume less effective energy that a shorter link with higher receiver noise. Real-life scenarios, however, impose both an upper bound on the maximum possible transmission power as well as a minimum energy threshold for successful packet reception — if the received power level is below this threshold, no reception is possible even in the absence of any receiver noise. Since any signal suffers channel attenuation $\propto D^4$, the transmission range is an alternative way of assuming that the received power level beyond a distance of 150 units is always lower than the minimum reception threshold, even if the transmitter operates at the maximum power level.

[2] We count each pair of nodes that are within the transmission range of each other as one wireless link.

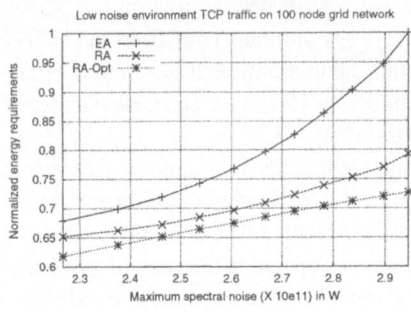
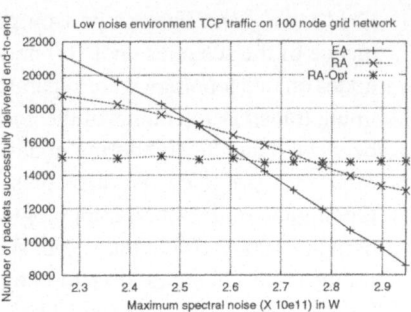

Fig. 2. Normalized energy costs for TCP flows (Low Noise Grid Topology).

Fig. 3. Comparison of throughput for TCP flows (Low Noise Grid Topology).

transmission algorithm (RA-Opt) appropriately chose a transmission power for each link so that the energy consumption for reliable data transfer across that link is minimized.

2. *High noise Environment:* In this scenario, we chose N_{min} to be $2.8 \times 10^{-11}W$; we varied N_{max} between $3.0 \times 10^{-11}W$ and $4.0 \times 10^{-11}W$.

Our results show that the RA-Opt scheme outperforms the other schemes in environments (other than zero noise environments). Additionally, our scheme shows significant benefits as the noise in the environment increases, as a comparison between these two environments show.

4.2 Metrics

To study the energy efficiency of the routing protocols, we observed two different metrics:

1. **Normalized energy:** We first compute the average energy per data packet by dividing the total energy expenditure (over all the nodes in the network) by the total number of unique packets received at any destination (sequence number for TCP and packets for UDP). We defined the normalized energy of a scheme, as the ratio of the average energy per data packet for that scheme to the average energy per data packet required by the maximum-energy experiment between all the schemes among all these experiments. This provides an easy representation for comparison of the different schemes with each other and with changing maximum spectral noise for different sets of studies.

2. **Effective Reliable Throughput:** This metric counts the number of packets that was reliably transmitted from the source to the destination, over the simulated duration. Since all the plots show results of runs of different schemes over the same time duration, we do not actually divide this packet count by the simulation duration. Different routing schemes will differ in the total number of packets that the underlying flows are able to transfer over an identical time interval.

4.3   Low Noise Environment for Grid Topologies

Figure 2 shows the normalized energy consumption for the different schemes for the TCP flows. For example, when N_{max} was set to $2.9 \times 10^{-11}W$, the relative energy requirements of RA-Opt, RA and EA were 0.72, 0.77 and 0.96 respectively. As expected, the energy requirements of all the schemes increase with increase in spectral noise. The EA scheme has the highest energy requirements among all the schemes when the maximum channel noise on links was high. Both the RA and the RA-Opt scheme performs significantly better than this scheme for the entire range of spectral noise. RA-Opt has the best performance among the three different schemes for the entire range of spectral noise. The EA scheme consumes about 10% to 33% more energy per packet, while the RA scheme consumes about 8% to 10% more energy per packet than the RA-Opt scheme,

It is interesting to note that in this low noise environment the energy costs of both the EA and RA schemes have a convexity property, while that of the RA-Opt scheme has a concavity property. This implies that the benefits of the RA-Opt scheme becomes more and more significant with increase in the spectral noise.

In Figure 3 we plot the effective reliable throughput metric for the different schemes in the same low noise experiments. The number of packets transmitted reliably over a fixed duration for the EA scheme falls rapidly with increase in spectral noise. This is expected because the EA scheme does not consider channel properties in choosing routes. In contrast, the number of packets reliably transferred by the RA scheme falls in a more gradual fashion. The decreasing trend in both these schemes is due to the increasing link error rates with the increase in spectral noise. As the link error rates increase, packets sees an increase in end-to-end delays, due to the delays spent in increased number of retransmissions necessary to ensure reliability.

However, the same metric stays relatively constant for the RA-Opt case. This is because, the RA-Opt scheme aggressively adapts the transmission power so as to minimize the energy costs for reliable packet delivery across a link. *The corresponding transmission power to achieve this optimal cost is such that the link error rate stays fairly stable across the entire range of spectral noise.*

4.4   High Noise Environment for Grid Topologies

Now we present results for the higher noise environment. Note that in this environment, the value of N_{min} is significantly larger than its corresponding value in the low noise environment. In Figure 4, we plot the normalized energy required per packet in the high noise environment for UDP flows. For example, when N_{max} is set to $3.375 \times 10^{-11}W$ the relative energy requirements of RA-Opt, RA and EA schemes are 0.16, 0.30 (i.e. 1.9 times that of RA-Opt) and 0.60 (i.e. 2.75 times that of RA-Opt) respectively. The benefits of the RA-Opt scheme is significantly higher than in the low noise environment (note that the scale of the Y-axis is much larger than the corresponding plots for low noise environments). The RA scheme consumes between 1.5 times to 2 times more energy per packet in this environment than the RA-Opt scheme, while the EA scheme consumes 6.67 times more energy in the worst case in these experiments.

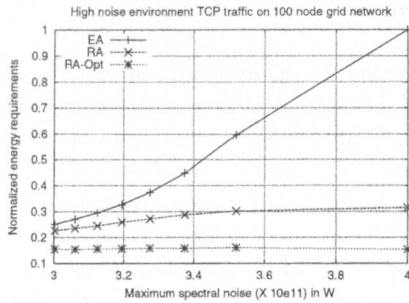

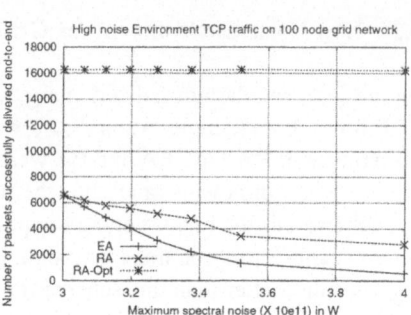

Fig. 4. Normalized energy costs for TCP flows (High Noise Grid Topology).

Fig. 5. Comparison of throughput for TCP flows (High Noise Grid Topology).

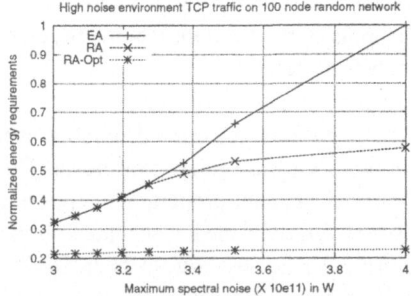

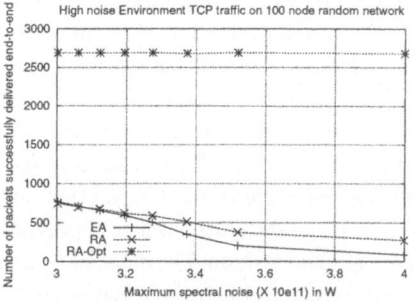

Fig. 6. Normalized energy costs for TCP flows (High Noise Random Topology).

Fig. 7. Comparison of throughput for TCP flows (High Noise Random Topology).

It is interesting to observe the behavior of the throughput achieved by TCP flows in this high noise environment. In trying to optimize the energy consumption, the RA-Opt scheme adapts the transmission power which suitably drives down the channel error rates in all environments. Therefore, the throughput achieved by the RA-Opt scheme is largely unaffected by the noise characteristics. In Figure 5 we plot the throughput for TCP flows in the high noise environment. We can observe that the RA-Opt scheme achieves the same throughput both in the low and high noise environments. Both the EA and RA schemes suffer in the high noise environment, as can be seen in the significant drop in their throughputs achieved.

Both the EA and RA schemes achieve similar throughputs when the maximum spectral noise is $3 \times 10^{-11}W$ with the minimum spectral noise being $2.8 \times 10^{-11}W$. This is because the range of error rates between different links are similar in this scenario, and so RA is unable to choose significantly better paths than EA. RA-Opt is able to make such a choice by increasing the transmission power at nodes to drive down the error rates significantly.

4.5   Random Topologies

We now present results of our studies with randomly generated topologies. Observing that the relative energy requirements for both TCP and UDP flows are similar in nature, we only present the results for the TCP flows in a high noise environment.

In Figure 6, we plot the energy requirements for the different schemes in the high noise environment. The benefits of the RA-Opt scheme is apparent in the plot. The relative energy costs of the RA-Opt, RA and EA schemes are 0.22, 0.49 (i.e. 123% more than RA-Opt) and 0.53 (i.e. 141% more than RA-Opt) respectively. At low noise levels (less than $3.28 \times 10^{-11} W$) the RA scheme and the EA scheme performs equally. However, by adapting the transmission power at the nodes, the RA-Opt scheme performs significantly better than these schemes.

In Figure 7 we show the throughput of the different schemes. Like before, the RA-Opt scheme maintains a stable throughput, while the other schemes see a significant degradation in performance.

5   Related Work

Typical energy-aware routing protocols (e.g. PAMAS [12] and PARO [6]) aim to minimize the total power consumed over the entire transmission path. However, these protocols do not address the problem of reliable data delivery, and hence, ignores the additional energy expenditure due to re-transmissions, if they become necessary. Such a formulation often leads to the formation of a path with a large number of hops. In [11], the authors extend this basic approach by including energy expenditure for packet reception in the link cost.

The basic idea of reducing communication energy costs for packet transmission over individual wireless links has also been explored in the literature. However, the focus has been on the use of intelligent *link scheduling* algorithms, rather than on transmission power control. For example, Zorzi and Rao [15] proposed the use of short, periodic probe packets to detect the condition of the wireless channel. Actual data packet transmissions were deferred when the channel is in a 'bad' state. A similar idea for energy-efficient scheduling of packets from a base station to a set of downstream wireless hosts has been explored in [3,10,4].

Ad-hoc routing protocols aim to compute minimum-cost paths; in contrast to generic (non ad-hoc) routing protocols, they contain special features to reduce the signaling overheads and convergence problems caused by node mobility and link failures. So, ad-hoc protocols, such as AODV [9] or DSR [7], can (in principle) be adapted, with suitable modifications, to yield minimum-energy paths by setting the link metric to be a function of the transmission energy.

Apart from minimum energy path problem, research in in energy-aware routing has also focused on other problems with related objectives. For example, *battery-aware routing* algorithms typically aim to extend the lifetime of all the ad-hoc nodes by distributing the transmission paths among nodes that currently possess greater battery resources [13, 14,8]. While minimum energy algorithms are most efficient, these network lifetime maximizing schemes are more "fair." A combination of both these approaches can therefore be useful as shown in [14,8].

Link error probabilities have been considered for single hop spread spectrum links in [5]. In contrast, we focus on end-to-end energy costs for multi-hop wireless networks.

6   Conclusions

We have defined an optimal energy routing scheme for reliable data transfer on a multi-hop wireless network. This scheme outperforms well-known existing routing schemes for a wide range of channel characteristics. This improvement in energy efficiency is achieved by explicitly considering the impact of receiver noise on packet errors, and by adjusting the transmission power to minimize the total energy spent in reliably forwarding a single bit. The scheme is general across different modulation techniques that can be employed for data delivery on wireless networks.

We investigated the issue existence of an optimal transmission energy for a given link. By employing information-theoretic bounds, we have shown how any link is associated with a fundamental lower bound on energy efficient reliable communication. Moreover, this fundamental bound is directly proportional to the channel attenuation rate ($\propto D^{\alpha}$), and is achieved by choosing arbitrary low transmission power. In contrast, practical communication systems are associated with a well-defined optimal transmission power, such that any decrease or increase from this optimal value results in a sharp increase in the total transmission energy spent in reliable data transfer.

We have studied the applicability of this technique in choosing optimal energy paths. By appropriately choosing the cost metric, it is possible to optimize other objective functions, e.g. end-to-end latencies, data delivery throughput, etc. We, therefore, believe that our scheme has a wider applicability to a range of operating modes depending on the the optimization objectives. Simulation studies indicate that performing adaptive power control based on the individual link conditions (error rates) can provide energy savings of $\approx 10\%$ in low-noise environments, and as much as $\approx 40\%$ in high-noise environments.

The analysis in this paper assumed that the use of retransmissions as the sole means of providing a reliable link layer. As discussed earlier, the fundamental technique can, however, also be applied to alternate reliability schemes such as forward error correcting codes through appropriate changes to the relationships in Equations 4 and 3. Since our power adaptation mechanism implicitly relies on relative stable variations in the packet error-rate, this technique is especially useful in static, or low-mobility, multi-hop networks, where link parameters such as distance or attenuation coefficients do not exhibit very rapid changes. (Of course, our formulation is applicable in the presence of typical wireless environment effects such as fading; the average bit error rate at the link-layer is typically a more stable statistical metric obtained by averaging over such physical layer variations). We have also assumed the existence of appropriate MAC-layer contention resolution mechanisms for common-channel networks, which present an abstraction of zero-interference to the higher layers. Such an abstraction is also provided by the use of distinct physical channels based on TDMA/FDMA/CDMA techniques. Finally, our approach to energy optimization is useful not just for forming energy-efficient routing paths, but also for independently optimizing the transmission energy on each individual link. The routing algorithm and individual link-layer power control techniques can operate on different time scales; while routes can be re-computed over longer time periods, individual transmitters can adjust their link transmission power over shorter time-scales.

References

1. S. Banerjee and A. Misra. Minimum energy paths for reliable communication in multi-hop wireless networks. In *Proceedings of ACM Mobihoc*, June 2002.
2. S. Banerjee and A. Misra. Power adaptation based optimization for energy efficient reliable wireless paths. Technical Report, Department of Computer Sciences, University of Wisconsin-Madison, WI 53706, USA, Available at http://www.cs.wisc.edu/~suman/pubs/optpower-tr.pdf, 2004.
3. P. Bhagwat, P. Bhattacharya, A. Krishna, and S.K. Tripathi. Using channel state dependent packet scheduling to improve tcp throughput over wireless lans. *Wireless Networks*, 3(1), 1997.
4. A. El Gamal, C. Nair, B. Prabhakar, E. Uysal-Biyikoglu, and S. Zahedi. Energy-efficient Scheduling of Packet Transmissions over Wireless Networks. In *Proceedings of IEEE Infocom*, June 2002.
5. J.H. Gass Jr., M.B. Pursley, H.B. Russell, and J.S. Wysocarski. An adaptive-transmission protocol for frequency-hop wireless communication networks. *Wireless Networks*, 7(5):487–495, September 2001.
6. J. Gomez and A. Campbell. Power-aware routing optimization for wireless ad hoc networks. In *High Speed Networks Workshop (HSN)*, June 2001.
7. D. Johnson and D. Maltz. Dynamic source routing in ad hoc wireless networks. In *Mobile Computing*, pages 153–181, 1996.
8. A. Misra and S. Banerjee. Mrpc: Maximizing network lifetime for reliable routing in wireless environments. In *Proceedings of IEEE Wireless Communications and Networking Conference*, March 2002.
9. C.E. Perkins and E.M. Royer. Ad-hoc on-demand distance vector routing. In *Proceedings of the 2nd IEEE Workshop on Mobile Computing Systems and Applications*, February 1999.
10. B. Prabhakar, E. Uysal-Biyikoglu, and A. El Gamal. Energy-efficient Transmission over a Wireless Link via Lazy Packet Scheduling. In *Proceedings of IEEE Infocom*, April 2001.
11. K. Scott and N. Bamboos. Routing and channel assignment for low power transmission in PCS. In *Proceedings of ICUPC*, October 1996.
12. S. Singh and C.S. Raghavendra. Pamas-power aware multi-access protocol with signaling for ad hoc networks. In *ACM Communications Review*, July 1998.
13. S. Singh, M. Woo, and C.S. Raghavendra. Power-aware routing in mobile ad-hoc networks. In *Proceedings of Mobicom*, October 1998.
14. C.K. Toh, H. Cobb, and D. Scott. Performance evaluation of battery-life-aware routing schemes for wireless ad hoc networks. In *Proceedings of ICC*, June 2001.
15. M. Zorzi and R. Rao. Error control and energy consumption in communications for nomadic computing. *IEEE Transactions on Computers*, 46(3), March 1997.

Adaptive Broadcast Consumption (ABC), a New Heuristic and New Bounds for the Minimum Energy Broadcast Routing Problem[*]

Ralf Klasing[1], Alfredo Navarra[1,2], Aris Papadopoulos[1,3], and
Stéphane Pérennes[1]

[1] MASCOTTE project, I3S-CNRS/INRIA/Univ. Nice, Sophia Antipolis, France.
{FirstName.LastName}@sophia.inria.fr
[2] Computer Science Department, University of L'Aquila, Italy.
navarra@di.univaq.it
[3] Department of Computing, Imperial College London, UK.
aris.papadopoulos@imperial.ac.uk

Abstract. In this paper we present a new heuristic called *Adaptive Broadcast Consumption* (ABC for short) for the *Minimum-Energy Broadcast Routing* (MEBR) problem. We first investigate the problem trying to understand which are the main properties not taken into account by the classic and well–studied MST and BIP heuristics, then we propose a new algorithm proving that it computes the MEBR with an approximation ratio less than or equal to MST, for which we prove an approximation ratio of at most 12.15 instead of the well–known 12 [10]. Finally we present experimental results supporting our intuitive ideas, comparing ABC with other heuristics presented in the literature and showing its good performance on random instances even compared to the optimum.

1 Introduction

The *Minimum-Energy Broadcast Routing* (MEBR) problem is one of the most intensively studied problems in recent years for the various applications in which it can be applied (see [3,4,5,9,10] for a survey). The best result presented in the literature concerning the approximation is a ratio of 12 thanks to [10], but we show in Section 3 that 12.15 is the actual value. In this paper, an analytical study of the performance of some basic heuristics such as Minimum Spanning Tree (MST), Shortest Path Tree (SPT) and Broadcast Incremental Power (BIP) is presented. According to the results we obtained, MST seems to be the most promising heuristic to reduce the gap between the best known lower and upper bounds. The Euclidean MST, in fact, has a lot of useful properties that can be applied to try to reduce the gap. On the other hand, MST seems to be very far

---

[*] Work supported by the European projects RTN ARACNE and IST FET CRESCCO, by the French MASCOTTE project I3S-CNRS/INRIA/University of Nice, Sophia Antipolis and by the Italian project REAL-WINE.

N. Mitrou et al. (Eds.): NETWORKING 2004, LNCS 3042, pp. 866–877, 2004.

from a good heuristic to solve MEBR due to the static properties of its construction. It is evident, in fact, that to refine a solution, a sort of backtracking is needed to adapt at each step the current solution to the newly considered configuration of nodes in the plane. In this paper, we investigate other kinds of heuristics, related to MST, adding a sort of backtracking in order to modify properly a current solution. We present a promising heuristic supported by experimental results and some basic theoretical results. In particular, after proving that our *Adaptive Broadcast Consumption* (ABC) algorithm outputs a solution whose value is at most equal to MST, we show the near optimal performance of ABC on random instances and the decrease, on average, of 5% on the total cost with respect to BIP.

The paper is organized as follows. In the next section, we introduce the problem of Minimum Energy Broadcast Routing with notations and the necessary definitions. In Section 3, we prove the mentioned upper bound 12.15 for MST refining the arguments of [10]. In Section 4, we present a new heuristic for the MEBR problem trying to trace some more specific property that is completely ignored by previous attempts. In Section 5, we present a formalization of the problem using integer linear programming to obtain optimal solutions. In Section 6, we provide experimental results obtained in various scenarios of random instances, comparing the developed heuristics. Finally, in Section 7, we give some conclusive remarks and discuss some open questions.

2   Minimum Energy Broadcast Routing

The Minimum Energy Broadcast Routing (MEBR) problem arises in the context of computing communication schedules in *Ad-Hoc* networks. These are usually special networks used in specific situations such as military operations, emergency disaster relief and so on. In contrast with *cellular, single-hop* networks, Ad Hoc wireless networks consist of a homogeneous network of mobile nodes and require no established, supporting backbone infrastructure. A communication session can be established through a series of wireless links involving any of the network mobile nodes and therefore Ad Hoc networks are *multi-hop* networks. In the problem we study, the nodes have the ability to adjust their transmission power as appropriate. Thus every mobile node is assigned a transmission range and every node inside this range receives its message. In simple terms, the aim is to assign ranges in such a way that the total energy that is consumed is minimum. Considering the fact that the mobile nodes operate using batteries of limited capability and the nature of the operations for which such kind of networks are used, it is easy to understand the importance of the energy conservation.

According to the mostly used power attenuation model [11] when a node that we call s transmits to a node r with power P_s, the power at the point that node r lies will be:

$$P_r = \frac{P_s}{\|s,r\|^\kappa} \tag{1}$$

where $\|s, r\|$ is the Euclidean distance between the source node s and the receiving node r, and κ is the *distance power gradient*. In the real world it holds that $2 \leq \kappa \leq 6$ according to the topology of the space, but for an even, flat surface $\kappa \approx 2$ (see [8]). Therefore for our analysis we assume that $\kappa = 2$. Also assuming normalization to 1, in order to establish communication between s and r, the following expression must be valid:

$$\frac{P_s}{\|s, r\|^\kappa} > 1 \tag{2}$$

Note that in order to send the message from s to r there is the possibility of using a node i as an intermediate node, if one exists, resulting in a decrement of the power needed. With the assumption that $\kappa = 2$ choosing the path $\{s, i, r\}$ is cheaper only if i lies inside the circle around the middle point of the line segment s, r with diameter $\|s, r\|$. Otherwise using i makes the solution more expensive.

As mentioned, the most common algorithms that are proposed in the literature to solve MEBR are MST and its improved variations, SPT and BIP (see [1,10,11]). In this paper we propose a new algorithm, *Adaptive Broadcast Consumption* (ABC for short), and we provide an analytical comparison of the algorithms through extensive experiments. We do that for all the mentioned algorithms except for SPT. The reason is that SPT is clearly more suitable for the *Minimum-Energy Unicast Routing Problem* (MEUR, see [7] for a survey) than for the MEBR. Moreover, the lower bound presented in [10] for SPT is very high. Although MST seems to suit the MEBR much better, the reality is that MEBR is different and much more challenging than the typical MST problem.

3   Upper Bound for the MST Heuristic

In this section, we prove the upper bound for the MST heuristic that will be useful also to bound the other heuristics. First of all, we briefly describe the technique used in the [10] where the authors claimed that the approximation ration of MST is at most 12. In that paper, the authors used some geometric properties of the Euclidean MST.

Informally speaking, they associated the so–called "diamond" to each edge of the produced MST to estimate its cost comparing it with the covered area. The diamond is a rhombus whose bigger diagonal coincides with an edge of the MST and the angles at its endpoints are 60 degree (see Figure 1). After proving that there are no superposition between each pair of diamonds associated to an MST computed over an instance of MEBR inside a circle of radius 1, they calculate the "sticking-out area" ($S(\alpha)$) of those diamonds, that is, the outgoing area from the circle where α is the angle between the two radii to the endpoints of a diamond placed over the circumference (an example can be seen in Figure 1). The sticking-out area of a diamond is $S(\alpha) = \frac{1}{2} \sin \alpha + \frac{\sqrt{3}}{6}(1 - \cos \alpha) - \frac{\alpha}{2}$, the maximum is obtained for $\alpha = \frac{\pi}{3}$ (see [10]). In their calculation they maximized properly the biggest sticking-out area of a single diamond but there is some problem with the global scenario. In the following we show the erroneous argumentation

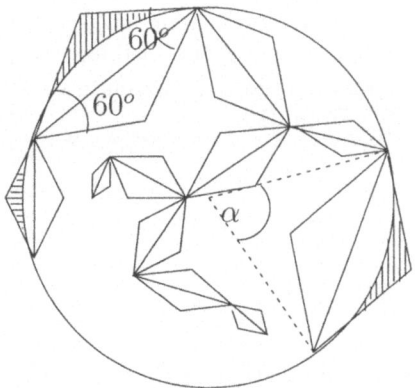

Fig. 1. Diamonds associated to an MST inside a circle.

of [10] that, in any case, remains a milestone for the MEBR problem. Actually, the quantity that must be maximized to obtain the upper bound is $\sum_{i=1}^{k} S(\alpha_i)$ with $\sum_{i=1}^{k} \alpha_i \leq 2\pi$. Roughly speaking, this quantity represents the maximum sticking-out area of k diamonds that can be placed around the circumference of the circle of radius 1 compatible with the Euclidean MST properties.

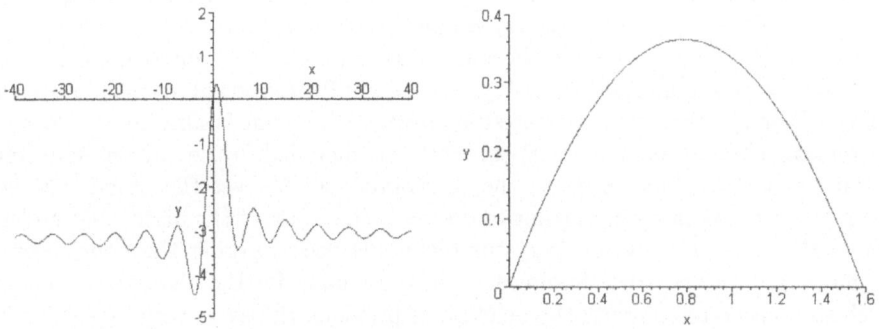

Fig. 2. $y = \frac{2\pi}{x} \left(\frac{1}{2} \sin x + \frac{\sqrt{3}}{6}(1 - \cos x) - \frac{x}{2} \right)$ and its enlargement.

Theorem 1. *The approximation ratio of the Euclidean minimum spanning tree heuristic for the minimum energy broadcast routing problem is at most 12.15.*

Proof.

$$\sum_{i=1}^{k} S(\alpha_i) = \sum_{i=1}^{k} \alpha_i \frac{S(\alpha_i)}{\alpha_i} \leq \sum_{i=1}^{k} \alpha_i \max_{0 \leq \alpha \leq 2\pi} \left\{ \frac{S(\alpha)}{\alpha} \right\} \leq 2\pi \max_{0 \leq \alpha \leq 2\pi} \left\{ \frac{S(\alpha)}{\alpha} \right\}$$

In Figure 2, we show the graph of the function $y = 2\pi \frac{S(x)}{x}$ whose maximum is $y_{\max} = 0.36324...$ and this maximum is obtained for $x = \frac{\pi}{4.00458...} \approx \frac{\pi}{4}$. Adding π (the area of the circle of radius 1) to the maximum sticking-out area, we obtain the upper bound for MST. From [10],

$$\sum_{e \in mst(V)} \|e\|^2 \leq \frac{\pi + y_{\max}}{\frac{\sqrt{3}}{6}} = 12.14110... < 12.15 .$$

$\square$

The proof of Theorem 1 shows that the quantity $\sum_{i=1}^{k} S(\alpha_i)$ (i.e. the maximum sticking-out area of k diamonds that can be placed around the circumference of the circle of radius 1 compatible with the Euclidean MST properties) is maximized by 8 diamonds of area $S(\frac{\pi}{4})$ instead of 6 of area $S(\frac{\pi}{3})$ (as was assumed in [10]). In fact, considering the maximum value for $S(\alpha)$ that is obtained for $\alpha = \frac{\pi}{3}$, only 6 diamonds can be placed around the circumference.

4   The ABC Algorithm

A first easy observation in order to improve the basic MST algorithm is that for each node we need at most only one outgoing edge, i.e. the longest one.

This technique is also used in [1,3,4,9,11] for the Broadcast Incremental Power (BIP) where the nodes are discovered one by one according to the improvement of the energy needed to cover the new node. But also this algorithm does not take into account another property that is really important from a practical point of view. What we really need is an algorithm that is able to decide how to remove some previous edge(s) (or circle(s)) included in the current solution, adding a new one according to the considered points over the plane that we want to cover. What is important to decide is the order of the discovered nodes. The BIP order or more precisely the technique used to order the nodes seems to be inappropriate, since finding the cheapest node for the algorithm, that at each step needs to compute the deletion of previous circles, is very expensive in terms of time and resources. A more appropriate order seems to be choosing at each step the closest node to the current solution. This intuition is supported by our experimental results. In practice, we order the nodes in Prim's MST order. Due to this property, we will show that in general the cost of ABC is lower than or equal to the one of MST. After having decided the order, we describe the algorithm that tries to add the newly discovered node, at each step, according to its Euclidean distance from the current solution, in the cheapest way. This means that we try to reach the new node from each node already present in the solution considering that we can remove previous circles that become useless because of the new one. Let V be a set of nodes and $s \in V$ the given source, denoting with i the i-th node discovered at step i ($s \equiv 0$), the two invariants that we need to guarantee a valid solution at each step i are:

1. Every node from 1 to $i - 1$ is covered by some circle.
2. Every node from 1 to $i - 1$ admits an induced path back to the source 0.

Now we can describe more precisely how to design the ABC algorithm:

procedure $ABC(s, V)$

1:   $S \leftarrow s$ * Set of covered nodes
2:   $V \leftarrow V \backslash s$ * Set of uncovered nodes
3:   $Sol \leftarrow null$ * Set of used circles
4: **while** $|S| < |V|$ **do**
5:      Let v be the closest node of V to S
6:      **if** v is not already covered by S **then**
7:        $Sol \leftarrow Sol + Cost(v, S, Sol)$
8:      **end if**
9:      $V \leftarrow V \backslash v$
10:     $S \leftarrow S + v$
11: **end while**

The Function $Cost(v, S, Sol)$ outputs the minimum cost circle needed to reach the current node v. As we said before this calculation is made according to the 2 previous invariants.

procedure $Cost(v, S, Sol)$

1: **for** each node $x \in S$ **do**
2:      $C_{xv} \leftarrow C_{xv} - C_{xy} - P$
3:      where $C_{xy} \in Sol$ is the previous circle
4:      used by x to reach some other node in S,
5:      unless it does not exists and
6:      P is the set of the circles that now
7:      can be removed due to C_{xv} coverage.
8: **end for**
    Return $\min_{x \in S}\{C_{xv}\}$

The most complicated part of the algorithm is how to compute the set P. In our implementation we associated a set of covered nodes to each circle, so that the first step to compute P is to verify if the set of covered nodes of a circle is a subset of the new circle. If not, we cannot remove that circle, otherwise we have to check the invariant number 2. This means, in practice that we have to verify if without the candidate circle to be removed, a path from the source 0 to the center of the new circle there exists.

By construction it is easy to see that the above algorithm is a refinement of Prim's MST algorithm in which at each step the closest node to the existing structure is chosen. The difference is that, due to the specific circle area coverage, the algorithm outputs a forest of paths refining the cost of the MST. Hence the following lemma holds.

Lemma 1. *Given a set of nodes V over the Euclidean 2−dimensional space and a source $s \in V$, $cost(ABC(s,V)) \leq cost(MST(s,V))$.*

ABC can also be considered as an extension of the BIP program since at each step we try to increase the power of each node, with the main difference that in ABC we can also reduce it due to some different kind of coverage. This process can be easily tested on the example of ten nodes presented in [11] for BIP in which, applying ABC, we obtain the optimal solution. Unfortunately we cannot claim a similar lemma for BIP (as for MST) because it can happen in some specific case that the BIP solution is cheaper. An example of this case is shown in Figure 3.

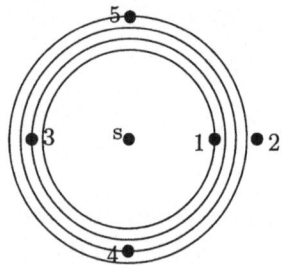

Fig. 3. Example in which ABC can lose with respect to BIP.

In this figure, ABC adds the circle $C_{1,2}$ just after $C_{s,1}$ because of the distance. This way, when it realizes that a bigger circle centered in s is needed, it does not cover also the node 2. BIP, instead, increasing the solution by the cost discovers node 2 at the end of its computation.

In terms of complexity, ABC takes $O(n^3)$ time, in fact the *while* loop inside the main procedure is executed exactly $n-1$ times, the *for* loop in the Procedure *Cost*() is executed at most $n-1$ times and to find the set P we need at most $n-1$ comparisons. Therefore in practice the complexity is the same as for BIP. Also for the approximation ratio we cannot say in general that ABC is better than BIP although the experimental results show better values on random instances but in general, we have as approximation ratio again 12.15 since it is bounded by the performances of MST.

The idea behind this algorithm is the result of our effort to solve all the practical problems presented in the literature. Every example and mainly worst-case example presented for each heuristic in [2,6,10,11] is optimally solved by ABC. This is due to the further property of this new algorithm for which, according to the order of the nodes, every path or subpath of the output solution is less expensive than the circle with radius equal to its endpoints when it covers the same set of nodes.

Before concluding the section, we show in Figure 4 the worst example for ABC that we found by studying the problem. Notice that in this example the

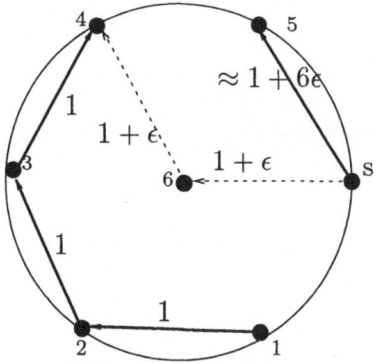

Fig. 4. Worst-case example for ABC.

approximation ratio obtained by ABC is 2 that is very far from 12.15. In the first four steps the ABC solution is composed by the path $\{s, 1, 2, 3, 4\}$. When node 5 is discovered ABC removes the edge $(s, 1)$ and adds $(s, 5)$, finally node 6 is already covered. The same solution is obtained by applying BIP. MST, instead, is even worse (its reduced version as well) as it outputs the path $\{s, 1, 2, 3, 4, 5, 6\}$. The optimal solution costs $2(1 + \epsilon)^2$ as it is composed by the two edges $(s, 6)$, $(6, 4)$.

The problem for ABC in the example of Figure 4 is that the central node is discovered at the end of the computation, so it does not test the circles centered in that node. On the other hand, the central node is discovered by BIP as the second node but the algorithm is not "smart" enough to realize that after it could cover all the remaining nodes with only one circle. As a consequence of the previous result and Theorem 1, the following corollary holds.

Corollary 1. *The approximation ratio ρ of ABC is $2 \leq \rho(ABC) \leq 12.15$.*

5   ILP Formulation

In this section we present an integer linear programming formulation to solve optimally the MEBR problem. We use it for our experiments to compare the results of ABC, MST and the other heuristics to the optimal solution in "small" instances. Formally we describe the problem as the research of the minimum cost of the sum of the squared radii of the used disks to perform a flow problem.

Given n nodes $\{s \equiv v_1, v_2, \ldots, v_n\}$, distributed on the 2-dimensional space, let $d_{i,j}$ be the Euclidean distance between nodes v_i and v_j; let $y_{i,j} \in \{0, 1\}$ be the variable equal to 1 if the solution contains the disk centered at v_i and radius $d_{i,j}$, 0 otherwise; let $x_{i,j} \in I\!N$ be the variable describing the number of units of flow going from v_i to v_j. We add a further node, t, as the destination node, setting $\forall i \neq s$, $d_{i,t} = 0$ and $x_{i,t} = 1$.

$$\min \sum_{i,j} d_{i,j}^2 y_{i,j}$$

$$\sum_{j \neq t} x_{s,j} = n - 1$$

$$\forall k, \sum_{i \neq k} x_{i,k} - \sum_{j \neq k} x_{k,j} = 0$$

$$\sum_{i \neq s} x_{i,t} = n - 1$$

$$\forall i \neq t \text{ and } j \neq s, \; x_{i,j} \leq (n-1) \sum_{k:d_{i,j} \leq d_{i,k}} y_{i,k}$$

$$\forall i,j \; \; y_{i,j} \in \{0,1\}, x_{i,j} \in \mathbb{N}$$

The first condition describes that the source s outputs $n-1$ units of flow; the second is the standard condition for the flow conservation. The third describes that each node, except s, sends a unit of flow to t. The last means that if a disk $y_{i,k}$ covers more than one node, then v_i is allowed to send units of flow to each covered node. This is not the first time the the linear programming method is used to formalize the MEBR problem (see [6] for instance) but, to the best of our knowledge, it is the first time that experimental results comparing several heuristics with the real optimal solution appear.

6 Some Experimental Results

In this section we briefly present a comparison between the heuristics that we took into consideration which are the MST, BIP, ABC and the optimal algorithm obtained by the ILP formulation, as part of the extensive experiments we ran. We show the good performance of the ABC algorithm compared to the others. The experiments were performed among 500 instances for each different size. The number of nodes is between 5 and 50. The distribution of the nodes is random in a 2-dimensional space inside a square of dimension 5×5, even the source node is randomly chosen. The metric used to perform the comparisons is the total power of the broadcast tree. In the following pictures the X-axis represents the instances and the Y-axis represents the correspondent costs of each heuristic.

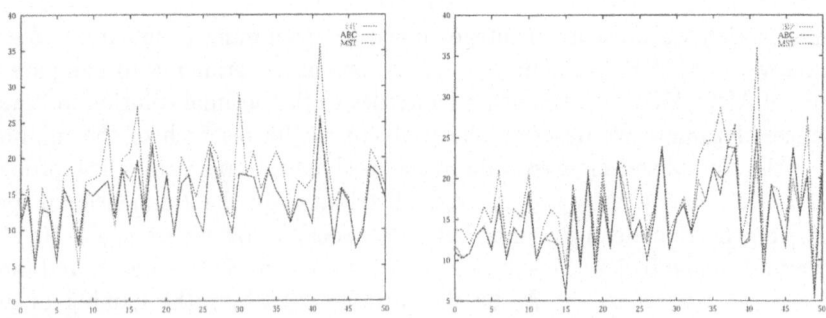

Fig. 5. ABC, BIP, MST with 5 and 7 nodes

In small instances, such as in Figure 5, of 5 and 7 nodes, the distance between the heuristics is on the average almost zero, mainly between ABC and BIP the solutions are almost the same.

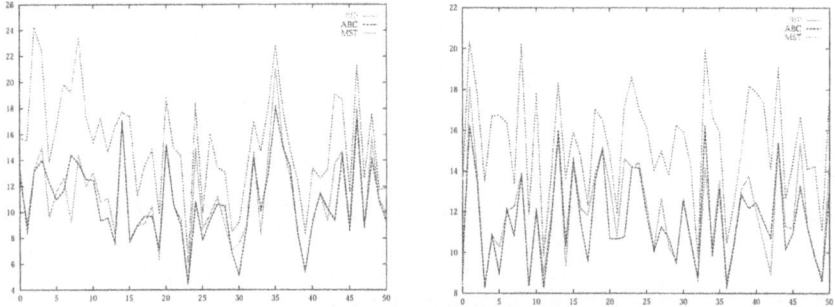

Fig. 6. ABC, BIP, MST with 10 and 15 nodes

Increasing the number of nodes we start observing some relevant difference between the heuristics. From Figure 6, using 10 and 15 nodes, the bad performance of the classic MST algorithm are already evident. This clearly shows again the intuition that MST is not the appropriate heuristic to solve the $MEBR$ problem. Anyway the properties of the Euclidean MST lead to the upper bound of Section 3 that remains the only valid upper bound for all the other heuristics. ABC is, on average, better than BIP but there are some peaks in which BIP is better.

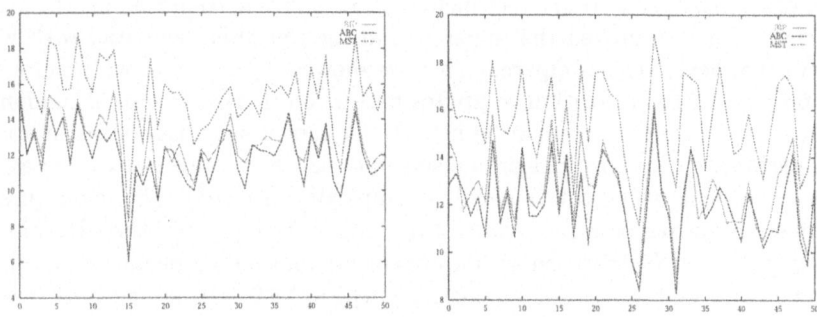

Fig. 7. ABC, BIP, MST with 25 and 30 nodes

Experimenting using 25 and 30 nodes (see Figure 7), the fact that ABC performs better than BIP becomes clear as it results in solutions that are, on the average 3.5% better than the ones calculated by BIP. Moreover, even if MST is far from the other heuristics, this distance is nothing compared to the known bounds.

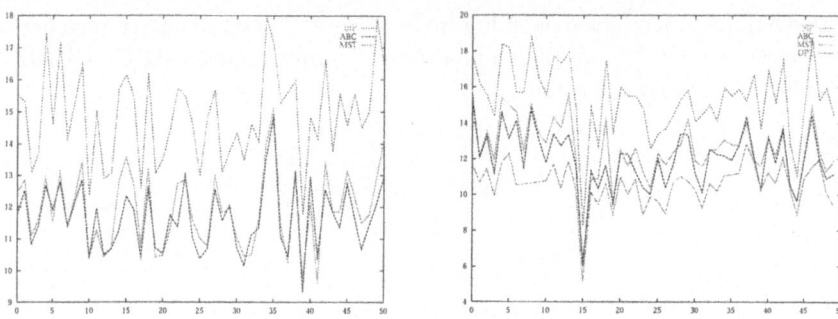

Fig. 8. ABC, BIP, MST with 45 nodes and ABC, BIP, MST, OPT with 50 nodes

Increasing further the number of nodes up to 45 and 50 (see Figure 8) the average of the power saved by ABC with respect to BIP reaches 5%. Moreover, in the last figure we also show the graph of the optimal solution obtained by the integer linear program proposed in Section 5. The optimal solution is, on average, 12% better than ABC so, in practice, on random instances we do not loose much with respect to the optimal and this is in contrast with the discussed upper bound of 12.15. By these experiments we can conclude that the advantage of the 5% obtained with ABC with respect to BIP can be considered to be very significant.

7   Conclusion

The aim of the paper was to produce a new heuristic able to solve some evident problems arising from experimental and analytical results about the MEBR problem. We first proved the upper bound for the MST heuristic, with which every other heuristic is compared. Then we showed, by experimental results, that the proposed ABC algorithm computes better solutions over several random instances of various sizes comparing it with the known heuristics of the literature. The performed computations produced solutions which values are on average 5% better than BIP using the same time complexity of $O(n^3)$. The main question left open in this paper is if it is possible to reduce the big gap of the ABC bounds. Anyway the new heuristic shows improvements on random instances and makes it difficult to find an ad hoc instance for which ABC outputs a solution worse than a 2-approximation.

We also developed an implementation of the optimal algorithm using integer linear programming. To the best of our knowledge this is the first work where the heuristics are compared with the real costs. With this method we showed that on random instances even the easy MST is not too far from the optimal and hence that our improvement of 5% is a very good result. Another way to try implementing other good approximation algorithms could be to study the relaxed formulation of the integer linear program. By our first tests there are good hopes in this direction. Another interesting issue could be to evaluate the heuristics for other kind of distributions such as the spatial Poisson one (see [2] for instance).

References

1. BANERJEE, S., MISRA, A., YEO, J., AND AGRAWALA, A. Energy-efficient broadcast and multicast trees for reliable wireless communication. In *Proceedings of the 19th Annual Joint Conference of the IEEE Computer and Communications Societies (INFOCOM)* (2000), IEEE Computer Society.

2. CAGALJ, M., HUBAUX, J. P., AND ENZ, C. Minimum-energy broadcast in all-wireless networks: NP-completeness and distribution issues. In *Proceedings of MOBICOM* (2002), pp. 172–182.

3. CLEMENTI, A., CRESCENZI, P., PENNA, P., ROSSI, G., AND VOCCA, P. On the complexity of computing minimum energy consumption broadcast subgraph. In *Proceedings of the 18th Annual Symposium on Theoretical Aspects of Computer Science* (2001), vol. 2010, Lecture Notes in Computer Science, Springer-Verlag, pp. 121–131.

4. CLEMENTI, A., HUIBAN, G., PENNA, P., ROSSI, G., AND VERHOEVEN, Y. C. On the approximation ratio of the mst-based heuristic for the energy-efficient broadcast problem in static ad-hoc radio networks. In *Proceedings of the 3rd IEEE IPDPS Workshop on Wireless, Mobile and Ad Hoc Networks (WMAN)* (2003), p. 222.

5. CLEMENTI, A., PENNA, P., AND SILVESTRI, R. On the power assignment problem in radio networks. *Mobile Networks and Applications (ACM MONET), Special Issue on Discrete Algorithms and Methods for Mobile Computing and Communications 9* (2004), 125–140.

6. DAS, A. K., MARKAS, R. J., EL-SHARKAWAI, M., ARABSHAHI, P., AND GRAY, A. Minimum energy broadcast trees for wireless networks: Integer programming formulations. In *Proceedings of the 22nd Annual Joint Conference of the IEEE Computer and Communications Societies (INFOCOM)* (2003), IEEE Computer Society.

7. KASHYAP, A., NISHAR, H., AND AGARWAL, P. Survey on unicast routing in mobile ad hoc networks, http://citeseer.nj.nec.com/481660.html, 2003.

8. RAPPAPORT, T. *Wireless communications: principles and practice.* Prentice-Hall, Englewood Cliffs, NY, 1996.

9. ROSSI, G. *The Range Assignment Problem in Static Ad-Hod Wireless Networks.* PhD thesis, Universitá degli Studi di Siena, 2002.

10. WAN, P. J., CALINESCU, G., LI, X., AND FRIEDER, O. Minimum energy broadcast routing in static ad hoc wireless networks. *Wireless Networks 8* (2002), 607–617.

11. WIESELTHIER, J. E., NGUYEN, G. D., AND EPHREMIDES, A. On the construction of energy-efficient broadcast and multicast trees in wireless networks. In *Proceedings of the 19th Annual Joint Conference of the IEEE Computer and Communications Societies (INFOCOM)* (2000), IEEE Computer Society, pp. 585–594.

Access Protocols to Support Different Service Classes in an Optical Burst Switching Ring*

Vishwas S. Puttasubbappa and Harry G. Perros

North Carolina State University
Computer Science Department
Raleigh NC 27695-7534 {vsputtas,hp}@csc.ncsu.edu

Abstract. Several access protocols are proposed to support different service classes in an optical burst switched ring. Their performance is evaluated through simulation. Various performance metrics such as throughput, utilization, burst loss rate, end-to-end delay and fairness are used to analyze the behaviour of each protocol.

Keywords: Optical burst switching, MAN, access protocols, service classes

1   Introduction

Optical Burst Switching (OBS) is a novel method currently under study that can be used to transport data over a Wavelength Division Multiplexing (WDM) optical network. Battestelli and Perros [1] provide a detailed survey on OBS and its variations. There is not much work done in the field of OBS over metropolitan-area rings. Xu et al [6] investigated access protocols for OBS rings based on the Just Enough Time (JET) scheme and a new scheme called the Only Destination Delay (ODD). Jong [4] proposed several access protocols for multicasting in such an environment. A new architecture called the LightRing has been proposed by Fumagalli and Krishnamoorthy [3] with multi-token protocol to prevent contention among bursts. Bouabdallah et al [2] proposed a collision avoidance MAC protocol for a metropolitan bus-based optical access network. Analytical models were developed to calculate the mean access delay of each node in such a shared-medium system. Fairness issues were also investigated.

The work done so far on OBS rings considered traffic to be best effort except in Fumagalli and Krishnamoorthy [3], where real-time and best-effort were considered. The aim of this paper is to investigate how an OBS ring can support different classes of traffic. In this study, we extend the OBS ring architecture proposed in Xu et al [6] in order to consider the following three different classes of traffic. The first class of traffic (Class 1) is a variable bit rate traffic with stringent end-to-end delay constraints, the second class (Class 2) is variable bit rate with no delay constraints, and the third class (Class 3) is non-real time

---

* This work was supported by ARDA under contracts MDA904-00-C-2133 and MDA904-02-C-0482

N. Mitrou et al. (Eds.): NETWORKING 2004, LNCS 3042, pp. 878–889, 2004.

variable bit rate best effort traffic. Several access protocols are proposed and their performance evaluated through simulation.

The paper is organized as follows: Section 2 presents the system architecture including the structure of the metro ring and the OBS nodes. The proposed protocols are presented in section 3, and in section 4 we describe the arrival processes used in the simulation model. The results of the simulation are discussed in 5. Finally, section 6 gives the conclusions from the study.

2 System Architecture

The metro ring uses the WDM architecture and it consists of several nodes which serve as concentration points to incoming and outgoing traffic from several access networks. The ring is divided into two co-existing rings as follows: Some of the wavelengths are used to host SONET/SDH rings, and the remaining wavelengths are used for optical burst switching. The metro ring has to carry all types of traffic, such as circuit switched traffic, ATM traffic and IP traffic. SONET/SDH rings can cater to circuit switched traffic and data traffic can be transported through the OBS ring. In this paper, we only investigate the OBS part of the metro ring.

The wavelengths allocated for optical burst switching are divided into S sets of N+1 wavelengths. Within a set of wavelengths, each of the N wavelengths is allocated to a different node. This wavelength is referred as the *home* wavelength of the node. The (N+1)th wavelength of the set is the control channel for the wavelengths in the set and it carries control frames. The control frames implement the signalling necessary for OBS. Since there are S sets of wavelengths, each node is allocated to S home wavelengths. A node can only transmit bursts on its home wavelengths.

Each OBS node in the ring has S transmitters each fixed-tuned to one of the S home wavelengths, and S tunable receivers one per wavelength set. These S pairs of transceivers are used for transmitting and receiving bursts. A node can transmit a burst on any free home wavelength. A free receiver can tune to receive a burst arriving on any wavelength in its corresponding wavelength set. Each node is additionally equipped with S transceivers, one set per control wavelength. The OBS node is equipped with a control module which performs its functions based on the information each control frame carries around the ring. Each control wavelength carries back to back control frames. The structure of a control frame is as shown in figure 1. Each node has its own slot into which it can write information during transmission. The control frames on the S control wavelengths travel around the ring in a synchronous manner. That is, the control frames in the (i+1)st control wavelength lag behind their corresponding ones in the ith control wavelength by the time the control module requires to process them. This arrangement ensures efficient usage of control frames for burst transmission. For instance, if a node cannot transmit on the first control channel, it has an opportunity to transmit immediately using the control frame

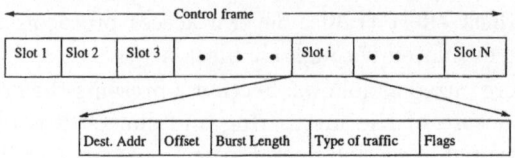

Fig. 1. Control Frame structure

in the second control channel without having to wait even for a small amount of time.

Each node serves a number of access networks. The incoming data from these networks is queued in the transmission queues of a node. Specifically, each node maintains N-1 transmission queues of each class, where N is the total number of nodes in the ring. Since we consider three classes, each node maintains 3(N-1) queues. In this paper we assume that class i has non-preemptive priority over class i+1, i=1,2. That is, the transmitter will always transmit a burst from a class 1 queue. If there is no class 1 traffic, it will transmit a burst from a class 2 queue, and if there is no class 1 or 2 traffic, it will transmit a burst from class 3. The N-1 class 1 transmission queues are assumed to form a single logical queue. Bursts in these N-1 class 1 queues are served in the order in which they arrive at this logical queue. The N-1 class 2 transmission queues are served in a round-robin fashion. Likewise, round robin is used to serve the N-1 class 3 transmission queues.

3   The Access Protocols

In this paper, we assume that the class 1 traffic consists of multiple HDTV streams. Each HDTV frame constitutes a single burst. In the case of class 2 and class 3 traffic, a burst is comprised of several data packets which may be IP packets, ATM cells etc. A class 2 and class 3 transmission queue is eligible to be served if there are enough packets in the queue whose aggregate size exceeds a minimum burst size.

We have defined and analyzed the following five protocols. *Destination-Reservation Free* which provides no guaranteed delivery to any traffic class, *Ack* and *Token* which provide guaranteed delivery to class 1 traffic and *Token-Token* and *Ack-Ack* which provide guaranteed service to both class 1 and class 2 traffic. All these protocols transmit class 3 bursts when bandwidth is available. These access protocols fall into two categories: *collision-free* and *collision* protocols. Collision-free protocols reserve resources at the destination and hence no burst loss at the destination and in the ring. None of these five protocols are collision-free for class 3 traffic. Token-Token and Ack-Ack are class 1 and class 2 collision-free. Token-Token uses tokens and Ack-Ack makes use of acknowledgements. Ack and Token are class 1 collision-free protocols. Each of the protocols is explained in detail in the following sub-sections.

3.1   Destination-Reservation Free Protocol (Dest-Resv-Free)

Nodes transmit bursts without making any reservations at the receiver node. This may result in multiple bursts arriving at the destination at the same time and hence collisions. This is the Tell-and-Go protocol currently used in OBS mesh networks.

Single transceiver case: Upon arrival of a control frame, the bursts are transmitted with the priority scheme as stated in section 2 in case the transmitter is not busy. On multiple bursts arriving at the same time, priority for reception is given to class 1 bursts. If multiple class 1 bursts arrive at the same time, one of them is randomly selected. There is pre-emption of class 2 and class 3 bursts on arrival of a class 1 burst. Class 2 bursts are given the next priority and in case many of them arrive at the same time, one of them is randomly chosen. Class 3 bursts are given the least priority and if many of them arrive at the same time, one of them is randomly selected.

Multiple transceivers case: In the case of multiple home wavelengths, the control module can choose any free home wavelength to transmit. The reception mechanism is identical in each of the different sets of wavelengths.

3.2   Token Protocol

The Token protocol uses the concept of tokens to resolve receiver collisions. Tokens are used only for class 1 bursts. Class 2 and class 3 is serviced through best-effort. Every node has a token circulating around the ring. If a source wants to transmit a class 1 burst to a particular destination, it has to have the token for that destination. All the nodes maintain a queue to hold tokens. The token is released after the transmission is completed. Since only the node that has possession of the token can transmit a burst to the appropriate destination, the destination can only receive a single burst at a time, and therefore the token protocol is a collision-free protocol, but only for class 1 traffic.

Single transceiver case: Each node monitors the control frames for tokens. If a control frame is carrying a token for destination k, then the flag field of slot k has the value 1. The node takes the token out of the control frame and queues it in its FIFO queue, provided that there is traffic for that destination node. The token has to be queued because the node may not be able to transmit the burst immediately because of busy transmitter. The token is released after the transmission is completed. This guarantees no collision at the destination between class 1 bursts. Class 2 and class 3 bursts do not have a collision-free reception mechanism. Bursts are simply transmitted following the Dest-Resv-Free protocol. On the receiver's side, class 1 bursts are given the highest priority. Since this is a class 1 collision-free protocol, there are no collisions among class 1 bursts. Priority of reception is similar to the Dest-Resv-Free protocol.

Multiple transceivers case: Each node maintains a separate token queue for each home wavelength. A node cannot use more than one home wavelength simultaneously to transmit bursts to a particular destination. This is achieved by making sure that only one token to a particular destination can be held in any of the multiple token queues. A class 1 burst can be transmitted on any home wavelength as long as the corresponding token queue has the token for the appropriate destination. Reception in any one set of wavelengths is independent and identical to the other sets.

3.3   Ack Protocol

This protocol also ensures guaranteed reception of class 1 bursts. It is different to the Token protocol and is based on the Tell-And-Wait (TAW) protocol proposed in OBS networks. The protocol uses a *Request* and *Acknowledgement* mechanism. A minimum of a round-trip delay is required for the node to transmit a burst after it can be formed. The acknowledgement mechanism is available only for class 1 traffic. Class 2 and class 3 bursts are transmitted as in the case of the Dest-Resv-Free protocol.

Single transceiver case: As soon as a HDTV frame arrives at any of the class 1 queues of the node, a request is sent out to the destination node requesting it to return an acknowledgement in which it indicates the earliest time it is free to receive this burst. The source is not allowed to send out a request to any destination as long as it has an outstanding request. This makes sure that there are no transmitter conflicts. The flag field of the control frame is used to indicate a *Request*. When the destination node sends an acknowledgement, the offset field in the destination's slot will be the earliest time by which the source can start transmitting. This is to make sure that by the time the source starts transmitting the burst, the destination is free to receive it. The transmission of class 2 and class 3 bursts is without any acknowledgement mechanism and is similar to the Dest-Resv-Free protocol. A destination node cannot send out an acknowledgement for a request that it has received until it starts receiving the burst from the previous acknowledgement that it sent. This is necessary because the source node to which the last acknowledgement was sent may not be able to transmit the burst immediately because of a busy transmitter. Reception is similar to Dest-Resv-Free protocol. The receiver has a queue to hold requests and the request with the earliest time-stamp will be served first. Class 2 bursts are given higher priority than class 3 bursts.

Multiple transceivers case: While sending a request on any of the control channels, the source node makes sure that no other home wavelength is being used for the same destination. If this is the case, other transmission queues are served which send out requests to different destinations. Reception is similar to the Token protocol.

3.4   Token-Token Protocol

This is a collision free protocol for class 1 and class 2 bursts. Nodes use the token mechanism.

Single transceiver case: The token for a particular destination is captured by the node only if it has bursts of class 1 or class 2 to be sent to that particular destination node. Priority for transmission is given for class 1 traffic. Unlike the token protocol, class 2 bursts require a token for transmission. Class 3 queues are served whenever the transmitter is free and the bursts can be formed. Reception is simpler in this case since there is no collision between class 1 and class 2 bursts and hence no priority between them. Class 3 bursts are subject to pre-emption in case either of the other classes' burst arrives.

Multiple transceivers case: The operation is similar to that of the Token protocol except that tokens are required for classes 1 and 2.

3.5   Ack-Ack Protocol

Guaranteed reception both for class 1 and class 2 bursts is provided by extending the acknowledgement scheme to cover both traffic classes. Class 3 is served through best effort.

Single transceivers case: The operation of the transmitter is similar to the Ack protocol, except that class 2 bursts need an acknowledgement before transmission. A request to transmit a class 2 burst is sent only if all the class 1 transmission queues are empty. Thus, this is a collision-free protocol for class 1 and class 2. Class 3 bursts are transmitted whenever the transmitter is free. The receiver exercises its decision to send out acknowledgements not in the FCFS manner, but based on priority. Once the receiver starts receiving a burst for the acknowledgement it last sent, it scans all the requests in its queue and sends out the next acknowledgement according to the following rules: (1). Requests which are for class 1 traffic are given priority. If there are none, then one of the class 2 requests is arbitrarily picked and served (2). If there are multiple class 1 requests, then the request that has the earliest time-stamp will be served first. Rule 2 makes sure that requests are served not in their order of arrival (which may be biased towards nodes close by) but by the earliest time-wise arrival of frames at their respective source nodes. Reception is similar to Token-Token protocol.

Multiple transceivers case: The operation is similar to that of the Ack protocol except that acknowledgements are required for classes 1 and 2.

4   The Simulation Model

An event-based simulation model was developed with a view to analyzing the performance of the proposed five protocols. For details, see Puttasubbappa and Perros [7].

For each node i, i=1,2,...,N, a number of HDTV streams are setup at the beginning of the simulation. Each of these streams originate at node i and terminate at destination node j. In each stream, frames are generated at a rate of 60 frames per second giving an inter-frame arrival time of 16.667 milliseconds. We assume that the frames follow the MPEG 2 Group Of Pictures (GOP) structure of IBB PBB PBB PBB. The size of each frame is generated using the autoregressive model, see Bragg [9], $S(t) - S(t-12) = e(t) - 0.69748 \times e(t-3)$, where $S(t)$ is the size of frame t, and $e(t) \sim N(0,\sigma^2)$ with $\sigma^2=4849.5$.

A class 2 source in our simulation is a variable bit rate source with no end-to-end time constraints. In our simulation experiments, we assume that the packets are generated from a storage area network (SAN), with the following packet-size distribution: 44 % of 64Kbytes, 18 % of 56K, 21 % of 40K, 4 % of 32K, 4 % of 24K and 6 % of 8Kbytes, see Trevitt [8]. The arrival process consists of packets arriving in succession with an exponentially distributed inter-packet delay. The time it takes for each packet to arrive is taken into account.

Finally the class 3 traffic arrival process is best effort traffic and it is modelled as in Xu et al [6] by a modified Interrupted Poisson Process. The ON and OFF periods are exponentially distributed. Packets arrive back to back during the ON period at the rate of 2.5 Gbps. The last packet that arrives when the ON period ends is truncated. During the OFF period, no packets are generated. The mean packet size is 500 bytes and any packet size above 5000 bytes is truncated to 5000 bytes. To calculate the ON and OFF periods, we use the coefficient of variation c^2, defined as the ratio of the variance of the packet inter-arrival time divided by the squared mean of the packet inter-arrival time. c^2 indicates the burstiness of the arrival process.

$$c_{IPP}^2 = 1 + 2\lambda\mu_1/(\mu_1 + \mu_2)^2 \tag{1}$$

where $\frac{1}{\lambda} = (500\text{bytes})/(2.5\text{Gbps}) = 1.6\mu s$, and $\frac{1}{\mu_1}$ and $\frac{1}{\mu_2}$ are the mean times of the ON and OFF periods. The arrival process of class 3 traffic is completed by the following equation:

$$\text{Average Arrival Rate} = 2.5 \text{ Gbps} \times \mu_1/(\mu_1 + \mu_2) \tag{2}$$

5   Discussion of Simulation Results

We simulated a ring consisting of 10 nodes and each node is separated by a distance of 5 km. Each wavelength was assumed to have a bandwidth of 2.5 Gbps and the control wavelength works at a rate of 622 Mbps. For each class i, i=1,2,3, the transmission queue in a node was assumed to have a buffer size of 1 MB. In the single transceiver case, 11 wavelengths are required for the OBS network. In the multiple transceivers case, the number of wavelengths used is an integral multiple of 11. The simulation results are plotted with 95% confidence interval estimated by the method of batch means, see Perros [5]. Each batch is completed when each node generates 10,000 bursts. The confidence intervals are very tight and are not discernible in the graphs.

The simulation model was used to evaluate the performance of each of the protocols discussed in section 3. For all the results obtained, the class 2 average arrival rate at each node was fixed to 0.8 Gbps and the average arrival rate of class 3 traffic to 0.5 Gbps. The x axis is always the number of HDTV streams originating at each node. Specifically, in each simulation experiment, the same number of HDTV streams originate at each node, and the destination node of each stream is randomly selected. Each stream contributes an average of 20 Mbps of the total traffic. The total average arrival rate is the sum of the average arrival rates of the 3 traffic classes. In all the experiments, the overall traffic a node transmits is less than the bandwidth of the home wavelength(s).

Most of the results are self-explanatory. For a detailed discussion, please refer our full paper Puttasubbappa and Perros [7]. Figure 2(a) plots the *Mean node overall throughput* versus the number of HDTV streams per node. The *mean node overall throughput* is defined as the average number of bits received (class 1, class 2 and class 3) by all the nodes in a unit time divided by the number of nodes in the ring. The Token-Token protocol has the highest mean node overall throughput followed by the Token protocol. Ack-Ack protocol provides less service to class 2 traffic as can be verified with figure 2(e). Token-Token is not as biased towards class 1 traffic as Ack-Ack. This can be confirmed from figure 2(c) where we note that even though the class 1 arrival rate increases, Token-Token cannot increase the throughput provided to class 1 traffic. The mean node overall throughput for the three transceivers case is plotted in figure 2(b). A notable feature is a better performance by Ack-Ack. It can be seen that Token-Token scales well.

Figure 2(c) plots the *Mean node class 1 throughput* versus the number of HDTV streams per node. The *mean node class 1 throughput* is defined as the average number of class 1 bits received by all the nodes in a unit time divided by the number of nodes in the ring. Ack-Ack performs very closely to Token and Ack despite the fact that it provides acknowledgement services to both class 1 and class 2 traffic. Token-Token does not perform as well as its counterpart. Due to more frequently available tokens, Token and Token-Token scale well for the mean node class 1 throughput when there are three transceivers per node as can be seen in the figure 2(d). The propagation delay limits the scalability of acknowledgement based protocols.

The *% Bandwidth utilization* is defined as the amount of time a home wavelength is busy transmitting bursts. Two graphs for the utilization of the home wavelength for node 1 are given: one depicting % bandwidth utilization for class 2 traffic and the other for class 1 traffic. Figure 2(e) plots the % bandwidth utilization for class 2 traffic for the 1 transceiver case. Token-Token supports class 2 traffic better than Ack-Ack. For the multiple transceivers case (figure not shown), Ack-Ack has higher utilization for class 2 traffic because of higher bandwidth available in the 3 transceivers case. Figure 2(f) plots the % bandwidth utilization of transmit wavelength of node 1 for class 1 traffic for the 1 transceiver case. From these plots, it can be seen that although Dest-Resv-Free protocol has a high % bandwidth utilization, it has a low class 1 throughput because of collisions. For the 3 transceivers case, Token and Token-Token scale up better than Ack and Ack-Ack (figure not shown).

The *Overall burst loss rate* is defined as the total number of bursts (class 1, class 2 and class 3) lost because of receiver collisions divided by the total number of bursts transmitted by all the nodes. Figure 2(g) plots the *overall burst loss rate* with varying number of HDTV streams per node.

Figure 2(h) plots the *hit ratio* for the five protocols. This is an important performance metric that describes how good a protocol is to support class 1

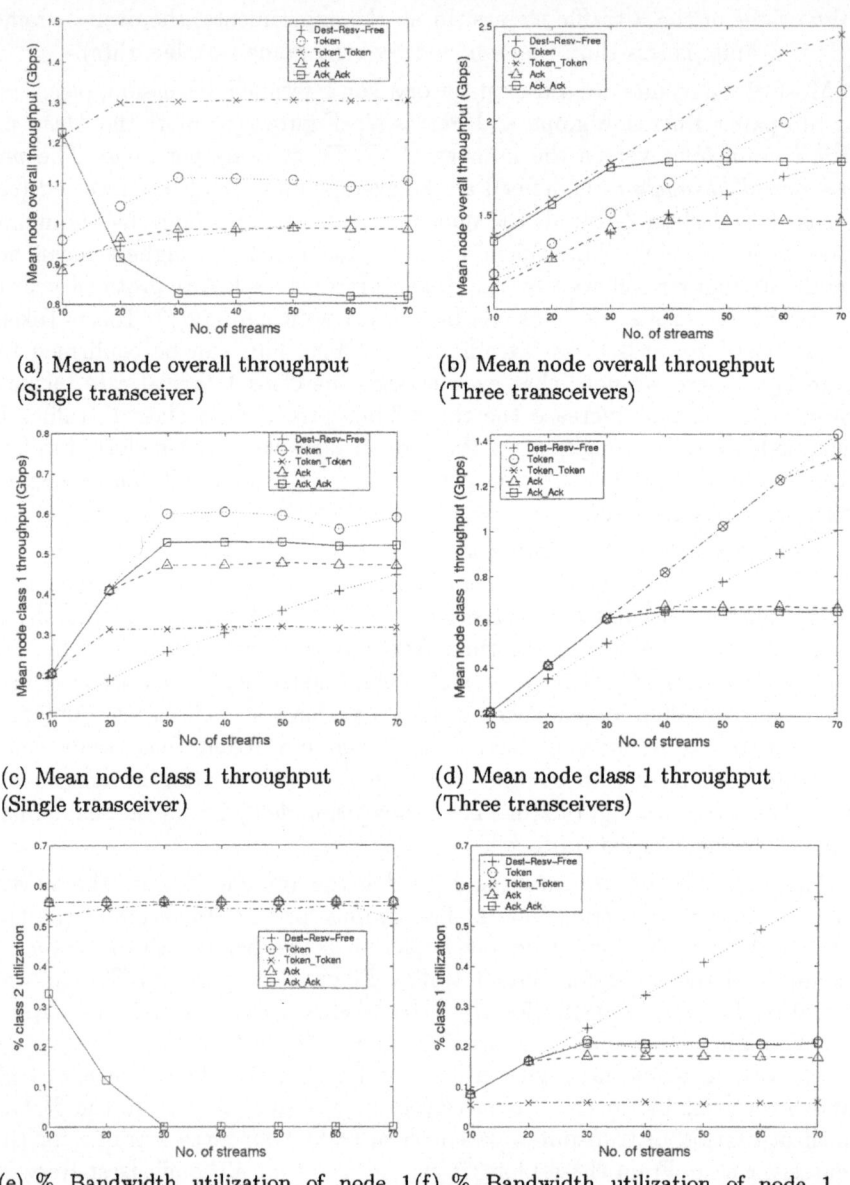

(a) Mean node overall throughput (Single transceiver)

(b) Mean node overall throughput (Three transceivers)

(c) Mean node class 1 throughput (Single transceiver)

(d) Mean node class 1 throughput (Three transceivers)

(e) % Bandwidth utilization of node 1 transmit wavelength for class 2 traffic (Single transceiver)

(f) % Bandwidth utilization of node 1 transmit wavelength for class 1 traffic (Single transceiver)

Fig. 2. Simulation results

traffic. Hit ratio is defined as the total number of class 1 bursts (i.e. HDTV frames) received in time by all the nodes, divided by the total number of frames sent by all the nodes. A frame is received in time if it arrives within 17 mil-

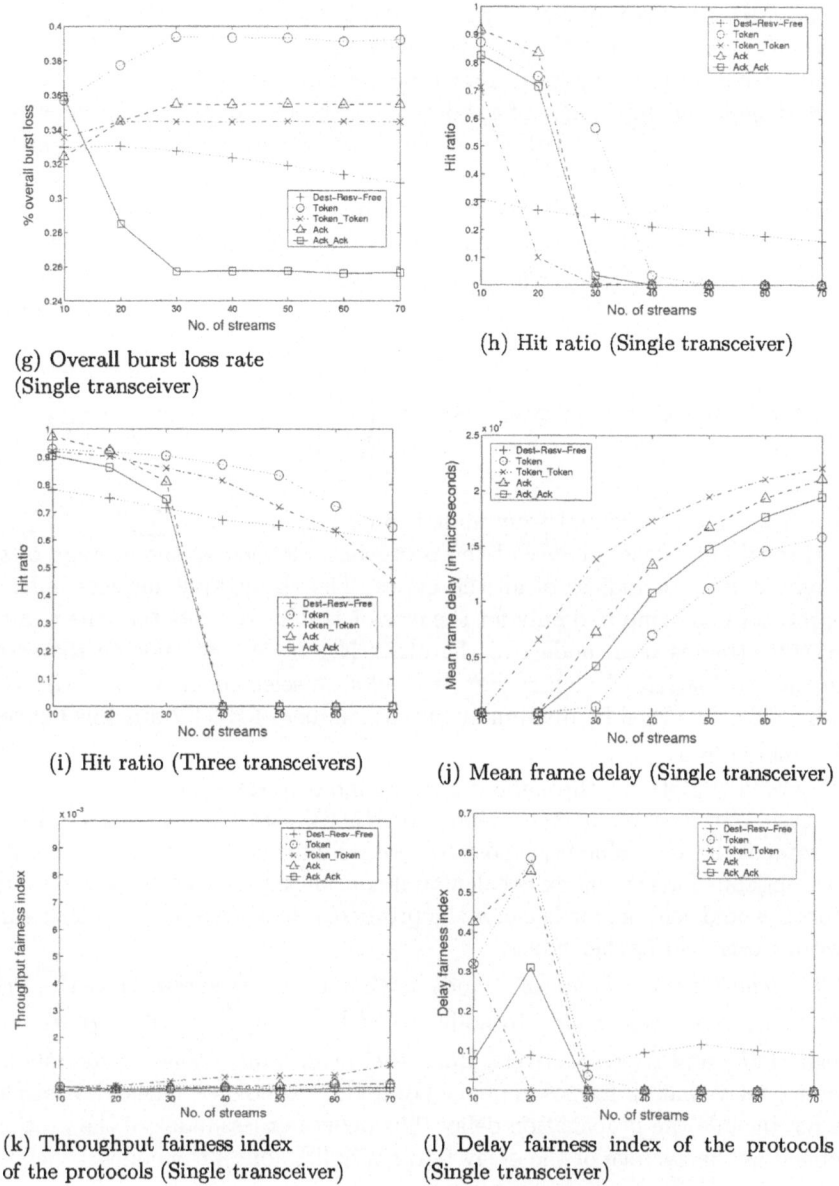

(g) Overall burst loss rate
(Single transceiver)

(h) Hit ratio (Single transceiver)

(i) Hit ratio (Three transceivers)

(j) Mean frame delay (Single transceiver)

(k) Throughput fairness index
of the protocols (Single transceiver)

(l) Delay fairness index of the protocols
(Single transceiver)

Fig. 2. Simulation results (continued)

liseconds of receiving the previous frame from the same stream, otherwise it is
a miss. A higher hit ratio for a larger number of streams per node is the desired
performance criterion. Token and Token-Token perform remarkably well when
more bandwidth is available as can be seen in figure 2(i) compared to Ack and
Ack-Ack.

The *fairness* of a protocol is an important criterion in ring networks, since it shows whether the positioning of a node in the ring has any effect on the protocol's performance metrics. Two types of fairness indices were calculated. The *throughput fairness index* of a protocol which determines how fair a protocol is with respect to the throughput of individual nodes, and the *delay fairness index* of a protocol which determines how fair a protocol is with respect to the delay in a node. The definition of the two indices metrics are the same as those used in Xu et al[6]. The throughput fairness index of a node i is defined as the c^2 of the throughput from node i to all other nodes.

$$\text{Throughput Fairness Index of Node i} = \left(\sum_{j=1, j\neq i}^{10} (H_{ij} - \overline{H_i})^2 \right) \times \frac{1}{\overline{H_i}^2} \quad (3)$$

where H_{ij} is the throughput from node i to node j, and $\overline{H_i} = (\sum_{j=1, j\neq i}^{10} H_{ij})/9$. The throughput fairness index of the protocol is defined as the average of the throughput fairness indices of all the nodes. The throughput fairness index of the protocol was computed only for the class 1 traffic. We note that the number of HDTV streams from node i to the other N-1 nodes may not be the same, since the destination of each stream is randomly selected. In view of this, the term H_{ij} is normalized by dividing it by the number of HDTV streams between node i and node j.

Figure 2(k) plots the throughput fairness index of the protocols (considering only class 1 traffic) versus the number of HDTV streams per node. Since all the protocols have a value very close to zero, all of them are throughput fair. If the throughput fairness index is calculated considering all the three classes, the protocols would still be fair because the protocols themselves do not distinguish between closer and farther nodes.

The *delay fairness index* of a node is similar to (3) except that W_{ij} (the mean frame delay from node i to node j) and $\overline{W_i} = (\sum_{j=1, j\neq i}^{10} W_{ij})/9$ are used instead of H_{ij} and $\overline{H_i}$ respectively. Here, the mean frame delay counts only the queueing delay that all frames in queue j of node i experience. The delay fairness does not include the propagation delay. The delay fairness index of the protocol is defined as the average of the delay fairness indices of all the nodes.

Figure 2(l) plots the delay fairness index of the protocols (considering only class 1 traffic) versus the number of HDTV streams per node. It shows that none of the protocols are delay fair. This is not due to the physical positioning of the nodes around the ring, because as mentioned earlier, the protocols do not distinguish between closer and farther nodes. Due to the asymmetric traffic pattern of class 1, there is difference in delays experienced by class 1 bursts, translating to a higher delay fairness index.

6   Conclusion

Five different protocols for the support of different service classes on an optical burst switched ring network were proposed and their performance was evaluated through simulation. Dest-Resv-Free protocol was based on the best-effort type of service. Ack and Token provided guaranteed delivery to class 1 bursts but do not ensure zero burst loss for classes 2 and 3. These two protocols can support more HDTV streams per node than other protocols because of guaranteed delivery only to class 1 bursts. Ack-Ack and Token-Token provided zero burst loss delivery for both classes 1 and 2 of traffic and they ensured that class 1 bursts are received in time and without too much jitter. Class 3 received best-effort service. For the single home wavelength case, Ack-Ack provides better service to class 1 traffic than Token-Token. But, the Token-Token and Token protocols perform better than their acknowledgement-based counterparts when multiple home wavelengths are available for transmission. Additional simulations experiments performed indicate that token based protocols are more scalable with ring size and number of nodes in the ring.

References

1. T. Battestilli and H. G. Perros. An introduction to optical burst switching. *IEEE communications optical magazine*, 41:510–515, 2003.
2. N. Bouabdallah, A-L. Beylot and G. Pujolle. Collision avoidance and fairness issues in metropolitan bus-based optical access networks. Technical report, Paris 6 university, 2003.
3. A. Fumagalli and P. Krishnamoorthy. A low-latency and bandwidth-efficient distributed optical burst switching architecture for metro ring. *In Proceedings of IEEE ICC*, 2:1340–1344, 2003.
4. James Jong. Multicast access protocols in an optical burst switched WDM ring network. Master's thesis, NC State University, 2002.
5. H. G. Perros. *Computer Simulation Techniques: The definitive introduction.* Available for free download from http://www.csc.ncsu.edu/faculty/perros/hp.html.
6. L. Xu, H. G. Perros and G. N. Rouskas. A simulation study of optical burst switching access protocols for WDM ring networks. *Computer Networks*, 41(2):143–160, January 2003.
7. V.S. Puttasubbappa and H.G. Perros. Access protocols to support different service classes in an optical burst switching ring, 2003. Available from World Wide Web: http://www.csc.ncsu.edu/faculty/perros/puttasubbappa-perros.pdf.
8. S. Trevitt. Traffic patterns in fiber channel fabrics, 2002. Available from World Wide Web: http://www.mcdata.com/downloads/mkt/wpaper/TrafficPatterns.pdf.
9. Bragg, A. W. *MPEG-4 Analysis.* private e-mail communication, March 2003.

Prioritized Multiplexing of Traffic Accessing an FSAN-Compliant GPON

John D. Angelopoulos, Helen-C. Leligou, Theodore Argyriou, and Stelios Zontos

National Technical University of Athens,
School of Electrical and Computer Engineering
Telecommunications Systems Laboratory
9 Heroon Polytechniou Street, Zographou, 15773, Athens, Greece
{jangel, nelly, teo, zontos}@telecom.ntua.gr

Abstract. The Gigabit per second PON, now under standardization by the FSAN group in ITU-T, constitutes the most promising approach for the photonisation of the local loop, an elusive target of telecom operators for more than a decade. The traffic multiplexing / concentration on which the system efficiency (and hence low cost) is based, relies upon the dynamic MAC protocol to allocate bandwidth according to momentary needs, following traffic fluctuations. To serve each traffic class according to its needs, four priority levels are employed along with a high number of logically separate queues. This paper presents and evaluates the FSAN access control algorithms and discusses the choice of suitable traffic service parameters that optimize system performance.

Keywords. FSAN, GPON, PON, EFM, Shared access, reservation MAC.

1 Introduction

The current initiative of FSAN (Full Services Access Network) consortium to standardize a Gigabit per second Passive Optical Network, constitutes the most promising approach to the photonisation of the local loop. The incentive for such technology lies in cost benefits stemming from the fact that they need less fiber and less costly optical interfaces at the central office (one optical interface serves the entire network) but also achieve high traffic concentration as appropriate for low cost residential access systems [1]. After a series of developments starting in the late 80s in BT, the need for a standard led in the mid-1990s a group of major network operators to establish the FSAN (Full Service Access Network) consortium with a mandate to define a common standard for PON equipment, so that vendors and operators could create a competitive market for the last mile. Already a 155 Mb/s PON system has been specified and later adopted as the ITU-T G.983 series of standards known as APONs/ BPONs [2], [3]. In APON, data is transported via fixed-size packets of 53 octets, implementing a segmentation and reassembly process, for variable length packets. The small, fixed-size packets are deemed essential for carrying delay-sensitive traffic, making APON appropriate for services like voice, video and data.

N. Mitrou et al. (Eds.): NETWORKING 2004, LNCS 3042, pp. 890–901, 2004.

On the other hand the provisioning of ATM virtual paths and virtual channels appears to add important complexity in the PON system, while the five-byte header added in each 48-byte payload consumes a large amount of upstream bandwidth. Trying to eliminate these drawbacks of APONs and following the major drive of the increasing packet traffic due to the omnipresence of Ethernet LANs and IP, the IEEE through the Ethernet in the First Mile (EFM) [5] group are also working on standardizing a new form of PON called EPON (Ethernet PON). In EPON data is transmitted in variable length packets of up to 1,518 bytes avoiding the frequent and time-consuming segmentation and reassembly procedure of APON.

In a parallel activity, but with emphasis in QoS and efficiency as well as IP support, the FSAN group, initiated in 2001 a new effort for standardizing PON networks operating at bit rates above 1 Gb/s and supporting multiple services, and operation, administration, maintenance and provisioning (OAM&P) functionality and scalability. The result is the Gigabit PON (GPON), intended to support all service needs of access networks by either packet or TDM transport.

The work presented in this paper was carried out in the framework of IST project GIANT (GIgaPON Access NeTwork) [4], which targets the design, implementation and demonstration of such an Ethernet-based GPON system. It will support all kind of services from very strict QoS down to plain best-effort. The studies of the GIANT project have heavily influenced FSAN through contributions of consortium members and particularly Alcatel Bell, the consortium leader, but in turn the GIANT system has been aligned to FSAN decisions as they become available resulting in an FSAN-compliant system.

2   GPON Operation

The PON access technology is based on a passive star fibre network, which connects a number of ONUs (Optical Network Units) at the subscriber side to one OLT (Optical Line Termination) in the local exchange, as shown in Figure 1. The traffic streams arriving at the ONUs from the customer premises are kept in logically separate queues identified by their Allocation Identifier (AllocID) in FSAN terminology. Depending on QoS requirements, the AllocIDs belong to one of five T-CONT (Traffic Container) types. In the downstream direction, the OLT broadcasts the packets and their destination ONU extracts them based on the knowledge of which AllocIDs are locally assigned. In upstream, an aggregate data flow is generated by the active ONU transmitters via a TDMA mechanism controlled by the MAC controller in the OLT, which allocates the channel to each ONU in turn, according to a MAC algorithm. As with all PONs, a ranging procedure is relied upon for eliminating the problems arising by the difference in distances due to the tree topology. The AllocID assignments take place during the activation and registration process of each ONU. The choice in FSAN was to handle each AllocID separately without multiplexing those with the same QoS in one queue, due to the will of operators to keep detailed control of the customer traffic so as to check compliance with SLAs.

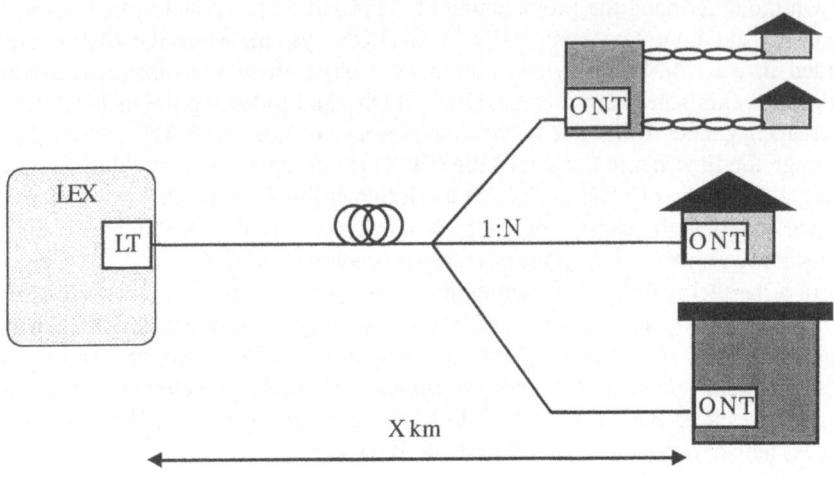

Fig. 1. PON structure

In the downstream direction frames of fixed size are transmitted every 125us consisting of a Physical Control Block downstream (PCBd) header of variable length and a payload field. The PCBd contains information about synchronization, bit error rate estimation, PLOAM messages [2] and a Bandwidth Map field, which is created by the GIANT MAC algorithm and contains the grants that enable upstream transmissions from the referenced ONU and AllocID. More specifically, the bandwidth map (shown in Figure 2) contains a number of allocation structures, which distribute the upstream bandwidth to the AllocIDs. Each allocation structure contains apart from the AllocID field two pointers, which indicate the starting and ending point of the upstream transmission in the channel (all with reference to the downstream frames, which play the role of time reference for the system). Thus, by the difference of the start and stop pointer each ONU learns the time that is granted to it for upstream transmission which can be compiled in the number of bytes it can transmit (Transmission Bytes, TB) avoiding, thus, collisions and at the same time exploiting in an efficient way the upstream bandwidth.

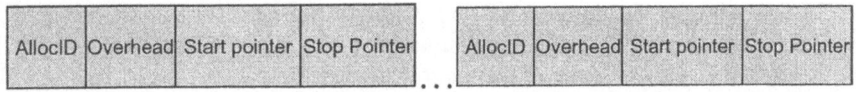

Fig. 2. BW Map structure

The upstream frame is created by a number of transmissions (shown in Figure 3) from one or more ONUs as dictated by the successive structures of the bandwidth map. Moreover, apart from user data, four PON overhead types can be inserted in the upstream frame. These types are:

• The physical layer overhead (PLO) which is always present at the start of an ONU upstream burst and contains the indispensable preamble, allowing proper physical layer operation (e.g. bit / byte alignment) of the burst mode upstream link.

• The physical layer operation, administration and management field (PLOAM), which is responsible for management functionalities, such as ranging, activation of ONU and alarm transfer.

• The power leveling sequence field (PLSu) which transfers information about the level of power at lasers of ONUs as been seen by the OLT.

• Finally, the Dynamic Bandwidth Report (DBR) field which reports to the OLT the queue length of each AllocID, allowing for the operation of the reservation based MAC algorithm according to the DBA (Dynamic Bandwidth Allocation) principle.

The inclusion of any of the above fields, except the first one, is controlled by appropriate flags sent by the OLT in the downstream direction.

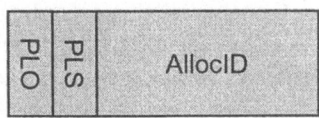

Fig. 3. Upstream frame format

In contrast to EPON, the GPON implements segmentation and reassembly for reasons of efficiency avoiding the overhead of reporting the boundaries of packets. So frames must be reconstructed in the egress of the system and the adopted method is called GEM (GPON Encapsulation Mode).

As mentioned, FSAN places emphasis on QoS support and seeks to control by means of the MAC protocol each traffic stream so as to be able to effect the SLA (service level agreement) and provide the required quality per user and stream. The quality class, and hence the service received, are determined by assigning each queue to one of the five Traffic Containers (T-CONTs) which follow different service policies.

The five traffic classes of FSAN are a legacy from the APON DBA specification G.983.4 [3] keeping the same term: T-CONT. Trying to offer different levels of QoS GPON determines four different T-CONTs corresponding to ATM's Class of Services:

• T-CONT-1 based on unsolicited periodic permits granting fixed payload allocations and intended for the emulation of leased line services,

• T-CONT-2 for VBR traffic and applications with both delay and throughput requirements, such as video and voice,

• T-CONT-3 for better than best effort services offering service at a guaranteed minimum rate while any surplus bandwidth is assigned only upon request and availability,

• T-CONT4 for purely best-effort services (browsing, FTP, SMTP, e.t.c.), and as such is serviced only upon bandwidth availability up to a provisioned maximum rate,

•   T-CONT5 is a combined class of two or more of the other four T-CONTs. It is now left to the ONU to choose which queue to service.

The MAC controller services the traffic belonging to each T-CONT starting from T-CONT 2, 3 and 4 with absolute priority i.e. no traffic of a lower priority is serviced while any requests for higher priority exist. The operation of the algorithm is presented in detailed next.

3   The MAC Algorithm

The MAC controller is responsible for the assignment of both the guaranteed and the surplus part of the bandwidth to the active queues. The service principle is a prioritised weighted round robin. Service of T-CONT1 is based on unsolicited periodic grants, which create a virtual leased line allocating a fraction of the link bandwidth inflexibly to each customer, (i.e. no variation according to the offered load, any excess is wasted as in a leased line). The priority order of the rest is: T-CONT2 first, then 3 and 4 while the weights follow SLA (Service Level Agreement) parameters. Unlike T-CONT 1, T-CONTs 2,3 and 4, use the reservation approach which regulates allocation according to fluctuating needs which in FSAN is called DBA (Dynamic Bandwidth Allocation). The information required for the proper operation of the MAC controller is based on the queue length reports sent for every AllocID, as well as the service level parameters negotiated during the activation phase by means of management tools.

Each flow is allocated at activation time two parameters: SDI (Successive Data Interval, i.e the time distance between two successive grants) and TB (Transmit Bytes, i.e how many bytes are to be sent with each allocation). Upper and lower bounds of these parameters are defined in the service agreement and are related to the specified peak and average rates as well as the allowed variation. This provides the tool to specify a guaranteed part (based on minTB, MaxSDI) allowing the surplus bandwidth to be assigned dynamically up to the peak rate (defined by MaxTB, MinSDI) by varying the actual values of TB and SDI in each allocation.

In more detail, for T-CONT1 the maximum and minimum TB and SDI values are equal (to keep delay variation zero). For T-CONT2, although the same relation between maximum and minimum TB and SDI values exists, the respective allocations are issued only on condition of outstanding requests i.e. unsatisfied reports. For T-CONT3, maximum and minimum values are different, resulting in the differentiation of guaranteed and surplus bandwidth assignments, while for T-CONT4, the maximum grant interval is theoretically infinite, providing no guarantees. The MAC controller in GIANT uses timers to enforce the extreme values while relying on DBA to decide each round of allocations by inspecting the "request" table where past unserviced requests are stored, reflecting the queue fill level. The examination of this table follows the round robin discipline within each T-CONT while absolute priority is observed among T-CONTs.

As expected, the polling frequency is a critical parameter for DBA performance, since nothing can be guaranteed a service faster than the polling cycle as the round trip for reservation and the processing times must also be taken into account. For example, in order to satisfy the maximum of 3ms of round trip delay budget for real-time services, a maximum polling interval of 500µs has to be adopted to guarantee an access delay below 1.5ms.

4 Performance versus Offered Load

To study the performance of the MAC algorithm, given the lack of analytical tools due to the high system complexity, a series of computer simulations have been carried out. In this section, the overall performance of the MAC is evaluated versus the total offered load under uniform loading among all sources.

It is worth noting that for T-CONT 1 the delay has a deterministic behaviour with well-defined limits so no simulation is needed. In contrast, the evaluation of T-CONT 2 access delay is very important since it is through this T-CONT that delay sensitive applications will be serviced based on a dynamic mechanism seeking greatly improved efficiency when compared with the rigid and wasteful T-CONT1 approach. For T-CONT 3 and 4 the metric of greater interest is the throughput rather than the delay, provided that the latter stays within reasonable limits (hundreds of ms).

The model consisted of 32 ONUs each supporting T-CONT 2, 3 and 4 with only one AllocID per T-CONT, i.e. totally 96 queues. The sources generating the traffic load were following the widely used for end-user data systems tri-modal length distribution model, which reflects IP data traffic length distribution from LANs. Exponential interarrival times were used. So, the packet length frequencies were about 60% of 64 byte long packets, 20% of 500 byte long and 20% of 1500 bytes while the load distribution among ONUs and T-CONTs was uniform. The polling period (maximum time between queue reports) was 1.25ms, i.e.10 frames.

The parameters that change from run to run were, first, the mean of time intervals between packet generation (decreasing for increasing load), in order to vary the total load, and second the negotiated parameter TB on which the MAC controller bases the allocations beside the AllocID queue report (implicit request).

Regarding T-CONT3 two options are investigated. As T-CONT 3 has in its specification a guarantied part and a surplus part, in the first scenario group the assured bandwidth for T CONT 3 is about 2/3 and in the other it is 1/3 of the provisioned bandwidth on the basis of its SLA.

As shown in Figure 4, as the total offered load increases, queues of T-CONT 4 suffer first the congestion at about 0.9Gbps due to the prioritized service. Hence, for total offered load below 0.9Gbps, all traffic is serviced and the observed delay remains in the order of ms for all types of traffic. When the total offered load is above 0.9Gbps, all traffic is protected except best-effort no-guarantee T-CONT 4 traffic which suffers all the congestion. It is worth noting that the sources did not contain any closed loop congestion control (i.e. TCP-like), which would in real life come into action to reduce the offered load. It was chosen to focus on the MAC mechanism and to exclude from the model interference from other network elements that are encountered by a flow in an end-to-end travel through a network since such an approach would involve many other assumptions about the rest of the network that do not play a role in the MAC evaluation.

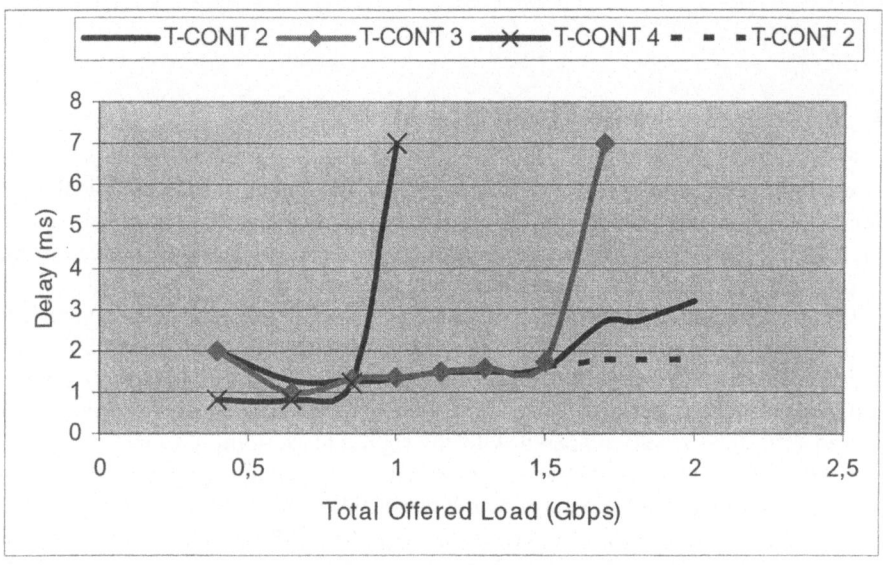

Fig. 4. Impact of increasing offered load

Queues of T-CONT 3 type are serviced to the demanded rate even when the total offered load is up to 1.6Gbps, i.e. above the nominal link rate. This of course is not expected to happen thanks to combined action of admission control and policing. So it can be considered a simulation of misbehaving T-CONT4 sources, which, as a result, have caused overflowing queues, (protecting within the designed limits the service received by T-CONT2 and 3 traffic). Of course, when the offered load reaches beyond 1.6Gbps, (at which time the offered load of T-CONT 2 and T-CONT 3 exceeds 1.06Gbps), the surplus bandwidth is not enough for the full service of T-CONT 3, which also gradually enters unstable conditions.

Focusing on the performance of T-CONT 2, this is as expected better than that of the other two, in both scenarios where the guaranteed part of T-CONT 3 is 2/3 or just 1/3 of the T-CONT 3 bandwidth. However in the scenario of high proportion of guar-antied BW to T-CONT3, the access delay of T-CONT 2 start increasing earlier (solid line) than when only 1/3 is guaranteed (dotted line), though not as much as the delay of T-CONT3 which has a lower priority. In the second case, the delay for T-CONT 2 is kept lower than 2 ms up to a total load of 2Gbps as shown by the dotted line. Despite the fact that such unrealistically high loads are to be prevented by SLAs, they have a value in evaluating the MAC under extreme conditions and checking the effect of instantaneous overloads, which cannot be excluded while they stay within the tolerance provided by the specified leaky bucket buffering of the policing unit.

5   The Impact of TB and SDI on Performance

To study the impact of the choice of TB and SDI on performance, five sets of simulation scenarios were carried out, changing in each scenario the values of TB and SDI while keeping constant their ratio TB/SDI, which is the service rate and must equal the provisioned source rate. The polling period (used for ONUs to inform the OLT about their queue lengths) also coincides with the SDI. Again only T-CONT2 type traffic was used in all scenarios.

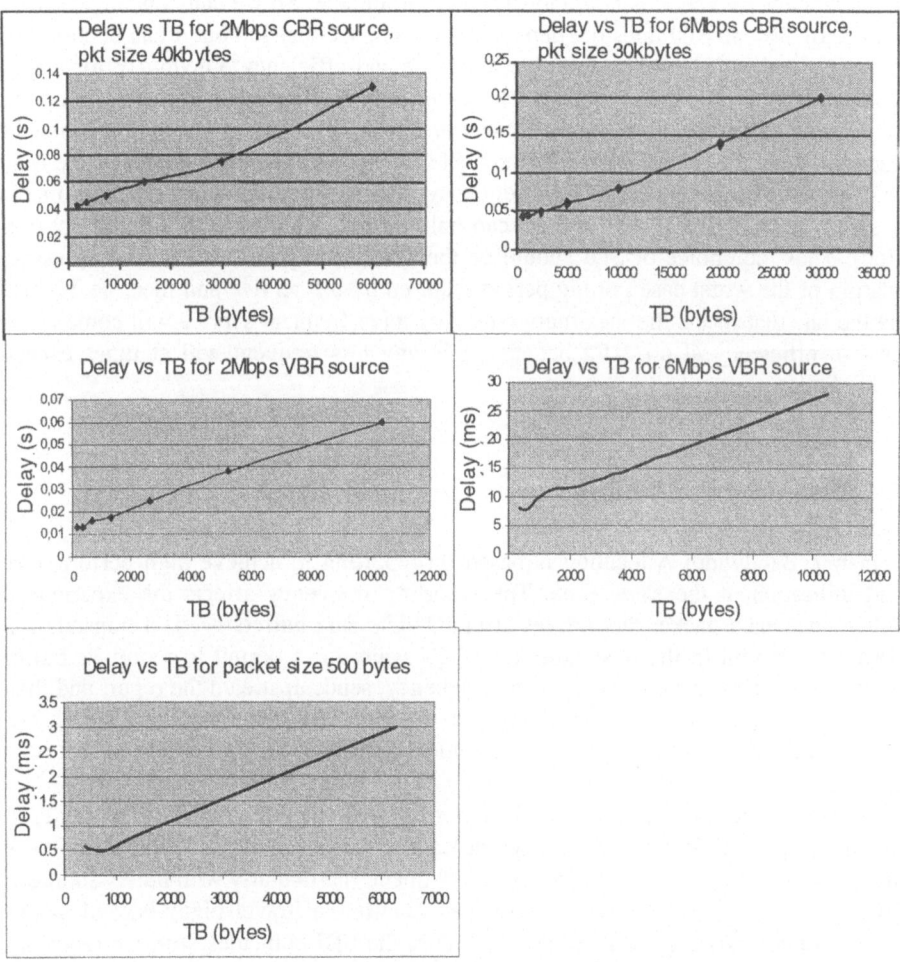

Fig. 5. Simulation results for testing TB and SDI

The sources employed in these five sets of scenarios were:
CBR: 25Mbps, packet size=500 bytes, interval =160µs,
CBR: 2Mbps, 10Kbytes every 40ms
CBR: 6Mbps, 30kbytes every 40ms

VBR: (mean rate 2Mbps) peak-rate=6Mbps, Ton=13ms, Toff=27ms, pkt size = 1300bytes

VBR: (mean rate 6Mbps) peak-rate=10Mbps, Ton=24ms, Toff=16ms, pkt size = 1300bytes

The results of these five sets of scenarios are shown in the figure 5.

It is obvious from the results that under all circumstances the system achieves better performance when the MAC serves each ONU with smaller TB, but more frequently (smaller SDI also). Even if longer packets take longer to have the last fragment reach the OLT, still the higher frequency of service ends in better performance. With smaller packets this is obviously more prominent. So we can conclude that the choice of TB and SDI does not depend upon complex decision but it can be based on a simple trade-off between delay performance and efficiency. Obviously for a high number of AllocIDs the overheads that accompany each upstream transmission make the choice of frequent service inefficient and thus indirectly affect the performance through the reduced net bandwidth available to payload. The other performance limit is of course dictated by the polling frequency, which affects the worst reaction time in a case of new arrivals (new arrivals can only be first reported with a polling allocation). Thus the choice of SDI should be the one that is just lower by the necessary margin of the worst case polling period required by the service and then the TB will be the one that meets the maximum rate. Any more frequent service will come at the cost of efficiency at no extra benefit while any less frequent will at times exceed delay limits.

6 The Impact of Polling

Dynamic Bandwidth Allocation is based on reporting to achieve high performance and utilisation at the same time. The reporting frequency affects the experienced delay since each packet that arrives at the ONU's data buffers needs a transmission permit to be sent in the upstream. The ONU waits for a permit to report its buffer status, which depends on the reporting frequency, sends upstream the report and then the OLT schedules the data permit. Hence, the reporting frequency has a strong impact on the worst-case delay, which is a critical parameter for the support of real-time applications.

The evident solution is to increase the reporting frequency regulated by the OLT, to minimise delay. However, such an action leads to efficiency degradation, since more often reports would travel announcing empty queues particularly with bursty sources.

To improve efficiency, the MAC also uses reports that travel piggy-backed in upstream frames carrying payload, in which case the SDI coincides with the reporting period. However, mainly because the value of the SDI affects the experienced delay, in the set of simulations run to demonstrate the impact of the reporting frequency in performance, the successive data interval and the reporting frequency are not assumed equal.

Table 2 presents the results from scenarios where both T-CONT 2 and 3 queues were activated and different traffic profiles were tested. The frequency that queues were

Table 1. The impact of the reporting frequency

Al-locID	Source type	T-CON T	Poll-ing Inter-val (frames)	Av. Delay (ms)	Number of Empty queue announce-ments	Max re-ported value
1	CBR	2	4	0.8	0.51%(20 of 3915)	48
2	CBR	2	8	1.5	0 of 2102	95
4	CBR	2	16	2.5	0 of 1135	161
6	Tri-modal/expo	2	4	0.85	1.19% (48 of 4022)	177
7	Tri-modal/expo	2	8	1.53	0.19% (4 of 2125)	192
9	Tri-modal/expo	2	16	2.6	1.18% (14 of 1190)	301
11	CBR	3	4	0.75	2.12% (86 of 4049)	48
12	CBR	3	8	1.13	0.19% (4 of 2122)	85
14	CBR	3	16	2	0 of 1208	170
16	Tri-modal/expo	3	4	0.83	3.74% (149 of 3980)	99
17	Tri-modal/expo	3	8	1.4	1.64% (35 of 2131)	99
19	Tri-modal/expo	3	16	2.3	0.27% (3 of 1092)	252

polled to report their lengths were either 4, 8 or 16 frames (remember frame duration is 125), and it was in general different to the service period (Max SDI).

In these runs, 20 queues with different service parameters were tested. CBR sources had a rate equal to 16 Mbps while sources with tri-modal traffic pattern a rate equal to 16.66 Mbps. So the system total offered load was about 600 Mbps. The traffic profile of the sources, the T-CONT type and the polling interval are given in columns 2, 3 and 4 respectively. The results include the average delay, the number of queue reports announcing empty queues and the maximum queue reported value, listed in columns 5, 6 and 7. The number of empty queue announcements allows for the estimation of efficiency degradation, which should be traded off with the performance amelioration.

A first important result is that the delay observed by sources of the same profile and serviced at the same rate (successive data interval) is increasing as the polling period increases. For example, compare sources corresponding to AllocIDs 1, 2 and 4. Should the polling frequency continue increasing, the delay would become greater than 3ms, which is the delay budget for the access domain for real-time applications (e.g. telephony).

Comparing T-CONT 2 to T-CONT 3 queues of the same profile and service parameters, (for example source 1 and 11), they seem to have identical performance, demonstrating the impact of the reporting period (polling). Even in the case that the sources have different traffic profiles, still the average delay stays at the same levels. For example, sources 1 and 6 which are of CBR and tri-modal traffic pattern respectively, generate packets at the same mean rate and their average delays differ only by 0.05 ms. This small difference is due to the fact that T-CONT 3 reports empty queue more frequently than T-CONT 2, leading to some inefficiency. The fact that the reporting frequency is the main factor that affects system performance, can be also deduced from the similar difference in average delay between the cases of 4, 8 and 16 frames DBA reporting period, regardless of traffic generation by tri-modal or CBR traffic pattern and T-CONT 2 or T-CONT 3 traffic classes.

The reporting period for all the sources included in the table is chosen so that performance remains acceptable for real-time services. Any increase of the reporting period to more than 20 frame times leads to considerably higher delay values.

As regards efficiency, the 6th column shows that the higher the polling frequency and the burstier the source, the higher the percentage of reports announcing empty queues, hence wasting resources. This leads to the recommendation that for services that do not require strict delay guaranties (e.g. for T-CONT4 or some T-CONT3 queues), the polling period should be much greater than 16 frames, particularly since such source types are expected to be much more bursty.

7 Conclusions

The cost-effective multiplexing of a variety of traffic in a GPON relies upon a dynamic MAC protocol that allows support of many services with response matching the fluctuating demand. The delay performance is dominated by the polling period so for services with strict delay requirements frequent polling below 16 frames should be chosen. For the non-real-time services, efficiency dictates larger polling values. The performance evaluation based on computer simulations shows that the FSAN GPON can satisfy any mix of service classes thanks to its prioritized MAC service policy with quite satisfactory efficiency.

Acknowledgement. The work of this paper was partly funded by the EU IST 2001-34523 GIANT project dealing with the design and development of an FSAN-compliant GPON system.

References

1. J.D. Angelopoulos, I.S. Venieris, G.I. Stassinopoulos, "A TDMA based Access Control Scheme for APON's," IEEE/OSA Journal of Lightwave Technology, Special Issue: Broadband Optical Networks, Vol. 11, No. 5/6, May/June 1993, pp. 1095-1103.
2. ITU G.983.1: "Broadband optical access systems based on passive optical networks (PON)".
3. ITU-T, G.983.4, Study Goup 15, "A Broadband Optical Access System with increased service capability using Dynamic Bandwidth Assignment".
4. A. Houghton, "Supporting the Rollout of Broadband in Europe: Optical Network Research in the IST Program", IEEE Communications Magazine, vol. 41, no 9, pp. 58-64, September 2003
5. G. Kramer, G. Pesavento, "Ethernet Passive Optical Network (EPON): Building a Next-Generation Optical Access Network", IEEE Communications Magazine, vol. 40, issue 9, pp. 66-73, February 2002

Traffic Regulation and Control in the Manhattan Street Network

O. Tayan and D. Harle

Department of Electronic and Electrical Engineering,
University Of Strathclyde, Glasgow, G1 1XW, U.K.
Tel: (+44) 141 548 2082 Fax: (+44) 141 552 4968
{otayan, d.harle}@eee.strath.ac.uk

Abstract.
The Manhattan Street Network (MSN) with the *Clockwork* Routing scheme has been proposed as a particularly suitable candidate for use in optical networks. However, without any medium access control (MAC), the success of the MSN with *Clockwork* Routing in optical networks would be inhibited by the destination-release protocol, since the protocol provides no guarantee of fair access or bounded traffic delay requirements of access nodes. Hence, new traffic regulation schemes providing medium access control are required to ensure fair access between network nodes and satisfy stringent quality of service (QoS) requirements of applications. This paper introduces a new scheme for medium access that supports fairness between competing access nodes while limiting the delay in the network. A ring model based on the "return-to-trail" property of the MSN is used to realise the logical behaviour of the MSN and connection services are regulated using a distributed view of the network loading.

Keywords: Medium Access Control, veto protocol, Manhattan Street Network

1 Introduction

The MSN is a self-routing virtual topology proposed for local and metropolitan area network applications. An MSN is characterised as a two-connected regular mesh with the nodes connected as rows and columns. Each node consists of a 2x2 cross-bar switch that connects incoming links to outgoing links. The *Clockwork* Routing Scheme has been proposed as a candidate for implementing optical packet switched networks. The *Clockwork* Routing Scheme is a time-slotted system that enhances the MSN by including a simple routing mechanism employed at intermediate nodes that prevents optical layer contentions, requires no re-sequencing of packets at the destination node and has comparable throughput to conventional routing schemes [1].

This paper focusses on the use of rings to model the logical operation of the MSN, allowing traffic regulation and performance to be monitored at the ring level. In [2], it was shown that an $n \times n$ MSN may be decomposed into n^2 independent rings by exploiting the return-to-trail property. The MSN with *Clockwork* Routing was viewed as a collection of independent rings, each operating as a

N. Mitrou et al. (Eds.): NETWORKING 2004, LNCS 3042, pp. 902–913, 2004.

slotted system while enforcing destination-release slot-reuse of the container at the destination node. A key advantage over single token passing protocols is evident in its support in allowing several nodes transmit simultaneously using different slots, one on each ring of the MSN. However, the drawback of using a destination-release based protocol was acknowledged: immediate reuse of the slot container at destination nodes may deprive other nodes of service. Such schemes are clearly restrictive in real time applications, whereby stringent and bounded delays must be guaranteed as part of the service. Therefore, traffic regulation schemes providing medium access control (MAC) have become an essential part of the research. In [2], the ring model was considered for purposes of deriving the stability criteria of each of the peripheral queues in the MSN and obtaining performance analyses of the ATM Ring medium access protocol. It was noted, however, that the ATMR scheme was initially proposed for bidirectional rings. Hence, it would be beneficial to study and compare alternative MAC protocols suitable for unidirectional rings. This study uses the ring approach to simulate and analyse an alternative MAC protocol; a protocol that regulates connection services to provide fair medium access by using a distributed view of the loading on the network.

The remainder of this paper is organised as follows. Section 2 explains the unfair medium access scheme associated with the destination-release based protocol. Section 3 explains how an $n \times n$ MSN with *Clockwork* Routing may be decomposed into n^2 independent rings. The behaviour of the proposed protocol in a single constituent ring is the focus of study in this paper. An alternative to the ATMR protocol (in respect of providing medium access control) is introduced in section 4, outlining the assumptions and objectives and explaining the methodology/implementation techniques used. Section 5 presents the results obtained for variations of the proposed model by altering design parameters. Assumed traffic metrics such as network load, are varied and the results are compared and discussed. Section 6 concludes the work.

2   Using the 'Return-to-Trail' Property to Model the MSN

This section exploits the 'return-to-trail' property to decompose an $n \times n$ MSN into n^2 constituent rings. In [2], it was shown that each ring is independent and can be used to model the logical behaviour of a *Clockwork*-Routed MSN. A 4×4 MSN is used as an example to explain how an $n \times n$ MSN with *Clockwork* Routing can be decomposed into 16 independent rings. Consider a single slot-container released from the horizontal output port of node 1 on timeslot 0. Any data on the currently selected horizontal buffer in node 1 (governed by the current timeslot) would consequently occupy the slot-container. The return-to-trail path of this slot-container is highlighted in Figure 1. After the first hop (timeslot 1), the slot-container visits node 2, allowing data selection from its horizontal buffers if the slot-container were unoccupied. On the second hop (timeslot 2), the slot-container visits node 3 where the horizontal queue associated with the current timeslot is selected to transmit data. An observation from Figure 1 is that the slot-container returns to its original node after a total of eight timeslots ($2n$ hops). It must be noted that each ring represents node 1 using two different nodes, whereby the first node would be the initial and returning position of the slot-container and the second node would be the midpoint intermediary node in the return-to-trail path. Additionally, horizontal and vertical peripheral queues are associated with the respective nodes. Hence, using a slot-container to traverse the MSN in a return-to-trail path has demonstrated the clockwise rotation of the slot-container in a predefined ring ($0 \rightarrow n\text{-}1$). Similarly, the decomposition obtained by considering node 1 at timeslot x for $\{0 \leq x \leq n\text{-}1\}$ can be extended for each MSN node by considering each node at timeslot x. In essence, this implies that an $n \times n$ MSN consists of n^2 independent rings each of which maintains the logical operation of the network. This paper focusses on a single ring to model the behaviour of the MSN.

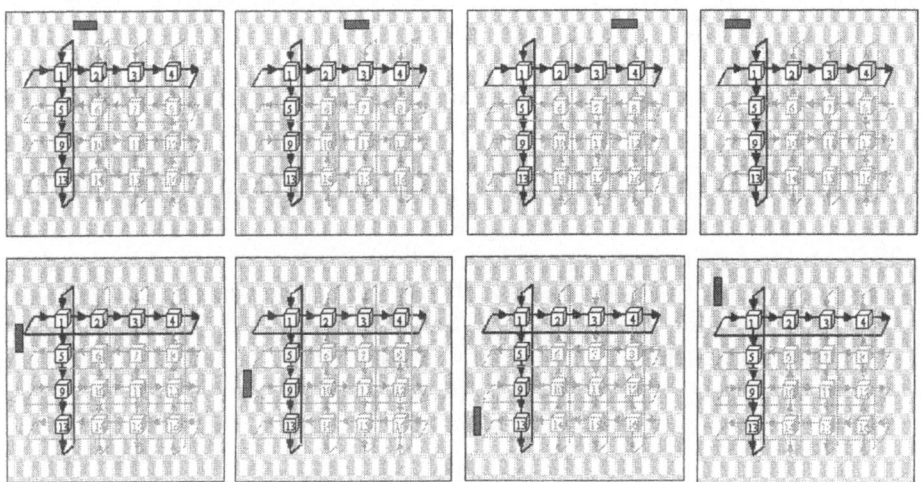

Fig. 1. Highlighting one possible sequence of a 'return-to-trail' path

3   Medium Access Unfairness in a Time-Slotted MSN

The ring based approach offers numerous advantages; providing a higher bandwidth utilisation than single-slotted systems and allowing for increased traffic to be offered to the network. A major drawback of the ring based approach, however, is that it provides no means of ensuring fair use of resources by access nodes. Traffic matrix parameters and positioning of nodes on the ring are influential in provisioning access to the shared medium. In a highly loaded network, destination-release of the slot-container may prevent some nodes from ever accessing the slot-container, therefore depriving these nodes of service [2]. For instance, Figure 2 shows two nodes (nodes 1 and 5) in isolation: neither of these nodes can ever access a free slot-container when upstream stations are continuously transmitting to downstream stations of the respective nodes. Consider when node 2 (upstream from node 5) continuously transmits to node 9 (downstream from node 5), where the slot-container is released. In this scenario, node 5 only ever sees in-transit packets and never any free slot-container and hence is unable to transmit data.

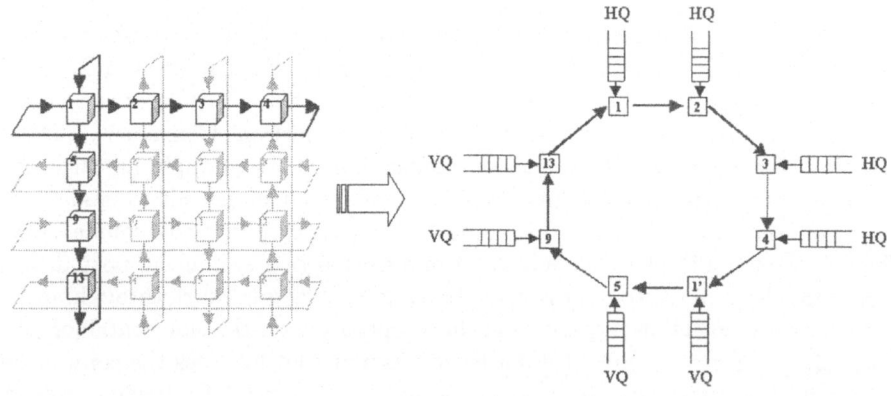

Fig. 2. Nodes deprived of service in the destination-release protocol

4   Medium Access Control for a Slotted-Ring Implementation of the MSN

4.1   Problem and Motivation

Sections 1 and 3 have stated the need for a medium access control scheme that ensures fair use of resources between access nodes. Without medium access control, the success of the MSN with *Clockwork* Routing in optical networks would be inhibited by the destination-release protocol since the protocol provides no guarantee of fair access or bounded traffic delay requirements of network nodes. Hence, new traffic regulation schemes providing medium access control are required to ensure fair medium access between network nodes and satisfy stringent quality of service (QoS) requirements of real time applications.

This section introduces a new scheme for medium access that supports fairness between competing access nodes. The MAC protocol uses a ring model to realise the logical behaviour of the MSN and connection services are regulated using a distributed view of the network loading.

4.2   Implementation and Methodology

Guaranteed QoS parameters in the MSN can be achieved by regulating traffic using a distributed view of the network loading, whereby all affected nodes are able to support/oppose required connection services. Hence, the ring model implementation must allow intermediate nodes to decide whether they support or oppose each proposed connection. The protocol would then accept or reject the proposed connection by considering its own local perception of the loading on the network and the stance taken by each affected node. Therefore, distributed protocols may be employed to establish connection request services, thereby removing the need for a central management node. A thorough explanation of the operational concepts of a simple "Veto" protocol is now discussed.

The 2N bit Connection Request Structure.   Initially, the veto protocol was realised using a $2n$ bit connection request field. Each node in a $2n$ node ring is allocated a single bit for use in the request field. The bit corresponding to each node is set when the node requests a connection. Conversely, the bit is reset if the node has not requested connection as for example initialisation or when an existing request has been 'vetoed' by an intermediary node that lies along the path of which the request circulates towards the source node. The veto protocol implies the behaviour of a slotted ring system supported by a token that circulates through the ring. To avoid reducing network throughput by transmitting connection requests and data separately on the slot container, the protocol operates on a single (uniform) cell format that includes the connection request field and the veto signal(s) in the header structure of a payload cell. A generic cell structure used in each version of the protocol is illustrated in Figure 3 and Figure 4 showing the connection request field used in this simulation.

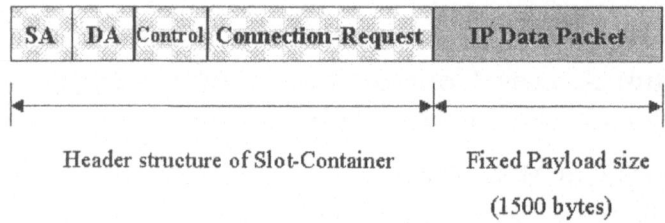

Fig. 3. Generic cell structure for each implementation of the veto protocol

The uniform cell format enables separate payload and connection request fields to be filled and emptied independently on each cell. Additionally, each traffic type must be stored separately before being transmitted on the slot container. Hence, each ring node must support an incoming connection request buffer and a payload data buffer.

The protocol initially circulates a connection-request cell around the ring (following a 'return-to-trail' path in the MSN), giving each intermediate node an opportunity to signal its opposition to the proposed connection. Any node wishing to block the proposed connection would alter the circulating connection

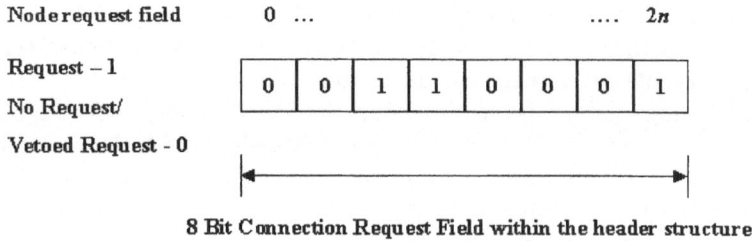

Node request field   0 ...   . $2n$

Request – 1
No Request/

Vetoed Request - 0

| 0 | 0 | 1 | 1 | 0 | 0 | 0 | 1 |

8 Bit Connection Request Field within the header structure

Fig. 4. A $2n$-bit Connection-Request structure

request to reflect this before passing it on through the network. A distributed view of the state of the network is obtained by circulating a connection request prior to connection establishment. Table 1 illustrates a possible sequence of operations on the cell as it circulates through the ring.

Table 1. A possible sequence of operations on the slot container.

Arrival at Node	Data Slot (Slot-Container)	Node C-R Field (Slot-Container)								Node State
		1	2	3	4	1'	5	9	13	
2	Free	S	R	R	S	R	R	S	R	Transmits Request
3	Free	S	S	R	S	R	R	S	R	Transmits Request
4	Free	S	S	S	S	R	R	S	R	Transmits Data Packet
1'	Occupied	S	S	S	R	R	R	S	R	Transmits Request
5	Occupied	S	S	S	R	S	R	S	R	Transmits Request
9	Occupied	S	S	S	R	S	S	S	R	Vetoes a single Request
13	Occupied	R	S	S	R	S	S	R	R	Sink packet/Transmit Data
1	Occupied	S	S	S	R	S	S	R	R	Veto/Transmits Request

R – node has its C-R field Reset
S – node has its C-R field Set

Consider the initial arrival of a connection request placed at the rear of the connection request queue. Each connection request is generated after the arrival of the ith data packet for transmission. An interesting comparison was to observe any impact on the performance for a variable connection length i. The buffered request progresses towards the head of the queue, where it awaits a free request slot on the cell. When a free request slot arrives, the node fills the free slot with a connection request and places the cell back onto the network then awaits its return after $2n$ timeslots. As each ring node receives the request, it must decide whether or not to accept the proposed connection. The decision to oppose (or accept) at each intermediate node can be based on any number of criteria including, instantaneous or average buffer occupancy or average service rate or the throughput into the node. If an intermediate node decides to oppose the connection, it resets a single bit in the connection request field before it is passed on towards the source node. Once the request has arrived at the source node, the source frees the request field for the following request and must decide whether or not to accept the connection. The decision of acceptance/rejection is based on the bit-value (within the request field) corresponding to the source node. If the corresponding bit-value was set, it would be deduced that no other node has vetoed this particular request. Hence, the data packet(s) associated with the request are accepted into data buffers awaiting transmission on a free payload slot within the circulating cell. A reset bit value, however, would indicate that either the node did not previously transmit a request or that the node sent a request which must have been vetoed by a ring node. The connection attempt would therefore be rejected and the associated data packets discarded. The protocol operates on the data fields independently of connection requests: when a node with an occupied data buffer sees a free data slot, it fills the empty slot with the packet at the head of the data queue. Once the cell has reached its destination, the data field becomes free again.

The protocol achieves simultaneous operation of data and connection requests by operating on the data and request fields independently of each other within the circulating cell.

The motivation in this initial veto scheme had considered simplicity in implementation and the idea of a simple request field was to ensure simple calculations and signalling operations were required and realised from simple optical logic. The veto scheme aims to provide medium access control in the MSN deployed in ultrafast optical networks.

4.3   Multi-queue Single Server Model

The veto scheme introduced in Section 4.2 can be described as a multi-queue single server (MQSS) model [3] in which a number of queues are served according to a certain discipline with the mean inter arrival rate λi at queue i.

The server visits each node sequentially, filling and freeing data and connection request slots independently as the cell circulates through the network. A cyclic-discipline governs the order by which connection-request queues are serviced by the server, whereas data packets are serviced using a stochastic mechanism[1]. The MQSS model provides support for cyclic connection-request services and stochastic data service schemes by inserting/removing connection-requests at fixed (deterministic) time intervals ($2n$ hops) and inserting/removing data randomly from the data slot when data packets are transmitted and received respectively. A polling system with Markovian server routing had been considered

---

[1] i.e. only occupied buffers at each station are serviced

in the literature [2][4] to model a destination-release based protocol. Several previous studies, including [2] were concerned with stochastic polling mechanisms to model arbitrary (data) packet arrivals and packet transmissions from access nodes. However, the model presented in this study uses a circulating slot-container, guaranteeing that each node is visited after $2n$ hops. The stochastic nature of data flows are maintained in the model described in Section 4.2 since the data-slot within the circulating cell serves data stations according to the destination-release protocol. Once the data-slot has been filled by a packet from a data station, it remains occupied until it reaches its destination, at which point the data-slot is freed and available to the destination station. Meanwhile, connection requests are inserted and removed at the designated fields within the request-slot on each cycle of the slot-container (i.e after a period known as the *cycle* time). Interestingly, the model introduced in Section 4.2 seems to combine the features of the strictly cyclic order discipline commonly found in token-ring based networks and the random polling scheme used in stochastic data traffic modelling [2][4]. Furthermore, the MQSS is classified as a 1-limited service model since on each cycle, the slot-container can only service a single packet at the head of each queue.

Arrival Process.   In [3], the system utilisation $(p0)$ was defined as the sum of the products between the arrival rate (λi) and service time (Si) for each node i, as expressed in Equation 1.

$$p0 = \sum_{i=1}^{n} \lambda_i.S_i \quad \forall i \epsilon[1, 2n] \tag{1}$$

An MQSS model operating in stability enforces the condition that $p0 \leq 1$. The load on each node in the return-to-trail path was assumed to be uniform. A uniform load for each node in the return-to-trail path is equivalent to a non-uniform load on the MSN. [2]

4.4   Packet Generation Process

Each node generates packets using Bernoulli trials and each packet is assigned an arbitrary destination from a uniform distribution. The veto protocol utilises a connection-length parameter[3], indicating the length of the call in constant size data cells (1500 bytes). A connection request is then generated for each call awaiting transmission and remains in the connection-request queue until arrival of the slot-container.

---

[2] A balanced MSN was shown to be equivalent to a heterogeneous loading on the nodes in the return-to-trail path [2].

[3] Different simulations were considered for each connection-length in order to demonstrate the impact of call length on performance.

5 Results

This section provides a performance evaluation of the veto scheme when used with a *Clockwork* Routed MSN.The two figures of merit investigated in this section are the mean call drop probability (i.e loss of discarded packets as a proportion of the offered load) and the mean call delay. The loss is an important network design parameter, as it indicates the point at which the network becomes overloaded and hence saturated. Mean call delay, on the other hand, is essential for monitoring packet delays and ensuring that they conform to the stringent requirements of intended applications. A number of variables including the call connection length, veto threshold and data buffer threshold were considered to obtain a detailed analysis of the veto protocol. Several assumption were made in order to model the performance of the veto protocol. These assumption include:

- An n x n MSN with Clockwork Routing may be decomposed into n^2 independent rings, each ring established by following the 'return-to-trail' path from each node at a chosen timeslot.
- Each ring is independent and equally loaded, hence a single ring was used to model the logical operation of the MSN.
- Uniform traffic generation was assumed at each node. In other words, the probability of success of Bernoulli trials was constant for all nodes. Additionally, source and destination nodes were chosen from a uniform distribution.
- All links between adjacent nodes were assumed to have 1 timeslot length. Moreover, it was assumed that the links were lossless and the propagation delay to traverse each link was considered negligible.
- The destination release protocol, allowing immediate reuse of the slot-container was considered. Essentially, this implies that a slot-container containing a packet destined for an arbitrary destination could be immediately reused at the destination if there was a packet at the head of the destination station.
- Processing times at the source and intermediate nodes were assumed negligible.

An analysis of the results obtained from a simulation must consider the degree of uncertainty introduced by random samples during simulation execution. This work uses a Student's t distribution to obtain a *small sample confidence interval* since the number of samples taken are typically below thirty.

A 95% confidence interval is assumed in all results presented in this section. All simulations have an initial warmup phase to eliminate any bias due to transient effects and obtain steady-state behaviour before recording statistics. Simulations were run for 10^7 trials and 10^3 warmup trials. A constant buffer threshold of 30 entries was used at each node. Consequently, exceeding the buffer threshold would result in a veto of the next connection request.

The experiments are based on a number of simulation runs of variable connection lengths in order to analyse the performance of a basic implementation of the veto protocol. Figures 5 and 6 show the probability of loss and the mean call delay, respectively, for a call length of 1, 5 and 20 packets in a 4x4 MSN.

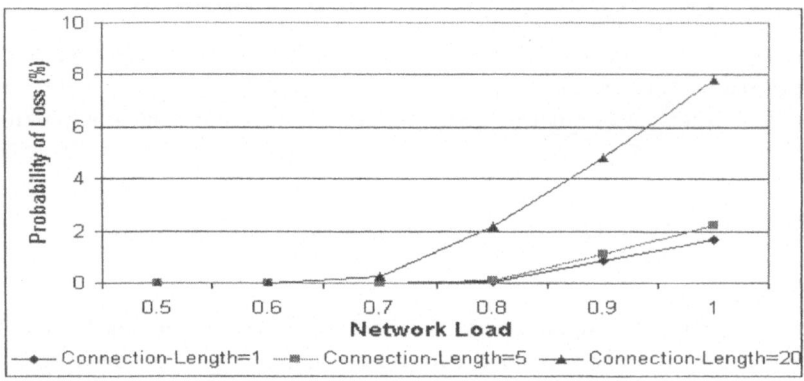

Fig. 5. Call Drop Probability for MAC scheme

Clearly, Figure 5 shows that the longest call duration results in the highest loss, since for each rejected call, 20 packets are dropped. It is expected that longer calls correspond to larger mean call delays, as shown in Figure 6 for a call length of 20. However, long calls require fewer connection requests, allowing for more efficient use of the bandwidth at higher loads. In contrast, call lengths of unity require a separate connection request for each packet, consequently occupying more time for connection-request passing and resulting in larger packet delays than longer calls at high loads. Figure 7 shows that at 0.5 load, the mean packet delay for short calls increases above the corresponding packet delay for longer calls, as a result of the additional time each node spends on connection request passing.

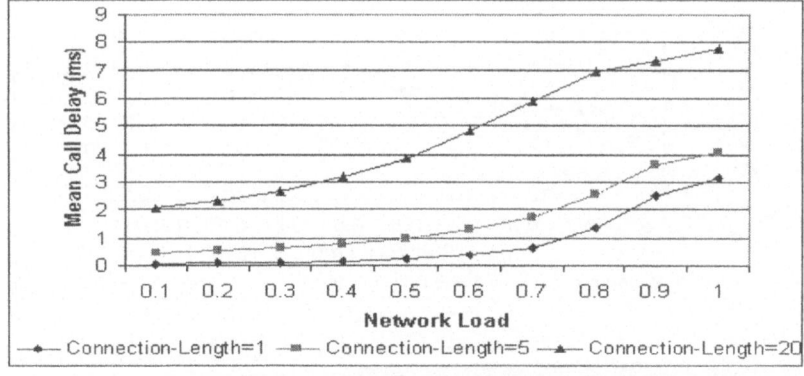

Fig. 6. Mean Call Delay for MAC scheme

Furthermore, it may be observed that for longer calls in particular, the mean packet delay levels off as the the load approaches unity, suggesting that the veto protocol limits the delay present in the network. The mean packet delay is obtained from the mean call delay and the connection-length, as defined in Equation 2.

$$MeanPacketDelay = MeanCallDelay/ConnectionLength \qquad (2)$$

The maximum throughput obtained was 0.248, in comparison to 0.31 from an MSN model that had avoided connection requests in [2]. However, [2] had derived the maximum theoretical throughput as the inverse of the mean path length in the MSN. Hence, the reduction in throughput can be explained, since in [2], a shortest path routing discipline was used, yielding a mean path length of 2.94 compared to a mean of 4.0 for the ring considered in this study. The theoretical maximum throughput for the ring topology was calculated as 0.25, for a mean path length of 4. Furthermore, a fair comparison of the loss performance with the losses obtained in [2] is possible when a connection length of 1 is considered. Figure 5 shows a maximum drop probability of 1.6% in comparison to a theoretical loss probability 0.7. However, Figure 5 shows that the loss probability remains negligible until 0.9 load, where the loss approaches the theoretical maximum value. The additional loss may be explained, since at very high loads the veto scheme must discard some packets in order to prevent high demand nodes starving other nodes of service. Maximum loading causes queue occupancies at nodes to exceed the buffer threshold, hence arising in vetoes from all affected nodes.

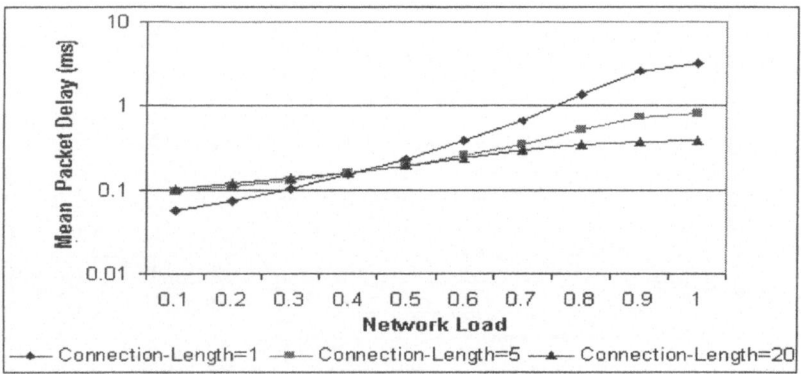

Fig. 7. Mean Packet Delay characteristics

6 Conclusions

This paper has investigated medium access control for the MSN with *Clockwork* Routing. In previous work, the MSN performance was inhibited by the destination-release protocol and clearly a new traffic regulation scheme was needed to prevent nodes being deprived of service. The significance of the "return-to-trail" property was evident, as it had enabled the MSN to be modelled as ring network with $2n$ nodes. A fair mechanism of controlling medium access and preventing node starvation, while limiting the delay in the network was achieved using the veto protocol. An analysis of the results had shown the significance of the veto protocol at high loads: nodes veto connections in order to keep their queue occupancy below the buffer threshold, thereby limiting the delay in the network with only the expense of a slight increase in loss as compared to the original model that lacks medium access control.

References

1. O.Tayan, J.Md.Desa, D.Harle, "A Practical Embedding Problem Exposing the Affect of Link Failures in the Manhattan Street Network", Proceedings of HET-NET's 2003, Ilkley, West Yorkshire, July 2003.
2. Chevalier, F., "Performance Evaluation Of The Clockwork Routing Scheme in Optical Packet-Switching Networks", PhD Thesis, 2000.
3. Akimaru H., Kawashima K., 'Teletraffic Theory and Applications', Published by Springer 1999.
4. Zafir-Vukotic M., "Wating Time Estimates in Symmetric ATM- Oriented Rings with the Destination Release of Used SLots", IEEE/ACM Transactions on Networking, Volume 7, Number 2, April 1999, pp251-261.
5. Hock NG C., "Queueing Modelling Fundamentals", Published by John Wiley and Sons 1996.
6. K. Watkins, "Discrete Event Simulation in C", Published by McGraw Hill Book Company, 1993.
7. F. Chevalier, D.Cotter and D.Harle, "Routing and Fairness Issues Associated with a Self Routing Control Strategy in Regular Mesh Photonic Networks", International Teletraffic Congress ITC-16, Edinburgh, Scotland, June 1999.

Fairness Issues in Bus-Based Optical Access Networks

Nizar Bouabdallah[1,3], Andre-Luc Beylot[2], and Guy Pujolle[1]

[1] LIP6, University of Paris 6, 8 rue du Capitaine Scott, F-75015 Paris, France
{nizar.bouabdallah, guy.pujolle}@lip6.fr
[2] ENSEEIHT - IRIT/TeSA Lab., 2 rue C. Camichel, BP7122, F-31071 Toulouse
Andre-Luc.Beylot@enseeiht.fr
[3] Alcatel Research & Innovation
Route de Nozay, F-91460 Marcoussis, France

Abstract. Packet-based optical ring becomes the standard access medium in metropolitan networks. Its performance depends mainly on how optical resource sharing, among different competing access nodes, takes place. This network architecture has mostly been explored in regard to synchronous transmission. However, in the present paper, we focus on the performance of asynchronous transmission-based metropolitan networks with variable packet sizes. An analytical model is presented in an attempt to provide explicit formulas that express the mean access delay of each node of the bus-based optical access network. In addition, we prove that in such a network, fairness problems are likely to arise between upstream and downstream nodes sharing a common data channel. Furthermore, we show that sharing the available bandwidth fairly and arbitrarily between access nodes, as in slotted WDM rings, doesn't resolve the fairness problem in asynchronous system.

1 Introduction

In next-generation metropolitan networks, internet traffic is deemed to be stamped by three important characteristics. In fact, packet-based data traffic of bursty nature will become prevalent. Moreover, it is believed that traffic will fluctuate heavily and on a random basis. Finally, internet traffic will keep on growing in the next few years up to, and eventually beyond, 1 Tbit/s. The architecture of next-generation metro networks must consequently evolve enabling them to tackle the new challenges, which are set by the aforementioned characteristics.

In the metropolitan segment, infrastructures are generally organized over a ring topology. We have proposed a new architecture named DBORN (Dual Bus Optical Ring Network), which satisfies all the requirements of next-generation metro networks. A detailed description of the DBORN architecture is out of this paper's scope, so for more information about this architecture the reader is invited to refer to [1]. Nonetheless, the work presented in this study, is more pertaining to the design of the media-access-control (MAC) protocol planned for DBORN. This protocol is designed for efficient transport of variable-sized IP packets, whereas it does not address the DBORN inherent fairness control issue, characteristic of shared medium networks.

N. Mitrou et al. (Eds.): NETWORKING 2004, LNCS 3042, pp. 914–925, 2004.

Generally, in order to avoid collisions on the individual WDM channels of such networks and arbitrate the bandwidth access, MAC protocols are needed. In the mean time, several access protocols for all-optical slotted WDM rings have been proposed in the literature [2], [3]. Most of them consider as many wavelength channels as nodes in the network, resulting in serious scalability issues, especially for MANs (Metropolitan Area Networks). In order to deal with the aforementioned limitations, we propose a novel access protocol for a packet-based optical metropolitan network supporting many more ring nodes than the available wavelengths in the network. The proposed MAC protocol addresses the case of non slotted WDM rings.

Since several source nodes share a common channel, one upstream node can grab all the available bandwidth, and possibly starve downstream nodes competing to access the same channel. Protocols at various levels (such as MAC or CAC – Call Admission Control) must be introduced to ensure good utilization of transmission resources and alleviate fairness problems. In general, fairness control mechanisms limit the transmission of upstream nodes in an attempt to leave enough bandwidth for downstream stations [4]. These schemes may be efficient in the case of slotted WDM rings (i.e. synchronous transmission). However, they don't perform well in the case of asynchronous transmission based architectures like DBORN. We present here an analytical model that aims to illustrate this issue. Despite its importance and up to now, the analytical study of asynchronous transmission in bus-based optical access networks has not been tackled.

The key behavior metric in such networks is the access delay at each node competing to access the shared data medium. By presenting a specific two-node bus as a first case study, we examine the average access delay of each node thanks to an exact analytical model. Afterwards, the model is extended to handle the general case, with many more nodes. The fairness issues are also dealt with in the proposed model. Simulation results show that the analytical model remains highly accurate under various traffic loads.

The remaining parts of this article are organized as follows. Section II focuses on the MAC context including a description of the network and node architectures along with the main features. The analysis of access delay of each ring node is presented in section III. Then, section IV discusses the effects of unfair access to the data channel, and at the same time it analyzes the corresponding numerical results. Finally, some conclusions are drawn in section V.

2 Network Architecture and MAC Design

This section describes the DBORN architecture and the proposed MAC protocol. DBORN can be described as a unidirectional fiber split into downstream and upstream channels spectrally disjoint (i.e. on different wavelengths). The downstream bus, initiated at the hub node, is a medium shared in reading, while the upstream bus, initiated in the ring nodes, is a multiple access-writing medium. Ring nodes consist mainly of optical couplers and burst mode transponder (i.e. transmitters and receivers) working in an asynchronous mode. This spectral separation allows the use of a simple passive structure for the optical part of ring nodes (right hand-side of Fig. 1). The

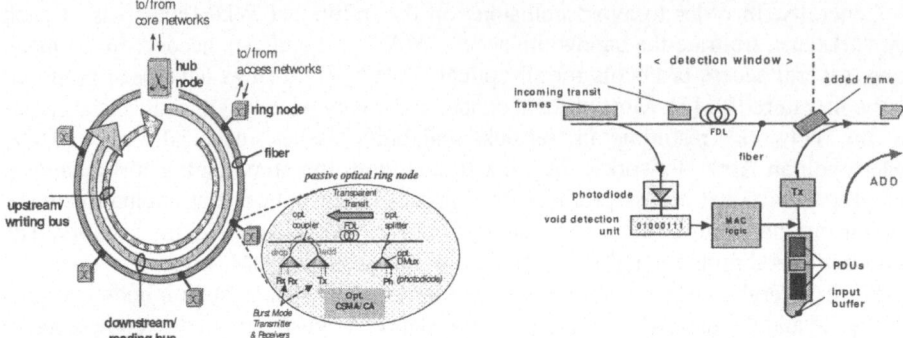

Fig. 1. Overview of DBORN network and node architecture

Fig. 2. Schema of the CSMA/CA based MAC of DBORN

network architecture (proposed in [1]) targets the simplicity of "old" shared Ethernet buses. It provides the necessary performance, flexibility, and capacity required for MANs via the usage of an optical asynchronous transport mode. Optical transparency is one of the key features that differentiates DBORN from existing solutions (such as NG-SONET/SDH and RPR). It relies on the optical transparency of the transit path in ring nodes and on shared optical wavelengths. Both aspects contribute in reducing the number of transponders required in the network:

• First, ring nodes need transponders for their local traffic only (no need to have O/E/O conversion for transit traffic going to the hub).

• Second, hub transponders are shared between several ring nodes.

In terms of logical performance, the main issue is related to the collision-free packet insertion on a shared writing bus. Since the transit path remains transparent and passive, no packet is dropped once transmitted on the ring (optical memory is still in the research stage). Hence, traffic control mechanisms are required at the electronic edge of the ring nodes to regulate data emission. In this regard, each DBORN ring node is equipped with void/null-detection mechanism in its upstream operating plane. This mechanism tends to retain the upstream traffic flow within the optical layer while monitoring the medium activity.

In a fixed-slotted ring system with fixed-packet size, void (i.e. slot) filling can be carried out immediately upon its detection, since the void duration is either one or multiple series of fixed-packet size duration. The detected void is therefore guaranteed to have a minimum duration of one fixed-packet length. However in non slotted ring systems with variable packet length and arbitrary void duration, it is very likely for a collision to occur if a packet is immediately transmitted upon detecting the edge of a void.

To avoid the abovementioned problem, a very simple collision avoidance system is implemented through photodiode power detection on each locally accessible upstream wavelength (Fig. 2). So, ring nodes first use an optical coupler to separate an incoming signal into two identical signals: the main transit signal and its copy used for control. A Fiber Delay Line (FDL) creates on the transit path a fixed delay between the control unit and the add function realized through a 2:1 coupler. With regard to the control part, as in [5], low bit rate photodiodes (ph) –typically 155 MHz– are used to monitor the activity on upstream wavelengths.

This way, voids are detected and a fixed length FDL – slightly larger than the MTU (Maximum Transmission Unit) size allowed on the network – ensures collision free packet insertion on the upstream bus from the add port. The introduction of a FDL delays the upstream flow by one maximum frame duration plus the information processing time, so that the MAC unit will have sufficient time to listen and measure the medium occupancy. The ring node will begin injecting a packet to fill the void only if the null period is large enough (i.e. at least equal to the size of the packet to be inserted). Undelivered data will remain buffered in the electronic memory of the ring node until a sufficient void space is detected.

However, considering only this basic mechanism, HOL (Head Of the Line) blocking and fairness issues arise. A directly resulting effect is performance degradation for ring nodes that are close to the hub node on the upstream bus. Additional flow control mechanisms have thus to be considered, both at the MAC layer and in upper layers at edge nodes.

3 Analytical Model

3.1 Framework

In this section, we will analyze the performance of the network in term of access delay. The proposed MAC protocol, which is based on CSMA/CA principle, avoids collision between transient packets and local ones that attempt to access the shared medium. It detects a gap between two packets on the optical channel. Then a local packet is inserted into the perceived gap. However, in such an environment, fairness issues could arise.

To simplify our analysis, we make the following assumptions: The network is composed of two ring nodes sharing a common medium (e.g. one wavelength) in order to contact the hub. Packets arrive to each node according to a Poisson process with an arrival rate λ. The traffic of the first node has a higher priority to access the medium. The head-of-line packet of the second queue can only access the channel if the medium is free for a sufficient period of time larger than its transmission time. So, the emission process of the second node depends on the activity of the first one. The first queue can be modeled simply by an M/G/1 system. Therefore in the remainder study, we will focus on the second queue analysis.

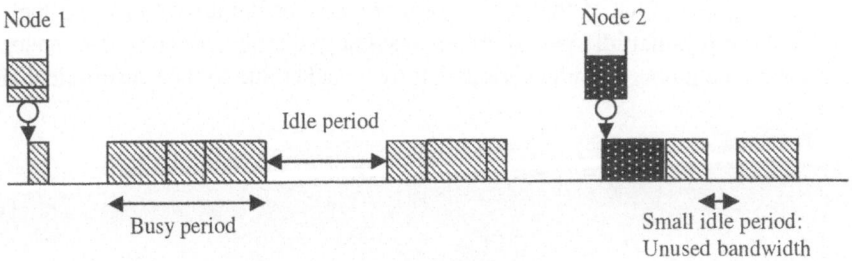

Fig. 3. Activity on the data channel

In this paper, "link state" refers to the state of the link when a packet from a ring node attempts to access the data channel. The wavelength channel can be in one of two states: free (idle) or occupied (busy). It is obvious that for packets from the upstream node, i.e. node 1, the channel is always idle. However, when packets from the downstream node, i.e. node 2, try to access the channel, it can be either free or occupied by upstream traffic. Since packet arrival in the first queue is assumed to follow a Poisson process, idle periods $I_j, (j = 1,2,\cdots)$ (Fig. 3) are iid and are exponentially distributed with parameter λ.

Let $B_j, (j = 1,2,\cdots)$ be the random variable representing the duration of the jth busy period encountered by a packet while trying to access the data channel. We make the observation that the state of the medium, as seen by node 2, alternates continuously between an idle and a busy period.

Let $S_i, (i = 1,2,\cdots)$ be the transmission time of packet i. We assume that S_i forms a sequence of iid random variables, distributed according to some common distribution f_S having a mean $E[S]$, a second moment $E[S^2]$ and a Laplace transform S^*. We assume that the length of the packets emitted by the different nodes has the same distribution.

Let us consider a packet entering the second queue. Two cases are to be distinguished:

a- When the packet arrives, the queue is empty. It has to wait for a suitable idle period to be transmitted (larger than its transmission time). When it arrives, it may encounter either an idle or a busy period on the medium (or of the first queue).

b- In the second case, the packet arrives while the second queue is not empty. In this case, the packet has to wait for its turn in order to be served. When the packet just in front of the arriving one leaves the buffer, the latter starts its service by waiting for an adequate idle period. It necessarily starts its service by an idle period which corresponds to the residual idle time resulting from the previous packet's service.

The second queue corresponds to a priority queue with pre-emptive repeat without re-sampling [6] (i.e. if a packet can not be sent because the idle period is not long enough, its size will not change).

Let X_i denote the service time of packet i. It is the time spent by packet i on the head of the buffer until the end of its transmission. It will correspond to a series of idle and busy periods. The expression of X_i depends on whether we consider the case (a) or (b).

Let us consider first the case (b). As explained earlier, the packet service starts by an idle period. The packet will remain on the head of the buffer until finding a suitable idle period. Due to the memoryless property of the arrival process, the idle period and the residual idle period are exponentially distributed with a parameter λ. Let $\tilde{n}$ denote the number of idle periods left by a packet due to their insufficient size. Then:

$$X_i = I_1 + B_1 + \cdots + I_{\tilde{n}} + B_{\tilde{n}} + S_i \tag{1}$$

Let us now consider case (a). As explained before, the arriving time of the packet can correspond to the progress of an idle or a busy period on the channel. If it corresponds to an idle period, X_i retains the same expression (1). Otherwise, i.e. packet i arrives in a busy period, we have to add another term corresponding to the residual time of the current busy period denoted $\tilde{B}_0$:

$$X_i = \tilde{B}_0 + I_1 + B_1 + \cdots + I_{\tilde{n}} + B_{\tilde{n}} + S_i \tag{2}$$

Note that $\tilde{B}_0$ does not have the same distribution as B_i, even when the service time is considered to be exponential. Finally, the framework for calculating the access delay of the second queue can be summarized as follows:

1- Analyze the second queue as a traditional M/G/1 queue with a service time X_i as defined in equations (1)-(2). The only difference in comparison with M/GI/1 is the particular distribution of the service time which depends on the state of the first queue.

2- Characterize the service distribution X_i by calculating its first two moments.

Below, we present the detailed analysis of the particular M/G/1 queue. We will first calculate with respect to the second queue the mean number of present packets, which is denoted at steady state by $E[q]$. Then, we get easily the expression of the response time $E[R]$.

Let X be the time that a packet spends at the head of the queue before the end of its transmission. Given the Little formula we get:

$$\rho = \lambda E[X] = \Pr[q > 0] \tag{3}$$

Where ρ represents the load of the second queue server and q denotes the number of packet in the system .

Let X_1 be the service time of a packet of type (b) (i.e. when the packet arrives the second queue is empty). Let X_0 be the service time of a packet of type (a) (i.e. when the packet arrives the second queue is not empty). The expression of X_1 and X_0 were presented respectively in equation (1) and (2).

Using nearly the same approach as the one presented by Kleinrock [7], we obtain the mean response time:

$$E[R] = E[X] + \frac{\lambda E[X^2]}{2(1 - \lambda E[X_1])} \tag{4}$$

Based on (4), we have to compute the first two moments of the general service time ($E[X]$ and $E[X^2]$) and the first moment of the service time for packets of type (b), that is $E[X_1]$, in order to calculate the mean response time. Below, we will discuss the calculation of $E[X_1]$, $E[X]$ and $E[X^2]$, respectively.

3.2 Determination of the "Mean Service Time"

Let us start first with the entity $E[X_1]$. As we have mentioned, X_1 is a series of idle and busy periods that begin with an idle period. Suppose that the transmission time of a packet is S and $\tilde{n}$ is the number of idle periods left by the packet due to their insufficient size. Based on (1), we obtain directly:

$$X_1 = I_1 + B_1 + ... + I_{\tilde{n}} + B_{\tilde{n}} + S \quad \text{Where } I_j < S, \forall \ j = 1, \tilde{n} \tag{5}$$

Further, we have:

$$E[X_1] = \int_{x=0}^{\infty} \sum_{n=0}^{\infty} E[X_1 | S = x, \tilde{n} = n] \Pr[\tilde{n} = n | S = x] f_s(x) \cdot dx \tag{6}$$

Where $E[X_1 | S = x, \tilde{n} = n]$ and $\Pr[\tilde{n} = n | S = x]$ are two conditional probabilities that need to be calculated first:
The mean "service time" of a case (b) packet, given the emission time S and the number of idle periods $\tilde{n}$, can be calculated as:

$$E[X_1 | S = x, \tilde{n} = n] = x + nE[B] + nE[I | S = x] \tag{7}$$

Let $\tilde{\rho}$ be the probability that the server of the first queue is busy. Applying the PASTA property [8], $\tilde{\rho}$ denotes also the probability that a packet entering the second queue encounters a medium in a busy state. Hence:

$\tilde{\rho} = \lambda E[S] = \dfrac{E[B]}{E[B] + E[I]}$, where $E[I] = 1/\lambda$ since the idle periods follow an exponential distribution of parameter λ. So the mean time of busy period is given by:

$$E[B] = \frac{E[S]}{1 - \tilde{\rho}} \tag{8}$$

Moreover,

$$E[I | S = x] = \frac{1}{\lambda} + \frac{xe^{-\lambda x}}{1 - e^{-\lambda x}} \tag{9}$$

Given the emission time S, the distribution of $\tilde{n}$ is clearly equal to:

$$\Pr[\tilde{n} = n | S = x] = (1 - e^{-\lambda x})^n e^{-\lambda x} \tag{10}$$

Finally by replacing (7), (8), (9) and (10) in (6) we get:

$$E[X_1] = \left(E[B] + \frac{1}{\lambda} \right) \left(S^*(-\lambda) - 1 \right) \tag{11}$$

We can obtain easily the expression of $E[X_0]$:

$$E[X_0] = E[X_1] + E[\tilde{B}]\tilde{\rho} \tag{12}$$

Where $\tilde{B}$ is the residual time of the first busy period if the arriving packet encounters the medium in a busy state.

We can now determine the expression of the mean service time of a packet in the second queue :

$$E[X] = \rho E[X_1] + (1 - \rho)E[X_0] \tag{13}$$

Where ρ, as mentioned earlier, is the probability that the server of the second queue is busy.

To summarize we have:

$$E[X] = \frac{E[X_0]}{1 + \lambda^2 E[S]E[\tilde{B}]} \tag{14}$$

3.3 Determination of the Second Moment of the Service Time

To complete the calculation of the overall mean response time of the second queue, it remains the expression of the second moment $E[X^2]$.

From (12) we have:

$$E[X_0^2] = E[X_1^2](1 - \tilde{\rho}) + \{E[\tilde{B}^2] + 2E[\tilde{B}]E[X_1] + E[X_1^2]\}\tilde{\rho} \tag{15}$$

Using (13) and (15) we get:

$$E[X^2] = E[X_1^2] + (1 - \lambda E[X])\tilde{\rho}\{E[\tilde{B}^2] + 2E[\tilde{B}]E[X_1]\} \tag{16}$$

From equation (16), we see that in order to obtain $E[X^2]$, we need to calculate the expression of the first and second moment of the busy period residual time ($E[\tilde{B}]$ and $E[\tilde{B}^2]$) and the second moment of the service time of packets of type (b) that is $E[X_1^2]$.

The first two moments of residual life $E[\tilde{B}]$ and $E[\tilde{B}^2]$ can be derived (see [7] for instance):

$$E[\tilde{B}] = \frac{E[S^2]}{2(1 - \tilde{\rho})^2 E[S]} \tag{17}$$

$$E[\tilde{B}^2] = \frac{\lambda E[S^2]^2}{(1 - \tilde{\rho})^4 E[S]} + \frac{E[S^3]}{3(1 - \tilde{\rho})^3 E[S]} \tag{18}$$

The basic idea for calculating $E[X_1^2]$ is nearly the same as for calculating $E[X_1]$. Therefore:

$$E[X_1^2] = \int_{x=0}^{\infty} \sum_{n=0}^{\infty} E[X_1^2 | S = x, \tilde{n} = n] \Pr[\tilde{n} = n | S = x] f_S(x) \cdot dx \qquad (19)$$

After some manipulations, we get:

$$E[X_1^2] = 2E[B]^2 + \frac{2E[B]}{\lambda} - E[B^2] + S^*(-\ ) \left\{ E[B^2] - \frac{2}{\lambda^2} - E[B]^2 - \frac{6E[B]}{\lambda} \right\} \qquad (20)$$

$$+ 2S^*(-2\ ) \left\{ \frac{1}{\lambda} + E[B] \right\}^2 + 2S'^*(-\ ) \left\{ \frac{2}{\lambda} + 2E[B] \right\}$$

Where $S'^*(-\lambda) = \dfrac{dS^*(s)}{ds} \bigg|_{s=-\lambda} = -\int_{x=0}^{\infty} x e^{\lambda x} f_s(x) \cdot dx$

Finally the calculation of overall response time $E[R]$ can be found easily through equation (4).

3.4 Example

Different packet length distributions can be considered. In the present paper, we consider packets of variable length (50, 500 and 1500 bytes) more or less representative of the peaks in packet size distribution in Ethernet. Let p_i be the probability of the different packet sizes and d_i the corresponding transmission time. The mean response time and the mean waiting time of the second queue (15) can be derived using the following parameters:

$$E[S^k] = \sum_i p_i d_i^k , \ S^*(s) = \sum_i p_i e^{-sd_i} \qquad (21)$$

4 Numerical Results

To evaluate the accuracy of the proposed analytical model, we compare the results of this model with those obtained from a simulation conducted on *network simulator 2*. In all our simulations, unless otherwise specified, we assume that (1) all the ring nodes share a common upstream wavelength modulated at 1 Gbit/s ; (2) the packets arrive according to a Poisson process; (3) the arrival rate of the packets to each node is the same; and (4) all the ring nodes transmit only to the hub. In all the figures depicting the simulation results, the traffic load on the x-axis denotes the average traffic load sourced from every node to the hub.

   The analysis results of access delay for each node are presented in Fig. 4, revealing a very good match with the simulation results. We consider packets of variable length

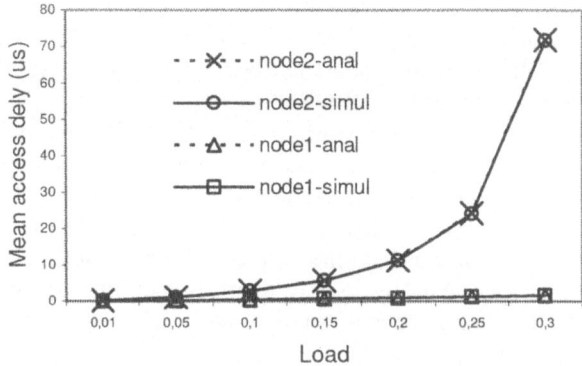

Fig. 4. Mean access delay of the first two nodes with variable-packet size traffic

(50, 500 and 1500 bytes) more or less representative of the peaks in packet size distribution in Ethernet. The total traffic volume comprises 50% of 1500 Bytes, 40% of 500 Bytes and 10% of 50 Bytes packets size. We observe that:
• Under light traffic load, the access delay of the downstream node is more important than upstream node access delay. As a result, the fairness issue is pronounced even under light traffic load.
• Under high traffic load, the difference between the performance of upstream and downstream nodes sharing the optical channel increases. The main reason is that upstream nodes grab more bandwidth thus leaving less capacity to the downstream node.

The analysis results in this special scenario are significant. We observe that even when the upstream node uses a small part of the available bandwidth, the downstream nodes' performance is strongly affected. The fairness issue is always present in shared medium networks. This is mainly due to the lack of organization of the emission in the network and the absence of control mechanisms. In fact, the mismatch between the idle period distribution resulting from upstream node utilisation and the packets' size distribution often leads to bandwidth waste as well as fairness problems with regard to resource access.

The poor performance is not due to the saturation of the medium: For instance, the fairness issue is pronounced even when the channel occupancy is no more than 20%. In fact, the available bandwidth is divided among unusable idle periods (voids) which can not be filled by new packets. This is due to the mismatch between the idle period distribution and the packet size distribution. A finer analysis of the problem shows that it results from the combination of two factors:
1- Wasted voids (i.e. voids smaller than the minimum packet size).
2- Inadequacy of voids with respect to the first packet at the head of the FIFO (First In First Out) insertion buffer of the ring node.

Once a packet of maximum size is at the head of the insertion buffer, it blocks the emission process until finding an adequate void: this is the well-known HOL blocking problem. Thus, sharing the bandwidth arbitrarily between nodes is not sufficient to ensure satisfactory results. The sharing process must thus be done smartly in order to preserve a maximum of useful bandwidth for downstream nodes. In general, fairness control mechanisms limit the transmission of upstream nodes to keep enough bandwidth for downstream stations. These schemes may be efficient in the case of slotted WDM rings. However, they don't perform well in the case of asynchronous transmission based architectures like DBORN.

Hence, we suggest preserving bandwidth (represented by idle periods) by upstream nodes in order to satisfy downstream nodes requirements in an organized way. A basic rule consists of avoiding random division of the resource which would lead to inadequacy between idle periods length and the layer 2 PDUs (Protocol Data Units) size. Therefore the control mechanism has to prevent greedy upstream stations from taking more than their fair share by forcing them to keep idle periods of sufficient size. In other words, the control mechanism required in this context has to regulate the distribution of idle and busy periods.

In this study, the analytical model handles the case of two-node ring for simplicity. However, it can be easily extended to more nodes without major modifications but still approximations have to be accounted for. Since the performance of an access node depends only on the upstream nodes of the ring, we assume all of them as M/G/1 system. Afterwards we can recursively apply exactly the same method. So, the method consists on aggregating all the upstream traffics in a single flow. The packets of the aggregated flow arrive according to a Poisson process. Then, we analyze system as a two-node ring.

The analysis results for the general case of six-node bus, depicted in the Fig. 5, emphasize the above mentioned results. The traffic load ρ sourced by each node is 0,05. The access delay of each node is found to increase monotonically when progressing towards the hub. Indeed, the closest nodes to the hub encounter relatively large delays, incompatible with the performance expected in metropolitan networks. In contrast, upstream nodes (numbers 1 to 4), which grab gradually the available bandwidth, are satisfied and benefit from short access delay. We insist that the performance degradation of downstream nodes is not due to the medium saturation since the medium occupation is not beyond 30%. This example exhibits the fairness issue already seen in the particular case of two nodes. The analytical model curve is very close to the simulation result curve. So, the approximate analytical model can achieve high accuracy. But, we make the observation that the model becomes less accurate for the closest nodes to the hub. This negligible difference, well below 5% , is expected since the approximations concern more and more nodes when coming near the hub.

The accuracy of the analytical model is studied further in the Fig. 6, which presents the evolution of the access delay of node 3 in function of the load ρ . This plot shows that approximate method is highly accurate when the load of each node is relatively low. However, this model becomes less accurate with the increase of the load.

In this study, an exact analytical model handling the case of two-node ring is studied revealing a good match with the simulation results. However, it can be easily extended to more nodes without major modifications but still approximations have to be accounted for. The approximate model is shown to be accurate but we point out that it loses slightly this accuracy with the increase of the load and the number of ring nodes.

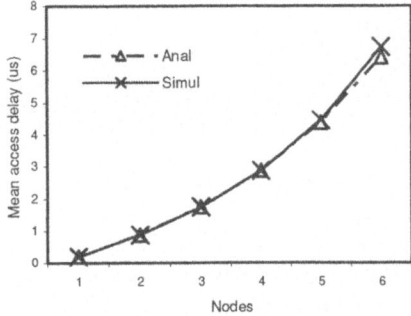

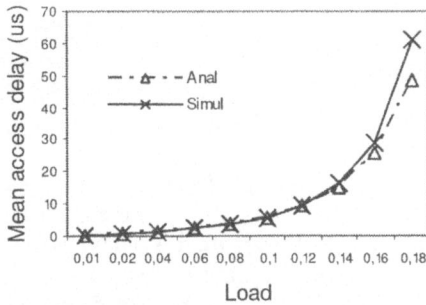

Fig. 5. Mean access delay of the six-node bus with variable-packet size traffic

Fig. 6. Mean access delay of the third node of the shared bus with variable-packet size traffic

5 Conclusion

This paper, to the author's knowledge, provides the first analysis of shared bus network behavior with asynchronous transmission. We analyzed the system performance in terms of access delay required by each node to inject a packet on the shared medium. The analysis results showed that fairness issues are likely to arise between upstream and downstream nodes even under light loads. We observed that sharing the available bandwidth fairly and arbitrarily between nodes does not resolve the fairness problem. Consequently, an additional flow control mechanism has to be considered, not only to limit the transmission of the upstream nodes but also to organize their emission process.

References

1. N. Le Sauze et al. : A novel, low cost optical packet metropolitan ring architecture. Proc. Of ECOC '01, Amsterdam, Netherlands, Vol. 4 (October 2001) 66-67
2. M. A. Marsan, A. Bianco, E. Leonardi, M. Meo, and F. Neri: MAC protocols and fairness control in WDM multirings with tunable transmitters and fixed receivers. IEEE/OSA J. Ligh. Tech., vol. 14 (June 1996) 1230-1244
3. A. Fumagalli, M. Johansson, and M. Roughan: A token-based protocol for integrated packet and circuit switching in WDM rings. Proc. of GLOBECOM '98, Sydney, Australia (November 1998) 2339-2344
4. M. A. Marsan, A. Bianco, E. Leonardi, F. Neri, and S. Toniolo: Metaring Fairness Control Schemes in All-Optical WDM Rings. Proc. of INFOCOM '97, Kobe, Japan, vol. 2 (April 1997) 752-760
5. R. Gaudino et al.: RINGO: a WDM Ring Optical Packet Network Demonstrator. Proc. of ECOC '01, Amsterdam, Netherlands, Vol. 4 (September 2001) 620-621
6. H. Takagi: Queueing Analysis Vol I: Vacation and Priority Systems Part I. North Holland (1991)
7. L. Kleinrock: Queueing Systems Volume I: Theory. John Wiley & Sons (1975)
8. R.W. Wolff: Poisson Arrivals See Time Averages. Operations Research, Vol. 20 (1982) 223-231

Loss Strategies for Competing TCP/IP Connections*

Eitan Altman[1], Rachid El-Azouzi[2], David Ros[3], and Bruno Tuffin[4]

[1] INRIA, 2004 route des Lucioles, 06902 Sophia Antipolis, France
Eitan.Altman@sophia.inria.fr
[2] Université d'Avignon et des Pays de Vaucluse (IUP), LIA-CERI,
339 chemin des Meinajariès, BP1228, 84911 Avignon Cedex 9, France
Rachid.Elazouzi@lia.univ-avignon.fr
[3] GET/ENST Bretagne, Rue de la Châtaigneraie, CS 17607,
35576 Cesson Sévigné Cedex, France
David.Ros@enst-bretagne.fr
[4] IRISA/INRIA, Campus Universitaire de Beaulieu, 35042 Rennes Cedex, France
Bruno.Tuffin@irisa.fr

Abstract. We study in this paper two competing TCP connections that share a common bottleneck link. When congestion occurs, one (or both) connections will suffer a loss that will cause its throughput to decrease by a multiplicative factor. The identity of the connection that will suffer a loss is determined by a randomized "loss strategy" that may depend on the throughputs of the connections at the congestion instant. We analyze several such loss strategies. After deriving some results for the general asymmetric case, we focus in particular on the symmetric case and study the influence of the strategy on the average throughput and average utilization of the link. As the intuition says, a strategy that assigns a loss to a connection with a higher throughput is expected to give worse performance since the total instantaneous throughput after a loss is expected to be lower with such a strategy. We show that, surprisingly, the average throughput and average link utilizations are *invariant*: they are the same under any possible strategy; the link utilization is 6/7 of the link capacity. We show, in contrast, that the second moment of the throughput does depend on the strategy.

1 Introduction

The mathematical analysis of the performance of TCP has been a major research area in networking. Different types of approaches have been suggested and validated. On the one hand, there have been models focusing on a single connection that is subject to some exogenous loss process (which does not depend on that connection), see e.g. [1]. This approach is appealing when there is a large amount of traffic, so that we can neglect the effect of the single connection on events that cause losses. An alternative approach is necessary when the window increase of a

---

* This work was partially supported by INRIA's "TCP" Cooperative Research Action.

N. Mitrou et al. (Eds.): NETWORKING 2004, LNCS 3042, pp. 926–937, 2004.
© IFIP International Federation for Information Processing 2004

connection is itself a central cause for losses. This occurs typically when a small number of connections compete over bandwidth, say, at a bottleneck link. A main mathematical approach for studying this situation has been to study several connections sharing a bottleneck, and then make the simplifying assumption that all connections reduce their windows simultaneously upon congestion [2,3, 4]. With this approach, it has been shown [4] that the throughput achieved by a TCP connection is inversely proportional to RTT^α with $1 < \alpha < 2$, where RTT is the two-way propagation delay of the connection. However, it turns out that in practice this assumption does not hold, except for drop tail buffers and connections with similar Round Trip Times (RTTs) [5]. Indeed, traces in [2] (e.g. Fig. 5) show that the synchronization assumption is invalid for asymmetric connections for a drop tail buffer.

Instead of considering synchronization, two modeling approaches have been developed for determining which connection will suffer a packet loss. In the model of Baccelli and Hong [6], the probability that a connection will lose a packet is a constant: it does not depend on its current throughput. As argued in [7], such an assumption is valid in describing AIMD protocols in which packet transmission rates are constant, and the throughput is varied by changing the packet size. An alternative model has been considered in [8] in which the probability that a connection loses a packet is proportional to the throughput at the congestion instant. This is called the "proportional strategy". As validated by simulations [9], this model is appropriate for standard TCP where packet size is constant.

Motivated by these two approaches, we raise the question of what is the throughput of an AIMD protocol as a function of the strategy that determines which connection loses a packet at a congestion instant. We focus on the simple scenario of two competing connections.

Our findings are as follows. We first study the constant probability model in [6]. In that paper, a linear set of stochastic recursive equations has been introduced for obtaining the throughput, in which the state variables correspond to the connections' throughputs after a loss. In this paper we present an alternative set of stochastic recursive equations in which the states correspond to the throughput *just before* the loss occurs. We show that our approach allows us to reduce the dimensionality of the system by one, so in particular, the case of two connections can be described by a one-dimensional state equation. This allows us to obtain an *explicit* expression for the throughput in the general asymmetric case for the constant probability model. As a corollary of this result, it is seen in the symmetric case that the link utilization is 6/7 of its capacity.

We then study a new strategy in which the connection with the larger instantaneous throughput is the one to lose a packet at congestion instants. Surprisingly, we obtain the same average throughput and link utilization in the symmetric case as for the constant probability model. Moreover, this is the same utilization also obtained for the proportional strategy. This motivated us to examine the behavior of an arbitrary strategy. Our main finding is that although the expectation of throughputs at loss instants depend on the strategy, the average throughput is an invariant quantity for the case of symmetric connections.

We finally derive a general expression for the second moment of the throughput and compare the performance of the three strategies mentioned above in the symmetric case, in order to find out which one has the smallest throughput variability.

The structure of the paper is as follows. In Sect. 2 we study the throughput of the constant loss strategy, whereas the Larger Throughput Loss (LTL) strategy is analyzed in Sect. 3. Section 4 then presents some numerical experimentations and comparisons between the strategies. Section 5 studies the average throughputs in the symmetric setting under an arbitrary strategy and obtains the invariance property. Section 6 then provides an expression for the second moment of the throughput under an arbitrary strategy and a comparison for the three aforementioned strategies. We end with a concluding section.

2   Fixed Loss Probabilities: Model and Analysis

2.1   Basic Definitions and Assumptions

This model is based on [6] where an additive increase, multiplicative decrease (AIMD) model is used to describe the joint throughput evolution of a set of TCP sessions sharing a common router bottleneck.

In full generality, let N be the number of TCP sessions competing for bandwidth, and C the capacity of the bottleneck router. Let T_n be the n-th congestion epoch and $\tau_{n+1} = T_{n+1} - T_n$. Let also η_i be the additive increase rate for session i and $\beta^{(i)}$ be its multiplicative decrease rate. Usually, $\beta^{(i)} = 1/2 \; \forall i$ and η_i is taken as the square inverse of the round trip time of session i. We consider here $Y_n^{(i)}$, the throughput of session i *before* the n-th congestion epoch, instead of $X_n^{(i)}$, the throughput after the n-th congestion epoch like in [6].

Denote by $\bar{Y}^{(i)}$ session i's mean throughput. As in [6], let $a_n^{(i)}$ be a Bernoulli random variable with value 1 if session i experiences a loss at the n-th congestion epoch, and 0 otherwise, so that $\mathbb{E}[a_n^{(i)}] = p^{(i)}$. Note that the $a_n^{(i)}$ ($1 \leq i \leq N$) are correlated to make sure that at least one packet is lost at each congestion time. We have

$$Y_{n+1}^{(i)} = \gamma_n^{(i)} Y_n^{(i)} + \tau_{n+1} \eta_i \tag{1}$$

where $\gamma_n^{(i)} = (1 - a_n^{(i)}) + \beta^{(i)} a_n^{(i)}$. As in [6], we assume here that there is a loss as soon as the router capacity is reached, i.e., as soon as

$$\sum_{i=1}^{N} \gamma_n^{(i)} Y_n^{(i)} + \tau_{n+1} \sum_{i=1}^{N} \eta_i = C \; . \tag{2}$$

This assumption will allow us to derive the throughput at the different congestion epochs.

2.2   Computation of the Average Throughput

The goal of this subsection is to derive the average throughput of a session in terms of the loss probabilities when the number of sessions is $N = 2$.

First, using (2), we get the time between the n-th and $(n+1)$-th congestion epochs

$$\tau_{n+1} = \frac{C - \sum_{i=1}^{N} \gamma_n^{(i)} Y_n^{(i)}}{\sum_{i=1}^{N} \eta_i} . \tag{3}$$

Using this relation we are able to derive a closed-form of the average throughput $\bar{Y}^{(1)}$ of session 1. The average throughput $\bar{Y}^{(2)}$ of session 2 can be obtained in the same way (or by switching the indexes 1 and 2 in the following formula).

Proposition 1. *Assume that $N = 2$. If we denote $p^{(12)} = \mathbb{E}(a^{(1)}a^{(2)})$, we obtain*

$$
\begin{aligned}
\bar{Y}^{(1)} = \frac{C}{2} \Bigg(& \Bigg(\frac{\eta_1^2 \left(1 - \beta^{(2)}\right)^2 p^{(2)}}{\left(\eta_1 + \eta_2\right)^2} + 2\eta_1^2 p^{(2)} \left(1 - \beta^{(2)}\right) \times \\
& \times \Bigg(\left(1 - \beta^{(2)}\right) p^{(2)} - \frac{\eta_1 \left(1 - \beta^{(2)}\right)^2 p^{(2)}}{\eta_1 + \eta_2} - \frac{\eta_2 \left(1 - \beta^{(1)}\right)\left(1 - \beta^{(2)}\right) p^{(12)}}{\eta_1 + \eta_2} \Bigg) \times \\
& \times \left(\left(1 - \beta^{(2)}\right)\eta_1 p^{(2)} + \left(1 - \beta^{(1)}\right)\eta_2 p^{(1)} \right)^{-1} \left(\eta_1 + \eta_2\right)^{-1} \Bigg) \times \\
& \times \Bigg(\frac{\eta_1 \left(1 - \beta^{(2)}\right)^2 p^{(2)}}{\eta_1 + \eta_2} - \left(1 + \frac{\eta_2}{\eta_1 + \eta_2}\right)\left(1 - \beta^{(1)}\right)^2 p^{(1)} + \\
& + 2\frac{\eta_2 \left(1 - \beta^{(1)}\right)\left(1 - \beta^{(2)}\right) p^{(12)}}{\eta_1 + \eta_2} - 2\left(1 - \beta^{(2)}\right) p^{(2)} + 2\left(1 - \beta^{(1)}\right) p^{(1)} \Bigg) \times \\
& \times \Bigg(2\frac{\left(1 - \beta^{(1)}\right)\eta_2 p^{(1)}}{\eta_1 + \eta_2} + 2\frac{\left(1 - \beta^{(2)}\right)\eta_1 p^{(2)}}{\eta_1 + \eta_2} - \frac{\eta_1^2 \left(1 - \beta^{(1)}\right)^2 p^{(1)}}{\left(\eta_1 + \eta_2\right)^2} \\
& - \frac{\eta_2^2 \left(1 - \beta^{(2)}\right)^2 p^{(2)}}{\left(\eta_1 + \eta_2\right)^2} - 2\frac{\eta_1 \left(1 - \beta^{(1)}\right) p^{(1)} \eta_2 \left(1 - \beta^{(2)}\right) p^{(12)}}{\left(\eta_1 + \eta_2\right)^2} \Bigg)^{-1} + \\
& + 2\eta_1 p^{(2)} \Bigg(\left(1 - \beta^{(2)}\right) p^{(2)} - \frac{\eta_1 \left(1 - \beta^{(2)}\right)^2 p^{(2)}}{\eta_1 + \eta_2} - \frac{\eta_2 \left(1 - \beta^{(1)}\right)\left(1 - \beta^{(2)}\right) p^{(12)}}{\eta_1 + \eta_2} \Bigg) \times \\
& \times \frac{1 - \beta^{(2)}}{\left(1 - \beta^{(2)}\right)\eta_1 p^{(2)} + \left(1 - \beta^{(1)}\right)\eta_2 p^{(1)}} + \frac{\eta_1 \left(1 - \beta^{(2)}\right)^2 p^{(2)}}{\eta_1 + \eta_2} \Bigg) \times \\
& \times \left(\frac{\eta_1 p^{(2)} + \eta_2 p^{(1)} - \beta^{(2)} \eta_1 p^{(2)} - \beta^{(1)} \eta_2 p^{(1)}}{\left(\eta_1 + \eta_2\right)\left(1 - \beta^{(1)}\right)\left(1 - \beta^{(2)}\right) p^{(1)} p^{(2)}} \right)
\end{aligned}
$$

The proof of this proposition is provided in [10].

Corollary 1. *Still assuming $N = 2$, the symmetric case yields*

$$
\begin{aligned}
\bar{Y}^{(1)} = \frac{C}{4} \times & \frac{2 p^{(1)} - p^{(12)} + p^{(12)} \beta^{(1)}}{p^{(1)2} \left(3 + \beta^{(1)} - p^{(12)} + p^{(12)} \beta^{(1)}\right)} \\
& \times \left(p^{(12)} \beta^{(1)} p^{(1)} + 2 p^{(1)} + 2 p^{(1)} \beta^{(1)} - p^{(1)} p^{(12)} + p^{(12)} - p^{(12)} \beta^{(1)} \right) .
\end{aligned} \tag{4}
$$

Proof. Just replace η_2 by η_1, $\beta^{(2)}$ by $\beta^{(1)}$ and $p^{(2)}$ by $p^{(1)}$ in Proposition 1. $\qquad \square$

2.3   Sampling the Loss Probabilities

The previous expressions of the average throughput are general in the sense that no special sampling structure has been used for the losses. In this section, we aim at studying how the losses can be sampled and how it impacts on the average throughput formula.

Independent Sampling. As in [6], we can assume that the $a_n^{(i)}$ are at first generated independently, such that $\mathbb{P}[a_n^{(i)} = 1] = \pi^{(i)}$, with $\pi^{(i)}$ given, but that the samples are restricted to the domain where at least one loss is experienced. This requires a derivation of $\pi^{(i)}$ in terms of the $p^{(j)}$.

Assuming $N = 2$, we have as in [6]

$$\begin{cases} p^{(1)} = \frac{\pi^{(1)}}{1-(1-\pi^{(1)})(1-\pi^{(2)})} \\ p^{(2)} = \frac{\pi^{(2)}}{1-(1-\pi^{(1)})(1-\pi^{(2)})} \end{cases}$$

where $\pi^{(i)}$ is for the loss probability for user i, sampled independently, but reduced to the domain such that a loss is actually experienced. This gives

$$\begin{cases} \pi^{(1)} = p^{(1)}(\pi^{(1)} + \pi^{(2)} - \pi^{(1)}\pi^{(2)}) \\ \pi^{(2)} = p^{(2)}(\pi^{(1)} + \pi^{(2)} - \pi^{(1)}\pi^{(2)}) \end{cases}$$

We obtain the relation

$$\pi^{(1)} = \frac{p^{(1)}}{p^{(2)}}\pi^{(2)}$$

which gives (assuming $\pi^{(2)} > 0$)

$$\pi^{(2)} = \frac{p^{(1)} + p^{(2)} - 1}{p^{(1)}} \quad \text{and then} \quad \pi^{(1)} = \frac{p^{(1)} + p^{(2)} - 1}{p^{(2)}} \ .$$

Then an assumption $p^{(1)} + p^{(2)} = 1$ cannot be used. Also, it seems difficult to make sure that $\pi^{(1)} \leq 1$ and $\pi^{(2)} \leq 1$ for every pair $(p^{(1)}, p^{(2)})$. Thus, this sampling procedure does not work in full generality.

A Single Loss at Congestion Epochs. The simplest way to sample is by using the relation

$$a_n^{(2)} = 1 - a_n^{(1)}$$

with $a_n^{(1)}$ Bernoulli random variable such that $\mathbb{P}[a_n^{(1)} = 1] = p^{(1)}$. This means that at each congestion epoch, one and only one session will see a decrease of its throughput. We then have

$$p^{(2)} = 1 - p^{(1)} \quad \text{and} \quad p^{(12)} = 0 \ .$$

Substituting these values of $p^{(2)}$ and $p^{(12)}$ in the equation given in Proposition 1, we can obtain a closed, explicit expression for $\bar{Y}^{(1)}$ (not shown for space reasons). The symmetric case (that is, taking $p^{(1)} = p^{(2)} = 1/2$ and $\beta^{(2)} = \beta^{(1)}$) yields

$$\bar{Y}^{(1)} = \frac{\left(1 + \beta^{(1)}\right) C}{3 + \beta^{(1)}} .$$

If $\beta^{(1)} = 1/2$, we obtain $\bar{Y}^{(1)} = \frac{3}{7}C$, like in [8] for the proportional loss strategy.

3   The Largest Throughput Loss (LTL) Strategy

Let us look at the case where the session that is penalized is systematically the one with the largest throughput. We call this the "Largest Throughput Loss" (LTL) strategy. Consider the n-th congestion epoch, with throughputs $Y_n^{(1)}$ and $Y_n^{(2)}$ such that $Y_n^{(1)} + Y_n^{(2)} = C$. Without loss of generality, assume $Y_n^{(1)} > Y_n^{(2)}$ and that the additive increase is 1.

3.1   The Symmetric Case: The Periodic Solution

We identify a periodic solution for the evolution of the system. In this regime, we assume (without loss of generality) that at time n, connection 1 has a larger throughput than connection 2. We seek for a regime in which at time $n+1$ the situation is reversed, and so on. This gives the following dynamics:

$$\begin{cases} Y_n^{(1)}/2 + \tau_{n+1} = Y_n^{(2)} \\ Y_n^{(2)} + \tau_{n+1} = Y_n^{(1)} \\ Y_n^{(1)} + Y_n^{(2)} = C , \end{cases}$$

leading to

$$\tau_{n+1} = \frac{1}{7}C, \quad Y_n^{(1)} = \frac{4}{7}C \quad \text{and} \quad Y_n^{(2)} = \frac{3}{7}C .$$

As in the proof of Proposition 1, but due to the periodicity of the system, the average throughput is given by $S/\mathbb{E}[2\tau]$ where S is the cumulative throughput of a session between congestion epochs n and $n+2$ (in one period, the throughput is going from $2C/7$ to $3C/7$ and in the other one from $3C/7$ to $4C/7$). This gives $S = \frac{12}{98}C^2$, leading again to $\bar{Y}^{(1)} = \bar{Y}^{(2)} = \frac{3}{7}C$ and an average utilization of $\frac{6}{7}$ as we obtained in the previous section and as is the case in the model in [8]. Obviously, $\mathbb{E}[Y_n^{(i)}]$ are also the same in all three cases (and equal to $C/2$). One could wonder whether in fact the distribution of the rates is independent of the way one chooses the connection to decrease the rate at T_n. Note however, that $\mathbb{E}[(Y_n^{(i)})^2] = 25C^2/98$ in our example, which is different than the value of $7C^2/26$ obtained in the regime considered in [8].

3.2   The Dynamic Equations for the Asymmetric Case

For each connection $i = 1, 2$ we have

$$Y_{n+1}^{(i)} = \begin{cases} Y_n^{(i)}/2 + \tau_{n+1}\eta_i & \text{if } Y_n^{(i)} > C/2, \\ Y_n^{(i)} + \tau_{n+1}\eta_i & \text{if } Y_n^{(i)} < C/2. \end{cases} \tag{5}$$

For the case that $Y_n^{(i)} = C/2$ any tie-breaking rule can be considered. Combining this with the relation $Y_n^{(2)} = C - Y_n^{(1)}$ as well as $Y_{n+1}^{(2)} = C - Y_{n+1}^{(1)}$ gives

$$
\tau_{n+1} = \begin{cases} \frac{Y_n^{(i)}}{2(\eta_1 + \eta_2)} & \text{if } Y_n^{(i)} > C/2, \\ \frac{C - Y_n^{(i)}}{2(\eta_1 + \eta_2)} & \text{if } Y_n^{(i)} < C/2. \end{cases}
$$

Substituting in (5) gives

$$
Y_{n+1}^{(i)} = \begin{cases} \frac{1}{2}\left(1 + \frac{\eta_i}{\eta_1 + \eta_2}\right) Y_n^{(i)} & \text{if } Y_n^{(i)} > C/2, \\ \left(1 - \frac{\eta_i}{2(\eta_1 + \eta_2)}\right) Y_n^{(i)} + \frac{C\eta_i}{2(\eta_1 + \eta_2)} & \text{if } Y_n^{(i)} < C/2. \end{cases}
$$

These equations can be used to obtain the exact transient behavior of the system. The average throughput can then be computed by

$$
\bar{Y}^{(i)} = \lim_{n \to \infty} \frac{\sum_{k=1}^{n} \tau_{k+1}(Y_{k+1}^{(i)} + \gamma_k^{(i)} Y_k^{(i)})/2}{\sum_{k=1}^{n} \tau_{k+1}} .
$$

3.3 The Case $\eta_2/\eta_1 \to 0$

We consider here the case of $x \to 0$ where $x := \eta_2/\eta_1$ and assume for simplicity that $\beta^{(i)} = 1/2$. We present a heuristic argument to compute the bandwidth sharing.

Connection 2 will increase its rate until it reaches $C/2$, so its trajectory at steady state will be periodic (with a period of duration of $C/(4\eta_2)$), linearly increasing between $C/4$ to $C/2$. Its average throughput is $3C/8$.

Connection 1 Fix $\Delta = \sqrt{x}/\eta_2$. We can view the problem as one with two time scales: connection 1 is much faster than connection 2, so during the interval $[n\Delta, (n+1)\Delta)$, the throughput of connection 2 can be approximated by a constant which we denote by $Y^{(2)}(n)$; assume that this constant is smaller than $C/2$. During that interval, the throughput of connection 1 will oscillate very quickly (between half of the remaining and all the remaining bandwidth) so that it will use in average over that interval $3/4$ of the remaining bandwidth. Thus its average bandwidth during the interval is $(3/4)(C - Y^{(2)}(n))$, and during the whole period of $C/(4\eta_2)$ it will be $(3/4)(C - 3C/8) = 15C/32$.

Thus as $x \to 0$ we see that the fast connection will get $5/4$ of the throughput of the slow connection under the LTL strategy.

4 Numerical Results for the Fairness in Bandwidth Sharing

We study in this section the fairness in throughput as a function of the round trip times. We recall that the square root formula of TCP as well as its refinements (see [1,11]) predict that the throughput of a connection should be inversely proportional to its RTT. We will compare this with the fairness obtained under our model of interacting connections.

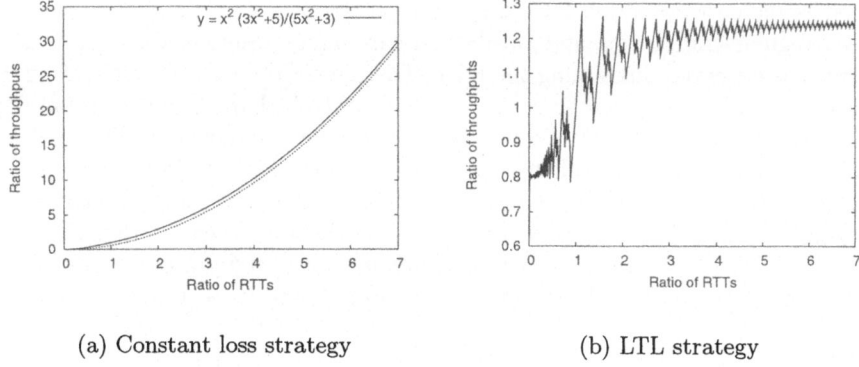

(a) Constant loss strategy   (b) LTL strategy

Fig. 1. The ratio $\bar{Y}_1/\bar{Y}_2$ as a function of the ratio $R^{(2)}/R^{(1)}$

4.1   Constant Loss Strategy

We now look at the ratio $\bar{Y}_1/\bar{Y}_2$ of average throughputs. To simplify the expressions, let us assume that $\beta^{(1)} = \beta^{(2)} = 1/2$ and that $p^{(1)} = p^{(2)} = p \geq 1/2$. We also assume that the linear growth rates are inversely proportional to the square of the round trip times, i.e., $\eta_i = 1/(R^{(i)})^2$ for $i = 1, 2$. (Indeed, the window increases by one each RTT, and since the throughput is given by the window size divided by the RTT, the increase rate of the throughput is $1/RTT^2$.)

We then obtain from (4) that

$$\frac{\bar{Y}_1}{\bar{Y}_2} = \left(\frac{R^{(2)}}{R^{(1)}}\right)^2 \frac{3p\frac{R^{(2)}}{R^{(1)}} + 5p - 2p^{(12)}}{5p\left(\frac{R^{(2)}}{R^{(1)}}\right)^2 + 3p - 2p^{(12)}\left(\frac{R^{(2)}}{R^{(1)}}\right)^2} .$$

If we further assume that exactly one flow will experience a loss, then we have $p^{(12)} = 0$ and $p = 1/2$ giving

$$\frac{\bar{Y}_1}{\bar{Y}_2} = \left(\frac{R^{(2)}}{R^{(1)}}\right)^2 \cdot \frac{3\left(\frac{R^{(2)}}{R^{(1)}}\right)^2 + 5}{5\left(\frac{R^{(2)}}{R^{(1)}}\right)^2 + 3} .$$

We show the fairness in throughputs for the fixed loss strategy in Fig. 1(a). Note that the ratio of average throughputs is very close to be linear in the square of the ratio of round trip times (the dotted line depicts the function $y = 3x^2/5$).

4.2   The LTL Strategy

In Fig. 1(b) we depict the throughput ratios as a function of the ratio of the inverse of the square of RTTs for the LTL strategy. The values are obtained by computing the throughput as in Sect. 3.2.

We observe that although in general the throughput has a tendency to increase as the corresponding RTT decreases, we see that the throughput curve is quite irregular and fractal, and locally there are many points where the opposite behavior is observed: increasing the RTT of a connection results in increasing its throughput. This can perhaps be explained in part by changes in the periodicity of the steady-state behavior and in other discrete nature behavior. The analysis of this phenomenon is beyond the scope of this paper. We note that other fractal aspects of AIMD connections in networks with several nodes have already been reported in [12]. We finally observe that as the RTT of a connection becomes negligible with respect to the other, its share of the throughput converges to 5/4 of the throughput of the other connection, as predicted in Sect. 3.3.

4.3   Comparisons

We first observe that the throughput sharing in the LTL strategy is much more fair than in the probabilistic sharing: it is much less sensible to the differences in RTT. Indeed, a connection with 3 times smaller RTT gets only 1.21 times more throughput in the LTL strategy, whereas it gets 6 times more throughput in the case of the constant probabilities strategy.

The fairness behavior of the proportional drop strategy has already appeared in [9], where the connection with 3 times smaller RTT gets 2.75 times more throughput. Comparing to these results we see that, in terms of fairness, the LTL strategy gives the best results whereas the worse performance is provided by the fixed loss probabilities strategy.

The behavior of the throughput as the ratio of RTTs goes to zero is in particular interesting. The throughput of the long connection and its share of the throughput tend to zero in the constant loss strategy, as well as with the proportional strategy [8, Sect. 7-8], whereas it tends to a positive constant under the LTL strategy.

Note that the fact that we obtain different average throughput sharing under different policies reflects the fact that, in contrast to the symmetric case, the throughput is not invariant with respect to the strategy in the general asymmetric case.

5   The Symmetric Case: Invariance of the Throughput for a General Strategy

Consider now a general strategy for deciding which connection will decrease its rate when capacity is reached. The decrease is by a constant β and the increase rate is η. We still restrict ourselves to the symmetric case of two connections, and assume that one and only one connection decreases its rate when the capacity is reached. At such a moment, connection 1 that transmits at a rate of y will decrease its rate with probability $f(y)$ and connection 2 will decrease its rate with probability $1 - f(y)$. We assume that the rate process of both connections is in a stationary ergodic regime. In particular we shall focus again on $Y_n^{(1)}$, the rate of connection 1 just before a rate decrease occurs.

Let us state one of the main results of the paper in the following proposition.

Proposition 2. *The average throughput $\bar{Y}$ of a connection in a symmetric network with two connections is given by*

$$\bar{Y} = \frac{1+\beta}{3+\beta}C \ , \tag{6}$$

independent of the sampling function f.

Proof. The proof is quite involved, so for space reasons we will roughly sketch its different steps; the full proof can be found in [10].

First, we focus in the throughput process $Y_n := Y_n^{(1)}$. In particular, we compute the cumulative throughput S between congestion epochs for two cases (y denotes the state of Y_n at time T_n):

- Connection 1 is the one to decrease its rate (this happens with probability $f(y)$), in which case we have: $S = \frac{1}{2}(\beta Y_n + Y_{n+1})\tau_{n+1} = y^2 \frac{(1+3\beta)(1-\beta)}{8\eta}$.
- Connection 2 is the one to decrease its rate (this occurs with probability $1 - f(y)$). In this case, $S = \frac{1}{2}(Y_n + Y_{n+1})\tau_{n+1} = \frac{1-\beta}{8\eta}(-(3+\beta)y^2 + 2C(1 + \beta)y + C^2(1-\beta))$.

Next, we compute the expected time interval between congestion epochs: $\mathbb{E}[\tau] = \mathbb{E}\left[Y\frac{1-\beta}{2\eta}f(Y) + \frac{(C-Y)(1-\beta)}{2\eta}(1 - f(Y))\right] = (1-\beta)\mathbb{E}[Yf(Y)]/\eta$, where Y denotes a random variable distributed like Y_n at steady-state. Remark that $\mathbb{E}[\tau]$ depends on the expectation $\mathbb{E}[Yf(Y)]$.

The average throughput $\bar{Y}$ is given by: $\bar{Y} = \mathbb{E}[S]/\mathbb{E}[\tau]$. Hence, we need to compute $\mathbb{E}[S]$, which can be expressed as: $\mathbb{E}[S] = \frac{1-\beta}{8\eta}\Big(-(3+\beta)\mathbb{E}[Y^2] + 4(1+\beta)\mathbb{E}[Y^2f(Y)] - 2C(1+\beta)\mathbb{E}[Yf(Y)] + C^2\frac{3+\beta}{2}\Big)$.

We then obtain three expressions relating both the three unknowns that appear in the formula of $\mathbb{E}[S]$, that is: $\mathbb{E}[Yf(Y)]$, $\mathbb{E}[Y^2]$, $\mathbb{E}[Y^2f(Y)]$, and the quantity $\mathbb{E}[Y^3]$ (hence, we have four unknowns and three equations). It happens that $\mathbb{E}[S]$ can be expressed as a function only of $\mathbb{E}[Yf(Y)]$, that is: $\mathbb{E}[S] = \frac{(1-\beta)(1+\beta)\mathbb{E}[Y(f(Y)]C}{\eta(3+\beta)}$. Therefore, the term $\mathbb{E}[Yf(Y)]$ cancels out when dividing $\mathbb{E}[S]$ by $\mathbb{E}[\tau]$, which gives (6).

$\square$

6   The Symmetric Case: Second Moment of the Throughput

Even if all possible loss strategies provide the same average throughput in steady-state in the symmetric case we can wonder about the variability of the throughput. In real-time applications that may use AIMD protocols in order to be TCP-friendly, it is clearly advantageous to have the lowest possible throughput variability.

The following Proposition gives a general expression for the second moment of the throughput. As will be seen, this expression is not invariant any more, in contrast to the first moment.

Proposition 3. *Let* $\mathbb{E}[S_2]$ *denote the mean cumulative of the square throughput between two loss epochs. The (average) second moment of throughput is*

$$\frac{\mathbb{E}[S_2]}{\mathbb{E}[\tau]} = \frac{1}{8}(3 + 2\beta + 3\beta^2)\frac{\mathbb{E}[Y^3 f(Y)]}{\mathbb{E}[Y f(Y)]} .$$

The proof of this proposition, which follows along the same lines as that of Proposition 2, is given in [10].

Since we still have two unknowns, one could argue that their ratio is constant. Actually, it is not the case from the following proposition where we compare the second order moment for the three loss strategies (constant, proportional or largest flow).

Proposition 4. *Let* $\beta = 1/2$. *Using the constant loss probability scheme, we get*

$$Q_{cst} = \frac{\mathbb{E}[S_2]}{\mathbb{E}[\tau]} = \frac{95}{448}C^2 \approx 0.21C^2$$

whereas when the loss is applied to the largest flow (LTL strategy), we have

$$Q_{ltl} = \frac{\mathbb{E}[S_2]}{\mathbb{E}[\tau]} = \frac{4}{21}C^2 \approx 0.19C^2$$

and the scheme with proportional losses gives

$$Q_{pro} = \frac{\mathbb{E}[S_2]}{\mathbb{E}[\tau]} = \frac{469373}{1467072}C^2 \approx 0.32C^2 .$$

We see from the proposition that in the symmetric case, the LTL strategy is to be preferred (in terms of lower second moment), whereas the strategy of losses proportional to the throughput has the worse performance. The proof of this proposition, omitted for space reasons, is given in [10].

7   Discussion and Future Research

We have introduced in this paper various loss strategies that determine which connection will lose a packet when a congestion occurs. We have shown that such loss strategies may have a considerable impact on the throughput variability (which may be an important performance measure in real-time applications that use AIMD protocols to be TCP-friendly) but that they all lead to the same average throughput in the special case of a symmetric network with two connections. Among three specific strategies that we introduced, we have shown in the above setting that the LTL strategy (i.e., the strategy that drops a packet from the connection with highest throughput) has the best performance in terms of throughput variability, and moreover, it guarantees a positive share of the throughput even when the RTT of one of the connections becomes arbitrarily large.

The mathematical study of the sharing of bandwidth under various loss strategies turns out to be quite involved. So far we have not been able to get explicit expressions for the asymmetric network with two connections when the LTL or the proportional loss strategies are used. We have provided however an (involved) explicit expression for the throughput for the case of constant loss strategy. For the symmetric case, however, we have obtained an explicit expression for the throughput under an arbitrary loss strategy.

Many open problems remain: 1. Is there any probabilistic argument that can explain the invariance of the average throughput in the loss strategy phenomenon in the case of two connections? 2. Does the invariance of the throughput holds for the case of more than two competing symmetric connections? 3. What is the reason for the fractal behavior of the throughput sharing under LTL? 4. How to implement LTL? Note that a desirable way of implementation should be stateless, and it should make use only of local information available at the bottleneck element.

References

1. Padhye, J., Firoiu, V., Kurose, J., Towsley, D.: Modeling TCP Throughput: A Simple Model and its Empirical Validation. In: Proceedings of ACM SIGCOMM 98. (1998) 303–314
2. Ait-Hellal, O., Altman, E., Elouadghiri, D., Erramdani, M., Mikou, N.: Performance of TCP/IP: The Case of Two Controlled Sources. In: Proceedings of ICCC'97, Cannes (1997) 469–477
3. Brown, P.: Resource Sharing of TCP Connections with Different Round Trip Times. In: IEEE Infocom. (2000)
4. Lakshman, T., Madhow, U.: The Performance of TCP/IP for Networks with High Bandwidth-Delay Products and Random Loss. IEEE/ACM Transactions on Networking (1997)
5. Zhang, L., Shenker, S., Clark, D.: Observations on the Dynamics of a Congestion Control Algorithm: The Effects of Two-Way Traffic. In: ACM SIGCOMM. (1991)
6. Baccelli, F., Hong, D.: AIMD, Fairness and Fractal Scaling of TCP Traffic. In: Proceedings of IEEE INFOCOM 02. (2002)
7. Vojnovic, M., Le-Boudec, J.Y.: Some Observations on Equation-Based Rate Control. In: Proceedings of ITC-17, Salvador da Bahia (2001)
8. Altman, E., Jiménez, T., Núñez-Queija, R.: Analysis of Two Competing TCP/IP Connections. Performance Evaluation 49 (2002) 43–56
9. Altman, E., Barakat, C., Laborde, E., Brown, P., Collange, D.: Fairness Analysis of TCP/IP. In: Proceedings of IEEE CDC'00, Sydney (2000)
10. Altman, E., El Azouzi, R., Ros, D., Tuffin, B.: Loss Strategies for Competing TCP/IP Connections. Research report RR-5100, INRIA (2004)
11. Altman, E., Avratchenkov, K., Barakat, C.: A Stochastic Model of TCP/IP with Stationary Random Losses. In: Proceedings of ACM SIGCOMM 2000. (2000)
12. Baccelli, F., Hong, D.: Interaction of TCP Flows as Billiards. Research Report RR-4437, INRIA (2002)

Modelling TCP Throughput and Fairness

Douglas J. Leith and Robert Shorten

Hamilton Institute, NUI Maynooth, Ireland

Abstract. Since TCP traffic is elastic, a fundamental issue is the behaviour of multiple TCP flows competing for bandwidth on a shared link. Motivated by the ubiquity of drop-tail queueing in current networks, our focus in this paper is on developing analytic models suited to characterising the throughput and fairness of competing TCP flows in drop-tail environments. Building on recent ideas from the theory of positive linear systems, we obtain simple, insightful closed-form expressions for throughput and fairness. The accuracy of these expressions is confirmed in extensive simulations across a range of network conditions. In particular, they are found to provide accurate estimate of mean fairness and throughput even when flows are not synchronised.

1 Introduction

TCP traffic continues to account for the majority of traffic on the internet. Since TCP traffic is elastic, a fundamental issue is the behaviour of multiple TCP flows competing for bandwidth on a shared link. A variety of fluid models have been studied, but while they have been found to provide useful insight and predictive power for active queueing disciplines such as RED (see, for example [1], [2]), their utility for drop-tail queuing has yet to be established. Our focus in this paper is on developing analytic models that characterise the throughput and fairness of competing TCP flows in drop-tail environments. This is motivated by the ubiquity of drop-tail queueing in current networks. With drop-tail queueing and a single TCP flow, the seminal work by Padhye et al [3] establishes a simple approximate model of the mean transmit rate as a function of network parameters (round-trip time, bandwidth, drop probability etc). However, while a number of empirical studies have been reported (e.g. [4], [5]), few analytic results for multiple flows sharing a link are available. Notable exceptions include the work of Chiu and Jain [6] in an abstract setting, and more recently that of Brown [7]. The latter is of most relevance to the the present paper.

While [7] adopts a fluid approach, here we take a modelling approach based on new results in [8] using ideas from the theory of positive systems to analyse network dynamics. In this paper we extend the results in [8] to account for buffering and flows with different round trip times. This allows us to obtain simple, insightful closed-form expressions for fairness and throughput. Both the model in [8] and that used in [7] assume synchronisation. The synchronisation

N. Mitrou et al. (Eds.): NETWORKING 2004, LNCS 3042, pp. 938–948, 2004.

assumption is that all flows experience a drop when the network "pipe" becomes full. Clearly, this is an unrealistic assumption in real networks. In this paper we show that the fairness and throughput predicted under the assumption of synchronisation, while inaccurate with respect to instantaneous values, provides a rather accurate estimate of the *mean* fairness and throughput even when flows are not synchronised.

The paper is organised as follows. We begin in Section 2 by briefly considering the case of a single TCP flow. In Section 3, we present our main result, an analytic expression for the throughput efficiency of TCP flows competing for shared bandwidth the accuracy of which is verified by extensive packet-level simulations. We conclude in Section 4 by summarising the results presented.

2   Preliminaries: Throughput of a Single TCP Flow in a Buffered Network

We begin by briefly considering the case of a single TCP flow. This was previously considered by Padhye et al [3] and others. Unlike this earlier work, here we explicitly take account of the influence of buffering on the throughput efficiency of a flow. The latter is defined here as the rate at which packets leave the bottleneck link normalised by the bandwidth of the bottleneck link; that is, the efficiency is 100% when the link operates continuously at its maximum bandwidth.

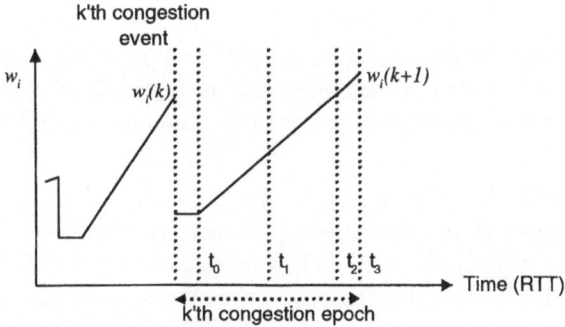

Fig. 1. Congestion window time history.

The single flow case with buffering may be analysed as follows. With reference to Figure 1, let $t_0(k)$ be the time at the beginning of the kth congestion epoch, $t_1(k)$ the time when queue starts to fill, $t_2(k)$ the time when a drop occurs and $t_3(k)$ the time when the source detects the drop. We use $w_i(k)$ to denote the source TCP congestion window immediately before backoff and note that for a single flow $w_i(k)$ equals the "pipe" size $P = BT + q_{max}$, with B the link bandwidth, T the propagation delay and q_{max} the queue size. Let α, β be the AIMD increase and decrease parameters respectively, and define the provisioning parameter $\gamma = q_{max}/BT$ where B is the bandwidth of bottleneck link, q_{max} the buffer size and T is the end-to-end propagation delay. For simplicity, assume that $1/(1+\gamma) \leq \beta$ i.e. that the queue empties at backoff. We immediately have that

$$t_1 - t_0 = (1/(1+\gamma) - \beta)P/\alpha \tag{1}$$
$$t_2 - t_1 = (1 - 1/(1+\gamma))P/\alpha \tag{2}$$
$$t_2 - t_0 = (1 - \beta)P/\alpha \tag{3}$$

That is,

$$(t_1 - t_0)/(t_2 - t_0) = \frac{(1/(1+\gamma) - \beta)}{(1-\beta)}, \quad (t_2 - t_1)/(t_2 - t_0) = \frac{(1 - 1/(1+\gamma))}{(1-\beta)} \tag{4}$$

The throughput (packets per second) averaged over one congestion epoch, y, is

$$y = 0.5(1/(1+\gamma) + \beta)(P/T)(t_1 - t_0)/(t_2 - t_0) + B(t_2 - t_1)/(t_2 - t_0) \tag{5}$$

(where, for simplicity, we have neglected the contribution from the one RTT interval between the drop time t_2 and the time t_3 that this is detected at the source). Substituting from (4) and normalising by B yields the efficiency

$$\eta = \frac{1}{(1-\beta)}[(0.5 + \gamma)/(1+\gamma) - 0.5\beta^2(1+\gamma)] \tag{6}$$

where we have made use of the fact that $P/BT = 1+\gamma$ to eliminate P from the expression. The accuracy of (6) can be seen from Figure 2, where the theoretical prediction (6) is compared against packet-level simulation results from ns-2 [9].

Comment: *RTT dependence.* In (6) the efficiency is determined by the AIMD backoff factor β and the queue provisioning parameter γ. Notice that $\gamma = q_{max}/BT$ is inversely proportional to propagation delay, T. Hence, by (6), for given bandwidth B and queue size q_{max} the efficiency decreases as the propagation delay T increases. Consider the data shown in Figure 3, which plots simulation results illustrating the change in efficiency as RTT varies. Re-plotting the same data against γ rather than RTT yields the points marked by o in Figure 2.

3 Fairness and Throughput of Competing TCP Flows

We extend our analysis to n flows with a shared bottleneck link as follows. Let $w_i(k)$ denote the congestion window of flow i (the total number of packets in

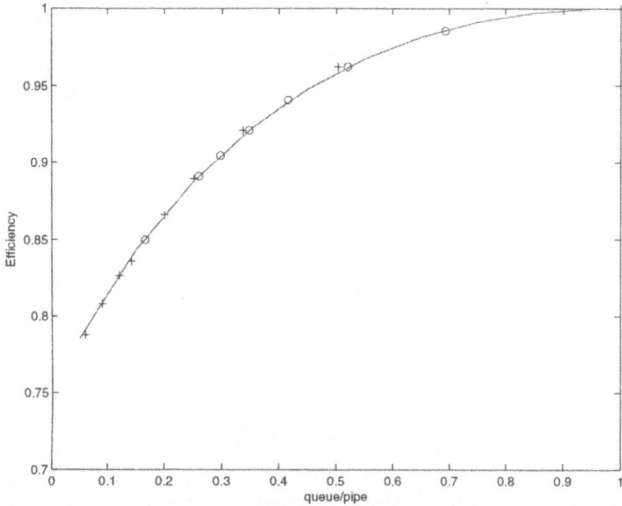

Fig. 2. Efficiency of TCP vs ratio of queue to delay-bandwidth product. The solid line is the theoretical efficiency curve given by (6). Key: + denotes simulation data as queue size varies (bandwidth 100Mb, RTT 40ms, queue is varied from 330 to 20 packets), o denotes data from Figure 3.

flight) immediately before the kth backoff event, α_i denote the AIMD increase parameter of flow i (number of extra packets added to the network per second) and β denote the backoff factor (which is assumed to be the same for all flows). With knowledge of these quantities, we can apply essentially the same analysis as in the single flow case to derive an expression for the efficiency of flow i. In the single flow case, $w(k)$ is simply the pipe size. With competing flows, to determine $w_i(k)$ we need to establish how the available bandwidth is divided up between the flows.

3.1   Distribution of Bandwidth between Flows

We proceed as follows. Let B denote the bandwidth of the bottleneck link, q_{max} the queue size and $\bar{T}$ the propagation delay of the bottleneck link. Let T_i denote the combined propagation delay of all other (non-congested) links traversed by flow i, see Figure 4. The queue at the bottleneck link does not begin to fill until

$$\sum_{i=1}^{n} cwnd_i(t_1)/(\bar{T} + T_i) = B \tag{7}$$

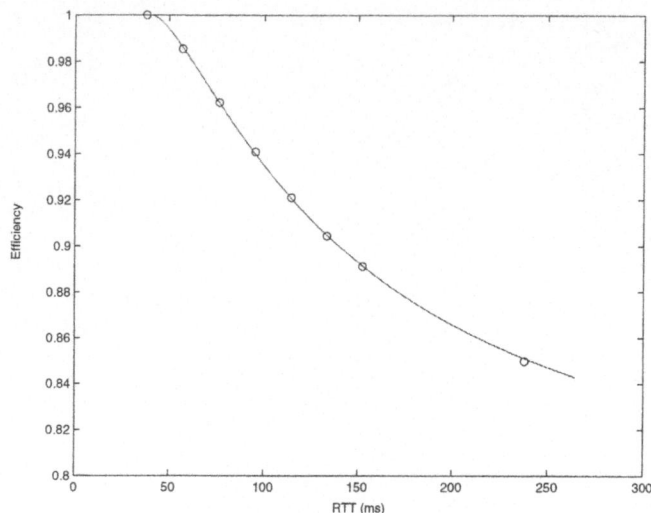

Fig. 3. Efficiency of TCP as RTT varies (bandwidth 100Mb, queue 330 packets). The solid line is the corresponding theoretical prediction (6).

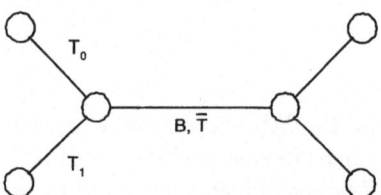

Fig. 4. Dumbell topology

where $cwnd_i(t)$ is the instantaneous value of the congestion window of flow i at time t and t_1 is the time when the queue just begins to fill. We re-write this as

$$\sum_{i=1}^{n} \overline{cwnd}_i(t_1) = B\bar{T} \tag{8}$$

with $\overline{cwnd}_i(t) = [\bar{T}/(\bar{T}+T_i)]cwnd_i(t)$; we might interpret $\overline{cwnd}_i(t)$ as the number of packets in flight in the congested link and $B\bar{T}$ is the delay-bandwidth product of this link. The pipe then becomes full at time t_2 after an additional q_{max} packets are added,

$$\sum_{i=1}^{n} \overline{cwnd}_i(t_2) = B\bar{T} + \bar{q}_{max} \tag{9}$$

where $\bar{q}_{max} = [\sum_{i=1}^{n} \bar{\alpha}_i / \sum_{i=1}^{n} \alpha_i] q_{max}$ and $\bar{\alpha}_i = [\bar{T}/(\bar{T}+T_i)]\alpha_i$. Observe that $\bar{\alpha}_i$ can be interpreted as the number of extra packets added to the *bottleneck* link by flow i every second.

Similarly to the single flow case, let $t_0^i(k)$ be the start of the kth congestion epoch for flow i, $t_2^i(k)$ be the time at which a drop occurs and $t_3^i(k)$ the time at which the source detects the loss. Let $w_i(k)$ denote the congestion window size of flow i immediately before backoff in the kth congestion epoch . Source i detects this packet loss one RTT, i.e. $T_i + \bar{T} + q_{max}/B$, later at time $t_3^i(k)$. Assuming for the moment (this assumption will be dropped later) that drops are synchronised, i.e. $t_2^i(k) = t_2(k) \forall i$, and neglecting the impact of RTT variation due to queueing,

$$\bar{w}_i(k+1) = \beta \bar{w}_i(k) + \bar{\alpha}_i(t_3^i(k+1) - t_0^i(k+1)) \tag{10}$$

and

$$t_0^i(k+1) = t_2(k) + 2RTT_i \tag{11}$$

$$t_2(k+1) = t_2(k) + 2\tau + \frac{\sum_{i=1}^{n}(1-\beta)\bar{w}_i(t_3^i(k)) - \sum_{i=1}^{n}\bar{\alpha}_i(2\tau - RTT_i)}{\sum_{i=1}^{n}\bar{\alpha}_i} \tag{12}$$

$$t_3^i(k+1) = t_2(k+1) + RTT_i \tag{13}$$

where $RTT_i = T_i + \bar{T} + q_{max}/B$, $\tau = \max_i(T_i + \bar{T} + q_{max}/B)$ and we have made use of

$$\bar{P} = B\bar{T} + \bar{q}_{max} = \sum_{i=1}^{n} \bar{w}_i(k) - \bar{\alpha}_i RTT_i \tag{14}$$

Hence,

$$t_3^i(k+1) - t_0^i(k+1) = \frac{\sum_{i=1}^{n}(1-\beta)\bar{w}_i(k)}{\sum_{i=1}^{n}\bar{\alpha}_i} + \epsilon \tag{15}$$

where $\epsilon = (2\tau - RTT_i) - \frac{\sum_{i=1}^{n}\bar{\alpha}_i(2\tau - RTT_i)}{\sum_{i=1}^{n}\bar{\alpha}_i}$. The bottleneck link population of packets therefore evolves according to

$$\bar{w}_i(k+1) = \beta \bar{w}_i(k) + \frac{\bar{\alpha}_i}{\sum_{i=1}^{n}\bar{\alpha}_i}[\sum_{i=1}^{n}(1-\beta)\bar{w}_i(k)] + \bar{\alpha}_i \epsilon \tag{16}$$

When ϵ is sufficiently small that it may be neglected (e.g. when the congestion epoch duration is sufficiently long), the dynamics (16) constitute a positive linear system and the analysis from [8] may be immediately applied. In particular, the fixed point is determined by the Perron eigenvector and we have $\bar{w}_i = [\bar{\alpha}_i / \sum_{j=1}^{n} \bar{\alpha}_j]\bar{P}$. Using $\alpha_i = 1/(\bar{T} + T_i)$, then we have that in steady-state

$$\bar{w}_i = \frac{(\bar{T} + T_i)^2}{\sum_{j=1}^{n}(\bar{T} + T_j)^2}\bar{P} \tag{17}$$

3.2 Throughput of Flow i

As before, define $t_1^i(k)$ to be the time during the kth congestion epoch when the queue just begins to fill. We have that

$$(t_1^i - t_0^i)/(t_2^i - t_0^i) = \frac{(1/(1+\bar{\gamma}) - \beta)}{(1-\beta)}, \ (t_2^i - t_1^i)/(t_2^i - t_0^i) = \frac{(1 - 1/(1+\bar{\gamma}))}{(1-\beta)} \quad (18)$$

where $\bar{\gamma}$ is the provisioning parameter of the bottleneck link,

$$\bar{\gamma} = \bar{q}_{max}/B\bar{T} \quad (19)$$

The throughput (packets per second) of flow i averaged over one congestion epoch is

$$y_i = 0.5(\beta + \frac{1}{1+\bar{\gamma}})\frac{\bar{w}_i}{\bar{T}}\frac{(t_1^i - t_0^i)}{(t_2^i - t_0^i)} + \frac{1}{1+\bar{\gamma}}\frac{\bar{w}_i}{\bar{T}}\frac{(t_2^i - t_1^i)}{(t_2^i - t_0^i)} \quad (20)$$

Hence, the efficiency is

$$\eta_i = \frac{1}{(1-\beta)(1+\bar{\gamma})}\frac{\bar{w}_i}{B\bar{T}}[\frac{0.5 + \gamma}{1+\bar{\gamma}} - 0.5\beta^2(1+\gamma)] \quad (21)$$

Observing that $B\bar{T} = (1 + \bar{\gamma})\bar{P}$ and substituting from (17) for $\bar{w}_i/B\bar{T}$ yields

$$\eta_i = \frac{\phi_i}{(1-\beta)}[\frac{0.5 + \bar{\gamma}}{1+\bar{\gamma}} - 0.5\beta^2(1+\bar{\gamma})] \quad (22)$$

where $\phi_i = \frac{(\bar{T}+T_i)^2}{\sum_{j=1}^{n}(\bar{T}+T_j)^2}$. It can be seen that this is simply the throughput expression (6) scaled by ϕ_i.

Comments

(i) *Fairness.* The factor ϕ_i captures the unfairness that can be introduced when AIMD flows with different effective increase and decrease factors compete for shared bandwidth. It is interesting to note that ϕ_i is quadratic in round-trip time and so the unfairness arising from competition between flows with different round-trip times is indicated as potentially much larger than that observed for isolated flows with different round-trip times (in which case throughput scales approximately linearly with RTT).

(ii) *Overall efficiency.* Since $\sum_{i=1}^{n}\phi_i = 1$, we have from (22) that the overall efficiency $\eta = \sum_{i=1}^{n}\eta_i$ achieved by multiple competing flows is identical to the efficiency (6) for a single flow, with γ replaced by $\bar{\gamma}$.

3.3 Comparison with Packet-Level Simulation Results

The foregoing analysis is based on a number of simplifying assumptions. In this section we compare the efficiency predicted using (22) with results obtained from *ns-2* packet-level simulations for a range of network conditions.

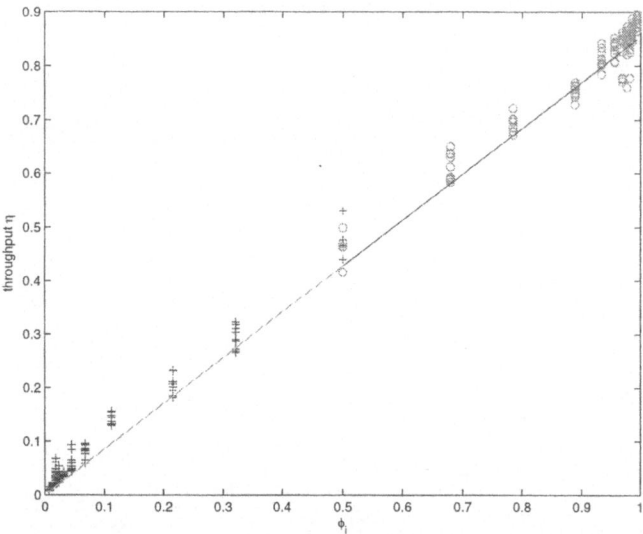

Fig. 5. Throughput efficiency η as RTT varies (B=100Mb, $\bar{T}$=20ms, T_1=2ms, T_2=2-252ms, queue 80 packets). Key: o denotes data for flow 1, + data for flow 2; the solid line is the analytic prediction given by (22)

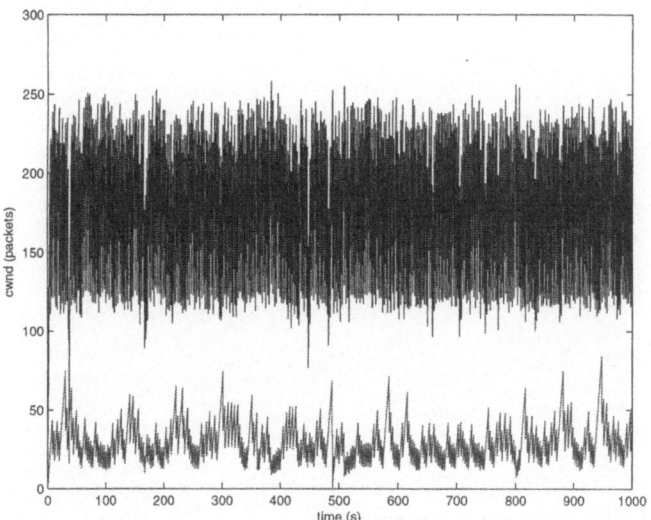

Fig. 6. Typical congestion window time histories of TCP (B=100Mb, $\bar{T}$=20ms, T_1=2ms, T_2=162ms, queue 80 packets).

Our analysis predicts that with multiple competing TCP flows the throughput efficiency of flow i is linear in ϕ_i (for given bottleneck buffer provisioning $\bar{\gamma}$). Figure 5 presents *ns-2* simulation results for competing TCP flows. The data is

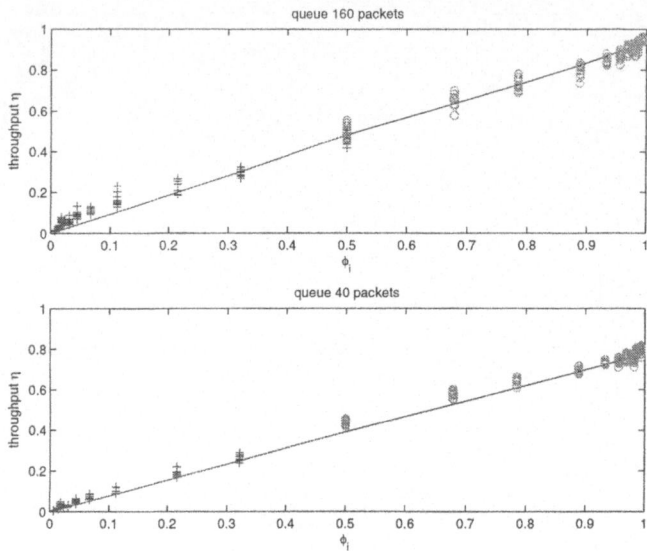

Fig. 7. Throughput efficiency as RTT and queue varies (B=100Mb, $\bar{T}$=20ms, T_1=2ms, T_2=2-252ms in 20ms steps, queue 40/160 packets). Key: o denotes data for flow 1, + data for flow 2; the solid line is the analytic prediction given by (22)

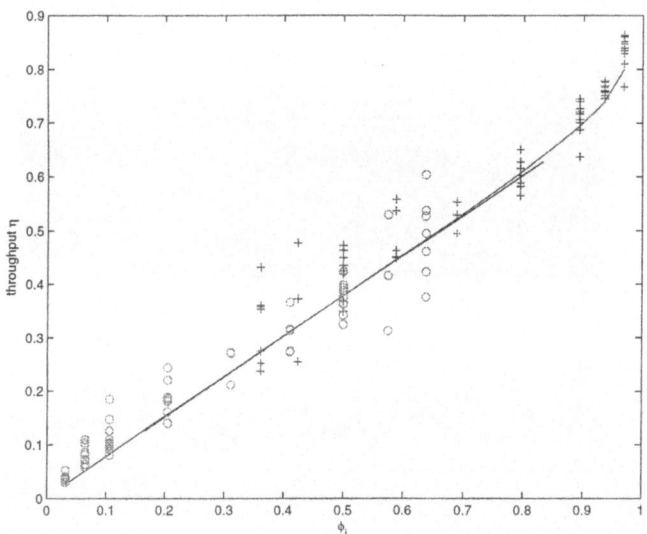

Fig. 8. Throughput efficiency as RTT varies (B=100Mb, $\bar{T}$=20ms, T_1=100ms, T_2=2-252ms in 20ms steps, queue 80 packets). Key: o denotes data for flow 1, + data for flow 2; the solid line is the analytic prediction given by (22)

for two flows, one flow with round-trip time fixed at 22ms and the other with round-trip varied from 22ms to 272ms in 20ms steps. In addition, a small amount (less than 1% of the bottleneck link bandwidth) of background web traffic was included to better capture realistic network conditions and to reduce artefacts associated with phase effects [10]. For each value of round-trip time, a total of 1000s of data was collected. The throughput binned over 100s intervals plotted in Figure 5. It can be seen that the data from the packet-level simulation does, indeed, exhibit a near linear dependence on ϕ_i. It is important to note that while the analytic expression (6) is derived under the assumption that the flows experience synchronised packet drops, it is readily verified that in these simulation results this assumption is significantly violated. See, for example, the congestion window time histories in Figure 6. The lack of synchronisation is reflected in the spread of throughput values observed.

The effect of changing the queue size is to alter $\bar{\gamma}$ thereby, according to (22), altering the slope of the line relating throughput to ϕ_i. Once again this analytic prediction is supported by simulation results: see Figure 7. These results are not specific to the network parameters used: see, for example, Figure 8.

4   Concluding Remarks

Since TCP traffic is elastic, a fundamental issue is the behaviour of multiple TCP flows competing for bandwidth on a shared link. Motivated by the ubiquity of drop-tail queueing in current networks, our focus in this paper is on developing analytic models suited to characterising the throughput and fairness of competing TCP flows in drop-tail environments. Our approach builds on recent results for synchronised networks [8], extending these earlier results to account for buffering and flows with different round trip times. This allows us to obtain simple, insightful closed-form expressions for fairness and throughput. The synchronisation assumption is that all flows experience a drop when the network "pipe" becomes full. Clearly, this is an unrealistic assumption in most real networks. Nevertheless, we find that the fairness and throughput predictions, while inaccurate with respect to instantaneous values, provide a rather accurate estimate of the *mean* fairness and throughput even when flows are not synchronised.

Acknowledgements. This work was supported by Science Foundation Ireland grant 00/PI.1/C067. This work was also partially supported by the European Union funded research training network *Multi-Agent Control*, HPRN-CT-1999-00107 and by the Enterprise Ireland grant SC/2000/084/Y.

References

1. E. Altman, C. Barakat, E. Laborde, P. Brown, and D. Collange, "Fairness analysis of tcp/ip," in *Proceedings IEEE Conference on Decision and Control, Sydney*, 2000.
2. S. Low, F. Paganini, and J. Doyle, "Internet congestion control," *IEEE Control Systems Magazine*, vol. 32, no. 1, pp. 28–43, 2002.
3. J. Padhye, V. Firoiu, D. Towsley, and J. Kurose, "Modeling tcp throughput: A simple model and its empirical validation," in *Proceedings SIGCOMM'98*, 1998.
4. R. Yang, M. Kim, and S. Lam, "Transient behaviors of tcp-friendly congestion control protocols," in *Proceedings INFOCOM 2001*, 2001.
5. S. Floyd, M. Handley, and J. Padhye, "A comparison of equation-based and aimd congestion control," in *Int. Computer Science Institute Technical Report*, 2000.
6. D. Chiu and R. Jain, "Analysis of the increase/decrease algorithms for congestion avoidance in computer networks," *Journal of Computer Networks*, vol. 17, no. 1, pp. 1–14, 1989.
7. P. Brown, "Resource sharing of tcp connections with different round trip times," in *Proceedings INFOCOMM 2000*, 2000.
8. R. Shorten, D. Leith, J. Foy, and R. Kilduff, "Towards an analysis and design framework for congestion control in communication networks," in *Proceedings of the 12th Yale Workshop on Adaptive and Learning Systems*, 2003.
9. "Network simulator, http://www.isi.edu/nsnam/ns/," tech. rep., University of California, Berkeley.
10. S. Floyd and V. Jacobson, "Traffic phase effects in packet-switched gateways," *ACM SIGCOMM Computer Communication Review*, vol. 21, no. 2, pp. 26–42, 1991.

The Undismissible Rats: How Do Short Connections Affect Long-Lived TCP Flows under Moderate Traffic Load?

Rong Pan

Stanford University
Stanford CA 94305, USA
rong@stanford.edu

Abstract. It is well-known that the distribution of file sizes in the Internet has a long tail, and that the traffic mainly consists of small flows - "the mice", while a large portion of the bytes are sent by large flows - "the elephants". However, it is not yet well understood regarding how the mice and the elephants interact with each other when they share a common link. A simplified modelling assumption is that the mice appear as an uncontrolled and random background load while the elephants take the remaining link bandwidth [7].

In this paper, we find that under moderate traffic load, short connections, particularly large-sized mice whom we refer to as "rats", have a significant impact on the throughput of long-lived TCP flows. The occurrence of rats, who are in the slow-start phase of the TCP protocol, impair the ability of long-lived TCP flows to grab a link's available bandwidth. Our analysis shows that the throughput of the long-lived TCP flows is a function of the rats' arrival rate; and since the file sizes in the Internet are Pareto-distributed, this arrival rate is actually a function of the traffic's shape parameter.

1 Introduction

Studies have found that the file size distribution of Internet traffic, 90% of which uses the TCP protocol, has a heavy-tail [2], [4], [9]. In simple terms, it means that most TCP connections are "mice" (with short lifetimes, still in the slow start phase), but a few large TCP flows referred to as "elephants" (with long lifetimes, in congestion avoidance phase) generate most of the traffic. This behavior, known as "the elephants and mice phenomenon" is considered to be one of the few invariants of Internet traffic.

The Pareto distribution is shown to be an accurate model to describe the heavy-tailed distribution of file sizes [4]. The Pareto distribution has a cumulative distribution function as follows:

$$F(x) = 1 - (\frac{b}{x})^\alpha \quad for \ \ x \geq b, \tag{1}$$

N. Mitrou et al. (Eds.): NETWORKING 2004, LNCS 3042, pp. 949–961, 2004.

where α is the shape parameter which determines the tail behavior and b is the scale parameter. The expected value of a Pareto-distributed random variable equals to $E(x) = b(\alpha - 1)^{-1}$ when $\alpha > 1$.

Separately, the performance of long-lived TCPs has also been well-studied [5], [7], [10], [11]. Previous work has shown that the performance of long-lived TCP flows depends on their round trip time (RTT) and packet loss probability. Assuming that the packet loss probability, p, is not too high and the receiver's window is not limited, the throughput $Th(p)$ of a long-lived TCP flow is given by

$$Th(p) \approx \frac{c}{RTT\sqrt{p}} \tag{2}$$

where c is a constant. This equation plays a fundamental role in determining a long-lived TCP flow's throughput when the traffic's arrival rate λ is relatively high. In this case, p is independent of any particular flow's behavior. Given p, a flow with a longer RTT will get lesser throughput than a flow with a shorter RTT. However, we note that p is not always independent of Th, especially when an elephant flow dominates a link's buffer space. For example, in the extreme case that only one long-lived flow is present on a link, Th will be equal to the link capacity C since the TCP protocol has the ability to fully utilize a link; and this is true regardless the value of RTT as long as there is enough buffering at the link. Here, Equation (2), while being true, only shows a secondary effect through which the drop probability can be obtained.

What kind of a role does Equation (2) play when the traffic mix consists of mice and elephants? Does the Pareto distribution play any role? It is not yet understood how the mice and the elephants interact with each other when they share a common link, i.e. how the bandwidth is distributed among the mice and the elephants. This task becomes more intricate because of the following fact: flows arrive and leave at random times; hence, the number of concurrent flows are varying over time. A modelling assumption, suggested by Kelly [7], is that the mice appear as an uncontrolled and random background load while the elephants take the remaining link bandwidth. This assumption is quite simplified; Kelly later indicates it might need more investigation [8].

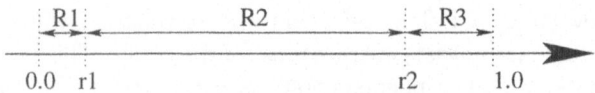

Fig. 1. Regions of Operations

This paper studies how the mice affect the performance of the elephants. We distinguish the following three regions that a long-lived TCP flow might operate in, shown in Figure 1. In Region $R1$, the link is lightly loaded. The boundary $r1$

could be around 10%.[1] Here, Kelly's assumption is valid. A long-lived TCP would dominate the behavior of the buffering queue and take most of link's bandwidth. Conversely in Region $R3$, the link's utilization is high. The boundary $r2$ could be around 70%.[2] In this region, since the aggregated arrival rate λ is high, no single flow will dominate. Hence, the drop probability only depends on the mixed arrival rate of mice and elephants. Given p, a long-lived TCP flow's throughput can be found using Equation (2). However, in Region $R2$, which is the typical region of operation for a network link, it is not yet understood how the mice and elephants share the link bandwidth. Our effort is focused on this normal, less-understood operating region.

The main contribution of this paper is to show that in Region $R2$ where a link is moderately loaded, the mice, especially those large-sized mice whom we refer to as "rats", play a rather negative role in sharing bandwidth with other TCP flows. These rats are big enough to just exit the slow-start phase of the TCP protocol, but not big enough to go through the saw-tooth like congestion avoidance. As a result, they undermine the ability of long-lived TCP flows to fill up a link's available bandwidth, one of the major strengths of the TCP protocol. Assuming that all flows share a common FIFO queue,[3] our analysis shows that the presence of rats causes extra delays and packet losses at the FIFO queue; and therefore force a long-lived TCP flow to back off and reduce its sending rate. As a result, the throughput of a long-lived TCP flow becomes a function of the rats' arrival rate λ_{rats} as follows

$$Th \approx \frac{1.5}{RTT^2 \lambda_{rats}}. \tag{3}$$

Since we know that the file sizes are Pareto-distributed and recent studies have shown that Internet traffic can be modeled as a stationary stochastic process [1], we can rewrite Equation (3) as a function of the traffic's shape parameter α:

$$Th \approx \frac{1.5}{RTT^2 \lambda} (\frac{b}{X})^{-\alpha}, \tag{4}$$

where λ is the traffic arrival rate and X is the minimum size of a rat. Equation (3) and (4), derived later in this paper, explain the role played by the mice in Region $R2$. Once again, they do not contradict the well-known Equation (2), which is valid in all regions. However, they do point to the fact that, in region $R2$, the drop probability becomes dependent on a flow's own behavior, i.e. Th, in this case.

We first study how mice and elephants interact with each other in Section 2, in which we introduce the notion of "rats". Section 3 uses a first-order mathe-

---

[1] We only use these boundaries as a general guideline, the numbers chosen might not be accurate.

[2] Network operators start considering an upgrade to their equipment when a link's average utilization is above 70%.

[3] There are proposals to buffer the mice and the elephants separately with some extra cost [3]. We do not study them in this paper.

matical model to analyze the role played by the rats. Then, we verify our findings using real trace statistics. We conclude in Section 4.

2   Interaction of Mice and Elephants

To gain insight into how short and long lived TCP flows share a link's bandwidth, we run a number of simulation experiments using the *ns2* simulator [13].

2.1   Simulation Study

The simulation setup consists of one bottleneck link shared by multiple flows. The link uses a buffer that can buffer up to 250ms worth of packets, which means any flow with an RTT of less than 250ms has the ability to fully utilize a link. Note this implies that the size of the buffer is different for different link speeds. All packets are 1000 bytes long. All connections have a maximum receiver window of 1000 packets so that a receiver window is not a limiting factor for a flow's throughput. Also, a recent version of the TCP protocol, TCP/Sack, is used throughout this paper.

The Basic Setup. In the basic setup, the bottleneck link bandwidth C is 15Mbps and the link uses a droptail buffer with a size of 450 packets. Web sessions have a Pareto distribution with parameters $\alpha = 1.25$ and $b = 6.25$. Hence, the average file length, $E(x) = b(\alpha - 1)^{-1}$, is 25 packets.

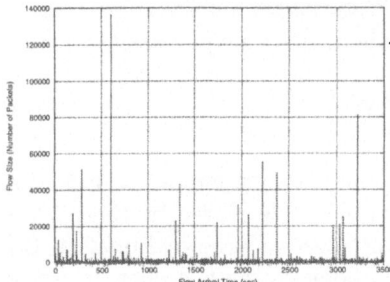
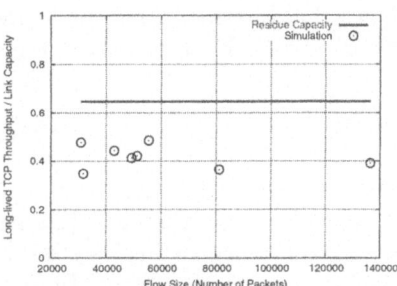

Fig. 2. Traffic

Fig. 3. Basic Setup: Long-lived TCP's Throughput

In the first experiment, the traffic of 100,000 web sessions consisting of mice and elephants, arrives randomly with an average demand of 5.3Mbps as shown in Figure 2. The mice have RTTs of either 30ms or 190ms (evenly distributed); and the elephants have an RTT of 190ms. The reason for the mice to have different RTTs is to reduce the phase effect under the droptail queue [6]. The rationale for the long-lived TCP flows having the same RTT is the following: since the link is moderately loaded, the degree of statistical multiplexing is not high. As a result, only a few elephants occur randomly in this simulation setup as shown in Figure 2.[4] It is easier and more accurate to calculate the average throughput of long-lived TCPs when they have the same RTT.

In this basic setup, the average traffic demand λ is 5.3Mbps. Hence the link load, $\rho = \frac{\lambda}{C}$, is around 35%. According to statistical bandwidth sharing [1], the throughput of long-lived TCP flows, which use all the capacity not used by other flows, should be approximately equal to the residual capacity $C(1 - \rho)$ at the bottleneck link. However, Figure 3 shows that the throughputs *actually* achieved by the long-lived TCP flows, whose sizes are over 30,000 packets, are below the expected value. This figure illustrates that while the link has 65% of its bandwidth available, the long-lived TCP flows fail to fully utilize it, yielding an average throughput of 6.2Mbps, only 41% of the link capacity.

If we vary the arrival rate of the traffic, the behavior of long-lived TCP flows remain similar. We plot the average link utilization *when at least one long-lived TCP flow is present* in Figure 4.[5] One would expect the link to be 100% utilized since with enough buffering a long-lived TCP flow has the ability to use up a link's available bandwidth regardless of its RTT and the link's traffic load ρ. However, as shown in Figure 4, the long-lived TCP flows fail to fill up the available bandwidth when the link is under a moderate load except when the RTT is very short, such as 30ms in the plot. The worst link utilization (when long-lived flows present) goes down quickly with increasing RTTs (from 90ms to 230ms).

Clearly, the link is not always fully-utilized in Region $R2$ when long-lived TCP flows are present: in the worst case, the link's utilization is only around 60%. The plot also points out an interesting behavior: the link utilization goes down initially when ρ starts to enter Region $R2$; stays low in Region $R2$; and increases again when the traffic demand is such that the link enters the congested region, Region $R3$. This figure demonstrates the fact that a long-lived TCP flow, under moderate traffic demands, could lose its strength to fully utilize a link.

Link with a Higher Speed. Suppose we increase the link speed to 20Mbps and the buffer space to 600 packets (still 250ms worth of buffering), while keeping the same traffic arrivals as in Section 2.1. Figure 5 depicts that, once again, the long-lived TCP flows can fail to capture the available link bandwidth in Region $R2$. The link utilization shows a similar, U-shaped behavior.

RED Queue instead of DropTail Queue. Suppose we alter the basic setup again by changing the queue management scheme to RED while keeping the

---

[4] Although some of them might overlap a bit during their life-time, they are not on and off at the exact same time.

[5] Note the link utilization plotted here is the average link utilization when at least one long-lived TCP flow is present. It is not the overall, average link utilization over the entire simulation time. With $\rho = 0.35$, the average link utilization over the entire simualtion is approximately 0.35 as well.

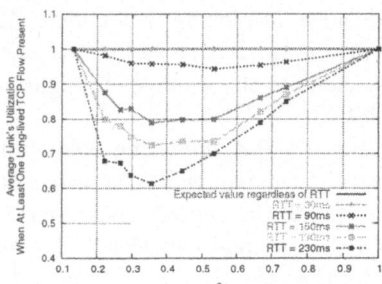

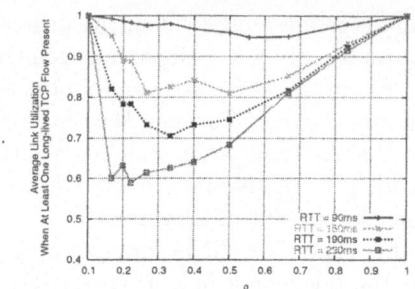

Fig. 4. Varying Traffic Demands: Average Link Utilization Under Long-lived TCP Flows

Fig. 5. With A Faster Link: Average Link Utilization Under Long-lived TCP Flows

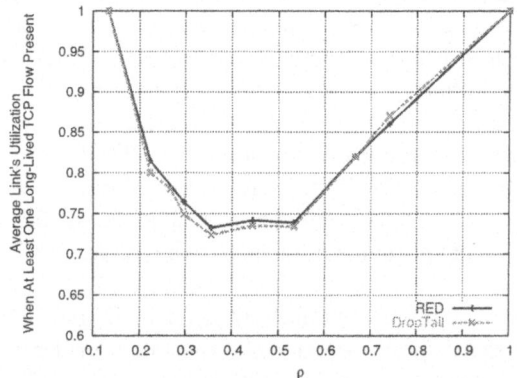

Fig. 6. With A RED-Queue: Average Link Utilization Under Long-lived TCP Flows

other parameters the same. The min_{th} and max_{th} are set to be 150 and 300 packets. Figure 6 shows that the dynamics observed in the previous simulations does not change much when we use RED as the queue management scheme. This indicates that the phenomenon we observed might not be related much to the average queue length, rather it is related to the instantaneous queue length.

Mice with Different Round Trip Times. If the mice have shorter or longer RTTs, would it affect the throughput of long-lived TCP flows? We now let the mice have two different round trip times with equal probability, RTT_m and 190ms, where RTT_m is a variable. We keep the RTT of 190ms for those flows whose sizes are above 30,000 packets. Table 1 shows that the average throughput of long-lived TCP flows depends little on the RTTs of mice. This behavior is rather different from that of a TCP flow operating in Region $R3$, where a flow can gain more bandwidth when competing against flows with shorter RTTs or lose bandwidth when competing against flows with longer RTTs as implied in

Table 1. Average Throughput of Long-lived TCP flows as a Function of Mice's RTTs

RTT_m	30ms	70ms	110ms	150ms	190ms	230ms
Throughput	0.418	0.419	0.440	0.437	0.464	0.454

Equation (2). This serves as another indication that the TCP protocol experiences a different throughput-limiting factor.

Summary. We have also run other simulations such as traffic with a different shape parameter or two-way traffic which are omitted here. All the simulation studies demonstrate that, under a variety of scenarios in Region $R2$, a long-lived TCP flow can fail to capture a link's available bandwidth even though it has the ability to fully utilize a link when it is alone, i.e. when no mice flows are present. Here, neither Kelly's assumption nor Equation (2) can help explain the behavior of long-lived flows. The studies also show that the rates of mice flows, not the round trip times of mice flows, play a role in limiting a long-lived TCP flow's throughput.

2.2   A Key Observation

In the congestion-avoidance phase, a long-lived TCP flow regulates its window size according to network conditions which are directly related to the traffic load on a link. To examine the effect of the Internet traffic arrivals on the window size, we run a simulation where the flows whose sizes are over 30000 packets in the basic setup are combined into one single FTP flow since this "long" FTP flow shows the evolution of the window size more visibly than those scattered long-lived flows in the original setup.

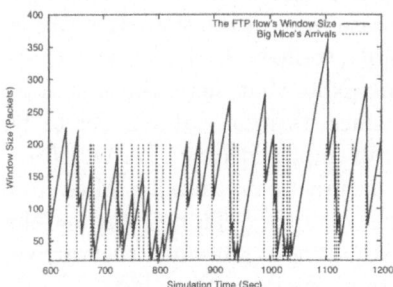 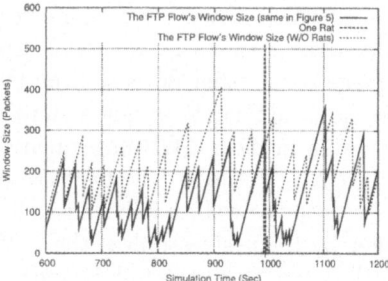

Fig. 7. Window Size vs. Traffic Arrivals   **Fig. 8.** Window Size Comparison

The solid line in Figure 7 represents the window size of the FTP flow. The dotted lines indicate the arrivals of mice whose sizes are above 750 packets.[6]

---

[6] Note that the heights of the dotted lines in Figure 7 have no meaning.

The plot illustrates that the time when the FTP flow cuts its window mostly coincides with the time when there is an arrival of a mouse whose size is above a few hundred packets. The arrivals of these big mice hinders the growth of the FTP flow's window. Hence, the throughput of the long-lived TCP flow becomes constrained. As a result, the FTP flow fails to capture the available link bandwidth.

2.3   The Notion of "Rats"

Since flows with a few hundred packets are significantly bigger than the common concept of "mice" - who send only a few packets, we introduce a new notion of "rats" to represent those flows whose sizes are big enough to disturb long-lived TCP flows, but small enough to avoid the saw-tooth type of window adjustment that a long-lived TCP flow would go through. The spike in Figure 8 shows an example of a rat's window size in the above simulation. If we remove rats and distribute their traffic load to flows under 600 packets (i.e. maintaining the same traffic arrival rate λ), the window size of the FTP flow becomes much larger, as shown in Figure 8. This serves as another evidence of the influence of rats. Since Internet traffic is Pareto-distributed, the rats are bound to occur. In other words, an elephant, whose life time is long, is certain to suffer from the disturbances caused by rats.

2.4   Discussion

One should be aware that the above observation does NOT indicate that long-lived TCP flows or small mice would not cause losses. As a matter of fact, they would. However, these drops occur less frequently as compared to the drops caused by rats. Long-lived TCP flows occur scarcely under a moderate traffic demand as shown in Figure 2. Although these flows may overlap over a small period of time, the effect they have on each other is less than the effect brought by rats which arrive much more frequently. These rats, in the slow start phase, send many packets in one RTT time, and will overflow the buffer and cause damage to all other flows sharing the same link. Note that, here, the slow start threshold, $ssthresh$, is not a limiting factor since its intial value is set to be very high in most TCP implementations. For example, in Linux kernel 2.5.73, $ssthresh$ is initialized to be 0x7fffffff, which is practically infinity. Mice may cause drops as well, but because their sizes are small they can not overflow the buffer by themselves unless the buffer is already almost full when they arrive, which is rare since the link is under a moderate traffic demand.

3   Modeling and Analysis of Rats

This section develops a first-order mathematical model for analyzing the effect of rats on long-lived TCP flows. The model shows that the throughput of a long-lived TCP flow depends on the arrival rate of rats, λ_{rats}, and since the Internet traffic is Pareto-distributed, λ_{rats} can be shown to be a function of the traffic's shape parameter α.

3.1   Model

In our model of long-lived TCP's throughput, we assume a network setup as follows: an infinite long TCP connection is sharing a link with rats which arrive regularly. We also assume that the data buffer overflows and packets are dropped whenever a rat arrives. Note that we don't take into account the losses that are caused by mice or other competing long-lived TCP flows because these losses are rare compared to the losess caused by rats as discussed in Section 2.4.

Although multiple packets might be dropped within a window, TCP/Sack only cuts down its window by half instead of going into timeout. Hence one rat's arrival is equivalent to a single drop event. After a drop event, the TCP sender increases its congestion window by one packet per round trip time, until the next rat arrives.

TCP throughput: a function of λ_{rats}. Suppose the inter-rat arrival time is λ_{rats}^{-1}, and the long-lived TCP connection has a rather constant round-trip time of RTT seconds. Each time a rat arrives, the TCP sender has a congestion window of W_{max} packets as shown in Figure 9.

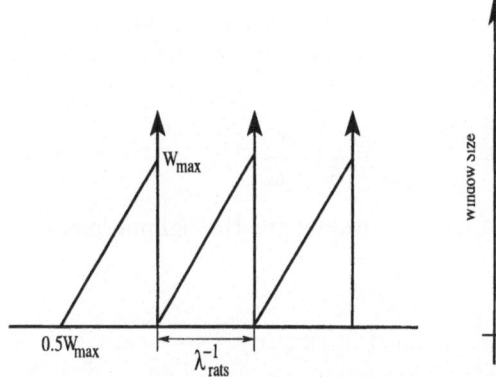

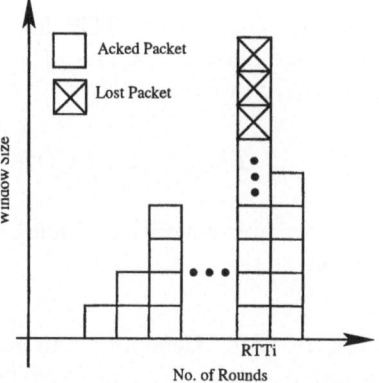

Fig. 9. Congestion Window Adjustment   **Fig. 10.** Evolution of Window Size

By cutting its window by half for each rat's arrival and increasing its window by one per round-trip time subsequently, the TCP sender goes through $0.5W_{max}$ round-trip times for each drop event, i.e. each rat's arrival. Hence, the following equation holds:

$$\frac{W_{max}}{2} * RTT = \lambda_{rats}^{-1} \Rightarrow W_{max} = \frac{2}{RTT\lambda_{rats}}.$$

Since the average window size in a saw-tooth period equals $0.75W_{max}$, the throughput of a long-lived TCP, Th_{tcp}, can be expressed as follows:

$$Th_{tcp} = \frac{0.75W_{max}}{RTT} = \frac{1.5}{RTT^2\lambda_{rats}}. \tag{5}$$

TCP throughput: a function of the shape parameter α. As discussed in Section 1, previous work has shown that sessions arrive in the Internet as a Poisson process [1] and the file sizes in the Internet are Pareto-distributed [4]. The probability that a flow is a rat equals

$$P_{rat} = (\frac{b}{S_{rat}})^\alpha - (\frac{b}{S_{ele}})^\alpha \approx (\frac{b}{S_{rat}})^\alpha, \tag{6}$$

where S_{rat} is the smallest size of a rat and S_{ele} is the smallest size of an elephant. Since the Pareto-distributed file sizes imply that the probability is much higher for a flow being a rat than being an elephant, the above Equation can approximated by $(\frac{b}{S_{rat}})^\alpha$.

Given that the traffic arrives as a Poisson process of rate λ, we can obtain that the rats also arrive as a Poisson process with a rate of $\lambda_{rats} \approx \lambda(\frac{b}{S_{rat}})^\alpha$. Combining Equation (5) and λ_{rats}, we obtain that

$$Th_{tcp} \approx \frac{1.5}{RTT^2\lambda}(\frac{b}{S_{rat}})^{-\alpha}.$$

Certainly, a long-lived TCP flow's throughput is also constrained by the available link bandwidth, $C(1-\rho)$. Hence we can find the throughput of a long-lived TCP flow in Region $R2$ as

$$Th_{tcp} \approx \min(\ C(1-\rho),\ \frac{1.5}{RTT^2\lambda}(\frac{b}{S_{rat}})^{-\alpha}\ ). \tag{7}$$

Note that the above model does not take into account the throughput losses due to time out.

The size of rats. Generally, rats are defined in this paper to be any connections which can cause the long-lived TCP flow to cut its congestion window size. Despite this loose definition of rats, we do follow a guideline in calculating the arrival rate of rats. We know that a rat, like any TCP flow, increases its congestion window geometrically in the slow start phase. For the ith round trip time, RTT_i, the congestion window size W_i is 2^i. The total number of packets sent up to RTT_i equals

$$\sum_{m=0}^{i} 2^m = 2^{i+1} - 1 \approx 2^{i+1}. \tag{8}$$

This equation means that, if a flow has N packets to send and it incurs no drop before the last round, then approximately $N/2$ are sent in the last round alone.

Given that $N/2$ is close to the buffer size B at an output link, the buffer is bound to overflow and some packets are lost. The lost packets are then resent. The evolution of the window size is illustrated in Figure 10. Although it normally takes a burst size of less than B packets to overflow a buffer since other flows take up buffer spaces as well, $N \approx 2B$ serves as an upper-bound estimate of a rat's minimum size. The theoretical analysis of a rat's minimum size is left for future work.

3.2 Verification

We apply the analytical model to the simulations in Section 2. Table 2 shows the comparison results. Recall that C is the bottleneck link capacity and λ is sessions' arrival rate. The traffic load, ρ, equals the traffic arrival rate divided by the link capacity, and U represents the link utilization when at least one long-lived TCP flow present. The buffer size, B, is set to be the bandwidth-delay product of the output link: 450 packets for a 15Mbps link and 600 packets for a 20Mbps link. The table clearly demonstrates that, under various traffic and network scenarios, our theoretical model is able to predict the performance of long-lived TCP flows accurately.

The model can not only predict a long-lived TCP flow's performance but also explain other phenomenon that we observed in Section 2: using a RED queue or varying mice's RTT times does not change the TCP flows' performance. Since a rat is in the slow-start phase of exponentially expanding its window size, the sudden increase of its packet burst size would not change much the average queue length of the buffer at the bottleneck link. Therefore, packets are dropped because the buffer overflows, not because the average queue length is over a certain threshold so that RED needs to preemptively drop packets. As a result, the RED queue behaves like the droptail queue. Separately, we know from Equation (7) that the arrivals of mice only contribute to the traffic load, ρ, and therefore their RTT values have little effect on the elephants' performance.

Another interesting fact that the model can help explain is how a flow's throughput depends on its RTT. When a link is congested, the throughput of a long-lived TCP flow is inversely proportional to its RTT as in Equation (2). When a link is moderately loaded, however, the dependency on RTT is quite different. When the RTT value is small, the long-lived TCP flow's throughput is limited by the available link bandwidth, $C(1 - \rho)$, as the model predicts, which is consistent with what we have seen in Figure 4 when RTT = 30ms. When the RTT value is big, the long tcp flow's throughput is inversely propotional to RTT^2. The drop in a flow's throughput is more drastic as shown in Figure 4.

3.3 Real Traffic Statistics

In this section, we demonstrate that the phenomenon indeed still exists when we use real traffic statistics. We obtained traces of traffic from backbone routers

Table 2. Theory vs. Simulation: $\alpha = 1.25$

C	λ	ρ	B	S_{rats}	$U(Simulation)$	$U(Model)$	RTT
15M	26.7	0.356	450	750	1.0	1.0	30ms
15M	26.7	0.356	450	750	1.0	0.96	90ms
15M	26.7	0.356	450	750	0.86	0.83	150ms
15M	26.7	0.356	450	900	0.74	0.72	190ms
15M	26.7	0.356	450	900	0.61	0.61	230ms
15M	33.3	0.444	450	900	0.76	0.74	190ms
15M	33.3	0.444	450	900	0.64	0.65	230ms
15M	50	0.667	450	900	0.84	0.82	190ms
15M	50	0.667	450	900	0.76	0.79	230ms
20M	26.7	0.267	600	1200	0.69	0.73	190ms
20M	33.3	0.333	600	1200	0.67	0.70	190ms
20M	50	0.5	600	1200	0.70	0.74	190ms

at two different locations: New York and Philadelphia.[7] It has been studied that these traces have a heavy tail and the arrival process of sessions is Poisson. We applied the sampled trace statistics to the access link of 15Mbps. The average arrival rate to the access link is 4.48Mbps, which is equivalent to a traffic load of 0.3.

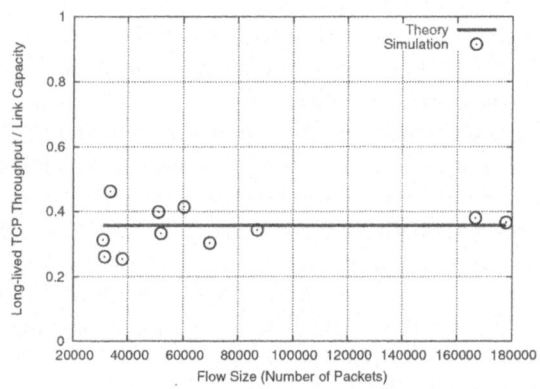

Fig. 11. Trace 1: Throughputs of Long-lived TCP flows

Figure 11 shows the simulation result using the traffic statistics of one trace. The circles in the figure represent the throughputs of long-lived TCP flows whose sizes are bigger than 30,000 packets. It is clear that these long-lived TCP flows fail to capture the available link bandwidth of 70%. The straight line in the plot shows our model prediction. Since there lacks enough data to fully characterize the shape parameter α for these traces, we measure the arrival rate of rats λ_{rats} directly from the sampled trace and apply Equation (5). The plot depicts that the simulation result using real traffic statistics, with $\lambda_{rats} = 0.062$, matches well with the analytical result. Most importantly, the plot shows that the phe-

---

[7] Due to the proprietary nature of these traces, we omit the details about them.

nomenon about the rats affecting elephants indeed exists for the real Internet traffic: an elephant would encounter multiple rats in its life-time and hence suffer from their presence.

4 Conclusion

In this paper, we have shown that short flows, especially "rats", have an important effect on the throughput of long-lived TCP flows. When a link is under moderate traffic demands, the occurrence of rats impairs the ability of the TCP protocol to fully utilize a link's available bandwidth, which leads to the link being under-utilized. We have also presented a theoretical model to explain the effect of rats. Our analysis shows that the throughput of a long-lived TCP flow is a function of the rats' arrival rate λ_{rats}. Since the file lengths of Internet traffic are Pareto-distributed and session arrivals are Poisson, λ_{rats} is in fact a function of the traffic shape parameter.

References

1. Ben Fredj, S., Bonald, T., Proutiere, A., Regnie, G., and Roberts, J., "Statistical Bandwidth Sharing: A Study of Congestion at Flow Level", *ACM SIGCOMM*, 2001.
2. CAIDA, "Traffic Workload Overview" and "Graphs of Ames Internet Exchange Traffic", *http://www.caida.org*, 2003.
3. Chen, X. and Heidemann, J. "Preferential Treatment for Short Flows to Reduce Web Latency", *Computer Networks*, 41(6), April, 2003.
4. Crovella, M. and Bestavros, A. "Self-Similarity in Worlf Wide Web Traffic: Evidence and Possible Causes", *IEEE/ACM Trans. on Networking*, vol.5, no.6, Dec. 1997, pp.835-846.
5. Floyd, S. and Fall, K., "Promoting the Use of End-to-End Congestion Control in the Internet", *IEEE/ACM Trans. on Networking*, August, 1999.
6. Floyd, S. and Jacobson, V., "On Traffic Phase Effects in Packet-Switched Gateways", *Internetworking: Research and Experience*, V.3 N.3, September 1992, p.115-156.
7. Kelly, F., "Mathematical modelling of the Internet", *Proceedings of the Fourth International Congress on Industrial and Applied Mathematics*, July, 1999.
8. Kelly, F., "Models for a self-managed Internet", *Philosophical Transactions of the Royal Society*, A358(2000), pp.2335-2348.
9. Leland, W., Taqqu, M., Willinger, W. and Wilson, D., "On the self-similar nature of Ethernet traffic", *ACM SIGCOMM*, 1993.
10. Misra, V., Gong, W. and Towsley, D., "A Fluid-based Analysis of a Network of AQM Routers Supporting TCP Flows with an Application to RED", *ACM SIGCOMM*, 2000.
11. Padhye J., Firoiu, V., Towsley, D. and Kurose, J., "Modeling TCP Throughput: A Simple Model and its Empirical Validation", *ACM SIGCOMM*, 1998.
12. Paxson, V. and Floyd, S., "Wide-area Traffic: The Failure of Poisson Modelling", *IEEE/ACM Trans. on Networking*, 3(3), June 1995, pp226-244.
13. Network Simulator, Version 2.1b8.

Theoretical Analysis of Performances of TCP/IP Congestion Control Algorithm with Different Distances

Tsuyoshi Ito and Mary Inaba

Department of Computer Science, The University of Tokyo
7-3-1 Hongo, Bunkyo-ku, Tokyo, Japan

Abstract. According to current TCP/IP implementations, the acceleration in additive-increase phase depends on the distance of connection. In this paper, the performance of Additive Increase and Multiplicative Decrease (AIMD) congestion control algorithm in TCP is analyzed in two ways, both focusing on effects of the *heterogeneity* or the mixture of different accelerations caused by different distances. First, we analyze flow time minimization, extending the competitive analysis by Edmonds et al. to heterogeneous case. We show (a) the performance loss of TCP/IP in Long Fat Pipe Networks (LFNs) is caused by the heterogeneity rather than long distance itself. Next, we step forward to more realistic *single-drop model*, where upon each congestion only one, instead of all, connection drops rate, and analyze asymptotic total and per-connection bandwidth utilizations. We show (b) increasing the number of connections makes total utilization better as opposed to common model, (c) in homogeneous environments, *victim policies* or choice of which connection drops do not affect total utilization, and (d) in heterogeneous two-connection environments, maximum total utilization is achieved by certain victim policy which leads to unfair share, whereas fair utilization is achieved by certain random victim policy.

1 Introduction

The Transmission Control Protocol (TCP) is used by most data transfer in the Internet. It has been widely known [1,2] that the current implementations of TCP do not perform well in long-distance high-bandwidth networks, or Long Fat Pipe Networks (LFNs). These days, the backbone network over gigabits per second such as Abilene and GÉANT is rapidly constructed, and the bandwidth of the links in the Internet, especially of the long-distance ones, is increasing. As a result, the Internet has become an LFN. However, the exact reason why the performance of TCP suffers in LFNs is not known. It is observed [3] that the negative impact of using faster network interface than the bottleneck capacity is more severe in long-latency connections than in short-latency connections.

At the same time, there are more and more needs for the transfer of various kinds of large data. For example, people will send e-mails with video images of tens or hundreds of megabytes length in near future. As an example where huge

N. Mitrou et al. (Eds.): NETWORKING 2004, LNCS 3042, pp. 962–973, 2004.

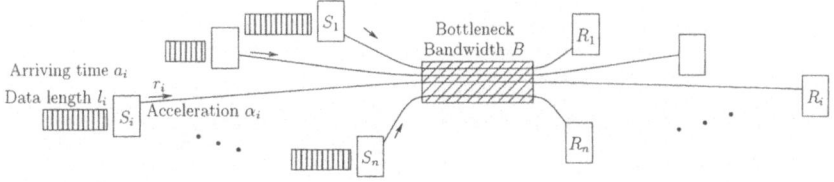

Fig. 1. A single-bottleneck network consisting of a bottleneck with bandwidth B and n connections with different distances. Each connection C_i has the acceleration α_i which is inversely proportional to the square of its RTT.

data is concerned, some research institutes currently receive data of terabytes produced by scientific measurement instruments by the physical transportation of Digital Linear Tapes (DLTs), but they can receive them online if the LFN problem is resolved [4]. This indicates the necessity of the analysis of the performance of the long-time transfer of very large data.

The congestion control in TCP makes a guess on the appropriate transmission rate by only using the data exchanged between the endpoints of the connection. The current congestion control algorithm increases the transmission rate by α in unit time while the transmission succeeds, and decreases it by multiplying $1 - \beta$ to the current rate. This algorithm is called Additive Increase and Multiplicative Decrease (AIMD) [5].

In this paper, in quest of the exact reason the current TCP does not perform well on LFNs, theoretical analyses are performed from various viewpoints on the most fundamental network model with a single bottleneck, as depicted in Figure 1. The performance is analyzed in the case that each of the connections with different distances transfers large data. As we focus on the transfer of large data, we consider only the AIMD congestion avoidance phase of TCP of sufficiently long period, ignoring the effect of the slow start phase which is relatively short period of time.

In the real world, the distance of a connection affects the behavior of the AIMD mainly in three ways. (1) Acceleration: In the AIMD algorithm, the transmission rate of a connection increases by $\alpha = c/T^2$ per unit time while the transmission succeeds, where c is Sender Maximum Segment Size (SMSS), which is a constant for usual case, and T is Round Trip Time (RTT), which reflects the distance of the connection. This α is called the *acceleration* of the connection. (2) Response time: After a node transmits its data, it takes the time amount of RTT to know whether the transmission has succeeded or failed. (3) The number of congestion points: Long-distance connections pass more congestion points such as routers and switches than short-distance connections.— We focus on the difference of (1) to isolate the effects of different distances of connections. We say the environment is *homogeneous* if all the connections have an equal acceleration, and *heterogeneous* otherwise.

Edmonds et al. [6] consider the single-bottleneck network and prove by theoretical analysis that the AIMD algorithm performs well when all the connections

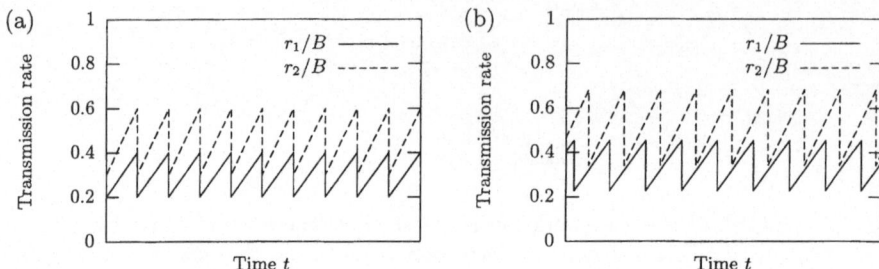

Fig. 2. Time evolution of the transmission rates r_1 and r_2 of the two connections with the accelerations $\alpha_1 : \alpha_2 = 2 : 3$ and the drop factor $\beta = 1/2$. (a) uses the all-drop model, and (b) uses the single-drop model and Periodic victim policy.

have a common acceleration, that is, in the homogeneous case. In section 4, we extend their result to the heterogeneous case and show a result that suggests the AIMD does not perform well when connections have different accelerations, thus explaining the low throughputs under the coexistence of short- and very-long-distance communication.

In section 5, we further analyze the total bandwidth utilization and the share of the available bandwidth in the stationary state. Many existing results, including the result by Edmonds et al. and our extension to it, assume that when congestion occurs at the bottleneck, all the connections drop their transmission rate at the same time as depicted in Figure 2 (a). With this assumption, it is shown that the total utilization does not depend on the number of connections. To fill a gap between this assumption and the reality, we consider another model of the drop as shown in Figure 2 (b). In the new model, when congestion occurs, one connection is chosen as *victim* and only the victim drops its transmission rate and the transmission rate of the other connections does not change. We call this model the *single-drop model* and refer to the previous model as the *all-drop model*. In the single-drop model, we have several choices of the order in which the connections are chosen as a victim. We refer to these choices as *victim policies*. Victim policy is an abstraction of the algorithm executed by the bottleneck router to choose which packet to discard when the network is congested. We investigate which victim policy maximizes the total utilization. In addition, we prove that in the all-drop model and some single-drop victim policies, the bandwidth is shared among the connections in proportion to their accelerations, hence unfairly in proportion to the inverse of the square of their RTTs, whereas one of the randomized victim policies results in fair per-connection utilization.

2   Related Works

TCP congestion control is an algorithm which works without knowledge about the bandwidth of links or information about other communication sharing the network. There are two approaches to the theoretical analysis of the performance of such incomplete-information algorithms. Probabilistic analysis is the analysis of the average case after assuming some probabilistic distribution of the unknown information, and competitive analysis is the analysis of the worst *competitive ratio* of the performance to the fictional case where the complete information were available to an algorithm.

Probabilistic analysis.   Several papers [7,8,9] analyze how the throughput of homogeneous TCP connections is affected by random packet losses under the assumption that every packet is dropped independently with a constant probability. De Vendictis et al. [10] consider the environment with two connections where one connection uses the current TCP and the other uses a different congestion control algorithm called TCP Vegas, and analyze the throughputs of the connections in the stationary state.

Competitive analysis.   At the top of our knowledge, the application of the competitive analysis to the performance evaluation of TCP congestion control was first proposed and performed by Karp et al. [11]. They formalized the congestion control as the algorithm to guess a secret available bandwidth which changes little by little over time. Edmonds et al. [6] consider the setting where multiple homogeneous connection jobs arrive and complete over time. They regard TCP as an online and distributed algorithm to share the available bandwidth among ongoing connections and compare it to scheduling algorithms which share the available processors among ongoing jobs in the centralized manner. They show that TCP achieves a constant competitive ratio independent of the number of connections by the competitive analysis against the optimal offline scheduling algorithm. However, the result holds only for the homogeneous case.

3   Definitions

Figure 1 illustrates the *single-bottleneck network* we consider. The network consists of one bottleneck with bandwidth B, n *senders* $S_1, \ldots, S_n$ on one side of the bottleneck, and n corresponding *receivers* $R_1, \ldots, R_n$ on the other side. Sender S_i sends its data to receiver R_i, together making a *connection* C_i. S_i sends data at the rate of r_i per unit time, where r_i, called the *transmission rate* of C_i, changes as time goes on. Any algorithm must control the transmission rates so that their sum $\sum_{i=1}^{n} r_i$ never exceed B. Here we use *fluid model*: r_i can be any nonnegative real value and the data can be sent as if it does not have the minimum unit such as a packet, an octet or even a bit.

Each connection C_i is associated with three constants: the *arriving time* a_i, the *data length* $l_i > 0$, and the *acceleration* $\alpha_i > 0$. The connection C_i starts at time a_i to send l_i amount of data. We consider both the case of $l_i < \infty$ and the case of $l_i = \infty$. The acceleration α_i is used by the AIMD algorithm as described later.

In this paper, the behavior of the AIMD congestion control algorithm is formalized as follows. A constant $0 < \beta \leq 1$ fixed. β is called *drop factor* and common to all the connections. Each C_i maintains its transmission rate $r_i \geq 0$ as follows. While $\sum_{i=1}^{n} r_i < B$, in other words, the sum of transmission rates of the n connections is less than the bottleneck bandwidth, each C_i transmits an infinitesimally small amount $r_i\,dt$ of data for an infinitesimally short time dt and increases r_i by $\alpha_i\,dt$. When $\sum_{i=1}^{n} r_i = B$, meaning that the sum of transmission rates hits the bandwidth, what happens depends on which *drop model* we adopt. (1) *All-drop model*: All the r_i's are multiplied by $(1 - \beta)$ instantly at the same time, as shown in Figure 2 (a). (2) *Single-drop model*: One connection C_i is chosen as *victim* and its transmission rate r_i is multiplied by $(1 - \beta)$ instantly. Note that in the single-drop model, the choice of victim is not unique, and we will discuss about *victim policies* in section 5. For example, Periodic victim policy defined in section 5.3 chooses every connection as victim in turn as shown in Figure 2 (b).

When $l_i < \infty$ and $\int_{a_i}^{t} r_i\,dt = l_i$, meaning that connection C_i has sent all of its data, then connection C_i terminates. In this case, the time elapsed since the arriving time a_i until the termination of C_i is called the *flow time* f_i of connection C_i, and the sum $F = \sum_{i=1}^{n} f_i$ is just called the *flow time*.

In the current TCP congestion control algorithm, β is fixed to $1/2$, and α_i is inversely proportional to the square of RTT of connection C_i. The case that α_i's are equal for all the connections is called *homogeneous* case, and the other case *heterogeneous* case.

4   Competitive Analysis of Flow Time in Heterogeneous Environments

In this section, we assume the all-drop model and we consider the case that $l_i < \infty$ for all i, that is, each sender sends a finite amount of data. In this setting, we consider the optimization problem of minimizing the flow time.

Now consider the arriving time a_i is not known until the request of data transfer of C_i arrives at time a_i. Similarly, consider the data length l_i is not known until the sender sends l_i amount of data, reaching the end of data. This situation is common, because it corresponds to the case that the congestion control algorithm is implemented as a protocol stack independent of the application which decides when and which data to send. The AIMD algorithm works without any problem in this situation, because it does not use any information given in future to work. In this sense, the AIMD algorithm is called an *online algorithm*.

Besides, the AIMD is a *distributed algorithm* in the following sense. Each connection C_i only requires the information about its own parameters, a_i, α_i and l_i, and does not need to know the bottleneck bandwidth B or the parameters of the other connections, provided the sender knows whether $\sum r_i < B$ or $\sum r_i = B$. In TCP, this last additional information is supplied by the presence or the absence of acknowledgment from the receiver.

In contrast to the online and distributed AIMD algorithm, we can consider fictional *offline* and *centralized* algorithms. This kind of algorithms know B, and a_i and l_i of all the n connections before any request arrives, and controls all the r_i's simultaneously. Because offline and centralized algorithms have more access to knowledge than online and distributed algorithms like the AIMD, the optimal offline and centralized algorithm achieves no longer flow time than the AIMD.

For the homogeneous case where $\alpha_1 = \cdots = \alpha_n = \alpha$, Edmonds et al. prove the following.

Theorem 1 ([6]). *The AIMD is competitive to the optimal offline and centralized algorithm with a limited bottleneck bandwidth in the following sense. Let $\varepsilon > 0$ and*

$$s = 2(2 + \varepsilon) \cdot \frac{1}{\beta} \cdot \frac{2}{2 - \beta}, \tag{1}$$

and suppose we compare the flow time $F(\mathcal{C})$ of the set $\mathcal{C} = \{C_1, \ldots, C_n\}$ of connections achieved by the AIMD with bottleneck bandwidth B and that achieved $F_{\mathrm{OPT}}(\mathcal{C})$ by the optimal offline and centralized algorithm with bottleneck bandwidth B/s. Then, for $D = 4n\beta B/(s\alpha)$, it holds that $\frac{F(\mathcal{C})}{F_{\mathrm{OPT}}(\mathcal{C}) + D} \leq 2 + \frac{4}{\varepsilon}$.

Now we consider the heterogeneous case. As we mentioned in section 3, the flow time $F(\mathcal{C})$ can be written by using the flow time f_i of individual connection as $F(\mathcal{C}) = \sum_{i=1}^{n} f_i$. In a similar way, we define *modified flow time* as: $F'(\mathcal{C}) = \sum_{i=1}^{n} \alpha_i f_i$, that is, the sum of the flow times of the connections weighted by the accelerations of the connections.

Then Theorem 1 is extended as follows.

Theorem 2. *Let $\mathcal{C} = \{C_1, \ldots, C_n\}$ be a set of connections. Suppose $\alpha_{\min} \leq \alpha_i \leq \alpha_{\max}$ for all i, and α_i be a multiple of α_{unit}. Let $\varepsilon > 0$, and define s is in equation (1). Let $F(\mathcal{C})$ be the flow time achieved by the AIMD with bottleneck bandwidth B and $F_{\mathrm{OPT}}(\mathcal{C})$ be that achieved by the optimal offline and centralized algorithm with bottleneck bandwidth B/s. Then, it holds that*

$$\frac{F'(\mathcal{C})}{F'_{\mathrm{OPT}}(\mathcal{C}) + \frac{\alpha_{\max}}{\alpha_{\mathrm{unit}}} D} \leq 2 + \frac{4}{\varepsilon} \quad and \quad \frac{F(\mathcal{C})}{F_{\mathrm{OPT}}(\mathcal{C}) + D} \leq \frac{\alpha_{\max}}{\alpha_{\min}} \left(2 + \frac{4}{\varepsilon} \right)$$

where $D = n \cdot \frac{\beta^2 (2-\beta)}{2+\varepsilon} \cdot \frac{B}{\alpha_{\mathrm{unit}}}$.

Proof (sketch). We make a new set $\mathcal{C}'$ of connections from the given set $\mathcal{C}$ so that all the connections in $\mathcal{C}'$ have the acceleration of α_{unit}. For each connection C_i in $\mathcal{C}$, let $n_i = \alpha_i / \alpha_{\mathrm{unit}}$ and divide C_i into n_i equal connections with arriving time a_i, data length l_i / n_i and acceleration α_{unit}. Then it holds $F'(\mathcal{C}') = F'(\mathcal{C})$. Because $\mathcal{C}'$ is made just by dividing the connections of $\mathcal{C}$ to smaller ones, it holds $F'_{\mathrm{OPT}}(\mathcal{C}') \leq F'_{\mathrm{OPT}}(\mathcal{C})$. The theorem is obtained by applying Theorem 1 to $\mathcal{C}'$.

This gives the same competitive ratio as the homogeneous case for the modified flow time, and $\alpha_{\max} / \alpha_{\min}$ times as worse competitive ratio as the homogeneous case for normal flow time.

Because the modified flow time attaches importance to the flow time of connections with large acceleration, or short-distance connections, the fact proven above that the modified flow time is near optimal explains that in heterogeneous case long-distance connections get less bandwidth, resulting worse competitive ratio of the normal flow time.

5   Analysis of Asymptotic Bandwidth Utilization

In this section, we consider the case that $l_i = \infty$ for all i, that is, all the senders have infinite data to transmit and the connections never terminate. As discussed in the introduction, this is an approximation of the case that all the connections continue for a long time. Under this assumption, we analyze the asymptotic bandwidth utilization.

Let us introduce some notations. Let $A = \sum_{i=1}^{n} \alpha_i$. The transmission rate at time t is denoted by $r_i[t]$. Let $\boldsymbol{r}[t] = (r_1[t], \ldots, r_n[t])^{\mathrm{T}}$.

For $t_1 \leq t_2$, the amount $W_i[t_1, t_2]$ of data transmitted in connection C_i between time t_1 and t_2 is $W_i[t_1, t_2] = \int_{t_1}^{t_2} r_i[t]\, dt$, and we let $W[t_1, t_2]$ be the total amount of data transmitted in n connections between the same period,

$$
W[t_1, t_2] = \sum_{i=1}^{n} W_i[t_1, t_2] = \int_{t_1}^{t_2} (r_1[t] + \cdots + r_n[t])\, dt.
$$

The (asymptotic) bandwidth utilization U_i of connection C_i and the (asymptotic) total utilization U are defined as the limit of time average of the proportion of transmission rate in available bandwidth[1]:

$$
U_i = \frac{1}{B} \lim_{T \to \infty} \frac{W_i[0, T]}{T} \quad \text{and} \quad U = \frac{1}{B} \lim_{T \to \infty} \frac{W[0, T]}{T}.
$$

A larger total utilization means the algorithm makes use of much bandwidth and that it is efficient. Besides, a small variation in the values of U_i means the algorithm is fair.

Most of the proofs are omitted due to space limitation.

5.1   Total and Per-connection Utilizations in All-Drop Case

Theorem 3. *In the all-drop model, the total and per-connection utilizations are* $U = 1 - \frac{\beta}{2}$ *and* $U_i = \frac{\alpha_i}{A} U$.

The proof of Theorem 3 uses the idea of "adjusted" and "unadjusted" bandwidths used in [6]. Theorem 3 says that in the all-drop model, the total utilization does not depend on the number of connections. This is different than the empirical fact. In the following sections, we consider the single-drop model.

5.2   Total Utilization in Homogeneous Single-Drop Case

In this section we consider the homogeneous single-drop case where $\alpha_1 = \cdots = \alpha_n = \alpha$.

Theorem 4. *In the homogeneous single-drop model, total bandwidth utilization* U *is* $U = \frac{(2-\beta)n}{(2-\beta)n+\beta}$ *regardless of how we choose victim of each drop.*

---

[1] U_i and U may not have limit values depending on the choice of victims. In such cases, U_i and U are not defined.

Proof (sketch). By using the potential function

$$\varphi(\boldsymbol{r}) = \frac{1}{2\alpha} \cdot \frac{(2-\beta)\{B^2 - (B - \sum r_i)^2\} - \beta \sum r_i^2}{(2-\beta)n + \beta}.$$

it can be proven that for any $t_1 \leq t_2$,

$$W[t_1, t_2] + \varphi(\boldsymbol{r}[t_2]) - \varphi(\boldsymbol{r}[t_1]) = B\frac{(2-\beta)n}{(2-\beta)n + \beta}(t_2 - t_1),$$

which proves the claim.

Theorem 4 shows that in the single-drop model, the total utilization U increases as n increases, which means dividing data into multiple streams gives better total throughput. This is different from the case of the all-drop model.

Here is an intuitive interpretation of Theorem 4

Suppose we want to achieve high total utilization by choosing appropriate victim. When the sum $\sum r_i$ of transmission rates hits the bandwidth B, we are forced to choose a victim C_v and decrease the sum $\sum r_i$ by βr_v. One choice is to choose a connection with small r_v as victim to keep $\sum r_i$ relatively high and achieve a high throughput for a moment. But this way the other r_i's will increase a little, meaning that when a connection $C_{v'}$ other than C_v is eventually chosen as victim, $\sum r_i$ will decrease by much. Because we cannot continue choosing C_v as victim for an arbitrarily long time, sooner or later we have to pay for the increased $C_{v'}$, canceling the short-term gain of total utilization.

5.3   Total and Per-connection Utilizations under Periodic Victim Policy

In this section, we consider Periodic victim policy as a typical example of a deterministic policy. This policy is similar to the all-drop model in that it chooses every connection C_i equal times.

Definition 1. Periodic victim policy *is the policy where connection C_1 is chosen as victim of the first drop, C_2 of the next drop, then $C_3, \ldots, C_n$, and this process is repeated infinitely. An example is shown in Figure 2 (b).*

Theorem 5. *Under Periodic victim policy, it holds*

$$U = \frac{2-\beta}{2 - \beta(1 - \sum_{i=1}^{n}(\alpha_i/A)^2)}, \quad U_i = U \cdot \frac{\alpha_i}{A}.$$

The proof of Theorem 5 is based on the fact that the operation on vector $\boldsymbol{r}$ in every period is represented as the multiplication of a matrix. The theorem is obtained by computing the eigenvector of the matrix.

Theorem 5 implies that under Periodic victim policy, the bandwidth is shared in proportion to α_i like the all-drop model, and α_i's with small deviation give better total utilization.

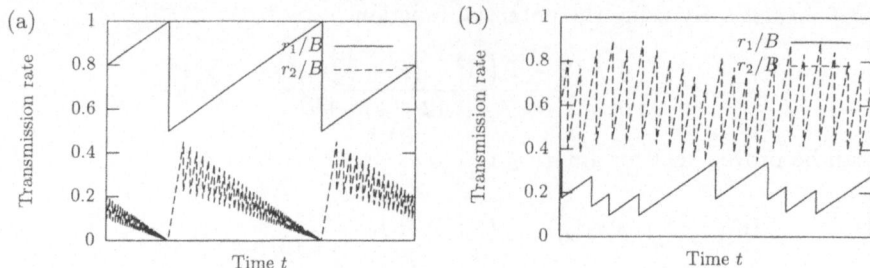

Fig. 3. Time evolution of the transmission rates r_1 and r_2 of the two connections with the accelerations $\alpha_1 : \alpha_2 = 1 : 9$ and the drop factor $\beta = 1/2$, under (a) Priority victim policy and (b) Share-Random victim policy.

5.4 Upper and Lower Bounds of Total Utilization in Heterogeneous Single-Drop Case

In this section, we consider Priority victim policy, which is the most unfair policy in some sense. Figure 3 (a) illustrates this policy. Intuitively, Priority victim policy chooses the connection C_i with the largest i that has nonzero transmission rate as victim. However, this informal definition is not accurate because the transmission rates are always nonzero. Instead, we define Priority victim policy as follows.

Definition 2. *Let* $0 < \varepsilon < 1/n$. ε-*Priority victim policy is the policy where on every drop, connection* C_i *with the largest* i *that satisfies* $r_i \geq \varepsilon B$ *is chosen as victim. Priority victim policy is the limit of* ε-*Priority policy as* $\varepsilon \to 0$.

Theorem 6. *Let* $A_0 = 0$ *and* $A_i = \alpha_1 + \cdots + \alpha_i$. *Under Priority victim policy,*

$$U = 1 - \prod_{i=1}^{n} \frac{(2-\beta)A_{i-1} + \beta\alpha_i}{(2-\beta)A_{i-1} + 2\alpha_i}, \quad U_i = \frac{(2-\beta)\alpha_i}{(2-\beta)A_{i-1} + 2\alpha_i} \prod_{j=1}^{i-1} \frac{(2-\beta)A_{j-1} + \beta\alpha_j}{(2-\beta)A_{j-1} + 2\alpha_j}.$$

When $n = 2$, Priority victim policy gives the maximum and the minimum of the total utilization as the following theorem implies.

Theorem 7. *Let* $n = 2$ *and* $\alpha_1 \leq \alpha_2$. *If the total utilization* U *converges to some value, it holds*

$$1 - \frac{\beta}{2} \cdot \frac{\beta\alpha_1 + (2-\beta)\alpha_2}{2\alpha_1 + (2-\beta)\alpha_2} \leq U \leq 1 - \frac{\beta}{2} \cdot \frac{(2-\beta)\alpha_1 + \beta\alpha_2}{(2-\beta)\alpha_1 + 2\alpha_2}.$$

The proof of Theorem 7 uses potential function method with the potential function

$$\varphi(r) = \frac{((2-\beta)\alpha_1 + \beta\alpha_2)r_1((2-\beta)B - r_1) + 2(2-\beta)\alpha_1(B - r_1)r_2 - 2\alpha_1 r_2^2}{2\alpha_1((2-\beta)\alpha_1 + 2\alpha_2)}.$$

This theorem indicates an interesting fact that as long as the total utilization is concerned, the router should discard the packet from the connection with the higher acceleration upon congestion. This strategy may also be useful to discourage the use of high acceleration by selfish connection, thus achieving high total utilization and penalty to selfish connection at the same time.

5.5 Total Utilization with Two Heterogeneous Connections under (p_1, p_2)-Random Victim Policy

In this section, we assume $n = 2$ and consider the following (p_1, p_2)-Random policy.

Definition 3. *Let $p_1, p_2 > 0$ and $p_1 + p_2 = 1$. (p_1, p_2)-Random victim policy is the policy where on every drop, C_1 is chosen as victim with probability p_1 and C_2 with p_2.*

When a randomized victim policy is used, the value of $W[t_1, t_2]$ varies depending on random choices. Therefore, we consider the expected total utilization $E[U]$ and $E[U_i]$.

Theorem 8. *Let $p_1 = \alpha_1/A$ and $p_2 = \alpha_2/A$. For any $s = (s_1, s_2)^{\mathrm{T}}$ with $s_1 + s_2 \leq B$, the expected total and per-connection utilization under the condition $r[t_0] = s$ are given by $E[U] = \frac{2(2-\beta)}{4-\beta}$ and $E[U_1] = E[U_2] = \frac{2-\beta}{4-\beta}$.*

The proof uses the potential function which is quadratic in r_1 and r_2.

This means that by choosing victim with probabilities proportional to α_i, the total utilization is equal to that in the two-connection homogeneous case and the two connections share the bandwidth in a fair manner, avoiding the inefficiency and unfairness caused by the heterogeneity.

5.6 Simulation of Two Heterogeneous Connections under Share-Random Victim Policy

Definition 4. *Share-Random victim policy is the policy where on every drop, each C_i is chosen as victim with probability r_i/B, as shown in Figure 3 (b).*

Share-Random victim policy is the policy which is most easily implemented by a router placed at the bottleneck. Provided all the packets are infinitesimally short and the same length, the number of packets received by the router for each connection C_i at some moment is in proportion to the transmission rate r_i. When the sum $\sum r_i$ exceeds the capacity B of the router, the router will discard one packet received at the moment, which is for the connection C_i with the probability r_i/B. This scenario assumed the drop-tail behavior of the router, but the same thing happens if the router uses the Random Early Detection (RED) [12] given the buffer in the router is small enough.

We performed the numerical simulation of the utilizations by two connections under Share-Random policy, with B and $\beta = 1/2$ fixed and α_1 and α_2 altered

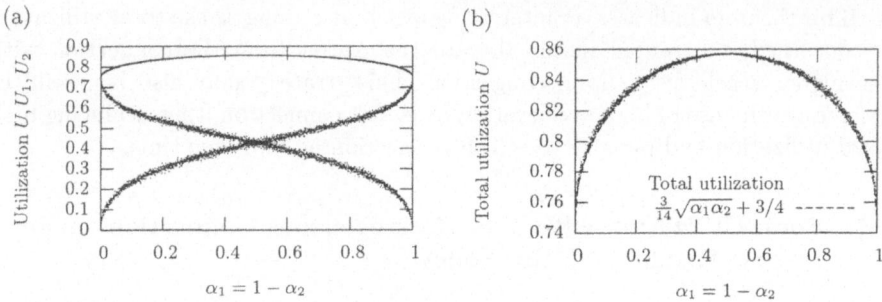

Fig. 4. Total utilization U and the utilizations U_i by each connection C_i of two connections under Share-Random victim policy with $\beta = 1/2$, $B = 1$ and different values for α_1 and α_2, while keeping $\alpha_1 + \alpha_2 = 1$.

while maintaining $A = \alpha_1 + \alpha_2 = 1$. Figure 4 shows the total and per-connection utilizations in this case. From Figure 4 (b) and the results with other values of β, we conjecture the following.

Conjecture. *In heterogeneous two-connection case under Share-Random victim policy, it holds*

$$E[U] = \left(1 - \frac{\beta}{2}\right)\left(1 + \frac{2\beta}{4 - \beta} \cdot \frac{\sqrt{\alpha_1 \alpha_2}}{A}\right).$$

In addition, Figure 4 (a) suggests that in Share-Random case, the sharing of bandwidth among the connections is closer to the fair sharing than the all-drop case and the single-drop Periodic case. It is nearly proportional to the square root of the acceleration, or inversely proportional to RTT. This can be interpreted that the Share-Random victim policy mitigates the unfairness caused by different accelerations by choosing the connection with higher throughput more often than the other connection.

6   Conclusion and Future Works

We performed analyses of the performance of AIMD congestion control algorithm focusing on the heterogeneity of accelerations α_i of connections. The competitive analysis of the total flow time showed that the performance loss was caused by the heterogeneity, suggesting that one of the causes of performance problem of TCP in LFNs is the mixture of connections with different distances, rather than just the long distances of connections. The analysis of bandwidth utilization in stationary state revealed that the victim policy greatly affected the performance. When two connections with different accelerations exist, the maximum total utilization is achieved by Priority victim policy and the fair utilization is achieved by (p_1, p_2)-Random policy.

To tackle the LFN problem, many alternative congestion control algorithms for TCP have been proposed [13,14,15]. The extension of our analyses to these new congestion control algorithms will be useful to compare them.

References

1. Jacobson, V., Braden, R.: TCP extensions for long-delay paths. RFC 1072 (1988) Obsoleted.
2. Jacobson, V., Braden, B., Borman, D.: TCP extensions for high performance. RFC 1323 (1992)
3. Nakamura, M., Inaba, M., Hiraki, K.: Fast Ethernet is sometimes faster than Gigabit Ethernet on LFN — observation of congestion control of TCP streams. In: Proceedings of the 15th IASTED International Conference on Parallel and Distributed Computing and Systems (PDCS), ACTA Press (2003) To appear.
4. Hiraki, K., Inaba, M., Tamatsukuri, J., Kurusu, R., Ikuta, Y., Koga, H., Zinzaki, A.: Data Reservoir: Utilization of multi-gigabit backbone network for data-intensive research. In: Proceedings of the IEEE/ACM SC2002 Conference. (2002)
5. Chiu, D.M., Jain, R.: Analysis of the increase and decrease algorithms for congestion avoidance in computer networks. Computer Networks and ISDN Systems **17** (1989) 1–14
6. Edmonds, J., Datta, S., Dymond, P.W.: TCP is competitive against a limited adversary. In: Proceedings of the Fifteenth Annual ACM Symposium on Parallel Algorithms and Architectures. (2003) 174–183
7. Mathis, M., Semke, J., Mahdavi, J., Ott, T.: The macroscopic behavior of the TCP congestion avoidance algorithm. Computer Communication Review **27** (1997)
8. Lakshman, T.V., Madhow, U.: The performance of networks with high bandwidth-delay products and random loss. IEEE/ACM Transactions on Networking **5** (1997) 336–350
9. Padhye, J., Firoiu, V., Towsley, D., Kurose, J.: Modeling TCP throughput: A simple model and its empirical validation. In: Proceedings of the ACM SIGCOMM '98. (1988) 303–314
10. De Vendictis, A., Baiocchi, A.: Modeling a mixed TCP Vegas and TCP Reno scenario. In: Networking 2002: Proceedings of 2nd International IFIP-TC6 Networking Conference. Volume 2345 of Lecture Notes in Computer Science. (2002) 612–623
11. Karp, R.M., Koutsoupias, E., Papadimitriou, C.H., Shenker, S.: Optimization problems in congestion control. In: 41st Annual Symposium on Foundations of Computer Science (FOCS), IEEE Computer Society (2000) 66–74
12. Floyd, S., Jacobson, V.: Random early detection gateways for congestion avoidance. IEEE/ACM Transactions on Networking **1** (1993) 397–413
13. Floyd, S.: HighSpeed TCP for large congestion windows. Internet Draft (work in progress) (2003)
    http://www.ietf.org/internet-drafts/draft-ietf-tsvwghighspeed-01.txt.
14. Kelly, T.: Scalable TCP: Improving performance in highspeed wide area networks. First International Workshop on Protocols for Fast Long-Distance Networks (PFLDnet 2003) (2003)
    http://datatag.web.cern.ch/datatag/pfldnet2003/papers/kelly.pdf.
15. Jin, C., Wei, D.X., Low, S.H.: FAST TCP for high-speed long-distance networks. Internet Draft (work in progress) (2003)
    http://netlab.caltech.edu/pub/papers/draft-jwl-tcp-fast-01.txt.

Last Encounter Routing
under Random Waypoint Mobility

Natasa Sarafijanovic-Djukic and Matthias Grossglauser

School of Computer and Communication Sciences
EPFL – Lausanne, Switzerland

Abstract. Last Encounter Routing (LER) algorithms for mobile ad hoc networks rely only on encounter histories at every node to route packets, and therefore do not need control traffic to track topology changes due to node mobility. LER exploits the fact that past information about a node's mobility helps to locate that node in the future. As we have pointed out in earlier work [1], the performance of LER algorithms depends on the mobility processes of nodes.

In this paper, we ask whether LER can work under the random waypoint (RWP) mobility model. This question is important for several reasons. First, as shown in [1], a good performance for the RWP model is harder to achieve than for another prominent mobility model, the random walk. This is because the RWP model has a much shorter relaxation time, i.e., a time-horizon over which past information is still useful. Also, the RWP model has a much less favorable ratio of number of encounters between nodes and the traveled distance. Second, in contrast to the random walk, the RWP model is predictable. This provides us with an opportunity to exploit additional information collected in an encounter (such as speed, direction, etc.) to improve routing.

We formally define the RWP model, and compute the optimal predictors for several observation sets, i.e., observed parameters of node mobility. We develop a new LER algorithm tuned to the RWP model called GREASE-RWP, and present simulation results that demonstrate that an efficient and scalable LER for the RWP model is possible.

1 Introduction

Last Encounter Routing (LER) [1,2,3] refers to a type of routing algorithm where the destination of a packet is located without the help of a location service, and without any control traffic to track topology changes due to node mobility (except for local discovery packets to detect immediate neighbors). Rather, a packet is routed using only the encounter histories at nodes it is forwarded through. In its basic form, the encounter history consists of the time and location when a node was a directly connected neighbor of another node.

In [1], the following model was considered. A set of nodes perform independent random walks on a square lattice. Two nodes are directly connected neighbors if they reside at the same lattice point. Every node remembers when and where it has encountered every other node, in a *last encounter table*. A very

N. Mitrou et al. (Eds.): NETWORKING 2004, LNCS 3042, pp. 974–988, 2004.
© IFIP International Federation for Information Processing 2004

simple algorithm called EASE was introduced in [1] to compute a route from a source node to a destination node, based only on LE history at every node. It was shown that the expected total cost of EASE routes is a small multiple of the expected shortest path length between a random source and destination. In other words, EASE is a scalable LER algorithm for the random walk mobility model, as the cost of routes relative to the shortest path does not blow up as the network size increases.

It is clear that the performance of LER algorithms is closely tied to the mobility pattern of the nodes in the network. To see this, consider an extreme scenario where there is no dependence between a node's position at different times, i.e., nodes "jump around" randomly in the network domain. In this case, history information is of no use, and any LER algorithm would perform as poorly as an exhaustive search for the destination.

The main question, then, is the interplay between the mobility model and the performance of LER algorithms. A complete answer to this question remains elusive, but it is possible to develop some intuition about features of the mobility model that are favorable to LER algorithms. In particular, we argue that the following three features of a mobility model help LER: locality, frequent intersections, and homogeneity. *Locality* means that a node's position at a time t correlates with its position at a certain time $t + \tau$ in the future; this ensures that information about past encounters is actually useful in locating a node. *Frequent intersections* mean that a node over a given time interval tends to encounter a large number of other nodes. This ensures that information about that node's location in the LE tables of other nodes is refreshed frequently. *Homogeneity* means that the statistical properties of each node mobility are similar. This ensures that the speed of diffusion of encounter histories due to movements of other nodes is matched to that of a destination node. In this paper, all the scenarios we consider are homogeneous.

Based on these three features, we can see that the random walk mobility model is quite advantageous to LER. First, as a node performs independent steps over time, the dependence between the current and a future position of the node decreases only slowly with time. Specifically, it follows from the central limit theorem that the difference between the two positions t seconds apart is a random variable with variance proportional to $\sqrt{t}$. Second, despite the locality, a node encounters other nodes frequently. Specifically, it was shown in [1] that over a time interval of length t, a node encounters $\Theta(t/\log t)$ other nodes. As a result, in the random walk mobility model, we observe that (i) a node's past location contains information about its current location over a relaxation time of $\Theta(n)$, where n is the network size, and (ii) when a node moves by a distance d, it encounters approximately $\Theta(d^2/\log d)$ other nodes. This means that information about a destination node's movement is quite dense around this node, and that such an algorithm can rely on fairly old information to locate a destination.

In this paper, we consider LER for the random waypoint (RWP) mobility model. Informally, this model is defined as follows. Every node moves independently of any other node. A node selects a random waypoint uniformly in the area of the network, and moves towards this waypoint in a straight line and at constant speed. Once it reaches this waypoint, it selects a new waypoint independently of the previous one, and starts moving towards it (possibly after some pause time), and so forth.

There are several reasons why we are interested in LER for the RWP model. First, the model is well studied and is very prominent in simulation studies of mobile ad hoc networks. Second, it was shown in [1] that the LER algorithms EASE and GREASE did not perform well with this model. Third, given the discussion of helpful mobility features above, the RWP model is much less favorable to LER than the random walk.

To see this, let us perform a back-of-the-envelope comparison with the random walk. Note that the average transition time between waypoints is $\Theta(\sqrt{n})$, proportional to the diameter of the network. Once a node has gone through a waypoint, the dependence with its current position drops very rapidly. A past position of a node is therefore only useful for $\Theta(\sqrt{n})$ time. Furthermore, when a node moves by a distance d, it encounters only $\Theta(d)$ other nodes. This means that information about a destination node's movement is much less dense around this node, and that an LER algorithm must rely on fairly recent information to catch up with a destination.

On the other hand, the RWP model has a feature that is in our favor: the node movement is highly predictable over a short time-scale, because a node moves at constant speed on a straight line between waypoints. This provides us with an opportunity to route a packet towards the predicted current position of its destination, rather than simply towards the location of the encounter with the destination. This is not possible in the random walk case, where the best predictor for a node's future position is simply its current position.

In this paper, we introduce a version of the GREASE algorithm in [1] that is tuned to the specific features of the RWP mobility model. GREASE-RWP takes into account the short relaxation time of random waypoint mobility by reducing the encounter age it searches for initially to obtain the first estimate of the destination's location. It then uses mobility prediction as the packet approaches the destination in order to move towards the destination in "shortcuts". We evaluate GREASE-RWP through simulation in networks with up to 1000 nodes, and we find that GREASE-RWP has significantly better performance than GREASE, and seems to scale with network size. Therefore, LER can be efficient even for much less favorable mobility models than the random walk.

The paper is structured as follows. In Section 2, we formally define the RWP model and discuss some of its properties that matter in the context of LER. In Section 3, we present an LER algorithm specifically designed for the RWP model. We present simulation results in Section 4. Section 5 concludes the paper.

2   The Random Waypoint Mobility Model

In this section we formally define the RWP model, and discuss the properties that play an important role in the performance of LER.

Nodes move on a square torus of side a. The origin is in the center of the square. The vector $c - b$ is the vector on the torus surface such that

$\|c - b\| \equiv d(b, c)$ is the shortest distance between the points b and c. Nodes move independently of each other, so it is enough to define the movement of one node.

We consider a special case of the RWP model for simplicity. Specifically, we assume that nodes move with constant speed v, and that nodes do not pause at waypoints.

This RWP model is formally defined as follows.

- The speed v.
- the sequence of independently and identically distributed (i.i.d.) random variables $\{P_w\}_{w \in \mathbb{N}}$, uniformly distributed in the torus.

The P_w is the w-th waypoint. If the node moves between the w-th and $w + 1$-st waypoints we say that it is on the w-th segment. The distance between these waypoints $L_w = d(P_w, P_{w+1})$ is the w-th segment length. The time for the node to traverse the w-th segment is the w-th segment duration. The time instant S_w when the node is at the w-th waypoint is equal to:

$$S_w = \sum_{j=1}^{w-1} \frac{L_j}{v},$$

where $S_1 = 0$. The speed vector between the w-th and $w + 1$-st waypoints is equal to:

$$V_w = v \frac{P_{w+1} - P_w}{L_w}.$$

Therefore, the node's position at time t can be formally expressed as:

$$X(t) = P_{w(t)} + V_{w(t)}(t - S_{w(t)}),$$

where $w(t)$ is an index such that $t \in [S_{w(t)}, S_{w(t)+1})$.

Note that on the torus the L_ws are i.i.d, thus the random process $\{S_w\}_{w \in \mathbb{N}}$ is a renewal process. This property comes from the fact that torus represents isotropic space where all points are equivalent.

In the remainder of this section, we assume that time $t = 0$ is the beginning of the observation period for the renewal process $\{S_w\}_{w \in \mathbb{N}}$ that has been operating long enough to be in steady state[1]. We enumerate the points of the renewal process as $\{S_w\}_{w \in \mathbb{Z}}$ with the convention that $S_0 \leq 0 < S_1$.

2.1   Relaxation Time

LER algorithms exploit a key property of realistic mobility processes, which is that the location of a node at a time t and a time $t + \tau$ are dependent, and that therefore information collected through an encounter at time t can be useful for a packet looking for its destination at time $t + \tau$. More specifically, consider

---

[1] The RWP model with random speed does not possess a steady state in terms of average speed [5]. Here, this problem does not arise because of constant speed.

a realistic mobility scenario where nodes have limited speed. Suppose that a source knows the destination position $X(t)$ and wants to route a packet to the destination at time $t + \tau$. If τ is relatively small, then the destination is in the small area around $X(t)$. Thus, information $X(t)$ is useful. If τ increases, then the area where the destination may be becomes larger and information $X(t)$ is less useful. For some large τ, the destination may be anywhere in the network area and information $X(t)$ is useless.

The relaxation time T_r is the minimum τ for which $X(t+\tau)$ does not depend on $X(t)$ any more (note that as we assume homogeneity, all nodes have the same relaxation time). The relaxation time is the maximum time after its observation that past information about a node's movement is still of use. It is therefore crucial to ensure that a LER algorithm does not rely on observations older than T_r, as this would lead to poor routes.

Let us consider the relaxation time in the RWP model. Let $Y_t = S_{w(t)+1} - t$ be the time the node travels to the next waypoint. After time $Y_t + L/v$ (L denotes the distribution of the segment lengths $L_w s$) the node will reach the second next waypoint. Hence, $X(t + \tau)$ for $\tau > Y_t + L/v$ depends only on $P_w s$, where $w \geq w(t) + 2$. The $X(t)$ depends only on $P_{w(t)}$ and $P_{w(t)+1}$. Therefore, the positions $X(t)$ and $X(t + \tau)$ are independent for $\tau > Y_t + L/v$.

2.2   Prediction

In the RWP model, a node moves in a sequence of lines. This results in fewer encounters with other nodes relative to the traveled distance, and hence a low density of nodes with recent encounters around this node. On the other hand, this movement also implies excellent predictability of a node's position a certain time into the future, given its current location and other observations, such as direction and speed. Our goal is to use prediction of the destination position in order to decrease the routing cost of LER.

The observations available to a node on its own mobility process depend on the scenario. At one extreme, a node might be able to determine only some basic parameters about its instantaneous movement, such as its direction and speed. This might arise when a node's movement is subject to external influences, and if the node lacks the capability to identify waypoints. At the other extreme, a node might be able to know its precise movements a long time into the future. This might arise when a node's movement is predetermined by the node itself (e.g., a person running errands in a city, or a doctor visiting patients). Obviously, the predictability improves as the observation set increases. As our focus here is on the "tough" cases, we only consider observation sets that do not look beyond the next waypoint. Specifically, we calculate mobility predictors for the following three different observation sets θ:

1. the node's current position and speed vector;
2. the node's current position, speed vector and previous waypoint;
3. the node's current position, speed vector and next waypoint.

The optimal mobility predictor is the one that minimizes in some sense the error between the node's predicted and true positions. We use the minimum mean square error criterion [7].

Definition 1. *The minimum mean square error (MMSE) predictor of the node's position $X(t + \tau)$ given an observation $\theta(t)$ at time t minimizes*

$$E[(X(t + \tau) - \widehat{X}(t + \tau))^2 \mid \theta(t)]. \tag{1}$$

and is given by

$$\widehat{X}(t + \tau) = E[X(t + \tau) \mid \theta(t)]. \tag{2}$$

We derive our predictors under the following assumptions. First, we assume that we make the observation at a random time t. Second, we neglect the direction the node is moving to. The length of the current interarrival interval is dependent on the direction; we ignore this for simplicity.

Theorem 1. *The MMSE predictor $\widehat{X}(t + \tau)$ given the following observations:*

- *the node's current position $X(t) = x(t)$*
- *the node's current speed vector $V(t) = v(t)$*

is given by:

$$\widehat{x}(t + \tau) = x(t) + \frac{v(t)}{v}(P(Z \geq v\tau)v\tau + E[Z1_{\{Z < v\tau\}}]), \tag{3}$$

where Z is a random variable with the pdf equal to:

$$f_Z(z) = \begin{cases} c(1 - \frac{z^2\pi}{a^2}) & \text{if } 0 \leq z \leq \frac{a}{2} \\ c(1 - \frac{z^2\pi}{a^2} + \frac{4z^2}{a^2}\arccos\frac{a}{2z} - \frac{2}{a}\sqrt{z^2 - \frac{a^2}{4}}) & \text{if } \frac{a}{2} \leq z \leq \frac{a\sqrt{2}}{2}, \\ 0 & \text{otherwise} \end{cases} \tag{4}$$

where $c = 6(a\sqrt{2} + a\ln(1 + \sqrt{2}))^{-1}$.

Theorem 2. *The optimal mobility predictor of $X(t + \tau)$ given the following observations:*

- *the node's current position $X(t) = x(t)$*
- *the node's current speed vector $V(t) = v(t)$*
- *the node's previous waypoint $P_{w(t)} = p$*

is given by:

$$\widehat{x}(t + \tau) = x(t) + \frac{v(t)}{v}(P(Z \geq v\tau)v\tau + E[Z1_{\{Z < v\tau\}}]), \tag{5}$$

where Z is a random variable with pdf

$$f_Z(z) = \frac{f_L(z + l_0)}{1 - F_L(l_0)}, \tag{6}$$

where $l_0 = d(p, x(t))$, and $f_L(l)$ and $F_L(l)$ are the pdf and the cdf of the segment length (see Lemma 1 in the Appendix).

Theorem 3. *The optimal mobility predictor of $X(t + \tau)$ in the defined RWP model if we know the following observations:*

- *the node's current position $X(t) = x(t)$*
- *the node's current speed vector $V(t) = v(t)$*
- *the node's next waypoint $P_{w(t)+1} = p_{next}$*

is given by:

$$\widehat{x}(t + \tau) = \begin{cases} x(t) + v(t)\tau & \text{if } v\tau \leq d(x(t), p_{next}) \\ p_{next} & \text{if } v\tau > d(x(t), p_{next}) \end{cases}. \qquad (7)$$

Next, we give the main idea on how we derive the predictors. The details can be found in the Appendix. To calculate the conditional expectation of the $X(t + \tau)$ given the observation set $\theta(t)$, we express $X(t + \tau)$ as

$$X(t + \tau) = X(t) + 1_{\{\tau \leq Y_t\}} V(t)\tau + 1_{\{\tau > Y_t\}}(V(t)Y_t + \triangle P(\tau - Y_t)), \qquad (8)$$

where $Y_t = S_{w(t)+1} - t$ is the time until the node hits the next waypoint, and $\triangle P(\tau - Y_t)$ is the displacement of the node after reaching the next waypoint until time $t + \tau$. Figure 1 explains (8). Expected value of the $\triangle P(\tau - Y_t)$ is equal to zero (Lemma 2 in the Appendix). Thus, we need to calculate only the conditional expectation of the Y_t given the $\theta(t)$. Since we observe the system at a random point in time, the Y_t is the residual time of the renewal process $\{S_w\}_{w \in \mathbb{N}}$. Note that the segment duration on which the node is at a random time t is not distributed according to the distribution of the intervals $S_{w+1} - S_w$. This is because a long segment is more likely to be "intercepted" by our observation than a short one [6]. We use elements of Palm calculus (Chapter 12,[8]) to relate these different viewpoints and to calculate the distribution of the residual time in the concrete case. Since the calculation for Theorem 3 is trivial, we give further calculations only for Theorems 1 and 2 in the Appendix.

We are unable to calculate the mobility predictors in Theorems 1 and 2 in a closed form. Instead, we use approximations to derive good closed-form predictor (cf. Approximations 1 and 2 in the Appendix).

3   A LER Algorithm for the Random Waypoint Model

We now introduce a new LER algorithm that takes into account the features of the RWP model described in the previous section. We first define some notation. Nodes are indexed by $1, 2, ..., n$ where n is the number of nodes. We focus on a single destination node with index 1 and assume w.l.g. that a packet is sent to the destination at time $t = 0$. The $X_i(t)$ is the position of the node i at time t. The $T_i(t)$ is the age of last encounter between node i and the destination. The $\theta_i(t)$ is the observation set of the node i at time t. The optimal mobility predictor of the destination position at time t is denoted by $\widehat{X_1}(t)$. Recall that $T_r = \Theta(\sqrt{n})$ is the relaxation time.

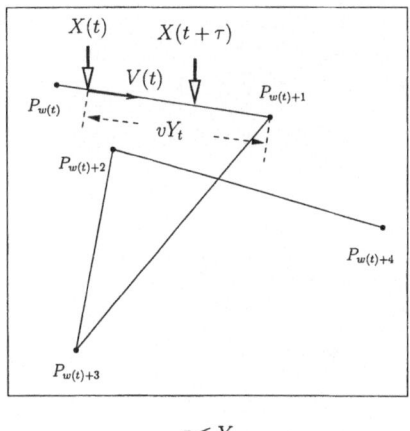 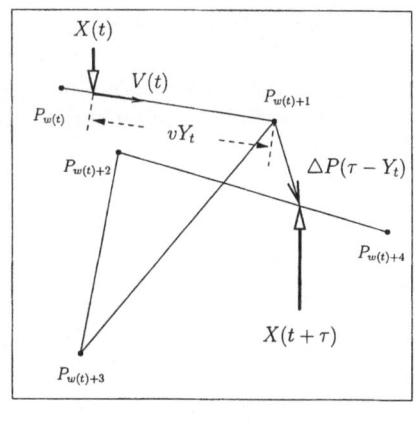

$$\tau \leq Y_t \qquad\qquad\qquad \tau > Y_t$$

Fig. 1. Explanation of (8): one possible movement of a node and its positions at times t and $t + \tau$ for different values of τ.

Algorithm 1: GREASE-RWP

1   Set $Y_0 := X_s(0)$, $T_0 := E[T_r]$, $k := 0$.
2   Repeat
3   Search the nodes around Y_k in the order of increasing distance until a node i is found such that $T_i(0) < T_k$.
4   Let $T_{k+1} = T_i(0)$, and $Y_{k+1} := \widehat{X_1}(0) = f(\theta_1(-T_{k+1}))$ be the new anchor point.
5   While not at Y_{k+1}
6   Route packet: find next hop j towards Y_{k+1} and forward packet to j.
7   If $T_j(0) < T_{k+1}$, then $T_{k+1} := T_j(0)$, $Y_{k+1} := \widehat{X_1}(0) = f(\theta_1(-T_{k+1}))$.
8   End while
9   k++.
10 Until $Y_k = X_1(0)$.

The main new features in GREASE-RWP in comparison with GREASE [1] are a more aggressive initial search and the use of prediction.

Aggressive initial search means that a source searches for an encounter with the destination younger than average relaxation time. As we saw in Sect. 2.1 average relaxation time is of the order of a segment time. This means that the packet immediately goes to the few last segments of the destination movement. Thus the packet avoids the useless walking over the network area.

Prediction is incorporated into GREASE by sending a packet towards the predicted position of the destination rather than towards the location of the encounter. Thus the packet tends to take "shortcuts" from segment to segment.

Of course, the use of prediction requires nodes to remember additional statistics about other nodes, in addition to the position of last encounter. This is incorporated into LER as follows. Every node tracks its own mobility. When two nodes encounter each other, they exchange observations about their own mobility (e.g., their current speed and direction) along with the "hello" messages. Each node records these observations in its LE table (cf. Fig. 2). They are used to calculate the optimal mobility predictor.

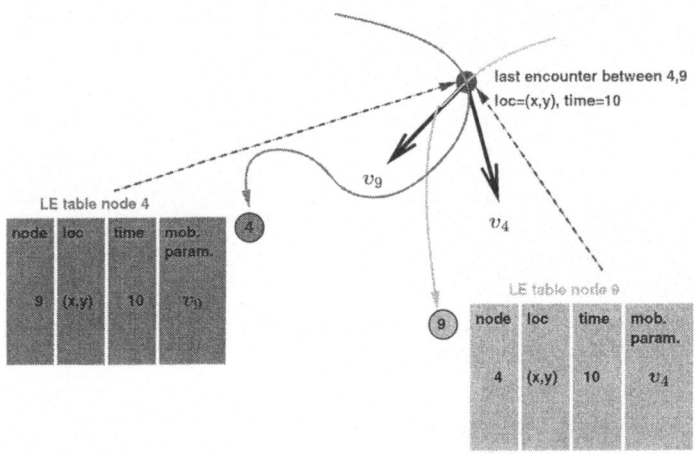

Fig. 2. A *last encounter table* in every node remembers both the location and time of the last encounter with every other node in the network, and in addition, some observations about every other node's movement at the time of the last encounter (in this example, speed vector).

4   Simulation Results

We have performed extensive simulations to evaluate the efficiency and scalability of GREASE-RWP. By efficiency, we mean the average routing cost relative to the shortest path cost. By scalability, we mean how the efficiency scales with increasing the network size.

Nodes move on a torus of surface n according to the RWP model. Initially the nodes are placed uniformly in the torus. We let them to move for a sufficient warm-up period so that a fair proportion of node pairs have encountered at least once. Then, we assume that the nodes are frozen for the time of the routing of the packet.

Routing of the packet is performed through the GREASE (Algorithm 2 in [1]), GREASE-RWP (Algorithm 1 in this paper), GREASE-M and GREASE-A algorithms. GREASE-M is GREASE with mobility prediction. We obtain it by changing the step 4 in GREASE with the step 4 in GREASE-RWP. GREASE-A is GREASE with aggressive initial search. We obtain it by changing the step 1 in GREASE with the step 1 in GREASE-RWP. The M_i ($i = 1, 2, 3$) denotes respectively the predictors from Theorems 1-3. If nothing is specified the predictor M2 is used.

At every time t, we assume that connectivity is given by the Delaunay graph generated by the set of points $\{X_i(t)\}$. This is equivalent to generating the Voronoi tessellation of the set of points $\{X_i(t)\}$, such that every node $X_i(t)$ is the center of a Voronoi cell, and is connected to the center nodes of its adjacent cells. Each node updates the entries in its LE table for its directly connected neighbors.

The advantage of this topology over other topologies (e.g., k nearest neighbors) is that we are guaranteed that a node always has a neighbor that is closer to the destination (except when that destination is already in the first node's Voronoi cell). Therefore, a packet can always make progress towards its anchor point, and we do not have to deal with backtracking, avoiding routing loops, etc., which are not the focus of this work. This allows us to focus on the main issue at hand, i.e., the quality of routes based on diffused information about last encounters.

The main metric we evaluate is the relative cost of the routes compared with the cost of the shortest path route. This metric therefore captures the relative penalty incurred for not having the exact position of the destination available.

Figure 3 shows an example of a route computed by GREASE and GREASE-RWP for the same source- destination pair. We see that GREASE-RWP achieves the shorter route than GREASE by increasing the cost of the initial search and by taking shortcuts with the help of mobility prediction.

In Fig. 4(a) and 4(b) we give the relative cost conditional on the distance between the source and destination. This provides an indication whether the relative quality of the routes increases or decreases as the routes get longer.

Figure 4(a) shows the benefit of mobility prediction if different observation sets are used. If more information about node mobility is available, improvement of GREASE is better.

Figure 4(b) shows that the penalty of the GREASE-RWP algorithm because of the uncertainty of the destination location is only 2.5 times greater than the shortest route (i.e., the ideal case where the destination location is known). Also, it shows that aggressive initial search or mobility prediction alone significantly increase efficiency.

In Fig. 5 we give dependence of the average relative cost of routes on the number of nodes n. This provides an indication whether the relative cost of the routes is scalable with the network size.

5   Discussion and Conclusion

In this paper, we have shown that efficient and scalable last encounter routing under random waypoint mobility is possible. We have achieved this by devising a new instance of the GREASE algorithm, which differs from the version reported in [1] in two respects. First, we have exploited the inherent predictability of nodal movement in the RWP model over short time-scales. When a packet looking for its destination picks up a more recent encounter, it can compute a predicted location for the destination that is better on average than the location of that encounter itself. Second, we account for the fact that the RWP model has a very

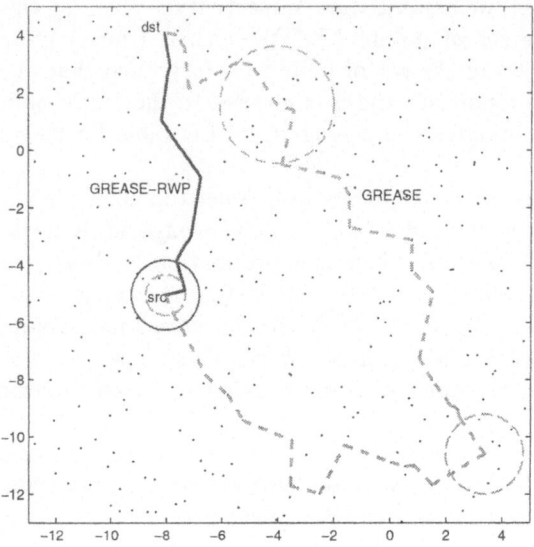

Fig. 3. A sample route computed by GREASE and GREASE-RWP for the same source-destination pair.

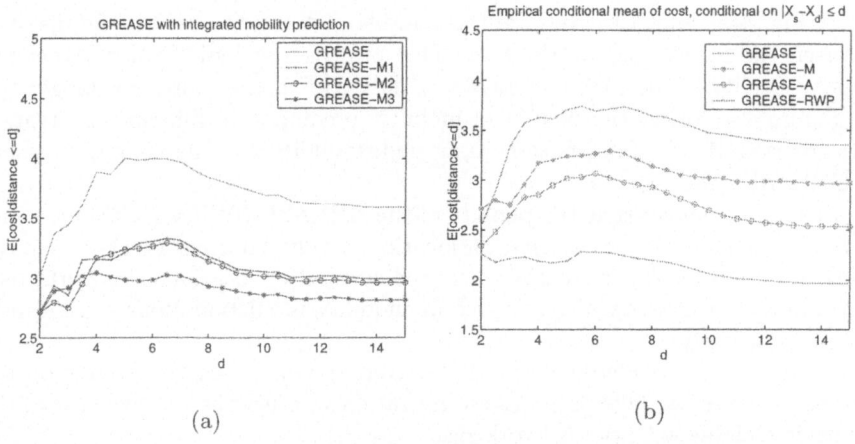

(a)   (b)

Fig. 4. a) Impact of the different observation sets in mobility prediction on the efficiency of GREASE. b) Efficiency of GREASE, GREASE-M (mobility prediction), GREASE-A (aggressive initial search) and GREASE-RWP (mobility prediction and aggressive initial search).

short relaxation time, by forcing a low target age for the initial search. Thus the packet avoids using outdated past encounters that are independent of the destination's actual position.

We have performed simulations that show the following results. First, the GREASE-RWP algorithm achieves drastically better performance than the "non-RWP" version reported in [1]. Specifically, the total average route cost

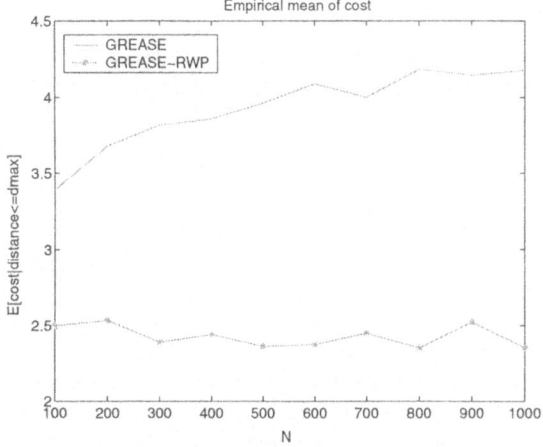

Fig. 5. Scalability of the GREASE-RWP algorithm.

is slightly more than twice as long as the shortest path. This is quite remarkable, given that no resources were invested to track the rapid change in the network topology due to RWP mobility. Second, our results show, as we would expect, that the benefit of prediction depends on what observations are available about a node's mobility. It should be pointed out that if we extended prediction beyond the next waypoint, the performance would be further improved. Third, the cost of GREASE-RWP routes relative to the shortest path does not seem to increase as we scale up the network size n. Therefore, we believe that a similar scaling result as shown in [1] for the random walk holds for the RWP model as well.

Although this paper shows that efficient and scalable LER is possible beyond the random walk mobility model, we are not yet able to fully characterize the space of mobility models for which this is true. However, as the RWP model can be considered unfavorable to LER, our results suggest that LER may be applicable to a large class of mobility processes. In our future work, we plan to find more general necessary conditions for a mobility process to be feasible.

Another open question concerns the generality of particular instances of LER algorithms. Note that GREASE-RWP would perform poorly with the random walk model, because the initial search criterion would be too aggressive; whereas, the original GREASE described in [1] has been shown to perform relatively poorly for the RWP model. Our ultimate goal, of course, would be to devise a LER algorithm that is *universal*, i.e., that would perform well over the entire class of feasible mobility models. We conjecture that such a model would require some a-priori knowledge about the mobility process, as it would be essentially impossible otherwise to decide what initial age to search for. Such a-priori knowledge might in turn be obtained through some estimation process that would run in the background as an "outer loop" (e.g., through a gossiping protocol) and make estimated mobility parameters available at every node.

References

1. M. Grossglauser and M. Vetterli. Locating Nodes with EASE: Last Encounter Routing through Mobility Diffusion. *IEEE Infocom*, April 2003, San Francisco, CA.
2. H. Dubois-Ferrière, M. Grossglauser and M. Vetterli. Age Matters: Efficient Route Discovery in Mobile Ad Hoc Networks Using Encounter Ages. In Proc. ACM MO-BIHOC 03, Maryland, June 2003.
3. H. Dubois-Ferrière, M. Grossglauser and M. Vetterli. Space-Time Routing in Ad Hoc Networks, Ad Hoc Now 03, MontrÃ©al, Canada, October 2003.
4. David B. Johnson and David A. Maltz. Dynamic Source Routing in Ad Hoc Wireless Networks. In Mobile Computing, edited by Tomasz Imielinski and Hank Korth, Chapter 5, pages 153-181, Kluwer Academic Publishers, 1996.
5. J. Yoon, M. Liu and B. Noble. Random Waypoint Considered Harmful. In Proc. IEEE INFOCOM, vol 2, pp 1312-1321, April 2003, San Francisco, CA.
6. L. Kleinrock. Queueing Systems, Volume 1, Jhon Wiley & Sons, 1975.
7. R. Gray and L. Davisson. An Introduction to Statistical Signal Processing. http://www-ee.stanford.edu/ gray/sp.html
8. Jean-Yves Le Boudec. Performance Evaluation Lecture Notes. EPFL, 2003. http://ica1www.epfl.ch/perfeval/lectureNotes

Appendix

Auxiliary Results

Lemma 1. *The sequence of the segment lengths L_ws has the following pdf:*

$$
f_L(l) = \begin{cases} \frac{2l\pi}{a^2} & \text{if } 0 \leq l \leq \frac{a}{2} \\ \frac{2l\pi}{a^2} - \frac{8l}{a^2} \arccos \frac{a}{2l} & \text{if } \frac{a}{2} \leq l \leq \frac{a\sqrt{2}}{2} \\ 0 & \text{otherwise} \end{cases} .
\tag{9}
$$

Lemma 2. *The expected displacement of a node from a waypoint is equal to zero, i.e., the expectation of the random variable $X(S_w + \triangle t) - X(S_w)$ is equal to zero, for every w and $\triangle t > 0$.*

Optimal Predictors

Proof of Theorems 1 and 2. Using (2) and (8), the optimal mobility predictor is:

$$
\hat{x}(t + \tau) = E[X(t + \tau) \mid \theta(t)] =
$$
$$
= x(t) + v(t)\tau P(Y_t \geq \tau \mid \theta(t)) + v(t)E[Y_t 1_{\{Y_t < \tau\}} \mid \theta(t)] +
$$
$$
+ E[\triangle P(\tau - Y_t) 1_{\{Y_t < \tau\}} \mid \theta(t)].
$$

The last term of this equation is equal to 0 (Lemma 2). Thus, we obtain:

$$
\hat{x}(t + \tau) = x(t) + v(t)\tau P(Z \geq v\tau) + \frac{v(t)}{v} E[Z 1_{\{Z > v\tau\}}],
$$

where Z is the random variable with the same distribution as vY_t given $\theta(t)$. Next, we calculate the distribution of Z separately for Theorems 1 and 2.

In Theorem 1, the set of observations is $\theta(t) = \{X(t) = x(t), V(t) = v(t)\}$. The residual time Y_t depends only on the renewal process S_w. The S_ws depend only on the segment lengths L_ws because of the constant speed. The L_ws do not depend on $X(t)$ since every point in the torus is the same. Therefore, the Y_t does not depend on $X(t)$. As mentioned previous, we neglect that the Y_t depends on the direction of the speed vector, thus the Y_t does not depend on $V(t)$. Hence, the Z has the same distribution as vY_t. The pdf of residual time Y_t is equal to [6]:

$$f_Y(y) = \frac{1 - F_R(y)}{m_R},$$

where $F_R(x)$ is the cdf of the intervals $R_w = S_{w+1} - S_w$ and m_R is the mean of R_w. Since $R_w = L_w/v$, the pdf of Z is:

$$f_Z(z) = \frac{1 - F_L(z)}{m_L},$$

where $F_L(z)$ is the cdf of L_ws (Lemma 1), and $m_L = a/6(\sqrt{2} + \ln(1 + \sqrt{2}))$ is its expected value. We obtain the pdf of Z given by (4).

In Theorem 2, the set of observations is $\theta(t) = \{X(t) = x(t), V(t) = v(t), P_{w(t)} = p\}$. As in the previous case, the residual time Y_t does not depend on $X(t)$ and $V(t)$. Knowing both $X(t) = x(t)$ and $P_{w(t)} = p$ we know that the previous waypoint was at time $t - d(p, x(t))/v = t - t_0$. This means that there is a point at time $t - t_0$ and that there is no point in interval $(t - t_0, t)$. This second condition is equivalent to $Y_{t-t_0} \geq t_0$. If we denote the conditional probability given that there exists a point at time t as P^t then the cdf of Z is equal to:

$$
\begin{aligned}
P(Z \leq z) &= P(vY_t \leq z \mid X(t) = x(t), V(t) = v(t), P_{w(t)} = p) \\
&= P(vY_t \leq z \mid \text{there exists a point at time } t - t_0, Y_{t-t_0} \geq t_0) \\
&= P^{t-t_0}(vY_t \leq z \mid Y_{t-t_0} \geq t_0) = P^0\left(Y_0 \leq \frac{z}{v} + t_0 \mid Y_0 \geq t_0\right) \\
&= P^0\left(S_1 \leq \frac{z}{v} + t_0 \mid S_1 \geq t_0\right).
\end{aligned}
$$

Using the result of Palm calculus that $P^0(S_1 \leq x) = P^0(R_1 \leq x) = F_R(x)$ (Chapter 12,[8]), we obtain:

$$P(Z \leq z) = \frac{F_R(\frac{z}{v} + t_0) - F_R(t_0)}{1 - F_R(t_0)}.$$

Since $R_w = L_w/v$, the cdf of Z can be expressed as:

$$P(Z \leq z) = \frac{F_L(z + d_0) - F_L(d_0)}{1 - F_L(d_0)},$$

and we obtain the pdf of Z given by (6).   □

Approximation 1. We are unable to compute in a closed form the predictor in Theorem 1. Therefore, we use the following mobility predictor:

$$\widehat{x}(t+\tau) = \begin{cases} x(t) + \frac{v(t)}{v}(v\tau - \frac{c(v\tau)^2}{2} + \frac{c\pi(v\tau)^4}{12a^2}) & \text{if } 0 \le v\tau \le \frac{a}{2} \\ x(t) + \frac{v(t)}{v}E[Z] & \text{if } v\tau > \frac{a\sqrt{2}}{2} \end{cases}$$

$$\widehat{x}(t+\tau) \approx x(t) + \frac{v(t)}{v}E[Z] \text{ if } \frac{a}{2} \le v\tau \le \frac{a\sqrt{2}}{2}.$$

The mean value of the Z is given by:

$$E[Z] = \frac{E[L]}{2} + \frac{Var(L)}{2E[L]},$$

where the mean $E[L]$ and the variance $Var(L)$ of the L_is are equal to $E[L] = a/6(\sqrt{2} + \ln(1 + \sqrt{2}))$ and $Var(L) = \frac{a^2}{6} - E[L]^2$, respectively. We make an approximation in the region $\frac{a}{2} < v\tau < \frac{a\sqrt{2}}{2}$ by neglecting the probability that there is not any waypoint in the period of $v\tau$.

Approximation 2. We are unable to compute in a closed form the predictor in Theorem 2. Therefore, we use the following mobility predictor:

$$\widehat{x}(t+\tau) = x(t) + \frac{v(t)}{v}\frac{\pi l_0^3 - \pi(v\tau + l_0)^3 + 3v\tau a^2}{3(a^2 - l_0^2 \pi)} \text{ if } (l_0, v\tau + l_0) \in [0, \frac{a}{2}]^2$$

$$\widehat{x}(t+\tau) \approx \begin{cases} x(t) + v\tau & \text{if } 0 \le l_0 \le \frac{a}{2}, \frac{a}{2} - l_0 \le v\tau \le E[Z] \\ x(t) + E[Z] & \text{if } 0 \le l_0 \le \frac{a}{2}, \frac{a}{2} - l_0 \le v\tau, v\tau > E[Z] \\ x(t) & \text{if } l_0 > \frac{a\sqrt{(2)}}{2} \end{cases}.$$

The mean value of the Z is given by:

$$E[Z] = \frac{3a^2(E[L] - l_0) + \pi l_0^3}{3(a^2 - l_0^2)},$$

where the mean value of the L_is is equal to $E[L] = \frac{a}{6}(\sqrt{2} + \ln(1 + \sqrt{2}))$. We make an approximation in the region $0 \le l_0 \le \frac{a}{2}, \frac{a}{2} - l_0 \le v\tau \le E[Z]$ by neglecting probability that there is a waypoint in the period of $v\tau$. We make also an approximation in the region $0 \le l_0 \le \frac{a}{2}, \frac{a}{2} - l_0 \le v\tau, v\tau > E[Z]$ by neglecting probability that there is not any waypoint in the period of $v\tau$. In the region $l_0 > a/2$ we do not make mobility prediction because we are unable to compute $E[Z]$. The probability of the appearance of this last case during the operation of the LER algorithm is small because the probability that $l_0 > a/2$ is small. Thus, this will have a small effect on the predictor.

A-STAR: A Mobile Ad Hoc Routing Strategy for Metropolis Vehicular Communications

Boon-Chong Seet[2], Genping Liu[1], Bu-Sung Lee[1], Chuan-Heng Foh[1],
Kai-Juan Wong[3], and Keok-Kee Lee[1]

[1] Centre for Multimedia and Network Technology, Nanyang Technological University,
Singapore. {asgpliu, ebslee, aschfoh,askklee}@ntu.edu.sg
[2] Network Technology Research Centre, Nanyang Technological University,
Singapore, ebcseet@ntu.edu.sg
[3] Institute for Computing Systems Architecture Informatics
University of Edinburgh, U.K., k.j.wong@sms.ed.ac.uk

Abstract. One of the major issues that affect the performance of Mobile Ad hoc NETworks (MANET) is routing. Recently, position-based routing for MANET is found to be a very promising routing strategy for inter-vehicular communication systems (IVCS). However, position-based routing for IVCS in a built-up city environment faces greater challenges because of potentially more uneven distribution of vehicular nodes, constrained mobility, and difficult signal reception due to radio obstacles such as high-rise buildings. This paper proposes a new position-based routing scheme called Anchor-based Street and Traffic Aware Routing (A-STAR), designed specifically for IVCS in a city environment. Unique to A-STAR is the usage of information on city bus routes to identify an anchor path with high connectivity for packet delivery. Along with a new recovery strategy for packets routed to a local maximum, the proposed protocol shows significant performance improvement in a comparative simulation study with other similar routing approaches.

1 Introduction

MANET is an autonomous system composed of mobile nodes communicating through wireless links in an environment without any fixed infrastructure support. Nodes in this network are self-organizing and rely on each other to relay messages to their correct destinations. As nodes are free to move randomly, the network topology may change rapidly and unpredictably. Thus, the routing protocol must be able to adapt and maintain routes in the face of changing network connectivity. Such networks are very useful in military and other tactical applications such as emergency rescue or exploration missions where an established (e.g. cellular) infrastructure is unavailable or unusable. Commercial applications are also likely where there is a need for ubiquitous communication services. Particularly in recent years, there is a growing commercial interest on the research and deployment of MANET technology for vehicular communications, e.g. FleetNet [1], VICS [2], CarNet 3 [3], etc.

N. Mitrou et al. (Eds.): NETWORKING 2004, LNCS 3042, pp. 989–999, 2004.

Existing MANET routing protocols work well in scenarios where nodes are uniformly distributed and moving freely in open space. However, these protocols do not work as well for IVCS in a city environment because of some additional inherent challenges. Generally, vehicular nodes are more unevenly distributed due to the fact that vehicles tend to concentrate more on some roads than others. Their constrained mobility by road patterns, along with more difficult signal reception in the presence of radio obstacles such as high-rise buildings, have contributed to greater fragility in the connectivity of the IVCS network, and the frequent formation of topology "holes", which could not be dealt with effectively by existing position-based routing protocols.

Recently, a project called BUSNet [4] was initiated to study the performance of MANET routing algorithms in the IVCS, based on a Metropolitan Grid model (M-Grid) [4][5]. It proposes using the regular network of buses to form a stable communication backbone for an otherwise fragile IVCS network. In [5], the performance of existing MANET routing protocols is found to be much lower in the M-Grid model than in the random waypoint model. This is because inter-node connectivity is much harder to establish with constrained mobility and obstacles in the M-Grid model.

For a large, metropolitan-scale IVCS network, the scalability of the routing protocol is very important. Position-based routing is known to be very scalable with respect to the size of the network. Thus, it is a good candidate for metropolitan-scale IVCS. However, applying position-based routing to IVCS may not be without any problems. An example is Greedy Perimeter Stateless Routing (GPSR) [6], one of the most well known position-based protocols in literature. It works best in a free open space scenario with evenly distributed nodes. But when applied to city scenarios [7][8], GPSR is found to suffer from several deficiencies, the details of which we will discuss in the next section.

This paper proposes a new position-based routing scheme called Anchor-based Street and Traffic Aware Routing (A-STAR), designed specifically for IVCS in a city environment. Unique to A-STAR is the usage of information on city bus routes to identify an anchor path with high connectivity for packet delivery. Along with a new recovery strategy for packets routed to a *local maximum*(to be explained in Section 2), the proposed protocol shows significant performance improvement in the M-Grid model. A-STAR is therefore proposed as a potential routing strategy for metropolis vehicular communications.

The remainder of the paper is organized as follows. Section 2 discusses with example the challenges faced by position-based routing in IVCS. Section 3 presents some works in literature related to this area. Section 4 describes the proposed A-STAR protocol. The mobility model and simulation setting are explained in Section 5. Performance results are presented in Section 6. Finally, the paper is concluded in Section 7.

2   Challenges of Position-Based Routing in IVCS

The challenges of position-based routing in a city environment have been discussed thoroughly in [7][8]. An example is given here to illustrate some main problems if typical GPSR is deployed directly to IVCS. Figure 1 shows a partial city environment.

Suppose node s wants to send a packet to node d. Greedy forwarding will fail in this case as there is no neighbor of s, which is nearer to d than s itself. Such a situation is what is commonly known as *local maximum*. Following the strategy in GPSR, the packet enters into perimeter-mode, using the right hand rule to travel through each node on the dotted route, including nodes a, b and c. At b, it is found that c is nearer to d than s, at which the packet enters into perimeter-mode. Thus, the packet switches back to greedy mode at b, and then reaches its destination d through c. It can be seen that this route is very long in terms of hop count. In fact, s can reach a, and a can reach b, both in one hop. This shows that the perimeter-mode which packet employs to recover from local maximum is very inefficient and time-consuming.

Another observation is that the packet can actually travel from s to d via a route that passes through e and f (shown as solid line), which is much shorter. However, this route is not exploited because the perimeter-mode of GPSR based on right hand rule is biased to a specific direction when selecting for the next hop.

It should be noted that in a city environment, the constrained mobility and frequently encountered obstacles can effectively force GPSR to run into perimeter-mode frequently. As a result, the performance of GPSR could deteriorate dramatically, and therefore may not be suitable for IVCS.

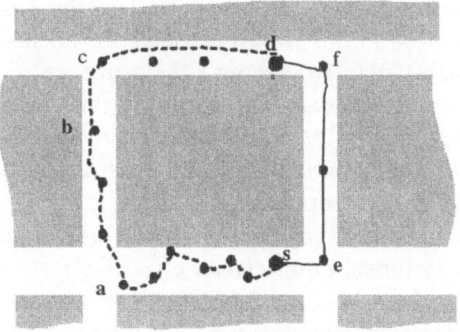

Fig. 1. Challenges of Position-Based Routing in IVCS

3   Related Work

3.1   Anchor-Based Routing

Anchor-based routing is analogous to the source routing of DSR [10]. In anchor-based routing, the source node includes into each packet a route vector composed of a list of anchors or fixed geographic points, through which packets must pass. Between anchors, the greedy position-based routing is employed. Both Termin-ode Remote Routing (TRR) [9] and Geographic Source Routing (GSR) [7][8] are examples of algorithms that employ anchor-based routing to forward packets to remote destinations.

3.2   Spatial Aware Routing

In spatial aware routing, spatial information such as streets map of a city or a description of how several towns are connected by highways, is utilized to assist in making routing decisions. The spatial information reflects the underlying node distribution and topology of the network. Spatial aware routing is usually used in conjunction with anchor-based routing, such as in TRR and GSR where anchored paths are computed using the spatial information.

4   Anchor-Based Street and Traffic Aware Routing (A-STAR)

Considering the challenges faced in a city environment, a new position-based routing scheme called A-STAR is proposed. Similar to GSR, A-STAR adopts the anchor-based routing approach with *street awareness*. The term "street aware-ness" is preferred over "spatial awareness" to describe more precisely the use of street map information in our routing scheme for anchor path computation. That is, using the street map to compute the sequence of junctions (anchors) through which a packet must pass to reach its destination. But unlike GSR, A-STAR computes the anchor paths with *traffic awareness*. "Traffic" herein refers to vehicular traffic, including cars, buses, and other roadway vehicles.

It is observed that in a metropolitan area, some streets are wider and accommodate more vehicular traffic than others. These are the major streets, served by a regular fleet of city buses. Connectivity on such streets can be higher due to higher density of vehicular nodes and more stable due to regular presence of city buses. With this observation, weight can be assigned to each street based on the number of bus lines by which it is served, i.e. the more bus lines by which a street is served, the less weight it is assigned, and vice-versa. The street map in use by the vehicle is assumed to be loaded with bus route information. An anchor path can thus be computed using Dijkstra's least-weight path algorithm. For such a map with pre-configured information, it is called a *statistically rated map*.

While bus route information can provide a reasonable estimate of the expected vehicular traffic on each street, the traffic conditions in a city area can be

---

Let R be a node receiving a packet p for destination D
Let N be the set of one-hop neighbors of R
Let AP represent the anchor path in the header of p
Let L represent the number of hops p has traversed
Let L_{max} represent the maximum hops p is allowed to traverse
Let LR represent the number of times p has been recovered
Let LR_{max} represent the maximum number of times p is allowed to be recovered
If $(R = $ source S of $p)$
        Initialize $AP = $ null
Else If $(L \geq L_{max})$ or $(LR > LR_{max})$
        Discard p
        Return
If "out of service" information present in the header of p
        Update local map with the "out-of-service" information
Forward:
If $(AP = $ null)
        //anchor path initialization
        Set $AP = $ least weight path from R to D with Dijkstra algorithm
        If $(AP = $ null)
                // no anchor path exists, drop the packet
                Discard p
                Return
//compute the next hop n along the anchor path
If $(\exists\, n \in N$: n resides on AP and has shortest distance along AP to $D)$
        Forward p to n
Else // local maximum occurs
        Mark the street where n resides as "out of service" for time period T
        Record the "out of service" information in the header of p
        Set $AP = $ null
        Goto ***Forward***

Fig. 2. Pseudo code of A-STAR algorithm

quite dynamic at times. A better weight assignment scheme is therefore one that dynamically monitors and assigns weight to a street based on its latest traffic condition, which can provide higher quality of anchor computation. It could be envisaged that future IVCS would be able to monitor the city traffic condition and distribute such information to every vehicle connected to the IVCS network. This information could then be used to re-compute the weight of each street on the map, e.g. more vehicles, less weight assigned, and vice-versa. Such a map with re-configurable information is called a ***dynamically rated map***.

4.1   Local Recovery

It has been shown that local recovery algorithm of GPSR using perimeter-mode is quite inefficient in a city area. Other recovery algorithms that rely on "right hand rule" such as face-1 or face-2 [11] also face a similar problem. GSR adopts a "switch back to greedy" approach for local recovery: when a packet reaches a local maximum along its anchor path, it switches back to greedy mode. This is not efficient at all as it has been shown that greedy forwarding does not perform well in a city environment.

Thus, a more efficient recovery strategy is proposed for A-STAR: a new anchor path is computed from the local maximum to which the packet is routed. The packet is salvaged by traversing the new anchor path. To prevent other packets from traversing through the same void area, the street at which local maximum occurred is marked as "out of service" temporarily, and this information is distributed to the network by piggybacking them onto the packets to be recovered. Nodes receiving these packets update their local map with the "out of service" information prior to making their forwarding decision. The "out of service" streets are not used for anchor computation or re-computation during the "out of service" duration and they resume "operational" after the time out duration. A maximum threshold value (LR_{max}) is also defined to limit the number of times a packet can be recovered to prevent the perpetual sending of outdated data and bandwidth wastage. Figure 2 presents the pseudo code of the A-STAR algorithm.

5   Mobility Model and Simulation Setting

5.1   M-Grid Mobility Model

Mobility model describes the movement of nodes in a certain environment. In this paper, the M-Grid mobility model [4][5] is used to describe the movement of vehicular nodes in a city area. M-Grid is a variant of the Manhattan model [12], which models the vehicular movement in a typical metropolis where streets are set out on a grid pattern. Key features which distinguish the M-Grid from Manhattan model, include:

- *Node heterogeneity*: Buses and cars are two types of vehicular nodes modeled in our M-Grid. Buses, which only travel along the bus routes, show higher regularity and lower mobility than cars. For the M-Grid in Figure 3, the bus routes are represented by bold lines in gray. It shows three loop lines (or service numbers), plying the streets in various parts of the city. Each line is bi-directional with buses running clockwise and anti-clockwise.
- *Preferential movements*: It is observed that in real life, some streets would attract more vehicles than others. More often than not, these are the main streets, which are bustling with people and therefore served by buses. In M-Grid, when a car reaches a junction, it would choose to move into another street with some preference. Given the observation above, the car at the junction shall give greater preference to a street which is on a bus route than one which is not.
- *Radio obstacles*: The blocking of signal transmissions by objects such as high-rise buildings in the city has been modeled in M-Grid. As Figure 4 shows, the gray areas represent obstacles, which are non-penetrable by the

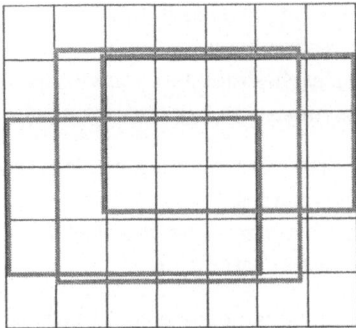

Fig. 3. M-Grid with bus routes

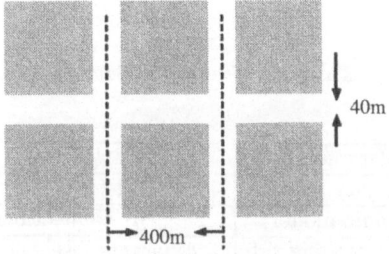

Fig. 4. M-Grid with obstacles

signals. Thus, for a node pair to communicate directly, they must have a "line-of-sight" to each other, in addition to being in range of one another.

5.2 Simulation Setting

Performance of A-STAR and other related protocols are evaluated using the ns-2 [13] simulator. Four protocols are implemented, namely: i) GPSR, ii) GSR, iii) A-STAR-SR, and iv) A-STAR-DR. Protocol iii and iv refer to the proposed A-STAR with *statistically rated* and *dynamically rated* maps respectively. The presence of an information system that provides location service (e.g. [14] offering information about the position of other network users) and current road traffic state is assumed. Table 1 summarizes the parametric settings used in our simulation.

Note that the number of vehicles (nodes) is varied to reflect different vehicle densities under which the performance of each protocol is evaluated. However, throughout the evaluation, the number of buses is a constant, with only the car density varying. Inter-bus distance is approximately 1 kilometer for each line in the same direction. With three bus lines for the M-Grid shown in Figure 3, a total of 37 buses will be running in the city: two with 12 buses, one with 13 buses. Moreover, cars at the junction would move into a street which is on a bus route with a probability three times that of which is not(to effect the preferential movements in the M-Grid as mentioned in Section 5.1). Speed limit of buses and cars are 50 and 70 km/h respectively.

Performance result for each simulated vehicle density (node number) is the average of five simulation runs. The key metrics of interest are:

- *Packet delivery ratio*: the ratio of packets delivered to the destinations to those generated by the sources.
- *End-to-end delay*: the average time it takes for a packet to traverse the network from its source to destination.

Results of the control overhead is not presented here because the overhead messages are predominantly beacon messages transmitted periodically by nodes to build up their neighbors' location information, the amount of which are the same for all position-based routing protocols considered in this study.

Table 1. Simulation Setting

Parameter	Setting
Mobility model	M-Grid
Traffic model	20 CBR connections
Packet sending rate	4 packets / second
Data packet size	64 bytes
Transmission range	350 meters
Map size	2800x2400m2 (7x6 grid)
Node number	200 to 500, in steps of 50
Simulation time	500 seconds
MAC protocol	IEEE 802.11 DCF

6   Simulation Results and Analysis

Recall that A-STAR differs from GSR and GPSR in two main aspects. Firstly, A-STAR incorporates traffic awareness by using statistically rated and dynamically rated maps. Secondly, A-STAR employs a new local recovery strategy that is more suitable for a city environment than the greedy approach of GSR, or the perimeter-mode of GPSR.

To investigate impacts of each aspect on the routing performance, protocols are evaluated initially without local recovery, and later with local recovery. Without local recovery, a packet is simply dropped when it encounters a local maximum. Figures 5 and 6 show the protocols performance (with 95% confidence intervals) without and with local recovery, respectively.

In Figure 5(a), it is observed that more packets are delivered as node number increases. This is expected since more nodes increases the probability of connectivity, which in turn reduces the number of packets dropped due to local

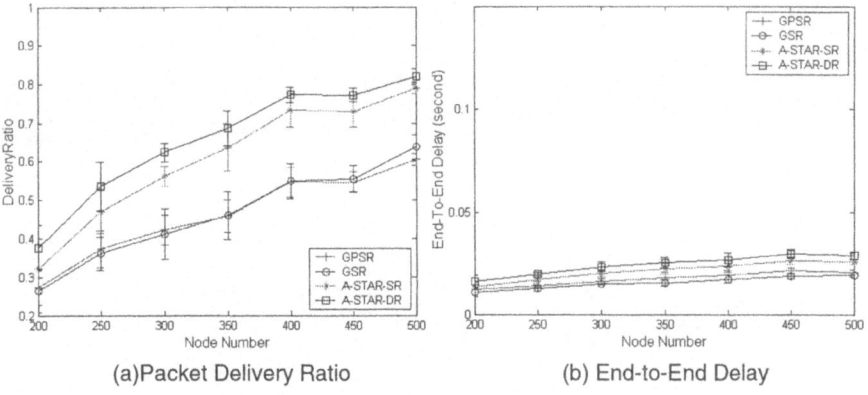

(a)Packet Delivery Ratio   (b) End-to-End Delay

Fig. 5. Performance without local recovery

maximum. It is also observed that GSR did not show a better performance than GPSR, possibly because the grid layout of streets did not pose as much problem to GPSR as did one with fork junctions in [8]. With traffic awareness, A-STAR shows the best performance because it can select paths with higher connectivity for packet delivery. As much as 40% more packets are delivered by A-STAR, compared to GSR. Between A-STAR-SR and A-STAR-DR, the latter performs better by using more precise vehicular traffic information. Figure 5(b) shows the result of end-to-end delay. Generally, no significant difference is observed between the protocols. A-STAR, however, shows slightly higher delay that may be attributed to possibly longer, but higher connectivity paths used for packet delivery.

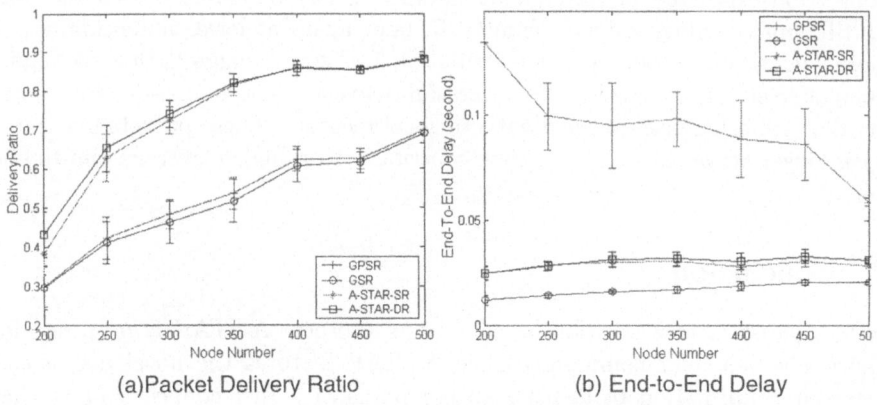

(a)Packet Delivery Ratio   (b) End-to-End Delay

Fig. 6. Performance with local recovery

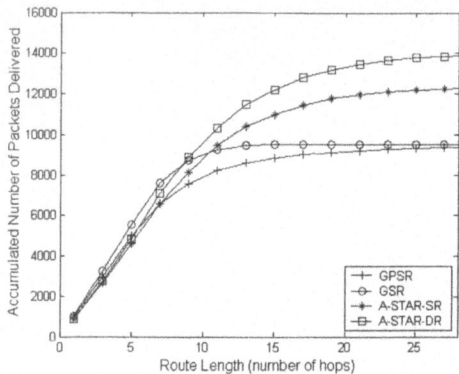

Fig. 7. Route length distribution (for 200 nodes)

With local recovery, packets that encounter local maximum can be rerouted and delivered instead of being dropped. Thus, more packets are delivered by each protocol as shown in Figure 6(a). The increase in packets delivered is more significant at lower node number where local maximum is encountered more frequently. For example, with local recovery, A-STAR-DR delivers 20% more packets at 250 nodes, while only 6% more at 400 nodes. It is also observed that local recovery allows A-STAR-SR to narrow its performance gap with A-STAR-DR. GSR and GPSR show improvement in packet delivery of not more than 15% with local recovery, which suggests that their recovery strategies may not be very effective in a city environment.

Figure 6(b) shows the corresponding result for end-to-end delay. A key observation is that GPSR with local recovery incurs significantly higher end-to-end delay. This is because of frequent attempts by GPSR to salvage packets from local maximum via perimeter-mode, which is generally inefficient and causes congestion especially at lower node number. Delay of A-STAR is lower than GPSR, but seemingly higher than GSR, once again at lower node number. A close analysis of its route length distribution in Figure 7 suggests that the higher delay is likely an artifact due to successful delivery of more long-distance packets that are otherwise dropped without local recovery. These packets inevitably have longer traversal time and thus contribute to a higher average end-to-end delay.

7   Conclusion

In this paper, a new position-based routing protocol A-STAR is proposed for metropolis vehicular communications. A-STAR features the novel use of city bus route information to identify anchor paths of higher connectivity so that more packets can be delivered to their destinations successfully. In our comparative simulation study with other position-based routing schemes, A-STAR demonstrates excellent improvement in packet delivery while maintaining reasonable end-to-end delay. As future work, the traffic awareness in A-STAR shall be extended to include data traffic to provide vehicular nodes with higher performance paths in terms of connectivity as well as delay. Another area that shall be looked into is how information on bus schedules, in addition to bus routes, can be utilized to further optimize the performance of our protocol.

References

1. W. Franz, R. Eberhardt, and T. Luckenbach, "FleetNet - Internet on the Road", Proc. 8th World Congress on Intelligent Transportation Systems, Sydney, Australia, Oct. 2001.
2. S. Yamada. "The Strategy and Deployment Plan for VICS", IEEE Communications, Vol. 34, No. 10, pp.94-97, 1996
3. R. Morris, J. Jannotti, F. Kaashoek, J. Li, and D. Decouto, "CarNet: A Scalable Ad Hoc Wireless Network System", Proc. 9th ACM SIGOPS European Workshop, Sept. 2000.
4. K. J. Wong, B. S. Lee, B. C. Seet, G. Liu, and L. Zhu, "BUSNet: Model and Usage of Regular Traffic Patterns in Mobile Ad Hoc Networks for Inter-Vehicular Communications", Proc. ICT 2003, Thailand, April, 2003
5. B. S. Lee, K. J. Wong, B. C. Seet, L. Zhu, and G. Liu, "Performance of Mobile Ad Hoc Network in Constrained Mobility Pattern", Proc. International Conference on Wireless Networks (ICWN'03), Las Vegas, USA, Jun. 2003.
6. B. Karp and H. T. Kung, "GPSR: Greedy Perimeter Stateless Routing for Wireless Networks", Proc. ACM/IEEE MobiCom, Boston, USA, Aug. 2000.
7. C. Lochert, H. Hartenstein, J. Tian, H. Füßler, D. Herrmann, and M. Mauve, "A Routing Strategy for Vehicular Ad Hoc Networks in City Environments", Proc. IEEE Intelligent Vehicles Symposium (IV2003), Ohio, USA, Jun. 2003.
8. J. Tian, I. Stepanov, and K. Rothermel, "Spatial Aware Geographic Forwarding for Mobile Ad Hoc Networks", Proc. MobiHoc, Lausanne, Switzerland, Jun. 2002.
9. L. Blazevic, S. Giordano, and J. Y. Le Boudec. "Self-Organizing Wide-Area Routing", Proc. SCI 2000/ISAS 2000, Orlando, USA, Jul. 2000.
10. D. B. Johnson and D. A. Maltz, "Dynamic Source Routing Protocol for Mobile Ad Hoc Networks", Mobile Computing, T. Imielinski and H. Korth, Eds., Kluwer, 1996, pp. 153-81.
11. P. Bose, P. Morin, I. Stojmenovic, and J. Urrutia, "Routing with Guaranteed Delivery in Ad Hoc Wireless Networks", Proc. 3rd ACM International Workshop on Discrete Algorithms and Methods for Mobile Computing and Communications (DIALM'99), Seattle, USA, Aug. 1999.
12. A. Kamat and R. Prakash. "Effects of Link Stability and Directionality of Motion on Routing Algorithms in MANETs". Proc. IEEE International Conference on Computer Communications and Networks (ICCCN), Las Vegas, USA, Oct 2000.
13. Network Simulator (ns-2), http://www.isi.edu/nsnam/ns/
14. M. Kasemann, H Fußler. Hartenstein, and M. Mauve. A Reactive Location Service for Mobile Ad Hoc Networks. Technical Report TR-02-014, Department of Computer Science, University of Mannheim, Nov. 2002.

Enhancing the Route Discovery Process of On-Demand Routing in Networks with Directional Antennas[*]

Marco Aurélio Spohn and Jose Joaquin Garcia-Luna-Aceves

University of California at Santa Cruz
Computer Science Department
Santa Cruz, CA 95064
{maspohn,jj}@cse.ucsc.edu

Abstract. The route discovery process of on-demand routing protocols is based on *route request* (RREQ) packets that are propagated using either an unrestricted broadcast (flooding) or an expanding ring search. We apply two enhancements to this basic process aimed at reducing the overhead incurred in the dissemination of *route request* (RREQs). First, we apply the Enhanced Dominant Pruning (EDP), which is a distributed connected dominating-set algorithm used for reducing the number of nodes that need to propagate RREQs transmitted on broadcast mode. Second, information regarding prior routes to a destination is used to unicast RREQs to a region close to the intended destination. Directional antennas are assumed, which provide higher spatial reuse than omni-directional antennas for unicast transmission. We show the benefits of our approach using the *Ad-hoc On-demand Distance Vector* (AODV) protocol as an example. Detailed simulation results show that using EDP and unicast RREQs provide substantial performance improvements over standard AODV running in ad hoc networks with directional antennas.

1 Introduction

Route discovery in on-demand routing protocols is based on *route request* (RREQ) and *route reply* (RREP) messages (e.g., AODV [1] and DSR [2]). A request is relayed until it reaches a node with a valid route to the destination or the destination itself, which triggers a reply message sent back to the originator. RREQs are propagated using either an unrestricted broadcast or an expanding ring search [3]. In either case, the resulting flooding operation causes considerable collisions of packets in wireless networks using contention-based channel access.

There are several techniques aimed at reducing the impact of broadcasting[4]. *Dominating sets* (DS) play an important role in some of these techniques. A DS is a set of nodes such that every node in the network is either in the set

---

[*] This work was supported in part by CNPq (Brazil), and the Baskin Chair of Computer Engineering

N. Mitrou et al. (Eds.): NETWORKING 2004, LNCS 3042, pp. 1000–1011, 2004.
© IFIP International Federation for Information Processing 2004

or is the neighbor of a node in the set. If the graph induced by the nodes in the dominating set is connected, then we have a *connected dominating set*. The problem of determining the *minimum connected dominating set* (MCDS) is known to be NP-complete [5]. Extensive work has been done on finding a good approximation of MCDS in terms of small approximation ratio. A protocol with a constant approximation ratio of eight has been proposed by Wan et. al. [6]. However, their approach requires that a spanning tree be constructed first in order to select the dominating nodes (forwarders), and only after that a broadcast can be performed. To improve the route discovery process we need an approach that is suitable for dynamic networks with mobile nodes, and is based on determining the CDS *on-the-fly*.

Lim and Kim [7] show that the MCDS problem can be reduced to the problem of building a *minimum cost flooding tree* (MCFT). Given that an optimal solution for the MCFT problem is not feasible, they propose heuristics for flooding trees, resulting in two algorithms: *self-pruning* and *dominant pruning* (DP). They show that both algorithms perform better than *blind flooding*, with which each node broadcasts a packet to its neighbors whenever it receives the packet along the shortest path from the source node, and that DP outperforms *self-pruning*.

A few enhancements to dominant pruning have been reported recently [8, 9]. Lou and Wu [8] propose two enhancements to DP: *total dominant pruning* (TDP), and *partial dominant pruning* (PDP). Spohn and Garcia-Luna-Aceves [9] propose *enhanced dominant pruning* (EDP), which is applied to AODV (using omni-directional antennas) to show its improvements compared to DP.

We apply two enhancements to the route discovery process of on-demand routing protocols aimed at reducing the overhead incurred in the dissemination of RREQs. First, we apply EDP to reduce the number of nodes that need to propagate RREQs transmitted on broadcast mode. Second, information regarding prior routes to a destination is used to unicast RREQs to a region close to the intended destination, so that broadcast RREQs are postponed as much as possible and occur only close to the destination, rather than on a network-wide basis.

Directional antennas are assumed, which provide higher spatial reuse [10] [11] than omni-directional antennas for unicast transmission. An advantage of using directional antennas is that they allow a larger number of simultaneous transmissions compared to omni-directional antennas .

The rest of this paper is organized as follows. Section 2 presents a review of dominant pruning and its enhancements. Section 3 presents our approach for improving the route discovery process in AODV. Section 4 presents detailed simulations showing the benefits of EDP when it is applied to AODV and compares it against AODV, and AODV with DP. In all cases we use directional antennas. The simulation results clearly show that AODV with EDP renders the best performance of all the AODV versions, which is a direct consequence of reducing packet collisions due to RREQs by means of EDP and the use of directional antennas. Section 5 concludes this work.

2   Dominant Pruning: Review and Enhancements

In DP [7] the sending node decides which adjacent nodes should relay the packet. The relaying nodes are selected using a distributed CDS algorithm, and the identifiers (IDs) of the selected nodes are piggybacked in the packet as the forwarder list. A receiving node that is requested to forward the packet again determines the forwarder list. The flooding ends when there is no more relaying nodes.

Nodes keep information about their two-hop neighborhood, which can be obtained by the nodes exchanging their adjacent node list with their neighbors. DP is a distributed algorithm that determines a set cover based on the partial knowledge of the two-hop neighborhood. Ideally, the number of forwarding nodes should be minimized to decrease the number of transmissions. However, the optimal solution is NP-complete and requires that nodes know the entire topology of the network. DP uses the *greedy set cover* (GSC) algorithm to determine the forwarder list of a packet (i.e., the list of nodes that should forward the packet) based just on partial knowledge of the network topology. GSC recursively chooses one-hop neighbors that cover the most two-hop neighbors, repeating the process until all two-hop neighbors are covered.

TDP [8] requires that the two-hop neighborhood of the sender be piggybacked in the header of the packet. This information reduces the size of the two-hop neighbor set that needs to be covered by the forwarders. The header size increases proportionally to the number of nodes in the two-hop neighborhood, which may become a problem in dense networks. PDP [8] enhances DP by eliminating the two-hop nodes advertised by a neighbor shared by both the sender and the receiver (forwarder). Simulation results assuming an ideal MAC layer with which no contention or collisions occur show that both TDP and PDP improve DP in a static environment. A dynamic scenario is also evaluated, and DP is shown to perform better than both TDP and PDP.

EDP [9] requires the second-to-previous (STP) list in addition to the forwarder list. EDP improves DP by reducing even more the number of redundant broadcasts, mainly due to the information provided by the STP list. Although EDP increases the packet header size with the STP list, EDP improves the performance of AODV when compared to standard DP.

3   Improvements to the Route Discovery Process of AODV

This section addresses the application of EDP to the route discovery process in AODV. The resulting variant of AODV is denoted by AODV-EDP. Our neighbor protocol uses hello packets to disseminate the one-hop neighborhood, which creates a picture of its two-hop neighborhood at any given node in the network.

To avoid pruning too many route requests in the presence of mobility and cross-traffic, we have chosen to implement the neighbor protocol as part of AODV. We extended the hello mechanism available in AODV to include the information about the one-hop neighborhood in hello messages, and we also rely on the AODV mechanisms for evaluating the link status to neighbors. A hello packet advertises the node's sequence number ($mySeqNum$), the identification of its known neighbors ($neighbors[]$), and the corresponding neighbors' sequence number ($neighSeqNum[]$). We have chosen a hello interval of $1.5s$. To reduce the number of broadcast messages, a RREQ also advertises the one-hop neighborhood information, and the transmission of a RREQ by a node reschedules its pending hello message.

Algorithm 1: RREQ algorithm

    Data   : n_i, destination D, $\mathcal{F}_S$, $\mathcal{F}_{S_S}$, U_i
    Result   : Unicast the RREQ, or Broadcast the RREQ
    begin

1   **if** *recently expired route to D and not retrying* **then**

2          $NextHop \longleftarrow previous\_nextHop(D)$

3   **if** *validRoute(NextHop)* **then**

4   ∟ $result \longleftarrow Unicast$

5   **else**

6   ∟ $result \longleftarrow Broadcast$

7   **else**

8   ∟ $result \longleftarrow Broadcast$

9      $\mathcal{F}_i \longleftarrow EDP(n_i, \mathcal{F}_S, \mathcal{F}_{S_S}, U_i)$

10   Update RREQ packet with $\mathcal{F}_i$

11   **if** *result == Unicast* **then**

12   ∟ Unicast the RREQ packet to $NextHop$

13   **else**

14   ∟ Broadcast the RREQ packet

    end

Algorithm 1 presents the pseudo-code for the modified RREQ. As defined in [9], $\mathcal{F}_S$, $\mathcal{F}_{S_S}$, and U_i, are respectively the forwarder list, the STP list, and the set of neighbors to be covered. A route request (RREQ) is handled as follows:

- If the source of a RREQ does not have any previous knowledge about the route to the destination or is retrying the RREQ, it calculates its forwarder list using EDP, and broadcasts the packet (Lines 8, 9, and 14).
- On the other hand, if the source of a RREQ has knowledge about a recently expired route to the destination, and there is a valid route to the next hop towards the destination (Lines 2, 3, and 4), the node calculates the forwarder list using EDP (Line 9), but instead of broadcasting the RREQ packet, the

node unicasts the packet to the last known next hop towards the destination (Line 12).

- Upon receiving a route request, a forwarder that cannot respond to this request calculates its own forwarder list using the information provided in the RREQ packet (i.e., forwarder list, second to previous forwarder list, and source node) and broadcasts or unicasts the packet (depending on which one of the two first cases apply) after updating it with its own forwarder list.

Eventually, the RREQ reaches a node with a route to the destination or the destination itself. Our approach attempts to reduce the number of collisions and the delay of the route discovery by unicasting a RREQ towards the region where the destination was previously located. The success of this approach depends on how fresh the previous known route to the destination is, and how fast the destination node is moving out of the previous known location. If an intermediate node has completely removed any route to the destination, the RREQ is then broadcasted. The intended effect is to postpone the broadcast of a RREQ to the region closest to the destination. In the case that the unicast approach fails, or there is no previous route to the destination, the source broadcasts by default.

Because of topology changes, nodes may not have correct two-hop neighborhood information, which may result in forwarding lists that do not cover all nodes in the neighborhood. However, this is not a major problem when the request is broadcasted, because a node incorrectly excluded from the forwarder list may also receive the request and is able to respond in the case it has a route to the destination.

4   Simulations and Performance Results

To compare AODV-EDP against other protocols, we use traffic and mobility models similar to those previously reported in [12]. We implemented AODV-EDP in *Qualnet* 3.5, and compare it against AODV-DP (AODV with *Dominant Pruning*), and standard AODV with no hello messages. In the simulation scenarios, we vary both the number of nodes in the network as well as the number of flows.

Qualnet provides two models for directional antennas: *switched beam* with multiple patterns (circular array with 8 patterns), and *steerable* with multiple steerable patterns (triangular array with 4 different beam widths). The antenna model is receiver side only due to the omni-directional MAC protocol. In our simulations we have used the *switched beam* model for all the simulations and routing protocols. The radio model used is a $2Mbps$ IEEE 802.11 device. Terrain size and radio range are adjusted for each particular scenario.

Traffic sources are continuous bit rate (CBR). Only 512-bytes data packets are used. The source-destination pairs are chosen randomly among the nodes in the network. Flows last in average for $50s$ (following an exponential distribution), unless otherwise mentioned. Source nodes keep active flows during all simulation time (new destinations are randomly selected as needed). Nodes begin transmitting at $50s$ plus an offset uniformly chosen over a $5s$ period to avoid synchronization in their initial transmissions. The simulation time is set to 600 seconds, and identical mobility and traffic scenarios are used for all protocols. Initially nodes are placed uniformly over a grid. Nodes move according to the random waypoint model with velocities between 0 and $20m/s$. Six pause times are tested: $0s$ (always moving), $50s$, $100s$, $300s$, $400s$, and $600s$.

Experiments are repeated for 10 trials with different random-number seeds. Results present a 95% confidence interval. Each data point represents the mean over the 10 runs discarding the lowest and largest results (quantile of one).

Four performance metrics are evaluated:

- *Packet delivery ratio*, the ratio of the data packets delivered to the destination to those generated by the CBR sources.
- *Average end-to-end delay* for data packets, including all possible delays caused by route discovery latency, queuing at the interface, retransmission delays at the MAC layer, and propagation and transfer times.
- *Routing load*, the number of routing packets transmitted per data packet delivered to the destination, where each hop traversed by the packet is counted as one transmission.
- *MAC collisions*, the number of collisions detected at the MAC layer.

Because the scenarios we have used to evaluate our approach differ from those presented in [8], and because we implemented our solution together with a neighbor and routing protocol, we do not know how our solution compares to TDP and PDP. The relation between the savings of pruning (too much, or too little) and the degree of broadcast redundancy achieved, can be different, depending on the physical environment under consideration. If we take into account that more packets being broadcasted translate into more contention and collisions, we could have a different picture, depending on the number of broadcasts that are avoided.

4.1   50-Node Scenario

The network is composed of 50 nodes spread over an area of $1500m$ x $300m$. The radio has a nominal transmission range of $250m$. The network is tested for three traffic models:

- 30 source nodes transmitting 4 *packets/s*, each flow lasting in average $50s$ (exponential distribution).
- 40 source nodes transmitting 3 *packets/s* (flows of $50s$ as in the previous scenario).
- 30 source nodes transmitting 4 *packets/s*, with very short flows (flows lasting in average $10s$ and $20s$).

In both scenarios, we have a total of 120 data packets being injected into the network every second. We show that, in all of the categories, AODV-EDP outperforms the other protocols. Figures 1 and 2 summarize the results for 30 and 40 flows, and Figures 3 and 4 summarize the results for 30 sources varying the flow duration.

30 flows. As pointed out in [12], the possibility of link failures is low with low mobility, but due to the node movement model (*random waypoint*) nodes usually get clustered. This situation is responsible for congestion in those regions in the presence of high traffic. This causes the link layer to report link failures even though the nodes are relatively static and a physical link still exists between the nodes. This is observed on Figure 1(a), where we notice a decreasing on the packet delivery ratio for some larger pause times.

Figure 1(b) shows the average end-to-end delay for 30 flows. AODV-EDP presents an almost constant mean latency, and is always the best for all pause times. Together with the packet delivery ratio, these results show that besides delivering more packets for most of the pause times, AODV-EDP delivers them faster than the other protocols. AODV-DP again shows that DP alone improves AODV. Clustering of nodes has a direct impact on the latency as well. Packets spend more time waiting on the queues, and usually need to be retransmitted due to increased congestion.

Figure 2(a) presents the routing load for 30 flows. As expected, AODV-EDP has a lower routing load compared to standard AODV, because it reduces the number of broadcast transmissions. AODV-DP reduces the control overhead compared to AODV, but not as much as AODV-EDP.

Figure 2(b) shows the number of collisions at the MAC layer for 30 flows. The number of collisions for standard AODV is noticeable larger than the other protocols, because a node always respond to the first received RREQ (if the TTL is valid, i.e., greater than zero). Because both AODV-EDP and AODV-DP reduce the number of necessary broadcasts, it translates in less collisions.

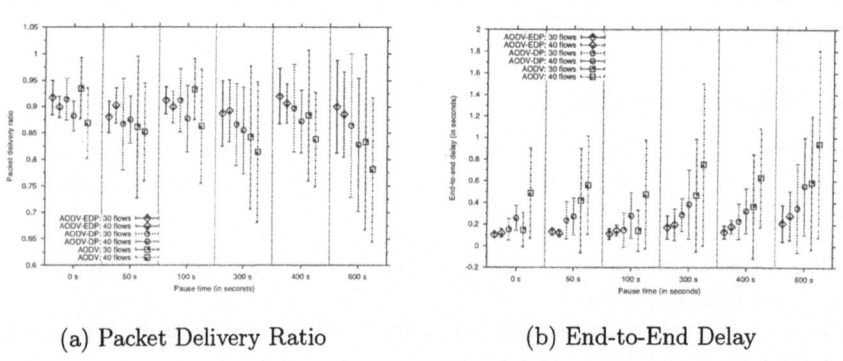

(a) Packet Delivery Ratio   (b) End-to-End Delay

Fig. 1. 50 Nodes, 30 and 40 flows: packet delivery ratio and end-to-end delay

40 flows. In this scenario we increase the number of flows but keep the same number of data packets being injected into the network (each source sends 3 packets/s). Figure 1(a) shows the packet delivery ratio. AODV-EDP presents

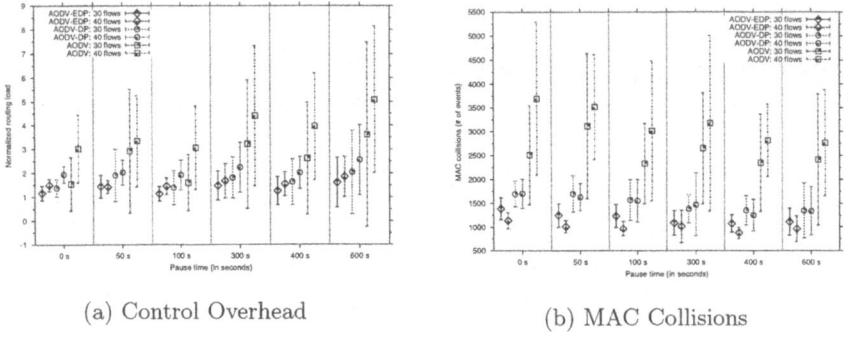

(a) Control Overhead   (b) MAC Collisions

Fig. 2. 50 Nodes, 30 and 40 flows: control overhead and MAC collisions

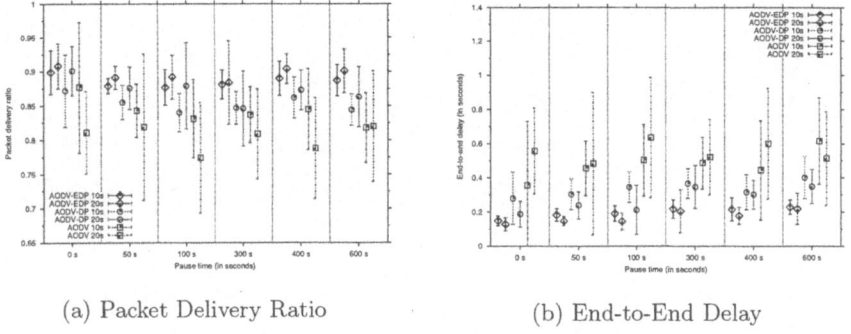

(a) Packet Delivery Ratio   (b) End-to-End Delay

Fig. 3. 50 Nodes, flows of 10s and 20s: packet delivery ratio and end-to-end delay

an almost constant packet delivery ratio for all pause times, and it has a higher delivery ratio for all the pause times. The effect of clustering is noticeable on Figure 1(a). This result shows that by increasing the number of flows, more nodes in the network participate in active communications, what translates in more replies coming from intermediate nodes during the route discovery process. In these circumstances, it helps even more when a request can be unicasted instead of broadcasted.

Figure 1(b) shows the average end-to-end delay for 40 flows. AODV-EDP presents an almost constant mean latency, and is always the best for all pause times. In this scenario AODV-EDP again deliver more packets, and doing it faster than the two other variants. AODV-DP again shows that DP alone improves AODV. For all the pause times, AODV-DP presents less than half the latency produced by AODV. On its turn, AODV-EDP reduces even more the end-to-end delay, having almost all the time half the latency produced by AODV-DP. The impact of clustering of nodes in the latency of data packets is more noticeable only for AODV.

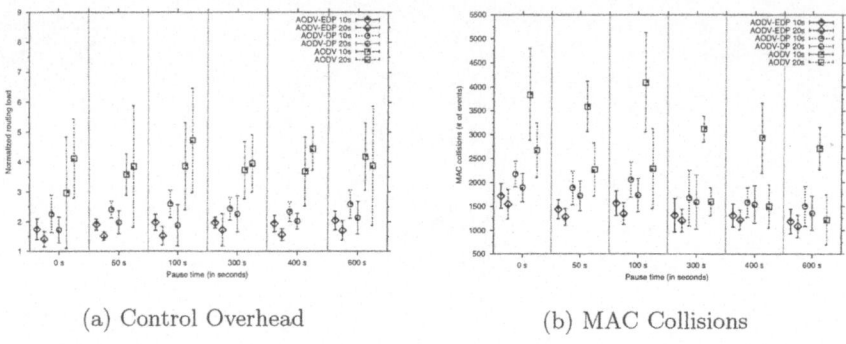

(a) Control Overhead

(b) MAC Collisions

Fig. 4. 50 Nodes, flows of 10s and 20s: control overhead and MAC collisions

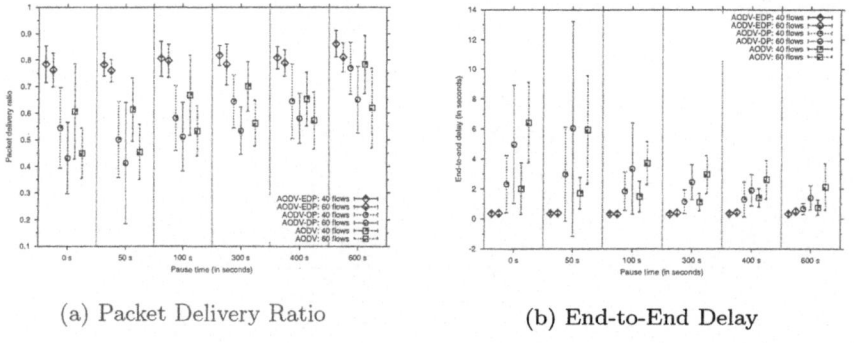

(a) Packet Delivery Ratio

(b) End-to-End Delay

Fig. 5. 100 Nodes, 40 and 60 flows: packet delivery ratio and end-to-end delay

Figure 2(a) presents the routing load for 40 flows. As expected, there is an increase in the routing load because there are more flows (and destinations) in the network. It is more noticeable the improvements introduced by both dominant pruning techniques, but AODV-EDP performs better for all pause times.

Figure 2(b) shows the number of collisions at the MAC layer for 40 flows. Although a larger number of flows, for both AODV-DP and AODV-EDP we notice only a slightly difference (sometimes even less collisions) compared to the 30 flows scenario. But AODV incurs on more collisions than on the previous scenario. In all situations AODV-EDP outperforms the two other variants.

30 sources: varying flow duration. In this set of simulations we play with the flow duration. At any given time, there are at least 30 active flows, and every node in the network has a chance to be the source of at least one session . In fact, because we are dealing with flows of short duration, every node participates as a sender and as a receiver on several different sessions during the simulation time. Flows last in average 10s and 20s (exponential distribution). As mentioned

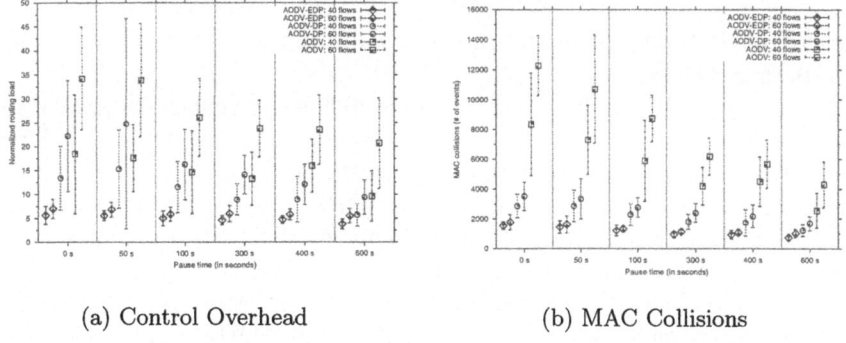

(a) Control Overhead   (b) MAC Collisions

Fig. 6. 100 Nodes, 40 and 60 flows: control overhead and MAC collisions

before, flows start at $50s$ of simulation time with a jitter of $5s$. For each flow duration, simulations are run for the same number of trials as in the previous scenarios.

The results presented in Figures 3 and 4 show that DP alone improves the performance of AODV for all pause times and for all flows. But AODV-EDP performs better than the other two protocols in all situations, and it also presents the smallest variance among the three protocols. Both AODV-DP and AODV-EDP present an almost constant performance for all pause times. As expected, we notice again a great reduction on the control overhead due to the pruning of redundant broadcasts. But we also notice that AODV performs as well as the other protocols regarding number of collisions in situations with large pause times and flows of $20s$.

4.2 100-Node Scenario

The network is composed of 100 nodes spread over an area of $2200m$ x $600m$. The radio has a nominal transmission range of $280m$. For traffic sources, we have two traffic models: 40 source nodes transmitting 3 *packets/s*, and 60 source nodes transmitting 2 *packets/s*. In both cases we have a total of 120 data packets being injected into the network every second. We show that, in most of the categories, AODV-EDP outperforms the other protocols. For all the metrics evaluated, AODV-EDP presents the smallest variance.

40 flows. Figure 5(a) shows the packet delivery ratio for 40 flows. AODV-EDP presents an almost constant packet delivery ratio for all pause times, as well as a higher delivery ratio for all pause times. AODV-DP performs worse than AODV specially in the high mobility scenarios, as the network gets more static the difference between AODV and AODV-DP becomes very small.

Figure 5(b) shows the average end-to-end delay for 40 flows. AODV-EDP presents an almost constant mean latency, and is always the best for all pause times. Together with the packet delivery ratio, these results show that besides delivering more packets, AODV-EDP delivers them faster than the other protocols. Although AODV-DP performs better than AODV, AODV-DP delivers less packets than AODV.

Figure 6(a) presents the routing load for 40 flows. As expected, AODV-EDP has a lower routing load in comparison to standard AODV, but the difference among the protocols is a way larger than in the 50 nodes scenario. AODV-DP reduces the control overhead compared to AODV, but not as much as AODV-EDP. AODV-DP shows that DP alone improves the control overhead, but it does not improve as much as EDP.

Figure 6(b) shows the number of collisions at the MAC layer for 40 flows. As expected, the number of collisions for standard AODV is noticeable larger than the other protocols. AODV-EDP incurs 4 to 5 times less collisions than AODV for most of the pause times, and almost half of the collisions incurred by AODV-DP. AODV-EDP also presents the smallest variance, and an almost constant number of collisions for all pause times.

60 flows. Figure 5(a) shows the packet delivery ratio for 60 flows. AODV-EDP presents an almost constant packet delivery ratio for all pause times, as well as a higher delivery ratio. AODV-DP performs worse than AODV but the difference is smaller compared to the 40 flows scenario.

Figure 5(b) shows the average end-to-end delay for 60 flows. AODV-EDP presents an almost constant mean latency, and is always the best for all pause times. As in the previous scenarios, besides delivering more packets, AODV-EDP delivers them faster. AODV-DP performs better than AODV, but it also delivers slightly less packets than AODV. As expected, the latency increases compared to the 40 flows scenario, but not as much for AODV-EDP.

Figure 6(a) presents the routing load for 60 flows. AODV-DP performs better than AODV, specially for larger pause times. AODV-EDP is the best again, and when comparing the results against the 40 flows scenario, we observe that only AODV-EDP does not increase the control overhead proportionally as observed in the two other protocols.

Figure 6(b) shows the number of collisions at the MAC layer for 60 flows. Compared to the 40 flows scenario, AODV increases about 50% the number of collisions, while both AODV-EDP and AODV-DP increase around 15% the number of collisions. Both pruning techniques show to be effective on reducing redundant broadcasts, but EDP outperforms DP in all aspects.

5   Conclusions

We applied two enhancements to the route discovery process of on demand routing protocols. Redundant broadcasts increase the number of packet collisions, and consequently delay the response for RREQ's in the route discovery process. EDP is shown to reduce the number of broadcast transmissions when compared to standard DP. To reduce the interference (and the number of packet collisions) we take advantage of directional antennas by unicasting RREQs to regions of the network where the destination may be located.

We show through extensive simulation results that AODV-EDP improves the performance in all aspects (i.e., the four metrics chosen) for all the pause times in the 50-node and the 100-node scenarios. The other protocols (standard AODV and AODV-DP) deliver fewer packets than AODV-EDP. AODV-EDP not only delivers more packets, but it does it faster than the other protocols. AODV-EDP also presents the smallest variance among the protocols, and almost constant results for all the metrics considered in the simulations (with some exceptions because of clustering of nodes due to the mobility model).

References

1. Perkins, C.: Ad-hoc on-demand distance vector routing. In: Second IEEE Workshop on Mobile Computing Systems and Applications. (1999)
2. Johnson, D.B., Maltz, D.A.: Dynamic source routing in ad hoc wireless networks. In Imielinski, Korth, eds.: Mobile Computing. Volume 353. Kluwer Academic Publishers (1996)
3. Segall, A.: Distributed network protocols. IEEE Transactions on Information Theory **29** (1983) 23–35
4. Williams, B., Camp, T.: Comparison of broadcasting techniques for mobile ad hoc networks. In: Proceedings of the ACM International Symposium on Mobile Ad Hoc Networking and Computing (MOBIHOC). (2002) 194–205
5. Garey, M.R., Johnson, D.S.: Computers and Intractability. Freeman, San Francisco, (1978)
6. Wan, P.J., Alzoubi, K.M., Frieder, O.: Distributed construction of connected dominating set in wireless ad hoc networks. In: Proceedings of INFOCOM. (2002)
7. H.Lim, Kim, C.: Flooding in wireless ad hoc networks. Computer Communications **24** (2001)
8. Lou, W., Wu, J.: On reducing broadcast redundancy in ad hoc wireless networks. IEEE Transactions on Mobile Computing **1** (2002)
9. Spohn, M.A., Garcia-Luna-Aceves, J.J.: Enhanced dominant pruning applied to the route discovery process of on-demand routing protocols. In: Proceedings of the 12th International Conference on Computer Communications and Networks (ICCCN). (2003)
10. Huang, Z., Shen, C.C.: A Comparison Study of Omnidirectional and Directional MAC Protocols for Ad Hoc Networks. In: IEEE Globecom 2002, Taipei, Taiwan (2002)
11. Wang, Y., Garcia-Luna-Aceves, J.J.: Collision Avoidance in Single-Channel Ad Hoc Networks Using Directional Antennas. In: Proceedings of ICDCS, Providence, Rhode Island, U.S.A. (2003)
12. Perkins, C., Royer, E., Das, S.R., Marina, M.K.: Performance comparison of two on-demand routing protocols for ad hoc networks. IEEE Personal Communications **8** (2001) 16–28

Hash-Based Dynamic Source Routing

Claude Castelluccia and Pars Mutaf

INRIA Rhône-Alpes
ZIRST - 655, Avenue de l'Europe
38334 Saint Ismier Cedex - France
{claude.castelluccia, pars.mutaf}@inria.fr
http://www.inrialpes.fr/planete.html

Abstract. This paper [1] presents and evaluates *Hash-Based DSR*, a *DSR* extension for large networks. This protocol reduces the per-packet control overhead of DSR by compressing the source-route with a Bloom filter. Simulations on large networks show that $HB - DSR$ increases the network capacity by a factor of up to 15. $HB - DSR$ is an attractive extension to DSR for large ad-hoc networks.

Another important property of $HB - DSR$ is that, as opposed to DSR, its performance is similar for IPv4 and IPv6. While IPv6 large addresses are prohibitive in DSR, we show by simulations that $HB - DSR$ performs as well for both IP versions. This is important contribution considering the growing interest of the wireless network community for IPv6.

1 Introduction

In an *ad hoc wireless network*, mobile nodes communicate with each other using multi-hop wireless links. Each node is also a router and is therefore part of the routing infrastructure. The bandwidth and the nodes' capacity (processing and power) of such networks are usually very limited. Therefore one of the biggest scientific challenges of this area is to design routing protocols that minimize the control overhead, i.e. the bandwidth overhead used to establish and maintain the routes, as much as possible.

Several protocols have been proposed recently. One of them is the *Dynamic Source Routing* (DSR) protocol [2]. The main characteristic of DSR is that it uses source-routing to route packets from the source to the destination. *DSR* is probably one of the most efficient protocols. However it has two important limitations:

1. *Scalability limitation*: *DSR* uses source-routing and, as a result, does not scale to large networks.
2. *IPv6 unfriendly*: IPv6's large address space limits considerably the performance of DSR. This is an important drawback considering the growing interest of the wireless community for IPv6.

---

[1] A full version of this paper is available as Technical Report INRIA-4784 [1].

N. Mitrou et al. (Eds.): NETWORKING 2004, LNCS 3042, pp. 1012–1023, 2004.

Our goal is to propose an extension to DSR that overcomes these two problems, while still preserving its current features. We propose *Hash-Based DSR*, a protocol that compresses the list of addresses in the source-route using a Bloom filter [3]. Instead of inserting a source-routing option in each packet, as in DSR, the source inserts the corresponding Bloom filter.

The rest of the paper is organized as follows: Section 2 presents the proposed *Hash-Based DSR* protocol. In Section 3, some simulation results are presented and analyzed. Section 4 presents some related work. Finally, the last section concludes the paper.

2 Hash-Based DSR

One of the biggest limitations of DSR for large network is that it uses source-routing to route packets. A DSR packet must carry the IP addresses of all the nodes that are on the path from the source to the destination. The generated bandwidth overhead is significant when the source and the destination are far from each other. Indeed as we will show later in our simulations (see Figure 5), this control overhead can use up to 95% of the network bandwith when the path length is about 100 nodes. This overhead is not acceptable and limits the use of DSR to small networks.

We propose, in this paper, *Hash-Based DSR* (HB-DSR) a DSR extension for large ad-hoc networks. The main idea of our scheme is to compress the list of addresses in the source-route using a Bloom filter. Instead of inserting a source-routing option in each packets as in DSR, the source inserts the corresponding Bloom filter. The resulting protocol is very similar to DSR and inherits from most of its features. Furthermore since the filter used in each packet is much smaller that the list of addresses, $HB - DSR$ is much more scalable than DSR for large networks.

2.1 Bloom Filters

A Bloom filter is a m-bit vector v that codes the membership of a set $A = \{a_1, a_2, .., a_n\}$ of n elements [3]. The idea is to allocate a vector v of m bits, initially all set to 0, and then to choose k independent hash functions, h_1, h_2,...,h_k, each with a range $\{1, ..., m\}$. For each element a of A, the k bits at positions $h_1(a)$, $h_2(a)$, ...,$h_k(a)$ in v are set to 1.

To verify whether a element b is the set A, it is enough to verify if all the k bits at positions $h_1(b)$, $h_2(b)$, ...,$h_k(b)$ in v are set to 1. If any of them is set to 0, b does clearly not belong to A. If all the bits are set to 1, then b is probably a member of A, although there is some probability that b is not a member of A. This is called a *false positive*. The probability a *false positive* can be calculated in a straightforward fashion. After all the elements of A are coded in the Bloom filter, the probability that a specific bit is still set to 0 is: $(1 - 1/m)^{kn}$ The probability of a false positive is then:

$$fp = (1 - (1 - 1/m)^{kn})^k \qquad (1)$$

It can be shown that for a given m and n, the optimal value of hash functions to used, k, is $k = ln2 \times m/n$. There is a clear trade-off between the size of the filter m and the probability of a false positive. For a given n, fp can be decreased by increasing m. However increasing m reduces the compression rate of the Bloom filter. The optimal value of m is application-specific. For some applications, the false-positive rate must be very small and as a result m must be large. Other applications can tolerate higher false positive rates and therefore can use smaller values of m.

2.2   Protocol Overview

In $HB - DSR$, the DSR source-routing option is *compressed* using a Bloom filter. The set A is therefore composed of the nodes' IP addresses of the path from the source to the destination. n is equal to the path length. The proposed protocol is the following:

1. A source S that wants to send packets to a destination D, invokes the DSR *route discovery* protocol. It then receives a *Route Reply* that contains the list of the nodes' addresses along the path from S to D.
2. S then computes from these addresses the corresponding Bloom filter as follows:

   1: **for** $(j = 1; j < plen; j + +)$ **do**
   2:    **for** $(i = 0; i < k; i + +)$ **do**
   3:       $BF[hash(i|Addr_j).mod(m)] = 1;$
   4:    **end for**
   5: **end for**

   where the Bloom filter, BF, is a bit-string of size m, k is the number of hashes used and $Addr_j$ are the $(plen - 1)$ addresses on the path from S to D [2].
3. When S sends packets to D, it inserts a (newly defined) hop-by-hop option that carries the bloom filter and the value of the parameter k (the number of hashes to use).
4. Upon receiving a packet, a node verifies whether the destination address is one of its addresses. If this is the case, the packet has reached its destination. If destination address is not one of its addresses and the TLL is zero, the packet is dropped, otherwise the node verifies if any of its neighbors' addresses (except the one it received the packet from) are contained in the Bloom filter carried in the packet using the following algorithm:

   1: **for** $(j = 1; j <= R; j + +)$ **do**
   2:    $IsMember[j] = 1;$
   3:    **for** $(i = 0; i < k; i + +)$ **do**
   4:       **if** $(BF[hash(i|addr_j).mod(m)] == 0)$ **then**
   5:          $IsMember[j] = 0;$
   6:       **end if**
   7:    **end for**

---

[2] The following sections describe how the parameters m and k are chosen.

8: **end for**

where R is the number of neighbors N_j defined by their address, $addr_j$.

If, when the algorithm terminates $IsMember[j]$ is set to 1 then the neighbor N_j belongs to the filter and is therefore a node on the path from S to D, otherwise N_j does not belong to the filter. If a neighbor belongs to the Bloom Filter, the packet's TTL is decremented and the packet is forwarded to that neighbor. If not, the packet is silently dropped (the packet was probably mis-routed from a false positive). If there are several neighbors that are contained in the packet's filter (this is the result of false positives), the packet is duplicated and forwarded to each of these neighbors.

As a result of this protocol, the packets are forwarded hop-by-hop until the destination.

The filter's size is obviously a crucial parameter of our protocol. In fact if the filter's size is too small, false positives will be frequent and the packets will be mis-routed. Although the packets will reach their destination, the bandwidth overhead will be quite large. On the other hand, if the filter's size is very large, false positives will be rare (and therefore no many packets will be mis-routed) but the size of the packets will be larger (since they carry the filter) and the benefit of the compression will be reduced.

2.3   Filter Size Computation

This section presents two different approaches to compute the size of the Bloom filter. In the first one (open-loop algorithm), a mobile host computes the "optimal" size using some cost function. This computation is performed locally without any feedback from the network. In the second approach (closed-loop algorithm), a host computes the filter size using some inputs he gets from the network. Both approaches can be used conjointly.

Notation. In this section, we define and use the following notation:
  - *plen:* the path length from source to destination.
- *R:* the average number of neighbors per intermediate node.
- *fp:* the false positive probability.
- *IP_addr_size:* the size of an IP address (32 bits for IPv4 and 128 bits for IPv6).
- *IP_head_size:* the size of an IP header.
- *IP_opt_size:* the size of the IP option header that carries a Bloom filter.
- *pkt_size:* the total size of a packet including its IP header, options and its data.
- *m:* the size of the Bloom filter.

Open-loop Algorithm. In DSR, the routing overhead resulting from the source routing is defined by:

$$dsr\_cost = plen \times [(plen - 1) \times IP\_addr\_size + IP\_opt\_size] \qquad (2)$$

In HB-DSR, the routing cost results from (1) the Bloom filter that each packet carries and (2) the false positives. If a packet creates a false positiveat a node,

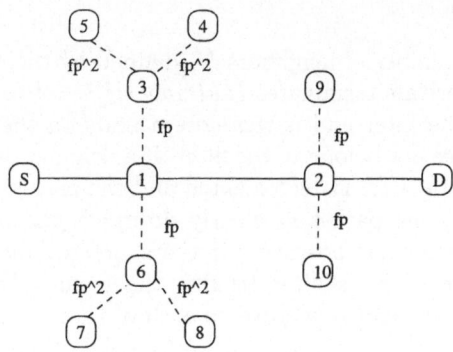

Fig. 1. HB-DSR routing

the packet is sent to one or several wrong paths. Let us consider the scenario displayed in Figure 1. In this figure, S is sending packets to D. The path length, *plen* is equal to 3.

At the first node, *N1*, the probability that a false positive happens towards N3 (resp. N6) is defined by *fp*. The resulting cost is $fp \times pkt\_size$. The probability that the forwarded packet is forwarded to N4 (resp. N8) or to N5 (resp. N7) is fp^2. The induced cost is then $fp^2 \times pkt\_size$. The probability that the forwarded packet is forwarded by N4 or N5 is zero because the TTL has then reached zero. The total cost at node N1 is therefore:

$$Cost_{N1} = 2 \times pkt\_size \times [fp + (R-1) \times fp^2] \qquad (3)$$

Similarly at the second node, N2, the probability that a false positive happens towards N9 (resp. N10) is *fp*. The resulting cost is $fp \times pkt\_size$. The probability that this packet is forwarded by N9 (resp. N10) is zero because the TTL is then 0. The total cost at node *N2* is then:

$$Cost_{N2} = 2 \times pkt\_size \times fp \qquad (4)$$

More generally, the cost at node Nx (with $0 < x < plen$) is defined by:

$$Cost_{Nx} = (R-2) \times pkt\_size \times fp \times \sum_{i=0}^{plen-x-1} ((R-1)^i \times fp^i) \qquad (5)$$

As a result, the total cost resulting from the false positives by a packet sent from a source to a destination is defined as follows:

$$fp\_cost = \sum_{j=1}^{plen-1} Cost_{Nj}$$

$$= (R-2) \times pkt\_size \times fp \times [\sum_{j=0}^{plen-2}((plen-j-1) \times (R-1)^j \times fp^j)] \qquad (6)$$

where *fp* is computed as follows:

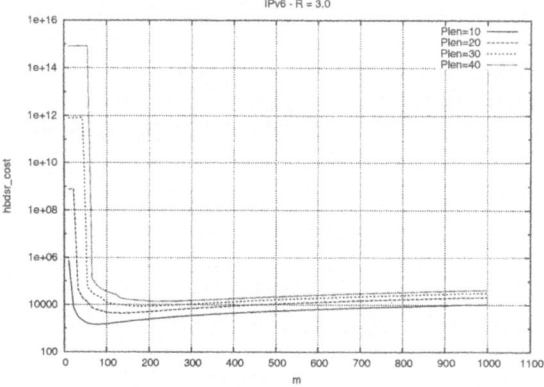

Fig. 2. HB-DSR cost vs m

$$fp = (1/2)^{ln(2) \times m/plen} \tag{7}$$

The total HB-DSR overhead is the cost resulting from the false positive and the cost of carrying the filter in each packet, i.e:

$$hbdsr\_cost = plen \times (IP\_opt\_size + m) + fp\_cost \tag{8}$$

Figure 2 plots the $hbdsr\_cost$ function according to m for several path lengths ($plen$). This figure shows that, for a given path length, there is only one value of m that minimizes the cost function.

In HB-DSR, when a source wants to send packet to a destination it performs a DSR route discovery. It then receives a Route Reply that contains the list of addresses along the path. It can deduce from this information the path length ($plen$) but still need to estimate the parameter R. We propose to extend the route request and reply messages with a *CN (Cumulative Neighbors) field*. The field is set to 0 in the route request by the source. Each intermediate node that forwards it increments this CN field by the number of its neighbors. When the destination receives the route request, it copies the CN value in the route reply and sends it to the source. The source can then compute an estimate of the parameter R by dividing the value CN by the path length ($plen$).

Once the source has an estimate of $plen$ and R, it computes for several m the value of the cost function $hbdsr\_cost$ and uses the value of m that minimizes it.

Closed-loop Algorithm. In the previous section, the source computes the filter size using a cost function. However in some scenarios, the source can still suffer from the cost of false positives. An closed-loop algorithm is often more efficient and practical. We therefore propose to use some feedbacks from the network to tune the filter size.

We define, *FP_DUP*, a new error message. When a node, N_i, detects one or several false positives (i.e. there are more than one neighbor, let's say r_i neighbors, in the filter), it returns a *FP_DUP* message to the source. This message contains the number of neighbors contained in the filter, i.e. r_i, at node N_i.

In order to avoid the explosion of *FP_DUP* messages and to avoid routing loop, we propose that nodes drop packets that have experienced two or more false positives. When a node detects a false positive upon reception of a packet, it sends a *FP_DUP* to the source and set a bit (the D bit) in the packet to 1. If this bit is already set, the packet is dropped and, instead of a *FP_DUP* message, a *FP_DROP* message is sent to the source.

The source executes the following algorithm:

- *Step1*: If the source receives one or several *FP_DROP* messages, it increases the filter size by one unit (i.e as shown in Section 3 by 64 bits for IPv6 or 32 bits for IPv4). The algorithm is then re-executed.
- *Step2*: If the source receives one or several *FP_DUP* (but no *FP_DROP* message) that specify N false positives, the source computes *FP_cost*, the cost resulting from the N false positives:

$$FP\_cost = N \times pkt\_size \qquad (9)$$

It also computes, *Delta_cost*, the cost resulting from increasing the filter size by one unit, *BF_unit*.

$$Delta\_cost = plen \times BF\_unit \qquad (10)$$

If *Delta_cost* < *FP_cost* then the filter's size is incremented by one unit and the algorithm is re-executed. Otherwise the filter size is kept to m and the algorithm terminates.
- *Step3*: If the source does not receive any *FP_DUP* nor any *FP_DROP* messages for a given value of m and if this m was not obtained from *Step2*, the source decrements m by one unit and the algorithm is re-executed. Otherwise the algorithm terminates.

3   Simulation

3.1   Simulation Model

The main goal of our simulations is to evaluate the gain of our approach over *DSR*. We mainly focus our simulations on the per-packet bandwidth gain achieved by compressing the source-routes with Bloom filters. We do not consider the control messages (such as RREP, RREQ or RRER) overhead since they are identical in *DSR* and *HB − DSR*. Furthermore, for simplicity, we do not consider mobility in our simulations. The main contribution of our scheme is to reduce the per-packet control overhead. All the rest, including mobility management, is similar to DSR. We expect the mobility management performance of DSR and HB-DSR to be very similar.

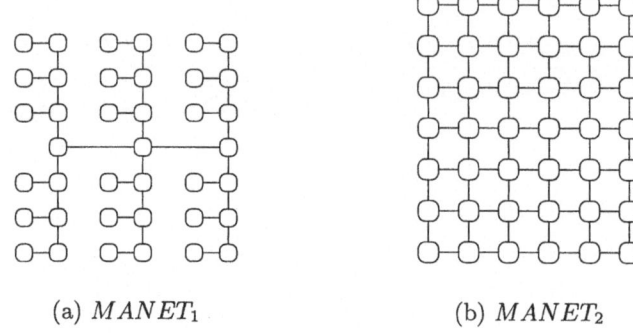

(a) $MANET_1$ (b) $MANET_2$

Fig. 3. Network topology

We considered two different network topologies (see Figure 3). Both of them contain 3600 nodes. The topology of the first one ($MANET_1$) is a tree and as a result there is only one possible path from a given source to a given destination. The second one ($MANET_2$) is highly connected (a node has on average 4 neighbors while in $MANET_1$ a node has on average 2.8 neighbors) and there are several paths for a given source to a given destination.

For each of the network, we randomly select the source and the destination. We then simulate the routing of packets from the source to the destination using DSR and the 3 following variants of $HB - DSR$:

1. $HB - DSR$: the filter size is constant (88 bits for IPv6).
2. $AHB - DSR_0$ (*A* stands for *Adaptive*): the filter size is computed using the cost function described in Section 2.3.
3. $AHB - DSR$: the filter size is computed using the cost function described in Section 2.3 and adjusted using feedbacks from the network (FP_DUP and FP_DROP messages), as described in Section 2.3.

We run these simulations 2500 times for IPv6 and IPv4.

3.2 Performance Results

Figure 4 displays the *bandwidth gain* of $HB - DSR$, $AHB - DSR_0$ and $AHB - DSR$ for IPv6 according to the path length between the source and the destination. We compute the *bandwidth gain* of a scheme S by dividing the bandwith used by DSR to transmit a packet containing 64 bytes of data by the bandwidth used by the scheme S to transmit the same data to the destination. Each hop-wise transmission is counted as one transmission. We assume here that the route discovery phase has been performed and that the source knows the route to the destination.

The results show that:

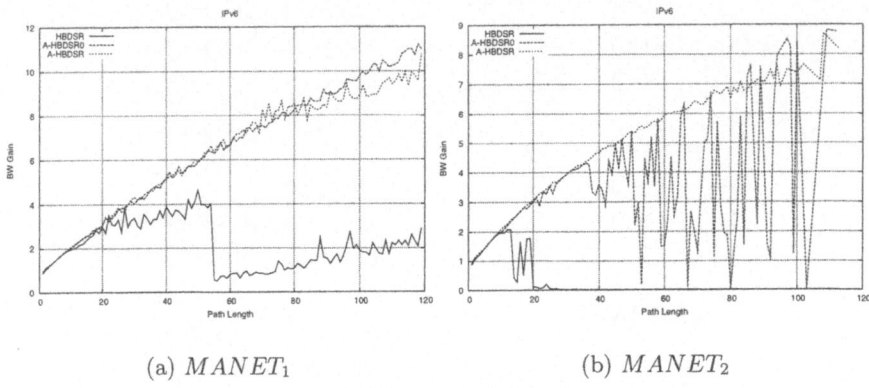

(a) $MANET_1$ (b) $MANET_2$

Fig. 4. Bandwidth gain (IPv6; data-size = 64 bytes)

- $HB-DSR$: When m is fixed and equal to 88, the gain decreases considerably with the path length. These results were expected because when the path length is large, the false positive rate is close to 1 and most of the transmitted packets are broadcast. For $MANET_1$, the gain is always greater than one because the network is not highly connected and the broadcast cost is not too high (and a least lower than source-routing). In contrast, for $MANET_2$, the gain converges to 0. $MANET_2$ is highly connected and the cost of broadcasting a packet is very large. It is more efficient to use source-routing, i.e. regular DSR, in this scenario.

- $AHB - DSR_0$: The performance for $MANET_1$ and $MANET_2$ are very different. With $MANET_1$ the gain is always greater than one and increases with the path length. When the path length is 100 nodes, the achieved gain is 10. This means that the network can accommodate 10 times more connections than if DSR was used. With $MANET_2$, $AHB - DSR_0$ does not perform well and its performance fluctuates a lot. These results are explained by the different costs of a false positive in each networks. In $MANET_1$, the cost of a false positive is not too high because the network is not very well connected and the mis-routed packets die out quickly. In $MANET_2$, the false positive cost is quite high. In fact, since each node has a higher number of neighbors, the probability of a false positive is larger and as a result, false positives are more frequent. Furthermore, if a packet is mis-routed twice in a row, it can reach the correct path again and enter a routing loop.

- $AHB - DSR$: $AHB - DSR$ corrects the problems of the $AHB - DSR_0$. By using feedbacks, the source can adjust the value of m accurately and minimizes the number of mis-routed packets. The gain increases with the path length. The gain obtained with $MANET_1$ is larger than the gain achieved with $MANET_2$. Indeed, since the connectivity is smaller, fewer bits are required per bloom filters in $MANET_1$.

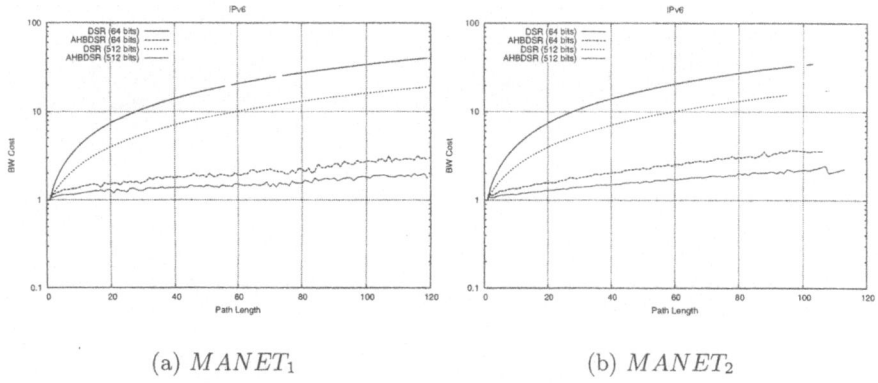

(a) $MANET_1$ (b) $MANET_2$

Fig. 5. DSR and AH-BDSR bandwidth cost (IPv6)

Figure 5 presents the *bandwidth cost* of $AHB - DSR$ and DSR. The bandwidth cost of a given proposal is computed by dividing the number of bytes necessary to transmit a given piece of data from the source to the destination using the proposal by the number of bytes necessary to transmit the same data with regular IP. This cost actually measures the control cost resulting from carrying in each packet a source-route or a Bloom Filter. We consider two data-sizes: 8 and 64 bytes. The results show that DSR cost increases drastically with the path length. In fact since a packet carries the addresses of all the nodes from the source to the destination, the control cost increases with the path length. When the path length is 100, the cost of DSR for a data size of 64 bytes is 20. This means that the cost of sending 64 bytes from the source to the destination is 20 times larger than the cost needed to transport the same data in a regular IP packet (i.e. without the source-routing option). The cost to transport these data with $AHB - DSR$ goes down to 3. This is the result of compressing the source-address with a Bloom filter. As we will see later in this section, this cost is much lower when the data size is larger.

Figures 6 compare the performance obtained with IPv4 and IPv6. They display the bandwidth (in bytes) used by DSR and $AHB - DSR$ to transmit 64 bytes of data from the source to the destination with IPv4 and IPv6. Each hop-wise transmission is counted as one transmission. These results show:

1. DSR is more expensive in IPv6 than in IPv4. In fact since addresses are much larger in IPv6 than in IPv4, the source-routing overhead is much larger. As a result, IPv6-DSR is more expensive than IPv4-DSR in term of bandwidth.
2. IPv4 and IPv6 $AHB - DSR$ costs are very similar. The filter's size used by $AHB - DSR$ is independent from the IP version. The resulting cost is the same despite large IPv6 address size. This result is very encouraging. While IPv6 is an handicap for DSR, it becomes very attractive for $AHB - DSR$.

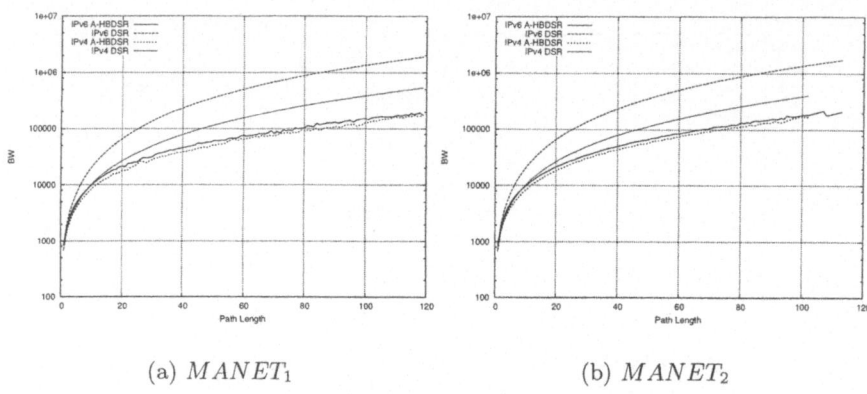

(a) $MANET_1$ (b) $MANET_2$

Fig. 6. IPv4 vs IPv6 bandwidth (data-size= 64 bytes)

4   Related Work

Bloom filters have recently received attention in the networking area. A complete survey of these proposals can be found in [4]. [5] introduces a longest prefix matching algorithm that deploys "counting Bloom filters", in order to efficiently narrow the scope of the search and hence increase the IP lookup speed. [6] proposes using Bloom filters to reduce the storage cost of interface lists held by routers for group communications. Instead of maintaining a list of interfaces for each group, for each interface a Bloom filter is maintained. If a group is active with respect to an interface, then the corresponding Bloom filter is inserted the group. False positives, in this case, lead to some packets being forwarded incorrectly, which will be eventually discarded by other nodes.

In another related work we have proposed using Bloom filters for reducing the broadcast cost of paging in IP-based cellular systems [7]. IP paging is an optimization that allows the mobile IP hosts to preserve energy by entering dormant mode within the boundaries of a wide paging area. Using "hash-based paging" the paging sub-system inserts the IP addresses of the called dormant hosts (in the same paging area) into a Bloom filter, and page them concurrently by broadcasting a single paging message. Similarly to $HB - DSR$, hash-based IP paging is more attractive for IPv6, since a paging message that contains a number of 128-bit IPv6 addresses would consume too much bandwidth.

The proposed optimization has some relevance with header compression, but in a more general sense. The term "header compression" mostly implies techniques such as [8][9][10]. These techniques are based on the observation that most TCP/IP header fields never or seldom change during a session, and can easily be compressed. The proposed optimization is more a compression technique that is specific to *a list of IP addresses*. Compression can be made in a stateless fashion. Decompression is replaced by membership query, which is also stateless.

This allows $HB - DSR$ to compress source-routes while avoiding routing state (or, routing context) at intermediate nodes.

5 Conclusions

We have presented and evaluated *Hash-Based DSR*, a *DSR* extension for large networks. This protocol reduces the per-packet control overhead of DSR by compressing the source-route with a Bloom filter. The simulation results on large networks (diameter of 100 nodes) show that $HB - DSR$ increases the network capacity by a factor of up to 15.

Another important benefit of $HB - DSR$ over DSR is that its performance is similar for IPv4 and IPv6. In fact while IPv6 large addresses are prohibitive in DSR, we showed by simulations that $HB - DSR$ performs as well for both IP versions. We expect this result to generate a lot of interest in the IPv6 community.

$HB - DSR$ can be extended to support *one-to-many* communication. In fact, if a source knows the source-routes to each of its destinations, it can build the Bloom filter that contains all the nodes of the delivery tree. By inserting such filter in a packet, the packet will be routed from the source to its destinations using the optimal delivery tree. We believe that this approach is promising for small groups.

References

1. Castelluccia, C.: Hash-Based Dynamic Source Routing. Technical Report 4784, INRIA (2003)
2. Johnson, D., Maltz, D., Broch, J.: The Dynamic Source Routing Protocol for Multihop Wireless Ad Hoc Networks. In Perkins, C., ed.: Ad Hoc Networking. Addison-Wesley (2001) 139–172
3. Bloom, B.: Space/time trade-offs in hash coding with allowable errors. Communications of the ACM **13** (1970) 422–426
4. Broder, A., Mitzenmacher, M.: Network Applications of Bloom Filters: A Survey. In: Proceedings of the 40th Annual Allerton Conference on Communication, Control, and Computing. (2002)
5. Dharmapurikar et al, S.: Longest Prefix Matching Using Bloom Filters. In: Proceedings of SIGCOMM'2003 Conference. (2003)
6. Gronvall, B.: Scalable multicast forwarding. SIGCOMM Poster (2001)
7. Mutaf, P., Castelluccia, C.: Hash-based Paging and Location Update Using Bloom Filters. to appear in ACM Mobile Networks and Applications (MONET) **10** (2005)
8. Jacobson, V.: Compressing TCP/IP Headers for Low-Speed Serial Links. RFC 1144, IETF (1990)
9. Degermark et al, M.: IP Header Compression. RFC 2507, IETF (1999)
10. Bormann et al, C.: RObust Header Compression (ROHC): Framework and four profiles: RTP, UDP, ESP, and uncompressed. RFC 3095, IETF (2001)

A New Method of Primary Routes Selection for Local Restoration

Krzysztof Walkowiak

Chair of Systems and Computer Networks, Faculty of Electronics, Wroclaw University of
Technology, Wybrzeze Wyspianskiego 27, 50-370 Wroclaw, Poland
Krzysztof.Walkowiak@pwr.wroc.pl

Abstract. We consider the problem of non-bifurcated multicommodity flows
rerouting in survivable connection-oriented networks. We focus on local restoration, in which the backup route is reroutes the flow only around the failed arc.
The origin node of the failed arc is responsible for rerouting. We introduce and
discuss a new convex function for optimisation of primary routes. We propose
a heuristic algorithm for assignment of primary routes employing the developed
function. During numerical experiments we examine performance of the new
function compared to other functions proposed previously in the literature.

1 Introduction

Recently, issues of network survivability become more important due to the growing
requirements for QoS and traffic engineering. Connection-oriented network technologies like Asynchronous Transfer Mode (ATM), MultiProtocol Label Switching
(MPLS) use comparable approaches to enable network survivability. The main idea
of this approach is as follows. Each circuit, i.e. virtual path in ATM or label switched
path in MPLS, has a primary route and a backup route. The primary route is used for
transmitting of data in normal, failure-free state of the network. After a failure of the
primary route the failed circuit is switched to the backup route. The process of
switching is easy, i.e. the circuit's identifier numbers are changed in network nodes.
All backup routes have zero bandwidth. After activation there are assigned with necessary bandwidth. In this work we focus on local restoration (called also rerouting or
repair) [1], [8]. The backup route is found only around the failed arc. The origin node
of the failed arc is responsible for rerouting.

In modern computer networks a single-link failure is the most common and frequently reported failure event [6]. Therefore, in most of optimization models a single-
link failure is considered as the basic occurrence. Spare capacity is computed to provide full restoration in case of a failure of any single-link. In networks having limited
resources of spare capacity 100% restoration is not always possible and routes are
designed to minimize effects of the failure, i.e. to minimize to amount of flow lost
due to a failure.

N. Mitrou et al. (Eds.): NETWORKING 2004, LNCS 3042, pp. 1024–1035, 2004.

For the context of this work we concentrate on an existing facility network, i.e. we do not consider facility capacity planning and topological design. Joint optimization of primary and backup routes must be carried out to find a globally optimal solution of the lost flow due to a network failure for a projected traffic demand. Since the optimization is conducted jointly over primary and backup routes, the complexity of the problem grows tremendously. The main idea of our approach is to partition the problem into two simpler problems: first optimize primary routes and next find backup routes for already established primary routes. Since there is mutual dependency between the primary routes and the backup routes assignment, the obtained solution cannot be claimed to be an optimum if these problems are treated separately. However, obtained results prove robustness of this approach [9], [10]. A key problem of our approach is to define an objective function for primary routes' assignment. Such a function must indicate preparation of the network to the rerouting process. Since there are many algorithms for convex multicommodity flow problem, it would be convenient if the function were convex.

In this work we define and discuss a new convex function for optimisation of primary routes in a network applying local rerouting. We also propose a heuristic algorithm for assignment of primary routes using this function. Furthermore, we present and discuss results of extensive simulations.

2 Definition of a New Function for Local Rerouting

Our objective in this section is twofold: Firstly, we want to define a function that reflects the preparation of the network to the local rerouting. Secondly, we plan to discuss main characteristics of this function.

Various network or transportation problems can be modeled as multicommodity (m.c.) flow problem [4], [5]. Multicommodity flows are of two types: bifurcated and non-bifurcated. In this work we focus on non-bifurcated m.c., in which each commodity flows along one route only.

We are given a network (G, c) where $G = (V, A)$ is a directed graph with n vertices representing routers or switches and m arcs representing links, $c : A \rightarrow R^+$ is a function that defines capacities of the arcs. We denote by $o : A \rightarrow V$ and $d : A \rightarrow V$ functions defining the origin and destination node of each arc. For each $a \in A$ we call $\text{in}(a) = \{k \in A \mid d(k) = d(a), k \neq a\}$ the set of incoming arcs of $d(a)$ except a, and $\text{out}(a) = \{k \in A \mid o(k) = o(a), k \neq a\}$ the set of outgoing arcs of $o(a)$ except a.

To mathematically represent the problem we introduce the following notations

f_a	represents the total flow on arc a.
c_a	capacity of arc a.
$g_v^{\text{out}} = \sum\limits_{i:o(i)=v} f_i$	aggregate flow of outgoing arcs of v.
$g_v^{\text{in}} = \sum\limits_{i:d(i)=v} f_i$	aggregate flow of incoming arcs of v.

$$e_v^{out} = \sum_{i:o(i)=v} c_i \qquad \text{aggregate capacity of outgoing arcs of } v.$$

$$e_v^{in} = \sum_{i:d(i)=v} c_i \qquad \text{The aggregate capacity of incoming arcs of } v.$$

Definition 1. *The global non-bifurcated m.c. flow denoted by* $\underline{f} = [f_1, f_2, ..., f_m]$ *is defined as a vector of flows in all arcs. We call a flow* $\underline{f}$ *feasible if for every arc* $a \in A$ *the following inequality holds*

$$\forall a \in A : f_a \le c_a \tag{1}$$

Inequality (1) ensures that in every arc flow is not greater then capacity. This inequality is called a capacity constraint.

For the sake of simplicity we introduce the following function

$$\varepsilon(x) = \begin{cases} 0 & \text{for} \quad x \le 0 \\ x & \text{for} \quad x > 0 \end{cases} \tag{2}$$

To analyze properties of the local restoration we consider an arc $k \in A$. We assume failure of k. Recall that in the local rerouting flow on the arc k must be rerouted by the source node of the arc k. Therefore, spare capacity of outgoing arcs of $o(k)$ except k is a potential bottleneck of the restoration process. Notice that if

$$f_k \le \sum_{i \in out(k)} (c_i - f_i) \tag{3}$$

then flow of the failed k can be restored using spare capacity of other links leaving the origin node of k. Otherwise if

$$f_k > \sum_{i \in out(k)} (c_i - f_i) \tag{4}$$

then some flow of the failed link k cannot be restored because spare capacity of other arcs leaving the origin node of k is too small. It means that those arcs block the 100% restoration and some flow of k is lost. In formulas (3-4) we assume that reroutable flow can be split to a number of different routes. It means that in order to define a function of lost flow we relax the non-bifurcated flow to bifurcated flow. It is, according to [6], a reasonable approach for backbone networks where vary large volume of various calls are transmitted. Furthermore, such assumption makes easier the analysis. Recalling definition of $g_{o(k)}^{out}$ and $e_{o(k)}^{out}$ and applying formulas (3-4) we define the function L_k^{out} of the arc k flow lost in the node $o(k)$ in the following way

$$L_k^{out}(\underline{f}) = \varepsilon \left(g_{o(k)}^{out} - (e_{o(k)}^{out} - c_k) \right) \tag{5}$$

Note that L_k^{out} denotes lost flow that cannot be restored using arcs leaving the node $o(k)$ due to limited spare capacity of these arcs. The L_k^{out} function depends on the flow $g_{o(k)}^{\text{out}}$ leaving the node $o(k)$. It is not dependent directly on the flow f_k. Correspondingly, we define the function L_k^{in} that denotes lost flow that cannot be restored using arcs entering the node $d(k)$.

$$L_k^{\text{in}}(\underline{f}) = \varepsilon\left(g_{d(k)}^{\text{in}} - (e_{d(k)}^{\text{in}} - c_k)\right) \tag{6}$$

Definition 2. *We call an arc* k adjacent *to an arc* a *if* $o(k)=o(a)$ *or* $d(k)=d(a)$. *If* k *is not adjacent to* a *we call it* remote *to* a.

Function L_k defined below is a linear combination of flow lost in arcs outgoing $o(k)$ and arcs incoming $d(k)$. This function considers only arcs adjacent to the failed arc a. Arcs remote to k, which can block some flow of the failed arc during the rerouting process, aren't taken into account. Therefore, L_k only estimates the flow of arc k lost after local rerouting. Function L_k is a lower bound of the flow of arc k lost due a failure of k.

$$L_k(\underline{f},\alpha) = \alpha L_k^{\text{in}}(\underline{f}) + (1-\alpha)L_k^{\text{out}}(\underline{f}) \qquad 0 \le \alpha \le 1 \tag{7}$$

Theorem 1. *Function* $L_k(\underline{f},\alpha)$ *(7) is a convex function for any feasible flow* $\underline{f}$ *and any* α *such that* $0 \le \alpha \le 1$.

Due to limited space of the paper we do not present the formal proof.

Using $L_k(\underline{f},\alpha)$ we can define a function $L(\underline{f},\alpha)$ that shows preparation of the whole network to the local rerouting after a failure of any single arc. We assume that probability of the arc failure is the same for all arcs. Therefore, probability is not included in this function.

$$L(\underline{f},\alpha) = \sum_{k \in A} L_k(\underline{f},\alpha) = \alpha \sum_{k \in A} L_k^{\text{in}}(\underline{f}) + (1-\alpha) \sum_{k \in A} L_k^{\text{out}}(\underline{f}) \qquad 0 \le \alpha \le 1 \tag{8}$$

Theorem 2. *Function* $L(\underline{f},\alpha)$ *(8) is a convex function for any feasible flow* $\underline{f}$ *and any* α *such that* $0 \le \alpha \le 1$.

Proof. It is sufficient to notice that the function $L(\underline{f},\alpha)$ is a sum of convex functions $L_k(\underline{f},\alpha)$ over all arcs $k \in A$. According to Theorem 1 the function $L_k(\underline{f},\alpha)$ is convex for any feasible flow $\underline{f}$ and any α such that $0 \le \alpha \le 1$. Consequently, the function $L(\underline{f},\alpha)$ is convex. This completes the proof.   □

Corollary 1. $L(\underline{f},\alpha)$ *is a continuous, non-decreasing, piece-wise linear function for any feasible flow* $\underline{f}$ *and any* α *such that* $0 \le \alpha \le 1$. *The function* $L(\underline{f},\alpha)$ *is differentiable except points for which one of the following condition holds*

$$g_{o(k)}^{\text{out}} = (e_{o(k)}^{\text{out}} - c_k) \qquad k \in A \tag{9}$$

$$g_{o(k)}^{in} = (e_{o(k)}^{in} - c_k) \quad k \in A \tag{10}$$

3 Related Work

In this section we present two functions proposed in the literature for optimization of primary routes using the local restoration and compare these functions with the function developed in previous section. In local restoration after a failure of the k-th arc all circuits using the failed arc must be rerouted around the arc k. In order to estimate the amount of the restored flow the maximum flow algorithm can be applied. The maximum flow criterion denotes the theoretical maximal rerouting capacity. The failed arc k is removed from the network. Next, the maximum flow between the origin and destination node of the failed arc is calculated taking into account spare capacity of network's arcs. Let $MF(k)$ denote flow of the failed arc a restored by the maximum flow method. The flow lost using the maximum flow rerouting is given by $\varepsilon(f_k - MF(k))$. If the $MF(k)$ is greater than f_k, no flow is lost. Otherwise, $(f_k - MF(k))$ flow is lost.

Another function applied for calculation of lost flow after the local rerouting is based on k-shortest paths (KSP) algorithm. The failed arc k is removed from the network. Next, the KSP algorithm finds k-successively shortest disjoint paths between the origin and the destination node of the failed arc. These paths one by one are saturated with flow of the failed arc and are used for restoration of flow f_k. The fraction of the flow f_k not restored during running KSP algorithm is lost. Let $KSP(k)$ denote flow of the failed arc a restored by the KSP method. The flow lost after failure of arc k using the KSP rerouting is given by $\varepsilon(f_k - KSP(k))$. Consequently, the lost flow after a failure of any single arc using KSP approach is calculated similarly to (8)

$$L_{KSP}(\underline{f}) = \sum_{k \in A} \varepsilon(f_k - KSP(k)) \tag{11}$$

Some previous authors have introduced similar approach for local rerouting of ATM network and formulated a problem of primary routes assignment with the objective function of lost flow using the KSP rerouting [6].

Authors of [3] compared maximum flow and KSP strategies using simulation methods. KSP restoration offers performance 99.9% of that from Max Flow. The advantage of KSP is time complexity of $O(nlogn)$ obtained for one link compared to maximum flow $O(n^3)$ using the centralized restoration by a single processor computation. Function L_k (7) requires only $O(n)$ time complexity. However, as mentioned above, the function (8) is a lower bound of the lost flow calculated using the maximum flow or KSP method. The function L takes into consideration only arcs adjacent to the failed arc k. Other arcs remote to k are not considered in this function.

Another advantage of the function $L(\underline{f}, \alpha)$ is convexity. There are many algorithms developed for nonlinear convex multicommodity flow problems. Since the function L is convex, it can be easily applied for optimization of network flows in survivable networks. According to [5], the most popular algorithm for optimization of nonlinear convex m.c. flow problems is Flow Deviation (FD) - method proposed in [4] and applied to comparable problem in [2]. A comprehensive list of other algorithms can be found in [7].

4 Algorithm for Flow Assignment

In order to solve the problem of static primary routes assignment using as objective function $L(\underline{f}, \alpha)$ we develop a heuristic algorithm based on the non-bifurcated FD algorithm proposed in [4].

With the purpose of making easier the consideration we define a new function

$$\varpi(x) = \begin{cases} 0 & \text{for} \quad x \le 0 \\ 1 & \text{for} \quad x > 0 \end{cases} \tag{12}$$

The FD algorithm uses an arc metric, which is a derivative of the objective function. Since according to the Corollary 1, the function $L(\underline{f}, \alpha)$ is not differentiable everywhere, we introduce the following metrics of arc k

$$l_k^{\text{out}}(\underline{f}) = \varpi \left(g_{o(k)}^{\text{out}} - (e_{o(k)}^{\text{out}} - c_k) \right) \tag{13}$$

$$l_k^{\text{in}}(\underline{f}) = \varpi \left(g_{o(k)}^{\text{in}} - (e_{o(k)}^{\text{in}} - c_k) \right) \tag{14}$$

$$l_k(\underline{f}, \alpha) = 1 + \alpha l_k^{\text{in}}(\underline{f}) + (1-\alpha) l_k^{\text{out}}(\underline{f}) \qquad 0 \le \alpha \le 1 \tag{15}$$

Note that $\alpha l_k^{\text{in}}(\underline{f}) + (1-\alpha) l_k^{\text{out}}(\underline{f})$ is a derivative of the function $L_k(\underline{f}, \alpha)$ (7) except points for which $g_{o(k)}^{\text{out}} = (e_{o(k)}^{\text{out}} - c_k)$ or $g_{o(k)}^{\text{in}} = (e_{o(k)}^{\text{in}} - c_k)$. In these points the function $l_k(\underline{f}, \alpha)$ is equal to the left-sided derivative of $L_k(\underline{f}, \alpha)$. Note that for less loaded networks metrics (13) and (14) are equal to 0. Therefore, we introduce to the formula (15) the hop number.

Algorithm FDP

Let $\underline{f}^1$ denote a feasible flow containing routes for all p circuits to be established. In order to find $\underline{f}^1$ we can apply an algorithm based on the initial phase of the FD algorithm [4]. Let $L(\underline{g}, \alpha)$ denote value of the L (8) function for a feasible flow $\underline{g}$. We start with $r := 1$.

<u>Step 1.</u> Find a flow $SR(f^r)$ defined as the set of shortest routes under the metric $l_k(f, \alpha)$ for all circuits. Set $i := 1$ and go to step 2.

<u>Step 2.</u> Let $g = f^r$.

a) Find v from g by deviating flow of circuit i to the shortest route given by $SR(f^r)$. Routes for other circuits except circuit i remain unchanged.

b) If v is a feasible flow and $L(v, \alpha) < L(g, \alpha)$ set $g = v$.

c) If $i = p$ go to step 3. Otherwise, set $i := i + 1$ and go to step 2a.

<u>Step 3.</u> If $g = f^r$ stop the algorithm, since the solution cannot be improved. Otherwise, set $r := r + 1$, $f^r = g$ and go to step 1.

Theorem 3. *Algorithm FDP converges in a finite number of steps and constitutes a feasible solution.*

Proof. The main idea of the FDP algorithm is as follows. We start with a feasible flow f^1. For each considered flow f^r we calculate $SR(f^r)$ containing the shortest routes according to the metric $l_k(f, \alpha)$ (Step 1). Next, we try to improve the solution by deviation of one selected circuit to another route (Step 2). Since there are a finite number of non-bifurcated flows, the algorithm converges in a finite number of steps. Repetitions of the same flow are impossible due to the stopping condition (Step 3). We assume that the initial flow f^1 is feasible. Next, in Step 2b we check whether the new flow v is feasible. If v is not feasible it is not analyzed further. Therefore, the algorithm FDP constitutes a feasible solution. □

We use the FDP algorithm also for optimization using other functions: lost flow using the KSP rerouting given by (11) and the overall network flow given by the sum of all arcs' flows. Clearly, we must modify the FDP according to these functions. For the former function we apply the following metric

$$l_k(f) = 1 + \omega(KSP(k)) \tag{16}$$

For the flow function we use the traditional hop number metric

$$l_k(f) = 1 \tag{17}$$

5 Results

The algorithm proposed in previous section was coded in C, and the program was run on an IBM-compatible PC with 2GHz Intel processor and 512 MB of RAM. Throughout the experiments, three objective functions were examined: lost flow in link given by (8), lost flow using the KSP rerouting (11) and the function of overall network flow. For the sake of simplicity in presentation of results, we refer to these functions in this section LFL, KSP and Hop, respectively. Also in tables and figures we use these names. The FDP algorithm is run for the same network and demand pattern three times with different objective functions. Labels FDP_LFL, FDP_KSP and FDP_Hop are used to denote the FDP algorithm applying functions LFL, KSP and Hop, respectively. In all cases the same starting solution found by the initial phase of the FD algorithm is applied. We assume that in function (8) $\alpha = 0.5$. We also show results given by the initial phase of the FD algorithm. Results presented in this section are obtained from simulations on 6 sample networks. Name of each network indicates the number of links in the network.

Table 1. Parameters of tested networks

Name of network	114	128	144	162	180	200
Number of nodes	36	36	36	36	36	36
Number of links	114	128	144	162	180	200
Node degree (average)	3.17	3.56	4.00	4.50	5.00	5.56
Node degree (minimum)	2	3	3	3	4	4
Node degree (maximum)	5	6	6	6	7	7
Number of tests	20	18	16	15	14	14
Flow requirement (minimum)	45	75	77	92	108	138
Flow requirement (maximum)	64	92	92	106	121	151

Table 1 summarizes the parameters of all sample networks. The first column specifies the name of the parameter, next columns includes values of these parameters for each network. Let bandwidth unit (BU) denote an arbitrary unit of bandwidth, for instance 1 Mb/s. We assume that for all networks capacity of each link is 5000 BU. Since, according to theoretical analysis presented above, the function of link lost flow depends on the node degree; we selected to numerical experiments networks with different values of the average node degree. In the experiment it is assumed that there is a requirement to set up a connection for each direction of every node pair. Thus, the total number of demands is 1260. Each demand is defined by: the source node, destination node and flow requirement. To make clear the effectiveness of function LFL in response to a varying traffic demand, several demand patterns are examined for each network. The flow requirement for all demands is the same. For instance, for network 114 we perform 20 simulations starting with flow requirement of each demand equal to 45 BU, the biggest value of flow requirement is 64 BU.

We introduce the following parameters to present results. Since the simplest performance indicator to show preparation of the network to the local rerouting is the KSP function given by (11) defined in section 4, in order to compare performance of three tested functions we use the *normalized loss* (NL) function calculated using the KSP function. NL is defined as a unit of a normalized flow where 100 NL is equal to the total flow in the network. For instance, if $L_{KSP}=100$ and the total flow in the network is 500, the normalized flow equals 20 NL. Also the concept of *average link utilization* (AVLU) is used to describe the simulation results. The AVLU parameter, which indicates the network load, is defined as the proportion of the total flow in the network summed over all links and the total capacity of all networks links.

Table 2. The aggregate normalized loss obtained for various functions

Networks	FDInit	LFL	KSP	Hop
All	12.47	12.19	11.99	13.36
114	12.68	10.88	10.97	13.80
128	28.16	27.95	26.81	30.33
144	15.80	15.72	15.59	16.90
162	12.52	12.44	12.38	13.49
180	7.55	7.55	7.55	7.70
200	6.71	6.71	6.71	6.96

In Table 2 we report performance of the initial phase of FD and three functions: LFL, KSP and Hop in terms of the normalized loss. Results are aggregated over all tests performed for a given network. Generally, we found the KSP approach to be superior to the other approaches. However, for the network 114 the LFL function gives better results. Summary for all networks shows that the difference between LFL and KSP is lower then 1.7%. FDInit and Hop yield much worse solutions.

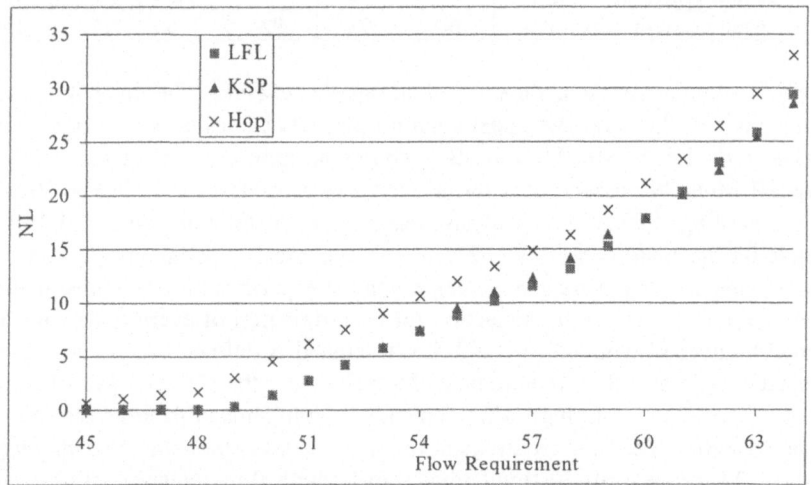

Fig. 1. Performance of various approaches in terms of the normalized flow for network 114

Fig. 1 shows the detailed performance of the three functions for network 114. The x-axis represents the flow requirement and the y-axis represents the normalized loss. The general trend is that both the LFL and KSP functions yield similar results while the Hop function provides much worse performance. Apparently, when the flow requirement increases, more flow is lost due to failure of any single arc.

In Table 3 we report CPU time taken for running algorithm FDP for all three functions. The time does not include time to do I/O for input of various files and design output. For each network we sum times over for all tested traffic patterns. Times are given in seconds and do not include the calculation time of the initial phase of the FD. We can see that the FDP_LFL performs 37 times faster then the FDP_KSP. This can be easily explained by the time complexity of both approaches discussed in Section 4. To calculate the KSP function we must find k-shortest paths, which is much more time consuming then the calculation of the LFL function.

Table 3. The aggregated decision times of the FDP algorithm obtained for various functions

Networks	FDP_LFL [s]	FDP_KSP [s]	FDP_Hop [s]
All	222	8119	207
114	60	1057	33
128	70	1585	39
144	30	1198	31
162	28	1432	34
180	17	1195	32
200	18	1653	39

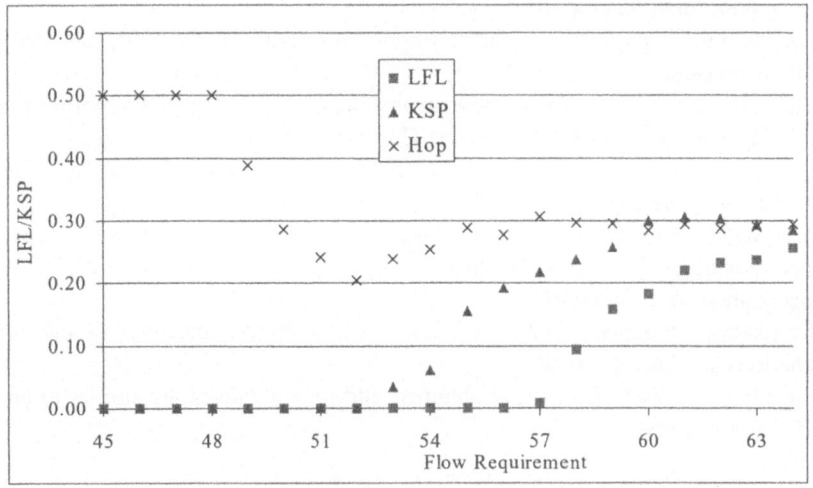

Fig. 2. The LFL/KSP ratio obtained for various functions for the network 114

One of the most interesting questions raised during the simulation on all networks was how the LFL function approximates the KSP function. As mentioned above, the LFL function given by (8) is a lower bound of the lost flow using the KSP rerouting (11). In Fig. 2 we compare performance of functions LFL, KSP and Hop. The x-axis represents the flow requirement. The y-axis represents the LFL/KSP ratio calculated for each traffic demand pattern. The data for each point in the figure are obtained by running the FDP algorithm using one of three functions for a given demand pattern in network 114. We studied the performance of the functions for increasing traffic load, examining the evolution of the network status toward a saturation condition. We can see that for the FDP_Hop algorithm values of LFL/KSP are the largest. It is due to the fact that the Hop function does not optimize the network in terms of the lost flow for local rerouting. Therefore, the FDP_HOP yields values of LFL and KSP functions far from minimal values. Since there is a correlation between these two functions, the ratio LFL/KSP for FDP_Hop function is relatively large. Under low load conditions, algorithm FDP_LFL and FDP_KSP give much lower values of LFL/KSP then the FDP_Hop. This implies that optimization of network flow according to LFL or KSP functions significantly reduces the lost flow, especially for low congested networks. Under high, near saturation, loads, all functions tested have similar performance. This is because for all functions, the FDP algorithm uses the same starting solution and for high loaded networks only a small part all possible solutions are feasible. Consequently, all three functions produce similar results.

Generally the experimental results are consistent with theoretical analysis. The LFL function is a lower bound of the KSP function. However, when the average node degree of tested network grows, the ratio LFL/KSP decreases. In networks 180 and 200 for all traffic demand patterns the LFL is 0, while values of KSP are larger than 0. Hence, the LFL function should be rather applied for networks with average node degree lower than 4.5. For highly connected networks, the LFL function does not estimate KSP function properly.

Due to limited paper space, other important results of our study are described briefly as follows:

1. Each function tested gives the best results calculated by the FDP algorithm applying this function. Values of KSP and Hop functions obtained for various versions of FDP are quite similar. However, the FDP_LFL can find values of LFL function 50% better in average than the FDP applying one of two other functions.
2. Minimizing overall network flow does not guarantee good restoration performance; sometimes demands should use longer routes in order to omit highly congested areas of the network.
3. If we split each demand to 2, 3 or 4 demands, the performance gain of any of three functions is relatively small.
4. The curves of the LFL function obtained during simulations are similar to analytically plotted curves.

6 Conclusion

The two main contributions in this paper are the definition of a new function for optimization of m.c. flows in survivable connection-oriented networks and experimental simulations performed to examine this function. We have studied how to simplify the optimization of m.c. flows in survivable networks using the LFL function given by (8). Moreover, we have compared this function with other functions proposed by previous authors.

We found the KSP approach to be superior to the LFL and Hop functions. However, the calculation time for the KSP function is much greater then the calculation time for two other functions. In addition, the gap between results of KSP and LFL is very small. Concluding, the function (8) developed in this work can be effectively applied for design of primary routes in order to prepare the network for local rerouting. The performance evaluation reveals that the LFL function yields results close to the KSP approach proposed in previous works. However, the time complexity is much lower and consequently, the calculation time is shorter.

There are also several shortcomings of the LFL that are worth to be mentioned. First, the presented approach relies on the information adjacent to the failed link. Therefore, it is a local metric that does not take into account global information on the network flows. Second, as shown in previous section, the LFL performs well for networks with the average node degree lower than 4.5 .

For design of computer networks we can use offline or online algorithms. The function $L(\underline{f}, \alpha)$ is applicable in both types of algorithms. The application for offline algorithms is shown above. In online algorithms, e.g. dynamic routing algorithms, we can use $l_k(\underline{f}, \alpha)$ (15) as an arc metric for computation of shortest routes.

References

1. Ayanoglu, E., Gitlin, R.: Broadband Network Restoration. IEEE Comm. Magazine, 7 (1996) 110-119
2. Burns, J., Ott, T., Krzesinski, A., Muller, K.,: Path selection and bandwidth allocation in MPLS networks. Performance Evaluation, 52 (2003) 133-152.
3. Dunn, A., Grover, W., MacGregor, M.: Comparison of k-Shortest Paths and Maximum Flow Routing for Network Facility Restoration. IEEE JSAC, 1 (1994) 88-99
4. Fratta, L., Gerla, M., Kleinrock, L.: The Flow Deviation Method: An Approach to Store-and-Forward Communication Network Design. Networks (1973) 97–133
5. Kasprzak, A.: Exact and Approximate Algorithms for Topological Design of Wide Area Networks with Non-simultaneous Single Commodity Flows. Lectures Notes In Computer Science, LNCS 2660, (2003) 799-808
6. Murakami, K., Kim, H.: Virtual Path Routing for Survivable ATM Networks. IEEE/ACM Transactions on Networking, 2 (1996) 22-39
7. Ouorou, A., Mahey, P., Vial, J.-Ph.: A survey of algorithms for convex multicommodity flow problems. Management Science, 1 (2000) 126-147
8. Sharma, V., Hellstrand, F. (ed.): Framework for MPLS-based Recovery. RFC 3469 (2003)
9. Walkowiak, K.: A New Approach to Survivability of Connection Oriented Networks. Lectures Notes In Computer Science, LNCS 2657, (2003) 501-510
10. Walkowiak, K.: A Branch and Bound Algorithm for Primary Routes Assignment in Survivable Connection Oriented Networks. Computational Optimization and Applications 2 (2004) 149-171

Multi-domain Diagnosis of End-to-End Service Failures in Hierarchically Routed Networks

Małgorzata Steinder[1] and Adarshpal S. Sethi[2]

[1] IBM T.J. Watson Research Center, Hawthorne, NY
steinder@us.ibm.com
[2] Computer and Information Sciences, University of Delaware, Newark, DE
sethi@cis.udel.edu

Abstract. This paper investigates an approach to improving the scalability and feasibility of probabilistic fault localization in communication systems by exploiting the domain semantics of computer networks. The proposed technique divides the computational effort and system knowledge among multiple, hierarchically organized managers. Each manager performs fault localization in the domain it manages and requires only the knowledge of its own domain. Since failures propagate among domains, domain managers cooperate with each other to find a consensus explanation of the observed disorder. We show through simulation that the proposed approach increases the effectiveness of probabilistic diagnosis and makes it feasible in networks of considerable size [1].

1 Introduction

End-to-end connectivity in a given protocol layer is provided through a sequence of intermediate nodes. Communication problems between a pair of these nodes, e.g., a malfunctioning interface, intermittent connectivity, etc., may disorder one or more end-to-end paths containing the failing link. These end-to-end problems propagate to higher system layers causing various application-level events, such as aborted transactions, session timeouts, or abnormal delays. The diagnosis of end-to-end network service failures [1,2] is a sub-task of fault localization [3,4,5] that isolates node-to-node services responsible for availability or performance problems experienced by end-to-end services. In the previous work [1,2], we investigated an application of probabilistic reasoning to end-to-end service failure diagnosis. The proposed approaches rely on a probabilistic *fault propagation model* (FPM), which represents causal relationships between end-to-end and node-to-node service failures. To solve the fault localization problem, in [1], an adaptation of Pearl's belief updating in belief networks [6] was used, and in [2], a novel algorithm was proposed, which is based on incremental hypothesis updating. The algorithms were shown effective in the diagnosis of end-to-end service failures in networks composed of tens of nodes.

[1] Prepared through collaborative participation in the Communications and Networks Consortium sponsored by the U. S. Army Research Laboratory under the Collaborative Technology Alliance Program, Cooperative Agreement DAAD19-01-2-0011. The U. S. Government is authorized to reproduce and distribute reprints for Government purposes notwithstanding any copyright notation thereon.

N. Mitrou et al. (Eds.): NETWORKING 2004, LNCS 3042, pp. 1036–1046, 2004.
© IFIP International Federation for Information Processing 2004

This paper introduces a distributed fault-localization technique [4,7], which increases the admissible network size by an order of magnitude by taking advantage of the domain semantics of communication systems. The technique divides the computational effort and system knowledge among hierarchically organized managers. Each manager is responsible for fault localization within the domain it governs, and reports to a higher-level manager that oversees and coordinates the fault-localization process of multiple domains. The technique is suitable for distributed diagnosis of end-to-end service failures in hierarchically routed networks such as the Internet. Although the technique is not expected to apply to the management of the entire Internet, we consider it applicable to failure diagnosis across a small subset of network domains that are used to provide a distributed service we want to manage.

The paper is structured as follows. Section 2 defines the problem of probabilistic end-to-end service failure diagnosis. In Section 3, an outline of a multi-domain fault localization technique for hierarchically routed networks is proposed. A distributed fault propagation model is proposed in Section 4, and a multi-domain fault localization algorithm is presented in Section 5. Section 6 evaluates the effectiveness of the proposed multi-domain techniques.

2 Probabilistic Diagnosis of End-to-End Service Failures

This paper adopts a service-oriented view of the network [8], in which end-to-end or node-to-node connectivity in a given protocol layer is considered a service provided by this layer to higher layers. The fault propagation model (FPM) for end-to-end service failure diagnosis is a bipartite causality graph in which parentless nodes (called *link* nodes) represent node-to-node service failures (faults) and childless nodes (called *path* nodes) represent end-to-end service failures (symptoms). Multiple *link* or *path* nodes may exist for every node-to-node or end-to-end service that correspond to different types of failures that may be experienced by the service. Since causal relationships between node-to-node and end-to-end service failures are difficult to determine due to their dynamic and unpredictable nature, the FPM is a probabilistic one, in which each *link* node is labeled with the probability of the corresponding fault's independent occurrence, and causal edges between *link* nodes and *path* nodes are weighted with the probability of the causal implication.

In our previous work, two approaches to solving this problem have been proposed. The first technique (referred to as **Alg. 1**) [1] adapts Pearl's belief updating [6] for polytrees to calculating the most probable explanation (MPE) of observed symptoms. The second approach (**Alg. 2**) [2,1] obtains the MPE by incrementally updating a set of alternative explanation hypotheses. Due to space limitations, we will not present these algorithms in this paper. Instead, we ask the reader to refer to the respective previous publications. We only state that both algorithms include an event-driven procedure called *inference* that analyzes an observed symptom and includes the results of this analysis in the fault-localization state. Also, both algorithms are capable of producing a conditional probability of a given fault's existence, $Prob\{f\}$, or non-existence, $Prob\{\neg f\}$ at any time in the process of fault localization.

Table 1. Basic notation

$n_k \rightarrow n_l$	A directed link from n_k to n_l, where n_k and n_l are node identifiers that are unique network-wide, e.g., IP addresses
$n_{p_1} \overset{*}{\rightarrow} n_{p_m}$	A directed, possibly multi-hop path from n_{p_1} to n_{p_m} consisting of links $n_{p_1} \rightarrow n_{p_2}$, $\ldots, n_{p_{m-1}} \rightarrow n_{p_m}$.
$s : n_k \overset{*}{\rightarrow} n_l$	A symptom indicating a failure of path $n_k \overset{*}{\rightarrow} n_l$
$f : n_k \rightarrow n_l$	A fault associated with link $n_k \rightarrow n_l$
$i \overset{*}{\rightarrow} j$	The set of all paths that begin in domain $\mathcal{D}_i$ and end in domain $\mathcal{D}_j$, i.e., $i \overset{*}{\rightarrow} j = \{n_k \overset{*}{\rightarrow} n_l \mid n_k \in \mathcal{D}_i$ and $n_l \in \mathcal{D}_j\}$, where i and j are unique domain identifiers, e.g., IP subnet masks.
$s : i \overset{*}{\rightarrow} j$	A symptom associated with the set of paths $i \overset{*}{\rightarrow} j$. We say that symptom $s : i \overset{*}{\rightarrow} j$ occurred when at least one $s : n_k \overset{*}{\rightarrow} n_l$ occurred such that $n_k \in \mathcal{D}_i$ and $n_l \in \mathcal{D}_j$.

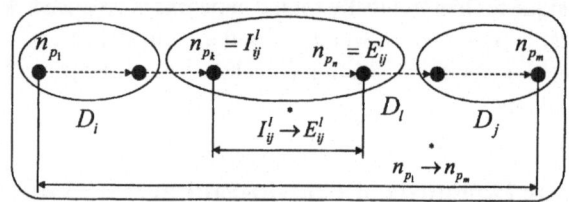

Fig. 1. Definition of a path segment, and ingress and egress gateways.

3   Multi-domain Approach to End-to-End Service Failure Diagnosis

The multi-domain approach to fault localization in hierarchically routed networks takes advantage of the domain semantics of communication systems. A management domain corresponds to a routing domain and may be identified, for example, by a subnet address. The technique proposed in this paper may be applied in networks with multiple levels of the routing hierarchy. However, for simplicity, we focus on a two-level architecture and use $\mathcal{N}$ and $\mathcal{D}_i$ to denote the entire network and its sub-domain, respectively. Domain $\mathcal{D}_i$ is managed by a separate manager, DM_i. At the root of the management hierarchy we place a network manager, NM, which coordinates the operation of managers DM_i.

We introduce the notation presented in Table 1. For an end-to-end path $n_{p_1} \overset{*}{\rightarrow} n_{p_m}$ consisting of links $n_{p_1} \rightarrow n_{p_2}, \ldots, n_{p_{m-1}} \rightarrow n_{p_m}$ we define the following concepts.

Definition 1. *Path $n_{p_1} \overset{*}{\rightarrow} n_{p_m}$ traverses $\mathcal{D}_i$ iff $\exists n_{p_j} | 1 \leq j \leq m, n_{p_j} \in \mathcal{D}_i$. Path $n_{p_1} \overset{*}{\rightarrow} n_{p_m}$ is an intra-domain path in $\mathcal{D}_i$ if $\forall n_{p_j} | 1 \leq j \leq m, n_{p_j} \in \mathcal{D}_i$. If $n_{p_1} \overset{*}{\rightarrow} n_{p_m}$ that traverses $\mathcal{D}_i$ is not an intra-domain path in $\mathcal{D}_i$, it is an inter-domain path with respect to $\mathcal{D}_i$.*

Definition 2. *Let $n_{p_1} \overset{*}{\rightarrow} n_{p_m}$ be an inter-domain path with respect to $\mathcal{D}_l$. Let $n_{p_1} \in \mathcal{D}_i$ and $n_{p_m} \in \mathcal{D}_j$. Node n_{p_k} such that $1 < k \leq m$, $n_{p_k} \in \mathcal{D}_l$, and $n_{p_{k-1}} \notin \mathcal{D}_l$ is an*

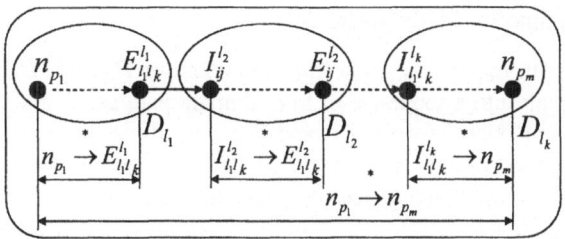

Fig. 2. Transformation of an end-to-end path into a sequence of inter-domain links and intra-domain path segments.

ingress gateway *from $\mathcal{D}_i$ to $\mathcal{D}_j$ in $\mathcal{D}_l$ and is denoted by* $I_{i,j}^l$. Similarly, n_{p_n} *such that* $1 \leq n < m$, $n_{p_n} \in \mathcal{D}_l$, *and* $n_{p_{n+1}} \notin \mathcal{D}_l$ *is an* egress gateway *from $\mathcal{D}_i$ to $\mathcal{D}_j$ in $\mathcal{D}_l$ and is denoted by* $E_{i,j}^l$ (Fig. 1).

Consequently, if a node in $\mathcal{D}_l$ is an ingress/egress gateway both from $\mathcal{D}_i$ to $\mathcal{D}_j$ and from $\mathcal{D}_{i'}$ to $\mathcal{D}_{j'}$, then it is denoted by both $I_{i,j}^l/E_{i,j}^l$ and $I_{i',j'}^l/E_{i',j'}^l$.

Definition 3. *Let $n_{p_1} \overset{*}{\rightarrow} n_{p_m}$ such that $n_{p_1} \in \mathcal{D}_i$ and $n_{p_m} \in \mathcal{D}_j$ be inter-domain with respect to $\mathcal{D}_l$. Path $I_{i,j}^l \overset{*}{\rightarrow} E_{i,j}^l$ is called an* intra-$\mathcal{D}_l$ segment *of $n_{p_1} \overset{*}{\rightarrow} n_{p_m}$ (Fig. 1).*

The solution proposed in this paper is based on the following assumptions, which are reasonable to make in hierarchically routed networks: (1) domains are disjoint, (2) no path enters the same domain more than once. In addition, to simplify the presentation of the methodology, we consider the case where, at a given point in time, all relevant traffic between two nodes is transfered using the same route (which is equivalent to single-path routing).

Each DM has the minimum knowledge necessary for fault diagnosis, i.e., it is aware of logical topology only in the domain it directly manages. DM_i is aware of link $n_k \rightarrow n_l$ iff both n_k and n_l belong to $\mathcal{D}_i$, whereas NM is aware of link $n_k \rightarrow n_l$ iff $n_k \rightarrow n_l$ is a link between $\mathcal{D}_i$ and $\mathcal{D}_j$, and n_k and n_l are egress and ingress gateways in $\mathcal{D}_i$ and $\mathcal{D}_j$, respectively. Consequently, NM is able to transform path $n_{p_1} \overset{*}{\rightarrow} n_{p_m}$ that traverses $\mathcal{D}_{l_1}, \ldots, \mathcal{D}_{l_k}$ into a sequence of intra-domain path segments and links $n_{p_1} \overset{*}{\rightarrow} E_{l_1,l_k}^{l_1}$, $E_{l_1,l_k}^{l_1} \rightarrow I_{l_1,l_k}^{l_2}$, $I_{l_1,l_k}^{l_2} \overset{*}{\rightarrow} E_{l_1,l_k}^{l_2}, \ldots, E_{l_1,l_k}^{l_{k-1}} \rightarrow I_{l_1,l_k}^{l_k}$, $I_{l_1,l_k}^{l_k} \overset{*}{\rightarrow} n_{p_m}$ (Fig. 2). DM_i is able to obtain a route for each path $n_k \overset{*}{\rightarrow} n_l$ such that $n_k, n_l \in \mathcal{D}_i$, but it cannot obtain the topology and routing information for any parts of the network located outside of $\mathcal{D}_i$.

4   Distributed Fault Propagation Model

In the multi-domain solution, the FPM of the entire network is distributed among DMs. Each manager maintains a part of the distributed FPM that represents the manager's knowledge of the system structure, i.e., it includes only faults of links that are located in its domain. Faults located in other domains that may propagate to the manager's domain are represented by proxy nodes, called $\mathcal{P}$-nodes.

4.1 Fault Propagation Model of the NM

Let us consider path $n_{p_1} \xrightarrow{*} n_{p_m}$ that traverses domains $\mathcal{D}_{l_1}, \ldots, \mathcal{D}_{l_k}$. Recall that NM transforms this path into a sequence of intra-domain path segments and links (Fig. 2). When only one path exists between $\mathcal{D}_{l_1}$ and $\mathcal{D}_{l_k}$ then all paths that begin in $\mathcal{D}_{l_1}$ and end in $\mathcal{D}_{l_k}$ are transformed into the same sequence of intra-domain path segments and links with the exception of the first and last segments. In this case, the FPM includes a single symptom node labeled $s : l_1 \xrightarrow{*} l_k$ that represents all paths that begin in $\mathcal{D}_{l_1}$ and end in $\mathcal{D}_{l_k}$.

In the FPM of NM, two types of fault nodes exist: (1) ordinary fault nodes, like ones in the centralized case, which represent failures of inter-domain links; these faults are directly isolated by NM, and (2) proxy fault nodes that represent path-segment failures, which cannot be isolated by NM alone because they are located in domains that are not directly managed by NM. For every $\mathcal{D}_i$, one or more $\mathcal{P}$-nodes are created as follows.

1. For every ingress gateway in $\mathcal{D}_i$, $I_{l,i}^i$, we create $\mathcal{P} : I_{l,i}^i \xrightarrow{*} *$ that represents all intra-$\mathcal{D}_i$ paths that begin in $I_{l,i}^i$.
2. For every egress gateway in $\mathcal{D}_i$, $E_{i,k}^i$, we create $\mathcal{P} : * \xrightarrow{*} E_{i,k}^i$ that represents all intra-$\mathcal{D}_i$ paths that end in $E_{i,k}^i$.
3. For each pair of gateways $I_{l,k}^i$ and $E_{l,k}^i$, we create $\mathcal{P} : I_{l,k}^i \xrightarrow{*} E_{l,k}^i$ that represents intra-$\mathcal{D}_i$ path $I_{l,k}^i \xrightarrow{*} E_{l,k}^i$.

In the FPM of NM, $s : l_1 \xrightarrow{*} l_k$ is connected to $\mathcal{P} : * \xrightarrow{*} E_{l_1,l_k}^{l_1}$, $f : E_{l_1,l_k}^{l_1} \to I_{l_1,l_k}^{l_2}$, $\mathcal{P} : I_{l_1,l_k}^{l_2} \xrightarrow{*} E_{l_1,l_k}^{l_2}, \ldots, f : E_{l_1,l_k}^{l_{k-1}} \to I_{l_1,l_k}^{l_k}, \mathcal{P} : I_{l_1,l_k}^{l_k} \xrightarrow{*} *$. The FPM of NM contains multiple such symptom nodes for all pairs of domains in $\mathcal{N}$.

The approach chosen in this paper assigns all conditional probabilities between $\mathcal{P}$-nodes and symptom nodes to 1. Prior failure probabilities associated with $\mathcal{P}$-nodes must be calculated by the multi-domain technique based on the state of the fault localization process in domains they represent.

4.2 Fault Propagation Model of DM

As it was stated at the beginning of this section, the FPM built by DM_i includes all intra-$\mathcal{D}_i$ paths and links, i.e., all the information contained in the centralized model of $\mathcal{D}_i$. Such model is sufficient for the diagnosis of symptoms observed in $\mathcal{D}_i$ but is not sufficient for the diagnosis of symptoms DM_i receives from NM. In particular, NM may delegate to DM_i a part of a task involved in the diagnosis of path $n_{p_1} \xrightarrow{*} n_{p_m}$ that traverses $\mathcal{D}_i$. In this case, DM_i will be notified about a failure of its intra-domain path that constitutes the intra-$\mathcal{D}_i$ segment of $n_{p_1} \xrightarrow{*} n_{p_m}$. This notification only indicates a possibility of the segment's failure, since the failure of $n_{p_1} \xrightarrow{*} n_{p_m}$ could have been caused by its path-segment or link that is not located in domain $\mathcal{D}_i$. Thus, symptoms received by DM from NM are typically associated with a high degree of uncertainty, i.e., they are likely to be spurious. To deal with spurious symptoms, we use ideas from previous work [1] as follows.

Let $s : n_r \xrightarrow{*} n_t$ be an intra-$\mathcal{D}_i$ symptom received by DM_i from NM in the process of diagnosing a failure of inter-domain path $n_{p_1} \xrightarrow{*} n_{p_m}$. To model the possibility that

$s : n_r \overset{*}{\to} n_t$ is spurious in the FPM of DM_i, we create a $\tilde{\mathcal{P}}$-node that represents all possible causes of $s : n_r \overset{*}{\to} n_t$ that are not located in $\mathcal{D}_i$. Observe that, since $n_r \overset{*}{\to} n_t$ constitutes a segment of an inter-domain path, at least one of n_r, n_t is a gateway in $\mathcal{D}_i$. Let l and k be identifiers of domains that contain n_{p_1} and n_{p_m}, respectively. Let us consider three cases.

1. $i = l$; We create $\tilde{\mathcal{P}} : * \overset{*}{\to} E^i_{i,k}$ and connect it to $s : n_r \overset{*}{\to} n_t$.
2. $i = k$; We create $\tilde{\mathcal{P}} : I^i_{l,i} \overset{*}{\to} *$ and connect it to $s : n_r \overset{*}{\to} n_t$.
3. $i \neq l$ and $i \neq k$; We create $\tilde{\mathcal{P}} : I^i_{l,k} \overset{*}{\to} E^i_{l,k}$ and connect it to $s : n_r \overset{*}{\to} n_t$.

Similar to the FPM of NM, conditional probabilities on edges between $\tilde{\mathcal{P}}$-nodes and symptom nodes in the FPM of DM_i are set to 1, while prior failure probabilities assigned to $\tilde{\mathcal{P}}$-nodes in the FPM of DM_i are calculated by NM and sent to DM_i together with reported symptoms.

5   Multi-domain Fault Localization Algorithm

In this section, we present an outline of a multi-domain fault localization algorithm (Alg. 3) based on the distributed FPM described in Section 4, which may be refined to create multi-domain versions of Algs. 1 and 2. In the pseudo-code of Alg. 3, which is presented on page 1043, sections of the algorithm that are specialized for different probabilistic reasoning mechanisms are underlined. The multi-domain fault localization algorithm proceeds in two phases performed by every DM and NM: (1) symptom analysis and (2) fault selection. Initially, the model is reset by assigning prior failure probabilities to proxy nodes. In our implementation, these probabilities are set to 0 in the FPM of NM. In the FPM of DM, no $\tilde{\mathcal{P}}$-nodes exist at the beginning, and therefore no assignment is needed. Symptom-analysis and fault-selection phases progress by traversing the hierarchy of managers in a bottom-up or top-down manner.

5.1   Symptom Analysis Phase

The symptom analysis phase is executed for every received alarm that indicates a failure of an end-to-end path. This alarm can be received either by the NM or a DM. A DM can start symptom analysis only if the entire failed path belongs to its domain. If the DM is not able to diagnose the symptom it forwards it to the NM, which initiates the symptom diagnosis (function *analyze_internal*).

Symptom Processing by NM: In the process of diagnosing $s : n_{p_1} \overset{*}{\to} n_{p_m}$ (see function *analyze_internal* in the pseudo-code of Alg. 3), the NM first maps it into node $s : l_1 \overset{*}{\to} l_k$ in its FPM, such that $n_{p_1} \in \mathcal{D}_{l_1}$ and $n_{p_m} \in \mathcal{D}_{l_k}$. Then, it splits $n_{p_1} \overset{*}{\to} n_{p_m}$ into path segments and links. Failures of path segments are then interpreted as symptoms $s_1, s_2, \ldots, s_k$ that will be reported to $DM_{l_1}, DM_{l_2}, \ldots, DM_{l_k}$, respectively. Note that in the FPM of DM_{l_j}, all causes of s_j that are not located in $\mathcal{D}_{l_j}$ are represented by a $\tilde{\mathcal{P}}$-node that is attached to node s_j. To indicate that s_j may be spurious in $\mathcal{D}_{l_j}$, NM calculates the prior probability associated with this $\tilde{\mathcal{P}}$-node in the FPM of DM_{l_j}. Suppose that $s_j = s : n_r \xrightarrow{*} n_t$. Then the probability that s_j is spurious is obtained as follows:

$$p_{\text{spurious}}(s : n_t \xrightarrow{*} n_r) = \prod_{\mathcal{P} \in \mathcal{P}(n_t \xrightarrow{*} n_r)} \underline{Prob\{\neg\mathcal{P}\}} \tag{1}$$

$$\mathcal{P}(n_t \xrightarrow{*} n_r) = \begin{cases} \{\mathcal{P} : n_r \xrightarrow{*} n_t, \mathcal{P} : * \xrightarrow{*} n_t, \mathcal{P} : n_r \xrightarrow{*} *\} \\ \qquad \text{if } n_r \text{ and } n_t \text{ are ingress and egress gateways} \\ \{\mathcal{P} : * \xrightarrow{*} n_t\} \text{ if } n_r \text{ is an ingress gateway} \\ \{\mathcal{P} : n_r \xrightarrow{*} *\} \text{ if } n_t \text{ is an egress gateway} \end{cases}$$

After calculating $p_{\text{spurious}}(s_j)$, NM delegates the diagnosis of s_j to DM_{l_j}, for $j = 1 \dots k$ by invoking *analyze_external*. As a result of the diagnosis performed by DM_j, the NM obtains $p(\mathcal{P}_j)$, where $\mathcal{P}_j$ is the $\mathcal{P}$-node representing $\mathcal{D}_i$ that is connected to $s : l_1 \xrightarrow{*} l_k$ in the FPM of NM. Then NM updates its FPM. Finally, NM analyzes $s : l_1 \xrightarrow{*} l_k$ using the symptom-analysis procedure of either Alg. 1 or 2 (function *inference*).

To limit duplicate delegations of the same symptom to DM_{l_j}, NM marks nodes as either UNOBSERVED or OBSERVED_INTERNAL. While analyzing $s : n_{p_1} \xrightarrow{*} n_{p_m}$, when $s : l_1 \xrightarrow{*} l_k$ is marked OBSERVED_INTERNAL, the NM does not delegate symptoms to DM_{l_j}s for $j = 2 \dots k - 1$. It does, however, delegate the analysis to DM_{l_1} and DM_{l_k}, since paths represented by $s : l_1 \xrightarrow{*} l_k$ differ in their segments located in $\mathcal{D}_{l_1}$ and $\mathcal{D}_{l_k}$.

Symptom processing by DM: DM_i may start the processing of $s_r = s : n_{p_1} \xrightarrow{*} n_{p_m}$ when (1) it observes a failure of intra-$\mathcal{D}_i$ path $n_{p_1} \xrightarrow{*} n_{p_m}$ or (2) s_r is delegated to DM_i by NM. In the former case, s_r is an internal symptom; in the latter case it is called an external symptom. To distinguish between different observations of the same symptom, DM_i marks symptom nodes as either UNOBSERVED, OBSERVED_INTERNAL, and OBSERVED_EXTERNAL when they are not processed, processed as a result of internal observation, and processed as a result of a delegation by NM, respectively.

Internal symptoms are processed by function *analyze_internal*. First, the association between the observed symptom and its $\tilde{\mathcal{P}}$-node (if one exists) is removed, as the symptom can no longer be explained by external causes. Then, a symptom-analysis procedure is executed.

The processing of external symptoms is done by function *analyze_external*. Assume that $s_r = s : n_{p_1} \xrightarrow{*} n_{p_m}$ has been delegated to DM_i as a result of a failure of a path between domains $\mathcal{D}_l$ and $\mathcal{D}_k$. DM_i also receives two parameters from NM: $\mathcal{P}^i_{l,k}$ and p_{spurious}, where $\mathcal{P}^i_{l,k}$ is a description of a $\mathcal{P}$-node that is connected to node $s : l \xrightarrow{*} k$ in the FPM of NM, and p_{spurious} is the probability that s_r is spurious. DM_i first updates its FPM by assigning p_{spurious} as the prior probability to the $\tilde{\mathcal{P}}$-node connected to symptom s_r. If the symptom has been previously analyzed, DM_i returns the stored value of $p(\mathcal{P}^i_{l,k})$. Otherwise, it updates the FPM by connecting s_r to its corresponding $\tilde{\mathcal{P}}$-node, and updates the state of fault localization to reflect the modified value of its prior failure probability. Then, a probabilistic reasoning mechanism is used to analyze the symptom. Finally, $p(\mathcal{P}^i_{l,k})$ is calculated as follows:

$$p(\mathcal{P}^i_{l,k}) = \begin{cases} 0 & \mathcal{S}^i_{l,k} = \emptyset \\ \prod_{s_i \in \mathcal{S}^i_{l,k}} bel(s_i) & \text{otherwise} \end{cases} \tag{2}$$

$$\mathcal{S}^i_{l,k} = \{s_k = s : n_r \xrightarrow{*} n_t | n_r \xrightarrow{*} n_t \in \mathcal{P}^i_{l,k} \text{ and } s_k \text{is not UNOBSERVED}\} \tag{3}$$

$$bel(s_i) = \begin{cases} 1 & \text{if } s_i \text{ is OBSERVED\_INTERNAL} \\ 1 - \prod_{f_j \in \mathcal{F}} (1 - p(s_i|f_j)\underline{Prob\{f_j\}}) & \text{otherwise} \end{cases} \tag{4}$$

5.2   Fault Selection Phase

In the fault selection phase, DMs and NM have to synchronize their FPMs by updating prior failure probabilities associated with their proxy nodes. Afterward, DMs and NM choose the most likely hypotheses.

Algorithm 3: Multi-domain algorithm

Symptom analysis phase:
DM: FOR *every observed symptom* $s : n_{p_1} \xrightarrow{*} n_{p_m}$ DO
    IF *internal symptom analyze_internal*$(s : n_{p_1} \xrightarrow{*} n_{p_m})$
    ELSE NM→*analyze_internal*$(s : n_{p_1} \xrightarrow{*} n_{p_m})$
NM: FOR *every observed symptom* $s : n_{p_1} \xrightarrow{*} n_{p_m}$ DO *analyze_internal*$(s : n_{p_1} \xrightarrow{*} n_{p_m})$
DM_i: FUNCTION *analyze_internal*(s_r)
    IF s_r *is not marked* OBSERVED_INTERNAL THEN *update the model* AND *run* <u>*inference*(s_r)</u>
NM: FUNCTION *analyze_internal*$(s : n_{p_1} \xrightarrow{*} n_{p_m})$
    map $s : n_{p_1} \xrightarrow{*} n_{p_m}$ *to* $s : l_1 \xrightarrow{*} l_k$ *such that* $n_{p_1} \xrightarrow{*} n_{p_m} \in l_1 \xrightarrow{*} l_k$
    transform $n_{p_1} \xrightarrow{*} n_{p_m}$ *into* $n_{p_1} \to E^{l_1}_{l_1,l_k}, E^{l_1}_{l_1,l_k} \to I^{l_2}_{l_1,l_k}, I^{l_2}_{l_1,l_k} \xrightarrow{*} E^{l_2}_{l_1,l_k}, \ldots, I^{l_k}_{l_1,l_k} \xrightarrow{*} n_{p_m}$
    set $s_1 = s : n_{p_1} \to E^{l_1}_{l_1,l_k}, s_2 = s : I^{l_2}_{l_1,l_k} \xrightarrow{*} E^{l_2}_{l_1,l_k}, \ldots, s_k = s : I^{l_k}_{l_1,l_k} \xrightarrow{*} n_{p_m}$
    find $\mathcal{P}$-*nodes connected to* $s : l_1 \xrightarrow{*} l_k : \mathcal{P}_1 = \mathcal{P} : * \xrightarrow{*} E^{l_1}_{l_1,l_k}, \ldots, \mathcal{P}_k = \mathcal{P} : I^{l_k}_{l_1,l_k} \xrightarrow{*} *$
    FOR $1 \le j \le k$ DO
        IF $s : l_1 \xrightarrow{*} l_k$ *is marked* UNOBSERVED OR $j = 1$ OR $j = k$ THEN
            $p(\mathcal{P}_j) = \text{DM}_{l_j} \to$*analyze_external*$(s_j, \mathcal{P}_j, p_{\text{spurious}}(s_j))$
        IF $s : l_1 \xrightarrow{*} l_k$ *is not marked* OBSERVED_INTERNAL THEN
            update the model AND *run* <u>*inference*$(s : l_1 \xrightarrow{*} l_k)$</u>
DM_i: FUNCTION *analyze_external*$(s_r, \mathcal{P}^i_{l,k}, p_{\text{spurious}})$
    IF s_r *is not marked* UNOBSERVED THEN *return* $p(\mathcal{P}^i_{l,k})$
    ELSE *update the model, run* <u>*inference*(s_r)</u>, AND RETURN $p(\mathcal{P}^i_{l,k})$
Fault selection phase:
NM: FOR *every* $\mathcal{P}^i_{l,k}$ DO *obtain* $p(\mathcal{P}^i_{l,k})$ *from* DM_i AND *update the model*
    FOR *every* $\mathcal{P}^i_{l,k}$ DO *send* $Prob\{\neg\mathcal{P}^i_{l,k}\}$ *to* DM_i
    FOR *every* DM_i DO *obtain the most likely set of faults from* DM_i
    obtain the most likely set of faults in NM

It is not difficult to refine Alg. 3 to create multi-domain versions of Algs. 1 and 2. The resultant multi-domain versions of Algorithms 1 and 2 are labeled **Alg. 3A** and **Alg. 3B**, respectively [9]. The computational complexities of fault localization performed by a single manager using Algs. 3A and 3B in a network domain composed of n nodes or domains are $\mathcal{O}(n^5)$ and $\mathcal{O}(n^4)$, respectively.

In the entire algorithm, the messaging overhead is $\mathcal{O}(\max(|\mathcal{S}_O|, n^3))$ per domain, where n is a number of nodes or sub-domains in the domain.

6 Simulation Study

We evaluate the performance of Algs. 3A and 3B through simulation. The study uses sets of fault localization scenarios in which faults and symptoms are randomly generated based on the conditional probability distribution that describes non-deterministic causal relationships between faults and symptoms.

The simulation study uses network topologies similar to those of the Internet. The generation of random graphs resembling the topology of real-life networks has been a widely studied research area [10,11,12,13]. This study uses a generator based on Barabasi-Albert power-law model [11], because its implementation (BRITE [14]) is available in public domain, and because topologies built based on this model are representative of the Internet topology [15].

Using the topology generator we create a random network composed of $N = 10$ domains and n nodes in each domain, where n varies between 5 and 70. We determine routes between any source and destination using the shortest-path policy for intra-domain routes. We choose inter-domain routes such that the number of visited domains is minimized. Then, we generate prior failure probabilities for inter-domain and intra-domain links, which are uniformly distributed over the range $[0.0001, 0.001]$. For each intra-domain link l and path p, we randomly choose the probability that p fails if l fails from set $\{0.25, 0.5, 0.75\}$. In the FPM of the NM, the conditional probabilities are all equal to 1. We randomly generate a subset of symptoms observable in every domain to include 50% of all intra-domain paths. The observability ratio [1,2] for inter-domain paths is 2%.

We distinguish three types of experiments: those involving only intra-domain link failures, inter-domain link failures, and both types of failures. In every study, two performance metrics are calculated: detection rate, DR, defined as a percentage of faults occurring in the network which are isolated by the technique, and false positive rate, FPR, defined as a percentage of faults reported by the technique that are not occurring in the network [1].

In Figs. 3a-3(b), we show the accuracy of Alg. 3A in a ten-domain network, in which each domain is composed of up to 70 nodes. Thus the entire network consists of up to 700 nodes. Figs. 4(a)-4(b) present the results of the same experiment executed using Alg. 3B.

The figures compare the accuracy achievable in scenarios involving only inter-domain, only intra-domain, and both types of faults. Clearly, the mixed-failure scenarios are the most difficult to diagnose since they always involve at least two concurrent faults located in different network domains. The interpretation of the faults' symptoms, which may overlap, leads to ambiguity. This results in a lower fault-localization accuracy of

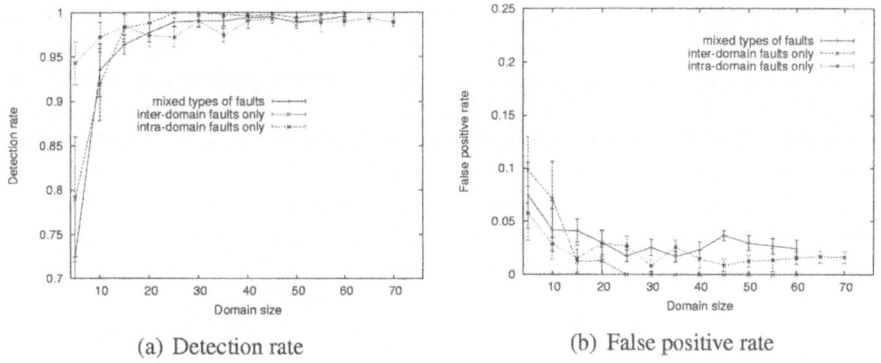

(a) Detection rate   (b) False positive rate

Fig. 3. Accuracy of Algorithm 3A in a ten-domain network.

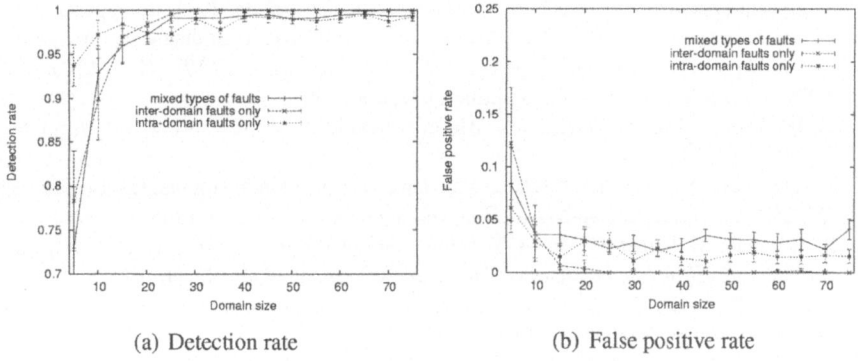

(a) Detection rate   (b) False positive rate

Fig. 4. Accuracy of Algorithm 3B in a ten-domain network.

mixed-fault scenarios compared to that of other types of scenarios, which is conspicuous in networks of small size. Scenarios involving only inter-domain symptoms are the easiest to solve as the number of suspect faults is usually small compared to the amount of available evidence. In intra-domain- and mixed-fault scenarios, increasing the domain size also increases the frequency of multi-fault scenarios. For example, in mixed scenarios the number of simultaneous faults exceeds 3 in a 700-node network. In inter-domain scenarios, the number of faults is equal to 1.02 on average, and does not change when the network size increases, while the number of received symptoms grows with the increasing network size.

We repeated the same set of experiments using networks composed of 50 domains. The results, which are not shown in this paper, confirm the results obtained using ten-domain networks (see [1] for details).

7   Conclusion

The paper identifies two main difficulties of fault management in multi-domain networks: failure propagation among domains and a lack of global information about the system structure and state. To address these challenges, the paper proposes a multi-domain algorithm, which is shown to provide high accuracy while increasing the admissible network size by an order of magnitude. Clearly, the biggest challenge in applying the fault localization technique proposed in this paper to real-life problems is obtaining the probabilistic FPM. To build an FPM for end-to-end service failure diagnosis a knowledge of network logical topology and communication protocols is needed. The problem of building FPMs is beyond the scope of this paper. Our previous publications in this subject survey some of the techniques that can be used to build an FPM for end-to-end service failure diagnosis [9]. [2]

References

1. M. Steinder and A. S. Sethi, "Non-deterministic fault localization in communication systems using belief networks," *IEEE/ACM Transactions on Networking*, 2004, (to appear).
2. M. Steinder and A. S. Sethi, "Non-deterministic event-driven fault diagnosis through incremental hypothesis updating," in *Integrated Network Management VIII*, G. Goldszmidt and J. Schoenwaelder, Eds., Colorado Springs, CO, Mar. 2003.
3. G. Jakobson and M. D. Weissman, "Alarm correlation," *IEEE Network*, vol. 7, no. 6, pp. 52–59, Nov. 1993.
4. I. Katzela and M. Schwartz, "Schemes for fault identification in communication networks," *IEEE/ACM Transactions on Networking*, vol. 3, no. 6, pp. 733–764, 1995.
5. S. A. Yemini, S. Kliger, E. Mozes, Y. Yemini, and D. Ohsie, "High speed and robust event correlation," *IEEE Communications Magazine*, vol. 34, no. 5, pp. 82–90, 1996.
6. J. Pearl, *Probabilistic Reasoning in Intelligent Systems: Networks of Plausible Inference*, Morgan Kaufmann Publishers, 1988.
7. A. T. Bouloutas, S. B. Calo, A. Finkel, and I. Katzela, "Distributed fault identification in telecommunication networks," *J. of Network and Systems Management*, vol. 3, no. 3, 1995.
8. P. Hasselmeyer, "An infrastructure for the management of dynamic service networks," *IEEE Communications Magazine*, vol. 41, no. 4, pp. 120–126, 2003.
9. M. Steinder and A. S. Sethi, "Multi-domain diagnosis of end-to-end service failures in hierarchically routed networks," Tech. Rep. 2003-10, CIS, University of Delaware, 2003.
10. W. Aiello, F. Chung, and L. Lu, "A random graph model for massive graphs," in *ACM Symposium on Theory of Computing*, Portland, OR, May 2000, pp. 171–180.
11. A. Barabasi and R. Albert, "Emergence of scaling in random networks," *Science*, pp. 509–512, Oct. 1999.
12. K. Calvert, M. Doar, and E. Zegura, "Modeling Internet topology," *IEEE Transactions on Communications*, pp. 160–163, Dec. 1997.
13. C. Jin, Q. Chen, and S. Jamin, "Inet: Internet topology generator," Tech. Rep. CSE-TR443-00, EECS, University of Michigan, 2000.
14. A. Medina, A. Lakhina, I. Matta, and J. Byers, "BRITE:universal topology generation from a user's perspective," Tech. Rep. BUCS-TR-2001-003, CS Dept, Boston Univ., 2001.
15. T. Bu and D. Towsley, "On distringuishing between Internet power law topology generators," in *Proc. of IEEE INFOCOM*, New York, NY, Jun. 2002.

---

[2] The views and conclusions contained in this document are those of the authors and should not be interpreted as representing the official policies, either expressed or implied of the Army Research Lab or the U.S. Government.

Detecting Traffic Anomalies through Aggregate Analysis of Packet Header Data

Seong Soo Kim[1], A.L. Narasimha Reddy[1], and Marina Vannucci[2]

[1] Department of Electrical Engineering, [1] TX 77843-3128
{skim, reddy}@ee.tamu.edu
[2] Statistics, Texas A&M University, TX 77843-3143, USA, College Station,
mvannucci@stat.tamu.edu

Abstract. If efficient network analysis tools were available, it could become possible to detect the attacks, anomalies and to appropriately take action to contain the attacks. In this paper, we suggest a technique for traffic anomaly detection based on analyzing correlation of destination IP addresses in outgoing traffic at an egress router. This address correlation data are transformed through discrete wavelet transform for effective detection of anomalies through statistical analysis. Our techniques can be employed for postmortem and real-time analysis of outgoing network traffic at a campus edge. Results from trace-driven evaluation suggest that proposed approach could provide an effective means of detecting anomalies close to the network. We also present data analyzing the correlation of port numbers as a means of detecting anomalies.

1 Introduction

At present, attacks on Internet infrastructure, in the form of denial of service (DoS) attacks and worms, have become one of the most serious threats to the network security. If efficient analysis tools for analyzing and monitoring traffic were available, it could become possible to detect the attacks, anomalies and to appropriately take action to mitigate them before they have had much time to propagate across the network. In this paper*, we study the possibilities of traffic-analysis based mechanisms for attack and anomaly detection.

Traffic is monitored at regular intervals to obtain a signal that can be analyzed through statistical techniques and compared to historical norms to detect anomalies. By observing the traffic and correlating it to previous states of traffic, it may be possible to see whether the current traffic is behaving in a similar/correlated manner.

Our methodology to detecting anomalies envisions two kinds of detection mechanisms: postmortem and real-time modes.

---

* This work is supported by a NSF grant ANI-0087372, Texas Higher Education Board, Texas Information Technology and Telecommunications Taskforce, Intel Corp, and a NSF CAREER award DMS-0093208.

N. Mitrou et al. (Eds.): NETWORKING 2004, LNCS 3042, pp. 1047–1059, 2004.

Recently, statistical analysis of aggregate traffic data has been studied [1, 3, 9]. Our previous work [1] and the work in [3] have studied traffic volume as a signal for wavelet analysis and these earlier works have considerably motivated our current study here. Traditionally, various forms of signatures have been utilized for representing the contents or certain identities. Traffic analysis signatures have been proposed for detecting anomalies. For example, disproportion of bi-directional flows can be used as a signature of anomalistic traffic [4]. The changing ratios (i.e., the rate of decrease) between the flow numbers of neighboring specific bit-prefix aggregate flows can be calculated and used for detecting peculiarities [5].

2 Our Approach

2.1 Traffic Analysis at the Source

We focus on analyzing the traffic at an egress router. A traffic monitoring at a source network enables a detector to detect attacks early and is able to control hijacking of AD (administrative domain, e.g., campus) machines. Outbound filtering has been advocated for limiting the possibility of address spoofing i.e., to make sure that source addresses correspond to the designated addresses for the campus. With such filtering in place, we can focus on destination addresses and port numbers of the outgoing traffic for analysis purposes.

Our approach is based on the following observations: the outbound traffic from an AD is likely to have a strong correlation with itself over time. Recent studies have shown the traffic can have strong patterns of behavior over several timescales [3]. We hypothesize that the destination addresses will have a high degree of correlation for a number of reasons: (i) popular web sites are shown to receive a significant portion of the traffic, (ii) individual users are shown to access similar web sites over time due to their habits, and (iii) long-term flows, such as ftp download and video accesses, tend to correlate addresses over longer timescales. If this is the case, sudden changes in correlation of outgoing addresses can be used to detect anomalies in traffic behavior.

2.2 General Mechanism of the Detector and Traces

Our detection mechanisms can be explained in three major steps shown in Fig. 1. The first step is traffic parser, in which a network traffic signal is generated from packet header traces or NetFlow records as input. The second step involves data transformation for statistical analysis. In this paper, we employ wavelet transforms to study the

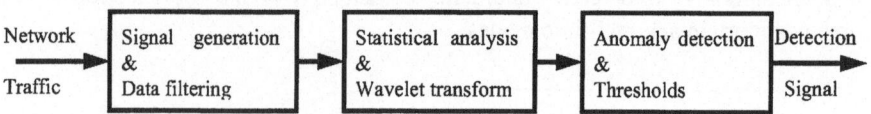

Fig. 1. The block diagram of our detector

address and port number correlation over several timescales. The final is detection, in which attacks and anomalies are checked using thresholds. The analyzed information will be compared with historical thresholds to see whether the traffic's characteristics are out of regular norms. This comparison will lead to some form of a detection signal that could be used to alert the network administrator of the potential anomalies in the network traffic as explained in section 5, 6 and 7.

To verify the validity of our approach, we run our algorithm on two kinds of traffic traces. First, we examine the detector on KREONet2 traces from July 21, 2003 to July 28, 2003 which contain real worm attacks. Currently KREONET member institutions are over 230 organizations, which include 50 government research institutes, 72 universities, 15 industrial research laboratories, and is connecting with 155Mbps international ATM link [7]. Additionally we employ the packet traces from the NLANR [2], which are later superimposed with simulated virtual attacks. We employ Auckland-IV traces which are transmitted about 5000 connections at the rate of 5Mbps and 1500 packets/second. These traces were anonymized, but preserved IP prefix relationships.

3 Signal Generation

Individual fields in the packet header are analyzed to observe anomalies in the traffic. Individual fields in the traffic header data take discrete values and show discontinuities in the sample space. For example, IP address space can span 2^{32} possible addresses and addresses in a sample are likely to exhibit many discontinuities over this space making it harder to analyze the data over the address space. In order to overcome such discontinuities over a discrete space, we convert packet header data into a continuous signal through correlation of samples over successive samples. To investigate the sequence of a random process, we employ a simplified correlation of time-series for computational efficiency without compromising performance.

For each address, a_m, in the traffic, we count the number of packets, p_{mn}, sent in the sampling instant, s_n. For computing address correlation signal, we consider two adjacent sampling instants. We define address correlation signal in sampling point n as

$$C(n) = \sum_m p_{mn-1} * p_{mn} \Big/ \sum_m p_{mn} \qquad (1)$$

If an address a_m spans the two sampling points $n-1$ and n, we will obtain a positive contribution to $C(n)$.

In order to minimize storage and processing complexity, we employ a simple but powerful data structure used in our previous work [8]. A location $count [i][j]$ is used to record the packet count for the address j in i^{th} field of the IP address through scaling. This provides a concise description of the address instead of 2^{32} locations that would be required to store the address occurrence uniquely. We filter this signal by computing a correlation of the address in two success samples, i.e., by computing

$$C_{in} = \sum_{j=0}^{255} \frac{count[i][j][n-1]}{\sum_{j=0}^{255} count[i][j][n-1]} * \frac{count[i][j][n]}{\sum_{j=0}^{255} count[i][j][n]}, i = 1,2,3,4 \qquad (2)$$

Consequently four correlation signals are calculated as C_{1n} through C_{4n}. The employment of this approximate representation of addresses allows us to reduce the computational and storage demands by a factor of 2^{22}. In order to generate the address correlation signal $S(n)$ at the end of sampling point n, we multiply each segment correlation C_{in} with scaling factors α_i and generate $S(n)$ as

$$S(n) = A * (\alpha_1 * C_{1n} + \alpha_2 * C_{2n} + \alpha_3 * C_{3n} + \alpha_4 * C_{4n}) + B \qquad (3)$$
$$where, \alpha_1 + \alpha_2 + \alpha_3 + \alpha_4 = 1$$

Our approach could introduce errors when the addresses segments match even though addresses themselves don't match. In normal traffic without attacks, we compared the full-32 bit address correlation with the correlation signal generated by our approach. The upper two sub-pictures and bottom two sub-pictures in Fig. 2 show the weighted signal computed with the full-32 bit address correlation and our data structure with respect to Auckland-IV traces. From the figure, we see that the differences are negligible i.e., our approach does not add significant noise. From a statistical standpoint, they have an approximately same mean ($\cong 50$) and dispersion (standard deviation $\cong 12.4 \sim 12.6$), and have $\rho_{XY} \approx 0.77$ as cross-correlation coefficient.

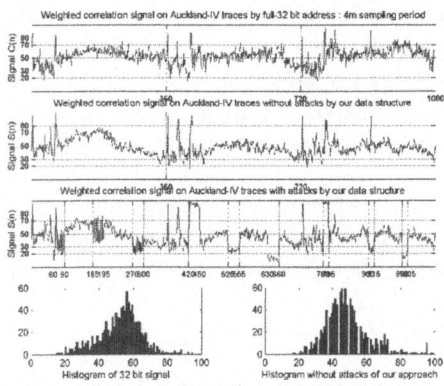

Fig. 2. Comparison of full-32 bit correlation and data structure

3.1 Attacks

Besides the actual attacks observed in the KREONet2 traces, we construct virtual attacks on the Auckland-IV traces. This allows us to test the proposed technique under different conditions. As shown in Table I, these attacks cover a diversity of behaviors and allow us to deterministically test the efficacy of proposed mechanisms. These are classified by following criteria.

- **Persistency:** The first 3 attacks send malicious packets for 3 minutes and pause for 3 minutes. Such intermittent pattern is intended to model crafty attackers that attempt to dilute their trails. The remnant attacks persistently assault.
- **IP address:** The 1st attack among every 3 attacks targets for a single destination IP address. The 2nd attack style composes the IP address in which a portion of addresses preserve the class-A and a partition of addresses preserve class-B for the infiltration efficiency. The 3rd type is randomly generated address
- **Protocol:** The three major protocols, ICMP, TCP and UDP, are exploited in turn
- **Port:** The 1st port among every 3 attacks is a representative #80 that stands for the reserved ports for well-known services. The 2nd port targets for randomly generated destination ports that is used to probe port-scan. The 3rd port is a #1434 that acts for the ephemeral client ports, which was exploited in SQL Slammer worm
- **Size:** The three denominations are random size, 4K Bytes and 404 Bytes [10].

The third-top sub-picture in Fig. 2 represents the weighted correlation signal of IP address in 3-day Auckland-IV traces with attacks. The simulated attacks are staged between the vertical lines, shown in the figure.

4 Data Transform

The generated signal can be, in general, analyzed by employing techniques such as FFT (Fast Fourier Transform) and wavelet transforms. The analysis carried out on the signal may exploit the statistical properties of the signal such as correlation over several timescales and its distribution properties.

Since wavelet analysis can reveal scaling properties of the temporal and frequency dynamics simultaneously unlike Fourier Transform used in [6], we compute a wavelet transform of the generated address correlation signal over several sampling points. Through signal can be detected in certain timescales that imply frequency components, and in certain positions of the timescales that mean temporal information, we can induce the frequency and temporal components respectively.

Table 1. The Nine Kinds of Simulated Attacks

	1 (2,I,SD)	2 (2,I,SR)	3 (2,I,R)	4 (2,P,SD)	5 (2,P,SR)	6 (2,P,R)	7 (1,P,SD)	8 (1,P,SR)	9 (1,P,R)
Duration	2 hours	2h	2h	2h	2h	2h	1 hour	1h	1h
Persistency	int.	int.	int.	per.	per.	per.	per.	per.	per.
IP	single	semi-random	random	single	semi-random	random	single	semi-random	random
Protocol	ICMP	TCP	UDP	ICMP	TCP	UDP	ICMP	TCP	UDP
Port	#80	random	#1434	#80	random	#1434	#80	random	#1434
Size	random	4KB	404B	random	4KB	404B	random	4KB	404B

Discrete Wavelet Transform (DWT) consists of decomposition and reconstruction. We iterate a multilevel one-dimensional wavelet analysis up to 8 levels in case of the postmortem analysis, so our final analysis (approximation and detail) coefficients are $[cA_8, cD_8, cD_7, cD_6, cD_5, cD_4, cD_3, cD_2, cD_1]$. We employ a daubechies-6 two-band filter. The filtered signal is down-sampled by 2 at each level of the analysis procedure; the signal of each level has an effect that sampling interval extends 2 times. Consequently it means that the wavelet transform identifies the changes in the signal over several timescales. When we use t minutes as sampling interval, the time range at level j spans $t*2^j$ minutes. These time range can independently sample and restore frequency components of $1/t*2^{j+1}$ by the Nyquist sampling theorem.

5 Detection in Postmortem Analysis

5.1 Selective Reconstruction in DWT

Our postmortem analysis allows the administrator to choose the timescales over which attacks/anomalies detection is desired. The network operator can analyze the traffic successively at different sampling times or choose to analyze the traffic at multiple timescales at the same time. Because of the time-scaled decomposition of the wavelets we are able to detect changes in the behavior of the network traffic that may appear at some resolution but go un-noticed at others.

The first three attacks described in (*,I,*) have an ON/OFF timing of 3 minutes. This signal could be effectively detected by only the 1st coefficient in case of 1-minute sampling period. The last six attacks expressed in (*,P,*) are persistent attacks. Attacks last for 1 hour at a minimum. It means that we could choose the cD_5, cD_6 and cD_7 among all the coefficients for reconstruction that are equivalent to 32 minutes, 1 hour 4 minutes and 2 hour 8 minutes respectively.

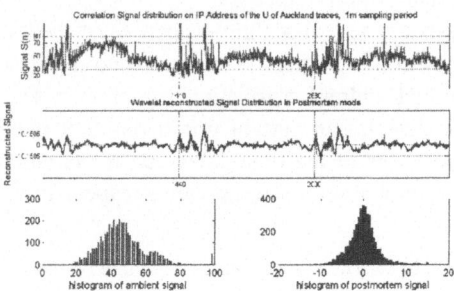

Fig. 3. The distribution of the ambient traces

The network operators can select reconstructed levels that they wish to be captured. We assume that the network administrators are interested in detecting shorter anomalies of sufficient intensity and anomalies of more than 30-minute duration. In order to detect these attacks, we extract only the 1st, 5th, 6th and 7th levels in decomposition and reconstruct the signal based only on coefficients at these levels.

5.2 Thresholds Setting through Statistical Analysis

We develop a theoretical basis for deriving thresholds for anomaly detection. The right-bottom sub-picture in Fig. 3 shows the histogram of the reconstructed signal of the ambient Auckland-IV traces in postmortem mode. We verify normality of the Fri/Sun data in Table II through the Lilliefors test for goodness of fit to a normal distribution with unspecified mean and variance. The postmortem transformed data have a normal distribution at 5% significance level, namely $X \sim N(0, 3.38^2)$. The original weighted correlation data fail to pass the null hypothesis of normality; however, the DWT converts it to normal distribution. By selecting some of the levels, we have removed some of the features from the signal that were responsible for the non-normality in the original signal.

When we set the thresholds to -10.15 and 10.15 respectively, these figures are equivalent to $\pm 3.0\sigma$ confidence interval for random process X. This interval corresponds to 99.7% confidence level. With such thresholds, we can detect attacks with error rate of 0.3%.

5.2.1 Statistical Consideration of Thresholds: Wide-Sense Stationary

If statistical parameters of network traffic, such as mean and standard deviation, are stationary distributed under given traffic, parameters of specific day could be applied to other days. We gather the 4-week traces and analyze their statistical summary measures. Table II shows the distribution in other days. We could infer wide-sense stationary (WSS) regarding these traces from the following: (i) the average is not dependent on time, and (ii) autocorrelation function is a function of time difference regardless of sample path. From the viewpoint of communication, the postmortem analysis of the ambient trace could be considered as WSS Gaussian white noise, on the other hand, the attack and anomaly could be considered as signal of interest. It illustrates that the thresholds could remain approximately the same over several days.

Table 2. The Statistical Parameters

	ambient traffic		postmortem analysis	
	$\overline{x}$	s	$\overline{x}$	s
1st week	53.0	13.5	-0.0	3.3
Mon/Tue	58.1	13.0	-0.2	3.8
Wed/Thu	55.3	13.2	-0.2	3.5
Fri/Sun	48.2	12.4	-0.0	3.4
2nd week	51.5	14.5	-0.0	3.9
3rd week	50.6	14.1	+0.1	3.3
4th week	47.8	13.5	-0.0	4.1

5.3 Detection of Anomalies Using the Real Attack Traces

Detection results of our composite approach with respect to 7-day KREONet2 traces are shown in Fig. 4. The top-most sub-picture illustrates a weighted correlation signal of IP addresses that is used for wavelet transform with real attacks. The second sub-picture is the wavelet-transformed and reconstructed signal in postmortem and its detection results. The actual attacks assail between the vertical lines, and the detection signal is shown with dots at the bottom of the second sub-picture.

A 7-day wide DWT window and a 20-minute wide DETECTION window are used for DWT analysis and detection, respectively. To evaluate the reconstructed signal we use $\pm 4.0\sigma$ as statistical threshold in second sub-picture of Fig 4. Overall, our results show that our approach may provide a good detector of attacks.

First 2 attacks attempted to attack web-server, which sequentially generated source port and targeted for 80 TCP port. A single source machine sent 48 byte-sized packets to (semi) single destination IP addresses in /24 address which preserved first 3 bytes of IP and randomly changed the last byte.

The last attack is the SQL slammer worm attack which generated random IP addresses at a specific port. A few compromised machines enormously sent 404 byte-sized packets to randomly generated destination IP addresses and 1434 UDP port.

As the bottom 2 sub-pictures shown, except the first attack, the remaining 2 attacks didn't set off any distinguishable variance in volume. It shows that the approach using traffic volume itself doesn't appropriately detect the bandwidth attacks.

Table 3. The Detectionability of the IP Correlation Signal and the DWT signal

	confidence level	DWT	1	2	3	4	5	6	7	8	9	false positive	false negative
1.0σ	68 %	IP[a]	c	.	.	.	.	.	.	.	.	5	0
		DWT[b]	.	.	.	.	.	.	.	.	.	6	0
1.5σ	86 %	IP	.	.	.	.	.	.	.	.	.	4	0
		DWT	.	.	.	.	.	.	.	.	.	5	0
2.0σ	95.5 %	IP	.	x[d]	.	.	.	.	.	.	.	3	1
		DWT	.	.	.	.	.	.	.	.	.	3	0
2.5σ	98.5 %	IP	.	x	x	.	x	.	.	x	.	1	4
		DWT	.	.	.	.	.	.	.	.	.	2	0
3.0σ	99.7 %	IP	.	x	x	.	x	.	.	x	x	0	5
		DWT	.	.	.	.	.	.	.	.	.	0	0
3.5σ	99.95 %	IP	x	x	x	.	x	x	.	x	x	0	7
		DWT	.	x	.	.	x	.	.	x	.	0	2
4.0σ	99.99 %	IP	x	x	x	x	x	x	x	x	x	0	9
		DWT	.	x	x	.	x	.	.	x	.	0	4

a.   IP means the original IP correlation signal without applying of DWT
b.   DWT means the DWT transformed signal
c.   . means a detection
d.   x means a non-detection

5.4 Effectiveness of DWT

For evaluating the effectiveness of employing DWT, we compare the detection results of our scheme employing DWT with a scheme that directly employs statistical analysis of the correlation signal. The anomaly detection results are shown in Table III. At low confidence levels (below 90%), DWT doesn't offer any advantage. However, when confidence levels of most interest (90% ~ 99.7%) are considered, DWT provides significantly better detection results than the simpler statistical analysis. This clearly shows that DWT offers significant improvement in the detection of anomalies.

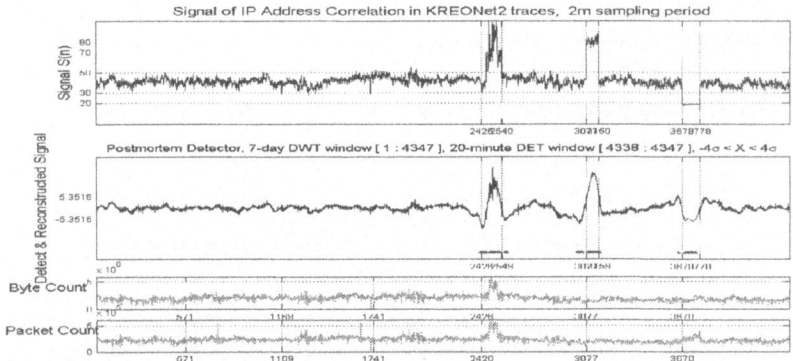

Fig. 4. Address based detection results using real attack traces in postmortem

Table 4. The Latencies in Nine Kinds of Attacks in real-time mode

	1 (2,I,SD)	2 (2,I,SR)	3 (2,I,R)	4 (2,P,SD)	5 (2,P,SR)	6 (2,P,R)	7 (1,P,SD)	8 (1,P,SR)	9 (1,P,R)	f p^c	f n
1.0σ	0[a]	0	0	0	0	0	0	0	0	11	0
1.5σ	0	0	0	0	0	0	0	0	0	7	0
2.0σ	0	0	0	0	0	0	0	0	0	5	0
2.5σ	0	0	0	0	0	1	0	0	0	3	0
3.0σ	0	0	0	0	0	2	0	2	0	2	0
3.5σ	0	0	1	0	20	9	0	3	2	2	0
4.0σ	1	0	1	0	X[b]	11	0	5	3	1	1

a.   Latency is measured by minute unit
b.   X means non-detection
c.   false positive is counted a series of relevant signal as 1

6   Detection in Real-Time Analysis

6.1   Individual Reconstruction in DWT

In real-time analysis, the network administrator may not have the luxury to selectively analyze the traffic at different timescales since attacks and anomalies need to be detected as they occur. Due to this lack of a priori knowledge of timescales of attacks or anomalies, real-time analysis requires analysis of data at all the time scales. Because of these two needs of analyzing data at all timescales, and the need to have lower latencies of attack/anomaly detection, real-time analysis is much more challenging. Because the number of the transformable samples is closely connected with the size of DWT window, the maximum allowable levels are restricted at $\log_2 n$, where n is the number of samples. If we want to investigate a specific level j, it requires 2^j samples for reconstruction at least. In our analysis here, we employed the most recent 2-hour data of traffic for prompt response and robustness. Detecting anomalies through all individual levels will have a number of advantages: (i) By setting a high threshold at each level, anomalies can be detected with high confidence, (ii) Depending on network administrator's filtering criteria, he/she can adjust the threshold between accuracy and flexibility as shown in Table IV, and (iii) the attributes of attacks, such as the frequency and pattern, can be straightforwardly determined.

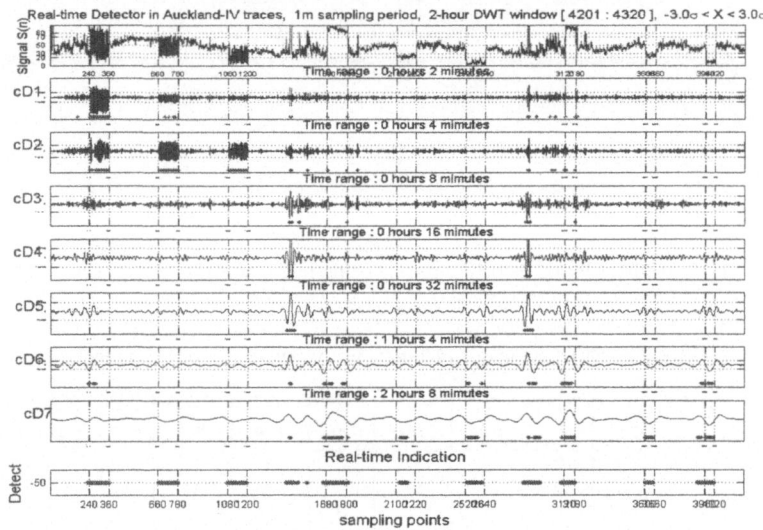

Fig. 5. Address based detection results using simulated attack traces in real-time

6.2 Detection of Anomalies Using the Simulated Attack Trace

We employed a 2-hour DWT window in 1-minute sampling interval. It can be decomposed up to level 7. The results of our real-time analysis are shown in Fig. 5. The DWT signal at each timescales is shown along with the horizontal detector (an anomaly detected over successive samples at the same level). The bottom-most picture shows the composite detector that employs two-dimensional mechanism using horizontal and vertical detection simultaneously. The results indicate that the real-time analysis detects all the attacks along with a few anomalies present in the base signal.

Table IV shows the overall timing relationship between detection latency and the confidence level of our attacks in real-time mode. As we expect, the higher the confidence level, the higher the detection latency. According to the network administrator's security standard, the appropriate confidence level could be established.

7 Multidimensional Indicators

It seems feasible to carry out a similar correlation and wavelet-based analysis of network packets based on their port numbers. Is it possible to combine several indicators to build a more robust anomaly detector that is less prone to false alarms? Fig. 6 shows the comprehensive anomaly detector based on a combination of addresses and port numbers. The two kinds of dots at the bottom of the picture show detection results. The dots located on top are marked when both the address and port methods detect anomalies simultaneously. The dots located on the bottom are displayed when only one of the two detection methods detects anomalies. It can be understood that the above markings imply very high confidence and the lower dots imply probable detections.

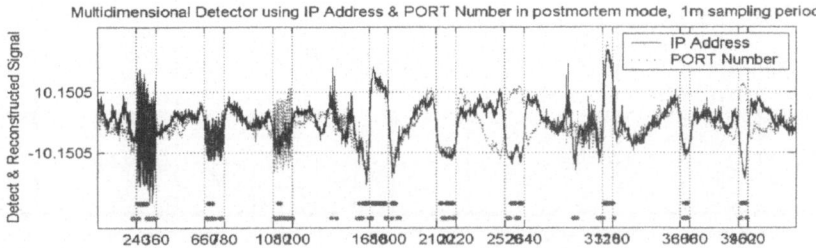

Fig. 6. The multidimensional detection results

7.1 Attack Volume

We carried out similar analysis of traffic to study the sensitivity of our detectors to the relative volume of attack traffic in Auckland-IV traces. We varied the ratio of attack traffic to normal traffic volume from 1:2, to 1:5 to 1:10. The results of this study are shown in Table V and VI. The results show that the proposed schemes are effective even when the attack traffic volume as low as 10% of the normal traffic. The latencies for real-time detection get longer with smaller attack traffic volume as to be expected. The results indicate that the DWT analysis of address correlation signal is useful over a wide range of attack traffic volumes.

8 Future Work and Conclusion

As a further research, the relation between sampling rate and latency should be investigated from statistical point of view. We also plan to study the effectiveness of the analysis of traffic header data at various points in the network.

We studied the feasibility of analyzing packet header data through wavelet analysis for detecting traffic anomalies. Specifically, we proposed the use of correlation of destination IP addresses and port numbers in the outgoing traffic at an egress router.

Table 5. The Detectionability Of the various mixture ratios in postmortem mode

	Mix. Ratio	1 (2,I,SD)	2 (2,I,SR)	3 (2,I,R)	4 (2,P,SD)	5 (2,P,SR)	6 (2,P,R)	7 (1,P,SD)	8 (1,P,SR)	9 (1,P,R)	f_p	f_n
1.0σ	1:2	.	.	.	.	.	.	.	.	.	6	0
	1:5	.	.	.	.	.	.	.	.	.	6	0
	1:10	.	.	.	.	.	.	.	.	.	6	0
1.5σ	1:2	.	.	.	.	.	.	.	.	.	5	0
	1:5	.	.	.	.	.	.	.	.	.	5	0
	1:10	.	.	.	.	.	.	.	.	.	4	0
2.0σ	1:2	.	.	.	.	.	.	.	.	.	3	0
	1:5	.	.	.	.	.	.	.	.	.	4	0
	1:10	.	.	.	.	.	.	.	.	.	2	0
2.5σ	1:2	.	.	.	.	.	.	.	.	.	2	0
	1:5	.	.	.	.	.	.	.	.	.	3	0
	1:10	.	.	.	.	.	.	.	.	.	2	0
3.0σ	1:2	.	.	.	.	.	.	.	.	.	0	0
	1:5	.	.	X	.	.	.	.	.	.	2	1
	1:10	.	.	X	.	.	.	.	.	.	1	1
3.5σ	1:2	.	.	X	.	.	.	.	X	.	0	2
	1:5	.	.	X	.	.	.	.	.	.	1	1
	1:10	.	X	X	.	X	X	X	.	.	1	5
4.0σ	1:2	.	X	X	.	X	.	.	X	.	0	4
	1:5	.	.	X	.	.	.	X	X	.	0	3
	1:10	.	X	X	.	X	X	X	X	.	1	6

Table 6. The Detection Latency Of the various mixture ratios in Real-time mode

	Mix. Ratio	1 (2,I,SD)	2 (2,I,SR)	3 (2,I,R)	4 (2,P,SD)	5 (2,P,SR)	6 (2,P,R)	7 (1,P,SD)	8 (1,P,SR)	9 (1,P,R)	f_p	f_n
1.0σ	1:2	0	0	0	0	0	0	0	0	0	11	0
	1:5	0	0	0	0	0	0	0	0	0	9	0
	1:10	0	0	0	0	0	0	0	0	0	9	0
1.5σ	1:2	0	0	0	0	0	0	0	0	0	7	0
	1:5	0	0	1	0	0	0	0	0	0	7	0
	1:10	0	2	2	0	0	2	0	0	0	6	0
2.0σ	1:2	0	0	0	0	0	0	0	0	0	5	0
	1:5	0	0	2	0	6	0	0	0	0	5	0
	1:10	0	4	2	0	7	10	0	0	9	5	0
2.5σ	1:2	0	0	0	0	0	1	0	0	0	3	0
	1:5	0	0	2	0	8	14	0	2	4	3	0
	1:10	0	5	34	0	24	32	0	2	12	2	0
3.0σ	1:2	0	0	0	0	0	2	0	2	0	2	0
	1:5	0	0	5	1	8	30	0	5	6	2	0
	1:10	0	7	38	0	28	32	0	5	16	2	0
3.5σ	1:2	0	0	1	0	20	9	0	3	2	2	0
	1:5	0	2	34	1	10	40	0	10	9	1	0
	1:10	0	8	40	1	28	X	0	6	20	1	1
4.0σ	1:2	1	0	1	0	X	11	0	5	3	1	1
	1:5	0	2	40	3	50	X	4	12	12	1	1
	1:10	0	10	X	1	X	X	0	15	24	1	3

We studied the effectiveness of our approach in postmortem and real-time analysis of network traffic. The results of our analysis are encouraging and point to a number of interesting directions for future research.

Acknowledgment. We are very grateful to Deukwoo Kwon for his comments on statistical analysis, to Man Hee Lee and Dr. Okhwan Byeon at Kisti for their help in accessing traces.

References

1. Anu Ramanathan, "WADeS: A Tool for Distributed Denial of Service Attack Detection", TAMU-ECE-2002-02, Master of Science Thesis, August 2002.
2. National Laboratory for Applied Network Research (NLANR), measurement and operations analysis team, "NLANR network traffic packet header traces", accessed in August 2002.
3. P. Barford, J. Kline, D. Plonka and A. Ron, "A Signal Analysis of Network Traffic Anomalies," in *Proc. of ACM SIGCOMM IMW*, Marseille, France, November 2002.
4. T. M. Gil and M. Poletto, "MULTOPS: A Data-Structure for Bandwidth Attack Detection", in *Proc. of the 10th USENIX Security Symposium*, Washington, D.C., USA, August 2001.
5. E. Kohler, J. Li, V. Paxson and S. Shenker, "Observed Structure of Addresses in IP Traffic, in *Proc. of ACM SIGCOMM IMW*, Marseille, France, November 2002.
6. Chen-Mou Cheng, H. T. Kung and Koan-Sin Tan, "Use of spectral analysis in defense against DoS attacks", in *Proc. of IEEE Globecom*, 2002.
7. KREONet2 (Korea Research Environment Open NETwork2), www.kreonet2.net
8. Seong Soo Kim, A. L. Narasimha Reddy and Marina Vannucci, "Detecting Traffic Anomalies using Discrete Wavelet Transform", in *Proc. of ICOIN 2004*, Busan, Korea, Feb 2004
9. Anja Feldmann, Anna Gilbert, Polly Huang and Walter Willinger, "Dynamics of IP traffic: A study of the role of variability and the impact of control", *Computer Communication Review, Vol. 29, No. 4 (Proc. of the ACM Sigcomm'99, Cambridge, MA), pp. 301-313*, 1999.
10. CERT Coordination Center (CERT/CC), "CERT Advisory CA-2003-04 MS-SQL Server Worm", January 2003. http://www.cert.org/advisories/CA-2003-04.html

Lazy Agent Replication and Asynchronous Consensus for the Fault-Tolerant Mobile Agent System

Taesoon Park[1], Ilsoo Byun[1], and Heon Y. Yeom[2]

[1] Department of Computer Engineering, Sejong University,
Seoul 143-747, KOREA
{tspark,widepis}@sejong.ac.kr
[2] Department of Computer Science, Seoul National University,
Seoul 151-742, KOREA
yeom@snu.ac.kr

Abstract. In this paper, we propose a low overhead replication scheme for the fault-tolerant mobile agent system. In the proposed lazy replication scheme, execution of a primary agent and migration of its replicas are concurrently processed. Also, the primary agent performs asynchronous consensus with fixed consensus agents so that the consensus step and the replica migration step can concurrently be processed. As a result, the primary agent should not wait for the completion of the replica migration step unless any of the consensus agents fails. The proposed scheme has been implemented on top of the Aglet system and its performance has been measured.

1 Introduction

A mobile agent is a software program which moves from a site to another site to execute a task assigned by a user [1]. As the mobile agent system has drawn attention as a new distributed computing paradigm, the importance of reliable agent execution is more emphasized. Reliable execution of a mobile agent is to guarantee the exactly-once execution of an agent even in case of a system failure [11]. Many fault-tolerance schemes for the mobile agent system have been proposed and they are categorized into the replication schemes [3,7,8,11] and the checkpointing schemes [2,10,12].

Replication schemes show a high degree of fault tolerance since a replication scheme with $2k + 1$ replicas can tolerate up to k failures. On the other hand, checkpointing schemes may cause a severe delay in recovery even after a single failure. However, considering the execution time, checkpointing does not require much overhead, while the time to replicate an agent and migrate the replicas and the time to perform the consensus among replicas are not negligible.

To reduce the replication cost, we have suggested asynchronous agent replication schemes and measured the performance [5,6]: In the asynchronous replication scheme, agent replicas are migrated to the designated sites in an asynchronous manner so that the primary can begin its execution without waiting

N. Mitrou et al. (Eds.): NETWORKING 2004, LNCS 3042, pp. 1060–1071, 2004.

for the migration of other replicas. With this optimization, we have achieved up to 37% reduction of the replication cost compared to the synchronous agent replication. However, there is still a performance gap between the systems using the asynchronous replication scheme and no fault-tolerance scheme.

In this paper, to fill the performance gap, we propose a scheme for lazy replication and asynchronous consensus. In the proposed scheme, execution and consensus of an agent may proceed asynchronously with the agent replication. For a new stage, the first replica of a current agent is migrated to the new execution site and begins the execution, while the rest of replicas are sent to the designated sites. Since an agent begins its execution without waiting for the other replicas, it is sometimes possible that the agent ends the stage before its replicas are ready for the consensus. In order to proceed the consensus without waiting for the late replicas, we use fixed consensus agents. Since an agent performs the consensus with fixed consensus agents instead of waiting for the late replicas, there is no delay in consensus unless the primary replica or one of consensus agents fails. To validate the correctness of the proposed scheme and evaluate its performance, we have implemented the lazy replication with asynchronous consensus on top of the Aglet system and measured the performance.

The rest of this paper is organized as follows: Section 2 describes the Aglet system and the failure model. Existing replication and consensus schemes are presented in Section 3 and Section 4 presents the proposed lazy replication scheme with asynchronous consensus. Section 5 describes the experimental environment and discusses the experimental results. Section 6 concludes the paper.

2   The Aglet System

A mobile agent system consists of a number of system sites connected by the communication network. Each of the sites, to support execution and migration of agents, provides one or more *places*. An agent executes its task *on the place* and migrates *between the places*. While residing in a place, the agent performs an assigned task. The execution of an agent in a place and the migration of the agent into the next place are called a *stage*. In other words, the computation of an agent is denoted by a sequence of stages. Figure 1 shows the execution of a mobile agent, MA_i, consisting of four stages. In the figure, $SG_{i,\alpha}$ denotes the α-th stage of MA_i. The task execution and the migration of a stage, $SG_{i,\alpha}$, are denoted by $E_{i,\alpha}$ and $M_{i,\alpha}$, respectively.

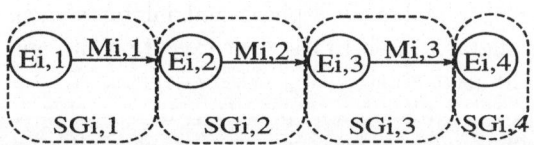

Fig. 1. Stages of a Mobile Agent, MA_i

The *Aglet* [4] is a Java-based mobile agent system. To support execution, migration and communication of agents, the system provides the *AgletContext* environment. The agents in the Aglet system inherit the properties and the methods from the *AgletClass* and perform event-driven activities. For the inter-agent communication, a message-passing mechanism is used in the Aglet system and the *AgletProxy* is provided to support the location transparency of the agent. The AgletProxy is an interface to an Aglet object and every message is sent to the Aglet object through the AgletProxy, regardless of its location.

Failures considered in the system are the *agent failure*, the *place failure* and the *system failure*. For all of these failure types, the fail-stop [9] model is assumed; that is, once a component fails, it stops its execution and does not perform any malicious actions.

3   Replication and Consensus

3.1   Synchronous Replication

For the fault-tolerant execution of an agent, one execution stage consists of three steps, which are the task execution step, the replication step, and the consensus step. Figure 2 shows an example of these three steps where one primary agent and two replicas are used.

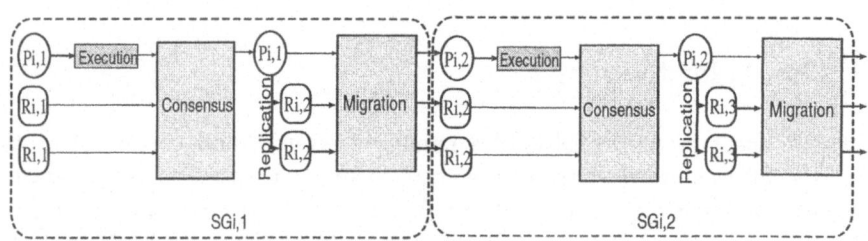

Fig. 2. Synchronous Agent Replication

• Agent Replication: Before a stage, $SG_{i,\alpha}$, begins, the primary agent, $P_{i,\alpha-1}$, of the previous stage, $SG_{i,\alpha-1}$, makes its replicas. A primary agent is the one responsible for the initial task execution of a stage and at the end of the stage, it is also responsible for the agent replication for the next stage. During the replication step, $P_{i,\alpha-1}$ makes $2k$ replicas and migrates them to $2k$ different sites. $P_{i,\alpha-1}$ then migrates itself to the next execution site and becomes a new primary $P_{i,\alpha}$ for the next stage, $SG_{i,\alpha}$. Every replica, $R_{i,\alpha}$, observes the task execution step and the replication step of $P_{i,\alpha}$. In case that a replica suspects the failure of a primary, it may become a new primary after a consensus step.

• Task Execution and Consensus: The primary, $P_{i,\alpha}$, begins the task execution step for the stage, $SG_{i,\alpha}$, as soon as it arrives in a new execution site. When the primary successfully completes the task, it begins the consensus step. The consensus step is to confirm the task completion of the primary and to prevent any redundant execution by false failure detection of a replica. For example, due to slow execution or communication, a replica may suspect the failure of a primary and try to become a new primary. In such a case, two primary agents may execute the same task, which violates the exactly-once execution property.

The primary begins the consensus by sending out the *consensus_begin* message to every replica, $R_{i,\alpha}$. Every $R_{i,\alpha}$ replies with the *consensus_ack* message unless it has already sent out the message for the same stage. When the primary receives the majority of *consensus_ack* messages, it sends out *consensus_confirm* messages to the replicas and completes the consensus step. If the primary fails to obtain the majority of *consensus_ack* messages, it gives up the current stage and undoes the executed task. In the consensus step, only one primary can obtain the majority votes for a stage and complete the task.

• Failure Handling: A replica $R_{i,\alpha}$ is made to detect any failure of the primary for the stage $SG_{i,\alpha}$. For the failure detection, the time-out is often used. If $R_{i,\alpha}$ cannot receive any *consensus_begin* message within a time-out period, it suspects the failure of the primary during the task execution step. Also, if $R_{i,\alpha}$ cannot receive the majority of the *successful_migration* messages within a time-out period, it suspects the failure of the primary during the replication step. For this, the *successful_migration* of every replica $R_{i,\alpha+1}$ is acknowledged not only to the primary $P_{i,\alpha}$ but also to every $R_{i,\alpha}$. When a failure of the primary is suspected, the replica $R_{i,\alpha}$ becomes a new primary through its own consensus and begins an alternative task execution step for the stage $SG_{i,\alpha}$.

Any replica which first detects the failure of the primary can initiate the consensus step and take over the execution as proposed in [11]. Or as in [8], the priority of every replica is predetermined and only the replica with the highest priority can initiate the consensus step when a primary fails. In this scheme, the $(j+1)$-th replica can take over the task execution step of the current stage, when up to the j-th replica fails. In either case, the agent can keep alive as long as $k+1$ replicas survive failures. Among them, one replica is for the task execution and k replicas are for the majority voting. Therefore, the replication scheme with one primary and $2k$ replicas can tolerate up to k failures.

3.2   Asynchronous Replication

To reduce the replication and migration cost of the synchronous scheme, we have proposed an asynchronous agent replication scheme [5], in which the replica migration proceeds asynchronously with the primary execution.

• Asynchronous Agent Replication: In the replication step, the primary, $P_{i,\alpha}$, makes $2k+1$ replicas and transfers them as in the synchronous replication scheme. However, every replica in the asynchronous replication scheme has a predetermined priority and the replica with the highest priority becomes the new primary $P_{i,\alpha+1}$ when it arrives in the next site for the stage, $SG_{i,\alpha+1}$. Since the new primary, $P_{i,\alpha+1}$, is the first replica migrated by the previous primary,

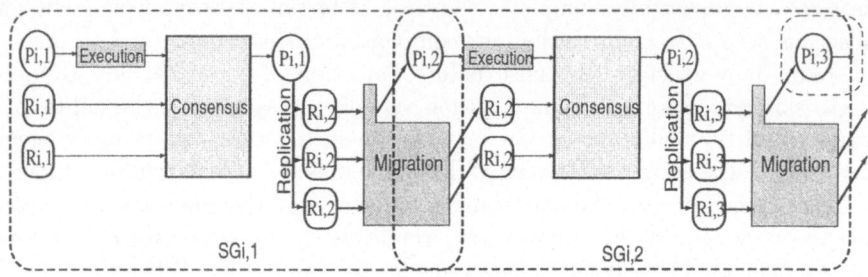

Fig. 3. Asynchronous Agent Replication

$P_{i,\alpha}$, and $P_{i,\alpha}$ may still process the migration of the other replicas, the task execution step of $P_{i,\alpha+1}$ and the replication step by $P_{i,\alpha}$ may proceed concurrently as shown in Figure 3. $P_{i,\alpha}$ terminates itself when it completes the replication step. Because of this asynchronous replication, the time for the replication step by $P_{i,\alpha}$ can fully or partially be masked by the task execution step of $P_{i,\alpha+1}$. As a result, the total execution time of the agent can be reduced.

• Failure Handling: The failure detection during the task execution step and the replication step is processed in the same way as in the synchronous replication scheme. One difference is that the primary $P_{i,\alpha}$ in the asynchronous scheme may finish the task execution step before some replicas arrive in their execution sites. In this case, $P_{i,\alpha}$ should wait so that every replica can participate in the consensus step. Another problem may happen when $P_{i,\alpha}$ fails before replicating and migrating the majority of the replicas for the next stage. Note that in this case, any replica $R_{i,\alpha}$ becomes a new primary and re-processes the task execution step of the stage $SG_{i,\alpha}$, while the next primary $P_{i,\alpha+1}$ may have already begun its task execution step for the stage $SG_{i,\alpha+1}$. However, even in this case, $P_{i,\alpha+1}$ cannot obtain the majority of votes during the consensus step and therefore it can be discarded anyway.

4   Lazy Replication and Asynchronous Consensus

Asynchronous agent replication can reduce the total execution time of an agent as much as the overlapped time of the replication step of a primary, $P_{i,\alpha}$, and the task execution step of the next primary, $P_{i,\alpha+1}$. However, $P_{i,\alpha+1}$ still has to wait before the consensus step unless the execution step of $P_{i,\alpha+1}$ is longer than the replication step of $P_{i,\alpha}$. In the Aglet system, for a primary agent to communicate with its replicas, it has to obtain the proxy of the replicas first. In the synchronous scheme, the primary migrates with the proxy information of its replicas, since it migrates after all the replicas. However, in the asynchronous scheme, the primary, $P_{i,\alpha}$, should send the proxy information of the replicas, $R_{i,\alpha+1}$s, to the next primary, $P_{i,\alpha+1}$, after it completes the replication step. As a result, $P_{i,\alpha+1}$ cannot begin the consensus step until the replica proxy information arrives.

• Fixed Consensus Agents: To eliminate the waiting time for late replicas, we propose to use fixed consensus agents. The replication scheme with one primary and $2k$ replicas is designed to tolerate up to k failures. This means that only the majority of replicas have a chance for the alternative task execution and the others are made just for attending the consensus step. Assuming the consensus algorithm using the predetermined priority [8], the k replicas with lower priorities are sure to attend only the consensus step. Therefore, in the proposed scheme, we replace the k lower priority replicas with fixed consensus agents.

A consensus agent contains simple codes to perform the consensus step. In the first stage of agent execution, k consensus agents are created and sent to k different sites. Therefore, the primary agent in any stage is sure to know the location of consensus agents and have their proxy information. Now, for the consensus, a primary or any replica suspecting the failure of the primary sends consensus messages to the consensus agents. To differentiate the consensus messages of different stages, consensus related messages should carry the stage number. Consensus agents then reply with the *consensus_ack* or *consensu_nak* messages according to the consensus algorithm.

Using fixed consensus agents, we can take the following advantages: First, a primary makes only $k + 1$ replicas and the time to make k more replicas can be eliminated. Also, a primary no longer waits for the late replicas. As soon as the primary completes the task execution step, it can begin the consensus step with fixed consensus agents. As a result, there is virtually no blocking of primary agent execution when there is no failure.

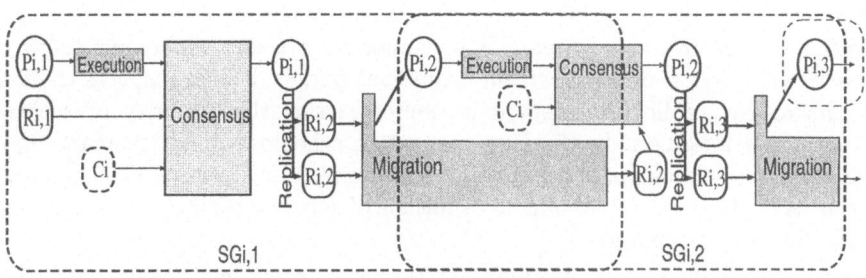

Fig. 4. Lazy Agent Replication and Asynchronous Consensus

• Lazy Replication and Asynchronous Consensus: In the proposed scheme, a primary $P_{i,\alpha}$ performs the asynchronous replication as described in the previous section. Therefore, while $P_{i,\alpha}$ processes the replication step, the next primary, $P_{i,\alpha+1}$, may complete the task execution step for the stage, $SG_{i,\alpha+1}$. When $P_{i,\alpha+1}$ completes the task execution step, it can begin the consensus step by sending out the *consensus_begin* messages to the consensus agents. In case that all the consensus agents are alive, $P_{i,\alpha+1}$ can obtain the majority of *consensus_ack* messages from fixed consensus agents and then proceed the replication step for the next stage, $SG_{i,\alpha+2}$. Using fixed consensus agents whose location information is known, the primary agent can complete the consensus step without waiting for the migration of all the other replicas.

Figure 4 shows an example of lazy replication and asynchronous consensus with one primary, one replica and one fixed consensus agent, C_i. As it can be noticed from the figure, a primary $P_{i,\alpha+2}$ can begin the task execution step while the previous primaries, $P_{i,\alpha+1}$ and $P_{i,\alpha}$, still perform the replication step. However, for the proposed scheme to be complete, the replication step of each stage should be performed in a very lazy manner. One responsibility of a primary, $P_{i,\alpha}$, is to notify every replica, $R_{i,\alpha}$, of the successful completion of the consensus. Also, each of the next replicas, $R_{i,\alpha+1}$, made by $P_{i,\alpha}$ should know the location information of their previous replicas, $R_{i,\alpha}$s, to inform the $successful\_migration$ message.

To handle these processes without blocking the agent execution, the primary, $P_{i,\alpha}$, proceeds with the replication step and migrates the first replica for the next stage, $SG_{i,\alpha+1}$, after it successfully completes the consensus step with consensus agents. $P_{i,\alpha}$ then waits for the location information of the late replicas of the current stage. The rest of the consensus step and the replication step can be continued when it receives the necessary location information of the other replicas. To complete the consensus step, $P_{i,\alpha}$ sends the $consensus\_confirm$ message to every $R_{i,\alpha}$ instead of $consensus\_begin$ message and continues the migration of the replicas for the next stage. As a result, in the proposed scheme, the consensus step of a stage can be overlapped with the replication step for the next stage and also the replication step of several stages can be overlapped.

• Failure Handling: The role of the replica agent in the lazy replication scheme is the same as in the other replication schemes. A replica $R_{i,\alpha}$ may suspect the failure of its primary if it cannot receive any $consensus\_begin$ or $consensus\_confirm$ message within a time-out period. It also suspects the failure during the replication step, if it cannot receive the majority of $successful\_migration$ messages. In the lazy replication scheme, when a primary, $P_{i,\alpha}$, ends the replication step, it informs the fixed consensus agents of the beginning of the next stage. The location information of current replicas, $R_{i,\alpha}$s, is also carried in that information.

On the receipt of that information, consensus agents act as the replicas of the next stage, $R_{i,\alpha+1}$, and send $successful\_migration$ messages to the previous replicas, $R_{i,\alpha}$. Therefore, even if any new replica may fail on the arrival of a new site, previous replicas can receive the majority of $successful\_migration$ messages as long as more than $k+1$ agents among the primary, replicas and consensus agents survive. The failure of a consensus agent may affect the performance of the agent execution, since a primary cannot asynchronously complete the consensus step if a consensus agent fails. However, in such a case, the primary can wait for the location information of any replica and process the consensus step with that replica. Therefore, even in case of the consensus agent failure, the primary can complete the consensus step, while the execution time may slightly be longer.

5   Performance Study

5.1   Experimental Setup

To validate the correctness of the lazy replication scheme and evaluate its performance, we have implemented the lazy replication and asynchronous consensus
scheme (the *LazyRep* scheme) on top of the *Aglet* system. The Aglet system is a
Java-based mobile agent system and for our experiments, *Aglet DSK 1.1b2* has
been used. The synchronous replication scheme (the *SyncRep* scheme) and the
asynchronous replication scheme (the *AsyncRep* scheme) have also been implemented for the performance comparison.

A cluster of five Pentium IV 1 GHz PCs connected by a 100 Mbps Ethernet
was used for experiments. Each machine supported one place. An agent traversed
the places in a predetermined order and the replicas were also sent to the pre-
determined places. To obtain a stable performance, we used an agent consisting
of twenty stages. For each experimental data, ten runs of the agent execution
were measured and then eight measured values were averaged out, excluding the
lowest and the highest ones. In each stage, the agent sleeps for T milliseconds,
instead of performing any task. To observe the influence of the agent size, an
NXN integer array is included in the agent so that we can control the size of
the agent by varying the size of the array.

5.2   Experimental Results

Figure 5(a) first compares the agent execution time of three schemes when the
replica number is 3, 5 and 7. This number includes the primary, replicas and
consensus agents of one stage. The size of the agent is 100 KBytes and the task
execution time of one stage is 1000 milliseconds.

Considering the time for the replication step and the consensus step, the
SyncRep scheme shows 184% increase as the replica number changes from 3 to
7. Compared to this, the LazyRep scheme shows only 75% increase when the
replica number changes. Since the increase of the replication time in the SyncRep
scheme is proportional to the number of replicas, the total execution time sharply
increases when the replica number is increased. However, the replication time of
the LazyRep scheme is not affected by the replica number. Only the consensus
time can be longer when the replica number is large, since the primary has to
wait for *consensus_ack* messages from more number of consensus agents.

Compared to the LazyRep scheme, the performance of the AsyncRep scheme
is somewhat disappointing. When the replica number is three, the AsyncRep
scheme achieves 31.6% reduction of the replication and the consensus time, compared to the SyncRep scheme. However, it does not show much reduction when
the number of replicas are seven and the performance is even worse for five
replicas. One possible explanation is that the stage execution time of 1000 milliseconds may be too short to mask the replication time for a large number of
replicas.

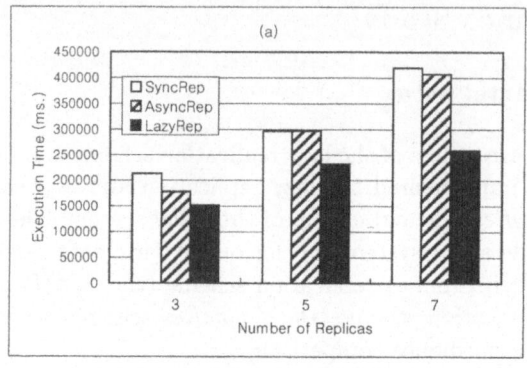

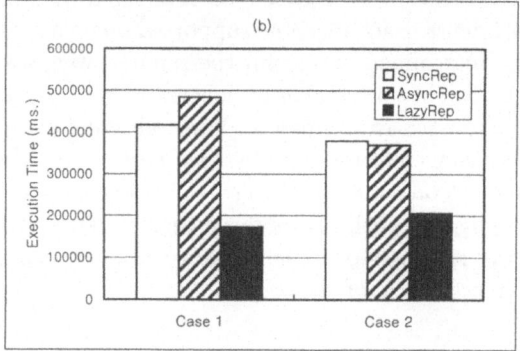

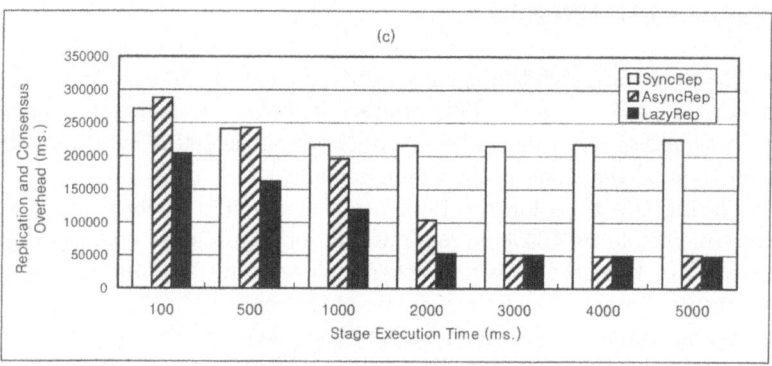

Fig. 5. Experimental Results I

To confirm this, we have measured the performance of two extreme cases as shown in Figure 5.(b). For the Case 1, we have reduced the task execution time of one stage into 100 milliseconds where the agent size is 100 KBytes and the replica number is five. In this case, the AsyncRep scheme is much worse than the SyncRep scheme and this worse performance is due to the time to make replicas. In the AsyncRep scheme, a primary makes $2k+1$ replicas instead of $2k$ replicas. Since copying of one agent takes about 20 milliseconds in this case, the marginal performance gain obtained by asynchronous replication should be lost to copy one more replica.

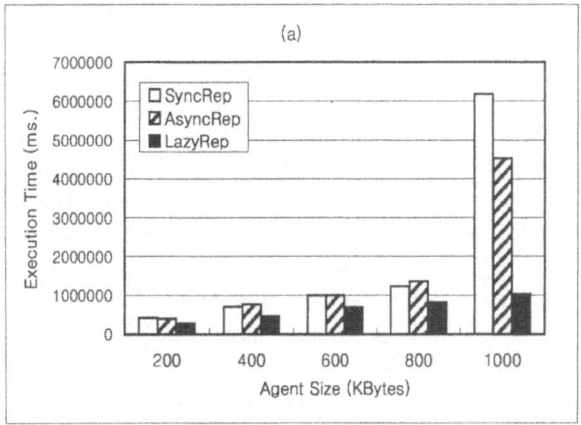

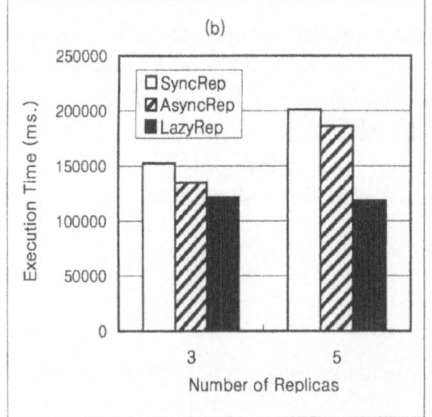
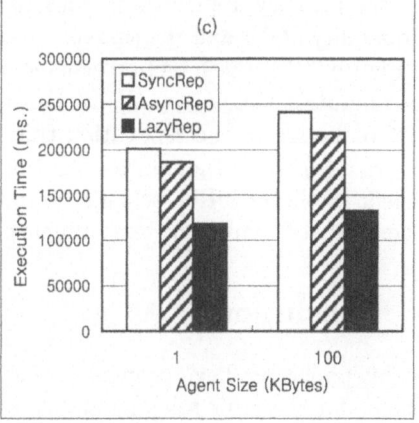

Fig. 6. Experimental Results II

For the performance of the Case 2, we have reduced the agen size to 10 KBytes where the task execution time of one stage is 1000 milliseconds and the replica number is five. By reducing the agent size, the time to make one more replica can be reduced and also the replica migration time can be reduced. Because of this reduction, the AsyncRep scheme can show the slight performance gain compared to the SyncRep scheme. This disappointing performance of the AsyncRep scheme is only for the cases where the replication time is much longer than the task execution time. When the task execution time becomes long enough to mask the replication step and the consensus step, the performance of the AsyncRep scheme is close to the LazyRep scheme as shown in Figure 5.(c).

The performance of Figure 5.(c) was obtained when the agent size is 100 KBytes and the replica number is five. As shown in the figure, the time for the replication step and the consensus step of the SyncRep scheme is not affected by the task execution time of one stage. However, as the task execution time increases, the replication step and the consensus step of the AsyncRep and the LazyRep schemes can concurrently be processed with the task execution step. As a result, the total execution time of an agent can significantly be reduced.

Figure 6.(a) shows the execution time of three schemes, when the agent size varies from 200 KBytes to 1000 KBytes. The replica number is five and the task execution time of one stage is 1000 milliseconds for this result. A large agent requires more time to make replicas and to migrate them. Therefore, the performance of the SyncRep scheme is heavily influenced by the agent size. The performance of the AsyncRep scheme is also affected a lot by the agent size, however, nonnegligible performance gain can be achieved with asynchronous replication. Compared to this, the LazyRep scheme shows very desirable performance and also very stable performance.

Figure 6.(b) and 6.(c) show the impact of failures on the agent execution time of three schemes. For these results, it is assumed that any agent, including the primary, replicas and consensus agents, can fail on each stage with the probability of 0.1 and the time-out to detect a failure is three seconds. As shown in Figure 6.(b), the LazyRep scheme now achieves only 30%–54.6% reduction of the migration time under the various number of replicas. It also achieves 54.6%–56% reduction of the replication time under the various agent size as shown in Figure 6.(c). Even though re-execution caused by a failure partially takes off the benefit of the LazyRep scheme, Figure 6.(b) and 6.(c) show that the LazyRep scheme is still under the less influence.

6   Conclusions

In this paper, we have proposed a lazy replication and asynchronous consensus scheme for the fault tolerant mobile agent system. In the proposed scheme, the replication step, the consensus step and the task execution step of an agent proceeds asynchronously. To process the consensus step without waiting for the late replicas, we have introduced fixed consensus agents and an agent can perform the consensus step with fixed consensus agents. As a result, there is practically no delay in the consensus step unless the primary or one of consensus agents fails. To evaluate the performance of the proposed scheme, we have implemented the proposed scheme on top of the Aglet system. The performance results show that the lazy replication with asynchronous consensus scheme can achieve much reduction of the replication and the consensus cost, compared to earlier replication schemes. Also, the lazy replication scheme shows very stable performance.

Acknowledgments. This work was supported by grant No. R04-2002-000-20102-02003 from the Basic Research Program of the Korea Science & Engineering Foundation.

References

1. Baumann, J., Hohl, F., Rothermel, K., Strasser, M.: Mole - Concepts of a Mobile Agent System. World Wide Web Journal, Vol. 1, No. 3 (1998) 12–137
2. Gendelman, E., Bic, L.F., Dillencourt, M.B.: An Application-Transparent, Platform-Independent Approach to Rollback-recovery for Mobile Agent Systems. Proc. of the 20th Int'l Conf. on Distributed Computing Systems (2000)
3. Johansen, D., Marzullo, K., Schneider, F.B., Jacobsen, K.: NAP: Practical Fault-Tolerance for Itinerant Computations. Proc. of the 10th Int'l Conf. on Distributed Computing Systems (1999)
4. Karjoth, G., Lange, D.B., Oshima, M.: A Security Model for Aglets. IEEE Internet Computing (1997)
5. Park, T., Byun, I., Kim, H., Yeom, H.Y.: The Performance of Checkpointing and Replication Schemes for Fault Tolerant Mobile Agent Systems. Proc. of the 21st Symp. on Reliable Distributed Systems (2002) 256–261
6. Park, T., Byun, I.: Low Overhead Agent Replication for the Reliable Mobile Agent System. Lecture Notes in Computer Science, Vol. 2790. Springer-Verlag, Berlin Heidelberg New York (2003) 1170–1179
7. Pleisch, S., Schiper, A.: Modeling Fault-Tolerant Mobile Agent Execution as a Sequence of Agreement Problems. Proc. of the 19th Symp. on Reliable Distributed Systems (2000) 11-20
8. Pleisch, S., Schiper, A.: FATOMAS - A Fault-Tolerant Mobile Agent System Based on the Agent-Dependent Approach. Proc. of the Int'l Conf. on Dependable Systems and Networks (2001) 215–224
9. Schlichting, R.D., Schneider, F.B.: Fail-stop Processors: An Approach to Designing Fault-tolerant Computing Systems. ACM Transactions on Computer Systems, Vol. 1, No. 3 (1983) 222–238
10. Silva, L., Batista, V., Silva, J.G.: Fault-Tolerant Execution of Mobile Agents. Proc. of the Int'l Conf. on Dependable Systems and Networks (2000)
11. Strasser, M., Rothermel, K.: Reliability Concepts for Mobile Agents. International Journal of Cooperative Information Systems, Vol. 7, No. 4 (1998) 355-382
12. Strasser, M., Rothermel, K.: System Mechanism for Partial Rollback of Mobile Agent Execution. Proc. of the 20th Int'l Conf. on Distributed Computing Systems (2000)

LMPS: Localized Multi-path Selection for QoS Routing in VoIP Networks

Khaled M.F. Elsayed, Hassan Fadel, and Amin M. Nassar

Department of Electronics and Communications Engineering
Faculty of Engineering, Cairo University, Giza, Egypt 12613
khaled@ieee.org, hassan.fadel@egti.com.eg, anassar@eng.cu.edu.eg

Abstract. Localized QoS routing techniques were proposed to achieve acceptable performance without exchanging global state information over the network. In such techniques, the ingress nodes estimate the network state and make the routing decision locally, hence reducing the signaling traffic at core routers. However, using only locally collected information to make routing decision is not a completely reliable technique to achieve trusted QoS performance. The Localized Multi-Path Selection (*LMPS*) scheme is proposed in this paper for connection-oriented bandwidth-aware flows to overcome the common drawbacks of the localized techniques. The *LMPS* technique is a multi-path selection algorithm, which selects paths that are capable of satisfying the requested bandwidth requirement of an incoming call, while at the same time trying to avoid the overloaded links to minimize the overall blocking ratio in the network. We demonstrate through extensive simulations and comparisons with a fully localized approach known as virtual capacity routing (*VCR*) and a shortest-widest multipath scheme based on QOSPF that the LMPS scheme provides good performance.

1 Introduction

Most QoS routing approaches require network routers to periodically exchange link state information in order to obtain a global view of the network state. As network resource availability changes continuously with traffic dynamics, maintaining accurate network state information for provisioning of acceptable service quality requires frequent information exchange in the network. The fast change in network state and the associated regular state updates could lead to large amount of signaling traffic especially in large scale networks. Furthermore, the core routers, whose main function is to forward packets as fast as possible, could be overburdened by tasks such as collecting and exchanging the information that is usually needed to perform quality-of-service (QoS) routing. Also, the network QoS state information maintained by the network nodes can become out-of-date if the QoS state update interval is large rela-

N. Mitrou et al. (Eds.): NETWORKING 2004, LNCS 3042, pp. 1072–1083, 2004.

tive to the time scale of traffic dynamics. Furthermore, path selection based on short-est path algorithms such as Dijkstra's algorithm does not seem to be well-justified in the presence of expired state information. Another well-known problem is path oscil-lations which may occur when a node discovers that a certain link is congested and after the next route selection will avoid routing any traffic on these links resulting in their under-utilization. The links then seem attractive again and re-selected by the routing algorithm. This oscillatory behavior can have a negative impact on the system performance.

As an alternative to the global QoS routing schemes, localized approaches [2,7,8,10] were introduced to maintain QoS routing without exchanging global state information. In these approaches, nodes infer the network QoS state using locally collected information based on flow blocking statistics, and perform flow routing using this localized view of the network QoS state. The localized QoS routing ap-proaches have several advantages. The communication overhead involved is minimal, with no global information exchange. Also the core routers are not involved with interpreting QoS signaling messages and do not need to keep or update any QoS state database necessary for global QoS routing.

Nelakuditi, and Zhang [2] proposed an adaptive proportional QoS routing method which makes local routing decisions based on locally collected QoS state information (with no global network information exchange among routers). They compared the performance of these localized routing methods with that of a global optimal propor-tional model that has knowledge of the global network QoS state. They demonstrated that an appropriately designed localized QoS routing scheme that makes routing deci-sions solely based on local information can potentially achieve such global objectives as a stable network system with good overall system performance.

Extensions to support QoS routing based on OSPF have been proposed in the QOSPF [1] protocol. The QOSPF is considered as an extension to OSPF to support QoS by flooding the network with information about the available and used link re-sources. In this proposal the routing decisions are being done based on topology, link resources available and traffic requirements. The QOSPF framework signals ingress routers to send the QoS requirements for incoming traffic. If a QoS route can be computed and a path reserved, a message is sent back, reserving the resource and accepting the request. In [3] Lakkakorpi describes the extensions suggested to the OSPF protocol to support QoS routes. An overview for QoS routing and examples of source routing protocols are discussed in [4]. More information about QoS extension to OSPF can be found in [1], [5] and [6]

In this paper, we propose the Localized Multi-Paths Selection (LMPS) technique to overcome the common drawbacks of the localized approaches like the inflexibility in bandwidth assignment on some paths and the inaccurate traffic parameters calcula-tions. We apply LMPS to VoIP networks where all VoIP calls have the same band-width. Information about the used paths are exchanged between the ingress/egress nodes without involving the core routers. We can classify LMPS to lie somewhere in between the global link-state QoS schemes and the localized fully protocols. The simulations indicate that the LMPS is capable of achieving good performance com-pared to the mentioned localized protocols.

2 System Model

We consider a carrier network for transporting of VoIP calls as depicted in Fig. 1. Ingress/egress IP routers (abbreviated as IE nodes or routers) are connected to a traditional PSTN-based local exchanges and/or PBXs. Core routers are used to interconnect the IE routers. The IE routers contain the necessary functionality to process the PSTN signaling and format conversion. All PSTN calls can be optionally transcoded at the IE nodes for bandwidth reduction.

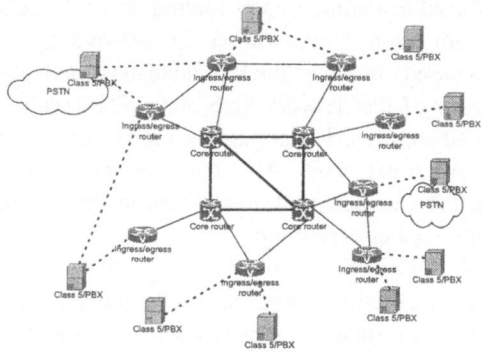

Fig. 1. The system model

A link-state protocol such as OSPF is employed within the network. It is used to collect information about network state and topology updates. All techniques discussed here use the information collected by the link-state protocol to select candidate paths.

The offered load for a generic source-destination (SD) IE node pair is given by V = $\lambda \times$ AHT, where λ presents the call arrival rate (calls/sec) for the SD pair, and AHT is the average holding time for the calls. Let the capacity of a generic link be C_l bits/sec and let B be the required bandwidth for (an optionally transcoded) phone call in bits/sec.

The objective is to route calls for each SD pair in the most efficient and scalable manner possible. More precisely, it is required to reduce the blocking probability for establishing a new call and to maximize the potential revenue and network utilization. While such a network can be used in a multiservice setting where VoIP calls are mixed with other traffic types such as best-effort data-oriented traffic, out concentration will be a VoIP dedicated network or where a fixed portion of bandwidth is allocated for VoIP calls.

3 Localized Quality of Service Routing Approaches

The localized QoS routing techniques of [2,8] were developed based on the concept of the "*virtual capacity*". This concept takes into account that typically the paths between an SD pair could traverse links shared with other SD pairs. A bottleneck link of a path is defined as the link offering the path the minimum (virtual) capacity. Furthermore, as traffic patterns across a network change, the bottleneck link of a path and its capacity could also change. The virtual capacity is a function of the offered load to the links belonging to the path and the perceived blocking probability at the IE node. Assume that a path r between an SD pair carries offered load v_r, and the corresponding blocking probability observed by the IE node is b_r. The blocking probability on this path can be presented using the Erlang's loss formula as function of both the path load v_r and the capacity c_r as follows:

$$b_r = E(v_r, c_r) \tag{1}$$

If the quantities v_r and b_r are measured, an estimate of the virtual capacity of the path vc_r can be obtained by the inverse of the Erlang function:

$$vc_r = E^{-1}(v_r, b_r) \tag{2}$$

If m paths $(r_1, r_2, \ldots, r_i, \ldots, r_m)$ share a link with capacity c, then the virtual capacity vc_i of path r_i represents its "capacity share" of the link. Let v_i denote the offered load on the link from path r_i. The blocking probability on the link is given by $b = E(\sum_{i=1}^{m} v_i, c)$. In the case when all flows have the same bandwidth requirement, and the equalizing of the blocking probability for all paths sharing a given link is a target, the virtual capacity of path r_i is given by:

$$vc_i = E^{-1}(v_i, b_i) = E^{-1}(v_i, b) \tag{3}$$

where, b_i denotes the observed blocking probability of path r_i. We can see from (3) that for path r_i the larger the offered load v_i is, the larger is its virtual capacity vc_i.

The virtual capacity routing (VCR) model [2,8] is based on load balancing across multiple paths (it is to be noted that the selected paths are fixed). Initially, a set of paths are identified using a route selection scheme such as shortest-hop algorithm. In the VCR scheme, the IE nodes estimate a certain virtual capacity for each path as a private share in the shared links traversed by the path. Let the overall offered traffic to SD pair (s,d) be V, and let there be m available paths between (s,d), then the load balancing problem is finding the set of flow proportions $\{\alpha_i\}_{i=1}^{m}$, with the constraint $\sum_{i=1}^{m} \alpha_i = 1$. The offered load to path i is then given by $\alpha_i V$. The values $\{\alpha_i\}_{i=1}^{m}$ are dynamically updated to shift traffic away from paths observing high blocking. The algorithm uses the equalizing blocking probability (EBP) strategy to simplify and solve the minimization problem of the blocking probability of the paths.

The VCR algorithm depends on an observation period to measure the quantities of interest. The observation period is divided into two intervals. In the first interval (blocking observation interval) the IE nodes collect local statistics regarding the blocking on each path, to be able to adapt its traffic parameters to minimize the blocking probability on this path. The adaptation is being done during the second interval (traffic parameters recalculation interval), where each IE node recalculates some traffic parameters to be able to minimize the blocking probability on this path during the next cycle. More details about QoS routing without exchanging global information can be found in [7], and about the VCR model in [2,8].

4   The Localized Multipath Selection (LMPS) Scheme

One of the drawbacks of localized techniques of [2,8] is that the set of multiple paths is fixed. If the bottleneck links of the paths are overloaded, then there is not much that could be done to reduce the blocking as shifting traffic may also lead to a path with high blocking probability. Another drawback is the possibility of reaching inaccurate decisions due to errors in the blocking observation and load proportions calculations.

The LMPS technique is based on identifying the currently best possible set of paths satisfying the bandwidth and other QoS constraints to be used by each SD pair, in order to minimize blocking probability on each path. In LMPS, network state information is collected and exchanged only by the IE nodes on an on-demand basis when replacement of current paths is needed. This is done in conjunction with the localized approach. This hybrid of global and local information helps to compensate the drawbacks of the traditional localized technique and lead to better QoS performance without the common drawbacks of the global state protocols (such as QOSPF).

The observation mechanism: The observation process in the LMPS technique depends on the observation period T_g. The period T_g is the time interval between two successive instants at which a node examines its paths. The period T_g consists of two intervals: the path checking interval T_{pc} recalculation interval T_{rc}. During and period T_{pc} each node checks the status of its paths, and may decide to create new ones. It then sends updates to all other IE nodes regarding its new set of paths. The other nodes update their conceived network state according to the updates. The above process occurs once during T_g (within T_{pc}) at any time, and it does not matter if every node executes it in different time therefore storms of signaling traffic are not likely to occur. The second interval is the traffic parameters recalculation interval T_{rc} where all IE nodes are required to start the second time interval at approximately the same time, and exit it also at the same time. The interval T_{rc} is relatively small (few seconds), and within it each IE node recalculates its new paths parameters according to the new set of paths selected by each IE node and the observed blocking of its paths. The path updates tables will be discussed later in details.

The LMPS algorithm: Each IE node picks m (typically $m=3$) paths satisfying the fixed QoS constraints (propagation delay, reliability, ..) to each destination from the network topology information collected by the link-state protocol. Each IE node broadcasts its own set of paths and their respective links to other IE nodes. An IE node also keeps a table of its own paths and the paths of other IE nodes. Using this information, each IE node can identify how many paths share a specific link and then determines its own capacity share in each physical link, and the most shared link (bottleneck link) in each path. If the number of paths using link l is given by N_l, then the link with the smallest C_l/BN_l is the bottleneck link. Here, the bottleneck is defined as function of number of paths rather than load, since the node exchange the path information with the assumption that links are fairly shared among the paths.

After each IE node determines the bottleneck link(s) of its own paths, it calculates the virtual capacity of each path. The virtual capacity of a path in LMPS is given by the maximum number of calls that can be used by the path and is equal to C_l/BN_l at the bottleneck link. This is based on the assumption that traffic between all SD pairs is uniform and all calls request identical amount of bandwidth. However, in the case of non-uniform traffic or call bandwidths, the virtual capacity of each path should be weighted by the service provider as part of a service level agreement (SLA). LMPS restricts the number of active calls on any path to be less than or equal to its virtual capacity. In contrast to the VCR approach, we note that there is no need to calculate the virtual capacity using inverse Erlang formula. The IE nodes use the resulting virtual capacity to calculate the related path proportions α_i, for each path. If we have m active paths with LMPS virtual capacities $\{\overline{vc_i}\}_{i=1}^{m}$, then the flow proportions are simply given by $\alpha_i = \overline{vc_i} / \sum_{i=1}^{m} \overline{vc_i}, i = 1, \cdots, m$. The offered load to path i is then $v_i = V \times \alpha_i$, where V is total offered load to a given SD pair. A counter C_i records the number of calls routed on path i.

At the start of the period T_{pc}, each IE node checks for paths satisfying the condition $C_i \quad \psi \times vc_i$, where ψ is configurable ratio with default value one. However, the ratio ψ may be set to a value less than one (e.g. 0.8) as a protective procedure to discover the blocked paths before the blocking occurs. If blocked paths are found, the node marks the bottleneck links of these paths as forbidden links, and selects as many new paths as needed to replace the blocked paths with the constraint of avoiding the forbidden links. After identifying the new set of paths, the nodes with newly created paths broadcast their new paths information to all other IE nodes. In the mean time, the IE nodes receive information from other nodes about their paths. By the end of period T_{pc}, all nodes assume no further updates will be broadcast and they start calculating the parameters of the new paths (virtual capacity, flow proportions, flow load, path counter initial value...) to be able to route calls via these new paths. It is so important that all IE nodes start the interval T_{rc} and leave it at the same time (at least with reasonable time differences) to start using the new paths simultaneously. It is very important to note that currently existing calls are kept on whatever paths are chosen and new routes are used by the new incoming calls, i.e. no rerouting in involved. The core routers have no role to play in this process (except forwarding the messages between the IE nodes), thereby simplifying the design of such routers and reducing the processing overhead.

LMPS signaling information: LMPS is based on exchanging some global information between the IE nodes to achieve reliable QoS performance. This global information is considered as signaling traffic, and can be classified as: 1) path updates messages which are sent by an IE node to all other IE nodes to announce a new paths set; 2) acknowledgment messages with which an IE node confirms reception of a path update message; 3) link state requests: when an IE node changes its paths, it needs to know the current load on each new path, so it sends request to the routers located on the new path to get this information; and 4) link state update: as a response for the (link state request), each node sends an information about single physical link to the requesting IE node.

The amount of signaling information in LMPS (during a fixed period) is a function of factors such as the number of IE nodes in the network, the number of blocked paths reported every T_g , the average path length in the given topology, and the number of parallel paths used for each SD pair. The OSPF protocol can be tailored to transport the signaling information used by LMPS. However, the signaling traffic in the LMPS technique is not periodic and is only generated when an IE node needs to change its paths.

Time synchronization: In the LMPS technique, IE nodes are required to maintain time synchronization among themselves to be able to proceed in the time interval T_{rc} simultaneously, or at least with acceptable reasonable time differences. The worst case performance occurs when T_{rc} intervals from different IE nodes are completely unsynchronized. This causes an IE node to repeat its traffic parameters calculations several times during the period T_g due to the frequent changes. This problem can be solved by using network time synchronization protocol like the Simple Network Time Protocol sNTP [9].

5   Performance Evaluation

Since the focus in this paper is routing of VoIP calls, the blocking probability is considered as the main metric to evaluate the performance of the different routing schemes. The Network Simulator NS-2 is used to simulate the mentioned models. NS-2 is a discrete-event packet level multi-protocol simulator.

The voice CODEC G.729 (voice rate 8 kb/s voice) is used to generate voice traffic during the simulation. The call bandwidth is 14.8 kbps after using the RTP header compression (cRTP). Two network topologies are used during the simulation. The first one called NSFNET (Fig. 2a), with 14 nodes, 21 links, average node connectivity degree = 3, and five IE nodes. The nodes (0, 1, 2, 3 and 4) are selected to act as IE nodes (the choice is mainly for evaluation purposes not for specific reasons with these nodes). The second topology is called LCTNET (Fig. 2b), with 12 nodes, 13 links, average node connectivity degree = 2.167, and 4 IE nodes. The nodes (0, 1, 2 and 3) are selected to act as IE nodes. We select the NS-2 class based queuing (CBQ) as the queue management technique, and the weighted round robin (WRR) queuing discipline. A constraint of 250 msec is assumed on the end-to-end delay (including both propagation and processing delays). Three parallel paths are used to route calls for each SD pair. For the two network topologies, we set all links capacities to be equal to two Mbps (unchannelized E1).

The simulation considers using two levels of call generation rates (λ_1, λ_2). The rate λ_1 presents the normal call arrival rate for most SD pairs in the network, while the rate λ_2 presents the arrival rate for the hot pairs in the network (higher rates). The factor β presents the degree of the non-uniformity in the network $\beta = \dfrac{x\lambda_2}{x\lambda_2 + (1-x)\lambda_1}$ where x is the percentage of hot pairs, and $(1-x)$ is the percentage of normal pairs from the total number of SD pairs in the network. The average call arrival rate for each SD pair is presented by λ given by: $\lambda = \dfrac{\lambda_T}{n(n-1)} = (1-x)\lambda_1 + x\lambda_2$, where n is number of IE nodes, and $n(n-1)$ is number of SD pairs. λ_T presents the total rate of call generation in the network (by all IE nodes) every second: $\lambda_T = n(n-1)(1-x)\lambda_1 + n(n-1)x\lambda_2$. The rate of the normal pair can be expressed as:

$$\lambda_1 = (1-\beta)\lambda_T / n(n-1)(1-x) = (1-\beta)\lambda /(1-x) \tag{4}$$

and of the hot pair as:

$$\lambda_2 = \beta\lambda_T / n(n-1)x = \beta\lambda / x \tag{5}$$

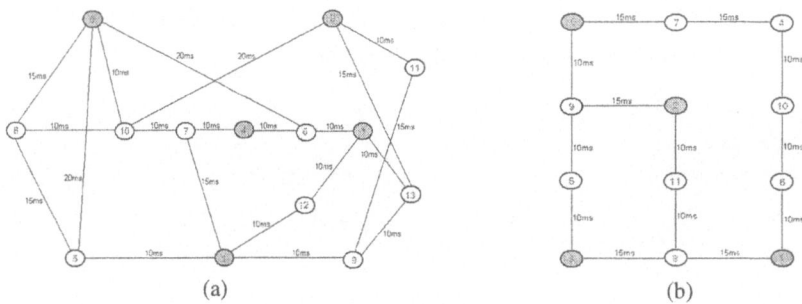

(a)   (b)

Fig. 2. The network topologies used in the experiments

The simulation compares three models to evaluate the QoS routing performance. The first model is the LMPS routing algorithm, the second is the pure localized approach multi-path VCR, and the third model is the multipath selection based on SWP and QOSPF protocol signaling (explained below). All algorithms are evaluated for the case of three parallel paths for each SD pair. All models consider two QoS constraints, the bandwidth availability and end-to-end delay restrictions. Paths with delay exceeds 250 msec are rejected.

5.1   Multipath Routing with Global QoS Information

To assess the performance of LMPS against routing protocols with global QoS information, we utilize a multipath routing model with QOSPF-based signaling. The
model works by selecting m paths for each SD pair using a shortest-path algorithm.
These paths are fixed. When a call arrives, the shortest-widest path (SWP) policy is
used to select the best path (from the m paths) to establish the incoming call. The
QOSPF updates are sent regularly every T_{upd}, to report the changes in the capacity
status on each link in the network. In QOSPF, there are two types of signaling messages exchanged, the RES-LSA messages, and the Acknowledgment messages. The
amount of signaling information in QOSPF (during fixed period) is being affected by
some factors like the update period T_{upd}, and the connectivity degree of each node.

5.2   Performance with Uniform Traffic Loading

Fig. 3 and Fig. 4 compare the blocking behavior for the three models using the
NSFNET and LCTNET topologies respectively. The common parameter is $\beta = 0.333$.
The graphs show that the LMPS model has better performance than the traditional
VCR, and less than the performance of the QOSPF model. The QOSPF-based multipath SWP has the best performance (if used with tight update periods), but costs
other disadvantages that were discussed earlier. In case of the LCTNET, the LMPS
model has performance closer to the VCR performance. The reason is that the topology of the LCTNET (low node connectivity degree) doesn't offer many alternative
paths to recover from blocked paths when they occur.

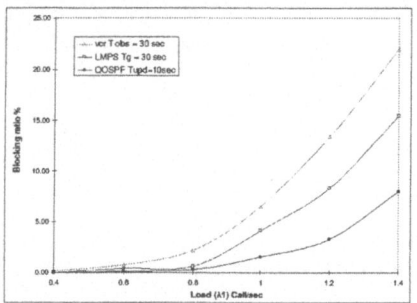
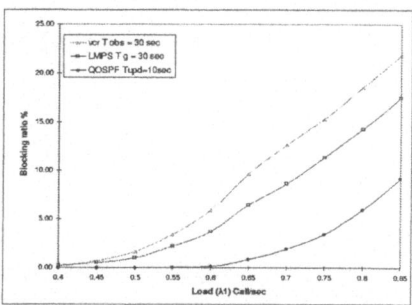

Fig. 3. The blocking behavior (NSFNET)   **Fig. 4.** The blocking behavior (LCTNET)

5.3   Performance with Non-uniform Traffic Loading

The factor β reflects the non-uniformity of the traffic in the network. As β increases,
the network suffers from hotspots (paths with hot traffic load λ_2), while the total load
in the network is always constant. The percentage x (percentage of hot pairs) was
configured to be 20% of the SD pairs in the network (hot pairs with rate λ_2), while the

remaining pairs (80%) carry normal load with rate λ_1. For each value of β, the values of λ_1 and λ_2 are adjusted using equations (4) and (5) while keeping the value of the average and total loads respectively λ and λ_T constant. We set $\lambda=0.96$ calls/sec while changing the value of β and show the result in Fig. 5. It shows that changing the traffic uniformity and distribution affects the overall blocking ratio directly. The most affected model as is the VCR model. The QOSPF-based multipath SWP and the LMPS models are more capable of coping with the hot spots occurring on some the links.

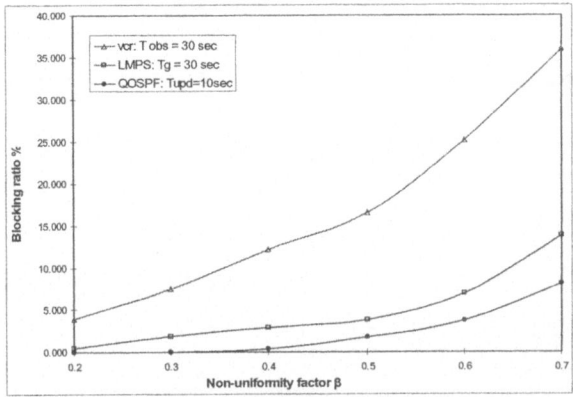

Fig. 5. Effect of changing the non-uniformity factor β

5.4 Network Utilization and Link Over-Utilization

In the utilization calculation, the number of hops used by each call is being considered. The network utilizations for three models are compared using the NSFNET topology with $\beta = 0.333$. Fig. 6 shows that the traditional localized proportional approach (VCR) has the highest network utilization values compared to the LMPS and the QOSPF models. As an indication for the algorithm efficiency, the load of the links is monitored during the simulation to record any overflow. The overflow indicates that some calls are accepted on certain physical links although that no physical capacity is available on these links. The *over utilization* is not accepted in networks with QoS guarantees as this reflect the incapability of the network to provide guaranteed QoS. Fig. 7 shows the maximum overflow for each model during its operation. The overflow load is being monitored against the call arrival rate λ_1 with fixed non-uniform factor $\beta = 0.333$ at the LCTNET network with 2Mbps links (135 calls as integer capacity). The dashed horizontal line (at 135 calls) indicates the restriction on each physical link (maximum capacity 135 calls). The relatively long update period (T_{upd}) in the QOSPF model result in accepting calls while no capacity available in some links. LMPS (with its capability to change paths) does not allow the existence of the blocked paths, hence reduces the overflow on physical links.

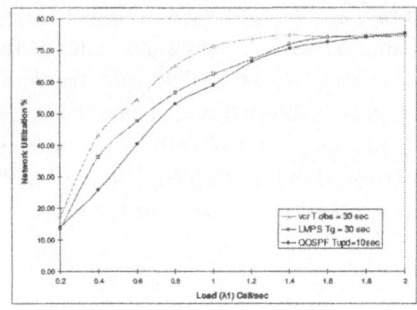

Fig. 6. Network utilization

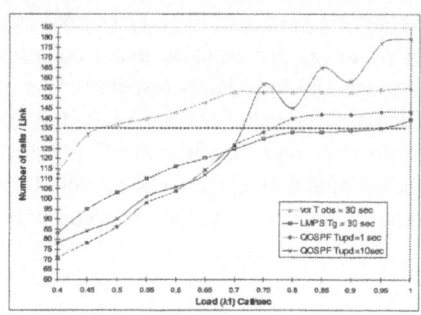

Fig. 7. The link over-utilization

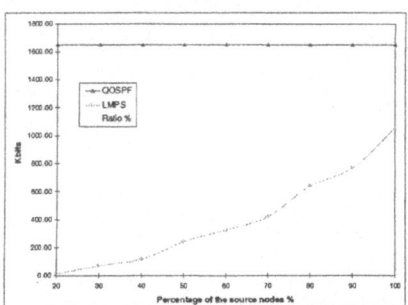

Fig. 8. The effect of changing the percentage of source nodes in the network

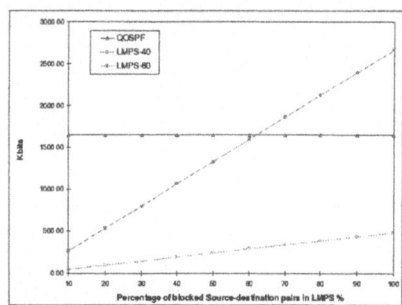

Fig. 9. The effect of changing the percentage of the blocked source-destination pairs in LMPS model

5.5 Signaling Overhead

The amount of the signaling information exchanged by the nodes in the network reflects the efficiency of the QoS algorithm and its capability to save the network resources. In this section, a comparison between both QOSPF and LMPS algorithms shows the amount of signaling information exchanged by each one. Increasing the percentage of the IE nodes in the network affect the amount of signaling information exchanged by the LMPS model, where the signaling information is being exchanged only between the IE nodes. In QOSPF the signaling information is being exchanged between all nodes, hence increasing the IE nodes in the network does not affect the QOSPF signaling information directly. This is illustrated in Fig. 8.

Another important factor to consider in the LMPS model is the percentage of SD pairs suffering from blocked paths during a single update period. This affects the amount of path updates exchanged by the IE nodes. Fig. 9 shows the effect of changing the percentage of SD pairs with blocked paths. The LMPS-40 and LMPS-80 curves represent the cases when 40% respectively 80% of the nodes in the NSFNET are IE nodes.

6 Conclusions

We proposed LMPS as a multipath routing protocol that uses local decisions to load balance traffic across multiple paths. It uses global information that approximates the network state and is only exchanges between the IE nodes and only triggered when saturated paths are identified. The LMPS approach proved superior to fully localized approaches such as (VCR). It should be noticed that the operation of the LMPS approach depends strongly on the availability alternate paths between the IE nodes in the network. Using the LMPS approach in networks with low connectivity (nodes have low connectivity degrees), and as a result a small number of paths between IE nodes exist, is not advisable. In the worst cases, the LMPS will act similar to the traditional localized techniques (e.g. the adaptive proportional approach). The best topologies are when a plethora disjoint paths for each SD pair exist.

Further evidence of LMPS stability and susceptibility to path oscillation is needed. Also, packet-level performance and assessment of typical signaling information overhead should be investigated. Another important area for investigation is effect pf exchanging actual path load between the IE nodes rather in addition to the set of links comprising the path.

References

1.   G. Apostolopoulos et al., RFC 2676 QoS Routing Mechanisms and OSPF Extensions.
2.   S. Nelakuditi, Z.-L. Zhang, R.P. Tsang, Adaptive Proportional Routing: A Localized QoS Routing Approach, Proc. of IEEE INFOCOM , pp. 1566-1575, 2000.
3.   J. Lakkakorpi, QoS Routing Extensions to OSPF, technical report Helsinki University of Technology, available from http://keskus.hut.fi/tutkimus/ipana/paperit/, 2000.
4.   S. Chen and K. Nahrstedt, An Overview of Quality of Service Routing for Next-Generation High-Speed Networks: Problems and Solutions, IEEE Network, Vol. 12, pp. 64 -79, Nov./Dec. 1998.
5.   G. Apostolopoulos, R. Guerin and S. Kamat, Implementation and Performance Measurements of QoS Routing Extensions to OSPF, Proc. IEEE INFOCOM, pages 680-688, 1999.
6.   Z. Zhang, C. Sanchez, B. Salkewicz, E. Crawley, QoS Extension to OSPF, IETF Internet draft, Sep 1997.
7.   S. Nelakuditi, R. P. Tsang, and Z.-L. Zhang, Quality-of-Service Routing without Global Information Exchange, Proc. of IWQOS 1999.
8.   S. Nelakuditi, S. Varadarajan, and Z.-L. Zhang, On Localized Control in Quality-of-Service Routing, IEEE Transactions on Automatic Control, Systems and control Methods for Communication Networks, June 2002.
9.   D. L. Mills, RFC 1769 Simple Network Time Protocol V3.
10. I. Gojmerac, T. Ziegler, P. Reichl: Adaptive Multipath Routing Based on Localized Distribution of Link Load Information. Proc. QofIS'03, Stockholm, Oct. 2003.

Link-Based Fair Aggregation: A Simple Approach to Scalable Support of Per-Flow Service Guarantees

Yuming Jiang

Center for Quantifiable QoS in Communication Systems (Q2S)
Norwegian University of Science and Technology (NTNU)
ymjiang@ieee.org

Abstract. To support service guarantees in packet-switched networks, three approaches have been proposed. They are the Stateless Core (SCORE) approach, the Integrated Services (IntServ) approach, and the Differentiated Services (DiffServ) approach. The granularities of service guarantees provided by these approaches at each router are respectively *packet level*, *flow level*, and *class level*. In this paper, we propose a novel approach, called *Link-Based Fair Aggregation* (LBFA) approach to scalable support of service guarantees. While the granularity of service guarantees supported by LBFA is *link level* at each router, we show through analysis that the proposed LBFA approach can achieve as good as or even better *per-flow* service guarantees than the current three approaches.

1 Introduction

The Internet was initially designed to provide one simple service: best-effort datagram delivery. Such a design allows routers to be *stateless* and to forward packets in a First-In-First-Out (FIFO) manner. As a consequence, today's Internet is highly *scalable* in the sense that router complexity does not increase with the number of flows in the network. However, with the development and deployment of multimedia and network technologies, multimedia has become an indispensable feature of the Internet. Unlike traditional applications such as file transfer, many multimedia applications such as Internet telephony are delay-sensitive. Thus, there is a demand for introducing a service in the Internet with which both bandwidth and delay guarantees are provided.

Chronologically, three approaches have been proposed in the literature to provide such services in addition to best-effort service, which are the Integrated Services (IntServ) approach [4], Differentiated Services (DiffServ) approach [3], and Stateless Core (SCORE) approach [20]. Specifically, the Guaranteed service [19] in IntServ, the Expedited Forwarding service [7] in DiffServ, and the feature of providing guaranteed services in SCORE [20] are for this purpose. These approaches have important differences in achieving these services. In particular, while the IntServ approach can provide *end-to-end flow level* service guarantees, it is *stateful* in the sense that every router needs to maintain per-flow states.

N. Mitrou et al. (Eds.): NETWORKING 2004, LNCS 3042, pp. 1084–1095, 2004.

The DiffServ approach is *core-stateless* since per-flow states are only maintained at edge routers. However, service guarantees under the DiffServ approach are provided only to aggregate *class level* and such guarantees are mainly defined for the *per-hop case*. Additional effort is needed to make DiffServ support end-to-end per-flow service guarantees. Like the DiffServ approach, the SCORE approach is also *core-stateless* in providing scalable support of guaranteed services, but it achieves this in a different way by letting each packet carry *packet state* and each router forward the packet by a deadline calculated based on the carried state. Hence, *packet level* service guarantees at each router and consequently end-to-end per-flow service guarantees are provided by the SCORE approach.

In this paper, we propose a novel approach, the Link-Based Fair Aggregation (LBFA) approach, to provide scalable support of per-flow service guarantees. This approach is inspired by a natural phenomenon in the network: a flow is aggregated with other flows on each link along its transit path. The idea behind LBFA comes from the conjecture that if a flow is properly aggregated with other flows of the same traffic class on each link along the path, it is possible to preserve service guarantees end-to-end to the flow by aggregating the corresponding aggregates on each link properly and providing service guarantees to the formed *link level* aggregate at each router along the path.

In the proposed LBFA approach, except for a couple of fixed or pre-configured parameters associated with each router, no per-flow information is maintained in the core as is the case for the core-stateless DiffServ and SCORE approaches. In addition, as opposed to the SCORE and DiffServ approaches, the LBFA approach does not mandate maintaining per-flow states at edge routers. Hence, the LBFA approach can be *stateless*. In addition, as opposed to the SCORE approach, no additional information needs to be added to each packet in the LBFA approach. In the LBFA approach, service guarantees at each router are provided to the aggregate of flows of the same traffic class from the same incoming link. In this sense, the LBFA approach is said to provide *link level* service guarantees at the router. In the paper, we show through analysis that end-to-end per-flow service guarantees can be provided if LBFA is implemented in the network. These guarantees are as good as or even better than those provided by the three existing approaches.

2   Network Model and Scheduling Model

2.1   Network Model

We consider a single multi-service network domain with feedforward routing. Routers are building blocks of the network. In the network, flows are partitioned based on the service classes that they belong to. The service discipline at each output port of a router allocates resources of the corresponding output link among different service classes in a link-sharing manner [9]. For each scheduler at the scheduling hierarchy of such link-sharing [9], its buffer, if there is any, is assumed to operate in FIFO manner and its size is large enough to ensure no packet drop.

When entering the network, every flow is shaped at the edge before releasing into the network. Packets of the flow are transmitted in the network along a single path, which is modeled as a list of link servers. These packets traverses the path in the FIFO order so that the ordering of packets in the flow is preserved at every router along the path. In this paper, we are interested in the end-to-end service guarantees provided to the flow.

All routers in the network are assumed to be output-buffered and implement aggregate class-based scheduling. Fig. 1 depicts the architecture of a typical router. The router possibly has multiple input links and multiple output links. For each packet that arrives on an input link, the router determines the next hop on its path and transmits the packet on the corresponding output link. At each output link, a certain percentage of bandwidth is reserved for each traffic class according to some link-sharing policy. Throughout the rest of the paper, we use *link server* to represent the whole link-sharing scheduling hierarchy. Specifically, the link server includes both the class-based flow aggregation part and the class-based scheduling part shown in Fig. 1.

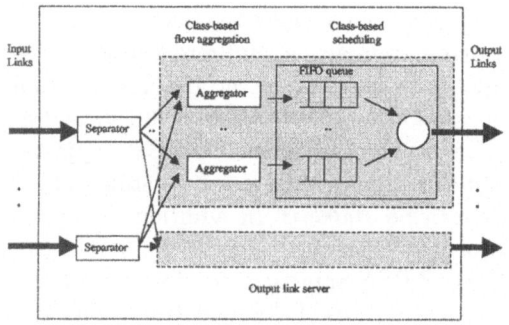

Fig. 1. Architecture of a typical router

Consider a flow f of a certain traffic class c. Let $\mathcal{F}_s$ denote the set of all class c flows constituting the class c aggregate on link server s along the path of flow f; I_s the number of incoming links to the router with output link s; R_s the reserved bandwidth for class c traffic on link s. Assume f is shaped before entering the network to conform to a token bucket with parameters (r^f, σ^f) where r^f is the token arrival rate and σ^f is the bucket size.

We further make the following assumptions for later analysis in the paper. 1) For ease of exposition, the delay experienced by a packet at the edge traffic shaper and the propagation delay are excluded from the end-to-end delay. 2) The amount of class c traffic on any link s does not exceed a certain ratio $\alpha_s (\leq 1)$ of the reserved rate R_s. Specifically, we require that for any link s in the network $\sum_{f \in \mathcal{F}_s} r^f \leq \alpha_s R_s$. 3) For any link s, let $\beta_s = \frac{1}{R_s} \sum_{f \in \mathcal{F}_s} \sigma^f$ and β be a bound on β_s. 4) Any flow in the network is assumed to traverse at most H hops. In other words, the network diameter is H. 5) There exists a bound on the size of any packet in the network, which is denoted by L. 6) No packet delay is introduced by traffic separator. 7) Finally, we adopt the convention that a packet has been received/transmitted if and only if its last bit has been received/transmitted.

2.2   Scheduling Model

Each scheduler is assumed to be Guaranteed Rate (GR) server [10], which includes the class-based scheduler in Fig. 1 and the corresponding link server.

GR server is defined based on the *guaranteed rate clock* (GRC) value of a packet [10]. Let $p^{f,j}$ be the jth packet of flow f, $l^{f,j}$ be its length, and r_s^f be the bandwidth allocated to the flow. Then, the GRC value for packet $p^{f,j}$ at server s, denoted by $F_s^{f,j}$, is iteratively defined as:

$$F_s^{f,j} = max\{a_s^{f,j}, F_s^{f,j-1}\} + \frac{l^{f,j}}{r_s^f}, \tag{1}$$

where $F_s^{f,0} = 0$, and $a_s^{f,j}$ denotes the arrival time of packet $p^{f,j}$ to the server.

Similar to $F_s^{f,j}$, we define a virtual time function for flow f and denote it by $F^{f,j}$ which is iteratively obtained by replacing r_s^f with r^f as:

$$F^{f,j} = max\{a_1^{f,j}, F^{f,j-1}\} + \frac{l^{f,j}}{r^f}. \tag{2}$$

The difference between $F_s^{f,j}$ and $F^{f,j}$ is that while the former is server-dependent, the latter is server-independent. In fact, if we view the whole network as a blackbox, then $F^{f,j}$ is the GRC for the blackbox. In this sense, $F^{f,j}$ can be considered as the end-to-end GRC function for the flow f across the network.

A server s is said to be GR server to flow f with rate r_s^f and error term E_s^f, iff it guarantees that any packet $p^{f,j}$ of the flow is transmitted by [10]

$$d_s^{f,j} \leq F_s^{f,j} + E_s^f \tag{3}$$

where E_s^f is a constant that depends on the scheduling algorithm.

Similarly, we say the network provides *per-domain rate guarantee* to flow f with rate r^f and error term E^f, iff for any packet $p^{f,j}$, it guarantees that

$$d_H^{f,j} \leq F^{f,j} + E^f \tag{4}$$

where E^f is a constant that depends on the link servers along its path and $d_H^{f,j}$ denotes the departure time of packet $p^{f,j}$ from the network.

For every GR scheduling algorithm, a corresponding *Core-Stateless GR* (CSGR) algorithm can be defined [15] [16]. In such a CSGR algorithm, it assigns the following *GRCore* values to packets of any flow f and schedules packets in the increasing order of their *GRCore* values at each link server s:

$$GRCore_1^{f,j} = F_1^{f,j}, \quad and, \tag{5}$$

$$GRCore_{s+1}^{f,j} = GRCore_s^{f,j} + \frac{L}{r^f} + E_{s+1}^f, \tag{6}$$

for $s \geq 1$. Conversely, the GR scheduler is said to be the corresponding GR scheduler of the CSGR scheduler [15] [16].

3   Current Approaches

3.1   The IntServ Approach

In the IntServ approach [4], service guarantees are provided end-to-end on a per-flow basis. In order to do so, this approach uses end-to-end signaling to set up flow classification and reservation states on each router along the path. Usually, the router implements per-flow scheduling for resource allocation among flows. In particular, per-flow fair queueing (PFFQ) is commonly adopted by the router for the class-based flow aggregation part shown in Fig. 1. We call the resulted IntServ *PFFQ-based IntServ*. The granularity of service guarantees provided by the PFFQ-based IntServ approach at each router is hence *flow level*.

The following Theorems 1 to 3 summarize the per-flow service guarantees provided by the PFFQ-based IntServ approach. Due to space limitation, their proofs are omitted and can be found from [14].

Theorem 1. [Rate Guarantee] *If in a network, the link server at each router s along the path of a flow f belongs to GR with rate $r_s^f(\geq r^f)$ and error term E_s^f, then the network guarantees that*

$$d_H^{f,j} \leq F^{f,j} + \sum_{s=1}^{H-1} \frac{L}{r_s^f} + \sum_{s=1}^{H} E_s^f \qquad (7)$$

where H is the number of routers on the path.

Theorem 2. [Bounded Delay] *Under the same condition as Theorem 1, the end-to-end delay of any packet in the flow is bounded by:*

$$D^f = \frac{\sigma^f}{r^f} + \sum_{s=1}^{H-1} \frac{L}{r_s^f} + \sum_{s=1}^{H} E_s^f. \qquad (8)$$

Theorem 3. [Throughput Guarantee] *Under the same condition as Theorem 1, if the source of flow f transmits packets at least at rate r^f, the network guarantees to the flow*

$$W^f(t_1, t_2) \geq r^f \left(t_2 - t_1 - \sum_{s=1}^{H} \frac{L}{r^f} - \sum_{s=1}^{H} E_s^f \right)^+, \qquad (9)$$

where $W^f(t_1, t_2)$ denotes the work done by the network to the flow in any interval $[t_1, t_2]$ with $0 \leq t_1 \leq t_2$, and $(x)^+ \equiv max\{0, x\}$.

3.2   The DiffServ Approach

In the DiffServ approach [3], packets of flows are classified into a small fixed number of classes, such as the Expedited Forwarding (EF) class [7] and Assured Forwarding (AF) groups [11]. Complex per-flow classification is implemented only at edge routers. In the core, routers provide service guarantees only on a class basis rather than a per-flow basis. The granularity of such a service guarantee at each router is hence *class level*. Since only edge routers need to maintain per-flow states and core routers do not, the DiffServ approach is *core-stateless*.

As oppose to end-to-end service guarantees for individual flows in the IntServ approach, the current DiffServ mainly supports per-hop service guarantees, i.e. Per-Hop Behaviors (PHBs), to each traffic class aggregate. The extension of per-hop guarantees to per-domain or end-to-end guarantees is still an undergoing work [13] [18].

To date, a lot of DiffServ PHBs implementations have emerged. A typical implementation is to aggregate packets of the same class in a single FIFO queue and service them in the order of their arrival times. In such implementations, the aggregator shown in Fig. 1 is actually virtual, since the buffer of the class-based scheduler has been assumed to operate in the FIFO manner. Clearly, FIFO aggregation results in a very simple implementation of DiffServ. Throughout the rest of the paper, we shall focus on the DiffServ approach with FIFO aggregation and call it *FIFO-based DiffServ*. In the following, we present per-flow service guarantees provided by the FIFO-based DiffServ approach.

As discussed above, the link-server of a router under FIFO-based DiffServ is indeed the class-based scheduling part shown in Fig. 1. Assume that each link-server s guarantees rate R_s to the corresponding class aggregate of flow f with error term E_s. Then, we have the following theorem which presents the bounded delay guarantee supported by the FIFO-based DiffServ approach. It has been proved in a previous work [12].

Theorem 4. [Bounded Delay] *If the following condition on link utilization is satisfied*

$$\alpha < min_s \frac{P_s}{(H-1)(P_s - R_s)^+ + R_s},\tag{10}$$

then a bound on end-to-end delay for any flow f exists and is

$$D^f = \frac{H}{1 - (H-1)u\alpha}(u\beta + E'),\tag{11}$$

where P_s denotes the capacity sum of all input links that have the considered traffic class input, $u_s = \frac{(P_s - R_s)^+}{P_s - \alpha R_s}$, $u = max_s\{u_s\}$, and

$$E' = max_s\left\{\frac{(1-u_s)L_s}{R_s} + E_s\right\}.\tag{12}$$

Conversely with Theorem 4, we can get the following results for the FIFO-based DiffServ approach. Their proofs can be found from [14].

Theorem 5. [Rate Guarantee] *If condition (10) is satisfied, then the network guarantees that*

$$d_H^{f,j} \le F^{f,j} + D^f\tag{13}$$

where D^f is determined by (11).

Theorem 6. [Throughput Guarantee] *Under the same condition (10), if the source of flow f transmits packets at least at rate r^f, then the network guarantees to the flow that*

$$W^f(t_1, t_2) \geq r^f(t_2 - t_1 - D^f - \frac{L}{r^f})^+. \tag{14}$$

where D^f is determined by (11).

3.3   The SCORE Approach

Like the DiffServ approach, the SCORE approach is also *core-stateless* in providing service guarantees [20]. However, it achieves this in a different way from the DiffServ approach. The key construct in the SCORE approach is the notion of *packet state* (PS) and the main ideal behind PS is to have packets carry per-flow states instead of having routers maintain the per-flow states [20].

The packet state is inserted by ingress edge routers which, as in the DiffServ approach, maintain per-flow states. In the core, a router processes each incoming packet based on the state carried by the packet and the router's internal state. Before forwarding the packet to the next hop, the core router may update both the packet state and its internal state. In such a way, PS coordinates actions of edge and core routers along the path of a flow to provide service guarantees to the flow. In fact, a router schedules packets even unaware of each individual flow. The router guarantees that each packet is forwarded to the next hop within a certain time limit that is computed from its carried PS. In this sense, the SCORE approach provides *packet level* service guarantees at each router. Like the DiffServ approach, the SCORE approach does not maintain per-flow states at core routers and hence achieves scalability of core routers.

The following results show that the same service guarantees are provided by a SCORE network as by a PFFQ-based IntServ network if each router in the SCORE network implements the corresponding core-stateless version of the GR algorithm used in the IntServ network [15]. Here, Theorem 7 has been proved in [15]. Theorem 8 and Theorem 9 can be easily proved using the same method as for Theorem 2 and Theorem 3.

Theorem 7. [Rate Guarantee] *A SCORE network of CSGR servers provides the same rate guarantee (7) as a PFFQ-based IntServ network of the corresponding GR servers.*

Theorem 8. [Bounded Delay] *A SCORE network of CSGR servers provides the same bounded delay (8) as a PFFQ-based IntServ network of the corresponding GR servers.*

Theorem 9. [Throughput Guarantee] *A SCORE network of CSGR servers provides the same throughput guarantee (9) as a PFFQ-based IntServ network of the corresponding GR servers.*

4   The Link-Based Fair Aggregation Approach

4.1   The Approach

The idea of LBFA was motivated by the following observation. Let us look at Fig. 2, which shows that when a flow passes through each link of its path, it is aggregated with other flows. As shown in Fig. 2, other flows may join and leave the path. In other words, the considered flow is aggregated with other flows of the same traffic class on each link along the path. Observing this, the idea of LBFA comes from the conjecture that since flows traversing the same link have already been aggregated, it may be possible to preserve service guarantees to each individual flow by aggregating those aggregates on each link properly.

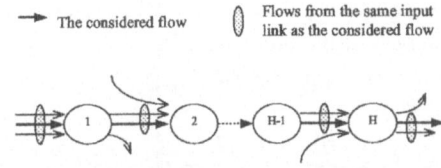

Fig. 2. Aggregation nature of an end-to-end flow

In particular, we treat the aggregation of all flows of the same traffic class, which are from the same input link and destined to the same output link, as a single link-level aggregate. We let this link-level aggregate first pass through a traffic shaper, and then use a fair queueing (FQ) scheduler to multiplex such shaped link-level aggregates from different input links to generate the class-based aggregate fed into the class-based scheduler. We call the resulted rate-controlled scheduler [21], including both the shaper and the normal FQ scheduler, *Fair Queueing with Shaping* (FQ-S) scheduler and call the normal FQ scheduler the corresponding FQ scheduler of the FQ-S scheduler.

More specifically, we require the shaper for any input link i at router s to be a greedy shaper [17] that works like a link with fixed capacity $r_s^{f_i}$. Such a shaper can be implemented by a leaky bucket shaper with leaking rate $r_s^{f_i}$. It is easy to verify that such a shaper has a shaping curve $r_s^{f_i} \cdot t + L$ (e.g. see [12]). Here, f_i denotes the corresponding link-level aggregate from input link i of flow f; $r_s^{f_i}$ denotes the sum of allocated rates of all constituent flows of the link-level aggregate f_i. In addition, we require that both the class-based scheduler and the FQ scheduler belong to the Latency-Rate Worst-case Service Guarantee (LR-WSG) server class [13]. This requirement is (to some extent) necessary since for analyzing hierarchical schedulers, to the best of our knowledge, [13] and [2] are the two main available references. Particularly, the class-based scheduler is an LR-WSG server to the considered traffic class aggregate with rate R_s and error term E_s; the corresponding FQ scheduler provides LR-WSG to each link-level aggregate with rate $r_s^{f_i}$ and error term $e_s^{f_i}$.

The following are some properties of the link server with LBFA implemented, which consists of the shaper, the FQ scheduler and the class-based scheduler. Their proofs can be found from [14].

Lemma 1. *(i) The class-based scheduler is GR server to the class aggregate with rate R_s and error term E_s. (ii) The hierarchical scheduler made of the class-based scheduler and the FQ scheduler in FQ-S is GR server to the link-level aggregate with rate $r_s^{f_i}$ and error term $E_s^{f_i} (= e_s^{f_i} + E_s)$.*

Lemma 2. *The link server implementing LBFA is GR server to flow f_i with rate $r_s^{f_i}$ and error term $E_s^{f_i} + \frac{L}{r_s^{f_i}}$.*

Remarks: With Lemma 2, it can be shown that various service guarantees are provided by router s to the link-level aggregate f_i. In this sense, we say *link level* service guarantees are provided by the LBFA approach at each router.

Lemma 3. *The link server implementing LBFA guarantees to f_i*

$$G_s^{g,k} \leq F^{f_i,j} + \mathcal{E}_s \tag{15}$$

with

$$\mathcal{E}_s = \frac{L}{r_s^{f_i}} + \frac{I_s \cdot L}{R_s} + E_s + E_s^{f_i}. \tag{16}$$

Here in Lemma 3, g denotes the corresponding class-based aggregate of f_i, $p^{g,k} = p^{f_i,j}$ and with $G_s^{g,0} = 0$, $G_s^{g,k}$ is iteratively defined as

$$G_s^{g,k} = max[d_s^{g,k}, G_s^{g,k-1}] + \frac{l^{g,k}}{R_s}. \tag{17}$$

4.2 Per-Flow Service Guarantees

Having introduced the idea of LBFA and some properties of the link server implementing LBFA, we now show through analysis that end-to-end per-flow service guarantees are provided by the network if LBFA is implemented even though each node is unaware of individual flows. We shall see that these guarantees are independent of link utilization level (as long as the link is not overloaded) as in the IntServ and SCORE approaches.

In the following analysis, we assume that in the network, for flow f, the allocated rate R_s to the corresponding class-based aggregate at any router s along its path is not less than the sum of allocated rates to the link-level aggregates forming the class-based aggregate at their pervious hops. It is worth highlighting that this assumption makes sense for real networks. An example is IntServ networks with aggregation of end-to-end reservation [1]. Another example is MPLS networks with LSP merging [8]. In these networks, when bandwidth guarantee is required, it is usually assumed that in cases of reservation aggregation or LSP merging, the reserved rate at downstream routers of the merged point is sufficient to carry the sum of merged traffic [1] [8]. The above assumption could be relaxed based on the idea of enforcing spacing between packets at the edge which has been used in [5].

Lemma 4. *For any packet $p^{f,j}$, the network guarantees that*

$$d_H^{f,j} \leq F^{f_1,k} + \sum_{s=1}^{H} \mathcal{E}_s, \tag{18}$$

where f_1 denotes the corresponding link-level aggregate of flow f at the first router, $\mathcal{E}_s$ is determined by (16), and $p^{f,j} = p^{f_1,k}$.

With Lemma 4, the following can be further derived. (See [14] for proofs.)

Theorem 10. **[Rate Guarantee]** *For the same network as in Lemma 4, it guarantees to flow f that, with $\sigma^{\overline{f}_1} = \sum_{v \in f_1, \neq f} \sigma^v$,*

$$d_H^{f,j} \leq F^{f,j} + \sum_{s=1}^{H} \mathcal{E}_s + \frac{\sigma^{\overline{f}_1}}{r^{f_1}}. \tag{19}$$

Theorem 11. **[Bounded Delay]** *For the same network as in Lemma 4, the end-to-end delay of any packet in flow f is bounded by:*

$$d_H^{f,j} \leq \frac{\sigma^{f_1}}{r^{f_1}} + \sum_{s=1}^{H} \mathcal{E}_s. \tag{20}$$

Theorem 12. **[Throughput Guarantee]** *For the same network as in Lemma 4, if flow f transmits packets at least at rate r^f, then the network guarantees to the flow that, with $\sigma^{\overline{f}} = \sum_{v \in f_1, \neq f} \sigma^v$,*

$$W^f(t_1, t_2) \geq r^f \left(t_2 - t_1 - \sum_{s=1}^{H} \mathcal{E}_s - \frac{\sigma^{\overline{f}}}{r^{f_1}} - \frac{L}{r^f} \right)^{+}. \tag{21}$$

Remarks: By expending $\mathcal{E}_s$ in (19), we get $d_H^{f,j} \leq F^{f,j} + \sum_{s=1}^{H} \frac{L}{r_s^{f_i}} +$ $\sum_{s=1}^{H}(E_s + E_s^{f_i}) + \frac{\sigma^{\overline{f}_1}}{r^{f_1}} + \sum_{s=1}^{H} \frac{I_s \cdot L}{R_s}$, comparing which with (7), it is not difficult to verify that in general (19) is comparable with (7). In addition, if the edge router maintains per-flow states as in SCORE and DiffServ, a per-flow FQ scheduler can be used to aggregate constituent flows of f_1 to form the link-level aggregate. As a result, the forth term can be removed. Consequently, the resulted network is only *core-stateless* while the guarantee can be better than (7).

A similar approach as LBFA can be found from [6], in which flows are aggregated at path level using fair aggregator. (The author would like to thank the anonymous reviewers for pointing out this.) In [6], a fair aggregator is implemented by limiting the total output rate of a fair queueing scheduler. Hence, there are two major differences between LBFA and the approach in [6]. One is that the levels of service guarantees provided by them at each router are different; the other is their suggested implementations of flow aggregators are different. Nevertheless, one might view LBFA as an extension of the approach in [6]. In fact, LBFA may be applied to aggregating flows at path level and the fair aggregator suggested in [6] may be used to perform link-level flow aggregation.

Table 1. Comparison of approaches to service guarantees

Approach	PFFQ-based IntServ	FIFO-based DiffServ	SCORE	LBFA
Granualarity	flow level	class level	packet level	link level
Guarantees	Th. 1 - 3	Th. 4 - 6	Th. 7 - 9	Th. 10 - 12
Limitation	no	yes	yes³	no
Flow states	stateful	core-stateless	core-stateless	(core-)stateless
# of states	$O(\mathcal{N})$	no	no	no
Complexity	$O(log(\mathcal{N}))^1/O(1)^2$	straightforward	$O(log(\mathcal{M}))$	$O(\mathcal{I})^1 / O(1)^2$
Packet state	no	no	yes	no

[1]: use deadline-based FQ. [2]: use RR-based FQ. [3]: "No" only if PS is stored finestly.
$\mathcal{N}$: # of active flows $\mathcal{I}$: # of input links $\mathcal{M}$: # of packets in queue

5   Comparison

Table 1 presents a brief comparison of the existing and proposed approaches. More detailed discussion can be found from [14]. From the table, it can be seen that PFFQ-based IntServ provides powerful service guarantees but has serious scalability problem. FIFO-based DiffServ is scalable but cannot provide comparable per-flow service guarantees as IntServ and has a limitation on link utilization level. While SCORE is both scalable and has the same ability as IntServ in providing per-flow service guarantees, it introduces additional implementation requirements that could limit its use in the current Internet infrastructure. The proposed LBFA approach, which is simple to implement, provides another option for providing per-flow service guarantees that are comparable to those provided by PFFQ-based IntServ. LBFA can be considered as a compromise between the PFFQ-based IntServ approach and the FIFO-based DiffServ approach. In fact, if we simply treat LBFA as a flow aggregation method, it could be used under both the IntServ architecture and the DiffServ architecture. In addition, LBFA may be used with IntServ to solve its scalability problem or with DiffServ to avoid its link utilization problem.

6   Conclusion

In this paper, we discussed from the service guarantee granularity perspective the three widely studied approaches for providing service guarantees, which are IntServ, DiffServ and SCORE. We also summarized per-flow service guarantees provided by them. In addition, we proposed a new approach, called LBFA approach to scalable support of per-flow service guarantees. While the granularity of service guarantees provided by LBFA at each route is *link level*, we showed that the per-flow service guarantees provided by LBFA are comparable to those by PFFQ-based IntServ and SCORE, and are better than those provided by FIFO-based DiffServ. Moreover, we compared briefly the three current approaches and the LBFA approach. The comparison showed that the LBFA approach is as scalable as the FIFO-based DiffServ and SCORE approaches and provides comparable per-flow service guarantees as the PFFQ-based IntServ and SCORE approaches. We believe that LBFA is a simple yet effective approach and it could be used with IntServ or DiffServ to solve their specific problems.

References

1. F. Baker et al. Aggregation of RSVP for IPv4 and IPv6 reservation. *IETF RFC 3175*, Sept. 2001.
2. J. C. R. Bennett and H. Zhang. Hierarchical packet fair queueing algorithms. *IEEE/ACM Trans. Networking*, 5(5):675 – 689, Oct 1997.
3. S. Blake et al. An architecture for Differentiated Services. *IETF RFC 2475*, 1998.
4. R. Braden, D. Clark, and S. Shenker. Integrated services in the Internet architecture: An overview. *IETF RFC1633*, 1994.
5. D. Chlamtac et al. A deterministic approach to the end-to-end analysis of packet flows in connection-oriented networks. *IEEE/ACM ToN*, 6(4):422–431, Aug. 1998.
6. J. A. Cobb. Preserving quality of service guarantees in spite of flow aggregation. *IEEE/ACM Trans. Networking*, 10(1):43–53, Feb. 2002.
7. B. Davie et al. An Expedited Forwarding PHB. *IETF RFC 3246*, March 2002.
8. F. L. Faucheur and et al. Multiprotocol label switching (MPLS) support of Differentiated Services. *IETF RFC 3270*, May 2002.
9. S. Floyd and V. Jacobson. Link-sharing and resource management models for packet networks. *IEEE/ACM Trans. Networking*, 3(4), 1995.
10. P. Goyal, S. S. Lam, and H. M. Vin. Determining end-to-end delay bounds in heterogeneous networks. *Springer Multimedia Systems*, 5:157–163, 1997.
11. J. Heinanen, F. Baker, W. Weiss, and J. Wroclawski. Assured forwarding PHB group. *IETF RFC 2597*, June 1999.
12. Y. Jiang. Delay bounds for a network of Guaranteed Rate servers with FIFO aggregation. *Computer Networks*, 40(6):683–694, Dec. 2002.
13. Y. Jiang. Per-domain packet scale rate guarantee for Expedited Forwarding. Proc. IWQoS 2003, LNCS 2707, pp. 422–439, 2003
14. Y. Jiang. Link-Based Fair Aggregation: A Simple Approach to Scalable Support of Per-Flow Service Guarantees. Technical Report, Q2S, NTNU, 2004.
15. J. Kaur and H. M. Vin. Core-stateless guaranteed rate scheduling algorithms. In *Proc. INFOCOM'01*, 2001.
16. J. Kaur and H. M. Vin. Core-stateless guaranteed throughput networks. In *Proc. INFOCOM'03*, 2003.
17. J.-Y. Le Boudec and P. Thiran. *Network Calculus: A Theory of Deterministic Queueing Systems for the Internet*. Springer-Verlag, 2001.
18. K. Nichols and B. Carpenter. Definition of differentiated services per domain behaviors and rules for their specification. *IETF RFC 3086*, April 2001.
19. S. Shenker, C. Partridge, and R. Guerin. Specification of guaranteed quality of service. *IETF RFC 2212*, Sept 1997.
20. I. Stoica and H. Zhang. Providing guaranteed services without per flow management. In *Proc. SIGCOMM'99*, 1999.
21. H. Zhang and D. Ferrari. Rate-controlled service disciplines. *J. High Speed Networks*, 3(4), 1994.

Reducing Packet-Loss by Taking Long-Range Dependences into Account

J. Ignacio Alvarez-Hamelin* and Pierre Fraigniaud

Laboratoire de Recherche en Informatique, Bât 490 Université Paris Sud,
91405 Orsay CEDEX, France, {ihameli,pierre}@lri.fr

Abstract. We show that the "fractal" behavior of Internet traffic can be efficiently and practically employed to significantly reduce packet-loss. Thanks to recent advances in the theory of self-similar processes, we define the *probabilistic congestion* of a link, based on an estimated computation of the packet-loss probability over that link. This congestion parameter allows valid predictions on the future behavior of the network, on which one can base efficient routing strategies. We show how to implement the computation of the probabilistic congestion, and we illustrate several applications for improving unicast and multicast protocols.

Keywords: Self-Similar Traffic, Routing, Multicast, IP Networks.

1   Introduction

Starting in the early 90's, there has been a number of empirical studies that provide evidence of the prevalence of self-similar traffic patterns in packet networks such as the Internet (see, e.g., [9,17]). This "fractal" behavior is very different both from conventional telephone traffic, and from standard models for packet traffic (e.g., Poisson). In particular, significant traffic variance (burstiness) is present on a wide range of time scales in self-similar traffic, and hence such traffic exhibits long-range dependencies (i.e., values at any instant are correlated with values at future instants). This paper aims to explore situations for which routing can take advantage of the long-range dependence nature of the traffic, and to provide effective solutions for such situations.

Self-similarity was observed in Local Area Networks [9] as well as in Wide Area Networks [17]. In [19,22] it is shown that the self-similarity at the LAN level can result from the superposition of on/off sources with strictly alternating on- and off-periods, and whose on-periods or off-periods have high variability. User related traffic, as World-Wide-Web (WWW) traffic [5,4], WWW workload [18], and Variable-Bit-Rate (VBR) video data [2,7] present characteristics that are consistent with self similarity. On the other hand, [16] shows that transport

---

* Also from the Facultad de Ingeniería of Universidad de Buenos Aires. Supported by the ALPHA program CORDIAL.

N. Mitrou et al. (Eds.): NETWORKING 2004, LNCS 3042, pp. 1096–1107, 2004.
© IFIP International Federation for Information Processing 2004

mechanisms are important factors in translating self-similarity from the application layer to the link layer. In particular, the flow control mechanisms of TCP seem to maintain the long-range dependency structure induced by heavy-tailed file size distributions, a phenomenon which is not observed to the same extent when using the non-flow-controlled UDP transport protocol (see [15]). In fact, it is shown in [21] that TCP congestion control "propagates" self-similarity. Roughly speaking, if a TCP stream meets another TCP stream which exhibits large time-scale fluctuations, then TCP will react in a way that causes the former traffic to inherit the self-similarity nature from the latter. In addition to that, [20] concludes that the congestion control of TCP itself, as a deterministic process, creates chaos, which generates self-similarity. There are hence many reasons why traffic in IP exhibits a self-similar behavior. This paper shows that this traffic property can be effectively employed to reduce packet-loss.

Packet-loss is the cause of important performance degradations in packet networks, for it is intrinsically related to all "standard" Quality of Services (QoS) measures: latency, bandwidth, jitter, etc. Using TCP, it is necessary to resends packets, with time-consuming effects. Using UDP, packet-loss has a significant impact on the user-perception for certain applications. IPv4 allows the use of a Type-of-Service (ToS) field in the IP-headers, for choosing the route with minimum packet-loss. However, this facility has not been implemented, or rarely used, because of the uncertainty of foreseeing the future behavior of a system with short-range dependencies, as Internet was assumed to be. On the other hand, recent advances in the theory of self-similar variation allow the anticipation of the future behavior of systems with long-range dependencies, as the Internet really is. Anticipating the traffic behavior allows efficient routing strategies to be developed, while preserving the best effort nature of IP, and in particular no resource reservation is required.

Various routing protocols include QoS in their definition, in the sense that they are able to set up the best routes according to pre-defined parameters such as number of hops, maximum delay, or available bandwidth. The route is chosen among a set of alternative paths proposed by the protocol. This is typically the case of the QoSMIC [6] and YAM [3] multicast routing protocols, as well as of any Dijkstra-like unicast routing protocol. Basically, we propose to perform these choices according to the probability of packet-loss.

We define the *probabilistic congestion c* of a link as a logarithmic transformation of the packet-loss probability p along that link. (The probabilistic congestion of a route is thus simply the sum of the probabilistic congestion of each of its links.) The probability p is estimated using the theory developed in [11,12,13], which proposes an explicit formula for the probability that a given buffer contains more than x bytes, assuming a plausible model of self-similar traffic. The computation requires the values of three parameters: the mean input rate m, the "variance coefficient" a, and the Hurst parameter H. The latter two parameters characterize the "quality" of the traffic, in contrast to the long run mean rate m which characterizes its "quantity" alone [12]. We show that these three parameters can be computed on-line at every router, using locally accessible traffic variables. They are estimated at time t based on the traffic observed during the last time period $[t - \Delta, t]$, for some Δ. Assuming a traffic with large time-scale fluctuations, we expect that the traffic will offer the same characteristics during the time interval $[t, t + \delta]$ as it did during $[t - \Delta, t]$, at least for a sufficiently small $\delta < \Delta$. In particular, the packet-loss probability computed based on parameters specific to the traffic observed during $[t - \Delta, t]$ is expected to not vary too much in the near future. We have grounds to believe in such a behavior of the packet-loss probability since we observed, based on real traffic observations on Abilene's backbone, that long-range dependences imply smooth variations of the probabilistic congestion c, as opposed to, e.g., the available bandwidth, which experiences abrupt variations.

In this paper, we demonstrate the interest of the probabilistic congestion by improving QoS-sensitive multicast routing protocols. More precisely, we compare the performance (in terms of packet-loss) of QoSMIC using probabilistic congestion vs. QoSMIC using standard dynamic criteria. Simulations were performed on Abilene's topology, as well as on UUNET's topology, for several group configurations. The simulated traffic was based on real traffic samples from Abilene, and hence offers realistic characteristics. In all cases, we observed a significant improvement for QoSMIC using probabilistic congestion compared to QoSMIC using any other dynamic criteria. Note that the multicast protocol itself is not modified, only the information collected by the nodes for the construction of the multicast tree differ: they collect probabilistic congestion rather than current delays or available bandwidth. The remaining of the protocol is exactly the same.

We also demonstrate the interest of the probabilistic congestion for improving performances of unicast routing. In particular, in a network supporting different classes of services, the routing paths of the priority classes can be chosen as the ones with minimum probabilistic congestion. Simulations on the Abilene's backbone demonstrate that this approach yields significant improvements in term of packet losses.

2   Probabilistic Congestion

We follow the model extensively explored by I. Norros in a series of papers (see, e.g., [11,12,13]). The self-similar variation of a packet-network traffic is modeled by a Gaussian self-similar process. A normalized fractional Brownian motion with self-similar parameter $H \in [\frac{1}{2}, 1)$ is a stochastic process Z_t, $t \in \mathbb{R}$, characterized by the following properties: (1) Z_t is Gaussian, (2) with probability 1, $t \mapsto Z_t$ is continuous, (3) $\mathbf{E}Z_t^2 = |t|^{2H}$ for all t, (4) $Z_0 = 0$, and $\mathbf{E}Z_t = 0$ for all t, and (5) for all t, and $s_1 < s_2 < \ldots < s_k$, the distribution of $(Z_{t+s_2} - Z_{t+s_1}, \ldots, Z_{t+s_k} - Z_{t+s_{k-1}})$ is independent of t. Note that, for $H = \frac{1}{2}$, Z_t is the standard Brownian motion [10]. The parameter H is the Hurst parameter. Let us denote by A_t the amount of traffic (in bytes, say) offered to a link of the network during time interval $[0, t]$. Norros models A_t as a so called fractional

Brownian *traffic*, that is

$$A_t = mt + \sqrt{am}Z_t \ , \tag{1}$$

where $m > 0$ is the mean input rate, and $a > 0$ is a variance coefficient. The self-similarity is captured by the Brownian scaling relation stating that, for any $t > 0$, $\{Z_{st}, s \geq 0\}$ and $\{t^H Z_s, s \geq 0\}$ have the same finite-dimensional distributions. The traffic fluctuations buffering of a fractional Brownian traffic of parameters m, a, and H offered to a link of capacity $C > m$ with infinite queue length is then defined as

$$X_t = \sup_{s \leq t} \Big(A_t - A_s - C(t - s) \Big).$$

Norros [12] derived a lower bound of the probability that the local storage exceeds a certain limit:

$$\mathbf{Pr}(X_t > x) \geq \bar{\varPhi}\Big(\frac{(C - m)^H x^{1-H}}{H^H (1 - H)^{1-H}\sqrt{am}} \Big) \ , \tag{2}$$

where $\bar{\varPhi}(y) = \mathbf{Pr}(Z_1 > y)$, that is the residual distribution function of the standard Gaussian distribution. Since $\bar{\varPhi}(y) \sim e^{-y^2/2}$, $\mathbf{Pr}(X_t > x)$ can hence be lower bounded by a Weibull distribution, in particular with regards to the tail behavior. Norros performed simulation to check the accuracy of the bound in Eq. (2). The queue length process was generated by the usual formula

$$X_{k\tau} = \Big(X_{(k-1)\tau} - C\tau + A_{k\tau} - A_{(k-1)\tau} \Big)^+ \ , \tag{3}$$

where τ is the resolution of the experimental sample, and $x^+ = x$ if $x > 0$, and 0 otherwise. Interestingly, one observation coming out from the simulations is that the Weibull approximation $\mathbf{Pr}(X_t > x) \sim Y$ where Y is the right side of Eq. 2 is a satisfactory accurate approximation. This motivates us to define the following parameter:

Definition. The *probabilistic congestion* c of a link with buffer size b supporting self-similar traffic of parameters m, a and H, is defined by $c =$

$$- \ln \Big(1 - \exp \Big(- \frac{(C - m)^{2H}}{2H^{2H}(1 - H)^{2-2H} am} b^{2-2H} \Big) \Big) \ . \tag{4}$$

According to the Weibull approximation, we have $c \sim - \ln(1 - \mathbf{Pr}(X_t > b))$, and hence, up to the logarithmic rescaling, the probabilistic congestion follows the same behavior as the packet-loss probability. In particular, given a path $P = \{e_1, \ldots, e_k\}$, with probabilistic congestion of e_i equal to c_i, the probabilistic congestion of P is naturally defined as $\sum_i c_i$.

The probabilistic congestion c of any link can be approximated by its tail router. For each of its outgoing link e, every router counts the number of bytes that were sent through e during specific time-intervals of length t_u. At time t, let A_i, $1 \leq i \leq N$, be the number of bytes offered to link e during time-interval

$I_i = (t - it_u, t - (i-1)t_u]$. Then let $B_i = A_i/t_u$ be the average throughput of link e during I_i. The mean rate m is simply estimated by $\overline{B} = \frac{1}{N}\sum_{i=1}^{N} B_i$, and the variance coefficient a is estimated by $\sigma^2(B)/m$. The auto-covariance of the B_i's is $\dot{B}_k = \sum_{i=1}^{N}(B_i - \overline{B})(B_{i-k} - \overline{B})$, for $k = -N+1,\ldots,N-1$. The $\dot{B}_k$'s, $k \geq 0$, plot is a second order self-similar process $x \mapsto \alpha e^{-\beta x}$, and the Hurst parameter characterizing the traffic observed through e between times $t - Nt_u$ and t is $H = 1 - \beta/2$. (Here $\Delta = Nt_u$.) $\dot{B}$ can be computed rapidly thanks to a constant number of Fast Fourier Transforms (FFT): $\dot{B} = FFT^{-1}(\widehat{B} \cdot \overline{\widehat{B}})$ where $\widehat{B} = FFT(B)$, and $\overline{x}$ is the conjugate of x. Then H can be easily obtained by a linear regression applied on $\log \dot{B}$. Once m, a, and H have been estimated, the probabilistic congestion c of link e is computed according to its definition in Eq. (4). This probabilistic congestion will be used during the next time-interval $(t, t + \delta]$. It is worth mentioning that although the N-point FFT requires $O(N \log N)$ arithmetic operations, it can be performed in logarithmic time in parallel, and the sequential computation of the FFT is certainly doable for reasonably large N without overloading current routers. Moreover, the estimation of m, a, and H is performed every δ time units. In practice, one can set δ equal to a couple of minutes (say 10mn), for $\Delta \simeq$ 1h and $t_u \simeq$ 3.5s. This yields the computation of FFT's on $\Delta/t_u = 1024$ points, which is reasonably small. Still the estimation of the long-scale behavior of the traffic will remain good enough, as shown in the next sections.

3   Application to Multicast

Several protocols have been proposed to support group communications. Among them, YAM [3], QoSMIC [6] and MORF [24] take into account QoS for the construction of the multicast tree. A common feature of these three protocols is to offer multiple routes between a new member and the current tree. This allows them to select the "best" route with regards to some QoS requirements. MORF and YAM use static parameters (e.g., link capacity, link delay, or reliability), but QoSMIC uses dynamic parameters (e.g., available bandwidth, current delay). Hence, we focussed our attention to this latter protocol, and show that using probabilistic congestion offers better performances than using the available bandwidth or the current delay, as far as packet-loss is concerned. QoSMIC can create shared trees or source-based trees. In this paper, we assume a behavior in the shared tree mode. A new router, aiming to join the group, searches for a router in the tree where to connect. There are two search procedures in QoSMIC: *local search* and *multicast tree search*. The local search is similar to the one proposed in YAM: the new router performs exploration of its neighborhood at successive distance $1, 2, \ldots$, until the Time To Live field of the IP header is exhausted. The multicast tree search is performed via a call to a Manager router, which starts exploration of the tree to find appropriate candidates in the tree for the new router to connect with. How much the simultaneous use of these two types of search allows to reduce the complexity of joining is discussed in [6]. In our experiments, we consider all possible paths constructed by the unicast routing from the new router to all routers in the current tree. Therefore, the search performed from a router x aiming to join the group results in a set $y_1, \ldots, y_k$ of candidate routers in the tree. Then x selects the appropriate candidate by comparing the characteristics of the routes between itself and every y_i. It is underlined in [6] that, although the routes are restricted by static information in the routing information base, the new router selects among these routes using dynamic routing information. The use of the probabilistic congestion ideally fits with this setting.

In order to compare the QoSMIC selection using available bandwidth vs. the selection using probabilistic congestion, we also use the Minimal λ-Tree multicast protocol (MλT) briefly described as follows [1]. MλT is built upon a multiple-path routing protocol which maintains a table D_v at each router v, such that $D_v[i, x]$ is the length of the shortest path connecting node v to node x when leaving v through the i-th interface. MλT explores, for every node x of the group, all paths from the source s to x that are of length at most $\varrho \cdot d(s, x) + r$ where ϱ and r are constants fixed a priori, and $d(s, x)$ is the distance from s to x (measured in #hops). Then MλT selects, for every router x, the path from s to x whose QoS is the best. Finally, it constructs a multicast tree by merging all selected paths. Therefore, MλT constructs a tree T rooted at the source s such that, for every node x of the group, the route from s to x in T has the best QoS among all routes from s to x of length at most $\varrho \cdot d(s, x) + r$. Obviously, there is tradeoff between the control overhead, and the efficiency of MλT. The construction of the tree by MλT is indeed costly if ϱ or r is large. On the other hand, the efficiency (in terms of packet-loss) of the tree is expected to be good if many routes are considered for the selection, i.e., if ϱ or r is large. In some sense, the performances of MλT are those of an ideal multicast protocol that would be allowed to explore a large portion of the network to select the routes.

In total, we have considered four tree-construction protocols: Reverse Shortest Path (RSP), Greedy, QoSMIC, and MλT. Recall that RSP sets up the tree simply as the union of the (unicast driven) paths from the source to every destination. This technique is used in CBT and PIM-SM. In Greedy, a new node aiming to join the group connects to the closest node in the current tree. This technique is used in YAM. We compare the efficiency of using available bandwidth vs. probabilistic congestion for the selection of the routes in QoSMIC. MλT was implemented using probabilistic congestion only. On one hand RSP and Greedy allows to compare the two versions of QoSMIC with (sort of) basic approaches for tree-construction. On the other hand, MλT allows to compare the same two versions of QoSMIC with a (sort of) ideal tree-construction protocol.

3.1   Experimental Results

We have performed simulations based on realistic traffic patterns obtained from the traffic samples observed on Abilene's backbone. Actually, since Abilene is underutilized, we assumed links with 15% capacities of those of the real Abilene network. This rescaling yields a load of up to 98% for the most occupied links.

We have divided the time in 389 non-overlapping windows of 4h10 each. Since the network is observed at the period of 5mn, every window contains 50 samples. For each window, we computed the three corresponding characteristic parameters m, a, and H. Therefore, we based our simulations on 389 sets $\{(m_e, a_e, H_e), e \in E\}$, where E is the set of the 28 links of Abilene's backbone, and each triple is computed from a real IP traffic. Therefore, we obtained more than 10,000 experimental triples in total. Each of them is rescaled for a link of unit capacity (i.e., m and a are rescaled to m/C and a/C where C is the link capacity). We have performed experiments on the UUNET backbone topology. UUNET is a world-wide private network with 129 nodes. We did not had access to the buffer size b of the links of the network, therefore we set $b = 4800$ bytes for all links, which is a bit more than the maximum size of an FDDI packet. We used UUNET's backbone with all its original link capacities, which vary from few tens Mb/s for some links, to roughly 10 Gb/s for others.

Each experiment is performed according to the following protocol. For each link e of the network, we choose one triple (m, a, H) among the 10,000 experimental triples, and re-scale m and a to fit with the capacity C of link e (i.e., m and a are rescaled to m C and a C). Then the probabilistic congestion of each link is computed according to Eq. 4. The probabilistic congestion of a path is the sum of the probabilistic congestion of its links. Similarly, the available bandwidth $C - w$ of each link is computed, where C is the physical capacity of the link, and w is the used capacity. The used capacity is computed as the average number of bytes crossing the link during some predefined interval of time. The available bandwidth of a path is the minimum of the available bandwidth of its links.

For a considered multicast group $\{x_1, \ldots, x_k\}$, and a source s, we simulate the execution of RSP, Greedy, and the two versions of QoSMIC, assuming members arrive successively in the order $x_1, \ldots, x_k$. Hence, we get four trees T_{RSP}, T_{Greedy}, $T_{\text{QoSMIC}}^{\text{prob cong}}$, and $T_{\text{QoSMIC}}^{\text{avail bdw}}$. For the ideal protocol MλT, a tree $T_{\text{M}\lambda\text{T}}$ spanning all group members, with minimum probabilistic congestion, is computed globally (although MλT can also be implemented in a distributed greedy manner). Actually, to make MλT more realistic, we didn't explore all paths, but only those of bounded length. More precisely, we have run MλT with $\varrho = 1$ and $r = 2$.

Then, for each link, we have performed simulations by injection of a self-similar traffic A_t through the link during 512s (roughly 8mn30), where A_t is described in Eq. (1), and where m, a, and H are the parameters chosen for that link. That is, for every link with buffer size b, we run the iterative process

$$Y_{k\tau} = (m - C)\tau + \sqrt{am} \left(Z_{k\tau} - Z_{(k-1)\tau} \right)$$

and

$$X_{k\tau} = \begin{cases} 0 & \text{if } X_{(k-1)\tau} + Y_{k\tau} \leq 0; \\ b & \text{if } X_{(k-1)\tau} + Y_{k\tau} \geq b; \\ X_{(k-1)\tau} + Y_{k\tau} & \text{otherwise.} \end{cases}$$

The resolution of the simulation is $\tau = 1/512$s. The Z_t's were generated using the simulator of Norros [14]. For each link e, we computed the ratio between the

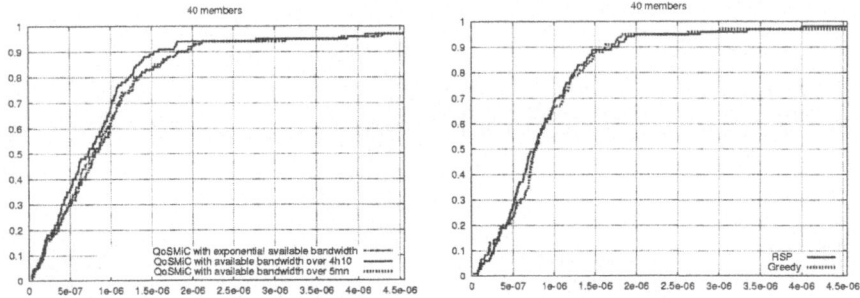

Fig. 1. Cumulative distributions of the packet-loss probability

total number of bytes lost by link e during the 8mn30 of the simulation, and the total number of bytes sent over link e during the same period of time. This ratio is therefore the probability p_e for an application using link e to lose packets. For each tree-construction protocol, we then computed the probability for a multicast application using the resulting tree $T = (V, E)$ to not lose packets, estimated by $\mathbf{Pr}(T) = \Pi_{e \in E}(1 - p_e)$. For the same multicast group, we repeated the experiment 100 times (i.e., with 100 different triples (m, a, H) for each link). We performed experiments for different multicast groups, of different sizes. Results are displayed using the cumulative distributions F of the packet-loss probability. That is, the vertical axis is a percentage, and the horizontal axis is a probability. Given a probability p, $F(p)$ is the percentage of experiments which returned a tree for which the packet-loss probability was $\leq p$.

Fig. 1 (left) displays the performances of three versions of QoSMIC, all using predictions based on the available bandwidth, for a group of size 40. All versions average the used capacity w of each link. The first version uses exponential averaging [8]. The last two versions differ on the interval of time during which the used capacity w of each link was averaged. We considered either average over the last 4h10, i.e., $w = m$ where m is the mean input rate of the link, or average over 5mn, i.e., $w = B$ where B is the average throughput of the link during the last 5mn. Fig. 1 (left) shows that, compared to the two others type of averaging, predictions based on an average over a long period of time (here 4h10) offer slightly better results. For instance, using an average over 4h10, 35% of our experiments return a tree for which the packet-loss probability is $\leq 5 \cdot 10^{-7}$. Using an average over 5mn, only 30% of our experiments return a tree for which the packet-loss probability is $\leq 5 \cdot 10^{-7}$. This latter packet-loss probability is roughly the same for exponential averaging. Indeed, exponential averaging is similar to short time averaging because the exponential decays very quickly. All our experiments have confirmed this behavior. Therefore, for comparisons with tree-constructions based on other methods, we considered only the version of QoSMIC averaging the available bandwidth over 4h10.

Fig. 1 (right) shows the performances of RSP and Greedy for a group of size 40. As far as packet-loss is concerned, the two strategies perform roughly

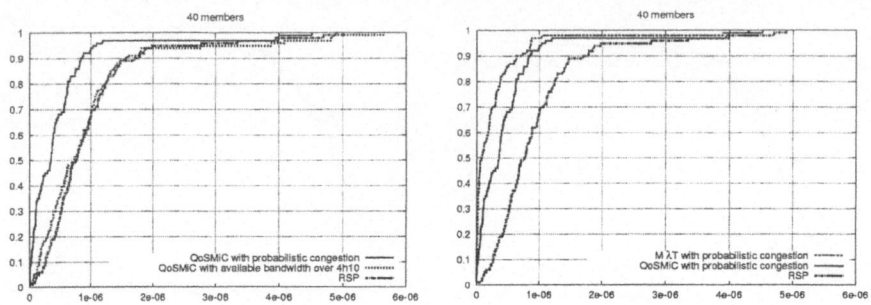

Fig. 2. Cumulative distributions of the packet-loss probability (continued)

the same, with perhaps a little advantage to RSP. All our other experiments show the same relative behavior of the two protocols. In the following, we will hence consider RSP only, as a representative of popular protocols such as PIM-SM or CBT.

Fig. 2 (left) presents a first set of significant results. It displays the performances of RSP, QoSMIC with available bandwidth , and QoSMIC with probabilistic congestion, for a group of size 40. QoSMIC with available bandwidth performs better than RSP in general, and hence better than the Greedy protocol. That is, although the traffic is subject to bursts at all time scales, which makes average bandwidth a poor predictor for the future behavior of the system, QoSMIC takes in fact some benefits from the selection of the routes according to the available bandwidth. More interestingly, QoSMIC with probabilistic congestion performs significantly better than QoSMIC with available bandwidth, bringing supports to the main claim of this paper. For instance, almost 70% of our experiments on QoSMIC with probabilistic congestion return a tree for which the packet-loss probability is $\leq 5 \cdot 10^{-7}$ for the 40-node group. In comparison, only 35% of our experiments on QoSMIC with available bandwidth return a tree with the same QoS. In 90% of our experiments, QoSMIC with probabilistic congestion has packet-loss probability $\leq 8.6 \cdot 10^{-5}$, and QoSMIC with available bandwidth has packet-loss probability $\leq 16 \cdot 10^{-5}$. Actually, the improvement of using QoSMIC with probabilistic congestion compared to QoSMIC with available bandwidth is by far bigger than the improvement of using QoSMIC with available bandwidth compared to RSP.

Finally, Fig. 2 (right) displays comparisons between QoSMIC using probabilistic congestion with, on one hand, the standard protocol RSP, and on the other hand, the ideal protocol MλT. Of course, MλT outperforms the two other protocols. The gain obtained from using MλT compared to QoSMIC with probabilistic congestion is roughly the same as the gain of using the latter compared to RSP. In particular, in terms of average packet-loss, QoSMIC with probabilistic congestion performs almost 2 times better than RSP.

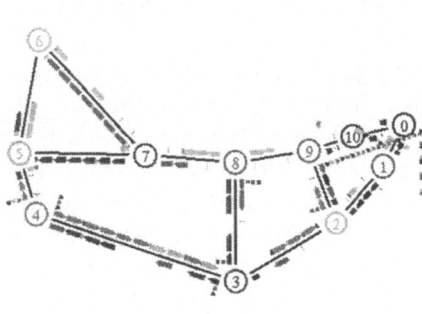

# flows	% l. P_1	# t.l.P_1	% l. P_2	# t.l.P_2
1	2.87	12930	0	4014
2	2.94	14207	0	3462
3	2.93	13819	0	3590
5	3.06	13371	0	3284
10	2.92	11925	0	2625
20	4.83	14218	0.01	3866
30	6.27	15068	0.05	4668
40	6.96	14968	0.14	3927
50	7.98	14968	0.27	3920
60	9.37	15829	0.64	5145

Fig. 3.

4 Application to Unicast

Probabilistic congestion could also be used for unicast routing. Consider for instance a network supporting two classes of services: a priority class (i.e., voice transmission), and a non-priority class. Assume moreover that the amount of priority class traffic is small in front of the non-priority class traffic. In this context, one could use probabilistic congestion to construct and maintain the unicast tables of the priority class. A straightforward approach consists of setting up the weight of each link as its probabilistic congestion, and using a standard Dijkstra-like unicast routing protocols. Then the priority class traffic would benefit from a significant decrease in packet-loss. We illustrate this approach by using the following scenario[1]. We used again the backbone of the Abilene network (c.f. Fig. 3 left), in which we injected exponential on-off TCP traffic between (almost) every pair of nodes. More precisely, we took the pairs $\{3,4\}$, $\{3,5\}$, $\{3,6\}$, $\{4,5\}$, $\{4,6\}$, $\{5,6\}$, $\{9,2\}$, $\{9,3\}$, $\{9,4\}$, $\{10,2\}$, $\{10,3\}$, $\{10,4\}$, $\{10,9\}$, $\{6,9\}$, $\{5,7\}$, $\{3,8\}$, $\{0,2\}$, and $\{0,10\}$, and connect a source and a receptor at each node or router of each pair. We used the NS2 [23] simulator to inject and route the traffic. In this simulation we observed a self-similar traffic traversing every link, because of TCP. Figure 3 (left) shows a screen capture of NAM [23], the network animator of NS2. In this figure, it is possible to observe the traffic on all links, and the buffers occupation. We selected two nodes, $x = 2$ and $y = 10$ arbitrarily. The path set up by the routing tables of Abilene between x and y is denoted by $P_1 = \{2, 9, 10\}$, through one intermediate node. There is an alternative longer path, through two intermediate nodes, denoted by $P_2 = \{2, 1, 0, 10\}$. We have simulated 60 seconds of traffic, and we have computed the probabilistic congestion (using the natural logarithm) of P_1 and P_2 ($c_1 = 48.8 \cdot 10^{-2}$ and $c_2 = 26.5 \cdot 10^{-2}$) as well as the available bandwidth of these two paths ($b_1 = 4.07$Mb/s and $b_2 = 0.579$Mb/s). Clearly, the choice between P_1 and P_2 depends on whether one uses probabilistic congestion or available bandwidth. We show that probabilistic congestion is the best criterion. For that purpose,

---

[1] The complete description of the our scenario is available on request to the authors.

we again performed simulations during 60 seconds of TCP traffic, with several CBR voice flows from node x to node y, at 23Kb/s. The table in Figure 3 (right) displays the number of voice flows from x to y, the percentage of voice-packet lost on path P_1, the total number of packets lost on P_1 (during the 60 s), the percentage of voice-packet lost on path P_2, and the total number of packets lost on P_2 (during the 60 s). Using path P_2, the total number of packets losses is approximately $1/3$ of the number of packet losses when using path P_1, independently of the number of flows. Moreover, looking at the voice flows packets only, we observed that the percentage of packet losses in these flows is significantly smaller using path P_2 than using path P_1. For instance, if there are 60 aggregated voice flows, then the percentage of packet losses in these flows is 9.37% for path P_1, whereas it is only 0.64% for path P_2. Since voice is very sensitive to packet losses, the benefit in term of QoS of using a prediction criteria such as probabilistic congestion would be significant in comparison with other criteria such as available bandwidth.

5   Conclusion and Future Works

A challenging problem consists of checking whether the probabilistic congestion could be efficiently used for the whole unicast traffic. We have shown that probabilistic congestion captures simultaneously qualitative and quantitative traffic characteristics, and, from that perspective, it offers better properties than standard QoS parameters such as delays or bandwidth. However, it is unclear *how much* the global traffic would improve in terms of packet-loss. Indeed, the behavior of the network traffic depends heavily on TCP which dynamically adapts to the congestion of the routes. We leave this question open for future investigations.

References

[1] Alvarez-Hamelin, J.I., Fraigniaud, P.: MλT: A Multicast Protocol with QoS Support. In: The 12th International Conference on Computer Communications and Networks, ICCCN 2003, IEEE (2003) 264–269
[2] Beran, J., Sherman, R., Taqqu, M., Willinger, W.: Long-range dependence in Variable-Bit-Rate video traffic. IEEE Transaction on Communications **43** (1995) 1566–1579
[3] Carlberg, K., Crowcroft, J.: Building Shared Trees using a one-to-many joining mechanism. In: ACM SIGCOMM Computer Communication Review. (1997) 5–11
[4] Crovella, M., Taqqu, M., Bestavros, A.: Heavy-tailed probability distributions in the World Wide Web. chapter 1, pages 3-26, Chapman & Hall, New York (1998)
[5] Crovella, M.E., Bestavros, A.: Self-similarity in World Wide Web traffic: evidence and possible causes. IEEE/ACM Transactions on Networking **5** (1997) 835–846
[6] Faloutsos, M., Banerjea, A., Pankaj, R.: QoSMIC: Quality of Service sensitive Multicast Internet protoCol. In: SIGCOMM. (1998) Vancouver BC.
[7] Garrett, M., Willinger, W.: Analysis, modeling and generation of self-similar vbr video traffic. In: ACM SIGCOMM. (1994) 269–280

[8] Jain, R.: Congestion control and traffic management in ATM networks: Recent
     advances and A survey. Computer Networks and ISDN Systems **28** (1996) 1723–
     1738
[9] Leland, W., Taqqu, M., Willinger, W., Wilson, D.: On the self-similar nature of
     the Ethernet traffic. IEEE/ACM Transaction on Networking **2** (1994) 1–15
[10] Mandelbrot, B., Ness, J.V.: Fractional Brownian motion, fractional noises and
     applications. SIAM Review **10** (1968) 422–437
[11] Norros, I.: A storage model with self-similar input. Queueing Systems **16** (1994)
     387–396
[12] Norros, I.: On the Use of Fractional Brownian Motion in the Theory of Connec-
     tionless Networks. IEEE Journal of Selected Areas in Communications **13** (1995)
     953–962
[13] Norros, I.: Busy periods of fractional Brownian storage: a large deviations ap-
     proach. Adv. Perf. Anal. **2** (1999) 1–19
[14] Norros, I., Mannersalo, P., Wang, J.: Simulation of fractional Brownian motion
     with conditionalized random midpoint displacement. Adv. Perf. Anal. **2** (1999)
     77–101
[15] Park, K., Kim, G., Crovella, M.: On the relationship between file sizes, transport
     protocols, and self-similar network traffic. In: 4th Int. Conference on Network
     Protocols (ICNP '96). (1996) 171–180
[16] Park, K., Kim, G., Crovella, M.: On the Effect and Control of Self-Similar Network
     Performance. In: SPIE Conf. on Performance and Control of Network Systems.
     (1997)
[17] Paxson, V., Floyd, S.: Wide-Area Traffic: The Failure of Poisson Modeling.
     IEEE/ACM Transaction on Networking **3** (1995) 226–244
[18] Richard, O., Cappello, F.: Sur la nature auto-similaire de l'activité de stations
     de travail et de serveurs HTTP. Technique et Science informatiques **17** (1998)
     635–658
[19] Taqqu, M.S., Willinger, W., Sherman, R.: Proof of a fundamental result in self-
     similar traffic modeling. ACMCCR: Computer Communication Review **27** (1997)
     5–23
[20] Veres, A., Boda, M.: The chaotic nature of TCP congestion control. In: IEEE
     INFOCOM. (2000) 1715–1723
[21] Veres, A., Kenesi, Z., Molnár, S., Vattay, G.: On the propagation of Long-Range
     Dependence in the Internet. In: ACM SIGCOMM. (2000) 243–254
[22] Willinger, W., Taqqu, M.S., Sherman, R., Wilson, D.V.: Self-similarity through
     high-variability: statistical analysis of Ethernet LAN traffic at the source level.
     IEEE/ACM Transactions on Networking **5** (1997) 71–86
[23] : http://www.isi.edu/nsnam/ns/.
[24] Zappala, D., Estrin, D., Shenker, S.: Alternate path routing and pinning for inter-
     domain multicast routing. Technical Report USC-CS-TR-97-655, University of
     Southern California (1997)

A Family of Performance Bounds for QoS Measures in Packet-Based Networks*

József J. Bíró, Zalán Heszberger, and Mátyás Martinecz

Budapest University of Technology and Economics
Department of Telecommunications and Media Informatics
H-1117, Magyar tudósok körútja 2., Budapest, Hungary
biro@tmit.bme.hu

Abstract. In characterizing statistical multiplexing models of network nodes important QoS measures are often to be estimated such as saturation probability and workload loss ratio. In the two main streams of multiplexer models (bufferless and buffered statistical multiplexers) the meaning of these measures and their possible estimations are different. In this paper, we set up a family of closed-form performance bounds which can be used for estimating QoS measures in both multiplexing models. The underlying approximation technique in the framework of bufferless fluid flow multiplexing model (*bffm*) is systematically described, and besides the resultant new QoS measure estimates previously known ones are also identified. The bounds in *bffm* have been analyzed and compared based on numerical investigations. The applicability of the performance bounds for buffered multiplexers is also briefly discussed.

1 Introduction

Setting and estimating QoS measures play central role in traffic flow control and management functions in QoS guaranteed packet-based networks. As regards the available resources (e.g. transmission link capacities, buffers, processing capacities) at the nodes in the network, an important question is how often such resources are overloaded due to high volume of packets arrived. The class of QoS measures which quantify these overloading phenomena is often referred to as resource-based QoS measures [1]. One representative measure within this family is the link saturation probability which corresponds to the fraction of time when the sum of the instantaneous (or average over a sufficiently small time interval) arrival rate of traffic flows exceeds the transmission link capacity. The use of this measure assumes the bufferless fluid flow multiplexing framework which turned out to be powerful dimensioning tool in either the case of elastic traffic in certain access network scenarios [2] or the case of stream-like traffic [3].

Another characteristic member of the family of resource-based QoS measures is the buffer saturation (or buffer overflow) probability. The analysis of this apparently requires a buffered statistical multiplexing model. Several asymptotic

---

* This work was supported by the Inter University Centre for Telecommunications and Informatics, Hungary.

N. Mitrou et al. (Eds.): NETWORKING 2004, LNCS 3042, pp. 1108–1119, 2004.

and approximate results have been formulated for buffer overflow probability under different assumptions. Recent result in [4] incorporates bounds for buffer overflow probability provided the arrival traffic flows are regulated (the regulation is characterized by arrival curves) and the service offered to the traffic by the nodes is described by a so-called service curves. These bounds are based on Hoeffding's result on the tail probability estimation of sum of partial backlogs as bounded random variables [5].

Besides the fraction of resource overload periods, it is also important to identify quantities (measures) which are based on the amount of traffic becoming unconformant due to the resource overload. The class of these measures is referred to as stream-based[1] measures. Although the unconformant packets can be either downgraded into lower level QoS class (e.g. best effort) or simply discarded, for simplicity, the ratio between the unconformant traffic and the whole offered traffic is called as workload loss ratio (*WLR*). In the *bffm* modeling framework *WLR* corresponds to the fraction of traffic which can not be transmitted due to the link saturation. In buffered statistical multiplexers *WLR* means the fraction of packets which can not be placed into the buffer due to buffer overflow.[2]

In this paper, we set up a family of conservative upper bounds for saturation probability and workload loss ratio, which can directly be applied in both (bufferless and buffered) multiplexing framework. For this purpose, the so-called Chernoff-Hoeffding bounding method as an approximation technique has been used which enables to treat the estimates for saturation probability and workload loss ratio in a common way. This underlying approximation method in the framework of bufferless fluid flow multiplexing model (*bffm*) is systematically described, and besides the resultant new QoS measure estimates previously known ones are also identified. The very attractive properties of the bounds that they are expressed in closed-form formulae and use few characteristic information (known a priori and/or measurable) on the traffic flows.

In Section 2 we introduce the general concept of Chernoff-Hoeffding bounding technique in the context of bufferless fluid flow multiplexing model, which originally motivates the design of the bounds on the related QoS measures. In Section 3 the underlying probability generating function approximations have been performed. Previously known and newly developed bounds has been presented under a common framework in Section 4. After that analysis and comparisons based on extensive numerical investigations have been highlighted. The applicability of the performance bounds for buffered multiplexers is also briefly discussed.

---

[1] It does not necessarily mean that these measures can be applied only for streaming traffic.

[2] In buffered multiplexing models the buffer size is usually assumed to be infinite, hence, *WLR* is often identified as the fraction of workload in the buffer being above a certain threshold.

2   The Chernoff-Hoeffding Bounding Method

Bufferless fluid flow multiplexing is often used in the literature to analyze QoS measures, e.g., packet loss probability in a multiplexer [4,1,3]. Because this approach assumes no buffer at burst time scales, it is able to provide conservative estimates for the QoS measures under question. For modeling purposes under *bffm*, let us assume that we have n fluid flows to be multiplexed on a communication link with transmission capacity C. Let the instantaneous stationary (that is time dependence can be eliminated) arrival rate of flow i be noted by X_i, as a random variable. Because every flow has a peak rate p_i we also have $0 \leq X_i \leq p_i$. Further, let the aggregate flow arrival rate be $X = \sum_{i=1}^{n} X_i$.

The link saturation probability can now be defined as

$$P_{\text{sat}} \stackrel{\text{def}}{=} \mathbb{P}(X > C).$$

This probability reflects the fraction of time when the link is overloaded (provided the system is ergodic), i.e. the frequency that the combined arrival rate exceeds the link capacity. This resource-based congestion measure could be important from network operation point of view. The workload loss ratio can be identified as

$$WLR \stackrel{\text{def}}{=} \frac{\mathbb{E}[(X - C)^+]}{\mathbb{E}[X]} , \tag{1}$$

where $\mathbb{E}[.]$ stands for the expectation value operator and $(X - C)^+ = \max(X - C, 0)$. The estimation of this quantity can provide more accurate loss performance analysis. This measure better characterizes the expected loss rate and could also contribute to determining the users' satisfaction. From traffic management (e.g. connection admission control) point of view an important question can arise: Whether the ongoing session (possibly together with a newcomer) satisfies a predefined QoS constraint related to some quality of service measure. In a more formal way, the inequalities

$$\mathbb{P}(X > C) \leq e^{-\gamma} , \quad \frac{\mathbb{E}[(X - C)^+]}{\mathbb{E}[X]} \leq e^{-\gamma} \tag{2}$$

represent the fulfillment of the constraint on saturation probability and workload loss ratio, respectively.

The Chernoff bound of P_{sat} and WLR are as follows [6]:

$$\mathbb{P}(X > C) \leq \inf_{s>0} \frac{G_X(s)}{e^{sC}} = \inf_{s>0} \exp\left(\Lambda_X(s) - sC\right) , \tag{3}$$

$$WLR \leq \exp\left(\Lambda_X(s^*) - s^*C - \log(s^*M)\right) , \tag{4}$$

where

$$s^* = \operatorname{arginf}_s(\Lambda_X(s) - sC) , \tag{5}$$

and $G_X(s) \stackrel{\text{def}}{=} \mathbb{E}[\exp(sX)]$ and $\Lambda_X(s) \stackrel{\text{def}}{=} \log G_X(s)$ are the probability generating function (PGF) and the cumulant generating function (CGF) of X, respectively.

The computation of these bounds is usually not possible, because the underlying generating functions would require all the moments of X to be known. Instead, the CGF's are to be further bounded based on the available information (moments) on X and embedded into the Chernoff bound. This is called the Chernoff-Hoeffding bounding method.

3 Approximations of Probability Generating Functions (PGF)

In the previous section we have seen that the cumulant generating function plays important role in QoS measure approximation through the Chernoff bounding method. In this section, we provide three conservative bounds of the PGF of aggregate traffic rate distribution, provided only the following pieces of information are available on X: the number of traffic flows multiplexed (n), the peak rates of the traffic flows (p_i) and the aggregate mean arrival rate ($M \overset{\text{def}}{=} \mathbb{E}[X]$).

3.1 Approximations Based on Hoeffding's Result

The following lemmas due to Hoeffding (1963), is on the PGF approximation of bounded random variables.

Lemma 1 ([5]).
   Let X_i, $i = 1 \ldots n$ be independent random variables with $X = \sum_{i=1}^{n} X_i$, $M = \mathbb{E}[X]$ and $0 \le X_i \le p$. Then, for $s > 0$

$$G_X(s) \le \left(1 - \frac{M}{np} + \frac{M}{np} \exp(sp) \right)^n \qquad (6)$$

Lemma 2 ([5]). Let Y be a random variable with $\mathbb{E}[Y] = 0$, $a \le Y \le b$. Then for $s > 0$, $\mathbb{E}[\exp(sY)] \le \exp\left(\frac{s^2 (b-a)^2}{8} \right)$

Based on this lemma it can be easily constructed an upper bound for the PGF of sums of independent and bounded random variables.

Corollary 1. Let X_i, $i = 1 \ldots n$ be independent random variables with $X = \sum_{i=1}^{n} X_i$, $M = \mathbb{E}[X]$ and $0 \le X_i \le p_i$. Then for $s > 0$,

$$\mathbb{E}[\exp(sX)] \le \exp(sM) \exp\left(\frac{s^2 \sum_{i=1}^{n} p_i^2}{8} \right) \qquad (7)$$

The PGF bound in (6) is applicable for random variable bounded uniformly (but with not necessarily identical distribution), while that in (7) covers a more general case with non-uniformly bounded random variables. Nevertheless, the latter one does not coincide the former one in the special case of $p_1 = p_2 = \ldots = p_n$. This fact motivated the construction of the following upper bound.

1112 J.J. Bíró, Z. Heszberger, and M. Martinecz

3.2 An Improved Hoeffding-Type Approximation

Applying the Chernoff-Hoeffding bounding method on X in a different way, we have obtained the following conservative bound for $\mathbb{E}[\exp(sX)]$.

Theorem 1 ([1]). *Let X_i be independent bounded random variables with $0 \leq X_i \leq p_i$, $X = \sum_{i=1}^n X_i$ and $M = \mathbb{E}[X]$. Then for $s > 0$,*

$$E\left[e^{sX}\right] \leq \prod_{i=1}^n \left(\frac{e^{sp_i} - 1}{p_i}\right) \left(\frac{M + \sum_{k=1}^n \frac{p_k}{e^{sp_k} - 1}}{n}\right)^n. \tag{8}$$

The proof of this theorem is not detailed here, it can be found in [1]. Nevertheless, an important step in the chain of bounding formulae is worth repeating here:

$$G_X(s) = \mathbb{E}[e^{sX}] \leq \prod_{i=1}^n \left(1 + m_i \frac{e^{sp_i} - 1}{p_i}\right), \tag{9}$$

where $m_i \overset{\text{def}}{=} \mathbb{E}[X_i]$.

Corollary 2. *The PGF bound on the right-hand side in (9) is the exact generating function of the sum of heterogeneous on-off random variables with the distribution*

$$\mathbb{P}\left(X_i^{\text{onoff}} = p_i\right) = \frac{m_i}{p_i}, \quad \mathbb{P}\left(X_i^{\text{onoff}} = 0\right) = 1 - \frac{m_i}{p_i}.$$

This is because

$$\mathbb{E}\left[e^{sX_i^{\text{onoff}}}\right] = \left(1 - \frac{m_i}{p_i} + \frac{m_i}{p_i}e^{sp_i}\right).$$

It can also be seen that the formula on the right hand side in (8) gives back the bound in (6) in the case of uniformly bounded random varibales. In this way, this improved Hoeffding-type approximation is a consistent extension of the result presented in Lemma 1.

3.3 A PGF Approximation Based on Stochastic Ordering

In this subsection let us recall the essential definitions and properties of a certain type of stochastic ordering of random variables to be applied for PGF approximation.

Definition 1 ([4]). *Given two random variables X and Y with distribution function F_X and F_Y, respectively. Then, X is said to be smaller than Y with respect to increasing convex ordering, written as*

$$X <_{icx} Y,$$

if the condition

$$\int_{-\infty}^{\infty} \phi(x)\mathrm{d}F_X(x) \leq \int_{-\infty}^{\infty} \phi(x)\mathrm{d}F_Y(x)$$

holds for all increasing convex function ϕ, for which the integral exists.

An important consequence of this definition for probability generating functions of random variables is the following:

Lemma 3. *Let X and Y be two random variables with the relation $X <_{icx} Y$. Then for $s > 0$, $G_X(s) \leq G_Y(s)$.*

This can be justified by the substitution $\phi(x) = \exp(sx)$.

The following results presented in [4] leads us to construct a new PGF bound.

Lemma 4. *Let the random variables $X_1^{\text{onoff}}, \ldots, X_n^{\text{onoff}}$ represent n independent heterogeneous on-off sources with peak rates $p_1, \ldots, p_n$ and mean rates $m_1, \ldots, m_n$. Let $Y_1^{\text{onoff}}, \ldots, Y_{n_Y}^{\text{onoff}}$ be n_Y independent homogeneous on-off sources with the identical peak rate $p = \max(p_i, i = 1, \ldots, n)$, $n_Y = \lceil \sum_{i=1}^n p_i / p \rceil$, and identical mean rate $m = \sum_{i=1}^n m_i / n_y$. Then $X_{\text{onoff}} <_{icx} Y_{\text{onoff}}$, where*

$$X_{\text{onoff}} \overset{def}{=} \sum_{i=1}^n X_i^{\text{onoff}} \quad \text{and} \quad Y_{\text{onoff}} \overset{def}{=} \sum_{i=1}^{n_Y} Y_i^{\text{onoff}}. \tag{10}$$

For a proof of Lemma 4 see [4].

Now, the PGF bound based on increasing convex stochastic ordering can be formulated in the following theorem [7]:

Theorem 2. *Let $X_1, \ldots, X_n$ indicate n independent random variables with $0 \leq X_i \leq p_i$, $X = \sum_{i=1}^n X_i$ and $M = \mathbb{E}[X]$. Then for $s > 0$,*

$$G_X(s) \leq \left(1 - \frac{M}{n_Y p} + \frac{M}{n_Y p} e^{sp}\right)^{n_Y} \tag{11}$$

Proof of Theorem 2: By Corollary 2 we have $G_X(s) \leq G_{X_{\text{onoff}}}(s)$, $\forall\, s > 0$. Further, by combining Lemma 4 and Lemma 3 the following relation also holds: $G_{X_{\text{onoff}}}(s) \leq G_{Y_{\text{onoff}}}(s)$, $\forall\, s > 0$. The two inequalities above give the statement of the theorem, because $G_{Y_{\text{onoff}}}(s) = \left(1 - \frac{M}{n_Y p} + \frac{M}{n_Y p} e^{sp}\right)^{n_Y}$. Q.E.D.

Let the PGF approximations presented in (7), (8) and (11), be designated by $\widetilde{G}_{X,\text{hoe}}(s)$, $\widetilde{G}_{X,\text{ih}}(s)$ and $\widetilde{G}_{X,\text{so}}(s)$, respectively. The corresponding cumulant generation functions (CGF's) are $\widetilde{\Lambda}_{X,\text{hoe}}(s)$, $\widetilde{\Lambda}_{X,\text{ih}}(s)$ and $\widetilde{\Lambda}_{X,\text{so}}(s)$.

4   Conservative Upper Bounds Based on the PGF Approximations

4.1   Bounds for the Saturation Probability P_{sat}

Applying the Chernoff bound for the saturation probability of uniformly bounded random variables with the CGF bound in (6) the following bound can be obtained [5]:

Theorem 3. *Let X_i, $i = 1 \ldots n$ be independent random variables with $X = \sum_{i=1}^{n} X_i$, $M = \mathbb{E}[X]$ and $0 \le X_i \le p$. Then, for $C > M$*

$$\mathbb{P}(X > C) \le \left(\frac{M}{C}\right)^{\frac{C}{p}} \left(\frac{np - M}{np - C}\right)^{n - \frac{C}{p}} \tag{12}$$

In this case the optimal s parameter can also be expressed as

$$s^* = \frac{1}{p} \log \frac{C}{M} \frac{np - M}{np - C}. \tag{13}$$

In the more interesting case of non-uniformly bounded random variables the following bounds can be obtained by the substitution of $\tilde{\Lambda}_{\text{hoe}}(s)$, $\tilde{\Lambda}_{\text{ih}}(s)$ and $\tilde{\Lambda}_{\text{so}}(s)$ into the Chernoff bound of P_{sat} (3).

Theorem 4 ([5]). *Let X_i be independent bounded random variables with $0 \le X_i \le p_i$, $X = \sum_{i=1}^{n}$ and $M = \mathbb{E}[X]$, then*

$$\mathbb{P}(X > C) \le \exp\left(\frac{-2(C - M)^2}{\sum_{i=1}^{n} p_i^2}\right) \tag{14}$$

The optimizing parameter s can be formulated here as

$$s^* = \frac{4(C - M)}{\sum_{i=1}^{n} p_i^2}. \tag{15}$$

Theorem 5 ([1]). *If $X_1, X_2, \ldots, X_n$ are independent (and not necessarily identically distributed) random variables, for which $0 \le X_i \le p_i$ holds, then*

$$\mathbb{P}(X \ge C) \le e^{-s^* C} \left(\frac{M + \sum_{j=1}^{n} \frac{p_j}{e^{s^* p_j} - 1}}{n}\right)^n \prod_{k=1}^{n} \frac{e^{s^* p_k} - 1}{p_k}, \tag{16}$$

where s^* is the solution of the following equation.

$$\sum_{k=1}^{n} \frac{e^{s p_k} p_k}{e^{s p_k} - 1} - \frac{n \sum_{j=1}^{n} \frac{e^{s p_j} p_j^2}{(e^{s p_j} - 1)^2}}{M + \sum_{j=1}^{n} \frac{p_j}{e^{s p_j} - 1}} - C = 0. \tag{17}$$

Unfortunately, in this case neither the optimizing parameter s^* nor the bound of P_{sat} can not be expressed in closed form. In [1] closed-form solutions have been developed for P_{sat} through finding closed-form suboptimal solutions of the equation above with respect to s. One of them is repeated here:

$$\mathbb{P}(X \ge C) \le e^{-\tilde{s}^* C} \left(\frac{M + \sum_{j=1}^{n} \frac{p_j}{e^{\tilde{s}^* p_j} - 1}}{n}\right)^n \prod_{k=1}^{n} \frac{e^{\tilde{s}^* p_k} - 1}{p_k}, \tag{18}$$

where

$$\tilde{s}^* = \frac{C - M}{\frac{1}{4} \sum_{i=1}^{n} p_i^2 - \frac{1}{n} \left(M - \frac{1}{2} \sum_{i=1}^{n} p_i\right)^2} \tag{19}$$

As a further new contribution of this paper, applying the CGF approximation $\widetilde{\Lambda}_{\text{so}}(s)$ based on stochastic ordering, inherently a closed form upper bound can be obtained for P_{sat}:

Theorem 6. *Let X_i be independent bounded random variables with $0 \leq X_i \leq p_i$, $X = \sum_{i=1}^{n} X_i$ and $M = \mathbb{E}[X]$. Further, let $p = \max(p_i, i = 1, \ldots, n)$, $n_Y = \lceil \sum_{i=1}^{n} p_i / p \rceil$, and $m = \sum_{i=1}^{n} m_i / n_y$, then*

$$\mathbb{P}(X > C) \leq \left(\frac{M}{C}\right)^{\frac{C}{p}} \left(\frac{n_Y p - M}{n_Y p - C}\right)^{n_Y - \frac{C}{p}} \tag{20}$$

Proof sketch of Theorem 6: Combining the result of Theorem 2 and Theorem 3 gives the required statement.

In this case the optimal s parameter can also be expressed as

$$s^* = \frac{1}{p} \log \frac{C}{M} \frac{n_Y p - M}{n_Y p - C} . \tag{21}$$

4.2   Bounds for the Workload Loss Ratio *WLR*

Turning to the *WLR* approximation, here it is worth using formula (4)[3], because in this case the optimizing parameter s and hence the resulted conservative upper bounds can be expressed directly in closed-form (like in the previously presented P_{sat} bounds) when the CGF approximations $\widetilde{\Lambda}_{X,\text{hoe}}(s)$ and $\widetilde{\Lambda}_{X,\text{so}}(s)$ are embedded in (4). When the CGF bound $\widetilde{\Lambda}_{X,\text{ih}}(s)$ is used similar sub-optimal solutions can be obtained as in the corresponding P_{sat} bound in (18).

In the following theorem we summarize these new closed form conservative bounds:

Theorem 7. *Let X_i be independent bounded (and not necessarily identically distributed) random variables with $0 \leq X_i \leq p_i$, $X = \sum_{i=1}^{n} X_i$ and $M = \mathbb{E}[X]$. Further, let $K = \frac{1}{4} \sum_{i=1}^{n} p_i^2 - \frac{1}{n} \left(M - \frac{1}{2} \sum_{i=1}^{n} p_i\right)^2$, $p = \max(p_i, i = 1, \ldots, n)$, $n_Y = \lceil \sum_{i=1}^{n} p_i / p \rceil$, and $m = \sum_{i=1}^{n} m_i / n_y$, then the following three inequalities hold for WLR:*

$$WLR \leq \frac{\sum_{i=1}^{n} p_i^2}{4(C - M)M} \exp\left(\frac{-2(C - M)^2}{\sum_{i=1}^{n} p_i^2}\right) , \tag{22}$$

$$WLR \leq \left(\frac{M + \sum_{j=1}^{n} \frac{p_j}{e^{\frac{C-M}{K} p_j} - 1}}{n}\right)^{n} \frac{K e^{\frac{(M-C)C}{K}}}{(C - M)M} \prod_{k=1}^{n} \frac{e^{\frac{C-M}{K} p_k} - 1}{p_k} , \tag{23}$$

$$WLR \leq \left(\frac{M}{C}\right)^{C/p} \left(\frac{n_Y p - M}{n_Y p - C}\right)^{n_Y - C/p} \frac{p}{M \log \frac{C}{M} \frac{n_Y p - M}{n_Y p - C}} . \tag{24}$$

---

[3] A slight improvement of formula (4) would be $\inf_{s>0} \exp\left(\Lambda_X(s) - sC - \log(sM)\right)$, but the use of this would require further approximations to arrive closed-form expression.

Proof sketch of Theorem 7 : Substituting the three CGF approximations $\tilde{\Lambda}_{X,\text{hoe}}(s)$, $\tilde{\Lambda}_{X,\text{ih}}(s)$ and $\tilde{\Lambda}_{X,\text{so}}(s)$ and the corresponding optimization parameters s^* performed in equations (15), (19), (21) into the Chernoff bound of *WLR* (4), the three bounds above are obtained.

Finally, it is worth again emphasizing that in the set of bounds presented above the ones in (14) and (18) are already known from [5] and [1], but, to the authors best knowledge, the bounds in (20), (22), (23) and (24) are neither presented nor analyzed previously.

5   Performance Analysis

In this section the performance of the bounds are analyzed and illustrated through numerical examples. For this purpose a simple two-class on-off traffic mix has been defined. The number of sources within the classes are represented by n_1 and n_2, respectively. The mean arrival rate and the peak rate of a source within a class are assumed to be identical and indicated by m_i, p_i, $i = \{1, 2\}$. The representative traffic scenarios considered in the paper for illustrating the numerical investigations are summarized in Table 1. The first traffic mix (Mix 1) resembles the aggregation of uncompressed voice and compressed video flows. The second (Mix 2) and third one (Mix 3) represent the multiplexing of uncompressed and compressed voice traffic with low and high peak to mean ratio, respectively.

Table 1. Traffic Scenarios

	n_1	m_1 [kbit/s]	p_1 [kbit/s]	n_2	m_2 [kbit/s]	p_2 [kbit/s]	$\frac{P}{M}$
Mix 1	100	51	64	10	2×10^3	5×10^3	2.24
Mix 2	100	51	64	1000	4.8	5.8	1.34
Mix 3	100	25	64	1000	2.3	5.8	2.54

In the figures the 10-based logarithm of the exact values of saturation probability and workload loss ratio and their bounds are drawn in the function of the transmission capacity C. Since the bounds presented give reasonable values when $M < C < P$ ($P \overset{\text{def}}{=} \sum_{i=1}^{n} p_i$), parts of the interval (M, C) is considered in the drawing in such a way that the exact values of P_{sat} and *WLR* should be no smaller than 10^{-8}. The exact values are drawn with continuous lines, while the bounds based on the Hoeffding, improved Hoeffding and stochastic ordering-based CGF bounds are represented by dotted, dash-dotted and dash-dot-dotted lines, respectively.

Common observations and remarks based on our extensive numerical analysis are given, which are partly illustrated by the numerical examples.

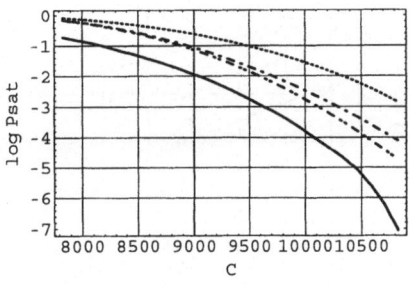

Fig. 1. Bounds on P_{sat}, Mix 1

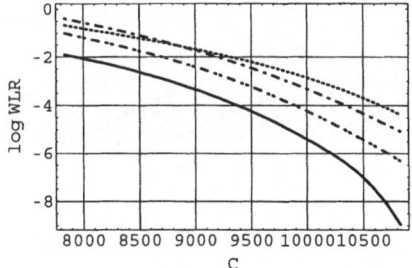

Fig. 2. Bounds on WLR, Mix 1

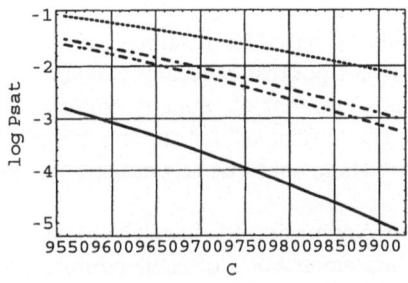

Fig. 3. Bounds on P_{sat}, Mix 2

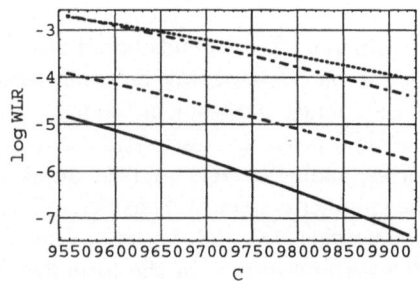

Fig. 4. Bounds on WLR, Mix 2

Observations:

- The bounds (14), (22) based on the CGF approximation (7) has usually the poorest performance, due to the underlying coarse bound on the cumulant generating function.
- The differences between the improved Hoeffding and stochastic ordering-based P_{sat} bounds are usually small, furthermore, it turned out to be negligible when the number of sources are higher than 100 in each traffic class and the peak rates of the traffic classes are in similar order of magnitude (e.g. in Mix2 and Mix 3).
- The superiority of the stochastic ordering-based WLR bound can be observed in several cases, especially when the aggregate peak to mean ratio (P/M) is small (e.g. in Mix 2).
- In case of high peak to mean ratio and high differences between the peak rates of the traffic classes, the improved Hoeffding-based P_{sat} bound can outperform (such figures not presented) the stochastic ordering-based P_{sat} bound.
- The horizontal and vertical distances between the curves are usually increases with increasing γ (with tightening the QoS constraint).

Remarks: Although all the bounds presented require the same amount of information on the the traffic flows, the complexity of their closed-form formulae

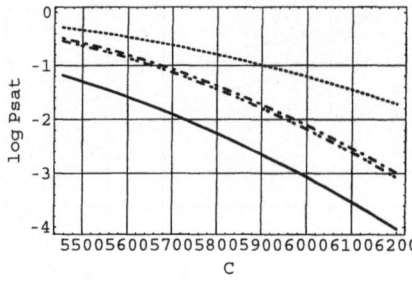

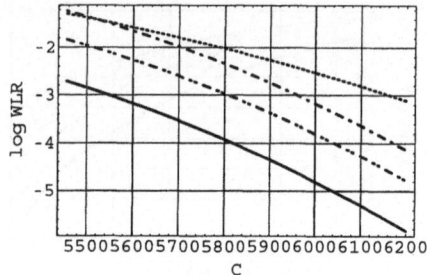

Fig. 5. Bounds on P_{sat}, Mix 3   **Fig. 6.** Bounds on WLR, Mix 3

are different. The bounds based on the Hoeffding-based CGF approximation appear in the simplest way, however, these have the poorest accuracy. Nevertheless, these bounds can result in saving considerable amount of capacity compared to the plain peak rate reservation scheme (see the horizontal distances between the corresponding curves), and the application of them could be recommended when the simplicity is an exclusive criterion.

The bounds (18), (23) based on the CGF approximation (8) have the most complicated appearance in the formulae, the implementation of their computation might encounter serious problems due to the presence of the several exponential-like terms. The improved Hoeffding-based saturation probability bound can have better performance in some cases (not seen in the figures) than the stochastic ordering-based one (see *Observations*), but the gain in capacity savings is not in proportion to the higher complexity of implementation.

The formulae of the stochastic ordering-based bounds are relatively simple, they seem to be implementable (especially the logarithm of the bounds) in a straightforward manner. Consequently, the application of these bounds (especially the *WLR* bound) is strongly encouraged, also because of the good performance in accuracy.

6   Conclusion

In this paper a family of bounds on saturation probability and workload loss ratio has been set up under the bufferless fluid flow multiplexing framework. This family comprises previously known as well as newly developed bounds. According to the analysis the stochastic ordering-based bounds have the best performance, especially in the case of workload loss ratio approximation. Nevertheless, the simple Hoeffding-based bounds could still form viable alternative from implementation point of view.

References

1. Heszberger, Z., Zátonyi, J., Bíró, J.: Efficient chernoff-based resource assessment techniques in multi-service networks. Telecommunication Systems **20** (2002) 59–80

2. Charzinski, J.: Fun factor dimensioning for elastic traffic. In: ITC Specialist Seminar on Internet Traffic Measurement, Modeling and Management, Monterey, CA, USA (2000)
3. Grossglauser, M., Tse, D.: A framework for robust measurement-based admission control. In: ACM SIGCOMM Symposium on Communications Architectures and Protocols, Cannes, France (1997) 237–248
4. Mao, G., Habibi, D.: Loss performance analysis for heterogeneous on-off sources with application to connection admission control. IEEE/ACM Transactions on Networking (TON) **10** (2002) 125–138
5. Hoeffding, W.: Probability inequalities for sums of bounded random variables. Journal of the American Statistical Association **58** (1963) 13–30
6. Hsu, I., Walrand, J.: Admission control for ATM networks. Stochastic Networks (1995) 413–429 Springer Verlag, New York.
7. Bíró, J.J., Heszberger, Z., Martinecz, M.: Equivalent capacity estimators for bufferless fluid flow multiplexing. In: IEEE Globecom, San Fransisco, CA (2003)
8. Kesidis, G., Tassiulas, L.: Traffic shaping for a loss system. IEEE Communication Letters (Dec. 2000) 417–419
9. Chang, C.S.: On the performance of multiplexing independent regulated inputs. In: Sigmetrics, Cambridge, Massachusetts, USA (2001)
10. Vojnovic, M., Le Boudec, J.Y.: Bounds for independent regulated inputs multiplexed in a service curve network element. IEEE Trans. on Communications **to appear** (2003)

Appendix: Improved Bounds for Independent Regulated Flows Multiplexed into a Service Curve Network Element

In [8] and [9] performance bounds (bounds on buffer overflow probability) have been derived when independent and regulated traffic flows have been multiplexed into a buffer with fixed service rate. These bounds have been extended in [10] to the case of a more general multiplexer model, i.e. the service is characterized by a general service curve [4] .

All the buffer overflow bounds presented in [8], [9] and [10] relies on the use of Hoeffding's inequalities [5], also presented in this paper in equation (12) and (14) for the homogeneous and heterogeneous case, respectively. Our closed form bounds presented in (18) and (20) are also based on and improve (see [1] and Section 4) one of the Hoeffding's inequalities (14), and use the same amount of information. Therefore, they can apparently be used for further improving the results of Vojnovic and Le Boudec in [10] in the case of heterogeneously regulated traffic multiplexed into a service curve network element. For the same reason, improved bounds on buffer workload loss ratio can also be set up by the use of (23) and (24). The performance evaluation of the improved bounds in the buffered multiplexer context is a matter of future work.

---

[4] In this model the multiplexer is referred to as service curve network element.

Impact of Frequency-Based Cache Management Policies on the Performance of Segment Based Video Caching Proxies

Anna Satsiou and Michael Paterakis

Laboratory of Information and Computer Networks
Department of Electronic and Computer Engineering
Technical University of Crete,
73100 Chania, Crete, Greece.
Tel: +30-28210-37343, 37225, Fax: +30-28210-37542
{anna, pateraki}@telecom.tuc.gr

Abstract. This paper examines a novel cache management policy, which we call LRLFU and is applied to groups of blocks of the media files. This cache management policy is capable of capturing the changing popularities of the various videos by attaching a caching value to every video according to how recently and how frequently the video was requested, and decides to cache the most 'valuable' videos. Our event-driven simulations have shown that the frequency considerations we have introduced in the caching values attached to the videos (i) improve the byte-hit ratio and (ii) significantly reduce the fraction of user requests with delayed starts.

Keywords: Multimedia Streaming, Video Proxy Caching, Segment-Based Video Caching, Frequency Based Cache Management Policies.

1 Introduction

The widespread use of the Internet, and a mature digital video technology have led to an increase in various streaming media applications such as corporate communications, distance education, Internet TV broadcasting and entertainment video distribution. However, the high bandwidth requirements and the long duration of digital video are two major limitations in supporting high quality streaming applications at a large scale over the Internet. Particularly, during periods of network congestion and media server overload, start-up latency and interrupts of video streams are very common. In order to alleviate the above situation, research efforts have been undertaken in two directions: a) use of multicast or broadcast connections to stream a popular video to groups of clients, and b) video caching at proxies located closer to the end-user.

Recently, there has been research in a combination of the above areas (see [1]-[4]), where partial caching and i) periodic broadcast [1],[2], or ii) techniques like batching, patching and stream merging [3],[4], are studied together. However multicast is considered impractical as its deployment in the Internet is still very limited. As far as sole proxy caching is concerned, research has been focusing in two directions: i) the effective maximization of delivered quality to heterogeneous clients (see [5],[6]) and

N. Mitrou et al. (Eds.): NETWORKING 2004, LNCS 3042, pp. 1120–1131, 2004.
© IFIP International Federation for Information Processing 2004

ii) the reduction of server loads, network traffic and user access latencies (see [7]-[11]).

Most of the above video caching schemes do not dynamically cache the files in response to individual client requests, rather they employ replication of parts of the videos according to a pre-estimated access pattern of each video. In reality, the request rate for a particular video may vary with time, and the relative popularities of the videos may vary from proxy to proxy. Proxies must be flexible to the particular preferences of different client populations and should be able to dynamically change their content accordingly. In [9] dynamic caching mechanisms, which are robust for evolving client workloads, are examined. Partial caching is used by segmenting video files in segments of variable size forming a pyramid and a portion of the cache capacity is dedicated to cache only the prefixes of each object, while the remaining part of the cache is used for the later segments. Caching a prefix of the video relieves clients from delays and jitter on the server-proxy path, while a proper cache admission and replacement policy is used to achieve a high byte-hit-ratio (BHR) even when popularity changes occur frequently. Video segments are ejected from the cache according to a caching value, which is based on how recently they have been requested, and their distance from the beginning of the video file.

Our work is based on some ideas from [9], it adopts the idea of using a portion of the cache referred to as part A to cache the prefixes of the videos and the rest referred to as part B for the latter segments, while using a different than in [9] cache management policy and examining a new segmentation scheme. Our work aims at i) improving the byte-hit ratio (BHR) at the proxy, thus reducing the average backbone load, irrespectively of the transmission scheme that is used and ii) reducing the fraction of user requests with delayed starts.

The remainder of the paper is organized as follows. Section 2 presents the system model and describes the segmentation schemes and cache management policies that we use. Section 3 presents our event-driven simulation model, parameters and results. The same section contains the discussion of the results. Finally, section 4 concludes the paper and describes ideas of ongoing work.

2 System Description

2.1 Network Topology

Figure 1 illustrates the system architecture we are considering. Our system consists of a far-distant origin server which stores the video files, proxy servers located close to the client populations which cache the files or parts of the files, and client devices. User requests for videos are directed to the nearest proxy. If the proxy has the whole or part of the media file, it transmits it to the client, while it contacts the content server for the rest of the file. If the video is popular enough, the proxy server decides to cache a part of it so that it can satisfy subsequent requests for the same video. In order for the traffic volume in the server-proxy path to be reduced, the most popular videos must be cached in the proxy. It is assumed that the bandwidth between the proxy and the clients is sufficient to support video streaming with negligible latency, while the latency between the content server and the proxy is quite significant.

Therefore, the idea is to cache enough initial segments of each video to mask the latency between the content and the proxy server.

2.2 Segmentation of Video Files

Here we introduce and briefly discuss two different segmentation schemes. In the first scheme, a media object consists of multiple equal-sized blocks, which is the smallest unit of transfer. Blocks of a media file received by the proxy server are then grouped into variable-sized, distance-sensitive segments. The segment size increases exponentially from the beginning segment, forming a pyramid. Assume that b is a constant and $b > 1$, S_i^p denotes segment i of a media stream under pyramid segmentation and N_i^p denotes the number of media blocks contained in S_i^p then

$$N_i^p = 1 \qquad\qquad \text{if } i = 0$$
$$N_i^p = (b - 1)b^{i-1} \quad \text{if } i > 0$$

Segment S_i^p contains media blocks $b^{i-1}, b^{i-1} + 1, \cdots b^i - 1$, if $i > 0$. Segment S_0^p contains block 0.

In the second segmentation scheme, we consider that the first segment consists of a small number of blocks Bp, forming the prefix of the video and later segments consist of a larger fixed number of blocks. Bp is chosen in a way to guarantee that enough blocks will be in the cache to mask the latency in the server-proxy path in order to guarantee continuous streaming of the video once it is started. We refer to this segmentation scheme as the 'fixed segmentation with prefix'.

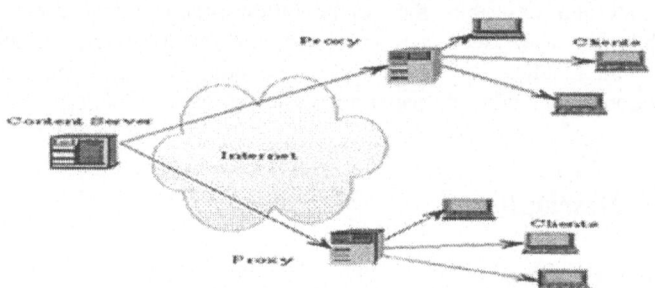

Fig. 1. Streaming video in the Internet.

2.3 Cache Management Policies

We have adopted from [9] the idea of dividing the cache in two parts. The size of the first part is a small portion of the total cache capacity and is dedicated to store only the initial blocks Bp (prefix) of each video while the rest is used to cache later segments. We refer to the first part as Cache A and to the second as Cache B. Different replacement policies are used in Cache A and Cache B, that is, different replacement policies are applied to prefixes than to later segments. When a media object that is not in the cache is requested, its Bp blocks are always eligible for caching in Cache A and form the prefix of the video. The later segments of the video will be considered for caching in Cache B in subsequent requests for the same video. Once an objects' prefix is cached in Cache A, information is maintained for this object, such as the object ID, the timestamp of the last request to this object T' (timestamp records the last time the object has been requested), and the number i of the last segment of the object inside the cache. We further keep the number of the requests for this object, RF. This information is updated at every request of the particular video.

In summary, the replacement policy in Cache A decides which videos should be cached in the proxy and the replacement and admission policies in Cache B determine the portion of these videos that should be cached.

2.3.1 Replacement Policy in Cache A

Bp blocks are cached as an entity in Cache A and can only be replaced by other initial Bp blocks of another video. In [9], in order to find space to cache the Bp blocks of a video, a simple LRU policy is applied to all the videos in Cache A that are not currently played. We however use here another replacement policy. We consider that every prefix in Cache A has a caching value given by $\dfrac{RF}{T - T'}$, where T is the current time. In order to cache the Bp blocks (the prefix) of a requested video, the caching values of the prefixes of all the videos in Cache A that are not currently played are examined and the one with the smallest caching value is removed from the cache. Every time a prefix of a video is moved out from Cache A the remaining segments of the video (the suffix), if any, that are cached in Cache B are also moved out from Cache B and all the information maintained for this video is deleted. The prefix of the requested video is cached, and as previously mentioned, its later segments will be considered for caching in subsequent requests for the particular video.

Notice that by applying this caching value to each prefix, the video that is ejected from the cache is not the least recently used but the least valued according to how frequently and how recently it was requested.

2.3.2 Replacement and Admission Policies in Cache B

Once a video is requested and its Bp blocks are already cached in cache A, the proxy checks to see if it must cache another segment X of that video in Cache B. At every subsequent request, only one segment is considered for caching. The authors in [9] assigned a caching value to every segment of a video, given by $\dfrac{1}{(T - T')i}$, where i is the number of the segment. According to this scheme, initial segments have larger caching values. Furthermore in [9], a **small number** of videos in Cache B that are not currently played are examined from the bottom of an LRU stack and the caching values of their last cached segments are compared. Then, the caching value of the least valued segment Y is compared with the caching value of the segment X that is considered for caching. If it is smaller than the caching value of X, the proxy removes segment Y. The same procedure is repeated until there is enough free space in Cache B to cache segment X. If not enough segments with smaller caching values than the caching value of X can be found in order to free space in Cache B, segment X is not cached in the proxy. Segments are always removed from the end of the cached video files.

In contrast to the above, we use a caching value for every video and not for individual video segments as this would simplify the cache management operation. The caching value of every video participating in the cache admission and replacement policy in Cache B is the same with the one used in Cache A, given by $\dfrac{RF}{T-T'}$. We would like to add here that we also tried to use the number of segment i in the denominator, as in [9], and assign a different caching value to each segment of a video file given by $\dfrac{RF}{(T-T')i}$. However, our simulation results were no sensitive at all to the presence of the segment number i in the denominator.

The procedure followed by our scheme in order to cache segment X of the requested video is the same with the one described above with the only differences that we examine **all** the videos in Cache B that are not currently played in order to select our victim segments and that we use a different caching value, which is the same for all the segments of the same video. Similarly, at every subsequent request only one segment of the video is considered for caching and video segments are always removed from the end of a cached video file.

In order to compare our scheme with the one in [9], we simulated their scenario but in order to find segment victims from Cache B, we examined the last segments of **all** the videos in Cache B that are not currently played and this explains the observation that their scenario appears to work better in our simulations than in theirs (according to the results reported in [9]). We compared our scheme with the one in [9], using pyramid and fixed segmentation with prefix schemes and we have shown that our scheme achieves much better results in all the performance metrics. We have concluded that this is attributed to the use of the Reference Frequency RF, which plays important role both in the replacement policy we use in Cache A and the replacement and admission policies we use in Cache B. In this way we attain a combination of an LRU and an LFU replacement policy. According to this LRLFU policy, the more frequently and the most recently a video file is requested the larger its caching value, and therefore the more difficult for it to be removed from the cache.

When a video is requested for the first time, the proxy always cache its initial Bp blocks as no information is available for that video to compare it with other videos in the cache. Once this video is cached in Cache A, we know ist RF and timestamp T' and therefore we can compute its caching value and update it at every subsequent request. If this video is popular enough, it can gradually bring its whole content inside the cache in subsequent requests, as in every request only one segment is considered for caching, dependent on its caching value. If it is not popular enough its Bp blocks will quickly be discarded from Cache A and the wrong decision to cache a small part of this video does not significantly affect the performance of the system. Popularity of each video is well captured by RF and timestamp, two metrics that adjust their values according to the client preferences.

3   Performance Evaluation

3.1   Performance Metrics

The main goal of the proxy server is to efficiently manage the cache capacity in order to reduce the required backbone rate. Byte hit ratio (BHR) is the primary metric that provides a direct measure of the savings in remote network and server bandwidth. It is defined as the fraction of total bytes that can be served directly from the proxy over the total bytes of media objects requested.

Playback delay is a very annoying effect to the clients and for that reason another important performance metric is the percentage of requests with delayed start. If a request for a video does not find the first initial Bp blocks in the cache A, it has a delayed start.

Another performance metric that we examined is the number of segment replacements during the simulation. The smaller the number of segment replacements is, the less the cache management overhead and the more effective the cache management policies are. However, due to loss of space we do not present results for this metric in the paper.

3.2   Simulation Model

We conducted event-driven simulations to evaluate the performance of a proxy with the cache management policy that was previously described. In order to compare our scheme with the one in [9], we have used the same system parameters, but we have also examined some additional scenarios .The Bp initial blocks cached for a video (prefix) was set to 32 blocks both in the pyramid and the fixed segmentation with prefix scheme.

We assume that client requests for videos arrive according to a Poisson process, therefore the inter-arrival times are exponentially distributed with mean λ. The default value of λ is 60.0 seconds. Videos are selected from a total of V distinct videos that are stored in the content server. The size of the videos is assumed uniformly distributed between 0.5M and 1.5M blocks, where M is the mean video size. The default value for M is 2,000. The playing time for a block is assumed to be 1.8 seconds, which means that the default playing time for a video is between 30 and 90 minutes, and that the playing time for a video prefix is almost one minute.

The popularity of each of the V videos is assumed to follow a Zipf-like distribution Zipf (s,V), where s corresponds to the degree of skew and V to the total number of videos in the content server. Every video x, x $\in$ {1,....,V} has a probability given by $p_x = c / x^{1-s}$, where $c = 1 / \sum_{x=1}^{V} 1 / x^{1-s}$ is a normalization constant. For s=0 the distribution is highly skewed, while for s=1 the distribution is uniform with no skew. The default value for s is 0.2. We also assume that the popularity of each video changes over time, so that we examine the behavior of the cache when popularity changes occur. In particular, as in [9], it is assumed that the popularity distribution changes every R requests in our simulations. When it does, another well-correlated Zipf-like distribution with the same parameters s, and V, is used. The correlation between the two Zipf-like distributions is modeled by using a single parameter k that can take any integer value between 1 and V. First, the most popular video in Zipf-like distribution 1 is made to correspond to the r1-th most popular video in Zipf-like distribution 2, where r1 is chosen randomly between 1 and k. Then, the second most popular video in distribution 1 is made to correspond to the r2-th most popular video in distribution 2, where r2 is chosen randomly between 1 and min (V, k +1), except that r1 is not allowed, and so on.

Thus, k represents the maximum position in popularity a video title may shift from one distribution to the next. Hence, k = 1 corresponds to perfect correlation, and k = V to the random case or no correlation. In most of our simulation scenarios, we consider that k=10 and R=200. This means that we adopted a scenario where the popularities of videos change progressively, approximately every 3 hours. The change is slight, but happens very frequently. We have also examined scenarios with larger or steeper popularity changes.

The cache size is expressed in terms of the number of media blocks. The default size is assumed equal to 400,000 blocks. This means that in the default scenario, we consider a cache capacity equal to 10% of the video repository. The portion of the cache capacity that is used for the initial Bp blocks of videos (Cache A) is denoted by pr and its default value is 10%. All the parameter default values are shown in Table 1.

Table 1. System parameters and default values

Notation	Definition (Default values)
Bp	Initial blocks cached for a video (prefix=32 blocks)
λ	Mean request inter-arrival time (60 sec)
V	Number of distinct video titles (2,000)
M	Mean number of blocks per video (2,000 blocks)
s	Parameter s in Zipf distribution (0.2)
k	Maximum shifting distance for a hot video (10)
R	Number of requests between shifting of video Popularity distribution (200)
C	Total cache capacity (400,000 blocks)
pr	Portion of cache capacity used for storing initial blocks (10%)
Runs	Number of requests in a simulation run (100,000 requests)

3.3 Simulation Results

In our simulations, we compared our cache management policy with the cache management policy used in [9] when i) pyramid segmentation and ii) fixed segmentation with prefix is used. The prefix was chosen to be 32 blocks long in both segmentation schemes to facilitate result comparisons.

We examined the byte-hit ratio for a variety of fixed segment sizes (all multiples of 64 blocks) in the fixed segmentation with prefix scheme and we observed that the scheme works better with a larger segment size. This is due to the fact that an attentive cache admission policy is used, in order to cache a segment of a video.

Therefore, when most of the decisions to cache a segment are right, it is preferable to cache a large segment. Throughout our simulations of the fixed segmentation with prefix scheme, we used the 32-960 case as our default case, where 32 is the size of the prefix in blocks and 960 is the size of the fixed segment also in blocks. We also examined the 32-960 case in our simulations, as the default fixed segmentation with prefix scheme for the cache management policy used in [9] which for brevity we denote by LRU-i because of the contribution of the number of segment i in an LRU policy. In our simulations of the 32-960 case and for video length smaller than 960 blocks, we considered that such videos are divided in two parts: the prefix which is 32 blocks and is cached in cache A and the suffix which consists of the remaining video and is cached in Cache B as one segment.

3.3.1 The Impact of Cache Size, Number of Videos, and Mean Video Length

Initially, we studied the performance of our scheme and the scheme in [9] with pyramid and fixed segmentation with prefix for different cache sizes, or different number of videos in the content server or different mean video lengths. For all these differentiations in the above parameters we observe that the pyramid and fixed segmentation with prefix schemes perform equally well in terms of byte-hit ratio and percentage of delayed starts in our scenario, but the pyramid segmentation scheme outperforms the fixed segmentation with prefix scheme in the scenario of [9] in terms of byte-hit ratio.

The percentage of delayed starts for the pyramid and fixed segmentation with prefix scheme is the same for a given cache management policy, as it depends only on the presence or not of the initial 32 blocks of the requested videos in Cache A, which is determined by the cache management policy and not by the segmentation scheme. Hence, our figures present the fraction of requests with delayed start only when the pyramid segmentation scheme is used.

It is remarkable that the LRLFU method performs better than LRU-i in terms of all the performance metrics for all the different cases examined. For the pyramid segmentation scheme, there is an 8% improvement in byte-hit ratio and 17% improvement in percentage of delayed starts when using LRLFU instead of LRU-i. For the fixed segmentation with prefix scheme the improvement is even larger. We note here that this improvement in byte-hit ratio is observed when the comparison is done between our scheme and the scheme in [9], simulated by us with an exhaustive search of segment victims in Cache B. Otherwise, direct comparison with the results in [9] shows that our scheme is 11% better in terms of byte-hit ratio.

From the results in Figure 2 we conclude that byte-hit ratio is better for larger cache sizes, smaller number of videos and smaller video lengths for all the scenarios examined, as it was expected. When we vary the size of Cache A, that is, the percentage of cache capacity used for storing initial blocks, we notice from Figure 2 (b) that for pr=5%, pyramid segmentation in LRU-i attains a much lower byte-hit ratio than in LRLFU. 32-960 scheme in LRU-i behaves better than pyramid in LRU-i but not as well as pyramid and 32-960 in LRLFU. These results reveal the superiority of our replacement policy in Cache A which decides smartly which videos to cache. When the size of Cache A is very small, not many prefixes can be cached and this

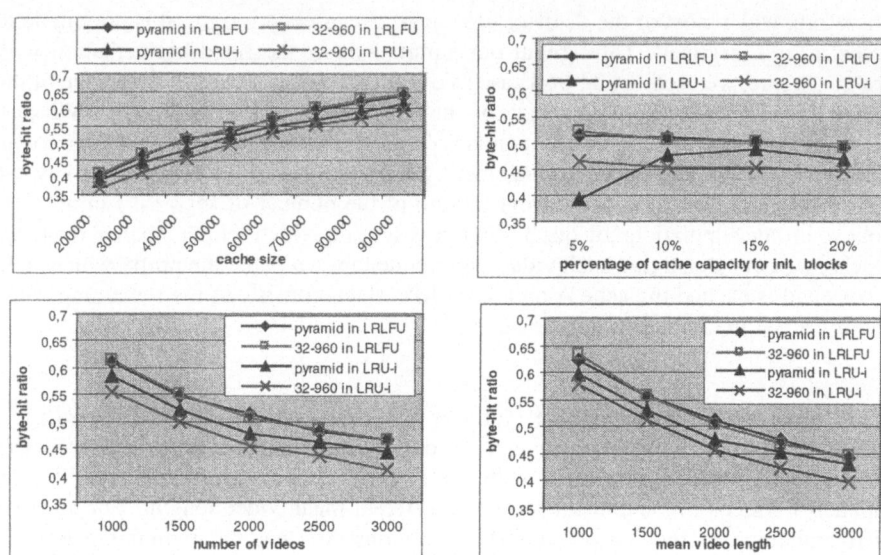

Fig. 2. Byte-hit ratio versus a) cache size b) percentage of cache capacity used for cache A c) number of total videos d) mean video length.

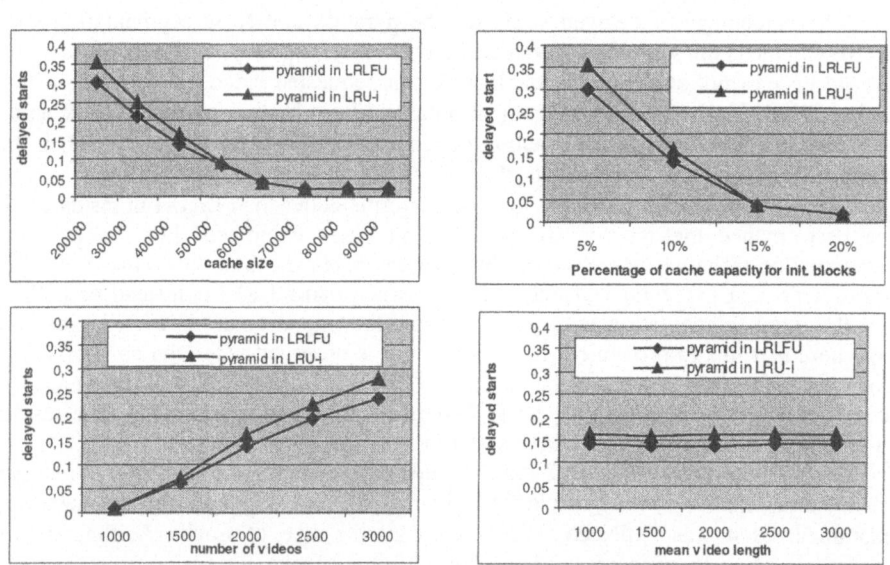

Fig. 3. Fraction of requests with delayed starts versus a) cache size b) percentage of cache capacity used for Cache A c) number of total videos d) mean video length.

also restricts the number of videos in general that can be cached in the proxy. As it has already been mentioned, only videos that are cached in Cache A can be cached in Cache B. 32-960 scheme in LRU-i performs better than pyramid in LRU-i for pr=5%, however, because the cache admission policy in Cache B succeeds to cache larger portion of popular videos with 32-960 scheme than with the pyramid scheme, before a change in popularity occurs and different videos are cached in Cache A.

The fraction of requests with delayed start depends only on the replacement policy in Cache A and the results presented in Figure 3 exhibit once again the superiority of our replacement policy in Cache A. When this fraction is near zero, Cache A has enough size to cache the prefixes of all the videos in the content server.

3.3.2 The Impact of the Parameters of the Popularity Distribution

The results in Figure 4 show the impact of the degree of skew in video popularity on the byte-hit ratio and percentage of delayed starts. Small values of the Zipf parameter s correspond to a more skewed popularity distribution, i.e., more clients are interested in fewer videos. As the parameter s increases, i.e., client's preferences are dispersed among a plethora of videos, byte-hit ratio decreases while the percentage of delayed starts increase. However, LRLFU performs better than LRU-i even for large values of the parameter s.

The results in Figure 5 show the impact of maximum shifting distance k on byte-hit ratio and on delayed start. Maximum shifting distance k determines the extent of the popularity change once such a change occurs. We can see from Figure 5(a) that as k increases from k=10 to k=50, LRLFU and LRU-i perform almost identically in terms of byte-hit ratio. However LRLFU still provides better results in terms of delayed starts (Figure 5(b)).We have also examined cases where either a steeper change in popularity distribution occurs or large but less frequent popularity changes occur and we concluded that LRLFU is preferable when there are small and regular popularity changes, i.e., a progressive change in the popularities of videos, or large but less frequent popularity changes. For large and frequent changes in the popularities of the videos, an uncommon case in real systems, the performance of all the scenarios deteriorates, but among them the 32-960 in LRU-i scenario provides the best results (the corresponding results are not shown here due to space considerations).

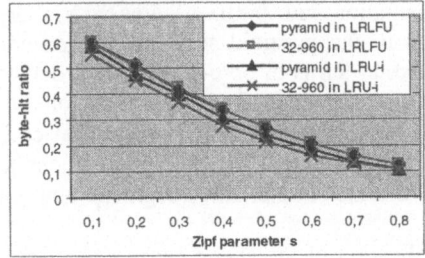 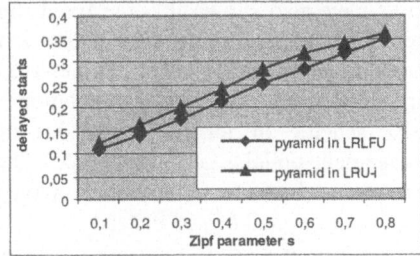

Fig. 4. The impact of skew in video popularity on (a) byte-hit ratio (b) delayed starts.

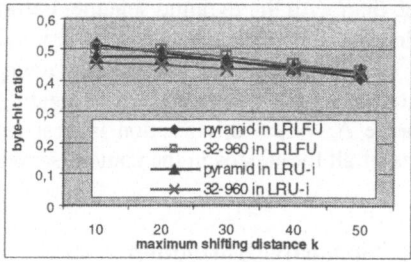 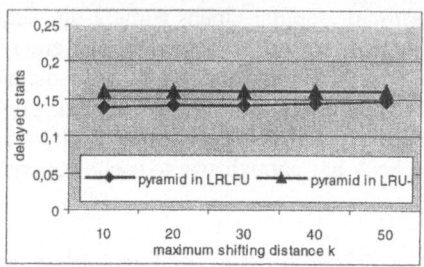

Fig. 5. The impact of maximum shifting distance on (a) byte-hit ratio (b) delayed starts.

4   Conclusions and Ongoing Work

In this paper, we have studied caching strategies for media objects in order to reduce the bandwidth requirements in the backbone links of the network and shield clients from delays and jitter on the server-proxy path. We considered the division of the cache into two parts, part A and part B. Replacement policy in part A decides which videos should be evicted from the cache in order for newly-requested videos to be cached, while replacement and admission policies in part B control the portion of the cached videos that should be inside the cache. All the above decisions are based on the caching value of each video which reflects its popularity. Popularity of each video is better captured when apart from the timestamp T', which records the last time that a video has been requested, the number of requests for that video once it is cached, RF, is also considered in the caching value. Our simulation results have shown that the frequency considerations we have introduced in the caching value of the videos and the caching strategies that we used compared to the one in [9], improve the byte-hit ratio and significantly reduce the fraction of requests with delayed starts. he performance of our scheme deteriorates only for large and frequent changes in the popularity distribution, cases which are not expected to be common. Our simulation results have also shown that the fixed segmentation with prefix scheme with a large fixed segment size performs equally well with the pyramid segmentation scheme, when our cache management policies are used, in terms of byte-hit ratio and delayed starts. However pyramid segmentation outperforms the fixed segmentation with prefix scheme in terms of byte-hit ratio and delayed starts when the cache management policies of [9] are used. Therefore, when frequency considerations are used, pyramid segmentation can be avoided and the simpler fixed segmentation with prefix scheme can be used instead.

Our current work is focusing on the application of the above cache management policies and segmentation schemes in a collaborative environment of more than one proxy servers which serve homogeneous or even heterogeneous client communities. More specifically, a hierarchical tree topology system of proxies is considered where the prefixes of the videos are stored into small size proxy caches each located very close to the corresponding client community and larger caches located further away from the client communities. In such a collaborative system caches A and B are implemented in different proxy servers in a way that reduces the overall cache capacity needed to achieve certain performance goals.

References

[1]   D.Eager, M.Ferris, M.Vermon,"Optimized regional caching for on-demand data delivery" in Proc. of Multimedia Computing and Networking Jan. 1999.

[2]   Y. Guo D. Towsley, "Prefix caching assisted periodic broadcast: Framework and techniques to support streaming for popular videos", Technical report, UM-CS-2001-022, Dept. of Computer Science, University of Massachusetts, May 2001.

[3]   B. Wang, S. Sen, M. Adler, D. Towsley, "Optimal proxy cache allocation for efficient streaming media distribution" , in Proc. of the 2002 IEEE INFOCOM Conf., New York, NY, pages 1726–1735.

[4]   C. Venkatramani, O. Verscheure, P. Frossard, K. Lee, "Optimal Proxy Management for Multimedia Streaming in - Content Distribution Networks", NOSSDAV'02, May 12-14, 2002, Miami, Florida, USA.

[5]   J. Kangasharju, F. Hartanto, M. Reisslein, K. W. Ross. "Distributing layered encoded video through caches", in Proc. of the 2001 IEEE INFOCOM Conf., Anchorage, Alaska, April 2001, pages 1791–1800.

[6]   R. Rejaie J. Kangasharju, "Mocha: A quality adaptive multimedia proxy cache for internet streaming", in Proc. of the International Workshop on Network and Operating Systems Support for Digital Audio and Video (NOSSDAV'01), June 2001, pages 3-10.

[7]   S. Sen, J. Rexford, D. Towsley, "Proxy prefix caching for multimedia streams", in Proc. of the 1999 IEEE INFOCOM Conf., pages 1310–1319.

[8]   M. Almeida, D.L.Eager, M.K. Vernon,"A Hybrid Caching Strategy for Streaming Media Files" in Proc. of the MMCN '01, San Jose, CA, Jan. 2001.

[9]   K.-L. Wu, P. S. Yu, J. L. Wolf, "Segment-based proxy caching of multimedia streams", in Proc. of the 10th International WWW Conference, Hong Kong, 2001. An extended version titled "Segmentation of Multimedia Streams for proxy caching" will appear in IEEE Trans. on Multimedia.

[10]   E. Balafoutis, A. Panagakis, N. Laoutaris, I. Stavrakakis, "The impact of replacement granularity on video caching", in Proc. of the 2002 IFIP Networking Conf., Pisa, Italy, pages 214–225.

[11]   Z.-L. Zhang, Y. Wang, D. H. C. Du, D. Su, "Video staging: A proxy-server-based approach to end--to--end video delivery over wide--area networks", IEEE/ACM Transactions on Networking, vol. 8, no. 4, pages 429–442, Aug. 2000.

Traffic Conscious Distribution of Service Components

Miltiades E. Anagnostou[1] and Maria A. Lambrou[2]

[1] National Technical University of Athens, School of Electrical and Computer Engineering, GR-15780, Athens, Greece, miltos@central.ntua.gr
[2] University of the Aegean,Business School, Department of Shipping, Trade and Trasport, GR-82000, Chios, Greece, mlambrou@aegean.gr

Abstract. A service is commonly realized by a set of components distributed over different nodes. For example, Internet based applications are orchestrated across a large scale distributed computing infrastructures and underlying resource elements; Similarly, in TINA-like approaches intelligence for control and management of services and resource, in particular, is distributed among network nodes and user/terminal nodes. In fact a service can be seen as a set of interacting components with a common purpose, being it application-oriented or of a support nature . The placement of components in different machines is more or less empirically determined at the design phase, by loosely (and occasionally subconsciously) taking into account "reasonable" predictions of the component usage. Code mobility has added complexity to the distribution problem. The aim of this paper is to present a methodology of dealing with component distribution, to explore its limitations, and to present its effect on service and network design.

1 Introduction

The service-oriented computing paradigm considers services as the fundamental elements for constructing applications. Composite services as resulting by basic service components aggregation are utilized by service providers as commercial solutions to be offered to a diverse customer base. Thus, services may be viewed as open, self-contained software components that support efficient configuration as well quality of service composition of distributed applications. Services may be supplied by different business stakeholders and comprise a distributed computing infrastructure in support of intra- and cross-enterprise application integration and collaboration [1].

Quality of service (QoS) considerations, are seen as an integral part of the service design lifecycle, taking into account important functional and non-functional service properties, such as performance, security, reliability, transactional integrity and services overall cost. Today's dominant manifestation of service-oriented computing implementation is realized in terms of emerging web technologies [2] Current service design frameworks merely focus on service capabilities, interface and behavior models and notations, which are expressed in a

N. Mitrou et al. (Eds.): NETWORKING 2004, LNCS 3042, pp. 1132–1142, 2004.

universal format, independent of a particular modeling tool and implementation platforms. In this paper, the component distribution analysis activities, in particular, are examined as an integrated phase of a performance-centered and thus quality of service-centered design process. The argument here made is that the service design philosophy and methodology can be explicitly extended to incorporate component allocation considerations in terms of combined mathematical programming and formal specifications activities.

The relationship between information mobility, which is supported by a network infrastructure, and the network itself, has so far been rather fuzzy; design has been based on empirical decisions, which aim at adapting network design to the needs it is assumed to serve. Ideally, given a set of services, a geographical distribution of users, and a demand pattern for the services requested by each user, one would design a minimum cost network, which would satisfy the quality requirements of all services, and, of course, the services themselves. Consider now that a service can be designed in a network independent manner: It consists of components, whose types and interactions are network topology independent, while only their distribution depends on the network. In this respect component based service design becomes an isolated problem, which can be separated from the aforementioned general problem. Therefore, service design (excluding the component distribution phase) becomes an input to this holistic design problem. Even after this reduction, the problem remains very ambitious, as its solution should produce both (a) the network topology, including node and channel capacities, and (b) the placement of service components over the network. The realisation of this grand objective cannot be achieved unless the relationship between the different factors of this problem becomes explicit. This paper takes us at least halfway to the solution of the problem, as it addresses subproblem (b) of the design, and partially subproblem (a) in the following sense: Although the network topology is taken as granted, node and link capacities can be determined or corrected by using the model presented in this paper; they can also be taken as given, which is plausible when new services are deployed over an existing network, and actually this is the most common situation. Therefore the main contributions of this paper are (i) that it presents a new design methodology, and (ii) it describes a number of steps, which are necessary to make the relationship between demand distribution and component distribution explicit.

2   Component Distribution Issues

To illustrate some of the issues, which will be explored in this paper, a simple example is useful: Assume that a new mobile device is in the design phase. A user directory application, which will of course include phone numbers, will be offered to the device user. The designer's dilemma is where to place the main component, i.e. in the device or in the network. Her choice of preference might be to make the device as light as possible, not only because this will give her less development pain, but also because it will be mass produced and it must be as cheap as possible. However, if the whole application is in the network, each time a user asks for a person's number, a directory application interface (and proxy) in the phone must send a message to the main application component, thereby consuming communications resources, including expensive air-interface time. Driven be these thoughts she might after all decide to spend the extra effort and try to put the main application in the phone. Storage, then, is likely to become an issue, but the price of memory is constantly sinking and today's phones are equipped with a memory, which well surpasses the storage capacity of yesterday's computers. Yet, again, each time a user related piece of information changes, all phone resident copies of this information must also be updated, sooner or later. The obvious reaction of the overwhelmed designer would be to ask a colleague to prepare for her a cost-benefit analysis of the alternatives.

What is a common objective behind these and similar considerations in a designer's mind? Is it not to produce the cheapest possible product, which satisfies a set of quality criteria? In the previous example, putting the directory in the network may produce a cheaper device, but the response time in retrieving a person's number must be kept under an acceptable limit. In a cost conscious design, the total communication cost should be minimized. Short sighted design may ignore such considerations or pursue a partial cost optimization, but market competition is likely to punish such policies in the long term. In the specific example, a designer may choose a network based user directory on the grounds of creating a lower market price product. On the other hand, a misinformed product buyer will discover that she pays higher communication bills than her neighbour, just because of the directory application. A healthy market will sooner or later reject the specific phone design.

From this discussion it should have become clear that communication cost is an important factor in assigning components to *physical entities*, i.e. physical containers, such as nodes, terminals, and other devices. Other important factors in the realm of service provision over network infrastructures are development costs of devices and applications, and, more recently, the cost of acquiring and processing information. In this paper we mainly deal with the problem of component distribution over a given network infrastructure, based on communication cost minimization.

A service can be seen as a collection of concerted components, which should produce a desired outcome. Components exchange messages in order to establish cooperation and communicate results. The following observations add to the complexity of the component distribution problem:

- There can be common components between services. For example, a charging component may be used by different services.
- Certain components can be attached only to specific physical entities or specific types of physical entities (due to functional or administrative reasons). For example, a module, which captures human voice and converts it to a digital signal can only exist in the mobile phone.
- Communication volume reduction or load sharing may favor the distribution of copies of the same component over a network. For example, a user

directory database may be implemented in multiple copies, each serving the demand of a major city.

A general and loose formulation of the problems, which are explored in this paper is the following:

Problem 1. Assuming

1. a set of services,
2. a set of components for each service (including common service components),
3. an estimate of the volume of interaction between components produced by invoking a service,
4. a network topology,
5. a population of terminals attached to network nodes,
6. an estimate of service demand (per different service) created at a terminal, and
7. a charging scheme (i.e. an algorithm, which transforms traffic volume to cost),

find the assignment of components to nodes (and terminals), which minimizes the total communication cost.

3   Related Work

The problem of distributing a set of components with known mutual interaction volumes reduces to the *multiterminal cut* problem [3], when network node pairs are equidistant (or the charging scheme is flat with respect to distance). In [4] the objective is to minimize the total running time of program modules rather than communication. The problems in [4] also reduce to the multiterminal cut problem [3].

A binary program formulation of the problem can be found in [5]. The development and results of a software tool, which implemented the methodology of [5] for networks with fixed and mobile nodes has been described in [6].

Other related work concerns modeling agent mobility and performance. [7] considers the problem of optimally scheduling a single mobile agent that is assigned to perform a certain task in a computer network. The cost to be minimized is the overall response time, which consists of the time spent in the network nodes plus the time spent for the migration between nodes. [8] compares two possible implementations (static vs. mobile) of a particular service component of the TINA Service Architecture, i.e. the User Agent (UA). There are no optimization concepts involved. Mobile component optimization is explored in [9].

The problems discussed in the aforementioned papers are partly related with the popular *file allocation problem*: Individual files are allowed to replicate in order to reduce communication cost by bringing information closer to the programs that access it, but at the expense of increasing update costs. Papers [10, 11] discuss various optimization models for the distribution of files in a computer network. The costs considered include communication and delay. Additional requirements refer to parallelism, availability and security. The problem of file migration, that is the reorganization of the file allocation scheme, is also discussed in [10]. Reference [12] discusses and compares various file migration and dynamic file allocation problems. Both adaptive and non-adaptive models are discussed for both types of problems. References [13,14] consider the problem of reallocating a single file. For this purpose a stochastic control problem is formulated. Whereas in [13] the decision for the location of the file is made centrally, in [14] various nodes decide independently. Finally [15] presents an online algorithm for the dynamic replication of a single file.

4   Problem Formulation

The core of the general problem is a mathematical problem, which will henceforth be called the *fixed component distribution problem*. This core problem has been presented and solved in papers [5,6], where several theoretical and practical examples can be found as well (while the preconditions to use this core and its consequences are presented in this paper). To make the present paper as self-contained as possible, we outline the fixed component distribution problem, but we have omitted the examples for obvious reasons. However, a better familiarisation with certain technical aspects requires a reading of the examples.

Problem 2. Given

- a network topology graph $G(V, E)$ (where $V = \{v_1, v_2, \ldots, v_n\}$ is the set of nodes and $E = \{e_1, e_2, \ldots, e_r\}$ is the set of links),
- a set of components C,
- a collection of N service topology graphs $G(C^k, F^k)$ ($k = 1, \ldots, N$), where $C^k = \{c_1^k, \ldots\}$ is the set of components of service k ($C^k \subseteq C$) and edges $F = \{f_1^k, f_2^k, \ldots, f_{m_k}^k\}$ represent the interaction between components according to service k, for each service a set of labels $\Lambda^k = \{\lambda_1^k, \lambda_2^k, \ldots, \lambda_{m_k}^k\}$ (that denote the traffic exchanged between components for each unit of traffic offered by a user to the service triggering component c_1^k,
- a collection of N functions $t^k : V \to \mathcal{R}$ ($k = 1, \ldots, N$) that describe the total volume of k service demand due to users attached to a node,
- a routing scheme (i.e. a collection of paths $P = \{p_{ij}\}_{i \in V, j \in V}$, where p_{ij} is a path for each pair of nodes (i, j) to be used by the traffic exchanged between them) and
- certain link and node capacity constraints

find the allocation of service components to nodes that minimises a given communication cost function.

The linear program is based on initially placing copies of all components in each node. The program variables are the traffic variables $x_{ij}^{k;mn}$, where $x_{ij}^{k;mn}$ is the traffic generated by service k on edge (i, j) between a copy of component

c_m, which has been placed in node v_i, and a copy of component c_n, which has been placed in node v_j.

It is assumed that the communication cost is a (preferably linear) known function of the information traffic volumes $x_{ij} = \sum_{k,m,n} x_{ij}^{k;mn}$, where x_{ij} is the total traffic over link (v_i, v_j). The exact form and validity of this assumption depends on the charging scheme imposed by the network operator and on network protocols.

Linear programming versions of Problem 2, examples and numerical results have been presented in [5,9]. The technique used in [5] is to place a component copy in each network node and to calculate the traffic served by each component for each service. Components producing zero traffic are finally removed. Note that the problem cannot be decomposed into independent subproblems, one for each service, because of the existence of common components between services. Additional features to the problem may include node setup costs or component installation costs. Such additions make the problem non-linear.

The single component copy version of this problem, i.e. when each component is unique in the network, is not necessarily easy in terms of complexity. This problem is easily proven to be NP-complete. Even a simplified single copy problem with only three equidistant nodes and a single service is NP-complete. Actually it can be easily shown to be equivalent to the *multiterminal cut problem* [3] (as already mentioned in the previous section), which is NP complete. Fortunately, if only two nodes exist, the problem can be solved in polynomial time by using the max-low min-cut theorem and flow maximization techniques. The mobile version of the component placement problem is treated in [9]

5   The Component Distribution Methodology

Effectively a new importand phase in the service design methodology has evolved. After a service has been designed to the point that its components and their mutual interactions are known, an estimation of the traffic generated between components can be performed. Then by using and solving the distribution problem for sets of services with common components, components can optimally be assigned to nodes and possibly to terminals, if the latter are also in the design phase or if they are reconfigurable.

6   Factors Determining Component Distribution

Problem 2 formulation is enlightening in the sense that it can reveal the effect of various factors on component distribution. We take up these factors one by one in the following few paragraphs.

Service demand distribution: The solution to Problem 2 is partly based on the estimation of service demand. Service demand is created at specific nodes or (groups of) terminals. Demand can follow a quite complex pattern; consequently the traffic source related literature is extensive. If the demand volume is modeled as a random process, the problem of finding the appropriate distribution for such a process is largely open. However, in fixed network topology and capacity problems demand usually assumes the form of an average over a suitable time period, or it is a compromise between an average a peak traffic volume. In reconfigurable networks, which are by design able to respond to quasi-static traffic, i.e. traffic, which has different statistical properties in different time intervals, but its characteristics remain the same within each interval, the component distribution problem can also be formulated in a piecewise manner. In this case, the designer should cater for the transition between successive configurations.

In general, service components are "pulled" by a node with a high volume of service demand in an effort to reduce communication cost. If the capacity of such a node is not a limiting factor, if the node, component installation and maintenance costs are negligible, and if all components are movable (i.e. they are not assigned to particular nodes for particular reasons), there is a trivial solution to the distribution problem: All service components are likely to be copied to this node.

Is service demand and its distribution predictable? In general the answer is negative, and from time to time totally unpredictable and revolutionary changes may happen. The emergence of WWW is a typical example. The answer depends on the collection of services, which are likely to be offered, on customer profiles and distribution. In certain networks, e.g. in cellular mobile networks, basic services remain constant or at least predictable for a significant period of time. Often network operators try to contain change by hindering the spreading of certain technologies and services: A typical example is voice over IP. The quasi-stability assumption remains valid in special purpose networks, e.g. in the private network of a company. Internet type networks are less predictable. Obviously, the answer to the lack of totally predictable traffic patterns is the creation of reconfigurable networks. From time to time an operator should run the component distribution algorithm again and adjust the configuration of the network and its components.

Network topology and charging: The objective function of Problem 2 expresses the total communication cost in terms of traffic variables. It may contain a sum of terms of the form $d_{ij} \times x_{ij}$, where d_{ij} is the distance between nodes i and j if the communication cost is a linear function of the distance. In general d_{ij} is influenced by the physical distance between nodes, but it may also reflect the operator's charging policy. Flat charging, which depends only on volume, can be modeled by setting all distances to the same value. A quantized charging scheme will usually require a non-linear formulation. Flat charging is likely to favor component concentration in a smaller number of nodes, which offer cheaper processing and storage. Flat charging has become popular with the success of the Internet, but it is becoming increasingly attractive even in the PSTN.

To some extent charging, and the subsequent accuracy of the objective function, will also depend on packet lengths, retransmission protocols, failure recovery protocols, error rates, compression algorithms, and any other factor, which contributes to an increased number of transmitted bits for the same number of information (i.e. service generated) bits. However, for the purposes of component distribution the influence of these factors can be seen as a second order effect.

Processing and storage capacity: In general the larger the capacity of a node, the more components it can host. Since communication cost is paid only for the interaction of pairs of components, which reside in machines separated by a physical distance, the total elimination of communication cost is ideally achieved by putting all components in the same machine, or at least in machines accommodated under the same roof. An extreme version of this idea has been realized in *computer farms.* This would be the case if users could somehow be transported in negligible time to the farms and use the services there. However, distance in our world is a factor, and a major mission of a network is to make up for distance. In other words, certain service components, which at least include a user interface, must be in the user's premises. Exactly these components pull other components towards them if the user equipment has host them in a cost effective manner. If communication becomes cheaper and cheaper, component concentration is favored again. There are some notable "exceptions": For example, a processing intensive problem can be solved by a large number of geographically separated machines, if installation and processing are also cheap. This is the SETI (Search for ExtraTerrestrial Intelligence) case and the security related decomposition of a long integer to a product of prime numbers. In both cases large numbers of machines have been volunteered by their owners, while low communication volumes and cheap Internet prices have made the communication cost negligible.

Channel capacities: Capacities have an indirect influence on the distribution problem, as they may have an impact on charges. However, a capacity constraint directly delimits the traffic, which is allowed to pass through a link. A cheap link tends to absorb more traffic. When the link is saturated, traffic is diverted to the next cheapest path.

7   Service Internal Traffic Estimation

In Problem 2 for each service k a graph $G(C^k, F^k)$ has been defined, together with a set of labels on its edges. A label λ_{ij}^k on an edge aiming from component i to component j denotes the traffic generated on this edge for data sent from i to j for a unit traffic from a service user to the service interface component. The existence of this graph is based on the assumption that the traffic generated between pairs of service components can be evaluated or estimated. The rest of this section is devoted to the description of techniques, which allow for the estimation of service internal traffic.

The obvious solution in internal traffic estimation is to monitor the traffic between pairs of components for a certain test period. While this approach is conceptually simple and the test can run on a single machine, it requires one or more monitoring components, depending on the testing architecture. If service execution varies and depends on user input, a statistically stable result must be pursued by letting different users interact with the service interface. Also, this approach does not solve the problem of accurately determining the actual volume of the traffic generated between two components, due to the omission of the network protocols. A more realistic approach would be to install the components in different machines, which are separated by a network similar to the target network.

Simulation is another interesting choice. Occasionally the required programming effort may be close to the service development effort. Modern popular simulation packages, like OPNET and NS2, are capable of capturing the required protocols. However, they will require extra effort in modeling the service components. The problem of the statistical stability of the results and the associated user behavior model must be taken care of by the simulation designer.

A relatively recent approach is to use formal specification techniques like SDL or UML. This approach is suitable only if the service creation process contains a formal definition phase.

7.1   SDL Based Estimation

The existence of formal specification techniques determine certain features of a service before its actual implementation. So far formal specification has been reserved for the functional characteristics of services. In this section a non functional usage is introduced.

SDL tools are capable of producing a simulation run, but usually this will not entail network protocol details. Also, the specification may not have modeled all those aspects of the service, which are necessary for accurate traffic estimation purposes. For example, it may not use the actual messages, which are used in communication between service components. The statistical stability of the results is also a problem.

SDL trains of events can altermatively be derived from Message Sequence Charts (MSC). MSCs do not offer any particular advantage over simulation, nevertheless they might become available before the existence of an SDL specification (which may or may not appear), e.g. as a part of a preliminary definition of the service. In this sense a collection of representative MSC (or even informal event and message exchange sequences) may be a lazy or a hurried developer's last refuge.

The observation that real messages may differ from their representation in a model, holds for MSCs as well. While the MSC user may decide to make do with the virtual message names or even with a simple message count, certain events are definitely different than simple messages in terms of volume and must be treated accordingly. Such events are packet transmissions and file tranfers in general. Note that such events are of major importance in multimedia services.

7.2   UML Based Estimation

UML being a standard notation for formal analysis and design of software system, offers several diagrams for separating concerns of different system views, and arguably this approach makes feasible to derive early performance models by taking into account combined data from these diagrams. In UML, a use case diagram (UCD) provides a functional description of a system, by means of its major use cases and its external users or actors. Sequence diagrams (SD) depict a number of software components and the messages that are exchanged between them in a given scenario (generally a single use case can be described by a set of scenarios, i.e., a set of sequence diagrams).Thus, sequence diagrams provide specific information about the order in which events occur and the interactions required for each event. Consequently, estimates on component traffic generation can be obtained similar to to SDL-MSC based estimates that were treated in the previous section. A deployment diagram (DD) is a graph of computing nodes connected by communication links. Nodes may contain component instances (indicating that the component lives on the node) so it shows the mapping of components on processing nodes. It is apparent that, for example, SDs alone can directly support the traffic conscious specifications of service systems, since they depict in a straightforward manner the timed sequence and generated traffic of service events. Also, DD where the mapping of software components to hardware nodes is described can support the modeling of service distribution. Nevertheless, the level of modeling detail acquired does not directly stem from the set of diagrams and their refined semantics and constructs adopted to describe the hardware/software system; it rather depends on the depth of system knowledge and the designers intuition. Besides, extracting combined information from other UML diagrams would be helpful in order to keep into a performance model relevant characteristics of the system that are not explicitly captured from the SD and DD diagrams being considered. In [16] a comprehensive survey on using UML diagrams for performance modeling and a particular performance estimation algorithm are given.

8   Conclusions

In this paper ample evidence and methodological directions are given in support of service design activities, which will deal with component distribution in such a way, as to minimize communication cost. Communication cost has been considered the main factor of component distribution in this work. Other important and indicative factors, as security and reliability, have not been considered, and can be explored in future research on component distribution. The overall design philosophy should tend to an integrated and QoS aware service-oriented design methodology that explicitly incorporates non functional requirements modeling such as cost, performance, security, availability and reliability in the established service analysis and design methodology phases.

References

1. Papazoglou M.P. et al. "Service-Oriented Computing", *Communications of the ACM*, Vol.46, No3, pp. 25-28, October 2003.
2. Yian Yang. "Web Service Componentization", *Communications of the ACM*, Vol.46, No3, pp. 35-40,October 2003.

3. E. Dahlhaus et al. "The Complexity of Multiterminal Cuts", *SIAM J. Comput.*, Vol. 23 No. 4, pp. 864-894, August 1994.

4. H. S. Stone, " Multiprocessor Scheduling with the Aid of Network Flow Algorithms", IEEE Transactions on Software Engineering, Vol. SE-3, No. 1, Jan. 1977, pp. 85-93.

5. M. Anagnostou, "Optimal Distribution of Service Components", *Lecture Notes in Comp. Science*, No. 1430, pp. 17-30, Springer, 1998.

6. M.Anagnostou, A.Rouskas, S.Trigila, "The DOLMEN Component Distribution Methodology and Tool", *IS&N Book 1999: On the Way to the Information Society - 5 Years of European ACTS IS&N Research*, IOS Press, Amsterdam, Netherlands, 2000.

7. K. Moizumi and G. Cybenko, "The Travelling Agent Problem", *Mathematics of Control, Signals and Systems*, Jan. 1998.

8. A. Kuepper and A. S. Park, "Stationary vs. Mobile User Agents in Future Mobile Telecommunication Networks", *Proceedings of the Second International Workshop MA '98*, Stuttgart, Germany, Sept. 1998, pp. 112-123.

9. I. Avramopoulos , M. Anagnostou , "Optimal Component Configuration and Component Routing," *IEEE Transactions on Mobile Computing*, Vol. 1, No. 4, pp. 303-312, 2002.

10. B. W. Wah, "File Placement on Distributed Computer Systems", *IEEE Computer Magazine*, Jan. 1984, pp. 23-30.

11. L. W. Dowdy and D. V. Foster, "Comparative Models of the File Assignment Problem", *Computing Surveys*, Vol. 14, No. 2, June 1982, pp. 287-313.

12. B. Gavish and O. R. L. Sheng, "Dynamic File Migration in Distributed Computer Systems", *Communications of the ACM*, Vol.33, No. 2, Feb. 1990, pp. 177-189.

13. A. Segall, "Dynamic File Assignment in a Computer Network", *IEEE Transactions on Automatic Control*, Vol. AC-21, No. 2, April 1976, pp. 161-173.

14. A. Segall and N. R. Sandell, "Dynamic File Assignment in a Computer Network - Part 2: Decentralized Control", *IEEE Transactions on Automatic Control*, Vol. AC-24, No.5, Oct. 1979, pp. 709-715.

15. O. Wolfson et al., "An Adaptive Data copytion Algorithm", *ACM Transactions on Database Systems*, Vol. 22, No. 2, June 1997, pp. 255-314.

16. Cortelessa V. et al, "PRIMA-UML: a performance validation incremental methodology on early UML diagrams", *Science of Computer Programming* No. 44, pp 101-129, 2002.

Network-Embedded Programmable Storage and Its Applications

Sumeet Sobti[1], Junwen Lai[1], Yilei Shao[1], Nitin Garg[1],
Chi Zhang[1], Ming Zhang[1], Fengzhou Zheng[1],
Arvind Krishnamurthy[2]*, and Randolph Y. Wang[1]*

[1] Department of Computer Science, Princeton University, Princeton, NJ 08544, USA.
{sobti, lai, yshao, nitin, chizhang, mzhang, zheng,
rywang}@cs.princeton.edu.
[2] Department of Computer Science, Yale University, New Haven, CT 06520, USA.
arvind@cs.yale.edu

Abstract. We consider the utility of two key properties of network-embedded storage: programmability and network-awareness. We describe two extensive applications, whose performance and functionalities are significantly enhanced through innovative combination of the two properties. One is an incremental file-transfer system tailor-made for low-bandwidth conditions. The other is a "customizable" distributed file system that can assume very different personalities in different topological and workload environments. The applications show how both properties are necessary to exploit the full potential of network-embedded storage. We also discuss the requirements of a general infrastructure to support easy and effective access to network-embedded storage, and describe a prototype implementation of such an infrastructure.

1 Introduction

For wide-area distributed services, network-embedded storage offers optimization opportunities that are not available when storage resides only at the edges of the network. A prime example of this is content-distribution networks, such as Akamai, that place storage servers at strategic locations inside the network and direct client requests to servers that are "close" to them, thus achieving reduced access latency for the clients and better load balance at the servers.

Given the desirability of network-embedded storage, a natural question to ask is this: What is a good "access model" for network-embedded storage that allows services to realize its full potential? By access model, we mean mechanisms through which diverse services can use the network-embedded storage resources to satisfy their diverse needs.

One simple access model is what can be referred to as the *fixed-interface* model. In this model, each embedded storage element exports a fixed set of high-level operations (such as caching operations). Service-specific code is executed

---

* Krishnamurthy is supported by NSF grants CCR-9985304, ANI-0207399, and CCR-0209122, and Wang is supported by NSF grants CCR-9984790 and CCR-0313089.

N. Mitrou et al. (Eds.): NETWORKING 2004, LNCS 3042, pp. 1143–1155, 2004.

only at edge-nodes. This code manufactures service-specific messages and sends them into the network to manipulate the embedded storage elements through the fixed interface. An example of this model is the Internet Backplane Protocol (IBP) proposed in the "Logistical Networking" approach [1].

Although the fixed-interface model does benefit a certain class of services, it has two main limitations. First, it does not have sufficient flexibility. Due to the extremely diverse needs of distributed services, it may be difficult to arrive at an interface that caters well to all present and future services. Second, the restriction that service-specific code executes only at the edges of the network, and not at the embedded storage elements, imposes a severe limitation, both on the functionalities provided by the services and the optimization opportunities available to them. For example, for application code executing at the edges, it is often difficult to gather information about changes in the load and network conditions around an embedded storage element, and then to respond to such changes in a timely fashion.

These limitations point to the need for the following properties. (1) *Programmability:* the services should be able to execute service-specific code of some form at the embedded storage elements. (2) *Network-awareness:* the code executing at these elements should be able to use dynamic information about the resources at and around them. We do not claim that any of these properties is novel by itself. We, however, do believe that it is the combination of the two that is necessary to realize the full potential of embedded storage.

To support this hypothesis, this paper presents qualitative and quantitative evidence in the form of two applications of network-embedded storage. One is an incremental file-transfer service tailor-made for low-bandwidth conditions (Section 2). The other is a "customizable" distributed file system that can assume very different personalities in different topological and workload environments (Section 3). In these applications, we explicitly point out how the absence of any one of the two properties would significantly limit their power, both in terms of functionality and performance. These applications also show that the combination of programmability and network-awareness is useful in a diverse set of environments, including both local and wide area networks. A general theme of our work is that in any system configuration or service, if a storage element is in a position to exploit its location advantage intelligently, it should be programmed to do so.

We also discuss the requirements of a general infrastructure to support easy and effective access to programmable network-embedded storage, and describe a prototype implementation (Section 4). We refer to such an infrastructure as a *Prognos* (PROGrammable Network Of Storage), and to each embedded storage element in it as a *Stone* (STOrage Network Element). As long as the Stones have access to network information, the making of the Stones and the links among them can be quite flexible. One possibility is to construct a Prognos on top of an overlay network. The overlay links used should approximate the underlying physical topology and the Stones can simply be general-purpose computers. The

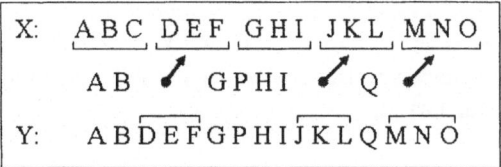

Fig. 1. A simple rsync example.

other potentially more efficient possibility is to co-locate a Stone with a router and the links among the Stones would largely be physical.

We refer to the systems-support module of a Prognos as SOS (Stone Operating System). SOS is responsible for managing the physical resources at the participating Stones, and for allowing services to inject service-specific code into the Stones in a secure fashion. A collaboratively supported platform, such as PlanetLab [2] (www.planet-lab.org), can be turned into a Prognos by loading the participating machines (also referred to as Stones) with the SOS module.

2   Incremental File Transfer

We now describe a service intended to facilitate transfer of incrementally changing, large files. An example usage scenario of this service is one where a producer periodically releases new versions of the Linux kernel file, and multiple consumers update their versions at different times.

The basic idea is to use network-embedded storage elements (or Stones) to optimize these file transfers. As data flows through a sequence of Stones during a file transfer, there is an obvious caching opportunity to benefit subsequent transfers. If, however, the Stones are capable of executing complex service-specific code, more sophisticated optimizations become possible. Our service, which we call "Prognos-based rsync" (or Prsync), programs the Stones to use the rsync protocol to propagate files.

2.1   The rsync Protocol

The rsync protocol [3] (rsync.samba.org) is a tool for updating an old version of a file with a remotely-located new version. The protocol seeks to reduce network usage by not transferring those portions of the new version that are already present in the old version. A checksum-search algorithm is used to identify such portions when the two versions are not located on the same machine.

As a simple example, suppose that nodes X and Y have two versions of a file with contents shown in the top and bottom rows of Figure 1, and X wants to get Y's version. X first partitions its version into fixed size blocks and sends the checksums of those blocks to Y. In the example shown, X sends five checksums to Y. Using the checksums, Y is able to identify portions that are common between the two versions. Y then sends to X a description of its version referencing the blocks at X wherever possible. The middle row of letters shows the description Y sends to X. X is then able to reconstruct Y's version from this description. If the two versions share several blocks, then there is significant saving in the number of bytes transferred.

2.2   Prsync

We examine four aspects of Prsync relating to the programmability and network-awareness of the Stones. First, we show how programmability of Stones enables rapid deployment of Prsync-like services, even when one does not have full co-operation of edge machines. Second, we describe how Stones can themselves use pair-wise rsync exchanges to improve end-to-end performance. Third, we describe how Prsync adapts to its environment by exploiting the network-awareness of Stones. Fourth, we describe how network information can be combined with service-specific state in a service-specific manner to achieve good performance.

Interaction with Legacy Protocols. Consider a scenario where a producer and a consumer want to engage in a file update, but they lack the ability to participate in rsync exchanges. Assume that the Stones have been programmed to cache files, execute checksum-search algorithms, and participate in the Prsync protocol. The system can still be used to transfer files efficiently. The file is first copied from the producer to a nearby Stone using a legacy protocol. The file is then efficiently propagated using Prsync to a Stone that is located close to the consumer. As the last step, the file is copied from this Stone to the consumer using a legacy protocol. This is an example of an end-to-end legacy protocol that benefits from programmable network-embedded storage.

Hop-by-Hop Interaction. In the above scenario, the Prsync protocol is executed between two Stones that are potentially separated by a weak wide-area connection. The performance could be further improved if we were to enlist intermediate Stones to decompose a long-distance rsync into a sequence of short-distance hop-by-hop rsyncs. First, intermediate Stones may already have a version that is very close to the fresh version being propagated. In such cases, fewer bytes will have to be transferred along some portions of the path. Second, after a sequence of hop-by-hop rsync exchanges, all the intermediate Stones also end up receiving the fresh version and can satisfy future requests without requiring end-to-end interactions. The hop-by-hop protocol demonstrates that simple caching in particular, or any hardwired storage interface in general, on intermediate Stones is not sufficient—instead, the programmability of Stones is needed to allow them to participate in a sophisticated protocol.

Adapting to Changing Environments. The rsync program employs a computationally expensive checksum and compression algorithm. Its use may in fact be counterproductive in cases of abundant link bandwidth, drastic file content changes, or high CPU load on participating nodes. In order for Prsync to adapt to these environmental factors in a timely fashion, the programmability and the network-awareness of Stones become indispensable. When an upstream node X starts to send fresh data to a downstream node Y, the two nodes begin with the checksum-based rsync algorithm. Node X monitors two quantities dynamically: (1) the ratio (r) between the number of bytes that has been actually transferred and the size of the content that has been synchronized, and (2) the physical bandwidth achieved (B). If r exceeds a threshold, which in turn is a pre-determined function of B (implemented as an empirical table lookup), then the communicating nodes would abandon the checksum-based rsync and revert to simply transmitting the literal bytes of the fresh file. Note that such adaptive optimizations need to be performed on a hop-by-hop basis within the network—they are difficult, if not impossible, to replicate at the edge. An additional optimization to further reduce rsync overhead is to compute the per-block checksums off-line and store them along with the file in the Stone's persistent store.

Selecting Propagation Paths. In scenarios where there exists path diversity and pairs of Stones are connected by multiple paths (as in overlay networks), Prsync can select propagation paths for hop-by-hop synchronization based on application-specific metrics. We have experimented with two specific methods of doing this. In the *tree-based* method, an overlay tree spanning all the Stones is constructed. The tree is constructed using a minimum-spanning tree algorithm on a graph where the nodes are Stones and the edges are weighted with the inverse of pair-wise bandwidth. The tree construction uses heuristics for constraining the node degree and diameter of the resulting tree. The resulting tree thus contains high bandwidth paths between all pairs of Stones, and only these paths are used for hop-by-hop rsync exchanges. The *mesh-based* method maintains an overlay graph in which each Stone is adjacent to a certain number of other Stones to which it has high-bandwidth links. When selecting a path between a pair of Stones, all paths in this overlay graph are considered. Note that the time taken for a pair-wise rsync exchange is determined by the link bandwidth and the difference between the file versions at the two Stones. Prsync can maintain estimates of the differences between the file versions at different stones and also monitor pair-wise bandwidths. By using these estimates, a *best* path (i.e., one for which the expected time for hop-by-hop propagation of data is minimized) can be selected in the mesh. This is an instance where information about the network characteristics is combined with service-specific state in a service-specific manner to improve performance.

2.3   Summary of Prsync Experimental Results

We have experimentally validated the Prsync design. The experiments were performed on two testbeds – one constructed in our laboratory, and another implemented on a set of PlanetLab machines distributed across the wide-area. Due to lack of space, we refer the reader to [4] for details.

Here, we only provide a brief summary of the results as they relate to the four aspects of Prsync described in the previous section. (1) The rsync protocol is observed to perform more than 5× better than simpler legacy protocols for copying files, especially in low bandwidth conditions. (2) Hop-by-hop use of rsync can improve upon end-to-end rsync by an additional factor of 2. These results demonstrate the utility of executing complex service-specific code (e.g., rsync) at the embedded storage elements for functionality and performance gains. (3) The adaptive nature of Prsync allows it to perform well in a diverse range of network conditions. Lack of adaptivity can degrade performance by as much as 2×. (4) In a PlanetLab experiment, the mesh-based method of selecting propagation paths performs 30% better than the tree-based method, which in turn performs about 30% better than a simple end-to-end rsync. These results demonstrate the kind of performance benefits that a service can get by being network-aware, and by intelligently using network information in a service-specific manner.

3   A Customizable Distributed File System

Today, we build cluster-based distributed file systems [5,6,7] that are very different from wide-area storage systems [8,9,10]. Life would be simpler if we only had to build two stereotypical file systems: one for LAN and one for WAN. The reality, however, is more complicated than just two mythical "representative" extremes: we face an increasingly diverse continuum, often with users and servers distributed across a complex interconnection of subnets.

Prognosfs is a "meta file system" in the sense that its participating Stones can be customized to allow the resulting system to exhibit different personalities in different environments. Prognosfs software has two parts: (1) a fixed framework that is common, and (2) a collection of injectable components that run on participating Stones and may be tailored for different workloads, and network topologies and characteristics. (In the near future, we envision injectable Prognosfs parts to be compiled from high-level specifications of the workload and the physical environment.)

3.1   Architecture and Component Details

Unlike several existing wide-area storage systems that support only immutable objects and loose coherence semantics [11,8], Prognosfs is a read/write file system with strong coherence semantics: when file system update operations are involved, users on different client machines see their file system operations strictly serialized. Of course, we are not advocating that this is the only coherence semantics that one should implement—it just happens to be one of the desirable semantics that makes collaboration easy.

Figure 2 shows the Prognosfs parts in greater detail. The fixed part is similar to that of the Petal/Frangipani systems [6,7]. For each file system call, a Prognosfs client kernel module translates it into a sequence of a lock acquisition, block reads/writes, and a lock release. This sequence is forwarded to a Prognosfs client user module via the Linux NBD pseudo disk driver. The read and write locks provide serialization at the granularity of a user-defined "volume" and they are managed by the Distributed Lock Manager. If a client fails without holding

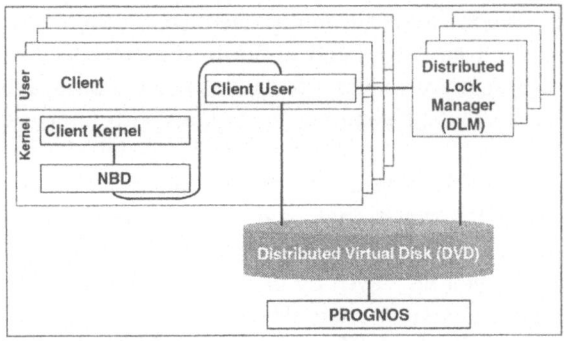

Fig. 2. Components of Prognosfs.

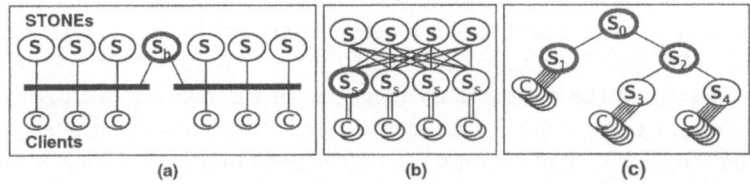

Fig. 3. Example topologies connecting client machines with their Stones.

a write lock, no recovery action is required. If a client fails while holding the write lock of a volume, a recovering client inherits the write lock and runs `fsck` on the failed volume. These components of Prognosfs are fixed.

The customizable part of Prognosfs lies within the Distributed Virtual Disk (DVD). Externally, the interface to the DVD is very much like existing distributed virtual disks such as Petal [6]. The difference is that, internally, while all Petal servers are identical, the DVD consists of a number of peer Stones, each of which can run a specialized piece of code to perform functions such as selective caching, active forwarding, replication, and distribution of data to other Stones. These decisions can be made based on network topology, network condition, Stone load, and Stone capacity information that is typically either unavailable or difficult to determine accurately and responsively at the edge.

Figure 3 shows several example topologies. In Figure 3(a), clients on each of the two subnets can read data served by Stones on either subnet. If, for example, the clients of the right subnet repeatedly read data from Stones on the left, they might increase the load on the left subnet. As the "bridge Stone" S_b detects this access pattern, due to its awareness of the topology, S_b can take several possible actions to reduce the load: (1) S_b could cache data from the left subnet in its own persistent store. (2) If S_b itself becomes a bottleneck, S_b could forward a copy of the data to a Stone in the right subnet and this Stone would absorb future reads. (3) As reply data flows from the left subnet to a client in the right subnet, S_b could distribute the data across multiple Stones in the right subnet.

In Figure 3(b), the Stones in the middle layer (S_s) form a "switching fabric"—
they accept requests from clients and perform functions such as load-balancing
and striping as they forward requests to the next tier Stones. The role played by
an S_s is analogous to that played by a μproxy, an NFS interposition agent [12].
Such interposition agents are just an example of the kind of functionalities
that Prognosfs can enable. (Unlike a μproxy, the switching fabric is fully programmable, can have its own storage, and is not limited to the NFS protocol.)

In Figure 3(c), we replace a number of wide-area routers with their Stone
counterparts. To see the role played by network-awareness, consider an example
where S_4, on its clients' behalf, reads data stored at S_1. As data flows back on
the path $S_1 \rightarrow S_0 \rightarrow S_2 \rightarrow S_4$, S_0 does not need to cache the data, S_2 may
cache the data in the hope that S_3 may demand it later, and S_4 may cache the
data in the hope that its own clients may demand it again. Once S_3 does read
the cached data at S_2 and caches it itself, S_2 may choose to discard it.

In each of these examples, the function executed by a Stone is intimately
associated with its often unique position in the network. Furthermore, although
we have described the above Stone functions in the context of Prognosfs, the
concepts are more generally applicable to other Prognos applications.

While the Prsync application relies on the combination of a known producer
to ensure that a requester receives an up-to-date copy of the desired data, the
presence of multiple readers and writers and the presence of multiple copies in
Prognosfs demand a data location service from the underlying Prognos infrastructure. Given an object ID, the location service is responsible for locating *a*
replica for a read request, and for locating *all* obsolete replicas to invalidate (or
update) for a write request. This service is briefly described in Section 4.

We have implemented an initial prototype Prognosfs, along with a few of
its incarnations that are customized to work for some different topologies. Existing applications on multiple Linux client machines are able to transparently
read/write-share Prognosfs volumes.

3.2   Summary of Prognosfs Experimental Results

Detailed experimental results from both local area and wide area configurations
are described in [4]. Here, we only present some observations from our experiments with the topology of Figure 3(a). The main role of the bridge Stone S_b
is to forward blocks from one side to the other. In addition, its behavior can be
customized in at least two ways. In one case, it is programmed to "cache" any
data blocks that flow through it, so that it may be able to satisfy any subsequent requests for those blocks. In another case, it is programmed to actively
"distribute" blocks flowing through it among the Stones on the destination side
in a round-robin fashion. These "cache" and "distribute" strategies pay the cost
of replication the first time a block flows through S_b for potential benefit during
subsequent accesses to that block. This illustrates the fact that the benefits of
any given strategy may be highly workload- and application-dependent. Therefore, the ability to dynamically adapt the behavior of embedded storage is often

important. In some cases, it may be possible to execute these functions by issuing commands from the edges of the network, but this often incurs overheads and limits the ability to quickly adapt to the workload.

Prognosfs is an example that illustrates some of the extremely diverse customizations made possible by programmable embedded storage. The example strategies, such as those mentioned in the context of Figure 3, serve to show that a fixed interface for embedded storage may not always be sufficient. Different strategies suit different system configurations, and one needs both programmability and network-awareness of embedded storage to tailor application behavior to prevailing conditions.

4 Prototype Prognos

Resource Management and Security. The three key players in resource management are: the Stone Operating System (SOS), the application-specific service running on a Prognos, and the user of the service. In general, the user trusts the service, which in turn trusts the SOS. The SOS must protect different services from each other on a Stone; the distributed participants implementing the same service on multiple Stones must be able to authenticate each other; and the service must implement its own application-specific protection to protect its users from each other. We discuss each of these issues in turn.

One simple way of insulating the multiple services that run on a Stone simultaneously from each other is to employ one process per service per allocated Stone. Such a daemon is present as long as the service is up. Code specific to each service is executed within its own separate address space, thus isolating it from other services running concurrently on the same Stone. The service daemons request resources from the SOS, which is currently implemented as a simple Linux user-level process. Prognos could benefit from resource accounting abstractions that are more precise than the process model, such as "resource containers" [13], but our prototype does not support such fine-grained mechanisms. More efficient alternatives than the process model, such as software-based fault isolation and safe language-based extensions, also exist. One of the chief aims of building this prototype is to have a vehicle with which we can experiment with several Prognos-based applications and demonstrate the utility of the Prognos approach. To this end, we have not started with a potentially more efficient kernel-based and/or language-based implementations.

All the participants that collaborate in a Prognos to implement a particular service, such as Stones allocated to this service and the processes on edge machines belonging to the service provider, must be able to authenticate each other. Existing cryptographic techniques for authentication, secure booting, and secure links can be used for this purpose [14,15].

The codes that implement different services can choose their own means of authenticating their users. Application-specific access control and resource management is entirely left to individual services.

Code Injection. Service-specific code is injected into the Prognos at service launch time. (Updating code requires re-starting the service.) The Prognos supports an interface to allow services to inject code in native binary format. The code fragments injected into different Stones might be different because they may be tailor-made for Stones at different locations in the network.

Persistent Storage. Each service is allocated a separate storage partition on each participating Stone at service launch time. At each Stone, storage is available in three alternative forms, and a service is free to choose one or even switch among them. The alternatives are: (1) A raw disk partition interface that is essentially the Linux /dev/raw/ interface. (2) A logical disk interface that is similar to several existing ones [16]. A user of this interface can read and write blocks that are keyed by their 64-bit logical addresses. This interface is useful for those who desire a block-level interface but do not care to explicitly manage their own storage layout. Our implementation is log-structured. Prognosfs uses this interface. (3) A subset of the Linux local file system interface. Prsync uses this interface.

Connectivity. The communication links between Stones can be either physical or virtual. The current SOS implementation enforces no resource arbitration mechanisms such as proportional bandwidth sharing[17], which we plan to add. The SOS also needs to be able to provide local connectivity information in the form of, for example, the set of neighboring Stones, and estimates of pair-wise bandwidth, latency and loss-rate.

Location Service. Our prototype includes an efficient, network-aware object location service to track copies of objects in a set of participating Stones. We refer to it as Canto (Coherent And Network-aware Tracking of Objects). Canto is heavily used by Prognosfs. It is designed as a network-aware generalization of the manager-based approach commonly used in cluster-based systems [5,6,7]. Due to lack of space, we refer the reader to [18] for further details on Canto.

5   Related Work

Many active network prototypes have been built [19,20,21,22]. Prognos shares their goal of allowing new services to be loaded into the infrastructure on demand. Most active networking efforts to date, however, have consciously avoided tackling persistent storage inside the network. This decision typically limits the injected intelligence to those related to low-level forwarding decisions. By embracing embedded storage, Prognos makes it possible for services to inject high-level intelligence that is qualitatively different and more sophisticated.

In a DARPA proposal [23], Nagle proposes "Active Storage Nets," which are active networks applied to network-attached storage. In this proposal, active routers may implement storage functions such as striping, caching, and prefetching of storage objects, and quality-of-service responsibilities of I/O operations. "Logistical Networking", a system proposed in a recent SIGCOMM position paper [1], argues for an IP-like embedded storage infrastructure that allows arbitrary packets to manipulate the embedded storage using a fixed low-level interface. In our experience, applications such as Prsync and Prognosfs can fully benefit from the embedded storage only when application-specific intelligence, which could be more sophisticated than conventional caching of objects, is co-located with embedded storage.

Active technologies have been successfully applied to applications such as web caching [24] and media transcoding [25]. We hope to generalize these approaches for a wider array of applications that can benefit from network-embedded programmable storage. Active technologies have also been successfully realized in the context of "Active Disks" [26,27]. One important difference between Active Disks and Prognos is that the intelligence in the former is at the "ends" of the network while in the latter case, it is embedded "inside" the network.

The applications, Prsync and Prognosfs, represent extensions to previous work that is either limited to client-server settings or lacks customizability. LBFS [28] is a client/server file system that employs a checksum-based algorithm to reduce network bandwidth consumption in a way that is analogous to rsync. By using the Prognos infrastructure, Prsync extends this approach to fully exploit multiple peer Stones and their network-awareness. Prognosfs is similar to Petal/Frangipani [6,7] in its break down of the file system into three components: clients, a distributed lock manager, and a distributed virtual disk (DVD), but it improves upon existing cluster file systems that possess little network awareness [5,6,7]. The most novel part of Prognosfs lies within its DVD—the DVD consists of a number of peer Stones, each of which can be customized for a specific environment.

6   Conclusion

We describe two applications that gain significant performance and functionality benefits by using a clever combination of the programmability and network-awareness of network-embedded storage. These applications qualitatively and quantitatively show that such combination is necessary to exploit the full power of embedded storage. They are also evidence to support our belief that the benefits of such combination are not limited to content-distribution networks, but extend to many conventional applications too. The applications run on our prototype Prognos system that currently works on LAN clusters and wide-area PlanetLab-like overlay networks.

References

1. Beck, M., Moore, T., Plank, J.S.: An End-to-End Approach to Globally Scalable Network Storage. In: Proc. of ACM SIGCOMM 2002. (2002)

2. Peterson, L., Anderson, T., Culler, D., Roscoe, T.: A Blueprint for Introducing Disruptive Technology into the Internet. In: Proc. First ACM Workshop on Hot Topics in Networking (HotNets). (2002)

3. Tridgell, A.: Efficient Algorithms for Sorting and Synchronization. PhD thesis, Australian National University (1999)

4. Sobti, S., Lai, J., Shao, Y., Garg, N., Zhang, C., Zhang, M., Zheng, F., Krishnamurthy, A., Wang, R.Y.: Network-Embedded Programmable Storage and Its Applications. Technical report, CS Dept., Princeton University (2004)

5. Anderson, T., Dahlin, M., Neefe, J., Patterson, D., Roselli, D., Wang, R.: Serverless Network File Systems. ACM Transactions on Computer Systems **14** (1996)

6. Lee, E.K., Thekkath, C.E.: Petal: Distributed Virtual Disks. In: Conference on Architectural Support for Programming Languages and Operating Systems. (1996)

7. Thekkath, C.A., Mann, T., Lee, E.K.: Frangipani: A Scalable Distributed File System. In: Proc. ACM Symposium on Operating Systems Principles. (1997)

8. Dabek, F., Kaashoek, M.F., Karger, D., Morris, R., Stoica, I.: Wide-Area Cooperative Storage with CFS. In: Proc. of SOSP. (2001)

9. Kubiatowicz, J., Bindel, D., Chen, Y., Czerwinski, S., Eaton, P., Geels, D., Gummadi, R., Rhea, S., Weatherspoon, H., Weimer, W., Wells, C., Zhao, B.: OceanStore: An Architecture for Global-Scale Persistent Storage. In: Proc. of ASPLOS. (2000)

10. Rowstron, A., Druschel, P.: Storage Management and Caching in PAST, a Large-Scale, Persistent Peer-to-Peer Storage Utility. In: Proc. of SOSP. (2001)

11. Clarke, I., Sandberg, O., Wiley, B., Hong, T.: Freenet: A Distributed Anonymous Information Storage and Retrieval System. In: Workshop on Design Issues in Anonymity. (2000)

12. Anderson, D., Chase, J., Vahdat, A.: Interposed Request Routing for Scalable Network Storage. In: Proc. of Operating Systems Design and Implementation. (2000)

13. Banga, G., Druschel, P., Mogul, J.C.: Resource Containers: A New Facility for Resource Management In Server Systems. In: Operating Systems Design and Implementation. (1999)

14. Wobber, E., Abadi, M., Burrows, M., Lampson, B.: Authentication in the Taos operating system. ACM Transactions on Computer Systems **12** (1994) 3–32

15. Gibson, G., Nagle, D., Amiri, K., Chang, F., Feinberg, E., Gobioff, H., Lee, C., Ozceri, B., Riedel, E., Rochberg, D., Zelenka, J.: File Server Scaling with Network-Attached Secure Disks. In: Proc. of the 1997 SIGMETRICS. (1997)

16. de Jonge, W., Kaashoek, M.F., Hsieh, W.C.: The Logical Disk: A New Approach to Improving File Systems. In: Proc. Symposium on Operating Systems Principles. (1993)

17. Zhang, M., Wang, R.Y., Peterson, L., Krishnamurthy, A.: Probabilistic Packet Scheduling: Achieving Proportional Share Bandwidth Allocation for TCP Flows. In: Proc. IEEE Infocom 2002. (2002)

18. Zhang, C., Lai, J., Garg, N., Sobti, S., Zheng, F., Krishnamurthy, A., Wang, R.: Coherent and Network-aware Tracking of Objects. Technical Report TR-672-03, CS Dept., Princeton University (2003)

19. Alexander, D.S., Shaw, M., Nettles, S., Smith, J.M.: Active Bridging. In: Proc. of ACM SIGCOMM '97. (1997) 101–111

20. Decasper, D., Dittia, Z., Parulkar, G.M., Plattner, B.: Router Plugins: A Software Architecture for Next Generation Routers. In: Proc. of ACM SIGCOMM '98. (1998)

21. Nygren, E.L., Garland, S.J., Kaashoek, M.F.: PAN: A High-Performance Active Network Node Supporting Multiple Mobile Code Systems. In: Proc. of OpenArch'99. (1999)
22. Wetherall, D.: Active Network Vision and Reality: Lessons from a Capsule-Based System. In: Proc. of the ACM Seventeenth Symposium on Operating Systems Principles. (1999)
23. Nagle, D.: Active Storage Nets. http://www.ece.cmu.edu/~ asn/old/pubs/-Active%20Storage%20Nets%20Intro.pdf (1998)
24. Cao, P., Zhang, J., Beach, K.: Active Cache: Caching Dynamic Contents on the Web. In: Intl. Conf. on Distributed Systems Platforms and Open Distributed Processing. (1998)
25. Amir, E., McCanne, S., Katz, R.H.: An Active Service Framework and Its Application to Real-Time Multimedia Transcoding. In: Proc. of ACM SIGCOMM '98. (1998)
26. Acharya, A., Uysal, M., Saltz, J.: Active Disks: Programming Model, Algorithms and Evaluation. In: Proc. of ASPLOS. (1998)
27. Riedel, E., Gibson, G.A., Faloutsos, C.: Active Storage For Large-Scale Data Mining and Multimedia. In: Proc. of International Conference on Very Large Data Bases. (1998)
28. Muthitacharoen, A., Chen, B., Mazieres, D.: A Low-bandwidth Network File System. In: Proc. ACM Symposium on Operating Systems Principles. (2001)

Distributed Algorithm for Service Replication in Service Overlay Network

Kevin Y.K. Liu[1], John C.S. Lui[1], and Zhi-Li Zhang[2]*

[1] Computer Science & Engineering, The Chinese University of Hong Kong
{ykliu,cslui}@cse.cuhk.edu.hk
[2] Department of Computer Science, University of Minnesota zhzhang@cs.umn.edu

Abstract. The service overlay network (SON) is an effective mean to deploy end-to-end QoS guaranteed content delivery services on the current Internet. We model the content delivery service on a SON as a service delivery tree (SDT). Previous studies have addressed the optimal distribution tree formulation issues. In this paper, we focus on the problem of maximizing the total operation profit of the SON. In [1], authors introduced a cost model for optimal bandwidth provisioning in SON. In this paper, we extended this concept and propose an alternative approach to maximize the total effective throughput of SDT as well as to minimize the QoS violation penalty of the SON by service replication. We present both centralized and distributed algorithms for the placement of replicated servers on the SDT. Experiments are carried to quantify the merit, effectiveness and the scalability of the proposed service replication algorithm. In particular, the performance gain is very close to the exhaustive search. The algorithm performs well when we scale up the service overlay networks. Finally, we show that one only needs to perform a small number of replications to attain the optimal gain.

Keywords: Overlay networks, QoS, replication algorithm

1 Introduction

The Internet is being used for many different user activities, including emails, software distribution, video/audio entertainment, e-commerce, and real-time games. Although some of these applications are designed to be adaptive to available network resources, they still expect good level of services from the network, for example, low latency and low packet loss, so as to deliver the desired performance at the application layer. However, the primary service provided by the Internet is the *best-effort* service model which does not perform any service differentiation, therefore, end-to-end quality-of-service (QoS) guarantees are difficult to maintain. Another reason for the difficulty in providing end-to-end QoS guarantees is that the Internet is organized as many different autonomous systems (**ASs**) wherein each AS manages its own traffic, performance guarantees and internal routing decisions. These autonomous systems also have various

* John C.S Lui was supported in part by the RGC Research Grant 4420/01E. Zhi-Li Zhang was supported in part by the National Science Foundation (NSF)under the Grant ITR-0085824. Any opinions, findings, conclusions or recommendations expressed in this paper are those of the authors and do not necessarily reflect the views the NSF.

bilateral business relationships (e.g., peering and provider-customer) for traffic exchange so as to maintain the Internet global connectivity. For many network applications, the data traffic usually traverses across multiple autonomous systems, and it is difficult to establish a *"multi-lateral"* business relationship which spans many autonomous systems. Therefore, network services which need end-to-end QoS guarantees are still far from realization and the above mentioned problems hinder the deployment of many QoS sensitive services on the Internet.

In [1], authors advocate the notion of *service overlay network* (SON) as an effective mean to address problems of providing end-to-end services. A SON is an overlay network that spans many autonomous systems. In general, a SON purchases bandwidth with certain QoS guarantees from all ASs that the overlay network spans. This way, a logical end-to-end service delivery infrastructure can be built on top of the existing network infrastructure. On this logical network, one can provide different types of time sensitive services, such as video-on-demand, Internet radio, VoIP, etc. SON offers these services to users who pay the SON for using these value-added services.

The ultimate goal of the service overlay network is to maximize its revenue and minimize the operating cost. In some previous works [1,2], authors formulate this problem as bandwidth provisioning model, wherein the revenue of SON comes from the fee paid by users and the costs consist of bandwidth provisioning cost and the QoS violation penalties. However, one important point to observe is that once the bandwidth provisioning is carried out, the overlay network is *committed* to a topology wherein each link in the overlay network has a *fixed* bandwidth capacity. This capacity of each link remains unchanged until the next bandwidth provisioning instant.

In general, the time scale of bandwidth provisioning can be in terms of weeks or months. Since traffic/service demand is time varying and stochastic in nature, it is possible that there will be a sudden surge on traffic due to some unexpected event (e.g., a popular pay-per-view sport or musical event). This type of traffic surge may not be well-represented or characterized in the original measured traffic distribution that was used for the bandwidth provisioning process. In this case, the allocated bandwidth for the SON may not be sufficient to provide the end-to-end QoS guarantees. This translates to lower profit for the SON operator since the operator needs to pay for the penalty for these QoS violations.

Note that many time-sensitive services provided by the SON are in the one-to-many format, for example, services such as video-on-demand and multi-players on-line games, wherein one "logical" server needs to support many users of the overlay network. As shown in [3], to deliver this type of service, a data delivery process is usually in form of a tree topology. When the user demands increase, some links of the delivery tree could be overloaded or even congested. Instead of delivering a low quality of service over these congested links, (i.e. reduction in profit of SON), we propose to *dynamically replicate* services on the service gateways of SON so as to reduce the QoS penalty as well as increase the effective throughput of the SON. The problem of service replication along the delivery tree is to choose among a set of service gateways to place the additional server for service replication such that the total profit can be maximized.

2 Background on Service Overlay Network

In this section, we provide the necessary background on service overlay networks and its bandwidth provisioning problem.

A SON $\mathcal{G}$ is a logical overlay network with a set of nodes $\mathcal{N}$ and a set of links $\mathcal{L}$. Each node in $\mathcal{N}$ is a *service gateway* which performs service-specific data forwarding and control functions. One can view a service gateway as a physical end host on the Internet. A link in $\mathcal{L}$ is a *logical* connection between two service gateways and the link is an IP level path provided by the underlying autonomous systems. The advantages of the SON architectural framework are : 1) one can purchase different bandwidth for different links in the SON and, 2) one can bypass congested peering points among ASs and thereby provide end-to-end QoS guarantees. When a user requests for a specific QoS guaranteed service, it will connect to the SON through its own network domain and its request will be forwarded to the proper service gateway.

The advantage of the SON architecture is that it decouples the application services from the network services and thereby reduces the complexity of network control and management. Meanwhile, the SON can provide more diverse end-to-end QoS guaranteed services to satisfy the needs of its users.

The *"bandwidth provisioning problem"*[1,2,4] for a SON is to determine the appropriate amount of bandwidth to purchase for each link in $\mathcal{L}$ from the underlying ASs, so that the QoS sensitive traffic demand for any source-destination pair in $\mathcal{R}$ can be satisfied and at the same time, the total net profit of the SON is maximized.

The formal mathematical framework for performing the bandwidth provisioning can be described as follows. Given a network topology $\mathcal{G}$, the source-destination (SD) path requirements in $\mathcal{R}$, the stochastic traffic demand $\{\rho_r\}$ for each $r \in \mathcal{R}$, and the routing method, one can provide a *lower bound* of the expected net profit (or income) for the service overlay network. Let r denote a path in the source-destination path set $\mathcal{R}$. Assume that the traffic demand distribution on path r is known[1] and traffic of all paths in $\mathcal{R}$ are described by the stochastic traffic demand matrix $\{\rho_r\}$, the total net income for the SON, denoted by the random variable W, can be expressed as:

$$W(\{\rho_r\}) = \sum_{r \in \mathcal{R}} e_r \rho_r - \sum_{l \in \mathcal{L}} \Phi_l(c_l) - \sum_{r \in \mathcal{R}} \pi_r \rho_r B_r(\{\rho_r\}). \tag{1}$$

where $\sum_{r \in \mathcal{R}} e_r \rho_r$ is the total revenue received by a SON for carrying $\{\rho_r\}$ traffic along the SD path $r \in \mathcal{R}$; $\sum_{l \in \mathcal{L}} \Phi_l(c_l)$ is the total bandwidth cost that a SON must purchase from all its underlying autonomous systems; $\sum_{r \in \mathcal{R}} \pi_r \rho_r B_r(\{\rho_r\})$ is the total *penalty* that a SON suffered when the QoS guarantees for those traffic demands are violated. The variable B_r represents the probability that QoS guarantees for the SD pair r is violated. The problem of bandwidth provisioning can thus be formulated as the optimization of the average total net profit $E(W)$, or:

$$\max_{c_l} E(W). \tag{2}$$

In other words, determining the appropriate amount of capacity $\{c_l\}$ for each link $l \in \mathcal{L}$.

---

[1] This traffic demand distribution can be obtained through long-term observation or measurement of past traffic history.

Note that the above mentioned bandwidth provisioning method is only practical in an *off-line manner*. That is, once bandwidth is provisioned, it cannot be changed until the next bandwidth provisioning instant. It is possible that there may be a surge in traffic demand due to some unexpected events, e.g., a popular pay-per-view sport or musical event that attracts many users. The variation of traffic flow will increase the QoS violation probability B_r. Therefore, it is crucial for the SON to have the adaptive capability to traffic flow fluctuation. In this paper, we propose to *dynamically replicate services* within a SON so as to reduce the traffic demands on "overloaded" links and to maximize the net income of an SON operator.

3   Mathematical Formulation for Service Replication

Real-time content delivery is one of the major applications of SON, many QoS sensitive services can be deployed on the SON's infrastructure. As illustrated in [3], the optimal data delivery topology for these applications is a tree topology. In this paper, we call the topology as a *service delivery tree* (SDT). The root node of SDT is an application level service gateway. All the leaf nodes are called client nodes. They are access points for users within the same network domain. The formation of delivery tree can be different for different applications [3,5,6,7], however, our model is generic for any tree formation.

To formally define the service replication problem, we use the following notations:

$\mathcal{T}$:	the service delivery tree.
T_u:	subtree of $\mathcal{T}$ rooted at node u.
S_u:	all children nodes of u.
ρ_u:	traffic demand from node u.
D_u:	total average demand from all client nodes under the subtree T_u.
c_u:	the allocated capacity of the uplink of node u.
$q_u(D_u, c_u)$:	denoting the probability of QoS guarantee on the uplink of node u, given that the traffic demand D_u and capacity c_u of the uplink.
$F(T_u)$:	total effective throughput of subtree T_u.

In the original SON bandwidth provisioning model [1,4], the QoS violation on link l is defined as $\rho_l B_l(\rho_l)$ where B_l is the QoS "*violation*" probability. In our service replication problem on SDT, we use an alternative metric. In the original model, the first two terms of the objective function in Eq. (1) are the total revenue (total income leveraged from all users) and the total bandwidth cost (total cost paid to purchase the bandwidth from underlying ASs) of the SON. Note that for the service replication process, the values of these two terms will not change. Therefore, in formulation of the service replication problem, we only need to focus on the third term of Eq. (1), namely, the total QoS violation penalty.

As stated in Section 2, the derivation of the expression of QoS violation is difficult due to the functional dependency on the joint traffic distribution and the violation probability B. Instead of directly evaluating the QoS violation penalty, we define a new function $F(T_u)$ to evaluate the effective throughput, which in fact quantifies the level of QoS guarantee of any subtree T_u rooted at the node u.

First, we denote the generic link QoS guarantee probability function as following:

$$q_u(D_u, c_u) = 1 - B_l, \qquad \text{where } l \text{ is uplink of } u. \tag{3}$$

This probability function is independent of any particular form of QoS violation function B. Similar to the QoS violation penalty in Eq.(1), $F(T_u)$ can then be defined as:

$$F(T_u) = \sum_{v \in L_u} \rho_v \prod_{i \in path(u,v)} q_i(D_i, c_i), \tag{4}$$

where L_u denotes the set of leaf nodes of the subtree T_u and $path(u, v)$ denotes all the nodes along the path from u to v. $F(T_u)$ can also be expressed in a recursive form:

$$F(T_u) = \begin{cases} \rho_u & \text{if node } u \text{ is a leaf node,} \\ \sum_{v \in S_u} F(T_v) \cdot q_v(D_v, c_v) & \text{otherwise.} \end{cases} \tag{5}$$

Using the above recursive function, we can compute $F(T_r)$, i.e. the total effective throughput of the SDT with the root node r. [2]

Finally, given a SDT T_r, the *service replication problem* is formally defined as :

$$\max_{v \in \mathcal{D}_r} \{F(T_r - T_v) + F(T_v)\}. \tag{6}$$

where $\mathcal{D}_r$ is the set representing the descendant nodes of the root node r. In other words, find a node v under the SDT T_r to maximize the gain in the effective throughput.

3.1   Distributed Approach to Evaluate the Effective Throughput of SDT

One way to find the optimal solution to the above problem in Eq. (6) is to perform an exhaustive evaluation at every nodes in the tree T_r and choose the node which maximizes the objective function in Eq. (6). However, since runtime of this approach is $O(n^2)$, it is computational prohibitive when the size of SDT is large. Another disadvantage of this exhaustive evaluation approach is that it requires a centralized entity which has the view of the whole network topology, as well as all the traffic information and probability of QoS guarantees of all the links and nodes of the SDT. Thus, this approach suffers from the potential of single point failure and it is not scalable as the network size grows.

In the following, we propose a *distributed* approach to solve the service replication problem. In our approach, each node only maintains *three* variables that summarize the characteristics of the subtree which rooted at that node. This way, the information can be *recursively evaluated* from the leaf nodes up to the root node. Since only a small amount of information is maintained at each node, the decision making can be carried out very efficiently in a top-down evaluation method.

We require that each node u of the SDT $\mathcal{T}$ maintains three variables, namely, D_u, q_u and F_u. The first variable D_u represents the total traffic demand of the subtree T_u, and it can be recursively evaluated using the following expression:

$$D_u = \begin{cases} \rho_u & \text{if } u \text{ is leaf node} \\ \sum_{v \in S_u} D_v & \text{otherwise} \end{cases} \tag{7}$$

---

[2] We have omitted an example here due to page limit. It is available at [8].

The variable q_u is the probability of QoS guarantee on the uplink of node u to its parent node. It is computed at node u. It is defined as an generic QoS guarantee function of the traffic D_u and the capacity c_u. It is important to point out that our approach can be applied to any specific form of this function, as long as it is an increasing function of D_u. Lastly, the variable F_u is the total effective throughput of the subtree T_u. One can use the recursive expression in Eq. (5) to update these three variables and thereby obtain the effective throughput of the SDT.

Our evaluation scheme begins with all leaf nodes. Each leaf node, say u, will send the values of $\{D_u, q_u, F_u\}$ to its parent node v. The node v, upon receiving all the information from all its children nodes, will then update its own variables $\{D_v, q_v, F_v\}$ accordingly, and then send them to its parent. All the other nodes are updated accordingly in a bottom-up manner. This process will continue until the root node r computes its effective throughput F_r.

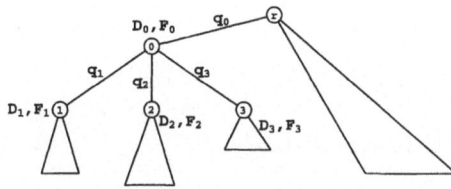

Fig. 1. Illustration on the evaluation of SDT

Consider an example illustrated in Fig.1. After receiving the updated values from all its children, node 0 will then update its own values as follows: $D_0 = D_1 + D_2 + D3$, $q_0 = q_u(D_0, c_0)$ and $F_0 = q_1 F_1 + q_2 F_2 + q_3 F3$.

The above distributed approach is used to evaluate the effective throughput of a SDT only. To find the proper node for service replication efficiently, we require each node, say node u, to maintain an extra variable G_u This G_u represents the gain of total effective throughput by placing the additional server at node u. It is defined as:

$$G_u = F(T_r - T_u) + F(T_u) - F(T_r) \tag{8}$$

In other words, after placing the additional server at node u, the additional server will serve all the users of the subtree T_u only; while the original server r will serve all users from the remaining tree $T_r - T_u$. Therefore, the sum of the first two terms in Eq. (8) is the total effective throughput after service replication at node u and G_u represents the *gain* in the effective throughput if the replicated server is placed at node u.

Note that during the updating process, the update of the G function at node u is not so easy as the updating the D or F values. For each node u, we need to evaluate the $F(T_r - T_u)$ value. However, placing the additional server at node u will *affect* the QoS guarantee probability along all the links between root node r and node u. Thus we need to re-evaluate the F value of each node along that path. To address this difficulty, we take the following approach. We calculate the G function at node u simply by using the uplink probability only, i.e. $G_u = F(T_u)(1 - q_u)$ instead of calculate $F(T_r - T_u)$, because the $F(T_u)$ and q_u are directly available information at node u. In other words, G_u is the *minimum* guaranteed gain of the total effective throughput of SDT $\mathcal{T}$.

4   Service Replication Algorithms

In this section, we present the algorithm for selecting a node for service replication. To enhance the readers' understanding, we first present a centralized service replication algorithm, then we extend the concept to a distributed approach of service replication.

4.1   Centralized Service Replication Algorithm

The centralized algorithm has two phases, namely, *preprocessing* and *searching*. The preprocessing phase can be carried in a recursive manner. Fig.2 illustrates the pseudocode of the recursive update of node u.

```
UPDATE-NODE (u)
1  if S_u = ∅                              /* if node u is leaf node */
2       D_u ← ρ_u
3       F_u ← ρ_u
4       G_u ← ρ_u · (1 − q_u(D_u, c_u))
5  else                                    /* node u is not leaf node */
6       for v ∈ S_u do UPDATE-NODE (v)     /* for each child v of node u */
7       D_u ← Σ_{v∈S_u} D_v
8       F_u ← Σ_{v∈S_u} F_v q_v(D_v, c_v)
9       G_u ← max{G_v}, (∀v ∈ S_u)
10      if parent(u) ≠ ∅                    /* if node u is not root node */
11          G_u ← max{G_u, F_u · (1 − q_u(D_u, c_u))}
```

Fig. 2. Recursive update of node u of SDT $\mathcal{T}$

Lemma 1. *Assuming the average degree of SDT is constant, the runtime complexity of preprocessing phase is $O(n)$.* [3]

At the end of preprocessing phase, each node will obtain the updated values of $\{D_u, F_u, G_u\}$. Then one can search for the optimal server placement in a top-down manner starting at the root node of SDT $\mathcal{T}$. Fig.3 illustrates the procedures. The searching algorithm will output the node that maximizes the gain of effective throughput.

Lemma 2. *The average runtime complexity of searching phase is $O(\log(n))$.*

4.2   Distributed Service Replication Algorithm

Though the above centralized algorithm is simple to implement, it requires a centralized entity in the SON for execution. This requires extra resources and also has the potential of a single-point-failure problem. These problems will become significant when the size of the SON is large. We propose the following distributed algorithm, which can be concurrently executed on each node inside the SDT. Thus, no centralized management

---

[3] The proof of the lemma is omitted here due to page limit. Please refer to [8].

```
FIND-REP-NODE (T)
1  u ← root(T)
2  v ← max_v{G_v}, ∀v ∈ S_u          /* get the child v with maximum gain */
3  while G_v ≥ G_u and S_u ≠ ∅ do     /* while G_v ≥ G_u and u is not leaf node */
4      u ← v                          /* proceed to the next node */
5      v ← max_v{G_v}, ∀v ∈ S_u
6  return u
```

Fig. 3. Finding the node to place the replicated server

```
DISTRIBUTED-NODE-UPDATE (u)
1  upon receiving ⟨request_update⟩
2  if S_u = ∅                                          /* if u is not leaf node */
3      D_u ← ρ_u, F_u ← ρ_u
4      G_u ← ρ_u(1 − q_u(D_u, c_u))
5      send ⟨reply_update : D_u, F_u, G_u⟩ to parent(u)
6  else send ⟨request_update⟩ to all v ∈ S_u            /* ask all children to update */

7  upon receiving ⟨reply_update : D_v, F_v, G_v⟩ from child v
8  D_u ← D_u + D_v
9  F_u ← F_u + F_v q_v(D_v, c_v)
10 G_u ← max{G_u, G_v}
11 if received ⟨reply_update⟩ message from all children   /* wait until all children are updated */
12     if parent(u) ≠ ∅                                   /* if not root node */
13         G_u ← max{G_u, F_u(1 − q_u(D_u, c_u))}
14         send ⟨reply_update : D_u, F_u, G_u⟩ to parent(u)  /* send updated information to parent */
15     else send⟨exec_search⟩ to self                       /* start searching phase */

16 upon receiving ⟨exec_search⟩
17 if S_u = ∅ output u                                    /* if leaf node, output u */
18 else send ⟨request_G⟩ to all v ∈ S_u                   /* else ask each child v to send G_v */

19 upon receiving ⟨request_G⟩
20 send ⟨reply_G : G_u⟩ to parent(u)                      /* reply G value to parent node */

21 upon receiving ⟨reply_G : G_v⟩ from child v
22 if G_v ≥ G_u send ⟨exec_search⟩ to node v
23 else if received ⟨reply_G⟩ messages from all children
24     output u
```

Fig. 4. Distributed algorithm running at each node u

is required and the server replication can be carried more efficiently. The distributed algorithm achieves the same result as the centralized algorithm by sending messages among the nodes of SDT. Fig.4 illustrates the distributed service replication algorithm. It is divided into five parts. The first two parts (lines 1-15), correspond to the preprocessing phase, while the rest three parts (lines 16-24) correspond to the searching phase.

The DISTRIBUTED-NODE-UPDATE() procedure can be implemented as an event driven program running at each node. The information exchange between nodes can be implement as a simple protocol with the following set of of messages ⟨request_update⟩, ⟨reply_update⟩, ⟨exec_search⟩, ⟨request_G⟩, ⟨reply_G⟩.

The root node will initiate the distributed algorithm by sending the ⟨request_update⟩ to all its child nodes. Upon receiving this message, these nodes will send the same message to their children (line 6), and this message will be propagated till the leaf nodes. The leaf nodes will then send the ⟨reply_request⟩ to their parents with the updated values of D, F, G (line 5). Each node, upon receiving ⟨reply_request⟩ message will then update its own D, F, G values (line 8-10). When it receives the updates from all its children, the processing phase on that node is finished, and it will send the $\langle reply\_request \rangle$ message to its parent (line 14). When the root node finally receives all the updates from its children and updates its own D, F, G, the whole preprocessing phase is terminated.

The root node will then start the searching phase (line 15). It will ask the G values of all its children by sending the $\langle request\_G \rangle$ message. Upon receiving the reply, it will pick the child node with the G value not less than the G value of itself, and then send the $\langle exec\_search \rangle$ message (line 22). This process will stop when there is one node in which the G value of all its children are less than itself (line 23). At this moment, the searching phase is terminated and that node will be picked to place the replication.

For the distributed service replication algorithm, the preprocessing phase can be executed in a *parallel* fashion, in which case the total running time of the *preprocessing phase* can be improved to $O(\log(n))$ (proportional to the height of the tree). Therefore, the total running time of our algorithm is also improved to $O(\log(n))$. It is much faster than the exhaustive searching method $(O(n^2))$.

4.3 Improved Distributed Algorithm

The centralized and distributed algorithms discussed above are easy to implement on top of SON. However, one may provide a better solution (e.g., in terms of finding a closer-to-optimal gain in the effective throughput) if each node is allowed to stored more information. In the following, we provide an improved version of the distributed algorithm which can find a better solution at the cost of extra computational resources.

In the previous algorithms, to determine the minimum possible gain in the total effective throughput (G_u) of placing a replicated server at node u, we consider the uplink QoS guarantee probability (q_u) only. However, in this improved distributed algorithm, we use the total QoS guarantee probability *along the path* from the root node r to the node u, and we denote this total probability to be Q_u for each node u. We can define Q_u recursively as:

$$Q_u = \begin{cases} 1 & u \text{ is root node} \\ q_u \cdot q_{parent(u)} & \text{otherwise} \end{cases} \qquad (9)$$

Therefore, we redefine the G_u to be:

$$G_u = F(T_u)(1 - Q_u) \qquad (10)$$

To deploy this new algorithm, each node needs to maintain an extra variable Q_u, and the following procedure NODE-IMPROVE which served as an add-on module to the basic distributed algorithm, can be invoked, if necessary, after the *preprocessing phase* and before the *searching phase*. To use this add-on module, we only need to modify the (line 15) of DISTRIBUTED-NODE-UPDATE to:

15   **else send** $\langle improve\_Q : 1 \rangle$ **to self**

Then the root node, before start the searching phase, will first initiate the updating of Q_u as well G_u value of each node. Fig.5 illustrate the add-on module where line (1-7) updates the Q_u of each node, and line (8-15) updates the G_u of each node.

```
NODE-IMPROVE (u)
1  upon receiving ⟨improve_Q : Q_p⟩
2     Q_u ← q_u(D_u, c_u) · Q_p
3     if S_u = ∅                                          /* if u is leaf node */
4         G_u ← F_u(1 − Q_u)
5         send ⟨improve_G : G_u⟩ to parent(u)             /* ask the parent node to update the G value */
6     else                                                /* if u is internal node */
7         send ⟨improve_Q : Q_u⟩ to all v ∈ S_u           /* ask the children to update Q */

8  upon receiving ⟨improve_G : G_v⟩ from child v
9     G_u ← max{G_u, G_v}
10    if received ⟨improve_G⟩ message from all children   /* wait till all children updated G value */
11        if parent(u) ≠ ∅                                /* if u is not root node */
12            G_u ← max{G_u, F_u · (1 − Q_u)}
13            send ⟨improve_G : G_u⟩ to parent(u)         /* ask parent node to update G */
14        else                                            /* if the root receives all updates */
15            send ⟨exec_search⟩ to self                  /* the root node start searching phase */
```

Fig. 5. Add-on module for improved distributed algorithm

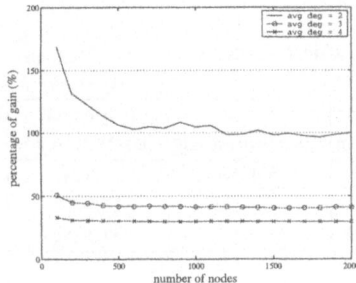

Fig. 6. Illustration on the performance gain of our service replication algorithm when the size of the tree grows from 100 to 2000 nodes.

5 Experiments

In this section, we perform two experiments [4] so as to evaluate the performance and effectiveness of our service replication algorithm. The first experiment evaluates the quality of the results obtained by our algorithm as compares to random selection and exhaustive selection the replication. The second experiment illustrates the scalability of the service replication algorithm when we increase the size of the SDT.

Experiment 1: (Comparing the quality of the distributed service replication algorithm with random selection and exhaustive search): In this experiment, we show the quality of our distributed algorithm comparing with the other two algorithms. The random selection algorithm will arbitrarily pick an internal node of SDT for service replication. Obviously, it has the least computational overhead as compare to other algorithms. The exhaustive search algorithm will search throughout the whole SDT tree and find the optimal node for replication. This algorithm has the largest computational complexity and is not scalable.

---

[4] There are two more experiment in [8] to demonstrate our replication algorithm, which are omitted here due to page limit.

In this experiment, we randomly generate 100 instances of SDTs of 500 nodes each. The average number of children of each internal node is set to 3. Each client node has a random traffic demand uniformly distributed within a range from 1 to 1000 unit. The link capacity is provisioned in the way such that the loading on each link (D/c) is a constant. We compare the gain of placing a replicated server at variable link loading.

Table 1. Comparison of our algorithm with random placement and optimal placement, when $q_u = 1 - D_u/c_u$.

D/c	Random selection	Basic Distributed Alg	Improved Distributed Alg	Exhaustive selection
0.4	2.21%	19.45%	21.29%	22.33%
0.6	7.90%	35.71%	40.56%	42.61%
0.8	46.34%	84.51%	87.80%	91.76%

Table 2. When $q_u = 1 - (D_u/c_u)^2$.

D/c	Random selection	Basic Distributed Alg	Improved Distributed Alg	Exhaustive selection
0.4	0.61%	6.43%	7.10%	7.79%
0.6	1.99%	16.93%	18.93%	20.97%
0.8	11.27%	40.33%	47.17%	53.62%

Table 3. When $q_u = 1 - (D_u/c_u)^4$.

D/c	Random selection	Basic Distributed Alg	Improved Distributed Alg	Exhaustive selection
0.4	0.19%	0.94%	1.05%	1.37%
0.6	0.60%	5.11%	5.69%	6.73%
0.8	3.10%	20.09%	22.90%	27.95%

Table 1- 3 illustrate the result of our experiments. From these tables, we can conclude that our improved distributed algorithm, which has a much lower computational complexity than the exhaustive search, has a performance very closed to the optimal. Another observation can be made from these tables is that when the average loading on each link (D/c) is high, it is more beneficial to perform the service replication.

Experiment 2: (Illustration of the scalability of our algorithm): In this experiment, we illustrate the performance of our algorithm when the size of the SDT grows from 100 nodes to 2000 nodes. For each size of the SDT, we generate 100 instances of SDT and compute the average performance gain. All the link capacity are set to the value such that the D/c is 0.8. As shown in Fig.6, although there is a little fluctuation when the SDT size is small, the average gain of our replication algorithm still remains at a certain percentage even the network size grows.

We also test on SDTs of different average degrees. When the average degree increases, the average gain decrease. The reason is that when the average degree of tree is small, the height of tree is larger, i.e. the average path length from client node to the root node is longer. This means the QoS guarantee are much harder to preserve for the client nodes, therefore doing replication at SDT of small average degree will have more benefit. Meanwhile, because of the constraint of resources at each service gateway of SON, the average degree of SDT in real situation will not be a big number, so our service replication is suitable for SDT.

6 Conclusions

Previous works have studied the bandwidth provisioning problems and optimal distribution tree formulation on SON. However, since the bandwidth is fixed after provisioning and the topology is static, the SON is inflexible to traffic demand variation.

In this paper, we proposed to resolve this problem by service replication in the service delivery tree. We have presented both centralized and distributed algorithms to find the placement of a replicated server, which maximize the total effective throughput of SDT. The distributed algorithm requires very little resource at each node, and can be implemented as a simple protocol among all the service gateways of SON. The complexity of the algorithm is much lower than the brute-force exhaustive search method, but still achieve a near-optimal result. Furthermore, it has a good scalability and can be deployed in large scale SON networks.

References

1. Z. Duan, Z.-L. Zhang, and Y. T. Hou, "Service Overlay Networks: SLAs, QoS and Bandwidth Provisioning," in *IEEE 10th International Conference on Network Protocols (ICNP'02)*, (Paris, France), Nov. 2002.
2. D. Mitra and Q. Wang, "Stochastic traffic engineering, with applications to network revenue management," in *IEEE Infocom 2003*, (San Francisco, USA), 2003.
3. M. S. Kim, S. S. Lam, and D.-Y. Lee, "Optimal Distribution Tree for Internet Streaming Media," in *23rd IEEE ICDCS*, May 2003.
4. Z. Duan, Z.-L. Zhang, and Y. T. Hou, "Service Overlay Networks: SLAs, QoS and bandwidth provisioning," tech. rep., Computer Science Department, University of Minnesota, Feb. 2002.
5. J. Jannotti, D. K. Gifford, K. L. Johnson, M. F. Kaashoek, and J. W. O'Toole, Jr., "Overcast: Reliable Multicasting with an Overlay Network," in *the Fourth Symposium on Operating System Design and Implementation (OSDI)*, pp. 197–212, Oct. 2000.
6. Y. Chu, S. G. Gao, S. Seshan, and H. Zhang, "Enabling conferencing applications on the internet using an overlay multicast architecture," in *ACM SIGCOMM 2001*, Apr. 2001.
7. Y. Cui, Y. Xue, and K. Nahrstedt, "Optimal resource allocation in overlay multicast," in *"IEEE 11th International Conference on Network Protocols (ICNP'03)"*, Nov. 2003.
8. K. Y. Liu, J. C. Lui, and Z.-L. Zhang, "Distributed algorithm for serivce replication in service overlay network." unpublished. http://www.cse.cuhk.edu.hk/~ykliu/research/dason.ps.

Routing in Turn-Prohibition Based Feed-Forward Networks

Markus Fidler and Gerrit Einhoff

Aachen University, Department of Computer Science,
Ahornstr. 55, 52074 Aachen
fidler@i4.informatik.rwth-aachen.de
gerrit@einhoff.com

Abstract. The application of queuing theory to communications systems often requires that the respective networks are of a feed-forward nature, that is they have to be free of cyclic dependencies. An effective way to ensure this property is to identify a certain set of critical turns and to prohibit their use. A turn is a concatenation of two adjacent, consecutive links.

Unfortunately, current routing algorithms are usually not equipped to handle forbidden turns and the required extensions are nontrivial. We discuss the relevant issues for the example of the widely deployed Dijkstra algorithm. Then, we address the general case and introduce the Turnnet concept, which supports arbitrary combinations of routing algorithms with turn-prohibiting feed-forward mechanisms.

1 Introduction

Classical queuing theory has been investigated for a long time to better understand many qualities of communication systems [7]. It has recently been complemented by Network Calculus [10,2], which extends known queuing theory by means of a worst-case analysis to provide deterministic performance bounds. A field of application of Network Calculus are Quality of Service (QoS) enabling architectures, like the Differentiated Services framework [1], where it allows to efficiently compute delay bounds [6,14] for a so-called Premium Service [3].

1.1 The Feed-Forward Property

Unfortunately, a variety of methods from the field of classical queuing theory, as well as the direct application of Network Calculus have one important prerequisite, namely the network has to be of a feed-forward nature.

Definition 1 (Feed-Forward Property). *A feed-forward queuing network is a network, in which all queues can be ordered in such a way that whenever a traffic flow traverses from queue i to queue j, this implies that $i < j$ [7], or in a more verbatim way: the links of a feed-forward network cannot form any cycles, i.e. it is impossible for traffic flows to create cyclic dependencies on each other [2].*

N. Mitrou et al. (Eds.): NETWORKING 2004, LNCS 3042, pp. 1168–1179, 2004.

Dependencies occur, for example in case of Network Calculus, if two flows use the same queuing and scheduling unit on an outgoing link. In this scenario the service offered to each of the flows individually depends on the service that is consumed by the respective other flow. Now, consider a network consisting of three nodes a, b, and c and three links (a, b), (b, c), and (c, a). Assume two flows use the network, whereby the path of flow 1 is $a \to b \to c$ and the path of flow 2 is $b \to c \to a$. The service that remains for flow 2 at link (b, c) depends on the service that is consumed by flow 1 at the same link, which in turn depends on the output of flow 1 from link (a, b). Fortunately, the output of flow 1 from link (a, b) does not depend on flow 2. Thus, the dependency is not cyclic and the system can be solved in an inductive manner. However, adding a third flow that traverses the path $c \to a \to b$ creates a cyclic dependency. The output of flow 1 from link (a, b) depends on the output of flow 3 from link (c, a). Flow 3's output from link (c, a) depends, however, on flow 2's output from link (b, c), and again on flow 1's output from link (a, b), which completes the cycle.

1.2 Feed-Forward Mechanisms

Obviously, real-world networks are not necessarily of a feed-forward nature, unless they are for example star-shaped. One way to nevertheless realize Network Calculus based QoS offerings [6] is to take provisions to prevent from the creation of cyclic dependencies between different flows.

To ensure the feed-forward property in an arbitrary network, the usual approach is to restrict the usage in a certain way that makes it impossible for flows to create a cyclic dependency. The simplest way to do so is to build a spanning tree covering all nodes and to prohibit the use of all links not belonging to that tree. Since a spanning tree cannot contain any circles by definition, the feed-forward property is ensured. On the other hand this approach can disable large parts of the network, potentially causing a big performance impact [5].

A more intelligent approach is not to prohibit the use of complete links, but only of certain turns. A turn is a triple of three nodes connected by two links. For example a prohibited turn (a, b, c) would disallow a flow to utilize the path $a \to b \to c$, but it could still use $a \to b \to d$, provided the link $b \to d$ exists and the turn (a, b, d) is permitted.

Two possible algorithms that determine a set of turns, which have to be prohibited to make a network feed-forward compliant, are Up/Down Routing [13] and Turn Prohibition [15]. Both algorithms return a set of turns that have to be prohibited within a given network topology. As expected the performance impacts of the two turn-prohibiting algorithms on the routing performance are a lot smaller than with the link-prohibiting spanning tree approach [5].

2   Routing in Networks with Prohibited Turns

Using a turn-prohibiting mechanism creates a problem though. Routing algorithms are usually not equipped to handle forbidden turns. In difference to link-prohibiting mechanisms that return a smaller, but still valid network consisting only of nodes, links, and metrics that routing algorithms can work with[1], turn-prohibiting mechanisms require that the routing algorithm takes the forbidden turns into account and does not use them. Obviously, commonly used routing schemes do usually not fulfill this requirement.

2.1   The Challenge of Routing with Prohibited Turns

One option to apply a routing algorithm to a network with prohibited turns, is to adapt the algorithm to honor the forbidden turns. In [15] an example is provided for the Bellman-Ford algorithm. However, a solution that is applicable to arbitrary routing algorithms is not self-evident and to our knowledge missing in current literature.

An algorithm used by a lot of routing schemes [5] is Dijkstra's shortest path algorithm. Examples include Shortest Path First (SPF) and its enhancements [11,12,17,5]. Yet, Dijkstra's algorithm is not aware of prohibited turns, although it does not seem to hard to extend it by just cancelling the consideration of a new path as soon as it includes a prohibited turn. However, figure 1 gives a motivating example, why this approach does not work.

Part (a) shows a simple network with four nodes. The number at each link specifies the additive link-costs. When searching a least-cost path $1 \rightarrow 4$ using Dijkstra's algorithm, it finds the correct path $1 \rightarrow 3 \rightarrow 4$ with a cost of 2 as shown in (b). Assuming that the turn $1 \rightarrow 3 \rightarrow 4$ is prohibited (see (c)), the correct least-cost path from 1 to 4 is now $1 \rightarrow 2 \rightarrow 3 \rightarrow 4$ with a cost of 5 as can be seen in (d). However, using Dijkstra's algorithm extended in the way described above, it would find the path $1 \rightarrow 2 \rightarrow 4$ (see (e)) with a cost of 6, which is *not* the path with the least costs.

The reason for this wrong result is that the algorithm finds the shortest path to each node in an incremental way, i.e. once it has found the shortest path to a node, that path is fixed. In the given example the first node that is examined is node 3. The correct shortest path $1 \rightarrow 3$ is identified and the backpointer of node three is set to point to node 1. With prohibited turns, however, the shortest path to a node depends on the next node of the path. Unfortunately, Dijkstra's algorithm does not consider the following node in its local shortest path decision as it is not designed to do so. With prohibited turns, a node may need more than one backpointer, depending on the destination node. For the request $1 \rightarrow 3$, the backpointer of node 3 should point to node 1, but for the request $1 \rightarrow 4$ it should point to 2 (see (f)).

It is obvious that extending Dijkstra's algorithm to work with prohibited turns is far from trivial and also would not constitute a general solution for all routing schemes. For example the Maximum Disjoint Paths [16] and the Minimum Interference Routing [9] algorithm–although making use of Dijkstra's algorithm–would need individual adaptations to honor prohibited turns. Thus, a general purpose concept that allows for an arbitrary combination of routing algorithms with feed-forward mechanisms is needed.

---

[1] Note, that routing in a spanning tree is trivial, since there exists only a single path between any two nodes.

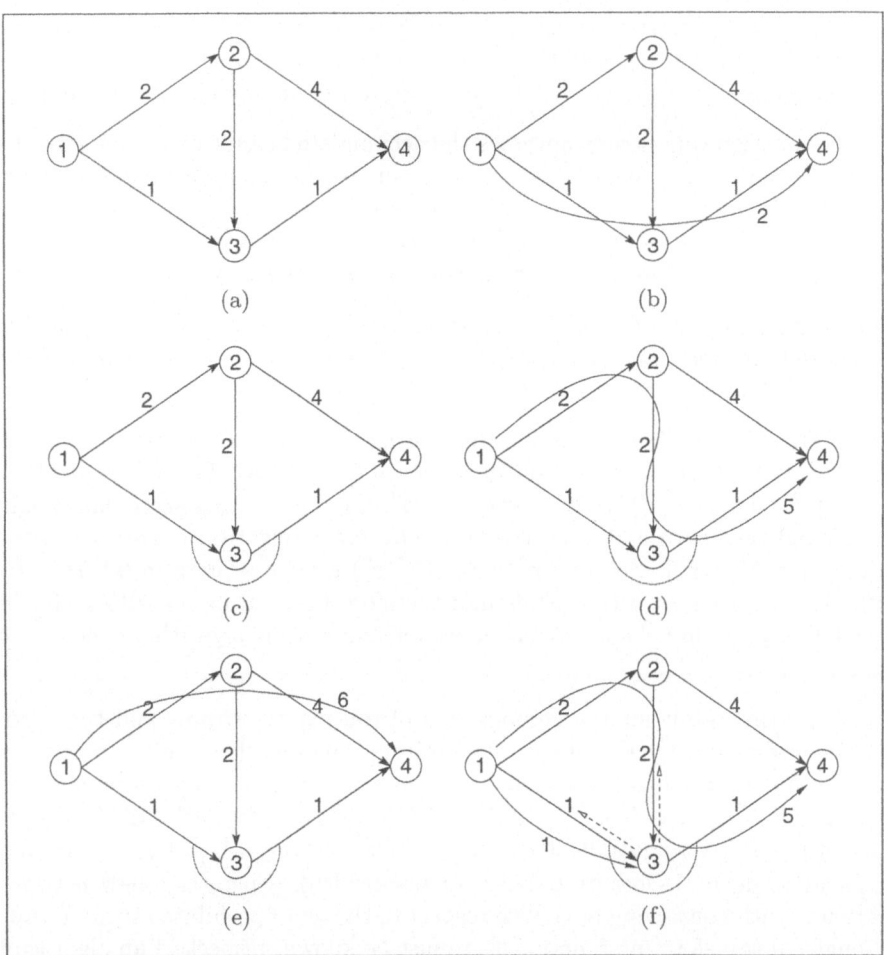

Fig. 1. Dijkstra's algorithm does not work with forbidden turns.

2.2 Formal Requirements Specification

Routing algorithms expect a network to consist of nothing else but nodes, links, and link metrics, which can be used without any restrictions. To work with prohibited turns one solution is to transform a network with a given set of prohibited turns into another network without prohibited turns, with the constraint that routing results can be transferred back to the original network without impacts on their correctness.

The following definition helps in the formal specification of this requirement.

Definition 2 (Path-Conserving). *Given two networks $G^1 = (N^1, E^1)$ and $G^2 = (N^2, E^2)$ consisting of nodes N^i and edges E^i. Select two nodes $s^1, d^1 \in N^1$ with $s^1 \neq d^1$ and define P_{s^1, d^1} to be the set of possible paths between s^1 and d^1.*

G^2 is called path-conserving to G^1, if there exist $s^2, d^2 \in N^2$ and a bijective function f between P_{s^1,d^1} and P_{s^2,d^2} so that all defined path metrics are the same for p and $f(p)$, i.e. $m(p) = m(f(p))$ with $m()$ being the metrics for a path p.

The function $m()$ hereby specifies the accumulated result of the metrics of a path. For an additive metric this means the sum of all link metrics on that path. For example the path p in Fig. 1 (d) would result in $m(p) = 5$. If a network has a vector of different metrics for each link, $m()$ results in a vector also.

From this definition, the following corollary can be extracted immediately.

Corollary 1. *If a network G^2 is path-conserving to a network G^1 with function f, any routing algorithm that finds optimal paths by link metrics and produces a path p^2 in G^2 would produce a path p^1 in G^1 with $m(p^1) = m(f^{-1}(p^2)) = m(p^2)$.*

Proof. By contradiction. Assuming the routing algorithm would produce a path p^1 in G^1 and a path p^2 in G^2 and $m(p^1) \neq m(f^{-1}(p^2))$ would hold. If $m(p^1) > m(f^{-1}(p^2))$, i.e. $f^{-1}(p^2)$ is a better path in G^1, then the routing algorithm should have found that path and is therefore no optimizing algorithm contrary to the assumption. If $m(p^1) < m(f^{-1}(p^2))$, i.e. $f^{-1}(p^2)$ is a worse path in G^1, then by definition $m(f(p^1)) < m(p^2)$ holds and therefore the routing algorithm should have found $f(p^1)$ in G^2 and is thus no optimizing routing algorithm contrary to the assumption.   □

To use this result for the combination of routing algorithms and turn prohibiting feed-forward mechanisms, an algorithm is needed that, if given a network $G = (N, E)$, a source and a destination node $s, d \in N$, a set of prohibited turns $T \subseteq \{(i, j, k) : i, j, k \in N \wedge (i, j) \in E \wedge (j, k) \in E\}$, and for each link $(i, j) \in E$ a set of m additive, multiplicative[2], or concave[3] metrics $m_{(i,j)}^t, 1 \leq t \leq m$ (e.g. propagation delay, $1-$loss probability, or bandwidth), generates a new network G^2 that is path-conserving to G with respect to the set of prohibited turns T. Additionally, a transforming function f^{-1} must be known, respective an algorithm that produces $f^{-1}(p)$ for the input p.

The Turnnet algorithm provides exactly that and is described in the following section.

3   The Turnnet Concept

The basic idea behind Turnnet is that a routing algorithm should not look at a path node-by-node but link-by-link, i.e. it should shift its focus from the visited nodes to the visited links. Going from one node to the next includes crossing a link, but going from one link to the next includes crossing a turn around a node. So by focusing on the link-steps in a path, rather than the node-steps, the turns are included in the observations.

---

[2] A multiplicative metric can be converted into an additive metric applying the logarithmic transformation.

[3] A concave metric is a metric that is accumulated by forming the minimum, i.e. $m = \min(m_1, m_2, ..., m_n)$.

3.1   The Algorithm

To achieve this, Turnnet transforms an arbitrary network with prohibited turns into a new one without prohibited turns with the following steps:

1. Add two special nodes to the original network, one connected to the source node, the other connected from the destination node. Set the link metrics of the new links to neutral, that is zero for additive metrics and infinity for concave metrics.
2. For each link in the original network, generate a node in the new network.
3. For each turn in the original network, generate a link in the new network connecting the nodes corresponding to the two links of the original turn.
4. Set the link metrics of the new links to be the same as the metrics of the second link of the corresponding turn in the original network.
5. Delete all links from the new network, whose corresponding turns in the original network are prohibited.

The routing algorithm is then run on the new network using the nodes corresponding to the newly added special links in the original network as source and destination nodes.

Transforming a path from the new to the original network can be done efficiently by cycling through the nodes of the new path and replacing them with corresponding nodes in the original network like this:

1. Cut the last link from the path (which is the link to the node corresponding to the second special node in the original network).
2. Cycle through the nodes of the path and append the destination node of the corresponding link in the original network to the new path.

3.2   The Initial Example Revisited

Figure 2 continues the previous example from figure 1, by applying the Turnnet algorithm to the network and showing that Dijkstra's algorithm produces optimal results this time. In part (a) the original network is shown, with the new special nodes connected to the source and destination nodes. In (b) the network has been transformed as described in steps 2-4. The new nodes are marked with the labels of the source and destination nodes of the corresponding links in the original network in (a). In the following they will be labelled with the "$\rightsquigarrow$" symbol. Step 5 of the algorithm is shown in (c). The original network has only one prohibited turn, namely $1 \rightarrow 3 \rightarrow 4$, i.e. $T = \{(1,3,4)\}$. This turn corresponds to the link $(1 \rightsquigarrow 3) \rightarrow (3 \rightsquigarrow 4)$ in the new network. Thus, in compliance with step 5, the use of that link is prohibited. After running Dijkstra's algorithm on the new network, two paths are found as shown in (d), the best path being $(-1 \rightsquigarrow 1) \rightarrow (1 \rightsquigarrow 2) \rightarrow (2 \rightsquigarrow 3) \rightarrow (3 \rightsquigarrow 4) \rightarrow (4 \rightsquigarrow -2)$ with a cost of 5. Transforming this path back into the original network one gets the path $1 \rightarrow 2 \rightarrow 3 \rightarrow 4$ which is identical to the path shown in part (d) of figure 1 and indeed the correct shortest path.

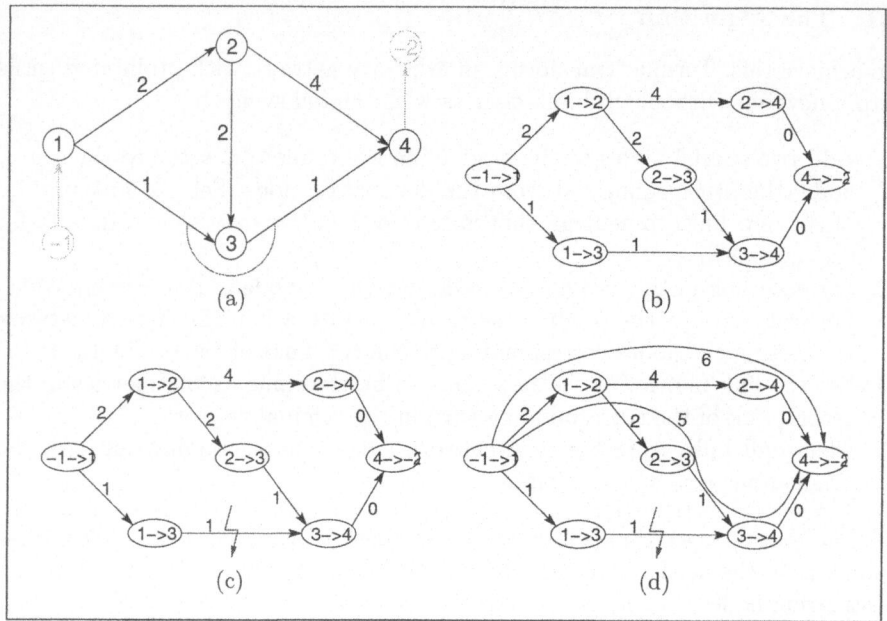

Fig. 2. Example for the transformation of a network with the Turnnet concept.

3.3   Formal Definition and Proof of Correctness

In the following a formal definition of the Turnnet concept is given and its path-conserving property is proven.

Definition 3 (Turnnet algorithm). *Given a network $G = (N, E)$, a source and a destination node $s, d \in N$, a set of prohibited turns $T \subseteq \{(i, j, k) : i, j, k \in N \wedge (i, j) \in E \wedge (j, k) \in E\}$, and for each link $(i, j) \in E$ a set of m additive, multiplicative, or concave metrics $m_{(i,j)}^t, 1 \leq t \leq m$, the Turnnet algorithm produces two special nodes n_s and n_d, a network $G^{TN} = (N^{TN}, E^{TN})$, two nodes $s^{TN}, d^{TN} \in N^{TN}$, and a set of link metrics m^{TN} as follows:*

$$N^{TN} = \{(i \rightsquigarrow j)\} : (i, j) \in E\} \cup \tag{1}$$

$$\{(n_s \rightsquigarrow s), (d \rightsquigarrow n_d)\} \tag{2}$$

$$E^{TN} = \{\{((i \rightsquigarrow j), (j \rightsquigarrow k)) : (i, j), (j, k) \in E\} \cup \tag{3}$$

$$\{((n_s \rightsquigarrow s), (s \rightsquigarrow i)) : i \in N \wedge (s, i) \in E\} \cup \tag{4}$$

$$\{((j \rightsquigarrow d), (d \rightsquigarrow n_d)) : j \in N \wedge (j, d) \in E\}\} \setminus \tag{5}$$

$$\{((i \rightsquigarrow j), (j \rightsquigarrow k)) : (i, j, k) \in T\} \tag{6}$$

$$s^{TN} = (n_s \rightsquigarrow s) \tag{7}$$

$$d^{TN} = (d \rightsquigarrow n_d) \tag{8}$$

$$m_{((i \rightsquigarrow j), (j \rightsquigarrow k))}^{TN,t} = m_{(j,k)}^t \quad \forall ((i \rightsquigarrow j), (j \rightsquigarrow k)) \in E^{TN} \tag{9}$$

The two nodes $(n_s \rightsquigarrow s)$ and $(d \rightsquigarrow n_d)$ in (2) emerge from the addition of the two special nodes n_s and n_d to the original graph G in step 1 of the algorithm. Equation (9) only holds for the links in E^{TN} that have a corresponding link in E. For the new links leading to the second special node (specified in (5)), the metrics have to be set to "neutral", i.e. they should not influence the total path metric. This is given in (10).

$$m^{TN,t}_{((j \rightsquigarrow d),(d \rightsquigarrow n_d))} = \begin{cases} 0, & \text{if } m^t \text{ is an additive metric} \\ 1, & \text{if } m^t \text{ is a multiplicative metric} \\ \infty, & \text{if } m^t \text{ is a concave metric} \end{cases} \qquad (10)$$

Definition 4 (Transformation Function). *Given a path $p^{TN} = ((i \rightsquigarrow j)_1, ...,$
$(i \rightsquigarrow j)_q)$ with $(i \rightsquigarrow j)_x \in N^{TN}$, the transformation function tr to the corresponding path p in G is defined as given in (11):*

$$tr(p^{TN}) = (j_1, j_2, ..., j_{q-1}) \qquad (11)$$

Given these definitions the following theorem shows that the Turnnet algorithm indeed provides a method to combine arbitrary routing algorithms with turn-prohibiting feed-forward mechanisms.

Theorem 1. *Given a network G, source and destination nodes s, d, a set of prohibited turns T, and for each link a set of metrics $m^t_{(i,j)}$, the Turnnet algorithm produces a network G^{TN} that is path-conserving to G with respect to the prohibited turns.*

Proof. The proof is divided into two parts. First, it is shown that tr is a bijective function between G and G^{TN} and then $m(p) = m(tr^{-1}(p))$ is proven.

Given a path p in G from the source node s to the destination node d. The path is defined as $p = (p_1, ..., p_q)$ with $p_1 = s$, $p_q = d$, and $p_x \in N$ for $1 < x < q$. The inverse transformation function tr^{-1} transforms this path p to a path p^{TN} in G^{TN} like this:

$$p^{TN} = (p_1^{TN}, ..., p_{q+1}^{TN}) \qquad (12)$$

$$p_x^{TN} = \begin{cases} s^{TN}, & \text{if } x = 1 \\ (p_{x-1} \rightsquigarrow p_x), & \text{if } 1 < x < q+1 \\ d^{TN}, & \text{if } x = q+1 \end{cases} \qquad (13)$$

To prove that p^{TN} is a valid path in G^{TN} it suffices to show that all its nodes are valid, i.e. $p_x^{TN} \in N^{TN}$ for $1 \le x \le q+1$, and that they are connected to each other, i.e. $(p_{x-1}^{TN} \to p_x^{TN}) \in E^{TN}$ for $2 \le x \le q+1$.

Clearly, all p_x^{TN} are valid nodes in N^{TN}, because all $p_{x-1} \to p_x$ are links in G (otherwise p would not be a path) and N^{TN} includes all links from G (see (1)).

Since every pair of nodes (p_{x-1}^{TN}, p_x^{TN}) in p^{TN} corresponds to the two links $((p_{x-2} \to p_{x-1}), (p_{x-1} \to p_x))$ in E, according to (3) there is also a link $p_{x-1}^{TN} \to p_x^{TN}$ in E^{TN}. Therefore, every node in p^{TN} is connected to its predecessor node and thus, p^{TN} is a valid path in G^{TN}.

The other way around, i.e. given p^{TN}, the tr function from definition 4 produces p, because $p_x^{TN} = (p_{x-1} \rightsquigarrow p_x)$ for $1 < x < q+1$ and $p_1^{TN} = s^{TN} = (n_s \rightsquigarrow s)$ (see (7)) still holds and therefore $p = (p_1, ..., p_q)$ according to (11).

Thus, tr^{-1} produces an unique, valid path in G^{TN} and $p = tr(tr^{-1}(p))$ holds and consequently tr is a bijective function.

Since tr is a bijective function, it suffices to show that $m(p) = m(tr^{-1}(p)) = m(p^{TN})$ to prove that G^{TN} is path-conserving to G.

Because $p = (p_1, ..., p_q)$, $p^{TN} = (p_1^{TN}, ..., p_{q+1}^{TN})$ and, with (9), $m_{(p_{x-1}^{TN} \to p_x^{TN})}^{TN,t} = m_{(p_{x-1} \to p_x)}^{t}$ for $2 \le x \le q$, for the path metric $m(p) = m((p_1^{TN}, ..., p_q^{TN}))$ holds, i.e. the path metrics for p in G and the first q nodes of $p^{TN} = tr^{-1}(p)$ in G^{TN} is the same.

Now, since the metrics for the last link of p^{TN}, namely $p_q^{TN} \to p_{q+1}^{TN}$ which is $p_q^{TN} \to (d \rightsquigarrow n_d)$, are set to neutral according to (10), they have no influence of the accumulated path metrics. Thus, $m(p) = m((p_1^{TN}, ..., p_q^{TN})) = m((p_1^{TN}, ..., p_{q+1}^{TN})) = m(p^{TN})$ holds and G^{TN} is path-conserving to G. ☐

3.4   Application to the G-WiN Topology

This section provides a real-world scenario applying the G-WiN topology of the German Research Network (DFN) as of 2000 [8] that is shown in the left of figure 3. It consists of a dense level one mesh, which allows for multiple alternative paths, thus achieving redundancy. The level two sites are each connected to a single level one site only, however, using two links in parallel, thereby providing backup capabilities.

In the right of figure 3 the level one mesh is reproduced, including a set of forbidden turns that are derived by Turn Prohibition [15]. The level one nodes have been visited by the Turn Prohibition algorithm in the order of their numbering. The star-shaped level two components are excluded here, because all routing decisions are already determined. Further on, the star structure assures that the routes are feed-forward compliant anyway.

Figure 4 illustrates the Turnnet that corresponds to the G-WiN level one topology excluding the prohibited turns from figure 3. It can be immediately seen, that the Turnnet graph provides a valid order for an inductive application of Network Calculus. However, adding any of the prohibited turns that are shown in figure 3, for example $(7, 1, 2)$, will render an inductive approach impossible. The Turnnet graph represents the dependencies that exist between the links in the original network. A related structure is also known as channel dependency graph from [4], where it is used to analyze deadlock conditions.

Adding special nodes as described in section 3.1 and applying Dijkstra's algorithm to the Turnnet in figure 4 allows to derive shortest paths without introducing cycles that can be transformed backwards to the original network.

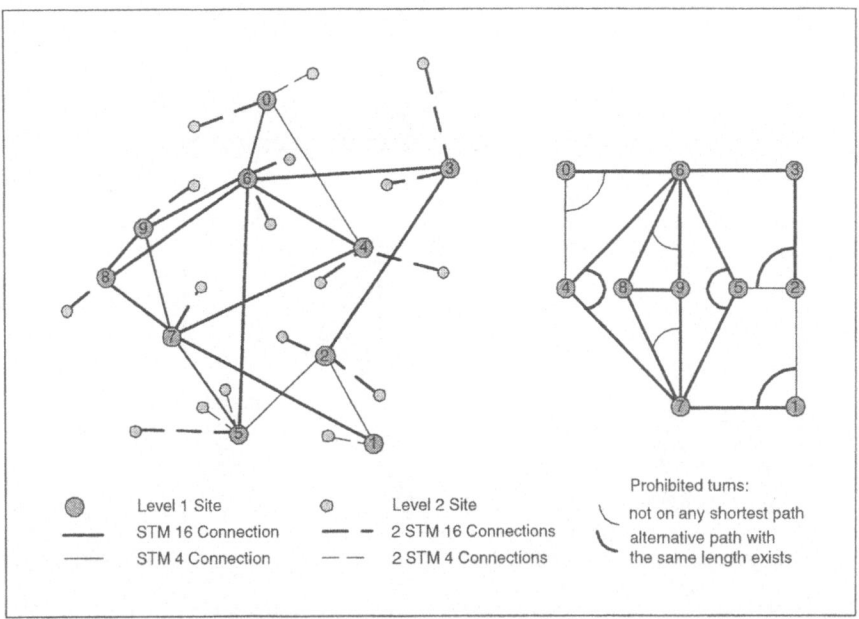

Fig. 3. G-WiN topology and Turn Prohibition example.

For example to derive the shortest path from node 5 to node 3 two special nodes -1 and -2 have to be connected to the original network by links $(-1, 5)$ and $(3, -2)$. These are then transformed to the Turnnet and become nodes $(-1 \rightsquigarrow 5)$ and $(3 \rightsquigarrow -2)$ and links $((-1 \rightsquigarrow 5), (5 \rightsquigarrow 2))$, $((-1 \rightsquigarrow 5), (5 \rightsquigarrow 6))$, $((-1 \rightsquigarrow 5), (5 \rightsquigarrow 7))$, $((2 \rightsquigarrow 3), (3 \rightsquigarrow -2))$, and $((6 \rightsquigarrow 3), (3 \rightsquigarrow -2))$. The shortest path that is found in the Turnnet is $(-1 \rightsquigarrow 5) \rightarrow (5 \rightsquigarrow 6) \rightarrow (6 \rightsquigarrow 3) \rightarrow (3 \rightsquigarrow -2)$, which becomes $5 \rightarrow 6 \rightarrow 3$ after backwards transformation.

An analysis of the paths that can be derived with or without Turn Prohibition allows to classify the turns as shown in the right of figure 3. Three of the seven prohibited turns do not impact any shortest paths and the remaining four forbid potential shortest paths, for which alternatives with the same hop count exist.

3.5   Discussion

From theorem 1 and corollary 1 it immediately follows that any routing results found in the Turnnet network are also valid in the original network, when transformed back with function tr. Thus, it is proven that the Turnnet algorithm solves the problems described in section 2 and allows for the combination of arbitrary routing algorithms and turn-prohibiting feed-forward mechanisms.

This flexibility, however, comes at the price of increased complexity. According to the definition, the Turnnet algorithm has to be executed for each source/destination pair. Fortunately, it is very easy to implement Turnnet in a way that allows for a single execution at initialization time and a very simple

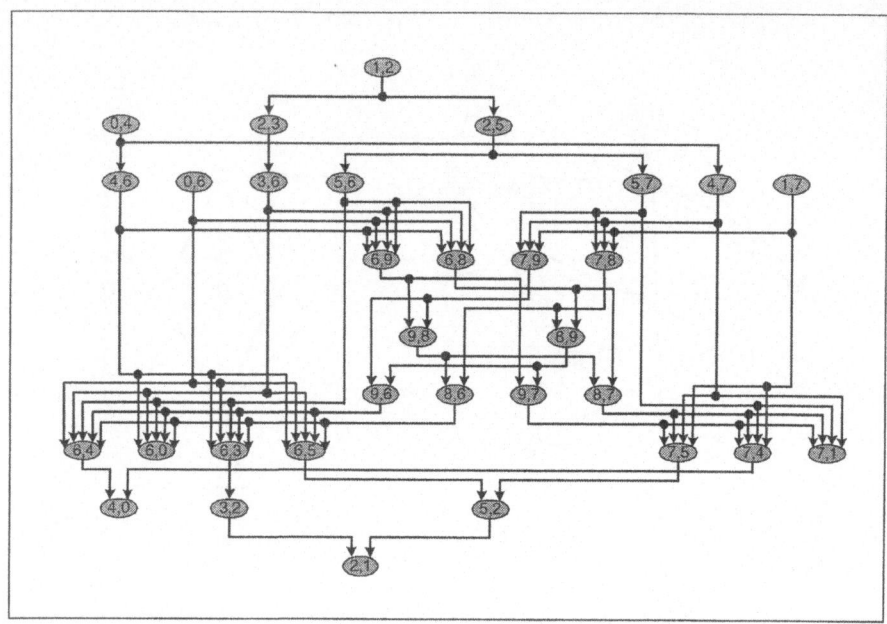

Fig. 4. G-WiN Turnnet excluding prohibited turns.

adjustment of the source/destination nodes for each request. The complexity of the initial Turnnet computation is in $O(|E|^2)$. However, the resulting network G^{TN} is bigger than the original one, i.e. $|N^{TN}| = (|E| + 2)$ and $|E^{TN}|$ depends on the number of turns in the original network. Therefore, the computational complexity of the applied routing algorithm may increase. For routing schemes based on Dijkstra's algorithm the complexity rises from $O(n^2)$ to $O^{TN}(|E|^2)$.

A big advantage of the Turnnet concept is that routing algorithms do not have to be aware of it. For a routing algorithm it makes no difference if the network it operates on is a Turnnet or not, which can be used efficiently for practical implementations.

4   Conclusions

The application of feed-forward mechanisms to data networks is relatively new and the problem of applying conventional routing algorithms to networks with prohibited turns has to our knowledge not been investigated in detail so far. By developing the Turnnet concept, we have evolved a general-purpose solution, which allows to use arbitrary routing schemes with prohibited turns.

The Turnnet algorithm is not very complicated. It does not raise the routing complexity in an unacceptable manner and can be easily implemented. Thus, offering a service with delay guarantees based on Network Calculus and the application of a feed-forward mechanism in conjunction with the use of Turnnet for routing, is a viable and recommendable option for network operators.

Acknowledgments. This work was supported in part by the Path Allocation in Backbone Networks (PAB) project funded by the German Research Network (DFN) and the Federal Ministry of Education and Research (BMBF) and in part by the German Research Community (DFG) under grant GRK (Graduate School) 643.

References

1. S. Blake, D. Blake, M. Carlson, E. Davies, Z. Wang, and W. Weiss. An Architecture for Differentiated Services. *RFC 2475*, December 1998.
2. C.-S. Chang. *Performance Guarantees in Communication Networks.* Springer, 2000.
3. B. Davie, A. Charny, J.C.R. Bennett, K. Benson, J.Y. Le Boudec, W. Courtney, S. Davari, V. Firoiu, and D. Stiliadis. An Expedited Forwarding PHB (Per-Hop Behavior). *RFC 3246*, March 2002.
4. J. Duato, S. Yalamanchili, and N. Lionel. *Interconnection Networks: An Engineering Approach.* Morgan Kaufmann, 2003.
5. Gerrit Einhoff. Quality of Service Routing for an IP Premium Service based on MPLS Traffic Engineering. Master's Thesis, Aachen University, June 2003.
6. M. Fidler, and V. Sander. A Parameter Based Admission Control for Differentiated Services Networks, *Elsevier Computer Networks*, 44(4):463–479, 2004.
7. B. R. Haverkort. *Performance of Computer Communication Systems: A Model-Based Approach.* John Wiley & Sons, January 1999.
8. G. Hoffmann. G-WiN - the Gbit/s Infrastructure for the German Scientific Community. *Proceedings of Terena Networking Conference*, 2000.
9. M. S. Kodialam and T. V. Lakshman. Minimum Interference Routing with Applications to MPLS Traffic Engineering. *Proccedings of IEEE INFOCOM (2)*, pages 884–893, 2000.
10. J.-Y. Le Boudec and P. Thiran. *Network Calculus: A Theory of Deterministic Queueing Systems for the Internet.* Number 2050 in LNCS. Springer, July 2002.
11. Q. Ma and P. Steenkiste. On Path Selection for Traffic with Bandwidth Guarantees. *Proceedings of IEEE International Conference on Network Protocols*, October 1997.
12. Q. Ma, P. Steenkiste, and H. Zhang. Routing High-Bandwidth Traffic in Max-Min Fair Share Networks. *Proceedings of ACM SIGCOMM*, pages 206–217, 1996.
13. M. D. Schroeder et al. Autonet: A High-speed, Self-configuring Local Area Network Using Point-to-point Links. *IEEE Journal on Selected Areas in Communications*, 9(8):1318–1335, October 1991.
14. V. Sander. *Design and Evaluation of a Bandwidth Broker that Provides Network Quality of Service for Grid Applications*, volume 16 of *NIC*. February 2003. PhD Thesis, Aachen University.
15. D. Starobinski, M. Karpovsky, and L. Zakrevski. Application of Network Calculus to General Topologies using Turn-Prohibition. *IEEE/ACM Transactions on Networking*, June 2003.
16. N. Taft-Plotkin, B. Bellur, and R. Ogier. Quality-of-Service Routing Using Maximally Disjoint Paths. *Proceedings of IEEE/IFIP IWQoS*, pages 119–128, June 1999.
17. Z. Wang and J. Crowcroft. Quality-of-Service Routing for Supporting Multimedia Applications. *IEEE Journal of Selected Areas in Communications*, 14(7):1228–1234, 1996.

On the Representability of Arbitrary Path Sets as Shortest Paths: Theory, Algorithms, and Complexity

Gábor Rétvári, Róbert Szabó, and József J. Bíró

High Speed Networks Laboratory, QoSIT Laboratory
Department of Telecommunications and Media Informatics
Budapest University of Technology and Economics*
H-1117, Magyar Tudósok körútja 2., Budapest, Hungary
{retvari,robert.szabo,biro}@tmit.bme.hu

Abstract. The question, whether an optional set of routes can be represented as shortest paths, and if yes, then how, has been a rather scarcely investigated problem up until now. In turn, an algorithm that, given an arbitrary set of traffic engineered paths, can efficiently compute OSPF link weights as to map the given paths to shortest paths may be of huge importance in today's IP networks, which still rely on legacy shortest-path-first routing protocols. This article establishes the fundamental theory and algorithms of shortest path representability, and concludes that in general it is much more difficult task to compute shortest path representable paths than to actually calculate link weights for such paths.

Keywords: Traffic engineering, routing, linear programming, OSPF

1 Introduction

Most of today's Traffic Engineering (TE, [1]) proposals require the deployment of expensive routing and traffic forwarding hardware and software. On the other hand, ISPs have huge installation base of routers running legacy routing protocols like OSPF (Open Shortest Path First, [2]) or IS-IS (Intermediate-System-to-Intermediate-System). Both OSPF and IS-IS rely on shortest-path-first routing, i.e., there is an administrative weight associated with network links, and, for a given destination IP address prefix, the routing protocol uses the shortest aggregate cost path to that destination. The network operator manipulates routing by setting the administrative link weights appropriately. Usually, optional load balancing by ECMP (Equal-Cost-MultiPath) is also available, where traffic is split roughly evenly amongst multiple shortest paths, if such paths exist. Depending on the choice of the manufacturer, ECMP may implement per packet, per

---

* This work was supported by the Ministry of Education, Hungary under the reference No. IKTA-0092/2002.

N. Mitrou et al. (Eds.): NETWORKING 2004, LNCS 3042, pp. 1180–1191, 2004.

destination or per source-destination pair load distribution using round-robin or some hashing technique.

Hence, it is an easy-to-deploy and overly cost-effective solution to implement traffic engineering on top of OSPF while retaining existing routing equipment. In such an architecture, a suitable Traffic Engineer (*i*) participates in OSPF signaling to learn routing information, (*ii*) assigns paths for each session, (*iii*) computes link weights as to assure that the link weights reflect the assignment of paths (i.e., all paths, which are assigned for a particular session are shortest paths for that session) and (*iv*) distributes the selected link weights back to OSPF routers. However, this solution is almost certainly sub-optimal due to the inherent limitation of ECMP, which restricts load balancing to equal splitting.

The foundations of OSPF traffic engineering are laid down by [3] and [4]. An unpublished work [5] of the same authors shows that it is NP hard to compute link weights, as to assure that the resultant set of shortest paths fulfill some useful traffic engineering criteria. Therefore, the authors propose a local search heuristic achieving nearly optimal routing in some cases. However, the applicability of the algorithms proposed is restricted to the long-term process of network dimensioning. This is because of the running time of these algorithms, which may amount to hours even in a middle sized network. In contrast, on-line traffic engineering requires rapid algorithms to assure quick adaptation to topology changes or management controls. The authors also missed to identify, whether in path selection or in shortest path representation hides the real origin of exponential complexity.

To the best of our knowledge, the outstanding paper of Wang *et al.* [6] has been the only work dealing with shortest path representability up until now. According to their definition, a set of paths is *shortest path reproducible* if there exists a positive valued weight set based on which all the paths in the set are shortest paths. They establish the sufficient and necessary condition of shortest path representability and conclude that a set of paths is either loopy, and therefore is of negligible interest to traffic engineering, or it is shortest path reproducible. This precious work (and some derivatives, e.g., [7]) disproved the common belief of many researchers that shortest-path-first routing is, by nature, useless to traffic engineering.

It is of extreme importance to understand that the definition of shortest path representability and the implied linear programming solution can only guarantee that the selected paths are reproduced as shortest paths. Though, it claims nothing about other paths. Therefore, a path that was originally not designated for data forwarding may be given small cost and so, be introduced into routing. The traffic engineer does not have total control over routes and may experience unwanted interference caused by the additional paths, which he or she did not even consider to use. This may very well amortize the overall performance of the network. In fact, we can show that depending on the actual topology of the network and distribution of source-destination pairs the worst case performance might degrade to an arbitrary small fraction of the optimal performance due to unintended interference. This happens regardless of using ECMP or not.

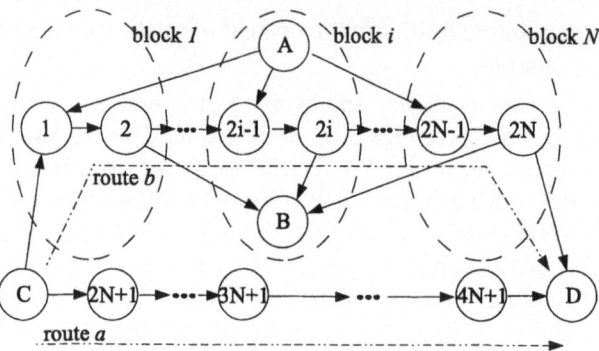

Fig. 1. Sample graph topology. All link capacities and weights equal to 1 and there are N and 1 units of demand for $A \to B$ and $C \to D$, respectively

Consider Figure 1. The network consists of N identical blocks with an alternate path circumventing these blocks. All link capacities and weights equal to 1. There are two sessions (A, B) and (C, D) communicating over the network, willing to use N and 1 units of bandwidth, respectively. In an optimal setting, the traffic engineer may select the circumventing $C \to D$ path (route a in the figure) of hop count $2N+2$, and let (A, B) demands flow through the N blocks, one unit of demand per block. The optimal throughput is therefore $N + 1$. An obvious way to achieve this would be to set the weight of link $(C, 1)$ to a high value. However, traditional shortest path representation methods can not guarantee this, therefore, route b (also of length $2N+2$) is often introduced unintentionally into data forwarding[1]. In the worst case, all (C, D) communication may use route b completely blocking (A, B). This causes the performance to degrade to some $\frac{1}{N+1}$ fraction of the optimal value. If ECMP is not used, then, depending on the actual implementation of the shortest-path-first routing engine, in around half of the cases route a and in the other half route b would be selected, In average, this causes half of (C, D) traffic to interfere with (A, B) traffic. Observe that the very same situation rises with ECMP, too. Asymptotically, the total throughput degrades to the half of the optimal performance in the average case.

Yet another important thing to know about the definition of shortest path representability proposed in [6] is that it provides no means to avoid multi-path routing. Deciding whether or not multi-path routing is a beneficial feature to have is completely beyond the scope of this paper. On the one hand, multi-path routing promises load balancing and may yield higher performance and network utilization. On the other hand, multi-path routing introduces a huge amount of uncertainty compared to single-path routing, ranging from the actual implementation of the ECMP splitting algorithm to the fact that in the presence of equal-cost paths, it is totally impossible to predict, which particular path a certain connection will take. This is completely undesirable in some cases (e.g.,

---

[1] Note that as of our favorite open source linear programming toolkit, GLPK, the optimal solution of ILP-SPR almost always yields this "bad" configuration.

it precludes call admission control to assess the route of a connection prior to actually instantiating it in the network). By all means, it may be advantageous to set link weights as to avoid multiple paths of a source-destination pair to have the same cost. To this end, [8] extends the local search heuristic proposed in [3], however, the proposed method is neither exact nor rapid.

In this paper we introduce the notions of explicit and unique shortest path representation to avoid the use of unintended paths. In Section 2 we give the basic mathematical formulation and definitions. As far as we know, this is the first time that strictly combinatorial algorithms to verify explicit shortest path representability are defined. The algorithms and some theoretical background is discussed in Section 3. Section 4 reveals the complexity of path assignment and gives simple approximations to the NP hard problem. Section 5 briefly outlines related simulation studies, and finally, Section 6 concludes our work.

2   Mathematical Formulation

Let $G(V, E)$ be a directed graph, formed by the set of nodes N ($|N| = n$) and the set of edges E ($|E| = m$). Let K denote the set of source-destination pairs (s_k, d_k), which are referred to as *sessions* for short. Let $\mathcal{P}^{s_k \to d_k}$ be the set of all paths that connect a particular source-destination pair (s_k, d_k). Our task then is to explicitly represent a given subset $\mathcal{P}^k \subseteq \mathcal{P}^{s_k \to d_k}$ as shortest paths. A path $P \in \mathcal{P}^k$ of length L_P is defined by its consecutive edges: $P := \{(v_i, v_{i+1}) \in E : i = 1 \ldots L_P - 1, v_1 = s_k, v_{L_P} = d_k\}$. We assume that there is a positive valued weight $w_{ij} \in \mathcal{Z}^+$ associated with each edge (i, j), and we let the aggregate cost of a path P be $W(P) = \sum_{(i,j) \in P} w_{ij}$. Furthermore, let p_k denote the number of paths for session k (i.e., $p_k = |\mathcal{P}^k|$), let $\mathcal{P}$ be the set of all designated paths (i.e., $\mathcal{P} = \bigcup_{k \in K} \mathcal{P}^k$) and let the aggregate cost of a path P over link weights w_{ij} be $W(P) = \sum_{(i,j) \in P} w_{ij}$.

Now we introduce the notion of path-graphs, which will be heavily used throughout this paper. Let n_{ij}^k be the number of paths of session k traversing link (i, j) and $n_{ij} = \sum_{k \in K} n_{ij}^k$ be the number of all paths using that link. Then, the path-graph $G_{\mathcal{P}}$ induced by a path set $\mathcal{P}$ is a special network, which includes all edges of all paths of $\mathcal{P}$ and the capacity of the edges equals to the number of paths in $\mathcal{P}$ using that link. We also let the demand t_k for session k be $t_k = p_k$. Formally, a path-graph $G_{\mathcal{P}}$ is a network on $G(V, E)$, such that the capacity of a link (i, j) is given by $u_{ij} = n_{ij}$ and all zero capacity links and zero degree nodes are removed from the network. Observe that a path set unambiguously determines the corresponding path-graph, though, the reverse is not necessarily true. This is because a path-graph may contain additional paths, which are formed by the concatenation of some sub-paths of the original path set $\mathcal{P}$. This parallels with the property of shortest paths that if $a \to b$ and $b \to c$ are shortest paths, then $a \to b \to c$ is also a shortest path. Note that in this case, we consider these additional paths to belong to $\mathcal{P}$, too. In other words, a set of paths is said to include a particular path if it includes all edges of that path.

According to [6], a path set $\mathcal{P}$ is shortest path reproducible, if there exists a positive weight setting $\mathcal{W} = \{w_{ij}\} : w_{ij} > 0$, such that all paths in $\mathcal{P}$ are shortest paths over $\mathcal{W}$:

Definition 1 (SPR). *A path set $\mathcal{P}$ is shortest path representable, if there exists a positive weight setting $\mathcal{W}$, such that for all $P' \in \mathcal{P}^k$ it holds that*

$$\forall P \in \mathcal{P}^{s_k \to d_k} \setminus \mathcal{P}^k : W(P') \leq W(P) \tag{1}$$

for every session $k \in K$.

In this context, verification of shortest path representability of a particular path set $\mathcal{P}$ and the actual link weights implementing $\mathcal{P}$ can be given by solving the following integer linear program (ILP-SPR) over the path-graph:

$$\max \sum_{k \in K} p_k \pi_{d_k}^k - \sum_{(i,j) \in E} n_{ij} w_{ij} \tag{2}$$

$$\text{s.t.} \quad \pi_j^k - \pi_i^k \leq w_{ij} \qquad \forall (i,j) \in E, \forall k \in K \tag{3}$$

$$w_{ij} \geq 1 \qquad \forall (i,j) \in E \tag{4}$$

Observe that this definition tells nothing about paths outside of $\mathcal{P}^k$. Such paths are either shortest paths or not. In other words, the representation is not explicit. As shown in the previous section, this may lead to undesirable interference amortizing the overall performance of the network. To eliminate this shortcoming we introduce the notion of *explicit shortest path representation* in the following way:

Definition 2 (eSPR). *A path set $\mathcal{P}$ is explicitly shortest path representable, if there exists a positive weight setting $\mathcal{W}$, such that for all $P' \in \mathcal{P}^k$ it holds that*

$$\forall P \in \mathcal{P}^{s_k \to d_k} \setminus \mathcal{P}^k : W(P') < W(P) \tag{5}$$

for every session $k \in K$ (note the strict inequality!).

What makes the difference is that eSPR explicitly prohibits a path outside of the desired path set $\mathcal{P}^k$ to become shortest path. Observe that any eSPR path set also fulfills Definition 1.

In certain situations, it is important to avoid having multiple different parallel routes for any sessions. Therefore, it is a plausible idea to define some sorts of *uniqueness* of routing by definitely precluding the existence of multiple equal-cost shortest paths between any source-destination pairs:

Definition 3 (uSPR). *A path set $\mathcal{P}$ is uniquely reproducible as shortest paths, if there exists a positive weight setting $\mathcal{W}$, such that for every session $k \in K$ there is exactly one path P_k', for which it holds that*

$$\forall P \in \mathcal{P}^{s_k \to d_k} \setminus \{P_k'\} : W(P_k') < W(P) . \tag{6}$$

Wang *et al.* propose a linear program to verify and perform shortest path representation [6], though, they do not seem to recognize that the resultant representation will be neither explicit nor easily implementable in network devices. In the sequel, we focus on the more difficult problem of eSPR instead of SPR and introduce some simple sufficient conditions to both explicit and unique shortest path representation.

3   Explicit and Unique Shortest Path Representation

In order to provide further insight into the relation of SPR and eSPR, first, we borrow some basic results of network flow theory and linear programming [9] [10]. Paths in a path set $\mathcal{P}$ define shortest paths if and only if there exist node potentials $\pi_n : n \in V$ and positive link weights $w_{ij} : (i,j) \in E$, such that:

$$\forall P \in \mathcal{P}, \forall (i,j) \in P : \pi_j - \pi_i = w_{ij} \tag{7}$$

$$\forall P \notin \mathcal{P}, \exists (i,j) \in P : \pi_j - \pi_i < w_{ij} \tag{8}$$

The cost of an $s \to d$ shortest path is given as $W(P) = \sum_{(i,j)\in P} w_{ij} = \pi_d - \pi_s$.

An intriguing question to investigate is to assess, under which conditions a shortest path representable path set is also explicitly shortest path representable. As it turns out, the two concepts of shortest path representability are identical under some surprisingly mild assumptions. This is formulated in the following important result:

Theorem 1. *Let $\mathcal{P}$ be a path set, such that for all $k \in K$ there exists an $s_k \to d_k$ path in the path-graph $G_\mathcal{P}$ of $\mathcal{P}$. Then, $\mathcal{P}$ is shortest path representable if and only if it is explicitly shortest path representable.*

Proof. eSPR $\Rightarrow$ SPR is obvious in light of the fact that an explicit shortest path representation immediately conforms to the SPR definition.

SPR $\Rightarrow$ eSPR: we give a constructive proof by presenting an algorithm, which turns a SPR into an eSPR in polynomial time. Given a path set $\mathcal{P}$ let the path-graph induced by $\mathcal{P}$ be $G_\mathcal{P}(V_\mathcal{P}, E_\mathcal{P})$. From Definition 1 we have that over a proper SPR weight set $\mathcal{W}$ and node potentials π_n, all $s_k \to d_k$ paths in $G_\mathcal{P}$ are shortest paths. What we need to assure is that no paths outside of $G_\mathcal{P}$ are also shortest paths. Consider the following simple modification of $\mathcal{W}$: For some edge $(i,j) \notin E_\mathcal{P}$ let $w_{ij} = W_{MAX}$, where W_{MAX} is defined as:

$$W_{MAX} = \max_{k \in K}(\pi_{d_k} - \pi_{s_k}) + 1 \tag{9}$$

Now, we make the following observations:

- Let $n \in V_\mathcal{P}$ be a node in the path-graph. By assumption, we have that for any session $k \in K$ there exists at least one $s_k \to n$ path, which lies completely inside $G_\mathcal{P}$. Since this path is a shortest path and its length is not affected by the weight of any $(i,j) \notin E_\mathcal{P}$, the node potentials $\pi_n : n \in V_\mathcal{P}$ are invariant with respect to the above modification of the link weights. The same applies to the value of W_{MAX}, itself.
- Furthermore, (9) is constructed as to assure that W_{MAX}, and as such, any path containing at least one edge of weight W_{MAX} is longer than the longest one of all $s_k \to d_k$ shortest paths.

Then, one can set the weight of all the links outside of $E_\mathcal{P}$ to W_{MAX} to obtain an explicit shortest path representation. This can be done in $O(m)$ time. To prove the correctness of this algorithm, let P_1 be any optional $s_k \to d_k$ path for some session $k \in K$, such that $P_1 \notin \mathcal{P}$. Hence, P_1 traverses some edge $(i,j) \notin E_{\mathcal{P}}$ (otherwise, it would lie completely in $G_{\mathcal{P}}$ and therefore, by definition, it would belong to $\mathcal{P}$). So $w_{ij} = W_{MAX}$. Furthermore, let P_2 be a $s_k \to d_k$ path inside $G_{\mathcal{P}}$, i.e., $\forall(u,v) \in P_2 : (u,v) \in E_{\mathcal{P}}$. Note that, by assumption, one can always find such P_2 path. Additionally, the algorithm leaves the node potentials, and as such, the length of any path in $G_{\mathcal{P}}$ intact. Thus, for the length of paths P_1 and P_2 we have that $W(P_2) < W_{MAX} \leq W(P_1)$. Since the strict inequality holds for any P_1 outside of $G_{\mathcal{P}}$ and any P_2 inside $G_{\mathcal{P}}$ we conclude that the modified weight set implements explicit shortest path representation (cf. (7)-(8)). This completes the second part of the proof.   □

The significance of the above theorem is two-fold, First, as shall be shown by the simulation results in Section 5, explicit shortest path representation generally improves the performance of OSPF traffic engineering as it avoids unnecessary and adverse interference caused by unintentional paths. Moreover, proving SPR is usually simpler than proving eSPR. Hence, Theorem 1 assures that the resultant weight set can be transformed in polynomial time into one, which implements explicit shortest path representation. Henceforward, we restrict our discussions to path sets, which satisfy the assumption of Theorem 1, i.e., which contain at least one path for each session.

The linear program ILP-SPR does not provide easy way to determine, whether or not a particular set of paths is shortest path representable. It requires the rapid solution of a potentially large scale linear program, which may very well fall beyond the capabilities of today's network devices. Therefore, in the sequel, we show some easy-to-check conditions to test for eSPR and uSPR, respectively.

Lemma 1 (Sufficient condition for uSPR). *A path set $\mathcal{P}$ is uniquely representable as shortest paths if the path graph $G_{\mathcal{P}}$ induced by $\mathcal{P}$ is a directed forest.*

Proof. Given that $G_{\mathcal{P}}$ is connected with respect to source-destination pairs (s_k, d_k) and it forms a directed forest, by definition, there is only one path between any two nodes. Hence, arbitrary positive setting of the link weights will conform to (8) as long as the weight setting W_{MAX} defined in (9) is respected for the links, which do not reside in $G_{\mathcal{P}}$. This can be done in $O(m)$ time.   □

Lemma 2 (Sufficient condition for eSPR 1.). *A path set $\mathcal{P}$ is explicitly representable as shortest paths if the path graph $G_{\mathcal{P}}$ induced by $\mathcal{P}$ is acyclic (i.e., it does not contain any directed circles).*

Proof. We construct a simple shortest path representation of $\mathcal{P}$, which then, according to Theorem 1 can be easily converted to an explicit shortest path representation. From graph theory, we know that every acyclic graph $G(V,E)$ possesses one or more *topological ordering*. A topological ordering is a labeling $order(n)$ of the nodes $n \in V$, such that every edge joins a lower-labeled node to a higher-labeled node, i.e., $\forall(i,j) \in E : order(i) < order(j)$. It is fairly easy to see that any topological ordering of the path graph $G_{\mathcal{P}}$ induced by $\mathcal{P}$ defines suitable node potentials, and hence, link weights in the form: $w_{ij} = order(j) - order(i)$. A weight set w_{ij}, defined in this way is, by definition, integer and positive valued. Thus, proper eSPR link weights can be computed in $O(m)$ time in this case.   □

Note that verifying any of the above conditions takes $O(m)$ steps. This is the lower bound on the complexity of any link weight setting algorithm, since at least $O(m)$ steps are necessary to walk through all the links in the network. Also note that a set of paths may very well be shortest path representable even if its path graph contains directed circles. Therefore, the above condition is obviously not a necessary one. In order to catch a larger class of path graphs than acyclic graphs, we present yet another sufficient condition of eSPR, which is of considerably broader scope:

Lemma 3 (Sufficient condition for eSPR 2.). *Consider the single commodity flow problem (the so called mass-flow problem) derived from the original K-commodity flow problem in the following way. For every node $n \in V$, let the imbalance of n be $e(n) = \sum_{k \in K : n = s_k} p_k - \sum_{k \in K : n = d_k} p_k$. Find a minimum cost mass-flow instance that satisfies $e(n)$. This can be done in polynomial time by some combinatorial algorithm, e.g., by minimum mean-cycle cancellation in $O(n^2 m^3 \log n)$ time [9]. Then, a path set $\mathcal{P}$ is explicitly representable as shortest paths if for the aggregate cost C_{mass} of the optimal mass-flow:*

$$C_{mass} = \sum_{P \in \mathcal{P}} L_P , \tag{10}$$

where L_P is the length of path P.

Proof. Consider the the dual linear program instance I of ILP-SPR over $G_{\mathcal{P}}$. It is fairly easy to show that I is a minimum cost multi-commodity flow problem. From [6] we know that a path set $\mathcal{P}$ is shortest path representable if and only if the aggregate length $\sum_{P \in \mathcal{P}} L_P$ of $\mathcal{P}$ equals to the optimal objective function of I. Solve I and let the optimal objective value be C_I. It is straightforward that $C_{mass} \leq C_I$. Hence, if (10) holds, then $C_I = \sum_{P \in \mathcal{P}} L_P$ and $\mathcal{P}$ is optimal.   □

In general, given a set of paths, one can either conclude that the path set is loopy or otherwise provide an explicit shortest path representation in strictly polynomial time. This implies that this is not the shortest path representation problem, but rather the determination of optimal paths, which hides the exponential complexity of OSPF traffic engineering. In fact, the next section confirms just this claim.

4   Complexity of Optimal Path Assignment

Now, we move on to investigate the complexity of optimal path selection and to show that in general, it is a hard task to compute eSPR path sets with respect to some reasonable traffic engineering optimization criteria. The reader is advised to the following discussion.

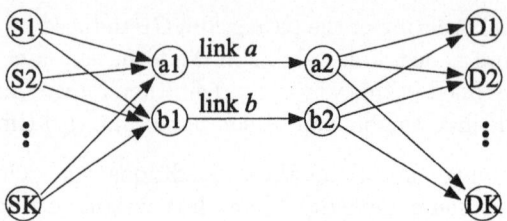

Fig. 2. Sample configuration with K sessions and demands $0 < t_k < 1$

Theorem 2. *Given a demand set $t_k : k \in K$ it is NP hard to compute a uniquely shortest path representable path set, such that all t_k demands are satisfied. In fact, it is also NP hard to even decide, whether or not the demand set can be satisfied along an uSPR path set.*

Proof. The transformation is from *2 bin packing*. Consider the sample configuration depicted in Figure 2. There are K sessions with demands $0 < t_k < 1$ and two link disjoint paths of capacity 1 from each source to each destination, one through link a and another through link b. Individual demands must be routed without any sorts of splitting to form a directed forest of a uSPR. Therefore, all $s_k \to d_k$ traffic t_k is either packed into link a or link b, alternatively. Thus, any uSPR path set in this setting also solves 2 bin packing. It is also NP hard to even decide, whether a particular demand set can be packed into the two bins or not. □

Note that it is also NP hard to determine the maximum number of sessions that can be satisfied, which would be the objective if one was to maximize the throughput of the network. The network dimensioning case is also NP hard (how many links to deploy in order to assure proper uSPR routing), since it maps to the minimum bin packing problem. Also note that the proof of Theorem 2 remains to be valid, if we let individual demands to be split between the two paths evenly, i.e., let half of the demand flow through link a and the other half through link b. Observe that such a path set is acyclic, and as such, it conforms to the condition of explicit shortest path representability given in Lemma 2.

Corollary 1. *In general, determining a path set that is explicitly shortest path representable and optimizes OSPF ECMP routing is NP hard.*

Despite of the intractable complexity of eSPR path selection, there are certain relaxations of the full-fledged problem, which are both easy to solve and may prove to be of substantial interest in some realistic scenario. The NP complete nature of OSPF-ECMP routing is closely coupled with the requirement that either flows are unsplittable (uSPR) or can only be split evenly (eSPR). Relaxing this requirement of unsplittable flows immediately yields polynomial approximate algorithms for eSPR path selection, such as the optimal routing or the minimum cost maximum throughput linear programs [9], [10].

uSPR path selection is a more difficult problem, because one must avoid any branching of the optimal paths in this case. Therefore, some more relaxation is necessary. First, it is plausible to let $\forall k \in K : t_k = 1$, since most of today's networking architectures do not provide means for a user to specify his or her demand size. In addition, the scope of the routing information retrievable from OSPF link state information is currently limited to the actual topology of the network. This gives rise to a unit-demand-unit-capacity relaxation of the original problem. In this setting, the integer linear program below (if solvable) provides a path set $\mathcal{P}$ that can be uniquely represented as shortest paths:

$$\max \sum_{k \in K} t_k - \alpha \sum_{k \in K} \sum_{(i,j) \in E} X_{ij}^k \tag{11}$$

$$\sum_{j:(i,j) \in E} X_{ij}^k - \sum_{j:(j,i) \in E} X_{ji}^k = \begin{cases} t_k & \text{if } i = s_k \\ -t_k & \text{if } i = d_k \\ 0 & \text{otherwise} \end{cases} \quad \forall k \in K, \forall i \in V \tag{12}$$

$$\sum_{k \in K} \sum_{j:(i,j) \in E} X_{ij}^k \leq 1 \qquad\qquad \forall i \in V \tag{13}$$

$$X_{ij}^k \in [0,1], \ t_k \in [0,1] \qquad\qquad \forall k \in K, \forall (i,j) \in E \tag{14}$$

The objective function (11) maximizes the overall throughput of the network, while minimizing the aggregate flow to avoid loops (α is a suitably small constant). (12) requires flow conservation. It is assured that the resultant path set (which comes in the form $\exists k : X_{ij}^k > 0 \Rightarrow (i,j) \in \mathcal{P}$) is uSPR, since (13) lets only one unit of flow to emanate from any node. Thus, the path graph consists of isolated paths for some sessions, which together form a directed forest. Finally, (14) keeps per-session traffic t_k and the link flow X_{ij}^k integer.

5   Simulation Studies

In this section, we briefly outline some results of our simulation studies that demonstrate the benefits of rendering the representation explicit. We used the *BRITE* tool [11] with the *router-level Waxman-model* ($\alpha = 0.15, \beta = 0.2, m = 3$) to generate a sequence of increasing sized realistic random graphs. We tried to keep the load constant throughout the sequence by setting the number of sessions, request intensity, average demand size and average holding time as to assure that the generated traffic keeps track with the growing capacity of the consecutively increasing networks. We used a call level OSPF-ECMP simulator to compare the performance of distance-vector routing (minimum hop-count routing, *MINHOP*), shortest path representation of the optimal paths generated by the maximum throughput relaxation (*SPR*) and the *explicit* representation (using Theorem 1) of the very same paths (*eSPR*). The results presented below are averaged over 30 graph sequences.

Figure 3 depicts the average number of ECMP routes as the function of the network size. MINHOP may only accidently create ECMP paths, while the solution of the maximum throughput relaxation intentionally, though, being just a relaxation of the NP hard path selection problem, sub-optimally forms ECMP

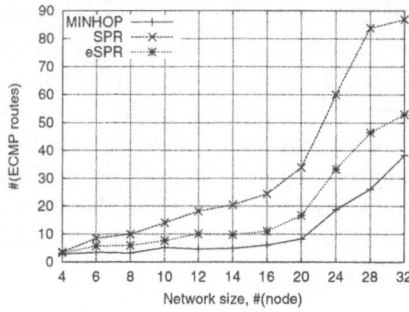 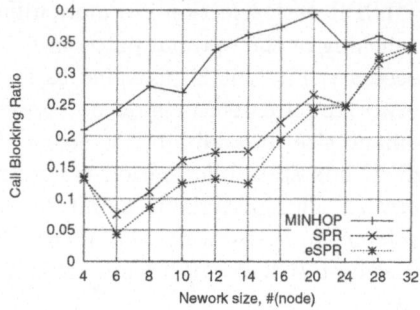

Fig. 3. Average number of ECMP routes as the function of network size

Fig. 4. Average call blocking ratio as the function of the network size

paths. However, compared to eSPR, which exclusively implements these paths, almost every second path formed by SPR is unintended. The average call blocking ratio depicted in Figure 4 insists that this nature of SPR is indeed leading to interference owing to the additional paths. The difference amounts to some 5-8% in average, which, in individual cases may be highly significant gain implied by eSPR. The figure also underlines the superiority of OSPF traffic engineering in comparison to traditional MINHOP routing.

6   Conclusions

This paper focuses on shortest path representation, a question of crucial importance in the majority of today's IP networks, which still rely on legacy shortest-path-first routing protocols. Our most important contribution to the groundbreaking work of [6] in this field comes from the recognition that it is not enough to blindly map the desired paths to shortest paths. We provided strong theoretical and practical evidence that if one can not exclude unintended paths from becoming shortest paths, he or she risks substantial amortization of the network revenue. We showed that, under reasonable assumptions, every non-explicit shortest path representation can be turned into an explicit one in polynomial time. As the explicit representation is a stronger and more useful one, we propose to use it instead of the non-explicit case. To ease this, we gave some novel sufficient conditions to test for SPR, which are, in contrast with prior work, strictly combinatorial. We also introduced uniqueness and proposed an exact method to compute uSPR link weights. Finally, we dealt with the problem of the selection of traffic engineered paths subject to eSPR or uSPR, and concluded that this problem is NP hard. As our major contribution, we concluded that actual shortest path representation is an easily tractable problem, and instead, optimal eSPR/uSPR path selection hides the real origin of exponential complexity. Therefore, we plan to take further efforts in this field.

References

1. D. Awduche, A. Chiu, A. Elwalid, I. Widjaja, and X. Xiao, "Overview and principles of Internet traffic engineering." RFC 3272, May 2002.
2. J. Moy, "OSPF Version 2." RFC 2328, April 1998.
3. B. Fortz, J. Rexford, and M. Thorup, "Traffic engineering with traditional IP routing protocols," *IEEE Communications Magazine*, vol. 40, pp. 118–124, Oct 2002.
4. B. Fortz and M. Thorup, "Optimizing OSPF/IS-IS weights in a changing world," *IEEE Journal of Selected Areas in Communications*, vol. 20, pp. 756–767, May 2002.
5. B. Fortz and M. Thorup, "Increasing internet capacity using local search," 2000. unpublished manuscript, http://www.research.att.com/~mthorup/PAPERS/or_ospf.ps.
6. Z. Wang, Y. Wang, and L. Zhang, "Internet traffic engineering without full-mesh overlaying," in *Proceedings of INFOCOM 2001*, April 2001.
7. A. Sridharan, C. Diot, and R. Guérin, "Achieving near-optimal traffic engineering solutions for current OSPF/IS-IS networks," in *Proceedings of INFOCOM 2003*, March 2003.
8. M. Thorup, "Avoiding ties in shortest path first routing," 2001. unpublished manuscript, http://www.research.att.com/~mthorup/PAPERS/ties_ospf.ps.
9. R. K. Ahuja, T. L. Magnanti, and J. B. Orlin, *Network Flows: Theory, Algorithms, and Applications*. Prentice-Hall, Englewood Cliffs, NJ, 1993.
10. M. S. Bazaraa, J. J. Jarvis, and H. D. Sherali, *Linear Programming and Network Flows*. John Wiley & Sons, January 1990.
11. A. Medina, A. Lakhina, I. Matta, and J. Byers, "BRITE: Universal topology generation from a user's perspective," Tech. Rep. 2001-003, 1 2001.

Precomputation of Constrained Widest Paths in Communication Networks

Stavroula Siachalou and Leonidas Georgiadis

Aristotle Univ. of Thessaloniki, Faculty of Engineering, School of Electrical and
Computer Engineering, Telecommunications Dept. Thessaloniki, 54124, GREECE.
{ssiachal,leonid}@auth.gr

Abstract. We consider the problem of precomputing constrained
widest paths in a communication network. Precomputing and storing
of all relevant paths minimizes the computational overhead required
to determine an optimal path when a new connection request arrives.
We present three algorithms that precompute paths with maximal
bandwidth (widest paths), which in addition satisfy given end-to-end
delay constraints. We analyze and compare the algorithms both in worst
case and through simulations using a wide variety of networks.

Keywords: Precomputation, QoS Routing, Widest Paths, Bottleneck
Paths, Graph Theory.

1   Introduction

In today's communication networks, transmission of multimedia traffic with
varying performance requirements (bandwidth, end-to-end delay, packet loss,
etc.), collectively known as Quality of Service (QoS) requirements, introduces
many challenges. In such an environment, where a large number of new requests
with widely varying QoS requirements arrive per unit of time, it is important to
develop algorithms for the identification of paths that satisfy the QoS requirements (i.e. feasible paths) of a given connection request, with minimal computational overhead. Minimization of the computational overhead per request can
be achieved by computing a priori (precomputing) and storing all relevant paths
in a data base.

While a large number of studies addressed the Constrained Path Routing
Problem (see [2], [4], [10], [12], [17] and the references therein) there are relatively
few works dealing with the specific issues related to precomputing paths with
QoS constraints [6], [8], [14]. In [8], the problem of precomputing optimal paths
under hop-count constraints is investigated. They propose an algorithm that
has superior performance than Bellman Ford's algorithm in terms of worst case
bounds. In [14], by considering the hierarchical structure which is typical in large
scale networks, an algorithm which offers substantial improvements in terms of
computational complexity is presented. These studies concentrated on the hop-
count path constraint.

N. Mitrou et al. (Eds.): NETWORKING 2004, LNCS 3042, pp. 1192–1203, 2004.

In [9] Guerin, Orda and Williams presented the link available bandwidth metric as one of the information on which path selection may be based. They mentioned that the leftover minimum bandwidth on the path links after connection acceptance must be as large as possible in order to accept as many requests as possible. In this paper we focus on the problem of precomputing paths with maximal bandwidth (path bandwidth is the minimal of the path link bandwidths), which in addition must satisfy given end-to-end delay requirements which become known upon the arrival of a new request. We present three algorithms that provide all relevant paths. The first algorithm is an application in the specific context of the algorithm developed in [17] for the Constrained Path Routing Problem. The second is based on an implementation of the basic algorithmic steps in [17], where we introduce new data structures that take advantage of useful properties of the problem at hand. The third algorithm is based on an approach whereby iteratively relevant paths are determined and links that are not needed for further computation are eliminated. We analyze and compare the algorithms both in worst case and through simulations. The analysis considers both computation times and memory requirements and shows the trade-offs involved in the implementation of each of the algorithms.

The rest of the paper is organized as follows. The problem is formulated in Section 2. We present the three algorithms in Section 3 and in Section 4 we examine the algorithms in terms of worst case running time and memory requirements. Section 5 presents numerical experiments that evaluate the performance of the proposed algorithms. Conclusions of the work are presented in Section 6. Due to space limitation proofs are omitted. We refer the interested reader to the site [20] for a version containing proofs.

2 Model and Problem Formulation

In this section we formulate the problem related to the precomputation of constrained widest paths and define some notation that will be used in the rest of the paper.

A network is represented by a directed graph $G = (V, E)$, where V is the set of nodes and E is the set of edges (links). Let $N = |V|$ and $M = |E|$. A link l with origin node u and destination node v is denoted by (u, v). A path is a sequence of nodes $p = (u_1, u_2, ..., u_k)$, such that $u_i \neq u_j$ for all $1 \leq i, j \leq k$, $i \neq j$, and $k - 1$ is the number of hops of p. By p we also denote the set of links on the path, i.e., all links of the form (u_i, u_{i+1}), $i = 1, ..., k - 1$. By $V_{in}(u)$ and $V_{out}(u)$ we denote respectively the set of incoming and outgoing neighbors to node u, that is

$$V_{in}(u) = \{v \in V : (v, u) \in E\}, \quad V_{out}(u) = \{v \in V : (u, v) \in E\}.$$

With each link $l = (u, v)$, $u, v \in V$ there is an associated width $w_l \geq 0$ and a delay $\delta_l \geq 0$. We define the width and the delay of the path p respectively,

$$W(p) = \min_{l \in p} \{w_l\}, \ D(p) = \sum_{l \in p} \delta_l.$$

The set of all paths with origin node s, destination node u and delay less than or equal to d is denoted by $P_u(d)$. The set of all paths from s to u is denoted by P_u.

In a computer network environment, w_l may be interpreted as the free bandwidth on link l and δ_l as the link delay. Assume that a connection request has bandwidth requirements b and end-to-end delay requirement d. Upon the arrival of a new connection request with origin node s and destination node u, a path must be selected that joins the source to the destination, such that the connection bandwidth is smaller than the free bandwidth on each link on the path, and the end-to-end delay of connection packets is smaller than the path delay. It is often desirable to route the connection through the path with the largest width in $P_u(d)$; this ensures that the bandwidth requirements of the connection will be satisfied, if at all possible, and the delay guarantees will be provided. Moreover, the leftover minimum bandwidth on the path links after connection acceptance will be as large as possible. We call such a path "constrained widest path".

According to the previous discussion, upon the arrival of a new connection request with end-to-end delay requirement d, we must select a path $p^* \in P_u(d)$ that solves the following problem.

Problem I: Given a source node s, a destination node u and a delay requirement d, find a path $p_u^* \in P_u(d)$ that satisfies

$$W(p_u^*) \geq W(p) \text{ for all } p \in P_u(d).$$

Note that when $\delta_l = 1$ for all $l \in E$, Problem I reduces to the problem addressed in [8], i.e., the problem of finding a widest path with hop count at most d. Let us assume that the source node s is fixed. In principle, in order to be able to select the appropriate path for any delay requirement one must precompute for each destination u and each delay d, an appropriate optimal path $p_u^*(d)$. At first this may seem rather formidable, both in terms of running time and in terms of space requirements. However, the situation is greatly simplified by the observation that one needs to precompute the paths $p_u^*(d)$ for only a subset of the delays. Indeed, let $W_u^*(d)$ be the value of the solution to Problem I (if no solution exists set $W_u^*(d) = -\infty$). It can be easily seen using similar arguments as in [17] that $W_u^*(d)$ is a piecewise constant, left continuous, non-decreasing function with a finite number of discontinuities. Hence, to determine the function $W_u^*(d)$, we only need to know the values of $W_u^*(d)$ at these discontinuities (we also need the paths that cause these discontinuities - see Section 3.1). A discontinuity of $W_u^*(d)$ will also be referred to as a discontinuity of node u.

In fact, from the route designer's perspective, the pairs $(d_k, W_u^*(d_k))$, where d_k is a discontinuity point of $W_u^*(d)$ are the most interesting ones, even if one takes into account routing requirements different than those considered in Problem I. Specifically, under our interpretation of path width and delay, among pairs $(D(p_i), W(p_i))$, $p_i \in P_u$, $i = 1, 2$, there is a natural "preference relation". That is, we would like to obtain paths that have as small delay as possible and as large width as possible. We are thus lead to the following natural definition of dominance

> **Definition I (Dominance Relation):** We say that pair $(D(p_1), W(p_1))$ dominates pair $(D(p_2), W(p_2))$ (or that path p_1 dominates path p_2) if either $\{W(p_1) > W(p_2) \text{ and } D(p_1) \leq D(p_2)\}$, or $\{W(p_1) \geq W(p_2) \text{ and } D(p_1) < D(p_2)\}$.

Hence, the pairs of interest under our setup are those for which no other dominating pair can be found for the same origin-destination nodes. This set of paths is generally known as the non-dominated or the Pareto-optimal set [3], [12]. From a precomputation perspective, it is desirable to determine for each destination u, the non-dominated set of pairs (and the associated paths). It can be shown that this set is exactly the set of discontinuities of $W_u^*(d)$, $u \in V$.

In the next section we present three algorithms for precomputing the discontinuities of the functions $W_u^*(d)$, $u \in V$.

3   Algorithm Description

The problem of determining the function discontinuities when link widths and delays are both additive costs (i.e., the cost of a path is the sum of its link costs) has been addressed in [17]. In the current setup, the main difference is that the path width is the minimum of its link widths (rather than the sum). However, the general algorithms in [17] can be adapted to the problem under consideration with minor modifications, as outlined in Section 3.1. In Sections 3.2 and 3.3 we present two additional algorithms that take into account the particular form of the problem under consideration. The first is an implementation of the algorithm in [17] that uses efficient data structures. The second uses a "natural" approach that eliminates successively unneeded graph edges and uses a dynamic version of Dijkstra's algorithm to determine all function discontinuities. Our intent is to compare these algorithms in terms of worst case, average running times and space requirements.

3.1   Algorithm I (ALG I)

The algorithms proposed in [17] are based on the following facts, which carry over to the situation at hand. In the discussion that follows we assume for convenience that $W_u^*(d)$ is defined for any real d, $W_u^*(d) = -\infty$, $d < 0$, and $W_s^*(d) = \infty$, $d \geq 0$. Hence by convention the source node s has a discontinuity at zero.

– For any $u \in V - \{s\}$, if $W_u^*(d)$ is discontinuous at d, then there is a $v \in V_{in}(u)$ such that $W_v^*(d)$ is discontinuous at $d - \delta_{vu}$ and $W_u^*(d) = \min \{W_v^*(d - \delta_{vu}), w_{vu}\}$. We call the pair $(d, W_u^*(d))$ the successor discontinuity of $(d - \delta_{vu}, W_v^*(d - \delta_{vu}))$. Also, $(d - \delta_{vu}, W_v^*(d - \delta_{vu}))$ is called the predecessor discontinuity of $(d, W_u^*(d))$. If it is known that the pair $(d, W_v^*(d))$ is a discontinuity point, then its "possible" successor discontinuities are pairs of the form

$$(d + \delta_{vu}, \min\{W_v^*(d), w_{vu}\}), \ u \in V_{out}(v).$$

- If $W_u^*(d)$ is discontinuous at d then there is a path $p^*(d) \in P_u(d)$ such that

$$W(p^*(d)) = W_u^*(d), \ D(p^*(d)) = d.$$

- Suppose that we impose a lexicographic order relation between discontinuity pairs $(\dot{d}_i, W_i)$, $i = 1, 2$, as follows:

$$(d_1, W_1) \prec (d_2, W_2) \text{ iff either } d_1 < d_2 \text{ or } (d_1 = d_2 \text{ and } W_1 > W_2).$$

Suppose also that among all the discontinuities of the functions $W_u^*(d), u \in V$ we know the set of the k smallest ones (with respect to the lexicographic order). Call this set $\widehat{D}$. Let $\widehat{D}(u)$ be the discontinuities in $\widehat{D}$ that belong to node function $W_u^*(d)$. Hence $\widehat{D} = \cup_{u \in E} \widehat{D}(u)$. The set of possible successor discontinuities of those in $\widehat{D}$ is denoted by $\widehat{P}$. Let (d, W) be a smallest element of $\widehat{P}$ and let u be the node to which this possible discontinuity belongs. Then (d, W) is a real discontinuity for node u if and only if

$$W > \max\left\{W_m : (d_m, W_m) \in \widehat{D}(u)\right\}.$$

Based on these facts, we can construct an algorithm for determining all the node discontinuities as described below. In the following we will need to know the node u to which a real or possible discontinuity (d, W) belongs. For clarity we denote this discontinuity by (d, W, u). For initialization purposes we set $\widehat{D}(u) = \{(-\infty, -\infty, u)\}$, $u \in V$ and $\widehat{D}(s) = \{(0, \infty, s)\}$. The generic algorithm is presented below.

Generic Algorithm I

Input: Graph G with link widths w_{uv} and delays δ_{uv}. **Output:** The queues $\widehat{D}(u)$, $\forall u \epsilon V$.

1. /* Begin Initialization
2. $\widehat{D}(u) = \{(-\infty, -\infty, u)\}$; $u \in V$, $\widehat{P} = \varnothing$;
3. $\widehat{D}(s) = \{(0, \infty, s)\}$; $(d, W, u) = (0, \infty, s)$;
4. /*End Initialization*/
5. Create all possible successor discontinuities of (d, W, u)
   (i.e., the set $\{(d + \delta_{uv}, \min\{W, w_{uv}\}, v), \ v \in V_{out}(u)\}$ and add them to $\widehat{P}$);
6. If $\widehat{P}$ is empty, then stop;
7. Among the elements $\widehat{P}$ (possible successor discontinuities), find and extract (i.e., delete from $\widehat{P}$) the minimum one in the lexicographic order. Denote this element (d, W, u);
8. If $W \leq \max\left\{w_m : (d_m, W_m, u) \in \widehat{D}(u)\right\}$, then go to step 6. Else,

9. $\widehat{D}(u) \leftarrow \widehat{D}(u) \cup \{(d, W, u)\}$;
10. go to step 5;

In [17] two implementations of the generic algorithm were proposed, which differ mainly in the manner in which the set $\widehat{P}$ is organized. In the current work we pick the implementation that was shown to be more efficient both in worst case and average case analysis. For our purposes, it is important to note that the sets $\widehat{D}(u)$ are implemented as FIFO queues, and that the elements (d, W, u) in these queues are generated and stored in increasing order of both d and W as the algorithm proceeds. Furthermore, in our implementation of Algorithm I, we introduce an additional optimization that is based on the following observation in [8]: whenever a real discontinuity (d, W, u) is found and the possible discontinuities caused by (d, W, u) are created, then links (v, u), $v \in V_{in}(u)$ with $w_{vu} \leq W$ can be removed from further consideration. This is so, since these links cannot contribute to the creation of new discontinuities for node u. Indeed, any newfound discontinuity (d_1, W_1, v) at node v, will create a possible discontinuity $(d_1 + \delta_{vu}, \min(W_1, w_{vu}), u)$. But $\min(W_1, w_{vu}) \leq W$ and hence this possible discontinuity cannot be a real one for node u.

As usual, in order to be able to find by the end of the algorithm not only the discontinuities, but paths that correspond to these discontinuities, one must keep track of predecessor discontinuities as well. That is, in the implementation we keep track of $(d, w, u, predecessor\_disc)$, where for the source node s, $predecessor\_disc = null$, and for any other node u, $predecessor\_disc$ is a pointer to the predecessor discontinuity of (d, w, u). To simplify the notation, in the description of all algorithms we do not explicitly denote $predecessor\_disc$, unless it is needed for the discussion.

3.2   Algorithm II (ALG II)

The Generic Algorithm in Section 3.1 works also when lexicographic order is defined as

$$(d_1, W_1) \preccurlyeq (d_2, W_2) \text{ if either } W_1 > W_2 \text{ or } (W_1 = W_2 \text{ and } d_1 < d_2).$$

In this case, the elements (d, W, u) in the FIFO queues $\widehat{D}(u)$ are generated and stored in decreasing order of both d and W as the algorithm proceeds.

Algorithm II uses the lexicographic order $\preccurlyeq$, and is based on an extension of ideas presented in [7] to speedup computations. The basic observations are the following.

- Suppose that link widths take $K \leq M$ different values $g_1 < g_2 < \dots < g_K$. Let $r(w_l)$ be the ranking order of w_l, i.e. if for link l it holds $w_l = g_i$, set $r(w_l) = i$. If one uses $r(w_l)$ instead of the link's actual width in the calculations, the resulting discontinuities occur at the same delays and for the same paths as if the actual widths were used.

– Path widths always take one of the values in the set $\{w_{vu}, (v, u) \in E\}$ i.e., they take at most K different values. Hence the same holds for the values of $W_u^*(d)$ and the widths of all possible discontinuities.

We use these observations to speed up the computations of Generic Algorithm I as follows. First, we use $r(w_l)$ in place of the link widths. Next we organize the set of possible discontinuities $\widehat{P}$ as follows. We create an array $A[u, k]$, $1 \leq u \leq N$, $1 \leq k \leq K$, where $A[u, k]$, if nonnull, denotes a possible discontinuity of the form (d, k, u). We also create K heaps $H[k]$, $1 \leq k \leq K$. Heap $H[k]$ contains the nonnull elements of $A[u, k]$, $1 \leq u \leq N$ and uses as key the delay d of a possible discontinuity. Reference [5] contains various descriptions of heap structures. For our purposes we need to know that the following operations can be performed on the elements of a heap structure.

– create_heap(H): creates an empty heap H.
– insert(e, H): inserts element e to H.
– get_min(e, H): removes and returns an element e in H with the smallest key.
– decrease_key(e_{new}, e, H): replaces in H element e with e_{new}, where element e_{new} has smaller key than e.

With these data structures, we implement steps 5 and 7 of Generic Algorithm I as follows. For an element $e = (d, W, u)$ we denote $e.delay = d$, $e.width = W$.

– **Step 5:** *Create all possible successor discontinuities of (d, W, u) and add them to $\widehat{P}$.*
  /* let $k' = r(W)$, hence we have available the discontinuity (d, k', u) */
  1. For $v \in V_{out}(u)$ do
     a) $e_{new} = (d + \delta_{uv}, \min\{k', r(w_{uv})\}, v)$; $k = e_{new}.width$;
     b) If $A[v, k]$ is null $\{A[v, k] = e_{new};$ insert($e_{new}, H[k]$)$\}$. Else $\{$
     c) If $e_{new}.delay < e.delay$ then$\{$
        i. $e = A[v, k]$; $A[v, k] = e_{new};$
        ii. decrease_key($e_{new}, e, H[k]$)$\}\}$;
  2. end do
  In step 1b, if $A[v, k]$ is null, there is no possible discontinuity for node v with width k. Hence a new possible discontinuity for node v with width k is created and placed both in $A[v, k]$ and $H[k]$. In step 1c, when $e_{new}.delay < e.delay$ we know that the old possible discontinuity for node v cannot be a real discontinuity since e_{new} dominates e and therefore in step 1(c)i we replace the e with e_{new} both in $A[v, k]$ and $H[k]$. These last two steps avoid inserting unnecessary elements in the heap $H[k]$, thus decreasing the time that the get_min operation takes in step 7 of Generic Algorithm I. The trade-off is extra memory space requirements due to array $A[v, k]$. We discuss this issue further in Sections 4 and 5.
– **Step 7:** *Among the elements $\widehat{P}$, find and extract the minimum one in the lexicographic order. Denote this element (d, W, u).*
  /* let $k' = r(W)$, hence we have available the discontinuities (d, k', u) */

The heaps $H[k]$ are scanned starting from the largest index and moving to the smallest. The index of the heap currently scanned is stored in the variable L which is initialized to K.

1. Find the largest $k' \leq L$ such that the heap $H[k']$ is nonempty;
2. get_min$(e, H[k'])$; $(d, k', u) = e$;
3. Set $A[u, k']$ to null;
4. $L = k'$;

The scanning process (largest to smallest) works since whenever a possible discontinuity (d, k, u) is removed from $\hat{P}$, any possible discontinuities that already exist or might be added later to $\hat{P}$ are larger (with respect to $\preccurlyeq$) than (d, k, u) and thus will have width at most k. Notice that this would not be true if the order $\prec$ were used. The real discontinuities $\hat{D}(u)$, $u \in V$ are again implemented as FIFO queues.

3.3   Algorithm III (ALG III)

The third algorithm we consider is based on the idea of identifying discontinuities, eliminating links that are not needed to identify new discontinuities and repeating the process all over again. Specifically, the algorithm performs iterations of the basic steps shown below. Again $\hat{D}(u)$, $u \in V$ are implemented as FIFO queues.

Algorithm III

Input: Graph G with link widths w_{uv} and delays δ_{uv}. **Output:** The queues $\hat{D}(u)$, $\forall u \epsilon V$.

1. Find the widest-shortest paths from s to all nodes in G. That is, for any node $u \in V$, among the shortest-delay paths find one, say p_u, that has the largest width.
2. Let W^* be the minimum among the widths of the paths p_u, $u \in V - \{s\}$. For any $u \in V$, if $W(p_u) = W^*$, add $(D(p_u), W(p_u))$ at the end of queue $\hat{D}(u)$.
3. Remove from G all links with width at most W^*.
4. If s has no outgoing links, stop. Else go to step 1.

This algorithm produces all discontinuities in $\hat{D}$ as the next theorem shows.

Theorem 1. *Algorithm III produces all discontinuities in $\hat{D}$.*

Proof. Proof can be found in [20].

The widest-shortest path problem can be solved by a modification of Dijkstra's algorithm [15]. In fact, after the removal of the links of G in step 3, paths whose width is larger than W^* will still remain the widest-shortest paths when the algorithm returns to step 1. Hence the computations in the latter step can be reduced by taking advantage of this observation. Algorithms that address this issue have been presented in [13] and we pick for our implementation the one that was shown to be the most efficient.

4   Worst Case Analysis

In this section we examine the three algorithms proposed in Section 3 in terms of worst case running time and memory requirements, the analysis of the algorithms is presented in [20]. In all three algorithms we assume a Fibonacci heap implementation [5]. In such implementation of a heap H, all operations except get_min(e, H) take $O(1)$ time. Operation get_min(e, H) takes $O(\log L)$ time, where L is the number of elements in the heap.

The worst case running times of the algorithms are, $O(MN \log N + M^2 \log N)$ for the first algorithm and $O(MN \log N + M^2)$ for ALG II and III. All three algorithms have the same worst case memory requirements equal to $O(NM)$. ALG II and ALG III have the same worst case running time, which is slightly better than the worst case running time of ALG I. Hence based on these metrics, all three algorithms have similar performance. However, worst case analysis alone is not a sufficient indicator of algorithm performance. The simulation results in Section 5 reveal the performance difference of the algorithms in several networks of interest.

5   Simulation Results

We run the following set of experiments. We generate:

Power Law Networks: This is one of the methods that attempt to generate network topologies that are "Internet like". We choose a number of N nodes and a number of M links ($M = \alpha N, \alpha > 1$). The links are used to connect nodes randomly with each other in such a manner that the node degrees follow a power law [18].

Real Internet Networks: These networks were taken from [19] and are based on network topologies observed on the dates 20/09/1998, 01/01/2000 and 01/02/2000.

For each experiment the delay of a link is picked randomly with uniform distribution among the integers [1, 100]. For the generation of the link widths we use the following method.

Widths are generated in such a manner that they are correlated to their delays. Thus, for each link l a parameter β_l is generated randomly among the integers [1, 10]. The width of link l will then be $w_l = \beta_l(101 - d_l)$.

We also run experiments using link widths uncorrelated to their delays, thus w_l is picked randomly with uniform distribution among the integers [1, 100]. For a given algorithm and for fixed number of nodes and edges we notice that the running time is much smaller when the width values are uncorrelated to delays and therefore are not presented here. This is due to the fact that when widths are correlated to delays, the number of discontinuities is increased.

We generate Power Law Networks with 400, 800 and 1200 nodes and with ratios $\alpha = M/N$ equal to 4, 8, 16. For each N and α we generate 10 different networks and for each network we generate the link widths according to the method previously described (correlated to delays).

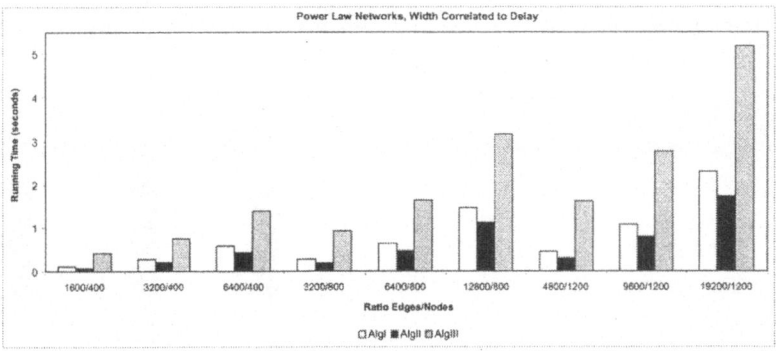

Fig. 1. Running Time for Power Law Networks with width correlated to delays

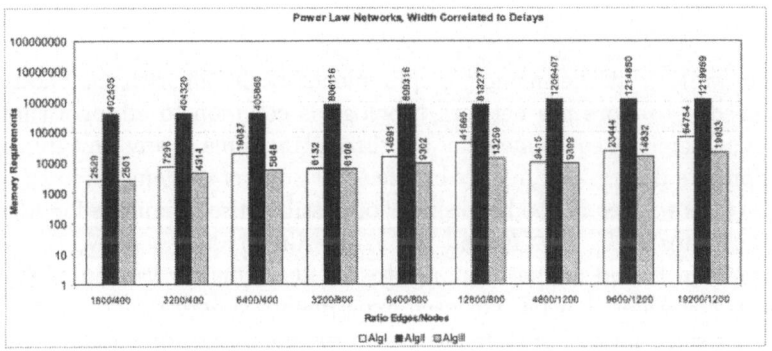

Fig. 2. Memory Requirements for Power Law Networks with width correlated to delays.

The experiments were run on a Pentium PC IV, 1.7GHz, 256MB RAM. In Figure 1 we present the average running times (in seconds) of the three algorithms for Power Law Networks. We make the following observations.

- Algorithm II has the best running time performance, and Algorithm III the worst.
- Compared to Algorithm II, the running times of Algorithm I and Algorithm III are found to be up to 1.5 times and 6 times larger, respectively.
- Algorithm II performs better than Algorithm I and III for all experiments and especially for large networks.

The additional optimization (removal of unneeded links) in Algorithm I improves its running time but not by much.

The Real Internet Networks have $N = 2107, 4120, 6474$ nodes and $M = 9360, 16568, 27792$ links respectively. In these networks we also performed 10 experiments, where in each experiment we picked randomly a source node. Figure 3(a) shows the average running time of the three algorithms. We notice again that Algorithm II has the best running time performance and Algorithm III the worst. The running time of Algorithm III has been found to be 20 times larger than that of Algorithm II in some experiments. The performance of Algorithm I is worse, but comparable to that of Algorithm II.

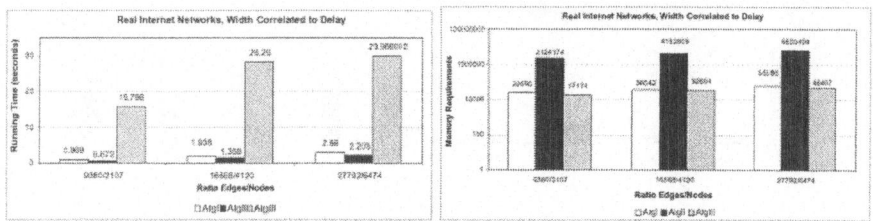

Fig. 3. (a)Running Time and (b)Memory Requirements for Real Internet Networks with width correlated to delays.

Next we look at the memory requirements of the algorithms. The memory space needed to store the network topology is common to all algorithms. The additional memory requirements of the three algorithms at any time during their execution, are determined mainly by the total number of elements in the queues $\widehat{D}(u)$, $u\epsilon V$ as well as: a) the heap size $\widehat{P}$ of possible discontinuities for Algorithm I, b) the heaps $H[k]$, $k \in K$ and the array $A[u,k]$, $1 \leq u \leq N$, $1 \leq k \leq K$ for Algorithm II and c) the heap size to run the dynamic version of Dijkstra's algorithm for Algorithm III. For each experiment we determined the maximum of memory space needed to store the previously mentioned quantities. This space depends on the particular network topology for Algorithm I and III, while for Algorithm II it is already of order $O(KN)$ due to the array $A[u,k]$. As a result, the memory requirements of Algorithm II are significantly larger than those of the other two algorithms. This is indicated in Figure 2-3(b) where we present the memory requirements of the three algorithms for Power Law and Real Internet Networks. Algorithm III has the smallest memory, followed by Algorithm I whose memory requirements are comparable to those of Algorithm III. Due to the need of array $A[u,k]$, Algorithm II has significantly larger memory requirements.

Summarizing our observations, Algorithm II has the best running time, however its memory requirements are significantly worse than those of the other two algorithms. At the other end, Algorithm III has the best memory space requirements, however its running time is significantly worse than that of the other two. Algorithm I represents a compromise between running time and space requirements, as its performance with respect to these measures, while not the best, is comparable to the best.

6   Conclusions

We presented three algorithms for precomputing constrained widest paths in a communication network. We analyzed the algorithms in terms of worst case running time and memory requirements. We also presented simulation results indicating the performance of the algorithms in networks of interest. The worst case analysis showed that all three algorithms have similar performance, with Algorithm I being slightly worse in case of worst case running time. However, the simulations revealed significant performance differences and indicated the conditions under which each algorithm is appropriate to be used.

References

1. Claude Berge, *Graphs*, North-Holland Mathematical Library, 1991.
2. D. Blokh, G. Gutin, "An Approximation Algorithm for Combinatorial Optimization Problems with Two Parameters", *IMADA* preprint PP-1995-14, May 1995.
3. K. Deb, *Multi-Objective Optimization using Evolutionary Algorithms*, Wiley, 2001.
4. S. Chen, K. Nahrstedt, "On Finding Multi-Constrained Paths", *in Proc. of IEEE International Conference on Communications (ICC'98)*, pp. 874-879, Atlanta, GA, June 1998.
5. T. H. Cormen, C. E. Leiserson, R. L. Rivest, *Introduction to Algorithms*, Mc Graw Hill, 1990.
6. Yong Cui, Ke Xu, Jianping Wu, "Precomputation for Multi-Constrained QoS Routing in High Speed Networks", *IEEE INFOCOM 2003*.
7. L. Georgiadis, "Bottleneck Multicast Trees in Linear Time", to be published in *IEEE Communications Letters*.
8. R. Guerin, A. Orda, "Computing Shortest Paths for Any Number of Hops", *IEEE/ACM Transactions on Networking*, vol. 10, no. 5, October 2002.
9. R. Guerin, A. Orda and Williams D., "QoS Routing Mechanisms and OSPF Extensions", 2nd IEEE Global Internet Mini-Conference, Phoenix, AZ, November 1997.
10. T. Korkmaz, M. Krunz and S. Tragoudas, "An Efficient Algorithm for Finding a Path Subject to Two Additive Constraints", *Computer Communications Journal*, vol. 25, no. 3, pp. 225-238, Feb. 2002.
11. K. Mehlhorn, S. Naher, *Leda:A Platform for Combinatorial and Geometric Computing*, Cambridge University Press, 2000.
12. P. Van Mieghem, H. De Neve and F.A. Kuipers, "Hop-by-hop Quality of Service Routing", *Computer Networks*, vol. 37/3-4, pp. 407-423, November 2001.
13. P. Narvaez, Kai-Yeung Siu, and Hong-Yi Tzeng, "New Dynamic Algorithms for Shortest Path Tree Computation ", *IEEE/ACM Transactions on Networking*, vol. 8, no. 6, December 2000.
14. A. Orda and A. Sprintson, "QoS Routing: The Precomputation Perspective", *IEEE INFOCOM 2000*, vol. 1, pp. 128-136, 2000.
15. J. L. Sobrino, "Algebra and Algorithms for QoS Path Computation and Hop-by-Hop Routing in the Internet", *IEEE INFOCOM 2001*, Anchorage, Alaska, April 22-26, 2001.
16. A. Orda and A. Sprintson, "A Scalable Approach to the Partition of QoS Requirements in Unicast and Multicast", *IEEE INFOCOM 2002*.
17. S. Siachalou, L. Georgiadis, "Efficient QoS Routing", *Computer Networks Journal*,vol.43/3, pp351-367, October 2003.
18. The Power Law Simulator, *http://*www.cs.bu.edu/brite.
19. The Real Networks, http://moat.nlanr.net/Routing/raw-data.
20. http://genesis.ee.auth.gr/georgiadis/english/public/networking04full.pdf

A Traffic Engineering-Aware Shortest-Path Routing Algorithm in IP Networks[*]

Youngseok Lee[1] and Biswanath Mukherjee[2]

[1] Department of Computer Science and Engineering,
Chungnam National University, Daejon, 305-764, Korea
yslee@cs.cnu.ac.kr
[2] Department of Computer Science,
University of California, Davis, CA 95616, USA
mukherje@cs.ucdavis.edu

Abstract. Single shortest-path routing is known to perform poorly for Internet traffic engineering (TE) where the typical optimization objective is to minimize the maximum link load. Splitting traffic uniformly over equal-cost multiple shortest paths in OSPF and IS-IS does not always minimize the maximum link load when multiple paths are not carefully selected for the global traffic demand matrix. However, a *TE-aware shortest path* among all the equal-cost multiple shortest paths between each ingress-egress pair can be selected such that the maximum link load is significantly reduced. IP routers can use the TE-aware shortest path without any change to existing routing protocols and without any serious configuration overhead. While calculating TE-aware shortest paths, the destination-based forwarding constraint at a node should be satisfied, because an IP router will forward a packet to the next-hop towards the destination by looking up the destination prefix. In this paper, we present a mathematical problem formulation for finding a set of TE-aware shortest paths for the given network as an integer linear program (ILP), and we propose a simple heuristic for solving large instances of the problem. The proposed algorithm is evaluated through simulations in IP networks.

Keywords: Shortest-path routing, Traffic engineering, IP, ILP, Simulations, Optimization.

1 Introduction

1.1 Traffic Engineering

In recent years, the pervasive usage of Internet services has significantly increased the volume of Internet traffic, which has overtaken that of voice traffic.

[*] This work was supported in parts by a Korean Science & Engineering Foundation (KOSEF) Postdoctoral Fellowship Program and by the US National Science Foundation (NSF) Grant No. ANI-02-07864. This work was performed when Dr. Youngseok Lee was a visiting scholar at University of California, Davis.

N. Mitrou et al. (Eds.): NETWORKING 2004, LNCS 3042, pp. 1204–1215, 2004.

Due to the popularity of World Wide Web (WWW), audio and video streaming applications, and bandwidth-intensive services such as peer-to-peer applications, the network will become much more congested in spite of increased capacity unless appropriate techniques are designed. Aside from the rapid increase in traffic demand, congestion also happens even in the over-provisioned IP backbone network when traffic is detoured because of link failures. Traffic engineering (TE) is an effective solution to control the network congestion and optimize network performance. As stated in [1,2], traffic engineering encompasses the application of technology and scientific principles to the measurement, modeling, characterization, and control of Internet traffic. The purpose of traffic engineering is to facilitate efficient and reliable network operations while simultaneously optimizing network-resource utilization and traffic performance. The typical optimization objective of traffic engineering is to minimize the maximum link load in the network which represents the overall congestion in the network [3].

Traffic engineering in Internet Protocol (IP) networks is usually done by routing. The Interior Gateway Protocol (IGP) routing protocol in IP networks is based on the shortest-path routing algorithm where the shortest path is computed using link weights. An IP router maintains a routing table consisting of a next-hop (shortest path) and a destination routing prefix. An incoming packet is forwarded to the next-hop after the variable-sized destination IP prefix of the packet is looked up in the routing table, which is called the destination-based forwarding rule. Although the shortest-path routing algorithm is practical and easy to implement, it potentially limits the throughput of the network because only a single path between a source and a destination is used. Theoretically, optimal routing [3] can supply bifurcation of any source-destination traffic at strategic points with alternate routes. Optimal routing is based on the multi-commodity-flow problem formulation, and is also known as minimum-delay routing [4], because it can minimize the average end-to-end delay. Optimal routing can be realized through a form of multipath routing. In the shortest-path algorithm used by Internet routing protocols, multiple shortest paths can be found when the destination can be reached via multiple paths with the same cost from the source. Multi-path routing is employed by some Internet routing protocols in order to increase the total network utilization and the end-to-end throughput. IGPs such as Open Shortest Path First (OSPF) [5] and Intermediate System-Intermediate System (IS-IS) [6] explicitly support the option of multi-path routing, which is known as Equal-Cost Multi-Path (ECMP) routing, when there are multiple paths with the same cost. Some router implementations allow equal-cost multipath with Routing Information Protocol (RIP) and other routing protocols. Optimized Multi-Path (OMP) [7] which improves the performance beyond that of ECMP has been proposed for OPSF, IS-IS, and Multi-Protocol Label Switching (MPLS) [8] networks in the Internet Engineering Task Force (IETF). Although ECMP is widely used in OSPF and IS-IS for splitting traffic equally to multiple paths, it does not always minimize the maximum link load if multiple paths are not carefully selected for the global traffic demand matrix.

On the other hand, a recent network technology has been developed, called MPLS, where the path between ingress and egress routers is established by signaling protocols such as Resource ReSerVation Protocol (RSVP) [9] and a fixed-sized label is attached to IP packets by the ingress router. MPLS provides the functions of fast switching and explicit path setup to IP networks. Intermediate Label-Switched Routers (LSRs) forward packets by looking up not the variable IP prefix but the fixed-sized MPLS label. Thus, traffic engineering in MPLS networks is more efficient than in traditional IP networks due to the explicit Labeled-Switched Path (LSP). However, since MPLS plays the role of the control plane for IP routing, the maintenance cost may increase because of imposing an additional layer to the IP network.

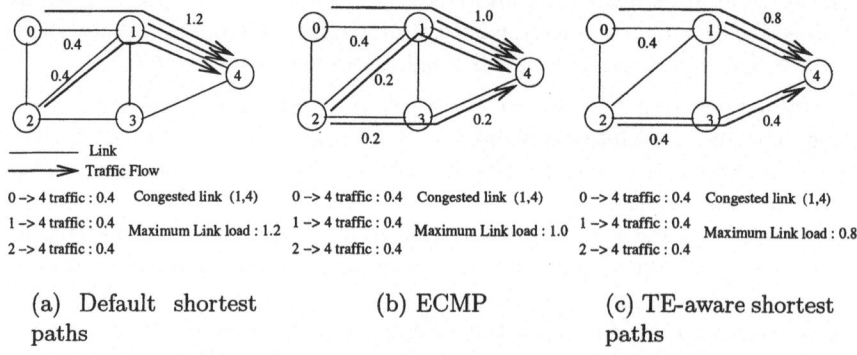

(a) Default shortest paths

(b) ECMP

(c) TE-aware shortest paths

Fig. 1. An example of TE-aware shortest paths.

1.2   TE-Aware Shortest-Path Routing

It is well known that single shortest-path routing performs poorly for achieving the traffic-engineering goal. Usually, it is assumed that, when equal-cost multiple shortest paths are available for the routing prefix, IP routers will randomly select one of the possible candidates for the routing prefix without considering the link load. When the network topology and the traffic demand matrix are given, however, a *TE-aware shortest path* can be selected among all possible shortest paths for each node pair such that the maximum link load is significantly reduced even when a single shortest path is used for routing. TE-aware shortest-path routing is simple and easily realized in current networks. It also mitigates the complexity of maintaining multipath routing, in which, for example, the number of multiple paths should be carefully configured at each router for global optimization. In addition, when traffic is distributed per routing prefix to avoid packet reordering, load balancing may not be achieved due to different traffic load per routing prefix. Although TE-aware shortest-path routing uses a single path, it is shown by experiments that it achieves a near-optimal solution as does multipath routing in a typical network topology with a typical traffic demand matrix. It is likely that TE-aware shortest paths will be more viable as the number of equal-cost shortest paths and the number of destination routing prefixes become large. However, it is shown by experiments that, even under the case of a small number of equal-cost multiple paths, TE-aware shortest paths are useful in reducing the maximum link load. Since it was shown in [10] that small routing prefixes which take a large fraction of traffic volume are enough for achieving the traffic-engineering goal, we assume that the destination routing prefixes are the egress node addresses for simplicity. Hence, when IP routers use TE-aware shortest paths instead of randomly chosen ones, the traffic-engineering goal will be easily achieved without any change to existing routing protocols and without any serious configuration overhead.

Figure 1 shows an example of the different behavior patterns of default shortest-path routing[1], ECMP, and TE-aware shortest-path routing for a given set of traffic demands. In Fig. 1(a), default shortest-path routing, where node 2 selects node 1 as the next-hop for traffic from source node 2 to destination node 4, results in the high maximum link load of 1.2. If splitting traffic evenly across two paths between nodes 2 and 4 is allowed (Fig. 1(b)), the maximum link load is reduced, but it is not the optimal solution. When the alternate shortest path (2-3-4) is used rather than 2-1-4 from node 2 to node 4 in Fig. 1(c), the maximum link load can be minimized. Therefore, the TE-aware shortest path between source node 2 and destination node 4 is 2-3-4. In this example, optimal shortest-path routing which allows traffic bifurcation is same as TE-aware shortest-path routing. In this paper, we present the TE-aware shortest-path routing problem as an ILP for the optimal solution, and we propose a heuristic algorithm to solve the problem in polynomial time.

2 Related Work

For the MPLS network, in [11], the traffic bifurcation problem is formulated in linear program (LP) and heuristics for the non-bifurcating problem are proposed. Although the LP problem formulated in [11] minimizes the maximum of link utilization, its solution does not consider total network resources, Therefore, the routing loop or unnecessary long paths may be generated for the LP solution. Wang, et al., have included the total network resources as the optimization objective in [12] to prevent the unnecessary long paths. In [13], the dynamic routing algorithm for MPLS networks is proposed where the path for each request is selected to prevent the interference among paths for the future demands. It considers only single-path routing for simplicity. Reference [14] proposes a traffic-partitioning mechanism for differentiation in MPLS networks, where an expedited forwarding (EF) traffic demand is divided into multiple LSPs to minimize the average end-to-end delay. However, it needs recalculation of load-splitting ratios whenever new requests arrive or the average rate of requests changes.

On the other hand, Internet traffic-engineering schemes without the MPLS feature have been recently studied. In [12], it is shown that the optimal routing

---

[1] In this paper, it is assumed that, in default shortest-path routing, a next-hop is randomly chosen for the destination node if multiple next-hops exist.

problem can be transformed into the shortest-path routing problem by choosing appropriate link metrics. Traffic engineering in IP networks can be achieved by changing link weights appropriately [15]. Although link-weight-based traffic engineering is simple, it may cause slow convergence time because of flooding of Link State Advertisement (LSA) messages throughout the network and recalculating all the shortest paths. During the convergence period, unstable network status such as load fluctuation and routing loops may occur. In general, it is not easy for a network operator to determine which link metric should be modified and to predict its impact on the network performance. A recent study [10] showed that traffic bifurcation over equal-cost multiple shortest paths can achieve the near optimal solution when the subset of next-hops are intelligently chosen for each routing prefix.

3   Problem Statement for TE-Aware Shortest-Path Routing

In order to calculate the optimal TE-aware shortest paths, we formulate and solve the ILP formulation in two steps: the first step calculates the weights of the shortest paths by minimizing the sum of the integer link weights; and the second step minimizes the maximum link load among the possible sets of shortest paths which satisfy the minimum weights of the shortest paths found in the first step. The reasons behind choosing this two-step ILP formulation are as follows: the first reason is that two dependent objectives need to be minimized; and the second reason is that, when one ILP formulation with the combination of two objectives is used, as in [12], an inappropriately-selected coefficient for combining the two objectives sometimes does not find the lower bound for one objective. For the calculation of TE-aware shortest paths, we assume a centralized TE server which has all the information on the network topology, the traffic demand matrix, the link-load measurement data, and all the routing tables of the network.

The following constants regarding the network parameters are given for the problem statement.

- $G = (N, E)$: a graph consisting of a set of nodes, N, and a set of links, E.
- $\Lambda = \{\Lambda_{sd}\}$: traffic demand matrix.
- C_{ij} : capacity of each link (i, j).
- $w(i, j)$: weight function of link (i, j) as an integer value, which is set to 1 for each $w(i, j)$ in this investigation, but could be generalized, if necessary.

Given the above constants, we need to determine the following variables.

- L_{max} : the maximum link load (≥ 0).
- X_{ij}^{sd} : the binary integer variable for flow assignment of traffic demand from source s to destination d on the link (i, j).
- Y_{ij}^{d} : the binary integer variable for aggregated flow assignment of traffic demand to destination d on the link (i, j), which means the next-hop of destination d at node i is node j.

The final objective of the problem is to minimize the maximum link load, L_{max}, while all paths are the shortest ones and the destination-based forwarding constraint is preserved at each node. The problem formulation of the first step is as follows.

$$Minimize \quad \sum_{i,j,s,d} X_{ij}^{sd} \cdot w(i,j) \tag{1}$$

$$\sum_{j:(i,j)\in E} X_{ij}^{sd} - \sum_{j:(j,i)\in E} X_{ji}^{sd} = \begin{cases} 1, i = s, \forall i, s, d \\ 0, i \neq s, d, \forall i, s, d \\ -1, i = d, \forall i, s, d \end{cases} \tag{2}$$

$$\sum_{s,d} \Lambda_{sd} \cdot X_{ij}^{sd} \leq L_{max} \cdot C_{ij}, \forall i, j \tag{3}$$

$$X_{ij}^{sd} \leq Y_{ij}^{d}, \forall i, j, s, d \tag{4}$$

$$\sum_{j} Y_{ij}^{d} \leq 1, \forall i, d \tag{5}$$

$$Int \quad X_{ij}^{sd}, Y_{ij}^{d} \in \{0,1\} \tag{6}$$

The objective of the first-step in the formulation, i.e., Eqn. (1), is to minimize the sum of the weights on links, which guarantees that all the traffic are assigned to the shortest path. Constraint (2) represents the flow-conservation rules for source, intermediate nodes, and destination. Constraint (3) explains that traffic assignment on each link will be constrained by the link capacity multiplied by L_{max} which indicates the maximum link load. Constraints (4) and (5) state that, if the traffic destined to d uses next-hop j at node i (Y_{ij}^{d}), then all the traffic towards the destination d at node i should have the same next-hop j regardless of their sources (using the destination-based forwarding rule).

After finding the sum of weights of the shortest paths in the first step, which will be given as a constant value denoted by H_{total} to the second-step formulation, we turn to the second step in order to minimize the maximum link load, L_{max}, by using the following objective (7) and adding the constraint (8) to the first-step formulation.

$$Minimize \quad L_{max} \tag{7}$$

$$\sum_{i,j,s,d} X_{ij}^{sd} \cdot w(i,j) \leq H_{total} \tag{8}$$

The optimization objective in Eqn. (7) is to minimize the maximum link load, L_{max}. Constraint (8) states that all the paths for traffic assignment will be the shortest paths.

As the above problem formulation contains integer variables, its computation complexity is not scalable as the network grows. The problem without the destination forwarding constraint has been shown to be $NP\text{-}hard$ [11]. Even though the number of equal-cost multiple paths between source and destination, $|P_{sd}|$, is known, the size of the search space is exponential ($\prod_{i=1}^{|N|} \prod_{j=1, j\neq i}^{|N|} |P_{ij}|$). Therefore, we propose a simple heuristic algorithm for solving large instances of this problem in the next section.

4   Heuristic Algorithm for TE-Aware Shortest-Path Routing

The proposed algorithm is based on the information of all the available equal-cost multiple shortest paths. Given the shortest-path information, we try to find a set of shortest paths which minimizes the maximum link load while the destination-based forwarding rule is satisfied at each node.

[**Step 1**] Calculate ECMPs for all (s, d) pairs.
[**Step 2**] Initialize a random set of single shortest paths satisfying the destination-based forwarding rule.
[**Step 3**] Find the set of flows, $\{F_i\}$, over the most congested link, l_c, with the maximum link load, L_{max}.
[**Step 4**] For each candidate flow, F_i, select an alternate path, and determine "induced flows" which also have to change their paths to meet the destination-based forwarding rule.
[**Step 5**] If swapping paths of F_i and induced flows determined in [Step 4] reduces L_{max}, exchange the current shortest path with the alternate one.
[**Step 6**] Mark F_i as examined.
[**Step 7**] If the new congested link is different from l_c, go to [Step 3].
[**Step 8**] If flows to be examined remain in $\{F_i\}$, go to [Step 4]; otherwise, terminate.

In order to satisfy the destination-based forwarding rule, [Step 4] enumerates the "induced flows" affected by F_i instead of all flows to be examined for fast computation. The induced flows for F_i are ones that share the "branched-sub-shortest path" of F_i as shown in Fig. 2(a). Thus, the shortest paths of the induced flows will also be changed when the primary shortest path for the flow, F_i, is swapped with the alternate one.

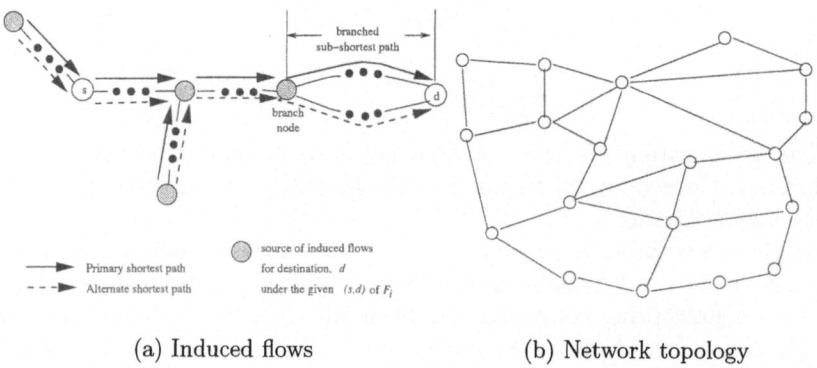

(a) Induced flows   (b) Network topology

Fig. 2. (a) "Induced flows" affected by F_i flow from node s to node d and (b) sample network topology.

In [Step 4], various ways of deciding a candidate flow, F_i, to be swapped are possible such as random, maximum flow-preferred, or minimum flow-preferred selections. In addition, when the alternate shortest path for the candidate flow is examined, different selection criteria such as random, maximum residual capacity, and minimum residual capacity may be possible. However, these variations do not yield much difference on the performance, because any selection method may generate the induced flows to be swapped so that the difference between selection methods will be compensated. Therefore, randomly selecting a candidate flow and an alternate path is used for the experiments.

The complexity of the heuristic algorithm is $O(|N|^3 + |E||N|(|N| - 1)^2)$. The computation complexity of the initialization step for finding all the equal-cost shortest paths using linear *Dijkstra* algorithm is $O(|N|^3)$, and for each flow $(O(|N|(|N|-1)))$ on the congested link $(O(|E|))$, the induced flows for the same destination $(O(|N| - 1))$ will be examined in the worst case.

5   Performance Evaluation

5.1   Case I: Sample Network

For our simulation experiments, we use a simple network topology (Fig. 2(b)) with 19 nodes and 62 directed links, and we use 24 sample traffic matrices for one day [16] where a traffic model is derived based on the observations on several backbones' link loads. Backbone traffic is the aggregation of several end systems' traffic, and the aggregation process filters out the short-term variations. On the other hand, long-term variations (on a scale of hours) remain and repeat their pattern in one-day periods. Hence, we believe that it is reasonable to use the representative sampled traffic matrices for the simulation of the periodic TE operations. To obtain a realistic model, 24 representative link traffic rates are sampled from real networks over a 24-hour period. These traffic rates are used to generate 24 traffic samples between any node pair in the network. We use a scaling factor in the range [0.2, 1.2] to create discrepancies in the volume of traffic flowing between different node pairs (values on the main diagonal remain zero). It is assumed that the capacity of each link is 310 Mbps (2 OC-3's).

Dynamic TE-Aware Shortest-Path Routing. In the dynamic TE-aware shortest-path routing algorithm, the routing table may be changed if the better shortest path is known. When the TE process is performed every one hour for one-day traffic matrix, we compare the heuristic for TE-aware shortest-path routing with default shortest-path routing, ECMP[2], optimal TE-aware shortest-path routing, and optimal multipath routing for each sampled traffic matrix. Figure 3(a) shows that the proposed heuristic reduces the maximum link load by 25% when compared with default shortest-path routing, whereas the maximum link load of ECMP is less than that of shortest-path routing by 8.5%. When compared with the results of optimal TE-aware shortest-path routing (solved by

---

[2] The number of equal-cost shortest paths is set to 2, because its result was found to be the best among all the possible cases in this experiment.

CPLEX[3] [17]), the maximum link load of the heuristic has increased only by 4.9% on average. It is interesting to note that, even when compared with the results of optimal multipath routing which are obtained by LP after the integer variables are relaxed to be real and the destination-based forwarding rule is removed, the maximum link load of the heuristic increases only by 6.6%.

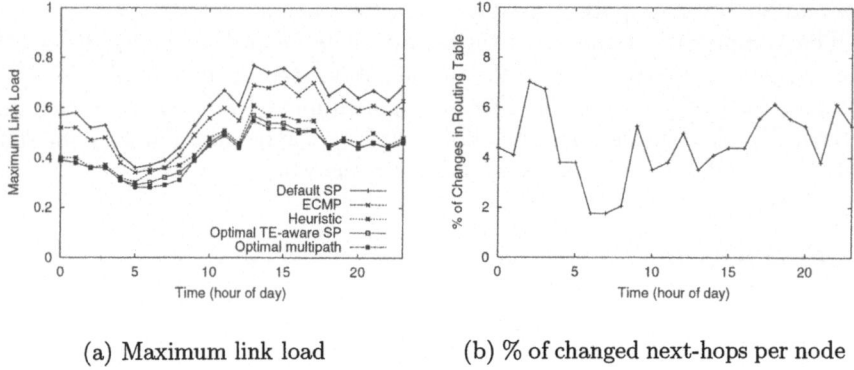

(a) Maximum link load   (b) % of changed next-hops per node

Fig. 3. Maximum link load and the percentage of changed next-hops per node in the sample network for one-day traffic demand matrix.

If the TE-aware shortest paths calculated by the TE server are to be utilized by IP routers, the next-hops for the destination may be changed, if they are different from those of the default shortest paths. Swapping the next-hops for the destination can be completed by configuring the new next-hop and deleting the old next-hop sequentially, while ECMP with two paths is enabled temporarily to prevent traffic disruption. Therefore, during the transition period, packets are distributed to two paths due to ECMP. The number of changed next-hops is considered as the configuration overhead of using the TE-aware shortest paths.

In Fig. 3(b), the percentage of changed next-hops over all the possible routing tables at each node $(= \frac{Next-hop\ Changes}{N*(N-1)} * 100)$ is shown. In this network topology which provides 1.56 equal-cost multiple paths on average, the percentage of next-hop changes over all the possible routing tables at each node is about 4.5%, and nine nodes experience a routing-table change. On average, two changes of next-hops per node are enough to reduce the maximum link load in this experiment.

Static TE-Aware Shortest-Path Routing. In order to avoid the configuration overhead and traffic disruption for every TE period, the static TE-aware shortest path routing algorithm uses the pre-configured routing table. We ran the simulation after fixing the routing table for the representative traffic demand matrix among 24 sample traffic matrices for one day [16], and the corresponding

---

[3] CPLEX 6.5.2 was run on a Linux 2.4.2-2 machine using a Pentium PC running at 1 GHz with 512 MB RAM.

results are shown in Fig. 4(a). In the best case (the first sample among 24 traffic demand tables), the maximum link load does not increase much on average (by 1.7%), compared with dynamic TE-aware shortest-path routing. However, in the worst case (the 7th traffic demand table), it increases on average by 26.5%, which is greater than that of ECMP, but still less than that of default shortest-path routing. Therefore, when the traffic demand matrix is known, the fixed routing table for the best performance case may be used to reduce the maximum link load without any configuration overhead.

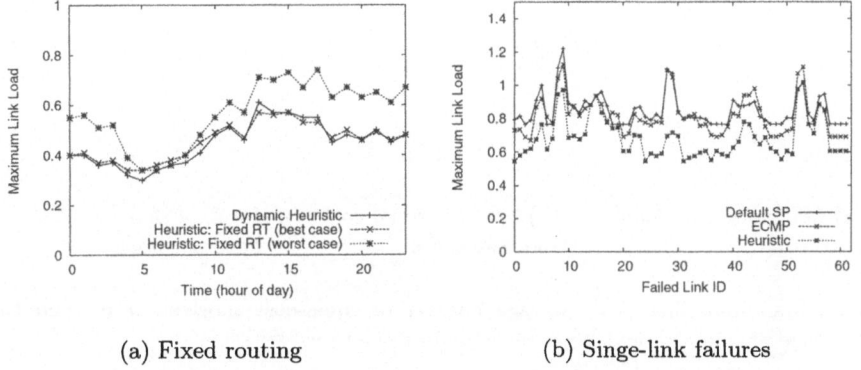

(a) Fixed routing   (b) Singe-link failures

Fig. 4. (a) Maximum link load of the heuristic with fixed routing table for one-day traffic demand matrix and (b) maximum link load under single-link failures.

Single-Link Failures. Noting that single-link failures are the dominant form of failures in the network, the effects on the performance of every link failure are shown in Fig. 4(b). For a single-link failure, the maximum link load over 24-hour traffic demand is plotted. From the simulation results, it is observed that this network needs to be over-provisioned with at least 1.3 times of the current capacity to protect traffic against the worst case failure, namely link 9 failure, if default shortest path routing is used. In some cases, the maximum link load of ECMP is greater than that of default shortest-path routing because ECMP splits traffic equally without considering the congested link globally. Under the critical link 9 failure, the maximum link load of the heuristic is still less than that of default shortest-path routing by 20%.

5.2   Case II: Random Networks

Random network topologies and traffic demand matrices are generated to compare default single shortest-path routing, ECMP, optimal multipath routing, and our proposed heuristic[4]. The network topology consist of nodes with the size of

---

[4] Optimal TE-aware shortest-path routing was omitted because a solution to the ILP for a large network cannot be achieved within reasonable time.

$\{10, 20, 30, 40, 50\}$ and links with the size of $\{30, 60, 90, 120, 150\}$, assuming that a node has three interfaces on average. The nodes are randomly distributed on the Euclidean plane, and the nodes are randomly connected until the connected graph has the given number of links. The capacity of each link is uniformly distributed over the range of [155, 310] Mbps, and the traffic demand is uniformly distributed over [0, 10] Mbps. We carried out 30 runs of experiments for each random graph with the given nodes and links.

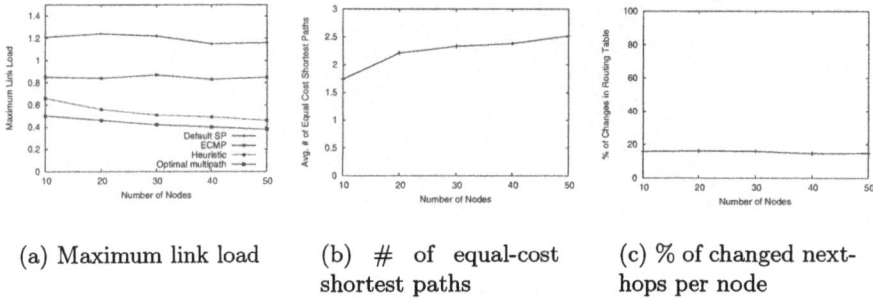

(a) Maximum link load   (b) # of equal-cost shortest paths   (c) % of changed next-hops per node

Fig. 5. Maximum link load, average number of equal-cost shortest paths, and the percentage of changed next-hops per node in random networks.

In Fig. 5(a), it is seen that the proposed heuristic performs better than shortest-path routing and ECMP. When shortest-path routing is used, the maximum link load becomes greater than 1, which means that traffic is lost because of severe congestion. Although the ECMP way of utilizing up to two paths reduces the maximum link load by 30%, the maximum link load is still high (about 0.85). The proposed heuristic maintains the maximum link load at the level of 0.54. Even when compared with optimal multipath routing with the capability of splitting traffic unequally whose result was obtained by CPLEX, its marginal difference is about 20% on average. The performance of our proposed heuristic gets better as the number of equal-cost multiple paths becomes large (Fig. 5(b)).

The difference between next-hops of TE-aware shortest-path routing and those of default shortest-path routing is shown in Fig. 5(c) which describes the percentage of the next-hop changes over all the possible routing tables at each node. Overall, less than 20% of the routing-table entries have been changed in this experiment.

6   Conclusion

In this paper, we presented the mathematical problem formulation for TE-aware shortest-path routing as an ILP. We proposed a heuristic algorithm which can minimize the maximum link load without any serious modification overhead to the Internet routing protocols. TE-aware shortest paths found by the proposed heuristic (from among several available shortest paths between any node pair) were shown to be useful for the global optimization goal of traffic engineering. Traffic engineering can be performed at IP routers by changing the next-hop for the destination with the optimally calculated TE-aware shortest paths. Using our simulation experiments, it was shown that the maximum link load can be minimized when IP routers can change the next-hops for the TE-aware shortest path. Even though the next-hops for the TE-aware shortest paths are fixed for the known traffic demand matrix and the network topology, it was also observed that the maximum link load can be lower than those of default shortest-path routing and ECMP. For the case of single-link failures, TE-aware shortest-path routing was found to perform better than default shortest-path routing and ECMP.

References

1. D. Awduche, A. Chiu, A. Elwalid and X. Xiao, "Overview and Principles of Internet Traffic Engineering," IETF RFC3272, 2002.
2. D. Awduche, J. Malcolm, J. Agogbua, M. O'Dell, and J. McManus, "Requirements for Traffic Engineering Over MPLS," IETF RFC2702, 1999.
3. D. Bertsekas, and R. Gallager, Data Networks, Prentice Hall, 1992.
4. S. Vutukury and J. J. Garcia-Luna-Aceves, "A Traffic Engineering Approach based on Minimum Delay Routing," Proc., IEEE ICCCN, 2000.
5. J. Moy, "OSPF Version 2," IETF RFC2328, 1998.
6. R. Callon, "Use of OSI IS-IS for Routing in TCP/IP and Dual Environments," IETF RFC1195, 1990.
7. Optimized Multipath, http://www.fictitious.org/omp
8. E. Rosen, A. Viswanathan, and R. Callon, "Multiprotocol Label Switching Architecture," IETF RFC3031, 2001.
9. D. Awduche, L. Berger, D. Gan, T. Li, V. Srinivasan, and G. Swallow, "RSVP-TE: Extensions to RSVP for LSP Tunnels," IETF RFC 3209, 2001.
10. A. Sridharan, R. Guerin and C. Diot, "Achieving Near-Optimal Traffic Engineering Solutions for Current OSPF/IS-IS Networks," Proc., IEEE INFOCOM, 2003.
11. Y. Wang, and Z. Wang, "Explicit Routing Algorithms for Internet Traffic Engineering," Proc., IEEE ICCCN, 1999.
12. Z. Wang, Y. Wang, and L. Zhang, "Internet Traffic Engineering without Full Mesh Overlaying," Proc., IEEE INFOCOM, 2001.
13. M. Kodialam and T. V. Lakshman, "Minimum Interference Routing with Applications to MPLS Traffic Engineering," Proc., IEEE INFOCOM, 2000.
14. E. Dinan, D. O. Awduche, and B. Jabbari, "Analytical Framework for Dynamic Traffic Partitioning in MPLS Networks," Proc., IEEE ICC, 2000.
15. B. Fortz and M. Thorup, "Internet Traffic Engineering by Optimizing OSPF Weights," Proc., IEEE INFOCOM, 2000.
16. A. Gencata and B. Mukherjee, "Virtual-Topology Adaptation for WDM Mesh Networks Under Dynamic Traffic," Proc., IEEE INFOCOM, 2002.
17. CPLEX, http://www.ilog.com

Performance of Directional Collision Avoidance in Ad Hoc Networks*

Yu Wang and Jose Joaquin Garcia-Luna-Aceves

Department of Computer Engineering
University of California at Santa Cruz
Santa Cruz, CA 95064, U.S.A.
{ywang,jj}@cse.ucsc.edu

Abstract. This paper analyzes the performance of directional collision avoidance schemes, in which antenna systems are used to direct the transmission and reception of control and data packets in channel access protocols based on four-way collision avoidance handshake. We present an analytical model that considers both directional reception and the possible difference in gain between omni-directional and directional transmissions. Analytical results show that, when the directional collision avoidance scheme in which all transmissions are directional is augmented with directional receiving, one-hop throughput does not decrease due to the increased spatial reuse, even when the number of competing nodes within a region increases as long as the directional transmission/reception beamwidth is narrow. It is also shown that, as expected, the performance of directional collision avoidance schemes degrades when directional transmissions have much higher gain than omni-directional transmissions. However, this degradation is relatively small. Simulations of the IEEE 802.11 protocol and its directional variants validate the results predicted in the analysis; and show that side lobes affect little on throughput if they are reasonably low and the carrier sensing threshold is raised to make nodes less sensitive to channel activities. It is argued that the all-directional scheme in which both transmission and reception of packets is directional is the most attractive collision avoidance approach.

1 Introduction

Collision avoidance is very important in ad hoc networks to combat the "hidden terminal" problem that can degrade throughput significantly. The usual four-way collision avoidance handshake (RTS-CTS-data-ACK), as deployed in the popular IEEE 802.11 MAC protocol [1] and some other protocols, performs much better than the simple carrier sense multiple access (CSMA) protocols in ad hoc networks. However, its performance still degrades rapidly when the number of nodes competing for a shared channel increases moderately due to the much reduced spatial reuse [2]. Recently several MAC schemes that use directional antennas have been proposed to enhance the performance of existing omni-directional collision-avoidance schemes [3,4,5,6,7], which we call *directional collision*

---

* This work was supported in part by the Defense Advanced Research Projects Agency (DARPA) under Grant No. DAAD19-01-C-0026 and by the US Air Force/OSR under Grant No. F49620-00-1-0330.

N. Mitrou et al. (Eds.): NETWORKING 2004, LNCS 3042, pp. 1216–1227, 2004.

avoidance protocols. However, the majority of the performance analyses of directional collision avoidance schemes have been done via simulations [3, 4, 5, 6, 7], and there is little prior work on the analytical modeling of directional collision avoidance protocols. Wang and Garcia-Luna-Aceves [8] extended the model by Takagi and Kleinrock [9] to analyze three directional collision-avoidance schemes based on omni-directional packet reception, together with omni-directional and directional transmissions. The limitation of that work is that it assumes that the gain for omni-directional transmissions is the same as that for directional transmissions, and that all packet reception is omni-directional. In current systems, it is possible to have directional packet reception, and the range of directional transmissions can be longer than the range of omni-directional transmissions. Such limitation motivates our work reported in this paper and the rest of the paper is organized as follows.

Section 2 outlines the directional collision avoidance schemes that we study analytically and by simulation. Section 3 presents the first analytical study of directional collision avoidance in ad hoc networks that considers (a) the effect of directional transmitting and receiving on spatial reuse and collision avoidance, and (b) the effect of the differences in gains between omni-directional and directional transmissions. Our model generalizes the analytical model adopted in [8]. Section 4 presents the results of simulations carried out to validate the results from the analytical model, as well as to analyze the effect of side lobes in directional transmissions. The IEEE 802.11 MAC protocol and its variants that implement directional collision avoidance are investigated. Section 5 summarizes our results and outlines directions for future work.

2   Directional Collision Avoidance Schemes

In omni-directional MAC schemes, all packet transmissions and receptions are omni-directional. The IEEE 802.11 MAC protocol is an example of a protocol based on this scheme. We call this scheme *OTOR* (for "omni-directional transmission and omni-directional reception").

We consider two directional collision-avoidance schemes in our analytical study. In both schemes, RTS, CTS, data packets and ACK are transmitted directionally. When a node is transmitting in one direction, it appears "deaf" to other directions and cannot sense any channel activity at all.

One scheme uses omni-directional receiving mode whenever it is not transmitting; we call such a scheme *DTOR* (for "directional transmission and omni-directional reception"). In the other scheme, which we call *DTDR* (for "directional transmission and directional reception"), a node directs its antenna to the neighbor from which it expects to receive a packet. A node that is not expecting a packet from a neighbor listens to the channel omni-directionally.

It is also possible to establish schemes that combine both omni-directional and directional transmissions and receptions. For example, in one scheme, only CTS packets are transmitted omni-directionally, while all the other types of packets are transmitted directionally. Omni-directional or directional reception can be applied to this approach. However, directional reception is intuitively more attractive, and results in the *MTDR* scheme (for "mixed-directional transmission and directional reception").

To implement any of the directional collision avoidance schemes, nodes have to know the relative locations of their neighbors. In this paper, we assume that broadcast beacons are used by nodes to determine who their neighbors are and their relative locations. These broadcast beacons are transmitted omni-diretionally in the same channel as other packets and nodes communicate directly only with nodes within its omni-directional transmission range to avoid the complexity of maintaining additional neighbors that are only reachable through directional transmissions.

3   Approximate Analysis

In this section, we present the analysis of the *DTOR* and *DTDR* schemes and compare their performance with the existing omni-directional *OTOR* scheme. The *MTDR* scheme can be analyzed using the approach presented here, but its analysis is omitted for brevity. However, it is still compared with the other schemes later in simulations.

3.1   Assumptions

For simplicity, we assume that directional transmissions and receptions have equal beamwidth. Most of the assumptions made in our analysis of directional collision avoidance schemes have been used successfully in the performance evaluation of several MAC protocols [10, 9, 11, 2, 8] to obtain tractable analytical models. Below we just highlight the key assumptions.

We assume a multi-hop network in which nodes are two dimensionally Poisson distributed with density λ, i.e., the probability $p(i, S)$ of finding i nodes in an area of S is given by:

$$p(i, S) = \frac{(\lambda S)^i}{i!} e^{-\lambda S}.$$

Each node has omni-directional and directional transmission ranges of R and R' respectively. We have $R' = \gamma R$, where $\gamma \geq 1$. Suppose N is the average number of nodes within a circular region of radius R; therefore, we have $N = \lambda \pi R^2$. Similarly we have $N' = \lambda \pi R'^2 = \gamma^2 N$.

We use a time-slotted system as an approximation in which each time slot lasts τ and the transmission times of RTS, CTS, data, and ACK packets are normalized with regard to τ and are denoted by l_{rts}, l_{cts}, l_{data}, and l_{ack}, respectively.

We assume that a silent node begins transmission with probability p at each time slot which is a protocol-specific parameter but is slot independent. As advocated in [8], we use a three-state Markov chain to derive the throughput which is largely decided by p.

The node model is shown in Fig. 1(from Fig. 1 in [8]), where *wait* is the state when the node defers for other nodes or backs off, *succeed* is the state when the node can complete a successful four-way handshake with other nodes, and *fail* is the state when the node initiates a handshake that is unsuccessful or cannot be completed due to collisions.

In our analytical modeling, we assume complete signal attenuation outside the main transmission beamwidth. In reality, beamforming antennas can generate side lobes whose interference range can be modeled by a distance parameter d. However, we reason that

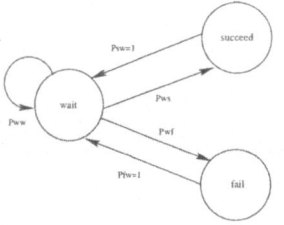

Fig. 1. Markov chain model for a node

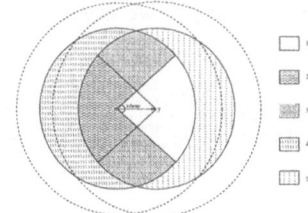

Fig. 2. The *DTOR* scheme

the influence of side lobes on throughput is insignificant if the side lobe level is low enough and the carrier sensing (CS) threshold is raised high enough so that nodes are less sensitive to channel activities. Simulations are used to validate our conjecture and are presented in Section 4.

In addition, in our analysis and simulations, we also assume that a node communicates directly only with other nodes that are within its omni-directional transmission range R, and will not communicate directly with nodes outside R and inside its directional transmission range R'.

3.2 Throughput

The throughput Th of each directional collision avoidance scheme can be calculated by the proportion of time that a node spends transmitting data packets successfully in the average. With some simplifications, all the directional collision avoidance schemes we address in this paper can be analyzed using the same node model of Fig. 1, and differ only in the duration of certain states and the transition probabilities among these states. Let π_s, π_w and π_f denote the steady-state probability of states *succeed*, *wait* and *fail* respectively. From the node model of Fig. 1 we have:

$$Th = \frac{\pi_s \cdot l_{data}}{\pi_w T_w + \pi_s T_s + \pi_f T_f} \tag{1}$$

where T_s, T_f and T_w are the duration of states *succeed*, *fail* and *wait*, respectively.

We first derive those steady-state probabilities, transition probabilities and times spent at different states that are common to *DTOR* and *DTDR*, and then derive the results that are particular to each scheme.

It is easy to shown [8] that for all schemes

$$T_s = l_{rts} + l_{cts} + l_{data} + l_{ack} + 4 \quad \text{and} \quad T_w = 1.$$

Given that a node in the *wait* state listens omni-directionally, the transition probability P_{ww} that node x continues to stay in *wait* state in a slot equals the probability that it does not initiate any transmission and there is no node around it initiating a transmission in the direction towards node x. Because these two events are independent, we have that

$$P_{ww} = (1 - p)e^{-p'N'} \tag{2}$$

where $p' = p\theta/(2\pi)$ and N' is the average number of nodes within the directional transmission range (R') of a node.

According to [8], we have

$$\pi_w = \frac{1}{2 - P_{ww}} = \frac{1}{2 - (1-p)e^{-p'N'}}$$

$$\pi_s = \pi_w P_{ws} = \frac{P_{ws}}{2 - (1-p)e^{-p'N'}}$$

$$\pi_f = 1 - \pi_w - \pi_s.$$

(3)

To derive the transition probability P_{ws} from *wait* to *succeed*, we need to calculate the probability $P_{ws}(r)$ that node x successfully initiates a four-way handshake with node y at a given time slot when the two nodes are at a distance r apart. The configuration is shown in Fig. 2, where θ is the beamwidth of transmissions and receptions. In Fig. 2, solid circles indicate omni-directional transmission ranges of nodes, while dashed circles indicate directional transmission ranges.

The success of the handshake between nodes x and y depends on the nodes for which y is within their omni-directional transmission range and those nodes for which y is within their directional transmission range. Fig. 2 indicates in dashed lines the area around nodes x and y that may contain nodes whose directional transmissions can reach x or y. To simplify our computation of throughput, we assume that there are, in effect, N' nodes around a node's omni-directional transmission range, though no node is assumed to communicate directly with any other node that is only reachable from directional transmissions. In fact, this simplifying assumption avoids the complexity of calculating interference directly from those nodes that are between the solid and dashed circles and instead such interference is taken into account by increasing the number of nodes within omni-directional range from N to N'.

From Fig. 2, we can see that the region around nodes x and y can be divided into five areas. The sizes of the five areas have been calculated in Ref. [8] and are omitted for brevity.

With the above definitions, $P_{ws}(r)$ equals the probability that x transmits in a given time slot, y does not transmit in the same time slot, and none of the nodes in the five areas defined above interfere with the handshake between x and y. Given that transmissions are independent, we have:

$$P_{ws}(r) = p_x \cdot p_y \cdot \prod_{i=1}^{5} p_i$$

where

$$p_x = \text{Prob.}\{x \text{ transmits in the time slot}\} = p,$$

$$p_y = \text{Prob.}\{y \text{ does not transmit in the time slot}\} = 1 - p,$$

$$p_i = \text{Prob.}\{\text{none of the nodes within Area } i \text{ interfere}$$

$$\text{with the handshake between } x \text{ and } y \}.$$

The next two subsections compute the remaining periods and probabilities needed to compute the throughput of the *DTOR* and *DTDR* schemes.

3.3   The DTOR Scheme

The DTOR scheme has been analyzed in Ref. [8] and we just cite the results here.

$$T_f = \frac{1-p}{1-p^{T_2-T_1+1}} \sum_{i=0}^{T_2-T_1} p^i (T_1 + i)$$

$$p_1 = e^{-pS_1 N'}$$

$$p_2 = e^{-p' S_2 N' (2l_{rts})} \cdot e^{-pS_2 N'}$$

$$p_3 = e^{-p' S_3 N' (2l_{rts} + l_{cts} + l_{data} + l_{ack} + 4)}$$

$$p_4 = e^{-p' S_4 N' (2l_{rts} + l_{cts} + l_{ack} + 2)}$$

$$p_5 = e^{-p' S_5 N' (3l_{rts} + l_{data} + 2)}$$

$$P_{ws} = \int_0^1 2r \left(p \cdot (1-p) \cdot p_1 \cdot p_2 \cdot p_3 \cdot p_4 \cdot p_5 \right) dr$$

Please note that we need to substitute N' for N when considering the possible higher gain in directional transmissions.

3.4   The DTDR Scheme

Referring to Fig. 2, it is clear that, for the *DTDR* scheme, only nodes in Areas 1 and 2 can interfere with the handshake between nodes x and y. However, in the *DTDR* scheme, nodes are more vulnerable to the transmissions from other nodes in these areas than they are in the *DTOR* scheme, because nodes receive omni-directionally only if they are in the *wait* state. To take the higher vulnerability into account, we use $l_{rts} + l_{cts} + 2$ as the lower bound for the distribution of T_f which still follows a truncated geometric distribution [8].

On the other hand, because nodes x and y are immune to the transmissions from nodes in Areas 3, 4, and 5, and because concurrent transmissions can go on unobstructed in these areas, we introduce a spatial reuse factor $\gamma(r)$ for this scheme in the calculation of $P_{ws}(r)$. The parameter $\gamma(r)$ is defined to be the number of possible concurrent transmissions in the combined region covered by nodes x and y, i.e., $\gamma(r) = \min(\gamma_1(r), \gamma_2(r))$. In the expression, $\gamma_1(r)$ is the ratio between the total region covered nominally by nodes x and y and the actual region covered by the handshake between nodes x and y. If there is one handshake in Areas 1 and 2, then in theory there can be possibly $\gamma_1(r)$ concurrent handshakes in the total area of Areas 1 to 5. Hence, $\gamma_1(r) = (S_1 + S_2 + S_3 + S_4 + S_5)/(S_1 + S_2)$. On the other hand, $\gamma_2(r)$ can be explained as follows: There are on average $N \cdot (S_3 + S_4 + S_5)$ nodes in the area of Areas 3 to 5 and in theory they can form a maximum of $\gamma_2(r) = N \cdot (S_3 + S_4 + S_5)/2$ pairs of concurrent handshakes. To be conservative, we take the smaller value of $\gamma_1(r)$ and $\gamma_2(r)$ to estimate the spatial reuse benefit enabled by the *DTDR* scheme.

It should be noted that the above is a very crude estimation of the gain in spatial reuse for the *DTDR* scheme. Still, for this scheme, P_{ws} can be adjusted as follows:

$$P_{ws} = \int_0^1 2r\gamma(r) P_{ws}(r) dr.$$

The other quantities needed to derive the throughput are the same as those for the *DTOR* scheme.

3.5   Analytical Results

We compare the performance of the *OTOR*, *DTOR* and *DTDR* schemes. For the *OTOR* scheme, we use the results reported by Wang and Garcia-Luna-Aceves [8], who assume that correct collision avoidance is enforced.

We present the results of two typical configurations for the three schemes; similar results can be readily obtained for other configurations. In these configurations, τ denotes the duration of one slot and RTS, CTS, and ACK packets last 5τ, while a data packet lasts 100τ. In configuration one, both omni-directional and directional transmissions have the same gain and thus we have $N' = N$. In configuration two, directional transmissions have higher gain than omni-directional transmissions and $\gamma = 1.5$. In this case, $N' = 1.5^2 N = 2.25N$.

For each configuration, we derive the maximum achievable throughput when the antenna beamwidth changes from $\theta = 15°(\frac{\pi}{12})$ to $\theta = 120°(2\pi/3)$ in increment of $\theta = 15°(\frac{\pi}{12})$. The results for configurations one and two are shown in Figs. 3 and 4, respectively.

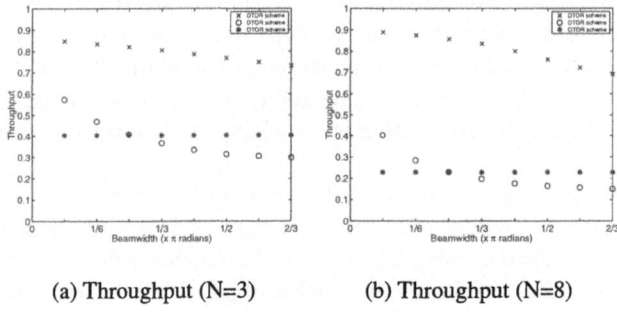

(a) Throughput (N=3)   (b) Throughput (N=8)

Fig. 3. Throughput comparison when omni-directional and directional transmissions have equal gain ($l_{rts} = l_{cts} = l_{ack} = 5\tau, l_{data} = 100\tau, \gamma = 1$).

Figs. 3 and 4 clearly show that the *DTDR* scheme maintains the highest throughput among the three schemes, even with the increase of transmission and receiving beamwidth. Two factors contribute to the superiority of the *DTDR* scheme. One is the significant increase in spatial reuse, because only a small area is covered by the transmissions between two nodes engaged in a handshake according to the analysis. The other is the much reduced interference from those nodes that are not aware of the handshake because of directional receiving.

Even though the *DTDR* scheme does not ensure perfect collision avoidance, the directional reception capability makes the receiving node immune to the transmissions from many other nodes in Areas 3, 4, and 5 after it transmits a CTS packet. Hence, in terms of avoiding collisions, the *DTDR* scheme is as good as or even better than the *OTOR* scheme, which silences all the neighbors around both a sender and a receiver.

Another significant advantage of the *DTDR* scheme is that its performance does not degrade with the increase of competing nodes within a neighborhood when antenna beamwidth is narrow. Instead, it even has a slight increase in throughput. This can be explained as follows: When the number of nodes is small, spatial reuse may be not utilized to its full advantage because some nodes may have to stay idle when all of their neighbors are engaged. This is not due to collision avoidance, but due to the scarity of nodes. Hence, when more nodes are around, the effect of spatial reuse is more conspicuous and one-hop throughput increases accordingly. However, when antenna beamwidth increases, spatial reuse is reduced and throughput still degrades with the increase of N as people usually expect.

The results in Figs. 3 and 4 also show that, as expected, the performance of the *DTOR* and *DTDR* schemes degrades when directional transmissions have higher gain than omni-directional transmissions. This is a direct consequence of the fact that the higher gain of directional transmissions leads to more interference at nodes receiving in omni-directional mode. However, it is clear that the *DTDR* scheme is superior to the *OTOR* scheme in all cases, and the same conclusion can be derived from the results of simulation experiments described in Section 4. This helps to justify our approach of having each node consider as its neighbors those nodes that it hears through their omni-directional beacon transmissions.

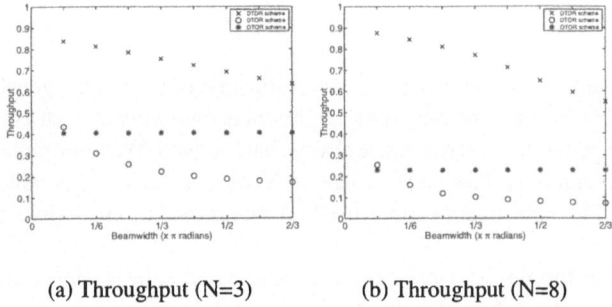

(a) Throughput (N=3)   (b) Throughput (N=8)

Fig. 4. Throughput comparison when the gain of directional transmissions equals one and a half times the gain of omni-directional transmissions ($l_{rts} = l_{cts} = l_{ack} = 5\tau, l_{data} = 100\tau, \gamma = 1.5$).

4   Simulation Results

This section describes the results of computer simulations used to investigate the performance of the popular IEEE 802.11 DFWMAC protocol, which is labeled as *OTOR* in this section, and its variants corresponding to three directional collision avoidance schemes. The directional schemes considered are the *DTOR*, *MTDR* and *DTDR* schemes.

We use GloMoSim 2.0 as the network simulator and implement the directional collision avoidance schemes under the assumption that there is a neighbor protocol that maintains a list of neighbors as well as their locations by means of beacons transmitted omni-directionally and periodically.

In addition to evaluating the performance of the directional collision avoidance schemes with ideal directional antennas, we have also evaluated the performance of these schemes with directional antennas that generate side lobes in directional transmissions.

4.1   Performance Evaluation with Ideal Directional Antennas

Direct sequence spread spectrum (DSSS) parameters are used throughout the simulations, which are shown in Table 1. The raw channel bit rate is 2Mbps. We use a uniform distribution to approximate the Poisson distribution used in our network model. In this network model, we place nodes in concentric circles or rings and focus on the performance of the innermost N nodes in networks with radius of $3R$. The network model has been described in detail in [8] and is omitted here for brevity.

Table 1. IEEE 802.11 protocol configuration parameters

RTS	CTS	data	ACK	DIFS	SIFS
20-byte	14-byte	1460-byte	14-byte	50μsec	10μsec

contention window	slot time	sync. time	prop. delay
31–1023	20μsec	192μsec	1μsec

In our simulation, each node has a constant-bit-rate (CBR) traffic generator with data packet size of 1460 bytes, and one of its neighbors is randomly chosen as the destination for each packet generated. All nodes are always backlogged. We run simulation programs with $N = 3$, 5, and 8, and for each choice of N we use beamwidth values of $\theta = 30°$, $90°$, and $150°$. The same beamwidth is used for directional transmissions and receptions for simplicity.

Fifty random topologies were generated that satisfy the uniform distribution and the average of the throughput for the N nodes in the innermost circle of radius R was computed for each configuration.

The results for the case in which omni-directional and directional transmissions have equal gain are shown in Fig. 5. The results for the case in which directional transmissions have higher gain than omni-directional transmissions and $\gamma = 1.5$ are shown in Fig. 6.

In Figs. 5–6, the vertical lines show the range of throughput achieved by each scheme, i.e., *mean ± standard variance*. The lines are shifted a bit for clarity. The *DTDR* scheme performs the best among all these schemes and its performance does not degrade even for large values of N as predicted in the analysis when antenna beamwidth is narrow. The results also show that the *MTDR* scheme outperforms the *DTOR* scheme, which indicates that the directional receiving capability can boost performance significantly.

Without directional receiving, a scheme with mixed transmissions (*MT* scheme) performs worse than a scheme with only directional transmissions (*DT* scheme). This is

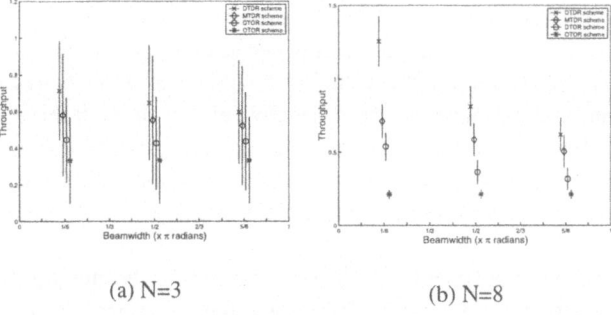

(a) N=3   (b) N=8

Fig. 5. Throughput comparison – equal gain

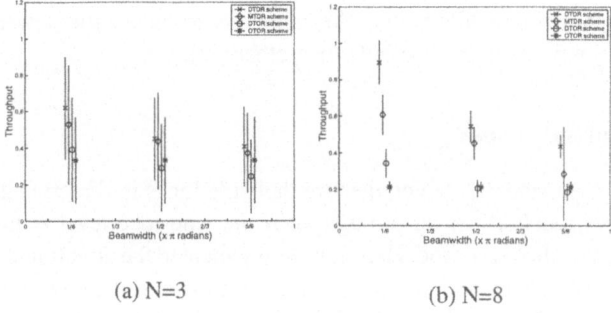

(a) N=3   (b) N=8

Fig. 6. Throughput comparison – higher gain ($\gamma = 1.5$)

because omni-directionally transmitted CTS packets make almost all the nodes around the receiver defer their access to the shared channel or interfere with the ongoing handshake around the nodes that transmit CTS packets. Such conservative collision avoidance can largely nullify the benefits of spatial reuse and an all-directional scheme such as *DT* is shown to perform much better than *MT* when both schemes use only directional transmission capability of antenna systems.

However, when directional receiving is used, even though CTS is transmitted omnidirectionally, the handshakes of those nodes that have turned their receiving to other directions are not affected. Hence, the *MTDR* scheme can outperform the *DTOR* scheme in this case, although its performance is still inferior to the *DTDR* scheme because of the reduced spatial reuse.

It is also clear that, when beamwidth becomes wider, the performance of the *DTDR* scheme degrades faster when *N* becomes larger. This shows that when networks are dense, the performance of a directional scheme is more influenced by the transmission/reception beamwidth.

It should be noted again that, because correct collision avoidance is not enforced in the IEEE 802.11 MAC protocol, collisions of data packets can still occur and hence the

OTOR scheme cannot achieve the same performance predicted in the analysis, which assumes correct collision avoidance. It is for this reason that the *DTOR* scheme performs better than the *OTOR* scheme, even when wider beamwidths are used.

When comparing the results shown in Fig. 5 with those in Fig. 6, it is clear that higher directional transmission gains can have negative effects on both throughput and delay (which is not shown here due to limited space). This is because a node's directional transmissions interfere with more nodes, which translates into a reduction in spatial reuse, given that more nodes spend more time in the wait state after perceiving the channel busy.

We also find that the schemes with narrow receiving beamwidth have far smaller data packet collision ratios than the schemes without directional receiving. Hence, with directional receiving, the adverse effects of hidden terminals are almost completely canceled, leading to much higher throughput. It can also be noted that higher directional transmission gain leads to higher data collision ratio due to the increased interference it introduces. Therefore, directional antenna systems that are able to transmit and receive with narrow beamwidth while having the capability to reduce the power of directional transmissions are much more desirable than other variants.

4.2 Impact of Side Lobes

As discussed in Section 3.1, we conjecture that side lobes in directional transmissions should not have much effect on throughput if the side lobe level is low enough and carrier sensing threshold is raised. Hence, we implemented the directional antenna model in GloMoSim and conducted some simulations. In our simulations, we also set the interference range of side lobes $d = R/\gamma$. Other configurations remain the same as the case when directional transmissions have higher gain, i.e., $R' = \gamma R$. Due to the limited space, we just summarize the results. We find that the presence of side lobes has negligible effect on throughput and it only causes larger variance for access delay when N is small.

5   Conclusion

We have presented an analytical modeling of directional collision avoidance schemes that takes into account directional transmission and reception capabilities and the possibility of having different gains in omni-directional and directional transmissions. The analytical results show that the scheme in which all transmitting and receiving are done directionally can achieve much higher throughput than any other hybrid scheme that combines directional and omni-directional transmissions or receptions. Furthermore, the one-hop throughput of the all-directional scheme does not degrade with the increase of competing nodes within a region, which shows that the all-directional scheme is also much more scalable in dense ad hoc networks. It is also shown that higher directional transmission gain can have negative effects on the performance of directional collision avoidance schemes due to the increased interference range and reduced spatial reuse.

Extensive simulations of the popular IEEE 802.11 MAC protocol and its directional variants validate the analytical results. Simulation results also show that side lobes have little effect on throughput if side lobes are reasonably suppressed and carrier sensing threshold is raised to make nodes less sensitive to channel activities. Augmented with prior work that shows broadcast traffic does not degrade much the performance of directional collision avoidance schemes [12], it is argued that an all-directional scheme is very attractive and practical for ad hoc networks.

In practice, some form of power control to achieve similar gains for both omni-directional and directional transmissions is desirable to take full advantage of the antenna systems. It is also possible to use power control to reduce both interference and energy consumption. Interesting areas of future research include analyzing the impact of eliminating omni-directional transmissions and receptions altogether by means of a directional beaconing mechanism, and comparing the performance of such a scheme against schemes that rely on omni-directional beaconing.

References

1. IEEE, *IEEE Standard for Wireless LAN Medium Access Control (MAC) and Physical Layer (PHY) Specifications*. IEEE Std 802.11-1997, IEEE, 1997.
2. Y. Wang and J. J. Garcia-Luna-Aceves, "Performance of Collision Avoidance Protocols in Single-Channel Ad Hoc Networks," in *Proc. of IEEE ICNP '02*, (Paris, France), Nov. 2002.
3. Y.-B. Ko, V. Shankarkumar, and N. H. Vaidya, "Medium Access Control Protocols Using Directional Antennas in Ad Hoc Networks," in *IEEE INFOCOM 2000*, Mar. 2000.
4. A. Nasipuri, S. Ye, J. You, and R. E. Hiromoto, "A MAC Protocol for Mobile Ad Hoc Networks Using Directional Antennas," in *Proc. of the IEEE WCNC 2000*, (Chicago, IL, USA), Sept. 2000.
5. R. Ramanathan, "On the Performance of Ad Hoc Networks with Beamforming Antennas," in *ACM MobiHoc '01*, (Long Beach, CA, U.S.A.), Oct. 2001.
6. M. Takai, J. Martin, A. Ren, and R. Bagrodia, "Directional Virtual Carrier Sensing for Directional Antennas in Mobile Ad Hoc Networks," in *ACM MobiHoc '02*, (Lausanne, Switzerland), June 2002.
7. R. R. Choudhury, X. Yang, R. Ramanathan, and N. Vaidya, "Medium Access Control in Ad Hoc Networks Using Directional Antennas," in *ACM MobiCom '02*, (Atlanta, GA, USA), Sept. 2002.
8. Y. Wang and J. J. Garcia-Luna-Aceves, "Collision Avoidance in Single-Channel Ad Hoc Networks Using Directional Antennas," in *Proc. of IEEE ICDCS '03*, (Providence, RI, U.S.A.), May 2003.
9. H. Takagi and L. Kleinrock, "Optimal Transmission Range for Randomly Distributed Packet Radio Terminals," *IEEE Trans. on Comm.*, vol. 32, no. 3, pp. 246–57, 1984.
10. F. Cali, M. Conti, and E. Gregori, "Dynamic Tuning of the IEEE 802.11 Protocol to Achieve a Theoretical Throughput Limit," *IEEE/ACM Trans. on Net.*, vol. 8, pp. 785–799, Dec. 2000.
11. L. Wu and P. Varshney, "Performance Analysis of CSMA and BTMA Protocols in Multihop Networks (I). Single Channel Case," *Information Sciences, Elsevier Sciences Inc.*, vol. 120, pp. 159–77, 1999.
12. Y. Wang and J. J. Garcia-Luna-Aceves, "Broadcast Traffic in Ad Hoc Networks with Directional Antennas," in *IEEE Globecom 2003*, (San Francisco, CA, U.S.A.), Dec. 2003.

On the Performance of the European LMDS System

Michel Marot*, Monique Becker**, and Paul-Vincent Marboua***

G.E.T./I.N.T.; U.M.R. 5157 S.A.M.O.V.A.R.
Institut National des Télécommunications
9 rue Charles Fourier
91011 EVRY-FRANCE
Phone : (33) 1-60-76-47-81
Fax : (33) 1-60-76-47-80
{Michel.Marot,Monique.Becker,Paul-Vincent.Marboua}@int-evry.fr

Abstract. This paper investigates the performance of the European LMDS system. The collision rate and the Laplace transform of the access delay to the shared medium, are analytically computed and used to derive numerically the mean of the response time (the access delay to the shared medium). At last, the performance of the LMDS system when the contention mode is used for dynamic reservation request is analytically studied.

Among several results, we prove that the use of the max parameter in the LMDS collision recovery mechanism prevents the system from unstability, whatever the collision rate is. It differs from results derived in other papers for binary exponential backoff algorithms without max parameter. It appears also that stability has to be carefully defined. Several definitions of stability may be chosen and the derived conclusions are then different.

Keywords: Contention access mode, fixed point, exponential backoff, collision rate, analytical study

1 Introduction

Local Multipoint Delivery Systems (LMDS [2]) is designed to be used to implement interactive services in a DVB environment, providing a bi-directional communication path between the user terminal and the service provider. A DVB downstream channel is broadcasting video, audio and data from a base station to clients. LMDS can be used to provide wireless access from the user to the base station, for instance to provide Internet access. Data are sent downstream from the base station to the clients in a FIFO manner, each client looking at the downstram traffic to see whether he is concerned with the received packets

---

* Michel Marot is an Assistant Professor at I.N.T.
** Monique Becker is a Professor at I.N.T.
*** Paul-Vincent Marboua is a Master student at I.N.T.

N. Mitrou et al. (Eds.): NETWORKING 2004, LNCS 3042, pp. 1228–1239, 2004.

or not. The upstream channel is a slotted TDMA multiple access channel. It can be shared in a fixed mode, where each client has a dedicated sequence of slots, in a contention mode, or in a dynamical mode where clients wanting to send data send reservation requests to the base station which temporarily reserves data slots to the user. The reservation requests can be sent in contention mode. Contention mode, which use a collision recovery mechanism similar to the standard Binary Backoff Exponential (BEB) mechanism used in ethernet [8] or wireless LAN [9] standards, and mixed contention reservation modes are difficult to analyse and are the subject of this paper.

For thirty years, there has been a lot of studies about BEB (e.g. [1], [3]). An excellent survey on this area is given in [4]. The authors reference several papers which have proved that BEB is unstable for an infinite node model while it is stable for a finite node model, if the system arrival rate is small enough but unstable if the arrival rate is too large. Of course, when the throughput of each of the sources increases, or for a given throughput when the number of sources increases, the expected number of packets waiting to be transmitted or the expected delay before transmission increases. Of course LMDS is unstable in that way. But for a given number of users and a given throughput of the requests of each of them, in case the throughput of the successfully transmitted packets tends to zero when the requests throughputs is increasing, the network will be said to be unstable and stable otherwise In [4], an interesting new approach is proposed where the collision rate is numerically derived as the solution of a problem of fixed point. Actually, the collision rate depends on the emission rate of the sources but the emission rate also depends on the collision rate: there is an equation to invert. The authors also study the stability of the BEB. They prove that if the collision rate is larger than 0.5 the system is unstable.

In the present paper we study the collision rate as a fixed point, but not only in saturation condition as the authors of [4] do and we propose a method to analytically analyze the variations of the fixed point of the collision rate as a function of the parameters of the system (section 2). We prove that with max and min values bounding backoff exponent, the LMDS system can be stable in some cases even when the collision rate is greater than 0.5. Moreover, we also derive (in section 3) the Laplace transform of the access delay to the share medium which permits us to numerically obtain the moments of the service and response times. At last (section 4), we consider the mixed scenario where data are sent with dynamical reservation mechanism, the dynamical reservation requests being sent through the contention mode. The distribution of the reservation queue on the base station and the total mean response time are analytically derived. We show that there is an optimal value for the amount of contention slots compared to the amount of data slots in this scenario, which is a tradeoff between the collision rate and the amount of the data slots.

2   The Collision Rate

The system is constituted by N clients sending ATM cells to queues implementing the MAC protocol. Each client has such a single queue. Each of these queues sends traffic through the upstream shared channel to the base station. The time is divided in periods of 3ms and each period is divided in $m = 18$ ATM slots. A queue can send at most one cell per period. When it just sends a cell, it waits during three periods for an acknowledgment to know if a collision occured or if the cell has been successfully transmitted. Actually, it is possible that simultaneous transmissions occur in a single slot, which is called a collision. If a collision has occured, the packet may be retransmitted. A counter records the number of collisions encountered by the packet and generates a $backoff\_exponent$ which is included between a min and a max value. The retransmission will be scheduled in a slot whose number is uniformly distributed between 1 and $2^{backoff\_exponent}$. The $backoff\_exponent$ is encreased by one when a new collision happens, but when it reaches its max value. If the counter reaches the maximum number, it remains at this value regardless of the number of subsequent collisions.

For clarity of the description, let us call applications the clients generating traffic, which are modeled as Poisson processes, and sources the queues below the applications and above the shared medium. All applications are assumed to be Poisson with identical rate. Let us define the following notations:

- λ: the rate of each Poisson source
- N: the number of sources
- U: the utilisation factor of the sources
- A_{ck}: number of periods a source waits to receive the acknowledgement.
- min, max: parameters of the LMDS collision resolution algorithm
- m: number of slots per period
- n_r: number of retransmissions for a given cell to be transmitted successfully
- X: random variable representing the time in periods necessary for a source to transmit successfully a given cell (the source service time)
- Y: random variable representing the time in periods between two transmissions, whatever they are successful or not (i.e.: with collision), when the queue of the source is not empty
- p_c: the collision rate
- p_e: effective emission rate of the sources ($p_e \geq \lambda$ because of retransmissions).

$Z \hookrightarrow Uniform\left(\llbracket A; B \rrbracket\right)$ means that the random variable Z is discrete and uniformly distributed between A and B and $\lceil r \rceil$ denotes the upper integer part.

2.1   The Equation of the Collision Rate

In each period slot, the collision rate depends on the emission rate p_e of the sources. It depends also of the number of emitting sources k (see [6] for details):

$$p_c = 1 - \left(1 - \frac{1}{m}p_e\right)^{N-1} \tag{1}$$

Moreover, $p_e = \dfrac{U}{\mathbb{E}\left[Y\right]}$ and, each source being an M/G/1 queue, $U = \rho = \lambda \mathbb{E}\left[X\right]$. We may determine $\mathbb{E}[X]$ and $\mathbb{E}[Y]$.

First approximation: Theoretically, when a collision occurs, the source has to wait for a given number of slots before retransmitting. Because we choose the period as time unit, and because this number of slots is generated according to a discrete uniform distribution between 1 and 2^{min}, 2^i or 2^{max}, we consider that the corresponding time to wait before retransmitting is generated according to a discrete uniform distribution between 1 and $\left\lceil \dfrac{2^{min}}{m} \right\rceil$, 1 and $\left\lceil \dfrac{2^j}{m} \right\rceil$ or 1 and $\left\lceil \dfrac{2^{max}}{m} \right\rceil$ (it is an approximation).

The number of periods necessary to transmit successfully a cell is:

$$X = A_{ck}(1 + n_r) + \sum_{j=1}^{n_r} i_j \tag{2}$$

where,
$$i_j \hookrightarrow Uniform\left(\left[\!\left[1; \left\lceil \frac{2^{min}}{m} \right\rceil\right]\!\right]\right) \ for \ 1 \le i_j \le min$$
$$i_j \hookrightarrow Uniform\left(\left[\!\left[1; \left\lceil \frac{2^j}{m} \right\rceil\right]\!\right]\right) \quad for \ min \le i_j \le max$$
$$i_j \hookrightarrow Uniform\left(\left[\!\left[1; \left\lceil \frac{2^{max}}{m} \right\rceil\right]\!\right]\right) \ for \ max \le i_j$$

If $P(n_r)$ denotes the probability that the number of retransmissions is n_r,

$$\mathbb{E}\left[X\right] = A_{ck} + A_{ck}\mathbb{E}[n_r] + \sum_{n_r=1}^{+\infty} P(n_r)\mathbb{E}\left[\sum_{j=1}^{n_r} i_j \bigg/ n_r\right] \tag{3}$$

Since n_r is geometrically distributed $P(n_r) = p_c^{n_r}(1 - p_c)$, $\mathbb{E}[n_r] = \frac{p_c}{1-p_c}$ and

$$\sum_{n_r=1}^{+\infty} P(n_r)\mathbb{E}\left[\sum_{j=1}^{n_r} i_j \bigg/ n_r\right] = \sum_{n_r=1}^{min}\left[(1-p_c)p_c^{n_r}\sum_{j=1}^{n_r}\frac{1}{2}\left(\left\lceil\frac{2^{min}}{m}\right\rceil + 1\right)\right]$$

$$+ \sum_{n_r=min+1}^{max}(1-p_c)p_c^{n_r}\left[\sum_{j=1}^{min}\frac{1}{2}\left(\left\lceil\frac{2^{min}}{m}\right\rceil + 1\right) + \sum_{j=min+1}^{n_r}\frac{1}{2}\left(\left\lceil\frac{2^j}{m}\right\rceil + 1\right)\right]$$

$$+ \sum_{n_r=max+1}^{+\infty}(1-p_c)p_c^{n_r}\left[\sum_{j=1}^{min}\frac{1}{2}\left(\left\lceil\frac{2^{min}}{m}\right\rceil + 1\right) + \sum_{j=min+1}^{max}\frac{1}{2}\left(\left\lceil\frac{2^j}{m}\right\rceil + 1\right)\right.$$

$$\left. + \sum_{n_r=max+1}^{n_r}\frac{1}{2}\left(\left\lceil\frac{2^{max}}{m}\right\rceil + 1\right)\right] \tag{4}$$

Second approximation: by approximating $\left\lceil \dfrac{2^j}{m} \right\rceil$ by $\dfrac{2^j}{m}$, the expression (4) can be simplified in

$$\mathbb{E}\left[X\right] = \frac{1}{1-p_c}\left[A_{ck} + \frac{1}{2}p_c\left(\frac{2^{min}}{m} + 1\right) + \frac{\left((2p_c)^{min+1} - (2p_c)^{max+1}\right)}{4m(1-2p_c)}\right] \tag{5}$$

Similarly, the expectation of the time between two transmissions of a source when it is not empty is

$$\mathbb{E}\left[Y\right] = \mathbb{E}\left[A_{ck} + i_{n_r}\right] = \sum_{n_r=1}^{+\infty} \mathbb{E}\left[A_{ck} + i_{n_r}/n_r\right] P(n_r) + A_{ck} P(0)$$

$$= \left[A_{ck} + \frac{1}{2}p_c\left(\frac{2^{min}}{m}+1\right) + \frac{\left((2p_c)^{min+1} - (2p_c)^{max+1}\right)}{4m(1-2p_c)}\right] \qquad (6)$$

So, the effective emission probability (probability of a cell emission including retransmission) is:

$$p_e = \lambda \frac{\mathbb{E}[X]}{\mathbb{E}[Y]} = \frac{\lambda}{1-p_c} \qquad (7)$$

And the collision probability becomes:

$$p_c = 1 - \left(1 - \frac{1}{m}\frac{\lambda}{1-p_c}\right)^{N-1} \qquad (8)$$

This result is similar to the one found in [4], except that in our case the sources are not saturated but are fed by Poisson processes. It can be noticed that, contrary to the results found in [4], due to the existence of the max parameter, the infinite sums in (4) and (6) always converge, even when $p_c \geq \frac{1}{2}$. There are no more stability problem. The drawback will be that there can be less sources in our study than in [4].

2.2   The Collision Rate as a Function of λ

We may study the variations of p_c as a function of λ. The collision rate can be found by inverting the equation (8). A possible parametrisation is:

$$\begin{cases} p_c = 1 - t^{N-1} \\ \lambda = mt^{N-1}(1-t) \end{cases} \qquad (9)$$

The variations of p_c and λ are given in Fig. 3 and Fig. 4 (for $m = 18$ and $A_{ck} = 3$).

These curves represent all the solutions of equation 8. The interesting part of this curve is the portion increasing from 0 to $1 - \left(\frac{N-1}{N}\right)^{N-1}$ when λ increases from 0 to $\frac{m}{N}\left(\frac{N-1}{N}\right)^{N-1}$ if it is less than 1 or to 1 otherwise. The other parts are due to the fact that equation (8) is derived with a sequence of implications and not with a sequence of equivalences, so p_c is solution of (8), but all solutions of (8) are not necessarely p_c.

When N is enough large, the maximum value of the collision rate is $1 - \left(\frac{N-1}{N}\right)^{N-1}$ and is obtained for the maximum possible value of the application

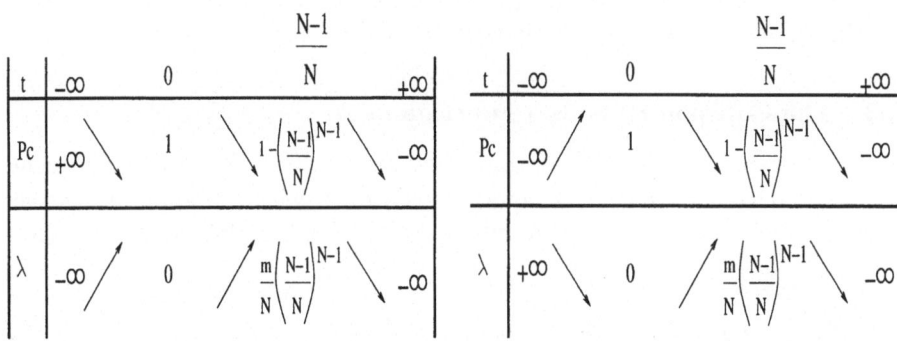

Fig. 1. p_c and λ variations when N is even   **Fig. 2.** p_c and λ variations when N is odd

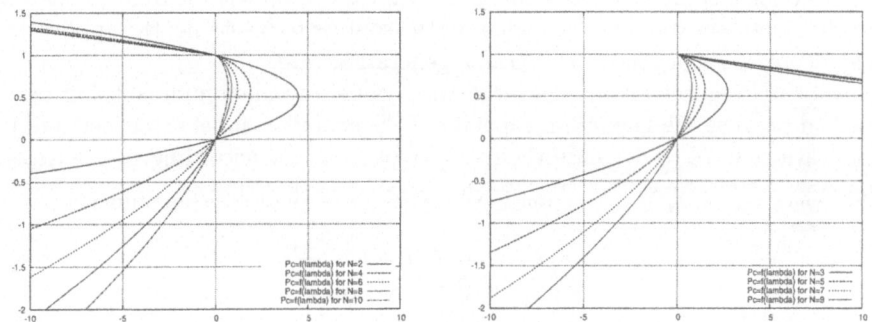

Fig. 3. p_c as a function of λ when N is   **Fig. 4.** p_c as a function of λ when N is
even   odd

rate $\lambda_{max} = \frac{m}{N}\left(\frac{N-1}{N}\right)^{N-1}$. The maximum happens when $N\frac{\lambda_{max}}{1-p_{c_{max}}} = m$, i.e. when the shared medium is saturated (the product of the total throughput by the expected value of the number of transmissions and by the mean service time is one). Due to the fact that a source can send at most one cell per period, this case may only appear when there is a sufficiently large number of sources N. In this case,

$$\lim_{N \to +\infty} p_c = 1 - \frac{1}{e} \tag{10}$$

Contrary to what stands in [4], the maximum value of the collision rate is $1 - \frac{1}{e}$ and thus is greater that 0.5; it is explained by the use of the *max* parameter of the collision recovery algorithm.

p_c, derived by simulation, is plotted in Fig. 6 as a function of λ for different number of sources. Results are obtained with confidence intervals less or equal to 5%. The curves have been truncated at a value of λ corresponding to he maximal possible load of the queues upstream the shared channel. There is an excellent match between simulations and analytical results.

2.3   The Collision Rate as a Function of N

The maximum possible rate per application is $\lambda_{max} = \frac{1}{E[X]}$, otherwise queues are saturated. The study of p_c as a function of N when all applications send traffic at their maximum possible rate is derived from the following expression:

$$p_c = 1 - \left(1 - \frac{1}{m} \frac{1}{A_{ck} + \frac{1}{2}p_c \left(\frac{2min}{m} + 1 \right) + \frac{\left((2p_c)^{min+1} - (2p_c)^{max+1} \right)}{4m(1-2p_c)}} \right)^{N-1} \tag{11}$$

When N increases, p_c increases and so the service time of the sources, that is the time for a source to successfully transmit a cell on the medium to the base station, increases too. For the queues of the sources to be stable, the applications may decrease accordingly their traffic generation rate.

When N reaches an enough large N_{lim} value, the maximum collision rate is reached because the maximum capacity of the shared channel is reached and the applications have to decrease their rate hyperbolically according to the expression $\lambda_{max} = \frac{m}{N} \left(\frac{N-1}{N} \right)^{N-1}$. N_{lim} is the solution of the following equation:

$$\frac{1}{E[X]} = \frac{m}{N} \left(\frac{N-1}{N} \right)^{N-1} \tag{12}$$

with $p_c = 1 - \left(\frac{N-1}{N} \right)^{N-1}$. When N is large, (12) can be solved with a limited expansion. The solution is:

$$N_{lim} = m \left[A_{ck} + \frac{1}{2} \left(1 + \frac{2^{min}}{m} \right) \left(1 - \frac{1}{e} \right) + \frac{\left(2 - \frac{2}{e} \right)^{min+1} - \left(2 - \frac{2}{e} \right)^{max+1}}{4m \left(\frac{2}{e} - 1 \right)} \right.$$

$$+ \frac{1}{A_{ck} + \frac{1}{2} \left(\frac{2^{min}}{m} + 1 \right) \left(1 - \frac{1}{e} \right) + \frac{1}{4m(\frac{2}{e}-1)} \left(\left(2 - \frac{2}{e} \right)^{min+1} - \left(2 - \frac{2}{e} \right)^{max+1} \right)}$$

$$\times \left[\frac{1}{2e} \left(1 + \frac{2^{min}}{m} \right) - \frac{1}{4m \left(\frac{2}{e} - 1 \right)} \times \left[\left(2 - \frac{2}{e} \right)^{min+1} \left(\frac{2(min+1)}{2e-2} \right) \right. \right.$$

$$\left. \left. \left. + \frac{2}{2-e} \right) - \left(2 - \frac{2}{e} \right)^{max+1} \left(\frac{2(max+1)}{2e-2} + \frac{2}{2-e} \right) \right] \right] \tag{13}$$

In figure 5 the maximal value of λ is plotted as a function of N, for different values of max, and the corresponding collision rate. The limit values of p_c ($1 - \left(\frac{N-1}{N} \right)^{N-1}$) and of λ ($\frac{m}{N} \left(\frac{N-1}{N} \right)^{N-1}$) are also plotted. For N around 70, which corresponds to N_{lim}, λ_{max} and $p_{c_{max}}$ reach their limit values. For $N > N_{lim}$, those curves may not be considered but instead their limit values.

It can be noticed that the influence of the max parameter is very marginal.

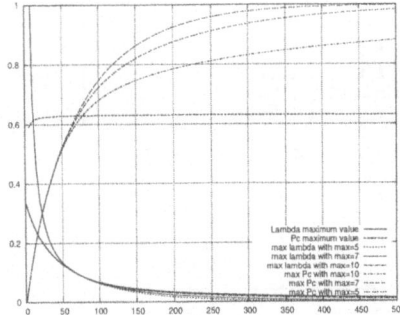
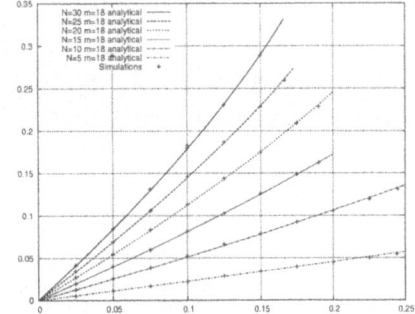

Fig. 5. λ_{max} and p_c as a function of N, for different values of max

Fig. 6. Comparison between simulations and analytical results fo p_c

3   Laplace Transform of the Service and Response Times

In this section, the Laplace transform of the service time is presented. The unit (or the *granularity*) of the service time is the period. It is justified by the fact that first the MAC layer can send the next cell only when the current cell has been successfully received by the base station and second by the fact that in this case the client must wait for the next period before sending the next cell. The Laplace transform of the service time is:

$$
S_X^*(t) = (1 - p_c)\, e^{-A_{ck}t}
\left[
\frac{
1 - \left[p_c e^{-A_{ck}t} \dfrac{e^{-t}}{\left\lceil \frac{2^{min}}{m} \right\rceil} \dfrac{1 - e^{-t\left\lceil \frac{2^{min}}{m} \right\rceil}}{1 - e^{-t}} \right]^{min+1}
}{
1 - \left[p_c e^{-A_{ck}t} \dfrac{e^{-t}}{\left\lceil \frac{2^{min}}{m} \right\rceil} \dfrac{1 - e^{-t\left\lceil \frac{2^{min}}{m} \right\rceil}}{1 - e^{-t}} \right]
}
\right]
$$

$$
+ \left[\frac{e^{-t}}{\left\lceil \frac{2^{min}}{m} \right\rceil} \frac{1 - e^{-t\left\lceil \frac{2^{min}}{m} \right\rceil}}{1 - e^{-t}} \right]^{min}
$$

$$
\sum_{n_r = min+1}^{max} \left[\left(p_c e^{-A_{ck}t} \right)^{n_r} \prod_{j=min+1}^{n_r} \frac{e^{-t}}{\left\lceil \frac{2^j}{m} \right\rceil} \frac{1 - e^{-t\left\lceil \frac{2^j}{m} \right\rceil}}{1 - e^{-t}} \right]
$$

$$
+\left[\frac{e^{-t}}{\left\lceil\frac{2^{min}}{m}\right\rceil}\frac{1-e^{-t\left\lceil\frac{2^{min}}{m}\right\rceil}}{1-e^{-t}}\right]^{min}\prod_{j=min+1}^{max}\left[\frac{e^{-t}}{\left\lceil\frac{2^{j}}{m}\right\rceil}\frac{1-e^{-t\left\lceil\frac{2^{j}}{m}\right\rceil}}{1-e^{-t}}\right]
$$

$$
\times\left(p_{c}e^{-A_{ck}t}\right)^{max+1}\frac{\dfrac{e^{-t}}{\left\lceil\frac{2^{max}}{m}\right\rceil}\dfrac{1-e^{-t\left\lceil\frac{2^{max}}{m}\right\rceil}}{1-e^{-t}}}{1-p_{c}e^{-A_{ck}t}\dfrac{e^{-t}}{\left\lceil\frac{2^{max}}{m}\right\rceil}\dfrac{1-e^{-t\left\lceil\frac{2^{max}}{m}\right\rceil}}{1-e^{-t}}}\right] \tag{14}
$$

The reader can refer to [6] for details on the calculations and for references about Laplace transform inversion. The response time (because the system under study can be modeled as an M/G/1 queue) and the moments of the service time are obtained from the following formulae (where $\rho = \lambda \times \mathbb{E}[X]$):

$$
R = \mathbb{E}[X] + \frac{\lambda\mathbb{E}[X^2]}{2(1-\rho)}, \quad \mathbb{E}[X^k] = (-1)^k\left(\frac{d^k S_X^*(s)}{ds^k}\right)_{s=0} \tag{15}
$$

Since the above Laplace transform has been calculated by using only the first approximation mentionned p.1231, and not the second one, by comparing $\mathbb{E}[X]$ obtained numerically from the Laplace transform with the expression (5), we can validate the second approximation mentionned p.1231.

We can numerically compute as a function of p_c the response time by using the derivative (15), or as a function of λ by using also the parametrisation given in (9). In fig. 8, the response time is plotted as a function of λ, for $m = 18$, $A_{ck} = 3$ and different values for min, max and N.

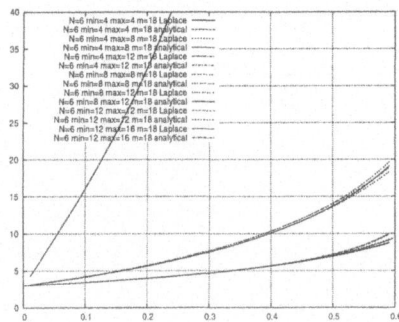
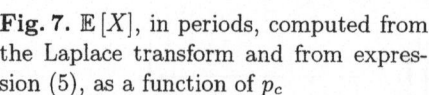
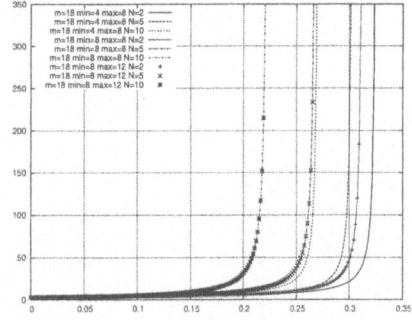

Fig. 7. $\mathbb{E}[X]$, in periods, computed from the Laplace transform and from expression (5), as a function of p_c

Fig. 8. R, in periods, as a function of λ

4   Performance of the Dynamical Reservation Mechanism

We assume now that the contention mode is used to transmit dynamical reservation requests. The time is still divided in periods, each period being divided in slots, but there are m slots per period dedicated to reservation requests transmissions and m' slots dedicated to data transmisson. When a user wants to send data, he sends one or several *messages* containing an appropriate number of *requests* of m' data slots by using the contention mechanim. The base station receives the requests, it inserts them in a FIFO queue with constant service time, m' slots for a request of m' slots, and allocates the required number of slots. The answer to the request is sent to the user who can then send his data by using the allocated data slots. Such a system can be modeled by fig. 9.

During one period, messages are received during the first m slots and they are served during the last m' slots. Let q_n be the length of the reservation queue in *requests* of m' slots during the n^{st} period, v_n be the number of arrivals of *requests* during the n^{st} period, Ξ_i be the number of *requests* in the i^{st} message. We assume that all Ξ_i are independant and identically distributed (i.i.d.) with a geometric distribution with parameter ξ (denoted in the following Geo(ξ)). Let $Q(z)$ be the z-transfom of q_n and $V(z)$ the one of v_n. We assume that the departure process of *messages* from the users are Poisson with rate λ. It is a reasonable hypothesis since studies have proved that arrivals of user sessions are Poisson (cf. [7] and [5]).

Moreover, because of the collision recovery algorithm, the sources are not independent, but the *message* arrival process is a binomial process with parameters $Min(m, N)$ and $\zeta = \frac{N\lambda}{Min(m,N)}$, where $Min(m, N)$ denotes the smallest integer from m and N. Actually, at most one arrival can occur per contention slot.

The stationnary limits of q_n and v_n being $\bar{q}$ and $\bar{v}$, and since Ξ_i are independant and Geo(ξ), it can be schown that (see [6]):

$$Q(z) = \frac{V(z)\,P(\bar{q}=0)\,(z-1)}{z - V(z)}, V(z) = \left(\frac{(1-\xi)\zeta}{1-\xi z} + 1 - \zeta\right)^{Min(N,m)} \qquad (16)$$

By writing $Q(z)$ under the form $\sum_{i=0}^{+\infty} P(\bar{q}=i)z^i$ and we obtain:

$$\begin{cases} P(\bar{q}=0) & = 1 - \frac{\xi}{1-\xi}\zeta Min(N,m) \\ P(\bar{q}=1) & = \left(1 - \frac{\xi}{1-\xi}\zeta Min(N,m)\right)(a_1 - 1) \\ \forall k \geq 2, P(\bar{q}=k) & = \left(1 - \frac{\xi}{1-\xi}\zeta Min(N,m)\right)(a_k - a_{k-1}) \end{cases} \qquad (17)$$

where, for all $k \geq 1$, a_k is (C_n^p denotes the combination of order p from n elements):

$$\frac{1}{(1-\zeta)^{Min(N,m)k}} + \sum_{n=1}^{+\infty}\sum_{l=0}^{k-1} [(1-\xi)\zeta]^n \, \xi^l C_{l+n-1}^{n-1} \frac{(-1)^n C_{n+Min(N,m)(k-l)-1}^{Min(N,m)(k-l)-1}}{(1-\zeta)^{n+Min(N,m)(k-l)}} \qquad (18)$$

The mean length L of the reservation queue, by differentiating $Q(z)$ and making z going to 1, and the response time R' of the reservation queue (Little's law: $R' = L/\Lambda$) are:

$$L = \frac{1}{P(\overline{q}=0)}\left[1 - P(\overline{q}=0) + Min(N,m)\frac{\zeta\xi^2(2-\xi)}{1-\xi}\right], R' = \frac{L(1-\xi)}{\lambda N\xi} \quad (19)$$

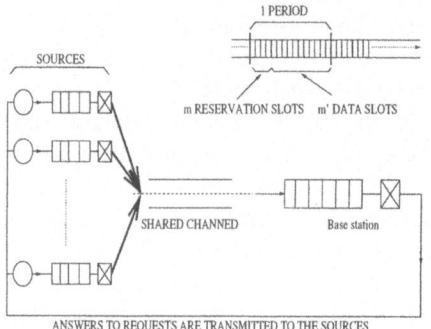

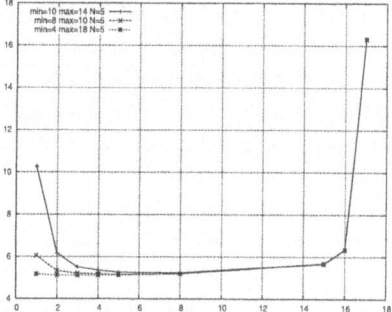

Fig. 9. The reservation model and the way the time is organised

Fig. 10. $R + R'$ as a function of m at a fixed load

The total response time is $R + R'$. For a given load, that is when λ is fixed and when $\lambda N \frac{\xi}{1-\xi} m' = Ct$ (where Ct is a constant value), and when m increases but $m + m'$ remains constant (i.e.: when the ratio of slots dedicated to send reservation messages over the number of slots dedicated for data transport increases with $m + m'$ constant), the number of collisions decreases but the number of requests arriving on the reservation queue is more important compared to the amount of data slots. So, the response time R decreases but R' increases: there is a tradeoff to find. Figure 10 presents the total response time as a function of m, for different values for min and max, and with $N = 5$, $\frac{\xi}{1-\xi} m' = 100$ and $\lambda \simeq 0.0016$ which represents an average load of about 0.8.

5 Conclusion

In this paper, the performance of the European LMDS system is investigated. The characteristic of this system is that it can combine an aloha-kind access mode with a dynamical reservation one. This makes the performance of the system a non-linear function of the parameters of the system. We propose a method to derive the value of the collision rate. The Laplace transform of the service time is also derived. Its inversion remains for future work, but it can already be used to compute numerically the response and service times. At last, the response time of the system when the contention mode is used to send messages for data slot reservation is calculated. The distribution of the length of the reservation queue is also analytically computed.

The stability has to be precisely defined. LMDS networks are of course unstable if we consider the expected delay before transmission or the number of messages waiting in the queue. Those two performance criteria tend to infinity when the number of sources increases. But, for a given number of users and a given throughput of the requests of each of them it is interesting to determine the throughput of successfully transmitted packets. Because of *max* parameter, it appears that when the load increases the successfully transmitted flow do not tend to zero and that is why we say that because of *max* parameter LMDS networks are stable.

In the future work, the performance of TCP on such a system may be investigated. It seems to be difficult: multi-layer models could be profitably used.

Acknowledgement. This work has been driven within the ERASME project. ERASME is a project funded by the French Réseau National de la Recherche en Télécommunication. The goal of ERASME is to set up an LMDS wireless loop in Limoges (France).

The authors would also like to thank Pr. G. Hébuterne for fruitfull discussions and remarks.

References

1. D.J. Aldous, *Ultimate instability of exponential back-off protocol for acknowledgement-based transmission control of random access communication channels*, IEEE Trans. Information Theory, vol. 33, no. 2, 1987.
2. ETSI European Standard (Telecommunications series), *Digital Video Broadcasting(DVB); Interaction Channel for Local Multi-point Distribution Systems (LMDS)* ETSI EN 301 199 V1.2.1 (1999-06).
3. F.P. Kelly, I.M. MacPhee, *The number of packets transmitted by collision detect random access schemes*, Annals of Probability, vol. 15 pp. 1557-1568, 1987.
4. B.-J. Kwak, N.O. Song, L. E. Miller, *Analysis of the Stability and Performance of Exponential Backoff*. J. Research of NIST, vol. 108, July-August 2003.
5. Z. Liu, N. Niclausse, C. Jalpa-Villanueva, *WAGON: A Web Server Benchmarking Tool*. Poster proceedings of the 8th World Wide Web Conference, Toronto, Canada, May 1999.
6. M. Marot, M. Becker, P.-V. Marboua, *On the performance of the European LMDS system*, Collection des rapports de recherche de l'Institut National des Télécommunications, rapport 03 013 RST, 21 octobre 2003.
7. M. Marot, M. Becker, P. Vincent, *Modelling a WEB user, application to performance comparisons of UDP versus TCP for WEB traffic transport*, 8th International Conference on Telecommunication Systems, 9-12 March 2000, Nashville, Tennessee, USA.
8. R. M. Metcalfe, D. R. Boggs, *Ethernet: Distributed packet switching for local computer networks*, Communication of the ACM, vol. 19, no. 7, July 1976.
9. *P802.11, IEEE standard for wireless lan medium access control (MAC) and physical layer (PHY) specifications*, november 1997.

Computing Call Dropping and Waiting Probabilities in LEO Satellite Systems for Voice Communications

Sebastià Galmés and Ramon Puigjaner

Dept. de Ciències Matemàtiques i Informàtica, Universitat de les Illes Balears
Cra. de Valldemossa, km. 7.5, 07122 Palma, Illes Balears, Spain
{dmisgo0, putxi}@uib.es

Abstract. In this paper we propose and analyze a call management architecture for a cell in a satellite cellular network, where handoff arrivals are prioritized by using a number of guard channels and only new calls are temporarily buffered if necessary. The idea behind this is to avoid any extra delay during handoff procedures, in addition to the intrinsic round trip propagation time, since it might cause severe user dissatisfaction. The analysis is based on a Markov chain model of a single spot (cell) and yields exact expressions for the dropping and waiting probabilities of handoff and new arrivals respectively. Then, we derive an efficient algorithm to evaluate these probabilities, which is based on the well-known Erlang-B formula and its recursion. Finally, we use the expressions for the waiting and dropping probabilities to formulate and solve an optimization problem in the context of network planning. Numerical results are provided.

1 Introduction

Within the context of cellular wireless networks, several prioritized handoff procedures have been proposed in the literature, usually in one or both of the two following ways: by reserving a number of channels (called guard channels) for handoff connections, or by allocating some buffer space to them. Examples can be found in [1-6].

One way to support voice communications with mobile and personal terminals is by means of a cellular wireless network implemented as a constellation of LEO (Low Earth Orbit) satellites [7-11]. In essence, these systems are similar to the terrestrial ones, but they have to cope with the specific problems that arise when base stations are orbiting at a non-fixed apparent position around the earth. Particularly, there is also a need to manage the incoming handoff traffic into a cell, caused in this case either by the user mobility (Earth Fixed Coverage – EFC - systems) or by the satellite mobility (Satellite Fixed Coverage – SFC - systems).

More specifically, with regards to handoff management, there are some interesting suggestions in the literature. For instance, in [12-13] a buffer with a time deadline is used to store handoff connections that find all channels busy. Other works assume that a number of guard channels are reserved for handoff attempts [14]. In general, most of works rely on the combination of guard channels and buffer space as a way of prioritizing handoff connections. From our point of view, the presence of a buffer to temporally store handoff calls in a LEO system, adds an extra delay to the intrinsic

N. Mitrou et al. (Eds.): NETWORKING 2004, LNCS 3042, pp. 1240–1251, 2004.

round trip propagation time that users may not tolerate. Thus, to decrease the time consumption incurred by handoff procedures, we propose a model without buffering for this type of traffic, but with the necessary increment on the number of guard and/or common channels to keep the desired dropping probability. At the same time, since new calls are more tolerant to delays, we propose the use of a buffer exclusively for them, which in turn contributes to increase the total traffic carried by the cell.

The paper is organized as follows. In Section 2, we propose and analyze a Markov model for a single spot beam (cell) of a LEO satellite. In Section 3, we obtain exact analytical expressions for the dropping probability of handoff calls and the waiting probability of new calls. In Section 4, we discuss some computational aspects with regards to the formulas obtained in the previous section, and we provide an efficient algorithm to evaluate them. In Section 5, we test the proposed algorithm and obtain some numerical results. In Section 6, we propose a method to determine the feasible region of solutions for the common and guard channels, when the total number of channels is fixed by the satellite frequency plan, and certain bounds on the dropping and waiting probabilities are imposed. Finally, in Section 7, we draw the main conclusions and suggestions for further research.

2   The Markov Model

We consider the performance model of a single cell in a satellite cellular system. There is a Poisson arrival stream of new calls at the rate λ_1 and a Poisson arrival stream of handoff calls at the rate λ_2. Handoff calls may be caused either by the user mobility (EFC coverage) or by the satellite mobility (SFC coverage). There is a total amount of N channels (servers), where each one can hold a voice communication. An ongoing call (new or handoff) leaves a channel when its service is completed (call holding time) at a rate μ_1, or when it is handed off, at a rate μ_2. In order to prioritize handoff arrivals, a number of g guard channels are reserved for them. The rest of channels ($c=N-g$), called common or shared channels, are available to both arrival classes. This means that a new call is accepted only if $g+1$ or more channels are in idle state. Otherwise, the new call is queued on a FCFS buffer. We assume that the buffer has always space to allocate new calls, so from the mathematical point of view we model it as of infinite capacity. On the other hand, as we said before, there is no buffer space for handoff arrivals, in order to avoid extra delays.

Within the context of LEO satellite systems, the proposed model is quite general, since it can be adapted to represent EFC or SFC mechanisms, FCA or DCA schemes and both beam-to-beam and satellite-to-satellite handoffs.

The model so far described was already analyzed in other contexts in [15] and some references therein, but following different formulations and objectives. Particularly, in [15] the model was motivated by its possible application to cellular radio systems, as a way of increasing the total traffic supported by a cell while improving the perceived quality of service by users. This paper focused on the application of a novel methodology to obtain closed-form expressions for the steady-state probabilities. In the present paper, we re-formulate that analysis in the context of LEO satellite systems, by introducing a significant system parameter and by referring it to the well-known Erlang-B formula. The new formulation is simpler and facilitates the treatment of some multiobjective optimization problems that arise when planning the network.

According to the definitions given before, the global arrival rate is $\lambda = \lambda_1 + \lambda_2$. Since any arrival (new or handoff) may leave the cell as a terminated call or after a handoff procedure, the total service rate is $\mu = \mu_1 + \mu_2$.

Let b be the number of busy channels and $S(t)$ the state of the system at time t. From the specification of the model, we propose the following state description:

$$S(t) = \begin{cases} 0,1,\ldots,c-1\,, b < c \\ \\ \{(c+i,j): i = 0,\ldots,g \; ; \; j = 0,\ldots,\infty\}, b \geq c \end{cases} \tag{1}$$

Here, variables i and j denote respectively the number of busy guard channels and the number of new calls present in the queue. The state transition diagram for the case $b < c$ is represented in Figure 1.

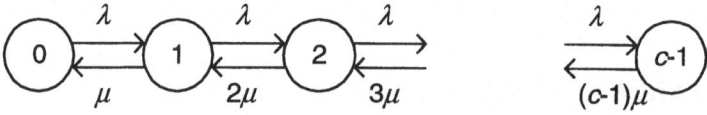

Fig. 1. State transition diagram for $b < c$

On the other hand, if $b \geq c$ the diagram takes the form of a meshed structure, as it is shown in Figure 2. Note that state $S(t) = c\text{-}1$ has been depicted again, since it links both the linear and meshed structures of the global state transition diagram. Since it in fact corresponds to an irreducible homogeneous continuous time Markov chain, an ergodic (stable) steady-state solution may exist. Then, if we assume that stability conditions are fulfilled, we can easily obtain the following solution for the steady-state probabilities p_k associated with the linear diagram, where $A = \lambda/\mu$ is the total traffic offered to the cell (in Erlangs):

$$p_k = p_0 \cdot \frac{A^k}{k!}, k = 1,\ldots,c-1 \tag{2}$$

On the other hand, let $p_{c+i,j}$ be the steady-state probability of state $(c+i,j)$, with $i = 0$, ..., g and $j = 0, \ldots, \infty$. Let also define $A_1 = \lambda_1/\mu, A_2 = \lambda_2/\mu$, so that $A = A_1 + A_2$, and the following aggregate probabilities on a per-column basis:

$$q_{c+i} = \sum_{j=0}^{\infty} p_{c+i,j} \, , i = 0,\ldots,g \tag{3}$$

Then, by following standard procedures based on applying balance equations to state aggregations [16], we can obtain the following result (see [15] for a more detailed derivation with another notation):

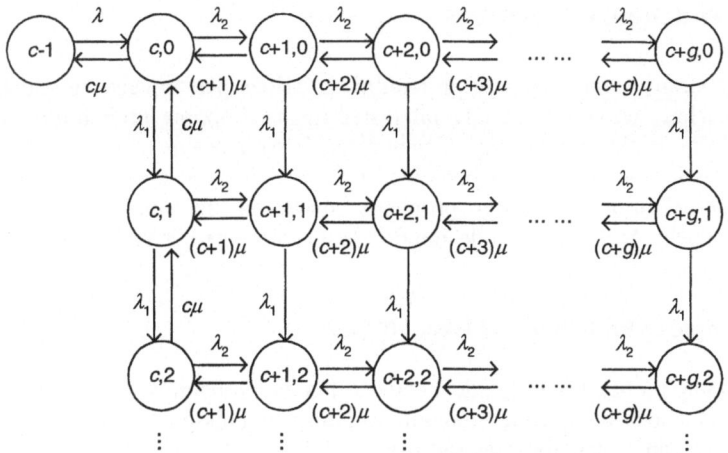

Fig. 2. State transition diagram for $b \geq c$

$$q_{c+i} = \frac{A_2^i}{(c+i)!} \frac{A^c}{1 - \frac{A_1 H}{c}} p_0 \, , i = 0,...,g \tag{4}$$

where

$$H = \sum_{i=0}^{g} A_2^i \frac{c!}{(c+i)!} \tag{5}$$

Here, $H \geq 1$ is an auxiliary parameter that it is directly proportional to the offered handoff traffic (A_2) and the number of guard channels (g). In fact, $H = 1$ only if $A_2 = 0$ and/or $g = 0$. Also, H increases as the number of common channels decreases. Therefore, since all contributions to an increment on the value of H are always associated with an increment of the buffer occupancy, we call H the *congestion parameter*. On the other hand, the denominator in expression (4) reveals the unique stability condition for the system, which imposes an upper bound to H:

$$\frac{A_1 H}{c} < 1 \tag{6}$$

Finally, by applying the normalizing condition to all probabilities, we obtain p_0:

$$\sum_{k=0}^{c-1} p_k + \sum_{i=0}^{g} q_{c+i} = 1 \Rightarrow p_0 = \left(\sum_{k=0}^{c-1} \frac{A^k}{k!} + \frac{A^c}{c!} \cdot \frac{H}{1 - \frac{A_1 H}{c}} \right)^{-1} \tag{7}$$

3   Performance Measures

From the steady-state aggregate probabilities, various performance measures can be obtained. Here, we are especially interested in the dropping probability of handoff calls and the waiting probability of new calls. These measures are directly related to the quality or grade of service perceived by users, and can be used as the basis for several optimization and design problems concerning the number of common and guard channels. An example of design problem is presented in Section 7.

3.1   Dropping Probability of Handoff Calls

The dropping probability of handoff calls (P_d) is the probability that a handoff arrival finds all channels busy. Since handoff arrivals are Poisson, the PASTA argument holds and P_d can be calculated as follows:

$$P_d = P_d(c,g) = q_{c+g} = \frac{\dfrac{A^c \cdot A_2^g}{(c+g)!} \cdot \dfrac{1}{1 - \dfrac{A_1 \cdot H}{c}}}{\displaystyle\sum_{k=0}^{c-1} \dfrac{A^k}{k!} + \dfrac{A^c}{c!} \cdot \dfrac{H}{1 - \dfrac{A_1 \cdot H}{c}}} \tag{8}$$

This expression becomes simpler when there are no guard channels in the system, in which case the congestion parameter is 1, and even simpler when additionally the handoff traffic is zero $(A_1 = A)$. In the latter case, the resulting model is nothing else but a pure M/M/c queue, and therefore the dropping probability becomes the Erlang-C formula.

3.2   Waiting Probability of New Calls

The waiting probability of new calls (P_w) is the probability that an arriving new call cannot capture a channel immediately and has to wait in the queue. Obviously, this occurs whenever the number of busy channels is not less than the number of common channels:

$$P_w = P_w(c,g) = \sum_{i=0}^{g} q_{c+i} = \frac{\dfrac{A^c}{c!} \cdot \dfrac{H}{1 - \dfrac{A_1 \cdot H}{c}}}{\displaystyle\sum_{k=0}^{c-1} \dfrac{A^k}{k!} + \dfrac{A^c}{c!} \cdot \dfrac{H}{1 - \dfrac{A_1 \cdot H}{c}}} \tag{9}$$

Again, we can easily obtain particular simplifications of this result when there are no guard channels in the system or handoff traffic does not exist. For instance, if there are no guard channels, obviously the waiting and blocking probabilities coincide. If additionally the handoff traffic is zero, we again obtain the Erlang-C formula. From (8) and (9) we can derive the following relation between both probabilities:

$$P_w(c,g) = P_d(c,g) \frac{H(c+g)!}{A_2^g \cdot c!} \tag{10}$$

4 Computational Aspects

Usually the total number of voice channels available in the satellite is relatively large, leading to numerical difficulties in the evaluation of the performance measures. Thus, it is necessary to introduce recursive procedures in order to avoid overflow problems caused by the direct computation of large powers and factorials. Since both probabilities are related through expression (10), it is enough to find a recursion for one of them. Particularly, expression (9) for the waiting probability is simpler, and in fact it can be easily rewritten in terms of the standard Erlang-B formula as follows:

$$P_w(c,g) = \frac{B(c,A)}{\left(\dfrac{1}{H} - \dfrac{A_1}{c}\right) + \left[1 - \left(\dfrac{1}{H} - \dfrac{A_1}{c}\right)\right]B(c,A)} \tag{11}$$

Then, the waiting probability can be calculated by using the well known recursion for the Erlang-B formula, traditionally used in the context of wired telephone networks [17]:

$$B(c,A) = \frac{\dfrac{A}{c} B(c-1,A)}{1 + \dfrac{A}{c} B(c-1,A)} \, ; \, B(0,A) = 1.0 \tag{12}$$

On the other hand, expression (5) for the congestion parameter can be written in the following way for computational purposes:

$$H = \sum_{i=g}^{0} h_i \, , \, h_i = h_{i-1} \frac{A_2}{c+i} \, , i = 1,\ldots,g \, , h_0 = 1.0 \tag{13}$$

The idea behind (13) is to evaluate the congestion parameter by starting with the least significant term so as to reduce round-off errors.

Once the Erlang-B formula and the congestion parameter have been evaluated separately, the waiting probability can be easily derived from (11) and, next, the dropping probability through (10) following this practical implementation:

$$P_d(c,g) = \frac{P_w(c,g)}{H} \prod_{i=1}^{g} \frac{A_2}{c+i} \tag{14}$$

The following algorithm summarizes the proposed methodology:

```
Algorithm 1:
Step 0.   Initialization phase:
          Set A₁, A₂, c, g
          A:=A₁+A₂
Step 1.   Compute partial results:
          Evaluate the Erlang-B formula (E_B):
            E_B:=1.0
            for i:=1 to c do
              E_B:=((A/i)*E_B)/(1+(A/i)*E_B)
          Evaluate the congestion parameter (H):
            m[0]:=1.0
            for i:=1 to g do m[i]:=m[i-1]*A₂/(c+i)
            H:=0
            for i:=g down to 0 do H:=H+m[i]
Step 2.   Check for stability:
          if A₁*H/c>1 then STOP: "system unstable"
          else go to Step 3.
Step 3.   Compute performance parameters:
          Waiting probability (Pw):
          Pw:=E_B/((1/H-A₁/c)+(1-(1/H-A₁/c))*E_B)
          Dropping probability (Pd):
          Pd:=Pw/H
          for i:=1 to g do Pd:=Pd*A₂/(c+i)
```

5   Numerical Results

In this section, we provide the results obtained from applying the proposed algorithm to different sets of input data. Particularly, we varied the number of shared channels (c) between 100 and 120, and the number of guard channels (g) between 0 and 10. For the remaining parameters, we assumed with no loss of generality $A_1 = 15$ and $A_2 = 90$.

As an intermediate result, Figure 3 shows the evolution of the congestion parameter (H) as a function of g, for different values of c. For reasons of stability, the congestion parameter cannot achieve arbitrarily high values; in fact, condition (6) imposes the following upper bound to it:

$$H < \frac{c}{A_1} \tag{15}$$

Figure 4 plot the evolution of the waiting probability in terms of the number of common channels, with the number of guard channels as a parameter. As it is shown, the waiting probability decreases as the number of common channels increases, since more channels are available for new calls. On the other hand, an improvement of the service offered to handoff traffic, by increasing the amount of guard channels, causes more congestion to new calls. This is because more handoff calls are accepted in the system and it takes more time until all busy guard channels become idle.

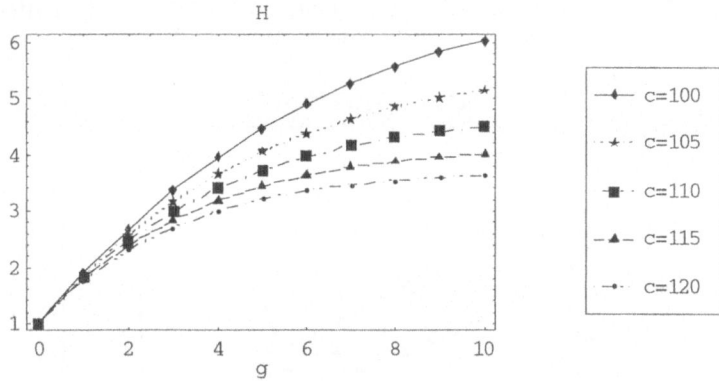

Fig. 3. The congestion parameter as a function of the number of guard channels, for different values of the number of common channels (the offered traffics are $A_1=15$ and $A_2=90$)

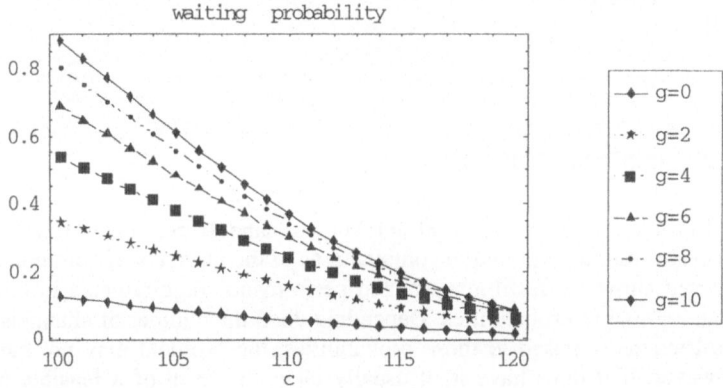

Fig. 4. The waiting probability as a function of both the common and guard channels ($A_1=15$ and $A_2=90$)

With regards to Figure 5, we see that an increment on the number of any type of channels always causes a benefit to handoff traffic.

Also, we have conducted some experiments to compare the buffered case studied in the present paper with the unbuffered case analyzed in [2]. In general, we have observed that for given values of the number of common and guard channels, the unbuffered case always behaves better than the buffered one with regards to dropping probability. This is because more new calls are accepted into the system when a buffer is present, in detriment of the handoff traffic (take into account that both traffic classes compete for common channels). Thus, in order to keep the quality of service offered to handoff traffic, the buffered configuration requires a larger number of guard and/or common channels. However, when this requirement is fulfilled, the buffered configuration supports a larger value of total carried traffic than the unbuffered one.

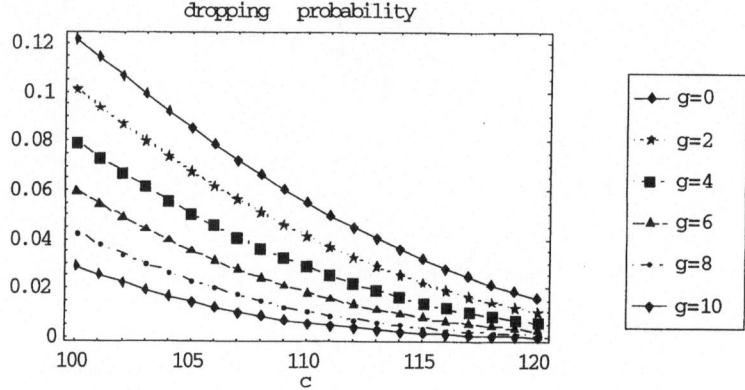

Fig. 5. The dropping probability as a function of both the common and guard channels (A_1=15 and A_2=90)

6 Optimization Problem

A typical problem is when the total number of channels per cell is fixed by the operator, and we want to explore the combinations of the numbers of common and guard channels that allow for fulfilling prefixed performance requirements in terms of the waiting and dropping probabilities. Depending upon the number of channels available or the restrictions imposed to those probabilities, the problem may not have a solution. However, if it does have it, it usually takes the form of a feasible region expressed as an interval [cmin, cmax] for the number of common channels, or as [gmin, gmax] for the number of guard channels (recall that $N = c + g$ is now fixed).

One way of facing this problem is by determining the feasible region for one performance measure, and then intersecting it with the feasible region of the other one. Since the total amount of channels is prefixed by the experiment, the numbers of both common and guard channels will follow opposite evolutions. Then, under this hypothesis, the behavior of the waiting probability becomes easier to predict than that of the dropping probability, as we can deduce from Figures 4 and 5. More formally, we can check the monotonic properties of the waiting probability by reformulating expression (9) as follows:

$$P_w(c,g) = \frac{1}{1 + \left(\dfrac{1}{H} - \dfrac{A_1}{c}\right)\left[\dfrac{1}{B(c,A)} - 1\right]} \tag{16}$$

Since both the congestion parameter and the Erlang-B formula decrease as the number of common channels increases, the waiting probability also decreases. On the other hand, if the number of guard channels increases, the congestion parameter and consequently the waiting probability also increase.

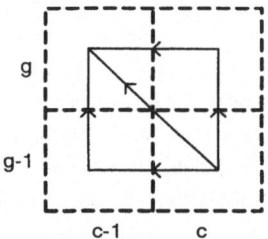

Fig. 6. Monotonic properties of the waiting probability

Figure 6 summarizes the monotonic properties of the waiting probability, and explains its adequateness to the design problem formulated above. Thus, we first investigate the feasible region of the waiting probability, and then evaluate both probabilities at each point belonging to it (on a point-per-point basis). This will provide the final feasible region, expressed as the interval [*cmin, cmax*] mentioned before.

Suppose the total number of channels N is fixed (by the operator), and we want to check the combinations of values of c and g such that $P_w(c,g) < P_{w0}$ and $P_d(c,g) < P_{d0}$, where P_{w0} and P_{d0} are certain given bounds. Algorithm 2 proposed below is designed to efficiently determine the minimum value of the number of common channels (*cmin*) necessary for the waiting probability to fulfill its requirement:

```
Algorithm 2:
Set A₁, A₂, N, Pw0
A:=A₁+A₂
if Pw(N,0)>Pw0 then STOP: "no feasible region"
else
          max:=N
          min:=1
          mid:=[(max+min)/2]
          while (max-min>1) do
            compute Pw(mid,N-mid) by using "Algorithm 1"
            if Pw(mid,N-mid)<= Pw0 then max:=mid
            else min:=mid
            mid:=[(max+min)/2]
          cmin:=max
```

Note that the algorithm first checks if the requirement is satisfied at the extreme point $c = N$ and $g = 0$. Because of the monotonic properties of the waiting probability described in Figure 6, if the test fails at this point, then there is no possible feasible region and the algorithm can conclude. Finally, the third algorithm evaluates both probabilities for $c \in$ [*cmin,N*], taking into account that the number of guard channels must satisfy the condition $g = N-c$:

```
Algorithm 3:
for c:=cmin to N do
            compute Pw(c,N-c) by using "Algorithm 1"
            compute Pd(c,N-c) by using "Algorithm 1"
```

To work out with some practical examples, we tested Algorithms 2 and 3 for $A_1=90$, $A_2=10$, $P_{w0}=10^3$, $P_{d0}=10^4$ and different values of N. For instance, for $N=145$ Algorithm 2 yielded $cmin=134$, that is, $c \in [134,145]$ as the feasible region for the waiting probability. Then, we applied Algorithm 3 to evaluate both probabilities along to the latter interval, and observed that in all cases the dropping probability was also falling below its upper bound. So definitely $c \in [134,145]$ is the final feasible region.

Something slightly different happened when $N=138$. In this case, the resulting feasible region for the waiting probability was [134,138], but the values obtained for the dropping probability by applying Algorithm 3 restricted this interval to [134,137].

In general, by varying N between 135 and 150, we obtained a constant value of 134 for $cmin$, and a quasi linear increase of the feasible region with $cmax \cong N$.

Also, we conclude from all experiments that usually the dropping probability either does not cause severe restriction on the feasible region of the waiting probability, or it cancels it completely. The reason is that the dropping probability decays very slowly as the number of common channels decreases while N is fixed, and thus typically this probability falls almost completely below or above its prefixed upper bound.

7   Conclusions

In this paper, we have proposed an alternative architecture for a single cell in a LEO satellite cellular network aimed to support voice communications, and analyzed its performance. Based on a Markovian characterization of the model, we have derived exact closed form expressions for the waiting and dropping probabilities of new and handoff arrivals respectively, in terms of a system parameter and the Erlang-B formula. These expressions allow for the development of an efficient algorithm to compute both probabilities, as well as for the formulation and efficient implementation of a multiobjective optimization problem in the context of network planning.

Further investigation can focus on different problems:

– First, other multiobjective optimization problems can be formulated to account for alternative design requirements.
– Second, the analysis performed in this paper can be extended to more realistic scenarios consisting of networks of cells. In fact, when dealing with a network of cells, the handoff arrival rate should be determined as a function of the new arrival rate, and call blocking probabilities do not longer depend only on the availability of channels at the originating cell, but also on the availability of channels at the intersatellite links and the destination cell. Some related works can be found in [18-20].

References

1. Hong, D., Rappaport, S. S.: Traffic Model and Performance Analysis for Cellular Mobile Radio Telephone Systems with Prioritized and Nonprioritized Handoff Procedures. IEEE Transactions on Vehicular Technology, Vol. VT-35, No. 3 (1986) 77-92.
2. Haring, G., Marie, R., Puigjaner, R., Trivedi, K.: Loss Formulas and Their Application to Optimization for Cellular Networks. IEEE Transactions on Vehicular Technology, Vol. 50, No. 3 (2001) 664-673.
3. McMillan, D.: Delay Analysis of a Cellular Mobile Priority Queueing System. IEEE/ACM Transactions on Networking, Vol. 3, No. 3 (1995) 310-319.
4. Sato, T., Mori, M.: An Application of the Lumping Method to a Loss System with Two Types of Customers. Journal of the Operations Research Society of Japan, Vol. 26, No. 1 (1983) 51-59.
5. Kawashima, K.: An Approximation of a Loss System with Two Heterogeneous Types of Calls. Journal of the Operation Research Society of Japan, Vol. 28, No. 2 (1985) 163-176.
6. Yue, W., Matsumoto, Y.: Performance Analysis of Multi-Channel and Multi-Traffic on Wireless Communication Networks. Kluwer Academic Publishers, Boston (2002).
7. Gavish, B.: LEO/MEO Systems – Global Mobile Communication Systems. Telecommunication Systems, Vol. 8, pp. 99-141, 1997.
8. Stallings, W.: Wireless Communications and Networks. Prentice-Hall, 2002.
9. Dimitrijevic, D. D., Vuceti, J.: Design and Performance Analysis of the Algorithms for Channel Allocation in Cellular Networks. IEEE Transactions on Vehicular Technology, Vol. 42, No. 4, November 1993.
10. Boukhatem, L., Beylot, A. L., Gaïti, D., Pujolle, G.: Performance Analysis of Dynamic and Fixed Channel Allocation Techniques in a LEO Constellation with an Earth-Fixed Cell System. Globecom'00, San Francisco, November 2002.
11. Boukhatem, L., Beylot, A. L., Gaïti, D., Pujolle, G.: Channel Assignment with Handover Queuing in LEO Satellite Systems based on an "Earth-Fixed Cell" Coverage. ITC 17, Salvador da Bahia (Brazil), December 2001.
12. Del Re, E., Fantacci, R., Giambene, G.: Efficient Dynamic Channel Allocation Techniques with Handover Queuing for Mobile Satellite Networks. IEEE Journal on Selected Areas in Communications, Vol. 13, No. 2, February 1995.
13. Del Re, E., Fantacci, R., Giambene, G.: Handover Queuing Strategies with Dynamic and Fixed Channel Allocation Techniques in Low Earth Orbit Mobile Satellite Systems. IEEE Transactions on Communications, Vol. 47, No. 21, January 1999.
14. Ruiz, G., Doumi, T. L., Gardiner, J. G.: Teletraffic Analysis and Simulation of Mobile Satellite Systems. IEEE Transactions on Vehicular Technology, Vol. 47, No. 1, February 1998.
15. Guérin, R.: Queueing-Blocking System with Two Arrival Streams and Guard Channels. IEEE Transactions on Communications, Vol. 36, No. 2, February 1988.
16. Harrison, P. G., Patel, N. M.: Performance Modelling of Communication Networks and Computer Architectures. Addison-Wesley (1993).
17. Akimaru, H., Kawashima, K.: Teletraffic. Theory and Applications. Springer-Verlag, Germany (1993).
18. Zaim, A. H., Rouskas, G. N., Perros, H. G.: Computing Call-Blocking Probabilities in LEO Satellite Networks: The Single-Orbit Case. IEEE Transactions on Vehicular Technology, Vol. 51, No. 2, March 2002.
19. Zaim, A. H., Perros, H. G., Rouskas, G. N.: Computing Call-Blocking Probabilities in LEO Satellite Constellations. IEEE Transactions on Vehicular Technology, Vol. 52, No. 3, May 2003.
20. Ganz, A. et al.: Performance Study of Low Earth Orbit Satellite Systems. IEEE Transactions on Communications, Vol. 42, 1994.

Performance Modelling of a Wireless GSM/GPRS Cell under Partial Sharing Scheme*

Demetres D. Kouvatsos, Yue Li, and Irfan Awan

Department of Computing, School of Informatics, University of Bradford
BD7 1DP, Bradford, West Yorkshire, England, UK
{D.D.Kouvatsos, Y.Li5, I.Awan}@bradford.ac.uk

Abstract. Novel analytic and simulation methodologies are devised for the performance modelling and evaluation of a wireless GSM/GPRS cell with bursty multiple class flows of voice calls and data packets subject to a traffic handling partial sharing scheme (PSS). In this context, a delay network model is proposed consisting of two interacting multiple class GE-type delay systems, namely a GSM GE/GE/c/c loss system with c servers and a GPRS GE/GE/1/N/PS delay system with finite capacity N and discriminatory processor share (PS) transfer rule. The principle of maximum entropy (ME) is used to characterize analytic solutions, subject to appropriate GE-type delay theoretic mean value constraints and closed form expressions for state and blocking probability distributions are obtained. Furthermore, related simulation programs in Java are developed. Typical numerical examples are included to compare the ME solutions against simulation and to study the effect of bursty multiple class traffics upon the performance of the cell.

Keywords: Global System for Mobile Telecommunication (GSM), General Packet Radio Service (GPRS), wireless GSM/GPRS cell, partial sharing scheme (PSS), quality-of-service(QoS), maximum entropy (ME) principle, performance evaluation, generalized exponential (GE) distribution.

1   Introduction

Cost-effective algorithms for queueing and delay network models under various traffic handling schemes are widely recognized as powerful and realistic tools for the performance evaluation and prediction of complex mobile networks with ever increasing volumes of multimedia traffic.

Most of the published performance studies in the field are based on simulation modelling and numerical solution of Markov models covering different traffic scenarios, mostly at voice call level, with single or multiple service classes(e.g.,[1-3]). Notably, Foh et al [3] proposed a single server infinite capacity queue for modelling GPRS data packet traffic in a Markovian environment and applied

---

* This work is supported by the EU IST project IASON under grant IST-2000-32392.

N. Mitrou et al. (Eds.): NETWORKING 2004, LNCS 3042, pp. 1252–1262, 2004.

matrix geometric methods for the computation of the performance metrics. More recently, Kouvatsos et al [4] presented an analytic framework, based on the principle of maximum entropy (ME) [5], for the performance modelling and evaluation of a wireless GSM/GPRS cell with a generalized exponential (GE) type of bursty traffic under a complete partitioning scheme (CPS).

Simulation modelling is an efficient tool for studying detailed system behaviour but it becomes costly, particularly as the system size increases. Markov models on the other hand provide greater flexibility. However, associated numerical solutions may suffer from several drawbacks, such as restrictive assumptions of Poisson arrival process and/or state space explosion, limiting the analysis to small mobile systems.

In this paper novel analytic and simulation delay network methodologies are devised for the performance evaluation of a delay network model (DNM) representing a wireless GSM/GPRS cell with bursty flows of voice calls and data packets subject to a traffic handling partial sharing scheme (PSS). The network model consists of two interacting multiple class GE-type delay systems, namely, a GSM GE/GE/c/c loss system with c (> 1) servers and a GPRS single sever GE/GE/1/N/PS delay system with finite capacity N (> 1) and processor share (PS) transfer rule. The principle of ME is used to characterize analytic solutions, subject to appropriate GE-type delay theoretic mean value constraints, leading into decomposition of the DNM into two individual but interacting systems each of which can be analyzed in isolation. Subsequently, closed form expressions for state and blocking probability distributions are obtained. Furthermore, the Java programming language is used to carry out detailed network model simulations at 95% confidence intervals.

Section 2 presents a GSM/GPRS architecture with potential spots of congestion together with a proposed DNM under PSS. An overview of the ME analysis of the GE/GE/c/c and GE/GE/1/N/PS delay systems is highlighted in Section 3. Typical numerical validation results are included in Section 4. Concluding remarks follow in Section 5.

Remarks on the GE-Type Distribution

The GE-type distribution is of the form [5]

$$F(t) = P(X \leq t) = 1 - \tau e^{-\tau v t}, \ \tau = 2/(1 + C^2), \ t \geq 0 \qquad (1)$$

where X is an interevent time random variable and $\{1/v, C^2\}$ are the mean and squared coefficient of variation (SCV) of the interevent times, respectively.

The GE distribution has a counting compound Poisson process (CPP) with geometrically distributed batch sizes with mean $1/\tau$. It may be meaningfully used to model the inter-arrival times of bursty multiple class mobile connections with different minimum capacity demands. Note that an IP packet length distribution is known to be non-exponential and should at least be described by the mean, $1/v$, and SCV, C^2. This is because IP packets are restricted by the underlying physical network, such as Ethernet and ATM, and thus, they have different packet lengths, typically 1500 bytes and 53 bytes, respectively.

The GE distribution may also be employed to model short range dependence (SRD) traffic with small error. For example, an SRD process may be approximated by an ordinary GE distribution whose first two moments of the count distribution match the the corresponding first two SRD moments. This approximation of a correlated arrival process by an uncorrelated GE traffic process may facilitate (under certain conditions) problem tractability with a tolerable accuracy and, thus, the understanding of the performance behaviour of external SRD traffic in the interior of the network. It can be further argued that, for a given buffer size, the shape of the autocorrelation curve, from a certain point onwards, does not influence system behaviour (c.f., [3]). Thus, in the context of system performance evaluation, an SRD model may be used to approximate accurately long range dependence (LRD) real traffic.

2   A DNM for a GSM/GPRS Cell under PSS

Efficient traffic handling schemes for wireless GSM/GPRS cells are of crucial importance for the support of mobile multimedia applications with quality-of-service (QoS) guarantees. Resources for GPRS traffic can be reserved statically or dynamically whereas a combination of both is possible. Different cell capacity partitioning schemes for handling the transmission of voice calls/data packets can be defined. In such environment, partitions of the available bandwidth may be created for GSM and GPRS traffics. For GPRS traffic, a complete partition is used for different data services. However, some data packets may be allocated higher priority and, therefore, they can be given higher share of the available bandwidth. Whenever voice calls and data packets share bandwidth, voice calls are always given preemptive priority over data packets. Two main GSM/GPRS call/data handling schemes are described below.

- *Complete partitioning scheme (CPS)* divides the total cell capacity to serve simultaneously GSM and GPRS traffics. As a consequence, the GSM and GPRS systems can be analyzed separately.
- *Partial sharing scheme (PSS)* allocates a fixed number of channels for data traffic and the remaining channels are shared by both voice calls and data packets with preemptive service priority for voice calls.

Focusing on the PSS, a simplified GSM/GPRS wireless cell architecture can be seen in Fig. 1. The User Equipment (UE) can generate traffic consisting of GSM voice calls and GPRS data packets which go through the same Base Station (BS) but each of them will have its own Um and Gb radio interfaces. A joint bandwidth management procedure under PSS is in operation. Potential spots of congestion and delays are often attributed to the constraints imposed by the Base Station Controller (BSC) and the Serving GPRS Support Node (SGSN) dealing with GSM voice and GPRS data traffics, respectively.

A number of bufferless channels may be allocated to the GSM partition which can be clearly seen as a loss system for multiple class voice calls. An admitted voice call requires the assignment of a single channel for its entire duration. Moreover, the GPRS traffic is packet based and, therefore, each connection will attempt to

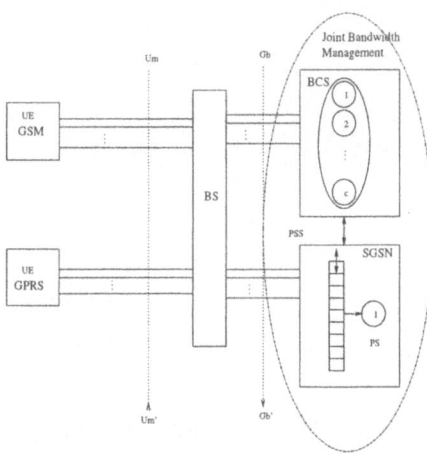

Fig. 1. A GSM/GPRS wireless cell architecture with a PSS system of queues

use the complete available bandwidth. As soon as there are more active channels, it is the responsibility of the SGSN to ensure that bandwidth is shared fairly among them. Thus, the GPRS data partition can be modelled by a PS delay system where all data connections may belong to different classes and share the total available capacity. These classes may have their own characteristics such as maximum or minimum data rates, delay sensitivity, service discrimination, arrival rates, inter-arrival-time variability and transferable file (data) length. Under PSS, there is a joint bandwidth management procedure according to which free GSM channels are acquired dynamically to increase the transmission capacity of the GPRS partition until the arrival of new voice calls.

A DNM of a GSM/GPRS cell under PSS can be seen in Fig. 2. The CPP with geometrically distributed batches is used to represent bursty GSM and GPRS multiple class arrival processes (or, equivalently GE-type interarrival times) of voice calls and data packets, respectively. Moreover, GE distributions are used to describe the call durations and packet channel transmission times. In the context of this work, only up link traffic streams are considered.

The GSM partition may be generally modelled by a classical pseudo-birth death GE/GE/c/c loss system with multiple voice call streams. On the other hand, the GPRS partition can be modelled by a single server GE/GE/1/N/PS system with a finite capacity, N, for the transmission of multiple class data packets under discriminatory PS rule. This is feasible as all admitted data connections will share simultaneously the available GPRS bandwidth according to assigned priorities. Physically, the GPRS system is capable of allocating all available channels to one connection (subject to some battery restrictions). Also, in case of multiple connections, one time slot (channel) can be shared by eight different connections [6].

The joint bandwidth management protocol of the PSS is implemented under the following traffic handling operational conditions: At any given time, free

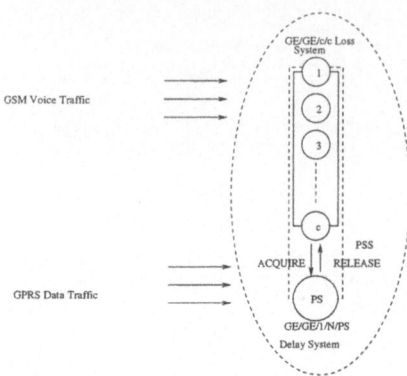

Fig. 2. A DNM of a wireless GSM/GPRS cell under PSS

GSM channels belonging to the GE/GE/c/c loss system are used (temporarily) to increase the transmission capacity of the GPRS GE/GE/1/N/PS delay system. However, new arrivals of voice calls, with a preemptive priority over the data packets, will cause the immediate release of some or all GSM channels, as appropriate, from the GPRS partition to the loss system for their own immediate transmission. Subsequently, the transmission capacity of the GPRS system will be progressively either reduced or increased, as appropriate. Note that, in the absence of available capacity under PSS, both voice calls and data packets will be lost on arrival.

3 Delay Theoretic Performance Analysis of a Wireless GSM/GPRS Cell under PSS

The proposed analytic DNM of Fig. 2 consists of the multiple class GE/GE/c/c loss system interacting under PSS with the multiple class GE/GE/1/N/PS delay system. An overview of the ME methodology, as applied to the analysis of the aforementioned systems, is presented below. Details of mathematical proofs associated with key analytic GE-type results can be found in [7].

Notation

For each class i $(i = 1, 2, \ldots, R, R > 1)$ let
$\{1/\lambda_i, C_{ai}^2\}, \{1/\mu_i, C_{si}^2\}$ be the mean and SCV of the interarrival and service time distributions, respectively.

Moreover, let at any given time at equilibrium

n_i be the number of either voice calls at the GE/GE/c/c loss system or data packets at the GE/GE/1/N/PS delay system;

$\mathbf{n} = (n_1, n_2, \ldots, n_R)$ be joint state of either GE/GE/c/c or GE/GE/1/N/PS systems, (n.b., $\mathbf{0} = (0, \ldots, 0)$);

Ω be the set of all feasible states $\mathbf{n}$;

π_i be the blocking probability that an arrival of class i will find either the GE/GE/c/c or GE/GE/1/N/PS systems at the full capacity.

The form of the state probability distribution, $P(\mathbf{n}), \mathbf{n} \in \Omega$, can be characterized by maximizing the entropy functional $H(\mathbf{P}) = -\sum_{\mathbf{n}} P(\mathbf{n}) \log P(\mathbf{n})$, subject to prior information expressed in terms of the normalization and, for each class $i(i = 1, 2, \ldots, R)$, the marginal constraints of either server utilization for a GE/GE/1/N/PS delay system under discriminatory PS rule or, state probabilities $P(0), P(1), \ldots, P(c-1)$ for a GE/GE/c/c loss system, mean number of voice calls and/or data packets, as appropriate and full buffer state probabilities for either system satisfying the flow balance equations, namely

$$\lambda_i(1 - \pi_i) = \mu_i U_i, \; i = 1, \ldots, R \tag{2}$$

where U_i is the steady state probability of having customers of class i in the system. By employing Lagrange's method of undetermined multipliers, the following solutions are obtained:

$$P(\mathbf{n}) = \begin{cases} \dfrac{1}{Z} \dfrac{\left(\sum_{j=1}^{R} n_j - 1\right)!}{\prod_{j=1}^{R} n_j!} \left(\prod_{j=1}^{R} x_j^{n_j}\right) \left(\sum_{i=1}^{R} n_i g_i y_i^{f_i(\mathbf{n})}\right), & \text{GE/GE/1/N/PS} \\[4mm] \dfrac{1}{Z} \prod_{i=1}^{R} \left\{\prod_{k=1}^{c} g_{ik}^{h_{ik}(\mathbf{n})} y_{ik}^{f_{ik}(\mathbf{n})}\right\} x_i^{n_i}, & \text{GE/GE/c/c} \end{cases} \tag{3}$$

where $Z = 1/P(0)$, is the normalizing constant, $\{g_i, x_i, y_i, i = 1, 2, \ldots, R\}$ and $\{g_{ik}, y_{ik}, i = 1, 2, \ldots, R; k = 1, 2, \ldots, c\}$ are the Lagrangian coefficients corresponding to the aforementioned constraints per class, respectively and $h_{ik}(\mathbf{n})$, $f_{ik}(\mathbf{n})$ and $f_i(\mathbf{n})$ are suitable auxiliary functions defined as: $h_{ik}(\mathbf{n}) = 1$ if $(n_i \geq k$ and $k \leq c - \sum_{\ell=1}^{i-1} n_i, k = 1, 2, \ldots, c)$ or 0 otherwise; $f_{ik}(\mathbf{n}) = 1$ if $(h_{ik}(\mathbf{n}) = 1$ and $\sum_{i=1}^{R} n_i = c, k = 1, 2, \ldots, c)$ or 0 otherwise and $f_i(\mathbf{n}) = 1$ if $\sum_{i=1}^{R} n_i = N, i = 1, 2, \ldots, R$ or 0 otherwise. Note that, the Lagrangian coefficients $\{x_i, g_i, g_{ik}, \forall i, k\}$ can be approximated analytically by making asymptotic connections to the corresponding GE-type infinite capacity systems [7]. Moreover, the Lagrangian coefficients $\{y_i, y_{ik} \, \forall i, k\}$ can be determined via flow balance equation (2).

From ME solution (3), and after some manipulation, the aggregate state probabilities can be obtained by

$$P(n) = \begin{cases} \dfrac{1}{W} \left(\sum_{i=1}^{R} g_i x_i y_i^{f_i(\mathbf{n})}\right) X^{n-1}, & n \in [1, N], \; GE/GE/1/N/PS \\[4mm] \dfrac{1}{W} \sum_{\mathbf{n} \in A_n} \prod_{i=1}^{R} \left\{\prod_{k=1}^{c} g_{ik}^{h_{ik}(\mathbf{n})} y_{ik}^{f_{ik}(\mathbf{n})}\right\} x_i^{n_i}, & GE/GE/c/c \end{cases} \tag{4}$$

where $W = 1/P(0)$, $X = \sum_{i=1}^{R} x_i$ and $A_n = \{\mathbf{n} : \sum_{i=1}^{R} n_i = n\}$. Moreover, focusing on a tagged voice call or data packet, as approriate, within an arriving bulk and applying GE-type probabilistic arguments, the blocking probabilities

$\{\pi_i, i = 1, 2, \ldots, R\}$ can be approximated by

$$\pi_i = \begin{cases} \sum_{n=0}^{N} \delta_i(n)(1 - \sigma_i)^{N-n} P(n), \ GE/GE/1/N/PS \\ \sum_{n=0}^{c} \delta_i(n)(1 - \sigma_i)^{c-n} P(n), \ GE/GE/c/c \end{cases} \tag{5}$$

where $\delta_i(0) = \frac{r_i}{r_i(1-\sigma_i)+\sigma_i}$, $\delta_i(n) = 1(\forall n > 0)$, $\sigma_i = 2/(1 + C_{a\,i}^2)$ and $r_i = 2/(1 + C_{s\,i}^2)$.

By using the blocking probabilities $\{\pi_i, i = 1, 2, \ldots, R\}$ and the flow balance condition (2), after some manipulation, recursive relationships for the computation of Lagrangian coefficients $\{y_i, y_{ik} \forall i, k\}$, can be obtained (c.f., [7]).

GPRS Weighted Average Performance Measures

GPRS data packets will receive under PSS variable transmission capacity depending on the availability of free GSM channels. Thus, an average performance statistic for data packet of class i, say, S_{GPRS}^i (e.g., class utilisation, mean number of data packets per class, etc) of the $GE/GE/1/N/PS$ delay system can be clearly determined by the weighted average measure S_{GPRS}^i, given by

$$S_{GPRS}^i = \sum_{l=0}^{c} S_{GPRS}^{l,i} P_{GSM}(n = l), i = 1, 2, \ldots, R \tag{6}$$

where $P_{GSM}(.)$ is the aggregate steady state probability of the number of voice calls in the GE/GE/c/c loss system and $\{S_{GPRS}^{l,i}, l = 0, 1, 2, \ldots, c\}$ are estimated values of statistic S_{GPRS}^i corresponding to the maximum available mean GPRS transfer rates $\{\mu_{GPRS}^i + (c-l)\mu_{GSM}, l = 0, 1, 2, \ldots, c\}$, under PSS with $\{\mu_{GSM}, \mu_{GPRS}^i\}$ being the initially allocated aggragate class i transfer rates to the GSM GE/GE/c/c loss system and the GPRS GE/GE/1/N/PS delay system, respectively.

Remarks

The proposed analytic GE/GE/1/N/PS delay model differs from and, in some respect, extends overall the MMPP/M/1 delay model suggested by Fol et al. [3]. Although the latter model incorporates a Markov modulated Poisson arrival process (MMPP), nevertheless it is only applicable to a single class of data packets, assumes exponential transmission times and, being an infinite capacity delay model, does not capture the adverse blocking effect on system performance. Moreover, the GE-type delay model can be solved via closed-form expressions as opposed to those requiring computationally demanding matrix geometric methods.

4   Numerical Results

This section presents typical numerical experiments to illustrate the credibility of the ME methodology against simulation and also to demonstrate the applicability of ME solutions, as simple and cost-effective performance evaluation tools,

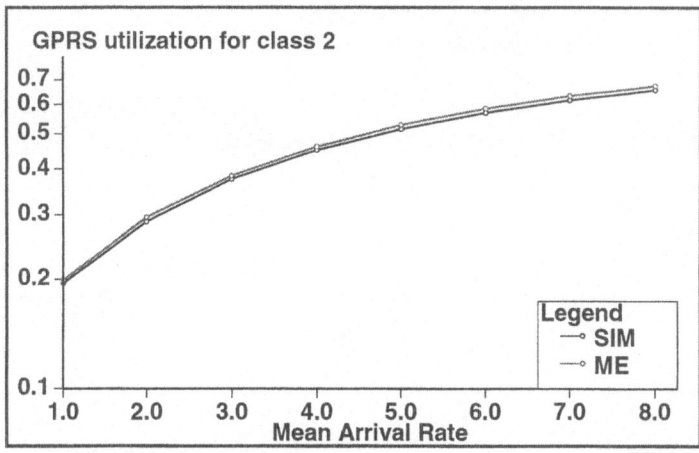

Fig. 3. GPRS class 2 utilization for GE/GE/1/N/PS delay system at overall GSM utilization of 40 % (Experiment 1-Input Data: $\{\lambda = 1.0\ \mu = 0.58,\ c = 6,\ Ca^2 = 2,\ Cs^2 = 1\}$ for GSM and $\{N = 20,\ \lambda_1 = 0.3,\ \mu_1 = 2.92,\ \mu_2 = 0.58,\ Ca_1^2 = 2,\ Ca_2^2 = 3,\ Cs_1^2 = 4,\ Cs_2^2 = 4\}$ for GPRS).

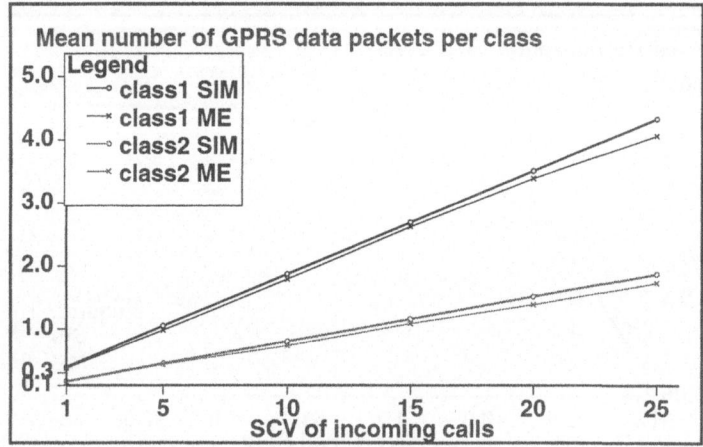

Fig. 4. Effect of varying degrees of SCV on mean number of GPRS data packets at overall GSM utilization of 50 % (Experiment 1-Input Data: $\{\lambda = 1.0,\ \mu = 0.58,\ c = 6,\ Ca^2 = 2,\ Cs^2 = 1\}$ for GSM and $\{N = 20,\ \lambda_1 = 0.3,\ \lambda_2 = 0.05,\ \mu_1 = 2.92,\ \mu_2 = 0.58,\ Cs_1^2 = 4,\ Cs_2^2 = 4\}$ for GPRS).

for assessing the effect of bursty multiple class traffics upon the performance of the wireless cell. Note that the Java programming language was used to carry out the simulation analysis of the DNM of Fig. 2 at 95 % confidence intervals.

Two experiments have been devised. Experiment-1 (c.f., Figs. 3-5) deals with an aggregate class of GSM voice calls and two GPRS classes of data packets having different mean sizes, namely, 62.5 KBytes (class 1, e.g., web browsing) and

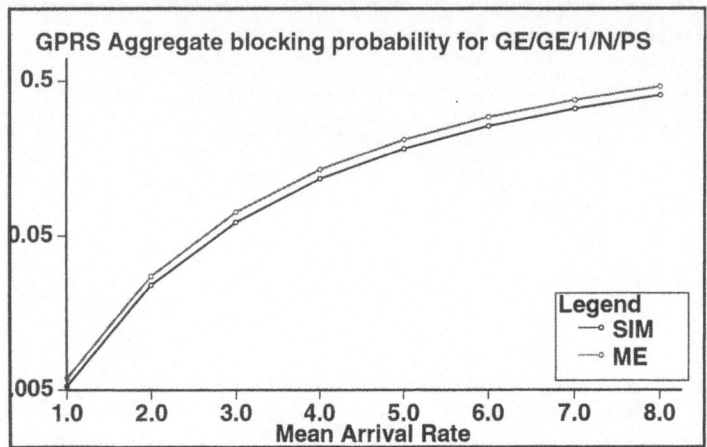

Fig. 5. Effect of varying degrees of arrival rate λ_1 on the aggregate GPRS data packets blocking probability at overall GSM utilization of 40 % (Experiment 1-Input Data: $\{\lambda = 1.0,\ \mu = 0.58,\ c = 6,\ Ca^2 = 2,\ Cs^2 = 1\}$ for GSM and $\{N = 20,\ \lambda_2 = 0.05,$ $\mu_1 = 2.92,\ \mu_2 = 0.58,\ Ca_1^2 = 2,\ Ca_2^2 = 3,\ Cs_1^2 = 4,\ Cs_2^2 = 4\}$ for GPRS).

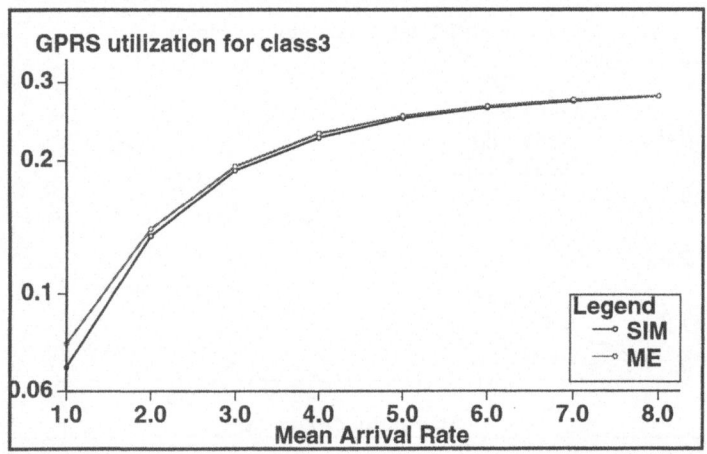

Fig. 6. GPRS class 3 of utilization for GE/GE/1/N/PS delay system at overall GSM utilization of 40 % (Experiment 2-Input Data: $\{\lambda = 1.0\ \mu = 0.58,\ c = 6,\ Ca^2 = 2,$ $Cs^2 = 1\}$ for GSM and $\{N = 20,\ \lambda_1 = 0.3,\ \lambda_2 = 0.0,\ \mu_1 = 2.92,\ \mu_2 = 0.58,\ \mu = 14.01,$ $Ca_1^2 = 2,\ Ca_2^2 = 3,\ Ca_3^3 = 2,\ Cs_1^2 = 4,\ Cs_2^2 = 4,\ Cs_3^2 = 4\}$ for GPRS).

12.5 KBytes (class 2, e.g., email). Experiment-2 (c.f., Figs. 6,7) includes a study involving aggregate class GSM voice calls and three GPRS classes of data packets with different mean sizes given by 300 KBytes (class 1, e.g., video stream), 62.5 KBytes (class 2, e.g., web browsing), 12.5 KBytes (class 3, e.g., email). It is assumed that both GSM and GPRS partitions consist of one frequency providing total capacity of 171.2 Kbps. Note that the comparative study makes use of

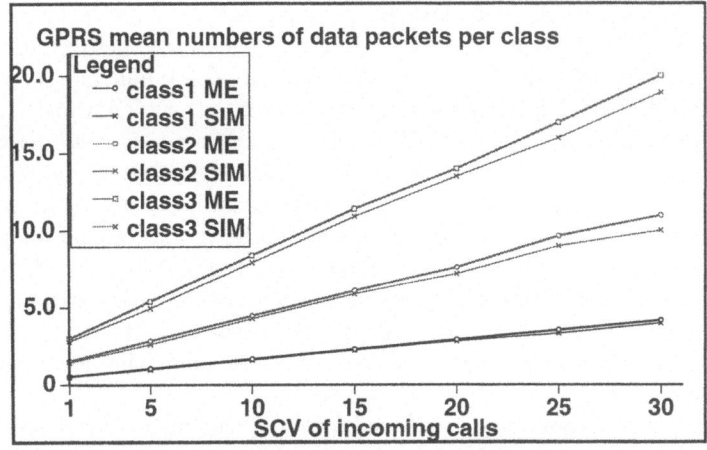

Fig. 7. Effect of varying degrees of SCV on mean number of GPRS data packets at overall GSM utilization of 50 % (Experiment 2-Input Data: $\{\lambda = 1.0 \ \mu = 0.58, c = 6, Ca^2 = 2, Cs^1 = 1\}$ for GSM and $\{N = 20, \lambda_1 = 0.2 \ \lambda_2 = 0.05, \lambda_3 = 2.0, \mu_1 = 2.92, \mu_2 = 0.58, \mu = 14.01, Cs_1^2 = 4, Cs_2^2 = 4, Cs_3^2 = 4\}$ for GPRS).

varying GPRS traffic values of mean arrival rates and SCV of the interarrival times.

As the GE/GE/c/c loss system is unaffected, under PSS, by the GPRS patition, without the loss of generality, the comparative study focuses on marginal performance metrics of GPRS utilization, mean number of data packets per class together and the aggregate GPRS blocking probability of the GE/GE/1/N/PS delay system. It can be observed (c.f., Figs. 3-7) that, over a wide range of parameterisation, the ME results are slightly pessimistic but very comparable to those obtained via simulation. Moreover, it can be seen that the interarrival time SCV has an inimical effect on the performance metrics of GPRS data packets (c.f., Figs. 4,7). Note that additional validation tests involving a multiple GPRS class GE/GE/1/N/PS delay system with a discriminatory PS rule together with a multiple GSM class GE/GE/c/c loss system can be seen in [7].

Remarks

The credibility of the ME results, as reliable GE-type or generalized geometric (GGeo)-type approximations [5], against those obtained from the simulation of more complex correlated traffic models, such as Markov Modulated Batch Bernoulli Process (MMBBP) and Batch Renewal Process (BRP), is beyond the scope of this paper and will be the subject of further study.

5   Conclusions

Novel analytic and simulation methodologies are devised for the performance modelling and evaluation of a wireless GSM/GPRS cell with bursty multiple class flows of voice calls and data packets subject to a traffic handling PSS. In this context, a DNM is proposed consisting of two interacting mutiple class systems, namely, a GSM GE/GE/c/c loss system with c multiple servers and a GPRS GE/GE/1/N/PS delay system with discriminatory PS transfer rule. The principle of maximum entropy (ME) is used to characterize new analytic solutions, subject to appropriate GE-type delay theoretic mean value constraints. Subsequently, closed form expressions for state probability distributions and related performance metrics are determined. Furthermore, simulation programs developed in Java are employed to validate the credibility of the ME solutions against simulation results at 95% confidence internals and also verify the adverse effect of bursty traffics upon the performance of the cell. It was observed that the ME analytic results are comparable to those of simulation for a wide range of parametrization.

The analytic methodology of ME, subject to queueing and/or delay theoretic mean value coustraints, provides simple and robust analytic building block tools for the performance modelling and prediction of wireless cells and networks.

The proposed DNM can be extended to specify alternative operational interpretations of the joint management bandwidth protocol of PSS. Current work in progress focuses on the GSM/GPRS enhancement of the Universal Mobile Telecommunication System (UMTS) network architecture.

Acknowledgements. We wish to thank our industrial partners of the EU IST project IASON for their valuable advice towards the specification of the GSM/GPRS analytic and delay models of this paper.

References

1. K. Begain, G. Bolch, M. Telek, Scalable Schemes for Call Admission and Handover Handling in Cellular Networks with Multiple Services. *Journal on Wireless Personal Communications*, Volume 15, No. 2, Kluwer Academic Publishers, 2000, pp. 125-144.
2. R. Litjens, R. Boucherie, Radio Resource Sharing in GSM/GPRS Network. em ITC Specialist Seminar on Mobile Systems and Mobility, Lillehammer, Norway, March 22 - 24, 2000. pp. 261-274.
3. C.H.Foh, B.Meini, B. Wydrowski and M.Zuerman, Modeling and Performance Evaluation of GPRS, Proc. of IEEE VTC, 2001, Rhodes, Greece, pp. 2108-2112, May 2001.
4. D.Kouvatsos, I.Awan and K.Al-Begain, Performance Modelling of GPRS with bursty multiclass traffic, IEE Proc.Comput.Digit.Tech., Vol. 150, No. 2, pp. 75-85, 2003.
5. D.D. Kouvatsos, Entropy Maximisation and Queueing Network Models, *Annals of Operation Research*, Vol. 48, pp. 63-126, 1994.
6. Rappaport, T.S.: *Wireless Communications*, Prentice Hall, NJ, 1996.
7. I.U. Awan and D.D.Kouvatos, Maximum Entropy Analysis of Arbitrary Queueing Network Models with Multiple Servers, PR Service Priority and finite capacity queues *Research Report* RS-05-02, Performancè Modelling and Engineering Research Group, Department of Computing, University of Bradford, May (2003).

An Efficient Probabilistic Packet Marking Scheme for IP Traceback

Basheer Duwairi, Anirban Chakrabarti, and Govindarasu Manimaran

Department of Electrical and Computer Engineering
Iowa State University, Ames, IA 50011, USA
{dbasheer,anirban,gmani}@iastate.edu

Abstract. Denial of Service (DoS) attacks represent a major threat to the availability of Internet services. Identifying the sources of these attacks is considered an important step toward a DoS-free Internet. In this paper, we propose a new scheme, called *Distributed Link-List Traceback*, which combines the good features of probabilistic packet marking [6] and Hash-based traceback [9]. The main idea used in the scheme is to preserve the marking information at intermediate routers in such a way that it can be collected in an efficient manner. We evaluate the effectiveness of the proposed scheme for various performance metrics through combination of analytical and simulation studies. Our studies show that the proposed scheme requires small number of packets, adjustable amount of memory. At the same time, offers high attack source detection percentage.

1 Introduction

In DDoS attacks, the attacker's machine (the master) instructs previously compromised innocent machines (the slaves) to aggressively overwhelm the victim by high volume streams of flooding packets with faked IP source addresses, leaving the victim with no clue about the true sources of these packets. This distributed anonymous nature of the attack helps the attacker to stay behind the scenes. Attack traceback, which can be defined as the process of identifying the true physical sources of attack packets, has emerged as a promising solution to DoS attacks. This has the following benefits: first, isolating or even shutting down the attack facility, which greatly reduces the impact of the ongoing attack or stopping it completely. Second, holding attackers responsible for abusing the Internet. Personal identification of attackers can be done by further investigation and analysis of the compromised systems discovered by the attack traceback process.

The stateless nature of the Internet combined with the destination oriented IP routing increases the difficulty of tracing attacks back to their sources. This problem is also complicated by the fact of having millions of hosts connected to the Internet, which implies a huge search space. The imminent threats imposed by DoS attacks call for efficient and fast traceback schemes. A good traceback scheme should provide accurate information about routers near the attack source rather than those near to the victim, recognize and exclude false information injected by the attacker, avoid using large amount of attack packets to construct the attack path or attack tree, avoid imposing high processing and storage overhead at intermediate routers, and if packet information is to be maintained at intermediate routers then collecting this information must be efficient.

N. Mitrou et al. (Eds.): NETWORKING 2004, LNCS 3042, pp. 1263–1269, 2004.
© IFIP International Federation for Information Processing 2004

In this paper, we develop a novel concept called *Distributed link-list* (DLL), which refers to the process of keeping track of a selected set of routers that were involved in forwarding certain packet by establishing a temporary link between them in a distributed manner. We utilize this concept and develop a novel traceback scheme, called *Distributed Link-List Traceback* (DLLT) that combines the desirable features of PPM [6] and hash-based traceback [9]. The rest of this paper is organized as follows. In the next section, we discuss the related work. In section 3, we present the proposed work: distributed link-list traceback. In section 4, we provide theoretical analysis. In section 5, we describe the simulation studies. Finally, conclusions are drawn in section 6.

2   Related Work

Traceback schemes [6][9] [7][4] [2][8][3] usually rely on router assistance to determine the path followed by attack packets and eventually identify the attack source. For example, in PPM [6], routers mark forwarded packets (i.e., write their own IP addresses into the packets) probabilistically, such that the victim can reconstruct the attack path after receiving huge amount of packets. In hash-based traceback [9], bloom filters [1] were used to save packet digests at intermediate routers to be collected and searched when attack is detected.

PPM requires very large number of packets to be collected before starting the traceback process. This is due to the fact of allowing routers to overwrite marking information written by previous routers. Also, the ability of the attackers to spoof the marking information represents a major weakness of PPM [5]. In Hash-based scheme, processing of every packet passing through, imposes significant router overhead. Also, the method employed to download packet information from network routers is inefficient and requires special resources. Moreover, a major concern in Hash-based traceback is the small window of time through which packets can be successfully traced.

The main contribution of this paper is a novel concept called *Distributed Link-List (DLL)* and using which we developed a new traceback scheme, called *Distributed Link-List Traceback* (DLLT). DLLT exhibits the features of PMM [6] in the sense that routers probabilistically mark forwarded packets. Also, it exhibits the features of Hash-based scheme [9] in the sense that processing and storage at intermediate routers are necessary. The significance of DLLT is due to drastic reduction of the number of packets required in the traceback process compared to PPM, and the adjustable memory requirement and efficient marking information collection compared to Hash-based traceback.

3   Proposed Solution: Distributed Link-List Traceback

Distributed Link-List Concept: The main idea of DLL is to keep track of some of the routers that were involved in forwarding certain packet by establishing a temporary link between them in a distributed manner. DLL is based on "store, mark and forward" approach. A single marking field is allocated in each packet. Any router that decides to mark the packet, stores the current IP address found in the marking field along with the packet ID in a special data structure called *Marking Table* maintained at the router, then marks the packet by overwriting the marking field by its own IP address, and then forwards the packet as usual. Any router that decides not to mark the packet just forwards it.

A link list is inherently established because the marking field serves as a pointer to the last router that did the marking for the given packet and the marking table of that router contains a pointer (i.e., the IP address) of the previous marking router and so on. Therefore, each packet received by the destination contains the start point of a link list that is part of the packet path. We call it distributed link-list because each router decides by its own to be on the list or not according to certain marking probability.

Details of Distributed Link-List Traceback: Distributed Link-List Traceback (DLLT) uses DLL concept to keep track of the routers that have been involved in forwarding malformed packets toward the victim. DLLT employs a probabilistic marking and storage scheme. When a router receives a packet, it makes a decision based on certain marking probability q of whether to mark the packet (i.e., write some information, called the marking information, into the packet) or not. Whenever a router decides to mark a packet it has to store the marking information found in the packet before remarking it. Therefore, packet marking and storage is an integrated procedure. Before going into details of this procedure, we show the main data structure used for storing packet information.

Logging packet information at intermediate routers is not a new idea. Storing packet digests was considered in [9]. However, our storage scheme is probabilistic in nature, which means that only fraction of the traffic is to be logged at each router. Also, we store this information in such a way as to ensure that it can be collected in a predetermined manner. We borrow the idea of using bloom filters [1] from [9], and we modify it to satisfy our requirements:

- Storing the packet digests to be able to verify that a given packet has been actually forwarded by the router.
- Mapping the digests of a given packet to certain memory location where the marking information of that particular packet can be stored.

The first requirement can be achieved exactly the same way as in [9], where a bloom filter computes j distinct packet digests for each marked packet using j independent uniform hash functions, and uses the n-bit result to index the 2^n-sized bit Digests Array (DA). The array is initialized to all zeros, and bits are set to one as packets are received.

The second requirement can be achieved by storing the marking information of a given packet in the Marking Information Table (MIT) at the memory location indexed by the first hash function that maps to zero bit in the digests array. Fig. 1 depicts both the DA and MIT with j hash functions. It also shows the marking information of a given packet before and after being marked. The marking fields reserved in each packet and in the MIT are shown also.

It can be realized that *probabilistic edge marking* (an edge is composed of two adjacent routers on the packet path) is simple to implement in our scheme. Whenever a router decides to mark a packet we enforce the subsequent router to mark the same packet. This can be achieved by maintaining a 1-bit field called *marking flag* as part of the marking information to be held in the packet. This flag is used to enforce deterministic marking when it is on. When it is off the marking becomes probabilistic. With this flag, the probabilistic edge marking in DLLT can be implemented as follows: When a router

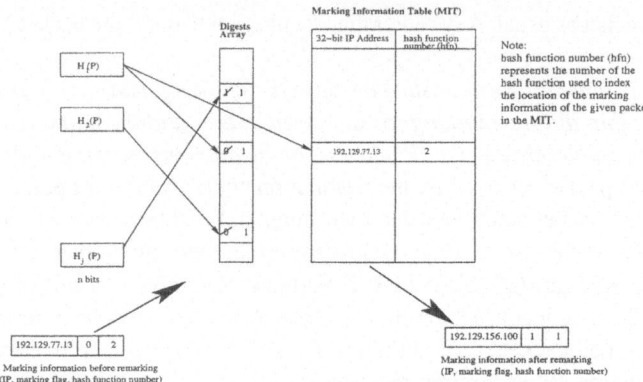

Fig. 1. Digests array (DA) and marking information table (MIT) at router R. The marking information of a given packet before and after being marked at router R which has the IP address of 192.129.156.100

receives a packet, it checks the marking flag. If it is on, it has to do the marking and storage procedure and then reset the flag. Otherwise (i.e., when the flag is off), it takes the decision based on some probability q. If the decision outcome is to mark the packet it will do so, and then set the flag such that the next adjacent router will do the marking deterministically.

After detecting an attack, the victim has to collect marking information that belongs to k of the received attack packets. This information can be retrieved from intermediate routers by following the link list associated with each of the chosen k packets. The attack sources are then determined by inferring the relative ordering of routers based on the retrieved marking information.

4   Analysis

Storage Analysis: The amount of storage that needs to be allocated to a traceback scheme is a critical issue. In this section, we quantify the amount of memory required in both DLLT and Hash-based scheme [9]. First, we review some characteristics of the bloom filters (or what we call digests arrays) that would be necessary in our analysis. A bloom filter is characterized by its size s bits, the number of hash functions used j, and its capacity factor f. A bloom filter of size s and capacity factor f can be used to store the digests of at most s/f packets. The effective false positive rate of a bloom filter is directly dependent on the previous parameters. Please refer to [1] and [9] for theoretical and experimental bounds on the false positive rate. What follows is a quantification of memory requirement at each router in both schemes.

Let b denotes the number of bits required to store the marking information of one packet in the MIT (i.e., this includes 32-bit IP address plus $\lceil \lg(j) \rceil$ bits for the hash function number). To store marking information of x packets, we need an MIT of size xb bits to be *shared* among f Digests Arrays each of size x bits. Therefore, the total memory requirement (M_{dllt}) to store x packets information is given by: $M_{dllt} = x(b+f)$.

Assuming an aggregate incoming link capacity of p packets/sec and marking probability of value q at each router, x can be replaced in the previous equation by qp. Therefore, the amount of memory required to store a second's worth of digests can be rewritten as: $M_{dllt} = qp(b + f)$. In Hash based scheme, To store digests of x packets, we just need a digests array of size fx bits. Meaning that the amount of memory required to store a second's worth of digests assuming an aggregate link capacity of p packets/sec can be expressed as: $M_{hash}=fp$. Expressing M_{dllt} as a function of q makes it adjustable to meet the limitations imposed by current memory technology.

Number of Attack Packets Required to Identify the Attacker: Our objective is to find a bound on the minimum number of packets that has to be received by the victim such that every router on the path from attacker to victim is involved in marking at least one of these packets with high confidence probability u. Let k represents this lower bound. Let the marking probability at router R be q. Let P_f be the probability that R fails to mark any packet out of the k packets. Clearly, $P_f = (1 - q)^k$. Therefore, the probability that R will succeed in marking (we call it the success probability) at least one packet is given by:

$$P_s = 1 - P_f = 1 - (1 - q)^k \qquad (1)$$

To obtain the desired bound we can safely assume that the success probability for all routers a long a path of length l is the same and equal to that of the farthest router (i.e., as given in equation 1)

If we define X to be a random variable that represents the number of routers out of l that were successful in the marking process (l is path length), then X follows the binomial distribution with success probability P_s given in equation 1. We need to find k such that: $P(X = l) \geq u$ (i.e., the probability that each router succeeds in marking at least one packet is larger than u. But, $P(X = l) = (1 - (1 - q)^k)^l \geq u$, solving for k, we obtain:

$$k \geq \frac{\lg(1 - u^{1/l})}{\lg(1 - q)} \qquad (2)$$

For example, for an attack path of length 15 and marking probability of 0.3 and confidence probability 0.95 the number of packets required by DLLT is 16 compared to 1340 in PPM [6]. This significant reduction is due to the mechanism employed by DLLT to maintain marking information of routers far away from the victim.

5 Simulation Studies

We have carried out several simulation experiments to evaluate the proposed scheme. *Detection Percentage (DP)* defined as the percentage of exactly detected attack sources, is the metric used to evaluate our scheme. For example, if a attack sources are exactly located out of m attack sources, then we express the detection percentage as $\frac{a \times 100}{m}$ %.

In each simulation experiment we generated a random attack tree with m attackers and one victim. The attack path length l was the same for all attackers. Packets were marked according to a specific probability q. Attackers were instructed to inject their packets simultaneously with a rate of 1000 packets/attacker. k attack packets were used to conduct the traceback process. The default values for m, q, l, and k were set to be 100,

0.15, 20, and 1000 respectively. Each of the following results represents the average of 500 independent simulation runs based on the default parameter values unless otherwise specified. The simulation experiments were designed to study the effect of the above parameters on detection percentage. The results are shown in Fig. 2.

We measured the detection percentage of DLLT under different circumstances. Fig. 2 shows the detection percentage as a function of the marking probability q for different values of k, l, and m, respectively. In all these experiments we used the default parameter values mentioned above except for the parameter under investigation which made equal to the values shown in the figure. From theses figures we can make the following observations:

In all cases, the detection percentage of DLLT increases by increasing the marking probability. This is expected since marking routers would have better chance to appear in the information collected by the victim. Therefore, attack source identification becomes more accurate by increasing q. However, we should not forget the effect of increasing q on the amount of storage required by DLLT, and we should limit the marking probability to low values. Therefore, higher number of packets must be used to initiate marking information retrieval.

Increasing the number of packets used by the victim to identify attack sources results in better detection percentage. This can be observed in Fig.2 (left). As can be seen in Fig. 2 (middle), the attack path length seems to have negligible effect on the detection percentage. This can be explained by recalling that in short attack paths there is a low chance for any of the routers to mark a given packet, while in long attack paths there is a low chance for most of the routers to mark the given packet.

Fig. 2 (right) depicts the effect of increasing the number of attackers m while fixing the number of attack packets used by the victim k. It is clear that the detection percentage is affected negatively by this increase. In fact, the detection percentage in this case can not be increased without increasing the number of packets k used by the victim.

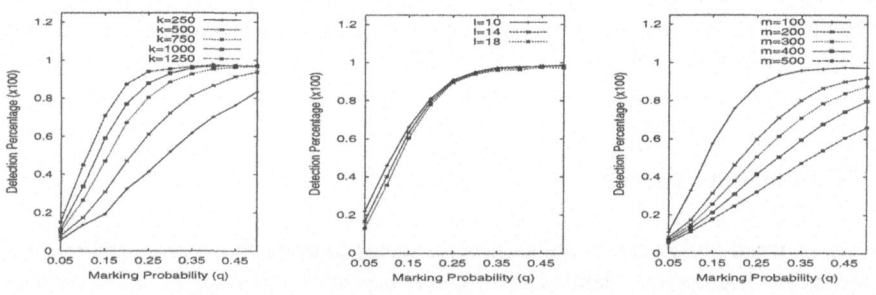

Fig. 2. Left: Effect of number of attack packets k used by the victim on the detection percentage. Middle: Effect of attack path length l on the detection percentage. Right: Effect of number of attackers m on the detection percentage.

6   Conclusion

An efficient traceback scheme is necessary to identify the sources of denial of service attacks which impose an imminent threat to the availability of Internet services. In this paper we proposed an efficient traceback scheme called DLLT. In this scheme, the probabilistic nature of marking and storage offers the advantage of minimizing router and storage overhead. Also, storing the packet digests at intermediate routers provides an authentic way to verify that a given router has actually forwarded certain packet. This prevents attackers from passing spoofed marking information to the victim even if the marking probability is very low. DLLT employs an efficient scheme to collect marking information from intermediate routers. Moreover, we showed that the number of packets required to identify the attack sources is low. Simulation studies show that DLLT offers high attack source detection percentage.

References

1. B. H. Bloom, "Space/time trade-offs in hash coding with allowable errors," in *Communications of ACM 13*, July 1970, 422-426.
2. H. Burch and B. Cheswick, "Tracing anonymous packets to their approximate source," in *Proc. 2000 USENIX LISA Conf.*, Dec. 2000, pp.319-327.
3. D. Dean, M. Franklin, and A. Stubblefield, "An algebraic approach to IP traceback," in *Network and Distributed System Security Symposium (NDSS '01)*, Feb. 2001.
4. M. T. Goodrich, "Efficient Packet Marking for Large-Scale IP Traceback," in *Proc. of ACM CCS 2002*, Nov. 2002.
5. K. Park and H. Lee, "On the Effectiveness of Probabilistic Packet Marking for IP Traceback under Denial of Service Attack," in *Proc. of IEEE INFOCOM 2001*, Mar. 2001.
6. S. Savage, D. Wetherall, A. Karlin and T. Anderson, "Practical network support for IP traceback," in *Proc. of ACM SIGCOMM*, Aug. 2000, pp. 295-306.
7. D. Song and A. Perrig, "Advanced and authenticated marking schemes for IP traceback," in *Proc. of IEEE INFOCOMM 2001*, April 2001.
8. R. Stone, "Centertrack: An IP overlay network for tracking DoS floods," in *Proc. of 9_{th} USENIX Security Symposium*, Aug. 2000.
9. A. C. Snoeren, C. Partiridge, L. A. Sanchez, C. E. Jones, F. Tchhakountio, S. T. Kent, and W. T. Strayer, "Hash-Based IP TraceBack," in *Proc. of ACM SIGCOMM*, Aug. 2001.

A Geomulticast Architecture and Analysis Model for Ad-Hoc Networks

Beongku An[1] and Dohyeon Kim[2]

[1]School of Electronic, Electrical and Computer Engineering, Hongik University
Jochiwon, Chungnam, Korea, 339-701, Tel.:041-860-2243
beongku@wow.hongik.ac.kr
[2]School of Information and Communications, Cheonan University
Cheonan-City, Chungnam, Korea, 330-180, Tel.:041-620-9418
dhkim@cheonan.ac.kr

Abstract. In this paper, we propose a geomulticast architecture and analysis model for ad-hoc networks. We define and formulate the probability of geomulticast error and geomulticast error cost. Based on the geomulticast architecture and analysis model, we present how to evaluate the performance of the geomulticast architectures and protocols, and we obtain some insight about the impact of the design and operational parameters on the system performance.

1 Introduction

Geomulticast is a specialized location-dependent multicasting technique, where messages are multicast to some specific user groups within a specific geographical zone. While conventional multicast protocols define a multicast group as a set of nodes with a multicast address and geocast[1,2] defines a geocast group as all the nodes within a specified zone at a given time, a geomulticast group is defined as a set of nodes of some specific groups within a specified zone. In general, geomulticast combines geocast and multicast for dynamic message delivery to a specific group within a geomulticast zone. If a specific geographic zone is very large and there are many different kind of groups within this zone, we do not need to broadcast to all nodes within the specific zone. In this paper, we propose a geomulticast architecture and analysis model to support geomulticast services for performance analysis and evaluation in the challenging environment of mobile ad-hoc wireless networks. We define and formulate the probability of geomulticast error and geomulticast error cost in order to evaluate the performance of the design alternatives and protocols used for the support of geomulticast services in mobile ad-hoc wireless networks.

2 Geomulticast Architecture and Analysis Model

To support geomulticast, it is first needed to define the geomulticast zone formulation method and geomulticast zone representation method. We use several zone formulation methods that range from simple specific zone formulation techniques (i.e., circle

N. Mitrou et al. (Eds.): NETWORKING 2004, LNCS 3042, pp. 1270–1275, 2004.

with given radius) to the most general cases, while we use several approximation methods for geomulticast zone representation to reduce routing overhead as follows: circular representation, elliptical representation, and rectangular representation. Fig. 1 presents a high level description of the proposed network architecture for supporting geomulticast using clustering[3] as underlying structure in mobile ad-hoc wireless networks. The key idea of the proposed network structure is the extension of cellular network concepts into mobile ad-hoc wireless networks to support geomulticast services in a stable and cost-effective way. Our network structure for supporting geomulticast services consists of the following elements and components: Geomulticast Control Office (GeoCO), clusterheads (CHs), mobile nodes (MNs). The functions performed by CH and GeoCO are similar to those performed by BS (base station) and MSC (mobile switching center) respectively in cellular networks. In our structure, the source clusterhead (CH) plays the role of GeoCO. The main function of GeoCO is the membership management (i.e., geomulticast group construction and maintenance).

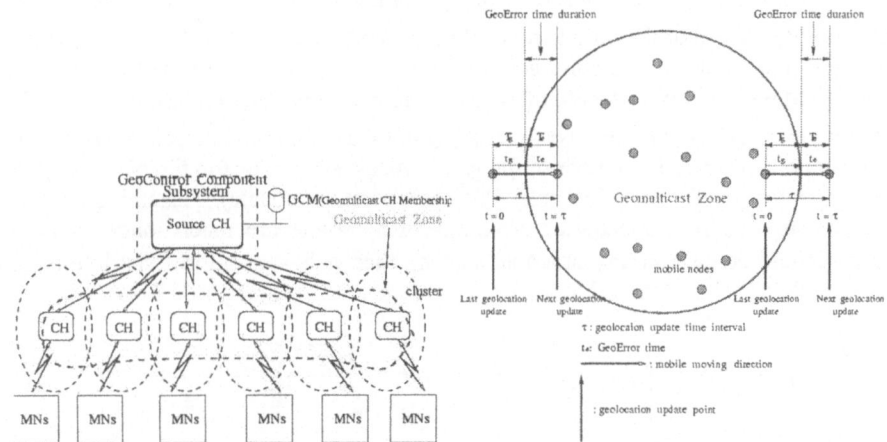

Fig. 1. Network architecture   **Fig. 2.** The model

Next, we define and formulate a framework model in order to evaluate the performance of the design alternatives, algorithms and implementations used for geomulticast services in mobile ad-hoc wireless networks. Since the main objective of our efforts is to design a modeling framework for performance analysis and evaluation, we introduce and define the probability of geomulticasting error (GeoError), Pge, and the Geomulticast Packet Delivery Ratio(GPDR) as performance measures. The probability of GeoError (Fig. 2), Pge, refers to the probability of occurrence of either of the following events: a) a mobile that is inside the geomulticast zone does not receive the geomulticast message for some time, and b)a mobile that is not in the geomulticast zone does receive the geomulticast message for some time. Let us denote by τ the geolocation update time interval and by T_e the GeoError time interval for a specific mobile due to either one of the two events described above (see fig. 2).

From the above discussion, it is obvious that the rate of geomulticast zone boundarycrossing, λ_c, before the geomulticast membership list is updated is one of the most significant contributors to the total probability of GeoError, especially when the geomulticast membership decision is made at the network level. Here, we develop a framework in order to evaluate the effects and associated trade offs, and we gain some insight about the optimum location update interval through a numerical study. Specifically, we study the effect of the mobile node geolocation periodic update interval on the packet delivery accuracy and the associated overhead (signaling load). This model assumes that the mobile nodes move with constant velocity during time interval τ in a random direction θ that is uniformly distributed in $[0, 2\pi]$. Considering a geomulticast area with perimeter S where ρ mobile terminals per unit area are located, the rate of geomulticast zone boundary crossing per unit time (average number of mobile nodes crossing the area border per unit of time), denoted by λ_c is where β denotes the activity ratio of mobiles (in the following without loss of generality, we may assume $\beta = 1$), and V is the average mobile node velocity. For example, if the geomulticast zone is represented by a circle with radius R then $\lambda_c = 2\rho VR$. In general, λ_c depends on the actual size of the geomulticast zone, the network density, and the mobility pattern of the mobile nodes (e.g. velocity). We assume that the number of messages that arrive at a mobile node during a time t is a Poisson random variable with λ_a. Moreover, let us denote by te the random variable that represents the GeoError time interval (i.e. the time between the point that a mobile is erroneously considered (or not considered) member of the geomulticast session due to the two events described before, and the point until the next position update). Assuming that the interval between two boundaries crossings of a mobile are much longer than τ, which should be case in any practical system, then te has a uniform distribution in interval $[0, \tau]$. Thus, the probability of k message arrival during time Te is

$$P_e[n = k] = \int_0^\tau P[n = k \,|\, T_e = t] f_e(t) dt = \int_0^\tau \frac{e^{-\lambda_a t}(\lambda_a t)^k}{k!} \frac{1}{\tau} dt \qquad (1)$$

Since as mentioned before te follows a uniform distribution between $[0, \tau]$, we can easily show based on the properties of conditional expectation the average rate of error ($E[error\ packet\ rate]$) message arrival. Then, we can calculate the probability of GeoError(P_{ge}) as follows:

$$P_{ge} = \frac{E[error\ packet\ rate]}{E[desired\ packet\ rate]} = \frac{2*\lambda_a*\lambda_c*\frac{\tau}{2}*h}{\lambda_a*B} = \frac{\lambda_c*\tau*h}{B} \qquad (2)$$

where B denotes the average number of nodes within the geomulticast zone that are geomulticast members. Based on the model under consideration, the total number of nodes within a zone of area F is given by $\rho *F$, and therefore $h = \frac{B}{\rho*F}$. Substituting this in the previous relation, we obtain: $P_{ge} = \frac{\lambda_c*\tau}{\rho*F}$. The Geomulticast Packet Delivery Ratio(GPDR) can be defined as: $GPDR = 1 - P_{ge}$.

In the following, we develop a cost function to quantify the trade off between the penalty associated with the occurrence of errors in packet delivery and the associated cost for supporting certain degree of accuracy in a geomulticast architecture[6]. We consider only the cost directly related to the geomulticast session setup, the update of the geolocation and the loss of packets due to the inaccuracy of position information as a result of the node mobility and the gelocation update time interval. Let us denote by Csetup the cost for geomulticast session setup, by Cgeo-update the cost of geolocation update, and by Cerror the GeoError cost per time unit. Then, the corresponding average cost per time unit, Ctotal, is: Ctotal = Csetup + Cgeo-update + Cerror.. . . We denote by a the rate at which geolocation updates are generated by a mobile node(e.g. updates per time unit). Let us denote by c_n, n=1,2,..., the cost of loosing n calls. Then, the corresponding average GeoError cost per time unit, Cerror is:

$$C_{error} = 2 * \lambda_c * h * \sum_{n=1}^{\infty} c_n P_e[n] \tag{3}$$

For the sake of simplicity in the following, we can reasonably assume that the cost of loosing the n_{th} message arrival after geomulticast zone boundary crossing by a mobile node is independent of n. Thus, $\forall\ n,\ c_n = n\ c_1$. Consequently:

$$C_{error} = 2 * \lambda_c * h * c_1 \sum_{n=1}^{\infty} n P_e[n] = c_1 * 2 * \lambda_c * h * \lambda_a * \frac{\tau}{2} \tag{4}$$

Finally, we get:

$$C_{total} = C_{setup} + C_{geo\text{-}update} + c_1 * 2 * \lambda_c * h * \lambda_a * \frac{\tau}{2} \tag{5}$$

3 Performance Evaluation and Discussion

In this section, we present some numerical results that we obtained based on the analytical framework as well as via modeling and simulation using the Optimized Network Engineering Tool (OPNET).

A mobile ad-hoc network consisting of 50 mobile nodes that are placed randomly within a rectangular region of 1000 m x 1000 m is modeled for the purposes of this study. The results presented in the following section correspond to the use of circular geomulticast zone with radius R (150 m - 300 m). The mobile nodes are assumed to have constant radio range of Z=250 m. Throughout our study, we assume that a link fails, or reappears, as a node goes out or in transmission range of another node, due to the mobility of the nodes. Mobile nodes are assumed to be moving around throughout the network. The speed and the direction of each move are uniformly distributed, with speed range [0, 80 km/h] and direction range [0, 2π], respectively. We use one geomulticast source node (sender) while the gomulticast members are selected randomly within the geomulticast zone.

One of the main objectives of this numerical evaluation is to compare the geomulticast packet delivery ratio between simulation results and results of analysis using the proposed model. We implement a geomulticast routing protocol[6], namely DGR

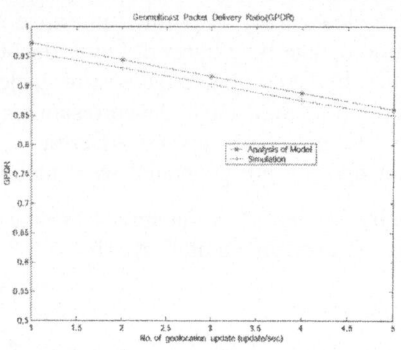

Fig. 3. GPDR as a function of update interval

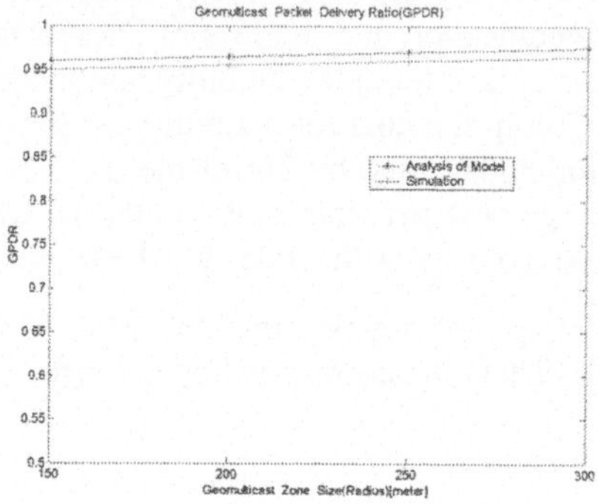

Fig. 4. GPDR as function of geomulticast zone size

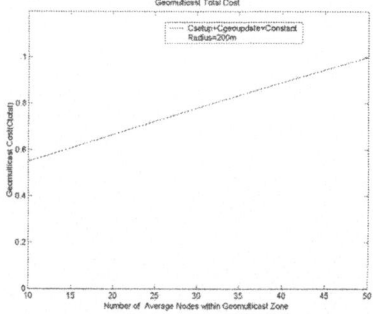

Fig. 5. Ctotal as a function of average number of nodes within geomulticast zone

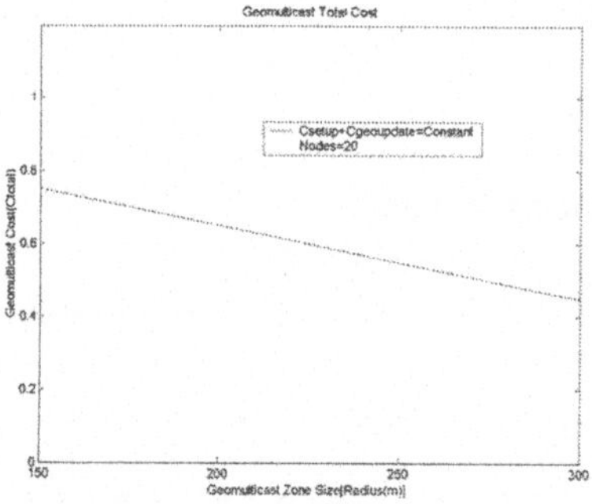

Fig. 6. Ctotal as a function geomultict zone size(Radius)

(direction-guided routing). Fig. 5 presents the geomulticast packet delivery ratio as a function of geolocation update time interval (i.e., τ) for mobility 40 km/hr, packet arrival rate(PAR) 5 packets/sec and geomulticast zone with Radius 250 m. From this figure we can see that geomulticast packet delivery ratio(GPDR) is decreasing with the increase of τ. However, as we discussed before, this is accompanied with a decrease in signaling load due to the position update generation process. Specifically, fig. 4 presents the GPDR as a function of the geomulticast zone size (i.e., radius R) for mobility 40 km/hr, PAR 5 packets/sec and geolocation update time interval τ =1 sec. As we can see from this figure, the geomulticast packet delivery ratio remains relatively constant with the change of the geomulticast zone size. The main goal of the fig. 5 and figure 6 is the evaluation of the error cost (Cerror) for supporting geomulticast services. Therefore, we assume Csetup and Cgeo-update are constant while Ctotal is normalized as Ctotal = 1 when Radius is 200m and nodes within geomulticast zone is 50. Fig. 5 presents the total cost as a function of the average number of nodes within the geomulticast zone while fig. 6 presents the total cost as a function of geomulticast zone size. As can be see from these figures(fig. 5 and fig. 6), the total cost for geomulticast services is linearly decreased as the geomulticast zone size is increased while the total cost is linearly decreased as the average number of nodes within the geomulticast zone is increased.

The results in figure3-figure6 show the proposed model can be efficiently used as measures of performance analysis and evaluation for supporting geomulticast services in mobile ad-hoc wireless networks.

4 Conclusion

In this paper, we have presented a geomulticast architecture and analysis model in order to analyze and evaluate the performance of the design alternatives for algorithms and implementations used for geomulticast services. The probability of GeoError, the geomulticast packet delivery ratio, and GeoError cost are derived and the tradeoff between the geolocation update interval is discussed. The performance evaluation of the modeling framework has demonstrated their efficiency in terms of geomulticast packet delivery ratio(GPDR) as a function of geomulticast zone size, geolocation update time interval. The results in figure3-figure6 show the proposed model can be efficiently used as measures of performance analysis and evaluation for supporting geomulticast services in mobile ad-hoc wireless networks.

References

1. J.C. Navas and T. Imielinski, "GeoCast - Geographic Addressing and Routing", Proc. of MOBICOM'97, pp.151-160, Budapest, Hungary, September 26-30, 1997.
2. Y.-B. Ko, N.H. Vaidya, "Geocasting in mobile ad-hoc networks: location-based multicast algorithms", Proc. of IWMC-Sa'99, New Orleans, USA, 1999.
3. Beongku An, Symeon Papavassiliou, "A Mobility-Based Approach to Support Mobility Management and Multicast Routing in Mobile Ad-hoc Wireless Networks", The International Journal of Network Management(JNM), vol.11, no.6, pp.387-395, December 2001.
4. I.F. Akyildiz and J.S.M. Ho, "Dynamic mobile user location update for wireless PCS networks", Wireless Networks 1 (1995) 187-196.
5. A. Bar-Noy, I. Kessler and M. Sidi, "Mobile users: To update or not to update ?", Wireless Networks 1 (1995) 175-186.
6. Beongku An, Symeon Papavassiliou, "An architecture for supporting geomulticast services in mobile ad-hoc wireless networks", Proc. of MILCOM2001, USA, October 2001.

An Architecture to Support QoS Multicast Routing for Ad-Hoc Networks

Beongku An[1], Do Hyeon Kim[2], and Nam-Soo Kim[3]

[1]School of Electrical, Electronic & Computer Engineering, Hongik University
Jochiwon, Chungnam, Korea, 339-701,Tel.: +82-41-860-2243
beongku@wow.hongik.ac.kr
[2]School of Information and Communications, Cheonan University, Cheonan-City,
Chungnam, Korea, 330-180 Tel.: +82-41-620-9418, Fax: +82-41-620-9507
dhkim@cheonan.ac.kr
[3]Dept. of Computer and Communication Engineering, Chongju University, Chongju-City,
Chungbook, Korea, 360-764 Tel.: +82-43-229-8459
nskim@chongju.ac.kr

Abstract. In this paper, we present an architecture for supporting QoS multicast routing in ad-hoc networks. The proposed architecture consists of three parts as follows. The first part is a clustering as underlying structure for supporting stable multicast services. In the second part, a framework which can support and evaluate the stability of route and network for supporting QoS routing is presented. In the third part, we describe a method which uses two structures of the first and second parts for supporting QoS multicast routing services.

1 Introduction

The goal of mobile ad-hoc wireless networking is to extend mobility into the realm of a set of wireless mobile nodes, where they form the network infrastructure in an ad-hoc fashion. Management functions, routing and multicasting in mobile ad-hoc wireless networks present more complex problems than in wired or last-hop networks, due to the random movement of nodes, the bandwidth and power limitations, and the lack of fixed infrastructure. In this paper, we present an architecture for supporting QoS multicast routing services for mobile ad-hoc wireless networks in those challenging environments as follows. First, a mobility-based clustering approaches and group based hierarchical structures are presented in order to facilitate the implementation of efficient and scalable multicasting techniques and mobility management functions. Second, we present a modeling framework to support and evaluate route and network stability for QoS routing services in mobile ad-hoc wireless networks. To support stable QoS routing services, the support and evaluation of the stability of route and network are very important. For example, if we can select most stable route among multiple routes between source node and destination node, we can save much bandwidth and reduce delay and overload. Third, we describe a method which can support QoS multicast routing services using two structures presented above.

The remaining of this paper is organized as follows. Section 2 describes the architecture for supporting QoS multicast routing services in mobile ad-hoc wireless net

N. Mitrou et al. (Eds.): NETWORKING 2004, LNCS 3042, pp. 1276–1281, 2004.
© IFIP International Federation for Information Processing 2004

works. Section 3 contains the performance evaluation of our proposed methods, while section 4 concludes the paper.

2 The Architecture

First, we introduce a mobility-based clustering structure to support a stable multicast service. The main idea of the clustering is to combine both physical and logical partitions of the network as well as the concept of relative mobility in order to improve the stability in the clustered topology. The proposed mobility-based hierarchical clustering algorithm[1] may result in variable-size clusters depending on the mobility characteristics of the nodes. A group may consist of clusters that present similar mobility characteristics. Several groups can be hierarchically merged into one group depending on the mobility of each group. Fig. 1 shows the basic concepts of the proposed clustering.

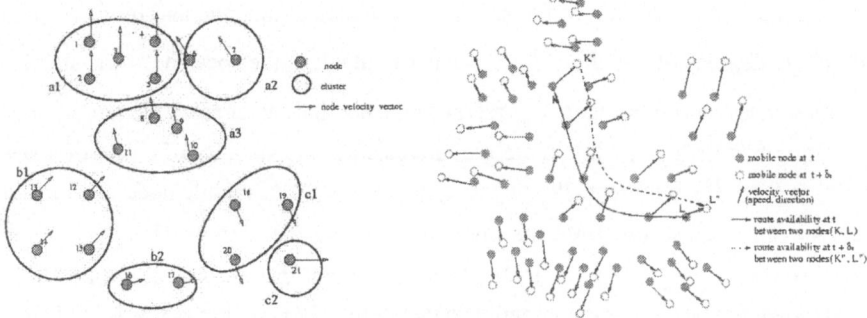

Fig. 1. Basic concepts of the clustering   **Fig. 2.** Route stability for supporting QoS routing

Second, we present a modeling framework for supporting QoS routing. The main goal of the presented modeling framework is to select the most stable route between a source and a destination, in an environment where multiple paths are available, as well as to create a convenient performance measure to be used for the evaluation of the stability and connectivity in a mobile ad-hoc wireless networks. The basic motivations of the proposed modeling approach stem from the commonality observed in the location uncertainty in mobile ad-hoc wireless networks and the concept of entropy [2]. We also associate each node m with a set of variable features denoted by $a_{m,n}$ where node n is a neighbor of node m. In this paper, two nodes are considered neighbors if they can reach each other in one hop (e.g. direct communication). These variable features $a_{m,n}$ represent a measure of the relative speed among two nodes and are defined rigorously later in this section. Any change of the system can be described as a change of variable values $a_{m,n}$ in the course of time t such as $a_{m,n}(t) \rightarrow a_{m,n}(t+\Delta_t)$. Let us also denote by v(m,t) the velocity vector of node m and by v(n,t) the velocity vector of node n at time t. Please note that velocity vectors v(m,t) and v(n,t) have two parameters, namely speed and direction. The relative velocity v(m,n,t) between nodes m and n at time t is defined as: $v(m,n,t) = v(m,t) - v(n,t)$. Then, the relative mobility between any pair (m, n) of nodes during some time interval is defined as their absolute relative speed averaged over time. Therefore, we have:

$$a_{m,n} = \frac{1}{N}\sum_{i=1}^{N}|v(m,n,t_i)|$$

. where N is the number of discrete times t_i that velocity information can be calculated and disseminated to other neighboring nodes within time interval. Based on this, we can define the entropy $H_m(t,\Delta_t)$ at mobile m during time interval. The entropy can be defined either within the whole neighboring range of node m, or for any subset of neighboring nodes of interest. In general, the entropy $H_m(t,\Delta_t)$ at mobile m is calculated as follows:

$$H_m(t,\Delta_t) = \frac{-\sum_{k \in F_m} P_k(t,\Delta_t)\log P_k(t,\Delta_t)}{\log C(F_m)}$$, where $$P_k(t,\Delta_t) = \frac{a_{m,k}}{\sum_{i \in F_m} a_{m,i}}$$. In this relation by F_m, we

denote the set (or any subset) of the neighboring nodes of node m, and by $C(F_m)$ the cardinality (degree) of set F_m. If we want to calculate the local network stability (with reference to node m), then F_m refers to the set that includes all the neighboring nodes of mobile node m , while if we are interested in the stability of a part of a specific route then F_m represents the two neighboring nodes of mobile node m over that route. As can be observed from the previous relation, the entropy $H_m(t,\Delta_t)$ is normalized so that $0 \quad H_m(t,\Delta_t) \quad 1$. It should be noted that the entropy, as defined here, is small when the change of the variable values in the given region is severe and large when the change of the values is small [2]. Let us present the route stability (RS) between two nodes k and l during some interval Δ_t as $\gamma = RS_{k,l}(t,\Delta_t)$. We also define and evaluate two different measures to estimate and quantify end to end route stability, denoted by $\gamma^1 = RS^1_{k,l}(t,\Delta_t)$ and $\gamma^2 = RS^2_{k,l}(t,\Delta_t)$ and defined as follows respectively:

$$\gamma^1 = RS^1_{k,l}(t, \Delta_t) = \prod_{t=1}^{N_r}[H_i(t,\Delta_t)] \qquad (1)$$

$$\gamma^2 = RS^2_{k,l}(t,\Delta_t) = \min_{i=[1,2,3,...,N_r]} [H_i(t, \Delta_t)] \qquad (2)$$

where Nr denotes the number of intermediate mobile nodes over a route between the two end nodes (k, l). Parameter $\gamma^1(\gamma^2)$ can be used to measure the route availability and stability.

Third, we present a method which uses two structures of the first and second parts for supporting QoS multicast routing services. The main objective for supporting QoS multicast routing is to reduce the unnecessary routes and the redundancy for bandwidth saving as well as delay and overload. Fig. 3 describes the steps for supporting QoS multicast routing on an example multicast routing protocol, namely MHMR[3], for mobile ad-hoc wireless networks. In first step (Fig.3(a)), MHMR uses a limited mesh structure over the clusterheads only that are created by the underlying mobility-based clustering. In second step (Fig.3(b)), although the MHMR creates a mesh infra-structure for multicast routing, we are trying to use a source-based tree for multicast packet forwarding to support QoS multicast routing. Then, we can send multicast message data over the most stable single route among the mesh routes depending on the stability of routes. Therefore, we can support QoS multicast routing services(i.e., reduce redundancy and unnecessary routes for bandwidth saving, reduce delay and overload, increase throughput) in mobile ad-hoc wireless networks.

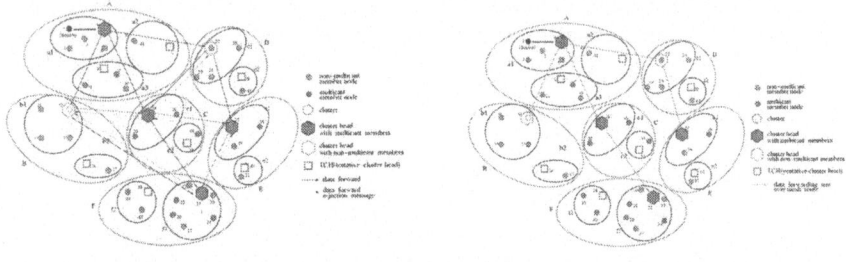

<table>
<tr><td>(a) Mesh structure</td><td>(b)Tree structure supporting route stability</td></tr>
</table>

Fig. 3. Supporting QoS multicast routing

3   Evaluation and Discussion

The performance evaluation of our protocol is accomplished via modeling and simu-lation using the Optimized Network Engineering Tool (OPNET). A mobile ad-hoc network consisting of 200 nodes that are placed randomly within a rectangular region of 2 km x 2 km is modeled in the simulation. Each node is modeled as an infinite-buffer, store-and-forward queuing station, and is assumed to be aware of its position with the aid of a reliable position location system (i.e., GPS). The mobile nodes are assumed to have constant radio range of $Z= 250$m. Two different mobility scenarios are considered in this study. In the first one (in the following, we refer to it as mobil-ity 1), the speed and the direction of each move are uniformly distributed with speed range [0, Vmax km/h] and direction range [0, 2π] respectively. In the second one (in the following, we refer to it as mobility 2), a group-based mobility pattern[4] is mod-eled. Specifically, nodes are grouped into several groups, where we assume that nodes in the same group have similar mobility characteristics (speed and direction). The speed and direction of each group are selected randomly at the start point of the simulation within the speed range [0, Vmax km/h] and the direction range [0, 2π] and is assumed that the group holds these speed and direction for the duration of the simulation. Initially, each group consists of 10 nodes. If a mobile arrives at the boundary of the given network coverage area, the node reenters into network.

In the first part simulation, we measure and evaluate the stability of the created clustered topology for supporting multicast. The used metrics is number of cluster changes that represent a measure of the rate of clusterhead changes (or construction of

new clusters). Fig. 4 presents the cluster stability for individual random mobility scenario (i.e., mobility 1) under three different clustering algorithms: lowest-ID algorithm[5], highest-connectivity (degree) algorithm[6], and MBC algorithm. Similarly,
Figure 5 shows the corresponding cluster stability under the group mobility scenario
(i.e. mobility 2). As can be seen by the first set of fig. 4, MBC algorithm outperforms
both of them by improving the cluster stability metrics. The improvement is considerably higher in the experiment using group mobility scenario (Fig. 5) where MBC
algorithm reduces the clusterhead change ratio by approximately several times compared to the lowest-ID algorithm and the highest-connectivity algorithm. This is
achieved because MBC algorithm takes advantage of the similarity that several nodes
demonstrate in this set of experiment due to the group mobility pattern assumed.

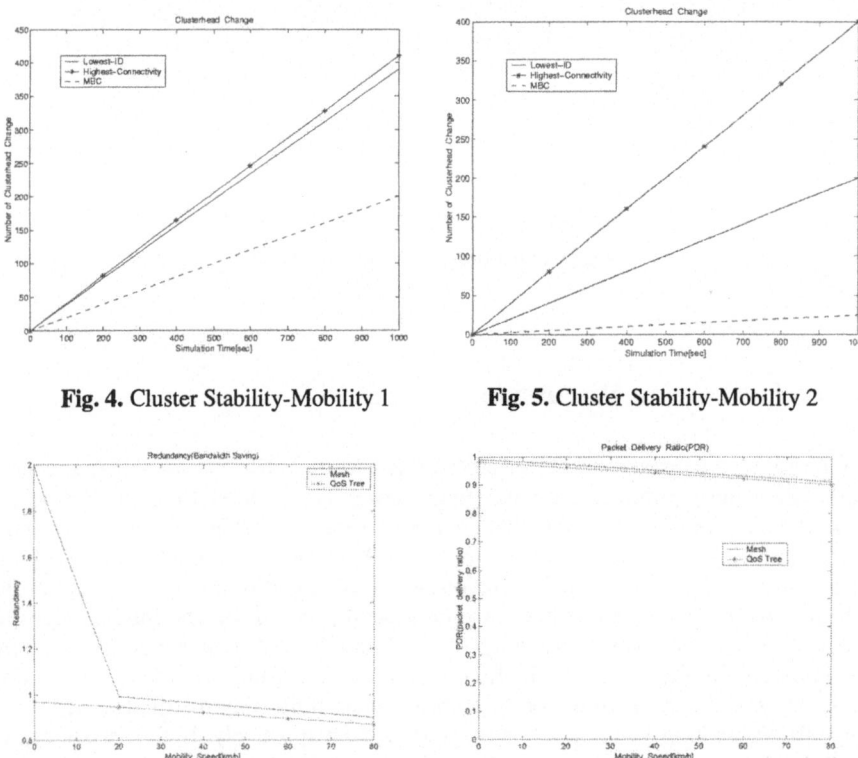

Fig. 4. Cluster Stability-Mobility 1   **Fig. 5.** Cluster Stability-Mobility 2

Fig. 6. Redundancy(Bandwidth Saving)-   **Fig. 7.** Packet delivery ratio-Mobility2
Mobility2

In the next part simulation, we are trying to measure and evaluate the degree of
QoS multicast routing on the implemented multicast routing protocol, namely MHMR
[3]. The performance metrics, redundancy & bandwidth saving, are studied. Fig. 6
present the degree of redundancy for group mobility at one destination node per one
packet sending from one source as mobility function, while fig. 7 present the packet
delivery ratio as mobility function group mobility. As can be seen by figure 6 and fig.

7, even though the redundancy using QoS tree is less than the redundancy using just mesh concept, as can be seen by figure 6, the packet delivery ratio is similar for mesh by fig. 7. The reason is that QoS tree can select the most stable single route among the multiple routes between source node and destination node.

4 Conclusions

In this paper, we present an architecture for supporting QoS multicast routing in mobile ad-hoc wireless networks. The proposed architecture consists of three parts as follows. First, we present a clustering structure which can support stable multicast routing services. Second, a framework which can support and evaluate the stability of route and network for supporting QoS routing is presented. Third, we present a method for supporting QoS multicast routing services in mobile ad-hoc wireless networks. The performance evaluation of our proposed methods is accomplished via modeling and simulation. The performance results demonstrate that our proposed architecture can efficiently support QoS multicast routing services in mobile ad-hoc wireless networks.

References

1. R. Ramanathan and M. Steenstrup, "Hierarchically-Oragnized, Multihop Mobile Wireless Networks for Quality-of-Service Support", ACM/Baltzer Mobile Networks and Applications, vol. 3, no. 1, p.p. 101-119, 1998.
2. Akira Shiozaki, "Edge Extraction Using Entropy Operator", Computer Vision, Graphics, and Image Processing 36, 1-9, 1986.
3. Beongku An and Symeon Papavassiliou, "MHMR: A Mobility-Baed Hybrid Multicast Routing Protocol in Mobile Ad-hoc Wireless Networks", Wireless Communications and Mobile Computing (WCMC), vol.3, issue 2, pp.255-270, 2003.
4. Xiaoyan Hong, Mario Gerla, Guangyu Pei and Ching-Chuan Chiang, "A Group Mobility Model for Ad-Hoc Wireless networks", Proc. of ACM/IEEE MSWiM'99, August 1999.
5. Anthony Ephremides, Jeffrey E. Wieselthier, and Dennis J. Baker, "A design concept for reliable mobile radio networks with frequency hopping signaling", Proc. of IEEE 75(1), pp.56-73, 1987.
6. Abhay K. Parekh, "Selecting routers in ad-hoc wireless networks", in ITS, 1994.

Linkcast: Fast and Scalable Multicast Routing Protocol

Mozafar Bag-Mohammadi, Siavash Samadian-Barzoki, and Nasser Yazdani

Router Lab., Dept. of EE & Computer Eng., Univ. of Tehran, Tehran, Iran
{mozafarb, s.samadian}@ece.ut.ac.ir, yazdani@ut.ac.ir

Abstract. The current multicast routing protocols require state maintenance in the on-tree routers in order to forward multicast packet properly. Therefore, the routers memory may be easily saturated when there are very large number of low to moderate size multicast groups. In contrast, the explicit multicast protocols offer a stateless design using header space of multicast data packets. In this paper, we introduce new stateless scheme called Linkcast that efficiently eliminates processing overhead of explicit multicast protocols like Xcast. The Linkcast represents the multicast tree by encoding its counterpart links. Simulation results show that Linkcast completely eradicates the required unicast lookups in explicit multicast protocols with less header size overhead.

1 Introduction

IP multicast significantly enhances bandwidth utilization eliminating duplicate packets crossing the network links. Traditional multicast routing protocols like DVMRP [11] and PIM-SM [12] require state maintenance in the on-tree routers, commonly known as Multicast Forwarding Table (MFT), in order to forward multicast packet properly. The state maintenance is performed in a per-group basis at on-tree routers. Therefore, the routers may easily run out of memory when there are very large number of low to moderate size multicast groups.

Although current multicast routing protocols are able to support small number of large multicast groups, they don't scale well when supporting very large number of low to moderate size multicast groups [1]. There are many small to moderate size multi-party applications such as video and audio conferencing, IP telephony and network games, which can not be well serviced in large scale by the current model. Xcast [1] [2] [3] [4], its variations Xcast+ [5] [6], Bcast [8] and ERM [7] are designed to serve this class of applications in a scalable manner.

In Xcast and Xcast+, a list of destination IP addresses is sent with each multicast data packet. In Bcast and ERM, IP addresses of the receivers and branching points of multicast tree are sent with each Bcast packet. These protocols trade-off the header size and processing power to obtain scalability and simplicity. The processing overhead consists of two parts: the number of required unicast lookups and the header processing. In Linkcast, we achieve the same goal with very small processing overhead.

N. Mitrou et al. (Eds.): NETWORKING 2004, LNCS 3042, pp. 1282–1287, 2004.

In Linkcast, multicast sender encodes the tree listing tree links in a proper way. We present two complementary coding schemes, which can be used interchangeably in different circumstances. The first encoding scheme is appropriate when distances between the branching routers are short. Otherwise, the other scheme is suitable. Simulation results for various network graphs show that the header size overhead in Linkcast is comparable with Bcast and Xcast. This is more noteworthy if one considers negligible Linkcast forwarding cost.

In section two, we briefly describe the related work. Then, the key components of Linkcast are discussed in section three. Section four deals with the simulation results. Finally, we conclude the paper in section five.

2   Related Work

Xcast [1] [2] [3] [4] was originally designed to overcome scalability and deployment problems of the current multicast routing protocols. In Xcast, multicast sender simply encodes the list of receiver IP addresses in a special header and sends it with all data packets. Therefore, at each router, each destination in the Xcast header of the packet requires a unicast lookup. This processing overhead limits the number of possible destinations to a very small number (less than 10 [14]). Xcast+ [5] [6] introduces a simple modification to Xcast combining it with the well-known IGMP (Internet Group Management Protocol) [13] protocol.

Sender Initiated Multicast (SIM) [9] is another Xcast like scheme which reduces Xcast forwarding cost. SIM capable routers construct an MFT-like table to forward multicast packets. This state maintenance severely limits SIM scalability. Simple Explicit Multicast (SEM) [10] uses the receivers' list to construct the multicast distribution tree. SEM also suffers from scalability issues due to the state maintenance at branching routers of the multicast tree.

Bcast [8] efficiently removes unnecessary lookups that take part in Xcast and Xcast+ forwarding mechanism. In Bcast, IP addresses of the receivers and branching points are encoded in Bcast header. Furthermore, destination IP address field in IP header is filled with address of the next branching point. This enables Bcast to use unicast forwarding between each branching point pair.

3   Linkcast

3.1   Tree Encoding

The main objective of tree encoding method is to minimize size of generated code. We believe that the arrangement of the tree nodes is the most important parameter that affects the code size. The node in a multicast tree can be partitioned into three categories based on the node degree [15] : 1- Member nodes which have degree 1, 2- Relay nodes which have degree 2, 3- Branching nodes which have degree 3 or more. We introduce two encoding schemes that consider different mix of relay and branching nodes. The first scheme is appropriate when there is large number of relay nodes in the tree. We call this mode Sparse

Branching Mode or SBM for short. The second is suitable for a tree with high average node degree. The second scheme is called Dense Branching Mode or DBM shortly. In either case, the generated code is placed in a special header between IP header and transport layer header.

We explain SBM and DBM encodings through an example. The example tree and its corresponding codes are shown in Fig. 1. The number on each link is the local link ID (see subsection 3.2). In both method, we differentiate between different types of link by means of link ID coding.

Each link in SBM may have one or more pointers which point(s) to the next link(s) in the tree. Obviously, the member links e.g. m2 need not to have any pointer. The incoming link of a relay node e.g. f has only one next link. We arrange the tree links so that the pointer value for this type of links is always one eliminating the need of storing a pointer for relay links. For incoming link of a branching node, the branching factor must be determined. In addition, we need to store a pointer for each of its branches.

In the DBM, each links has a pointer which points to the end node of the link. For member links, the end node is not required to be in the code. The outgoing link(s) of each node begin from the node itself to the next node in the code. The nodes are represented with a special value in the code. The number of required bytes in the SBM and DBM coding can be calculated as:

SBM: $(\#oftreebranches) + total\_br + (\#ofbranchingnodes)$ (1)

DBM: $(\#oftreebranches - 1) + total\_br + (\#ofrelaynodes)$ (2)

where total_br is total branching factor of branching nodes. In the above example the code sizes are 30B and 34B for SBM and DBM coding respectively.

The tree has a general pointer P1 which points to current link in the code. When a router receives a Linkcast packet, it examines the value of P1. Then, it finds the next link(s) by interpreting the tree code. Finally, it modifies P1 value accordingly and forwards the packet.

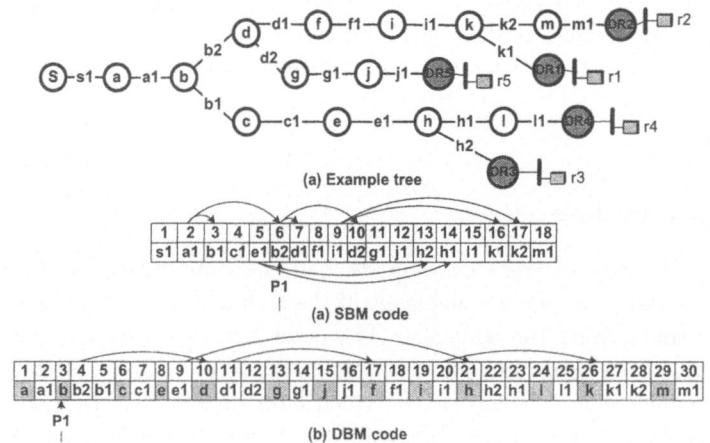

Fig. 1. An example tree and resulted tree codes

3.2   Gathering Tree Information

The sender collects the tree information from the receivers using their *Join* messages. Each new member sends a *Join* message to the source of the multicast session. All routers on the path must examine the message and append incoming link ID, outgoing link ID and their IP addresses to the packet. Having the path information from receivers to the source, the source can construct reverse shortest path tree. The receivers must repeat their *Join* messages periodically in order to refresh their state in sender. Thus, the sender can repair the multicast distribution tree against temporary route changes. If the sender misses three consecutive *Join* messages from a receiver, it will remove the receiver from the tree code. The receivers can also immediately depart the multicast session by sending *Leave* messages to the sender.

3.3   Multi-access Links

In a multi-access link, the broadcast nature of the link is source of ambiguity. Since possible end nodes of the link are more than one, it is not clear that what node actually belongs to the tree. To solve this ambiguity, we decompose a multi-access link into $n * (n-1)/2$ virtual links connecting each pair of nodes where n is the number of nodes on the link. Each virtual link has a unique ID. All nodes must know the ID of each virtual link. Therefore, the nodes must run a simple protocol to agree on the ID of the resulted virtual links.

3.4   Source Branching

Since the sender generates the encoded tree, it is possible to produce different codes for each sub-tree rooted at the sender when the sender itself is a branching point. This significantly reduces size of the encoded tree. Bcast has the same property as well. This allows them to support larger number of receivers compared to Xcast+. It is worth noting that formula 1 and 2 in subsection 3.1 are applied to largest sub-tree rooted at the sender. Simulation results show that this capability of Linkcast results in lower header size than Xcast+ in most cases.

4   Simulation Result

We have evaluated the Linkcast header overhead in comparison with Xcast and Bcast using NS-2 environment [19]. We performed two sets of experiments, one for small networks and another for large networks. Any router with at least one member is counted as a single receiver. Therefore, the actual number of manageable receivers is slightly more.

4.1   Small Networks

Small networks are generated based on Doar-Leslie random graph model using GT-ITM network topology generator [16]. We changed the network size from 20

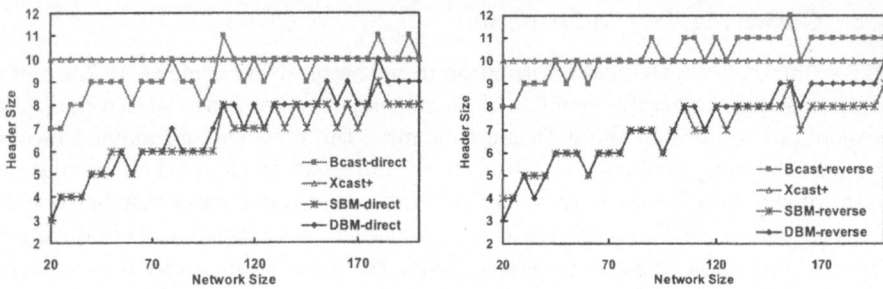

Fig. 2. The header size comparison based on network size in Doar-Leslie flat random topology model.

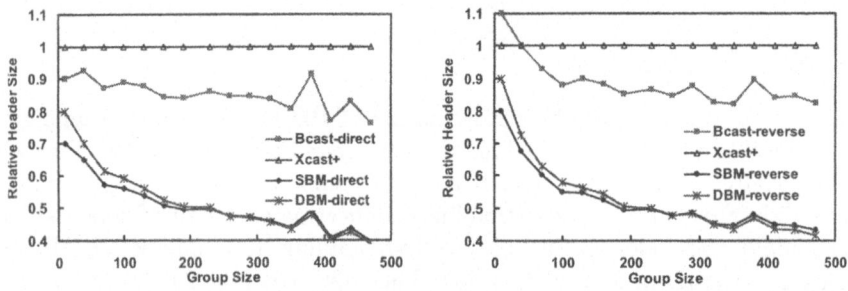

Fig. 3. The header size comparison based on group size in Barabasi-Albert flat random topology model.

to 200 while the average node degree is fixed approximately at 3.5. The group size is fixed at 10 because Xcast+ is not supposed to support more than 10 receivers [14]. For each generated network, we intentionally introduced 50% asymmetry in the network links. Then, we considered the reverse and forward shortest path trees in Linkcast and Bcast methods, which resulted in five different graphs. Finally, all plots are normalized to Xcast+. As Fig. 2 suggests, the header size of Linkcast is always smaller than Xcast and Bcast header. Furthermore, the differences between the direct and reverse versions of Linkcast are negligible.

4.2   Large Networks

For the large networks, we generated random graphs based on Barabasi-Albert [17] topology model using BRITE [18] network topology generator. The Barabasi-Albert model takes the power law relationship of Internet into account. We fixed the network size at 1204 nodes and performed the simulations with various group sizes ranging from 10 to 500. The average node degree is fixed at 4. The simulation results with Barabasi-Albert graph model is shown in Fig. 3. As can be seen in this figure, Linkcast has significantly less header overhead than Xcast and Bcast. We believe that Linkcast can support very large number of moderate size group (less than 50). Linkcast header size for 50 receivers is 120B.

5 Conclusion

Traditional multicast protocols fail to support very large number of any size multicast groups mainly due to their state-full design. Although Xcast and Bcast benefit from stateless design, they have two main difficulties to support moderate size multicast groups. First, their header size grows rapidly. Second, they need more unicast lookups in intermediate on-tree nodes when number of multicast members increase. Linkcast solve the header size problem of Xcast and Bcast without having to perform any form of table lookup. We believe that Linkcast is more appropriate than Xcast and Bcast in supporting huge amount of moderate and fairly large size multicast groups (less than 70). Linkcast fill the gap between Bcast, Xcast and traditional multicast for moderate size multicast groups.

References

1. R. Boivie, et al, "Explicit Multicast (Xcast) Basic Specification, IETF Internet-Draft, 2003
2. R. Boivie, N. Feldman, "Small Group Multicast", IETF Internet-Draft, July 2000.
3. R. Boivie, N. Feldman, C. Metz, "Small Group Multicast: A New Solution for Multicasting on the Internet", Internet Computing, Vol. 4, No. 3, May/June 2000.
4. D. Ooms, W. Livens, "Connectionless Multicast", IETF Internet-Draft, April 2000.
5. M.K. Shin, Y.J. Kim, K.S Park, S.H Kim, "Explicit Multicast Extension (Xcast+) Supporting Receiver Initiated Join", IETF Internet-Draft, October, 2002.
6. M.K. Shin, Y.J Kim, K.S. Park, S.H. Kim, "Explicit Multicast Extension (Xcast+) for Efficient Multicast Packet Delivery", ETRI journal, Vol. 23, No. 4, Dec. 2001.
7. J. Bion, D. Farinacci, M. Shand, A. Tweedly, "Explicit Route Multicast (ERM)", IETF Internet-Draft, June 2000.
8. M. Bag-Mohammadi, S. Samadian-Barzoki, N. Yazdani, "Using Branching Points for Multicast Data Distribution", submitted to SIGCOMM 2004.
9. V. Visoottiviseth, H. Kido, Y. Kadobayashi, S. Yamaguchi, "Sender-Initiated Multicast Forwarding Scheme", Proc. of IEEE ICT'2003, Tahiti, Feb. 2003.
10. A. Boudani, B. Cousin, "SEM: A New Small Group Multicast Routing Protocol", Proc. of IEEE ICT2003, Tahiti, Feb. 2003.
11. D. Waitzman, C. Partridge, S. Deering, "Distance Vector Multicast Routing Protocol", RFC 1075, Nov.1988
12. S. Deering, et al, "The PIM architecture for wide-area multicast routing", IEEE/ACM Trans. on Networking, Vol.4, No.2, April 1996
13. B. Cain, S. Deering, I. Kouvelas, B. Fenner, A. Thyagarajan, "Internet Group Management Protocol, Version 3" RFC 3376, October 2002.
14. O. Paridaens, D. Ooms, "Security Framework for Explicit Multicast", IETF Internet-Draft, November 2000.
15. J. Pansiot and D. Grad, "On routes and multicast trees in the Internet", ACM Computer Communication Review, vol. 28, no. 1, pp. 41-50, Jan.1998.
16. E. W. Zegura, K. Calvert, S. Bhattacharjee, "How to model an Internetwork.", Proc. of IEEE Infocom'96, San Francisco, CA
17. A.L. Barabasi, R. Albert, "Emergence of Scaling in Random Networks". Science, 286:509-512, October 1999.
18. A. Medina, A. Lakhina, I. Matta, J. Byers, "BRITE: An Approach to Universal Topology Generation" , In Proc. of MASCOTS'01, Cincinnati, Ohio, August 2001.
19. "The Network Simulator - ns - 2", http://www.isi.edu/nsnam/ns/

Towards an Erlang-Like Formula for the Performance Evaluation of GPRS/EDGE Networks with Finite-Length Sessions

Bruno Baynat[1], Khaled Boussetta[1], Pierre Eisenmann[2], and Nidham Ben Rached[2]

[1] Université de Paris VI, Laboratoire d'Informatique de Paris 6, Paris, France
{Bruno.Baynat, Khaled.Boussetta}@rp.lip6.fr
[2] Nortel Networks, Wireless Network Engineering, Chateaufort, France
{pierree, nbenrac}@nortelnetworks.com

Abstract. In this paper, we focus on a simple yet important objective: to provide Erlang-like models for GPRS/EDGE radio engineering. To this end, we develop a constructive model based on an original discrete-time Markov process that captures the detailed behavior of the radio allocator when a finite number of users are performing ON/OFF sessions in the cell. After a preliminary focus on infinite-length sessions, which provide a simpler analytical framework, we include finite-length sessions and study their specific effects.

1 Introduction

Several papers have been published on performance evaluation of GPRS or EDGE systems. A number of them were based on simulations (see e.g. [8,6]). Performance evaluation using analytical models were proposed in [7,5]. In these papers, GPRS session duration was supposed to be infinite. Finite GPRS sessions were investigated in [4,2,3]. All these studies rely on a number of relevant hypotheses and modeling assumptions, and quite usefully contribute to the understanding of the GPRS/ EDGE system. However none of them fully allowed us to reach our goal which was to develop an Erlang-like law for GPRS/EDGE network engineering relying on the most possible accurate description of existing GPRS infrastructure. So, we have developed a model based on an original constructive description of GPRS/EDGE with a discrete-time Markovian model at the granularity of the GPRS radio block (note that all previously mentioned works have developed continuous-time models). This first model is presented in [1]. The original contribution of our work in [1] is the detailed description of the GPRS/EDGE system in a discrete-time model which has allowed the derivation of a formula that has the simplicity of Erlang's B and C laws for voice but which applies to GPRS/EDGE.

In this paper, we extend our work presented in [1] by considering finite-length sessions. Our aim is to obtain, as in [1], closed form expressions of system performance parameters that could easily be used for GPRS/EDGE engineering, with a same focus on capturing the detailed behavior of the GPRS/EDGE system.

N. Mitrou et al. (Eds.): NETWORKING 2004, LNCS 3042, pp. 1288–1293, 2004.

2   Model Description

Our study is focused on the analysis of the bottleneck i.e. the radio downlink, studied in a particular cell. Our GPRS/EDGE system is characterized by the following parameters:

- T: the fixed number of time-slots in the cell that are dedicated to GPRS.
- t_B: the radio block duration which is equal to 20 ms.
- x_B: the number of data bytes that are transferred during t_B. The value of x_B depends on the radio coding scheme (e.g. in GPRS CS2, $x_B = 30$ bytes).
- d: the maximum number of time-slots that can be simultaneously used by a mobile for the downlink traffic (all mobiles have same d value). Today, most mobiles have $d = 3$ or 4.
- tbf_{max}: the maximum number of mobiles that can simultaneously have an active downlink TBF, because of the GPRS system limitations.

We also assume that there is a fixed number M of GPRS mobiles on the cell. Their traffic is made of finite-length sessions: each mobile alternates between Session and Inter-session periods. During an Inter-session, a mobile doesn't generate any traffic while during a Session, a mobile generates an ON/OFF traffic:

- ON periods correspond to the download of an element (e.g. a www page, an email, a file, etc.). Its size in bytes is characterized by a discrete random variable X_{on}, with an average value of x_{on}. Note that the time-scale of the system requires the ON characterization to be discretized. Therefore, the X_{on} bytes of an ON period are divided into N_{on} blocks of x_B bytes each. In this study N_{on} is assumed to be geometric with parameter p, where p is the probability for a given block to be the last one of the considered packet. According to this modeling assumption[1], $p = 1 - e^{-\frac{x_B}{x_{on}}}$.
- OFF periods correspond to the reading time, which is modeled as a continuous random variable T_{off}, with an average value of t_{off}. The T_{off} seconds of an OFF period are divided into B_{off} units of t_B seconds. B_{off} is also assumed to be geometric with parameter q, where q is the probability for the current unit of time to be the last one of the OFF period. By equating the average OFF time, we obtain the value of $q = 1 - e^{-\frac{t_B}{t_{off}}}$.

Let us denote by N the number of mobiles in active session. N lies between 0 and M. Let us finally emphasize that there is a limitation $n_{max} = min(tbf_{max}, M)$ on the number of mobiles that can simultaneously be on active transfer in the cell. We chose to describe the succession of Session and Inter-session periods for each mobile independently from each others as follows:

- A Session period is characterized by its number of ON periods. For instance, for a WAP or a WEB traffic, each ON period corresponds to the download of a given page. This number is described by a discrete random variable

---

[1] The memory-less hypotheses on the traffic characteristics are strong assumptions that are discussed and validated in details in [1].

N_{pages}, with an average value of n_{pages}. In this study, N_{pages} is assumed to be geometric with parameter v, where $v = n_{pages}^{-1}$ is the probability for a given ON period to be the last one of the considered session.

- An Inter-session period is modeled by a continuous random variable T_{is}, with an average value of t_{is}. Again, the number of units of t_B seconds in an Inter-session period is assumed to be geometric with parameter r, where $r = 1 - e^{-\frac{t_B}{t_{is}}}$ is the probability for the current unit of time to be the last one of the Inter-session.

Finally, we also suppose that the allocator fairly allocates bandwidth between all active mobiles (no QoS is modeled so far). Thus, if there are n mobiles in an active downlink process, each one receives an average bandwidth equal to the total cell bandwidth divided by n as soon as its capacity (given by d) allows it.

3   The Infinite-Length Session Markovian Model

In this section, we first recall the simpler case of infinite-length sessions [1]. Every mobile in the cell is thus doing an ON/OFF traffic with an infinite number of pages (i.e. $t_{is} = 0$ or $n_{pages} = +\infty$). Therefore, $N = M$ anytime.

3.1   The Erlang-Like Model

The smallest time-scale of the system, namely t_B, will be accounted for in the modeling process, by associating with it a discrete-time Markovian model of equal time step. Thus, the state of the system will only be described at the end of each radio block period. As we assume no QoS and a fair allocation among all mobiles, we choose to describe the state of the model by the number n of mobiles in active transfer. Since there is a physical limitation n_{max} on n then the state space is given by the set $\{0, n_{max}\}$. We then define the following probabilities:

- $a_i^n =$ probability of i arrivals[2], provided that the state of the system is n;
- $d_j^n =$ probability of j departures[3], provided that the state of the system is n.

Considering the fact that there are at least two orders of magnitude between the time scales of the block duration (20 ms) and the ON/OFF duration (seconds), we make the assumption that between two steps of the process (i.e. 20 ms), there can be at most one arrival and one departure. The probabilities a_i^n and d_j^n can thus only take non zero values when i or j are equal to 0 and 1. In other words, $a_0^n + a_1^n = 1$ and $d_0^n + d_1^n = 1$.

In order to derive departure probabilities remember that each mobile can use a maximum of d time-slots in downlink. As a consequence if $nd < T$, all the mobiles in active transfer can get the maximum capacity with regard to their downlink capability. Otherwise, all the T time-slots of the cell are used, and each mobile receives a reduced downloading bandwidth. Let us also recall that, if n

---

[2] i.e. i new mobiles among $N - n$ ones entering an ON period
[3] i.e. j mobiles among the n ones in active transfer that ends up their download.

mobiles are currently in an ON period, $N - n$ are in an OFF period. Moreover, under realistic conditions $q \ll 1$ and $p \ll 1$. Therefore [1],

$$\forall n \in \{0, n_{max}\}, \ a_0^n \approx 1 - (N - n)\, q \ \text{and} \ d_0^n \approx 1 - \min\,(nd, T)\, p \qquad (1)$$

According to these assumptions, the resulting Markov chain is linear and its transition probabilities can be calculated as:

$$
\begin{aligned}
p_{n\,n+1} &= a_1^n d_0^n && \text{if } 0 \leq n < n_{max} \\
p_{n\,n-1} &= a_0^n d_1^n && \text{if } 0 < n \leq n_{max} \\
p_{n\,n} &= 1 - p_{n\,n+1} - p_{n\,n-1}
\end{aligned}
\qquad (2)
$$

Because of the birth-death structure of the Markov chain the steady-state probabilities $p(n)$, $\forall n \in \{0, n_{max}\}$, are given [1] by the following closed form:

$$
p(n) =
\begin{cases}
\dfrac{N!}{n!\, d^n\, (N-n)!} \left(\dfrac{q}{p}\right)^n p(0) & \text{if } 0 \leq n \leq n_0 \\[2ex]
\dfrac{N!}{n_0!\, d^{n_0}\, T^{n-n_0}\, (N-n)!} \left(\dfrac{q}{p}\right)^n p(0) & \text{if } n_0 < n \leq n_{max}
\end{cases}
\qquad (3)
$$

where $p(0)$ is obtained by normalization. The expression 3 has the simplicity of Erlang's B formula yet it applies to GPRS with two quality criteria, blocking and throughput. It has the same attributes: it provides generic reference curves and fruitful qualitative insight on the dimensioning of GPRS system.

3.2   Performance Evaluation

From the stationnary probabilities given by equation 3, we can compute the average utilization $\overline{U}$ of the cell as follow:

$$\overline{U} = \sum_{n=0}^{n_{max}} p(n) \min\left(\frac{nd}{T}, 1\right) \qquad (4)$$

Furthermore, we can deduce from the Markov chain the mean number of mobiles ending up their transfers per unit of time as well as the mean number of mobiles in active transfer. From Little's law, we can thus derive the average duration of an active transfer. Then, we can calculate the average throughput $\overline{X}$ obtained by each mobile in active transfer as

$$\overline{X} = \frac{x_{on} \sum_{n=1}^{n_{max}} p(n)\, d_1^n}{t_B \sum_{n=1}^{n_{max}} np(n)} \qquad (5)$$

Finally, we can derive the last important performance parameter namely the so-called "*blocking*" (or "*reject*") probability P_r. P_r is the probability that an arrival is rejected, *i.e.* that a mobile that wants to switch to an ON state cannot do it because the limit of n_{max} mobiles in the cell is reached. As shown in [1],

$$P_r = \frac{\sum_{n=0}^{n_{max}} p(n)\, (a_1^n - d_1^n)}{\sum_{n=0}^{n_{max}} p(n)\, a_1^n} \qquad (6)$$

4   Erlang-Like Law for Finite-Length Sessions

We now come back to the finite-length sessions model as described in Section 2. Focusing on our goal to obtain a closed-form Erlang-like formula, we transform our finite-length sessions traffic into an "equivalent" infinite-length sessions traffic as follows: the inter-session period is "shared" among the inter-page OFF periods of an approximate infinite-length sessions model. So the equivalent OFF period duration is given by 7, all the other system parameters being equal.

$$t'_{off} = t_{off} + \frac{t_{is}}{n_{pages}} = t_{off} + v\,t_{is} \qquad (7)$$

This is obviously an approximation as the distribution of the OFF period in the equivalent linear model are geometric and will not respect the variability of the actual succession of OFF and OFF+Inter-session. However, assuming the approximation holds, all the Erlang-like formulas of Sections 3.1 and 3.2 can be applied by performing the above transform on t_{off}. We will see in the next section that this approximation results in a very accurate estimation of all the average performance parameters.

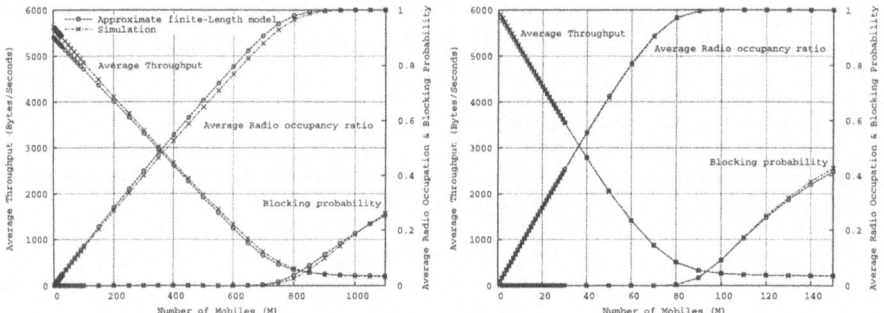

Fig. 1. Validation of Erlang-like law for finite-length sessions against simulations

4.1   Validation

We have compared the performance parameters obtained under the Erlang-like law for finite-length sessions against those obtained by simulations. To that end we have considered several test scenarios. In a first set of tests, we have considered the following parameters: $T = d = 4$ slots, $t_{off}= 7$ seconds, $t_{is} = 20$ minutes and $n_{pages} = 10$. Curves in the left side of figure 1 show average throughput, average radio occupation ratio and blocking probability obtained by both the Erlang-Like finite-length sessions model and simulations when $x_{on} = 1$ kB and M varies form 1 until 1100 mobiles. These curves show clearly that the values obtained by the approximate finite-length sessions model are extremely close to those obtained by simulations. In fact in these test scenarios the error remains less than 3%. In other comparison test scenarios we have observed that the error gets even smaller when x_{on} grows or when t_{is} decreases. We also have investigated the influence of the average number of pages to be transmitted during a session on the equivalence between the finite-length sessions model and the equivalent infinite one. We ran the set of tests previously described while taking now n_{pages} = 3 and t_{is} = 2 minutes. The curves in the right side of figure 1 illustrate the average performance parameters obtained for x_{on} = 4 kB. These curves confirm the same observations we made before. They indicates that even for a small inter-session period of 2 minutes and a modest average number of transmitted pages set to 3 a good fit is obtained between simulation and the approximate finite-length sessions model. Other similar results[4], allows us to conclude that an Erlang-like law can be utilized for GPRS/EDGE network engineering with finite-length sessions traffic models. A framework quite similar to the standard voice dimensioning methods is thus available.

5   Conclusion

In this paper, we have developed an Erlang-like model for GPRS/EDGE network dimensioning with finite-length sessions. To that end, we have first derived a simple Erlang-like formula for the infinite-length sessions case. Then we have included finite-length sessions in our model. The simulation results demonstrated that finite-length sessions models could also be solved by a simple Erlang-like law with a very good accuracy: this is achieved by performing the appropriate transform in the input traffic, namely by increasing the OFF periods by a portion of the inter-session period.

References

1. B. Baynat and P. Eisenmann. Towards an Erlang-like formula for GPRS/EDGE network engineering. In *Proc. of IEEE ICC*, Paris, France, June 2004.
2. B. Wydrowski C. H. Foh, B. Meini and M. Zukerman. Modeling and Performance Evaluation of GPRS. In *Proc. of IEEE VTC*, pages 2108–2112, Rhodes, Greece, May 2001.
3. X. Fang and D. Ghosal. Performance Modeling and QoS Evaluation of MAC/RLC Layer in GSM/GPRS Networks. In *Proc. of IEEE ICC*, Alaska, USA, May 2003.
4. C. Lindemann and A. Thummler. Performance Analysis of the General Packet Radio Service. *Computer Networks*, 41(1):1–17, January 2003.
5. S. Ni and S. Haggman. GPRS performance estimation in GSM voice and GPRS shared resource system. In *Proc. of IEEE WCNC*, pages 1417–1421, New Orleans, USA, September 1999.
6. M. Oliver and C. Ferrer. Overview and Capacity of the GPRS (General Packet Radio Service). In *Proc. of IEEE PIMRC*, Boston, MA, USA, June 1998.
7. J. Romero S. Pedraza and J. Munoz. (E)GPRS Hardware Dimensioning Rules with Minimum Quality Criteria. In *Proc. of IEEE VTC Spring*, pages 391–395, Birmingham, Al, May 2002.
8. P. Stuckmann and O. Paul. Dimensioning Rules for GSM/GPRS Networks. In *Proc. of the Aachen Symposium on Signal Theory*, pages 169–174, Aachen, Germany, September 2001.

---

[4] Which we can not show here due to space limitation.

Multicast Transport Protocol Analysis:
Self-Similar Sources*

Mine Çağlar[1] and Öznur Özkasap[2]

[1]Koç University, Department of Mathematics, Istanbul, Turkey
[2]Koç University, Department of Computer Engineering, Istanbul, Turkey
{mcaglar,oozkasap}@ku.edu.tr

Abstract. We study the traffic that scalable multicast protocols generate in terms of message delays over the network as well as traffic counts at the link level in the case of self-similar sources. In particular, we study Bimodal Multicast and Scalable Reliable Multicast protocols proposed for scalable reliable multicasting. These protocols are based on different mechanisms for recovering from message losses and providing scalability. We discuss the protocol mechanisms as the main underlying factor in our empirical results. Our results can be considered as a contribution to the general problem of integration of multicast communication to large scale.

Keywords: Scalable multicast; loss recovery; self-similarity; long-range dependence.

1 Introduction

Analyses of fine-grained measurements over the last decade reveal that long-range dependence exists in the link level network traffic as well as self-similarity [1]. Self-similarity in the presence of long-range dependence has adverse consequences on network performance. We focus on the traffic that scalable multicast protocols generate. These protocols have been compared with respect to several performance measures such as scalability, reliability and congestion control. However, the nature of the traffic stream generated by each type of protocol particularly with respect to self-similarity has not been studied extensively.

We consider multicast communication traffic in the case of self-similar sources. In particular, we study Bimodal Multicast [2] and Scalable Reliable Multicast (SRM) [3] protocols proposed for scalable reliable multicasting. These protocols are based on different mechanisms for recovering message losses and providing scalability, namely epidemic dissemination for Bimodal Multicast and nonhierarchical feedback control for SRM. Our previous studies demonstrate that epidemic approach of Bimodal Multicast generates a more desirable traffic than SRM with lower overhead traffic and transport delays in the case of a constant bit rate source [4,5]. In this study, we consider a self-similar source, namely an on/off sender that transmits with Pareto on

---

* This work is supported by TUBITAK and COST Action 279 "Analysis and Design of Advanced Multiservice Networks supporting Mobility, Multimedia, and Internetworking".

N. Mitrou et al. (Eds.): NETWORKING 2004, LNCS 3042, pp. 1294–1299, 2004.

and off times as an extension of [6]. The delays and traffic at the link level are studied and compared with constant bit rate case. Bimodal Multicast generates shorter delays; however long-range dependence arises in the link level. For SRM, the traffic becomes worse in terms of both delays and at the link level.

We elaborate on the protocol mechanisms as the main underlying factor in our empirical results. The intrinsic relation of these mechanisms to traffic characteristics is explored. Our results can be considered toward the general problem of integration of multicast communication to the Internet. The ultimate aim is to discover and develop multicast protocols that not only feed well-behaved traffic discretely into the existing networks, but also can cope with the existing self-similar traffic and its adverse consequences. Next section describes comparative simulation results for protocols that we investigate. Then, we give our conclusions and point at future directions.

2 Simulations, Analysis, and Results

The implementation of Bimodal Multicast that we developed over ns-2 [2] and the available ns-2 model of SRM are used [7]. The simulation scenario is transit-stub topology with 60, 80, 100 and 120 nodes where every node is a group member. Transit-stub topologies approximate the structure of the Internet that can be viewed as a collection of interconnected routing domains where each domain can be classified as either a stub or a transit domain [8]. The sender is located on a central node and the receiver that we analyze for transport delays is located as far as possible from the sender. We also monitor a central link that is heavily loaded. A 1% drop rate set on every link forms a system-wide noise. We obtain our results from a sequence of 35000 (approximately 2^{15}) multicast data messages each with size 210 bytes. We have a single on/off source which multicasts to all receivers continuously. On and off times follow a Pareto distribution with shape parameter 1.5 and mean 500 milliseconds each. On the average, 50 messages per second are transmitted like the cbr simulations of [5]. In order to estimate the Hurst parameter H from the delay of these messages or from traffic counts at the link level, we apply the wavelet estimation method as given in [9] using Daubechies wavelets with three vanishing moments.

Bimodal Multicast [2] which is based on an epidemic loss recovery mechanism is a novel option in the spectrum of multicast protocols. It has been shown to impose constant loads on links and routers if configured correctly and to exhibit stable throughput under failure scenarios that are common on real large-scale networks. In contrast, this kind of behavior can cause other reliable multicast protocols to yield unstable throughput. Bimodal Multicast consists of two sub-protocols, namely an optimistic dissemination protocol and a two-phase anti-entropy protocol. The former is a best-effort, hierarchical multicast used to efficiently deliver a multicast message to its destinations. This phase is unreliable and does not attempt to recover a possible message loss. When available, IP multicast, or a randomized dissemination protocol can be used. The second stage is responsible for message loss recovery. It is based on an anti-entropy protocol that detects and corrects inconsistencies in a system by continuous gossiping. We refer the interested reader to [2] for further details on Bimodal Multicast and the theory behind the protocol.

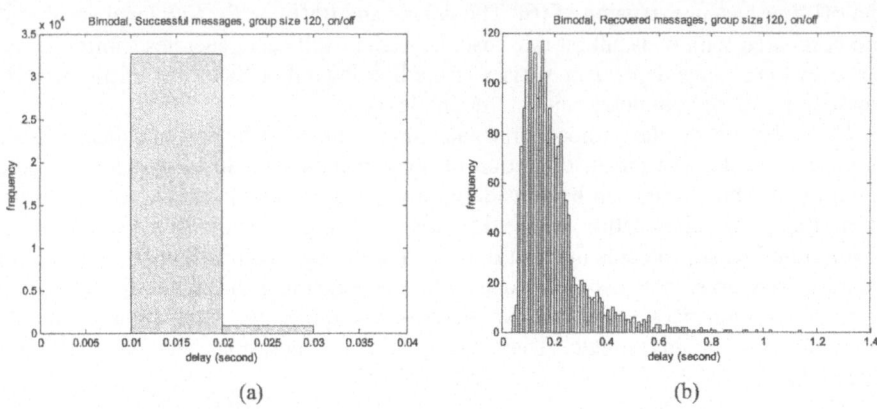

Fig. 1. Marginal delay distribution of (a) successful (b) recovered messages with Bimodal Multicast for group size 120.

In view of the simulation results in [5], Bimodal Multicast has an exponential marginal delay distribution and a Markovian recovery mechanism. As a result, LRD is not an expected phenomenon. Indeed, the delay sequence obtained with a cbr source shows no LRD in the case of Bimodal Multicast whereas the delay of SRM is long range dependent for larger group sizes 100 and 120 [5]. The transport layer mainly translates LRD from application level to link level while recent studies show that it can also cause LRD and self-similarity up to the scale of minutes, not only at small scales. This has been shown for TCP, which is the prevalent protocol over the Internet. This leads us to the question: what happens if the source does not multicast with cbr but mimics an on/off source, which is typical in current networks? It is well known that when sufficiently many of traffic streams from such sources are aggregated, LRD arises at the link level [10].

The marginal delay distributions for both Bimodal Multicast and SRM do not change in general with an on/off source. Figures 1 and 2 show these distributions separately for messages that are successful in the initial multicast and for those recovered through loss recovery mechanism. Bimodal Multicast has an exponential delay distribution as given in Fig. 1 (b). For the group size 120, LRD has been detected for SRM with a cbr source where the successful messages follow a normal distribution [5]. In the on/off case, this distribution becomes lognormal with a pronounced right tail as shown in Fig. 2 (a). The mean delays are slightly higher in SRM for group sizes 100 and 120 with an on/off source, but not significantly. The means are plotted for all group sizes in Fig. 3 comparatively with cbr results. As in the cbr case, SRM delay is much higher than Bimodal delay.

The correlations in delay are measured through the Hurst parameter H. If the delay sequence is self-similar with LRD, then H takes values in $(0.5,1)$. It has been found that neither the delay sequence nor the link level traffic of Bimodal Multicast show LRD. However SRM shows LRD for larger group sizes even with a cbr source. The TCP-like approach of SRM induces LRD and self-similarity. SRM necessitates the basic IP delivery model and forms reliability on an end-to-end basis. Similar to TCP that adaptively sets timers or congestion control windows, SRM algorithms

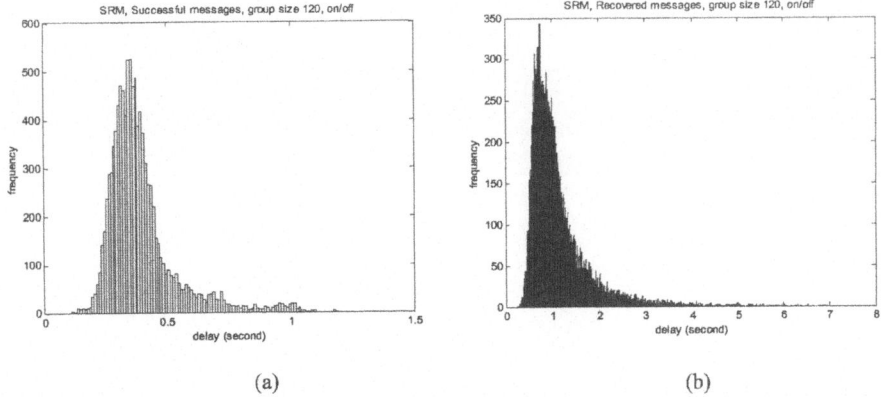

Fig. 2. Marginal delay distribution of (a) successful (b) recovered messages with SRM for group size 120.

dynamically regulate their control parameters such as request and repair timers, based on the observed performance within a session. Repair requests and retransmissions are multicast to the whole group. A lost packet ideally triggers only a single request from a host just downstream of the point of failure.

With an on/off source, the Hurst parameter estimates are given in Fig. 4 where H is simply estimated as 1 in group sizes 100 and 120 for SRM. In fact, the scaling diagrams in these sizes do not show a linear scaling even for larger scales and indicate a multifractal sequence. Therefore, a linear fit yields H greater than 1. The same value is obtained for the delay of recovered messages. This was not the case with a cbr source; we had estimated H to be around 0.8 for both group sizes with a clearly self-similar delay sequence. On the other hand, Bimodal Multicast delays are scalable with respect to group size as H remains around 0.5 all throughout.

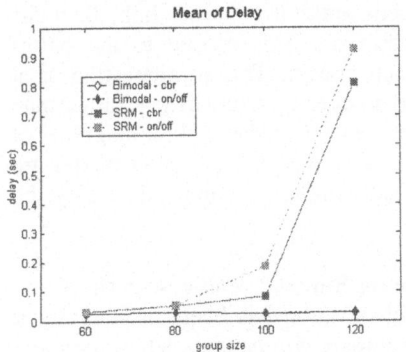

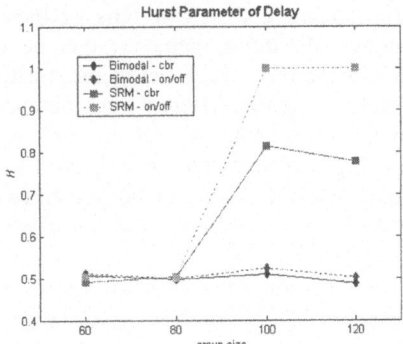

Fig. 3. Mean delay of a message

Fig. 4. Hurst parameter obtained from delay sequence

We measure goodput as the ratio of the total number of messages successfully received to the total number received including protocol overhead. Fig. 5 shows that goodput remains relatively constant in Bimodal Multicast whereas it decreases

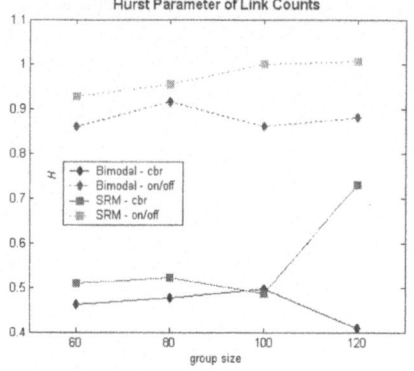

Fig. 5. Goodput versus group size

Fig. 6. Hurst parameter obtained from link counts

significantly in SRM as the group size increases. These results are very close to those with a cbr source [5]. In view of Hurst parameter calculations, we can conclude that the correlations for several performance indicators worsen with an on/off source whereas the means do not get much affected.

In comparison to the transport level, the link level results are different for Bimodal Multicast but similar for SRM in terms of LRD. Fig. 6 gives the Hurst parameters estimated from a monitored link incoming to our fixed receiver. The results from the reverse direction are similar with slightly lower H values and significantly lower mean. As the receiver is chosen to be one of the end nodes, the latter direction is less loaded. Bimodal Multicast shows LRD at the link level with an on/off source. In fact, this is not surprising. Although there is a single on/off source, at the link level there is an aggregation arising from the recovery process of all receivers in the network. If the traffic streamed to a few tens of these members multiplex in this link, then the limiting self-similar behavior can be easily observed. This is due to the on/off behavior of the source being propagated by each stream. This is interesting as it indicates a slightly different aggregation of on/off sources by multicasting, rather than many sources being present in a unicast environment [10]. The scaling diagrams of the delay sequence and the link level packet counts which yield the H estimates are similar in SRM. This can be explained due to self-similarity being induced by the protocol itself in addition to an on/off source.

The most important effect of an on/off source on Bimodal Multicast traffic is the emergence of LRD at the link level. Another observation has been an increase in message loss. There is a significant probability that there will be high delays for some messages with a Pareto on/off source although these messages are few. Therefore, the mean delay does not increase significantly but we can detect more message loss. The number of rounds for holding a lost message in the buffer is a parameter of the Bimodal Multicast simulation and is called the *stability threshold*. By default, this parameter is set to 10. We have observed that increasing the stability threshold decreases loss.

We have already shown that SRM induces self-similarity on both delays at the transport level and the traffic counts at the link level even in the case of a cbr source. The immediately observed effect of an on/off source is that LRD gets more pronounced. In fact, a monofractal model of self-similarity does not hold in this case. In the scaling diagram of the delay sequence in the case of a cbr source, the linear scaling over larger scales is evident [5]. However, for an on/off source, the scaling is much different indicating a multifractal character. The analysis of the coupling of an on/off source with SRM's loss recovery mechanism to yield such results remains as future work.

3 Conclusions and Future Directions

In this study, we focus on the traffic that scalable multicast protocols generate. We consider a self-similar source, namely an on/off sender that transmits with Pareto on and off times. The delays and traffic at the link level are studied and compared with constant bit rate case. Bimodal Multicast generates desirable delays; however long-range dependence arises in the link level. For SRM, the traffic becomes worse in terms of both delays and at the link level. Our results can be considered toward the general problem of integration of multicast communication to the Internet. We have demonstrated that Bimodal Multicast is a good candidate for both generating well-behaved traffic and also coping with the existing self-similar traffic.

References

1. Leland, W. E., Taqqu, M. S., Willinger, W. and Wilson, D. V.: On the Self-Similar Nature of Ethernet Traffic (Extended Version), IEEE/ACM Trans. On Networking, 2 (1994) 1-15
2. Birman, K.P., Hayden, M., Ozkasap, O., Xiao, Z., Budiu, M. and Minsky, Y.: Bimodal Multicast, ACM Transactions on Computer Systems, 17 (1999) 41-88
3. Floyd, S., Jacobson, V., Liu, C., McCanne, S. and Zhang, L.: A Reliable Multicast Framework for Light-weight Sessions and Application Level Framing, IEEE/ACM Transactions on Networking 5 (1997) 784-803
4. Özkasap, Ö, Çağlar, M.: Traffic Behavior of Scalable Multicast: Self-similarity and Protocol Dependence. Proceedings, International Teletraffic Congress 18, Berlin, Germany, 31 Aug.- 5 Sep. (2003)
5. Çağlar, M., Özkasap, O.: Traffic Properties of Scalable Multicast Communication: Comparison of Bimodal Multicast and SRM. Submitted (2002)
6. Özkasap, Ö, Çağlar, M.: Traffic Characterization of Scalable Multicasting in the case of a Self-Similar Source (poster), ACM SIGCOMM, Karlsruhe, Germany, 25-29 Aug. (2003)
7. Bajaj, S., Breslau, L., Estrin, D., et al.: Improving Simulation for Network Research, USC Computer Science Dept. Technical Report 99-702 (1999)
8. Calvert, K., Doar, M., and Zegura, E.W.: Modeling Internet Topology. IEEE Communications Magazine, June (1997)
9. Veitch, D. and Abry, P.: A Wavelet Based Joint Estimator of the Parameters of Long-Range Dependence, IEEE Trans. on Information Theory, 45 (1999) 878-897
10. Taqqu, M.S., Willinger, W. and Sherman, R.: Proof of a Fundamental Result in Self-Similar Traffic Modeling, Computer Communication Review, 27 (1997) 5-23

SPEED: Scalable Protocols for Efficient Event Delivery in Sensor Networks

Tassos Dimitriou[1], Ioannis Krontiris[1], Fotios Nikakis[1], and Paul Spirakis[2]

[1] Athens Information Technology, Markopoulo Ave.,
190 02 Peania, Athens, Greece
{tassos, ikro, fnik}@ait.gr
[2] Computer Technology Institute, Greece
spirakis@cti.gr

Abstract. One of the most eminent problems in sensor networks is the routing of data to a central destination in a robust and efficient manner. In this work we propose a new scalable protocol for propagating information about a sensed event towards a receiving center. Using only local information and total absence of coordination between sensors our protocol achieves to propagate the sensed data to a receiving center by activating only those nodes that lie very close to the optimal path between the source of the event and the destination, resulting in low activation of the network's sensors. Thus the protocol is very energy efficient. Furthermore, our protocol is robust as it manages to propagate the information even when sensors fail with certain probability.

1 Introduction

Sensor networks[1] have attracted much scientific interest during the past few years. These networks use hundreds to thousands of inexpensive wireless sensor nodes over an area for the purpose of monitoring certain phenomena and capture geographically distinct measurements over a long period of time.

These networks differ from wireless ad hoc networks in the sense that their nodes are characterized by limited resources such as storage, computational and communication capabilities. The power of sensor networks, however, lies exactly in the fact that their nodes are so small and cheap to build that a large number of them can be used to cover an extended geographical area, gather information in-site and process it in parallel enabling an accurate and reliable monitoring process that was previously hard or infeasible to get. Due to the limited resources available to sensors, expensive routing protocols, costly flooding mechanisms, or complex algorithms that don't scale to large number of nodes cannot be used. Furthermore, random distribution of nodes in the physical environment, node failure during their deployment and dynamic change of nodes' power supply make the design of communication protocols a very challenging task.

In this work we focus on the efficient propagation of a sensed event towards some receiving center, assuming an event-driven data delivery model[6]. The need for communication between a regular sensor (the source) and some base

N. Mitrou et al. (Eds.): NETWORKING 2004, LNCS 3042, pp. 1300–1305, 2004.

station (called the destination or the sink) can arise at any time, possibly triggered by unexpected changes in the environment. It is exactly this change in the environment (i.e. a fire, a person entering a restricted area, etc.) that we feel it is important to reach the base station as quickly as possible without depleting the network from its resources through the use of complicated protocols.

There is an abundance of routing protocols for sensor networks (see for example [3] and [4] for two distinct representatives) that attempt to optimize a variety of different measures including efficiency, robustness, number of activated particles, etc. In our setting, however, where sensed data need to be sent to a receiving center, these approaches seem to be somewhat of overkill as we only need to send a single message or packet back to the base station. Taking into account the small communication throughput and the limited memory and computational capabilities of sensor networks a simple flooding approach seems to be the best alternative.

Flooding is the most computationally efficient protocol due to its simplicity as every node broadcasts every new incoming packet. Therefore, data are bound to reach their destination, assuring correctness, and the protocol is immune to node failures, assuring robustness. Although this protocol can be integrated even in the most simplistic implementations of sensor nodes, it is extremely energy consuming as all nodes must receive and transmit the message at least once. Gossiping or wandering approaches [2] seem to alleviate this problem, at the cost, however, of increasing path lengths or failing to reach destination.

Our Contribution: In this work we propose a new light-weight protocol that tries maximizing efficiency and minimizing energy consumption by favoring certain paths of local data transmission towards the sink. Just like flooding, the protocol is very easy to implement, even by the simplest of sensor nodes, as it only requires nodes to decide whether or not to forward the message to their neighbors. Unlike flooding protocols, however, it avoids depleting the network from its resources by restricting the nodes that receive and hence retransmit the message. The mechanism that controls this propagation of information is very simple; every node decides to retransmit according to the value of a single parameter that essentially captures the node's distance from the optimal path between source and sink. By carefully tuning the retransmission threshold one obtains a trade-off between robustness (the fraction of times the message reaches the destination) and load incurred in the network (measured in terms of the ratio of activated nodes).

2   Description of the Protocol

In what follows we assume that the number of sensor nodes in the network is N and there is a single destination point D that represents the center where data should be sent. We denote the node that sensed the event by S. Our focus is the design of a simple algorithm that solves this communication problem between S and D in an efficient and robust way by adhering to the following design principles that are particular to sensor networks: (i) low energy requirements and utilization of the sensors, (ii) scalability and use of local information only, and (iii) robustness under node failures.

The goal of the algorithm is to activate only those sensors that lie very close to the optimal line (in terms of hops) between S and D. The thinner this zone of transmitting nodes is, the more efficient will be the protocol in terms of the ratio of activated nodes. The algorithm is shown below:

Angle Protocol

Let P be the sensor that needs to decide whether to forward a received message or not. Denote by P_{prev} the sensor that transmitted the message to P and by $\omega = (P_{prev}PD)$ the angle defined by the lines $P_{prev}P$ and PD.
Then P deterministically forwards the message according to whether

$$\omega \geq \omega_{threshold}$$

where $\omega_{threshold}$ is some predefined parameter of the protocol.

In order to be able to implement the proposed algorithm, we assume that each sensor node has the ability to estimate the Direction of arrival (DoA) of incoming packets received from nearby nodes, and it can estimate the relative direction of the sink D, provided that the sink sends out beacon messages in regular intervals, to account for network mobility, and these can be captured by all the nodes in the network.

It is noted that the algorithm does not need any distance information (see the previous line of work [5] that analyzes a similar protocol under a stronger set of assumptions). Protocols requiring distance or exact location information can be implemented either with the use of GPS receivers that are unsuitable for sensor applications, by triangulation methods implied in neighboring nodes, which require severe bandwidth overhead, or by the deployment of acoustic sensors that add up to the overall node cost. Moreover, the proposed algorithm does not require nodes to have the same co-ordinates system. Such an approach would require the deployment of magnetic sensors (electronic compasses) that would again result in additional sensor cost. The algorithm is executed locally, and takes into account only local DoA measurements.

3   Experimental Analysis

In order to analyze the performance of the algorithm described above, we performed a set of large scale experiments whose goal was to test the protocol's effectiveness under the following measures:

1. *Efficiency*: A small number (compared to the total number) of nodes must be activated for each data transmission towards the sink.

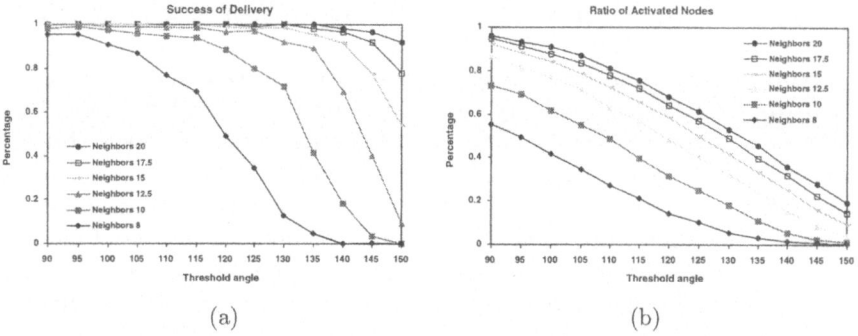

Fig. 1. (a) Ratio of activated nodes. (b) Success ratio

2. *Robustness or Fault tolerance*: The protocol should be able to deliver data to the destination, even when a large number of nodes is not responding (due to failure at deployment time)

3. *Scalability*: The routing algorithm should adapt easily to changes in the size of the network.

4. *Simplicity*: The algorithm must be able to run in an 8-bit microprocessor with minimal data memory (i.e. 4Kbytes).

It is obvious that the algorithm is extremely simple and since it does not depend on the network's size but only on its density (details omitted) it is also scalable so we will concentrate on proving its efficiency and robustness. We used the following setup for our experiments: $N = 2500$ sensors were spread uniformly at random in a square field, where all sensors have the same communication range R. For each simulation run we choose the sensor with the smallest and the largest $x - y$ coordinates to be the source and the sink, respectively. Hence, transmission will occur along the "diagonal" of the square field. Finally, to obtain valid statistical results all experiments were repeated a 1000 times.

Figure 1(a) shows the ratio r of activated sensors over the total number of sensors N as a function of the threshold angle, for $\omega_{threshold}$ equal to 90 up to 150 degrees. The six curves correspond to different communication ranges, each one resulting in a different average number μ of neighbors (8, 10, 12.5, 15, 17.5 and 20 neighbors, respectively).

As it can be seen in this figure, for small values of $\omega_{threshold}$ (around 90°) a large percentage of the sensors gets activated but the protocol is still better than flooding, especially for smaller values of μ. More importantly, however, the number of activated sensors decreases with the value of $\omega_{threshold}$ and becomes extremely small for suitable values of this quantity.

While the information shown on Figure 1(a) indicates that the protocol manages to activate only a few sensors for large values of $\omega_{threshold}$, it is important to know at the same time the success ratio of the protocol, i.e. the number of times the propagated information reached the sink. In particular, we would like to know whether there exists a value for $\omega_{threshold}$ where both the number of activated particles is small and the success ratio is high. The answer to this question is shown on Figure 1(b).

It is obvious from this figure that the larger the average number of neighbors is the better the success probability becomes for any given threshold angle. Hence if we want to achieve a success ratio of 90%, we can either choose a threshold angle of about 95° when $\mu_{neighbors} = 8$, an angle of 115° when $\mu_{neighbors} = 10$, or an angle of 140° when $\mu_{neighbors} = 15$. So, one may ask: are all these settings equivalent? The answer of course depends on the number of activated particles. If we go back to Figure 1(a) and look at the corresponding curves, we will see that for these angles the number of activated particles becomes 50%, 40% and 25%, respectively. Hence using a forwarding angle of 140° in the $\mu_{neighbors} = 15$ setting, we manage to reach the destination 90% of the time while activating less than 25% of all the nodes.

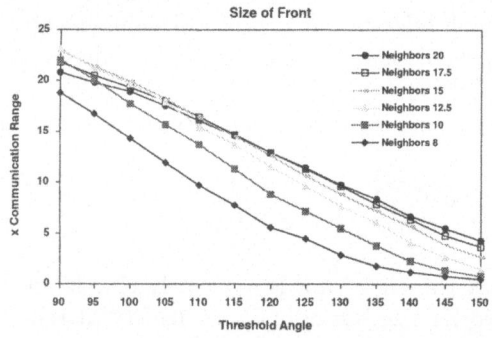

Fig. 2. Size of propagating front of activated sensors

We have also been able to compute the average size of the propagating front of particles during transmission. The front is simply the particles that lie at the edge of the transmission zone towards the sink and is depicted in Figure 2. As can be seen from the figure the average front size decreases almost linearly with respect to $\omega_{threshold}$. When $\omega_{threshold} = 90°$, the front size is less than $23R$, while at $\omega_{threshold} = 140°$, the front size is less than $7R$ for all neighborhood distributions. Hence at this forwarding angle about 7 particles get activated on the average.

Finally, we investigated the fault-tolerance nature of our protocol when sensors die with various probabilities. The invariance of our algorithm under changes in the network size suggests the following approach: when we know that sensors may die with certain probability we can either plant more nodes or increase the communication range slightly to counteract the effect of dead nodes. In any case, using the results of Figure 1 we can optimize the algorithm's performance and obtain the required robustness (details omitted due to space restrictions).

4   Conclusions and Future Research

It is known from ancient times that the closest route between two points (at least in the Euclidean setting) is a straight line. In this work we have presented

SPEED, a Scalable Protocol for Efficient Event Delivery in sensor networks, where sensed data is sent to a receiving center using only local information and total absence of coordination. Our protocol is suited for those cases where unexpected changes to the environment (i.e. a fire, a person entering a restricted area, etc.) must be propagated quickly back to the base station without the use of complicated protocols that may deplete the energy of the network.

Our protocol is very easy to be implemented as it only requires nodes to decide whether or not to forward the message according to whether the angle formed between the previous node P_{prev}, the current node P and the destination D is greater than a predefined threshold angle $\omega_{threshold}$. We show that by carefully tuning this value one can obtain a tradeoff between the number of activated sensors and the success ratio of the protocol.

There are, however, some issues that need to be addressed. First, we need to overcome the problems related to the use of beacon frames for determining the relative direction of the sink as nodes in shadowed areas will be unable to receive the beacon frames. An initialization protocol is therefore required to ensure this information is shared by all nodes. Another issue is to increase the robustness of the protocol by assuring that data may reach the destination using more than one simple paths. This would make the protocol resistant to obstacles blocking communications between sensors. Finally, it is important to avoid depleting the energy of the sensors that lie in the optimal path between S and D (in the hypothetical scenario where S must send data to D frequently). Since the protocol is deterministic, the same sensors will be used continuously and hence they will be quickly drained of their energy. A solution to this problem is for the source S to choose a "random" path (not necessarily optimal) so that data is routed using this path. Of course, for this to be of any value, the protocol must again use only local information and no coordination between sensors.

References

1. I. F. Akyildiz, W. Su, Y. Sankarasubramaniam, and E. Cayirci, "Wireless sensor networks: a survey," *Computer Networks*, vol. 38, pp. 393–422, March 2002.
2. Z. Haas, J. Halpern, and L. Li, "Gossip-based ad hoc routing," in *IEEE InfoCom Proceedings 2002*, vol. 3, pp. 1707–1716, June 2002.
3. J. Kulik, W. R. Heinzelman, and H. Balakrishnan, "Negotiation-based protocols for disseminating information in wireless sensor networks," *Wireless Networks*, vol. 8, pp. 169–185, March 2002.
4. C. Intanagonwiwat, R. Govindan, and D. Estrin, "Directed diffusion: a scalable and robust communication paradigm for sensor networks," in *Proc. 6th International Conference on Mobile Computing and Networking*, pp. 56–67, 2000.
5. I. Chatzigiannakis, T. Dimitriou, M. Mavronicolas, S. Nikoletseas, and P. Spirakis, "A comparative study of protocols for efficient data propagation in smart dust networks," in *International Conference on Parallel and Distributed Computing (EUROPAR)*, August 2002.
6. S. Tilak, N. B. Abu-Ghazaleh, and W. Heinzelman, "A taxonomy of wireless microsensor network models," *ACM SIGMOBILE Mobile Computing and Communications Review*, vol. 6, pp. 28–36, April 2002.

Improved Position Estimation in Wireless Heterogeneous Networks

Erwan Ermel[1,2], Anne Fladenmuller[1], Guy Pujolle[1], and André Cotton[2]

[1] Université Pierre et Marie Curie, Laboratoire d'Informatique de Paris 6 (LIP6),
8 rue du Capitaine Scott, 75005 Paris, France
tel:+33(0)144278879 - fax: +33(0)144277495
{Erwan.Ermel,Anne.Fladenmuller,Guy.Pujolle}@lip6.fr
[2] Thales Communication, BGCOM/TCF/SAS/SEA/TAI,
160 bld de Valmy - BP 82, 92704 Colombes Cedex, France
tel:+33(0)146132216 - fax: +33(0)146132668
Andre.Cotton@fr.thalesgroup.com

Abstract. This paper addresses the problem of nodes localization in wireless ad hoc networks. Two types of nodes are considered: nodes with self-locating capability like GPS and nodes with no self-locating capability. For the last ones it is thus important to infer a position which will be retrieved from the position of the neighbor's nodes. The precision of this information clearly depends on the environment and may not be very accurate. We propose a method which consists in selecting and processing only nodes that are likely to enhance the accuracy of an estimated position. We focus our approach on defining a hull, made up of neighboring nodes, as a key element of position accuracy enhancement. The improvements of using such a method are then validated by a set of simulations.

1   Introduction

Ad hoc networks consist of wireless heterogeneous nodes that communicate with each other without any pre-established infrastructure. These autonomous networks focus on providing self-configuring networks that are easily and quickly deployable. Ad hoc networks are of great interest and have a wide range of applications, as for in war theatres or disasters relief.

Due to the wide diversity of wireless devices available on the market and the various capabilities they offer, it seems restrictive to suppose that all nodes will possess a localization system such as GPS [1] or Galileo [2]. Thus, nodes without such type of equipment will have to retrieve a geographical position. Such information can useful for many purposes: geographical routing, location services or more generally in ambient networks.

The localization process can be fully distributed, based on the information a node can get from its neighbors position, whether this position has itself been inferred by a localization process or obtained thanks to dedicated devices. The purpose of this paper is thus to propose a method to improve the precision

N. Mitrou et al. (Eds.): NETWORKING 2004, LNCS 3042, pp. 1306–1311, 2004.

of the position estimation in the most generic environments such as a network composed of basic laptops with simple WiFi cards.

Most position estimation techniques are based on geometrical computations like triangulation or trilateration. To evaluate the distance between two nodes, several methods can be used. Four classes of position estimation methods can be defined: the first one consists in determining the *time-of-flight* of a signal between two anchors (Time Of Arrival [3,4] and Time Difference Of Arrival [5,6]). The second class is based on the signal strength [7,8]: when the emission power is known, the distance to the anchor can be approximated. The third class is based on triangulation like the Angle of Arrival (AoA) estimates the direction of an incoming signal from several anchors, and then estimates the position. As for the fourth, it gathers all remaining position estimation methods like connectivity based approach [9,10,11]. Our approach belongs to the fourth and last class.

We define two classes of nodes: *self-locating* nodes, which are embedded with self-locating capability, such as, and *simple* nodes with no self-locating capability. We call anchor or landmark a node that knows its position. In this paper, we investigate a simple method to select nodes with the aim to enhance the accuracy of estimated positions. Our technique can be implemented and used by every position estimation methods stated above as it is a step just before the real position estimation process.

The structure of this paper is done as follow. We first present the assumptions and the definition made in this paper. Section 3 details the hull method to select anchors within neighboring nodes, followed in Section 4 by our simulation results. Section 5 concludes the paper.

2   Assumptions and Definitions

We limit our approach to select only one-hop anchors but this technique is also feasible for n-hops nodes selection. No distance measurement is to be used to estimate the position of a simple node. Thus a node only exploits its neighbor's nodes connectivity.

Let S be a *simple* node. Let S_{est} be the estimated position of the node S, and S_{real} be the coordinates of its real location. Note that S_{real} information is only used by simulations to evaluate the precision of our algorithm. We can not possess such information in reality as it represents what we are looking for. Let R_{max} be the maximum theoretical transmission range of S. We also define the accuracy of the node position C_{acc}, as a function of the localization error represented by the distance between S_{real} and S_{est}:

$$C_{acc} = 1 - \frac{\|(S_{real}, S_{est})\|_2}{R_{max}} \tag{1}$$

By definition, $0 \leq C_{acc} \leq 1$. Self-locating nodes with accurate coordinates, like the ones given by a GPS have a position accuracy of 1. On the other side, simple nodes, which have to estimate their position, have a position accuracy $0 \leq C_{acc} < 1$. Note that these simple nodes don't only have to retrieve their position coordinates but also to evaluate the precision of this information. In order to do so, several methods such as statistical or area computing approaches are detailed in [12].

3   Convex Hull Selection

Our main goal in this paper is to enhance the accuracy of an estimated position by selecting only anchors that are likely to improve the position estimation process. We detail in this section a simple approach to select anchors amount the one-hop nodes in a wireless networks: the convex hull selection.

Computational geometry deals with geometrical problems. These problems are for example convex hull among a list of nodes, Voronoï diagrams, geometric searching. These algorithms are well detailed in [13,14,15,16].

The main idea of using a convex hull as a selection method among nodes is to choose only nodes which are at the greatest distance from anchors. As the position estimation process is based on trilateration, the further apart the anchors are, the better will be the accuracy of the estimated position.

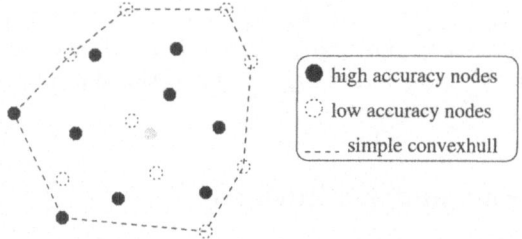

Fig. 1. Convex hull: simple convex hull considers the distance metric to elect hull nodes.

We choose the convex hull approach as our nodes selection method. Only the hull nodes are taken into account in the position estimation process. The remaining nodes are simply discarded. A convex hull example is shown in Fig. 1.

The convex hull of a set of points S in n dimensions is the intersection of all convex sets containing S. For N nodes $n_1, ..., n_N$, the convex hull C is then given by the expression:

$$C \equiv \left\{ \sum_{j=1}^{N} \lambda_j n_j : \lambda_j \geq 0 \text{ for all } j \text{ and } \sum_{j=1}^{N} \lambda_j = 1 \right\}.$$

The simple convex hull selects only the nodes for their physical position whatever their position accuracy C_{acc} is. We will study in Section 4 the performances of this simple hull selection method.

4 Evaluation of the Hull Selection

The simulations were performed under Java. 50 nodes were randomly placed in a 1000m x 1000m square. Self-locating nodes and simple nodes were also randomly elected. The maximum theoretical transmission range R_{max} was set to 170m.

Fig.2 shows the impact of our simple hull selection method on the accuracy of an estimated position. To compare our selection method, we choose as the reference model the greedy scheme: all the neighbors nodes are selected in the position estimation process. None are discarded. The estimated position is obtained by a simple centroid formula, where all the nodes have got the same weight.

We want here to compare the impact of choosing carefully some nodes from our neighborhood (simple hull method) or of getting as much information as we can (greedy method) to enter the position estimation process. The simulations provide us the accuracy of the estimated position by monitoring the distance between the real and the estimated positions. For each selection method (hull or greedy), we keep track of the number of neighbors and their type: self-locating and simple nodes. It seems important to differentiate both types as self-locating nodes have accurate position information, whereas simple nodes already have an inferred position with a much less accurate value. We thus want to evaluate the impact on the accuracy of the position to estimate another position. By running a large number of simulations, we then obtain an average accuracy for each combination of neighbors (number of simple and self-locating nodes).

In Fig. 2, the graphs plots in (a),(c),(e) (respectively (b),(d),(f)) the average accuracy of the estimated position as a function of the number of simple nodes (resp. self-locating nodes) in the neighborhood. The number of self-locating nodes (respectively simples nodes) in the direct neighborhood is set to 0 (resp. 2 and 6) for (a),(b) (resp. (c),(d) and (e),(f)).

It is obvious that the more precise the position information retrieved from the neighborhood is, the better will be the accuracy of the estimated position. Nevertheless it also appears clearly from these plots that the hull method, which consists of selecting the position information from the nodes the further apart from each other, gives better results. As for in Fig. 2 (a), where we can see for instance that selecting 3 simple nodes gives on average similar performances than taking the all 6 simple nodes in the greedy approach.

We also notice that the more sln nodes are used bye the estimation process, the better is the position accuracy.

In every case, the simple selection gives a better position accuracy than the greedy approach does. The selection enhances the position accuracy up to 20%.

5 Conclusion

We present and compare in this paper a simple method to select anchors in a wireless network to enhance the position estimation of simple nodes, nodes with no self-locating capabilities.

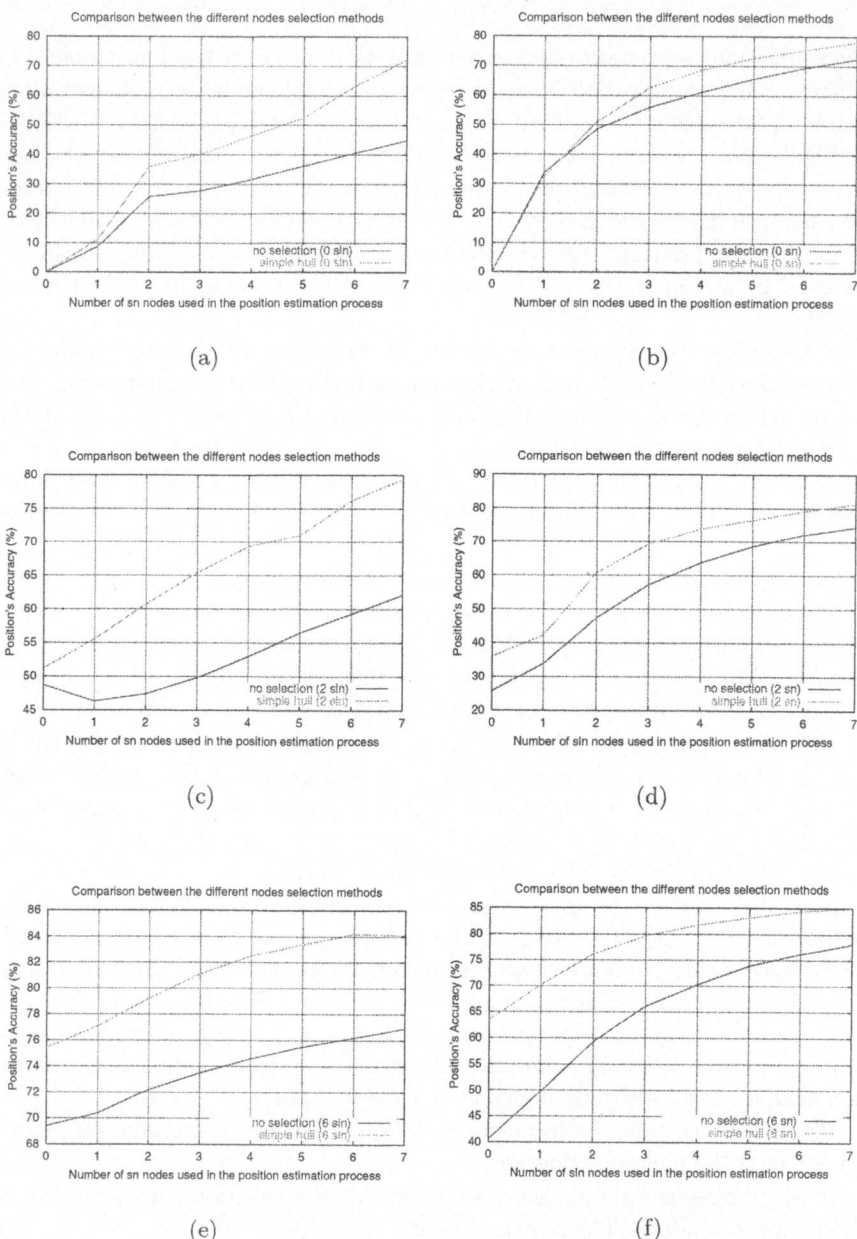

Fig. 2. Impact of nodes selection method on the average accuracy of the estimated position.

Our nodes selection approach deals with the definition of a convex hull among neighbor nodes. The resulting position accuracy using such nodes selection method is enhancing from a greedy scheme up to 20%. We also show that the accuracy of an estimated position only take advantage of the distance between the hull nodes, whatever their position accuracy is.

Our next step consists of implementing such selection algorithm in a global geographical routing protocol in a heterogeneous network under NS-2.

References

1. Hofmann-Wellenhof, B., Lichtenegger, H., Collins, J.: Global Positioning System : Theory and Practice. Springer-Verlag (1997)
2. (Galileo) http://europa.eu.int/comm/dgs/energy_transport/galileo/.
3. Capkun, S., Hamdi, M., Hubaux, J.P.: Gps-free positioning in mobile ad-hoc networks. Cluster Computing 5 (2002)
4. Werb, J., Lanzl, C.: A positioning system for finding things indoors. IEEE Spectrum 35 (1998) 71–78
5. Savvides, A., Han, C.C., Strivastava, M.B.: Dynamic fine-grained localization in ad-hoc networks of sensors. (2001)
6. Ward, A., Jones, A., Hopper, A.: A new location technique for the active office. IEEE Personal Communications 4 (1997) 42–47
7. Savarese, C., Rabaey, J.M., Beutel, J.: Localization in distributed ad-hoc wireless sensor networks. Proceedings of the ICASSP (2001)
8. Beutel, J.: Geolocalisation in a picoradio environment. Master's thesis, ETH Zurich, Electronics Lab (1999)
9. Doherty, L., Pister, K.S.J., Ghaoui, L.E.: Convex optimization methods for sensor node position estimation. In: Proceedings of IEEE INFOCOM'2001, Anchorage (2001)
10. Niculescu, D., Nath, B.: Dv based positioning in ad hoc networks. Telecommunication Systems 22 (2003) 267–280
11. Bulusu, N., Heidemann, J., Estrin, D.: Gps-less low cost outdoor localization for very small devices. IEEE Personal Communication, Special Issue on Smart Spaces and Environment 7 (2000) 28–34
12. Ermel, E., Fladenmuller, A., Pujolle, G., Cotton, A.: Estimation de positions dans des réseaux sans-fil hybrides. In: CFIP 2003. (2003)
13. Preparata, F.P., Shamos, M.I.: Computational Geometry: An Introduction. Springer Verlag (1991)
14. de Berg, M., Kreveld, M., Overmars, M., Scharzkopf, O.: Computational Geometry, Algoritms and Application. Springer (1997)
15. O'Rourke, J.: Computational Geometry in C. Cambridge University Press (1998)
16. Lemaire, C.: Triangulatation de Delaunay et arbres multidimensionnels. PhD thesis, Ecole des mines de Saint-Etienne (1997)

A Geometric Derivation of the Probability of Finding a Relay in Multi-rate Networks[*]

Laura Marie Feeney[1], Daniel Hollos[2], Martin Kubisch[2], Seble Mengesha[2], and Holger Karl[2]

[1] Swedish Institute of Computer Science, lmfeeney@sics.se
[2] Telecommunication Networks Group, TU Berlin,
{hollos, kubisch, mengesha, karl}@tkn.tu-berlin.de

Abstract. Relaying can improve performance of a wireless network, especially when different transmission modes with distance/cost tradeoffs are available. Examples of such modes include data rates or transmission power. This paper geometrically analyzes the probability that a high-cost direct transmission can be replaced by a combination of low-cost relay transmissions. The main result of the analysis is a technology-agnostic characterization of a communication system's amenability to relaying strategies and some recommendations for how to structure such systems.

1 Introduction

One way to improve the performance of a wireless system is to replace high-cost transmissions to distant terminals with a sequence of lower-cost transmissions, using intermediate terminals as relays. An example of "cost" is the data transmit rate. In an IEEE 802.11b system, a single, long-distance 2 Mbit/s transmission might be divided into two 11 Mbit/s transmissions, if a relay is available in the right place, resulting in an (ideal) effective data rate of 5.5 Mbit/s.

Obviously, whether or not a relay is available is a probabilistic question that depends on the distribution of terminals in a cell and on the relationship between the cost of a specific transmission mode and the distances over which it can be used. In the example above, an 11 Mbit/s transmission can only be used over short distances, limiting the choice of relays.

In practice, whether or not relaying results in performance improvement depends on the cost structure for a particular link technology. For example, there is no benefit to replace a 1 Mbit/s transmission with a pair of 2 Mbit/s transmissions. In fact, the overhead cost of transmitting headers twice would certainly result in performance degradation. Accounting for such fixed costs and per-transmission overhead is essential, although analysis of specific link technologies is outside the scope of this paper.

The primary contribution of this work is a geometry-based, analytic model for bounding the performance improvement that can be obtained using intra-cell relaying. In particular, we develop a closed form expression for the expected

[*] Corresponding author: Holger Karl. This paper has been partially supported by the German Federal Ministry of Education and Research (BMBF) IBMS2 project.

N. Mitrou et al. (Eds.): NETWORKING 2004, LNCS 3042, pp. 1312–1317, 2004.

proportion of terminals in a cell that can take advantage of relaying, given the terminal density and ranges of two transmission modes. Because it is based on geometric construction, the result generalizes to any link technology and transmission cost metrics, including data rate and energy consumption.

The model makes no assumption about protocol mechanisms for relaying, in effect, assuming that relay terminals are selected without protocol overhead. This idealized result provides a useful basis for evaluating the performance results of some specific relaying protocol. We conclude this paper by proposing a rate adaptive relaying protocol for IEEE 802.11 that exploits the probabilistic insights gained from our analytic work.

2   Geometric Analysis

We derive the probability that, for a given terminal, there is another terminal located such that using it as a relay reduces the total cost of communication.

2.1   Relaying and Network Model

Consider a wireless cell consisting of an access point and N terminals. Each terminal supports a small number, k, of transmission modes $M_1, \ldots, M_k$. Each mode supports a maximum transmit radius $d_1, \ldots, d_k$, where $d_i < d_j$ for $i < j$; d_k defines the cell radius. Each transmission mode M_i also has associated with it some transmission "cost" C_i. We assume that there is a direct relationship between distance and cost, thus $C_i < C_j$ for $i < j$. (The higher the cost, the longer distance can be covered.)

In the IEEE 802.11 rate adaptation example above, "cost" is the time required to transmit a packet. Energy cost, based on variable transmit power, is another example. The discrete cost model is suited to any link technology which provides a small number of discrete transmission modes. Assigning costs for a specific link technology is outside the scope of this work. We assume that transmission mode combinations for which relaying reduces the total cost have been identified.

We do not take into account the impact of relaying on end-to-end packet error rate (PER). As long as the maximum transmit radius for a mode is defined in terms of a relatively low PER and there is only a single relay, we believe that this is an acceptable simplification.

In this work, we only consider the case of a single relay and two transmission modes ($k = 2$). It is fairly straightforward to extend the analysis to larger k, although it becomes somewhat cumbersome as the number of cases to be considered grows rapidly; we leave this to forthcoming work. For this restricted problem, we can simplify the notation by normalizing the smaller (low-cost) *relay radius* d_1 to 1 and defining the larger (high-cost) transmit radius and cell size $D = d_2/d_1 > 1$. We assume that for the link technology in question, it is known that relaying is (at least potentially) cost efficient i.e. that $C_1 + C_1 < C_2$.

2.2   Derivation of the Probability of Relay Utilization

First, we compute the area $A(r)$ of the region in which a terminal must be located in order to act as a relay for a transmitter located at distance r from the access point. Assuming terminals are uniformly distributed in the cell, it is then straightforward to compute the probability that there is at least one terminal in a region of area $A(r)$. Finally, we compute for each terminal in a cell its expected probability of being able to take advantage of a relay terminal.

Area of the Relay Region. We define $R_t(r)$, the *relay region* of a transmitter t located at distance r from the access point, as the region in which a terminal must be located in order to act as a relay for t. The relay region is non-empty only for $1 < r \le \min(D, 2)$. For $0 < r \le 1$, the transmitter is already able to communicate directly with the access point using the lower cost transmission mode. For $r > min(2, D)$, the transmitter is either too far away to reach the access point via a single relay, or it is outside the cell. [1]

In order for a terminal to act as a relay for a transmitter, the distance between the relay and the transmitter must be less than the relay radius, which is 1; otherwise, the transmitter cannot communicate with the relay using the lower cost (smaller transmission radius) transmission mode. Similarly, the distance between the relay and the access point must also be less than the relay radius; otherwise the relay cannot communicate with the access point at the lower cost transmission mode. Figure 1 illustrates this requirement.

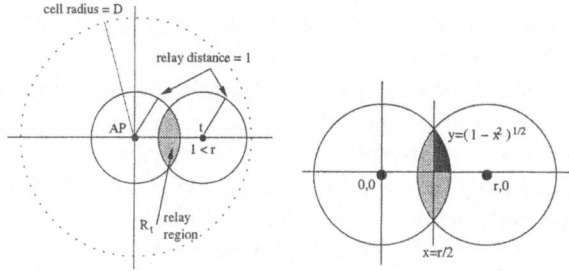

Fig. 1. Relay region $R_t(r)$(left). Detail (right). Without loss of generality, we set the access point at the origin and the transmitter at $(r, 0)$.

We can now compute $A(r)$, the area of the relay region $R_t(r)$ for transmitter t. Integrating to find the intersection of the unit disks (Figure 1 detail) gives:

$$A(r) = 4 \int_{x=r/2}^{1} \int_{y=0}^{\sqrt{1-x^2}} \mathrm{d}y \mathrm{d}x = \pi - r\sqrt{1 - \left(\frac{r}{2}\right)^2} - 2\arcsin\frac{r}{2}. \qquad (1)$$

---

[1] The analysis is easily extended to address the use of relays for coverage extension.

Probability of finding a relay. The probability that a transmitter t is able to take advantage of a relay is equivalent to the probability that there is at least one terminal in its relay region $R_t(r)$. (Here we assume an optimal relay protocol.)

Any relay region is, by definition, contained within the cell radius D. Assuming terminals are uniformly distributed within the cell, the probability p that a terminal in the cell is in a relay region of area $A(r)$ is therefore $p(r) = \frac{A(r)}{\pi D^2}$.

We define $P_{n \geq 1}(r)$ to be the probability that, for some transmitter t, at least one of the other $N-1$ terminals is in its relay region $R_t(r)$. This is the complement of the probability that all $N-1$ terminals are not in its relay region.

$$P_{n \geq 1}(r) = 1 - (1-p)^{N-1} = 1 - \left(1 - \frac{A(r)}{\pi D^2}\right)^{N-1} \tag{2}$$

Expected proportion of terminals finding a relay. To find $P(N, D)$, the expected proportion of terminals in a cell which can take advantage of relaying, we compute the expected value of the probability $P_{n \geq 1}(r)$ for each terminal, which is its expected value over all positions in the cell. The equivalent polar formulation integrates over r the probability that a terminal is located at distance r from the access point and there is at least one terminal in its relay region.

$$P(N, D) = \lim_{\Delta r \to 0} \sum_{i=0}^{\lfloor \frac{\min(D,2)}{\Delta r} \rfloor} \left(P(\text{terminal in ring from } i\Delta r, \Delta r \text{ wide}) \, P_{n \geq 1}(i\Delta r) \right)$$

$$= \int_{r=1}^{\min(D,2)} \frac{2r}{D^2} P_{n \geq 1}(r) \mathrm{d}r = \int_{r=1}^{\min(D,2)} \frac{2r}{D^2} \left(1 - \left(1 - \frac{1}{\pi D^2} A(r)\right)^{N-1} \right) \mathrm{d}r$$

2.3 Numeric Solution of $P(N, D)$

Figure 2 shows the behavior[2] of $P(N, D)$ for various values of N and D. The proportion of terminals that can take advantage of a relay grows rapidly as the Terminal density increases – precisely the situation in which relaying is most needed. There is also a narrow range of optimal values for D, which is the ratio between the range of the high and low cost transmission modes.

For the case of rate adaptation in IEEE 802.11 networks, realistic values of N and D ($D \approx 1.25 - 2.25$, $N \approx 8 - 32$) suggest that between 15 and 55 percent of terminals in a cell can use a relay to increase their effective transmission rate.

3 Protocol Outline

We now outline a relaying strategy for rate adaptation in an IEEE 802.11 cell. The proposed technique retains the basic operation of the IEEE 802.11 DCF

---

[2] Confirmed using Matlab to analyze 10000 randomly generated topologies.

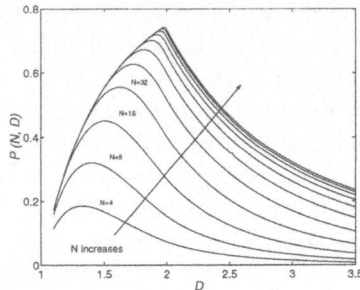

Fig. 2. Behavior of $P(N, D)$ for various D as N becomes large (graph shows curves for $N = 4, 8, 16, \ldots, 4196$)

protocol: RTS and CTS messages are transmitted at the base (lowest) transmission bit rate of the BSS and reserve the channel for the end-to-end duration of the transmission. A terminal that acts as a relay is thus permitted to transmit during this interval without first performing media reservation.

We build upon opportunistic rate adaptation. A terminal transmitting at a low bit rate occasionally attempts a higher bit rate. The terminal will use the RTS/CTS mechanism to reserve the channel for a duration that assumes a sequence of two transmissions at a higher bit rate, plus some backoff interval. The originating terminal then transmits the DATA frame at the higher bit rate.

Any terminal that successfully receives the high-bit-rate frame is a potential relay. It can determine the minimum date rate at which the frame must be retransmitted by comparing the frame duration with the date rate at which the frame was transmitted. For the relayed transmission to succeed only one of the potential relays should actually retransmit the frame and that relay terminal must be able to communicate with the access point at the required data rate.

Each potential relay performs a short random backoff procedure, after which it checks for an already ongoing relay transmission. Thus, there will be a collision only if two relays select the same slot in which to begin their transmissions. The backoff interval can be tuned to reflect the probability of finding a relay, as derived above. The backoff procedure can also be biased such that a relay terminal that is more likely to succeed in communicating with the access point at the higher data rate (e.g. because it has recently done so) selects a shorter backoff interval. The maximum length of the backoff interval must be included in the duration specified in the RTS, allowing a tradeoff between collisions and overhead.

If any of the higher bit rate transmissions fail or if there is no relay available, the originating terminal will not receive an ACK. The terminal simply retransmits at the lower bit rate and continues to transmit at the lower bit rate until it makes another opportunistic attempt to communicate at a higher bit rate.

This approach has the advantage of simplicity and minimal modification to the existing IEEE 802.11 framework. The technique applies directly only to traffic on the up-link, although bi-directional down-link traffic can be supported by caching up-link relay information at the access point. The overhead of the backoff interval and risk of collision among relays are potential disadvantages.

4   Related Work

The capacity of multi-hop networks has been widely studied. More closely related to the present work are a number of probabilistic studies of the distributions of link distance and path lifetime in multi-hop networks. In particular, Miller derives the probability of a relay connection in a random network as a function of node dispersion [1].

Relaying is frequently proposed to minimize energy consumption in wireless networks. In [2], a geometric model is used to determine for each node which of its neighbors are most efficiently reached directly and which are most efficiently reached via a relay. A distributed algorithm is also presented.

The combination of relaying with rate adaptation has been a motivating example throughout this paper. While this problem has not been extensively studied, a scheme proposed in [3] indicates a significant capacity increase. The impact of multiple transmission rates in an IEEE 802.11 cell is examined in [4].

5   Conclusion and Future Work

We have developed a geometric analysis of the probability of finding a suitable relay, using a generic formulation for cost/distance-adaptive transmission modes. For the case of two transmission modes, we determined the expected proportion of terminals in a cell that can take advantage of a relay, as a function of the ratio between the two transmission ranges and the transmitter density in the cell.

We believe that this analysis provides a good general tool for capacity investigations. Further results have already provided insights into the relative suitability for various, multi-rate capable physical layers (IEEE 802.11b and Hiper-LAN/2). We are also looking at the impact of several transmission modes and a further specification of the relaying protocol outlined here. We expect that the probabilistic understanding of relaying developed here will assist in the design and performance analysis aspects of that work.

References

1. Miller, L.E.: Probability of a two-hop connection in a random mobile network. In: Proc. 35th Conf. on Information Sciences and Systems (CISS 2001). (2001)
2. Rodoplu, V., Meng, T.H.: Minimum energy mobile wireless networks. IEEE J. on Selected Areas on Communication **17** (1999) 1333–1344
3. Mengesha, S., Karl, H.: Relay routing and scheduling for capacity improvement in cellular WLANs. In: Proc. of Modeling and Optimization in Mobile, Ad Hoc and Wireless Networks (WiOpt'03), Sophia-Antipolis, France (2003)
4. Heusse, M., Rousseau, F., Berger-Sabbatel, G., Duda, A.: Performance anomaly of 802.11b. In: Proc. INFOCOM, San Francisco, CA (2003)

Metadata Design for Introspection-Capable Reconfigurable Systems

Vangelis Gazis, Nancy Alonistioti, and Lazaros Merakos

Communication Networks Laboratory, Department of Informatics & Telecommunications,
University of Athens, 157 84, Athens, Greece,
{gazis, nancy, merakos}@di.uoa.gr

Abstract. Global vision consensus on the next generation of wireless mobile communications, broadly termed 4G, sketches a hybrid infrastructure, comprising different wireless access systems in a complementary manner and vested with reconfiguration capabilities that facilitate a flexible and dynamic adaptation of the wireless infrastructure to meet the ever-changing service requirements. We identify essential metadata classes to support the reconfiguration of communication systems, introducing a respective object-oriented UML model. We elaborate on the design rationale that underpins the UML model, describing its classes and associations and discussing the possible metadata representation technologies and encoding formats. We proceed to identify existing metadata standards that are candidate for the representation of reconfiguration metadata, discussing and evaluating their suitability. Ultimately, we present a developed reconfiguration metadata description vocabulary and illustrate its application with an example.

1 Introduction

Over the last decade, the mobile industry has developed into a breeding ground for innovative wireless access technologies. In addition to second (2G) and third (3G) generation mobile communication systems, broadband WLAN type systems such as HIPERLAN/2, IEEE 802.11 and broadcast systems like DAB and DVB-T are becoming available and short range connectivity systems like Bluetooth are being developed rapidly. Considering that the observed proliferation of wireless access technologies is likely to persist and that future mobile devices will need to support multiple dissimilar wireless access standards, the mobile communication industry has been focusing on the reconfigurability concept as a technological enabler of future (multi-standard) mobile systems and radio resource management across different wireless standards. The now widely accepted vision for reconfigurable systems and networks sketches a seamless ubiquitous computing and communication infrastructure where mobile and immobile devices may proactively and/or reactively adapt their own communication capabilities by dynamically discovering, selecting, downloading and activating software implementations for the communication personalities they wish to assume in any given time instance.

N. Mitrou et al. (Eds.): NETWORKING 2004, LNCS 3042, pp. 1318–1325, 2004.

The rest of the paper proceeds as follows: The next section highlights the fundamental concepts of reconfiguration, introducing key definitions and providing an overview of current approaches as well as the related standardization status. Next we focus in the realm of each individual reconfigurable system and introduce a generic object-oriented UML model that facilitates discovery of reconfiguration options (i.e., the reconfiguration space) to support the application of reconfiguration within and across communication standards. We go on to elaborate on the design rationale of the UML model, followed by a discussion and brief evaluation of instrumentation options. Finally, we conclude the paper and highlight directions for future work.

2 Reconfiguration – Basic Definitions and Standardization Issues

In general, the term reconfiguration refers to the (dynamic) instantiation, parameterization and inter-connection of protocols (i.e., communication-related functional entities) within the user, control and management planes of a collection of operating communication systems in a manageable, consistency-preserving and – preferably – transparent fashion. For the rest of the paper, the term reconfiguration will refer to the dynamic adaptation of implementation mappings of internal (communication) equipment components [1] that does not compromise their consistency or their ability to provide their services. Leveraging the work of from early software radio projects in the military domain [2], SDR Forum has pioneered in exploring reconfiguration in the domain of wireless communications. However, being the vanguard of reconfiguration developments and the first to define a software radio architecture [3], seems to have come at the expense of a rather restricted view on reconfiguration that focuses primarily on the radio domain (e.g., RF processing, down-conversion, IF processing, A/D conversion, etc) [4]. Soon it was realized that, restricting the concept of reconfiguration solely to radio-dependent communication functionality under control of mobile network operators or equipment manufacturers only, would severely limit its application domain and undercut its beneficial impact on the long run. Support grew on the viewpoint that the full potential of reconfiguration would best be served by opening up reconfigurable device capabilities to the wider service provision process and leveraging technical expertise in third-parties (e.g., software developers) [5], creating an open market for (software) implementations of reconfigurable equipment components that will propel the development of universally reconfigurable mobile systems.

The (now joint) Parlay/OSA standardization initiative has been a major step forward towards the openness of the mobile value chain and the participation of multiple players in mobile service provision. However, it did not anticipate the case of reconfigurable systems; Parlay/OSA consider the network infrastructure as immutable and specify logical interfaces for invoking the particular functionality it supports. Although not precluded by their logical architecture, the case of reconfigurable wireless networks and mobile systems capable of dynamically adapting their internal instrumentation is beyond the scope of the current standard which does not include reconfiguration-supporting interfaces. This shortcoming has been identified in [6] along with the need for appropriate Parlay/OSA extensions to support third-party driven reconfiguration actions upon the mobile network infrastructure.

Another paramount issue not yet identified in the literature that should be addressed by standardization, and which is explicitly identified here, concerns the specification of an appropriate object model for reconfigurable communication systems. Architectures supporting reconfiguration will require a suitable object-oriented information model to capture and express the internal organization and structure of reconfigurable equipment in an abstract, implementation neutral way that effectively provides the unified view necessary to start specifying a generic reconfiguration capability. Through object orientation and inheritance, common parts can be factored out and reused as an abstract foundation model from which wholly different instrumentation inherit, thereby allowing a fine-grain mix of standardized behavior with innovative, performance-focused, proprietary instrumentations. Object orientation does not necessarily restrict the granularity of structural analysis to individual classes and objects; use of more coarse-grain analysis modules (i.e., components) is also possible (and to a large extent desirable).

3   Designing for a Generic Reconfiguration Capability

Reconfigurable systems must be adaptable at two different levels: the base level that includes the (software-based) instrumentations of communication-related functionality and the so-called meta-level comprising the (abstract) specifications of that functionality. That will allow development of architectures supporting adaptation between different (software-based) instrumentations of communication personality and across disparate communication personalities (e.g., ad-hoc, cellular, broadcast, etc) in a uniform way. From that viewpoint, generic support structures like architectural frameworks for flexibly expressing and circulating reconfiguration-related metadata become of paramount importance. The next section introduces a UML model designed to provide suitable metadata abstractions for the development of manageable reconfigurable communication systems in beyond 3G mobile networks.

4   Modeling Reconfiguration Metadata

4.1   Metadata Classes

Product, the root abstract class in our model, specifies a 'marketable' item (i.e., a resource that may constitute the subject of an exchange in an economic system), which can be identified through a textually represented name. It includes a single (URI-convertible) URL attribute that provides a unique identifier of each individual product instance as a Web-identifiable resource, thereby streamlining it to the Semantic Web model and its Resource Description Framework (RDF) [7].

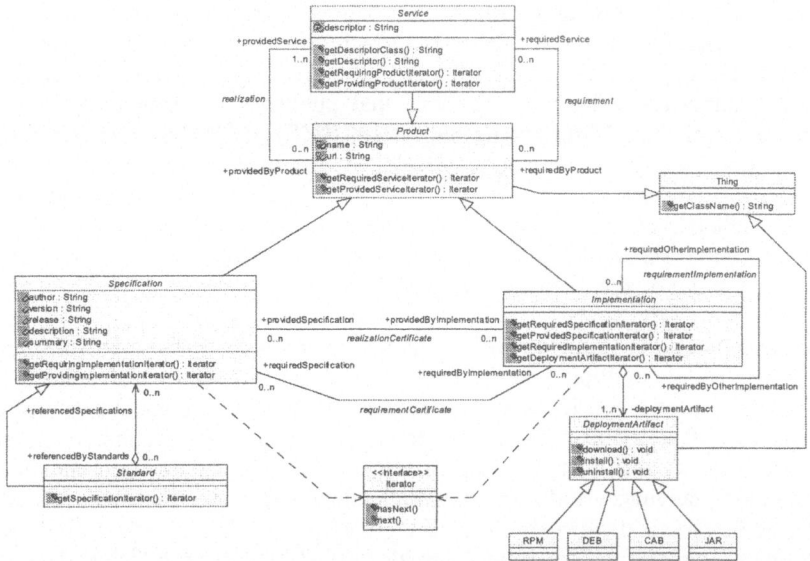

Fig. 1. The object-oriented information model for reconfiguration metadata.

Service is a subclass of Product that refers to some precisely defined functionality and has a textual description property. It is meant to provide an abstract yet unambiguous placeholder for a service's definition accompanied by a textual descriptor that might be associated with arbitrary formal semantics, provided those semantics support a textual representation. It is not particularly important whether a unique formal format is employed for the service descriptor, since generic adaptation mechanisms may be used to identify the appropriate handler for each available format. However, it is of paramount importance that the service descriptor identifies the service unambiguously, an overlooked issue that is further elaborated on in the subsection entitled "Metadata encoding".

Specification is a subclass of Product with additional (textual) attributes, namely author, version, release, description and summary. Specification provides an abstract class for commonly representing behavioral and/or functional specifications (e.g., the specification of a authentication protocol). It is meant to provide a first-class abstraction for standards developed and published by authoritative bodies, such as the Universal Mobile Telecommunication System (UMTS) specifications developed and published by the 3rd Generation Partnership Project (3GPP). Currently, such specifications are recorded in a documentation system in various human-readable formats, such as the IETF Request For Comments (RFC) textual system. The lack of a common (machine-interpretable) format for specifications published by different authoritative bodies rules out the possibility of having those specifications parsed, understood and exploited by an intelligent agent in control of reconfigurable communication capabilities.

Standard is a subclass of specification designed to provide a generic container for related specification instances, in order to facilitate modeling of specifications that reference (as opposed to specialize) other specifications, possibly published by a different authoritative body (to the one that publishes the standard). The 3GPP specification of the IP Multimedia Subsystem (IMS) in UMTS is an example of a standard that leverages specifications developed by a different authoritative body (i.e., the IETF SIP specification). We stress that, through the Specification and Standard classes, inheritance-based as well as composition-based modeling of actual communication standards is supported, thereby rendering the full spectrum of modeling options available to the designer [8].

Implementation is a subclass of Product that refers to a real-life (software) artifact, which may realize multiple specifications. It is meant to model the real-life software instrumentation of a specification but may also be used to represent software-based functionality that is not associated to a particular specification (e.g., utility functionality). Given that an implementation may be developed in different programming languages and supporting technologies (e.g., C, C++, Java, .NET) and packaged in various deployment formats (e.g., Microsoft CAB, RedHat Linux RPM), modeling of implementations should provide unified support for different deployment artifacts through a common base class, such as the DeploymentArtifact abstract class included in Fig. 1.

4.2 Metadata Associations

A particular specification may depend on the availability of multiple services much as it may render multiple services. Similarly, a particular implementation, in addition to the set of services that its associated specifications collectively require and realize, may depend on the availability of additional services to function properly and may realize additional services during operation. Because they apply to Specification and Implementation instances alike, these concerns are expressed through a pair of associations between the Product and Service classes named requirement and realization, respectively.

Access to the aforementioned associations is supported based on an application of the Iterator design pattern [8] that abstracts the implementation details of the association from client entities. An agent may navigate these associations through an Iterator instance returned by any of the (getRequiredServiceIterator, getProvidedServiceIterator) and (getRequiringProductIterator, getProvidingPro-ductIterator) method pairs of the Product and Service classes, respectively, rendering client implementations dependent solely on the Iterator interface, while the supporting implementation of the navigation facility may vary arbitrarily from a local database row set to an hyperlinked knowledgebase distributed over the Internet or any suitable combination. Finally, an implementation may de dependent upon the availability of other implementations to function properly (e.g., object libraries), a concern expressed through the requirementImplementation association.

Regarding the relation between Specification and Implementation instances, we should note that it is not mandatory that an Implementation instance be associated to a

Specification instance; it might as well be an implementation of utility functionality not subject to standardization yet required by other implementations. Thus, the case of an Implementation unassociated to a Specification instance is considered valid. In the typical case, however, the association between a Specification and an Implementation is expressed via the realizationCertificate and requirementCertificate named (multilateral) associations. The former signifies that the Implementation instance realizes the behavior of the set of Specification instances, while the latter marks the dependence of the Implementation instance upon a set of Specification instances. Agents may navigate the realizationCertificate and requirementCertificate associations through an Iterator instance returned by any of the (getRequiredSpecificationIterator, getRequiringImplementationIterator) and (getProvidedSpecificationIterator, getProvidingImplementationIterator) method of the Implementation and Specification classes, respectively.

4.3 Metadata Encoding

The aforementioned UML model provides a common information model for expressing reconfiguration metadata that may be exploited by a reconfiguration management process. Considering that reconfiguration metadata may be subject to processing and exchange in different administrative domains, it should be represented in an instrumentation-independent format that ensures interoperability. Two recommendations of the World-Wide-Web Consortium, XML [9] and RDF [10] are considered as prime candidates for this task. In general, XML is easier to use and manipulate, while RDF has greater capabilities for expressing semantically rich information. However, only RDF is capable of unambiguous semantic representation, since there is an explicit unique interpretation of any RDF data, based on the RDF Model Theory [11]. Consequently, a certain piece of information can be represented in RDF in exactly one unique way, while in XML many different representations with the same meaning are possible [12]. This advantage of RDF comes at the cost of being more verbose and significantly more complex, making it less attractive for the vast majority of users and developers [13].

In our approach, all reconfiguration metadata are represented in RDF, while the vocabulary employed by the RDF representation is a combination of W3C-standard RDF vocabulary, industry-used vocabularies and an extension vocabulary defined in an RDF Schema document, all using XML as a serialization format. An extension vocabulary named RCM that is derived from an isomorphic mapping [14] of the aforementioned UML model to an RDF Schema document has been developed and used to describe the UML model classes and associations. To ease prototype implementation, we chose the widely used the Red Hat Package Manager (RPM) vocabulary [15], a superset of the Linux Standard Base Specification [16], for representing the metadata of the DeploymentArtifact class. Reconfiguration metadata are represented using the RCM extension vocabulary, which, thanks to the namespace extensibility mechanism of RDF, provides also for integration to the standard RDF and RPM vocabularies. The text below serves as an illustrative example of our RDF Schema applied for the case of the 3GPP GTP specification, which is dependent upon an ITU service identified via its RDF URI.

```
<?xml version="1.0"?>
<rdf:RDF xmlns:RDF="http://www.w3.org/TR/WD-rdf-syntax#"
         xmlns:rdf="http://www.w3.org/1999/02/22-rdf-syntax-ns#"
         xmlns:RPM="http://www.rpm.org/"
         xmlns:rcm="http://cnl.di.uoa.gr/People/Gazis/RCM/1.0/"
  <rcm:Specification rdf:about="GTP610">
          <rcm:Name>GTP</rcm:Name>
          <rcm:URL>
http://www.3gpp.org/ftp/Specs/archive/29_series/29.060/29060-610.zip"
          </rcm:URL>
          <rcm:Author>3GPP</rcm:Author>
          <rcm:Version>6.1.0</rcm:Version>
          <rcm:Release>6</rcm:Release>
          <rcm:Description>
GTP provides user data tunnelling within and across GPRS domains
          </rcm:Description>
          <rcm:Summary>
          GPRS Tunnelling Protocol (GTP) across the Gn & Gp interfaces specification
          </rcm:Summary>
          <rcm:Provides rdf:parseType="Collection">
            <rdf:Description
rdf:about="http://www.itu.ch/specifications/services/transport/reliable.rdf"/>
            <rdf:Description
rdf:about="http://www.itu.ch/specifications/services/general/tunnelling.rdf"/>
          </rcm:Provides>
          <rcm:Requires rdf:parseType="Collection">
            <rdf:Description
rdf:about="http://www.itu.ch/specifications/services/transport/unreliable.rdf"/>
          </rcm:Requires>
  </rcm:Specification>
</rdf:RDF>
```

The primary reason for preferring RDF over XML for metadata representation is that RDF has been specifically designed for unambiguous representation. Considering that RDF models can be serialized in XML, RDF provides an ideal instrument for unambiguously representing reconfiguration metadata whilst supporting their serialization into an interoperable, machine-interpretable textual format that can be widely circulated across different administrative domains without alteration of semantics. Naturally, the higher complexity associated with RDF is the price to pay for *semantic univocality* – although we feel that other significant benefits, such as seamless plug-in to the Semantic Web infrastructure and laying the foundation for a reconfiguration knowledge base upon which to build self-aware, cognitive communication systems, offset the cost in the long run.

5 Conclusions

In the forthcoming future, mobile communication devices will be vested with a cognitive introspective intelligence that monitors its operational context as well as its own instrumentation, adapting it whenever and wherever it deems necessary and in any way it sees fit through the dynamic download and assembly of software components into standard-compliant operating instrumentations. Availability of appropriate reconfiguration metadata is a prerequisite to the advent of introspective cognition capabilities and a facilitator of efficient reconfigurations, an issue that has not been at the focus of mobile communication research. Similarly, efficiency concerns dealing with the optimality of different metadata representation standards for the representation of reconfiguration metadata have met little attention in the literature. On the Parlay/OSA initiative front, reconfiguration is yet to be included in the standardization agenda and the issue of object-oriented models as reconfiguration enabling frameworks remains in research twilight. We address these issues by introducing a generic object-oriented model to express reconfiguration metadata that will enable future systems to evolve not just the instrumentation of their behavior but the behavior itself, thus facilitating reconfigurations across disparate network architectures (e.g., ad-hoc, cellular) and deployment topologies. In addition to design issues, we have discussed and evaluated the potential of existing technologies and related standards for representing and encoding reconfiguration metadata in a machine-interpretable format that can be circulated across different administrative domains without semantic losses. Future extensions of our work will focus on the development of appropriate algorithms to support service-driven reconfiguration of mobile communication devices, both for inter- and intra- standard scenarios in order to assess algorithm complexity and to conduct comparative performance evaluations.

References

1. Tang, Z.: Dynamic reconfiguration of component-based applications in Java, M.Sc. thesis, MIT, September 2000.
2. Cox, M. C.: Joint tactical radio system (JTRS), presentation available from http://www.jtrs.sarda.army.mil/.
3. Bickle, J.: Software radio architecture (SRA) 2.0 overview, OMG TC, December 11, 2000, Orlando, Florida.
4. Blust, S. M.: SDR definitions, SDR Forum Plenary & Technical Committee, September 1, 2000.
5. Pereira, J.: Beyond software radio, VTC Fall 1999, Amsterdam, Netherlands, September 22, 1999.
6. Alonistioti, A., Houssos, N., Panagiotakis, S.: A framework for reconfigurable provisioning of services in mobile networks, International Symposium on Communications Theory & Applications (ISCTA), Ambleside Cumbria UK (2001).
7. Manola, F., Miller, E.: RDF Primer, see http://www.w3.org/TR/rdf-primer/.
8. Gamma, E., Helm, R., Johnson, R., Vlissides, J.: Design Patterns: Elements of Reusable Object Oriented Software, Addison Wesley Longman (1995).
9. XML: Extensible Markup Language home page, see http://www.w3.org/XML/.
10. RDF: Resource Description Framework home page, see http://www.w3.org/RDF/.
11. Hayes, P.: RDF Semantics, see http://www.w3.org/TR/rdf-mt/.
12. Berners Lee, T.: Why RDF model is different from the XML model, W3C discussion note, see http://www.w3.org/DesignIssues/RDF-XML.html.
13. Butler, M.: Barriers to the real world adoption of Semantic Web technologies, HP Labs Technical Report, HPL-2002-333, see http://www.hp.com/.
14. Chang, W.: A discussion of the relationship between RDF-Schema and UML, W3C discussion note, see http://www.w3.org/TR/1998/NOTE-rdf-uml-19980804.
15. Red Hat Package Management format, see http://www.rpm.org/.
16. Linux Standard Base Specification, see http://www.linuxbase.org/.

Security Clustering: A Network-Wide Secure Computing Mechanism in Pervasive Computing[*]

Jabeom Gu[1], Sehyun Park[1][**], Jaehoon Nah[2], Sungwon Sohn[2], and Ohyoung Song[1]

[1] School of Electrical and Electronics Engineering,
Chung-Ang University, Seoul 156-756, Korea
`jabeom@ms.cau.ac.kr`, {`shpark, song`}`@cau.ac.kr`
[2] Electronics and Telecommunications Research Institute (ETRI)
{`jhnah, swsohn`}`@etri.re.kr`

Abstract. In this paper, we introduce a new security paradigm, called security clustering, for pervasive computing environment that enables network-wide defend against increasing evolutionary attacks on the heterogeneous network and hosts. Security clustering make use of dynamic security context exchange between cluster members and distributed information sharing to achieve scalable and efficient cooperation.

1 Introduction

The pervasive computing can be envisioned as an open network with a high degree of heterogeneity, providing advanced Internet services to mobile users [1-3]. The openness of the network may well be the most important feature that the success of many future mobile applications rests on. However, in dealing with interwork of huge set of heterogeneous components, the absence of appropriate mechanism to actively detect and put down various attacks will result in a liability to the open environment. Mobile users will face increased possibility of unwilling exposure to the significant security hazards caused by various types of attacks conducted on the network or on the device. In this paper, we propose *security clustering*, a network-wide defending mechanism against attacks. Security clustering make use of dynamic security context exchange between cluster members and distributed information sharing to achieve scalable and efficient cooperation.

The rest of this paper is organized as follows: Sect. 2 describes the environmental changes in pervasive computing and evolutionary threats. In Sect. 3 and 4, we present security clustering mechanisms, security context exchange and cooperation protocol for information sharing. We conclude the paper in Sect. 5.

---

[*] This work was supported by Korea Research Foundation Grant (KRF-2003-003-D00441).

[**] The corresponding author

N. Mitrou et al. (Eds.): NETWORKING 2004, LNCS 3042, pp. 1326–1331, 2004.

2   Environmental Changes

On the basis of the paradigm shift of the mobile Internet, the network is expected form a loosely coupled and highly dynamic environments. The mobile users will be able to move around the network while connected to the environment: directly to other users or devices in their vicinity or indirectly (through the backbone) to the external. In this environment, the importance of user's location, service context, and various contents will be more evident than ever. But the management of such networks will have many dimensions in service provisioning, customization, and personalization, which will lead to a more complicated network revolution. Consequently, the network will face new security challenges because of the heterogeneity of the network, lack of centralized control, and presence of foreign users. Many vulnerabilities and weaknesses that have existed in the wired environment can easily be exploited in the new environment [4, 5].

The Presence of foreign users adds vulnerability on top of that openness and heterogeneity. The migration of foreign users especially have important implications on the network because they might have no pre-established secure association nor been authenticated and authorized to access the network through a decent mechanism. Users are potentially insecure in that they might conduct some kind of attacks intentionally or be victims of such attack and would act as slave for subsequent attacks. Furthermore, because they are basically mobile, the infected victims migrated into local network will be the security glitch while they are connected. This vulnerability introduces many risks to the network: First, the network becomes vulnerable to theft of data and DoS attacks. Second, network entities are effectively exposing the data on every remote system and creating thousands of unprotected entry points to the local network.

Security services in this open environment can be discussed in two different domains: *trust management* between communication entities and *system security* from various attacks such as distributed DoS (DDoS) or Interent worms. In this paper, we focus on how the pervasive computing environment can have network-wide defending mechanism against attacks. We propose *security clustering*, a distributed cooperation mechanism, to actively detect and put down various attacks conducted on the open network.

3   Security Clustering

3.1   Security Agent

The proposed secure clustering extends the agent-based management features to enable network-wide, reliable, and timely response to various and evolving attacks. The *security agent* plays a key role in managing the communication channel. The Fig. 1 shows a generalized pervasive computing environment in which we have shown three *security clusters* (C1, C2, and C3) that comprises one or more security agents and mobile users. The major function of the security agent is to establish secure inter-cluster communication channel and to exchange the *security context* with mobile users. The inter-cluster communication uses secure

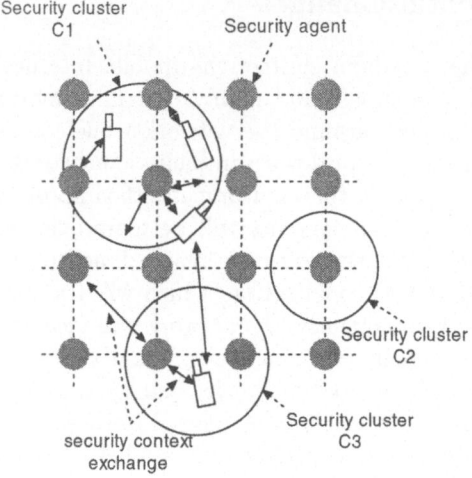

Fig. 1. Secure cluster

multicast protocol [6-8] to provide efficient and reliable information exchange mechanism to the security cluster.

3.2   Security Context Exchange

Since a lot of information for various attacks such as the records of suspicious behaviors and network events is required for constructing new signatures and uncovering relevant attacks, immediate countermeasure may not be guaranteed in many cases. The openness of pervasive computing environment seems to worsen the problem further in that there exist higher possibilities of widely conducted attacks on the open network.

The consequent main drawback of the current detection system would be the size of signature database that hinders real-time detection for various attacks. To increase the chance for successful detection, the detection system will need to collect more signatures for various attacks and its size will be much bigger that the mobile terminals can afford. Though the probability of successful detection is increased, the delayed detection procedure will lose QoS and user's interest.

To address these problems, canonical countermeasure architecture for pervasive computing should include these requirements:

— Cooperation of many network entities in scalable and robust manner
— Active detection for various attacks
— Management of attack signatures
— Authenticity of the exchanged context

To achieve these requirements, security clustering makes use of security context exchange mechanism between network entities.

The *security context* is a data structure that defines the sets of attributes or rules to describe the signatures of various attacks. The security context is differ from the *usual context* discussed in the pervasive computing, where the context means the environment, status, situation, and surroundings of a system or a user. The security context, on the other hand, is the security-specific information including the description of the on going attacks, status of the security alarm on the network, and security knowledge of each host. With the security context, one can detect the malicious activities of specific transaction or data. As shown in Fig. 1, users and agents exchange security context with each other. The secrecy of the context exchange relies on the secure multicast session that the security agents and users are involved.

4   Cooperation Protocol

4.1   Distributed Information Sharing

We use the *cooperation protocol* for exchanging security context and cooperating between multiple clusters. Because the security cluster should be able to respond timely to attacks, the volume of the security context database that each entity possess should also be minimized so that the local database scanning consume minimum horsepower and generate search result in time. For this purpose, each user that participates in the clustering possesses *differentiated* context for small amount of attack signatures. Each entity has specific policies that define the type and amount of the security context that it possesses. Therefore the security context database can be differentiated between participating entities. As a result, a host possess the *common security context*, which is the information of on going attacks or most recent security update, and small amount of the *differentiated security context* specific to that host.

A mobile user can request to the entire network for context for specific attack through the cooperation protocol. One who have proper context responses to the request.

4.2   Cooperation Protocol

We designed the *cooperation protocol* on the basis of secure multicast (shown in Fig. 2). The proposed protocol is as following.

Message Notations

-   *REQ (Request)*: requests for a security context for specific fingerprints
-   *RES (Response)*: response to the request with appropriate security context

Cooperation protocol

i.   An initiator (a host) multicasts *REQ* message (Fig. 2(a)).
ii. Other entities of security cluster immediately perform detection procedures (Fig. 2(b)) with their security context database.

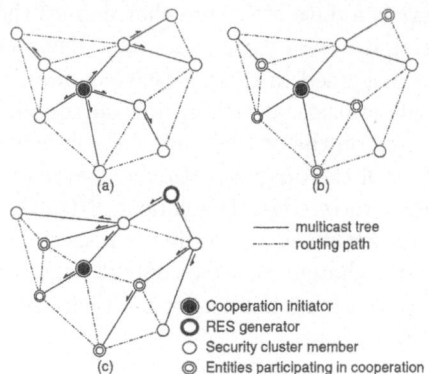

Fig. 2. Cooperation protocol

iii. Those who have no match silently ignores the request and does not respond.
iv. If one encounters appropriate context, then it multicasts the *RES* (Fig. 2(c)).
 v. If one has no sufficient resources for the detection, just ignore and do not reply.

Main procedures of the cooperation protocol are as follows:

```
Collaboration_request :=
    if trigger(i) == TRUE then {
        if i ∈ common_context_pool /* common context */
        then countermeasure(i)
        else if i ∈ local_context_pool /* local differentiated context
*/
        then countermeasure(i)
        else {
        req = build_request(i)
        collaborative_request(req)
        if receive_response(r) then
            r -> common_context_pool /* update common context*/
            countermeasure(i) }}
Migration :=
    if migrating() == TRUE then {
        req = build_request(null) /* use null to indicate common context
*/
        collaborative_request(req) /* request for common context */
        if receive_response(res)
        then res -> common_context_pool } /* update common context */
Collaboration_response :=
    if receive_request(req)==TRUE then {/* is it a cooperation request?
*/
        if req.ctx ∈ common_context_pool then
            res = build_response(req.ctx, common_context_pool)
            send_response(res)
    } else if common_context_req() == TRUE then { /* or is it a common
context
        res = build_response(common_context_pool)         request? */
        send_response(res) }
```

In the event of suspected operation or access, a host performs the *cooperation_request* procedure. The procedure check if the event corresponds to the common security context. If it does, then call *countermeasure* function. It it doesn't, the host call the *cooperate_request* function to initiate cooperation. On successful detection, the *receive_response* returns with proper context. Mobile users migrated into the local network performs the *migration* procedure to receive the security context specific to the local network and to adapt to new environment. The hosts that participate in the cooperation performs the *cooperation_response* procedure. In this procedure, each cooperating host use their differentiated security context to generate proper response.

5   Conclusion

The purpose of the security clustering is to timely preempt the attack and quickly recover the systems on the basis of network-wide cooperative interwork. With the help of network-wide cooperation, relatively low powered mobile systems can have equivalent or higher level of security services than the ordinary single host or server that carry out self-reliant countermeasures. This will be very important feature for the highly mobile and heterogeneous environment of pervasive computing. Network-wide cooperation diminishes the migration of infected users and prevents the outbreak of attacks or viruses effectively. The continuous security context exchange and cooperation enables timely response to various attacks.

Although much work still remains to be done to design the security clustering in pervasive computing, the advantages of the security clustering suggests that it can be applied as a real-time countermeasure to the active attacks such as distributed DoS attacks and Internet worms.

References

1. M. Satyanarayanan: Pervasive computing: vision and challenges. IEEE Personal Communications, Vol. 8, Issue 4, August 2001.
2. L. Kagal, T. Finin, A. Joshi: Trust-Based Security in Pervasive Computing Environments. IEEE Computer, December 2001.
3. F. Stajano: Security for whom? The shifting security assumptions of pervasive computing. Proceedings of International Security Symposium 2002, LNCS 2609, (c) Springer-Verlag.
4. D. Moor, V. Paxson, S. Savage, C. Shannon, S. Staniford, N. Weaver: The Spread of the Sapphire/Slammer Worm. Technical Report, http://www.caida.org/analysis/security/sapphire, 2003.
5. S. R. White: Open Problems in Computer Virus Research. Virus Bulletin Conference, Munich Germany, Oct 22, 1998.
6. M.J. Moyer, J.R. Rao, P. Rohotgi: A Survey of Security Issues in Multicast Communications. IEEE Network, Vol. 13 Issue 6, Nov.-Dec. 1999.
7. C.K. Wong, M. Gouda, S.S. Lam: Secure Group Communication Using Key Graphs. Proceedings of ACM SIGCOMM'98, pp.68-99, September 1998.
8. R. Canetti, J. Garay, G. Itkis, D. Micciancio, M. Naor, B. Pinkas: Multicast security: A Taxonomy and Some Efficient Constructions. Proceedings of the IEEE INFOCOM'99, pp.708-716, 1999.

A Game Theoretic Approach to Web Caching

Stathes Hadjiefthymiades, Yiannis Georgiadis, and Lazaros Merakos

Communication Networks Laboratory, Department of Informatics and Telecommunications,
University of Athens, Panepistimioupolis, Ilisia, Athens 15784, Greece
shadj@di.uoa.gr

Abstract. In this paper, the Game Theoretic framework is applied to Web caching. The interaction of multiple clients with a caching subsystem is viewed as a non-cooperative game. Some clients may continuously request resources, occupy a large segment of the cache disk space and thus, enjoy high hit rates. Owing to this situation, the remaining clients may suffer the removal of their "important" resources from the cache, and, subsequently, experience numerous cache misses. A utility function is introduced and calculated by clients in a de-centralized fashion to avoid such monopolizing scenarios and guarantee similar performance levels for all users.

1 Introduction

Since the early '90s Game Theory has been extensively used in networking problems. Seminal papers like [3] and [4] provide a very insightful consideration of problems like bandwidth allocation and datagram switch operation disciplines. In this paper, we discuss the application of Game Theory in Web caching. The breathtaking increase in the volume of Web content world-wide renders the caching of resources a very important and promising area of research. Web caching has been extensively used to expedite users' queries by shortening the request-response chain. In this paper, our objective is to avoid having a single client monopolizing the allocated disk space in the Web caching proxy and thus, achieve high hit rates in contrast to the performance achieved by other users. As discussed in [10], the performance (hit rate) achieved in known Web cache servers is characterized by increased variance; in some cases, the variance in hit rates exceeds the average hit rate. Such statistical evidence clearly indicates a very wide range of performance levels seen by the users of the caching service. The hit rate variance is a decreasing function of the number of requests but persists even at high numbers of requests. To avoid such situations, we introduce a game theoretic mechanism that takes into account the disk space already allocated to a specific client and the actual benefit obtained by the retrieval of resources. Our scheme is based on a concave utility function, which secures the existence of Nash equilibrium points (NEP), in the considered game. The paper is structured as follows. In Section 2, we elaborate on the details of the utility function. Section 3 discusses the simulation set-up that we have adopted for evaluating the performance of the suggested solution and the respective results. Other Game Theoretic studies of networking problems are discussed in Section 4. Section 5 concludes the paper.

N. Mitrou et al. (Eds.): NETWORKING 2004, LNCS 3042, pp. 1332–1337, 2004.
© IFIP International Federation for Information Processing 2004

2 Problem Statement and Game Theoretic Solution

As discussed above, our objective is to avoid having a single client monopolizing the disk space in the Web cache and thus, achieve high hit rates in contrast to the performance achieved by other users. Continuous requests by some clients will cause the reservation of a constantly increasing disk segment, force out the popular resources of other users and enjoy high hit rates. Users that do not interact in such a systematic way with the cache, suffer very low hit rates, cache misses and high response times. To cope with the problem, we introduce a utility function that takes into account the disk space already allocated to the specific client and the actual benefit obtained by the retrieval of resources. Such utility function is structured so as to secure the existence of NEP in the considered game. The interacting client calculates the value of the utility function (U) based on feedback received by the proxy cache (piggybacked in the HTTP responses delivered by the proxy). Whenever the marginal utility (ΔU) drops below a certain positive threshold (Eq.1), the cost for the client for the specific resource retrieval increases very rapidly and the client instructs the proxy to cease caching the retrieved objects.

$$\Delta U = U_{t+\Delta t} - U_t < \varepsilon,\ \varepsilon > 0,\ \Delta t > 0 \tag{1}$$

Hence, the disk space allocated to the user rests below a threshold and a stable convergence is secured. The utility function U_k for user k has as follows:

$$U_k = P_k - C_k \qquad P_k = \frac{\alpha \cdot P_{hit}(k) \cdot R_k \cdot S}{[1 - P_{hit}(k)] \cdot R_k \cdot S} = \frac{\alpha \cdot P_{hit}(k)}{1 - P_{hit}(k)}$$

$$C_k = \frac{P_{hit}(k) \cdot R_k \cdot S}{CacheSize - \sum_i P_{hit}(i) \cdot R_i \cdot S} = \frac{P_{hit}(k) \cdot R_k \cdot S}{CacheFreeSpace} \tag{2}$$

The term $P_{hit}(k)$ denotes the caching performance achieved by user k. The term α is a normalizing constant (in our simulations, $\alpha=1.1$). Term R_k denotes the number of requests made by user k towards the caching system. S is the average size of a Web resource file and can be used to determine the mean retrieval/storage cost associated with the resource. CacheSize denotes the total size of the cache disk space. The CacheFreeSpace term denotes the free cache disk capacity and, practically, represents the strategies of all involved players (i.e., shows the cache disk space reserved by the competing players). Term P_k represents the benefit (profit) owing to the caching capability of the proxy subsystem. The term C_k represents the cost owing to the retrieval of resources and the allocation of disk capacity. The term $P_{hit}(k)$ can be approximated by the Web client (agent) as a function of the number of requests issued by the client. Specifically, $P_{hit}(k)$ can be calculated by the R_k number of requests as follows [1].

$$P_{hit}(k) = \gamma \cdot \ln(R_k) - \delta \tag{3}$$

Constants γ and δ are non-negative real numbers. The concavity of the utility function (Eq. 2) guarantees the existence of a NEP. As proven in [5], an equilibrium point exists for every concave n-person game. For a game with multiple players, say two, the payoff function of each player (i.e., $\varphi_1(x)$, $\varphi_2(x)$) is dependent upon the strategies of all players (i.e., the point $x=\{x_1, x_2\}$ belongs to the space S of feasible strategies which, in turn, is a subset of the Cartesian product $E_1 \times E_2$ of the domains of definition for strategy coordinates x_i). The function $\varphi_i(x)$ should be continuous in S and concave in x_i for fixed values of the other coordinate(s) $(x_j, j \neq i)$. In the considered game, all payoff functions are identical, provided by (Eq.2). The strategy variable of user k is the number of requests issued by the user (R_k). The strategy variables of other users are taken into account in the formulation of the denominator of P_k. The strategy space E_i represents the number of requests issued by the Web user during interaction with the cache. Practically, this space is a bounded interval of $\aleph$, $[0, u]$, where u is a fairly large number. Typically, u can increase as high as infinity. However, this is practically infeasible as Web sessions typically last 30 minutes. Additionally, an infinite increase of u would prevent us from applying the findings in [5] for establishing NEP existence. It can be easily shown that the utility function in (Eq.2) is concave. Since, the considered game is a two-players game with identical utility functions, the space of acceptable allocations is a convex, closed and bounded area in $\aleph^2$. The concave utility function guarantees the existence of a NEP.

Below, we examine the main characteristics of the utility function. The formulation of the utility function is based on the deduction of two terms. The first term (P_k) denotes the gain for the involved player from the caching interactions. It is an increasing function of the number of client requests. The second term (C_k) denotes the cost induced to the client through the retrieval and caching of resources. Such cost increases very rapidly as the remaining cache disk space approaches zero. It is known that P_{hit} can range to as high as 50% [8]. Analyses of extensive access traces from second and third level proxy caches, presented in [1], show that hit rates vary up to 45%. In [9] a maximum hit rate of 49% is reported for infinite cache size. Based on the above, the term P_k may increase from 0 up to the α constant. The structure of the P_k term favors the intensive user activity (the obtained benefit is higher when P_{hit} increases). Another interesting issue is how the term C_k varies as a function of R_k. Should the cache free space be very limited the C_k term demonstrates a very rapid increase. The utility function allows a "new" user (i.e., with a limited history of requests) to interact freely with the cache (i.e., cache all the fetched files). A "new" user tries to exploit the structure of P_k and increase his benefit through intensive behavior. Conversely, the interaction of a user with increased number of requests is restricted by the availability of cache disk space. Hence, a user tries to reach a balance between increased benefit and increased cost. As discussed above, the Web proxy returns to the interacting client (supported by an agent) the free disk space that is currently available (piggypacked in HTTP responses). Such information is exploited by the client to decide whether the object subsequently requested by the user should be cached or not (if not found in the cache). HTTP/1.1 provides the means for such, selective, caching through the cache-control header directive and a "no-store" value in specific. To calculate the value of the utility function, the interacting Web client needs also to estimate the average size of the retrieved resources (S in Eq.2). The client maintains an estimate of the average resource size through a low pass filter (Eq.4).

$$S_{new} = w \cdot S_{old} + (1-w) \cdot S_{resource} \qquad 0 < w < 1 \qquad (4)$$

The term w is a smoothing factor, $S_{resource}$ is the size of the recently retrieved resource while S is the estimate of the average resource size.

3 Simulation Framework and Results

We have to tried to assess the impact of the proposed game theoretic mechanism on a Web caching setting involving the intensive interaction of 1500 users (n) with a single cache for a period of 10 days. We have also simulated the non-game theoretic scenario. User interaction patterns followed the Web traffic model reported in [6] and [7]. The metrics recorded throughout the simulation were the number of requests made by each user (R), the number of observed cache hits (CH), the number of cache misses that were affected by the game theoretic mechanism (i.e., were not found in cache, were retrieved but not, subsequently, cached) (GT), the number of invocations to the LRU cache replacement mechanism, the percentage of available cache free space. In this paper, the cache hit rate (H) is calculated as follows.

Game theoretic scenario	Unregulated, non-game theoretic scenario
$H=CH/(R+GT)$	$H=CH/R$

Measurements were collected for all simulated user objects every 120 minutes. We have adopted the coefficient of variation of the H_i measurements as an indicator of the fairness achieved by the caching scheme. A high value of this fairness criterion (FC) means that different users do not enjoy the same benefits from caching and some monopolize the disk space. On the contrary, a low FC value implies that the behaviour experienced by the majority of users is almost identical and all have been allocated an almost equal disk share. Specifically, the fairness metric is defined as follows.

$$FC = s/\overline{H} = \left[\sqrt{\frac{1}{n-1}\sum_{i=1}^{n}(H_i - \overline{H})^2}\right]\cdot\left[\frac{1}{n}\sum_{i=1}^{n}H_i\right]^{-1} \qquad (5)$$

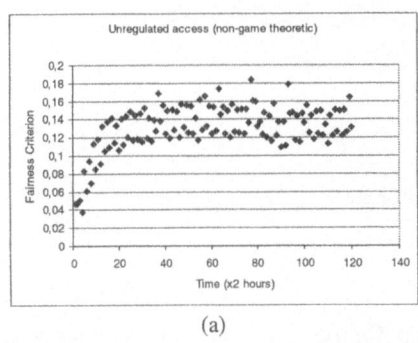
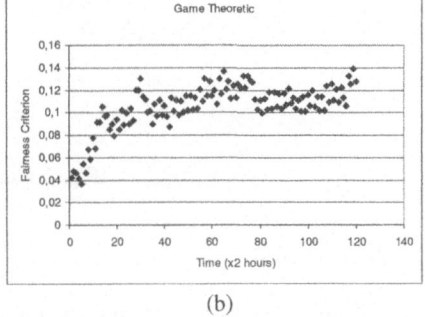

(a)   (b)

Fig. 1. Fairness Criterion

As shown in Fig.1, the game theoretic mechanism achieves (a) lower values for the coefficient of variation of H (i.e., the fairness criterion) and (b) a more predictable caching behaviour since the plot in Fig.1.b is much more condensed than in Fig.1.a. Quite similar observations have been made for the standard deviation of H. The proposed mechanism managed to drastically reduce the number of LRU (Least Recently Used) replacements in the Web cache. Specifically, the number of replacements in the game theoretic scenario was reduced to the 1.17% of the unregulated case. The LRU scheme removed the least recently used items of the cache to free the 15% of the allocated disk space. We have also observed how the cache completeness (or, reversely, the free cache space) varies as time progresses. The game theoretic solution achieves completeness levels between 90% and 99%, in contrast to the non-game theoretic scenario where completeness levels are uniformly distributed between 85% and 99%. The observed number of LRU replacements indicates that the disk space usage remains close to 85% for much more time in the unregulated scenario. If the allocated disk space is the resource that users are charged for, then the game theoretic case entails more total revenue to the cache operator. Lastly, it is important to assess the penalty that users have to pay for the de-monopolizing policy enforced by the game theoretic mechanism. The game theoretic solution achieves an average cache hit rate of 17-17,5 % (Fig.2). The unregulated ("laissez-faire") solution achieves higher cache hit rates, in the order of 21-22%. However, the game theoretic solution appears much more predictable, since all the relevant points lie very close to each other (Fig.2). During our simulations, a very important observation was that equilibrium was reached very rapidly since the behaviour of all clients was governed by the same traffic model. To monitor the performance of the suggested solution over an extended time period, a 20% of the population of users were initialised every 2 hours (i.e., the number of requests of each client were reduced to a very low level). It is implied, that this part of the client population are "new" users that are allowed to interact freely with the caching system.

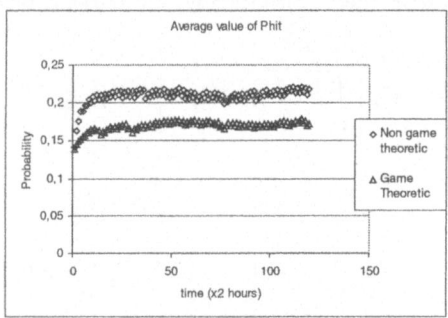

Fig. 2. Cache Hit Rate

4 Prior Work

Game Theory has been extensively employed for resolving networking problems (e.g., flow control, routing). Nevertheless, to our knowledge, Game Theory has not been applied in Web systems engineering. In [2], the authors demonstrate that despite the non-cooperative decisions of network users, there is, still, room for network performance improvement. The authors in [3] discuss the available bandwidth distribution to ATM virtual paths controlled by different selfish users. The work in [4] focuses on the proper design of the disciplines followed by network switches that could drive the network system to optimal conditions despite the selfish nature of the involved users.

5 Conclusions

In the context of Web caching, clients typically reserve more disk space in order to improve the observed cache performance. Such performance is denoted by the cache hit rate. As the cache disk space is a finite resource, a social interaction problem is formulated. Study of this problem is based on game theory. Specifically, we adopt the study of a non-cooperative game where the existence of a NEP is investigated. Users compete with each other trying to selfishly improve a utility function until a NEP is reached. We propose a utility function consisting of profit and cost components. Users has to determine a rational course of interaction taking into account the performance advantages and associated resource retrieval/storage costs. The cost component is dependent upon the strategies that different players assume. An extensive simulation of the game theoretic mechanism has been conducted. Our findings indicate considerable improvement in the adopted fairness criterion metric. The performance seen by different users is comparable and more predictable. At the game theoretic scenario, the number of cache replacement operations is drastically reduced and the cache enjoys higher utilization. The average hit rate seen by users is degraded at the game theoretic scenario and that is the penalty users have to pay for the demonopolizing policy of operation.

References

1. B.M. Duska, D. Marwood, and M.J. Feeley, "The Measured Access Characteristics of WWW Client Proxy Caches", proceedings of USENIX Symposium on Internet Technologies and Systems, December 1997.
2. Y. Korilis, A. Lazar, and A. Orda, "Architecting Noncooperative Networks" IEEE JSAC, Vol. 13, No. 8, 1995.
3. A. Lazar, A. Orda and D. Pendarakis, "Virtual Path Bandwidth Allocation in Multiuser Networks", IEEE/ACM Transactions on Networking, Vol. 5, No. 6, December 1997.
4. S.J. Shenker, "Making Greed Work in Networks: A Game Theoretic Analysis of Switch Service Disciplines", IEEE/ACM Trans. Networking, Vol.3, No.6, December 1995.
5. J.B. Rosen, "Existence and Uniqueness of Equilibrium Points for Concave N-Person Games," Econometrica, Vol.33, No.3, 1965.
6. M. Crovella, and A. Bestavros, "Self-Similarity in World Wide Web Traffic: Evidence and Possible Causes," IEEE/ACM Trans. Networking, Vol. 5, No. 6, December 1997.
7. P. Barford, and M. Crovella, "Generating Representative Web Workloads for Network and Server Performance Evaluation," Proceedings of ACM SIGMETRICS, July 1998.
8. E. Markatos, and C. E. Chronaki, "A Top-10 Approach to Pre-fetching the web", Proceedings of INET '98 Geneva, Switzerland, July 1999.
9. L. Fan, P. Cao, J. Almeida, and A.Z. Broder, "Summary Cache: A Scalable Wide-Area Web Cache Sharing Protocol", IEEE/ACM Trans. Networking, Vol.8, No.3, June 2000.
10. C. Roadknight and I. Marshall, "Variations in cache behaviour", in proceedings of 7th International WWW Conference (WWW7), Brisbane, Australia, April 1998.

Transient Analysis of the D-BMAP/G/1 Queue with an Application to the Dimensioning of a Playout Buffer for VBR Video*

Tom Hofkens, Kathleen Spaey, and Chris Blondia

University of Antwerp, Department of Mathematics and Computer Science
Performance Analysis of Telecommunication Systems Research Group
Middelheimlaan 1, BE-2020 Antwerpen - Belgium
{tom.hofkens, kathleen.spaey, chris.blondia}@ua.ac.be

Abstract. In this paper the D-BMAP/G/1 queue is considered. The goal is to derive an explicit expression for the transform of the queueing delay of the nth arriving customer, based on a transient analysis. While deriving this transform, intermediate results such as an explicit expression for the transform of the probability of having an empty system at the nth departure, are also obtained. These results are then applied to the dimensioning of a playout buffer for variable bit rate video traffic.

1 Introduction

In this paper the D-BMAP/G/1 queue is considered. This is a discrete-time single-server queue of infinite capacity with general service times. The arrival process is a discrete-time batch Markovian arrival process (D-BMAP), a quite general traffic model for discrete-time Markov sources [1,2]. In [1] and [2], a steady state analysis of queueing systems with a D-BMAP as input is performed. The goal of this paper is to derive an explicit expression for the transform of the queueing delay of the nth arriving customer of a D-MAP, based on a transient analysis. The paper is based on results presented in [3] about the transient analysis of the continuous-time BMAP/G/1 queue. While deriving the transform of the queueing delay of the nth arrival, intermediate results such as an explicit expression for the transform of the probability of having an empty system at the nth departure, are also obtained. The transform of the queueing delay of the nth arrival is used to dimension a playout buffer for a video application. The time the video application needs to keep the first packet of a video stream in the buffer before starting to playout is determined such that underflow is avoided.

The structure of the paper is as follows. Section 2 introduces the D-BMAP arrival process as well as the queueing model considered in this paper. It also summarizes the transient analysis of the queueing system and presents an expression for the transform of the queueing delay of the nth arrival in the D-MAP/G/1

---

* This work was carried out within the framework of the project CoDiNet sponsored by the Flemish Institute for the promotion of Scientific and Technological Research in the Industry (IWT).

N. Mitrou et al. (Eds.): NETWORKING 2004, LNCS 3042, pp. 1338–1343, 2004.

queueing system. The obtained results are then applied in Section 3 to dimension a playout buffer for a video application. Finally, Section 4 concludes the paper.

2   The D-BMAP/G/1 Queue

2.1   The Discrete-Time Batch Markovian Arrival Process

A discrete-time batch Markovian arrival process (D-BMAP) is a general traffic model for discrete-time Markov sources. Consider a two-dimensional discrete-time Markov chain $\{N(k), J(k)|k \in \mathbb{N}\}$ on the state space $\mathbb{N} \times \{1, \ldots, M\}$. $N(k)$ is a counting variable representing the number of arrivals that have occurred since time 0 until time k (not including the possible arrivals at time k), and $J(k)$ is the phase of the arrival process immediately before the possible arrivals of time k occur. The transition matrix of the process has the following structure:

$$
\mathbf{T} = \begin{pmatrix} \mathbf{D}_0 & \mathbf{D}_1 & \mathbf{D}_2 & \cdots \\ 0 & \mathbf{D}_0 & \mathbf{D}_1 & \cdots \\ 0 & 0 & \mathbf{D}_0 & \cdots \\ \vdots & \vdots & \vdots & \ddots \end{pmatrix},
$$

where the $\mathbf{D}_n$, $n \geq 0$, are $M \times M$ matrices. The matrices $\mathbf{D}_n$ govern the phase transitions of the arrival process for a batch arrival of size n. The matrix $\mathbf{D} = \sum_{n=0}^{\infty} \mathbf{D}_n$ is the transition matrix of the underlying Markov chain. Define the matrix generating function of the D-BMAP as $\mathbf{D}(z) = \sum_{n=0}^{\infty} \mathbf{D}_n z^n$, $|z| \leq 1$. Let $\boldsymbol{\pi}$ be the stationary probability vector of this Markov chain, i.e., $\boldsymbol{\pi} \mathbf{D} = \boldsymbol{\pi}$, $\boldsymbol{\pi} \mathbf{e} = 1$, where $\mathbf{e}$ is a column vector of 1's. The fundamental arrival rate λ of this process is then given by $\lambda = \boldsymbol{\pi} \left(\sum_{n=1}^{\infty} n \, \mathbf{D}_n \right) \mathbf{e}$.

More details and properties about D-BMAPs can be found in [1,2].

2.2   The Queueing Model

Consider a discrete-time single-server queue of infinite capacity with a D-BMAP $(\mathbf{D}_n)_{n \in \mathbb{N}}$ as arrival process. Call the underlying time unit of the D-BMAP a slot, where slot l is the time unit between time instants $l - 1$ and l. Let the service time have an arbitrary distribution H with z-transform $h(z) = \sum_{k=1}^{\infty} H(k) \, z^k$, where $H(k)$ is the probability that the service time equals k slots.

2.3   The Embedded Process at Departures

Define $\left[\hat{\mathbf{A}}_n(m) \right]_{i,j}$ as the probability that, given a departure at time 0 leaving at least one customer in the system and the phase of the arrival process is i, the next departure occurs at time m, at that time the phase of the arrival process is j, and there have been n arrivals since time 0. Define $\left[\hat{\mathbf{B}}_n(m) \right]_{i,j}$ as the probability that, given a departure at time 0 leaving the system empty and the phase of the arrival process is i, the next departure occurs at time m, at that time the phase of the arrival process is j, and there have been $n+1$ arrivals since time 0.

Consider the queueing system at departure instants $t_0, t_1, t_2, \ldots$. Let $L(t_k)$ be the number of customers in the system at instant t_k (after the departure), and let $J(t_k)$ be the phase of the arrival process at time t_k. Then the process $\{(L(t_k), J(t_k), t_{k+1} - t_k)|k \geq 0\}$ is a semi-Markov chain with state space $\mathbb{N} \times \{1, \ldots, M\}$. The transition matrix of the semi-Markov chain is given by

$$\mathbf{Q}(k) = \begin{pmatrix} \hat{\mathbf{B}}_0(k) & \hat{\mathbf{B}}_1(k) & \hat{\mathbf{B}}_2(k) & \cdots \\ \hat{\mathbf{A}}_0(k) & \hat{\mathbf{A}}_1(k) & \hat{\mathbf{A}}_2(k) & \cdots \\ 0 & \hat{\mathbf{A}}_0(k) & \hat{\mathbf{A}}_1(k) & \cdots \\ \vdots & \vdots & \vdots & \ddots \end{pmatrix}, \quad k \geq 0,$$

which shows that the system has an embedded Markov chain of M/G/1-type.

2.4   The Delay of the nth Arrival in the D-MAP/G/1 Queue

Let $\left[\hat{\mathbf{G}}^{(r)}(k, m)\right]_{i,j}$ be the probability that the first passage from state $(l+r, i)$ to state (l, j), with $1 \leq i, j \leq M$, $l \geq 0$ and $r \geq 1$, occurs in k transitions during m slots and (l, j) is the first state visited in the set $\{(l, l')|1 \leq l' \leq M\}$. Define $\mathbf{G}(z) = \sum_{k=1}^{\infty} \sum_{m=k}^{\infty} \hat{\mathbf{G}}^{(1)}(k, m) y^m$ with $|y| \leq 1$. Then from [4], $\mathbf{G}(z)$ satisfies

Property 1. $\mathbf{G}(z) = z\, h(\mathbf{D}(\mathbf{G}(z)))$.

Define the n-step transition probability matrices $\mathbf{P}_{i,j}^{(n)}$ as

$$\left[\mathbf{P}_{i,j}^{(n)}\right]_{k,l} = \mathrm{P}\left[L(t_n) = j, J(t_n) = l|L(t_0) = i, J(t_0) = k\right],$$

and the transform matrix $\tilde{\mathbf{P}}_{i,j}(w) = \sum_{n=0}^{\infty} \mathbf{P}_{i,j}^{(n)} w^n$, $|w| \leq 1$. Then it is proven in [4] that

Property 2. $\tilde{\mathbf{P}}_{i,0}(w) = [\mathbf{G}(w)]^i \left[\mathbf{I} - (\mathbf{I} - \mathbf{D}_0)^{-1}[\mathbf{D}[\mathbf{G}(w)] - \mathbf{D}_0]\right]^{-1}$.

This result can then be used to derive an expression for the transform of the delay for the D-MAP/G/1 queue. A D-MAP is a D-BMAP in which no batch arrivals occur, i.e., $\mathbf{D}_n = \mathbf{0}$ for $n \geq 2$.

Let $[\mathbf{W}_n(k)]_{i,j}$ be the probability that, given a departure at time 0 and the phase of the arrival process is i, the queueing delay of the nth arrival is k slots and the phase of the arrival process immediately after the nth arrival is j. Denote its z-transform as $\mathbf{w}_n(z) = \sum_{k=0}^{\infty} \mathbf{W}_n(k) z^k$, and let $\mathbf{w}(y, z) = \sum_{n=1}^{\infty} \mathbf{w}_n(z) y^n$, $|y| \leq 1, |z| \leq 1$. Define the matrix $\mathbf{U}$ as $\mathbf{U} = (\mathbf{I} - \mathbf{D}_0)^{-1} \mathbf{D}_1$. Then the following theorem holds [4]:

Theorem 1. $\mathbf{w}(y, z) = y(\mathbf{w}_1(z) + \sum_{l=0}^{\infty} \mathbf{W}_1(l) \mathbf{D}(\mathbf{G}(y))^l \mathbf{G}(y)(\mathbf{I} - \mathbf{U}\mathbf{G}(y))^{-1}$
$(\mathbf{U} - (z\mathbf{I} - \mathbf{D}_0)^{-1}\mathbf{D}_1)) (\mathbf{I} - h(z)y(z\mathbf{I} - \mathbf{D}_0)^{-1}\mathbf{D}_1)^{-1}.$

$\mathbf{w}_1(z)$ gives the transform of the queueing delay of the first arrival and is set to the initial conditions of the system when the first customer arrives.

More details about the transient analysis can be found in [4].

3   Application to the Dimensioning of a Playout Buffer

In this section the results are applied to the dimensioning of a playout buffer for a video application. This is achieved by numerically inverting the two-dimensional transform of the queueing delay of the nth arrival in Theorem 1 using [5].

Consider a scenario in which the traffic of a variable bit rate video source is sent towards a video player. Because of varying delays within the network caused by the random queueing delays in the routers in the network, the end-to-end delay between source and receiver can fluctuate from packet to packet. This phenomenon is called jitter. To compensate for the jitter, the video player uses a playout buffer. The player waits a fixed amount of time Δ after the first packet has arrived before starting the video playout. In order to avoid underflow, it is important to carefully choose the initial delay Δ. In [4] it is derived that this is achieved if the delay of the jth packet d_j satisfies $d_j \leq d_1 + \Delta$ $(j \geq 1)$.

Recent measurements [6] revealed that video streamers generate video traffic in bursts of multiple video frames. The duration of these bursts can vary from 1.5-2 ms for short bursts and 5-7 ms for long bursts. The silences between the bursts may be much longer than the bursts themselves, implying that the instantaneous bit rate during a burst is much higher than the average bit rate of the stream.

D-MAPs and D-BMAPs are good models for bursty traffic sources (e.g., VBR video) [1,7]. In this paper the traffic generated by a video streamer is modelled by a 4-state D-MAP which incorporates the typical characteristics of the video traffic as described above. The transition matrix $\mathbf{D}$ of the D-MAP is given by

$$
\begin{pmatrix}
1-\alpha & \alpha & 0 & 0 \\
\beta & 1-\beta-\gamma & \gamma & 0 \\
0 & 0 & 1-\delta & \delta \\
\phi & 0 & \epsilon & 1-\epsilon-\phi
\end{pmatrix} . \quad
\mathbf{D}_1 =
\begin{pmatrix}
\lambda_1(1-\alpha) & \lambda_1\alpha & 0 & 0 \\
0 & 0 & 0 & 0 \\
0 & 0 & \lambda_2(1-\delta) & \lambda_2\delta \\
0 & 0 & 0 & 0
\end{pmatrix} ,
$$

and $\mathbf{D}_0 = \mathbf{D} - \mathbf{D}_1$. Note that this D-MAP is a kind of on/off source, with two on states (state 1 and state 3) during which packets are generated in a slot with probabilities λ_1 and λ_2 respectively, and two off states during which no packets are generated. A period in the first, respectively second on state is always followed by a period in the first, respectively second off state, while a period in an off state is always followed by a period in an on state. So this D-MAP mimics the bursty character of a video source. By carefully choosing the values of the parameters, properties such as the mean burst and silence durations, the average arrival rate and the instantaneous arrival rates during the bursts can be tuned.

In this example, the parameters are set as follows: $\alpha = 1/50$, $\beta = (1/950) - \gamma$, $\gamma = 10^{-3}$, $\delta = 3/50$, $\phi = 19\gamma/59$, $\epsilon = (3/2950) - \phi$, and $\lambda_1 = \lambda_2 = 0.6$. Assuming that the video traffic enters the network over a link of 100 Mbit/s in packets of 1500 bytes, this means that the average bit rate of the source is 1.5 Mbit/s. Packets are generated during bursts which have an average duration of respectively 6 ms and 2 ms and a standard deviation of respectively 5.94 ms and 1.94 ms, and these bursts are followed by silence periods of on average 114 ms or 118 ms respectively and a standard deviation of respectively 113.94 ms and

117.94 ms. 75% of the bursts are 'long' bursts, the remaining bursts are 'short' bursts. During a burst, packets are generated at a rate of 60 Mbit/s.

The transport of the video stream through the network and the introduction of delay and delay jitter by the network is modelled by the D-MAP/G/1 queue. The general service time distribution H follows a shifted binomial distribution $B(9, 1/3)$, i.e., if $H_B \sim B(9, 1/3)$, then $H(k) = H_B(k-1)$ for $k > 0$, where $H(k)$ is the probability that the service time of a packet equals k slots. This distribution has a mean of 4 slots and a standard deviation of $\sqrt{2}$ slots.

Using the theory developed before, values Δ_n are determined such that with probability 1-p all of the first n generated packets arrive before their scheduled playout time. It is assumed that at time 0 the D-MAP/G/1 queueing system is empty, and the phase of the arrival process is i with probability π_i, where $\pi = (\pi_1, \ldots, \pi_4)$ is the stationary probability vector of the D-MAP. Figure 1

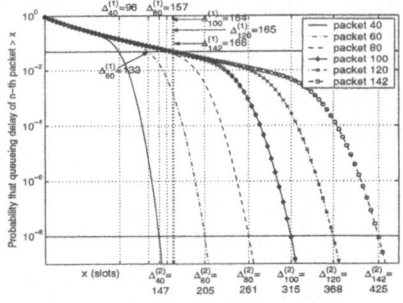

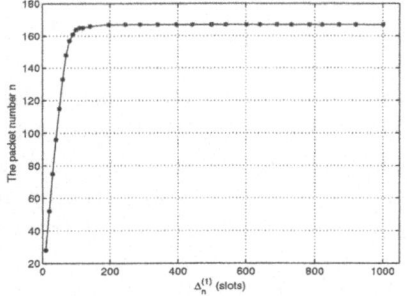

Fig. 1. The complementary cumulative distribution of the queueing delay.

Fig. 2. The different values of $\Delta_n^{(1)}$ ($p = 0.05$).

shows the complementary cumulative distributions of the queueing delay of the nth packet, for $n \in \{40, 60, 80, 100, 120, 142\}$ with $p = 0.05$ and $p = 10^{-8}$. The values for Δ_n are denoted by $\Delta_n^{(1)}$ for $p = 0.05$ and by $\Delta_n^{(2)}$ for $p = 10^{-8}$. Figure 2 shows the different values of $\Delta_n^{(1)}$ for increasing n. Both the queueing delay and Δ_n are measured in slots. The horizontal lines on Figure 1 are positioned at probability p. It is the intersection of these lines and the delay curve for packet n that gives the corresponding values for $\Delta_n^{(1)}$ and $\Delta_n^{(2)}$. Because of the bursty nature of the traffic, subsequent packets in a burst have larger delays with a higher probability. Thus the values for Δ_n need to increase with increasing n since the probability that a packet needs more time to arrive at the playout buffer than a previous packet also increases. Note the influence of the high variance of the burst length on the delay curves. For a long burst the average burst length is 6 ms, which corresponds to sending on average 30 packets. Because of the high variance of the burst length however, the actual number of packets that are sent in a burst can be much higher, hence the values for Δ_n increase with increasing n for values of n much larger than 30. This increase will however not continue indefinitely because bursts are followed by silence periods during which the buffer of the D-MAP/G/1 system is able to empty again. Therefore, Δ_n will stabilize to a fixed Δ, as is shown in Figure 2 for $p = 0.05$.

All of the delay curves first follow a common straight line and then drop relatively fast. As n increases, the common portion of the curves becomes larger because of the increasing delays. Therefore Δ_n no longer increases when this common portion crosses the horizontal line indicating that the required condition of having a probability of $1 - p$ that packets arrive on time, is satisfied.

When $p = 10^{-8}$ a stronger demand is imposed on the system, i.e., a higher probability that packets arrive before their scheduled playout time is required. Where for $p = 0.05$, $\Delta_n^{(1)}$ stabilizes to $\Delta^{(1)} = 167$ slots $= 20.04$ ms, $\Delta^{(2)}$ will take a much larger value for $p = 10^{-8}$, as is confirmed by the different values of $\Delta_n^{(1)}$ and $\Delta_n^{(2)}$ in Figure 1.

4   Conclusion

In this paper the D-BMAP/G/1 queue was considered. For this queueing system a transient analysis was done in order to derive an explicit expression for the transform of the queueing delay of the nth arriving customer of a D-MAP. These results were then applied to dimension a playout buffer for a video application. A simple model was proposed to model the bursty nature of variable bit rate video and used as traffic source into a network. The transport of the video stream and the introduction of delay and delay jitter by the network was modelled by the D-MAP/G/1 queue. Using the developed theory, values for the time Δ_n the video application needs to keep the first packet of a video stream in the playout buffer were determined, such that with probability $1 - p$ all of the first n packets arrive before their scheduled playout time in order to avoid buffer underflow.

References

1. Blondia, C., Casals, O.: Statistical multiplexing of VBR sources: A matrix-analytic approach. Performance Evaluation **1** (1992) 5–20
2. Blondia, C.: A discrete-time batch Markovian arrival process as B-ISDN traffic model. Belgian Journal of Operations Research, Statistics and Computer Science **32** (1993) 3–23 http://www.pats.ua.ac.be/chris-personal.html.
3. Lucantoni, D.: Further Transient Analysis of the BMAP/G/1 Queue. Stochastic Models **14** (1998) 461–478
4. Hofkens, T., Spaey, K., Blondia, C.: Transient analysis of the D-BMAP/G/1 queue with an application to the dimensioning of a playout buffer for VBR video (extended version). (http://www.pats.ua.ac.be/publications.html)
5. Choudhury, G.L., Lucantoni, D.M., Whitt, W.: Multidimensional transform inversion with applications to the transient M/G/1 queue. Ann. Appl. Prob. **4** (1994)
6. Balint, Z., Truyts, B.: Traffic characteristics: measurements. (Internal report of the CoDiNet project)
7. Spaey, K., Blondia, C.: Circulant matching method for multiplexing ATM traffic applied to video sources. In: Proceedings IFIP PICS'98, Lund, Sweden (1998)

A Kerberos-Based Authentication Architecture for Wireless LANs

Mohamed Ali Kâafar[1], Lamia Benazzouz[1], Farouk Kamoun[1], and Davor Males[2]

[1] Ecole Nationale des Sciences de l'Informatique, Université de la Manouba. Tunisia
{Medali.kaafar, Lamia.benazzouz, Farouk.kamoun}@ensi.rnu.tn
[2] Laboratoire d'Informatique de Paris 6 Université Pierre et Marie Curie 8, rue du capitaine
Scott 75015 Paris. France
davor.males@lip6.fr

Abstract. This work addresses the issues related to authentication in wireless LAN environments, with emphasis on the IEEE 802.11 standard. It proposes an authentication architecture for Wireless networks. This architecture called Wireless Kerberos (W-Kerberos), is based on the Kerberos authentication server and the IEEE 802.1X-EAP model, in order to satisfy both security and mobility needs. It then, provides a mean of protecting the network, assuring mutual authentication, thwarts cryptographic attack risks via a key refreshment mechanism and manages fast and secure Handovers between access points.

1 Introduction

Over recent years, wireless communication has enjoyed enormous growth, becoming popular in both public and private sectors. Wireless Local Area Network (WLAN) technology is capable of offering instant, high-speed and mobile connectivity. While this technology offers a lot of advantages, it does also introduce issues related to authentication, access control, confidentiality and data integrity. Today, wireless products are being developed that do not address all of the security services related to this technology. Although the IEEE 802.11i framework is proposing a "Robust Security Nework" architecture (RSN) to deal with the security wireless networks limitations, actually there is not a complete set of standards available that solves all the issues related to Wireless security [1].

While the Kerberos approach has been proposed as a standard for enhanced security in IEEE TGe [2], currently there is no valid proposals using a Kerberos-like mechanism to provide authentication in a WLAN environment, preventing from cryptographic attacks and handling fast and secure handovers. In this paper, we propose a mobility aware authentication architecture for the IEEE 802.11 networks, based on the IEEE 802.11i works and exploiting the Kerberos protocol to overcome the RSN limitations and provide a global framework. We first begin by introducing the Kerberos protocol, and concepts related to the RSN architecture such as the EAP-802.1X model. This is followed by a description of the proposed architecture (called

N. Mitrou et al. (Eds.): NETWORKING 2004, LNCS 3042, pp. 1344–1353, 2004.

W-Kerberos) and the authentication process. Next, we describe the implementation of the system and conclude with perspectives of this work.

2   The Kerberos Protocol

The following subsections present the Kerberos protocol and the authentication process in a Kerberos-based system.

2.1   Presentation

Kerberos was developed as an open software at the Massachusetts Institute of Technology (MIT) as part of its Athena project [3]. Since its version 4, Kerberos is under the IETF Common Authentication Technology Working Group responsibility[4].
The Kerberos architecture defines three entities: the client wanting to reach resources of a certain server, the service supplier or server, and the authentication Kerberos server. The latter is based on two distinct logical entities: An AS server (Authentication Server), responsible for the identification of clients, and a TGS server (Ticket Granting Service) which provides clients with access authorizations on the basis of an AS identification. These two entities are regrouped under the name of KDC to mean Key Distribution Center [5].

2.2   The Kerberos Authentication Process

The Kerberos authentication takes place in a set of steps as shown in Figure 1 and described below:

1. Before the client attempts to use any service of the network, he must be authenticated by a Kerberos Authentication Server AS. This authentication consists in an initial ticket request: Ticket Granting Ticket (TGT). The TGT is used subsequently to get credentials for several services.
2. When the client wants to communicate with a particular server, he sends a request to the TGS asking for credentials for this server. The TGS answers with these credentials encrypted by the user's key. The credentials consist of a temporary session key Sk and a ticket for the service supplier called Service Ticket ST, containing the identity of the client and the session key, all of them encoded with the server's key.
3. The client, wanting to reach a server's ressources, transmits the ticket to this server.
4. The session key, now shared by the client and the server, can be used to encrypt the next communications.

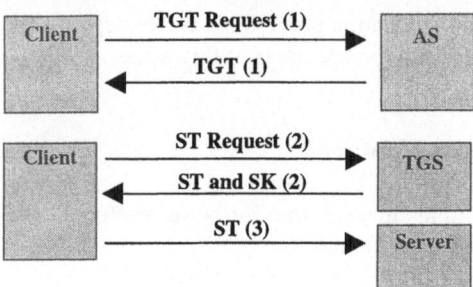

Fig. 1. Kerberos service ticket request.

3   W-KERBEROS or Kerberos for the 802.11 Networks

The proposed authentication process is based on tickets delivered by a W-Kerberos server. These tickets are going to direct the access points either to allow or not the traffic of a particular client. In the same way, it exploits the notion of dual ports of the IEEE 802.1X framework and the Extensible Authentication Protocol.

We present in the following the IEEE 802.1X framework as a pillar of the IEEE 802.11i architecture, and the EAP protocol as a generic authentication methods transporter. We describe then the proposed Kerberos authentication architecture called Wireless Kerberos: W-Kerberos.

3.1   The IEEE 802.1X Framework

The IEEE standard 802.1X [7] defines a port-based network access control using the physical characteristics of LAN (IEEE 802) infrastructures. This can be used to authenticate and authorize network access to certain physical devices. This access control is performed at the data link layer. The IEEE 802.1X standard abstracts three entities (Figure 2).

- The *supplicant*: that wishes to access services, usually the client.

- The *authenticator*: which is the entity that wishes to enforce authentication before allowing access to its services, usually within the device the supplicant connects to.

- The *authentication server*: which the role is to authenticate supplicants on behalf of the authenticator.

The IEEE 802.1X framework does not specify one particular authentication mechanism; it rather uses the Extensible Authentication Protocol (EAP) [8] as its authentication framework. EAP is a protocol that supports exchange of information for multiple authentication mechanisms. The authenticator is responsible for relaying this information between the supplicant and the authentication server.

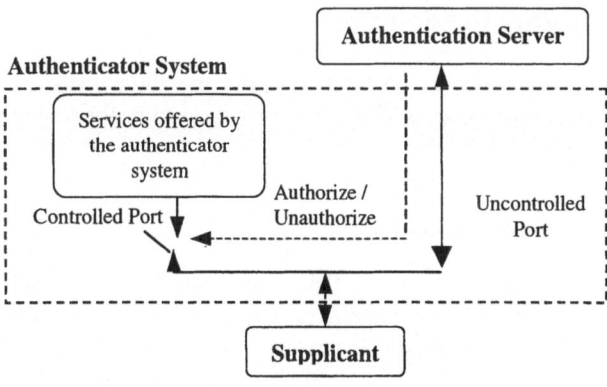

Fig. 2. The IEEE 802.1X Setup

The authenticator's port-based access control defines two logical ports via a single physical LAN port. These are controlled and uncontrolled ports. The uncontrolled port allows uncontrolled exchange (typically information for the authentication mechanism) between the authenticator and other entities on the LAN, irrespective of the authentication state of the system. Any other exchange between the supplicant and servers takes place via the controlled port.

3.2   The W-Kerberos Architecture

The W-Kerberos system is composed of three main entities:

- The *client* trying to have access to the network.
- The *access points* considered as the Kerberos service suppliers, offering the service of access to the network.
- The *W-Kerberos server* allowing identification, tickets transmission, key refreshment and secured Handovers.

In this architecture, the authentication process takes place only once for the user. The principle of "Single Sign-on ", a principle according to which the user identifies himself only one time to the network to reach its different resources is applied. This transparency provides both security and convenience which palliates to certain EAP methods limitations, such as certificate-based methods [9]. Moreover, Mobility, a major asset in the Wireless networks, is handled by the proposed architecture. In fact, the authentication of the Handover phase, during which a client terminal should associate to a new access point, takes place without the exchange of any security context between access points and avoids an initialisation of the authentication process.

4   The Authentication Process

In the following subsections, we will describe the three main phases of the W-Kerberos authentication process: the initial authentication, the key refreshment or re-authentication and the Handover phase.

4.1   The Initial Authentication

This phase is typically initiated by the client terminal, which achieved a 802.11 association. In a first step, the client, receiving an EAP Request Identity from the access point, sends an EAP Response message, encapsulating an initial Service Ticket request (KRB-AS-REQ). The key used to encode the KRB messages is shared between the client and the Kerberos server and derived from the password provided by the client[1].

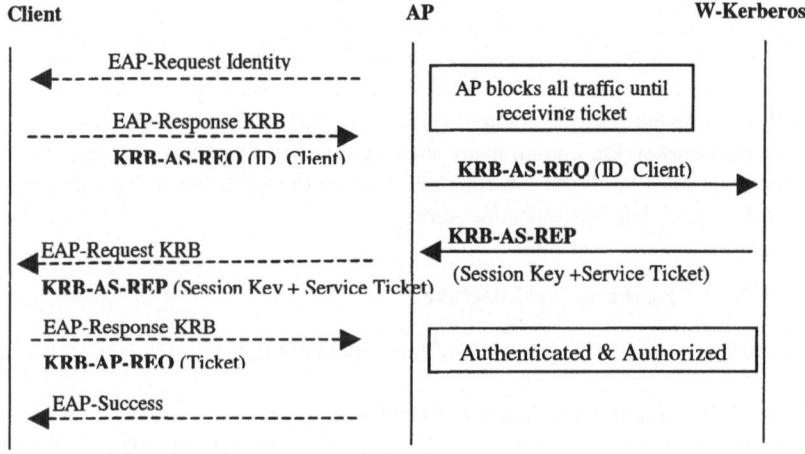

Fig. 3. Initial authentication phase.

After receiving the EAP Response, a Kerberos authentication request is sent from the access point to the W-Kerberos authentication server on the non controlled IEEE 802.1X port. The authentication server consults then the basis of principals, fixes the session time (needed for key refreshment), and generates a session key. An answer message KRB-AS-REP containing the session key, the ticket encoded with the AP secret key, and some authentication information is sent to the client via the access point. Data transmitted in this message is encrypted with the client key. To have access to the network resources, the client issues the ticket to the access point as a KRB-AP-REQ message encapsulated in an EAP Response packet. Thus, the client is now authenticated and authorised by the access point.

---

[1] For more details on key generation see [4].

4.2   The Key Refreshment Phase

W-Kerberos offers a secure channel for communications via encryption mechanisms where key exchange is dynamic. This avoids the possibility of passive attacks to retrieve encryption keys. Hence, in addition to the ticket validity time, a key refreshment mechanism based on a session time out, sent in the initial authentication ticket, is specified by our architecture (see Figure 4). For this purpose, after having received an initial ticket, the access point calculates two time values:

- The *TTSR* (or Time To Send Re-authentication key): defines the instant when the access point must renew the session key and after which he is waiting for a receipt notification from the client.
- The *TTSN* (or Time To Send Notification) which is the instant when the client is considered as no longer authenticated. The access point will then send a Client Reject message (KRB-Cl-Rej) to the client station and the authentication server.

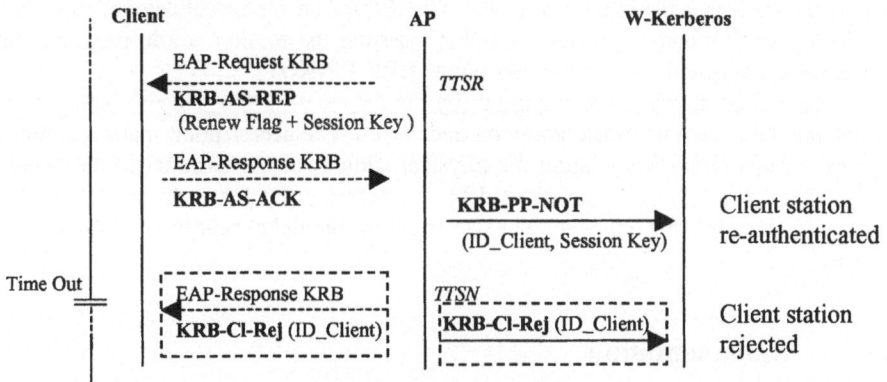

Fig. 4. Key refreshment phase.

While receiving the KRB-AS-REP message with the Renew Flag set (indicating that it is a key refreshment), the client station sends a receipt notification (KRB-AS-ACK), using the new session key, to indicate that the key update has been done. The access point has then to mention this to the W-Kerberos server by sending a notification message (KRB-PP-NOT) containing the new session key.

4.3   The Handover Phase

This phase is completely transparent to the client in a way that no new authentication does take place. The client terminal will transparently handle all the actions needed to perform a fast, efficient and transparent Handover (Figure 5).

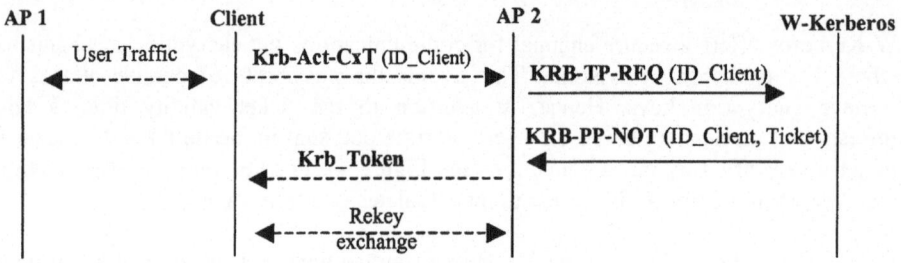

Fig. 5. Handover phase.

As soon as the client terminal performs the IEEE 802.11 reassociation, it has to send a context activation message Krb-Act-CxT to the new access point, in order to move to the authenticated status. The access point contacts the W-Kerberos authentication server by sending a ticket request (KRB-TP-REQ). The server validates if this client is already authenticated, generates a ticket inserting the session key in progress, and sends back a notification to the access point (KRB-PP-NOT).

Once its context activation is acquitted (by the means of a Krb_Token message), the client can then, have its traffic going on once again. The access point maintains now a context of this client, associating the physical address of the terminal to the session key. A Rekeying exchange is initiated by the access point for each HandOver. The time out values for this exhange are extracted from the ticket sent by the W-Kerberos server.

5   Implementation

This section describes the W-Kerberos architecture implementation. This architecture is mainly composed of the client (W-Kerberos Client), the W-Kerberos authenticator (the access point) and the authentication server (W-Kerb). In the following, we will present each architecture component.

5.1   W-Kerberos Client

The W-Kerberos client is composed of two modules : WClient and WXsupplicant.

WClient. This module handles the different Kerberos messages from and to the W-Kerberos server. This entity is in charge of the Kerberos messages encryption, the authenticity check, tickets transmission. The GSS-API library [10] has been used for this client's implementation.

WXsupplicant. WXsupplicant is an extension of the client side open source implementation of the IEEE 802.1X standard called XSupplicant [11]. Some useful

Kerberos authentication functions have been added to the Xsupplicant source and a new authentication type has been implemented.

5.2   W-Kerberos Authenticator

A physically secured computer bridges the Wireless network (IEEE 802.11) to the wired network (Ethernet IEEE 802.3). The W-Kerberos authenticator was implemented using HostAP [12], which is a Linux driver for wireless LAN cards supporting the Host AP mode, i.e. it takes care of IEEE 802.11 management frames and acts as an access point. Over HostAP, a layer was implemented to take in charge the Kerberos service at the access point level. Our authenticator will then act as a "Kerberized[2]"server offering the service of network access. It is composed of three main components (Figure 6).

HostAP-802.1X. This entity is responsible for the IEEE 802.1X client authorizations. Besides the MAC addresses control carried out to either allow or not the traffic, it will also check the EAP packets authenticity[3].

EAP / W-Kerberos. The 802.1X implementation, present within the HostAP module, supports a RADIUS authentication server. We have then, implemented an EAP kerberos method that manages on the one hand, the EAP packets from the client terminal, and on the other hand, Kerberos packets sent to the W-Kerberos authentication server.

AP Server. It represents the Kerberos service of access to the network. It is under this server's name that we must record the access point close to the W-Kerberos server. This layer is in charge of tickets validity verification, message authenticity check, context activation in Handover phase, etc.

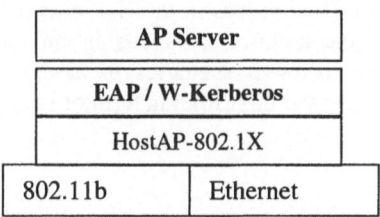

Fig. 6. The W-Kerberos authenticator.

5.3   The Wireless Kerberos Server (W-Kerb)

The authentication server must be physically secured. This server's physical violation could compromise the entire system. On the other hand, W-Kerberos being basically based on the Kerberos architecture, it was necessary to define a W-Kerberos messages management module within the Kerberos server. This module is called W-Kerb. Its

---

[2] The term Kerberized is used for applications that include Kerberos authentication as a feature.

[3] For this purpose, we have added an authenticator attribute at the level of EAP packets [6].

main task is to deal with the W-Kerberos messages that are sent by hosts and access points.

6   Conclusions and Further Works

In this paper, we have proposed a kerberos-based Wireless authentication architecture that conforms to the IEEE 802.11i standard and is mobility aware. Mobility has been the major virtue of Wireless computing and as Wireless networks are deployed, users will expect secure mobility support even when presented with effective access control. The ticket concept existing in the Kerberos protocol is well adapted to such needs.

Although the Kerberos protocol is known to be vulnerable to dictionnary attacks, several works could address this vulnerabily [14], [15]. On the other hand, the W-Kerberos architecture provides transparent authentication of users and access points, as well as a secure channel for communications via encryption mechanisms where key exchange is dynamic and changes periodically via a key refreshment mechanism and a secure Handover phase. The specified architecture provides also, an effective means of protecting the network from unauthorized users and rogue access points, making then the possibility to steal valuable information ruled out, due to the fact that Kerberos provides mutual authentication, i.e., clients and access points ascertain that they are communicating with authentic counterparts. Finally, this architecture is highly customizable, allowing the use of different available encryption mechanism and maintaining thus ability to plug-in different cryptographic algorithms.

The main goal of this work has consisted in analysing and defining the security level within the IEEE 802.11 networks, with a security architecture proposal trying to satisfy both security and mobility needs. Future activities will expand this work, considering: public-key based techniques in the Kerberos model, the implementation of further components of the architecture focusing on the Handover phase, and performances evaluation in different scenarios to assess best values for various parameters (session time, ticket validity time) in term of security and overhead.

References

1.   M. Casole, "WLAN security–Status, Problems and Perspective", in Proceedings of European Wireless 2002, Florence Italy, February 2002. Available from: http://www.ing.unipi.it/ew2002/proceedings/sec002.pdf
2.   IEEE. 802.11 "TGe Security Baseline Draft", March 2001
3.   The MIT Kerberos distribution. Available from: http://www.mit.edu/ afsnet.mit.edu/project/krb5 /.f/kerberosindex.html
4.   J. Kohl, C. Neuman, "The Kerberos Network Authentication Service (V5)", September 1993. Available from: http://www.ietf.org/rfc/rfc1510.txt
5.   N. Fischbach, "Kerberos en environnement ISP", January 2003. Available from: http://www.securite.org/ presentations/Krb5/ OSSIR2001-krb5_1.13.ppt
6.   M. Mishra, W.Arbaugh, "An initial Security Analysis of the IEEE 802.1X Standard", February 2002. Available from: http://www.cs.umd.edu/~waa/ 1x.pdf

7. IEEE. Standards for local and metropolitan area networks: Standard for port based network access control. IEEE Draft P802.1X/D11, March 2001

8. Blunk, J. Vollbrecht, "PPP Extensible Authentication Protocol (EAP)", March 1998. Available from: http://www.faqs.org/rfcs/rfc2284.html

9. F. Moioli, "Security in Public Access Wireless LAN Networks", M.Sc. Thesis, Royal Institute of Technology, Stockholm, June 2000. Available from: http://downloads.securityfocus.com/library/fabio-thesis.pdf

10. J. Linn, "Generic Security Service Application Program Interface", September 1993. Available from: http://www.ietf.org/rfc/rfc1508.txt

11. The open1x project, Web site: http://www.open1x.org

12. A linux wireless card driver, software access point. Available from: http://www.hostap.epitest.fi/

13. A. Chickinsky, Litton/TASC, "Wireless LAN Security Threats", IEEE 802.11-01/258, May 2001. Available from: http://grouper.ieee.org/groups/802/11/ Documents/DocumentHolder/1-258.zip

14. T. Wu, "The Secure Remote Password Protocol", In proceedings of the fifth Annual Symposium on Network and Distributed System Security, San Diego, March 1998. Available from: http://www.isoc.org/isoc/conferences/ndss/98/ wu_sl.pdf

15. B.Tung, et al., "Public Key Cryptography for initial authentication in Kerberos", Interbet Draft, 2001. Available from: http://www.ietf.org/internet-drafts/draft-ietf-cat-kerberos-pk-init-18.txt

A Preliminary Study of Scalability of TCP/IP Based Clusters under Database Workloads

Krishna Kant

Enterprise Technology Labs
Intel Corporation

Abstract. In this paper we study the scalability of non-partitioned, clustered database management systems as a function of inter-process communication (IPC) latency and number of nodes. It is assumed that the clustered DBMS has a fully shared IO subsystem and multiversion concurrency control over the data in various buffer caches. The cluster interconnect fabric is assumed to be TCP/IP over Ethernet with and w/o hardware offload. The main contribution of the paper is to shed some light on the scalability of DBMS workloads in a scaleout environment as a function of number of nodes and interconnect latencies.

1 Introduction

In the e-business environment, mid-tier and backend applications have traditionally been implemented on SMPs (symmetric multiprocessors) because of its easier programming model and efficient inter-process communication (IPC). However, with the emergence of high bandwidth, low-latency cluster interconnect technologies, there is a move afoot towards clustered implementations. In particular, the availability of cost effective 10 Gb/sec Ethernet networking solutions along with hardware offloaded TCP/IP could make clustered implementations even more attractive. In this paper we examine the scalability clustered implementations for DBMS systems since such systems are significantly impacted by IPC overhead and latency. We assume a shared disk type of clustered DBMS (such as the Oracle 9i/10g product) and a TPC-C like workload (http://www.tpc.org/tpcc/default.asp). In order to avoid complexities and idiosyncracies of actual systems, the modeling in this paper is not intended to project performance for any real system; its purpose is merely to study a simple model of clustered DBMS based on a limited set of measurements.

Although several papers in the literature have discussed IPC performance issues, much of the work is concentrated on the high performance computing (HPC) side rather than the commercial workloads. There are some industry papers that show substantial benefits of low latency interconnect technologies on application performance [1], but no performance models or sensitivity analyses are presented.

It is well known that end-to-end IPC latencies via the traditional TCP/IP over Ethernet stack can be almost an order of magnitude higher than other

N. Mitrou et al. (Eds.): NETWORKING 2004, LNCS 3042, pp. 1354–1359, 2004.

specialized fabrics such as Myrinet, QsNet, IBA, etc. [1]. A detailed discussion of these along with a study of performance benefits of HW TCP offload for front-end servers is contained in [4], and will be omitted here. Most of these inefficiencies can be addressed by using the Virtual Interface Architecture [VIA] like interface [3] and efficient fast-path processing. The former is supported by the RDMA (remote DMA) protocol [8] which is gaining widespread acceptance. We assume that RDMA/TCP offload is performed in a programmable engine which we call as *packet processing engine* (PPE). This PPE can be located in multiple places in the platform as reported in [4]; however, to avoid clutter, we shall consider only the "north-bridge" implementation that participates in the processor coherence protocol.

2   Clustered Database Overview

Oracle 9i/10g presents a premier example of clustered DBMS and is used as the representative clustering mechanism in this paper. In this architecture, also known as real application cluster (RAC), all nodes share a common disk subsystem that holds the entire database [6]. That is, no partitioning of the database among nodes is required for clustered operation. For efficient access, each node may cache portions of the data or indices in its main memory (normally called "buffer cache"). If a node requires data that is not present in its local buffer cache, it checks if the data is available in the buffer cache of another node and if not, it initiates the disk IO. Given a high BW, low latency and low overhead interconnection fabric, this technique can substantially reduce the IO overhead and thereby improve the scalability of the cluster.

RAC maintains a distributed directory indicating location and status of all database blocks available in various buffer caches. The directory information is migrated dynamically depending upon the data access pattern so that the directory entry is resident at the most frequently used node.

RAC uses *multiversion concurrency control* (MCC) to achieve high scalability [2] where each lockable entity (assumed to be a page here) carries with it the undo log and the version numbers. Thus, a read transaction does not need any locking since it can always get at the correct version of the page. Also, dirty data can be shared among nodes directly. The major additional cost of MCC is in heavier duty directory management and maintaining cascading undo logs that must be carried around. Of course, write locking is still required. In the model, the choice of values for the various RAC related parameters are somewhat arbitrary, since detailed measurements quantifying these are not available. This is adequate since the purpose of the paper is merely to illustrate scalability rather than do performance projections of an actual system.

For the workload, we used TPC-C, which is a popular benchmark for studying on-line transaction processing (OLTP). TPC-C is usually considered an inappropriate benchmark for clustering purposes since it is possible to partition TPC-C database such that the IPC traffic between nodes becomes negligible. However, note that in the RAC context, the database is *not* partitioned. We also assume that the query processing itself is not partitioned among nodes, although it is reasonable to assume intelligent transaction management/scheduling to take advantage of already cached data in various nodes. In this case, TPC-C becomes a reasonable clustered database workload. These assumptions also allow us to use existing non-clustered TPC-C measurements to calibrate our model.

We assume that all TCP connections used for IPC are persistent, so that connection setup/teardown overhead or latencies do not come into play for IPC. RDMA does require pre-registration, pre-pinning and exposure of user buffers. Registration and pinning require a Kernel call, which is expensive. Buffer exposure requires an explicit message exchange. We assume that control message buffers are registered and pinned at the time of thread creation and exposed to the directory node once per transaction. For the IPC data messages, we assume we assume that both pinning and exposure are done on the basis of adjustable but small "windows" explicitly exposed for every data transfer. Note that buffer exposure message are themselves IPC messages and need to be accounted for.

3   A Simple Performance Model

The major performance modelling exercise is to accurately estimate the impact of IO (IPC and disk) on the workload as a function of cluster size. To this end, we note that any IO has two major performance impacts:

1. Increased "path-length" (instruction count) of IO handling, which results in *increased utilization* of the host processor.
2. Increased data retrieval latency which leads to *increased stalls* on the host processor. Stalls happen whenever the communication latency cannot be hidden by using multiple threads/processes.

The cluster size primarily affects the shared content (and hence locking & synchronization frequency), amount of management information, and locality properties. Because the available measurements are only for small clusters, we have assumed certain functional behaviors with respect to number of nodes which remain to be validated.

In a typical benchmark performance context, one is usually interested in the achieved throughput when the CPU utilization is as close to 100% as possible. Assuming a multithreaded environment, the effective latency per transaction includes the following components.

1. Latency corresponding to basic per transaction path-length excluding disk IO and IPC, denoted L_{comp}.
2. IPC related latency. This includes code latency L_{eph} and the unhidable latency L_{uh} for each communication.
3. Disk IO related latency. This includes code latency L_{dkc} and the unhidable latency L_{uh} for each IO.

Here, the estimation of the unhidable latency L_{uh}, in turn, requires the estimation of the overall thread stall time, denoted L_{avg}. Let N_{ipc} and N_{disk} denote, respectively, the number of IPC communications and disk IOs per transaction. Also, let L_{ipc} and L_{disk} denote thread stall times per IPC and disk IO operation respectively. Then L_{avg} can be computed as the following weighted sum:

$$L_{avg} = (L_{ipc}N_{ipc} + L_{disk}N_{disk})/(N_{ipc} + N_{disk}) \qquad (1)$$

The estimation of the crucial parameters introduced above (including unhidable latencies) is omitted here due to lack of space and may be found in [5]. With this, the total per transaction latency L_{tot} can be estimated as:

$$L_{tot} = L_{comp} + N_{ipc}[L_{eph} + L_{uh}] + N_{disk}[L_{dkc} + L_{uh}] \qquad (2)$$

With this, the achievable throughput *per node* is given by simply $1/L_{tot}$.

Although absolute achievable throughput is interesting, a more important performance metric is *cluster efficiency*, which we define as ratio of nodal throughputs in the clustered and unclustered cases. The unclustered throughput can be obtained by independent (and much more mature) performance projection models; however, we express the unclustered throughput in the same framework as the clustered throughput. For brevity, the details of this calculation are omitted here.

4   Sample Modelling Results

The model calibration proved to be an ardous process because of the lack of consistent set of detailed measurements. The TCP offload related parameters were obtained using a prototype system which uses one processor in a SMP system as the TCP engine. Even here, RDMA related estimates are speculative since currently we do not have a working RDMA prototype. A further difficulty is that the modeled system (assumed to be a platform in 2005-2006 timeframe) is different from the measured system and a translation of parameters was required. The available TPC-C results were are also sketchy and were available only for 1, 2 and 4 node systems. In view of these deficiencies, the results must be treated as merely indicative of trends, rather than as actual achievable results.

For the results presented here, we assume a "rated utilization" model where the utilizations of various resources have been fixed at predetermined values. The practical interpretation of this is that the system uses just enough units (NICs, disk adapters, disk drives, etc.) to keep the utilization almost constant regardless of the cluster size or configuration. Apart from simplicity, the main motivation for this approach is that it does not color the cluster scalability results with system configuration issues. In particular, we assume a rated utilization of 60% for chipset, IO bus, disk adapter, NIC, and switch ports, and 30% for disk drives. All switches were assumed to be 16-port layer-2 and the topologies attempted to distribute the switch-port usage evenly. For the results shown here, we consider the following 3 cases with respect to the IPC transport implementation.

1. Kernel based software RDMA/TCP implementation. TCP parameters are calibrated based the study in [4], but RDMA calibration is speculative. This case represents the lower bound on cluster performance.

2. Memory control hub (MCH) based hardware RDMA/TCP engine, which was calibrated based on the current prototype and internal models of such an engine.

3. An "ideal" case characterized by (a) zero path-length for the PPE, (b) zero port-redirection latency in the switches, and (c) fully optimized host-PPE interface (i.e., no interrupts, no scheduling delays, etc.). This case represents an (almost unachievable) upper bound on performance.

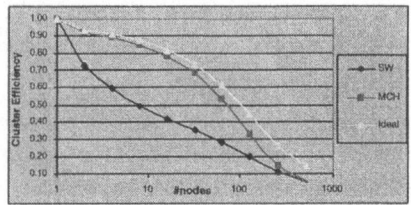

Fig. 1. Cluster Efficiency vs. cluster size   **Fig. 2.** 256B msg latency vs. cluster size

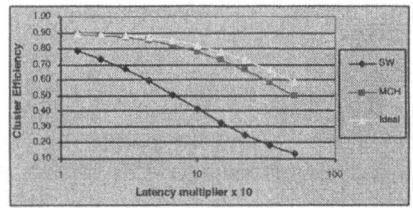
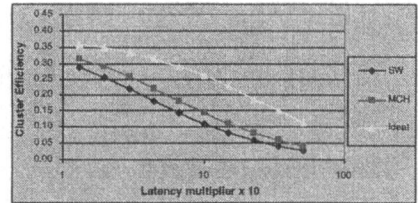

Fig. 3. Latency sensitivity w/ 16 nodes   **Fig. 4.** Latency sensitivity w/ 256 nodes

Figure 1 shows the cluster efficiency as a function of number of nodes (N_{nodes}). The efficiency starts out at 1.0 and generally decreases with N_{nodes}. The efficiency for Ideal and MCH cases remains high with small N_{nodes} but eventually rolls off due to the overhead of managing nodes, duplication in buffer contents of various nodes, and multiple switches in the path. Note that the SW TCP shows a steadily decreasing efficiency because the substantial IPC overhead cannot compensate for reduced IO. In fact, even a 2-node cluster shows an efficiency of only 73%.

It has been well recognized in the literature that IPC latency limits the *scalability*, i.e., the maximum cluster size that one could reasonably build. In the absence of a standardized definition, let's say that scalability refers to the size at which the efficiency drops to 50%. With this definition, Figure 1 shows that HW RDMA allows scalability to > 64 nodes whereas SW RDMA limits it to 8 nodes. The interesting point to note is that even the Ideal curve provides a scalability of only 100 nodes. That is, low IPC latency can only do so much for the scalability – at some point myriad issue of platform latencies, OS overhead, and application interface take over.

Figure 2 attempts to show end-to-end IPC latency $L_{ipc}^{(tot)}$ as a function of number of nodes. For this, we use a *reference message size* of 256B. Figure 2 shows that for small clusters, SW TCP provides a latency of 27 μs whereas HW TCP has a latency of only about 10 μs.[1] The Ideal case shows another factor of 3 reduction in latency (i.e., 3.5 μs). This last result is interesting since it shows that using very fast PPE's/switches still leaves various chipset, wire and OS latencies which may be substantial.

Figures 3 and 4 attempt to directly show the sensitivity of cluster efficiency to end-to-end IPC latency for $N_{nodes} = 16$ and 256. Here, the latency multiplier= 10 corresponds to normal latency and the latency for each successive point is 1.5 times that for the previous point. As expected, the latency sensitivity is small initially, increases in the middle, and then decreases eventually. The major differences between the 3 cases are due to the distinction between code and non-code latencies. Code latencies directly contribute to worse performance since they amount to a change in path-length, whereas non-code latencies can be hidden to large extent by multiple threads. This explains why SW TCP shows the highest sensitivity whereas the Ideal shows the least.

5   Conclusions and Open Issues

In this paper we studied the scalability of Ethernet based clustered non-partitioned DBMS with multiversion concurrency control. The results show that in such an environment, an end-to-end loaded latency of 10 μs is adequate to scale the cluster to 100 nodes and still achieve good cluster efficiency. Although the results are preliminary due to novelty of technologies and other practical challenges, we believe that our model provides a number of interesting insights into cluster performance and scalability that have hitherto been unavailable to the researchers. A more detailed setup and measurement work is currently underway.

References

1. B. Benton, "Infiniband's superiority over Myrinet and QsNet for high performance computing", whitepaper at www.FabricNetworks.com.
2. P.A. Bernstein and N. Goodman, "Multiversion concurrency control — theory and algorithms", ACM Trans on Database Systems ., 8(4):465–483, December 1983.
3. D. Dunning, G. Regnier, et. al., "The virtual interface architecture - a protected, zero-copy user-level interface to networks", IEEE Micro, March 1998, pp66-76.
4. K. Kant, "TCP offload performance for front-end servers", to appear in proc. of GLOBECOM 2003, Dec 2003, San Francisco, CA.
5. K. Kant, "Scalability of TCP/IP based clusters under Database Workloads", Full paper available at kkant.ccwebhost.com/download.html.
6. T. Lahiri, V. Srihari, et. al., "Cach Fusion: Extending shared disk clusters with shared caches", Proc. 27th VLDB conference, Rome, Italy 2001.
7. J. Liedtke, K. Elphinstone, et. al., "Achieved IPC performance", Proc. of 6th workshop on hot topics in operating systems, May 1997, Chatham, MA.
8. J. Pinkerton, www.rdmaconsortium.org/home/The_Case_for_RDMA020531.pdf

---

[1] Recall that these numbers are for 2006 platforms; for current platforms, SW TCP latencies are more like 90 μs and HW offloaded latencies estimated to be 30 μs.

Evaluation of Smart Antenna Algorithms for cdma2000 Reverse Link

Mustafa Karakoc [1] and Adnan Kavak [2]

[1] Kocaeli University, Dept. of Electronics and Computer Ed., 41100, Kocaeli, Turkey
[2] Kocaeli University, Dept. of Computer Engineering, 41040, Kocaeli, Turkey
{mkarakoc, akavak}@kou.edu.tr

Abstract. Adaptive algorithms which provide desired performance and have small computational load are key factors that enable the integration of smart antennas into 3G wireless communication systems. We evaluate non-blind (LMS, RLS), blind (CM), and semi-blind (CFA) type receive beamforming algorithms in terms of their performance, complexity, and convergence rate for cdma2000 systems. Simulations are performed for a 5-element uniform linear array operating in a wireless vector channel that consists of multipaths and multiple access interference. Results show that CFA has the highest SINR with minimum deviation in varying fading conditions. CM has the smallest SINR performance possibly due to not involving pilot information, but it has the computation load as low as LMS does, i.e. O(M) where M is the antenna number. RLS achieves SINR comparable to CFA with the fastest convergence rate.

1   Introduction

With the emergence of the third generation (3G) wireless systems, smart antenna systems (SAS) which are proven to provide capacity increase and coverage expansion becomes an attractive technology. The heart of the SAS is their advanced signal processing capability at the baseband, which enables them to adaptively adjust their beam pattern. In order to deploy SAS in 3G systems, it is, however, essential to employ beamforming algorithms which provide desired performance and have small computational complexity to be implemented in real time.

In this paper, we evaluate *blind* type Constant Modulus (CM) [1], *semi-blind* type Code Filtering Approach (CFA) [2], and *non-blind* type Least Mean Square (LMS) and Recursive Least Squares (RLS) [3,4] adaptive beamforming algorithms from various aspects for cdma2000 system in changing channel propagation conditions. The objective function common to these algorithms is that they try to maximize signal-to-interference plus noise ratio (SINR) at the output of the beamformer. However, they differ in the computation of beamforming weight vector that approaches Wiener solution. In obtaining optimum weight vector, while LMS, RLS, and CM algorithms use iterative approaches, CFA solves generalized eigenvalue problem using the estimated covariance matrices of the array output and post-correlation signal vector. In the non-blind type adaptive algorithms, a training signal is known to both the transmitter and receiver during the training period issued. On the

N. Mitrou et al. (Eds.): NETWORKING 2004, LNCS 3042, pp. 1360–1365, 2004.

other hand, the blind type adaptive algorithm does not require any training sequence and exploit some known properties of the desired signal, such as discrete-alphabet structure, constant-modulus property or direction-of-arrivals (DOAs) information. Semi-blind type adaptive algorithm overcomes the problem of insufficient training symbols of non-blind methods and the requirement of large amount of data as in blind methods.

Simulations are performed for the reverse link of cdma2000 under varying multipath and multiple access interference conditions. The results to be presented include the comparison of SINR variation, computation complexity, DOA estimation error, and convergence time for the above algorithms.

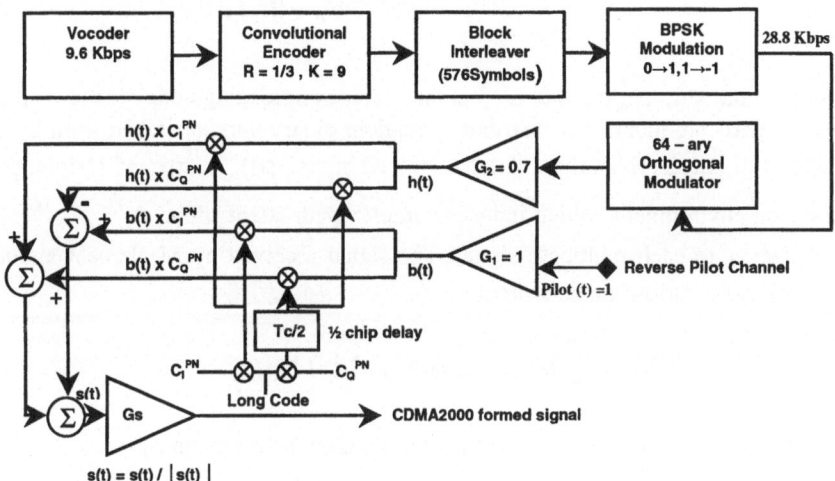

Fig. 1. Reverse link of the cdma2000 system for RC1

2   Uplink Signal Model

Cdma2000 is an approved 3G standard by ITU, which is based on the existing IS-95 CDMA system. [5, 6]. The cdma2000 reverse link baseband signal model in radio configuration 1 (RC1) is used in our simulations as shown in Figure 1. The transmitted baseband signal s(t) from a mobile unit can be expressed as,

$$s(t) = \sum_{k=1}^{\infty} d(k,t) \otimes c(k,t), \tag{1}$$

where k is the slot index, d(k,t) is the waveform before complex spreading, c(k,t) is the complex PN spreading sequence, $\otimes$ denotes the complex spread function. The waveform before complex spreading d(k,t) is composed of the traffic channel and pilot channel information symbols, which are given by

$$d(k,t) = d_I(k,t) + jd_Q(k,t), \tag{2}$$

where j represents the imaginary part of Eq. 2, and

$$d_I(k,t) = \sum_{i=1}^{N_C} G_1\, b((k-1)N_C + i)\,, \tag{3}$$

$$d_Q(k,t) = \sum_{i=1}^{N_C} G_2\, h((k-1)N_C + i)\,, \tag{4}$$

where $b(\cdot)$ is the bit stream of pilot channel which are all "1"s, $d(\cdot)$ is the bit stream of traffic channel and N_c is the length of spreading code per slot. The complex PN spreading sequence waveform is

$$c(k,t) = \sum_{j=0}^{N_c-1} C^{PN} p(t - [(k-1)N_c + j]T_c)\,, \tag{5}$$

where T_c is the chip period and C^{PN} is the complex spreading code whose real and imaginary parts are identically distributed, random binary numbers taken from the set $\{+1,-1\}$ with equal probability. The transmitted signal s(t) is exposed to multipath propagation environment, which induces complex path attenuation, $\alpha_l = \beta_l e^{j\phi_l}$ and time delay τ_ℓ to each multipath signal. The signal received by M element antenna array at the base station can be written as,

$$\mathbf{X}(t) = \sum_{\ell=1}^{L} \alpha_\ell s(t-\tau_\ell)\mathbf{a}(\theta_\ell) + \mathbf{I}(t) + \mathbf{N}(t)\,, \tag{6}$$

where $\mathbf{I}(t)$ is the multiple access interference (MAI) which is given by

$$\mathbf{I}(t) = \sum_{q=1}^{N-1} \sum_{\ell=1}^{L_q} \alpha_{q,\ell} s(t-\tau_{q,\ell})\mathbf{a}(\theta_{q,\ell}) \tag{7}$$

and $\mathbf{N}(t)$ is the Mx1 complex-valued additive spatially white Gaussian noise vector, $\mathbf{a}(\theta_\ell)$ is the Mx1 array response vector of the multipath arriving at DOA θ_ℓ. Finally, the array output is multiplied by a complex weight vector W which is determined according to aforementioned beamforming algorithms to result in

$$z(t) = \mathbf{w}^H(t)\mathbf{X}(t)\,, \tag{8}$$

where H denotes complex conjugation and transpose (Hermitian) operation.

3   SINR Calculation

SINR of the received signal at the beamformer output is calculated using the similar approach described in [2]. The post-correlation signal vector for the multipath signal of the desired user can be written as

$$\mathbf{y}_1(t) = \frac{1}{\sqrt{T_b}} \int\limits_{(t-1)T_c}^{tT_c} \mathbf{X}(t) c_1^*(t - \tau_1) dt , \qquad (9)$$

where, T_b is the symbol duration, c_1 is the code of desired user, τ_1 is the time delay for the desired multipath signal. In order to find the SINR, it is necessary to first estimate the pre-correlation $\mathbf{R}_{xx}$ and post-correlation $\mathbf{R}_{yy,1}$ covariance matrices of the signal vectors $\mathbf{X}(t)$ and $\mathbf{y}_1(t)$ in (7) and in (9), respectively. The interference-plus-noise covariance matrix $\mathbf{R}_{uu,1}$ can be estimated utilizing these matrices as given by

$$\mathbf{R}_{uu,1} = \frac{G}{G-1}\left(\mathbf{R}_{xx} - \frac{1}{G}\mathbf{R}_{yy,1}\right), \qquad (10)$$

where, G is the processing gain defined as $G = T_b/T_c$. Then, the covariance matrix for the desired signal component $s_1(t)$ is estimated from

$$\mathbf{R}_{ss} = \mathbf{R}_{yy,1} - \mathbf{R}_{uu,1} . \qquad (11)$$

Finally, using the optimum weight vector generated at the output of each beamforming algorithm, we can calculate the received SINR

$$\text{SINR} = \frac{\mathbf{w}_{opt}^* \mathbf{R}_{ss,1} \mathbf{w}_{opt}}{\mathbf{w}_{opt}^* \mathbf{R}_{uu,1} \mathbf{w}_{opt}} . \qquad (12)$$

4   Simulations and Results

We consider a multipath propagation scenario that has a direct path signal at 32° and two multipath signals at 44° and 85° for the desired user and an interference signal at 61°. The base station antenna configuration is 5-element uniform linear array (ULA). We perform repeated simulations (100 times) in order to compute the average SINR for each algorithm under varying channel fading conditions. These 100 runs represent that we take 100 spatial points as the mobile moves a small distance along a given direction. Therefore, as demonstrated in [7], DOAs (θ_l) are assumed to remain unchanged during this small movement. For each simulation, the antenna array receives 1000 signal samples and executes adaptive algorithms. We assume that Rayleigh random variable represents the corresponding amplitude fade (β_l) along the multipath. For the phase component (ϕ_l) of complex path attenuation at each multipath, uniformly distributed random variable is assumed, which represents path length difference and Doppler shift in the propagation medium. Fading parameters in the channel are adjusted such that direct path signal is minimum 3 dB above the multipaths and 5 dB above the interference. Multipath time delays are set as one chip period (T_c) for 85° and half chip period ($T_c/2$) for 44°. Convergence criteria for the LMS, RLS, and CM algorithms are determined from the norm of weight error vector given by

$$\|\Delta\mathbf{w}\| = \|\mathbf{w}(t+1) - \mathbf{w}(t)\| , \qquad (13)$$

which is 0,0001. Other relevant parameters for the simulations are chosen as following; step size parameter μ=0.01 for LMS and CM; weight factor ψ=0.99 for CM; forgetting factor δ_o=0.99 for RLS; code parameter η=1.5 and spreading gain G=4 for CFA.

During simulations, CFA, CM, RLS, and LMS algorithms are assumed to run parallel on the received signal vector $\mathbf{X}(t)$ and each generates a weight vector based on the calculations given in [1, 2, 3, 4]. Spatial spectrums formed with the weight vector resulting from these algorithms are plotted in Figures 2, 3, 4, and 5 for LMS, RLS, CM, and CFA, respectively. Note that these figures are the results of a single simulation run, which are representative of 100 simulation runs. In each figure, we see that main beam direction is steered towards to DOA of desired signal's direct path. This is because the direct path signal was the strongest and the algorithms lock on the signal which has the largest power,

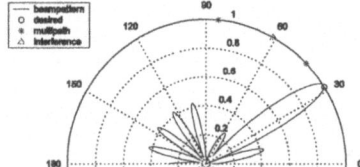

Fig. 2. Spatial spectrum via LMS alg.

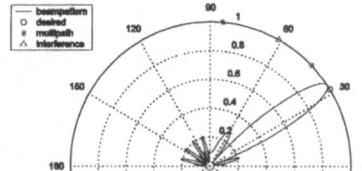

Fig. 3. Spatial spectrum via RLS alg.

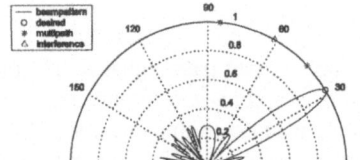

Fig. 4. Spatial spectrum via CM alg.

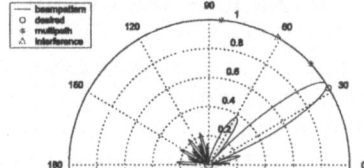

Fig. 5. Spatial spectrum via CFA alg.

Other results of the simulations are summarized in Table 1. Note that the mean (50% values) and standard deviation values of SINR are compiled from its cumulative distribution. CFA has the largest SINR among the four algorithms. It also has the smallest SINR deviation, implying the stability of this algorithm in changing channel conditions. RLS provides SINR value comparable to CFA, but it has the advantage of minimum convergence time. CFA is the most computationally complex algorithm with the order of $O(M^3)$ due to matrix inversion operation. The SINR performance of the CM is very low (5dB) as compared to other three algorithms. This may be due to not utilizing additional pilot information provided by cdma2000 system as a training signal in CM algorithm. The only advantage of CM is its simplicity in the computation load in the order of $O(M)$, which is as small as LMS algorithm.

Table 1. Simulation results.

ADAPTIVE ALGORITHMS		LMS	RLS	CM	CFA
Computational complexity	add	2M+1	$2M^2+M$	2M+4	M^3+2M^2-1
	mply	2M	$5M^2+3M+1$	2M	M^3+3M^2+2M+3
DOA estimation error (%)		1,5625	6,250	0,6250	3,125
Convergence time		283	24	546	------------[(1)]
SINR (dB)	mean	7,4565	11,8799	4,6018	12,056
	std. dev.	0,57	0,65	3,08	0,05

(1) Algorithm does not require iterative computation

5   Conclusion

We have found the following as a result of evaluating various receive beamforming algorithms for cdma2000 under changing channel conditions. We find that CFA is the robust algorithm providing the highest SINR (~12.1dB) among the four algorithms compared. The SINR performance of CFA is stable with minimum standard deviation. In terms of convergence time, RLS is the fastest algorithm. It also has the SINR performance (~11.9dB) as high as CFA does. The SINR performance of CM is the worst (~4.6dB) due to its blind type adaptation that possibly can not follow varying channel conditions. However, the results obtained herein need further verification for vehicular propagation conditions including the other beamforming methods such as DOA based or covariance matrix based beamforming.

Acknowledgement. This research was supported by TUBITAK (Scientific and Technical Research Council of Turkey) under contract EEEAG/102E015.

References

1.   Veen, A. J., Paulraj, A.: An analytical constant modulus algorithm. IEEE Trans. of Signal Proc., 44(5) (1996) 1-19.
2.   Naguib, A. F.: Adaptive antennas for CDMA wireless networks. Ph.D dissertation, Stanford Univ., (1996).
3.   Rong, Z.: Simulation of adaptive array algorithms for CDMA systems. MS. Thesis, Virginia PI & S Univ., (1996).
4.   Godara, L.: Application of antenna arrays to mobile communications, Part II: beamforming and direction-of-arrival considerations. Proc. IEEE, Vol. 85. (1997) 1195-1245.
5.   TIA/EIA Interim Standard, Physical Layer Standard for cdma2000 spread spectrum systems, TIA/EIA/S-2000-2.
6.   Wideband cdmaOne (TIA cdma2000) Radio Transmission Technology Proposal. International Telecommunication Union, Radio communication Study Groups (1998) http://www.cdg.org/frame_3giis.html.
7.   Kavak, A., Yang, W., Xu, G., Vogel, W. J.: Characteristics of vector propagation channels in dynamic mobile scenarios. IEEE Trans. on Antennas and Prop., 49(12) (2001) 1695-1703.

Control Theoretic Modelling and Design of Admission Control Mechanisms for Server Systems[*]

Maria Kihl[1], Anders Robertsson[2], and Björn Wittenmark[2]

[1]Department of Communication Systems, [2]Department of Automatic Control
Lund University, BOX 118, 221 00 Lund, Sweden
maria@telecom.lth.se, fax: +46 46 14 58 23, tel.no: +46 46 222 9010

Abstract. The admission control mechanism is an important part of many communication systems. In this paper we investigate load control mechanisms for server systems, that is systems that may be modelled as queueing systems. We show how control theory can be used when designing controllers for a *G/G/1*-system. We design a PI-controller for the system and compare the steady-state and transient behavior of this controller with the behavior of a static controller.

1 Introduction

One problem with all server systems, for example web servers or application servers in 3G networks, is that they are sensitive to overload. Therefore, admission control mechanisms can be implemented in the systems. The mechanism can either be static or dynamic. A static mechanism admits a predefined rate of calls whereas a dynamic mechanism contains a controller that, with periodic time intervals, calculates a new admission rate depending on some control objective.

The research concerning admission control has shown that the problem of optimally controlling the arrivals at a server system is a difficult task. The main problem comes from the fact that server systems usually are analyzed with queueing theory. However, there are no queueing theoretic methods that can be used when developing and designing controllers for the systems. Another solution is, therefore, to use control theory. Control theory has since long been used to analyze different types of automatic control systems. One well-known controller in automatic control is the PID-controller, which enables a stable control for many types of system (see, for example, [9]). The PID-controller uses three actions: one proportional, one integrating, and one derivative.

Very few papers have investigated admission control mechanisms for server systems with control theoretic methods. In [1] and [2] a web server was modelled as a static gain to find controller parameters for a PI-controller. A scheduling algorithm for an Apache web server was designed using system identification methods and linear control theory in [7]. In [4] a PI-controller is used in an admission control mechanism for a web server. However, no analysis is presented on how to design the controller parameters.

In [5] and [8], we analyzed queue length controllers for M/G/1-system. We developed a nonlinear fluid flow model and used this model when designing a PI-controller for the system. We demonstrated that linear models of this system are insufficient,

*. The work in this paper has partially been supported by the Swedish Research Council through the Multi Project Grant 621-2001-3020 and contract 621-2001-3053.

N. Mitrou et al. (Eds.): NETWORKING 2004, LNCS 3042, pp. 1366–1371, 2004.

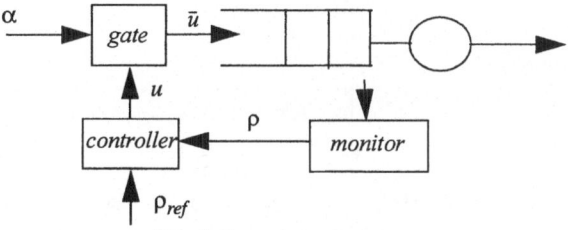

Fig. 1. Investigated system.

since the nonlinearities in the gate and queue introduce system dynamics that must be considered in the design process.

In this paper we instead analyze load control mechanisms. In [6], we developed and validated a control theoretic model of a G/G/1-system that can be used for the design of load control mechanisms. In [3] we show that the model is valid for an Apache web server. In this paper, we design and analyse a PI-controller.

2 System Model

The system model is shown in Fig. 1. We assume that the system may be modelled as a G/G/1-system with an admission control mechanism. The admission control mechanism consists of three parts: a *gate*, a *controller*, and a *monitor*. Continuous control is not possible in computer systems. Instead, time is divided into control intervals of length h seconds. Time interval $[kh-h, kh]$ is denoted interval kh.

The monitor measures the *control variable*, in this case the average server utilization during interval kh, $\rho(kh)$. At the end of interval kh, the controller calculates the desired admittance rate for interval $kh+h$, denoted $u(kh+h)$, from the measured average server utilization during interval kh, and the reference value, ρ_{ref}. The objective is to keep the server utilization as close as possible to the reference value. The gate rejects those requests that cannot be admitted. The variable representing the number of arrivals during control interval kh is denoted $\alpha(kh)$. Since the admittance rate may never be larger than the arrival rate, the actual admittance rate, $\bar{u}=\min[u, \alpha]$.

The gate uses a token bucket algorithm to reject those requests that cannot be admitted. Rejected requests are assumed to leave the system without retrials. An arriving request is only admitted if there is an available token. New tokens are generated at a rate of $u(kh)$ tokens per second during control interval kh.

3 Control Theoretic Model

We use the discrete-time control theoretic model shown in Fig. 2. This model has been validated in [6] for the single server queue in Section 2. The model is a flow or liquid model in discrete-time. The model is an averaging model in the sense that we are not considering the specific timing of different events, arrivals, or departures from the queue.

There are two stochastic traffic generators in the model. The *arrival generator* feeds the system with new requests. The number of new requests during interval kh is

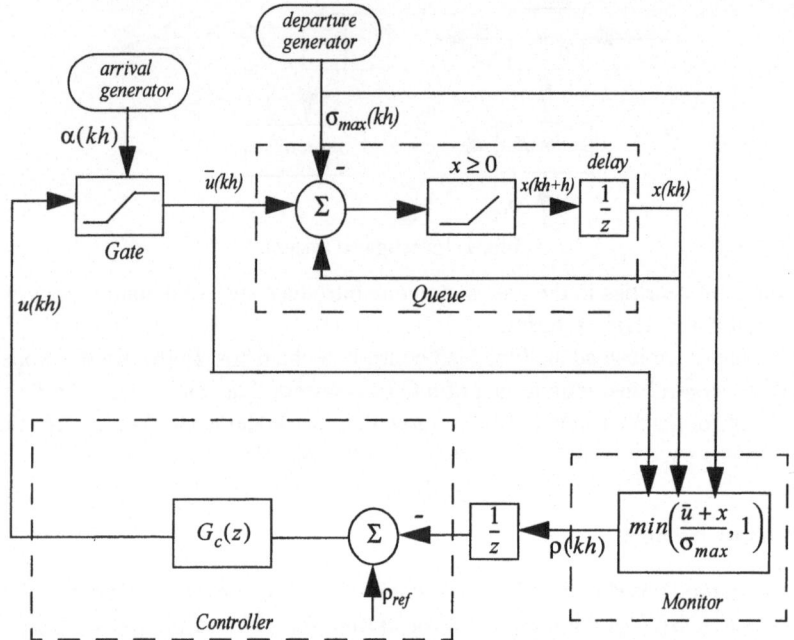

Fig. 2. A control theoretic model of a G/G/1-system with admission control.

denoted $\alpha(kh)$. $\alpha(kh)$ is an integrated stochastic process over one sampling period with a distribution obtained from the underlying interarrival time distribution. The *departure generator* decides the *maximum* number of departures during interval kh, denoted $\sigma_{max}(kh)$. $\sigma_{max}(kh)$ is also a stochastic process with a distribution given by the underlying service time distribution.

The *gate* is constructed as a saturation block that limits the number of admitted requests during interval kh, $\bar{u}(kh)$, to be zero when $u(kh) < 0$, $u(kh)$ when $0 \le u(kh) \le \alpha(kh)$, and $\alpha(kh)$ when $u(kh) > \alpha(kh)$.

The *queue* is represented by its state $x(kh)$, which corresponds to the number of requests in the system at the end of interval kh. The difference equation for the queue is given by $x(kh + h) = f(x(kh) + \bar{u}(kh) - \sigma_{max}(kh))$

where the limit function, $f(w)$, equals zero if $w<0$ and w otherwise. The limit function assures that $x(kh + h) \ge 0$. When the limit function is disregarded then the queue is a discrete-time integrator.

The *monitor* must estimate the server utilization since this is not directly measurable in the model. The server utilization during interval kh, $\rho(kh)$, is estimated as

$$\rho(kh) \;=\; min\!\left(\frac{\bar{u}(kh) + x(kh)}{\sigma_{max}(kh)},\, 1\right)$$

The objective of the *controller* is to minimize the difference between the server utilization during interval kh, $\rho(kh)$, and the reference value, ρ_{ref}. The control law is given by the transfer function, $G_c(z)$.

4 Controller Design

The system we investigated had an average service time of 0.02 seconds and the reference load, ρ_{ref}, was set to 0.8. We will use linear control design methods for finding parameters for a PI-controller. This means that we during the design consider a deterministic system with no active saturations. However, the real queueing system will for instance only allow positive queue lengths.

4.1 Static Controller

We used a static controller as a benchmark controller. A static controller uses a fixed acceptance rate, u_{fix}, that is set so that the average value of the control variable should be equal to the reference value. u_{fix} is in this case equal to 40 jobs per second.

4.2 PI-Controller

The PI-controller is a well-known controller in automatic control. It uses two actions: one proportional and one integrating. The control law for the PI-controller expressed in z-transform is given by

$$G_c(z) = K\left(1 + \frac{1}{T_i} \cdot \frac{h}{z-1}\right)$$

where the gain K and the integral time T_i are the controller parameters that are set so that the controlled system behaves as desired. The characteristic polynomial for the linear closed loop system will be

$$z \cdot \left(z^2 + \frac{K - 2\sigma}{\sigma} z + \frac{-KT_i + Kh + \sigma T_i}{\sigma T_i}\right) \quad (1)$$

where σ is the average value of σ_{max} and the pole at $z=0$ is cancelled in the transfer function from the input (the load reference) to the desired output (the load). Assume that the desired characteristic equation is

$$z(z^2 + a_1 z + a_2) = 0$$

The values of the controller parameters that gives this are

$$K = 2\sigma + a_1\sigma \qquad T_i = h \cdot \frac{2 + a_1}{1 + a_1 + a_2}$$

The controller parameters K and T_i influence the closed loop response for the system and need to be determined with respect to stability and robustness. The behavior of the PI-controller becomes better when the sampling period is short (should match desired dynamics). Therefore, for these investigations we used a sampling period of 0.2 seconds (h=0.2). This means that $\sigma = 10$, since σ is the average maximum number of departured jobs during a control interval. Choosing $\{K, T_i\}=\{12, 0.6\}$ the roots of the characteristic polynomial in eq. (1) will be 0.4 ± 0.2. This set of controller parameters can, therefore, be seen as a "good" choice.

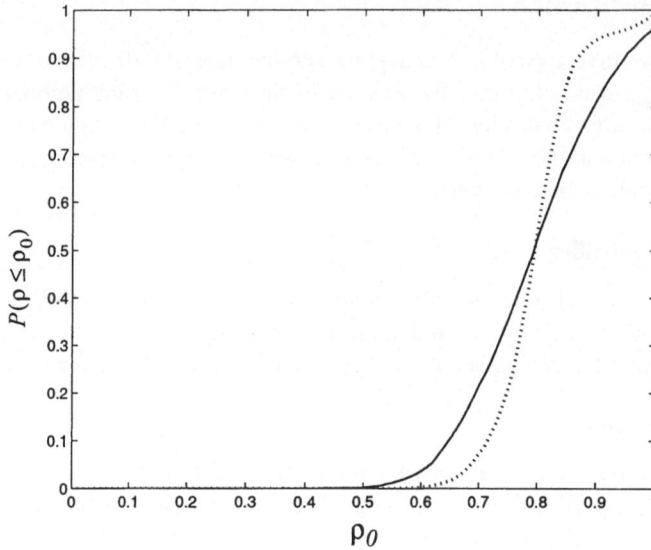

Fig. 3. Distribution function for an M/M/1-system, solid line: static controller , dotted line: PI-controller.

5   Numerical Investigations

The numerical investigations contain a comparison of the controllers described in the previous section. The queueing model was represented by a discrete-event simulation program implemented in C, and the control theoretic model was implemented with the Matlab Simulink package. During all investigations, the reference load was set to 0.8, and the average arrival rate was 150 jobs per second. We investigated an M/M/1-system, with average service time 0.02 seconds. Note that the controller design is independent of the type of arrival process and the service time distribution, since the system dynamics only depend on the average service time.

In Fig. 3, we show the *steady-state distribution* of the server utilization, by plotting the estimated distribution function, i.e. $P(\rho \le \rho_0)$ where $0 \le \rho_0 \le 1$. The distribution function was estimated from 5000 measurements of the server utilization for a specific parameter setting. The optimal distribution function is zero for $0 \le \rho \le 0.8$ and one for $0.8 \le \rho \le 1$. Each measurement is the average server utilization during one second. The distribution function shows how well the controller meets the first control objective. As can be seen, the systems with a PI-controller behave better than the system with a static controller. This phenomenon is due to that this controller can adapt to the stochastic variations in the system. This behavior requires a short sampling period. With a longer sampling period, for example one second, the PI-controller behave as the static controller.

The *step responses* for the load controlled M/M/1-system were also investigated. The step responses show the transient behavior of the controllers. The fastest controller is of course the static controller, since it already from start is set to an accurate admittance rate. However, the PI-controller has found a correct admittance rate only after a few seconds, which means that it can be regarded as fast enough.

Finally, we have investigated the *robustness* for the system. A good controller should maintain a good performance even when the system parameters change, that is the controller should be robust to modelling errors. In a real system, it is likely that the average service time will change slowly with time, for example due to changes in the user behavior. The results showed that the PI-controller was very robust to changes in the average service time. The static controller is, of course, dependent on a correct service time, which means that it cannot operate properly when the service times change.

6 Conclusions

In this paper, we have designed load control mechanisms for a G/G/1-system with control theoretic methods. We have designed a PI-controller. We have shown that, when considering transient and stationary behavior, the PI-controller behave well. One conclusion of this paper is that it is possible to use control theoretic methods when designing admission control mechanisms for server systems. The designs have been verified with simulations for discrete-event systems based on queuing theory.

References

[1] T.F. Abdelzaher and C. Lu, "Modeling and performance control of Internet servers", Proc. of the 39th IEEE Conference on Decision and Control, 2000, pp 2234-2239.

[2] T.F. Abdelzaher, K.G. Shin and N. Bhatti, "Performance guarantees for web server end-systems: a control theoretic approach", *IEEE Transactions on Parallel and Distributed Systems*, Vol. 13, No. 1, Jan 2002, pp 80-96.

[3] M. Andersson, M. Kihl, and A. Robertsson, "Modelling and design of admission control mechanisms for web servers using non-linear control theory", Proc. of SPIE ITCom, 2003.

[4] P. Bhoj, S. Ramanathan, and S. Singhal, "Web2K: Bringing QoS to web servers", HP Labs Technical report, HPL-2000-61, 2000.

[5] M. Kihl, A. Robertsson, and B. Wittenmark, "Analysis of admission control mechanisms using non-linear control theory", Proc. of IEEE International Symposium on Computer Communcations, 2003.

[6] M. Kihl, A. Robertsson, and B. Wittenmark, "Performance Modelling and Control of Server Systems using Non-linear Control Theory", Proc. of 18th International Teletraffic Congress, 2003.

[7] C. Lu, T.F. Abdelzaher, J.A. Stankovic and S.H. Son, "A feedback control approach for guaranteeing relative delays in web servers", Proc. of the 7th IEEE Real-Time Technology and Applications Symposium, 2001, pp 51-62.

[8] A. Robertsson, B. Wittenmark, and M. Kihl, "Analysis and design of admission control in web-server systems", American Control Conference, 2003.

[9] K.J. Åström and B. Wittenmark, *Computer-controlled systems, theory and design*, Prentice Hall International Editions, 3rd Edition, 1997.

A Fair Service Work Scheduling Approach for Differentiated Services in Optical Access Networks

NamUk Kim, HyunHo Yun, and Minho Kang

Optical Internet Research Center, Information and Communications University, 58-4, Hwaam-Dong, Yuseong-gu,Daejeon, 305-732, Korea
{niceguy, exhyho, mhkang}@icu.ac.kr

Abstract. We address the QoS (Quality of Services) bandwidth reservation and service work scheduling mechanism of how to fairly provide advance quality of service in polling based TDM networks like Ethernet PON (Passive Optical Networks). Unlike existing differentiated QoS supporting schemes, such as SP (Strict Priority) based DBA(Dynamic Bandwidth Allocation) or static work scheduling with min-max theory, proposed service quality pre-engagement (SQP) for prioritized packet scheduling and equal weighted fair bandwidth arbitration (EWF-BA) for fair services among access nodes reflect the dynamically changing service pattern of access node like ONU (Optical Network Unit). The performance results show that the service fairness and QoS are tightly coupled and proposed mechanisms achieve better QoS performances with high fairness and better network performances both in Poisson and Self-similar traffic environment.

1 Introduction

The polling based TDMA over the shared medium is recently adopted to many access networks such as Ethernet-PON (Passive Optical Network) and WDM optical access networks[1][2]. Especially, the upstream multiple access control is important because upstream channel must be shared efficiently and fairly by ONUs without channel collision while both high performances and relative QoS with differentiated services among classes are guaranteed. These objectives are mainly achieved by the service work scheduling, the controlled bandwidth allocation mechanism of OLT (Optical Line Termination), and the packet scheduling of ONU, generally the strict priority (SP) queuing or weighted fair queue (WFQ) service[4]. Up to now, some approaches, mainly be categorized into static bandwidth allocation (SBA) and DBA, are contributed to this topic for the optimized QoS performances[2]. These mechanisms guarantee the minimum service fairness with fixed service slot or min-max theory based dynamic slot allocation and provide high channel utilization, especially in DBA. But because these targets for the fairness of service slot not service thread, consistent service pattern of ONU, it restarts a new cycle although the unused bandwidth currently exists, which

N. Mitrou et al. (Eds.): NETWORKING 2004, LNCS 3042, pp. 1372–1377, 2004.

can be used for urgent heavy loaded ONUs. This can affect the fairness and effectiveness of services in short time epoch.

In this paper, we propose SQP which applies QoS bandwidth prediction and pre-reservation scheme based on SP. Up to now, most proposed scheduling methods only consider the prioritized services separated from DBA specific operation but we converge them for supporting efficient differentiated services in intra ONU. Then, we introduce EWF-BA, the policy based service work scheduling, to get the thread based pipeline service for high QoS performances as well as efficient service fairness in inter ONU domain. Our proposed schemes can be adapted to all kinds of generalized polling based TDMA network independent from physical media and topological difference.

2   The Fair Differentiated Services Supporting with SQP and EWF-BA

The SP can support QoS guaranteed services regardless of traffic changes but the bandwidth is easily monopolized by heavy traffic loaded high priority classes. Weight based methods like WFQ or DRR can not guarantee good performances by variable Ethernet frame size and high complexity[3][4]. Proposed SQP-DBA is a modified DBA mechanism based on SP scheduling and QoS bandwidth pre-reservation. For easy expressions, following notations for class i traffic are used when ONU_j and m_{th} cycle are assumed.

N,K: Number of ONUs and traffic classes. $Q_i^m(t)$:Queue size at time t.
D_i^m: Reported bandwidth for next cycle. A_i^m:Allocated bandwidth.
S_i^m: Total serviced work(bandwidth). V_l: Transmission speed of link.

We assumed that traffic is grouped into K service classes, which are ordered, such that class i is better than class i+1 for $0 < i < K - 1$, in terms of the queuing delay and packet losses. Generally, it is possible to assume that the size of class0 queue of ONU_j, $Q_{0,j}(t_m)$, where t_m is m_{th} service slot instance, does not rapidly change during several frame cycles because the aggregated CBR and real time application traffic shows nearly constant arrival pattern although their frame size varies. Based on this, in SQP-DBA, ONU reserves the service slot of next cycle as not $Q_{0,j}(t_m + A_j^m/V_l)$ but $Q_{0,j}(t_m + A_j^m/V_l)$ + QoS reservation bandwidth which is same size of current cycle's class0 serviced work. In addition, we assume the gated service scheme to minimize the long switching-over time between inter-classes and bandwidth monopolization of higher classes. Hence, the final report queue size and final service work is different to each class as follows.

$D_{0,j}^m$ = Q(late arrival)+Q(QoS reservation) = $Q_{0,j}^m(t_m + \sum_{i=0}^{K-1} A_{i,j}^m/V_l) + S_{0,j}^m$
$D_{0,j}^m(i \neq 0)$ = Q(late arrival or unfinished work) = $Q_{i,j}^m(t_m + \sum_{i=0}^{K-1} A_{i,j}^m/V_l)$
$A_{0,j}^m = min(Q_{0,j}^m(t_m), A_j^m) = S_{0,j}^m$, $A_{low,j}^m = A_j^m - min(Q_{0,j}^m(t_m), A_j^m)$
$S_{i+1}^m = min(Q_{i+1,j}^m(t_m), A_j^m - \sum_{n=0}^i Q_{N,j}^m(t_m)$, when $i < K - 2$.

Because the arrival pattern of lower priority ones such as aggregated VBR and nrt-VBR traffic rapidly changes, the efficiency of QoS bandwidth pre-reservation is relatively low. In this reason, we apply the general DBA's three step channel reservation mechanism to lower priority classes. With this, the high channel utilization of SQP-DBA is maintained by DBA and service quality reservation is done by SQP. Therefore, the QoS performance enhancement is expected to every class within the allocated bandwidth. But SQP client has no control authority to bandwidth allocation and its algorithm targets for QoS based differentiated services only in ONU. Hence, there must be a optimized inter ONU service scheduling algorithm in OLT, which is defined as bandwidth arbitration in this paper, for high QoS performances in network domain. General service work scheduling, SBA or DBA based, targets for the minimum service fairness by set limitation on slot size. However, these do not consider the service pattern of ONU, which can be important to QoS guaranteeing as well as network performances in short time epoch under heavy traffic environment. The service fairness shows how fairly and efficiently bandwidth is distributed to ONUs in long and even in short time period for full services. Because upstream packet service is done under non-exhaustive vacation service scheme, fairness is directly determined by two elements, T_j^m and A_j^m and these are tightly coupled with each other by arbitration. Proposed EWF-BA is a new service balancing mechanism that arbitrates the service slot size with real time information induced from traffic demands as well as the service thread of each ONU. The main idea is as follows. When the ideal cycle frame under SLA is limited to T_c excluding guard band, OLT gets three kinds of information in every arbitration points, sequentially expressed as $\sum_{n=0,n\neq j}^{N-1} A_n^m, D_j^m$, and $T_c^m V_l$. To effectively reflect the service thread or allocation pattern to A_j^m, we introduce two parameters. The one is the bandwidth occupation ratio C^m. It shows the serviced work ratio of each ONU normalized to total serviced work of ONUs in previous cycle and expressed as follows when $t_{j,R}^m$ is the m_{th} is the reporting instance.

$$C_j^{m+1} = A_j^{m+1} / \sum_{n=0}^{N-1} A_n^{m+1} = A_j^{m+1} / (t_{j,R}^{m+1} - t_{j,R}^m) V_l, 0 < j < N - 1.$$

C^m can be easily calculated from DBA operation. The other parameter is the bandwidth overflow ratio, $R_j^m = [(\sum_{n=0,n\neq j}^{N-1} A_n^m + D_j^m) - T_c V_l / T_c V_l]$, which is the excessively requested bandwidth ratio normalized by the SLA threshold value. It is meaningful only when the sum of service reserved and current demanding work is greater than $T_c V_l$. As mentioned, the over-reserved traffic service seriously degrades QoS performances and service fairness by the long service interval and bandwidth monopolization. To avoid these problems, OLT proportionally reflects the occupation ratio C_j^m to the final reduction weight value, W_j^{m+1}, while the thread of each ONU is kept as much as possible. OLT calculates final A_j with the reduction weight calculated from the occupation ratio and the overflow ratio as follows.

$$W_j^m = C_j^m R_j^m = A_j^m N[(\sum_{n=0,n\neq j}^{N-1} A_n^m + D_j^m) - T_c V_l / T_c V_l \sum_{n=0}^{N-1} A_n^m]$$

$$if \sum_{n=0,n\neq j}^{N-1} A_n{}^m + D_j^m > T_c V_l, A_j^{m+1} = D_j^m - D_j^m C_j^m R_j^m = D_j^m(1 - W_j^m).$$

else, $A_j^{m+1} = D_j^m$ by general DBA scheme.

Therefore, In EWF-BA, by the consideration of individual ONU's service thread in near real time, greedy ONU that has monopolized large bandwidth previously receives the reduced bandwidth with the same reduction weight to pervious occupation ratio in current cycle. The light loaded ONU rarely experiences arbitration effects due to low occupation ratio. It is efficient for minimum bandwidth guaranteeing under burst traffic environment. At the same time, because the heavy traffic load of some ONUs is efficiently distributed to every ONUs, There is no serious performance degradation of some ONUs. Moreover, the fairness guaranteed QoS service can be achieved since occupation ration which shows service pattern is effectively reflected to reduction weight and the cycle frame does not dynamically change by this. Below figure 1 shows the overall procedure of EWF-BA.

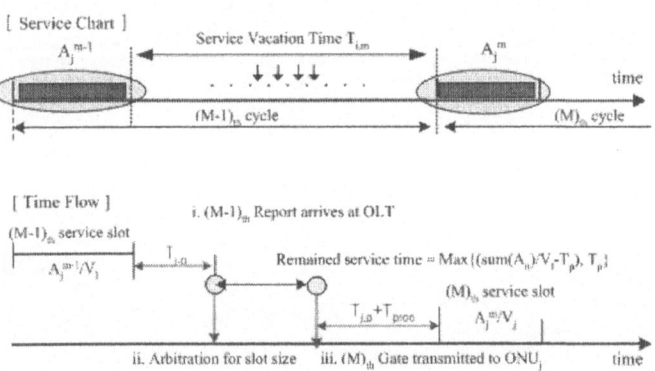

Fig. 1. Service work scheduling mechanism for EWF-BA.

3   Simulation Results and Performance Evaluation

For effective simulations, we assume that all incoming frames are classified into three traffic service classes in ONU as class0 for the CBR traffic and real time, class1 for non-real time and class2 for best effort service. The ON/OFF model is applied to the generation of Poisson and burst traffic and load is uniformly distributed over three classes. The frame size distribution follows that of the real traffic model in both Poisson and self similarity burst traffic analysis model [6,11]. In burst traffic environment, we applied Pareto distributed(Hurst parameter 0.8,0.9) traffic generation scenario.

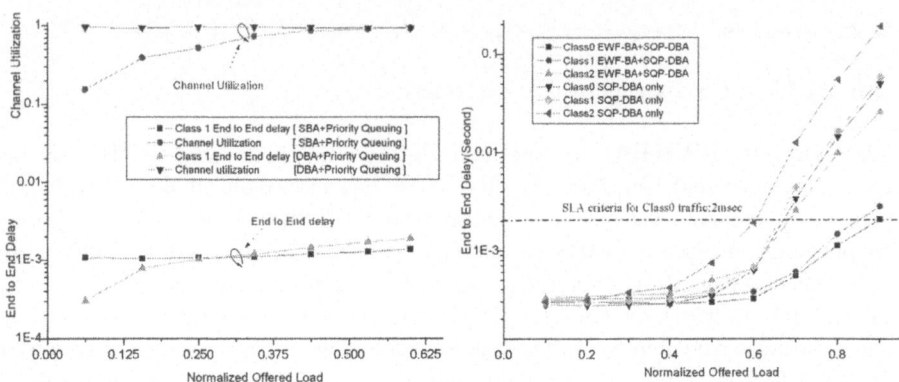

Fig. 2. (a)Average end to end delay and utilization of each class when Poisson traffic applied (b) Average end to end delay when Pareto traffic applied.

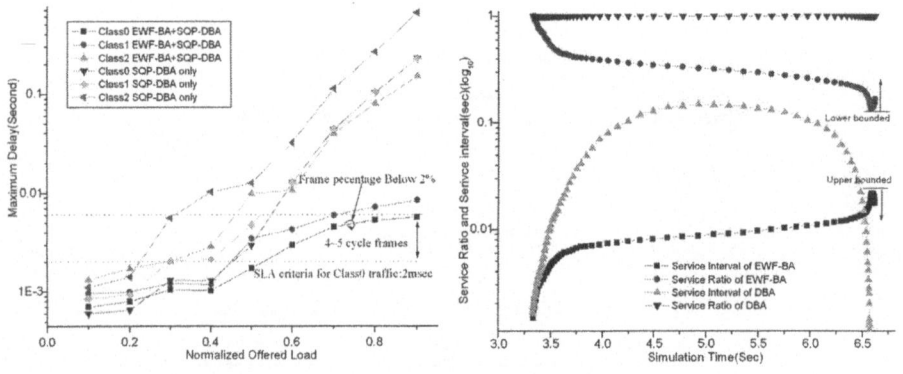

Fig. 3. (a)Maximum end to end delay of each class when Pareto traffic applied.(b) Service ratio and interval when Pareto traffic applied(load=0.8,H=0.9).

Figure 2(a) shows DBA achieves better utilization and lower delay in low traffic environment. But its delay gradually increases in high traffic environment due to the cycle frame change. Figure 2(b) shows that average delay of SQP-DBA plus EWF-BA is more enhanced than SQP-DBA only case in every classes. This result comes from the fact that EWF-BA efficiently distributes the heavy traffic load to all ONUs with the equal weighted reduction ratio until network goes into the stable service state. Moreover, figure 3(a) shows that the maximum delay of EWF-BA plus case is smaller than SQP-DBA only case and total numbers of longer delayed frames over targeted SLA value is relatively small. Finally, the congestion resolution period, time gap to overcome burst traffic, is just several cycle frames and this means that EWF-BA can service efficiently heavy loaded

ONUs while service fairness is maintained among ONUs. This is more obvious in figure 3(b). This result is derived from the same simulation scenario of above one but more burst environment. In DBA case, the service ratio that means serviced work normalized by the original demand bandwidth is always one by the DBA mechanism. But the service interval is too large to efficiently support the QoS guaranteed services. Moreover, the serious congestion duration is unpredictable and uncontrollable. Oppositely, EWF-BA guarantees the relatively short service interval and its value shows upper bounded pattern. This is possible by the load distribution and the service arbitration of that. The service ratio is generally lower than DBA but is also lower bounded. Therefore, the differentiated services for QoS can be guaranteed within minimum delay by SQP and under stable service pattern and fairness by EWF-BA. In this simulations, the average queue size of EWF-BA is generally lower than DBA but maximum queue size in short epoch is about 1.3 times larger than DBA case in very heavy traffic environment. This is the endurable performance degradation but there must be more improvement to this problem.

4   Conclusion

In this paper, we propose the SQP-DBA and EWF-BA for supporting QoS guaranteed fair service under the service work scheduling and reservation. The SQP-DBA achieves the minimum end-to-end delay of the highest priority traffic and high utilization with the minimum effects to lower classes. When EWF-BA scheme is applied, the network performances, especially the delay and service fairness among access nodes, are enhanced even if burst traffic happens in some ONUs. The bandwidth starvation by some heavy loaded ONUs can be also resolved within relatively short time epoch by the EWF-BA's thread based load distribution scheme although more buffer size is needed.

Acknowledgement. This work was supported in part by the Korea Science and Engineering Foundation (KOSEF) through OIRC project and ETRI.

References

1. Kramer, G., Mukherjee, B., Pesavento, G.: Ethernet passive optical network (EPON): building a next-generation optical access network. IEEE Comm. Mag, Vol. 40, issue. 2 (2002) 66-73
2. Kramer, G., Mukherjee, B., Pesavento, G.: IPACT-a dynamic protocol for an Ethernet PON (EPON). IEEE Comm. Mag, Vol. 40, issue. 2 (2002) 74-80
3. Rege, K., Dravida, S., Nanda, S., Narayan, S., Strombosky, J., Tandon, M., Gupta, D.: QoS Management in Trunk-and-Branch Switched Ethernet Network. IEEE Comm. Mag, Vol. 40, issue. 12 (2002) 30-36
4. Dovrolis, C., Stiliadis, D., Ramanathan, P.: Proportional Differentiated Services: Delay Differentiated and Packet Scheduling. IEEE/ACM Transaction on Networking, Vol.10, issue. 1 (2002) 12-26

Integrating (m,k)-Firm Real-Time Guarantees into the Internet QoS Model

Anis Koubaa, Ye-Qiong Song, and Jean-Pierre Thomesse

LORIA – INPL – UHP Nancy I
615 rue du jardin botanique, 54602 Villers-Les-Nancy
{akoubaa, song, thomesse}@loria.fr

Abstract. Having in mind that real-time streams tolerate some deadline misses according to (m,k)-firm constraints [1], this paper presents a solution that consists in integrating (m,k)-firm temporal requirements into the *guaranteed–rate* quality of service such as IntServ Model to provide both bandwidth and delay guarantees as required by real-time applications. Analytic study using Network Calculus gives the upper bound on delay guaranteed by the proposed approach and shows better behavior compared to the classical IntServ QoS model.

1   Introduction

Guaranteed QoS of IntServ model uses guaranteed-rate servers, such as Weighted Fair Queueing (WFQ) and its variants, to provide a share-driven scheduling policy to serve active streams according to their bandwidth requirements specified by the traffic parameter known as TSPEC, which defines an arrival curve for the real-time flow. Moreover, guaranteed end-to-end delay bound is also computed from the TSPEC parameter as well as the reserved bandwidth. This delay bound is affected by the bursty nature of the flow. In fact, for a given reserved bandwidth, when the burst size is large, the delay experienced by real-time packets gets higher and may exceed the required deadline. However, increasing service share to meet the required deadline leads to inefficient resource utilization since the reserved bandwidth would be higher than the actual need of the application. A solution proposed in [2], called PWFQ, consists in integrating static priority within WFQ scheduling algorithm in order to better manage the delay bounds for various sessions. The idea consists in serving with *Static Priority* policy, packets whose virtual finish tags [13] belong to a specified *sliding window*. This technique decouples the delay from service share and provides lower delay for low-share streams without degrading the delay of other sessions. The main critical point of this technique is to find the optimal window size and the priority assignment for the stream set.

Furthermore, congestion is an additional problem for streams with large burst size. In fact, when serving several concurrent streams, the guaranteed-rate router may suffer from congestion if its internal queue reaches its capacity. The basic solution to prevent such situation consists in performing stochastic dropping using RED mechanism; however, it could affect the QoS of real-time streams if packet drops are

N. Mitrou et al. (Eds.): NETWORKING 2004, LNCS 3042, pp. 1378–1383, 2004.

made inadequately. An elegant share-driven scheduling technique with drop was recently proposed by Koubâa and Song in [3], called (m,k)-WFQ, and resolves the problem of delay guarantees for low-share bursty streams as well as dropping process. This technique integrates (m,k)-firm temporal requirement [1] into WFQ scheduling algorithm and takes advantage of some deadline miss tolerance for real-time streams with the respect of (m,k)-firm timing constraints, , which consists in meeting the deadlines of at least m messages among any k consecutive message. (m,k)-WFQ is described in section 2.

In our previous work [3], we have only defined the (m,k)-WFQ algorithm and evaluate its performance using a basic simulation scenario. In this paper we propose a novel analytic study using Network Calculus to compute the delay bound for a given flow described with its TSPEC parameter and its (m,k)-firm requirement in the IntServ model. For this purpose, we introduce the (m,k)-filtering concept to integrate (m,k)-firm constraint into Network Calculus framework and derive the expression of the delay upper bound.

The remainder of this paper is structured as follows. Section 2 outlines the basic steps of (m,k)-WFQ scheduling algorithm. Section 3 presents mathematical background to adapt (m,k)-firm constraints to Network Calculus framework. In section 4, we propose to integrate (m,k)-firm constraints into IntServ QoS model using (m,k)-WFQ to provide bandwidth guarantee. Section 5 concludes this paper.

2   (m,k)-WFQ Scheduling Algorithm

In this section we briefly describe the (m,k)-WFQ scheduling algorithm. We refer the reader to our technical report [4] for more details.

(m,k)-WFQ is intended to fairly serve flows according to their bandwidth requirement and with respect to their (m,k)-firm constraints. A stream is said to have (m,k)-firm constraint if at least m packets inside any window of k consecutive packets must meet their required deadlines, otherwise the temporal QoS of the stream would be transgressed. The idea behind (m,k)-WFQ is to classify packets of incoming real-time flows into two parts - *Mandatory* and *Optional* - according to their (m,k)-firm constraints. This classification is made using the κ-pattern. The κ-pattern of a stream having (m,k)-firm deadline requirement is the succession of k elements from the alphabet $\Delta = \{O, M\}$ where:

$$\begin{cases} O & \text{Stands for an Optional packet} \\ M & \text{Stands for a Mandatory packet} \end{cases}$$

and contains exactly m 'M' symbols. κ(i) denotes the ith element of the κ-pattern for $1 \le i \le k$. The nth packet of a stream is classified as mandatory when $\kappa(n\%k) = 'M'$ for $n = 1, 2, \ldots$ where % is the modulus the operator.

The standard WFQ scheduling algorithm is based on the computation of virtual finish time to emulate the fluid GPS system. The virtual finish tag of a packet is defined as:

$$F_i^k = \max\left\{F_i^{k-1}, V(t)\right\} + \frac{L_i^k}{\Phi_i} \tag{1}$$

where F_i^k is the virtual finish time of the kth packet of ith stream. $V(t)$ is the virtual time when the kth packet arrives, L_i^k is the packet size of kth packet and Φ_i is the service share weight. This value is tagged into the packet. Then, the scheduler selects the packet with lowest finish tag. This tag doesn't consider any temporal constraint. It depends only on service share weight Φ_i and packet length L_i^k.

However, (m,k)-WFQ scheduling algorithm repairs this lack by further considering the (m,k)-firm constraint of each stream as well as the packet classification. The proposed algorithm fosters the transmission of mandatory packets to guarantee their deadline meet. In fact, (m,k)-WFQ scheduler makes the selection of the packet with lowest finish tag among mandatory packets present at the head of active queue of each served stream. Otherwise, *i.e.* -no mandatory packet is present at the head of queues- the optional packet with lowest F_i^k is then picked out for service. If the selected packet is *mandatory*, then the (m,k)-WFQ server sends it immediately. Else, if the selected packet is *optional*, the scheduler checks whether its deadline would be missed after being served. If it is the case, (m,k)-WFQ drops the selected optional packet, else, if the deadline would be met, the selected optional packet is transmitted.

3   The (m,k)-Filtering Concept

We introduce the (m,k)-filtering concept to adapt (m,k)-firm constraints to Network Calculus framework, and then derive the upper bound on delay in section 4. Further details are presented in [4].

Definition 1. *We define an (m,k)-filter as a device, that for an arrival function R(t), makes the output $\widetilde{R}(t)$ where only mandatory packets of the corresponding flow are sent according to its κ-pattern. Optional packets are discarded.*

The following theorem gives the arrival curve of a (σ,ρ)-shaped stream that crosses an (m,k)-filter. The proof is shown in [4].

Theorem 1. *Consider a stream S with arrival function R(t) upper constrained by the arrival curve $\alpha(t) = \sigma + \rho.t$ and crosses an (m,k)-filter device. Set λ_M the ratio of mandatory packets into the window of k consecutive packets according to its κ-pattern. The output produced by the (m,k)-filter is bounded by the arrival curve $\widetilde{\alpha}(t) = \widetilde{\sigma} + \widetilde{\rho}.t$ where:*

$$\begin{cases} \widetilde{\sigma} = \lambda_M \cdot \sigma \\ \widetilde{\rho} = \lambda_M \cdot \rho \end{cases}$$

and $t \in T = \{t_0, t_k, t_{2k}, ..., t_{nk}, ...\}$ where t_{nk} is $(nk)^{th}$ packet arrival time.

This curve is the **minimal arrival curve** of the stream.

4   Integrating (m,k)-Firm Constraints into IntServ QoS Model

In the IntServ QoS model, variable-bit rate flows are typically described by the quadruple TSPEC $= (M, p, b, r)$ where M is the maximum packet size of the flow, p is the peak rate, b is the maximum burst size and r is the average long-term rate.

The cumulative arrival function of the VBR traffic is then upper bounded by the arrival curve $\alpha(t) = \min(M + p.t, b + r.t)$. In this QoS model, routers use WFQ and its variants to provide bandwidth guarantee.

By making bandwidth reservation R and a maximum latency T, the maximum delay guaranteed for a flow using WFQ scheduling [5]:

$$D_{max} = \frac{M}{R} + \frac{b-M}{R}\left(\frac{p-R}{p-r}\right)^+ + T \tag{2}$$

This delay bound depends only on the traffic specification (M, p, b, r) and the reserved bandwidth. No temporal constraint is considered. As consequent, if the burst size b is large, the delay may exceed the required deadline of the real-time stream.

Using (m,k)-WFQ scheduler, the actual traffic transmitted would be slightly different from the TSPEC form since the scheduler would drop optional packet missing their deadlines. Hence, we need to estimate the arrival curve of the *effective flow* transmitted by the scheduler which includes all mandatory packets (bits) and the maximum number of optional packets (bits) transmitted by the scheduler. We denote by D_{req} the required delay for each packet of the flow. Then, the burst size of optional packets cannot be larger than $\sigma = D_{req} \cdot r$ since r, the average long-term rate, is the minimum reserved bandwidth. We denote λ_M (resp. λ_0) the ratio of mandatory (resp. optional) bits into the window of k consecutive packets according to its κ-pattern. Figure 1 presents the effective flow model.

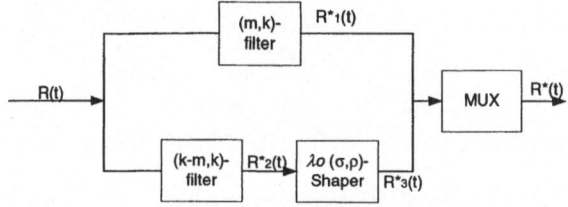

Fig. 1. The effective flow model is obtained using the (m,k)-filter.

The mandatory part of $R(t)$ is the output of the (m,k)-filter $R_1^-(t)$. The optional part of $R(t)$ is obtained when the flow crosses the *(k-m,k)-filter* according to the *reverse κ-pattern* of the stream. The output is denoted by $R_2^-(t)$. Finally, to get the maximum number of optional packets processed by the scheduler (not dropped), the flow $R_2^-(t)$ is shaped by a $\lambda_0(\sigma, r)$ leaky bucket controller to select only optional packets whose deadlines are lower than σ/r. The output $R^-(t)$ represents then the *effective flow*. Denote by $\alpha^-(t)$ its arrival curve. Using *Theorem 1*, we show that the curve of the traffic actually served by (m,k)-WFQ, called *smoothed curve* is (c.f figure 2):

$$\alpha^*(t) = \min\left(M + pt, (\lambda_M M + \lambda_0\sigma) + \rho t, (\lambda_M b + \lambda_0\sigma) + rt\right) \tag{3}$$

Where $\rho = \lambda_M p + \lambda_0 r$.

We denote by $\theta_\sigma = \dfrac{\sigma - M}{p - r}$, $\quad \theta_b = \dfrac{b - M}{p - r}$, $\quad \gamma_1 = \lambda_M b + \lambda_0 \sigma$, $\quad \gamma_2 = \lambda_M M + \lambda_0 \sigma$.

From the figure 2, the smoothed curve is made of three segments as expressed by equation 3. When time interval $\lfloor \theta_\sigma, \theta_b \rfloor$ is very short, a good approximation of this arrival curve is:

$$R^*(t) \sim \min\left(M + pt, (\lambda_M b + \lambda_0 \sigma) + rt\right) \tag{4}$$

Now, we assume that a service curve $\beta_{R,T}(t) = R.(t - T)$ is guaranteed to the effective flow. We propose to derive the delay bound experienced by the effective flow.

If we consider the approximation of equation 4, a direct result of the delay bound, when replacing the TSPEC curve by the approximated smoothed curve, is:

$$D_{max} = \frac{M}{R} + \frac{(\lambda_M b + \lambda_0 \sigma) - M}{R}\left(\frac{p - R}{p - r}\right)^+ + T \tag{5}$$

A finer bound is obtained by considering the three segment curve of equation 3. We show that the delay bound guaranteed by (m,k)-WFQ for a reserved bandwidth R and a latency T (*Figure 2*):

$$D_{max} = \max\left[\frac{M}{R} + \left(\frac{\sigma - M}{R}\right)\left(\frac{p - R}{p - r}\right)^+, \left(\frac{(\lambda_M M + \lambda_0 \sigma)}{R} + \left(\frac{b - M}{R}\right)\left(\frac{\rho - R}{p - r}\right)\right)\right] + T \tag{6}$$

All the details and proofs leading to this result could be found in [4].

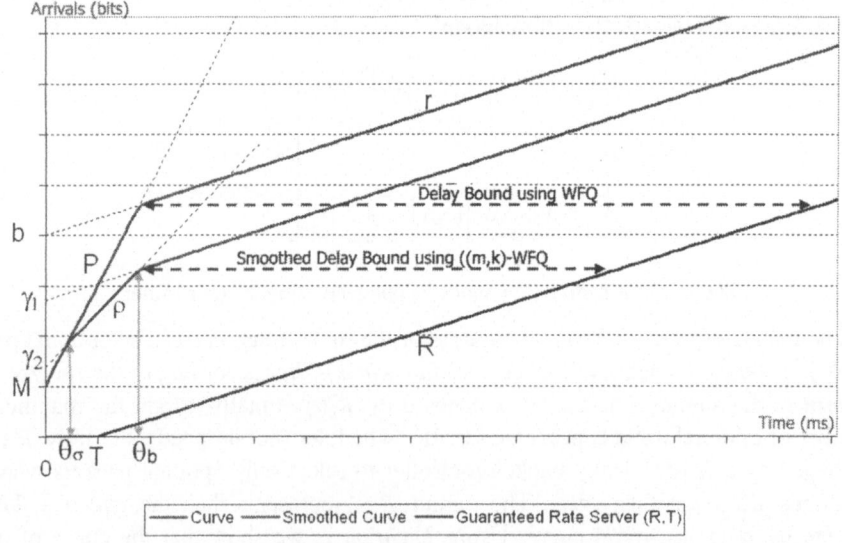

Fig. 2. This figure shows the delay bounds for a flow served by WFQ and (m,k)-WFQ with a bandwidth reservation R and a maximum latency T. The smoothed curve denotes the curve of the effective flow. The delay bound guaranteed by (m,k)-WFQ is lower than that guaranteed by WFQ due to the skipping of optional packets missing their deadlines.

The delay bound guaranteed by (m,k)-WFQ is always lower than that guaranteed by WFQ. Moreover, we can adjust the deadline to drop optional packets in order to make D_{max} equal to the required delay D_{req}. Hence to have $D_{max} = D_{req}$, according to equation 6, the maximum optional burst-size eligible for serving is:

$$\sigma = \min\left\{ \frac{R.(D_{req} - T) - M\left(1 - \frac{(p-R)^{+}}{(p-r)}\right)}{\frac{(p-R)^{+}}{(p-r)}}, \frac{R.(D_{req} - T) - \lambda_M M + (b - M)\left(\frac{\rho - R}{p - r}\right)}{\lambda_o} \right\} \quad (7)$$

Consequently, the maximum delay to serve an optional packet is not more than $D_{op} = \sigma / r$. Therefore, integrating (m,k)-firm timing constraints provides an important flexibility for real-time applications and makes guarantees on both bandwidth and delay. In fact, based on the TSPEC parameter and the (m,k)-firm requirement, an intermediate router makes the bandwidth reservation and adjusts, by using equation 7, the maximum allowed deadline to send an optional packet in order to guarantee the required delay for mandatory packets. Moreover, simulation study made in [4] shows the practical interest of using (m,k)-WFQ for guaranteeing loss-tolerant QoS for MPEG streams.

5   Conclusion

In this paper, we have presented a new approach of QoS that consists in integrating (m,k)-firm real-time guarantees into the Internet QoS model to provide delay guarantee for real-time applications that tolerate some deadline misses. Analytic study shows that (m,k)-WFQ provides lower delay guarantees than that provided by WFQ according to (m,k)-firm constraints.

We are currently working towards integrating (m,k)-firm guarantee into DiffServ QoS architecture to define loss-tolerant service classes using these timing constraints.

References

[1] M. Hamdaoui and P. Ramanathan. "A Dynamic Priority Assignment Technique for Streams with (m, k)-firm Deadline*s". *IEEE Trans. on Computers*, 44 (4), 1443–1451, Dec.1995.

[2] S. Wang, Y. Wang, K. Lin, "Integrating Priority with Share in the Priority-Based Weighted Fair Queueing Scheduler for Real-Time Networks" *Journal of RTS* pp. 119-149, Vol 22, 2002.

[3] A. Koubaa , Y.Q. Song "Amélioration des Délais dans les Réseaux à Débits Garantis pour des Flux Temps-Réel Sous Contrainte (m,k)-Firm" SETITE'03 Mars 2003 Sousse, Tunisie.

[4] A. Koubaa, Y.Q. Song "Integrating (m,k)-Firm Real-Time Guarantees in the Internet QoS Model" INRIA-LORIA Technical Report February 2004.

[5] J.Y. Le Boudec, P. Thiran, *Network Calculus: A Theory of Deterministic Queueing Systems for the Internet* Springer Verlag, July 2002.

Adaptive Channel Allocation in OFDM/SDMA Wireless LANs with Limited Transceiver Resources

Iordanis Koutsopoulos and Leandros Tassiulas

Department of Computer and Communications Engineering,
University of Thessaly, Greece

Abstract. Smart antennas increase capacity of wireless systems by allowing channel reuse by several users in the same cell through space division multiple access (SDMA). Orthogonal Frequency Division Multiplexing (OFDM) creates additional challenges in beam-forming when the beams that can be formed are limited due to transceiver unit limitations. We investigate the impact of smart antennas on channel allocation and propose meaningful heuristics to form beams and assign subcarriers and transceivers to users. Numerical results quantify performance benefits and provide useful design guidelines.

1 Introduction

A smart antenna enables intra-cell channel reuse by many spatially separable users by steering the main lobe of each beam to the desired direction and placing nulls in the directions of interfering users [1]. Smart antennas at the physical layer raise significant issues at higher layers that have not been addressed in literature. For instance, channel allocation has hitherto been studied separately from spatial separation and channel reuse. The latter are based on static cell sectorization or beam switching methods that do not capture channel dynamics, mobility and traffic load variations. Orthogonal frequency division multiplexing (OFDM) is a signaling and access technique for wireless broadband networks that provides immunity to inter-symbol interference (ISI) and high data rates [2]. The spectrum is divided into orthogonal narrow-band subcarriers and the user bit stream is split into subsymbols that are transmitted in parallel over subcarriers.

An unaddressed issue concerns transceiver limitations that are significant in indoor WLANs or WPANs, where implementation complexity and cost, physical space inadequacy or specifications on maximum induced interference may limit the number of beams that can be formed. A first category of studies consider single-channel multi-user systems [3], [4]. For M antennas, M transceiver modules are required to form a beam for each user in the co-channel set. This also holds for multi-channel multi-user systems with time division multiplexing [5], where co-channel sets change in different slots. In OFDM, each of the N subcarriers has different quality for a user due to different impact of frequency on user

N. Mitrou et al. (Eds.): NETWORKING 2004, LNCS 3042, pp. 1384–1389, 2004.

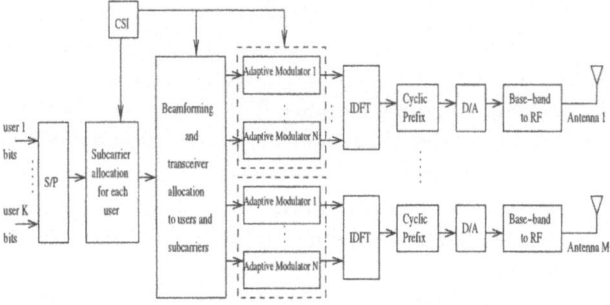

Fig. 1. Multi-user OFDM/SDMA transmitter with limited transceiver resources.

spatial and multi-path characteristics. In single-user systems, N transceivers are required [6]. However, in a multi-user OFDM system, a separate beam may be needed for each user in a co-channel set in each subcarrier and NM transceivers may be needed. Depending on N and M, this number can be of the order of several hundreds and may not comply with the limitations above. In [7], the existence of NM transceivers was implied. We investigate the impact of smart antenna systems with limited transceiver resources on channel allocation. This paper provides a brief overview of the topic. For a more detailed study, the reader is referred to [8].

We consider OFDM transmission with N subcarriers from an access point (AP) with a smart antenna of M elements to K users. The block diagram is depicted in figure 1. Each of the C transceivers can form a unit-power beam $\mathbf{u}_c = \left(u_c^1, \ldots, u_c^M\right)^T$. The expected SIR at the receiver of user k at subcarrier n that receives useful signal from transceiver c is $\mathrm{SINR}_{n,k}^c = (\mathbf{u}_c^H \mathcal{H}_{n,k} \mathbf{u}_c)/(\sum_{b=1,b\neq c}^{C} \mathbf{u}_b^H \mathcal{H}_{n,k} \mathbf{u}_b)$, where $\mathcal{H}_{n,k}$ is called spatial covariance matrix of user k and captures angular and multi-path characteristics such as path gains and delays and their impact on subcarrier n. Matrices $\mathcal{H}_{n,k}$ can be estimated by sending known pilot symbols in the up-link. The BER at the output of the detector in a subcarrier must satisfy BER $\leq \epsilon$. For one modulation level with b_0 bits/subsymbol, the minimum required SIR (in dB) for this BER specification is $\gamma = -(\ln(5\epsilon)/1.5)(2^{b_0} - 1)$.

2   Channel Allocation for OFDM/SDMA Systems with Limited Transceiver Resources

Each user receives useful signal from a transceiver and co-channel interference from beams of other transceivers that use the same subcarrier for other users. The co-channel set is *spatially separable* if there exist beams, one from each transceiver, so that a minimum SIR is ensured for each user. Spatial separability depends on user spatial covariance matrices, on the specific subcarrier and on beam-forming vectors. Clearly, users that are illuminated by one beam must use different subcarriers. Users that are served by different beams may or may not reuse a subcarrier, depending on user spatial and multi-path characteristics at the subcarrier, beam orientations and induced co-channel interference by beams. A user experiences co-channel interference from beams of other transceivers that use the same subcarrier. Therefore, subcarrier and transceiver assignment are coupled. Our algorithms consist of two stages. First, users are assigned to sub-carriers and beams are computed assuming there are no transceiver limitations. In the second stage, the beams are sequentially unified, until the desired number of C beams is reached.

In the first stage, the idea is to create large co-channel sets of spatially separable users in each subcarrier. An appropriate user is sequentially assigned to a subcarrier and beams of co-channel users are adjusted so that acceptable SIRs are ensured. Inserted users should cause least interference to users that are already assigned in the channel and should receive least interference from them. For each inserted user k, we evaluate a factor

$$\Phi_{n,k} = \frac{\mathbf{u}_{n,k}^{*H} \mathcal{H}_{n,k} \mathbf{u}_{n,k}^*}{\max\left\{\mathbf{u}_{n,k}^{*H}(\sum_{j\in\mathcal{U}^{(n)}} \mathcal{H}_{n,j})\mathbf{u}_{n,k}^*, \sum_{j\in\mathcal{U}^{(n)}} \mathbf{u}_{n,j}^{*H} \mathcal{H}_{n,k} \mathbf{u}_{n,j}^*\right\}}, \tag{1}$$

where beams maximize the ratio of created useful and interference signals. This factor captures the requirement of high useful signal power and least caused and received interference to or from other users in n. The assignment of a user with the maximum $\Phi_{n,k}$ is desired. Users are sequentially inserted in each subcarrier until one SIR is violated and this is performed for all subcarriers.

In the second stage, we need to reduce the number of beams to C while maintaining high subcarruer reuse. Clearly, only beams from different subcarriers can be unified to a new beam, since the new beam cannot serve users of the same subcarrier. At each iteration of the unification algorithm, we need to select an appropriate beam pair from different subcarriers n, m and replace it with a new beam. The rationale for selecting the beam pair is to *combine beams of different subcarriers with similar orientations*, so that desirable properties of old beams are maintained. Thue, the beam pair with minimum Euclidean distance is selected. For normalized beams, this is equivalent to $(k^*, \ell^*) = \arg\max_{(k,\ell)} \Re(\rho_{k\ell})$, where $\rho_{k,\ell}$ is the beam cross-correlation. The new beam $\mathbf{u}_c$ that replaces old beams $\mathbf{u}_{n,k}$ and $\mathbf{u}_{m,\ell}$ can be computed with one of the following methods.

Method A: Maximum new/old beam cross-correlation. New beam $\mathbf{u}_c^*$ should have maximum cross-correlation with old beams, subject to being normalized and is thus the solution of a constrained optimization problem. After computing $\mathbf{u}_c^*$, we replace $\mathbf{u}_{n,k}$ and $\mathbf{u}_{m,\ell}$ with $\mathbf{u}_c^*$ and evaluate user SIRs in k, ℓ and SIRs of other users in old beams which are affected by the replacement. If all SIRs exceed γ, we replace beams $\mathbf{u}_{n,k}$ and $\mathbf{u}_{m,\ell}$ with beam $\mathbf{u}_c^*$ and proceed to selecting the next beam pair. Otherwise, some beams (and thus users served by these beams) in subcarriers n and m must be removed, so that SIRs increase. We eliminate the user that maximizes the minimum SIR, since we intend to maintain a large number of users with acceptable SIRs. This continues until SIRs of all users are acceptable. Then, the algorithm proceeds to selecting the next beam pair and the procedure terminates when the number of beams is C.

Method B: Maximum signal strength/minimum induced interference. The new beam $\mathbf{u}_c^*$ must maximize the ratio of total useful power for users k and ℓ in initial beams and total interference to other users. After computing $\mathbf{u}_c^*$, SIRs of users are calculated and users are sequentially eliminated, until acceptable SIRs are ensured.

Note that as the algorithm progresses, one or both beams that are unified may contain several users in different subcarriers or a user with several subcarriers. All such users are included in the algorithm.

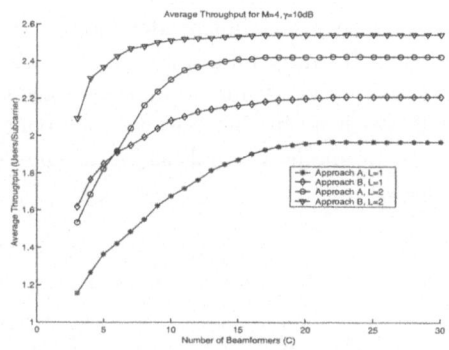

Fig. 2. Average throughput vs. number of transceivers for approaches A and B, for multi-path with $L = 1$ and $L = 2$ paths and $M = 4$ antennas.

3   Simulation Results

We simulate a system with an AP, $N = 10$ subcarriers and $K = 15$ users. In both of the considered approaches, the first stage is executed and the beam pairs for unification are selected with the cross-correlation criterion. The differences of the approaches are as follows. In approach A, the new beam is computed with method A. Next, beams are sequentially eliminated, until SIRs of remaining users exceed γ. In B, the new beam is computed with method B. Now, after beam elimination, a new beam is computed again with a ratio similar to that in method B. This iterative process of beam elimination and new beam computation stops when user SIRs are acceptable. We measure average subcarrier throughput, namely average number of assigned users per subcarrier. In figure 2, this is illustrated as a function of number of transceivers for $M = 4$ antennas, for different multi-path (number of paths, L) and $\gamma = 10\,\text{dB}$. We see that for the same multi-path conditions, approach B always performs better than A due to the iterative beam computations as opposed to A, where beams are computed once. Moreover, different criteria for computation of new beam were used in A and B. For relatively small C, approach B outperforms A by 20%, while for larger C, B is better than A by 4%. Note that the throughput with $L = 2$ is alaways larger than that for $L = 1$ due to the additive effect of multi-path.

The most significant observation is that performance improves as C increases until C reaches a limit value C^*, beyond which no further improvement occurs. Then, the system has reached spatial separability performance limits and cannot accommodate more users in the same subcarriers. Thus, for A and B and $L = 1$, we have $C_A^* = 17$ and $C_B^* = 13$ with corresponding throughput of 1.9 and 2.18 users per subcarrier. For $L = 2$, it is $C_A^* = 12$ and $C_B^* = 9$ with throughput of 2.4 and 2.5. It was also observed that the benefit of B over A increases as M increases and decreases as L increases. Furthermore, the number C^* is almost proportional to M. Although in a realistic system N and K are larger, performance is still determined by subcarrier reuse, which in turn depends on spatial and multi-path channel properties of users, beamforming and resource allocation policy. For a given BER requirement and a given value of M, there exists a crucial number of transceivers C^*, beyond which no further performance benefits are anticipated. Viewed differently, C can be made as small as C^* at the design stage with no incurred performance losses.

4   Discussion

In this study, we identified the particular features of the coupled resource allocation problem and demonstrated the impact of smart antenna systems with transceiver limitations on the design of MAC layer algorithms. There exist several directions for future study. A first issue is establishing a theoretical framework for determining the limiting system performance. A more general treatment could include power adaptation. Finally, the analogy between the addressed assignment problem and the scheduling problem at the packet level (where the transceiver is viewed as a server) is worth considering. The arising issue is the identification of eligible user activation sets for each subcarrier by considering spatial separability, number of subcarriers per user and queue lengths. Several existing ideas and stabilizing scheduling policies could be extended to such generalized scheduling problems.

References

1. K. Sheikh, D. Gesbert, D. Gore and A. Paulraj, "Smart antennas for broadband wireless access networks", *IEEE Commun. Mag.*, vol.37, no.11, pp.100-105, Nov. 1999.
2. T. Keller and L. Hanzo, "Adaptive multicarrier modulation: a convenient framework for time-frequency processing in wireless communications", *Proc. IEEE,* vol.88, no.5, pp.611-640, May 2000.
3. F. Rashid-Farrokhi, K.J.R. Liu and L. Tassiulas, "Transmit beamforming and power control for cellular wireless systems", *IEEE J. Select. Areas Commun.*, vol.16, no.8, pp.1437-1450, Oct. 1998.

4. C. Farsakh and J.A. Nossek, "Spatial Covariance based downlink beamforming in an SDMA mobile radio system", *IEEE Trans. Commun.*, vol.46, no.11, pp.1497-1506, Nov. 1998.
5. F. Shad, T.D. Todd, V. Kezys and J. Litva, "Dynamic slot allocation (DSA) in indoor SDMA/TDMA using a smart antenna basestation", *IEEE/ACM Trans. Networking*, vol.9, no.1, Feb. 2001.
6. K.-W. Wong, R.S.-K. Cheng, K. Ben Letaief and R.D. Murch, "Adaptive antennas at the mobile and base stations in an OFDM/TDMA system", *IEEE Trans. Commun.*, vol.49, no.1, pp.195-206, Jan. 2001.
7. I. Koutsopoulos and L. Tassiulas, "Adaptive resource allocation in wireless broadband networks with OFDM signaling", *Proc. IEEE INFOCOM 2002* vol.3, pp.1376-1385.
8. I. Koutsopoulos, "Resource Allocation issues in wireless broadband networks with OFDM signaling", Ph.D Dissertation, University of Maryland, College Park, December 2002.

Fast End-to-End Mobility Support Using SIP for Vertical Handoffs in 4G Wireless Communication Networks

Jung-Ho Lee, Sang-Hee Lee, Ki-Nam Seo, and Jae-Sung Lim

Graduate School of Information and Communication, Ajou University,
San 5, Wonchon-dong, Paldal-gu, Suwon 442-749, Korea
jackdl@hanmail.net, {dreami, knseo, jaslim}@ajou.ac.kr

Abstract. The mobility management is a significant issue in 4G heterogeneous networks. However, the mobility needs to be managed by upper layer in the networks because of the limitation of mobility management at link layer. Session Initiation Protocol (SIP) supports personal mobility so that a user can be found independent of location and network device. But it cannot support IP mobility and cannot maintain TCP connection when mobile node (MN) moves between different subnets during a session. In this paper, we propose a new mobility management scheme based on SIP to overcome mobility support problems of SIP. In the proposed scheme, an MN early gets an IP address that will be used in a new subnet and informs its correspondent node (CN) of the address. To minimize packet losses, the MN freezes data transmission during a handoff. The proposed scheme is compared with the previous mobility management scheme through computer simulations.

1 Introduction

With the start of 3G wireless services, people in both academia and industry have begun to show more and more interests in 4G wireless communications. The 4G mobile communication systems are characterized by heterogeneous access networks and IP-based transport technologies. The systems also provide voice and multi-media services as well as ultra-high speed Internet based on All-IP network. So, the systems have to support not only horizontal handoffs but also vertical handoffs, global roaming, and seamless services with service negotiation including mobility, security, and QoS. But, the global mobility between different access networks should be managed by upper layers because the mobility support of link layer has limitations.

SIP[1], a signaling protocol, was accepted for the next-generation standards in both cdma2000 and UMTS 3G wireless networks[2]. Also, it has been developed as a solution for mobility management of the application layer for real-time services.

[3] proposed to extend SIP with functions that would support the terminal mobility, alleviating some of the shortcomings associated with Mobile IP and its route optimization variants. It can apply over UDP, but it can not support TCP. Supporting terminal mobility for TCP with SIP requires a tracking agent on MN to maintain a record of its ongoing TCP connection, as well as, IP encapsulation capability on CN[4]. But, when an MN moves into another subnet, packet losses cannot be inevitable because of time delay that the MN is assigned with a new IP address and informs the CN of the

N. Mitrou et al. (Eds.): NETWORKING 2004, LNCS 3042, pp. 1390–1394, 2004.
© IFIP International Federation for Information Processing 2004

IP address. Also, the regular TCP tuned to perform well in traditional wireline networks leads to significant throughput degradation in the mobile networks because packet losses usually occur due to handoffs rather than congestions.

In this paper, we introduce a new mobility management scheme using SIP for vertical handoffs in wireless communication networks. The proposed scheme supports route optimization and minimizes network costs by providing fast SIP mobility control for movement of MN with a TCP connection. Also, the scheme can guarantee QoS by minimizing packet losses during the handoff because of being based on the TCP freeze scheme. As a result of the simulation, the proposed scheme showed better performance in terms of throughput than the previous scheme using SIP.

The rest of this paper is organized as follows. In the next section, we mainly present operations of fast end-to-end mobility support using SIP. Then we evaluate performance of the proposed scheme by computer simulations in section 3. Finally, concluding remarks are given in section 4.

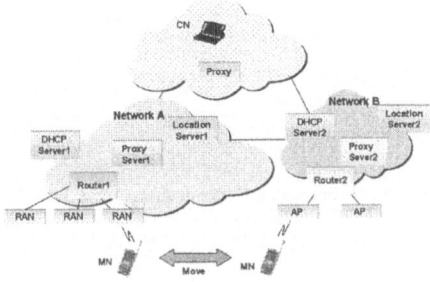

Fig. 1. Network model for proposed scheme

2 Fast End-to-End Mobility Support Using SIP

A network model for the proposed scheme is shown in Figure 1. This model allows for loosely coupled interworking. It also can be evolved into networks with tight coupled interworking in heterogeneous networks[5]. The MN should support dual mode that can detect two different frequencies. In the paper we don't consider Authentication, Authorization, and Accounting (AAA) between different network providers. A location server (LS) stores Access Network Identifiers (ANIDs) which are used as a layer 2 trigger to identify network areas and information of serving MNs. In this paper, SIP INFO[6] message is appropriately used to send handoff information.

As depicted in Figure 2, the proposed scheme is operated for vertical handoffs. At first, if an MN moves into a boundary of the domains, it receives ANID from the AP in the handoff zone. The MN sends this information to the proxy server 1 (PS1) in the current service area after putting the information in the body part of SIP INFO message. At this time, the SIP INFO message contains the SIP ID of the MN in From and To fields. Receiving this message, the PS1 asks the LS whether this ANID is one of areas which the PS1 manages. If the PS1 can not manage the area, it will reply with the information of other PSs around itself. Receiving the reply, PS1 forwards the

ANID to surrounding PSs for searching the network with the same ANID. If a PS which manages the area of the ANID receives it, the PS gets a new IP address which the MN will use in the area by a DHCP server and sends the IP address to the MN. At the time, the IP address is delivered to the MN being contained in body part of the SIP INFO message. As soon as the MN is assigned with the new IP address, the SIP module on the MN signals its TCP, the ACK is sent to the sender(CN) with the advertised window field set to zero, and the sender enters the persist mode. On sending the ACK with the advertised window field set to zero, the MN informs the CN of the new IP address by sending SIP INFO message. After the MN moves to the target network and establishes an air link, it sends an SIP INFO message representing completion of handoff to the CN. Then as the MN sends the ACK with advertised window field set to the previous value, the sender can come out from the persist mode.

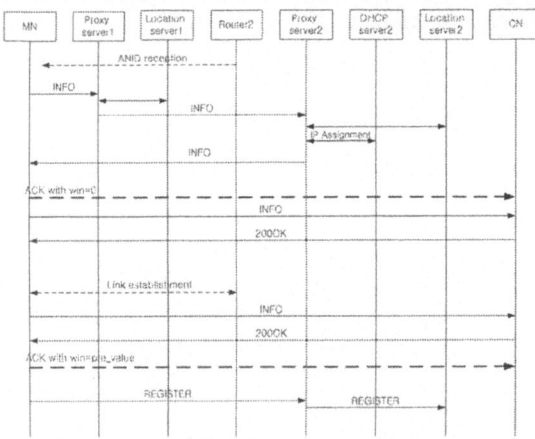

Fig. 2. Message flow in proposed scheme

3   Performance Analysis

In this paper, NS-2 was used for computer simulations[7]. We made a comparison between the proposed scheme and the previous mobility management scheme[4]. The simulation model is shown in Figure 3. The numbers beside a link indicate the bandwidth and delay of the link. We assume that an MN moves around between four domains, and one FTP connection from the CN to the MN is considered. It is also assumed that one of 4 domains is a hot spot area and others are cellular networks.

Figure 4 shows the throughput of the proposed scheme and the previous scheme according to average velocity of the MN. It is shown that the proposed scheme can overcome significant degradation in throughput of TCP because it freezes data transmission during vertical handoffs. As the average velocity is increased from 20km/h to 120km/h, the throughput is decreased because the faster the MN is, the more the MN crosses handoff zones.

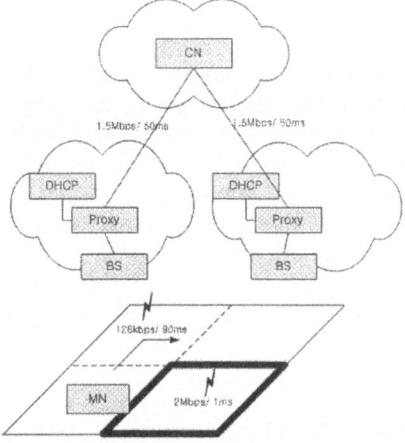

Fig. 3. Simulation Model

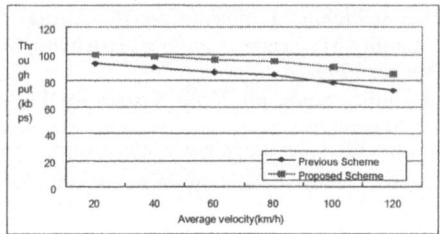

Fig. 4. Throughput according to average velocity of MN

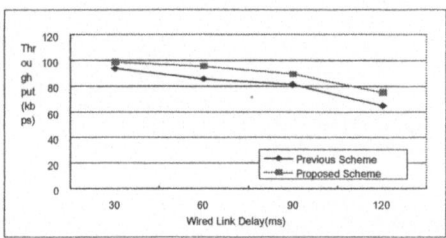

Fig. 5. Throughput according to wired link delay

Figure 5 shows the throughput according to increase of wired link delay when the average velocity of the MN is 50km/h. As the wired link is longer, the throughput is decreased because the time that the CN receives a new IP address from the MN becomes longer. As a result, the proposed scheme shows better throughput than the previous one.

4 Conclusion

In this paper, we have proposed a fast end-to-end mobility management scheme using SIP for vertical handoffs in 4G wireless communication networks. In the proposed scheme, an MN early obtains a new IP address that will be used after handoff and it makes TCP sender freeze data transmission to avoid timeout and slow start during vertical handoffs. So, the proposed scheme can guarantee QoS by minimizing packet loss. As results of computer simulations, it was shown that the scheme yielded better performance than the previous scheme using SIP.

References

1. J. Rosenberg et al., "SIP: Session Initiation Protocol", IETF RFC 3261, June 2002.
2. G. Patel and S. Dennett, "The 3GPP and 3GPP2 Movements Toward an All-IP Mobile Network", IEEE Personal Communications, vol. 7, no. 4, pp. 62-64, Aug. 2000.
3. E. Wedlund and H. Schulzrinne, "Mobility support using SIP", The second ACM International workshop on Wireless Mobile Multimedia, pp. 76-82, Aug. 1999.
4. F. Vakil et al., "Host Mobility Management Protocol Extending SIP to 3G-IP Networks", IETF Internet Draft, October 1999.
5. 3GPP TR 22.934, "Feasibility Study on 3GPP System to WLAN Interworking", Dec. 2002.
6. S. Donovan, "The SIP INFO Method", IETF RFC 2976, October 2000.
7. Network Simulator version 2 (NS-2). URL: http://www.isi.edu/nanam/ns/.

Optimal Monitoring Equipment Placement for Fault and Attack Location in Transparent Optical Networks

Carmen Mas Machuca and Ioannis Tomkos

Athens Information Technology (AIT) Center, Markopoulo Av, PO. BOX 68, 19002 Peania,
Athens, Greece
{cmas,itom}@ait.edu.gr

Abstract. Last decade, network security has become a very sensitive and important topic for equipment manufacturers and network operators. Physical layer security in opaque optical networks relies on the information obtained at the opaque nodes where the signal is electronically regenerated. However, in transparent optical networks, security is even more complex since the optical signals are not regenerated and, therefore, faults and attacks are more difficult to be detected and isolated. Our work deals with the study of the optical placement of monitoring equipment which may help operators to optimize the investment on their equipment while increasing the accuracy to locate the faults and attacks (so-called failures). For this purpose, we have extended our Transparent Failure Location Algorithm (TFLA), which is able to locate failure(s) in transparent optical networks in presence of false and/or lost alarms to propose an optimal location for new monitoring equipment. and tested it for the Pan-European network.

1 Introduction

Network management involves configuration, performance, fault, accounting and security management functionalities. Fault management relies on the information retrieved from the monitoring equipment in order to detect faults and react to them. Opaque networks allow supervising the signal at each opaque node where the optical signal is converted to the electrical domain. However, in transparent networks the data remains in the optical domain all along its path (i.e. without going through any optical-to-electrical conversion but through optical amplification and optical switching in the near future, and optical regeneration and conversion further in the future) and the optical signal is more exposed to degradation without being noticed by the network management system. On the other hand, the information received by the network management system is more limited in transparent networks as it relies on analog signal measurements at some points of the network, whereas in opaque networks per bit or BER performance based monitoring is available. However, we should point out that when the transmission speed increases, BER monitoring becomes very expensive as it requires high speed signal processing.

Fault management deals with the prevention, detection, and reaction to faults. Prevention deals with the component and network design so that it can prevent faults. When the fault has occurred, detection takes care of learning about the existence of

N. Mitrou et al. (Eds.): NETWORKING 2004, LNCS 3042, pp. 1395–1400, 2004.

the fault and to identify it. Finally, reaction manages to restore the connections that have been disturbed by the fault. All these functionalities become even more important in optical networks because of (i) the high bit rates that cause a huge amount of information to be lost (ii) the high latency of the network that allows a lot of data to get into the network when the fault occurs, (iii) the fault identification that should be efficient and exact in order to restore the connections and isolate the fault efficiently[1]. Fault identification is based on the alarms received by the network management system and it should cope with the existence of false and/or lost alarms. When there are two or more simultaneous faults, the number of alarms increases considerably, the alarms arrive intermingled to the management system, and the problem of locating the faults becomes even more difficult.

Fault management can be extended to also cover attacks. Attack can be defined as an intentional action against the ideal and secure functioning of the network. Attacks can be classified as eavesdropping or service disruption[2]. Hence, we can define as failure the set of faults and attacks that can interrupt the ideal functioning of the network. In this paper, we will present an algorithm to locate failures (both faults and attacks) in transparent optical network in presence of false and/or lost alarms and some results when the algorithm is applied to the Pan-European network.

The paper is organized as follows. Sect. 2 introduces transparent optical networks including an overview of their components and an example of a possible attack. Sect. 3 presents the Transparent Failure Location Algorithm (TFLA) which includes the methodology of the algorithm. Sect. presents some results on the study of the optimal location of new monitoring equipment. Finally, Sect. 5 concludes the paper.

2   Performance Monitoring in Transparent Optical Networks

In transparent optical networks the signal remains in the optical domain along its path without going through any optic-to-electric conversion. These networks are very promising as they reduce unnecessary, expensive optoelectronic conversions, offer high data-rate, provide flexible switching, and support multiple types of clients (different bit rates, modulation formats, protocols, etc.).

Transparent optical networks contain two classes of network components: (i) Optical components which take care of the optical signal transmission and are not able to generate alarms, and (ii) Monitoring equipment (ME) which is able to generate alarms and notifications when the optical signal is not the expected one. The alarms generated by monitoring equipment depend on the kind of equipment and its characteristics. The failure of the monitoring equipment does not interrupt/modify the data transmission and therefore their failure is not as relevant as the failure of an optical component. Moreover, when monitoring equipment fails, it may result in the loss of alarms which will be considered in the proposed algorithm as lost alarms.

As discussed previously, transparent optical networks are more vulnerable to failures than opaque networks because (i) the quality of the optical signal is not evaluated at each node and (ii) a single failure can affect more channels than in opaque networks, as there are no transparency boundaries supported by optoelectronic regenerators. An example of this is shown in Fig. 1. In this scenario, an attacker inserts optical power at a wavelength that is already used ($\lambda 2$). This attack will cause an increase of the optical power at that wavelength that will disturb neighbouring channels (e.g. when traversing an optical amplifier such as EDFA, the gain that $\lambda 2$ channel will experience will be greater than the gain of $\lambda 1$ channel (case a of Fig. 1)). Even after filtering channel $\lambda 1$ at the wavelength demultiplexer, there is some residual optical power at $\lambda 2$ higher than the one specified in the system, so it can degrade the performance of its neighbouring channels (case b of Fig. 1). When there are optical switches, crosstalk is very critical. In our example, $\lambda 2$ channel of Fibre Nf could be disturbed by $\lambda 2$ channel of Fibre 1 due to crosstalk (case c of Fig. 1). The degree of crosstalk is closely related to the optical power pumped by the attacker.

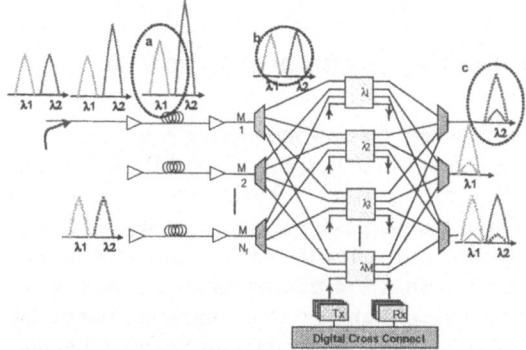

Fig. 1. Example of an attack on an Optical Cross-connect (OXC) with wavelength selective architecture and its propagation through different network components

The attacks that may occur in a transparent optical network can be classified into four categories[3] based on the effects they inflict on the signal: power drop (e.g. power decrease and cut), in-band jamming (including also intrachannel crosstalk), out-band jamming (including also interchannel crosstalk and non-linearities), and wavelength misalignment.

3   Transparent Failure Location Algorithm

The failure location algorithm has to be able to locate the optical component or set of optical components that cause the received alarms in case of failure. The problem of locating multiple failures has been shown to be NP-complete even in the ideal scenario of receiving each of the expected alarms[4]. The complexity increases further when lost and false alarms are considered. In order to minimize as much as possible the time to locate failure(s), the complexity of the proposed algorithm has been concentrated in a Pre-Computational Phase (PCP) so that the computation to be carried out when new alarms reach the manager is kept as small as possible. The second phase, which is called Core Phase (CP), consists just in traversing a simple binary tree when alarms reach the manager. The proposed PCP has been implemented on the basis of an algorithm used to locate multiple faults in non-ideal opaque networks[5]. This algorithm uses an important concept *Domain(component)* which is defined as the set of network components that will generate an alarm when this component fails. The result of the PCP is a binary tree with a depth equal to the number of monitoring components and whose leaves correspond to different failure scenarios with an accepted number of false and lost alarms (*mismatching threshold*).

We have extended this algorithm to the case of transparent networks (so-called Transparent Failure Location Algorithm or TFLA)[3]. In this case the algorithm computes as many domains for each component as types of failures (four in our case: power drop, in-band and out-band jamming, and wavelength misalignment). The extended algorithm includes the proposal of an optical location for new monitoring equipment as presented in next section.

4   Optimal Monitoring Equipment Location

The optimal location problem for new monitoring equipment (ME) has been studied. We define *optimal location* as the position of the ME that minimizes the number of network elements that are candidates to have a failure i.e. that minimizes the result given by the TFLA. For this purpose, the algorithm while computing the domains of all the optical components, it stores the series *(Xa,Yb)* with the highest number of components located between *Xa* (a transmitter or the first optical component right after a monitoring equipment), and *Yb* (the following monitoring equipment). This length is so-called MSL standing for Maximum Segment Length. By definition, the optimal position for a new ME will be the one that divides the series *(Xa,Yb)* in two series *(Xa,Xc)* and *(Xc+1,Yb)* with *Lac* and *Lc+1 b* as close as possible. In this way, after including this ME in this position, the network components that are candidates to be faulty will be certainly reduced.

4.1 Established Channels Based

For long term channels, the extended TFLA was run on the Pan-European Topology network[6] within a ring between Madrid, Barcelona, Lyon, Paris and Bordeaux (Fig. 2) that is assumed to be transparent. Optical Add/Drop Multiplexers (OADMs) are located at the cities of Madrid, Barcelona and Bordeaux, whereas Optical Cross-Connects (OXCs) are located at Paris and Lyon. The assumed architecture for the OXCs and OADMs is Wavelength Selective (as shown in Fig. 1).The number of amplifiers needed for each link depends on the distance between the cities. Due to the overall ring length, optical regeneration is needed in some nodes (Barcelona, Paris and Bordeaux). Three different channels have been considered: Ch. 1 from Barcelona to Madrid, Ch.2 from Barcelona to Bordeaux via Madrid, and Ch. 3 from Madrid to Paris via Bordeaux. Three cases have been compared:

Case 1: A single ME is installed at the end of each channel.

Case 2: One ME is installed at the location proposed by TFLA (at Madrid's node).

Case 3: One new ME (one more than Case 2) is installed at the location proposed by TFLA, which is at the output of Bordeaux's node.

The TFLA was run for the three cases considering that there are no false or lost alarms. The number of optical components that could be faulty was studied for two scenarios and plotted in Fig. 3:

Scenario I: when receiving an alarm from the receiver of Ch. 2 at Bordeaux and

Scenario II: when receiving two alarms issued by the monitoring equipment located when dropping Ch. 1 at Madrid and Ch. 2 at Bordeaux.

For both scenarios an important reduction on the number of candidates to be faulty is shown (e.g. 90% less in Scenario II when one ME was included).

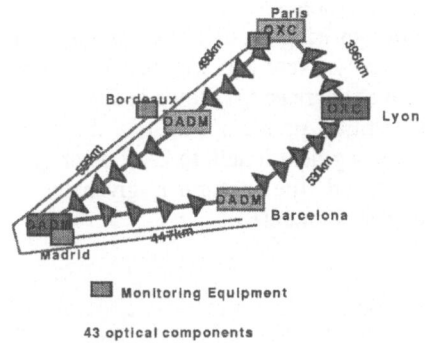

43 optical components

Fig. 2. Considered European transparent ring including the amplifiers and the regeneration nodes needed and the considered ME

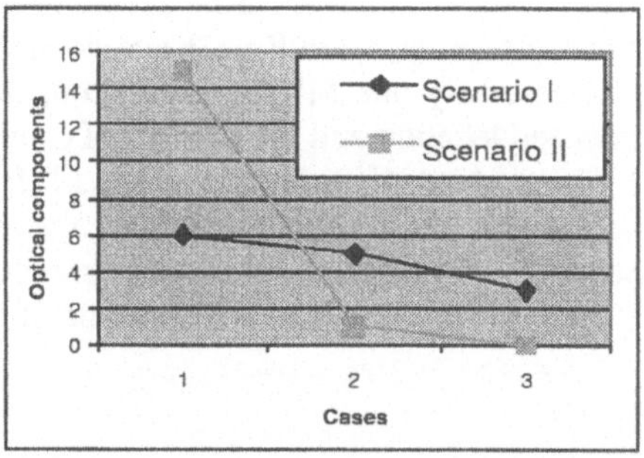

Fig. 3. Graph showing the decrease of the number of optical components that are candidate to have a failure when including new ME

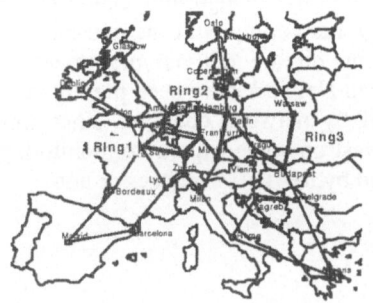

Fig. 4. Pan-European network with three interconnected rings

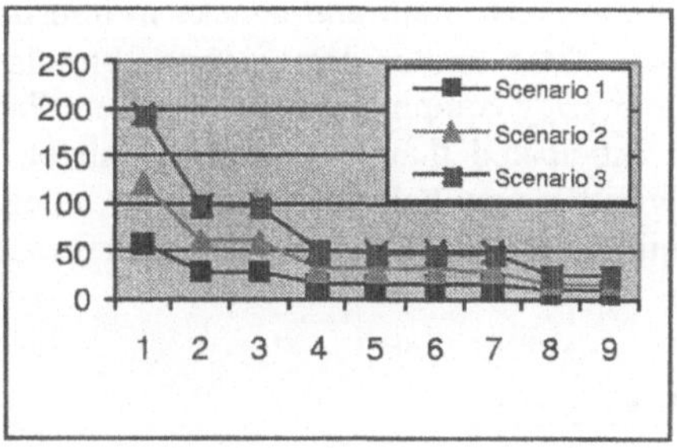

Fig. 5. Decrease of the maximum segment length as a function of the installed ME

4.2 Topology Based

The previous work shows the improvement on the failure location when new ME is located where the TFLA proposed based on the established channels. However, network operators may be more interested in the location of new ME based on the network topology rather than the channel based approach. The reason is that the established channels are not fixed and may change with the time, and hence, if we had optimized the ME location for a particular set of established channels, it won't be optimal if the set change.

The extension of the TFLA could be used for any topology if we are able to find the longest channel that can be established in the given topology. We focused on the case of interconnected Pan-European rings shown in Fig. 4 and tested three scenarios: Scenario 1 with a single ring (Ring 1) of 57 network elements, Scenario 2 with a double ring (Ring 1 and Ring 2) of 119 optical components, and Scenario 3 with a triple ring (Ring 1 and Ring 3 interconnected through Ring 2), of 193 components. In all cases, only one ME was been considered initially. New ME was installed in the proposed location and the number of candidates was decreasing (shown in Fig. 5).

The problem of minimizing the MSL is a partition problem with rate 2. For all these scenarios, in order to decrease the maximum segment length to 2^{-n} of its original value, $2^n - 1$ monitoring equipment should be installed (the location is given by the extended TFLA and it is in the middle of the longest segment).

5   Conclusions

This paper described the fault location problem in transparent optical networks and its extension to failure location and optimal placement of new monitoring equipment. Transparent networks are more vulnerable than opaque networks to failures due to the absence of electrical conversion of the optical signal and the cost of the optical monitoring equipment. Some work has been presented on attack location but it was limited to specific network components. We have proposed an algorithm called TFLA able to locate fault and attacks in transparent networks coping with the existence of some false and/or lost alarms. Simulation results on the optimal placement of monitoring equipment in a transparent Pan-European Network have been presented. It is anticipated that these results can be exploited by network operators since this algorithm may help them to decide whether to invest on some expensive monitoring equipment depending on whether the result returned by the TFLA reduces or not.

References

1. M. Medard et al., "Node wrappers for QoS monitoring in transparent optical nodes", Journal of High Speed Networks, Vol. 10, 2001, pp. 247-268
2. M. Medard, D. Marquis, R. A. Barry, and S. G. Finn "Security Issues in all-optical Networks" IEEE Network, May/June 1997, pp. 42-48.
3. C. Mas, I. Tomkos and O. K. Tonguz., "Optical Network Security: A Failure Management Framework" ITCom 2003, 5247 Session, pp. 230-241, Sept. 2003.
4. N. S. V. Rao. "Computational Complexity Issues in operative Diagnosis of graph-based systems", IEEE Transactions on Computers, 42(4), April 1993
5. C. Mas and P. Thiran "An efficient algorithm for locating soft and hard failures in WDM networks" JSAC Special Issue on protocols and architectures for next generation WDM optical networks, Vol. 18, Oct. 2000.
6. S. De Maesschalck et al., "Reference Scenario for a Pan-European Network" COST 266 Report, August 2002

Introducing Service Differentiation in a Bluetooth Piconet

Antoine Mercier[1] and Pascale Minet[2]

[1] Ecole Centrale d'Electronique, LACCSC, 53 rue de Grenelle, 75007 Paris, France
mercier@ece.fr
[2] INRIA, Rocquencourt, 78153 Le Chesnay, France
pascale.minet@inria.fr

Abstract. In this paper, we focus on the Bluetooth wireless network, analyzing its ability to support Quality of Service (QoS) requirements defined by the application. We focus on two QoS parameters : (i) an application constraint denoting the importance degree of a message, and (ii) an end-to-end delivery deadline. In a first step, we introduce a local scheduling accounting for the two QoS parameters in One-Round Robin (1-RR). We evaluate the performances of this enhanced version. In a second step, we define a new Bluetooth global scheduling, called Class-Based Earliest Deadline First (CB-EDF). It takes into account the two QoS parameters at the global level. Simulation results show that CB-EDF achieves a better service differentiation than enhanced 1-RR and allows the coexistence of messages with different application constraints.

1 Introduction

Bluetooth [1] is defined as a solution for ad-hoc networking using polling scheme as medium access control. Bluetooth devices are organized into piconets where a central device acts as the master, which solicits packets transmission from up to seven active slaves. In this paper, we study how Bluetooth devices account for Quality of Service (QoS) requirements, expressed by QoS sensitive applications. More precisely, we focus on the scheduling mechanisms used to support these QoS requirements. We distinguish two message scheduling mechanisms:

- Local scheduling refers to the mechanism selecting one message in the local waiting queues of the Bluetooth device considered.
- Intra-piconet scheduling refers to the polling scheme within a piconet. It is defined in [3] as the set of rules that determines when the piconet master switches from one slave to another. The combination of local and intra-piconet schedulings is referred to as global scheduling.

In order to support QoS applications, real-time constraints must be accounted for by both local and intra piconet schedulings. We report simulations of the native One-Round Robin (1-RR). We focus on the support of two application constraints: (i) the first one denoting the importance degree of a message, and (ii) the second referring to the message delivery deadline. As expected, 1-RR does not achieve service differentiation:

N. Mitrou et al. (Eds.): NETWORKING 2004, LNCS 3042, pp. 1401–1407, 2004.

- High importance messages are transmitted after low importance ones.
- Messages with close deadlines are transmitted after messages with the same importance and far deadlines.

To improve performances of 1-RR, we introduce a local scheduling that takes into account these two constraints by a combination of class based Priority Queuing (PQ) - a class contains all the messages of the same importance degree - and Earliest Deadline First (EDF). PQ is used between classes and EDF is used within a class. This local improvement allows a better service differentiation. However, it does not guarantee that the message belonging to the highest priority class is transmitted first. To achieve that, classes and deadlines have to be accounted for by the intra-piconet scheduling. We then design an optimal solution technically unrealizable, where the master has a complete knowledge of pending messages in the piconet. This solution is uniquely given as a reference for comparative evaluation. We finally define a global scheduling mechanism, called Class-Based Earliest Deadline First (CB-EDF), that supports the two previous application constraints. Performances of CB-EDF are evaluated by simulations.

This paper is organized as follows. In Section 2, we consider the local and intra-piconet problem and present a brief state of the art describing the main polling schemes. In section 3, we introduce into 1-RR a local scheduling accounting for classes and deadlines. Performances of this enhanced version are compared with those of the native one. In Section 4, we present our solution, CB-EDF and compare its performances with those of the optimal solution.

2   State of the Art

In this section, we present a brief state of the art of local and intra-piconet schedulings and give a classification of main intra-piconet schedulings. In this paper, we consider only Asynchronous ConnectionLess links, because they offer better flexibility [1]. We recall that the master maintains at least one queue per slave, denoted $M \to S$ and each slave maintains at least a slave-to-master queue, denoted $S \to M$.

2.1   Local Scheduling

Bluetooth specifications [1] recommend the use of FIFO for local scheduling. We preconize here the use of classes and deadlines to account for the message importance and deadline. The local scheduling is then class based. Moreover, we recommend the use of EDF algorithm within each class, as it has been proved optimal for a uniprocessor scheduling in both preemptive and non-preemptive context when the message arrival times are not known a priori[1].

---

[1] K. Jeffay, D. F. Stanat, C. U. Martel: On non-preemptive scheduling of periodic and sporadic tasks, IEEE Real-Time Systems Symposium (1991), San-Antonio, Texas.

2.2   Intra Piconet Scheduling

Polling schemes designed or extended for Bluetooth technology can be classified according to three main criteria. The first criterion allows determining which slaves are polled. In full polling mechanism such as 1-RR or ERR [3], all slaves in the piconet, even if they have nothing to transmit, are polled. In activity based polling (e.g. FEP in [4]), the polling cycle can be adjusted according to the activity of each slave. Slaves with empty queues are less frequently polled to prevent the bandwidth wastage. The second criterion corresponds to the slave polling order. This order can be determined to achieve fairness access between slaves or to maximize throughput. Static or dynamic priority can be assigned to each slave to determine the polling order. This priority can be mapped:

- In case of full polling:
    - arbitrarily: e.g. arrival order of link requests (e.g. 1-RR and ERR in [3]).
    - from slave parameters: e.g. the poll interval negotiated during the link establishment or the $M \to S$ queue length (e.g. EPM in [3]).
- In case of activity based polling:
    - arbitrarily: e.g. FEP in [4].
    - from slave parameters: bandwidth requirements (e.g. PFP in [5], FPQ in [6]), delivery deadline (eg. FPQ in [6]), $M \to S$ or $S \to M$ queue length (e.g. AFP and SAFP in [7]), size of the Head-Of-Line message (e.g. PP and KFP in [8]).

The third criterion is related to the use of information about the $S \to M$ queues. If used, such information is provided by a signalling scheme or a prediction algorithm. With a signalling scheme for $S \to M$ queue, slaves can communicate their queue status while transmitting packets. Such information can be included into a specific packet field (e.g. [7] uses the flow bit in the header). With a prediction algorithm, the master estimates the message presence in $S \to M$ queues. This prediction is based on the arrival laws of messages on a slave (e.g. PFP in [5], FPQ in [6]).

According to this classification, CB-EDF is based on an activity based polling; the priority of a slave is deduced from its message parameters (class and delivery deadline); CB-EDF uses a combination of signalling and prediction schemes.

3   Service Differentiation with 1-RR

In this section, we evaluate the performances by simulation of 1-RR with regard to the application constraints. We consider its native version and its version enhanced with local scheduling accounting for message importance and deadline.

3.1   Local Scheduling Algorithm Principles

The local scheduling combines class based PQ, a class contains all messages of the same importance degree, and EDF. PQ is used between classes and EDF is used within a class. This local scheduling is used by both master and slaves to manage $M \to S$ or $S \to M$ queues. Each $S \to M$ or $M \to S$ queue consists of N distinct sub-queues; each class i ($i \in [1, N]$) has its own sub-queue. According to PQ principle, messages from class $i+1$ are always transmitted before those from class i. In each sub-queue, the transmission order is determined by EDF. This local scheduling will be used with the enhanced version of 1-RR and CB-EDF.

3.2   Simulation Parameters and Scenario

Before running the simulations described hereafter, we have validated the simulation tool we have developed, by confrontation with the results already published in [3]. In all scenarios, we consider a piconet made up of 7 slaves. Four classes are managed (class 3 having the highest priority). Each slave generates 2 flows in each of the 4 classes. In the same way, the master has, for each slave, 4 classes with 2 flows per class. Each flow follows a Poisson arrival law. The size of flow messages follows an exponential law of parameter 8, representing the number of slots required for the message transmission. The relative deadlines associated with class 0 and 1 range from 600 to 1000ms, with class 2 from 400 to 600ms and with class 3 from 200 to 400ms. Each simulation provides the following results:

- For any class, the average response time of any $S \to M$ flow.
- For any class, the percentage of $S \to M$ messages missing their deadline.
- The efficiency of the polling scheme. It is computed as the percentage of useful slots over the total slot number.

To perform a comparative performance evaluation, we have considered a scenario representative of Bluetooth communications. The submitted load is equal to 80%. The traffic distribution varies as follows with n ranging from 0 to 10; when $n = 0$, the distribution is symmetric and uniform :

- $S \to M$ flows of slaves 1 to 3 are heavily loaded: the submitted load of these flows ranges from 17% ($n = 0$) to 80% ($n = 10$).
- The submitted load of $S \to M$ flows of slaves 4 to 7 and $M \to S$ flows of slaves 1 to 7 ranges from 63% ($n = 0$) to 0% ($n = 10$).
- Class 3 (resp. 2, 1, 0) contributes to 10% of the load (resp. 20%, 30%, 40%).

3.3   Evaluation of 1-RR

For 1-RR, the master polls successively each slave for one message. In this section, we evaluate the performances of 1-RR and enhanced 1-RR polling schemes. The average response times of slaves 1 and 4 are illustrated on Fig. 1.

As expected, the initial 1-RR does not provide service differentiation. For a given n, $n \in [0, 10]$, each class has the same average response time. For class 3, the percentage of messages missing their deadline is greater than in other classes; it ranges from 15% to 100% when n ranges from 0 to 10. It can be explained because the highest priority class has smaller deadlines. The efficiency is the same for 1-RR and enhanced 1-RR (eg. 79.6% for $n = 0$ and 47.7% for $n = 10$). Enhanced 1-RR only provides an acceptable service differentiation in case of traffic uniformly distributed over the slaves ($n = 0$).

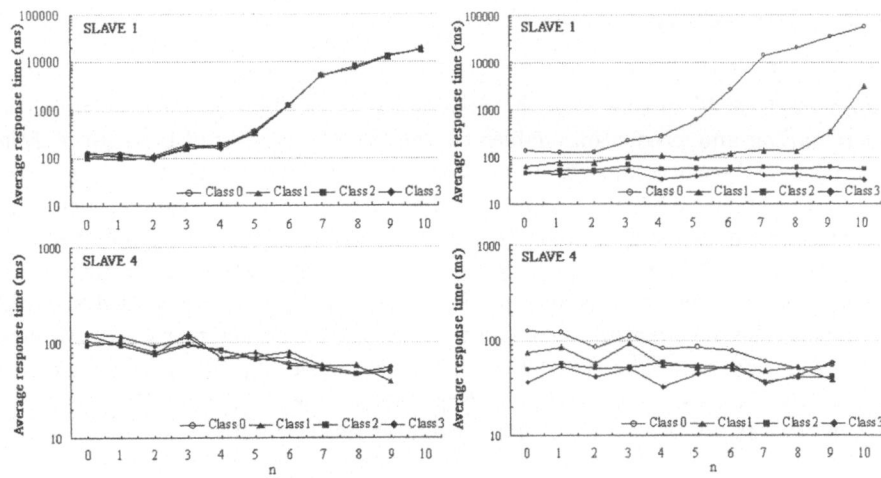

Fig. 1. Simulation for 1-RR and enhanced 1-RR for slaves 1 and 4.

4   A Class-Based EDF Scheduling

We present in this section our solution, called CB-EDF, to support QoS in the piconet. We give performance evaluations for the scenario used with 1-RR.

4.1   Principles

Our purpose is to define a global scheduling mechanism that supports service differentiation. Our solution accounts for two application constraints:

– the importance degree of a message that is mapped in a class,
– the end-to-end delivery deadline.

These constraints are accounted for and integrated within Bluetooth local and intra-piconet schedulings by a combination of PQ between classes and EDF within a class. We define two versions of this algorithm. The first one corresponds to an optimal solution, where a global and complete knowledge of the messages and their arrival times in each $M \rightarrow S$ and $S \rightarrow M$ queues is assumed. This version has no practical interest, except for comparative evaluation. In considering all $M \rightarrow S$ and $S \rightarrow M$ pending messages, the master always chooses the slave with the highest priority class. In case of equality, the master arbitrates in favor of the slave with the message having the earliest absolute deadline in this class. The second version corresponds to a realistic solution where the knowledge of $S \rightarrow M$ queues is obtained from an original scheme combining signalling and prediction. This version is called CB-EDF. It requires, for any slave, the knowledge of the message arrival laws and deadlines of any class present on this slave.

4.2   Message Presence Evaluation

We now describe the module of CB-EDF in charge of message presence evaluation in $S \to M$ queues of each active slave on the piconet. The combination of signalling and prediction enables to improve the quality of message presence evaluation. As we assume that messages in a given class can be generated by different application flows, L2CAP [1] mechanism signalling the presence of another segment in queue cannot be used. For that purpose, we propose a slight modification in Bluetooth specifications: we use 2 bits in the header of the Baseband packet to notify the master of pending message at the polled slave [2]. However, this mechanism is unable to signal the presence of messages arrived after the slave poll. That is why, we use a prediction algorithm. This algorithm is based on the knowledge of the message arrival law to estimate the presence of messages arrived in each class of a slave after its last poll.

4.3   Performance Evaluation of CB-EDF

In this section, we evaluate by simulation the performance of CB-EDF and compare them with those obtained by the optimal solution.

The results show that CB-EDF provides a good service differentiation in all simulated scenarios, better than enhanced 1-RR. Indeed, average response times for the two highest priority classes are shorter as well as the percentage of messages missing their deadline. The results obtained with CB-EDF are close to those obtained by the optimal solution. The efficiency is the same for CB-EDF and the optimal solution for any $n \in [0, 10]$ (eg. 79.8% for $n = 0$ and 76.0% for $n = 10$). CB-EDF approximates the global exact knowledge of the optimal solution by combining the signalling scheme with the prediction algorithm.

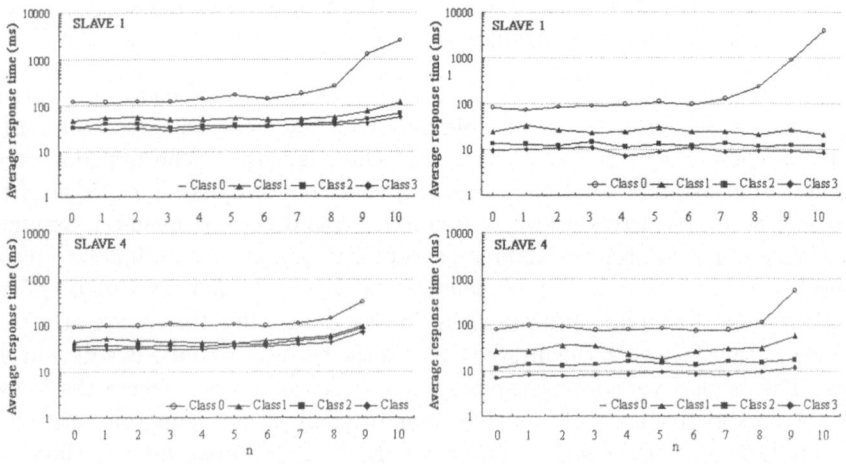

Fig. 2. Simulation for CB-EDF and the optimal solution for slaves 1 and 4.

5   Conclusion

In this paper, we have shown how to account for QoS in a piconet scheduling. Two QoS parameters, defined by the application, have been considered: the importance of a message and its delivery deadline. The simplest way to introduce service differentiation in a piconet consists in replacing the local scheduling, FIFO, by a scheduling managing classes and deadlines, while keeping the existing intra-piconet scheduling, 1-RR. This solution provides an acceptable service differentiation when the load is uniformly distributed. However, this solution does not account for classes and deadlines in the intra-piconet scheduling. Hence, in a second step, we have proposed CB-EDF that allows the coexistence of messages with different constraints. Simulations results show that in all simulated scenarios, CB-EDF provides service differentiation without efficiency loss and achieves good performances by the combination of signalling and prediction algorithms.

References

1. Specification of the Bluetooth system, version 1.b, www.bluetooth.com (1999).
2. A. Mercier, P. Minet, L. George: Introducing QoS support in Bluetooth Piconet with a Class-Based EDF Scheduling, INRIA Research report 5054 (2003).
3. A. Capone, M. Gerla, R. Kapoor: Efficient polling schemes for Bluetooth picocells, Proc. of IEEE ICC 2001, Helsinki, Finland.
4. N. J. Johansson, U. Korner, and P. Johansson: Performance evaluation of scheduling algorithms for Bluetooth, in Proc. of IFIP TC6 5th Int. Conf. on BC'99, Hong-Kong.
5. G. Heijenk, R. A. Yaiz: Predictive Fair Polling, Provisional Application for United States Letters Patent 60/241, 314 (2000).
6. J. B. Lapeyrie, T. Turletti: Adding QoS support for Bluetooth piconet, INRIA Research report 4514 (2002).
7. A. Das, A. Ghose, A. Razdan, H. Saran, R. Shorey: Enhancing performance of asynchronous data traffic over the Bluetooth wireless ad-hoc network, Proc. of IEEE INFOCOM 2001, Alaska, USA.
8. M. Kalia, D. Bansal, R. Shorey: MAC scheduling and SAR policies for Bluetooth - a master driven TDD pico-cellular wireless system, in Mobile Multimedia Communications (1999), IEEE Int. Workshop.

Call-Burst Blocking Probabilities of ON-OFF Traffic Sources under the Bandwidth Reservation Policy

Ioannis D. Moscholios and Michael D. Logothetis

University of Patras, Dept. of Electrical & Computer Engineering,
Wire Communications Laboratory, 265 00 Patras, Greece
{moscholios, m-logo}@wcl.ee.upatras.gr

Abstract. We calculate call and burst blocking probabilities (CBP and BBP respectively) in a single link loss system accommodating service-classes of ON-OFF traffic calls, either of high or low bandwidth requirements. To benefit high speed calls the bandwidth reservation (BR) policy is considered (ON-OFF/BR model). Poisson arriving calls enter the system via state ON and may alternate between states ON and OFF. When a call passes to state OFF it releases the bandwidth held in state ON which becomes available for new arrivals. When the call tries to return to state ON it re-requests its bandwidth. If it is available a new burst begins; otherwise, the burst is blocked and the call remains in state OFF. The CBP/BBP calculation is based on approximate formulas.

1 Introduction

Two bandwidth sharing policies are the most common at call-level: the complete sharing (CS) and the bandwidth reservation (BR) policy [1]. When calls of different service-classes compete for the available bandwidth of a link, under the CS policy (i.e. calls equally share the link bandwidth upon arrival, without any restriction), this leads to an unfair bandwidth allocation among the service-classes. Service-classes with higher bandwidth per call requirements will receive worse call-level QoS, namely, call blocking probability (CBP), than other service-classes accommodated in the link with fewer requirements. Usually an equal CBP among the service-classes is desirable; this is achieved with the BR policy. According to the BR policy, some fraction of the free bandwidth of the link is reserved to benefit the high speed calls.

We consider the ON-OFF model of [2] under the BR policy (ON-OFF/BR) and focus on the CBP equalization among the service-classes. According to the BR policy, the CBP equalization is achieved when the reserved link bandwidth for a service-class is defined so that the sum of it and the required bandwidth per call of the service-class is the same for all service-classes. The absence of a product form solution (PFS) [3] in the ON-OFF/BR model motivates the necessity for an approximate CBP calculation. To this end, we rely on the recursive formula of [2] and propose a similar-structured formula that achieves satisfactory CBP results. For the BBP calculation, we propose an approximate formula, which stands not only for the BR but also for the CS policy. For the derivation of this formula we follow the assumption that the population of calls of a service-class inside its reservation space (RS) is negligible [1]. The resultant BBP formula is based on the calculation of the average number of calls in the burst

N. Mitrou et al. (Eds.): NETWORKING 2004, LNCS 3042, pp. 1408–1413, 2004.

blocking state space. This calculation (and consequently the BBP) is improved by the method of Stasiak&Glabowski (S&G) [4] that takes into account the average population of calls inside the service-class RS that it is not negligible in reality.

In section 2 we propose the ON-OFF/BR model. We give an approximate but recursive CBP formula and derive a BBP formula. On the latter we apply the method of S&G in order to obtain better BBP results. In section 3 we present numerical and simulation CBP/BBP results for evaluation. We conclude in section 4.

2   The Proposed ON-OFF Model under the BR Policy

Consider a single link of capacity C, accommodating K service-classes of ON-OFF-type calls. Calls of a service-class k ($k=1,...,K$) arrive to the link according to a Poisson process with mean arrival rate λ_k (calls/sec) and require b_k bandwidth units (b.u.). The available link bandwidth is shared according to the CS policy. A call is accepted in the link in state ON, while the occupied link bandwidth is characterized real – real link. At the end of the ON-period a call of service-class k releases the b.u that holds and may begin an OFF-period with probability σ_k, or depart from the system with probability $1-\sigma_k$. While it is in state OFF, it is assumed that it seizes fictitious bandwidth (b_k) of a fictitious link with fictitious capacity C^*. At the end of the OFF-period a call of service-class k returns to state ON with probability 1. However, in order to return to state ON it requires again b_k b.u. If $C = C^*$, there is always available bandwidth for that call in state ON, i.e. no burst blocking occurs. If $C < C^*$, then if there is available bandwidth in the real link, i.e. if $j_1 + b_k \leq C$, (where j_1 is the occupied real link bandwidth), the call will return to state ON and a new burst will begin; otherwise the burst is blocked and the call will remain in state OFF for another OFF-period. When the BR policy is incorporated to the ON-OFF model, a new service-class k call is accepted in the system with b_k b.u., if it meets the following constraints:

$$j_1 + b_k \leq C - t_k \tag{1}$$

$$j_1 + j_2 + b_k \leq C^* - t_k \tag{2}$$

where t_k is the BR parameter of service-class k, and j_2 is the occupied fictitious link bandwidth.

The second constraint prevents the system from accepting new calls when most of the system calls are in state OFF. This model does not have a PFS; therefore the calculation of $G(\vec{j})$'s, is done in an approximate way, according to the recursive formula:

$$j_s G(\vec{j}) = \sum_{i=1}^{2} \sum_{k=1}^{K} D_{i,k,s}(\vec{j} - B_{i,k}) p_{ik} G(\vec{j} - B_{i,k})$$

where: $D_{i,k,s}(\vec{j} - B_{i,k}) = \begin{cases} b_{i,k,s} & \text{when } j_1 \leq C - t_k \text{ and } j_2 + j_1 \leq C^* - t_k \\ 0 & \text{otherwise} \end{cases}$ (3)

where, $i=1 \Rightarrow$ state ON, $i=2 \Rightarrow$ state OFF and $s=1 \Rightarrow$ real link, $s=2 \Rightarrow$ fictitious link.

$$b_{i,k,s} = \begin{cases} b_k & if \ s = i \\ 0 & if \ s \neq i \end{cases} \tag{4}$$

$B_{i,k}$: is the $(i, k)^{th}$ row of the $(2K \times 2)$ matrix B with entries $b_{i,k,s}$ and $B_{i,k} = (b_{i,k,1} \ b_{i,k,2})$;
p_{ik}: the utilization of the ith type link by service-class k,

$$p_{ik} = \frac{e_{ik}}{\mu_{ik}} = \begin{cases} \dfrac{\lambda_k}{(1-\sigma_\kappa)\mu_{1\kappa}} & for \ i = 1 \\[3mm] \dfrac{\lambda_k \sigma_\kappa}{(1-\sigma_\kappa)\mu_{2\kappa}} & for \ i = 2 \end{cases} \tag{5}$$

e_{ik}, μ_{ik}^{i} : total arrival rate and mean service time of service-class k calls in i^{th} state.
   The CBP of a service-class k, denoted as P_{b_k}, is calculated by:

$$P_{b_k} = \frac{\sum\limits_{\left\{\vec{j}\left|[(b_{1,k,1}+j_1)>C - t_k] \ \cup [(b_{1,k,1}+j_1+j_2)>C^* - t_k]\right\}\right.} G^{-1}G(\vec{j})}{} \qquad where \ G = \sum\limits_{j \in \Omega} G(\vec{j}) \tag{6}$$

The BBP of a service-class k, denoted as $P^*_{b_k}$, is calculated by:

$$P^*_{b_k} = \frac{\sum\limits_{(j \in \Omega^*)} y_{2k}(\vec{j})G(\vec{j})\mu_{2k}}{\sum\limits_{(j \in \Omega)} y_{2k}(\vec{j})G(\vec{j})\mu_{2k}} \tag{7}$$

where $\vec{j} \in \Leftrightarrow \left\{\left(j_1 \leq C \cap \left(\sum\limits_{s=1}^{2} j_s \leq C^*\right)\right)\right\}$, $y_{2k}(\vec{j})$, the average population of calls

of service-class k in state OFF (when the system state is $\vec{j}$) determined by ([5]):

$$y_{ik}(\vec{j}) = E(n_k^i|\vec{j}) = \frac{p_{ik}G(\vec{j} - B_{i,k})}{G(\vec{j})}, i = 1, 2 \tag{8}$$

and the G($\vec{j}$)'s are calculated by (3).
   To prove (7), if $\overline{n}_k^i$ is the average number of service-class k calls in i^{th} state ([2]):

$$\overline{n}_k^i = \sum\limits_{(j \in \Omega)} p_{ik}G(\vec{j} - B_{i,k}) \tag{9}$$

note the relation between (8) and (9):

$$\overline{n}_k^i = \sum\limits_{(j \in \Omega)} y_{ik}(\vec{j})G(\vec{j}) \tag{10}$$

   Multiplying $y_{2k}(\vec{j})$ by the corresponding $G(\vec{j})$ and the service rate in state OFF μ_{2k}
and summing up over the burst blocking space:

$$\vec{j} \in \Omega^* \Leftrightarrow \left\{ \left(C - b_k + 1 \leq j_1 \leq C \cap \left(\sum_{s=1}^{2} j_s \leq C^* \right) \right) \right\} \tag{11}$$

we obtain the following summation:

$$\sum_{(\vec{j} \in \Omega^*)} y_{2k}(\vec{j}) G(\vec{j}) \mu_{2k} \tag{12}$$

By normalizing it (taking into account the whole state space), we obtain (7).

The application of the BR policy to the ON-OFF model leads to $y_{2k}(\vec{j}) = 0$ when $j_1 > C - t_k$ or $j_2 > C^* - j_1 - t_k$. We have investigated that, depending on the values of t_k, equation (7) may result in a poor approximation of the BBP. A better approximation is obtained when one applies in (7) the method of S&G for the $y_{2k}(\vec{j})$ calculation, according to the following proposed procedure:

<u>Calculation of BBP of a service-class k with $t_k > 0$</u>

```
For the system state j⃗ = (j₁, j₂)

If j₁ ≤ C - tₖ   and   j₂ ≤ C* - j₁ - tₖ then

    Determine y₂ₖ (j⃗) according to Eq (8) and Eq. (3).

endif

If jreal > C - tₖ   then

    For all service-classes x (x = 1,...,K , but x ≠ k)

    If j₁ ≤ C - tₓ   and   j₂ ≤ C* - j₁ - tₓ then

        y₂ₖ (j⃗) = y₂ₖ (j⃗) + y₂ₖ (j₁-bₓ, j₂)

    endif

endif

Determine the BBP according to Eq. (7)
```

3 Numerical Examples – Evaluation

Consider a link of (real) capacity C=80 that accommodates three service-classes. The traffic description parameters of the service-classes are the following: $(\lambda_1, \lambda_2, \lambda_3)=$ (0.05, 0.06, 0.09), $(\mu^1_{11}, \mu^1_{12}, \mu^1_{13})=(0.5, 1.0, 1.2)$, $(\mu^1_{21}, \mu^1_{22}, \mu^1_{23})=(0.8, 1.9, 0.9)$, $(b_1, b_2, b_3)=(10, 8, 6)$, $(\sigma_1, \sigma_2, \sigma_3)=(0.85, 0.9, 0.95)$. CBP equalization is achieved by using the BR parameters: $(t_1, t_2, t_3)=(0, 2, 4)$. Two cases are examined: a) C=C*=80 and b) C=80, C*=100. In case (a) no burst blocking occurs, while in case (b) the increase of the fictitious capacity results in the burst blocking occurrence for all service-classes.

Each point in the horizontal axis of the figures entitled "arrival rate" has the following value: point 1 is $(\lambda_1, \lambda_2, \lambda_3)$=(0.05, 0.06, 0.09), point 2 is $(\lambda_1, \lambda_2, \lambda_3)$=(0.05, 0.06, 0.12),..., point 6 is $(\lambda_1, \lambda_2, \lambda_3)$=(0.05, 0.06, 0.24). Fig. 1 presents for the case (a), the analytical equalized CBP of all service-classes together with the analytical CBP of each service-class when the BR policy is not applied. As it was expected from the equalization of the CBP, Fig. 1 shows that the BR policy benefits the 1st and the 2nd service-class. Fig. 2 presents the analytical CBP for the case (b) for all service-classes either when the BR policy is applied or not. Simulation CBP results found to be quite close to the analytical CBP results for both cases and therefore they are not presented. The increase of the fictitious capacity from 80 to 100 decreases the CBP probabilities (more calls pass to state OFF, releasing bandwidth in state ON which is seized by new arrivals); the CBP of Fig. 2 are much lower than those of Fig. 1.

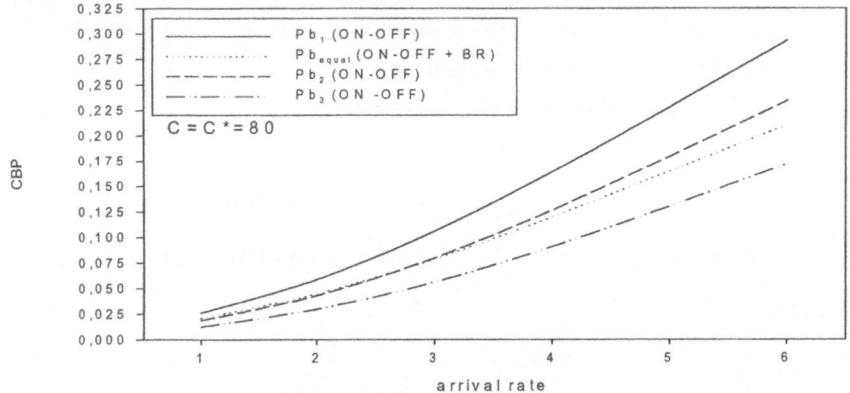

Fig. 1. Analytical CBP when C=C*=80 b.u

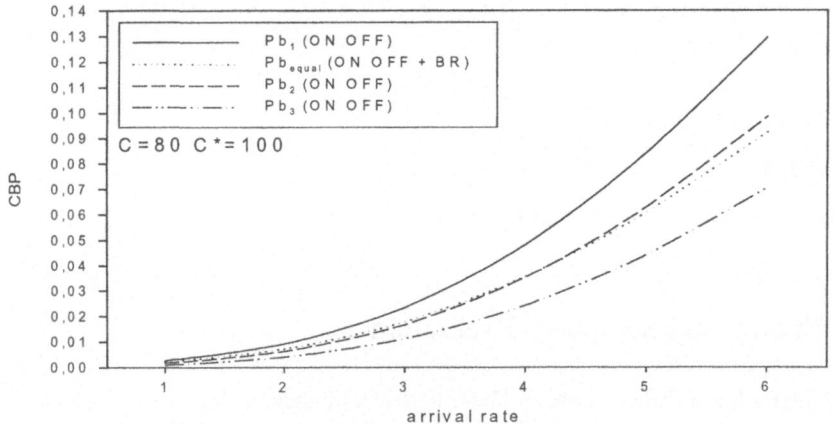

Fig. 2. Analytical CBP when C=80 and C*=100 b.u.

The increase of the fictitious capacity results in the appearance of BBP for all service-classes. In Table 1 we give both analytical and simulation BBP of all service-classes under the BR policy. Simulation results are mean values of 10 runs with 95%

confidence interval. The analytical BBP results of the 2nd and 3rd service-class are obtained by either the application of the method of S&G or not. For the 1st service-class, however, the application of the method of S&G in the BBP formula does not affect the BBP results, since there is no RS for the 1st service-class ($t_1=0$). For the 2nd service-class, the method of S&G results in slightly better BBP results compared to the corresponding results when the method of S&G is not applied. For the 3rd service-class, the method of S&G results in much better BBP results compared to the corresponding results when the method of S&G is not applied.

Table 1. Analytical and simulation BBP of the ON-OFF model under the BR policy

P	$P^*_{b_1}$		$P^*_{b_2}$			$P^*_{b_3}$		
	Analytical	Simulation	Analytical	Anal.S&G	Simulation	Analytical	Anal.S&G	Simulation
1	9.06×10^{-5}	9.59×10^{-5} $\pm 1.45 \times 10^{-5}$	4.67×10^{-5}	4.69×10^{-5}	5.62×10^{-5} $\pm 8.90 \times 10^{-6}$	2.00×10^{-5}	2.66×10^{-5}	3.06×10^{-5} $\pm 7.05 \times 10^{-6}$
2	2.87×10^{-4}	2.77×10^{-4} $\pm 4.90 \times 10^{-5}$	1.48×10^{-4}	1.49×10^{-4}	1.58×10^{-4} $\pm 3.22 \times 10^{-5}$	6.44×10^{-5}	8.33×10^{-5}	9.24×10^{-5} $\pm 9.05 \times 10^{-6}$
3	6.86×10^{-4}	6.35×10^{-4} $\pm 8.51 \times 10^{-5}$	3.54×10^{-4}	3.56×10^{-4}	3.71×10^{-4} $\pm 3.80 \times 10^{-5}$	1.56×10^{-4}	1.96×10^{-4}	2.14×10^{-4} $\pm 1.35 \times 10^{-5}$
4	1.34×10^{-3}	1.15×10^{-3} $\pm 1.32 \times 10^{-4}$	6.83×10^{-4}	6.87×10^{-4}	7.48×10^{-4} $\pm 6.10 \times 10^{-5}$	3.01×10^{-4}	3.71×10^{-4}	4.35×10^{-4} $\pm 1.89 \times 10^{-5}$
5	2.25×10^{-3}	2.07×10^{-3} $\pm 1.16 \times 10^{-4}$	1.12×10^{-3}	1.12×10^{-3}	1.27×10^{-3} $\pm 9.00 \times 10^{-5}$	4.87×10^{-4}	5.91×10^{-4}	7.68×10^{-4} $\pm 3.46 \times 10^{-5}$
6	3.38×10^{-3}	2.98×10^{-3} $\pm 1.52 \times 10^{-4}$	1.61×10^{-3}	1.62×10^{-3}	1.77×10^{-3} $\pm 8.61 \times 10^{-5}$	6.91×10^{-4}	8.25×10^{-4}	9.58×10^{-4} $\pm 4.40 \times 10^{-5}$

4 Conclusion

We investigate the call and burst level behavior of service-classes of Poisson arriving ON-OFF-type calls to a single link, under the BR policy. We propose the ON-OFF/BR model (recursive CBP and robust BBP formulas). The accuracy of the analytical CBP and BBP results is satisfactory taken as standard simulation results.

References

1. J.W. Roberts, "Teletraffic models for the Telecom 1 Integrated Services Network", ITC-10, 1983.
2. M. Mehmet Ali, "Call-burst blocking and call admission control in a broadband network with bursty sources", Performance Evaluation, 38 (1999), pp. 1-19.
3. K. W. Ross, *Multiservice Loss Models for Broadband Telecommunications Networks*, Springer Verlag London Limited, ISBN 3-540-19918-7, 1995.
4. M. Stasiak, M. Glabowski, "A simple approximation of the link model with reservation by a one-dimensional Markov chain", Performance Evaluation 41 (2000), pp. 195-208.
5. I. Moscholios, P. Nikolaropoulos, M. Logothetis, "Call level blocking of ON-OFF traffic sources with retrials under the complete sharing policy", ITC-18, Berlin, 2003, pp. 811-821.

The Connection Dependent Threshold Model for Finite Sources – A Generalization of the Engset Multirate Loss Model

Ioannis D. Moscholios and Michael D. Logothetis

University of Patras, Dept. of Electrical & Computer Engineering,
Wire Communications Laboratory, 265 00 Patras, Greece
{moscholios, m-logo}@wcl.ee.upatras.gr

Abstract. Call Blocking Probabilities (CBP) are the key index of the call-level QoS in multirate networks supporting either CBR or elastic traffic. We review the Engset Multirate Loss Model, in a single link, for CBR traffic, and for elastic traffic we propose the Connection Dependent Threshold Model with finite sources (f-CDTM). In f-CDTM, calls may adjust their traffic and bandwidth requirements according to sets of thresholds. Furthermore, we present the f-CDTM for a mixture of service-classes of finite and infinite sources. The proposed models don't have a product form solution; therefore the CBP calculation is based on approximate but recursive formulas used for the link occupancy distribution determination. The latter is complex since it requires enumeration and processing of the system state space. Simulation results validate our analysis.

1 Introduction

The classical Erlang Multirate Loss Model (EMLM) analyzes the call blocking behavior of service-classes with different bandwidth-per-call requirements when they are accommodated in a single link loss system. Calls of each service-class arrive to the link according to a Poisson process and compete for the available link bandwidth under the Complete Sharing (CS) policy. If the required bandwidth is available, calls are accepted and remain in the system for an arbitrarily distributed service time [1]; if not, calls are blocked and lost. The fact, that the EMLM is described by an efficient recursive formula ([1],[2]), not only simplifies the determination of Call Blocking Probability (CBP), but also serves as the basis in the analysis of other loss models [3]-[5].

In [3], single and multi retry models are proposed, in which blocked calls can retry (one or more times) to be connected in the system with reduced bandwidth and increased service time requirements. In [4], Single and Multi Threshold Models are proposed, in which the bandwidth requirement of a new call may depend on one or more thresholds, which indicate the occupied link bandwidth, j. In [5] the retry and threshold models, as well as the EMLM, are generalized to the Connection Dependent Threshold Model (CDTM); a threshold model, in which the state dependency is individualized among service-classes. In [6] the EMLM is extended to the Engset Multirate Loss Model (EnMLM) where the offered traffic of each service-class k, comes

N. Mitrou et al. (Eds.): NETWORKING 2004, LNCS 3042, pp. 1414–1419, 2004.

from a finite number of N_k sources. In EnMLM the CBP calculation is based on a recursive formula, similar to the EMLM. However, the determination of the link occupancy distribution, $G(j)$, which is essential for the CBP calculation, is complex. This is because the system state space needs enumeration and processing (in order for an equivalent system to be defined) prior to the $G(j)$ calculation.

In this paper, the EnMLM is extended to the CDTM for finite sources (f-CDTM), where each service-class has its own set of thresholds and calls may reduce their bandwidth and increase their service time requirements according to the value of j and the thresholds. We generalize the f-CDTM to include a mixture of service-classes with either finite or infinite sources. In the proposed models the determination of $G(j)$, is complex, since the state space requires enumeration and processing. We evaluate the accuracy of the proposed models by comparing analytical with simulation CBP results. The comparison is based on the Relative Approximation Errors (RAE); as a reference point, we use the RAE of the corresponding infinite source models.

In section 2 we propose the f-CDTM. We present the analytical model and prove the recursive formula used for the $G(j)$ calculation (not for the general case, due to lack of space). We proceed to the generalization to the CDTM to include a mixture of service-classes of either finite or infinite number of sources. In section 3 we present application examples. We conclude in section 4.

2 The Proposed CDTM for Finite Sources (f-CDTM)

A call of a service-class k is accepted in the system with its requirements $(b_{kc_t}, \mu_{kc_t}^{-1})$, $t =1, \ldots, T(k)$, when $J_{k_{t-1}} < j \leq J_{k_t}$, where $J_{T(k)} = C - b_{kc_{T(k)}}$ and $T(k)$ is the number of thresholds (J) and of contingency bandwidth requirements of call k (Fig. 1).

Eq. (1) is proposed for the $G(j)$'s calculation.

$$G(j) = \begin{cases} 1 & \text{for } j=0 \\ \dfrac{1}{j}\sum_{k=1}^{K}(N_k - n_k + 1)a_k b_k \delta_k(j)G(j-b_k) + \\ +\dfrac{1}{j}\sum_{k=1}^{K}\sum_{t=1}^{T}(N_k - (n_k + n_{kc_1} + \ldots + n_{kc_t} + \ldots + n_{kc_T}) + 1)a_{kc_t} b_{kc_t} \delta_{kc_t}(j)G(j - b_{kc_t}) & \text{for } j=1,\ldots,C \\ 0 & \text{otherwise} \end{cases} \quad (1)$$

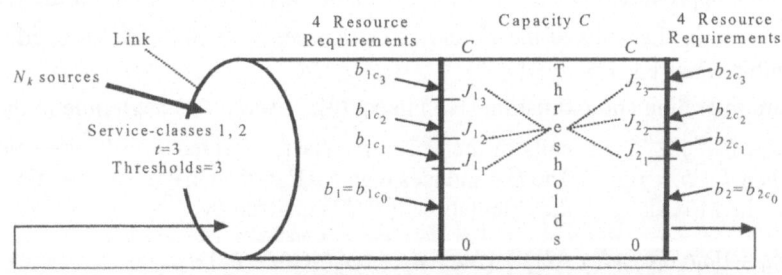

Fig. 1. Principle of the f-CDTM

- $\delta_k(j)=1$ when $1\leq j \leq C$ and $b_{kc}=0$, or, when $j\leq J_{k_l}+b_k$ and $b_{kc}>0$, otherwise $\delta_k(j)=0$.
- $\delta_{kc_l}(j)=1$ when $J_{k_l}+b_{kc_l}\geq j >J_{k_{l-1}}+b_{kc_l}$, otherwise $\delta_{kc_l}(j)=0$.

The CBP, $B_{kc_{T(k)}}$, that a call k is blocked with its last bandwidth requirement, $b_{kc_{T(k)}}$, is defined as $B_{kc_{T(k)}}=\text{Prob}\{\,j>b_{kc_{T(k)}}\}$ and is given by:

$$B_{kc_{T(k)}} = \sum_{j=C-b_{kc_{T(k)}}+1}^{C}G^{-1}G(j) \text{ where } G = \sum_{j=1}^{C}G(j) \tag{2}$$

To simplify the proof of (1), let us consider two service-classes only with the same single threshold, J_0, and bandwidth and service-time requirements (b_1,b_2), (μ_1^{-1}, μ_2^{-1}), respectively. Calls of the 1st service-class use the pair (b_{1c}, μ_{1c}^{-1}) when $j >J_0$. Although this model does not have a PFS, we assume that Local Balance (LB) equations are valid for calls of the 1st service-class:

$$[N_1 - n_1 +1]a_1b_1G(j-b_1) = E(n_1 b_1|j)G(j) \text{ for } j\text{-}b_1 \leq J_0 \tag{3}$$

$$[N_1 - (n_1 + n_{1c})+1]a_{1c}b_{1c}G(j-b_{1c}) = E(n_{1c}b_{1c}|j)G(j) \text{ for } j\text{-}b_{1c} > J_0 \tag{4}$$

where: a_1, a_{1c} are the offered traffic loads per idle source of the 1st service-class, such that $a_1b_1 = a_{1c} b_{1c}$, and n_{1c} is the number of in-service 1st service-class sources (calls) accepted in the system with b_{1c}.
As far as calls of the 2nd service-class are concerned, the L.B equation is of the form:

$$[N_2 - n_2 + 1]a_2b_2G(j-b_2) = E(n_2 b_2|j)G(j) \text{ for } 1\leq j \leq C \tag{5}$$

Equations (2) to (4) lead to a system of equations (6) to (8):

$$[N_1 - n_1 +1]a_1b_1G(j-b_1)+ [N_2 - n_2 +1]a_2b_2G(j-b_2) = E(n_1 b_1 + n_2b_2|j)G(j) \text{ for } 1\leq j \leq J_0+b_{1c} \tag{6}$$

$$[N_1-n_1+1]a_1b_1G(j-b_1)+[N_2-n_2+1]a_2b_2G(j-b_2)+[N_1-(n_1+n_{1c})+1]a_{1c}b_{1c}G(j-b_{1c})=jG(j) \text{ for } J_0+b_{1c}<j\leq J_0+b_1 \tag{7}$$

$$[N_2-n_2+1]a_2b_2G(j-b_2)+[N_1-(n_1+n_{1c})+1]a_{1c}b_{1c}G(j-b_{1c})= E(n_2b_2 +n_{1c}b_{1c})G(j) \text{ for } J_0+b_1<j\leq C \tag{8}$$

Eqs. (6)-(8) are combined into one for $G(j)$'s calculation, under two approximations:
- Migration approximation (M.A) in eq. (6): $E(n_{1c}b_{1c}|j)$ is negligible in the region $1\leq j \leq J_0+b_{1c}$, i.e. calls of the 1st service-class accepted with b_{1c} are assumed to be negligible when $1\leq j \leq J_0+b_{1c}$.
- Upward migration approximation (UA) in eq. (8): $E(n_1 b_1|j)$ is negligible in the region $J_0+b_1 < j \leq C$, i.e. calls of the 1st service-class, accepted with b_1, are negligible when $J_0+b_1<j \leq C$. The UA induces high RAE in the CBP results (section 3).
Based on the M.A and U.A the calculation of $G(j)$'s is given by:

$$[N_1 - n_1 + 1]a_1b_1\delta_1(j)G(j-b_1) + [N_2 - n_2 + 1]a_2b_2G(j-b_2)+ \\ [N_1 - (n_1 + n_{1c}) + 1]a_{1c}b_{1c}\delta_{1c}(j)G(j-b_{1c})= jG(j) \quad \text{for } 1<j \leq C \tag{9}$$

$\delta_2(j)=1$ for $1 \leq j \leq J_0+b_1$ otherwise $\delta_2(j)=0$, $\delta_{2c}(j)=1$ for $j>J_0+b_{1c}$ otherwise $\delta_{2c}(j)=0$.

Consider a single link that accommodates K_{fin} service-classes of finite sources and K_{inf} service-classes of infinite sources. Then, the calculation of $G(j)$'s is as follows

$$G(j)=\left\{ \begin{array}{l} 1 \quad \text{for } j=0 \\[4pt] \dfrac{1}{j}\sum_{k=1}^{K_{fin}}(N_k-n_k+1)a_k b_k \delta_k(j)G(j-b_k)+ \\[10pt] +\dfrac{1}{j}\sum_{k=1}^{K_{fin}}\sum_{t=1}^{T(k)}(N_k-(n_k+n_{kc_1}+...+n_{kc_t}+...+n_{kc_T})+1)a_{kc_t}b_{kc_t}\delta_{kc_t}(j)G(j-b_{kc_t}) \\[10pt] +\dfrac{1}{j}\sum_{k=1}^{K_{inf}}a_k b_k \delta_k(j)G(j-b_k)+\dfrac{1}{j}\sum_{k=1}^{K_{inf}}\sum_{t=1}^{T}a_{kc_t}b_{kc_t}\delta_{kc_t}(j)G(j-b_{kc_t}) \quad \text{for } j=1,...,C \\[10pt] 0 \quad \text{otherwise} \end{array} \right. \tag{10}$$

- $\delta_k(j)=1$ when $1 \leq j \leq C$ and $b_{kc}=0$, or, when $j \leq J_{k_t}+b_k$ and $b_{kc}>0$, otherwise $\delta_k(j)=0$.
- $\delta_{kc_t}(j)=1$ when $J_{k_t}+b_{kc_t} \geq j > J_{k_{t-1}}+b_{kc_t}$, otherwise $\delta_{kc_t}(j)=0$. This is the gener. f-CDTM.

3 Numerical Example – Evaluation

We present an application example to show the performance of the generalized f-CDTM. We compare the CBP approximation errors that appear in the f-CDTM, with that of CDTM (with infinite population). The analytical CBP results are compared with simulation results (mean values of 7 runs with 95% confidence interval). In graphs (Fig. 2), we compare the CBP results obtained by the infinite and finite models by using the Mean Relative Approximation Error (MRAE) of CBP:

$$MRAE = \frac{\sum_{k=1}^{K} RAE_k}{K}100\% \tag{11}$$

where RAE_k stands for the Relative Approximation Error for each service-class k:

$$RAE_k = \frac{\left| B_{k,an} - B_{k,sim} \right|}{B_{k,sim}} \tag{12}$$

where $B_{k,an}, B_{k,sim}$ are CBP obtained by analytical models and simulation, respectively.

Consider a link of capacity $C=50$ b.u. and two service-classes which require $b_1=10$ and $b_2=7$ b.u., respectively. Calls of the 1st service-class arrive to the link according to a Poisson process, whereas the offered traffic-load is α_1. Calls of the 2nd service-class arrive to the link according to a quasi-random process, whereas the offered traffic-load per idle source is α_2. Calls of the 1st service-class use their reduced bandwidth requirement $b_{1c_1}=8$ b.u when $j>J_{1_0}=30$, while $b_{1c_2}=6$ b.u when $j>J_{1_1}=35$. In the first case the offered traffic load is given by $\alpha_{1c_1}=\alpha_1 b_1/b_{1c_1}$, while in the second by $\alpha_{1c_2}=\alpha_1 b_1/b_{1c_2}$. Similarly, calls of the 2nd service-class use their reduced bandwidth requirement $b_{2c_1}=7$ b.u when $j>J_{1_0}=37$. In that case the offered traffic-load of calls of the second service-class is given by $\alpha_{2c_1}=\alpha_2 b_2 / b_{2c_1}$. Consider four values for the number of sources, N_2, of the 2nd service-class. Table 1 shows the various values of N_2 and the corresponding values of α_1, α_2, α_{1c_1}, α_{1c_2}, α_{2c_1}. The equivalent system used for the CBP calculation is: $C=50019$, $b_1=10000$, $b_2=7001$, $b_{1c_1}=8000$, $b_{1c_2}=6000$, $b_{2c_1}= 4005$, $J_{1_0}=$ 30000, $J_{1_1}=35000$ and $J_{2_0}=37001$. Fig. 2 shows the MRAE obtained from the generalized f-CDTM and the CDTM, for the four values of N_2. At each point in the horizontal axis entitled "offered traffic" α_1, α_{1c_1}, α_{1c_2} are constant, while α_2, α_{2c_1} are increased by $0.4/N_2$ and $0.7/N_2$ respectively, i.e. point 1 is $(\alpha_1,\alpha_2,\alpha_{1c_1},\alpha_{1c_2},\alpha_{2c_1})=(0.72,\ 2.4/N_2,\ 0.9,$ 1.2, $4.2/N_2)$, point 2:$(\alpha_1,\alpha_2,\alpha_{1c_1},\ \alpha_{1c_2},\alpha_{2c_1})=(0.72,\ 2.8/N_2,\ 0.9,\ 1.2,\ 4.9/N_2)$,..., point 6:$(\alpha_1,\alpha_2,\alpha_{1c_1},\alpha_{1c_2},\alpha_{2c_1})=(0.72,\ 4.4/N_2,\ 0.9,\ 1.2,\ 7.7/N_2)$. We present in Tables 2, 3, both analytical and simulation CBP results obtained from the generalized f-CDTM and the CDTM, for all points (P): 1,...,6.

According to Fig. 2 higher MRAE appear when $N_2=12$. As the number of N_2 increases the values of the MRAE tend to be reduced, while they approach those of the CDTM. A similar behavior is observed in the case of the CBP results; the CBP are increased, approaching the CBP results in the CDTM case, when N_2 increases.

Table 1. Service-classes characteristics

N_k	Model used	α_1 (erl)	α_2 (erl)	α_{1c_1} (erl)	α_{1c_2} (erl)	A_{2c_1} (erl)
$N_1=\infty,\ N_2=12$	Gener.f-CDTM	0.72	0.2 (=2.4 / N_2)	0.9	1.2	0.35 (=4.2/N_2)
$N_1=\infty,\ N_2= 60$	Gener. f-CDTM	0.72	0.04 (=2.4 / N_2)	0.9	1.2	0.07 (=4.2/N_2)
$N_1=\infty,N_2=600$	Gener. f-CDTM	0.72	0.004 (=2.4 / N_2)	0.9	1.2	0.007 (=4.2/N_2)
$N_1=N_2=\infty$	CDTM	0.72	2.4	0.9	1.2	4.2

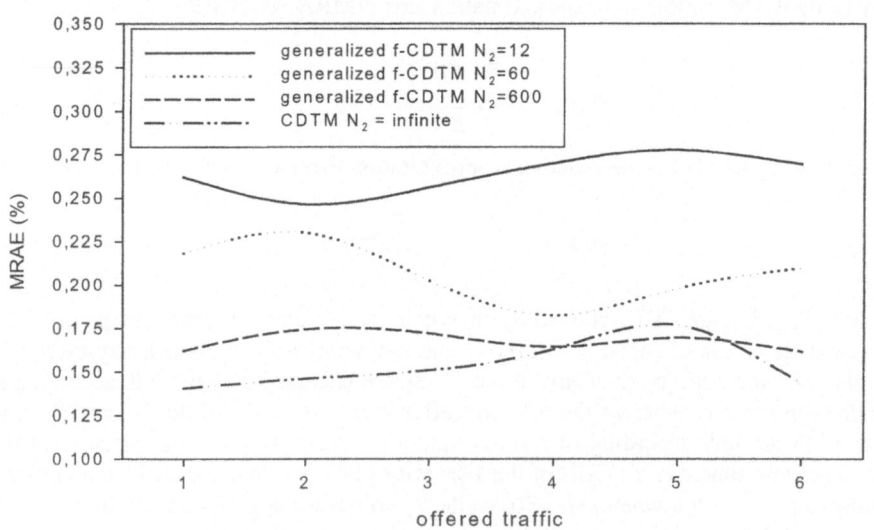

Fig. 2. MRAE of the generalized f-CDTM and the CDTM

Table 2. Analytical results for the generalized f-CDTM and the CDTM

P	$N_1=\infty, N_2=12$ (gener. f-CDTM)		$N_1=\infty, N_2=60$ (gener. f-CDTM)		$N_1=\infty, N_2=600$ (gener. f-CDTM)		$N_1=N_2=\infty$ (CDTM)	
	$B_{1c_2}(\%)$	$B_{2c_1}(\%)$	$B_{1c_2}(\%)$	$B_{2c_1}(\%)$	$B_{1c_2}(\%)$	$B_{2c_1}(\%)$	$B_{1c_2}(\%)$	$B_{2c_1}(\%)$
1	2.31	1.31	3.92	2.18	4.43	2.45	4.49	2.48
2	3.18	1.79	5.76	3.16	6.60	3.60	6.70	3.65
3	4.20	2.35	8.03	4.36	9.26	5.01	9.39	5.10
4	5.37	2.99	10.69	5.76	12.36	6.65	12.55	6.74
5	6.69	3.70	13.69	7.33	15.83	8.49	16.06	8.62
6	8.14	4.47	16.94	9.06	19.56	10.51	19.84	10.65

Table 3. Simulation results for the generalized f-CDTM and the CDTM

P	$N_1=\infty, N_2=12$ (gener. f-CDTM)		$N_1=\infty, N_2=60$ (gener. f-CDTM)		$N_1=\infty, N_2=600$ (gener. f-CDTM)		$N_1=N_2=\infty$ (CDTM)	
	$B_{1c_2}(\%)$	$B_{2c_1}(\%)$	$B_{1c_2}(\%)$	$B_{2c_1}(\%)$	$B_{1c_2}(\%)$	$B_{2c_1}(\%)$	$B_{1c_2}(\%)$	$B_{2c_1}(\%)$
1	1.98±0.09	0.97±0.05	3.10±0.27	1.86±0.04	3.66±0.18	2.20±0.11	3.81±0.34	2.25±0.18
2	2.66±0.15	1.38±0.05	4.34±0.11	2.79±0.09	5.22±0.17	3.32±0.08	5.29±0.38	3.55±0.27
3	3.44±0.26	1.82±0.07	6.17±0.22	3.95±0.08	7.38±0.25	4.60±0.17	7.46±0.57	4.87±0.15
4	4.35±0.17	2.29±0.07	8.34±0.37	5.32±0.19	9.66±0.40	6.34±0.09	9.80±0.55	6.45±0.23
5	5.32±0.17	2.85±0.04	10.22±0.18	6.94±0.22	12.11±0.31	8.23±0.15	11.87±0.29	8.61±0.67
6	6.49±0.23	3.48±0.10	12.66±0.29	8.38±0.23	14.86±0.40	10.57±0.16	15.44±0.45	10.63±0.56

4   Conclusion

We've proved the f-CDTM and shown its performance together with the generalized f-CDTM, by comparing their (numerical) CBP results with simulation results.

References

1. J.S. Kaufman, "Blocking in a Shared Resource Environment", IEEE Trans. Commun. 29 (10) pp. 1474-1481, 1981.
2. J.W. Roberts, "A service system with heterogeneous user requirements", in: G. Pujolle (Ed.), Performance of Data Communications systems and their applications, North Holland, Amsterdam, pp.423-431,1981.
3. J.S. Kaufman, "Blocking in a Completely Shared Resource Environment With State Dependent Resource And Residency Requirements", Proc. IEEE INFOCOM'92, pp. 2224-2232, 1992.
4. J.S. Kaufman, "Blocking with retrials in a completely shared resource environment", North-Holland, Performance Evaluation, 15, pp. 99-113, 1992.
5. I. Moscholios, M. Logothetis and G. Kokkinakis, "Connection Dependent Threshold Model: A Generalization of the Erlang Multiple Rate Loss Model", Performance Evaluation, Vol.48, Issue 1-4, pp. 177-200, May 2002.
6. G. Stamatelos and J. Hayes, "Admission control techniques with application to broadband networks", Comput. Commun.,Vol. 17, No. 9, pp. 663-673, September 1994.

Analysis of Protocol Operations and Scalability of COPS-SLS Negotiation System

Thi Mai Trang Nguyen[1,2], Nadia Boukhatem[2], and Guy Pujolle[1]

[1] Laboratoire d'Informatique de Paris 6, 8 rue du Capitaine Scott,
75015 Paris, France
{Thi-Mai-Trang.Nguyen, Guy.Pujolle}@lip6.fr
[2] GET-Télécom Paris, LTCI-UMR 5141 CNRS, 46 rue Barrault,
75013 Paris, France
{trnguyen, boukhatem}@enst.fr

Abstract. Dynamic service level negotiation is necessary to flexibly allocate resources according to QoS demands of a host or a domain. We have designed COPS-SLS as a service level negotiation protocol for IP-based networks. In this paper, we analyse a COPS-SLS-based negotiation system from the protocol design and system's scalability perspectives. This analysis is useful not only to the improvement of the COPS-SLS protocol design but also to the deployment of a service level negotiation system in general.

1 Introduction

To date, the relationship between customer and network as well as between network domains is based on static Service Level Agreement (SLA). The allocated network resources may be unused in low utilization periods or may be insufficient in high utilization periods. In addition, service levels in today's SLAs are offered only in one domain which is the domain of network provider. Once traffic is forwarded to another domain, the service level of the traffic is out of control. In the context of multiple domains, the offering of service levels is more complex because different domains can deploy different QoS architectures and apply different resource management policies. The motivation of our work is allowing SLA to be dynamically negotiated in order to enable the resource allocation on demand. Service levels can be negotiated and assured in any scope, one or multiple domains, specified by the customer, and independently of QoS architectures of individual domains while respecting the resource management policies of each domain.

We have defined the COPS-SLS (COPS usage for SLS negotiation) protocol for dynamic service level negotiation [1]. An SLS (Service Level Specification) can be considered as the technical part of the SLA which specifies the level of service that the traffic can obtain from the network. COPS-SLS extends the COPS protocol [2], which has been proposed for resource allocation in the context of Policy-based networking [3]. In this paper, we analyse a COPS-SLS based negotiation system from the protocol design and system's scalability perspectives. The organization of the

N. Mitrou et al. (Eds.): NETWORKING 2004, LNCS 3042, pp. 1420–1425, 2004.
© IFIP International Federation for Information Processing 2004

remainder of this paper is as follows. Section 2 provides an overview on the COPS-SLS protocol. The analysis of different operations of the COPS-SLS protocol impacting on the waiting time of a customer in a negotiation system is given in section 3. Section 4 presents a measurement of the scalability of a COPS-SLS negotiation system. Finally, section 5 concludes the paper.

2 Protocol Overview

We proposed COPS-SLS as an extension of the COPS protocol for the purpose of service level negotiation [1]. The SLS-PEP represents the customer which can be an end-host, a local network, an administrative domain, an ASP (Application Service Provider), or an ISP. The SLS-PDP represents the provider who provides service levels to customers. A provider can be an ASP or an ISP. SrNP (Service Negotiation Protocol) [5] and SIBBS (Simple Inter Bandwidth Broker Signaling) [6] are the protocols also designed for a similar purpose. SIBBS, SrNP and COPS-SLS are developed in a partially overlapping time periods during years 1999-2003. These three protocols have the common purpose of SLS negotiation, but each approach has its own particularities. The COPS-SLS protocol is organized in three phases as described in Fig. 1.

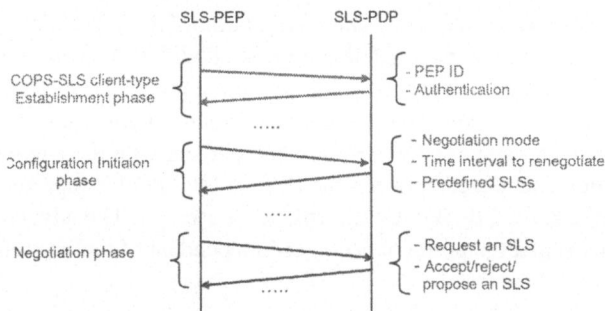

Fig. 1. COPS-SLS protocol

The system starts off with the *COPS-SLS client-type establishment phase*. The SLS-PEP must send its PEPID (PEP Identification) to the SLS-PDP for the authentication. If the SLS-PEP is authenticated, the COPS-SLS client-type is established and the system starts the *configuration initiation phase*. In this phase, the SLS-PEP and the SLS-PDP configure some parameters to prepare the negotiation process. Once the configuration of negotiation is successfully initiated, the system moves to the *negotiation phase*. In this phase, service levels can be negotiated and the configuration can also be changed if necessary.

In comparison with the alternative protocols such as SIBBS or SrNP, the configuration phase is the main difference distinguishing COPS-SLS from the others. This phase has been introduced in order for the negotiation to be able to flexibly adapt to a variety of customers, network providers, and QoS architectures. The customer and the provider are allowed to configure the manner of negotiation, and to change this configuration dynamically. The configuration parameters include *the set of negotiation parameters, the negotiation mode,* and *the renegotiation interval.* As different networks may provide guaranties of different QoS parameters and use different information to identify traffic, the agreement on these parameters allows the negotiation to be operational over heterogeneous environments. *The set of negotiation parameters* defines parameters which can be exchanged between the SLS-PEP and the SLS-PDP for the negotiation of a service level. For example, the SLS-PEP and the SLS-PDP can agree that only bandwidth and delay can be negotiated while the jitter parameter is not used in the negotiation. *The negotiation mode* defines how the customer can request a service level from the provider. COPS-SLS defines three negotiation modes and leaves the provider to choose the manner of negotiation which is most convenient to them. In the *predefined-SLS mode*, the provider advertises their customers of the service levels supported, using a set of predefined service levels. The customer can choose one of these predefined service levels to make a request. In the *non-predefined SLS with constraints mode*, the provider does not supply their customers all the supported service levels but only put some constraints on the values of negotiation parameters sent in the requests. These constraints must be locally checked by the customer before sending a request. The *non-predefined-SLS mode* neither puts any constraint on the negotiation request nor pre-defines service levels. The customer asks the provider any desired service level. *The renegotiation interval* reflects the time interval before that a service level negotiated cannot be modified. Indeed, there is a trade-off between the negotiation's dynamic level and the network performance. The modification of a negotiated service level may lead to some network reconfiguration.

Another characteristic of COPS-SLS in comparison with the alternatives is that the SLS information exchanged between the PEP and the PDP is represented by a named data structure, a.k.a. a PIB (Policy Information Base) [4]. The advantage of using the concept of PIB is that the protocol becomes independent of the information carried.

3   Analysis of Different Protocol Operations

In this section, we analyse the impact of different operations of the COPS-SLS protocol on *the starting time* and *the negotiation time*. These times have an important influence on the customer's satisfaction. Large starting times or negotiation times may irritate the customers participating in the negotiation. *The starting time* is defined as the time taken by the system to be ready for the negotiation of a service level. Relating to the COPS-SLS design described in Fig. 1, this time includes the TCP connection time, the COPS-SLS client-type establishment time, and the configuration initiation time. *The negotiation time* is defined as the time taken by a customer to perform a negotiation of a service level in the negotiation phase.

To analyse the impacts on the starting time, the times taken by the following operations have been measured. The *TCP connection time* is measured between the time that the SLS-PEP tries to connect to the SLS-PDP and the time that the TCP connection is established. The *COPS-SLS client-type establishment time* is measured between the time that the SLS-PEP sends an *OPN* (Client-Open) message (see [1] for COPS messages) to the SLS-PDP and the time that the SLS-PEP receives a *CAT* (Client-Accept) message from the SLS-PDP. The *configuration initiation time* is measured between the time that the SLS-PEP sends a *REQ* (Request) message with a *'context = configuration'* to the SLS-PDP to initiate the configuration of negotiation and the time that the SLS-PEP finishes the transmission of a *RPT* (Report) message reporting a successful installation.

To analyze the impacts on the negotiation time, the following times have been measured. The *negotiation time* is measured between the time that the SLS-PEP sends a *REQ* message with a *'context = resource-allocation'* to the SLS-PDP to negotiate a service level and the time that the SLS-PEP finishes the transmission of a *RPT* message reporting the final result of the negotiation. The *decision time* is measured between the time that the SLS-PDP receives a *REQ* message requesting a service level and the time that the SLS-PDP finishes making decision for this request.

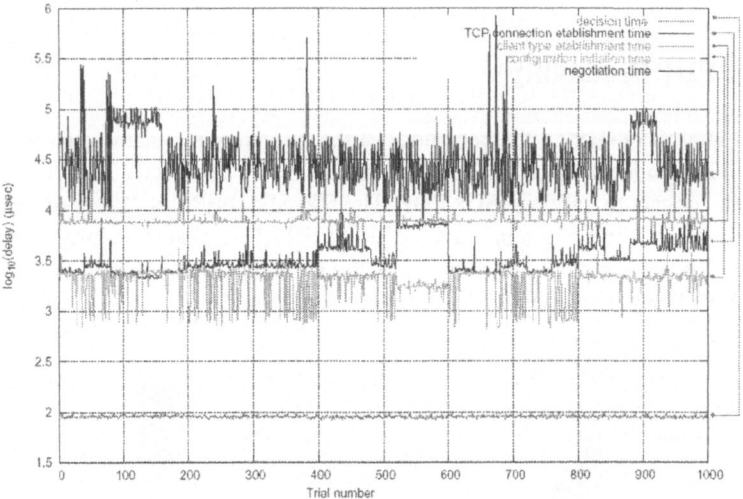

Fig. 2. Time measurements of different COPS-SLS protocol operations

One thousand trials have been realized from different SLS-PEPs to an SLS-PDP in a local network. In each trial, an SLS-PEP establishes the COPS-SLS connection with the SLS-PDP and realizes one negotiation. Fig. 2 presents the common logarithm of the times taken by the above-mentioned operations in each trial. The result shows that the operation that impacts most on the COPS-SLS starting time is the COPS-SLS client-type establishment, then the TCP connection establishment, and finally the configuration initiation. For the negotiation time, the time taken by the decision process of the SLS-PDP is minor in comparison with the transmission time of COPS-SLS messages. The analysis of this result leads to some conclusions on the design and the usage of a COPS-SLS based negotiation system. The *COPS-SLS client-type estab-*

lishment is mandatory because this is an operation of any COPS-based protocol. The time taken by this operation can be minimized by reducing the time taken by the authentication process of the SLS-PDP. The *TCP connection establishment* is also an operation of any COPS-based protocol. However, different COPS client-types can use the same TCP connection. If the TCP connection is already established by another COPS client-type, the impact of the TCP connection time on the starting time can be eliminated. The *configuration initiation time* depends on the complexity of the configuration to be initiated. The configuration can be simple and minimized by designing a light PIB. The *decision time* depends on the decision process of the SLS-PDP. This time can vary with the volume of negotiation policies that the SLS-PDP handles, and the number of SLS-PEPs requesting service levels simultaneously. The distance between the SLS-PDP and SLS-PEPs and the message size should be taken into account for both the starting time and the negotiation time because of the transmission delay. The maximum message size in this test is around 300 bytes. The negotiation PIB should be carefully designed to minimize the volume of information encapsulated into a message.

4   SLS-PDP Scalability

This test aims at observing the behaviour of the negotiation system when the number of SLS-PEPs connecting to a SLS-PDP increases. In this scenario, 500 SLS-PEPs attempt to connect to only one SLS-PDP. The 500 SLS-PEPs are launched one by one every five seconds from 25 terminals (i.e. 20 SLS-PEPs are launched from a terminal) to the SLS-PDP in a local network. Once the COPS-SLS connection is established, each SLS-PEP negotiates service levels every one second.

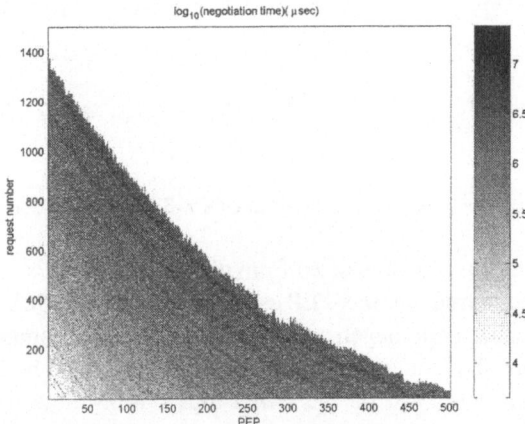

Fig. 3. Negotiation time measured in the scalability test of the negotiation system

The common logarithm of the negotiation time measured in the system is presented in Fig. 3. The horizontal axis identifies the ith SLS-PEP ($i = 1:500$) connecting to the SLS-PDP. The vertical axis identifies the kth negotiation request that an SLS-PEP can send to the SLS-PDP and obtain a decision. The colour represents the negotiation time of each request. This measurement shows that more the number of requests the SLS-PDP treats in a same lap of time, more the performance of the negotiation system is poor. When there are less than 50 SLS-PEPs in the system, the colour is light. The colour becomes darker as long as the number of connected SLS-PEPs increases. When the system is up to 450 - 500 simultaneously connected SLS-PEPs, the negotiation time can be up to $10^{6.5}$ and sometimes 10^7 microseconds.

A negotiation system should be carefully dimensioned. The mechanism of redirecting a PEP to another PDP provided by the COPS protocol [4] sounds an interesting solution to the scalability of the negotiation system. If the number of SLS-PEPs connecting to an SLS-PDP reaches a threshold, the SLS-PDP can redirect further SLS-PEPs to another SLS-PDP. The re-negotiation interval adjustment is also a useful mechanism to put the negotiation rhythm under control.

5 Conclusion

In this paper, we summarized the main characteristics of COPS-SLS and analyzed a negotiation system from the protocol design and scalability perspectives. The analysis of different operations of COPS-SLS gives a concrete picture on the impacts on the starting time and the negotiation time. The scalability of the negotiation system has been studied at different number of simultaneous connected SLS-PEPs. The design of negotiation PIBs, the ability of simplifying or customizing the configuration initiation phase, and the dimensioning of a negotiation system are to be investigated in future works.

References

1. Nguyen, T.M.T., Boukhatem, N., Gharmri-Doudarn, Y., and Pujolle, G., *COPS-SLS: A service level negotiation for the Internet.* IEEE Communications Magazine, May 2002. **40**(5): p. 158-165
2. Durham, D., et al., *The COPS (Common Open Policy Service) Protocol.* RFC 2748, January 2000
3. Yavatkar, R., D. Pendarakis, and R. Guerin, *A Framework for Policy Based Admission Control.* RFC 2753, January 2000
4. Chan, K., et al., *COPS Usage for Policy Provisioning (COPS-PR).* RFC 3084, March 2001
5. Goderis, D., et al., *A service-centric IP quality of service architecture for next generation networks.* in *IEEE/IFIP Network Operations and Management Symposium.* April 2002. Florence, Italy
6. Teitelbaum, B., et al., *Internet2 QBone: Building a Testbed for Differentiated Services.* IEEE Network, September/October 1999. **13**(5): p. 8-16

Outage Analysis for Multi-connection Multiclass Services in the Uplink of Wideband CDMA Cellular Mobile Networks

Chun Nie[1,2], Tung Chong Wong[1], and Yong Huat Chew[1]

[1]Institute for Infocomm Research, Agency for Science, Technology and Research
21 Heng Mui Keng Terrace, Singapore 119613
[2]Department of Electrical and Computer Engineering, National University of Singapore
10 Kent Ridge Crescent, Singapore 119260
{stuniec, wongtc, chewyh}@i2r.a-star.edu.sg

Abstract. In this paper, we address the data link layer quality of service (QoS) issue by investigating the outage probabilities for multi-connection multiclass services in the uplink of a wideband CDMA cellular mobile network. Four Universal Mobile Telecommunications System (UMTS) QoS classes are served within each mobile user simultaneously. Each class has different QoS constraints. Assuming perfect power control, a power allocation scheme is designed to fulfill the desired received powers for all traffic classes. The outage probability is formulated for each mobile user with its traffic classes in terms of bit error rate (BER) and signal-to-interference-plus-noise ratio (SINR). In addition, an admission region, satisfying the outage probability requirements of all mobile users, is computed.

1 Introduction

In UMTS network, wideband CDMA technology enables multimedia services with different QoS specifications. In [1], four different QoS traffic classes, including conversational, streaming, interactive and background classes, are defined for 3G systems. The network is responsible for providing different QoS guarantees to all traffic classes. QoS attribute at the data link layer, such as outage probability, attracts a lot of research interest. In [2], Gilhousen et al. studied the outage probability issue for a single class on/off source in a CDMA network. Recently, Wong et al. extended the analysis of outage probability from a single class sources to on/off multiclass sources, variable bit rate (VBR) multiclass sources and video multiclass sources in [3], [4] and [5], respectively. However, each mobile user can only have a single connection in these papers. This paper differs from the existing published work [3-5] by dealing with QoS provisioning for each mobile user having multiple traffic classes. In this analytical work, each user can have multiple connections to serve more than one traffic class. According to [1,6], voice, video, web-browsing and data services are chosen as typical examples of conversational, streaming, interactive and background classes in the UMTS QoS architecture, respectively. From [6], voice, web-browsing

N. Mitrou et al. (Eds.): NETWORKING 2004, LNCS 3042, pp. 1426–1432, 2004.

and data services are usually assumed to be on/off sources. Comparatively, video is usually modeled as a two-dimensional discrete-state, continuous-time Markov chain [7]. The main contribution of this paper is to analyze the outage probabilities for multi-connection multiclass services within all mobile users in the uplink of a wide-band CDMA cellular mobile network. In subsequent sections, the paper is organized as follows. The system model, power distribution, outage probability and feasible admission region will be investigated in sections 2, 3, 4 and 5, respectively.

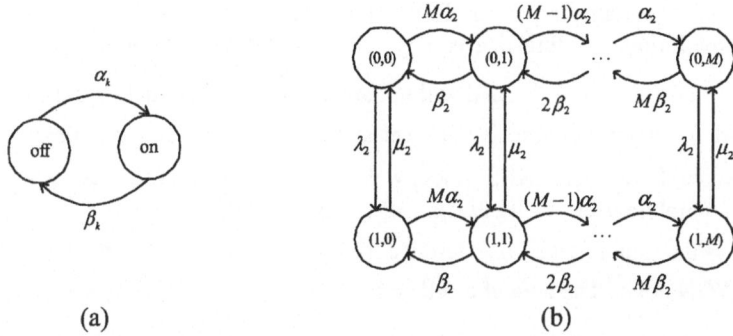

(a)   (b)

Fig. 1. Traffic Models with (a) a 2-state Markov chain for an on/off source and (b) a 2-dimensional Markov chain for a video source

2 System Model

As mentioned in section 1, voice, web-browsing and data services are usually modeled as on/off sources, which are shown as Markov chains in Fig. 1(a), $k \in \{1,3,4\}$, respectively. Thus, the activity factor, which is the probability that the process stays in the on state, for voice, web-browsing and data services, is given by

$$p_k = \alpha_k /(\alpha_k + \beta_k), \ k \in \{1,3,4\}. \tag{1}$$

On the other hand, a video service is usually referred to as Sen's model and its Markov chain is illustrated in Fig. 1(b) According to Sen's model [7], each video source can be decomposed into one high-bit-rate (HBR) and M low-bit-rate (LBR) mini-sources, which are identified by HBR and LBR spreading codes, respectively. Thus, the activity factors of LBR and HBR mini-source are given by

$$p_k = \alpha_2 /(\alpha_2 + \beta_2), \ k \in \{2l\}, \tag{2}$$

and

$$p_k = \lambda_2 /(\lambda_2 + \mu_2), \ k \in \{2h\}. \tag{3}$$

The followings are assumptions and system parameters used in this paper.

Assumptions
- All mobile users are uniformly located in each cell.
- An Additive White Guassian Noise (AWGN) channel is assumed.
- Perfect power control is assumed for each service.
- Convolutional coding is used for each traffic class, which is defined in [9].

System Parameter Definitions

- There exist N mobile users in each cell and the number of cells in the system is n.
- $n_{i,k}$ denotes the number of voice, video, web-browsing and data services within the ith ($1 \leq i \leq N$) mobile user, $k \in \{1,2,3,4\}$, respectively. p_k, G_k, γ_k^* and BER_k^*, denote the activity factors, spreading gains, SINR requirements and BER requirements for voice, video using LBR spreading codes, video using HBR spreading code, web-browsing and data services, $k \in \{1,2l,2h,3,4\}$, respectively.
- M denotes the maximum number of LBR spreading codes used by one video service.
- The convolutional coding rate is r_{coding}. The convolutional coding can result in a lower γ_k^*, $k \in \{1,2l,2h,3,4\}$, and will not appear in the mathematical formulation.
- $S_{i,k}$ and $l_{i,k}$, denote the received power and total number of active spreading codes used by voice, LBR video, HBR video, web-browsing and data services within the ith ($1 \leq i \leq N$) mobile user, $k \in \{1,2l,2h,3,4\}$, respectively.
- $I_{intercell}$ is the total intercell interference from neighbouring cells and η is the power of the AWGN noise. The area of a cell is A.

3 Power Distribution Algorithm

In the WCDMA network, the system capacity and QoS performance are directly associated with multiple access interference (MAI) which is contributed by interfering mobile users. Therefore, signal-to-interference-plus-noise ratio (SINR) is an important attribute at the data link layer. To attain good performance at the data link layer, it is necessary that the average SINR of each service should be maintained at a required level. Let us denote set V as $\{1,2l,2h,3,4\}$ and denote set V' as $\{1,2h,3,4\}$. In addition, let $n_{i,2l} = Mn_{i,2}$ and $n_{i,2h} = n_{i,2}$, ($1 \leq i \leq N$). Within the ith mobile user, the average SINR of voice, video using LBR spreading codes, video using HBR spreading code, web-browsing and data services are given by

$$S_{i,k}G_k / \{ \sum_{j=1;\, j \neq i}^{N} \sum_{k \in V} p_k n_{i,k} S_{i,k} + E[I_{intercell}] + \eta \} = \gamma_k^*, \qquad (4)$$

where $k \in \{1,2l,2h,3,4\}$, respectively.

Let $\Gamma_i = \sum_{k \in V} p_k n_{i,k} \gamma_k^* / G_k$ and $\bar{S}_i = \sum_{k \in V} p_k n_{i,k} S_{i,k}$. $\bar{S}_i$ denotes the average received power from the ith mobile user at the base station. Rearrange equation (4) algebraically and the following equation is satisfied.

$$[1+\Gamma_i]\bar{S}_i = \Gamma_i (\sum_{j=1}^{N} \bar{S}_j + E[I_{intercell}] + \eta), \ 1 \leq i \leq N \qquad (5)$$

$E[I_{intercell}]$ denotes the mean of the intercell interference. According to [2], the total intercell interference can be approximated by a Guassian distribution and a path loss exponent of 4 is assumed. Thus, the mean and variance of the intercell interference are given by

$$E[I_{intercell}] \le [\sum_{i=1}^{N} \sum_{k \in V} p_k n_{i,k} S_{i,k}] \int \int f(r_m / r_d) dA / A, \qquad (6)$$

and $$Var[I_{intercell}] \le \sum_{i=1}^{N} \{\sum_{k \in V'} S_{i,k}^2 n_{i,k} \int \int [p_k g(r_m / r_d) - p_k^2 f^2 (r_m / r_d)] dA / A$$

$$+ S_{i,2l}^2 n_{i,2} \int \int [Mp_{2l}[1 + (M-1)p_{2l}]g(r_m / r_d) - (Mp_{2l})^2 f^2 (r_m / r_d)] dA / A\}, \qquad (7)$$

where $f(r_m / r_d)$ and $g(r_m / r_d)$ are given by [2-5]

$$f(r_m / r_d) = (r_m / r_d)^4 e^{(\sigma \ln 10/10)^2} [1 - Q(40 \log(r_m / r_d) / \sqrt{2\sigma^2} - \sqrt{2\sigma^2} \ln 10/10)], \qquad (8)$$

and $$g(r_m / r_d) = (r_m / r_d)^8 e^{(\sigma \ln 10/5)^2} [1 - Q(40 \log(r_m / r_d) / \sqrt{2\sigma^2} - \sqrt{2\sigma^2} \ln 10/5)]. \qquad (9)$$

In equations (8) and (9), ε_m and ε_d are two independent Guassian random variables with zero mean and σ^2 variance. Let us suppose that r_m (r_d) denote the distance between an intercell service and its own base station (the intracell base station). Thus, based on equation (6), equation (5) is algebraically rearranged as follows.

$$(1 + \Gamma_i) \bar{S}_i / \Gamma_i = [1 + \int \int f(r_m / r_d) dA / A] \sum_{i=1}^{N} \bar{S}_i + \eta, \ 1 \le i \le N \qquad (10)$$

The power vector of the ith mobile user S_i is defined as $[S_{i,1}, S_{i,2l}, S_{i,2h}, S_{i,3}, S_{i,4}]$. Therefore, it is clear that the objective of the power distribution is to derive a positive solution for the vector S_i. Let $\varepsilon = 1 - \sum_{i=1}^{N} \Gamma_i [1 + \int \int f(r_m / r_d) dA / A]/(1 + \Gamma_i)$. According to [8], for equation (10), if and only if $0 \le \varepsilon \le 1$ is satisfied, $\bar{S}_i$ has a positive solution. The solution is easily given by

$$\bar{S}_i = \eta \Gamma_i / [\varepsilon(1 + \Gamma_i)]. \qquad (11)$$

Otherwise, it is impossible to find a positive solution for equation (10). From the definition of $\bar{S}_i$, if a feasible $\bar{S}_i$ is available, the positive power vector S_i exists and the desired received powers within the ith mobile user are formulated for each type of services. Thus, we have

$$S_{i,j} = \eta \gamma_j^* / [\varepsilon(1 + \Gamma_i) G_j], \ 1 \le i \le N, \ j = \{1, 2l, 2h, 3, 4\}. \qquad (12)$$

Accordingly, if the condition $0 \le \varepsilon \le 1$ holds, equation (12) satisfies the SINR requirements in equation (4). Obviously, if all positive received powers are increased by the same ratio, the achieved SINR will exceed the corresponding SINR requirements and the data link layer QoS of the system is improved. Therefore, in our calculation of outage probability, the positive power solutions of all services obtained from equation (12) are multiplied by a common factor, θ ($\theta > 1$).

4 Analysis of Outage Probability

In a WCDMA system, the outage probability refers to the probability that the achieved SINR is below the SINR requirement or the achieved BER is above the BER requirement. Within the ith mobile user, the outage probabilities for voice, video using a LBR spreading code, video using a HBR spreading code, web-browsing and data services are expressed as $P_{out,i,k}, 1 \le i \le N$, $k \in \{1, 2l, 2h, 3, 4\}$, respectively.

$$P_{out,i,k} = \sum_{l_{1,1}=0}^{n_{1,1}} \sum_{l_{1,2l}=0}^{Mn_{1,2}} \sum_{l_{1,2h}=0}^{n_{1,2}} \sum_{l_{1,3}=0}^{n_{1,3}} \sum_{l_{1,4}=0}^{n_{1,4}} \cdots \sum_{\substack{l_{j,1}=0 \\ j \neq i}}^{n_{j,1}} \sum_{\substack{l_{j,2l}=0 \\ j \neq i}}^{Mn_{j,2}} \sum_{\substack{l_{j,2h}=0 \\ j \neq i}}^{n_{j,2}} \sum_{\substack{l_{j,3}=0 \\ j \neq i}}^{n_{j,3}} \sum_{\substack{l_{j,4}=0 \\ j \neq i}}^{n_{j,4}} \cdots \sum_{l_{N,1}=0}^{n_{N,1}} \sum_{l_{N,2l}=0}^{Mn_{N,2}} \sum_{l_{N,2h}=0}^{n_{N,2}} \sum_{l_{N,3}=0}^{n_{N,3}} \sum_{l_{N,4}=0}^{n_{N,4}} \tag{13}$$

$$\{Q(\frac{\delta_{i,k} - \mu_i}{\sigma_i}) \prod_{\substack{j=1 \\ j \neq i}}^{N} \prod_{k \in V} \binom{n_{j,k}}{l_{j,k}} (p_k)^{l_{j,k}} (1-p_k)^{n_{j,k} - l_{j,k}} \},$$

where $\mu_i = \sum_{j=1; j \neq i}^{N} \sum_{k \in V} (l_{j,k} S_{j,k}) + E[I_{intercell}]$, $Q(x) = \int_x^{\infty} e^{-t^2/2} dt / \sqrt{2\pi}$,

$\delta_{i,k} = S_{i,k} G_k / \gamma_k^* - \eta$, $\sigma_i^2 = Var[I_{intercell}]$, $k \in \{1, 2l, 2h, 3, 4\}$.

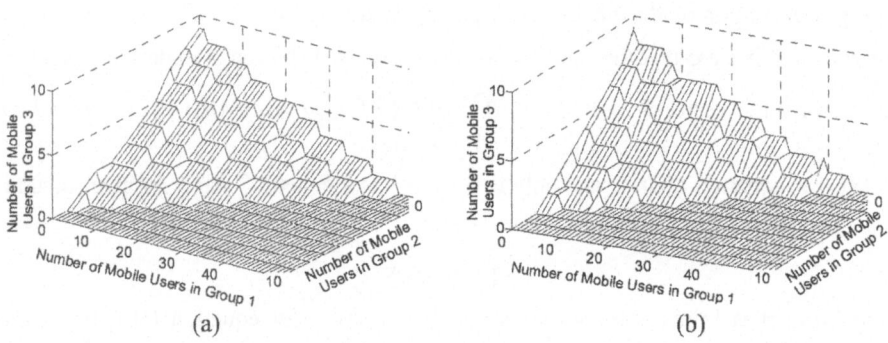

(a)   (b)

Fig. 2. System Admission Regions for (a) Analytical Admission Region, and (b) Simulation Admission Region

Table 1. Parameters value used

η	-103.2 dBm	M	8	θ	100.0
n	9	r_{coding}	1/2	$\iint f(r_m / r_d) dA / A$	0.456
p_k, $\{1, 2l, 2h, 3, 4\}$			{0.4, 0.3867, 0.5, 0.1, 0.2}		
γ_k^*, $\{1, 2l, 2h, 3, 4\}$			{2 dB, 2 dB, 2 dB, 3 dB, 3 dB}		
G_k, $\{1, 2l, 2h, 3, 4\}$			{64, 128, 64, 32, 16}		
BER_k^*, $\{1, 2l, 2h, 3, 4\}$			$\{10^{-2}, 10^{-2}, 10^{-2}, 10^{-3}, 10^{-3}\}$		

5   Numerical Results for Admission Region

In a WCDMA network, the admission control scheme must guarantee that all admitted mobile users are assured of the required QoS levels, which are referred to as the outage probability at the data link layer. In this paper, the admission region is provided by satisfying all services simultaneously. In this section, we present an admission region with three dimensions. Firstly, we assume all mobile users in the network can be divided into three groups. In the first group, each mobile user only serves one voice service. In the second group, each mobile user serves one voice service and one video service. In the third group, each mobile user serves one web-browsing service and one data service. The maximum acceptable outage probability is set at 10^{-2} for voice and video and is set at 10^{-3} for web-browsing and data. The admission region is obtained based on the following parameters in Table 1 and is given by Fig. 2.

In Fig. 2, a 3-dimensional feasible admission region is shown. The region on or below the surface indicates the admission region under the assumed conditions. Any set of mobile users with the given services can be guaranteed for their outage probability requirements at the data link layer. The parameters in Table 1 can be varied to achieve different admission regions. In order to verify the accuracy of our analytical work in sections 3 and 4, an analytical admission region and a simulation admission region are presented in Fig. 2(a) and Fig. 2(b), respectively. Clearly, the regions in Fig. 2(a) and Fig. 2(b) match well with each other. Thus, the proposed outage probability formulation is accurate and can be used to examine the QoS performance and system capacity at the data link layer of a wideband CDMA cellular mobile network.

6   Conclusion

This paper deals with the QoS performance in terms of the outage probability at the data link layer in a cellular WCDMA network. Multiple services are supported in each mobile user. The power allocation scheme is presented under the assumption of perfect power control and outage probabilities are formulated mathematically. As a criterion of call admission, the satisfaction of outage probabilities determines the admission region of the system at the data link layer. In this paper, a typical numerical admission region is shown in a three-dimensional graph in the Numerical Results section. The analytical formulation in this paper can be used to dimension the system capacity of multi-connection multiclass services in a WCDMA network.

References

1. 3GPP, TS 23.907, QoS concept and architecture, (2002)
2. Gilhousen, K., Jacobs, I., Padovani, R., Viterbi, A., Weaver, L., and Wheatley III, C.: On the capacity of a cellular CDMA system, *IEEE Transactions on Vehicular Technology*, Vol. 40, (1991) 303-312

3. Wong, T.C., Mark, J.W., Chua, K.C., Yao, J., and Chew, Y.H.: Performance analysis of multiclass services in the uplink of wideband CDMA, IEEE ICCS, (2002), 692 –696
4. Wong, T.C., Mark, J.W., Chua, K.C., and Kannan, B.: Performance analysis of variable bit rate multiclass services in the uplink of wideband CDMA, IEEE ICC, (2003), Vol. 1, 363 – 367
5. Wong, T.C., Mark, J.W., and Chua, K.C.: Performance evaluation of video services in a multirate DS-CDMA system, IEEE PIMRC, (2003), Vol. 2, 1490-1495
6. ETSI Technical Report, ETSI TR 101 112, Selection procedures for the choice of radio transmission technologies of the UMTS, (1998), V3.2.0
7. Sen, P., Maglaris, B., Rikli, N.E., and Anastassiou, D.: Model for packet switching of variable-bit-rate video sources, *IEEE Journal on Selected Areas in Communications*, (1989), Vol. 7, 865-869
8. Lee, S.J., Lee, H.W., and Sung, D.K.: Capacity calculation in DS-CDMA systems supporting multi-class services, IEEE PIMRC, (1997), Vol. 2, 297 –301
9. 3GPP TS 25.212, Multiplexing and channel coding (FDD), v5.5.0, (2003)

Experimental Analysis of the SABUL Congestion Control Algorithm*

Phoemphun Oothongsap, Yannis Viniotis, and Mladen Vouk

North Carolina State University, Raleigh NC 27606, USA

Abstract. Several new protocols such as RBUDP, User-Level UDP, Tsunami, and SABUL, have been proposed as alternatives to TCP for high-speed data transfer. The purpose of this paper is to analyze experimentally the effects of SABUL congeston control algorithm on SABUL and performance metrics such as bandwidth utilization, self-fairness, and aggressiveness. Our results confirm some expected behavior of SABUL and reveal some less expected one. Our experiments also indicate that SABUL implementation and design can result in an even more erratic behavior and degraded performance under high-congestion conditions.

1 Introduction

The high-performance networks being developed at present offer the promise of connectivity at speeds upto 40 Gbps or more. Such networks can enable new classes of high-performance applications, such as remote data analysis/visualization and high-performance grid-based computation. Although there is significant bandwidth available for such applications, the effective use of the available bandwidth is a challenge.

Several studies (e.g., [3] and references therein) have shown that, in practice, user-level distributed applications connected by a high-speed network (e.g., Abilene) cannot fully utilize the available bandwidth. The main reason for this subpar performance is the congestion control mechanisms of the transport protocol (e.g.,TCP). Thus, to improve bandwidth utilization, two alternatives are to:(i) improve the performance of TCP, and, (ii) develop new transport protocols that are suitable for a high-bandwidth environment.

An Example of new hybrid protocols is SABUL [4]. SABUL uses UDP to transfer the data and a TCP for signaling. SABUL has been evaluated empirically by simulation, and theoretically [1,4]. However, there appear be no published experiments that explore some more extreme behaviors of SABUL algorithms and implementations in a real environment. The purpose of this paper is to study SABUL behavior in situations where SABUL connections compete against each other in high congestion situations. We have also investigated how

---

* This work sponsored in part by the National Science Foundation grant No.NSF-9901004, DOE SciDAC grant DE-FC02-01ER25484, and IBM Corp. Shared University Research Program

N. Mitrou et al. (Eds.): NETWORKING 2004, LNCS 3042, pp. 1433–1439, 2004.

SABUL send-rate calculation (and its implementation) affect SABUL performance.

The remainder of this paper is divided as follows: Section II, experimental results are presented and discussed. Section III focuses on one specific observation, that of unusually high bandwidth oscillations found in high-congestions situations.

2   Experiments

To understand the general behavior of SABUL, experiments were performed in two environments: (i) a private local area network where the round trip time (RTT) is in microseconds (a short-haul network), (ii) Abilene network where the round trip time (RTT) is in milliseconds (a long-haul network). For the long-haul network, end hosts are located at three different Abilene end-point locations: North Carolina State University (NCSU), Georgia Institute of Technology (GT) and University of Washington (UW).

The purpose of this set of experiments was to study SABUL self-fairness, bandwidth utilization, and factors affecting these properties. Multiple SABUL connections were studied in both short- and long-haul networks. We emphasize the long-haul network part because the main purpose of SABUL is to aid file transfer in high-speed long RTT networks.

Table 1 shows bandwidth utilization of three SABUL connections ($Source_1$, $Source_2$, and $Source_3$). The fourth column in Table 1 is the receiver machine. The fifth to seventh columns represent the RTT from $Source_i$ to destination. The eighth to tenth columns represent the initial sending rate of each connection in Mb/s. The eleventh to thirteenth columns represent the rate control interval (round length) of each connection in msecs. The fourteenth to sixteenth columns represent the average sending rate of each connection in Mb/s and the last column represents the figure showing the instantaneous sending rate of each experiment.

Table 1. Average sending rate of three SABUL connections

Src_1	Src_2	Src_3	Dest	RTT_1	RTT_2	RTT_3	$Init_1$	$Init_2$	$Init_3$	T_1	T_2	T_3	$Rate_1$	$Rate_2$	$Rate_3$	Figure
n10	n11	n12	n20	0.204	0.204	0.203	320	260	210	200	200	200	320	320	320	1
n10	n11	n12	n20	0.204	0.204	0.203	320	260	210	600	400	200	295	310	361	2
fast1	fast2	fast3	localhost	23.3	23.2	23.3	293	291	279	200	200	200	270	270	270	3
fast1	fast2	fast3	fasttcp	57	57.5	57	298	293	258	200	200	200	298	310	295	4
fast1	fast3	fasttcp	localhost	23.3	23.2	57.5	273	282	286	200	200	200	220	280	300	5

Table 1 shows the three connections compete on the same bottleneck link. The results show that SABUL connections may or may not be fair to each other. In table 1, we can categorize the experiments into three cases:(i) same RTT and rate control interval, (ii) different RTT and same rate control interval, and (iii) same RTT and different rate control interval. We notice that all connections get the similar average sending rate when rate control interval is the same regardless of RTT and initial sending rate. All connections show an unfairness behavior when the rate control interval of all connections are different. This behavior can be explained as follows. SABUL sender recalculates a new sending rate every time it receives a SYN packet from the receiver and the receiver generates a SYN packet every constant rate control interval. For the connection having a short rate control interval, the sender will receive a signal to increase a sending rate more often than the connection having a longer rate control interval. Moreover, SABUL congestion control is a variant of Multiplicative Iincrease and Multiplicative Decrease algorithm. The sender increases sending rate aggressively. Then the connection with a short rate control interval increases the sending rate more aggressively than the connection with a longer rate control interval, causing unfairness.

Figures 3 to 5 show the instantaneous sending rate of each experiment. The x-axis represents the experimental time in seconds and the y-axis represents SABUL instantaneous throughput in Mbits/sec. In each figure, we notice that SABUL still exhibits the oscillation property. And also, we notice the "synchronized" behavior, i.e., sources oscillate in phase. Synchronized behavior is an unpleasant behavior since it can reduce the overall throughput of the system. Synchronized behavior occurs due to the drop-tail operations at the router, and the round trip time (RTT) effect. With drop-tail routers, each congestion period introduces global synchronization in the network as noted in [2]. When the queues overflow, packets from several connections are dropped and these connections decrease their sending rate at the same time. The consequence is loss of throughput at the router. The effect of the RTT on the send rate fluctuation was already mentioned.

Even though it is not apparent in figures 1 and 2 due to the log scale, the behavior is still the same.

3 Discussion

In this section, we discuss some of our results in more detail. Following are some of the observations one can make from the tables and accompanying figures.

The sharp increase and decrease in the sending rate in almost every experiment seems to be a problem with SABUL implementation, and not really a defect of the SABUL congestion control algorithm. From the implementation source code, one can find that

(1) SABUL sender calculates the sending rate at the application layer.

(2) SABUL average sending rate is equal to the number of packets sent in a particular interval divided by the time interval between two SYN packets. The number of packets sent in a particular interval is calculated as follows (i) the number of retransmitted packets plus the number of new packets or (ii) if the sum of the previous value is less than the number of ERR packets, the number of packet sent is equal to number of ERR packets.

(3) At the receiver side, SABUL receiver has to estimate the average round trip time once when the program starts.

(4) SABUL receiver will send the ERR packets back to the sender in two cases: (i) once it detects the gap in packet sequence number, and (ii) periodically every 1.5 * RTT second.

These actions can cause incorrect sending rate calculation which may cause sharp increase and decrease in the measured sending rate. As we know, SABUL sender calculates the sending rate from the number of packets sent divided by the time between two SYN packets. If the number of packets sent is less than the number of packet errors in that interval, the number of packets sent is set to the number of error packets. This way of computing can cause observed effects.

(i) When the number of packets that is actually sent to the network is less than the number of error packets, then this will cause the sending rate to have a higher value than its actaul one.

(ii) SABUL receiver uses TCP channel to transmit SYN packet. SABUL sender will process SYN packets upon receipt. However, if the SABUL sender CPU is busy, then the kernel will not pass the SYN packet up to the application layer right away. With this behavior, when SABUL sender see two SYN packets back to back, the time difference between two packets is either larger than rate control interval or smaller than rate control interval.(In effect, the time measured between two SYN packets is random.)

With such unavoidable randomness and errors in the number of packets sent and the time interval between two SYN packets, SABUL sending rate will increase sharply when the sender detects a large number of losses and the time interval between two SYN packets is very small. It will also decrease sharply when the sender detects small losses (this means the actual number of packets sent is also smaller than the number of packets lost). This behavior can be seen clearly in Figure 4.

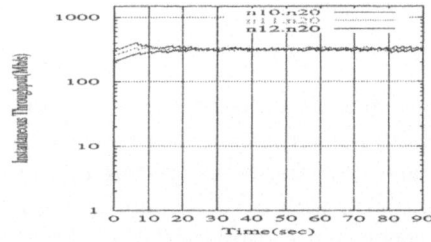

Fig. 1. Instantenous sending rate from n10, n11, and n12 to n20 with the same RTT, rate control interval and different initial sending rate

The number of error packets received from the receiver is sometimes higher than the actual number of packets lost. This phenomenon happens because the receiver periodically sends the ERR packets to the sender. When sender receives an ERR packet, it does not check whether the associated packet has already

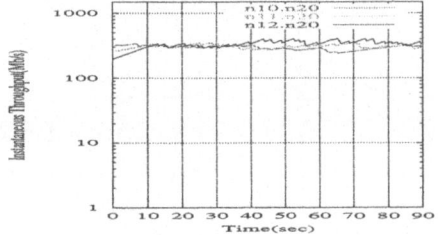

Fig. 2. Instantenous sending rate from n10, n11, and n12 to n20 with the same RTT, and different rate control interval and initial sending rate

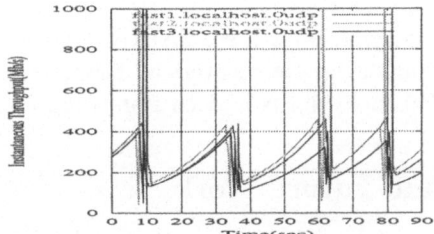

Fig. 3. Instantenous sending rate from fast1, fast2 and fast3 to localhost with the same RTT, rate control interval and initial sending rate

been retransmitted. It just adds the number of packets to the number of error packets and considers it loss total. This action, it can double the apparent error rate.

(i) the number of error packets is higher than the actual number of packets lost. This will cause a sharp increase/decrease of the sending rate as we explained previously.

(ii) the loss rate calculation is not accurate. When the number of packets sent is less than the number of error packets, SABUL sender sets the loss rate to one. This value of the loss rate will effect the new sending rate in the next interval.

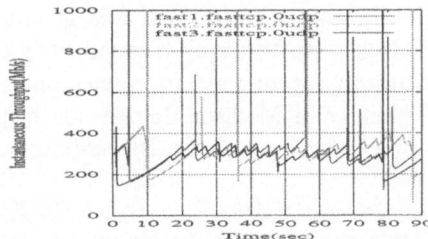

Fig. 4. Instantenous sending rate from fast1, fast2 and fast3 to fasttcp with the same RTT, rate control interval and initial sending rate

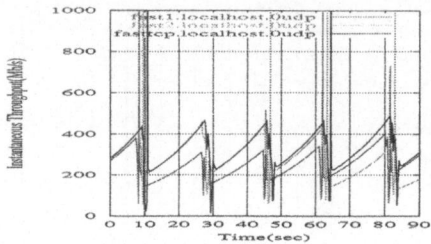

Fig. 5. Instantenous sending rate from fast1, fast3 and fasttcp to localhost with the same rate control interval, initial sending rate, and different RTT

Then in a congested network, we will see SABUL sender drop the sending rate to a smaller value. After the sender recovers all the losses, the sending rate will climb up gradually. This behavior is noticed from Figures 5.

4   Conclusion and Future Work

In this paper, we investigated performance of SABUL in both local area and wide-area network. We focused on SABUL bandwidth utilization, and SABUL self-fairness. The results of our experiments show that SABUL can utilize network bandwidth efficiently. As expected, the main factor having an effect on SABUL performance is rate control interval. SABUL shows a self-fairness property that depends on rate control interval, while RTT has no effect on SABUL self-fairness. However, we also noticed several less desirable SABUL traits: high large swing throughput oscillations (under certain conditions) and synchronization of these oscillations when several streams are involved. This degrades SABUL performance. We ascribe these behaviors not to the congestion control algorithm but to the implementation.

Acknowledgements. The authors wish to thank Dr. Steven Low and Raj Jayaram of FAST project for allowing us to use FAST project end-hosts in the experimetns, Stanilav Shalunov of Internet2 for scheduling the tests, and Scott Friedrich (of GT), Cas D'Angelo (of GT), David Richarson (of UW), Tommy Jacobson (of NCNI) for allowing us to transmit a large amount of data through their network. We would also like to thank to John Streck and John Moore of the NC ITEC and NC State Centaur Labs for advice and support given in this project, and Marhn Fullmer and Michael Bugaev for help. We would like to thank the Extreme Networks for the use of a Blackdiamond switch.

References

1. P. Oothongsap, M. Vouk, Y. Viniotis, CACC Technical Report, North Carolina State University, February 2003.

2. S. Floyd, V. Jacobson, Random Early Detection Gateways for Congestion Avoidance, *IEEE/ACM Transactions on Networking*, vol. 1, pp. 397-413, August 1993.
3. W. Feng, P. Tinnakornsrisuphap, The Failure of TCP in High Performance Computational Grids, *Proceedings of the Super Computing 2000*, (SC2000).
4. Y. Gu, M. Mazzucco, X. Hong, R. Grossman, Rate Based Congestion Control over High Bandwidth/Delay Links, Submitted to IEEE/ACM Transaction on Networking, http://www.rgrossman.com/faq/sabul-faq-03.htm, download on February 2003.
5. M. Mazzucco, A. Ananthanarayan, R. Grossman, J. Levera, G. Rao, Merging Multiple Data Streams on Common Keys over High Performance Networks, SuperComputing 2002, November 2002.

Rational Server Selection for Mobile Agents: Problem Solution and Performance Analysis

Carsten Pils, Jan Kritzner, and Stefan Diepolder

Informatik 4 (Communications Systems)
RWTH Aachen, 52056 Aachen, Germany
{pils, kritzner, diepolder}@informatik.rwth-aachen.de

Abstract. Since agents have the ability to migrate to outperforming resources they can potentially balance the load of heterogeneous systems. However, to balance resources efficiently agents must take the load into account. Thus, to support the agent migration strategy the application of server selection systems has been proposed recently. Server selection systems keep track of the load of network and host resources and hence predict the performance of different migration strategies. Yet, server selection comes at a cost and therefore agents must take care when applying it. This paper presents a decision strategy for the agent's decision problem. The performance of the approach is analysed with the help of a simple queuing model.

1 Introduction

Basically, the advantage of mobile agent technology is that it allows application designers to decide where an agent is processed. In that sense, developers can optimise the performance of an application by carefully selecting processing resources. Yet, approaches proposing an automatic selection of destination systems [1] [2] [3] have not gained much attention. A similar problem is server selection in the Internet. The deployment of mirror sites has motivated users to select a mirror offering the best performance. With gaining significance of bulk document, audio, and video file transfer, smart mirror site selection has become a compelling task and thus motivated numerous automatic server selection approaches. Lately, we discussed in [4] the application of server selection to mobile agents. It has been pointed out that due to the rather small resource requirements of mobile agents, these must be careful when applying server selection. That is, since agents are idle while the server selection system processes their requests, server selection comes at a cost. Thus, agents will only access a server selection system if the utility function U is positive:

$$U = d_\varnothing - d_{min} - \delta \qquad (1)$$

where $d_\varnothing$ is the average service time, d_{min} is the service time of a server recommended by a selection system, and δ is the agent idle time while the selection system processes its request. In the remainder of the paper, this problem will be

N. Mitrou et al. (Eds.): NETWORKING 2004, LNCS 3042, pp. 1440–1445, 2004.

referred to as the *rational server selection problem*. In [4] we presented a decision algorithm which solves the *rational server selection problem*. This paper briefly summarises our findings in [4] and analyses the performance of the algorithm with the help of a queuing model.

2   Rational Server Selection

The problem in developing a decision algorithm for the *server selection problem* is that the agents neither know the server resource capacities nor the waiting time δ. Yet, it is assumed that they have a basic knowledge of the system heterogeneity. Thus, let R be the resource capacity distribution describing the probability that a randomly selected server has a resource capacity $c \in \mathfrak{R}_{\geq 0}^m$, with $\bar{x} \leq c$ (where $\leq$ is a componentwise comparison and $\bar{x} \in \mathfrak{R}_{\geq 0}^m$). With the help of R, $d_\varnothing$ can be estimated by:

$$d_\varnothing(r) = \min\left\{d \in \mathfrak{R}_{\geq 0} \middle| r \, \nabla(E[R] \cdot d) = 0\right\}$$

where r is the resource requirement and ∇ is the resource consumption operator defined as: the mapping of a process's resource requirement to the requirements remaining after consuming a specified capacity is defined as $\nabla : \mathfrak{R}_{\geq 0}^m \times \mathfrak{R}_{\geq 0}^m \mapsto \mathfrak{R}_{\geq 0}^m$. That is, if capacity c is required to satisfy a resource requirement r equation $r\overline{\nabla}c = 0$ holds. According to the definition of R, its average value $E[R]$ is the resource capacity an agent expects when it selects a server randomly or the number of alternative servers is 1. To estimate d_{min}, the random distribution R_n giving the maximum resource capacity c, with $\bar{x} \leq c$ out of n randomly selected servers is required (again, $\leq$ is the componentwise comparison). Obviously, R_n is an order statistic distribution [5] and thus it is given by: $R_n(x) = R_1(x)^n = R(x)^n$ and $\frac{dR_n(x)}{dx} = n \cdot R(x)^{n-1} \cdot \frac{dR(x)}{dx}$.

$E[R_n]$ is the average maximum server capacity when a server is selected out of n. Consequently, d_{min} can be estimated by function $d_{min}(r, n)$ as follows:

$$d_{min}(r, n) = \min\left\{d \in \mathfrak{R}_{\geq 0} \middle| r \, \nabla(E[R_n] \cdot d) = 0\right\}$$

Finally, given the random distribution $Err(x)$ (prediction error distribution) that a prediction has a deviation of ξ with $\xi \leq \bar{\xi}$ for an interval of length h the estimated utility function is:

$$E[U(r, n)] = d_\varnothing(r) - d_{min}(r, n) - \delta - E[Err(d_\varnothing(r) - d_{min}(r, n))] \quad (2)$$

3   Performance Analysis

Given that rational server selection is widely deployed it does not only improve the performance of individual agents. Just as much it influences the performance of the overall system: As server selection systems struggle to assign agents to the

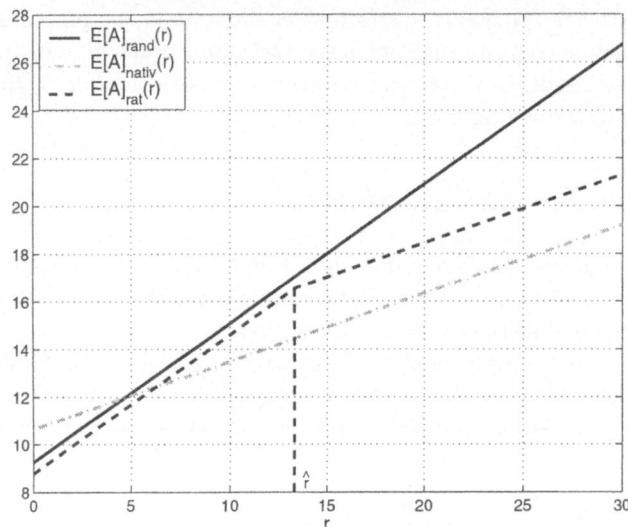

Fig. 1. Customer response time versus customer resource requirement

Table 1. Queuing model: Parameters

Para.	Description	Para.	Description
m	Number of servers (2)	μ_i	deterministic service rate of server i ($\mu_1 = 1, \mu_2 = 6$)
λ	customer arrival rate (exponentially distributed, $\lambda = 1$)	$\frac{1}{\alpha}$	Average service requirement (exponentially distributed, $\alpha = 0.55$)
$\hat{r}$	break-even resource requirement	δ	latency of server selection system interaction ($\delta = 10$)

best performing destination system, they effectively balance the server load and thus increase the throughput rate of the whole system. In general, load-balancing is just oriented at a single resource. Therefore, only one resource is considered in the analysis which is shared by the agents as is the case for network or processor resources. Thus, to compare random, native (server selection without application of the decision algorithm), and rational server selection (i.e. application of the decision algorithm) a simple $M/M/m$ queuing model is used: The service rates μ_i are deterministic, yet the customer's resource requirements are exponentially distributed with rate α. Table 1 summarises all model parameters and their settings in this analysis. To ease the analysis, it is assumed that customers, i.e. agents respectively, compete for the same kind of resource and that the server selection time is constant. Moreover, all servers are able to satisfy the customer's service requirements. Thus, on each request, the server selection system evaluates all servers. Finally, errors of the selection system have not been considered. Next a performance model for rational server selection is developed. This model will finally be compared with random and native server selection.

If only a single resource is considered equation 2 can be simplified to:

$$E[U(r,n)] = \frac{r}{E[R]} - \frac{r}{E[R_n]} - \delta - E\left[Err\left(\frac{r}{E[R]} - \frac{r}{E[R_n]}\right)\right] \quad (3)$$

Consequently, for a fixed n there exists a break-even resource requirement $\hat{r}$ which meets $U(\hat{r},n) = 0$. That is, customers having resource requirement smaller than $\hat{r}$ select servers randomly; on the contrary, customers applying server selection have a requirement greater than $\hat{r}$. The break-even resource requirement is given as:

$$\hat{r} = \frac{\left(\delta + E\left[Err\left(\frac{\hat{r}}{E[R]} - \frac{\hat{r}}{E[R_n]}\right)\right]\right) \cdot E[R] \cdot E[R_n]}{E[R_n] - E[R]} \quad (4)$$

Thus, to model rational server selection two customer classes must be distinguished, namely customers which apply random selection and those which apply server selection. Looking at a server i the fraction of customers with resource requirement greater than $\hat{r}$ depends on its load share. To model the ratio between the two customer classes the divided exponential distribution is derived (see appendix). Its density function f is:

$$f(\hat{r},\omega,\alpha,x) = \frac{1}{\omega \cdot \left(1 - e^{-\alpha \cdot \hat{r}}\right) + (1-\omega) \cdot e^{-\alpha \cdot \hat{r}}} \cdot \begin{cases} \omega \cdot \alpha \cdot e^{-\alpha \cdot x} & x < \hat{r} \\ (1-\omega) \cdot \alpha \cdot e^{-\alpha \cdot x} & x \geq \hat{r} \end{cases} \quad (5)$$

where ω reflects the ratio between the customer classes. Thus, the mean $E[S_i](\hat{r})$ of a system's i service time is given by:

$$E[S_i](\hat{r}) = \int_0^\infty f(\hat{r},\omega_i,\alpha,x) \cdot \left(\frac{x}{\mu_i}\right) dx$$

With the help of density function f the ratio between two customer classes can be modelled. Yet, this approach requires a model transformation, i.e. an adaptation of the individual arrival rates. However, at first the weights ω_i must be derived. Preconditioned that none of the servers is overloaded when random selection is applied, customers with a requirement smaller than $\hat{r}$ are equally distributed among the servers. The others are fairly distributed among the servers where each server i receives a share of $\frac{\mu_i}{\sum_{j=1}^m \mu_j}$. Thus the relation ω_i at server i is:

$$\omega_i = \frac{1}{m \cdot \left(\frac{1}{m} + \frac{\mu_i}{\sum_{j=1}^m \mu_j}\right)}$$

The arrival rate of customers at server i, λ_i, is:

$$\lambda_i = \lambda \cdot \left(\frac{1}{m} \cdot \left(1 - e^{-\alpha \cdot \hat{r}}\right) + \frac{\mu_i}{\sum_{j=1}^m \mu_j} e^{-\alpha \cdot \hat{r}}\right)$$

where the first summand is the fraction of customers with resource requirements lower than $\hat{r}$ arriving at i and the second the fraction of those with requirements greater or equal than $\hat{r}$. By applying the Pollaczek-Kinchin formula [6], the average response time of a customer with resource requirement $r \in \mathfrak{R}_{\geq 0}$, $E[R]_{rat}(\hat{r}, r)$ is:

$$
E[R]_{rat}(\hat{r}, r) = \sum_{i=1}^{m} \begin{cases} \frac{1}{m} \cdot \left(E[W_i]_{rat}(\hat{r}) + \frac{r}{\mu_i} \right) & r < \hat{r} \\ \frac{\mu_i}{\sum_{j=1}^{m} \mu_j} \cdot \left(E[W_i]_{rat}(\hat{r}) + \frac{r}{\mu_i} \right) + \delta & r \geq \hat{r} \end{cases} \tag{6}
$$

where the average service time at server i is given by:

$$
E[W_i]_{rat}(\hat{r}) = \frac{\lambda_i^2 \cdot E[S_i^2](\hat{r})}{2 \cdot (1 - \rho_{rat})}
$$

$$
\rho_{rat}(\hat{r}) = \sum_{i=1}^{m} \frac{\lambda}{m^2} \cdot E[S_i](\hat{r})
$$

$E[S_i^2](\hat{r})$ is the second moment of the service time at server i. Apparently, at the break-even point $\hat{r}$ there is no difference between selecting customers randomly or server selection. Therefore, it can easily calculated by solving the equation:

$$
\sum_{i=1}^{m} \frac{1}{m} \cdot \left(E[W_i]_{rat}(\hat{r}) + \frac{\hat{r}}{\mu_i} \right) = \sum_{i=1}^{m} \frac{\mu_i}{\sum_{j=1}^{m} \mu_j} \cdot \left(E[W_i]_{rat}(\hat{r}) + \frac{\hat{r}}{\mu_i} \right) + \delta
$$

The random and native server selection performance models are special cases of the rational server selection model. That is, random selection corresponds to a rational server selection setting where the break-even point is infinite. Likewise, native server selection corresponds to rational server selection with a break-even point of zero. Thus, based on equation 6 derivation of the average response times of random server selection $E[R]_{rand}(r)$ and native server selection $E[R]_{native}(r)$ are straightforward and are given by:

$$
E[R]_{rand} = \lim_{\hat{r} \to \infty} E[R]_{rat}(\hat{r}, r) \quad E[R]_{nativ} = E[R]_{rat}(0, r)
$$

The performance evaluation has been restricted to a queuing system comprising only two servers. Though this scenario is quite simple, it is sufficient to illustrate the characteristics of rational server selection. Figure 1 shows the average customer response time versus customer resource requirements of the random, native and rational server selection approach. Apparently, the server selection approaches outperform random selection if a customer's resource requirement exceeds the breakeven point. However, those customers which use server selection even though their resource requirements are less than the breakeven point perform poor. Comparison of rational and native server selection shows: If rational selection is used customer's having less resource requirements perform well at the costs of those customers having significant requirements. But if native server selection is used, customer's having considerable requirements perform well at the costs of those having small.

4   Conclusions and Future Work

With the help of a simple performance model the performance characteristics of rational server selection have been discussed. According to this analysis, a relaxed load-balancing results in improved agent performance. Future work will focus on implementation of the decision algorithm in a server selection system and its evaluation in a real world scenario.

References

1. Gray, R.S., Kotz, D., Nog, S., Rus, D., Cybenko, G.: Mobile agents for mobile computing. Technical Report PCS-TR96-285, Dartmouth College, Computer Science, Hanover, NH (1996)
2. Theilmann, W., Rothermel, K.: Efficient dissemination of mobile agents. In: Proceedings. 19th IEEE International Conference on Distributed Computing Systems. Workshops on Electronic Commerce and Web-based Applications, Austin, TX, USA, IEEE Comput. Soc (1999) 9–14
3. Brewington, B., Gray, R., Moizumi, K., Kotz, D., Cybenko, G., Rus, D.: Mobile agents in distributed information retrieval. In Klusch, M., ed.: Intelligent Information Agents. Springer-Verlag, Germany (1999) 355–395
4. Pils, C., Diepolder, S.: Rational server selection for mobile agents. In Stefani, J.B., Demeure, I., Hagimont, D., eds.: 4th IFIP International Conference on Distributed Applications and Interoperable Systems (DAIS03). Volume 2893 of Lecture Notes in Computer Science (LNCS)., Paris, France, Springer-Verlag, Germany (2003) 61–72
5. Ogawa, J.: Distribution and moments of order statistics. In Sarhan, A.E., Greenberg, B.G., eds.: Contributions to order statistics. Wiley publications in statistics. John Wiley and Sons, Inc. (1962) 11–19
6. Kleinrock, L.: Queueing systems volume 1: Theory. John Wiley and Sons (1975)

Appendix

The divided exponential distribution models two customer streams arriving at a server A which are distinguished by different resource requirements. Basically, the resource requirements are exponentially distributed. However, dividing the customers in a number of streams according to weights $P(A|r < \hat{r})$ and $P(A|r \geq \hat{r})$ their resource requirements result in the divided distribution. Preconditioned, that the fraction of customers with resource requirement $r < \hat{r}$ and $r \geq \hat{r}$ are known, i.e. $P(A|r < \hat{r})$ and $P(A|r \geq \hat{r})$ respectively, the probability that a customer arriving at A has a resource requirement s is:

$$P(r|A) = \frac{P(r,A)}{P(A)} = \begin{cases} \frac{P(r,A,r<\hat{r})}{P(I)} & r < \hat{r} \\ \frac{P(r,A,r\geq\hat{r})}{P(I)} & r \geq \hat{r} \end{cases} \tag{7}$$

Since the streams descend from an exponential distribution with rate α, $P(A) = P(A|r < \hat{r}) \cdot \int_0^{\hat{r}} \alpha \cdot e^{-\alpha \cdot x} dx + P(A|r \geq \hat{r}) \cdot \int_{\hat{r}}^\infty \alpha \cdot e^{-\alpha \cdot x} dx$. Moreover, $P(r, A, r < \hat{r}) = P(A|r < \hat{r}) \cdot \alpha \cdot e^{-\alpha \cdot x}$, $P(r, A, r \geq \hat{r}) = P(A|r \geq \hat{r}) \cdot \alpha \cdot e^{-\alpha \cdot x}$. Finally, giving that $\omega = P(A|r < \hat{r})$ the density function of $P(r|A)$ is f (see equation 5).

Collaborative Trust-Based Secure Routing in Multihop Ad Hoc Networks

Niki Pissinou, Tirthankar Ghosh, and Kia Makki

Florida International University, Miami, Florida
{pissinou, tirthankar.ghosh, makkik}@fiu.edu

Abstract. In this paper we have proposed a secure routing protocol based on AODV for multihop ad hoc networks. Our protocol is unique in the sense that it is capable of finding a secure end-to-end route free of any malicious entity, thus resisting an internal attack within the network either in the form of compromised or disloyal nodes. We propose to find a secure and efficient route to a destination based on collaborative effort of all the nodes.

1 Introduction

Most of the research on ad hoc networks, so far, has been done in the area of routing protocols, though, in recent years, security issues have also been addressed. As the ad hoc nodes are characterized by minimum trust for each other, finding a secure end-to-end route is truly challenging. Most of the work on routing security focus on the efficient use of digital signatures or shared secret keys to authenticate and confide the data and routing headers. However, they always tend to find the shortest path between source and destination irrespective of the presence of malicious nodes in between.

In this paper we propose a secure routing algorithm to find a secure end-to-end route based on collaborative effort of all the nodes in an ad hoc network, which can withstand the attack of any malicious entity, like a compromised or a disloyal node. Our protocol is robust against any internal attack within the network tending to inject malicious routing information and disrupting the network operation.

2 Related Work

Not much work has been done to find an end-to-end secured route in ad hoc networks. Some work has been done to design some secured routing protocols in a public key infrastructure. [8,7,4]. The protocol proposed in [8] works under the assumption of a trusted certificate server, which itself violates the basic paradigm of MANET. The security of the protocol proposed in [7] is based on the assumption of the existence of an efficient key management system that enables the nodes to obtain public key information of other nodes. The authors, however, did not consider the threat from compromised node.

Some symmetric key solutions are also proposed to secure ad hoc networks [5,6,9]. In [5] the authors have proposed a protocol whose security depends upon the

N. Mitrou et al. (Eds.): NETWORKING 2004, LNCS 3042, pp. 1446–1451, 2004.

assumption of an efficient distribution of shared secrets between the nodes. This itself is a burning research problem. A similar approach is proposed in [6]. The security here is based on the efficient use of one-way hash function and works under the assumption that some secure means of distributing the elements of the hash chain is already there.

A token based approach is proposed in [9] where the use of threshold cryptography is suggested to distribute the tokens securely among the nodes. Another protocol, proposed in [12], aims at isolating the misbehaving nodes, thus making noncooperation unattractive. The monitoring mechanism is implemented by a neighborhood watch concept where the no-forwarding behavior of the nodes are monitored and reported.

In [4], the authors have proposed a protocol for securely discovering the network topology in a public key infrastructure. The protocol is responsible for securing the discovery and distribution of link state information. Another protocol to achieve a similar goal is proposed in [3], which works under the assumption of an already established shared secret between the source and destination.

All the above solutions tend to find the shortest path from source to destination irrespective of some malicious nodes in between. In [1], the authors have proposed a secured routing protocol based upon the trust level of the nodes. Although their approach is unique, the protocol fails under the attack of a compromised or a disloyal node.

3 Our Protocol

3.1 Goals and Assumptions

Our design is based on the following assumptions which we think are justified. First, there is a prior distribution of trust level[1] of all the nodes. Second, all the nodes communicate via a shared wireless channel and all communication channels are bi-directional. Third, all the nodes operate in a promiscuous mode. Fourth, we do not consider here physical layer or MAC layer security. Instead, we concentrate on the network layer. Our proposed routing protocol is actually a secure extension of AODV. Last but not least, we assume that all the nodes are identical in their physical characteristics, i.e., if node A is within the transmission range of B, then B is also within the transmission range of A.

3.2 Protocol Description

Essentially all routing protocols in the ad hoc community tend to find the shortest path to the destination irrespective of the presence of any malicious node in that path.

---

[1] defined in line with the organizational hierarchy of the specific application of the network deployment which is not discussed in this paper. This metric can be dynamically changed depending upon the history of the past behavior of the nodes. We are currently working on that.

We can argue that, as internal threat[2] in the network in the form of a compromised[3] or disloyal[4] node is of significant concern, a path free of malicious node is more important than the shortest path. The protocol that we propose here is an extension of the Ad hoc On Demand Distance Vector (AODV) routing protocol. The protocol works as follows:

When a node wants to find a route to another node, it initiates a route discovery. The RREQ packet header contains a *trust_level* field, in addition to the other fields in AODV RREQ. When an intermediate node receives the RREQ packet, it rebroadcasts it after modifying the *trust_level* field to include the trust level of the node that sends it the RREQ. Every node checks back the rebroadcasted RREQ packet from its next node to see whether it has provided the proper information. If not, it immediately broadcasts a warning message questioning the sanctity of that node. Our protocol does not encourage any intermediate node to send a route reply. The final route selection is based upon the *trust_level* metric. H*op_count* plays a role in deciding the final route only when more than one packet has same *trust_level*. The RREP packet has the next hop information. This is in line with the solution given in [11] to counter the black hole problem. When the source node gets back the first RREP, it waits for a specified amount of time before using that route. If within that time another RREP comes, the source node queries the next hops of the two RREPs. The next hop of the malicious RREP will obviously not have the same route to the destination. Thus, malicious route injection into the network can be prevented.

The pseudocode below shows the action of a node after it receives a route request packet.

```
// when a node receives a Route Request packet
   Receice_RREQ( ) {
// check whether it is the destination of the route
// request
     if destination {
      compute_highest_trust_level( )
      // in case more than one RREQ has same trust_level
      // decides on the basis of lowest hop_count
      sends_RREP_to_source( ) }
     else (not destination) {
        if duplicate packet {
          cross_checks_trust_level( )
          if found ok
            drops the packet
          else
            broadcasts roure_warning message( ) }
        else (not duplicate) {
```

---

[2] defined as an active attack by a compromised or a disloyal node which actively takes part in the ongoing communication.

[3] defined as a node which has been physically taken over by an intruder thus giving access to all its stored secrets and system codes.

[4] defined as a node which has ended its loyalty to the network and has decided to disrupt the network operation by non-cooperation of some means.

```
modifies trust_level
increments hop_count
rebroadcasts RREQ }}}
```

The pseudocode below shows the detailed action of the source node after it receives the first route reply.

```
// when the source node gets back the first Route Reply
Receive_RREP( ) {
    waits for a specified period
    if receives another RREP {
        queries next_hop( ) }
    else {
        sends data( ) }}
```

The function *cross_checks_trust_level* can be implemented in two ways. When an intermediate node receives a duplicate route request packet, it checks back the *hop_count* field to find out from which node it is receiving the packet. The following algorithm implements the function.

```
if (current ->hop_count = = hop_count - 1) {
cross_checks_trust_level( ) }
else {
    // the node is trying to put malicious information
    // finds out which node is malicious
    broadcasts roure_warning message( ) }
```

The above algorithm works under our assumption that all the nodes are identical in their radio range. If they are not, a node can receive a duplicate route request from any other node which it cannot reach directly and wrongly assume that the later is trying to act malicious. This will generate false warning messages in the network.

The second possible implementation takes care of this. An intermediate node, on receiving a duplicate route request packet, extracts the address stored in the *lastaddr* field (the *lastaddr* field contains the address of the node from which the next node receives a route request packet) and checks from the neighbor table whether it is from any of its neighbor. The algorithm works as follows:

```
if (lastaddr = = neighbortable->addr) {
    cross_checks_trust_level( ) }
else {
    drops the packet( ) }
```

The above implementation can actually increase the computational overhead in each node. The complexity can however be reduced by efficient searching of the neighbor table.

4 Simulation and Results

We have used Glomosim for our simulation. Glomosim is a scalable simulation software used for mobile ad hoc networks. We defined a region of 2 Km by 2 Km with random node placement. The nodes move with uniform speed chosen between 0 to 10 meters/sec. The pause between each successive movement is 30 seconds.

The results that we got confirm the efficiency of our protocol. We have benchmarked our protocol, which we call *Trust-embedded AODV (T-AODV)* with the original AODV protocol. The small percentage increase in overhead (Fig.1) can be traded off with the incorporation of security into the protocol. This increase is due to retransmission of some route request packets because of delayed receipt of route reply by the source nodes (as we do not encourage intermediate nodes to send route reply in our protocol). Actually, this overhead can be brought down by increasing the NET_TRAVERSAL time. We can also see from Fig.1 that the percentage variation in overhead decreases with increasing number of nodes. This is because, in AODV, as more and more nodes join the network, the probability of sending route replies by intermediate nodes increases, which is not the case in T-AODV as no intermediate node can send route replies.

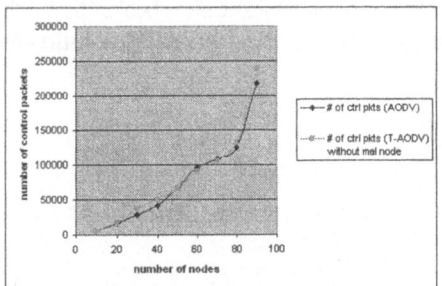

Fig. 1. Routing overhead

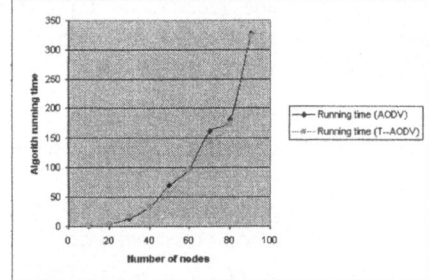

Fig. 2. Algorithm Running times

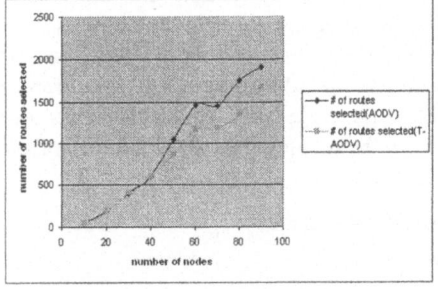

Fig. 3. Numbers of Routes selected

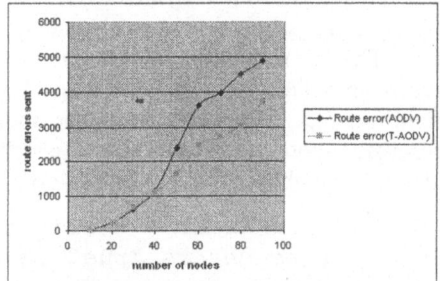

Fig. 4. Route Errors sent

The comparison of the algorithm running times in Fig.2 also shows that our protocol is efficient which runs more efficiently with increased number of nodes. A probable explanation of this is, in AODV, as we increase the number of nodes in the network, more nodes tend to find a route from its cache. In our protocol, as we do not encourage intermediate nodes to send route replies, the running times becomes lower than AODV as more nodes join the network. The time can go up by a small percentage if we increase the NET_TRAVERSAL time to lower the routing overhead.

As no intermediate node is encouraged to come up with route replies, we obviously have lesser number of routes selected in our protocol that that in AODV (Fig.3). However, this should not give any misconception that some of the routes are not properly selected. In fact, our protocol has lesser number of route errors reported than that in AODV (Fig.4). Lesser number of routes selected, in effect, renders lower processing overhead for the source nodes, as they do not have to process all the route replies from the intermediate nodes.

5 Conclusion

Currently we are working on two extensions of the protocol. We are developing a dynamic trust model instead of a predistributed static one. We are also working to make the protocol robust enough to withstand the attack from multiple malicious nodes colluding to disrupt the network, which is not currently incorporated.

References

1. Yi, S., Naldurg, P., Kravets, R.: Security-Aware Ad hoc Routing for Wireless Networks. Report No. UIUCDCS-R- 2001-2241, UILU-ENG-2001-1748, August 2001.
2. Perkins, C., Royer, E.: Ad hoc On-Demand Distance Vector Routing. In Proc. IEEE Workshop on Mobile Computing Systems and Applications, 1999.
3. Papadimitratos, P., Haas, Z. J.: Secure Routing for Mobile Ad hoc Networks. In Proc. SCS Communication Networks and Distributed Systems Modeling and Simulation Conference (CNDS 2002), San Antonio, TX, January 27-31, 2002.
4. Papadimitratos, P., Haas, Z. J.: Secure Link State Routing for Mobile Ad hoc Networks. In Proc. IEEE Workshop on Security and Assurance in Adhoc Networks, in conjunction with the 2003 International Symposium on Applications and the Internet, Orlando, FL, January 28, 2003.
5. Hu, Yih-Chun, Perrig. A., Johnson, D. B.: Ariadne: A Secure On-Demand Routing Protocol for Ad hoc Networks. MobiCom '02, September 23-26, 2002, Atlanta, Georgia, USA.
6. Hu, Yih-Chun, Perrig. A., Johnson, D. B.: SEAD: Secure Efficient Distance Vector Routing for Mobile Wireless Ad hoc Networks. In Fourth IEEE Workshop on Mobile Computing Systems and Applications (WMCSA '02), June 2002, pages 3-13, June 2002.
7. Zapata, M. G., Asokan, N.: Securing Ad hoc Routing Protocols. WiSe '02, September 28, 2002, Atlanta, Georgia, USA.
8. Sanzgiri, K., et al,: A Secure Routing Protocol for Ad hoc Networks. In Proc of the 10th IEEE International Conference on Network Protocols (ICNP'02).
9. Yang, H., Meng, X., Lu, S.: Self-Organized Network Layer Security in Mobile Ad hoc Networks. WiSe '02, September 28, 2002, Atlanta, Georgia, USA.
10. Zhou, L., Haas, Z. J.: Securing Ad hoc Networks. IEEE Network, November/December 1999.
11. Deng, H., Li, W., Agrawal, D. P.: Routing Security in Wireless Ad Hoc Networks. IEEE Communications Magazine, October 2002.
12. Buchegger, S., Le Boudec, Jean-Yves.: Performance Analysis of the CONFIDANT Protocol (Cooperation Of Nodes: Fairness In Dynamic Ad-hoc Networks), MOBIHOC '02, June 9-11, 2002, Switzerland.

Comparative Evaluation of Two Scalable QoS Architectures

Rui Prior, Susana Sargento, Pedro Brandão, and Sérgio Crisóstomo

DCC & LIACC, Faculty of Sciences, University of Porto
Rua do Campo Alegre, 823, 4150-180 Porto, Portugal
{rprior, ssargento, pbrandao, slc}@ncc.up.pt

Abstract. This paper performs a comparative evaluation of two QoS architectures, RSVP Reservation Aggregation and Scalable Reservation-Based QoS, aimed at providing QoS levels similar to the ones provided by the well-known RSVP/IntServ architecture, but scalable enough for use in high traffic core networks. The comparative analysis, based on each model's characteristics and on simulation results, shows that our Scalable Reservation-Based QoS architecture is able to support the same QoS guarantees provided by RSVP Reservation Aggregation, with significantly increased resource utilisation and a small penalty in signalling processing overhead.

Keywords: QoS, scalability, performance evaluation

1 Introduction

With the goal of benefiting from the virtues of both IntServ [1] and DiffServ [2] architectures and mitigating their problems, several architectures have been proposed in the literature. None of these architectures, however, ensures simultaneously the strict and differentiated QoS support and the maximisation of the usage of network resources without scalability concerns.

One of the most promising architectures [3] is based on aggregation of per-flow reservations, where the RSVP protocol [4] is extended to allow RSVP signalling messages to be hidden inside an aggregate. In the simplest case, all edge routers reserve bandwidth between ingress and egress routers of a network domain; these reservations can be updated in bulks much larger than the individual flow's bandwidth. When a flow requests admission in an aggregate region, the edge routers of the region check if there is enough bandwidth to accept the flow on the aggregate. If there is, the flow is accepted without signalling the core routers. Otherwise, the core routers will be signalled in an attempt to increase the aggregate's bandwidth in bulk quantities. If this attempt succeeds, the flow is admitted; otherwise, it is rejected. This architecture benefits from the fact that signalling messages are only exchanged when the aggregate's bandwidth needs to be updated. Unfortunately, the decrease in signalling rate is accompanied by a decrease in resource utilisation.

In order to address the requirements of end-to-end QoS support without resource utilisation and scalability concerns, we developed a new architecture [5]

N. Mitrou et al. (Eds.): NETWORKING 2004, LNCS 3042, pp. 1452–1457, 2004.

based on scalable per-flow signalling and resource reservations at both transit (core) and access networks with aggregate packet classification and scheduling. Several techniques and algorithms have been developed aiming at the minimisation of the computational complexity and, therefore, the improvement of the signalling scalability. More specifically, a label switching mechanism was developed with the goal of avoiding expensive lookups in flow reservation tables. Moreover, a scalable implementation of expiration timers for soft reservations, with a complexity that is low and independent from the number of flows, was also developed. In terms of QoS guarantees, [6] showed that our architecture is able to support strict and soft QoS guarantees to each flow, irrespectively of the behaviour of the other flows in the same and in different classes, with resource utilisation similar to the one obtained with IntServ, but increased scalability.

In this paper we qualitatively and quantitatively compare both architectures in terms of QoS guarantees, resource utilisation, and scalability, in order to evaluate their relative merits and shortcomings, as well as their suitability to replace the reference RSVP/IntServ architecture, which suffers from scalability problems that disallow its usage in high traffic core networks. The results indicate that both the Scalable Reservation-Based QoS (SRBQ) and the RSVP Reservation Aggregation (RSVPRAgg) models are able to provide adequate QoS levels and may, therefore, be used in place of RSVP/IntServ.

This paper is organised as follows. Section 2 consists on a comparative analysis of these two models, and is divided in two subsections, where the models are qualitatively compared based on their main characteristics (2.1) and quantitatively compared based on simulation results (2.2). Section 3 presents the most important conclusions from this work and points out some topics for future work.

2   Comparison between SRBQ and RSVPRAgg

In this section we address both qualitative and quantitative comparisons of SRBQ and RSVPRAgg. The qualitative comparison is based on the nature of the architectures, drawing their advantages and disadvantages. The quantitative comparison is based on simulation results.

A mapping between SRBQ and RSVPRAgg needs to be performed for the comparison to be fair. The aggregation regions of RSVPRAgg and the non-aggregated RSVP regions correspond to the core and access domains of SRBQ; the aggregators and deaggregators in RSVPRAgg to edge routers in SRBQ. The network topology used for the simulations is the same given this mapping.

2.1   Qualitative Comparison

The core routers in the RSVPRAgg architecture only need to store the state of aggregates, whereas in SRBQ per-flow state is stored. This is not a limiting factor of SRBQ, however, considering the routers available nowadays: in a core router handling an average of a hundred thousand flows, if each flow's information occupies 100 bytes, only 10 Mbytes are required. The most complex task for the core routers is the lookup of the flow information, based on the

5-tuple parameters that specify the flow, when the number of flows is very high. Since SRBQ uses the labels to overcome this problem, the existence of per-flow reservation structures is not a limitation of the architecture.

The scalability of the classification and scheduling procedures at the core nodes is similar in both architectures, since they are performed on a per-aggregate basis according to the DSCP of the flows. At the edge routers classification is much lighter in the SRBQ model, since no flow to aggregate mapping is performed. As scheduling is based on the DSCP on both architectures, the efficiency is comparable. At the access routers, packet classification is per-flow based in both models. Therefore, the processing load of classification is similar, but may be much reduced in SRBQ if labels are used in data packets. Packet scheduling is more efficient in SRBQ, since in RSVPRAgg usually Weighted Fair Queueing (WFQ) or a similar discipline is used outside the aggregation regions.

The approach to reducing the signalling processing load is radically different. In SRBQ end-to-end reservations are used, and scalability is achieved by making use of highly efficient techniques and algorithms (like labels and efficient timers). In RSVPRAgg end-to-end reservations are aggregated, reducing the signalling processing at the core to that needed to maintain and update aggregate reservations in bulk quantities. This approach has two disadvantages: (1) the number of signalling messages processed at the edge nodes of the aggregation region, which may be a high-traffic transit domain, is even higher than in the case of regular RSVP, since both end-to-end and aggregate messages must be processed; worse, packet classification at the edge is per-flow; and (2) the reduction in signalling is highly dependent on the bulk size, but large bulk sizes lead to a very poor utilisation of network resources. Due to the different approaches, although the number of signalling messages processed at core nodes is much higher in SRBQ, the processing load is comparable.

2.2   Quantitative Comparison

Both the SRBQ and the RSVPRAgg models were implemented in the ns-2 simulator. An existing implementation of RSVP for this simulator was also used. It is important to keep in mind that ns-2 has some limitations, the most significant of which is the inability to simulate and measure processing delays.

The simulated scenario is depicted in figure 1. It includes 1 transit (TD) and 6 access (AD) domains. Each terminal in the access domains simulates a set of terminals. The bandwidth of the connections in the transit domain, and in the interconnections between the transit and the access domains, is 10 Mbps. The propagation delay is 2 ms in the transit domain connections and 1 ms in the interconnections between the access and the transit domain. The transit domain corresponds to the aggregation region in the RSVPRAgg architecture.

The simulated scenario contains a class for signalling traffic, CL and BE classes. At each referred connection, the bandwidth assigned to the signalling traffic is 1 Mbps. Note that, although this seems very high, the unused signalling bandwidth is used for BE traffic. The bandwidth assigned to the CL class is 7 Mbps. The remaining bandwidth, as well as unused CL and signalling bandwidth, is used for BE traffic.

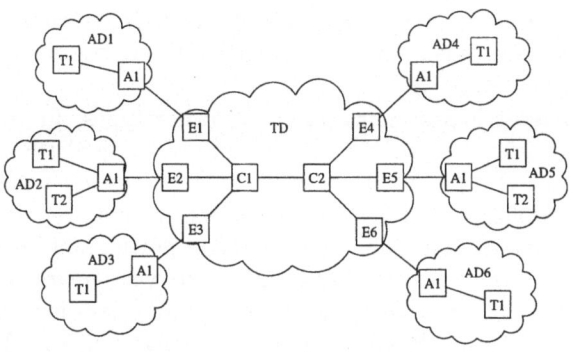

Fig. 1. Simulation topology

Each terminal of the access domains on the left side generates a set of flows belonging to the CL and BE classes. Each source may generate traffic to all destinations; the destination of each flow is randomly chosen in the set of the terminals in the right side access domains. With 3 source and 3 destination edge routers in the core domain, the number of required end-to-end aggregates in the domain is 9. Traffic belonging to the CL class is a mixture of different types of flows: CBR, exponential on-off and Pareto on-off. These flows are initiated according to a Poisson process with a certain mean time interval between calls (MTBC), and flows' durations are exponentially distributed. Filler traffic in the BE class is composed by on-off Pareto and FTP flows. All simulations were run for 5,400 simulation seconds, discarding the data for the first 1,800 seconds. All values presented are an average of 5 simulation runs with different random seeds.

Table 1 shows the flow types used in the simulations, along with their characteristics. Reservations for RSVPRAgg and RSVP are of token-bucket type; in SRBQ they are based on 3 water-marks. The mean time between calls (MTBC) is adjusted so that all flow types have, on average, the same amount of reserved bandwidth; The amount of offered load is varied from 0.8 to 1.2 (load factor) times the CL bandwidth at the core by adjusting the MTBC in all flows. The MTBC value presented in the table corresponds to a load factor of 1.

Table 1. Flow characteristics

Type	Avg. rate (kbps)	Pkt. size (Bytes)	On (ms)	Off (ms)	Pk. rate (kbps)	Token Bucket R (kbps)	Token Bucket B (Bytes)	Watermarks (kbps) 1	Watermarks (kbps) 2	Watermarks (kbps) 3	MTBC (s)	Avg. dur. (s)
cbr48cl	48	500				48	1500	48	48.048	56	13.2	120
cbr64cl	64	500				64	1500	64	64.064	72	17.5	120
exp1cl	48	500	200	200	96	64	15000	32	64	96	17.5	120
pareto1cl	48	500	200	200	96	64	15000	32	64	96	17.5	120

Figure 2 contains the most relevant results. The mean delay is shown for all flow types in SRBQ and RSVPRAgg with a bulk size of 300 kbps. As can be seen, all flows in SRBQ suffer similar average delays, which is an obvious result since they share all the queues. In RSVPRAgg the average delay inflicted to different flow types is different due to the use of WFQ outside the aggregation region; Pareto flows in this model have significantly higher queueing delays than the other flows.

Regarding packet losses, contrary to all other types, Pareto flows have a very significant packet loss ratio of about 10% in the RSVPRAgg model. It is worth noting that the Pareto distribution is a heavy-tailed one, with infinite variance. This implies that Pareto on-off flows are not well suited for token-bucket characterisation, since unless we use disproportionately large bucket sizes, packet losses will always be high. The three rate water-marks characterisation used in SRBQ for the CL class is much more appropriate for this kind of flow: losses for Pareto flows in SRBQ are always less than 0.003%. Packet losses for exponential on-off flows are very low in RSVPRAgg (about 0.003%) and null in SRBQ. All CBR flows have no packet losses in both models.

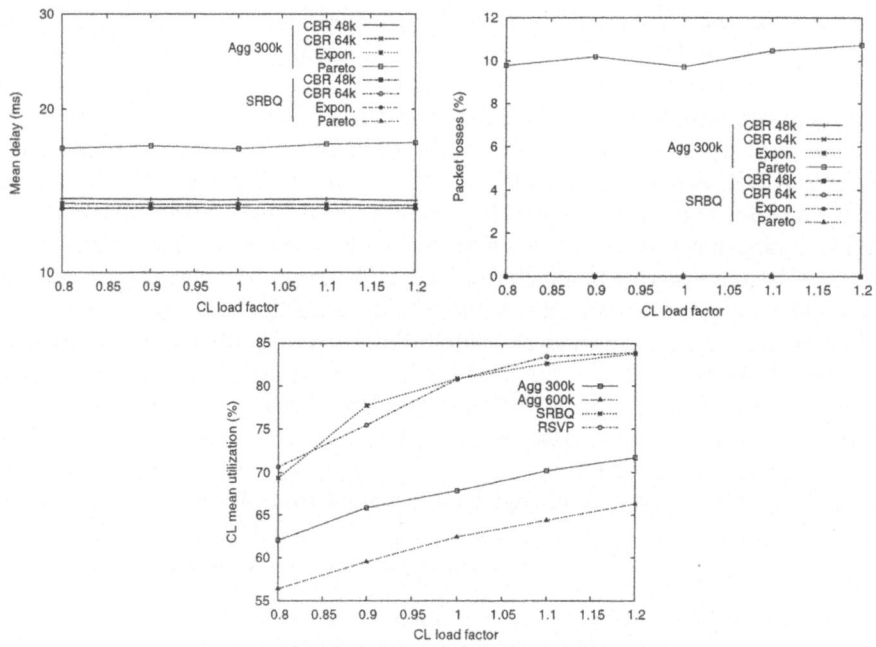

Fig. 2. Delay, packet loss and CL class utilization results

SRBQ and RSVP have similar utilisation figures, showing link saturation around an offered load factor of 1; RSVPRAgg has lower utilisation values which decrease with increasing bulk size, and does not exhibit saturation.

From the previous results we may conclude that both models provide adequate QoS, except for Pareto flows in RSVPRAgg which are not well suited for the token bucket type reservations used in that model.

3   Conclusions and Future Work

In this paper we performed a comparative evaluation of two QoS architectures, RSVPRAgg and SRBQ, aimed at providing QoS levels similar to those provided by the RSVP/IntServ architecture, but scalable enough for use in high traffic core networks. The comparison was based on implementations of both architectures in the ns-2 simulator.

From the results presented in the previous section, we may state that both the RSVPRAgg and SRBQ models provide adequate QoS levels and flow isolation in the CL class. In terms of packet classification and scheduling, both models perform aggregation of end-to-end flows and make use of the DSCP field. The number of signalling messages processed at core nodes is much lower in RSVPRAgg due to the fact that signalling scalability in SRBQ is not obtained by performing reservations at an aggregate level; instead, it makes use of highly efficient techniques and algorithms while keeping the end-to-end character of signalling. The signalling gain in RSVPRAgg is dependent on the bulk size used in aggregate bandwidth management, but the use of large bulks leads to gross under-utilisation of network resources. Based on end-to-end reservations, the SRBQ model does not have this limitation, making good use of network resources under all conditions. In addition to this, edge nodes of aggregation regions in RSVPRAgg must perform per-flow signalling (and in some cases also packet classification and scheduling). In high traffic transit networks, the number of flows even at the edge of transit domains may be huge, imposing a scalability limit to RSVPRAgg which must not be overlooked.

As future work, we plan to evaluate the possibilities for interoperability between SRBQ and other QoS architectures. We also plan to perform simulations with more realistic flows, based on data collected in real networks, and to implement prototypes of both architectures in order to evaluate performance parameters not provided by ns-2, namely regarding processing power required.

References

1. Braden, R., Clarck, D., Shenker, S.: Integrated Services in the Internet Architecture: an Overview. RFC 1633, Internet Engineering Task Force (1994)
2. Blake, S., Blake, D., Carlson, M., Davies, E., Wang, Z., Weiss, W.: An Architecture for Differentiated Services. RFC 2475, Internet Engineering Task Force (1998)
3. Baker, F., Iturralde, C., Faucheur, F.L., Davie, B.: Aggregation of RSVP for IPv4 and IPv6 Reservations. RFC 3175, Internet Engineering Task Force (2001)
4. Braden, R., Zhang, L., Berson, S., Herzog, S., Jamin, S.: Resource Reservation Protocol (RSVP) - Version 1 Functional Specification. RFC 2205, Internet Engineering Task Force (1997)
5. Prior, R., Sargento, S., Cris stomo, S., Brand o, P.: End-to-end Quality of Service with Scalable Reservations. In: Proceedings of the 11th International Conference on Telecommunication System, Modeling and Analysis. (2003)
6. Prior, R., Sargento, S., Brand o, P., Cris stomo, S.: Efficient Reservation-Based QoS Architecture. In: Interactive Multimedia on Next Generation Networks. Volume 2899 of Lecture Notes in Computer Science., Springer-Verlag (2003) 161–181

An Efficient ID-Based Authenticated Key Agreement Protocol from Pairings

Eun-Kyung Ryu, Eun-Jun Yoon, and Kee-Young Yoo

Department of Computer Engineering, Kyungpook National University,
Daegu 702-701, South Korea
{ekryu,ejyoon}@infosec.knu.ac.kr, yook@knu.ac.kr

Abstract. In this paper, we describe a new ID-based authenticated key agreement protocol that makes use of bilinear pairings. We then discuss the security properties of our scheme, including known-key security, perfect forward secrecy and no key control. It is also able to withstand both passive and active attacks. An important advantage of our scheme is that it preserves the perfect forward secrecy even though the long-term secret key of a trusted key generation center is compromised. We also show that it is more efficient than Chen and Kudla's protocol with same security properties as ours.

1 Introduction

Key agreement is one of the fundamental problems considered in cryptography. The best-known protocol for key agreement is the Diffie-Hellman protocol, which allows two parties to establish a shared secret by exchanging messages over an open channel without the need for any prior communication. However, the basic Diffie-Hellman protocol is susceptible to a man-in-the-middle attack because it does not authenticate the communicating parties.

Many solutions to this vulnerability in the Diffie-Hellman scheme have been developed over the years; recently, the identity-based (ID-based) approach has been the subject of much interest. In ID-based schemes, a public key is calculated directly from the user's identity rather than being extracted from a certificate that is issued by a trusted third-party. Such schemes can potentially provide the benefits of public key cryptography without the need for certificates and their attendant public key infrastructure.

One of the first feasible solutions for ID-based encryption was Boneh and Franklin's scheme[3], which is based on pairings on elliptic curves. Other feasible ID-based key agreements based on the pairing technique were then developed; in particular, Smart[8] proposed an ID-based authenticated key agreement protocol based on a combination of the ideas from [1] and [2].

All ID-based key agreement protocols require a Key Generation Center (KGC) that is relied upon to create and deliver private keys to entities and to not abuse its knowledge of those keys. However, a property that should be required of ID-based protocols is that if two entities are communicating, then the KGC cannot derive the established session key. In addition, if at any stage the

N. Mitrou et al. (Eds.): NETWORKING 2004, LNCS 3042, pp. 1458–1463, 2004.

KGC's key is compromised, this should not compromise the previously established session keys. This property is called *full forward secrecy* or *perfect forward secrecy*, which should be an important consideration when designing ID-based authenticated key agreement protocols.

However, Shim[6] pointed out that Smart's scheme does not have the property of perfect forward secrecy, which we believe to be an important security requirement for authenticated key agreement protocols. Shim proposed an alternative ID-based authenticated key agreement protocol, which is claimed to be efficient and to provide many security properties such as known-key security, perfect forward secrecy, key compromise impersonation resilience, and unknown key-share resilience. Nonetheless, Shim's protocol still suffers from an important security flaw because it is not protected from a man-in-the-middle attack, as described in [5]. After that Chen and Kudla in [7] introduced a ID-based authenticated key agreement protocol which includes the property of the perfect forward secrecy by increasing communication and computation overhead.

In this paper, we describe a new ID-based authenticated key agreement protocol in which computation and communication overheads for computing a session key are significantly reduced, while it provides same security propertis with Chen and Kulda's protocol. This new protocol combines the idea of ID-based cryptosystems from pairing on elliptic curve with the basic Diffie-Hellman key agreement scheme.

2 Bilinear Pairings

In this section, we briefly describe the basic definition of the bilinear pairing that is necessary for the description of our protocol. Let G_1 be a cyclic additive group generated by P whose order is a prime number q, and let G_2 be a cyclic multiplicative group of the same order q. Typically G_1 will be a subgroup of the group of points on an elliptic curve over a finite field, and G_2 will be a subgroup of the multiplicative group of a related finite field. A mapping

$$\hat{e} : G_1 \times G_1 \to G_2$$

is called a *bilinear pairing* which has the following properties:

- *Bilinearity*: $\hat{e}(P_1 + P_2, Q) = \hat{e}(P_1, Q) \cdot \hat{e}(P_2, Q)$ and $\hat{e}(P, Q_1 + Q_2) = \hat{e}(P, Q_1) \cdot \hat{e}(P, Q_2)$, or $\hat{e}(aP, bQ) = \hat{e}(P, Q)^{ab}$;
- *Non-degeneracy*: If P is a generator of G_1, then $\hat{e}(P, P)$ is a generator of G_2. In other words, $\hat{e}(P, P) \neq 1$;
- *Computability*: There is an efficient algorithm to compute $\hat{e}(P, Q)$ for all $P, Q \in G_1$.

We note that the Weil and Tate pairings associated with supersingular elliptic curves or Abelian varieties can be modified to create such bilinear maps; details can be found in [3,4].

For the remainder of the paper, we will use G_1 to refer to an additive group and G_2 to refer to a multiplicative group. We assume that the Discrete Logarithm Problem in both G_1 and G_2 is hard.

3   The Protocol

In this section we demonstrate a new ID-based authenticated key agreement protocol. This protocol consists of two phases: system setup and authenticated key agreement.

3.1   System Setup

As stated previously, an ID-based system requires a Key Generation Center(KGC) that is relied upon to create and deliver private keys to entities and to not abuse its knowledge of those keys. A KGC constructs two groups, G_1 and G_2, and a mapping $\hat{e} : G_1 \times G_1 \rightarrow G_2$ that is a bilinear pairing, as described in the preceding section. The KGC publishes $\{G_1, G_2, \hat{e}, P, H_1, H\}$, where P is a primitive root of G_1 and H_1 is a cryptographic hash function $H_1 : \{0,1\}^* \rightarrow G_1$ that maps a message of arbitrary length into a nonzero point of G_1, as described in [8]. H is a key derivation function, typically a secure hash function. The KGC then chooses a random integer $s \in Z_q^*$ as the secret key. Note that without the use of the key derivation function H, an adversary might be able to gain partial information about the session key despite the hardness of the underlying problem.

For an entity with identity information ID, the public key is given by $Q_{ID} = H_1(ID)$, and the KGC computes the private key as $S_{ID} = sQ_{ID}$. The KGC then issues S_{ID} to the entity via a secure channel. Thus, an ID-based key pair is defined as (Q_{ID}, S_{ID}), where $Q_{ID}, S_{ID} \in G_1$.

3.2   Authenticated Key Agreement

Suppose two communication entities, Alice and Bob want to establish a secret session key. To achieve this, they perform an instance of the protocol run. We denote their respective private keys as

$$S_A = sQ_A \text{ and } S_B = sQ_B$$

that have been obtained from the KGC.

Alice(A) and Bob(B) each randomly choose an ephemeral private key $a, b \in Z_q^*$, and compute the values of corresponding public keys, $T_A = aP$ and $T_B = bP$. Then they exchange the public keys as follows:

1. A → B: T_A
2. B → A: T_B

After that, Alice computes the session key $K_{AB} = H(A, B, K_A, V_A)$, where $K_A = a \cdot T_B$ and $V_A = \hat{e}(S_A, Q_B)$. Bob also computes the session key $K_{BA} = H(A, B, K_B, V_B)$, where $K_B = b \cdot T_A$ and $V_B = \hat{e}(S_B, Q_A)$.

Note, that both parties have the secret key $K_{AB} = K_{BA} = H(A, B, abP, \hat{e}(Q_A, Q_B)^s)$. Therefore the share secret key depends on the identities Q_A, Q_B of two parties, the secret key s of the key generation center and the two ephemeral keys a, b. For the process of key confirmation it can easily be added to our protocol in the same manner as described in [8,2].

The strength of our protocol depends only on the difficulty of the well-known Discrete Logarithm Problem in G_1 and on the classical Diffie-Hellman assumption. The protocol makes use of the bilinearity property, but this does not require any additional assumptions to be made.

4 Security Analysis

In this section, we argue that our scheme has the following security properties.

- **Passive attack:** If an adversary who eavesdrops on a successful protocol run can compute a session key using only information obtainable over network, then the adversary could also break the Diffie-Hellman Problem(DHP) in G_1. This is because computing the session key involves deriving the keying material abP from the values $T_A = aP$ and $T_B = bP$. Thus, we claim that it is no less difficult to break the DHP in G_1 even though the adversary knows the long-term secret key s of the KGC. Therefore our protocol resists passive attack at least as well as the Diffie-Hellman scheme.
- **Man-in-the-middle attack:** A man-in-the-middle attack, which requires an adversary to fool both sides of a legitimate conversation, cannot be carried out by an adversary who does not know Alice or Bob's private key. For example, suppose that an adversary, Eve, wants to fool Bob into thinking he is talking to Alice. First, Eve can compute $A' = a'P$ and send A' to Bob. Conversely, Bob computes $B = bP$ and send them to Eve, believing her to be Alice. The adversary must then compute $\hat{e}(Q_A, Q_B)^s$ to derive a correct session key. Therefore, it is argued that an adversary with no knowledge of S_A or S_B, is not in a position to launch a classical man-in-the-middle attack against it.
- **Known-key security:** Suppose that an adversary learned a key $K_{AB} = H(A, B, abP, \hat{e}(Q_A, Q_B)^s)$ from a past session. The adversary does not gain any additional information from combining the past key with publicly visible data for the purpose of deducing future session keys. This is true since each run of the protocol computes a unique session key that depends on the ephemeral private keys a and b. There does not appear to be any easier way for him to carry out an expensive brute-force attack. It means that the adversary, having obtained some past session keys, gains no advantage toward computing future session keys. Thus, it the protocol resists the known-key attack.
- **Perfect Forward secrecy:** Suppose that an adversary has learned a long-term private key, either S_A or S_B, or both of the entities involved in a conversation. To extract the past session keys, the adversary must compute abP from aP and bP. However, this is assumed to be a hard problem equivalent to solving the DHP in G_1. In our scheme, any previous session key will not be compromised even if the long-term key s of the KGC may be corrupted. Therefore, it preserves the property of perfect forward secrecy. This property is one advantage of our scheme over the Smart's protocol[8] in which the

Table 1. Efficiencies of ID-based authenticated key agreement protocols

Protocol	Pairng	Point multiplication	Large blocks
Smart's protocol	2	2	2
Chen and Kudla's protocol	1	4	4
Ours	1	2	2

compromise of the long-term private keys or the KGC's secret key allows past session keys to be computed.

- **Key-compromise impersonation attack:** Suppose that Alice's long-term private key S_A is revealed to an adversary, Eve. Then, Eve can of course impersonate Alice in any protocol in which Alice is identified by this key. However, in our protocol, the compromise of one entity's long-term private key does not imply that the private key of the other entity will also be compromised. That is, possession of this key does not allow Eve to impersonate Bob to Alice, nor can she impersonate any entities besides Alice to Bob. To achieve this goal, the adversary would have to solve the Discrete Logarithm Problem in G_1. Thus, our protocol resists the key-compromise impersonation attack.

- **No key control:** The session keys in our protocol are determined jointly by both parties, so that neither party alone can control the outcome of the session key by restricting it to lie in some predetermined small set. Therefore, there is no key control in our protocol.

5 Efficiency

The proposed protocol is role symmetric, meaning both communication entities execute the same operations. We compare our protocol with Smart's protocol[8] and Chen and Kudla's[7], which are also role-symmetric ID-based schemes.

The factors that most affect the overall performance of authenticated key agreement protocols include the number of rounds, the communication overhead, and the computational overhead; therefore it is desirable to minimize these properties of the protocol used. In this section, we thus compare our protocol with them in terms of computation overhead and exchanged large message blocks except the number of rounds. Since message flows in our protocol are identical with the message flows of the two pass elliptic curve based unauthenticated Diffie-Hellman protocol as well as Smart's protocol and Chen and Kudla's.

Table 1 shows efficiencies of ID-based authenticated key agreement protocols for each user. In our protocol, each user requires to compute only one pairing and two elliptic curve point multiplications for establishing session key. The calculation of a bilinear pairing is a computationally expensive process; therefore reducing the number of pairing operations in a pairing-based protocol leads to significantly greater efficiency.

As we see from Table 1, our scheme and Chen and Kudla's protocol require only one pairing while Smart's needs two pairings. Furthermore, Smart's protocol does not preserve the security property of perfect forward secrecy. In Chen and Kudla's protocol, each entity needs more two elliptic curve point multiplications and two large data blocks exchanged than ours by allowing their scheme to include the perfect forward secrecy. Therefore, the proposed scheme can be expected as the most efficient one in terms of computation and communication overhead.

6   Conclusion

Recently, many cryptographic schemes from pairings have been proposed. In this paper, we presented a new ID-based authenticated key agreement protocol that makes use of bilinear pairings. The security of our scheme is based on the difficulty of the well-known Discrete Logarithms Problem over an elliptic curve and on the classical Diffie-Hellman assumption. We argued that the proposed scheme has the properties of known-key security, perfect forward secrecy and no key control; it is also able to withstand both passive and active attacks, including key compromise impersonation and man-in-the-middle. We have also shown that our protocol is more efficient than Chen and Kudla's protocol with same security properties as ours.

Acknowledgement. We would like to thank anonymous reviewers for the helpful comments.This work was supported by the Brain Korea 21 Project in 2003.

References

1. A. Joux: A one-round protocol for tripartite Diffie-Hellman. Algorithm Number Theory Symposium, Lecture Notes in Computer Science 1838 (2000), pp. 385-394
2. A. Menezes, M. Qu, J. Solinas and S. Vanstone: Some new key agreement protocols providing mutual implicit authentication. In proceedings of the second workshop on Selected Area in Cryptography. (1995), pp. 22-32
3. D. Boneh, and M. Franklin: Identity-based encryption from the Weil pairing. Advances in Cryptology(Crypto'2001), Lecture Notes in Computer Science 2139 (2001), pp. 213-229
4. D. Boneh, B. Lynn, and H. Shacham: Short signatures from the Weil pairing. Advances in Cryptology(Asiacrypt'2001), Lecture Notes in Computer Science 2248 (2002), pp. 514-532
5. H. Sun and B. Hsieh: Security Analysis of Shim's Authenticated Key Agreement Protocols from Pairings. (available on eprint.iacr.org)
6. K. Shim: Efficient ID-based authenticated key agreement protocol from the Weil pairing. Electronics Letters 39 (2003), pp. 653-654
7. L. Chen and C. Kudla: Identity based authenticated key agreement protocols from pairings, Computer Security Foundations Workshop, (2003), pp.219-233
8. N.P. Smart: An identity based authenticated key agreement protocol based on the Weil pairing. Electronics Letters 38 (2002), pp. 630-632

Fast Flooding in Power Save Mode of IEEE 802.11 DCF Based Mobile Ad Hoc Networks

Myung-Hwan Seo[1], Hyeon-Ju Yoon[2], and Joong-Soo Ma[1]

[1] Information and Communications University,
103-6, Munji-Dong, Yusong-Gu, Daejeon 305-714, S. Korea
http://mm.icu.ac.kr
[2] Korea Advanced Institute of Science and Technology,
373-1, Gusong-Dong, Yusong-Gu, Daejeon 305-701, S. Korea
{mhseo, jsma}@icu.ac.kr, juyoon@kaist.ac.kr

Abstract. We propose Fast Flooding in Power Save Mode (FFPSM), a modification of IEEE 802.11 power save mode operation for Mobile Ad hoc NETworks (MANET). The previous works for the power save mode operation usually focused on shortening the duration of the *awake* state to minimize the power consumption of mobile nodes. We observed that the longer sleeping period results in seriously long delivery latency because the packets can move forward only one hop for a fixed interval. Our FFPSM scheme allows the flooding packets to be forwarded several hops in a transmission period. Each node propagates the announcements for next flooding to nodes of several hops away, thus the packets can travel multiple hops during one interval. Simulation results show that the proposed approach significantly reduces the end-to-end delay of flooding without extra energy consumption.

1 Introduction

Mobile ad hoc network (MANET) is a network of self-organized, wireless, mobile hosts with limited capacity of battery power. In the literature of MANET research, energy efficient techniques have been very important issues. IEEE 802.11 DCF (Distributed Coordination Function) [1] is a fully distributed MAC protocol for MANETs. It specifies a power saving mechanism, PSM (Power Save Mode), which consists of 3 states of power consumption, *off*, *awake*, and *doze* state. Very little power is consumed in the *doze* state in which the wireless interface is not able to transmit or receive.

In DCF, time is divided into fixed intervals, beacon intervals (BI), and each BI is divided into ATIM (Ad hoc traffic Indication Message) window and data transmission window (Fig. 1). During the *ATIM Window*, each node wakes up and exchanges the announcements for next data transmission with neighbors. The nodes that send or receive ATIMs stay awake for data transmission and others go into the *doze* state. The data packets usually move forward hop by hop, taking one BI per hop.

N. Mitrou et al. (Eds.): NETWORKING 2004, LNCS 3042, pp. 1464–1469, 2004.

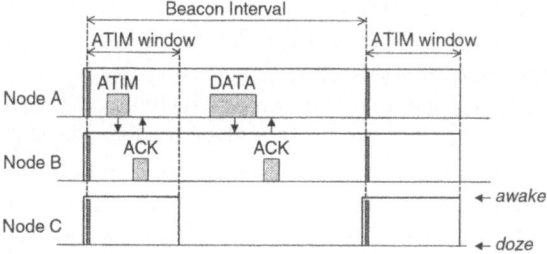

Fig. 1. Operation in IEEE 802.11 DCF Power Save Mode

Some modification of PSM have been developed to achieve more energy saving, by reducing the *awake* interval. However, the packet delivery latency from source to destination has not been an important issue, though it may generate significant user inconvenience and even more power consumption overhead.

We propose a new approach, FFPSM (Fast Flooding in Power Save Mode), which can significantly reduce the end-to-end delay of flooding and maintain other performance measures at the acceptable levels. The main idea is to maximize the utilization of the compulsory *awake* state interval, *ATIM Window*, not to shrink or adjust its size. FFPSM makes many nodes awake at the same time, so the packet flooding speed can be enhanced.

2   Related Works

In the IEEE 802.11 DCF-PSM, the different sizes of *BI* and *ATIM Window* have significant impact on energy saving and throughput achieved by the nodes. [2] recommended 95ms of *BI* and about 25% of *BI* as *ATIM Window* size.

DPSM (Dynamic PSM) by Jung et al.[3] dynamically adjusts the ATIM window size based on observed network conditions. It improved energy efficiency without degrading throughput, and they revealed that highly loaded network may cause low throughput per energy unit because of the extra use of channel capacity for *ATIM Window*. TIPS (Traffic Indication-based Power Saving)[4] is another approach to prolong the *doze* state. It allows the beacon transmission to indicate the presence of following ATIM, thus nodes to go to sleep early. Both [3] and [4], the modifications of IEEE 802.11 DCF-PSM, have focused on reducing the duration of the *awake* state, thus achieving more energy saving. However, they didn't pay attention to the long latency caused by the periodic sleeping and intermittent packet forwarding, and the effect of multi-hop transmission.

Span [5] is a power saving technique improving the latency. In span, only a small number of nodes called coordinator make a forwarding backbone and remain awake at any time to forward traffic for active connections, and others work in PSM. Coordinators are elected based on local decision and change periodically. With the geographic routing which uses excessive broadcasts, Span showed more energy saving and shorter latency than those of IEEE 802.11 PSM.

However, it is customized for geographic routing, so the distributed coordinator election overhead would be large and usually intensive use of specific nodes may result in shorter network lifetime under other conditions.

On the other hand, some power saving techniques focused on the clock synchronization[6][7]. In the standard, nodes synchronize their clocks by distributed method. In real environment, due to the mobility and the radio interference, there is possibility not to recognize each other for a long time even though they are within the communication range. [6] and [7] proposed asynchronous PSM, in which each node repeats *awake* and *doze* state by a specific pattern devised to meet others without fail. However, packet delivery rate may degrade because of the asynchrouns and fewer coincidence, and become more critical in the case of broadcasting. In this paper, we assume that the nodes are fully synchronized as assumed in [2][3][4].

3   Fast Flooding in Power Save Mode

We propose a fast flooding scheme for the power save mode (FFPSM) of IEEE 802.11 DCF to reduce the end-to-end delay. Each node has to wake up periodically and stay awake for a fixed duration (*ATIM Window*), regardless of participating in the data transmission. If the upcoming packets can be announced to receivers of several hops ahead during the current *ATIM Window*, the receivers stay awake for the data transmission and the packets can move forward several hops at an interval.

This idea inevitably involves the upper layer protocols and requires the modification of the standard. The MAC protocol for wireless networks is responsible for only one-hop communication. To propagate the announcement for unicast messages through the network, the MAC frames should contain the routing information, or network layer operation should be performed at the step of ATIM exchange. In the case of broadcasting, network layer information or operation is not required because the packets have to be delivered to all nodes. Also, the broadcasting is a fundamental communication method in many essential protocols, because of the ever-changing topology in MANETs. Therefore, the proposed scheme considers only the broadcast packets by *flooding*, one of the simplest and reliable broadcasting algorithm [8]. We surmise that the improvement of flooding performance makes the overall network performance better and our scheme can be adapted to other broadcast algorithms.

ATIM frames in IEEE 802.11 DCF may be classified into 2 types by the destination address (DA). If DA is a group address, all single-hop neighbors which listen the announcement stay awake for the subsequent broadcast transmission. If not, the only designated node sends an ACK and stays awake, while others go to sleep. We refer to them as B-ATIM and U-ATIM for each, and introduce another class, F-ATIM. The F-ATIM contains a group address of DA and additional information to tell if it is for multi-hop broadcasting.

The ATIM and frame transmission of FFPSM is as follows. Unspecified procedures conform to the standard specification [1].

(a) If a mobile node has flooding packets to send, it transmits a F-ATIM frame during the *ATIM Window*. After the *ATIM window*, it remains awake and transmits the buffered flooding packets.

(b) The nodes that receive a F-ATIM frame also transmit a F-ATIM frame during the rest of the current *ATIM Window* and remain awake, though they don't have buffered packets yet. After they receive the flooding packet announced by the F-ATIM, they forward it to neighbors during the rest of the *BI*. If a node receives a F-ATIM frame but cannot transmit it during the current *ATIM Window*, the node doesn't forward the flooding packets until the beginning of next *BI*.

(c) A node transmits only one F-ATIM frame, though it receives several F-ATIMs.

(d) The node that transmits a F-ATIM shall not transmit B-ATIMs, though it has buffered broadcast packets.

(e) The nodes that receive F-ATIMs or B-ATIMs shall not transmit U-ATIMs to the originators of received ATIMs, determining that the nodes will remain awake afterwards.

We propose a tentative solution to indicate ATIM types. ATIM frame is a 802.11 management frame, of which type and subtype value in Frame Control field is 1001. U-ATIM and B-ATIM are classified by the DA field. To indicate the F-ATIM, we use the DA field combined with the Frame Body field, not used in the original ATIM frame. The value of DA field of F-ATIM is a common broadcast IP or a special multicast IP assigned by applications, and the Frame Body field contains a TTL (time-to-live) value copied from the IP header. The responsibility to set the address and TTL field is upon the applications or routing protocols in the source node. Other intermediate nodes check and copy the TTL value from the received frames. Additional use of Frame Body field is simpler than changing the definition or function of other fields, and makes the later improvement or modification of ATIM frame easier.

4   Simulation Results and Discussion

We simulated the proposed FFPSM under various conditions, and compared the performance in terms of end-to-end delay and power consumption with IEEE 802.11 Active mode (AM) and Power Save mode (PSM). We implemented IEEE 802.11 PSM and our FFPSM modules into the ns-2.26 network simulator[9] with wireless extensions from the CMU Monarch project[10]. Multi-hop transmission was actuated by AODV routing protocol.

A simulation area of 1000m × 1000m is used, and we varied the number of nodes from 20 to 100, randomly placed in the area. For each number of nodes, 10 different node deployment scripts are generated and the results are averaged. We use 100ms and 20ms as value of *BI* and *ATIM Window* size, based on the results of [2] and [3]. To measure energy consumption, we use 1.327W, 0.976W, 0.843W, and 0.066W as value of power, consumed by the network interface in transmit, receive, idle, and doze state.

Fig. 2 shows the results of the case of 60 nodes. 10 sources generate unicast CBR (constant bit rate) flows starting randomly between 0 and 80 sec. Each flow lasts for 20 seconds and total simulation time is 100 sec. We varied the traffic load of each flow as 10, 20, 30, 40, and 50kbps. On the other hand, total 30 nodes are randomly chosen for flooding flows. Each node generates one flooding packet at random time.

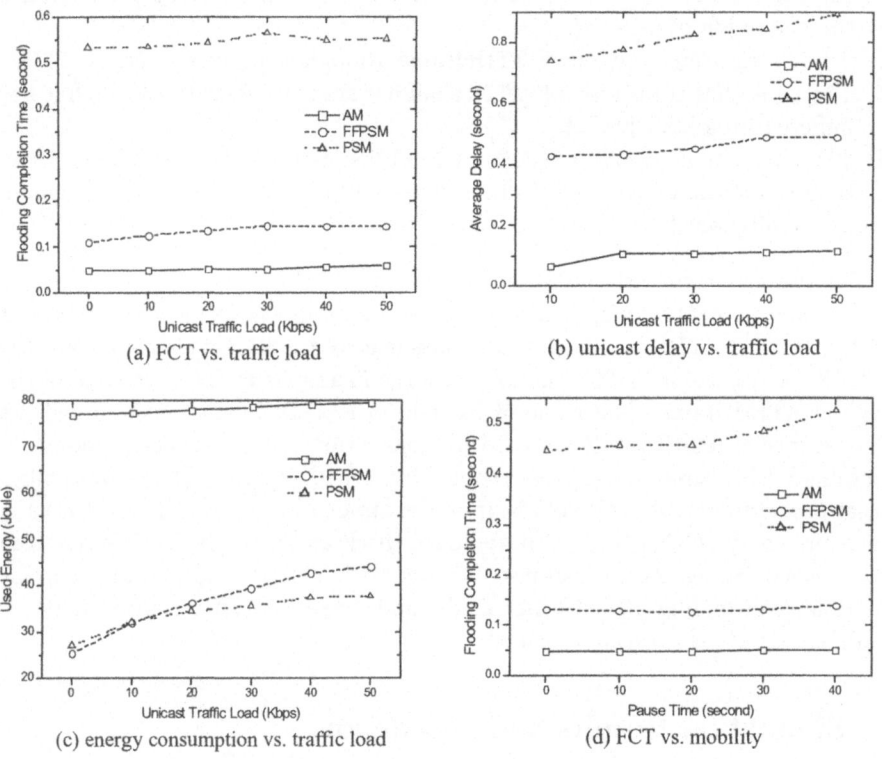

Fig. 2. Performance comparison

In the Fig. 2(a), end-to-end delay is represented as flooding completion time (FCT), which is the time difference between the departure time from source node and the arrival time at the latest destination. FFPSM significantly improves the latency problem of PSM. In PSM, flooding packets can travel the network one hop per *BI*, so the FCT of PSM is determined by the longest hops to reach the farthest node, while FFPSM consumes just 2 *BI*s. FFPSM does not provide any method to reduce unicast transmission delay, but FFPSM shows better performance than PSM in unicasting, too (Fig. 2(b)). The speedup of unicast messages is explained with 2 reasons: 1) by multi-hop announcement, many nodes are awake in the same *BI*, and 2) routing protocol also uses flooding to discovery routes.

Fig. 2(c) shows that FFPSM consumes similar amount of energy with PSM, though its transmission latency is much smaller than that of PSM.

We also evaluated the effect of mobility on the performance of FFPSM and others (Fig. 2(d)). The random waypoint model parameterized by *pause time* was used for node mobility. The mobility varies from constant (0 *pause time*) to low (40 *pause time*), and the maximum node speed is 10m/s. The FCT values are almost same except that the FCT of PSM is more sensitive to the level of mobility.

5   Conclusion

In this paper, we presented Fast Flooding in Power Save Mode (FFPSM), a modification of IEEE 802.11 power save mode operation. Power saving mechanisms can reduce the power consumption, but they suffer from long latency of packet delivery due to the periodic sleeping and intermittent packet forwarding. FFPSM does not reduce the duration of the *awake* state for energy efficiency, but maximizes its utilization. Each node propagates the announcements of next flooding to nodes of several hops ahead, thus the packets can travel multiple hops during one beacon interval.

We evaluated the performance by simulation in terms of end-to-end delay and power consumption compared with IEEE 802.11 PSM. Simulation results showed that the proposed approach significantly reduces the end-to-end delay of flooding. For various traffic load and mobility, FFPSM showed much less delay of both flooding and unicast, with similar level of energy consumption.

References

1. IEEE Computer Society LAN MAN Standards Committee: IEEE Std 802.11-1999, Wireless LAN Medium Access Control and Physical Layer specifications (1999)
2. Woesner, H. et al.: Power-Saving Mechanisms in Emerging Standards for Wireless LANs: The MAC Level Perspective. IEEE Personal Communication (1998)
3. Jung, E.-S., Vaidya, N. H.: An Energy Efficient MAC Protocol for Wireless LANs. IEEE Infocom'02, (2002)
4. Choi, J.-M., Ko, Y.-B., Kim, J.-H.: Enhanced Power Saving Scheme for IEEE 802.11 DCF based Wireless Networks. IFIP PWC 2003, Venice, Italy (2003)
5. Chen, B. et al.: Span: An Energy-Efficient Coordination Algorithm for Topology Maintenance in Ad Hoc Wireless Networks. ACM MobiCom'01 (2001)
6. Tseng, Y.-C., Hsu, C.-S. and Hsieh, T.-Y.: Power-Saving Protocols for IEEE 802.11-Based Multi-Hop Ad Hoc Networks. IEEE Infocom'02 (2002)
7. Zheng, R., Hou, J. C., and Sha, L.: Asynchronous Wakeup for Ad Hoc Networks. ACM MobiHoc'03 (2003)
8. Williams, B., Camp, T.: Comparison of Broadcasting Techniques for Mobile Ad Hoc Networks. ACM MobiHoc'02 (2002)
9. http://www.isi.edu/nsnam/ns/index.html
10. http://www.monarch.cs.cmu.edu/cmu-ns.html

Seamless Congestion Control over Wired and Wireless IEEE 802.11 Networks

Vasilios A. Siris* and Despina Triantafyllidou

Institute of Computer Science (ICS)
Foundation for Research and Technology - Hellas (FORTH)
P.O. Box 1385, GR 711 10 Heraklion, Crete, Greece
vsiris@ics.forth.gr

Abstract. We present a new approach for seamless congestion control over heterogeneous networks containing wired and wireless IEEE 802.11 links. The approach uses ECN (Explicit Congestion Notification) as a common signalling mechanism for conveying congestion information from both wired and wireless links. Two additional novel aspects of the approach are that ECN marking for a wireless link, due to the way resources are shared, is performed for both the uplink and the downlink based on measurements of the aggregate traffic in both directions, and the marking mechanism dynamically adapts to varying traffic and load conditions. Simulation results demonstrate that our approach achieves higher fairness compared to drop-tail queueing, and can effectively control the average packet delay over the wireless link.

1 Introduction

The number of users accessing the Internet and enterprise intranets through wireless links, and IEEE 802.11 wireless LANs (WLANs) in particular, is expected to grow dramatically with the proliferation of wireless hotspots and enterprise WLANs. Hence, there is a need for efficient and fair congestion control over heterogeneous networks that include both wired and wireless links.

In this paper we propose a new approach that enables TCP, in conjunction with Explicit Congestion Control (ECN), to operate seamlessly over heterogeneous networks. The approach combines three key ideas: First, it uses ECN as the common end-to-end signalling mechanism for conveying congestion information from both wired and wireless links; second, marking for the wireless link is performed using a load-based marking (LBM) algorithm, where the marking probability is a function of the aggregate utilization; third, the load-based marking algorithm dynamically adapts to varying traffic and load conditions in order to achieve an average packet delay over the wireless link within a target range.

Although the application of ECN to wireless networks is not new, e.g. see [1, 2], its application as a common signalling mechanism for conveying congestion information in wired and wireless networks, in a way that takes into account the

---

* The authors are also with the Department of Computer Science, University of Crete

N. Mitrou et al. (Eds.): NETWORKING 2004, LNCS 3042, pp. 1470–1475, 2004.
© IFIP International Federation for Information Processing 2004

particular characteristics of the underlying wireless technology was first proposed in [3], for the case of 3G networks based on Wideband CDMA. However, as we discuss in this paper, IEEE 802.11 WLANs differ from 3G WCDMA based cellular networks, hence the marking procedure for each should be different.

It is interesting to position our approach with respect to the cross-layer paradigm, which has emerged as an effective way for designing efficient network protocols over wireless link technologies [4]. Cross-layer design departs from the strict layer separation, which has been the traditional approach for network protocol design. Our approach follows the cross-layer design paradigm in the sense that the proposed ECN marking procedure takes into account the particular characteristics and the resource sharing model of IEEE 802.11 WLANs, hence differs from the marking procedure for wired links. Interestingly, our approach maintains TCP's end-to-end operation and semantics, hence adheres to the end-to-end argument stated in [5]: congestion control is performed at the end-systems, which is where aggregate information on the level of congestion for the whole end-to-end path exists. The particular characteristics of the underlying link technology are hidden from the TCP layer through appropriate design of the ECN marking procedure.

The rest of the paper is structured as follows. In Section 2 we present our approach for seamless end-to-end congestion control, in Section 3 we present and discuss simulation results, and finally in Section 4 we conclude the paper.

2   Seamless Congestion Control

2.1   ECN as a Common Signalling Mechanism

Explicit Congestion Notification (ECN) has been approved as an IETF proposed standard [6]. With ECN, congestion of a network link is explicitly signaled by having routers set the CE (Congestion Experienced) bit located in the IP header, rather than implicitly signaled through lost packets as is the case with TCP's current operation. ECN can thus provide an early warning of incipient congestion, before packets start to be dropped, thus avoiding their retransmission. Hence, ECN can, to a large extent, avoid packet drops due to congestion and the corresponding overhead of retransmitting lost packets.

Our approach for using ECN goes one step further, and proposes to use ECN to convey congestion information from both the wired and the wireless links. For wired networks, marking is performed at the output link of routers, whereas for a wireless link, marking is performed at the access point, Figure 1. Indeed, the marking procedure for wired and wireless links need not be, and as we argue in this paper, should not be the same.

2.2   Load-Based Marking (LBM)

For WLANs based on IEEE 802.11, both the uplink and the downlink share the same resource (wireless spectrum). Hence, the aggregate throughput in both

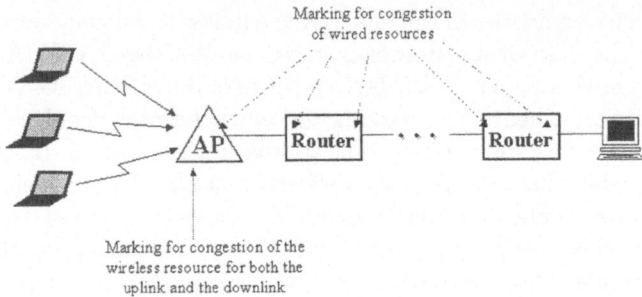

Fig. 1. Routers are responsible for marking based on the congestion level at each of their wired links, whereas the access point (AP) is responsible for marking based on the congestion level of the wireless resource (spectrum).

directions should be taken as an indication of the utilization, and hence of the level of congestion of the wireless resource. Moreover, since there is no single shared buffer that is used for the packets flowing in both directions, a RED (Random Early Detection)-like marking algorithm, where the packet marking probability is a function of an average queue length, cannot be applied.

Based on the above discussion, we propose that the probability of marking a packet, flowing either in the uplink or the downlink direction the wireless link, is a function of the aggregate utilization over some time interval t_{avg}, taking into account traffic flowing in both directions. The marking probability can have a piecewise linear dependence on the aggregate utilization: the marking probability is zero when the average utilization is less than ρ_0. For utilization values ρ larger than ρ_0, the marking probability is given by $\min\{\alpha(\rho - \rho_0), 1\}$.

It is interesting to note that in the case of 3G cellular networks based on Wideband CDMA, unlike the case of WLANs based on IEEE 802.11, the uplink and downlink directions use different frequency bands. Furthermore, in the downlink of WCDMA networks there is a shared buffer located at the base station; this is not the case in the uplink direction, where each mobile has its own local buffer. Hence, similar to 802.11, a RED-like mechanism cannot be applied in the uplink direction of WCDMA networks.

2.3   LBM Adaptation

Recall that the LBM algorithm has three parameters: the time interval t_{avg} over which the average utilization is measured, the minimum utilization ρ_0, and the slope parameter α. The time interval t_{avg} determines how quickly the algorithm adjusts the marking probability to changes of the aggregate utilization, and the timescale over which congestion is detected. Typically, t_{avg} will be set to some number of round trip times, in order to obtain stable measurements of the load. The slope parameter α affects the reactivity and the stability of the marking algorithm [7]: A higher slope would yield a more reactive algorithm since a small change of the utilization would give a large change of the marking probability. Finally, for a fixed slope parameter α, the minimum utilization ρ_0 determines the marking probability for a given aggregate utilization, hence the utilization achieved in the steady state.

Based on the above discussion, we propose the following procedure for adjusting the minimum utilization parameter ρ_0: The average delay for transmitting a packet over the wireless link is measured in a non-intrusive manner, i.e. the delay is measured for actual data packets transmitted over the wireless link. The parameter ρ_0 is adaptively adjusted so that the average delay is within a target interval $[d_{min}, d_{max}]$. In particular, ρ_0 is increased when the average delay is less than d_{min}, and is decreased when the average delay is greater than d_{max}. The magnitude of the change in each increase or decrease step is determined by the minimum utilization step size $\Delta\rho_0$. The values of d_{min} and d_{max} are related to the target packet delay requirements over the wireless link.

3 Simulation Results and Discussion

In this section we present and discuss simulation results comparing the proposed marking approach with drop tail queueing. Our experiments were conducted using the ns-2 simulator. The topology simulated is the one shown in Figure 1. Traffic flows from the fixed host to the wireless hosts, i.e. from right to left. In the experiments the IEEE 802.11 MAC layer performs retransmission of corrupted packets; losses due to corruption are assumed to be independent (non-bursty). We consider ftp flows that transfer files whose sizes follow a pareto distribution with average 500 KBytes.

3.1 Fairness and Throughput

The graphs we present next show the average and 95% confidence interval, from 10 independent runs of the same experiment. As a measure of fairness we consider

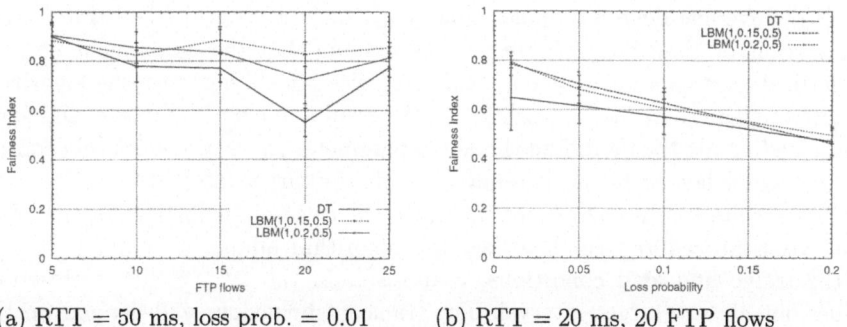

(a) RTT = 50 ms, loss prob. = 0.01   (b) RTT = 20 ms, 20 FTP flows

Fig. 2. Fairness for different number of FTP flows and different loss probabilities. LBM parameters: $\alpha = 1$, $\rho_0 = 0.15, 0.2$, $t_{avg} = 500$ ms. Average file size = 500 KBytes.

Table 1. Average and standard deviation of packet delay (in ms) over the wireless
link. File size = 500 KBytes, RTT = 50 ms, loss prob. = 0.01, LBM parameters:
$\alpha = 1, \rho_0 = 0.2, t_{avg} = 500$ ms

# of flows	DT		LBM	
	avg	std.dev.	avg	std.dev.
5	39.1	16.4	15.9	4.3
10	65.9	17.3	32.9	14.3
15	92.9	19.1	57.4	8.7
20	101.1	10.7	71.9	15.0

the fairness index given by Fairness Index $= \dfrac{\left(\sum_{i=1}^{N} x_i\right)^2}{N \sum_{i=1}^{N} x_i^2}$, where x_i is the rate of
flow i and N is the total number of flows. The fairness index takes values in the
interval $(0,1]$, with a higher value indicating higher fairness.

Figure 2(a) shows that LBM achieves better fairness compared to DT; moreover, the difference between the fairness achieved by LBM and DT is larger
for a larger number of flows. Figure 2(b) shows the fairness for different packet
loss probabilities over the wireless link. Observe that the difference between the
fairness achieved with LBM and DT is larger for smaller loss probabilities. The
utilization achieved by both DT and LBM, with $\rho_0 = 0.2$, is identical. The fact
that the use of ECN does not result in higher utilization compared to DT should
not be that surprising, since experiments for wired networks also show that, for
an appropriately dimensioned network, TCP with ECN does not achieve higher
throughput compared to TCP with drop tail queueing [8].

3.2   Packet Delay and LBM Adaptation

Table 1 shows the average delay and the standard deviation of the delay over
the wireless link. Observe that LBM achieves a smaller average delay and delay
jitter, as indicated by the smaller values of the standard deviation.

Next we investigate the dynamic adaptation feature of the proposed approach. Assume that the goal is to maintain the average delay within the interval
$[d_{min}, d_{max}] = [10, 15]$ milliseconds. To achieve this, the value of ρ_0 is increased
when the average delay is smaller than d_{min}, or is decreased when the average
delay is larger than d_{max}. Moreover, we assume that the adjustment of ρ_0 is
performed in steps of $\Delta\rho_0 = 0.01$; this step size affects how fast the algorithm
adapts to changes in the network load. The results appear in Table 2, and show
that by adjusting the minimum threshold parameter ρ_0 we can effectively control
the average delay, such that it remains inside the target interval.

The results in Table 2 do not illustrate the dynamic behaviour of the LBM
adaptation procedure, e.g. how fast the algorithm adapts to sudden changes
in the traffic and load conditions, that result in the average delay obtaining
values outside the target interval. The transient behaviour will depend on the
step parameter $\Delta\rho_0$ and the difference $d_{max} - d_{min}$; results for such dynamic
behaviour will be included in an extended version of this paper.

Table 2. Minimum utilization threshold ρ_0 for a different number of flows, when the target delay interval is $[10, 15]$ ms.

# of flows	avg delay	std.dev.	throughput (Mbps)	ρ_0
5	13.7	3.4	1.7	0.19
10	13.8	2.5	2.2	0.13
15	14.5	2.4	2.3	0.07
20	14.0	2.4	2.2	0.05

4 Conclusions

We presented an approach for seamless congestion control over heterogeneous networks containing wired and wireless IEEE 802.11 links. The approach combines three key features: First, it uses ECN as the common end-to-end signalling mechanism for conveying congestion information from both wired and wireless links; second, marking for the wireless link is performed using a load-based marking algorithm, where the marking probability is a function of the aggregate utilization; third, the load-based marking algorithm dynamically adapts to varying traffic and load conditions.

Further experimental work is investigating the transient behaviour of the proposed algorithm, in a dynamic environment where flows arrive and leave, different shapes of the marking probability curve (convex and concave, rather than piecewise-linear that we consider in this paper), and different measures of congestion of the wireless medium, such as the delay to access the wireless medium and the throughput measured in packets per time unit. Other interesting areas are the application of the proposed approach to multi-hop wireless networks and its combination with more advanced link-layer retransmission mechanisms.

References

1. Montenegro, G., Dawkins, S., Kojo, M., Magret, V., Vaidya, N.: Long thin networks. RFC 2757 (2000)
2. Peng, F., Cheng, S.D., Ma, J.: A proposal to apply ECN to wireless and mobile networks. In: Proc. INET'00. (2000)
3. Siris, V.A.: Resource control for elastic traffic in CDMA networks. In: Proc. of ACM MOBICOM'02. (2002)
4. Shakkottai, S., Rappaport, T.S., Karlsson, P.C.: Cross-layer design for wireless networks. IEEE Commun. Mag. (2003) 74–80
5. Saltzer, J., Reed, D., Clark, D.: End-to-end arguments in system design. ACM Trans. Comp. Sys. **2** (1984) 277–288
6. Ramakrishnan, K.K., Floyd, S., Black, D.: The Addition of Explicit Congestion Notification (ECN) to IP. RFC 3168 (2001)
7. Kelly, F.P., Maulloo, A., Tan, D.: Rate control in communication networks: shadow prices, proportional fairness and stability. Journal of the Operational Research Society **49** (1998) 237–252
8. Pentikousis, K., Badr, H.: An evaluation of TCP with Explicit Congestion Notification. To appear in *Annals of Telecommunications*. (2003)

Shared Protection by Concatenated Rings in Optical WDM Networks[*]

Hyunseung Choo, Minhan Son, Min Young Chung, and Tae-Jin Lee

School of Information and Communication Engineering
Sungkyunkwan University
440-746, Suwon, Korea +82-31-290-7145
choo@ece.skku.ac.kr

Abstract. The design of survivable all-optical mesh networks based on bidirectional wavelength division multiplexing (WDM) self-healing rings (SHRs) to supply complete protection against any single link failure requires an efficient solution. This paper proposes a shared protection algorithm called shared protection by concatenated rings (SPCR) using self-healing capability and a pre-configuration mechanism. Our algorithm protects and recovers any single link failure with very little capacity, faster detection, and recovery time on failures in mesh networks. The comprehensive computer simulation shows that the protection cost is reduced up to about three times and the protection delay is improved up to about five times comparing to short leap shared protection (SLSP) scheme which is known to be effective.

1   Introduction

Explosive growth of hosts, users and services connected to the Internet requires very high speed network technology such as optical networks. This trend is accelerated by ever increasing data traffic. Recently it has already exceeded voice traffic. In this context research on optical networks based on wavelength division multiplexing (WDM) technology is drawing much interest for future high speed network infrastructure. In such high speed network environment, protection of user service becomes increasingly important since even very short interruption of service due to link or node failure will cause huge data loss and incur tremendous restoration cost. Thus fast and efficient protection and restoration is one of the most important issues to be addressed.

Protection and restoration schemes can be categorized into P-Cycle[1,2], SLSP[3,4] and schemes using other type of cycles, e.g., Eulerian tour and Hamilton cycle[5,6,7]. Cycles are pre-configured in P-Cycle so that it can provide fast protection. If the length of a protection cycle is long, the protection delay may be increased. SLSP is a scalable end-to-end service-guaranteeing shared protection scheme, which accommodates the characteristics of both path-based protection

---

[*] This work was supported in part by Brain Korea 21 and University ITRC. Dr. T.-J. Lee is the corresponding author.

N. Mitrou et al. (Eds.): NETWORKING 2004, LNCS 3042, pp. 1476–1482, 2004.

and link-based protection. The main idea is to divide the working path into several overlapping segments (domains), each of which is assigned a protection domain ID (PDID). The formation of the protection path of each protection domain starts in the first node of each protection domain. If the diameter of a protection domain is d, we denote the protection method as SLSPd (we investigate SLSP3, SLSP4, SLSP5, and SLSP6 in this work).

In this paper, we propose a shared protection algorithm, shared protection by concatenated rings (SPCR), which utilized the concept of self-healing ring (SHR) in ring networks, and P-Cycle in mesh networks. We compare performance of SPCR and SLSP via simulations. Protection costs, protection delay and failure detection time are evaluated and our proposed SPCR is shown to demonstrate better performance.

2   The Proposed Protection Scheme

In contrast to dedicated protection such as 1+1[8], shared protection that allows several working paths to make use of the same protection resources can yield better capacity efficiency. For shared protection the speed of recovering service availability after the occurrence of failure may be relatively slower than in dedicated protection, and thus shared protection resources must be configured to speed up before the optical traffic flow can be switched to them. We employ a basic concept of pre-configuration method for a failed edge along with the shared protection to have reasonable protection performance in terms of time in a practical networking environment. Here the pre-configuration means pre-determined concatenated rings as shown in Fig. 1 depending on the dimension of the mesh network topology. Each ring covers four nodes and adjacent two rings share exactly one node.

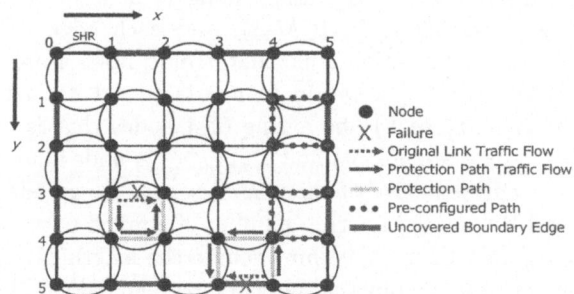

Fig. 1. Concatenated rings on $mesh(6,6)$ for path protections.

Another protection mechanism we consider in this work is the SHR. The motivation behind a ring-based backup configuration in mesh networks is that only rings can contribute to find alternative paths in a graph which represents a network topology. In addition, carefully designed rings can provide high sharability of backup paths and spare capacity along with the inherently simple and fast recovery operation for fault tolerant networks. This paper is focused to propose a new and efficient shared protection algorithm in a full mesh topology. The proposed scheme called SPCR is based on the SHR and the pre-configured protection path algorithm. Characteristics of the SPCR is as follows. The concatenated rings are basically SHRs in a full mesh topology and uncovered boundary edges that are not included in any SHR exist in the mesh boundary. A protection path for any uncovered boundary edge is determined by the pre-arrangement. If a failure occurs on a SHR, then the failed edge is protected by a path consisted of remaining edges in the SHR. A failure in a uncovered boundary edge is recovered by three nearest rings, especially an edge from each ring makes a protection path with length 3.

```
Algorithm DESIGN of CR(N,M)
  y=0; k=0;
  The ring number in each edge is initialized to -1;
  While (y < M - 1) do
    x=0;
    If ((y + 1) mod 2 = 0) then x = x + 1;
    While (x < N - 1) do
      ring_k = {(x, y), (x + 1, y), (x + 1, y + 1), (x, y + 1)};
      The ring number for each edge in ring _k become k;
      k = k + 1; x = x +2;
    y = y + 1;
```

Fig. 2. Pseudo code for Design of CR(N, M).

For consistency, we employ the similar definitions and notation for mesh networks in computer architecture[9]. A two-dimensional mesh, $mesh(N, M)$, is an $N \times M$ square grid of $N \cdot M$ nodes where N and M represent the length and width of the mesh, respectively. Each node in a mesh is represented by a coordinate (x, y) ($0 \leq x \leq N-1$, $0 \leq y \leq M-1$), and each edge $e_{(a,b),(c,d)}$ corresponds to a direct communication link that traffic flows from a node (a, b) to an adjacent node $(c, d) \in \{(a + 1, b), (a, b - 1), (a - 1, b), (a, b + 1)\}$. Let a $ring_k = \{(a, b), (c, d), (e, f), (g, h)\}$ be a ring of 4 nodes that is composed of 4 edges $e_{(a,b),(c,d)}$, $e_{(c,d),(e,f)}$, $e_{(e,f),(g,h)}$, and $e_{(g,h),(a,b)}$. A node in upper-left corner of a ring denotes a representative of the ring. Let e_{fail} be an edge which has a detected failure and RN(e_{fail}) be a ring number of a failed edge. Let D be a set of demands. The inputs to our algorithm include the length(N) and width(M) of a full mesh topology for the physical networks, a demand set D, and a failure detected edge e_{fail} if it exists. The output of the algorithm is an edge protection, and thus a link protection in contrast to a path protection. Fig. 2 shows how to make a layout of concatenated rings for a given mesh dimension. It starts from the upper-left corner node of the mesh as a representative at the $ring_0$ and finds the next representative (x, y) for the next ring. It is repeated from the left to the right and then from the top to the bottom of the mesh as you see in Fig. 1. The algorithm Design of CR(N, M) operates as follows. Variables x and y are used for coordinates of the mesh and k is for a ring number. The ring number for each edge in the mesh is initialized to -1. A ring$_k$ is a ring of 4 nodes which has 4 edges with the ring number k.

Table 1. The representative (x, y) of ring$_k$ based on N and k. $(k' = k \bmod (N - 1))$

	$N(= even)$	$N(= odd)$
$k' < \lfloor \frac{N}{2} \rfloor$	$z = 2k' + 2N\lfloor \frac{k}{N-1} \rfloor$	$z = 2k' + 2N\lfloor \frac{k}{N-1} \rfloor$
$k' \geq \lfloor \frac{N}{2} \rfloor$	$z = 2k' + 1 + 2N\lfloor \frac{k}{N-1} \rfloor$	$z = 2k' + 2 + 2N\lfloor \frac{k}{N-1} \rfloor$

$$x = z \bmod N \quad and \quad y = \lfloor \tfrac{z}{N} \rfloor$$

The algorithm Protection for Failure(N, M, e_{fail}) basically protects a failed edge in the following manner. Let us assume that k stores the ring number RN(e_{fail}). Then we can find the representative of the ring$_k$ based on N and k. Refer to Table 1. If k is positive, the algorithm protects the failed edge e_{fail} by the ring$_k$. Otherwise, e_{fail} is protected by a pre-configured protection path as you see in the circle dotted line at Fig. 1. This algorithm discussed is summarized in Fig. 3. Table 1 shows a method to calculate the representative (x, y) of ring$_k$ based on N and k. After identifying the representative in ring$_k$ which contains the failed edge, the original link is protected by the default protection path.

Algorithm PROTECTION for FAILURE (N, M, e_{fail})
  k = RN(e_{fail});
  If $(k$!= -1$)$ then the e_{fail} is protected by ring$k$ in mesh(N, M)
  Else uncovered boundary edge e_{fail} is protected by a pre-configured path for the e_{fail}

Fig. 3. Pseudo code for Protection for Failure(N, M, e_{fail}).

Fig. 4 shows the main algorithm. Step 1 generates SHRs for $mesh(N, M)$ as in Fig. 1 by Design of CR(N, M), and step 2 routes paths for all demands by any RWA algorithm. If a failure edge is detected in step 3, it is protected by the algorithm PROTECTION for FAILURE(N, M, e_{fail}) in step 4 (refer to Fig. 3).

      Algorithm SPCR (N, M, D)
  1   DESIGN of CR(N, M);
  2   Run any RWA algorithm for all demands in D;
  3   If (a failure detected in an edge e_{fail}) then
  4   PROTECTION for FAILURE(N, M, e_{fail});

Fig. 4. Pseudo code for SPCR(N, M, D).

3 Simulation Results

We now describe some numerical results of the proposed SPCR with which we compare the performance of the SLSP. We evaluate two schemes in terms of the protection costs required to reroute a given set of lightpaths, protection delay, and the time for failure detection. The details of our experiments are as follows. Our experimental topology is $mesh(N, M)$ what is formed of bidirectional edges. An edge used in any lightpath costs 1 unit, and thus five lightpaths for five different demands passing an edge cost 5 units. The experiment is performed with five protection methods (SLSP3, SLSP4, SLSP5, SLSP6, and SPCR) on four topologies (6×6, 6×7, 7×7, and 8×8). In this simulation, we try 500 times for each method and topology, and measure its performance on $mesh(N, M)$ with a set of 150 to 1000 random demands. The customized simulation is done at Pentium IV-1.7GHz PC with windows XP which is equipped with 256MB RAM. We now show bar graphs for each performance measure.

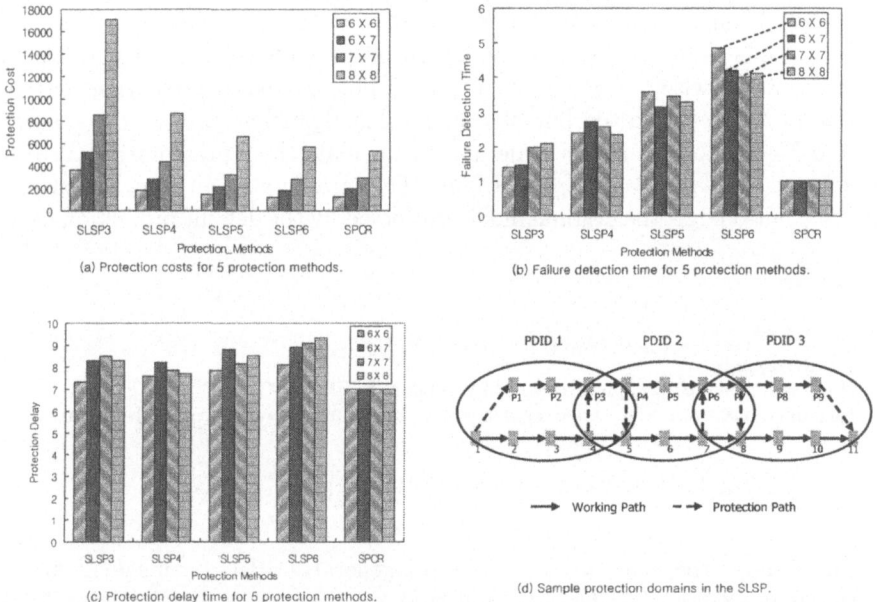

(a) Protection costs for 5 protection methods.

(b) Failure detection time for 5 protection methods.

(c) Protection delay time for 5 protection methods.

(d) Sample protection domains in the SLSP.

Fig. 5. Simulation Results(a),(b),(c) and sample protection domains in the SLSP (d).

Fig. 5(a) shows the protection costs for 5 protection methods mentioned earlier. The protection cost is determined by the maximum value of edge costs on a SHR when a failure occurs on the SHR. This is because the failure can be occurred at the maximum valued edge, and thus can be protected by the proposed scheme. If a failure occurs on an uncovered boundary edge, the protection cost is just the value of the edge cost. The protection cost of SLSP that is composed of protection domains is equal to the number of edges for protection domains. Fig. 5(a) shows that the SLSP requires much more protection costs than the SPCR.

Variables for the protection time are defined as follows. α is the time that a node detects a failure on the working path. β is the processing time for each node. γ is the transmission time for an edge. For instance in SLSP, if a failure occurs between nodes 2 and 3 in Fig. 5(d), the failure detection time is calculated as follows. Detection delay from a failure to node 2 is α, processing time at node 2 is β and transmission time for the edge between node 2 and node 1 is γ. And thus the failure detection time is $\alpha+\beta+\gamma$. For the protection time, processing time at nodes 1, P1, P2, P3, P4, and 5 is $6 \cdot \beta$ and transmission time for those related 5 edges is $5 \cdot \gamma$. And thus protection time is $6 \cdot \beta + 5 \cdot \gamma$. The failure detection time added to protection time makes total protection delay which is $\alpha+7 \cdot \beta+6 \cdot \gamma$.

The protection delay of SPCR is as follows. The failure detection time is always α. For the protection time, processing time at nodes (1,3), (1,4), (2,4), and (2,3) requires $4 \cdot \beta$ and transmission time for those related 3 edges needs $3 \cdot \gamma$. And thus the protection time is $4 \cdot \beta + 3 \cdot \gamma$, therefore the protection delay including the failure detection time is $\alpha + 4 \cdot \beta + 3 \cdot \gamma$ in Fig. 1. Figs 5(b) and (c) show results of the failure detection time and the protection delay, respectively, when α, β, and γ are 1. Fig. 5(b) shows that the failure detection time is improved up to about five times and Fig. 5(c) shows that the protection delay is improved a little comparing to SLSP scheme.

4   Conclusion

In this paper we have presented a shared protection algorithm based on self-healing capability a pre-configuration mechanism for $mesh(N, M)$. The SPCR provide shared protection against any single link failure using self-healing capability a pre-configuration mechanism. The comprehensive computer simulation shows that the protection cost is reduced up to about three times and the protection delay time is improved up to about five times comparing to the SLSP. This method shows its simplicity and better performance for various measures. We evaluate our proposed scheme for the physical topology instead of $mesh(N, M)$ with virtual nodes.

References

1. W. D. Grover and D. Stamatelakis, "Cycle-oriented Distributed Preconfiguration: Ring-like Speed with Mesh-like Capacity for Self-planning Network Restoration," *Proc. of IEEE International Conference on Communications,* vol.1, pp.537-543, Jun. 1998.
2. D. A. Schupke, C. G. Gruber, and A. Autenrieth, "Optimal Configuration of p-Cycles in WDM Networks," *Proc. of IEEE International Conference on Communications,* vol.5, pp.2761-2765, 2002.
3. P.-H. Ho and H. T. Mouftah, "A Framework of a Survivable Optical Internet Using Short Leap Shared Protection (SLSP)," *Proc. of IEEE Workshop on High Performance Switching and Routing,* pp. 21-25, 2001.

4. P.-H. Ho and H. T. Mouftah, "A Framework for Service-Guaranteed Shared Protection in WDM Mesh Networks," *IEEE Communication Magazine*, vol.40, pp.97-103, Feb. 2002.
5. H. Zhang and O. Yang, "Finding Protection Cycles in DWDM Networks," *Proc. of IEEE International Conference on Communications*, vol.5, pp.2756-2760, 2002.
6. H. Hwang, S. Ahn, Y. Yoo, and C. S. Kim, "Multiple Shared Backup Cycles for Survivable Optical Networks," *Proc. of International Conference on Computer Communications and Networks*, pp.284-289, 2001.
7. A. Sen, B. Hao, B. H. Shen, and G. Lin, "Survivable Routing in WDM Networks-Logical Ring in Arbitrary Physical Topology," *Proc. of IEEE International Conference on Communications*, vol.5, pp.2771-2775, 2002.
8. S. Ramamurthy and B. Mukherjee, "Survivable WDM Mesh Networks, Part II-Restoration," *Proc. of IEEE International Conference on Communications*, vol.3, pp.6-10, Jun. 1999
9. K. Hwang, Advenced Computer Architecture: Papallelism, Scalability, Programmability. New York: McGraw-Hill, 1993.

Distributed Scheduling of Recording Tasks with Interconnected Servers*

Sergios Soursos[1], George D. Stamoulis[1], and Theodoros Bozios[2]

[1] Department of Informatics, Athens University of Economics and Business
{sns, gstamoul}@aueb.gr
[2] New Technologies Department, INTRACOM S.A.
tmpo@intracom.gr

Abstract. We consider a system with multiple interconnected video servers storing TV programs that are received through satellite antennas. Users, equipped with set-top boxes, submit requests for TV programs, to each of which they assign a utility value according to their preferences. We develop a distributed scheduling algorithm that selects the programs to be recorded and the servers to store them, so that a high total utility is generated to the users' population. Our scheduling algorithm is based on the programs' broadcasting information, the users' preferences, the constraints regarding the capabilities of simultaneous recordings and storage, and the system's topology. In fact, servers belonging to the same cluster co-operate in order to attain increased efficiency by exchanging content through streaming or replication. The efficient performance of our scheduling algorithm is shown by means of experiments. The algorithm constitutes a practically applicable solution, already implemented and integrated in the testbed of the IST project UP-TV.

1   Introduction

The technology of digital television offers new possibilities for personalization and optimization of services that cannot be provided by analog broadcasting technologies. Additional information (i.e. metadata) concerning classification of content, starting and ending times etc. can be included in the digital broadcast stream and can greatly help in scheduling the recording and broadcast of the various programs so as to optimize certain performance indices. These issues were investigated by IST project UP-TV (**U**biquitous **P**ersonalized Interactive Multimedia **TV** System and Services, IST-1999-20751) [6]. The purpose of UP-TV was to create advanced and expandable architectures and systems for TV Anytime applications, thus providing personalized access to broadband information in an interactive way.

---

* This work is partly funded by the project IST-1999-20751 UP-TV whose partners are: INTRACOM S.A.(prime contractor), Universitat Paderborn, Invonova Gmdh, Pixelpark, PPS (Press Programm Service), Grundig, ERT (Hellenic Broadcasting Corporation), TUC (Technical Univ. of Crete), NV TV Limburg, and AUEB (Athens Univ. of Economics and Business) as a subcontractor of INTRACOM S.A.

N. Mitrou et al. (Eds.): NETWORKING 2004, LNCS 3042, pp. 1483–1488, 2004.

In this paper, we present the algorithm employed by the distributed Scheduler module for the purpose of scheduling the recording, the replication and the streaming of TV programs within the UP-TV network of servers, taking into account the preferences of the users and the various limitations regarding the capabilities of simultaneous recordings and storage, and the system's topology. The objective of the module is to generate a schedule leading to a high total utility for the users' population. This problem is very complicated to solve optimally for a realistic system with multiple servers, many different channels and numerous users. Therefore, we resort to a heuristic approach and develop a practically applicable algorithm. In fact, our approach does not only apply to the "simple" one-shot case, where we have to decide on the recording schedule once and for all. On the contrary, information on future TV programs can be received (as input) by the Scheduler *at any time instant*.

The problem addressed in this paper is similar to that of filling caches and scheduling content placement in Content Distribution Networks. For example, certain replication heuristics are presented and evaluated in [1], while several replica placement algorithms are developed and evaluated in [2]. However, contrary to our paper, such works are not dealing with content placement in conjunction with scheduling of program recordings.

2   The UP-TV Environment

The UP-TV distributed environment consists of several *clusters* which, in general, are interconnected through low bandwidth connections (e.g. through the Internet). A cluster is formed by a high-bandwidth network of UP-TV enabled servers (see Fig. 1). A number of users within a certain geographic area or a certain organization are served by the cluster's servers, through ADSL or CATV connections. Every server in the cluster has the following components, each of which poses restrictions to the Scheduler's algorithm: i) one or more hard disks for storing TV programs, with total capacity C, ii) a number R of satellite receiver cards that can record simultaneously different programs from different channels, iii) a number A of satellite antennas, each receiving the broadcast signal from one satellite, and iv) a number L of Low-Noise Block downconverters (LNBs) that can lower a signal's frequency and at the same time isolate the desired quarter of the signal's band, which then feeds a satellite receiver card. Note that it only makes sense to select L so that $L \leq 4A$ and $L \leq R$.

3   The Distributed Scheduling Algorithm

Within the cluster, all the servers run a local Scheduler module and one of them runs both the local Scheduler and the cluster's Central Scheduler (CS). The distributed scheduling algorithm consists of three parts. In Part 1, each local Scheduler decides which program will be selected for recording based on the preferences of the local users and the local technological limitations. In Part 2, each server sends its decisions to the CS; then, to each program which is to be

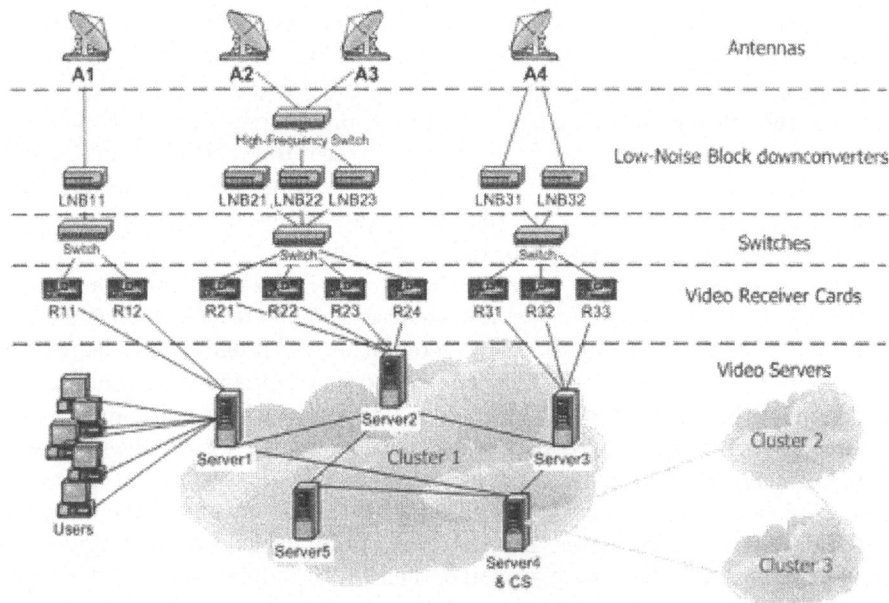

Fig. 1. The UP-TV environment

recorded by multiple servers the CS assigns a server which should certainly record it. Finally, in Part 3, the decisions of the CS are sent back to the local Schedulers, along with the information about the cluster's topology; each Scheduler then checks if it is beneficial to store locally programs that are assigned to other servers (in Part 2) or retrieve them when needed by means of streaming.

In our approach, each program is assigned a utility value that represents the total satisfaction to be generated to all users to which the program is of interest. Note that the system gives each user the opportunity to select his preferred programs and assign to them a utility value. (For each program, the utilities assigned thereto by the various users are summed.) The objective of the local Scheduler (i.e. Part 1 of the algorithm) is to compute recording schedules that will lead to a high *total satisfaction* (i.e. total utility) of the users, subject to the aforementioned limitations on the basis of the program's broadcasting information and the users' preferences. Such a maximization problem is very hard to solve exactly. (Even simplified versions of the problem reduce to the knapsack problem, which is NP-hard [3].) Therefore, we resort to a heuristic approach, the main ideas of which are outlined below. The key idea of our algorithm is that the profitability of a program depends on both the utility that it will generate and on the size of disk space it will occupy. Therefore, a proxy of the profitability of a program expressing the aforementioned trade-off, is the *Utility per Mbyte*(UpM); i.e., the ratio of the utility and the required disk space. The algorithm then considers for recording the programs *one-by-one* in decreasing order of UpM, while conforming to the constraints on availability of disk space, satellite receivers and LNBs. This approach can be motivated by considering each program as an individual bidder in a *sealed-bid auction*, who is bidding an amount of money equal to its own utility; e.g. see [4]. Another critical issue is the efficient exploitation of LNBs. A proxy representing the trade-off between the value introduced by a program's recording and the "cost" due to the time of occupying an LNB is the Utility per Second (UpS). However, this is almost equivalent to UpM, since the coding rate of the TV channels' program will be the same (or almost so) for different channels.

An important feature of our algorithm is that it does not only apply to the "simple" one-shot case, where we have to decide on the recording schedule once and for all. On the contrary, information on future TV programs can be received (as input) by the Scheduler *at any time instant* and previous information (i.e. programs' utilities) is updated. Future recording decisions are ignored, while recording decisions in progress are not subject to change, for simplicity reasons. Already recorded programs may have to be deleted, in order to save disk space for recording more profitable programs. The utility of each such program diminishes exponentially as time elapses. If a new program is decided to be recorded but there is not enough disk space available, then the Scheduler checks if it can gain the required capacity by deleting one or more recorded programs, if beneficial.

In Part 2 of the distributed algorithm, the clsuter's Central Scheduler (CS) receives the local decisions of all servers. Except from their local decisions (i.e. a sorted program list in decreasing order of UpM), all the servers of the cluster send also a few high-value programs that could not be selected for recording, due to satellite receiver or LNB unavailability. In cases where more than one cluster's servers have decided to record the *same* program, the CS decides which one of the servers should certainly record the program, based on the ranking that each server gave to the particular program. Thus, the CS gives the other servers the option of either i) storing locally the program or ii) streaming it from the server decided by the CS. Finally, the CS sends its own decisions back to the local Schedulers of the other servers together with the cluster's topology, which includes all the links of the cluster along with the link's capacity. Also, the CS sends the location of the high-value programs that couldn't be recorded, giving the option of duplication to the interested servers. In Part 3, each local Scheduler must decide how to handle the programs that were initially selected and the CS has assigned their storage to other servers. Motivated by [5], we use two metrics for evaluating the cost for storing a TV program to the disk and the cost for streaming it from another server via the network; namely, the *storage cost per unit* and the *streaming cost per unit*, which can be assigned proper values on the basis of current prices of hard disks and leased lines. The total storage and streaming costs for a specific TV program can be calculated as follows:

- *total storage cost =*
  (storage cost per unit) ∗ (program storage time) ∗ (program storage size).
- *total streaming cost =*
  (streaming cost per unit) ∗ hops ∗ (program storage size) ∗ requests ∗ 0.6,
  where the factor 0.6 accounts for the fact that due to caching, not every new

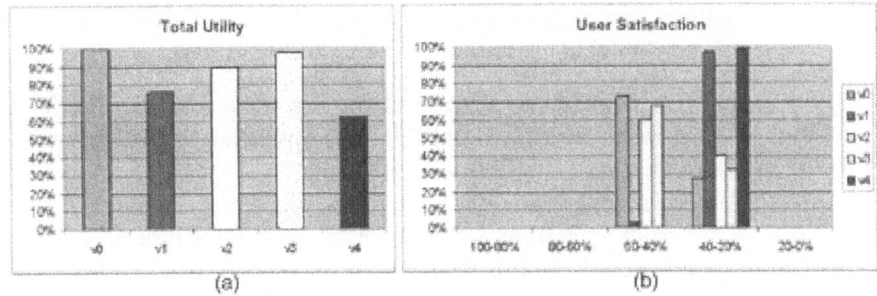

Fig. 2. Comparison of variations. v0 stands for the original algorithm, v1 for its first variation etc.

local request for a specific program causes a new streaming session thereof, and thus it does not add to its total streaming cost.

Each local Scheduler finds the shortest path and computes the *total streaming cost* and the *total storage cost* for each program that was initially decided to be stored locally and the CS has assigned its storage to another server. Comparing the above costs will lead the local Scheduler to a decision of streaming, if the former cost is lower than the later and at the same time the amount of bandwidth necessary for streaming is available in all links of the path.

After making all of the above decisions for leaving or not programs for streaming, each local Scheduler checks if any disk space has been released due to cancellation of recording decision(s). Then, it checks one-by-one (in decreasing order of UpM) the high-value programs that couldn't be recorded due to unavailability of satellite receivers or LNBs, and decides which ones to replicate from other servers. Finally, it re-executes Part 1 of the distributed algorithm, in order to take advantage of any unused disk space by recording additional programs.

4   Experimental Results

As already explained, the problem dealt is extremely complicated to solve exactly. Thus, in order to assess how efficient the proposed heuristic is, we must provide considerable evidence that other less complicated heuristic methods provide inferior results. Therefore, for the evaluation of the Scheduler's algorithm, we have defined the following simpler variations of the algorithm:

- Under the first two variations, the programs are sorted in Part 1 according to a different metric than UpM: i) sorted in descending order of the total utility and ii) sorted in ascending order of the required disk capacity.
- The third variation does not comprise Parts 2 and 3 of the original algorithm.
- The fourth variation is a totally greedy algorithm where the programs are sorted in ascending order of their starting times, while Parts 2 and 3 of the original algorithm and other optimization heuristics are not included either.

We have defined certain performance metrics that are appropriate for the comparison of the original algorithm with the above variations:

- *Total Utility:* the sum of the utilities of the recorded programs for all the servers in the cluster; streamed and duplicated programs are also included for the variations comprising Parts 2 and 3.
- *User Preferences:* the percentage of users of the cluster that have access (either locally or via streaming) to their first choice of a program, to their first *and* second choices, and to their first *or* second choices.
- *User Satisfaction:* the percentage of users of the cluster belonging to each specific level-of-satisfaction group (namely, 100%-80% group, 80%-60% group, etc.), where for each user we compute the fraction of the utility of all programs to which the user has access divided by the utility of all programs requested by him.

A realistic broadcast schedule was employed in our experiments. Users' preferences were generated randomly. A star topology was taken for the UP-TV cluster, with a router at the center and three servers at the edges. The bandwidth of each link was taken 100 Mbps. We have conducted many experiments, with different setup. In Fig. 2, we present some indicative results. In particular, our algorithm out-performs the rest variations in terms of total utility, with the total-greedy algorithm having the lowest one [see Fig. 2(a)]. The use of simpler metrics for sorting definitely leads to performance degradation. Considering the users' preferences (not depicted here), only the first variation is close in performance with the original algorithm. But, when it comes to the users' satisfaction [see Fig. 2(b)], it is clear that our algorithm has the best performance since it satisfies to a relatively high degree a high percentage of users.

In general, experiments have revealed that the distributed scheduling algorithm outperforms its variations with respect to the various performance criteria. Given also its relatively low computational complexity, it appears that this algorithm constitutes a promising and practically applicable solution.

References

1. J. Kangasharju, J. Roberts, K.W. Ross, "Object Replication Strategies in Content Distribution Networks," Proceedings of WCW '01: Web Caching and Content Distribution Workshop, Boston, MA, June 2001.
2. L. Qiu, V.N. Padmanabhan, G.M. Voelker, "On the Placement of Web Server Replicas," Proceedings of the 20th IEEE INFOCOM, 2001.
3. M.R. Garey, D.S. Johnson, "Computers and Intractability - A Guide to the Theory of NP-Completeness," W.H. Freeman and Company, ISBN: 0-7167-1045-5.
4. L.M. Ausubel, P. Cramton, "Demand Reduction and Inefficiency in Multi-Unit Auctions," working paper, University of Maryland, 2002.
5. S.-H. G. Chan, F. Tobagi, "Caching Schemes for Distributed Video Services," Proceedings of the 1999 IEEE ICC, Vancouver, Canada, June 1999.
6. Project IST-1999-20751 UP-TV, URL: http://www.up-tv.com

Impairment Constrained Based Routing for Managed Reach WDM Optical Networks

Dimitrios Vogiatzis, Carmen Mas Machuca, and Ioannis Tomkos

Athens Information Technology (AIT) Center, Markopoulo Av. PO Box 68, 19002 Peania,
Athens, Greece
{dvog,cmas,itom}@ait.edu.gr

Abstract. Managed reach networks are networks that allow placing of regenerators in the main cities and at requested distances. These networks minimize the number of regenerators but need a routing approach able to establish the connections through regenerating nodes when the signal quality deteriorates. For this purpose, Impairment Constrained Based Routing (ICBR) is used. Considering that impairment related information is exchanged through signaling, the ICBR should be able to compute paths that guarantee a certain level of optical signal quality at the receiver. In this paper we propose an ICBR that is combined with a Routing and Wavelength Assignment algorithm. The proposed ICBR will be presented and used with two different impairments: the Polarization Mode Dispersion (PMD) of the fibers and the Amplified Spontaneous Emission (ASE) noise of the amplifiers. Savings in the number of regenerators is shown when the proposed ICBR is used instead of traditional routing algorithms such as minimum hop and shortest path. Dependence of the blocking probability on different parameters such as span length and $OSNR_{min}$ is also demonstrated.

1 Introduction

Today's optical networks are *opaque*, i.e. regeneration of all the channels is done at each network node. This solution allows the signal to travel and reach large distances; however it is quite expensive due to the number of regenerators needed in the network and the connection bit-rate dependency. One proposed solution is *transparent* optical networks [1]. In transparent optical networks the signal is transported end-to-end all optically, without being converted to the electrical domain along its path. Another solution called *managed reach* networks was recently proposed [2] and is a compromise between transparent and opaque networks. This scenario proposes a network that allows placing of regenerators in the main cities and at requested distances. This approach minimizes the number of regenerators but needs a routing approach able to route the connections through regenerating nodes when the signal quality deteriorates.

Routing in opaque networks assumes optical link engineering on a per link basis, whereas in transparent network is done on a per connection basis [3]. Transparent and managed reach networks require routing approaches that take into account the physical limitations of the optical layer: in the former it will limit the transparency reach, whereas in the latter it will determine the number of regenerators needed. The reason

N. Mitrou et al. (Eds.): NETWORKING 2004, LNCS 3042, pp. 1489–1494, 2004.

is that traditional routing protocols use simple cost parameter (e.g. link length) and the resulting paths do not guarantee that the received optical signal, which has been transparently transported through the network, has enough quality. These physical impairments should be taken into account by the routing protocol. For this purpose, impairment information should be exchanged between the network components through signaling. MPLS and GMPLS platforms are able to support the signaling and the routing required to perform Impairment Constraint Based Routing (ICBR)[4].

In order to deal with this problem, this study has been focused on the analysis of the number of regenerators that can be saved when the routing approach used in the network takes into account the physical impairments.

2   Routing in Transparent and Managed Reach Networks

In opaque optical networks the signal is electrically regenerated at each node, i.e. any impairment of the signal can be eliminated. In these networks, the used routing algorithm gives for each connection request a path that may use different wavelengths (electrical conversion is available nowadays). Existing routing approaches find a path minimizing a certain cost parameter such as the number of hops or the fiber length. Routing in transparent and managed reach networks is more complicated due to the absence of optical conversion and regeneration. In this case, routing is in charge of finding a path with a common available wavelength through all the links of the path. The routing algorithm should take into account the physical limitations of the path in order to guarantee that the optical signal will reach the receiver with acceptable quality. This happens because the optical signal in a transparent network cannot be easily monitored and can be easily deteriorated since signal impairments accumulate along the all-optical path. The routing algorithms that take into account the physical impairment are known as Impairment Constrained Based Routing (ICBR).

In transparent and managed reach networks, not only the path but also the wavelength should be determined. This problem is identified as Routing and Wavelength Assignment (RWA). The logical connections between any pair of edge nodes will be established using lightpaths. Previously proposed RWA schemes assume that once the path and the wavelengths have been identified, the connection is feasible to be established. However, in real world networks this assumption is not necessarily true. Certain physical layer optical impairments may prevent the connection establishment through the proposed path because the signal quality will be severely affected.

3   Impairment Constrained Based Routing and Wavelength Assignment in All-Optical Networks

The proposed algorithm performs impairment constrained based routing and wavelength assignment taking into account defined physical constraints. In this work we will present how the algorithm takes into account the PMD and the ASE constraints. The inputs of the algorithm are:
- End-to-end connection requests: The connection requests, given in Gbps represent the traffic between two cities.

- Physical layer parameters: Several parameters related with the physical layer characteristics should be given such as the optical fiber attenuation, gain of the amplifiers, bit rate, launched optical power, noise figure of the amplifiers, $OSNR_{min}$, span and link lengths.

The output of the algorithm is for a given connection request, a path and an assigned wavelength, whenever the connection is not blocked.

Initially (preprocessing phase), the source-destination nodes of all the connection requests are identified and the bit rates of the connections are specified, based on traffic models that predict the traffic increase on a year per year basis. Additionally, the link costs are computed based on the physical layer and the *'cost impairment'* parameters which are related to the impairments that should be taken into account when finding the paths. When ASE is under consideration, either on its own or in combination with the PMD, the cost parameter is the number of amplifiers per link, whereas when only the PMD constraint is examined, the corresponding cost parameter is the D_{PMD} of the link along with the link length. Once the link costs are found, the routing and wavelength assignment (RWA phase) may begin. For this purpose, k shortest paths for each source-destination pair are computed. The k parameter is set before running the algorithm and in our simulations k has been set to 3. These k candidate paths are used for the RWA problem. Since the only parameter that captures the real performance of a network at the end of the day is the blocking probability, we have tried to reduce the blocking probability by evenly distributing the traffic in the network, so that load balancing is achieved. For this reason, the algorithm that we have used for the RWA problem attempts to minimize the cost induced by all flows in the links of the network, by a proper selection of a monotonically increasing function, as a flow cost function. The RWA module [6] selects the 'best' paths with the available wavelengths, which guarantee minimization of the blocking probability. This module constantly interacts with the Impairment Aware Constrained Based Routing module. Starting with the shortest path delivered from the RWA module, and continuing with the others, it is checked whether the path accomplishes the considered impairment requirements. If so, the connection is established; otherwise and if none of the paths delivered from the RWA module satisfies the physical impairment requirements, the connection is blocked.

Some of the novelties of the proposed algorithm are:

— It is suited for all types of topologies. It has been shown that the RWA problem can be partitioned into two sub-problems: the routing and the wavelength assignment. However, this partitioning is possible with star and ring topologies but not in mesh networks [4]. Our approach merges both sub-problems and is able to find optimal solution in mesh networks.

— It is also well known [5], that dividing the RWA problem into two distinct sub-problems, it yields only to sub-optimal results. Optimality can be only achieved, if the two sub-problems are treated as one joint optimization problem. Our algorithm guarantees this, by the proper selection of the k-shortest paths, which give a good representation of the space where the optimal solution lies in.

— The algorithm is recursive, that is, if the first solution does not accomplish the impairment constrains, a second solution may be found.

4 Simulation Scenario

As basis for our simulations we used the pan-European network shown in Fig. 1. This network consists of 8 nodes interconnected by 16 bidirectional links. The cumulative distribution of the fibre link lengths between the nodes is shown in Fig. 2. The average link length is about 700 km and each link is able to support up to 80 wavelengths. The fibers that are considered can be either 'new' (PMD factor equal to 0.1 ps/√km) or 'old' (PMD factor equal to 0.5 ps/ km). The parameter a gives the percentage of new fibers in the network.

The considered traffic matrix is static and consists of 56 end-to-end connections (full connectivity). The connections matrix has been considered to be symmetric and with values of 10, 20 and 40 Gbps bit rates (i.e. each connection between a pair of cities will be mapped into a single wavelength of the abovementioned capacity).

Fig. 1. Topology of the reference network

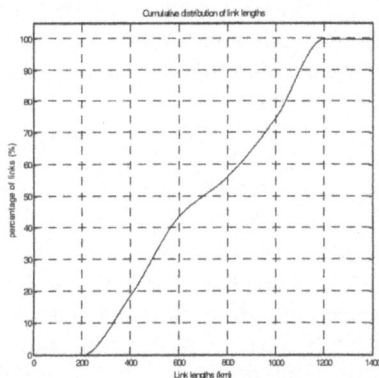

Fig. 2. Link length cumulative percentage

5 Simulation Results

5.1 ICBR Studies Considering the PMD Impairment as a Constraint

In this scenario connections are routed with ICBR taking into account the PMD parameter of the fibers. As mentioned earlier, our study focuses on managed reach networks, which are networks that locate regeneration where is needed due to the effect of impairments. In particular, this study has looked at the number of regenerators that are required when different routing algorithms are used. These algorithms are: the minimum hop, the shortest length and the proposed Impairment Constrained Based Routing (ICBR). The number of regenerators required in the 10 Gbps connections routed by the different routing schemes for different percentage of new fibers has been plotted in the graph of Fig. 3. This graph clearly shows that when the percentage of new fibers (x-axis) increases, the number of regenerators (y-axis) needed when using ICBR decreases much faster than with the other two routing algorithms. The percentage of fewer regenerators needed with ICBR compared to minimum hop increases as the percentage of new fibers increases. For example for a=0.4, ICBR requests 75% less regenerators than minimum hop and for a=0.6 ICBR requests 90% less regenerators than minimum hop. In the particular case when the percentage of new fibers is 0.8, no regeneration is needed if ICBR is used.

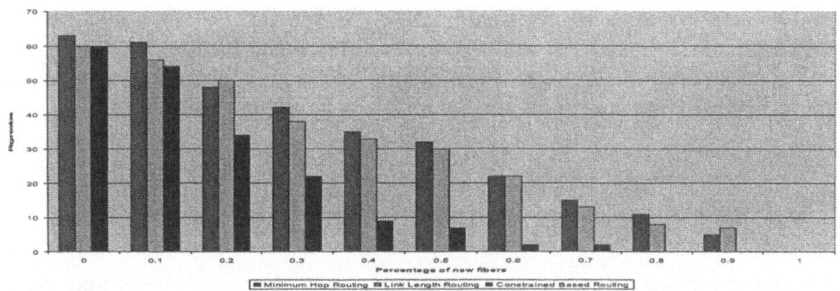

Fig. 3. Number of regenerators vs. the percentage of new fibers for B=10 Gbps

5.2 ICBR Studies Considering the ASE Impairment as a Constraint

The connections are now routed based on the ASE noise. In the simulations a homogeneous network has been considered and therefore, the results of minimum hop and shortest length routing, which are the same have been compared with the proposed ICBR. In case of non-homogeneity, the results of minimum hop and shortest length will differ but still ICBR uses fewer resources and outperforms the other two.

Let us consider the core network shown in Fig. 1 to be transparent. This section is focused on the study of the blocking probability versus the span length in function of the bit rate and the $OSNR_{min}$. Different $OSNR_{min}$ have been considered (each of them related to different FEC schemes). Fig. 4 shows the increase of the blocking probability with the $OSNR_{min}$ value and also with the bit rate. The blocking probability also increases for longer span lengths since although there will be fewer amplifiers, the cumulative noise will be higher and therefore more connections will not be able to accomplish the minimum OSNR required and will be blocked.

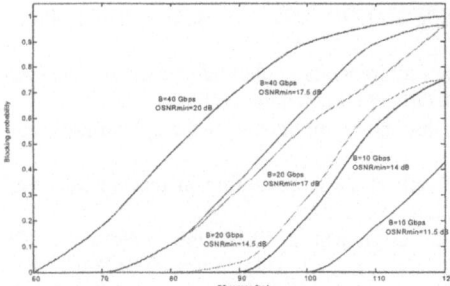 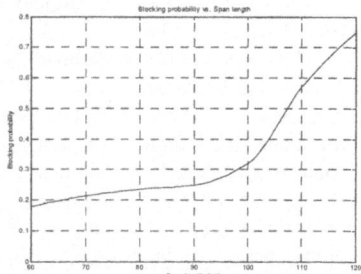

Fig. 4. Blocking probability vs. fiber span length wrt. $OSNR_{min}$ and connection bit rates

Fig. 5. Blocking probability vs. span length taking into account PMD and ASE

5.3 ICBR Studies Considering the PMD and ASE Impairments as Constraint

The two constraints have been considered as parameters of our proposed ICBR. In our case, we looked at the scenario of span lengths of 70 km, bit rates of 10 Gbps, PMD parameter equal to 0.25 ps/ km. The blocking probability versus the span length has been plotted in Fig. 5. This figure shows that there is a smooth increase of the blocking probability for span lengths shorter than 90 km and that after this value, the blocking probability increases much abruptly. This is due to the fact that for spans shorter than 90 km, the dominant impairment is the PMD, whereas for longer spans, both impairments contribute to the increase of blocking probability.

6 Conclusions

This paper described the most important issues related to transparent and managed reach optical networks. It has been shown that in these networks Impairment Constrained Based Routing should be used to guarantee that, by the proper selection of the paths, the quality of received optical signal is within acceptable limits. The presented results take into account as impairment constraints the fiber PMD and the cumulative ASE noise. One of the results is the reduction in the required number of regenerators when the proposed ICBR is used instead of the tradition minimum hop and shortest length routing algorithms. For example, when considering the PMD as constraint, the number of regenerators that could be saved increases with the percentage of new fiber installed in the network, since ICBR tries to route the connections through more new links. In some scenarios the number of regenerators needed with ICBR is null. Other results study the blocking probability and its dependency based on different parameters such as span length, percentage of new fibers, or $OSNR_{min}$. In the conference, details on the algorithm and the simulations will be presented.

References

1. R. E. Wagner et al., 'MONET: Multiwavelength optical networking', *Journal of Lightwave Technology*, 14(6), 1349–1355 (1996).
2. R. Wagner "Evolution of Optical Networking", *LEOS 2000 Proceedings*, TuC1, Puerto Rico, November 2000.
3. Frank Chang and Mazda Salmanian "Routing requirements in photonic networks" *National Fiber Optic Engineers Conference (NFOEC)*, Orlando, September 2003.
4. A.Chiu, J. Strand and R. Tkach "Unique features and requirements for the optical layer control plane" IETF draft, May 2001.
5. John Strand, Angela L. Chiu, and Robert Tkach "Issues for routing in the optical layer" *IEEE Communications Magazine*, February 2001, pp.81-87.
6. Dimitrios Vogiatzis "Routing and Wavelength Assignment in WDM Optical Networks" Diploma thesis, Computer Engineering and Informatics Department, University of Patras, June 2002.

Performance Analysis of Variable Bit Rate Multiclass Services in the Uplink of a Dynamic Complete Partitioning Round-Robin Carrier-Hopping Multirate Multi-carrier DS-CDMA System

Tung Chong Wong[1], Jon W. Mark[2], and Kee Chaing Chua[3]

[1] Institute for Infocomm Research, 21 Heng Mui Keng Terrace, Singapore 119613, Singapore
wongtc@i2r.a-star.edu.sg
[2] Center for Wireless Communications, University of Waterloo,
Waterloo, Ontario, Canada N2L 3G1
jwmark@bbcr.uwaterloo.ca
[3] Electrical and Computer Engineering, National University of Singapore,
10 Kent Ridge Crescent, Singapore 119260, Singapore
chuakc@nus.edu.sg

Abstract. An analytical formulation of the outage probability in terms of bit error rate specification for variable bit rate (VBR) multiclass services in the uplink of a dynamic complete partitioning (DCP) round-robin carrier-hopping multirate Multi-Carrier (MC) DS-CDMA cellular system is presented. The analytical framework is formulated for the general case in which different traffic classes have different spreading gains in each of the subcarriers. The analytical work leads to the determination of the capacity region of this cellular system for VBR traffic. Numerical results show that the system capacity with DCP can be larger than that with complete sharing (CS) of the subcarriers.

1 Introduction

3G/4G cellular mobile networks are expected to support integrated multimedia services with different qualities of service (QoS) for different service classes. Ref. [1] considers K traffic classes in 3G/4G cellular systems. In [1], the traffic source is assumed to be an on/off process which is valid only for voice or data traffic but not for video or other traffic which can be modeled as a multirate traffic [2]. Ref. [3] considers a multirate traffic model for *variable bit rate* (VBR) traffic in the uplink of a round-robin carrier-hopping multirate Multi-Carrier (MC) DS-CDMA system with *multiclass* traffic using complete sharing (CS) of all the subcarriers in the system.

Recently, MC-CDMA systems have been receiving a lot of attention as they can promise *high data rate* required by 4G mobile cellular systems and be effective in mitigating multipath fading and rejecting narrowband interference [4]. This paper considers a multirate traffic model for VBR *multiclass* traffic to study the capacity of the uplink of a dynamic complete partitioning (DCP) round-robin carrier-hopping multirate MC-DS-CDMA system. In traditional resource allocation schemes at the

N. Mitrou et al. (Eds.): NETWORKING 2004, LNCS 3042, pp. 1495–1501, 2004.

connection level, a CS scheme will give lower blocking probabilities than those from a CP scheme as they have "hard capacity." On the other hand, CDMA-based systems have "soft capacity" depending on intra- and inter-cell interference. A salient feature of this paper is the use of DCP to dynamically allocate subcarriers to different traffic classes to eliminate intra- and inter-cell interference from other classes. This helps to increase system capacity. The outages in the subcarriers of each class are minimized by spreading the subcarriers' signal-to-interference ratio evenly among all its subcarriers by using a dynamic round-robin carrier-hopping allocation scheme. This will also help to increase system capacity. The activity factor of a user in a subcarrier is modeled as though it is an on/off source in a single-carrier CDMA system as in [1].

2 Dynamic Complete Partitioning Round-Robin Carrier-Hopping Multirate MC-DS-CDMA

A single-carrier DS-CDMA system has a spread-spectrum bandwidth of W at a center frequency of f. On the other hand, a multi-carrier DS-CDMA system has a spread-spectrum bandwidth of W/N_c at frequency f_c in each of its subcarriers, where N_c is the total number of subcarriers, $c=1,2,...,N_c$, $N_c = \sum_{i=1}^{K} N_{ci}$, and N_{ci} is the number of class i subcarriers. The same spreading code can be used in each of the subcarriers for a user. A VBR source with discrete bit rate levels as modeled in Section 3 is considered. Each level has a bit rate of R_i, where $i=1,2,...,K$ and K is the number of traffic classes. The highest level is M_i and it has a bit rate of M_iR_i. Each level of bit rate is transmitted through one subcarrier. Thus a variable bit rate source is transmitted through a number of subcarriers, each using the same spreading code. Every user in the same class transmits at the same bit rate, R_i, in each group of subcarriers that are dynamically allocated to that class.

Consider the case of two traffic classes. To evenly spread the usage of the subcarriers and outage probabilities in the subcarriers and to increase the support of the number of users in the system, we propose to use a round-robin carrier-hopping multirate MC-DS-CDMA. This round-robin carrier-hopping subcarrier allocation scheme works as follows. Assume that the first level of bit rate is transmitted first in one of the N_{c1} subcarriers and the other levels of bit rate are transmitted in the next few subcarriers for a class 1 user. Similarly, the first level of bit rate for a class 2 user is transmitted first in one of the $(N_{c1}+1)$th to (N_c)th subcarriers. The usages of the subcarriers are shown in Table 1. The symbol, ô, represents that a reference subcarrier is used, while the symbol, o, represents that the subsequent subcarriers are used. The reference subcarrier of each class is moved in a round-robin manner over all its subcarriers. The number of subcarriers dynamically allocated to class 1 users is N_{c1}, while that to class 2 users is $N_{c2}=N_c-N_{c1}$. As shown in Table 1, the goal of evenly spreading the usage of the subcarriers is achieved. Thus the outage probabilities in the subcarriers are lowered and the number of users that can be supported is increased. Furthermore, the complete partitioning of the subcarriers to the different classes is dynamically adjusted to maximize the number of users that can be supported by using the DCP admission region curves in Section 4. The DCP admission region can be stored in a lookup table and be referenced to execute the dynamic complete partitioning allocation of subcarriers to each class when DCP can support more users than CS.

Table 1. Subcarriers used for VBR sources in a dynamic complete partitioning round-robin carrier-hopping MC-DS-CDMA system with 2 classes

Time epoch	Subcarrier(s) used									
	Class 1				Class 2					
	1	2	3	N_{cl}	$N_{cl}+1$	$N_{cl}+2$	$N_{cl}+3$	$N_{cl}+4$	$N_{cl}+5$	N_c
t_1	ô				ô					
t_2		ô	o			ô	o			
t_3	o		ô	o			ô	o	o	
t_4	o		ô				ô	o		
t_5	ô							ô		
t_6		ô	o		o					ô
t_7	o		ô	o	ô	o	o			

3 Analytical Model

From [2], a VBR source can be modeled by a continuous-time Markov chain with finite states. Each state represents the discrete level of bit rate generated by a single source. We assume that the highest level is state M_i, which is also matched to the maximum number of subcarriers used by a class i user. This means that each level has a data rate of R_i corresponding to the use one class i spreading code in one subcarrier. Every user uses the same spreading code in its subcarriers. If $M_i = 1$, the source is an on/off source. Each level can be modeled by a two-state mini-source with increase and decrease rates of α_i and β_i, respectively. Thus the continuous-time Markov chain for a single source at state m has increase and decrease rates of $(M_i - m)\alpha_i$ and $m\beta_i$, respectively. This Markov chain is shown in Fig. 1.

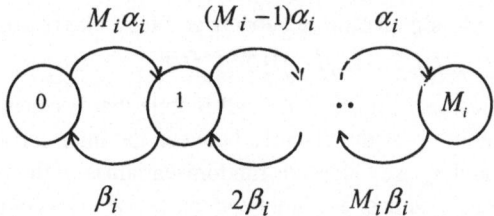

Fig. 1. Continuous-time Markov chain for a single variable bit rate source

The steady-state probability of being in state m, denoted by P_m, is given by

$$P_m = \binom{M_i}{m}(p_i)^m (1-p_i)^{M_i-m}, \ m = 0,1,2,...,M_i, \tag{1}$$

where $p_i = \alpha_i/(\alpha_i + \beta_i)$. The probability that subcarrier c is used for transmission given that the source is in state m, denoted by $P_{c|m}$, is given by

$$P_{c|m} = m/N_{ci}, \ m = 0,1,2,...,M_i, c = 1,2,...,N_{ci}, M_i \le N_{ci}. \tag{2}$$

Unconditioning the dependence on state m, the probability that subcarrier c is used for transmission, denoted by P_c, is given by

$$P_c = \sum_{m=0}^{M_i} P_{c|m}P_m = \sum_{m=0}^{M_i}\left(\frac{m}{N_{ci}}\right)\binom{M_i}{m}(p_i)^m(1-p_i)^{M_i-m} = \frac{M_i p_i}{N_{ci}}, M_i \le N_{ci}. \tag{3}$$

Note that if $M_i = 1$, the source is an on/off source and the probability that subcarrier c is used for transmission is simply equal to the source activity factor, p_i, divided by its number of subcarriers, N_{ci}. This is expected due to the round-robin carrier-hopping allocation scheme used. This allocation scheme evenly spreads the subcarriers' usage by a source. The E_b/I_0 for class i in each of its subcarriers, γ_i, is given by

$$\gamma_i = G_i\Big/\Big(\sum_{j=1}^{n_i-1}\psi_{ij} + (I_i/S_i) + (\eta/S_i)\Big), i = 1,2,...K, \tag{4}$$

where $\psi_{ij} \in \{0,1\}$ is a Bernoulli random variable indicating the activity factor of the spreading code used by the jth user of class i in each of its subcarriers. The probability that an active spreading code in a subcarrier is used by source i is given by $\Pr[\psi_{ij} = 1] = q_i = P_c, i = 1,2,...,K$. $G_i = (W/N_{ci})/R_i$ is the class i processing gain in subcarrier c. In the denominator of (4), n_i is the number of class i users, the first term is due to the intra-cell interference in subcarrier c from other users in class i, the second term is due to the inter-cell interference power, I_i, in the subcarrier from users from the same class, where S_i is the received power in the subcarrier for a class i user, and the last term is due to background noise power, η, in the subcarrier. Note that by using the proposed DCP scheme, the intra- and inter-cell interference in subcarrier c from other classes are eliminated.

Treating each subcarrier as *orthogonal* from other subcarriers, we can model each subcarrier as though it has on/off multiclass traffic with their multirate MC-DS-CDMA sources' subcarrier activity factors being treated like the on/off sources' activity factors. Thus we can make use of the approach and results in [1]. From [1], the inter-cell interference-to-signal ratio in subcarrier c for a class i user is given by

$$I_i/S_i = (r_m/r_d)^4 10^{(\varepsilon_d - \varepsilon_m)/10}, \tag{5}$$

where r_d is the distance between the inter-cell mobile that is causing interference and the intra-cell base station, r_m is the distance between the inter-cell mobile and its own base station, and ε_d and ε_m are Gaussian random variables with zero mean and standard deviation σ. Since ε_d and ε_m are independent, $(\varepsilon_d - \varepsilon_m)$ is a Gaussian random variable with zero mean and $2\sigma^2$ variance.

The mean and variance of I_i/S_i are upper bounded by [1]

$$E[I_i/S_i] = q_i \rho_i \iint f(r_m/r_d)dA \le \mu_{ii}, \tag{6}$$

and

$$Var[I_i/S_i] \le \rho_i \iint \Big[q_i g(r_m/r_d) - q_i^2 f^2(r_m/r_d)\Big]dA = \sigma_{ii}^2, \tag{7}$$

where μ_{ii} is the upper bound on the mean of I/S_i, σ_{ii}^2 is the upper bound on the variance of I/S_i, $\rho_i = 2n_i/\sqrt{3}$ is the density of class i users per unit area

$$f(r_m/r_d) = (r_m/r_d)^4 e^{(\sigma \ln 10/10)^2} \left[1 - Q\left(40\log(r_d/r_m)/\sqrt{2\sigma^2} - \sqrt{2\sigma^2} \ln 10/10 \right) \right], \quad (8)$$

$$g(r_m/r_d) = (r_m/r_d)^8 e^{(\sigma \ln 10/5)^2} \left[1 - Q\left(40\log(r_d/r_m)/\sqrt{2\sigma^2} - \sqrt{2\sigma^2} \ln 10/5 \right) \right], \quad (9)$$

and $Q(y) = \int_y^\infty e^{-x^2/2} dx / \sqrt{2\pi}$.

Let BER_i^* denote the BER requirement in a subcarrier for class i users and SIR_i^* denote the SIR requirement in a subcarrier for class i users. The system capacity is defined as the maximum $(n_1,...,n_i,...,n_K)$ that can be supported such that the achieved SIR in a subcarrier is greater than or equal to the required SIR_i^* in the subcarrier 99% of the time for all classes. That is, the outage probability in a subcarrier is defined as

$$\Pr[BER_i \geq BER_i^*] = \Pr[SIR_i \leq SIR_i^*] = \Pr\left[\sum_{j=1}^{n_i-1} \psi_{ij} + (I_i/S_i) \geq \delta_i \right], \quad (10)$$

where $\delta_i = (G_i/\gamma_i) - (\eta/S_i), i = 1,2,...,K$. Invoking the central limit approximation and solving (10), we have

$$\Pr[BER_i \geq BER_i^*] = \sum_{l_i=0}^{n_i-1} \Pr\left[\frac{I_i}{S_i} \geq \delta_i - l_i \mid \sum \psi_{ij} = l_i \right] \Pr[\sum \psi_{ij} = l_i]$$

$$= \sum_{l_i=0}^{n_i-1} Q\left(\frac{\delta_i - \mu_i}{\sigma_i} \right) \binom{n_i-1}{l_i} q_i^{l_i} (1-q_i)^{n_i-1-l_i}, \quad (11)$$

where $\mu_i = l_i + (q_i\rho_i/q_1\rho_1)\mu_{11}$, and $\sigma_i = \sqrt{\sigma_{ii}^2}$, $i = 1,2,...,K$. The analysis here can be used to determine the system capacity with the proposed DCP scheme for allocation of subcarriers in 4G MC-DS-CDMA mobile systems.

Table 2. Parameter Values Used

Symbol	Value	Symbol	Value
N_c	10 or 20	G_i	$(W/N_{ci})/R_i$
M_1	1	$BER_1^* = BER_2^*$	10^{-3}
M_2	3	σ	8 dB
α_1	1/0.625	$\gamma_1 = \gamma_2$	7 dB
β_1	1/0.375	S_1/η	-1 dB
α_2	0.9	S_2/η	$(S_1/\eta) \times (S_2/S_1)$, $\frac{S_2}{S_1} = \left(\frac{G_1}{\gamma_1} + q_1 \right) / \left(\frac{G_2}{\gamma_2} + q_2 \right)$ [3]
β_2	0.1	μ_{11}	$0.0247n_1$ or $0.01235n_1$
$R_1 = R_2$	80 kbps	σ_{11}^2	$0.01256n_1$ or $0.00641n_1$
W	50 MHz	σ_{22}^2	$0.01256n_2q_2/q_1$ or $0.00641n_2q_2/q_1$

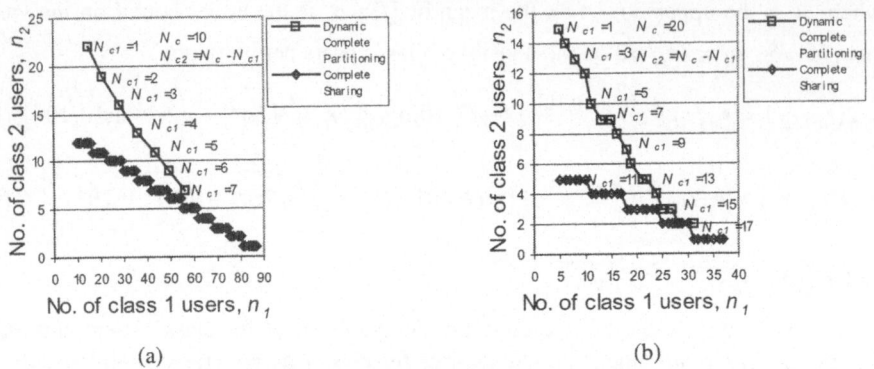

Fig. 2. System capacity with (a) $N_c = 10$, and (b) $N_c = 20$

4   Numerical Results

In this section we present results for the system capacity with 2 classes ($K=2$) as illustrative examples. The parameter values used in the numerical examples are tabulated in Table 2. The admissible regions for the system capacities for CS and DCP with $N_c = 10$ and $N_c = 20$ are shown in Fig. 2. The results for CS are obtained using the analytical formulation in [3]. Fig. 2 shows that the system capacities, (n_1, n_2), of DCP can be larger than those of CS, with these numbers of subcarriers. Admission to the system is possible only when the elements of the admitted doublet (n_1, n_2) are on or below these curves. Furthermore, the system capacities of CS and DCP for $N_c = 10$ is larger than those for $N_c = 20$. Note that the probability that an active spreading code in a subcarrier is used by source i, q_i, is increased by having a smaller number of subcarriers to use, $N_{ci} < N_c$, (see (4)) by using the DCP scheme for allocation of subcarriers to different classes. However, this is significantly offset, for low to above average values of N_{c1}, by eliminating the intra- and inter-cell interference from other classes in its own subcarriers. The admission region can be stored and retrieved from a look-up table to perform DCP of subcarriers to different classes when the number of users to be supported is larger than that of CS. The DCP round-robin carrier-hopping multirate MC-DS-CDMA cellular system introduced in this paper shows great promise in terms of capacity usage for multimedia services.

References

1. Wong, T.C., Mark, J.W., Chua, K.C., Yao, J., Chew, Y.H.: Performance Analysis of Multiclass Services in the Uplink of Wideband CDMA. IEEE ICCS (2000) 692-696
2. Maglaris, B., Anastassiou, D., Sen, P., Karlsson, G., Robbins, J.D.: Performance Models of Statistical Multiplexing in Packet Video Communications. IEEE Trans. on Comm., Vol. 36, No. 7, (1988) 834-844

3. Wong, T.C., Mark, J.W., Chua, K.C.: Performance Analysis of Variable Bit Rate Multiclass Services in the Uplink of a Round-Robin Carrier-Hopping Multirate Multi-Carrier DS-CDMA System. International Conference on CIC, (2003) CD-ROM
4. Simon, M.K., Alouini, M.-S.: Digital Communication over Fading Channels: A Unified Approach to Performance Analysis. John Wiley and Sons (2000)

ORBIS: A Reconfigurable Hybrid Optical Metropolitan Area Network Architecture

Yufeng Xin[1], Ilia Baldine[1], Mark Cassada[1], Daniel Stevenson[1],
Laura E. Jackson[2], and Harry Perros[2]

[1] ANR, MCNC-RDI, RTP NC 27709 USA
[2] Computer Science, NC State University, Raleigh NC 27609 USA

Abstract. In this paper, we propose a novel metropolitan area network (MAN) ring architecture, ORBIS, a multiple-service WDM platform that supports multiple transport techniques. The WDM ring is horizontally divided into multiple subnetworks, each of which supports one type of the traffic with a subset of the total wavelengths. The system and subnetwork control mechanism and reconfiguration issues are addressed in this paper.

1 Introduction

The challenges faced by metropolitan WDM networks are more complicated than those in the WDM backbones due to the existence of traffic from multiple transport protocols with variable bandwidth and QoS requirements [4].

Despite of the multitude of traffic, previous and current solutions are via a layered architecture on the top of a single transport platform such as SONET/SDH. This is a very expensive and not scalable approach because of the overhead of multiple layers and the complicated adaptation and encapsulation processes.

The WDM technique has provided another dimension to provide multiple services in an integrated network. By taking the advantage of the multiple wavelengths existing in a single link, we propose a novel hybrid MAN architecture, ORBIS, in which, multiple transport networks co-exist in a single WDM ring and each subnetwork is provisioned with a part of the total wavelengths. The merit behind this architecture is that different types of user traffic with different QoS and cost requirements can be simply dispatched to different optical transport subnetworks with different QoS guarantees and cost functions.

This is also a highly scalable network architecture. A piece of multitude traffic would first be admitted to an adequate subnetwork in a coarse QoS scale. Finer bandwidth and QoS provisioning can be provided inside each subnetwork. Dynamic network reconfiguration could be done via a simple way like re-allocating wavelengths to subnetworks. Finally, it allows a graceful deployment of new technologies in the same fiber infrastructure while protecting the investment on the old technologies like SONET/SDH.

N. Mitrou et al. (Eds.): NETWORKING 2004, LNCS 3042, pp. 1502–1507, 2004.

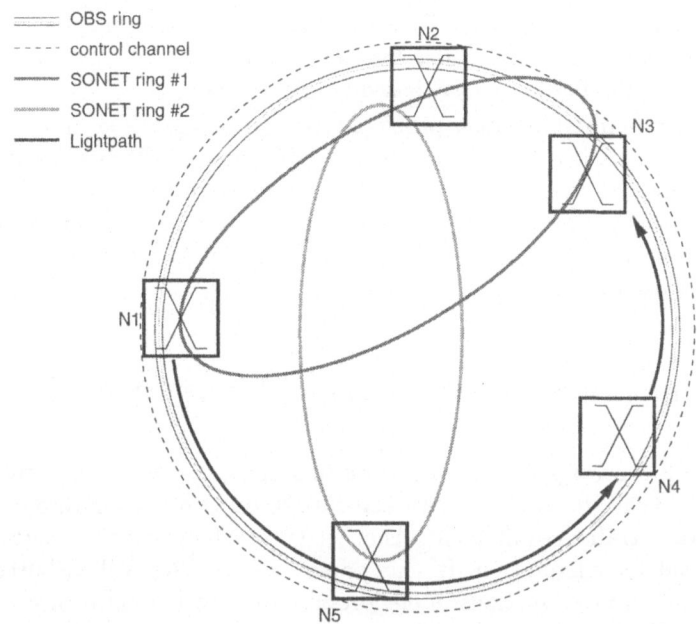

Fig. 1. ORBIS Optical Network Architecture

The ORBIS ring consists of three independent subnetworks: a SONET/SDH ring subnetwork (SONET), a wavelength switched subnetwork (LIGHTPATH), and an optical burst switched subnetwork (OBS). Circuit switching techniques like the SONET/SDH ring and setting up lightpaths are very inefficient to adopt data traffic, but are more suitable for real time traffic. Optical Burst Switching (OBS) has emerged as a more competitive choice for data traffic in the near future [Xu01].

2   System Model

The system is depicted as in the Fig. 1. Each node in the ring will have a minimum of 4 sets of optical transponders. The first is a transceiver fix-tuned to the control channel. The second one is a pair of fast-tunable transmitter and receiver for the BURST subnetwork, which can be tuned to any wavelength independently. The third set is dedicated to the LIGHTPATH and the transceiver can be slowly tuned to a specific wavelength to set up a lightpath between two nodes. The last one is an OADM integrating a set of SONET ADM for the SONET subnetwork. A part of nodes with their OADM tuned to a same wavelength will form a logic SONET/SDH ring. The network structure is depicted in Figure 1.

Each subnetwork will be independently allocated a subset of the total wavelengths, and the transponder for each subnetwork will be given the wavelength tuning range according to the allocated wavelengths. At every node in the network, there is an electrical buffer for each subnetwork and incoming traffic is dispatched to a buffer via a QoS-based traffic dispatcher.

There are many interesting problems need to be answered in order to make this architecture a feasible solution, for example, how to allocate the given wavelength resource to the subnetworks. We have developed a set of QoS-based optimal wavelength allocation algorithms based on a game theoretical model [2]. In this paper, we will focus on the control and reconfiguration protocols for the system and each subnetwork.

3   An Integrated System Control Plane Hierarchy

The network control plane has two main functions, one is to signal the network elements so that the correct route can be set up and proper resource can be reserved for a traffic requirement. Another is to achieve efficient network management and reconfiguration. In this paper, we consider a distributed control plane to fulfill above functions in our proposed network architecture.

The ring architectures feature simpler synchronization of nodes. This is usually accomplished by a control plan with a slotted control protocol. In the ring with N nodes, N control slots, one for each node, make up a control message frame. The control message will continuously circulate around the ring through the dedicated control wavelength.

Since our proposed network architecture consists of multiple subnetworks, a hierarchical control message format is a natural choice, in which, a piece of the message will be in charge of an individual subnetwork. Generally, a host node may write following control information into its slot:

1. Its SONET ring group membership information for the management and reconfiguration of the SONET/SDH subnetwork (details will be given in next section). 2. Lightpath setup, Acknowledgement, and release information. 3. Burst control information including the source and destination address, length, and offset time, etc. 4. Information about the availability of attached receiver(s). 5. Other necessary management information related to restoration and security.

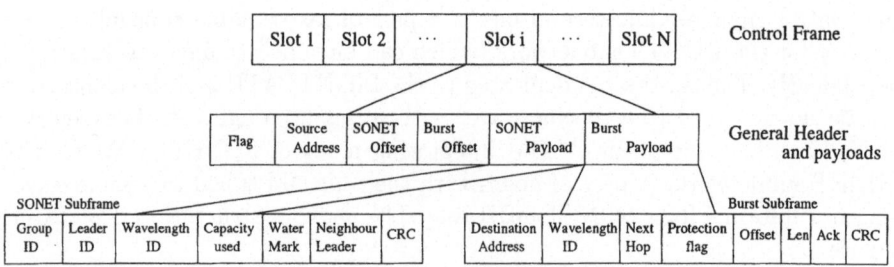

Fig. 2. ORBIS control frame

The proposed control message structure is depicted in Fig. 2. The general header of a slot provides the offset pointing to part of the payload containing the SONET and Lightpath/Burst sub-frame. The general header also contains the source address, the sub-frame payload length and a flag byte telling the receiving node which types of traffic is coming. The two sub-frames are variable in size. The SONET sub-frame is mainly used for the ring group management and reconfiguration. The Lightpath/Burst sub-frame is mainly used for resource reservation for integrated LIGHTPATH and BURST control.

4   SONET/WDM Rings

Since there are multiple rings in the SONET subnetwork, we need to address both ring construction and reconfiguration problem. We assume the BLSR/2 SONET ring structure, i.e. all traffic are routed via shortest paths and half of the capacity are initially reserved for protection/restoration purpose.

The ring construction problem in a WDM platform is usually modeled as a traffic grooming problem, i.e, assembling traffic elements from the given traffic matrix into different SONET rings. A recent survey on this problem can be found in [1]. All of existing studies try to minimize the network resource usage and treat the traffic among all the nodes equally. We argue that nodes with mutual traffic are better to be grouped in one ring due to the management and security concerns and customer requirement. Therefore we propose a novel ring grouping algorithm. It is based on the following ring grouping policies:

1. Those nodes requiring mutual interconnection should be better grouped into the same SONET/SDH ring.
2. Minimize the total number of rings.
3. The transition between different SONET rings should be minimized.

Our grouping problem can be proved to be equivalent the NP-Complet Maximum Clique Problem (MCP) in the graph theory by modeling the traffic matrix as a connected graph. The MCP tries to find the maximum fully connected subgraph (clique) in a simple graph [3]. Each clique found represents a logical SONET ring in the final solution.

To maintain a high utilization of the network resources, a reconfiguration mechanism should be able to change the ring groups by adding and releasing a ring, or migrating traffic between rings, upon the traffic SONET/SDH changes. A novel ring reconfiguration algorithm is given in Fig. 3. We use the backup capacity in the BLSR/2 ring to transfer the existing traffic during the reconfiguration to achieve hitless zero transition loss.

We develop a two-tier control protocol to achieve ring grouping and reconfiguration. The *group leader* is a special node that is the manager of a logical ring. The control plane for the SONET/WDM rings works in a semi-distributed manner with a two-tier structure. The first tier is for a individual ring under its group leader for the group membership management. The second tier is among

---

Input: The initial rings $C = (c_i)$, the watermarks W_H, W_L, the maximum traffic flow δ_i in every ring c_i
Output: A decision on the ring reconfiguration: adding, removing, or merging.

1.   begin
2.   Find the minimum ring load $\delta_{min} = min(\delta_i)$(ring c_{min}).
3.   If $\delta_{min} \leq W_L$ for the required timeout count.
4.   If there is another ring c_j having the required capacity and allowed membership.
5.   Switch the traffic from the work ring to the backup ring.
6.   Release the work ring and merge the ring c_{min} with ring c_j.
7.   Release the backup ring.
8.   End If
9.   End If
10.   Find the maximum ring load $\delta_{max} = max(\delta_i)$(ring c_{max}).
11.   If $\delta_m ax \geq W_H$ for the required timeout count.
12.   Find the nodes with minimum traffic in c_{max} untill $\delta_m ax \leq W_H$.
13.   Switch the according traffic to the backup ring and release the working load.
14.   Group these nodes into a new ring if the multiplexed traffic $\geq W_L$.
15.   If not, deliver these traffic to the non-SONET part of ORBIS.
16.   Remove the traffic from the backup ring.
17.   Update the ring membership.
18.   End If
19. end of the algorithm

Fig. 3. Algorithm for ring reconfiguration

all the group leaders. The group leaders communicate with each other in this control tier to make decisions on the ring reconfiguration.

For each type of the two groups, logical ring and the leader group, members can be added or removed dynamically. We propose a multicast-like signaling protocol for the operation of the 2-tier control planes. Each of above group, either the leaders' group or a logical ring group, is considered a multicast group. For each logical ring, the group leader plays the role of the multicast source node. For the leader group, we need to define a hub node to be the source node. We also assume a leaf-initiated joining mechanism.

In this control plane architecture, the two tiers of group membership management are closely cooperative. Initially the leader group is set up with only one member, the hub node. When a new ring is set up, a new group session is setup. Then the group leader is elected which will initiate a *ADD PARTY* message to join the leader group. When an existing ring needs to be removed, the leader will first initiate a *SESSION RELEASE* message to release the particular ring group. Then it will initiate a *DROP PARTY MESSAGE* to the leader group to get itself removed.

5   Lightpath and Burst-Switched Subnetwork

OBS uses a one-way resource reservation mechanism and a pre-sent control packet to set up the light path for a burst transmission session, while setting up a lightpath in the LIGHTPATH subnetwork requires to handle the two-way reservation message. It is obvious that these two tasks can be done by an integrated signaling protocol which will choose the resource reservation method according to the traffic requirement.

we have assumed both tunable transmitter and tunable receivers in each node. However, the tuning time of the receiver is usually much longer. This a fully dynamic case and both receiver and path wavelength assignment must be addressed to set up a lightpath between a pair of transceiver successfully.

The control packet will reserve the wavelength along the path and tune the receiver to the required wavelength. Since the data transmission follows the control packet, the tuning time needs to be included into the offset time. Here we propose a new wavelength reservation protocol, in which 2 control packets are sent out before the data transmission. The first one will not make any wavelength reservation along the path, but directly go to the selected receiver to tune the receiver to the required wavelength if it is free. The second control packet will follow the first one in a later time and will reserve the wavelength along the route. The reservation may be fail either due to the busy receiver (from control packet 1), or the occupied wavelength along the route (from control frame 2).

6   Conclusion

The focus of this paper is on the control and reconfiguration protocols for the proposed multiple-service WDM ring system and the 3 subnetworks. A set of novel algorithms for ring grouping and reconfiguration and a semi-distributed control protocol for ring membership management are given for the SONET ring subnetwork. For the lightpath and burst subnetworks, a distributed control protocol based on a mixed one-way and two-way reservation protocol is described.

References

1. Rudra Dutta and George N. Rouskas. Traffic grooming in wdm networks: past and future. *IEEE Network* 16(6), Nov./Dec. 2002.
2. Yufeng Xin *et al.* Optimal wavelength allocation in a novel hybrid optical network architecture with a game theoretical approach. In *Submitted for publication*, 2003.
3. Michael R. Garey and David S. Johnson. *Computers and Intractability: A Guide to the Theory of NP-Completeness.* W.H. Freeman Co, 1979.
4. N. Ghania, J.-Y. Pan, and X. Cheng. *Metropolitan optical networks.* Optical Fiber Telecommunications IVB, T. Li (Ed.), 2002.

A Layer Assignment and Resource Reservation Scheme for Hierarchical Cell Structures

Roland Zander and Johan M. Karlsson

Department of Communication Systems, Lund University, Sweden*
{rolandz, johan}@telecom.lth.se

Abstract. To provide for a high QoS in a cellular network, each call should be assigned to its preferred cell and priority given to already ongoing calls. Herein, a combined speed-sensitive layer selection method and dynamic guard channel scheme, called CLARR, for use in hierarchical networks, is introduced. To reduce the complexity, partially the same speed estimation and measurement procedures are applied for both tasks. The main features of CLARR are the use of dwell time feedback control in the layer selection method and the grouping of calls depending on previous movement patterns. The measurements needed for making mobility predictions are performed separately for calls of different groups.

1 Introduction

In a cellular network, micro cells can out of efficiency reasons be inserted to cover hotspots such as airports and shopping malls. A micro cell causes usually less interference than a normal (macro) cell due to the low transmission power required for covering its smaller cell area. Another advantage with a hierarchical structure is the possibility to overflow traffic to alternative layers/cells. The layers are ordered according to cell size, where a layer consisting of small/large cells is denoted a lower/higher layer.

Cell/layer assignment is usually triggered by the location, carried service and/or mobility of the call. By assigning all calls to cells according to largest received signal strength, the average transmission power can be lowered, enabling a smaller reuse distance. The carried services must be considered due to differences in delay constraints and bandwidth demands [1]. Speed-sensitive methods assign slow/fast-moving calls to cells in the lower/higher layers, reducing the number of handovers, $E[H]$. There exist different procedures for estimating the subscriber speed. These include to quantify the power adjustments needed for power control [2], to continuously make subscriber location updates [3] and to measure the dwell times, T_d, in previous visited cells [4].

A subscriber finds it more annoying to have an ongoing call dropped than a new call blocked, meaning that prioritizing ongoing calls is a good idea for the operators. This can be accomplished by having some channels, called Guard Channels (GCs), exclusively reserved for handovers. The introduction of GCs reduces the handover dropping probabilities, P_d, at the expense of larger blocking probabilities, P_b, for new calls. The number of GCs can be either fixed or time-varying (dynamic).

* This work was supported by the Swedish Foundation for Strategic Research under the Personal Computing and Communication (PCC) grant.

N. Mitrou et al. (Eds.): NETWORKING 2004, LNCS 3042, pp. 1508–1513, 2004.
© IFIP International Federation for Information Processing 2004

In this paper, a combined speed-sensitive layer selection method and GC scheme for use in hierarchical cellular networks, called *Combined Layer Assignment and Resource Reservation* (CLARR), is introduced. In contrast to other layer assignment methods, CLARR does not only put the emphasis on the past speed of a call, but also on predicting its future T_d-values, using feedback control. A subscriber having previously moved with high/low speed will not necessarily have a short/long T_d in its next cell. For example, the speed of a subscriber traveling by car to a major city will probably decrease due to speed limits. CLARR applies dynamic GCs to give higher priority to ongoing calls. The movements of previous calls are used for predicting the future whereabouts of the ongoing calls, determining the GC requirements in the different cells.

The outline of this paper is as follows. Section 2 includes a description of the CLARR method. The simulation model and the obtained numerical results can be found in section 3. Finally, conclusions are drawn in section 4.

2   CLARR

In CLARR, predictions of T_d are used to determine to which cell/layer the ongoing calls shall be assigned (large/small predicted $T_d \Rightarrow$ lower/higher layer). When a handover is required, T_d in the coverage area of the potential target cell belonging to the lowest covering layer is predicted. Previously measured T_d-values in this cell are used as input through a feedback procedure. By optimizing the assignment to the lower cellular layers where handovers are frequent, a handover rate reduction can be obtained.

Personalized subscriber treatment may be applied in the prediction procedure. Since this solution is of high complexity and mostly suffer from data shortage, T_d of other subscribers ought to be considered. The accuracy of the predictions can be improved by grouping the T_d-values. This means that only measurements of calls of the same group as the call to be predicted are considered. In CLARR, the calls are grouped according to past speed since slow-moving subscribers usually have longer T_d-values than fast-moving subscribers. Further, the calls are grouped according to previous traveled paths (movement groups) since this helps anticipate route-dependent speed alterations (e.g., traffic jams) and traveled distances in the next cell (e.g., road paths).

When a call leaves the area of a cell belonging to its lowest covering layer and/or requires a handover, a mean of the most recent performed speed estimations (using an arbitrary speed estimation method) of the call is calculated. The subscriber speed in the cell is characterized by a number between 0 (slowest) and 1 (fastest), denoted v, where a value of 0.3 means that 70% of the previous calls in the cell had a higher speed. Exponential smoothing is applied to give the more recent data a larger weight. In (1), V_n denotes the speed value obtained after n iterations and α the smoothing parameter ($0 \leq \alpha \leq 1$). V_n is the parameter used to group the calls according to speed.

$$V_n = \alpha V_{n-1} + (1 - \alpha)v \tag{1}$$

Following a call departure from the area of a cell of the lowest covering layer, its T_d-value is compared with those of all other calls measured in the cell. A percentile is calculated for T_d and is sent to the location where V_{n-1} for the call was calculated. There, the data are used for making predictions of T_d for the next visited cell, T_{next}, for calls

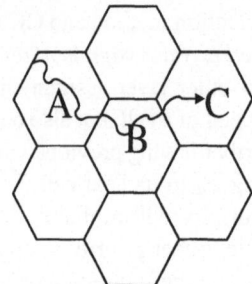

Fig. 1. Subscriber movements in a cellular network

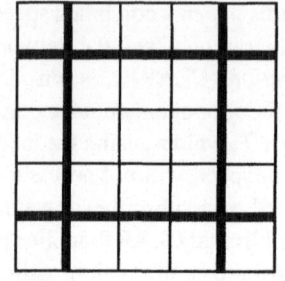

Fig. 2. Simulated network

of its movement and speed group. A movement group consists of calls having similar entry (identical previous cell) and exit points (identical next cell). T_{next} is calculated as a weighted mean of the most recent T_d-percentiles derived for calls of its group. For example, if the subscriber in Fig. 1 requires a handover from its current cell, C, its T_d-percentile in C is sent to its latest visited cell, B. There, it is considered when determining T_{next} for calls of a certain speed and movement group. Since the call arrived from A and then moved to C, the movement group in question is (A,C). The speed group is the group in which the call was placed following its handover from cell B.

When a call requires a handover, the T_{next}-value of its group is compared to a number of time-varying thresholds, equal to the number of cellular layers present in the region minus one. The call is assigned to the cell that corresponds to the number of thresholds that is larger/smaller than T_{next} (high/low $T_{next} \Rightarrow$ lower/higher layer). If the load in any of the candidate cells is very high, the thresholds are adjusted so that fewer calls are assigned to the heavily loaded cells [4,5].

CLARR includes a dynamic GC scheme measuring movement patterns of previous calls. The probability for a call residing in a certain cell to require a handover, P_h, and the probability for the handover to occur to cell y, $P_{hd}(y)$, where y can be a cell of any hierarchical layer [6], are continuously measured. From these probabilities and the number of ongoing calls in the cell, C, the number of ongoing calls expected to require a handover to cell y, $G(y)$, can be calculated, see (2). By summation of all $G(y)$-values, an estimation of the current handover arrival rate in cell y, $G_{tot}(y)$, is derived. Since all expected handover call arrivals to a cell will not occur simultaneously, the number of required GCs in cell y is significantly less than $G_{tot}(y)$. The number of allocated GCs is derived through a multiplication of $G_{tot}(y)$ and a time-varying parameter ψ ($\psi < 1$).

$$G(y) = CP_hP_{hd}(y) \qquad (2)$$

To improve the efficiency, separate measurements are performed for calls considered probable to have different movement patterns. Subscribers moving in opposite directions on a road have for instance completely different P_{hd}-values. Hence, the calls are divided into measurement groups, called source cell groups, according to latest visited cell. Since fast-moving subscribers often are more probable to require a handover than slow-moving subscribers, the calls in a source cell group are arranged into speed groups [6]. Here, the principles applied in the layer assignment part are reused.

3   Simulation Model and Numerical Results

The network consists of two layers with micro and macro cells, respectively. It is out of simplicity reasons chosen to give the cells a quadratic shape although a hexagonal shape is much more realistic. The advantage with using quadratic cells is that it enables the area of each micro cell to be entirely covered by one macro cell. The lower layer consists of 81 micro cells in a 9*9 ring topology with wrapped round edges. The micro cells are arranged into clusters of nine, 3*3, where a macro cell provides coverage over the entire cluster area, see Fig. 2. A thin line corresponds to a border between two micro cells and a thick line to a border between both adjacent micro and macro cells.

The number of channels in a micro/macro cell is set to 20 and 60, respectively, and the call length is exponentially distributed (mean of 10 minutes). A new call is assigned to a micro cell or overflowed to the overlapping macro cell. The layer assignment thresholds are adjusted to achieve similar load ratios between the layers for all scenarios. A handover is attempted at the instant a call crosses a cell border. In case of a handover failure, the call is immediately dropped. Every single call is assigned an uniformly distributed number between 1 and 10, indicating the subscriber speed, through a randomization process at call arrival. This subscriber speed value is multiplied with a speed constant, unique for its current covering micro cell, resulting in a value describing how fast it is possible for the subscriber to move in its current cell. The mean dwell time in the covering micro cell is derived through multiplication between the inverse of this number and the distance to be traveled by the subscriber. The latter parameter indicates how much the subscribers tend to move around in the cell. This distance measure is dependent on the entry and exit points of the subscriber in the cell. The subscriber speed is estimated from past T_d-values mainly due to the low modeling complexity of this technique, but CLARR can operate under an arbitrary speed estimation procedure.

First, CLARR is compared to a reference method with identical features except for the division of calls into movement groups (set 1). The bandwidth reservation and load balancing features are discarded. Four different scenarios (SCs), I-IV in Table 1, are investigated. The T_d-distribution and the speed constants applied for the different micro cells separate the SCs. T_d is assumed to be either uniformly or exponentially distributed. Three different parameter settings for the cell speed parameter are used, namely 1 for all cells, 2 for all cells, and 0.8 for 1/3 of the cells and 1.1 for the remaining 2/3 of the cells. When a call leaves its covering micro cell, it moves to each of the adjacent micro cells in a non-diagonal direction with an equal probability, namely 1/4.

In simulation set 2, it is investigated how beneficial it is to apply load balancing in CLARR. Still, the bandwidth reservation feature is discarded. The reference method is identical except for the load balancing feature. Two SCs are investigated (I and II). Cells with a momentary number of available channels smaller than a threshold, T_{load}, are classified as being heavily loaded. If a micro/macro cell is heavily loaded at the time of a call arrival and its overlapping macro/micro cell is not, the threshold determining to which cell the call should be assigned is increased/reduced. The size of the reduction/increase, ΔQ, is derived through subtracting T_{load} by the number of available channels in the heavily loaded cell, N_{av}, and then multiply the result by a constant, C.

Simulation set 1 and 2 are investigated for three different traffic loads. The call failure probability, P_f, is the percentage of calls that are either blocked or dropped. In Table 3,

Table 1. Simulation scenarios (SCs)

SC	T_d-distribution	P_{hd}	Cell Speed
I	Uniform	$P_{1-4}=1/4$	1/1/1
II	Uniform	$P_{1-4}=1/4$	2/2/2
III	Uniform	$P_{1-4}=1/4$	0,8/1,1/1,1
IV	Exponential	$P_{1-4}=1/4$	1/1/1
V	Uniform	$P_{1-3}=1/3, P_4=0$	1/1/1

Table 2. Handover directions (hd)

hd	Explanation
1	Micro cell straight ahead
2	Back to the latest visited cell
3	Micro cell to the left
4	Micro cell to the right

Table 3. Numerical results for simulation sets 1 and 2

Set	SC	P_f(%) CLARR			P_f(%) Reference			Gain (%)		
1	I	1,010	1,997	3,000	1,177	2,194	3,182	14,2	9,0	5,7
1	II	1,007	2,004	2,987	1,261	2,354	3,366	20,1	14,9	11,3
1	III	0.999	2,006	3,014	1,220	2,289	3,309	18,1	12,4	8,9
1	IV	0,999	2,011	2,995	1,185	2,276	3,261	15,7	11,6	8,2
2	I	0,839	1,765	2,636	1,011	1,997	3,000	17,0	11,6	12,1
2	II	0,791	1,649	2,548	1,007	2,004	2,987	21,4	17,7	14,7

P_f is shown for the CLARR and reference methods. The gain column shows the gain in percent obtained from dividing the calls into movement groups (set 1) or introducing load balancing (set 2). Further, $E[H]$ is also marginally larger for the reference scheme.

The introduction of movement groups increase the accuracy of the obtained speed estimations when using T_d as a speed indicator, because calls of the same group have had similar straight path distances. The obtained gain of the grouping feature is larger at low traffic load because the T_d-predictions used for layer assignment only matters if there are available channels in the covering cells of both layers. Further, the gain is increased in case of a larger $E[H]$, see SC II. For every performed handover, a layer assignment selection is necessary. Thus, the importance of making accurate T_d-predictions increases with the number of handovers. Since the grouping feature places calls arriving from different cells into separate groups, differences in cell speed constants also increase the obtained gain, see SC III. When T_d is exponentially distributed (SC IV), it becomes more difficult to make accurate T_d-predictions. However, these difficulties seem to affect both the CLARR and the reference method in similar ways. The higher gain for SC IV in comparison with SC I can probably be explained by the larger $E[H]$ that appears when an exponentially distributed T_d is applied.

Introduction of load balancing in the CLARR method significantly reduces P_f, see set 2 in Table 3. The larger gain obtained for SC II is a result of an increased $E[H]$. However, it is somewhat difficult to make a comparison between the SCs because the speed threshold adjustments are not optimized but identical for both SCs. When the traffic load increases, the obtained gain is reduced due to a more evenly distributed momentary load between overlapping micro and macro cells.

In set 3, the efficiency of the complete CLARR method is investigated. The reference method is identical except for the division of calls into movement and source cell groups. The parameter separating the two investigated SCs, I and V, is the probability for a call

Table 4. Numerical results for simulation set 3

SC	$P_b(10^{-2})$ CLARR	$P_b(10^{-2})$ Reference	Gain (%)	$P_d(10^{-4})$ CLARR	$P_d(10^{-4})$ Reference	Gain (%)
I	1,579	2,129	25,8	8,34	8,75	4,7
V	1,534	1,819	15,7	8,02	8,35	4,0

to move to the coverage area of a specific adjacent micro cell, see Tables 1-2. For new calls, all adjacent micro cells are chosen with an equal probability.

The grouping feature significantly reduces P_b and P_d, see Table 4. Further, $E[H]$ is marginally smaller for CLARR. The obtained gain is reduced if the subscribers do not move back to their latest visited cell, see SC V. This alteration in P_{hd} increases the accuracy of the movement predictions when applying the grouping feature, which ought to increase the obtained gain. Simultaneously, the differences in traveling distances in a cell are reduced. Thus, the difference in speed estimation efficiency between CLARR and the reference method is lowered. Since the obtained gain is smaller for SC V, this latter argument must have had a greater impact on the chosen simulation settings.

4   Conclusion

A combined speed-sensitive layer selection method and dynamic guard channel scheme for hierarchical cellular networks, called CLARR, has been introduced. Partially the same speed estimation and measurement procedures are applied for both tasks. Cell dwell time predictions together with the momentary cell loads are used as input to the layer assignment procedure. The number of guard channels to be allocated is determined by the number of ongoing calls in adjacent cells and their predicted movement patterns. To improve the accuracy of the predictions performed in the CLARR method, the calls are grouped according to previous movements. It was shown using simulations that a substantial gain can be obtained through its load balancing and grouping features.

References

1. J. Ben-Othman and F. Valois. *Multiservice Allocation in Hierarchical Cellular Networks (MAHCN).* IEEE Int. Symposium on Computers and Communication 1999, pp. 80-86, 1999.
2. S. Zhicum, W. Ying, Z. Ping and L. Baolin. *Layer Selection Method using Power Control in Hierarchical Cellular Systems.* IEEE VTC 2001, vol. 4, no. 4. pp. 2407-2411, 2001.
3. Y. Zhao. *Standardization of Mobile Phone Positioning for 3G Systems.* IEEE Communications Magazine, vol. 40, no. 7, pp. 108-116, 2002.
4. K. Shum and C. W. Sung. *Fuzzy Layer Selection Method in Hierarchical Cellular Systems.* IEEE Transactions on Vehicular Technology, vol. 48, no. 6, pp. 1840-1849, 1999.
5. L. Jorguseski and D. Sparreboom. *HCS: Decision Procedure for an Interlayer Handover.* IEEE VTC 1999, vol. 3, pp. 1496-1500, 1999.
6. R. Zander and J. M. Karlsson. *A Measurement-Based Dynamic Guard Channel Scheme for Handover Prioritization in Cellular Networks.* 2nd International IFIP-TC6 Networking Conference "Networking 2002", pp. 1245-1251, 2002.

Author Index

Lecture Notes in Computer Science

For information about Vols. 1–2906

please contact your bookseller or Springer-Verlag

Vol. 2977: G. Di Marzo Serugendo, A. Karageorgos, O.F. Rana, F. Zambonelli (Eds.), Engineering Self-Organising Systems. X, 299 pages. 2004. (Subseries LNAI).

Vol. 2976: M. Farach-Colton (Ed.), LATIN 2004: Theoretical Informatics. XV, 626 pages. 2004.

Vol. 2973: Y. Lee, J. Li, K.-Y. Whang, D. Lee (Eds.), Database Systems for Advanced Applications. XXIV, 925 pages. 2004.

Vol. 2972: R. Monroy, G. Arroyo-Figueroa, L.E. Sucar, H. Sossa (Eds.), MICAI 2004: Advances in Artificial Intelligence. XVII, 923 pages. 2004. (Subseries LNAI).

Vol. 2971: J.I. Lim, D.H. Lee (Eds.), Information Security and Cryptology -ICISC 2003. XI, 458 pages. 2004.

Vol. 2970: F. Fernández Rivera, M. Bubak, A. Gómez Tato, R. Doallo (Eds.), Grid Computing. XI, 328 pages. 2004.

Vol. 2966: F.B. Sachse, Computational Cardiology. XVIII, 322 pages. 2004.

Vol. 2965: M.C. Calzarossa, E. Gelenbe, Performance Tools and Applications to Networked Systems. VIII, 385 pages. 2004.

Vol. 2964: T. Okamoto (Ed.), Topics in Cryptology – CT-RSA 2004. XI, 387 pages. 2004.

Vol. 2963: R. Sharp, Higher Level Hardware Synthesis. XVI, 195 pages. 2004.

Vol. 2962: S. Bistarelli, Semirings for Soft Constraint Solving and Programming. XII, 279 pages. 2004.

Vol. 2961: P. Eklund (Ed.), Concept Lattices. IX, 411 pages. 2004. (Subseries LNAI).

Vol. 2960: P.D. Mosses (Ed.), CASL Reference Manual. XVII, 528 pages. 2004.

Vol. 2958: L. Rauchwerger (Ed.), Languages and Compilers for Parallel Computing. XI, 556 pages. 2004.

Vol. 2957: P. Langendoerfer, M. Liu, I. Matta, V. Tsaoussidis (Eds.), Wired/Wireless Internet Communications. XI, 307 pages. 2004.

Vol. 2956: A. Dengel, M. Junker, A. Weisbecker (Eds.), Reading and Learning. XII, 355 pages. 2004.

Vol. 2954: F. Crestani, M. Dunlop, S. Mizzaro (Eds.), Mobile and Ubiquitous Information Access. X, 299 pages. 2004.

Vol. 2953: K. Konrad, Model Generation for Natural Language Interpretation and Analysis. XIII, 166 pages. 2004. (Subseries LNAI).

Vol. 2952: N. Guelfi, E. Astesiano, G. Reggio (Eds.), Scientific Engineering of Distributed Java Applications. X, 157 pages. 2004.

Vol. 2951: M. Naor (Ed.), Theory of Cryptography. XI, 523 pages. 2004.

Vol. 2949: R. De Nicola, G. Ferrari, G. Meredith (Eds.), Coordination Models and Languages. X, 323 pages. 2004.

Vol. 2948: G.L. Mullen, A. Poli, H. Stichtenoth (Eds.), Finite Fields and Applications. VIII, 263 pages. 2004.

Vol. 2947: F. Bao, R. Deng, J. Zhou (Eds.), Public Key Cryptography – PKC 2004. XI, 455 pages. 2004.

Vol. 2946: R. Focardi, R. Gorrieri (Eds.), Foundations of Security Analysis and Design II. VII, 267 pages. 2004.

Vol. 2943: J. Chen, J. Reif (Eds.), DNA Computing. X, 225 pages. 2004.

Vol. 2941: M. Wirsing, A. Knapp, S. Balsamo (Eds.), Radical Innovations of Software and Systems Engineering in the Future. X, 359 pages. 2004.

Vol. 2940: C. Lucena, A. Garcia, A. Romanovsky, J. Castro, P.S. Alencar (Eds.), Software Engineering for Multi-Agent Systems II. XII, 279 pages. 2004.

Vol. 2939: T. Kalker, I.J. Cox, Y.M. Ro (Eds.), Digital Watermarking. XII, 602 pages. 2004.

Vol. 2937: B. Steffen, G. Levi (Eds.), Verification, Model Checking, and Abstract Interpretation. XI, 325 pages. 2004.

Vol. 2936: P. Liardet, P. Collet, C. Fonlupt, E. Lutton, M. Schoenauer (Eds.), Artificial Evolution. XIV, 410 pages. 2004.

Vol. 2934: G. Lindemann, D. Moldt, M. Paolucci (Eds.), Regulated Agent-Based Social Systems. X, 301 pages. 2004. (Subseries LNAI).

Vol. 2930: F. Winkler (Ed.), Automated Deduction in Geometry. VII, 231 pages. 2004. (Subseries LNAI).

Vol. 2929: H. de Swart, E. Orlowska, G. Schmidt, M. Roubens (Eds.), Theory and Applications of Relational Structures as Knowledge Instruments. VII, 273 pages. 2003.

Vol. 2926: L. van Elst, V. Dignum, A. Abecker (Eds.), Agent-Mediated Knowledge Management. XI, 428 pages. 2004. (Subseries LNAI).

Vol. 2923: V. Lifschitz, I. Niemelä (Eds.), Logic Programming and Nonmonotonic Reasoning. IX, 365 pages. 2004. (Subseries LNAI).

Vol. 2919: E. Giunchiglia, A. Tacchella (Eds.), Theory and Applications of Satisfiability Testing. XI, 530 pages. 2004.

Vol. 2917: E. Quintarelli, Model-Checking Based Data Retrieval. XVI, 134 pages. 2004.

Vol. 2916: C. Palamidessi (Ed.), Logic Programming. XII, 520 pages. 2003.

Vol. 2915: A. Camurri, G. Volpe (Eds.), Gesture-Based Communication in Human-Computer Interaction. XIII, 558 pages. 2004. (Subseries LNAI).

Vol. 2914: P.K. Pandya, J. Radhakrishnan (Eds.), FST TCS 2003: Foundations of Software Technology and Theoretical Computer Science. XIII, 446 pages. 2003.

Vol. 2913: T.M. Pinkston, V.K. Prasanna (Eds.), High Performance Computing - HiPC 2003. XX, 512 pages. 2003. (Subseries LNAI).

Vol. 2911: T.M.T. Sembok, H.B. Zaman, H. Chen, S.R. Urs, S.H. Myaeng (Eds.), Digital Libraries: Technology and Management of Indigenous Knowledge for Global Access. XX, 703 pages. 2003.

Vol. 2910: M.E. Orlowska, S. Weerawarana, M.M.P. Papazoglou, J. Yang (Eds.), Service-Oriented Computing - ICSOC 2003. XIV, 576 pages. 2003.

Vol. 2909: R. Solis-Oba, K. Jansen (Eds.), Approximation and Online Algorithms. VIII, 269 pages. 2004.

Vol. 2908: K. Chae, M. Yung (Eds.), Information Security Applications. XII, 506 pages. 2004.

Vol. 2907: I. Lirkov, S. Margenov, J. Wasniewski, P. Yalamov (Eds.), Large-Scale Scientific Computing. XI, 490 pages. 2004.